Electronic Communications Systems

Fundamentals Through Advanced

Electronic Communications Systems

Fundamentals Through Advanced

Fifth Edition

Wayne Tomasi
DeVry University
Phoenix, Arizona

Upper Saddle River, New Jersey
Columbus, Ohio

Editor in Chief: Stephen Helba
Assistant Vice President and Publisher: Charles E. Stewart, Jr.
Assistant Editor: Mayda Bosco
Production Editor: Alexandrina Benedicto Wolf
Production Coordination: Carlisle Publishers Services
Design Coordinator: Diane Ernsberger
Cover Designer: Ali Mohrman
Cover art: Digital Vision
Production Manager: Matt Ottenweller
Marketing Manager: Ben Leonard

This book was set in Times Roman by Carlisle Communications, Ltd. It was printed and bound by Courier Kendallville, Inc. The cover was printed by Phoenix Color Corp.

Pearson Education Ltd.
Pearson Education Singapore Pte. Ltd.
Pearson Education Canada, Ltd.
Pearson Education—Japan
Pearson Education Australia Pty. Limited
Pearson Education North Asia Ltd.
Pearson Educación de Mexico, S.A. de C.V.
Pearson Education Malaysia Pte. Ltd.

10 9 8 7 6 5 4 3 2
ISBN 0-13-049492-5

To Cheryl
my best friend since high school
and my loving and faithful wife for the past 37 years

To our six children: Aaron, Pernell, Belinda, Loren, Tennille, and Marlis;
their wives and husbands: Kriket, Robin, Mark, Brent;
and of course, my five grandchildren: Avery, Kyren, Riley, Reyna, and Ethan

Preface

The purpose of this book is to introduce the reader to the basic concepts of traditional analog electronic communications systems and to expand the reader's knowledge of more modern digital, optical fiber, microwave, satellite, data, and cellular telephone communications systems. The book was written so that a reader with previous knowledge in basic electronic principles and an understanding of mathematics through the fundamental concepts of calculus will have little trouble understanding the topics presented. Within the text, there are numerous examples that emphasize the most important concepts. Questions and problems are included at the end of each chapter and answers to selected problems are provided at the end of the book.

This edition of *Electronic Communications Systems: Fundamentals Through Advanced* provides a modern, comprehensive coverage of the field of electronic communications. Although nothing has been omitted from the previous edition, there are several significant additions, such as three new chapters on telephone circuits and systems, two new chapters on cellular and PCS telephone systems, and three new chapters on the fundamental concepts of data communications and networking. In addition, numerous new figures have been added and many figures have been redrawn. The major topics included in this edition are as follows.

Chapter 1 introduces the reader to the basic concepts of electronic communications systems and includes a new section on power measurements using dB and dBm. This chapter defines modulation and demodulation and describes the electromagnetic frequency spectrum. Chapter 1 also defines bandwidth and information capacity and how they relate to one another, and provides a comprehensive description of noise sources and noise analysis.

Chapters 2 and 3 discuss signals, signal analysis, and signal generation using discrete and linear-integrated circuits. Chapter 3 gives a comprehensive coverage of phase-locked loops.

Chapters 4 through 8 describe analog communications systems, such as amplitude modulation (AM), frequency modulation (FM), phase modulation (PM), and single sideband (SSB). A comprehensive mathematical and theoretical description is given for each modulation technique and the basic components found in analog transmitters and receivers are described in detail.

Chapter 9 discusses the fundamental concepts of digital modulation, including comprehensive descriptions of amplitude-shift keying (ASK), frequency-shift keying (FSK), phase-shift keying (PSK), quadrature amplitude modulation (QAM), and differential phase-shift keying (DPSK). Chapter 9 introduces the student to trellis code modulation and gives a comprehensive description of probability of error, bit error rate, and error performance.

Chapters 10 and 11 describe the basic concepts of digital transmission and multiplexing. Chapter 10 describes pulse code modulation, while Chapter 11 describes time-division multiplexing of PCM-encoded signals and explains the North American Digital Hierarchy and the North American FDM Hierarchy. Wavelength division multiplexing of light waves is also introduced in Chapter 11.

Chapters 12 through 15 describe the fundamental concepts of electromagnetic waves, electromagnetic wave propagation, metallic and optical fiber transmission lines, free-space wave propagation, and antennas.

Chapters 16 through 18 give a comprehensive description of telephone instruments, signals, and wireline systems used in the public telephone network. Chapters 19 and 20 describe the basic concepts of wireless telephone systems, including cellular and PCS.

Chapters 21 through 23 introduce the fundamental concepts of data communications circuits and describe basic networking fundamentals, such as topologies, error control, protocols, hardware, accessing techniques, and network architectures.

Chapters 24 through 26 describe the fundamental concepts of terrestrial and satellite microwave-radio communications. Chapter 24 describes analog terrestrial microwave systems; Chapters 25 and 26 describe digital satellite systems.

Appendix A describes the Smith Chart.

ACKNOWLEDGMENTS

I would like to thank the following reviewers for their valuable feedback: Jeffrey L. Rankinen, Pennsylvania College of Technology; Walter Hedges, Fox Valley Technical College; Samuel A. Guccione, Eastern Illinois University; Costas Vassiliadis, Ohio University; and Siben Dasgupta, Wentworth Institute of Technology. I would also like to thank my project editor, Kelli Jauron, for her sincere efforts in producing the past two editions of this book and for being my friend for the past four years. I would also like to thank my assistant editor, Mayda Bosco, for all her efforts. The contributions from these people helped to make this book possible.

Wayne Tomasi

Brief Contents

Contents

CHAPTER 13 OPTICAL FIBER TRANSMISSION MEDIA 557

CHAPTER 14 ELECTROMAGNETIC WAVE PROPAGATION 603

CHAPTER 15 ANTENNAS AND WAVEGUIDES 631

CHAPTER 19 CELLULAR TELEPHONE CONCEPTS 773

CHAPTER 20 CELLULAR TELEPHONE SYSTEMS 795

CHAPTER 21 INTRODUCTION TO DATA COMMUNICATIONS AND NETWORKING 833

CHAPTER 22 FUNDAMENTAL CONCEPTS OF DATA COMMUNICATIONS 871

C H A P T E R 1

Introduction to Electronic Communications

CHAPTER OUTLINE

OBJECTIVES

- Define the fundamental purpose of an electronic communications system
- Describe analog and digital signals
- Define and describe the basic power units dB and dBm
- Define a basic electronic communications system
- Explain the terms *modulation* and *demodulation* and why they are needed in an electronic communications system
- Describe the electromagnetic frequency spectrum
- Describe the basic classifications of radio transmission
- Define *bandwidth* and *information capacity*
- Define *electrical noise* and describe the most common types
- Describe the prominent sources of electrical noise
- Explain *signal-to-noise* ratio and *noise figure* and describe their significance in electronic communications systems

1-1 INTRODUCTION

The fundamental purpose of an *electronic communications system* is to transfer information from one place to another. Thus, electronic communications can be summarized as the *transmission, reception,* and *processing* of information between two or more locations using electronic circuits. The original source information can be in *analog* form, such as the human voice or music, or in *digital* form, such as *binary-coded numbers* or *alphanumeric codes.* Analog signals are time-varying voltages or currents that are continuously changing, such as sine and cosine waves. An analog signal contains an infinite number of values. Digital signals are voltages or currents that change in discrete steps or levels. The most common form of digital signal is binary, which has two levels. All forms of information, however, must be converted to *electromagnetic energy* before being propagated through an electronic communications system.

Communications between human beings probably began in the form of hand gestures and facial expressions, which gradually evolved into verbal grunts and groans. Verbal communications using sound waves, however, was limited by how loud a person could yell. Long-distance communications probably began with smoke signals or tom-tom drums, and that using electricity began in 1837 when Samuel Finley Breese Morse invented the first workable telegraph. Morse applied for a patent in 1838 and was finally granted it in 1848. He used *electromagnetic induction* to transfer information in the form of dots, dashes, and spaces between a simple transmitter and receiver using a transmission line consisting of a length of metallic wire. In 1876, Alexander Graham Bell and Thomas A. Watson were the first to successfully transfer human conversation over a crude metallic-wire communications system using a device they called the *telephone.*

In 1894, Marchese Guglielmo Marconi successfully transmitted the first *wireless* radio signals through Earth's atmosphere, and in 1906, Lee DeForest invented the triode vacuum tube, which provided the first practical means of amplifying electrical signals. Commercial radio broadcasting began in 1920 when radio station KDKA began broadcasting *amplitude-modulated* (AM) signals out of Pittsburgh, Pennsylvania. In 1931, Major Edwin Howard Armstrong patented *frequency modulation* (FM). Commercial broadcasting of monophonic FM began in 1935. Figure 1-1 shows an electronic communications time line listing some of the more significant events that have occurred in the history of electronic communications.

1-2 POWER MEASUREMENTS (dB, dBm, AND Bel)

The *decibel* (abbreviated dB) is a logarithmic unit that can be used to measure ratios of virtually anything. For example, decibels are used to measure the magnitude of earthquakes. The Richter scale measures the intensity of an earthquake relative to a reference intensity, which is the weakest earthquake that can be recorded on a seismograph. Decibels are also used to measure the intensity of acoustical signals in dB-SPL, where SPL means sound pressure level. Zero dB-SPL is the threshold of hearing. The sound of leaves rustling is 10 dB-SPL, and the sound produced by a jet engine is between 120 and 140 dB-SPL. The threshold of pain is approximately 120 dB-SPL.

In the electronics communications field, the decibel originally defined only power ratios; however, as a matter of common usage, voltage or current ratios can also be expressed in decibels. The practical value of the decibel arises from its logarithmic nature, which permits an enormous range of power ratios to be expressed in terms of decibels without using excessively large or extremely small numbers.

The dB is used as a mere computational device, like logarithms themselves. In essence, the dB is a *transmission-measuring unit* used to express relative gains and losses of electronic devices and circuits and for describing relationships between signals and noise. Decibels compare one signal level to another. The dB has become the basic yardstick for calculating power relationships and performing power measurements in electronic communications systems.

1830: American scientist and professor Joseph Henry transmitted the first practical electrical signal.
1837: Samuel Finley Breese Morse invented the telegraph.
1843: Alexander Bain invented the facsimile.
1861: Johann Phillip Reis completed the first nonworking telephone.
1864: James Clerk Maxwell released his paper "Dynamical Theory of the Electromagnetic Field," which concluded that light, electricity, and magnetism were related.
1865: Dr. Mahlon Loomis became the first person to communicate wireless through Earth's atmosphere.
1866: First transatlantic telegraph cable installed.
1876: Alexander Graham Bell and Thomas A. Watson invent the telephone.
1877: Thomas Alva Edison invents the phonograph.
1880: Heinrich Hertz discovers electromagnetic waves.
1887: Heinrich Hertz discovers radio waves.
Marchese Guglielmo Marconi demonstrates wireless radio wave propagation.
1888: Heinrich Hertz detects and produces radio waves.
Heinrich Hertz conclusively proved Maxwell's prediction that electricity can travel in waves through Earth's atmosphere.
1894: Marchese Guglielmo Marconi builds his first radio equipment, a device that rings a bell from 30 feet away.
1895: Marchese Guglielmo Marconi discovered ground-wave radio signals.
1898: Marchese Guglielmo Marconi established the first radio link between England and France.
1900: American scientist Reginald A. Fessenden transmits first human speech through radio waves.
1901: Reginald A. Fessenden transmits the world's first radio broadcast using continuous waves.
Marchese Guglielmo Marconi transmits telegraphic radio messages from Cornwall, England, to Newfoundland.
First successful transatlantic transmission of radio signals.
1903: Valdemar Poulsen patents an arc transmission that generates continuous wave transmission of 100-kHz signal that is receivable 150 miles away.
John Fleming invents the two-electrode vacuum-tube rectifier.
1904: First radio transmission of music at Graz, Austria.
1905: Marchese Guglielmo Marconi invents the directional radio antenna.
1906: Reginald A. Fessenden invents amplitude modulation (AM).
First radio program of voice and music broadcasted in the United States by Reginald A. Fessenden.
Lee DeForest invents the triode (three-electrode) vacuum tube.
1907: Reginald A. Fessenden invents a high-frequency electric generator that produces radio waves with a frequency of 100 kHz.
1908: General Electric develops a 100-kHz, 2-kW alternator for radio communications.
1910: The Radio Act of 1910 is the first occurrence of government regulation of radio technology and services.
1912: The Radio Act of 1912 in the United States brought order to the radio bands by requiring station and operator licenses and assigning blocks of the frequency spectrum to existing users.
1913: The cascade-tuning radio receiver and the heterodyne receiver are introduced.
1914: Major Edwin Armstrong patents a radio receiver circuit with positive feedback.
1915: Vacuum-tube radio transmitters introduced.
1918: Major Edwin Armstrong develops the superheterodyne radio receiver.
1919: Shortwave radio is developed
1920: Radio station KDKA broadcasts the first regular licensed radio transmission out of Pittsburgh, Pennsylvania.
1921: Radio Corporation of America (RCA) begins operating Radio Central on Long Island. The American Radio League establishes contact via shortwave radio with Paul Godley in Scotland, proving that shortwave radio can be used for long-distance communications.
1923: Vladimir Zworykin invents and demonstrates television.
1927: A temporary five-member Federal Radio Commission agency was created in the United States.
1928: Radio station WRNY in New York City begins broadcasting television shows.
1931: Major Edwin Armstrong patents wide-band frequency modulation (FM).

FIGURE 1-1 Electronic communications time line *(Continued)*

1934: Federal Communications Commission (FCC) created to regulate telephone, radio, and television broadcasting.
1935: Commercial FM radio broadcasting begins with monophonic transmission.
1937: Alec H. Reeves invents binary-coded pulse-code modulation (PCM).
1939: National Broadcasting Company (NBC) demonstrates television broadcasting.
First use of two-way radio communications using walkie-talkies.
1941: Columbia University's Radio Club opens the first regularly scheduled FM radio station.
1945: Television is born. FM is moved from its original home of 42 MHz to 50 MHz to 88 MHz to 108 MHz to make room.
1946: The American Telephone and Telegraph Company (AT&T) inaugurated the first mobile telephone system for the public called MTS (Mobile Telephone Service).
1948: John von Neumann created the first stored program electronic digital computer.
Bell Telephone Laboratories unveiled the transistor, a joint venture of scientists William Shockley, John Bardeen, and Walter Brattain.
1951: First transcontinental microwave system began operation.
1952: Sony Corporation offers a miniature transistor radio, one of the first mass-produced consumer AM/FM radios.
1953: RCA and NBC broadcast first color television transmission.
1954: The number of radio stations in the world exceeds the number of newspapers printed daily.
Texas Instruments becomes the first company to commercially produce silicon transistors.
1956: First transatlantic telephone cable systems began carrying calls.
1957: Russia launches the world's first satellite (*Sputnik*).
1958: Kilby and Noyce develop first integrated circuits.
NASA launched the United States' first satellite.
1961: FCC approves FM stereo broadcasting, which spurs the development of FM.
Citizens band (CB) radio first used.
1962: U.S. radio stations begin broadcasting stereophonic sound.
1963: T1 (transmission 1) digital carrier systems introduced.
1965: First commercial communications satellite launched.
1970: High-definition television (HDTV) introduced in Japan.
1977: First commercial use of optical fiber cables.
1983: Cellular telephone networks introduced in the United States.
1999: HDTV standards implemented in the United States.
1999: Digital television (DTV) transmission begins in the United States.

FIGURE 1-1 (Continued) Electronic communications time line

If two powers are expressed in the same units (e.g., watts or microwatts), their ratio is a dimensionless quantity that can be expressed in decibel form as follows:

$$\text{dB} = 10 \log_{(10)}\left(\frac{P_1}{P_2}\right) \tag{1-1}$$

where P_1 = power level 1 (watts)
P_2 = power level 2 (watts)

Because P_2 is in the denominator of Equation 1-1, it is the reference power, and the dB value is for power P_1 with respect to power P_2. The dB value is the difference in dB between power P_1 and P_2.

When used in electronic circuits to measure a power gain or loss, Equation 1-1 can be rewritten as

$$A_{P(\text{dB})} = 10 \log_{(10)}\left(\frac{P_{\text{out}}}{P_{\text{in}}}\right) \tag{1-2}$$

Table 1-1 Decibel Values for Absolute Power Ratios Equal to or Greater Than One (i.e., Gains)

Absolute Ratio	$\log_{(10)}$[ratio]	10 $\log_{(10)}$[ratio]
1	0	0 dB
1.26	0.1	1 dB
2	0.301	3 dB
4	0.602	6 dB
8	0.903	9 dB
10	1	10 dB
100	2	20 dB
1000	3	30 dB
10,000	4	40 dB
100,000	5	50 dB
1,000,000	6	60 dB
10,000,000	7	70 dB
100,000,000	8	80 dB

where $A_{P(\text{dB})}$ = power gain (dB)
P_{out} = output power level (watts)
P_{in} = input power level (watts)
$\dfrac{P_{out}}{P_{in}}$ = absolute power gain (unitless)

Since P_{in} is the reference power, the power gain is for P_{out} with respect to P_{in}.

An absolute power gain can be converted to a dB value by simply taking the log of it and multiplying by 10:

$$A_{P(\text{dB})} = 10 \log_{(10)} (A_P) \tag{1-3}$$

The dB does not express exact amounts like the inch, pound, or gallon, and it does not tell you how much power you have. Instead, the dB represents the ratio of the signal level at one point in a circuit to the signal level at another point in a circuit. Decibels can be positive or negative, depending on which power is larger. The sign associated with a dB value indicates which power in Equation 1-2 is greater the denominator or the numerator. A positive (+) dB value indicates that the output power is greater than the input power, which indicates a power gain, where gain simply means amplification. A negative (−) dB value indicates that the output power is less than the input power, which indicates a power loss. A power loss is sometimes called *attenuation*. If $P_{out} = P_{in}$, the absolute power gain is 1, and the dB power gain is 0 dB. This is sometimes referred to as a *unity power gain*. Examples of absolute power ratios equal to or greater than 1 (i.e., power gains) and their respective dB values are shown in Table 1-1, and examples of absolute power ratios less than 1 (i.e., power losses) and their respective dB values are shown in Table 1-2.

Although Tables 1-1 and 1-2 list absolute ratios that range from 0.00000001 to 100,000,000 (a tremendous range), the dB values span a range of only 160 dB (−80 dB to +80 dB). From Tables 1-1 and 1-2, it can be seen that the dB indicates compressed values of absolute ratios, which yield much smaller values than the original ratios. This is the essence of the decibel as a unit of measurement and what makes the dB easier to work with than absolute ratios or absolute power levels. Power ratios in a typical electronic communications system can range from millions to billions to one, and power levels can vary from megawatts at the output of a transmitter to picowatts at the input of a receiver.

Properties of exponents correspond to properties of logarithms. Because logs are exponents (and the dB is a logarithmic unit), power gains and power losses expressed in decibels

Table 1-2 Decibel Values for Absolute Power Ratios Equal to or Greater Than One (i.e., Losses)

Absolute Ratio	$\log_{(10)}$[ratio]	$10 \log_{(10)}$[ratio]
0.79	−0.1	−1 dB
0.5	−0.301	−3 dB
0.1	−1	−10 dB
0.01	−2	−20 dB
0.001	−3	−30 dB
0.0001	−4	−40 dB
0.00001	−5	−50 dB
0.000001	−6	−60 dB
0.0000001	−7	−70 dB
0.00000001	−8	−80 dB

can be added or subtracted, whereas absolute ratios would require multiplying or dividing (in mathematical terms, these are called the *product rule* and the *quotient rule*).

Example 1-1

Convert the absolute power ratio of 200 to a power gain in dB.

Solution Substituting into Equation 1-3 yields

$$\begin{aligned} A_{P(\text{dB})} &= 10 \log_{(10)}[200] \\ &= 10(2.3) \\ &= 23 \text{ dB} \end{aligned}$$

The absolute ratio can be equated to:

$$200 = 100 \times 2$$

Applying the product rule for logarithms, the power gain in dB is:

$$\begin{aligned} A_{P(\text{dB})} &= 10 \log_{(10)}[100] + 10 \log_{10}(2) \\ &= 20 \text{ dB} + 3 \text{ dB} \\ &= 23 \text{ dB} \end{aligned}$$

or

$$200 = 10 \times 10 \times 2$$

and

$$\begin{aligned} A_{P(\text{dB})} &= 10 \log_{(10)}[10] + 10 \log_{10}(10) + 10 \log_{10}(2) \\ &= 10 \text{ dB} + 10 \text{ dB} + 3 \text{ dB} \\ &= 23 \text{ dB} \end{aligned}$$

Decibels can be converted to absolute values by simply rearranging Equations 1-2 or 1-3 and solving for the power gain.

Example 1-2

Convert a power gain $A_P = 23$ dB to an absolute power ratio.

Solution Substituting into Equation 1-2 gives

$$23 \text{ dB} = 10 \log_{1(10)}\left(\frac{P_1}{P_2}\right)$$

divide both sides by 10

$$2.3 = \log_{1(10)}\left(\frac{P_1}{P_2}\right)$$

take the antilog

$$10^{2.3} = \left(\frac{P_1}{P_2}\right)$$

$$200 = \left(\frac{P_1}{P_2}\right)$$

the absolute power ratio can be approximated as

$$23 \text{ dB} = 10 \text{ dB} + 10 \text{ dB} + 3 \text{ dB} = 10 \times 10 \times 2 = 200$$

or

$$23 \text{ dB} = 20 \text{ dB} + 3 \text{ dB} = 100 \times 2 = 200$$

Power gain can be expressed in terms of a voltage ratio as

$$A_{P(\text{dB})} = 10 \log_{(10)}\left[\frac{E_o^2/R_o}{E_i^2/R_i}\right] \quad \textbf{(1-4a)}$$

where A_P = power gain (dB)
E_o = output voltage (volts)
E_i = input voltage (volts)
R_o = output resistance (ohms)
R_i = input resistance (ohms)

When the input resistance equals the output resistance ($R_o = R_i$), Equation 1-4a reduces to

$$A_{P(\text{dB})} = 10 \log_{(10)}\left(\frac{E_o^2}{E_i^2}\right) \quad \textbf{(1-4b)}$$

or

$$A_{P(\text{dB})} = 10 \log_{(10)}\left(\frac{E_o}{E_i}\right)^2 \quad \textbf{(1-4c)}$$

Applying the power rule for exponents gives

$$A_{P(\text{dB})} = 20 \log_{(10)}\left(\frac{E_o}{E_i}\right) \quad \textbf{(1-4d)}$$

where $A_{P(\text{dB})}$ = power gain (dB)
E_o = output voltage (volts)
E_i = input voltage (volts)
$\left(\frac{E_o}{E_i}\right)$ = absolute voltage gain 1(unitless)

Equation 1-4d can be used to determine power gains in dB but only when the input and output resistances are equal. However, Equation 1-4d can be used to represent the dB voltage gain of a device regardless of whether the input and output resistances are equal. Voltage gain in dB is expressed mathematically as

$$A_{v(\text{dB})} = 20 \log_{(10)}\left(\frac{E_o}{E_i}\right) \quad \textbf{(1-5)}$$

where $A_{v(\text{dB})}$ = voltage gain (dB)

A dBm is a unit of measurement used to indicate the ratio of a power level with respect to a fixed reference level. With dBm, the reference level is 1 mW (i.e., dBm means decibel relative to 1 milliwatt). One milliwatt was chosen for the reference because it equals the average power produced by a telephone transmitter. The decibel was originally used to express sound levels (acoustical power). It was later adapted to electrical units and defined as 1 mW of electrical power measured across a 600-ohm load and was intended to be used on telephone circuits for voice-frequency measurements. Today, the dBm is the measurement

Table 1-3 dBm Values for Powers Equal to or Greater Than One mW

Power (P) in Watts	$10 \log_{(10)} (P/0.001)$
0.001	0 dBm
0.002	3 dBm
0.01	10 dBm
0.1	20 dBm
1	30 dBm
10	40 dBm
100	50 dBm
1000	60 dBm
10,000	70 dBm
100,000	80 dBm

Table 1-4 dBm Values for Powers Equal to or Less Than One mW

Power (P) in Milliwatts	$10 \log_{(10)} (P/0.001)$
1	0 dBm
0.5	− 3 dBm
0.1	− 10 dBm
0.01	− 20 dBm
0.001	− 30 dBm
0.0001	− 40 dBm
0.00001	− 50 dBm
0.000001	− 60 dBm
0.0000001	− 70 dBm
0.00000001	− 80 dBm

unit of choice for virtually all electromagnetic frequency bands from ultralow frequencies to light-wave frequencies terminated in a variety of impedances, such as 50-, 75-, 600-, 900-, 124-, and 300-ohm loads.

The dBm unit is expressed mathematically as

$$\text{dBm} = 10 \log_{1(10)} \frac{P}{0.001 \text{ W}} \qquad \textbf{(1-6)}$$

where 0.001 is the reference power of 1 mW
P is any power in watts

Tables 1-3 and 1-4 list power levels in both watts and dBm for power levels above and below 1 mW, respectively. As the tables show, a power level of 1 mW equates to 0 dBm, which means that 1 mW is 0 dB above or below 1 mW. Negative dBm values indicate power levels less than 1 mW, and positive dBm values indicate power levels above 1 mW. For example, a power level of 10 dBm indicates that the power is 10 dB above 1 mW, or 10 times 1 mW, which equates to 10 mW. A power level of 0.1 mW indicates a power level that is 10 dB below 1 mW, which equates to one-tenth of 1 mW.

Example 1-3

Convert a power level of 200 mW to dBm.

Solution Substituting into Equation 1-6

$$\begin{aligned} \text{dBm} &= 10 \log_{(10)}\left(\frac{200 \text{ mW}}{1 \text{ mW}}\right) \\ &= 10 \log_{(10)}(200) \\ &= 23 \text{ dBm} \end{aligned}$$

Example 1-4

Convert a power level of 23 dBm to an absolute power.

Solution Substitute into Equation 1-6 and solve for P:

$$23 \text{ dBm} = 10 \log_{(10)}\left(\frac{P}{0.001 \text{ W}}\right)$$

$$2.3 = \log_{(10)}\left(\frac{P}{0.001 \text{ W}}\right)$$

Take the antilog:

$$10^{2.3} = \left(\frac{P}{0.001 \text{ W}}\right)$$

$$200 = \left(\frac{P}{0.001 \text{ W}}\right)$$

$$P = 200\,(0.001 \text{ W})$$

$$P = 0.2 \text{ watts or } 200 \text{ mW}$$

The dBm value can be approximated as: 23 dBm is a power level 23 dB above 0 dBm (1 mW)

because 23 dB is an absolute power ratio of 200

then 23 dBm = 200 × 1 mW

23 dBm = 200 mW

Signal power can be referenced to powers other than 1 milliwatt. For example, dBμ references signal levels to 1 microwatt, dBW references signal levels to 1 watt, and dBkW references signals to 1 kilowatt.

The decibel originated as the Bel, named in honor of Alexander Graham Bell. The Bel is expressed mathematically as

$$\text{Bel} = \log_{(10)}\left(\frac{P_{out}}{P_{in}}\right) \tag{1-7}$$

From Equation 1-7, one can see that a Bel is one-tenth of a decibel. It was soon realized that the Bel provided too much compression. For example, the Bel unit compressed absolute ratios ranging from 0.00000001 to 100,000,000 to a ridiculously low range of only 16 Bel (−8 Bel to +8 Bel). This made it difficult to relate Bel units to true magnitudes of large ratios and impossible to express small differences with any accuracy. For these reasons, the Bel was simply multiplied by 10, thus creating the decibel.

1-2-1 Power Levels, Gains, and Losses

When power levels are given in watts and power gains are given as absolute values, the output power is determined by simply multiplying the input power times the power gains.

Example 1-5

Given: A three-stage system comprised of two amplifiers and one filter. The input power $P_{in} = 0.1$ mW. The absolute power gains are $A_{P_1} = 100$, $A_{P_2} = 40$, and $A_{P_3} = 0.25$. Determine (a) the input power in dBm, (b) output power (P_{out}) in watts and dBm, (c) the dB gain of each of the three stages, and (d) the overall gain in dB.

Solution **a.** The input power in dBm is calculated by substituting into Equation 1-6:

$$P_{in(dBm)} = 10 \log_{(10)}\left(\frac{0.0001}{0.001 \text{ W}}\right)$$

$$= -10 \text{ dBm}$$

b. The output power is simply the input power multiplied by the three power gains:

$$P_{out} = (0.1\text{ mW})(100)(40)(0.25) = 100\text{ mW}$$

To convert the output power to dBm, substitute into Equation 1-6:

$$P_{out(dBm)} = 10\log\left(\frac{100\text{ mW}}{1\text{ mW}}\right)$$

$$P_{out(dBm)} = 20\text{ dBm}$$

c. Since stages one and two have gains greater than 1, they provide amplifications. Stage three has a gain less than one and therefore represents a loss to the signal. The decibel value for the three gains are determined by substituting into Equation 1-3:

$$A_{P_1(dB)} = 10\log(100) = 20\text{ dB}$$

$$A_{P_2(dB)} = 10\log(40) = 16\text{ dB}$$

$$A_{P_3(dB)} = 10\log(0.25) = -6\text{ dB}$$

d. The overall or total power gain in dB $(A_{P_T(dB)})$ can be determined by simply adding the individual dB power gains

$$(A_{P_T(dB)}) = 20\text{ dB} + 16\text{ dB} + (-6\text{ dB}) = 30\text{ dB}$$

or by taking the log of the product of the three absolute power gains and then multiplying by 10:

$$(A_{P_T(dB)}) = 10\log[(100)(40)(0.25)] = 30\text{ dB}$$

The output power in dBm is the input power in dBm plus the sum of the gains of the three stages:

$$P_{out(dBm)} = P_{in(dBm)} + A_{P_1(dB)} + A_{P_2(dB)} + A_{P_3(dB)}$$
$$= -10\text{ dBm} + 20\text{ dB} + 16\text{ dB} + (-6\text{ dB})$$
$$= 20\text{ dBm}$$

When power levels are given in dBm and power gains are given as dB values, the output power is determined by simply adding the individual gains to the input power.

Example 1-6

For a three-stage system with an input power $P_{in} = -20$ dBm and power gains of the three stages as $A_{P_1} = 13$ dB, $A_{P_2} = 16$ dB, and $A_{P_3} = -6$ dB, determine the output power (P_{out}) in dBm and watts.

Solution The output power is simply the input power in dBm plus the sum of the three power gains in dB:

$$P_{out\,(dBm)} = -20\text{ dBm} + 13\text{ dB} + 16\text{ dB} + (-6\text{dB}) = 3\text{ dBm}$$

To convert dBm to watts, substitute into Equation 1-6:

$$\text{dBm} = 10\log\left(\frac{P_{out}}{1\text{ mW}}\right)$$

Therefore,

$$3 = 10\log\left(\frac{P_{out}}{1\text{ mW}}\right)$$

$$\frac{3}{10} = \log\left(\frac{P_{out}}{1\text{ mW}}\right)$$

$$0.3 = \log\left(\frac{P_{out}}{1\text{ mW}}\right)$$

$$10^{0.3} = \left(\frac{P_{out}}{1\text{ mW}}\right)$$

$$P_{out} = (1 \text{ mW})(10^{0.3})$$
$$= 2 \text{ mW}$$

To combine two power levels given in watts, you simply add the two wattages together. For example, if a signal with a power level of 1 mW is combined with another signal with a power level of 1 mW, the total combined power is 1 mW + 1 mW = 2 mW. When powers are given in dBm, however, they cannot be combined through simple addition. For example, if a signal with a power level of 0 dBm (1 mW) is combined with another signal with a power level of 0 dBm (1 mW), the total combined power is obviously 2 mW (3 dBm). However, if the two power levels are added in dBm, the result is 0 dBm + 0 dBm = 0 dBm. When a signal is combined with another signal of equal power, the total power obviously doubles. Therefore, 0 dBm + 0 dBm must equal 3 dBm. Why? Because doubling a power equates to a 3-dB increase in power, and 0 dBm + 3 dB = 3 dBm.

To combine two or more power levels given in dBm, the dBm units must be converted to watts, added together, and then converted back to dBm units. Table 1-5 shows a table that can be used to combine two power levels directly when they are given in dBm. The combining term is added to the higher of the two power levels to determine the total combined power level. As shown in the table, the closer the two power levels are to each other, the higher the combining term.

Table 1-5 Combining Powers in dBm

Difference between the Two dBm Quantities	Combining Term (dB)
0–0.1	+ 3
0.2–0.3	+ 2.9
0.4–0.5	+ 2.8
0.6–0.7	+ 2.7
0.8–0.9	+ 2.6
1.0–1.2	+ 2.5
1.3–1.4	+ 2.4
1.5–1.6	+ 2.3
1.7–1.9	+ 2.2
2.0–2.1	+ 2.1
2.2–2.4	+ 2.0
2.5–2.7	+ 1.9
2.8–3.0	+ 1.8
3.1–3.3	+ 1.7
3.4–3.6	+ 1.6
3.7–4.0	+ 1.5
4.1–4.3	+ 1.4
4.4–4.7	+ 1.3
4.8–5.1	+ 1.2
5.2–5.6	+ 1.1
5.7–6.1	+ 1.0
6.2–6.6	+ 0.9
6.7–7.2	+ 0.8
7.3–7.9	+ 0.7
8.0–8.6	+ 0.6
8.7–9.6	+ 0.5
9.7–10.7	+ 0.4
10.8–12.2	+ 0.3
12.3–14.5	+ 0.2
14.6–19.3	+ 0.1
19.4 and up	+ 0.0

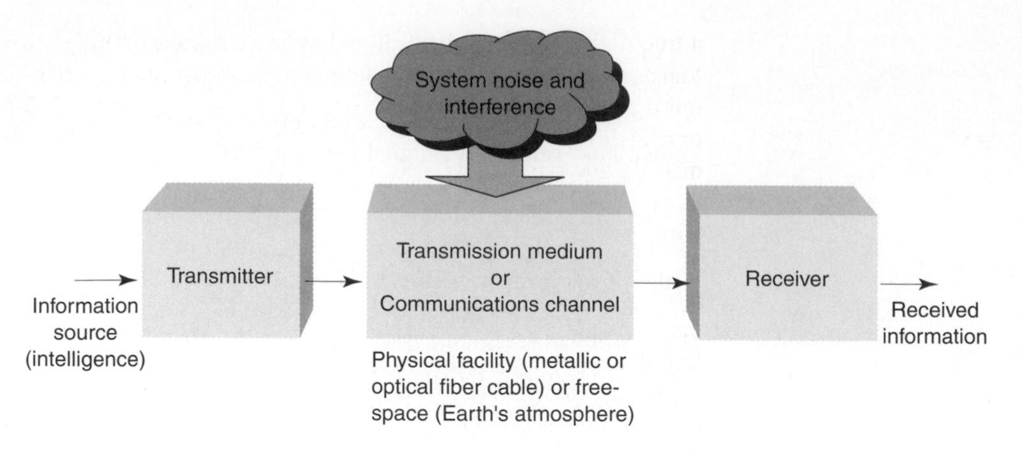

FIGURE 1-2 Simplified block diagram of an electronic communications system

Example 1-7

Determine the total power when a signal with a power level of 20 dBm is combined with a second signal with a power level of 21 dBm.

Solution The dB difference in the two power levels is 1 dB. Therefore, from Table 1-5, the combining term is 2.5 dB and the total power is

$$21 \text{ dBm} + 2.5 \text{ dB} = 23.5 \text{ dBm}$$

1-3 ELECTRONIC COMMUNICATIONS SYSTEMS

Figure 1-2 shows a simplified block diagram of an electronic communications system that includes a transmitter, a transmission medium, a receiver, and system noise. A transmitter is a collection of one or more electronic devices or circuits that converts the original source information to a form more suitable for transmission over a particular transmission medium. The transmission medium or communications channel provides a means of transporting signals between a transmitter and a receiver and can be as simple as a pair of copper wires or as complex as sophisticated microwave, satellite, or optical fiber communications systems. System noise is any unwanted electrical signals that interfere with the information signal. A receiver is a collection of electronic devices and circuits that accepts the transmitted signals from the transmission medium and then converts those signals back to their original form.

1-4 MODULATION AND DEMODULATION

Because it is often impractical to propagate information signals over standard transmission media, it is often necessary to modulate the source information onto a higher-frequency analog signal called a *carrier.* In essence, the carrier signal carries the information through the system. The information signal *modulates* the carrier by changing either its amplitude, frequency, or phase. *Modulation* is simply the process of changing one or more properties of the analog carrier in proportion with the information signal.

The two basic types of electronic communications systems are *analog* and *digital.* An *analog communications system* is a system in which energy is transmitted and received in analog form (a continuously varying signal such as a sine wave). With analog communications systems, both the information and the carrier are analog signals.

The term *digital communications,* however, covers a broad range of communications techniques, including *digital transmission* and *digital radio.* Digital transmission is a true digital system where digital pulses (discrete levels such as +5 V and ground) are transferred between two or more points in a communications system. With digital transmission, there is no analog carrier, and the original source information may be in digital or analog form. If it is in analog form, it must be converted to digital pulses prior to transmission and converted back to analog form at the receive end. Digital transmission systems require a physical facility between the transmitter and receiver, such as a metallic wire or an optical fiber cable.

Digital radio is the transmittal of digitally modulated analog carriers between two or more points in a communications system. With digital radio, the modulating signal and the demodulated signal are digital pulses. The digital pulses could originate from a digital transmission system, from a digital source such as a computer, or be a binary-encoded analog signal. In digital radio systems, digital pulses modulate an analog carrier. Therefore, the transmission medium may be a physical facility or free space (i.e., the Earth's atmosphere). Analog communications systems were the first to be developed; however, in recent years digital communications systems have become more popular.

Equation 1-8 is the general expression for a time-varying sine wave of voltage such as a high-frequency carrier signal. If the information signal is analog and the amplitude (V) of the carrier is varied proportional to the information signal, *amplitude modulation* (AM) is produced. If the frequency (f) is varied proportional to the information signal, *frequency modulation* (FM) is produced, and, if the phase (θ) is varied proportional to the information signal, *phase modulation* (PM) is produced.

If the information signal is digital and the amplitude (V) of the carrier is varied proportional to the information signal, a digitally modulated signal known as *amplitude shift keying* (ASK) is produced. If the frequency (f) is varied proportional to the information signal, *frequency shift keying* (FSK) is produced, and, if the phase (θ) is varied proportional to the information signal, *phase shift keying* (PSK) is produced. If both the amplitude and the phase are varied proportional to the information signal, *quadrature amplitude modulation* (QAM) results. ASK, FSK, PSK, and QAM are forms of digital modulation and are described in detail in Chapter 9.

$$v(t) = V\sin(2\pi ft + \theta), \tag{1-8}$$

where $v(t)$ = time-varying sine wave of voltage
V = peak amplitude (volts)
f = frequency (hertz)
θ = phase shift (radians).

A summary of the various modulation techniques is shown here:

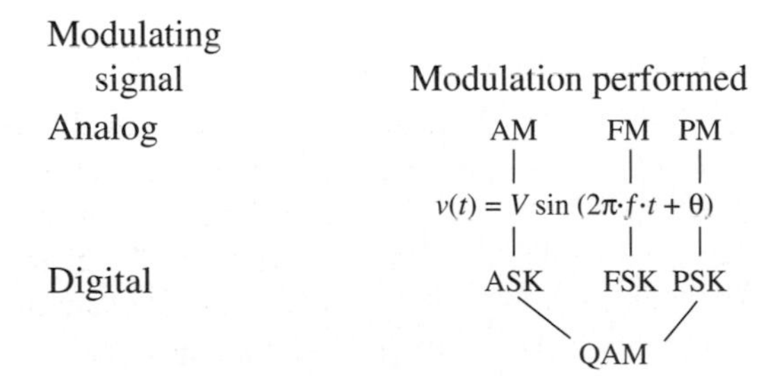

Modulation is performed in a transmitter by a circuit called a *modulator.* A carrier that has been acted on by an information signal is called a *modulated wave* or *modulated signal. Demodulation* is the reverse process of modulation and converts the modulated carrier back to the original information (i.e., removes the information from the carrier). Demodulation is performed in a receiver by a circuit called a *demodulator.*

There are two reasons why modulation is necessary in electronic communications: (1) It is extremely difficult to radiate low-frequency signals from an antenna in the form of electromagnetic energy, and (2) information signals often occupy the same frequency band and, if signals from two or more sources are transmitted at the same time, they would interfere with each other. For example, all commercial FM stations broadcast voice and music signals that occupy the audio-frequency band from approximately 300 Hz to 15 kHz. To avoid interfering with each other, each station converts its information to a different frequency band or channel. The term *channel* is often used to refer to a specific band of frequencies allocated a particular service. A standard voice-band channel occupies approximately a 3-kHz bandwidth and is used for transmission of voice-quality signals; commercial AM broadcast channels occupy approximately a 10-kHz frequency band, and 30 MHz or more of bandwidth is required for microwave and satellite radio channels.

Figure 1-3 is the simplified block diagram for an analog electronic communications system showing the relationship among the modulating signal, the high-frequency carrier, and the modulated wave. The information signal (sometimes called the intelligence signal) combines with the carrier in the modulator to produce the modulated wave. The information can be in analog or digital form, and the modulator can perform either analog or digital modulation. Information signals are *up-converted* from low frequencies to high frequencies in the transmitter and *down-converted* from high frequencies to low frequencies in the receiver. The process of converting a frequency or band of frequencies to another location in the total frequency spectrum is called *frequency translation.* Frequency translation is an intricate part of electronic communications because information signals may be up- and down-converted many times as they are transported through the system called a channel. The modulated signal is transported to the receiver over a transmission system. In the receiver, the modulated signal is amplified, down-converted in frequency, and then demodulated to reproduce the original source information.

1-5 THE ELECTROMAGNETIC FREQUENCY SPECTRUM

The purpose of an electronic communications system is to communicate information between two or more locations commonly called *stations.* This is accomplished by converting the original information into electromagnetic energy and then transmitting it to one or more receive stations where it is converted back to its original form. Electromagnetic energy can propagate as a voltage or current along a metallic wire, as emitted radio waves through free space, or as light waves down an optical fiber. Electromagnetic energy is distributed throughout an almost infinite range of frequencies.

Frequency is simply the number of times a periodic motion, such as a sine wave of voltage or current, occurs in a given period of time. Each complete alternation of the waveform is called a *cycle.* The basic unit of frequency is hertz (Hz), and one hertz equals one cycle per second (1 Hz = 1 cps). In electronics it is common to use metric prefixes to represent higher frequencies. For example, kHz (kilohertz) is used for thousands of hertz, and MHz (megahertz) is used for millions of hertz.

1-5-1 Transmission Frequencies

The total electromagnetic frequency spectrum showing the approximate locations of various services is shown in Figure 1-4. The useful electromagnetic frequency spectrum extends from approximately 10 kHz to several billions of hertz. The lowest frequencies are used only for special applications, such as communicating in water.

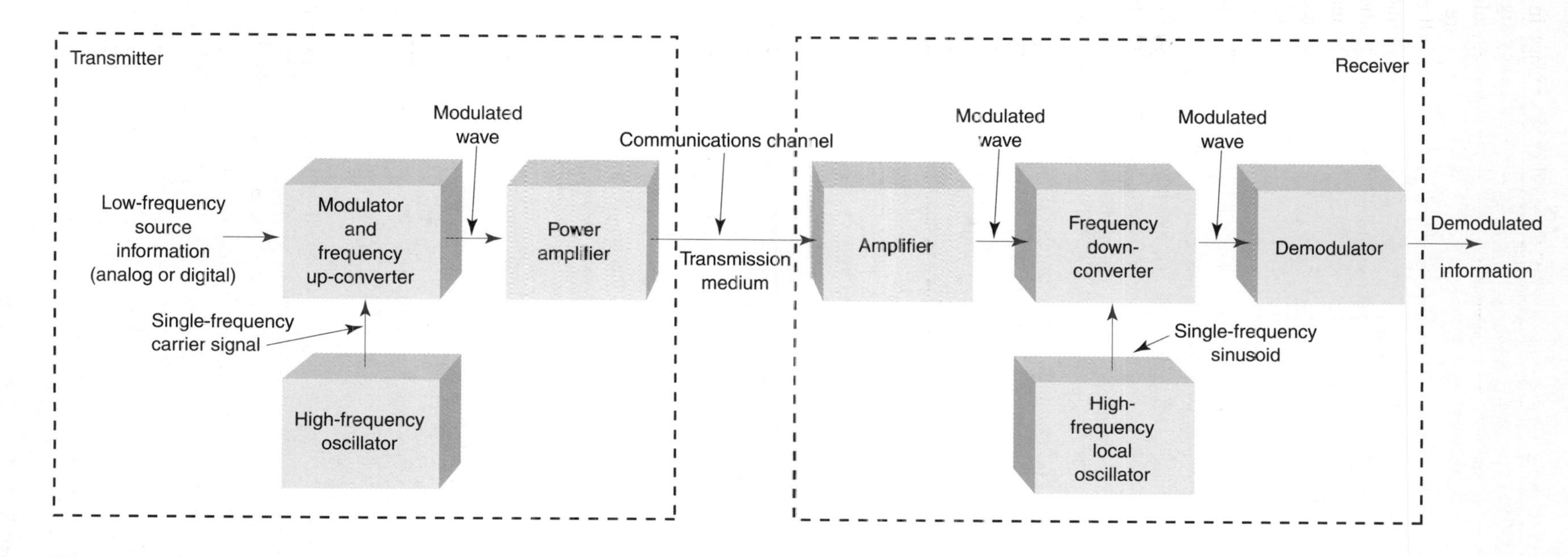

FIGURE 1-3 Simplified block diagram of an analog communications system

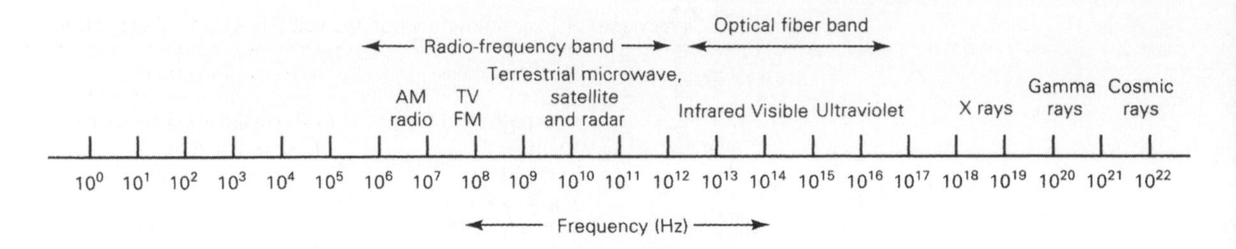

FIGURE 1-4 Electromagnetic frequency spectrum

The electromagnetic frequency spectrum is divided into *subsections,* or *bands,* with each band having a different name and boundary. The *International Telecommunications Union* (ITU) is an international agency in control of allocating frequencies and services within the overall frequency spectrum. In the United States, the *Federal Communications Commission* (FCC) assigns frequencies and communications services for free-space radio propagation. For example, the commercial FM broadcast band has been assigned the 88-MHz to 108-MHz band. The exact frequencies assigned a specific transmitter operating in the various classes of services are constantly being updated and altered to meet the world's communications needs.

The total usable *radio-frequency* (RF) spectrum is divided into narrower frequency bands, which are given descriptive names and band numbers, and several of these bands are further broken down into various types of services. The ITU's band designations are listed in Table 1-6. The ITU band designations are summarized as follows:

Extremely low frequencies. Extremely low frequencies (ELFs) are signals in the 30-Hz to 300-Hz range and include ac power distribution signals (60 Hz) and low-frequency telemetry signals.

Voice frequencies. Voice frequencies (VFs) are signals in the 300-Hz to 3000-Hz range and include frequencies generally associated with human speech. Standard telephone channels have a 300-Hz to 3000-Hz bandwidth and are often called *voice-frequency* or *voice-band channels.*

Very low frequencies. Very low frequencies (VLFs) are signals in the 3-kHz to 30-kHz range, which include the upper end of the human hearing range. VLFs are used for some specialized government and military systems, such as submarine communications.

Low frequencies. Low frequencies (LFs) are signals in the 30-kHz to 300-kHz range and are used primarily for marine and aeronautical navigation.

Medium frequencies. Medium frequencies (MFs) are signals in the 300-kHz to 3-MHz range and are used primarily for commercial AM radio broadcasting (535 kHz to 1605 kHz).

High frequencies. High frequencies (HFs) are signals in the 3-MHz to 30-MHz range and are often referred to as *short waves.* Most two-way radio communications use this range, and Voice of America and Radio Free Europe broadcast within the HF band. Amateur radio and citizens band (CB) radio also use signals in the HF range.

Very high frequencies. Very high frequencies (VHFs) are signals in the 30-MHz to 300-MHz range and are used for mobile radio, marine and aeronautical communications, commercial FM broadcasting (88 MHz to 108 MHz), and commercial television broadcasting of channels 2 to 13 (54 MHz to 216 MHz).

Table 1-6 International Telecommunications Union (ITU) Band Designations

Band Number	Frequency Range[a]	Designations
2	30 Hz–300 Hz	ELF (extremely low frequencies)
3	0.3 kHz–3 kHz	VF (voice frequencies)
4	3 kHz–30 kHz	VLF (very low frequencies)
5	30 kHz–300 kHz	LF (low frequencies)
6	0.3 MHz–3 MHz	MF (medium frequencies)
7	3 MHz–30 MHz	HF (high frequencies)
8	30 MHz–300 MHz	VHF (very high frequencies)
9	300 MHz–3 GHz	UHF (ultrahigh frequencies)
10	3 GHz–30 GHz	SHF (superhigh frequencies)
11	30 GHz–300 GHz	EHF (extremely high frequencies)
12	0.3 THz–3 THz	Infrared light
13	3 THz–30 THz	Infrared light
14	30 THz–300 THz	Infrared light
15	0.3 PHz–3 PHz	Visible light
16	3 PHz–30 PHz	Ultraviolet light
17	30 PHz–300 PHz	X rays
18	0.3 EHz–3 EHz	Gamma rays
19	3 EHz–30 EHz	Cosmic rays

[a] 10^0, hertz (Hz); 10^3, kilohertz (kHz); 10^6, megahertz (MHz); 10^9, gigahertz (GHz); 10^{12}, terahertz (THz); 10^{15}, petahertz (PHz); 10^{18}, exahertz (EHz).

Ultrahigh frequencies. Ultrahigh frequencies (UHFs) are signals in the 300-MHz to 3-GHz range and are used by commercial television broadcasting of channels 14 to 83, land mobile communications services, cellular telephones, certain radar and navigation systems, and microwave and satellite radio systems. Generally, frequencies above 1 GHz are considered microwave frequencies, which includes the upper end of the UHF range.

Superhigh frequencies. Superhigh frequencies (SHFs) are signals in the 3-GHz to 30-GHz range and include the majority of the frequencies used for microwave and satellite radio communications systems.

Extremely high frequencies. Extremely high frequencies (EHFs) are signals in the 30-GHz to 300-GHz range and are seldom used for radio communications except in very sophisticated, expensive, and specialized applications.

Infrared. Infrared frequencies are signals in the 0.3-THz to 300-THz range and are not generally referred to as radio waves. Infrared refers to electromagnetic radiation generally associated with heat. Infrared signals are used in heat-seeking guidance systems, electronic photography, and astronomy.

Visible light. Visible light includes electromagnetic frequencies that fall within the visible range of humans (0.3 PHz to 3 PHz). Light-wave communications is used with optical fiber systems, which in recent years have become a primary transmission medium for electronic communications systems.

Ultraviolet rays, X rays, gamma rays, and *cosmic rays* have little application to electronic communications and, therefore, will not be described.

When dealing with radio waves, it is common to use the units of wavelength rather than frequency. Wavelength is the length that one cycle of an electromagnetic wave occupies in space (i.e., the distance between similar points in a repetitive wave). Wavelength is inversely proportional to the frequency of the wave and directly proportional to the velocity of propagation (the velocity of propagation of electromagnetic energy in free space is assumed to be the speed of light, 3×10^8 m/s). The relationship among frequency, velocity, and wavelength is expressed mathematically as

$$\text{wavelength} = \frac{\text{velocity}}{\text{frequency}}$$

$$\lambda = \frac{c}{f} \quad \textbf{(1-9)}$$

where λ = wavelength (meters per cycle)
c = velocity of light (300,000,000 meters per second)
f = frequency (hertz)

The total *electromagnetic wavelength* spectrum showing the various services within the band is shown in Figure 1-5.

Example 1-8

Determine the wavelength in meters for the following frequencies: 1 kHz, 100 kHz, and 10 MHz.

Solution Substituting into Equation 1-9,

$$\lambda = \frac{300{,}000{,}000}{1000} = 300{,}000 \text{ m}$$

$$\lambda = \frac{300{,}000{,}000}{100{,}000} = 3000 \text{ m}$$

$$\lambda = \frac{300{,}000{,}000}{10{,}000{,}000} = 30 \text{ m}$$

Equation 1-9 can be used to determine the wavelength in inches:

$$\lambda = \frac{c}{f} \quad \textbf{(1-10)}$$

where λ = wavelength (inches per cycle)
c = velocity of light (11.8×10^9 inches per second)
f = frequency (hertz)

1-5-2 Classification of Transmitters

For licensing purposes in the United States, radio transmitters are classified according to their bandwidth, modulation scheme, and type of information. The *emission classifications* are identified by a three-symbol code containing a combination of letters and numbers as

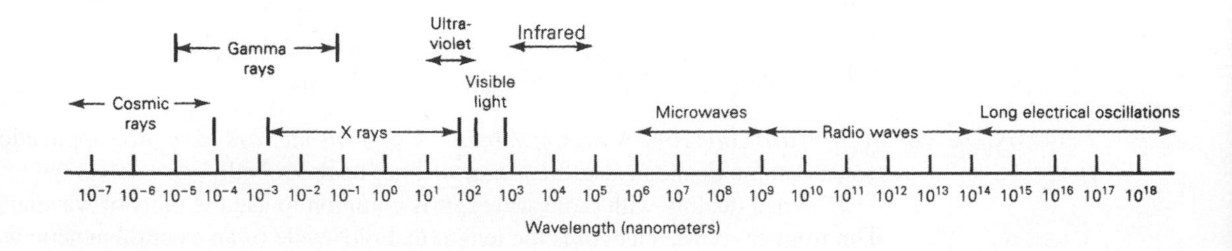

FIGURE 1-5 Electromagnetic wavelength spectrum

Table 1-7 Federal Communications Commission (FCC) Emission Classifications

Symbol	Letter	Type of Modulation
First	Unmodulated	
	N	Unmodulated carrier
	Amplitude modulation	
	A	Double-sideband, full carrier (DSBFC)
	B	Independent sideband, full carrier (ISBFC)
	C	Vestigial sideband, full carrier (VSB)
	H	Single-sideband, full carrier (SSBFC)
	J	Single-sideband, suppressed carrier (SSBSC)
	R	Single-sideband, reduced carrier (SSBRC)
	Angle modulation	
	F	Frequency modulation (direct FM)
	G	Phase modulation (indirect FM)
	D	AM and FM simultaneously or sequenced
	Pulse modulation	
	K	Pulse-amplitude modulation (PAM)
	L	Pulse-width modulation (PWM)
	M	Pulse-position modulation (PPM)
	P	Unmodulated pulses (binary data)
	Q	Angle modulated during pulses
	V	Any combination of pulse-modulation category
	W	Any combination of two or more of the above forms of modulation
	X	Cases not otherwise covered
Second	0	No modulating signal
	1	Digitally keyed carrier
	2	Digitally keyed tone
	3	Analog (sound or video)
	7	Two or more digital channels
	8	Two or more analog channels
	9	Analog and digital
Third	A	Telegraphy, manual
	B	Telegraphy, automatic (teletype)
	C	Facsimile
	D	Data, telemetry
	E	Telephony (sound broadcasting)
	F	Television (video broadcasting)
	N	No information transmitted
	W	Any combination of second letter

shown in Table 1-7. The first symbol is a letter that designates the type of modulation of the main carrier. The second symbol is a number that identifies the type of emission, and the third symbol is another letter that describes the type of information being transmitted. For example, the designation A3E describes a double-sideband, full-carrier, amplitude-modulated signal carrying voice or music telephony information.

1-6 BANDWIDTH AND INFORMATION CAPACITY

1-6-1 Bandwidth

The two most significant limitations on the performance of a communications system are *noise* and *bandwidth.* Noise is discussed later in this chapter. The bandwidth of an information signal is simply the difference between the highest and lowest frequencies contained in the information, and the bandwidth of a communications channel is the difference between the highest and lowest frequencies that the channel will allow to pass through it (i.e., its *passband*). The bandwidth of a communications channel must be large (wide) enough to pass all significant information frequencies. In other words, the bandwidth of the communications channel must be equal to or greater than the bandwidth of the information. For example, voice frequencies contain signals between 300 Hz and 3000 Hz. Therefore, a voice-frequency channel must have a bandwidth equal to or greater than 2700 Hz (300 Hz–3000 Hz). If a cable television transmission system has a passband from 500 kHz to 5000 kHz, it has a bandwidth of 4500 kHz. As a general rule, a communications channel cannot propagate a signal that contains a frequency that is changing at a rate greater than the bandwidth of the channel.

1-6-2 Information Capacity

Information theory is a highly theoretical study of the efficient use of bandwidth to propagate information through electronic communications systems. Information theory can be used to determine the *information capacity* of a data communications system. Information capacity is a measure of how much information can be propagated through a communications system and is a function of bandwidth and transmission time.

Information capacity represents the number of independent symbols that can be carried through a system in a given unit of time. The most basic digital symbol used to represent information is the *binary digit* or *bit.* Therefore, it is often convenient to express the information capacity of a system as a *bit rate.* Bit rate is simply the number of bits transmitted during one second and is expressed in *bits per second* (bps).

In 1928, R. Hartley of Bell Telephone Laboratories developed a useful relationship among bandwidth, transmission time, and information capacity. Simply stated, Hartley's law is

$$I \propto B \times t \qquad \textbf{(1-11)}$$

where I = information capacity (bits per second)
B = bandwidth (hertz)
t = transmission time (seconds)

From Equation 1-11, it can be seen that information capacity is a linear function of bandwidth and transmission time and is directly proportional to both. If either the bandwidth or the transmission time changes, a directly proportional change occurs in the information capacity.

In 1948, mathematician Claude E. Shannon (also of Bell Telephone Laboratories) published a paper in the *Bell System Technical Journal* relating the information capacity of a communications channel to bandwidth and *signal-to-noise ratio.* The higher the signal-to-noise ratio, the better the performance and the higher the information capacity. Mathematically stated, *the Shannon limit for information capacity* is

$$I = B \log_2\left(1 + \frac{S}{N}\right) \qquad \textbf{(1-12a)}$$

or

$$I = 3.32\,B \log_{10}\left(1 + \frac{S}{N}\right) \qquad \textbf{(1-12b)}$$

where I = information capacity (bps)
B = bandwidth (hertz)
$\frac{S}{N}$ = signal-to-noise power ratio (unitless)

Example 1-9

For a standard telephone circuit with a signal-to-noise power ratio of 1000 (30 dB) and a bandwidth of 2.7 kHz, determine the Shannon limit for information capacity.

Solution The Shannon limit for information capacity is determined by substituting into Equation 1-12b:

$$I = (3.32)(2700) \log_{10} (1 + 1000)$$
$$= 26.9 \text{ kbps}$$

Shannon's formula is often misunderstood. The results of the preceding example indicate that 26.9 kbps can be propagated through a 2.7-kHz communications channel. This may be true, but it cannot be done with a binary system. To achieve an information transmission rate of 26.9 kbps through a 2.7-kHz channel, each symbol transmitted must contain more than one bit.

1-7 NOISE ANALYSIS

Electrical noise is defined as any undesirable electrical energy that falls within the passband of the signal. For example, in audio recording, any unwanted electrical signals that fall within the audio frequency band of 0 Hz to 15 kHz will interfere with the music and therefore be considered noise. Figure 1-6 shows the effect that noise has on an electrical signal. Figure 1-6a shows a sine wave without noise, and Figure 1-6b shows the same signal except in the presence of noise. The grassy-looking squiggles superimposed on the sine wave in Figure 1-6b are electrical noise, which contains a multitude of frequencies and amplitudes that can interfere with the quality of the signal.

Noise can be divided into two general categories: *correlated* and *uncorrelated.* Correlation implies a relationship between the signal and the noise. Therefore, correlated noise exists only when a signal is present. Uncorrelated noise, on the other hand, is present all the time whether there is a signal or not.

1-7-1 Uncorrelated Noise

Uncorrelated noise is present regardless of whether there is a signal present or not. Uncorrelated noise can be further subdivided into two general categories: external and internal.

External noise. *External noise* is noise that is generated outside the device or circuit. The three primary sources of external noise are atmospheric, extraterrestrial, and man-made.

Atmospheric noise. Atmospheric noise is naturally occurring electrical disturbances that originate within Earth's atmosphere. Atmospheric noise is commonly called

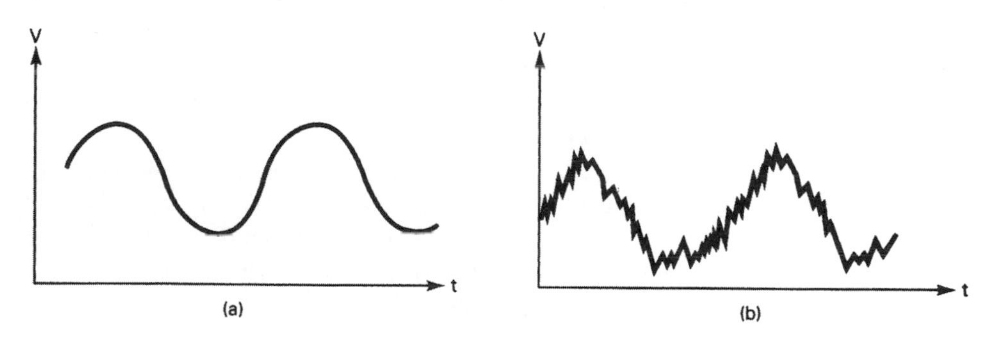

FIGURE 1-6 Effects of noise on a signal: (a) signal without noise; (b) signal with noise

static electricity and is the familiar sputtering, crackling, and so on often heard from a speaker when there is no signal present. The source of most static electricity is naturally occurring electrical conditions, such as lightning. Static electricity is often in the form of impulses that spread energy throughout a wide range of frequencies. The magnitude of this energy, however, is inversely proportional to its frequency. Consequently, at frequencies above 30 MHz or so, atmospheric noise is relatively insignificant.

Extraterrestrial noise. Extraterrestrial noise consists of electrical signals that originate from outside Earth's atmosphere and is therefore sometimes called *deep-space noise.* Extraterrestrial noise originates from the Milky Way, other galaxies, and the sun. Extraterrestrial noise is subdivided into two categories: solar and cosmic.

Solar noise is generated directly from the sun's heat. There are two parts to solar noise: a *quiet* condition, when a relatively constant radiation intensity exists, and *high intensity,* sporadic disturbances caused by *sunspot* activity and *solar flare-ups.* The magnitude of the sporadic noise caused by sunspot activity follows a cyclic pattern that repeats every 11 years.

Cosmic noise sources are continuously distributed throughout the galaxies. Because the sources of galactic noise are located much farther away than our sun, their noise intensity is relatively small. Cosmic noise is often called *black-body noise* and is distributed fairly evenly throughout the sky.

Man-made noise. Man-made noise is simply noise that is produced by mankind. The predominant sources of man-made noise are spark-producing mechanisms, such as commutators in electric motors, automobile ignition systems, ac power-generating and switching equipment, and fluorescent lights. Man-made noise is impulsive in nature and contains a wide range of frequencies that are propagated through space in the same manner as radio waves. Man-made noise is most intense in the more densely populated metropolitan and industrial areas and is therefore sometimes called *industrial noise.*

Internal noise. *Internal noise* is electrical interference generated within a device or circuit. There are three primary kinds of internally generated noise: shot, transit time, and thermal.

Shot noise. Shot noise is caused by the random arrival of carriers (holes and electrons) at the output element of an electronic device, such as a diode, field-effect transistor, or bipolar transistor. Shot noise was first observed in the anode current of a vacuum-tube amplifier and was described mathematically by W. Schottky in 1918. The current carriers (for both ac and dc) are not moving in a continuous, steady flow, as the distance they travel varies because of their random paths of motion. Shot noise is randomly varying and is superimposed onto any signal present. When amplified, shot noise sounds similar to metal pellets falling on a tin roof. Shot noise is sometimes called *transistor noise* and is additive with thermal noise.

Transit-time noise. Any modification to a stream of carriers as they pass from the input to the output of a device (such as from the emitter to the collector of a transistor) produces an irregular, random variation categorized as *transit-time noise.* When the time it takes for a carrier to propagate through a device is an appreciable part of the time of one cycle of the signal, the noise becomes noticeable. Transit-time noise in transistors is determined by carrier mobility, bias voltage, and transistor construction.

Carriers traveling from emitter to collector suffer from emitter-time delays, base transit-time delays, and collector recombination-time and propagation-time delays. If transit delays are excessive at high frequencies, the device may add more noise than amplification to the signal.

Thermal noise. Thermal noise is associated with the rapid and random movement of electrons within a conductor due to thermal agitation. The English botanist Robert Brown first noted this random movement. Brown first observed evidence for the moving-particle nature of matter in pollen grains and later noticed that the same phenomenon occurred with smoke particles. J. B. Johnson of Bell Telephone Laboratories first recognized random movement of electrons in 1927. Electrons within a conductor carry a unit negative charge, and the mean-square velocity of an electron is proportional to the absolute temperature. Consequently, each flight of an electron between collisions with molecules constitutes a short pulse of current that develops a small voltage across the resistive component of the conductor. Because this type of electron movement is totally random and in all directions, it is sometimes called random noise. With random noise, the average voltage in the substance due to electron movement is 0 V dc. However, such a random movement does produce an ac component.

Thermal noise is present in all electronic components and communications systems. Because thermal noise is uniformly distributed across the entire electromagnetic frequency spectrum, it is often referred to as *white noise* (analogous to the color white containing all colors, or frequencies, of light). Thermal noise is a form of additive noise, meaning that it cannot be eliminated, and it increases in intensity with the number of devices in a circuit and with circuit length. Therefore, thermal noise sets the upper bound on the performance of a communications system.

The ac component produced from thermal agitation has several names, including *thermal noise,* because it is temperature dependent; *Brownian noise,* after its discoverer; *Johnson noise,* after the man who related Brownian particle movement of electron movement; and *white noise* because the random movement is at all frequencies. Hence, thermal noise is the random motion of free electrons within a conductor caused by thermal agitation.

Johnson proved that thermal noise power is proportional to the product of bandwidth and temperature. Mathematically, noise power is

$$N = KTB \tag{1-13}$$

where N = noise power (watts)
B = bandwidth (hertz)
K = Boltzmann's proportionality constant (1.38×10^{-23} joules per kelvin)
T = absolute temperature (kelvin) (room temperature = 17°C, or 290 K)

To convert °C to kelvin, simply add 273°; thus, T = °C + 273°.

Example 1-10

Convert the following temperatures to kelvin: 100°C, 0°C, and −10°C

Solution The formula T = °C + 273° is used to convert °C to kelvin.

$$T = 100°\text{C} + 273° = 373\text{ K}$$
$$T = 0°\text{C} + 273° = 273\text{ K}$$
$$T = -10°\text{C} + 273° = 263\text{ K}$$

Noise power stated in dBm is a logarithmic function and equal to

$$N_{(\text{dBm})} = 10\log\frac{KTB}{0.001} \tag{1-14}$$

Equations 1-13 and 1-14 show that at absolute zero (0 K, or $-273°$ C), there is no random molecular movement, and the product KTB equals zero.

Rearranging Equation 1-14 gives

$$N_{(\mathrm{dBm})} = 10 \log \frac{KT}{0.001} + 10 \log B \qquad \textbf{(1-15)}$$

and for a 1-Hz bandwidth at room temperature,

$$N_{(\mathrm{dBm})} = 10 \log \frac{(1.38 \times 10^{-23})(290)}{0.001} + 10 \log 1$$

$$= -174 \text{ dBm}$$

Thus, at room temperature, Equation 1-14 can be rewritten for any bandwidth as

$$N_{(\mathrm{dBm})} = -174 \text{ dBm} + 10 \log B \qquad \textbf{(1-16)}$$

Random noise results in a constant power density versus frequency, and Equation 1-13 indicates that the available power from a thermal noise source is proportional to bandwidth over any range of frequencies. This has been found to be true for frequencies from 0 Hz to the highest microwave frequencies used today. Thus, if the bandwidth is unlimited, it appears that the available power from a thermal noise source is also unlimited. This, of course, is not true, as it can be shown that at arbitrarily high frequencies thermal noise power eventually drops to zero. Because thermal noise is equally distributed throughout the frequency spectrum, a thermal noise source is sometimes called a *white noise source,* which is analogous to white light, which contains all visible-light frequencies. Therefore, the rms noise power measured at any frequency from a white noise source is equal to the rms noise power measured at any other frequency from the same noise source. Similarly, the total rms noise power measured in any fixed bandwidth is equal to the total rms noise power measured in an equal bandwidth anywhere else in the total noise spectrum. In other words, the rms white noise power present in the band from 1000 Hz to 2000 Hz is equal to the rms white noise power present in the band from 1,001,000 Hz to 1,002,000 Hz.

Thermal noise is random and continuous and occurs at all frequencies. Also, thermal noise is predictable, additive, and present in all devices. This is why thermal noise is the most significant of all noise sources.

1-7-2 Noise Voltage

Figure 1-7 shows the equivalent circuit for a thermal noise source where the internal resistance of the source (R_I) is in series with the rms noise voltage (V_N). For the worst-case condition and maximum transfer of noise power, the load resistance (R) is made equal to R_I. Thus, the noise voltage dropped across R is equal to half the noise source ($V_R = V_N/2$),

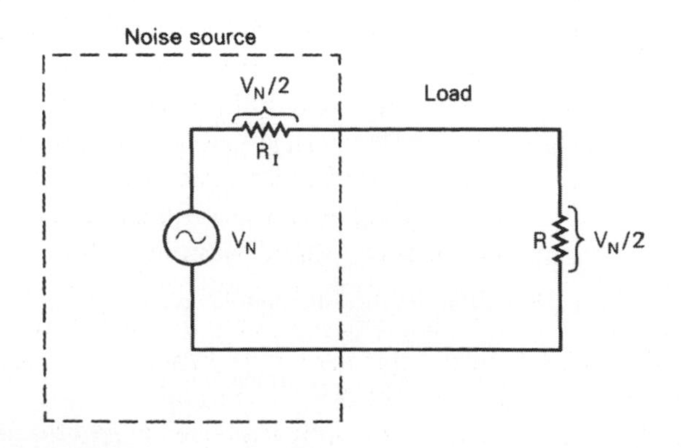

FIGURE 1-7 Noise source equivalent circuit

and from Equation 1-13 the noise power (N) developed across the load resistor is equal to KTB. The mathematical expression for V_N is derived as follows:

$$N = KTB = \frac{(V_N/2)^2}{R} = \frac{V_N^2}{4R}$$

Thus,

$$V_N^2 = 4RKTB$$

and

$$V_N = \sqrt{4RKTB} \tag{1-17}$$

Example 1-11

For an electronic device operating at a temperature of 17°C with a bandwidth of 10 kHz, determine

a. thermal noise power in watts and dBm

b. rms noise voltage for a 100-Ω internal resistance and a 100-Ω load resistance

Solution **a.** The thermal noise power is found by substituting into Equation 1-13 (N = KTB).

$$T\,(\text{kelvin}) = 17^\circ\text{C} + 273 = 290\text{ K}$$
$$N = (1.38 \times 10^{-23})(290)(1 \times 10^4) = 4 \times 10^{-17}\text{ W}$$

Substituting, Equation 1-14 gives the noise power in dBm:

$$N_{(\text{dBm})} = 10 \log \frac{[4 \times 10^{-17}]}{0.001} = -134\text{ dBm}$$

Or substitute into Equation 1-16:

$$\begin{aligned} N_{(\text{dBm})} &= -174\text{ dBm} + 10 \log 10{,}000 \\ &= -174\text{ dBm} + 40\text{ dB} \\ &= -134\text{ dBm} \end{aligned}$$

b. The rms noise voltage is found by substituting into Equation 1-17:

$$\begin{aligned} V_N &= \sqrt{4RKTB} \quad \text{where } KTB = 4 \times 10^{-17} \\ &= \sqrt{(4)(100)(4 \times 10^{-17})} = 0.1265\ \mu V \end{aligned}$$

1-7-3 Correlated Noise

Correlated noise is a form of internal noise that is correlated (mutually related) to the signal and cannot be present in a circuit unless there is a signal—simply stated, *no signal, no noise!* Correlated noise is produced by *nonlinear amplification* and includes *harmonic* and *intermodulation distortion,* both of which are forms of *nonlinear distortion.* All circuits are nonlinear to some extent; therefore, they all produce nonlinear distortion. Nonlinear distortion creates unwanted frequencies that interfere with the signal and degrade performance.

Harmonic distortion occurs when unwanted *harmonics* of a signal are produced through nonlinear amplification (*nonlinear mixing*). Harmonics are integer multiples of the original signal. The original signal is the *first harmonic* and is called the *fundamental frequency.* A frequency two times the original signal frequency is the *second harmonic,* three times is the *third harmonic,* and so forth. *Amplitude distortion* is another name for harmonic distortion.

There are various degrees of harmonic distortion. Second-order harmonic distortion is the ratio of the rms amplitude of the second harmonic to the rms amplitude of the fundamental. Third-order harmonic distortion is the ratio of the rms amplitude of the third harmonic to the rms value of the fundamental. A more meaningful measurement is total harmonic distortion (TDH), which is the ratio of the quadratic sum of the rms values of all the higher harmonics to the rms value of the fundamental.

Figure 1-8a show the input and output frequency spectrums for a nonlinear device with a single input frequency (f_1). As the figure shows, the output spectrum contains the

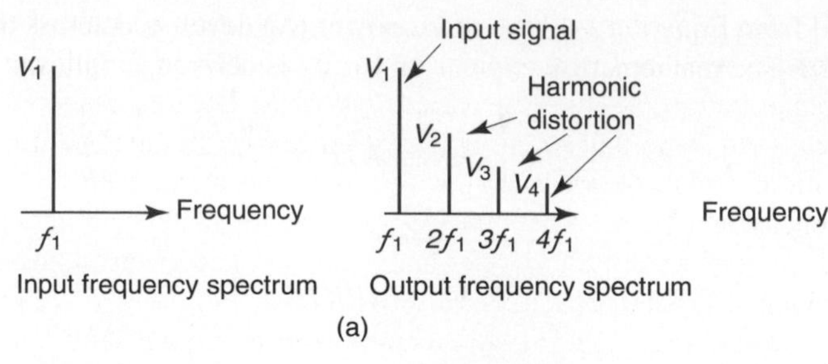

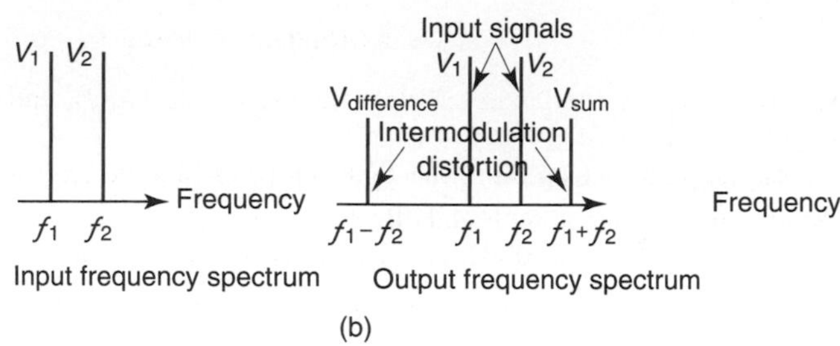

FIGURE 1-8 Correlated noise: (a) Harmonic distortion; (b) Intermodulation distortion

original input frequency plus several harmonics ($2f_1$, $3f_1$, $4f_1$) that were not part of the original signal.

Mathematically, total harmonic distortion (THD) is

$$\% \text{ THD} = \frac{v_{\text{higher}}}{v_{\text{fundamental}}} \times 100 \qquad \textbf{(1-18)}$$

where $\% \text{ THD}$ = percent total harmonic distortion

v_{higher} = quadratic sum of the rms voltages of the harmonics above the fundamental frequency, $\sqrt{v_2^2 + v_3^2 + v_n^2}$

$v_{\text{fundamental}}$ = rms voltage of the fundamental frequency

Example 1-12

Determine

a. 2nd, 3rd, and 12th harmonics for a 1-kHz repetitive wave.

b. Percent second-order, third-order, and total harmonic distortion for a fundamental frequency with an amplitude of 8 *V*rms, a second harmonic amplitude of 0.2 *V*rms, and a third harmonic amplitude of 0.1 *V*rms.

Solution **a.** Harmonic frequencies are simply integer multiples of the fundamental frequency:

$$\text{2nd harmonic} = 2 \times \text{fundamental} = 2 \times 1 \text{ kHz} = 2 \text{ kHz}$$
$$\text{3rd harmonic} = 3 \times \text{fundamental} = 3 \times 1 \text{ kHz} = 3 \text{ kHz}$$
$$\text{12th harmonic} = 12 \times \text{fundamental} = 12 \times 1 \text{ kHz} = 12 \text{ kHz}$$

b.

$$\% \text{ 2nd order} = \frac{V_2}{V_1} \times 100 = \frac{0.2}{8} \times 100 = 2.5\%$$

$$\% \text{ 3rd order} = \frac{V_3}{V_1} \times 100 = \frac{0.1}{8} \times 100 = 1.25\%$$

$$\% \text{ THD} = \frac{\sqrt{(0.2)^2 + (0.1)^2}}{8} = 2.795\%$$

Intermodulation distortion is the generation of unwanted *sum* and *difference* frequencies produced when two or more signals mix in a nonlinear device. The sum and difference frequencies are called *cross products*. The emphasis here is on the word *unwanted* because in communications circuits it is often desirable to produce harmonics or to mix two or more signals to produce sum and difference frequencies. Unwanted cross-product frequencies can interfere with the information signals in a circuit or with the information signals in other circuits. Cross products are produced when harmonics as well as fundamental frequencies mix in a nonlinear device.

Mathematically, the sum and difference frequencies are

$$\text{cross products} = mf_1 \pm nf_2 \tag{1-19}$$

where f_1 and f_2 are fundamental frequencies, where $f_1 > f_2$, and m and n are positive integers between one and infinity.

Figure 1-8b shows the input and output frequency spectrums for a nonlinear device with two input frequencies (f_1 and f_2). As the figure shows, the output spectrum contains the two original frequencies plus their sum and difference frequencies ($f_1 - f_2$ and $f_1 + f_2$). In actuality, the output spectrum would also contain the harmonics of the two input frequencies ($2f_1$, $3f_1$, $2f_2$, and $3f_2$), as the same nonlinearities that caused the intermodulation distortion would also cause harmonic distortion. The harmonics have been eliminated from the diagram for simplicity.

Example 1-13

For a nonlinear amplifier with two input frequencies, 3 kHz and 8 kHz, determine

a. First three harmonics present in the output for each input frequency.

b. Cross-product frequencies produced for values of m and n of 1 and 2.

Solution **a.** The first three harmonics include the two original frequencies, 3 kHz and 8 kHz; two times each of the original frequencies, 6 kHz and 16 kHz; and three times each of the original frequencies, 9 kHz and 24 kHz.

b. The cross products for values of m and n of 1 and 2 are determined from Equation 1-19 and are summarized as follows:

m	n	Cross Products
1	1	8 kHz ± 3 kHz = 5 kHz and 11 kHz
1	2	8 kHz ± 6 kHz = 2 kHz and 14 kHz
2	1	16 kHz ± 3 kHz = 13 kHz and 19 kHz
2	2	16 kHz ± 6 kHz = 10 kHz and 22 kHz

1-7-4 Impulse Noise

Impulse noise is characterized by high-amplitude peaks of short duration in the total noise spectrum. As the name implies, impulse noise consists of sudden bursts of irregularly shaped pulses that generally last between a few microseconds and several milliseconds, depending on their amplitude and origin. The significance of impulse noise (hits) on voice communications is often more annoying than inhibitive, as impulse hits produce a sharp popping or crackling sound. On data circuits, however, impulse noise can be devastating.

Common sources of impulse noise include transients produced from electromechanical switches (such as relays and solenoids), electric motors, appliances, electric lights (especially fluorescent), power lines, automotive ignition systems, poor-quality solder joints, and lightning.

Table 1-8 Electrical Noise Source Summary

Correlated noise (internal)
Nonlinear distortion
Harmonic distortion
Intermodulation distortion
Uncorrelated noise
External
Atmospheric
Extraterrestrial
Solar
Cosmic
Man-made
Impulse
Interference
Internal
Thermal
Shot
Transient time

1-7-5 Interference

Interference is a form of external noise and, as the name implies, means "to disturb or detract from." Electrical interference is when information signals from one source produce frequencies that fall outside their allocated bandwidth and interfere with information signals from another source. Most interference occurs when harmonics or cross-product frequencies from one source fall into the passband of a neighboring channel. For example, CB radios transmit signals in the 27-MHz to 28-Mhz range. Their second harmonic frequencies (54–55 MHz) fall within the band allocated to VHF television (channel 3 in particular). If one person transmits on a CB radio and produces a high-amplitude second harmonic component, it could interfere with other people's television reception. Most interference occurs in the radio-frequency spectrum and is discussed in more detail in later chapters of this book.

1-7-6 Noise Summary

Table 1-8 summarizes the electrical noise sources described in this chapter.

1-7-7 Signal-to-Noise Power Ratio

Signal-to-noise power ratio (S/N) is the ratio of the signal power level to the noise power level. Mathematically, signal-to-noise power ratio is expressed as

$$\frac{\mathrm{S}}{\mathrm{N}} = \frac{P_s}{P_n} \tag{1-20}$$

where P_s = signal power (watts)
P_n = noise power (watts)

The signal-to-noise power ratio is often expressed as a logarithmic function with the decibel unit:

$$\frac{\mathrm{S}}{\mathrm{N}}(\mathrm{dB}) = 10 \log \frac{P_s}{P_n} \tag{1-21}$$

Example 1-14

For an amplifier with an output signal power of 10 W and an output noise power of 0.01 W, determine the signal-to-noise power ratio.

Solution The signal-to-noise power ratio is found by substituting into Equation 1-20:

$$\frac{S}{N} = \frac{P_s}{P_n} = \frac{10}{0.01} = 1000$$

To express in dB, substitute into Equation (1-21):

$$\frac{S}{N}(\text{dB}) = 10 \log \frac{P_s}{P_n} = 10 \log \frac{10}{0.01} = 30 \text{ dB} \quad \textbf{(1-22)}$$

Signal-to-noise power ratio can also be expressed in terms of voltages and resistances as shown here:

$$\frac{S}{N}(\text{dB}) = 10 \log\left(\frac{V_s^2/R_{\text{in}}}{V_n^2/R_{\text{out}}}\right) \quad \textbf{(1-23)}$$

where $\frac{S}{N}$ = *signal-to-noise power ratio* (decibels)

R_{in} = input resistance (ohms)

R_{out} = output resistance (ohms)

V_s = signal voltage (volts)

V_n = noise voltage (volts)

If the input and output resistances of the amplifier, receiver, or network being evaluated are equal, then Equation 1-23 reduces to

$$\frac{S}{N}(\text{dB}) = 10 \log\left(\frac{V_s^2}{V_n^2}\right)$$

$$= 10 \log\left(\frac{V_s}{V_n}\right)^2$$

$$\frac{S}{N}(\text{dB}) = 20 \log \frac{V_s}{V_n} \quad \textbf{(1-24)}$$

Example 1-15

For an amplifier with an output signal voltage of 4 V, an output noise voltage of 0.005 V, and an input and output resistance of 50 Ω, determine the signal-to-noise power ratio.

Solution The signal-to-noise power ratio is found by substituting into Equation 1-23:

$$\frac{S}{N}(\text{dB}) = 20 \log \frac{V_s}{V_n} = 20 \log \frac{4}{0.005} = 58.06 \text{ dB}$$

1-7-8 Noise Factor and Noise Figure

Noise factor (F) and *noise figure* (NF) are figures of merit used to indicate how much the signal-to-noise ratio deteriorates as a signal passes through a circuit or series of circuits. Noise factor is simply a ratio of input signal-to-noise power ratio to output signal-to-noise power ratio. In other words, it is a ratio of ratios. Mathematically, noise factor is

$$F = \frac{\text{input signal-to-noise power ratio}}{\text{output signal-to-noise power ratio}} \text{(unitless ratio)} \quad \textbf{(1-25)}$$

Noise figure is simply the noise factor stated in dB and is a parameter commonly used to indicate the quality of a receiver. Mathematically, noise figure is

$$\text{NF(dB)} = 10 \log \frac{\text{input signal-to-noise power ratio}}{\text{output signal-to-noise power ratio}} \quad \textbf{(1-26)}$$

or

$$\text{NF(dB)} = 10 \log F$$

In essence, noise figure indicates how much the signal-to-noise ratio deteriorates as a waveform propagates from the input to the output of a circuit. For example, an amplifier with a noise figure of 6 dB means that the signal-to-noise ratio at the output is 6 dB less than it was at the input. If a circuit is perfectly noiseless and adds no additional noise to the signal, the signal-to-noise ratio at the output will equal the signal-to-noise ratio at the input. For a perfect, noiseless circuit, the noise factor is 1, and the noise figure is 0 dB.

An electronic circuit amplifies signals and noise within its passband equally well. Therefore, if the amplifier is ideal and noiseless, the input signal and noise are amplified the same, and the signal-to-noise ratio at the output will equal the signal-to-noise ratio at the input. In reality, however, amplifiers are not ideal. Therefore, the amplifier adds internally generated noise to the waveform, reducing the overall signal-to-noise ratio. The most predominant noise is thermal noise, which is generated in all electrical components. Therefore, all networks, amplifiers, and systems add noise to the signal and thus reduce the overall signal-to-noise ratio as the signal passes through them.

Figure 1-9a shows an ideal noiseless amplifier with a power gain (A_P), an input signal power level (S_i), and an input noise power level (N_i). The output signal level is simply $A_P S_i$, and the output noise level is $A_P N_i$. Therefore, the input and output S/N ratios are equal and are expressed mathematically as

$$\frac{S_{out}}{N_{out}} = \frac{A_P S_i}{A_P N_i} = \frac{S_i}{N_i}$$

where A_P equals amplifier power gain.

Figure 1-9b shows a nonideal amplifier that generates an internal noise (N_d). As with the ideal noiseless amplifier, both the input signal and noise are amplified by the circuit gain. However, the circuit adds the internally generated noise to the waveform. Consequently, the output signal-to-noise ratio is less than the input signal-to-noise ratio by an amount proportional to N_d. Mathematically, the S/N ratio at the output of a nonideal amplifier is expressed mathematically as

$$\frac{S_{out}}{N_{out}} = \frac{A_P S_i}{A_P N_i + N_d} = \frac{S_i}{N_i + N_d/A_P}$$

where A_P = amplifer power gain
N_d = internal noise

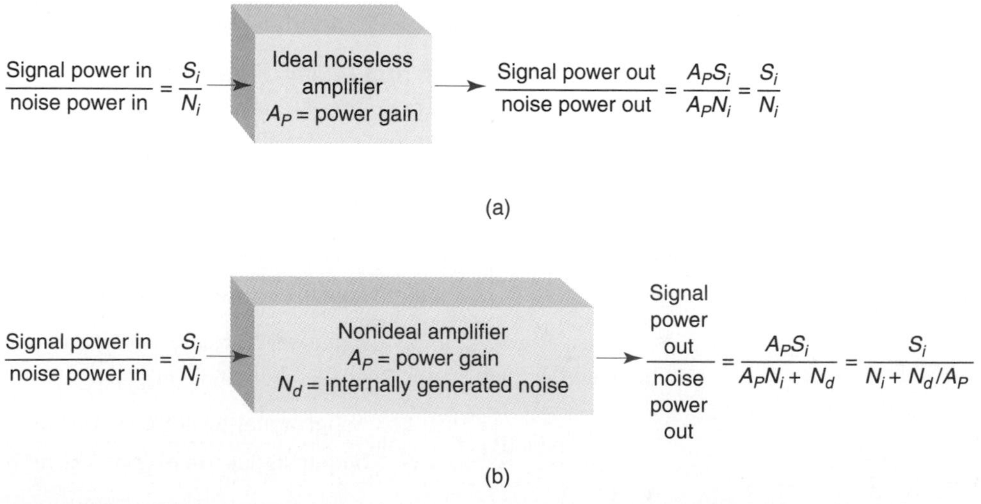

Signal power in / noise power in $= \frac{S_i}{N_i}$ → Ideal noiseless amplifier, A_P = power gain → Signal power out / noise power out $= \frac{A_P S_i}{A_P N_i} = \frac{S_i}{N_i}$

(a)

Signal power in / noise power in $= \frac{S_i}{N_i}$ → Nonideal amplifier, A_P = power gain, N_d = internally generated noise → Signal power out / noise power out $= \frac{A_P S_i}{A_P N_i + N_d} = \frac{S_i}{N_i + N_d/A_P}$

(b)

FIGURE 1-9 Noise figure: (a) ideal, noiseless device; (b) amplifier with internally generated noise

Example 1-16

For a nonideal amplifier and the following parameters, determine

a. Input S/N ratio (dB).
b. Output S/N ratio (dB).
c. Noise factor and noise figure.

Input signal power $= 2 \times 10^{-10}$ W
Input noise power $= 2 \times 10^{-18}$ W
Power gain $= 1{,}000{,}000$
Internal noise $(N_d) = 6 \times 10^{-12}$ W

Solution a. For the input signal and noise power levels given and substituting into Equation 1-22, the input S/N is

$$\frac{S}{N} = \frac{2 \times 10^{-10}\ \text{W}}{2 \times 10^{-18}\ \text{W}} = 100{,}000{,}000$$

$$10 \log(100{,}000{,}000) = 80\ \text{dB}$$

b. The output noise power is the sum of the internal noise and the amplified input noise.

$$N_{\text{out}} = 1{,}000{,}000(2 \times 10^{-18}) + 6 \times 10^{-12} = 8 \times 10^{-12}\ \text{W}$$

The output signal power is simply the product of the input power and the power gain.

$$P_{\text{out}} = 1{,}000{,}000(2 \times 10^{-10}) = 200\ \mu\text{W}$$

For the output signal and noise power levels calculated and substituting into Equation 1-22, the output S/N is

$$\frac{S}{N} = \frac{200 \times 10^{-6}\ \text{W}}{8 \times 10^{-12}\ \text{W}} = 25{,}000{,}000$$

$$10 \log(25{,}000{,}000) = 74\ \text{dB}$$

c. The noise factor is found by substituting the results from steps (a) and (b) into Equation 1-25,

$$F = \frac{100{,}000{,}000}{25{,}000{,}000} = 4$$

and the noise figure is calculated from Equation 1-26,

$$NF = 10 \log 4 = 6\ \text{dB}$$

When two or more amplifiers are cascaded as shown in Figure 1-10, the total noise factor is the accumulation of the individual noise factors. *Friiss' formula* is used to calculate the total noise factor of several cascaded amplifiers. Mathematically, Friiss' formula is

$$F_T = F_1 + \frac{F_2 - 1}{A_1} + \frac{F_3 - 1}{A_1 A_2} + \frac{F_n - 1}{A_1 A_2 \cdots A_n} \tag{1-27}$$

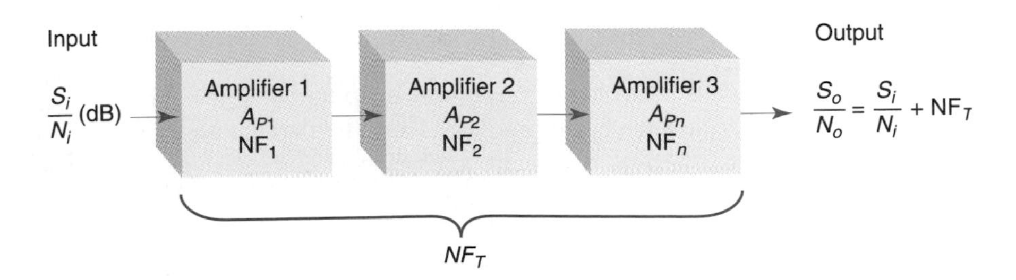

FIGURE 1-10 Noise figure of cascaded amplifiers

where F_T = total noise factor for n cascaded amplifiers
F_1 = noise factor, amplifier 1
F_2 = noise factor, amplifier 2
F_3 = noise factor, amplifier 3
F_n = noise factor, amplifier n
A_1 = power gain, amplifier 1
A_2 = power gain, amplifier 2
A_n = power gain, amplifier n

Note that to use Friiss' formula, the noise figures must be converted to noise factors. The total noise figure is simply

$$\text{NF}_T(\text{dB}) = 10 \log \text{F}_T \tag{1-28}$$

Example 1-17

For three cascaded amplifier stages, each with noise figures of 3 dB and power gains of 10 dB, determine the total noise figure.

Solution The noise figures must be converted to noise factors, then substituted into Equation 1-27, giving us a total noise factor of

$$\begin{aligned} \text{F}_T &= \text{F}_1 + \frac{\text{F}_2 - 1}{A_1} + \frac{\text{F}_3 - 1}{A_1A_2} + \frac{\text{F}_n - 1}{A_1A_2 \cdots A_n} \\ &= 2 + \frac{2-1}{10} + \frac{2-1}{100} = 2.11 \end{aligned}$$

Thus, the total noise figure is

$$\text{NF}_T = 10 \log 2.11 = 3.24 \text{ dB}$$

Several important observations can be made from Example 1-17. First, the overall noise figure of 3.24 dB was not significantly larger than the noise figure of the first stage (3 dB). From Equation 1-27, it can be seen that the first stage in a series of amplifiers, such as found in audio amplifiers and radio receivers, contributes the most to the overall noise figure. This is true as long as the gain of the first stage is sufficient to reduce the effects of the succeeding stages. For example, if A_1 and A_2 in Example 1-17 were only 3 dB, the overall noise figure would be 4.4 dB, a significant increase. Worse yet, if the first stage were passive and had a loss of 3 dB ($A = 0.5$), the overall noise figure would increase to 7.16 dB.

Figure 1-11 shows how signal-to-noise ratio can be reduced as a signal passes through a two-stage amplifier circuit. As the figure shows, both the input signal and the input noise are amplified 10 dB in amplifier 1. Amplifier 1, however, adds an additional 1.5 dB of noise (i.e., a noise figure of 1.5 dB), thus reducing the signal-to-noise ratio at the output of amplifier 1 to 28.5 dB. Again, the signal and noise are both amplified by 10 dB in amplifier 2. Amplifier 2, however, adds 2.5 dB of additional noise (i.e., a noise figure of 2.5 dB), thus reducing the signal-to-noise ratio at the output of amplifier 2 to 26 dB. The overall reduction in the signal-to-noise ratio from the input of amplifier 1 to the output of amplifier 2 is 4 dB; thus, the total noise figure for the two amplifiers is 4 dB.

1-7-9 Equivalent Noise Temperature

Because the noise produced from thermal agitation is directly proportional to temperature, thermal noise can be expressed in degrees as well as watts or dBm. Rearranging Equation 1-13 yields

$$T = \frac{N}{KB} \tag{1-29}$$

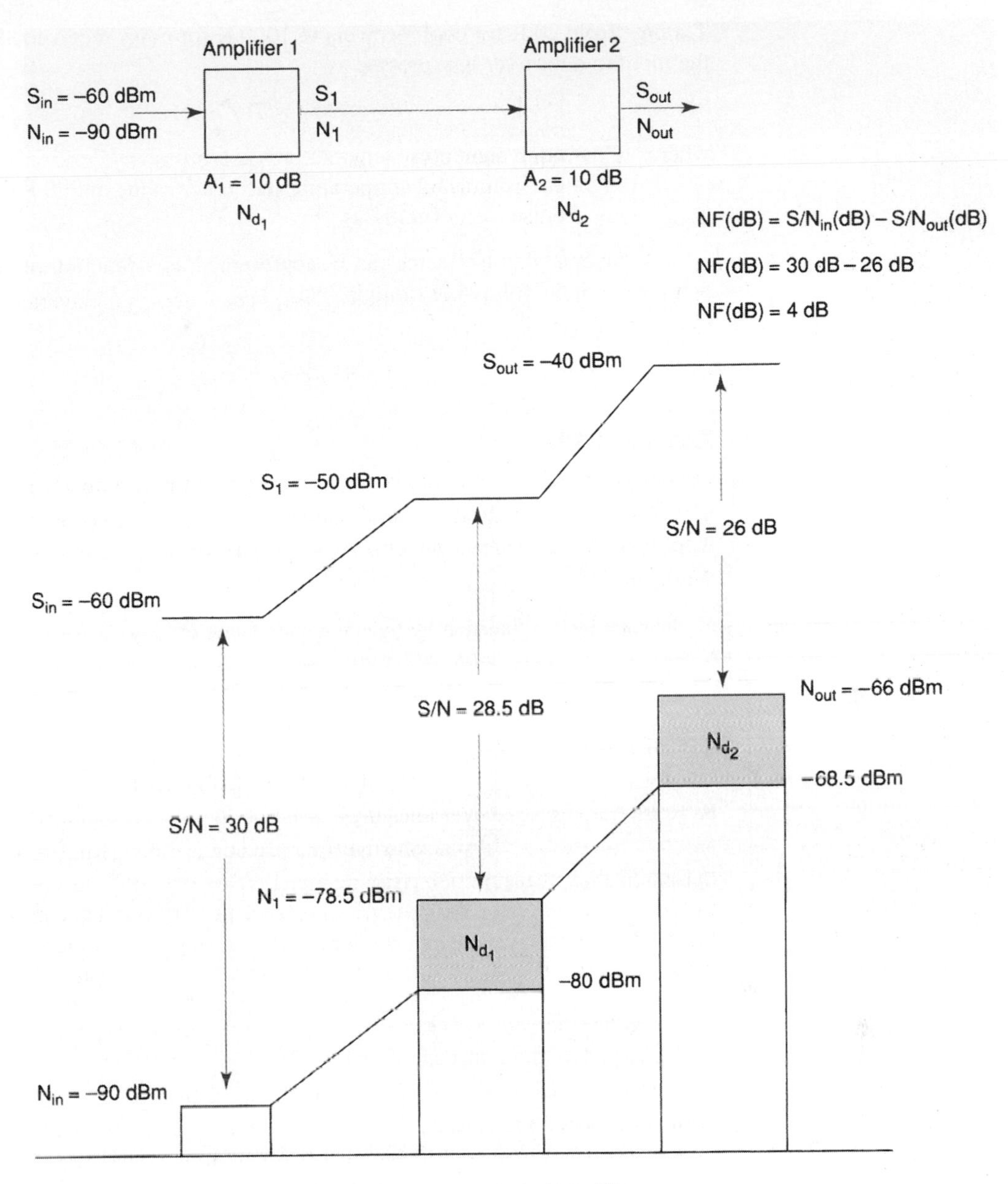

FIGURE 1-11 Noise figure degradation in cascaded amplifiers

where T = environmental temperature (kelvin)
N = noise power (watts)
K = Boltzmann's constant (1.38×10^{-23} J/K)
B = bandwidth (hertz)

Equivalent noise temperature (T_e) is a hypothetical value that cannot be directly measured. T_e is a convenient parameter often used rather than noise figure in low-noise, sophisticated VHF, UHF, microwave, and satellite radio receivers. T_e, as is noise factor, indicates the reduction in the signal-to-noise ratio a signal undergoes as it propagates through a receiver. The lower the equivalent noise temperature is, the better the quality of a receiver. A noise figure of 1 dB corresponds to an equivalent noise temperature of 75 K, and a noise figure of 6 dB corresponds to an equivalent noise temperature of 870 K. Typical values for

T_e range from 20 K for cool receivers to 1000 K for noisy receivers. Mathematically, T_e at the input to a receiver is expressed as

$$T_e = T(\mathrm{F} - 1) \tag{1-30}$$

where T_e = equivalent noise temperature (kelvin)
T = environmental temperature (reference value of 290 K)
F = noise factor (unitless)

Conversely, noise factor can be represented as a function of equivalent noise temperature with the following formula:

$$\mathrm{F} = 1 + \frac{T_e}{T} \tag{1-31}$$

Example 1-18

Determine

a. Noise figure for an equivalent noise temperature of 75 K (use 290 K for the reference temperature).
b. Equivalent noise temperature for a noise figure of 6 dB.

Solution

a. Substituting into Equation 1-31 yields a noise factor of

$$\mathrm{F} = 1 + \frac{T_e}{T} = 1 + \frac{75}{290} = 1.258$$

and noise figure is simply

$$\mathrm{NF} = 10 \log(1.258) = 1 \text{ dB}$$

b. Noise factor is found by rearranging Equation 1-26:

$$\mathrm{F} = \text{antilog}\ (\mathrm{NF}/10) = \text{antilog}\ (6/10) = (10)^{0.6} = 4$$

Substituting into Equation 1-30 gives

$$T_e = T(\mathrm{F} - 1) = 290(4 - 1) = 870 \text{ K}$$

QUESTIONS

1-1. Define *electronic communications.*
1-2. When did radio communications begin?
1-3. Define *decibel.*
1-4. What does dB represent?
1-5. What is the difference between a positive and a negative decibel?
1-6. Define *dBm* and describe what the term means.
1-7. What are the three primary components of an electronic communications system?
1-8. Define the terms *modulation* and *demodulation.*
1-9. What are the two basic types of electronic communications systems?
1-10. Describe the following terms: *carrier signal, modulating signal,* and *modulated wave.*
1-11. What are the three properties of a sine wave that can be varied? Name the types of modulation that result from each.
1-12. Describe two reasons why modulation is necessary in electronic communications.
1-13. Describe the terms *frequency up-conversion* and *frequency down-conversion.*
1-14. Define the following terms: *frequency, cycle, wavelength,* and *radio frequency.*
1-15. Briefly describe the ITU's frequency band designations.
1-16. What are the two most significant limitations on the performance of an electronic communications system?

1-17. Define *information capacity.*

1-18. Relate *Hartley's law* and the *Shannon limit for information capacity* to the performance of an electronic communications system.

1-19. Define *electrical noise.*

1-20. What are the two *general categories* of noise?

1-21. What is meant by the terms *external noise* and *internal noise*?

1-22. List several sources of external noise and give a brief description of each.

1-23. What is meant by the term *man-made noise*? Give several examples.

1-24. What is the most significant form of internal noise?

1-25. Define *thermal noise* and describe its relationship to temperature and bandwidth.

1-26. Explain the difference between *correlated* and *uncorrelated* noise.

1-27. List and describe the two most significant forms of correlated noise.

1-28. Describe the term *impulse noise* and list several sources.

1-29. Describe the term *interference* and list several sources.

1-30. Describe *signal-to-noise power ratio.*

1-31. Define noise factor and *noise figure* and describe their significance.

1-32. Describe the term *equivalent noise temperature* and describe its significance.

PROBLEMS

1-1. Convert the following absolute power ratios to dB:

a. 5
b. 15
c. 25
d. 125
e. 2000
f. 10,000
g. 100,000

1-2. Convert the following absolute power ratios to dB:

a. 0.1
b. 0.04
c. 0.008
d. 0.0001
e. 0.00002
f. 0.000005

1-3. Convert the following decibel values to absolute ratios:

a. 26 dB
b. 2 dB
c. 43 dB
d. 56 dB

1-4. Convert the following decibel values to absolute ratios:

a. −3 dB
b. −9 dB
c. −23 dB
d. −36 dB

1-5. Convert the following powers to dBm:

a. 0.001 μW
b. 1 pW, 2×10^{-15} W
c. $1.4 \times 10_{-16}$ W

1-6. Convert the following dBm values to watts:

a. -110 dBm
b. -50 dBm
c. -13 dBm
d. 26 dBm
e. 60 dBm

1-7. Given a three-stage system comprised of two amplifiers and one filter with an input power of $P_{in} = 0.01$ mW and absolute power gains of $A_{p_1} = 200$, $A_{p_2} = 0.1$, and $A_{p_3} = 1000$, determine

a. The input power in dBm
b. Output power (P_{out}) in watts and dBm
c. The dB gain of each of the three stages
d. The overall gain in dB

1-8. Given a three-stage system with an input power $P_{in} = -26$ dBm and power gains of the three stages of $A_{p_1} = 23$ dB, $A_{p_2} = -3$ dB, $A_{p_3} = 16$ dB, determine the output power (P_{out}) in dBm and watts.

1-9. Determine the combined power when a signal with a power level of 10 dBm is combined with a second signal with a power level of 8 dBm.

1-10. What it the ITU's designation for the following frequency ranges?

a. 3–30 kHz
b. 0.3–3 MHz
c. 3–30 GHz

1-11. Determine the wavelengths for the following frequencies:

a. 50 MHz
b. 400 MHz
c. 4 GHz
d. 100 GHz

1-12. Determine the information capacity for a communications channel with a bandwidth of 50 kHz and a signal-to-noise ratio of 40 dB.

1-13. What is the effect on the information capacity of a communications channel if the bandwidth is halved? Doubled?

1-14. What is the effect on the information capacity of a communications channel if the transmission time is doubled?

1-15. Convert the following temperatures to Kelvin:

a. 17°C
b. 27°C
c. -17°C
d. -50°C

1-16. Calculate the thermal noise power in watts and dBm for the following bandwidths and temperatures:

a. $B = 100$ Hz, $T = 17$°C
b. $B = 100$ kHz, $T = 100$°C
c. $B = 1$ MHz, $T = 500$°C

1-17. Determine the bandwidth necessary to produce 8×10^{-17} watts of thermal power at a temperature of 17°C.

1-18. Determine the second, third, and total harmonic distortion for a repetitive wave with a fundamental frequency amplitude of 10 V_{rms}, a second harmonic amplitude of 0.2 V_{rms}, and a third harmonic amplitude of 0.1 V_{rms}.

1-19. For a nonlinear amplifier with sine wave input frequencies of 3 kHz and 5 kHz, determine the first three harmonics present in the output for each input frequency and the cross-product frequencies produced for values of m and n of 1 and 2.

1-20. Determine the power ratios in dB for the following input and output powers:

a. $P_{in} = 0.001$ W, $P_{out} = 0.01$ W
b. $P_{in} = 0.25$ W, $P_{out} = 0.5$ W
c. $P_{in} = 1$ W, $P_{out} = 0.5$ W
d. $P_{in} = 0.001$ W, $P_{out} = 0.001$ W
e. $P_{in} = 0.04$ W, $P_{out} = 0.16$ W
f. $P_{in} = 0.002$ W, $P_{out} = 0.0002$ W
g. $P_{in} = 0.01$ W, $P_{out} = 0.4$ W

1-21. Determine the voltage ratios in dB for the following input and output voltages (assume equal input and output resistance values):

a. $v_{in} = 0.001$ V, $v_{out} = 0.01$ V
b. $v_{in} = 0.1$ V, $v_{out} = 2$ V
c. $v_{in} = 0.5$ V, $v_{out} = 0.25$ V
d. $v_{in} = 1$ V, $v_{out} = 4$ V

1-22. Determine the overall noise factor and noise figure for three cascaded amplifiers with the following parameters:

$A_1 = 10$ dB
$A_2 = 10$ dB
$A_3 = 20$ dB
$NF_1 = 3$ dB
$NF_2 = 6$ dB
$NF_3 = 10$ dB

1-23. Determine the overall noise factor and noise figure for three cascaded amplifiers with the following parameters:

$A_1 = 3$ dB
$A_2 = 13$ dB
$A_3 = 10$ dB
$NF_1 = 10$ dB
$NF_2 = 6$ dB
$NF_3 = 10$ dB

1-24. If an amplifier has a bandwidth $B = 20$ kHz and a total noise power $N = 2 \times 10^{-17}$ W, determine the total noise power if the bandwidth increases to 40 kHz. Decreases to 10 kHz.

1-25. For an amplifier operating at a temperature of 27°C with a bandwidth of 20 kHz, determine

a. The total noise power in watts and dBm.
b. The rms noise voltage (V_N) for a 50-Ω internal resistance and a 50-load resistor.

1-26. a. Determine the noise power in watts and dBm for an amplifier operating at a temperature of 400°C with a 1-MHz bandwidth.

b. Determine the decrease in noise power in decibels if the temperature decreased to 100°C.
c. Determine the increase in noise power in decibels if the bandwidth doubled.

1-27. Determine the noise figure for an equivalent noise temperature of 1000 K (use 290 K for the reference temperature).

1-28. Determine the equivalent noise temperature for a noise figure of 10 dB.

1-29. Determine the noise figure for an amplifier with an input signal-to-noise ratio of 100 and an output signal-to-noise ratio of 50.

1-30. Determine the noise figure for an amplifier with an input signal-to-noise ratio of 30 dB and an output signal-to-noise ratio of 24 dB.

1-31. Calculate the input signal-to-noise ratio for an amplifier with an output signal-to-noise ratio of 16 dB and a noise figure of 5.4 dB.

1-32. Calculate the output signal-to-noise ratio for an amplifier with an input signal-to-noise ratio of 23 dB and a noise figure of 6.2 dB.

1-33. Determine the thermal noise voltages for components operating at the following temperatures, bandwidths, and equivalent resistances:

a. $T = -50°\text{C}$, $B = 50$ kHz, and $R = 50\ \Omega$

b. $T = 100°\text{C}$, $B = 10$ kHz, and $R = 100\ \Omega$

c. $T = 50°\text{C}$, $B = 500$ kHz, and $R = 72\ \Omega$

1-34. Determine the 2nd, 5th, and 15th harmonics for a repetitive wave with a fundamental frequency of 2.5 kHz.

CHAPTER 2

Signal Analysis and Mixing

CHAPTER OUTLINE

OBJECTIVES

- Define *analog* and *digital signals* and describe the differences between them
- Define *signal analysis*
- Define what is meant by a periodic wave
- Describe time and frequency domain
- Define a complex wave
- Define *wave symmetry* and describe what is meant by even, odd, and half-wave symmetry
- Describe *frequency spectrum* and *bandwidth* and how they relate to each other
- Describe the spectral content of a square wave and a rectangular wave
- Describe the relationship between power and energy spectra
- Explain the differences between discrete and fast Fourier transforms
- Explain linear and nonlinear mixing and the differences between them

2-1 INTRODUCTION

Electrical signals can be in *analog* or *digital* form. With analog signals, the amplitude changes continuously with respect to time with no breaks or discontinuities. Figure 2-1a shows a sine wave, which is the most basic analog signal. As the figure shows, the amplitude of the sine wave varies continuously with time between its maximum value V_{max} and its minimum value V_{min}. With a sine wave, the time the waveform is above its average

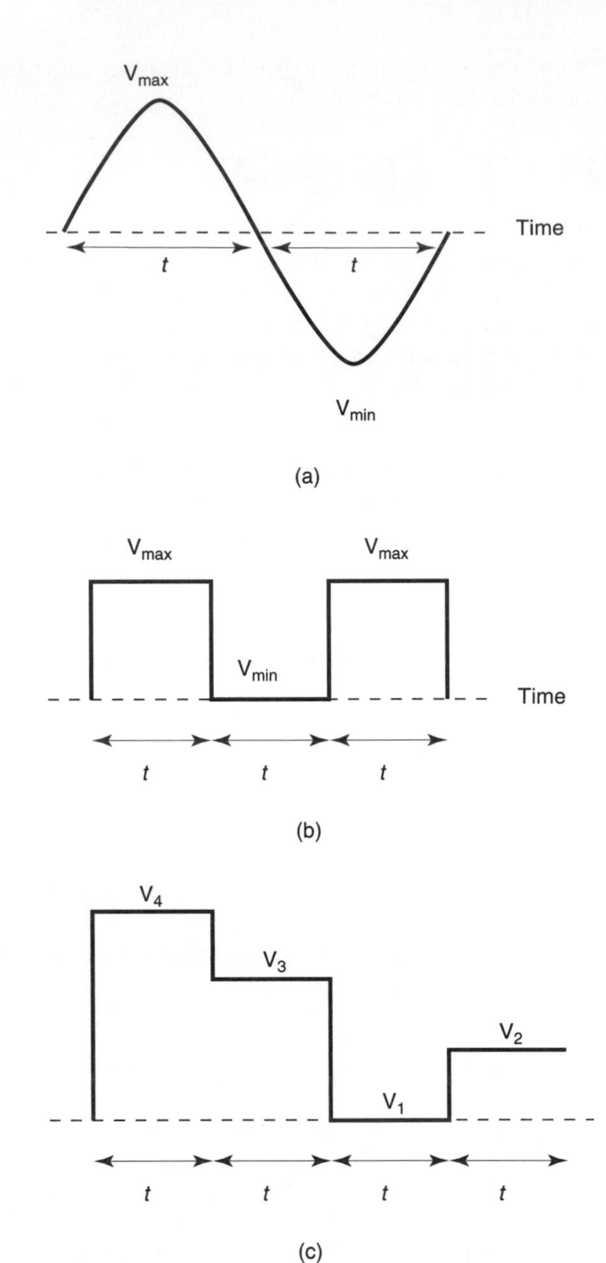

FIGURE 2-1 Electrical signals: (a) sine wave; (b) binary digital signal; (c) quaternary digital signal

amplitude equals the time it is below its average value, and the time of the positive half cycle and time of the negative half cycle are equal (t).

Digital signals are described as discrete; their amplitude maintains a constant level for a prescribed period of time and then it changes to another level. If there are only two levels possible, it is called a *binary signal.* All binary signals are digital, but all digital signals are not necessarily binary. Figure 2-1b shows a binary digital signal (called a pulse). As the figure shows, the amplitude of the waveform is at its maximum value V_{max} for t seconds, then changes to its minimum value V_{min} for t seconds. The only time the signal is not at either V_{max} or V_{min} is when it is transitioning between the maximum and minimum values. However, the voltage is only stable (constant) when it is at V_{max} or V_{min}. Figure 2-1c shows a four-level digital signal. Because there are four levels (V_1, V_2, V_3, and V_4), the signal is called a *quaternary* digital signal. Again, the voltage is constant for t seconds, then it changes to one of the three other values.

2-2 SIGNAL ANALYSIS

When designing electronic communications systems, it is often necessary to analyze and predict the performance of the circuit based on the voltage distribution and frequency composition of the information signal. This is done with mathematical signal analysis. Although all signals in electronic communications systems are not single-frequency sine or cosine waves, many of them are, and the signals that are not can often be represented by combinations of sine and cosine waves.

2-2-1 Sinusoidal Signals

In essence, signal analysis is the mathematical analysis of the frequency, bandwidth, and voltage level of a signal. Electrical signals are voltage- or current-time variations that can be represented by a series of sine or cosine waves. Mathematically, a single-frequency voltage or current waveform is

$$v(t) = V\sin(2\pi ft + \theta) \quad \text{or} \quad v(t) = V\cos(2\pi ft + \theta)$$
$$i(t) = I\sin(2\pi ft + \theta) \quad \text{or} \quad i(t) = I\cos(2\pi ft + \theta)$$

where
$v(t)$ = time-varying voltage sine wave
$i(t)$ = time-varying current sine wave
V = peak voltage (volts)
f = frequency (hertz)
θ = phase shift (radians)
I = peak current (amperes)
$2\pi f$ = angular velocity (radians per second)

Whether a sine or a cosine function is used to represent a signal is purely arbitrary and depends on which is chosen as the reference. However, it should be noted that $\sin\theta = \cos(\theta - 90°)$. Therefore, the following relationships hold true:

$$v(t) = V\sin(2\pi ft + \theta) = V\cos(2\pi ft + \theta - 90°)$$
$$v(t) = V\cos(2\pi ft + \theta) = V\sin(2\pi ft + \theta + 90°)$$

The preceding formulas are for a single-frequency, repetitive waveform. Such a waveform is called a *periodic* wave because it repeats at a uniform rate. In other words, each successive cycle of the signal takes exactly the same length of time and has exactly the same amplitude variations as every other cycle—each cycle has the same shape. A series of sine, cosine, or square waves are examples of periodic waves. Periodic waves can be analyzed in either the *time domain* or the *frequency domain.* In fact, it is often necessary when analyzing system performance to switch from the time domain to the frequency domain and vice versa.

2-2-1-1 Time domain. A description of a signal with respect to time is called a *time-domain representation.* A standard oscilloscope is a time-domain instrument. The display on the cathode ray tube (CRT) is an amplitude-versus-time representation of the signal and is commonly called a *signal waveform.* Essentially, a signal waveform shows the shape and instantaneous magnitude of the signal with respect to time but does not directly indicate its frequency content. With an oscilloscope, the vertical deflection is proportional to the amplitude of the input signal, and the horizontal deflection is a function of time (sweep rate). Figure 2-2 shows the signal waveform for a single-frequency sinusoidal signal with a peak amplitude of V volts and a frequency of $f = 1/T$ hertz.

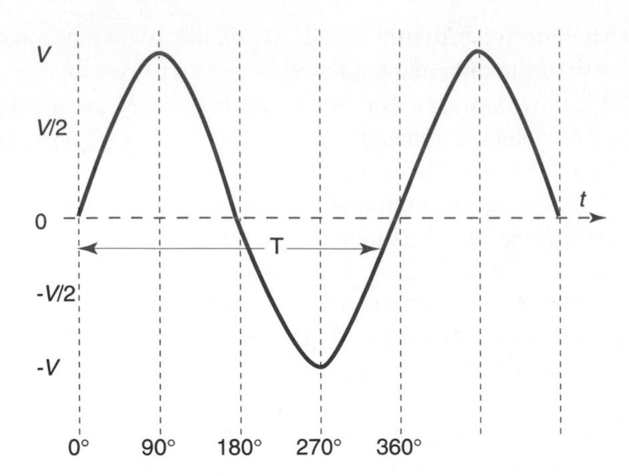

FIGURE 2-2 Time domain representation of a single-frequency sine wave

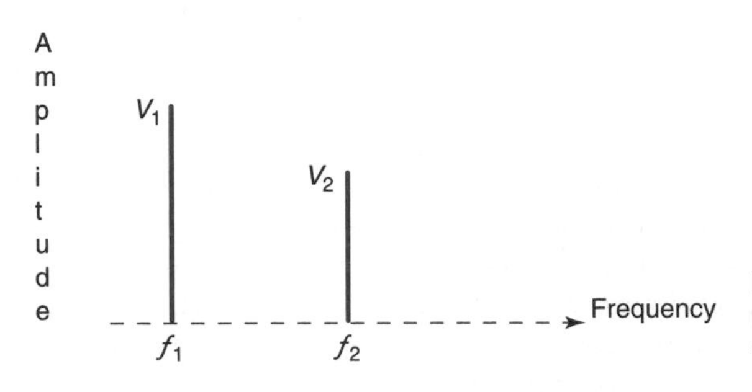

FIGURE 2-3 Frequency spectrum (frequency domain representation) of two sine waves

2-2-1-2 Frequency domain. A description of a signal with respect to its frequency is called a *frequency-domain representation.* A *spectrum analyzer* is a frequency-domain instrument. Essentially, no waveform is displayed on the CRT. Instead, an amplitude-versus-frequency plot is shown (this is called a *frequency spectrum*). With a spectrum analyzer, the horizontal axis represents frequency and the vertical axis amplitude. Therefore, there is a vertical deflection for each frequency present in the waveform. The vertical deflection (height) of each line is proportional to the amplitude of the frequency that it represents. Effectively, the input waveform is swept with a variable-frequency, high-Q bandpass filter whose center frequency is synchronized to the horizontal sweep rate of the CRT. Each frequency present in the input waveform produces a vertical line on the CRT (these are called spectral components). The vertical deflection (height) of each line is proportional to the amplitude of the frequency that it represents. A frequency-domain representation of a wave shows the frequency content but does not necessarily indicate the shape of the waveform or the combined amplitude of all the input components at any specific time. Figure 2-3 shows the frequency spectrum for a two sinusoidal signal with peak amplitudes of V_1 and V_2 volts and frequencies of f_1 and f_2 hertz, respectively.

2-3 COMPLEX WAVES

Essentially, any repetitive waveform that is comprised of more than one harmonically related sine or cosine wave is a *nonsinusoidal, complex* wave. Thus, a complex wave is any periodic (repetitive) waveform that is not a sinusoid, such as square waves, rectangular waves, and triangular waves. To analyze a complex periodic wave, it is necessary to use a mathematical series developed in 1826 by the French physicist and mathematician Baron Jean Fourier. This series is appropriately called the *Fourier series.*

2-3-1 The Fourier Series

Fourier analysis is a mathematical tool that allows us to move back and forth between the time and frequency domains. The Fourier series is used in signal analysis to represent the sinusoidal components of nonsinusoidal periodic waveforms (i.e., to change a time-domain signal to a frequency-domain signal). In general, a Fourier series can be written for any periodic function as a series of terms that include trigonometric functions with the following mathematical expression:

$$f(t) = A_0 + A_1 \cos \alpha + A_2 \cos 2\alpha + A_3 \cos 3\alpha + \cdots A_n \cos n\alpha$$
$$+ A_0 + B_1 \sin \beta_1 + B_2 \sin 2\beta + B_3 \sin 3\beta + \cdots B_n \sin n\beta \qquad \textbf{(2-1)}$$

where $\alpha = \beta$

Equation 2-1 states that the waveform $f(t)$ comprises an average (dc) value (A_0) and either sine or cosine functions in which each successive term has a frequency that is an integer multiple of the frequency of the first cosine term in the series and a series of sine functions in which each successive term has a frequency that is an integer multiple of the frequency of the first sine term in the series. There are no restrictions on the values or relative values of the amplitudes for the sine or cosine terms. Equation 2-1 is expressed in words as follows: Any *periodic waveform* is comprised of an average dc component and a series of harmonically related sine or cosine waves. A *harmonic* is an integral multiple of the *fundamental frequency.* The fundamental frequency is the first harmonic and is equal to the frequency (*repetition rate*) of the waveform. The second multiple of the fundamental frequency is called the *second harmonic,* the third multiple is called the *third harmonic,* and so forth. The fundamental frequency is the minimum frequency necessary to represent a waveform. Therefore, Equation 2-1 can be rewritten as

$$f(t) = \text{dc} + \text{fundamental} + \text{2nd harmonic} + \text{3rd harmonic} + \cdots \text{nth harmonic}$$

2-3-2 Wave Symmetry

Simply stated, wave symmetry describes the symmetry of a waveform in the time domain, that is, its relative position with respect to the horizontal (time) and vertical (amplitude) axes.

2-3-3 Even Symmetry

If a periodic voltage waveform is *symmetric* about the vertical axis, it is said to have *axes,* or *mirror, symmetry* and is called an *even function.* For all even functions, the β coefficients in Equation 2-1 are zero. Therefore, the signal simply contains a dc component and the cosine terms (note that a cosine wave is itself an even function). Even functions satisfy the condition

$$f(t) = f(-t) \qquad \textbf{(2-2)}$$

Equation 2-2 states that the magnitude and polarity of the function at $+t$ is equal to the magnitude and polarity at $-t$. A waveform that contains only the even functions is shown in Figure 2-4a.

2-3-4 Odd Symmetry

If a periodic voltage waveform is symmetric about a line midway between the vertical axis and the negative horizontal axis (i.e., the axes in the second and fourth quadrants) and passing through the coordinate origin, it is said to have *point,* or *skew, symmetry* and is called an *odd function.* For all odd functions, the α coefficients in Equation 2-1 are zero. Therefore, the signal simply contains a dc component and the sine terms (note that a sine wave is itself an odd function). Odd functions must be mirrored first in the *Y*-axis and then in the *X*-axis for superposition. Thus,

$$f(t) = -f(-t) \qquad \textbf{(2-3)}$$

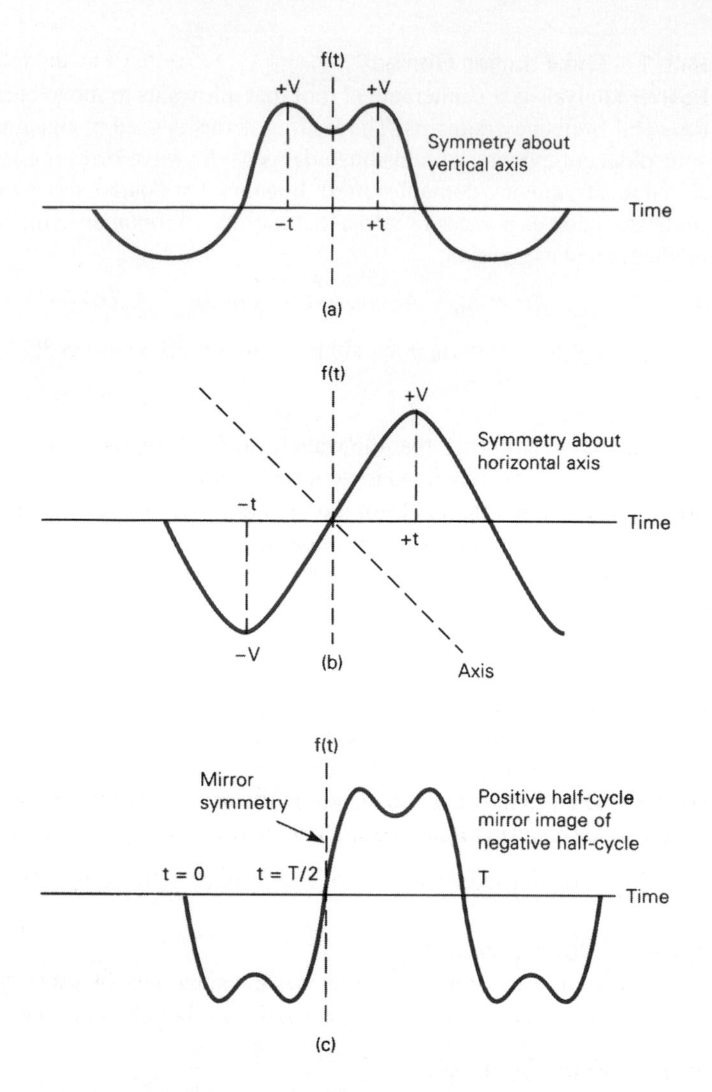

FIGURE 2-4 Wave symmetries: (a) even symmetry; (b) odd symmetry; (c) half-wave symmetry

Equation 2-3 states that the magnitude of the function at $+t$ is equal to the negative of the magnitude at $-t$ (i.e., equal in magnitude but opposite in sign). A periodic waveform that contains only the odd functions is shown in Figure 2-4b.

2-3-5 Half-Wave Symmetry

If a periodic voltage waveform is such that the waveform for the first half cycle ($t = 0$ to $t = T/2$) repeats itself except with the opposite sign for the second half cycle ($t = T/2$ to $t = T$), it is said to have *half-wave symmetry.* For all waveforms with half-wave symmetry, the even harmonics in the series for both the sine and cosine terms are zero. Therefore, half-wave functions satisfy the condition

$$f(t) = \frac{-f(T + t)}{2} \tag{2-4}$$

A periodic waveform that exhibits half-wave symmetry is shown in Figure 2-4c. It should be noted that a waveform can have half wave as well as either odd or even symmetry at the same time. The coefficients A_0, B_1, B_n and A_1 to A_n can be valuated using the following integral formulas:

$$A_0 = \frac{1}{T}\int_0^T f(t)\, dt \tag{2-5}$$

$$A_n = \frac{2}{T}\int_0^T f(t) \cos n\omega t\, dt \tag{2-6}$$

$$B_n = \frac{2}{T}\int_0^T f(t) \sin n\omega t\, dt \tag{2-7}$$

Solving Equations 2-5 to 2-7 requires integral calculus, which is beyond the intent of this book. Therefore, in subsequent discussions, the appropriate solutions are given.

Table 2-1 is a summary of the Fourier series for several of the more common nonsinusoidal periodic waveforms.

Example 2-1

For the train of square waves shown in Figure 2-5,

a. Determine the peak amplitudes and frequencies of the first five odd harmonics.

b. Draw the frequency spectrum.

c. Calculate the total instantaneous voltage for several times and sketch the time-domain waveform.

Solution **a.** From inspection of the waveform in Figure 2-5, it can be seen that the average dc component is 0 V and that the waveform has both odd and half-wave symmetry. Evaluating Equations 2-5 to 2-7 yields the following Fourier series for a square wave with odd symmetry:

$$v(t) = V_0 + \frac{4V}{\pi}\left[\sin \omega t + \frac{1}{3}\sin 3\omega t + \frac{1}{5}\sin 5\omega t + \frac{1}{7}\sin 7\omega t + \frac{1}{9}\sin 9\omega t + \cdots\right] \tag{2-8}$$

where
$v(t)$ = time-varying voltage
V_0 = average dc voltage (volts)
V = peak amplitude of the square wave (volts)
$\omega = 2\pi f$ (radians per second)
T = period of the square wave (seconds)
f = fundamental frequency of the square wave ($1/T$) (hertz)

The fundamental frequency of the square wave is

$$f = \frac{1}{T} = \frac{1}{1 \text{ ms}} = 1 \text{ kHz}$$

From Equation 2-8, it can be seen that the frequency and amplitude of the nth odd harmonic can be determined from the following expressions:

$$f_n = n \times f \tag{2-9}$$

$$V_n = \frac{4V}{n\pi} \qquad n = \text{odd positive integer value} \tag{2-10}$$

where
n = nth harmonic (odd harmonics only for a square wave)
f = fundamental frequency of the square wave (hertz)
V_n = peak amplitude of the nth harmonic (volts)
f_n = frequency of the nth harmonic (hertz)
V = peak amplitude of the square wave (volts)

Table 2-1 Fourier Series Summary

	Waveform	Fourier Series
odd	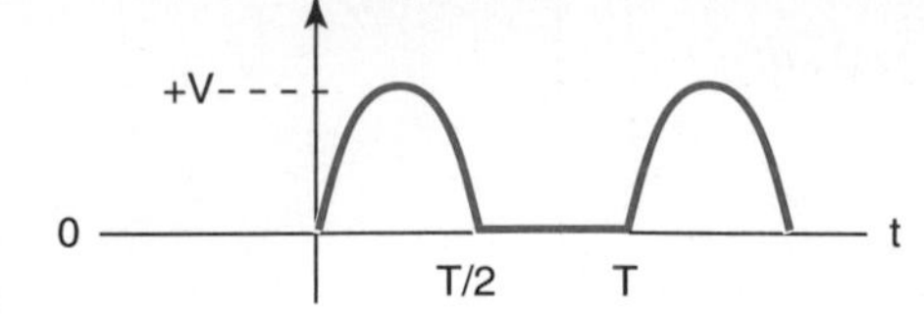	$v(t) = \frac{V}{\pi} + \frac{V}{2}\sin\omega t - \frac{2V}{3\pi}\cos 2\omega t - \frac{2V}{15\pi}\cos 4\omega t + \cdots$ $v(t) = \frac{V}{\pi} + \frac{V}{2}\sin\omega t + \sum_{N=2}^{\infty}\frac{V[1+(-1)^N]}{\pi(1-N^2)}\cos N\omega t$
even	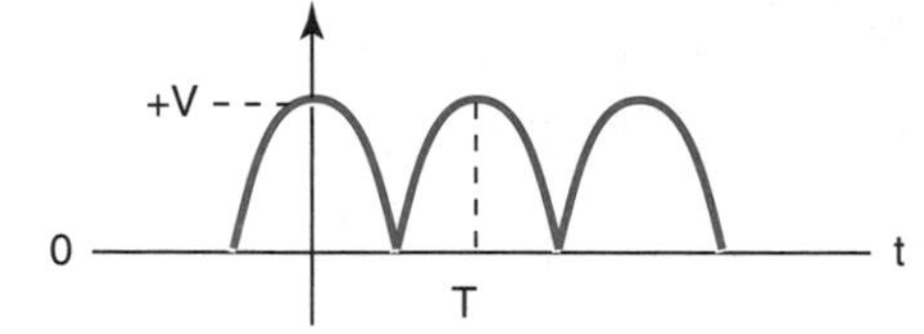	$v(t) = \frac{2V}{\pi} + \frac{4V}{3\pi}\cos\omega t - \frac{4V}{15\pi}\cos 2\omega t + \cdots$ $v(t) = \frac{2V}{\pi} + \sum_{N=1}^{\infty}\frac{4V(-1)^N}{\pi[1-(2N)^2]}\cos N\omega t$
odd	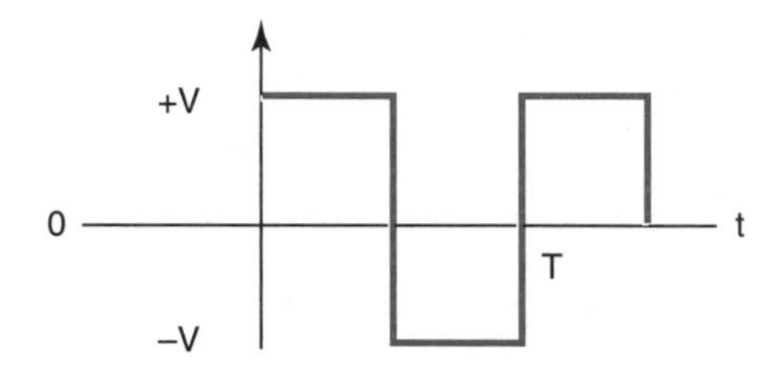	$v(t) = \frac{4V}{\pi}\sin\omega t + \frac{4V}{3\pi}\sin 3\omega t + \cdots$ $v(t) = \sum_{N=\text{odd}}^{\infty}\frac{4V}{N\pi}\sin N\omega t$
even	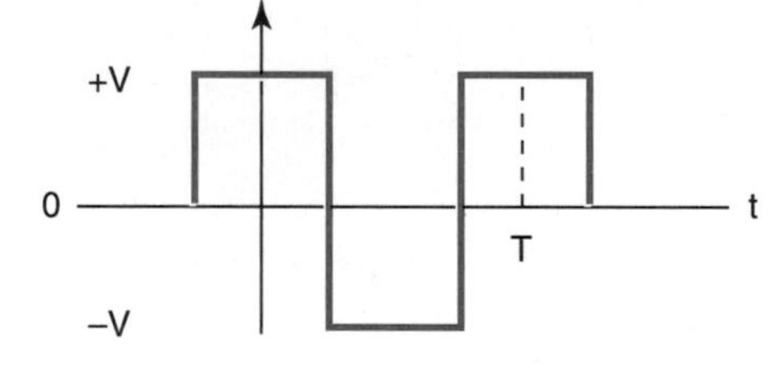	$v(t) = \frac{4V}{\pi}\cos\omega t - \frac{4V}{3\pi}\cos 3\omega t + \frac{4V}{5\pi}\cos 5\omega t + \cdots$ $v(t) = \sum_{N=\text{odd}}^{\infty}\frac{V\sin N\pi/2}{N\pi/2}\cos N\omega t$
even	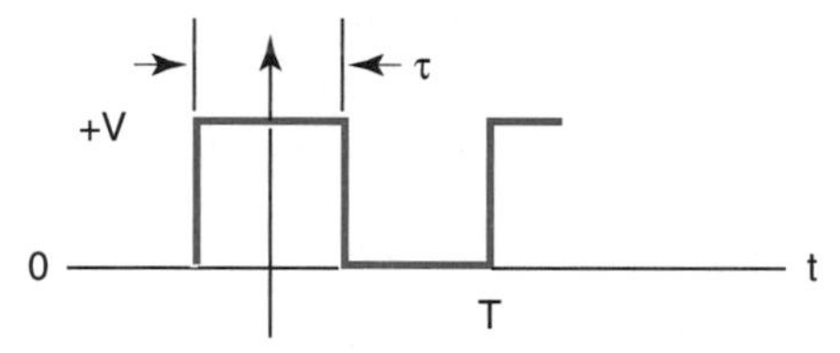	$v(t) = \frac{v\tau}{T} + \sum_{N=1}^{\infty}\left(\frac{2V\tau}{T}\,\frac{\sin N\omega t/T}{N\pi t/T}\right)\cos N\pi t$
even	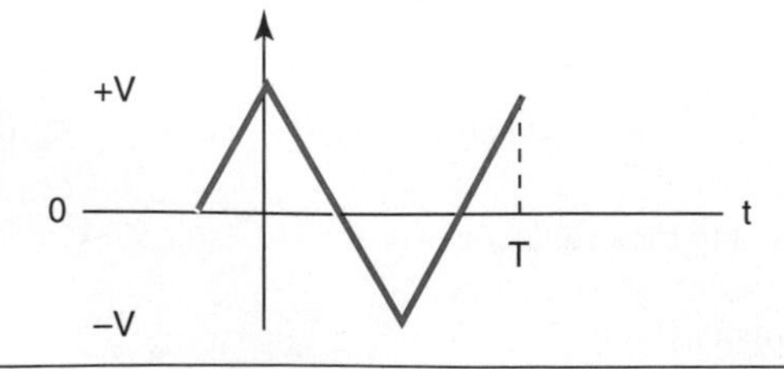	$v(t) = \frac{8V}{\pi^2}\cos\omega t + \frac{8V}{(3\pi)^2}\cos 3\omega t + \frac{8V}{(5\pi)^2}\cos 5\omega t + \cdots$ $v(t) = \sum_{N=\text{odd}}^{\infty}\frac{8V}{(N\pi)^2}\cos N\omega t$

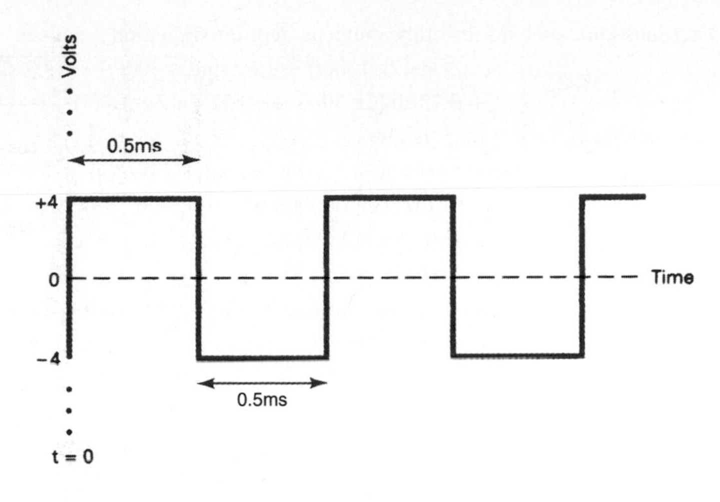

FIGURE 2-5 Waveform for Example 2-1

Substituting $n = 1$ into Equations 2-9 and 2-10 gives

$$V_1 = \frac{4(4)}{\pi} = 5.09 \text{ V}_p \qquad f_1 = 1 \times 1000 = 1000 \text{ Hz}$$

Substituting $n = 3, 5, 7$, and 9 into Equations 2-9 and 2-10 gives

n	Harmonic	Frequency (Hz)	Peak Voltage (V_p)
1	First	1000	5.09
3	Third	3000	1.69
5	Fifth	5000	1.02
7	Seventh	7000	0.73
9	Ninth	9000	0.57

b. The frequency spectrum is shown in Figure 2-6.

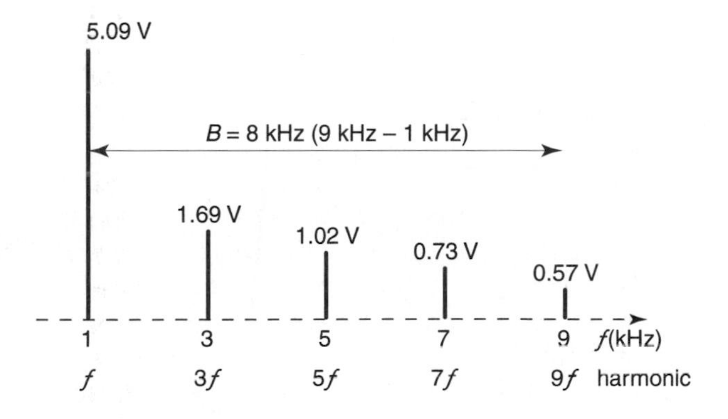

FIGURE 2-6 Frequency spectrum for Example 2-1

c. Substituting the results of the previous steps into Equation 2-8 gives

$$v(t) = 5.09 \sin[2\pi\, 1000t] + 1.69 \sin[2\pi\, 3000t] + 1.02 \sin[2\pi\, 5000t] + 0.73 \sin[2\pi\, 7000t] + 0.57 \sin[2\pi\, 9000t]$$

Solving for $v(t)$ at $t = 62.5\ \mu s$ gives

$$v(t) = 5.09 \sin[2\pi\, 1000(62.5\ \mu s)] + 1.69 \sin[2\pi\, 3000(62.5\ \mu s)] + 1.02 \sin[2\pi\, 5000(62.5\ \mu s)] + 0.73 \sin[2\pi\, 7000(62.5\ \mu s)] + 0.57 \sin[2\pi\, 9000(62.5\ \mu s)]$$

$$v(t) = 4.51\ V$$

Solving for $v(t)$ for several additional values of time gives the following table:

Time (μs)	$v(t)$ (Volts Peak)
0	0
62.5	4.51
125	3.96
250	4.26
375	3.96
437.5	4.51
500	0
562.5	− 4.51
625	− 3.96
750	− 4.26
875	− 3.96
937.5	− 4.51
1000	0

The time-domain signal is derived by plotting the times and voltages calculated above on graph paper and is shown in Figure 2-7. Although the waveform shown is not an exact square wave, it does closely resemble one. To achieve a more accurate time-domain waveform, it would be necessary to solve for $v(t)$ for more values of time than are shown in this diagram.

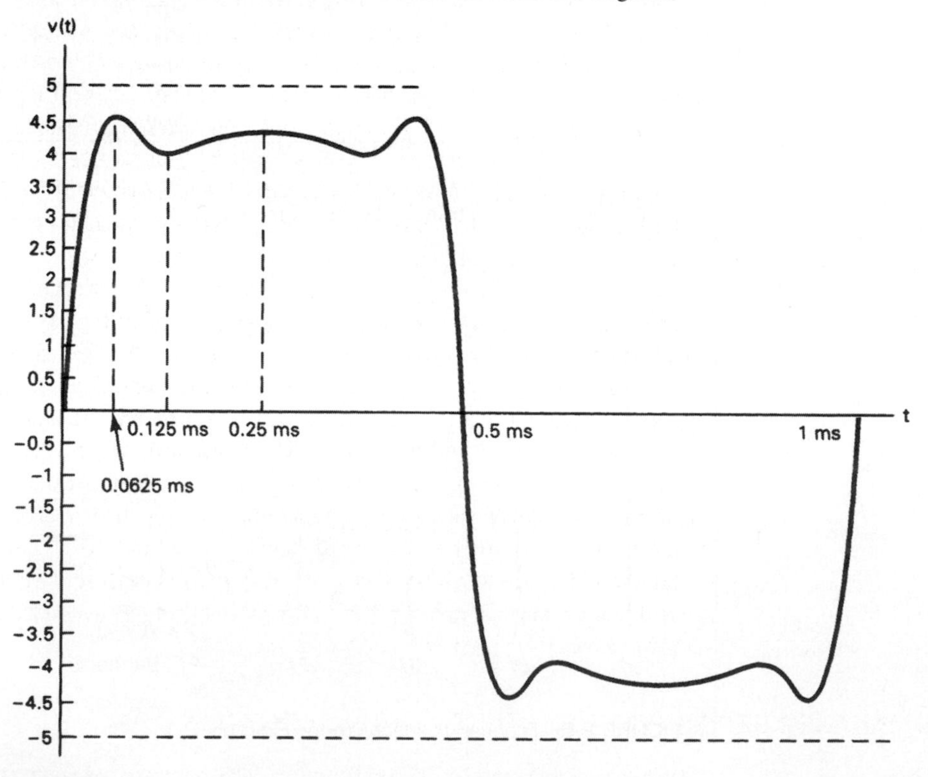

FIGURE 2-7 Time-domain signal for Example 2-1

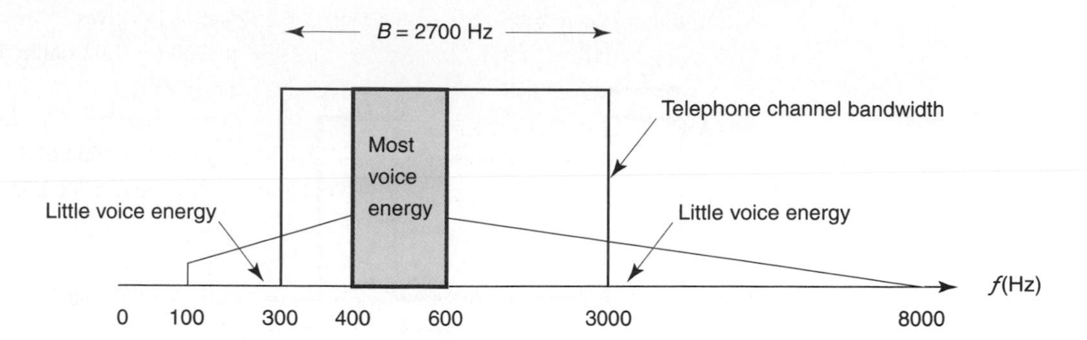

FIGURE 2-8 Voice-frequency spectrum and telephone circuit bandwidth

2-4 FREQUENCY SPECTRUM AND BANDWIDTH

The *frequency spectrum* of a waveform consists of all the frequencies contained in the waveform and their respective amplitudes plotted in the frequency domain. Frequency spectrums can show absolute values of frequency-versus-voltage or frequency-versus-power level, or they can plot frequency-versus-some relative unit of measurement, such as decibels (dB).

The term *bandwidth* can be used in several ways. The bandwidth of a frequency spectrum is the range of frequencies contained in the spectrum. The bandwidth is calculated by subtracting the lowest frequency from the highest. The bandwidth of the frequency spectrum shown in Figure 2-6 for Example 2-1 is 8000 Hz (9000 − 1000).

The bandwidth of an information signal is simply the difference between the highest and lowest frequencies contained in the information, and the bandwidth of a communications channel is the difference between the highest and lowest frequencies that the channel will allow to pass through it (i.e., its *passband*). The bandwidth of a communications channel must be sufficiently large (wide) to pass all significant information frequencies. In other words, the bandwidth of a communications channel must be equal to or greater than the bandwidth of the information signal. Speech contains frequency components ranging from approximately 100 Hz to 8 kHz, although most of the energy is distributed in the 400-Hz to 600-Hz band with the fundamental frequency of typical human voice about 500 Hz. However, standard telephone circuits have a passband between 300 Hz and 3000 Hz, as shown in Figure 2-8, which equates to a bandwidth of 2700 Hz (3000 − 300). Twenty-seven hundred hertz is well beyond what is necessary to convey typical speech information. If a cable television transmission system has a passband from 500 kHz to 5000 kHz, it has a bandwidth of 4500 kHz (4.5 MHz). As a general rule, a communications channel cannot propagate a signal through it that is changing at a rate that exceeds the bandwidth of the channel.

In general, the more complex the information signal, the more bandwidth required to transport it through a communications system in a given period of time. Approximately 3 kHz of bandwidth is required to propagate one voice-quality analog telephone conversation. In contrast, it takes approximately 32 kHz of bandwidth to propagate one voice-quality digital telephone conversation. Commercial FM broadcasting stations require 200 kHz of bandwidth to propagate high-fidelity music signals, and almost 6 MHz of bandwidth is required for broadcast-quality television signals.

2-5 FOURIER SERIES FOR A RECTANGULAR WAVEFORM

When analyzing electronic communications circuits, it is often necessary to use *rectangular pulses.* A waveform showing a string of rectangular pulses is given in Figure 2-9. The *duty*

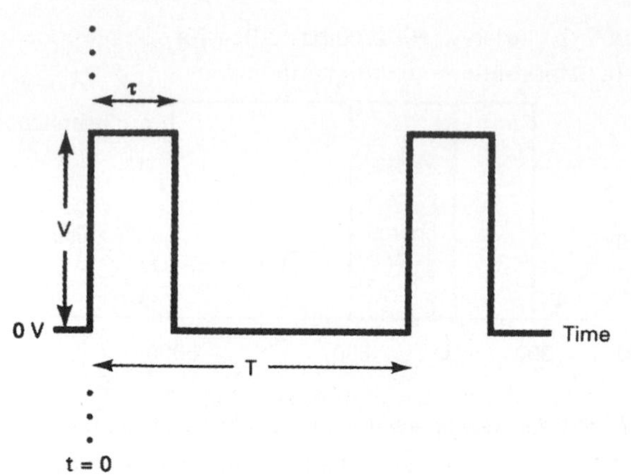

FIGURE 2-9 Rectangular pulse waveform

cycle (DC) for the waveform is the ratio of the active time of the pulse to the period of the waveform. Mathematically, duty cycle is

$$DC = \frac{\tau}{T} \tag{2-11}$$

$$DC(\%) = \frac{\tau}{T} \times 100 \tag{2-12}$$

where DC = duty cycle as a decimal
DC(%) = duty cycle as a percent
τ = pulse width of the rectangle wave (seconds)
T = period of the rectangular wave (seconds)

Regardless of the duty cycle, a rectangular waveform is made up of a series of harmonically related sine waves. However, the amplitude of the spectral components depends on the duty cycle. The Fourier series for a rectangular voltage waveform with even symmetry is

$$v(t) = \frac{V_\tau}{T} + \frac{2V\tau}{T}\left[\frac{\sin x}{x}(\cos \omega t) + \frac{\sin 2x}{2x}(\cos 2\omega t) + \cdots + \frac{\sin nx}{nx}(\cos n\omega t)\right] \tag{2-13}$$

where $v(t)$ = time-varying voltage wave
τ = pulse width of the rectangular wave (seconds)
T = period of the rectangular wave (seconds)
$x = \pi\,(\tau/T)$
n = nth harmonic and can be any positive integer value
V = peak pulse amplitude (volts)

From Equation 2-13, it can be seen that a rectangular waveform has a 0-Hz (dc) component equal to

$$V_0 = V \times \frac{\tau}{T} \text{ or } V \times DC \tag{2-14}$$

where V_0 = dc voltage (volts)
DC = duty cycle as a decimal
τ = pulse width of rectangular wave (seconds)
T = period of rectangular wave (seconds)

The narrower the pulse width is, the smaller the dc component will be. Also, from Equation 2-13, the amplitude of the nth harmonic is

$$V_n = \frac{2V\tau}{T} \times \frac{\sin nx}{nx} \qquad \textbf{(2-15)}$$

or

$$V_n = \frac{2V\tau}{T} \times \frac{\sin[(n\pi\tau)/T]}{(n\pi\tau)/T} \qquad \textbf{(2-16)}$$

where V_n = peak amplitude of the nth harmonic (volts)
n = nth harmonic (any positive integer)
π = 3.14159 radians
V = peak amplitude of the rectangular wave (volts)
τ = pulse width of the rectangular wave (seconds)
T = period of the rectangular wave (seconds)

The $(\sin x)/x$ function is used to describe repetitive pulse waveforms. Sin x is simply a sinusoidal waveform whose instantaneous amplitude depends on x and varies both positively and negatively between its peak amplitudes at a sinusoidal rate as x increases. With only x in the denominator, the denominator increases with x. Therefore, a $(\sin x)/x$ function is simply a damped sine wave in which each successive peak is smaller than the preceding one. A $(\sin x)/x$ function is shown in Figure 2-10.

Figure 2-11 shows the frequency spectrum for a rectangular pulse with a pulse width-to-period ratio of 0.1. It can be seen that the amplitudes of the harmonics follow a damped sinusoidal shape. At the frequency whose period equals $1/\tau$ (i.e., at frequency $10f$ hertz), there is a 0-V component. A second null occurs at $20f$ hertz (period $= 2/\tau$), a third at $30f$ hertz (period $= 3/\tau$), and so on. All spectrum components between 0 Hz and the first null frequency are considered in the first lobe of the frequency spectrum and are positive. All spectrum components between the first and second null frequencies are in the second lobe and are negative, components between the second and third nulls are in the third lobe and positive, and so on.

The following characteristics are true for all repetitive rectangular waveforms:

1. The dc component is equal to the pulse amplitude times the duty cycle.
2. There are 0-V components at frequency $1/\tau$ hertz and all integer multiples of that frequency providing $T = n\tau$, where n = any odd integer.
3. The amplitude-versus-frequency time envelope of the spectrum components take on the shape of a damped sine wave in which all spectrum components in odd-numbered lobes are positive and all spectrum components in even-numbered lobes are negative.

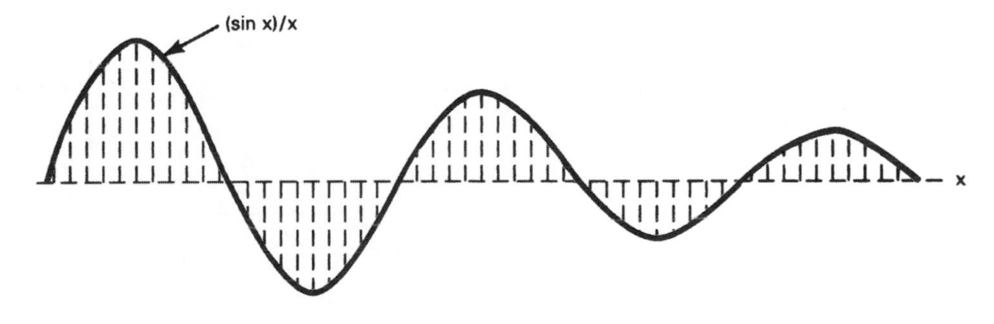

FIGURE 2-10 (sin x)/x function

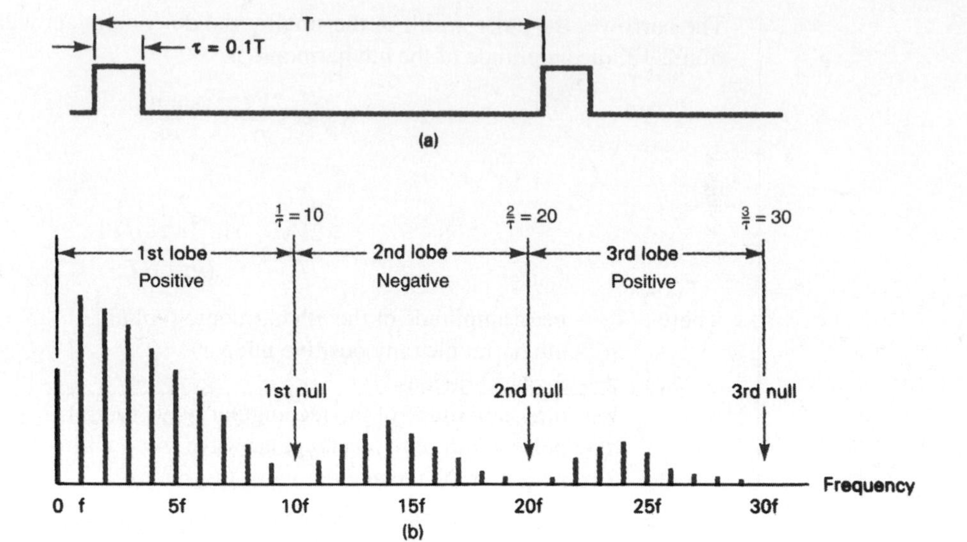

FIGURE 2-11 (sin *x*)/*x* function: (a) rectangular pulse waveform; (b) frequency spectrum

Example 2-2

For the pulse waveform shown in Figure 2-12,

a. Determine the dc component.

b. Determine the peak amplitudes of the first 10 harmonics.

c. Plot the (sin *x*)/*x* function.

d. Sketch the frequency spectrum.

Solution **a.** From Equation 1-16, the dc component is

$$V_0 = \frac{1(0.4 \text{ ms})}{2 \text{ ms}} = 0.2 \ V$$

b. The peak amplitudes of the first 10 harmonics are determined by substituting the values for τ, T, V, and n into Equation 2-16, as follows:

$$V_n = 2(1)\left(\frac{0.4 \text{ ms}}{2 \text{ ms}}\right)\left\{\frac{\sin[(n\pi)(0.4 \text{ ms}/2 \text{ ms})]}{(n\pi)(0.4 \text{ ms}/2 \text{ ms})}\right\}$$

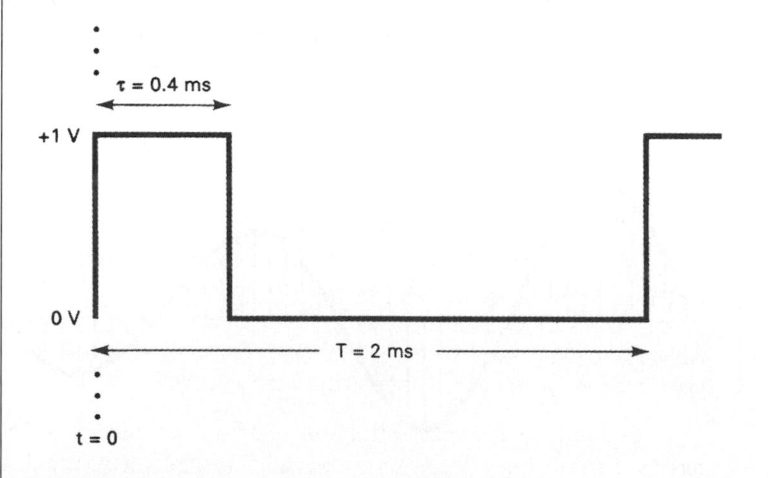

FIGURE 2-12 Pulse waveform for Example 2-2

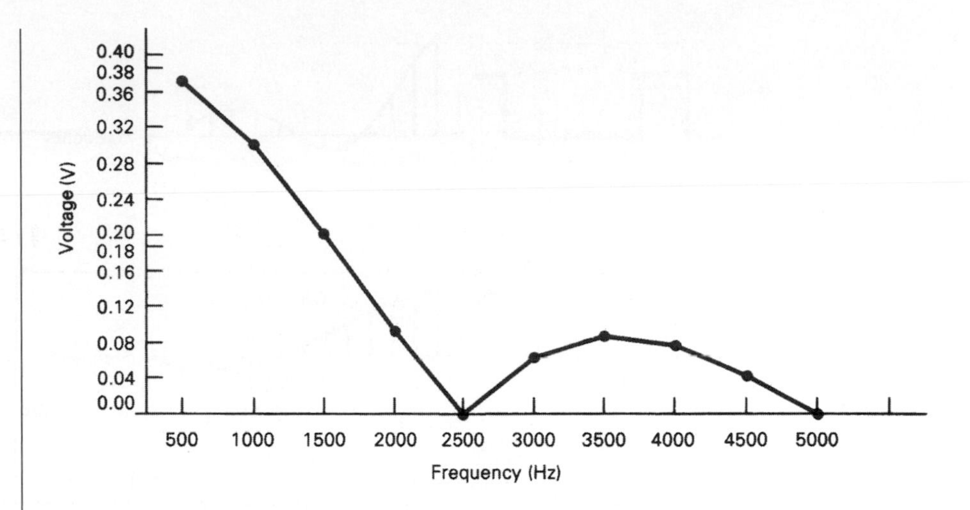

FIGURE 2-13 (sin *x*)/*x* function for Example 2-2

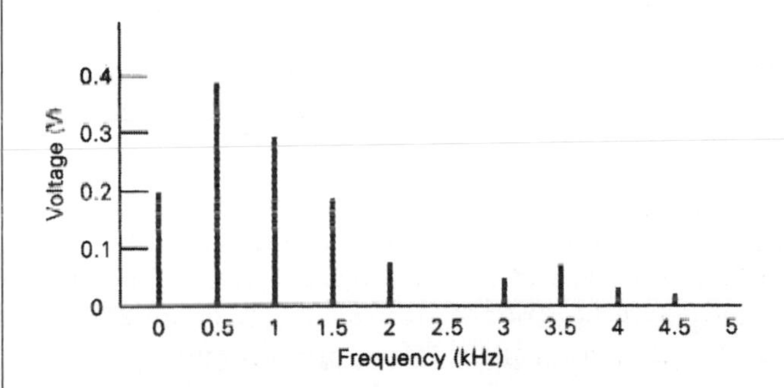

FIGURE 2-14 Frequency spectrum for Example 2-2

n	Frequency (Hz)	Amplitude (Volts)
0	0	0.2 V dc
1	500	0.374 V_p
2	1000	0.303 V_p
3	1500	0.202 V_p
4	2000	0.094 V_p
5	2500	0.0 V
6	3000	− 0.063 V_p
7	3500	− 0.087 V_p
8	4000	− 0.076 V_p
9	4500	− 0.042 V_p
10	5000	0.0 V

c. The (sin *x*)/*x* function is shown in Figure 2-13.
d. The frequency spectrum is shown in Figure 2-14.
Although the frequency components in the even lobes are negative, it is customary to plot all voltages in the positive direction on the frequency spectrum.

Figure 2-15 shows the effect that reducing the duty cycle (i.e., reducing the τ/T ratio) has on the frequency spectrum for a nonsinusoidal waveform. It can be seen that narrowing the pulse width produces a frequency spectrum with a more uniform amplitude. In fact,

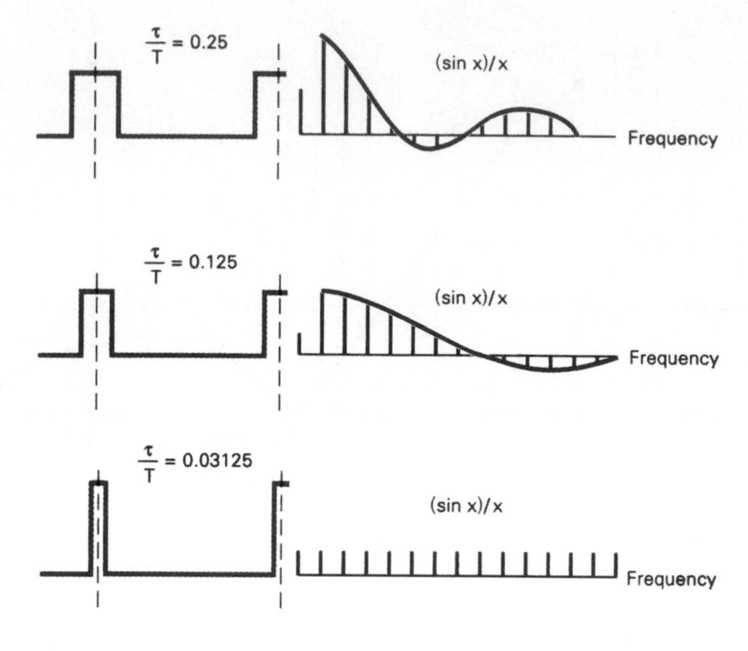

FIGURE 2-15 Effects of reducing the *t/T* ratio (either decreasing *t* or increasing *T*)

for infinitely narrow pulses, the frequency spectrum comprises an infinite number of harmonically related frequencies of equal amplitude. Such a spectrum is impossible to produce, let alone to propagate, which explains why it is difficult to produce extremely narrow pulses. Increasing the period of a rectangular waveform while keeping the pulse width constant has the same effect on the frequency spectrum.

2-5-1 Power and Energy Spectra

In the previous sections, we used the Fourier series to better understand the frequency- and time-domain representation of a complex signal. Both the frequency and the time domain can be used to illustrate the relationship of signal voltages (magnitudes) with respect to either frequency or time for a time-varying signal.

However, there is another important application of the Fourier series. The goal of a communications channel is to transfer electromagnetic energy from a source to a destination. Thus, the relationship between the amount of energy transmitted and the amount received is an important consideration. Therefore, it is important that we examine the relationship between energy and power versus frequency.

Electrical power is the rate at which energy is dissipated, delivered, or used and is a function of the square of the voltage or current ($P = E^2/R$ or $P = I^2 \times R$). For power relationships, in the Fourier equation, $f(t)$ is replaced by $[f(t)]^2$. Figure 2-16 shows the power spectrum for a rectangular waveform with a 25% duty cycle. It resembles its voltage-versus-frequency spectrum except it has more lobes and a much larger primary lobe. Note also that all the lobes are positive because there is no such thing as negative power.

From Figure 2-16, it can be seen that the power in a pulse is dispersed throughout a relatively wide frequency spectrum. However, note that most of that power is within the primary lobe. Consequently, if the bandwidth of a communications channel is sufficiently wide to pass only the frequencies within the primary lobe, it will transfer most of the energy contained in the pulse to the receiver.

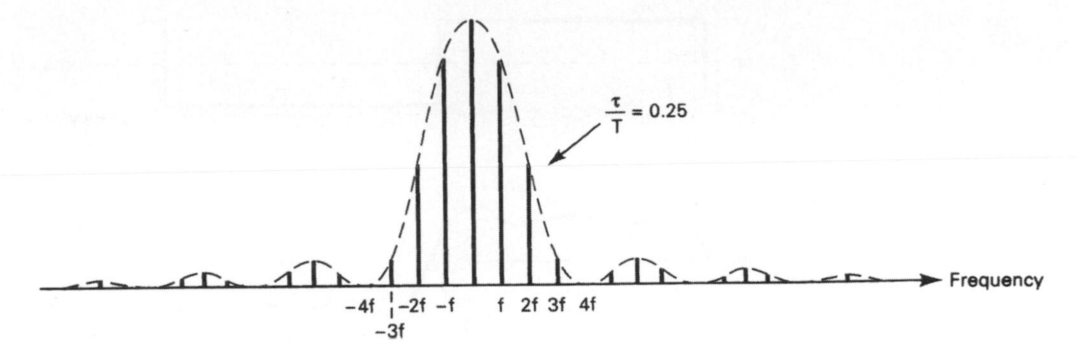

FIGURE 2-16 Power spectrum of a 25% duty cycle rectangular pulse

2-5-2 Discrete and Fast Fourier Transforms

Many waveforms encountered in typical communications systems cannot be satisfactorily defined by mathematical expressions; however, their frequency-domain behavior is of primary interest. Often there is a need to obtain the frequency-domain behavior of signals that are being collected in the time domain (i.e., in real time). This is why the *discrete Fourier transform* was developed. With the discrete Fourier transform, a time-domain signal is sampled at discrete times. The samples are fed into a computer where an algorithm computes the transform. However, the computation time is proportional to n^2, where n is the number of samples. For any reasonable number of samples, the computation time is excessive. Consequently, in 1965 a new algorithm called the *fast Fourier transform* (FFT) was developed by Cooley and Tukey. With the FFT the computing time is proportional to $n \log 2n$ rather than n^2. The FFT is now available as a subroutine in many scientific subroutine libraries at large computer centers.

2-5-3 Effects of Bandlimiting on Signals

All communications channels have a limited bandwidth and, therefore, have a limiting effect on signals that are propagated through them. We can consider a communications channel to be equivalent to an ideal *linear-phase filter* with a finite bandwidth. If a nonsinusoidal repetitive waveform passes through an ideal low-pass filter, the harmonic frequency components that are higher in frequency than the upper cutoff frequency of the filter are removed. Consequently, both the frequency content and the shape of the waveform are changed. Figure 2-17a shows the time-domain waveform for the square wave used in Example 2-1. If this waveform is passed through a low-pass filter with an upper cutoff frequency of 8 kHz, frequencies above the eighth harmonic (9 kHz and above) are cut off, and the waveform shown in Figure 2-17b results. Figures 2-17c, d, and e show the waveforms produced when low-pass filters with upper cutoff frequencies of 6 kHz, 4 kHz, and 2 kHz are used, respectively.

It can be seen from Figure 2-17 that *bandlimiting* a signal changes the frequency content and, thus, the shape of its waveform and, if sufficient bandlimiting is imposed, the waveform eventually comprises only the fundamental frequency. In a communications system, bandlimiting reduces the information capacity of the system, and, if excessive bandlimiting is imposed, a portion of the information signal can be removed from the composite waveform.

2-5-3-1 Mixing. *Mixing* is the process of combining two or more signals and is an essential process in electronic communications. In essence, there are two ways in which signals can be combined or mixed: linearly and nonlinearly.

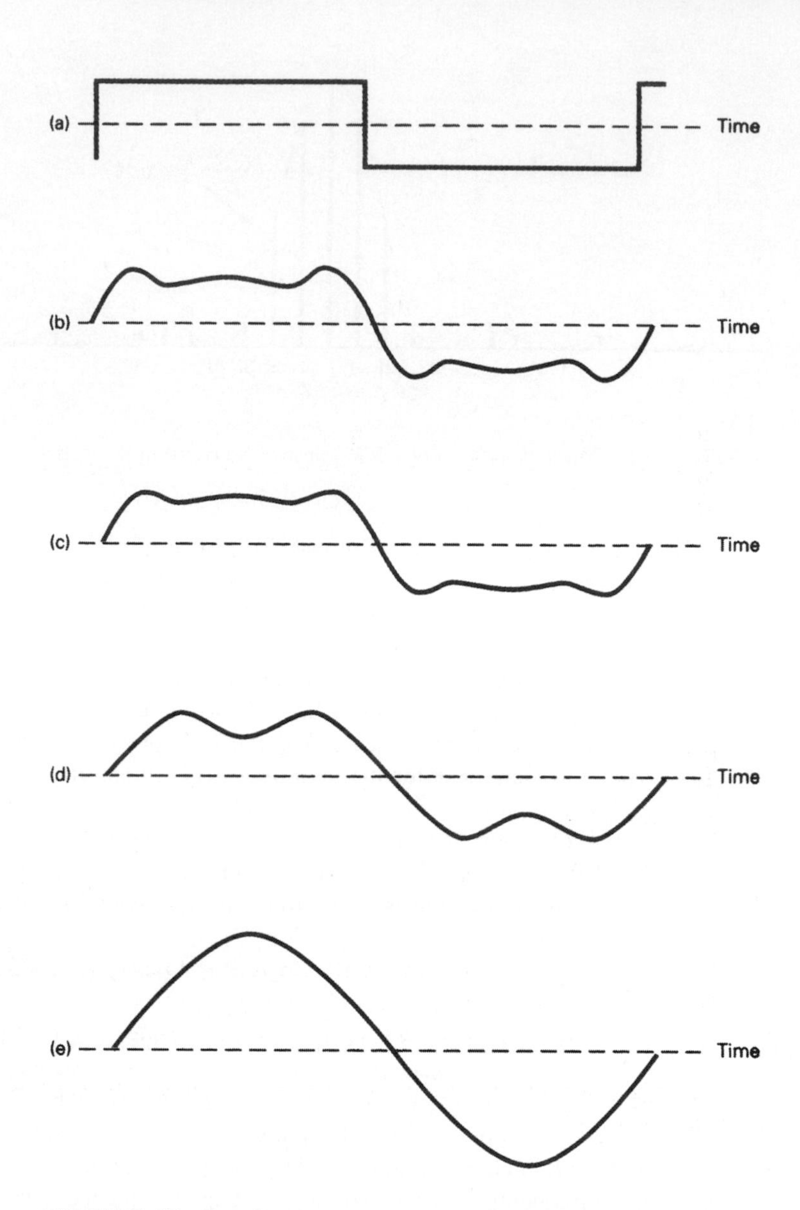

FIGURE 2-17 Bandlimiting signals: (a) 1-kHz square wave; (b) 1-kHz square wave bandlimited to 8 kHz; (c) 1-kHz square wave bandlimited to 6 kHz; (d) 1-kHz square wave bandlimited to 4 kHz; (e) 1-kHz square wave bandlimited to 2 kHz

2-6 LINEAR SUMMING

Linear summing occurs when two or more signals combine in a linear device, such as a passive network or a small-signal amplifier. The signals combine in such a way that no new frequencies are produced, and the combined waveform is simply the linear addition of the individual signals. In the audio recording industry, linear summing is sometimes called linear *mixing;* however, in radio communications, mixing almost always implies a nonlinear process.

2-6-1 Single-Input Frequency

Figure 2-18a shows the amplification of a single-input frequency by a linear amplifier. The output is simply the original input signal amplified by the gain of the amplifier (A). Figure 2-18b

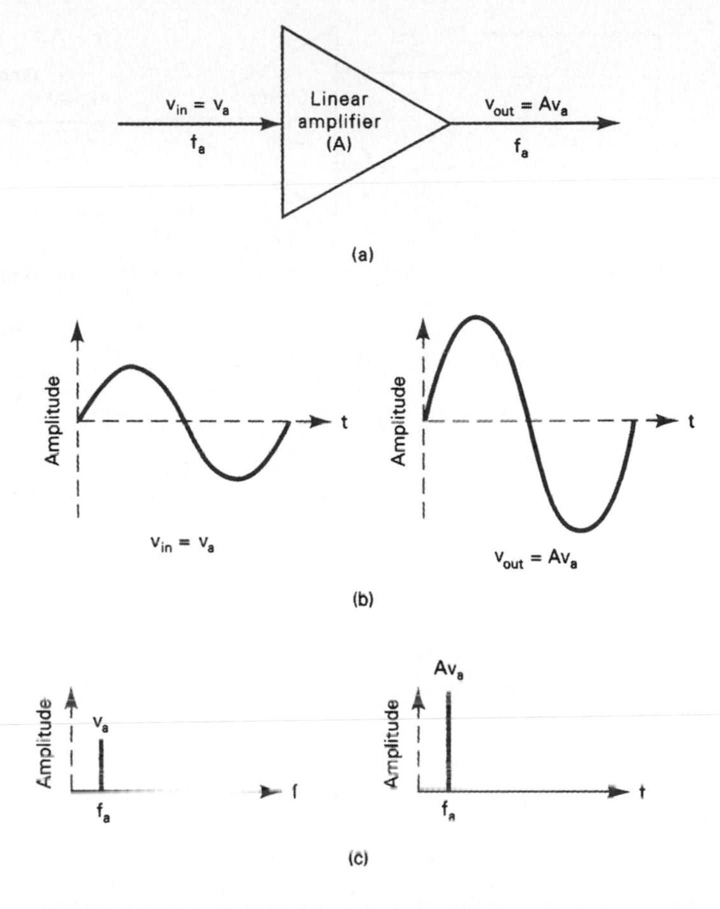

FIGURE 2-18 Linear amplification of a single-input frequency: (a) linear amplification; (b) time domain; (c) frequency domain

shows the output signal in the time domain, and Figure 2-18c shows the frequency domain. Mathematically, the output is

$$v_{out} = Av_{in} \tag{2-17}$$

or

$$v_{in} = V_a \sin 2\pi f_a t$$

Thus,

$$v_{out} = AV_a \sin 2\pi f_a t$$

2-6-2 Multiple-Input Frequencies

Figure 2-19a shows two input frequencies combining in a small-signal amplifier. Each input signal is amplified by the gain (A). Therefore, the output is expressed mathematically as

$$v_{out} = Av_{in}$$

where

$$v_{in} = V_a \sin 2\pi f_a t + V_b \sin 2\pi f_b t$$

Therefore,

$$v_{out} = A(V_a \sin 2\pi f_a t + V_b \sin 2\pi f_b t) \tag{2-18}$$

or

$$v_{out} = A(V_a \sin 2\pi f_a t + AV_b \sin 2\pi f_b t) \tag{2-19}$$

v_{out} is simply a complex waveform containing both input frequencies and is equal to the algebraic sum of v_a and v_b. Figure 2-19b shows the linear summation of v_a and v_b in the time domain, and Figure 2-19c shows the linear summation in the frequency domain. If additional input frequencies are applied to the circuit, they are linearly summed with v_a and

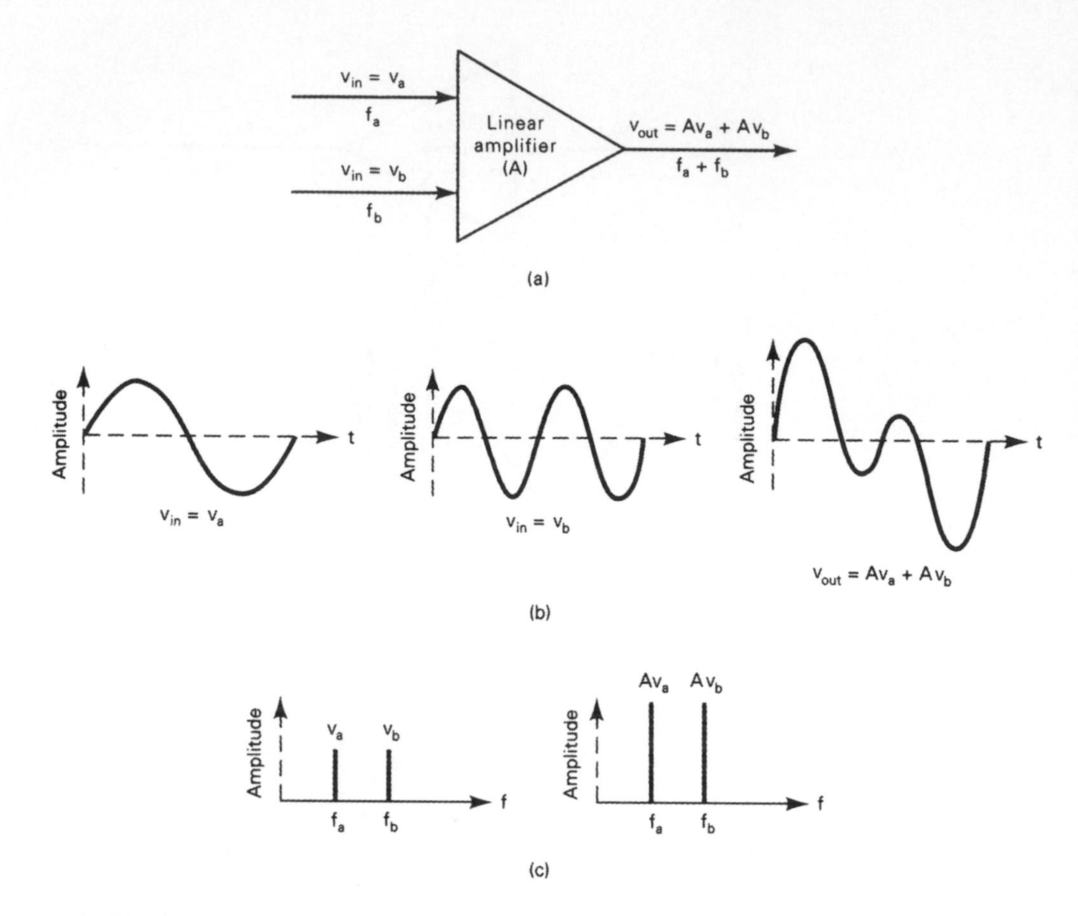

FIGURE 2-19 Linear mixing: (a) linear amplification; (b) time domain; (c) frequency domain

v_b. In high-fidelity audio systems, it is important that the output spectrum contain only the original input frequencies; therefore, linear operation is desired. However, in radio communications where modulation is essential, nonlinear mixing is often necessary.

2-7 NONLINEAR MIXING

Nonlinear mixing occurs when two or more signals are combined in a nonlinear device such as a diode or large-signal amplifier. With nonlinear mixing, the input signals combine in a nonlinear fashion and produce additional frequency components.

2-7-1 Single-Input Frequency

Figure 2-20a shows the amplification of a single-frequency input signal by a nonlinear amplifier. The output from a nonlinear amplifier with a single-frequency input signal is not a single sine or cosine wave. Mathematically, the output is in the infinite power series

$$v_{out} = Av_{in} + Bv_{in}^2 + Cv_{in}^3 \tag{2-20}$$

where $$v_{in} = V_a \sin 2\pi f_a t$$

Therefore, $$v_{out} = A(V_a \sin 2\pi f_a t) + B(V_a \sin 2\pi f_a t)^2 + C(V_a \sin 2\pi f_a t)^3 \tag{2-21}$$

where Av_{in} = linear term or simply the input signal (f_a) amplified by the gain (A)
Bv_{in}^2 = quadratic term that generates the second harmonic frequency ($2f_a$)
Cv_{in}^3 = cubic term that generates the third harmonic frequency ($3f_a$)

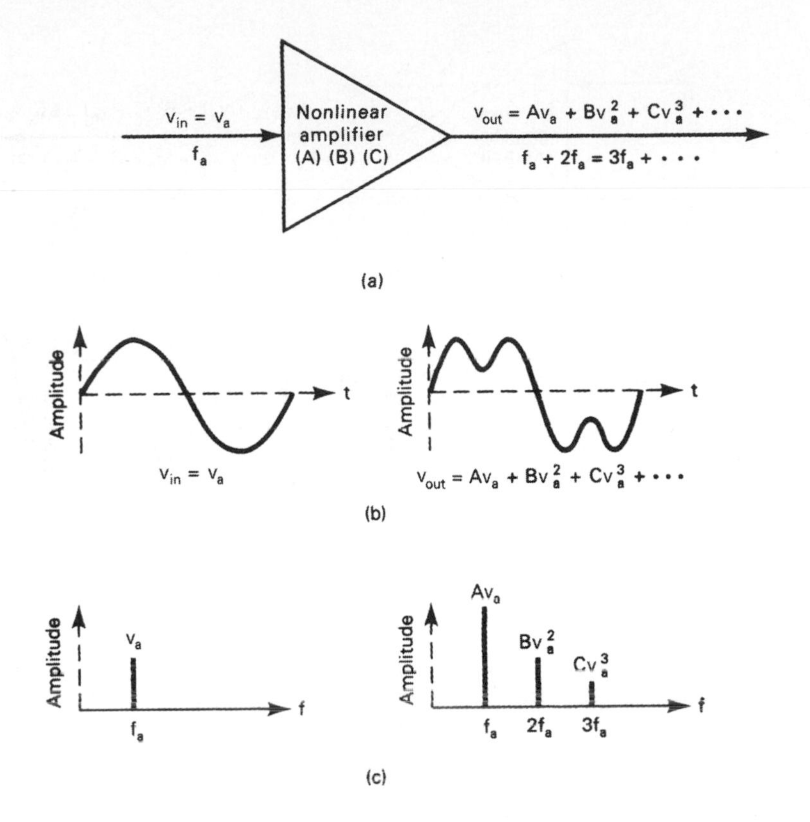

FIGURE 2-20 Nonlinear amplification of a single-input frequency: (a) nonlinear amplification; (b) time domain; (c) frequency domain

v_{in}^n produces a frequency equal to n times f. For example, Bv_{in}^2 generates a frequency equal to $2f_a$. Cv_{in}^3 generates a frequency equal to $3f_a$ and so on. Integer multiples of a *base* frequency are called *harmonics.* As stated previously, the original input frequency (f_a) is the first harmonic or the fundamental frequency, $2f_a$ is the second harmonic, $3f_a$ is the third, and so on. Figure 2-20b shows the output waveform in the time domain for a nonlinear amplifier with a single-input frequency. It can be seen that the output waveform is simply the summation of the input frequency and its higher harmonics (multiples of the fundamental frequency). Figure 2-20c shows the output spectrum in the frequency domain. Note that adjacent harmonics are separated in frequency by a value equal to the fundamental frequency, f_a.

Nonlinear amplification of a single frequency results in the generation of multiples or harmonics of that frequency. If the harmonics are undesired, it is called *harmonic distortion.* If the harmonics are desired, it is called *frequency multiplication.*

A JFET is a special-case nonlinear device that has characteristics that are approximately those of a square-law device. The output from a square-law device is

$$v_{\text{out}} = Bv_{\text{in}}^2 \tag{2-22}$$

The output from a square-law device with a single-input frequency is dc and the second harmonic. No additional harmonics are generated beyond the second. Therefore, less harmonic distortion is produced with a JFET than with a comparable BJT.

2-7-2 Multiple-Input Frequencies

Figure 2-21 shows the nonlinear amplification of two input frequencies by a large-signal (nonlinear) amplifier. Mathematically, the output of a large-signal amplifier with two input frequencies is

$$v_{\text{out}} = Av_{\text{in}} + Bv_{\text{in}}^2 + Cv_{\text{in}}^3$$

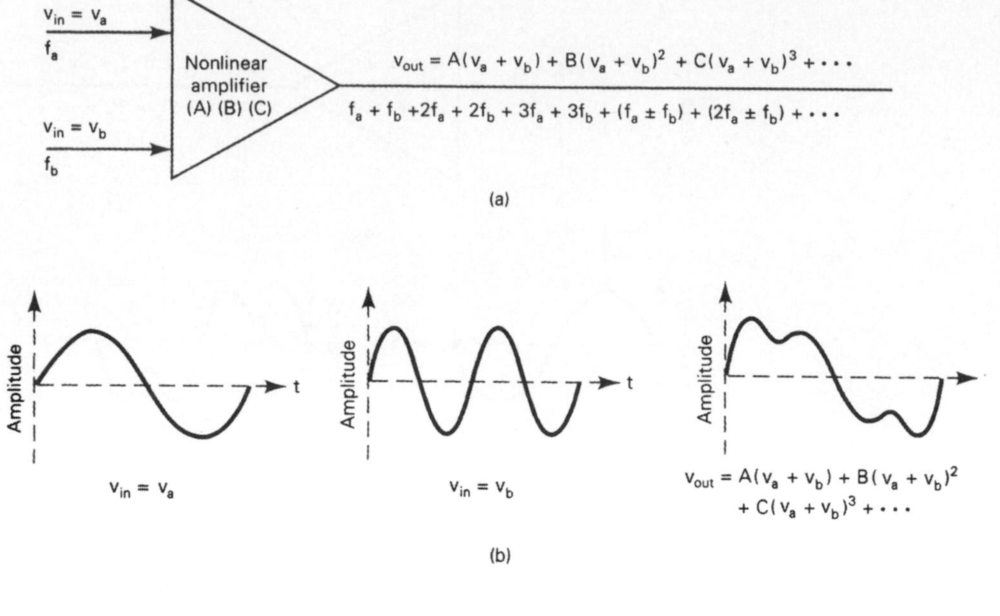

FIGURE 2-21 Nonlinear amplification of two sine waves: (a) nonlinear amplification; (b) time domain; (c) frequency domain

where

$$v_{\text{in}} = V_a \sin 2\pi f_a t + V_b \sin 2\pi f_b t$$

Therefore,

$$v_{\text{out}} = A(V_a \sin 2\pi f_a t + V_b \sin 2\pi f_b t) + B(V_a \sin 2\pi f_a t + V_b \sin 2\pi f_b t)^2 + C(V_a \sin 2\pi f_a t + V_b \sin 2\pi f_b t)^3 + \dots \tag{2-23}$$

The preceding formula is an infinite series, and there is no limit to the number of terms it can have. If the binomial theorem is applied to each higher-power term, the formula can be rearranged and written as

$$v_{\text{out}} = (Av'_a + Bv'^2_a + Cv'^3_a + \cdots) + (Av'_b + Bv_b{}^2 + Cv'^3_b + \cdots) + (2Bv'_a v'_b + 3Cv'^2_a v'_b + 3Cv'_a v'^2_b + \cdots) \tag{2-24}$$

where $v'_a = V_a \sin 2\pi f_a t$
$v'_b = V_b \sin 2\pi f_b t$

The terms in the first set of parentheses generate harmonics of f_a($2f_a$, $3f_a$, and so on). The terms in the second set of parentheses generate harmonics of f_b($2f_b$, $3f_b$, and so on). The terms in the third set of parentheses generate the cross products ($f_a + f_b$, $f_a - f_b$, $2f_a + f_b$, $2f_a - f_b$, and so on). The cross products are produced from intermodulation among the

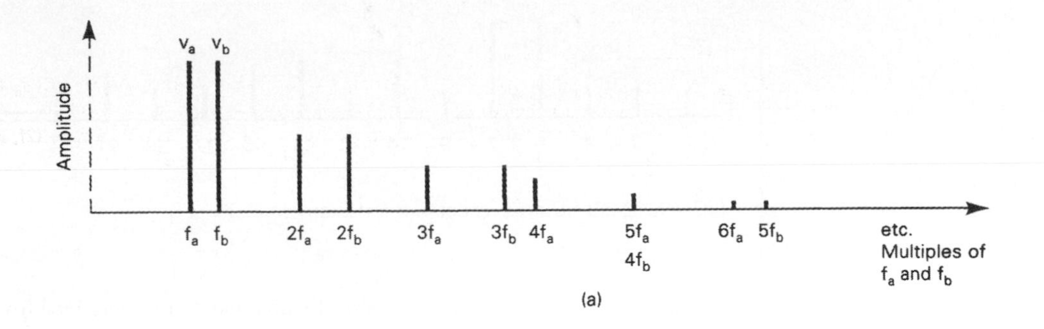

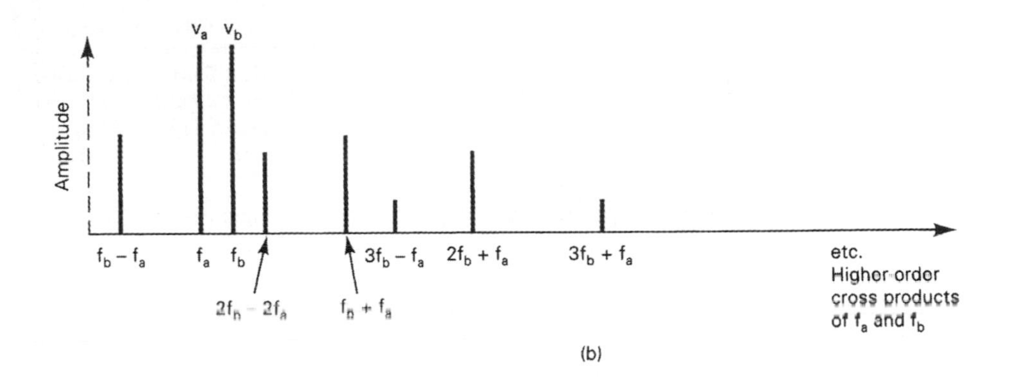

FIGURE 2-22 Output spectrum from a nonlinear amplifier with two input frequencies: (a) harmonic distortion; (b) intermodulation distortion

two original frequencies and their harmonics. The cross products are the *sum* and *difference* frequencies; they are the sum and difference of the two original frequencies, the sums and differences of their harmonics, and the sums and differences of the original frequencies and all the harmonics. An infinite number of harmonic and cross-product frequencies are produced when two or more frequencies *mix* in a nonlinear device. If the cross products are undesired, it is called *intermodulation distortion.* If the cross products are desired, it is called *modulation.* Mathematically, the sum and difference frequencies are

$$\text{cross products} = mf_a \pm nf_b \tag{2-25}$$

where m and n are positive integers between one and infinity. Figure 2-22 shows the output spectrum from a nonlinear amplifier with two input frequencies.

Intermodulation distortion is the generation of any unwanted cross-product frequency when two or more frequencies are mixed in a nonlinear device. Consequently, when two or more frequencies are amplified in a nonlinear device, both harmonic and intermodulation distortions are present in the output.

Example 2-3

For a nonlinear amplifier with two input frequencies, 5 kHz and 7 kHz,

a. Determine the first three harmonics present in the output for each input frequency.

b. Determine the cross products produced in the output for values of m and n of 1 and 2.

c. Draw the output frequency spectrum for the harmonics and cross-product frequencies determined in steps (a) and (b).

Solution **a.** The first three harmonics include the two original input frequencies of 5 kHz and 7 kHz; two times each of the original input frequencies, 10 kHz and 14 kHz; and three times each of the original input frequencies, 15 kHz and 21 kHz.

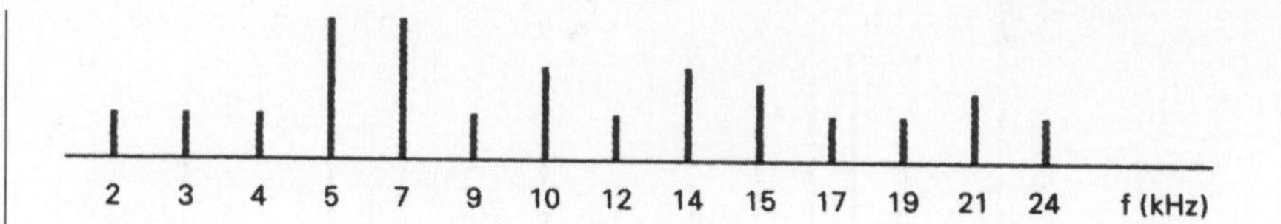

FIGURE 2-23 Output spectrum for Example 2-3

b. The cross products for values of m and n of 1 and 2 are determined from Equation 2-25 and are summarized next.

m	n	Cross Products
1	1	7 kHz ± 5 kHz = 2 kHz and 12 kHz
1	2	7 kHz ± 10 kHz = 3 kHz and 17 kHz
2	1	14 kHz ± 5 kHz = 9 kHz and 19 kHz
2	2	14 kHz ± 10 kHz = 4 kHz and 24 kHz

c. The output frequency spectrum is shown in Figure 2-23.

QUESTIONS

2-1. Briefly describe the differences between *analog* and *digital* signals.

2-2. Describe *signal analysis* as it pertains to electronic communications.

2-3. Describe *time domain* and give an example of a time-domain instrument.

2-4. Describe *frequency domain* and give an example of a frequency-domain instrument.

2-5. Describe a *complex signal.*

2-6. What is the significance of the *Fourier series*?

2-7. Describe the following wave symmetries: *even, odd,* and *half wave.*

2-8. Define frequency *spectrum* and *bandwidth.*

2-9. Describe the frequency content of a *square wave.* A *rectangular wave.*

2-10. Explain the terms *power spectra* and *energy spectra.*

2-11. Describe the differences between *discrete* and *fast Fourier transforms.*

2-12. Briefly describe the effects of *bandlimiting* electrical signals.

2-13. Describe what is meant by *linear summing.*

2-14. Describe what is meant by *nonlinear mixing.*

PROBLEMS

2-1. Determine the fundamental frequency for the square wave shown below:

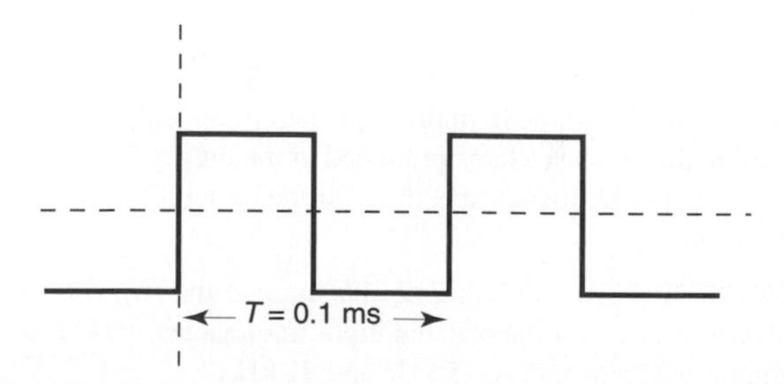

2-2. What kind of symmetry(ies) does the waveform below have?

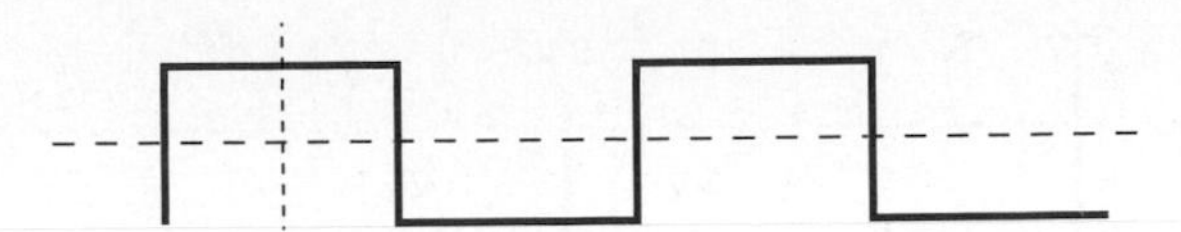

2-3. Determine the first three harmonics of the waveform shown below:

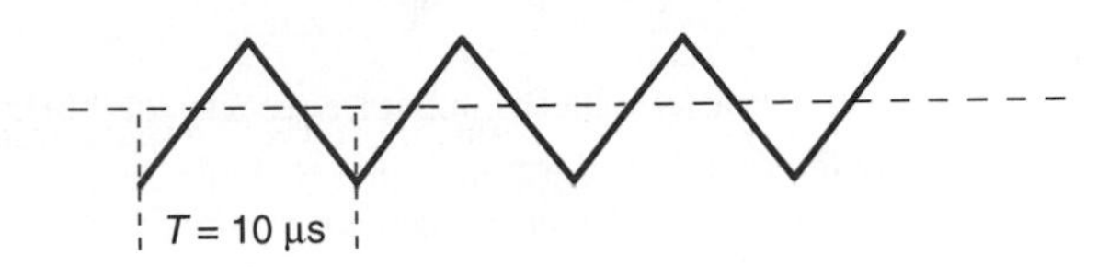

2-4. Determine the bandwidth for the frequency spectrum shown below:

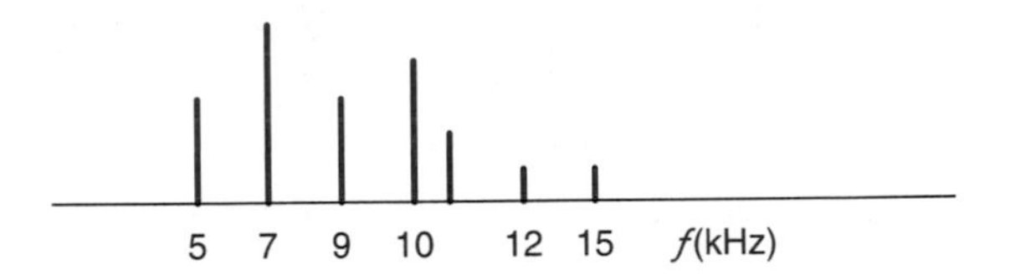

2-5. For the train of square waves shown below,

a. Determine the amplitudes of the first five harmonics.
b. Draw the frequency spectrum.
c. Sketch the time-domain signal for the frequency components, including the first five harmonics.

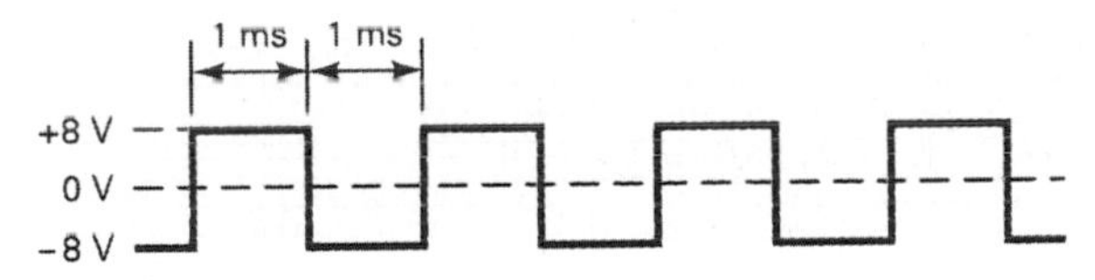

2-6. For the pulse waveform shown,

a. Determine the dc component.
b. Determine the peak amplitudes of the first five harmonics.
c. Plot the $(\sin x)/x$ function.
d. Sketch the frequency spectrum.

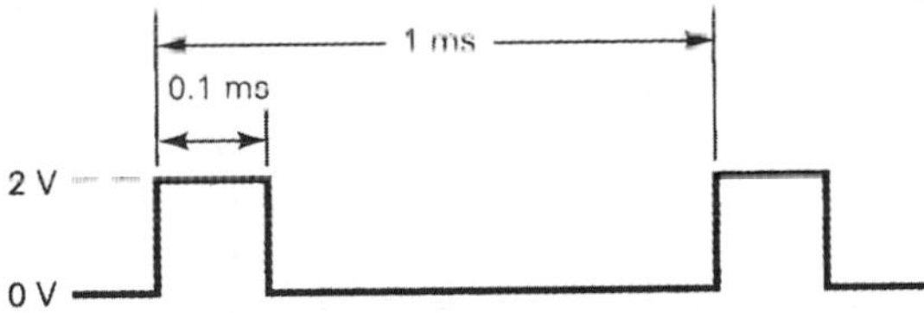

2-7. Describe the spectrum shown below. Determine the type of amplifier (linear or nonlinear) and the frequency content of the input signal:

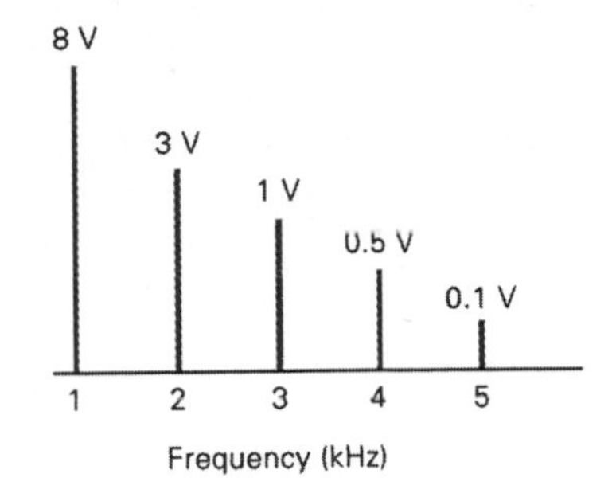

2-8. Repeat Problem 2-7 for the spectrum shown below:

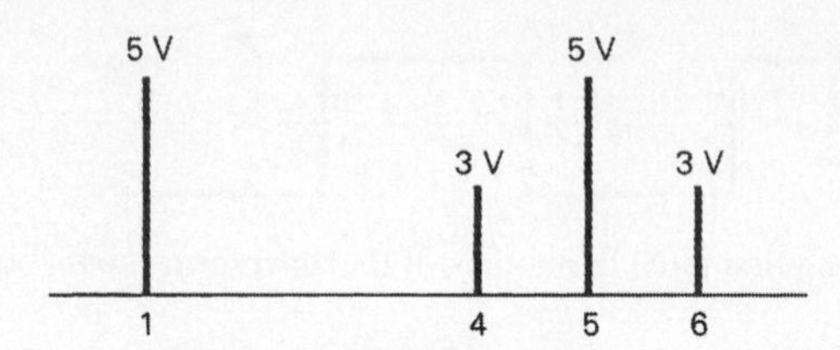

2-9. For a nonlinear amplifier with two input frequencies of 7 kHz and 4 kHz,

a. Determine the first three harmonics present in the output for each frequency.

b. Determine the cross-product frequencies produced in the output for values of *m* and *n* of 1 and 2.

c. Draw the output spectrum for the harmonics and cross-product frequencies determined in steps (a) and (b).

2-10. Determine the percent second-order, third-order, and total harmonic distortion for the output spectrum shown below:

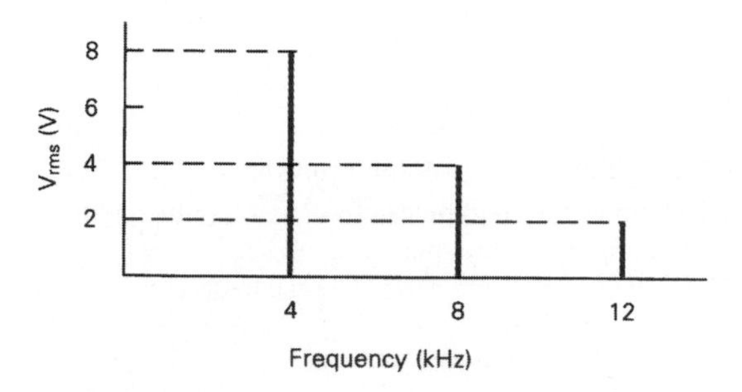

CHAPTER 3

Oscillators, Phase-Locked Loops, and Frequency Synthesizers

CHAPTER OUTLINE

OBJECTIVES

- Define *oscillate.*
- Outline the difference between self-sustaining and one-shot oscillators
- Describe the basic requirements of feedback oscillators
- Explain untuned and tuned oscillators
- Describe the term *frequency stability*
- Explain the piezoelectric effect
- Describe the basic operation of crystal oscillators
- Explain temperature coefficient
- Describe the basic operation of large-scale integration oscillators
- Introduce the basic operation of a phase-locked loop
- Define *lock* and *capture range*
- Describe the operation of a voltage-controlled oscillator
- Explain the operation of a phase detector
- Describe the operation of an integrated-circuit phase-locked loop
- Define *direct frequency synthesis*
- Describe the operation of a direct frequency synthesizer
- Define *indirect frequency synthesis*
- Describe the operation of an indirect frequency synthesizer

3-1 INTRODUCTION

Modern electronic communications systems have many applications that require stable, repetitive waveforms (both sinusoidal and nonsinusoidal). In many of these applications, more than one frequency is required, and very often these frequencies must be synchronized to each other. Therefore, *signal generation, frequency synchronization,* and *frequency synthesis* are essential parts of an electronic communications system. The purpose of this chapter is to introduce the reader to the basic operation of oscillators, phase-locked loops, and frequency synthesizers and to show how these circuits are used for signal generation.

3-2 OSCILLATORS

The definition of *oscillate* is to fluctuate between two states or conditions. Therefore, to oscillate is to vibrate or change, and *oscillating* is the act of fluctuating from one state to another. An *oscillator* is a device that produces oscillations (i.e., generates a repetitive waveform). There are many applications for oscillators in electronic communications, such as high-frequency carrier supplies, pilot supplies, clocks, and timing circuits.

In electronic applications, an oscillator is a device or circuit that produces electrical oscillations. An electrical oscillation is a repetitive change in a voltage or current waveform. If an oscillator is *self-sustaining,* the changes in the waveform are *continuous* and *repetitive;* they occur at a periodic rate. A self-sustaining oscillator is also called a *free-running* oscillator. Oscillators that are not self-sustaining require an external input signal or *trigger* to produce a change in the output waveform. Oscillators that are not self-sustaining are called *triggered* or *one-shot* oscillators. The remainder of this chapter is restricted to explaining self-sustaining oscillators, which require no external input other than a dc supply voltage. Essentially, an oscillator converts a dc input voltage to an ac output voltage. The shape of the output waveform can be a sine wave, a square wave, a sawtooth wave, or any other waveform shape as long as it repeats at periodic intervals.

3-3 FEEDBACK OSCILLATORS

A *feedback oscillator* is an amplifier with a *feedback loop* (i.e., a path for energy to propagate from the output back to the input). Free-running oscillators are feedback oscillators. Once started, a feedback oscillator generates an ac output signal of which a small portion is fed back to the input, where it is amplified. The amplified input signal appears at the output and the process repeats; a *regenerative* process occurs in which the output is dependent on the input and vice versa.

According to the *Barkhausen criterion,* for a feedback circuit to sustain oscillations, the net voltage gain around the feedback loop must be unity or greater, and the net phase shift around the loop must be a positive integer multiple of 360°.

There are four requirements for a feedback oscillator to work: *amplification, positive feedback, frequency determination,* and a *source* of electrical power.

1. *Amplification.* An oscillator circuit must include at least one active device and be capable of voltage amplification. In fact, at times it may be required to provide an infinite gain.
2. *Positive feedback.* An oscillator circuit must have a complete path for a portion of the output signal to be returned to the input. The feedback signal must be *regenerative,* which means it must have the correct phase and amplitude necessary to sus-

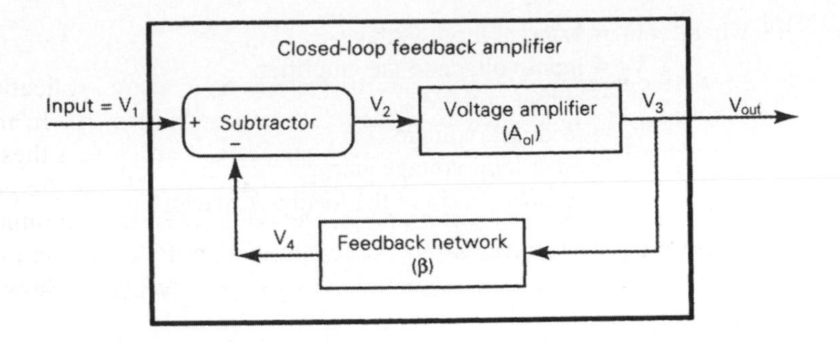

FIGURE 3-1 Model of an amplifier with feedback

tain oscillations. If the phase is incorrect or if the amplitude is insufficient, oscillations will cease. If the amplitude is excessive, the amplifier will saturate. *Regenerative feedback* is called *positive feedback,* where "positive" simply means that its phase aids the oscillation process and does not necessarily indicate a positive (+) or negative (−) polarity. *Degenerative feedback* is called *negative feedback* and supplies a feedback signal that inhibits oscillations from occurring.

3. *Frequency-determining components.* An oscillator must have frequency-determining components such as resistors, capacitors, inductors, or crystals to allow the frequency of operation to be set or changed.
4. *Power source.* An oscillator must have a source of electrical energy, such as a dc power supply.

Figure 3-1 shows an electrical model for a *feedback oscillator* circuit (i.e., a voltage amplifier with regenerative feedback). A feedback oscillator is a *closed-loop* circuit comprised of a voltage amplifier with an *open-loop voltage gain* (A_{ol}), a frequency-determining regenerative feedback path with a *feedback ratio* (β), and either a summer or a subtractor circuit. The open-loop voltage gain is the voltage gain of the amplifier with the feedback path open circuited. The *closed-loop voltage gain* (A_{cl}) is the overall voltage gain of the complete circuit with the feedback loop closed and is always less than the open-loop voltage gain. The feedback ratio is simply the transfer function of the feedback network (i.e., the ratio of its output to its input voltage). For a passive feedback network, the feedback ratio is always less than 1.

From Figure 3-1, the following mathematical relationships are derived:

$$\frac{V_{\text{out}}}{V_{\text{in}}} = \frac{V_3}{V_1}$$

$$V_2 = V_1 - V_4$$

$$V_3 = A_{ol}V_2$$

$$A_{ol} = \frac{V_3}{V_2}$$

$$V_4 = \beta V_3$$

$$\beta = \frac{V_4}{V_3}$$

where V_1 = external input voltage
V_2 = input voltage to the amplifier
V_3 = output voltage
V_4 = feedback voltage
A_{ol} = open-loop voltage gain
β = feedback ratio of the feedback network

Substituting for V_4 gives us $$V_2 = V_1 - \beta V_3$$

Thus, $$V_3 = (V_1 - \beta V_3)A_{ol}$$

and $$V_3 = V_1 A_{ol} - V_3 \beta A_{ol}$$

Rearranging and factoring yields

$$V_3 + V_3 \beta A_{ol} = V_1 A_{ol}$$

Thus, $$V_3(1 + \beta A_{ol}) = V_1 A_{ol}$$

and $$\frac{V_{out}}{V_{in}} = \frac{V_3}{V_1} = \frac{A_{ol}}{1 + \beta A_{ol}} = A_{cl} \qquad \textbf{(3-1)}$$

where A_{cl} is closed-loop voltage gain.

$A_{ol}/(1 + \beta A_{ol})$ is the standard formula used for the closed-loop voltage gain of an amplifier with feedback. If at any frequency βA_{ol} goes to -1, the denominator in Equation 3-1 goes to zero and V_{out}/V_{in} is infinity. When this happens, the circuit will oscillate, and the external input may be removed.

For self-sustained oscillations to occur, a circuit must fulfill the four basic requirements for oscillation outlined previously, meet the criterion of Equation 3-1, and fit the basic feedback circuit model shown in Figure 3-1. Although oscillator action can be accomplished in many different ways, the most common configurations use *RC* phase shift networks, *LC* tank circuits, quartz crystals, or integrated-circuit chips. The type of oscillator used for a particular application depends on the following criteria:

1. Desired frequency of operation
2. Required frequency stability
3. Variable or fixed frequency operation
4. Distortion requirements or limitations
5. Desired output power
6. Physical size
7. Application (i.e., digital or analog)
8. Cost
9. Reliability and durability
10. Desired accuracy

3-3-1 Untuned Oscillators

The Wien-bridge oscillator is an untuned *RC* phase shift oscillator that uses both positive and negative feedback. It is a relatively stable, low-frequency oscillator circuit that is easily tuned and commonly used in signal generators to produce frequencies between 5 Hz and 1 MHz. The Wien-bridge oscillator is the circuit that Hewlett and Packard used in their original signal generator design.

Figure 3-2a shows a simple lead–lag network. At the frequency of oscillation (f_o), $R = X_C$, and the signal undergoes a $-45°$ phase shift across Z_1 and a $+45°$ phase shift across Z_2. Consequently, at f_o, the total phase shift across the lead–lag network is exactly 0°. At frequencies below the frequency of oscillation, the phase shift across the network leads and for frequencies above the phase shift lags. At extreme low frequencies, C_1 looks like an open circuit, and there is no output. At extreme high frequencies, C_2 looks like a short circuit, and there is no output.

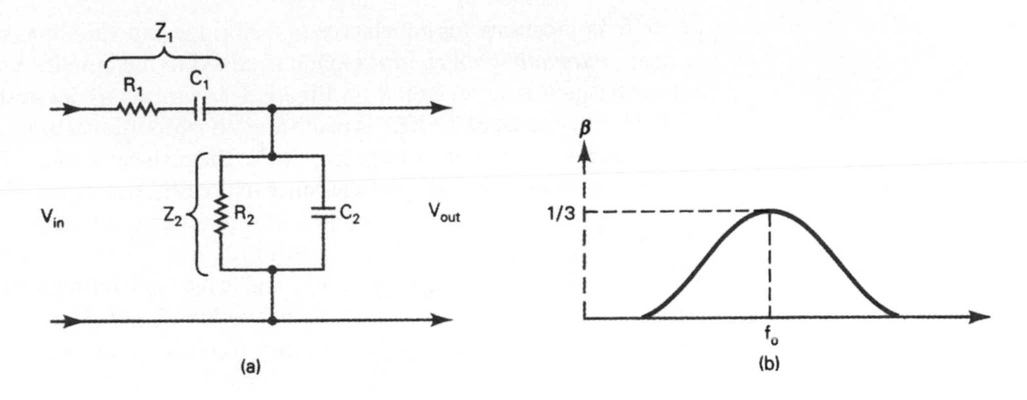

FIGURE 3-2 Lead-lag network: (a) circuit configuration; (b) input-versus-output transfer curver (β)

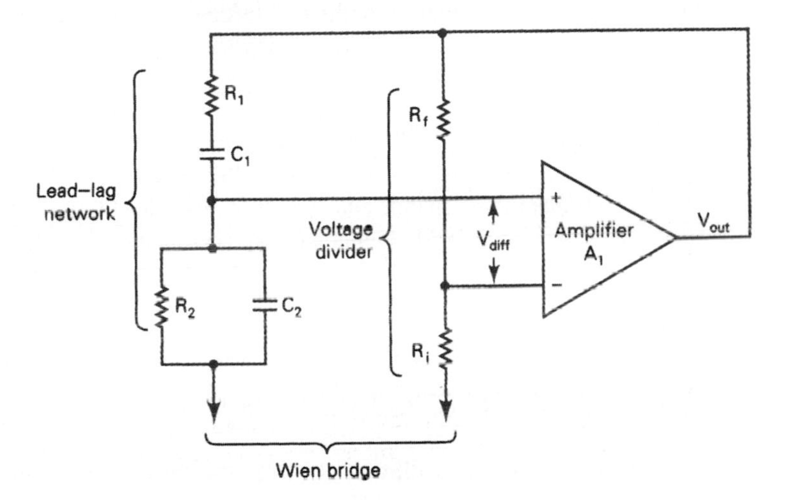

FIGURE 3-3 Wien-bridge oscillator

A lead–lag network is a reactive voltage divider in which the input voltage is divided between Z_1 (the series combination of R_1 and C_1) and Z_2 (the parallel combination of R_2 and C_2). Therefore, the lead–lag network is frequency selective, and the output voltage is maximum at f_o. The transfer function for the feedback network (β) equals $Z_2/(Z_1 + Z_2)$ and is maximum and equal to 1/3 at f_o. Figure 3-2b shows a plot of β versus frequency when $R_1 = R_2$ and $C_1 = C_2$. Thus, f_o is determined from the following expression:

$$f_o = \frac{1}{2\pi RC}$$

where $R = R_1 = R_2$
$C = C_1 = C_2$

Figure 3-3 shows a Wien-bridge oscillator. The lead–lag network and the resistive voltage divider make up a Wien bridge (hence the name *Wien-bridge oscillator*). When the bridge is balanced, the difference voltage equals zero. The voltage divider provides negative or degenerative feedback that offsets the positive or regenerative feedback from the lead–lag network. The ratio of the resistors in the voltage divider is 2:1, which sets the noninverting voltage gain of amplifier A_1 to $R_f/R_i + 1 = 3$. Thus, at f_o, the signal at the output of A_1 is reduced by a factor of 3 as it passes through the lead–lag network ($\beta = 1/3$) and then amplified by 3 in amplifier A_1. Thus, at f_o the loop voltage gain is equal to $A_{ol}\beta$ or $3 \times 1/3 = 1$.

To compensate for imbalances in the bridge and variations in component values due to heat, *automatic gain control* (AGC) is added to the circuit. A simple way of providing automatic gain is to replace R_i in Figure 3-3 with a variable resistance device such as a FET. The resistance of the FET is made directly proportional to V_{out}. The circuit is designed such that, when V_{out} increases in amplitude, the resistance of the FET increases, and when V_{out} decreases in amplitude, the resistance of the FET decreases. Therefore, the voltage gain of the amplifier automatically compensates for changes in amplitude in the output signal.

The operation of the circuit shown in Figure 2-3 is as follows. On initial power-up, noise (at all frequencies) appears at V_{out} and is fed back through the lead–lag network. Only noise at f_o passes through the lead–lag network with a 0° phase shift and a transfer ratio of 1/3. Consequently, only a single frequency (f_o) is fed back in phase, undergoes a loop voltage gain of 1, and produces self-sustained oscillations.

3-3-2 Tuned Oscillators

LC oscillators are oscillator circuits that utilize tuned *LC tank circuits* for the frequency-determining components. Tank-circuit operation involves an exchange of energy between *kinetic* and *potential.* Figure 3-4 illustrates *LC* tank-circuit operation. As shown in Figure 3-4a, once current is injected into the circuit (time t_1), energy is exchanged between the inductor and capacitor, producing a corresponding ac output voltage (times t_2 to t_4). The output voltage waveform is shown in Figure 3-4b. The frequency of operation of an *LC* tank circuit is simply the resonant frequency of the parallel *LC* network, and the bandwidth is a function of the circuit Q. Mathematically, the resonant frequency of an *LC* tank circuit with a $Q \geq 10$ is closely approximated by

$$f_o = \frac{1}{2\pi\sqrt{(LC)}} \quad \textbf{(3-2)}$$

LC oscillators include the Hartley and Colpitts oscillators.

3-3-2-1 Hartley oscillator. Figure 3-5a shows the schematic diagram of a *Hartley oscillator.* The transistor amplifier (Q_1) provides the amplification necessary for a loop voltage gain of unity at the resonant frequency. The coupling capacitor (C_C) provides the path for regenerative feedback. L_{1a}, L_{1b}, and C_1 are the frequency-determining components, and V_{CC} is the dc supply voltage.

Figure 3-5b shows the dc equivalent circuit for the Hartley oscillator. C_C is a blocking capacitor that isolates the dc base bias voltage and prevents it from being shorted to ground through L_{1b}. C_2 is also a blocking capacitor that prevents the collector supply voltage from being shorted to ground through L_{1a}. The *radio-frequency choke* (RFC) is a dc short.

Figure 3-5c shows the ac equivalent circuit for the Hartley oscillator. C_C is a coupling capacitor for ac and provides a path for regenerative feedback from the tank circuit to the base of Q_1. C_2 couples ac signals from the collector of Q_1 to the tank circuit. The RFC looks open to ac, consequently isolating the dc power supply from ac oscillations.

The Hartley oscillator operates as follows. On initial power-up, a multitude of frequencies appear at the collector of Q_1 and are coupled through C_2 into the tank circuit. The initial noise provides the energy necessary to charge C_1. Once C_1 is partially charged, oscillator action begins. The tank circuit will only oscillate efficiently at its resonant frequency. A portion of the oscillating tank circuit voltage is dropped across L_{1b} and fed back to the base of Q_1, where it is amplified. The amplified signal appears at the collector 180° out of phase with the base signal. An additional 180° of phase shift is realized across L_1; consequently, the signal fed back to the base of Q_1 is amplified and shifted in phase 360°. Thus, the circuit is regenerative and will sustain oscillations with no external input signal.

The proportion of oscillating energy that is fed back to the base of Q_1 is determined by the ratio of L_{1b} to the total inductance ($L_{1a} + L_{1b}$). If insufficient energy is fed back, os-

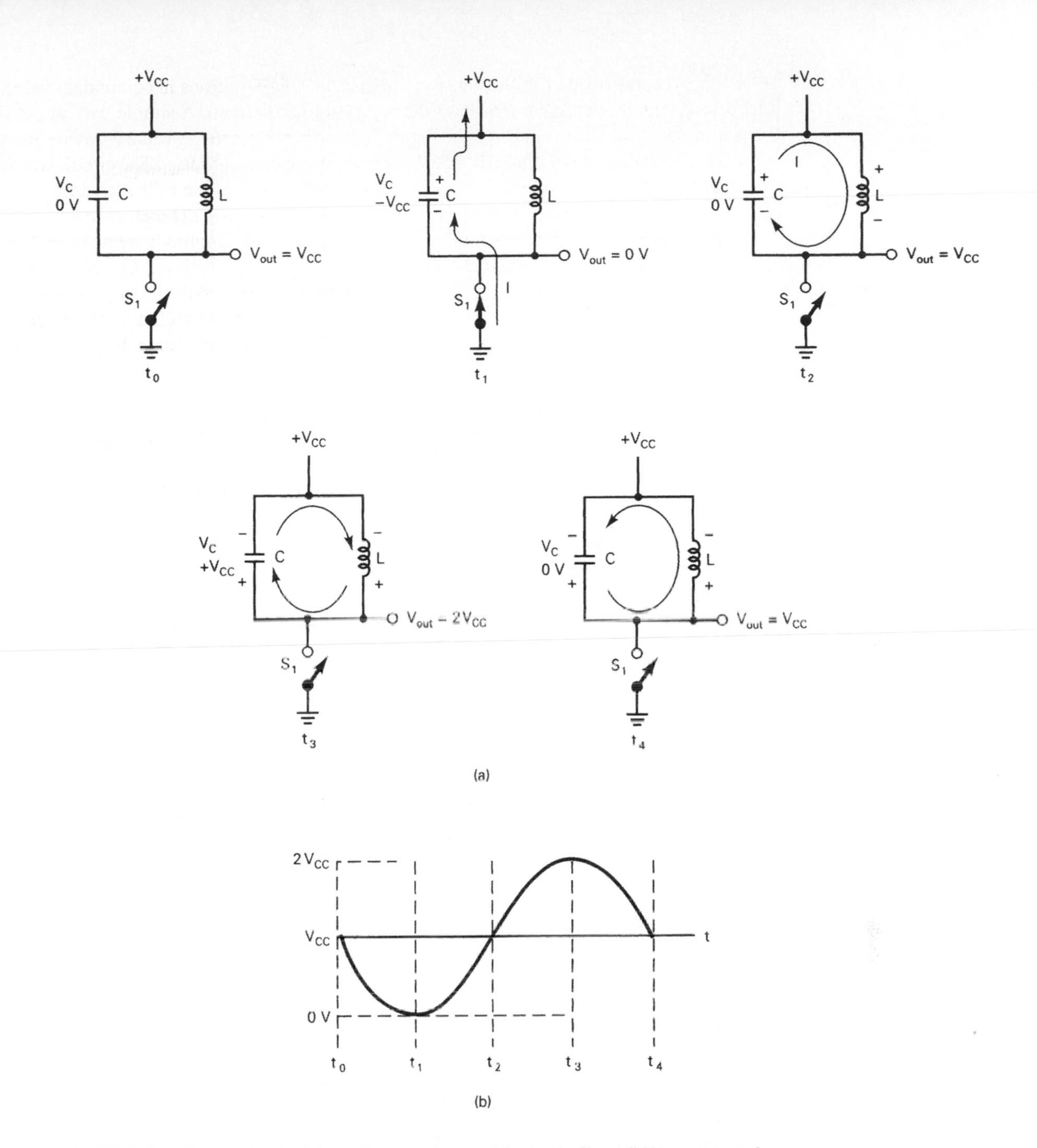

FIGURE 3-4 *LC* tank circuit; (a) oscillator action and flywheel effect; (b) output waveform

cillations are damped. If excessive energy is fed back, the transistor saturates. Therefore, the position of the wiper on L_1 is adjusted until the amount of feedback energy is exactly what is required for a unity loop voltage gain and oscillations to continue.

The frequency of oscillation for the Hartley oscillator is closely approximated by the following formula:

$$f_o = \frac{1}{2\pi\sqrt{(LC)}} \tag{3-3}$$

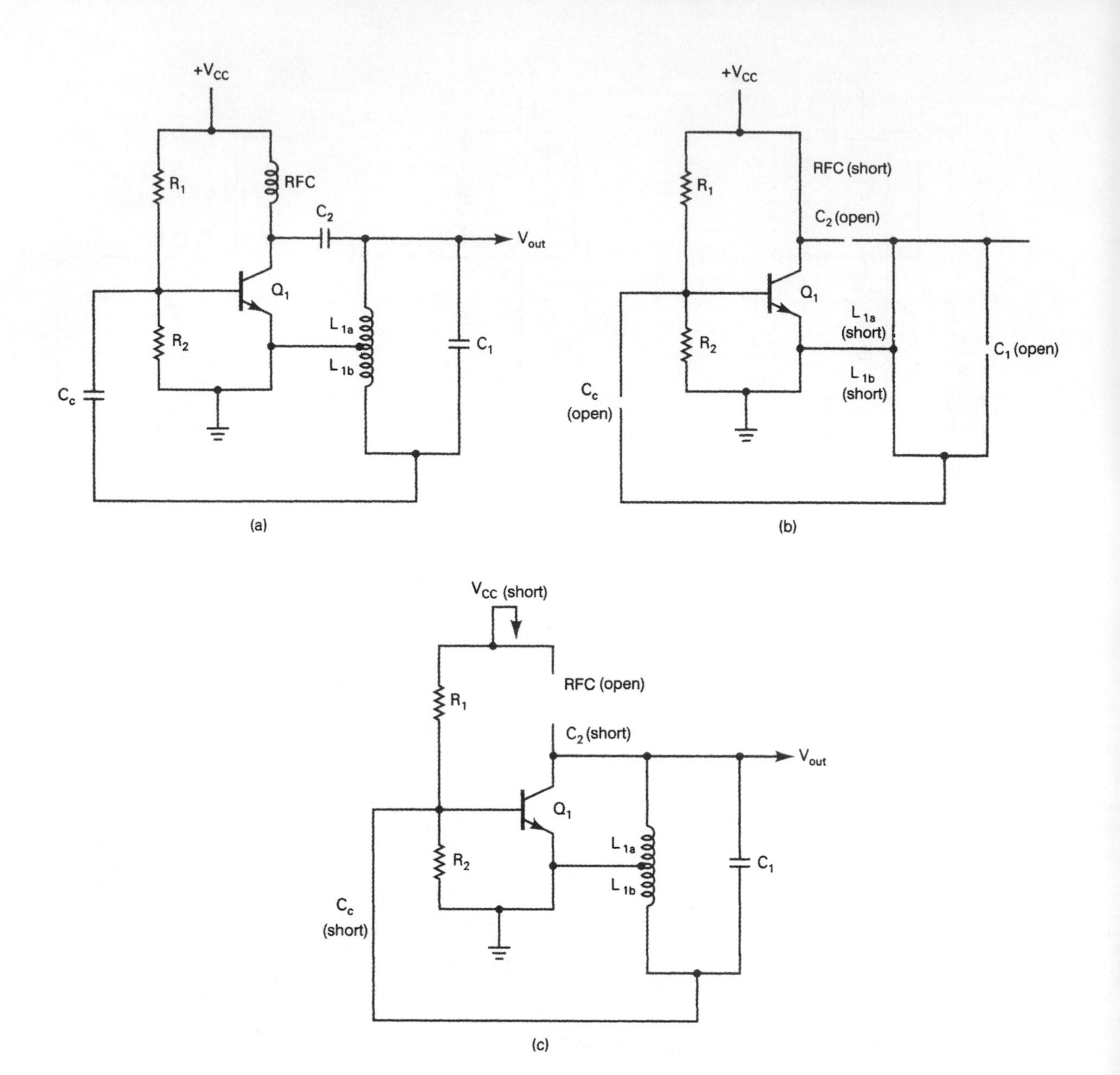

FIGURE 3-5 Hartley oscillator: (a) schematic diagram; (b) dc equivalent circuit; (c) ac equivalent circuit

where $L = L_{1a} + L_{1b}$
$C = C_1$

3-3-2-2 Colpitts oscillator. Figure 3-6a shows the schematic diagram of a *Colpitts oscillator.* The operation of a Colpitts oscillator is very similar to that of the Hartley except that a capacitive divider is used instead of a tapped coil. Q_1 provides the amplification; C_C provides the regenerative feedback path; L_1, C_{1a}, and C_{1b} are the frequency-determining components; and V_{CC} is the dc supply voltage.

Figure 3-6b shows the dc equivalent circuit for the Colpitts oscillator. C_2 is a blocking capacitor that prevents the collector supply voltage from appearing at the output. The RFC is again a dc short.

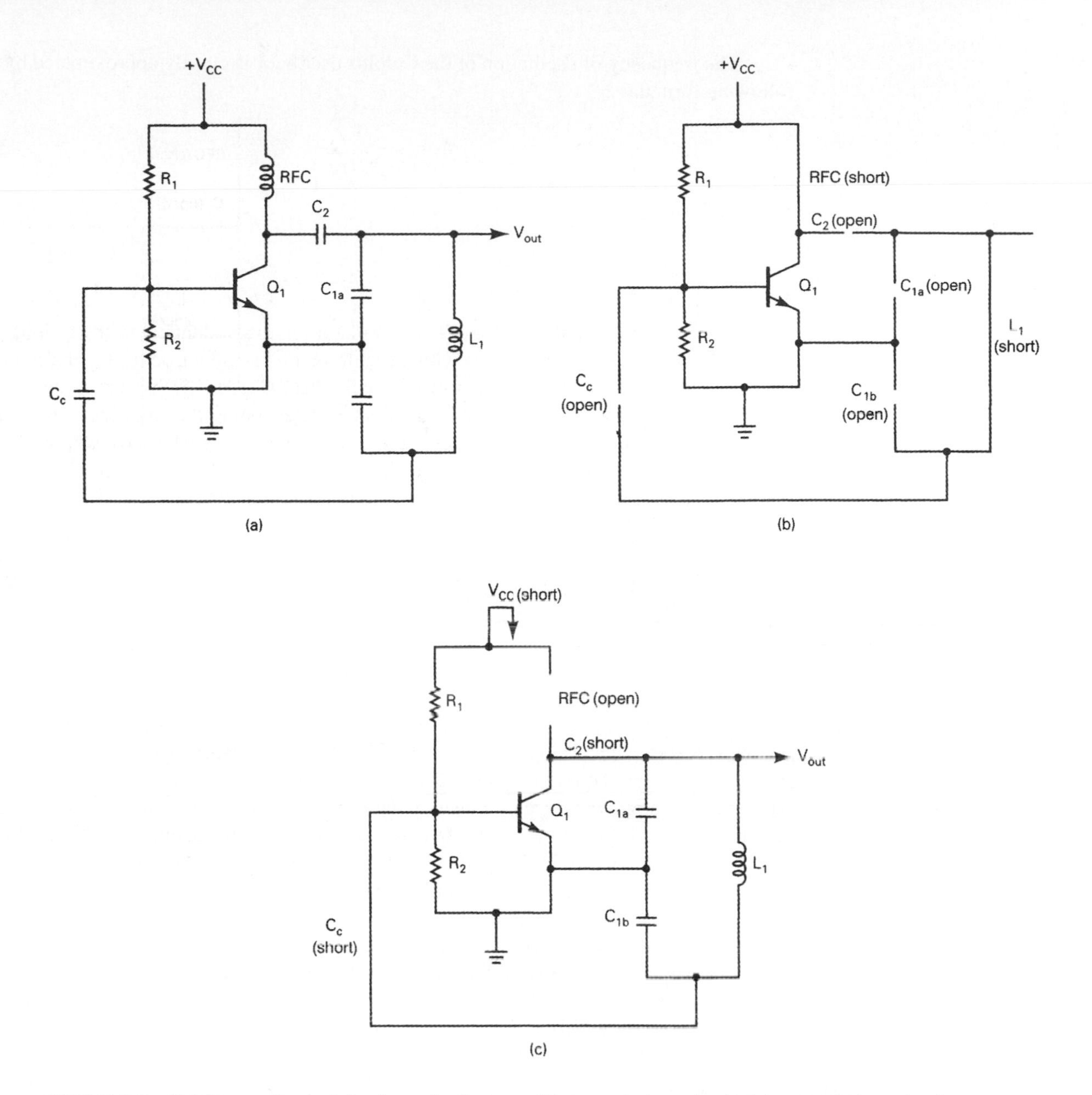

FIGURE 3-6 Colpitts oscillator: (a) schematic diagram; (b) dc equivalent circuit; (c) ac equivalent circuit

Figure 3-6c shows the ac equivalent circuit for the Colpitts oscillator. C_C is a coupling capacitor for ac and provides the feedback path for regenerative feedback from the tank circuit to the base of Q_1. The RFC is open to ac and decouples oscillations from the dc power supply.

The operation of the Colpitts oscillator is almost identical to that of the Hartley oscillator. On initial power-up, noise appears at the collector of Q_1 and supplies energy to the tank circuit, causing it to begin oscillating. C_{1a} and C_{1b} make up an ac voltage divider. The voltage dropped across C_{1b} is fed back to the base of Q_1 through C_C. There is a 180° phase shift from the base to the collector of Q_1 and an additional 180° phase shift across C_1. Consequently, the total phase shift is 360°, and the feedback signal is regenerative. The ratio of C_{1a} to $C_{1a} + C_{1b}$ determines the amplitude of the feedback signal.

The frequency of oscillation of the Colpitts oscillator is closely approximated by the following formula:

$$f_o = \frac{1}{2\pi\sqrt{(LC)}} \tag{3-4}$$

where $L = L_1$

$$C = \frac{(C_{1a}C_{1b})}{(C_{1a} + C_{1b})}$$

3-3-2-3 Clapp oscillator. A Clapp oscillator circuit is identical to the Colpitts oscillator shown in Figure 3-6a except with the addition of a small capacitor C_S placed in series with L_1. The capacitance of C_S is made smaller than C_{1a} or C_{1b}, thus providing a large reactance. Consequently, C_S has the most effect in determining the frequency of the tank circuit. The advantage of a Clapp oscillator is that C_{1a} and C_{1b} can be selected for an optimum feedback ratio, while C_S can be variable and used for setting the frequency of oscillation. In some applications C_S incorporates a negative temperature coefficient that improves the oscillator's frequency stability.

3-4 FREQUENCY STABILITY

Frequency stability is the ability of an oscillator to remain at a fixed frequency and is of primary importance in communications systems. Frequency stability is often stated as either short or long term. *Short-term stability* is affected predominantly by fluctuations in dc operating voltages, whereas *long-term stability* is a function of component aging and changes in the ambient temperature and humidity. In the *LC* tank-circuit and *RC* phase shift oscillators discussed previously, the frequency stability is inadequate for most radio communications applications because *RC* phase shift oscillators are susceptible to both short- and long-term variations. In addition, the *Q*-factors of the *LC* tank circuits are relatively low, allowing the resonant tank circuit to oscillate over a wide range of frequencies.

Frequency stability is generally given as a percentage of change in frequency (tolerance) from the desired value. For example, an oscillator operating at 100 kHz with a $\pm 5\%$ stability will operate at a frequency of 100 kHz $\pm$ 5 kHz or between 95 kHz and 105 kHz. Commercial FM broadcast stations must maintain their carrier frequencies to within ± 2 kHz of their assigned frequency, which is approximately a 0.002% tolerance. In commercial AM broadcasting, the maximum allowable shift in the carrier frequency is only ± 20 Hz.

Several factors affect the stability of an oscillator. The most obvious are those that directly affect the value of the frequency-determining components. These include changes in inductance, capacitance, and resistance values due to environmental variations in temperature and humidity and changes in the quiescent operating point of transistors and field-effect transistors. Stability is also affected by ac ripple in dc power supplies. The frequency stability of *RC* or *LC* oscillators can be greatly improved by regulating the dc power supply and minimizing the environmental variations. Also, special temperature-independent components can be used.

The FCC has established stringent regulations concerning the tolerances of radio-frequency carriers. Whenever the airway (free-space radio propagation) is used as the transmission medium, it is possible that transmissions from one source could interfere with transmissions from other sources if their transmit frequency or transmission bandwidths overlap. Therefore, it is important that all sources maintain their frequency of operation within a specified tolerance.

3-5 CRYSTAL OSCILLATORS

Crystal oscillators are feedback oscillator circuits in which the *LC* tank circuit is replaced with a crystal for the frequency-determining component. The crystal acts in a manner similar to the *LC* tank, except with several inherent advantages. Crystals are sometimes called crystal resonators, and they are capable of producing precise, stable frequencies for frequency counters, electronic navigation systems, radio transmitters and receivers, televisions, videocassette recorders (VCRs), computer system clocks, and many other applications too numerous to list.

Crystallography is the study of the form, structure, properties, and classifications of crystals. Crystallography deals with lattices, bonding, and the behavior of slices of crystal material that have been cut at various angles with respect to the crystal's axes. The mechanical properties of crystal lattices allow them to exhibit the piezoelectric effect. Sections of crystals that have been cut and polished vibrate when alternating voltages are applied across their faces. The physical dimensions of a crystal, particularly its thickness and where and how it was cut, determine its electrical and mechanical properties.

3-5-1 Piezoelectric Effect

Simply stated, the *piezoelectric effect* occurs when oscillating mechanical stresses applied across a *crystal lattice structure* generate electrical oscillations and vice versa. The stress can be in the form of squeezing (compression), stretching (tension), twisting (torsion), or shearing. If the stress is applied periodically, the output voltage will alternate. Conversely, when an alternating voltage is applied across a crystal at or near the natural resonant frequency of the crystal, the crystal will break into mechanical oscillations. This process is called *exciting* a crystal into *mechanical vibrations.* The mechanical vibrations are called *bulk acoustic waves* (BAWs) and are directly proportional to the amplitude of the applied voltage.

A number of natural crystal substances exhibit piezoelectric properties: *quartz, Rochelle salt,* and *tourmaline* and several manufactured substances such as ADP, EDT, and DKT. The piezoelectric effect is most pronounced in Rochelle salt, which is why it is the substance commonly used in crystal microphones. Synthetic quartz, however, is used more often for frequency control in oscillators because of its *permanence,* low *temperature coefficient,* and high *mechanical Q.*

3-5-2 Crystal Cuts

In nature, complete quartz crystals have a hexagonal cross section with pointed ends, as shown in Figure 3-7a. Three sets of axes are associated with a crystal: *optical, electrical,* and *mechanical.* The longitudinal axis joining points at the ends of the crystal is called the *optical* or *Z-axis.* Electrical stresses applied to the optical axis do not produce the piezoelectric effect. The *electrical* or *X-axis* passes diagonally through opposite corners of the hexagon. The axis that is perpendicular to the faces of the crystal is the *Y*- or *mechanical axis.* Figure 3-7b shows the axes and the basic behavior of a quartz crystal.

If a thin flat section is cut from a crystal such that the flat sides are perpendicular to an electrical axis, mechanical stresses along the *Y*-axis will produce electrical charges on the flat sides. As the stress changes from compression to tension and vice versa, the polarity of the charge is reversed. Conversely, if an alternating electrical charge is placed on the flat sides, a mechanical vibration is produced along the *Y*-axis. This is the piezoelectric effect and is also exhibited when mechanical forces are applied across the faces of a crystal cut with its flat sides perpendicular to the *Y*-axis. When a crystal wafer is cut parallel to the *Z*-axis with its faces perpendicular to the *X*-axis, it is called an *X*-cut crystal. When the faces are perpendicular to the *Y*-axis, it is called a *Y*-cut crystal. A variety of cuts can be obtained by rotating the plane of the cut around one or more axes. If the *Y* cut is made at a 35° 20′ angle from the vertical axis (Figure 3-7c), an *AT* cut is obtained. Other types of crystal cuts include the *BT, CT, DT, ET, AC, GT, MT, NT,* and *JT* cuts. The *AT* cut is the most popular for high-frequency and very-high-frequency crystal resonators. The type, length, and thickness of a

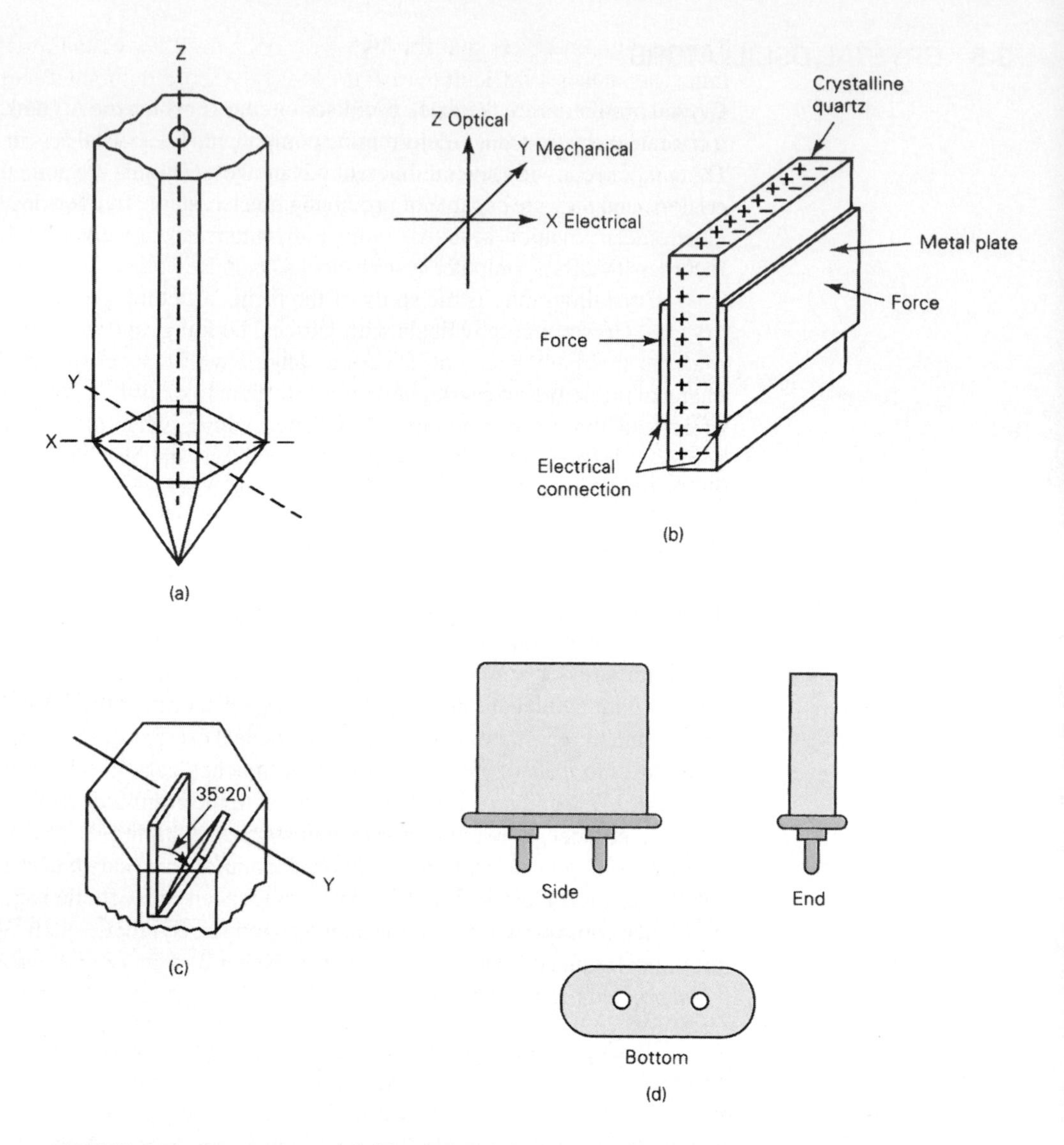

FIGURE 3-7 Quartz crystal: (a) basic crystal structure; (b) crystal axes; (c) crystal cuts; (d) crystal mountings

cut and the mode of vibration determine the natural resonant frequency of the crystal. Resonant frequencies for *AT*-cut crystals range from approximately 800 kHz up to approximately 30 MHz. *CT* and *DT* cuts exhibit low-frequency shear and are most useful in the 100-kHz to 500-kHz range. The *MT* cut vibrates longitudinally and is useful in the 50-kHz to 100-kHz range, and the *NT* cut has a useful range under 50 kHz.

Crystal *wafers* are generally mounted in *crystal holders,* which include the mounting and housing assemblies. A *crystal unit* refers to the holder and the crystal itself. Figure 3-7d shows a common crystal mounting. Because a crystal's stability is somewhat temperature dependent, a crystal unit may be mounted in an oven to maintain a constant operating temperature.

The relationship between a crystal's operating frequency and its thickness is expressed mathematically as

$$h = \frac{65.5}{f_n} \tag{3-5}$$

where h = crystal thickness (inches)
f_n = crystal natural resonant frequency (hertz)

This formula indicates that for high-frequency oscillations the quartz wafer must be very thin. This makes it difficult to manufacture crystal oscillators with fundamental frequencies above approximately 30 MHz because the wafer becomes so thin that it is exceptionally fragile, and conventional cutting and polishing can be accomplished only at extreme costs. This problem can be alleviated by using chemical etching to achieve thinner slices. With this process, crystals with fundamental frequencies up to 350 MHz are possible.

3-5-3 Overtone Crystal Oscillator

As previously stated, to increase the frequency of vibration of a quartz crystal, the quartz wafer is sliced thinner. This imposes an obvious physical limitation: The thinner the wafer, the more susceptible it is to damage and the less useful it becomes. Although the practical limit for fundamental-mode crystal oscillators is approximately 30 MHz, it is possible to operate the crystal in an overtone mode. In the overtone mode, harmonically related vibrations that occur simultaneously with the fundamental vibration are used. In the overtone mode, the oscillator is tuned to operate at the third, fifth, seventh, or even the ninth harmonic of the crystal's fundamental frequency. The harmonics are called overtones because they are not true harmonics. Manufacturers can process crystals such that one overtone is enhanced more than the others. Using an overtone mode increases the usable limit of standard crystal oscillators to approximately 200 MHz.

3-5-4 Temperature Coefficient

The natural resonant frequency of a crystal is influenced somewhat by its operating temperature. The ratio of the magnitude of frequency change (Δf) to a change in temperature (ΔC) is expressed in hertz change per megahertz of crystal operating frequency per degree Celsius (Hz/MHz/°C). The fractional change in frequency is often given in parts per million (ppm) per °C. For example, a temperature co-efficient of +20 Hz/MHz/°C is the same as +20 ppm/°C. If the direction of the frequency change is the same as the temperature change (i.e., an increase in temperature causes an increase in frequency and a decrease in temperature causes a decrease in frequency), it is called a *positive temperature coefficient.* If the change in frequency is in the direction opposite to the temperature change (i.e., an increase in temperature causes a decrease in frequency and a decrease in temperature causes an increase in frequency), it is called a *negative temperature coefficient.* Mathematically, the relationship of the change in frequency of a crystal to a change in temperature is

$$\Delta f = k(f_n \times \Delta C) \tag{3-6}$$

where Δf = change in frequency (hertz)
k = temperature coefficient (Hz/MHz/°C)
f_n = natural crystal frequency (megahertz)
ΔC = change in temperature (degrees Celsius)

and

$$f_o = f_n + \Delta f \tag{3-7}$$

where f_o is frequency of operation.

The temperature coefficient (k) of a crystal varies depending on the type of crystal cut and its operating temperature. For a range of temperatures from approximately +20°C to +50°C, both *X*- and *Y*-cut crystals have a temperature coefficient that is nearly constant. *X*-cut crystals are approximately 10 times more stable than *Y*-cut crystals. Typically, *X*-cut crystals have a temperature coefficient that ranges from −10 Hz/MHz/°C to −25 Hz/MHz/ C. *Y*-cut crystals have a temperature coefficient that ranges from approximately −25 Hz/MHz/°C to +100 Hz/MHz/°C.

Today, zero-coefficient (*GT*-cut) crystals are available that have temperature coefficients as low as −1 Hz/MHz/°C to +1 Hz/MHz/°C. The *GT*-cut crystal is almost a perfect zero-coefficient crystal from freezing to boiling but is useful only at frequencies below a few hundred kilohertz.

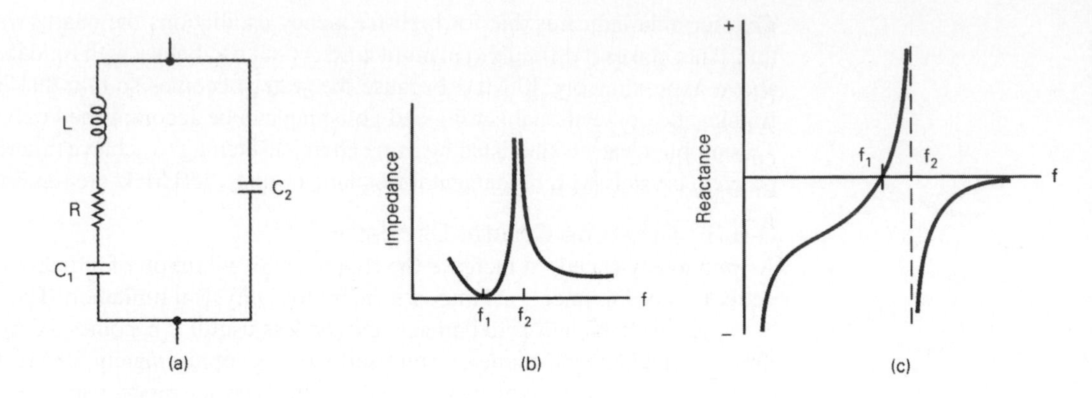

FIGURE 3-8 Crystal equivalent circuit: (a) equivalent circuit; (b) impedance curve; (c) reactance curve

Example 3-1

For a 10-MHz crystal with a temperature coefficient $k = +10$ Hz/MHz/°C, determine the frequency of operation if the temperature

a. Increases 10°C.

b. Decreases 5°C.

Solution **a.** Substituting into Equations 3-6 and 3-7 gives us

$$\Delta f = k(f_n \times \Delta C)$$
$$= 10(10 \times 10) = 1 \text{ kHz}$$
$$f_o = f_n + \Delta f$$
$$= 10 \text{ MHz} + 1 \text{ kHz} = 10.001 \text{ MHz}$$

b. Again, substituting into Equations 3-6 and 3-7 yields

$$\Delta f = 10[10 \times (-5)] = -500 \text{ Hz}$$
$$f_o = 10 \text{ MHz} + (-500 \text{ Hz})$$
$$= 9.9995 \text{ MHz}$$

3-5-5 Crystal Equivalent Circuit

Figure 3-8a shows the electrical equivalent circuit for a crystal. Each electrical component is equivalent to a mechanical property of the crystal. C_2 is the actual capacitance formed between the electrodes of the crystal, with the crystal itself being the dielectric. C_1 is equivalent to the mechanical compliance of the crystal (also called the resilience or elasticity). L is equivalent to the mass of the crystal in vibration, and R is the mechanical friction loss. In a crystal, the mechanical *mass-to-friction ratio* (L/R) is quite high. Typical values of L range from 0.1 H to well over 100 H; consequently, Q-factors are quite high for crystals. Q-factors in the range from 10,000 to 100,000 and higher are not uncommon (as compared with Q-factors of 100 to 1000 for the discrete inductors used in LC tank circuits). This provides the high stability of crystal oscillators as compared to discrete LC tank-circuit oscillators. Values for C_1 are typically less than 1 pF, and values for C_2 range between 4 pF and 40 pF.

Because there is a series and a parallel equivalent circuit for a crystal, there are also two equivalent impedances and two resonant frequencies: a series and a parallel. The series impedance is the combination of R, L, and C_1 (i.e., $Z_s = R \pm jX$, where $X = |X_L - X_C|$). The parallel impedance is approximately the impedance of L and C_2 (i.e., $Z_p = [X_L \times X_{C2}]/[X_L + X_{C2}]$). At extreme low frequencies, the series impedance of L, C_1, and R is very high and capacitive (−). This is shown in Figure 3-8c. As the frequency is increased, a point is reached where $X_L = X_{C1}$. At this frequency (f_1), the series impedance is minimum, resistive, and equal to R. As the frequency is increased even further (f_2), the series impedance becomes high and inductive (+). The parallel combination of L and C_2 causes the crystal to act as a parallel resonant circuit (maximum impedance at resonance). The difference between f_1 and f_2 is usually quite small (typically about 1% of the crystal's natural frequency). A crystal can operate at either its series or parallel resonant frequency, depending on the circuit configuration in which it is used. The relative steepness of the impedance curve shown in Figure 3-8b also attributes to the stability and accuracy of a crystal. The series resonant frequency of a quartz crystal is simply

$$f_1 = \frac{1}{2\pi\sqrt{(LC_1)}} \tag{3-8}$$

and the parallel resonant frequency is

$$f_2 = \frac{1}{2\pi\sqrt{(LC)}} \tag{3-9}$$

where C is the series combination of C_1 and C_2.

3-5-6 Crystal Oscillator Circuits

Although there are many different crystal-based oscillator configurations, the most common are the discrete and integrated-circuit Pierce and the *RLC* half-bridge. If you need very good frequency stability and reasonably simple circuitry, the discrete Pierce is a good choice. If low cost and simple digital interfacing capabilities are of primary concern, an IC-based Pierce oscillator will suffice. However, for the best frequency stability, the *RLC* half-bridge is the best choice.

3-5-7 Discrete Pierce Oscillator

The discrete Pierce crystal oscillator has many advantages. Its operating frequency spans the full fundamental crystal range (1 kHz to approximately 30 MHz). It uses relatively simple circuitry requiring few components (most medium-frequency versions require only one transistor). The Pierce oscillator design develops a high output signal power while dissipating very little power in the crystal itself. Finally, the short-term frequency stability of the Pierce crystal oscillator is excellent (because the in-circuit loaded Q is almost as high as the crystal's internal Q). The only drawback to the Pierce oscillator is that it requires a high-gain amplifier (approximately 70). Consequently, you must use a single high-gain transistor or possibly even a multiple-stage amplifier.

Figure 3-9 shows a discrete 1-MHz Pierce oscillator circuit. Q_1 provides all the gain necessary for self-sustained oscillations to occur. R_1 and C_1 provide a 65° phase lag to the feedback signal. The crystal impedance is basically resistive with a small inductive component. This impedance combined with the reactance of C_2 provides an additional 115° of phase lag. The transistor inverts the signal (180° phase shift), giving the circuit the necessary 360° of total phase shift. Because the crystal's load is primarily nonresistive (mostly the series combination of C_1 and C_2), this type of oscillator provides very good short-term frequency stability. Unfortunately, C_1 and C_2 introduce substantial losses and, consequently, the transistor must have a relatively high voltage gain; this is an obvious drawback.

3-5-8 Integrated-Circuit Pierce Oscillator

Figure 3-10 shows an IC-based Pierce crystal oscillator. Although it provides less frequency stability, it can be implemented using simple digital IC design and reduces costs substantially over conventional discrete designs.

To ensure that oscillations begin, RFB dc biases inverting amplifier A_1's input and output for class A operation. A_2 converts the output of A_1 to a full rail-to-rail swing (cutoff to saturation), reducing the rise and fall times and buffering A_1's output. The output resistance of A_1 combines with C_2 to provide the RC phase lag needed. Complementary metal-oxide

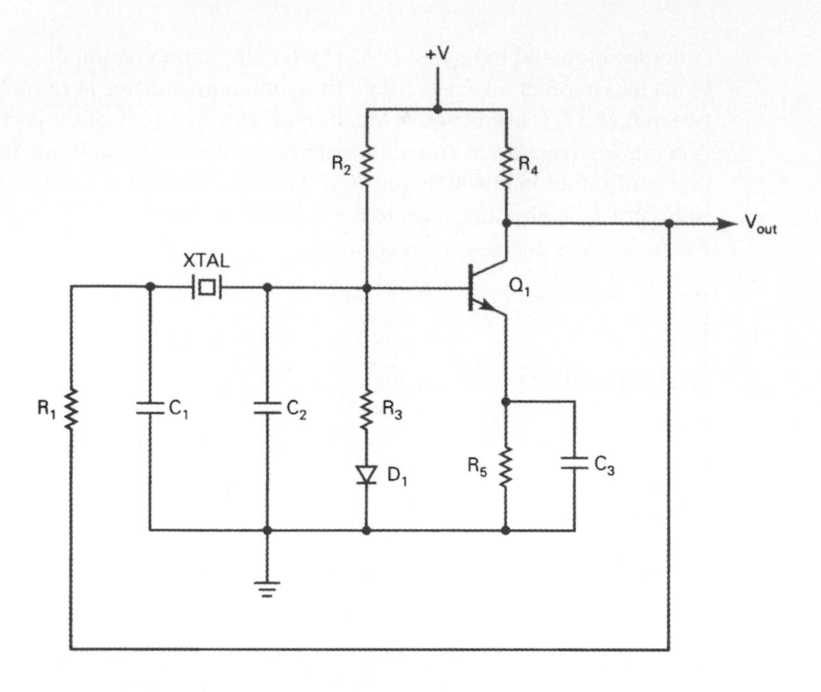

FIGURE 3-9 Discrete Pierce crystal oscillator

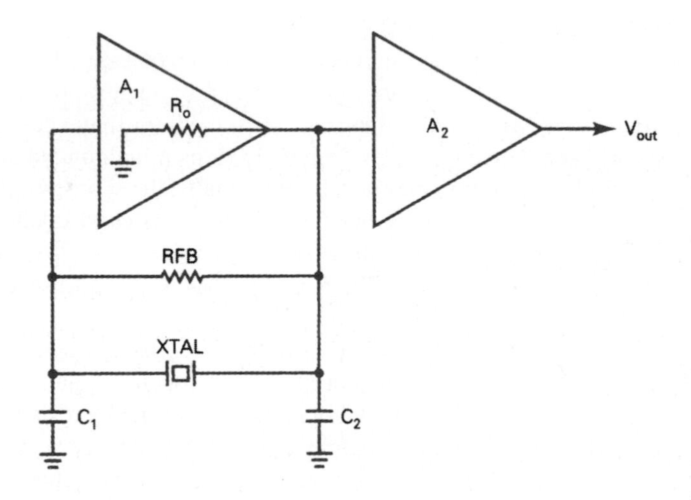

FIGURE 3-10 IC Pierce crystal oscillator

semiconductor (CMOS) versions operate up to approximately 2 MHz, and emitter-coupled logic (ECL) versions operate as high as 20 MHz.

3-5-9 RLC Half-Bridge Crystal Oscillator

Figure 3-11 shows the Meacham version of the *RLC* half-bridge crystal oscillator. The original Meacham oscillator was developed in the 1940s and used a full four-arm bridge and a negative-temperature-coefficient tungsten lamp. The circuit configuration shown in Figure 3-11 uses only a two-arm bridge and employs a negative-temperature-coefficient thermistor. Q_1 serves as a phase splitter and provides two 180° out-of-phase signals. The crystal

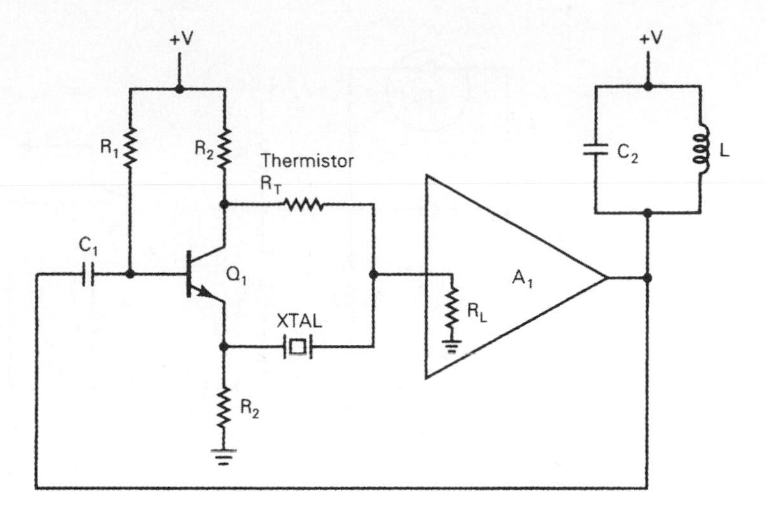

FIGURE 3-11 *RLC* half-bridge crystal oscillator

must operate at its series resonant frequency, so its internal impedance is resistive and quite small. When oscillations begin, the signal amplitude increases gradually, decreasing the thermistor resistance until the bridge almost nulls. The amplitude of the oscillations stabilizes and determines the final thermistor resistance. The *LC* tank circuit at the output is tuned to the crystal's series resonant frequency.

3-5-10 Crystal Oscillator Module

A *crystal oscillator module* consists of a crystal-controlled oscillator and a voltage-variable component such as a *varactor diode*. The entire oscillator circuit is contained in a single *metal can*. A simplified schematic diagram for a Colpitts crystal oscillator module is shown in Figure 3-12a. X_1 is a crystal itself, and Q_1 is the active component for the amplifier. C_1 is a shunt capacitor that allows the crystal oscillator frequency to be varied over a narrow range of operating frequencies. VC_1 is a voltage-variable capacitor (*varicap* or *varactor diode*). A varactor diode is a specially constructed diode whose internal capacitance is enhanced when reverse biased, and by varying the reverse-bias voltage, the capacitance of the diode can be adjusted. A varactor diode has a special depletion layer between the *p*- and *n*-type materials that is constructed with various degrees and types of doping material (the term *graded junction* is often used when describing varactor diode fabrication). Figure 3-12b shows the capacitance versus reverse-bias voltage curves for a typical varactor diode. The capacitance of a varactor diode is approximated as

$$C_d = \frac{C}{\sqrt{(1 + 2|V_r|)}} \quad \textbf{(3-10)}$$

where C = diode capacitance with 0-V reverse bias (farads)
$|V_r|$ = magnitude of diode reverse-bias voltage (volts)
C_d = reverse-biased diode capacitance (farads)

The frequency at which the crystal oscillates can be adjusted slightly by changing the capacitance of VC_1 (i.e., changing the value of the reverse-bias voltage). The varactor diode, in conjunction with a temperature-compensating module, provides instant frequency compensation for variations caused by changes in temperature. The schematic

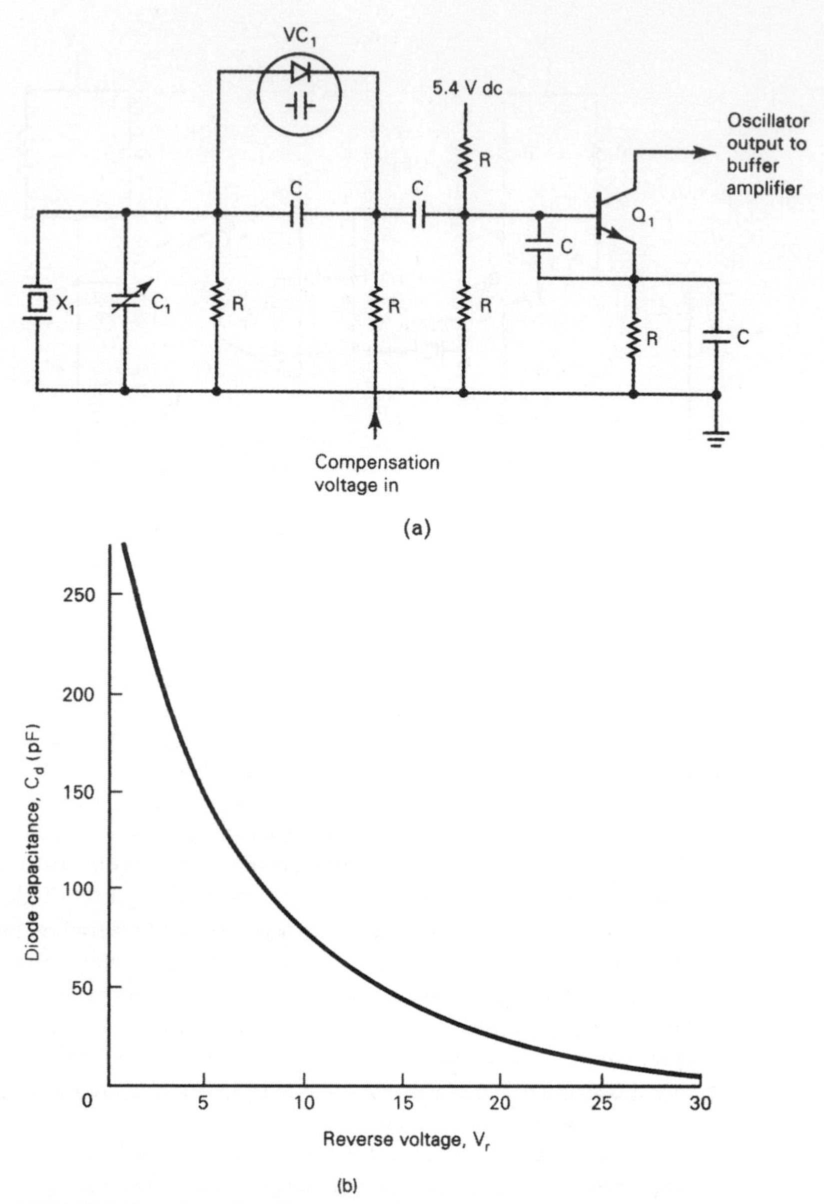

FIGURE 3-12 Crystal oscillator module: (a) schematic diagram; (b) varactor diode characteristics

diagram of a temperature-compensating module is shown in Figure 3-13. The compensation module includes a buffer amplifier (Q_1) and a temperature-compensating network (T_1). T_1 is a negative-temperature-coefficient thermistor. When the temperature falls below the threshold value of the thermistor, the compensation voltage increases. The compensation voltage is applied to the oscillator module, where it controls the capacitance of the varactor diode. Compensation modules are available that can compensate for a frequency stability of 0.0005% from −30°C to +80°C.

3-6 LARGE-SCALE INTEGRATION OSCILLATORS

In recent years, the use of *large-scale integration* (LSI) integrated circuits for frequency and waveform generation has increased at a tremendous rate because integrated-circuit oscillators have excellent frequency stability and a wide tuning range and are easy to use.

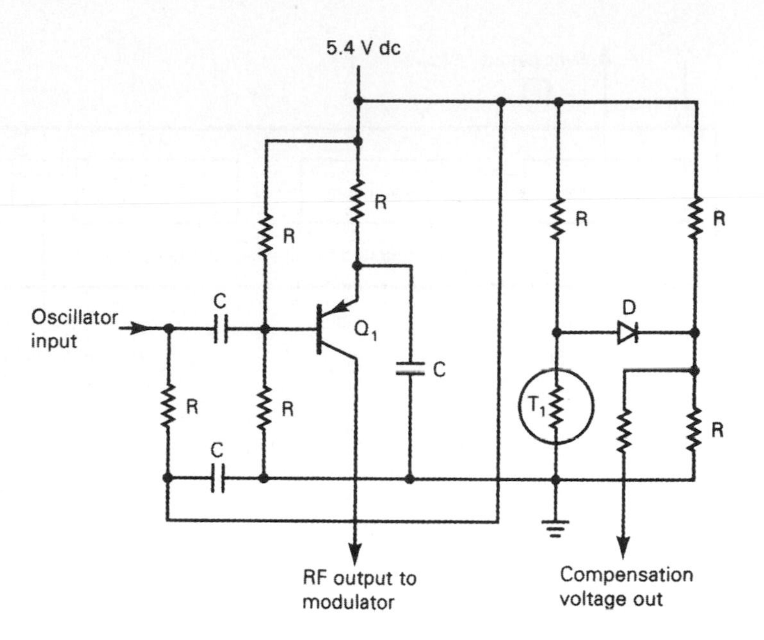

FIGURE 3-13 Compensation circuit

Waveform and *function generators* are used extensively in communications and telemetry equipment as well as in laboratories for test and calibration equipment. In many of these applications, commercial monolithic integrated-circuit oscillators and function generators are available that provide the circuit designer with a low-cost alternative to their noninte-grated-circuit counterparts.

The basic operations required for waveform generation and shaping are well suited to monolithic integrated-circuit technology. In fact, *monolithic linear integrated circuits* (LICs) have several inherent advantages over discrete circuits, such as the availability of a large number of active devices on a single chip and close matching and thermal tracking of component values. It is now possible to fabricate integrated-circuit waveform generators that provide a performance comparable to that of complex discrete generators at only a fraction of the cost.

LSI waveform generators currently available include function generators, timers, programmable timers, voltage-controlled oscillators, precision oscillators, and waveform generators.

3-6-1 Integrated-Circuit Waveform Generation

In its simplest form, a waveform generator is an oscillator circuit that generates well-defined, stable waveforms that can be externally modulated or swept over a given frequency range. A typical waveform generator consists of four basic sections: (1) an oscillator to generate the basic periodic waveform, (2) a waveshaper, (3) an optional AM modulator, and (4) an output buffer amplifier to isolate the oscillator from the load and provide the necessary drive current.

Figure 3-14 shows a simplified block diagram of an integrated-circuit waveform generator circuit showing the relationship among the four sections. Each section has been built separately in monolithic form for several years; therefore, fabrication of all four sections onto a single monolithic chip was a natural extension of a preexisting technology. The oscillator section generates the basic oscillator frequency, and the wave-shaper circuit converts the output from the oscillator to either a sine-, square-, triangular-, or ramp-shaped waveform. The modulator, when used, allows the circuit to produce

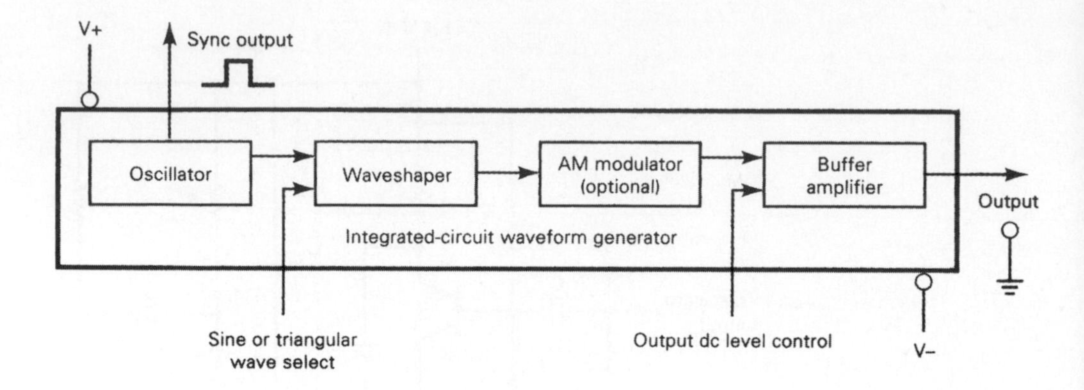

FIGURE 3-14 Integrated-circuit waveform generator

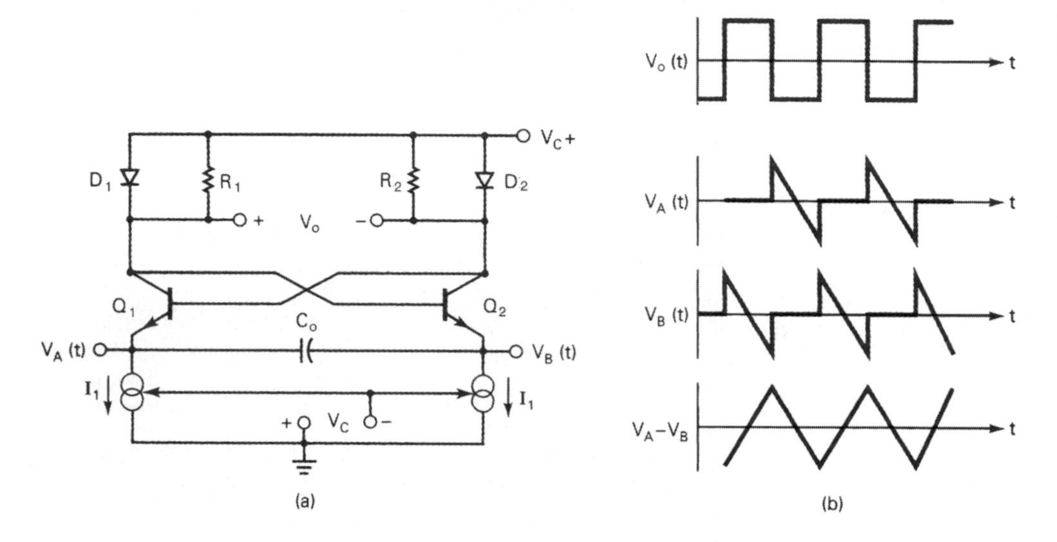

FIGURE 3-15 Simplified integrated-circuit waveform generator: (a) schematic diagram; (b) waveforms

amplitude-modulated signals, and the output buffer amplifier isolates the oscillator from its load and provides a convenient place to add dc levels to the output waveform. The sync output can be used either as a square-wave source or as a synchronizing pulse for external timing circuitry.

A typical IC oscillator circuit utilizes the constant-current charging and discharging of external timing capacitors. Figure 3-15a shows the simplified schematic diagram for such a waveform generator that uses an emitter-coupled multivibrator, which is capable of generating square waves as well as triangle and linear ramp waveforms. The circuit operates as follows. When transistor Q_1 and diode D_1 are conducting, transistor Q_2 and diode D_2 are off and vice versa. This action alternately charges and discharges capacitor C_o from constant current source I_1. The voltage across D_1 and D_2 is a symmetrical square wave with a peak-to-peak amplitude of $2V_{BE}$. V_A is constant when Q_1 is on but becomes a linear ramp with a slope equal to $-I_1/C_o$ when Q_1 goes off. Output $V_B(t)$ is identical to $V_A(t)$, except it is delayed by a half-cycle. Differential output, $V_A(t) - V_B(t)$ is a triangle wave. Figure 3-15b shows the output voltage waveforms typically available.

3-6-1-1 Monolithic function generators. The XR-2206 is a monolithic function generator integrated circuit manufactured by EXAR Corporation that is capable of producing high-quality sine, square, triangle, ramp, and pulse waveforms with both a high degree

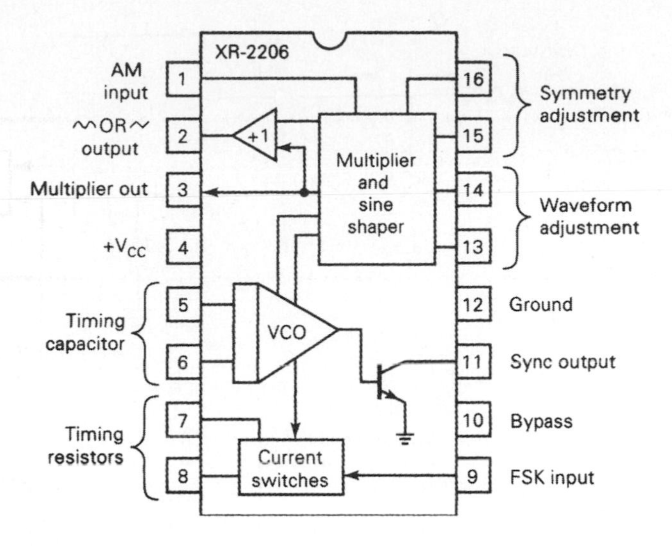

FIGURE 3-16 Block diagram for the XR-2206 monolithic function generator

of stability and accuracy. The output waveforms from the XR-2206 can be both amplitude and frequency modulated by an external modulating signal, and the frequency of operation can be selected externally over a range from 0.01 Hz to more than 1 MHz. The XR-2206 is ideally suited to communications, instrumentation, and function generator applications requiring sinusoidal tone, AM, or FM generation. The XR-2206 has a typical frequency stability of 20 ppm/°C and can be linearly swept over a 2000:1 frequency range with an external control voltage.

The block diagram for the XR-2206 is shown in Figure 3-16. The function generator is comprised of four functional blocks: a voltage-controlled oscillator (VCO), an analog multiplier and sineshaper, a unity-gain buffer amplifier, and a set of input current switches. A *voltage-controlled oscillator* is a free-running oscillator with a stable frequency of oscillation that depends on an external timing capacitance, timing resistance, and control voltage. The output from a VCO is a frequency, and its input is a bias or control signal that can be either a dc or an ac voltage. The VCO actually produces an output frequency that is proportional to an input current that is produced by a resistor from the timing terminals (either pin 7 or 8) to ground. The current switches route the current from one of the timing pins to the VCO. The current selected depends on the voltage level on the frequency shift keying input pin (pin 9). Therefore, two discrete output frequencies can be independently produced. If pin 9 is open circuited or connected to a bias voltage ≥2 V, the current passing through the resistor connected to pin 7 is selected. Similarly, if the voltage level at pin 9 is ≤1 V, the current passing through the resistor connected to pin 8 is selected. Thus, the output frequency can be keyed between f_1 and f_2 by simply changing the voltage on pin 9. The formulas for determining the two frequencies of operation are

$$f_1 = \frac{1}{R_1 C} \quad f_2 = \frac{1}{R_2 C} \tag{3-11}$$

where R_1 = resistor connected to pin 7
R_2 = resistor connected to pin 8

The frequency of oscillation is proportional to the total timing current on either pin 7 or 8. Frequency varies linearly with current over a range of current values between 1 μA to 3 μA. The frequency can be controlled by applying a control voltage, V_C, to the

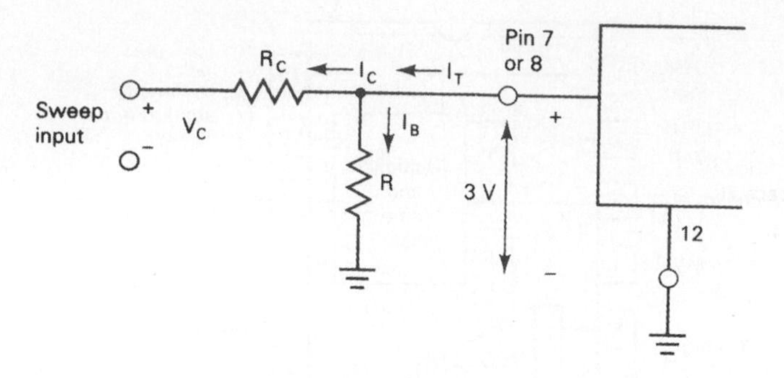

FIGURE 3-17 Circuit connection for control voltage frequency sweep of the XR-2206

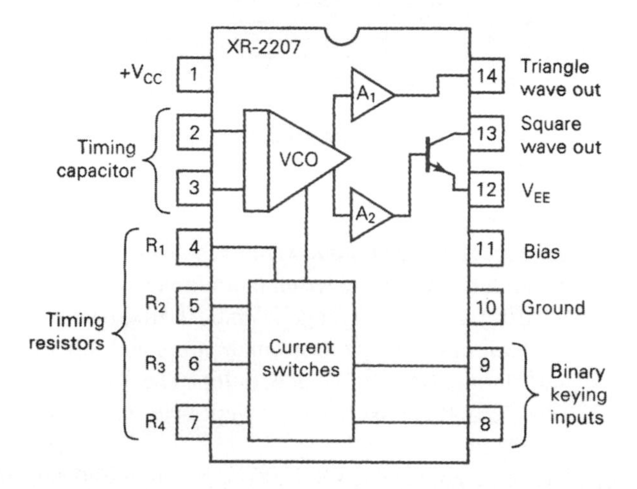

FIGURE 3-18 Block diagram for the XR-2207 monolithic voltage-controlled oscillator (VCO)

selected timing pin, as shown in Figure 3-17. The frequency of oscillation is related to V_C by

$$f = \frac{1}{RC}\left[1 + \frac{R}{R_C}\frac{(1 - V_C)}{3}\right]\text{Hz} \tag{3-12}$$

The voltage-to-frequency conversion gain K is given as

$$K = \frac{\Delta f}{\Delta V_C} = \frac{-0.32}{R_C C}\ \text{Hz/V} \tag{3-13}$$

3-6-1-2 Monolithic voltage-controlled oscillators. The XR-2207 is a monolithic voltage-controlled oscillator (VCO) integrated circuit featuring excellent frequency stability and a wide tuning range. The circuit provides simultaneous triangle- and square-wave outputs over a frequency range of from 0.01 Hz to 1 MHz. The XR-2207 is ideally suited for FM, FSK, and sweep or tone generation as well as for phase-locked-loop applications. The XR-2207 has a typical frequency stability of 20 ppm/°C and can be linearly swept over a 1000:1 frequency range with an external control voltage. The duty cycle of the triangular- and square-wave outputs can be varied from 0.1% to 99.9%, generating stable pulse and sawtooth waveforms.

The block diagram for the XR-2207 is shown in Figure 3-18. The circuit is a modified emitter-coupled multivibrator that utilizes four main functional blocks for frequency

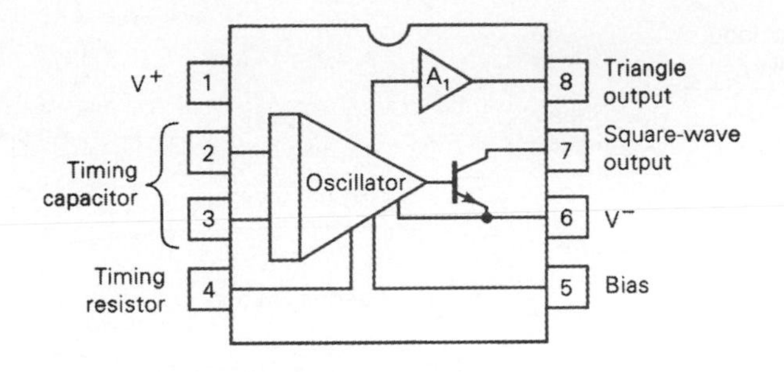

FIGURE 3-19 Block diagram for the XR-2209 monolithic precision oscillator

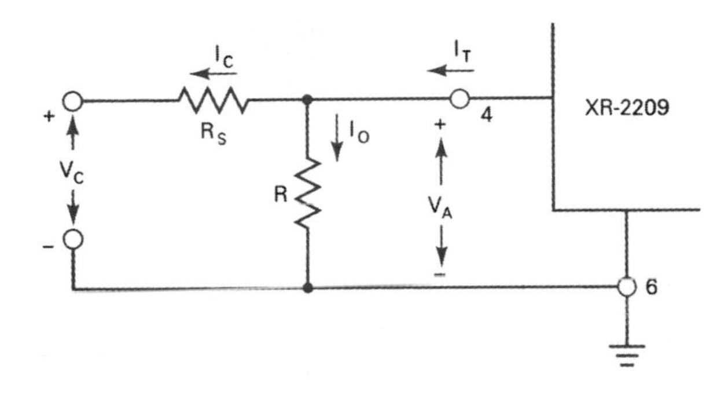

FIGURE 3-20 Circuit connection for control voltage frequency sweep of the XR-2209

generation: a VCO, four current switches that are activated by binary keying inputs, and two buffer amplifiers. Two binary input pins (pins 8 and 9) determine which of the four timing currents are channeled to the VCO. These currents are set by resistors to ground from each of the four timing input terminals (pins 4 through 7). The triangular output buffer provides a low-impedance output (10 Ω typical), while the square-wave output is open collector.

3-6-1-3 Monolithic precision oscillators. The XR-2209 is a monolithic variable-frequency oscillator circuit featuring excellent temperature stability and a wide linear sweep range. The circuit provides simultaneous triangle- and square-wave outputs, and the frequency is set by an external *RC* product. The XR-2209 is ideally suited for frequency modulation, voltage-to-frequency conversion, and sweep or tone generation as well as for phase-locked-loop applications when used in conjunction with an appropriate phase comparator.

The block diagram for the XR-2209 precision oscillator is shown in Figure 3-19. The oscillator is comprised of three functional blocks: a variable-frequency oscillator that generates the basic periodic waveforms and two buffer amplifiers for the triangular- and square-wave outputs. The oscillator frequency is set by an external capacitor and timing resistor. The XR-2209 is capable of operating over eight frequency decades from 0.01 Hz to 1 MHz. With no external sweep signal or bias voltage, the frequency of oscillation is simply equal to $1/RC$.

The frequency of operation for the XR-2209 is proportional to the timing current drawn from the timing pin. This current can be modulated by applying a control voltage, V_C, to the timing pin through series resistor R_S as shown in Figure 3-20. If V_C is negative with respect to the voltage on pin 4, an additional current, I_O, is drawn from the timing pin, causing the total input current to increase, thus increasing the frequency of oscillation. Conversely, if V_C is higher than the voltage on pin 4, the frequency of oscillation is decreased.

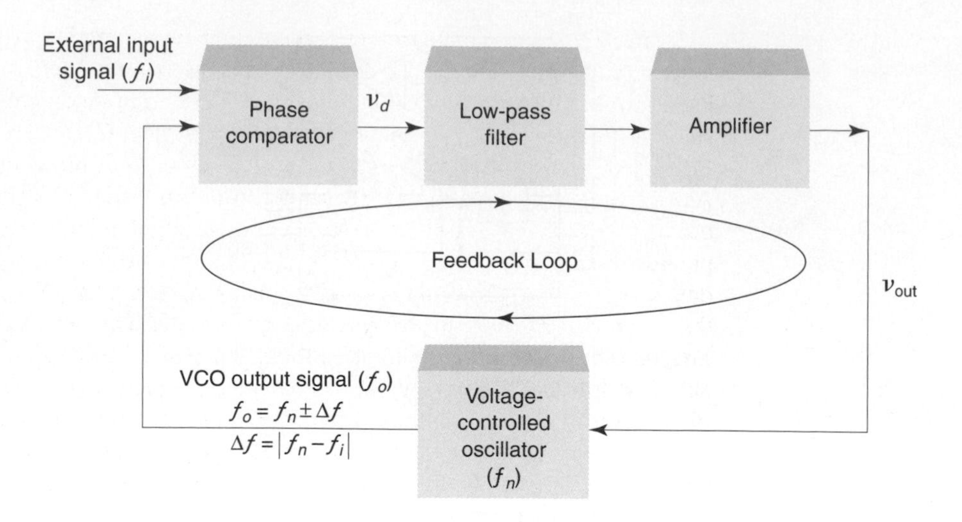

FIGURE 3-21 Phase-locked loop block diagram

3-7 PHASE-LOCKED LOOPS

The *phase-locked loop* (PLL) is an extremely versatile circuit used extensively in modern electronic communications systems for performing a wide variety of functions, including modulation, demodulation, signal processing, carrier and clock recovery, frequency generation, frequency synthesis, and a wide variety of other electronic communications applications. PLLs are used in transmitters and receivers using both analog and digital modulation and with the transmission of digital pulses.

PLLs were first used in 1932 for synchronous detection and demodulation of radio signals, instrumentation circuits, and space telemetry systems. However, for many years circuit designers avoided using PLLs because of their large size, necessary complexity, narrow bandwidth, and high cost. However, with the advent of large-scale integration, PLLs can now provide reliable, high-performance operation and at the same time be extremely small and easy to use and dissipate little power. Therefore, PLLs have changed from a specialized design technique to a general-purpose, universal building block with an abundance of applications. Today, dozens of integrated-circuit PLL products are available from a wide assortment of manufacturers. Some of these products are designated as general-purpose circuits suitable for many uses, while others are intended to be optimized and used for special applications, such as tone detection, stereo demodulation, and frequency synthesis. A PLL can provide precision frequency selective tuning and filtering without the need of bulky coils or inductors.

In essence, a PLL is a *closed-loop feedback control system* in which either the frequency or the phase of the feedback signal is the parameter of interest rather than the magnitude of the signal's voltage or current. The basic block diagram for a phase-locked loop circuit is shown in Figure 3-21. As the figure shows, a PLL consists of four primary blocks: (1) a phase comparator or phase detector, (2) a low-pass filter (LPF), (3) a low-gain operational amplifier, and (4) a VCO. The four circuits are modularized and placed on an integrated circuit, with each circuit provided external input and output pins, allowing users to interconnect the circuits as needed and to set the break frequency of the low-pass filter, the gain of the amplifier, and the frequency of the VCO.

3-7-1 PLL Loop Operation

A PLL ultimately uses phase lock to perform its intended function. However, before phase lock can occur, a PLL must be frequency locked. After frequency lock has occurred, the phase comparator produces an output voltage that is proportional to the difference in phase between the VCO output frequency and the external input frequency.

For a PLL to operate properly, there must be a complete path around the feedback loop, as shown in Figure 3-21. When there is no external input signal or when the feedback loop is open, the VCO operates at a preset frequency called its *natural* or *free-running frequency* (f_n). The natural frequency is the VCO's output frequency when the PLL is not locked. The VCO's natural frequency is determined by external components. As previously stated, before a PLL can perform its intended function, frequency lock must occur. When an external input signal (f_i) is initially applied to the PLL, the phase comparator compares the frequency of the external input signal to the frequency of the VCO output signal. The phase comparator produces an error voltage (v_d) that is proportional to the difference in frequency between the two signals. The error voltage is filtered, amplified, and then applied to the input to the VCO. If the frequency of the external input signal (f_i) is sufficiently close to the VCO natural frequency (f_n), the feedback nature of the PLL causes the VCO to *synchronize* or *lock* onto the external input signal. Therefore, the VCO output frequency is the VCO natural frequency plus or minus the difference between the external input frequency and the VCO's natural frequency. Mathematically, the VCO output frequency (f_o) is

$$f_o = f_n \pm \Delta f \tag{3-14}$$

where f_o = VCO output frequency (hertz)
f_n = VCO natural frequency (hertz)
$\Delta f = f_i - f_n$ (hertz)
f_i = external input frequency (hertz)

In essence, a PLL has three operating states: free running, capture, and lock. In the *free-running state,* either there is no external input frequency or the feedback loop is open. When in the free-running state, the VCO oscillates at its natural frequency determined by external components. To be in the *capture state,* there must be an external input signal, and the feedback loop must be complete. When in the capture state, the PLL is in the process of acquiring frequency lock. In the *lock state,* the VCO output frequency is *locked* onto (equal to) the frequency of the external input signal. When in the lock state, the VCO output frequency *tracks* (follows) changes in the frequency of the external input signal.

3-7-2 Loop Acquisition

When an external input signal ($V_i \sin[2\pi ft + \theta i]$) enters the phase comparator shown in Figure 3-22a, it mixes with the VCO output signal ($V_o \sin[2\pi f_o t + \theta_o]$). Initially, the two frequencies are not equal ($f_o \neq f_i$), and the loop is *unlocked.* Because the phase comparator is a nonlinear device, the external input signal and VCO output signal mix and generate cross-product frequencies (i.e., sum and difference frequencies). Therefore, the primary output frequencies from the phase comparator are the external input frequency (f_i), the VCO output frequency (f_o), and their sum ($f_o + f_i$) and difference ($f_o - f_i$) frequencies.

The low-pass filter blocks the two input frequencies (f_i and f_o) and the sum frequency ($f_o + f_i$). Thus, the only signal allowed to pass through to the output of the LPF is the relatively low difference frequency ($f_d = f_o - f_i$), which is sometimes called the *beat frequency.* The beat frequency is amplified and then applied to the input to the voltage-controlled oscillator, where it changes the VCO output frequency by an amount proportional to its polarity and amplitude. As the VCO output frequency changes, the amplitude and frequency of the beat frequency changes proportionately. Figure 3-22b shows the beat frequency produced when the VCO is swept by the difference frequency (f_d). After several cycles around the loop, the VCO output frequency equals the external input frequency, and the loop is said to have acquired frequency lock. Once frequency lock has occurred, the beat frequency at the output of the LPF is 0 Hz (a dc voltage), and its magnitude and polarity is proportional to the difference in phase between the external input signal and the VCO output signal. The

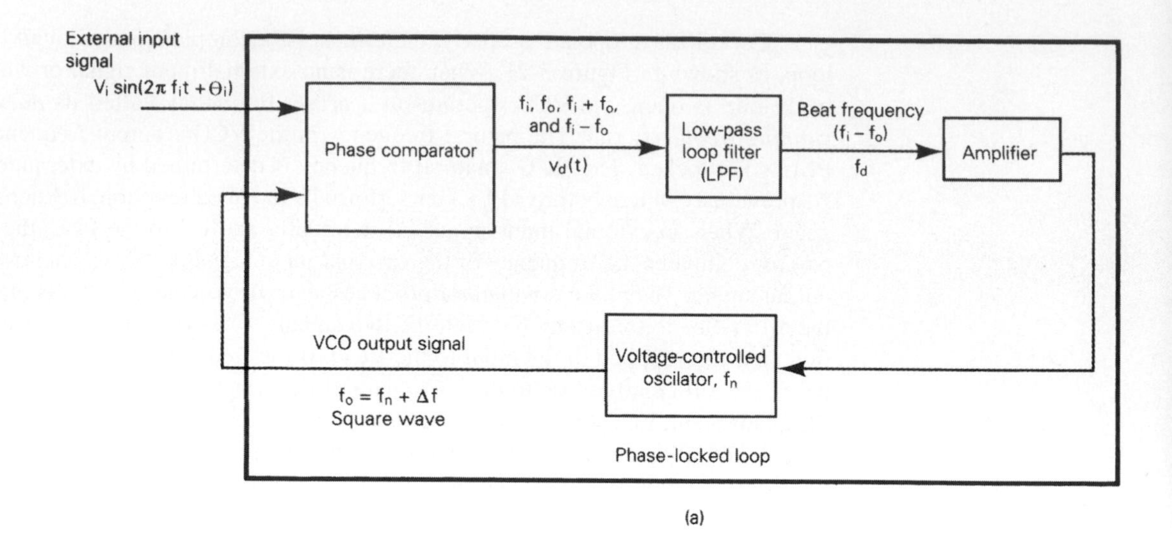

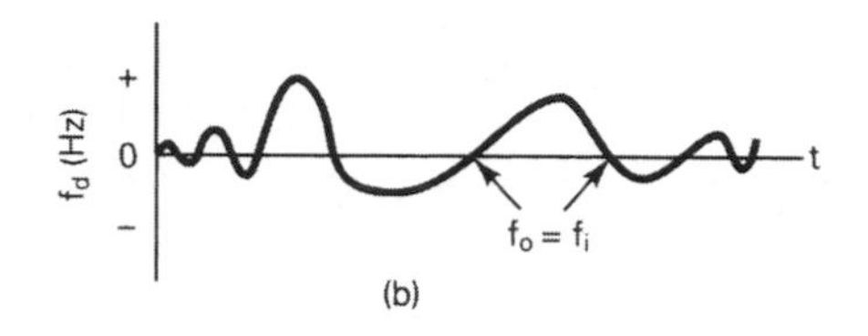

FIGURE 3-22 PLL operation: (a) block diagram; (b) beat frequency

dc voltage provides an input bias to the VCO, keeping it locked onto the frequency of the external input signal.

In essence, the phase comparator is a frequency comparator until frequency acquisition (zero beat) is achieved, then it becomes a phase comparator. Once the loop is frequency locked, the phase difference between the external input and VCO output frequency is converted to a dc bias voltage (v_d) in the phase comparator, filtered, amplified, and then fed back to the VCO to hold lock. Therefore, to maintain frequency lock, it is necessary that a phase error be maintained between the external input signal and the VCO output signal. The time required to achieve lock is called the *acquisition time* or *pull-in time.*

3-8 PLL CAPTURE AND LOCK RANGES

Two key parameters of a phase-locked loop that indicate its useful frequency range are capture and lock range.

3-8-1 Capture Range

Capture range is defined as the band of frequencies centered around the VCO natural frequency (f_n) where the PLL can initially establish or acquire frequency lock with an external input signal from an unlocked condition. The capture range is generally between 0.5 and 1.7 times the VCO's natural frequency, depending on PLL design, the bandwidth of the low-pass filter, and the gain of the feedback loop. Capture range is sometimes called *acquisition range. Pull-in range* is the capture range expressed as a peak value (i.e., capture range = 2 times pull-in range). Capture range and pull-in range are shown in frequency-diagram form in Figure 3-23. The lowest frequency the PLL can lock onto is called the lower capture limit (f_{cl}), and the highest frequency the PLL can lock onto is called the upper capture limit (f_{cu}).

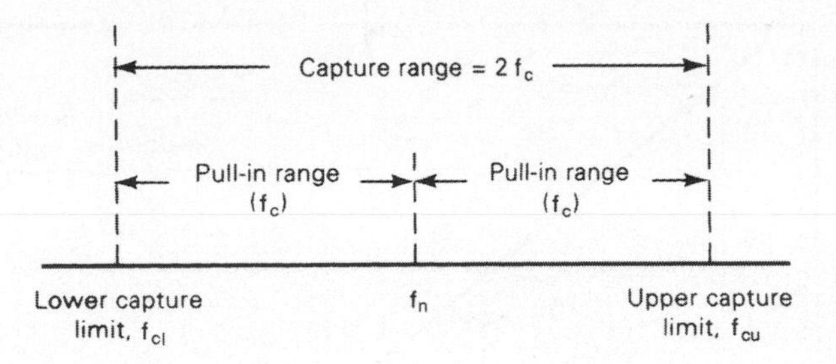

FIGURE 3-23 PLL capture range

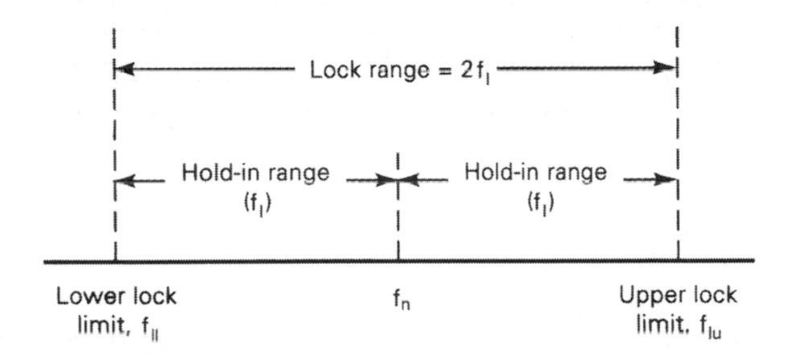

FIGURE 3-24 PLL lock range

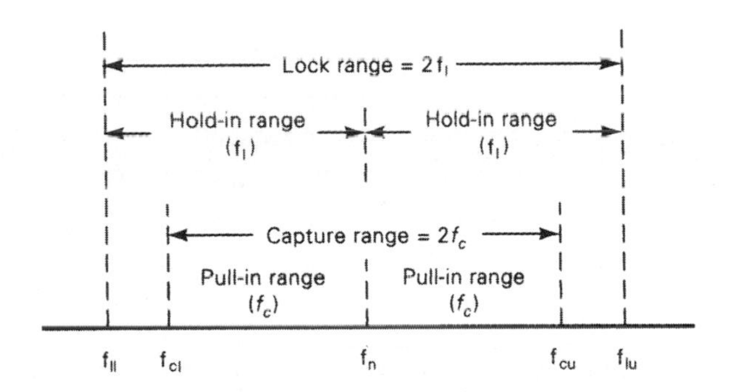

FIGURE 3-25 PLL capture and lock ranges

3-8-2 Lock Range

Lock range is defined as the band of frequencies centered on the VCO's natural frequency over which a PLL can maintain frequency lock with an external input signal. Lock range presumes that the PLL has initially captured and locked onto the external input signal. Lock range is also known as *tracking range.* Lock range is the range of frequencies over which the PLL will accurately track or follow the frequency of the external input signal after frequency lock has occurred. *Hold-in range* is the lock range expressed as a peak value (i.e., lock range = 2 times hold-in range). The relationship between lock and hold-in range is shown in frequency-diagram form in Figure 3-24. The lowest frequency a PLL will track is called the lower lock limit (f_{ll}), and the highest frequency that a PLL will track is called the upper lock limit (f_{lu}).

Capture and lock range are directly proportional to the dc gain of the PLL's feedback loop. The capture range is never greater than, and is almost always less than, the lock range. The relationship among capture, lock, hold-in, and pull-in range is shown in frequency-diagram form in Figure 3-25. Note that the lock range is greater than or equal to the capture range and that the hold-in range is greater than or equal to the pull-in range.

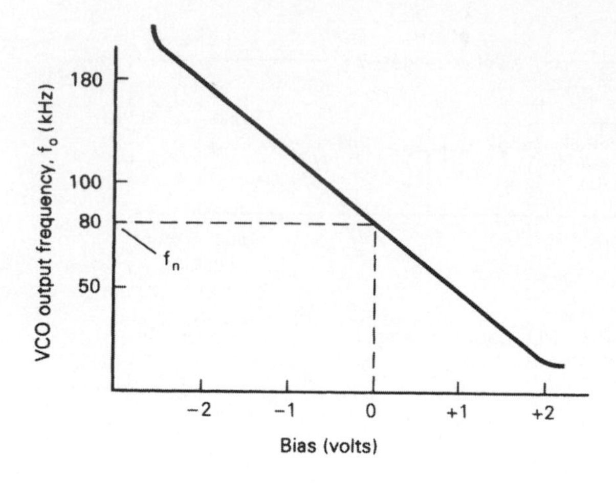

FIGURE 3-26 Voltage controlled oscillator output frequency-versus-input bias voltage characteristics

3-9 VOLTAGE-CONTROLLED OSCILLATOR

A *voltage-controlled oscillator* (VCO) is an oscillator (more specifically, a free-running multivibrator) with a stable frequency of oscillation that depends on an external bias voltage. The output from a VCO is a frequency, and its input is a bias or control signal that may be a dc or an ac voltage. When a dc or slowly changing ac voltage is applied to the VCO input, the output frequency changes or deviates proportionally. Figure 3-26 shows a transfer curve (output frequency-versus-input bias voltage characteristics) for a typical VCO. The output frequency (f_o) with 0-V input bias is the VCO's natural frequency (f_n), which is determined by an external RC network, and the change in the output frequency caused by a change in the input voltage is called frequency deviation (Δf). Consequently, $f_o = f_n + \Delta f$, where f_o = VCO output frequency. For a symmetrical Δf, the natural frequency of the VCO should be centered within the linear portion of the input-versus-output curve. The transfer function for a VCO is

$$K_o = \frac{\Delta f}{\Delta V} \tag{3-15}$$

where K_o = input-versus-output transfer function (hertz per volt)
ΔV = change in the input control voltage (volts)
Δf = change in the output frequency (hertz)

3-10 PHASE COMPARATOR

A phase comparator, sometimes called a *phase detector,* is a nonlinear device with two input signals: an external input frequency (f_i) and the VCO output signal (f_o). The output from a phase comparator is the product of the two signals of frequencies f_i and f_o and, therefore, contains their sum and difference frequencies ($f_i \pm f_o$). Figure 3-27a shows the schematic diagram for a simple phase comparator. v_o is applied simultaneously to the two halves of input transformer T_1. D_1, R_1, and C_1 make up a half-wave rectifier, as do D_2, R_2, and C_2 (note that $C_1 = C_2$ and $R_1 = R_2$). During the positive alternation of v_o, D_1 and D_2 are forward biased and *on,* charging C_1 and C_2 to equal values but with opposite polarities. Therefore, the average output voltage is $V_{out} = V_{C1} + (-V_{C2}) = 0$ V. This is shown in Figure 3-27b. During the negative half-cycle of v_o, D_1 and D_2 are reverse biased and *off.* Therefore, C_1 and C_2 discharge equally through R_1 and R_2, respectively, keeping the output voltage equal to 0 V. This is shown in Figure 3-27c. The two half-wave rectifiers produce equal-magnitude, opposite-polarity output voltages. Therefore, the output voltage due to v_o is constant and equal to 0 V. The corresponding input and output waveforms for a square-wave VCO signal are shown in Figure 3-27d.

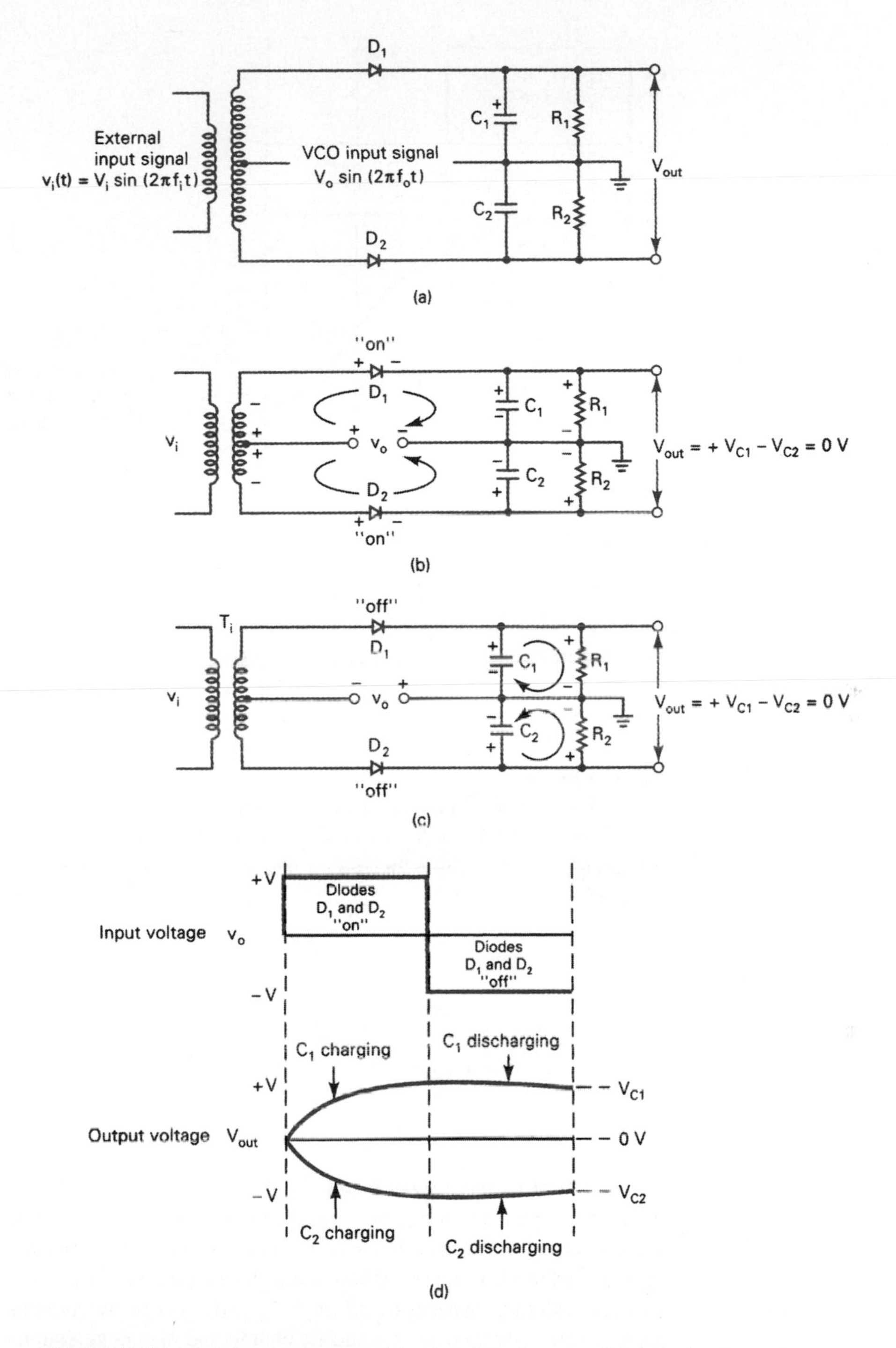

FIGURE 3-27 Phase comparator: (a) schematic diagram; (b) output voltage due to positive half-cycle of v_o; (c) output voltage due to negative half-cycle of v_o; (d) input and output voltage waveforms

3-10-1 Circuit Operation

When an external input signal ($v_{in} = V_i \sin[2\pi f_i t]$) is applied to the phase comparator, its voltage adds to v_o, causing C_1 and C_2 to charge and discharge, producing a proportional change in the output voltage. Figure 3-28a shows the unfiltered output waveform shaded when $f_o = f_i$ and v_o leads v_i by 90°. For the phase comparator to operate properly, v_o must be much larger than v_i. Therefore, D_1 and D_2 are switched *on* only during the positive

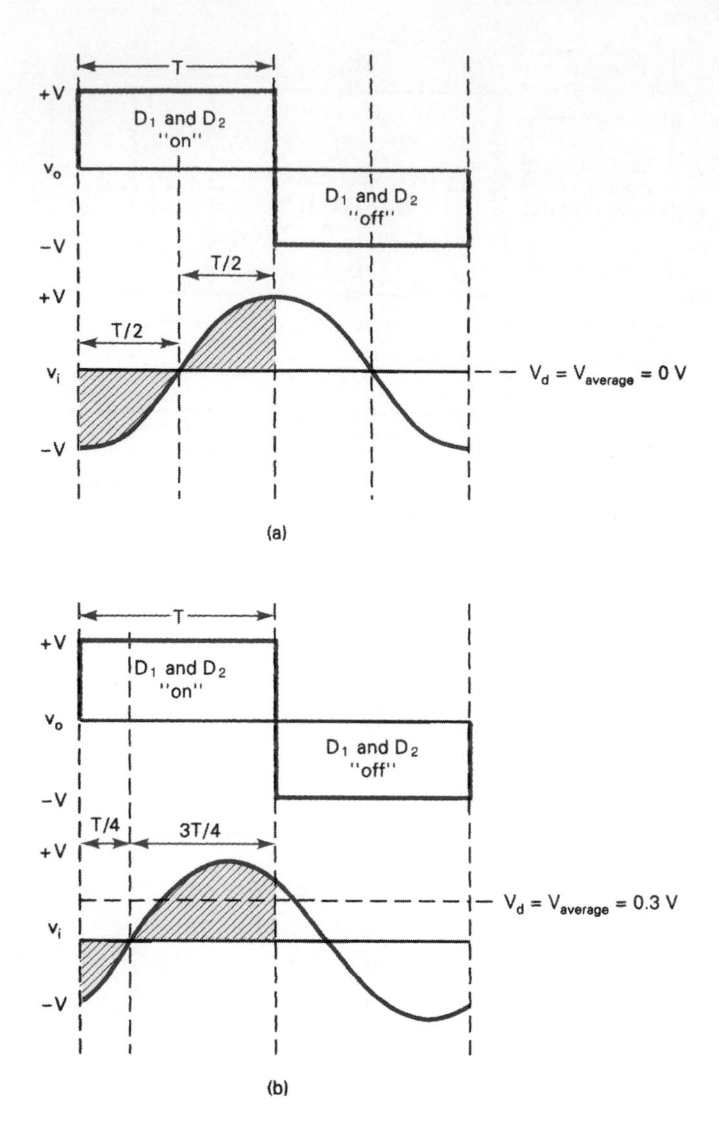

FIGURE 3-28 Phase comparator output voltage waveforms: (a) v_o leads v_i by 90°; (b) v_o leads v_i by 45°; *(Continued)*

alternation of v_o and are *off* during the negative alternation. During the first half of the *on* time, the voltage applied to $D_1 = v_o - v_i$, and the voltage applied to $D_2 = v_o + v_i$. Therefore, C_1 is discharging while C_2 is charging. During the second half of the *on* time, the voltage applied to $D_1 = v_o + v_i$, the voltage applied to $D_2 = v_o - v_i$, and C_1 is charging while C_2 is discharging. During the *off* time, C_1 and C_2 are neither charging nor discharging. For each complete cycle of v_o, C_1 and C_2 charge and discharge equally, and the average output voltage remains at 0 V. Thus, the average value of V_{out} is 0 V when the input and VCO output signals are equal in frequency and 90° out of phase.

Figure 3-28b shows the unfiltered output voltage waveform shaded when v_o leads v_i by 45°. v_i is positive for 75% of the *on* time and negative for the remaining 25%. As a result, the average output voltage for one cycle of v_o is positive and approximately equal to 0.3 V, where V is the peak input voltage. Figure 3-28c shows the unfiltered output waveform when v_o and v_i are in phase. During the entire *on* time, v_1 is positive. Consequently, the output voltage is positive and approximately equal to 0.636 V. Figures 3-28d and e show

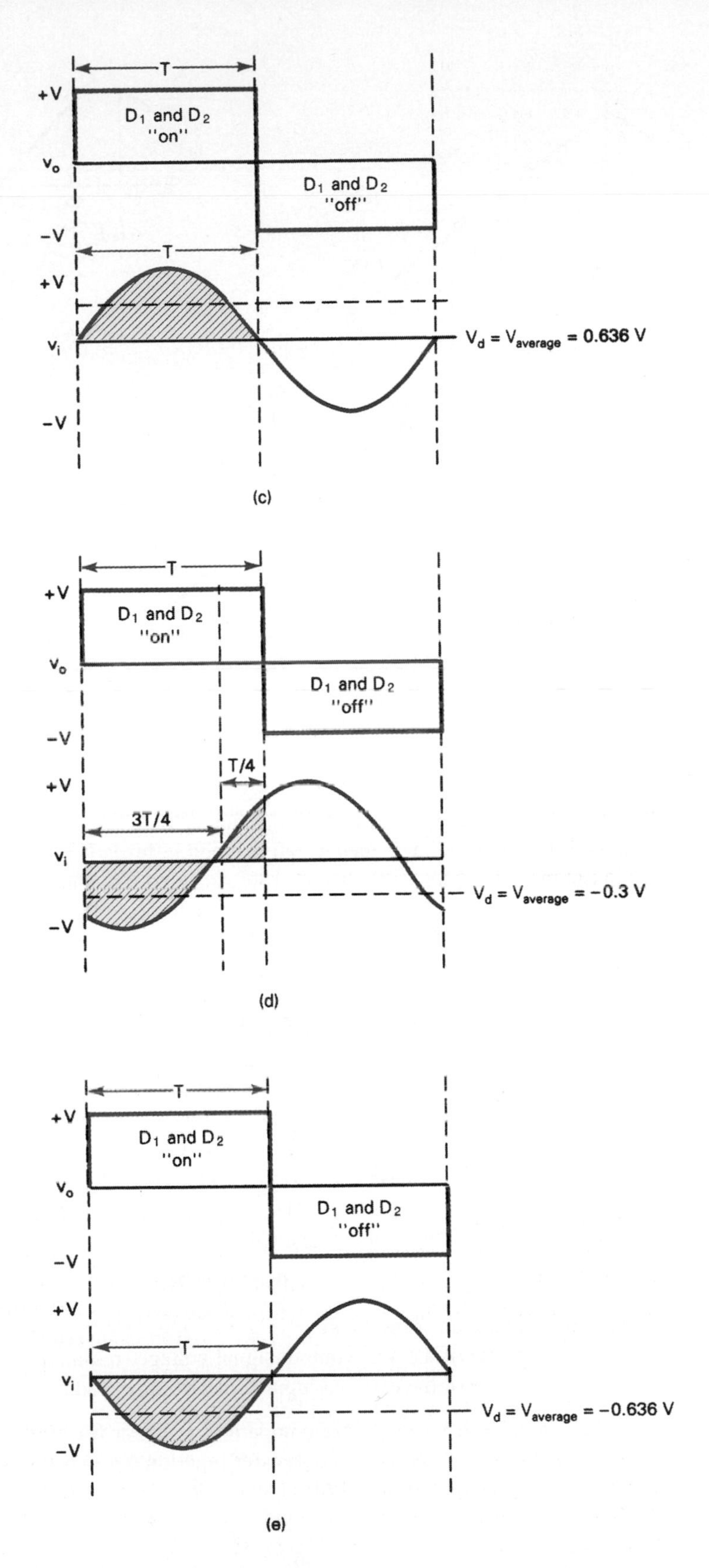

FIGURE 3-28 (Continued) (c) v_o and v_i in phase; (d) v_o leads v_i by 135°; (e) v_o leads v_i by 180°

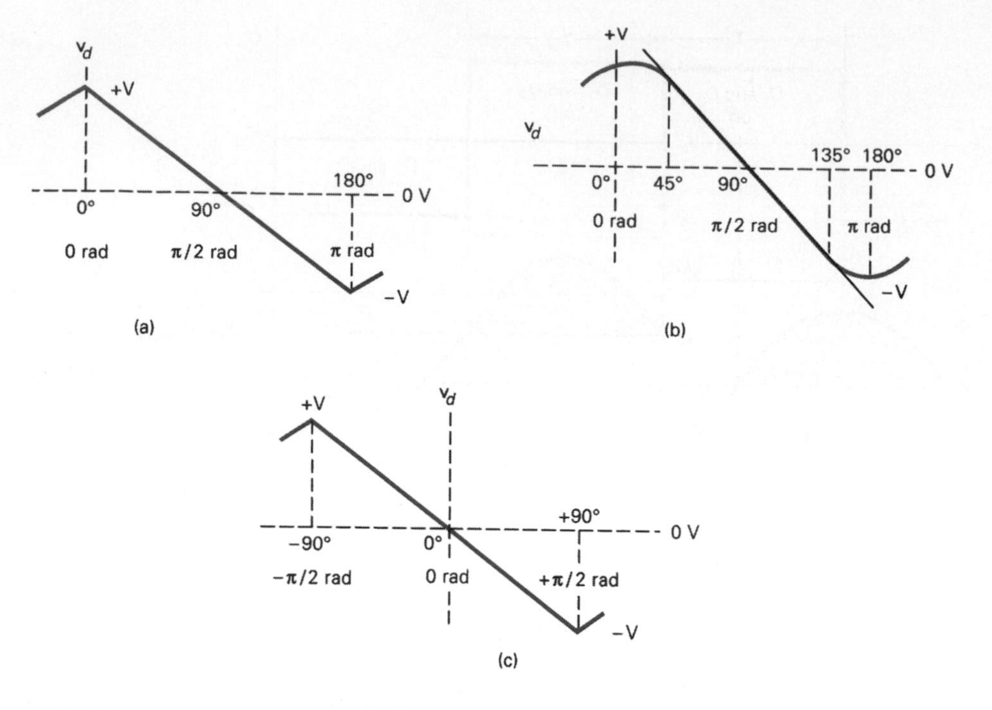

FIGURE 3-29 Phase comparator output voltage (*Vd*) versus phase difference (θ_e) characteristics: (a) square-wave inputs; (b) sinusoidal inputs; (c) square-wave inputs, phase bias reference

the unfiltered output waveform when v_o leads v_i by 135° and 180°, respectively. It can be seen that the output voltage goes negative when v_o leads v_i by more than 90° and reaches its maximum value when v_o leads v_i by 180°. In essence, a phase comparator rectifies v_i and integrates it to produce an output voltage that is proportional to the difference in phase between v_o and v_i.

Figure 3-29 shows the output voltage-versus-input phase difference characteristics for the phase comparator shown in Figure 3-27a. Figure 3-29a shows the curve for a square-wave phase comparator. The curve has a triangular shape with a negative slope from 0° to 180°. V_{out} is maximum positive when v_o and v_i are in phase, 0 V when v_o leads v_i by 90°, and maximum negative when v_o leads v_i by 180°. If v_o advances more than 180, the output voltage become less negative and, if v_o lags behind v_i, the output voltage become less positive. Therefore, the maximum phase difference that the comparator can track is 90° ± 90° or from 0° to 180°. The phase comparator produces an output voltage that is proportional to the difference in phase between v_o and v_i. This phase difference is called the *phase error.* The phase error is expressed mathematically as

$$\theta_e = \theta_i - \theta_o \tag{3-16}$$

where θ_e = phase error (radians)
θ_o = phase of the VCO output signal voltage (radians)
θ_i = phase of the external input signal voltage (radians)

The output voltage from the phase comparator is linear for phase errors between 0° and 180° (0 to π radians). Therefore, the transfer function for a square-wave phase comparator for phase errors between 0° and 180° is given as

$$K_d = \frac{V_d}{\theta_e} = \frac{2V_d}{\pi} \tag{3-17}$$

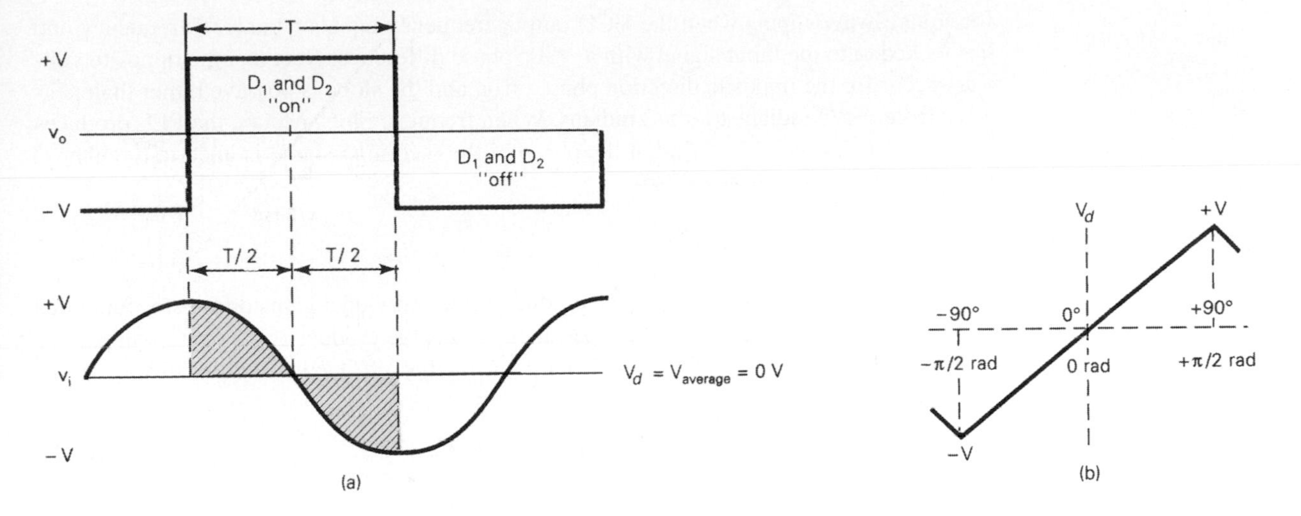

FIGURE 3-30 Phase comparator output voltage: (a) unfiltered output voltage waveform when v_i leads v_o by 90°; (b) output voltage-versus-phase difference characteristics

where K_d = transfer function or gain (volts per radian)
V_d = phase comparator output voltage (volts)
θ_e = phase error $(\theta_i - \theta_o)$(radians)
π = 3.14 radians

Figure 3-29b shows the output voltage-versus-input phase difference curve for an analog phase comparator with sinusoidal characteristics. The phase error versus output is nearly linear only from 45° to 135°. Therefore, the transfer function is given as

$$K_d = \frac{V_d}{\theta_e} \text{ volts per radian} \tag{3-18}$$

where K_d = transfer function or gain (volts per radian)
θ_e = phase error $(\theta_i - \theta_o)$ (radians)
V_d = phase comparator output voltage (volts)

Equation 3-16 can be rearranged to solve for V_d as follows:

$$V_d = K_d\theta \tag{3-19}$$

From Figures 3-29a and b, it can be seen that the phase comparator output voltage $V_{out} = 0$ V when $f_o = f_i$ and v_o and v_i are 90° out of phase. Therefore, if the input frequency (f_i) is initially equal to the VCO's natural frequency (f_n), a 90° phase difference is required to keep the phase comparator output voltage at 0 V and the VCO output frequency equal to its natural frequency ($f_o = f_n$). This 90° phase difference is equivalent to a bias or offset phase. Generally, the phase bias is considered as the reference phase, which can be deviated $\pm\pi/2$ radians (±90°). Therefore, V_{out} goes from its maximum positive value at $-\pi/2$ radians (−90°) and to its maximum negative value at $+\pi/2$ radians (+90°). Figure 3-29c shows the phase comparator output voltage-versus-phase error characteristics for square-wave inputs with the 90° phase bias as the reference.

Figure 3-30a shows the unfiltered output voltage waveform when v_i leads v_o by 90°. Note that the average value is 0 V (the same as when v_o led v_i by 90°). When frequency lock occurs, it is uncertain whether the VCO will lock onto the input frequency with a + or −90° phase difference. Therefore, there is a 180° phase ambiguity in the phase of VCO output frequency. Figure 3-30b shows the output voltage-versus-phase difference characteristics for square-wave inputs when the VCO output frequency equals its natural frequency and has locked onto the input signal with a $-90°$ phase difference. Note that the opposite voltages occur for the opposite direction phase error, and the slope is positive rather than negative from $-\pi/2$ radians to $+\pi/2$ radians. When frequency lock occurs, the PLL produces a coherent frequency ($f_o = f_i$), but the phase of the incoming signal is uncertain (either f_o leads f_i by $90° \pm \theta_e$ or vice versa).

3-11 PLL LOOP GAIN

The *loop gain* for a PLL is simply the product of the individual gains or transfer functions around the loop. In Figure 3-31, the open-loop gain is the product of the phase comparator gain, the low-pass filter gain, the amplifier gain, and the VCO gain. Mathematically, open-loop gain is

$$K_L = K_d K_f K_a K_o \tag{3-20}$$

where K_L = PLL open-loop gain (hertz per radian)
K_d = phase comparator gain (volts per radian)
K_f = low-pass filter gain (volts per volt)
K_a = amplifier gain (volts per volt)
K_o = VCO gain (hertz per volt)

and

$$K_L = \frac{(\text{volt})(\text{volt})(\text{volt})(\text{hertz})}{(\text{rad})(\text{volt})(\text{volt})(\text{volts})} = \frac{\text{hertz}}{\text{rad}}$$

or PLL open-loop gain (K_v) in radian/second (s^{-1}) is

$$K_v = \frac{\text{cycles/s}}{\text{rad}} = \frac{\text{cycles}}{\text{rad-s}} \times \frac{2\pi \text{ rad}}{\text{cycle}} = 2\pi\, K_L \tag{3-21}$$

Expressed in decibels, this gives us

$$K_{v(dB)} = 20 \log K_v \tag{3-22}$$

From Equations 3-15 and 3-19, the following relationships are derived:

$$V_d = (\theta_e)(K_d) \text{ volts} \tag{3-23}$$

$$V_{\text{out}} = (V_d)(K_f)(K_a) \text{ volts} \tag{3-24}$$

$$\Delta f = (V_{\text{out}})(K_o) \text{ hertz} \tag{3-25}$$

As previously stated, the hold-in range for a PLL is the range of input frequencies over which the PLL will remain locked. This presumes that the PLL was initially locked. The hold-in range is limited by the peak-to-peak swing in the phase comparator output voltage (ΔV_d) and depends on the phase comparator, amplifier, and VCO transfer functions. From Figure 3-29c, it can be seen that the phase comparator output voltage (V_d) is corrective for $\pm\pi/2$ radians ($\pm 90°$). Beyond these limits, the polarity of V_d reverses and actually chases the VCO frequency away from the external input frequency. Therefore, the maximum phase error (θ_e) that is allowed is $\pm\pi/2$ radians, and the maximum phase comparator output voltage is

$$\pm V_{d(\max)} = [\theta_{e(\max)}](K_d) \tag{3-26}$$

$$= \pm\left(\frac{\pi}{2}\text{rad}\right)(K_d) \tag{3-27}$$

where $\pm V_{d(\max)}$ = maximum peak change at the phase comparator output voltage
K_d = phase comparator transfer function

Consequently, the maximum change in the VCO output frequency is

$$\pm\Delta f_{max} = \pm\left(\frac{\pi}{2}\text{rad}\right)(K_d)(K_f)(K_a)(K_o) \tag{3-28}$$

where $\pm\Delta f_{max}$ is the hold-in range (maximum peak change in VCO output frequency).

Substituting K_v for $K_d K_f K_a K_o$ yields

$$\pm\Delta f_{max} = \pm\left(\frac{\pi}{2}\text{rad}\right)K_v \tag{3-29}$$

Example 3-2

For the PLL shown in Figure 3-31, a VCO natural frequency f_n = 200 kHz, an external input frequency f_i = 210 kHz, and the transfer functions K_d = 0.2 V/rad, K_f = 1, K_a = 5, and K_o = 20 kHz/V determine

a. PLL open-loop gain in Hz/rad and rad/s.
b. Change in VCO frequency necessary to achieve lock (Δf).
c. PLL output voltage (V_{out}).
d. Phase detector output voltage (V_d).
e. Static phase error (θ_e).
f. Hold-in range (Δf_{max}).

Solution **a.** From Equations 3-20 and 3-21,

$$K_L = \frac{0.2\text{ V}}{\text{rad}}\frac{1\text{ V}}{\text{V}}\frac{5\text{ V}}{\text{V}}\frac{20\text{ kHz}}{\text{V}} = \frac{20\text{ kHz}}{\text{rad}}$$

$$K_v = 2\pi K_L = \frac{20\text{ kHz}}{\text{rad}} = \frac{20\text{ kilocycles}}{\text{rad-s}} \times \frac{2\pi\text{ rads}}{\text{cycle}} = 125{,}600\text{ rad/s}$$

$$K_{v(dB)} = 20\log 125{,}600\text{ k} = 102\text{ dB}$$

b.

$$\Delta f = f_i - f_n = 210\text{ kHz} - 200\text{ kHz} = 10\text{ kHz}$$

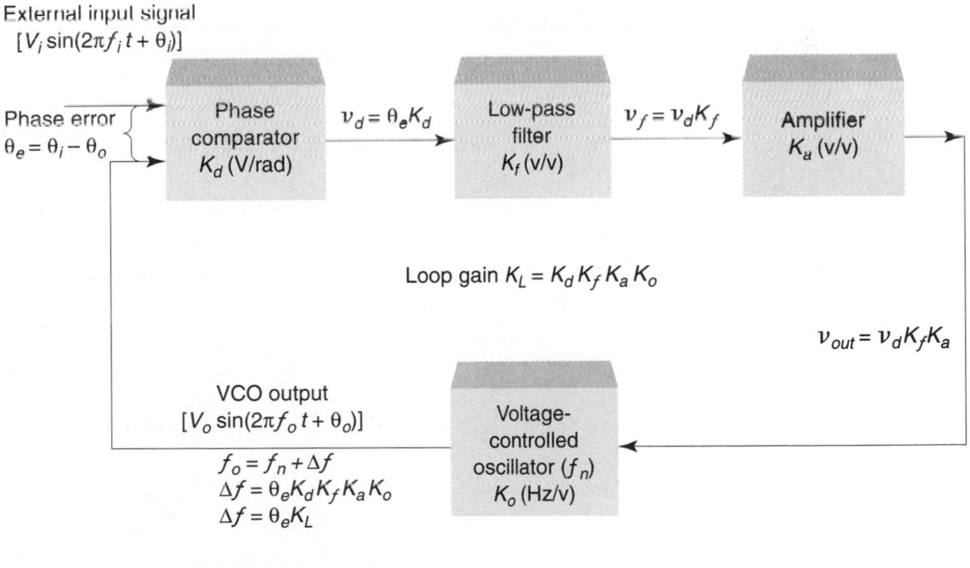

External input signal
$[V_i \sin(2\pi f_i t + \theta_i)]$

Phase error
$\theta_e = \theta_i - \theta_o$

Phase comparator
K_d (V/rad)

$v_d = \theta_e K_d$

Low-pass filter
K_f (v/v)

$v_f = v_d K_f$

Amplifier
K_a (v/v)

Loop gain $K_L = K_d K_f K_a K_o$

$v_{out} = v_d K_f K_a$

VCO output
$[V_o \sin(2\pi f_o t + \theta_o)]$

$f_o = f_n + \Delta f$
$\Delta f = \theta_e K_d K_f K_a K_o$
$\Delta f = \theta_e K_L$

Voltage-controlled oscillator (f_n)
K_o (Hz/v)

FIGURE 3-31 PLL for Example 3-2

c. Rearranging Equation 3-15 gives us

$$V_{out} = \frac{\Delta f}{K_o} = \frac{10 \text{ kHz}}{20 \text{ kHz/V}} = 0.5 \text{ V}$$

d. $$V_d = \frac{V_{out}}{(K_f)(K_a)} = \frac{0.5}{(1)(5)} = 0.1 \text{ V}$$

e. Rearranging Equation 3-18 gives us

$$\theta_e = \frac{V_d}{K_d} = \frac{0.1 \text{ V}}{0.2 \text{ V/rad}} = 0.5 \text{ rad or } 28.65°$$

f. Substituting into Equation 3-29 yields

$$\Delta f_{max} = \frac{(\pm\pi/2 \text{ rad})(20 \text{ kHz})}{\text{rad}} = \pm 31.4 \text{ kHz}$$

Lock range is the range of frequencies over which the loop will stay locked onto the external input signal once lock has been established. Lock range is expressed in rad/s and is related to the open-loop gain K_v as

$$\text{lock range} = 2\Delta f_{max} = \pi K_L \qquad \textbf{(3-30)}$$

where $K_L = (K_d)(K_f)(K_o)$ for a simple loop with a LPF, phase comparator, and VCO or $K_L = (K_d)(K_f)(K_a)(K_o)$ for a loop with an amplifier.

The lock range in radians per second is π times the dc loop voltage gain and is independent of the LPF response. The capture range depends on the lock range and on the LPF response, so it changes with the type of filter used and with the filter cutoff frequency. For a simple single-pole *RC* LPF, it is given by

$$\text{capture range} = \frac{2\sqrt{\Delta f_{max}}}{RC} \qquad \textbf{(3-31)}$$

Once the loop is locked, any change in the input frequency is seen as a phase error, and the comparator produces a corresponding change in its output voltage, V_d. The change in voltage is amplified and fed back to the VCO to reestablish lock. Thus, the loop dynamically adjusts itself to follow input frequency changes.

Mathematically, the output from the phase comparator is (considering only the fundamental frequency for V_o and excluding the 90° phase bias)

$$\begin{aligned} V_d &= [V \sin(2\pi f_o t + \theta_o) \times V \sin(2\pi f_i t + \theta_i)] \\ &= \frac{V}{2}\cos(2\pi f_o t + \theta_o - 2\pi f_i t - \theta_i) - \frac{V}{2}\cos(2\pi f_o t + \theta_o + 2\pi f_i t - \theta_i) \end{aligned} \qquad \textbf{(3-32)}$$

where V_d = the phase detector output voltage (volts)
$V = V_o V_i$ (peak volts)

When $f_o = f_i$,

$$\begin{aligned} V_d &= \frac{V}{2}\cos(\theta_i + \theta_o) \\ &= \frac{V}{2}\cos\theta_e \end{aligned} \qquad \textbf{(3-33)}$$

where $\theta_i + \theta_o = \theta_e$ (phase error). θ_e is the phase error required to change the VCO output frequency from f_n to f_i (a change $= \Delta f$) and is often called the *static phase error.*

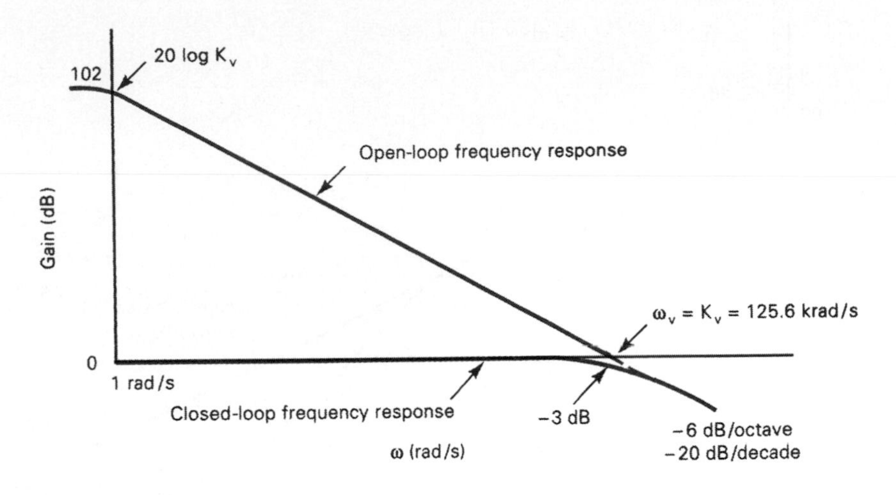

FIGURE 3-32 Frequency response for an uncompensated phase-locked loop

3-12 PLL CLOSED-LOOP FREQUENCY RESPONSE

The *closed-loop frequency response* for an *uncompensated* (unfiltered) PLL is shown in Figure 3-32. The open-loop gain of a PLL for a frequency of 1 rad/s = K_v. The frequency response shown in Figure 3-32 is for the circuit and PLL parameters given in Example 3-2. It can be seen that the open-loop gain (K_v) at 1 rad/s = 102 dB, and the open-loop gain equals 0 dB at the loop cutoff frequency (ω_v). Also, the closed-loop gain is unity up to ω_v, where it drops to -3 dB and continues to roll off at 6 dB/octave (20 dB/decade). Also, $\omega_v = K_v =$ 125.6 krad/s, which is the single-sided bandwidth of the uncompensated closed loop.

From Figure 3-32, it can be seen that the frequency response for an uncompensated PLL is identical to that of a single-pole (first-order) low-pass filter with a break frequency of $\omega_c = 1$ rad/s. In essence, a PLL is a low-pass tracking filter that follows input frequency changes that fall within a bandwidth equal to $\pm K_v$.

If additional bandlimiting is required, a low-pass filter can be added between the phase comparator and amplifier as shown in Figure 3-31. This filter can be either a single- or a multiple-pole filter. Figure 3-33 shows the loop frequency response for a simple single-pole *RC* filter with a cutoff frequency of $\omega_c = 100$ rad/s. The frequency response follows that of Figure 3-32 up to the loop filter break frequency, then the response rolls off at 12 dB/octave (40 dB/decade). As a result, the compensated unity-gain frequency (ω_c) is reduced to approximately ± 3.5 krad/s.

Example 3-3

Plot the frequency response for a PLL with a loop gain of $K_L = 15$ kHz/rad ($\omega_v = 94.3$ krad/s). On the same log paper, plot the response with the addition of a single-pole loop filter with a cutoff frequency $\omega_c = 1.59$ Hz/rad (10 rad/s) and a two-pole loop filter with the same cutoff frequency.

Solution The specified frequency response curves are shown in Figure 3-34. It can be seen that with the single-pole filter, the compensated loop response = $\omega'_v = 1$ krad/s and with the two-pole filter, $\omega''_v = 200$ rad/s.

The bandwidth of the loop filter (or for that matter, whether a loop filter is needed) depends on the specific application.

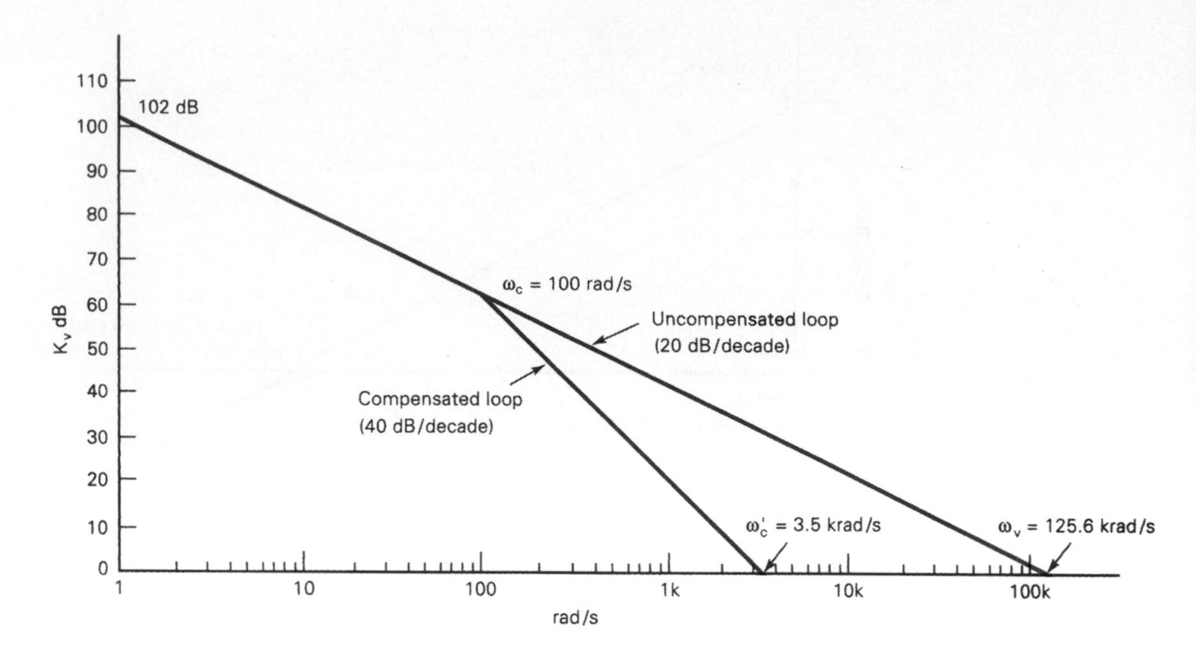

FIGURE 3-33 PLL frequency response for a single-pole RC filter

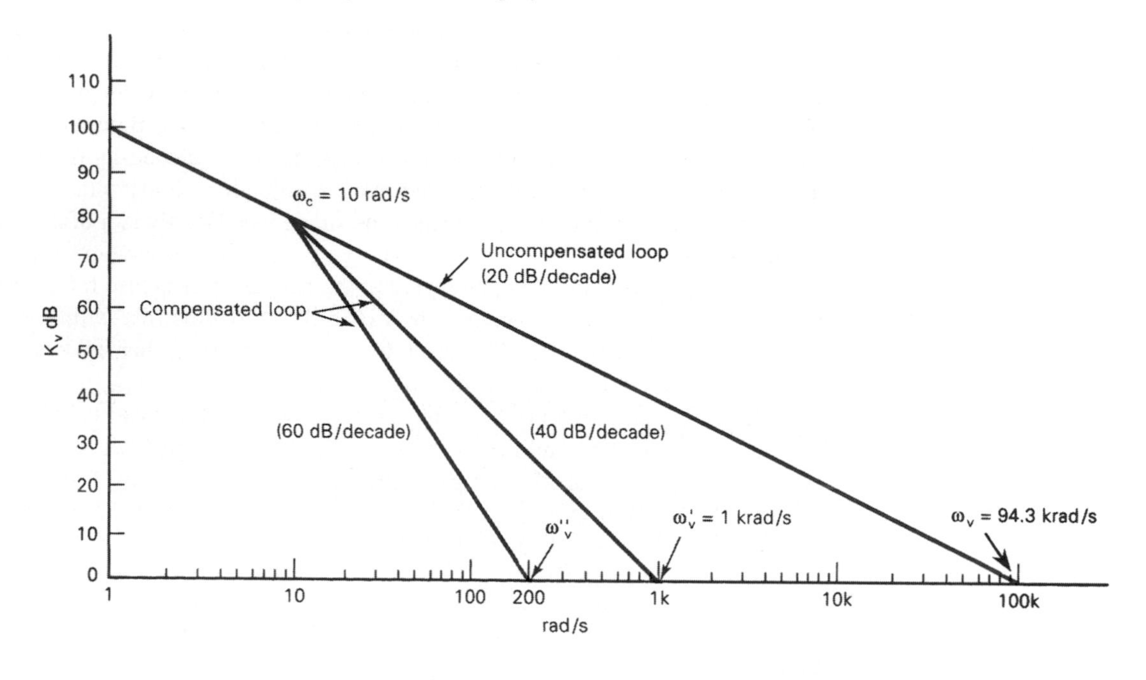

FIGURE 3-34 PLL frequency response for Example 2-3

3-13 INTEGRATED-CIRCUIT PRECISION PHASE-LOC,ED LOOP

The XR-215 is an ultrastable monolithic phase-locked-loop system designed by EXAR Corporation for a wide variety of applications in both analog and digital communications systems. It is especially well suited for FM or FSK demodulation, frequency synthesis, and tracking filter applications. The XR-215 can operate over a relatively wide frequency range from 0.5 Hz to 35 MHz and can accommodate analog input voltages between 300 μV and 3 V. The XR-215 can interface with conventional DTL, TTL, and ECL logic families.

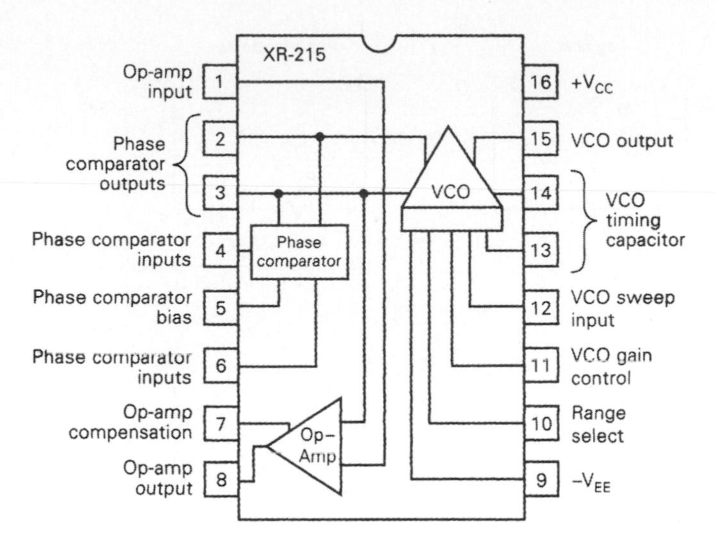

FIGURE 3-35 Block diagram for the XR-215 monolithic phase-locked loop

The block diagram for the XR-215 is shown in Figure 3-35 and consists of three main sections: a balanced phase comparator, a highly stable VCO, and a high-speed operational amplifier (op-amp). The phase comparator outputs are internally connected to the VCO inputs and to the noninverting amplifier of the op-amp. A self-contained PLL system is formed by simply ac coupling the VCO output to either of the phase comparator inputs and adding a low-pass filter to the phase comparator output terminals.

The VCO section has frequency sweep, on-off keying, sync, and digital programming capabilities. Its frequency is highly stable and determined by a single external capacitor. The op-amp can be used for audio preamplification in FM detector applications or as a high-speed sense amplifier (or comparator) in FSK demodulation.

3-13-1 Phase Comparator

One input to the phase comparator (pin 4) is connected to the external input signal, and the second input (pin 6) is ac coupled to the VCO output pin. The low-frequency ac (or dc) voltage across the phase comparator output pins (pins 2 and 3) is proportional to the phase difference between the two signals at the phase comparator inputs. The phase comparator outputs are internally connected to the VCO control terminals. One output (pin 3) is internally connected to the operational amplifier. The low-pass filter is achieved by connecting an *RC* network to the phase comparator outputs as shown in Figure 3-36. A typical transfer function (conversion gain) for the phase detector is 2 V/rad for input voltages ≥ 50 mV.

3-13-2 VCO

The VCO free-running or natural frequency (f_n) is inversely proportional to the capacitance of a timing capacitor (C_o) connected between pins 13 and 14. The VCO produces an output signal with a voltage amplitude of approximately 2.5 V_{p-p} at pin 15 with a dc output level of approximately 2 V. The VCO can be swept over a broad range of output frequencies by applying an analog sweep voltage (V_s) to pin 12 as shown in Figure 3-37. Typical sweep characteristics are also shown. The frequency range of the XR-215 can be extended by connecting an external resistor between pins 9 and 10. The VCO output frequency is proportional to the sum of currents I_1 and I_2 flowing through two internal transistors.

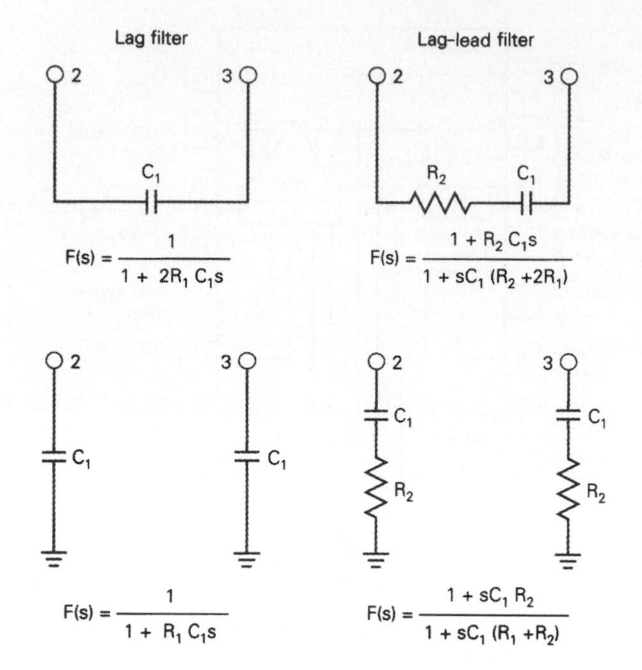

FIGURE 3-36 XR-215 low-pass filter connections

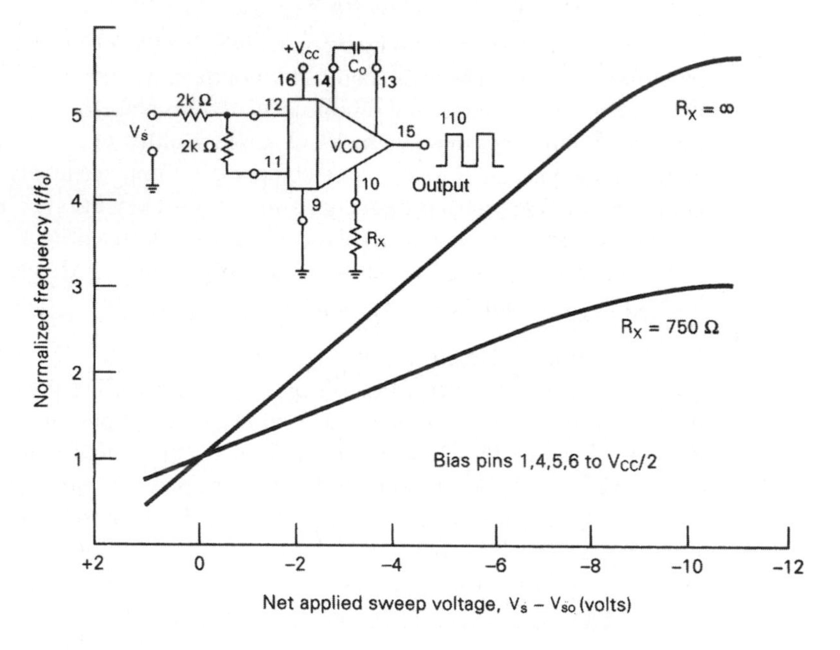

FIGURE 3-37 Typical frequency sweep characteristics as a function of applied sweep voltage

Current I_1 is set internally, whereas I_2 is set by an external resistor, R_x. Thus, for any value of C_o, the VCO free-running frequency can be expressed as

$$f_n = f\left(1 + \frac{0.6}{R_x}\right) \tag{3-34}$$

where f_n = VCO free-running frequency (hertz)
f = VCO output frequency with pin 10 open circuited (hertz)
R_x = external resistance (kilohms)

or
$$f_n = \frac{200}{C_o}\left(1 + \frac{0.6}{R_x}\right) \tag{3-35}$$

where C_o = external timing capacitor (microfarads)
R_x = external resistance (kilohms)

The VCO voltage-to-frequency conversion gain (transfer function) is determined by the choice of timing capacitor C_o and gain control resistor R_o connected externally across pins 11 and 12. Mathematically, the transfer function is expressed as

$$K_o = \frac{700}{C_o R_o}\text{(rad/s)/V} \tag{3-36}$$

where K_o = VCO conversion gain (radians per second per volt)
C_o = capacitance (microfarads)
R_o = resistance (kilohms)

3-13-3 Operational Amplifier

Pin 1 is the external connection to the inverting input of the operational amplifier section and is normally connected to pin 2 through a 10-kΩ resistor. The noninverting input is internally connected to one of the phase detector outputs. Pin 8 is used for the output terminal for FM or FSK demodulation. The amplifier voltage gain is determined by the resistance of feedback resistor R_f connected between pins 1 and 8. Typical frequency response characteristics for an amplifier are shown in Figure 3-38.

The voltage gain of the op-amp section is determined by feedback resistors R_f and R_p between pins (8 and 1) and (2 and 1), respectively, and stated mathematically as

$$A_v = \frac{-R_f}{R_s + R_p} \tag{3-37}$$

where A_v = voltage gain (volts per volt)
R_f = feedback resistor (ohms)
R_s = external resistor connected to pin 1 (ohms)
R_p = internal 6-kΩ impedance at pin 1 (ohms)

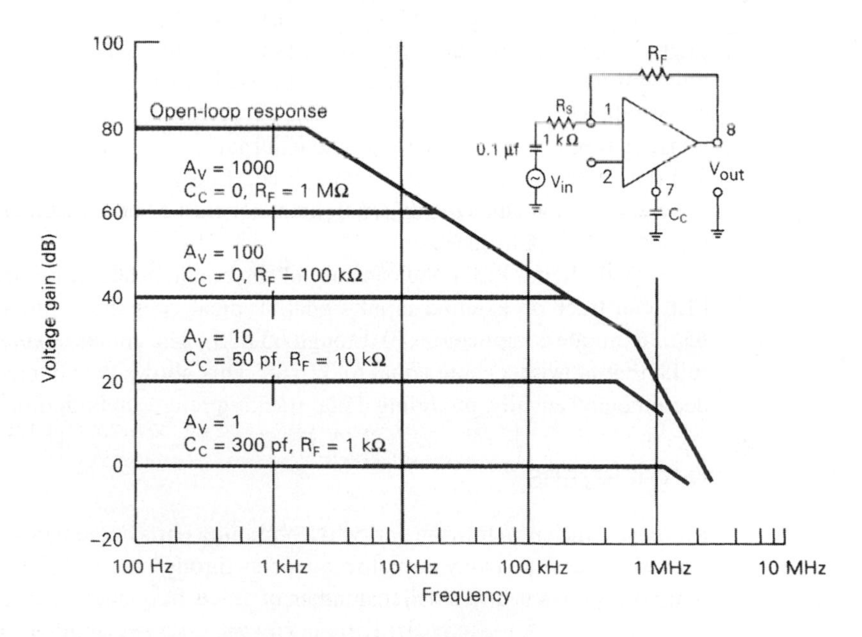

FIGURE 3-38 XR-215 Operational-amplifier frequency response

3-13-4 Lock Range

Lock range is the range of frequencies in the vicinity of the VCO's natural frequency over which the PLL can maintain lock with an external input signal. For the XR-215, if saturation or limiting does not occur, the lock range is equal to the open-loop gain or

$$\Delta\omega_L = K_v = (K_d)(K_o) \tag{3-38}$$

where $\Delta\omega_L$ = lock range (radians per second)
K_v = open-loop gain (radians per second)
K_d = phase detector conversion gain (volts per radian)
K_o = VCO conversion gain (radians per second per volt)

3-13-5 Capture Range

Capture range is the range of frequencies in the vicinity of the VCO's natural frequency where the PLL can establish or acquire lock with an input signal. For the XR-215, it can be approximated by a parametric equation of the form

$$\Delta\omega_C = \Delta\omega_L|F(j\Delta\omega_C)| \tag{3-39}$$

where $\Delta\omega_C$ = capture range (radians per second)
$\Delta\omega_L$ = lock range (radians per second)
$|F(j\Delta\omega_C)|$ = low-pass filter magnitude response at $\omega = \Delta\omega_C$

3-14 DIGITAL PLLs

Digital phase-locked loops are used to track digital pulses rather than analog signals, such as in clock recovery circuits. The goal of a digital PLL is to reproduce digital synchronization and timing signals rather than to extract information from an analog-modulated wave. With digital PLLs, only the rate of change of the external signal is of interest.

Digital PLLs are very similar to analog PLLs, except the VCO is replaced by a digital clock whose rate can be controlled, and the phase comparator is replaced with a incremental binary counter (i.e., a counter that can be incremented and decremented). The function of the counter is to control the clock frequency.

With a digital PLL, the counter is triggered by both the external input signal and the internal clock signal. The counter counts up one unit with each clock pulse and down one unit with each input pulse. When the external input and the internal clock are equal, the count remains unchanged, and the clock oscillates at a frequency. When the internal clock is oscillating at a higher rate than the external input signal, the counter increases in value and directs the internal clock to slow down. Likewise, when the internal clock is oscillating at a lower rate than the external input signal, the counter decreases in value and directs the internal clock to speed up.

With digital PLLs, the capacity of the counter determines the range over which the PLL can track an external input signal. Typical counters contain 16 bits, which affords 65,536 unique combinations (0 through 65,535). The counter in the PLL is initialized "half full," that is, with a count equal to 32,768. This allows the counter to be incremented and decremented equally, providing equal tracking ranges in both directions.

3-15 FREQUENCY SYNTHESIZERS

Synthesize means to form an entity by combining parts or elements. A *frequency synthesizer* is used to generate many output frequencies through the addition, subtraction, multiplication, and division of a smaller number of fixed frequency sources. Simply stated, a frequency synthesizer is a crystal-controlled variable-frequency generator. The objective of a synthesizer is twofold. It should produce as many frequencies as possible from a minimum number of sources, and each frequency should be as accurate and stable as every other frequency. The ideal frequency synthesizer can generate hundreds or even thousands of different frequencies from a single-crystal oscillator. A frequency synthesizer may be capable of simultaneously generating more than one output frequency, with each frequency being synchronous to a single reference or master oscillator frequency. Frequency synthesizers are used extensively in test and measurement equipment (audio and RF signal generators), tone-generating equipment (Touch-Tone), remote-control units (electronic tuners), multichannel communications systems (telephony), and music synthesizers.

In the early days of electronic communications, radio receivers were tuned manually by adjusting a tuned circuit. Manual tuning provided a continuous range of frequencies over the entire frequency band, and simply adjusting a knob on the front of the receiver could produce virtually any frequency. Although continuous manual tuning is flexible, it is also expensive to implement, susceptible to short- and long-term instability, relatively inaccurate, and difficult to interface with modern digital integrated circuits and microprocessors. Frequency synthesizers provide a modern alternative to continuous manually tuning.

Essentially, there are two methods of frequency synthesis: direct and indirect. With *direct frequency synthesis,* multiple output frequencies are generated by mixing the outputs from two or more crystal-controlled frequency sources or by dividing or multiplying the output frequency from a single-crystal oscillator. With *indirect frequency synthesis,* a feedback-controlled divider/multiplier (such as a PLL) is used to generate multiple output frequencies. Indirect frequency synthesis is slower and more susceptible to noise; however, it is less expensive and requires fewer and less complicated filters than direct frequency synthesis.

3-15-1 Direct Frequency Synthesizers

There are essentially two types of direct frequency synthesizers: multiple crystal and single crystal.

3-15-1-1 Multiple-crystal frequency synthesis. Figure 3-39 shows a block diagram for a *multiple-crystal frequency synthesizer* that uses nonlinear mixing (heterodyning) and filtering to produce 128 different frequencies from 20 crystals and two oscillator modules. For the crystal values shown, a range of frequencies from 510 kHz to 1790 kHz in 10-kHz steps is synthesized. A synthesizer such as this can be used to generate the carrier frequencies for the 106 AM broadcast-band stations (540 kHz to 1600 kHz). For the switch positions shown, the 160-kHz and 700-kHz oscillators are selected, and the outputs from the balanced mixer are their sum and difference frequencies (700 kHz ± 160 kHz = 540 kHz and 860 kHz). The output filter is tuned to 540 kHz, which is the carrier frequency for channel 1. To generate the carrier frequency for channel 106, the 100-kHz crystal is selected with either the 1700-kHz (difference) or the 1500-kHz (sum) crystal. The minimum frequency separation between output frequencies for a synthesizer is called *resolution.* The resolution for the synthesizer shown in Figure 3-39 is 10 kHz.

3-15-1-2 Single-crystal frequency synthesis. Figure 3-40 shows a block diagram for a *single-crystal frequency synthesizer* that again uses frequency addition, subtraction, multiplication, and division to generate frequencies (in 1-Hz steps) from 1 Hz to 999,999 Hz. A 100-kHz crystal is the source for the master oscillator from which all frequencies are derived.

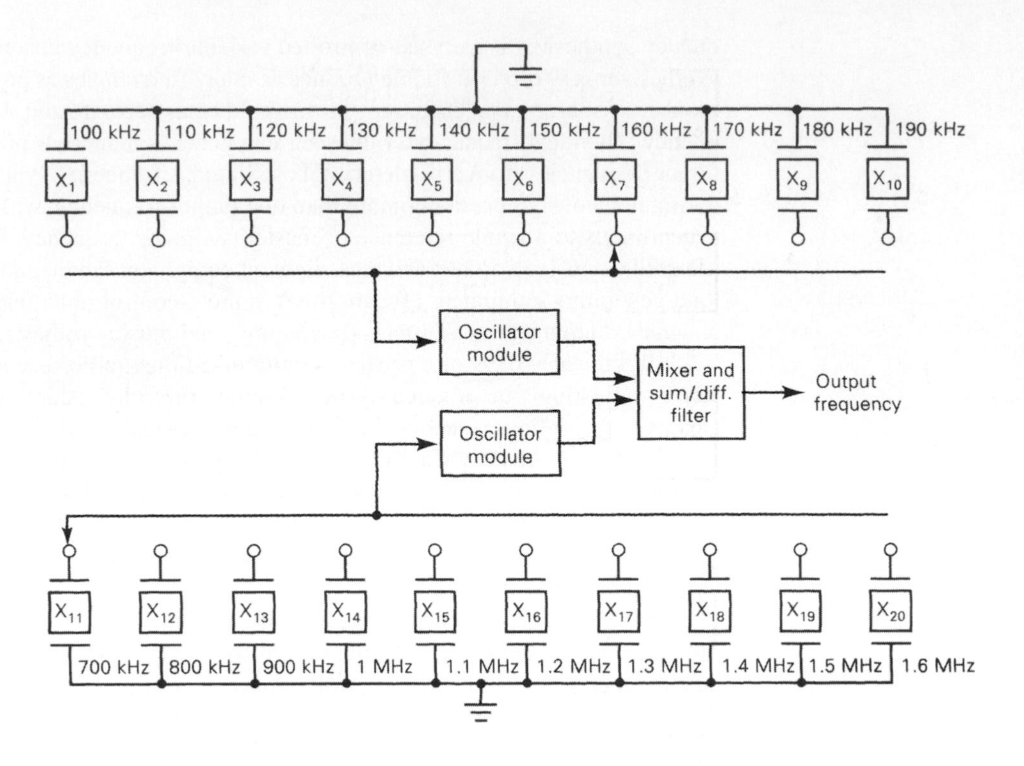

FIGURE 3-39 Multiple-crystal frequency synthesizer

The master oscillator frequency is a base frequency that is repeatedly divided by 10 to generate five additional subbase frequencies (10 kHz, 1 kHz, 100 Hz, 10 Hz, and 1 Hz). Each subbase frequency is fed to a separate harmonic generator (frequency multiplier), which consists of a nonlinear amplifier with a tunable filter. The filter is tunable to each of the first nine harmonics of its base frequency. Therefore, the possible output frequencies for harmonic generator 1 are 0 kHz to 900 kHz in 100-kHz steps; for harmonic generator 2, 10 kHz to 90 kHz in 10-kHz steps; and so on. The resolution for the synthesizer is determined by how many times the master crystal oscillator frequency is divided. For the synthesizer shown in Figure 3-40, the resolution is 1 Hz. The mixers used are balanced modulators with output filters that are tuned to the sum of the two input frequencies. For example, the harmonics selected in Table 3-1 produce a 246,313-Hz output frequency. Table 3-1 lists the selector switch positions for each harmonic generator and the input and output frequencies from each mixer. It can be seen that the five mixers simply sum the output frequencies from the six harmonic generators with three levels of mixing.

3-15-2 Indirect Frequency Synthesizers

3-15-2-1 PLL Frequency synthesizers. In recent years, PLL frequency synthesizers have rapidly become the most popular method for frequency synthesis. Figure 3-41 shows a block diagram for a simple *single-loop* PLL frequency synthesizer. The stable frequency reference is a crystal-controlled oscillator. The range of frequencies generated and the resolution depend on the divider network and the open-loop gain. The frequency divider is a divide-by-n circuit, where n is any integer number. The simplest form of divider circuit is a programmable digital *up–down counter* with an output frequency of $f_c = f_o/n$, where f_o = the VCO output frequency. With this arrangement, once lock has occurred, $f_c = f_{\text{ref}}$, and the VCO and synthesizer output frequency $f_o = nf_{\text{ref}}$. Thus, the synthesizer is essentially a times-n frequency multiplier. The frequency divider reduces the open-loop gain by a factor

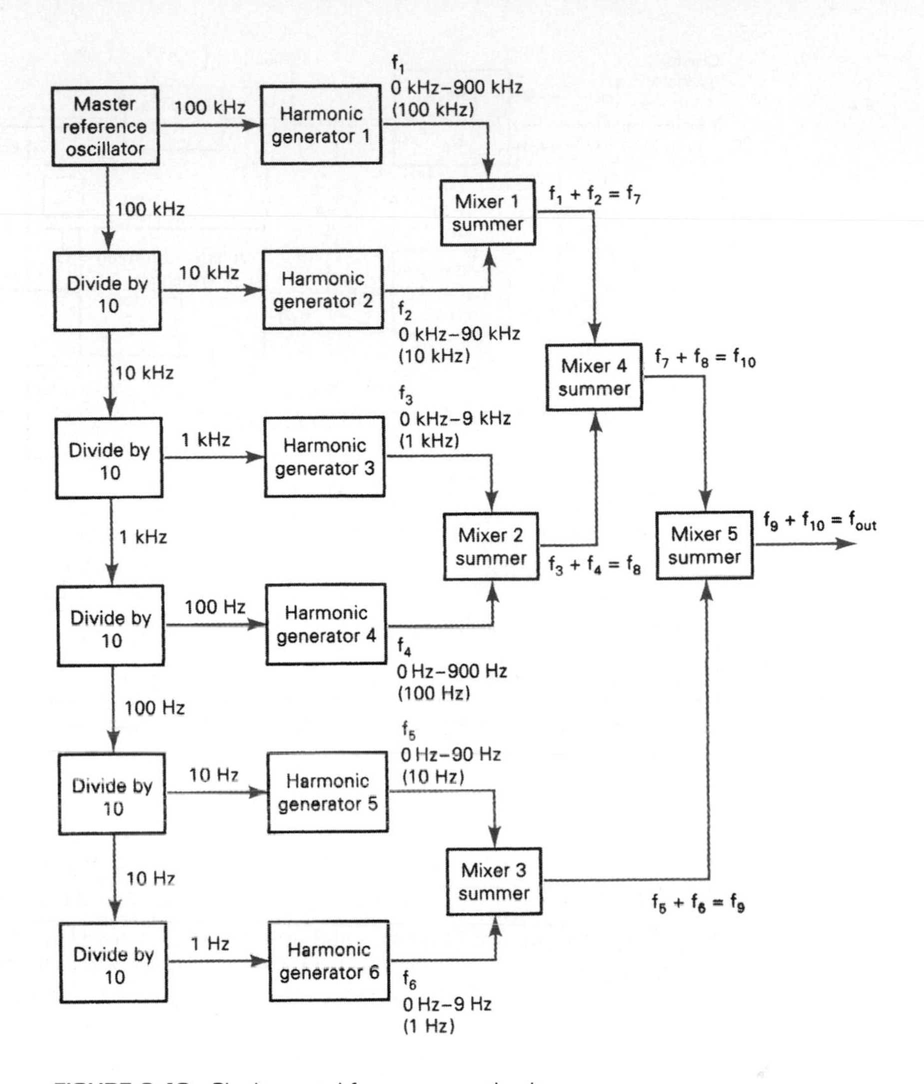

FIGURE 3-40 Single-crystal frequency synthesizer

Table 3-1 Switch Positions and Harmonics

Harmonic Generator	Selected Output Frequency	Mixer Output Frequency
1	200 kHz	Mixer 1 out = 200 kHz + 40 kHz = 240 kHz
2	40 kHz	
3	6 kHz	Mixer 2 out = 6 kHz + 0.3 kHz = 6.3 kHz
4	0.3 kHz	
5	10 Hz	Mixer 3 out = 10 Hz + 3 Hz = 13 Hz
6	3 Hz	
		Mixer 4 out = 240 kHz + 6.3 kHz = 246.3 kHz
		Mixer 5 out = 246.3 kHz + 13 Hz = 259.3 kHz

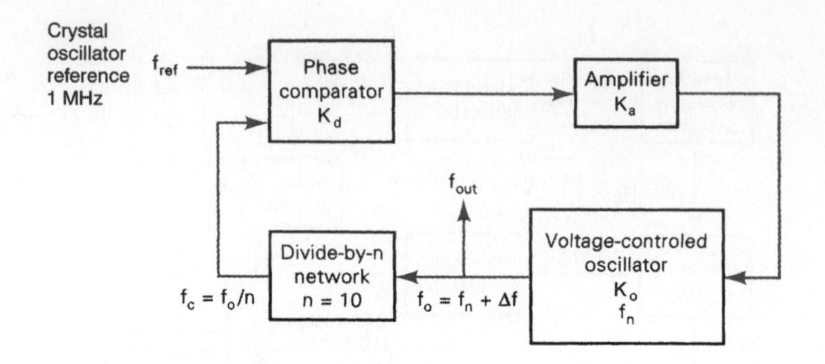

FIGURE 3-41 Single-loop PLL frequency synthesizer

of n. Consequently, the other circuits around the loop must have relatively high gains. The open-loop gain for the frequency synthesizer shown in Figure 3-41 is

$$K_v = \frac{(K_d)(K_a)(K_o)}{n} \tag{3-40}$$

From Equation 3-40, it can be seen that as n changes, the open-loop gain changes inversely proportionally. A way to remedy this problem is to program the amplifier gain as well as the divider ratio. Thus, the open-loop gain is

$$K_v = \frac{n(K_d)(K_a)(K_o)}{n} = (K_d)(K_a)(K_o) \tag{3-41}$$

For the reference frequency and divider circuit shown in Figure 3-41, the range of output frequencies is

$$\begin{aligned} f_o &= nf_{\text{ref}} \\ &= f_{\text{ref}} \text{ to } 10f_{\text{ref}} \\ &= 1 \text{ MHz to } 10 \text{ MHz} \end{aligned}$$

3-15-2-2 Prescaled frequency synthesizers. Figure 3-42 shows the block diagram for a frequency synthesizer that uses a PLL and a *prescaler* to achieve fractional division. Prescaling is also necessary for generating frequencies greater than 100 MHz because programmable counters are not available that operate efficiently at such high frequencies. The synthesizer in Figure 3-42 uses a *two-modulus* prescaler. The prescaler has two modes of operation. One mode provides an output for every input pulse (P), and the other mode provides an output for every $P + 1$ input pulse. Whenever the m register contains a nonzero number, the prescaler counts in the $P + 1$ mode. Consequently, once the m and n registers have been initially loaded, the prescaler will count down $(P + 1)$ m times until the m counter goes to zero, the prescaler operates in the P mode, and the n counter counts down $(n - m)$ times. At this time, both the m and n counters are reset to their initial values, which have been stored in the m and n registers, respectively, and the process repeats. Mathematically, the synthesizer output frequency f_o is

$$f_o = \left(n + \frac{m}{P}\right)Pf_i \tag{3-42}$$

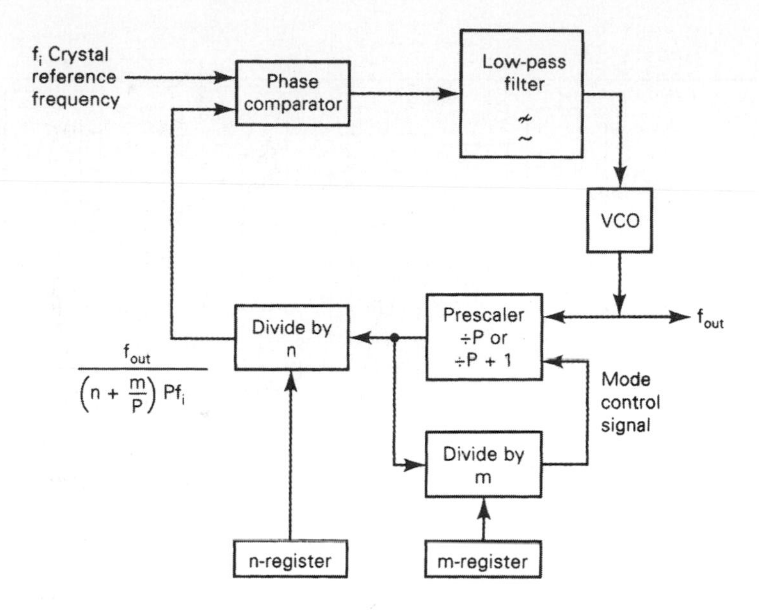

FIGURE 3-42 Frequency synthesizer using prescaling

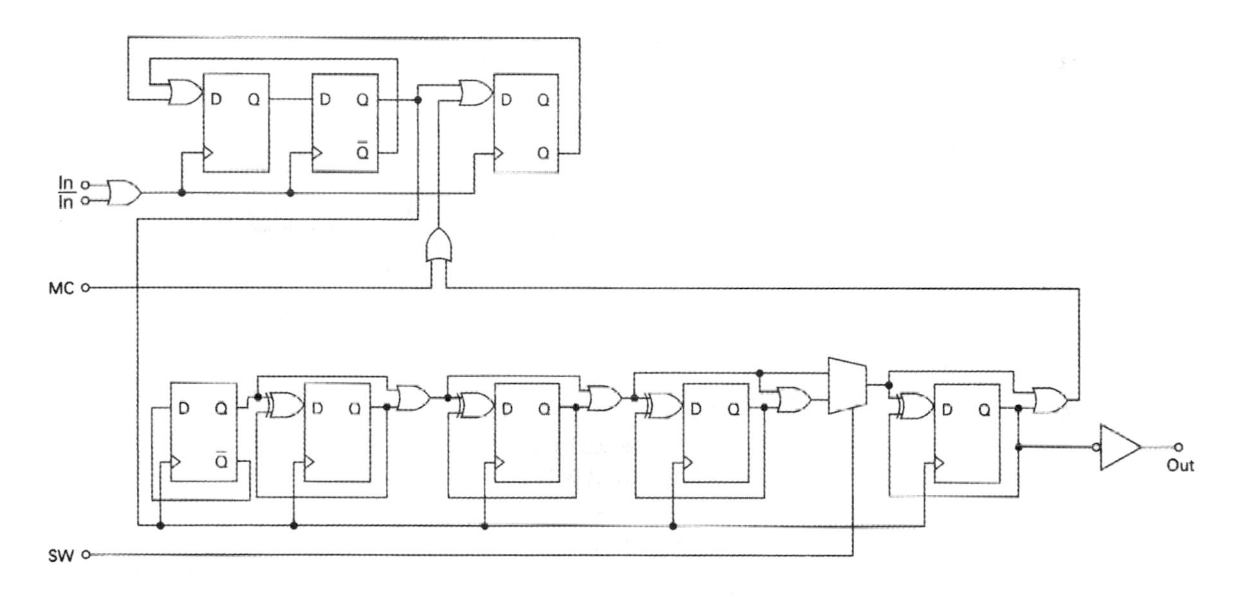

FIGURE 3-43 Block diagram of the NE/SA701 ECL prescaler

3-15-2-3 Integrated-circuit prescalers. Advanced ECL (emitter-coupled logic) integrated-circuit dual (divide by 128/129 or 64/65) and triple (divide by 64/65/72) modulus prescalers are now available that operate at frequencies from 1 Hz to 1.3 GHz. These integrated-circuit prescalers feature small size, low-voltage operation, low-current consumption, and simplicity. Integrated-circuit prescalers are ideally suited for cellular and cordless telephones, RF LANs (local area networks), test and measurement equipment, military radio systems, VHF/UHF mobile radios, and VHF/UHF handheld radios.

Figure 3-43 shows the block diagram for the NE/SA701 prescaler manufactured by Signetics Company. The NE701 is an advanced dual-modulus (divide by 128/129 or 64/65),

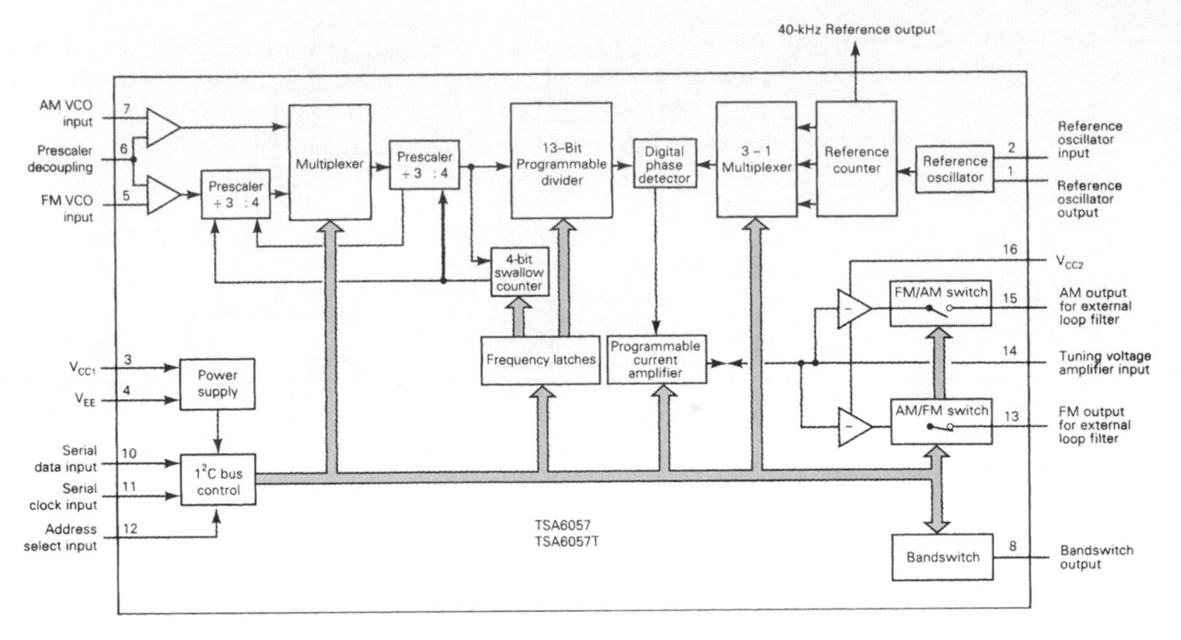

FIGURE 3-44 Block diagram of the TSA6057 radio-tuning PLL frequency synthesizer

low-power, ECL prescaler. It will operate with a minimum supply voltage of 2.5 V and has a maximum current drain of 2.8 mA, allowing application in battery-operated, low-power equipment. The maximum input signal frequency is 1.2 GHz for cellular and other land mobile applications. The circuit is implemented in ECL technology on the HS4+ process. The circuit is available in an 8-pin SO package.

The NE701 comprises a frequency divider implemented by using a divide-by-4 or -5 synchronous prescaler followed by a fixed five-stage synchronous counter. The normal operating mode is for the SW (modulus set switch) input to be low and the MC (modulus control) input to be high, in which case it functions as a divide-by-128 counter. For divide-by-129 operation, the MC input is forced low, causing the prescaler to switch into divide-by-5 operation for the last cycle of the synchronous counter. Similarly, for divide-by-64/65, the NE701 will generate those respective moduli with the SW signal forced high, in which the fourth stage of the synchronous divider is bypassed. With SW open circuited, the divide-by-128/129 mode is selected, and with SW connected to V_{CC}, divide-by-64/65 is selected.

3-15-2-4 Integrated-circuit radio-tuning frequency synthesizer. Figure 3-44 shows the block diagram for the Signetics TSA6057/T radio-tuning PLL frequency synthesizer. The TSA6057 is a bipolar, single-chip frequency synthesizer manufactured in SUBILO-N technology (components laterally separated by oxide). It performs all the tuning functions of a PLL radio-tuning system. The IC is designed for applications in all types of radio receivers and has the following features:

1. Separate input amplifiers for the AM and FM VCO signals.
2. On-chip, high-input sensitivity AM (3:4) and FM (15:16) prescalers.
3. High-speed tuning due to a powerful digital memory phase detector.
4. On-chip high-performance, one-input (two-output) tuning voltage amplifier. One output is connected to the external AM loop filter and the other output to the external FM loop filter.
5. On-chip, two-level current amplifier that consists of a 5-μA and 450-μA current source. This allows adjustment of the loop gain, thus providing high-current, high-speed tuning, and low-current stable tuning.

6. One reference oscillator (4 MHz) for both AM and FM followed by a reference counter. The reference frequency can be 1 kHz, 10 kHz, or 25 kHz and is applied to the digital memory phase detector. The reference counter also outputs a 40-kHz reference frequency to pin 9 for cooperation with the FM/IF system.
7. Oscillator frequency ranges of 512 kHz to 30 MHz and 30 MHz to 150 MHz.

QUESTIONS

3-1. Define *oscillate* and *oscillator.*

3-2. Describe the following terms: *self-sustaining, repetitive, free-running,* and *one-shot.*

3-3. Describe the regenerative process necessary for self-sustained oscillations to occur.

3-4. List and describe the four requirements for a *feedback oscillator* to work.

3-5. What is meant by the terms *positive* and *negative feedback?*

3-6. Define *open-* and *closed-loop gain.*

3-7. List the four most common oscillator configurations.

3-8. Describe the operation of a Wien-bridge oscillator.

3-9. Describe oscillator action for an *LC* tank circuit.

3-10. What is meant by a *damped oscillation?* What causes it to occur?

3-11. Describe the operation of a Hartley oscillator; a Colpitts oscillator.

3-12. Define *frequency stability.*

3-13. List several factors that affect the frequency stability of an oscillator.

3-14. Describe the *piezoelectric effect.*

3-15. What is meant by the term *crystal cut?* List and describe several crystal cuts and contrast their stabilities.

3-16. Describe how an overtone crystal oscillator works.

3-17. What is the advantage of an overtone crystal oscillator over a conventional crystal oscillator?

3-18. What is meant by *positive temperature coefficient? Negative temperature coefficient?*

3-19. What is meant by a *zero coefficient* crystal?

3-20. Sketch the electrical equivalent circuit for a crystal and describe the various components and their mechanical counterparts.

3-21. Which crystal oscillator configuration has the best stability?

3-22. Which crystal oscillator configuration is the least expensive and most adaptable to digital interfacing?

3-23. Describe a crystal oscillator module.

3-24. What is the predominant advantage of crystal oscillators over *LC* tank-circuit oscillators?

3-25. Describe the operation of a varactor diode.

3-26. Describe a phase-locked loop.

3-27. What types of LSI waveform generators are available?

3-28. Describe the basic operation of an integrated-circuit waveform generator.

3-29. List the advantages of a monolithic function generator.

3-30. List the advantages of a monolithic voltage-controlled oscillator.

3-31. Briefly describe the operation of a monolithic precision oscillator.

3-32. List the advantages of an integrated-circuit PLL over a discrete PLL.

3-33. Describe the operation of a voltage-controlled oscillator.

3-34. Describe the operation of a phase detector.

3-35. Describe how loop acquisition is accomplished with a PLL from an initial unlocked condition until frequency lock is achieved.

3-36. Define the following terms: *beat frequency, zero beat, acquisition time,* and *open-loop gain.*

3-37. Contrast the following terms and show how they relate to each other: *capture range, pull-in range, closed-loop gain, hold-in range, tracking range,* and *lock range.*

3-38. Define the following terms: *uncompensated PLL, loop cutoff frequency,* and *tracking filter.*

3-39. Define *synthesize.* What is a frequency synthesizer?

3-40. Describe direct and indirect frequency synthesis.

3-41. What is meant by the resolution of a frequency synthesizer?

3-42. What are some advantages of integrated-circuit prescalers and frequency synthesizers over conventional nonintegrated-circuit equivalents?

PROBLEMS

3-1. For a 20-MHz crystal with a negative temperature coefficient of $k = -8$ Hz/MHz/°C, determine the frequency of operation for the following temperature changes:

a. Increase of 10°C
b. Increase of 20°C
c. Decrease of 20°C

3-2. For the Wien-bridge oscillator shown in Figure 3-3 and the following component values, determine the frequency of oscillation: $R_1 = R_2 = 1$ kΩ; $C_1 = C_2 = 100$ pF.

3-3. For the Hartley oscillator shown in Figure 3-5a and the following component values, determine the frequency of oscillation: $L_{1a} = L_{1b} = 50$ μH; $C_1 = 0.01$ μF.

3-4. For the Colpitts oscillator shown in Figure 3-6a and the following component values, determine the frequency of oscillation: $C_{1a} = C_{1b} = 0.01$ μF; $L_1 = 100$ μH.

3-5. Determine the capacitance for a varactor diode with the following values: $C = 0.005$ μF; $V_r = -2$ V.

3-6. For the VCO input-versus-output characteristic curve shown, determine

a. Frequency of operation for a −2-V input signal.
b. Frequency deviation for a ±2-V_p input signal.
c. Transfer function, K_o, for the linear portion of the curve (−3 to +3 V).

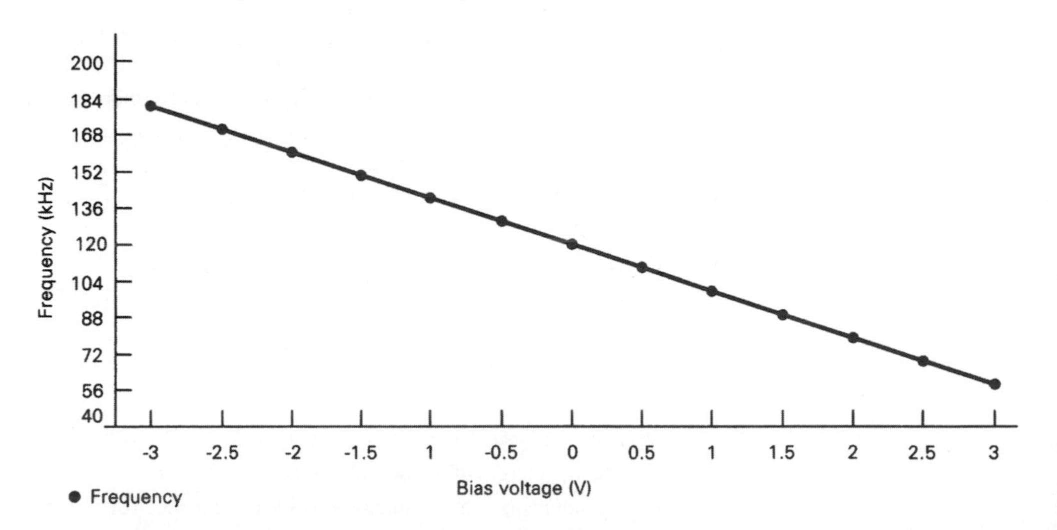

3-7. For the output voltage-versus-phase difference (θ_e) characteristic curve shown, determine

a. Output voltage for a −45° phase difference.
b. Output voltage for a +60° phase difference.
c. Maximum peak output voltage.
d. Transfer function, K_d.

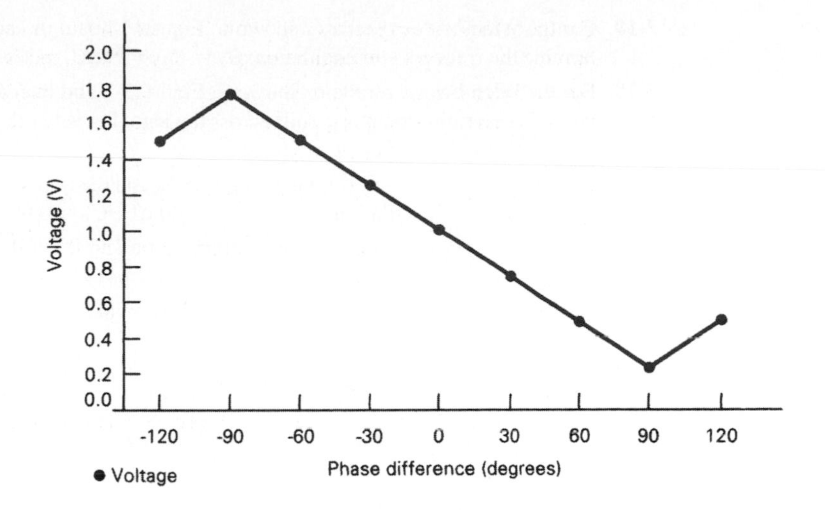

3-8. For the PLL shown in Figure 3-22a, a VCO natural frequency of f_n = 150 kHz, an input frequency of f_i = 160 kHz, and the circuit gains K_d = 0.2 V/rad, K_f = 1, K_a = 4, and K_o = 15 kHz/V, determine

a. Open-loop gain, K_L.
b. Δf.
c. V_{out}.
d. V_d.
e. θ_e.
f. Hold-in range, Δf_{max}.

3-9. Plot the frequency response for a PLL with an open-loop gain of K_L = 20 kHz/rad. On the same log paper, plot the response with a single-pole loop filter with a cutoff frequency of ω_c = 100 rad/s and a two-pole filter with the same cutoff frequency.

3-10. Determine the change in frequency (Δf) for a VCO with a transfer function of K_o = 2.5 kHz/V and a dc input voltage change of ΔV = 0.8 V.

3-11. Determine the voltage at the output of a phase comparator with a transfer function of K_d = 0.5 V/rad and a phase error of θ_e = 0.75 rad.

3-12. Determine the hold in range (Δf_{max}) for a PLL with an open-loop gain of K_L = 20 kHz/rad.

3-13. Determine the phase error necessary to produce a VCO frequency shift of Δf = 10 kHz for an open-loop gain of K_L = 40 kHz/rad.

3-14. Determine the output frequency from the multiple-crystal frequency synthesizer shown in Figure 3-39 if crystals X_8 and X_{18} are selected.

3-15. Determine the output frequency from the single-crystal frequency synthesizer shown in Figure 3-40 for the following harmonics.

Harmonic Generator	Harmonic	Harmonic Generator	Harmonic
1	6	4	1
2	4	5	2
3	7	6	6

3-16. Determine f_c for the PLL shown in Figure 3-41 for a natural frequency of f_n = 200 kHz, Δf = 0 Hz, and n = 20.

3-17. For a 10-MHz crystal with a negative temperature coefficient k = 12 Hz/MHz/°C, determine the frequency of operation for the following temperature changes:

a. Increase of 20°C
b. Decrease of 20°C
c. Increase of 10°C

3-18. For the Wien-bridge oscillator shown in Figure 3-3 and the following component values, determine the frequency of oscillation: $R_1 = R_2 = 2$ k; $C_1 = C_2 = 1000$ pF.

3-19. For the Wien-bridge oscillator shown in Figure 3-3 and the component values given in Problem 3-2, determine the phase shift across the lead–lag network for frequencies an octave above and below the frequency of oscillation.

3-20. For the Hartley oscillator shown in Figure 3-5a and the following component values, determine the frequency of oscillation: $L_{1a} = L_{1b} = 100$ μH; $C_1 = 0.001$ μF.

3-21. For the Colpitts oscillator shown in Figure 3-6a and the following component values, determine the frequency of oscillation: $C_{1a} = 0.0022$ μF, $C_{1b} = 0.022$ μF, and $L_1 = 3$ mH.

3-22. Determine the capacitance for a variactor diode with the following values: $C = 0.001$ μF and $v_r = -1.5$ V.

3-23. For the VCO input-versus-output characteristic curve shown below, determine

a. The frequency of operation for a -1.5 V input signal.
b. The frequency deviation for a 2 $V_{p\text{-}p}$ input signal.
c. The transfer function, K_o, for the linear portion of the curve (-2 V to $+2$ V).

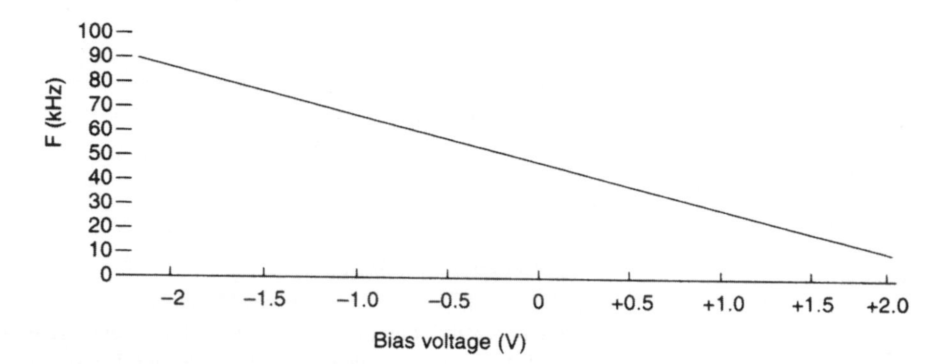

3-24. For the phase detector output voltage-versus-phase difference (θ_e) characteristic curve shown below, determine

a. The output voltage for a $-45°$ phase difference.
b. The output voltage for a $+60°$ phase difference.
c. The maximum peak output voltage.
d. The transfer function, K_d.

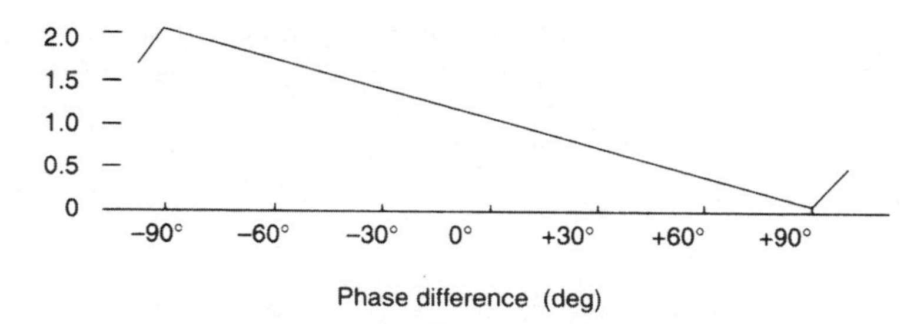

3-25. For the PLL shown in Figure 3-22a, a VCO natural frequency $f_n = 120$ kHz, an input frequency $f_i = 125$ kHz, and the following circuit gains $K_d = 0.2$ V/rad, $K_f = 1$, $K_a = 5$, and $K_o = 12$ kHz/V, determine

a. The open-loop gain, K_L.
b. Δf.
c. V_{out}.
d. V_d.
e. θ_e.
f. The hold-in range, Δf_{max}.

3-26. Plot the frequency response for a PLL with an open-loop gain $K_L = 30$ kHz/rad. On the same log paper, plot the response with a single-pole filter with a cutoff frequency $\omega_c = 200$ rad/s and a two-pole filter with the same cutoff frequency.

3-27. Determine the change in frequency for a VCO with a transfer function $K_o = 4$ kHz/V and a dc input voltage change $\Delta V = 1.2\ V_p$.

3-28. Determine the voltage at the output of a phase comparator with a transfer function $K_d = 0.4$ V/rad and a phase error $\theta_e = 0.55$ rad.

3-29. Determine the hold-in range for a PLL with an open-loop gain $K_v = 25$ kHz/rad.

3-30. Determine the phase error necessary to produce a VCO frequency shift of 20 kHz for an open-loop gain $K_L = 50$ kHz/rad.

CHAPTER 4

Amplitude Modulation Transmission

CHAPTER OUTLINE

OBJECTIVES

- Define *amplitude modulation*
- Describe the AM envelope
- Describe the AM frequency spectrum and bandwidth
- Explain the phasor representation of an AM wave
- Define and explain the following terms: *coefficient of modulation* and *percent modulation*
- Derive AM voltage distribution
- Analyze AM in the time domain
- Derive AM power distribution
- Explain AM current calculations
- Describe AM with a complex information signal
- Describe AM modulator circuits and the difference between low- and high-level modulation
- Describe linear-integrated circuit AM modulators
- Describe AM transmitters and the difference between low- and high-level transmitters
- Describe trapezoidal patterns and their significance in analyzing AM waveforms
- Define *carrier shift*
- Describe quadrature amplitude modulation

4-1 INTRODUCTION

Information signals are transported between a transmitter and a receiver over some form of transmission medium. However, the original information signals are seldom in a form that is suitable for transmission. Therefore, they must be transformed from their original form into a form that is more suitable for transmission. The process of impressing low-frequency information signals onto a high-frequency *carrier signal* is called *modulation. Demodulation* is the reverse process where the received signals are transformed back to their original form. The purpose of this chapter is to introduce the reader to the fundamental concepts of *amplitude modulation* (AM).

4-2 PRINCIPLES OF AMPLITUDE MODULATION

Amplitude modulation is the process of changing the amplitude of a relatively high frequency carrier signal in proportion with the instantaneous value of the modulating signal (information). Amplitude modulation is a relatively inexpensive, low-quality form of modulation that is used for commercial broadcasting of both audio and video signals. Amplitude modulation is also used for two-way mobile radio communications, such as citizens band (CB) radio.

AM modulators are nonlinear devices with two inputs and one output. One input is a single, high-frequency carrier signal of constant amplitude and the second input is comprised of relatively low-frequency information signals that may be a single frequency or a complex waveform made up of many frequencies. Frequencies that are high enough to be efficiently radiated by an antenna and propagated through free space are commonly called *radio frequencies,* or simply RFs. In the modulator, the information acts on or modulates the RF carrier producing a modulated waveform. The information signal may be a single frequency or more likely consist of a range of frequencies. For example, typical voice-grade communications systems utilize a range of information frequencies between 300 Hz and 3000 Hz. The modulated output waveform from an AM modulator is often called an AM envelope.

4-2-1 The AM Envelope

Although there are several types of amplitude modulation, AM *double-sideband full carrier* (DSBFC) is probably the most commonly used. AM DSBFC is sometimes called *conventional* AM or simply AM. Figure 4-1 illustrates the relationship among the carrier ($V_c \sin[2\pi f_c t]$), the modulating signal ($V_m \sin[2\pi f_m t]$), and the modulated wave ($V_{am}[t]$) for conventional AM. The figure shows how an AM waveform is produced when a single-frequency modulating signal acts on a high-frequency carrier signal. The output waveform contains all the frequencies that make up the AM signal and is used to transport the information through the system. Therefore, the shape of the modulated wave is called the *AM envelope.* Note that with no modulating signal, the output waveform is simply the carrier signal. However, when a modulating signal is applied, the amplitude of the output wave varies in accordance with the modulating signal. Note that the repetition rate of the envelope is equal to the frequency of the modulating signal and that the shape of the envelope is identical to the shape of the modulating signal.

4-2-2 AM Frequency Spectrum and Bandwidth

An AM modulator is a nonlinear device. Therefore, nonlinear mixing occurs, and the output envelope is a complex wave made up of a dc voltage, the carrier frequency, and the sum ($f_c + f_m$) and difference ($f_c - f_m$) frequencies (i.e., the cross products). The sum and difference frequencies are displaced from the carrier frequency by an amount equal to the modulating signal frequency. Therefore, an AM signal spectrum contains frequency components spaced f_m Hz on either side of the carrier. However, it should be noted that the modulated wave does not contain a frequency component that is equal to the modulating

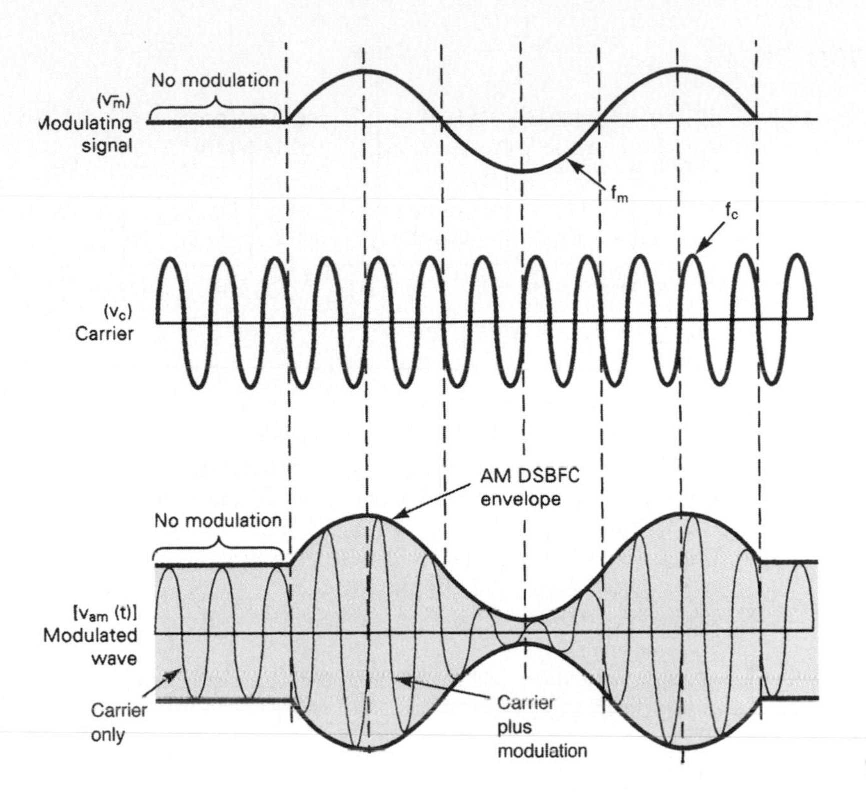

FIGURE 4-1 AM generation

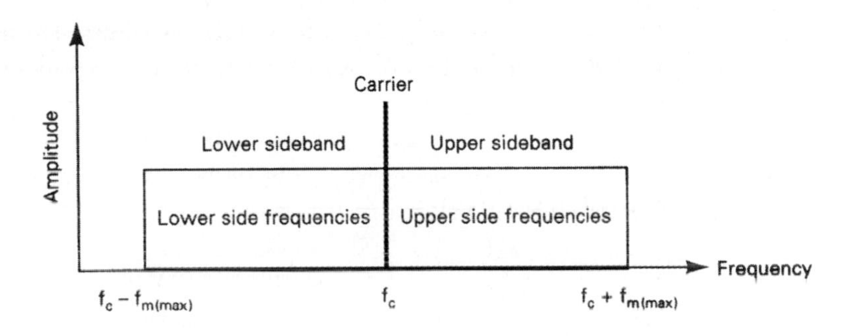

FIGURE 4-2 Frequency spectrum of an AM DSBFC wave

signal frequency. The effect of modulation is to translate the modulating signal in the frequency domain so that it is reflected symmetrically about the carrier frequency.

Figure 4-2 shows the frequency spectrum for an AM wave. The AM spectrum extends from $f_c - f_{m(\max)}$ to $f_c + f_{m(\max)}$, where f_c is the carrier frequency and $f_{m(\max)}$ is the highest modulating signal frequency. The band of frequencies between $f_c - f_{m(\max)}$ and f_c is called the *lower sideband* (LSB), and any frequency within this band is called a *lower side frequency* (LSF). The band of frequencies between f_c and $f_c + f_{m(\max)}$ is called the *upper sideband* (USB), and any frequency within this band is called an *upper side frequency* (USF). Therefore, the bandwidth (B) of an AM DSBFC wave is equal to the difference between the highest upper side frequency and the lowest lower side frequency, or two times the highest modulating signal frequency (i.e., $B = 2f_{m(\max)}$). For radio wave propagation, the carrier and all the frequencies within the upper and lower sidebands must be high enough to be sufficiently propagated through Earth's atmosphere.

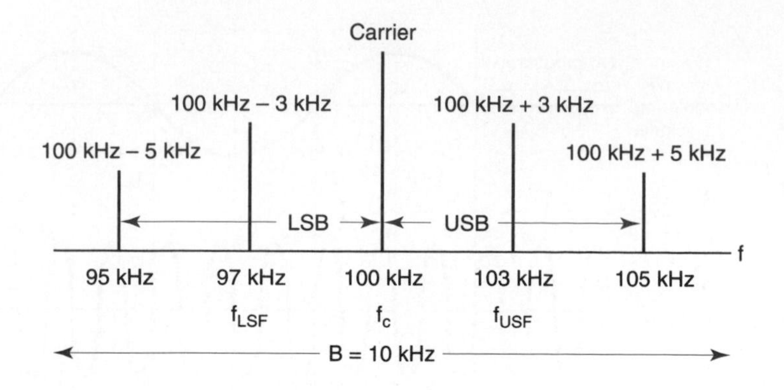

FIGURE 4-3 Output spectrum for Example 3-1

Example 4-1

For an AM DSBFC modulator with a carrier frequency $f_c = 100$ kHz and a maximum modulating signal frequency $f_{m(\text{max})} = 5$ kHz, determine

a. Frequency limits for the upper and lower sidebands.

b. Bandwidth.

c. Upper and lower side frequencies produced when the modulating signal is a single-frequency 3-kHz tone.

Then

d. Draw the output frequency spectrum.

Solution **a.** The lower sideband extends from the lowest possible lower side frequency to the carrier frequency or

$$\begin{aligned} \text{LSB} &= [f_c - f_{m(\text{max})}] \text{ to } f_c \\ &= (100 - 5) \text{ kHz to } 100 \text{ kHz} = 95 \text{ kHz to } 100 \text{ kHz} \end{aligned}$$

The upper sideband extends from the carrier frequency to the highest possible upper side frequency or

$$\begin{aligned} \text{USB} &= f_c \text{ to } [f_c + f_{m(\text{max})}] \\ &= 100 \text{ kHz to } (100 + 5) \text{ kHz} = 100 \text{ kHz to } 105 \text{ kHz} \end{aligned}$$

b. The bandwidth is equal to the difference between the maximum upper side frequency and the minimum lower side frequency or

$$\begin{aligned} B &= 2f_{m(\text{max})} \\ &= 2(5 \text{ kHz}) = 10 \text{ kHz} \end{aligned}$$

c. The upper side frequency is the sum of the carrier and modulating frequency or

$$f_{\text{usf}} = f_c + f_m = 100 \text{ kHz} + 3 \text{ kHz} = 103 \text{ kHz}$$

The lower side frequency is the difference between the carrier and the modulating frequency or

$$f_{\text{lsf}} = f_c - f_m = 100 \text{ kHz} - 3 \text{ kHz} = 97 \text{ kHz}$$

d. The output frequency spectrum is shown in Figure 4-3.

4-2-3 Phasor Representation of an Amplitude-Modulated Wave

For a single-frequency modulating signal, an AM envelope is produced from the vector addition of the carrier and the upper and lower side frequencies. The two side frequencies combine and produce a resultant component that combines with the carrier vector. Figure 4-4a shows this phasor addition. The phasors for the carrier and the upper and lower side frequencies all rotate in a counterclockwise direction. However, the upper side frequency rotates faster than the carrier ($\omega_{\text{usf}} > \omega_c$), and the lower side frequency rotates slower ($\omega_{\text{lsf}} < \omega_c$). Consequently, if the phasor for the carrier is held stationary, the phasor for the upper side frequency will continue to rotate in a counterclockwise direction relative

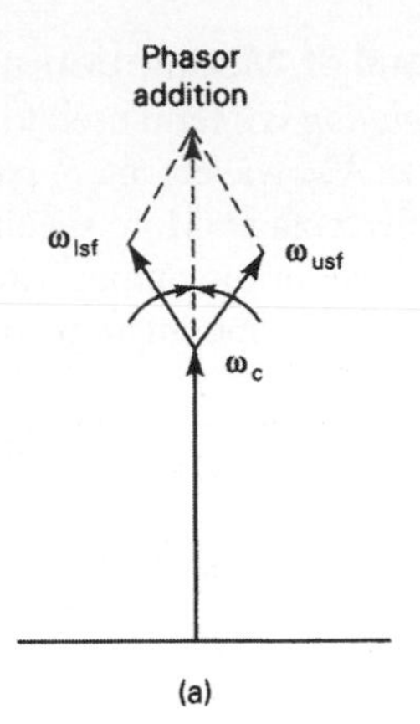

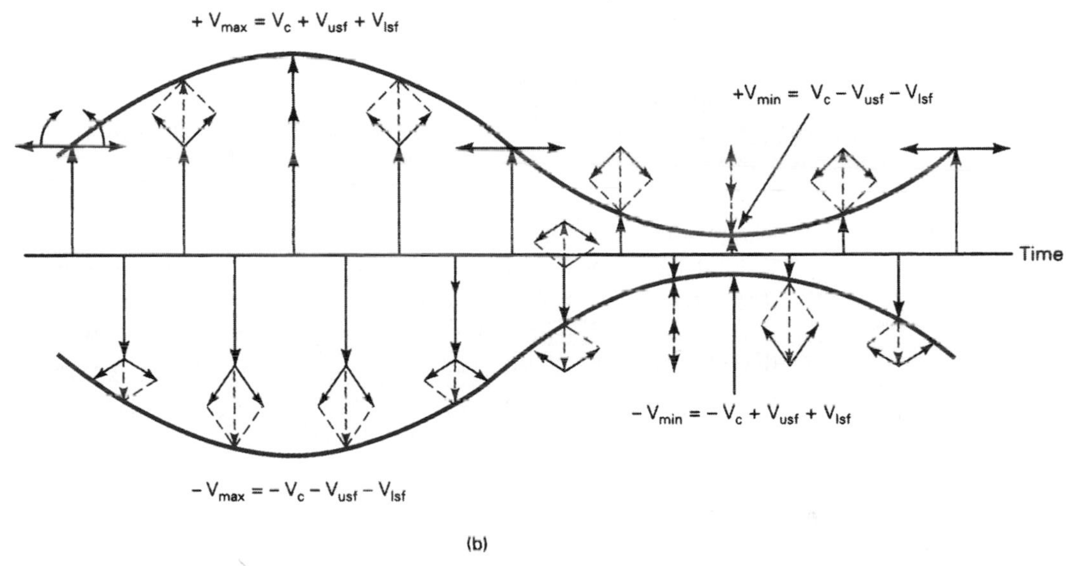

FIGURE 4-4 Phasor addition in an AM DSBFC envelope: (a) phasor addition of the carrier and the upper and lower side frequencies; (b) phasor addition producing an AM envelope

to the carrier, and the phasor for the lower side frequency will rotate in a clockwise direction. The phasors for the carrier and the upper and lower side frequencies combine, sometimes in phase (adding) and sometimes out of phase (subtracting). For the waveform shown in Figure 4-4b, the maximum positive amplitude of the envelope occurs when the carrier and the upper and lower side frequencies are at their maximum positive values at the same time ($+V_{max} = V_c + V_{usf} + V_{lsf}$). The minimum positive amplitude of the envelope occurs when the carrier is at its maximum positive value at the same time that the upper and lower side frequencies are at their maximum negative values ($+V_{min} = V_c - V_{usf} - V_{lsf}$). The maximum negative amplitude occurs when the carrier and the upper and lower side frequencies are at their maximum negative values at the same time ($-V_{max} = -V_c - V_{usf} - V_{lsf}$). The minimum negative amplitude occurs when the carrier is at its maximum negative value at the same time that the upper and lower side frequencies are at their maximum positive values ($-V_{min} = -V_c + V_{usf} + V_{lsf}$).

4-2-4 Coefficient of Modulation and Percent Modulation

Coefficient of modulation is a term used to describe the amount of amplitude change (modulation) present in an AM waveform. *Percent modulation* is simply the coefficient of modulation stated as a percentage. More specifically, percent modulation gives the percentage change in the amplitude of the output wave when the carrier is acted on by a modulating signal. Mathematically, the modulation coefficient is

$$m = \frac{E_m}{E_c} \tag{4-1}$$

where m = modulation coefficient (unitless)
E_m = peak change in the amplitude of the output waveform voltage (volts)
E_c = peak amplitude of the unmodulated carrier voltage (volts)

Equation 4-1 can be rearranged to solve for E_m and E_c as

$$E_m = mE_c \tag{4-2}$$

and

$$E_c = \frac{E_m}{m} \tag{4-3}$$

and percent modulation (M) is

$$M = \frac{E_m}{E_c} \times 100 \text{ or simply } m \times 100 \tag{4-4}$$

The relationship among m, E_m, and E_c is shown in Figure 4-5.

If the modulating signal is a pure, single-frequency sine wave and the modulation process is symmetrical (i.e., the positive and negative excursions of the envelope's amplitude are equal), then percent modulation can be derived as follows (refer to Figure 4-5 for the following derivation):

$$E_m = \frac{1}{2}(V_{max} - V_{min}) \tag{4-5}$$

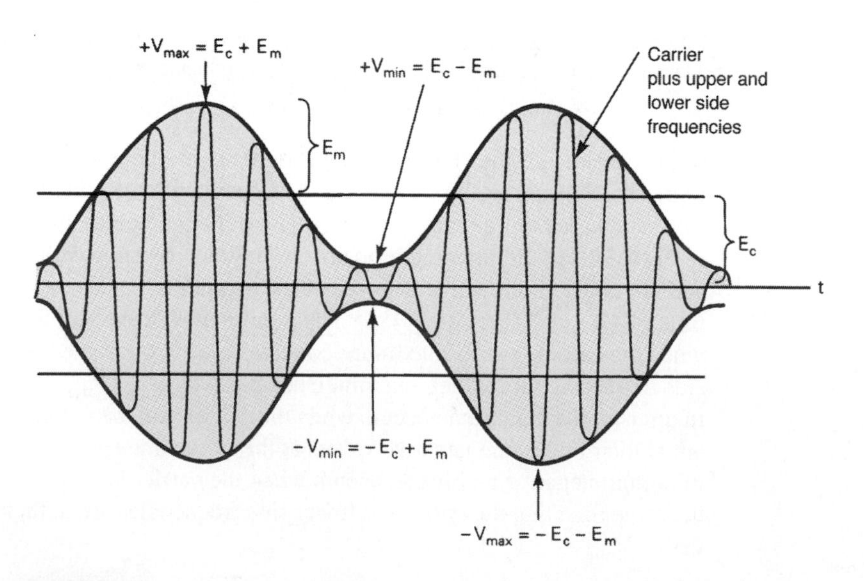

FIGURE 4-5 Modulation coefficient, E_m, and Ec

and
$$E_c = \frac{1}{2}(V_{max} + V_{min}) \tag{4-6}$$

Therefore,
$$M = \frac{1/2(V_{max} - V_{min})}{1/2(V_{max} + V_{min})} \times 100$$
$$= \frac{(V_{max} - V_{min})}{(V_{max} + V_{min})} \times 100 \tag{4-7}$$

where $V_{max} = E_c + E_m$
$V_{min} = E_c - E_m$

The peak change in the amplitude of the output wave (E_m) is the sum of the voltages from the upper and lower side frequencies. Therefore, since $E_m = E_{usf} + E_{lsf}$ and $E_{usf} = E_{lsf}$, then

$$E_{usf} = E_{lsf} = \frac{E_m}{2} = \frac{1/2(V_{max} - V_{min})}{2} = \frac{1}{4}(V_{max} - V_{min}) \tag{4-8}$$

where E_{usf} = peak amplitude of the upper side frequency (volts)
E_{lsf} = peak amplitude of the lower side frequency (volts)

From Equation 4-1, it can be seen that the percent modulation goes to 100% when $E_m = E_c$. This condition is shown in Figure 4-6d. It can also be seen that at 100% modulation, the minimum amplitude of the envelope V_{min} = 0 V. Figure 4-6c shows a 50% modu-

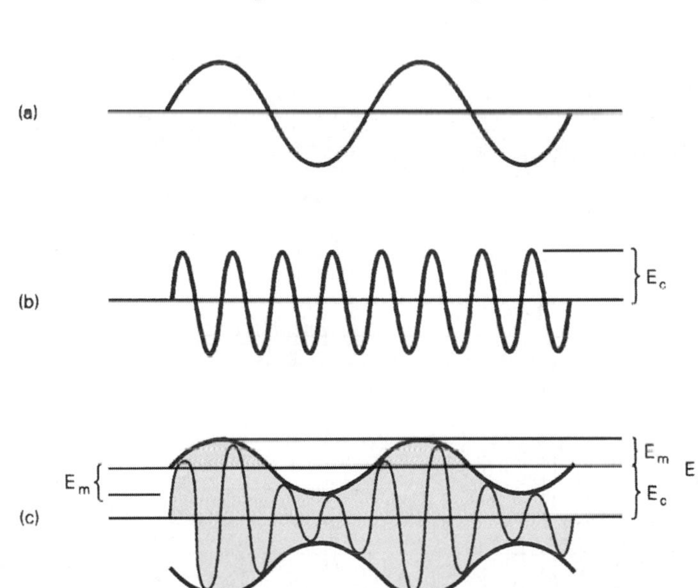

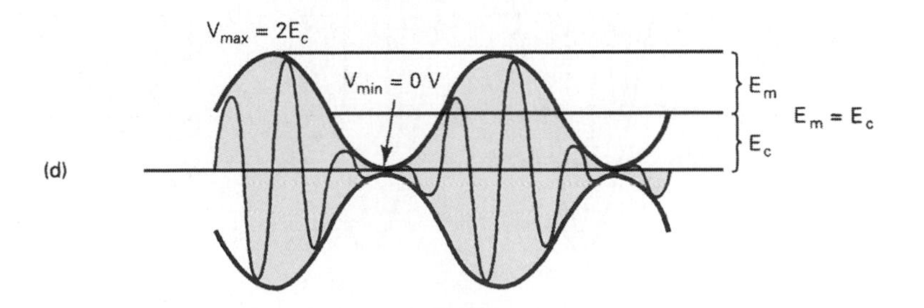

FIGURE 4-6 Percent modulation of an AM DSBFC envelope: (a) modulating signal; (b) unmodulated carrier; (c) 50% modulated wave; (d) 100% modulated wave

lated envelope; the peak change in the amplitude of the envelope is equal to one-half the amplitude of the unmodulated wave. The maximum percent modulation that can be imposed without causing excessive distortion is 100%. Sometimes percent modulation is expressed as the peak change in the voltage of the modulated wave with respect to the peak amplitude of the unmodulated carrier (i.e., percent change $= \Delta E_c/E_c \times 100$).

Example 4-2

For the AM waveform shown in Figure 4-7, determine

a. Peak amplitude of the upper and lower side frequencies.
b. Peak amplitude of the unmodulated carrier.
c. Peak change in the amplitude of the envelope.
d. Coefficient of modulation.
e. Percent modulation.

Solution **a.** From Equation 4-8,

$$E_{usf} = E_{lsf} = \frac{1}{4}(18 - 2) = 4\text{ V}$$

b. From Equation 4-6,

$$E_c = \frac{1}{2}(18 + 2) = 10\text{ V}$$

c. From Equation 4-5,

$$E_m = \frac{1}{2}(18 - 2) = 8\text{ V}$$

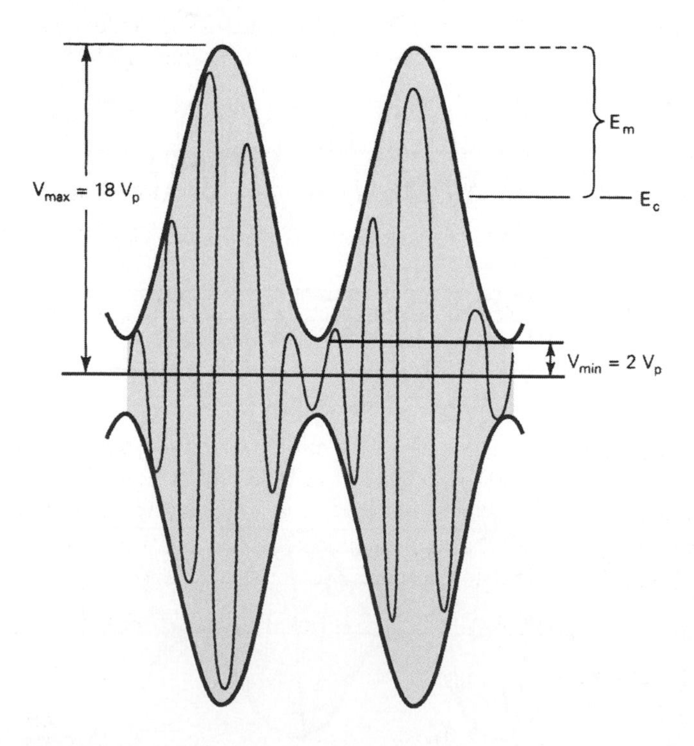

FIGURE 4-7 AM envelope for Example 3-2

d. From Equation 4-1,

$$m = \frac{8}{10} = 0.8$$

e. From Equation 4-4,

$$M = 0.8 \times 100 = 80\%$$

and from Equation 4-7,

$$M = \frac{18 - 2}{18 + 2} \times 100 = 80\%$$

4-2-5 AM Voltage Distribution

An unmodulated carrier can be described mathematically as

$$v_c(t) = E_c \sin(2\pi f_c t) \quad \textbf{(4-9)}$$

where $v_c(t)$ = time-varying voltage waveform for the carrier
E_c = peak carrier amplitude (volts)
f_c = carrier frequency (hertz)

In a previous section, it was pointed out that the repetition rate of an AM envelope is equal to the frequency of the modulating signal, the amplitude of the AM wave varies proportional to the amplitude of the modulating signal, and the maximum amplitude of the modulated wave is equal to $E_c + E_m$. Therefore, the instantaneous amplitude of the modulated wave can be expressed as

$$v_{am}(t) = [E_c + E_m \sin(2\pi f_m t)][\sin(2\pi f_c t)] \quad \textbf{(4-10)}$$

where $[E_c + E_m \sin(2\pi f_m t)]$ = amplitude of the modulated wave
E_m = peak change in the amplitude of the envelope (volts)
f_m = frequency of the modulating signal (hertz)

If mE_c is substituted for E_m,

$$v_{am}(t) = [(E_c + mE_c \sin(2\pi f_m t)][\sin(2\pi f_c t)] \quad \textbf{(4-11)}$$

where $[E_c + mE_c \sin(2\pi f_m t)]$ equals the amplitude of the modulated wave.

Factoring E_c from Equation 4-11 and rearranging gives

$$v_{am}(t) = [1 + m \sin(2\pi f_m t)][E_c \sin(2\pi f_c t)] \quad \textbf{(4-12)}$$

where $[1 + m \sin(2\pi f_m t)]$ = constant + modulating signal
$[E_c \sin(2\pi f_c t)]$ = unmodulated carrier

In Equation 4-12, it can be seen that the modulating signal contains a constant component (1) and a sinusoidal component at the modulating signal frequency ($m \sin[2\pi f_m t]$). The following analysis will show how the constant component produces the carrier component in the modulated wave and the sinusoidal component produces the side frequencies. Multiplying out Equation 4-11 or c yields

$$v_{am}(t) = E_c \sin(2\pi f_c t) + [mE_c \sin(2\pi f_m t)][\sin(2\pi f_c t)] \quad \textbf{(4-13)}$$

Therefore,

$$v_{am}(t) = E_c \sin(2\pi f_c t) - \frac{mE_c}{2}\cos[2\pi(f_c + f_m)t] + \frac{mE_c}{2}\cos[2\pi(f_c - f_m)t] \quad \textbf{(4-14)}$$

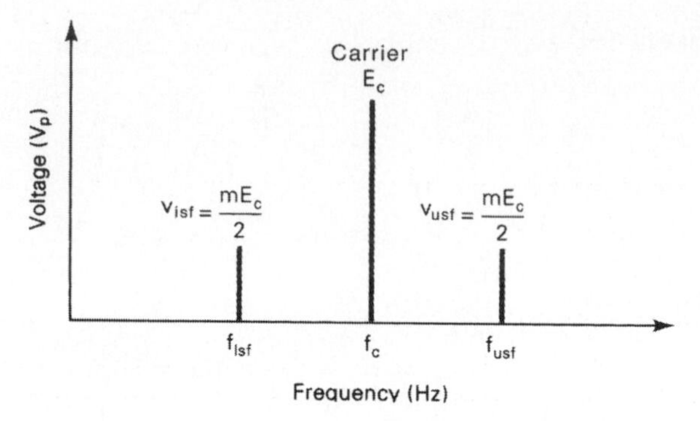

FIGURE 4-8 Voltage spectrum for an AM DSBFC wave

where

$$\begin{aligned} E_c \sin(2\pi f_c t) &= \text{carrier signal (volts)} \\ -(mE_c/2)\cos[2\pi (f_c + f_m)t] &= \text{upper side frequency signal (volts)} \\ +(mE_c/2)\cos[2\pi (f_c - f_m)t] &= \text{lower side frequency signal (volts)} \end{aligned}$$

Several interesting characteristics about double-sideband full-carrier amplitude modulation can be pointed out from Equation 4-14. First, note that the amplitude of the carrier after modulation is the same as it was before modulation (E_c). Therefore, the amplitude of the carrier is unaffected by the modulation process. Second, the amplitude of the upper and lower side frequencies depends on both the carrier amplitude and the coefficient of modulation. For 100% modulation, $m = 1$, and the amplitudes of the upper and lower side frequencies are each equal to one-half the amplitude of the carrier ($E_c/2$). Therefore, at 100% modulation,

$$V_{(\max)} = E_c + \frac{E_c}{2} + \frac{E_c}{2} = 2E_c$$

and

$$V_{(\min)} = E_c - \frac{E_c}{2} - \frac{E_c}{2} = 0 \text{ V}$$

From the relationships shown above and using Equation 4-14, it is evident that, as long as we do not exceed 100% modulation, the maximum peak amplitude of an AM envelope $V_{(\max)} = 2E_c$, and the minimum peak amplitude of an AM envelope $V_{(\min)} = 0$ V. This relationship was shown in Figure 4-6d. Figure 4-8 shows the voltage spectrum for an AM DSBFC wave (note that all the voltages are given in peak values).

Also, from Equation 4-14, the relative phase relationship between the carrier and the upper and lower side frequencies is evident. The carrier component is a + sine function, the upper side frequency a − cosine function, and the lower side frequency a + cosine function. Also, the envelope is a repetitive waveform. Thus, at the beginning of each cycle of the envelope, the carrier is 90° out of phase with both the upper and lower side frequencies, and the upper and lower side frequencies are 180° out of phase with each other. This phase relationship can be seen in Figure 4-9 for $f_c = 25$ Hz and $f_m = 5$ Hz.

Example 4-3

One input to a conventional AM modulator is a 500-kHz carrier with an amplitude of 20 V_p. The second input is a 10-kHz modulating signal that is of sufficient amplitude to cause a change in the output wave of ± 7.5 V_p. Determine

a. Upper and lower side frequencies.
b. Modulation coefficient and percent modulation.
c. Peak amplitude of the modulated carrier and the upper and lower side frequency voltages.
d. Maximum and minimum amplitudes of the envelope.

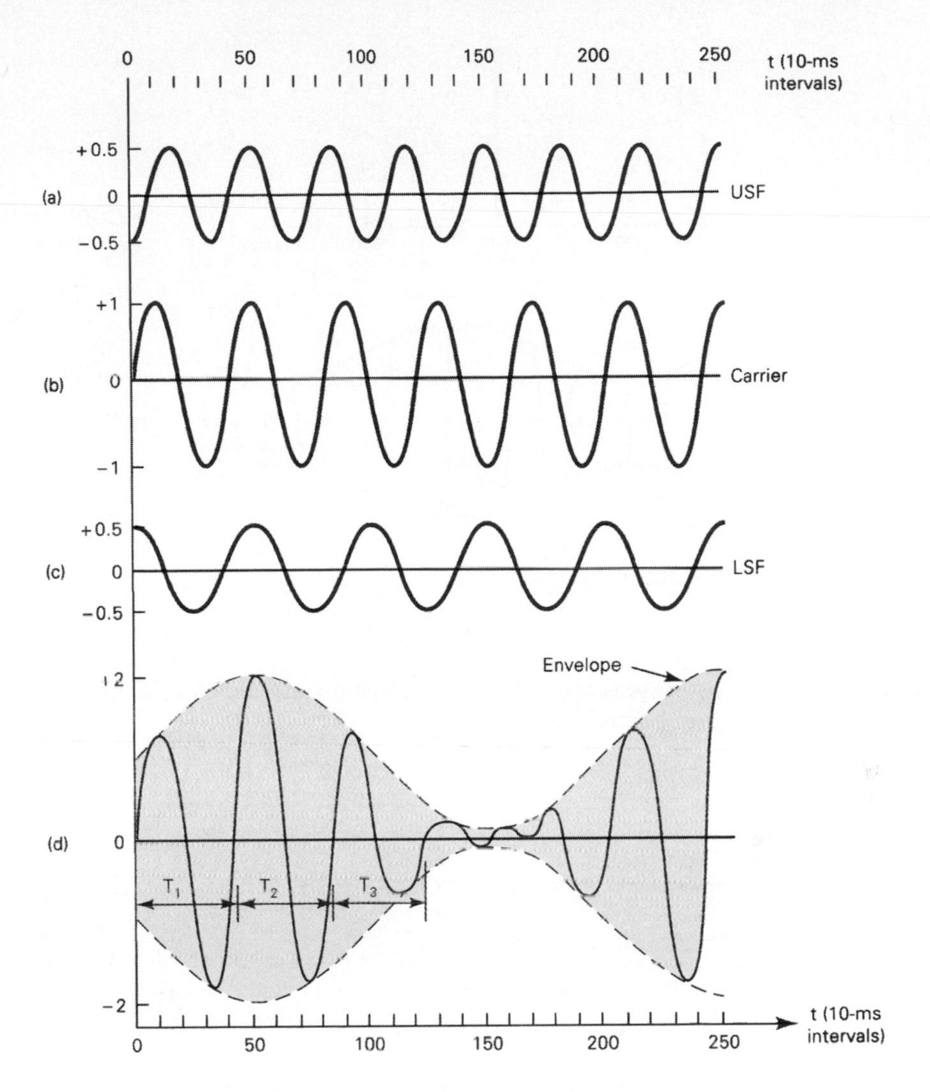

FIGURE 4-9 Generation of an AM DSBFC envelope shown in the time domain: (a) $-\frac{1}{2}\cos(2\pi 30t)$; (b) $\sin(2\pi 25t)$; (c) $+\frac{1}{2}\cos(2\pi 20t)$; (d) summation of (a), (b), and (c)

e. Expression for the modulated wave.

Then

f. Draw the output spectrum.

g. Sketch the output envelope.

Solution **a.** The upper and lower side frequencies are simply the sum and difference frequencies, respectively:

$$f_{usf} = 500\text{ kHz} + 10\text{ kHz} = 510\text{ kHz}$$
$$f_{lsf} = 500\text{ kHz} - 10\text{ kHz} = 490\text{ kHz}$$

b. The modulation coefficient is determined from Equation 4-1:

$$m = \frac{7.5}{20} = 0.375$$

Percent modulation is determined from Equation 4-2:

$$M = 100 \times 0.375 = 37.5\%$$

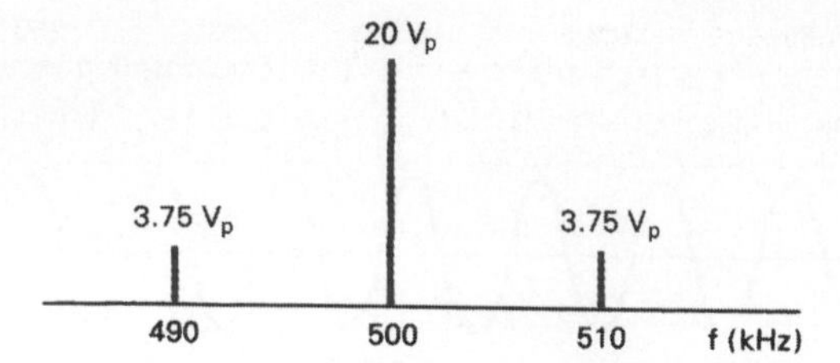

FIGURE 4-10 Output spectrum for Example 4-3

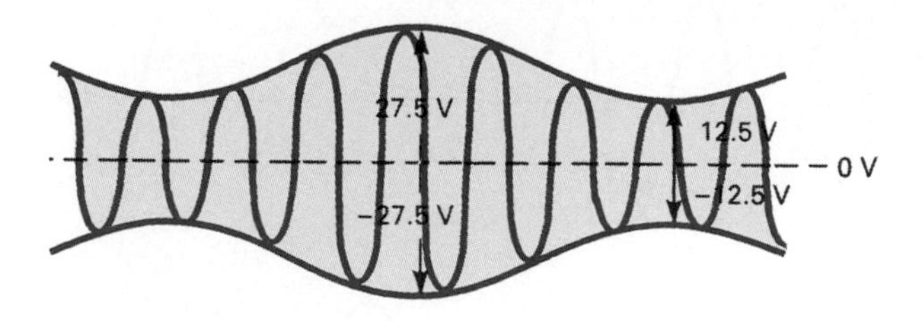

FIGURE 4-11 AM envelope for Example 4-3

c. The peak amplitude of the modulated carrier and the upper and lower side frequencies is

$$E_c(\text{modulated}) = E_c(\text{unmodulated}) = 20\ \text{V}_\text{p}$$

$$E_{\text{usf}} = E_{\text{lsf}} = \frac{mE_c}{2} = \frac{(0.375)(20)}{2} = 3.75\ \text{V}_\text{p}$$

d. The maximum and minimum amplitudes of the envelope are determined as follows:

$$V_{(\max)} = E_c + E_m = 20 + 7.5 = 27.5\ \text{V}_\text{p}$$
$$V_{(\min)} = E_c - E_m = 20 - 7.5 = 12.5\ \text{V}_\text{p}$$

e. The expression for the modulated wave follows the format of Equation 4-14:

$$v_{am}(t) = 20\sin(2\pi 500kt) - 3.75\cos(2\pi 510kt) + 3.75\cos(2\pi 490kt)$$

f. The output spectrum is shown in Figure 4-10.

g. The modulated envelope is shown in Figure 4-11.

4-2-6 AM Time-Domain Analysis

Figure 4-9 shows how an AM DSBFC envelope is produced from the algebraic addition of the waveforms for the carrier and the upper and lower side frequencies. For simplicity, the following waveforms are used for the modulating and carrier input signals:

$$\text{carrier} = v_c(t) = E_c\sin(2\pi 25t) \tag{4-15}$$

$$\text{modulating signal} = v_m(t) = E_m\sin(2\pi 5t) \tag{4-16}$$

Substituting Equations 4-15 and 4-16 into Equation 4-14, the expression for the modulated wave is

$$v_{am}(t) = E_c\sin(2\pi 25t) - \frac{mE_c}{2}\cos(2\pi 30t) + \frac{mE_c}{2}\cos(2\pi 20t) \tag{4-17}$$

where
$E_c\sin(2\pi 25t)$ = carrier (volts)
$-(mE_c/2)\cos(2\pi 30t)$ = upper side frequency (volts)
$+(mE_c/2)\cos(2\pi 20t)$ = lower side frequency (volts)

Table 4-1 lists the values for the instantaneous voltages of the carrier, the upper and lower side frequency voltages, and the total modulated wave when values of t from 0 to 250 ms,

Table 4-1 Instantaneous Voltages

USF, $-\frac{1}{2}\cos(2\pi 30t)$	Carrier, $\sin(2\pi\ 25t)$	LSF, $+\frac{1}{2}\cos(2\pi 20t)$	Envelope, v_{am}lthnl(t)	Time, t(ms)
−0.5	0	+0.5	0	0
+0.155	+1	+0.155	+1.31	10
+0.405	0	−0.405	0	20
−0.405	−1	−0.405	−1.81	30
−0.155	0	+0.155	0	40
+0.5	+1	+0.5	2	50
−0.155	0	+0.155	0	60
−0.405	−1	−0.405	−1.81	70
+0.405	0	−0.405	0	80
+0.155	+1	+0.155	+1.31	90
−0.5	0	+0.5	0	100
+0.155	−1	+0.155	−0.69	110
+0.405	0	−0.405	0	120
−0.405	+1	−0.405	+0.19	130
−0.155	0	+0.155	0	140
+0.5	−1	+0.5	0	150
−0.155	0	+0.155	0	160
−0.405	+1	−0.405	+0.19	170
+0.405	0	−0.405	0	180
+0.155	−1	+0.155	−0.69	190
−0.5	0	+0.5	0	200
+0.155	+1	+0.155	+1.31	210
+0.405	0	−0.405	0	220
−0.405	−1	−0.405	−1.81	230
+0.405	0	−0.405	0	240
+0.155	+1	+0.155	+1.31	250

in 10-ms intervals, are substituted into Equation 4-17. The unmodulated carrier voltage $E_c = 1\ V_p$, and 100% modulation is achieved. The corresponding waveforms are shown in Figure 4-9. Note that the maximum envelope voltage is 2 V ($2E_c$) and that the minimum envelope voltage is 0 V.

In Figure 4-9, note that the time between similar zero crossings within the envelope is constant (i.e., $T_1 = T_2 = T_3$, and so on). Also note that the amplitudes of successive peaks within the envelope are not equal. This indicates that a cycle within the envelope is not a pure sine wave and, thus, the modulated wave must be comprised of more than one frequency: the summation of the carrier and the upper and lower side frequencies. Figure 4-9 also shows that the amplitude of the carrier does not vary but, rather, that the amplitude of the envelope varies in accordance with the modulating signal. This is accomplished by the addition of the upper and lower side frequencies to the carrier waveform.

4-2-7 AM Power Distribution

In any electrical circuit, the power dissipated is equal to the voltage squared divided by the resistance. Thus, the average power dissipated in a load by an unmodulated carrier is equal to the rms carrier voltage squared divided by the load resistance. Mathematically, power in an unmodulated carrier is

$$P_c = \frac{(0.707E_c)^2}{R}$$

$$= \frac{(E_c)^2}{2R} \tag{4-18}$$

where P_c = carrier power (watts)
E_c = peak carrier voltage (volts)
R = load resistance (ohms)

The upper and lower sideband powers are expressed mathematically as

$$P_{usb} = P_{lsb} = \frac{(mE_c/2)^2}{2R}$$

where $mE_c/2$ is the peak voltage of the upper and lower side frequencies. Rearranging yields

$$P_{usb} = P_{lsb} = \frac{m^2E_c^2}{8R} \quad \textbf{(4-19)}$$

where P_{usb} = upper sideband power (watts)
P_{lsb} = lower sideband power (watts)

Rearranging Equation 4-19 gives

$$P_{usb} = P_{lsb} = \frac{m^2}{4}\left\{\frac{E_c^2}{2R}\right\} \quad \textbf{(4-20)}$$

Substituting Equation 4-18 into Equation 4-19 gives

$$P_{usb} = P_{lsb} = \frac{m^2P_c}{4} \quad \textbf{(4-21)}$$

It is evident from Equation 4-21 that for a modulation coefficient $m = 0$, the power in the upper and lower sidebands is zero, and the total transmitted power is simply the carrier power.

The total power in an amplitude-modulated wave is equal to the sum of the powers of the carrier, the upper sideband, and the lower sideband. Mathematically, the total power in an AM DSBFC envelope is

$$P_t = P_c + P_{usb} + P_{lsb} \quad \textbf{(4-22)}$$

where P_t = total power of an AM DSBFC envelope (watts)
P_c = carrier power (watts)
P_{usb} = upper sideband power (watts)
P_{lsb} = lower sideband power (watts)

Substituting Equation 4-21 into Equation 4-22 yields

$$P_t = P_c + \frac{m^2P_c}{4} + \frac{m^2P_c}{4} \quad \textbf{(4-23)}$$

Combining terms gives

$$P_t = P_c + \frac{m^2P_c}{2} \quad \textbf{(4-24)}$$

where $(m^2 P_c)/2$ is the total sideband power.

Factoring P_c gives us

$$P_t = P_c\left(1 + \frac{m^2}{2}\right) \quad \textbf{(4-25)}$$

From the preceding analysis, it can be seen that the carrier power in the modulated wave is the same as the carrier power in the unmodulated wave. Thus, it is evident that the

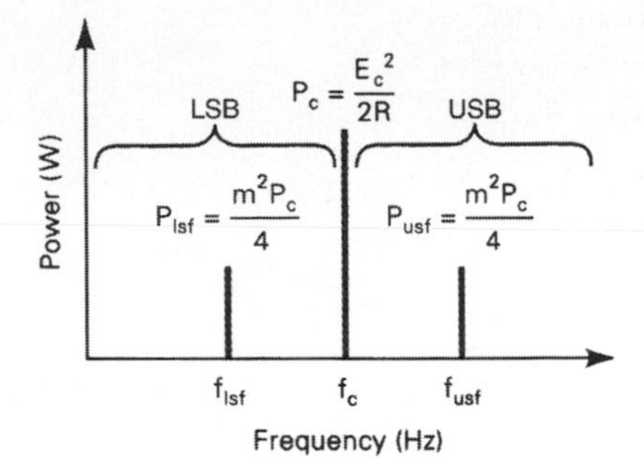

FIGURE 4-12 Power spectrum for an AM DSBFC wave with a single-frequency modulating signal

power of the carrier is unaffected by the modulation process. Also, because the total power in the AM wave is the sum of the carrier and sideband powers, the total power in an AM envelope increases with modulation (i.e., as m increases, P_t increases).

Figure 4-12 shows the power spectrum for an AM DSBFC wave. Note that with 100% modulation the maximum power in the upper or lower sideband is equal to only one-fourth the power in the carrier. Thus, the maximum total sideband power is equal to one-half the carrier power. One of the most significant disadvantages of AM DSBFC transmission is the fact that the information is contained in the sidebands although most of the power is wasted in the carrier. Actually, the power in the carrier is not totally wasted because it does allow for the use of relatively simple, inexpensive demodulator circuits in the receiver, which is the predominant advantage of AM DSBFC.

Example 4-4

For an AM DSBFC wave with a peak unmodulated carrier voltage $V_c = 10\ \text{V}_\text{p}$, a load resistance $R_L = 10\ \Omega$, and a modulation coefficient $m = 1$, determine

a. Powers of the carrier and the upper and lower sidebands.
b. Total sideband power.
c. Total power of the modulated wave.

Then

d. Draw the power spectrum.
e. Repeat steps (a) through (d) for a modulation index $m = 0.5$.

Solution **a.** The carrier power is found by substituting into Equation 4-18:

$$P_c = \frac{10^2}{2(10)} = \frac{100}{20} = 5\ \text{W}$$

The upper and lower sideband power is found by substituting into Equation 4-21:

$$P_{usb} = P_{lsb} = \frac{(1^2)(5)}{4} = 1.25\ \text{W}$$

b. The total sideband power is

$$P_{sbt} = \frac{m^2P_c}{2} = \frac{(1^2)(5)}{2} = 2.5\ \text{W}$$

c. The total power in the modulated wave is found by substituting into Equation 4-25:

$$P_t = 5\left[1 + \frac{(1)^2}{2}\right] = 7.5\ \text{W}$$

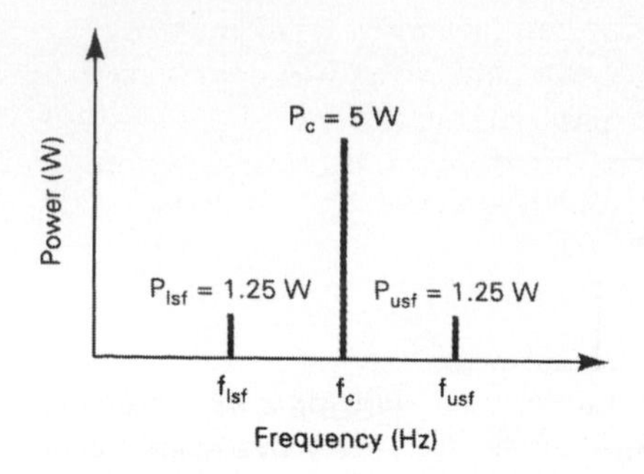

FIGURE 4-13 Power spectrum for Example 4-4d

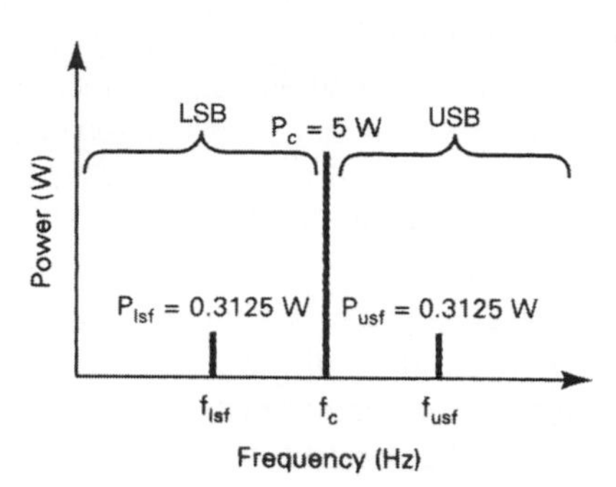

FIGURE 4-14 Power spectrum for Example 4-4

d. The power spectrum is shown in Figure 4-13.
e. The carrier power is found by substituting into Equation 4-18:

$$P_c = \frac{10^2}{2(10)} = \frac{100}{20} = 5\text{ W}$$

The upper and lower sideband power is found by substituting into Equation 4-21:

$$P_{\text{usb}} = P_{\text{lsb}} = \frac{(0.5)^2(5)}{4} = 0.3125\text{ W}$$

The total sideband power is

$$P_{\text{sbt}} = \frac{m^2 P_c}{2} = \frac{(0.5)^2(5)}{2} = 0.625\text{ W}$$

The total power of the modulated wave is found by substituting into Equation 4-25:

$$P_t = 5\left[1 + \frac{(0.5)^2}{2}\right] = 5.625\text{ W}$$

The power spectrum is shown in Figure 4-14.

From Example 4-4, it can be seen why it is important to use as high a percentage of modulation as possible while still being sure not to overmodulate. As the example shows, the carrier power remains the same as m changes. However, the sideband power was reduced dramatically when m decreased from 1 to 0.5. Because sideband power is proportional to the square of the modulation coefficient, a reduction in m of one-half results in a reduction in the sideband power of one-fourth (i.e., $0.5^2 = 0.25$). The relationship between modulation coefficient and power can sometimes be deceiving because the total transmitted power consists primarily of carrier power and is, therefore, not dramatically affected by changes in m. However, it should be noted that the power in the intelligence-carrying portion of the transmitted signal (i.e., the sidebands) is affected dramatically by changes in m. For this reason, AM DSBFC systems try to maintain a modulation coefficient between 0.9 and 0.95 (90% to 95% modulation) for the highest-amplitude intelligence signals.

4-2-8 AM Current Calculations

With amplitude modulation, it is very often necessary and sometimes desirable to measure the current of the carrier and modulated wave and then calculate the modulation index from these measurements. The measurements are made by simply metering the transmit antenna current with and without the presence of a modulating signal. The relationship between carrier current and the current of the modulated wave is

$$\frac{P_t}{P_c} = \frac{I_t^2 R}{I_c^2 R} = \frac{I_t^2}{I_c^2} = 1 + \frac{m^2}{2} \tag{4-26}$$

where P_t = total transmit power (watts)
P_c = carrier power (watts)
I_t = total transmit current (ampere)
I_c = carrier current (ampere)
R = antenna resistance (ohms)

and

$$\frac{I_t}{I_c} = \sqrt{1 + \frac{m^2}{2}} \tag{4-27}$$

Thus,

$$I_t = I_c\sqrt{1 + \frac{m^2}{2}} \tag{4-28}$$

4-2-9 Modulation by a Complex Information Signal

In the previous sections of this chapter, frequency spectrum, bandwidth, coefficient of modulation, and voltage and power distribution for double-sideband full-carrier AM were analyzed for a single-frequency modulating signal. In practice, however, the modulating signal is very often a complex waveform made up of many sine waves with different amplitudes and frequencies. Consequently, a brief analysis will be given of the effects such a complex modulating signal would have on an AM waveform.

If a modulating signal contains two frequencies (f_{m1} and f_{m2}), the modulated wave will contain the carrier and two sets of side frequencies spaced symmetrically about the carrier. Such a wave can be written as

$$v_{am}(t) = \sin(2\pi f_c t) + \frac{1}{2}\cos[2\pi(f_c - f_{m1})t] - \frac{1}{2}\cos[2\pi(f_c + f_{m1})t]$$
$$+ \frac{1}{2}\cos[2\pi(f_c - f_{m2})t] - \frac{1}{2}\cos[2\pi(f_c + f_{m2})t] \tag{4-29}$$

When several frequencies simultaneously amplitude modulate a carrier, the combined coefficient of modulation is the square root of the quadratic sum of the individual modulation indexes as follows:

$$m_t = \sqrt{m_1^2 + m_2^2 + m_3^2 + m_n^2} \tag{4-30}$$

where m_t = total coefficient of modulation
$m_1, m_2, m_3,$ *and* m_n = coefficients of modulation for input signals 1, 2, 3, and n

The combined coefficient of modulation can be used to determine the total sideband and transmit powers as follows:

$$P_{usbt} = P_{lsbt} = \frac{P_c m_t^2}{4} \quad \textbf{(4-31)}$$

and

$$P_{sbt} = \frac{P_c m_t^2}{2} \quad \textbf{(4-32)}$$

Thus,

$$P_t = P_c\left(1 + \frac{m_t^2}{2}\right) \quad \textbf{(4-33)}$$

where P_{usbt} = total upper sideband power (watts)
P_{lsbt} = total lower sideband power (watts)
P_{sbt} = total sideband power (watts)
P_t = total transmitted power (watts)

In an AM transmitter, care must be taken to ensure that the combined voltages of all the modulating signals do not overmodulate the carrier.

Example 4-5

For an AM DSBFC transmitter with an unmodulated carrier power $P_c = 100$ W that is modulated simultaneously by three modulating signals with coefficients of modulation $m_1 = 0.2$, $m_2 = 0.4$, and $m_3 = 0.5$, determine

a. Total coefficient of modulation.
b. Upper and lower sideband power.
c. Total transmitted power.

Solution **a.** The total coefficient of modulation is found by substituting into Equation 4-30:

$$m_t = \sqrt{0.2^2 + 0.4^2 + 0.5^2}$$
$$= \sqrt{0.04 + 0.16 + 0.25} = 0.67$$

b. The total sideband power is found by substituting the results of step (a) into Equation 4-32:

$$P_{sbt} = \frac{(0.67^2)100}{2} = 22.445 \text{ W}$$

c. The total transmitted power is found by substituting into Equation 4-33:

$$P_t = 100\left(1 + \frac{0.67^2}{2}\right) = 122.445 \text{ W}$$

4-3 AM MODULATING CIRCUITS

The location in a transmitter where modulation occurs determines whether the circuit is a *low-* or a *high-level transmitter.* With low-level modulation, the modulation takes place prior to the output element of the final stage of the transmitter, in other words, prior to the collector of the output transistor in a transistorized transmitter, prior to the drain of the output FET in a FET transmitter, or prior to the plate of the output tube in a vacuum-tube transmitter.

An advantage of low-level modulation is that less modulating signal power is required to achieve a high percentage of modulation. In high-level modulators, the modulation takes place in the final element of the final stage where the carrier signal is at its maximum amplitude and, thus, requires a much higher amplitude modulating signal to achieve a reasonable percent modulation. With high-level modulation, the final modulating signal amplifier must supply all the sideband power, which could be as much as 33% of the total transmit power. An obvious disadvantage of low-level modulation is in high-power applications when all the amplifiers that follow the modulator stage must be linear amplifiers, which is extremely inefficient.

4-3-1 Low-Level AM Modulator

A small signal, class A amplifier, such as the one shown in Figure 4-15a, can be used to perform amplitude modulation; however, the amplifier must have two inputs: one for the carrier signal and the second for the modulating signal. With no modulating signal present, the circuit operates as a linear class A amplifier, and the output is simply the carrier amplified by the quiescent voltage gain. However, when a modulating signal is applied, the amplifier operates nonlinearly, and signal multiplication as described by Equation 4-10 occurs. In Figure 4-15a, the carrier is applied to the base and the modulating signal to the emitter. Therefore, this circuit configuration is called *emitter modulation.* The modulating signal varies the gain of the amplifier at a sinusoidal rate equal to the frequency of the modulating signal. The depth of modulation achieved is proportional to the amplitude of the modulating signal. The voltage gain for an emitter modulator is expressed mathematically as

$$A_v = A_q\,[1 + m\sin(2\pi f_m t)] \qquad \textbf{(4-34)}$$

where A_v = amplifier voltage gain with modulation (unitless)
A_q = amplifier quiescent (without modulation) voltage gain (unitless)

$\sin(2\pi f_m t)$ goes from a maximum value of $+1$ to a minimum value of -1. Thus, Equation 4-35 reduces to

$$A_v = A_q(1 \pm m) \qquad \textbf{(4-35)}$$

where m equals the modulation coefficient. At 100% modulation, $m = 1$, and Equation 4-35 reduces to

$$A_{v(\max)} = 2A_q$$
$$A_{v(\min)} = 0$$

Figure 4-15b shows the waveforms for the circuit shown in Figure 4-15a. The modulating signal is applied through isolation transformer T_1 to the emitter of Q_1, and the carrier is applied directly to the base. The modulating signal drives the circuit into both saturation and cutoff, thus producing the nonlinear amplification necessary for modulation to occur. The collector waveform includes the carrier and the upper and lower side frequencies as well as a component at the modulating signal frequency. Coupling capacitor C_2 removes the modulating signal frequency from the AM waveform, thus producing a symmetrical AM envelope at V_{out}.

With emitter modulation, the amplitude of the output signal depends on the amplitude of the input carrier and the voltage gain of the amplifier. The coefficient of modulation depends entirely on the amplitude of the modulating signal. The primary disadvantage of emitter modulation is the amplifier operates class A, which is extremely inefficient. Emitter modulators are also incapable of producing high-power output waveforms.

Example 4-6

For a low-level AM modulator similar to the one shown in Figure 4-15 with a modulation coefficient $m = 0.8$, a quiescent voltage gain $A_q = 100$, an input carrier frequency $f_c = 500$ kHz with an amplitude $V_c = 5$ mV, and a 1000-Hz modulating signal, determine

a. Maximum and minimum voltage gains.
b. Maximum and minimum amplitudes for V_{out}.

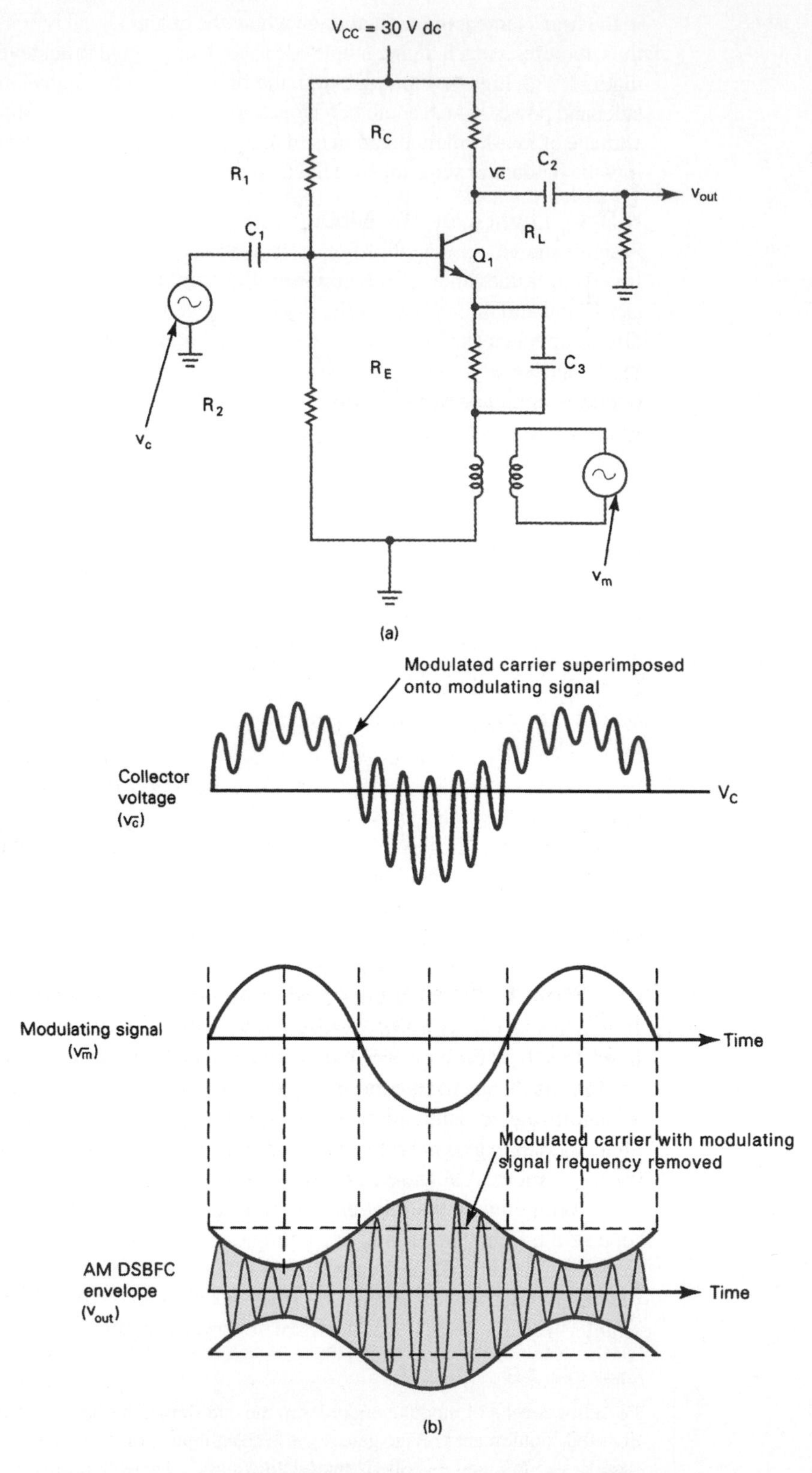

FIGURE 4-15 (a) Single transistor, emitter modulator; (b) output waveforms

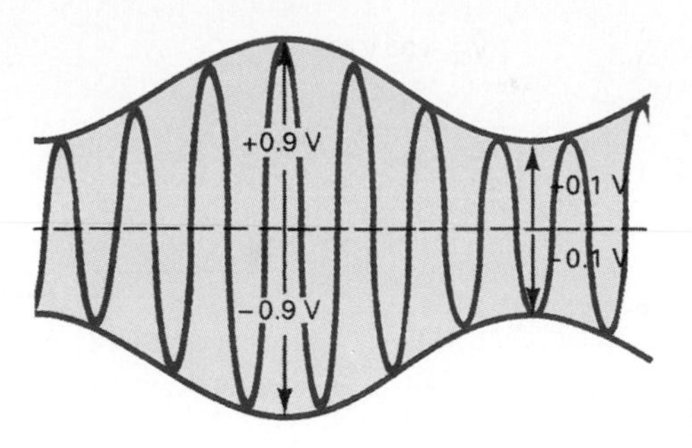

FIGURE 4-16 AM envelope for Example 4-6

Then

c. Sketch the output AM envelope.

Solution a. Substituting into Equation 4-34,

$$A_{max} = 100(1 + 0.8) = 180$$
$$A_{min} = 100(1 - 0.8) = 20$$

b.

$$V_{out(max)} = 180(0.005) = 0.9 \text{ V}$$
$$V_{out(min)} = 20(0.005) = 0.1 \text{ V}$$

c. The AM envelope is shown in Figure 4-16.

4-3-2 Medium-Power AM Modulator

Early medium- and high-power AM transmitters were limited to those that used vacuum tubes for the active devices. However, since the mid-1970s, solid-state transmitters have been available with output powers as high as several thousand watts. This is accomplished by placing several final power amplifiers in parallel such that their output signals combine in phase and are, thus, additive.

Figure 4-17a shows the schematic diagram for a single-transistor medium-power AM modulator. The modulation takes place in the collector, which is the output element of the transistor. Therefore, if this is the final active stage of the transmitter (i.e., there are no amplifiers between it and the antenna), it is a high-level modulator.

To achieve high power efficiency, medium- and high-power AM modulators generally operate class C. Therefore, a practical efficiency of as high as 80% is possible. The circuit shown in Figure 4-17a is a class C amplifier with two inputs: a carrier (v_c) and a single-frequency modulating signal (v_m). Because the transistor is biased class C, it operates nonlinear and is capable of nonlinear mixing (modulation). This circuit is called a *collector modulator* because the modulating signal is applied directly to the collector. The RFC is a radio-frequency choke that acts as a short to dc and an open to high frequencies. Therefore, the RFC isolates the dc power supply from the high-frequency carrier and side frequencies while still allowing the low-frequency intelligence signals to modulate the collector of Q_1.

4-3-2-1 Circuit operation. For the following explanation, refer to the circuit shown in Figure 4-17a and the waveforms shown in Figure 4-17b. When the amplitude of the carrier exceeds the barrier potential of the base–emitter junction (approximately 0.7 V for a silicon transistor), Q_1 turns on, and collector current flows. When the amplitude of the carrier drops below 0.7 V, Q_1 turns off, and collector current ceases. Consequently, Q_1 switches between saturation and cutoff controlled by the carrier signal, collector current flows for less than 180° of each carrier cycle, and class C operation is achieved. Each successive cycle of the carrier turns Q_1 on for an instant and allows current to flow for a short time, producing a negative-going waveform at the collector. The collector current and voltage waveforms are shown in Figure 4-17b. The collector voltage waveform resembles a repetitive half-wave rectified signal with a fundamental frequency equal to f_c.

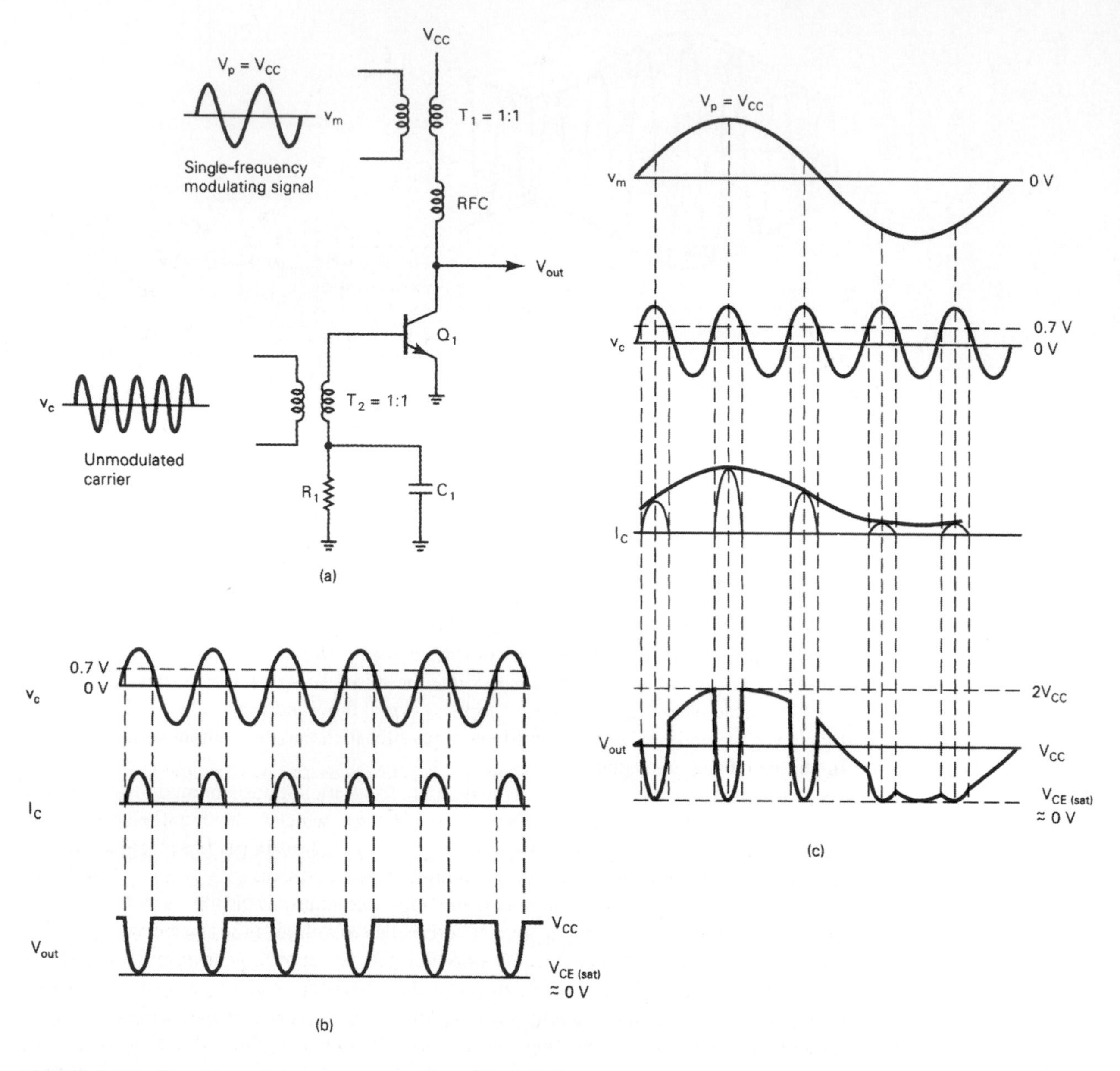

FIGURE 4-17 Simplified medium-power transistor AM DSBFC modulator: (a) schematic diagram; (b) collector waveforms with no modulating signal; (c) collector waveforms with a modulating signal

When a modulating signal is applied to the collector in series with the dc supply voltage, it adds to and subtracts from V_{CC}. The waveforms shown in Figure 4-17c are produced when the maximum peak modulating signal amplitude equals V_{CC}. It can be seen that the output voltage waveform swings from a maximum value of $2V_{CC}$ to approximately 0 V ($V_{CE(\text{sat})}$). The peak change in collector voltage is equal to V_{CC}. Again, the waveform resembles a half-wave rectified carrier superimposed onto a low-frequency ac intelligence signal.

Because Q_1 is operating nonlinear, the collector waveform contains the two original input frequencies (f_c and f_m) and their sum and difference frequencies ($f_c \pm f_m$). Because the output waveform also contains the higher-order harmonics and intermodulation components, it must be bandlimited to $f_c \pm f_m$ before being transmitted.

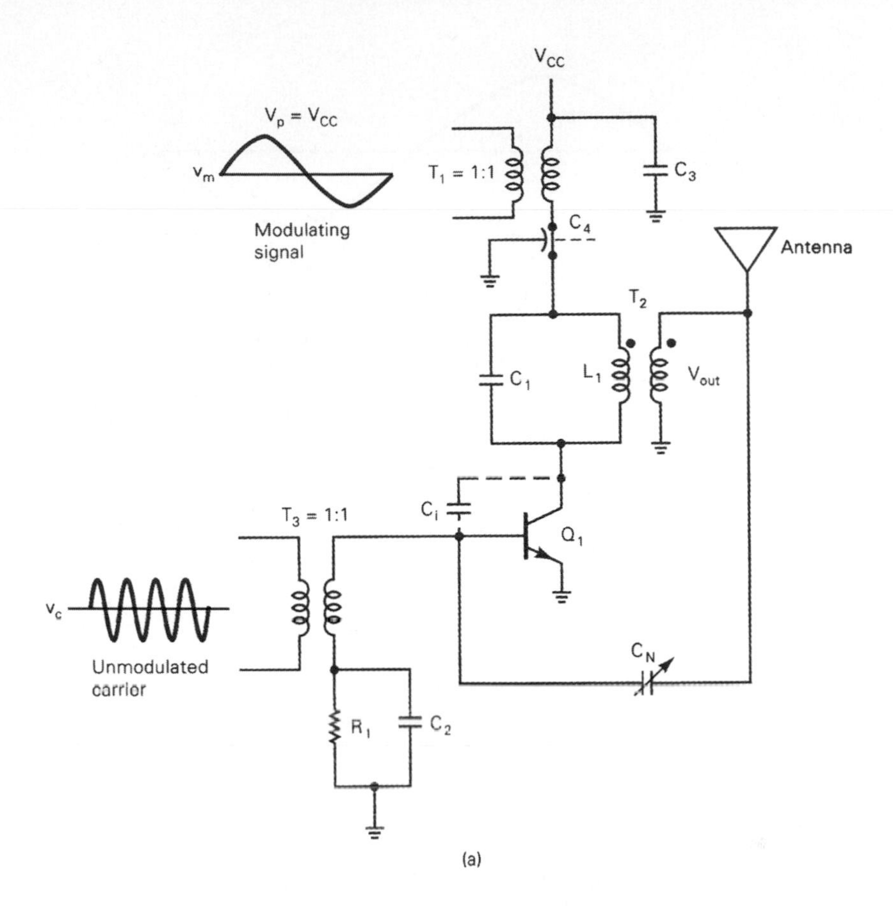

FIGURE 4-18 Medium-power transistor AM DSBFC modulator: (a) schematic diagram *(Continued)*

A more practical circuit for producing a medium-power AM DSBFC signal is shown in Figure 4-18a, with corresponding waveforms shown in Figure 4-18b. This circuit is also a collector modulator with a maximum peak modulating signal amplitude $V_{m(\max)} = V_{CC}$. Operation of this circuit is almost identical to the circuit shown in Figure 4-17a except for the addition of a tank circuit (C_1 and L_1) in the collector of Q_1. Because the transistor is operating between saturation and cutoff, collector current is not dependent on base drive voltage. The voltage developed across the tank circuit is determined by the ac component of the collector current and the impedance of the tank circuit at resonance, which depends on the quality factor (Q) of the coil. The waveforms for the modulating signal, carrier, and collector current are identical to those of the previous example. The output voltage is a symmetrical AM DSBFC signal with an average voltage of 0 V, a maximum positive peak amplitude equal to $2V_{CC}$, and a maximum negative peak amplitude equal to $-2V_{CC}$. The positive half-cycle of the output waveform is produced in the tank circuit by the *flywheel effect.* When Q_1 is conducting, C_1 charges to $V_{CC} + V_m$ (a maximum value of $2V_{CC}$), and when Q_1 is off, C_1 discharges through L_1. When L_1 discharges, C_1 charges to a minimum value of $-2V_{CC}$. This produces the positive half-cycle of the AM envelope. The resonant frequency of the tank circuit is equal to the carrier frequency, and the bandwidth extends from $f_c - f_m$ to $f_c + f_m$. Consequently, the modulating signal, the harmonics, and all the higher-order cross products are removed from the waveform, leaving a symmetrical AM DSBFC wave. One hundred percent modulation occurs when the peak amplitude of the modulating signal equals V_{CC}.

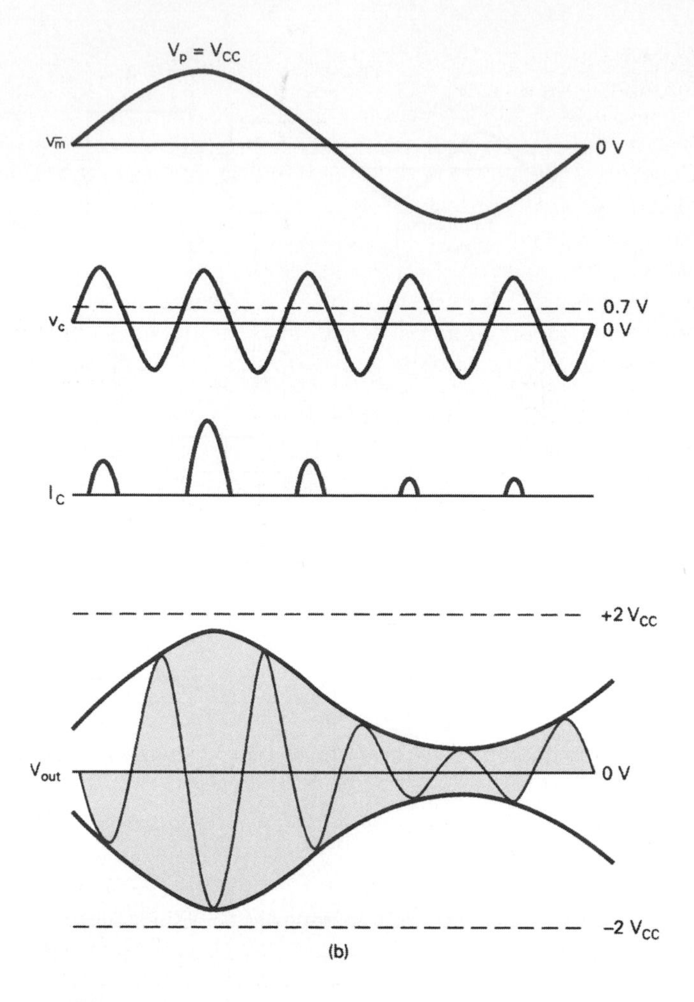

FIGURE 4-18 *(Continued)* (b) collector and output waveforms

Several components shown in Figure 4-18a have not been explained. R_1 is the bias resistor for Q_1. R_1 and C_2 form a clamper circuit that produces a reverse "self" bias and, in conjunction with the barrier potential of the transistor, determines the turn-on voltage for Q_1. Consequently, Q_1 can be biased to turn on only during the most positive peaks of the carrier voltage. This produces a narrow collector current waveform and enhances class C efficiency.

C_3 is a bypass capacitor that looks like a short to the modulating signal frequencies, preventing the information signals from entering the dc power supply. *C_{bc} is the base-to-collector junction capacitance of Q_1*. At radio frequencies, the relatively small junction capacitances within the transistor are insignificant. If the capacitive reactance of C_{bc} is significant, the collector signal may be returned to the base with sufficient amplitude to cause Q_1 to begin oscillating. Therefore, a signal of equal amplitude and frequency and 180° out of phase must be fed back to the base to cancel or *neutralize* the *interelectrode capacitance feedback. C_N* is a *neutralizing capacitor.* Its purpose is to provide a feedback path for a signal that is equal in amplitude and frequency but 180° out of phase with the signal fed back through C_{bc}. C_4 is a RF bypass capacitor. Its purpose is to isolate the dc power supply from radio frequencies. Its operation is quite similar; at the carrier frequency, C_4 looks like a short circuit, preventing the carrier from *leaking* into the power supply or the modulating signal circuitry and being distributed throughout the transmitter.

4-3-3 Simultaneous Base and Collector Modulation

Collector modulators produce a more symmetrical envelope than low-power emitter modulators, and collector modulators are more power efficient. However, collector modulators require a higher amplitude-modulating signal, and they cannot achieve a full saturation-to-cutoff output voltage swing, thus preventing 100% modulation from occurring. Therefore, to achieve symmetrical modulation, operate at maximum efficiency, develop a high output power, and require as little modulating signal drive power as possible, emitter and collector modulations are sometimes used simultaneously.

4-3-3-1 Circuit operation. Figure 4-19 shows an AM modulator that uses a combination of both base and collector modulations. The modulating signal is simultaneously fed into the collectors of the push–pull modulators (Q_2 and Q_3) and to the collector of the driver amplifier (Q_1). Collector modulation occurs in Q_1; thus, the carrier signal on the base of Q_2 and Q_3 has already been partially modulated, and the modulating signal power can be reduced. Also, the modulators are not required to operate over their entire operating curve to achieve 100% modulation.

4-4 LINEAR INTEGRATED-CIRCUIT AM MODULATORS

Linear integrated-circuit function generators use a unique arrangement of transistors and FETs to perform signal multiplication, which is a characteristic that makes them ideally suited for generating AM waveforms. Integrated circuits, unlike their discrete counterparts, can precisely match current flow, amplifier voltage gain, and temperature variations. Linear integrated-circuit AM modulators also offer excellent frequency stability, symmetrical modulation characteristics, circuit miniaturization, fewer components, temperature immunity, and simplicity of design and troubleshooting. Their disadvantages include low output power, a relatively low usable frequency range, and susceptibility to fluctuations in the dc power supply.

The XR-2206 *monolithic function generator* is ideally suited for performing amplitude modulation. Figure 4-20a shows the block diagram for the XR-2206, and Figure 4-20b shows the schematic diagram. The XR-2206 consists of four functional blocks: a voltage-controlled oscillator (VCO), an analog multiplier and sineshaper, a unity-gain buffer, and a set of current switches. The VCO frequency of oscillation f_c is determined by the external timing capacitor (C_1) between pins 5 and 6 and by timing resistor (R_1) connected between either pin 7 or 8 and ground. Whether pin 7 or 8 is selected is determined by the voltage level on pin 9. If pin 9 is open circuited or connected to an external voltage ≥ 32 V, pin 7 is selected. If the voltage on pin 9 is ≤ 1 V, pin 8 is selected. The oscillator frequency is given by

$$f_c = \frac{1}{R_1 C_1} \text{ Hz} \tag{4-36}$$

The output amplitude on pin 2 can be modulated by applying a dc bias and a modulating signal to pin 1. Figure 4-20c shows the normalized output amplitude-versus-dc bias. A normalized output of 1 corresponds to maximum output voltage, a normalized value of 0.5 corresponds to an output voltage equal to half the maximum value, and a normalized value of 0 corresponds to no output signal. As the figure shows, the output amplitude varies linearly with input bias for voltages within ± 4 volts of $V^+/2$. An input voltage equal to $V^+/2$ causes the output amplitude to go to 0 V, and an input voltage either 4 V above or below $V^+/2$ produces maximum output amplitude.

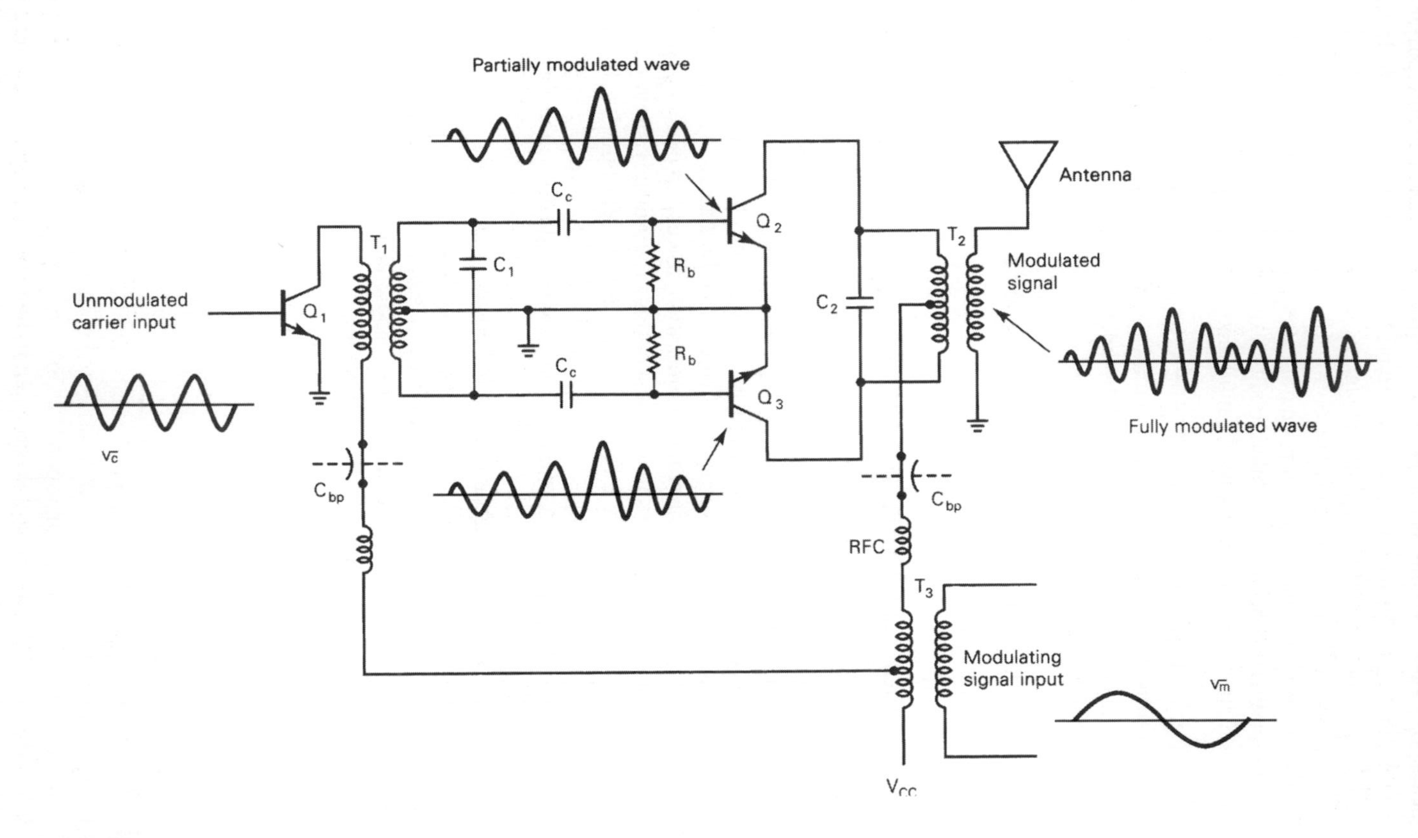

FIGURE 4-19 High-power AM DSBFC transistor modulator

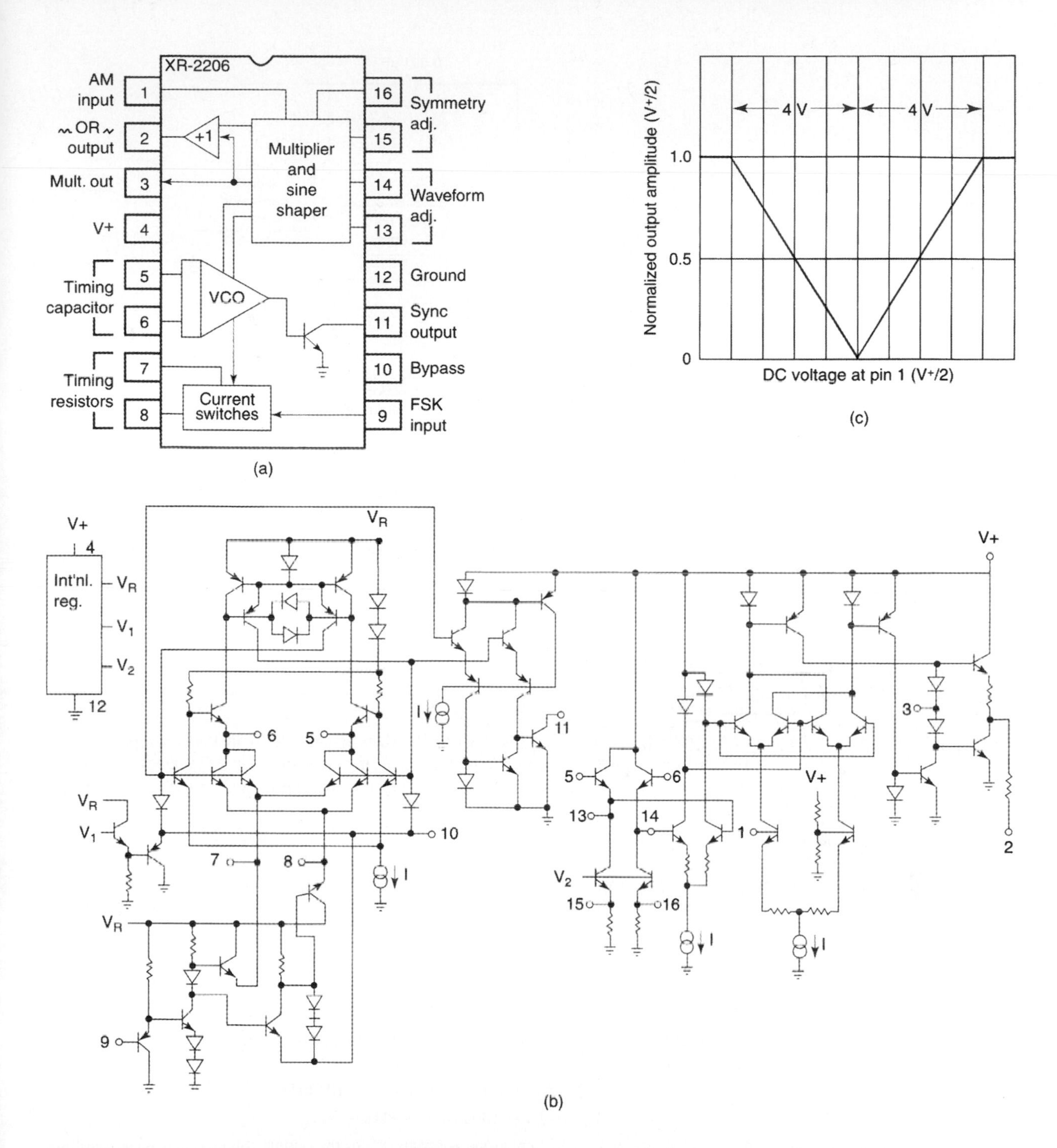

FIGURE 4-20 XR-2206: (a) Block diagram; (b) schematic diagram; (c) output voltage-versus-input voltage curve

Figure 4-21 shows the schematic diagram for a linear integrated-circuit AM modulator using the XR-2206. The VCO output frequency is the carrier signal. The modulating signal and bias voltage are applied to the internal multiplier (modulator) circuit through pin 1. The modulating signal mixes with the VCO signal producing an AM wave at V_{out}. The output wave is a symmetrical AM envelope containing the carrier and the upper and lower side frequencies.

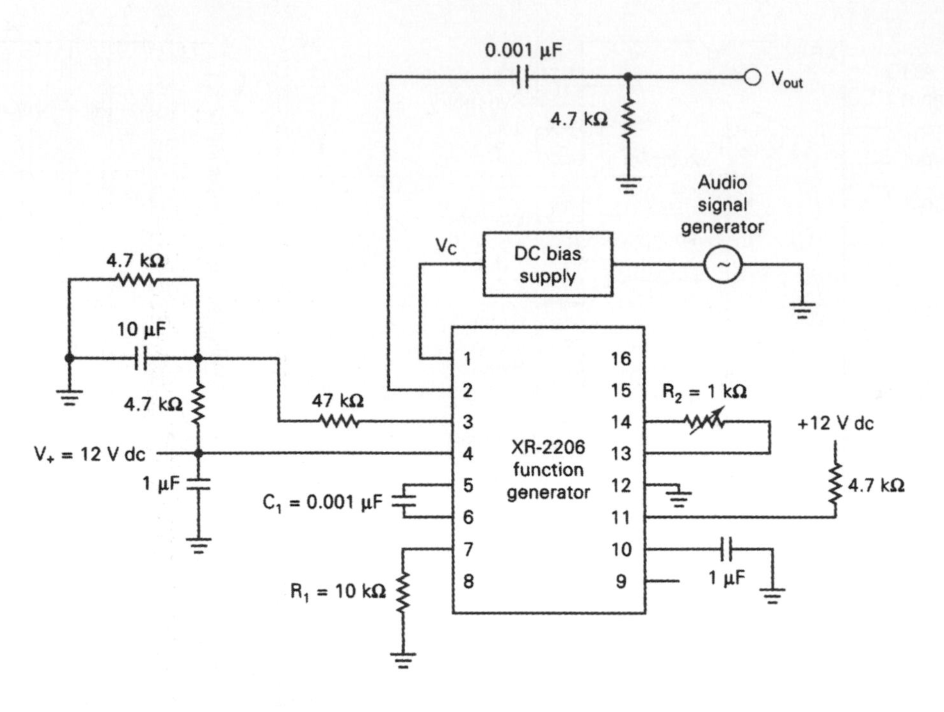

FIGURE 4-21 Linear integrated-circuit AM modulator

Example 4-7

For an XR-2206 LIC modulator such as the one shown in Figure 4-21 with a power supply voltage $V^+ = 12$ V dc, a modulating signal amplitude $V_m = 2\ V_p$, a modulating signal frequency $f_m = 4$ kHz, a dc bias $V_{bias} = +4$ V dc, timing resistor $R_1 = 10$ kΩ, and timing capacitor $C_1 = 0.001$ μF, determine

a. Carrier frequency.

b. Upper and lower side frequencies.

Then

c. Sketch the output wave.

d. From the output waveform, determine the coefficient of modulation and percent modulation.

Solution a. The carrier frequency is determined from Equation 4-36:

$$f_c = \frac{1}{(10\text{ k}\Omega)(0.001\,\mu\text{F})} = 100\text{ kHz}$$

b. The upper and lower side frequencies are simply the sum and difference frequencies between the carrier and the modulating signal.

$$f_{usf} = 100\text{ kHz} + 4\text{ kHz} = 104\text{ kHz}$$
$$f_{lsf} = 100\text{ kHz} - 4\text{ kHz} = 96\text{ kHz}$$

c. Figure 4-22 shows how an AM envelope is produced for the output voltage-versus-input voltage characteristics of the XR-2206.

d. The percent modulation is determined from the AM envelope shown in Figure 4-22 using Equation 4-7:

$$V_{max} = 10\text{ V}_p \quad + V_{min} = 0\text{ V}$$

$$m = \frac{10 - 0}{10 + 0} = 1$$

$$M = 1 \times 100 = 100\%$$

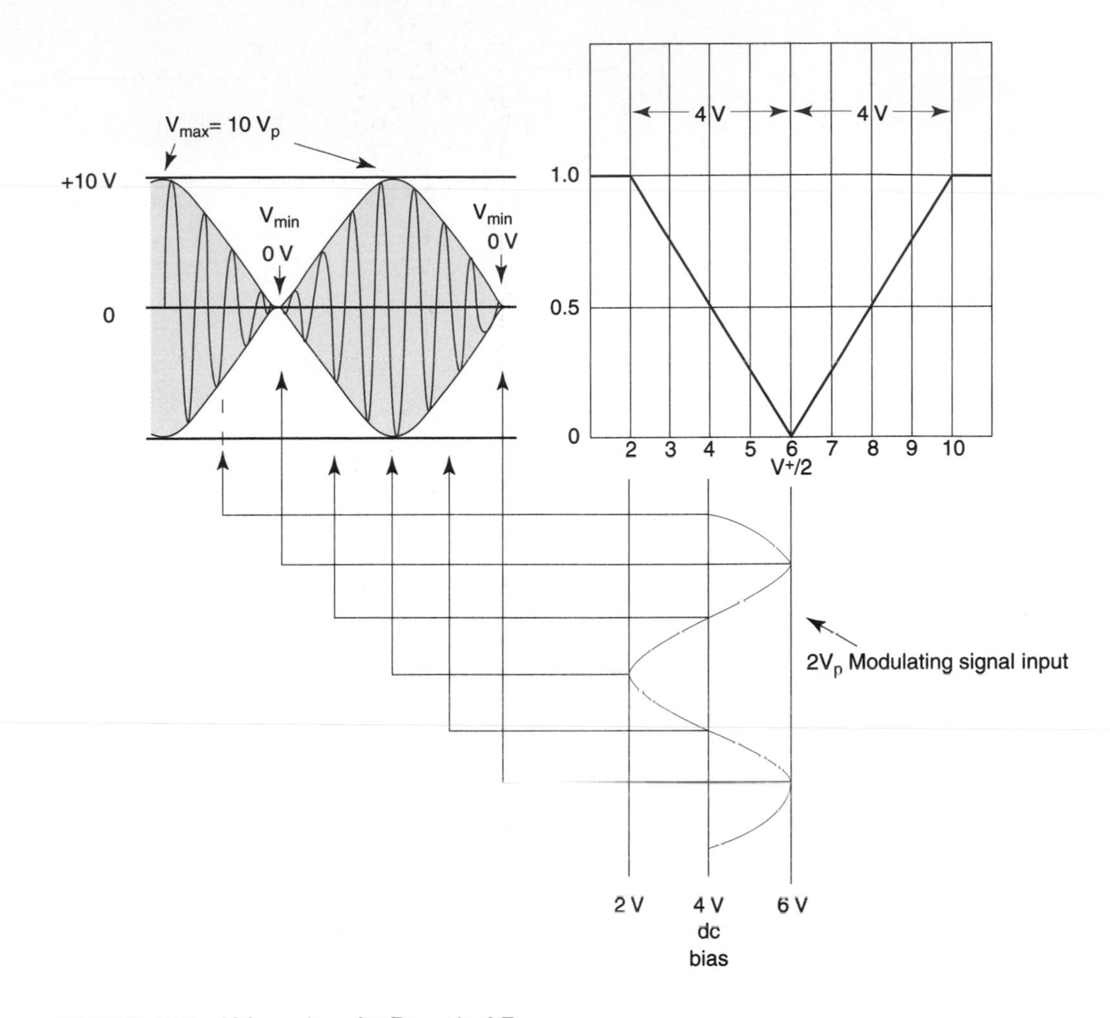

FIGURE 4-22 AM envelope for Example 4-7

4-5 AM TRANSMITTERS

4-5-1 Low-Level Transmitters

Figure 4-23 shows a block diagram for a low-level AM DSBFC transmitter. For voice or music transmission, the source of the modulating signal is generally an acoustical transducer, such as a microphone, a magnetic tape, a CD, or a phonograph record. The *preamplifier* is typically a sensitive, class A linear voltage amplifier with a high input impedance. The function of the preamplifier is to raise the amplitude of the source signal to a usable level while producing minimum nonlinear distortion and adding as little thermal noise as possible. The driver for the modulating signal is also a linear amplifier that simply amplifies the information signal to an adequate level to sufficiently drive the modulator. More than one drive amplifier may be required.

The RF *carrier oscillator* can be any of the oscillator configurations discussed in Chapter 3. The FCC has stringent requirements on transmitter accuracy and stability; therefore, crystal-controlled oscillators are the most common circuits used. The *buffer amplifier* is a low-gain, high-input impedance linear amplifier. Its function is to isolate the oscillator from the high-power amplifiers. The buffer provides a relatively constant load to the oscillator, which helps to reduce the occurrence and magnitude of short-term frequency variations. Emitter followers or integrated-circuit op-amps are often used for the buffer. The modulator

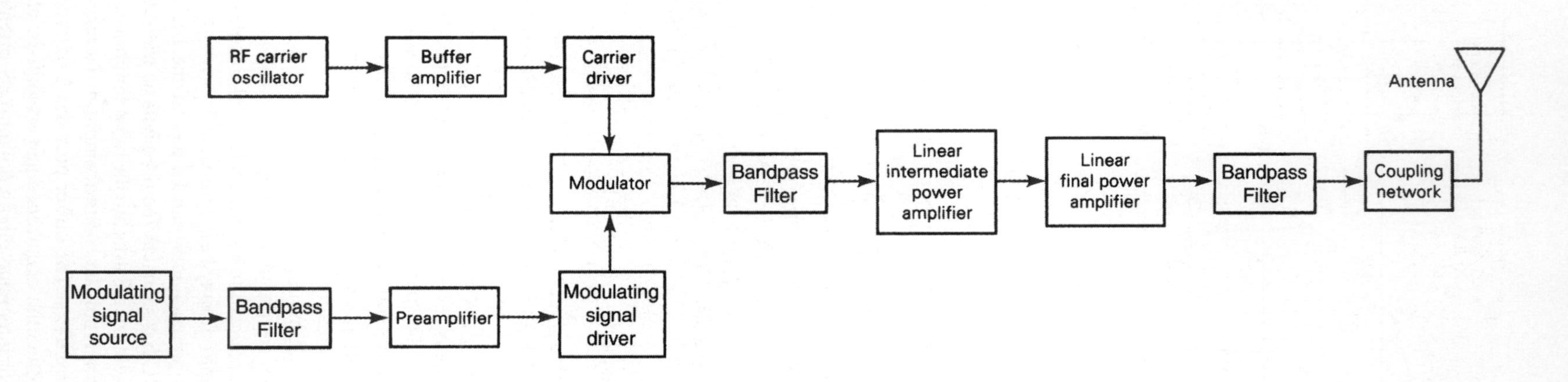

FIGURE 4-23 Block diagram of a low-level AM DSBFC transmitter

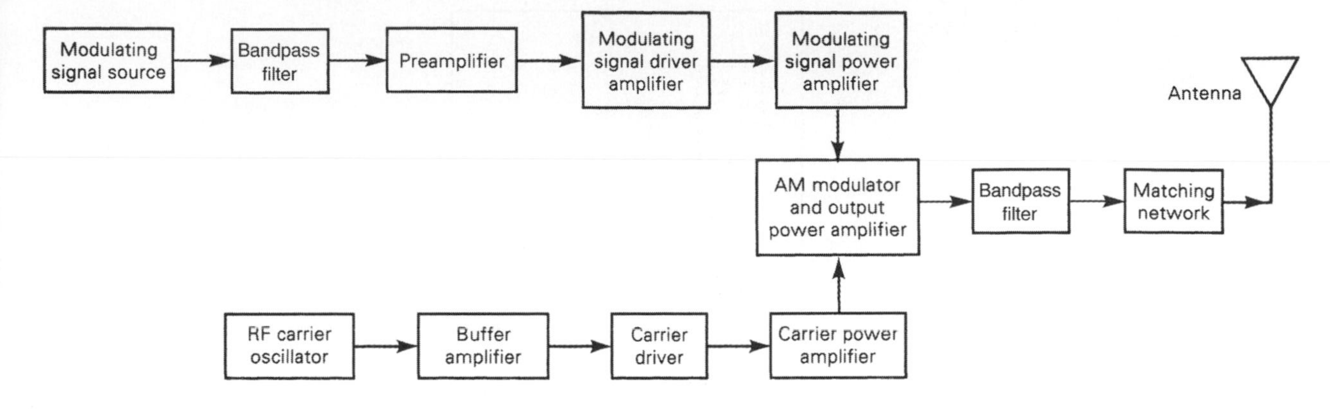

FIGURE 4-24 Block diagram of a high-level AM DSBFC transmitter

can use either emitter or collector modulation. The intermediate and final power amplifiers are either linear class A or class B push–pull modulators. This is required with low-level transmitters to maintain symmetry in the AM envelope. The antenna coupling network matches the output impedance of the final power amplifier to the transmission line and antenna.

Low-level transmitters such as the one shown in Figure 4-23 are used predominantly for low-power, low-capacity systems, such as wireless intercoms, remote-control units, pagers, and short-range walkie-talkies.

4-5-2 High-Level Transmitters

Figure 4-24 shows the block diagram for a high-level AM DSBFC transmitter. The modulating signal is processed in the same manner as in the low-level transmitter except for the addition of a power amplifier. With high-level transmitters, the power of the modulating signal must be considerably higher than is necessary with low-level transmitters. This is because the carrier is at full power at the point in the transmitter where modulation occurs and, consequently, requires a high-amplitude modulating signal to produce 100% modulation.

The RF carrier oscillator, its associated buffer, and the carrier driver are also essentially the same circuits used in low-level transmitters. However, with high-level transmitters, the RF carrier undergoes additional power amplification prior to the modulator stage, and the final power amplifier is also the modulator. Consequently, the modulator is generally a drain-, plate-, or collector-modulated class C amplifier.

With high-level transmitters, the modulator circuit has three primary functions. It provides the circuitry necessary for modulation to occur (i.e., nonlinearity), it is the final power amplifier (class C for efficiency), and it is a frequency up-converter. An up-converter simply translates the low-frequency intelligence signals to radio-frequency signals that can be efficiently radiated from an antenna and propagated through free space.

4-6 TRAPEZOIDAL PATTERNS

Trapezoidal patterns are used for observing the modulation characteristics of AM transmitters (i.e., coefficient of modulation and modulation symmetry). Although the modulation characteristics can be examined with an oscilloscope, a trapezoidal pattern is more easily and accurately interpreted. Figure 4-25 shows the basic test setup for producing a trapezoidal pattern on the CRT of a standard oscilloscope. The AM wave is applied to the vertical input of the oscilloscope, and the modulating signal is applied to the external horizontal input with the internal horizontal sweep disabled. Therefore, the horizontal sweep rate is determined by the modulating signal frequency, and the magnitude of the horizontal deflection is proportional to the amplitude of the modulating signal. The vertical deflection is totally dependent on the

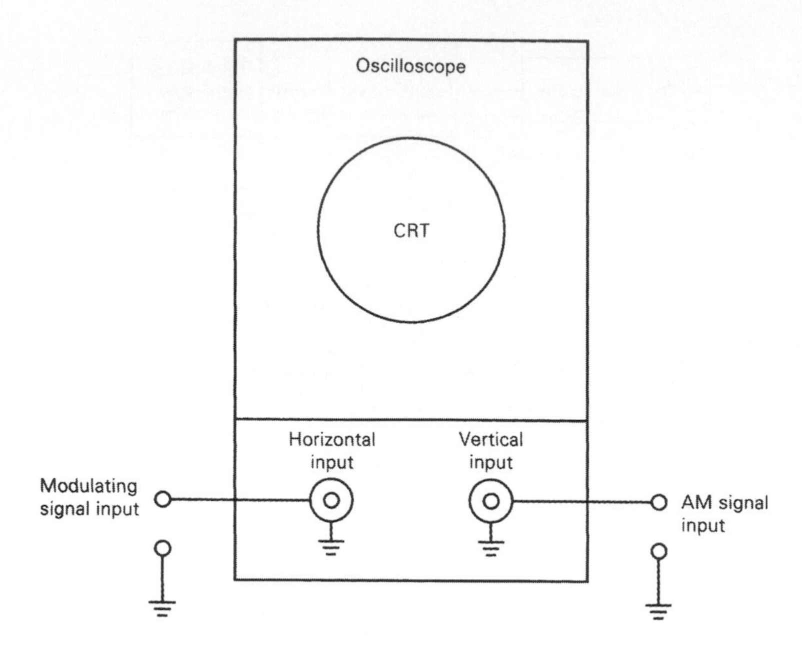

FIGURE 4-25 Test setup for displaying a trapezoidal pattern on an oscilloscope

amplitude and rate of change of the modulated signal. In essence, the electron beam emitted from the cathode of the CRT is acted on simultaneously in both the horizontal and vertical planes.

Figure 4-26 shows how the modulated signal and the modulating signal produce a trapezoidal pattern. With an oscilloscope, when 0 V is applied to the external horizontal input, the electron beam is centered horizontally on the CRT. When a voltage other than 0 V is applied to the vertical or horizontal inputs, the beam will deflect vertically and horizontally, respectively. If we begin with both the modulated wave and the modulating signal at 0 V (t_0), the electron beam is located in the center of the CRT. As the modulating signal goes positive, the beam deflects to the right. At the same time, the modulated signal is going positive, which deflects the beam upward. The beam continues to deflect to the right until the modulating signal reaches its maximum positive value (t_1). While the beam moves toward the right, it is also deflected up and down as the modulated signal alternately swings positive and negative. Notice that on each successive alternation, the modulated signal reaches a higher magnitude than the previous alternation. Therefore, as the CRT beam is deflected to the right, its peak-to-peak vertical deflection increases with each successive cycle of the modulated signal. As the modulating signal becomes less positive, the beam is deflected to the left (toward the center of the CRT). At the same time, the modulated signal alternately swings positive and negative, deflecting the beam up and down, except now each successive alternation is lower in amplitude than the previous alternation. Consequently, as the beam moves horizontally toward the center of the CRT, the vertical deflection decreases. The modulating signal and the modulated signal pass through 0 V at the same time, and the beam is again in the center of the CRT (t_2). As the modulating signal goes negative, the beam is deflected to the left side of the CRT. At the same time, the modulated signal is decreasing in amplitude on each successive alternation. The modulating signal reaches its maximum negative value at the same time as the modulated signal reaches its minimum amplitude (t_3). The trapezoidal pattern shown between times t_1 and t_3 folds back on top of the pattern displayed during times $t-$ and t_1. Thus, a complete trapezoidal pattern is displayed on the screen after both the left-to-right and right-to-left horizontal sweeps are complete.

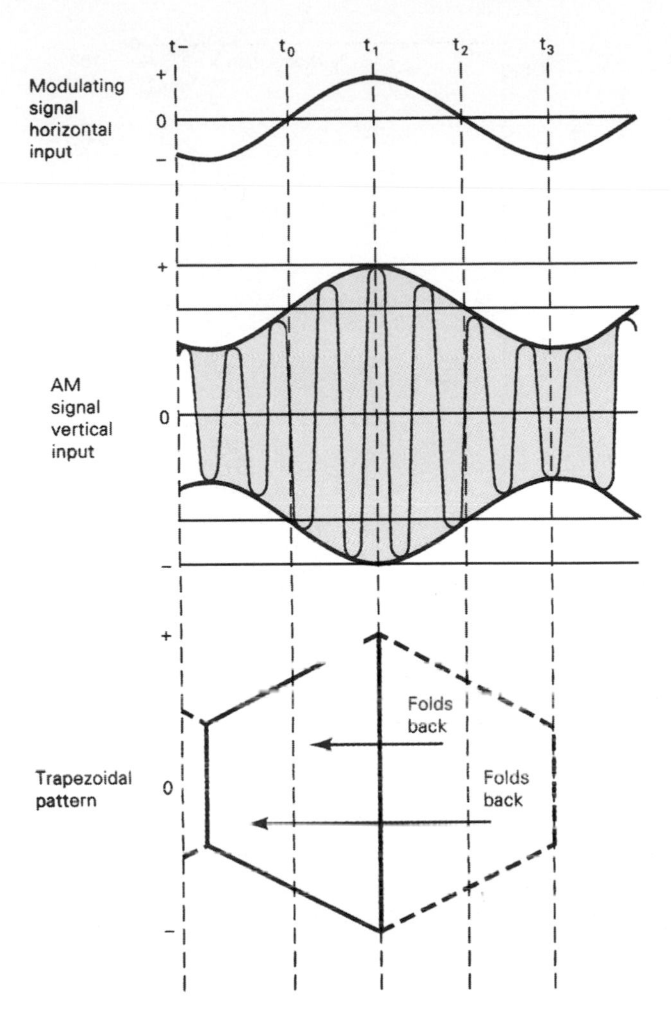

FIGURE 4-26 Producing a trapezoidal pattern

If the modulation is symmetrical, the top half of the modulated signal is a mirror image of the bottom half, and a trapezoidal pattern, such as the one shown in Figure 4-27a, is produced. At 100% modulation, the minimum amplitude of the modulated signal is zero, and the trapezoidal pattern comes to a point at one end as shown in Figure 4-27b. If the modulation exceeds 100%, the pattern shown in Figure 4-27c is produced. The pattern shown in Figure 4-27a is a 50% modulated wave. If the modulating signal and the modulated signal are out of phase, a pattern similar to the one shown in Figure 4-27d is produced. If the magnitude of the positive and negative alternations of the modulated signal are not equal, the pattern shown in Figure 4-27e results. If the phase of the modulating signal is shifted 180° (inverted), the trapezoidal patterns would simply point in the opposite direction. As you can see, percent modulation and modulation symmetry are more easily observed with a trapezoidal pattern than with a standard oscilloscope display of the modulated signal.

4-7 CARRIER SHIFT

Carrier shift is a term that is often misunderstood or misinterpreted. Carrier shift is sometimes called *upward* or *downward modulation* and has absolutely nothing to do with the frequency of the carrier. Carrier shift is a form of amplitude distortion introduced when the

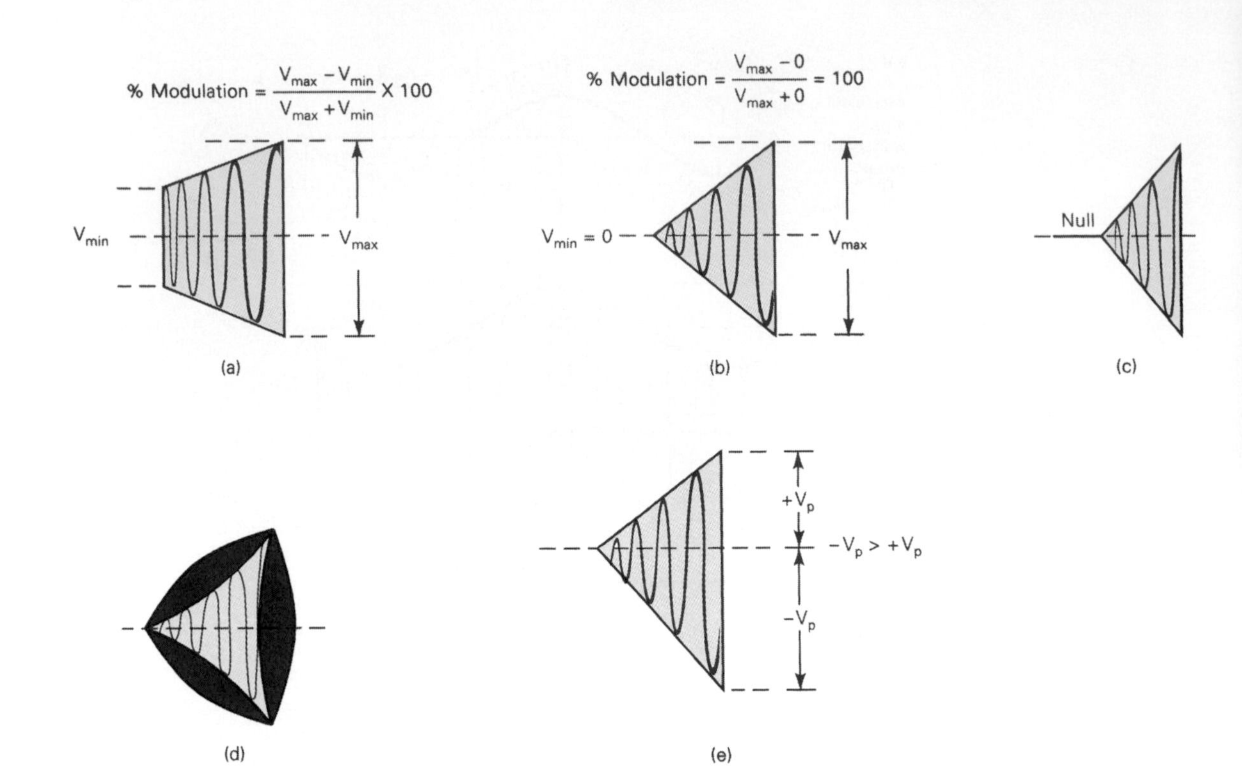

FIGURE 4-27 Trapezoidal patterns: (a) linear 50% AM modulation; (b) 100% AM modulation; (c) more than 100% AM modulation; (d) improper phase relationship; (e) nonsymmetrical AM envelope

positive and negative alternations in the AM modulated signal are not equal (i.e., nonsymmetrical modulation). Carrier shift may be either positive or negative. If the positive alternation of the modulated signal has a larger amplitude than the negative alternation, positive carrier shift results. If the negative alternation is larger than the positive, negative carrier shift occurs.

Carrier shift is an indication of the average voltage of an AM modulated signal. If the positive and negative halves of the modulated signal are equal, the average voltage is 0 V. If the positive half is larger, the average voltage is positive, and if the negative half is larger, the average voltage is negative. Figure 4-28a shows a symmetrical AM envelope (no carrier shift); the average voltage is 0 V. Figures 4-28b and c show positive and negative carrier shifts, respectively.

4-8 AM ENVELOPES PRODUCED BY COMPLEX NONSINUSOIDAL SIGNALS

Nonsinusoidal signals are complex waveforms comprised of two or more frequencies. Complex repetitive waveforms are complex waves made up of two or more harmonically related sine waves and include square, rectangular, and triangular waves. Complex modulating signals can also contain two or more unrelated frequencies, such as voice signals originating from different sources. When signals other than pure sine or cosine waves amplitude modulate a carrier, the modulated envelope contains upper and lower sideband frequencies commensurate with those contained in the modulating

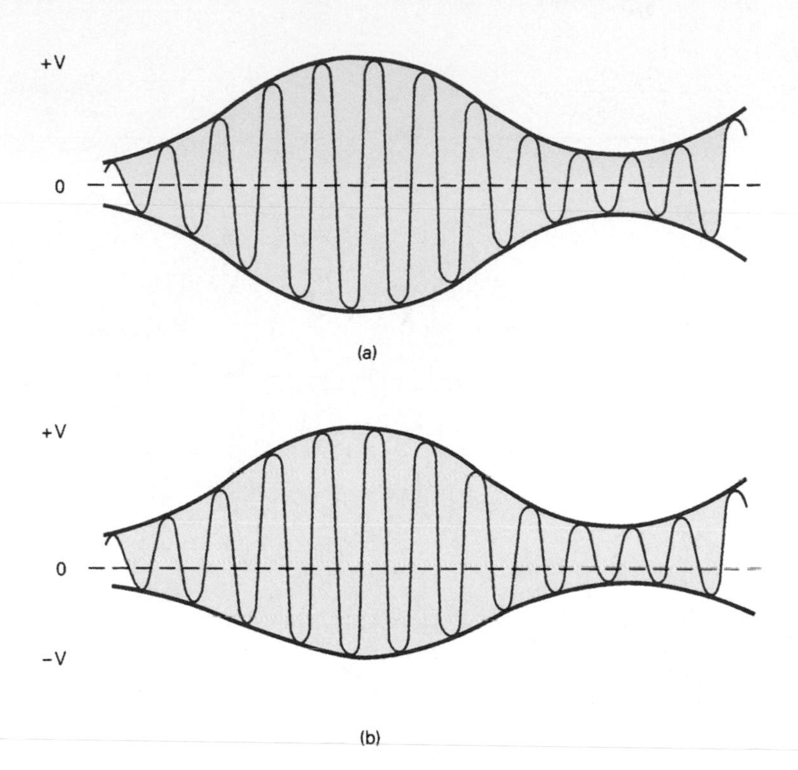

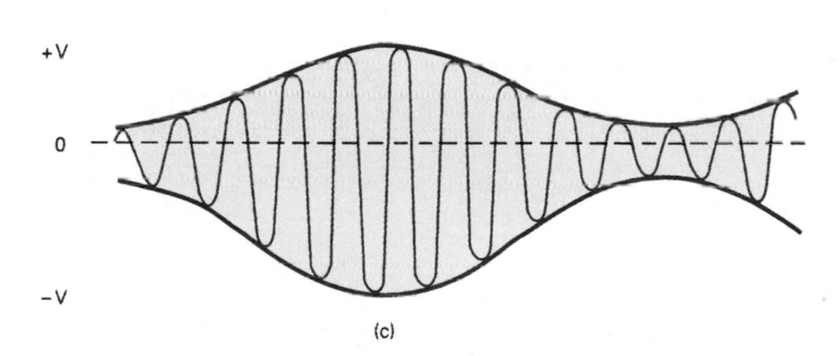

FIGURE 4-28 Carrier shift: (a) linear modulation; (b) positive carrier shift; (c) negative carrier shift

signal and thus, the shape of the envelope resembles the shape of the original modulating waveform.

Figure 4-29 shows three complex modulating signals and their respective AM envelopes.

4-9 QUADRATURE AMPLITUDE MODULATION

Quadrature amplitude modulation is a form of amplitude modulation where signals from two separate information sources (i.e., two channels) modulate the same carrier frequency at the same time without interfering with each other. The information sources modulate the same carrier after it has been separated into two carrier signals that are 90° out of phase with each other. This scheme is sometimes called *quadrature AM* (QUAM or QAM).

A simplified block diagram of a quadrature AM modulator is shown in Figure 4-30a. As the figure shows, there is a single carrier oscillator that produces an in-phase

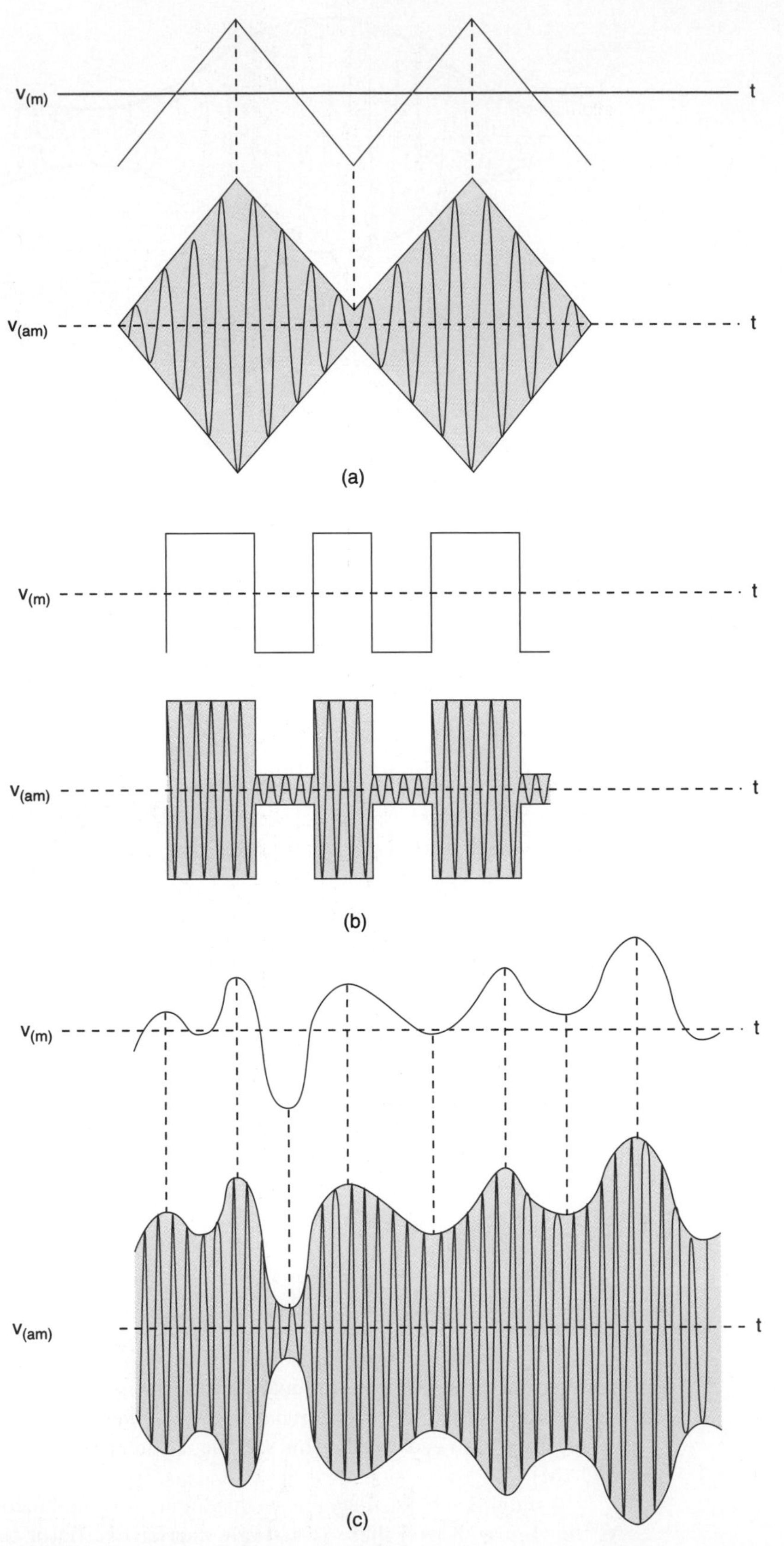

FIGURE 4-29 AM with complex modulating signal: (a) triangular wave modulation; (b) rectangular wave modulation; and (c) voice modulation

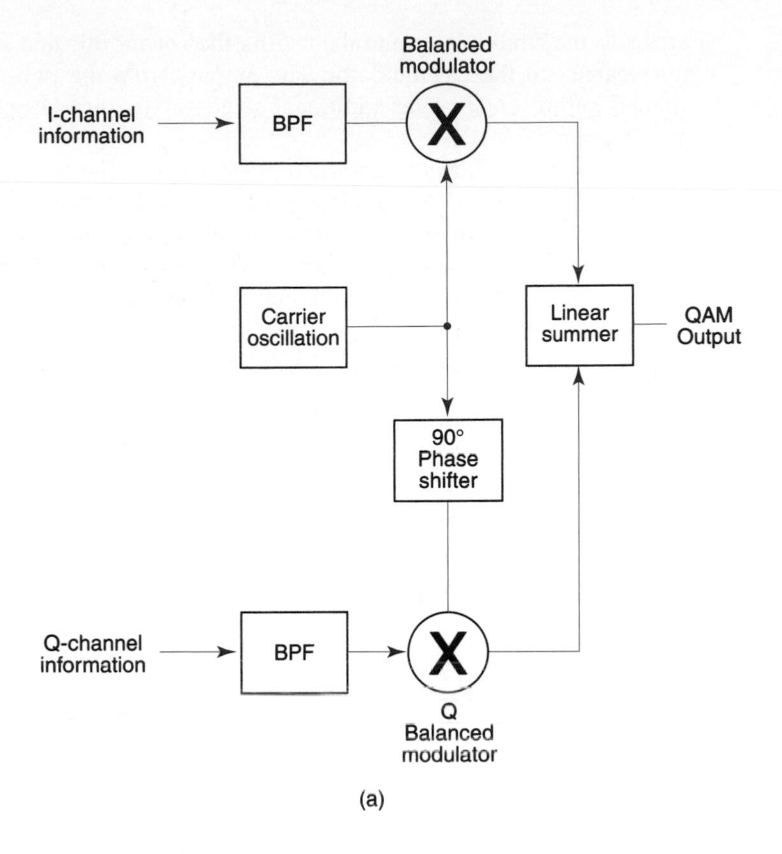

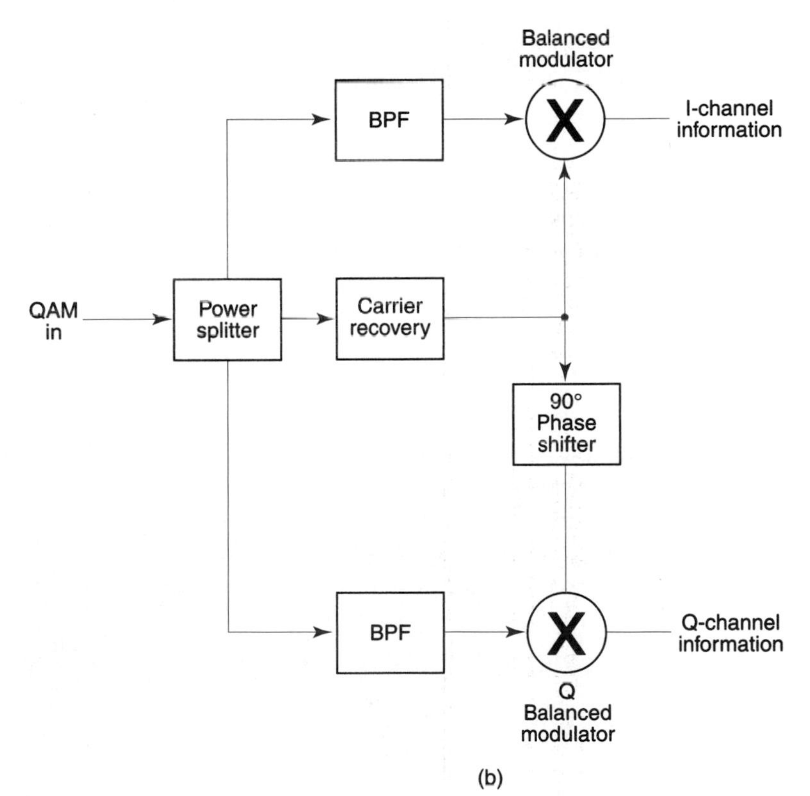

FIGURE 4-30 Quadrature AM: (a) modulator; (b) demodulator

carrier to the *I*-modulator and then shifts the carrier 90° and supplies a second quadrature carrier to the *Q*-modulator. The outputs from the two modulators are linearly summed before undergoing additional stages of frequency up-conversion and power amplification.

Figure 4-30b shows a simplified block diagram for a quadrature AM demodulator. As the figure shows, demodulating quadrature AM signals requires a carrier recovery circuit to reproduce the original carrier frequency and phase and two balanced modulators to actually demodulate the signals. This is called *synchronous detection,* and the process makes demodulating quadrature AM signals quite expensive when compared with conventional AM demodulator circuits. As you can see, quadrature AM is much more complex than conventional AM, costs more to implement, and produces approximately the same-quality demodulated signal. The primary advantage, however, of quadrature AM is conservation of bandwidth. Quadrature AM requires only half as much bandwidth as conventional AM, and two separate channels can modulate the same carrier. QAM is sometimes called *phase-division multiplexing* and was one of the modulation techniques considered for stereo broadcasting of AM signals. For now, quadrature AM is the modulation scheme used for encoding color signals in analog television broadcasting systems.

Today, quadrature AM is used almost exclusively for digital modulation of analog carriers in data modems to convey data through the public telephone network. Quadrature AM is also used for digital satellite communications systems. Digital quadrature modulation and synchronous detection are topics covered in more detail in later chapters of this book.

QUESTIONS

4-1. Define *amplitude modulation.*

4-2. Describe the basic operation of an *AM modulator.*

4-3. What is meant by the term *RF?*

4-4. How many inputs are there to an *amplitude modulator?* What are they?

4-5. In an AM communications system, what is meant by the terms *modulating signal, carrier, modulated wave,* and *AM envelope?*

4-6. What is meant by the *repetition rate* of the AM envelope?

4-7. Describe *upper* and *lower sidebands* and the *upper* and *lower side frequencies.*

4-8. What is the relationship between the *modulating signal frequency* and the *bandwidth* in a *conventional AM system?*

4-9. Define *modulation coefficient* and *percent modulation.*

4-10. What is the highest modulation coefficient and percent modulation possible with a conventional AM system without causing excessive distortion?

4-11. For 100% modulation, what is the relationship between the voltage amplitudes of the side frequencies and the carrier?

4-12. Describe the meaning of the following expression:

$$v_{am}(t) = E_c \sin(2\pi f_c t) - \frac{mE_c}{2}\cos[2\pi(f_c + f_m)t] + \frac{mE_c}{2}\cos[2\pi(f_c - f_m)t]$$

4-13. Describe the meaning of each term in the following expression:

$$v_{am}(t) = 10 \sin(2\pi\, 500kt) - 5 \cos(2\pi\, 515kt) + 5 \cos(2\pi 485kt)$$

4-14. What effect does modulation have on the amplitude of the carrier component of the modulated signal spectrum?

4-15. Describe the significance of the following formula:

$$P_t = P_c\left(1 + \frac{m^2}{2}\right)$$

4-16. What does *AM DSBFC* stand for?

4-17. Describe the relationship between the *carrier* and *sideband* powers in an AM DSBFC wave.

4-18. What is the predominant disadvantage of AM DSBFC?

4-19. What is the predominant advantage of AM DSBFC?

4-20. What is the primary disadvantage of low-level AM?

4-21. Why do any amplifiers that follow the modulator circuit in an AM DSBFC transmitter have to be linear?

4-22. Describe the differences between *low-* and *high-level modulators.*

4-23. List the advantages of low-level modulation and high-level modulation.

4-24. What are the advantages of using *linear-integrated circuit modulators* for AM?

4-25. What is the advantage of using a *trapezoidal pattern* to evaluate an AM envelope?

PROBLEMS

4-1. For an AM DSBFC modulator with a carrier frequency f_c = 100 kHz and a maximum modulating signal $f_{m(\max)}$ = 5 kHz, determine

a. Frequency limits for the upper and lower sidebands.
b. Bandwidth.
c. Upper and lower side frequencies produced when the modulating signal is a single-frequency 3-kHz tone.

Then

d. Sketch the output frequency spectrum.

4-2. What is the maximum modulating signal frequency that can be used with an AM DSBFC system with a 20-kHz bandwidth?

4-3. If a modulated wave with an average voltage of 20 V_p changes in amplitude ± 5 V, determine the minimum and maximum envelope amplitudes, the modulation coefficient, and the percent modulation.

4-4. Sketch the envelope for Problem 4-3 (label all pertinent voltages).

4-5. For a 30-V_p carrier amplitude, determine the maximum upper and lower side frequency amplitudes for an AM DSBFC envelope.

4-6. For a maximum positive envelope voltage of +12 V and a minimum positive envelope amplitude of +4 V, determine the modulation coefficient and percent modulation.

4-7. Sketch the envelope for Problem 4-6 (label all pertinent voltages).

4-8. For an AM DSBFC envelope with a $+V_{\max}$ = 40 V and $+V_{\min}$ = 10 V, determine

a. Unmodulated carrier amplitude.
b. Peak change in amplitude of the modulated wave.
c. Coefficient of modulation and percent modulation.

4-9. For an unmodulated carrier amplitude of 16 V_p and a modulation coefficient m = 0.4, determine the amplitudes of the modulated carrier and side frequencies.

4-10. Sketch the envelope for Problem 4-9 (label all pertinent voltages).

4-11. For the AM envelope shown below, determine

a. Peak amplitude of the upper and lower side frequencies.
b. Peak amplitude of the carrier.
c. Peak change in the amplitude of the envelope.
d. Modulation coefficient.
e. Percent modulation.

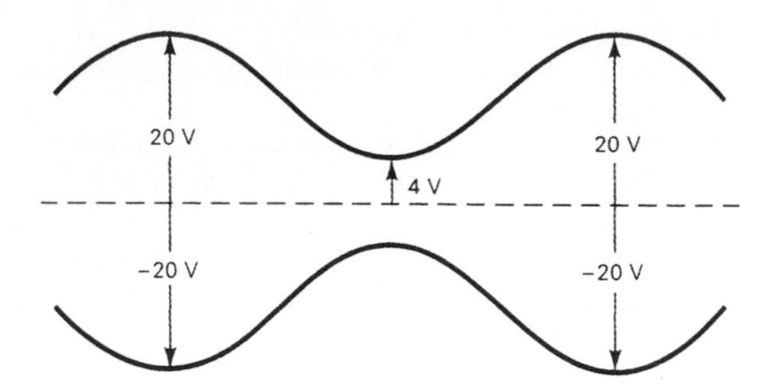

4-12. One input to an AM DSBFC modulator is an 800-kHz carrier with an amplitude of 40 V_p. The second input is a 25-kHz modulating signal whose amplitude is sufficient to produce a ±10-V change in the amplitude of the envelope. Determine

a. Upper and lower side frequencies.
b. Modulation coefficient and percent modulation.
c. Maximum and minimum positive peak amplitudes of the envelope.
Then
d. Draw the output frequency spectrum.
e. Draw the envelope (label all pertinent voltages).

4-13. For a modulation coefficient $m = 0.2$ and an unmodulated carrier power $P_c = 1000$ W, determine

a. Total sideband power.
b. Upper and lower sideband power.
c. Modulated carrier power.
d. Total transmitted power.

4-14. Determine the maximum upper, lower, and total sideband power for an unmodulated carrier power $P_c = 2000$ W.

4-15. Determine the maximum total transmitted power (P_t) for the AM system described in Problem 4-14.

4-16. For an AM DSBFC wave with an unmodulated carrier voltage of 25 V_p and a load resistance of 50 Ω, determine

a. Power in the unmodulated carrier.
b. Power of the modulated carrier, upper and lower sidebands, and total transmitted power for a modulation coefficient $m = 0.6$.

4-17. For a low-power transistor modulator with a modulation coefficient $m = 0.4$, a quiescent voltage gain $A_q = 80$, and an input carrier amplitude of 0.002 V, determine

a. Maximum and minimum voltage gains.
b. Maximum and minimum voltages for v_{out}.
Then
c. Sketch the modulated envelope.

4-18. For the trapezoidal pattern shown below, determine

a. Modulation coefficient.
b. Percent modulation.
c. Carrier amplitude.
d. Upper and lower side frequency amplitudes.

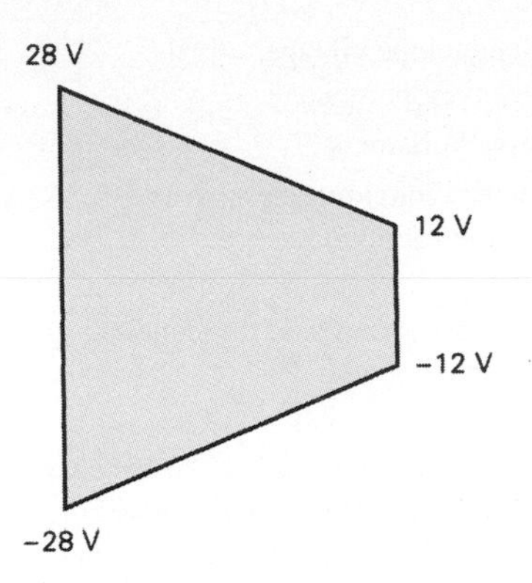

4-19. Sketch the approximate trapezoidal patterns for the following percent modulations and modulation conditions:

a. 100%
b. 50%
c. > 100%
d. Improper phase relationship.
e. Nonsymmetrical AM modulation.

4-20. For an AM modulator with a carrier frequency $f_c = 200$ kHz and a maximum modulating signal frequency $f_{m(max)} = 10$ kHz, determine

a. Frequency limits for the upper and lower sidebands.
b. Upper and lower side frequencies produced when the modulating signal is a single-frequency 7-kHz tone.
c. Bandwidth necessary to pass the maximum modulating signal frequency

Then

d. Draw the output spectrum.

4-21. For an unmodulated carrier voltage of 10 V_p and a ±4-V change in amplitude of the envelope, determine

a. Modulation coefficient.
b. Percent modulation.

4-22. For a maximum positive envelope voltage $V_{max} = +20$ V and a minimum positive envelope amplitude of +6 V, determine

a. Modulation coefficient.
b. Percent modulation.
c. Carrier amplitude.

4-23. For an envelope with $+V_{max} = +30\ V_p$ and $+V_{min} = +10\ V_p$, determine

a. Unmodulated carrier amplitude.
b. Modulated carrier amplitude.
c. Peak change in the amplitude of the envelope.
d. Modulation coefficient.
e. Percent modulation.

4-24. Write the expression for an AM voltage wave with the following values:

Unmodulated carrier = 20 V_p
Modulation coefficient = 0.4
Modulating signal frequency = 5 kHz
Carrier frequency = 200 kHz

4-25. For an unmodulated carrier amplitude of 12 V_p and a modulation coefficient of 0.5, determine the following:

a. Percent modulation.
b. Peak voltages of the carrier and side frequencies.

c. Maximum positive envelope voltage.
d. Minimum positive envelope voltage.

4-26. Sketch the envelope for Problem 4-25.

4-27. For an AM envelope with a maximum peak voltage of 52 V and a minimum peak-to-peak voltage of 24 V, determine the following:

a. Percent modulation.
b. Peak voltages of the carrier and side frequencies.
c. Maximum positive envelope voltage.
d. Minimum positive envelope voltage.

4-28. One input to an AM DSBFC modulator is a 500-kHz carrier with a peak amplitude of 32 V. The second input is a 12-kHz modulating signal whose amplitude is sufficient to produce a 14-V_p change in the amplitude of the envelope. Determine the following:

a. Upper and lower side frequencies.
b. Modulation coefficient and percent modulation.
c. Maximum and minimum amplitudes of the envelope.
Then
d. Draw the output envelope.
e. Draw the output frequency spectrum.

4-29. For a modulation coefficient of 0.4 and a carrier power of 400 W, determine

a. Total sideband power.
b. Total transmitted power.

4-30. For an AM DSBFC wave with an unmodulated carrier voltage of 18 V_p and a load resistance of 72 Ω, determine

a. Unmodulated carrier power.
b. Modulated carrier power.
c. Total sideband power.
d. Upper and lower sideband powers.
e. Total transmitted power.

4-31. For a low-power AM modulator with a modulation coefficient of 0.8, a quiescent gain of 90, and an input carrier amplitude of 10 mV_p, determine

a. Maximum and minimum voltage gains.
b. Maximum and minimum envelope voltages.
Then
c. Sketch the AM envelope.

CHAPTER 5

Amplitude Modulation Reception

CHAPTER OUTLINE

5-1 Introduction
5-2 Receiver Parameters
5-3 AM Receivers
5-4 AM Receiver Circuits
5-5 Double-Conversion AM Receivers
5-6 Net Receiver Gain

OBJECTIVES

- Define *AM demodulation*
- Define and describe the following receiver parameters: selectivity, bandwidth improvement, sensitivity, dynamic range, fidelity, insertion loss, and equivalent noise temperature
- Describe the operation of a tuned radio-frequency (TRF) receiver
- Explain the functions of the stages of a superheterodyne receiver
- Describe the operation of a superheterodyne receiver
- Describe local oscillator tracking
- Define and describe image frequency and image frequency rejection
- Describe the operation of the following AM receiver stages: RF amplifiers, mixer/converters, IF amplifiers, and AM detectors
- Describe the operation of single- and double-tuned transformer coupling circuits
- Describe the operation of a peak detector circuit
- Define *automatic gain control* and describe simple, delayed, and forward AGC
- Describe the purpose of squelch circuits
- Explain the operation of noise limiters and blankers
- Describe the operation of a double-conversion AM receiver
- Describe net receiver gain

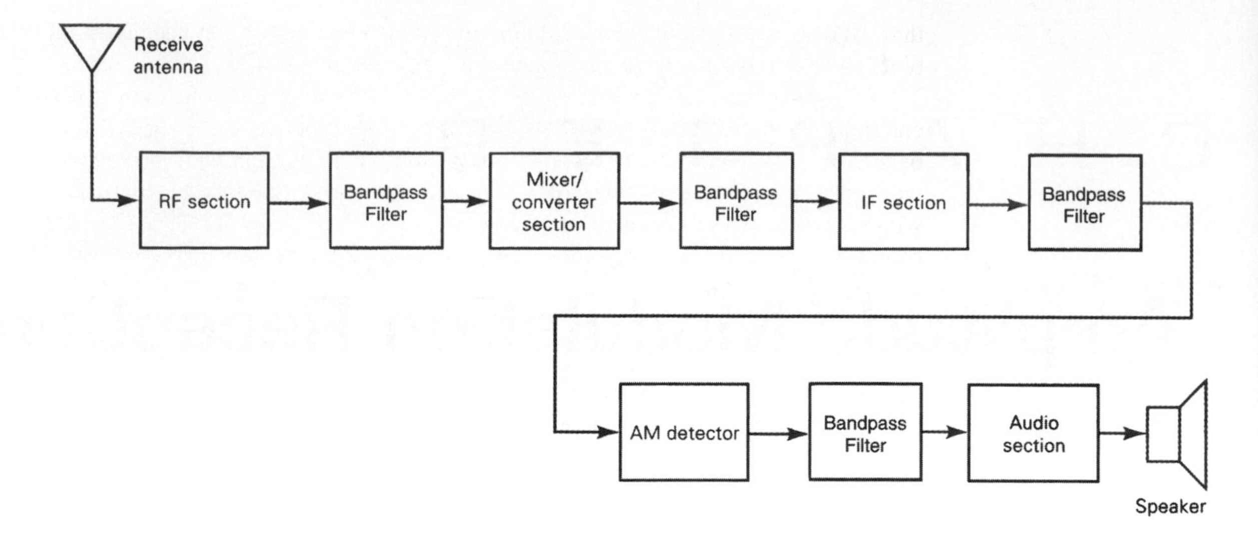

FIGURE 5-1 Simplified block diagram of an AM receiver

5-1 INTRODUCTION

AM demodulation is the reverse process of AM modulation. A conventional double-sideband AM receiver simply converts a received amplitude-modulated wave back to the original source information. To do this, a receiver must be capable of receiving, amplifying, and demodulating an AM wave. It must also be capable of bandlimiting the total radio-frequency spectrum to a specific desired band of frequencies. The selection process is called *tuning the receiver.*

To completely understand the demodulation process, first it is necessary to have a basic understanding of the terminology commonly used to describe radio receivers and their characteristics. Figure 5-1 shows a simplified block diagram of a typical AM receiver. The *RF section* is the first stage of the receiver and is therefore often called the *receiver front end.* The primary functions of the RF section are detecting, bandlimiting, and amplifying the received RF signals. The *mixer/converter section* is the next stage. This section down-converts the received RF frequencies to *intermediate frequencies* (IFs), which are simply frequencies that fall somewhere between the RF and information frequencies, hence the name *intermediate.* The primary functions of the *IF section* are amplification and selectivity. The *AM detector* demodulates the AM wave and converts it to the original information signal, and the *audio section* simply amplifies the recovered information.

5-2 RECEIVER PARAMETERS

There are several parameters commonly used to evaluate the ability of a receiver to successfully demodulate a radio signal. The most important parameters are selectivity and sensitivity, which are often used to compare the quality of one radio receiver to another.

5-2-1 Selectivity

Selectivity is a receiver parameter that is used to measure the ability of the receiver to accept a given band of frequencies and reject all others. For example, with the commercial AM broadcast band, each station's transmitter is allocated a 10-kHz bandwidth. Therefore, for a receiver to select only those frequencies assigned a single channel, the receiver must limit its bandwidth to 10 kHz. If the passband is greater than 10 kHz, more than one channel may be received and demodulated simultaneously. If the passband of a receiver is less than 10 kHz, a portion of the modulating signal information for that channel is rejected or blocked from entering the demodulator and, consequently, lost.

There are several acceptable ways to describe the selectivity of a radio receiver. One common way is to simply give the bandwidth of the receiver at the −3-dB points. This bandwidth, however, is not necessarily a good means of determining how well the receiver will reject unwanted frequencies. Consequently, it is common to give the receiver bandwidth at two levels of attenuation, for example, −3 dB and −60 dB. The ratio of these two bandwidths is called the *shape factor* and is expressed mathematically as

$$SF = \frac{B_{(-60\text{ dB})}}{B_{(-3\text{ dB})}} \tag{5-1}$$

where SF = shape factor (unitless)
$B_{(-60\text{ dB})}$ = bandwidth 60 dB below maximum signal level
$B_{(-3\text{ dB})}$ = bandwidth 3 dB below maximum signal level

Ideally, the bandwidth at the −3-dB and −60-dB points would be equal, and the shape factor would be 1. This value, of course, is impossible to achieve in a practical circuit. A typical AM broadcast-band radio receiver might have a −3-dB bandwidth of 10 kHz and a −60-dB bandwidth of 20 kHz, giving a shape factor of 2. More expensive and sophisticated satellite, microwave, and two-way radio receivers have shape factors closer to the ideal value of 1.

In today's overcrowded radio-frequency spectrum, the FCC makes adjacent channel assignments as close together as possible, with only 10 kHz separating commercial broadcast-band AM channels. Spacing for adjacent commercial broadcast-band FM channels is 200 kHz, and commercial television channels are separated by 6 MHz. A radio receiver must be capable of separating the desired channel's signals without allowing interference from an adjacent channel to spill over into the desired channel's passband.

5-2-2 Bandwidth Improvement

As stated in Chapter 1 and given in Equation 1-13, thermal noise is the most prevalent form of noise and is directly proportional to bandwidth. Therefore, if the bandwidth can be reduced, the noise will also be reduced by the same proportion, thus increasing the signal-to-noise power ratio, improving system performance. There is, of course, a system performance limitation as to how much the bandwidth can be reduced. The bottom line is that the circuit bandwidth must exceed the bandwidth of the information signal; otherwise, the information power and/or the frequency content of the information signal will be reduced, effectively degrading system performance. When a signal propagates from the antenna through the RF section, mixer/converter section, and IF section, the bandwidth is reduced, thus reducing the noise. The theoretical problem is how much the bandwidth should be reduced, and the practical problem is in the difficulty of constructing stable, narrow-band filters.

The input signal-to-noise ratio is calculated at a receiver input using the RF bandwidth for the noise power measurement. However, the RF bandwidth is generally wider than the bandwidth of the rest of the receiver (i.e., the IF bandwidth is narrower than the RF bandwidth for reasons that will be explained in subsequent sections of this chapter). Reducing the bandwidth is effectively equivalent to reducing (improving) the noise figure of the receiver. The noise reduction ratio achieved by reducing the bandwidth is called *bandwidth improvement* (BI) and is expressed mathematically as

$$\text{BI} = \frac{B_{\text{RF}}}{B_{\text{IF}}} \tag{5-2}$$

where BI = bandwidth improvement (unitless)
B_{RF} = RF bandwidth (hertz)
B_{IF} = IF bandwidth (hertz)

The corresponding reduction in the noise figure due to the reduction in bandwidth is called *noise figure improvement* and is expressed mathematically in dBas:

$$NF_{improvement} = 10 \log BI \tag{5-3}$$

Example 5-1

Determine the improvement in the noise figure for a receiver with an RF bandwidth equal to 200 kHz and an IF bandwidth equal to 10 kHz.

Solution Bandwidth improvement is found by substituting into Equation 5-2:

$$BI = \frac{200 \text{ kHz}}{10 \text{ kHz}} = 20$$

and noise figure improvement is found by substituting into Equation 5-3:

$$NF_{improvement} = 10 \log 20 = 13 \text{ dB}$$

5-2-3 Sensitivity

The *sensitivity* of a receiver is the minimum RF signal level that can be detected at the input to the receiver and still produce a usable demodulated information signal. What constitutes a usable information signal is somewhat arbitrary. Generally, the signal-to-noise ratio and the power of the signal at the output of the audio section are used to determine the quality of a received signal and whether it is usable. For commercial AM broadcast-band receivers, a 10-dB-or-more signal-to-noise ratio with 1/2 W (27 dBm) of power at the output of the audio section is considered to be usable. However, for broadband microwave receivers, a 40-dB-or-more signal-to-noise ratio with approximately 5 mW (7 dBm) of signal power is the minimum acceptable value. The sensitivity of a receiver is usually stated in microvolts of received signal. For example, a typical sensitivity for a commercial broadcast-band AM receiver is 50 μV, and a two-way mobile radio receiver generally has a sensitivity between 0.1 μV and 10 μV. Receiver sensitivity is also called receiver *threshold.* The sensitivity of an AM receiver depends on the noise power present at the input to the receiver, the receiver's noise figure (an indication of the noise generated in the front end of the receiver), the sensitivity of the AM detector, and the bandwidth improvement factor of the receiver. The best way to improve the sensitivity of a receiver is to reduce the noise level. This can be accomplished by reducing either the temperature or the bandwidth of the receiver or improving the receiver's noise figure.

5-2-4 Dynamic Range

The *dynamic range* of a receiver is defined as the difference in decibels between the minimum input level necessary to discern a signal and the input level that will overdrive the receiver and produce distortion. In simple terms, dynamic range is the input power range over which the receiver is useful. The minimum receive level is a function of front-end noise, noise figure, and the desired signal quality. The input signal level that will produce overload distortion is a function of the net gain of the receiver (the total gain of all the stages in the receiver). The high-power limit of a receiver depends on whether it will operate with a single- or a multiple-frequency input signal. If single-frequency operation is used, the *1-dB compression point* is generally used for the upper limit of usefulness. The 1-dB compression point is defined as the output power when the RF amplifier response is 1 dB less than the ideal linear-gain response. Figure 5-2 shows the linear gain and 1-dB compression point for a typical amplifier where the linear gain drops off just prior to saturation. The 1-dB compression point is often measured directly as the point where a 10-dB increase in input power results in a 9-dB increase in output power.

A dynamic range of 100 dB is considered about the highest possible. A low dynamic range can cause a desensitizing of the RF amplifiers and result in severe intermodulation distortion of the weaker input signals. Sensitivity measurements are discussed later in this chapter.

5-2-5 Fidelity

Fidelity is a measure of the ability of a communications system to produce, at the output of the receiver, an exact replica of the original source information. Any frequency, phase, or

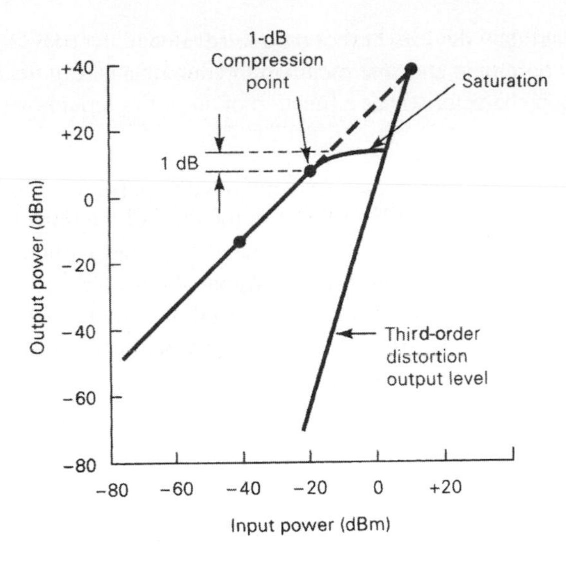

FIGURE 5-2 Linear gain, 1-dB compression point, and third-order intercept distortion for a typical amplifier

amplitude variations that are present in the demodulated waveform that were not in the original information signal are considered distortion.

Essentially, there are three forms of distortion that can deteriorate the fidelity of a communications system: *amplitude, frequency,* and *phase.* Phase distortion is not particularly important for voice transmission because the human ear is relatively insensitive to phase variations. However, phase distortion can be devastating to data transmission. The predominant cause of phase distortion is filtering (both wanted and unwanted). Frequencies at or near the break frequency of a filter undergo varying values of phase shift. Consequently, the cutoff frequency of a filter is often set beyond the minimum value necessary to pass the highest frequency information signals (typically the upper cutoff frequency of a low-pass filter is approximately 1.3 times the minimum value). *Absolute phase shift* is the total phase shift encountered by a signal and can generally be tolerated as long as all frequencies undergo the same amount of phase delay. *Differential phase shift* occurs when different frequencies undergo different phase shifts and may have a detrimental effect on a complex waveform, especially if the information is encoded into the phase of the carrier as it is with phase shift keying modulation. If phase shift versus frequency is linear, delay is constant with frequency. If all frequencies are not delayed by the same amount of time, the frequency-versus-phase relationship of the received waveform is not consistent with the original source information, and the recovered information is distorted.

Amplitude distortion occurs when the amplitude-versus-frequency characteristics of a signal at the output of a receiver differ from those of the original information signal. Amplitude distortion is the result of *nonuniform gain* in amplifiers and filters.

Frequency distortion occurs when frequencies are present in a received signal that were not present in the original source information. Frequency distortion is a result of harmonic and intermodulation distortion and is caused by nonlinear amplification. *Second-order products* ($2F_1 \pm F_2$, $F_1 \pm 2F_2$, and so on) are usually only a problem in broadband systems because they generally fall outside the bandwidth of a narrowband system. However, *third-order products* often fall within the system bandwidth and produce a distortion called *third-order intercept distortion.* Third-order intercept distortion is a special case of intermodulation distortion and the predominant form of frequency distortion. Third-order intermodulation components are the cross-product frequencies produced when the second harmonic of one signal is added to the fundamental frequency of another signal (i.e., $2f_1 \pm f_2, f_1 \pm 2f_2$, and so on). Frequency distortion can be reduced by using a *square-law device,* such as a FET, in the front end of a receiver. Square-law devices have a unique advantage over BJTs in that they produce only second-order harmonic and intermodulation components. Figure 5-2 shows a typical third-order distortion characteristic as a function of amplifier input power and gain.

5-2-6 Insertion Loss

Insertion loss (IL) is a parameter associated with the frequencies that fall within the passband of a filter and is generally defined as the ratio of the power transferred to a load with a filter in the circuit to the power transferred to a load without the filter. Because filters are generally constructed from lossy components, such as resistors and imperfect capacitors, even signals that fall within the passband of a filter are attenuated (reduced in magnitude). Typical filter insertion losses are between a few tenths of a decibel to several decibels. In essence, insertion loss is simply the ratio of the output power of a filter to the input power for frequencies that fall within the filter's passband and is stated mathematically in decibels as

$$\mathrm{IL}_{(\mathrm{dB})} = 10 \log \frac{P_{\mathrm{out}}}{P_{\mathrm{in}}} \tag{5-4}$$

5-2-7 Noise Temperature and Equivalent Noise Temperature

Because thermal noise is directly proportional to temperature, it stands to reason that noise can be expressed in degrees as well as watts or volts. Rearranging Equation 1-13 yields

$$T = \frac{N}{KB} \tag{5-5}$$

where T = environmental temperature (kelvin)
N = noise power (watts)
K = Boltzmann's constant (1.38×10^{-23} J/K)
B = bandwidth (hertz)

Equivalent noise temperature (T_e) is a hypothetical value that cannot be directly measured. T_e is a parameter that is often used in low-noise, sophisticated radio receivers rather than noise figure. T_e is an indication of the reduction in the signal-to-noise ratio as a signal propagates through a receiver. The lower the equivalent noise temperature, the better the quality of the receiver. Typical values for T_e range from 20° for *cool* receivers to 1000° for *noisy* receivers. Mathematically, T_e at the input to a receiver is expressed as

$$T_e = T(\mathrm{F} - 1) \tag{5-6}$$

where T_e = equivalent noise temperature (kelvin)
T = environmental temperature (kelvin)
F = noise factor (unitless)

Table 5-1 lists several values of noise figure, noise factor, and equivalent noise temperature for an environmental temperature of 17°C (290°K).

Table 5-1 (T = 17°C)

NF (dB)	F (unitless)	T_e (°K)
0.8	1.2	58
1.17	1.31	90
1.5	1.41	119
2.0	1.58	168

5-3 AM RECEIVERS

There are two basic types of radio receivers: *coherent* and *noncoherent.* With a coherent, or *synchronous,* receiver, the frequencies generated in the receiver and used for demodulation are synchronized to oscillator frequencies generated in the transmitter (the receiver must have some means of recovering the received carrier and synchronizing to it). With noncoherent, or *asynchronous,* receivers, either no frequencies are generated in the receiver or the frequencies used for demodulation are completely independent from the transmitter's carrier frequency. *Noncoherent detection* is often called *envelope detection* because the information is recovered from the received waveform by detecting the shape of the modulated envelope. The receivers described in this chapter are noncoherent. Coherent receivers are described in Chapter 6.

5-3-1 Tuned Radio-Frequency Receiver

The *tuned radio-frequency* (TRF) *receiver* was one of the earliest types of AM receivers. TRF receivers are probably the simplest designed radio receiver available today; however, they have several shortcomings that limit their use to special applications. Figure 5-3 shows the block diagram of a three-stage TRF receiver that includes an RF stage, a detector stage, and an audio stage. Generally, two or three RF amplifiers are required to filter and amplify the received signal to a level sufficient to drive the detector stage. The detector converts RF signals directly to information, and the audio stage amplifies the information signals to a usable level.

Although TRF receivers are simple and have a relatively high sensitivity, they have three distinct disadvantages that limit their usefulness to single-channel, low-frequency applications. The primary disadvantage is their bandwidth is inconsistent and varies with center frequency when tuned over a wide range of input frequencies. This is caused by a phenomenon called the *skin effect.* At radio frequencies, current flow is limited to the outermost area of a conductor; thus, the higher the frequency, the smaller the effective area and the greater the resistance. Consequently, the *quality factor* ($Q = X_L/R$) of the tank circuits remains relatively constant over a wide range of frequencies, causing the bandwidth (f/Q) to increase with frequency. As a result, the selectivity of the input filter changes over any appreciable range of input frequencies. If the bandwidth is set to the desired value for low-frequency RF signals, it will be excessive for high-frequency signals.

The second disadvantage of TRF receivers is instability due to the large number of RF amplifiers all tuned to the same center frequency. High-frequency, multistage amplifiers are susceptible to breaking into oscillations. This problem can be reduced somewhat by tuning each amplifier to a slightly different frequency, slightly above or below the desired center frequency. This technique is called *stagger tuning.* The third disadvantage of TRF receivers is their gains are not uniform over a very wide frequency range because of the nonuniform L/C ratios of the transformer-coupled tank circuits in the RF amplifiers.

With the development of the *superheterodyne receiver,* TRF receivers are seldom used except for special-purpose, single-station receivers and therefore do not warrant further discussion.

Example 5-2

For an AM commercial broadcast-band receiver (535 kHz to 1605 kHz) with an input filter Q-factor of 54, determine the bandwidth at the low and high ends of the RF spectrum.

Solution The bandwidth at the low-frequency end of the AM spectrum is centered around a carrier frequency of 540 kHz and is

$$B = \frac{f}{Q} = \frac{540\text{ kHz}}{54} = 10\text{ kHz}$$

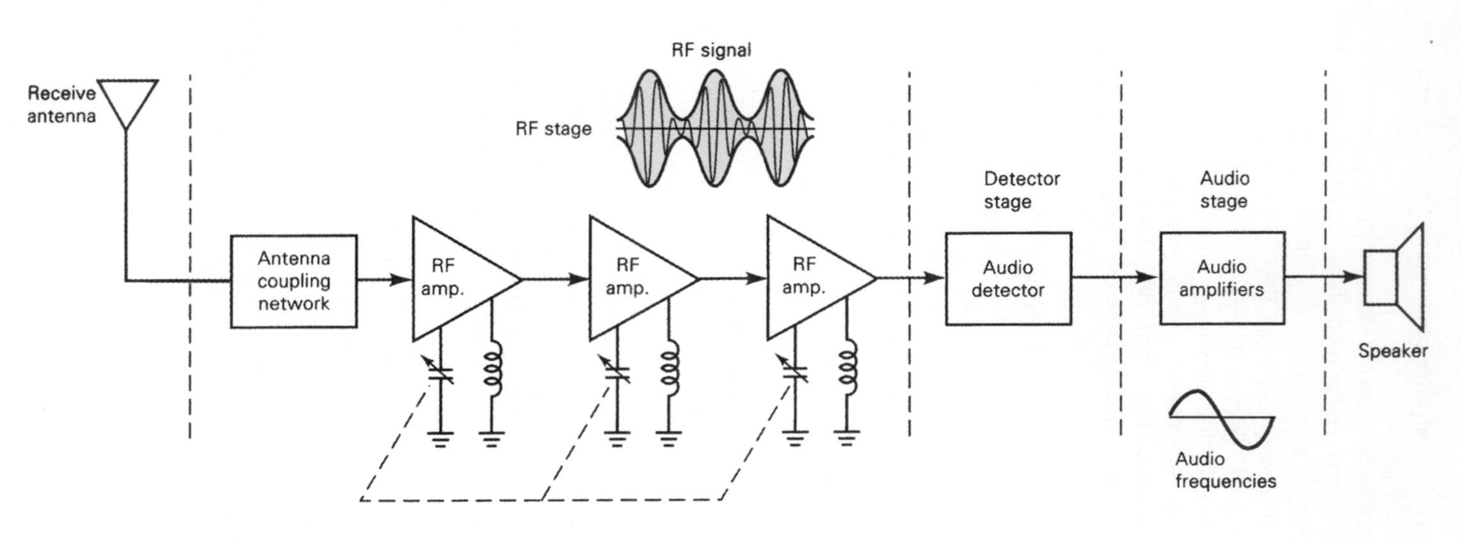

FIGURE 5-3 Noncoherent tuned radio frequency receiver block diagram

The bandwidth at the high-frequency end of the AM spectrum is centered around a carrier frequency of 1600 kHz and is

$$B = \frac{1600 \text{ kHz}}{54} = 29{,}630 \text{ Hz}$$

The −3-dB bandwidth at the low-frequency end of the AM spectrum is exactly 10 kHz, which is the desired value. However, the bandwidth at the high-frequency end is almost 30 kHz, which is three times the desired range. Consequently, when tuning for stations at the high end of the spectrum, three stations would be received simultaneously.

To achieve a bandwidth of 10 kHz at the high-frequency end of the spectrum, a Q of 160 is required (1600 kHz/10 kHz). With a Q of 160, the bandwidth at the low-frequency end is

$$B = \frac{540 \text{ kHz}}{160} = 3375 \text{ Hz}$$

which is obviously too selective because it would block approximately two-thirds of the information bandwidth.

5-3-2 Superheterodyne Receiver

The nonuniform selectivity of the TRF led to the development of the *superheterodyne receiver* near the end of World War I. Although the quality of the superheterodyne receiver has improved greatly since its original design, its basic configuration has not changed much, and it is still used today for a wide variety of radio communications services. The superheterodyne receiver has remained in use because its gain, selectivity, and sensitivity characteristics are superior to those of other receiver configurations.

Heterodyne means to mix two frequencies together in a nonlinear device or to translate one frequency to another using nonlinear mixing. A block diagram of a noncoherent superheterodyne receiver is shown in Figure 5-4. Essentially, there are five sections to a

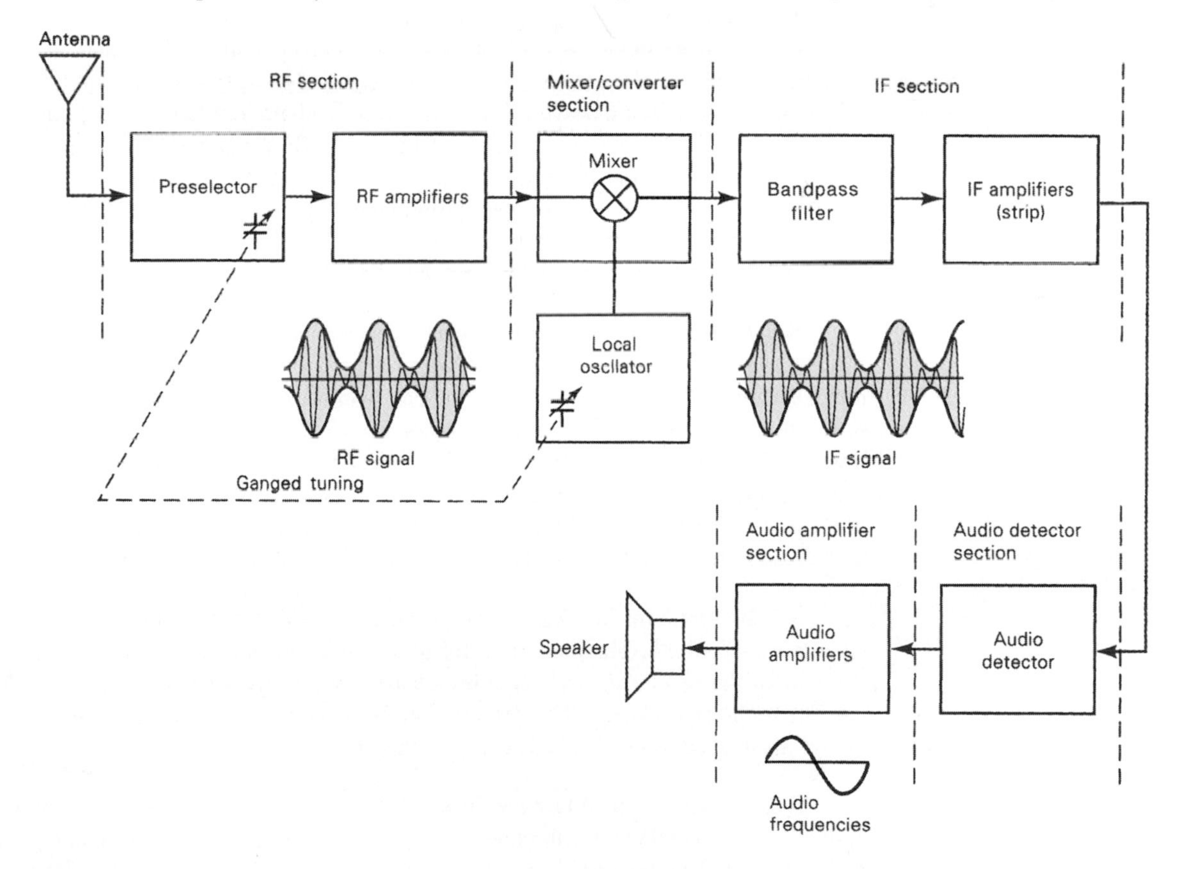

FIGURE 5-4 AM superheterodyne receiver block diagram

superheterodyne receiver: the RF section, the mixer/converter section, the IF section, the audio detector section, and the audio amplifier section.

5-3-2-1 RF section. The RF section generally consists of a preselector and an amplifier stage. They can be separate circuits or a single combined circuit. The preselector is a broad-tuned bandpass filter with an adjustable center frequency that is tuned to the desired carrier frequency. The primary purpose of the preselector is to provide enough initial bandlimiting to prevent a specific unwanted radio frequency, called the *image frequency,* from entering the receiver (image frequency is explained later in this section). The preselector also reduces the noise bandwidth of the receiver and provides the initial step toward reducing the overall receiver bandwidth to the minimum bandwidth required to pass the information signals. The RF amplifier determines the sensitivity of the receiver (i.e., sets the signal threshold). Also, because the RF amplifier is the first active device encountered by a received signal, it is the primary contributor of noise and, therefore, a predominant factor in determining the noise figure for the receiver. A receiver can have one or more RF amplifiers, or it may not have any, depending on the desired sensitivity. Several advantages of including RF amplifiers in a receiver are as follows:

1. Greater gain, thus better sensitivity
2. Improved image-frequency rejection
3. Better signal-to-noise ratio
4. Better selectivity

5-3-2-2 Mixer/converter section. The mixer/converter section includes a radio-frequency oscillator stage (commonly called a *local oscillator*) and a mixer/converter stage (commonly called the *first detector*). The local oscillator can be any of the oscillator circuits discussed in Chapter 2, depending on the stability and accuracy desired. The mixer stage is a nonlinear device and its purpose is to convert radio frequencies to intermediate frequencies (RF-to-IF frequency translation). Heterodyning takes place in the mixer stage, and radio frequencies are down-converted to intermediate frequencies. Although the carrier and sideband frequencies are translated from RF to IF, the shape of the envelope remains the same and, therefore, the original information contained in the envelope remains unchanged. It is important to note that although the carrier and upper and lower side frequencies change frequency, the bandwidth is unchanged by the heterodyning process. The most common intermediate frequency used in AM broadcast-band receivers is 455 kHz.

5-3-2-3 IF section. The IF section consists of a series of IF amplifiers and bandpass filters and is often called the *IF strip.* Most of the receiver gain and selectivity is achieved in the IF section. The IF center frequency and bandwidth are constant for all stations and are chosen so that their frequency is less than any of the RF signals to be received. The IF is always lower in frequency than the RF because it is easier and less expensive to construct high-gain, stable amplifiers for the low-frequency signals. Also, low-frequency IF amplifiers are less likely to oscillate than their RF counterparts. Therefore, it is not uncommon to see a receiver with five or six IF amplifiers and a single RF amplifier or possibly no RF amplification.

5-3-2-4 Detector section. The purpose of the detector section is to convert the IF signals back to the original source information. The detector is generally called an *audio detector* or the *second detector* in a broadcast-band receiver because the information signals are audio frequencies. The detector can be as simple as a single diode or as complex as a phase-locked loop or balanced demodulator.

5-3-2-5 Audio amplifier section. The audio section comprises several cascaded audio amplifiers and one or more speakers. The number of amplifiers used depends on the audio signal power desired.

5-3-3 Receiver Operation

During the demodulation process in a superheterodyne receiver, the received signals undergo two or more frequency translations: First, the RF is converted to IF, then the IF is converted to the source information. The terms RF and IF are system dependent and are often misleading because they do not necessarily indicate a specific range of frequencies. For example, RF for the commercial AM broadcast band are frequencies between 535 kHz and 1605 kHz, and IF signals are frequencies between 450 kHz and 460 kHz. In commercial broadcast-band FM receivers, intermediate frequencies as high as 10.7 MHz are used, which are considerably higher than AM broadcast-band RF signals. Intermediate frequencies simply refer to frequencies that are used within a transmitter or receiver that fall somewhere between the radio frequencies and the original source information frequencies.

5-3-3-1 Frequency conversion. *Frequency conversion* in the mixer/converter stage is identical to frequency conversion in the modulator stage of a transmitter except that, in the receiver, the frequencies are down-converted rather than up-converted. In the mixer/converter, RF signals are combined with the local oscillator frequency in a nonlinear device. The output of the mixer contains an infinite number of harmonic and cross-product frequencies, which include the sum and difference frequencies between the desired RF carrier and local oscillator frequencies. The IF filters are tuned to the difference frequencies. The local oscillator is designed such that its frequency of oscillation is always above or below the desired RF carrier by an amount equal to the IF center frequency. Therefore, the difference between the RF and the local oscillator frequency is always equal to the IF. The adjustment for the center frequency of the preselector and the adjustment for the local oscillator frequency are *gang tuned.* Gang tuning means that the two adjustments are mechanically tied together so that a single adjustment will change the center frequency of the preselector and, at the same time, change the local oscillator frequency. When the local oscillator frequency is tuned above the RF, it is called *high-side injection* or *high-beat injection.* When the local oscillator is tuned below the RF, it is called *low-side injection* or *low-beat injection.* In AM broadcast-band receivers, high-side injection is always used (the reason for this is explained later in this section). Mathematically, the local oscillator frequency is

For high-side injection: $$f_{lo} = f_{RF} + f_{IF} \quad (5\text{-}7)$$

For low-side injection: $$f_{lo} = f_{RF} - f_{IF} \quad (5\text{-}8)$$

where f_{lo} = local oscillator frequency (hertz)
f_{RF} = radio frequency (hertz)
f_{IF} = intermediate frequency (hertz)

Figure 5-5 illustrates the frequency conversion process for an AM broadcast-band superheterodyne receiver using high-side injection. The input to the receiver could contain any of the AM broadcast-band channels, which occupy the bandwidth between 535 kHz and 1605 kHz. The preselector is tuned to channel 2, which operates on a 550-kHz carrier frequency and contains sidebands extending from 545 kHz to 555 kHz (10-kHz bandwidth). The preselector is broadly tuned to a 30-kHz passband, allowing channels 1, 2, and 3 to pass through it into the mixer/converter stage, where they are mixed with a 1005-kHz local oscillator frequency. The mixer output contains the same three channels except, because high-side injection is used, the heterodyning process causes the sidebands to be inverted (i.e., the upper and lower sidebands of each channel are flipped over). In addition, channels 1 and 3 switch places in the frequency domain with respect to channel 2 (i.e., channel 1 is now above channel 2 in the frequency spectrum, and channel 3 is now below channel 2).

The heterodyning process converts channel 1 from the 535-kHz to 545-kHz band to the 460-kHz to 470-kHz band, channel 2 from the 545-kHz to 555-kHz band to the 450-kHz to 460-kHz band, and channel 3 from the 555-kHz to 565-kHz band to the 440-kHz to 450-kHz

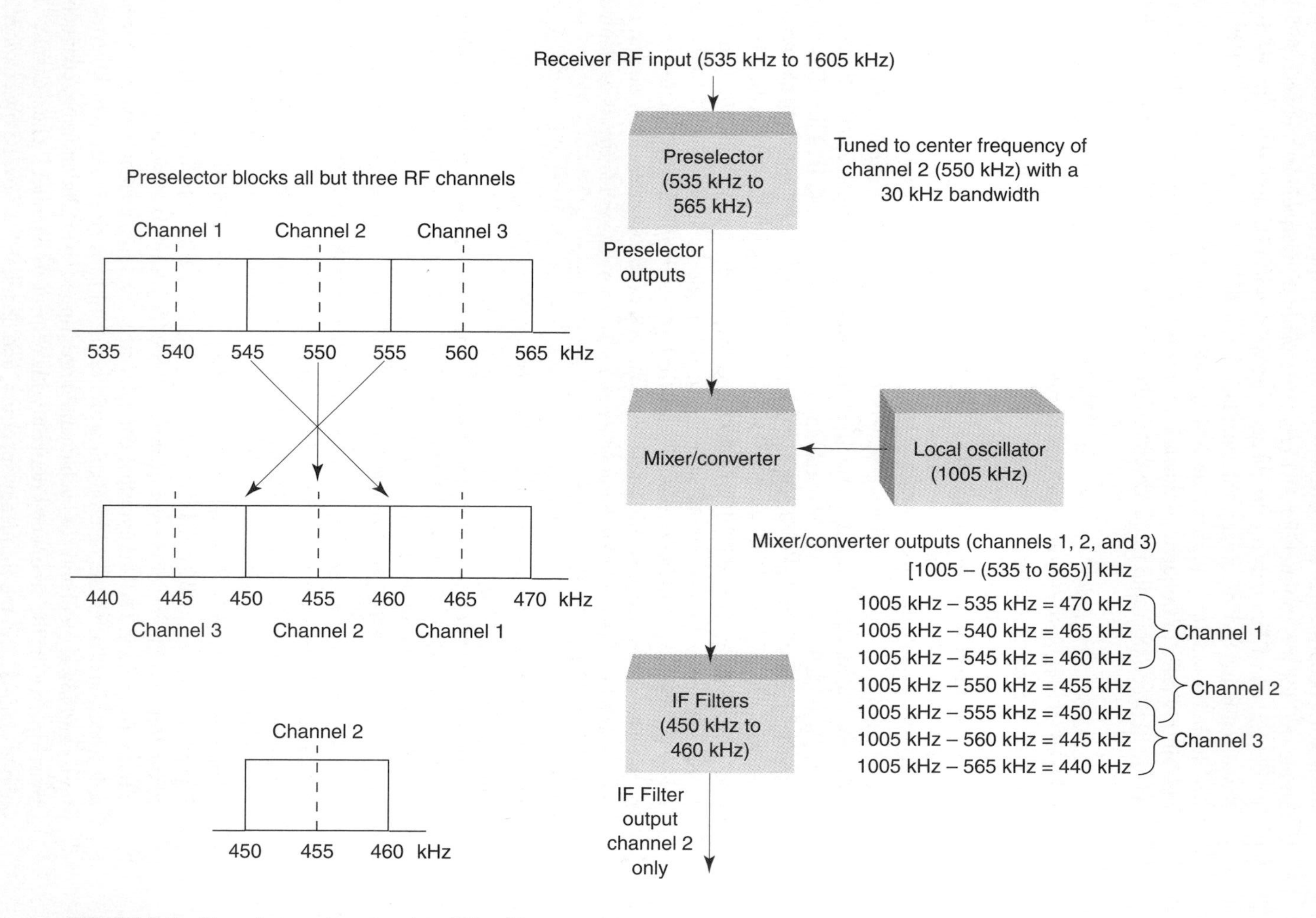

Receiver RF input (535 kHz to 1605 kHz)
Preselector (535 kHz to 565 kHz)
Tuned to center frequency of channel 2 (550 kHz) with a 30 kHz bandwidth
Preselector blocks all but three RF channels
Channel 1
Channel 2
Channel 3
535
540
545
550
555
560
565 kHz
Preselector outputs
Mixer/converter
Local oscillator (1005 kHz)
440
445
450
455
460
465
470 kHz
Channel 3
Channel 2
Channel 1
Mixer/converter outputs (channels 1, 2, and 3)
[1005 – (535 to 565)] kHz
1005 kHz – 535 kHz = 470 kHz
1005 kHz – 540 kHz = 465 kHz
1005 kHz – 545 kHz = 460 kHz
1005 kHz – 550 kHz = 455 kHz
1005 kHz – 555 kHz = 450 kHz
1005 kHz – 560 kHz = 445 kHz
1005 kHz – 565 kHz = 440 kHz
Channel 1
Channel 2
Channel 3
IF Filters (450 kHz to 460 kHz)
Channel 2
450
455
460 kHz
IF Filter output channel 2 only

FIGURE 5-5 Superheterodyne receiver RF-to-IF conversion

band. Channel 2 (the desired channel) is the only channel that falls within the bandwidth of the IF filters (450 kHz to 460 kHz). Therefore, channel 2 is the only channel that continues through the receiver to the IF amplifiers and eventually the AM demodulator circuit.

Example 5-3

For an AM superheterodyne receiver that uses high-side injection and has a local oscillator frequency of 1355 kHz, determine the IF carrier, upper side frequency, and lower side frequency for an RF wave that is made up of a carrier and upper and lower side frequencies of 900 kHz, 905 kHz, and 895 kHz, respectively.

Solution Refer to Figure 5-6. Because high-side injection is used, the intermediate frequencies are the difference between the radio frequencies and the local oscillator frequency. Rearranging Equation 5-7 yields

$$f_{IF} = f_{lo} - f_{RF} = 1355 \text{ kHz} - 900 \text{ kHz} = 455 \text{ kHz}$$

The upper and lower intermediate frequencies are

$$f_{IF(usf)} = f_{lo} - f_{RF(lsf)} = 1355 \text{ kHz} - 895 \text{ kHz} = 460 \text{ kHz}$$

$$f_{IF(lsf)} = f_{lo} - f_{RF(usf)} = 1355 \text{ kHz} - 905 \text{ kHz} = 450 \text{ kHz}$$

Note that the side frequencies undergo a sideband reversal during the heterodyning process (i.e., the RF upper side frequency is translated to an IF lower side frequency, and the RF lower side frequency is translated to an IF upper side frequency). This is commonly called *sideband inversion.* Sideband inversion is not detrimental to conventional double-sideband AM because exactly the same information is contained in both sidebands.

5-3-3-2 Local oscillator tracking. *Tracking* is the ability of the local oscillator in a receiver to oscillate either above or below the selected radio frequency carrier by an amount equal to the intermediate frequency throughout the entire radio frequency band. With high-side injection, the local oscillator should track above the incoming RF carrier by a fixed frequency equal to $f_{RF} + f_{IF}$, and with low-side injection, the local oscillator should track below the RF carrier by a fixed frequency equal to $f_{RF} - f_{IF}$.

Figure 5-7a shows the schematic diagram of the preselector and local oscillator tuned circuit in a broadcast-band AM receiver. The broken lines connecting the two tuning capacitors indicate that they are *ganged* together (connected to a single tuning control). The tuned circuit in the preselector is tunable from a center frequency of 540 kHz to 1600 kHz (a ratio of 2.96 to 1), and the local oscillator is tunable from 995 kHz to 2055 kHz (a ratio of 2.06 to 1). Because the resonant frequency of a tuned circuit is inversely proportional to the square root of the capacitance, the capacitance in the preselector must change by a factor of 8.8 whereas, at the same time, the capacitance in the local oscillator must change by a factor of only 4.26. The local oscillator should oscillate 455 kHz above the preselector center frequency over the entire AM frequency band, and there should be a single tuning control. Fabricating such a circuit is difficult if not impossible. Therefore, perfect tracking over the entire AM band is unlikely to occur. The difference between the actual local oscillator frequency and the desired frequency is called *tracking error.* Typically, the tracking error is not uniform over the entire RF spectrum. A maximum tracking error of ± 3 kHz is about the best that can be expected from a domestic AM broadcast-band receiver with a 455-kHz intermediate frequency. Figure 5-7b shows a typical tracking curve. A tracking error of $+3$ kHz corresponds to an IF center frequency of 458 kHz, and a tracking error of -3 kHz corresponds to an IF center frequency of 452 kHz.

The tracking error is reduced by a technique called *three-point tracking.* The preselector and local oscillator each have a trimmer capacitor (C_t) in parallel with the primary tuning capacitor (C_o) that compensates for minor tracking errors at the high end of the AM spectrum. The local oscillator has an additional padder capacitor (C_p) in series with the tuning coil that compensates for minor tracking errors at the low end of the AM spectrum. With three-point tracking, the tracking error is adjusted to 0 Hz at approximately 600 kHz, 950 kHz, and 1500 kHz.

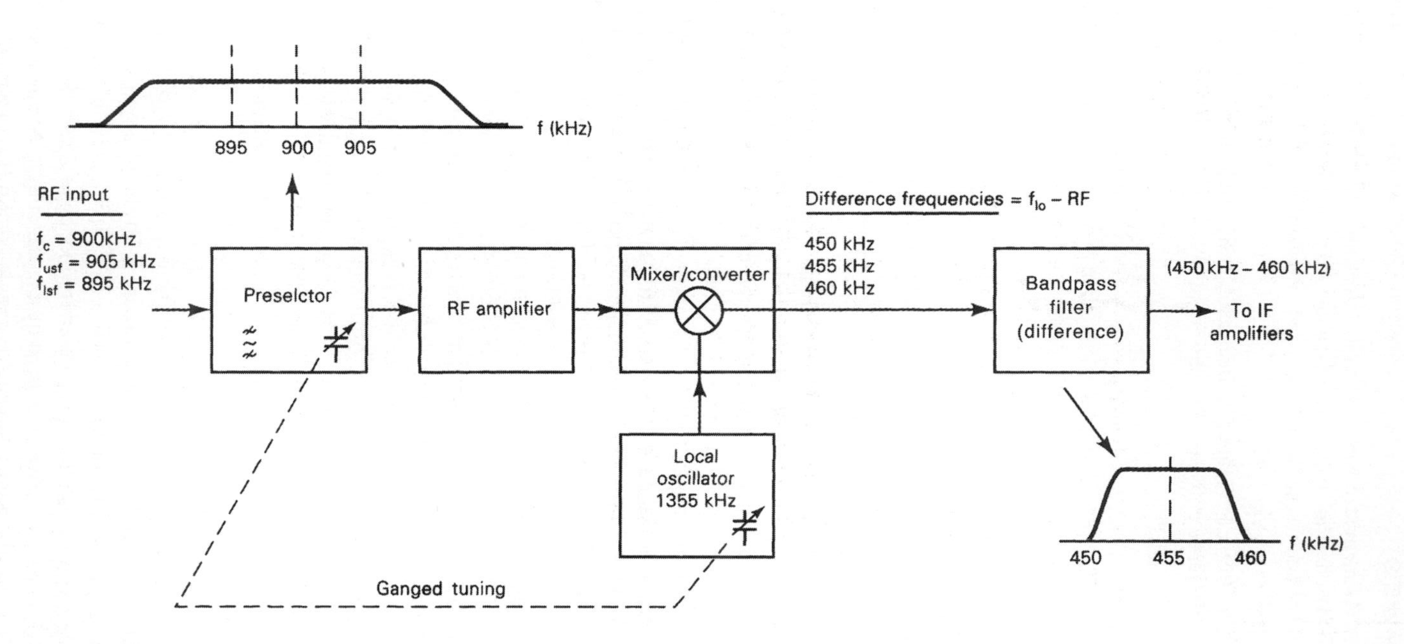

FIGURE 5-6 Figure for Example 5-3

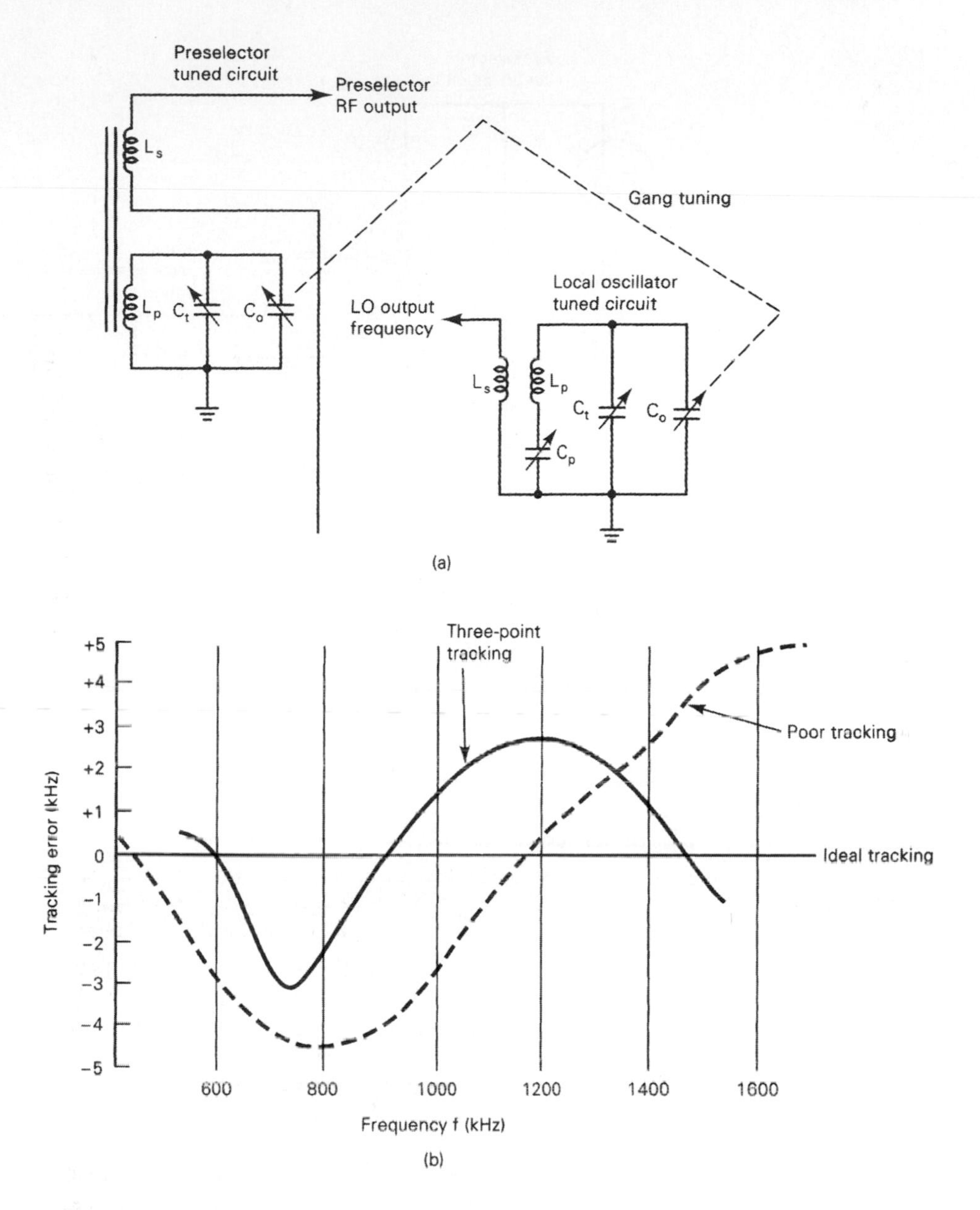

FIGURE 5-7 Receiver tracking: (a) preselector and local oscillator schematic; (b) tracking curve

With low-side injection, the local oscillator would have to be tunable from 85 kHz to 1145 kHz (a ratio of 13.5 to 1). Consequently, the capacitance must change by a factor of 182. Standard variable capacitors seldom tune over more than a 10 to 1 range. This is why low-side injection is impractical for commercial AM broadcast-band receivers. With high-side injection, the local oscillator must be tunable from 995 kHz to 2055 kHz, which corresponds to a capacitance ratio of only 4.26 to 1.

Ganged capacitors are relatively large, expensive, and inaccurate, and they are somewhat difficult to compensate. Consequently, they are being replaced by solid-state electronically tuned circuits. Electronically tuned circuits are smaller, less expensive, more accurate, relatively immune to environmental changes, more easily compensated, and more easily adapted to digital remote control and push-button tuning than their mechanical counterparts.

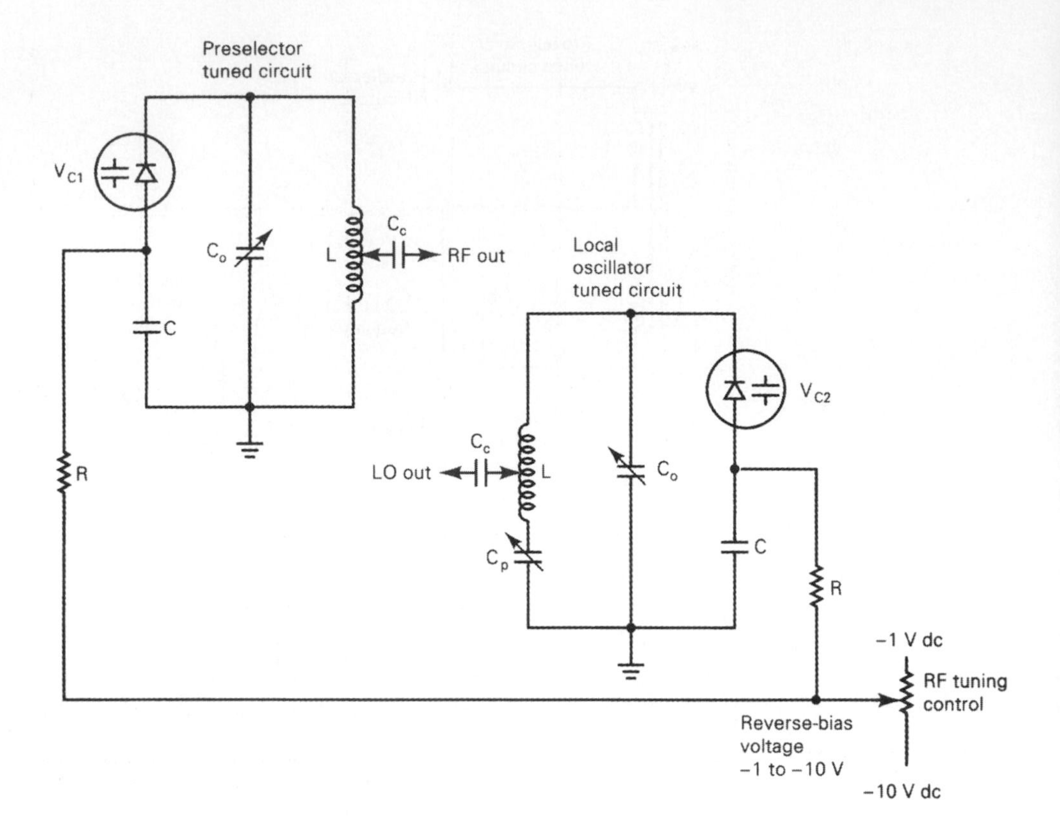

FIGURE 5-8 Electronic tuning

As with the crystal oscillator modules explained in Chapter 3, electronically tuned circuits use solid-state variable-capacitance diodes (varactor diodes). Figure 5-8 shows a schematic diagram for an electronically tuned preselector and local oscillator. The −1-V to −10-V reverse-biased voltage comes from a single tuning control. By changing the position of the wiper arm on a precision variable resistor, the dc reverse bias for the two tuning diodes (V_{C1} and V_{C2}) is changed. The diode capacitance and, consequently, the resonant frequency of the tuned circuit vary with the reverse bias. Three-point compensation with electronic tuning is accomplished the same as with mechanical tuning.

In a superheterodyne receiver, most of the receiver's selectivity is accomplished in the IF stage. For maximum noise reduction, the bandwidth of the IF filters is equal to the minimum bandwidth required to pass the information signal, which with double-sideband transmission is equal to two times the highest modulating signal frequency. For a maximum modulating signal frequency of 5 kHz, the minimum IF bandwidth with perfect tracking is 10 kHz. For a 455-kHz IF center frequency, a 450-kHz to 460-kHz passband is necessary. In reality, however, some RF carriers are tracked as much as ±3 kHz above or below 455 kHz. Therefore, the IF bandwidth must be expanded to allow the IF signals from the off-track stations to pass through the IF filters.

Example 5-4

For the tracking curve shown in Figure 5-9a, a 455-kHz IF center frequency, and a maximum modulating signal frequency of 5 kHz, determine the minimum IF bandwidth.

Solution A double-sideband AM signal with a maximum modulating signal frequency of 5 kHz would require 10 kHz of bandwidth. Thus, a receiver with a 455-kHz IF center frequency and ideal tracking would produce IF signals between 450 kHz and 460 kHz. The tracking curve shown in

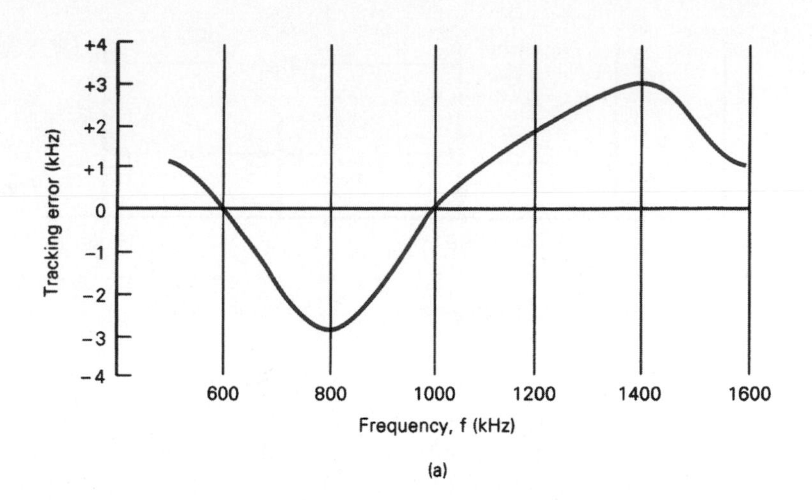

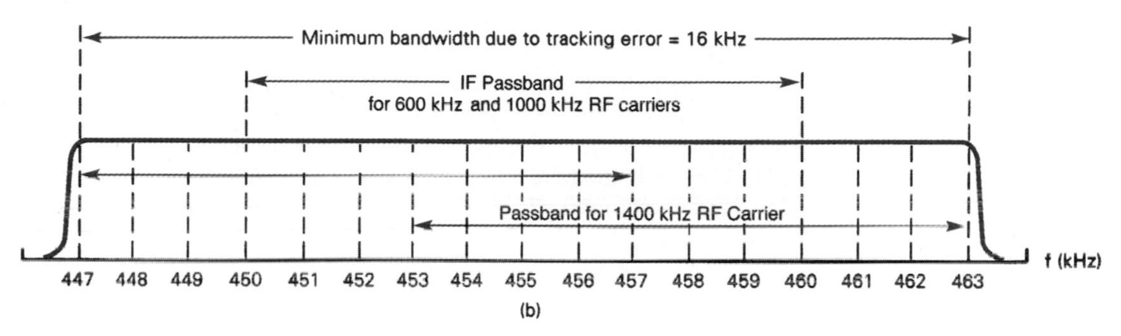

FIGURE 5-9 Tracking error for Example 5-4: (a) tracking curve; (b) bandpass characteristics

Figure 5-9a is for a receiver where perfect tracking occurs only for RF carrier frequencies of 600 kHz and 1000 kHz. The ideal IF passband is shown in Figure 5-9b.

The tracking curve in Figure 5-9a also shows that the maximum negative tracking error of −3 kHz occurs for a RF carrier frequency of 800 kHz and that the maximum positive tracking error of +3 kHz occurs for a RF carrier frequency of 1400 kHz. Consequently, as shown in Figure 5-9b, the IF frequency spectrum produced for an 800-kHz carrier would extend from 447 kHz to 457 kHz, and the IF frequency spectrum produced for a 1400-kHz carrier would extend from 453 kHz to 463 kHz.

Thus, the maximum intermediate frequency occurs for the RF carrier with the most positive tracking error (1400 kHz) and a 5-kHz modulating signal:

$$f_{IF(max)} = f_{IF} + \text{tracking error} + f_{m(max)}$$
$$= 455 \text{ kHz} + 3 \text{ kHz} + 5 \text{ kHz} = 463 \text{ kHz}$$

The minimum intermediate frequency occurs for the RF carrier with the most negative tracking error (800 kHz) and a 5-kHz modulating signal:

$$f_{IF(min)} = f_{IF} + \text{tracking error} - f_{m(max)}$$
$$= 455 \text{ kHz} + (-3 \text{ kHz}) - 5 \text{ kHz} = 447 \text{ kHz}$$

The minimum IF bandwidth necessary to pass the two sidebands is the difference between the maximum and minimum intermediate frequencies, or

$$B_{min} = 463 \text{ kHz} - 447 \text{ kHz} = 16 \text{ kHz}$$

Figure 5-9b shows the IF bandpass characteristics for Example 5-4.

5-3-3-3 Image frequency. An *image frequency* is any frequency other than the selected radio frequency carrier that, if allowed to enter a receiver and mix with the local oscillator, will produce a cross-product frequency that is equal to the intermediate frequency. An image frequency is equivalent to a second radio frequency that will produce an IF that

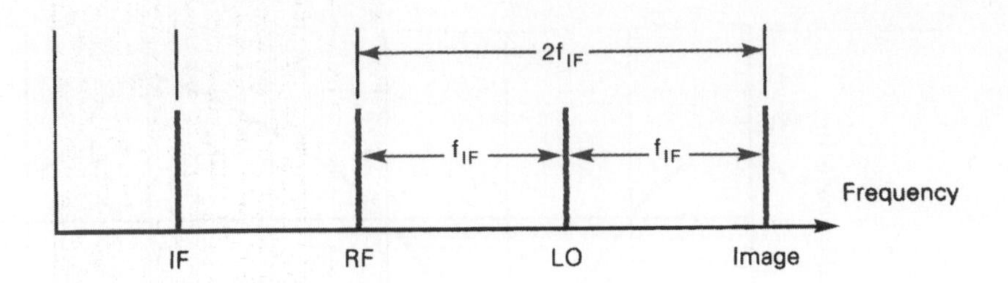

FIGURE 5-10 Image frequency

will interfere with the IF from the desired radio frequency. Once an image frequency has been mixed down to IF, it cannot be filtered out or suppressed. If the selected RF carrier and its image frequency enter a receiver at the same time, they both mix with the local oscillator frequency and produce difference frequencies that are equal to the IF. Consequently, two different stations are received and demodulated simultaneously, producing two sets of information frequencies. For a radio frequency to produce a cross product equal to the IF, it must be displaced from the local oscillator frequency by a value equal to the IF. With high-side injection, the selected RF is below the local oscillator by an amount equal to the IF. Therefore, the image frequency is the radio frequency that is located in the IF frequency above the local oscillator. Mathematically, for high-side injection, the image frequency (f_{im}) is

$$f_{im} = f_{lo} + f_{IF} \tag{5-9}$$

and, because the desired RF equals the local oscillator frequency minus the IF,

$$f_{im} = f_{RF} + 2f_{IF} \tag{5-10}$$

Figure 5-10 shows the relative frequency spectrum for the RF, IF, local oscillator, and image frequencies for a superheterodyne receiver using high-side injection. Here we see that the higher the IF, the farther away in the frequency spectrum the image frequency is from the desired RF. Therefore, for better *image-frequency rejection,* a high intermediate frequency is preferred. However, the higher the IF, the more difficult it is to build stable amplifiers with high gain. Therefore, there is a trade-off when selecting the IF for a radio receiver between image-frequency rejection and IF gain and stability.

5-3-3-4 Image-frequency rejection ratio. The *image-frequency rejection ratio* (IFRR) is a numerical measure of the ability of a preselector to reject the image frequency. For a single-tuned preselector, the ratio of its gain at the desired RF to the gain at the image frequency is the IFRR. Mathematically, IFRR is

$$\text{IFRR} = \sqrt{(1 + Q^2\rho^2)} \tag{5-11}$$

where $\rho = (f_{im}/f_{RF}) - (f_{RF}/f_{im})$.

If there is more than one tuned circuit in the front end of a receiver (perhaps a preselector filter and a separately tuned RF amplifier), the total IFRR is simply the product of the two ratios.

Example 5-5

For an AM broadcast-band superheterodyne receiver with IF, RF, and local oscillator frequencies of 455 kHz, 600 kHz, and 1055 kHz, respectively, refer to Figure 5-11 and determine

a. Image frequency.
b. IFRR for a preselector Q of 100.

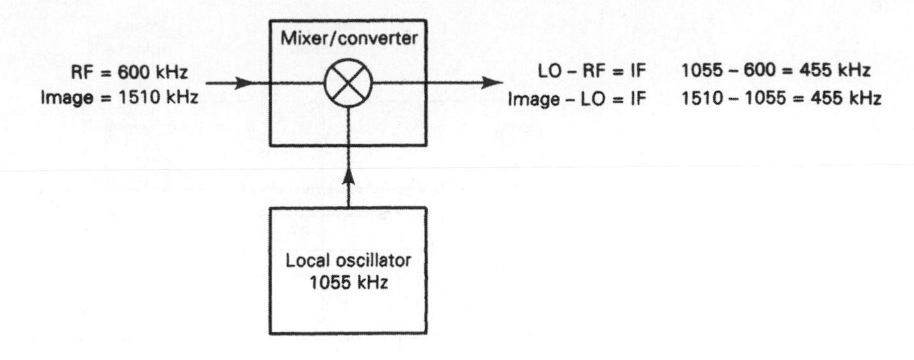

FIGURE 5-11 Frequency conversion for Example 5-5

Solution a. From Equation 5-9,

$$f_{im} = 1055 \text{ kHz} + 455 \text{ kHz} = 1510 \text{ kHz}$$

or from Equation 5-10,

$$f_{im} = 600 \text{ kHz} + 2(455 \text{ kHz}) = 1510 \text{ kHz}$$

b. From Equation 5-11,

$$\rho = \frac{1510 \text{ kHz}}{600 \text{ kHz}} - \frac{600 \text{ kHz}}{1510 \text{ kHz}} = 2.51 - 0.397 = 2.113$$

$$\text{IFRR} = \sqrt{1 + (100^2)(2.113^2)} = 211.3$$

Once an image frequency has been down-converted to IF, it cannot be removed. Therefore, to reject the image frequency, it has to be blocked prior to the mixer/converter stage. Image-frequency rejection is the primary purpose for the RF preselector. If the bandwidth of the preselector is sufficiently narrow, the image frequency is prevented from entering the receiver. Figure 5-12 illustrates how proper RF and IF filtering can prevent an image frequency from interfering with the desired radio frequency.

The ratio of the RF to the IF is also an important consideration for image-frequency rejection. The closer the RF is to the IF, the closer the RF is to the image frequency.

Example 5-6

For a citizens band receiver using high-side injection with an RF carrier of 27 MHz and an IF center frequency of 455 kHz, determine

a. Local oscillator frequency.
b. Image frequency.
c. IFRR for a preselector Q of 100.
d. Preselector Q required to achieve the same IFRR as that achieved for an RF carrier of 600 kHz in Example 5-5.

Solution a. From Equation 5-7,

$$f_{lo} = 27 \text{ MHz} + 455 \text{ kHz} = 27.455 \text{ MHz}$$

b. From Equation 5-9,

$$f_{im} = 27.455 \text{ MHz} + 455 \text{ kHz} = 27.91 \text{ MHz}$$

c. From Equation 5-11,

$$\text{IFRR} = 6.77$$

d. Rearranging Equation 5-11,

$$Q = \sqrt{\frac{(\text{IFRR}^2 - 1)}{\rho^2}} = 3167$$

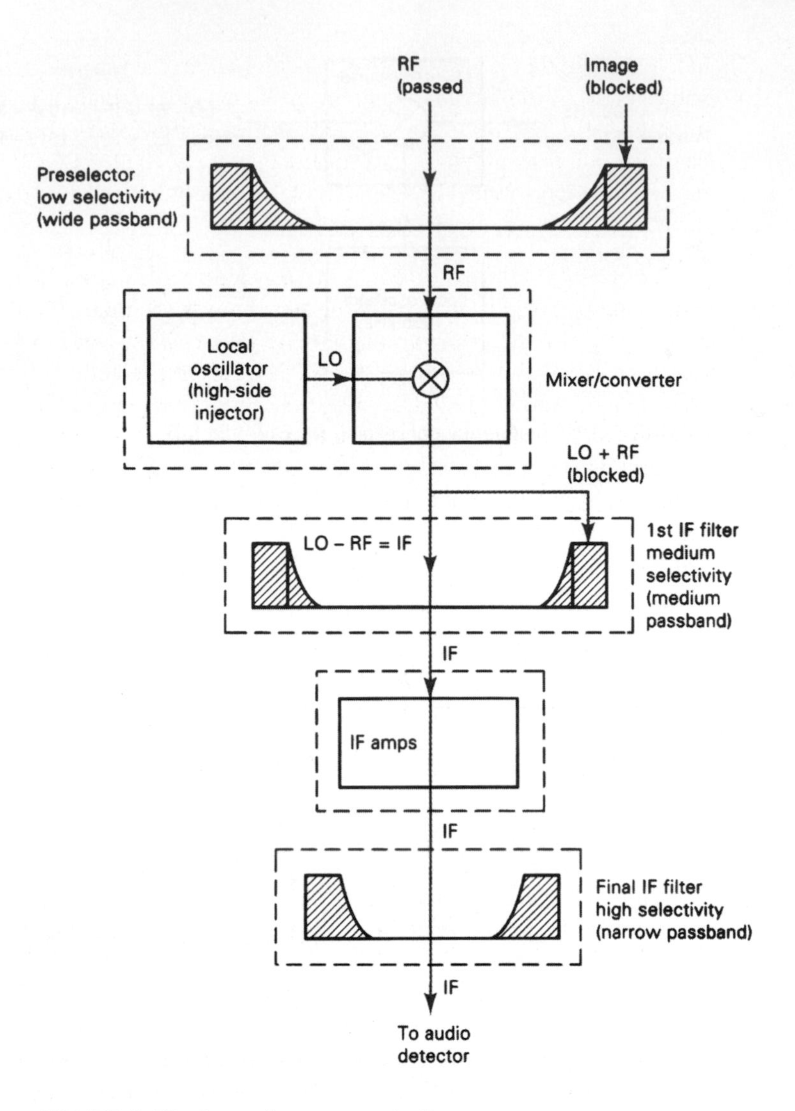

FIGURE 5-12 Image-frequency rejection

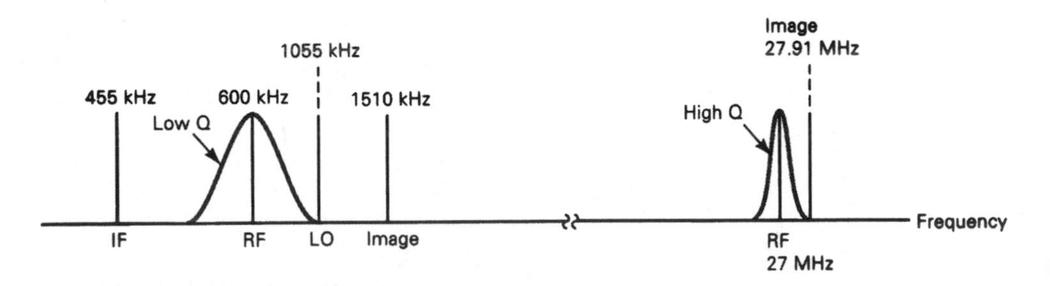

FIGURE 5-13 Frequency spectrum for Example 5-6

From Examples 5-5 and 5-6, it can be seen that the higher the RF carrier, the more difficult it is to prevent the image frequency from entering the receiver. For the same IFRR, the higher RF carriers require a much higher-quality preselector filter. This is illustrated in Figure 5-13.

From Examples 5-5 and 5-6, it can be seen that for a 455-kHz IF, it is more difficult to prevent the image frequency from entering the receiver with high RF carrier frequencies than with low RF carrier frequencies. For the same IFRR, the higher RF carriers require a much more higher-quality preselector—in fact, in many cases unrealistic or unachievable values of Q. A simpler solution to the problem is to use higher IF frequencies when receiving higher RF carrier frequencies. For example, if a 5-MHz IF were used for the citizens band receiver in Example 5-6, the image frequency would be 37 MHz, which is sufficiently far away from the 27-MHz carrier to allow a preselector with a realistic Q to easily prevent the image frequency from entering the receiver.

5-3-3-5 Double spotting. *Double spotting* occurs when a receiver picks up the same station at two nearby points on the receiver tuning dial. One point is the desired location, and the other point is called the *spurious point.* Double spotting is caused by poor front-end selectivity or inadequate image-frequency rejection.

Double spotting is harmful because weak stations can be overshadowed by the reception of a nearby strong station at the spurious location in the frequency spectrum. Double spotting may be used to determine the intermediate frequency of an unknown receiver because the spurious point on the dial is precisely two times the IF center frequency below the correct receive frequency.

5-4 AM RECEIVER CIRCUITS

5-4-1 RF Amplifier Circuits

An RF amplifier is a high-gain, low-noise, tuned amplifier that, when used, is the first active stage encountered by the received signal. The primary purposes of an RF stage are selectivity, amplification, and sensitivity. Therefore, the following characteristics are desirable for RF amplifiers:

1. Low thermal noise
2. Low noise figure
3. Moderate to high gain
4. Low intermodulation and harmonic distortion (i.e., linear operation)
5. Moderate selectivity
6. High image-frequency rejection ratio

Two of the most important parameters for a receiver are amplification and noise figure, which both depend on the RF stage. An AM demodulator (or detector as it is sometimes called) detects amplitude variations in the modulated wave and converts them to amplitude changes in its output. Consequently, amplitude variations that were caused by noise are converted to erroneous fluctuations in the demodulator output, and the quality of the demodulated signal is degraded. The more gain a signal experiences as it passes through a receiver, the more pronounced the amplitude variations at the demodulator input and the less noticeable the variations caused by noise. The narrower the bandwidth, the less noise propagated through the receiver and, consequently, the less noise demodulated by the detector. From Equation 1-17 ($V_N = \sqrt{4RKTB}$), noise voltage is directly proportional to the square root of the temperature, bandwidth, and equivalent noise resistance. Therefore, if these three parameters are minimized, the thermal noise is reduced. The temperature of an RF stage can be reduced by artificially cooling the front end of the receiver with air fans or even liquid helium in the more expensive receivers. The bandwidth is reduced by using tuned amplifiers and filters, and the equivalent noise resistance is reduced by using specially constructed solid-state components for the active devices. Noise figure is essentially a measure of the noise added by an amplifier. Therefore, the noise figure is improved (reduced) by reducing the amplifier's internal noise.

Intermodulation and harmonic distortion are both forms of nonlinear distortion that increase the magnitude of the noise figure by adding correlated noise to the total noise spectrum. The more linear an amplifier's operation, the less nonlinear distortion produced and the better the receiver's noise figure. The image-frequency reduction by the RF amplifier combines with the image-frequency reduction of the preselector to reduce the receiver input bandwidth sufficiently to help prevent the image frequency from entering the mixer/converter stage. Consequently, moderate selectivity is all that is required from the RF stage.

Figure 5-14 shows several commonly used RF amplifier circuits. Keep in mind that RF is a relative term and simply means that the frequency is high enough to be efficiently

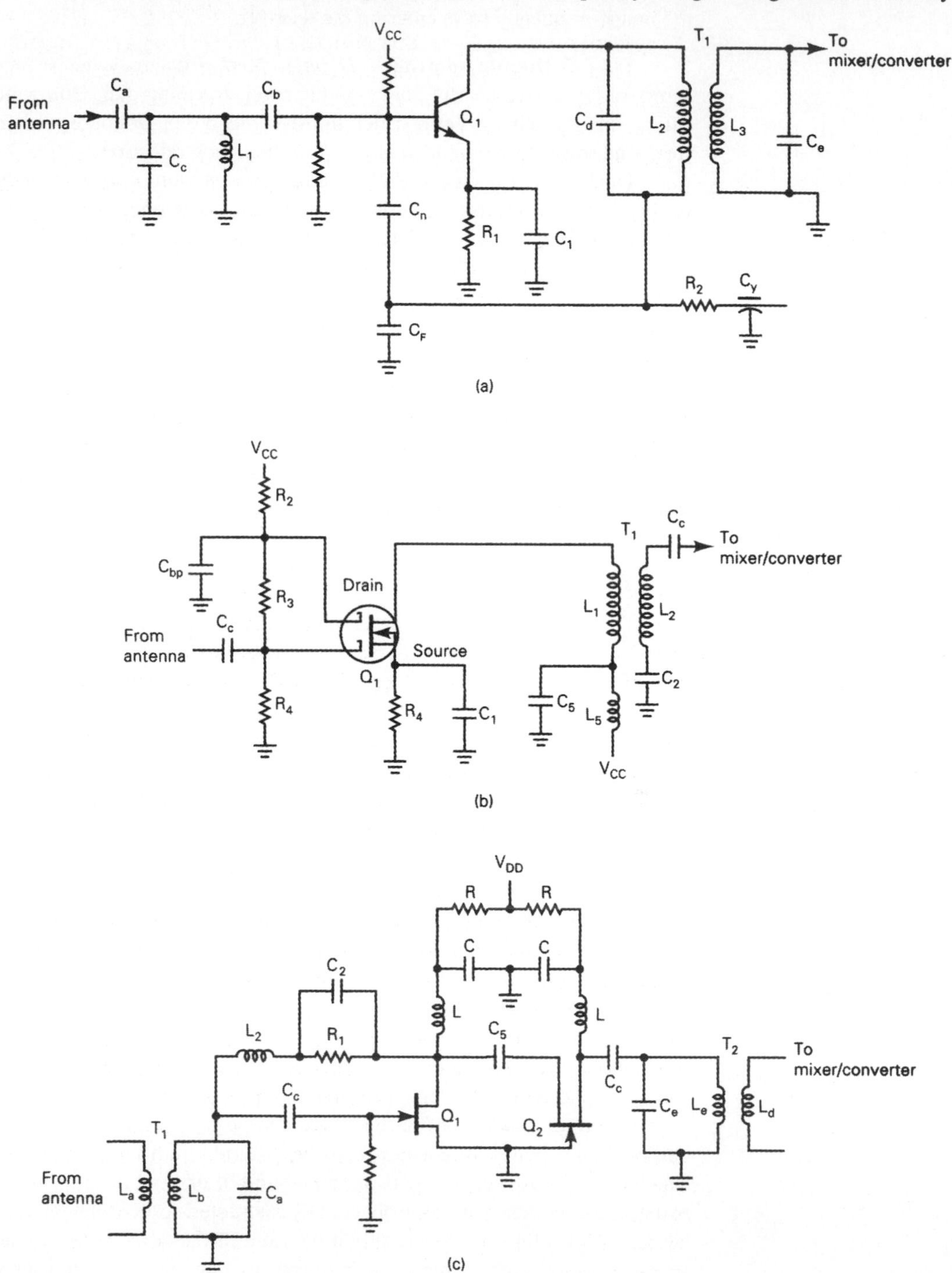

FIGURE 5-14 RF amplifier configurations: (a) bipolar transistor RF amplifier; (b) DEMOS-FET RF amplifier; (c) cascoded RF amplifier

radiated by an antenna and propagated through free space as an electromagnetic wave. RF for the AM broadcast band is between 535 kHz and 1605 kHz, whereas RF for microwave radio is in excess of 1 GHz (1000 MHz). A common intermediate frequency used for FM broadcast-band receivers is 10.7 MHz, which is considerably higher than the radio frequencies associated with the AM broadcast band. RF is simply the radiated or received signal, and IF is an intermediate signal within a transmitter or receiver. Therefore, many of the considerations for RF amplifiers also apply to IF amplifiers, such as neutralization, filtering, and coupling.

Figure 5-14a shows a schematic diagram for a bipolar transistor RF amplifier. C_a, C_b, C_c, and L_1 form the coupling circuit from the antenna. Q_1 is class A biased to reduce nonlinear distortion. The collector circuit is transformer coupled to the mixer/converter through T_1, which is double tuned for more selectivity. C_x and C_y are RF bypass capacitors. Their symbols indicate that they are specially constructed *feedthrough* capacitors. Feedthrough capacitors offer less inductance, which prevents a portion of the signal from radiating from their leads. C_n is a *neutralization* capacitor. A portion of the collector signal is fed back to the base circuit to offset (or neutralize) the signal fed back through the transistor collector-to-base lead capacitance to prevent oscillations from occurring. C_f, in conjunction with C_n, forms an ac voltage divider for the feedback signal. This neutralization configuration is called *off-ground* neutralization.

Figure 5-14b shows an RF amplifier using dual-gate field-effect transistors. This configuration uses DEMOS (depletion-enhancement metal-oxide semiconductor) FETs. The FETs feature high input impedance and low noise. A FET is a square-law device that generates only second-order harmonic and intermodulation distortion components, therefore producing less nonlinear distortion than a bipolar transistor. Q_1 is again biased class A for linear operation. T_1 is single tuned to the desired RF carrier frequency to enhance the receiver's selectivity and to improve the IFRR. L_5 is a radio frequency choke and, in conjunction with C_5, decouples RF signals from the dc power supply.

Figure 5-14c shows the schematic diagram for a special RF amplifier configuration called a *cascoded* amplifier. A cascoded amplifier offers higher gain and less noise than conventional cascaded amplifiers. The active devices can be either bipolar transistors or FETs. Q_1 is a common-source amplifier whose output is impedance coupled to the source of Q_2. Because of the low input impedance of Q_2, Q_1 does not need to be neutralized; however, neutralization reduces the noise figure even further. Therefore, L_2, R_1, and C_2 provide the feedback path for neutralization. Q_2 is a common-gate amplifier and because of its low input impedance requires no neutralization.

5-4-2 Low-Noise Amplifiers

High-performance microwave receivers require a *low-noise amplifier* (LNA) as the input stage of the RF section to optimize their noise figure. Equation 1-27 showed that the first amplifier in a receiver is the most important in determining the receiver's noise figure. The first stage should have low noise and high gain. Unfortunately, this is difficult to achieve with a single amplifier stage; therefore, LNAs generally include two stages of amplification along with impedance-matching networks to enhance their performance. The first stage has moderate gain and minimum noise, and the second stage has high gain and moderate noise.

Low-noise RF amplifiers are biased class A and usually utilize silicon bipolar or field-effect transistors up to approximately 2 GHz and gallium arsenide FETs above this frequency. A special type of gallium arsenide FET most often used is the MESFET (MEsa Semiconductor FET). A MESFET is a FET with a metal-semiconductor junction at the gate of the device, called a Schottky barrier. Low-noise amplifiers are discussed in more detail in a later chapter of this book.

5-4-2-1 Integrated-circuit RF amplifiers. The NE/SA5200 (Figure 5-15) is a wideband, unconditionally stable, low-power, dual-gain linear integrated-circuit RF amplifier

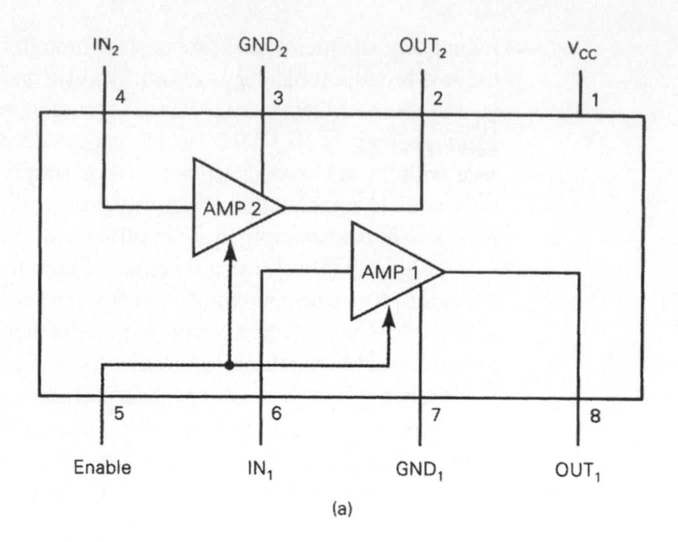

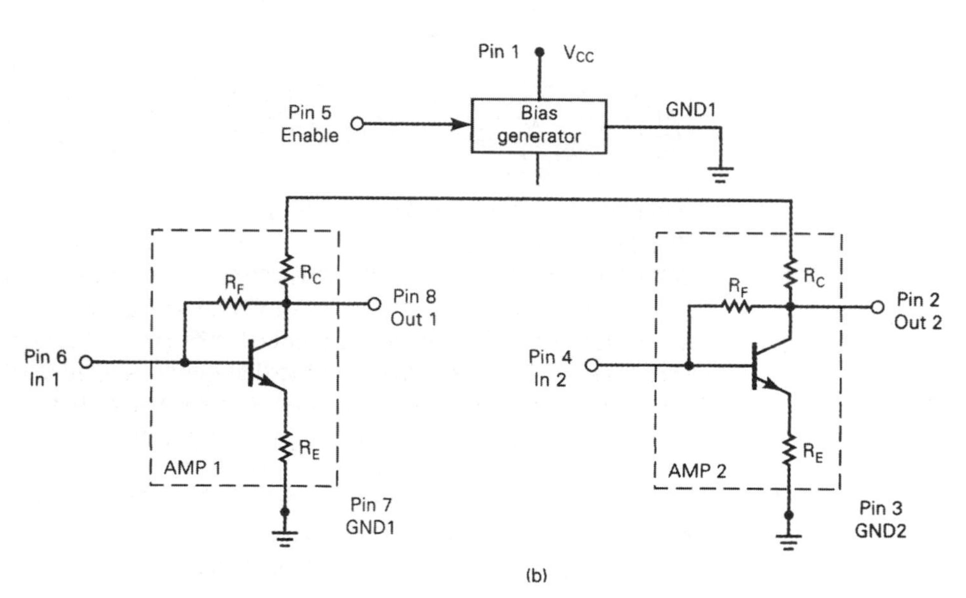

FIGURE 5-15 NE/SA5200 RF dual-gain stage: (a) block diagram; (b) simplified schematic diagram

manufactured by Signetics Corporation. The NE/SA5200 will operate from dc to approximately 1200 MHz and has a low-noise figure. The NE/SA5200 has several inherent advantages over comparable discrete implementations; it needs no external biasing components, it occupies little space on a printed circuit board, and the high level of integration improves its reliability over discrete counterparts. The NE/SA5200 is also equipped with a power-down mode that helps reduce power consumption in applications where the amplifiers can be disabled.

The block diagram for the SA5200 is shown in Figure 5-15a, and a simplified schematic diagram is shown in Figure 5-15b. Note that the two wideband amplifiers are biased from the same bias generator. Each amplifier stage has a noise figure of about 3.6 dB and a gain of approximately 11 dB. Several stages of NE/SA5200 can be cascaded and used as an IF strip, and the enable pin can be used to improve the dynamic

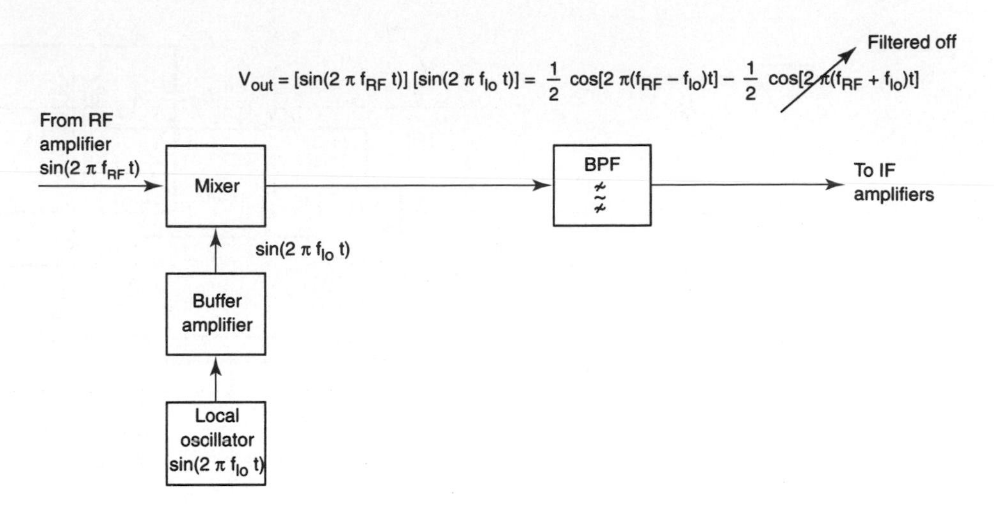

FIGURE 5-16 Mixer/converter block diagram

range of the receiver. For extremely high input levels, the amplifiers in the NE/SA5200 can be disabled. When disabled, the input signal is attenuated 13 dB, preventing receiver overload.

5-4-3 Mixer/Converter Circuits

The purpose of the mixer/converter stage is to down-convert the incoming radio frequencies to intermediate frequencies. This is accomplished by mixing the RF signals with the local oscillator frequency in a nonlinear device. In essence, this is heterodyning. A mixer is a nonlinear amplifier similar to a modulator, except that the output is tuned to the difference between the RF and local oscillator frequencies. Figure 5-16 shows a block diagram for a mixer/converter stage. The output of a balanced mixer is the product of the RF and local oscillator frequencies and is expressed mathematically as

$$V_{out} = (\sin 2\pi f_{RF}t)(\sin 2\pi f_{lo}t) \tag{5-12}$$

where f_{RF} = incoming radio frequency (hertz)
f_{lo} = local oscillator frequency (hertz)

Therefore, using the trigonometric identity for the product of two sines, the output of a mixer is

$$V_{out} = \frac{1}{2}\cos[2\pi(f_{RF} - f_{lo})t] - \frac{1}{2}\cos[2\pi(f_{RF} + f_{lo})t] \tag{5-13}$$

The absolute value of the difference frequency $(|f_{RF} - f_{lo}|)$ is the intermediate frequency.

Although any nonlinear device can be used for a mixer, a transistor or FET is generally preferred over a simple diode because it is also capable of amplification. However, because the actual output signal from a mixer is a cross-product frequency, there is a net loss to the signal. This loss is called *conversion loss* (or sometimes *conversion gain*) because a frequency conversion has occurred, and, at the same time, the IF output signal is lower in amplitude than the RF input signal. The conversion loss is generally 6 dB (which corresponds to a conversion gain of −6 dB). The conversion gain is the difference between the level of the IF output with an RF input signal to the level of the IF output with an IF input signal.

Figure 5-17 shows the schematic diagrams for several common mixer/converter circuits. Figure 5-17a shows what is probably the simplest mixer circuit available (other than a single diode mixer) and is used exclusively for inexpensive AM broadcast-band receivers.

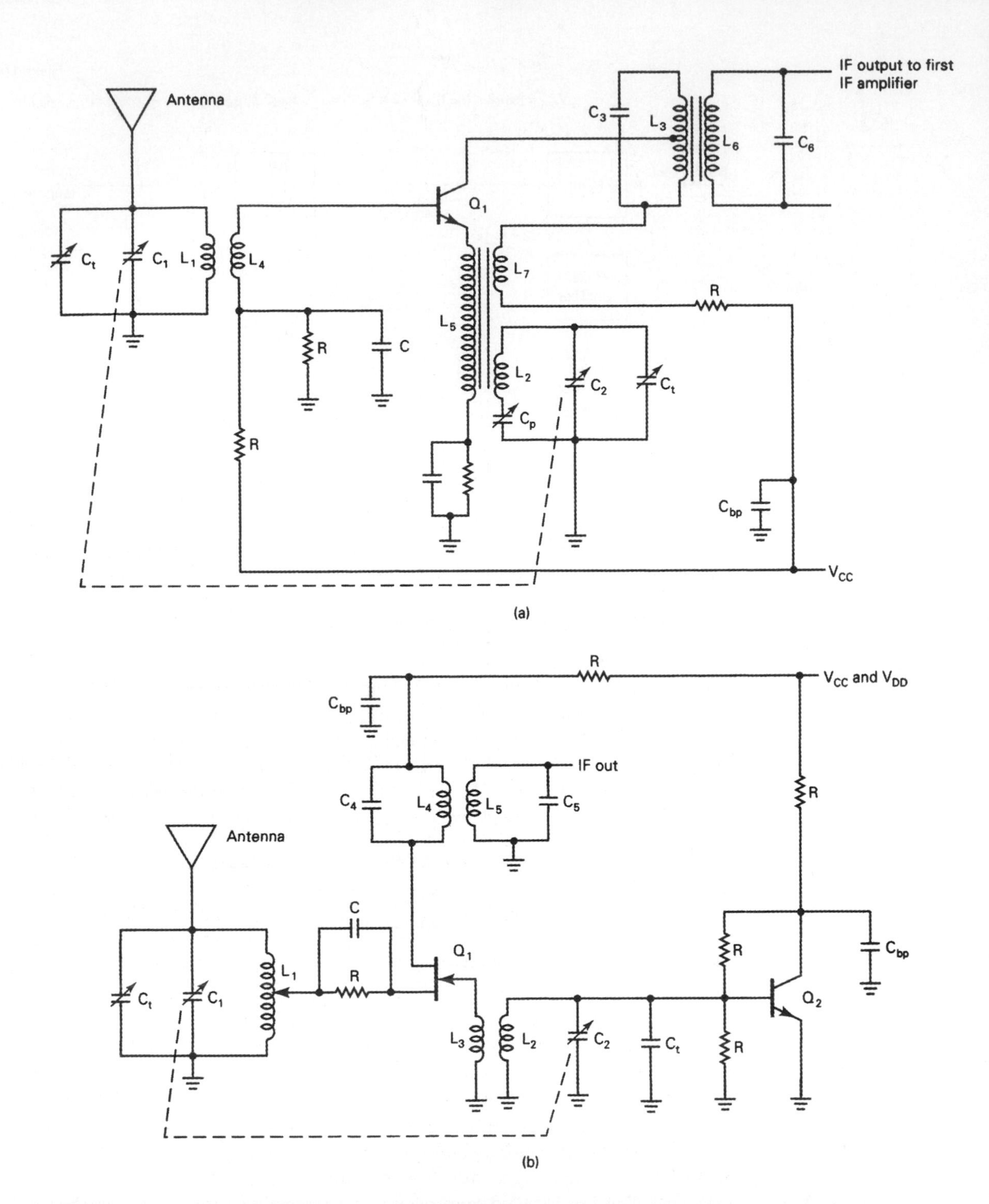

FIGURE 5-17 Mixer-converter circuits: (a) self-excited mixer; (b) separately excited mixer; (*Continued*)

Radio frequency signals from the antenna are filtered by the preselector tuned circuit (L_1 and C_1) and then transformer coupled to the base of Q_1. The active device for the mixer (Q_1) also provides amplification for the local oscillator. This configuration is commonly called a *self-excited* mixer because the mixer excites itself by feeding energy back to the local oscillator tank circuit (C_2 and L_2) to sustain oscillations. When power is initially applied, Q_1

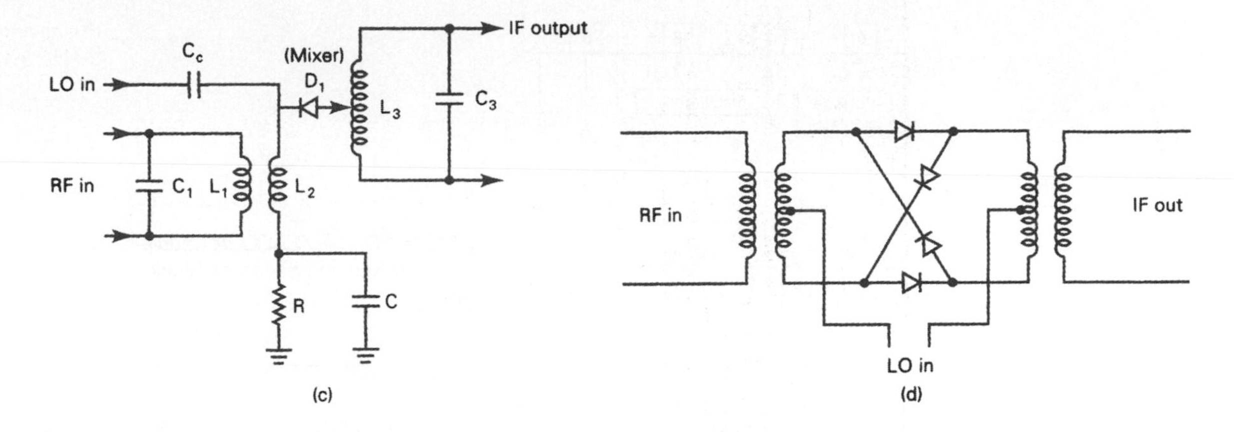

FIGURE 5-17 (*Continued*) (c) diode mixer; (d) balanced diode mixer

amplifies both the incoming RF signals and any noise present and supplies the oscillator tank circuit with enough energy to begin oscillator action. The local oscillator frequency is the resonant frequency of the tank circuit. A portion of the resonant tank-circuit energy is coupled through L_2 and L_5 to the emitter of Q_1. This signal drives Q_1 into its nonlinear operating region and, consequently, produces sum and difference frequencies at its collector. The difference frequency is the IF. The output tank circuit (L_3 and C_3) is tuned to the IF band. Therefore, the IF signal is transformer coupled to the input of the first IF amplifier. The process is regenerative as long as there is an incoming RF signal. The tuning capacitors in the RF and local oscillator tank circuits are ganged into a single tuning control. C_p and C_t are for three-point tracking. This configuration has poor selectivity and poor image-frequency rejection because there is no amplifier tuned to the RF signal frequency and, consequently, the only RF selectivity is in the preselector. In addition, there is essentially no RF gain, and the transistor nonlinearities produce harmonic and intermodulation components that may fall within the IF passband.

The mixer/converter circuit shown in Figure 5-17b is a *separately excited* mixer. Its operation is essentially the same as the self-excited mixer except that the local oscillator and the mixer have their own gain devices. The mixer itself is a FET, which has nonlinear characteristics that are better suited for IF conversion than those of a bipolar transistor. Feedback is from L_2 to L_3 of the transformer in the gate of Q_1. This circuit is commonly used for high-frequency (HF) and very-high-frequency (VHF) receivers.

The mixer converter circuit shown in Figure 5-17c is a *single-diode* mixer. The concept is quite simple: The RF and local oscillator signals are coupled into the diode, which is a nonlinear device. Therefore, nonlinear mixing occurs, and the sum and difference frequencies are produced. The output tank circuit (C_3 and L_3) is tuned to the difference (IF) frequency. A single-diode mixer is inefficient because it has a net loss. However, a diode mixer is commonly used for the audio detector in AM receivers and to produce the audio subcarrier in television receivers.

Figure 5-17d shows the schematic diagram for a *balanced diode* mixer. Balanced mixers are one of the most important circuits used in communications systems today. Balanced mixers are also called *balanced modulators, product modulators,* and *product detectors.* The phase detectors used in phase-locked loops and explained in Chapter 3 are balanced modulators. Balanced mixers are used extensively in both transmitters and receivers for AM, FM, and many of the digital modulation schemes, such as PSK and QAM. Balanced mixers have two inherent advantages over other types of mixers: noise reduction and carrier suppression.

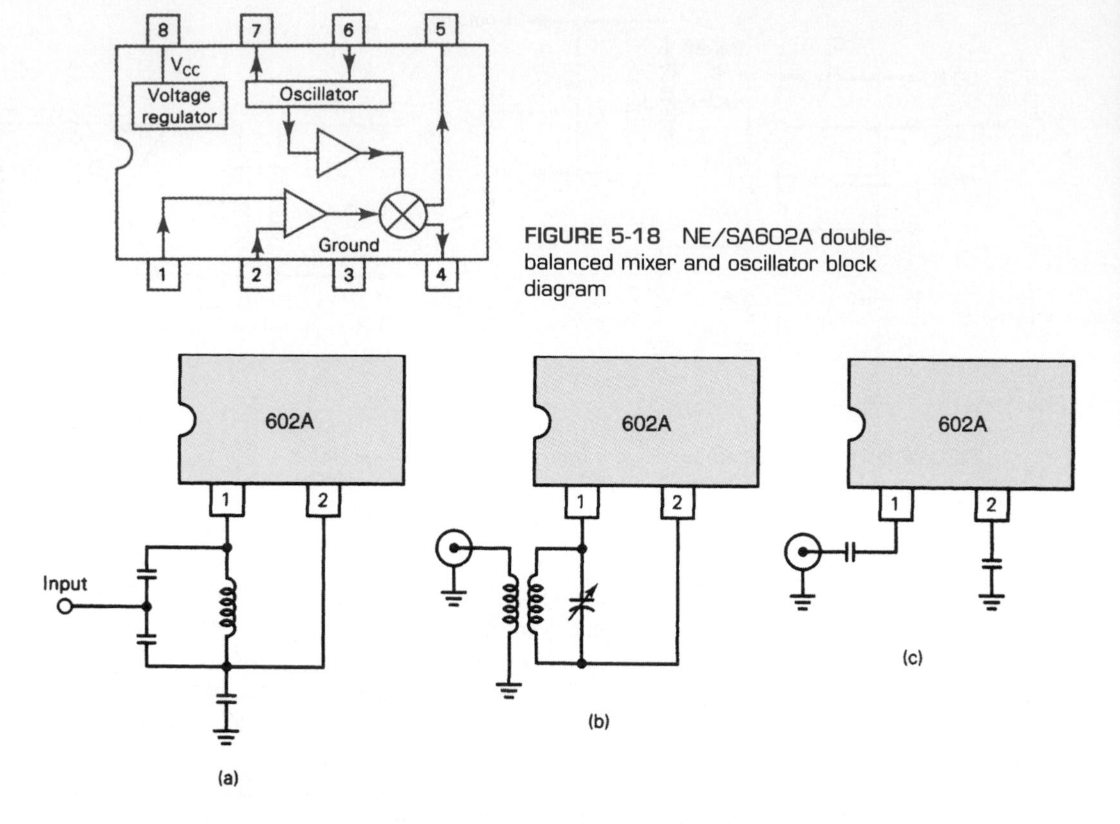

FIGURE 5-18 NE/SA602A double-balanced mixer and oscillator block diagram

FIGURE 5-19 NE/SA602A. Typical input configurations: (a) single-ended tuned input; (b) balanced input; (c) single-ended untuned input

5-4-3-1 Integrated-circuit mixer/oscillator. Figure 5-18 shows the block diagram for the Signetics NE/SA602A *double-balanced mixer and oscillator.* The NE/SA602A is a low-power VHF monolithic double-balanced mixer with input amplifier, on-board oscillator, and voltage regulator. It is intended to be used for high-performance, low-power communications systems; it is particularly well suited for *cellular radio* applications. The mixer is a *Gilbert cell* multiplier configuration, which typically provides 18 dB of gain at 45 MHz. A Gilbert cell is a differential amplifier that drives a balanced switching cell. The differential input stage provides gain and determines the noise figure and signal-handling performance of the system. The oscillator will operate up to 200 MHz and can be configured as a crystal or tuned LC tank-circuit oscillator of a buffer amplifier for an external oscillator. The noise figure for the NE/SA602A at 45 MHz is typically less than 5 dB. The gain, third-order intercept performance, and low-power and noise characteristics make the NE/SA602A a superior choice for high-performance, battery-operated equipment. The input, RF mixer output, and oscillator ports can support a variety of input configurations. The RF inputs (pins 1 and 2) are biased internally and are symmetrical. Figure 5-19 shows three typical input configurations: single-ended tuned input, balanced input, and single-ended untuned.

5-4-4 IF Amplifier Circuits

Intermediate frequency (IF) amplifiers are relatively high-gain tuned amplifiers that are very similar to RF amplifiers, except that IF amplifiers operate over a relatively narrow, fixed frequency band. Consequently, it is easy to design and build IF amplifiers that are stable, do not radiate, and are easily neutralized. Because IF amplifiers operate over a fixed frequency band, successive amplifiers can be inductively coupled with *double-tuned* circuits (with

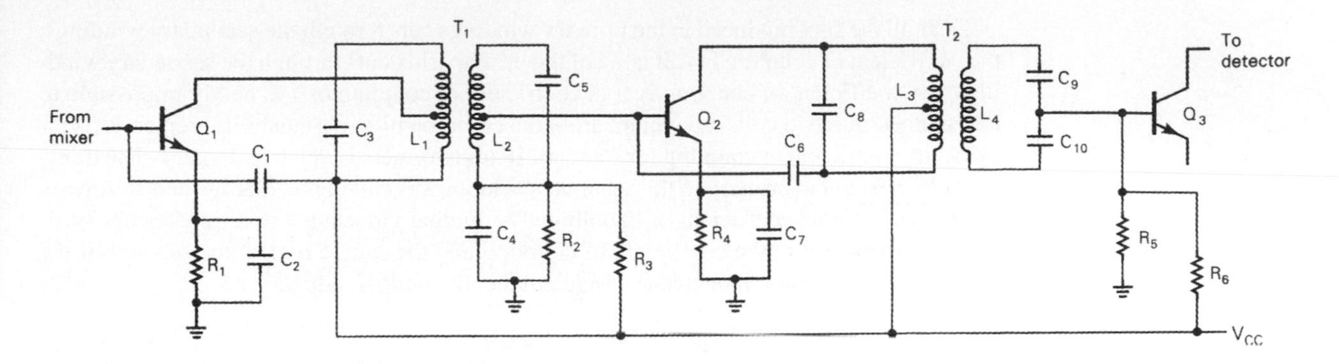

FIGURE 5-20 Three-stage IF section

double-tuned circuits, both the primary and secondary sides of the transformer are tuned tank circuits). Therefore, it is easier to achieve an optimum (low) shape factor and good selectivity. Most of a receiver's gain and selectivity is achieved in the IF amplifier section. An IF stage generally has between two and five IF amplifiers. Figure 5-20 shows a schematic diagram for a three-stage IF section. T_1 and T_2 are double-tuned transformers, and L_1, L_2, and L_3 are tapped to reduce the effects of loading. The base of Q_3 is fed from the tapped capacitor pair, C_9 and C_{10}, for the same reason. C_1 and C_6 are neutralization capacitors.

5-4-5 Inductive Coupling

Inductive or *transformer coupling* is the most common technique used for coupling IF amplifiers. With inductive coupling, voltage that is applied to the primary windings of a transformer is transferred to the secondary windings. The proportion of the primary voltage that is coupled across to the secondary depends on the number of turns in both the primary and secondary windings (the turns ratio), the amount of *magnetic flux* in the primary winding, the *coefficient of coupling,* and the speed at which the flux is changing (angular velocity). Mathematically, the magnitude of the voltage induced in the secondary windings is

$$E_s = \omega M I_p \tag{5-14}$$

where E_s = voltage magnitude induced in the secondary winding (volts)
ω = angular velocity of the primary voltage wave (radians per second)
M = mutual inductance (henrys)
I_p = primary current (amperes)

The ability of a coil to induce a voltage within its own windings is called *self-inductance* or simply *inductance* (*L*). When one coil induces a voltage through *magnetic induction* into another coil, the two coils are said to be *coupled* together. The ability of one coil to induce a voltage in another coil is called *mutual inductance* (*M*). Mutual inductance in a transformer is caused by the *magnetic lines of force* (*flux*) that are produced in the primary windings and cut through the secondary windings and is directly proportional to the coefficient of coupling. Coefficient of coupling is the ratio of the secondary flux to the primary flux and is expressed mathematically as

$$k = \frac{\phi_s}{\phi_p} \tag{5-15}$$

where k = coefficient of coupling (unitless)
ϕ_p = primary flux (webers)
ϕ_s = secondary flux (webers)

If all the flux produced in the primary windings cuts through the secondary windings, the coefficient of coupling is 1. If none of the primary flux cuts through the secondary windings, the coefficient of coupling is 0. A coefficient of coupling of 1 is nearly impossible to attain unless the two coils are wound around a common high-permeability iron core. Typically, the coefficient of coupling for standard IF transformers is much less than 1. The transfer of flux from the primary to the secondary windings is called *flux linkage* and is directly proportional to the coefficient of coupling. The mutual inductance of a transformer is directly proportional to the coefficient of coupling and the square root of the product of the primary and secondary inductances. Mathematically, mutual inductance is

$$M = k\sqrt{L_s L_p} \tag{5-16}$$

where M = mutual inductance (henrys)
L_s = inductance of the secondary winding (henrys)
L_p = inductance of the primary winding (henrys)
k = coefficient of coupling (unitless)

Transformer-coupled amplifiers are divided into two general categories: single and double tuned.

5-4-5-1 Single-tuned transformers. Figure 5-21a shows a schematic diagram for a *single-tuned inductively coupled amplifier.* This configuration is called *untuned primary–tuned secondary.* The primary side of T_1 is simply the inductance of the primary winding, whereas a capacitor is in parallel with the secondary winding, creating a tuned secondary. The transformer windings are not tapped because the loading effect of the FET is insignificant. Figure 5-21b shows the response curve for an untuned primary–tuned secondary transformer. E_s increases with frequency until the resonant frequency (f_o) of the secondary is reached; then E_s begins to decrease with further increases in frequency. The peaking of the

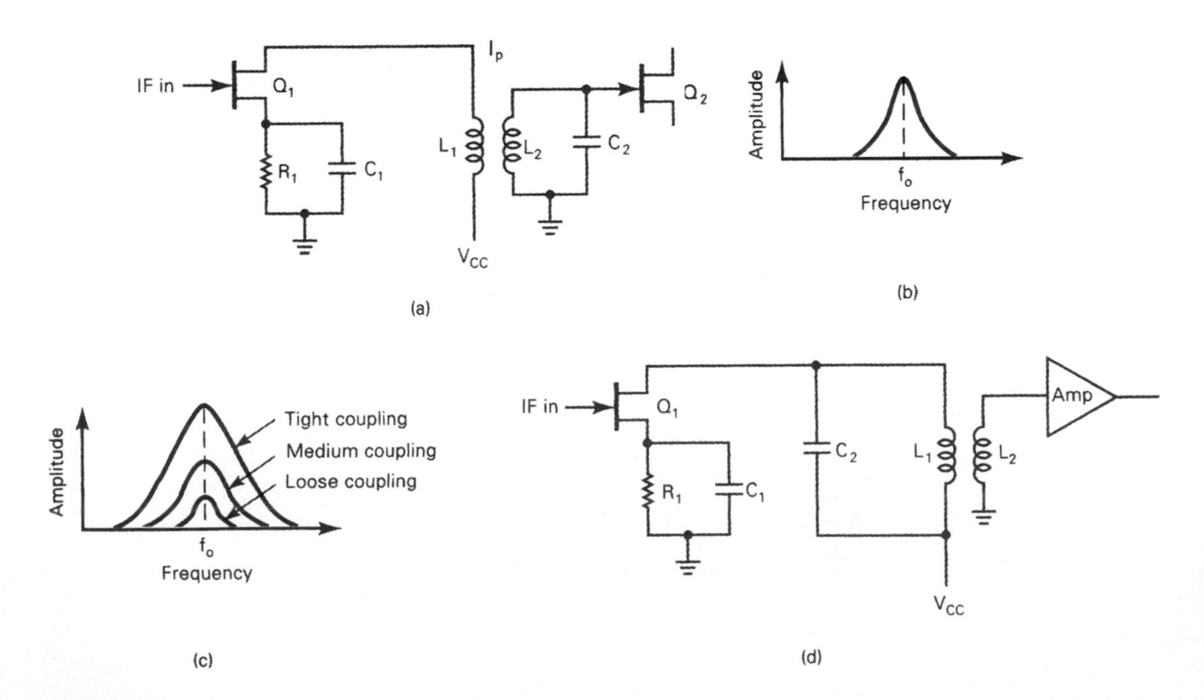

FIGURE 5-21 Single-tuned transformer: (a) schematic diagram; (b) response curve; (c) effects of coupling; (d) tuned primary–untuned secondary

response curve at the resonant frequency is caused by the reflected impedance. The impedance of the secondary is reflected back into the primary because of the mutual inductance between the two windings. For frequencies below resonance, the increase in ωM is greater than the decrease in I_p; therefore, E_s increases. For frequencies above resonance, the increase in ωM is less than the decrease in I_p; therefore, E_s decreases.

Figure 5-21c shows the effect of coupling on the response curve of an untuned primary–tuned secondary transformer. With *loose coupling* (low coefficient of coupling), the secondary voltage is relatively low, and the bandwidth is narrow. As the degree of coupling increases (coefficient of coupling increases), the secondary induced voltage increases, and the bandwidth widens. Therefore, for a high degree of selectivity, loose coupling is desired; however, signal amplitude is sacrificed. For high gain and a broad bandwidth, *tight coupling* is necessary. Another single-tuned amplifier configuration is the *tuned primary–untuned secondary,* which is shown in Figure 5-21d.

5-4-5-2 Double-tuned transformers. Figure 5-22a shows the schematic diagram for a *double-tuned inductively coupled amplifier.* This configuration is called *tuned primary–tuned secondary* because both the primary and secondary windings of transformer T_1 are tuned tank circuits. Figure 5-22b shows the effect of coupling on the response curve for a double-tuned inductively coupled transformer. The response curve closely resembles that of a single-tuned circuit for coupling values below *critical coupling* (k_c). Critical coupling is the point where the reflected resistance is equal to the primary resistance and the Q of the primary tank circuit is halved and the bandwidth doubled. If the coefficient of coupling is increased beyond the critical point, the response at the resonant frequency decreases, and two new peaks occur on either side of the resonant frequency. This *double peaking* is caused by the reactive element of the reflected impedance being significant enough to change the resonant frequency of the primary tuned circuit. If the coefficient of coupling is increased further, the dip at resonance becomes more pronounced,

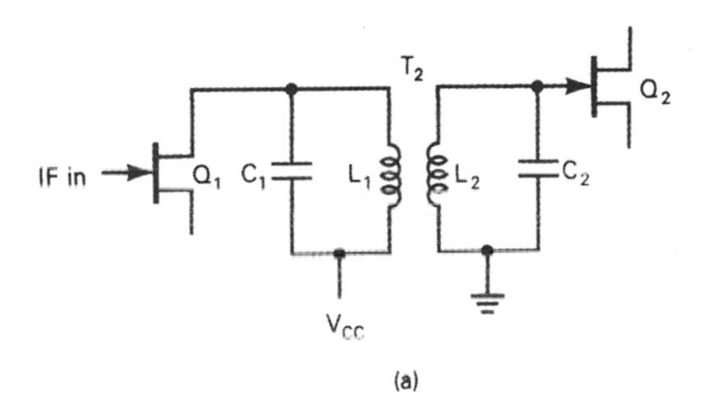

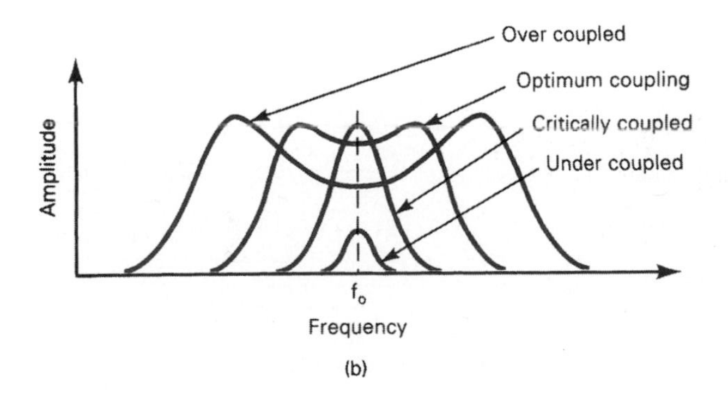

FIGURE 5-22 Double-tuned transformer: (a) schematic diagram; (b) response curve

and the two peaks are spread even farther away from the resonant frequency. Increasing coupling beyond the critical value broadens the bandwidth but, at the same time, produces a ripple in the response curve. An ideal response curve has a rectangular shape (a flat top with vertical skirts). From Figure 5-22b, it can be seen that a coefficient of coupling approximately 50% greater than the critical value yields a good compromise between flat response and steep skirts. This value of coupling is called *optimum coupling* (k_{opt}) and is expressed mathematically as

$$k_{opt} = 1.5k_c \tag{5-17}$$

where k_{opt} = optimum coupling

k_c = critical coupling $= 1/\sqrt{Q_p Q_s}$

where Q_p and Q_s are uncoupled values.

The bandwidth of a double-tuned amplifier is

$$B_{dt} = kf_o \tag{5-18}$$

5-4-5-3 Bandwidth reduction. When several tuned amplifiers are cascaded together, the total response is the product of the amplifiers' individual responses. Figure 5-23a shows a response curve for a tuned amplifier. The gain at f_1 and f_2 is 0.707 of the gain at f_o. If two identical tuned amplifiers are cascaded, the gain at f_1 and f_2 will be reduced to 0.5 (0.707 × 0.707), and if three identical tuned amplifiers are cascaded, the gain at f_1 and f_2 is reduced to 0.353. Consequently, as additional tuned amplifiers are added, the overall shape of the response curve

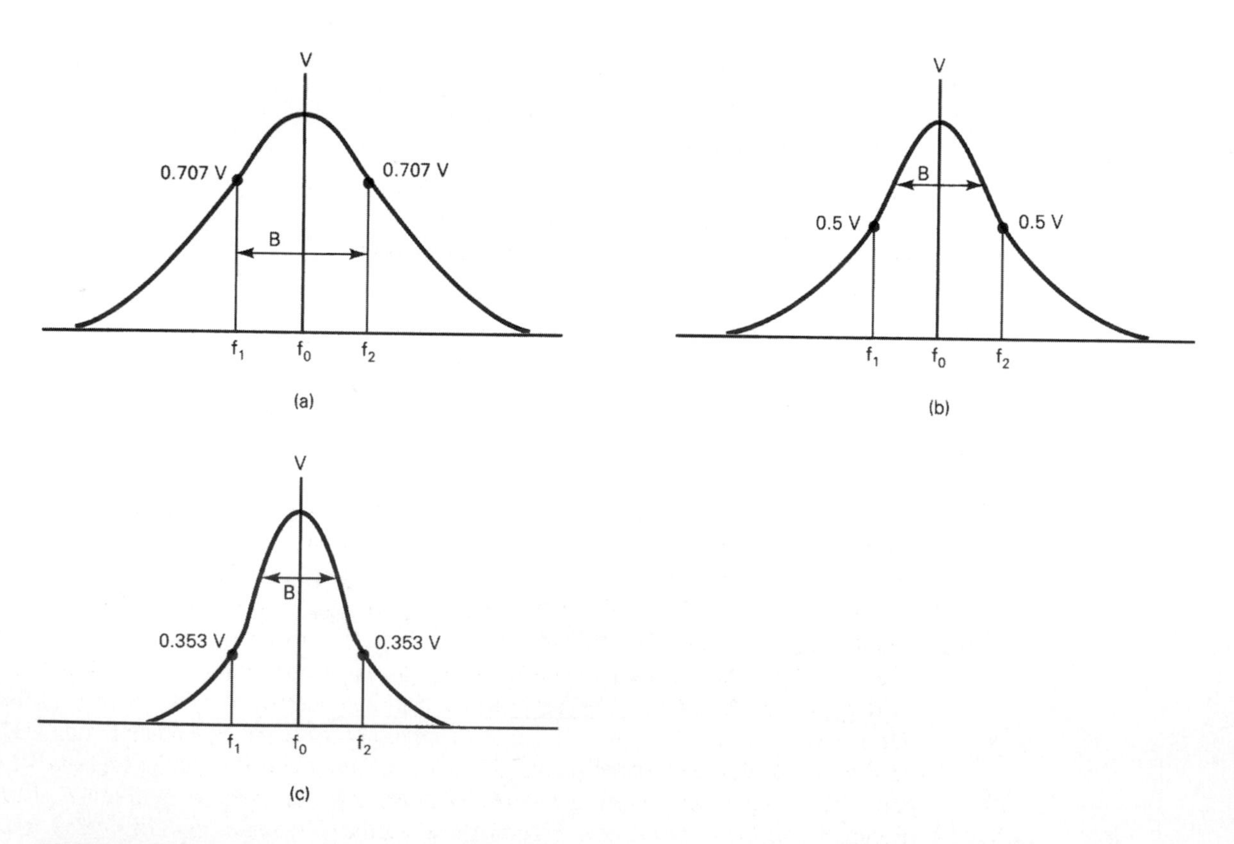

FIGURE 5-23 Bandwidth reduction: (a) single-tuned stage; (b) two cascaded stages; (c) three cascaded stages

narrows, and the bandwidth is reduced. This bandwidth reduction is shown in Figures 5-23b and c. Mathematically, the overall bandwidth of n single-tuned stages is given as

$$B_n = B_1(\sqrt{2^{1/n} - 1}) \tag{5-19}$$

where B_n = bandwidth of n single-tuned stages (hertz)
B_1 = bandwidth of one single-tuned stage (hertz)
n = number of stages (any positive integer)

The bandwidth for n double-tuned stages is

$$B_{ndt} = B_{1dt}[2^{1/n} - 1]^{1/4} \tag{5-20}$$

where B_{ndt} = overall bandwidth of n double-tuned amplifiers (hertz)
B_{1dt} = bandwidth of double-tuned amplifier (hertz)
n = number of double-tuned stages (any positive integer)

Example 5-7

Determine the overall bandwidth for

a. Two single-tuned amplifiers each with a bandwidth of 10 kHz.
b. Three single-tuned amplifiers each with a bandwidth of 10 kHz.
c. Four single-tuned amplifiers each with a bandwidth of 10 kHz.
d. A double-tuned amplifier with optimum coupling, a critical coupling of 0.02, and a resonant frequency of 1 MHz.
e. Repeat parts (a), (b), and (c) for the double-tuned amplifier of part (d).

Solution **a.** From Equation 5-19,

$$B_2 = 10\text{ kHz}(\sqrt{2^{1/2} - 1}) = 6436\text{ Hz}$$

b. Again from Equation 5-19,

$$B_3 = 10\text{ kHz}(\sqrt{2^{1/3} - 1}) = 5098\text{ Hz}$$

c. Again from Equation 5-19,

$$B_4 = 10\text{ kHz}(\sqrt{2^{1/4} - 1}) = 4350\text{ Hz}$$

d. From Equation 5-17, $K_{opt} = 1.5(0.02) = 0.03$
From Equation 5-18, $B_{dt} = 0.03(1\text{ MHz}) = 30\text{ kHz}$
e. From Equation 5-20,

n	B (Hz)
2	24,067
3	21,420
4	19,786

IF transformers come as specially designed tuned circuits in groundable metal packages called *IF cans.* Figures 5-24a and b show the physical and schematic diagrams for a typical IF can. The primary winding comes with a shunt 125-pF capacitor. The vertical arrow shown between the primary and secondary windings indicates that the ferrite core is tunable with a nonmetallic screwdriver or tuning tool. Adjusting the ferrite core changes the mutual inductance, which controls the magnitude of the voltage induced into the secondary windings. The tap in the primary winding can be used to increase the Q of the collector circuit of the driving transistor. If the tap is not used, the equivalent circuit is shown in Figure 5-24b; the effective $Q = R_L/X_L$ and the bandwidth $B = f/Q$, where f is the resonant frequency. If ac ground is connected to the tap, the equivalent circuit is shown in Figure 5-24c. With the tap at ac ground, the effective Q increases and the overall response is more selective.

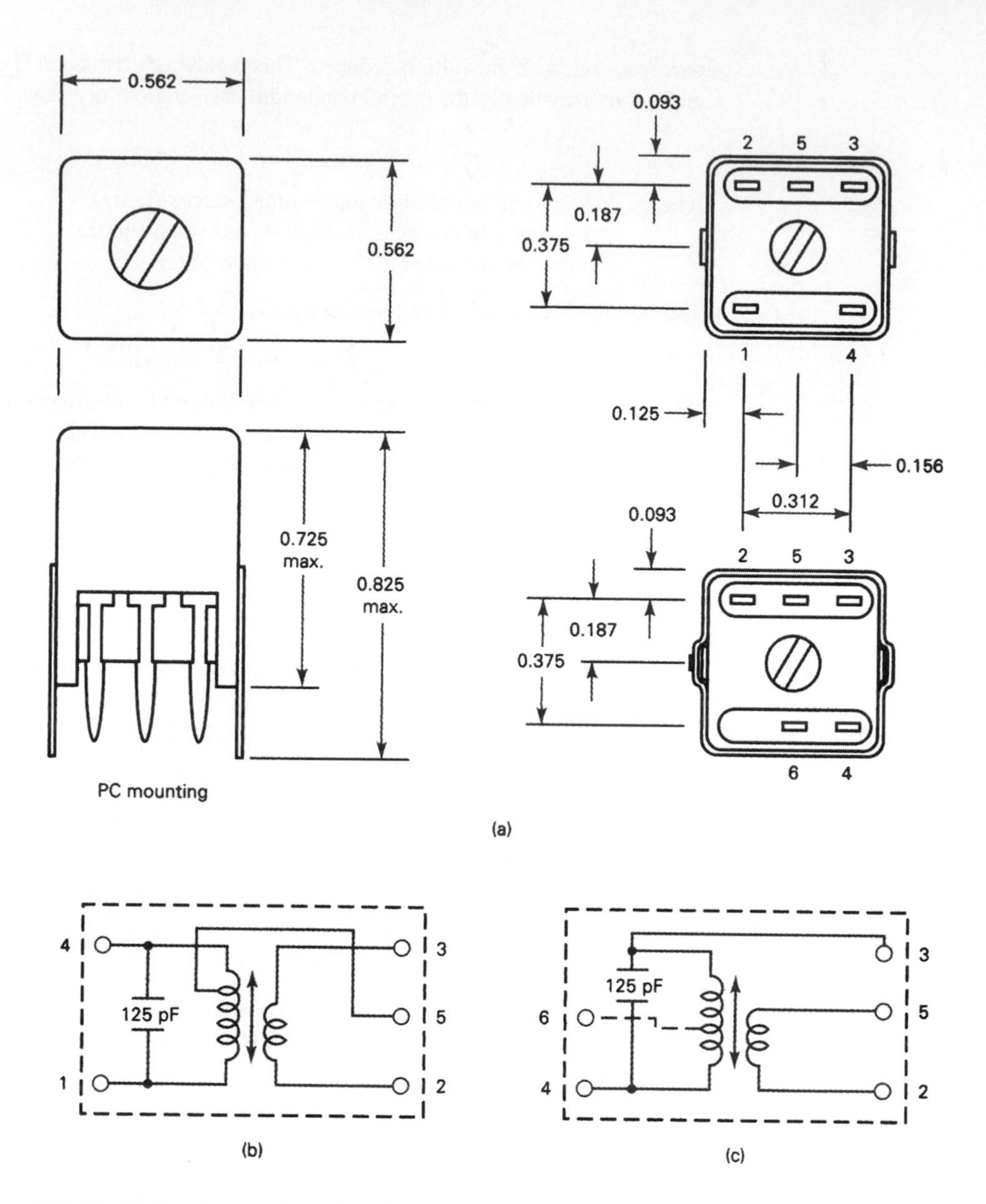

FIGURE 5-24 IF can: (a) physical diagram; (b) schematic diagram untapped coil; (c) schematic diagram tapped coil

5-4-6 Integrated-Circuit IF Amplifiers

In recent years, integrated-circuit IF amplifiers have seen universal acceptance in mobile radio systems such as two-way radio. Integrated circuits offer the obvious advantages of small size and low power consumption. One of the most popular IC intermediate-frequency amplifiers is the CA3028A. The CA3028A is a differential cascoded amplifier designed for use in communications and industrial equipment as an IF or RF amplifier at frequencies from dc to 120 MHz. The CA3028A features a controlled input offset voltage, offset current, and input bias current. It uses a balanced differential amplifier configuration with a controlled constant-current source and can be used for both single- and dual-ended operation. The CA3028A has balanced-AGC (automatic gain control) capabilities and a wide operating-current range.

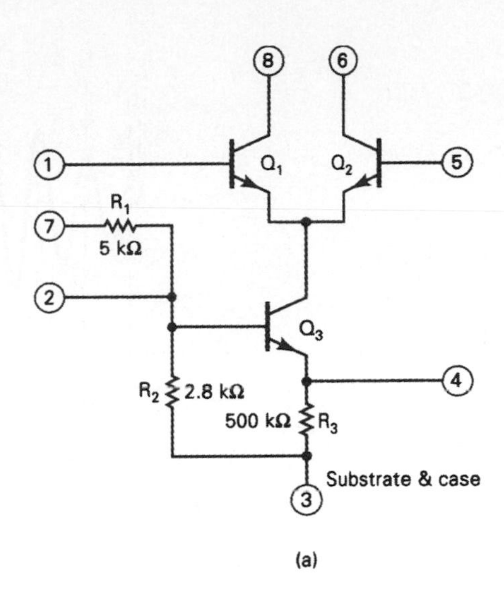

(a)

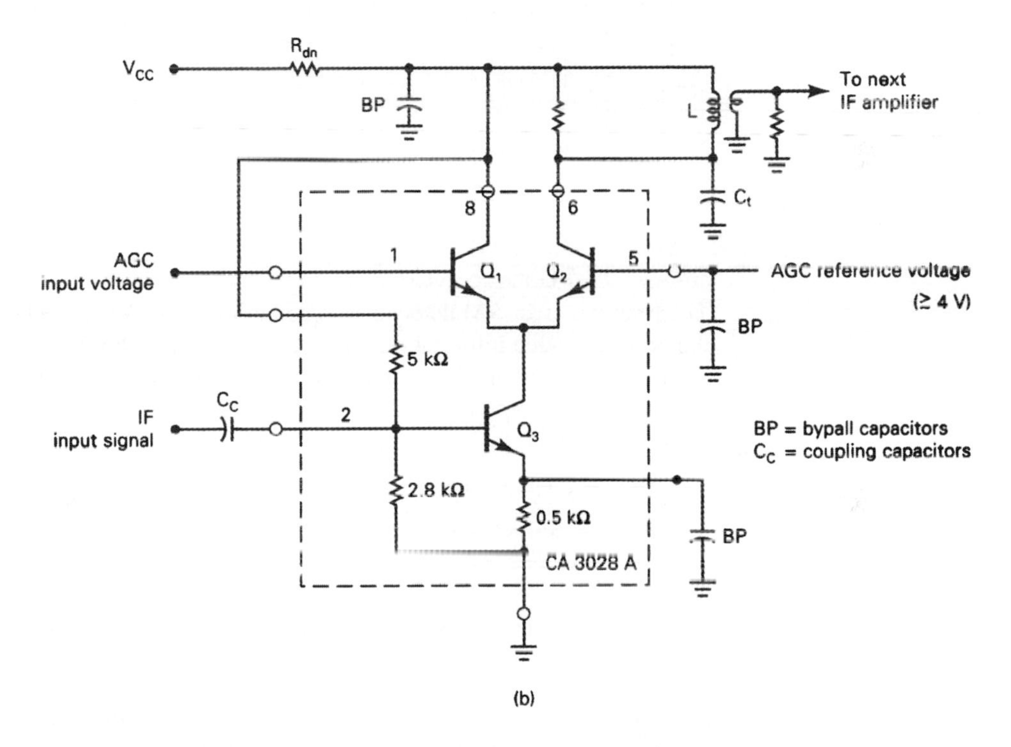

(b)

FIGURE 5-25 CA3028A linear integrated-circuit differential/cascoded amplifier: (a) schematic diagram; (b) cascoded amplified configuration

Figure 5-25a shows the schematic diagram for the CA3028A, and Figure 5-25b shows the CA3028A used as a cascoded IF amplifier. The IF input is applied to pin 2, and the output is taken from pin 6. When the AGC voltage on pin 1 is equal to the AGC reference voltage on pin 5, the emitter currents flowing in Q_1 and Q_2 are equal, and the amplifier has high gain. If the AGC voltage on pin 1 increases, Q_2 current decreases, and the stage gain decreases.

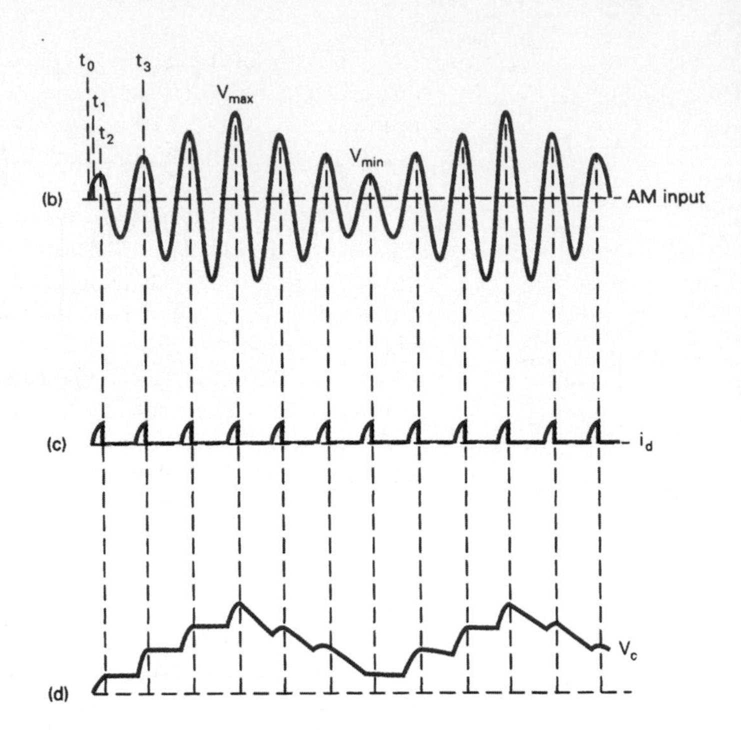

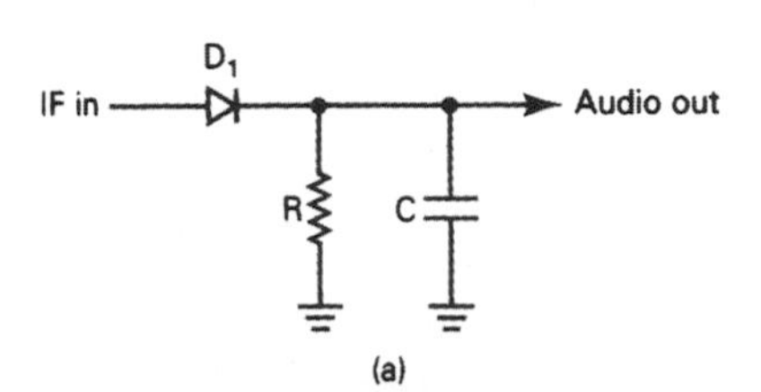

FIGURE 5-26 Peak detector: (a) schematic diagram; (b) AM input waveform; (c) diode current waveform; (d) output voltage waveform

5-4-7 AM Detector Circuits

The function of an AM detector is to demodulate the AM signal and recover or reproduce the original source information. The recovered signal should contain the same frequencies as the original information signal and have the same relative amplitude characteristics. The AM detector is sometimes called the *second detector,* with the mixer/converter being the first detector because it precedes the AM detector.

5-4-7-1 Peak detector. Figure 5-26a shows a schematic diagram for a simple noncoherent AM demodulator, which is commonly called a *peak detector.* Because a diode is a nonlinear device, nonlinear mixing occurs in D_1 when two or more signals are applied to its input. Therefore, the output contains the original input frequencies, their harmonics, and their cross products. If a 300-kHz carrier is amplitude modulated by a 2-kHz sine wave, the modulated wave is made up of a lower side frequency, carrier, and upper side frequency of 298 kHz, 300 kHz, and 302 kHz, respectively. If the resultant signal is the input to the AM detector shown in Figure 5-26a, the output will comprise the three input frequencies, the harmonics of all three frequencies, and the cross products of all possible combinations of the three frequencies and their harmonics. Mathematically, the output is

$$V_{\text{out}} = \text{input frequencies} + \text{harmonics} + \text{sums and differences}$$

Because the *RC* network is a low-pass filter, only the difference frequencies are passed on to the audio section. Therefore, the output is simply

$$\begin{aligned} V_{\text{out}} &= 300 - 298 = 2 \text{ kHz} \\ &= 302 - 300 = 2 \text{ kHz} \\ &= 302 - 298 = 4 \text{ kHz} \end{aligned}$$

Because of the relative amplitude characteristics of the upper and lower side frequencies and the carrier, the difference between the carrier frequency and either the upper or lower side frequency is the predominant output signal. Consequently, for practical purposes, the original modulating signal (2 kHz) is the only component that is contained in the output of the peak detector.

In the preceding analysis, the diode detector was analyzed as a simple mixer, which it is. Essentially, the difference between an AM modulator and an AM demodulator is that the output of a modulator is tuned to the sum frequencies (up-converter), whereas the output of a demodulator is tuned to the difference frequencies (down-converter). The demodulator circuit shown in Figure 5-26a is commonly called a *diode detector* because the nonlinear device is a diode, a *peak detector* because it detects the peaks of the input envelope, or a *shape* or *envelope detector* because it detects the shape of the input envelope. Essentially, the carrier signal *captures* the diode and forces it to turn on and off (rectify) synchronously (both frequency and phase). Thus, the side frequencies mix with the carrier, and the original baseband signals are recovered.

Figures 5-26b, c, and d show a detector input voltage waveform, the corresponding diode current waveform, and the detector output voltage waveform. At time t_0, the diode is reverse biased and off ($i_d = 0$ A), the capacitor is completely discharged ($V_C = 0$ V), and thus, the output is 0 V. The diode remains off until the input voltage exceeds the barrier potential of D_1 (approximately 0.3 V). When V_{in} reaches 0.3 V (t_1), the diode turns on, and diode current begins to flow, charging the capacitor. The capacitor voltage remains 0.3 V below the input voltage until V_{in} reaches its peak value. When the input voltage begins to decrease, the diode turns off, and i_d goes to 0 A (t_2). The capacitor begins to discharge through the resistor, but the RC time constant is made sufficiently long so that the capacitor cannot discharge as rapidly as V_{in} is decreasing. The diode remains off until the next input cycle, when V_{in} goes 0.3 V more positive than V_C (t_3). At this time, the diode turns on, current flows, and the capacitor begins to charge again. It is relatively easy for the capacitor to charge to the new value because the RC charging time constant is R_dC, where R_d is the *on* resistance of the diode, which is quite small. This sequence repeats itself on each successive positive peak of V_{in}, and the capacitor voltage follows the positive peaks of V_{in} (hence the name *peak detector*). The output waveform resembles the shape of the input envelope (hence, the name *shape detector*). The output waveform has a high-frequency ripple that is equal to the carrier frequency. This is due to the diode turning on during the positive peaks of the envelope. The ripple is easily removed by the audio amplifiers because the carrier frequency is much higher than the highest modulating signal frequency. The circuit shown in Figure 5-26 responds only to the positive peaks of V_{in} and, therefore, is called a *positive peak detector.* By simply turning the diode around, the circuit becomes a negative peak detector. The output voltage reaches its peak positive amplitude at the same time that the input envelope reaches its maximum positive value (V_{max}), and the output voltage goes to its minimum peak amplitude at the same time that the input voltage goes to its minimum value (V_{min}). For 100% modulation, V_{out} swings from 0 V to a value equal to $V_{max} - 0.3$ V.

Figure 5-27 shows the input and output waveforms for a peak detector with various percentages of modulation. With no modulation, a peak detector is simply a filtered half-wave rectifier, and the output voltage is approximately equal to the peak input voltage minus the 0.3 V. As the percent modulation changes, the variations in the output voltage increase and decrease proportionately; the output waveform follows the shape of the AM envelope. However, regardless of whether modulation is present, the average value of the output voltage is approximately equal to the peak value of the unmodulated carrier.

5-4-7-2 Detector distortion. When successive positive peaks of the detector input waveform are increasing, it is important that the capacitor hold its charge between

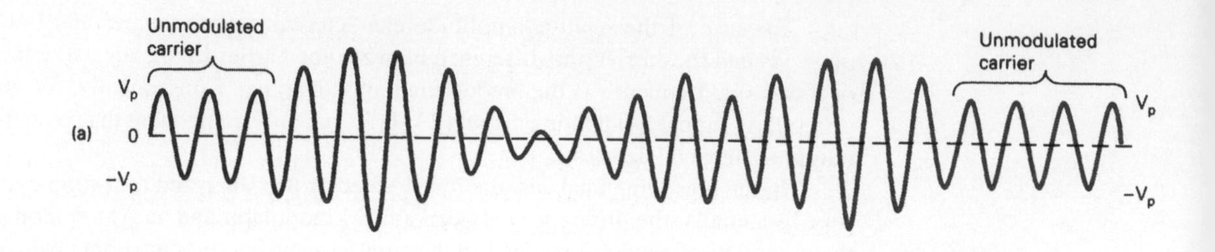

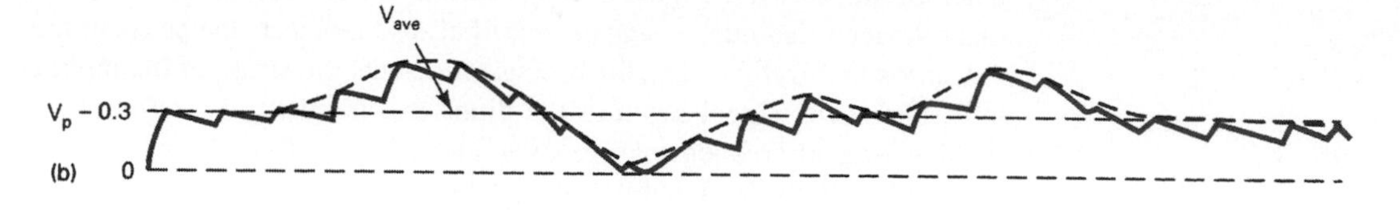

FIGURE 5-27 Positive peak detector: (a) input waveform; (b) output waveform

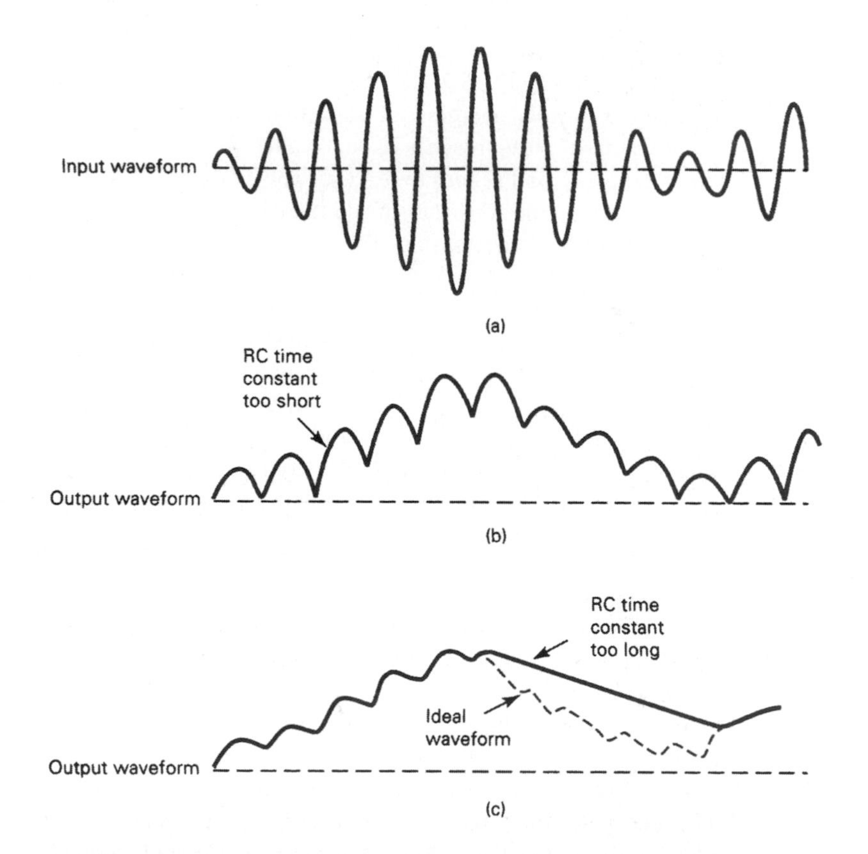

FIGURE 5-28 Detector distortion: (a) input envelope; (b) rectifier distortion; (c) diagonal clipping

peaks (i.e., a relatively long *RC* time constant is necessary). However, when the positive peaks are decreasing in amplitude, it is important that the capacitor discharge between successive peaks to a value less than the next peak (a short *RC* time constant is necessary). Obviously, a trade-off between a long- and a short-time constant is in order. If the *RC* time constant is too short, the output waveform resembles a half-wave rectified signal. This is sometimes called *rectifier distortion* and is shown in Figure 5-28b. If the *RC* time constant is too long, the slope of the output waveform cannot follow the trailing slope of the envelope. This type of distortion is called *diagonal clipping* and is shown in Figure 5-28c.

The RC network following the diode in a peak detector is a low-pass filter. The slope of the envelope depends on both the modulating signal frequency and the modulation coefficient (m). Therefore, the maximum slope (fastest rate of change) occurs when the envelope is crossing its zero axis in the negative direction. The highest modulating signal frequency that can be demodulated by a peak detector without attenuation is given as

$$f_{m(\max)} = \frac{\sqrt{(1/m^2) - 1}}{2\pi RC} \qquad \textbf{(5-21)}$$

where $f_{m(\max)}$ = maximum modulating signal frequency (hertz)
m = modulation coefficient (unitless)
RC = time constant (seconds)

For 100% modulation, the numerator in Equation 5-21 goes to zero, which essentially means that all modulating signal frequencies are attenuated as they are demodulated. Typically, the modulating signal amplitude in a transmitter is limited or compressed such that approximately 90% modulation is the maximum that can be achieved. For 70.7% modulation, Equation 5-21 reduces to

$$f_{m(\max)} = \frac{1}{2\pi RC} \qquad \textbf{(5-22)}$$

Equation 5-22 is commonly used when designing peak detectors to determine an approximate maximum modulating signal.

5-4-8 Automatic Gain Control Circuits

An *automatic gain control* (AGC) circuit compensates for minor variations in the received RF signal level. The AGC circuit automatically increases the receiver gain for weak RF input levels and automatically decreases the receiver gain when a strong RF signal is received. Weak signals can be buried in receiver noise and, consequently, be impossible to detect. An excessively strong signal can overdrive the RF and/or IF amplifiers and produce excessive nonlinear distortion and even saturation. There are several types of AGC, which include direct or simple AGC, delayed AGC, and forward AGC.

5-4-8-1 Simple AGC. Figure 5-29 shows a block diagram for an AM superheterodyne receiver with simple AGC. The automatic gain control circuit monitors the received signal level and sends a signal back to the RF and IF amplifiers to adjust their gain automatically. AGC is a form of degenerative or negative feedback. The purpose of AGC is to allow a receiver to detect and demodulate, equally well, signals that are transmitted from different stations whose output power and distance from the receiver vary. For example, an AM radio in a vehicle does not receive the same signal level from all the transmitting stations in the area or, for that matter, from a single station when the automobile is moving. The AGC circuit produces a voltage that adjusts the receiver gain and keeps the IF carrier power at the input to the AM detector at a relatively constant level. The AGC circuit is not a form of *automatic volume control* (AVC); AGC is independent of modulation and totally unaffected by normal changes in the modulating signal amplitude.

Figure 5-30 shows a schematic diagram for a simple AGC circuit. As you can see, an AGC circuit is essentially a peak detector. In fact, very often the AGC correction voltage is taken from the output of the audio detector. In Figure 5-27, it was shown that the dc voltage at the output of a peak detector is equal to the peak unmodulated carrier amplitude minus the barrier potential of the diode and is totally independent of the depth of modulation.

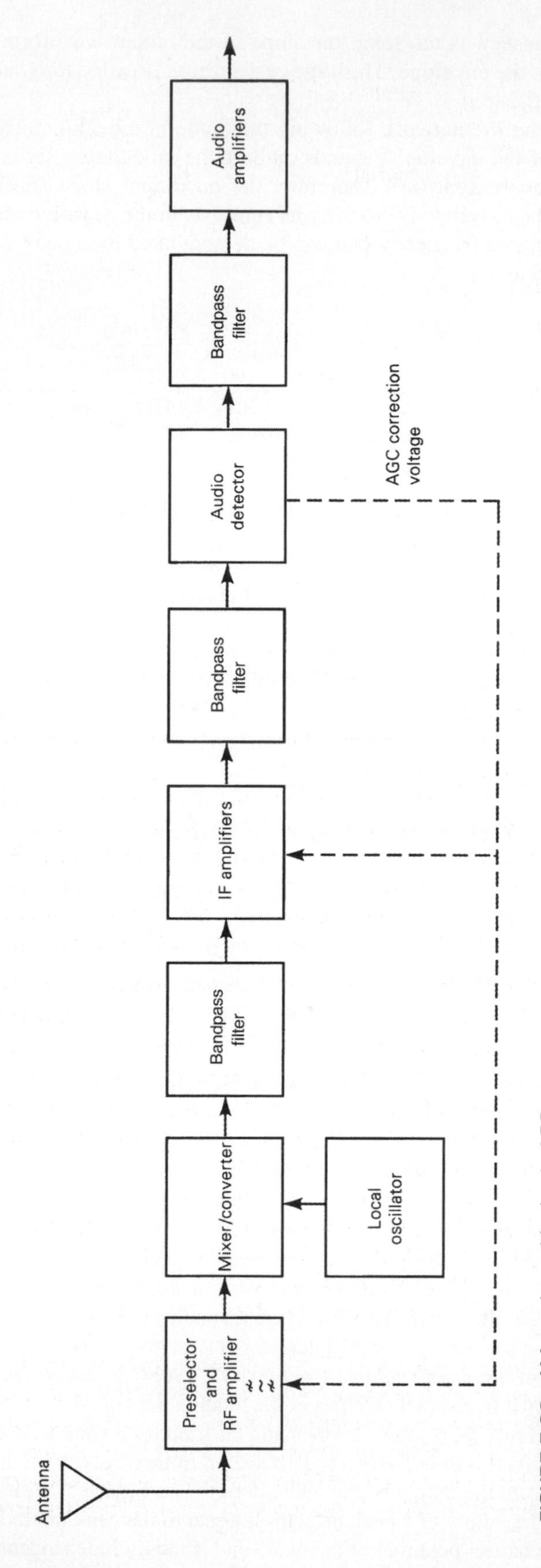

FIGURE 5-29 AM receiver with simple AGC

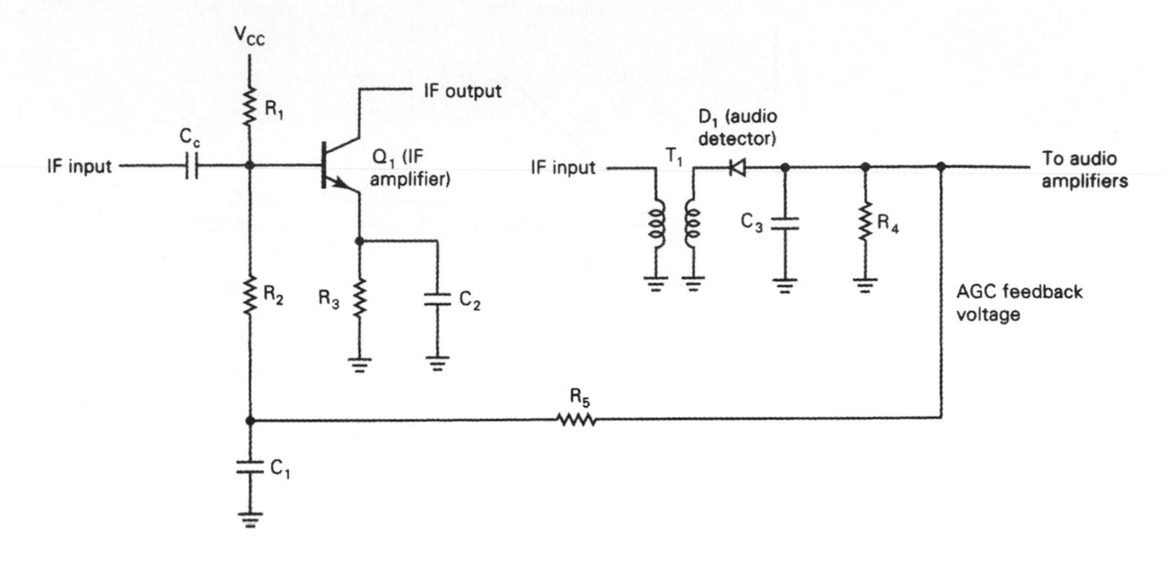

FIGURE 5-30 Simple AGC circuit

If the carrier amplitude increases, the AGC voltage increases, and if the carrier amplitude decreases, the AGC voltage decreases. The circuit shown in Figure 5-30 is a negative peak detector and produces a negative voltage at its output. The greater the amplitude of the input carrier, the more negative the output voltage. The negative voltage from the AGC detector is fed back to the IF stage, where it controls the bias voltage on the base of Q_1. When the carrier amplitude increases, the voltage on the base of Q_1 becomes less positive, causing the emitter current to decrease. As a result, r'_e increases and the amplifier gain (r_c/r'_e) decreases, which in turn causes the carrier amplitude to decrease. When the carrier amplitude decreases, the AGC voltage becomes less negative, the emitter current increases, r'_e decreases, and the amplifier gain increases. Capacitor C_1 is an audio bypass capacitor that prevents changes in the AGC voltage due to modulation from affecting the gain of Q_1.

5-4-8-2 Delayed AGC. Simple AGC is used in most inexpensive broadcast-band radio receivers. However, with simple AGC, the AGC bias begins to increase as soon as the received signal level exceeds the thermal noise of the receiver. Consequently, the receiver becomes less sensitive (which is sometimes called *automatic desensing*). *Delayed AGC* prevents the AGC feedback voltage from reaching the RF or IF amplifiers until the RF level exceeds a predetermined magnitude. Once the carrier signal has exceeded the threshold level, the delayed AGC voltage is proportional to the carrier signal strength. Figure 5-31a shows the response characteristics for both simple and delayed AGC. It can be seen that with delayed AGC, the receiver gain is unaffected until the AGC threshold level is exceeded, whereas with simple AGC, the receiver gain is immediately affected. Delayed AGC is used with more sophisticated communications receivers. Figure 5-31b shows IF gain-versus-RF input signal level for both simple and delayed AGC.

5-4-8-3 Forward AGC. An inherent problem with both simple and delayed AGC is the fact that they are both forms of *post-AGC* (after-the-fact) compensation. With post-AGC, the circuit that monitors the carrier level and provides the AGC correction voltage is located after the IF amplifiers; therefore, the simple fact that the AGC voltage changed indicates that it may be too late (the carrier level has already changed and propagated through the receiver). Therefore, neither simple nor delayed AGC can accurately compensate for

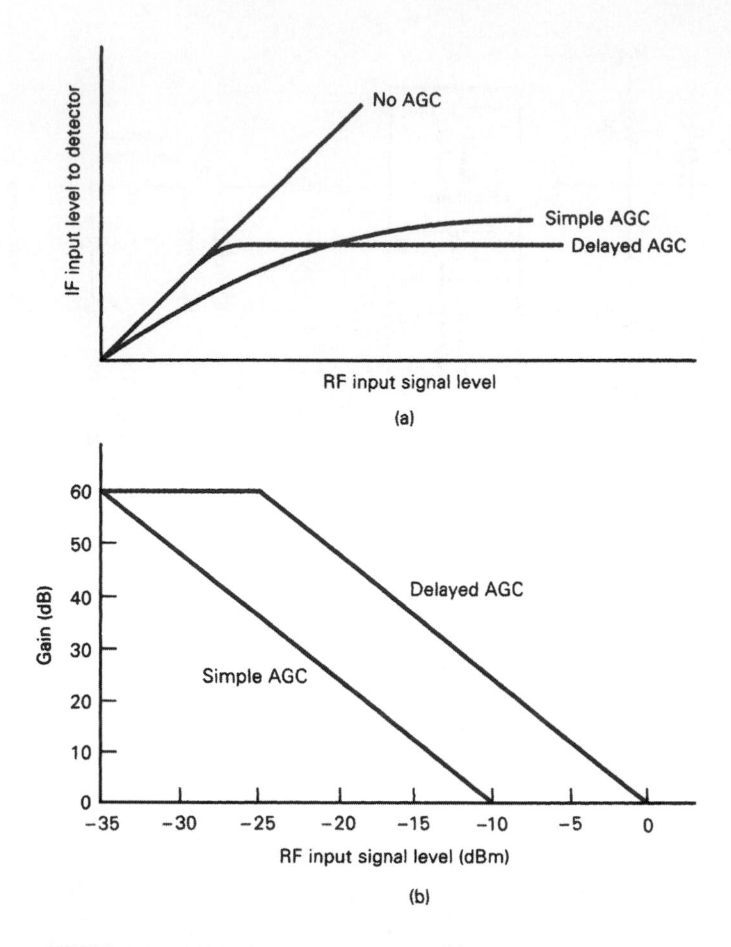

FIGURE 5-31 Automatic gain control (AGC): (a) response characteristics; (b) IF gain-versus-RF input signal level

rapid changes in the carrier amplitude. *Forward AGC* is similar to conventional AGC except that the receive signal is monitored closer to the front end of the receiver and the correction voltage is fed forward to the IF amplifiers. Consequently, when a change in signal level is detected, the change can be compensated for in succeeding stages. Figure 5-32 shows an AM superheterodyne receiver with forward AGC.

5-4-9 Squelch Circuits

The purpose of a *squelch circuit* is to *quiet* a receiver in the absence of a received signal. If an AM receiver is tuned to a location in the RF spectrum where there is no RF signal, the AGC circuit adjusts the receiver for maximum gain. Consequently, the receiver amplifies and demodulates its own internal noise. This is the familiar crackling and sputtering heard on the speaker in the absence of a received carrier. In domestic AM systems, each station is continuously transmitting a carrier regardless of whether there is any modulation. Therefore, the only time the idle receiver noise is heard is when tuning between stations. However, in two-way radio systems, the carrier in the transmitter is generally turned off except when a modulating signal is present. Therefore, during idle transmission times, a receiver is simply amplifying and demodulating noise. A squelch circuit keeps the audio section of the receiver turned off or *muted* in the absence of a received signal (the receiver is squelched). A disadvantage of a squelch circuit is weak RF signals will not produce an audio output.

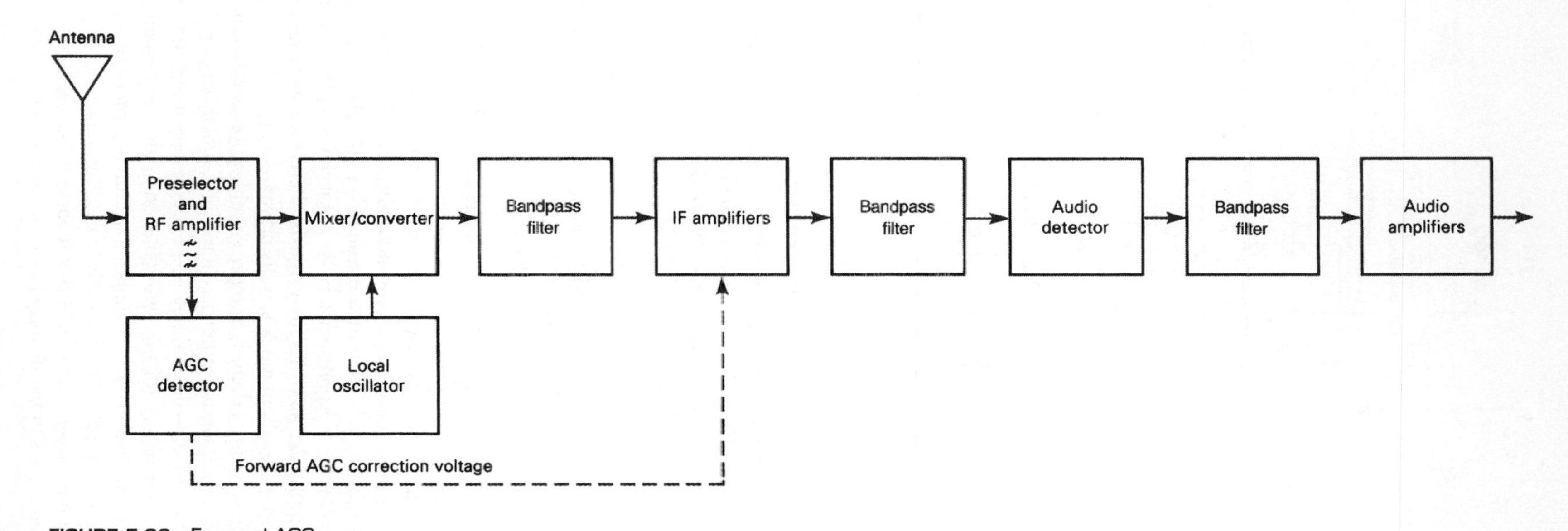

FIGURE 5-32 Forward AGC

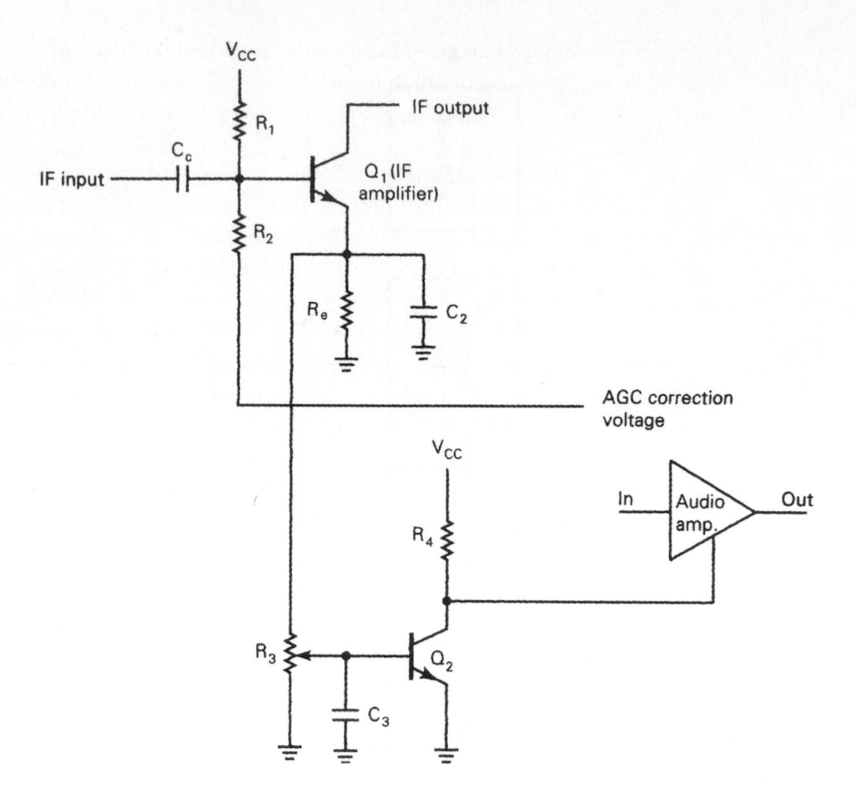

FIGURE 5-33 Squelch circuit

Figure 5-33 shows a schematic diagram for a typical squelch circuit. This squelch circuit uses the AGC voltage to monitor the received RF signal level. The greater the AGC voltage, the stronger the RF signal. When the AGC voltage drops below a preset level, the squelch circuit is activated and disables the audio section of the receiver. In Figure 5-33, it can be seen that the squelch detector uses a resistive voltage divider to monitor the AGC voltage. When the RF signal drops below the squelch threshold, Q_2 turns on and shuts off the audio amplifiers. When the RF signal level increases above the squelch threshold, the AGC voltage becomes more negative, turning off Q_2 and enabling the audio amplifiers. The squelch threshold level can be adjusted with R_3.

5-4-10 Noise Limiters and Blankers

Sporadic, high-amplitude noise transients of short duration, such as impulse noise, can often be removed using diode *limiters* or *clippers* in the audio section of a receiver. The limiting or clipping threshold level is normally established just above the maximum peak level of the audio signal. Therefore, the signal is virtually unaffected by them, but noise pulses will be limited to approximately the same level as the signal. Noise pulses are generally large-amplitude, short-duration signals; therefore, limiting them removes much of their energy and leaves them much less disturbing.

A *blanking circuit* is another circuit option commonly used for reducing the effects of high-amplitude noise pulses. In essence, a blanking circuit detects the occurrence of a high-amplitude, short-duration noise spike, then mutes the receiver by shutting off a portion of the receiver for the duration of the pulse. For example, when a noise pulse is detected at the input to the IF amplifier section of a receiver, the blanking circuit shuts off the IF amplifiers for the duration of the pulse, thus quieting the receiver. Shutting off the IF amplifiers has proven more effective than shutting off the audio section because the wider bandpass filters in the IF stage have a tendency to broaden the noise pulse.

Limiter and blanker circuits have little effect on white noise, however, because white noise power is generally much lower than the signal power level, and limiters and blankers work only when the noise surges to a level above that of the signal.

5-4-11 Alternate Signal-to-Noise Measurements

Sensitivity has little meaning unless it is accompanied by a signal-to-noise ratio and, because it is difficult to separate the signal from the noise and vice versa, sensitivity is often accompanied by a *signal plus noise-to-noise reading* (S+ N)/N. The sensitivity of an AM receiver is generally given as the minimum signal level at the input of the receiver with 30% modulation necessary to produce at least 500 mW of audio output power with a 10-dB (S+ N)/N ratio.

To measure (S+ N)/N, an RF carrier modulated 30% by a 1-kHz tone is applied to the input of the receiver. Total audio power is measured at the output of the receiver, which includes the audio signal and any noise present within the audio bandwidth. The modulation is then removed from the RF signal, and the total audio power is again measured. This time, however, only noise signals are present. The purpose of leaving the carrier on rather than terminating the input to the receiver is that the carrier is necessary to prevent the AGC circuit from detecting an absence of the carrier and turning the gain of the receiver up to maximum. The noise reading when the receiver is operating at maximum gain would amplify the internally generated noise well beyond its normal level, thus yielding a meaningless noise reading.

Another method of measuring signal strength relative to noise strength is called the *signal-to-notched noise* ratio. Again, an RF carrier modulated 30% by a 1-kHz tone is applied to the input of the receiver. Total audio power plus noise is measured at the output of the receiver. A narrowband 1-kHz notch filter is then inserted between the receiver output and the power meter, and another power measurement is taken. Again, the power reading will include only noise. This time, however, the entire receiver operated under near-normal signal conditions because it was receiving and demodulating a modulated carrier. Signal-to-notched noise ratios are meaningful only if the notch filter has an extremely narrow bandwidth (a few hertz) and introduces 40 dB or more of attenuation to the signal.

5-4-12 Linear Integrated-Circuit AM Receivers

Linear integrated circuits are now available from several manufacturers that perform all receiver functions except RF and IF filtering and volume control on a single chip. Figure 5-34 shows the schematic diagram of an AM receiver that uses the National Semiconductor Corporation LM1820 linear integrated-circuit AM radio chip. The LM1820 has onboard RF amplifier, mixer, local oscillator, and IF amplifier stages. However, RF and IF selectivity is accomplished by adjusting tuning coils in externally connected tuned circuits or cans. Also, an LIC audio amplifier, such as the LM386, and a speaker are necessary to complete a functional receiver.

LIC AM radios are not widely used because the physical size reduction made possible by reducing the component count through integration is offset by the size of the external components necessary for providing bandlimiting and channel selection. Alternatives to *LC* tank circuits and IF cans, such as ceramic filters, may be integrable in the near future. Also, new receiver configurations (other than TRF or superheterodyne) may be possible in the future using phase-locked-loop technology. Phase-locked-loop receivers would need only two external components: a volume control and a station tuning control.

5-5 DOUBLE-CONVERSION AM RECEIVERS

For good image-frequency rejection, a relatively high intermediate frequency is desired. However, for high-gain selective amplifiers that are stable and easily neutralized, a low intermediate frequency is necessary. The solution is to use two intermediate frequencies. The

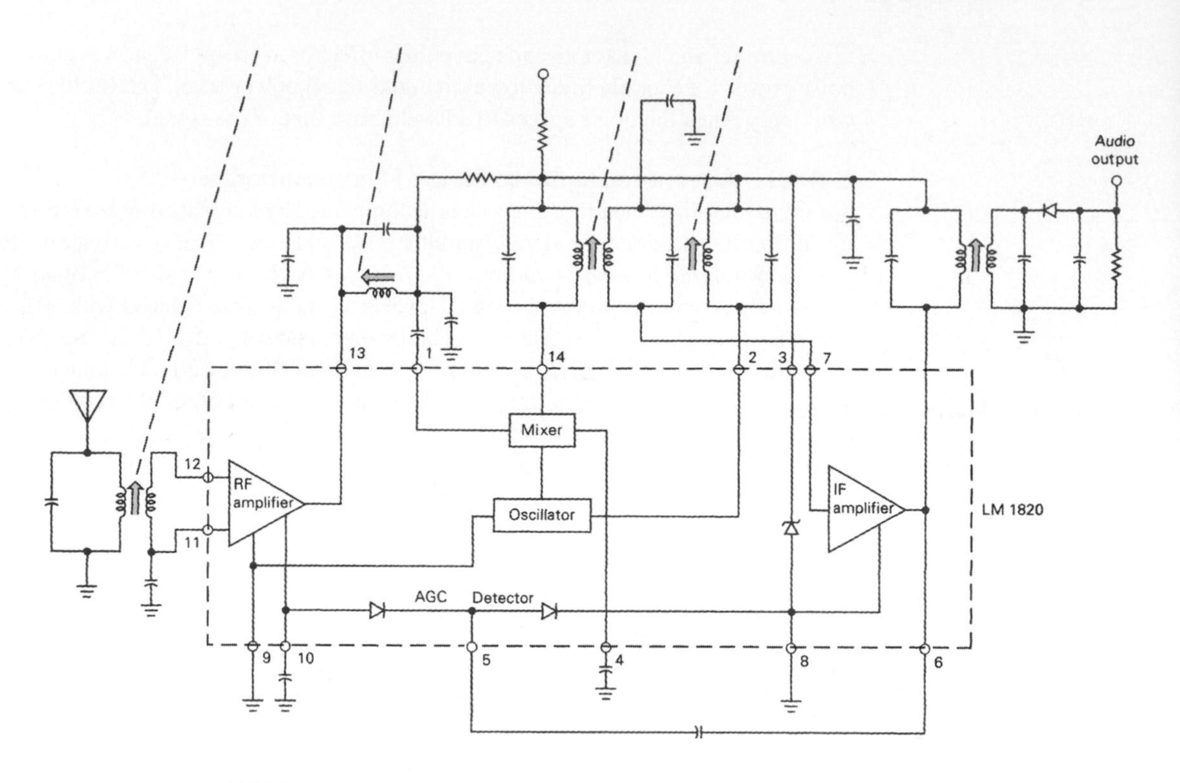

FIGURE 5-34 LM1820 linear integrated-circuit AM radio system

first IF is a relatively high frequency for good image-frequency rejection, and the *second IF* is a relatively low frequency for easy amplification. Figure 5-35 shows a block diagram for a *double-conversion* AM receiver. The first IF is 10.625 MHz, which pushes the image frequency 21.25 MHz away from the desired RF. The first IF is immediately down-converted to 455 kHz and fed to a series of high-gain IF amplifiers. Figure 5-36 illustrates the filtering requirements for a double-conversion AM receiver.

5-6 NET RECEIVER GAIN

Thus far, we have discussed RF gain, conversion gain, and IF gain. However, probably the most important gain is *net receiver gain.* The net receiver gain is simply the ratio of the demodulated signal level at the output of the receiver (audio) to the RF signal level at the input to the receiver, or the difference between the audio signal level in dBm and the RF signal level in dBm.

In essence, net receiver gain is the dB sum of all the gains in the receiver minus the dB sum of all the losses. Receiver losses typically include preselector loss, mixer loss (i.e., conversion gain), and detector losses. Gains include RF gain, IF gain, and audio-amplifier gain. Figure 5-37 shows the gains and losses found in a typical radio receiver.

Mathematically, net receiver gain is

$$G_{\mathrm{dB}} = \mathrm{gains}_{\mathrm{dB}} - \mathrm{losses}_{\mathrm{dB}}$$

where gains = RF amplifier gain + IF amplifier gain + audio-amplifier gain
losses = preselector loss + mixer loss + detector loss

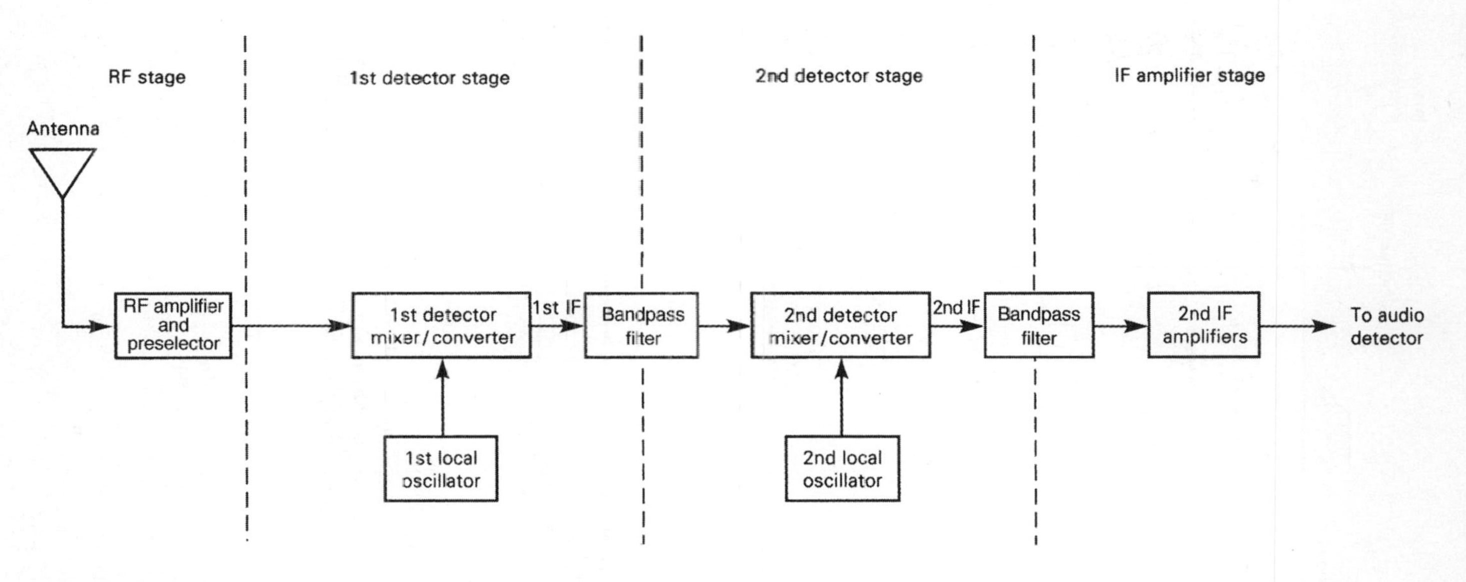

FIGURE 5-35 Double-conversion AM receiver

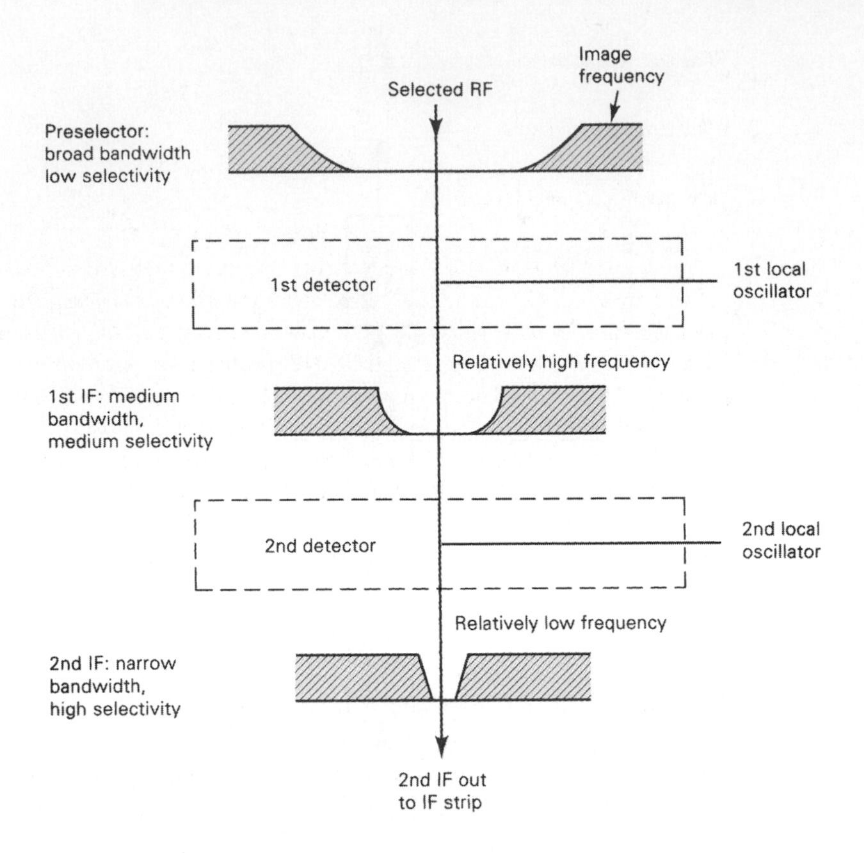

FIGURE 5-36 Filtering requirements for the double-conversion AM receiver shown in Figure 5-34

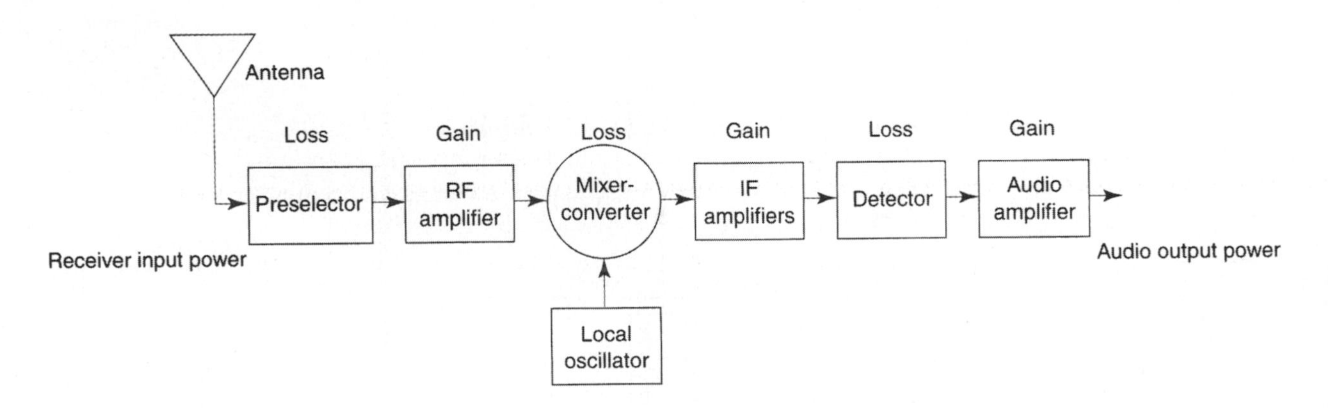

FIGURE 5-37 Receiver gains and losses

Example 5-8

For an AM receiver with a −80-dBm RF input signal level and the following gains and losses, determine the net receiver gain and the audio signal level:

Gains: RF amplifier = 33 dB, IF amplifier = 47 dB, audio amplifier = 25 dB

Losses: preselector loss = 3 dB, mixer loss = 6 dB, detector loss = 8 dB

Solution The sum of the gains is

$$33 + 47 + 25 = 105 \text{ dB}$$

The sum of the losses is

$$3 + 6 + 8 = 17 \text{ dB}$$

Thus, net receiver gain

$$G = 105 - 17 = 88 \text{ dB}$$

and the audio signal level is

$$-80 \text{ dBm} + 88 \text{ dB} = 8 \text{ dBm}$$

It is important to note, however, that because of the effects of AGC, the IF and/or RF gain of a receiver could be stated as a maximum, minimum, or average value. For example, the IF gain for the receiver in Example 5-8 could vary because of AGC between 20 dB and 60 dB, depending on the input signal strength. Therefore, the net receiver gain could vary between a maximum value of 101 dB, a minimum value of 61 dB, with an average value of 81 dB.

Net receiver gain should not be confused with overall *system gain.* Net receiver gain includes only components within the receiver beginning at the input to the preselector. System gain includes all the gains and losses incurred by a signal as it propagates from the transmitter output stage to the output of the detector in the receiver and includes antenna gain and transmission line and propagation losses. System gain is discussed in detail in Chapter 23.

QUESTIONS

5-1. What is meant by the *front end* of a receiver?

5-2. What are the primary functions of the front end of a receiver?

5-3. Define *selectivity* and *shape factor.* What is the relationship between receiver noise and selectivity?

5-4. Describe *bandwidth improvement.* What is the relationship between bandwidth improvement and receiver noise?

5-5. Define *sensitivity.*

5-6. What is the relationship among receiver noise, bandwidth, and temperature?

5-7. Define *fidelity.*

5-8. List and describe the three types of distortion that reduce the fidelity of a receiver.

5-9. Define *insertion loss.*

5-10. Define *noise temperature* and *equivalent noise temperature.*

5-11. Describe the difference between a *coherent* and a *noncoherent* radio receiver.

5-12. Draw the block diagram for a TRF radio receiver and briefly describe its operation.

5-13. What are the three predominant disadvantages of a TRF receiver?

5-14. Draw the block diagram for an AM superheterodyne receiver and describe its operation and the primary functions of each stage.

5-15. Define *heterodyning.*

5-16. What is meant by the terms *high-* and *low-side injection?*

5-17. Define *local oscillator tracking* and *tracking error.*

5-18. Describe *three-point tracking.*

5-19. What is meant by *gang* tuning?

5-20. Define *image frequency.*

5-21. Define *image-frequency rejection ratio.*

5-22. List six characteristics that are desirable in an RF amplifier.

5-23. What advantage do FET RF amplifiers have over BJT RF amplifiers?

5-24. Define *neutralization.* Describe the neutralization process.

5-25. What is a *cascoded* amplifier?

5-26. Define *conversion gain.*

5-27. What is the advantage of a relatively high-frequency intermediate frequency; a relatively low-frequency intermediate frequency?

5-28. Define the following terms: *inductive coupling, self-inductance, mutual inductance, coefficient of coupling, critical coupling,* and *optimum coupling.*

5-29. Describe *loose* coupling; *tight* coupling.

5-30. Describe the operation of a *peak detector.*

5-31. Describe *rectifier distortion* and its causes.

5-32. Describe *diagonal clipping* and its causes.

5-33. Describe the following terms: *simple AGC, delayed AGC,* and *forward AGC.*

5-34. What is the purpose of a *squelch* circuit?

5-35. Explain the operation of a double-conversion superheterodyne receiver.

PROBLEMS

5-1. Determine the IF bandwidth necessary to achieve a bandwidth improvement of 16 dB for a radio receiver with an RF bandwidth of 320 kHz.

5-2. Determine the improvement in the noise figure for a receiver with an RF bandwidth equal to 40 kHz and IF bandwidth of 16 kHz.

5-3. Determine the equivalent noise temperature for an amplifier with a noise figure of 6 dB and an environmental temperature $T = 27°C$.

5-4. For an AM commercial broadcast-band receiver with an input filter Q-factor of 85, determine the bandwidth at the low and high ends of the RF spectrum.

5-5. For an AM superheterodyne receiver using high-side injection with a local oscillator frequency of 1200 kHz, determine the IF carrier and upper and lower side frequencies for an RF envelope that is made up of a carrier and upper and lower side frequencies of 600 kHz, 604 kHz, and 596 kHz, respectively.

5-6. For a receiver with a ±2.5-kHz tracking error, a 455-kHz IF, and a maximum modulating signal frequency $f_m = 6$ kHz, determine the minimum IF bandwidth.

5-7. For a receiver with IF, RF, and local oscillator frequencies of 455 kHz, 900 kHz, and 1355 kHz, respectively, determine

a. Image frequency.
b. IFRR for a preselector Q of 80.

5-8. For a citizens band receiver using high-side injection with an RF carrier of 27.04 MHz and a 10.645 MHz first IF, determine

a. Local oscillator frequency.
b. Image frequency.

5-9. For a three-stage double-tuned RF amplifier with an RF carrier equal to 800 kHz and a coefficient of coupling $k_{opt} = 0.025$, determine

a. Bandwidth for each individual stage.
b. Overall bandwidth for the three stages.

5-10. Determine the maximum modulating signal frequency for a peak detector with the following parameters: $C = 1000$ pF, $R = 10$ kΩ, and $m = 0.5$. Repeat the problem for $m = 0.707$.

5-11. Determine the bandwidth improvement for a radio receiver with an RF bandwidth of 60 kHz and an IF bandwidth of 15 kHz.

5-12. Determine the equivalent noise temperature for an amplifier with a noise figure $F = 8$ dB and an environmental temperature $T = 122°C$.

5-13. For an AM commercial broadcast-band receiver with an input filter Q-factor of 60, determine the bandwidth at the low and high ends of the RF spectrum.

5-14. For an AM superheterodyne receiver using high-side injection with a local oscillator frequency of 1400 kHz, determine the IF carrier and upper and lower side frequencies for an RF envelope that is made up of a carrier and upper and lower side frequencies of 800 kHz, 806 kHz, and 794 kHz, respectively.

5-15. For a receiver with a ± 2800-Hz tracking error and a maximum modulating signal frequency $f_m = 4$ kHz, determine the minimum IF bandwidth.

5-16. For a receiver with IF, RF, and local oscillator frequencies of 455 kHz, 1100 kHz, and 1555 kHz, respectively, determine

a. Image frequency.
b. Image-frequency rejection ratio for a preselector $Q = 100$.
c. Image-frequency rejection ratio for a $Q = 50$.

5-17. For a citizens band receiver using high-side injection with an RF carrier of 27.04 MHz and a 10.645 MHz IF, determine

a. Local oscillator frequency.
b. Image frequency.

5-18. For a three-stage, double-tuned RF amplifier with an RF equal to 1000 kHz and a coefficient of coupling $k_{opt} = 0.01$, determine

a. Bandwidth for each individual stage.
b. Bandwidth for the three stages cascaded together.

5-19. Determine the maximum modulating signal frequency for a peak detector with the following parameters: $C = 1000$ pF, $R = 6.8$ kΩ, and $m = 0.5$. Repeat the problem for $m = 0.707$.

5-20. Determine the net receiver gain for an AM receiver with an RF input signal power of -87 dBm and an audio signal power of 10 dBm.

5-21. Determine the net receiver gain for an AM receiver with the following gains and losses:

Gains: RF amplifier = 30 dB, IF amplifier = 44 dB, audio amplifier = 24 dB

Losses: Preselector loss = 2 dB, mixer loss = 6 dB, detector loss = 8 dB

5-22. Determine the minimum RF input signal power necessary to produce an audio signal power of 10 dBm for the receiver described in Problem 5-21.

5-23. Determine the net receiver gain for an AM receiver with the following gains and losses:

Gains: RF amplifier = 33 dB, IF amplifier = 44 dB, audio amplifier = 22 dB

Losses: Preselector loss = 3.5 dB, mixer loss = 5 dB, detector loss = 9 dB

C H A P T E R 6

Single-Sideband Communications Systems

CHAPTER OUTLINE

OBJECTIVES

- Define and describe AM single-sideband full carrier
- Define and describe AM single-sideband suppressed carrier
- Define and describe AM single-sideband reduced carrier
- Define and describe AM independent sideband
- Define and describe AM vestigial sideband
- Compare single-sideband transmission to conventional double-sideband AM
- Explain the advantages and disadvantages of single-sideband transmission
- Develop the mathematical expression for single-sideband suppressed carrier
- Describe the operation of a balanced ring modulator
- Describe the operation of a FET push–pull balanced modulator
- Describe the operation of a balanced bridge modulator
- Describe the operation of a linear integrated-circuit balanced modulator
- Explain the operation of single-sideband transmitters using the filter and phase-shift methods
- Describe the operation of the following filters: crystal, ceramic, mechanical, and SAW
- Describe the operation of an independent-sideband transmitter

- Explain the operation of the following single-sideband receivers: noncoherent BFO, coherent BFO, envelope detection, and multichannel pilot carrier
- Describe how single-sideband suppressed carrier is used with frequency-division multiplexing
- Describe how double-sideband suppressed carrier is used with quadrature multiplexing
- Define and describe single-sideband measurement units

6-1 INTRODUCTION

Conventional AM double-sideband communications systems, such as those discussed in Chapters 3 and 4, have two inherent disadvantages. First, with conventional AM, carrier power constitutes two-thirds or more of the total transmitted power. This is a major drawback because the carrier contains no information; the sidebands contain the information. Second, conventional AM systems utilize twice as much bandwidth as needed with single-sideband systems. With double-sideband transmission, the information contained in the upper sideband is identical to the information contained in the lower sideband. Therefore, transmitting both sidebands is redundant. Consequently, conventional AM is both power and bandwidth inefficient, which are the two most predominant considerations when designing modern electronic communications systems.

The purpose of this chapter is to introduce the reader to several single-sideband AM systems and explain the advantages and disadvantages of choosing them over conventional double-sideband full-carrier AM.

The most prevalent use of single-sideband suppressed-carrier systems is with multichannel communications systems employing frequency-division multiplexing (FDM) such as long-distance telephone systems. Frequency-division multiplexing is introduced later in this chapter, then discussed in more detail in Chapter 11.

6-2 SINGLE-SIDEBAND SYSTEMS

Single sideband was mathematically recognized and understood as early as 1914; however, not until 1923 was the first patent granted and a successful communications link established between England and the United States. There are many different types of *sideband* communications systems. Some of them conserve bandwidth, some conserve power, and some conserve both. Figure 6-1 compares the frequency spectra and relative power distributions for conventional AM and several of the more common single-sideband (SSB) systems.

6-2-1 AM Single-Sideband Full Carrier

AM *single-sideband full carrier* (SSBFC) is a form of amplitude modulation in which the carrier is transmitted at full power but only one of the sidebands is transmitted. Therefore, SSBFC transmissions require only half as much bandwidth as conventional double-sideband AM. The frequency spectrum and relative power distribution for SSBFC are shown in Figure 6-1b. Note that with 100% modulation, the carrier power (P_c) constitutes four-fifths (80%) of the total transmitted power (P_t), and only one-fifth (20%) of the total power is in the sideband. For conventional double-sideband AM with 100% modulation, two-thirds (67%) of the total transmitted power is in the carrier, and one-third (33%) is in the sidebands. Therefore, although SSBFC requires less total power, it actually utilizes a smaller percentage of that power for the information-carrying portion of the signal.

Figure 6-2 shows the waveform for a 100%-modulated SSBFC wave with a single-frequency modulating signal. The 100%-modulated single-sideband, full-carrier envelope looks identical to a 50%-modulated double-sideband, full-carrier envelope. Recall from Chapter 4 that the maximum positive and negative peaks of an AM DSBFC wave occur when the carrier and both sidebands reach their respective peaks at the same time, and the peak change in the envelope is equal to the sum of the amplitudes of the upper and lower side frequencies. With single-sideband transmission, there is only one sideband (either the

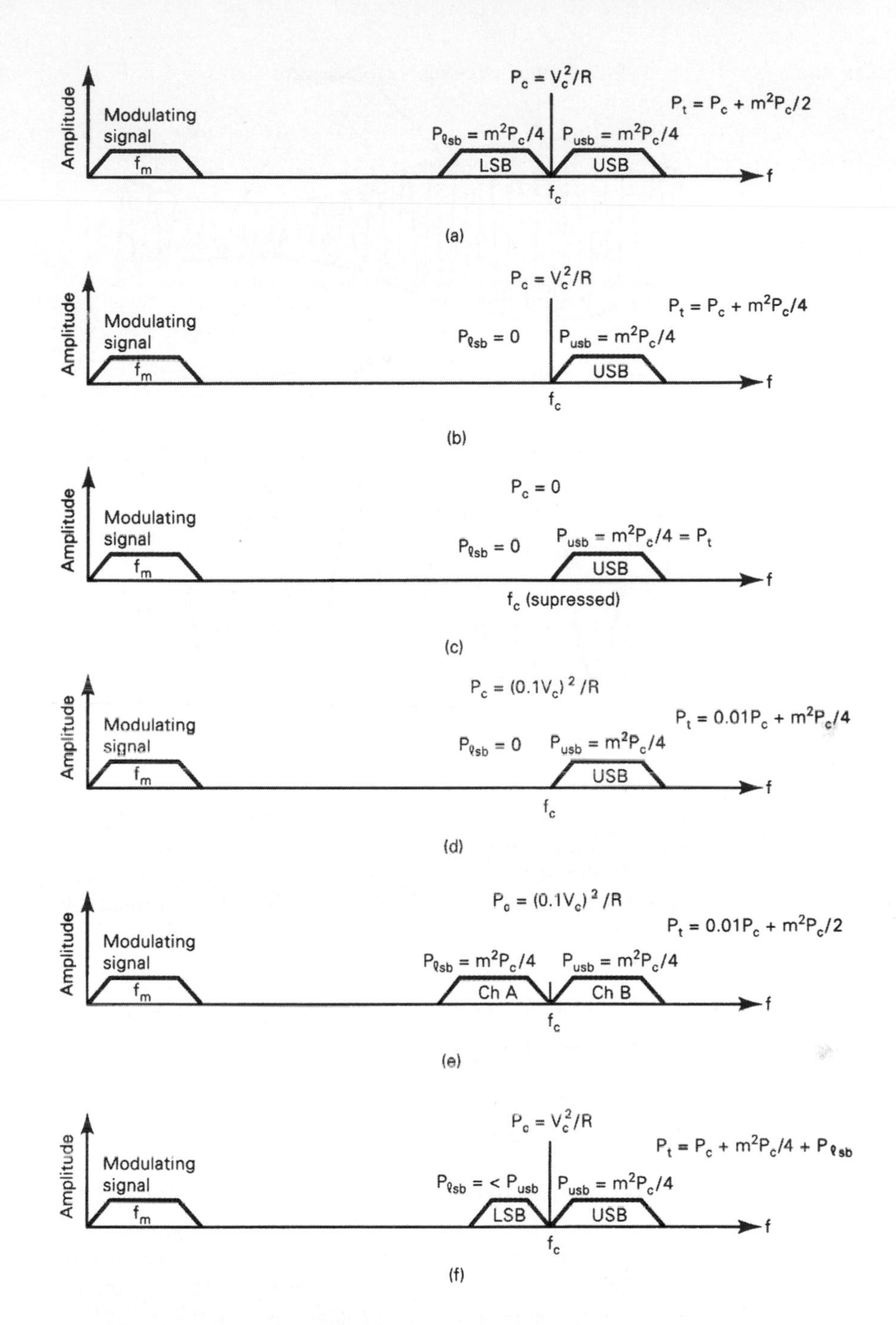

FIGURE 6-1 Single-banded systems: (a) conventional DSBFC AM; (b) full-carrier single sideband; (c) suppressed-carrier single sideband; (d) reduced-carrier single sideband; (e) independent sideband; (f) vestigial sideband

upper or the lower) to add to the carrier. Therefore, the peak change in the envelope is only half of what it is with double-sideband transmission. Consequently, with single-sideband full-carrier transmission, the demodulated signals have only half the amplitude of a double-sideband demodulated wave. Thus, a trade-off is made. SSBFC requires less bandwidth than DSBFC but also produces a demodulated signal with a lower amplitude. However, when the bandwidth is halved, the total noise power is also halved (i.e., reduced by 3 dB); and if one

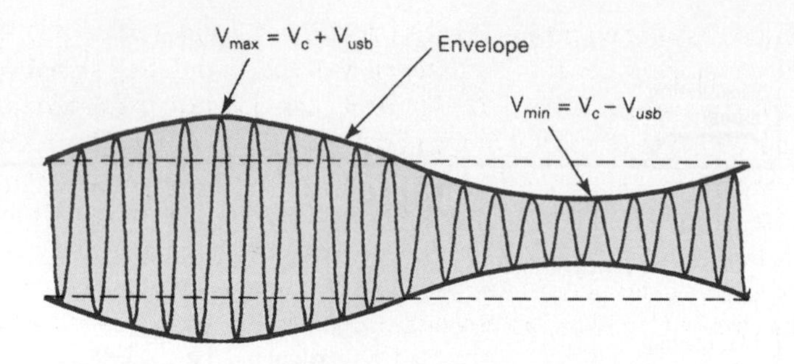

FIGURE 6-2 SSBFC waveform, 100% modulation

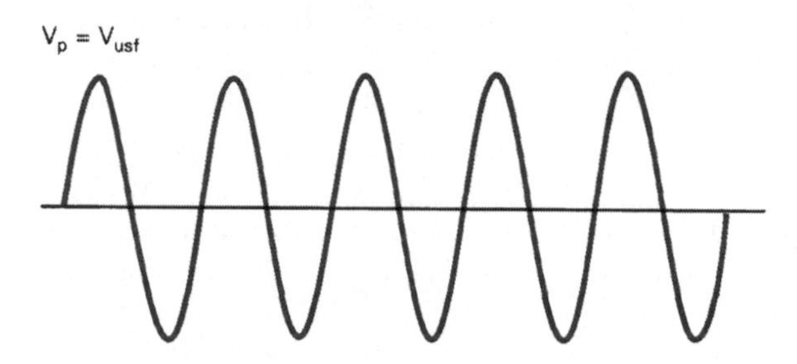

FIGURE 6-3 SSBSC waveform

sideband is removed, the power in the information portion of the wave is also halved. Consequently, the signal-to-noise ratios for single and double sideband are the same.

With SSBFC, the repetition rate of the envelope is equal to the frequency of the modulating signal, and the depth of modulation is proportional to the amplitude of the modulating signal. Therefore, as with double-sideband transmission, the information is contained in the envelope of the full-carrier modulated signal.

6-2-2 AM Single-Sideband Suppressed Carrier

AM *single-sideband suppressed carrier* (SSBSC) is a form of amplitude modulation in which the carrier is totally suppressed and one of the sidebands removed. Therefore, SSBSC requires half as much bandwidth as conventional double-sideband AM and considerably less transmitted power. The frequency spectrum and relative power distribution for SSBSC with upper sideband transmission are shown in Figure 6-1c. It can be seen that the sideband power makes up 100% of the total transmitted power. Figure 6-3 shows a SSBSC waveform for a single-frequency modulating signal. As you can see, the waveform is not an envelope; it is simply a sine wave at a single frequency equal to the carrier frequency plus the modulating-signal frequency or the carrier frequency minus the modulating-signal frequency, depending on which sideband is transmitted.

6-2-3 AM Single-Sideband Reduced Carrier

AM *single-sideband reduced carrier* (SSBRC) is a form of amplitude modulation in which one sideband is totally removed and the carrier voltage is reduced to approximately 10% of its unmodulated amplitude. Consequently, as much as 96% of the total power transmitted is in the unsuppressed sideband. To produce a reduced carrier component, the carrier is totally suppressed during modulation and then reinserted at a reduced amplitude. Therefore, SSBRC is sometimes called single-sideband *reinserted* carrier. The reinserted carrier is often called a pilot carrier and is reinserted for demodulation purposes, which is explained later in this chapter. The frequency spectrum and relative power distribution for SSBRC are shown in Figure 6-1d. The figure shows that the sideband power constitutes almost 100%

of the transmitted power. As with double-sideband, full-carrier AM, the repetition rate of the envelope is equal to the frequency of the modulating signal. To demodulate a reduced carrier waveform with a conventional peak detector, the carrier must be separated, amplified, and then reinserted at a higher level in the receiver. Therefore, reduced-carrier transmission is sometimes called *exalted* carrier because the carrier is elevated in the receiver prior to demodulation. With exalted-carrier detection, the amplification of the carrier in the receiver must be sufficient to raise the level of the carrier to a value greater than that of the sideband signal. SSBRC requires half as much bandwidth as conventional AM and, because the carrier is transmitted at a reduced level, also conserves considerable power.

6-2-4 AM Independent Sideband

AM *independent sideband* (ISB) is a form of amplitude modulation in which a single carrier frequency is independently modulated by two different modulating signals. In essence, ISB is a form of double-sideband transmission in which the transmitter consists of two independent single-sideband suppressed-carrier modulators. One modulator produces only the upper sideband, and the other produces only the lower sideband. The single-sideband output signals from the two modulators are combined to form a double-sideband signal in which the two sidebands are totally independent of each other except that they are symmetrical about a common carrier frequency. One sideband is positioned above the carrier in the frequency spectrum and one below. For demodulation purposes, the carrier is generally reinserted at a reduced level as with SSBRC transmission. Figure 6-1e shows the frequency spectrum and power distribution for ISB.

ISB conserves both transmit power and bandwidth, as two information sources are transmitted within the same frequency spectrum, as would be required by a single source using conventional double-sideband transmission. ISB is one technique that is used in the United States for stereo AM transmission. One channel (the left) is transmitted in the lower sideband, and the other channel (the right) is transmitted in the upper sideband.

6-2-5 AM Vestigial Sideband

AM *vestigial sideband* (VSB) is a form of amplitude modulation in which the carrier and one complete sideband are transmitted, but only part of the second sideband is transmitted. The carrier is transmitted at full power. In VSB, the lower modulating-signal frequencies are transmitted double sideband, and the higher modulating-signal frequencies are transmitted single sideband. Consequently, the lower frequencies can appreciate the benefit of 100% modulation, whereas the higher frequencies cannot achieve more than the effect of 50% modulation. Consequently, the low-frequency modulating signals are emphasized and produce larger-amplitude signals in the demodulator than the high frequencies. The frequency spectrum and relative power distribution for VSB are shown in Figure 6-1f. Probably the most widely known VSB system is the picture portion of a commercial television broadcasting signal, which is designated A5C by the FCC.

6-3 COMPARISON OF SINGLE-SIDEBAND TRANSMISSION TO CONVENTIONAL AM

From the preceding discussion and Figure 6-1, it can be seen that bandwidth conservation and power efficiency are obvious advantages of single-sideband suppressed- and reduced-carrier transmission over conventional double-sideband full-carrier transmission (i.e., conventional AM). Single-sideband transmission requires only half as much bandwidth as double sideband, and suppressed- and reduced-carrier transmissions require considerably less total transmitted power than full-carrier AM.

The total power transmitted necessary to produce a given signal-to-noise ratio at the output of a receiver is a convenient and useful means of comparing the power requirement and relative performance of single-sideband to conventional AM systems. The signal-to-noise ratio determines the degree of intelligibility of a received signal.

Figure 6-4 summarizes the waveforms produced for a given modulating signal for three of the more common AM transmission systems: double-sideband full carrier (DSBFC), double-sideband suppressed carrier (DSBSC), and single-sideband suppressed carrier (SSBSC). As the figure shows, the repetition rate of the DSBFC envelope is equal to the modulating signal frequency, the repetition rate of the DSBSC envelope is equal to twice the modulating signal frequency, and the SSBSC waveform is not an envelope at all but rather a single-frequency sinusoid equal in frequency to the unsuppressed sideband frequency (i.e., either the upper or the lower side frequency).

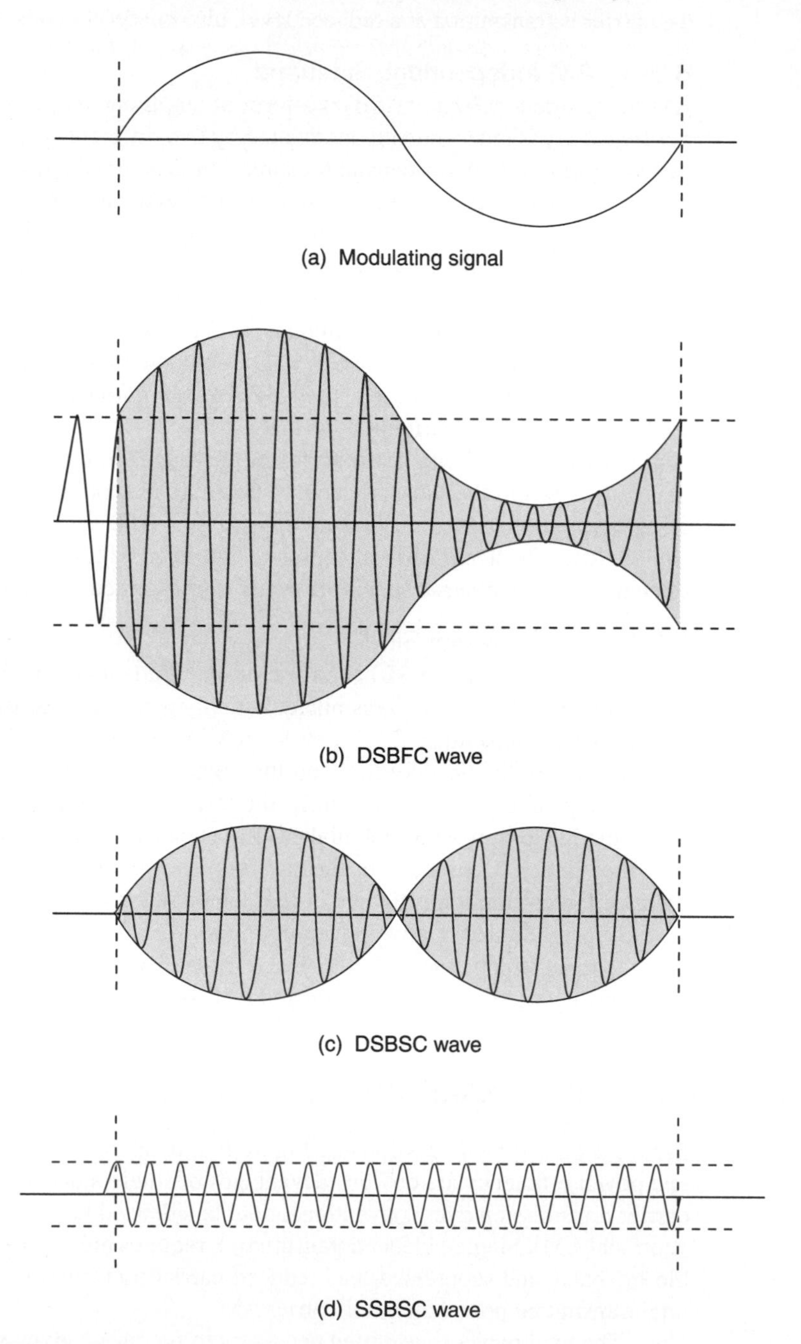

FIGURE 6-4 Comparison of three common AM transmission systems: (a) modulating signal; (b) DSBFC wave; (c) DSBSC wave; (d) SSBSC wave

A conventional AM wave with 100% modulation contains 1 unit of carrier power and 0.25 unit of power in each sideband for a total transmitted peak power of 1.5 units. A single-sideband transmitter rated at 0.5 unit of power will produce the same S/N ratio at the output of a receiver as 1.5 units of carrier plus sideband power from a double-sideband full-carrier signal. In other words, the same performance is achieved with SSBSC using only one-third as much transmitted power and half the bandwidth. Table 6-1 compares conventional AM to single-sideband suppressed carrier for a single-frequency modulating signal. *Peak envelope power* (PEP) is the rms power developed at the crest of the modulation envelope (i.e., when the modulating-signal frequency components are at their maximum amplitudes).

The voltage vectors for the power requirements stated are also shown. It can be seen that it requires 0.5 unit of voltage per sideband and 1 unit for the carrier with conventional AM for a total of 2 PEV (peak envelope volts) and only 0.707 PEV for single sideband. The RF envelopes are also shown, which correspond to the voltage and power relationships previously outlined. The demodulated signal at the output from a conventional AM receiver

Table 6-1 Conventional AM versus Single Sideband

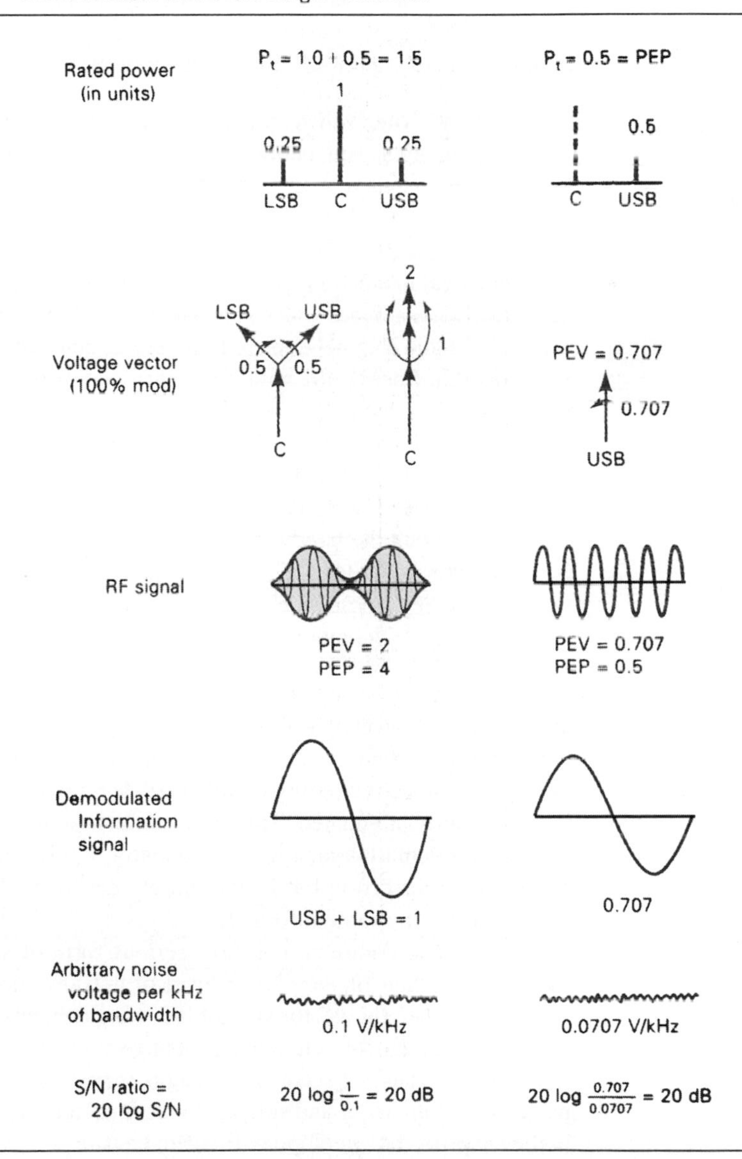

is proportional to the quadratic sum of the voltages from the upper and lower sideband signals, which equals 1 PEV unit. For single-sideband reception, the demodulated signal is $0.707 \times 1 = 0.707$ PEV. If the noise voltage for conventional AM is arbitrarily chosen as 0.1 V/kHz, the noise voltage for single-sideband signal with half the bandwidth is 0.0707 V/kHz. Consequently, the S/N performance for SSBSC is equal to that of conventional AM.

6-3-1 Advantages of Single-Sideband Transmission

Following are four predominant advantages of single-sideband suppressed- or reduced-carrier transmission over conventional double-sideband full-carrier transmission.

6-3-1-1 Power conservation. Normally, with single-sideband transmission, only one sideband is transmitted, and the carrier is either suppressed or reduced significantly. As a result, much less total transmitted power is necessary to produce essentially the same-quality signal in the receiver as is achieved with double-sideband, full-carrier transmission. At least two-thirds of the power in a standard double-sideband, full-carrier AM signal is contained in the carrier, and the maximum power contained in either sideband is only one-sixth of the total power. Thus, eliminating the carrier would increase the power available for the sidebands by at least a factor of 3, providing a signal power advantage of 10 log (3), or approximately a 4.8 dB improvement in the signal-to-noise ratio.

6-3-1-2 Bandwidth conservation. Single-sideband transmission requires half as much bandwidth as conventional AM double-sideband transmission. This advantage is especially important today with an already overcrowded radio-frequency spectrum. Eliminating one sideband actually reduces the required bandwidth by more than a factor of 2 because most modulating signals, including audio signals, rarely extend all the way down to 0 Hz (dc). A more practical lower frequency limit for audio signals is 300 Hz; thus, a 3-kHz audio channel actually has a bandwidth of approximately 2700 Hz (300 Hz to 3000 Hz). Consequently, a 2700-Hz audio channel transmitted over a double-sideband AM system would require 6 kHz of bandwidth, whereas the same audio information would require only 2700 Hz of bandwidth using a single-sideband system. Hence, the single-sideband system described appreciates a bandwidth improvement of 10 log (6000/2700), or a 3.5-dB reduction in the noise power. A safe, general approximation is a 50% reduction in bandwidth for single sideband compared to double sideband, which equates to an improvement in the signal-to-noise ratio of 3 dB.

Combining the bandwidth improvement achieved by transmitting only one sideband and the power advantage of removing the carrier, the overall improvement in the signal-to-noise ratio using single-sideband suppressed carrier is approximately 7.8 dB ($3 + 4.8$) better than double-sideband full carrier.

6-3-1-3 Selective fading. With double-sideband transmission, the two sidebands and carrier may propagate through the transmission media by different paths and, therefore, experience different transmission impairments. This condition is called *selective fading.* One type of selective fading is called *sideband fading.* With sideband fading, one sideband is significantly attenuated. This loss results in a reduced signal amplitude at the output of the receiver demodulator and, consequently, a 3-dB reduced signal-to-noise ratio. This loss causes some distortion but is not entirely detrimental to the signal because the two sidebands contain the same information.

The most common and most serious form of selective fading is *carrier-amplitude fading.* Reduction of the carrier level of a 100%-modulated wave will make the carrier voltage less than the vector sum of the two sidebands. Consequently, the envelope resembles an overmodulated envelope, causing severe distortion to the demodulated signal.

A third cause of selective fading is carrier or sideband phase shift. When the relative positions of the carrier and sideband vectors of the received signal change, a decided change in the shape of the envelope will occur, causing a severely distorted demodulated signal.

When only one sideband and either a reduced or totally suppressed carrier are transmitted, carrier phase shift and carrier fading cannot occur, and sideband fading changes only the amplitude and frequency response of the demodulated signal. These changes do not generally produce enough distortion to cause loss of intelligibility in the received signal. With single-sideband transmission, it is not necessary to maintain a specific amplitude or phase relationship between the carrier and sideband signals.

6-3-1-4 Noise reduction. Because a single-sideband system utilizes half as much bandwidth as conventional AM, the thermal noise power is reduced to half that of a double-sideband system. Taking into consideration both the bandwidth reduction and the immunity to selective fading, SSB systems enjoy approximately a 12-dB S/N ratio advantage over conventional AM (i.e., a conventional AM system must transmit a 12-dB more powerful signal to achieve the same performance as a comparable single-sideband system).

6-3-2 Disadvantages of Single-Sideband Transmission

Following are two major disadvantages of single-sideband reduced- or suppressed-carrier transmission as compared to conventional double-sideband, full-carrier transmission.

6-3-2-1 Complex receivers. Single-sideband systems require more complex and expensive receivers than conventional AM transmission because most single-sideband transmissions include either a reduced or a suppressed carrier; thus, envelope detection cannot be used unless the carrier is regenerated at an exalted level. Single-sideband receivers require a carrier recovery and synchronization circuit, such as a PLL frequency synthesizer, which adds to their cost, complexity, and size.

6-3-2-2 Tuning difficulties. Single-sideband receivers require more complex and precise tuning than conventional AM receivers. This is undesirable for the average user. This disadvantage can be overcome by using more accurate, complex, and expensive tuning circuits.

6-4 MATHEMATICAL ANALYSIS OF SUPPRESSED-CARRIER AM

An AM modulator is a *product modulator;* the output signal is the product of the modulating signal and the carrier. In essence, the carrier is multiplied by the modulating signal. Equation 4-12 was given as

$$v_{am}(t) = [1 + m\sin(2\pi f_m t)]\,[E_c\sin(2\pi f_c t)]$$

where $1 + m\sin(2\pi f_m t)$ = constant + modulating signal
$E_c\sin(2\pi f_c t)$ = unmodulated carrier

If the constant component is removed from the modulating signal, then

$$v_{am}(t) = [m\sin(2\pi f_m t)]\,[E_c\sin(2\pi f_c t)] \quad \textbf{(6-1)}$$

Multiplying yields

$$v_{am}(t) = -\frac{mE_c}{2}\cos[2\pi(f_c + f_m)t] + \frac{mE_c}{2}\cos[2\pi(f_c - f_m)t] \quad \textbf{(6-2)}$$

where $-(mE_c/2)\cos[2\pi(f_c + f_m)t]$ = upper side frequency component
$+(mE_c/2)\cos[2\pi(f_c - f_m)t]$ = lower side frequency component

From the preceding mathematical operation, it can be seen that, if the constant component is removed prior to performing the multiplication, the carrier component is removed from the modulated wave, and the output signal is simply two cosine waves, one at the sum frequency ($f_c + f_m = f_{usf}$) and the other at the difference frequency ($f_c - f_m = f_{lsf}$).

The carrier has been suppressed in the modulator. To convert to single sideband, simply remove either the sum or the difference frequency.

6-5 SINGLE-SIDEBAND GENERATION

In the preceding sections it was shown that with most single-sideband systems the carrier is either totally suppressed or reduced to only a fraction of its original value, and one sideband is removed. To remove the carrier from the modulated wave or to reduce its amplitude using conventional notch filters is extremely difficult, if not impossible, because the filters simply do not have sufficient Q-factors to remove the carrier without also removing a portion of the sideband. However, it was also shown that removing the constant component suppressed the carrier in the modulator itself. Consequently, modulator circuits that inherently remove the carrier during the modulation process have been developed. Such circuits are called *double-sideband suppressed-carrier (DSBSC) modulators.* It will be shown later in this chapter how one of the sidebands can be removed once the carrier has been suppressed.

A circuit that produces a double-sideband suppressed-carrier signal is a *balanced modulator.* The balanced modulator has rapidly become one of the most useful and widely used circuits in electronic communications. In addition to suppressed-carrier AM systems, balanced modulators are widely used in frequency and phase modulation systems as well as in digital modulation systems, such as phase shift keying and quadrature amplitude modulation.

6-5-1 Balanced Ring Modulator

Figures 6-5 and 6-6 show the schematic diagrams and waveforms for a *balanced ring modulator.* The schematic in Figure 6-5a is constructed with diodes and transformers. Semiconductor diodes are ideally suited for use in balanced modulator circuits because they are stable, require no external power source, have a long life, and require virtually no maintenance. The balanced ring modulator is sometimes called a *balanced lattice modulator* or simply *balanced modulator.* A balanced modulator has two inputs: a single-frequency carrier and the modulating signal, which may be a single frequency or a complex waveform. For the balanced modulator to operate properly, the amplitude of the carrier must be sufficiently greater than the amplitude of the modulating signal (approximately six to seven times greater). This ensures that the carrier and not the modulating signal controls the on or off condition of the four diode switches (D_1 to D_4).

6-5-1-1 Circuit operation. Essentially, diodes D_1 to D_4 are electronic switches that control whether the modulating signal is passed from input transformer T_1 to output transformer T_2 as is or with a 180° phase shift. With the carrier polarity as shown in Figure 6-5b, diode switches D_1 and D_2 are forward biased and on, while diode switches D_3 and D_4 are reverse biased and off. Consequently, the modulating signal is transferred across the closed switches to T_2 without a phase reversal. When the polarity of the carrier reverses, as shown in Figure 6-5c, diode switches D_1 and D_2 are reverse biased and off, while diode switches D_3 and D_4 are forward biased and on. Consequently, the modulating signal undergoes a 180° phase reversal before reaching T_2. Carrier current flows from its source to the center taps of T_1 and T_2, where it splits and goes in opposite directions through the upper and lower halves of the transformers. Thus, their magnetic fields cancel in the secondary windings of the transformer and the carrier is suppressed. If the diodes are not perfectly matched or if the transformers are not exactly center tapped, the circuit is out of balance, and the carrier is not totally suppressed. It is virtually impossible to achieve perfect balance; thus, a small carrier component is always present in the output signal. This is

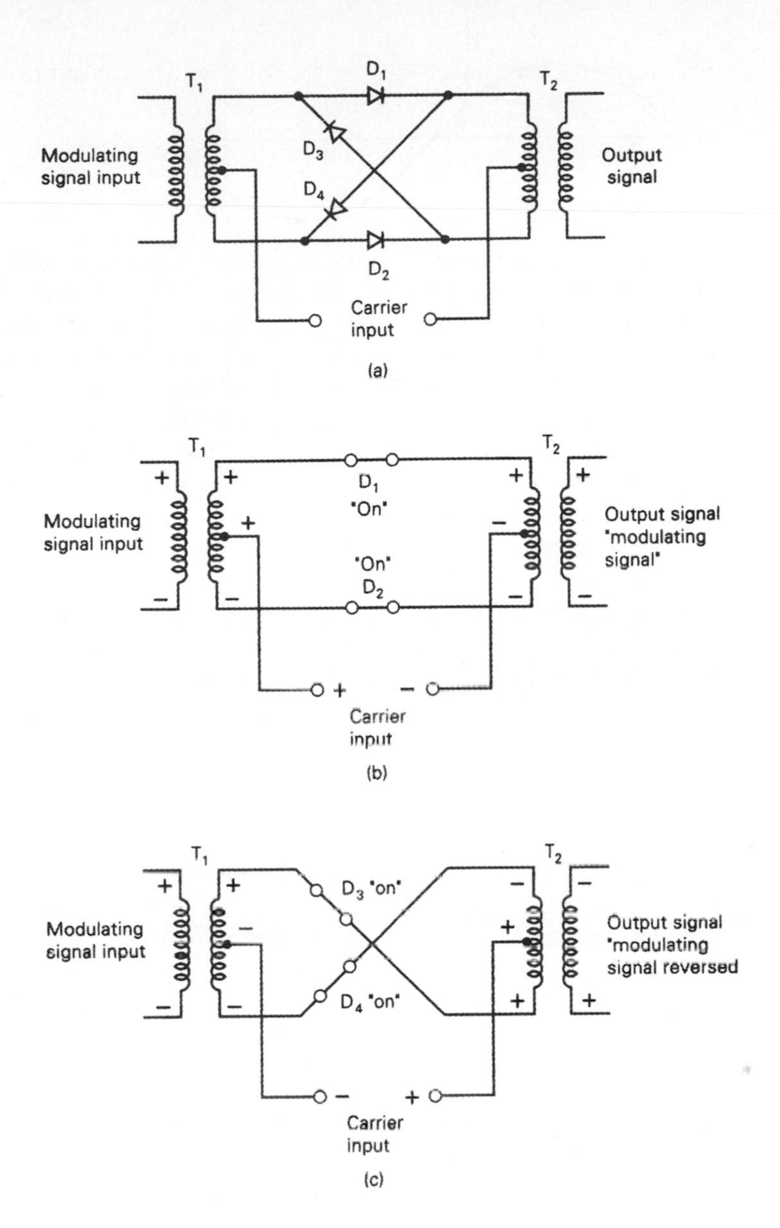

FIGURE 6-5 Balanced ring modulator: (a) schematic diagram; (b) D_1 and D_2 biased *on;* (c) D_3 and D_4 biased *on*

commonly called *carrier leak.* The amount of carrier suppression is typically between 40 dB and 60 dB.

Figure 6-6 shows the input and output waveforms associated with a balanced modulator for a single-frequency modulating signal. It can be seen that D_1 and D_2 conduct only during the positive half-cycles of the carrier input signal, and D_3 and D_4 conduct only during the negative half-cycles. The output from a balanced modulator consists of a series of RF pulses whose repetition rate is determined by the RF carrier switching frequency, and amplitude is controlled by the level of the modulating signal. Consequently, the output waveform takes the shape of the modulating signal, except with alternating positive and negative polarities that correspond to the polarity of the carrier signal.

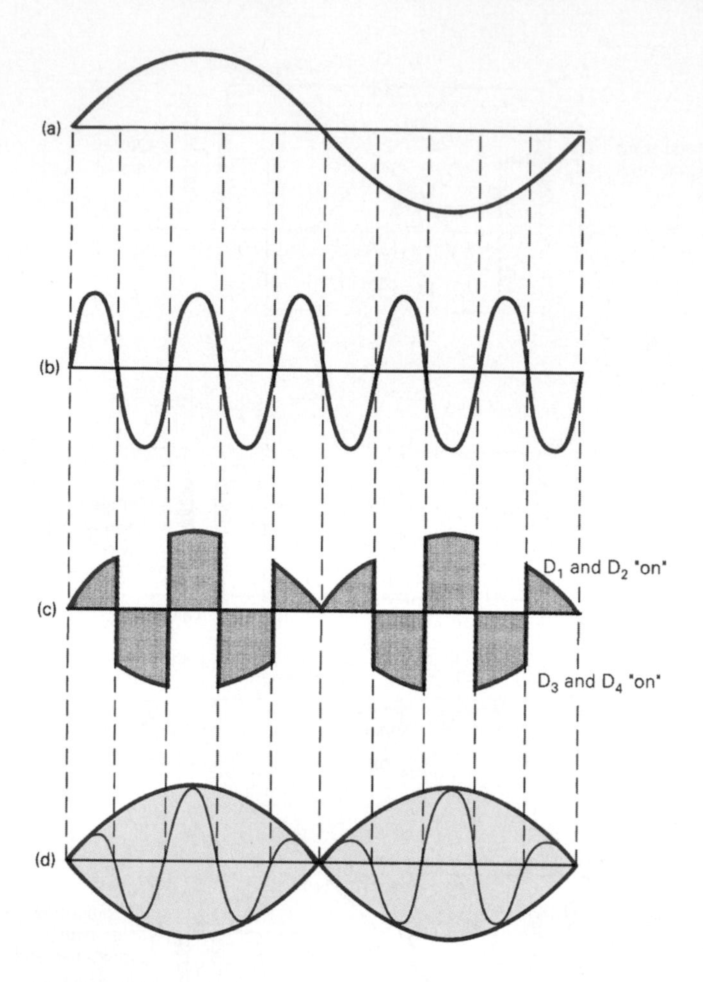

FIGURE 6-6 Balanced modulator waveforms: (a) modulating signal; (b) carrier signal; (c) output waveform before filtering; (d) output waveform after filtering

6-5-2 FET Push–Pull Balanced Modulator

Figure 6-7 shows a schematic diagram for a balanced modulator that uses FETs rather than diodes for the nonlinear devices. A FET is a nonlinear device that exhibits square-law properties and produces only second-order cross-product frequencies. As is the diode balanced modulator, a FET modulator is a product modulator and produces only the sidebands at its output and suppresses the carrier. The FET balanced modulator is similar to a standard push–pull amplifier except that the modulator circuit has two inputs (the carrier and the modulating signal).

6-5-2-1 Circuit operation. The carrier is fed into the circuit in such a way that it is applied simultaneously and in phase to the gates of both FET amplifiers (Q_1 and Q_2). The carrier produces currents in both the top and the bottom halves of output transformer T_3 that are equal in magnitude but 180° out of phase. Therefore, they cancel, and no carrier component appears in the output waveform. The modulating signal is applied to the circuit in such a way that it is applied simultaneously to the gates of the two FETs 180° out of phase. The modulating signal causes an increase in the drain current in one FET and a decrease in the drain current in the other FET.

Figure 6-8 shows the phasor diagram for the currents produced in the output transformer of a FET balanced modulator. Figure 6-8a shows that the quiescent dc drain currents from Q_a and Q_b (I_{qa} and I_{qb}) pass through their respective halves of the primary winding of

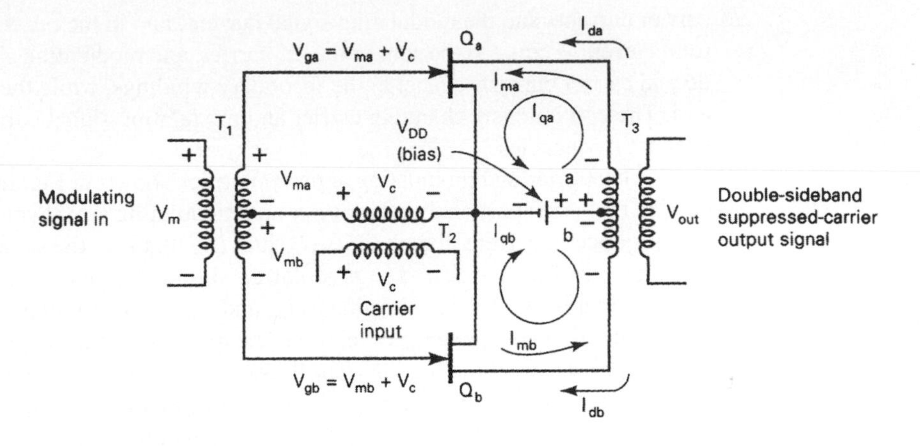

FIGURE 6-7 FET balanced modulator for the polarities shown

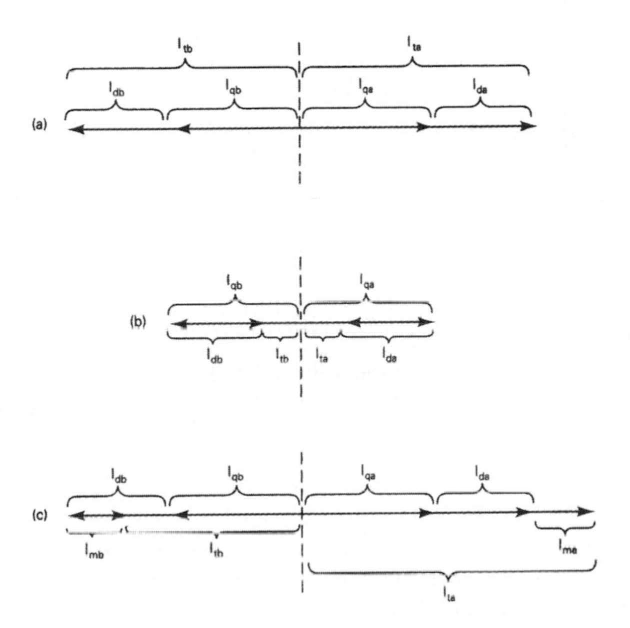

FIGURE 6-8 FET balanced modulator phasor diagrams: (a) in-phase sum of dc and carrier currents; (b) out-of-phase sum of dc and carrier currents; (c) sum of dc, carrier, and modulating-signal currents

T_3 180° out of phase with each other. Figure 6-8a also shows that an increase in drain current due to the carrier signal (I_{da} and I_{db}) adds to the quiescent current in both halves of the transformer windings, producing currents (I_{qa} and I_{qb}) that are equal and simply the sum of the quiescent and carrier currents. I_{qa} and I_{qb} are equal but travel in opposite directions; consequently, they cancel each other. Figure 6-8b shows the phasor sum of the quiescent and carrier currents when the carrier currents travel in the opposite direction to the quiescent currents. The total currents in both halves of the windings are still equal in magnitude, but now they are equal to the difference between the quiescent and carrier currents. Figure 6-8c shows the phasor diagram when a current component is added because of a modulating signal. The modulating signal currents (I_{ma} and I_{mb}) produce in their respective halves of the output transformer currents that are in phase with each other. However, it can be seen that in one-half of the windings, the total current is equal to the difference between the dc and carrier currents and the modulating signal current, and in the other half of the winding, the total current is equal to the sum of the dc, carrier, and modulating signal currents. Thus, the dc and carrier currents cancel in the secondary windings, while the difference components add. The continuously changing carrier and modulating signal currents produce the cross-product frequencies.

The carrier and modulating signal polarities shown in Figure 6-7 produce an output current that is proportional to the carrier and modulating signal voltages. The carrier signal (V_c) produces a current in both FETs (I_{da} and I_{db}) that is in the same direction as the quiescent currents (I_{qa} and I_{qb}). The modulating signal (V_{ma} and V_{mb}) produces a current in Q_a (I_{ma}) that is in the same direction as I_{da} and I_{qa} and a current in $Q_b(I_{mb})$ that is in the opposite direction as I_{db} and I_{qb}. Therefore, the total current through the *a* side of T_3 is $I_{ta} = I_{da} + I_{qa} + I_{ma}$, and the total current through the *b* side of T_3 is $I_{tb} = -I_{db} - I_{qb} + I_{mb}$. Thus, the net current through the primary winding of T_3 is $I_{ta} + I_{tb} = I_{ma} + I_{mb}$. For a modulating signal with the opposite polarity, the drain current in Q_b will increase and the drain current in Q_a will decrease. Ignoring the quiescent dc current (I_{qa} and I_{qb}), the drain current in one FET is the sum of the carrier and modulating signal currents ($I_d + I_m$), and the drain current in the other FET is the difference ($I_d - I_m$).

T_1 is an audio transformer, whereas T_2 and T_3 are radio-frequency transformers. Therefore, any audio component that appears at the drain circuits of Q_1 and Q_2 is not passed on to the output. To achieve total carrier suppression, Q_a and Q_b must be perfectly matched, and T_1 and T_3 must be exactly center tapped. As with the diode balanced modulators, the FET balanced modulator typically adds between 40 dB and 60 dB of attenuation to the carrier.

6-5-3 Balanced Bridge Modulator

Figure 6-9a shows the schematic diagram for a *balanced bridge modulator.* The operation of the bridge modulator, as the balanced ring modulator, is completely dependent on the switching action of diodes D_1 through D_4 under the influence of the carrier and modulating signal voltages. Again, the carrier voltage controls the on or off condition of the diodes and, therefore, must be appreciably larger than the modulating signal voltage.

6-5-3-1 Circuit operation. For the carrier polarities shown in Figure 6-9b, all four diodes are reverse biased and off. Consequently, the audio signal voltage is transferred directly to the load resistor (R_L). Figure 6-9c shows the equivalent circuit for a carrier with the opposite polarity. All four diodes are forward biased and on, and the load resistor is bypassed (i.e., *shorted out*). As the carrier voltage changes from positive to negative and vice versa, the output waveform contains a series of pulses that is comprised mainly of the upper and lower sideband frequencies. The output waveform is shown in Figure 6-9d. The series of pulses is shown as the shaded area in the figure.

6-5-4 Linear Integrated-Circuit Balanced Modulators

Linear integrated-circuit (LIC) balanced modulators are available up to 100 MHz, such as the LM1496/1596, that can provide carrier suppression of 50 dB at 10 MHz and up to 65 dB at 500 kHz. The LM1496/1596 balanced modulator integrated circuit is a *double-balanced modulator/ demodulator* that produces an output signal that is proportional to the product of its input signals. Integrated circuits are ideally suited for applications that require balanced operation.

6-5-4-1 Circuit operation. Figure 6-10 shows a simplified schematic diagram for a differential amplifier, which is the fundamental circuit of an LIC balanced modulator because of its excellent *common-mode rejection ratio* (typically 85 dB or more). When a carrier signal is applied to the base of Q_1, the emitter currents in both transistors will vary by the same amount. Because the emitter current for both Q_1 and Q_2 comes from a common

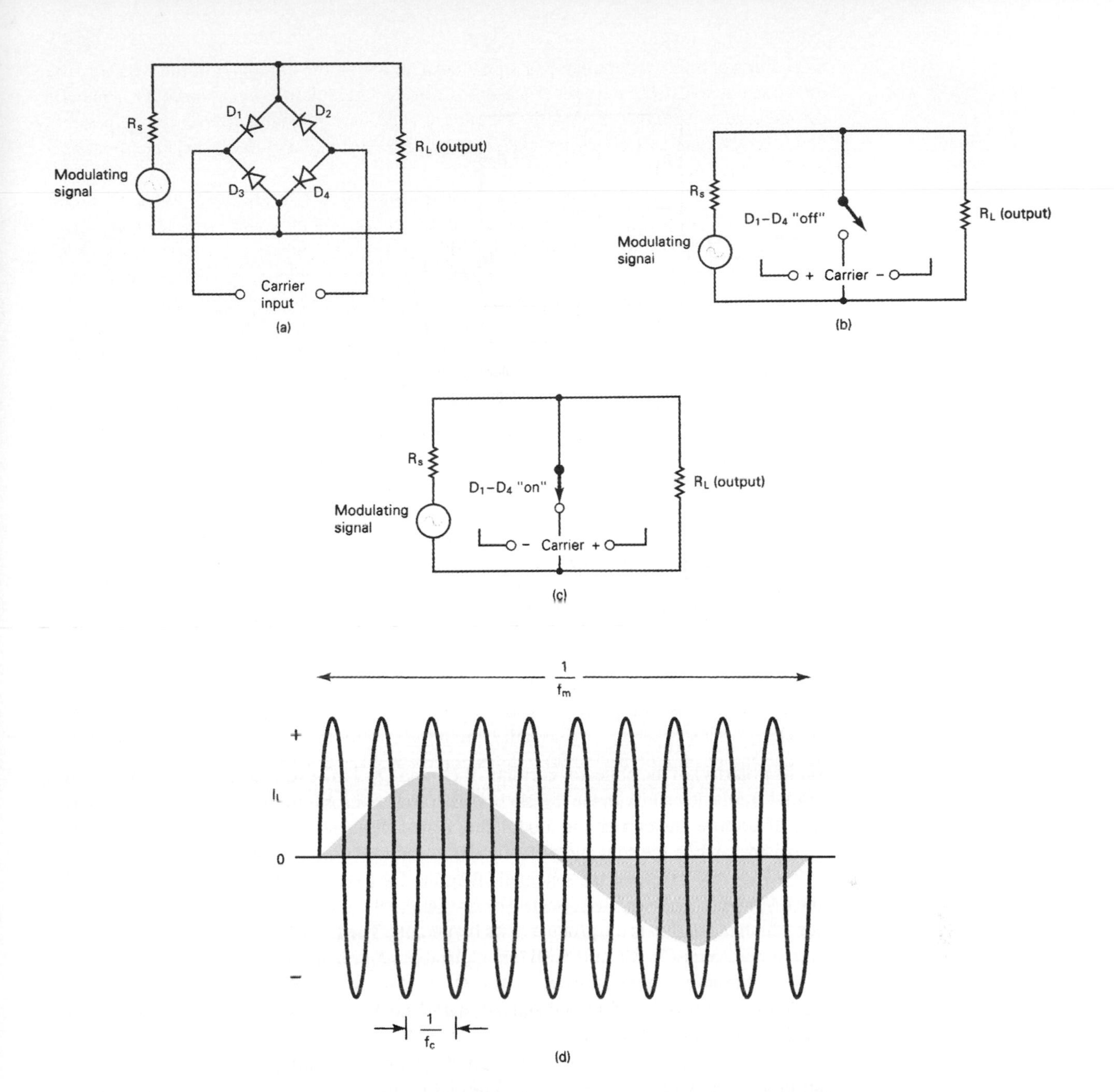

FIGURE 6-9 Balanced bridge modulator: (a) schematic diagram; (b) diodes biased *off;* (c) diodes biased *on;* (d) output waveform

constant-current source (Q_4), any increase in Q_1's emitter current results in a corresponding decrease in Q_2's emitter current and vice versa. Similarly, when a carrier signal is applied to the base of Q_2, the emitter currents of Q_1 and Q_2 vary by the same magnitude, except in opposite directions. Consequently, if the same carrier signal is fed simultaneously to the bases of Q_1 and Q_2, the respective increases and decreases are equal and, thus, cancel. Therefore, the collector currents and output voltage remain unchanged. If a modulating signal is applied to the base of Q_3, it causes a corresponding increase or decrease (depending

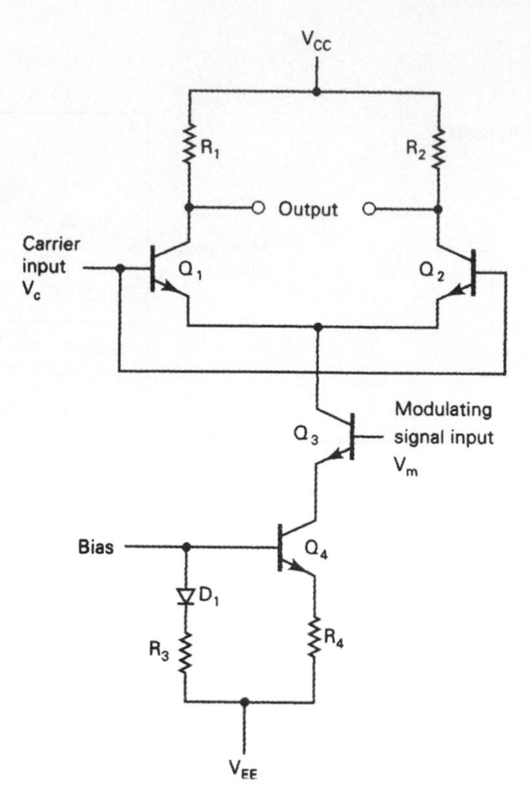

FIGURE 6-10 Differential amplified schematic

on its polarity) in the collector currents of Q_1 and Q_2. However, the carrier and modulating signal frequencies mix in the transistors and produce cross-product frequencies in the output. Therefore, the carrier and modulating signal frequencies are canceled in the balanced transistors, while the sum and difference frequencies appear in the output.

Figure 6-11 shows the schematic diagram for a typical AM DSBSC modulator using the LM1496/1596 integrated circuit. The LM1496/1596 is a balanced modulator/demodulator for which the output is the product of its two inputs. The LM1496/1596 offers excellent carrier suppression (65 dB at 0.5 MHz), adjustable gain, balanced inputs and outputs, and a high common-mode rejection ratio (85 dB). When used as a product detector, the LM1496/1596 has a sensitivity of 3.0 μV and a dynamic range of 90 dB when operating at an intermediate frequency of 9 MHz.

The carrier signal is applied to pin 10 which, in conjunction with pin 8, provides an input to a quad cross-coupled differential output amplifier. This configuration is used to ensure that full-wave multiplication of the carrier and modulating signal occurs. The modulating signal is applied to pin 1 which, in conjunction with pin 4, provides a differential input to the current driving transistors for the output difference amplifier. The 50-kΩ potentiometer, in conjunction with V_{EE} (−8 V dc), is used to balance the bias currents for the difference amplifiers and null the carrier. Pins 6 and 12 are single-ended outputs that contain carrier and sideband components. When one of the outputs is inverted and added to the other, the carrier is suppressed, and a double-sideband suppressed-carrier wave is produced. Such a process is accomplished in the op-amp subtractor. The subtractor inverts the signal at the inverting (−) input and adds it to the signal at the noninverting (+) input. Thus, a double-sideband suppressed-carrier wave appears at the output of the op-amp. The 6.8-kΩ resistor connected to pin 5 is a bias resistor for the internal constant-current supply.

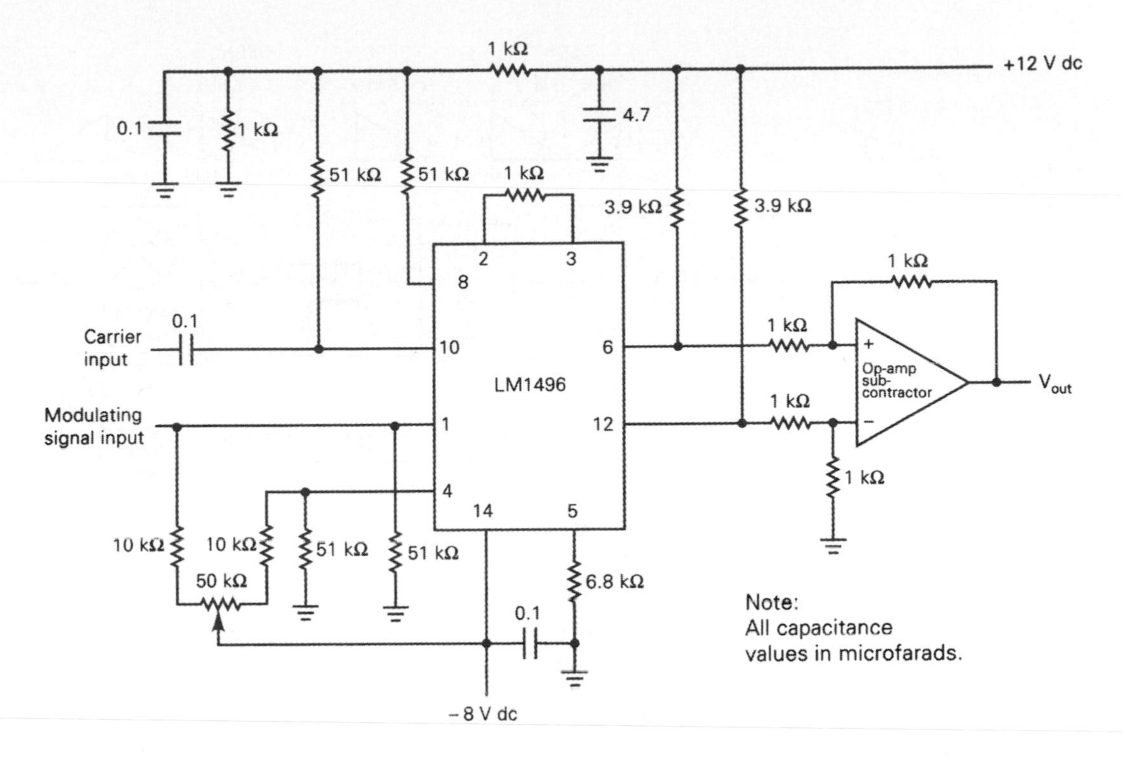

FIGURE 6-11 AM DSBSC modulator using the LM1496/1596 linear integrated circuit

The XR-2206 linear integrated-circuit AM DSBFC modulator described in Chapter 4 and shown in Figure 4-20a can also be used to produce a double-sideband suppressed-carrier wave by simply setting the dc bias to $V^+/2$ and limiting the modulating-signal amplitude to $\pm 4\ V_p$. As the modulating signal passes through its zero crossings, the phase of the carrier undergoes a 180° phase reversal. This property also makes the XR-2206 ideally suited as a phase shift modulator. The dynamic range of amplitude modulation for the XR-2206 is approximately 55 dB.

6-6 SINGLE-SIDEBAND TRANSMITTERS

The transmitters used for single-sideband suppressed- and reduced-carrier transmission are identical except that the reinserted carrier transmitters have an additional circuit that adds a low-amplitude carrier to the single-sideband waveform after suppressed-carrier modulation has been performed and one of the sidebands has been removed. The reinserted carrier is called a *pilot carrier.* The circuit where the carrier is reinserted is called a *linear summer* if it is a resistive network and a *hybrid coil* if the SSB waveform and pilot carrier are inductively combined in a transformer bridge circuit. Three transmitter configurations are commonly used for single-sideband generation: the filter method, the phase-shift method, and the so-called *third method.*

6-6-1 Single-Sideband Transmitter: Filter Method

Figure 6-12 shows a block diagram for a SSB transmitter that uses balanced modulators to suppress the unwanted carrier and filters to suppress the unwanted sideband. The figure shows a transmitter that uses three stages of frequency up-conversion. The modulating signal is an audio spectrum that extends from 0 kHz to 5 kHz. The modulating signal mixes with a low-frequency (LF) 100-kHz carrier in balanced modulator 1 to produce a double-sideband frequency spectrum centered around the suppressed 100-kHz IF carrier. Bandpass filter 1 (BPF 1) is tuned to a 5-kHz bandwidth centered around 102.5 kHz, which

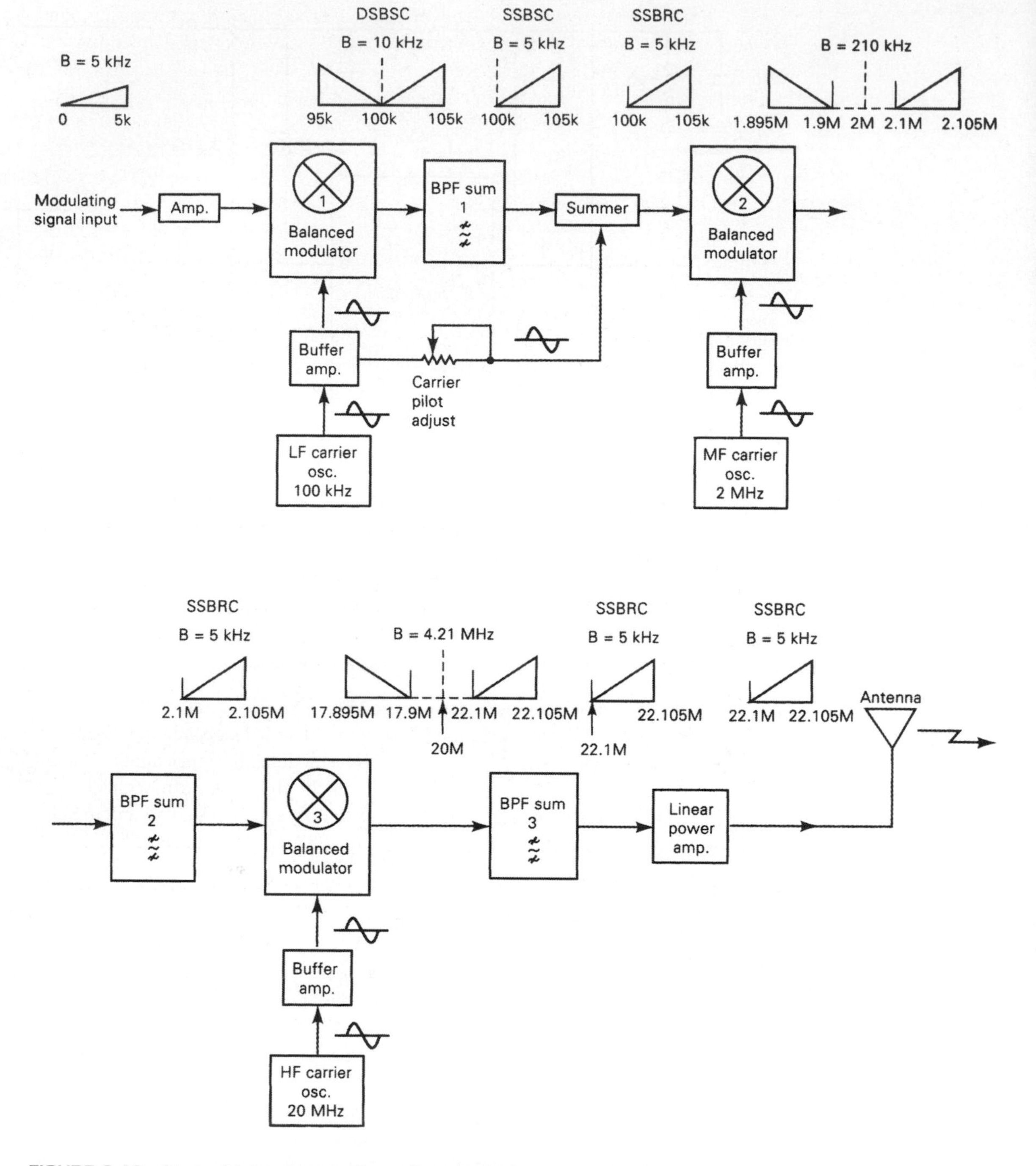

FIGURE 6-12 Single-sideband transmitter: filter method

is the center of the upper sideband frequency spectrum. The pilot or reduced-amplitude carrier is added to the single-sideband waveform in the carrier reinsertion stage, which is simply a linear summer. The summer is a simple adder circuit that combines the 100-kHz pilot carrier with the 100-kHz to 105-kHz upper sideband frequency spectrum. Thus, the output of the summer is a SSBRC waveform. (If suppressed-carrier transmission is desired, the carrier pilot and summer circuit can be omitted.)

The low-frequency IF is converted to the final operating frequency band through a series of frequency translations. First, the SSBRC waveform is mixed in balanced modulator 2 with a 2-MHz medium-frequency (MF) carrier. The output is a double-sideband suppressed-carrier signal in which the upper and lower sidebands each contain the original SSBRC frequency spectrum. The upper and lower sidebands are separated by a 200-kHz frequency band that is void of information. The center frequency of BPF 2 is 2.1025 MHz with a 5-kHz bandwidth. Therefore, the output of BPF 2 is once again a single-sideband reduced-carrier waveform. Its frequency spectrum comprises a reduced 2.1-MHz second IF carrier and a 5-kHz-wide upper sideband. The output of BPF 2 is mixed with a 20-MHz high-frequency (HF) carrier in balanced modulator 3. The output is a double-sideband suppressed-carrier signal in which the upper and lower sidebands again each contain the original SSBRC frequency spectrum. The sidebands are separated by a 4.2-MHz frequency band that is void of information. BPF 3 is centered on 22.1025 MHz with a 5-kHz bandwidth. Therefore, the output of BPF 3 is once again a single-sideband waveform with a reduced 22.1-MHz RF carrier and a 5-kHz-wide upper sideband. The output waveform is amplified in the linear power amplifier and then transmitted.

In the transmitter just described, the original modulating-signal frequency spectrum was up-converted in three modulation steps to a final carrier frequency of 22.1 MHz and a single upper sideband that extended from the carrier to 22.105 MHz. After each up-conversion (frequency translation), the desired sideband is separated from the double-sideband spectrum with a BPF. The same final output spectrum can be produced with a single heterodyning process: one balanced modulator, one bandpass filter, and a single HF carrier supply. Figure 6-13a shows the block diagram and output frequency spectrum for a single-conversion transmitter. The output of the balanced modulator is a double-sideband frequency spectrum centered around a suppressed-carrier frequency of 22.1 MHz. To separate the 5-kHz-wide upper sideband from the composite frequency spectrum, a multiple-pole BPF with an extremely high Q is required. A BPF that meets this criterion is in itself difficult to construct, but suppose that this were a multichannel transmitter and the carrier frequency were tunable; then the BPF must also be tunable. Constructing a tunable BPF in the megahertz frequency range with a passband of only 5 kHz is beyond economic and engineering feasibility. The only BPF in the transmitter shown in Figure 6-12 that has to separate sidebands that are immediately adjacent to each other is BPF 1. To construct a 5-kHz-wide, steep-skirted BPF at 100 kHz is a relatively simple task, as only a moderate Q is required. The sidebands separated by BPF 2 are 200 kHz apart; thus, a low Q-filter with gradual roll-off characteristics can be used with no danger of passing any portion of the undesired sideband. BPF 3 separates sidebands that are 4.2 MHz apart. If multiple channels are used and the HF carrier is tunable, a single broadband filter can be used for BPF 3 with no danger of any portion of the undesired sideband leaking through the filter. For single-channel operation, the single conversion transmitter is the simplest design, but for multi-channel operation, the three-conversion system is more practical. Figures 6-13b and c show the output spectrum and filtering requirements for both methods.

6-6-1-1 Single-sideband filters. It is evident that filters are an essential part of any electronic communications system and especially single-sideband systems. Transmitters as well as receivers have requirements for highly selective networks for limiting both the signal and noise frequency spectrums. The quality factor (Q) of a single-sideband filter depends on the carrier frequency, the frequency separation between sidebands, and the desired attenuation level of the unwanted sideband. Q can be expressed mathematically as

$$Q = \frac{f_c(\log^{-1} S/20)^{1/2}}{4\Delta f} \tag{6-3}$$

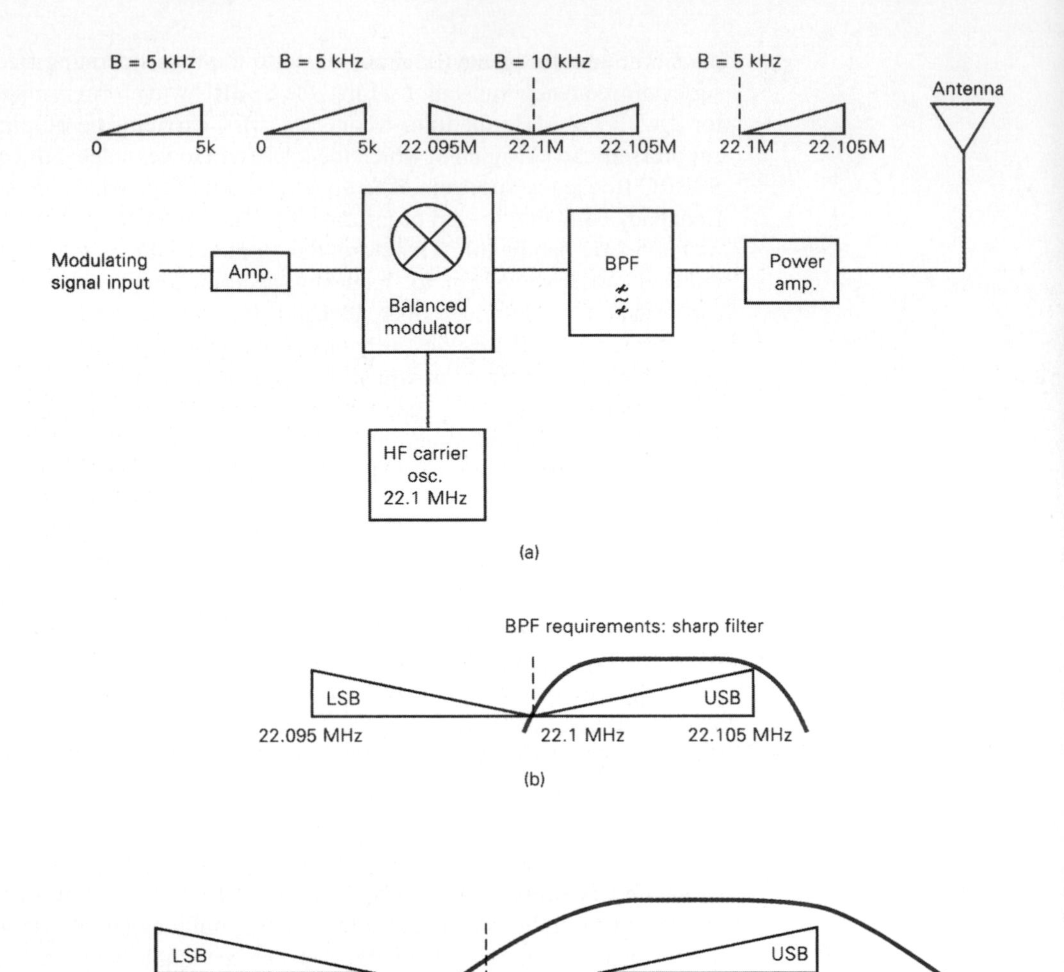

FIGURE 6-13 Single conversion SSBSC transmitter, filter method: (a) block diagram; (b output spectrum and filtering requirements for a single-conversion transmitter; (c) output spectrum and filtering requirements for a three-conversion transmitter

where Q = quality factor
f_c = center or carrier frequency
S = dB level of suppression of unwanted sideband
Δf = frequency separation between the highest lower sideband frequency and the lowest upper sideband frequency

Example 6-1

Determine the quality factor (Q) necessary for a single-sideband filter with a 1-MHz carrier frequency, 80-dB unwanted sideband suppression, and the following frequency spectrum:

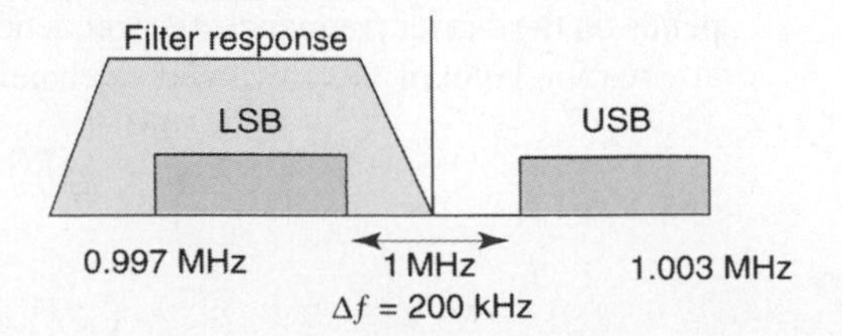

Solution Substituting into Equation 6-3 gives

$$Q = \frac{1 \text{ MHz}(\log^{-1} 80/20)^{1/2}}{4(200)} = 125{,}000$$

Conventional *LC* filters have relatively low *Q*s and are, therefore, not selective enough for most single-sideband applications. Therefore, filters used for single-sideband generation are usually constructed from either crystal or ceramic materials, mechanical filters, or surface acoustic wave (SAW) filters.

6-6-1-2 Crystal filters. The *crystal lattice filter* is commonly used in single-sideband systems. The schematic diagram for a typical crystal lattice bandpass filter is shown in Figure 6-14a. The lattice comprises two sets of matched crystal pairs (X_1 and X_2, X_3 and X_4) connected between tuned input and output transformers T_1 and T_2. Crystals X_1 and X_2 are series connected, whereas X_3 and X_4 are connected in parallel. Each pair of crystals is matched in frequency within 10 Hz to 20 Hz. X_1 and X_2 are cut to operate at the filter lower cutoff frequency, and X_3 and X_4 are cut to operate at the upper cutoff frequency. The input and output transformers are tuned to the center of the desired passband, which tends to spread the difference between the series and parallel resonant frequencies. C_1 and C_2 are used to correct for any overspreading of frequency difference under matched crystal conditions.

The operation of the crystal filter is similar to the operation of a bridge circuit. When the reactances of the bridge arms are equal and have the same sign (either inductive or capacitive), the signals propagating through the two possible paths of the bridge cancel each

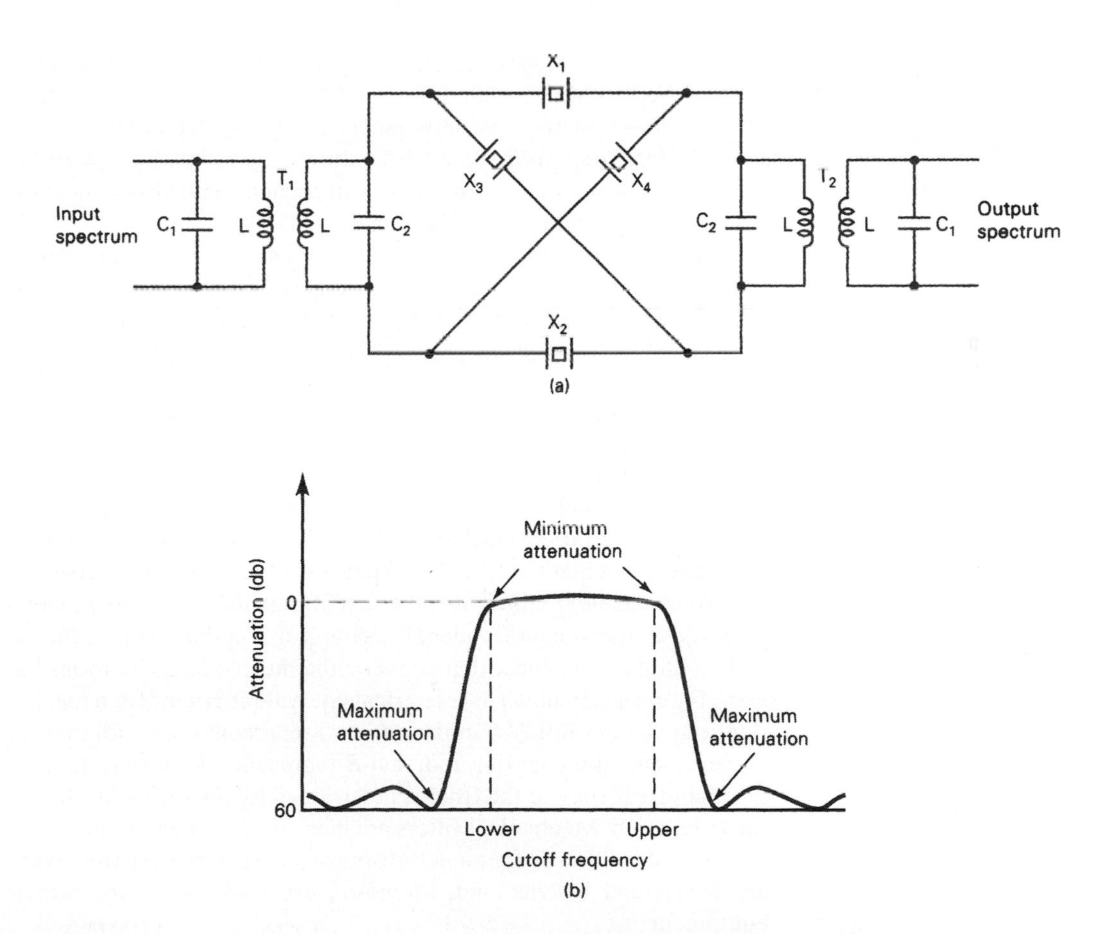

FIGURE 6-14 Crystal lattice filter: (a) schematic diagram; (b) characteristic curve

other out. At the frequency where the reactances have equal magnitudes and opposite signs (one inductive and the other capacitive), the signal is propagated through the network with maximum amplitude.

Figure 6-14b shows a typical characteristic curve for a crystal lattice bandpass filter. Crystal filters are available with a Q as high as 100,000. The filter shown in Figure 6-14a is a single-element filter. However, for a crystal filter to adequately pass a specific band of frequencies and reject all others, at least two elements are necessary. Typical insertion losses for crystal filters are between 1.5 dB and 3 dB.

6-6-1-3 Ceramic filters. *Ceramic filters* are made from lead zirconate-titanate, which exhibits the piezoelectric effect. Therefore, they operate quite similar to crystal filters except that ceramic filters do not have as high a Q-factor. Typical Q values for ceramic filters go up to about 2000. Ceramic filters are less expensive, smaller, and more rugged than their crystal lattice counterparts. However, ceramic filters have more loss. The insertion loss for ceramic filters is typically between 2 dB and 4 dB.

Ceramic filters typically come in one-element, three-terminal packages; two-element, eight-terminal packages; and four-element, 14-terminal packages. Ceramic filters feature small size, low profile, symmetrical selectivity characteristics, low spurious response, and excellent immunity to variations in environmental conditions with minimum variation in operating characteristics. However, certain precautions must be taken with ceramic filters, which include the following:

1. ***Impedance matching and load conditions.*** Ceramic filters differ from coils in that their impedance cannot readily be changed. When using ceramic filters, it is very important that impedances be properly matched.
2. ***Spurious signals.*** In practically all cases where ceramic filters are used, spurious signals are generated. To suppress these responses, impedance matching with IF transformers is the simplest and most effective way.
3. ***Matching coils.*** When difficulties arise in spurious response suppression or for improvement in selectivity or impedance matching in IF stages, use of an impedance matching coil is advised.
4. ***Error in wiring input and output connections.*** Care must be taken when connecting the input and output terminals of a ceramic filter. Any error will cause waveform distortion and possibly frequency deviation of the signal.
5. ***Use of two ceramic filters in cascade.*** For best performance, a coil should be used between two ceramic filter units. When cost is a factor and a direct connection is necessary, a suitable capacitor or resistor can be used.

6-6-1-4 Mechanical filters. A *mechanical filter* is a *mechanically resonant transducer.* It receives electrical energy, converts it to mechanical vibrations, and then converts the vibrations back to electrical energy at its output. Essentially, four elements comprise a mechanical filter: an input transducer that converts the input electrical energy to mechanical vibrations, a series of mechanical resonant metal disks that vibrate at the desired resonant frequency, a coupling rod that couples the metal disks together, and an output transducer that converts the mechanical vibrations back to electrical energy. Figure 6-15 shows the electrical equivalent circuit for a mechanical filter. The series resonant circuits (LC combinations) represent the metal disks, coupling capacitor C_1 represents the coupling rod, and R represents the matching mechanical loads. The resonant frequency of the filter is determined by the series LC disks, and C_1 determines the bandwidth. Mechanical filters are more rugged than either ceramic or crystal filters and have comparable frequency-response characteristics. However, mechanical filters are larger and heavier and, therefore, are impractical for mobile communications equipment.

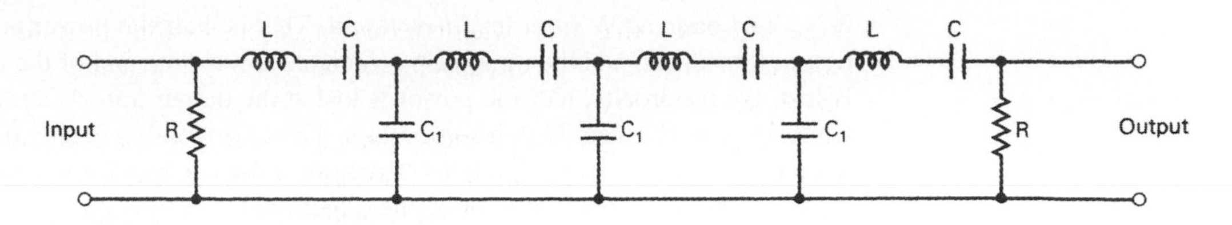

FIGURE 6-15 Mechanical filter equivalent circuit

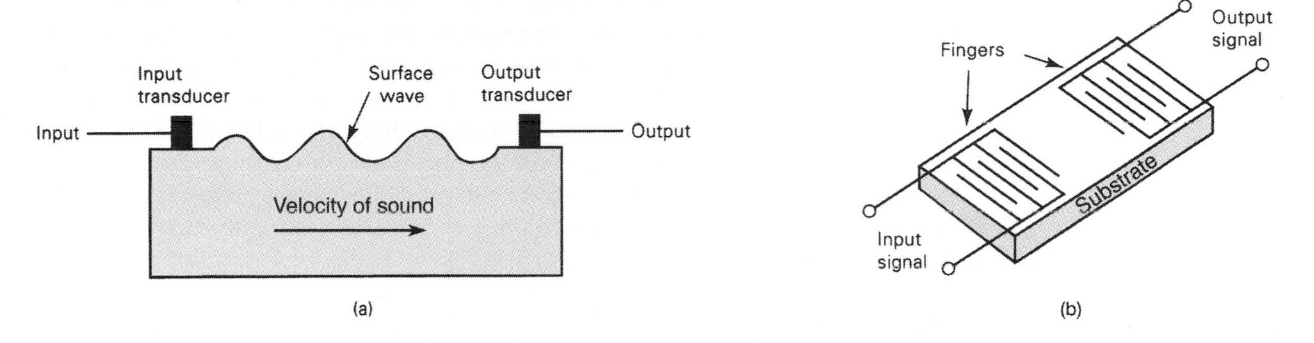

FIGURE 6-16 SAW filter: (a) surface wave; (b) metallic fingers

6-6-1-5 Surface acoustic wave filters. *Surface acoustic wave (SAW) filters* were first developed in the 1960s but did not become commercially available until the 1970s. SAW filters use acoustic energy rather than electromechanical energy to provide excellent performance for precise bandpass filtering. In essence, SAW filters trap or guide acoustical waves along a surface. They can operate at center frequencies up to several gigahertz and bandwidths up to 50 MHz with more accuracy and reliability than their predecessor, the mechanical filter, and they do it at a lower cost. SAW filters have extremely steep roll-off characteristics and typically attenuate frequencies outside their passband between 30 dB and 50 dB more than the signals within their passband. SAW filters are used in both single- and multiple-conversion superheterodyne receivers for both RF and IF filters and in single-sideband systems for a multitude of filtering applications.

A SAW filter consists of transducers patterned from a thin aluminum film deposited on the surface of a semiconductor crystal material that exhibits the piezoelectric effect. This results in a physical deformation (rippling) on the surface of the substrate. These ripples vary at the frequency of the applied signal but travel along the surface of the material at the speed of sound. With SAW filters, an oscillating electrical signal is applied across a small piece of semiconductor crystal that is part of a larger, flat surface, as shown in Figure 6-16a. The piezoelectric effect causes the crystal material to vibrate. These vibrations are in the form of acoustic energy that travels across the surface of the substrate until it reaches a second crystal at the opposite end, where the acoustic energy is converted back to electrical energy.

To provide filter action, a precisely spaced row of metallic *fingers* is deposited on the flat surface of the substrate, as shown in Figure 6-16b. The finger centers are spaced at either a half- or a quarter-wavelength of the desired center frequency. As the acoustic waves travel across the surface of the substrate, they reflect back and forth as they impinge on the fingers. Depending on the acoustical wavelength and the spacing between the fingers, some of the reflected energy cancels and attenuates the incident wave energy (this is called *destructive interference*), while some of the energy aids (*constructive interference*). The exact frequencies of acoustical energy that are canceled depend on the spacing between the fingers. The bandwidth of the filter is determined by the thickness and number of fingers.

The basic SAW filter is *bidirectional.* That is, half the power is radiated toward the output transducer, and the other half is radiated toward the end of the crystal substrate and is lost. By reciprocity, half the power is lost at the output transducer. Consequently, SAW filters have a relatively high insertion loss. This shortcoming can be overcome to a certain degree by using a more complex structure called a *unidirectional transducer,* which launches the acoustic wave in only one direction.

SAW filters are inherently very rugged and reliable. Because their operating frequencies and bandpass responses are set by the photolithographic process, they do not require complicated tuning operations, nor do they become detuned over a period of time. The semiconductor wafer processing techniques used in manufacturing SAW filters permit large-volume production of economical and reproducible devices. Finally, their excellent performance capabilities are achieved with significantly reduced size and weight when compared to competing technologies.

The predominant disadvantage of SAW filters is their extremely high insertion loss, which is typically between 25 dB and 35 dB. For this reason, SAW filters cannot be used to filter low-level signals. SAW filters also exhibit a much longer delay time than their electronic counterparts (approximately 20,000 times as long). Consequently, SAW filters are sometimes used for *delay lines.*

6-6-2 Single-Sideband Transmitter: Phase-Shift Method

With the phase-shift method of single-sideband generation, the undesired sideband is canceled in the output of the modulator; therefore, sharp filtering is unnecessary. Figure 6-17 shows a block diagram for a SSB transmitter that uses the phase-shift method to remove the upper sideband. Essentially, there are two separate double-sideband modulators (balanced modulators 1 and 2). The modulating signal and carrier are applied directly to one of the modulators, then both are shifted 90° and applied to the second modulator. The outputs from the two balanced modulators are double-sideband suppressed-carrier signals with the proper phase such that, when they are combined in a linear summer, the upper sideband is canceled.

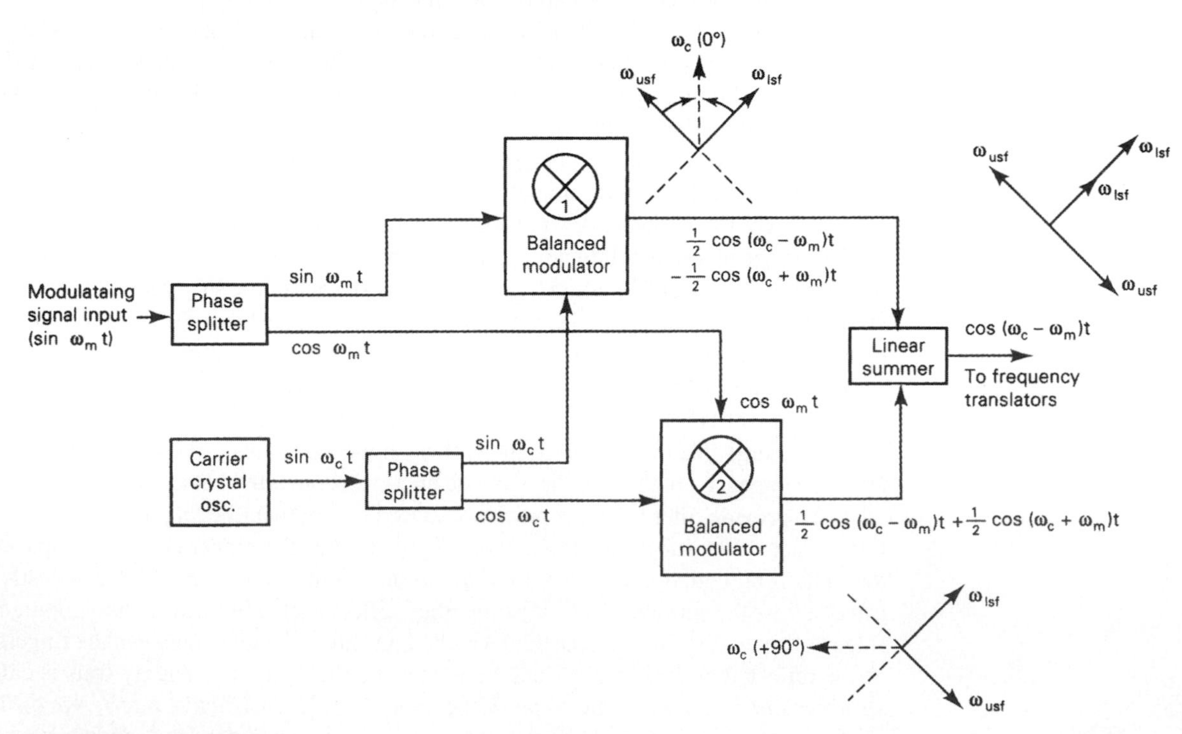

FIGURE 6-17 SSB transmitter: phase shift method

6-6-2-1 Phasor representation. The phasors shown in Figure 6-17 illustrate how the upper sideband is canceled by rotating both the carrier and the modulating signal 90° prior to modulation. The output phase from balanced modulator 1 shows the relative position and direction of rotation of the upper (ω_{usf}) and lower (ω_{lsf}) side frequencies to the suppressed carrier (ω_c). The phasors at the output of balanced modulator 2 are essentially the same except that the phase of the carrier and the modulating signal are each rotated 90°. The output of the summer shows the sum of the phasors from the two balanced modulators. The two phasors for the lower sideband are in phase and additive, whereas the phasors for the upper sideband are 180° out of phase and, thus, cancel. Consequently, only the lower sideband appears at the output of the summer.

6-6-2-2 Mathematical analysis. In Figure 6-17, the input modulating signal ($\sin \omega_m t$) is fed directly to balanced modulator 1 and shifted 90° ($\cos \omega_m t$) and fed to balanced modulator 2. The low-frequency carrier ($\sin \omega_c t$) is also fed directly to balanced modulator 1 and shifted 90° ($\cos \omega_c t$) and fed to balanced modulator 2. The balanced modulators are product modulators, and their outputs are expressed mathematically as

$$\begin{aligned}\text{output from}&\\ \text{balanced modulator 1} &= (\sin \omega_m t)(\sin \omega_c t)\\ &= \frac{1}{2}\cos(\omega_c - \omega_m)t - \frac{1}{2}\cos(\omega_c + \omega_m)t\end{aligned}$$

$$\begin{aligned}\text{output from}&\\ \text{balanced modulator 2} &= (\cos \omega_m t)(\cos \omega_c t)\\ &= \frac{1}{2}\cos(\omega_c - \omega_m)t + \frac{1}{2}\cos(\omega_c + \omega_m)t\end{aligned}$$

and the output from the linear summer is

$$\begin{array}{rr} \frac{1}{2}\cos(\omega_c - \omega_m)t & -\ \frac{1}{2}\cos(\omega_c + \omega_m)t \\ +\frac{1}{2}\cos(\omega_c - \omega_m)t & +\ \frac{1}{2}\cos(\omega_c + \omega_m)t \\ \hline \cos(\omega_c - \omega_m)t & \text{canceled} \\ \text{lower sideband} & \\ \text{(difference signal)} & \end{array}$$

6-7 INDEPENDENT SIDEBAND

Figure 6-18 shows a block diagram for an *independent sideband* (ISB) transmitter with three stages of modulation. The transmitter uses the filter method to produce two independent single-sideband channels (channel A and channel B). The two channels are combined, then a pilot carrier is reinserted. The composite ISB reduced-carrier waveform is up-converted to RF with two additional stages of frequency translation. There are two 5-kHz-wide information signals that originate from two independent sources. The channel A information signals modulate a 100-kHz LF carrier in balanced modulator A. The output from balanced modulator A passes through BPF A, which is tuned to the lower sideband (95 kHz to 100 kHz). The channel B information signals modulate the same 100-kHz LF carrier in balanced modulator B. The output from balanced modulator B passes through BPF B, which is tuned to the upper sideband (100 kHz to 105 kHz). The two single-sideband frequency spectrums are combined in a hybrid network to form a composite ISB suppressed-carrier spectrum (95 kHz to 105 kHz). The LF carrier (100 kHz) is reinserted in the linear

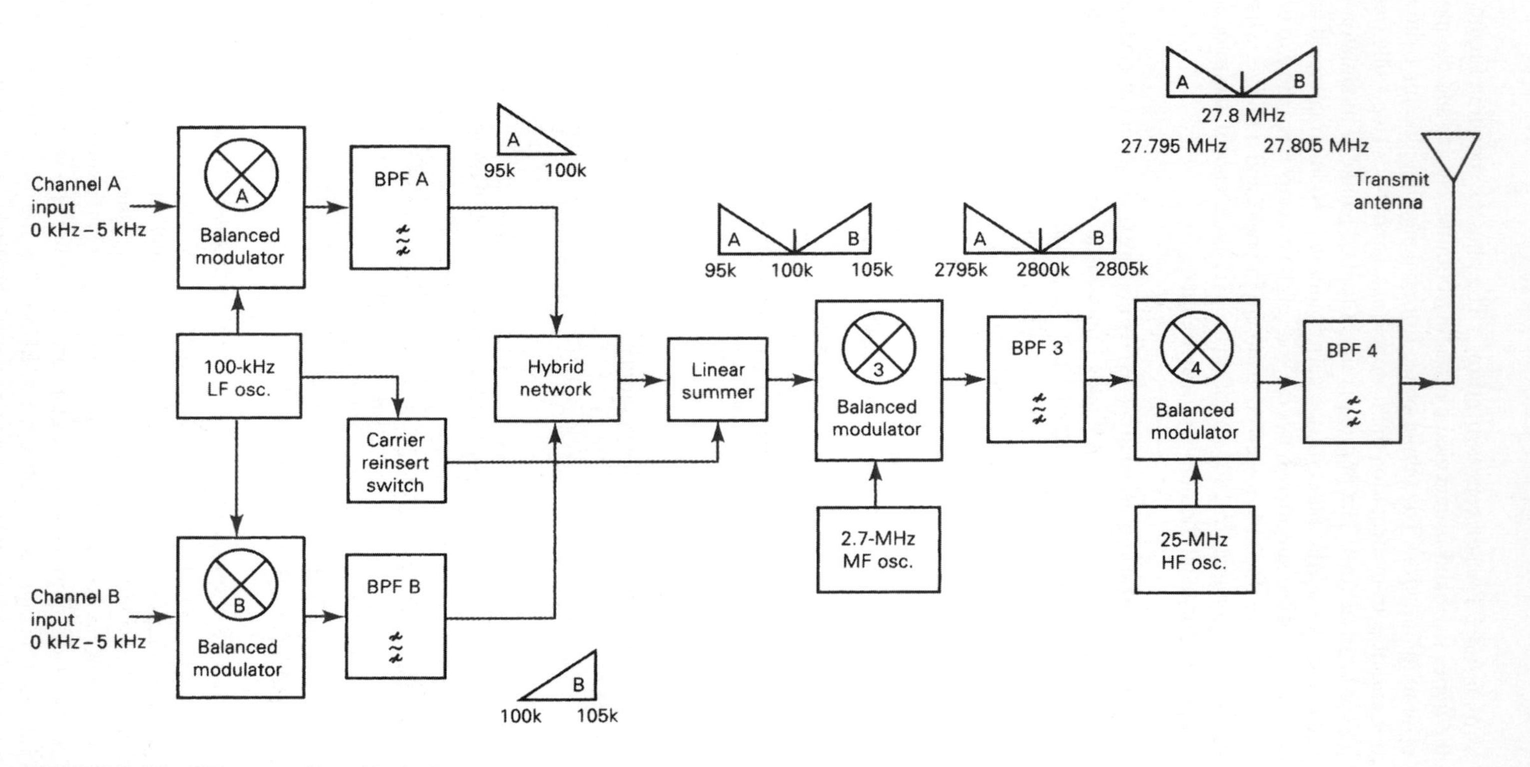

FIGURE 6-18 ISB transmitter: block diagram

summer to form an ISB reduced-carrier waveform. The ISB spectrum is mixed with a 2.7-MHz MF carrier in balanced modulator 3. The output from balanced modulator 3 passes through BPF 3 to produce an ISB reduced-carrier spectrum that extends from 2.795 MHz to 2.805 MHz with a reduced 2.8-MHz pilot carrier. Balanced modulator 4, BPF 4, and the HF carrier translate the MF spectrum to an RF band that extends from 27.795 MHz to 27.8 MHz (channel A) and 27.8 MHz to 27.805 MHz (channel B) with a 27.8-MHz reduced-amplitude carrier.

6-8 SINGLE-SIDEBAND RECEIVERS

6-8-1 Single-Sideband BFO Receiver

Figure 6-19 shows the block diagram for a simple noncoherent single-sideband *BFO receiver.* The selected radio-frequency spectrum is amplified and then mixed down to intermediate frequencies for further amplification and band reduction. The output from the IF amplifier stage is heterodyned (beat) with the output from a *beat frequency oscillator* (BFO). The BFO frequency is equal to the IF carrier frequency; thus, the difference between the IF and the BFO frequencies is the information signal. Demodulation is accomplished through several stages of mixing and filtering. The receiver is noncoherent because the RF local oscillator and BFO signals are not synchronized to each other or to the oscillators in the transmitter. Consequently, any difference between the transmit and receive local oscillator frequencies produces a frequency offset error in the demodulated information signal. For example, if the receive local oscillator is 100 Hz above its designated frequency and the BFO is 50 Hz above its designated frequency, the restored information is offset 150 Hz from its original frequency spectrum. Fifty hertz or more offset is distinguishable by a normal listener as a tonal variation.

The RF mixer and second detector shown in Figure 6-19 are product detectors. As with the balanced modulators in the transmitter, their outputs are the product of their inputs. A product modulator and product detector are essentially the same circuit. The only difference is that the input to a product modulator is tuned to a low-frequency modulating signal and the output is tuned to a high-frequency carrier, whereas with a product detector, the input is tuned to a high-frequency modulated carrier and the output is tuned to a low-frequency information signal. With both the modulator and detector, the single-frequency carrier is the switching signal. In a receiver, the input signal, which is a suppressed or reduced RF carrier and one sideband, is mixed with the RF local oscillator frequency to produce an intermediate frequency. The output from the second product detector is the sum and difference frequencies between the IF and the beat frequency. The difference frequency band is the original input information.

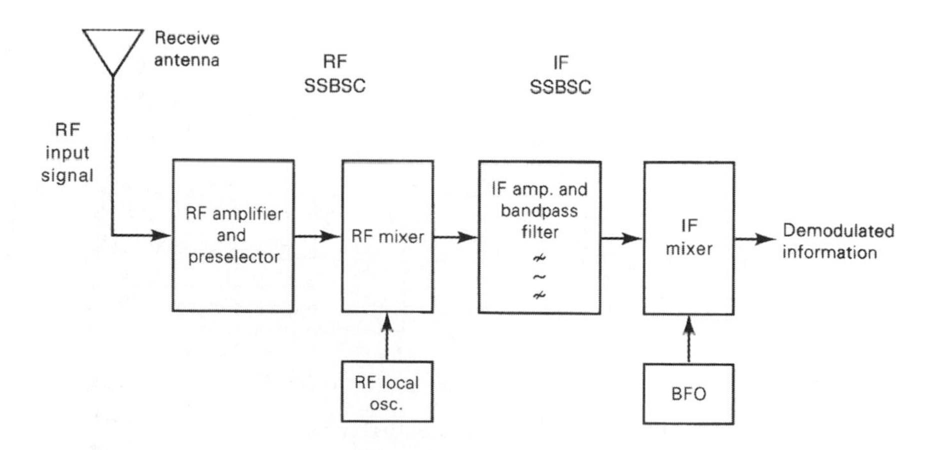

FIGURE 6-19 Noncoherent BFO SSB receiver

Example 6-2

For the BFO receiver shown in Figure 6-19, a received RF frequency band of 30 MHz to 30.005 MHz, an RF local oscillator frequency of 20 MHz, an IF frequency band of 10 MHz to 10.005 MHz, and a BFO frequency of 10 MHz, determine

a. Demodulated first IF frequency band and demodulated information frequency band.

b. Demodulated information frequency band if the RF local oscillator frequency drifts down 0.001%.

Solution **a.** The IF output from the RF mixer is the difference between the received signal frequency and the RF local oscillator frequency, or

$$f_{\text{IF}} = (30 \text{ MHz to } 30.005 \text{ MHz}) - 20 \text{ MHz} = 10 \text{ MHz to } 10.005 \text{ MHz}$$

The demodulated information signal spectrum is the difference between the intermediate frequency band and the BFO frequency, or

$$f_m = (10 \text{ MHz to } 10.005 \text{ MHz}) - 10 \text{ MHz} = 0 \text{ kHz to } 5 \text{ kHz}$$

b. A 0.001% drift would cause a decrease in the RF local oscillator frequency of

$$\Delta f = (0.00001)(20 \text{ MHz}) = 200 \text{ Hz}$$

Thus, the RF local oscillator frequency would drift down to 19.9998 Hz, and the output from the RF mixer is

$$\begin{aligned} f_{\text{IF}} &= (30 \text{ MHz to } 30.005 \text{ MHz}) - 19.9998 \text{ MHz} \\ &= 10.0002 \text{ MHz to } 10.0052 \text{ MHz} \end{aligned}$$

The demodulated information signal spectrum is the difference between the intermediate frequency band and the BFO, or

$$\begin{aligned} f_m &= (10.0002 \text{ MHz to } 10.0052 \text{ MHz}) - 10 \text{ MHz} \\ &= 200 \text{ Hz to } 5200 \text{ Hz} \end{aligned}$$

The 0.001% drift in the RF local oscillator frequency caused a corresponding 200-Hz error in the demodulated information signal spectrum.

6-8-2 Coherent Single-Sideband BFO Receiver

Figure 6-20 shows a block diagram for a coherent single-sideband BFO receiver. This receiver is identical to the BFO receiver shown in Figure 6-19 except that the LO and BFO frequencies are synchronized to the carrier oscillators in the transmitter. The carrier *recovery circuit* is a narrowband PLL that tracks the pilot carrier in the composite SSBRC receiver signal and uses the recovered carrier to regenerate coherent local oscillator fre-

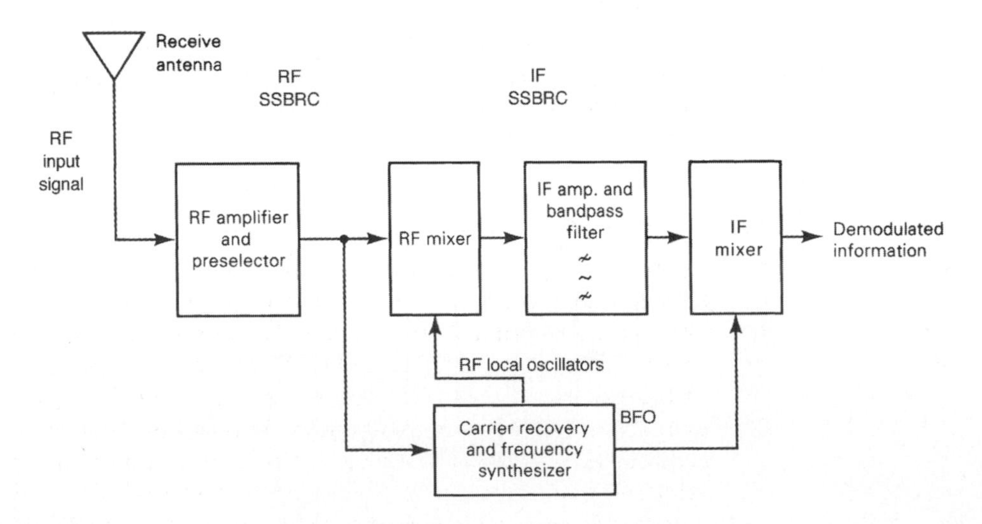

FIGURE 6-20 Coherent SSB BFO receiver

quencies in the synthesizer. The synthesizer circuit produces a coherent RF local oscillator and BFO frequency. The carrier recovery circuit tracks the received pilot carrier. Therefore, minor changes in the carrier frequency in the transmitter are compensated for in the receiver, and the frequency offset error is eliminated. If the coherent receiver shown in Figure 6-20 had been used in Example 6-2, the RF local oscillator would not have been allowed to drift independently.

Example 6-3

For the coherent single-sideband BFO receiver shown in Figure 6-20, an RF reduced carrier frequency of 30 MHz with an upper sideband that extends from just above 30 MHz to 30.005 MHz, an RF local oscillator frequency of 20 MHz, an IF center frequency of 10 MHz, and a BFO output frequency of 10 MHz, determine

a. Demodulated first IF frequency band and demodulated information frequency band.

b. Demodulated information frequency band if the reduced RF carrier input frequency drifted upward 60 Hz producing an RF carrier frequency of 30.0006 MHz and an upper sideband that extends to 30.00056 MHz.

Solution **a.** The solution is identical to that provided in Example 6-2. The only difference is the method in which the RF local oscillator and BFO frequencies are produced. In the coherent receiver, the RF local oscillator and BFO frequencies are produced in the carrier recovery circuit and are, therefore, synchronized to the received RF carrier:

$$f_{\text{IF}} = (30\text{ MHz to }30.005\text{ MHz}) - 20\text{ MHz} = 10\text{ MHz to }10.005\text{ MHz}$$

The demodulated information signal spectrum is simply the difference between the intermediate frequency band and the BFO frequency:

$$f_m = (10\text{ MHz to }10.005\text{ MHz}) - 10\text{ MHz} = 0\text{ Hz to }5\text{ kHz}$$

b. Because the RF local oscillator and BFO frequencies are synchronized to the received RF carrier signal, the RF local oscillator will shift proportionally with the change in the RF input signal. Therefore, the RF local oscillator frequency will automatically adjust to 20.0004 MHz, producing an IF frequency spectrum of

$$f_{\text{IF}} = (30.0006\text{ MHz to }30.00056\text{ MHz}) - 20.0004\text{ MHz} = 10.0002\text{ MHz to }10.0052\text{ MHz}$$

The BFO output frequency will also automatically adjust proportionally to 10.0002 MHz, producing a demodulated information signal of

$$f_m = (10.0002\text{ MHz to }10.0052\text{ MHz}) - 10.0002\text{ MHz} = 0\text{ Hz to }5\text{ kHz}$$

From Example 6-3 it can be seen that the coherent single-sideband receiver automatically adjusts to frequency drifts in the transmitted carrier frequency. Therefore, the coherent receiver is immune to carrier drift as long as the magnitude of the drift is within the limits of the carrier recovery circuit.

6-8-3 Single-Sideband Envelope Detection Receiver

Figure 6-21 shows the block diagram for a single-sideband receiver that uses synchronous carriers and envelope detection to demodulate the received signals. The reduced carrier pilot is detected, separated from the demodulated spectrum, and regenerated in the carrier recovery circuit. The regenerated pilot is divided and used as the stable frequency source for a frequency synthesizer, which supplies the receiver with frequency coherent local oscillators. The receive RF is mixed down to IF in the first detector. A regenerated IF carrier is added to the IF spectrum in the last linear summer, which produces a SSB full-carrier envelope. The envelope is demodulated in a conventional peak diode detector to produce the original information signal spectrum. This type of receiver is often called an *exalted carrier receiver.*

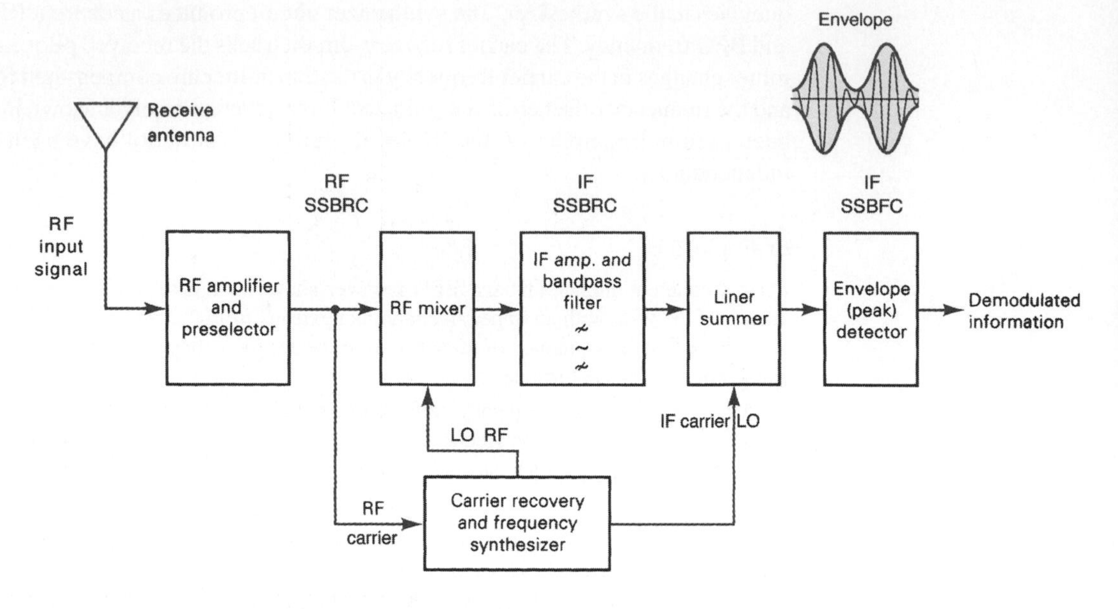

FIGURE 6-21 Single-sideband envelope detection receiver

6-8-4 Multichannel Pilot Carrier Single-Sideband Receiver

Figure 6-22 shows a block diagram for a multichannel pilot carrier SSB receiver that uses a PLL carrier recovery circuit and a frequency synthesizer to produce coherent local and beat frequency oscillator frequencies. The RF input range extends from 4 MHz to 30 MHz, and the VCO natural frequency is coarsely adjusted with an external channel selector switch over a frequency range of 6 MHz to 32 MHz. The VCO frequency tracks above the incoming RF by 2 MHz, which is the first IF. A 1.8-MHz beat frequency sets the second IF to 200 kHz.

The VCO frequency is coarsely set with the channel selector switch and then mixed with the incoming RF signal in the first detector to produce a first IF difference frequency of 2 MHz. The first IF mixes with the 1.8-MHz beat frequency to produce a 200-kHz second IF. The PLL locks onto the 200-kHz pilot and produces a dc correction voltage that fine-tunes the VCO. The second IF is beat down to audio in the third detector, which is passed on to the audio preamplifier for further processing. The AGC detector produces an AGC voltage that is proportional to the amplitude of the 200-kHz pilot. The AGC voltage is fed back to the RF and/or IF amplifiers to adjust their gains proportionate to the received pilot level and to the squelch circuit to turn the audio preamplifier off in the absence of a received pilot. The PLL compares the 200-kHz pilot to a stable crystal-controlled reference. Consequently, although the receiver carrier supply is not directly synchronized to the transmit oscillators, the first and second IFs are, thus compensating for any frequency offset in the demodulated audio spectrum.

6-9 AMPLITUDE-COMPANDORING SINGLE SIDEBAND

Amplitude-compandoring single-sideband (ACSSB) systems provide narrowband voice communications for land-mobile services with nearly the quality achieved with FM systems and do it using less than one-third the bandwidth. With ACSSB, the audio signals are compressed before modulation by amplifying the higher-magnitude signals less than the lower-magnitude signals. After demodulation in the receiver, the audio signals are expanded by amplifying the higher-magnitude signals more than the lower-magnitude

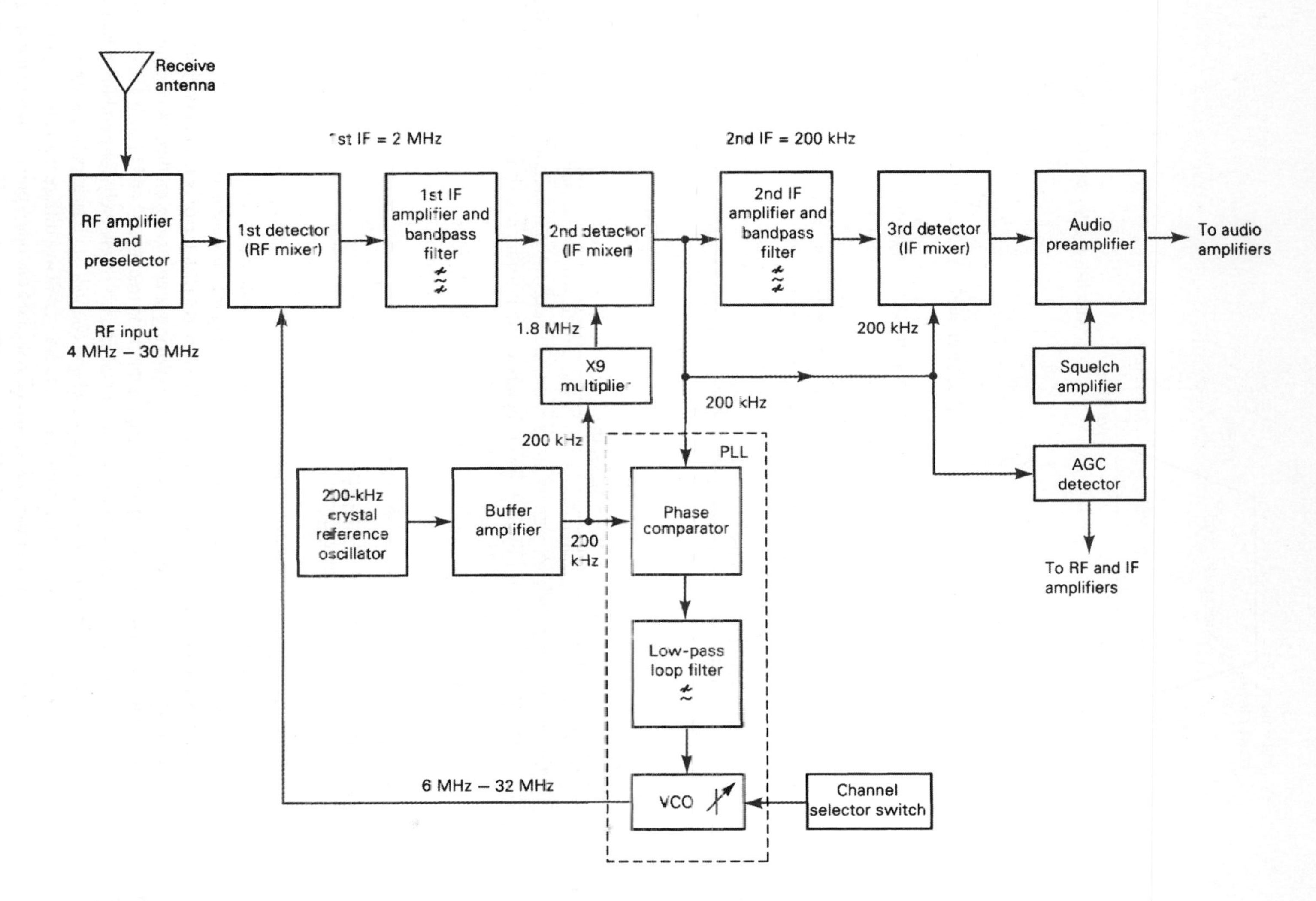

FIGURE 6-22 Multichannel pilot carrier SSB receiver

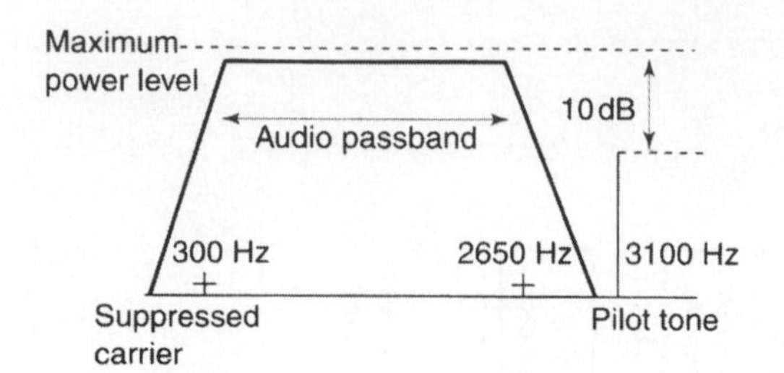

FIGURE 6-23 ACSSB signal

signals. A device that performs compression and expansion is called a *compandor* (*comp*ressor-exp*ander*).

Companding an information signal increases the dynamic range of a system by reducing the dynamic range of the information signals prior to transmission and then expanding them after demodulation. For example, when companding is used, information signals with an 80-dB dynamic range can be propagated through a communications system with only a 50-dB dynamic range. Companding slightly decreases the signal-to-noise ratios for the high-amplitude signals while considerably increasing the signal-to-noise ratios of the low-amplitude signals.

ACSSB systems require that a pilot carrier signal be transmitted at a reduced amplitude along with the information signals. The pilot is used to synchronize the oscillators in the receiver and provides a signal for the AGC that monitors and adjusts the gain of the receiver and silences the receiver when no pilot is received.

ACSSB significantly reduces the dynamic range allowing the lower-level signals to be transmitted with greater power while remaining within the power ratings of the transmitter when higher-level signals are present. Consequently, the signal-to-noise ratio is significantly improved for the low-level signals while introducing an insignificant increase in the noise levels for the higher-level signals.

Figure 6-23 shows the relative location and amplitudes of the audio passband and pilot tone for an ACSSB system. It can be seen that the pilot is transmitted 10 dB below the maximum power level for signals within the audio passband.

6-10 SINGLE-SIDEBAND SUPPRESSED CARRIER AND FREQUENCY-DIVISION MULTIPLEXING

Because of the bandwidth and power efficiencies inherent with single-sideband suppressed carrier, the most common application for it is in *frequency-division multiplexing* (FDM). *Multiplexing,* in general, is the process of combining transmissions from more than one source and transmitting them over a common facility, such as a metallic or optical fiber cable or a radio-frequency channel. Frequency-division multiplexing is an analog method of combining two or more analog sources that originally occupied the same frequency band in such a manner that the channels do not interfere with each other. FDM is used extensively for combining many relatively narrowband sources into a single wideband channel, such as in public telephone systems.

With FDM, each narrowband channel is converted to a different location in the total frequency spectrum. The channels are essentially stacked on top of one another in the frequency domain. Figure 6-24 shows a simple FDM system where four 5-kHz channels are frequency-division multiplexed into a single 20-kHz combined channel. As Figure 6-24a shows, channel 1 signals modulate a 100-kHz carrier in a balanced modulator, which inherently suppressed the 100-kHz carrier. The output of the balanced modulator is a double-sideband suppressed waveform with a bandwidth of 10 kHz. The DSBSC wave passes through a bandpass filter (BPF) where it is converted to a SSBSC signal. For this example, the lower sideband is blocked; thus,

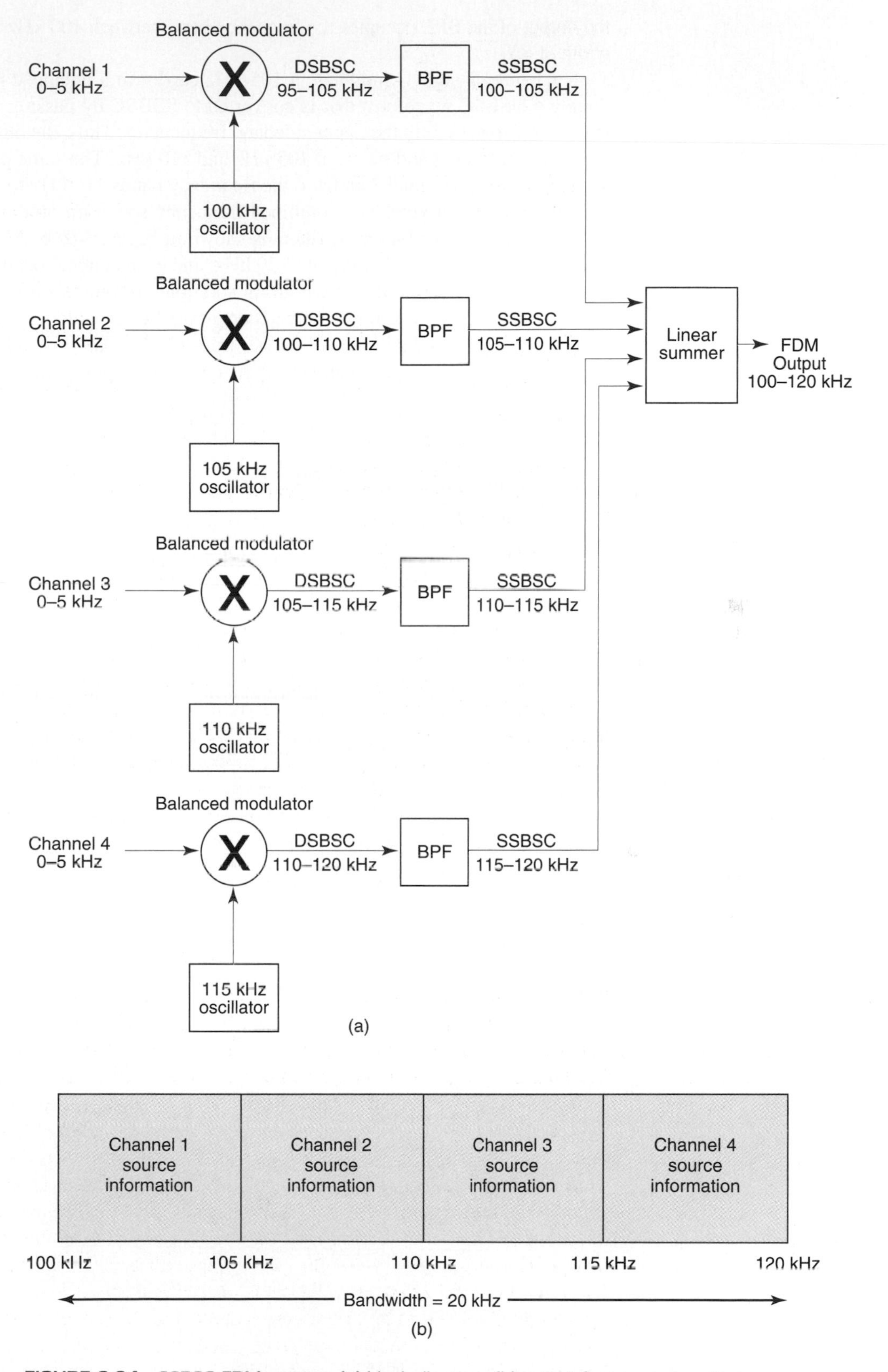

FIGURE 6-24 SSBSC FDM system: (a) block diagram; (b) output frequency spectrum

the output of the BPF occupies the frequency band between 100 kHz and 105 kHz (a bandwidth of 5 kHz).

Channel 2 signals modulate a 105-kHz carrier in a balanced modulator, again producing a DSBSC waveform that is converted to SSBSC by passing it through a bandpass filter tuned to pass only the upper sideband frequencies. Thus, the output from the BPF occupies a frequency band between 105 kHz and 110 kHz. The same process is used to convert signals from channel 3 and 4 to the frequency bands 110 kHz to 115 kHz and 115 kHz to 120 kHz, respectively. The combined frequency spectrum produced by combining the outputs from the four bandpass filters is shown in Figure 6-24b. As the figure shows, the total combined bandwidth is equal to 20 kHz, and each channel occupies a different 5-kHz portion of the total 20-kHz bandwidth. In addition, all four carriers have been suppressed, enabling all the available power to be concentrated in the sideband signals.

Single-sideband suppressed-carrier transmission can be used to combine hundreds or even thousands of narrowband channels (such as voice or low-speed data circuits) into a single, composite wideband channel without the channels interfering with each other. For a more detailed description of frequency-division multiplexing, refer to Chapter 11.

6-11 DOUBLE-SIDEBAND SUPPRESSED CARRIER AND QUADRATURE MULTIPLEXING

Quadrature multiplexing (QM) is a multiplexing method that uses double-sideband suppressed-carrier transmission to combine two information sources into a single composite waveform that is then transmitted over a common facility without the two channels interfering with each other.

Figure 6-25 shows how two information sources are combined into a single communications channel using quadrature multiplexing. As the figure shows, each channel's

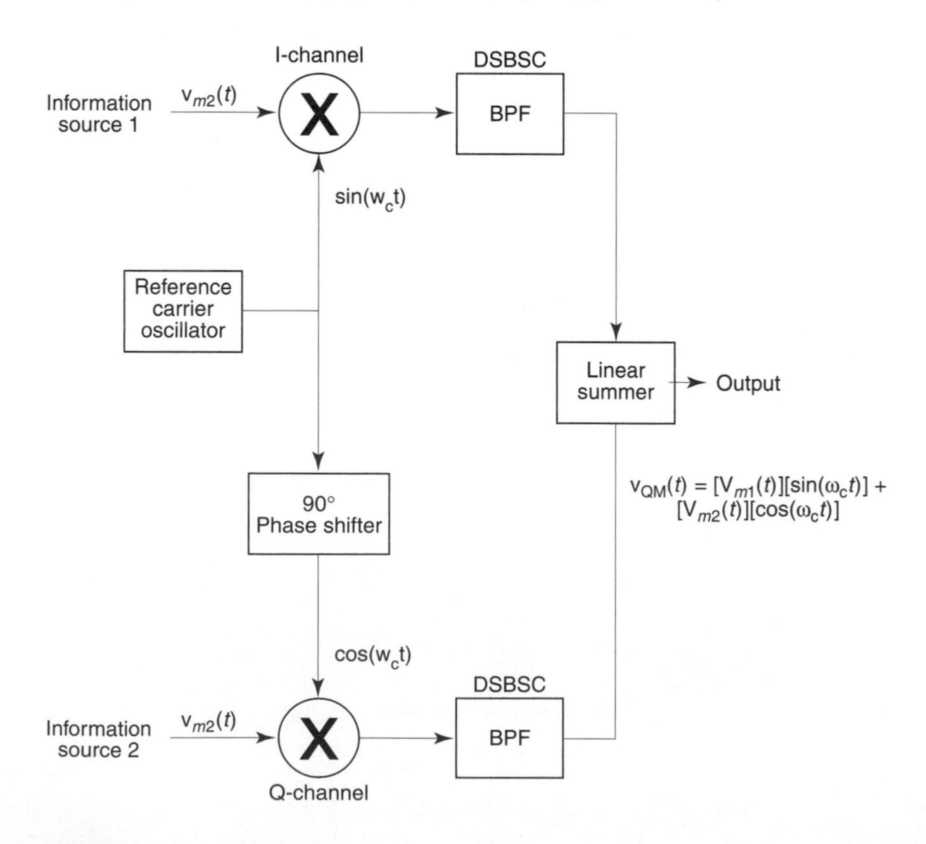

FIGURE 6-25 DSBSC QM system block diagram

information modulates the same carrier signal in balanced modulators. The primary difference between the two modulators is that the Q-channel carrier has been shifted in phase 90° from the I-channel carrier. The two carriers are said to be *in quadrature* with each other. Thus, the output from the channel I modulator is the product of the information signals from source 1 ($v_{m1}[t]$) and the in-phase carrier signal ($\sin[\omega_c t]$). The output from the channel 2 modulator is the product of the information signals from source 2 ($v_{m2}[t]$) and a carrier signal that has been shifted 90° in phase from the reference oscillator ($\cos[\omega_c t]$). The outputs from the two bandpass filters are combined in a linear summer producing a composite waveform consisting of the two orthogonal (90°) double-sideband signals symmetrical about the same suppressed carrier, $v_{QM}(t) = (v_{m1}[t])(\sin[\omega_c t]) + (v_{m2}[t])(\cos[\omega_c t])$. The two channels are thus separated in the phase domain. Quadrature multiplexing is typically used to multiplex information channels in data modems (Chapter 21) and to multiplex color signals in broadcast-band television (Chapter 26).

6-12 SINGLE-SIDEBAND MEASUREMENTS

Single-sideband transmitters are rated in peak envelope power (PEP) and peak envelope voltage (PEV) rather than simply rms power and voltage. For a single-frequency modulating signal, the modulated output signal with single-sideband suppressed-carrier transmission is not an envelope but rather a continuous, single-frequency signal. A single frequency is not representative of a typical information signal. Therefore, for test purposes, a *two-frequency* test signal is used for the modulating signal for which the two tones have equal amplitudes. Figure 6-26a shows the waveform produced in a SSBSC modulator with a two-tone modulating signal. The waveform is the vector sum of the two equal-amplitude side frequencies and is similar to a conventional AM waveform except that the repetition rate is equal to the difference between the two modulating-signal frequencies. Figure 6-26b shows the envelope for a two-tone test signal when a low-amplitude pilot carrier is added. The

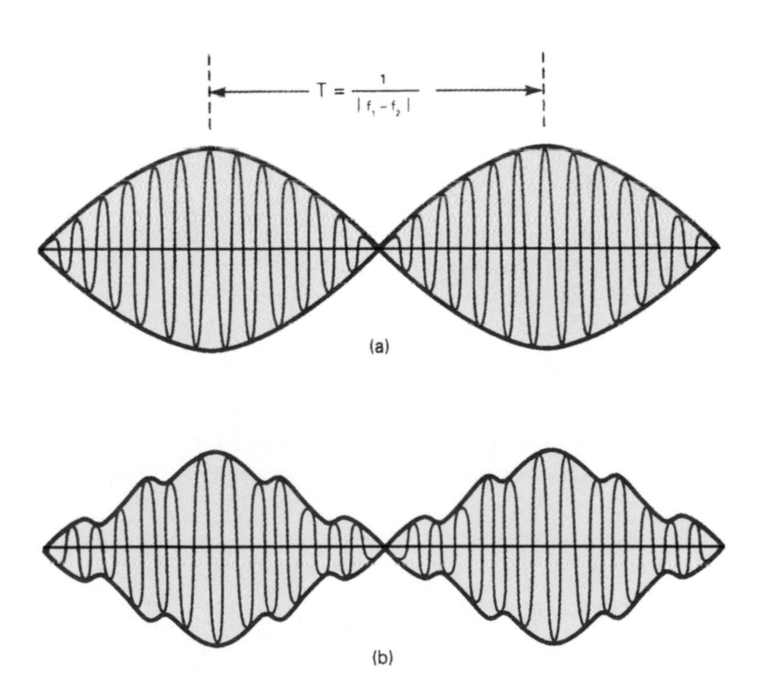

FIGURE 6-26 Two-tone SSB test signal: (a) without reinserted carrier; (b) with reinserted carrier

envelope has basically the same shape except with the addition of a low-amplitude sine-wave ripple at the carrier frequency.

The envelope out of two-tone SSB is an important consideration because it is from this envelope that the output power for a SSB transmitter is determined. The PEP for a SSBSC transmitter is analogous to the total output power from a conventional double-sideband full-carrier transmitter. The rated PEP is the output power measured at the peak of the envelope when the input is a two-tone test signal and the two tones are equal in magnitude. With such an output signal, the actual power dissipated in the load is equal to half the PEP. Therefore, the voltage developed across the load is

$$e_{total} = \sqrt{E_1^2 + E_2^2} \tag{6-4}$$

where E_1 and E_2 are the peak voltages of the two test tones. Therefore,

$$\text{PEP} = \frac{(\sqrt{E_1^2 + E_2^2})^2}{R} \tag{6-5}$$

and because $E_1 = E_2$,

$$\text{PEP} = \frac{2E^2}{R} \tag{6-6}$$

However, the average power dissipated in the load is equal to the sum of the powers of the two tones:

$$P_{ave} = \frac{E_1^2}{2R} + \frac{E_2^2}{2R} = \frac{2E^2}{2R} = \frac{E^2}{R} \tag{6-7}$$

which simplifies to

$$P_{ave} = \frac{\text{PEP}}{2} \tag{6-8}$$

Two equal-amplitude test tones are used for the test signal for the following reasons:

1. One tone produces a continuous single-frequency output that does not produce intermodulation.
2. A single-frequency output signal is not analogous to a normal information signal.
3. More than two tones make analysis impractical.
4. Two tones of equal amplitude place a more demanding requirement on the transmitter than is likely to occur during normal operation.

Example 6-4

For a two-tone test signal of 1.5 kHz and 3 kHz and a carrier frequency of 100 kHz, determine for a single-sideband suppressed-carrier transmission

a. Output frequency spectrum if only the upper sideband is transmitted.
b. For $E_1 = E_2 = 5$ V and a load resistance of 50 Ω, the PEP and average output power.

Solution **a.** The output frequency spectrum contains the two upper side frequencies:

$$f_{usf1} = 100 \text{ kHz} + 1.5 \text{ kHz} = 101.5 \text{ kHz}$$
$$f_{usf2} = 100 \text{ kHz} + 3 \text{ kHz} = 103 \text{ kHz}$$

b. Substituting into Equation 6-6 yields

$$\text{PEP} = \frac{2(0.707 \times 5)^2}{50} = 0.5 \text{ W}$$

Substituting into Equation 6-8 yields

$$P_{ave} = \frac{\text{PEP}}{2} = \frac{0.5}{2} = 0.25 \text{ W}$$

QUESTIONS

6-1. Describe AM SSBFC. Compare SSBFC to conventional AM.
6-2. Describe AM SSBSC. Compare SSBSC to conventional AM.
6-3. Describe AM SSBRC. Compare SSBRC to conventional AM.
6-4. What is a *pilot carrier?*
6-5. What is an *exalted* carrier?
6-6. Describe AM ISB. Compare ISB to conventional AM.
6-7. Describe AM VSB. Compare VSB to conventional AM.
6-8. Define *peak envelope power.*
6-9. Describe the operation of a balanced ring modulator.
6-10. What is a product modulator?
6-11. Describe the operation of a FET push–pull balanced modulator.
6-12. Describe the operation of a balanced bridge modulator.
6-13. What are the advantages of an LIC balanced modulator over a discrete circuit?
6-14. Describe the operation of a filter-type SSB transmitter.
6-15. Contrast crystal, ceramic, and mechanical filters.
6-16. Describe the operation of a phase-shift-type SSB transmitter.
6-17. Describe the operation of the "third type" of SSB transmitter.
6-18. Describe the operation of an independent sideband transmitter.
6-19. What is the difference between a product modulator and a product detector?
6-20. What is the difference between a coherent and a noncoherent receiver?
6-21. Describe the operation of a multichannel pilot carrier SSBRC receiver.
6-22. Why is a two-tone test signal used for making PEP measurements?

PROBLEMS

6-1. For the balanced ring modulator shown in Figure 6-5a, a carrier input frequency f_c = 400 kHz, and a modulating-signal frequency range f_m = 0 kHz to 4 kHz, determine

a. Output frequency spectrum.
b. Output frequency for a single-frequency input f_m = 2.8 kHz.

6-2. For the LIC balanced modulator shown in Figure 6-11, a carrier input frequency of 200 kHz, and a modulating-signal frequency range f_m = 0 kHz to 3 kHz, determine

a. Output frequency spectrum.
b. Output frequency for a single-frequency input f_m = 1.2 kHz.

6-3. For the SSB transmitter shown in Figure 6-12, a low-frequency carrier of 100 kHz, a medium-frequency carrier of 4 MHz, a high-frequency carrier of 30 MHz, and a modulating-signal frequency range of 0 kHz to 4 kHz,

a. Sketch the frequency spectrums for the following points: balanced modulator 1 out, BPF 1 out, summer out, balanced modulator 2 out, BPF 2 out, balanced modulator 3 out, and BPF 3 out.
b. For a single-frequency input f_m = 1.5 kHz, determine the translated frequency for the following points: BPF 1 out, BPF 2 out, and BPF 3 out.

6-4. Repeat Problem 6-3, except change the low-frequency carrier to 500 kHz. Which transmitter has the more stringent filtering requirements?

6-5. For the SSB transmitter shown in Figure 6-13a, a modulating input frequency range of 0 kHz to 3 kHz, and a high-frequency carrier of 28 MHz,

a. Sketch the output frequency spectrum.
b. For a single-frequency modulating-signal input of 2.2 kHz, determine the output frequency.

6-6. Repeat Problem 6-5, except change the audio input frequency range to 300 Hz to 5000 Hz.

6-7. For the SSB transmitter shown in Figure 6-17, an input carrier frequency of 500 kHz, and a modulating-signal frequency range of 0 kHz to 4 kHz,

a. Sketch the frequency spectrum at the output of the linear summer.

b. For a single modulating-signal frequency of 3 kHz, determine the output frequency.

6-8. Repeat Problem 6-7, except change the input carrier frequency to 400 kHz and the modulating-signal frequency range to 300 Hz to 5000 Hz.

6-9. For the ISB transmitter shown in Figure 6-18, channel A input frequency range of 0 kHz to 4 kHz, channel B input frequency range of 0 kHz to 4 kHz, a low-frequency carrier of 200 kHz, a medium-frequency carrier of 4 MHz, and a high-frequency carrier of 32 MHz,

a. Sketch the frequency spectrums for the following points: balanced modulator A out, BPF A out, balanced modulator B out, BPF B out, hybrid network out, linear summer out, balanced modulator 3 out, BPF 3 out, balanced modulator 4 out, and BPF 4 out.

b. For an A-channel input frequency of 2.5 kHz and a B-channel input frequency of 3 kHz, determine the frequency components at the following points: BPF A out, BPF B out, BPF 3 out, and BPF 4 out.

6-10. Repeat Problem 6-9, except change the channel A input frequency range to 0 kHz to 10 kHz and the channel B input frequency range to 0 kHz to 6 kHz.

6-11. For the SSB receiver shown in Figure 6-19, a RF input frequency of 35.602 MHz, a RF local oscillator frequency of 25 MHz, and a 2-kHz modulating signal frequency, determine the IF and BFO frequencies.

6-12. For the multichannel pilot carrier SSB receiver shown in Figure 6-22, a crystal oscillator frequency of 300 kHz, a first IF frequency of 3.3 MHz, an RF input frequency of 23.303 MHz, and modulating signal frequency of 3 kHz, determine the following: VCO output frequency, multiplication factor, and second IF frequency.

6-13. For a two-tone test signal of 2 kHz and 3 kHz and a carrier frequency of 200 kHz,

a. Determine the output frequency spectrum.

b. For $E_1 = E_2 = 12\ V_p$ and a load resistance $R_L = 50\ \Omega$, determine PEP and average power.

6-14. For the balanced ring modulator shown in Figure 6-5a, a carrier input frequency $f_c = 500$ kHz, and a modulating input signal frequency $f_m = 0$ kHz to 5 kHz, determine

a. Output frequency range.

b. Output frequency for a single input frequency $f_m = 3.4$ kHz.

6-15. For the LIC balanced modulator circuit shown in Figure 6-11, a carrier input frequency $f_c = 300$ kHz, and a modulating input signal frequency $f_m = 0$ kHz to 6 kHz, determine

a. Output frequency range.

b. Output frequency for a single-input frequency $f_m = 4.5$ kHz.

6-16. For the SSB transmitter shown in Figure 6-12, LF carrier frequency $f_{LF} = 120$ kHz, MF carrier frequency $f_{MF} = 3$ MHz, HF carrier frequency $f_{HF} = 28$ MHz, and an audio input frequency spectrum $f_m = 0$ kHz to 5 kHz,

a. Sketch the frequency spectrums for the following points: BPF 1 out, BPF 2 out, and BPF 3 out.

b. For a single-input frequency $f_m = 2.5$ kHz, determine the translated frequency at the following points: BPF 1 out, BPF 2 out, and BPF 3 out.

6-17. Repeat Problem 6-16 except change the LF carrier frequency to 500 kHz. Which transmitter has the more stringent filtering requirements?

6-18. For the SSB transmitter shown in Figure 6-13a, an audio input frequency $f_m = 0$ kHz to 4 kHz, and an HF carrier frequency $f_{HF} = 27$ MHz,

a. Sketch the output frequency spectrum.

b. For a single-frequency input signal $f_m = 1.8$ kHz, determine the output frequency.

6-19. Repeat Problem 6-18, except change the audio input frequency spectrum to $f_m = 300$ Hz to 4000 Hz.

6-20. For the SSB transmitter shown in Figure 6-17, a carrier frequency $f_c = 400$ kHz, and an input frequency spectrum $f_m = 0$ kHz to 5 kHz,

a. Sketch the frequency spectrum at the output of the linear summer.

b. For a single audio input frequency $f_m = 2.5$ kHz, determine the output frequency.

6-21. Repeat Problem 6-20, except change the carrier input frequency to 600 kHz and the input frequency spectrum to 300 Hz to 6000 Hz.

6-22. For the ISB transmitter shown in Figure 6-18, channel A input frequency f_a = 0 kHz to 5 kHz, channel B input frequency f_b = 0 kHz to 5 kHz, LF carrier frequency f_{LF} = 180 kHz, MF carrier frequency f_{MF} = 3 MHz, and HF carrier frequency f_{HF} = 30 MHz,

a. Sketch the frequency spectrums for the following points: BPF A out, BPF B out, BPF 3 out, and BPF 4 out.

b. For an A-channel input frequency f_a = 2.5 kHz and a B-channel input frequency f_b = 2 kHz, determine the frequency components at the following points: BPF A out, BPF B out, BPF 3 out, and BPF 4 out.

6-23. Repeat Problem 6-22, except change the channel A input frequency spectrum to 0 kHz to 8 kHz and the channel B input frequency spectrum to 0 kHz to 6 kHz.

6-24. For the SSB receiver shown in Figure 6-22, RF input frequency f_{RF} = 36.803 MHz, RF local oscillator frequency f_{lo} = 26 MHz, and a 3-kHz modulating-signal frequency, determine the following: BFO output frequency and detected information frequency.

6-25. For the multichannel pilot carrier SSB receiver shown in Figure 6-22, crystal oscillator frequency f_{co} = 400 kHz, first IF frequency f_{IF} = 4.4 MHz, RF input frequency f_{RF} = 23.403 MHz, and modulating-signal frequency f_m = 3 kHz, determine the following: VCO output frequency, multiplication factor, and second IF frequency.

6-26. For a two-tone test signal of 3 kHz and 4 kHz and a carrier frequency of 400 kHz, determine

a. Output frequency spectrum.

b. For E_1 and E_2 = 20 V_p and a load resistor R_L = 100 Ω, determine the PEP and average power.

C H A P T E R 7

Angle Modulation Transmission

CHAPTER OUTLINE

OBJECTIVES

- Define and mathematically describe *angle modulation*
- Explain the difference between frequency and phase modulation
- Describe direct and indirect frequency modulation
- Describe direct and indirect phase modulation
- Define *deviation sensitivity*
- Describe FM and PM waveforms
- Define *phase deviation* and *modulation index*
- Explain frequency deviation and percent modulation
- Analyze the frequency content of an angle-modulated waveform
- Determine the bandwidth requirements for frequency and phase modulation

- Describe deviation ratio
- Describe the commercial FM broadcast band
- Develop the phasor representation of an angle-modulated wave
- Describe the power distribution of an angle-modulated wave
- Explain how noise affects angle-modulated waves
- Describe preemphasis and deemphasis and explain how they affect the signal-to-noise ratio of an angle-modulated wave
- Describe the operation of frequency and phase modulators
- Explain of operation of direct FM modulators
- Explain the operation of direct PM modulators
- Describe the operation of linear integrated-circuit FM modulators
- Describe two methods of up-converting the frequency of angle-modulated waves
- Describe the operation of direct FM transmitters
- Explain the function and operation of an AFC loop
- Describe the operation of a phase-locked loop direct FM transmitter
- Describe the operation of indirect FM transmitters
- Compare the advantages and disadvantages of angle-modulation versus amplitude modulation

7-1 INTRODUCTION

As previously stated, there are three properties of an analog signal that can be varied (modulated) by the information signal. Those properties are amplitude, frequency, and phase. Chapters 4 to 6 described amplitude modulation. This chapter and Chapter 8 describe *frequency modulation* (FM) and *phase modulation* (PM), which are both forms of *angle modulation.* Unfortunately, both frequency and phase modulation are often referred to as simply FM, although there are actual distinctions between the two. Angle modulation has several advantages over amplitude modulation, such as noise reduction, improved system fidelity, and more efficient use of power. However, angle modulation also has several disadvantages when compared to AM, including requiring a wider bandwidth and utilizing more complex circuits in both the transmitters and the receivers.

Angle modulation was first introduced in 1931 as an alternative to amplitude modulation. It was suggested that an angle-modulated wave was less susceptible to noise than AM and, consequently, could improve the performance of radio communications. Major E. H. Armstrong (who also developed the superheterodyne receiver) developed the first successful FM radio system in 1936, and in July 1939, the first regularly scheduled broadcasting of FM signals began in Alpine, New Jersey. Today, angle modulation is used extensively for commercial radio broadcasting, television sound transmission, two-way mobile radio, cellular radio, and microwave and satellite communications systems.

The purposes of this chapter are to introduce the reader to the basic concepts of frequency and phase modulation and how they relate to each other, to show some of the common circuits used to produce angle-modulated waves, and to compare the performance of angle modulation to amplitude modulation.

7-2 ANGLE MODULATION

Angle modulation results whenever the phase angle (θ) of a sinusoidal wave is varied with respect to time. An angle-modulated wave is expressed mathematically as

$$m(t) = V_c \cos[\omega_c t + \theta(t)] \tag{7-1}$$

where $m(t)$ = angle-modulated wave
V_c = peak carrier amplitude (volts)
ω_c = carrier radian frequency (i.e., angular velocity, $2\pi f_c$ radians per second)
$\theta(t)$ = instantaneous phase deviation (radians)

With angle modulation, it is necessary that $\theta(t)$ be a prescribed function of the modulating signal. Therefore, if $v_m(t)$ is the modulating signal, the angle modulation is expressed mathematically as

$$\theta(t) = F[v_m(t)] \tag{7-2}$$

where $v_m(t) = V_m \sin(\omega_m t)$
ω_m = angular velocity of the modulating signal ($2\pi f_m$ radian per second)
f_m = modulating signal frequency (hertz)
V_m = peak amplitude of the modulating signal (volts)

In essence, the difference between frequency and phase modulation lies in which property of the carrier (the frequency or the phase) is directly varied by the modulating signal and which property is indirectly varied. Whenever the frequency of a carrier is varied, the phase is also varied and vice versa. Therefore, FM and PM must both occur whenever either form of angle modulation is performed. If the frequency of the carrier is varied directly in accordance with the modulating signal, FM results. If the phase of the carrier is varied directly in accordance with the modulating signal, PM results. Therefore, direct FM is indirect PM, and direct PM is indirect FM. Frequency and phase modulation can be defined as follows:

Direct frequency modulation (*FM*): Varying the frequency of a constant-amplitude carrier directly proportional to the amplitude of the modulating signal at a rate equal to the frequency of the modulating signal

Direct phase modulation (*PM*): Varying the phase of a constant-amplitude carrier directly proportional to the amplitude of the modulating signal at a rate equal to the frequency of the modulating signal

Figure 7-1 shows an angle-modulated signal ($m[t]$) in the frequency domain. The figure shows how the carrier frequency (f_c) is changed when acted on by a modulating signal ($v_m[t]$). The magnitude and direction of the frequency shift (Δf) is proportional to the amplitude and polarity of the modulating signal (V_m), and the rate at which the frequency changes are occurring is equal to the frequency of the modulating signal (f_m). For this example, a positive modulating signal produces an increase in frequency, and a negative modulating signal produces a decrease in frequency, although the opposite relationship could occur depending on the type of modulator circuit used.

Figure 7-2a shows in the time domain the waveform for a sinusoidal carrier for which angle modulation is occurring. As the figure shows, the phase (θ) of the carrier is changing proportional to the amplitude of the modulating signal ($v_m[t]$). The relative angular displacement (shift) of the carrier phase in radians in respect to the reference phase is called

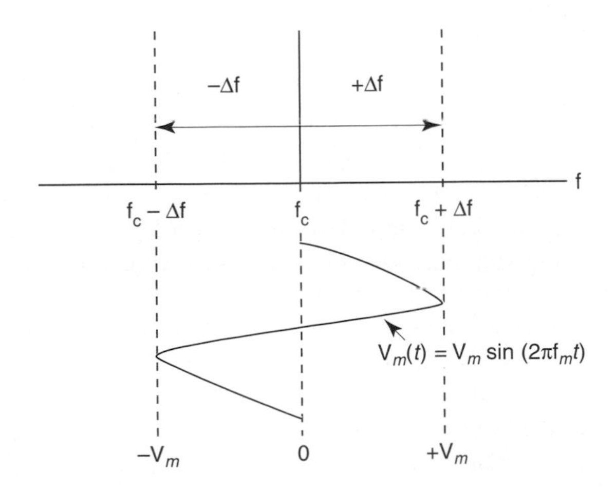

FIGURE 7-1 Angle-modulated wave in the frequency domain.

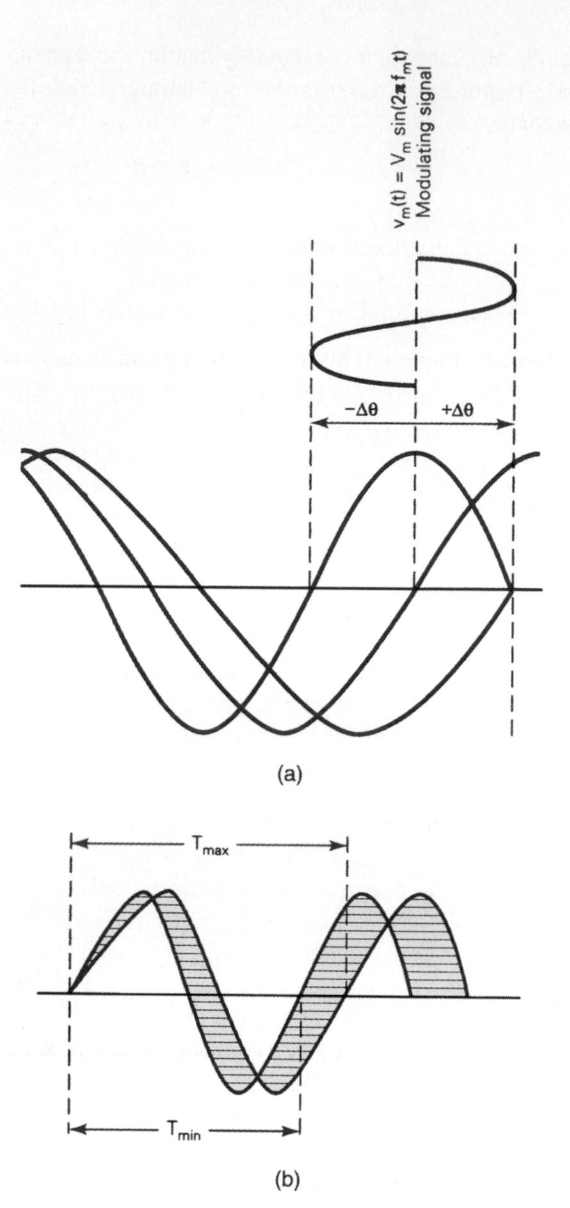

FIGURE 7-2 Angle modulation in the time domain: (a) phase changing with time; (b) frequency changing with time

phase deviation ($\Delta\theta$). The change in the carrier's phase produces a corresponding change in frequency. The relative displacement of the carrier frequency in hertz in respect to its unmodulated value is called *frequency deviation* (Δf). The magnitude of the frequency and phase deviation is proportional to the amplitude of the modulating signal (V_m), and the rate at which the changes are occurring is equal to the modulating signal frequency (f_m).

Figure 7-2b shows a sinusoidal carrier in which the frequency (f) is changed (*deviated*) over a period of time. The fat portion of the waveform corresponds to the peak-to-peak change in the period of the carrier (ΔT). The minimum period (T_{min}) corresponds to the maximum frequency (f_{max}), and the maximum period (T_{max}) corresponds to the minimum frequency (f_{min}). The peak-to-peak frequency deviation is determined by simply measuring the difference between the maximum and minimum frequencies ($\Delta f_{p\text{-}p} = 1/T_{min} - 1/T_{max}$).

Whenever the period (T) of a sinusoidal carrier changes, its frequency and phase also change, and if the changes are continuous, the wave is no longer a single frequency. It will be shown that the resultant angle-modulated waveform comprises the original unmodulated carrier frequency (often called the *carrier rests frequency*) and an infinite number of pairs of side frequencies are displaced on either side of the carrier by an integral multiple of the modulating-signal frequency.

7-3 MATHEMATICAL ANALYSIS

The difference between FM and PM is more easily understood by defining the following four terms with reference to Equation 7-1: instantaneous phase deviation, instantaneous phase, instantaneous frequency deviation, and instantaneous frequency.

1. *Instantaneous phase deviation.* The *instantaneous phase deviation* is the instantaneous change in the phase of the carrier at a given instant of time and indicates how much the phase of the carrier is changing with respect to its reference phase. Instantaneous phase deviation is expressed mathematically as

$$\text{instantaneous phase deviation} = \theta(t) \text{ rad} \tag{7-3}$$

2. *Instantaneous phase.* The *instantaneous phase* is the precise phase of the carrier at a given instant of time and is expressed mathematically as

$$\text{instantaneous phase} = \omega_c t + \theta(t) \text{ rad} \tag{7-4}$$

where $\omega_c t$ = carrier reference phase (radians)
$= [2\pi \text{ (rad/cycle)}][f_c\text{(cycles/s)}][t(s)] = 2\pi f_c t \text{ (rad)}$
f_c = carrier frequency (hertz)
$\theta(t)$ = instantaneous phase deviation (radians)

3. *Instantaneous frequency deviation.* The *instantaneous frequency deviation* is the instantaneous change in the frequency of the carrier and is defined as the first time derivative of the instantaneous phase deviation. Therefore, the instantaneous phase deviation is the first integral of the instantaneous frequency deviation. In terms of Equation 7-3, the instantaneous frequency deviation is expressed mathematically as

$$\text{instantaneous frequency deviation} = \theta'(t) \text{ rad/s} \tag{7-5}$$

or

$$= \frac{\theta'(t) \text{ rad/s}}{2\pi \text{ rad/cycle}} = \frac{\text{cycles}}{\text{s}} = \text{Hz}$$

The prime (′) is used to denote the first derivative with respect to time.

4. *Instantaneous frequency.* The *instantaneous frequency* is the precise frequency of the carrier at a given instant of time and is defined as the first time derivative of the instantaneous phase. In terms of Equation 7-4, the instantaneous frequency is expressed mathematically as

$$\text{instantaneous frequency} = \omega_i(t) = \frac{d}{dt}[\omega_c t + \theta(t)] \tag{7-6}$$

$$= \omega_c + \theta'(t) \text{ rad/s} \tag{7-7}$$

Substituting $2\pi f_c$ for ω_c gives

$$\text{instantaneous frequency} = f_i(t)$$

and

$$\omega_i(t) = \left(2\pi\frac{\text{rad}}{\text{cycle}}\right)\left(f_c\frac{\text{cycles}}{\text{s}}\right) + \theta'(t) = 2\pi f_c + \theta'(t) \text{ rad/s}$$

$$\text{or} \qquad f_i(t) = \frac{2\pi f_c + \theta(t) \text{ rad/s}}{2\pi \text{ rad/cycle}} = f_c + \frac{\theta'(t) \text{ cycles}}{2\pi \text{ s}} = f_c + \frac{\theta'(t)}{2\pi} \text{ Hz} \qquad \textbf{(7-8)}$$

7-4 DEVIATION SENSITIVITY

Phase modulation can then be defined as angle modulation in which the instantaneous phase deviation, $\theta(t)$, is proportional to the amplitude of the modulating signal voltage and the instantaneous frequency deviation is proportional to the slope or first derivative of the modulating signal. Similarly, frequency modulation is angle modulation in which the instantaneous frequency deviation, $\theta'(t)$, is proportional to the amplitude of the modulating signal and the instantaneous phase deviation is proportional to the integral of the modulating-signal voltage.

For a modulating signal $v_m(t)$, the phase and frequency modulation are

$$\text{phase modulation} = \theta(t) = Kv_m(t) \text{ rad} \qquad \textbf{(7-9)}$$

$$\text{frequency modulation} = \theta'(t) = K_1 v_m(t) \text{ rad/s} \qquad \textbf{(7-10)}$$

where K and K_1 are constants and are the *deviation sensitivities* of the phase and frequency modulators, respectively. The deviation sensitivities are the output-versus-input transfer functions for the modulators, which give the relationship between what output parameter changes in respect to specified changes in the input signal. For a frequency modulator, changes would occur in the output frequency in respect to changes in the amplitude of the input voltage. For a phase modulator, changes would occur in the phase of the output frequency in respect to changes in the amplitude of the input voltage.

The deviation sensitivity for a phase modulator is

$$K = \frac{\text{rad}}{\text{V}}\left(\frac{\Delta\theta}{\Delta \text{V}}\right)$$

and for a frequency modulator

$$K_1 = \frac{\text{rad/s}}{\text{V}}\left(\frac{\Delta\omega}{\Delta \text{V}}\right)$$

Phase modulation is the first integral of the frequency modulation. Therefore, from Equations 7-9 and 7-10,

$$\begin{aligned} \text{phase modulation} &= \theta(t) = \int \theta'(t)\, dt \\ &= \int K_1\, v_m(t)\, dt \\ &= K_1 \int v_m(t)\, dt \end{aligned} \qquad \textbf{(7-11)}$$

Therefore, substituting a modulating signal $v_m(t) = V_m \cos(\omega_m t)$ into Equation 7-1 yields,

for phase modulation,
$$\begin{aligned} m(t) &= V_c \cos[\omega_c t + \theta(t)] \\ &= V_c \cos[\omega_c t + KV_m \cos(\omega_m t)] \end{aligned} \qquad \textbf{(7-12)}$$

for frequency modulation,
$$\begin{aligned} m(t) &= V_c \cos[\omega_c t + \int \theta'(t)] \\ &= V_c \cos[\omega_c t + \int K_1 v_m(t)\, dt] \\ &= V_c \cos[\omega_c t + K_1 \int V_m \cos(\omega_m t)\, dt] \\ &= V_c \cos\left[\omega_c t + \frac{K_1 V_m}{\omega_m} \sin(\omega_m t)\right] \end{aligned} \qquad \textbf{(7-13)}$$

The preceding mathematical relationships are summarized in Table 7-1. Also, the expressions for the FM and PM waves that result when the modulating signal is a single-frequency cosinusoidal wave are shown.

Table 7-1 Equations for Phase- and Frequency-Modulated Carriers

Type of Modulation	Modulating Signal	Angle-Modulated Wave, $m(t)$
(a) Phase	$v_m(t)$	$V_c \cos[\omega_c t + K v_m(t)]$
(b) Frequency	$v_m(t)$	$V_c \cos[\omega_c t + K_1 \int v_m(t)\, dt]$
(c) Phase	$V_m \cos(\omega_m t)$	$V_c \cos[\omega_c t + K V_m \cos(\omega_m t)]$
(d) Frequency	$V_m \cos(\omega_m t)$	$V_c \cos\left[\omega_c t + \dfrac{K_1 V_m}{\omega_m} \sin(\omega_m t)\right]$

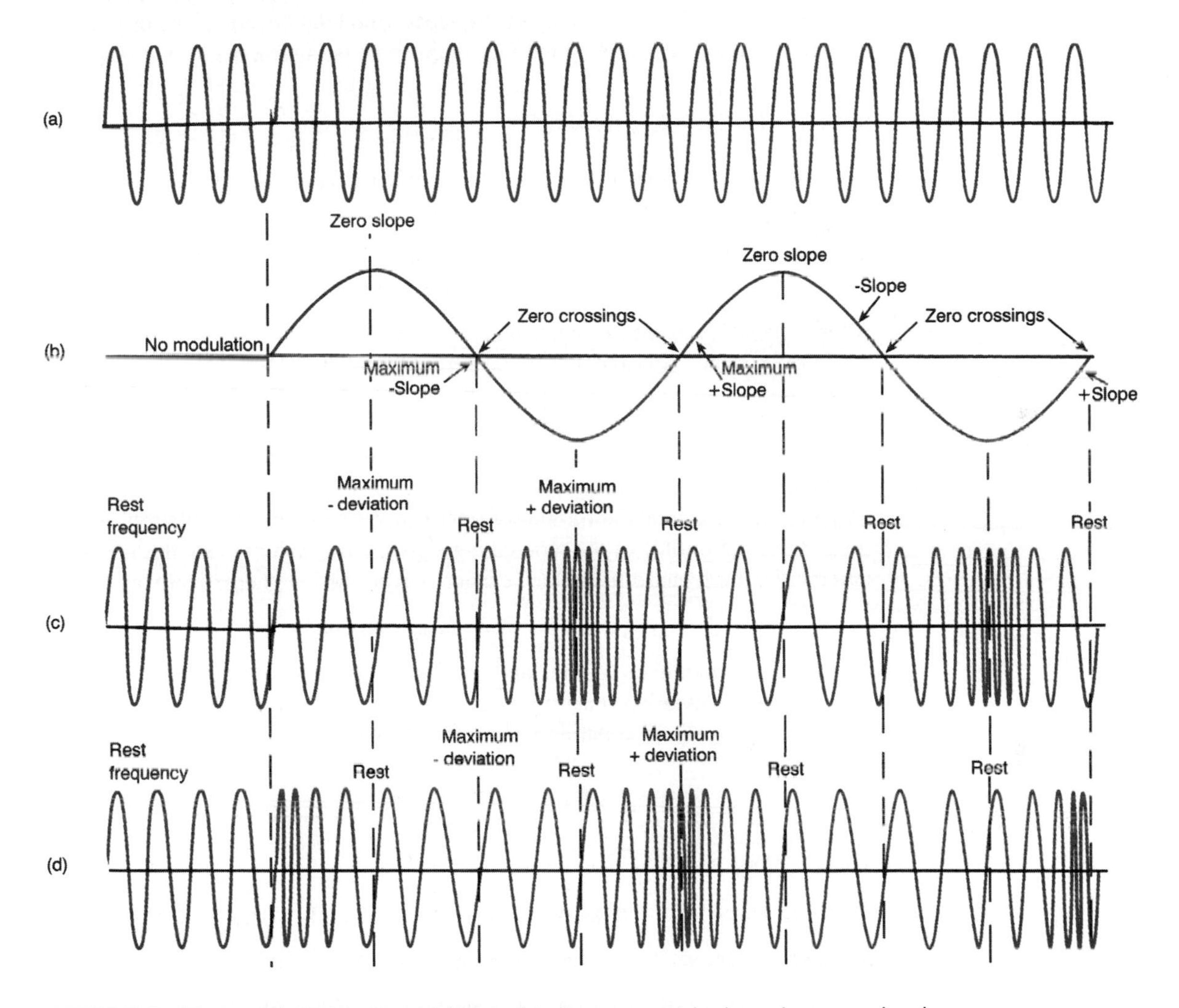

FIGURE 7-3 Phase and frequency modulation of a sine-wave carrier by a sine-wave signal: (a) unmodulated carrier; (b) modulating signal; (c) frequency-modulated wave; (d) phase-modulated wave

7-5 FM AND PM WAVEFORMS

Figure 7-3 illustrates both frequency and phase modulation of a sinusoidal carrier by a single-frequency modulating signal. It can be seen that the FM and PM waveforms are identical except for their time relationship (phase). Thus, it is impossible to distinguish an FM waveform from a PM waveform without knowing the dynamic characteristics of the modulating signal. With FM, the maximum frequency deviation (change in the carrier frequency) occurs during the maximum positive and negative peaks of the modulating signal (i.e., the frequency deviation is proportional to the amplitude of the modulating signal). With PM, the maximum frequency deviation occurs during the zero crossings of the modulating signal (i.e., the frequency deviation is proportional to the slope or first derivative of the modulating signal). For both frequency and phase modulation, the rate at which the frequency changes occur is equal to the modulating-signal frequency.

Similarly, it is not apparent from Equation 7-1 whether an FM or PM wave is represented. It could be either. However, knowledge of the modulating signal will permit correct identification. If $\theta(t) = Kv_m(t)$, it is phase modulation, and if $\theta'(t) = K_1 v_m(t)$, it is frequency modulation. In other words, if the instantaneous frequency is directly proportional to the amplitude of the modulating signal, it is frequency modulation, and if the instantaneous phase is directly proportional to the amplitude of the modulating frequency, it is phase modulation.

7-6 PHASE DEVIATION AND MODULATION INDEX

Comparing expressions (c) and (d) for the angle-modulated carrier in Table 7-1 shows that the expression for a carrier that is being phase or frequency modulated by a single-frequency modulating signal can be written in a general form by modifying Equation 7-1 as follows:

$$m(t) = V_c \cos[\omega_c t + m \cos(\omega_m t)] \qquad \textbf{(7-14)}$$

where $m \cos(\omega_m t)$ is the instantaneous phase deviation, $\theta(t)$. When the modulating signal is a single-frequency sinusoid, it is evident from Equation 7-14 that the phase angle of the carrier varies from its unmodulated value in a simple sinusoidal fashion.

In Equation 7-14, m represents the *peak phase deviation* in radians for a phase-modulated carrier. Peak phase deviation is called the *modulation index* (or sometimes *index of modulation*). One primary difference between frequency and phase modulation is the way in which the modulation index is defined. For PM, the modulation index is proportional to the amplitude of the modulating signal, independent of its frequency. The modulation index for a phase-modulated carrier is expressed mathematically as

$$m = KV_m \text{ (radians)} \qquad \textbf{(7-15)}$$

where m = modulation index and peak phase deviation ($\Delta\theta$, radians)
K = deviation sensitivity (radians per volt)
V_m = peak modulating-signal amplitude (volts)

thus,

$$m = K\left(\frac{\text{radians}}{\text{volt}}\right)V_m(\text{volts}) = \text{radians}$$

Therefore, for PM, Equation 7-1 can be rewritten as

$$m(t) = V_c \cos[\omega_c t + KV_m \cos(\omega_m t)] \qquad \textbf{(7-16)}$$

or

$$m(t) = V_c \cos[\omega_c t + \Delta\theta \cos(_m t)] \qquad \textbf{(7-17)}$$

or

$$m(t) = V_c \cos[\omega_c t + m \cos(\omega_m t)] \qquad \textbf{(7-18)}$$

For a frequency-modulated carrier, the modulation index is directly proportional to the amplitude of the modulating signal and inversely proportional to the frequency of the modulating signal. Therefore, for FM, modulation index is expressed mathematically as

$$m = \frac{K_1 V_m}{\omega_m} \text{(unitless)} \qquad \textbf{(7-19)}$$

where m = modulation index (unitless)
K_1 = deviation sensitivity (radians per second per volt or radians per volt)
V_m = peak modulating-signal amplitude (volts)
ω_m = radian frequency (radians per second)

thus,

$$m = \frac{K_1\left(\frac{\text{radians}}{\text{volt}-\text{s}}\right)V_m(\text{volt})}{\omega_m(\text{radians/s})} = (\text{unitless})$$

From Equation 7-19, it can be seen that for FM the modulation index is a unitless ratio and is used only to describe the depth of modulation achieved for a modulating signal with a given peak amplitude and radian frequency.

Deviation sensitivity can be expressed in hertz per volt allowing Equation 7-19 to be written in a more practical form as

$$m = \frac{K_1 V_m}{f_m}(\text{unitless}) \tag{7-20}$$

where m = modulation index (unitless)
K_1 = deviation sensitivity (cycles per second per volt or hertz per volt)
V_m = peak modulating-signal amplitude (volts)
f_m = cyclic frequency (hertz per second)

thus,

$$m = \frac{K_1\left(\frac{\text{hertz}}{\text{volt}}\right)V_m(\text{volt})}{f_m(\text{hertz})} = (\text{unitless})$$

7-7 FREQUENCY DEVIATION AND PERCENT MODULATION

7-7-1 Frequency Deviation

Frequency deviation is the change in frequency that occurs in the carrier when it is acted on by a modulating-signal frequency. Frequency deviation is typically given as a peak frequency shift in hertz (Δf). The peak-to-peak frequency deviation ($2\Delta f$) is sometimes called *carrier swing*.

For an FM, the deviation sensitivity is often given in hertz per volt. Therefore, the peak frequency deviation is simply the product of the deviation sensitivity and the peak modulating-signal voltage and is expressed mathematically as

$$\Delta f = K_1 V_m \text{ (Hz)} \tag{7-21}$$

Equation 7-21 can be substituted into Equation 7-20, and the expression for the modulation index in FM can be rewritten as

$$m = \frac{\Delta f(\text{Hz})}{f_m(\text{Hz})}(\text{unitless}) \tag{7-22}$$

Therefore, for FM, Equation 7-1 can be rewritten as

$$m(t) = V_c \cos\left[\omega_c t + \frac{K_1 V_m}{f_m}\sin(\omega_m t)\right] \tag{7-23}$$

or

$$m(t) = V_c \cos\left[\omega_c t + \frac{\Delta f}{f_m}\sin(\omega_m t)\right] \tag{7-24}$$

or

$$m(t) = V_c \cos[\omega_c t + m\sin(\omega_m t)] \tag{7-25}$$

From examination of Equations 7-19 and 7-20, it can be seen that the modulation indices for FM and PM relate to the modulating signal differently. With PM, both the modulation index and the peak phase deviation are directly proportional to the amplitude of the

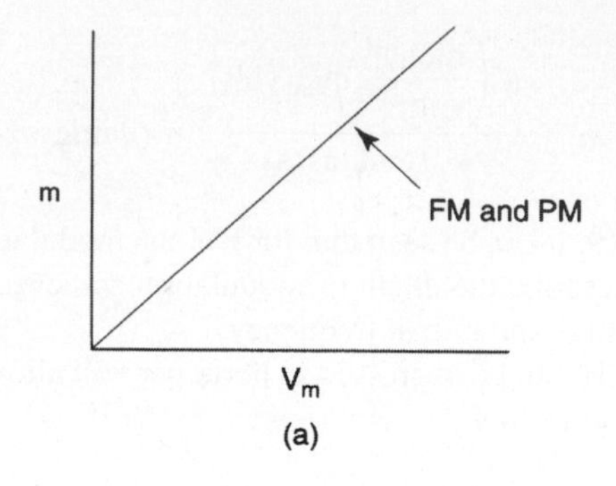

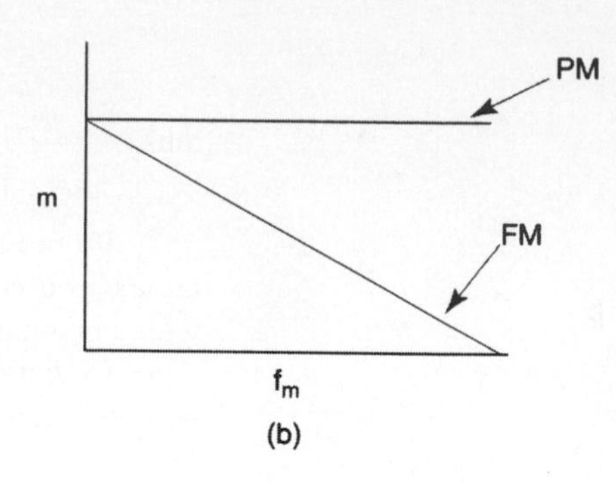

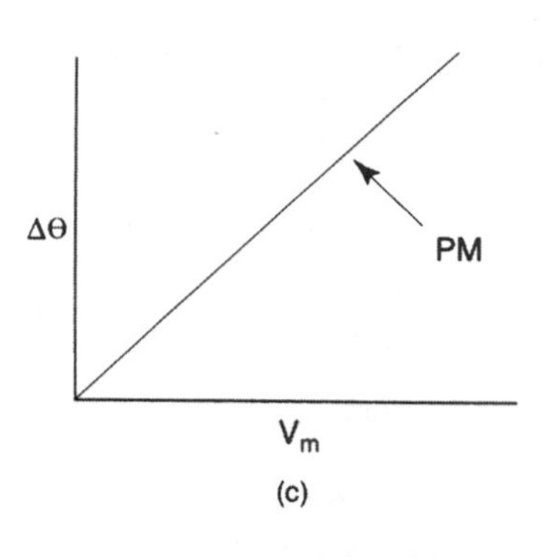

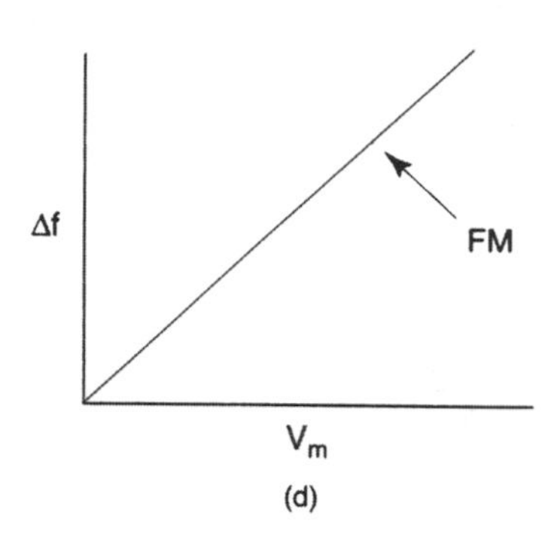

FIGURE 7-4 Relationship between modulation index, frequency deviation, and phase deviation in respect to modulations signal amplitude and frequency: (a) modulation index versus amplitude; (b) frequency deviation versus modulating frequency; (c) phase deviation versus amplitude; (d) frequency deviation versus amplitude

modulating signal and unaffected by its frequency. With FM, however, both the modulation index and the frequency deviation are directly proportional to the amplitude of the modulating signal, and the modulation index is inversely proportional to its frequency. Figure 7-4 graphically shows the relationship among modulation index and peak phase deviation for PM and the modulation index and peak frequency deviation for FM in respect to the modulating signal amplitude and frequency.

The preceding mathematical relationships are summarized in Table 7-2.

Example 7-1

a. Determine the peak frequency deviation (Δf) and modulation index (m) for an FM modulator with a deviation sensitivity $K_1 = 5$ kHz/V and a modulating signal $v_m(t) = 2\cos(2\pi 2000t)$.

b. Determine the peak phase deviation (m) for a PM modulator with a deviation sensitivity $K = 2.5$ rad/V and a modulating signal $v_m(t) = 2\cos(2\pi 2000t)$.

Solution

a. The peak frequency deviation is simply the product of the deviation sensitivity and the peak amplitude of the modulating signal, or

$$\Delta f = \frac{5\text{ kHz}}{\text{V}} \times 2\text{ V} = 10\text{ kHz}$$

The modulation index is determined by substituting into Equation 7-22:

$$m = \frac{10\text{ kHz}}{2\text{ kHz}} = 5$$

Table 7-2 Angle Modulation Summary

	FM	PM
Modulated wave	$m(t) = V_c \cos\left[\omega_c t + \frac{K_1 V_m}{f_m}\sin(\omega_m t)\right]$	$m(t) = V_c \cos[\omega_c t + KV_m \cos(\omega_m t)]$
or	$m(t) = V_c \cos[\omega_c t + m\sin(\omega_m t)]$	$m(t) = V_c \cos[\omega_c t + m\cos(\omega_m t)]$
or	$m(t) = V_c \cos\left[\omega_c t + \frac{\Delta f}{f_m}\sin(\omega_m t)\right]$	$m(t) = V_c \cos[\omega_c t + \Delta\theta \cos(\omega_m t)]$
Deviation sensitivity	K_1 (Hz/V)	K (rad/V)
Deviation	$\Delta f = K_1 V_m$ (Hz)	$\Delta\theta = KV_m$ (rad)
Modulation index	$m = \frac{K_1 V_m}{f_m}$ (unitless)	$m = KV_m$ (rad)
or	$m = \frac{\Delta f}{f_m}$ (unitless)	$m = \Delta\theta$ (rad)
Modulating signal	$v_m(t) = V_m \sin(\omega_m t)$	$v_m(t) = V_m \cos(\omega_m t)$
Modulating frequency	$\omega_m = 2\pi f_m$ rad/s	$\omega_m = 2\pi f_m$ rad/s
or	$\omega_m/2\pi = f_m$ (Hz)	$\omega_m/2\pi = f_m$ (Hz)
Carrier signal	$V_c \cos(\omega_c t)$	$V_c \cos(\omega_c t)$
Carrier frequency	$\omega_c = 2\pi f_c$ (rad/s)	$\omega_c = 2\pi f_c$ (rad/s)
or	$\omega_c/2\pi = f_c$ (Hz)	$\omega_c/2\pi = f_c$ (Hz)

b. The peak phase shift for a phase-modulated wave is the modulation index and is found by substituting into Equation 7-15:

$$m = \frac{2.5 \text{ rad}}{\text{V}} \times 2 \text{ V} = 5 \text{ rad}$$

In Example 7-1, the modulation index for the frequency-modulated carrier was equal to the modulation index of the phase-modulated carrier (5). If the amplitude of the modulating signal is changed, the modulation index for both the frequency- and the phase-modulated waves will change proportionally. However, if the frequency of the modulating signal changes, the modulation index for the frequency-modulated wave will change inversely proportional, while the modulation index of the phase-modulated wave is unaffected. Therefore, under identical conditions, FM and PM are indistinguishable for a single-frequency modulating signal; however, when the frequency of the modulating signal changes, the PM modulation index remains constant, whereas the FM modulation index increases as the modulating-signal frequency decreases and vice versa.

7-7-2 Percent Modulation

The percent modulation for an angle-modulated wave is determined in a different manner than it was with an amplitude-modulated wave. With angle modulation, percent modulation is simply the ratio of the frequency deviation actually produced to the maximum frequency deviation allowed by law stated in percent form. Mathematically, percent modulation is

$$\%\text{ modulation} = \frac{\Delta f_{(\text{actual})}}{\Delta f_{(\text{max})}} \times 100 \qquad \textbf{(7-26)}$$

For example, in the United States, the Federal Communications Commission (FCC) limits the frequency deviation for commercial FM broadcast-band transmitters to ±75 kHz.

If a given modulating signal produces ±50-kHz frequency deviation, the percent modulation is

$$\% \text{ modulation} = \frac{50 \text{ kHz}}{75 \text{ kHz}} \times 100 = 67\%$$

7-8 PHASE AND FREQUENCY MODULATORS AND DEMODULATORS

A *phase modulator* is a circuit in which the carrier is varied in such a way that its instantaneous phase is proportional to the modulating signal. The unmodulated carrier is a single-frequency sinusoid and is commonly called the *rest* frequency. A *frequency modulator* (often called a *frequency deviator*) is a circuit in which the carrier is varied in such a way that its instantaneous phase is proportional to the integral of the modulating signal. Therefore, with a frequency modulator, if the modulating signal $v(t)$ is differentiated prior to being applied to the modulator, the instantaneous phase deviation is proportional to the integral of $v(t)$ or, in other words, proportional to $v(t)$ because $\int v'(t) = v(t)$. Similarly, an FM modulator that is preceded by a differentiator produces an output wave in which the phase deviation is proportional to the modulating signal and is, therefore, equivalent to a phase modulator. Several other interesting equivalences are possible. For example, a frequency demodulator followed by an integrator is equivalent to a phase demodulator. Four commonly used equivalences are as follows:

1. PM modulator = differentiator followed by an FM modulator
2. PM demodulator = FM demodulator followed by an integrator
3. FM modulator = integrator followed by a PM modulator
4. FM demodulator = PM demodulator followed by a differentiator

7-9 FREQUENCY ANALYSIS OF ANGLE-MODULATED WAVES

With angle modulation, the frequency components of the modulated wave are much more complexly related to the frequency components of the modulating signal than with amplitude modulation. In a frequency or phase modulator, a single-frequency modulating signal produces an infinite number of pairs of side frequencies and, thus, has an infinite bandwidth. Each side frequency is displaced from the carrier by an integral multiple of the modulating signal frequency. However, generally most of the side frequencies are negligibly small in amplitude and can be ignored.

7-9-1 Modulation by a Single-Frequency Sinusoid

Frequency analysis of an angle-modulated wave by a single-frequency sinusoid produces a peak phase deviation of m radians, where m is the modulation index. Again, from Equation 7-14 and for a modulating frequency equal to ω_m, $m(t)$ is written as

$$m(t) = V_c \cos[\omega_c t + m \cos(\omega_m t)]$$

From Equation 7-14, the individual frequency components that make up the modulated wave are not obvious. However, *Bessel function identities* are available that may be applied directly. One such identity is

$$\cos(\alpha + m \cos \beta) = \sum_{n=-\infty}^{\infty} J_n(m) \cos\left(\alpha + n\beta + \frac{n\pi}{2}\right) \quad \textbf{(7-27)}$$

$J_n(m)$ is the Bessel function of the first kind of nth order with argument m. If Equation 7-28 is applied to Equation 7-15, $m(t)$ may be rewritten as

$$m(t) = V_c \sum_{n=-\infty}^{\infty} J_n(m) \cos\left(\omega_c t + n\omega_m t + \frac{n\pi}{2}\right) \quad \textbf{(7-28)}$$

Expanding Equation 7-28 for the first four terms yields

$$m(t) = V_c\Big\{J_0(m) \cos \omega_c t + J_1(m) \cos\left[(\omega_c + \omega_m)t + \frac{\pi}{2}\right]$$

$$- J_1(m) \cos\left[(\omega_c - \omega_m)t - \frac{\pi}{2}\right] + J_2(m) \cos[(\omega_c + 2\omega_m t)] \quad \textbf{(7-29)}$$

$$+ J_2(m) \cos[(\omega - 2\omega_m t)] + \cdots J_n(m) \cdots\Big\}$$

where $m(t)$ = angle-modulated wave
m = modulation index
V_c = peak amplitude of the unmodulated carrier
$J_0(m)$ = carrier component
$J_1(m)$ = first set of side frequencies displaced from the carrier by ω_m
$J_2(m)$ = second set of side frequencies displaced from the carrier by $2\omega_m$
$J_n(m)$ = nth set of side frequencies displaced from the carrier by $n\omega_m$

Equations 7-28 and 7-29 show that with angle modulation, a single-frequency modulating signal produces an infinite number of sets of side frequencies, each displaced from the carrier by an integral multiple of the modulating signal frequency. A sideband set includes an upper and a lower side frequency ($f_c \pm f_m, f_c \pm 2f_m, f_c \pm nf_m$, and so on). Successive sets of sidebands are called first-order sidebands, second-order sidebands, and so on, and their magnitudes are determined by the coefficients $J_1(m)$, $J_2(m)$, and so on, respectively.

To solve for the amplitude of the side frequencies, J_n, Equation 7-29 can be converted to

$$J_n(m) = \left(\frac{m}{2}\right)^n\left[\frac{1}{n} - \frac{(m/2)^2}{1!(n+1)!} + \frac{(m/2)^4}{2!(n+2)!} - \frac{(m/2)^6}{3!(n+1)!} + \cdots\right] \quad \textbf{(7-30)}$$

where ! = factorial ($1 \times 2 \times 3 \times 4$, etc.)
n = J or number of the side frequency
m = modulation index

Table 7-3 shows the Bessel functions of the first kind for several values of modulation index. We see that a modulation index of 0 (no modulation) produces zero side frequencies, and the larger the modulation index, the more sets of side frequencies produced. The values shown for J_n are relative to the amplitude of the unmodulated carrier. For example, $J_2 = 0.35$ indicates that the amplitude of the second set of side frequencies is equal to 35% of the unmodulated carrier amplitude ($0.35\ V_c$). It can be seen that the amplitude of the higher-order side frequencies rapidly becomes insignificant as the modulation index decreases below unity. For larger values of m, the value of $J_n(m)$ starts to decrease rapidly as soon as $n = m$. As the modulation index increases from zero, the magnitude of the carrier $J_0(m)$ decreases. When m is equal to approximately 2.4, $J_0(m) = 0$ and the carrier component go to zero (this is called the *first carrier null*). This property is often used to determine the modulation index or set the deviation sensitivity of an FM modulator. The carrier reappears as m increases beyond 2.4. When m reaches approximately 5.4, the carrier component

Table 7-3 Bessel Functions of the First Kind, $J_n(m)$

Modulation Index	Carrier	Side Frequency Pairs													
m	J_0	J_1	J_2	J_3	J_4	J_5	J_6	J_7	J_8	J_9	J_{10}	J_{11}	J_{12}	J_{13}	J_{14}
0.00	1.00	—	—	—	—	—	—	—	—	—	—	—	—	—	—
0.25	0.98	0.12	—	—	—	—	—	—	—	—	—	—	—	—	—
0.5	0.94	0.24	0.03	—	—	—	—	—	—	—	—	—	—	—	—
1.0	0.77	0.44	0.11	0.02	—	—	—	—	—	—	—	—	—	—	—
1.5	0.51	0.56	0.23	0.06	0.01	—	—	—	—	—	—	—	—	—	—
2.0	0.22	0.58	0.35	0.13	0.03	—	—	—	—	—	—	—	—	—	—
2.4	0	0.52	0.43	0.20	0.06	0.02	—	—	—	—	—	—	—	—	—
2.5	−0.05	0.50	0.45	0.22	0.07	0.02	0.01	—	—	—	—	—	—	—	—
3.0	−0.26	0.34	0.49	0.31	0.13	0.04	0.01	—	—	—	—	—	—	—	—
4.0	−0.40	−0.07	0.36	0.43	0.28	0.13	0.05	0.02	—	—	—	—	—	—	—
5.0	−0.18	−0.33	0.05	0.36	0.39	0.26	0.13	0.05	0.02	—	—	—	—	—	—
5.45	0	−0.34	−0.12	0.26	0.40	0.32	0.19	0.09	0.03	0.01	—	—	—	—	—
6.0	0.15	−0.28	−0.24	0.11	0.36	0.36	0.25	0.13	0.06	0.02	—	—	—	—	—
7.0	0.30	0.00	−0.30	−0.17	0.16	0.35	0.34	0.23	0.13	0.06	0.02	—	—	—	—
8.0	0.17	0.23	−0.11	−0.29	−0.10	0.19	0.34	0.32	0.22	0.13	0.06	0.03	—	—	—
8.65	0	0.27	0.06	−0.24	−0.23	0.03	0.26	0.34	0.28	0.18	0.10	0.05	0.02	—	—
9.0	−0.09	0.25	0.14	−0.18	−0.27	−0.06	0.20	0.33	0.31	0.21	0.12	0.06	0.03	0.01	—
10.0	−0.25	0.05	0.25	0.06	−0.22	−0.23	−0.01	0.22	0.32	0.29	0.21	0.12	0.06	0.03	0.01

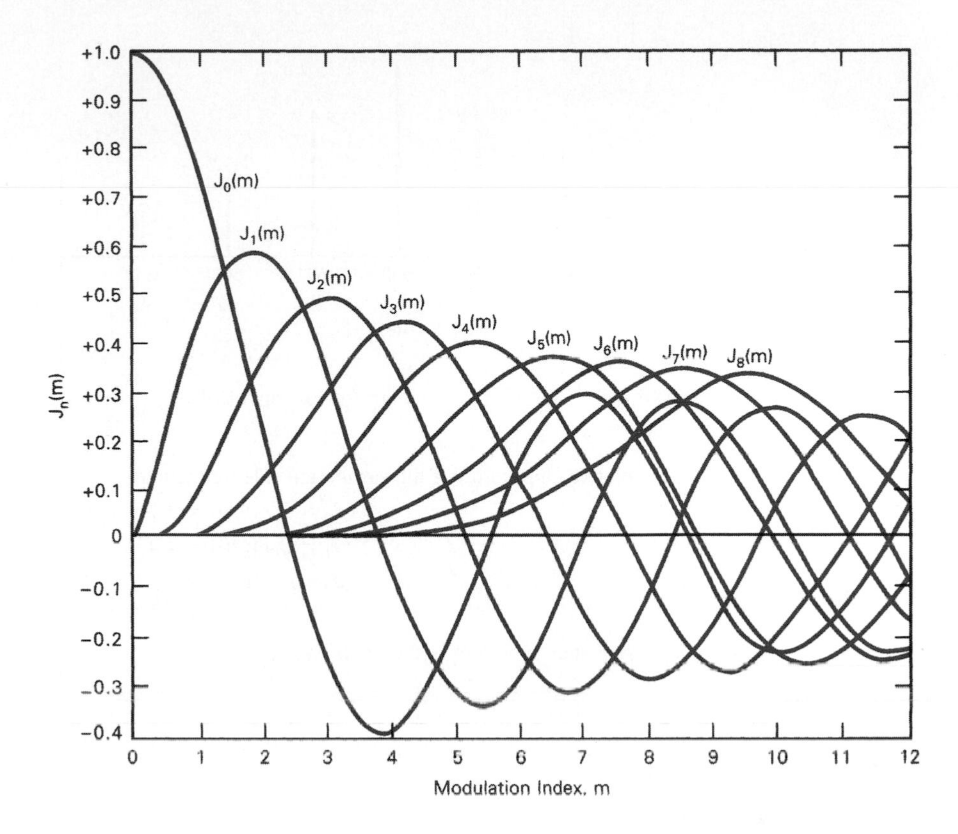

FIGURE 7-5 $J_n(m)$ versus m

once again disappears (this is called the *second carrier null*). Further increases in the modulation index will produce additional carrier nulls at periodic intervals.

Figure 7-5 shows the curves for the relative amplitudes of the carrier and several sets of side frequencies for values of m up to 10. It can be seen that the amplitudes of both the carrier and the side frequencies vary at a periodic rate that resembles a damped sine wave. The negative values for $J(m)$ simply indicate the relative phase of that side frequency set.

In Table 7-3, only the significant side frequencies are listed. A side frequency is not considered significant unless it has an amplitude equal to or greater than 1% of the unmodulated carrier amplitude ($J_n \geq 0.01$). From Table 7-3, it can be seen that as m increases, the number of significant side frequencies increase. Consequently, the bandwidth of an angle-modulated wave is a function of the modulation index.

Example 7-2

For an FM modulator with a modulation index $m = 1$, a modulating signal $v_m(t) = V_m \sin(2\pi 1000t)$, and an unmodulated carrier $v_c(t) = 10 \sin(2\pi 500kt)$, determine

a. Number of sets of significant side frequencies.

b. Their amplitudes.

Then

c. Draw the frequency spectrum showing their relative amplitudes.

Solution **a.** From Table 7-3, a modulation index of 1 yields a reduced carrier component and three sets of significant side frequencies.

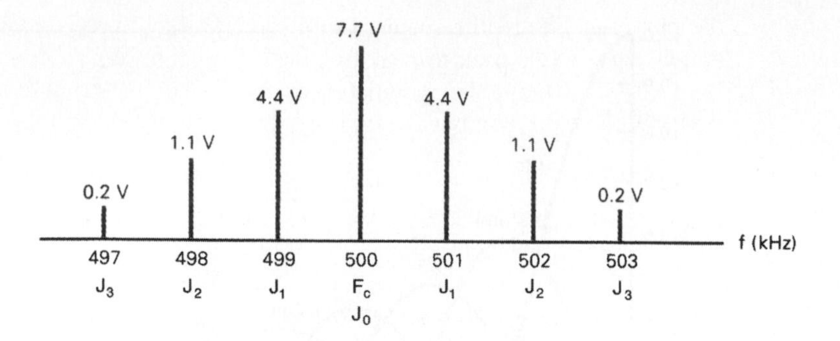

FIGURE 7-6 Frequency spectrum for Example 6-2

b. The relative amplitudes of the carrier and side frequencies are

$$J_0 = 0.77(10) = 7.7\text{ V}$$
$$J_1 = 0.44(10) = 4.4\text{ V}$$
$$J_2 = 0.11(10) = 1.1\text{ V}$$
$$J_3 = 0.02(10) = 0.2\text{ V}$$

c. The frequency spectrum is shown in Figure 7-6.

If the FM modulator used in Example 7-2 were replaced with a PM modulator and the same carrier and modulating signal frequencies were used, a peak phase deviation of 1 rad would produce exactly the same frequency spectrum.

7-10 BANDWIDTH REQUIREMENTS OF ANGLE-MODULATED WAVES

In 1922, J. R. Carson mathematically proved that for a given modulating-signal frequency, a frequency-modulated wave cannot be accommodated in a narrower bandwidth than an amplitude-modulated wave. From the preceding discussion and Example 7-2, it can be seen that the bandwidth of an angle-modulated wave is a function of the modulating signal frequency and the modulation index. With angle modulation, multiple sets of sidebands are produced, and, consequently, the bandwidth can be significantly wider than that of an amplitude-modulated wave with the same modulating signal. The modulator output waveform in Example 7-2 requires 6 kHz of bandwidth to pass the carrier and all the significant side frequencies. A conventional double-sideband AM modulator would require only 2 kHz of bandwidth and a single-sideband system only 1 kHz.

Angle-modulated waveforms are generally classified as either *low, medium,* or *high index.* For the low-index case, the modulation index is less than 1, and the high-index case occurs when the modulation index is greater than 10. Modulation indices greater than 1 and less than 10 are classified as medium index. From Table 7-3, it can be seen that with low-index angle modulation, most of the signal information is carried by the first set of sidebands, and the minimum bandwidth required is approximately equal to twice the highest modulating-signal frequency. For this reason, low-index FM systems are sometimes called *narrowband FM.* For a high-index signal, a method of determining the bandwidth called the *quasi-stationary* approach may be used. With this approach, it is assumed that the modulating signal is changing very slowly. For example, for an FM modulator with a deviation sensitivity $K_1 = 2$ kHz/V and a 1-V_p modulating signal, the peak frequency deviation $\Delta f = 2000$ Hz. If the frequency of the modulating signal is very low, the bandwidth is determined by the peak-to-peak frequency deviation. Therefore, for large modulation indexes, the minimum bandwidth required to propagate a frequency-modulated wave is approximately equal to the peak-to-peak frequency deviation ($2\Delta f$).

Thus, for low-index modulation, the frequency spectrum resembles double-sideband AM, and the minimum bandwidth is approximated by

$$B = 2f_m \text{ Hz} \tag{7-31}$$

and for high-index modulation, the minimum bandwidth is approximated by

$$B = 2\Delta f \text{ Hz} \tag{7-32}$$

The actual bandwidth required to pass all the significant sidebands for an angle-modulated wave is equal to two times the product of the highest modulating-signal frequency and the number of significant sidebands determined from the table of Bessel functions. Mathematically, the rule for determining the minimum bandwidth for an angle-modulated wave using the Bessel table is

$$B = 2(n \times f_m) \text{ Hz} \tag{7-33}$$

where n = number of significant sidebands
f_m = modulating-signal frequency (hertz)

In an unpublished memorandum dated August 28, 1939, Carson established a general rule to estimate the bandwidth for all angle-modulated systems regardless of the modulation index. This is called *Carson's rule*. Simply stated, Carson's rule approximates the bandwidth necessary to transmit an angle-modulated wave as twice the sum of the peak frequency deviation and the highest modulating-signal frequency. Mathematically stated, Carson's rule is

$$B = 2(\Delta f + f_m) \text{ Hz} \tag{7-34}$$

where Δf = peak frequency deviation (hertz)
f_m = modulating-signal frequency (hertz)

For low modulation indices, f_m is much larger than Δf, and Equation 7-34 reduces to Equation 7-31. For high modulation indices, Δf is much larger than f_m, and Equation 7-34 reduces to Equation 7-32.

Carson's rule is an approximation and gives transmission bandwidths that are slightly narrower than the bandwidths determined using the Bessel table and Equation 7-33. Carson's rule defines a bandwidth that includes approximately 98% of the total power in the modulated wave. The actual bandwidth necessary is a function of the modulating signal waveform and the quality of transmission desired.

Example 7-3

For an FM modulator with a peak frequency deviation Δf = 10 kHz, a modulating-signal frequency f_m = 10 kHz, V_c = 10 V, and a 500-kHz carrier, determine

a. Actual minimum bandwidth from the Bessel function table.
b. Approximate minimum bandwidth using Carson's rule.
Then
c. Plot the output frequency spectrum for the Bessel approximation.

Solution **a.** Substituting into Equation 7-22 yields

$$m = \frac{10 \text{ kHz}}{10 \text{ kHz}} = 1$$

From Table 7-3, a modulation index of 1 yields three sets of significant sidebands. Substituting into Equation 7-33, the bandwidth is

$$B = 2(3 \times 10 \text{ kHz}) = 60 \text{ kHz}$$

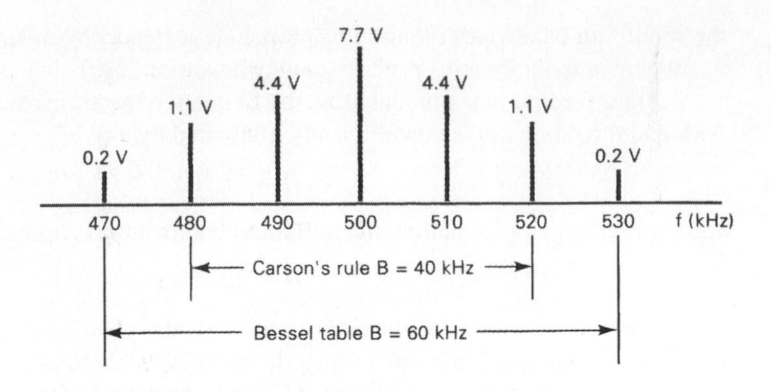

FIGURE 7-7 Frequency spectrum for Example 7-3

b. Substituting into Equation 7-34, the minimum bandwidth is

$$B = 2(10 \text{ kHz} + 10 \text{ kHz}) = 40 \text{ kHz}$$

c. The output frequency spectrum for the Bessel approximation is shown in Figure 7-7.

From Example 7-3, it can be seen that there is a significant difference in the minimum bandwidth determined from Carson's rule and the minimum bandwidth determined from the Bessel table. The bandwidth from Carson's rule is less than the actual minimum bandwidth required to pass all the significant sideband sets as defined by the Bessel table. Therefore, a system that was designed using Carson's rule would have a narrower bandwidth and, thus, poorer performance than a system designed using the Bessel table. For modulation indexes above 5, Carson's rule is a close approximation to the actual bandwidth required.

7-11 DEVIATION RATIO

For a given FM system, the minimum bandwidth is greatest when the maximum frequency deviation is obtained with the maximum modulating-signal frequency (i.e., the highest modulating frequency occurs with the maximum amplitude allowed). By definition, *deviation ratio* (DR) is the *worst-case* modulation index and is equal to the maximum peak frequency deviation divided by the maximum modulating-signal frequency. The worst-case modulation index produces the widest output frequency spectrum. Mathematically, the deviation ratio is

$$\text{DR} = \frac{\Delta f_{(\text{max})}}{f_{m(\text{max})}} \qquad \textbf{(7-35)}$$

where DR = deviation ratio (unitless)
$\Delta f_{(\text{max})}$ = maximum peak frequency deviation (hertz)
$f_{m(\text{max})}$ = maximum modulating-signal frequency (hertz)

For example, for the sound portion of a commercial TV broadcast-band station, the maximum frequency deviation set by the FCC is 50 kHz, and the maximum modulating-signal frequency is 15 kHz. Therefore, the deviation ratio for a television broadcast station is

$$\text{DR} = \frac{50 \text{ kHz}}{15 \text{ kHz}} = 3.33$$

This does not mean that whenever a modulation index of 3.33 occurs, the widest bandwidth also occurs at the same time. It means that whenever a modulation index of 3.33 occurs for a maximum modulating-signal frequency, the widest bandwidth occurs.

Example 7-4

a. Determine the deviation ratio and bandwidth for the worst-case (widest-bandwidth) modulation index for an FM broadcast-band transmitter with a maximum frequency deviation of 75 kHz and a maximum modulating-signal frequency of 15 kHz.
b. Determine the deviation ratio and maximum bandwidth for an equal modulation index with only half the peak frequency deviation and modulating-signal frequency.

Solution **a.** The deviation ratio is found by substituting into Equation 7-36:

$$\text{DR} = \frac{75\text{ kHz}}{15\text{ kHz}} = 5$$

From Table 7-3, a modulation index of 5 produces eight significant sidebands. Substituting into Equation 7-33 yields

$$B = 2(8 \times 15{,}000) = 240\text{ kHz}$$

b. For a 37.5-kHz frequency deviation and a modulating-signal frequency $f_m = 7.5$ kHz, the modulation index is

$$m = \frac{37.5\text{ kHz}}{7.5\text{ kHz}} = 5$$

and the bandwidth is $$B = 2(8 \times 7500) = 120\text{ kHz}$$

From Example 7-4, it can be seen that although the same modulation index (5) was achieved with two different modulating-signal frequencies and amplitudes, two different bandwidths were produced. An infinite number of combinations of modulating-signal frequency and frequency deviation will produce a modulation index of 5. However, the case produced from the maximum modulating-signal frequency and maximum frequency deviation will always yield the widest bandwidth.

At first, it may seem that a higher modulation index with a lower modulating-signal frequency would generate a wider bandwidth because more sideband sets are produced; but remember that the sidebands would be closer together. For example, a 1-kHz modulating signal that produces 10 kHz of frequency deviation has a modulation index of $m = 10$ and produces 14 significant sets of sidebands. However, the sidebands are displaced from each other by only 1 kHz and, therefore, the total bandwidth is only 28,000 Hz (2[14 × 1000]).

With Carson's rule, the same conditions produce the widest (worst-case) bandwidth. For the maximum frequency deviation and maximum modulating-signal frequency, the maximum bandwidth using Carson's rule for Example 7-4a is

$$\begin{aligned} B &= 2[\Delta f_{(\max)} + f_{m(\max)}] \\ &= 2(75\text{ kHz} + 15\text{ kHz}) \\ &= 180\text{ kHz} \end{aligned}$$

Example 7-5

Given FM and PM modulators with the following parameters:

FM modulator
deviations sensitivity $K_1 = 1.5$ kHz/v
carrier frequency $f_c = 500$ kHz
modulating signal $v_m = 2\sin(2\pi\, 2kt)$

PM modulator
deviations sensitivity $K = 0.75$ rad/v
carrier frequency $f_c = 500$ kHz
modulating signal $v_m = 2\sin(2\pi\, 2kt)$

a. Determine the modulation indexes and sketch the output spectrums for both modulators.
b. Change the modulating signal amplitude for both modulators to 4 V_p and repeat step (a).
c. Change the modulating signal frequency for both modulators to 1 kHz and repeat step (a).

Solution

a. *FM modulator* | *PM modulator*

$$m = \frac{v_m k_1}{f_m} \qquad\qquad m = v_m k$$

$$m = \frac{2(1.5 \text{ kHz/v})}{2 \text{ kHz}} = 1.5 \qquad\qquad m = 2(0.75 \text{ rad/v}) = 1.5$$

Since the modulation index and modulating frequency are the same for both modulators, they have the exact same output frequency spectrum (shown in Figure 7-8a), and it is impossible to distinguish the FM spectrum from the PM spectrum.

b. *FM modulator* | *PM modulator*

$$m = \frac{v_m k_1}{f_m} \qquad\qquad m = v_m k$$

$$m = \frac{4(1.5 \text{ kHz/v})}{2 \text{ kHz}} = 3 \qquad\qquad m = 4(0.75 \text{ rad/v}) = 3$$

Again, the modulation index and modulating signal frequency are the same for both modulators. Therefore, the FM and PM waveforms have the same output frequency spectrum (shown in Figure 7-8b), and it is still impossible to distinguish the FM spectrum from the PM spectrum.

c. *FM modulator* | *PM modulator*

$$m = \frac{v_m k_1}{f_m} \qquad\qquad m = v_m k$$

$$m = \frac{2(1.5 \text{ kHz/v})}{1 \text{ kHz}} = 3 \qquad\qquad m = 2(0.75 \text{ rad/v}) = 1.5$$

When the modulating signal frequency changes, the modulation index for the FM modulator changes inversely proportional. Figure 7-8c shows the output frequency spectrum for the FM modulator. From the figure, it can be seen that changing the modulating signal frequency changed the number of sidebands, the frequency separation between the sidebands, the amplitude of the sideband components, and the bandwidth. Figure 7-8d shows the output frequency spectrum for the PM modulator. From the figure, it can be seen that changing the modulating signal frequency had no effect on the modulation index for the PM modulator. Consequently, changing the modulating signal frequency had no effect on the number of sidebands produced or their relative amplitudes. However, changing the modulating signal frequency did change the frequency separation between the sidebands and the bandwidth.

From Example 7-5, it can be seen that from the frequency spectrums, it is impossible to distinguish FM from PM by changing the modulating signal amplitude. To see the difference between the two forms of angle modulation, the modulating signal frequency must be changed.

7-12 COMMERCIAL BROADCAST-BAND FM

The FCC has assigned the commercial FM broadcast service a 20-MHz band of frequencies that extends from 88 MHz to 108 MHz. The 20-MHz band is divided into 100, 200-kHz wide channels beginning at 88.1 MHz (i.e., 88.3 MHz, 88.5 MHz, and so on). To provide high-quality, reliable music, the maximum frequency deviation allowed is 75 kHz with a maximum modulating-signal frequency of 15 kHz.

Using Equation 7-33, the worst-case modulation index (i.e., the deviation ratio) for a commercial broadcast-band channel is 75 kHz/15 kHz = 5. Referring to the Bessel table, eight pairs of significant side frequencies are produced with a modulation index of 5. Therefore,

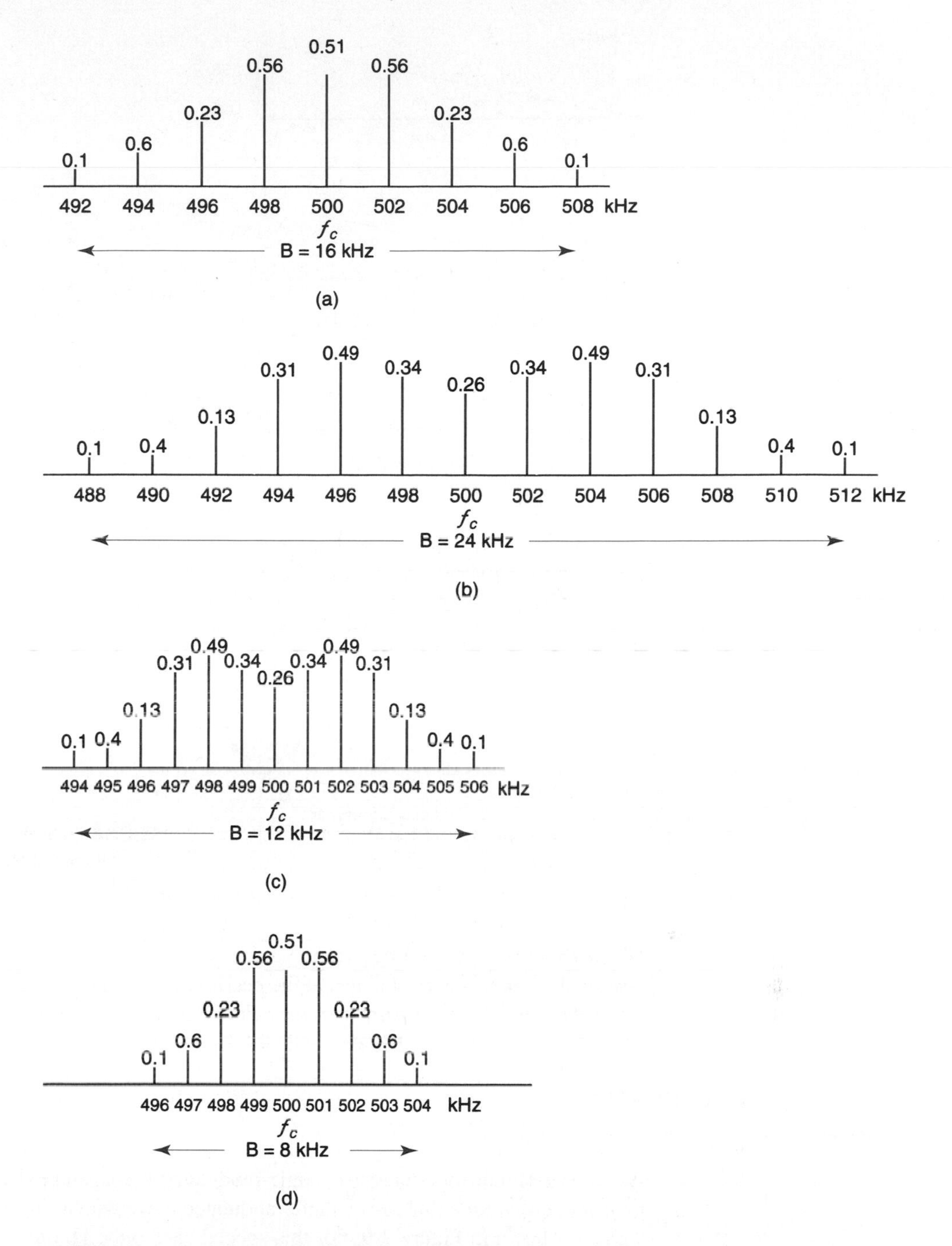

FIGURE 7-8 Output frequency spectrums for Example 7-5: (a) frequency spectrums for FM and PM; (b) frequency spectrums for FM and PM; (c) frequency spectrum for FM; (d) frequency spectrum for PM

from Equation 7-33, the minimum bandwidth necessary to pass all the significant side frequencies is $B = 2(8 \times 15\text{ kHz}) = 240\text{ kHz}$, which exceeds the allocated FCC bandwidth by 40 kHz. In essence, this says that the highest side frequencies from one channel are allowed to spill over into adjacent channels, producing an interference known as *adjacent channel interference.* This is generally not a problem, however, because the FCC has historically assigned only every other channel in a given geographic area. Therefore, a 200-kHz-wide guard band is usually on either side of each assigned channel. In addition, the seventh and eighth sets

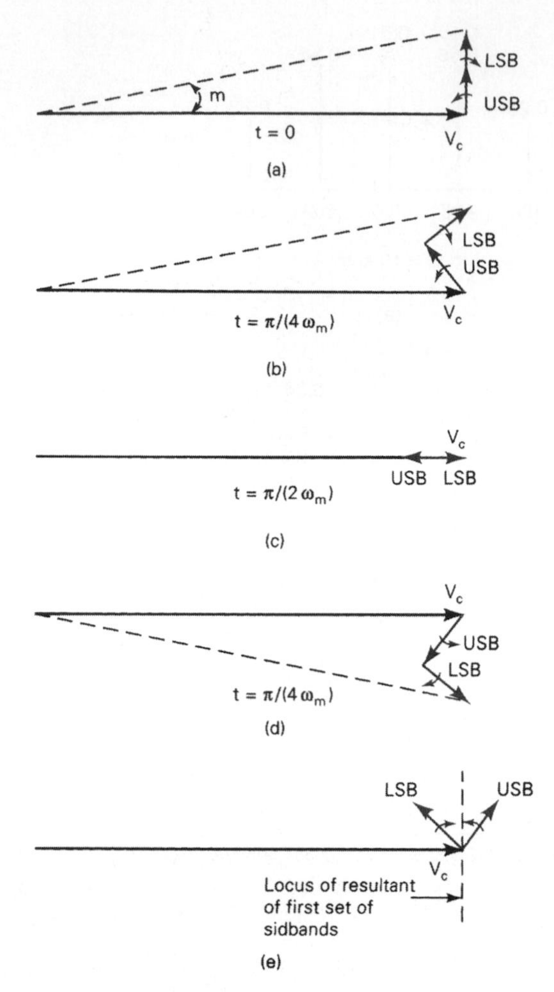

FIGURE 7-9 Angle modulation phasor representation, low modulation index

of side frequencies have little power in them, and it is also highly unlikely that maximum frequency deviation is ever obtained at the maximum modulating-signal frequency. Ironically, if you use Carson's approximation, the bandwidth for commercial broadcast-band channels is 2(75 kHz + 15 kHz) = 180 kHz, which is well within the band limits assigned by the FCC.

7-13 PHASOR REPRESENTATION OF AN ANGLE-MODULATED WAVE

As with amplitude modulation, an angle-modulated wave can be shown in phasor form. The phasor diagram for a low-index angle-modulated wave with a single-frequency modulating signal is shown in Figure 7-9. For this special case ($m < 1$), only the first set of sideband pairs is considered, and the phasor diagram closely resembles that of an AM wave except for a phase reversal of one of the side frequencies. The resultant vector has an amplitude that is close to unity at all times and a peak phase deviation of m radians. It is important to note that if the side frequencies from the higher-order terms were included, the vector would have no amplitude variations. The dashed line shown in Figure 7-9e is the locus of the resultant formed by the carrier and the first set of side frequencies.

Figure 7-10 shows the phasor diagram for a high-index, angle-modulated wave with five sets of side frequencies (for simplicity only the vectors for the first two sets are shown). The resultant vector is the sum of the carrier component and the components of the significant side frequencies with their magnitudes adjusted according to the Bessel table. Each

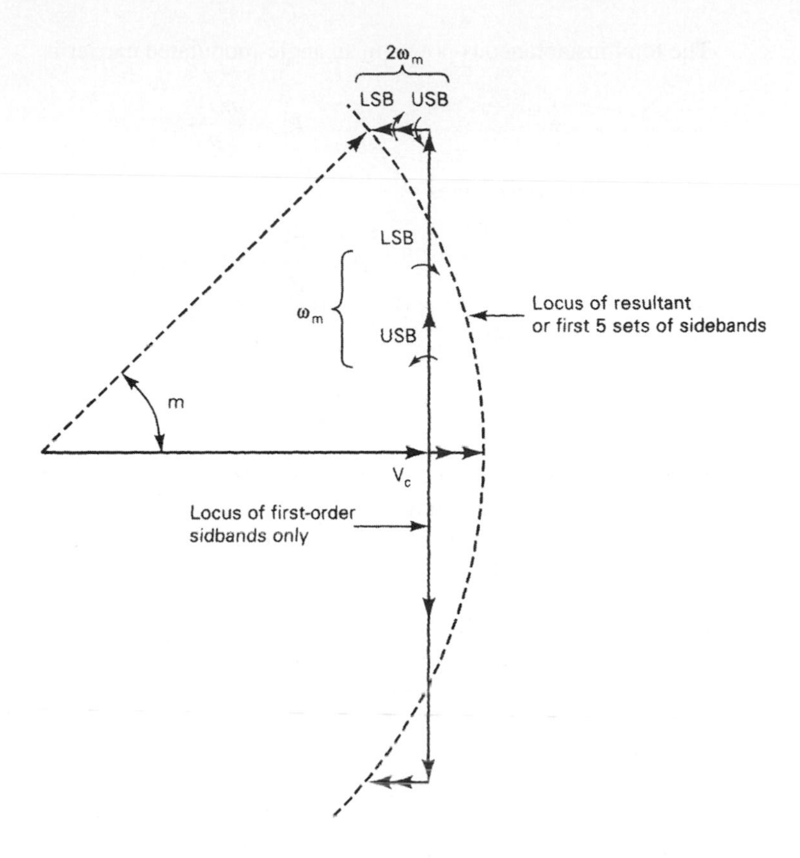

FIGURE 7-10 Angle modulation phasor representation, high modulation index

side frequency is shifted an additional 90° from the preceding side frequency. The locus of the resultant five-component approximation is curved and closely follows the signal locus. By definition, the locus is a segment of the circle with a radius equal to the amplitude of the unmodulated carrier. It should be noted that the resultant signal amplitude and, consequently, the signal power remain constant.

7-14 AVERAGE POWER OF AN ANGLE-MODULATED WAVE

One of the most important differences between angle modulation and amplitude modulation is the distribution of power in the modulated wave. Unlike AM, the total power in an angle-modulated wave is equal to the power of the unmodulated carrier (i.e., the sidebands do not add power to the composite modulated signal). Therefore, with angle modulation, the power that was originally in the unmodulated carrier is redistributed among the carrier and its sidebands. The average power of an angle-modulated wave is independent of the modulating signal, the modulation index, and the frequency deviation. It is equal to the average power of the unmodulated carrier, regardless of the depth of modulation. Mathematically, the average power in the unmodulated carrier is

$$P_c = \frac{V_c^2}{2R}\text{W} \tag{7-36}$$

where P_c = carrier power (watts)
V_c = peak unmodulated carrier voltage (volts)
R = load resistance (ohms)

The total instantaneous power in an angle-modulated carrier is

$$P_t = \frac{m(t)^2}{R} \text{W} \tag{7-37}$$

Substituting for $m(t)$ gives

$$P_t = \frac{V_c^2}{R} \cos^2[\omega_c t + \theta(t)] \tag{7-38}$$

and expanding yields

$$= \frac{V_c^2}{R}\left\{\frac{1}{2} + \frac{1}{2}\cos[2\omega_c t + 2\theta(t)]\right\} \tag{7-39}$$

In Equation 7-39, the second term consists of an infinite number of sinusoidal side frequency components about a frequency equal to twice the carrier frequency ($2\omega_c$). Consequently, the average value of the second term is zero, and the average power of the modulated wave reduces to

$$P_t = \frac{V_c^2}{2R} \text{W} \tag{7-40}$$

Note that Equations 7-36 and 7-40 are identical, so the average power of the modulated carrier must be equal to the average power of the unmodulated carrier. The modulated carrier power is the sum of the powers of the carrier and the side frequency components. Therefore, the total modulated wave power is

$$P_t = P_0 + P_1 + P_2 + P_3 + P_n \tag{7-41}$$

$$P_t = \frac{V_c^2}{2R} + \frac{2(V_1)^2}{2R} + \frac{2(V_2)^2}{2R} + \frac{2(V_3)^2}{2R} + \frac{2(V_n)^2}{2R} \tag{7-42}$$

where P_t = total power (watts)
P_0 = modulated carrier power (watts)
P_1 = power in the first set of sidebands (watts)
P_2 = power in the second set of sidebands (watts)
P_3 = power in the third set of sidebands (watts)
P_n = power in the nth set of sidebands (watts)

Example 7-6

a. Determine the unmodulated carrier power for the FM modulator and conditions given in Example 7-2 (assume a load resistance $R_L = 50\Omega$).
b. Determine the total power in the angle-modulated wave.

Solution

a. Substituting into Equation 7-36 yields

$$P_c = \frac{10^2}{2(50)} = 1 \text{ W}$$

b. Substituting into Equation 7-41 gives us

$$P_t = \frac{7.7^2}{2(50)} + \frac{2(4.4)^2}{2(50)} + \frac{2(1.1)^2}{2(50)} + \frac{2(0.2)^2}{2(50)}$$

$$= 0.5929 + 0.3872 + 0.0242 + 0.0008 = 1.0051 \text{ W}$$

The results of (a) and (b) are not exactly equal because the values given in the Bessel table have been rounded off. However, the results are close enough to illustrate that the power in the modulated wave and the unmodulated carrier are equal.

7-15 NOISE AND ANGLE MODULATION

When thermal noise with a constant spectral density is added to an FM signal, it produces an unwanted deviation of the carrier frequency. The magnitude of this unwanted frequency deviation depends on the relative amplitude of the noise with respect to the carrier. When this unwanted carrier deviation is demodulated, it becomes noise if it has frequency components that fall within the information-frequency spectrum. The spectral shape of the demodulated noise depends on whether an FM or a PM demodulator is used. The noise voltage at the output of a PM demodulator is constant with frequency, whereas the noise voltage at the output of an FM demodulator increases linearly with frequency. This is commonly called the FM *noise triangle* and is illustrated in Figure 7-11. It can be seen that the demodulated noise voltage is inherently higher for the higher-modulating-signal frequencies.

7-15-1 Phase Modulation Due to an Interfering Signal

Figure 7-12 shows phase modulation caused by a single-frequency noise signal. The noise component V_n is separated in frequency from the signal component V_c by frequency f_n. This is shown in Figure 7-12b. Assuming that $V_c > V_n$, the peak phase deviation due to an interfering single-frequency sinusoid occurs when the signal and noise voltages are in quadrature and is approximated for small angles as

$$\Delta\theta_{peak} \simeq \frac{V_n}{V_c} \text{ rad} \tag{7-43}$$

Figure 7-12c shows the effect of *limiting* the amplitude of the composite FM signal on noise. (Limiting is commonly used in angle-modulation receivers and is explained in Chapter 8.) It can be seen that the single-frequency noise signal has been transposed into a noise sideband pair each with amplitude $V_n/2$. These sidebands are coherent; therefore, the peak phase deviation is still V_n/V_c radians. However, the unwanted amplitude variations have been removed, which reduces the total power but does not reduce the interference in the demodulated signal due to the unwanted phase deviation.

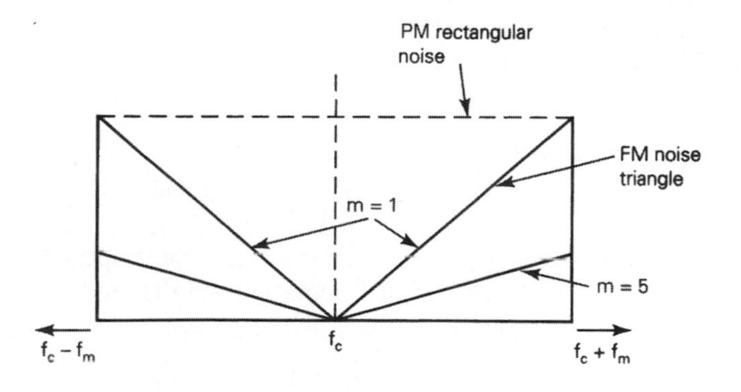

FIGURE 7-11 FM noise triangle

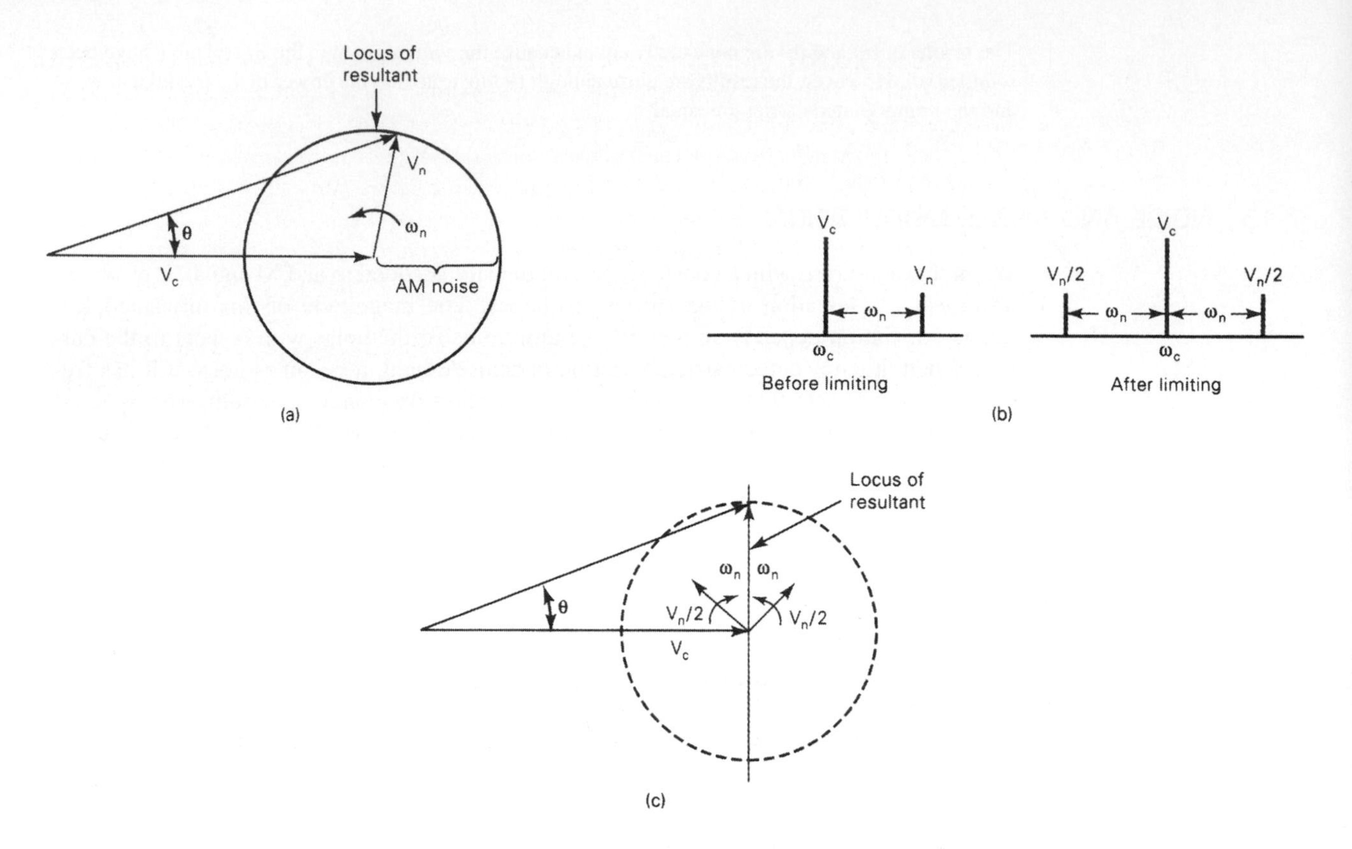

FIGURE 7-12 Interfering sinusoid of noise: (a) before limiting; (b) frequency spectrum; (c) after limiting

7-15-2 Frequency Modulation Due to an Interfering Signal

From Equation 7-5, the instantaneous frequency deviation $\Delta f(t)$ is the first time derivative of the instantaneous phase deviation $\theta(t)$. When the carrier component is much larger than the interfering noise voltage, the instantaneous phase deviation is approximately

$$\theta(t) = \frac{V_n}{V_c} \sin(\omega_n t + \theta_n) \text{ rad} \tag{7-44}$$

and, taking the first derivative, we obtain

$$\Delta\omega(t) = \frac{V_n}{V_c} \omega_n \cos(\omega_n t + \theta_n) \text{ rad/s} \tag{7-45}$$

Therefore, the peak frequency deviation is

$$\Delta\omega_{\text{peak}} = \frac{V_n}{V_c} \omega_n \text{ rad/s} \tag{7-46}$$

$$\Delta f_{\text{peak}} = \frac{V_n}{V_c} f_n \text{ Hz} \tag{7-47}$$

Rearranging Equation 7-22, it can be seen that the peak frequency deviation (Δf) is a function of the modulating-signal frequency and the modulation index. Therefore, for a noise modulating frequency f_n, the peak frequency deviation is

$$\Delta f_{\text{peak}} = m f_n \text{ Hz} \tag{7-48}$$

where m equals modulation index ($m << 1$).

From Equation 7-48, it can be seen that the farther the noise frequency is displaced from the carrier frequency, the larger the frequency deviation. Therefore, noise frequencies that produce components at the high end of the modulating-signal frequency spectrum produce more frequency deviation for the same phase deviation than frequencies that fall at the low end. FM demodulators generate an output voltage that is proportional to the frequency deviation and equal to the difference between the carrier frequency and the interfering signal frequency. Therefore, high-frequency noise components produce more demodulated noise than do low-frequency components.

The signal-to-noise ratio at the output of an FM demodulator due to unwanted frequency deviation from an interfering sinusoid is the ratio of the peak frequency deviation due to the information signal to the peak frequency deviation due to the interfering signal:

$$\frac{S}{N} = \frac{\Delta f_{\text{due to signal}}}{\Delta f_{\text{due to noise}}} \qquad \textbf{(7-49)}$$

Example 7-7

For an angle-modulated carrier $V_c = 6\cos(2\pi 110\text{ MHz } t)$ with 75-kHz frequency deviation due to the information signal and a single-frequency interfering signal $V_n = 0.3\cos(2\pi 109.985\text{ MHz } t)$, determine

a. Frequency of the demodulated interference signal.
b. Peak phase and frequency deviations due to the interfering signal.
c. Voltage signal-to-noise ratio at the output of the demodulator.

Solution **a.** The frequency of the noise interference is the difference between the carrier frequency and the frequency of the single-frequency interfering signal:

$$f_c - f_n = 110\text{ MHz} - 109.985\text{ MHz} = 15\text{ kHz}$$

b. Substituting into Equation 7-43 yields

$$\Delta\theta_{\text{peak}} = \frac{0.3}{6} = 0.05\text{ rad}$$

Substituting into Equation 7-47 gives us

$$\Delta f_{\text{peak}} = \frac{0.3 \times 15\text{ kHz}}{6} = 750\text{ Hz}$$

c. The voltage S/N ratio due to the interfering tone is the ratio of the carrier amplitude to the amplitude of the interfering signal, or

$$\frac{6}{0.3} = 20$$

The voltage S/N ratio after demodulation is found by substituting into Equation 7-49:

$$\frac{S}{N} = \frac{75\text{ kHz}}{750\text{ Hz}} = 100$$

Thus, there is a voltage signal-to-noise improvement of $100/20 = 5$, or $20\log 5 = 14$ dB.

7-16 PREEMPHASIS AND DEEMPHASIS

The noise triangle shown in Figure 7-11 shows that, with FM, there is a nonuniform distribution of noise. Noise at the higher-modulating-signal frequencies is inherently greater in amplitude than noise at the lower frequencies. This includes both single-frequency interference and thermal noise. Therefore, for information signals with a uniform signal level, a nonuniform signal-to-noise ratio is produced, and the higher-modulating-signal frequencies have a lower signal-to-noise ratio than the lower frequencies. This is shown in Figure 7-13a. It can be seen that the S/N ratio is lower at the high-frequency ends of the triangle. To compensate for this, the high-frequency modulating signals are emphasized or

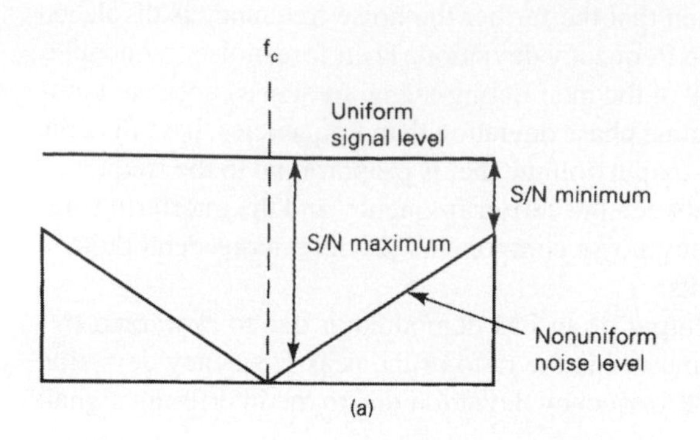

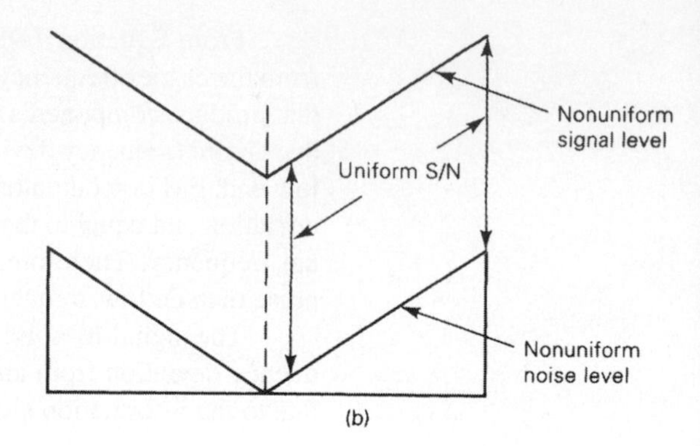

FIGURE 7-13 FM signal-to-noise: (a) without preemphasis; (b) with preemphasis

boosted in amplitude in the transmitter prior to performing modulation. To compensate for this boost, the high-frequency signals are attenuated or deemphasized in the receiver after demodulation has been performed. Deemphasis is the reciprocal of preemphasis and, therefore, a deemphasis network restores the original amplitude-versus-frequency characteristics to the information signals. In essence, the preemphasis network allows the high-frequency modulating signals to modulate the carrier at a higher level and, thus, cause more frequency deviation than their original amplitudes would have produced. The high-frequency signals are propagated through the system at an elevated level (increased frequency deviation), demodulated, and then restored to their original amplitude proportions. Figure 7-13b shows the effects of pre- and deemphasis on the signal-to-noise ratio. The figure shows that pre- and deemphasis produce a more uniform signal-to-noise ratio throughout the modulating-signal frequency spectrum.

A preemphasis network is a high-pass filter (i.e., a differentiator) and a deemphasis network is a low-pass filter (an integrator). Figure 7-14a shows the schematic diagrams for an active preemphasis network and a passive deemphasis network. Their corresponding frequency-response curves are shown in Figure 7-14b. A preemphasis network provides a constant increase in the amplitude of the modulating signal with an increase in frequency. With FM, approximately 12 dB of improvement in noise performance is achieved using pre- and deemphasis. The break frequency (the frequency where pre- and deemphasis begins) is determined by the *RC* or *L/R* time constant of the network. The break frequency occurs at the frequency where X_C or X_L equals *R*. Mathematically, the break frequency is

$$f_b = \frac{1}{2\pi RC} \tag{7-50}$$

$$f_b = \frac{1}{2\pi L/R} \tag{7-51}$$

The networks shown in Figure 7-14 are for the FM broadcast band, which uses a 75-μs time constant. Therefore, the break frequency is approximately

$$f_b = \frac{1}{2\pi 75\ \mu\text{s}} = 2.12\ \text{kHz}$$

The FM transmission of the audio portion of commercial television broadcasting uses a 50-μs time constant.

As shown in Figure 7-14, an active rather than a passive preemphasis network is used because a passive preemphasis network provides loss to all frequencies with more loss introduced at the lower modulating-signal frequencies. The result of using a passive network

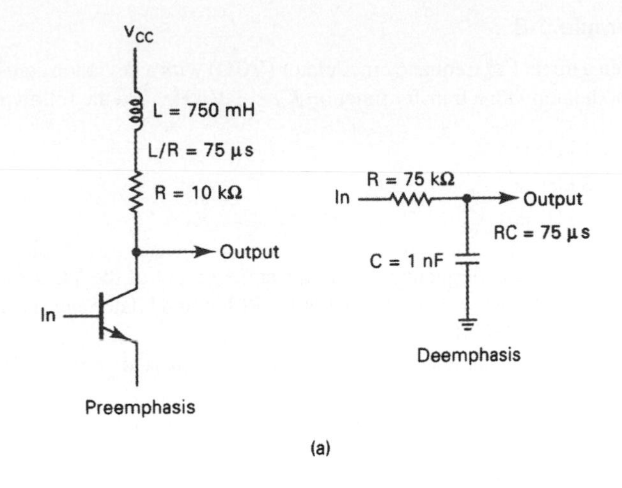

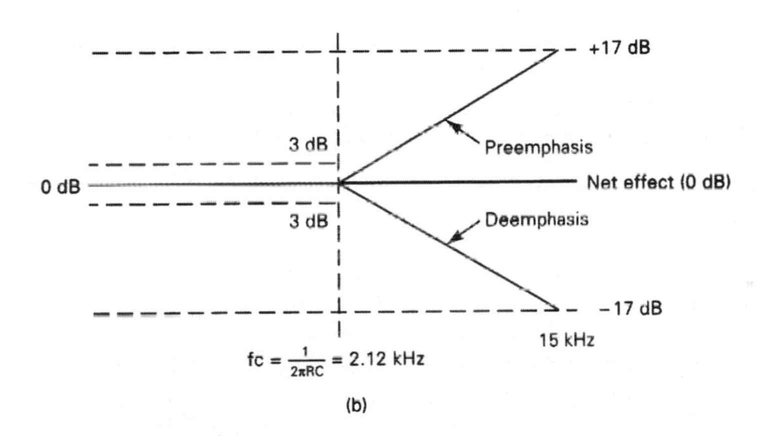

FIGURE 7-14 Preemphasis and deemphasis: (a) schematic diagrams; (b) attenuation curves

would be a decrease in the signal-to-noise ratio at the lower modulating-signal frequencies rather than an increase in the signal-to-noise ratio at the higher modulating-signal frequencies.

From the preceding explanation and Figure 7-14, it can be seen that the output amplitude from a preemphasis network increases with frequency for frequencies above the break frequency. Referring back to Equation 7-20, it can be seen that if changes in the frequency of the modulating signal (f_m) produce corresponding changes in its amplitude (V_m), the modulation index (m) remains constant with frequency. This, of course, is a characteristic of phase modulation. Consequently, with commercial broadcast-band modulators, frequencies below 2112 Hz produce frequency modulation, and frequencies above 2112 Hz produce phase modulation. Converting FM to PM is not the function of a preemphasis network, however, but rather a consequence.

The noise generated internally in FM demodulators inherently increases with frequency, which produces a nonuniform signal-to-noise ratio at the output of the demodulator. The signal-to-noise ratios are lower for the higher modulating-signal frequencies than for the lower modulating-signal frequencies. Using a preemphasis network in front of the FM modulator and a deemphasis network at the output of the FM demodulator improves the signal-to-noise ratio for the higher modulating-signal frequencies, thus producing a more uniform signal-to-noise ratio at the output of the demodulator. The benefits of using pre- and deemphasis are best appreciated with an example.

Example 7-8

Given a direct FM frequency modulator (VCO) with a deviation sensitivity $K_1 = 1$ kHz/V, a PLL FM demodulator with a transfer function $K_d = 1$ V/kHz, and the following input signals:

1 kHz at 4 V_p
2 kHz at 2 V_p
3 kHz at 1 V_p

a. Determine the frequency deviations at the output of the VCO for the three input signals and the demodulated voltages at the output of the PLL demodulator and sketch the frequency spectrum at the output of the demodulator.
b. For the following internally generated noise signals, determine the signal-to-noise ratios at the output of the demodulator:

1 kHz at 0.1 V_p
2 kHz at 0.25 V_p
3 kHz at 0.5 V_p

c. Determine the frequency spectrum at the output of the preemphasis network, the frequency deviations at the output of the modulator, the demodulator output voltages, the frequency spectrums at the output of the PLL demodulator and at the output of the deemphasis network, and the signal-to-noise ratios at the output of the PLL demodulator and the deemphhasis network.

Solution **a.** Figure 7-15a shows the frequency deviations at the output of the modulator, the demodulated voltages, and the demodulated frequency spectrum.
b. The signal-to-noise ratios at the output of the FM demodulator are shown in Figure 7-15a.
c. The frequency spectrum at the output of the preemphasis network, the frequency deviations at the output of the modulator, the demodulator output voltages, the frequency spectrums at the output of the PLL demodulator and at the output of the deemphasis network, and the signal-to-noise ratios at the output of the PLL demodulator and deemphasis network are shown in Figure 7-15b.

From Example 7-8, it can be seen that the signal-to-noise ratios achieved for the 3 kHz-signal improved from 2 to 8 when the pre- and deemphasis networks were used. This improvement equates to a 4:1 (12-dB) improvement. It can also be seen that the signal-to-noise ratios achieved using pre- and deemphasis are more uniform than when not using them.

7-17 FREQUENCY AND PHASE MODULATORS

In essence, the primary difference between frequency and phase modulators lies in whether the frequency or the phase of the carrier is directly changed by the modulating signal and which property is indirectly changed. When the frequency of the carrier oscillator is modulated by the information signal, direct FM (indirect PM) results. When the phase of the carrier signal is modulated by the information signal, direct PM (indirect FM) results.

The primary disadvantage of direct FM is that relatively unstable LC oscillators must be used to produce the carrier frequency, which prohibits using crystal oscillators. Thus, direct FM requires the addition of some form of automatic frequency control circuitry to maintain the carrier frequency within the FCC's stringent frequency-stability requirements. The obvious advantage of direct FM is that relatively high-frequency deviations and modulation indices are easily obtained because the oscillators are inherently unstable.

The primary advantage of direct PM is that the carrier oscillator is isolated from the actual modulator circuit and, therefore, can be an extremely stable source such as a crystal oscillator. The obvious disadvantage of direct PM is that crystal oscillators are inherently stable and, therefore, it is more difficult for them to achieve high phase deviations and modulation indices.

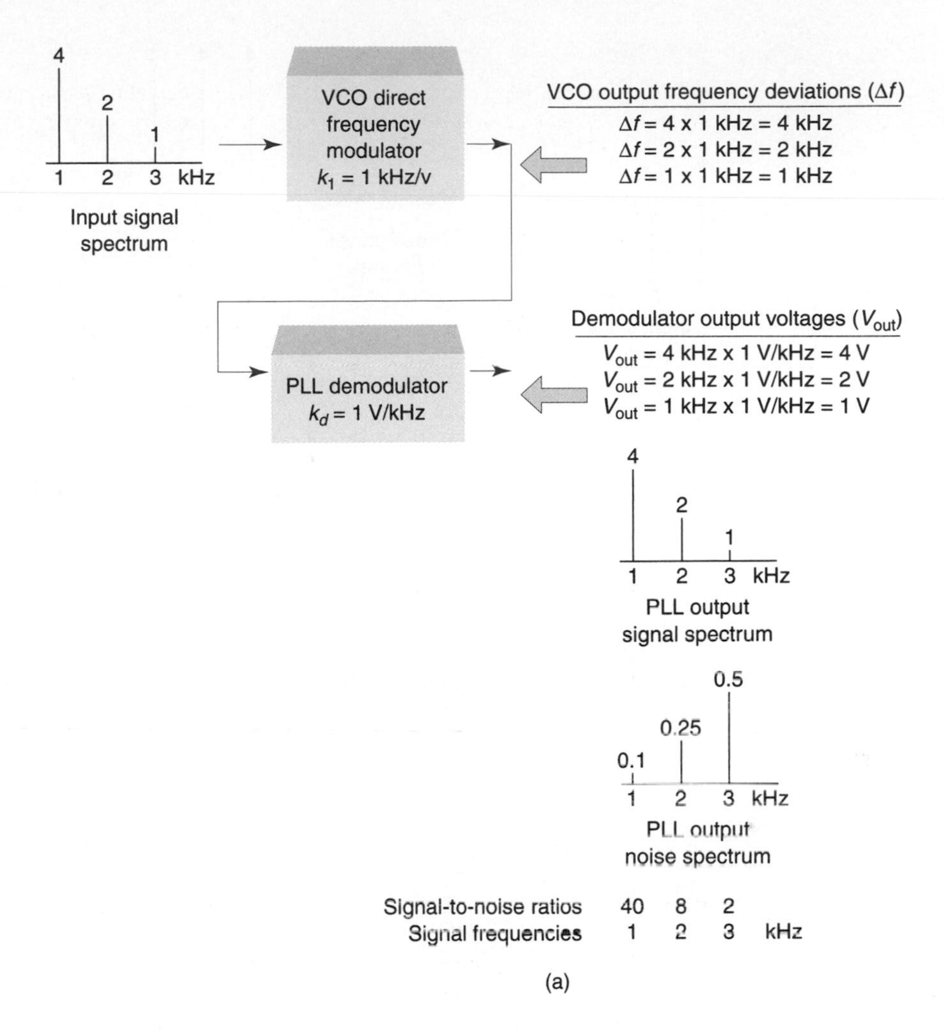

FIGURE 7-15 Figure for Example 7-8: (a) S/N without preemphasis and deemphasis; *(Continued)*

7-17-1 Direct FM Modulators

Direct FM is angle modulation in which the frequency of the carrier is varied (deviated) directly by the modulating signal. With direct FM, the instantaneous frequency deviation is directly proportional to the amplitude of the modulating signal. Figure 7-16 shows a schematic diagram for a simple (although highly impractical) direct FM generator. The tank circuit (L and C_m) is the frequency-determining section for a standard LC oscillator. The capacitor microphone is a transducer that converts acoustical energy to mechanical energy, which is used to vary the distance between the plates of C_m and, consequently, change its capacitance. As C_m is varied, the resonant frequency is varied. Thus, the oscillator output frequency varies directly with the external sound source. This is direct FM because the oscillator frequency is changed directly by the modulating signal, and the magnitude of the frequency change is proportional to the amplitude of the modulating-signal voltage. There are three common methods for producing direct frequency modulation: varactor diode modulators, FM reactance modulators, and linear integrated-circuit direct FM modulators.

7-17-1-1 Varactor diode modulators. Figure 7-17 shows the schematic diagram for a more practical direct FM generator that uses a varactor diode to deviate the frequency

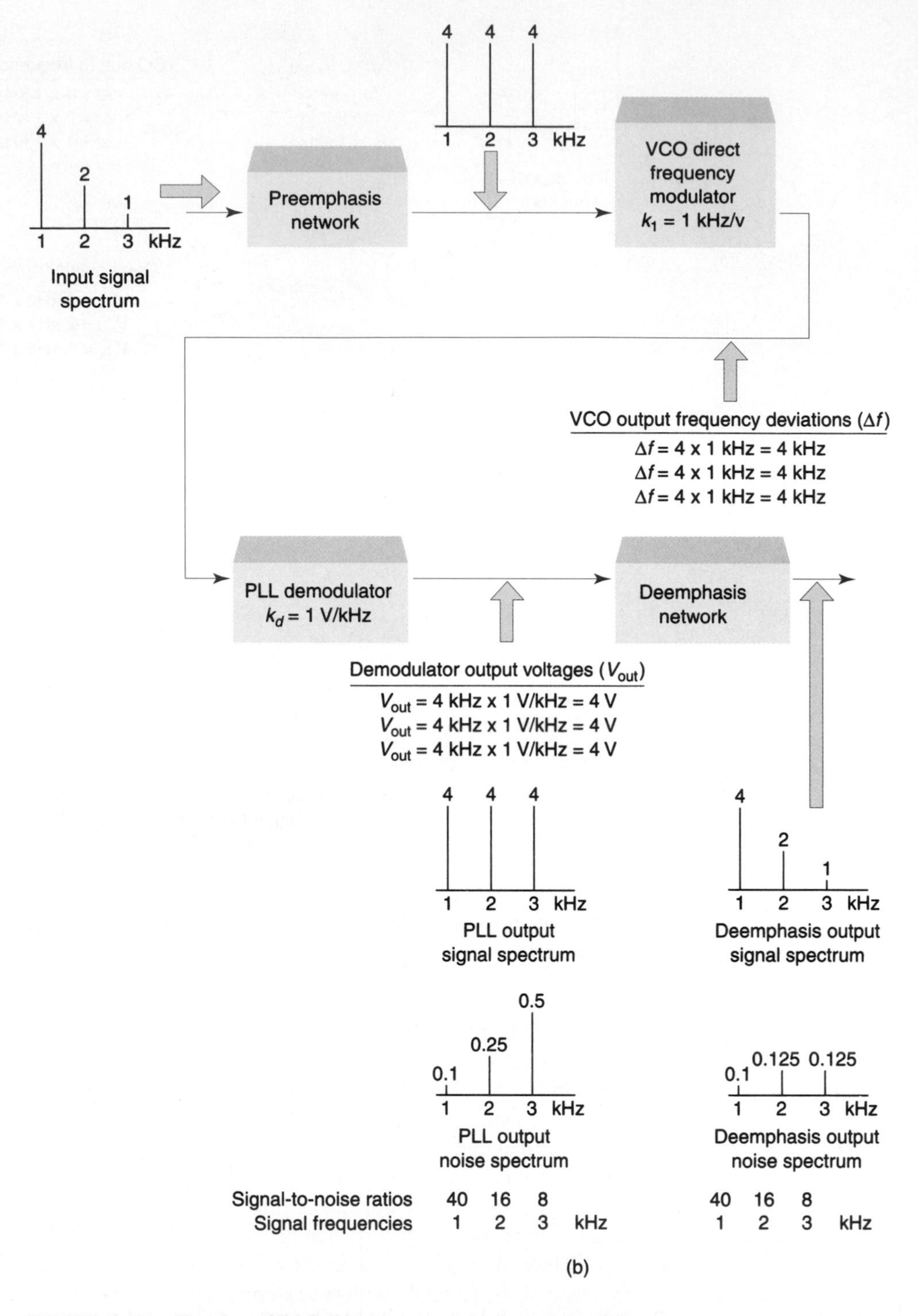

FIGURE 7-15 *(Continued)* (b) S/N with preemphasis and deemphasis

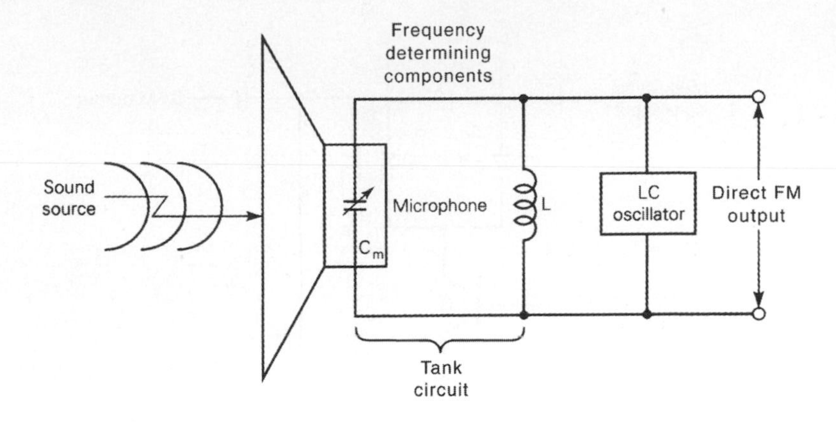

FIGURE 7-16 Simple direct FM modulator

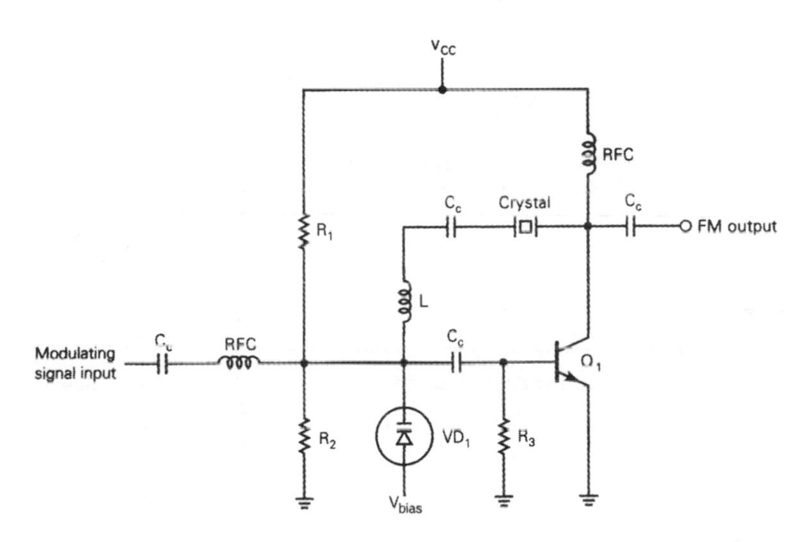

FIGURE 7-17 Varactor diode direct FM modulator

of a crystal oscillator. R_1 and R_2 develop a dc voltage that reverse biases varactor diode VD_1 and determines the rest frequency of the oscillator. The external modulating-signal voltage adds to and subtracts from the dc bias, which changes the capacitance of the diode and, thus, the frequency of oscillation (see Chapter 2 for more detailed description of varactor diodes). Positive alternations of the modulating signal increase the reverse bias on VD_1, which decreases its capacitance and increases the frequency of oscillation. Conversely, negative alternations of the modulating signal decrease the frequency of oscillation. Varactor diode FM modulators are extremely popular because they are simple to use and reliable and have the stability of a crystal oscillator. However, because a crystal is used, the peak frequency deviation is limited to relatively small values. Consequently, they are used primarily for low-index applications, such as two-way mobile radio.

Figure 7-18 shows a simplified schematic diagram for a voltage-controlled oscillator (VCO) FM generator. Again, a varactor diode is used to transform changes in the modulating-signal amplitude to changes in frequency. The center frequency for the oscillator is determined as follows:

$$f_c = \frac{1}{2\pi\sqrt{LC}} \text{ Hz} \tag{7-52}$$

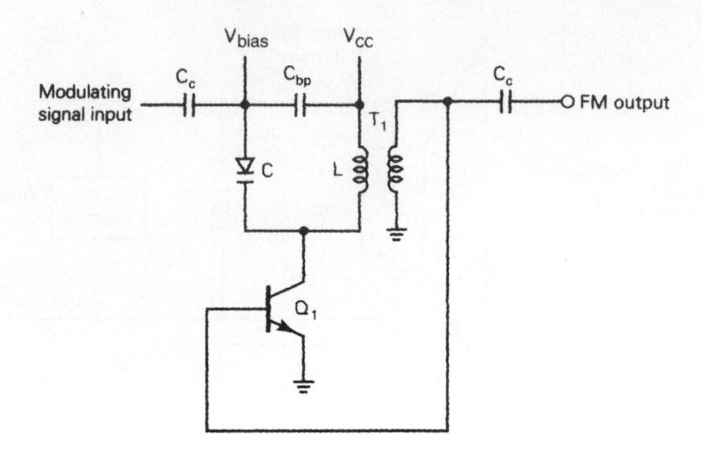

FIGURE 7-18 Varactor diode VCO FM modulator

where f_c = carrier rest frequency (hertz)
L = inductance of the primary winding of T_1 (henries)
C = varactor diode capacitance (farads)

With a modulating signal applied, the frequency is

$$f = \frac{1}{2\pi\sqrt{L(C + \Delta C)}} \text{ Hz} \tag{7-53}$$

where f is the new frequency of oscillation and ΔC is the change in varactor diode capacitance due to the modulating signal. The change in frequency is

$$\Delta f = |f_c - f| \tag{7-54}$$

where Δf = peak frequency deviation (hertz)

7-17-1-2 FM reactance modulators. Figure 7-19a shows a schematic diagram for a reactance modulator using a JFET as the active device. This circuit configuration is called a reactance modulator because the JFET looks like a variable-reactance load to the *LC* tank circuit. The modulating signal varies the reactance of Q_1, which causes a corresponding change in the resonant frequency of the oscillator tank circuit.

Figure 7-19b shows the ac equivalent circuit. R_1, R_3, R_4, and R_C provide the dc bias for Q_1. R_E is bypassed by C_c and is, therefore, omitted from the ac equivalent circuit. Circuit operation is as follows. Assuming an ideal JFET (gate current $i_g = 0$),

$$v_g = i_g R \tag{7-55}$$

where

$$i_g = \frac{v}{R - jX_C} \tag{7-56}$$

Therefore,

$$v_g = \frac{v}{R - jX_C} \times R \tag{7-57}$$

and the JFET drain current is

$$i_d = g_m v_g = g_m\left(\frac{v}{R - jX_C}\right) \times R \tag{7-58}$$

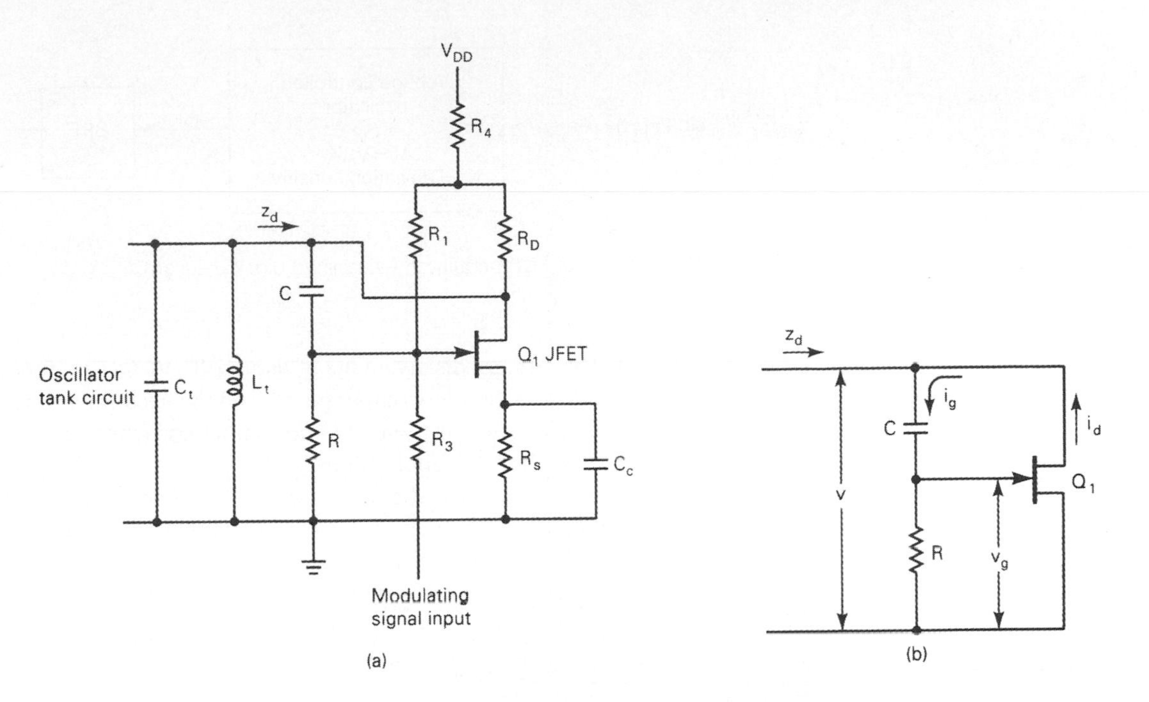

FIGURE 7-19 JFET reactance modulator: (a) schematic diagram; (b) ac equivalent circuit

where g_m is the transconductance of the JFET, and the impedance between the drain and ground is

$$z_d = \frac{v}{i_d} \tag{7-59}$$

Substituting and rearranging gives us

$$z_d = \frac{R - jX_C}{g_m R} = \frac{1}{g_m}\left(1 - \frac{jX_C}{R}\right) \tag{7-60}$$

Assuming that $R <<< X_C$,

$$z_d = -j\frac{X_C}{g_m R} = \frac{-j}{2\pi f_m g_m RC} \tag{7-61}$$

$g_m RC$ is equivalent to a variable capacitance, and z_d is inversely proportional to resistance (R), the angular velocity of the modulating signal ($2\pi f_m$), and the transconductance (g_m) of Q_1, which varies with the gate-to-source voltage. When a modulating signal is applied to the bottom of R_3, the gate-to-source voltage is varied accordingly, causing a proportional change in g_m. As a result, the equivalent circuit impedance (z_d) is a function of the modulating signal. Therefore, the resonant frequency of the oscillator tank circuit is a function of the amplitude of the modulating signal, and the rate at which it changes is equal to f_m. Interchanging R and C causes the variable reactance to be inductive rather than capacitive but does not affect the output FM waveform. The maximum frequency deviation obtained with a reactance modulator is approximately 5 kHz.

7-17-1-3 Linear integrated-circuit direct FM modulators. *Linear integrated-circuit voltage-controlled oscillators* and *function generators* can generate a direct FM output waveform that is relatively stable, accurate, and directly proportional to the input

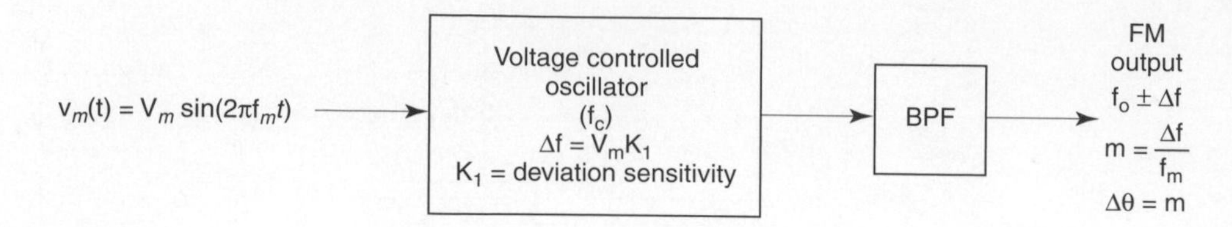

FIGURE 7-20 LIC direct FM modulator—simplified block diagram

modulating signal. The primary disadvantage of using LIC VCOs and function generators for direct FM modulation is their low output power and the need for several additional external components for them to function, such as timing capacitors and resistors for frequency determination and power supply filters.

Figure 7-20 shows a simplified block diagram for a linear integrated-circuit monolithic function generator that can be used for direct FM generation. The VCO center frequency is determined by external resistor and capacitor (R and C). The input modulating signal ($v_m[t] = V_m \sin[2\pi f_m t]$) is applied directly to the input of the voltage-controlled oscillator where it deviates the carrier rest frequency (f_c) and produces an FM output signal. The peak frequency deviation is determined by the product of the peak modulating-signal amplitude (V_m) and the deviation sensitivity of the VCO (K_1). The modulator output is

$$FM_{(out)} = f_c + \Delta f \tag{7-62}$$

where f_c = carrier rest frequency (VCO's natural frequency $-$ $1/RC$ hertz)
Δf = peak frequency deviation ($V_m K_1$ hertz)
K_1 = deviation sensitivity (hertz per volt)

Figure 7-21a shows the schematic diagram for the Motorola MC1376 monolithic FM transmitter. The MC1376 is a complete FM modulator on a single 8-pin DIP integrated-circuit chip. The MC1376 can operate with carrier frequencies between 1.4 MHz and 14 MHz and is intended to be used for producing direct FM waves for low-power applications, such as cordless telephones. When the auxiliary transistor is connected to a 12-V supply voltage, output powers as high as 600 mW can be achieved. Figure 7-21b shows the output frequency-versus-input voltage curve for the internal VCO. As the figure shows, the curve is fairly linear between 2 V and 4 V and can produce a peak frequency deviation of nearly 150 kHz.

7-17-2 Direct PM Modulators

7-17-2-1 Varactor diode direct PM modulator. Direct PM (i.e., *indirect FM*) is angle modulation in which the frequency of the carrier is deviated indirectly by the modulating signal. Direct PM is accomplished by directly changing the phase of the carrier and is, therefore, a form of direct phase modulation. The instantaneous phase of the carrier is directly proportional to the modulating signal.

Figure 7-22 shows a schematic diagram for a direct PM modulator. The modulator comprises a varactor diode VD_1 in series with an inductive network (tunable coil L_1 and resistor R_1). The combined series–parallel network appears as a series resonant circuit to the output frequency from the crystal oscillator. A modulating signal is applied to VD_1, which changes its capacitance and, consequently, the phase angle of the impedance seen by the carrier varies, which results in a corresponding phase shift in the carrier. The phase shift is directly proportional to the amplitude of the modulating signal. An advantage of indirect FM is that a buffered crystal oscillator is used for the source of the carrier signal. Consequently, indirect FM transmitters are more frequency stable than their direct counterparts. A disadvantage is that the capacitance-versus-voltage characteristics of a varactor diode

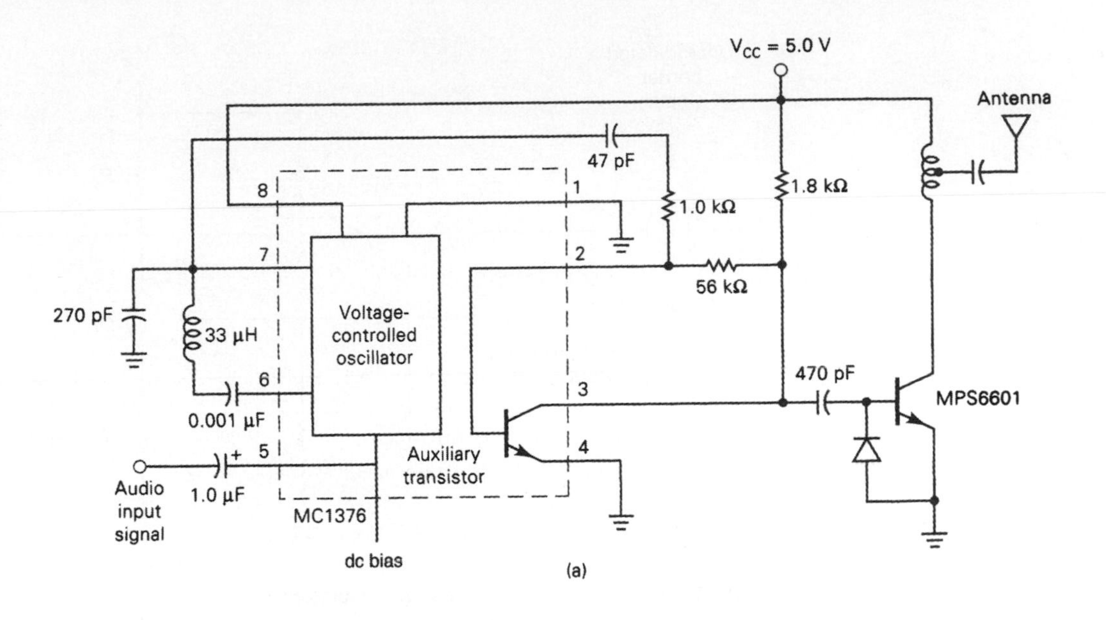

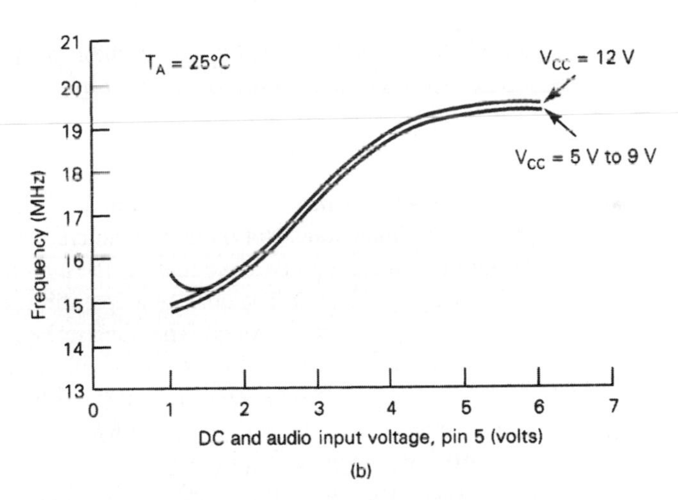

FIGURE 7-21 MC1376 FM transmitter LIC: (a) schematic diagram; (b) VCO output-versus-input frequency-response curve

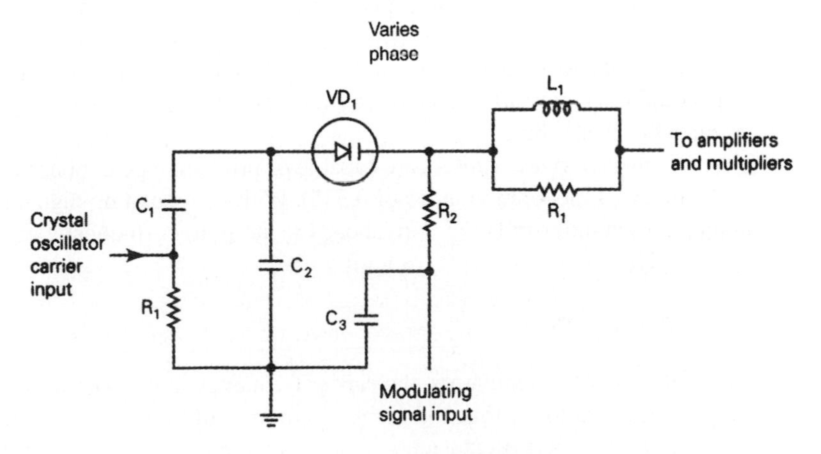

FIGURE 7-22 Direct PM modulator schematic diagram

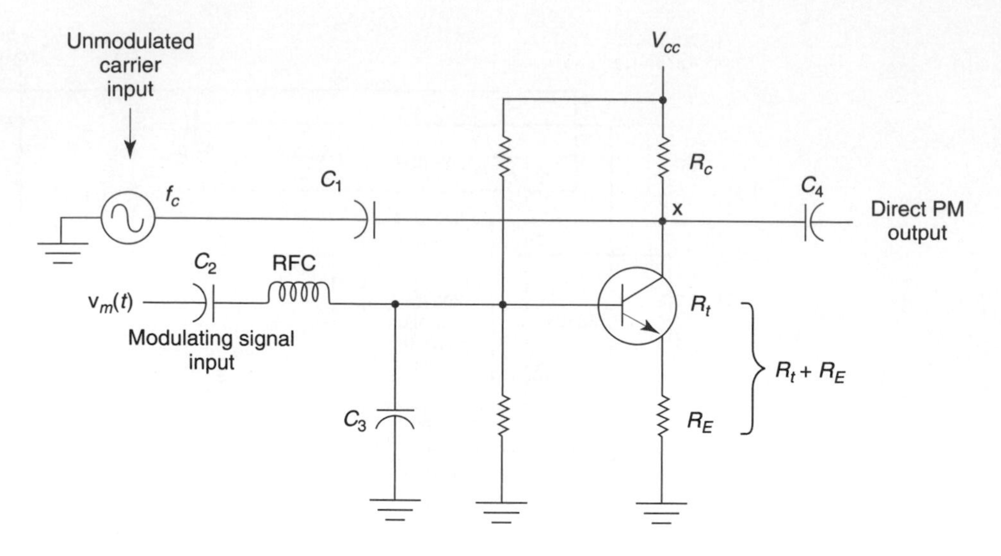

FIGURE 7-23 Transistor direct PM modulator

are nonlinear. In fact, they closely resemble a square-root function. Consequently, to minimize distortion in the modulated waveform, the amplitude of the modulating signal must be kept quite small, which limits the phase deviation to rather small values and its uses to low-index, narrowband applications.

7-17-2-2 Transistor direct PM modulator. Figure 7-23 shows the schematic diagram for a simple transistor direct PM modulator. The circuit is a standard class A common-emitter amplifier with two external inputs: a modulating-signal input ($v_m[t]$) and an external carrier input ($v_c[t]$). The quiescent operating conditions cause the transistor to act like a resistor from point x to ground. The transistor emitter-to-collector resistance (R_t) is part of a phase shifter consisting of C_1 in series with R_t and emitter resistor R_E. The output is taken across the series combination of R_t and R_E. If the circuit is designed such that at the carrier input frequency (f_c) the sum of R_t and R_E equals the capacitive reactance of C_1 (i.e., X_{C1}), the carrier input signal is shifted 45°.

When the modulating signal is applied, its voltage adds to and subtracts from the dc base bias, producing corresponding changes in the collector current. The changes in collector current dynamically change the transistor emitter-to-collector resistance, producing changes in the phase shift that the carrier undergoes as it passes through the phase shifting network. The phase shift is directly proportional to the amplitude of the modulating signal and occurs at a rate equal to the modulating-signal frequency. The higher the amplitude of the modulating input signal, the greater the change in emitter-to-collector resistance and the greater the phase shift.

Transistor phase shifters are capable of producing peak phase shifts as high as 0.375 radians (i.e., a modulation index of 0.375). With a modulating-signal frequency of 15 kHz, a modulation index of 0.375 corresponds to an indirect frequency shift of $15{,}000 \times 0.375 = 5625$ Hz.

7-18 FREQUENCY UP-CONVERSION

With FM and PM modulators, the carrier frequency at the output of the modulator is generally somewhat lower than the desired frequency of transmission. Therefore, with FM and PM transmitters, it is often necessary to up-convert the frequency of the modulated carrier after modulation has been performed. There are two basic methods of performing frequency

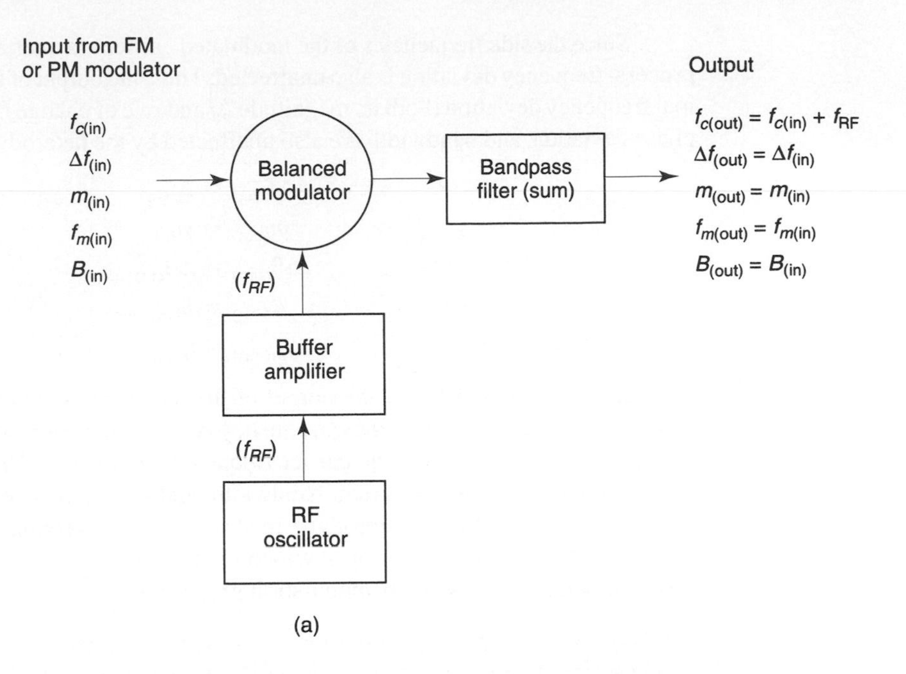

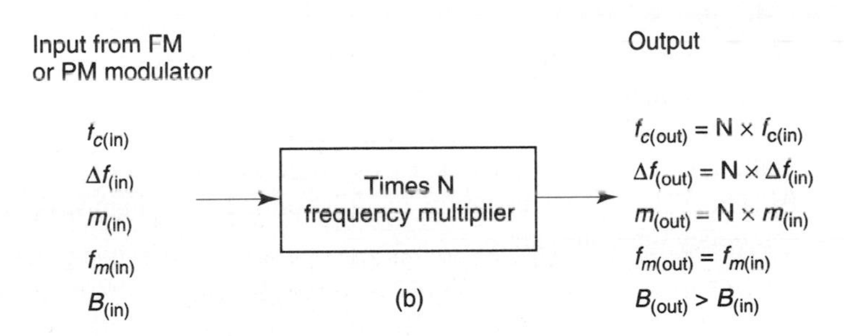

FIGURE 7-24 Frequency up-conversion: (a) heterodyne method; (b) multiplication method

up-conversion. One is by using a heterodyning process, and the other is through frequency multiplication. Figure 7-24 shows the two methods of performing frequency up-conversion.

7-18-1 Heterodyne Method of Frequency Up-Conversion

Figure 7-24a shows the heterodyne method of frequency up-conversion. With the heterodyne method, a relatively low-frequency, angle-modulated carrier along with its side frequencies are applied to one input of a balanced modulator. The second input is a relatively high-frequency, unmodulated RF carrier signal. In the balanced modulator, the two inputs mix nonlinearly, producing sum and difference frequencies at its output. Sum and difference frequencies are also produced between the side frequencies of the modulated signal and the RF carrier. The bandpass filter (BPF) is tuned to the sum frequency with a passband wide enough to pass the carrier plus the upper and lower side frequencies. Thus, the BPF passes the sum of the modulated and the unmodulated carriers while the difference frequencies are blocked. The output from the bandpass filter is

$$f_{c(\text{out})} = f_{c(\text{in})} + f_{\text{RF}} \tag{7-63}$$

where $f_{c(\text{out})}$ = up-converted modulated signal
$f_{c(\text{in})}$ = input modulated signal
f_{RF} = RF carrier

Since the side frequencies of the modulated carrier are unaffected by the heterodyning process, frequency deviation is also unaffected. Thus, the output of the BPF contains the original frequency deviation (both its magnitude Δf and rate of change f_m). The modulation index, phase deviation, and bandwidth are also unaffected by the heterodyne process. Therefore,

$$\Delta f_{(\text{out})} = \Delta f_{(\text{in})}$$

$$m_{(\text{out})} = m_{(\text{in})}$$

$$\Delta\theta_{(\text{out})} = \Delta\theta_{(\text{in})}$$

$$B_{(\text{out})} = B_{(\text{in})}$$

$$f_{m(\text{out})} = f_{m(\text{in})}$$

If the bandpass filter at the output of the balanced modulator in Figure 7-24a were tuned to the difference frequency, frequency down-conversion would occur. As with frequency up-conversion, only the carrier frequency is affected. The modulation index, frequency deviation, phase deviation, bandwidth, and rate of change would be unchanged.

In essence, with the heterodyne method of up-conversion, a low-frequency modulated carrier can be either up- or down-converted to a different location in the frequency spectrum without changing its modulation properties.

7-18-2 Multiplication Method of Up-Conversion

Figure 7-24b shows the multiplication method of frequency up-conversion. With the multiplication method of frequency up-conversion, the modulation properties of a carrier can be increased at the same time that the carrier frequency is up-converted. With the multiplication method, the frequency of the modulated carrier is multiplied by a factor of N in the frequency multiplier. In addition, the frequency deviation is also multiplied. The rate of deviation, however, is unaffected by the multiplication process. Therefore, the output carrier frequency, frequency deviation, modulation index, and phase deviation at the output of the frequency multiplier are

$$f_{c(\text{out})} = Nf_{c(\text{in})}$$

$$\Delta f_{(\text{out})} = \text{N}\Delta f_{(\text{in})}$$

$$m_{(\text{out})} = \text{N}m_{(\text{in})}$$

$$\Delta\theta_{(\text{out})} = \text{N}\Delta\theta_{(\text{in})}$$

$$f_{m(\text{out})} = f_{m(\text{in})}$$

N equals the multiplication factor.

Since the frequency deviation and modulation index are multiplied in the frequency multiplier, the number of side frequencies also increased, producing a corresponding increase in bandwidth. For modulation indices greater than 10 (i.e., high index modulation), Carson's rule can be applied and the bandwidth at the output of the multiplier can be approximated as

$$B_{(\text{out})} = \text{N}(2\Delta f)$$

$$= \text{N}B_{(\text{in})} \qquad (7\text{-}64)$$

It is important to note, however, that the frequency deviation occurs at the modulating-signal frequency (f_m), which remains unchanged by the multiplication process. Therefore, the separation between adjacent side frequencies remains unchanged (i.e., $\pm f_m$, $\pm 2f_m$, $\pm 3f_m$, and so on).

In subsequent sections of this book, you will see that sometimes it is advantageous to use the heterodyne method of frequency up-conversion in FM and PM transmitters, sometimes it is advantageous to use the multiplication method, and sometimes both techniques are used in the same transmitter.

Example 7-9

For the carrier frequency and modulation properties listed, determine the carrier frequency and modulation properties at the output of

a. A balanced modulator with a bandpass filter tuned to the sum frequencies and an RF carrier input frequency of 99.5 MHz.

b. A times 10 frequency multiplier.

$$\Delta f = 3 \text{ kHz}, f_m = 10 \text{ kHz}, m = 0.3, \text{ and } f_c = 500 \text{ kHz}$$

Solution **a.** The output carrier frequency is simply the sum of the input modulated carrier and the RF carrier:

$$f_{c(\text{out})} = 0.5 \text{ MHz} + 99.5 \text{ MHz}$$
$$= 100 \text{ MHz}$$

The output modulation properties are identical to the input modulation properties:

$$\Delta f(\text{in}) = 3 \text{ kHz}, f_{m(\text{in})} = 10 \text{ kHz}, m = 0.3, \text{ and } f_{c(\text{in})} = 500 \text{ kHz}$$

b. The output carrier frequency, frequency deviation, modulation index, and modulating frequency are simply

$$f_{c(\text{out})} = 10(500 \text{ kHz}) = 5 \text{ MHz}$$
$$\Delta f_{(\text{out})} = 10(3 \text{ kHz}) = 30 \text{ kHz}$$
$$m_{(\text{out})} = 10(0.3) = 3$$
$$f_{m(\text{out})} = 10 \text{ kHz}$$

7-19 DIRECT FM TRANSMITTERS

Direct FM transmitters produce an output waveform in which the frequency deviation is directly proportional to the modulating signal. Consequently, the carrier oscillator must be deviated directly. Therefore, for medium- and high-index FM systems, the oscillator cannot be a crystal because the frequency at which a crystal oscillates cannot be significantly varied. As a result, the stability of the oscillators in direct FM transmitters often cannot meet FCC specifications. To overcome this problem, *automatic frequency control* (AFC) is used. An AFC circuit compares the frequency of the noncrystal carrier oscillator to a crystal reference oscillator and then produces a correction voltage proportional to the difference between the two frequencies. The correction voltage is fed back to the carrier oscillator to automatically compensate for any drift that may have occurred.

7-19-1 Crosby Direct FM Transmitter

Figure 7-25 shows the block diagram for a commercial broadcast-band transmitter. This particular configuration is called a *Crosby direct FM transmitter* and includes an *AFC loop*. The frequency modulator can be either a reactance modulator or a voltage-controlled oscillator. The carrier rest frequency is the unmodulated output frequency from the master oscillator (f_c). For the transmitter shown in Figure 7-25, the center frequency of the master oscillator $f_c = 5.1$ MHz, which is multiplied by 18 in three steps ($3 \times 2 \times 3$) to produce a final transmit carrier frequency, $f_t = 91.8$ MHz. At this time, three aspects of frequency conversion should be noted. First, when the frequency of a frequency-modulated carrier is multiplied, its frequency and phase deviations are multiplied as well. Second, the rate at which the carrier is deviated (i.e., the modulating signal frequency, f_m) is unaffected by the multiplication process. Therefore, the modulation index is also multiplied. Third, when an angle-modulated carrier is heterodyned with another frequency in a nonlinear mixer, the carrier can be either up- or down-converted, depending on the passband of the output filter. However, the frequency deviation, phase deviation, and rate of change are unaffected by the heterodyning process. Therefore, for the transmitter shown in Figure 7-25, the frequency and phase deviations at the output of the modulator are

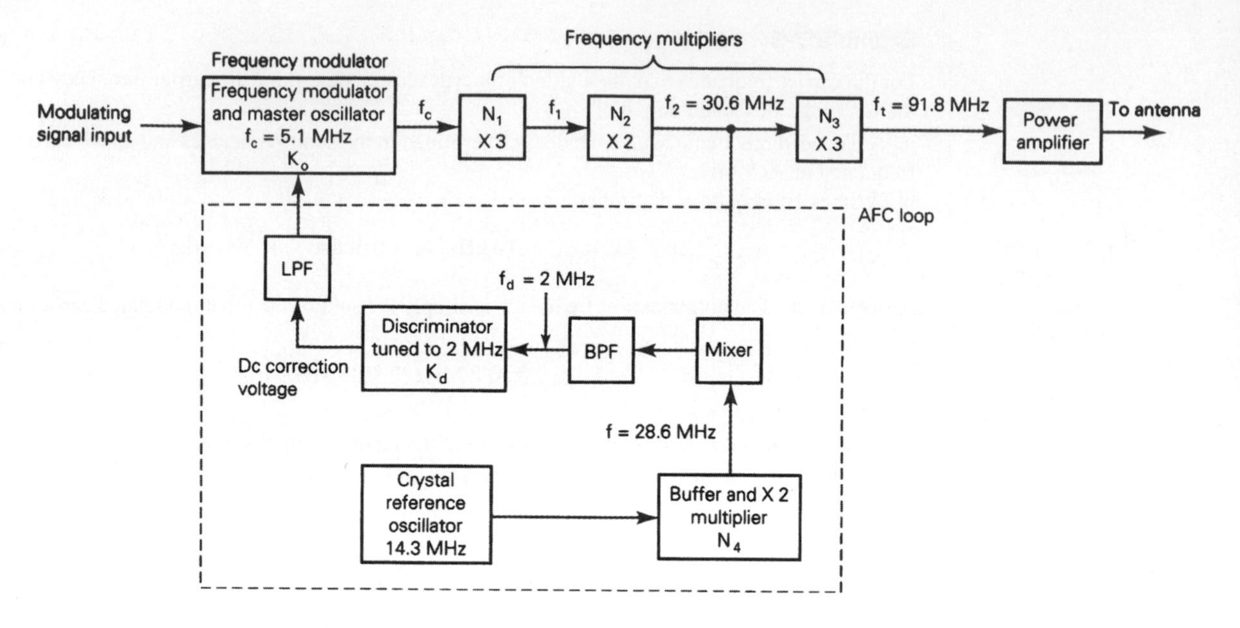

FIGURE 7-25 Crosby direct FM transmitter

also multiplied by 18. To achieve the maximum frequency deviation allowed FM broadcast-band stations at the antenna (75 kHz), the deviation at the output of the modulator must be

$$\Delta f = \frac{75\text{ kHz}}{18} = 4166.7\text{ Hz}$$

and the modulation index must be

$$m = \frac{4166.7\text{ Hz}}{f_m}$$

For the maximum modulating-signal frequency allowed, $f_m = 15$ kHz,

$$m = \frac{4166.7\text{ Hz}}{15{,}000\text{ Hz}} = 0.2778$$

Thus, the modulation index at the antenna is

$$m = 0.2778(18) = 5$$

which is the deviation ratio for commercial FM broadcast transmitters with a 15-kHz modulating signal.

7-19-1-1 AFC loop. The purpose of the *AFC loop* is to achieve near-crystal stability of the transmit carrier frequency without using a crystal in the carrier oscillator. With AFC, the carrier signal is mixed with the output signal from a crystal reference oscillator in a nonlinear device, down-converted in frequency, and then fed back to the input of a *frequency discriminator.* A frequency discriminator is a frequency-selective device whose output voltage is proportional to the difference between the input frequency and its resonant frequency (discriminator operation is explained in Chapter 8). For the transmitter shown in Figure 7-25, the output from the doubler $f_2 = 30.6$ MHz, which is mixed with a crystal-controlled reference frequency $f_r = 28.6$ MHz to produce a difference frequency $f_d = 2$ MHz. The discriminator is a relatively high-Q (narrowband) tuned circuit that reacts only to frequencies near its center frequency (2 MHz in this case). Therefore, the discriminator responds to long-term, low-frequency changes in the carrier center frequency because of master oscillator frequency drift and because low-pass filtering does not respond to the frequency deviation produced by the modulating signal. If the discriminator responded to the frequency deviation, the feedback loop would cancel the deviation and, thus, remove the modulation from the FM wave (this effect is called *wipe off*). The dc correction voltage is added to the modulating signal to automatically adjust the master oscillator's center frequency to compensate for the low-frequency drift.

Example 7-10

Use the transmitter model shown in Figure 7-25 to answer the following questions. For a total frequency multiplication of 20 and a transmit carrier frequency f_t = 88.8 MHz, determine

a. Master oscillator center frequency.
b. Frequency deviation at the output of the modulator for a frequency deviation of 75 kHz at the antenna.
c. Deviation ratio at the output of the modulator for a maximum modulating-signal frequency f_m = 15 kHz.
d. Deviation ratio at the antenna.

Solution **a.**

$$f_c = \frac{f_t}{N_1N_2N_3} = \frac{88.8 \text{ MHz}}{20} = 4.43 \text{ MHz}$$

b.

$$\Delta f = \frac{\Delta f_t}{N_1N_2N_3} = \frac{75 \text{ kHz}}{20} = 3750 \text{ Hz}$$

c.

$$\text{DR} = \frac{\Delta f_{(\max)}}{f_{m(\max)}} = \frac{3750 \text{ Hz}}{15 \text{ kHz}} = 0.25$$

d.

$$\text{DR} = 0.25 \times 20 = 5$$

7-19-1-2 Automatic frequency control. Because the Crosby transmitter uses either a VCO, a reactance oscillator, or a linear integrated-circuit oscillator to generate the carrier frequency, it is more susceptible to frequency drift due to temperature change, power supply fluctuations, and so on than if it were a crystal oscillator. As stated in Chapter 3, the stability of an oscillator is often given in parts per million (ppm) per degree Celsius. For example, for the transmitter shown in Figure 7-25, an oscillator stability of ±40 ppm could produce ±204 Hz (5.1 MHz × ±40 Hz/million) of frequency drift per degree Celsius at the output of the master oscillator. This would correspond to a ±3672-Hz drift at the antenna (18 × ±204), which far exceeds the ±2-kHz maximum set by the FCC for commercial FM broadcasting. Although an AFC circuit does not totally eliminate frequency drift, it can substantially reduce it. Assuming a rock-stable crystal reference oscillator and a perfectly tuned discriminator, the frequency drift at the output of the second multiplier without feedback (i.e., open loop) is

$$\text{open-loop drift} = df_{o1} = N_1N_2df_c \tag{7-65}$$

where d denotes drift. The closed-loop drift is

$$\text{closed-loop drift} = df_{c1} = df_{o1} - N_1N_2k_dk_odf_{c1} \tag{7-66}$$

Therefore,

$$df_{c1} + N_1N_2k_dk_odf_{c1} = df_{o1}$$

and

$$df_{c1}(1 + N_1N_2k_dk_o) = df_{o1}$$

Thus,

$$df_{c1} = \frac{df_{o1}}{1 + N_1N_2k_dk_o} \tag{7-67}$$

where k_d = discriminator transfer function (volts per hertz)
k_o = master oscillator transfer function (hertz per volt)

From Equation 7-67, it can be seen that the frequency drift at the output of the second multiplier and, consequently, at the input to the discriminator is reduced by a factor of $1 + N_1N_2k_dk_o$ when the AFC loop is closed. The carrier frequency drift is multiplied by the AFC loop gain and fed back to the master oscillator as a correction voltage. The total frequency error cannot be canceled because then there would be no error voltage at the output of the discriminator to feed back to the master oscillator. In addition, Equations 7-65 to 7-67 were derived assuming that the discriminator and crystal reference oscillator were perfectly stable, which of course they are not.

Example 7-11

Use the transmitter block diagram and values given in Figure 7-25 to answer the following questions. Determine the reduction in frequency drift at the antenna for a transmitter without AFC compared to a transmitter with AFC. Use a VCO stability = +200 ppm, k_o = 10 kHz/V, and k_d = 2 V/kHz.

Solution With the feedback loop open, the master oscillator output frequency is

$$f_c = 5.1\text{ MHz} + (200\text{ ppm} \times 5.1\text{ MHz}) = 5{,}101{,}020\text{ Hz}$$

and the frequency at the output of the second multiplier is

$$f_2 = N_1N_2f_c = (5{,}101{,}020)(6) = 30{,}606{,}120\text{ Hz}$$

Thus, the frequency drift is

$$df_2 = 30{,}606{,}120 - 30{,}600{,}000 = 6120\text{ Hz}$$

Therefore, the antenna transmit frequency is

$$f_t = 30{,}606{,}120(3) = 91.81836\text{ MHz}$$

which is 18.36 kHz above the assigned frequency and well out of limits.

With the feedback loop closed, the frequency drift at the output of the second multiplier is reduced by a factor of $1 + N_1N_2k_dk_o$ or

$$1 + \frac{(2)(3)(10\text{ kHz})}{\text{V}}\,\frac{2\text{V}}{\text{kHz}} = 121$$

Therefore,

$$df_2 = \frac{6120}{121} = 51\text{ Hz}$$

Thus,

$$f_2 = 30{,}600{,}051\text{ Hz}$$

and the antenna transmit frequency is

$$f_t = 30{,}600{,}051 \times 3 = 91{,}800{,}153\text{ Hz}$$

The frequency drift at the antenna has been reduced from 18,360 Hz to 153 Hz, which is now well within the ±2-kHz FCC requirements.

The preceding discussion and Example 7-11 assumed a perfectly stable crystal reference oscillator and a perfectly tuned discriminator. In actuality, both the discriminator and the reference oscillator are subject to drift, and the worst-case situation is when they both drift in the same direction as the master oscillator. The drift characteristics for a typical discriminator are on the order of ±100 ppm. Perhaps now it can be seen why the output frequency from the second multiplier was mixed down to a relatively low frequency prior to being fed to the discriminator. For a discriminator tuned to 2 MHz with a stability of ±100 ppm, the maximum discriminator drift is

$$df_d = \pm 100\text{ ppm} \times 2\text{ MHz} = \pm 200\text{ Hz}$$

If the 30.6-MHz signal were fed directly into the discriminator, the maximum drift would be

$$df_d = \pm 100\text{ ppm} \times 30.6\text{ MHz} = \pm 3060\text{ Hz}$$

Frequency drift due to discriminator instability is multiplied by the AFC open-loop gain. Therefore, the change in the second multiplier output frequency due to discriminator drift is

$$df_2 = df_d N_1 N_2 k_d k_o \tag{7-68}$$

Similarly, the crystal reference oscillator can drift and also contribute to the total frequency drift at the output of the second multiplier. The drift due to crystal instability is multiplied by 2 before entering the nonlinear mixer; therefore,

$$df_2 = N_4 df_o N_1 N_2 k_d k_o \tag{7-69}$$

and the maximum open-loop frequency drift at the output of the second multiplier is

$$df_{(\text{total})} = N_1 N_2 (df_c + k_o k_d f_d + k_o k_d N_4 df_o) \tag{7-70}$$

7-19-2 Phase-Locked-Loop Direct FM Transmitter

Figure 7-26 shows a *wideband* FM transmitter that uses a phase-locked loop (PLL) to achieve crystal stability from a VCO master oscillator and, at the same time, generate a high-index, wideband FM output signal. The VCO output frequency is divided by *N* and fed back to the PLL phase comparator, where it is compared to a stable crystal reference frequency. The phase comparator generates a correction voltage that is proportional to the difference between the two frequencies. The correction voltage is added to the modulating signal and applied to the VCO input. The correction voltage adjusts the VCO center frequency to its proper value. Again, the low-pass filter prevents changes in the VCO output frequency due to the modulating signal from being converted to a voltage, fed back to the VCO, and wiping out the modulation. The low-pass filter also prevents the loop from locking onto a side frequency.

7-19-2-1 PM from FM. An FM modulator preceded by a differentiator generates a PM waveform. If the transmitters shown in Figures 7-25 and 7-26 are preceded by a preemphasis network, which is a differentiator (high-pass filter), an interesting situation occurs. For a 75-μs time constant, the amplitude of frequencies above 2.12 kHz is emphasized through differentiation. Therefore, for modulating frequencies below 2.12 kHz, the output

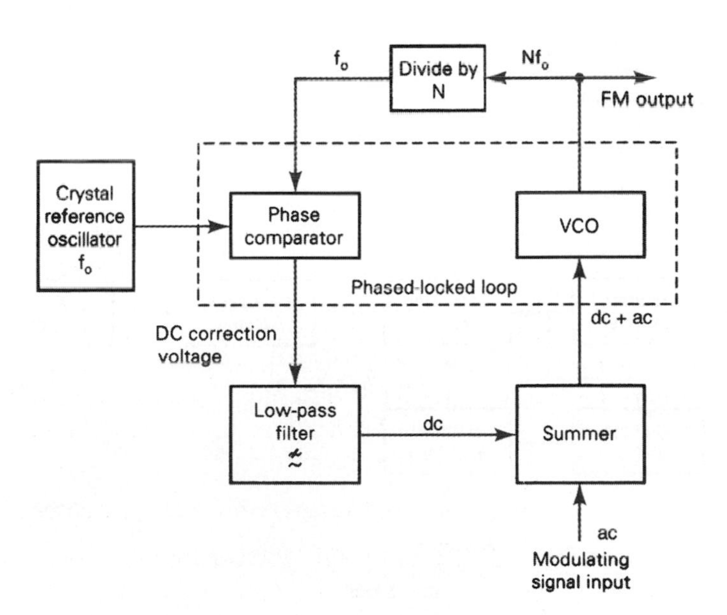

FIGURE 7-26 Phase-locked-loop FM transmitter

waveform is proportional to the modulating signal, and for frequencies above 2.12 kHz, the output waveform is proportional to the derivative of the input signal. In other words, frequency modulation occurs for frequencies below 2.12 kHz, and phase modulation occurs for frequencies above 2.12 kHz. Because the gain of a differentiator increases with frequency above the break frequency (2.12 kHz) and because the frequency deviation is proportional to the modulating-signal amplitude, the frequency deviation also increases with frequencies above 2.12 kHz. From Equation 7-23, it can be seen that if Δf and f_m increase proportionately, the modulation index remains constant, which is a characteristic of phase modulation.

7-20 INDIRECT FM TRANSMITTERS

Indirect FM transmitters produce an output waveform in which the phase deviation is directly proportional to the modulating signal. Consequently, the carrier oscillator is not directly deviated. Therefore, the carrier oscillator can be a crystal because the oscillator itself is not the modulator. As a result, the stability of the oscillators with indirect FM transmitters can meet FCC specifications without using an AFC circuit.

7-20-1 Armstrong Indirect FM Transmitter

With indirect FM, the modulating signal directly deviates the phase of the carrier, which indirectly changes the frequency. Figure 7-27 shows the block diagram for a wideband *Armstrong*

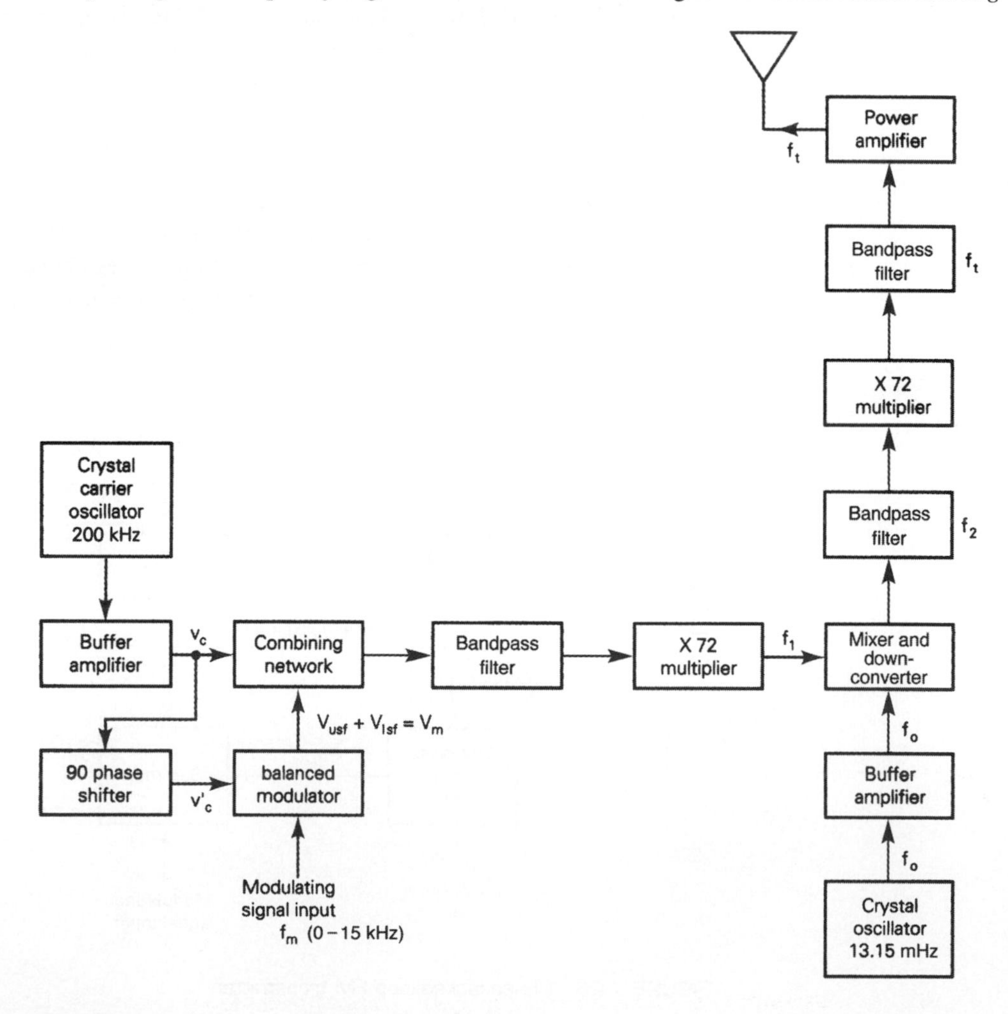

FIGURE 7-27 Armstrong indirect FM transmitter

indirect FM transmitter. The carrier source is a crystal; therefore, the stability requirements for the carrier frequency set by the FCC can be achieved without using an AFC loop.

With an Armstrong transmitter, a relatively low-frequency subcarrier (f_c) is phase shifted 90° (f_c) and fed to a balanced modulator, where it is mixed with the input modulating signal (f_m). The output from the balanced modulator is a double-sideband, suppressed-carrier wave that is combined with the original carrier in a combining network to produce a low-index, phase-modulated waveform. Figure 7-28a shows the phasor for the original carrier (V_c), and Figure 7-28b shows the phasors for the side frequency components of the suppressed-carrier wave (V_{usf} and V_{lsf}). Because the suppressed-carrier voltage (V_c') is 90° out of phase with V_c, the upper and lower sidebands combine to produce a component (V_m) that is always in quadrature (at right angles) with V_c. Figures 7-28a through f show the progressive phasor addition of V_c, V_{usf}, and V_{lsf}. It can be seen that the output from the combining network is a signal whose phase is varied at a rate equal to f_m and whose magnitude is directly proportional to the magnitude of V_m. From Figure 7-28, it can be seen that the peak phase deviation (modulation index) can be calculated as follows:

$$\theta = m = \arctan \frac{V_m}{V_c} \tag{7-71}$$

For very small angles, the tangent of the angle is approximately equal to the angle; therefore,

$$\theta = m = \frac{V_m}{V_c} \tag{7-72}$$

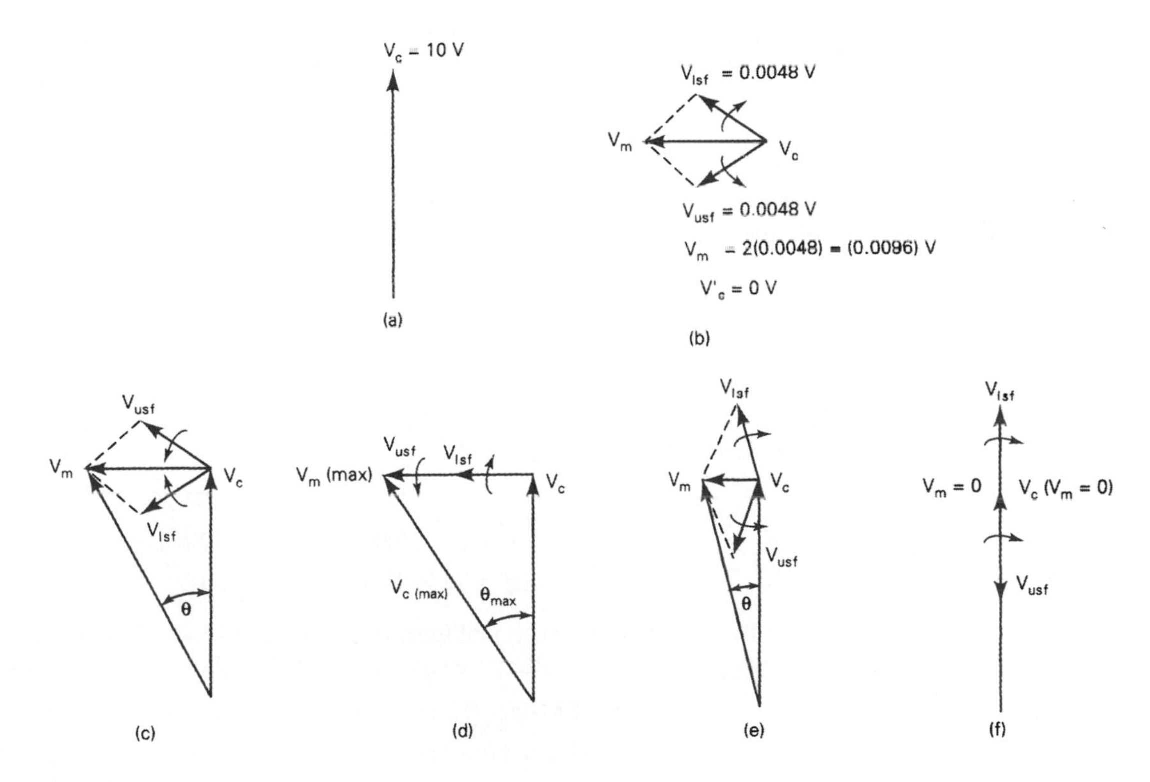

FIGURE 7-28 Phasor addition of V_c, V_{usf}, and V_{lsf}: (a) carrier phasor; (b) sideband phasors; (c)–(f) progressive phasor addition. Part (d) shows the peak phase shift.

Example 7-12

For the Armstrong transmitter shown in Figure 7-27 and the phase-shifted carrier (V'_c), upper side frequency (V_{usf}), and lower side frequency (V_{lsf}) components shown in Figure 7-28, determine

a. Peak carrier phase shift in both radians and degrees.
b. Frequency deviation for a modulating-signal frequency $f_m = 15$ kHz.

Solution a. The peak amplitude of the modulating component is

$$V_m = V_{usf} + V_{lsf}$$
$$= 0.0048 + 0.0048 = 0.0096$$

Peak phase deviation is the modulation index and can be determined by substituting into Equation 7-71:

$$\theta = m = \arctan\frac{0.0096}{10} = 0.055°$$

$$= 0.055° \times \frac{\pi \text{ rad}}{180°} = 0.00096 \text{ rad}$$

b. Rearranging Equation 7-22 gives us

$$\Delta f = mf_m = (0.00096)(15 \text{ kHz}) = 14.4 \text{ Hz}$$

From the phasor diagrams shown in Figure 7-28, it can be seen that the carrier amplitude is varied, which produces unwanted amplitude modulation in the output waveform, and $V_{c(\max)}$ occurs when V_{usf} and V_{lsf} are in phase with each other and with V_c. The maximum phase deviation that can be produced with this type of modulator is approximately 1.67 milliradians. Therefore, from Equation 7-22 and a maximum modulating-signal frequency $f_{m(\max)} = 15$ kHz, the maximum frequency deviation possible is

$$\Delta f_{\max} = (0.00167)(15{,}000) = 25 \text{ Hz}$$

From the preceding discussion, it is evident that the modulation index at the output of the combining network is insufficient to produce a wideband FM frequency spectrum and, therefore, must be multiplied considerably before being transmitted. For the transmitter shown in Figure 7-27, a 200-kHz phase-modulated subcarrier with a peak phase deviation $m = 0.00096$ rad produces a frequency deviation of only 14.4 Hz at the output of the combining network. To achieve 75-kHz frequency deviation at the antenna, the frequency must be multiplied by approximately 5208. However, this would produce a transmit carrier frequency at the antenna of

$$f_t = 5208 \times 200 \text{ kHz} = 1041.6 \text{ MHz}$$

which is well beyond the frequency limits for the commercial FM broadcast band. It is apparent that multiplication by itself is inadequate. Therefore, a combination of multiplying and mixing is necessary to develop the desired transmit carrier frequency with 75-kHz frequency deviation. The waveform at the output of the combining network is multiplied by 72, producing the following signal:

$$f_1 = 72 \times 200 \text{ kHz} = 14.4 \text{ MHz}$$
$$m = 72 \times 0.00096 = 0.06912 \text{ rad}$$
$$\Delta f = 72 \times 14.4 \text{ Hz} = 1036.8 \text{ Hz}$$

The output from the first multiplier is mixed with a 13.15-MHz crystal-controlled frequency (f_o) to produce a difference signal (f_2) with the following characteristics:

$$f_2 = 14.4 \text{ MHz} - 13.15 \text{ MHz} = 1.25 \text{ MHz (down-converted)}$$
$$m = 0.6912 \text{ rad (unchanged)}$$
$$\Delta f = 1036.8 \text{ Hz (unchanged)}$$

Note that only the carrier frequency is affected by the heterodyning process. The output from the mixer is once again multiplied by 72 to produce a transmit signal with the following characteristics:

$$f_t = 1.25 \text{ MHz} \times 72 = 90 \text{ MHz}$$
$$m = 0.06912 \times 72 = 4.98 \text{ rad}$$
$$\Delta f = 1036.8 \times 72 = 74{,}650 \text{ Hz}$$

In the preceding example with the use of both the multiplying and heterodyning processes, the carrier was increased by a factor of 450; at the same time, the frequency deviation and modulation index were increased by a factor of 5184.

With the Armstrong transmitter, the phase of the carrier is directly modulated in the combining network through summation, producing indirect frequency modulation. The magnitude of the phase deviation is directly proportional to the amplitude of the modulating signal but independent of its frequency. Therefore, the modulation index remains constant for all modulating-signal frequencies of a given amplitude. For example, for the transmitter shown in Figure 7-28, if the modulating-signal amplitude is held constant while its frequency is decreased to 5 kHz, the modulation index remains at 5, while the frequency deviation is reduced to $\Delta f = 5 \times 5000 = 25{,}000$ Hz.

7-20-2 FM from PM

A PM modulator preceded by an integrator produces an FM waveform. If the PM transmitter shown in Figure 7-27 is preceded by a low-pass filter (which is an integrator), FM results. The low-pass filter is simply a $1/f$ filter, which is commonly called a *predistorter* or *frequency correction network*.

7-20-3 FM versus PM

From a purely theoretical viewpoint, the difference between FM and PM is quite simple: The modulation index for FM is defined differently than for PM. With PM, the modulation index is directly proportional to the amplitude of the modulating signal and independent of its frequency. With FM, the modulation index is directly proportional to the amplitude of the modulating signal and inversely proportional to its frequency.

Considering FM as a form of phase modulation, the larger the frequency deviation, the larger the phase deviation. Therefore, the latter depends, at least to a certain extent, on the amplitude of the modulating signal, just as with PM. With PM, the modulation index is proportional to the amplitude of the modulating signal voltage only, whereas with FM, the modulation index is also inversely proportional to the modulating-signal frequency. If FM transmissions are received on a PM receiver, the bass frequencies would have considerably more phase deviation than a PM modulator would have given them. Because the output voltage from a PM demodulator is proportional to the phase deviation, the signal appears excessively bass-boosted. Alternatively (and this is the more practical situation), PM demodulated by an FM receiver produces an information signal in which the higher-frequency modulating signals are boosted.

7-21 ANGLE MODULATION VERSUS AMPLITUDE MODULATION

7-21-1 Advantages of Angle Modulation

Angle modulation has several inherent advantages over amplitude modulation.

7-21-1-1 Noise immunity. Probably the most significant advantage of angle modulation transmission (FM and PM) over amplitude modulation transmission is noise immunity. Most noise (including man-made noise) results in unwanted amplitude variations in the modulated wave (i.e., AM noise). FM and PM receivers include limiters that remove most of the AM noise from the received signal before the final demodulation process occurs—a process that cannot be used with AM receivers because the information is also contained in amplitude variations, and removing the noise would also remove the information.

7-21-1-2 Noise performance and signal-to-noise improvement. With the use of limiters, FM and PM demodulators can actually reduce the noise level and improve the signal-to-noise ratio during the demodulation process (a topic covered in more detail in Chapter 8). This is called *FM thresholding.* With AM, once the noise has contaminated the signal, it cannot be removed.

7-21-1-3 Capture effect. With FM and PM, a phenomenon known as the *capture effect* allows a receiver to differentiate between two signals received with the same frequency. Providing one signal at least twice as high in amplitude as the other, the receiver will capture the stronger signal and eliminate the weaker signal. With amplitude modulation, if two or more signals are received with the same frequency, both will be demodulated and produce audio signals. One may be larger in amplitude than the other, but both can be heard.

7-21-1-4 Power utilization and efficiency. With AM transmission (especially DSBFC), most of the transmitted power is contained in the carrier while the information is contained in the much lower-power sidebands. With angle modulation, the total power remains constant regardless if modulation is present. With AM, the carrier power remains constant with modulation, and the sideband power simply adds to the carrier power. With angle modulation, power is taken from the carrier with modulation and redistributed in the sidebands; thus, you might say, angle modulation puts most of its power in the information.

7-21-2 Disadvantages of Angle Modulation

Angle modulation also has several inherent disadvantages over amplitude modulation.

7-21-2-1 Bandwidth. High-quality angle modulation produces many side frequencies, thus necessitating a much wider bandwidth than is necessary for AM transmission. Narrowband FM utilizes a low modulation index and, consequently, produces only one set of sidebands. Those sidebands, however, contain an even more disproportionate percentage of the total power than a comparable AM system. For high-quality transmission, FM and PM require much more bandwidth than AM. Each station in the commercial AM radio band is assigned 10 kHz of bandwidth, whereas in the commercial FM broadcast band, 200 kHz is assigned each station.

7-21-2-2 Circuit complexity and cost. PM and FM modulators, demodulators, transmitters, and receivers are more complex to design and build than their AM counterparts. At one time, more complex meant more expensive. Today, however, with the advent of inexpensive, large-scale integration ICs, the cost of manufacturing FM and PM circuits is comparable to their AM counterparts.

QUESTIONS

7-1. Define *angle modulation.*

7-2. Define *direct FM* and *indirect FM.*

7-3. Define *direct PM* and *indirect PM.*

7-4. Define *frequency deviation* and *phase deviation.*

7-5. Define *instantaneous phase, instantaneous phase deviation, instantaneous frequency,* and *instantaneous frequency deviation.*

7-6. Define *deviation sensitivity* for a frequency modulator and for a phase modulator.

7-7. Describe the relationship between the instantaneous carrier frequency and the modulating signal for FM.

7-8. Describe the relationship between the instantaneous carrier phase and the modulating signal for PM.

7-9. Describe the relationship between frequency deviation and the amplitude and frequency of the modulating signal.

7-10. Define *carrier swing.*

7-11. Define *modulation index* for FM and for PM.

7-12. Describe the relationship between modulation index and the modulating signal for FM; for PM.

7-13. Define *percent modulation* for angle-modulated signals.

7-14. Describe the difference between a direct frequency modulator and a direct phase modulator.

7-15. How can a frequency modulator be converted to a phase modulator; a phase modulator to a frequency modulator?

7-16. How many sets of sidebands are produced when a carrier is frequency modulated by a single input frequency?

7-17. What are the requirements for a side frequency to be considered significant?

7-18. Define a *low,* a *medium,* and a *high* modulation index.

7-19. Describe the significance of the *Bessel* table.

7-20. State *Carson's general rule* for determining the bandwidth for an angle-modulated wave.

7-21. Define *deviation ratio.*

7-22. Describe the relationship between the power in the unmodulated carrier and the power in the modulated wave for FM.

7-23. Describe the significance of the FM *noise triangle.*

7-24. What effect does *limiting* have on the composite FM waveform?

7-25. Define *preemphasis* and *deemphasis.*

7-26. Describe a preemphasis network; a deemphasis network.

7-27. Describe the basic operation of a varactor diode FM generator.

7-28. Describe the basic operation of a reactance FM modulator.

7-29. Describe the basic operation of a linear integrated-circuit FM modulator.

7-30. Draw the block diagram for a Crosby direct FM transmitter and describe its operation.

7-31. What is the purpose of an AFC loop? Why is one required for the Crosby transmitter?

7-32. Draw the block diagram for a phase-locked-loop FM transmitter and describe its operation.

7-33. Draw the block diagram for an Armstrong indirect FM transmitter and describe its operation.

7-34. Compare FM to PM.

PROBLEMS

7-1. If a frequency modulator produces 5 kHz of frequency deviation for a 10-V modulating signal, determine the deviation sensitivity. How much frequency deviation is produced for a 2-V modulating signal?

7-2. If a phase modulator produces 2 rad of phase deviation for a 5-V modulating signal, determine the deviation sensitivity. How much phase deviation would a 2-V modulating signal produce?

7-3. Determine (a) the peak frequency deviation, (b) the carrier swing, and (c) the modulation index for an FM modulator with deviation sensitivity $K_1 = 4$ kHz/V and a modulating signal $v_m(t) = 10\sin(2\pi 2000t)$. What is the peak frequency deviation produced if the modulating signal were to double in amplitude?

7-4. Determine the peak phase deviation for a PM modulator with deviation sensitivity $K = 1.5$ rad/V and a modulating signal $v_m(t) = 2\sin(2\pi 2000t)$. How much phase deviation is produced for a modulating signal with twice the amplitude?

7-5. Determine the percent modulation for a television broadcast station with a maximum frequency deviation $\Delta f = 50$ kHz when the modulating signal produces 40 kHz of frequency deviation at the antenna. How much deviation is required to reach 100% modulation of the carrier?

7-6. From the Bessel table, determine the number of sets of sidebands produced for the following modulation indices: 0.5, 1.0, 2.0, 5.0, and 10.0.

7-7. For an FM modulator with modulation index $m = 2$, modulating signal $v_m(t) = V_m \sin(2\pi 2000t)$, and an unmodulated carrier $v_c(t) = 8 \sin(2\pi 800kt)$,

a. Determine the number of sets of significant sidebands.
b. Determine their amplitudes.
c. Draw the frequency spectrum showing the relative amplitudes of the side frequencies.
d. Determine the bandwidth.
e. Determine the bandwidth if the amplitude of the modulating signal increases by a factor of 2.5.

7-8. For an FM transmitter with 60-kHz carrier swing, determine the frequency deviation. If the amplitude of the modulating signal decreases by a factor of 2, determine the new frequency deviation.

7-9. For a given input signal, an FM broadcast-band transmitter has a frequency deviation $\Delta f = 20$ kHz. Determine the frequency deviation if the amplitude of the modulating signal increases by a factor of 2.5.

7-10. An FM transmitter has a rest frequency $f_c = 96$ MHz and a deviation sensitivity $K_1 = 4$ kHz/V. Determine the frequency deviation for a modulating signal $v_m(t) = 8 \sin(2\pi 2000t)$. Determine the modulation index.

7-11. Determine the deviation ratio and worst-case bandwidth for an FM signal with a maximum frequency deviation $\Delta f = 25$ kHz and a maximum modulating signal $f_{m(\max)} = 12.5$ kHz.

7-12. For an FM modulator with 40-kHz frequency deviation and a modulating-signal frequency $f_m = 10$ kHz, determine the bandwidth using both the Bessel table and Carson's rule.

7-13. For an FM modulator with an unmodulated carrier amplitude $V_c = 20$ V, a modulation index $m = 1$, and a load resistance $R_L = 10\ \Omega$, determine the power in the modulated carrier and each side frequency, and sketch the power spectrum for the modulated wave.

7-14. For an angle-modulated carrier $v_c(t) = 2 \cos(2\pi 200 \text{ MHz } t)$ with 50 kHz of frequency deviation due to the modulating signal and a single-frequency interfering signal $V_n(t) = 0.5 \cos(2\pi 200.01 \text{ MHz } t)$, determine

a. Frequency of the demodulated interference signal.
b. Peak phase and frequency deviation due to the interfering signal.
c. Signal-to-noise ratio at the output of the demodulator.

7-15. Determine the total peak phase deviation produced by a 5-kHz band of random noise with a peak voltage $V_n = 0.08$ V and a carrier $v_c(t) = 1.5 \sin(2\pi 40 \text{ MHz } t)$.

7-16. For a Crosby direct FM transmitter similar to the one shown in Figure 7-25 with the following parameters, determine

a. Frequency deviation at the output of the VCO and the power amplifier.
b. Modulation index at the same two points.
c. Bandwidth at the output of the power amplifier.

$N_1 = \times 3$

$N_2 = \times 3$

$N_3 = \times 2$

Crystal reference oscillator frequency = 13 MHz

Reference multiplier = $\times 3$

VCO deviation sensitivity $K_1 = 450$ Hz/V

Modulating signal $v_m(t) = 3 \sin(2\pi 5 \times 10^3\ t)$

VCO rest frequency $f_c = 4.5$ MHz

Discriminator resonant frequency $f_d = 1.5$ MHz

7-17. For an Armstrong indirect FM transmitter similar to the one shown in Figure 7-27 with the following parameters, determine

a. Modulation index at the output of the combining network and the power amplifier.
b. Frequency deviation at the same two points.

c. Transmit carrier frequency.
Crystal carrier oscillator = 210 kHz
Crystal reference oscillator = 10.2 MHz
Sideband voltage $V_m = 0.018$ V
Carrier input voltage to combiner $V_c = 5$ V
First multiplier = ×40
Second multiplier = ×50
Modulating-signal frequency $f_m = 2$ kHz

7-18. If a frequency modulator produces 4 kHz of frequency deviation for a 10-V_p modulating signal, determine the deviation sensitivity.

7-19. If a phase modulator produces 1.5 rad of phase deviation for a 5-V_p modulating signal, determine the deviation sensitivity.

7-20. Determine (a) the peak frequency deviation, (b) the carrier swing, and (c) the modulation index for an FM modulator with a deviation sensitivity $K_1 = 3$ kHz/V and a modulating signal $v_m = 6 \sin(2\pi 2000t)$.

7-21. Determine the peak phase deviation for a PM modulator with deviation sensitivity $K = 2$ rad/V and a modulating signal $v_m = 4 \sin(2\pi 1000t)$.

7-22. Determine the percent modulation for a television broadcast station with a maximum frequency deviation $\Delta f = 50$ kHz when the modulating signal produces 30 kHz of frequency deviation.

7-23. From the Bessel table, determine the number of side frequencies produced for the following modulation indices: 0.25, 0.5, 1.0, 2.0, 5.0, and 10.

7-24. For an FM modulator with modulation index $m = 5$, modulating signal $v_m = 2 \sin(2\pi 5kt)$, and an unmodulated carrier frequency $f_c = 400$ kHz, determine

a. Number of sets of significant sidebands.
b. Sideband amplitudes.
Then c. Draw the output frequency spectrum.

7-25. For an FM transmitter with an 80-kHz carrier swing, determine the frequency deviation. If the amplitude of the modulating signal decreases by a factor of 4, determine the new frequency deviation.

7-26. For a given input signal, an FM broadcast transmitter has a frequency deviation $\Delta f = 40$ kHz. Determine the frequency deviation if the amplitude of the modulating signal increases by a factor of 4.3.

7-27. An FM transmitter has a rest frequency $f_c = 94$ MHz and a deviation sensitivity $K_1 = 5$ kHz/V. Determine the frequency deviation for a modulating signal $v_m(t) = 4\ V_p$.

7-28. Determine the deviation ratio and worst-case bandwidth for an FM system with a maximum frequency deviation of 40 kHz and a maximum modulating-signal frequency $f_m = 10$ kHz.

7-29. For an FM modulator with 50 kHz of frequency deviation and a modulating-signal frequency $f_m = 8$ kHz, determine the bandwidth using both the Bessel table and Carson's rule.

7-30. For an FM modulator with an unmodulated carrier voltage $v_c = 12\ V_p$, a modulation index $m = 1$, and a load resistance $R_L = 12\ \Omega$, determine the power in the modulated carrier and each significant side frequency and sketch the power spectrum for the modulated output wave.

7-31. For an angle-modulated carrier $v_c = 4 \cos(2\pi 300 \text{ MHz } t)$ with 75 kHz of frequency deviation due to the modulating signal and a single-frequency interfering signal $v_n = 0.2 \cos(2\pi 300.015 \text{ MHz } t)$, determine

a. Frequency of the demodulated interference signal.
b. Peak and rms phase and frequency deviation due to the interfering signal.
c. S/N ratio at the output of the FM demodulator.

7-32. Determine the total rms phase deviation produced by a 10-kHz band of random noise with a peak voltage $V_n = 0.04$ V and a carrier with a peak voltage $V_c = 4.5\ V_p$.

7-33. For a Crosby direct FM transmitter similar to the one shown in Figure 7-25 with the following parameters, determine

a. Frequency deviation at the output of the VCO and the power amplifier.
b. Modulation index at the output of the VCO and the power amplifier.

c. Bandwidth at the output of the power amplifier.

$N_1 = \times 3$

$N_2 = \times 3$

$N_3 = \times 2$

Crystal reference oscillator frequency = 13 MHz

Reference multiplier = ×3

VCO deviation sensitivity k_1 = 250 Hz/V

Modulating-signal peak amplitude $v_m = 4\ V_p$

Modulating-signal frequency f_m = 10 kHz

VCO rest frequency f_c = 4.3 MHz

Discriminator resonant frequency f_d = 1.5 MHz

7-34. For an Armstrong indirect FM transmitter similar to the one shown in Figure 7-27 with the following parameters, determine

a. Modulation index at the output of the combining network and the power amplifier.

b. Frequency deviation at the same two points.

c. Transmit carrier frequency.

Crystal carrier oscillator = 220 kHz

Crystal reference oscillator = 10.8 MHz

Sideband voltage $V_m = 0.012\ V_p$

CHAPTER 8

Angle Modulation Reception and FM Stereo

CHAPTER OUTLINE

OBJECTIVES

- Give the block diagram for an FM receiver and describe the functions of each block
- Describe the process of FM demodulation
- Describe the operation of the following tuned-circuit FM demodulators: slope detector, balanced slope detector, Foster-Seeley discriminator, and ratio detector
- Explain the operation of a PLL FM demodulator
- Explain the operation of a quadrature FM demodulator
- Define *FM noise suppression*
- Describe the purpose and operation of amplitude limiters
- Define *FM thresholding*
- Describe the purpose of FM limiter circuits
- Define and explain the FM capture effect
- Compare frequency modulation to phase modulation
- Describe the operation of a linear integrated-circuit FM receiver
- Describe FM stereo broadcasting
- Explain FM stereo transmission
- Explain FM stereo reception
- Describe two-way mobile communications services
- Describe the operation of a two-way FM radio communications system

8-1 INTRODUCTION

Receivers used for angle-modulated signals are very similar to those used for conventional AM or SSB reception, except for the method used to extract the audio information from the composite IF waveform. In FM receivers, the voltage at the output of the audio detector is directly proportional to the frequency deviation at its input. With PM receivers, the voltage at the output of the audio detector is directly proportional to the phase deviation at its input. Because frequency and phase modulation both occur with either angle modulation system, FM signals can be demodulated by PM receivers and vice versa. Therefore, the circuits used to demodulate FM and PM signals are both described under the heading "FM Receivers."

With conventional AM, the modulating signal is impressed onto the carrier in the form of amplitude variations. However, noise introduced into the system also produces changes in the amplitude of the envelope. Therefore, the noise cannot be removed from the composite waveform without also removing a portion of the information signal. With angle modulation, the information is impressed onto the carrier in the form of frequency or phase variations. Therefore, with angle modulation receivers, amplitude variations caused by noise can be removed from the composite waveform simply by *limiting* (*clipping*) the peaks of the envelope prior to detection. With angle modulation, an improvement in the signal-to-noise ratio is achieved during the demodulation process; thus, system performance in the presence of noise can be improved by limiting. In essence, this is the major advantage of angle modulation over conventional AM.

The purposes of this chapter are to introduce the reader to the basic receiver configurations and circuits used for the reception and demodulation of FM and PM signals and to describe how they function and how they differ from conventional AM or single-sideband receivers and circuits. In addition, several FM communications systems are described, including FM stereo and two-way FM radio communications.

8-2 FM RECEIVERS

FM receivers, like their AM counterparts, are superheterodyne receivers. Figure 8-1 shows a simplified block diagram for a double-conversion superheterodyne FM receiver. As the figure shows, the FM receiver is similar to the AM receivers discussed in Chapter 5. The preselector, RF amplifier, first and second mixers, IF amplifier, and detector sections of an FM receiver perform almost identical functions as they did in AM receivers: The preselector rejects the image frequency, the RF amplifier establishes the signal-to-noise ratio and noise figure, the mixer/converter section down-converts RF to IF, the IF amplifiers provide most of the gain and selectivity of the receiver, and the detector removes the information from the modulated wave. Except for delayed AGC to prevent mixer saturation when strong RF signals are received, the AGC used with AM receivers is not used for FM receivers because with FM transmission there is no information contained in the amplitude of the received signal. Because of the inherent noise suppression characteristics of FM receivers, RF amplifiers are also often not required with FM receivers.

With FM receivers, a constant amplitude IF signal into the demodulator is desirable. Consequently, FM receivers generally have much more IF gain than AM receivers. In fact, with FM receivers, it is desirable that the final IF amplifier be saturated. The harmonics produced from overdriving the final IF amplifier are high enough that they are substantially reduced with bandpass filters that pass only the minimum bandwidth necessary to preserve the information signals. The final IF amplifier is specially designed for ideal saturation characteristics and is called a *limiter,* or sometimes *passband limiter* if the output is filtered.

The preselector, RF amplifiers, mixer/converter, and IF sections of an FM receiver operate essentially the same as they did in AM receivers; however, the audio detector stage used in FM receivers is quite different. The envelope (peak) detector common to AM

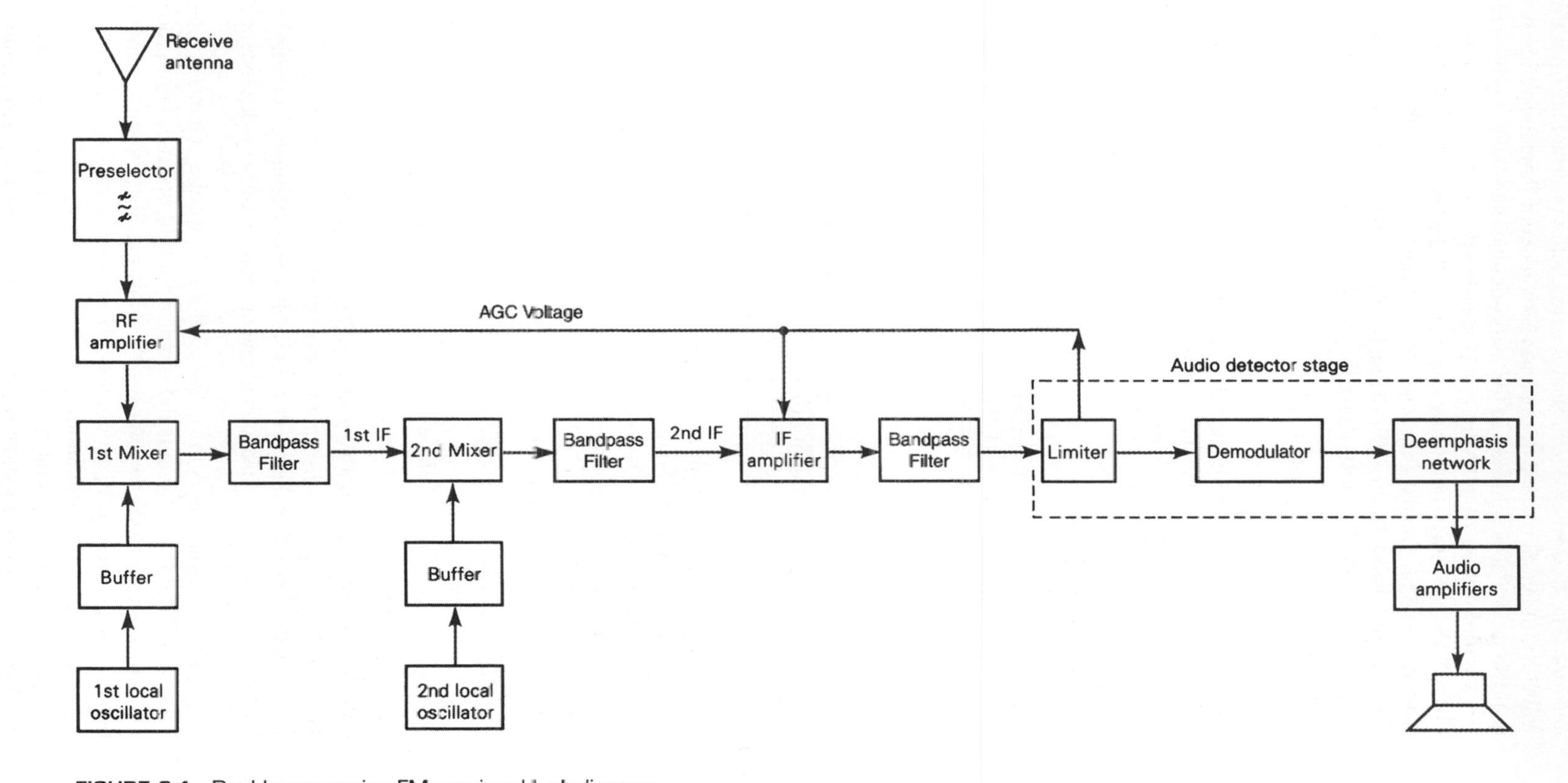

FIGURE 8-1 Double-conversion FM receiver block diagram

receivers is replaced in FM receivers by a *limiter, frequency discriminator,* and *deemphasis network.* The frequency discriminator extracts the information from the modulated wave, while the limiter circuit and deemphasis network contribute to an improvement in the signal-to-noise ratio that is achieved in the audio-demodulator stage of FM receivers.

For broadcast-band FM receivers, the first IF is a relatively high frequency (often 10.7 MHz) for good image-frequency rejection, and the second IF is a relatively low frequency (very often 455 kHz) that allows the IF amplifiers to have a relatively high gain and still not be susceptible to oscillating. With a first IF of 10.7 MHz, the image frequency for even the lowest frequency FM station (88.1 MHz) is 109.5 MHz, which is beyond the FM broadcast band.

8-3 FM DEMODULATORS

FM demodulators are frequency-dependent circuits designed to produce an output voltage that is proportional to the instantaneous frequency at its input. The overall transfer function for an FM demodulator is nonlinear but when operated over its linear range is

$$K_d = \frac{V\,(\text{volts})}{f\,(\text{Hz})} \tag{8-1}$$

where K_d equals transfer function.

The output from an FM demodulator is expressed as

$$v_{out}(t) = K_d \Delta f \tag{8-2}$$

where $v_{out}(t)$ = demodulated output signal (volts)
K_d = demodulator transfer function (volts per hertz)
Δf = difference between the input frequency and the center frequency of the demodulator (hertz)

Example 8-1

For an FM demodulator circuit with a transfer function $K_d = 0.2$ V/kHz and an FM input signal with 20 kHz of peak frequency deviation, determine the peak output voltage.

Solution Substituting into Equation 8-2, the peak output voltage is

$$v_{out}(t) = \frac{0.2\text{ V}}{\text{kHz}} \times 20\text{ kHz}$$

$$= 4\text{V}_\text{p}$$

Several circuits are used for demodulating FM signals. The most common are the *slope detector, Foster-Seeley discriminator, ratio detector, PLL demodulator,* and *quadrature detector.* The slope detector, Foster-Seeley discriminator, and ratio detector are forms of *tuned-circuit frequency discriminators.*

8-3-1 Tuned-Circuit Frequency Discriminators

Tuned-circuit frequency discriminators convert FM to AM and then demodulate the AM envelope with conventional peak detectors. Also, most frequency discriminators require a 180° phase inverter, an adder circuit, and one or more frequency-dependent circuits.

8-3-1-1 Slope detector. Figure 8-2a shows the schematic diagram for a *single-ended slope detector,* which is the simplest form of tuned-circuit frequency discriminator. The single-ended slope detector has the most nonlinear voltage-versus-frequency characteristics and, therefore, is seldom used. However, its circuit operation is basic to all tuned-circuit frequency discriminators.

In Figure 8-2a, the tuned circuit (L_a and C_a) produces an output voltage that is proportional to the input frequency. The maximum output voltage occurs at the resonant frequency of the tank circuit (f_o), and its output decreases proportionately as the input fre-

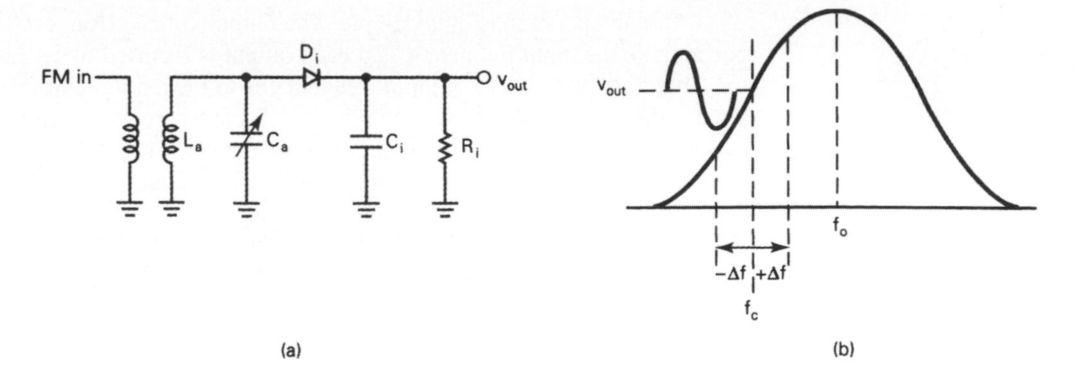

FIGURE 8-2 Slope detector: (a) schematic diagram; (b) voltage-versus-frequency curve

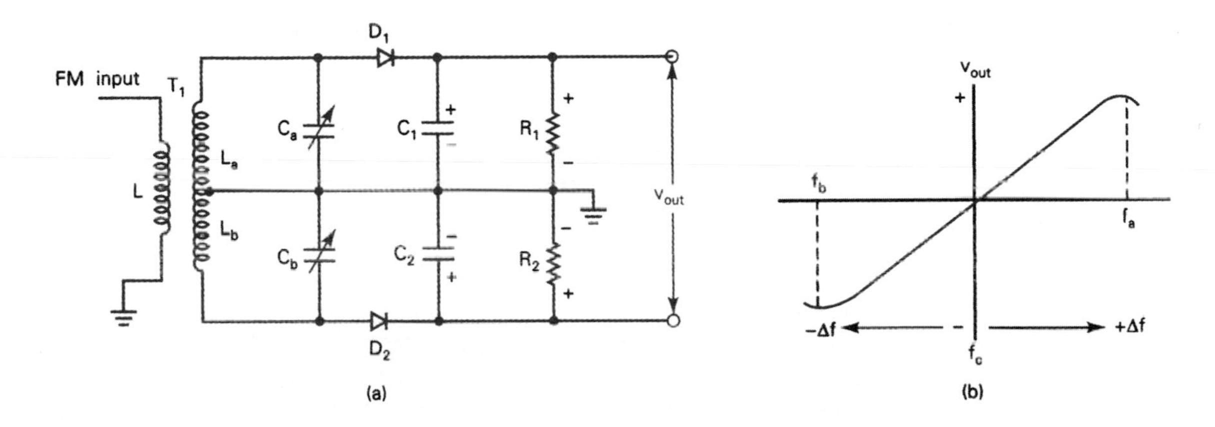

FIGURE 8-3 Balanced slope detector: (a) schematic diagram; (b) voltage-versus-frequency response curve

quency deviates above or below f_o. The circuit is designed so that the IF center frequency (f_c) falls in the center of the most linear portion of the voltage-versus-frequency curve, as shown in Figure 8-2b. When the intermediate frequency deviates above f_c, the output voltage increases; when the intermediate frequency deviates below f_c, the output voltage decreases. Therefore, the tuned circuit converts frequency variations to amplitude variations (FM-to-AM conversion). D_i, C_i, and R_i make up a simple peak detector that converts the amplitude variations to an output voltage that varies at a rate equal to that of the input frequency changes and whose amplitude is proportional to the magnitude of the frequency changes.

8-3-1-2 Balanced slope detector. Figure 8-3a shows the schematic diagram for a *balanced slope detector.* A single-ended slope detector is a tuned-circuit frequency discriminator, and a balanced slope detector is simply two single-ended slope detectors connected in parallel and fed 180° out of phase. The phase inversion is accomplished by center tapping the tuned secondary windings of transformer T_1. In Figure 8-3a, the tuned circuits (L_a, C_a, and L_b, C_b) perform the FM-to-AM conversion, and the balanced peak detectors (D_1, C_1, R_1, and D_2, C_2, R_2) remove the information from the AM envelope. The top tuned circuit (L_a and C_a) is tuned to a frequency (f_a) that is above the IF center frequency (f_o) by approximately $1.33 \times \Delta f$ (for the FM broadcast band, this is approximately 1.33×75 kHz $= 100$ kHz). The lower tuned circuit (L_b and C_b) is tuned to a frequency (f_b) that is below the IF center frequency by an equal amount.

Circuit operation is quite simple. The output voltage from each tuned circuit is proportional to the input frequency, and each output is rectified by its respective peak detector. Therefore, the closer the input frequency is to the tank-circuit resonant frequency, the greater the tank-circuit output voltage. The IF center frequency falls exactly halfway between the resonant frequencies of the two tuned circuits. Therefore, at the IF center frequency, the output voltages from the two tuned circuits are equal in amplitude but opposite in polarity. Consequently, the rectified output voltage across R_1 and R_2, when added, produce a differential output voltage $V_{out} = 0$ V. When the IF deviates above resonance, the top tuned circuit produces a higher output voltage than the lower tank circuit, and V_{out} goes positive. When the IF deviates below resonance, the output voltage from the lower tank circuit is larger than the output voltage from the upper tank circuit, and V_{out} goes negative. The output-versus-frequency response curve is shown in Figure 8-3b.

Although the slope detector is probably the simplest FM detector, it has several inherent disadvantages, which include poor linearity, difficulty in tuning, and lack of provisions for limiting. Because limiting is not provided, a slope detector produces an output voltage that is proportional to amplitude, as well as frequency variations in the input signal and, consequently, must be preceded by a separate limiter stage. A balanced slope detector is aligned by injecting a frequency equal to the IF center frequency and tuning C_a and C_b for 0 V at the output. Then frequencies equal to f_a and f_b are alternately injected while C_a and C_b are tuned for maximum and equal output voltages with opposite polarities.

8-3-1-3 Foster-Seeley discriminator. A *Foster-Seeley discriminator* (sometimes called a *phase shift discriminator*) is a tuned-circuit frequency discriminator whose operation is very similar to that of the balanced slope detector. The schematic diagram for a Foster-Seeley discriminator is shown in Figure 8-4a. The capacitance values for C_c, C_1, and C_2 are chosen such that they are short circuits for the IF center frequency. Therefore, the right side of L_3 is at ac ground potential, and the IF signal (V_{in}) is fed directly (in phase) across L_3 (V_{L3}). The incoming IF is inverted 180° by transformer T_1 and divided equally between L_a and L_b. At the resonant frequency of the secondary tank circuit (the IF center frequency), the secondary current (I_s) is in phase with the total secondary voltage (V_s) and 180° out of phase with V_{L3}. Also, because of loose coupling, the primary of T_1 acts as an inductor, and the primary current I_p is 90° out of phase with V_{in}, and, because magnetic induction depends on primary current, the voltage induced in the secondary is 90° out of phase with V_{in} (V_{L3}). Therefore, V_{La} and V_{Lb} are 180° out of phase with each other and in quadrature, or 90° out of phase with V_{L3}. The voltage across the top diode (V_{D1}) is the vector sum of V_{L3} and V_{La}, and the voltage across the bottom diode V_{D2} is the vector sum of V_{L3} and V_{Lb}. The corresponding vector diagrams are shown in Figure 8-4b. The figure shows that the voltages across D_1 and D_2 are equal. Therefore, at resonance, I_1 and I_2 are equal, and C_1 and C_2 charge to equal magnitude voltages except with opposite polarities. Consequently, $V_{out} = V_{C1} - V_{C2} = 0$ V. When the IF goes above resonance ($X_L > X_C$), the secondary tank-circuit impedance becomes inductive, and the secondary current lags the secondary voltage by some angle θ, which is proportional to the magnitude of the frequency deviation. The corresponding phasor diagram is shown in Figure 8-4c. The figure shows that the vector sum of the voltage across D_1 is greater than the vector sum of the voltages across D_2. Consequently, C_1 charges while C_2 discharges and V_{out} goes positive. When the IF goes below resonance ($X_L < X_C$), the secondary current leads the secondary voltage by some angle θ, which is again proportional to the magnitude of the change in frequency. The corresponding phasors are shown in Figure 8-4d. It can be seen that the vector sum of the voltages across D_1 is now less than the vector sum of the voltages across D_2. Consequently, C_1 discharges while C_2 charges and V_{out} goes negative. A Foster-Seeley discriminator is tuned by injecting a frequency equal to the IF center frequency and tuning C_o for 0 volts out.

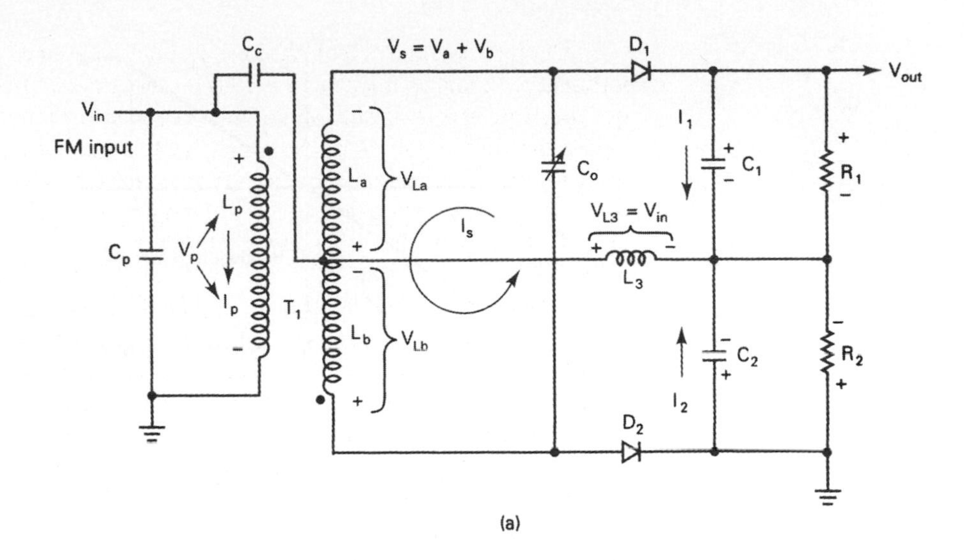

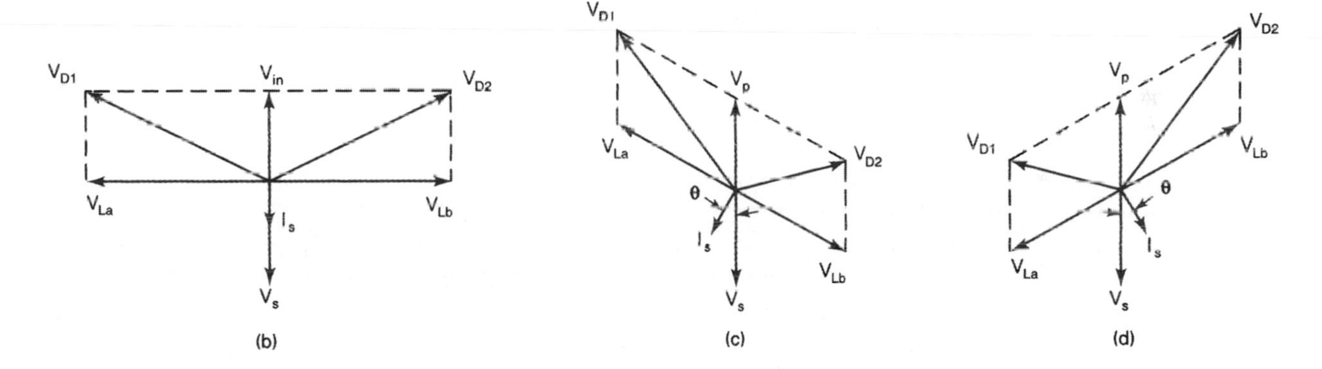

FIGURE 8-4 Foster-Seely discriminator: (a) schematic diagram; (b) vector diagram, $f_{in} = f_o$; (c) vector diagram, $f_{in} > f_o$; (d) vector diagram, $f_{in} < f_o$

The preceding discussion and Figure 8-4 show that the output voltage from a Foster-Seeley discriminator is directly proportional to the magnitude and direction of the frequency deviation. Figure 8-5 shows a typical voltage-versus-frequency response curve for a Foster-Seeley discriminator. For obvious reasons, it is often called an *S-curve.* It can be seen that the output voltage-versus-frequency deviation curve is more linear than that of a slope detector, and because there is only one tank circuit, it is easier to tune. For distortionless demodulation, the frequency deviation should be restricted to the linear portion of the secondary tuned-circuit frequency response curve. As with the slope detector, a Foster-Seeley discriminator responds to amplitude as well as frequency variations and, therefore, must be preceded by a separate limiter circuit.

8-3-1-4 Ratio detector. The *ratio detector* has one major advantage over the slope detector and Foster-Seeley discriminator for FM demodulation: A ratio detector is relatively immune to amplitude variations in its input signal. Figure 8-6a shows the schematic diagram for a ratio detector. As with the Foster-Seeley discriminator, a ratio detector has a single tuned circuit in the transformer secondary. Therefore, the operation of a ratio detector is similar to that of the Foster-Seeley discriminator. In fact, the voltage vectors for D_1 and D_2 are identical to those of the Foster-Seeley discriminator circuit shown in Figure 8-4.

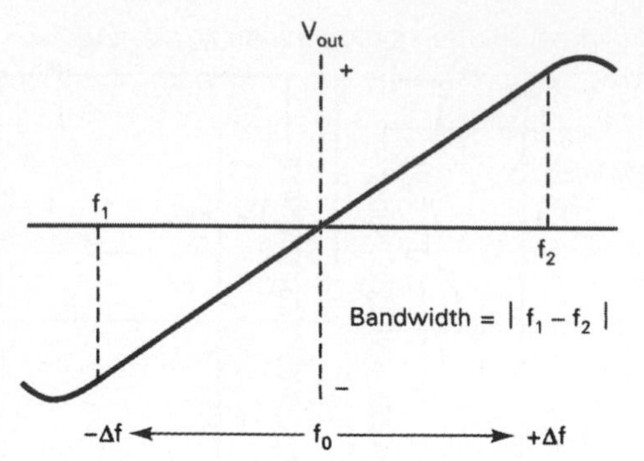

FIGURE 8-5 Discriminator voltage-versus-frequency response curve

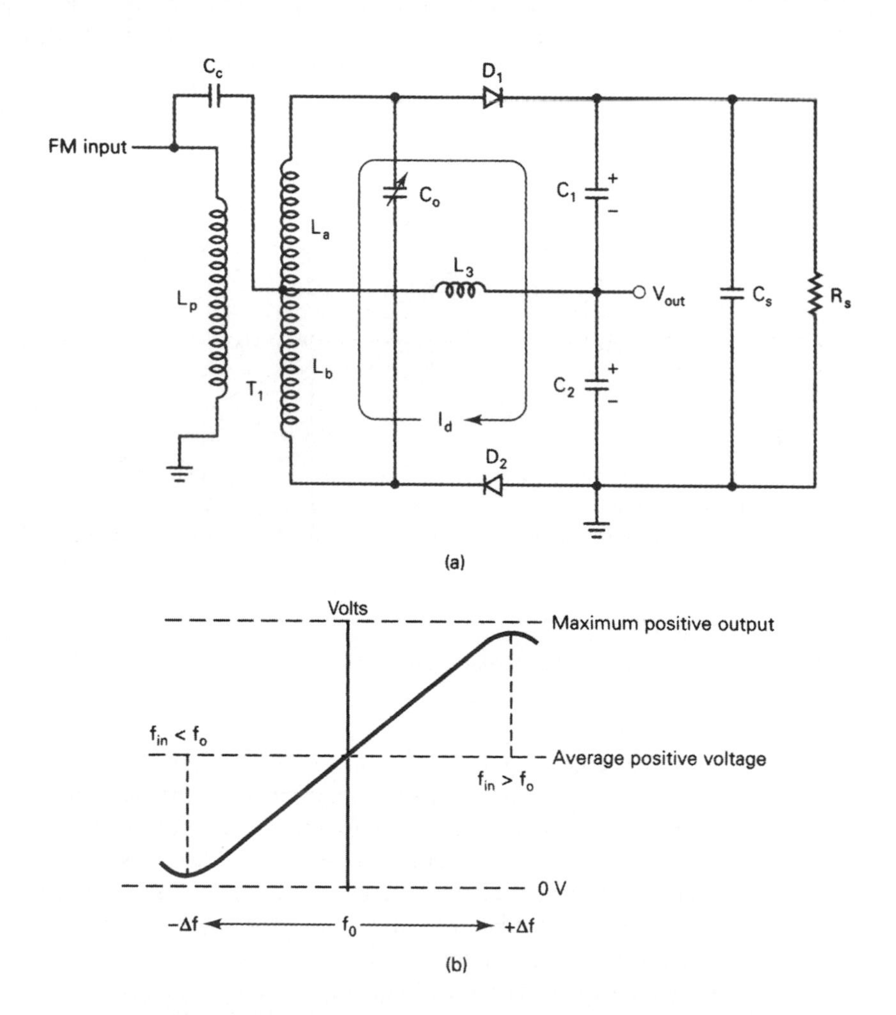

FIGURE 8-6 Ratio detector: (a) schematic diagram; (b) voltage-versus-frequency response curve

However, with the ratio detector, one diode is reversed (D_2), and current (I_d) can flow around the outermost loop of the circuit. Therefore, after several cycles of the input signal, shunt capacitor C_s charges to approximately the peak voltage across the secondary winding of T_1. The reactance of C_s is low, and R_s simply provides a dc path for diode current. Therefore, the time constant for R_s and C_s is sufficiently long so that rapid changes in the amplitude of the input signal due to thermal noise or other interfering signals are shorted to ground and have no effect on the average voltage across C_s. Consequently, C_1 and C_2 charge and discharge proportional to frequency changes in the input signal and are relatively immune to amplitude variations. Also, the output voltage from a ratio detector is taken with respect to ground, and for the diode polarities shown in Figure 8-6a, the average output voltage is positive. At resonance, the output voltage is divided equally between C_1 and C_2 and redistributed as the input frequency is deviated above and below resonance. Therefore, changes in V_{out} are due to the changing ratio of the voltage across C_1 and C_2, while the total voltage is clamped by C_s.

Figure 8-6b shows the output frequency response curve for the ratio detector shown in Figure 8-6a. It can be seen that at resonance, V_{out} is not equal to 0 V but, rather, to one-half of the voltage across the secondary windings of T_1. Because a ratio detector is relatively immune to amplitude variations, it is often selected over a discriminator. However, a discriminator produces a more linear output voltage-versus-frequency response curve.

8-4 PHASE-LOCKED-LOOP FM DEMODULATORS

Since the development of LSI linear integrated circuits, FM demodulation can be accomplished quite simply with a phase-locked loop (PLL). Although the operation of a PLL is quite involved, the operation of a *PLL FM demodulator* is probably the simplest and easiest to understand. A PLL frequency demodulator requires no tuned circuits and automatically compensates for changes in the carrier frequency due to instability in the transmit oscillator. Figure 8-7a shows the simplified block diagram for a PLL FM demodulator.

In Chapter 3, a detailed description of PLL operation was given. It was shown that after frequency lock had occurred the VCO would track frequency changes in the input signal by maintaining a phase error at the input of the phase comparator. Therefore, if the PLL input is a deviated FM signal and the VCO natural frequency is equal to the IF center frequency, the correction voltage produced at the output of the phase comparator and fed back to the input of the VCO is proportional to the frequency deviation and is, thus, the demodulated information signal. If the IF amplitude is sufficiently limited prior to reaching the PLL and the loop is properly compensated, the PLL loop gain is constant and equal to K_v. Therefore, the demodulated signal can be taken directly from the output of the internal buffer and is mathematically given as

$$V_{out} = \Delta f K_d K_a \tag{8-3}$$

Figure 8-7b shows a schematic diagram for an FM demodulator using the XR-2212. R_0 and C_0 are course adjustments for setting the VCO's free-running frequency. R_x is for fine tuning, and R_F and R_C set the internal op-amp voltage gain (K_a). The PLL closed-loop frequency response should be compensated to allow unattenuated demodulation of the entire information signal bandwidth. The PLL op-amp buffer provides voltage gain and current drive stability.

8-5 QUADRATURE FM DEMODULATOR

A *quadrature FM demodulator* (sometimes called a *coincidence detector*) extracts the original information signal from the composite IF waveform by multiplying two quadrature (90° out of phase) signals. A quadrature detector uses a 90° phase shifter, a single tuned circuit, and a product detector to demodulate FM signals. The 90° phase shifter produces a signal that is in quadrature with the received IF signals. The tuned circuit converts frequency variations to phase variations, and the product detector multiplies the received IF signals by the phase-shifted IF signal.

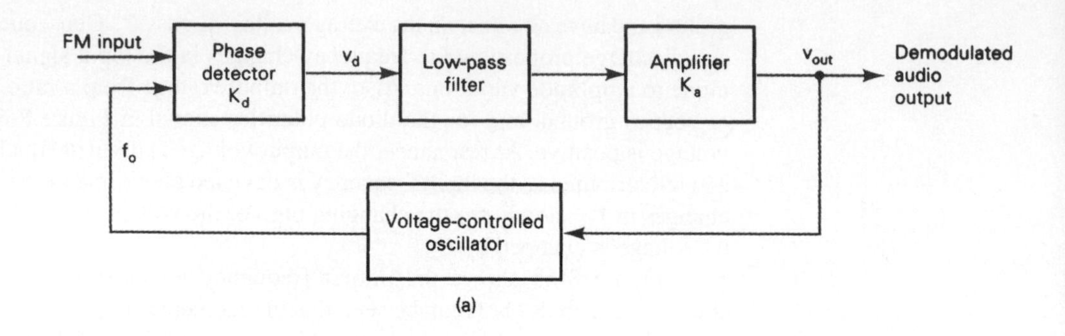

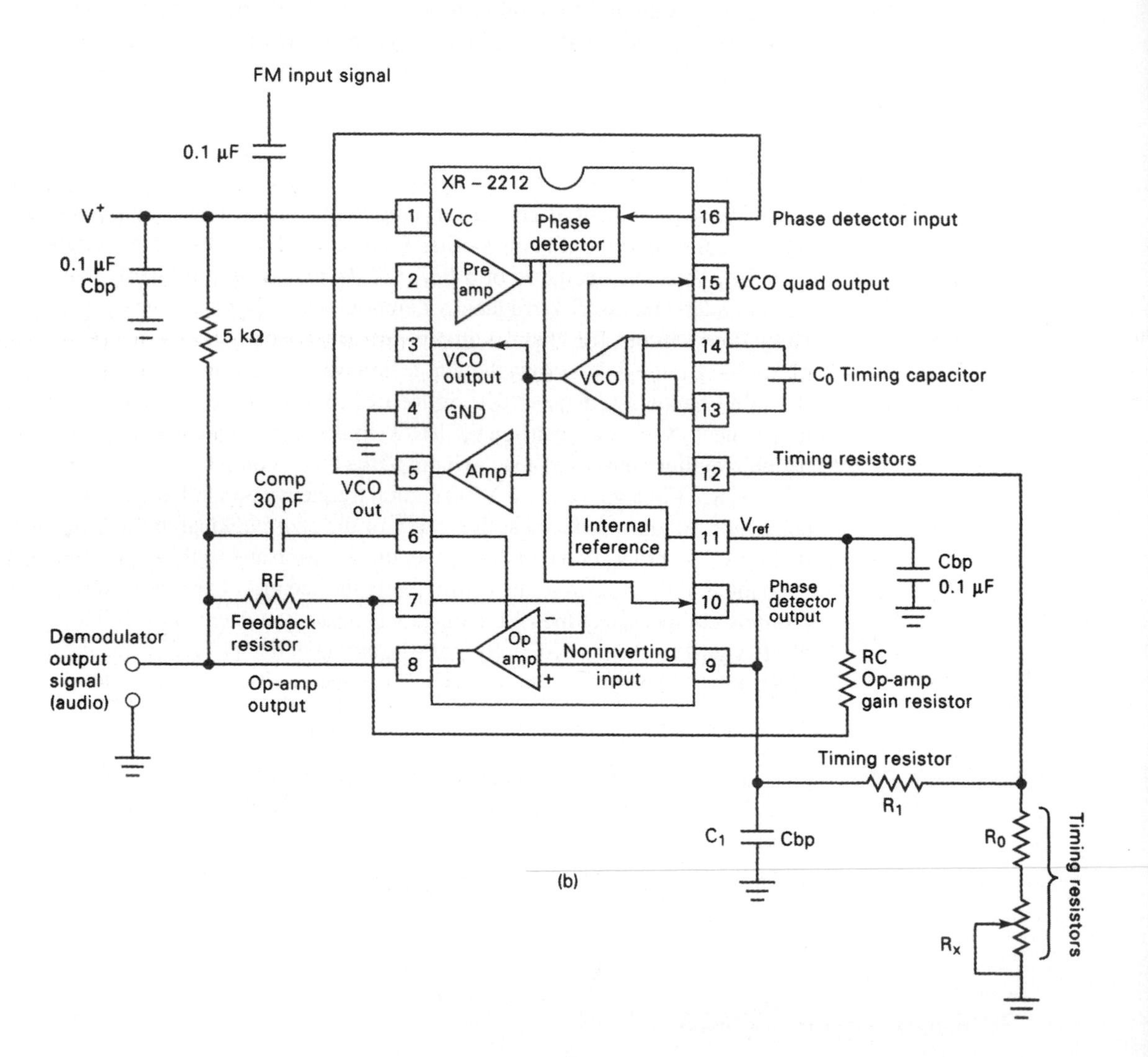

FIGURE 8-7 (a) Block diagram for a PLL FM demodulator; (b) PLL FM demodulator using the XR-2212 PLL

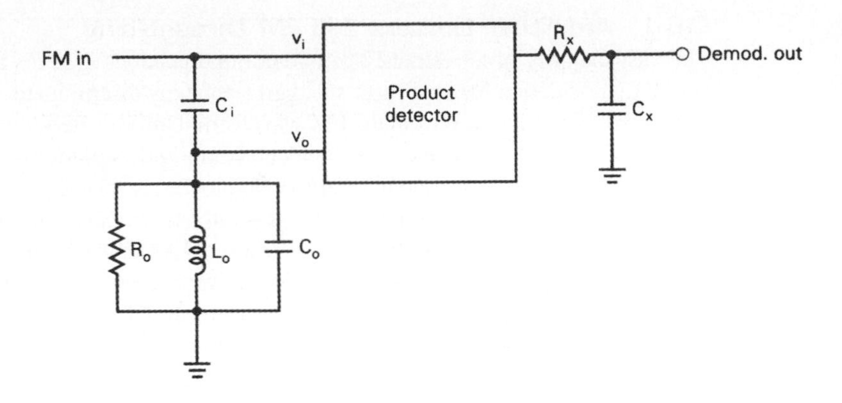

FIGURE 8-8 Quadrature FM demodulator

Figure 8-8 shows a simplified schematic diagram for an FM quadrature detector. C_i is a high-reactance capacitor that, when placed in series with tank circuit (R_o, L_o, and C_o), produces a 90° phase shift at the IF center frequency. The tank circuit is tuned to the IF center frequency and produces an additional phase shift (θ) that is proportional to the frequency deviation. The IF input signal (v_i) is multiplied by the quadrature signal (v_o) in the product detector and produces an output signal that is proportional to the frequency deviation. At the resonant frequency, the tank-circuit impedance is resistive. However, frequency variations in the IF signal produce an additional positive or negative phase shift. Therefore, the product detector output voltage is proportional to the phase difference between the two input signals and is expressed mathematically as

$$v_{\text{out}} = v_i v_o = [V_i \sin(\omega_i t + \theta)][V_o \cos(\omega_o t)] \tag{8-4}$$

Substituting into the trigonometric identity for the product of a sine and a cosine wave of equal frequency gives us

$$v_{\text{out}} = \frac{V_i V_o}{2}[\sin(2\omega_i t + \theta) + \sin(\theta)] \tag{8-5}$$

The second harmonic ($2\omega_i$) is filtered out, leaving

$$v_{\text{out}} = \frac{V_i V_o}{2}\sin(\theta) \tag{8-6}$$

where $\theta = \tan^{-1} \rho Q$
$\rho = 2\pi f/f_o$ (fractional frequency deviation)
Q = tank-circuit quality factor

8-6 FM NOISE SUPPRESSION

Probably the most important advantage of frequency modulation over amplitude modulation is the ability of FM receivers to suppress noise. Because most noise appears as amplitude variations in the modulated wave, AM demodulators cannot remove the noise without also removing some of the information. This is because the information is also contained in amplitude variations. With FM, however, the information is contained in frequency variations, allowing the unwanted amplitude variations to be removed with special circuits called *limiters.*

8-6-1 Amplitude Limiters and FM Thresholding

The vast majority of terrestrial FM radio communications systems use conventional noncoherent demodulation because most standard frequency discriminators use envelope detection to remove the intelligence from the FM waveform. Unfortunately, envelope detectors (including ratio detectors) will demodulate incidental amplitude variations as well as frequency variations. Transmission noise and interference add to the signal and produce unwanted amplitude variations. Also, frequency modulation is generally accompanied by small amounts of residual amplitude modulation. In the receiver, the unwanted AM and random noise interference are demodulated along with the signal and produce unwanted distortion in the recovered information signal. The noise is more prevalent at the peaks of the FM waveform and relatively insignificant during the zero crossings. A limiter is a circuit that produces a constant-amplitude output for all input signals above a prescribed minimum input level, which is often called the *threshold, quieting,* or *capture* level. Limiters are required in most FM receivers because many of the demodulators discussed earlier in this chapter demodulate amplitude as well as frequency variations. With amplitude limiters, the signal-to-noise ratio at the output of the demodulator (postdetection) can be improved by as much as 20 dB or more over the input (predetection) signal to noise.

Essentially, an amplitude limiter is an additional IF amplifier that is overdriven. Limiting begins when the IF signal is sufficiently large that it drives the amplifier alternately into saturation and cutoff. Figures 8-9a and b show the input and output waveforms for a

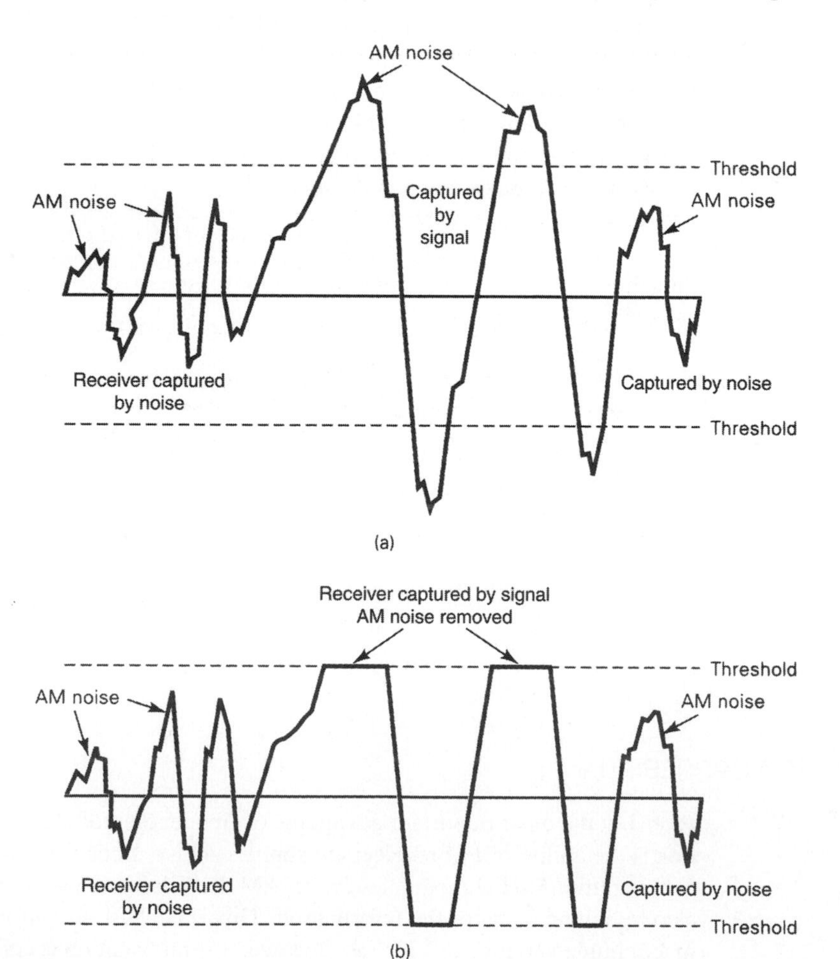

FIGURE 8-9 Amplitude limiter input and output waveforms; (a) input waveform; (b) output waveform

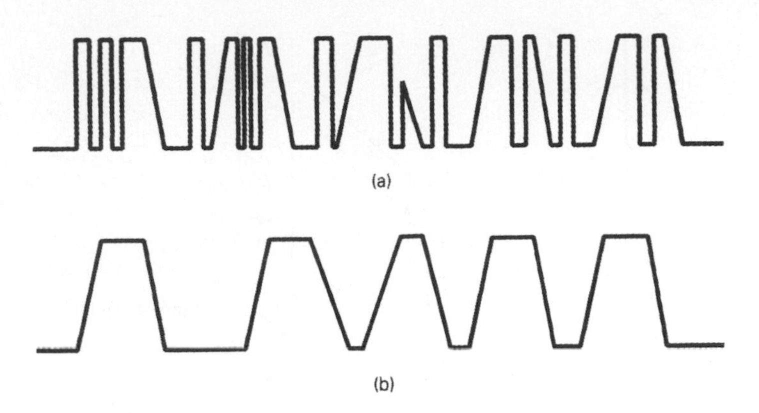

FIGURE 8-10 Limiter output: (a) captured by noise; (b) captured by signal

typical limiter. In Figure 8-9b, it can be seen that for IF signals that are below threshold, the AM noise is not reduced, and for IF signals above threshold, there is a large reduction in the AM noise level. The purpose of the limiter is to remove all amplitude variations from the IF signal.

Figure 8-10a shows the limiter output when the noise is greater than the signal (i.e., the noise has captured the limiter). The irregular widths of the serrations are caused by noise impulses saturating the limiter. Figure 8-10b shows the limiter output when the signal is sufficiently greater than the noise (the signal has captured the limiter). The peaks of the signal have the limiter so far into saturation that the weaker noise is totally eliminated. The improvement in the S/N ratio is called *FM thresholding, FM quieting,* or the *FM capture effect.* Three criteria must be satisfied before FM thresholding can occur:

1. The predetection signal-to-noise ratio must be 10 dB or greater.
2. The IF signal must be sufficiently amplified to overdrive the limiter.
3. The signal must have a modulation index equal to or greater than unity ($m \geq 1$).

Figure 8-11 shows typical FM thresholding curves for low ($m = 1$) and medium ($m = 4$) index signals. The output voltage from an FM detector is proportional to m^2. Therefore, doubling m increases the S/N ratio by a factor of 4 (6 dB). The quieting ratio for $m = 1$ is an input S/N = 13 dB and, for $m = 4$, 22 dB. For S/N ratios below threshold, the receiver is said to be captured by the noise, and for S/N ratios above threshold, the receiver is said to be captured by the signal. Figure 8-11 shows that IF signals at the input to the limiter with 13 dB or more S/N undergo 17 dB of S/N improvement. FM quieting begins with an input S/N ratio of 10 dB but does not produce the full 17-dB improvement until the input signal-to-noise ratio reaches 13 dB.

As shown in Figure 8-11, there is no signal-to-noise improvement with AM double-sideband or AM single-sideband transmission. With AM, the pre- and postdetection signal-to-noise ratios are essentially the same.

8-6-2 Limiter Circuits

Figure 8-12a shows a schematic diagram for a single-stage limiter circuit with a built-in output filter. This configuration is commonly called a *bandpass limiter/amplifier* (BPL). A BPL is essentially a class A biased tuned IF amplifier, and for limiting and FM quieting to occur, it requires an IF input signal sufficient enough to drive it into both saturation and cutoff. The output tank circuit is tuned to the IF center frequency. Filtering removes the harmonic and intermodulation distortion present in the rectangular pulses due to *hard limiting.* The effect of filtering is shown in Figure 8-13. If resistor R_2 were removed entirely, the amplifier would be biased for class C operation, which is also appropriate for this type

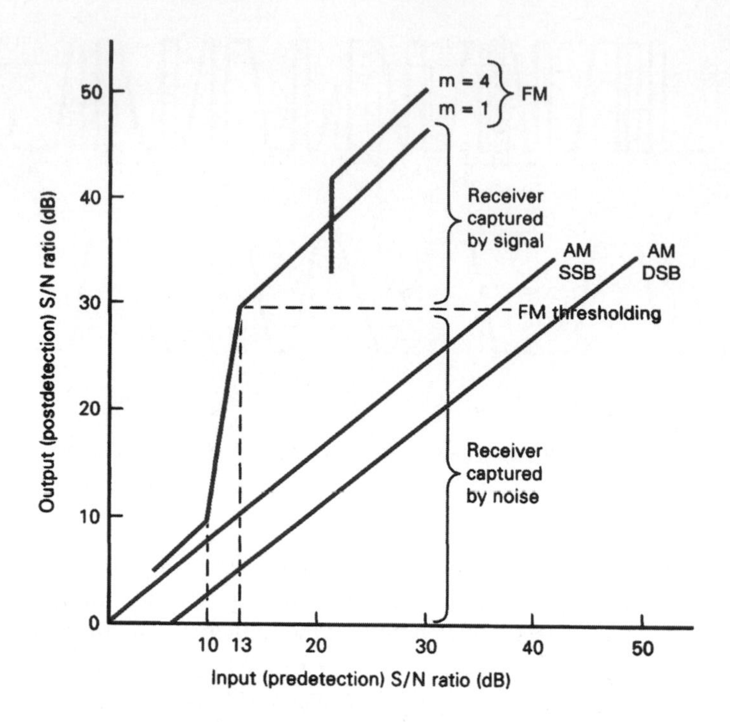

FIGURE 8-11 FM thresholding

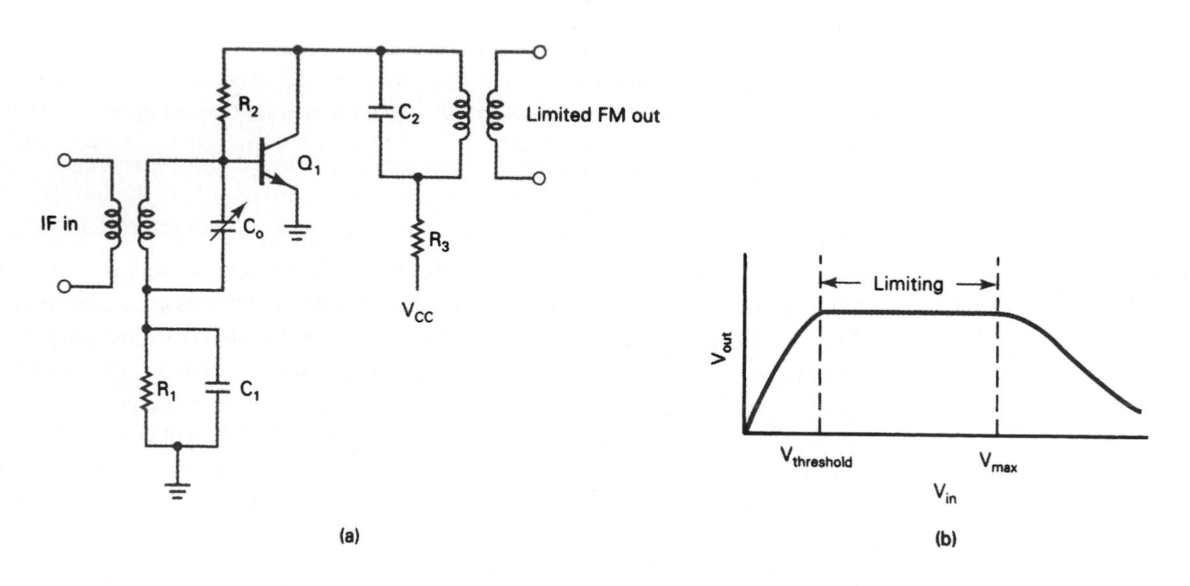

FIGURE 8-12 Single-stage tuned limiter: (a) schematic diagram; (b) limiter action

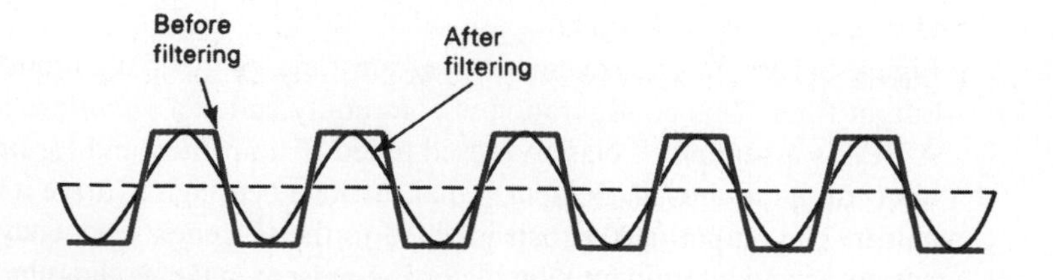

FIGURE 8-13 Filtered limiter output

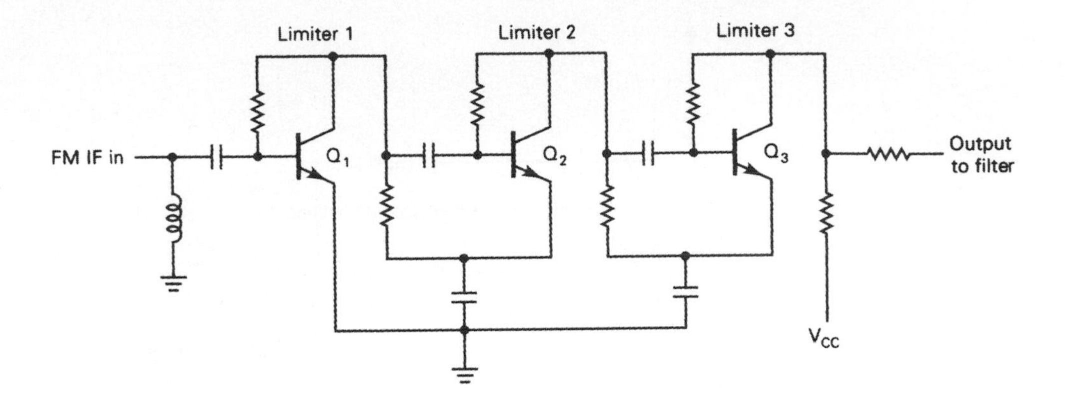

FIGURE 8-14 Three-stage cascaded limiter

of circuit, but requires more filtering. Figure 8-12b shows limiter action for the circuit shown in Figure 8-12a. For small signals (below the threshold voltage), no limiting occurs. When V_{in} reaches $V_{threshold}$, limiting begins, and for input amplitudes above V_{max}, there is actually a decrease in V_{out} with increases in V_{in}. This is because with high-input drive levels the collector current pulses are sufficiently narrow that they actually develop less tank-circuit power. The problem of overdriving the limiter can be rectified by incorporating AGC into the circuit.

8-6-3 FM Capture Effect

The inherent ability of FM to diminish the effects of interfering signals is called the *capture effect.* Unlike AM receivers, FM receivers have the ability to differentiate between two signals received at the same frequency. Therefore, if two stations are received simultaneously at the same or nearly the same frequency, the receiver locks onto the stronger station while suppressing the weaker station. Suppression of the weaker signal is accomplished in amplitude limiters in the same manner that AM noise is suppressed. If two stations are received at approximately the same signal level, the receiver cannot sufficiently differentiate between them and may switch back and forth. The *capture ratio* of an FM receiver is the minimum dB difference in signal strength between two received signals necessary for the capture effect to suppress the weaker signal. Capture ratios of 1 dB are typical for high-quality FM receivers.

When two limiter stages are used, it is called *double limiting;* three stages, *triple limiting;* and so on. Figure 8-14 shows a three-stage *cascaded limiter* without a built-in filter. This type of limiter circuit must be followed by either a ceramic or a crystal filter to remove the nonlinear distortion. The limiter shown has three *RC*-coupled limiter stages that are ac series connected to reduce the current drain. Cascaded amplifiers combine several of the advantages of common-emitter and common-gate amplifiers. Cascading amplifiers also decrease the thresholding level and, thus, improve the quieting capabilities of the stage. The effects of double and triple limiting are shown in Figure 8-15. Because FM receivers have sufficient gain to saturate the limiters over a relatively large range of RF input signal levels, AGC is usually unnecessary. In fact, very often AGC actually degrades the performance of an FM receiver.

Example 8-2

For an FM receiver with a bandwidth $B = 200$ kHz, a power noise figure NF $= 8$ dB, and an input noise temperature $T = 100$ K, determine the minimum receive carrier power necessary to achieve a postdetection signal-to-noise ratio of 37 dB. Use the receiver block diagram shown in Figure 8-1 as the receiver model and the FM thresholding curve shown in Figure 8-11 for $m = 1$.

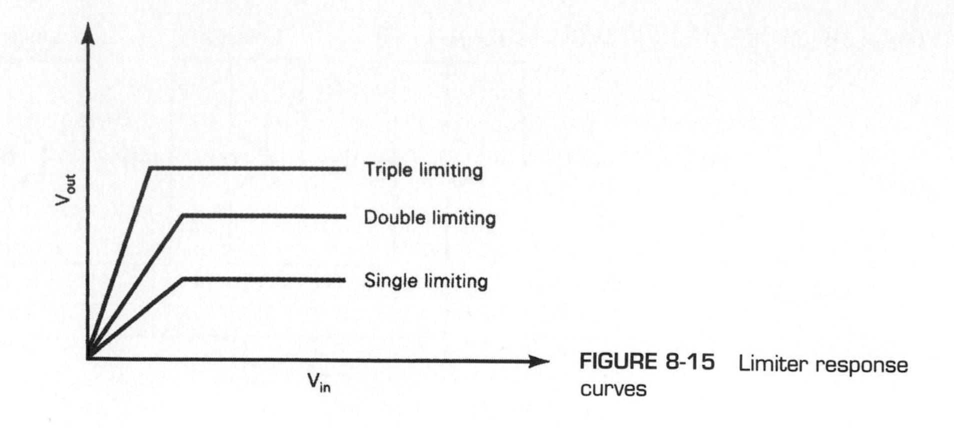

FIGURE 8-15 Limiter response curves

Solution From Figure 8-11, it can be seen that 17 dB of signal-to-noise improvement is evident in the detector, assuming the limiters are saturated and the input signal-to-noise is greater than 13 dB. Therefore, to achieve a postdetection signal-to-noise ratio of 37 dB, the predetection signal-to-noise ratio must be at least

$$37 \text{ dB} - 17 \text{ dB} = 20 \text{ dB}$$

Therefore, for an overall receiver noise figure equal to 8 dB, the S/N ratio at the input to the receiver must be at least

$$20 \text{ dB} + 8 \text{ dB} = 28 \text{ dB}$$

The receiver input noise power is

$$N_{(\text{dBm})} = 10 \log \frac{KTB}{0.001} = 10 \log \frac{(1.38 \times 10^{-23})(100)(200{,}000)}{0.001} = -125.6 \text{ dBm}$$

Consequently, the minimum receiver signal power for a 28-dB S/N ratio is

$$S = -125.6 \text{ dBm} + 28 \text{ dB} = -97.6 \text{ dBm}$$

Example 8-3

For an FM receiver with an input signal-to-noise ratio of 29 dB, a noise figure of 4 dB, and an FM improvement factor of 16 dB, determine the pre- and postdetection S/N ratios.

Solution The predetection signal-to-noise ratio is

input signal-to-noise ratio − noise figure

$$29 \text{ dB} - 4 \text{ dB} = 25 \text{ dB}$$

The postdetection signal-to-noise ratio is

predetection signal-to-noise + FM improvement

$$25 \text{ dB} + 16 \text{ dB} = 41 \text{ dB}$$

Example 8-4

For an FM receiver with an input noise level of −112 dBm, a postdetection S/N = 38 dB, an FM improvement factor of 17 dB, and a noise figure of 5 dB, determine the minimum receive signal level.

Solution The receiver input S/N is

postdetection S/N ratio − FM improvement + noise figure

$$38 \text{ dB} - 17 \text{ dB} + 5 \text{ dB} = 26 \text{ dB}$$

Therefore, the minimum receive signal level is

input noise level + minimum receiver S/N ratio

$$-112 \text{ dBm} + 26 \text{ dB} = -86 \text{ dBm}$$

8-7 FREQUENCY VERSUS PHASE MODULATION

Although frequency and phase modulation are similar in many ways, they do have their differences and, consequently, there are advantages and disadvantages of both forms of angle modulation. At one time for large-scale applications, such as commercial broadcasting, FM was preferred because PM requires coherent demodulation, usually using a PLL. Frequency modulation, on the other hand, can be demodulated using noncoherent demodulators. Today, however, PLLs are probably less expensive than their noncoherent counterparts mainly because they come as integrated circuits and require no transformers or *LC* tank circuits.

With PM, the modulation index is independent of the modulating-signal frequency. Therefore, PM offers better signal-to-noise performance than FM, and PM does not require a preemphasis network. One important advantage of PM is that phase modulation is performed in a circuit separate from the carrier oscillator. Therefore, highly stable crystal oscillators can be used for the carrier source. With FM, the modulating signal is applied directly to the carrier oscillator; thus, crystal oscillators cannot be used to produce the carrier signal. Therefore, FM modulators require AFC circuits to achieve the frequency stability required by the FCC.

One prominent advantage of FM over PM is that the VCOs used with FM can be directly modulated and produce outputs with high-frequency deviations and high modulation indices. PM modulators generally require frequency multipliers to increase the modulation index and frequency deviation to useful levels.

8-8 LINEAR INTEGRATED-CIRCUIT FM RECEIVERS

In recent years, several manufacturers of integrated circuits, such as Signetics, RCA, and Motorola, have developed reliable, low-power monolithic integrated circuits that perform virtually all the receiver functions for both AM and FM communications systems. These integrated circuits offer the advantages of being reliable, predictable, miniaturized, and easy to design with. The development of these integrated circuits is one of the primary reasons for the tremendous growth of both portable two-way FM and cellular radio communications systems that has occurred in the past few years.

8-8-1 Low-Power, Integrated-Circuit FM IF System

The NE/SA614A is an improved monolithic low-power FM IF system manufactured by Signetics Corporation. The NE/SA614A is a high-gain, high-frequency device that offers low-power consumption (3.3-mA typical current drain) and excellent input sensitivity (1.5 μV across its input pins) at 455 kHz. The NE/SA614A has an onboard temperature-compensated *received signal-strength indicator* (RSSI) with a logarithmic output and a dynamic range in excess of 90 dB. It has two audio outputs (one muted and one not). The NE/SA614A requires a low number of external components to function and meets cellular radio specifications. The NE/SA614A can be used for the following applications:

1. FM cellular radio
2. High-performance FM communications receivers
3. Intermediate-frequency amplification and detection up to 25 MHz
4. RF signal-strength meter
5. Spectrum analyzer applications
6. Instrumentation circuits
7. Data transceivers

The block diagram for the NE/SA614A is shown in Figure 8-16. As the figure shows, the NE/SA614A includes two limiting intermediate-frequency amplifiers, an FM quadrature

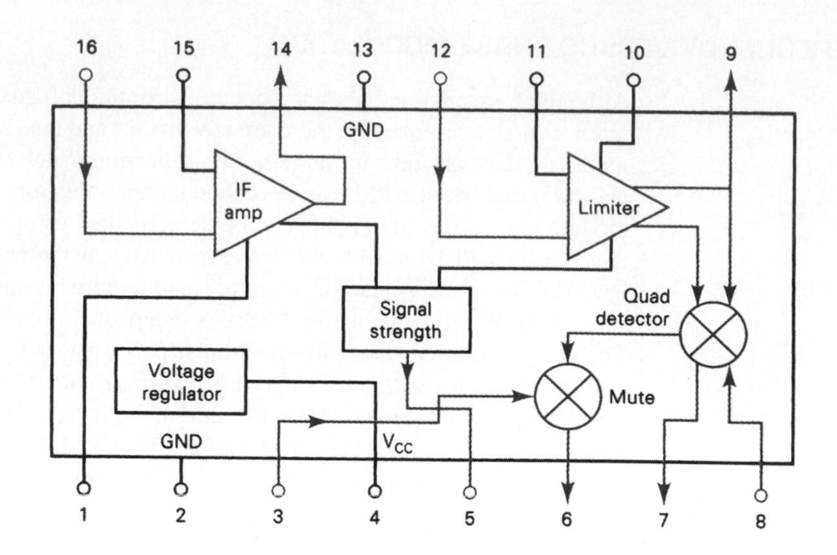

FIGURE 8-16 Block diagram for the Signetics NE/SA614A integrated-circuit, low-power FM IF system

detector, an audio muting circuit, a logarithmic received signal-strength indicator (RSSI), and a voltage regulator. The NE/SA614A is an IF signal-processing system suitable for frequencies as high as 21.4 MHz.

8-8-1-1 IF amplifiers. Figure 8-17 shows the equivalent circuit for the NE/SA614A. The IF amplifier section consists of two log-limiting amplifier stages. The first consists of two differential amplifiers with 39 dB of gain and a small-signal ac bandwidth of 41 MHz when driven from a 50-Ω source. The output of the first limiter is a low-impedance emitter follower with 1 kΩ of equivalent series resistance. The second limiting stage consists of three differential amplifiers with a total gain of 62 dB and a small-signal ac bandwidth of 28 MHz. The outputs of the final differential amplifier are buffered to the internal quadrature detector. One output is available to drive an external quadrature capacitor and *L/C* quadrature tank. Both limiting stages are dc biased with feedback. The buffered outputs of the final differential amplifier in each stage are fed back to the input of that stage through a 42-kΩ resistor. Because of the very high gain, wide bandwidth, and high input impedance of the limiters, the limiter stage is potentially unstable at IF frequencies above 455 kHz. The stability can be improved by reducing the gain. This is accomplished by adding attenuators between amplifier stages. The IF amplifiers also feature low phase shift (typically only a few degrees over a wide range of input frequencies).

8-8-1-2 Quadrature detector. Figure 8-18 shows the block diagram for the equivalent circuit for the quadrature detector in the NE/SA614A. A quadrature detector is a multiplier cell similar to a mixer stage, but instead of mixing two different frequencies, it mixes two signals with the same frequencies but with different phases. A constant-amplitude (amplitude-limited) signal is applied to the lower part of the multiplier. The same signal is applied single ended to an external capacitor connected to pin 9. There is a 90° phase shift across the plates of the capacitor. The phase shifted signal is applied to the upper port of the multiplier at pin 8. A quadrature tank (a parallel *LC* network) permits frequency selective phase shifting at the IF signal. The quadrature detector produces an output signal whose amplitude is proportional to the magnitude of the frequency deviation of the input FM signal.

8-8-1-3 Audio outputs. The NE/SA614A has two audio outputs. Both are PNP current-to-voltage converters with 55-kΩ nominal internal loads. The unmuted output is

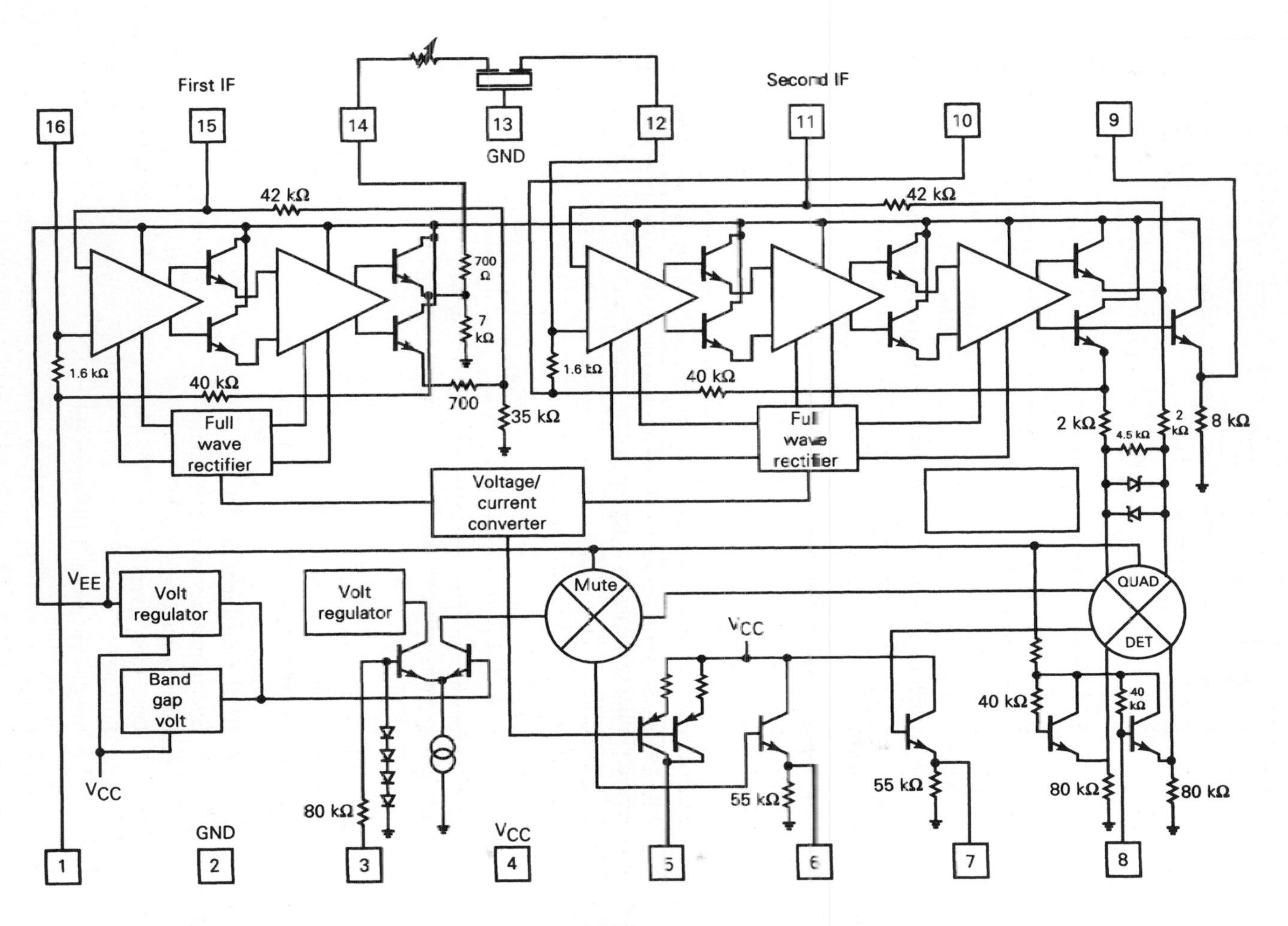

FIGURE 8-17 Equivalent circuit for the Signetics NE/SA614A integrated-circuit, low-power FM IF system

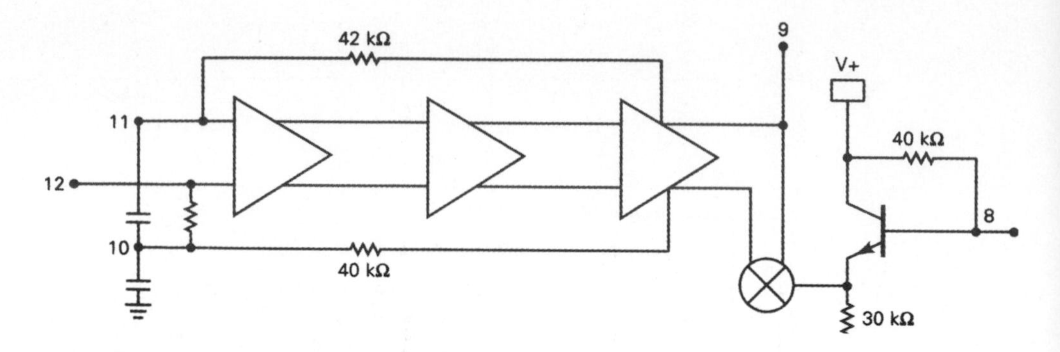

FIGURE 8-18 Quadrature detector block diagram

always active to permit the use of signaling tones, such as for cellular radio. The other output can be muted with 70-dB typical attenuation. The two outputs have an internal 180° phase difference and, therefore, can be applied to the differential inputs of an op-amp amplifier or comparator. Once the threshold of the reference frequency has been established, the two output amplitudes will shift in opposite directions as the input frequency shifts.

8-8-1-4 RSSI. The received signal-strength indicator demonstrates a monotonic logarithmic output over a range of 90 dB. The signal-strength output is derived from the summed stage currents in the limiting amplifiers. It is essentially independent of the IF frequency. Thus, unfiltered signals at the limiter input, such as spurious products or regenerated signals, will manifest themselves as an RSSI output. At low frequencies, the RSSI makes an excellent logarithmic ac voltmeter. The RSSI output is a current-to-voltage converter similar to the audio outputs.

8-8-2 Low-Voltage, High-Performance Mixer FM IF System

The NE/SA616 is a low-voltage, high-performance monolithic FM IF system similar to the NE/SA614A except with the addition of a mixer/oscillator circuit. The NE/SA616 will operate at frequencies up to 150 MHz and with as little as 2.7 V dc. The NE/SA616 features low power consumption, a mixer conversion power gain of 17 dB at 45 MHz, 102 dB of IF amplifier/limiter gain, and a 2-MHz IF amplifier/limiter small-signal ac bandwidth. The NE/SA616 can be used for the following applications:

1. Portable FM cellular radio
2. Cordless telephones
3. Wireless communications systems
4. RF signal-strength meter
5. Spectrum analyzer applications
6. Instrumentation circuits
7. Data transceivers
8. Log amps
9. Single-conversion VHF receivers

The block diagram for the NE/SA616 is shown in Figure 8-19. The NE/SA616 is similar to the NE/SA614A with the addition of a mixer and local oscillator stage. The input stage is a Gilbert cell mixer with an oscillator. Typical mixer characteristics include a noise figure of 6.2 dB, conversion gain of 17 dB, and input third-order intercept of −9 dBm. The oscillator will operate in excess of 200 MHz in an *LC* tank-circuit configuration. The

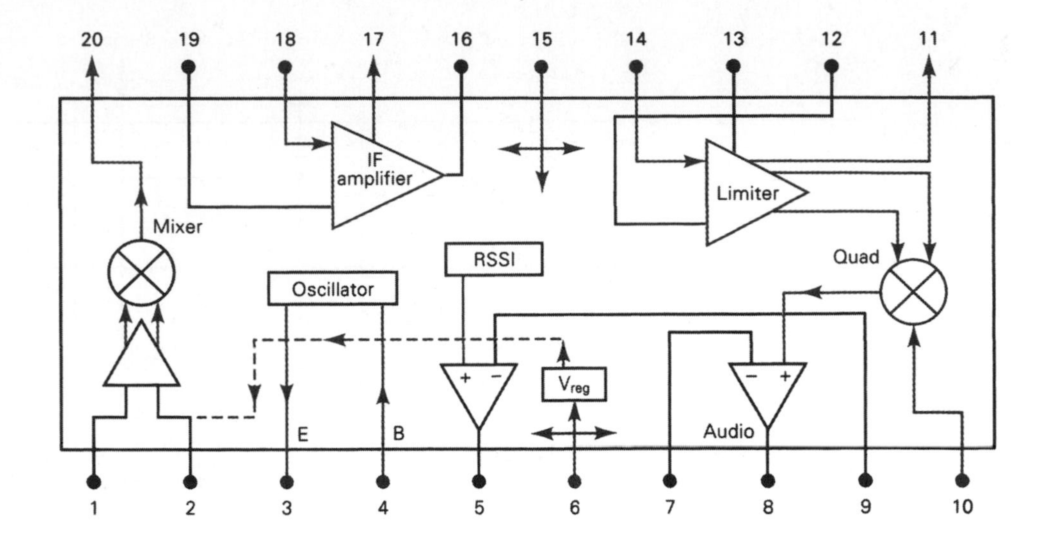

FIGURE 8-19 Block diagram for the Signetics NE/SA616 monolithic FM IF system

output impedance of the mixer is a 1.5-kΩ resistor, permitting direct connection to a 455-kHz ceramic filter. The IF amplifier has 43 dB of gain and a 5.5-MHz bandwidth. The IF limiter has 60 dB of gain and a 4.5-MHz bandwidth. The quadrature detector also uses a Gilbert cell. One port of the cell is internally driven by the IF signal, and the other output of the IF is ac coupled to a tuned quadrature network, where it undergoes a 90° phase shift before being fed back to the other port of the Gilbert cell. The demodulator output of the quadrature detector drives an internal op-amp. The op-amp can be configured as a unity-gain buffer or for simultaneous gain, filtering, and second-order temperature compensation if needed.

8-8-3 Single-Chip FM Radio System

The TDA7000 is a monolithic integrated-circuit FM radio system manufactured by Signetics Corporation for monophonic FM portable radios. In essence, the TDA7000 is a complete FM radio receiver on a single integrated-circuit chip. The TDA7000 features small size, lack of IF coils, easy assembly, and low power consumption. External to the IC is only one tunable *LC* tank circuit for the local oscillator, a few inexpensive ceramic plate capacitors, and one resistor. Using the TDA7000, a complete FM radio can be made small enough to fit inside a calculator, cigarette lighter, key-ring fob, or even a slim watch. The TDA7000 can also be used in equipment such as cordless telephones, radio-controlled models, paging systems, or the sound channel of a television receiver.

The block diagram for the TDA7000 is shown in Figure 8-20. The TDA7000 includes the following functional blocks: RF input stage, mixer, local oscillator, IF amplitude/limiter, phase demodulator, mute detector, and mute switch. The IC has an internal FLL (frequency-locked-loop) system with an intermediate frequency of 70 MHz. The FLL is used to reduce the total harmonic distortion (THD) by compressing the IF frequency swing (deviation). This is accomplished by using the audio output from the FM demodulator to shift the local oscillator frequency in opposition to the IF deviation. The principle is to compress 75 kHz of frequency deviation down to approximately 15 kHz. This limits the total harmonic distortion to 0.7% with ±22.5-kHz deviation and to 2.3% with ±75-kHz deviation. The IF selectivity is obtained with active *RC* Sallen-Key filters. The only function that needs alignment is the resonant circuit for the oscillator.

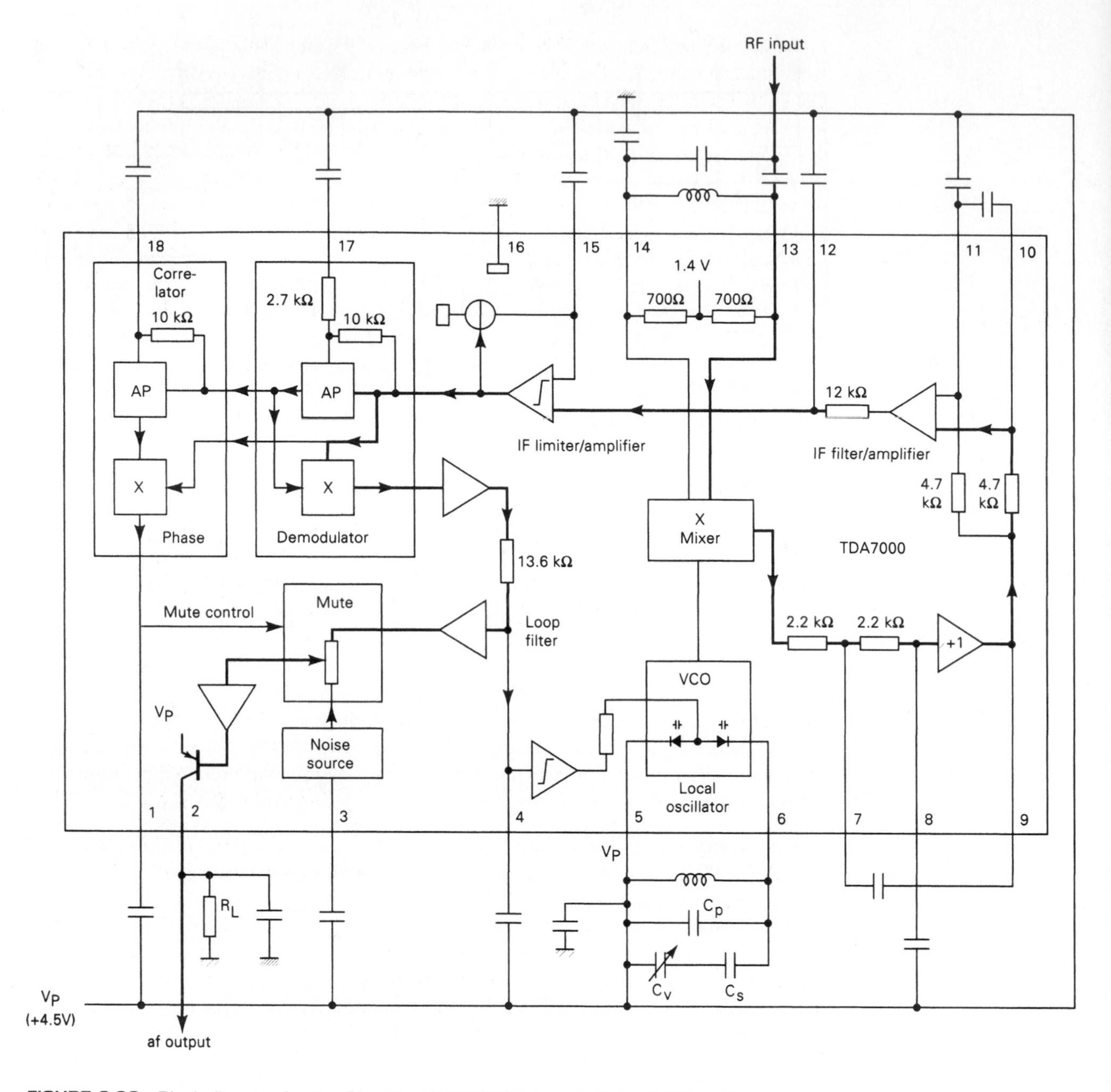

FIGURE 8-20 Block diagram for the Signetics TDA7000 integrated-circuit FM radio

8-9 FM STEREO BROADCASTING

Until 1961, all commercial FM broadcast-band transmissions were *monophonic.* That is, a single 50-Hz to 15-kHz audio channel made up the entire voice and music information frequency spectrum. This single audio channel modulated a high-frequency carrier and was transmitted through a 200-kHz-bandwidth FM communications channel. With *mono* transmission, each speaker assembly at the receiver reproduces exactly the same information. It is possible to separate the information frequencies with special speakers, such as *woofers* for low frequencies and *tweeters* for high frequencies. However, it is impossible to separate monophonic sound *spatially.* The entire information signal sounds as though it is coming from the same direction (i.e., from a *point source,* there is no directivity to the sound). In 1961, the FCC authorized *stereophonic* transmission for the commercial FM broadcast band. With stereophonic transmission, the information signal is spatially divided into two 50-Hz to 15-kHz audio channels (a left and a right). Music that originated on the left side is reproduced only on the left speaker, and music that originated on the right side is reproduced only on the right speaker. Therefore, with stereophonic transmission, it is possible to reproduce music with a unique directivity and spatial dimension that before was possible only with live entertainment (i.e., from an *extended* source). Also, with stereo transmission, it is possible to separate music or sound by *tonal quality,* such as percussion, strings, horns, and so on.

A primary concern of the FCC before authorizing stereophonic transmission was its compatibility with monophonic receivers. Stereo transmission was not to affect mono reception. Also, monophonic receivers must be able to receive stereo transmission as monaural without any perceptible degradation in program quality. In addition, stereophonic receivers were to receive stereo programming with nearly perfect separation (40 dB or more) between the left and right channels.

The original FM audio spectrum is shown in Figure 8-21a. The audio channel extended from 50 Hz to 15 kHz. In 1955, the FCC approved subcarrier transmission under the Subsidiary Communications Authorization (SCA). SCA is used to broadcast uninterrupted music to private subscribers, such as department stores, restaurants, and medical offices equipped with special SCA receivers. This is the music we sometimes cordially refer to as "elevator music." Originally, the SCA subcarrier ranged from 25 kHz to 75 kHz but has

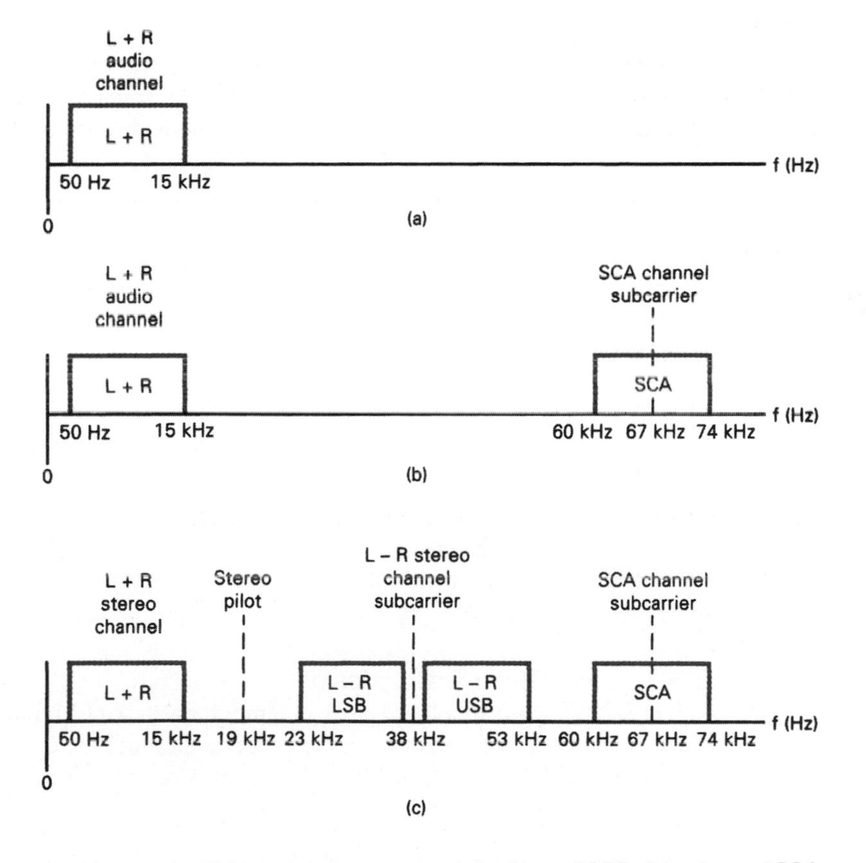

FIGURE 8-21 FM baseband spectrum: (a) prior to 1955; (b) prior to 1961; (c) since 1961

since been standardized at 67 kHz. The subcarrier and its associated sidebands become part of the total signal that modulates the main carrier. At the receiver, the subcarrier is demodulated along with the primary channel but cannot be heard because of its high frequency. The process of placing two or more independent channels next to each other in the frequency domain (stacking the channels), and then modulating a single high-frequency carrier with the combined signal is called *frequency division multiplexing* (FDM). With FM stereophonic broadcasting, three voice or music channels are frequency division multiplexed onto a single FM carrier. Figure 8-21b shows the total baseband frequency spectrum for FM broadcasting prior to 1961 (the composite baseband comprises the total modulating-signal spectrum). The primary audio channel remained at 50 Hz to 15 kHz, while an additional SCA channel is frequency translated to the 60-kHz to 74-kHz passband. The SCA subcarrier may be AM single- or double-sideband transmission or FM with a maximum modulating-signal frequency of 7 kHz. However, the SCA modulation of the main carrier is low-index narrowband FM and, consequently, is a much lower quality transmission than the primary FM channel. The total frequency deviation remained at 75 kHz with 90% (67.5 kHz) reserved for the primary channel and 10% (7.5 kHz) reserved for SCA.

Figure 8-21c shows the FM baseband frequency spectrum as it has been since 1961. It comprises the 50-Hz to 15-kHz stereo channel plus an additional stereo channel frequency division multiplexed into a composite baseband signal with a 19-kHz pilot. The three channels are (1) the left (L) plus the right (R) audio channels (the L − R stereo channel), (2) the left plus the inverted right audio channels (the L − R stereo channel), and (3) the SCA subcarrier and its associated sidebands. The L − R stereo channel occupies the 0-Hz to 15-kHz passband (in essence, the unaltered L and R audio information combined). The L − R audio channel amplitude modulates a 38-kHz subcarrier and produces the L − R stereo channel, which is a double-sideband suppressed-carrier signal that occupies the 23-kHz to 53-kHz passband, used only for FM stereo transmission. SCA transmissions occupy the 60-kHz to 74-kHz frequency spectrum. The information contained in the L + R and L − R stereo channels is identical except for their phase. With this scheme, mono receivers can demodulate the total baseband spectrum, but only the 50-Hz to 15-kHz L − R audio channel is amplified and fed to all its speakers. Therefore, each speaker reproduces the total original sound spectrum. Stereophonic receivers must provide additional demodulation of the 23-kHz to 53-kHz L − R stereo channel, separate the left and right audio channels, and then feed them to their respective speakers. Again, the SCA subcarrier is demodulated by all FM receivers, although only those with special SCA equipment further demodulate the subcarrier to audio frequencies.

With stereo transmission, the maximum frequency deviation is still 75 kHz; 7.5 kHz (10%) is reserved for SCA transmission, and another 7.5 kHz (10%) is reserved for a 19-kHz stereo pilot. This leaves 60 kHz of frequency deviation for the actual stereophonic transmission of the L + R and L − R stereo channels. However, the L + R and L − R stereo channels are not necessarily limited to 30-kHz frequency deviation each. A rather simple but unique technique is used to interleave the two channels such that at times either the L + R or the L − R stereo channel may deviate the main carrier 60 kHz by themselves. However, the total deviation will never exceed 60 kHz. This interleaving technique is explained later in this section.

8-9-1 FM Stereo Transmission

Figure 8-22 shows a simplified block diagram for a stereo FM transmitter. The L and R audio channels are combined in a matrix network to produce the L + R and L − R audio channels. The L − R audio channel modulates a 38-kHz subcarrier and produces a 23-kHz to 53-kHz L − R stereo channel. Because there is a time delay introduced in the L − R signal path as it propagates through the balanced modulator, the L + R stereo channel must be artificially delayed somewhat to maintain phase integrity with the L − R stereo channel for

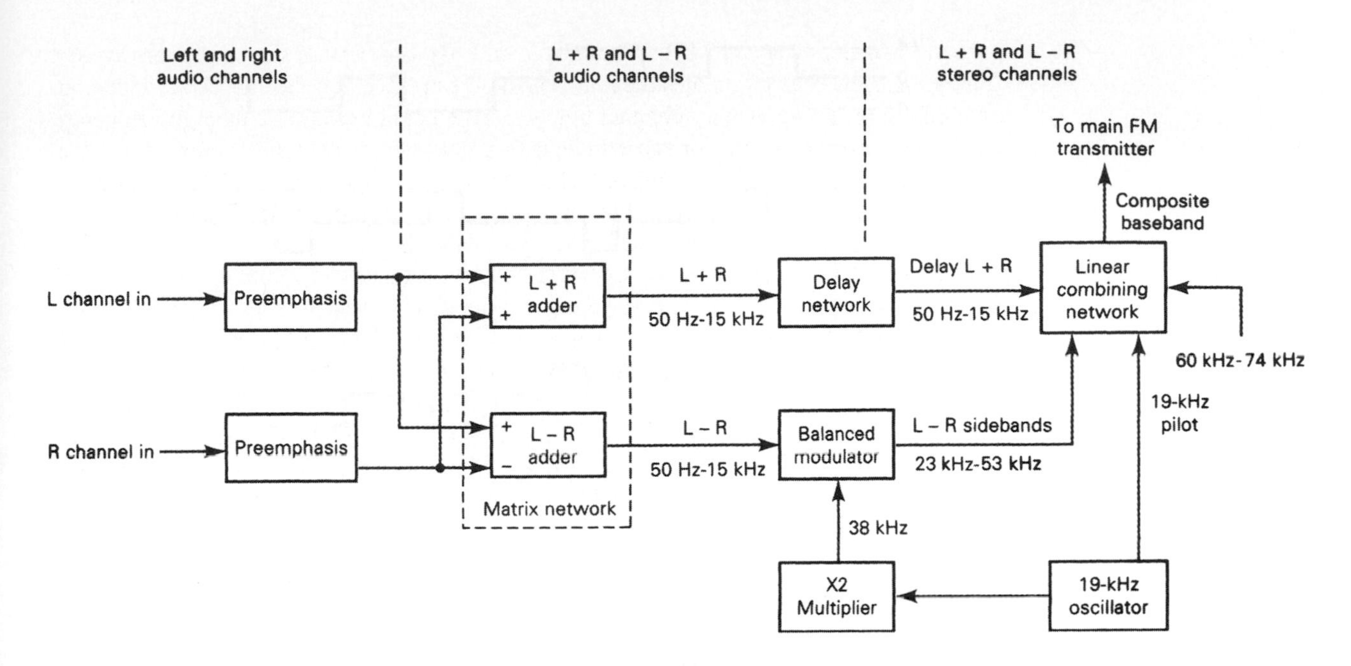

FIGURE 8-22 Stereo FM transmitter using frequency-division multiplexing

demodulation purposes. Also for demodulation purposes, a 19-kHz pilot is transmitted rather than the 38-kHz subcarrier because it is considerably more difficult to recover the 38-kHz subcarrier in the receiver. The composite baseband signal is fed to the FM transmitter, where it modulates the main carrier.

8-9-1-1 L + R and L − R channel interleaving. Figure 8-23 shows the development of the composite stereo signal for equal-amplitude L and R audio channel signals. For illustration purposes, rectangular waveforms are shown. Table 8-1 is a tabular summary of the individual and total signal voltages for Figure 8-23. Note that the L − R audio channel does not appear in the composite waveform. The L − R audio channel modulates the 38-kHz subcarrier to form the L − R stereo sidebands, which are part of the composite spectrum.

For the FM modulator in this example, it is assumed that 10 V of baseband signal will produce 75 kHz of frequency deviation of the main carrier, and the SCA and 19-kHz pilot polarities shown are for maximum frequency deviation. The L and R audio channels are each limited to a maximum value of 4 V; 1 V is for SCA, and 1 V is for the 19-kHz stereo pilot. Therefore, 8 V is left for the L + R and L − R stereo channels. Figure 8-23 shows the L, R, L + R, and L − R channels; the SCA and 19-kHz pilot; and

Table 8-1 Composite FM Voltages

L	R	L + R	L − R	SCA and Pilot	Total
0	0	0	0	2	2
4	0	4	4	2	10
0	4	4	−4	2	2
4	4	8	0	2	10
4	−4	0	8	2	10
−4	4	0	−8	−2	−10
−4	−4	−8	0	−2	−10

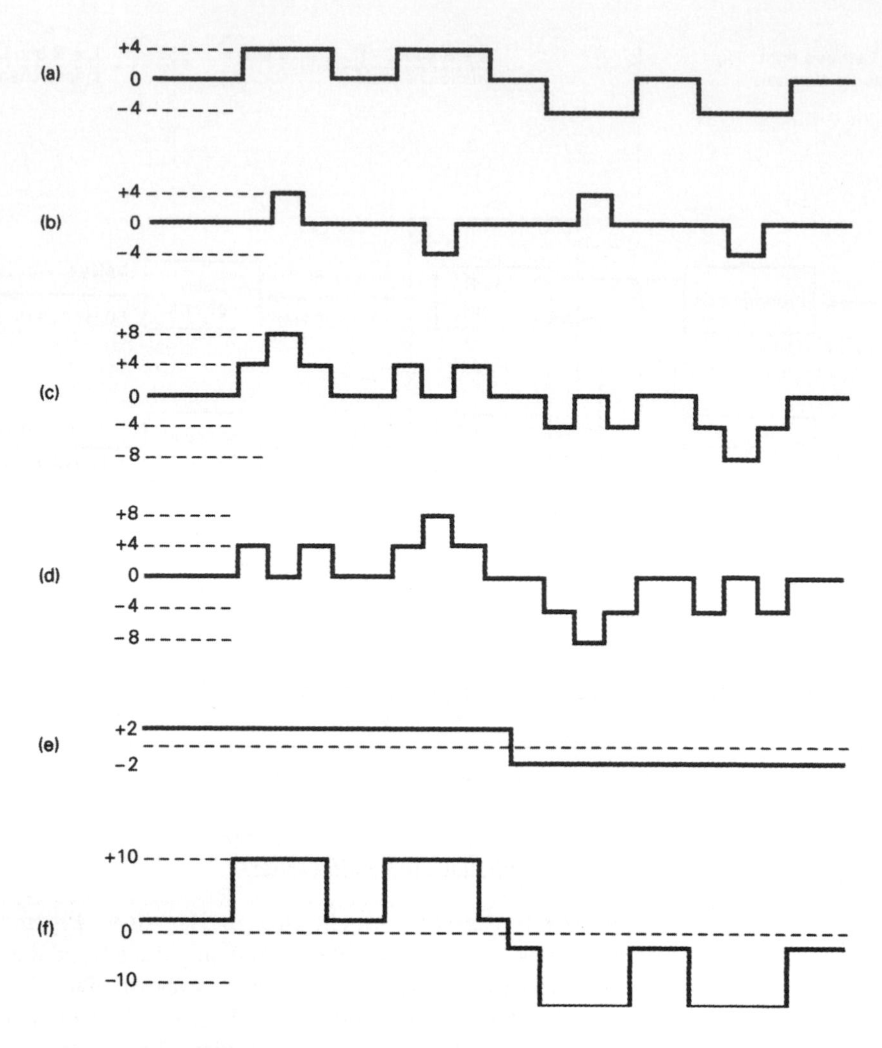

FIGURE 8-23 Development of the composite stereo signal for equal-amplitude L and R signals: (a) L audio signal; (b) R audio signal; (c) L + R stereo channel; (d) L − R stereo channel; (e) SCA + 19-kHz pilot; (f) composite baseband waveform

the composite stereo waveform. It can be seen that the L + R and L − R stereo channels interleave and never produce more than 8 V of total amplitude and, therefore, never produce more than 60 kHz of frequency deviation. The total composite baseband never exceeds 10 V (75-kHz deviation).

Figure 8-24 shows the development of the composite stereo waveform for unequal values for the L and R signals. Again, it can be seen that the composite stereo waveform never exceeds 10 V or 75 kHz of frequency deviation. For the first set of waveforms, it appears that the sum of the L + R and L − R waveforms completely cancels. Actually, this is not true; it only appears that way because rectangular waveforms are used in this example.

8-9-2 FM Stereo Reception

FM stereo receivers are identical to standard FM receivers up to the output of the audio detector stage. The output of the discriminator is the total baseband spectrum that was shown in Figure 8-21c.

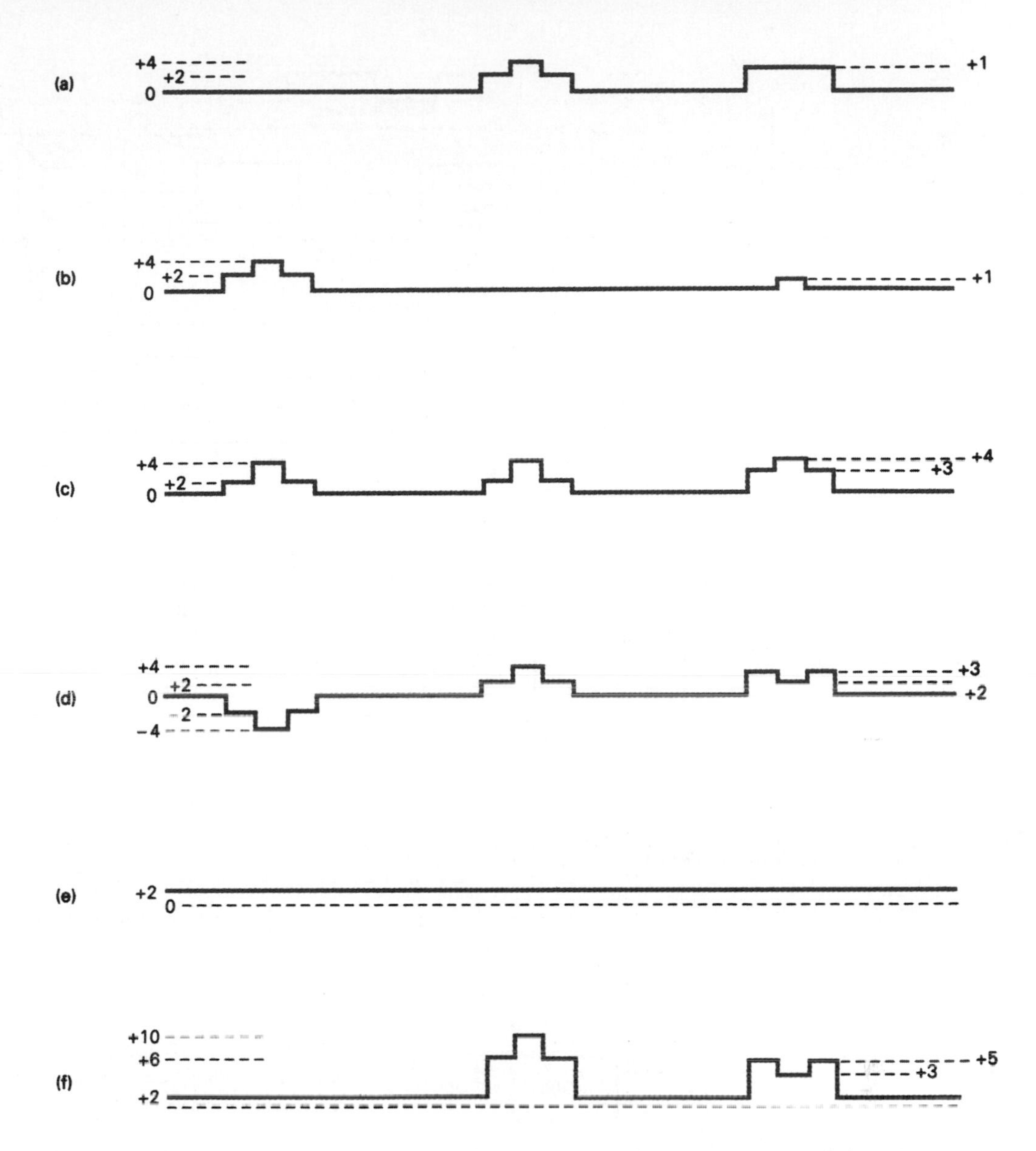

FIGURE 8-24 Development of the composite stereo signal for unequal amplitude L and R signals: (a) L audio signal; (b) R audio signal; (c) L + R stereo channel; (d) L − R stereo channel; (e) SCA + 19-kHz pilot; (f) composite baseband waveform

Figure 8-25 shows a simplified block diagram for an FM receiver that has both mono and stereo audio outputs. In the mono section of the signal processor, the L + R stereo channel, which contains all the original information from both the L and R audio channels, is simply filtered, amplified, and then fed to both the L and R speakers. In the stereo section of the signal processor, the baseband signal is fed to a stereo demodulator where the L and R audio channels are separated and then fed to their respective speakers. The L + R and L − R stereo channels and the 19-kHz pilot are separated from the composite baseband signal with filters. The 19-kHz pilot is filtered with a high-Q bandpass filter, multiplied by 2, amplified, and then fed to the L − R demodulator. The L + R stereo channel is filtered off by a low-pass filter with an upper cutoff frequency of 15 kHz. The L − R double-sideband signal is separated with a broadly tuned bandpass filter and then mixed with the recovered 38-kHz carrier in a balanced modulator to produce the L − R audio information. The matrix network

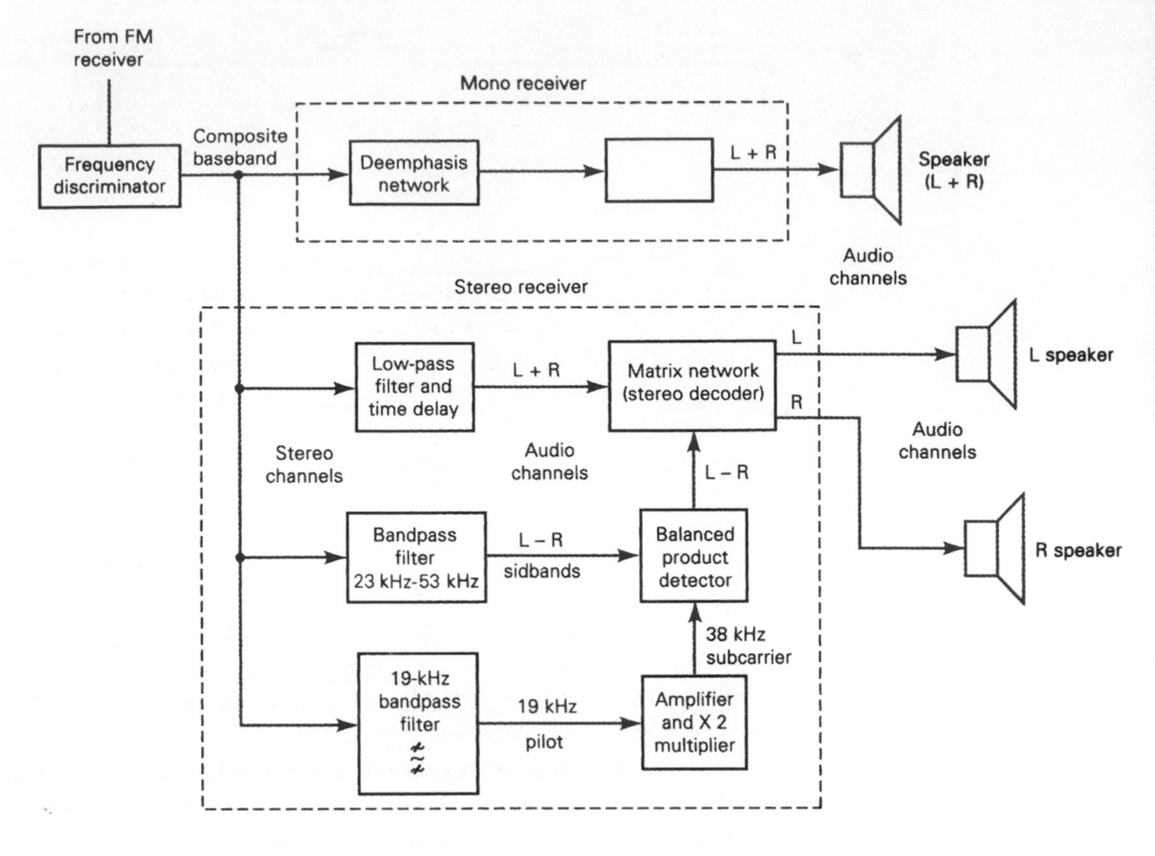

FIGURE 8-25 FM stereo and mono receiver

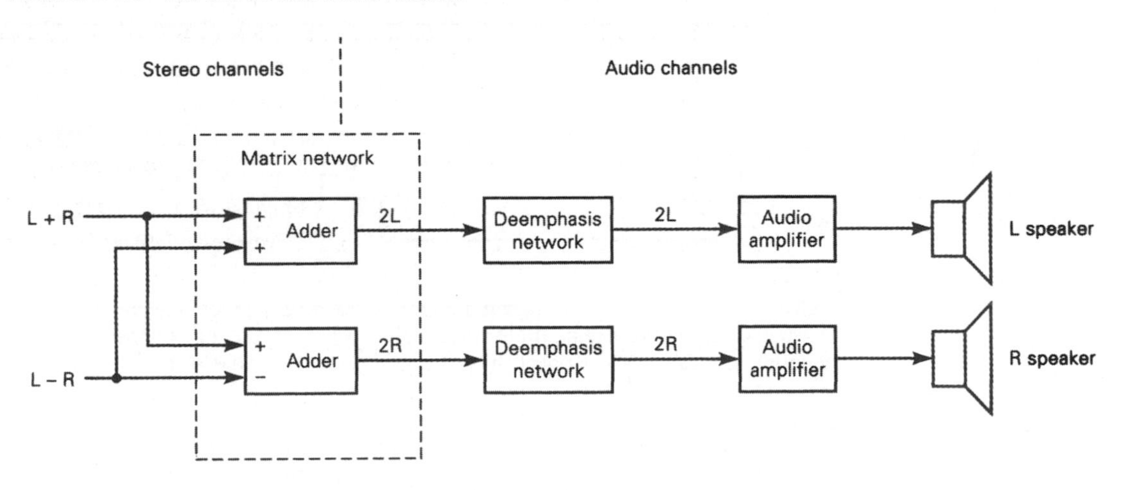

FIGURE 8-26 Stereo matrix network decoder

combines the L + R and L − R signals in such a way as to separate the L and R audio information signals, which are fed to their respective deemphasis networks and speakers.

Figure 8-26 shows the block diagram for a stereo matrix decoder. The L − R audio channel is added directly to the L + R audio channel. The output from the adder is

$$\begin{array}{r} L + R \\ + (L - R) \\ \hline 2L \end{array}$$

The L − R audio channel is inverted and then added to the L + R audio channel. The output from the adder is

$$\begin{array}{r} L + R \\ \underline{-\,(L - R)} \\ 2R \end{array}$$

The L + R and L − R stereo channels and the 19-kHz pilot are separated from the composite baseband signal with filters. The 19-kHz pilot is filtered with a high-Q bandpass filter, mulitiplied by 2, amplified, and then fed to the L − R demodulator. The L + R stereo channel is filtered off by a low-pass filter with an upper cutoff frequency of 15 kHz. The L − R double-sideband signal is separated with a broadly tuned bandpass filter and then mixed with the recovered 38-kHz carrier in a balanced modulator to produce the L − R audio information. The matrix network combines the L + R and L − R signals in such a way as to separate the L and R audio information signals, which are fed to their respective deemphasis networks and speakers.

8-9-2-1 Large-scale integration stereo demodulator. Figure 8-27 shows the specification sheet for the XR-1310 stereo demodulator/decoder. The XR-1310 is a monolithic FM stereo demodulator that uses PLL techniques to derive the right and left audio channels from the composite stereo signal. The XR-1310 uses a PLL to lock onto the 19-kHz pilot and regenerate the 38-kHz subcarrier. The XR-1310 requires no external LC tank circuits for tuning, and alignment is accomplished with a single potentiometer. The XR-1310 features simple noncritical tuning, excellent channel separation, low distortion, and a wide dynamic range.

8-10 TWO-WAY MOBILE COMMUNICATIONS SERVICES

There are many types of two-way radio communications systems that offer a wide variety of services, including the following:

1. Two-way mobile radio. Half-duplex, one-to-many radio communications with no dial tone.
   a. Class D citizens band (CB) radio. Provides 26.96 to 27.41 MHz (40, 10-kHz shared channels) public, noncommercial radio service for either personal or business use utilizing push-to-talk AM DSBFC and AM SSBFC. There are three other lesser known CB classifications (A, B, and C).
   b. Amateur (ham) radio. Covers a broad-frequency band from 1.8 MHz to above 300 MHz. Designed for personal use without pecuniary interest. Amateur radio offers a broad range of classes including CW, AM, FM, radio teleprinter (RTTY), HF slow scan still picture TV (SSTV), VHF or UHF slow- or fast-scan television and facsimile, and audio FSK (AFSK).
   c. Aeronautical Broadcasting Service (ABS). Provides 2.8 MHz to 457 MHz. ABS disseminates information for the purposes of air navigation and air-to-ground communications utilizing conventional AM and various forms of AM SSB in the HF, MF, and VHF frequency bands.
   d. Private land mobile radio services.
      i. Public safety radio, including two-way UHF and VHF push-to-talk FM systems typically used by police and fire departments, highway maintenance, forestry conservation, and local government radio services.
      ii. Special emergency radio including medical, rescue disaster relief, school bus, veterinarian, beach patrol, and paging radio services.
      iii. Industrial radio including power company, petroleum company, forest product, business, manufacturer, motion picture, press relay, and telephone maintenance radio service.

STEREO DEMODULATOR

March 1982

GENERAL DESCRIPTION

The XR-1310 is a unique FM stereo demodulator which uses phase-locked techniques to derive the right and left audio channels from the composite signal. Using a phase-locked loop to regenerate the 38 kHz subcarrier, it requires no external L-C tanks for tuning. Alignment is accomplished with a single potentiometer.

FEATURES

Requires No Inductors
Low External Part Count
Simple, Noncritical Tuning by Single Potentiometer Adjustment
Internal Stereo/Monaural Switch with 100 mA Lamp Driving Capability
Wide Dynamic Range: 600 mV (RMS) Maximum Composite Input Signal
Wide Supply Voltage Range: 8 to 14 Volts
Excellent Channel Separation
Low Distortion
Excellent SCA Rejection

FUNCTIONAL BLOCK DIAGRAM

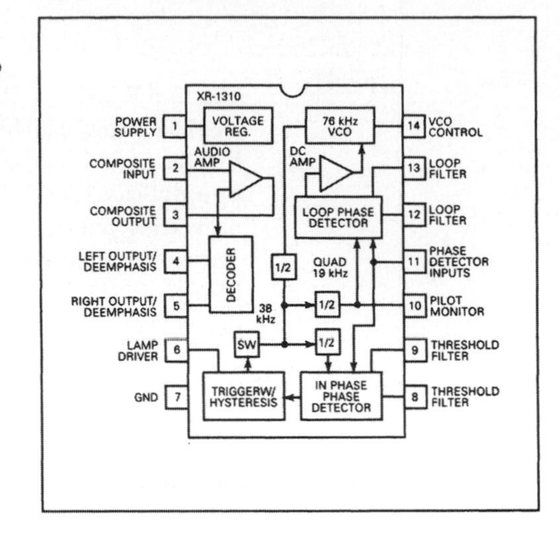

ORDERING INFORMATION

Part Number	Package	Operating Temperature
XR-1310CP	Plastic	–40°C to +85°C

APPLICATIONS

FM Stereo Demodulation

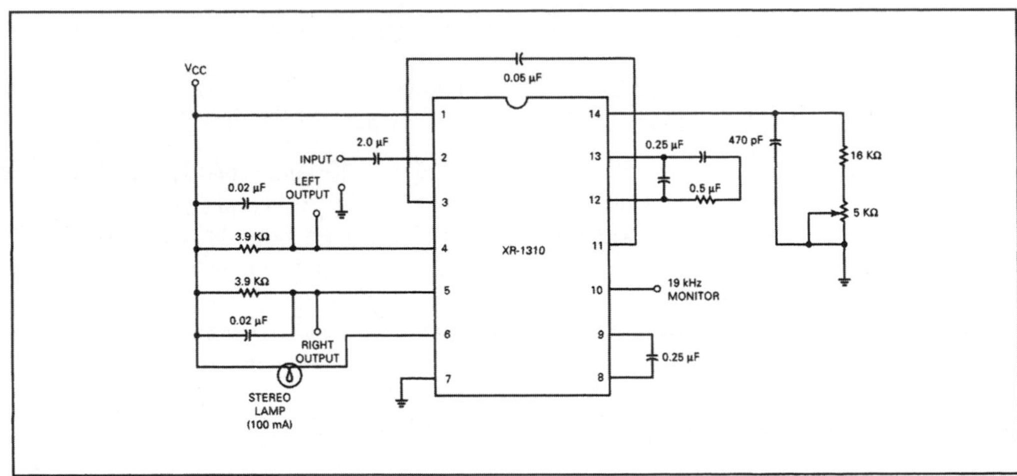

Figure 1. Typical Application

FIGURE 8-27 XR-1310 stereo demodulator

2. Mobile telephone service. Full-duplex, one-to-one radio telephone communications.
   a. Analog cellular radio. FM transmission using FDMA or TDMA.
   b. Digital cellular radio. Personal communications system (PCS). PSK transmission of PCM-encoded voice signals using TDMA, FDMA, and CDMA.
   c. Personal communications satellite service (PCSS). Provides worldwide telecommunications service using handheld telephones that communicate with each other through low earth-orbit satellite repeaters incorporating QPSK modulation and both FDMA and TDMA.

ELECTRICAL CHARACTERISTICS
Test conditions: Unless otherwise noted; V_{CC}* = +12Vdc,T_A = +25°C,560mV(RMS)(2.8Vp-p) standard multiplex composite signal with L or R channel only modulated at 1.0 kHz and with 100 mV (RMS) (10 % pilot level), using circuit of Figure 1.

PARAMETERS	MIN.	TYP.	MAX.	UNIT
Maximum Standard Composite Input Signal (0.5 % THD)	2.8			V (p-p)
Maximum Monural Input Signal (1.0 % THD)	2.8			V (p-p)
Input Impedance		50		kΩ
Stereo Channel Separation (50 Hz – 15 kHz)	30	40		dB
Audio Output Voltage (desired channel)		485		mV (rms)
Monaural Channel Balance (pilot tone "off")			1.5	dB
Total Harmonic Distortion		0.3		%
Ultrasonic Frequency Rejection 19 kHz	50	34.4		dB
38 kHz		45		
Inherent SCA Rejection (f = 67 kHz; 9.0 kHz beat note measured with 1.0 kHz modulation "off")		80		dB
Stereo Switch Level				
(19 kHz input for lamp "on")	13		20	mV (rms)
Hysteresis		6		dB
Capture Range (permissable tuning error of internal oscillator, reference circuit values of Figure 1)		±3.5		%
Operating Supply Voltage (loads reduced to 2.7 kΩ for 8.0-volt operation	8.0		14	V (dc)
Current Drain (lamp "off")		13		mA (dc)

*Symbols conform to JEDEC Engineering Bulletin No. 1 when applicable.

ABSOLUTE MAXIMUM RATINGS
(TA = +25°C unless otherwise noted)

Power Supply Voltage	14 V
Lamp Current (nominal rating, 12 V lamp)	75 mA
Power Dissipation (package limitation)	625 mW
Derate above TA = +25°C	5.0 mW/°C
Operating Temperature Range (Ambient)	−40 to +85°C
Storage Temperature Range	−65 to +150°C

FIGURE 8-27 *(Continued)* XR-1310 stereo demodulator

8-11 TWO-WAY FM RADIO COMMUNICATIONS

Two-way FM radio communication is used extensively for *public safety* mobile communications, such as police and fire departments and emergency medical services. Three primary frequency bands are allocated by the FCC for two-way FM radio communications: 132 MHz to 174 MHz, 450 MHz to 470 MHz, and 806 MHz to 947 MHz. The maximum frequency deviation for two-way FM transmitters is typically 5 kHz, and the maximum modulating-signal frequency is 3 kHz. These values give a deviation ratio of 1.67 and a maximum Bessel bandwidth of approximately 24 kHz. However, the allocated FCC channel spacing is 30 kHz. Two-way FM radio is half-duplex, which supports two-way communications but not simultaneously; only one party can transmit at a time. Transmissions are initiated by closing a *push-to-talk* (PTT) switch, which turns on the transmitter and shuts off the receiver. During idle conditions, the transmitter is shut off and the receiver is turned on to allow monitoring the radio channel for transmissions from other stations' transmitters.

8-11-1 Historical Perspective

Mobile radio was used as early as 1921 when the Detroit Police Department used a mobile radio system that operated at a frequency close to 2 MHz. In 1940, the FCC made available new frequencies for mobile radio in the 30-MHz to 40-MHz frequency band. However, not until researchers developed frequency modulation techniques to improve reception in the presence of electrical noise and signal fading did mobile radio become useful. The first commercial mobile telephone system in the United States was established in 1946 in St. Louis, Missouri, when the FCC allocated six 60-kHz mobile telephone channels in the 150-MHz frequency range. In 1947, a public mobile telephone system was established along the highway between New York City and Boston that operated in the 35-MHz to 40-MHz frequency range. In 1949, the FCC authorized six additional mobile channels to *radio common carriers,* which they defined as companies that do not provide public wireline telephone service but do interconnect to the public telephone network and provide equivalent *non wireline* telephone service. The FCC later increased the number of channels from 6 to 11 by reducing the bandwidth to 30 kHz and spacing the new channels between the old ones. In 1950, the FCC added 12 new channels in the 450-MHz band.

Until 1964, mobile telephone systems operated only in the *manual mode;* a special mobile telephone operator handled every call to and from each *mobile unit.* In 1964, *automatic channel selection systems* were placed in service for mobile telephone systems. This eliminated the need for push-to-talk operation and allowed customers to *direct dial* their calls without the aid of an operator. *Automatic call completion* was extended to the 450-MHz band in 1969, and *improved mobile telephone systems* (IMTS) became the United States' standard mobile telephone service. Presently, there are more than 200,000 *mobile telephone service* (MTS) subscribers nationwide. MTS uses FM radio channels to establish communication links between mobile telephones and central *base station* transceivers, which are linked to the local telephone exchange via normal metallic telephone lines. Most MTS systems serve an area approximately 40 miles in diameter, and each channel operates similarly to a *party line.* Each channel may be assigned to several subscribers, but only one subscriber can use it at a time. If the preassigned channel is busy, the subscriber must wait until it is idle before either placing or receiving a call.

The growing demand for the overcrowded mobile telephone frequency spectrum prompted the FCC to issue Docket 18262, which inquired into a means for providing a higher frequency-spectrum efficiency. In 1971, AT&T submitted a proposal on the technical feasibility of providing efficient use of the mobile telephone frequency spectrum. AT&T's report, titled *High Capacity Mobile Phone Service,* outlined the principles of cellular radio.

In April 1981, the FCC approved a licensing scheme for *cellular radio* markets. Each market services one *coverage area,* defined according to modified 1980 Census Bureau Standard Metropolitan Statistical Areas (SMSAs). In early 1982, the FCC approved a final plan for accepting cellular license applications beginning in June 1982 and a final round of applications by March 1983. The ensuing legal battles for cellular licenses between AT&T, MCI, GTE, and numerous other common carriers go well beyond the scope of this book.

8-11-2 Two-Way FM Radio Transmitter

The simplified block diagram for a *modular integrated-circuit* two-way indirect FM radio transmitter is shown in Figure 8-28. Indirect FM is generally used because direct FM transmitters do not have the frequency stability necessary to meet FCC standards without using AFC loops. The transmitter shown is a four-channel unit that operates in the 150-kHz to 174-MHz frequency band. The channel selector switch applies power to one of four crystal oscillator modules that operates at a frequency between 12.5 MHz and 14.5 MHz, depending on the final transmit carrier frequency. The oscillator frequency is temperature compensated by the compensation module to ensure a stability of $\pm 0.0002\%$. The phase modulator uses a varactor diode that is modulated by the audio signal at the output of the audio limiter. The audio signal amplitude is limited to ensure that the transmitter is not

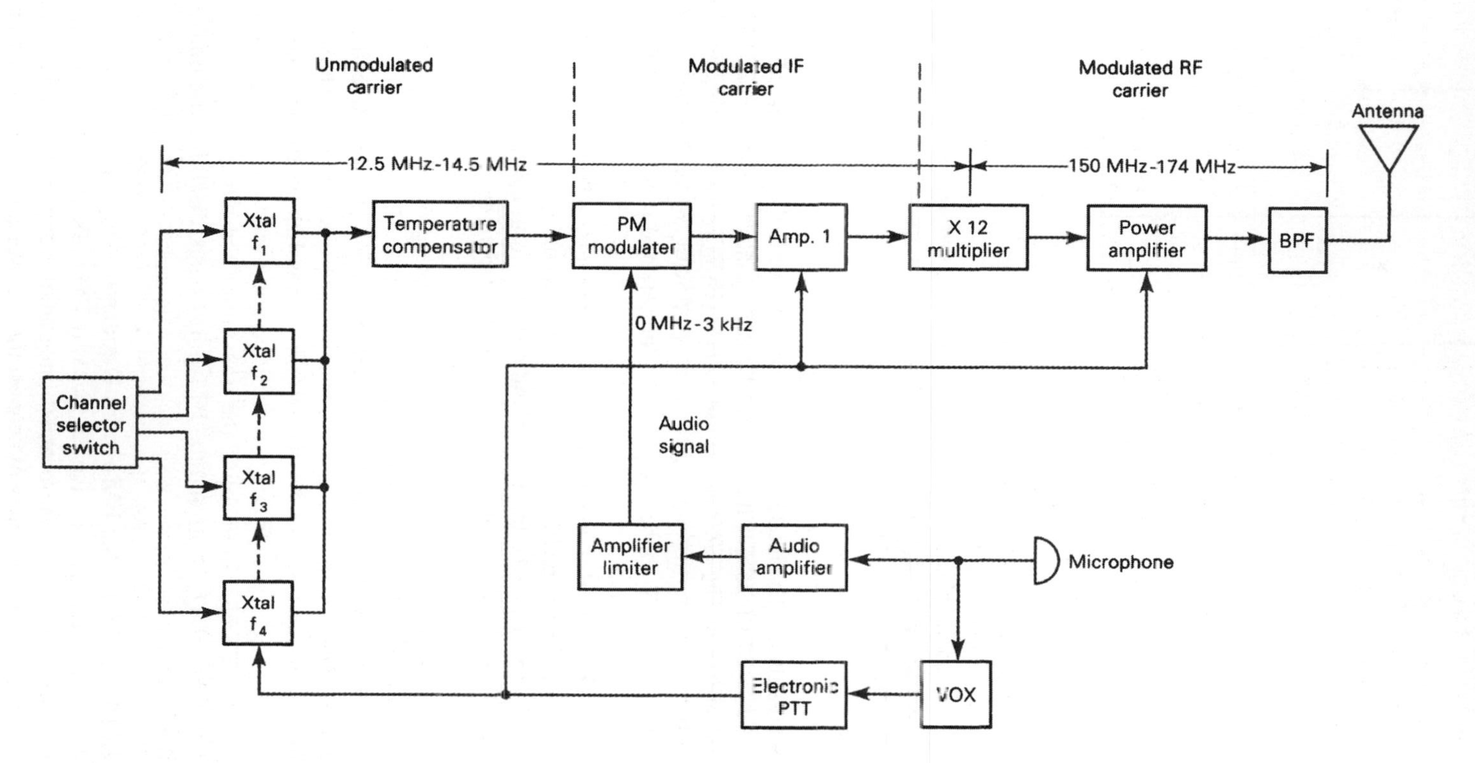

FIGURE 8-28 Two-way FM transmitter block diagram

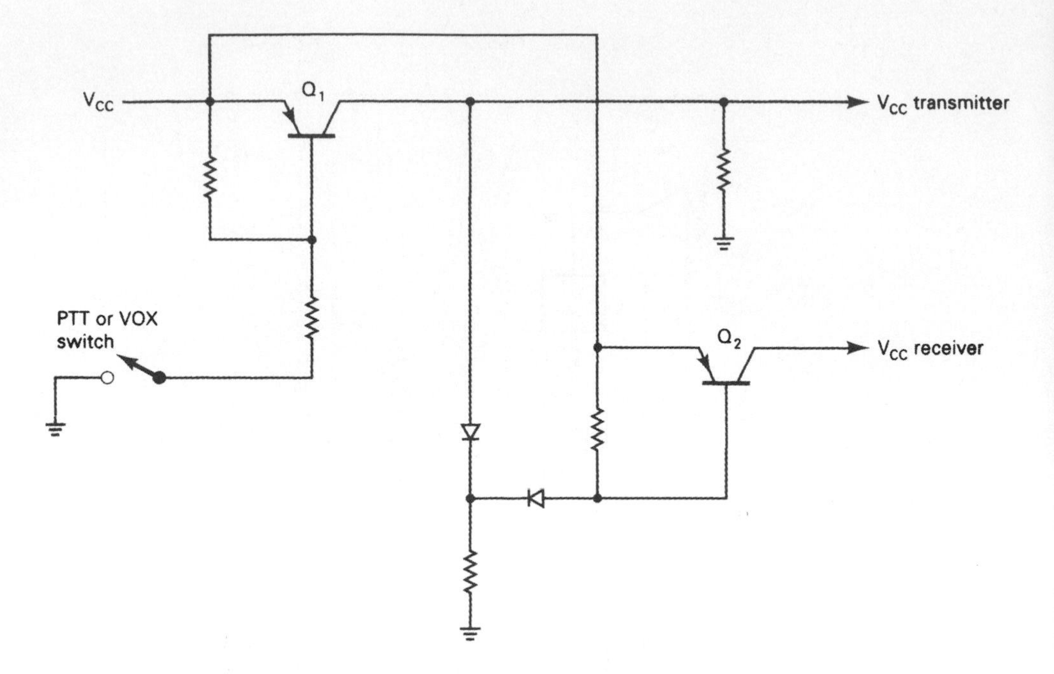

FIGURE 8-29 Electronic PTT schematic diagram

overdeviated. The modulated IF carrier is amplified and then multiplied by 12 to produce the desired RF carrier frequency. The RF signal is further amplified and filtered prior to transmission. The *electronic push-to-talk* (PTT) is used rather than a simple mechanical switch to reduce the static noise associated with *contact bounce* in mechanical switches. Keying the PTT applies dc power to the selected transmit oscillator module and the RF power amplifiers.

Figure 8-29 shows the schematic diagram for a typical electronic PTT module. Keying the PTT switch grounds the base of Q_1, causing it to conduct and turn off Q_2. With Q_2 off, V_{CC} is applied to the transmitter and removed from the receiver. With the PTT switch released, Q_1 shuts off, removing V_{CC} from the transmitter, turning on Q_2, and applying V_{CC} to the receiver.

Transmitters equipped with VOX (*voice-operated transmitter*) are automatically keyed each time the operator speaks into the microphone, regardless of whether the PTT button is depressed. Transmitters equipped with VOX require an external microphone. The schematic diagram for a typical VOX module is shown in Figure 8-30. Audio signal power in the 400-Hz to 600-Hz passband is filtered and amplified by Q_1, Q_2, and Q_3. The output from Q_3 is rectified and used to turn on Q_4, which places a ground on the PTT circuit, enabling the transmitter and disabling the receiver. With no audio input signal, Q_4 is off and the PTT pin is open, disabling the transmitter and enabling the receiver.

8-11-3 Two-Way FM Radio Receiver

The block diagram for a typical two-way FM radio receiver is shown in Figure 8-31. The receiver shown is a four-channel integrated-circuit modular receiver with four separate crystal oscillator modules. Whenever the receiver is on, one of the four oscillator modules is activated, depending on the position of the channel selector switch. The oscillator frequency is temperature compensated and then multiplied by 9. The output from the multiplier is applied to the mixer, where it heterodynes with the incoming RF signal to produce

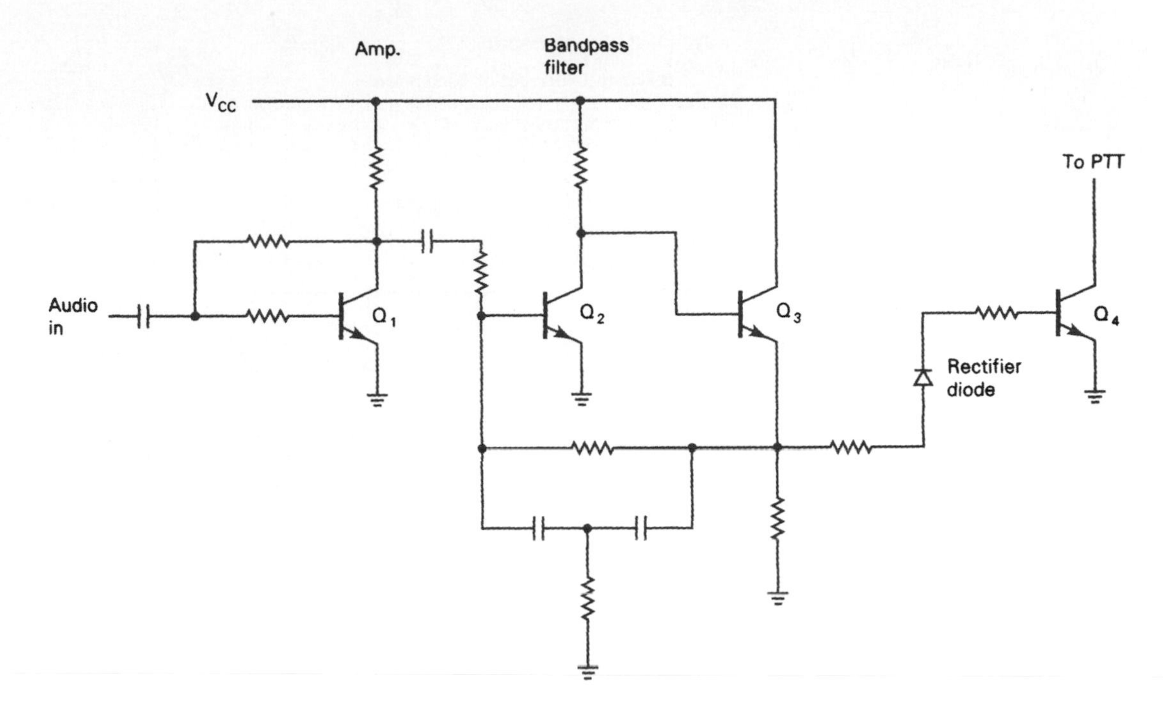

FIGURE 8-30 VOX schematic diagram

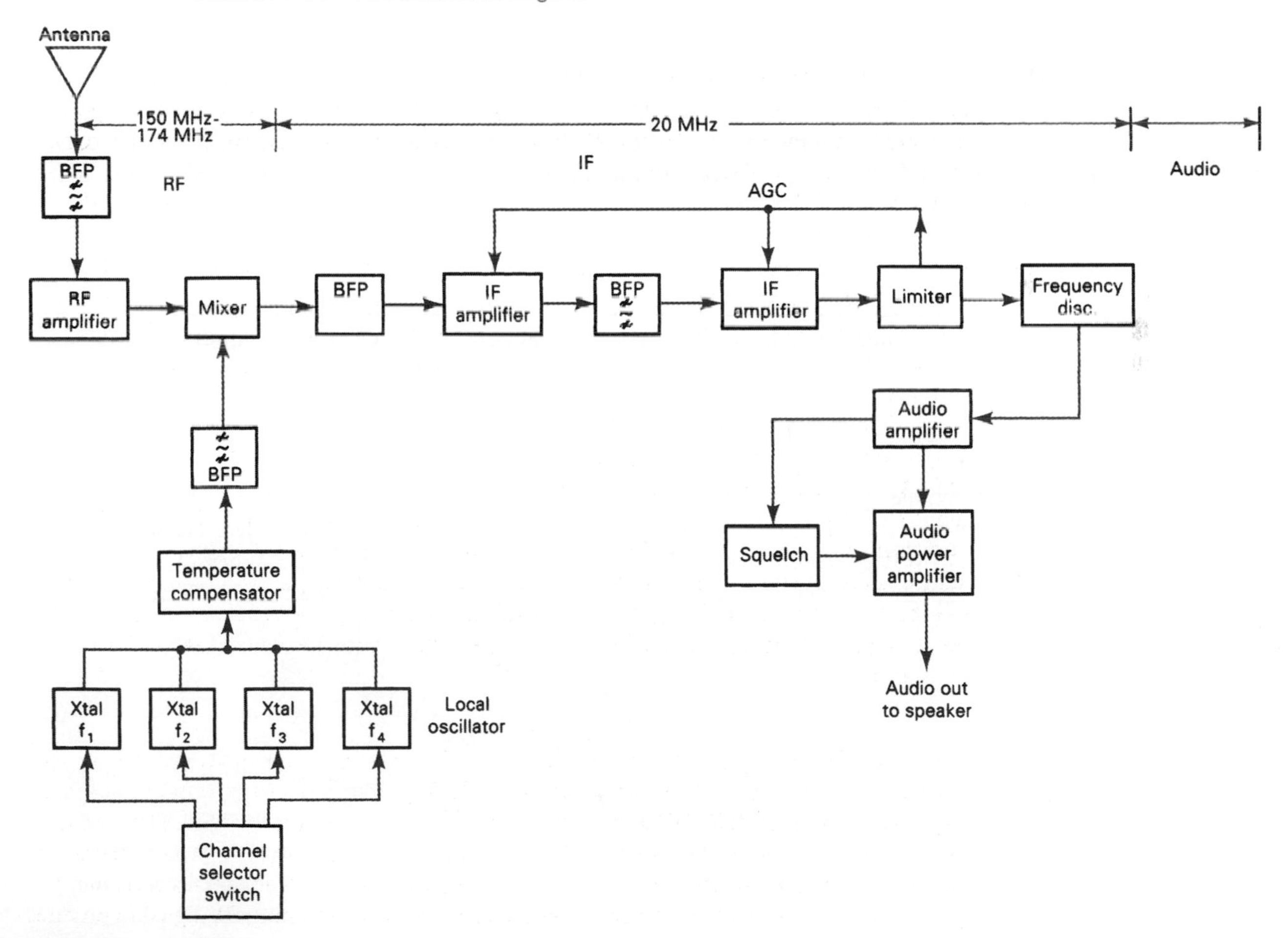

FIGURE 8-31 Two-way FM receiver block diagram

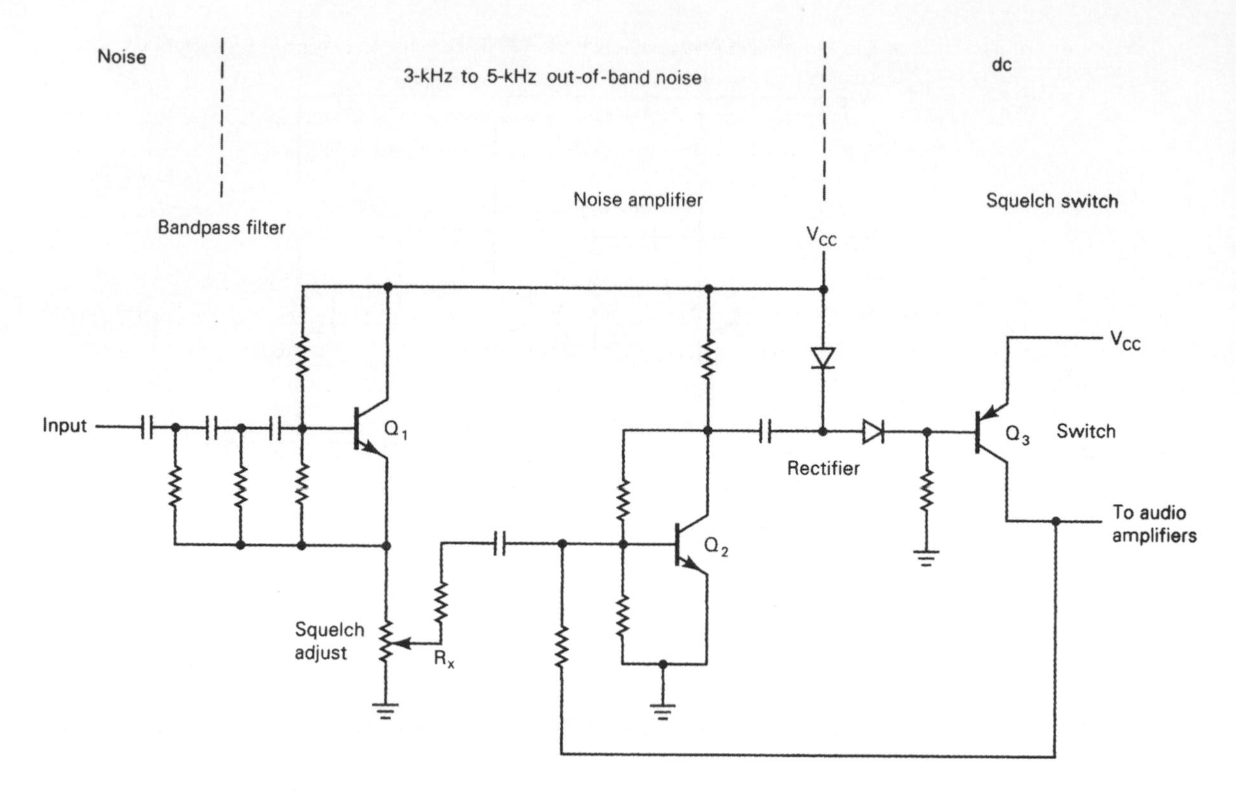

FIGURE 8-32 Squelch circuit

a 20-MHz intermediate frequency. This receiver uses low-side injection, and the crystal oscillator frequency is determined as follows:

$$\text{crystal frequency} = \frac{\text{RF frequency} - 20\text{ MHz}}{9} \tag{8-7}$$

The IF signal is filtered, amplified, limited, and then applied to the frequency discriminator for demodulation. The discriminator output voltage is amplified and then applied to the speaker. A typical noise amplifier/squelch circuit is shown in Figure 8-32. The squelch circuit is keyed by out-of-band noise at the output of the audio amplifier. With no receive RF signal, AGC causes the gain of the IF amplifiers to increase to maximum, which increases the receiver noise in the 3-kHz to 5-kHz band. Whenever excessive noise is present, the audio amplifier is turned off, and the receiver is quieted. The input bandpass filter passes the 3-kHz to 5-kHz noise signal, which is amplified and rectified. The rectified output voltage determines the off/on condition of squelch switch Q_3. When Q_3 is on, V_{CC} is applied to the audio amplifier. When Q_3 is off, V_{CC} is removed from the audio amplifier, quieting the receiver. R_x is a squelch sensitivity adjustment.

QUESTIONS

8-1. Describe the basic differences between AM and FM receivers.

8-2. Draw the schematic diagram for a *single-ended slope detector* and describe its operation.

8-3. Draw the schematic diagram for a *double-ended slope detector* and describe its operation.

8-4. Draw the schematic diagram for a *Foster-Seeley discriminator* and describe its operation.

8-5. Draw the schematic diagram for a *ratio detector* and describe its operation.

8-6. Describe the operation of a PLL FM demodulator.

8-7. Draw the schematic diagram for a *quadrature FM demodulator* and describe its operation.

8-8. Compare the advantages and disadvantages of the FM demodulator circuits discussed in Questions 8-1 through 8-7.

8-9. What is the purpose of a *limiter* in an FM receiver?

8-10. Describe *FM thresholding.*

8-11. Describe the operation of an FM *stereo transmitter;* an FM *stereo receiver.*

8-12. Draw the block diagram for a two-way FM radio transmitter and explain its operation.

8-13. Draw the block diagram for a two-way FM radio receiver and explain its operation.

8-14. Describe the operation of an electronic *push-to-talk circuit.*

8-15. Describe the operation of a *VOX circuit.*

8-16. Briefly explain how a composite *FM stereo* signal is produced.

8-17. What is meant by the term *interleaving* of L and R signals in stereo transmission?

8-18. What is the purpose of the 19-kHz *pilot* in FM stereo broadcasting?

8-19. What is the difference between *mobile radio* and *mobile telephone?*

PROBLEMS

8-1. Determine the minimum input S/N ratio required for a receiver with 15 dB of FM improvement, a noise figure NF = 4 dB, and a desired postdetection S/N = 33 dB.

8-2. For an FM receiver with a 100-kHz bandwidth, a noise figure NF = 6 dB, and an input noise temperature $T = 200°C$, determine the minimum receive carrier power to achieve a postdetection S/N = 40 dB. Use the receiver block diagram shown in Figure 8-1 as the receiver model and the FM thresholding curve shown in Figure 8-11.

8-3. For an FM receiver tuned to 92.75 MHz using high-side injection and a first IF of 10.7 MHz, determine the image frequency and the local oscillator frequency.

8-4. For an FM receiver with an input frequency deviation $\Delta f = 40$ kHz and a transfer ratio $K =$ 0.01 V/kHz, determine V_{out}.

8-5. For the balanced slope detector shown in Figure 8-3a, a center frequency $f_c = 20.4$ MHz, and a maximum input frequency deviation $\Delta f = 50$ kHz, determine the upper and lower cutoff frequencies for the tuned circuit.

8-6. For the Foster-Seeley discriminator shown in Figure 8-4, $V_{C1} = 1.2$ V and $V_{C2} = 0.8$ V, determine V_{out}.

8-7. For the ratio detector shown in Figure 8-6, $V_{C1} = 1.2$ V and $V_{C2} = 0.8$ V, determine V_{out}.

8-8. For an FM demodulator with an FM improvement factor of 23 dB and an input S/N = 26 dB, determine the postdetection S/N.

8-9. From Figure 8-11, determine the approximate FM improvement factor for an input S/N = 10.5 dB and $m = 1$.

8-10. Determine the minimum input S/N ratio required for a receiver with 15 dB of FM improvement, a noise figure NF = 6 dB, and a desired postdetection signal-to-noise ratio = 38 dB.

8-11. For an FM receiver with 200-kHz bandwidth, a noise figure NF = 8 dB, and an input noise temperature $T = 100°C$, determine the minimum receive carrier power to achieve a postdetection S/N = 40 dB. Use the receiver block diagram shown in Figure 8-1 as the receiver model and the FM thresholding curve shown in Figure 8-11.

8-12. For an FM receiver tuned to 94.5 MHz using high-side injection and a first IF of 10.7 MHz, determine the image frequency and the local oscillator frequency.

8-13. For an FM receiver with an input frequency deviation $\Delta f = 50$ kHz and a transfer ratio $K =$ 0.02 V/kHz, determine V_{out}.

8-14. For the balanced slope detector shown in Figure 8-3a, a center frequency f_c = 10.7 MHz, and a maximum input frequency deviation Δf = 40 kHz, determine the upper and lower cutoff frequencies for the circuit.

8-15. For the Foster-Seeley discriminator shown in Figure 8-4, V_{C1} = 1.6 V and V_{C2} = 0.4 V, determine V_{out}.

8-16. For the ratio detector shown in Figure 8-6, V_{C1} = 1.2 V and V_{C2} = 0.8 V, determine V_{out}.

8-17. For an FM demodulator with an FM improvement factor equal to 18 dB and an input (predetection) signal-to-noise S_i/N_i = 32 dB, determine the postdetection S/N.

8-18. From Figure 8-11, determine the approximate FM improvement factor for an input S/N = 11 dB and m = 1.

C H A P T E R 9

Digital Modulation

CHAPTER OUTLINE

OBJECTIVES

- Define *electronic communications*
- Define *digital modulation* and *digital radio*
- Define *digital communications*
- Define *information capacity*
- Define *bit, bit rate, baud,* and *minimum bandwidth*
- Explain Shannon's limit for information capacity
- Explain *M*-ary encoding
- Define and describe digital amplitude modulation
- Define and describe frequency-shift keying
- Describe continuous-phase frequency-shift keying
- Define *phase-shift keying*
- Explain binary phase-shift keying
- Explain quaternary phase-shift keying
- Describe 8- and 16-PSK
- Describe quadrature-amplitude modulation
- Explain 8-QAM
- Explain 16-QAM
- Define *bandwidth efficiency*
- Explain carrier recovery
- Explain clock recovery
- Define and describe differential phase-shift keying
- Define and explain trellis-code modulation
- Define *probability of error* and *bit error rate*
- Develop error performance equations for FSK, PSK, and QAM

9-1 INTRODUCTION

In essence, *electronic communications* is the transmission, reception, and processing of information with the use of electronic circuits. *Information* is defined as knowledge or intelligence that is communicated (i.e., transmitted or received) between two or more points. *Digital modulation* is the transmittal of digitally modulated analog signals (carriers) between two or more points in a communications system. Digital modulation is sometimes called *digital radio* because digitally modulated signals can be propagated through Earth's atmosphere and used in wireless communications systems. Traditional electronic communications systems that use conventional analog modulation, such as *amplitude modulation* (AM), *frequency modulation* (FM), and *phase modulation* (PM), are rapidly being replaced with more modern digital moduluation systems that offer several outstanding advantages over traditional analog systems, such as ease of processing, ease of multiplexing, and noise immunity.

Digital communications is a rather ambiguous term that could have entirely different meanings to different people. In the context of this book, digital communications include systems where relatively high-frequency analog carriers are modulated by relatively low-frequency digital information signals (*digital radio*) and systems involving the transmission of digital pulses (*digital transmission*). Digital transmission systems transport information in digital form and, therefore, require a physical facility between the transmitter and receiver, such as a metallic wire pair, a coaxial cable, or an optical fiber cable. Digital transmission is covered in Chapters 10 and 11. In digital radio systems, the carrier facility could be a physical cable, or it could be free space.

The property that distinguishes digital radio systems from conventional analog-modulation communications systems is the nature of the modulating signal. Both analog and digital modulation systems use analog carriers to transport the information through the system. However, with analog modulation systems, the information signal is also analog, whereas with digital modulation, the information signal is digital, which could be computer-generated data or digitally encoded analog signals.

Referring to Equation 9-1, if the information signal is digital and the amplitude (V) of the carrier is varied proportional to the information signal, a digitally modulated signal called *amplitude shift keying* (ASK) is produced. If the frequency (f) is varied proportional to the information signal, *frequency shift keying* (FSK) is produced, and if the phase of the carrier (θ) is varied proportional to the information signal, *phase shift keying* (PSK) is produced. If both the amplitude and the phase are varied proportional to the information signal, *quadrature amplitude modulation* (QAM) results. ASK, FSK, PSK, and QAM are all forms of digital modulation:

$$v(t) = \underset{\text{ASK}}{V} \sin (2\pi \cdot \underset{\text{FSK}}{f} t + \underset{\text{PSK}}{\theta}) \qquad \text{ASK, PSK} \rightarrow \text{QAM} \tag{9-1}$$

Digital modulation is ideally suited to a multitude of communications applications, including both cable and wireless systems. Applications include the following: (1) relatively low-speed voice-band data communications modems, such as those found in most personal computers; (2) high-speed data transmission systems, such as broadband *digital subscriber lines* (DSL); (3) digital microwave and satellite communications systems; and (4) cellular telephone *Personal Communications Systems* (PCS).

Figure 9-1 shows a simplified block diagram for a digital modulation system. In the transmitter, the precoder performs level conversion and then encodes the incoming data into groups of bits that modulate an analog carrier. The modulated carrier is shaped (fil-

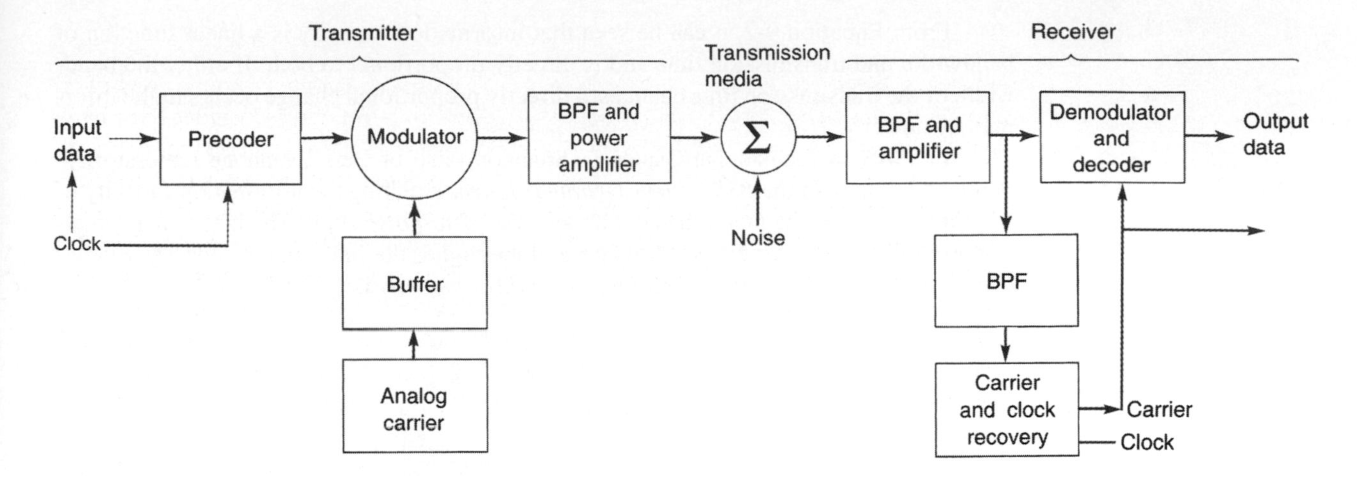

FIGURE 9-1 Simplified block diagram of a digital radio system

tered), amplified, and then transmitted through the transmission medium to the receiver. The transmission medium can be a metallic cable, optical fiber cable, Earth's atmosphere, or a combination of two or more types of transmission systems. In the receiver, the incoming signals are filtered, amplified, and then applied to the demodulator and decoder circuits, which extracts the original source information from the modulated carrier. The clock and carrier recovery circuits recover the analog carrier and digital timing (clock) signals from the incoming modulated wave since they are necessary to perform the demodulation process.

9-2 INFORMATION CAPACITY, BITS, BIT RATE, BAUD, AND *M*-ARY ENCODING

9-2-1 Information Capacity, Bits, and Bit Rate

Information theory is a highly theoretical study of the efficient use of bandwidth to propagate information through electronic communications systems. Information theory can be used to determine the *information capacity* of a data communications system. Information capacity is a measure of how much information can be propagated through a communications system and is a function of bandwidth and transmission time.

Information capacity represents the number of independent symbols that can be carried through a system in a given unit of time. The most basic digital symbol used to represent information is the *binary digit,* or *bit*. Therefore, it is often convenient to express the information capacity of a system as a *bit rate.* Bit rate is simply the number of bits transmitted during one second and is expressed in *bits per second* (bps).

In 1928, R. Hartley of Bell Telephone Laboratories developed a useful relationship among bandwidth, transmission time, and information capacity. Simply stated, Hartley's law is

$$I \propto B \times t \tag{9-2}$$

where I = information capacity (bits per second)
B = bandwidth (hertz)
t = transmission time (seconds)

From Equation 9-2, it can be seen that information capacity is a linear function of bandwidth and transmission time and is directly proportional to both. If either the bandwidth or the transmission time changes, a directly proportional change occurs in the information capacity.

In 1948, mathematician Claude E. Shannon (also of Bell Telephone Laboratories) published a paper in the *Bell System Technical Journal* relating the information capacity of a communications channel to bandwidth and *signal-to-noise ratio.* The higher the signal-to-noise ratio, the better the performance and the higher the information capacity. Mathematically stated, *the Shannon limit for information capacity* is

$$I = B\log_2\left(1 + \frac{S}{N}\right) \qquad \textbf{(9-3)}$$

or

$$I = 3.32B\log_{10}\left(1 + \frac{S}{N}\right) \qquad \textbf{(9-4)}$$

where I = information capacity (bps)
B = bandwidth (hertz)
$\frac{S}{N}$ = signal-to-noise power ratio (unitless)

For a standard telephone circuit with a signal-to-noise power ratio of 1000 (30 dB) and a bandwidth of 2.7 kHz, the Shannon limit for information capacity is

$$\begin{aligned} I &= (3.32)(2700)\log_{10}(1 + 1000) \\ &= 26.9 \text{ kbps} \end{aligned}$$

Shannon's formula is often misunderstood. The results of the preceding example indicate that 26.9 kbps can be propagated through a 2.7-kHz communications channel. This may be true, but it cannot be done with a binary system. To achieve an information transmission rate of 26.9 kbps through a 2.7-kHz channel, each symbol transmitted must contain more than one bit.

9-2-2 *M*-ary Encoding

M-ary is a term derived from the word *binary. M* simply represents a digit that corresponds to the number of conditions, levels, or combinations possible for a given number of binary variables. It is often advantageous to encode at a level higher than binary (sometimes referred to as *beyond binary* or *higher-than-binary encoding*) where there are more than two conditions possible. For example, a digital signal with four possible conditions (voltage levels, frequencies, phases, and so on) is an *M*-ary system where $M = 4$. If there are eight possible conditions, $M = 8$ and so forth. The number of bits necessary to produce a given number of conditions is expressed mathematically as

$$N = \log_2 M \qquad \textbf{(9-5)}$$

where N = number of bits necessary
M = number of conditions, levels, or combinations possible with N bits

Equation 9-5 can be simplified and rearranged to express the number of conditions possible with N bits as

$$2^N = M \qquad \textbf{(9-6)}$$

For example, with one bit, only $2^1 = 2$ conditions are possible. With two bits, $2^2 = 4$ conditions are possible, with three bits, $2^3 = 8$ conditions are possible, and so on.

9-2-3 Baud and Minimum Bandwidth

Baud is a term that is often misunderstood and commonly confused with bit rate (bps). Bit rate refers to the rate of change of a digital information signal, which is usually binary. Baud, like bit rate, is also a rate of change; however, baud refers to the rate of change of a signal on the transmission medium after encoding and modulation have occurred. Hence, baud is a unit of transmission rate, modulation rate, or symbol rate and, therefore, the terms *symbols per second* and *baud* are often used interchangeably. Mathematically, baud is the reciprocal of the time of one output *signaling element,* and a signaling element may represent several information bits. Baud is expressed as

$$\text{baud} = \frac{1}{t_s} \tag{9-7}$$

where baud = symbol rate (baud per second)
t_s = time of one signaling element (seconds)

A signaling element is sometimes called a *symbol* and could be encoded as a change in the amplitude, frequency, or phase. For example, binary signals are generally encoded and transmitted one bit at a time in the form of discrete voltage levels representing logic 1s (highs) and logic 0s (lows). A baud is also transmitted one at a time; however, a baud may represent more than one information bit. Thus, the baud of a data communications system may be considerably less than the bit rate. In binary systems (such as binary FSK and binary PSK), *baud* and *bits per second* are equal. However, in higher-level systems (such as QPSK and 8-PSK), bps is always greater than baud.

According to H. Nyquist, binary digital signals can be propagated through an ideal noiseless transmission medium at a rate equal to two times the bandwidth of the medium. The minimum theoretical bandwidth necessary to propagate a signal is called the minimum *Nyquist bandwidth* or sometimes the minimum *Nyquist frequency.* Thus, $f_b = 2B$, where f_b is the bit rate in bps and B is the *ideal Nyquist bandwidth.* The actual bandwidth necessary to propagate a given bit rate depends on several factors, including the type of encoding and modulation used, the types of filters used, system noise, and desired error performance. The ideal bandwidth is generally used for comparison purposes only.

The relationship between bandwidth and bit rate also applies to the opposite situation. For a given bandwidth (B), the highest theoretical bit rate is $2B$. For example, a standard telephone circuit has a bandwidth of approximately 2700 Hz, which has the capacity to propagate 5400 bps through it. However, if more than two levels are used for signaling (higher-than-binary encoding), more than one bit may be transmitted at a time, and it is possible to propagate a bit rate that exceeds $2B$. Using multilevel signaling, the Nyquist formulation for channel capacity is

$$f_b = B \log_2 M \tag{9-8}$$

where f_b = channel capacity (bps)
B = minimum Nyquist bandwidth (hertz)
M = number of discrete signal or voltage levels

Equation 9-8 can be rearranged to solve for the minimum bandwidth necessary to pass M-ary digitally modulated carriers

$$B = \left(\frac{f_b}{\log_2 M}\right) \tag{9-9}$$

If N is substituted for $\log_2 M$, Equation 9-9 reduces to

$$B = \frac{f_b}{N} \tag{9-10}$$

where N is the number of bits encoded into each signaling element.

If information bits are encoded (grouped) and then converted to signals with more than two levels, transmission rates in excess of $2B$ are possible, as will be seen in subsequent sections of this chapter. In addition, since baud is the encoded rate of change, it also equals the bit rate divided by the number of bits encoded into one signaling element. Thus,

$$\text{baud} = \frac{f_b}{N} \tag{9-11}$$

By comparing Equation 9-10 with Equation 9-11, it can be seen that with digital modulation, the baud and the ideal minimum Nyquist bandwidth have the same value and are equal to the bit rate divided by the number of bits encoded. This statement holds true for all forms of digital modulation except frequency-shift keying.

9-3 AMPLITUDE-SHIFT KEYING

The simplest digital modulation technique is *amplitude-shift keying* (ASK), where a binary information signal directly modulates the amplitude of an analog carrier. ASK is similar to standard amplitude modulation except there are only two output amplitudes possible. Amplitude-shift keying is sometimes called *digital amplitude modulation* (DAM). Mathematically, amplitude-shift keying is

$$v_{(ask)}(t) = [1 + v_m(t)]\left[\frac{A}{2}\cos(\omega_c t)\right] \tag{9-12}$$

where $v_{ask}(t)$ = amplitude-shift keying wave
$v_m(t)$ = digital information (modulating) signal (volts)
$A/2$ = unmodulated carrier amplitude (volts)
ω_c = analog carrier radian frequency (radians per second, $2\pi f_c t$)

In Equation 9-12, the modulating signal ($v_m[t]$) is a normalized binary waveform, where $+1$ V = logic 1 and -1 V = logic 0. Therefore, for a logic 1 input, $v_m(t) = +1$ V, Equation 9-12 reduces to

$$v_{(ask)}(t) = [1 + 1]\left[\frac{A}{2}\cos(\omega_c t)\right]$$

$$= A\cos(\omega_c t)$$

and for a logic 0 input, $v_m(t) = -1$ V, Equation 9-12 reduces to

$$v_{(ask)}(t) = [1 - 1]\left[\frac{A}{2}\cos(\omega_c t)\right]$$

$$= 0$$

Thus, the modulated wave $v_{ask}(t)$, is either $A\cos(\omega_c t)$ or 0. Hence, the carrier is either "*on*" or "*off*," which is why amplitude-shift keying is sometimes referred to as *on-off keying* (OOK).

Figure 9-2 shows the input and output waveforms from an ASK modulator. From the figure, it can be seen that for every change in the input binary data stream, there is one change in the ASK waveform, and the time of one bit (t_b) equals the time of one analog signaling element (t_s). It is also important to note that for the entire time the binary input is high, the output is a constant-amplitude, constant-frequency signal, and for the entire time the binary input is low, the carrier is off. The bit time is the reciprocal of the bit rate and the time of one signaling element is the reciprocal of the baud. Therefore, the rate of change of the

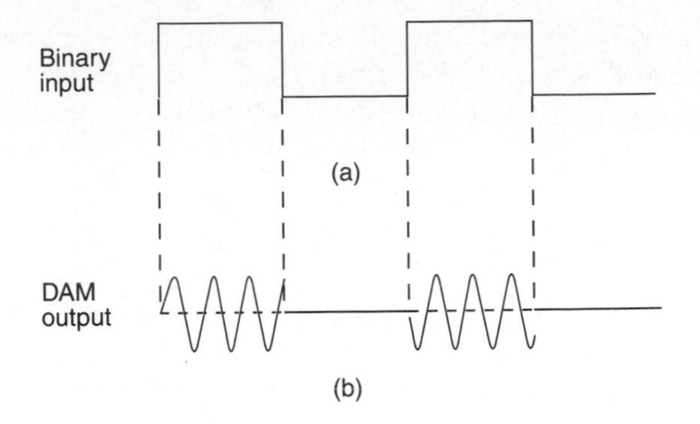

FIGURE 9-2 Digital amplitude modulation: (a) input binary; (b) output DAM waveform

ASK waveform (baud) is the same as the rate of change of the binary input (bps); thus, the bit rate equals the baud. With ASK, the bit rate is also equal to the minimum Nyquist bandwidth. This can be verified by substituting into Equations 9-10 and 9-11 and setting N to 1:

$$B = \frac{f_b}{1} = f_b \qquad \text{baud} = \frac{f_b}{1} = f_b$$

Example 9-1

Determine the baud and minimum bandwidth necessary to pass a 10 kbps binary signal using amplitude shift keying.

Solution For ASK, $N = 1$, and the baud and minimum bandwidth are determined from Equations 9-11 and 9-10, respectively:

$$B = \frac{10{,}000}{1} = 10{,}000$$

$$\text{baud} = \frac{10{,}000}{1} = 10{,}000$$

The use of amplitude-modulated analog carriers to transport digital information is a relatively low-quality, low-cost type of digital modulation and, therefore, is seldom used except for very low-speed telemetry circuits.

9-4 FREQUENCY-SHIFT KEYING

Frequency-shift keying (FSK) is another relatively simple, low-performance type of digital modulation. FSK is a form of constant-amplitude angle modulation similar to standard frequency modulation (FM) except the modulating signal is a binary signal that varies between two discrete voltage levels rather than a continuously changing analog waveform. Consequently, FSK is sometimes called *binary FSK* (BFSK). The general expression for FSK is

$$v_{fsk}(t) = V_c \cos\{2\pi[f_c + v_m(t)\,\Delta f]t\} \qquad \textbf{(9-13)}$$

where $v_{fsk}(t)$ = binary FSK waveform
V_c = peak analog carrier amplitude (volts)
f_c = analog carrier center frequency (hertz)
Δf = peak change (shift) in the analog carrier frequency (hertz)
$v_m(t)$ = binary input (modulating) signal (volts)

From Equation 9-13, it can be seen that the peak shift in the carrier frequency (Δf) is proportional to the amplitude of the binary input signal ($v_m[t]$), and the direction of the shift

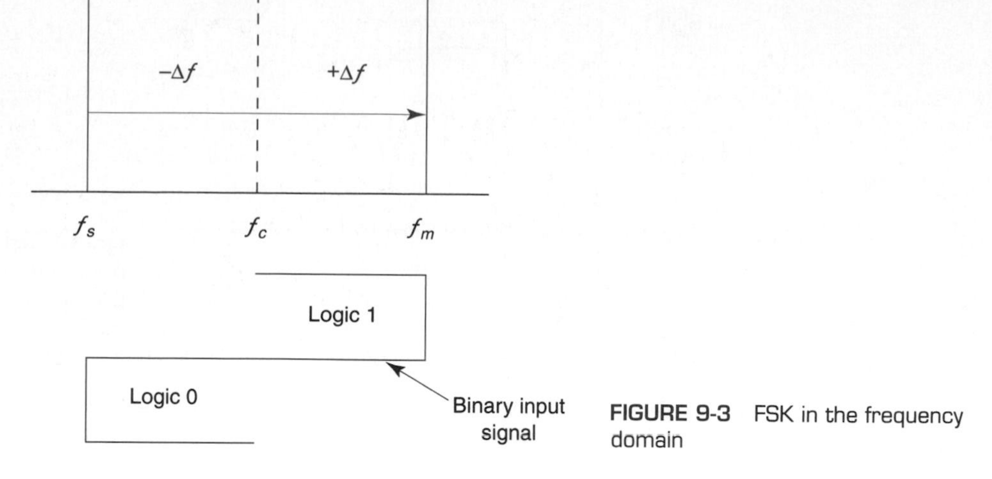

FIGURE 9-3 FSK in the frequency domain

is determined by the polarity. The modulating signal is a normalized binary waveform where a logic 1 = +1 V and a logic 0 = −1 V. Thus, for a logic 1 input, $v_m(t) = +1$, Equation 9-13 can be rewritten as

$$v_{fsk}(t) = V_c \cos[2\pi(f_c + \Delta f)t]$$

For a logic 0 input, $v_m(t) = -1$, Equation 9-13 becomes

$$v_{fsk}(t) = V_c \cos[2\pi(f_c - \Delta f)t]$$

With binary FSK, the carrier center frequency (f_c) is shifted (deviated) up and down in the frequency domain by the binary input signal as shown in Figure 9-3. As the binary input signal changes from a logic 0 to a logic 1 and vice versa, the output frequency shifts between two frequencies: a mark, or logic 1 frequency (f_m), and a space, or logic 0 frequency (f_s). The mark and space frequencies are separated from the carrier frequency by the peak frequency deviation (Δf) and from each other by 2 Δf.

With FSK, frequency deviation is defined as the difference between either the mark or space frequency and the center frequency, or half the difference between the mark and space frequencies. Frequency deviation is illustrated in Figure 9-3 and expressed mathematically as

$$\Delta f = \frac{|f_m - f_s|}{2} \tag{9-14}$$

where Δf = frequency deviation (hertz)
$|f_m - f_s|$ = absolute difference between the mark and space frequencies (hertz)

Figure 9-4a shows in the time domain the binary input to an FSK modulator and the corresponding FSK output. As the figure shows, when the binary input (f_b) changes from a logic 1 to a logic 0 and vice versa, the FSK output frequency shifts from a mark (f_m) to a space (f_s) frequency and vice versa. In Figure 9-4a, the mark frequency is the higher frequency ($f_c + \Delta f$), and the space frequency is the lower frequency ($f_c - \Delta f$), although this relationship could be just the opposite. Figure 9-4b shows the truth table for a binary FSK modulator. The truth table shows the input and output possibilities for a given digital modulation scheme.

9-4-1 FSK Bit Rate, Baud, and Bandwidth

In Figure 9-4a, it can be seen that the time of one bit (t_b) is the same as the time the FSK output is a mark of space frequency (t_s). Thus, the bit time equals the time of an FSK signaling element, and the bit rate equals the baud.

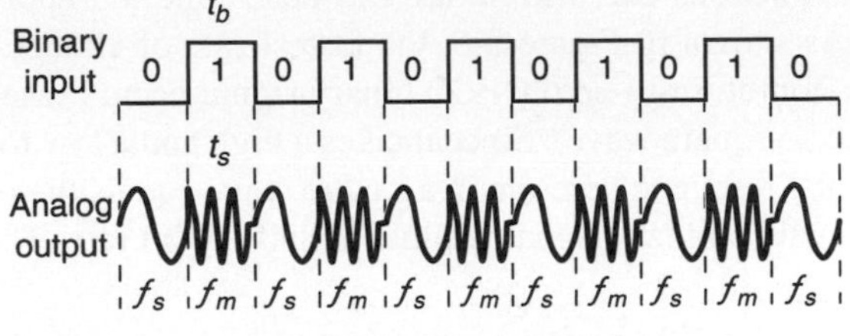

(a)

binary input	frequency output
0	space (f_s)
1	mark (f_m)

(b)

FIGURE 9-4 FSK in the time domain: (a) waveform; (b) truth table

The baud for binary FSK can also be determined by substituting $N = 1$ in Equation 9-11:

$$\text{baud} = \frac{f_b}{1} = f_b$$

FSK is the exception to the rule for digital modulation, as the minimum bandwidth is not determined from Equation 9-10. The minimum bandwidth for FSK is given as

$$B = |(f_s - f_b) - (f_m - f_b)|$$
$$= |f_s - f_m| + 2f_b$$

and since $|f_s - f_m|$ equals $2\Delta f$, the minimum bandwidth can be approximated as

$$B = 2(\Delta f + f_b) \tag{9-15}$$

where B = minimum Nyquist bandwidth (hertz)
Δf = frequency deviation ($|f_m - f_s|$) (hertz)
f_b = input bit rate (bps)

Note how closely Equation 9-15 resembles Carson's rule for determining the approximate bandwidth for an FM wave. The only difference in the two equations is that, for FSK, the bit rate (f_b) is substituted for the modulating-signal frequency (f_m).

Example 9-2

Determine (a) the peak frequency deviation, (b) minimum bandwidth, and (c) baud for a binary FSK signal with a mark frequency of 49 kHz, a space frequency of 51 kHz, and an input bit rate of 2 kbps.

Solution **a.** The peak frequency deviation is determined from Equation 9-14:

$$\Delta f = \frac{|49\text{kHz} - 51\text{kHz}|}{2}$$
$$= 1 \text{ kHz}$$

b. The minimum bandwidth is determined from Equation 9-15:

$$B = 2(1000 + 2000)$$
$$= 6 \text{ kHz}$$

c. For FSK, $N = 1$, and the baud is determined from Equation 9-11 as

$$\text{baud} = \frac{2000}{1} = 2000$$

Bessel functions can also be used to determine the approximate bandwidth for an FSK wave. As shown in Figure 9-5, the fastest rate of change (highest fundamental frequency) in a nonreturn-to-zero (NRZ) binary signal occurs when alternating 1s and 0s are occurring (i.e., a square wave). Since it takes a high and a low to produce a cycle, the highest fundamental frequency present in a square wave equals the repetition rate of the square wave, which with a binary signal is equal to half the bit rate. Therefore,

$$f_a = \frac{f_b}{2} \tag{9-16}$$

where f_a = highest fundamental frequency of the binary input signal (hertz)
f_b = input bit rate (bps)

The formula used for modulation index in FM is also valid for FSK; thus,

$$h = \frac{\Delta f}{f_a} \quad \text{(unitless)} \tag{9-17}$$

where h = FM modulation index called the h-factor in FSK
f_a = fundamental frequency of the binary modulating signal (hertz)
Δf = peak frequency deviation (hertz)

The worst-case modulation index (deviation ratio) is that which yields the widest bandwidth. The worst-case or widest bandwidth occurs when both the frequency deviation and the modulating-signal frequency are at their maximum values. As described earlier, the peak frequency deviation in FSK is constant and always at its maximum value, and the highest fundamental frequency is equal to half the incoming bit rate. Thus,

$$h = \frac{\dfrac{|f_m - f_s|}{2}}{\dfrac{f_b}{2}} \quad \text{(unitless)}$$

or

$$h = \frac{|f_m - f_s|}{f_b} \tag{9-18}$$

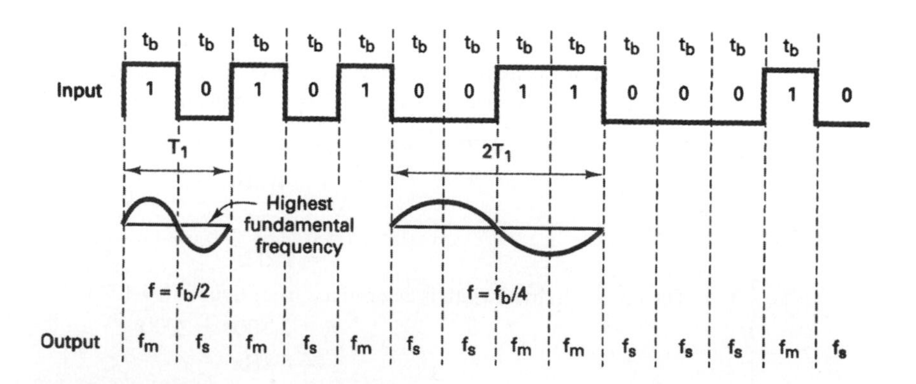

FIGURE 9-5 FSK modulator, t_b, time of one bit = $1/f_b$; f_m, mark frequency; f_s, space frequency; T_1, period of shortest cycle; $1/T_1$, fundamental frequency of binary square wave; f_b, input bit rate (bps)

where h = h-factor (unitless)
f_m = mark frequency (hertz)
f_s = space frequency (hertz)
f_b = bit rate (bits per second)

Example 9-3

Using a Bessel table, determine the minimum bandwidth for the same FSK signal described in Example 9-1 with a mark frequency of 49 kHz, a space frequency of 51 kHz, and an input bit rate of 2 kbps.

Solution The modulation index is found by substituting into Equation 9-17:

or

$$h = \frac{|49 \text{ kHz} - 51 \text{ kHz}|}{2 \text{ kbps}}$$

$$= \frac{2 \text{ kHz}}{2 \text{ kbps}}$$

$$= 1$$

From a Bessel table, three sets of significant sidebands are produced for a modulation index of one. Therefore, the bandwidth can be determined as follows:

$$B = 2(3 \times 1000)$$

$$= 6000 \text{ Hz}$$

The bandwidth determined in Example 9-3 using the Bessel table is identical to the bandwidth determined in Example 9-2.

9-4-2 FSK Transmitter

Figure 9-6 shows a simplified binary FSK modulator, which is very similar to a conventional FM modulator and is very often a voltage-controlled oscillator (VCO). The center frequency (f_c) is chosen such that it falls halfway between the mark and space frequencies. A logic 1 input shifts the VCO output to the mark frequency, and a logic 0 input shifts the VCO output to the space frequency. Consequently, as the binary input signal changes back and forth between logic 1 and logic 0 conditions, the VCO output shifts or deviates back and forth between the mark and space frequencies.

In a binary FSK modulator, Δf is the peak frequency deviation of the carrier and is equal to the difference between the carrier rest frequency and either the mark or the space frequency (or half the difference between the carrier rest frequency) and either the mark or the space frequency (or half the difference between the mark and space frequencies). A VCO-FSK modulator can be operated in the sweep mode where the peak frequency deviation is

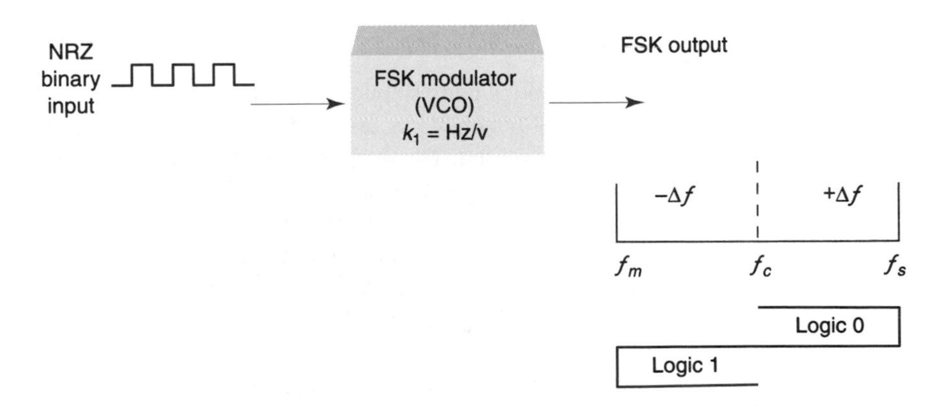

FIGURE 9-6 FSK modulator

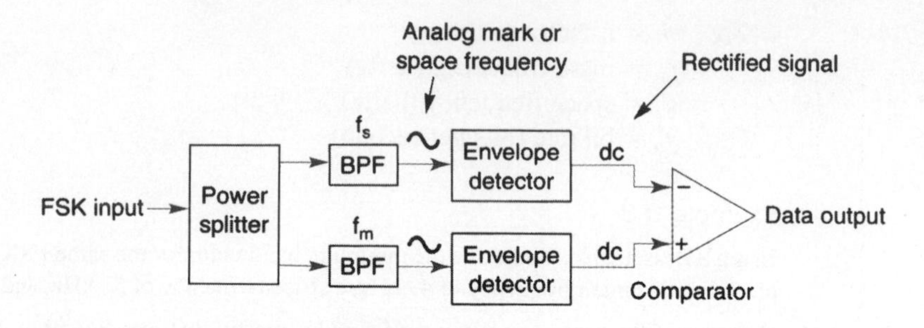

FIGURE 9-7 Noncoherent FSK demodulator

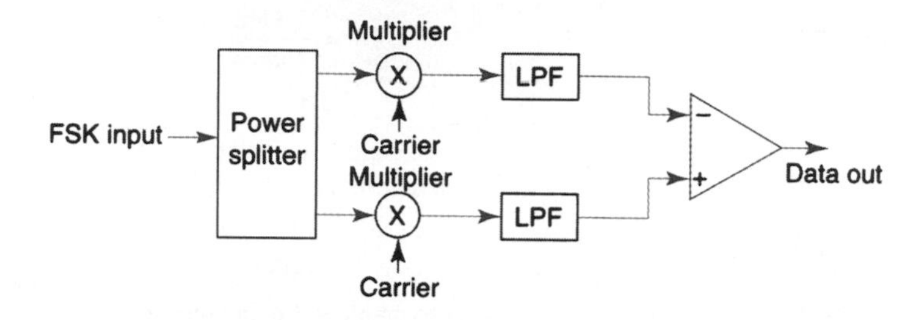

FIGURE 9-8 Coherent FSK demodulator

simply the product of the binary input voltage and the deviation sensitivity of the VCO. With the sweep mode of modulation, the frequency deviation is expressed mathematically as

$$\Delta f = v_m(t)k_l \qquad \text{(9-19)}$$

where Δf = peak frequency deviation (hertz)
$v_m(t)$ = peak binary modulating-signal voltage (volts)
k_l = deviation sensitivity (hertz per volt).

With binary FSK, the amplitude of the input signal can only be one of two values, one for a logic 1 condition and one for a logic 0 condition. Therefore, the peak frequency deviation is constant and always at its maximum value. Frequency deviation is simply plus or minus the peak voltage of the binary signal times the deviation sensitivity of the VCO. Since the peak voltage is the same for a logic 1 as it is for a logic 0, the magnitude of the frequency deviation is also the same for a logic 1 as it is for a logic 0.

9-4-3 FSK Receiver

FSK demodulation is quite simple with a circuit such as the one shown in Figure 9-7. The FSK input signal is simultaneously applied to the inputs of both bandpass filters (BPFs) through a power splitter. The respective filter passes only the mark or only the space frequency on to its respective envelope detector. The envelope detectors, in turn, indicate the total power in each passband, and the comparator responds to the largest of the two powers. This type of FSK detection is referred to as noncoherent detection; there is no frequency involved in the demodulation process that is synchronized either in phase, frequency, or both with the incoming FSK signal.

Figure 9-8 shows the block diagram for a coherent FSK receiver. The incoming FSK signal is multiplied by a recovered carrier signal that has the exact same frequency and phase as the transmitter reference. However, the two transmitted frequencies (the mark and space frequencies) are not generally continuous; it is not practical to reproduce a local reference that is coherent with both of them. Consequently, coherent FSK detection is seldom used.

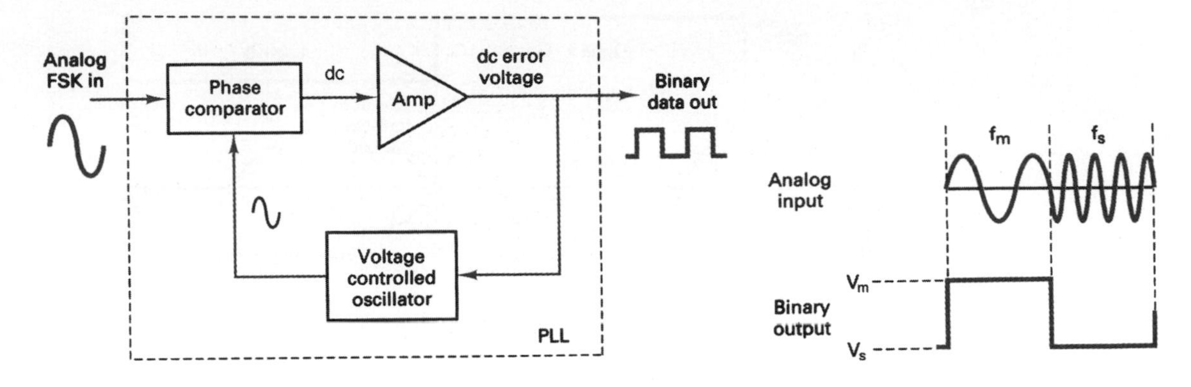

FIGURE 9-9 PLL-FSK demodulator

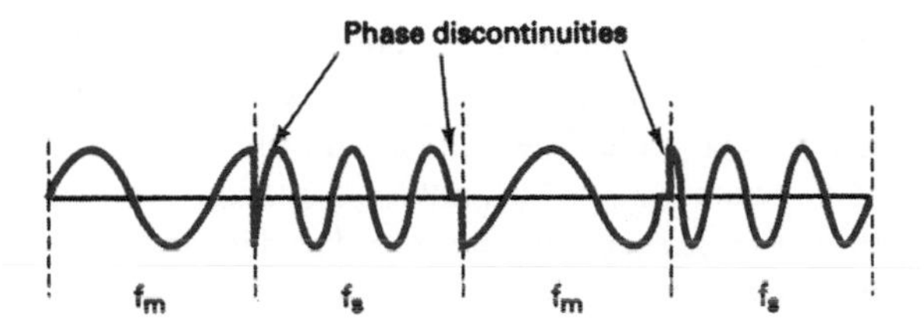

FIGURE 9-10 Noncontinuous FSK waveform

The most common circuit used for demodulating binary FSK signals is the *phase-locked loop* (PLL), which is shown in block diagram form in Figure 9-9. A PLL-FSK demodulator works similarly to a PLL-FM demodulator. As the input to the PLL shifts between the mark and space frequencies, the *dc error voltage* at the output of the phase comparator follows the frequency shift. Because there are only two input frequencies (mark and space), there are also only two output error voltages. One represents a logic 1 and the other a logic 0. Therefore, the output is a two-level (binary) representation of the FSK input. Generally, the natural frequency of the PLL is made equal to the center frequency of the FSK modulator. As a result, the changes in the dc error voltage follow the changes in the analog input frequency and are symmetrical around 0 V.

Binary FSK has a poorer error performance than PSK or QAM and, consequently, is seldom used for high-performance digital radio systems. Its use is restricted to low-performance, low-cost, asynchronous data modems that are used for data communications over analog, voice-band telephone lines.

9-4-4 Continuous-Phase Frequency-Shift Keying

Continuous-phase frequency-shift keying (CP-FSK) is binary FSK except the mark and space frequencies are synchronized with the input binary bit rate. Synchronous simply implies that there is a precise time relationship between the two; it does not mean they are equal. With CP-FSK, the mark and space frequencies are selected such that they are separated from the center frequency by an exact multiple of one-half the bit rate (f_m and $f_s = n[f_b/2]$), where n = any integer). This ensures a smooth phase transition in the analog output signal when it changes from a mark to a space frequency or vice versa. Figure 9-10 shows a noncontinuous FSK waveform. It can be seen that when the input changes from a logic 1 to a logic 0 and vice versa, there is an abrupt phase discontinuity in the analog signal. When this occurs, the demodulator has trouble following the frequency shift; consequently, an error may occur.

Figure 9-11 shows a continuous phase FSK waveform. Notice that when the output frequency changes, it is a smooth, continuous transition. Consequently, there are no phase discontinuities. CP-FSK has a better bit-error performance than conventional binary FSK for a given signal-to-noise ratio. The disadvantage of CP-FSK is that it requires synchronization circuits and is, therefore, more expensive to implement.

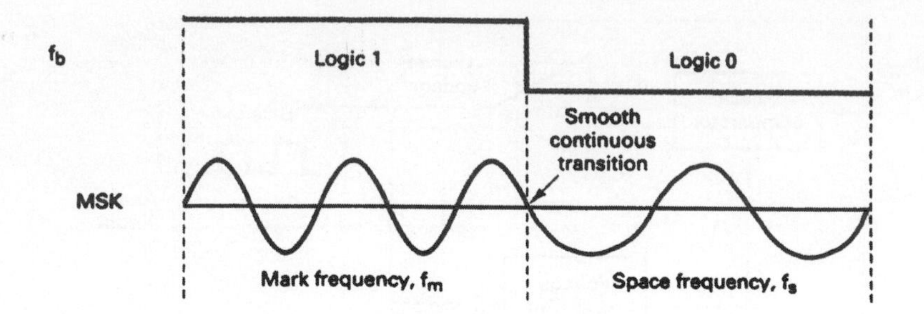

FIGURE 9-11 Continuous-phase MSK waveform

9-5 PHASE-SHIFT KEYING

Phase-shift keying (PSK) is another form of *angle-modulated, constant-amplitude* digital modulation. PSK is an M-ary digital modulation scheme similar to conventional phase modulation except with PSK the input is a binary digital signal and there are a limited number of output phases possible. The input binary information is encoded into groups of bits before modulating the carrier. The number of bits in a group ranges from 1 to 12 or more. The number of output phases is defined by M as described in Equation 9-6 and determined by the number of bits in the group (n).

9-5-1 Binary Phase-Shift Keying

The simplest form of PSK is *binary phase-shift keying* (BPSK), where $N = 1$ and $M = 2$. Therefore, with BPSK, two phases ($2^1 = 2$) are possible for the carrier. One phase represents a logic 1, and the other phase represents a logic 0. As the input digital signal changes state (i.e., from a 1 to a 0 or from a 0 to a 1), the phase of the output carrier shifts between two angles that are separated by 180°. Hence, other names for BPSK are *phase reversal keying* (PRK) and *biphase modulation*. BPSK is a form of square-wave modulation of a *continuous wave* (CW) signal.

9-5-1-1 BPSK transmitter. Figure 9-12 shows a simplified block diagram of a BPSK transmitter. The balanced modulator acts as a phase reversing switch. Depending on

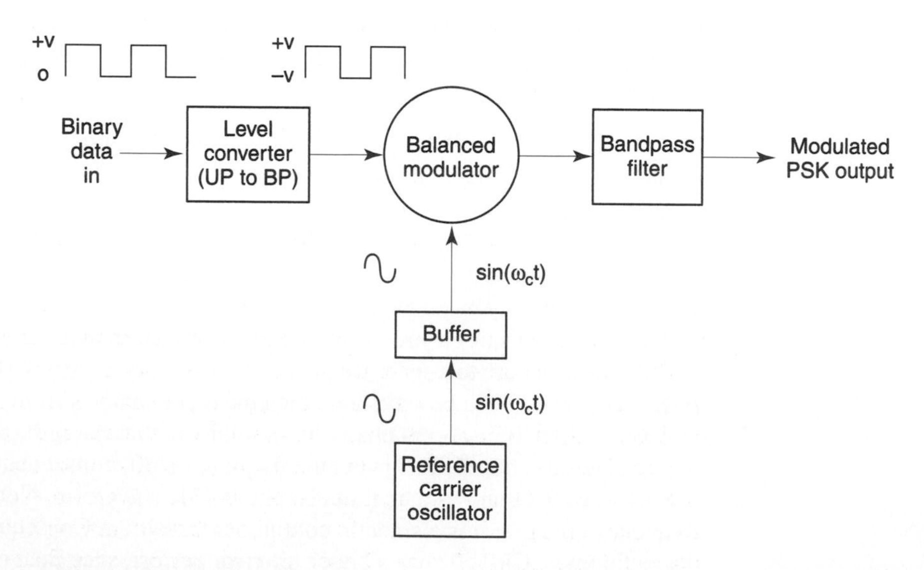

FIGURE 9-12 BPSK transmitter

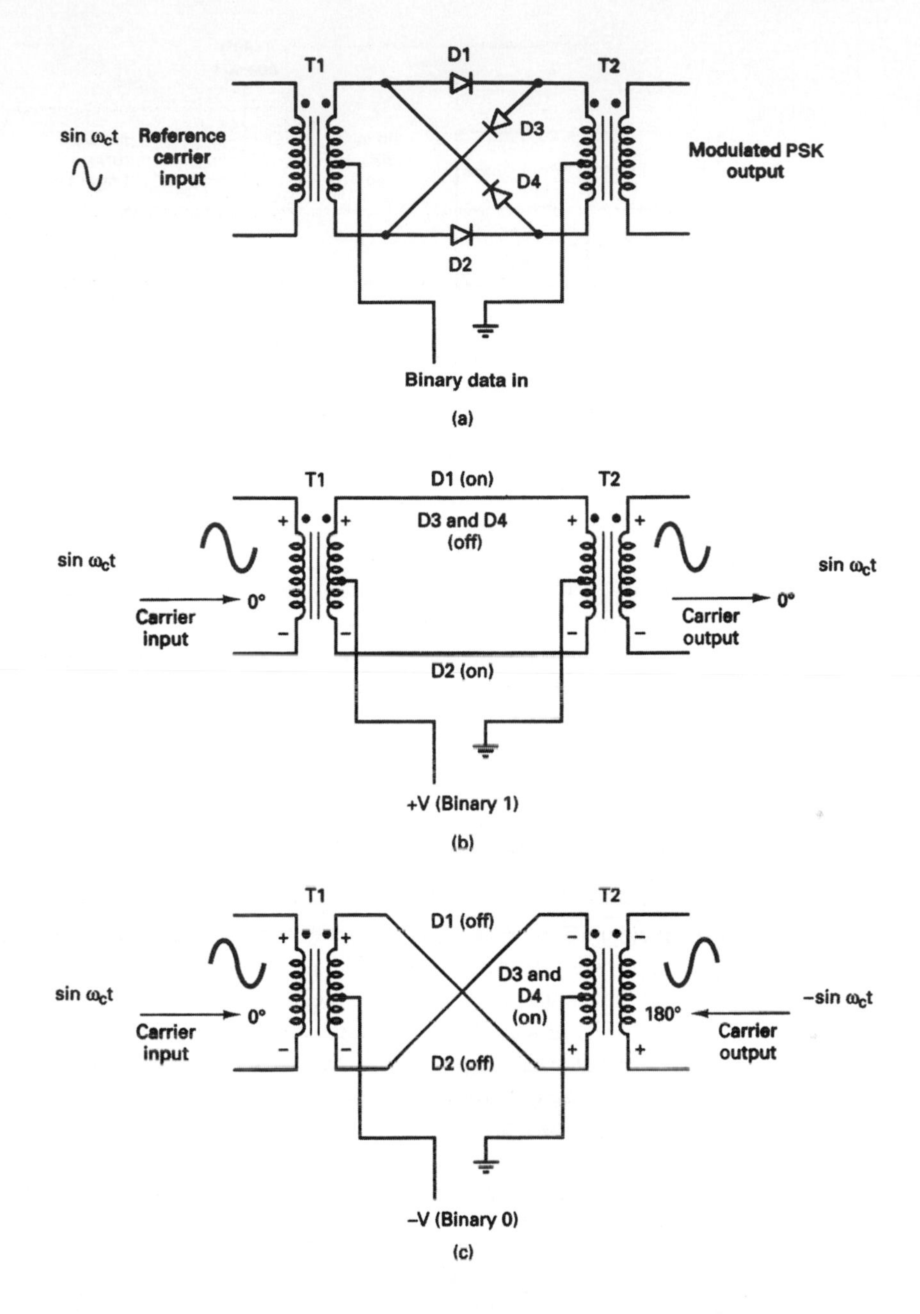

FIGURE 9-13 (a) Balanced ring modulator; (b) logic 1 input; (c) logic 0 input

the logic condition of the digital input, the carrier is transferred to the output either in phase or 180° out of phase with the reference carrier oscillator.

Figure 9-13 shows the schematic diagram of a balanced ring modulator. The balanced modulator has two inputs: a carrier that is in phase with the reference oscillator and the binary digital data. For the balanced modulator to operate properly, the digital input voltage must be much greater than the peak carrier voltage. This ensures that the digital input controls the on/off state of diodes D1 to D4. If the binary input is a logic 1 (positive voltage), diodes D1 and D2 are forward biased and on, while diodes D3 and D4 are reverse biased and off (Figure 9-13b). With the polarities shown, the carrier voltage is developed across

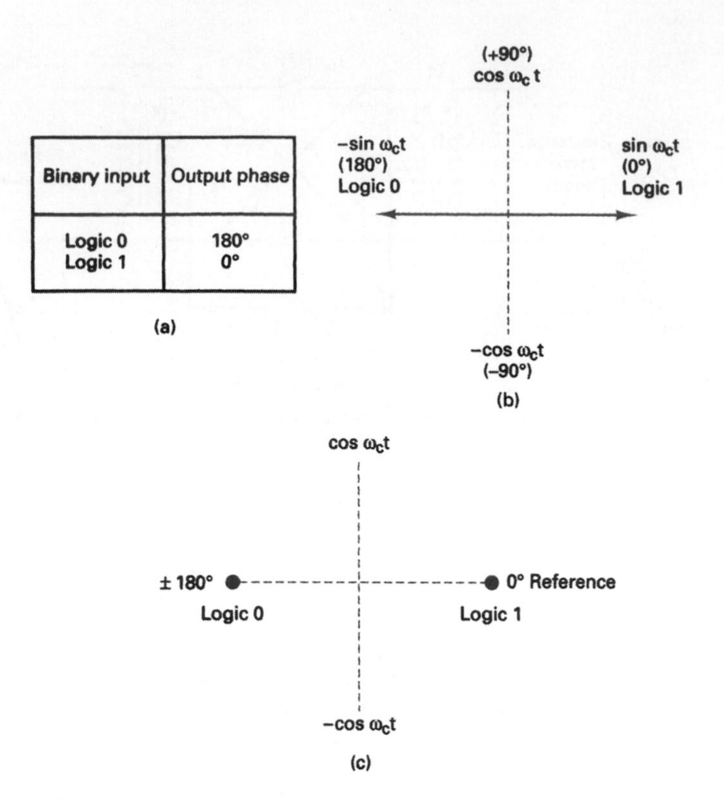

Binary input	Output phase
Logic 0	180°
Logic 1	0°

FIGURE 9-14 BPSK modulator: (a) truth table; (b) phasor diagram; (c) constellation diagram

transformer T2 in phase with the carrier voltage across T1. Consequently, the output signal is in phase with the reference oscillator.

If the binary input is a logic 0 (negative voltage), diodes D1 and D2 are reverse biased and off, while diodes D3 and D4 are forward biased and on (Figure 9-13c). As a result, the carrier voltage is developed across transformer T2 180° out of phase with the carrier voltage across T1. Consequently, the output signal is 180° out of phase with the reference oscillator. Figure 9-14 shows the truth table, phasor diagram, and constellation diagram for a BPSK modulator. A *constellation diagram,* which is sometimes called a *signal state-space diagram,* is similar to a phasor diagram except that the entire phasor is not drawn. In a constellation diagram, only the relative positions of the peaks of the phasors are shown.

9-5-1-2 Bandwidth considerations of BPSK. A balanced modulator is a *product modulator;* the output signal is the product of the two input signals. In a BPSK modulator, the carrier input signal is multiplied by the binary data. If +1 V is assigned to a logic 1 and −1 V is assigned to a logic 0, the input carrier ($\sin \omega_c t$) is multiplied by either a + or −1. Consequently, the output signal is either $+1 \sin \omega_c t$ or $-1 \sin \omega_c t$; the first represents a signal that is *in phase* with the reference oscillator, the latter a signal that is 180° out of phase with the reference oscillator. Each time the input logic condition changes, the output phase changes. Consequently, for BPSK, the output rate of change (baud) is equal to the input rate of change (bps), and the widest output bandwidth occurs when the input binary data are an alternating 1/0 sequence. The fundamental frequency (f_a) of an alternative 1/0 bit sequence is equal to one-half of the bit rate ($f_b/2$). Mathematically, the output of a BPSK modulator is proportional to

$$\text{BPSK output} = [\sin(2\pi f_a t)] \times [\sin(2\pi f_c t)] \qquad \text{(9-20)}$$

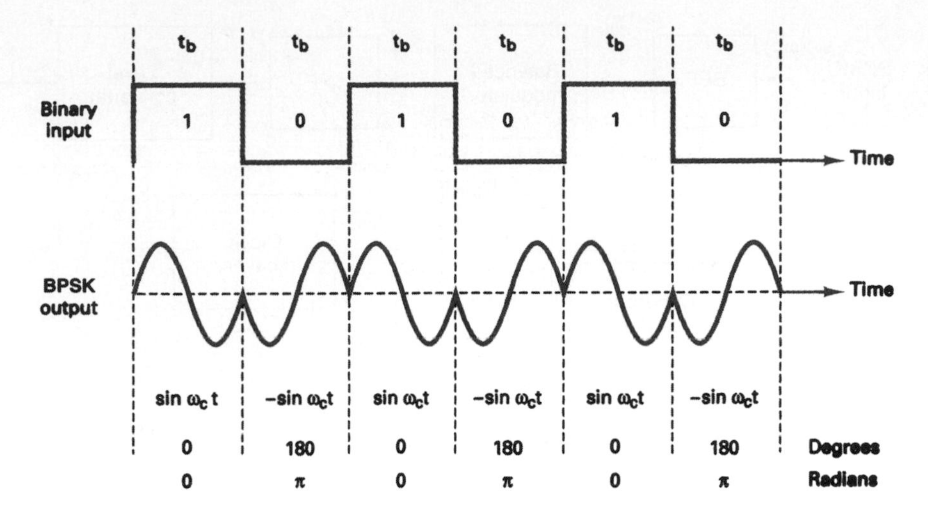

FIGURE 9-15 Output phase-versus-time relationship for a BPSK modulator

where f_a = maximum fundamental frequency of binary input (hertz)
f_c = reference carrier frequency (hertz)

Solving for the trig identity for the product of two sine functions,

$$\frac{1}{2}\cos[2\pi(f_c - f_a)t] - \frac{1}{2}\cos[2\pi(f_c + f_a)t]$$

Thus, the minimum double-sided Nyquist bandwidth (B) is

$$\begin{array}{r} f_c + f_a \\ -(f_c + f_a) \\ \hline \end{array} \quad \text{or} \quad \begin{array}{r} f_c + f_a \\ -f_c + f_a \\ \hline 2f_a \end{array}$$

and because $f_a = f_b/2$, where f_b = input bit rate,

$$B = \frac{2f_b}{2} = f_b$$

where B is the minimum double-sided Nyquist bandwidth.

Figure 9-15 shows the output phase-versus-time relationship for a BPSK waveform. As the figure shows, a logic 1 input produces an analog output signal with a 0° phase angle, and a logic 0 input produces an analog output signal with a 180° phase angle. As the binary input shifts between a logic 1 and a logic 0 condition and vice versa, the phase of the BPSK waveform shifts between 0° and 180°, respectively. For simplicity, only one cycle of the analog carrier is shown in each signaling element, although there may be anywhere between a fraction of a cycle to several thousand cycles, depending on the relationship between the input bit rate and the analog carrier frequency. It can also be seen that the time of one BPSK signaling element (t_s) is equal to the time of one information bit (t_b), which indicates that the bit rate equals the baud.

Example 9-4

For a BPSK modulator with a carrier frequency of 70 MHz and an input bit rate of 10 Mbps, determine the maximum and minimum upper and lower side frequencies, draw the output spectrum, determine the minimum Nyquist bandwidth, and calculate the baud.

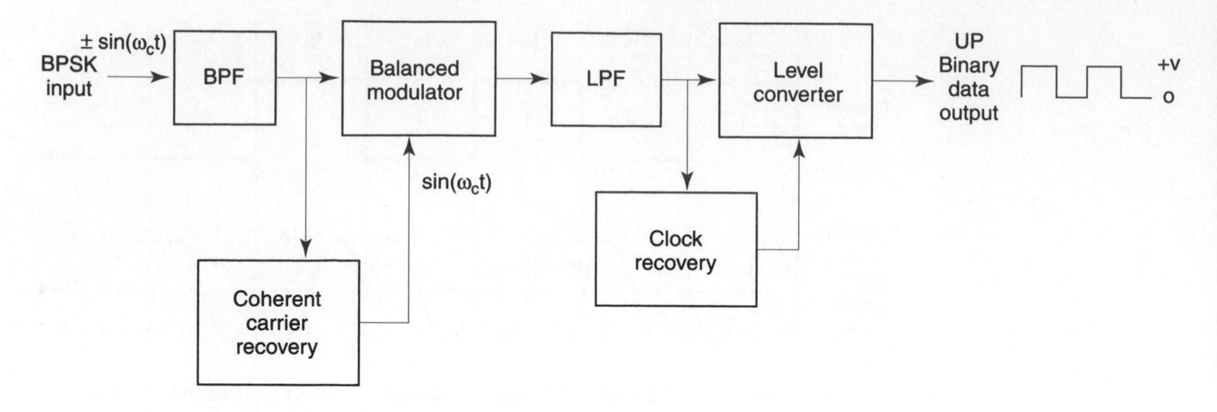

FIGURE 9-16 Block diagram of a BPSK receiver

Solution Substituting into Equation 9-20 yields

$$\text{output} = (\sin \omega_a t)(\sin \omega_c t)$$
$$= [\sin 2\pi(5\text{ MHz})t][\sin 2\pi(70\text{ MHz})t]$$
$$= \underbrace{\frac{1}{2}\cos 2\pi(70\text{ MHz} - 5\text{ MHz})t}_{\text{lower side frequency}} - \underbrace{\frac{1}{2}\cos 2\pi(70\text{ MHz} + 5\text{ MHz})t}_{\text{upper side frequency}}$$

Minimum lower side frequency (LSF):

$$\text{LSF} = 70\text{ MHz} - 5\text{ MHz} = 65\text{ MHz}$$

Maximum upper side frequency (USF):

$$\text{USF} = 70\text{ MHz} + 5\text{ MHz} = 75\text{ MHz}$$

Therefore, the output spectrum for the worst-case binary input conditions is as follows: The minimum Nyquist bandwidth (B) is

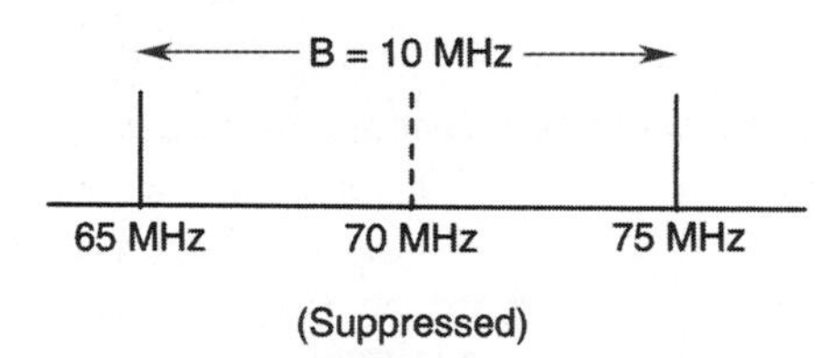

$$B = 75\text{ MHz} - 65\text{ MHz} = 10\text{ MHz}$$

and the baud = f_b or 10 megabaud.

9-5-1-3 BPSK receiver. Figure 9-16 shows the block diagram of a BPSK receiver. The input signal may be $+\sin \omega_c t$ or $-\sin \omega_c t$. The coherent carrier recovery circuit detects and regenerates a carrier signal that is both frequency and phase coherent with the original transmit carrier. The balanced modulator is a product detector; the output is the product of the two inputs (the BPSK signal and the recovered carrier). The low-pass filter (LPF) separates the recovered binary data from the complex demodulated signal. Mathematically, the demodulation process is as follows.

For a BPSK input signal of $+\sin \omega_c t$ (logic 1), the output of the balanced modulator is

$$\text{output} = (\sin \omega_c t)(\sin \omega_c t) = \sin^2 \omega_c t \qquad \textbf{(9-21)}$$

$$\text{or} \qquad \sin^2 \omega_c t = \frac{1}{2}(1 - \cos 2\omega_c t) = \frac{1}{2} - \underset{\text{(filtered out)}}{\frac{1}{2}\cos 2\,\omega_c t}$$

$$\text{leaving} \qquad \text{output} = +\frac{1}{2}\text{V} = \text{logic 1}$$

It can be seen that the output of the balanced modulator contains a positive voltage ($+[1/2]$V) and a cosine wave at twice the carrier frequency ($2\,\omega_c$) The LPF has a cutoff frequency much lower than $2\omega_c$ and, thus, blocks the second harmonic of the carrier and passes only the positive constant component. A positive voltage represents a demodulated logic 1.

For a BPSK input signal of $-\sin \omega_c t$ (logic 0), the output of the balanced modulator is

$$\text{output} = (-\sin \omega_c t)(\sin \omega_c t) = -\sin^2 \omega_c t$$

$$\text{or} \qquad -\sin^2 \omega_c t = -\frac{1}{2}(1 - \cos 2\omega_c t) = -\frac{1}{2} + \underset{\text{(filtered out)}}{\frac{1}{2}\cos 2\omega_c t}$$

$$\text{leaving} \qquad \text{output} = -\frac{1}{2}\text{V} = \text{logic 0}$$

The output of the balanced modulator contains a negative voltage ($-[1/2]$V) and a cosine wave at twice the carrier frequency ($2\omega_c$). Again, the LPF blocks the second harmonic of the carrier and passes only the negative constant component. A negative voltage represents a demodulated logic 0.

9-5-2 Quaternary Phase-Shift Keying

Quaternary phase shift keying (QPSK), or *quadrature* PSK as it is sometimes called, is another form of angle-modulated, constant-amplitude digital modulation. QPSK is an M-ary encoding scheme where $N = 2$ and $M = 4$ (hence, the name "quaternary" meaning "4"). With QPSK, four output phases are possible for a single carrier frequency. Because there are four output phases, there must be four different input conditions. Because the digital input to a QPSK modulator is a binary (base 2) signal, to produce four different input combinations, the modulator requires more than a single input bit to determine the output condition. With two bits, there are four possible conditions: 00, 01, 10, and 11. Therefore, with QPSK, the binary input data are combined into groups of two bits, called *dibits*. In the modulator, each dibit code generates one of the four possible output phases (+45°, +135°, −45°, and −135°). Therefore, for each two-bit dibit clocked into the modulator, a single output change occurs, and the rate of change at the output (baud) is equal to one-half the input bit rate (i.e., two input bits produce one output phase change).

9-5-2-1 QPSK transmitter. A block diagram of a QPSK modulator is shown in Figure 9-17. Two bits (a dibit) are clocked into the bit splitter. After both bits have been serially inputted, they are simultaneously parallel outputted. One bit is directed to the I channel and the other to the Q channel. The I bit modulates a carrier that is in phase with the reference oscillator (hence the name "I" for "in phase" channel), and the Q bit modulates a carrier that is 90° out of phase or in quadrature with the reference carrier (hence the name "Q" for "quadrature" channel).

It can be seen that once a dibit has been split into the I and Q channels, the operation is the same as in a BPSK modulator. Essentially, a QPSK modulator is two BPSK modulators combined in parallel. Again, for a logic 1 = +1 V and a logic 0 = −1 V, two phases are possible at the output of the I balanced modulator ($+\sin \omega_c t$ and $-\sin \omega_c t$), and two

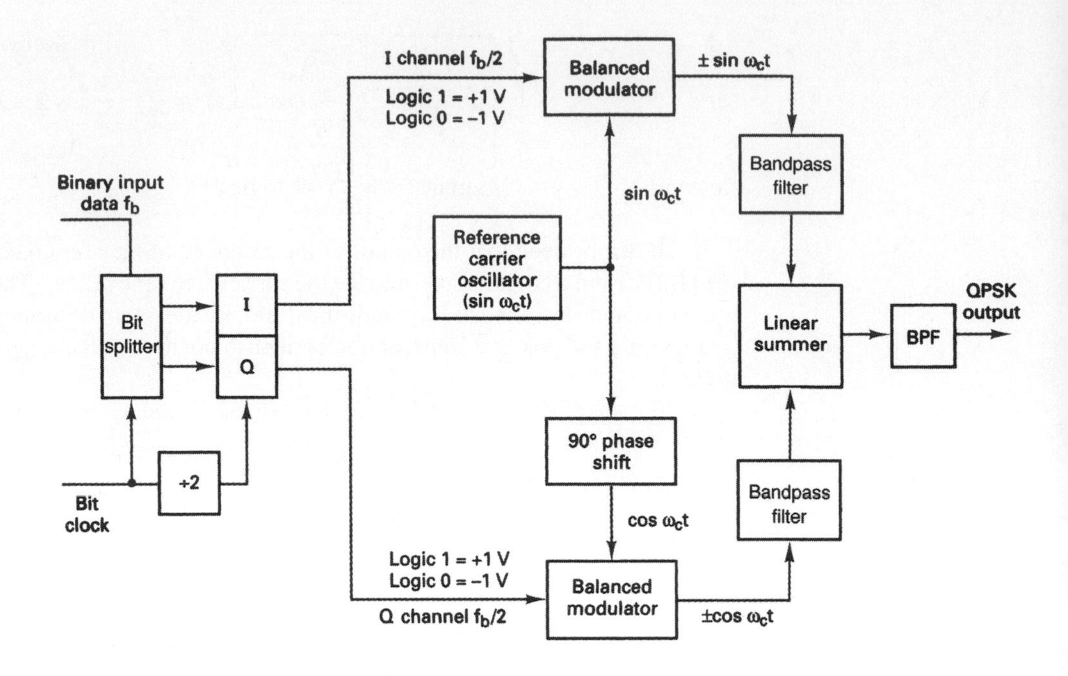

FIGURE 9-17 QPSK modulator

phases are possible at the output of the Q balanced modulator (+cos $\omega_c t$ and −cos $\omega_c t$). When the linear summer combines the two quadrature (90° out of phase) signals, there are four possible resultant phasors given by these expressions: + sin $\omega_c t$ + cos $\omega_c t$, + sin $\omega_c t$ − cos $\omega_c t$, −sin $\omega_c t$ + cos $\omega_c t$, and −sin $\omega_c t$ − cos $\omega_c t$.

Example 9-5

For the QPSK modulator shown in Figure 9-17, construct the truth table, phasor diagram, and constellation diagram.

Solution For a binary data input of Q = 0 and I = 0, the two inputs to the I balanced modulator are −1 and sin $\omega_c t$, and the two inputs to the Q balanced modulator are −1 and cos $\omega_c t$. Consequently, the outputs are

$$\text{I balanced modulator} = (-1)(\sin \omega_c t) = -1 \sin \omega_c t$$

$$\text{Q balanced modulator} = (-1)(\cos \omega_c t) = -1 \cos \omega_c t$$

and the output of the linear summer is

$$-1 \cos \omega_c t - 1 \sin \omega_c t = 1.414 \sin(\omega_c t - 135°)$$

For the remaining dibit codes (01, 10, and 11), the procedure is the same. The results are shown in Figure 9-18a.

In Figures 9-18b and c, it can be seen that with QPSK each of the four possible output phasors has exactly the same amplitude. Therefore, the binary information must be encoded entirely in the phase of the output signal. This constant amplitude characteristic is the most important characteristic of PSK that distinguishes it from QAM, which is explained later in this chapter. Also, from Figure 9-18b, it can be seen that the angular separation between any two adjacent phasors in QPSK is 90°. Therefore, a QPSK signal can undergo almost a +45° or −45° shift in phase during transmission and still retain the correct encoded information when demodulated at the receiver. Figure 9-19 shows the output phase-versus-time relationship for a QPSK modulator.

Binary input		QPSK output phase
Q	I	
0	0	−135°
0	1	−45°
1	0	+135°
1	1	+45°

(a)

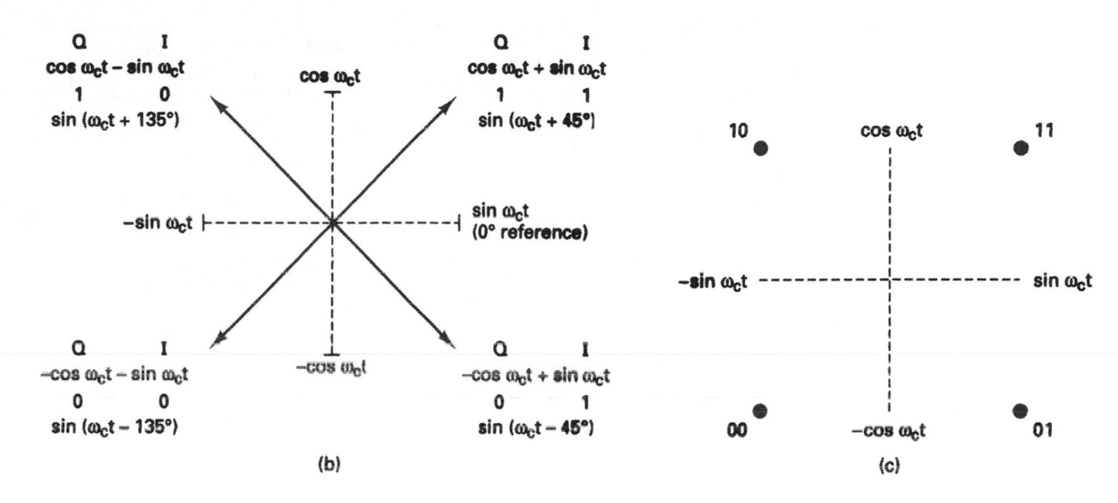

FIGURE 9-18 QPSK modulator: (a) truth table; (b) phasor diagram; (c) constellation diagram

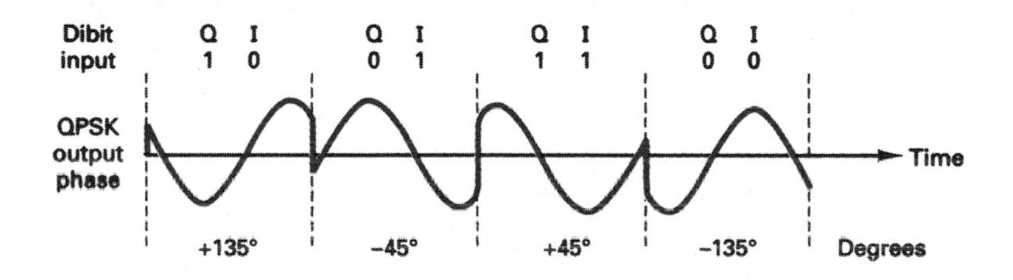

FIGURE 9-19 Output phase-versus-time relationship for a QPSK modulator

9-5-2-2 Bandwidth considerations of QPSK. With QPSK, because the input data are divided into two channels, the bit rate in either the I or the Q channel is equal to one-half of the input data rate ($f_b/2$). (Essentially, the bit splitter stretches the I and Q bits to twice their input bit length.) Consequently, the highest fundamental frequency present at the data input to the I or the Q balanced modulator is equal to one-fourth of the input data rate (one-half of $f_b/2 = f_b/4$). As a result, the output of the I and Q balanced modulators requires a minimum double-sided Nyquist bandwidth equal to one-half of the incoming bit rate (f_N = twice $f_b/4$ $= f_b/2$). Thus, with QPSK, a bandwidth compression is realized (the minimum bandwidth is less than the incoming bit rate). Also, because the QPSK output signal does not change phase until two bits (a dibit) have been clocked into the bit splitter, the fastest output rate of change (baud) is also equal to one-half of the input bit rate. As with BPSK, the minimum bandwidth and the baud are equal. This relationship is shown in Figure 9-20.

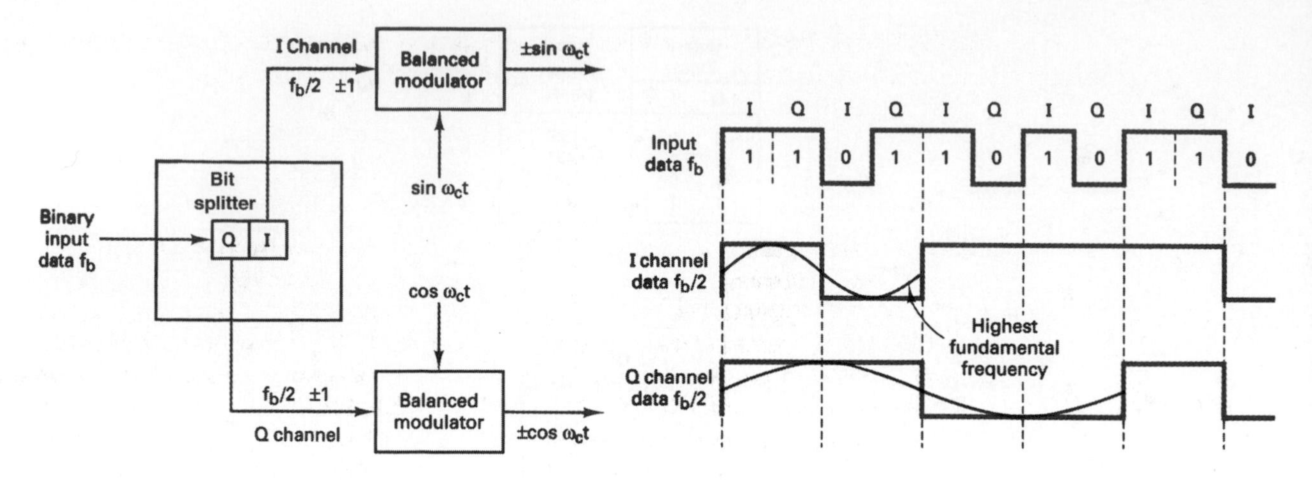

FIGURE 9-20 Bandwidth considerations of a QPSK modulator

In Figure 9-20, it can be seen that the worse-case input condition to the I or Q balanced modulator is an alternative 1/0 pattern, which occurs when the binary input data have a 1100 repetitive pattern. One cycle of the fastest binary transition (a 1/0 sequence) in the I or Q channel takes the same time as four input data bits. Consequently, the highest fundamental frequency at the input and fastest rate of change at the output of the balanced modulators is equal to one-fourth of the binary input bit rate.

The output of the balanced modulators can be expressed mathematically as

$$\text{output} = (\sin \omega_a t)(\sin \omega_c t) \tag{9-22}$$

where

$$\underbrace{\omega_a t = 2\pi \frac{f_b}{4} t}_{\text{modulating signal}} \quad \text{and} \quad \underbrace{\omega_c t = 2\pi f_c}_{\text{carrier}}$$

Thus,

$$\text{output} = \left(\sin 2\pi \frac{f_b}{4} t\right)(\sin 2\pi f_c t)$$

$$\frac{1}{2}\cos 2\pi\left(f_c - \frac{f_b}{4}\right)t - \frac{1}{2}\cos 2\pi\left(f_c + \frac{f_b}{4}\right)t$$

The output frequency spectrum extends from $f_c + f_b/4$ to $f_c - f_b/4$, and the minimum bandwidth (f_N) is

$$\left(f_c + \frac{f_b}{4}\right) - \left(f_c - \frac{f_b}{4}\right) = \frac{2f_b}{4} = \frac{f_b}{2}$$

Example 9-6

For a QPSK modulator with an input data rate (f_b) equal to 10 Mbps and a carrier frequency of 70 MHz, determine the minimum double-sided Nyquist bandwidth (f_N) and the baud. Also, compare the results with those achieved with the BPSK modulator in Example 9-4. Use the QPSK block diagram shown in Figure 9-17 as the modulator model.

Solution The bit rate in both the I and Q channels is equal to one-half of the transmission bit rate, or

$$f_{bQ} = f_{bI} = \frac{f_b}{2} = \frac{10 \text{ Mbps}}{2} = 5 \text{ Mbps}$$

The highest fundamental frequency presented to either balanced modulator is

$$f_a = \frac{f_{bQ}}{2} \text{ or } \frac{f_{bI}}{2} = \frac{5 \text{ Mbps}}{2} = 2.5 \text{ MHz}$$

The output wave from each balanced modulator is

$$(\sin 2\pi f_a t)(\sin 2\pi f_c t)$$

$$\frac{1}{2}\cos 2\pi(f_c - f_a)t - \frac{1}{2}\cos 2\pi(f_c + f_a)t$$

$$\frac{1}{2}\cos 2\pi[(70 - 2.5) \text{ MHz}]t - \frac{1}{2}\cos 2\pi[(70 + 2.5) \text{ MHz}]t$$

$$\frac{1}{2}\cos 2\pi(67.5 \text{ MHz})t - \frac{1}{2}\cos 2\pi(72.5 \text{ MHz})t$$

The minimum Nyquist bandwidth is

$$B = (72.5 - 67.5) \text{ MHz} = 5 \text{ MHz}$$

The symbol rate equals the bandwidth; thus,

$$\text{symbol rate} = 5 \text{ megabaud}$$

The output spectrum is as follows:

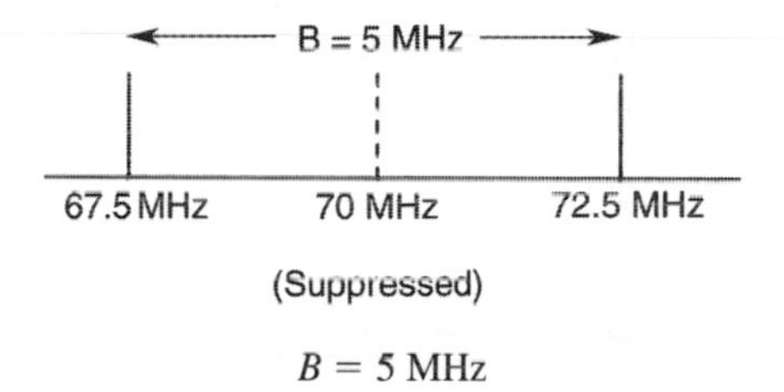

$B = 5$ MHz

It can be seen that for the same input bit rate the minimum bandwidth required to pass the output of the QPSK modulator is equal to one-half of that required for the BPSK modulator in Example 9-4. Also, the baud rate for the QPSK modulator is one-half that of the BPSK modulator.

The minimum bandwidth for the QPSK system described in Example 9-6 can also be determined by simply substituting into Equation 9-10:

$$B = \frac{10 \text{ Mbps}}{2}$$
$$= 5 \text{ MHz}$$

9-5-2-3 QPSK receiver. The block diagram of a QPSK receiver is shown in Figure 9-21. The power splitter directs the input QPSK signal to the I and Q product detectors and the carrier recovery circuit. The carrier recovery circuit reproduces the original transmit carrier oscillator signal. The recovered carrier must be frequency and phase coherent with the transmit reference carrier. The QPSK signal is demodulated in the I and Q product detectors, which generate the original I and Q data bits. The outputs of the product detectors are fed to the bit combining circuit, where they are converted from parallel I and Q data channels to a single binary output data stream.

The incoming QPSK signal may be any one of the four possible output phases shown in Figure 9-18. To illustrate the demodulation process, let the incoming QPSK signal be $-\sin \omega_c t + \cos \omega_c t$. Mathematically, the demodulation process is as follows.

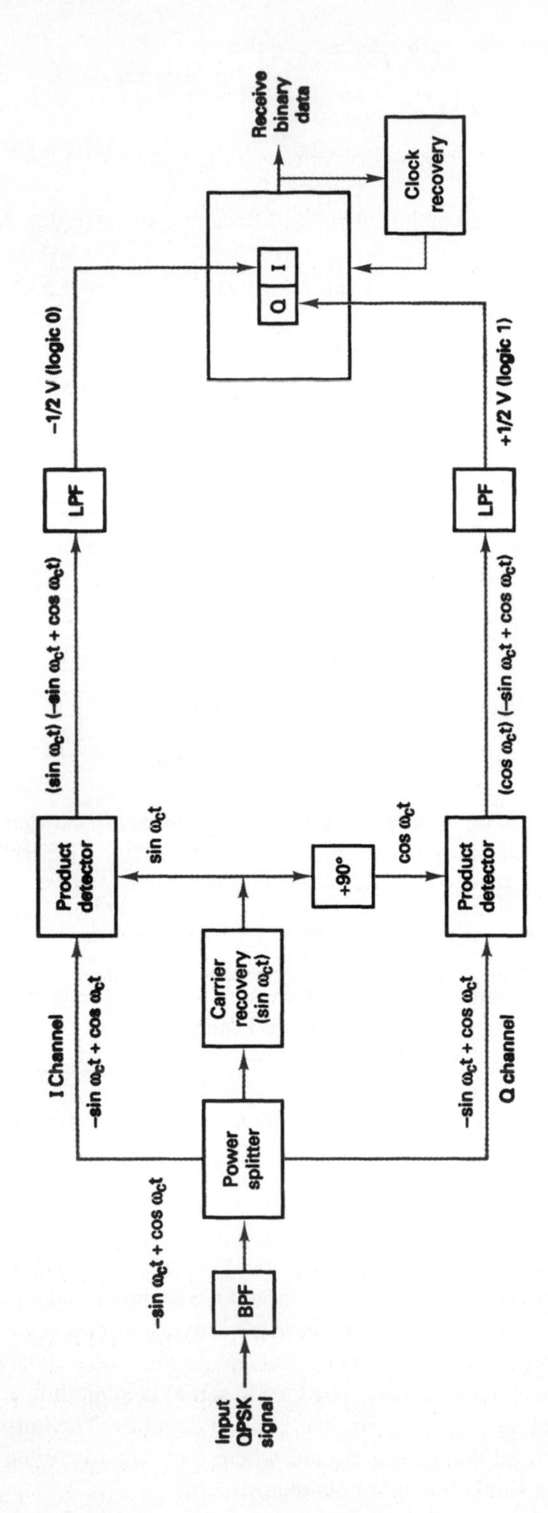

FIGURE 9-21 QPSK receiver

The receive QPSK signal ($-\sin \omega_c t + \cos \omega_c t$) is one of the inputs to the I product detector. The other input is the recovered carrier ($\sin \omega_c t$). The output of the I product detector is

$$I = \underbrace{(-\sin \omega_c t + \cos \omega_c t)}_{\text{QPSK input signal}}\underbrace{(\sin \omega_c t)}_{\text{carrier}} \tag{9-23}$$

$$= (-\sin \omega_c t)(\sin \omega_c t) + (\cos \omega_c t)(\sin \omega_c t)$$

$$= -\sin^2 \omega_c t + (\cos \omega_c t)(\sin \omega_c t)$$

$$= -\frac{1}{2}(1 - \cos 2\omega_c t) + \frac{1}{2}\sin(\omega_c + \omega_c)t + \frac{1}{2}\sin(\omega_c - \omega_c)t$$

(filtered out) (equals 0)

$$I = -\frac{1}{2} + \frac{1}{2}\cos 2\omega_c t + \frac{1}{2}\sin 2\omega_c t + \frac{1}{2}\sin 0$$

$$= -\frac{1}{2}\text{V (logic 0)}$$

Again, the receive QPSK signal ($-\sin \omega_c t + \cos \omega_c t$) is one of the inputs to the Q product detector. The other input is the recovered carrier shifted 90° in phase ($\cos \omega_c t$). The output of the Q product detector is

$$Q = \underbrace{(-\sin \omega_c t + \cos \omega_c t)}_{\text{QPSK input signal}}\underbrace{(\cos \omega_c t)}_{\text{carrier}} \tag{9-24}$$

$$= \cos^2 \omega_c t - (\sin \omega_c t)(\cos \omega_c t)$$

$$= \frac{1}{2}(1 + \cos 2\omega_c t) - \frac{1}{2}\sin(\omega_c + \omega_c)t - \frac{1}{2}\sin(\omega_c - \omega_c)t$$

(filtered out) (equals 0)

$$Q = \frac{1}{2} + \frac{1}{2}\cos 2\omega_c t - \frac{1}{2}\sin 2\omega_c t - \frac{1}{2}\sin 0$$

$$= \frac{1}{2}\text{V(logic 1)}$$

The demodulated I and Q bits (0 and 1, respectively) correspond to the constellation diagram and truth table for the QPSK modulator shown in Figure 9-18.

9-5-2-4 Offset QPSK. *Offset QPSK* (OQPSK) is a modified form of QPSK where the bit waveforms on the I and Q channels are offset or shifted in phase from each other by one-half of a bit time.

Figure 9-22 shows a simplified block diagram, the bit sequence alignment, and the constellation diagram for a OQPSK modulator. Because changes in the I channel occur at the midpoints of the Q channel bits and vice versa, there is never more than a single bit change in the dibit code and, therefore, there is never more than a 90° shift in the output phase. In conventional QPSK, a change in the input dibit from 00 to 11 or 01 to 10 causes a corresponding 180° shift in the output phase. Therefore, an advantage of OQPSK is the limited phase shift that must be imparted during modulation. A disadvantage of OQPSK is

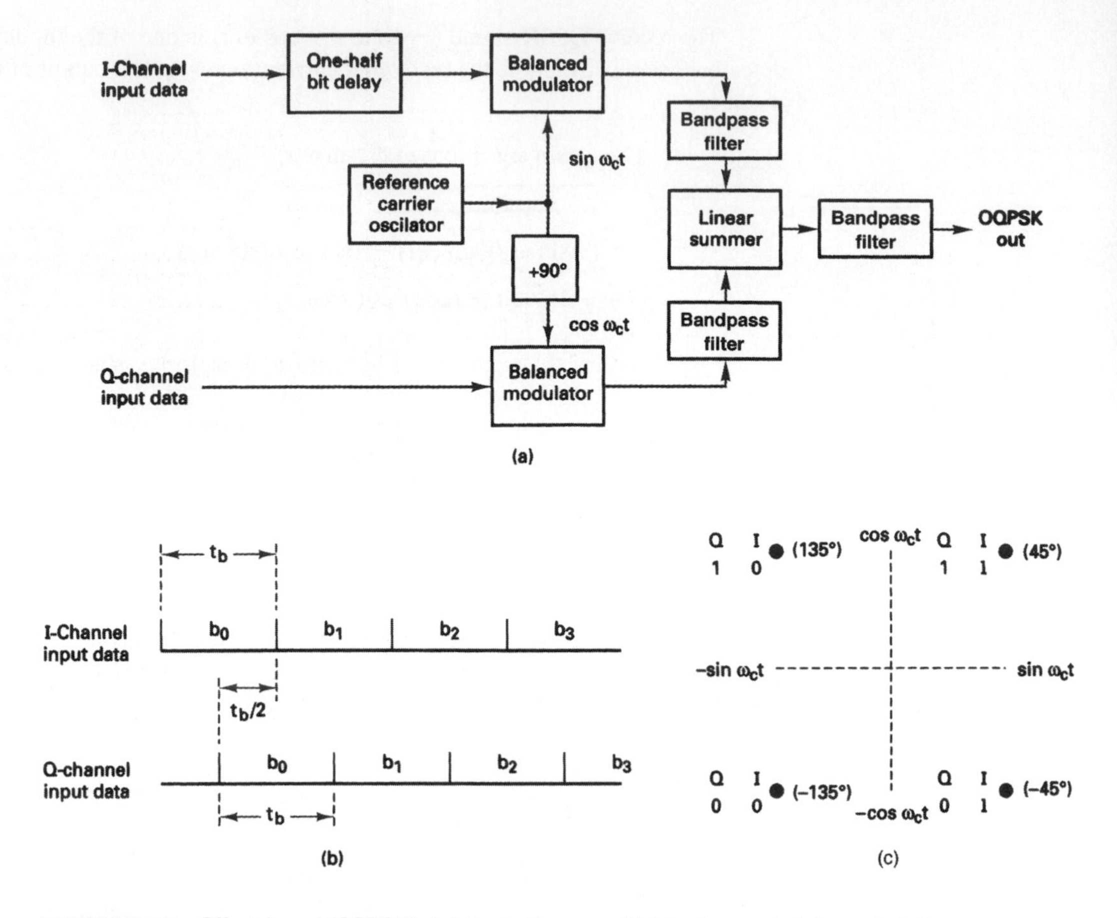

FIGURE 9-22 Offset keyed (OQPSK): (a) block diagram; (b) bit alignment; (c) constellation diagram

that changes in the output phase occur at twice the data rate in either the I or Q channels. Consequently, with OQPSK the baud and minimum bandwidth are twice that of conventional QPSK for a given transmission bit rate. OQPSK is sometimes called OKQPSK (*offset-keyed QPSK*).

9-5-3 8-PSK

With *8-PSK,* three bits are encoded, forming tribits and producing eight different output phases. With 8-PSK, $n = 3$, $M = 8$, and there are eight possible output phases. To encode eight different phases, the incoming bits are encoded in groups of three, called tribits ($2^3 = 8$).

9-5-3-1 8-PSK transmitter. A block diagram of an 8-PSK modulator is shown in Figure 9-23. The incoming serial bit stream enters the bit splitter, where it is converted to a parallel, three-channel output (the I or in-phase channel, the Q or in-quadrature channel, and the C or control channel). Consequently, the bit rate in each of the three channels is $f_b/3$. The bits in the I and C channels enter the I channel 2-to-4-level converter, and the bits in the Q and $\overline{C}$ channels enter the Q channel 2-to-4-level converter. Essentially, the 2-to-4-level converters are parallel-input *digital-to-analog converters* (DACs). With two input bits, four output voltages are possible. The algorithm for the DACs is quite simple. The I or Q bit determines the polarity of the output analog signal (logic 1 = +V and logic 0 = −V), whereas the C or $\overline{C}$ bit determines the magni-

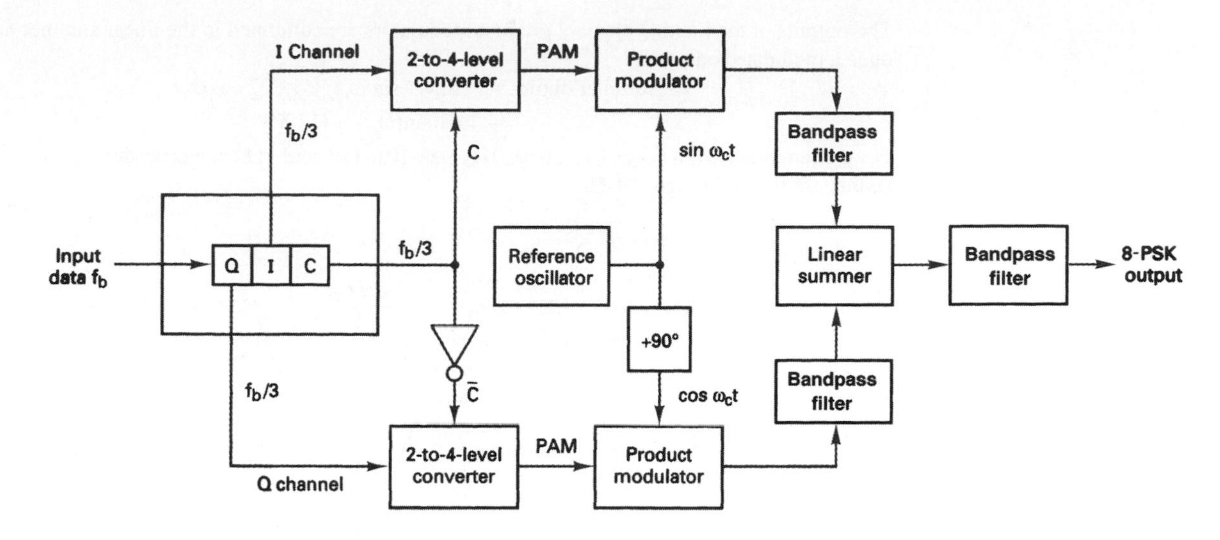

FIGURE 9-23 8-PSK modulator

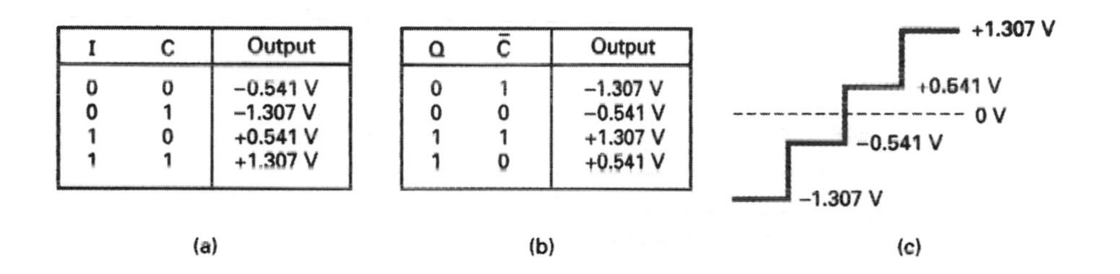

I	C	Output
0	0	−0.541 V
0	1	−1.307 V
1	0	+0.541 V
1	1	+1.307 V

(a)

Q	$\bar{C}$	Output
0	1	−1.307 V
0	0	−0.541 V
1	1	+1.307 V
1	0	+0.541 V

(b)

(c)

FIGURE 9-24 I- and Q-channel 2-to-4-level converters: (a) I-channel truth table; (b) Q-channel truth table; (c) PAM levels

tude (logic 1 = 1.307 V and logic 0 = 0.541 V). Consequently, with two magnitudes and two polarities, four different output conditions are possible.

Figure 9-24 shows the truth table and corresponding output conditions for the 2-to-4-level converters. Because the C and $\bar{C}$ bits can never be the same logic state, the outputs from the I and Q 2-to-4-level converters can never have the same magnitude, although they can have the same polarity. The output of a 2-to-4-level converter is an M-ary, *pulse-amplitude-modulated* (PAM) signal where $M = 4$.

Example 9-7

For a tribit input of Q = 0, 1 = 0, and C = 0 (000), determine the output phase for the 8-PSK modulator shown in Figure 9-23.

Solution The inputs to the I channel 2-to-4-level converter are I = 0 and C = 0. From Figure 9-24 the output is −0.541 V. The inputs to the Q channel 2-to-4-level converter are Q = 0 and $\bar{C}$ = 1. Again from Figure 9-24, the output is −1.307 V.

Thus, the two inputs to the I channel product modulators are −0.541 and sin $\omega_c t$. The output is

$$I = (-0.541)(\sin \omega_c t) = -0.541 \sin \omega_c t$$

The two inputs to the Q channel product modulator are −1.307 V and cos $\omega_c t$. The output is

$$Q = (-1.307)(\cos \omega_c t) = -1.307 \cos \omega_c t$$

The outputs of the I and Q channel product modulators are combined in the linear summer and produce a modulated output of

$$\text{summer output} = -0.541 \sin \omega_c t - 1.307 \cos \omega_c t$$
$$= 1.41 \sin(\omega_c t - 112.5°)$$

For the remaining tribit codes (001, 010, 011, 100, 101, 110, and 111), the procedure is the same. The results are shown in Figure 9-25.

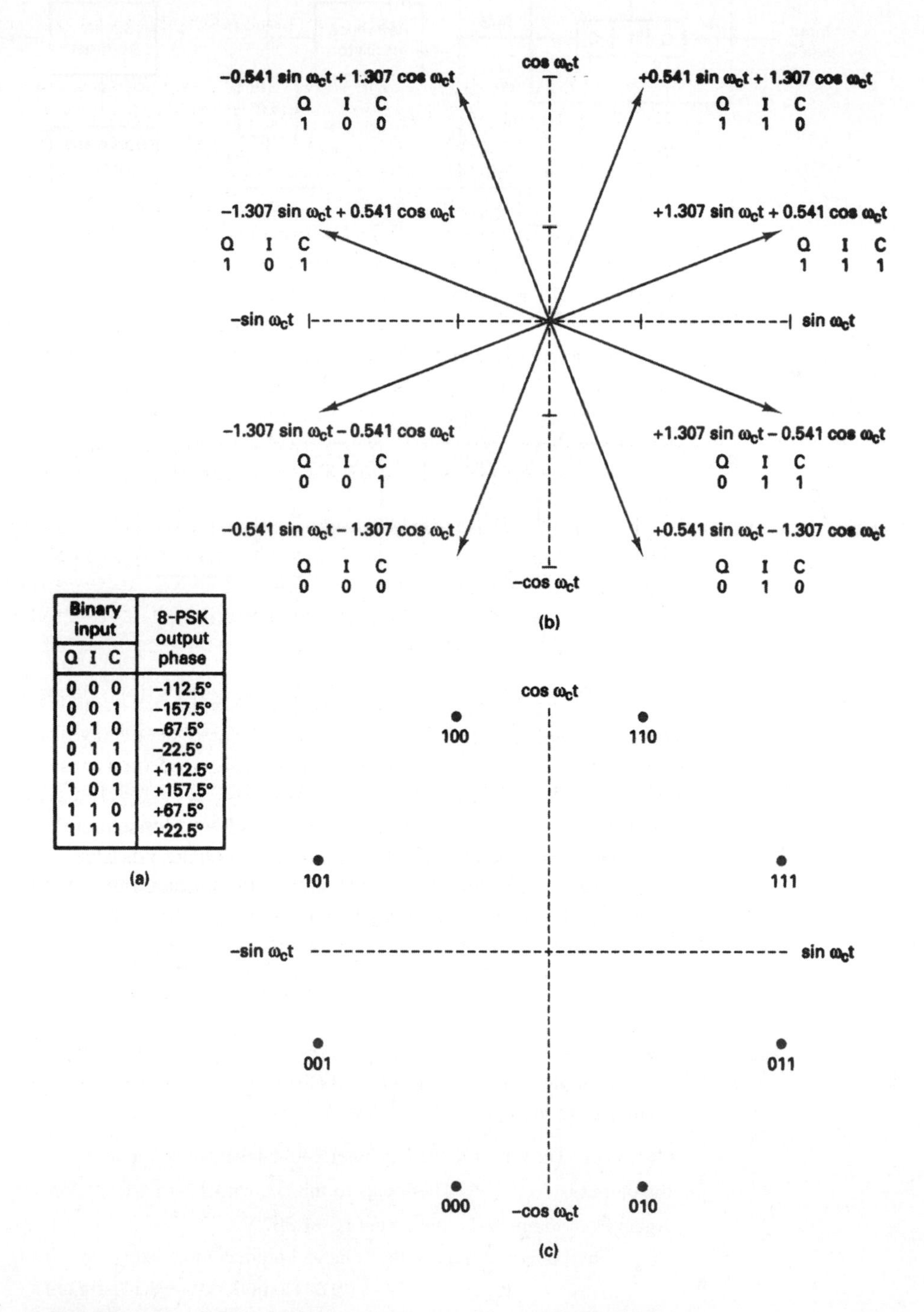

Binary input (Q I C)	8-PSK output phase
0 0 0	−112.5°
0 0 1	−157.5°
0 1 0	−67.5°
0 1 1	−22.5°
1 0 0	+112.5°
1 0 1	+157.5°
1 1 0	+67.5°
1 1 1	+22.5°

(a)

FIGURE 9-25 8-PSK modulator: (a) truth table; (b) phasor diagram; (c) constellation diagram

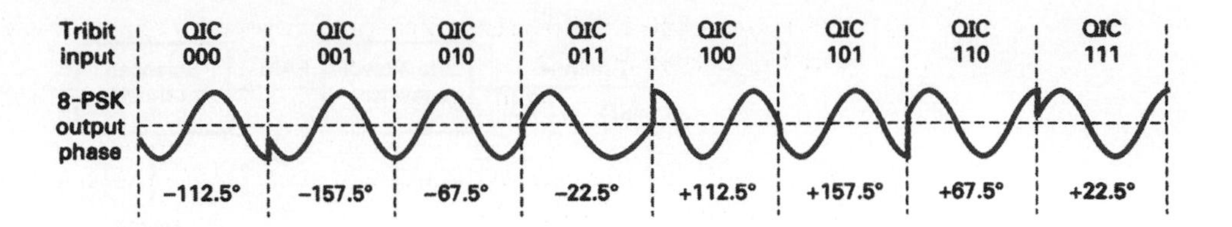

FIGURE 9-26 Output phase-versus-time relationship for an 8-PSK modulator

From Figure 9-25, it can be seen that the angular separation between any two adjacent phasors is 45°, half what it is with QPSK. Therefore, an 8-PSK signal can undergo almost a ±22.5° phase shift during transmission and still retain its integrity. Also, each phasor is of equal magnitude; the tribit condition (actual information) is again contained only in the phase of the signal. The PAM levels of 1.307 and 0.541 are relative values. Any levels may be used as long as their ratio is 0.541/1.307 and their arc tangent is equal to 22.5°. For example, if their values were doubled to 2.614 and 1.082, the resulting phase angles would not change, although the magnitude of the phasor would increase proportionally.

It should also be noted that the tribit code between any two adjacent phases changes by only one bit. This type of code is called the *Gray code* or, sometimes, the *maximum distance code*. This code is used to reduce the number of transmission errors. If a signal were to undergo a phase shift during transmission, it would most likely be shifted to an adjacent phasor. Using the Gray code results in only a single bit being received in error.

Figure 9-26 shows the output phase-versus-time relationship of an 8-PSK modulator.

9-5-3-2 Bandwidth considerations of 8-PSK. With 8-PSK, because the data are divided into three channels, the bit rate in the I, Q, or C channel is equal to one-third of the binary input data rate ($f_b/3$). (The bit splitter stretches the I, Q, and C bits to three times their input bit length.) Because the I, Q, and C bits are outputted simultaneously and in parallel, the 2-to-4-level converters also see a change in their inputs (and consequently their outputs) at a rate equal to $f_b/3$.

Figure 9-27 shows the bit timing relationship between the binary input data; the I, Q, and C channel data; and the I and Q PAM signals. It can be seen that the highest fundamental frequency in the I, Q, or C channel is equal to one-sixth the bit rate of the binary input (one cycle in the I, Q, or C channel takes the same amount of time as six input bits). Also, the highest fundamental frequency in either PAM signal is equal to one-sixth of the binary input bit rate.

With an 8-PSK modulator, there is one change in phase at the output for every three data input bits. Consequently, the baud for 8 PSK equals $f_b/3$, the same as the minimum bandwidth. Again, the balanced modulators are product modulators; their outputs are the product of the carrier and the PAM signal. Mathematically, the output of the balanced modulators is

$$\theta = (X \sin \omega_a t)(\sin \omega_c t) \tag{9-25}$$

where $\underbrace{\omega_a t = 2\pi \frac{f_b}{6} t}_{\text{modulating signal}}$ and $\underbrace{\omega_c t = 2\pi f_c t}_{\text{carrier}}$

and $X = \pm 1.307 \text{ or } \pm 0.541$

Thus,

$$\theta = \left(X \sin 2\pi \frac{f_b}{6} t\right)(\sin 2\pi f_c t)$$

$$= \frac{X}{2} \cos 2\pi\left(f_c - \frac{f_b}{6}\right)t - \frac{X}{2} \cos 2\pi\left(f_c + \frac{f_b}{6}\right)t$$

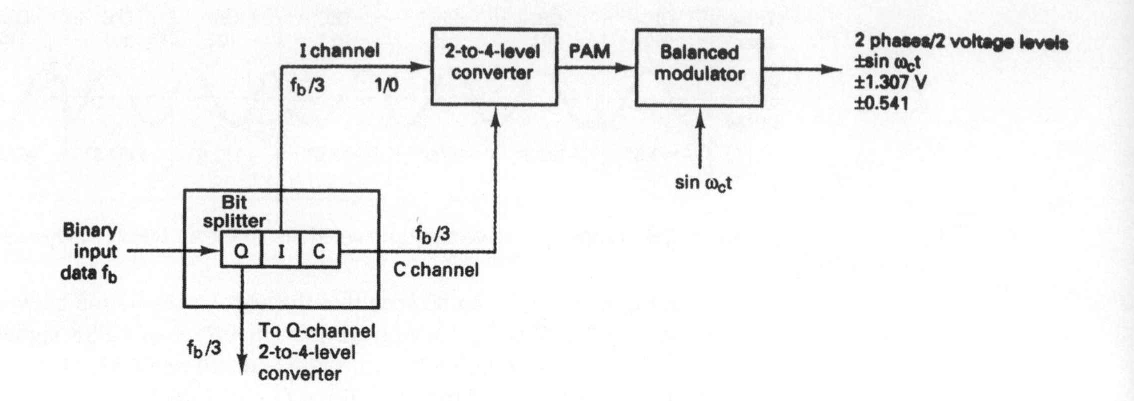

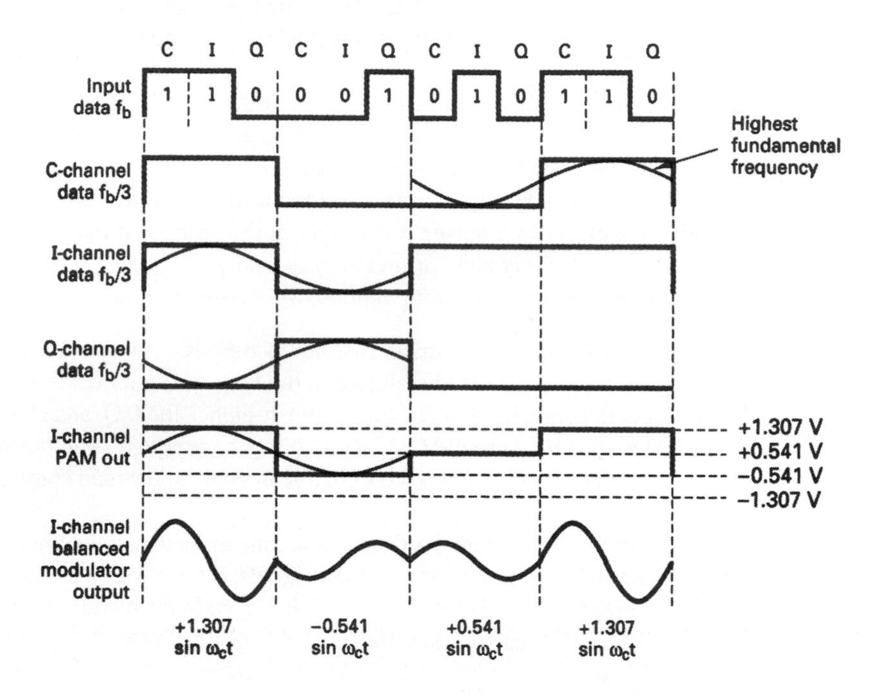

FIGURE 9-27 Bandwidth considerations of an 8-PSK modulator

The output frequency spectrum extends from $f_c + f_b/6$ to $f_c - f_b/6$, and the minimum bandwidth (f_N) is

$$\left(f_c + \frac{f_b}{6}\right) - \left(f_c - \frac{f_b}{6}\right) = \frac{2f_b}{6} = \frac{f_b}{3}$$

Example 9-8

For an 8-PSK modulator with an input data rate (f_b) equal to 10 Mbps and a carrier frequency of 70 MHz, determine the minimum double-sided Nyquist bandwidth (f_N) and the baud. Also, compare the results with those achieved with the BPSK and QPSK modulators in Examples 9-4 and 9-6. Use the 8-PSK block diagram shown in Figure 9-23 as the modulator model.

Solution The bit rate in the I, Q, and C channels is equal to one-third of the input bit rate, or

$$f_{bC} = f_{bQ} = f_{bI} = \frac{10 \text{ Mbps}}{3} = 3.33 \text{ Mbps}$$

Therefore, the fastest rate of change and highest fundamental frequency presented to either balanced modulator is

$$f_a = \frac{f_{bC}}{2} \text{ or } \frac{f_{bQ}}{2} \text{ or } \frac{f_{bI}}{2} = \frac{3.33 \text{ Mbps}}{2} = 1.667 \text{ Mbps}$$

The output wave from the balance modulators is

$$(\sin 2\pi f_a t)(\sin 2\pi f_c t)$$

$$\frac{1}{2}\cos 2\pi(f_c - f_a)t - \frac{1}{2}\cos 2\pi(f_c + f_a)t$$

$$\frac{1}{2}\cos 2\pi[(70 - 1.667) \text{ MHz}]t - \frac{1}{2}\cos 2\pi[(70 + 1.667) \text{ MHz}]t$$

$$\frac{1}{2}\cos 2\pi(68.333 \text{ MHz})t - \frac{1}{2}\cos 2\pi(71.667 \text{ MHz})t$$

The minimum Nyquist bandwidth is

$$B = (71.667 - 68.333) \text{ MHz} = 3.333 \text{ MHz}$$

The minimum bandwidth for the 8-PSK can also be determined by simply substituting into Equation 9-10:

$$B = \frac{10 \text{ Mbps}}{3}$$

$$= 3.33 \text{ MHz}$$

Again, the baud equals the bandwidth; thus,

$$\text{baud} = 3.333 \text{ megabaud}$$

The output spectrum is as follows:

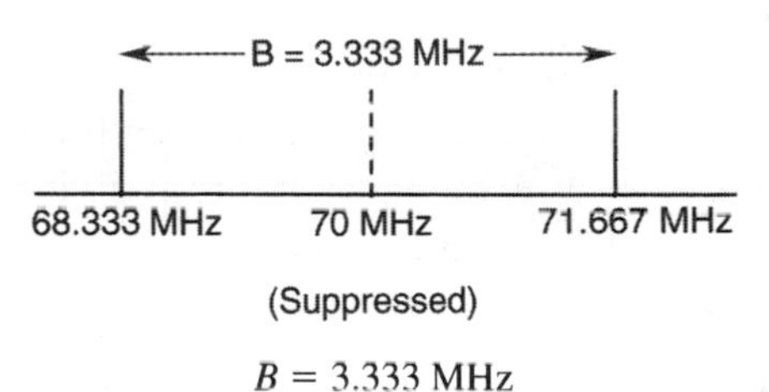

$$B = 3.333 \text{ MHz}$$

It can be seen that for the same input bit rate the minimum bandwidth required to pass the output of an 8-PSK modulator is equal to one-third that of the BPSK modulator in Example 9-4 and 50% less than that required for the QPSK modulator in Example 9-6. Also, in each case the baud has been reduced by the same proportions.

9-5-3-3 8-PSK receiver. Figure 9-28 shows a block diagram of an 8-PSK receiver. The power splitter directs the input 8-PSK signal to the I and Q product detectors and the carrier recovery circuit. The carrier recovery circuit reproduces the original reference oscillator signal. The incoming 8-PSK signal is mixed with the recovered carrier in the I product detector and with a quadrature carrier in the Q product detector. The outputs of the product detectors are 4-level PAM signals that are fed to the 4-to-2-level *analog-to-digital converters* (ADCs). The outputs from the I channel 4-to-2-level converter are the I and C bits, whereas the outputs from the Q channel 4-to-2-level converter are the Q and $\overline{C}$ bits. The parallel-to-serial logic circuit converts the I/C and Q/$\overline{C}$ bit pairs to serial I, Q, and C output data streams.

9-5-4 16-PSK

16-PSK is an *M*-ary encoding technique where $M = 16$; there are 16 different output phases possible. With 16-PSK, four bits (called *quadbits*) are combined, producing 16 different output phases. With 16-PSK, $n = 4$ and $M = 16$; therefore, the minimum bandwidth and

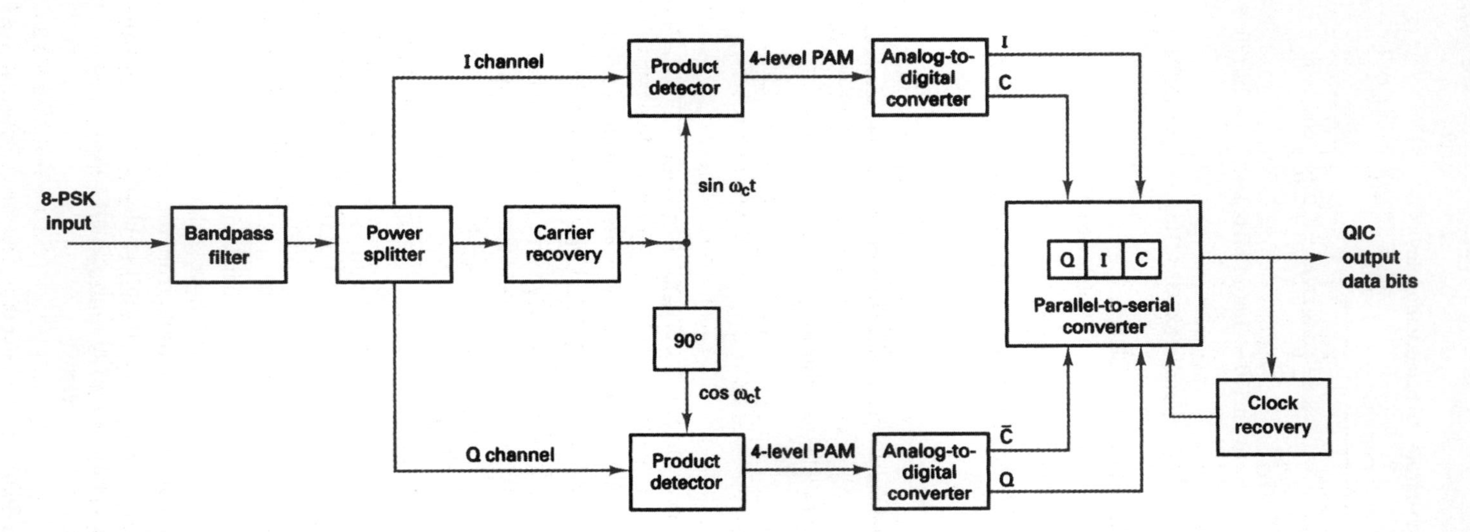

FIGURE 9-28 8-PSK receiver

Bit code	Phase	Bit code	Phase
0000	11.25°	1000	191.25°
0001	33.75°	1001	213.75°
0010	56.25°	1010	236.25°
0011	78.75°	1011	258.75°
0100	101.25°	1100	281.25°
0101	123.75°	1101	303.75°
0110	146.25°	1110	326.25°
0111	168.75°	1111	348.75°

(a)

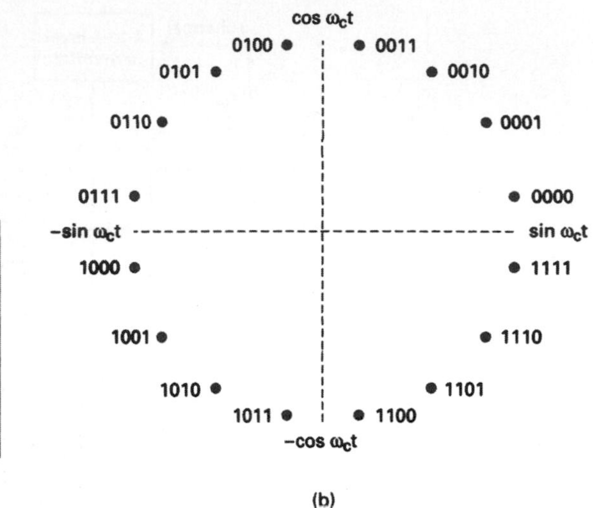

(b)

FIGURE 9-29 16-PSK: (a) truth table; (b) constellation diagram

baud equal one-fourth the bit rate ($f_b/4$). Figure 9-29 shows the truth table and constellation diagram for 16-PSK, respectively. Comparing Figures 9-18, 9-25, and 9-29 shows that as the level of encoding increases (i.e., the values of n and M increase), more output phases are possible and the closer each point on the constellation diagram is to an adjacent point. With 16-PSK, the angular separation between adjacent output phases is only 22.5°. Therefore, 16-PSK can undergo only a 11.25° phase shift during transmission and still retain its integrity. For an M-ary PSK system with 64 output phases ($n = 6$), the angular separation between adjacent phases is only 5.6°. This is an obvious limitation in the level of encoding (and bit rates) possible with PSK, as a point is eventually reached where receivers cannot discern the phase of the received signaling element. In addition, phase impairments inherent on communications lines have a tendency to shift the phase of the PSK signal, destroying its integrity and producing errors.

9-6 QUADRATURE-AMPLITUDE MODULATION

Quadrature-amplitude modulation (QAM) is a form of digital modulation similar to PSK except the digital information is contained in both the amplitude and the phase of the transmitted carrier. With QAM, amplitude and phase-shift keying are combined in such a way that the positions of the signaling elements on the constellation diagrams are optimized to achieve the greatest distance between elements, thus reducing the likelihood of one element being misinterpreted as another element. Obviously, this reduces the likelihood of errors occurring.

9-6-1 8-QAM

8-QAM is an M-ary encoding technique where $M = 8$. Unlike 8-PSK, the output signal from an 8-QAM modulator is not a constant-amplitude signal.

9-6-1-1 8-QAM transmitter. Figure 9-30a shows the block diagram of an 8-QAM transmitter. As you can see, the only difference between the 8-QAM transmitter and the 8-PSK transmitter shown in Figure 9-23 is the omission of the inverter between the C channel and the Q product modulator. As with 8-PSK, the incoming data are divided into groups of three bits (tribits): the I, Q, and C bit streams, each with a bit rate equal to one-third of

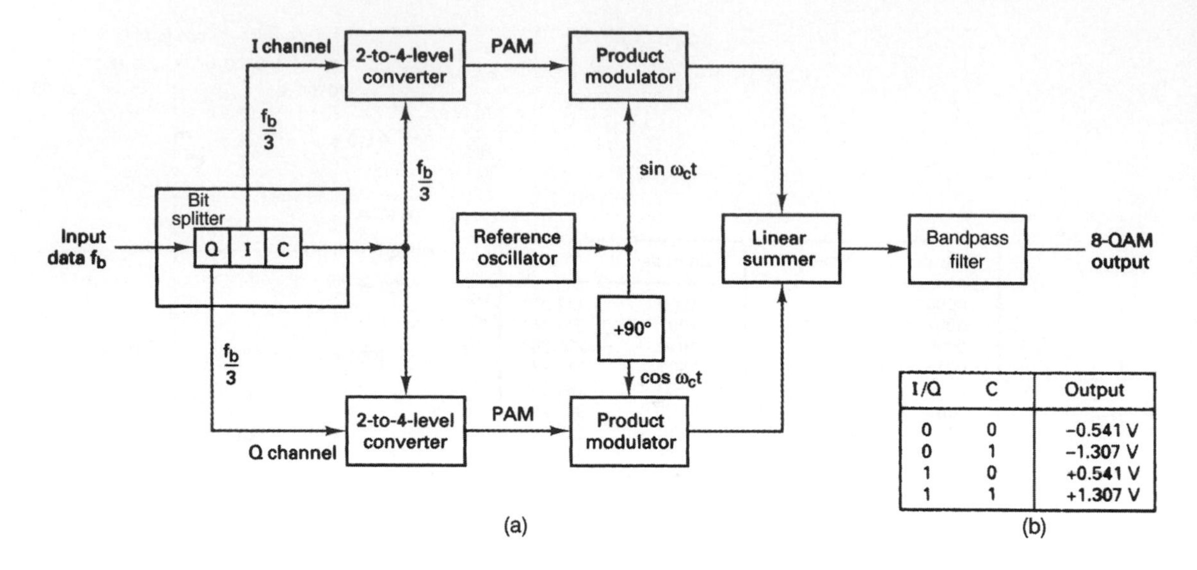

I/Q	C	Output
0	0	-0.541 V
0	1	-1.307 V
1	0	+0.541 V
1	1	+1.307 V

FIGURE 9-30 8-QAM transmitter: (a) block diagram; (b) truth table 2-4 level converters

the incoming data rate. Again, the I and Q bits determine the polarity of the PAM signal at the output of the 2-to-4-level converters, and the C channel determines the magnitude. Because the C bit is fed uninverted to both the I and the Q channel 2-to-4-level converters, the magnitudes of the I and Q PAM signals are always equal. Their polarities depend on the logic condition of the I and Q bits and, therefore, may be different. Figure 9-30b shows the truth table for the I and Q channel 2-to-4-level converters; they are identical.

Example 9-9

For a tribit input of Q = 0, I = 0, and C = 0 (000), determine the output amplitude and phase for the 8-QAM transmitter shown in Figure 9-30a.

Solution The inputs to the I channel 2-to-4-level converter are I = 0 and C = 0. From Figure 9-30b, the output is −0.541 V. The inputs to the Q channel 2-to-4-level converter are Q = 0 and C = 0. Again from Figure 9-30b, the output is −0.541 V.

Thus, the two inputs to the I channel product modulator are −0.541 and sin $\omega_c t$. The output is

$$I = (-0.541)(\sin \omega_c t) = -0.541 \sin \omega_c t$$

The two inputs to the Q channel product modulator are −0.541 and cos $\omega_c t$. The output is

$$Q = (-0.541)(\cos \omega_c t) = -0.541 \cos \omega_c t$$

The outputs from the I and Q channel product modulators are combined in the linear summer and produce a modulated output of

$$\text{summer output} = -0.541 \sin \omega_c t - 0.541 \cos \omega_c t$$
$$= 0.765 \sin(\omega_c t - 135°)$$

For the remaining tribit codes (001, 010, 011, 100, 101, 110, and 111), the procedure is the same. The results are shown in Figure 9-31.

Figure 9-32 shows the output phase-versus-time relationship for an 8-QAM modulator. Note that there are two output amplitudes, and only four phases are possible.

9-6-1-2 Bandwidth considerations of 8-QAM. In 8-QAM, the bit rate in the I and Q channels is one-third of the input binary rate, the same as in 8-PSK. As a result, the highest fundamental modulating frequency and fastest output rate of change in 8-QAM are the same as with 8-PSK. Therefore, the minimum bandwidth required for 8-QAM is $f_b/3$, the same as in 8-PSK.

Binary input			8-QAM output	
Q	I	C	Amplitude	Phase
0	0	0	0.765 V	−135°
0	0	1	1.848 V	−135°
0	1	0	0.765 V	−45°
0	1	1	1.848 V	−45°
1	0	0	0.765 V	+135°
1	0	1	1.848 V	+135°
1	1	0	0.765 V	+45°
1	1	1	1.848 V	+45°

(a)

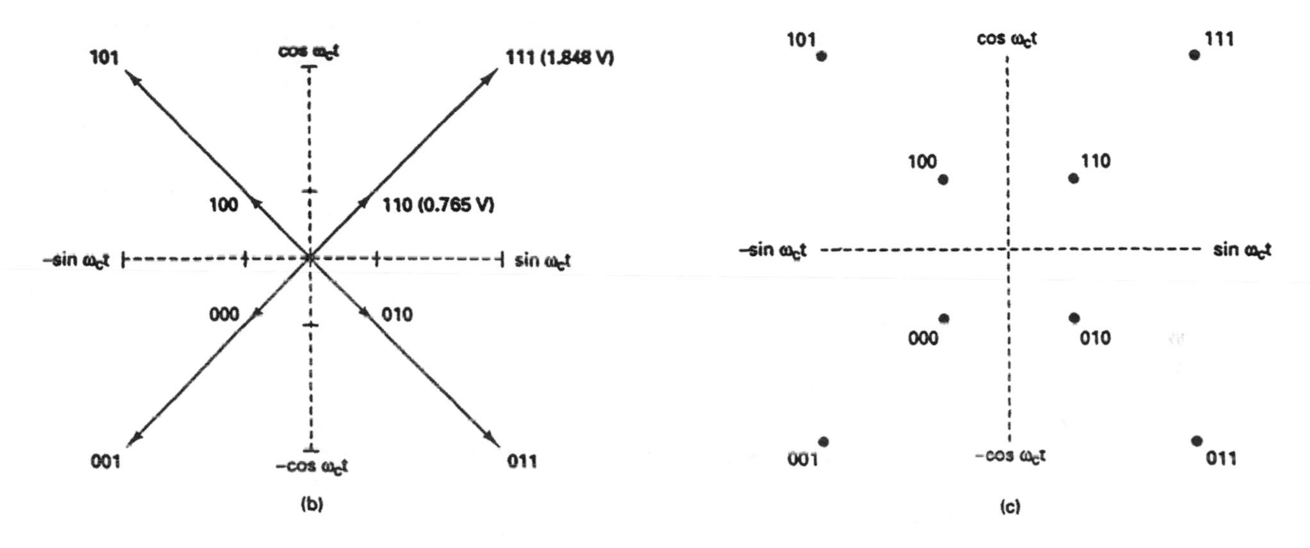

FIGURE 9-31 8-QAM modulator: (a) truth table; (b) phasor diagram; (c) constellation diagram

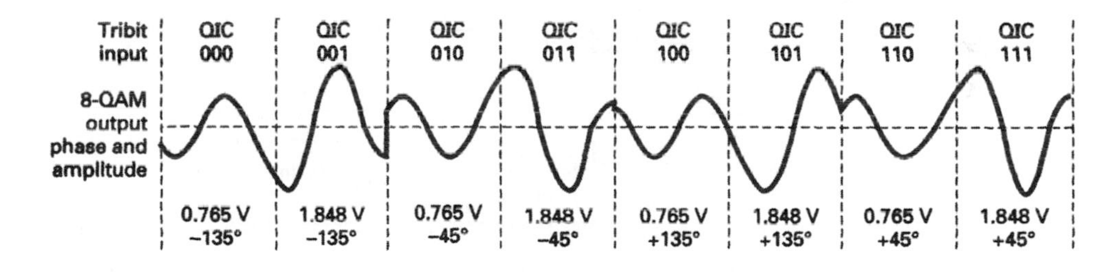

FIGURE 9-32 Output phase and amplitude-versus-time relationship for 8-QAM

9-6-1-3 8-QAM receiver. An 8-QAM receiver is almost identical to the 8-PSK receiver shown in Figure 9-28. The differences are the PAM levels at the output of the product detectors and the binary signals at the output of the analog-to-digital converters. Because there are two transmit amplitudes possible with 8-QAM that are different from those achievable with 8-PSK, the four demodulated PAM levels in 8-QAM are different from those in 8-PSK. Therefore, the conversion factor for the analog-to-digital converters must also be different. Also, with 8-QAM the binary output signals from the I channel analog-to-digital converter are the I and C bits, and the binary output signals from the Q channel analog-to-digital converter are the Q and C bits.

9-6-2 16-QAM

As with the 16-PSK, *16-QAM* is an *M*-ary system where $M = 16$. The input data are acted on in groups of four ($2^4 = 16$). As with 8-QAM, both the phase and the amplitude of the transmit carrier are varied.

9-6-2-1 QAM transmitter. The block diagram for a 16-QAM transmitter is shown in Figure 9-33. The input binary data are divided into four channels: I, I′, Q, and Q′. The bit rate in each channel is equal to one-fourth of the input bit rate ($f_b/4$). Four bits are serially clocked into the bit splitter; then they are outputted simultaneously and in parallel with the I, I′, Q, and Q′ channels. The I and Q bits determine the polarity at the output of the 2-to-4-level converters (a logic 1 = positive and a logic 0 = negative). The I′ and Q′ bits determine the magnitude (a logic I = 0.821 V and a logic 0 = 0.22 V). Consequently, the 2-to-4-level converters generate a 4-level PAM signal. Two polarities and two magnitudes are possible at the output of each 2-to-4-level converter. They are 0.22 V and ±0.821 V.

The PAM signals modulate the in-phase and quadrature carriers in the product modulators. Four outputs are possible for each product modulator. For the I product modulator, they are $+0.821 \sin \omega_c t$, $-0.821 \sin \omega_c t$, $+0.22 \sin \omega_c t$, and $-0.22 \sin \omega_c t$. For the Q product modulator, they are $+0.821 \cos \omega_c t$, $+0.22 \cos \omega_c t$, $-0.821 \cos \omega_c t$, and $-0.22 \cos \omega_c t$. The linear summer combines the outputs from the I and Q channel product modulators and produces the 16 output conditions necessary for 16-QAM. Figure 9-34 shows the truth table for the I and Q channel 2-to-4-level converters.

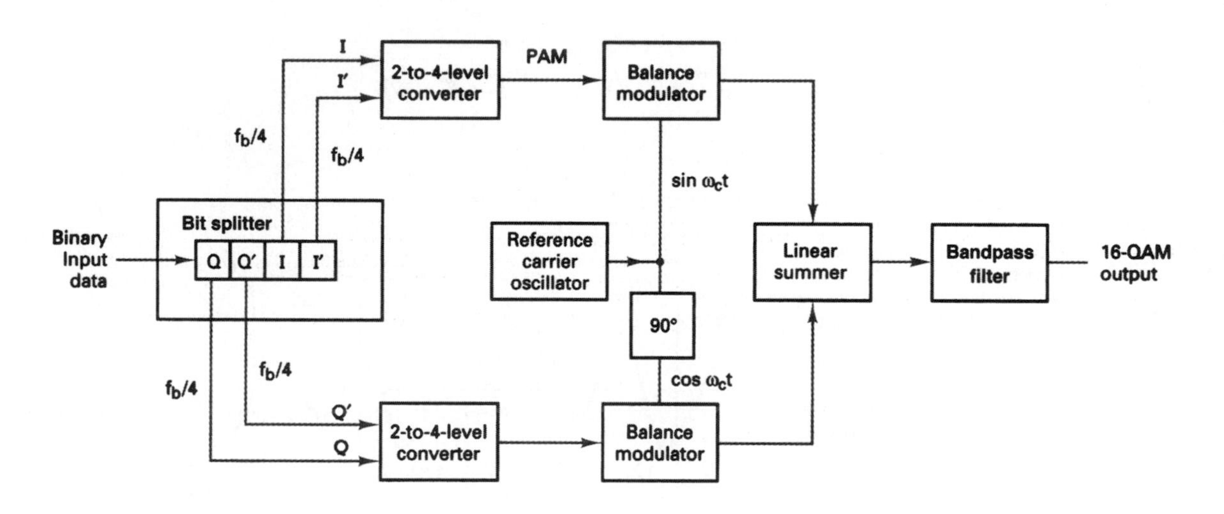

FIGURE 9-33 16-QAM transmitter block diagram

I	I′	Output
0	0	−0.22 V
0	1	−0.821 V
1	0	+0.22 V
1	1	+0.821 V

(a)

Q	Q′	Output
0	0	−0.22 V
0	1	−0.821 V
1	0	+0.22 V
1	1	+0.821 V

(b)

FIGURE 9-34 Truth tables for the I- and Q-channel 2-to-4-level converters: (a) I channel; (b) Q channel

Example 9-10

For a quadbit input of I = 0, I′ = 0, Q = 0, and Q′ = 0 (0000), determine the output amplitude and phase for the 16-QAM modulator shown in Figure 9-33.

Solution The inputs to the I channel 2-to-4-level converter are I = 0 and I′ = 0. From Figure 9-34, the output is −0.22 V. The inputs to the Q channel 2-to-4-level converter are Q = 0 and Q′ = 0. Again from Figure 9-34, the output is −0.22 V.

Thus, the two inputs to the I channel product modulator are −0.22 V and sin $\omega_c t$. The output is

$$I = (-0.22)(\sin \omega_c t) = -0.22 \sin \omega_c t$$

The two inputs to the Q channel product modulator are −0.22 V and cos $\omega_c t$. The output is

$$Q = (-0.22)(\cos \omega_c t) = -0.22 \cos \omega_c t$$

The outputs from the I and Q channel product modulators are combined in the linear summer and produce a modulated output of

$$\text{summer output} = -0.22 \sin \omega_c t - 0.22 \cos \omega_c t$$
$$= 0.311 \sin(\omega_c t - 135°)$$

For the remaining quadbit codes, the procedure is the same. The results are shown in Figure 9-35.

Binary input				16-QAM output	
Q	Q′	I	I′		
0	0	0	0	0.311 V	-135°
0	0	0	1	0.850 V	-165°
0	0	1	0	0.311 V	-45°
0	0	1	1	0.850 V	-15°
0	1	0	0	0.850 V	-105°
0	1	0	1	1.161 V	-135°
0	1	1	0	0.850 V	-75°
0	1	1	1	1.161 V	-45°
1	0	0	0	0.311 V	135°
1	0	0	1	0.850 V	165°
1	0	1	0	0.311 V	45°
1	0	1	1	0.850 V	15°
1	1	0	0	0.850 V	105°
1	1	0	1	1.161 V	135°
1	1	1	0	0.850 V	75°
1	1	1	1	1.161 V	45°

(a)

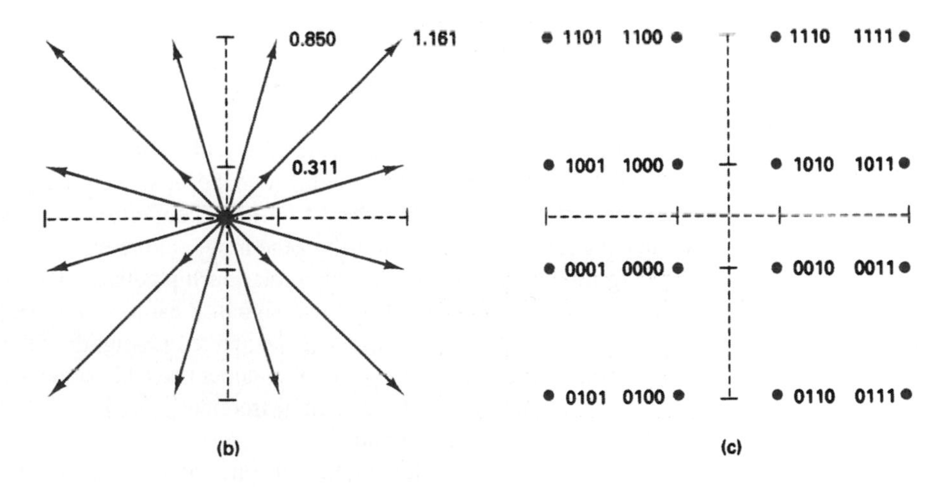

FIGURE 9-35 16-QAM modulator: (a) truth table; (b) phasor diagram; (c) constellation diagram

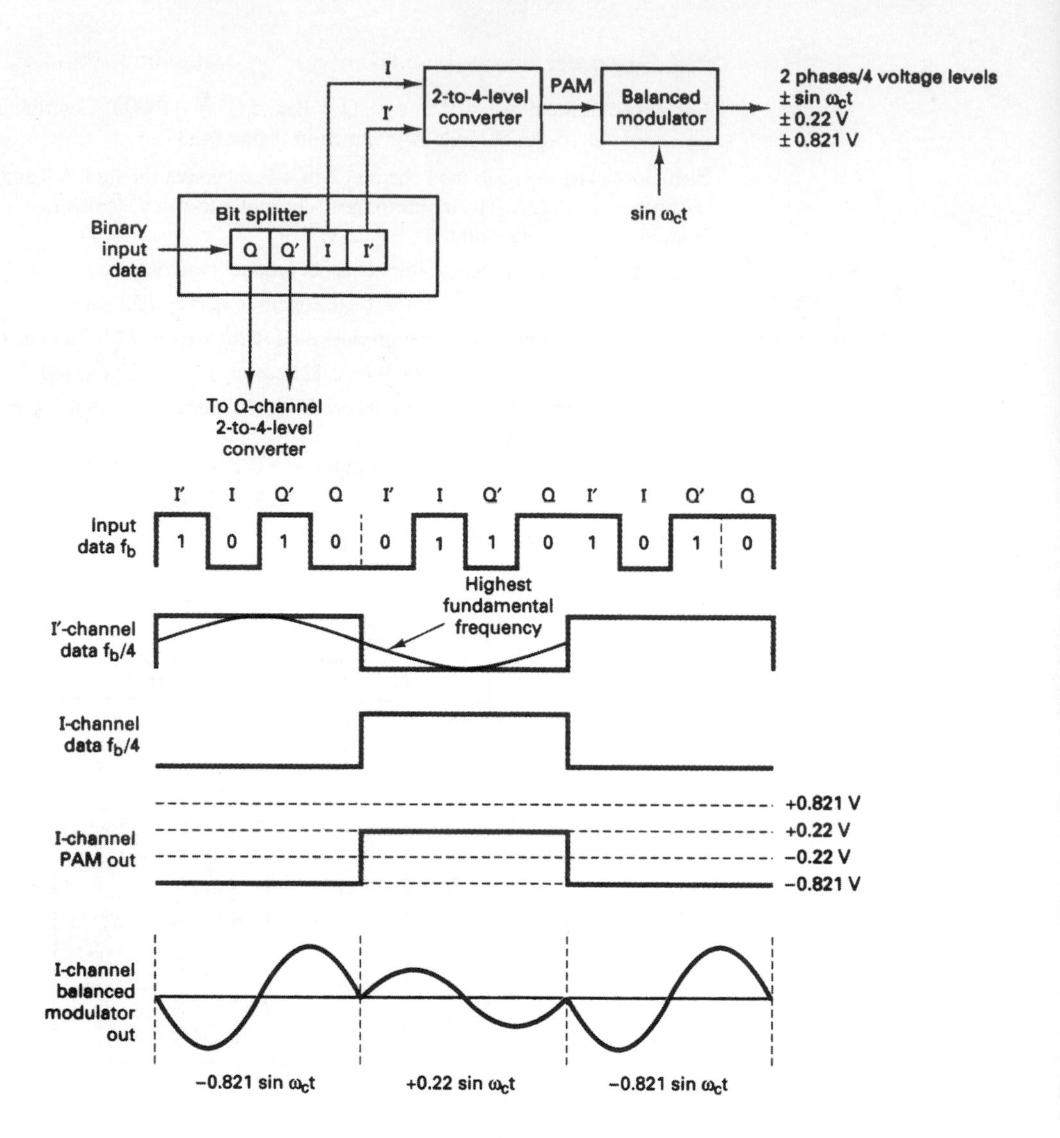

FIGURE 9-36 Bandwidth considerations of a 16-QAM modulator

9-6-2-2 Bandwidth considerations of 16-QAM. With a 16-QAM, because the input data are divided into four channels, the bit rate in the I, I′, Q, or Q′ channel is equal to one-fourth of the binary input data rate ($f_b/4$). (The bit splitter stretches the I, I′, Q, and Q′ bits to four times their input bit length.) Also, because the I, I′, Q, and Q′ bits are outputted simultaneously and in parallel, the 2-to-4-level converters see a change in their inputs and outputs at a rate equal to one-fourth of the input data rate.

Figure 9-36 shows the bit timing relationship between the binary input data; the I, I′, Q, and Q′ channel data; and the I PAM signal. It can be seen that the highest fundamental frequency in the I, I′, Q, or Q′ channel is equal to one-eighth of the bit rate of the binary input data (one cycle in the I, I′, Q, or Q′ channel takes the same amount of time as eight input bits). Also, the highest fundamental frequency of either PAM signal is equal to one-eighth of the binary input bit rate.

With a 16-QAM modulator, there is one change in the output signal (either its phase, amplitude, or both) for every four input data bits. Consequently, the baud equals $f_b/4$, the same as the minimum bandwidth.

Again, the balanced modulators are product modulators and their outputs can be represented mathematically as

$$\text{output} = (X \sin \omega_a t)(\sin \omega_c t) \quad \textbf{(9-26)}$$

where

$$\underbrace{\omega_a t = 2\pi \frac{f_b}{8} t}_{\text{modulating signal}} \quad \text{and} \quad \underbrace{\omega_c t = 2\pi f_c t}_{\text{carrier}}$$

and

$$X = \pm 0.22 \text{ or } \pm 0.821$$

Thus,

$$\text{output} = \left(X \sin 2\pi \frac{f_b}{8} t\right)(\sin 2\pi f_c t)$$

$$= \frac{X}{2} \cos 2\pi \left(f_c - \frac{f_b}{8}\right)t = \frac{X}{2} \cos 2\pi \left(f_c + \frac{f_b}{8}\right)t$$

The output frequency spectrum extends from $f_c + f_b/8$ to $f_c - f_b/8$, and the minimum bandwidth (f_N) is

$$\left(f_c + \frac{f_b}{8}\right) - \left(f_c - \frac{f_b}{8}\right) = \frac{2f_b}{8} = \frac{f_b}{4}$$

Example 9-11

For a 16-QAM modulator with an input data rate (f_b) equal to 10 Mbps and a carrier frequency of 70 MHz, determine the minimum double-sided Nyquist frequency (f_N) and the baud. Also, compare the results with those achieved with the BPSK, QPSK, and 8-PSK modulators in Examples 9-4, 9-6, and 9-8. Use the 16-QAM block diagram shown in Figure 9-33 as the modulator model.

Solution The bit rate in the I, I′, Q, and Q′ channels is equal to one-fourth of the input bit rate, or

$$f_{bI} = f_{bI'} = f_{bQ} = f_{bQ'} = \frac{f_b}{4} = \frac{10 \text{ Mbps}}{4} = 2.5 \text{ Mbps}$$

Therefore, the fastest rate of change and highest fundamental frequency presented to either balanced modulator is

$$f_a = \frac{f_{bI}}{2} \text{ or } \frac{f_{bI'}}{2} \text{ or } \frac{f_{bQ}}{2} \text{ or } \frac{f_{bQ'}}{2} = \frac{2.5 \text{ Mbps}}{2} = 1.25 \text{ MHz}$$

The output wave from the balanced modulator is

$$(\sin 2\pi f_a t)(\sin 2\pi f_c t)$$

$$\frac{1}{2}\cos 2\pi(f_c - f_a)t - \frac{1}{2}\cos 2\pi(f_c + f_a)t$$

$$\frac{1}{2}\cos 2\pi[(70 - 1.25) \text{ MHz}]t - \frac{1}{2}\cos 2\pi[(70 + 1.25) \text{ MHz}]t$$

$$\frac{1}{2}\cos 2\pi(68.75 \text{ MHz})t - \frac{1}{2}\cos 2\pi(71.25 \text{ MHz})t$$

The minimum Nyquist bandwidth is

$$B = (71.25 - 68.75) \text{ MHz} = 2.5 \text{ MHz}$$

The minimum bandwidth for the 16-QAM can also be determined by simply substituting into Equation 9-10:

$$B = \frac{10 \text{ Mbps}}{4}$$
$$= 2.5 \text{ MHz}$$

The symbol rate equals the bandwidth; thus,

$$\text{symbol rate} = 2.5 \text{ megabaud}$$

The output spectrum is as follows:

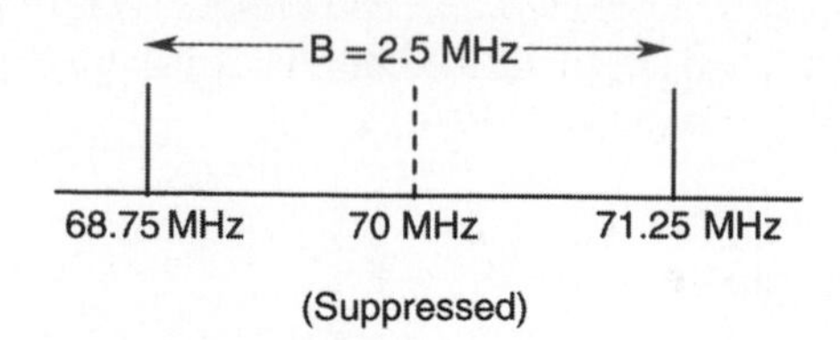

$$B = 2.5 \text{ MHz}$$

For the same input bit rate, the minimum bandwidth required to pass the output of a 16-QAM modulator is equal to one-fourth that of the BPSK modulator, one-half that of QPSK, and 25% less than with 8-PSK. For each modulation technique, the baud is also reduced by the same proportions.

Example 9-12

For the following modulation schemes, construct a table showing the number of bits encoded, number of output conditions, minimum bandwidth, and baud for an information data rate of 12 kbps: QPSK, 8-PSK, 8-QAM, 16-PSK, and 16-QAM.

Solution

Modulation	n	M	B (Hz)	baud
QPSK	2	4	6000	6000
8-PSK	3	8	4000	4000
8-QAM	3	8	4000	4000
16-PSK	4	16	3000	3000
16-QAM	4	16	3000	3000

From Example 9-12, it can be seen that a 12-kbps data stream can be propagated through a narrower bandwidth using either 16-PSK or 16-QAM than with the lower levels of encoding.

Table 9-1 summarizes the relationship between the number of bits encoded, the number of output conditions possible, the minimum bandwidth, and the baud for ASK, FSK, PSK, and QAM. Note that with the three binary modulation schemes (ASK, FSK, and

Table 9-1 ASK, FSK, PSK, and QAM Summary

Modulation	Encoding Scheme	Outputs Possible	Minimum Bandwidth	Baud
ASK	Single bit	2	f_b	f_b
FSK	Single bit	2	f_b	f_b
BPSK	Single bit	2	f_b	f_b
QPSK	Dibits	4	$f_b/2$	$f_b/2$
8-PSK	Tribits	8	$f_b/3$	$f_b/3$
8-QAM	Tribits	8	$f_b/3$	$f_b/3$
16-QAM	Quadbits	16	$f_b/4$	$f_b/4$
16-PSK	Quadbits	16	$f_b/4$	$f_b/4$
32-PSK	Five bits	32	$f_b/5$	$f_b/5$
32-QAM	Five bits	32	$f_b/5$	$f_b/5$
64-PSK	Six bits	64	$f_b/6$	$f_b/6$
64-QAM	Six bits	64	$f_b/6$	$f_b/6$
128-PSK	Seven bits	128	$f_b/7$	$f_b/7$
128-QAM	Seven bits	128	$f_b/7$	$f_b/7$

Note: f_b indicates a magnitude equal to the input bit rate.

BPSK), $n = 1$, $M = 2$, only two output conditions are possible, and the baud is equal to the bit rate. However, for values of $n > 1$, the number of output conditions increases, and the minimum bandwidth and baud decrease. Therefore, digital modulation schemes where $n > 1$ achieve *bandwidth compression* (i.e., less bandwidth is required to propagate a given bit rate). When data compression is performed, higher data transmission rates are possible for a given bandwidth.

9-7 BANDWIDTH EFFICIENCY

Bandwidth efficiency (sometimes called *information density* or *spectral efficiency*) is often used to compare the performance of one digital modulation technique to another. In essence, bandwidth efficiency is the ratio of the transmission bit rate to the minimum bandwidth required for a particular modulation scheme. Bandwidth efficiency is generally normalized to a 1-Hz bandwidth and, thus, indicates the number of bits that can be propagated through a transmission medium for each hertz of bandwidth. Mathematically, bandwidth efficiency is

$$B\eta = \frac{\text{transmission bit rate (bps)}}{\text{minimum bandwidth (Hz)}} \tag{9-27}$$

$$= \frac{\text{bits/s}}{\text{hertz}} = \frac{\text{bits/s}}{\text{cycles/s}} = \frac{\text{bits}}{\text{cycle}}$$

where $B\eta$ = bandwidth efficiency

Bandwidth efficiency can also be given as a percentage by simply multiplying $B\eta$ by 100.

Example 9-13

For an 8-PSK system, operating with an information bit rate of 24 kbps, determine (a) baud, (b) minimum bandwidth, and (c) bandwidth efficiency.

Solution a. Baud is determined by substituting into Equation 9-10:

$$\text{baud} = \frac{24{,}000}{3} = 8000$$

b. Bandwidth is determined by substituting into Equation 9-11:

$$B = \frac{24{,}000}{3} = 8000$$

c. Bandwidth efficiency is calculated from Equation 9-27:

$$B\eta = \frac{24{,}000 \text{ bps}}{8000 \text{ Hz}}$$

$$= 3 \text{ bits per second per cycle of bandwidth}$$

Example 9-14

For 16-PSK and a transmission system with a 10 kHz bandwidth, determine the maximum bit rate.

Solution The bandwidth efficiency for 16-PSK is 4, which means that four bits can be propagated through the system for each hertz of bandwidth. Therefore, the maximum bit rate is simply the product of the bandwidth and the bandwidth efficiency, or

$$\text{bit rate} = 4 \times 10{,}000$$

$$= 40{,}000 \text{ bps}$$

Table 9-2 ASK, FSK, PSK, and QAM Summary

Modulation	Encoding Scheme	Outputs Possible	Minimum Bandwidth	Baud	Bη
ASK	Single bit	2	f_b	f_b	1
FSK	Single bit	2	f_b	f_b	1
BPSK	Single bit	2	f_b	f_b	1
QPSK	Dibits	4	$f_b/2$	$f_b/2$	2
8-PSK	Tribits	8	$f_b/3$	$f_b/3$	3
8-QAM	Tribits	8	$f_b/3$	$f_b/3$	3
16-PSK	Quadbits	16	$f_b/4$	$f_b/4$	4
16-QAM	Quadbits	16	$f_b/4$	$f_b/4$	4
32-PSK	Five bits	32	$f_b/5$	$f_b/5$	5
64-QAM	Six bits	64	$f_b/6$	$f_b/6$	6

Note: f_b indicates a magnitude equal to the input bit rate.

9-7-1 Digital Modulation Summary

The properties of several digital modulation schemes are summarized in Table 9-2.

9-8 CARRIER RECOVERY

Carrier recovery is the process of extracting a phase-coherent reference carrier from a receiver signal. This is sometimes called *phase referencing.*

In the phase modulation techniques described thus far, the binary data were encoded as a precise phase of the transmitted carrier. (This is referred to as *absolute phase encoding.*) Depending on the encoding method, the angular separation between adjacent phasors varied between 30° and 180°. To correctly demodulate the data, a phase-coherent carrier was recovered and compared with the received carrier in a product detector. To determine the absolute phase of the received carrier, it is necessary to produce a carrier at the receiver that is phase coherent with the transmit reference oscillator. This is the function of the carrier recovery circuit.

With PSK and QAM, the carrier is suppressed in the balanced modulators and, therefore, is not transmitted. Consequently, at the receiver the carrier cannot simply be tracked with a standard phase-locked loop (PLL). With suppressed-carrier systems, such as PSK and QAM, sophisticated methods of carrier recovery are required, such as a *squaring loop,* a *Costas loop,* or a *remodulator.*

9-8-1 Squaring Loop

A common method of achieving carrier recovery for BPSK is the *squaring loop.* Figure 9-37 shows the block diagram of a squaring loop. The received BPSK waveform is filtered and then squared. The filtering reduces the spectral width of the received noise. The squaring circuit removes the modulation and generates the second harmonic of the carrier frequency. This harmonic is phase tracked by the PLL. The VCO output frequency from the PLL then is divided by 2 and used as the phase reference for the product detectors.

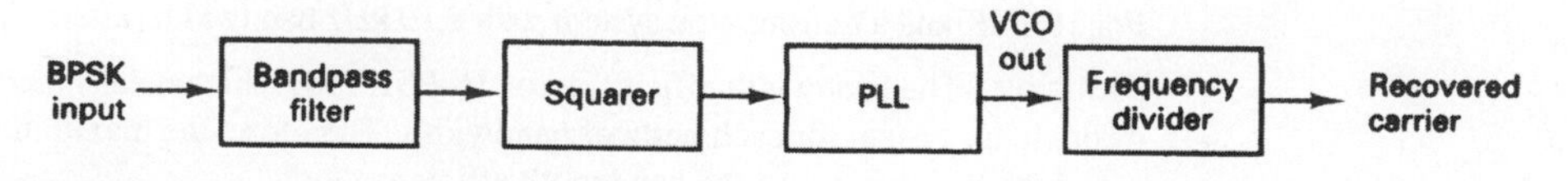

FIGURE 9-37 Squaring loop carrier recovery circuit for a BPSK receiver

With BPSK, only two output phases are possible: $+\sin \omega_c t$ and $-\sin \omega_c t$. Mathematically, the operation of the squaring circuit can be described as follows. For a receive signal of $+\sin \omega_c t$, the output of the squaring circuit is

$$\text{output} = (+\sin \omega_c t)(+\sin \omega_c t) = +\sin^2 \omega_c t$$

(filtered out)

$$= \frac{1}{2}(1 - \cos 2\omega_c t) = \frac{1}{2} - \frac{1}{2}\cos 2\omega_c t$$

For a received signal of $-\sin \omega_c t$, the output of the squaring circuit is

$$\text{output} = (-\sin \omega_c t)(-\sin \omega_c t) = +\sin^2 \omega_c t$$

(filtered out)

$$= \frac{1}{2}(1 - \cos 2\omega_c t) = \frac{1}{2} - \frac{1}{2}\cos 2\omega_c t$$

It can be seen that in both cases, the output from the squaring circuit contained a constant voltage (+1/2 V) and a signal at twice the carrier frequency ($\cos 2\omega_c t$). The constant voltage is removed by filtering, leaving only $\cos 2\omega_c t$.

9-8-2 Costas Loop

A second method of carrier recovery is the Costas, or quadrature, loop shown in Figure 9-38. The Costas loop produces the same results as a squaring circuit followed by an ordinary PLL in place of the BPF. This recovery scheme uses two parallel tracking loops (I and Q) simultaneously to derive the product of the I and Q components of the signal that drives the VCO. The in-phase (I) loop uses the VCO as in a PLL, and the quadrature (Q) loop uses a 90° shifted VCO signal. Once the frequency of the VCO is equal to the suppressed-carrier

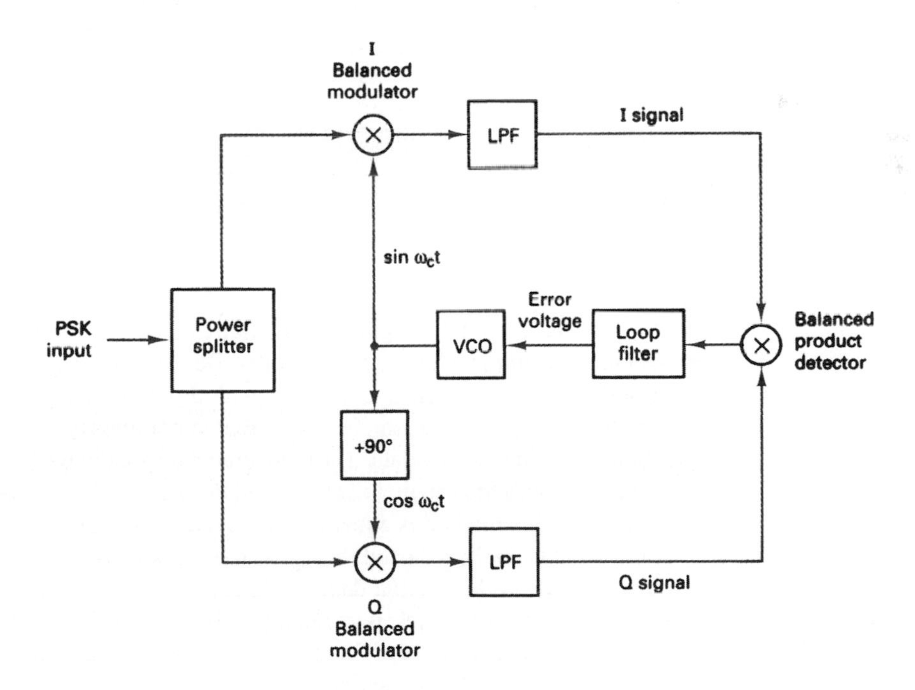

FIGURE 9-38 Costas loop carrier recovery circuit

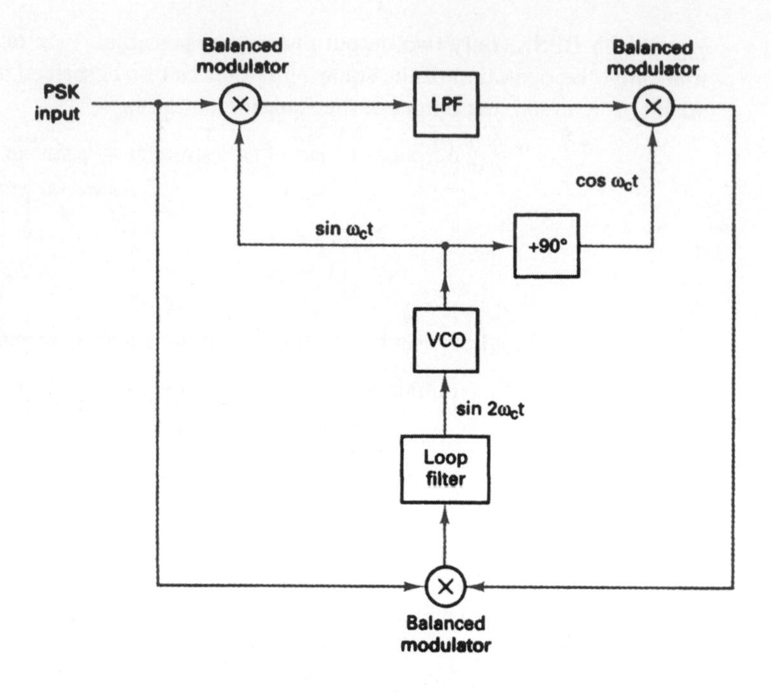

FIGURE 9-39 Remodulator loop carrier recovery circuit

frequency, the product of the I and Q signals will produce an error voltage proportional to any phase error in the VCO. The error voltage controls the phase and, thus, the frequency of the VCO.

9-8-3 Remodulator

A third method of achieving recovery of a phase and frequency coherent carrier is the remodulator, shown in Figure 9-39. The remodulator produces a loop error voltage that is proportional to twice the phase error between the incoming signal and the VCO signal. The remodulator has a faster acquisition time than either the squaring or the Costas loops.

Carrier recovery circuits for higher-than-binary encoding techniques are similar to BPSK except that circuits that raise the receive signal to the fourth, eighth, and higher powers are used.

9-9 CLOCK RECOVERY

As with any digital system, digital radio requires precise timing or clock synchronization between the transmit and the receive circuitry. Because of this, it is necessary to regenerate clocks at the receiver that are synchronous with those at the transmitter.

Figure 9-40a shows a simple circuit that is commonly used to recover clocking information from the received data. The recovered data are delayed by one-half a bit time and then compared with the original data in an XOR circuit. The frequency of the clock that is recovered with this method is equal to the received data rate (f_b). Figure 9-40b shows the relationship between the data and the recovered clock timing. From Figure 9-40b, it can be seen that as long as the receive data contain a substantial number of transitions (1/0 sequences), the recovered clock is maintained. If the receive data were to undergo an extended period of successive 1s or 0s, the recovered clock would be lost. To prevent this from occurring, the data are scrambled at the transmit end and descrambled at the receive end. Scrambling introduces transitions (pulses) into the binary signal using a prescribed algorithm, and the descrambler uses the same algorithm to remove the transitions.

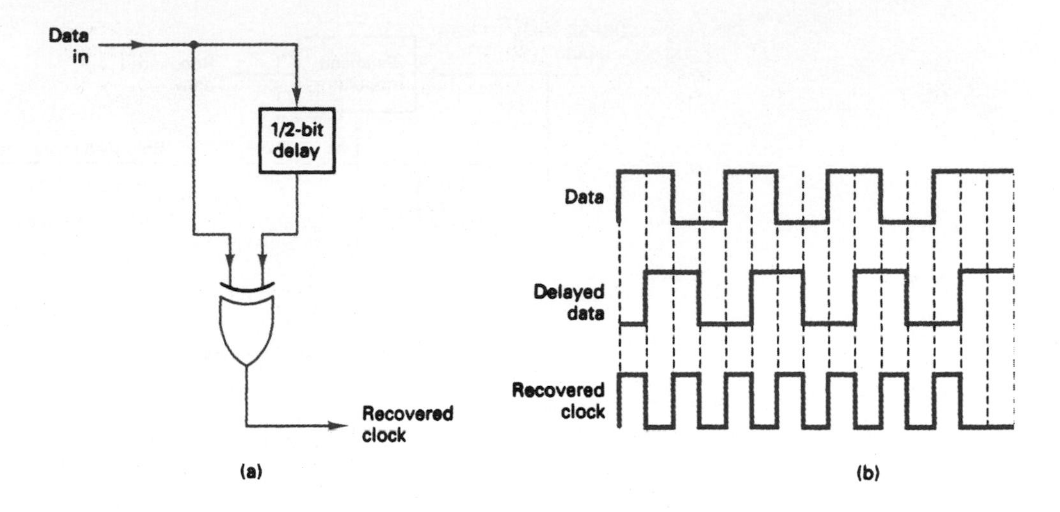

FIGURE 9-40 (a) Clock recovery circuit; (b) timing diagram

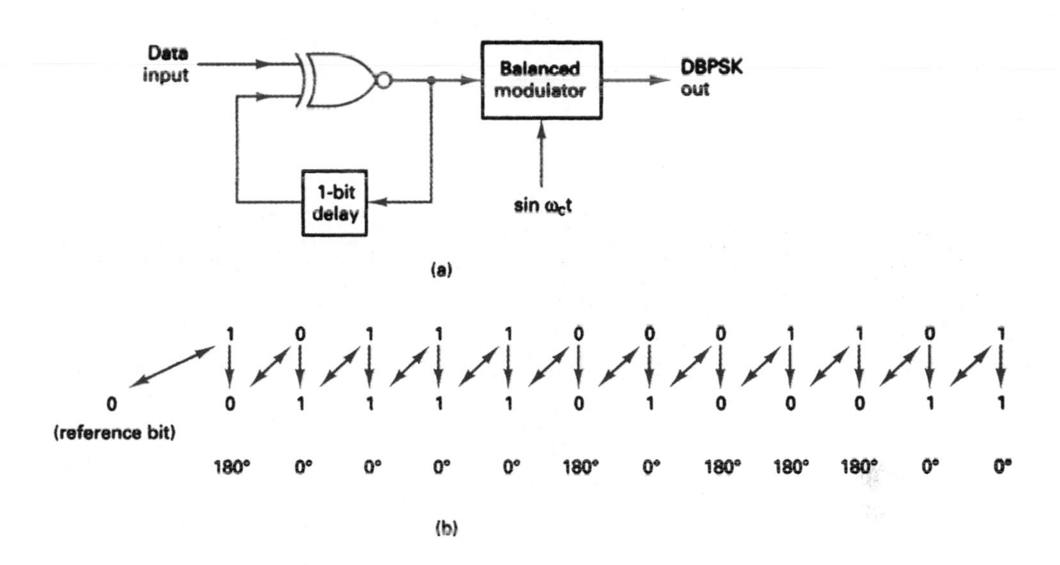

FIGURE 9-41 DBPSK modulator: (a) block diagram; (b) timing diagram

9-10 DIFFERENTIAL PHASE-SHIFT KEYING

Differential phase-shift keying (DPSK) is an alternative form of digital modulation where the binary input information is contained in the difference between two successive signaling elements rather than the absolute phase. With DPSK, it is not necessary to recover a phase-coherent carrier. Instead, a received signaling element is delayed by one signaling element time slot and then compared with the next received signaling element. The difference in the phase of the two signaling elements determines the logic condition of the data.

9-10-1 Differential BPSK

9-10-1-1 DBPSK transmitter. Figure 9-41a shows a simplified block diagram of a *differential binary phase-shift keying* (DBPSK) transmitter. An incoming information bit is

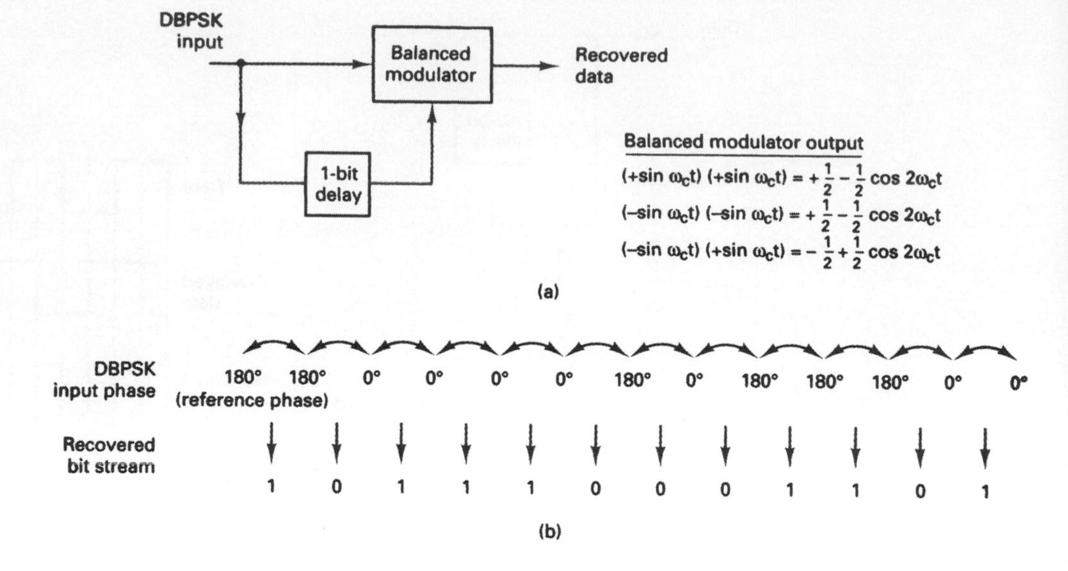

FIGURE 9-42 DBPSK demodulator: (a) block diagram; (b) timing sequence

XNORed with the preceding bit prior to entering the BPSK modulator (balanced modulator). For the first data bit, there is no preceding bit with which to compare it. Therefore, an initial reference bit is assumed. Figure 9-41b shows the relationship between the input data, the XNOR output data, and the phase at the output of the balanced modulator. If the initial reference bit is assumed a logic 1, the output from the XNOR circuit is simply the complement of that shown.

In Figure 9-41b, the first data bit is XNORed with the reference bit. If they are the same, the XNOR output is a logic 1; if they are different, the XNOR output is a logic 0. The balanced modulator operates the same as a conventional BPSK modulator; a logic 1 produces $+\sin \omega_c t$ at the output, and a logic 0 produces $-\sin \omega_c t$ at the output.

9-10-1-2 DBPSK receiver. Figure 9-42 shows the block diagram and timing sequence for a DBPSK receiver. The received signal is delayed by one bit time, then compared with the next signaling element in the balanced modulator. If they are the same, a logic 1 (+ voltage) is generated. If they are different, a logic 0 (− voltage) is generated. If the reference phase is incorrectly assumed, only the first demodulated bit is in error. Differential encoding can be implemented with higher-than-binary digital modulation schemes, although the differential algorithms are much more complicated than for DBPSK.

The primary advantage of DBPSK is the simplicity with which it can be implemented. With DBPSK, no carrier recovery circuit is needed. A disadvantage of DBPSK is that it requires between 1 dB and 3 dB more signal-to-noise ratio to achieve the same bit error rate as that of absolute PSK.

9-11 TRELLIS CODE MODULATION

Achieving data transmission rates in excess of 9600 bps over standard telephone lines with approximately a 3-kHz bandwidth obviously requires an encoding scheme well beyond the quadbits used with 16-PSK or 16-QAM (i.e., M must be significantly greater than 16). As might be expected, higher encoding schemes require higher signal-to-noise ratios. Using the Shannon limit for information capacity (Equation 9-4), a data transmission rate of 28.8 kbps through a 3200-Hz bandwidth requires a signal-to-noise ratio of

$$I(\text{bps}) = (3.32 \times B) \log(1 + \text{S/N})$$

therefore, $$28.8 \text{ kbps} = (3.32)(3200)\log(1 + S/N)$$

$$28{,}800 = 10{,}624 \log(1 + S/N)$$

$$\frac{28{,}800}{10{,}624} = \log(1 + S/N)$$

$$2.71 = \log(1 + S/N)$$

thus, $$10^{2.71} = 1 + S/N$$

$$513 = 1 + S/N$$

$$512 = S/N$$

in dB, $$S/N_{(dB)} = 10 \log 512$$

$$= 27 \text{ dB}$$

Transmission rates of 56 kbps require a signal-to-noise ratio of 53 dB, which is virtually impossible to achieve over a standard telephone circuit.

Data transmission rates in excess of 56 kbps can be achieved, however, over standard telephone circuits using an encoding technique called *trellis code modulation* (TCM). Dr. Ungerboeck at IBM Zuerich Research Laboratory developed TCM, which involves using *convolutional* (*tree*) codes, which combines encoding and modulation to reduce the probability of error, thus improving the bit error performance. The fundamental idea behind TCM is introducing controlled redundancy in the bit stream with a convolutional code, which reduces the likelihood of transmission errors. What sets TCM apart from standard encoding schemes is the introduction of redundancy by doubling the number of signal points in a given PSK or QAM constellation.

Trellis code modulation is sometimes thought of as a magical method of increasing transmission bit rates over communications systems using QAM or PSK with fixed bandwidths. Few people fully understand this concept, as modem manufacturers do not seem willing to share information on TCM. Therefore, the following explanation is intended not to fully describe the process of TCM but rather to introduce the topic and give the reader a basic understanding of how TCM works and the advantage it has over conventional digital modulation techniques.

M-ary QAM and PSK utilize a signal set of $2^N = M$, where N equals the number of bits encoded into M different conditions. Therefore, $N = 2$ produces a standard PSK constellation with four signal points (i.e., QPSK) as shown in Figure 9-43a. Using TCM, the number of signal points increases to two times M possible symbols for the same factor-of-M reduction in bandwidth while transmitting each signal during the same time interval. TCM-encoded QPSK is shown in Figure 9-43b.

Trellis coding also defines the manner in which signal-state transitions are allowed to occur, and transitions that do not follow this pattern are interpreted in the receiver as transmission errors. Therefore, TCM can improve error performance by restricting the manner in which signals are allowed to transition. For values of N greater than 2, QAM is the modulation scheme of choice for TCM; however, for simplification purposes, the following explanation uses PSK as it is easier to illustrate.

Figure 9-44 shows a TCM scheme using two-state 8-PSK, which is essentially two QPSK constellations offset by 45°. One four-state constellation is labeled 0-4-2-6, and the other is labeled 1-5-3-7. For this explanation, the signal point labels 0 through 7 are meant not to represent the actual data conditions but rather to simply indicate a convenient method of labeling the various signal points. Each digit represents one of four signal points permitted within each of the two QPSK constellations. When in the 0-4-2-6 constellation and a 0 or 4 is transmitted, the system remains in the same constellation. However, when either a 2 or 6 is transmitted, the system switches to the 1-5-3-7 constellation. Once in the 1-5-3-7

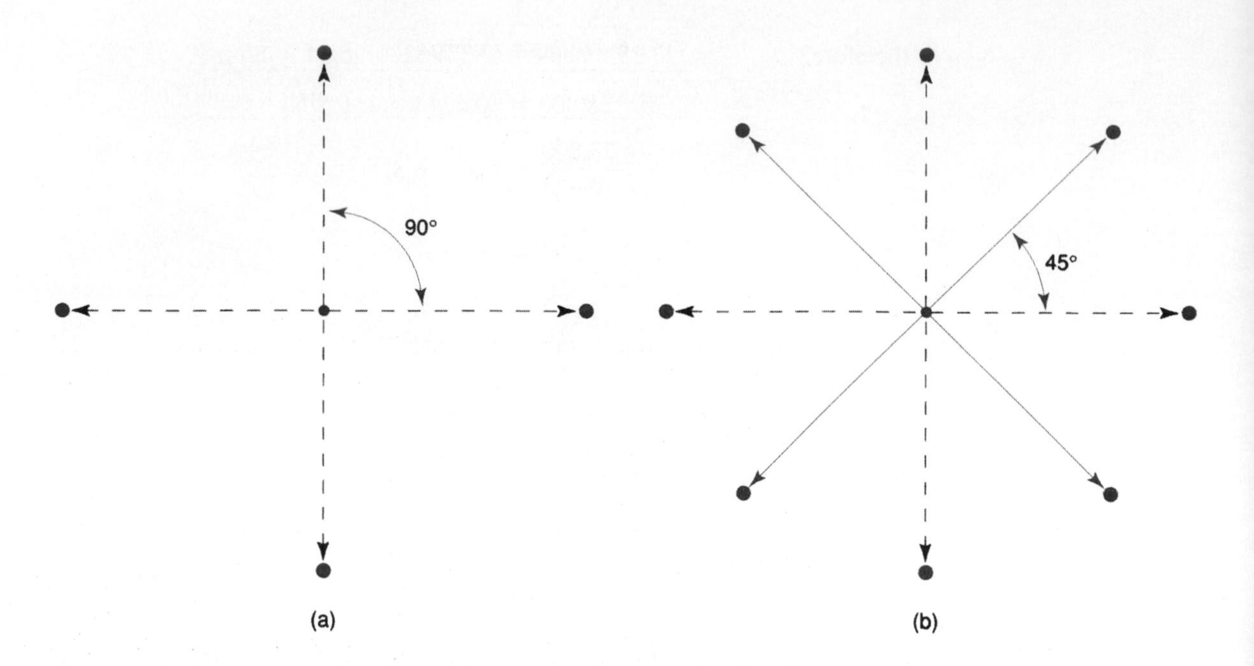

FIGURE 9-43 QPSK constellations: (a) standard encoding format; (b) trellis encoding format

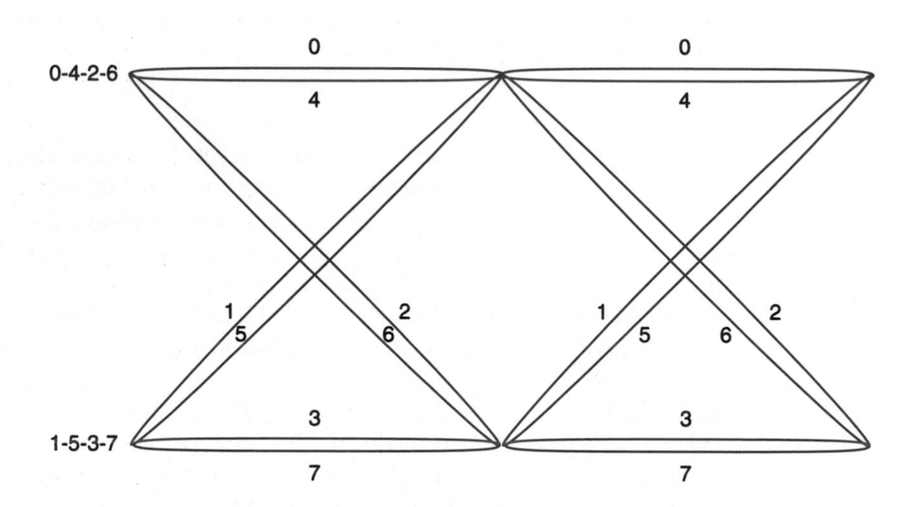

FIGURE 9-44 8-PSK TCM constellations

constellation and a 3 or 7 is transmitted, the system remains in the same constellation, and if a 1 or 5 is transmitted, the system switches to the 0-4-2-6 constellation. Remember that each symbol represents two bits, so the system undergoes a 45° phase shift whenever it switches between the two constellations. A complete error analysis of standard QPSK compared with TCM QPSK would reveal a coding gain for TCM of 2-to-1 1 or 3 dB. Table 9-3 lists the coding gains achieved for TCM coding schemes with several different trellis states.

The maximum data rate achievable using a given bandwidth can be determined by rearranging Equation 9-10:

$$N \times B = f_b$$

Table 9-3 Trellis Coding Gain

Number of Trellis States	Coding Gain (dB)
2	3.0
4	5.5
8	6.0
16	6.5
32	7.1
64	7.3
128	7.3
256	7.4

where N = number of bits encoded (bits)
B = bandwidth (hertz)
f_b = transmission bit rate (bits per second)

Remember that with M-ary QAM or PSK systems, the baud equals the minimum required bandwidth. Therefore, a 3200-Hz bandwidth using a nine-bit trellis code produces a 3200 baud signal with each baud carrying nine bits. Therefore, the transmission rate $f_b = 9 \times 3200 = 28.8$ kbps.

TCM is thought of as a coding scheme that improves on standard QAM by increasing the distance between symbols on the constellation (known as the *Euclidean distance*). The first TCM system used a five-bit code, which included four QAM bits (a quadbit) and a fifth bit used to help decode the quadbit. Transmitting five bits within a single signaling element requires producing 32 discernible signals. Figure 9-45 shows a 128-point QAM constallation.

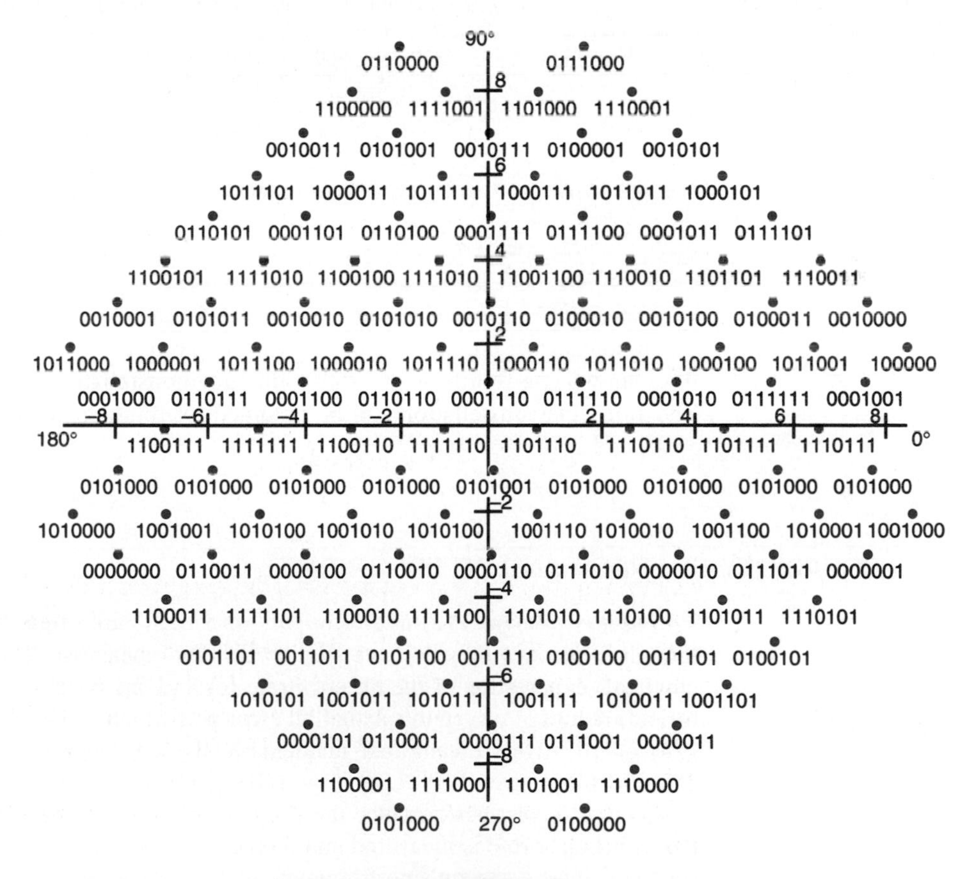

FIGURE 9-45 128-Point QAM TCM constellation

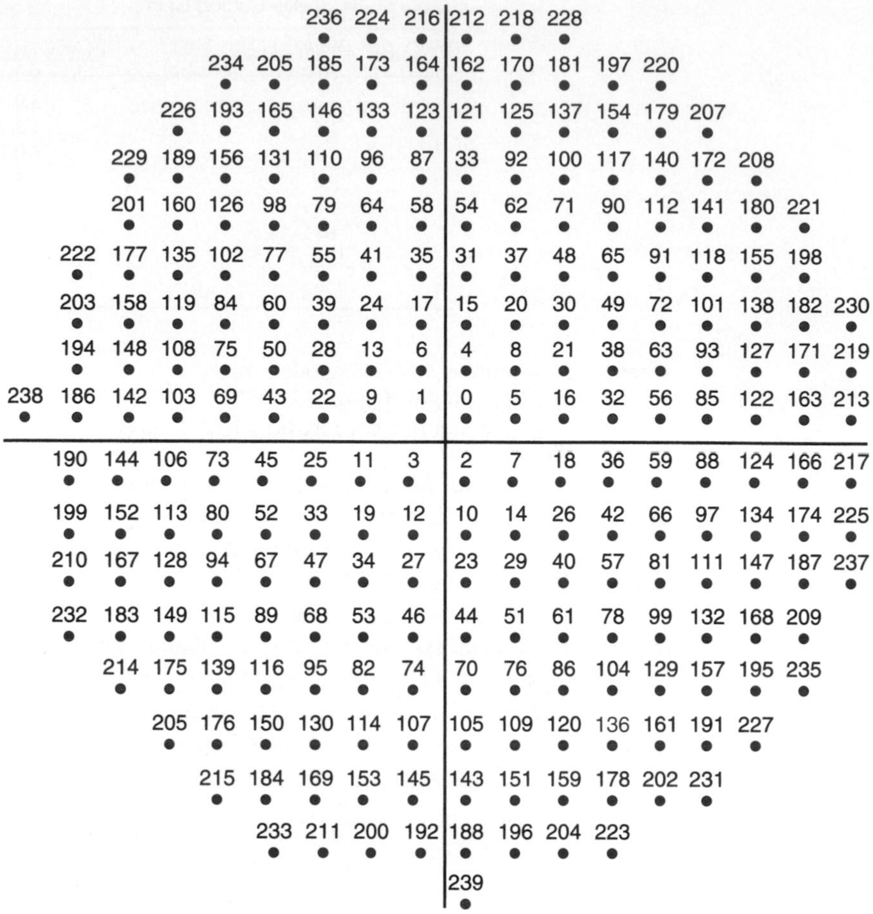

236 224 216 212 218 228
234 205 185 173 164 162 170 181 197 220
226 193 165 146 133 123 121 125 137 154 179 207
229 189 156 131 110 96 87 33 92 100 117 140 172 208
201 160 126 98 79 64 58 54 62 71 90 112 141 180 221
222 177 135 102 77 55 41 35 31 37 48 65 91 118 155 198
203 158 119 84 60 39 24 17 15 20 30 49 72 101 138 182 230
194 148 108 75 50 28 13 6 4 8 21 38 63 93 127 171 219
238 186 142 103 69 43 22 9 1 0 5 16 32 56 85 122 163 213
190 144 106 73 45 25 11 3 2 7 18 36 59 88 124 166 217
199 152 113 80 52 33 19 12 10 14 26 42 66 97 134 174 225
210 167 128 94 67 47 34 27 23 29 40 57 81 111 147 187 237
232 183 149 115 89 68 53 46 44 51 61 78 99 132 168 209
214 175 139 116 95 82 74 70 76 86 104 129 157 195 235
205 176 150 130 114 107 105 109 120 136 161 191 227
215 184 169 153 145 143 151 159 178 202 231
233 211 200 192 188 196 204 223
239

FIGURE 9-46 One-fourth of a 960-Point QAM TCM constellation

A 3200-baud signal using nine-bit TCM encoding produces 512 different codes. The nine data bits plus a redundant bit for TCM requires a 960-point constellation. Figure 9-46 shows one-fourth of the 960-point superconstellation showing 240 signal points. The full superconstellation can be obtained by rotating the 240 points shown by 90°, 180°, and 270°.

9-12 PROBABILITY OF ERROR AND BIT ERROR RATE

Probability of error P(e) and *bit error rate* (BER) are often used interchangeably, although in practice they do have slightly different meanings. $P(e)$ is a theoretical (mathematical) expectation of the bit error rate for a given system. BER is an empirical (historical) record of a system's actual bit error performance. For example, if a system has a $P(e)$ of 10^{-5}, this means that mathematically you can expect one bit error in every 100,000 bits transmitted ($1/10^{-5} = 1/100{,}000$). If a system has a BER of 10^{-5}, this means that in past performance there was one bit error for every 100,000 bits transmitted. A bit error rate is measured and then compared with the expected probability of error to evaluate a system's performance.

Probability of error is a function of the *carrier-to-noise power ratio* (or, more specifically, the average *energy per bit-to-noise power density ratio*) and the number of possible encoding conditions used (M-ary). Carrier-to-noise power ratio is the ratio of the average carrier power (the combined power of the carrier and its associated sidebands) to the *thermal noise power.* Carrier power can be stated in watts or dBm, where

$$C_{(\text{dBm})} = 10 \log \frac{C_{(\text{watts})}}{0.001} \tag{9-28}$$

Thermal noise power is expressed mathematically as

$$N = KTB \text{ (watts)} \tag{9-29}$$

where N = thermal noise power (watts)
K = Boltzmann's proportionality constant (1.38×10^{-23} joules per kelvin)
T = temperature (kelvin: 0 K = −273° C, room temperature = 290 K)
B = bandwidth (hertz)

Stated in dBm,

$$N_{(\text{dBm})} = 10 \log \frac{KTB}{0.001} \tag{9-30}$$

Mathematically, the carrier-to-noise power ratio is

$$\frac{C}{N} = \frac{C}{KTB} \text{ (unitless ratio)} \tag{9-31}$$

where C = carrier power (watts)
N = noise power (watts)

Stated in dB,

$$\frac{C}{N}(\text{dB}) = 10 \log \frac{C}{N}$$

$$= C_{(\text{dBm})} - N_{(\text{dBm})} \tag{9-32}$$

Energy per bit is simply the energy of a single bit of information. Mathematically, energy per bit is

$$E_b = CT_b \text{ (J/bit)} \tag{9-33}$$

where E_b = energy of a single bit (joules per bit)
T_b = time of a single bit (seconds)
C = carrier power (watts)

Stated in dBJ,

$$E_{b(\text{dBJ})} = 10 \log E_b \tag{9-34}$$

and because $T_b = 1/f_b$, where f_b is the bit rate in bits per second, E_b can be rewritten as

$$E_b = \frac{C}{f_b} \text{ (J/bit)} \tag{9-35}$$

Stated in dBJ,

$$E_{b\,(\text{dBJ})} = 10 \log \frac{C}{f_b} \tag{9-36}$$

$$= 10 \log C - 10 \log f_b \tag{9-37}$$

Noise power density is the thermal noise power normalized to a 1-Hz bandwidth (i.e., the noise power present in a 1-Hz bandwidth). Mathematically, noise power density is

$$N_0 = \frac{N}{B} \text{ (W/Hz)} \tag{9-38}$$

where N_0 = noise power density (watts per hertz)
N = thermal noise power (watts)
B = bandwidth (hertz)

Stated in dBm,
$$N_{0(\text{dBm})} = 10 \log \frac{N}{0.001} - 10 \log B \tag{9-39}$$

$$= N_{(\text{dBm})} - 10 \log B \tag{9-40}$$

Combining Equations 9-29 and 9-38 yields

$$N_0 = \frac{KTB}{B} = KT \text{ (W/Hz)} \tag{9-41}$$

Stated in dBm,
$$N_{0(\text{dBm})} = 10 \log \frac{K}{0.001} + 10 \log T \tag{9-42}$$

Energy per bit-to-noise power density ratio is used to compare two or more digital modulation systems that use different transmission rates (bit rates), modulation schemes (FSK, PSK, QAM), or encoding techniques (M-ary). The energy per bit-to-noise power density ratio is simply the ratio of the energy of a single bit to the noise power present in 1 Hz of bandwidth. Thus, E_b/N_0 normalizes all multiphase modulation schemes to a common noise bandwidth, allowing for a simpler and more accurate comparison of their error performance. Mathematically, E_b/N_0 is

$$\frac{E_b}{N_0} = \frac{C/f_b}{N/B} = \frac{CB}{Nf_b} \tag{9-43}$$

where E_bN_0 is the energy per bit-to-noise power density ratio. Rearranging Equation 9-43 yields the following expression:

$$\frac{E_b}{N_0} = \frac{C}{N} \times \frac{B}{f_b} \tag{9-44}$$

where E_b/N_0 = energy per bit-to-noise power density ratio
C/N = carrier-to-noise power ratio
B/f_b = noise bandwidth-to-bit rate ratio

Stated in dB,
$$\frac{E_b}{N_0}\text{(dB)} = 10 \log \frac{C}{N} + 10 \log \frac{B}{f_b} \tag{9-45}$$

or
$$= 10 \log E_b - 10 \log N_0 \tag{9-46}$$

From Equation 9-44, it can be seen that the E_b/N_0 ratio is simply the product of the carrier-to-noise power ratio and the noise bandwidth-to-bit rate ratio. Also, from Equation 9-44, it can be seen that when the bandwidth equals the bit rate, $E_b/N_0 = C/N$.

In general, the minimum carrier-to-noise power ratio required for QAM systems is less than that required for comparable PSK systems, Also, the higher the level of encoding used (the higher the value of M), the higher the minimum carrier-to-noise power ratio. In Chapter 24, several examples are shown for determining the minimum carrier-to-noise power and energy per bit-to-noise density ratios for a given M-ary system and desired $P(e)$.

Example 9-15

For a QPSK system and the given parameters, determine

a. Carrier power in dBm.
b. Noise power in dBm.

c. Noise power density in dBm.
d. Energy per bit in dBJ.
e. Carrier-to-noise power ratio in dB.
f. E_b/N_0 ratio.

$$C = 10^{-12}\ \text{W} \qquad f_b = 60\ \text{kbps}$$
$$N = 1.2 \times 10^{-14}\ \text{W} \qquad B = 120\ \text{kHz}$$

Solution **a.** The carrier power in dBm is determined by substituting into Equation 9-28:

$$C = 10 \log \frac{10^{-12}}{0.001} = -90\ \text{dBm}$$

b. The noise power in dBm is determined by substituting into Equation 9-30:

$$N = 10 \log \frac{1.2 \times 10^{-14}}{0.001} = -109.2\ \text{dBm}$$

c. The noise power density is determined by substituting into Equation 9-40:

$$N_0 = -109.2\ \text{dBm} - 10 \log 120\ \text{kHz} = -160\ \text{dBm}$$

d. The energy per bit is determined by substituting into Equation 9-36:

$$E_b = 10 \log \frac{10^{-12}}{60\ \text{kbps}} = -167.8\ \text{dBJ}$$

e. The carrier-to-noise power ratio is determined by substituting into Equation 9-34:

$$\frac{C}{N} = 10 \log \frac{10^{-12}}{1.2 \times 10^{-14}} = 19.2\ \text{dB}$$

f. The energy per bit-to-noise density ratio is determined by substituting into Equation 9-45:

$$\frac{E_b}{N_0} = 19.2 + 10 \log \frac{120\ \text{kHz}}{60\ \text{kbps}} = 22.2\ \text{dB}$$

9-13 ERROR PERFORMANCE

9-13-1 PSK Error Performance

The bit error performance for the various multiphase digital modulation systems is directly related to the distance between points on a signal state-space diagram. For example, on the signal state-space diagram for BPSK shown in Figure 9-47a, it can be seen that the two signal points (logic 1 and logic 0) have maximum separation (d) for a given power level (D). In essence, one BPSK signal state is the exact negative of the other. As the figure shows, a noise vector (V_N), when combined with the signal vector (V_S), effectively shifts the phase of the signaling element (V_{SE}) alpha degrees. If the phase shift exceeds ±90°, the signal element is shifted beyond the threshold points into the error region. For BPSK, it would require a noise vector of sufficient amplitude and phase to produce more than a ±90° phase shift in the signaling element to produce an error. For PSK systems, the general formula for the threshold points is

$$\text{TP} = \pm \frac{\pi}{M} \tag{9-47}$$

where M is the number of signal states.

The phase relationship between signaling elements for BPSK (i.e., 180° out of phase) is the optimum signaling format, referred to as *antipodal signaling,* and occurs only when two binary signal levels are allowed and when one signal is the exact negative of the other. Because no other bit-by-bit signaling scheme is any better, antipodal performance is often used as a reference for comparison.

The error performance of the other multiphase PSK systems can be compared with that of BPSK simply by determining the relative decrease in error distance between points

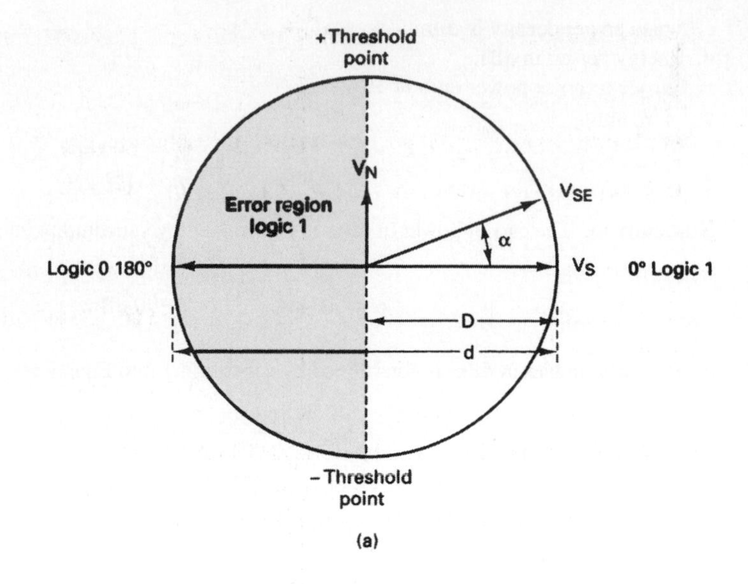

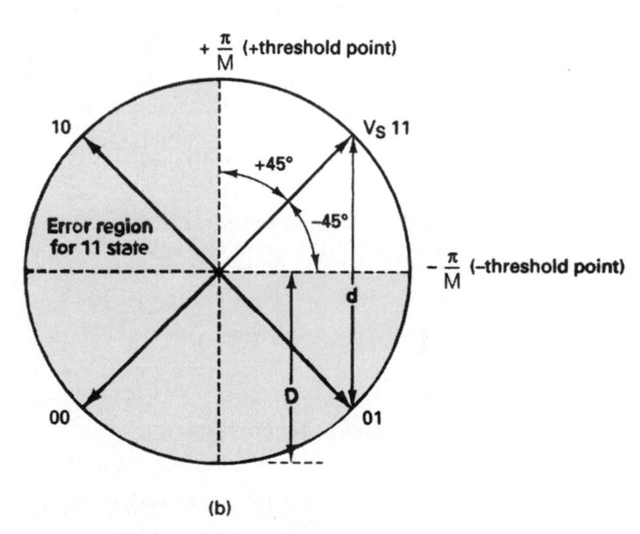

FIGURE 9-47 PSK error region: (a) BPSK; (b) QPSK

on a signal state-space diagram. For PSK, the general formula for the maximum distance between signaling points is given by

$$\sin\theta = \sin\frac{360°}{2M} = \frac{d/2}{D} \tag{9-48}$$

where d = error distance
M = number of phases
D = peak signal amplitude

Rearranging Equation 9-48 and solving for d yields

$$d = \left(2\sin\frac{180°}{M}\right) \times D \tag{9-49}$$

Figure 9-47b shows the signal state-space diagram for QPSK. From Figure 9-47 and Equation 9-48, it can be seen that QPSK can tolerate only a ±45° phase shift. From Equation 9-47, the maximum phase shift for 8-PSK and 16-PSK is ±22.5° and ±11.25°, respectively. Consequently, the higher levels of modulation (i.e., the greater the value of M) require a greater energy per bit-to-noise power density ratio to reduce the effect of noise interference. Hence, the higher the level of modulation, the smaller the angular separation between signal points and the smaller the error distance.

The general expression for the bit error probability of an M-phase PSK system is

$$P(e) = \frac{1}{\log_2 M}\,\mathrm{erf}(z) \tag{9-50}$$

where erf = error function

$$z = \sin(\pi/M)(\sqrt{\log_2 M})(\sqrt{E_b/N_0})$$

By substituting into Equation 9-50, it can be shown that QPSK provides the same error performance as BPSK. This is because the 3-dB reduction in error distance for QPSK is offset by the 3-dB decrease in its bandwidth (in addition to the error distance, the relative widths of the noise bandwidths must also be considered). Thus, both systems provide optimum performance. Figure 9-48 shows the error performance for 2-, 4-, 8-, 16-, and 32-PSK systems as a function of E_b/N_0.

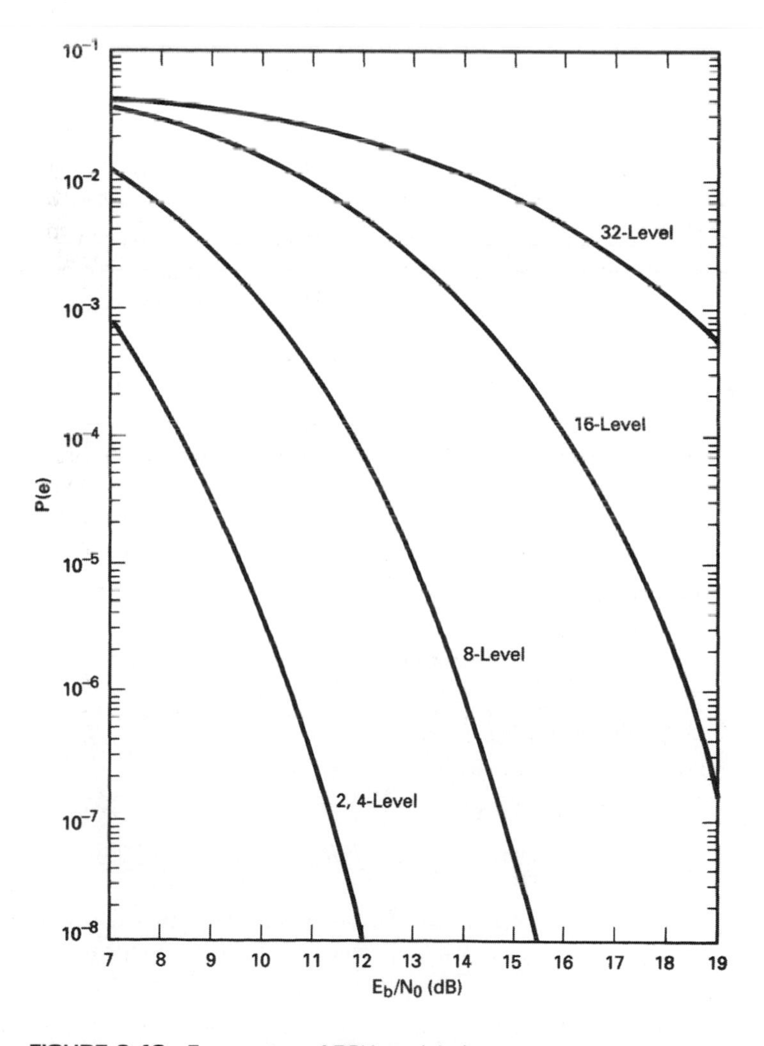

FIGURE 9-48 Error rates of PSK modulation systems

Example 9-16

Determine the minimum bandwidth required to achieve a $P(e)$ of 10^{-7} for an 8-PSK system operating at 10 Mbps with a carrier-to-noise power ratio of 11.7 dB.

Solution From Figure 9-48, the minimum E_b/N_0 ratio to achieve a $P(e)$ of 10^{-7} for an 8-PSK system is 14.7 dB. The minimum bandwidth is found by rearranging Equation 9-44:

$$\frac{B}{f_b} = \frac{E_b}{N_0} - \frac{C}{N}$$

$$= 14.7 \text{ dB} - 11.7 \text{ dB} = 3 \text{ dB}$$

$$\frac{B}{f_b} = \text{antilog } 3 = 2$$

$$B = 2 \times 10 \text{ Mbps} = 20 \text{ MHz}$$

9-13-2 QAM Error Performance

For a large number of signal points (i.e., M-ary systems greater than 4), QAM outperforms PSK. This is because the distance between signaling points in a PSK system is smaller than the distance between points in a comparable QAM system. The general expression for the distance between adjacent signaling points for a QAM system with L levels on each axis is

$$d = \frac{\sqrt{2}}{L-1} \times D \quad \textbf{(9-51)}$$

where d = error distance
L = number of levels on each axis
D = peak signal amplitude

In comparing Equation 9-49 to Equation 9-51, it can be seen that QAM systems have an advantage over PSK systems with the same peak signal power level.

The general expression for the bit error probability of an L-level QAM system is

$$P(e) = \frac{1}{\log_2 L}\left(\frac{L-1}{L}\right)\text{erfc}(z) \quad \textbf{(9-52)}$$

where erfc(z) is the complementary error function.

$$z = \frac{\sqrt{\log_2 L}}{L-1}\sqrt{\frac{E_b}{N_0}}$$

Figure 9-49 shows the error performance for 4-, 16-, 32-, and 64-QAM systems as a function of E_b/N_0.

Table 9-4 lists the minimum carrier-to-noise power ratios and energy per bit-to-noise power density ratios required for a probability of error 10^{-6} for several PSK and QAM modulation schemes.

Example 9-16

Which system requires the highest E_b/N_0 ratio for a probability of error of 10^{-6}, a four-level QAM system or an 8-PSK system?

Solution From Figure 9-49, the minimum E_b/N_0 ratio required for a four-level QAM system is 10.6 dB. From Figure 9-48, the minimum E_b/N_0 ratio required for an 8-PSK system is 14 dB. Therefore, to achieve a $P(e)$ of 10^{-6}, a four-level QAM system would require 3.4 dB less E_b/N_0 ratio.

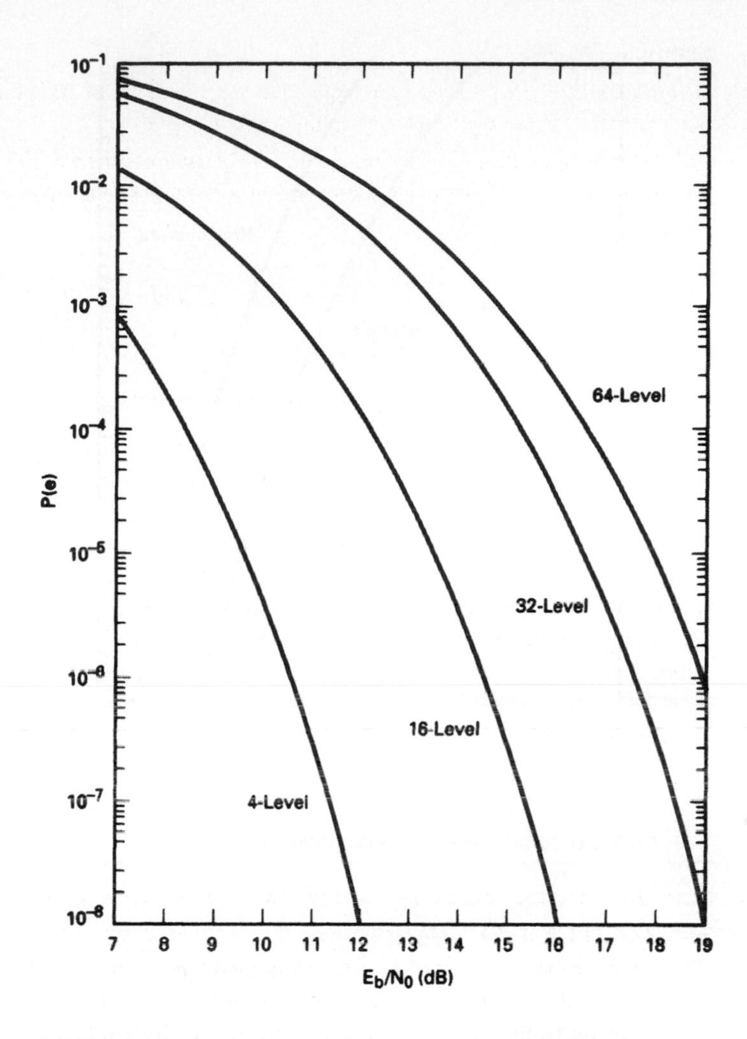

FIGURE 9-49 Error rates of QAM modulation systems

Table 9-4 Performance Comparison of Various Digital Modulation Schemes (BER = 10^{-6})

Modulation Technique	C/N Ratio (dB)	E_b/N_0 Ratio (dB)
BPSK	10.6	10.6
QPSK	13.6	10.6
4-QAM	13.6	10.6
8-QAM	17.6	10.6
8-PSK	18.5	14
16-PSK	24.3	18.3
16-QAM	20.5	14.5
32-QAM	24.4	17.4
64-QAM	26.6	18.8

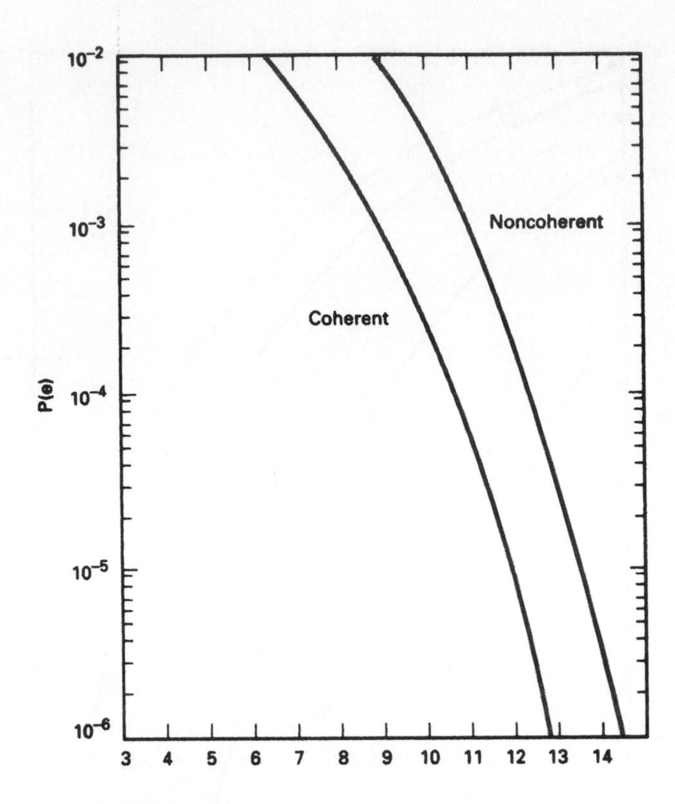

FIGURE 9-50 Error rates for FSK modulation systems

9-13-3 FSK Error Performance

The error probability for FSK systems is evaluated in a somewhat different manner than PSK and QAM. There are essentially only two types of FSK systems: noncoherent (asynchronous) and coherent (synchronous). With noncoherent FSK, the transmitter and receiver are not frequency or phase synchronized. With coherent FSK, local receiver reference signals are in frequency and phase lock with the transmitted signals. The probability of error for noncoherent FSK is

$$P(e) = \frac{1}{2}\exp\left(-\frac{E_b}{2N_0}\right) \tag{9-53}$$

The probability of error for coherent FSK is

$$P(e) = \operatorname{erfc}\sqrt{\frac{E_b}{N_0}} \tag{9-54}$$

Figure 9-50 shows probability of error curves for both coherent and noncoherent FSK for several values of E_b/N_0. From Equations 9-53 and 9-54, it can be determined that the probability of error for noncoherent FSK is greater than that of coherent FSK for equal energy per bit-to-noise power density ratios.

QUESTIONS

9-1. Explain *digital transmission* and *digital radio.*

9-2. Define *information capacity.*

9-3. What are the three most predominant modulation schemes used in digital radio systems?

9-4. Explain the relationship between bits per second and baud for an FSK system.

9-5. Define the following terms for FSK modulation: *frequency deviation, modulation index,* and *deviation ratio.*

9-6. Explain the relationship between (a) the minimum bandwidth required for an FSK system and the bit rate and (b) the mark and space frequencies.

9-7. What is the difference between standard FSK and MSK? What is the advantage of MSK?

9-8. Define *PSK.*

9-9. Explain the relationship between bits per second and baud for a BPSK system.

9-10. What is a constellation diagram, and how is it used with PSK?

9-11. Explain the relationship between the minimum bandwidth required for a BPSK system and the bit rate.

9-12. Explain *M*-ary.

9-13. Explain the relationship between bits per second and baud for a QPSK system.

9-14. Explain the significance of the I and Q channels in a QPSK modulator.

9-15. Define *dibit.*

9-16. Explain the relationship between the minimum bandwidth required for a QPSK system and the bit rate.

9-17. What is a coherent demodulator?

9-18. What advantage does OQPSK have over conventional QPSK? What is a disadvantage of OQPSK?

9-19. Explain the relationship between bits per second and baud for an 8-PSK system.

9-20. Define *tribit.*

9-21. Explain the relationship between the minimum bandwidth required for an 8-PSK system and the bit rate.

9-22. Explain the relationship between bits per second and baud for a 16-PSK system.

9-23. Define *quadbit.*

9-24. Define *QAM.*

9-25. Explain the relationship between the minimum bandwidth required for a 16-QAM system and the bit rate.

9-26. What is the difference between PSK and QAM?

9-27. Define *bandwidth efficiency.*

9-28. Define *carrier recovery.*

9-29. Explain the differences between absolute PSK and differential PSK.

9-30. What is the purpose of a clock recovery circuit? When is it used?

9-31. What is the difference between probability of error and bit error rate?

PROBLEMS

9-1. Determine the bandwidth and baud for an FSK signal with a mark frequency of 32 kHz, a space frequency of 24 kHz, and a bit rate of 4 kbps.

9-2. Determine the maximum bit rate for an FSK signal with a mark frequency of 48 kHz, a space frequency of 52 kHz, and an available bandwidth of 10 kHz.

9-3. Determine the bandwidth and baud for an FSK signal with a mark frequency of 99 kHz, a space frequency of 101 kHz, and a bit rate of 10 kbps.

9-4. Determine the maximum bit rate for an FSK signal with a mark frequency of 102 kHz, a space frequency of 104 kHz, and an available bandwidth of 8 kHz.

9-5. Determine the minimum bandwidth and baud for a BPSK modulator with a carrier frequency of 40 MHz and an input bit rate of 500 kbps. Sketch the output spectrum.

9-6. For the QPSK modulator shown in Figure 9-17, change the $+90°$ phase-shift network to $-90°$ and sketch the new constellation diagram.

9-7. For the QPSK demodulator shown in Figure 9-21, determine the I and Q bits for an input signal of $\sin \omega_c t - \cos \omega_c t$.

9-8. For an 8-PSK modulator with an input data rate (f_b) equal to 20 Mbps and a carrier frequency of 100 MHz, determine the minimum double-sided Nyquist bandwidth (f_N) and the baud. Sketch the output spectrum.

9-9. For the 8-PSK modulator shown in Figure 9-23, change the reference oscillator to cos $\omega_c t$ and sketch the new constellation diagram.

9-10. For a 16-QAM modulator with an input bit rate (f_b) equal to 20 Mbps and a carrier frequency of 100 MHz, determine the minimum double-sided Nyquist bandwidth (f_N) and the baud. Sketch the output spectrum.

9-11. For the 16-QAM modulator shown in Figure 9-33, change the reference oscillator to cos $\omega_c t$ and determine the output expressions for the following I, I′, Q, and Q′ input conditions: 0000, 1111, 1010, and 0101.

9-12. Determine the bandwidth efficiency for the following modulators:

a. QPSK, f_b = 10 Mbps
b. 8-PSK, f_b = 21 Mbps
c. 16-QAM, f_b = 20 Mbps

9-13. For the DBPSK modulator shown in Figure 9-40a, determine the output phase sequence for the following input bit sequence: 00110011010101 (assume that the reference bit = 1).

9-14. For a QPSK system and the given parameters, determine

a. Carrier power in dBm.
b. Noise power in dBm.
c. Noise power density in dBm.
d. Energy per bit in dBJ.
e. Carrier-to-noise power ratio.
f. E_b/N_0 ratio.

$C = 10^{-13}$ W $\qquad f_b = 30$ kbps
$N = 0.06 \times 10^{-15}$ W $\qquad B = 60$ kHz

9-15. Determine the minimum bandwidth required to achieve a $P(e)$ of 10^{-6} for an 8-PSK system operating at 20 Mbps with a carrier-to-noise power ratio of 11 dB.

9-16. Determine the minimum bandwidth and baud for a BPSK modulator with a carrier frequency of 80 MHz and an input bit rate f_b = 1 Mbps. Sketch the output spectrum.

9-17. For the QPSK modulator shown in Figure 9-17, change the reference oscillator to cos $\omega_c t$ and sketch the new constellation diagram.

9-18. For the QPSK demodulator shown in Figure 9-21, determine the I and Q bits for an input signal $-\sin \omega_c t + \cos \omega_c t$.

9-19. For an 8-PSK modulator with an input bit rate f_b = 10 Mbps and a carrier frequency f_c = 80 MHz, determine the minimum Nyquist bandwidth and the baud. Sketch the output spectrum.

9-20. For the 8-PSK modulator shown in Figure 9-23, change the +90° phase-shift network to a −90° phase shifter and sketch the new constellation diagram.

9-21. For a 16-QAM modulator with an input bit rate f_b = 10 Mbps and a carrier frequency f_c = 60 MHz, determine the minimum double-sided Nyquist frequency and the baud. Sketch the output spectrum.

9-22. For the 16-QAM modulator shown in Figure 9-33, change the 90° phase shift network to a − 90° phase shifter and determine the output expressions for the following I, I′, Q, and Q′ input conditions: 0000, 1111, 1010, and 0101.

9-23. Determine the bandwidth efficiency for the following modulators:

a. QPSK, f_b = 20 Mbps
b. 8-PSK, f_b = 28 Mbps
c. 16-PSK, f_b = 40 Mbps

9-24. For the DBPSK modulator shown in Figure 9-40a, determine the output phase sequence for the following input bit sequence: 11001100101010 (assume that the reference bit is a logic 1).

CHAPTER 10

Digital Transmission

CHAPTER OUTLINE

OBJECTIVES

- Define *digital transmission*
- List and describe the advantages and disadvantages of digital transmission
- Briefly describe pulse width modulation, pulse position modulation, and pulse amplitude modulation
- Define and describe pulse code modulation
- Explain flat-top and natural sampling
- Describe the Nyquist sampling theorem
- Describe folded binary codes
- Define and explain *dynamic range*
- Explain PCM coding efficiency
- Describe signal-to-quantization noise ratio
- Explain the difference between linear and nonlinear PCM codes
- Describe idle channel noise
- Explain several common coding methods
- Define *companding* and explain analog and digital companding
- Define *digital compression*

- Describe vocoders
- Explain how to determine PCM line speed
- Describe delta modulation PCM
- Describe adaptive delta modulation
- Define and describe differential pulse code modulation
- Describe the composition of digital pulses
- Explain intersymbol interference
- Explain eye patterns
- Explain the signal power distribution in binary digital signals

10-1 INTRODUCTION

As stated previously, *digital transmission* is the transmittal of digital signals between two or more points in a communications system. The signals can be binary or any other form of discrete-level digital pulses. The original source information may be in digital form, or it could be analog signals that have been converted to digital pulses prior to transmission and converted back to analog signals in the receiver. With digital transmission systems, a physical facility, such as a pair of wires, coaxial cable, or an optical fiber cable, is required to interconnect the various points within the system. The pulses are contained in and propagate down the cable. Digital pulses cannot be propagated through a wireless transmission system, such as Earth's atmosphere or free space (vacuum).

AT&T developed the first digital transmission system for the purpose of carrying digitally encoded analog signals, such as the human voice, over metallic wire cables between telephone offices. Today, digital transmission systems are used to carry not only digitally encoded voice and video signals but also digital source information directly between computers and computer networks. Digital transmission systems use both metallic and optical fiber cables for their transmission medium.

10-1-1 Advantages of Digital Transmission

The primary advantage of digital transmission over analog transmission is *noise immunity*. Digital signals are inherently less susceptible than analog signals to interference caused by noise because with digital signals it is not necessary to evaluate the precise amplitude, frequency, or phase to ascertain its logic condition. Instead, pulses are evaluated during a precise time interval, and a simple determination is made whether the pulse is above or below a prescribed reference level.

Digital signals are also better suited than analog signals for processing and combining using a technique called *multiplexing.* Digital signal processing (DSP) is the processing of analog signals using digital methods and includes bandlimiting the signal with filters, amplitude equalization, and phase shifting. It is much simpler to store digital signals than analog signals, and the transmission rate of digital signals can be easily changed to adapt to different environments and to interface with different types of equipment.

In addition, digital transmission systems are more resistant to analog systems to additive noise because they use signal *regeneration* rather than signal amplification. Noise produced in electronic circuits is additive (i.e., it accumulates); therefore, the signal-to-noise ratio deteriorates each time an analog signal is amplified. Consequently, the number of circuits the signal must pass through limits the total distance analog signals can be transported. However, digital regenerators sample noisy signals and then reproduce an entirely new digital signal with the same signal-to-noise ratio as the original transmitted signal. Therefore, digital signals can be transported longer distances than analog signals.

Finally, digital signals are simpler to measure and evaluate than analog signals. Therefore, it is easier to compare the error performance of one digital system to another digital system. Also, with digital signals, transmission errors can be detected and corrected more easily and more accurately than is possible with analog signals.

10-1-2 Disadvantages of Digital Transmission

The transmission of digitally encoded analog signals requires significantly more bandwidth than simply transmitting the original analog signal. Bandwidth is one of the most important aspects of any communications system because it is costly and limited.

Also, analog signals must be converted to digital pulses prior to transmission and converted back to their original analog form at the receiver, thus necessitating additional encoding and decoding circuitry. In addition, digital transmission requires precise time synchronization between the clocks in the transmitters and receivers. Finally, digital transmission systems are incompatible with older analog transmission systems.

10-2 PULSE MODULATION

Pulse modulation consists essentially of sampling analog information signals and then converting those samples into discrete pulses and transporting the pulses from a source to a destination over a physical transmission medium. The four predominant methods of pulse modulation include *pulse width modulation* (PWM), *pulse position modulation* (PPM), *pulse amplitude modulation* (PAM), and *pulse code modulation* (PCM).

PWM is sometimes called *pulse duration modulation* (PDM) or *pulse length modulation* (PLM), as the width (active portion of the duty cycle) of a constant amplitude pulse is varied proportional to the amplitude of the analog signal at the time the signal is sampled. PWM is shown in Figure 10-1c. As the figure shows, the amplitude of sample 1 is lower than the amplitude of sample 2. Thus, pulse 1 is narrower than pulse 2. The maximum analog signal amplitude produces the widest pulse, and the minimum analog signal amplitude produces the narrowest pulse. Note, however, that all pulses have the same amplitude.

With PPM, the position of a constant-width pulse within a prescribed time slot is varied according to the amplitude of the sample of the analog signal. PPM is shown in Figure 10-1d. As the figure shows, the higher the amplitude of the sample, the farther to the right the pulse is positioned within the prescribed time slot. The highest amplitude sample produces a pulse to the far right, and the lowest amplitude sample produces a pulse to the far left.

With PAM, the amplitude of a constant width, constant-position pulse is varied according to the amplitude of the sample of the analog signal. PAM is shown in Figure 10-1e, where it can be seen that the amplitude of a pulse coincides with the amplitude of the analog signal. PAM waveforms resemble the original analog signal more than the waveforms for PWM or PPM.

With PCM, the analog signal is sampled and then converted to a serial n-bit binary code for transmission. Each code has the same number of bits and requires the same length of time for transmission. PCM is shown in Figure 10-1f.

PAM is used as an intermediate form of modulation with PSK, QAM, and PCM, although it is seldom used by itself. PWM and PPM are used in special-purpose communications systems mainly for the military but are seldom used for commercial digital transmission systems. PCM is by far the most prevalent form of pulse modulation and, consequently, will be discussed in more detail in subsequent sections of this chapter.

10-3 PCM

Alex H. Reeves is credited with inventing PCM in 1937 while working for AT&T at its Paris laboratories. Although the merits of PCM were recognized early in its development, it was not until the mid-1960s, with the advent of solid-state electronics, that PCM became prevalent. In the United States today, PCM is the preferred method of communications within the public switched telephone network because with PCM it is easy to combine digitized voice and digital data into a single, high-speed digital signal and propagate it over either metallic or optical fiber cables.

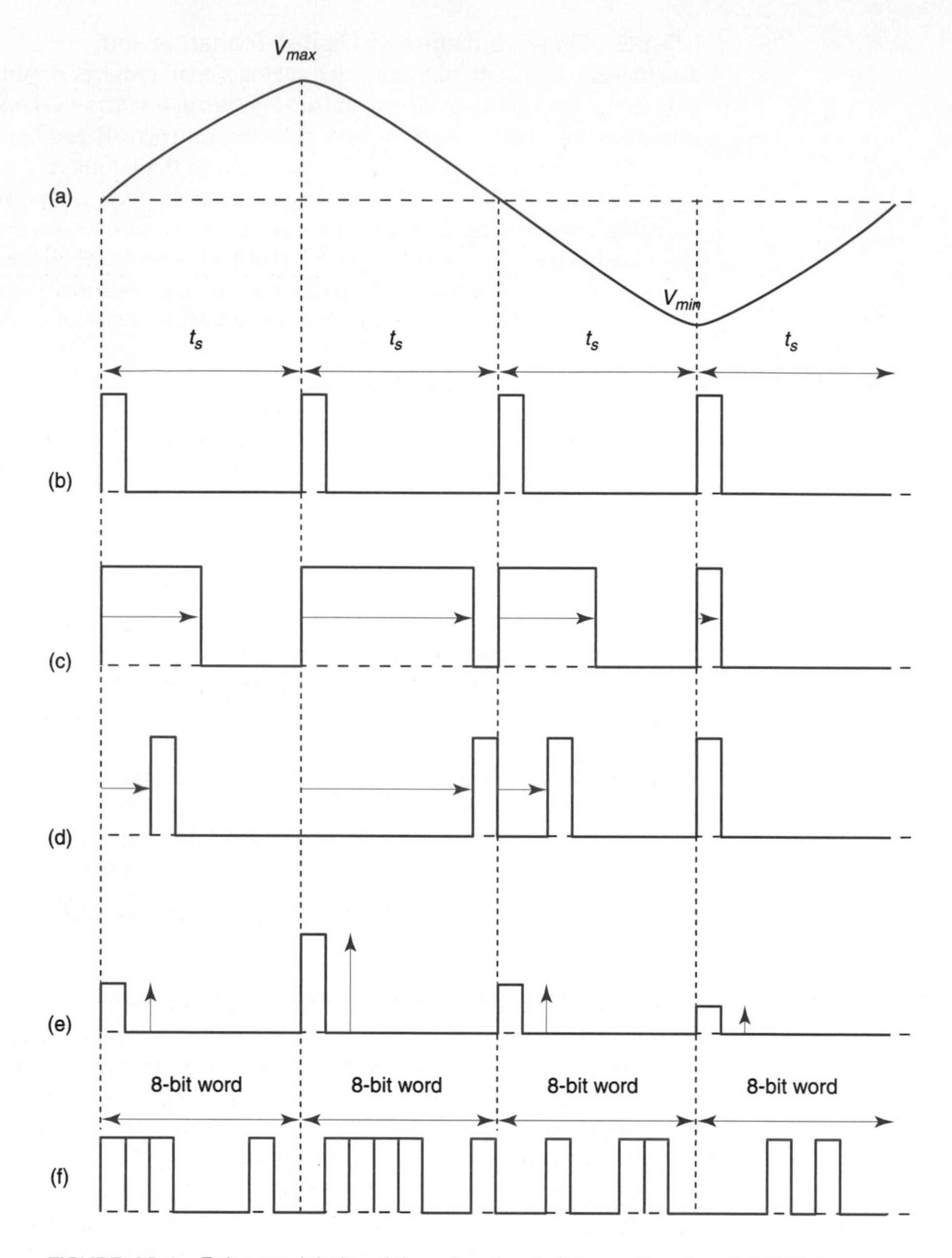

FIGURE 10-1 Pulse modulation: (a) analog signal; (b) sample pulse; (c) PWM; (d) PPM; (e) PAM; (f) PCM

PCM is the only digitally encoded modulation technique shown in Figure 10-1 that is commonly used for digital transmission. The term *pulse code modulation* is somewhat of a misnomer, as it is not really a type of modulation but rather a form of digitally coding analog signals. With PCM, the pulses are of fixed length and fixed amplitude. PCM is a binary system where a pulse or lack of a pulse within a prescribed time slot represents either a logic 1 or a logic 0 condition. PWM, PPM, and PAM are digital but seldom binary, as a pulse does not represent a single binary digit (bit).

Figure 10-2 shows a simplified block diagram of a single-channel, simplex (one-way only) PCM system. The bandpass filter limits the frequency of the analog input signal to the standard voice-band frequency range of 300 Hz to 3000 Hz. The *sample-and-hold* cir-

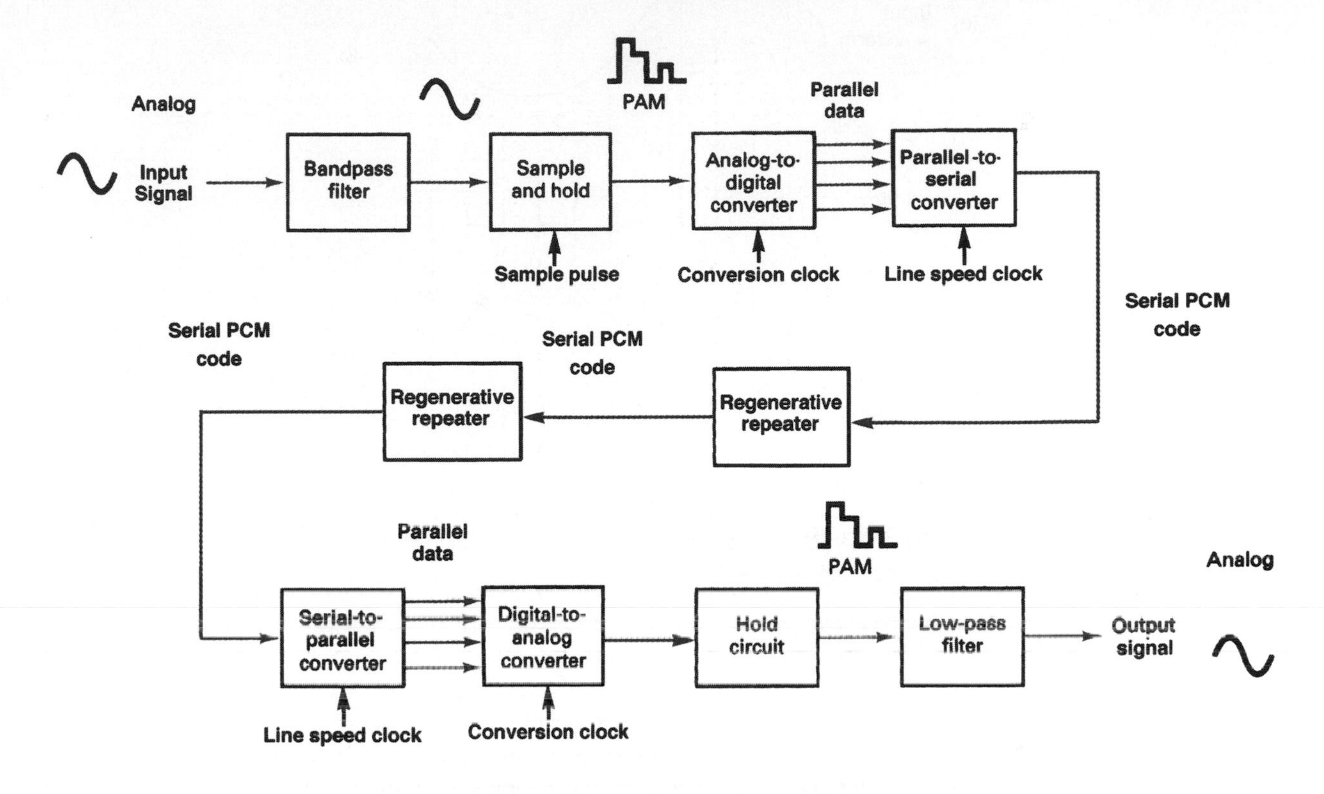

FIGURE 10-2 Simplified block diagram of a single-channel, simplex PCM transmission system

cuit periodically samples the analog input signal and converts those samples to a multilevel PAM signal. The *analog-to-digital converter* (ADC) converts the PAM samples to parallel PCM codes, which are converted to serial binary data in the *parallel-to-serial converter* and then outputted onto the transmission line as serial digital pulses. The transmission line *repeaters* are placed at prescribed distances to regenerate the digital pulses.

In the receiver, the *serial-to-parallel converter* converts serial pulses received from the transmission line to parallel PCM codes. The *digital-to-analog converter* (DAC) converts the parallel PCM codes to multilevel PAM signals. The hold circuit is basically a low-pass filter that converts the PAM signals back to its original analog form.

Figure 10-2 also shows several clock signals and sample pulses that will be explained in later sections of this chapter. An integrated circuit that performs the PCM encoding and decoding functions is called a codec (*co*der/*dec*oder), which is also described in a later section of this chapter.

10-4 PCM SAMPLING

The function of a sampling circuit in a PCM transmitter is to periodically sample the continually changing analog input voltage and convert those samples to a series of constant-amplitude pulses that can more easily be converted to binary PCM code. For the ADC to accurately convert a voltage to a binary code, the voltage must be relatively constant so that the ADC can complete the conversion before the voltage level changes. If not, the ADC would be continually attempting to follow the changes and may never stabilize on any PCM code.

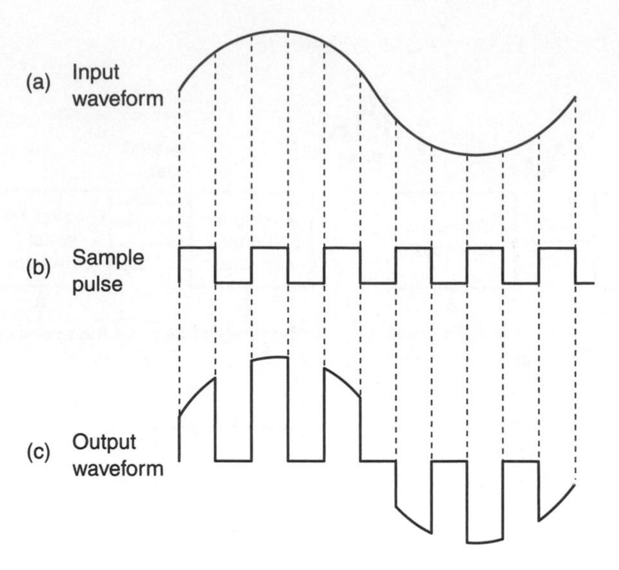

FIGURE 10-3 Natural sampling: (a) input analog signal; (b) sample pulse; (c) sampled output

Essentially, there are two basic techniques used to perform the sampling function: natural sampling and flat-top sampling. *Natural sampling* is shown in Figure 10-3. Natural sampling is when tops of the sample pulses retain their natural shape during the sample interval, making it difficult for an ADC to convert the sample to a PCM code. With natural sampling, the frequency spectrum of the sampled output is different from that of an ideal sample. The amplitude of the frequency components produced from narrow, finite-width sample pulses decreases for the higher harmonics in a (sin x)/x manner. This alters the information frequency spectrum requiring the use of frequency equalizers (compensation filters) before recovery by a low-pass filter.

The most common method used for sampling voice signals in PCM systems is *flat-top sampling*, which is accomplished in a *sample-and-hold circuit*. The purpose of a sample-and-hold circuit is to periodically sample the continually changing analog input voltage and convert those samples to a series of constant-amplitude PAM voltage levels. With flat-top sampling, the input voltage is sampled with a narrow pulse and then held relatively constant until the next sample is taken. Figure 10-4 shows flat-top sampling. As the figure shows, the sampling process alters the frequency spectrum and introduces an error called *aperture* error, which is when the amplitude of the sampled signal changes during the sample pulse time. This prevents the recovery circuit in the PCM receiver from exactly reproducing the original analog signal voltage. The magnitude of error depends on how much the analog signal voltage changes while the sample is being taken and the width (duration) of the sample pulse. Flat-top sampling, however, introduces less aperture distortion than natural sampling and can operate with a slower analog-to-digital converter.

Figure 10-5a shows the schematic diagram of a sample-and-hold circuit. The FET acts as a simple analog switch. When turned on, Q_1 provides a low-impedance path to deposit the analog sample voltage across capacitor C_1. The time that Q_1 is on is called the *aperture* or *acquisition time.* Essentially, C_1 is the hold circuit. When Q_1 is off, C_1 does not have a complete path to discharge through and, therefore, stores the sampled voltage. The *storage time* of the capacitor is called the A/D *conversion time* because it is during this time that the ADC converts the sample voltage to a PCM code. The acquisition time should be very short to ensure that a minimum change occurs in the analog signal while it is being deposited across C_1. If the input to the ADC is changing while it is performing the conversion, *aperture*

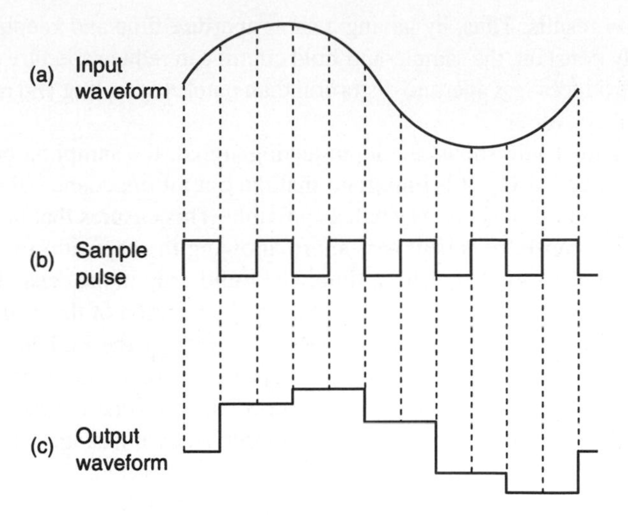

FIGURE 10-4 Flat-top sampling: (a) input analog signal; (b) sample pulse; (c) sampled output

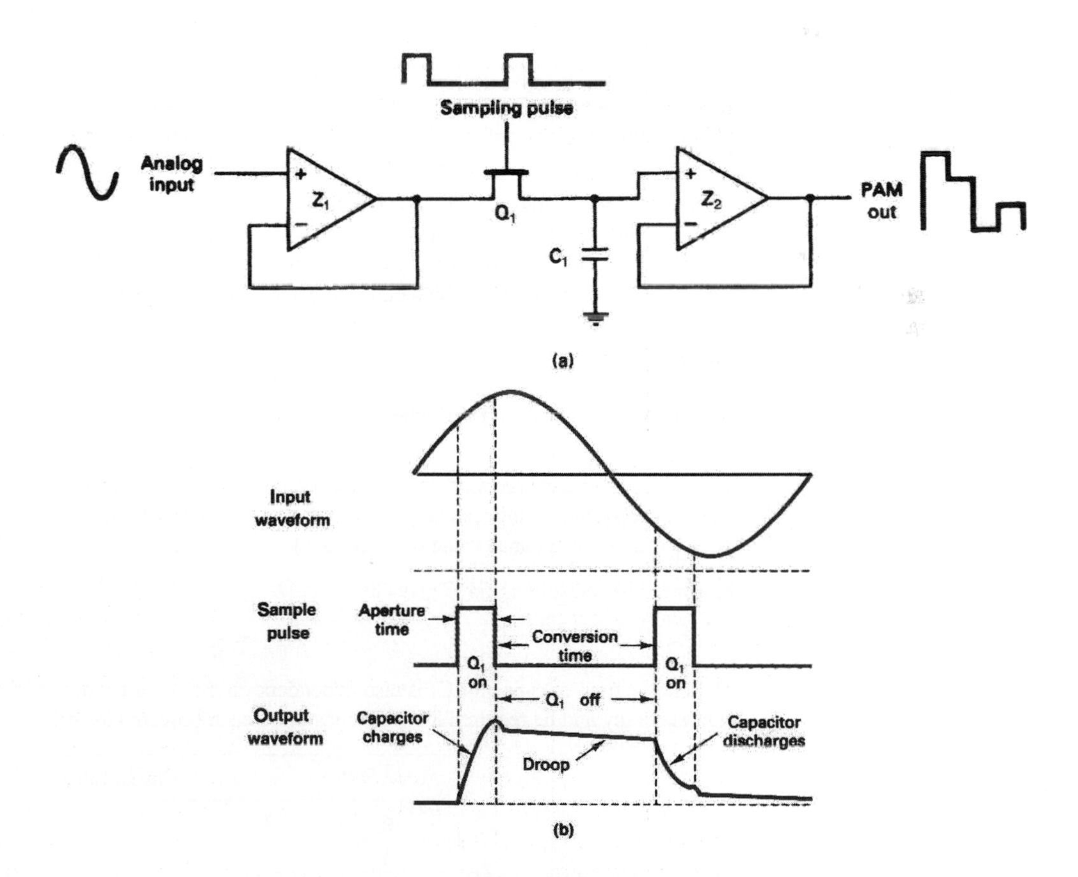

FIGURE 10-5 (a) Sample-and-hold circuit; (b) input and output waveforms

distortion results. Thus, by having a short aperture time and keeping the input to the ADC relatively constant, the sample-and-hold circuit can reduce aperture distortion. Flat-top sampling introduces less aperture distortion than natural sampling and requires a slower analog-to-digital converter.

Figure 10-5b shows the input analog signal, the sampling pulse, and the waveform developed across C_1. It is important that the output impedance of voltage follower Z_1 and the on resistance of Q_1 be as small as possible. This ensures that the *RC* charging time constant of the capacitor is kept very short, allowing the capacitor to charge or discharge rapidly during the short acquisition time. The rapid drop in the capacitor voltage immediately following each sample pulse is due to the redistribution of the charge across C_1. The interelectrode capacitance between the gate and drain of the FET is placed in series with C_1 when the FET is off, thus acting as a capacitive voltage-divider network. Also, note the gradual discharge across the capacitor during the conversion time. This is called *droop* and is caused by the capacitor discharging through its own leakage resistance and the input impedance of voltage follower Z_2. Therefore, it is important that the input impedance of Z_2 and the leakage resistance of C_1 be as high as possible. Essentially, voltage followers Z_1 and Z_2 isolate the sample-and-hold circuit (Q_1 and C_1) from the input and output circuitry.

Example 10-1

For the sample-and-hold circuit shown in Figure 10-5a, determine the largest-value capacitor that can be used. Use an output impedance for Z_1 of 10 Ω, an on resistance for Q_1 of 10 Ω, an acquisition time of 10 μs, a maximum peak-to-peak input voltage of 10 V, a maximum output current from Z_1 of 10 mA, and an accuracy of 1%.

Solution The expression for the current through a capacitor is

$$i = C\frac{dv}{dt}$$

Rearranging and solving for C yields

$$C = i\frac{dt}{dv}$$

where C = maximum capacitance (farads)
i = maximum output current from Z_1, 10 mA
dv = maximum change in voltage across C_1, which equals 10 V
dt = charge time, which equals the aperture time, 10 μs

Therefore,

$$C_{max} = \frac{(10\text{ mA})(10\text{ }\mu\text{s})}{10\text{ V}} = 10\text{ nF}$$

The charge time constant for C when Q_1 is on is

$$\tau = RC$$

where τ = one charge time constant (seconds)
R = output impedance of Z_1 plus the on resistance of Q_1 (ohms)
C = capacitance value of C_1 (farads)

Rearranging and solving for C gives us

$$C_{max} = \frac{\tau}{R}$$

The charge time of capacitor C_1 is also dependent on the accuracy desired from the device. The percent accuracy and its required *RC* time constant are summarized as follows:

Accuracy (%)	Charge Time
10	2.3τ
1	4.6τ
0.1	6.9τ
0.01	9.2τ

For an accuracy of 1%,

$$C = \frac{10\ \mu s}{4.6(20)} = 108.7\ \text{nF}$$

To satisfy the output current limitations of Z_1, a maximum capacitance of 10 nF was required. To satisfy the accuracy requirements, 108.7 nF was required. To satisfy both requirements, the smaller-value capacitor must be used. Therefore, C_1 can be no larger than 10 nF.

10-4-1 Sampling Rate

The *Nyquist sampling theorem* establishes the *minimum sampling rate* (f_s) that can be used for a given PCM system. For a sample to be reproduced accurately in a PCM receiver, each cycle of the analog input signal (f_a) must be sampled at least twice. Consequently, the minimum sampling rate is equal to twice the highest audio input frequency. If f_s is less than two times f_a, an impairment called *alias* or *foldover distortion* occurs. Mathematically, the minimum Nyquist sampling rate is

$$f_s \geq 2f_a \tag{10-1}$$

where f_s = minimum Nyquist sample rate (hertz)
f_a = maximum analog input frequency (hertz)

A sample-and-hold circuit is a nonlinear device (mixer) with two inputs: the sampling pulse and the analog input signal. Consequently, nonlinear mixing (heterodyning) occurs between these two signals.

Figure 10-6a shows the frequency-domain representation of the output spectrum from a sample-and-hold circuit. The output includes the two original inputs (the audio and the fundamental frequency of the sampling pulse), their sum and difference frequencies ($f_s \pm f_a$), all the harmonics of f_s and f_a ($2f_s$, $2f_a$, $3f_s$, $3f_a$, and so on), and their associated cross products ($2f_s \pm f_a$, $3f_s \pm f_a$, and so on).

Because the sampling pulse is a repetitive waveform, it is made up of a series of harmonically related sine waves. Each of these sine waves is amplitude modulated by the analog signal and produces sum and difference frequencies symmetrical around each of the harmonics of f_s. Each sum and difference frequency generated is separated from its respective center frequency by f_a. As long as f_s is at least twice f_a, none of the side frequencies from one harmonic will spill into the sidebands of another harmonic, and aliasing does not

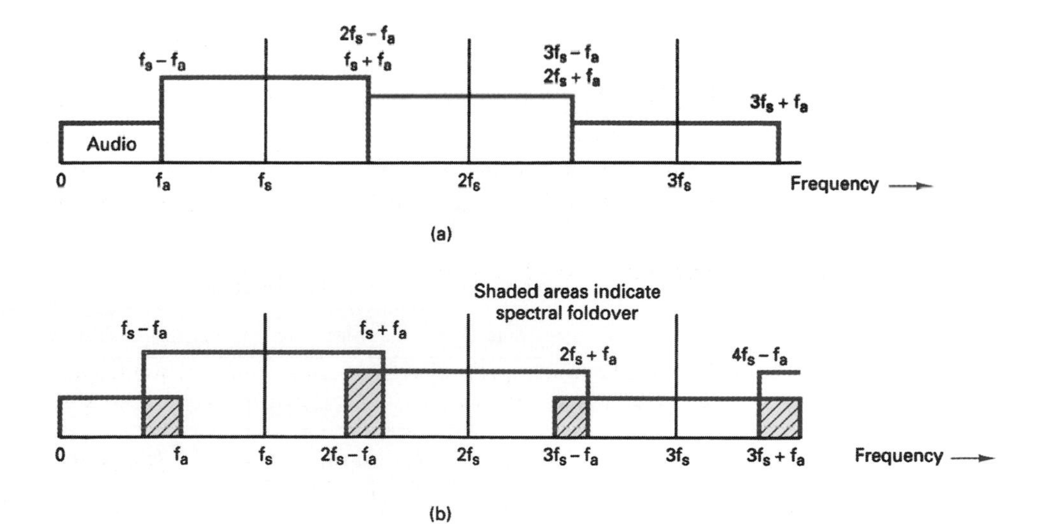

FIGURE 10-6 Output spectrum for a sample-and-hold circuit: (a) no aliasing; (b) aliasing distortion

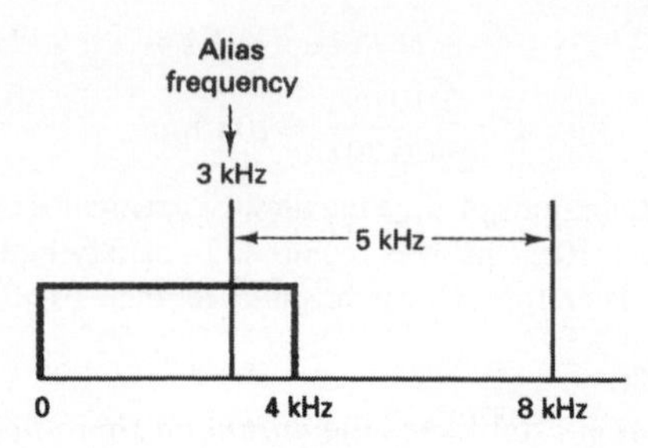

FIGURE 10-7 Output spectrum for Example 15-2

occur. Figure 10-6b shows the results when an analog input frequency greater than $f_s/2$ modulates f_s. The side frequencies from one harmonic fold over into the sideband of another harmonic. The frequency that folds over is an alias of the input signal (hence the names "aliasing" or "foldover distortion"). If an alias side frequency from the first harmonic folds over into the audio spectrum, it cannot be removed through filtering or any other technique.

Example 10-2

For a PCM system with a maximum audio input frequency of 4 kHz, determine the minimum sample rate and the alias frequency produced if a 5-kHz audio signal were allowed to enter the sample-and-hold circuit.

Solution Using Nyquist's sampling theorem (Equation 10-1), we have

$$f_s \geq 2f_a \quad \text{therefore,} \quad f_s \geq 8 \text{ kHz}$$

If a 5-kHz audio frequency entered the sample-and-hold circuit, the output spectrum shown in Figure 10-7 is produced. It can be seen that the 5-kHz signal produces an alias frequency of 3 kHz that has been introduced into the original audio spectrum.

The input bandpass filter shown in Figure 10-2 is called an *antialiasing* or *antifoldover filter.* Its upper cutoff frequency is chosen such that no frequency greater than one-half the sampling rate is allowed to enter the sample-and-hold circuit, thus eliminating the possibility of foldover distortion occurring.

With PCM, the analog input signal is sampled, then converted to a serial binary code. The binary code is transmitted to the receiver, where it is converted back to the original analog signal. The binary codes used for PCM are n-bit codes, where n may be any positive integer greater than 1. The codes currently used for PCM are *sign-magnitude codes,* where the *most significant bit* (MSB) is the sign bit and the remaining bits are used for magnitude. Table 10-1 shows an n-bit PCM code where n equals 3. The most significant bit is used to represent the sign of the sample (logic 1 = positive and logic 0 = negative). The two remaining bits represent the magnitude. With two magnitude bits, there are four codes possi-

Table 10-1 Three-Bit PCM Code

Sign	Magnitude	Decimal Value
1	1 1	+3
1	1 0	+2
1	0 1	+1
1	0 0	+0
0	0 0	−0
0	0 1	−1
0	1 0	−2
0	1 1	−3

ble for positive numbers and four codes possible for negative numbers. Consequently, there is a total of eight possible codes ($2^3 = 8$).

10-4-2 Quantization and the Folded Binary Code

Quantization is the process of converting an infinite number of possibilities to a finite number of conditions. Analog signals contain an infinite number of amplitude possibilities. Thus, converting an analog signal to a PCM code with a limited number of combinations requires quantization. In essence, quantization is the process of rounding off the amplitudes of flat-top samples to a manageable number of levels. For example, a sine wave with a peak amplitude of 5 V varies between +5 V and −5 V passing through every possible amplitude in between. A PCM code could have only eight bits, which equates to only 2^8, or 256 combinations. Obviously, to convert samples of a sine wave to PCM requires some rounding off.

With quantization, the total voltage range is subdivided into a smaller number of subranges, as shown in Table 10-2. The PCM code shown in Table 10-2 is a three-bit sign-magnitude code with eight possible combinations (four positive and four negative). The leftmost bit is the sign bit (1 = + and 0 = −), and the two rightmost bits represent magnitude. This type of code is called a *folded binary code* because the codes on the bottom half of the table are a mirror image of the codes on the top half, except for the sign bit. If the negative codes were folded over on top of the positive codes, they would match perfectly. With a folded binary code, each voltage level has one code assigned to it except zero volts, which has two codes, 100 (+0) and 000 (−0). The magnitude difference between adjacent steps is called the *quantization interval* or *quantum*. For the code shown in Table 10-2, the quantization interval is 1 V. Therefore, for this code, the maximum signal magnitude that can be encoded is +3 V (111) or −3 V (011), and the minimum signal magnitude is +1 V (101) or −1 V (001). If the magnitude of the sample exceeds the highest quantization interval, *overload distortion* (also called *peak limiting*) occurs.

Assigning PCM codes to absolute magnitudes is called quantizing. The magnitude of a quantum is also called the *resolution*. The resolution is equal to the voltage of the *minimum step size*, which is equal to the voltage of the *least significant bit* (V_{lsb}) of the PCM code. The resolution is the minimum voltage other than 0 V that can be decoded by the digital-to-analog converter in the receiver. The resolution for the PCM code shown in Table 10-2 is 1 V. The smaller the magnitude of a quantum, the better (smaller) the resolution and the more accurately the quantized signal will resemble the original analog sample.

In Table 10-2, each three-bit code has a range of input voltages that will be converted to that code. For example, any voltage between +0.5 and +1.5 will be converted to the code 101 (+1 V). Each code has a *quantization range* equal to + or − one-half the magnitude of a quantum except the codes for +0 and −0. The 0-V codes each have an input range equal to only one-half a quantum (0.5 V).

Table 10-2 Three-Bit PCM Code

	Sign	Magnitude		Decimal value	Quantization range
8 Sub ranges	1	1	1	+3	+2.5 V to +3.5 V
	1	1	0	+2	+1.5 V to +2.5 V
	1	0	1	+1	+0.5 V to +1.5 V
	1	0	0	+0	0 V to +0.5 V
	0	0	0	−0	0 V to −0.5 V
	0	0	1	−1	−0.5 V to −1.5 V
	0	1	0	−2	−1.5 V to −2.5 V
	0	1	1	−3	−2.5 V to −3.5 V

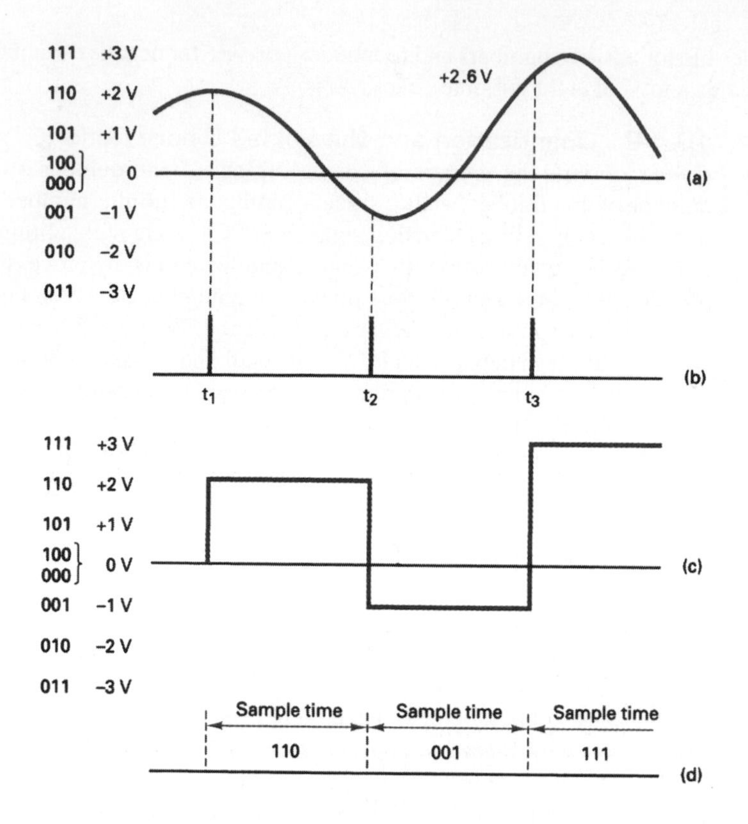

FIGURE 10-8 (a) Analog input signal; (b) sample pulse; (c) PAM signal; (d) PCM code

Figure 10-8 shows an analog input signal, the sampling pulse, the corresponding quantized signal (PAM), and the PCM code for each sample. The likelihood of a sample voltage being equal to one of the eight quantization levels is remote. Therefore, as shown in the figure, each sample voltage is rounded off (quantized) to the closest available level and then converted to its corresponding PCM code. The PAM signal in the transmitter is essentially the same PAM signal produced in the receiver. Therefore, any round-off errors in the transmitted signal are reproduced when the code is converted back to analog in the receiver. This error is called the *quantization error* (Q_e). The quantization error is equivalent to additive white noise as it alters the signal amplitude. Consequently, quantization error is also called *quantization noise* (Q_n). The maximum magnitude for the quantization error is equal to one-half a quantum (±0.5 V for the code shown in Table 10-2).

The first sample shown in Figure 10-8 occurs at time t_1, when the input voltage is exactly +2 V. The PCM code that corresponds to +2 V is 110, and there is no quantization error. Sample 2 occurs at time t_2, when the input voltage is −1 V. The corresponding PCM code is 001, and again there is no quantization error. To determine the PCM code for a particular sample voltage, simply divide the voltage by the resolution, convert the quotient to an n-bit binary code, and then add the sign bit. For sample 3 in Figure 10-9, the voltage at t_3 is approximately +2.6 V. The folded PCM code is

$$\frac{\text{sample voltage}}{\text{resolution}} = \frac{2.6}{1} = 2.6$$

There is no PCM code for +2.6; therefore, the magnitude of the sample is rounded off to the nearest valid code, which is 111, or +3 V. The rounding-off process results in a quantization error of 0.4 V.

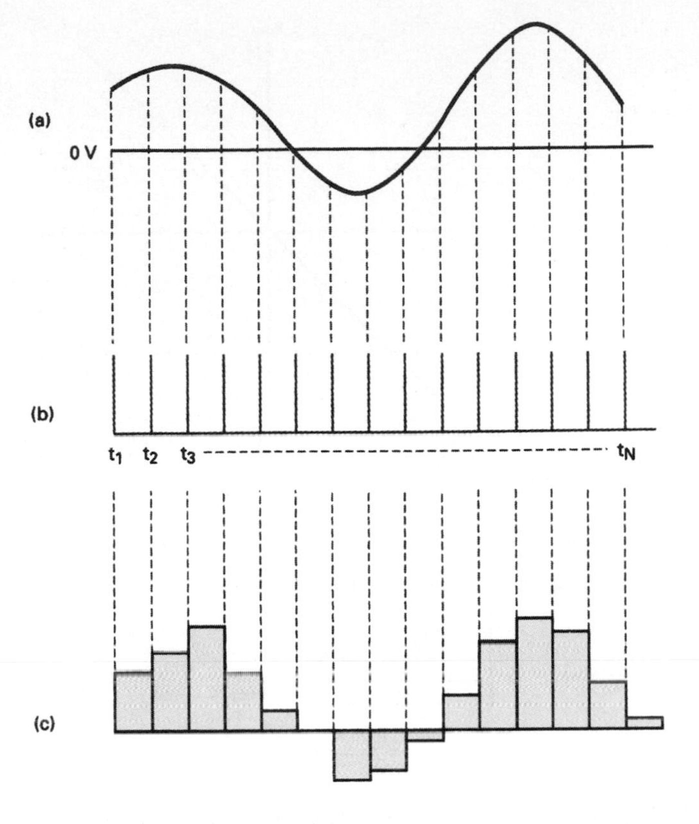

FIGURE 10-9 PAM: (a) input signal; (b) sample pulse; (c) PAM signal

The quantized signal shown in Figure 10-8c at best only roughly resembles the original analog input signal. This is because with a three-bit PCM code, the resolution is rather poor and also because there are only three samples taken of the analog signal. The quality of the PAM signal can be improved by using a PCM code with more bits, reducing the magnitude of a quantum and improving the resolution. The quality can also be improved by sampling the analog signal at a faster rate. Figure 10-9 shows the same analog input signal shown in Figure 10-8 except the signal is being sampled at a much higher rate. As the figure shows, the PAM signal resembles the analog input signal rather closely.

Figure 10-10 shows the input-versus-output transfer function for a linear analog-to-digital converter (sometimes called a linear quantizer). As the figure shows for a linear analog input signal (i.e., a ramp), the quantized signal is a staircase function. Thus, as shown in Figure 10-7c, the maximum quantization error is the same for any magnitude input signal.

Example 10-3

For the PCM coding scheme shown in Figure 10-8, determine the quantized voltage, quantization error (Q_e), and PCM code for the analog sample voltage of +1.07 V.

Solution To determine the quantized level, simply divide the sample voltage by resolution and then round the answer off to the nearest quantization level:

$$\frac{+1.07\ V}{1\ V} = 1.07 = 1$$

The quantization error is the difference between the original sample voltage and the quantized level, or

$$Q_e = 1.07 - 1 = 0.07$$

From Table 10-2, the PCM code for +1 is 101.

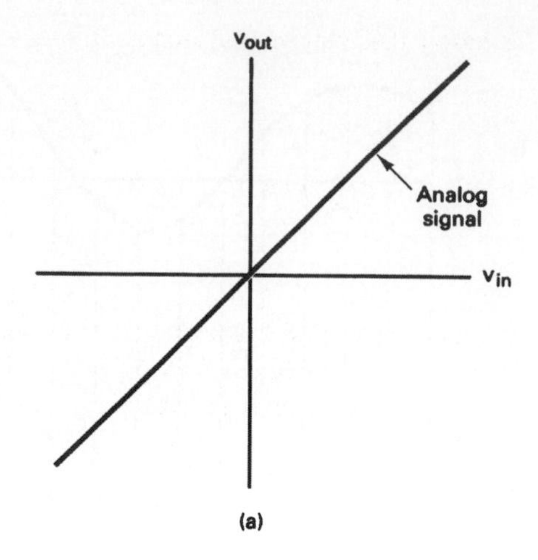

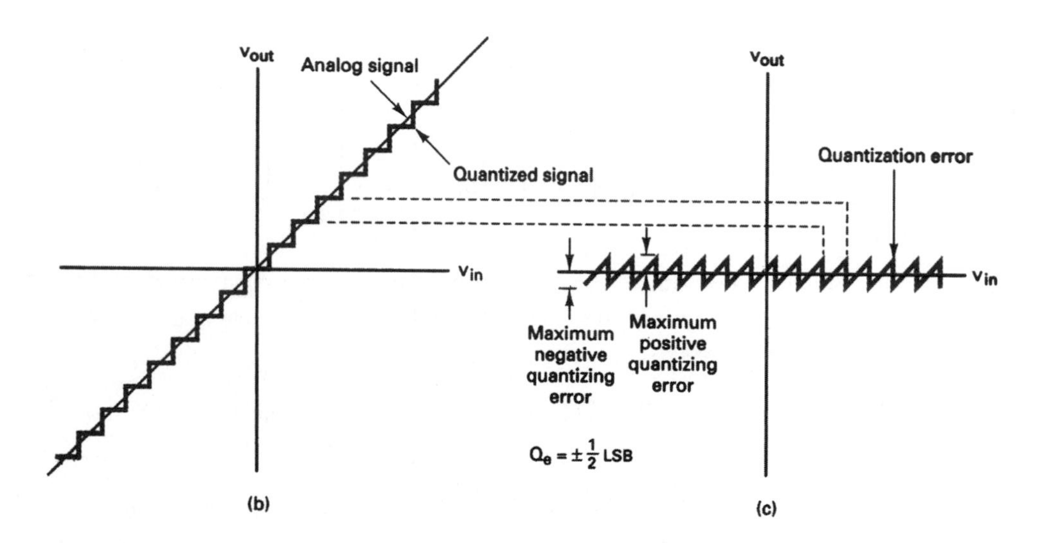

FIGURE 10-10 Linear input-versus-output transfer curve: (a) linear transfer function; (b) quantization; (c) Q_e

10-4-3 Dynamic Range

The number of PCM bits transmitted per sample is determined by several variables, including maximum allowable input amplitude, resolution, and dynamic range. *Dynamic range* (DR) is the ratio of the largest possible magnitude to the smallest possible magnitude (other than 0 V) that can be decoded by the digital-to-analog converter in the receiver. Mathematically, dynamic range is

$$DR = \frac{V_{max}}{V_{min}} \tag{10-2}$$

where DR = dynamic range (unitless ratio)
V_{min} = the quantum value (resolution)
V_{max} = the maximum voltage magnitude that can be discerned by the DACs in the receiver

Equation 10-2 can be rewritten as

$$DR = \frac{V_{max}}{\text{resolution}} \tag{10-3}$$

For the system shown in Table 10-2,

$$DR = \frac{3\ V}{1\ V} = 3$$

A dynamic range of 3 indicates that the ratio of the largest decoded voltage to the smallest decoded signal voltage is 3 to 1.

Dynamic range is generally expressed as a dB value; therefore,

$$DR = 20 \log \frac{V_{max}}{V_{min}} \tag{10-4}$$

For the system shown in Table 10-2,

$$DR = 20 \log 3 = 9.54 \text{ dB}$$

The number of bits used for a PCM code depends on the dynamic range. The relationship between dynamic range and the number of bits in a PCM code is

$$2^n - 1 \geq DR \tag{10-5a}$$

and for a minimum number of bits

$$2^n - 1 = DR \tag{10-5b}$$

where n = number of bits in a PCM code, excluding the sign bit
DR = absolute value of dynamic range

Why $2^n - 1$? One positive and one negative PCM code is used for 0 V, which is not considered for dynamic range. Therefore,

$$2^n = DR + 1$$

To solve for the number of bits (n) necessary to produce a dynamic range of 3, convert to logs,

$$\log 2^n = \log(DR + 1)$$
$$n \log 2 = \log(DR + 1)$$
$$n = \frac{\log(3 + 1)}{\log 2} = \frac{0.602}{0.301} = 2$$

For a dynamic range of 3, a PCM code with two bits is required. Dynamic range can be expressed in decibels as

$$DR_{(dB)} = 20 \log\left(\frac{V_{max}}{V_{min}}\right)$$

or

$$DR_{(dB)} = 20 \log(2^n - 1) \tag{10-6}$$

where n is the number of PCM bits. For values of $n > 4$, dynamic range is approximated as

$$DR_{(dB)} \approx 20 \log(2^n)$$
$$\approx 20n \log(2)$$
$$\approx 6n \tag{10-7}$$

Table 10-3 Dynamic Range versus Number of PCM Magnitude Bits

Number of Bits in PCM Code (n)	Number of Levels Possible ($M = 2^n$)	Dynamic Range (dB)
1	2	6.02
2	4	12
3	8	18.1
4	16	24.1
5	32	30.1
6	64	36.1
7	128	42.1
8	256	48.2
9	512	54.2
10	1024	60.2
11	2048	66.2
12	4096	72.2
13	8192	78.3
14	16,384	84.3
15	32,768	90.3
16	65,536	96.3

Equation 10-7 indicates that there is approximately 6 dB dynamic range for each magnitude bit in a linear PCM code. Table 10-3 summarizes dynamic range for PCM codes with n bits for values of n up to 16.

Example 10-4

For a PCM system with the following parameters, determine (a) minimum sample rate, (b) minimum number of bits used in the PCM code, (c) resolution, and (d) quantization error.

Maximum analog input frequency = 4 kHz
Maximum decoded voltage at the receiver = ±2.55 V
Minimum dynamic range = 46 dB

Solution a. Substituting into Equation 10-1, the minimum sample rate is

$$f_s = 2f_{a-} = 2(4\text{ kHz}) = 8\text{ kHz}$$

b. To determine the absolute value for dynamic range, substitute into Equation 10-4:

$$46\text{ dB} = 20\log\frac{V_{max}}{V_{min}}$$

$$2.3 = \log\frac{V_{max}}{V_{min}}$$

$$10^{2.3} = \frac{V_{max}}{V_{min}}$$

$$199.5 = DR$$

The minimum number of bits is determined by rearranging Equation 10-5b and solving for n:

$$n = \frac{\log(199.5 + 1)}{\log 2} = 7.63$$

The closest whole number greater than 7.63 is 8; therefore, eight bits must be used for the magnitude.

Because the input amplitude range is ±2.55, one additional bit, the sign bit, is required. Therefore, the total number of CM bits is nine, and the total number of PCM codes is $2^9 = 512$. (There are 255 positive codes, 255 negative codes, and 2 zero codes.)

To determine the actual dynamic range, substitute into Equation 10-6:

$$DR_{(dB)} = 20\log(2^n - 1)$$
$$= 20(\log 256 - 1)$$
$$= 48.13\text{ dB}$$

c. The resolution is determined by dividing the maximum positive or maximum negative voltage by the number of positive or negative nonzero PCM codes:

$$\text{resolution} = \frac{V_{max}}{2^n - 1} = \frac{2.55}{2^8 - 1} = \frac{2.55}{256 - 1} = 0.01 \text{ V}$$

The maximum quantization error is

$$Q_e = \frac{\text{resolution}}{2} = \frac{0.01 \, V}{2} = 0.005 \text{ V}$$

10-4-4 Coding Efficiency

Coding efficiency is a numerical indication of how efficiently a PCM code is utilized. Coding efficiency is the ratio of the minimum number of bits required to achieve a certain dynamic range to the actual number of PCM bits used. Mathematically, coding efficiency is

$$\text{coding efficiency} = \frac{\text{minimum number of bits (including sign bit)}}{\text{actual number of bits (including sign bit)}} \times 100 \quad \textbf{(10-8)}$$

The coding efficiency for Example 10-4 is

$$\text{coding efficiency} = \frac{8.63}{9} \times 100 = 95.89\%$$

10-5 SIGNAL-TO-QUANTIZATION NOISE RATIO

The three-bit PCM coding scheme shown in Figures 10-8 and 10-9 consists of *linear codes*, which means that the magnitude change between any two successive codes is the same. Consequently, the magnitude of their quantization error is also the same. The maximum quantization noise is half the resolution (quantum value). Therefore, the worst possible *signal voltage-to-quantization noise voltage ratio* (SQR) occurs when the input signal is at its minimum amplitude (101 or 001). Mathematically, the worst-case voltage SQR is

$$SQR = \frac{\text{resolution}}{Q_e} = \frac{V_{lsb}}{V_{lsb}/2} = 2$$

For the PCM code shown in Figure 10-8, the worst-case (minimum) SQR occurs for the lowest magnitude quantization voltage (±1 V). Therefore, the minimum SQR is

$$SQR_{(min)} = \frac{1}{0.5} = 2$$

or in dB

$$= 20 \log(2)$$
$$= 6 \text{ dB}$$

For a maximum amplitude input signal of 3 V (either 111 or 011), the maximum quantization noise is also equal to the resolution divided by 2. Therefore, the SQR for a maximum input signal is

$$SQR_{(max)} = \frac{V_{max}}{Q_e} = \frac{3}{0.5/2} = 6$$

or in dB

$$= 20 \log 6$$
$$= 15.6 \text{ dB}$$

From the preceding example, it can be seen that even though the magnitude of the quantization error remains constant throughout the entire PCM code, the percentage error does not; it decreases as the magnitude of the sample increases.

The preceding expression for SQR is for voltage and presumes the maximum quantization error; therefore, it is of little practical use and is shown only for comparison purposes and to illustrate that the SQR is not constant throughout the entire range of sample amplitudes. In reality and as shown in Figure 10-9, the difference between the PAM waveform and the analog input waveform varies in magnitude. Therefore, the SQR is not constant. Generally, the quantization error or distortion caused by digitizing an analog sample is expressed as an average signal power-to-average noise power ratio. For linear PCM codes (all quantization intervals have equal magnitudes), the *signal power-to-quantizing noise power ratio* (also called *signal-to-distortion ratio* or *signal-to-noise ratio*) is determined by the following formula:

$$SQR_{(\mathrm{dB})} = 10 \log \frac{v^2/R}{(q^2/12)/R} \qquad \textbf{(10-9a)}$$

where
R = resistance (ohms)
v = rms signal voltage (volts)
q = quantization interval (volts)
v^2/R = average signal power (watts)
$(q^2/12)/R$ = average quantization noise power (watts)

If the resistances are assumed to be equal, Equation 10-8a reduces to

$$SQR = 10 \log \left[\frac{v^2}{q^2/12} \right]$$

$$= 10.8 = 20 \log \frac{v}{q} \qquad \textbf{(10-9b)}$$

10-6 LINEAR VERSUS NONLINEAR PCM CODES

Early PCM systems used *linear codes* (i.e., the magnitude change between any two successive steps is uniform). With linear coding, the accuracy (resolution) for the higher-amplitude analog signals is the same as for the lower-amplitude signals, and the SQR for the lower-amplitude signals is less than for the higher-amplitude signals. With voice transmission, low-amplitude signals are more likely to occur than large-amplitude signals. Therefore, if there were more codes for the lower amplitudes, it would increase the accuracy where the accuracy is needed. As a result, there would be fewer codes available for the higher amplitudes, which would increase the quantization error for the larger-amplitude signals (thus decreasing the SQR). Such a coding technique is called *nonlinear* or *nonuniform encoding.* With nonlinear encoding, the step size increases with the amplitude of the input signal.

Figure 10-11 shows the step outputs from a linear and a nonlinear analog-to-digital converter. Note, with nonlinear encoding, there are more codes at the bottom of the scale than there are at the top, thus increasing the accuracy for the smaller-amplitude signals. Also note that the distance between successive codes is greater for the higher-amplitude signals, thus increasing the quantization error and reducing the SQR. Also, because the ratio of V_{max} to V_{min} is increased with nonlinear encoding, the dynamic range is larger than with a uniform linear code. It is evident that nonlinear encoding is a compromise; SQR is sacrificed for the higher-amplitude signals to achieve more accuracy for the lower-amplitude signals and to achieve a larger dynamic range. It is difficult to fabricate

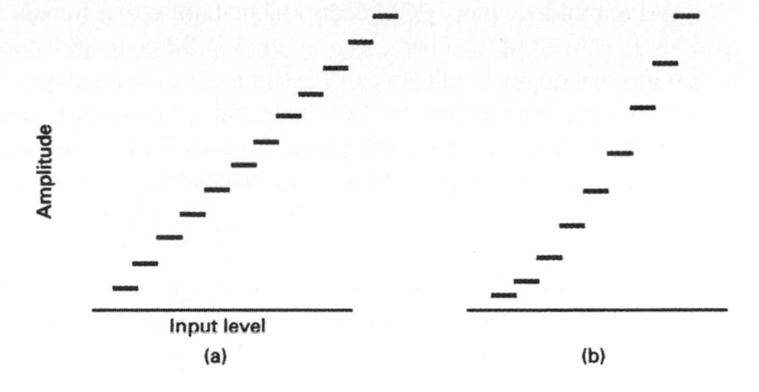

FIGURE 10-11 (a) Linear versus (b) nonlinear encoding

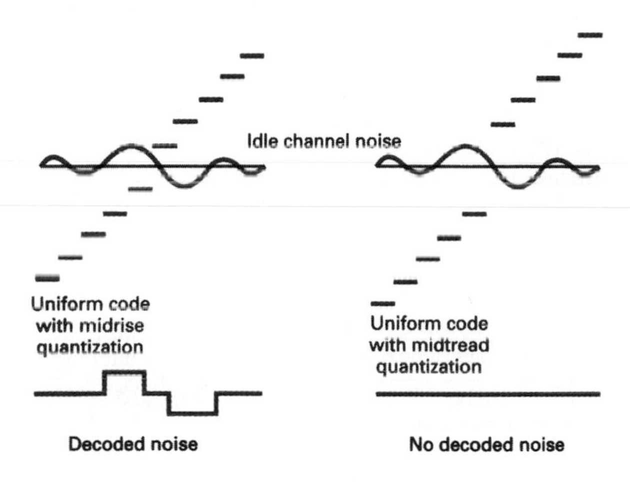

FIGURE 10-12 Idle channel noise

nonlinear analog-to-digital converters; consequently, alternative methods of achieving the same results have been devised.

10-7 IDLE CHANNEL NOISE

During times when there is no analog input signal, the only input to the PAM sampler is random, thermal noise. This noise is called *idle channel noise* and is converted to a PAM sample just as if it were a signal. Consequently, even input noise is quantized by the ADC. Figure 10-12 shows a way to reduce idle channel noise by a method called *midtread quantization.* With midtread quantizing, the first quantization interval is made larger in amplitude than the rest of the steps. Consequently, input noise can be quite large and still be quantized as a positive or negative zero code. As a result, the noise is suppressed during the encoding process.

In the PCM codes described thus far, the lowest-magnitude positive and negative codes have the same voltage range as all the other codes (+ or − one-half the resolution). This is called *midrise quantization.* Figure 10-12 contrasts the idle channel noise transmitted with a midrise PCM code to the idle channel noise transmitted when midtread quantization is used. The advantage of midtread quantization is less idle channel noise. The disadvantage is a larger possible magnitude for Q_e in the lowest quantization interval.

With a folded binary PCM code, residual noise that fluctuates slightly above and below 0 V is converted to either a + or − zero PCM code and, consequently, is eliminated. In systems that do not use the two 0-V assignments, the residual noise could cause the PCM encoder to alternate between the zero code and the minimum + or − code. Consequently, the decoder would reproduce the encoded noise. With a folded binary code, most of the residual noise is inherently eliminated by the encoder.

10-8 CODING METHODS

There are several coding methods used to quantize PAM signals into 2^n levels. These methods are classified according to whether the coding operation proceeds a level at a time, a digit at a time, or a word at a time.

10-8-1 Level-at-a-Time Coding

This type of coding compares the PAM signal to a ramp waveform while a binary counter is being advanced at a uniform rate. When the ramp waveform equals or exceeds the PAM sample, the counter contains the PCM code. This type of coding requires a very fast clock if the number of bits in the PCM code is large. Level-at-a-time coding also requires that 2^n sequential decisions be made for each PCM code generated. Therefore, level-at-a-time coding is generally limited to low-speed applications. Nonuniform coding is achieved by using a nonlinear function as the reference ramp.

10-8-2 Digit-at-a-Time Coding

This type of coding determines each digit of the PCM code sequentially. Digit-at-a-time coding is analogous to a balance where known reference weights are used to determine an unknown weight. Digit-at-a-time coders provide a compromise between speed and complexity. One common kind of digit-at-a-time coder, called a *feedback coder,* uses a successive approximation register (SAR). With this type of coder, the entire PCM code word is determined simultaneously.

10-8-3 Word-at-a-Time Coding

Word-at-a-time coders are flash encoders and are more complex; however, they are more suitable for high-speed applications. One common type of word-at-a-time coder uses multiple threshold circuits. Logic circuits sense the highest threshold circuit sensed by the PAM input signal and produce the approximate PCM code. This method is again impractical for large values of n.

10-9 COMPANDING

Companding is the process of *compressing* and then *expanding.* With companded systems, the higher-amplitude analog signals are compressed (amplified less than the lower-amplitude signals) prior to transmission and then expanded (amplified more than the lower-amplitude signals) in the receiver. Companding is a means of improving the dynamic range of a communications system.

Figure 10-13 illustrates the process of companding. An analog input signal with a dynamic range of 50 dB is compressed to 25 dB prior to transmission and then, in the receiver, expanded back to its original dynamic range of 50 dB. With PCM, companding may be accomplished using analog or digital techniques. Early PCM systems used analog companding, whereas more modern systems use digital companding.

10-9-1 Analog Companding

Historically, analog compression was implemented using specially designed diodes inserted in the analog signal path in a PCM transmitter prior to the sample-and-hold circuit.

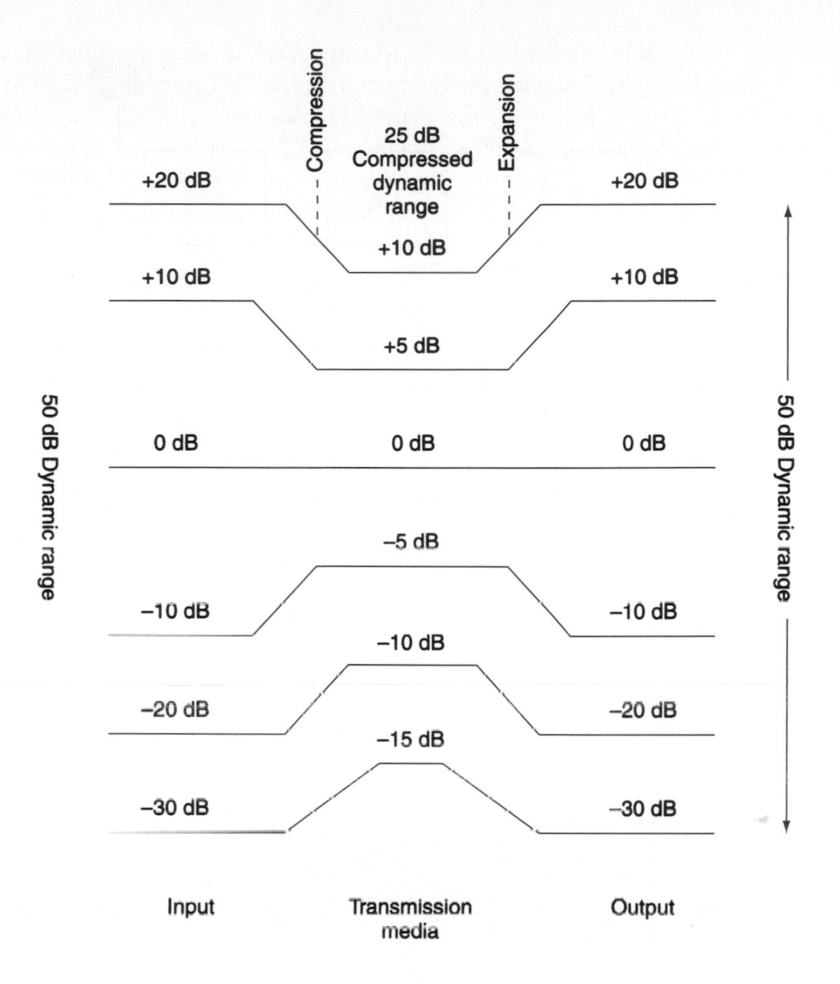

FIGURE 10-13 Basic companding process

Analog expansion was also implemented with diodes that were placed just after the low-pass filter in the PCM receiver.

Figure 10-14 shows the basic process of analog companding. In the transmitter, the dynamic range of the analog signal is compressed, sampled, and then converted to a linear PCM code. In the receiver, the PCM code is converted to a PAM signal, filtered, and then expanded back to its original dynamic range.

Different signal distributions require different companding characteristics. For instance, voice-quality telephone signals require a relatively constant SQR performance over a wide dynamic range, which means that the distortion must be proportional to signal amplitude for all input signal levels. This requires a logarithmic compression ratio, which requires an infinite dynamic range and an infinite number of PCM codes. Of course, this is impossible to achieve. However, there are two methods of analog companding currently being used that closely approximate a logarithmic function and are often called log-PCM codes. The two methods are μ-*law* and the *A-law* companding.

10-9-1-1 μ-Law companding. In the United States and Japan, μ-*law* companding is used. The compression characteristics for μ-*law* is

$$V_{out} = \frac{V_{max} \ln(1 + \mu V_{in}/V_{max})}{\ln(1 + \mu)} \quad \textbf{(10-10)}$$

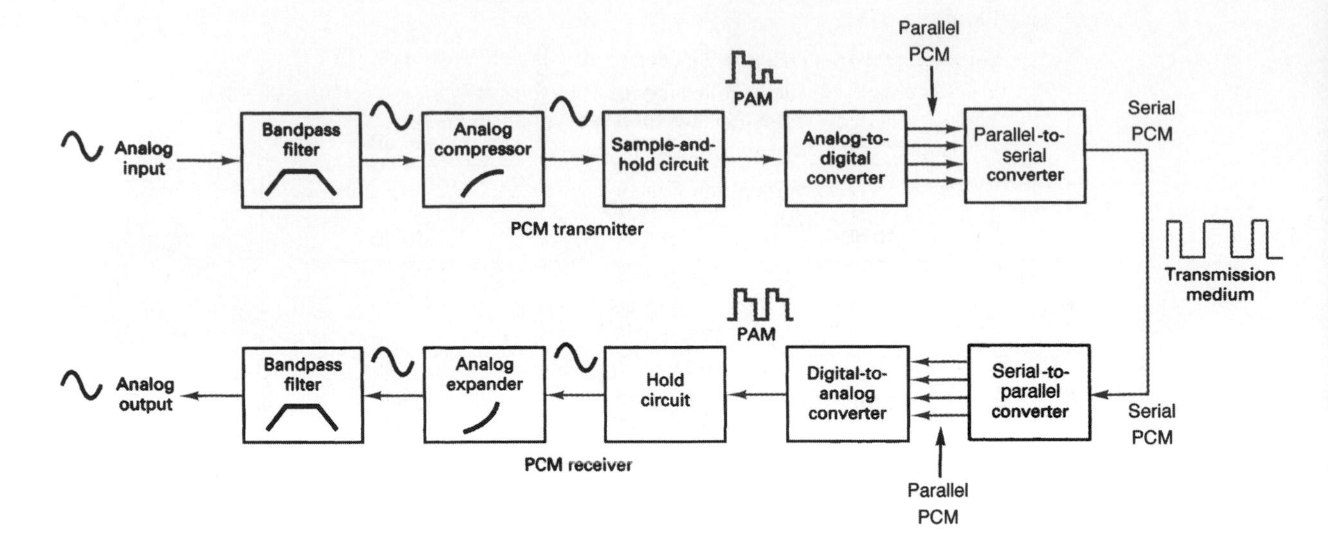

FIGURE 10-14 PCM system with analog companding

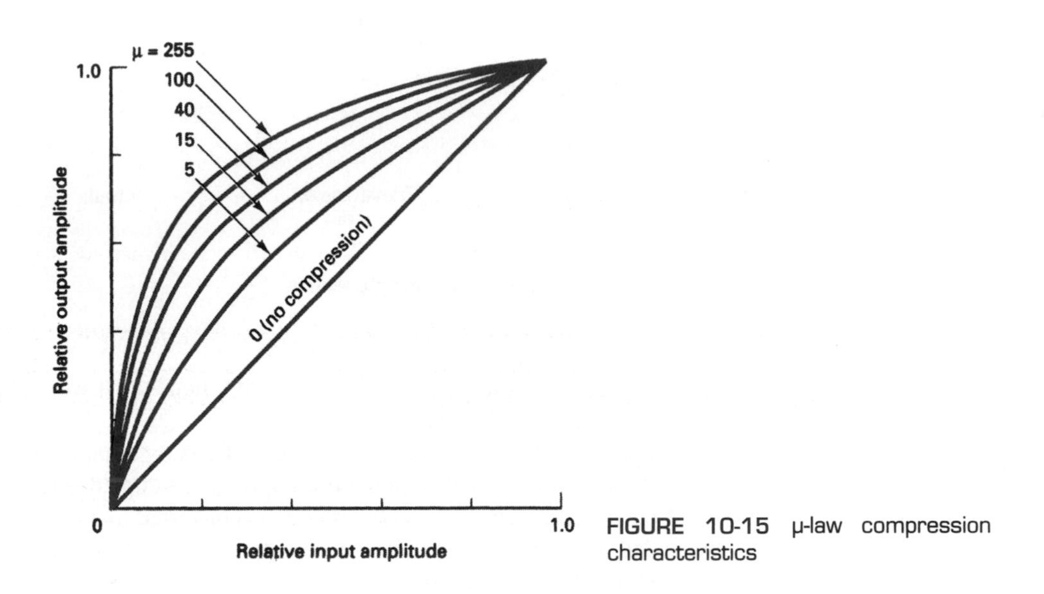

FIGURE 10-15 μ-law compression characteristics

where V_{max} = maximum uncompressed analog input amplitude (volts)
V_{in} = amplitude of the input signal at a particular instant of time (volts)
μ = parameter used to define the amount of compression (unitless)
V_{out} = compressed output amplitude (volts)

Figure 10-15 shows the compression curves for several values of μ. Note that the higher the μ, the more compression. Also note that for $\mu = 0$, the curve is linear (no compression).

The parameter μ determines the range of signal power in which the SQR is relatively constant. Voice transmission requires a minimum dynamic range of 40 dB and a seven-bit PCM code. For a relatively constant SQR and a 40-dB dynamic range, a $\mu \geq 100$ is required. The early Bell System PCM systems used a seven-bit code with a $\mu = 100$. However, the most recent PCM systems use an eight-bit code and a $\mu = 255$.

Example 10-5

For a compressor with a $\mu = 255$, determine

a. The voltage gain for the following relative values of V_{in}: V_{max}, $0.75\ V_{max}$, $0.5\ V_{max}$, and $0.25\ V_{max}$.
b. The compressed output voltage for a maximum input voltage of 4 V.
c. Input and output dynamic ranges and compression.

Solution **a.** Substituting into Equation 10-10, the following voltage gains are achieved for the given input magnitudes:

V_{in}	Compressed Voltage Gain
V_{max}	1.00
$0.75\ V_{max}$	1.26
$0.5\ V_{max}$	1.75
$0.25\ V_{max}$	3.00

b. Using the compressed voltage gains determined in step (a), the output voltage is simply the input voltage times the compression gain:

V_{in}	Voltage Gain	V_{out}
$V_{max} = 4\ V$	1.00	4.00 V
$0.75\ V_{max} = 3\ V$	1.26	3.78 V
$0.50\ V_{max} = 2\ V$	1.75	3.50 V
$0.25\ V_{max} = 1\ V$	3.00	3 00 V

c. Dynamic range is calculated by substituting into Equation 10-4:

$$\text{input dynamic range} = 20 \log\frac{4}{1} = 12\text{ dB}$$

$$\text{output dynamic range} = 20 \log\frac{4}{3} = 2.5\text{ dB}$$

$$\begin{aligned}\text{compression} &= \text{input dynamic range minus output dynamic range} \\ &= 12\text{ dB} - 2.5\text{ dB} = 9.5\text{ dB}\end{aligned}$$

To restore the signals to their original proportions in the receiver, the compressed voltages are expanded by passing them through an amplifier with gain characteristics that are the complement of those in the compressor. For the values given in Example 10-5, the voltage gains in the receiver are as follows:

V_{in}	Expanded Voltage Gain
V_{max}	1.00
$0.75\ V_{max}$	0.79
$0.5\ V_{max}$	0.57
$0.25\ V_{max}$	0.33

The overall circuit gain is simply the product of the compression and expansion factors, which equals one for all input voltage levels. For the values given in Example 10-5,

$$\begin{aligned}V_{in} &= V_{max} & 1 \times 1 &= 1 \\ V_{in} &= 0.75\ V_{max} & 1.26 \times 0.79 &\cong 1 \\ V_{in} &= 0.5\ V_{max} & 1.75 \times 0.57 &\cong 1 \\ V_{in} &= 0.25\ V_{max} & 3 \times 0.33 &\cong 1\end{aligned}$$

10-9-1-2 A-law companding. In Europe, the ITU-T has established *A-law* companding to be used to approximate true logarithmic companding. For an intended dynamic

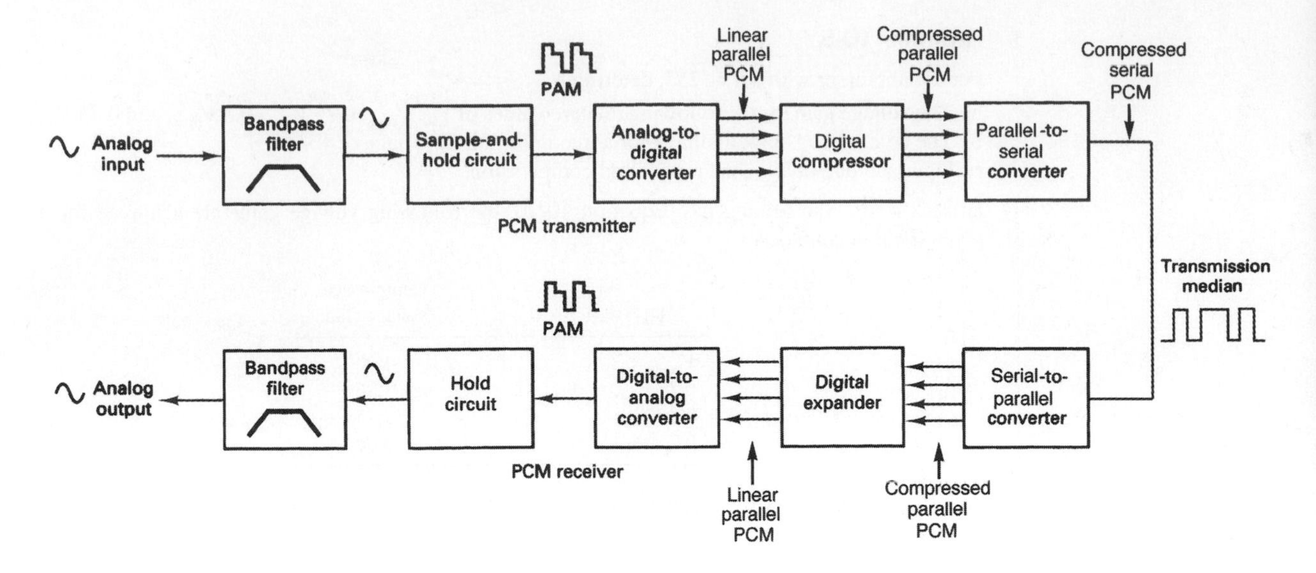

FIGURE 10-16 Digitally companded PCM system

range, *A-law* companding has a slightly flatter SQR than μ-law. *A-law* companding, however, is inferior to μ-law in terms of small-signal quality (idle channel noise). The compression characteristic for *A-law* companding is

$$V_{out} = V_{max}\frac{AV_{in}/V_{max}}{1 + \ln A} \qquad 0 \le \frac{V_{in}}{V_{max}} \le \frac{1}{A} \tag{10-11a}$$

$$= \frac{1 + \ln(AV_{in}/V_{max})}{1 + \ln A} \qquad \frac{1}{A} \le \frac{V_{in}}{V_{max}} \le 1 \tag{10-11b}$$

10-9-2 Digital Companding

Digital companding involves compression in the transmitter after the input sample has been converted to a linear PCM code and then expansion in the receiver prior to PCM decoding. Figure 10-16 shows the block diagram for a digitally companded PCM system.

With digital companding, the analog signal is first sampled and converted to a linear PCM code and then the linear code is digitally compressed. In the receiver, the compressed PCM code is expanded and then decoded (i.e., converted back to analog). The most recent digitally compressed PCM systems use a 12-bit linear PCM code and an eight-bit compressed PCM code. The compression and expansion curves closely resemble the analog μ-*law* curves with a μ = 255 by approximating the curve with a set of eight straight-line segments (segments 0 through 7). The slope of each successive segment is exactly one-half that of the previous segment.

Figure 10-17 shows the 12-bit-to-8-bit digital compression curve for positive values only. The curve for negative values is identical except the inverse. Although there are 16 segments (eight positive and eight negative), this scheme is often called *13-segment compression* because the curve for segments +0, +1, −0, and −1 is a straight line with a constant slope and is considered as one segment.

The digital companding algorithm for a 12-bit linear-to-8-bit compressed code is actually quite simple. The eight-bit compressed code consists of a sign bit, a three-bit segment identifier, and a 10-bit magnitude code that specifies the quantization interval within the specified segment (see Figure 10-18a).

In the μ255-encoding table shown in Figure 10-18b, the bit positions designated with an X are truncated during compression and subsequently lost. Bits designated A, B, C, and

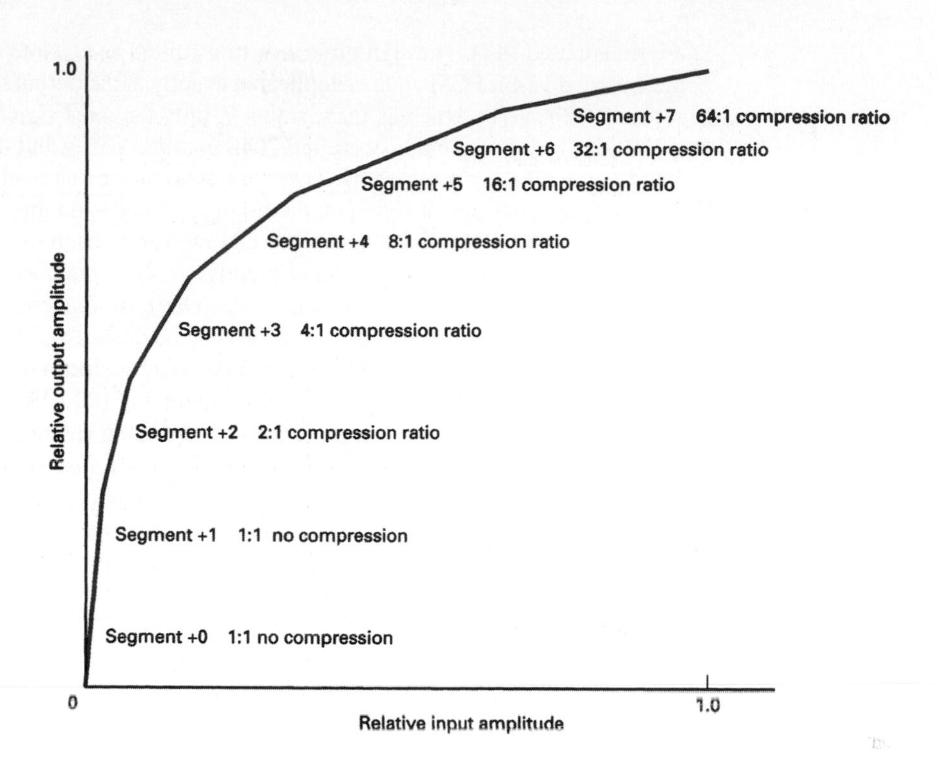

FIGURE 10-17 μ255 compression characteristics (positive values only)

Sign bit 1 = + 0 = −	3-Bit segment identifier	4-Bit quantization interval A B C D
	000 to 111	0000 to 1111

(a)

Transmission media — Transmit → Receive

Encoded PCM

Segment	12-Bit linear code	8-Bit compressed code
0	s0000000ABCD	s000ABCD
1	s0000001ABCD	s001ABCD
2	s000001ABCDX	s010ABCD
3	s0000ABCDXX	s011ABCD
4	s0001ABCDXXX	s100ABCD
5	s001ABCDXXXX	s101ABCD
6	s01ABCDXXXXX	s110ABCD
7	s1ABCDXXXXXX	s111ABCD

(b)

Decoded PCM

8-Bit compressed code	12-Bit recovered code	Segment
s000ABCD	s0000000ABCD	0
s001ABCD	s0000001ABCD	1
s010ABCD	s000001ABCD1	2
s011ABCD	s0000ABCD10	3
s100ABCD	s0001ABCD100	4
s101ABCD	s001ABCD1000	5
s110ABCD	s01ABCD10000	6
s111ABCD	s1ABCD100000	7

(c)

FIGURE 10-18 12-bit-to-8-bit digital companding: (a) 8-bit μ255 compressed code format; (b) μ255 encoding table; (c) μ255 decoding table

D are transmitted as is. The sign bit is also transmitted as is. Note that for segments 0 and 1, the encoded 12-bit PCM code is duplicated exactly at the output of the decoder (compare Figures 10-18b and c), whereas for segment 7, only the most significant six bits are duplicated. With 11 magnitude bits, there are 2048 possible codes, but they are not equally distributed among the eight segments. There are 16 codes in segment 0 and 16 codes in segment 1. In each subsequent segment, the number of codes doubles (i.e., segment 2 has 32 codes; segment 3 has 64 codes, and so on). However, in each of the eight segments, only 16 12-bit codes can be produced. Consequently, in segments 0 and 1, there is no compression (of the 16 possible codes, all 16 can be decoded). In segment 2, there is a compression ratio of 2:1 (of the 32 possible codes, only 16 can be decoded). In segment 3, there is a 4:1 compression ratio (64 codes to 16 codes). The compression ratio doubles with each successive segment. The compression ratio in segment 7 is 1024/16, or 64:1.

The compression process is as follows. The analog signal is sampled and converted to a linear 12-bit sign-magnitude code. The sign bit is transferred directly to an eight-bit compressed code. The segment number in the eight-bit code is determined by counting the number of leading 0s in the 11-bit magnitude portion of the linear code beginning with the most significant bit. Subtract the number of leading 0s (not to exceed 7) from 7. The result is the segment number, which is converted to a three-bit binary number and inserted into the eight-bit compressed code as the segment identifier. The four magnitude bits (A, B, C, and D) represent the quantization interval (i.e., subsegments) and are substituted into the least significant four bits of the 8-bit compressed code.

Essentially, segments 2 through 7 are subdivided into smaller subsegments. Each segment consists of 16 subsegments, which correspond to the 16 conditions possible for bits A, B, C, and D (0000 to 1111). In segment 2, there are two codes per subsegment. In segment 3, there are four. The number of codes per subsegment doubles with each subsequent segment. Consequently, in segment 7, each subsegment has 64 codes.

Figure 10-19 shows the breakdown of segments versus subsegments for segments 2, 5, and 7. Note that in each subsegment, all 12-bit codes, once compressed and expanded, yield a single 12-bit code. In the decoder, the most significant of the truncated bits is reinserted as a logic 1. The remaining truncated bits are reinserted as 0s. This ensures that the maximum magnitude of error introduced by the compression and expansion process is minimized. Essentially, the decoder guesses what the truncated bits were prior to encoding. The most logical guess is halfway between the minimum- and maximum-magnitude codes. For example, in segment 6, the five least significant bits are truncated during compression; therefore, in the receiver, the decoder must try to determine what those bits were. The possibilities include any code between 00000 and 11111. The logical guess is 10000, approximately half the maximum magnitude. Consequently, the maximum compression error is slightly more than one-half the maximum magnitude for that segment.

Example 10-6

Determine the 12-bit linear code, the eight-bit compressed code, the decoded 12-bit code, the quantization error, and the compression error for a resolution of 0.01 V and analog sample voltages of (a) +0.053 V, (b) −0.318 V, and (c) +10.234 V

Solution **a.** To determine the 12-bit linear code, simply divide the sample voltage by the resolution, round off the quotient, and then convert the result to a 12-bit sign-magnitude code:

$$\frac{+0.053\text{ V}}{+0.01\text{ V}} = +5.3, \text{ which is rounded off to 5 producing a quantization error}$$

$$Q_e = 0.3(0.01\text{ V}) = 0.003\text{ V}$$

```
                                             A B C D
12-bit linear code =   1         0 0 0 0 0 0 0 1 0 1
                       ↓       ←─────────────────────→
                    sign bit   11-bit magnitude bits 00000000101 = 5
                    (1 = +)
```

Segment	12-Bit linear code		12-Bit expanded code	Subsegment
7	s11111111111 . . . . s11111000000	64 : 1	s11111100000	15
7	s11110111111 . . . . s11110000000	64 : 1	s11110100000	14
7	s11101111111 . . . . s11101000000	64 : 1	s11101100000	13
7	s11100111111 . . . . s11100000000	64 : 1	s11100100000	12
7	s11011111111 . . . . s11011000000	64 : 1	s11011100000	11
7	s11010111111 . . . . s11010000000	64 : 1	s11010100000	10
7	s11001111111 . . . . s11001000000	64 : 1	s11001100000	9
7	s11000111111 . . . . s11000000000	64 : 1	s11000100000	8
7	s10111111111 . . . . s10111000000	64 : 1	s10111100000	7
7	s10110111111 . . . . s10110000000	64 : 1	s10110100000	6
7	s10101111111 . . . . s10101000000	64 : 1	s10101100000	5
7	s10100111111 . . . . s10100000000	64 : 1	s10100100000	4
7	s10011111111 . . . . s10011000000	64 : 1	s10011100000	3
7	s10010111111 . . . . s10010000000	64 : 1	s10010100000	2
7	s10001111111 . . . . s10001000000	64 : 1	s10001100000	1
7	s10000111111 . . . . s10000000000	64 : 1	s10000100000	0
	s1ABCD-------			

(a)

FIGURE 10-19 12-bit segments divided into subsegments: (a) segment 7; *(Continued)*

To determine the 8-bit compressed code,	1 1 sign bit (+)	<u>0 0 0 0 0 0 0</u> (7 − 7 = 0) unit identifier (segment 0)	<u>0 1 0 1</u> A B C D quantization interval (5)
8-bit compressed code	= 1	0 0 0	0 1 0 1
To determine the 12-bit recovered code, simply reverse the process:	1 s sign bit (+)	<u>0 0 0</u> (000 = segment 0) segment 0 has seven leading 0s	<u>0 1 0 1</u> A B C D quantization interval (0101 = 5)

12-bit recovered code = 1 0 0 0 0 0 0 0 0 1 0 1 = +5

recovered voltage = +5(0.01) = +0.05

Segment	12-Bit linear code		12-Bit expanded code	Subsegment
5	s00111111111 .... s00111110000	16 : 1	s00111111000	15
5	s00111101111 .... s00111100000	16 : 1	s00111101000	14
5	s00111011111 .... s00111010000	16 : 1	s00111011000	13
5	s00111001111 .... s00111000000	16 : 1	s001110010000	12
5	s00110111111 .... s00110110000	16 : 1	s00110111000	11
5	s00110101111 .... s00110100000	16 : 1	s00110101000	10
5	s00110011111 .... s00110010000	16 : 1	s00110011000	9
5	s00110001111 .... s00110000000	16 : 1	s00110001000	8
5	s00101111111 .... s00101110000	16 : 1	s00101111000	7
5	s00101101111 .... s00101100000	16 : 1	s00101101000	6
5	s00101011111 .... s00101010000	16 : 1	s00101011000	5
5	s00101001111 .... s00101000000	16 : 1	s00101001000	4
5	s00100111111 .... s00100110000	16 : 1	s00100111000	3
5	s00100101111 .... s00100100000	16 : 1	s00100101000	2
5	s00100011111 .... s00100010000	16 : 1	s00100011000	1
5	s00100001111 .... s00100000000	16 : 1	s00100001000	0
	s001ABCD----			

(b)

FIGURE 10-19 *(Continued)* (b) segment 5

As Example 10-6 shows, the recovered 12-bit code (+5) is exactly the same as the original 12-bit linear code (+5). Therefore, the decoded voltage (+0.05 V) is the same as the original encoded voltage (+0.5). This is true for all codes in segments 0 and 1. Thus, there is no compression error in segments 0 and 1, and the only error produced is from the quantizing process (for this example, the quantization error $Q_e = 0.003$ V).

b. To determine the 12-bit linear code,

$$\frac{-0.318\text{ V}}{+0.01\text{ V}} = -31.8, \text{ which is rounded off to } -32, \text{ producing a quantization error } Q_e = -0.2\ (0.01\text{ V}) = -0.002\text{ V}$$

$$\text{12-bit linear code} = \underset{\substack{\text{sign bit} \\ (0 = -)}}{0} \quad \underset{\longleftarrow \text{ 11-bit magnitude bits } \longrightarrow}{0\ 0\ 0\ 0\ 0\ \overset{A}{1}\ \overset{B}{0}\ \overset{C}{0}\ \overset{D}{0}\ 0\ 0}$$

Segment	12-Bit linear code		12-Bit expanded code	Subsegment
2	s00000111111 .... s00000111110	2 : 1	s00000111111	15
2	s00000111101 .... s00000111100	2 : 1	s00000111101	14
2	s00000111011 .... s00000111010	2 : 1	s00000111011	13
2	s00000111001 .... s00000111000	2 : 1	s00000111001	12
2	s00000110111 .... s00000110110	2 : 1	s00000110111	11
2	s00000110101 .... s00000110100	2 : 1	s00000110101	10
2	s00000110011 .... s00000110010	2 : 1	s00000110011	9
2	s00000110001 .... s00000110000	2 : 1	s00000110001	8
2	s00000101111 .... s00000101110	2 : 1	s00000101111	7
2	s00000101101 .... s00000101100	2 : 1	s00000101101	6
2	s00000101011 .... s00000101010	2 : 1	s00000101011	5
2	s00000101001 .... s00000101000	2 : 1	s00000101001	4
2	s000000100111 .... s00000100110	2 : 1	s00000100111	3
2	s00000100101 .... s00000100100	2 : 1	s00000100101	2
2	s000000100011 .... s00000100010	2 : 1	s00000100011	1
2	s00000100001 .... s00000100000	2 : 1	s00000100001	0
	s000001ABCD-			

(c)

FIGURE 10-19 *(Continued)* (c) segment 2

To determine the 8-bit compressed code,	0 0 sign bit (−)	0 0 0 0 0 1 (7 − 5 = 2) unit identifier (segment 2)	0 0 0 0 A B C D quantization interval (0)	0 X truncated
eight-bit compressed code	= 0	0 1 0	0 0 0 0	
Again, to determine the 12-bit recovered code, simply reverse the process:	0 sign bit (−)	0 1 0 (7 − 2 = 5) segment 5 has five leading 0s	0 0 0 0 A B C D quantization interval (0000 = 0)	

A B C D

12-bit recovered code = 0 (↑ s) 0 0 0 0 0 1 (↑ inserted) 0 0 0 0 1 (↑ inserted) = −33

decoded voltage = −33(0.1) = −0.33 V

Note the two inserted ones in the recovered 12-bit code. The least significant bit is determined from the decoding table shown in Figure 10-18c. As the figure shows, in the receiver the most significant of the truncated bits is always set (1), and all other truncated bits are cleared (0s). For segment 2 codes, there is only one truncated bit; thus, it is set in the receiver. The inserted 1 in bit position 6 was dropped during the 12-bit-to-8-bit conversion process, as transmission of this bit is redundant because if it were not a 1, the sample would not be in that segment. Consequently, for all segments except segments 0 and 1, a 1 is automatically inserted between the reinserted 0s and the ABCD bits.

For this example, there are two errors: the quantization error and the compression error. The quantization error is due to rounding off the sample voltage in the encoder to the closest PCM code, and the compression error is caused by forcing the truncated bit to be a 1 in the receiver. Keep in mind that the two errors are not always additive, as they could cause errors in the opposite direction and actually cancel each other. The worst-case scenario would be when the two errors were in the same direction and at their maximum values. For this example, the combined error was 0.33 V − 0.318 V = 0.012 V. The worst possible error in segments 0 and 1 is the maximum quantization error, or half the magnitude of the resolution. In segments 2 through 7, the worst possible error is the sum of the maximum quantization error plus the magnitude of the most significant of the truncated bits.

c. To determine the 12-bit linear code,

$$\frac{+10.234\text{ V}}{+0.01\text{ V}} = +1023.4, \text{ which is rounded off to 1023, producing a quantization error } Q_e = -0.4(0.01\text{ V}) = -0.004\text{ V}$$

		A B C D
12-bit linear code =	1	0 1 1 1 1 1 1 1 1 1 1
	↓ sign bit (1 = +)	←—— 11-bit magnitude bits ——→
To determine the 8-bit	1	0 1 1 1 1 1 1 1 1 1 1
compressed code,	1	A B C D X X X X X
		truncated
8-bit compressed code =	1	1 1 0 1 1 1 1
To determine the 12-bit	1	1 1 0 1 1 1 1
recovered code, simply	s	segment 6 A B C D
12-bit recovered code =	1 0	1 1 1 1 1 1 0 0 0 0 +1008
		↑ A B C D ↑
	s	inserted inserted

decoded voltage = +1008(0.01) = +10.08 V

The difference between the original 12-bit code and the decoded 12-bit code is

10.23 − 10.08 = 0.15

or

```
 1011 1111 1111
 1011 1111 0000
           1111  = 15(0.01) = 0.15 V
```

For this example, there are again two errors: a quantization error of 0.004 V and a compression error of 0.15 V. The combined error is 10.234 V − 10.08 V = 0.154 V.

10-9-3 Digital Compression Error

As seen in Example 10-6, the magnitude of the compression error is not the same for all samples. However, the maximum percentage error is the same in each segment (other than segments 0 and 1, where there is no compression error). For comparison purposes, the following formula is used for computing the percentage error introduced by digital compression:

$$\%\text{ error} = \frac{\text{12-bit encoded voltage} - \text{12-bit decoded voltage}}{\text{12-bit decoded voltage}} \times 100 \qquad \textbf{(10-12)}$$

Example 10-7

The maximum percentage error will occur for the smallest number in the lowest subsegment within any given segment. Because there is no compression error in segments 0 and 1, for segment 3 the maximum percentage error is computed as follows:

$$\begin{array}{lcccccccccccc} \text{transmit 12-bit code} & s & 0 & 0 & 0 & 0 & 1 & 0 & 0 & 0 & 0 & 0 & 0 \\ \text{receive 12-bit code} & \underline{s} & \underline{0} & \underline{0} & \underline{0} & \underline{0} & \underline{1} & \underline{0} & \underline{0} & \underline{0} & \underline{0} & \underline{1} & \underline{0} \\ & & 0 & 0 & 0 & 0 & 0 & 0 & 0 & 0 & 0 & 1 & 0 \end{array}$$

$$\% \text{ error} = \frac{|1000000 - 1000010|}{1000010} \times 100$$

$$= \frac{|64 - 66|}{66} \times 100 = 3.03\%$$

and for segment 7

$$\begin{array}{lcccccccccccc} \text{transmit 12-bit code} & s & 1 & 0 & 0 & 0 & 0 & 0 & 0 & 0 & 0 & 0 & 0 \\ \text{receive 12-bit code} & \underline{s} & \underline{1} & \underline{0} & \underline{0} & \underline{0} & \underline{0} & \underline{1} & \underline{0} & \underline{0} & \underline{0} & \underline{0} & \underline{0} \\ & & 0 & 0 & 0 & 0 & 0 & 1 & 0 & 0 & 0 & 0 & 0 \end{array}$$

$$\% \text{ error} = \frac{|10000000000 - 10000100000|}{10000100000} \times 100$$

$$= \frac{|1024 - 1056|}{1056} \times 100 = 3.03\%$$

As Example 10-7 shows, the maximum magnitude of error is higher for segment 7; however, the maximum percentage error is the same for segments 2 through 7. Consequently, the maximum SQR degradation is the same for each segment.

Although there are several ways in which the 12-bit-to-8-bit compression and 8-bit-to-12-bit expansion can be accomplished with hardware, the simplest and most economical method is with a lookup table in ROM (read-only memory).

Essentially every function performed by a PCM encoder and decoder is now accomplished with a single integrated-circuit chip called a *codec.* Most of the more recently developed codecs are called *combo* chips, as they include an antialiasing (bandpass) filter, a sample-and-hold circuit, and an analog-to-digital converter in the transmit section and a digital-to-analog converter, a hold circuit, and a bandpass filter in the receive section.

10-10 VOCODERS

The PCM coding and decoding processes described in the preceding sections were concerned primarily with reproducing waveforms as accurately as possible. The precise nature of the waveform was unimportant as long as it occupied the voice-band frequency range. When digitizing speech signals only, special voice encoders/decoders called *vocoders* are often used. To achieve acceptable speech communications, the short-term power spectrum of the speech information is all that must be preserved. The human ear is relatively insensitive to the phase relationship between individual frequency components within a voice waveform. Therefore, vocoders are designed to reproduce only the short-term power spectrum, and the decoded time waveforms often only vaguely resemble the original input signal. Vocoders cannot be used in applications where analog signals other than voice are present, such as output signals from voice-band data modems. Vocoders typically produce *unnatural* sounding speech and, therefore, are generally used for recorded information, such as "wrong number" messages, encrypted voice for transmission over analog telephone circuits, computer output signals, and educational games.

The purpose of a vocoder is to encode the minimum amount of speech information necessary to reproduce a perceptible message with fewer bits than those needed by a conventional encoder/decoder. Vocoders are used primarily in limited bandwidth applications. Essentially, there are three vocoding techniques available: the *channel vocoder,* the *formant vocoder,* and the *linear predictive coder.*

10-10-1 Channel Vocoders

The first channel vocoder was developed by Homer Dudley in 1928. Dudley's vocoder compressed conventional speech waveforms into an analog signal with a total bandwidth of approximately 300 Hz. Present-day digital vocoders operate at less than 2 kbps. Digital channel vocoders use bandpass filters to separate the speech waveform into narrower *subbands.* Each subband is full-wave rectified, filtered, and then digitally encoded. The encoded signal is transmitted to the destination receiver, where it is decoded. Generally speaking, the quality of the signal at the output of a vocoder is quite poor. However, some of the more advanced channel vocoders operate at 2400 bps and can produce a highly intelligible, although slightly synthetic sounding speech.

10-10-2 Formant Vocoders

A formant vocoder takes advantage of the fact that the short-term spectral density of typical speech signals seldom distributes uniformly across the entire voice-band spectrum (300 Hz to 3000 Hz). Instead, the spectral power of most speech energy concentrates at three or four peak frequencies called *formants.* A formant vocoder simply determines the location of these peaks and encodes and transmits only the information with the most significant short-term components. Therefore, formant vocoders can operate at lower bit rates and, thus, require narrower bandwidths. Formant vocoders sometimes have trouble tracking changes in the formants. However, once the formants have been identified, a formant vocoder can transfer intelligible speech at less than 1000 bps.

10-10-3 Linear Predictive Coders

A linear predictive coder extracts the most significant portions of speech information directly from the time waveform rather than from the frequency spectrum as with the channel and formant vocoders. A linear predictive coder produces a time-varying model of the *vocal tract excitation* and transfer function directly from the speech waveform. At the receive end, a *synthesizer* reproduces the speech by passing the specified excitation through a mathematical model of the vocal tract. Linear predictive coders provide more natural sounding speech than either the channel or the formant vocoder. Linear predictive coders typically encode and transmit speech at between 1.2 kbps and 2.4 kbps.

10-11 PCM LINE SPEED

Line speed is simply the data rate at which serial PCM bits are clocked out of the PCM encoder onto the transmission line. Line speed is dependent on the sample rate and the number of bits in the compressed PCM code. Mathematically, line speed is

$$\text{line speed} = \frac{\text{samples}}{\text{second}} \times \frac{\text{bits}}{\text{sample}} \qquad \textbf{(10-13)}$$

where line speed = the transmission rate in bits per second
samples/second = sample rate (f_s)
bits/sample = number of bits in the compressed PCM code

Example 10-8

For a single-channel PCM system with a sample rate f_s = 6000 samples per second and a seven-bit compressed PCM code, determine the line speed:

Solution

$$\text{line speed} = \frac{6000 \text{ samples}}{\text{second}} \times \frac{7 \text{ bits}}{\text{sample}}$$
$$= 42{,}000 \text{ bps}$$

10-12 DELTA MODULATION PCM

Delta modulation uses a single-bit PCM code to achieve digital transmission of analog signals. With conventional PCM, each code is a binary representation of both the sign and the magnitude of a particular sample. Therefore, multiple-bit codes are required to represent the many values that the sample can be. With delta modulation, rather than transmit a coded representation of the sample, only a single bit is transmitted, which simply indicates whether that sample is larger or smaller than the previous sample. The algorithm for a delta modulation system is quite simple. If the current sample is smaller than the previous sample, a logic 0 is transmitted. If the current sample is larger than the previous sample, a logic 1 is transmitted.

10-12-1 Delta Modulation Transmitter

Figure 10-20 shows a block diagram of a delta modulation transmitter. The input analog is sampled and converted to a PAM signal, which is compared with the output of the DAC. The output of the DAC is a voltage equal to the regenerated magnitude of the previous sample, which was stored in the up–down counter as a binary number. The up–down counter is incremented or decremented depending on whether the previous sample is larger or smaller than the current sample. The up–down counter is clocked at a rate equal to the sample rate. Therefore, the up–down counter is updated after each comparison.

Figure 10-21 shows the ideal operation of a delta modulation encoder. Initially, the up–down counter is zeroed, and the DAC is outputting 0 V. The first sample is taken, converted to a PAM signal, and compared with zero volts. The output of the comparator is a

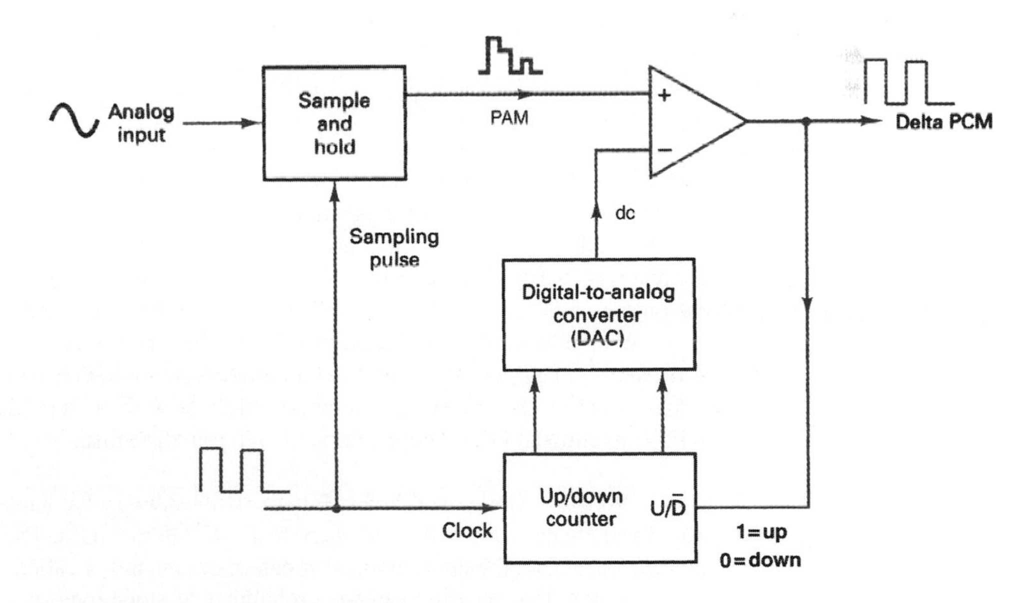

FIGURE 10-20 Delta modulation transmitter

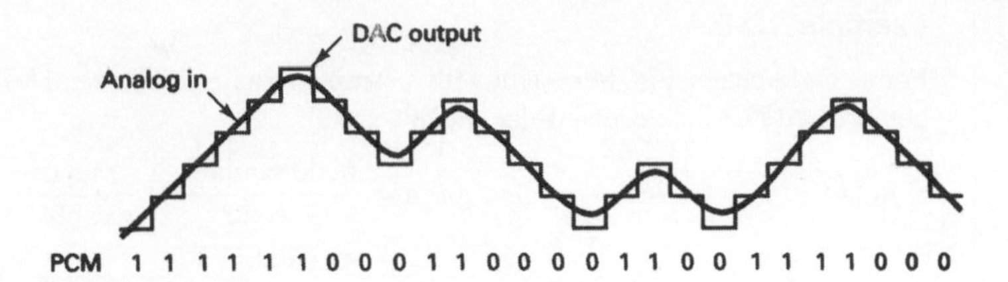

FIGURE 10-21 Ideal operation of a delta modulation encoder

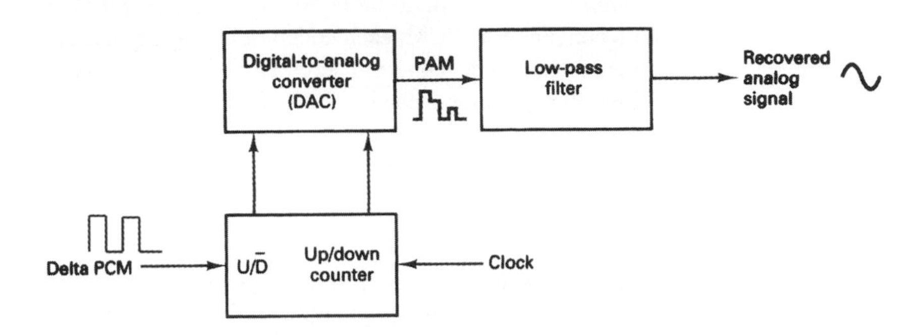

FIGURE 10-22 Delta modulation receiver

logic 1 condition (+V), indicating that the current sample is larger in amplitude than the previous sample. On the next clock pulse, the up–down counter is incremented to a count of 1. The DAC now outputs a voltage equal to the magnitude of the minimum step size (resolution). The steps change value at a rate equal to the clock frequency (sample rate). Consequently, with the input signal shown, the up–down counter follows the input analog signal up until the output of the DAC exceeds the analog sample; then the up–down counter will begin counting down until the output of the DAC drops below the sample amplitude. In the idealized situation (shown in Figure 10-21), the DAC output follows the input signal. Each time the up–down counter is incremented, a logic 1 is transmitted, and each time the up–down counter is decremented, a logic 0 is transmitted.

10-12-2 Delta Modulation Receiver

Figure 10-22 shows the block diagram of a delta modulation receiver. As you can see, the receiver is almost identical to the transmitter except for the comparator. As the logic 1s and 0s are received, the up–down counter is incremented or decremented accordingly. Consequently, the output of the DAC in the decoder is identical to the output of the DAC in the transmitter.

With delta modulation, each sample requires the transmission of only one bit; therefore, the bit rates associated with delta modulation are lower than conventional PCM systems. However, there are two problems associated with delta modulation that do not occur with conventional PCM: slope overload and granular noise.

10-12-2-1 Slope overload. Figure 10-23 shows what happens when the analog input signal changes at a faster rate than the DAC can maintain. The slope of the analog signal is greater than the delta modulator can maintain and is called *slope overload.* Increasing the clock frequency reduces the probability of slope overload occurring. Another way to prevent slope overload is to increase the magnitude of the minimum step size.

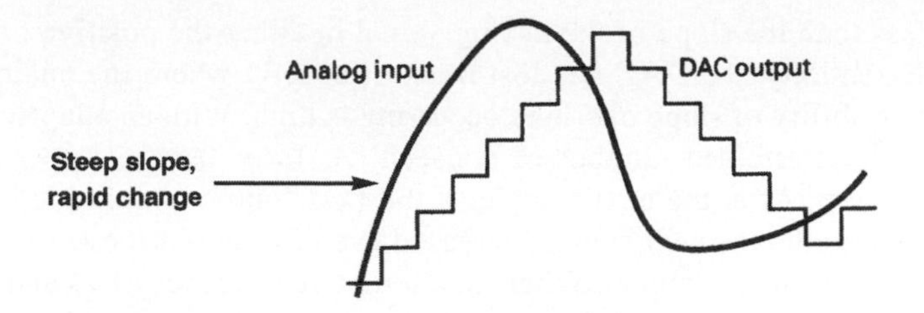

FIGURE 10-23 Slope overload distortion

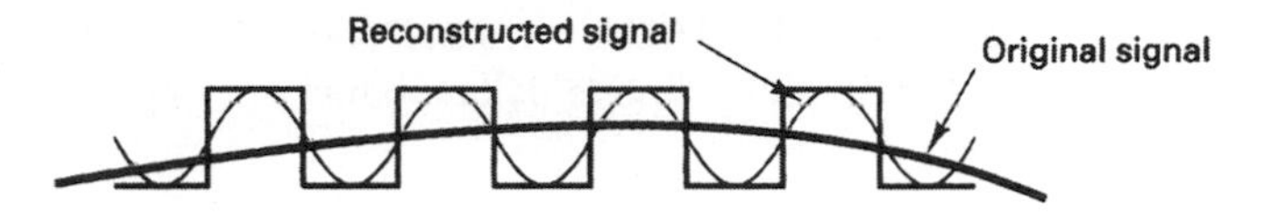

FIGURE 10-24 Granular noise

10-12-2-2 Granular noise. Figure 10-24 contrasts the original and reconstructed signals associated with a delta modulation system. It can be seen that when the original analog input signal has a relatively constant amplitude, the reconstructed signal has variations that were not present in the original signal. This is called *granular noise.* Granular noise in delta modulation is analogous to quantization noise in conventional PCM.

Granular noise can be reduced by decreasing the step size. Therefore, to reduce the granular noise, a small resolution is needed, and to reduce the possibility of slope overload occurring, a large resolution is required. Obviously, a compromise is necessary.

Granular noise is more prevalent in analog signals that have gradual slopes and whose amplitudes vary only a small amount. Slope overload is more prevalent in analog signals that have steep slopes or whose amplitudes vary rapidly.

10-13 ADAPTIVE DELTA MODULATION PCM

Adaptive delta modulation is a delta modulation system where the step size of the DAC is automatically varied, depending on the amplitude characteristics of the analog input signal. Figure 10-25 shows how an adaptive delta modulator works. When the output of the transmitter is a string of consecutive 1s or 0s, this indicates that the slope of the DAC output is

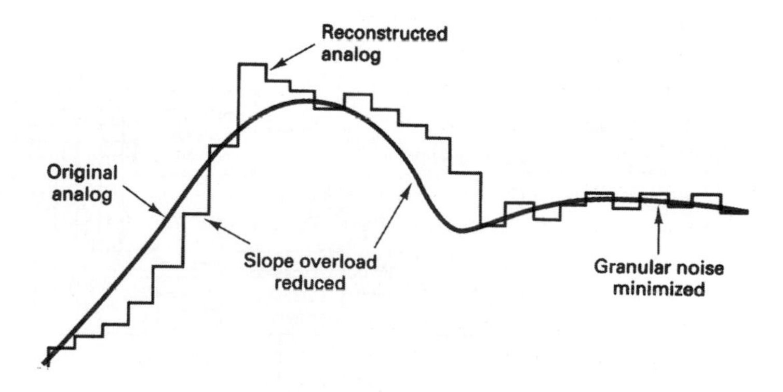

FIGURE 10-25 Adaptive delta modulation

less than the slope of the analog signal in either the positive or the negative direction. Essentially, the DAC has lost track of exactly where the analog samples are, and the possibility of slope overload occurring is high. With an adaptive delta modulator, after a predetermined number of consecutive 1s or 0s, the step size is automatically increased. After the next sample, if the DAC output amplitude is still below the sample amplitude, the next step is increased even further until eventually the DAC catches up with the analog signal. When an alternative sequence of 1s and 0s is occurring, this indicates that the possibility of granular noise occurring is high. Consequently, the DAC will automatically revert to its minimum step size and, thus, reduce the magnitude of the noise error.

A common algorithm for an adaptive delta modulator is when three consecutive 1s or 0s occur, the step size of the DAC is increased or decreased by a factor of 1.5. Various other algorithms may be used for adaptive delta modulators, depending on particular system requirements.

10-14 DIFFERENTIAL PCM

In a typical PCM-encoded speech waveform, there are often successive samples taken in which there is little difference between the amplitudes of the two samples. This necessitates transmitting several identical PCM codes, which is redundant. Differential pulse code modulation (DPCM) is designed specifically to take advantage of the sample-to-sample redundancies in typical speech waveforms. With DPCM, the difference in the amplitude of two successive samples is transmitted rather than the actual sample. Because the range of sample differences is typically less than the range of individual samples, fewer bits are required for DPCM than conventional PCM.

Figure 10-26 shows a simplified block diagram of a DPCM transmitter. The analog input signal is bandlimited to one-half the sample rate, then compared with the preceding accumulated signal level in the differentiator. The output of the differentiation is the difference between the two signals. The difference is PCM encoded and transmitted. The ADC operates the same as in a conventional PCM system, except that it typically uses fewer bits per sample.

Figure 10-27 shows a simplified block diagram of a DPCM receiver. Each received sample is converted back to analog, stored, and then summed with the next sample received. In the receiver shown in Figure 10-27, the integration is performed on the analog signals, although it could also be performed digitally.

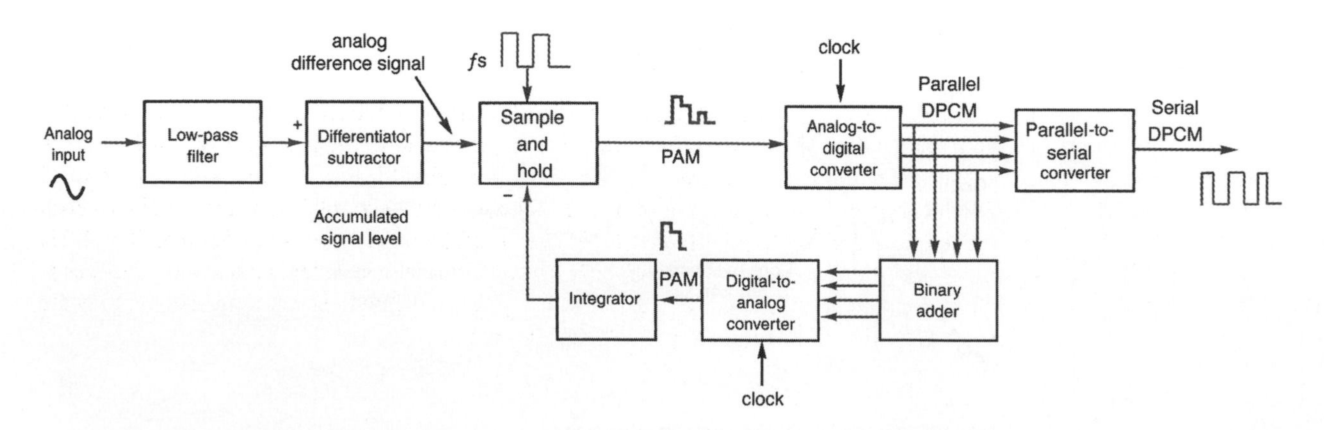

FIGURE 10-26 DPCM transmitter

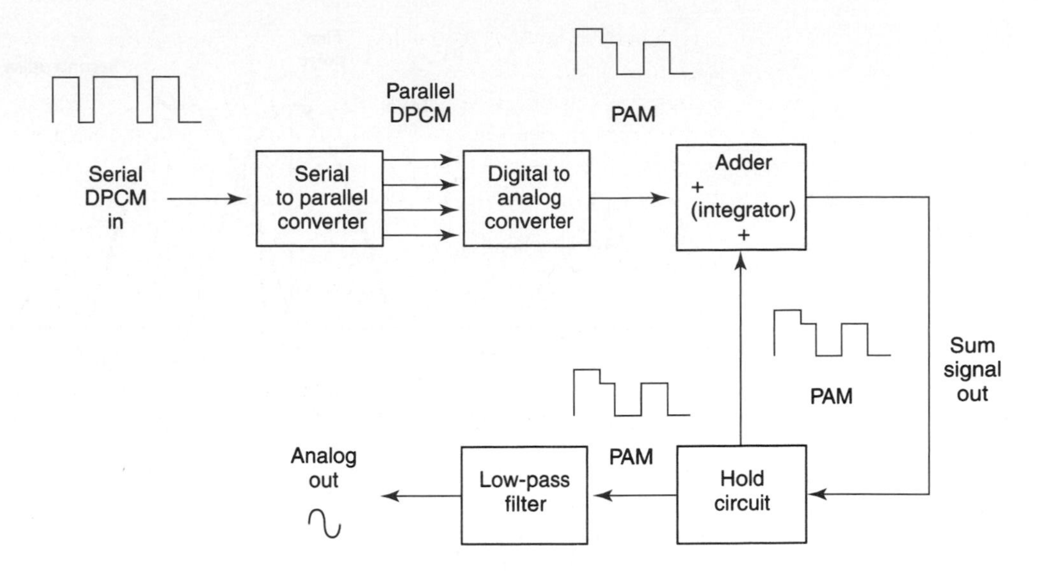

FIGURE 10-27 DPCM receiver

10-15 PULSE TRANSMISSION

All digital carrier systems involve the transmission of pulses through a medium with a finite bandwidth. A highly selective system would require a large number of filter sections, which is impractical. Therefore, practical digital systems generally utilize filters with bandwidths that are approximately 30% or more in excess of the ideal Nyquist bandwidth. Figure 10-28a shows the typical output waveform from a *bandlimited* communications channel when a narrow pulse is applied to its input. The figure shows that bandlimiting a pulse causes the energy from the pulse to be spread over a significantly longer time in the form of *secondary lobes*. The secondary lobes are called *ringing tails*. The output frequency spectrum corresponding to a rectangular pulse is referred to as a (sin x)/x response and is given as

$$f(\omega) = (T)\frac{\sin(\omega T/2)}{\omega T/2} \qquad \textbf{(10-14)}$$

where $\omega = 2\pi f$ (radians)
T = pulse width (seconds)

Figure 10-28b shows the distribution of the total spectrum power. It can be seen that approximately 90% of the signal power is contained within the first *spectral null* (i.e., $f = 1/T$). Therefore, the signal can be confined to a bandwidth $B = 1/T$ and still pass most of the energy from the original waveform. In theory, only the amplitude at the middle of each pulse interval needs to be preserved. Therefore, if the bandwidth is confined to $B = 1/2T$, the maximum signaling rate achievable through a low-pass filter with a specified bandwidth without causing excessive distortion is given as the Nyquist rate and is equal to twice the bandwidth. Mathematically, the Nyquist rate is

$$R = 2B \qquad \textbf{(10-15)}$$

where R = signaling rate = $1/T$
B = specified bandwidth

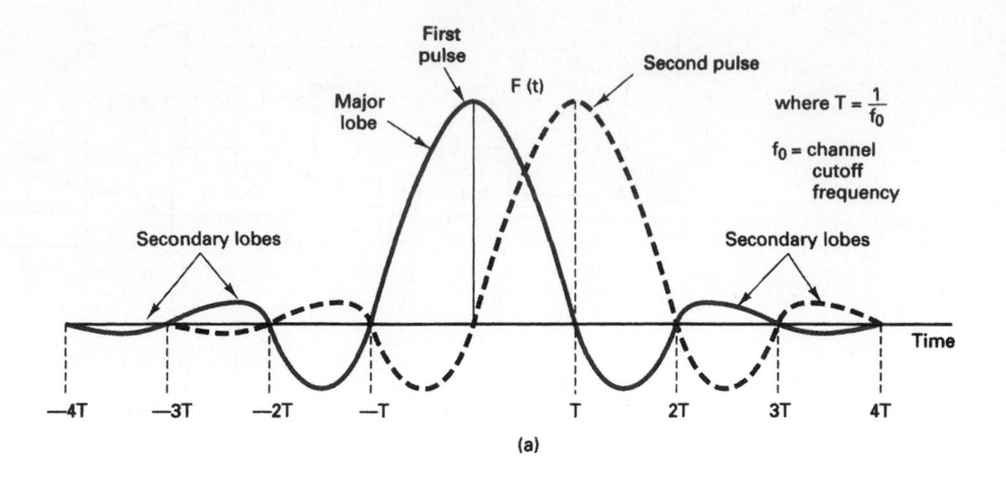

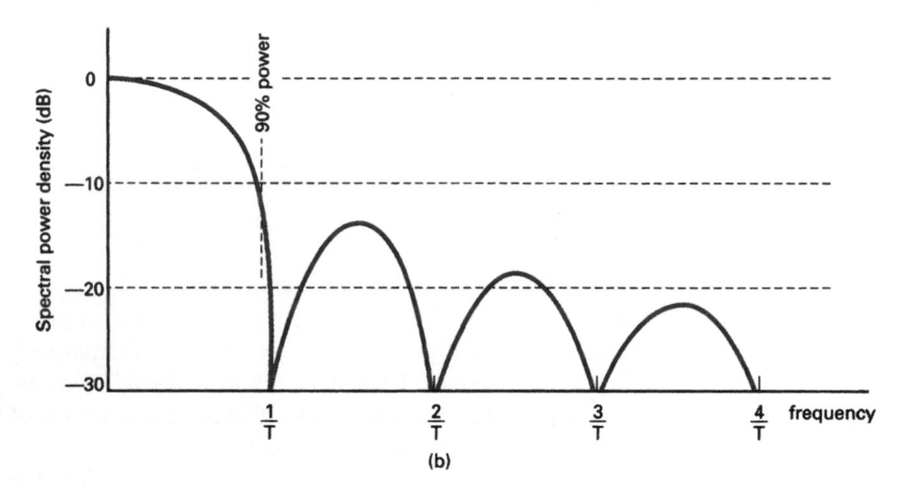

FIGURE 10-28 Pulse response: (a) typical pulse response of a bandlimited filter; (b) spectrum of square pulse with duration 1/T

10-15-1 Intersymbol Interference

Figure 10-29 shows the input signal to an ideal minimum bandwidth, low-pass filter. The input signal is a random, binary nonreturn-to-zero (NRZ) sequence. Figure 10-29b shows the output of a low-pass filter that does not introduce any phase or amplitude distortion. Note that the output signal reaches its full value for each transmitted pulse at precisely the center of each sampling interval. However, if the low-pass filter is imperfect (which in reality it will be), the output response will more closely resemble that shown in Figure 10-29c. At the sampling instants (i.e., the center of the pulses), the signal does not always attain the maximum value. The ringing tails of several pulses have *overlapped,* thus interfering with the *major pulse lobe.* Assuming no time delays through the system, energy in the form of spurious responses from the third and fourth impulses from one pulse appears during the sampling instant ($T = 0$) of another pulse. This interference is commonly called *intersymbol interference,* or simply ISI. ISI is an important consideration in the transmission of pulses over circuits with a limited bandwidth and a nonlinear phase response. Simply stated, rectangular pulses will not remain rectangular in less than an infinite bandwidth. The narrower the bandwidth, the more rounded the pulses. If the phase distortion is excessive, the pulse will *tilt* and, consequently, affect the next pulse. When pulses from more than

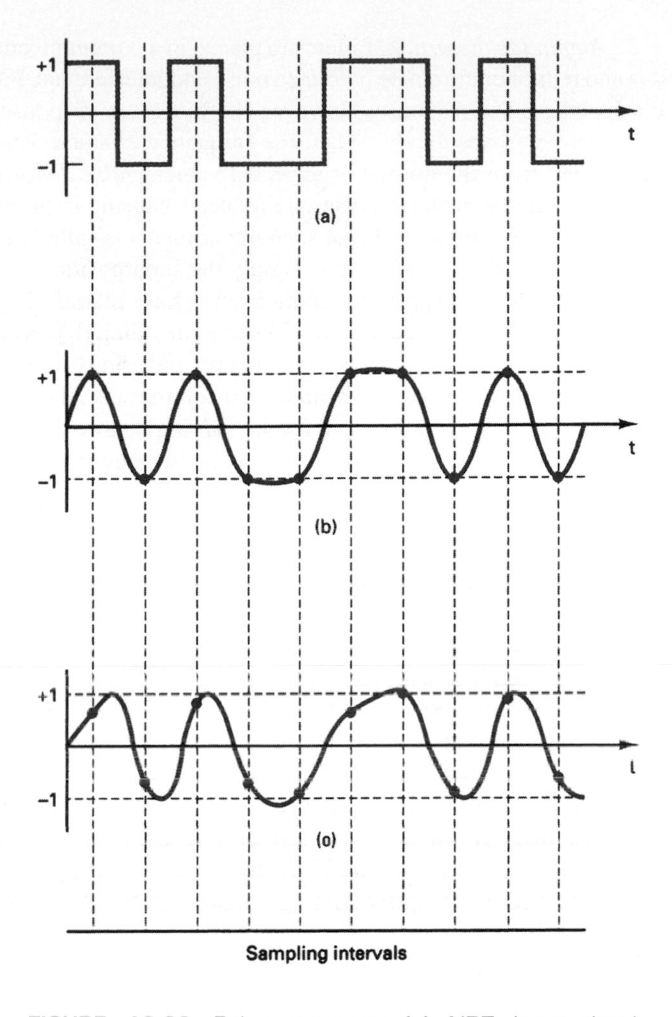

FIGURE 10-29 Pulse response: (a) NRZ input signal; (b) output from a perfect filter; (c) output from an imperfect filter

one source are multiplexed together, the amplitude, frequency, and phase responses become even more critical. ISI causes *crosstalk* between channels that occupy adjacent time slots in a time-division-multiplexed carrier system. Special filters called *equalizers* are inserted in the transmission path to "equalize" the distortion for all frequencies, creating a uniform transmission medium and reducing transmission impairments. The four primary causes of ISI are as follows:

1. *Timing inaccuracies.* In digital transmission systems, transmitter timing inaccuracies cause intersymbol interference if the rate of transmission does not conform to the *ringing frequency* designed into the communications channel. Generally, timing inaccuracies of this type are insignificant. Because receiver clocking information is derived from the received signals, which are contaminated with noise, inaccurate sample timing is more likely to occur in receivers than in transmitters.

2. *Insufficient bandwidth.* Timing errors are less likely to occur if the transmission rate is well below the channel bandwidth (i.e., the Nyquist bandwidth is significantly below the channel bandwidth). As the bandwidth of a communications channel is reduced, the ringing frequency is reduced, and intersymbol interference is more likely to occur.

3. *Amplitude distortion.* Filters are placed in a communications channel to bandlimit signals and reduce or eliminate predicted noise and interference. Filters are also used to produce a specific pulse response. However, the frequency response of a channel cannot always be predicted absolutely. When the frequency characteristics of a communications channel depart from the normal or expected values, *pulse distortion* results. Pulse distortion occurs when the peaks of pulses are reduced, causing improper ringing frequencies in the time domain. Compensation for such impairments is called *amplitude equalization.*

4. *Phase distortion.* A pulse is simply the superposition of a series of harmonically related sine waves with specific amplitude and phase relationships. Therefore, if the relative phase relations of the individual sine waves are altered, phase distortion occurs. Phase distortion occurs when frequency components undergo different amounts of time delay while propagating through the transmission medium. Special delay equalizers are placed in the transmission path to compensate for the varying delays, thus reducing the phase distortion. Phase equalizers can be manually adjusted or designed to automatically adjust themselves to varying transmission characteristics.

10-15-2 Eye Patterns

The performance of a digital transmission system depends, in part, on the ability of a repeater to regenerate the original pulses. Similarly, the quality of the regeneration process depends on the decision circuit within the repeater and the quality of the signal at the input to the decision circuit. Therefore, the performance of a digital transmission system can be measured by displaying the received signal on an oscilloscope and triggering the time base at the data rate. Thus, all waveform combinations are superimposed over adjacent signaling intervals. Such a display is called an *eye pattern* or *eye diagram.* An eye pattern is a convenient technique for determining the effects of the degradations introduced into the pulses as they travel to the regenerator. The test setup to display an eye pattern is shown in Figure 10-30. The received pulse stream is fed to the vertical input of the oscilloscope, and the symbol clock is fed to the external trigger input, while the sweep rate is set approximately equal to the symbol rate.

Figure 10-31 shows an eye pattern generated by a symmetrical waveform for *ternary* signals in which the individual pulses at the input to the regenerator have a cosine-squared shape. In an m-level system, there will be $m - 1$ separate eyes. The vertical lines labeled $+1$, 0, and -1 correspond to the ideal received amplitudes. The horizontal lines, separated by the signaling interval, T, correspond to the ideal *decision times.* The decision levels for the regenerator are represented by *crosshairs.* The vertical hairs represent the decision time, whereas the horizontal hairs represent the decision level. The eye pattern shows the quality of shaping and timing and discloses any noise and errors that might be present in the line equalization. The eye opening (the area in the middle of the eye pattern) defines a boundary within which no waveform *trajectories* can exist under any code-pattern condition. The eye

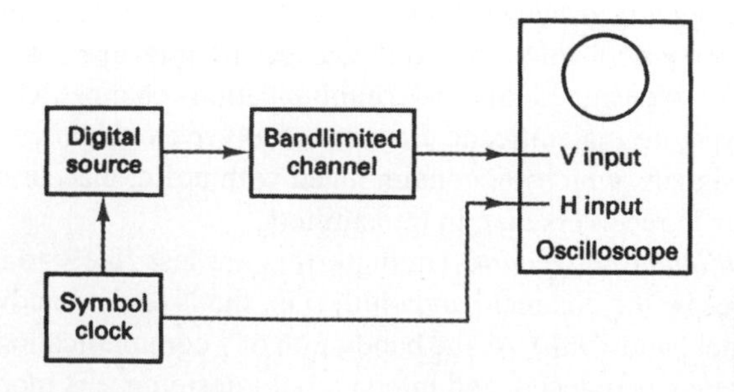

FIGURE 10-30 Eye diagram measurement setup

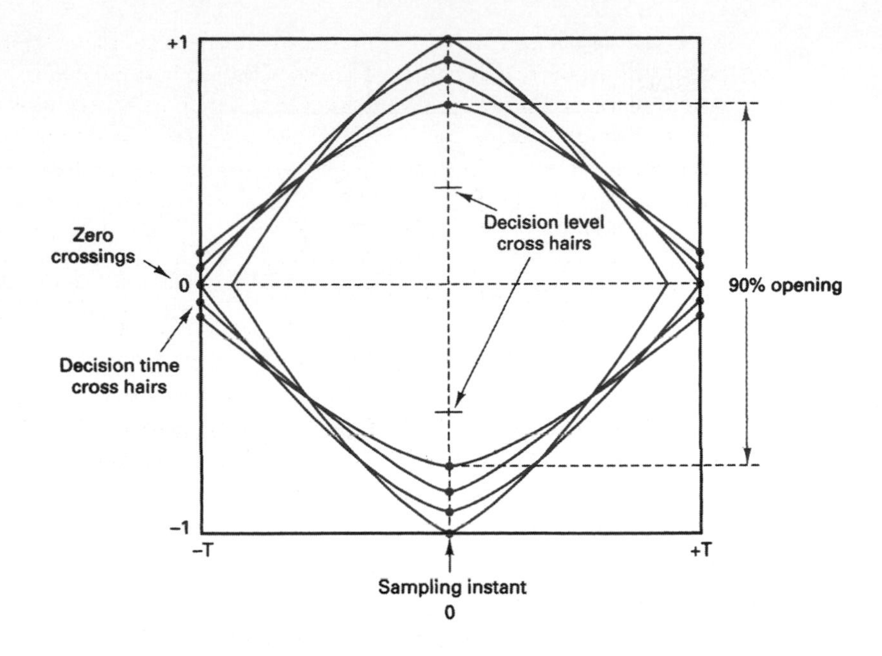

FIGURE 10-31 Eye diagram

opening is a function of the number of code levels and the intersymbol interference caused by the ringing tails of any preceding or succeeding pulses. To regenerate the pulse sequence without error, the eye must be open (i.e., a decision area must exist), and the decision crosshairs must be within the open area. The effect of pulse degradation is a reduction in the size of the ideal eye. In Figure 10-31, it can be seen that at the center of the eye (i.e., the sampling instant) the opening is about 90%, indicating only minor ISI degradation due to filtering imperfections. The small degradation is due to the nonideal Nyquist amplitude and phase characteristics of the transmission system. Mathematically, the ISI degradation is

$$\text{ISI} = 20 \log \frac{h}{H} \tag{10-16}$$

where H = ideal vertical opening (cm)
h = degraded vertical opening (cm)

For the eye diagram shown in Figure 10-31,

$$20 \log \frac{90}{100} = 0.915 \text{ dB (ISI degradation)}$$

In Figure 10-31, it can also be seen that the overlapping signal pattern does not cross the horizontal zero line at exact integer multiples of the symbol clock. This is an impairment known as *data transition jitter.* This jitter has an effect on the symbol timing (clock) recovery circuit and, if excessive, may significantly degrade the performance of cascaded regenerative sections.

10-16 SIGNAL POWER IN BINARY DIGITAL SIGNALS

Because binary digital signals can originate from literally scores of different types of data sources, it is impossible to predict which patterns or sequences of bits are most likely to occur over a given period of time in a given system. Thus, for signal analysis purposes, it is generally assumed that there is an equal probability of the occurrence of a 1 and a 0. Therefore,

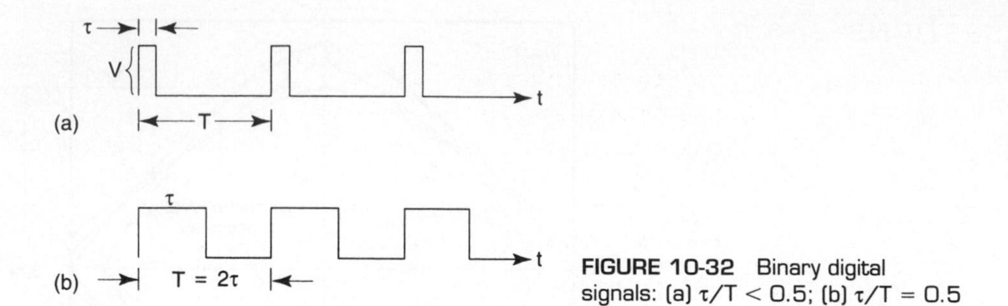

FIGURE 10-32 Binary digital signals: (a) $\tau/T < 0.5$; (b) $\tau/T = 0.5$

power can be averaged over an entire message duration, and the signal can be modeled as a continuous sequence of alternating 1s and 0s as shown in Figure 10-32. Figure 10-32a shows a stream of rectangularly shaped pulses with a pulse width-to-pulse duration ratio τ/T less than 0.5, and Figure 10-32b shows a stream of square wave pulses with a τ/T ratio of 0.5.

The normalized ($R - 1$) average power is derived for signal $f(t)$ from

$$\overline{P} = \lim_{T \to x} \frac{1}{T} \int_{-T/2}^{T/2} [f(t)]^2 \, dt \tag{10-17}$$

where T is the period of integration. If $f(t)$ is a periodic signal with period T_0, then Equation 10-17 reduces to

$$\overline{P} = \frac{1}{T_0} \int_{-T_0/2}^{T_0/2} [v(t)]^2 \, dt \tag{10-18}$$

If rectangular pulses of amplitude V with a τ/T ratio of 0.5 begin at $t = 0$, then

$$v(t) = \begin{cases} V & 0 \le t \le \tau \\ 0 & \tau < t \le T \end{cases} \tag{10-19}$$

Thus, from Equation 10-18,

$$\overline{P} = \frac{1}{T_0} \int_0^T (V)^2 \, dt = \frac{1}{T_0} V^2 t|_0^\tau$$

$$= \frac{\tau}{T_0} V^2 \tag{10-20}$$

and

$$\overline{P} = \left(\frac{\tau}{T}\right) \frac{V^2}{R}$$

Because the effective rms value of a periodic wave is found from $P = (V_{rms})^2/R$, the rms voltage for a rectangular pulse is

$$V_{rms} = \sqrt{\frac{\tau}{T}}(V) \tag{10-21}$$

Because $\overline{P} = (V_{rms})^2/R$, $\overline{P} = (\sqrt{\tau/T}V)^2/R = (\tau V^2)/(TR)$.

With the square wave shown in Figure 10-32, $\tau/T = 0.5$, therefore, $\overline{P} = V^2/2R$. Thus, the rms voltage for the square wave is the same as for sine waves, $V_{rms} = V/\sqrt{2}$.

QUESTIONS

10-1. Contrast the advantages and disadvantages of digital transmission.

10-2. What are the four most common methods of pulse modulation?

10-3. Which method listed in question 10-2 is the only form of pulse modulation that is used in a digital transmission system? Explain.

10-4. What is the purpose of the sample-and-hold circuit?

10-5. Define *aperture* and *acquisition time.*

10-6. What is the difference between natural and flat-top sampling?

10-7. Define *droop.* What causes it?

10-8. What is the Nyquist sampling rate?

10-9. Define and state the causes of foldover distortion.

10-10. Explain the difference between a magnitude-only code and a sign-magnitude code.

10-11. Explain overload distortion.

10-12. Explain quantizing.

10-13. What is quantization range? Quantization error?

10-14. Define *dynamic range.*

10-15. Explain the relationship between dynamic range, resolution, and the number of bits in a PCM code.

10-16. Explain coding efficiency.

10-17. What is SQR? What is the relationship between SQR, resolution, dynamic range, and the number of bits in a PCM code?

10-18. Contrast linear and nonlinear PCM codes.

10-19. Explain idle channel noise.

10-20. Contrast midtread and midrise quantization.

10-21. Define *companding.*

10-22. What does the parameter μ determine?

10-23. Briefly explain the process of digital companding.

10-24. What is the effect of digital compression on SQR, resolution, quantization interval, and quantization noise?

10-25. Contrast delta modulation PCM and standard PCM.

10-26. Define *slope overload* and *granular noise.*

10-27. What is the difference between adaptive delta modulation and conventional delta modulation?

10-28. Contrast differential and conventional PCM.

PROBLEMS

10-1. Determine the Nyquist sample rate for a maximum analog input frequency of

a. 4 kHz.
b. 10 kHz.

10-2. For the sample-and-hold circuit shown in Figure 10-5a, determine the largest-value capacitor that can be used. Use the following parameters: an output impedance for $Z_1 = 20\ \Omega$, an on resistance of Q_1 of $20\ \Omega$, an acquisition time of $10\ \mu s$, a maximum output current from Z_1 of 20 mA, and an accuracy of 1%.

10-3. For a sample rate of 20 kHz, determine the maximum analog input frequency.

10-4. Determine the alias frequency for a 14-kHz sample rate and an analog input frequency of 8 kHz.

10-5. Determine the dynamic range for a 10-bit sign-magnitude PCM code.

10-6. Determine the minimum number of bits required in a PCM code for a dynamic range of 80 dB. What is the coding efficiency?

10-7. For a resolution of 0.04 V, determine the voltages for the following linear seven-bit sign-magnitude PCM codes:

a. 0 1 1 0 1 0 1
b. 0 0 0 0 0 1 1
c. 1 0 0 0 0 0 1
d. 0 1 1 1 1 1 1
e. 1 0 0 0 0 0 0

10-8. Determine the SQR for a 2-v_{rms} signal and a quantization interval of 0.2 V.

10-9. Determine the resolution and quantization error for an eight-bit linear sign-magnitude PCM code for a maximum decoded voltage of 1.27 V.

10-10. A 12-bit linear PCM code is digitally compressed into eight bits. The resolution = 0.03 V. Determine the following for an analog input voltage of 1.465 V:

a. 12-bit linear PCM code
b. eight-bit compressed code
c. Decoded 12-bit code
d. Decoded voltage
e. Percentage error

10-11. For a 12-bit linear PCM code with a resolution of 0.02 V, determine the voltage range that would be converted to the following PCM codes:

a. 1 0 0 0 0 0 0 0 0 0 0 1
b. 0 0 0 0 0 0 0 0 0 0 0 0
c. 1 1 0 0 0 0 0 0 0 0 0 0
d. 0 1 0 0 0 0 0 0 0 0 0 0
e. 1 0 0 1 0 0 0 0 0 0 0 1
f. 1 0 1 0 1 0 1 0 1 0 1 0

10-12. For each of the following 12-bit linear PCM codes, determine the eight-bit compressed code to which they would be converted:

a. 1 0 0 0 0 0 0 0 1 0 0 0
b. 1 0 0 0 0 0 0 0 1 0 0 1
c. 1 0 0 0 0 0 0 1 0 0 0 0
d. 0 0 0 0 0 0 1 0 0 0 0 0
e. 0 1 0 0 0 0 0 0 0 0 0 0
f. 0 1 0 0 0 0 1 0 0 0 0 0

10-13. Determine the Nyquist sampling rate for the following maximum analog input frequencies: 2 kHz, 5 kHz, 12 kHz, and 20 kHz.

10-14. For the sample-and-hold circuit shown in Figure 10-5a, determine the largest-value capacitor that can be used for the following parameters: Z_1 output impedance = 15 Ω, an on resistance of Q_1 of 15 Ω, an acquisition time of 12 μs, a maximum output current from Z_1 of 10 mA, an accuracy of 0.1%, and a maximum change in voltage dv = 10 V.

10-15. Determine the maximum analog input frequency for the following Nyquist sample rates: 2.5 kHz, 4 kHz, 9 kHz, and 11 kHz.

10-16. Determine the alias frequency for the following sample rates and analog input frequencies:

f_a (kHz)	f_s (kHz)
3	4
5	8
6	8
5	7

10-17. Determine the dynamic range in dB for the following n-bit linear sign-magnitude PCM codes: n = 7, 8, 12, and 14.

10-18. Determine the minimum number of bits required for PCM codes with the following dynamic ranges and determine the coding efficiencies: DR = 24 dB, 48 dB, and 72 dB.

10-19. For the following values of μ, V_{max}, and V_{in}, determine the compressor gain:

μ	V_{max} (V)	V_{in} (V)
255	1	0.75
100	1	0.75
255	2	0.5

10-20. For the following resolutions, determine the range of the eight-bit sign-magnitude PCM codes:

Code	Resolution (V)
10111000	0.1
00111000	0.1
11111111	0.05
00011100	0.02
00110101	0.02
11100000	0.02
00000111	0.02

10-21. Determine the SQR for the following input signal and quantization noise magnitudes:

V_s	V_n (V)
1 v_{rms}	0.01
2 v_{rms}	0.02
3 v_{rms}	0.01
4 v_{rms}	0.2

10-22. Determine the resolution and quantization noise for an eight-bit linear sign-magnitude PCM code for the following maximum decoded voltages: V_{max} = 3.06 V_p, 3.57 V_p, 4.08 V_p, and 4.59 V_p.

10-23. A 12-bit linear sign-magnitude PCM code is digitally compressed into 8 bits. For a resolution of 0.016 V, determine the following quantities for the indicated input voltages: 12-bit linear PCM code, eight-bit compressed code, decoded 12-bit code, decoded voltage, and percentage error. $V_{in} = -6.592$ V, $+12.992$ V, and -3.36 V.

10-24. For the 12-bit linear PCM codes given, determine the voltage range that would be converted to them:

12-Bit Linear Code	Resolution (V)
100011110010	0.12
000001000000	0.10
000111111000	0.14
111111110000	0.12

10-25. For the following 12-bit linear PCM codes, determine the eight-bit compressed code to which they would be converted:

12-Bit Linear Code
100011110010
000001000000
000111111000
111111110010
000000100000

10-26. For the following eight-bit compressed codes, determine the expanded 12-bit code.

Eight-Bit Code
11001010
00010010
10101010
01010101
11110000
11011011

CHAPTER 11

Digital T-Carriers and Multiplexing

CHAPTER OUTLINE

OBJECTIVES

- Define *multiplexing*
- Describe the frame format and operation of the T1 digital carrier system
- Describe the format of the North American Digital Hierarchy
- Define *line encoding*
- Define the following terms and describe how they affect line encoding: *duty cycle, bandwidth, clock recovery, error detection,* and *detecting* and *decoding*
- Describe the basic T carrier system formats
- Describe the European digital carrier system
- Describe several methods of achieving frame synchronization
- Describe the difference between bit and word interleaving
- Define *codecs* and *combo chips* and give a brief explanation of how they work
- Define *frequency-division multiplexing*
- Describe the format of the North American FDM Hierarchy
- Define and describe baseband and composite baseband signals
- Explain the formation of a mastergroup
- Describe wavelength-division multiplexing
- Explain the advantages and disadvantages of wavelength-division multiplexing

11-1 INTRODUCTION

Multiplexing is the transmission of information (in any form) from one or more source to one or more destination over the same transmission medium (facility). Although transmissions occur on the same facility, they do not necessarily occur at the same time or occupy the same bandwidth. The transmission medium may be a metallic wire pair, a coaxial cable, a PCS mobile telephone, a terrestrial microwave radio system, a satellite microwave system, or an optical fiber cable.

There are several domains in which multiplexing can be accomplished, including space, phase, time, frequency, and wavelength.

Space-division multiplexing (SDM) is a rather unsophisticated form of multiplexing that simply constitutes propagating signals from different sources on different cables that are contained within the same trench. The trench is considered to be the transmission medium. QPSK is a form of *phase-division multiplexing* (PDM) where two data channels (the I and Q) modulate the same carrier frequency that has been shifted 90° in phase. Thus, the I-channel bits modulate a sine wave carrier, while the Q-channel bits modulate a cosine wave carrier. After modulation has occurred, the I- and Q-channel carriers are linearly combined and propagated at the same time over the same transmission medium, which can be a cable or free space.

The three most predominant methods of multiplexing signals are time-division multiplexing (TDM), frequency-division multiplexing (FDM), and the more recently developed wavelength-division multiplexing (WDM). The remainder of this chapter will be dedicated to time-, frequency-, and wavelength-division multiplexing.

11-2 TIME-DIVISION MULTIPLEXING

With *time division multiplexing* (TDM), transmissions from multiple sources occur on the same facility but not at the same time. Transmissions from various sources are *interleaved* in the time domain. PCM is the most prevalent encoding technique used for TDM digital signals. With a PCM-TDM system, two or more voice channels are sampled, converted to PCM codes, and then time-division multiplexed onto a single metallic or optical fiber cable.

The fundamental building block for most TDM systems in the United States begins with a DS-0 channel (digital signal level 0). Figure 11-1 shows the simplified block dia-

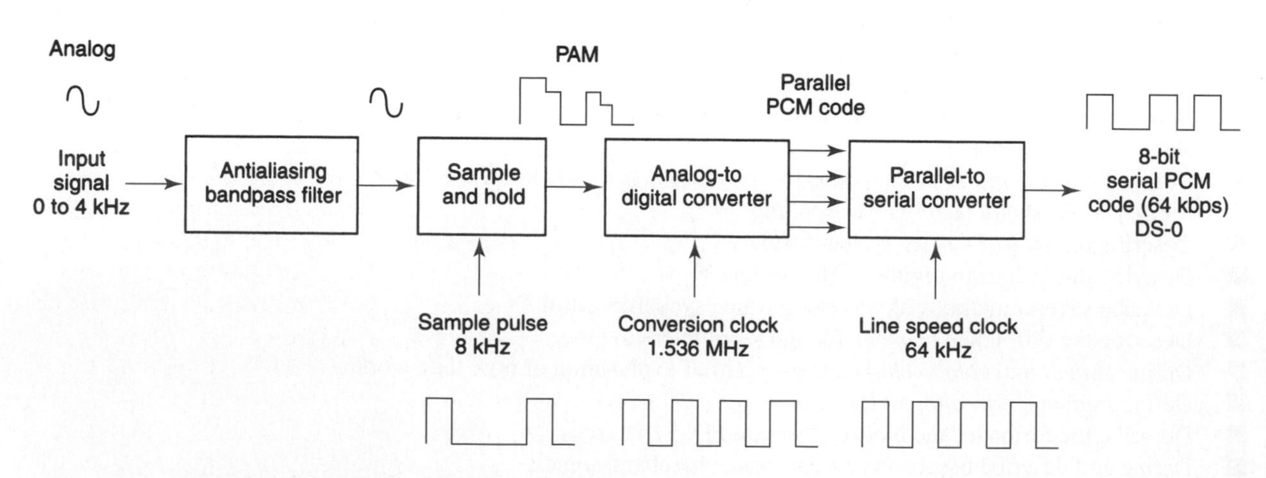

FIGURE 11-1 Single-channel (DS-0-level) PCM transmission system

gram for a DS-0 single-channel PCM system. As the figure shows, DS-0 channels use an 8-kHz sample rate and an eight-bit PCM code, which produces a 64-kbps PCM line speed:

$$\text{line speed} = \frac{8000 \text{ samples}}{\text{second}} \times \frac{8 \text{ bits}}{\text{sample}}$$

$$= 64{,}000 \text{ bps}$$

Figure 11-2a shows the simplified block diagram for a PCM carrier system comprised of two DS-0 channels that have been time-division multiplexed. Each channel's input is sampled at an 8-kHz rate and then converted to an eight-bit PCM code. While the PCM code for channel 1 is being transmitted, channel 2 is sampled and converted to a PCM code. While the PCM code from channel 2 is being transmitted, the next sample is taken from channel 1 and converted to a PCM code. This process continues, and samples are taken alternately from each channel, converted to PCM codes, and transmitted. The multiplexer is simply an electronically controlled digital switch with two inputs and one output. Channel 1 and channel 2 are alternately selected and connected to the transmission line through the multiplexer. One eight-bit PCM code from each channel (16 total bits) is called a TDM *frame,* and the time it takes to transmit one TDM frame is called the *frame time.* The frame time is equal to the reciprocal of the sample rate ($1/f_s$, or $1/8000 = 125\ \mu s$). Figure 11-2b shows the TDM frame allocation for a two-channel PCM system with an 8-kHz sample rate.

The PCM code for each channel occupies a fixed time slot (epoch) within the total TDM frame. With a two-channel system, one sample is taken from each channel during each frame, and the time allocated to transmit the PCM bits from each channel is equal to one-half the total frame time. Therefore, eight bits from each channel must be transmitted during each frame (a total of 16 PCM bits per frame). Thus, the line speed at the output of the multiplexer is

$$\frac{2 \text{ channels}}{\text{frame}} \times \frac{8000 \text{ frames}}{\text{second}} \times \frac{8 \text{ bits}}{\text{channel}} = 128 \text{ kbps}$$

Although each channel is producing and transmitting only 64 kbps, the bits must be clocked out onto the line at a 128-kHz rate to allow eight bits from each channel to be transmitted in a 1211-μs time slot.

11-3 T1 DIGITAL CARRIER

A digital carrier system is a communications system that uses digital pulse rather than analog signals to encode information. Figure 11-3a shows the block diagram for AT&T's T1 digital carrier system, which has been the North American digital multiplexing standard since 1963 and recognized by the ITU-T as Recommendation G.733. T1 stands for *transmission one* and specifies a digital carrier system using PCM-encoded analog signals. A T1 carrier system time-division multiplexes PCM-encoded samples from 24 voice-band channels for transmission over a single metallic wire pair or optical fiber transmission line. Each voice-band channel has a bandwidth of approximately 300 Hz to 3000 Hz. Again, the multiplexer is simply a digital switch with 24 independent inputs and one time-division multiplexed output. The PCM output signals from the 24 voice-band channels are sequentially selected and connected through the multiplexer to the transmission line.

Simply, time-division multiplexing 24 voice-band channels does not in itself constitute a T1 carrier system. At this point, the output of the multiplexer is simply a multiplexed first-level digital signal (DS level 1). The system does not become a T1 carrier until it is line encoded and placed on special conditioned cables called *T1 lines.* Line encoding is described later in this chapter.

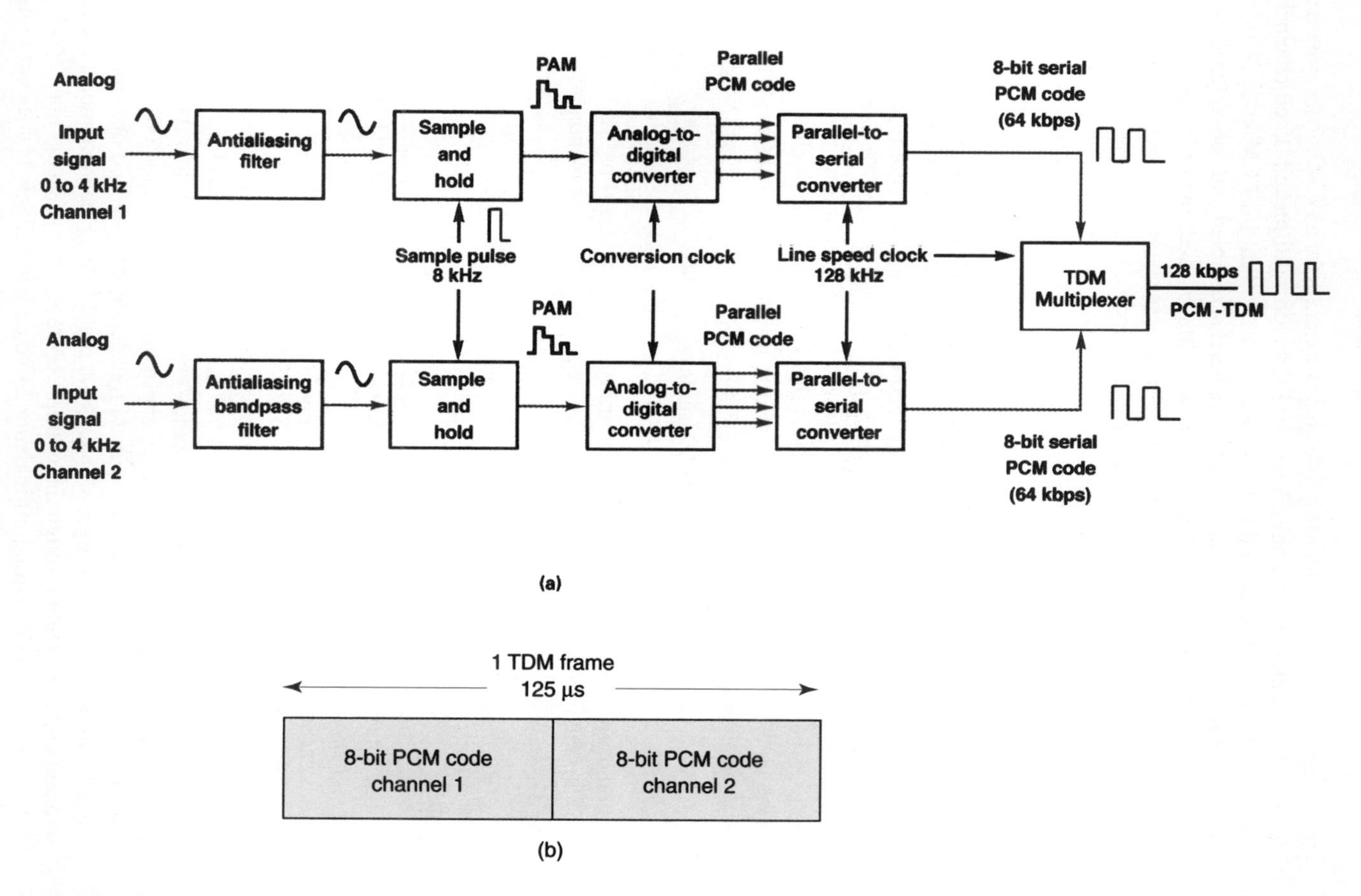

FIGURE 11-2 Two-channel PCM-TDM system: (a) block diagram; (b) TDM frame

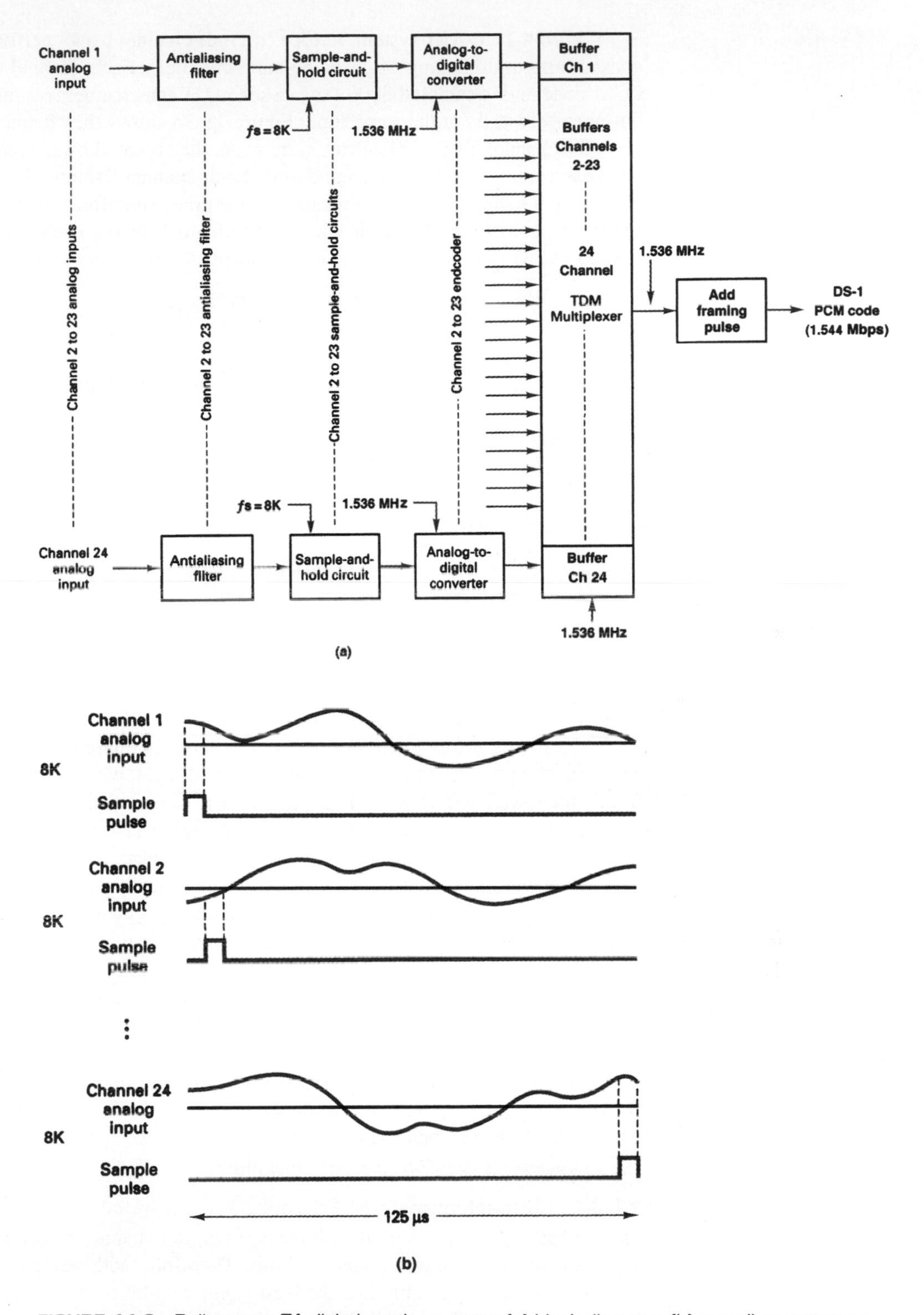

FIGURE 11-3 Bell system T1 digital carrier system: (a) block diagram; (b) sampling sequence

With a T1 carrier system, D-type (digital) channel banks perform the sampling, encoding, and multiplexing of 24 voice-band channels. Each channel contains an eight-bit PCM code and is sampled 8000 times a second. Each channel is sampled at the same rate but not necessarily at the same time. Figure 11-3b shows the channel sampling sequence for a 24-channel T1 digital carrier system. As the figure shows, each channel is sampled once each frame but not at the same time. Each channel's sample is offset from the previous channel's sample by 1/24 of the total frame time. Therefore, one 64-kbps PCM-encoded sample is transmitted for each voice-band channel during each frame (a frame time of 1/8000 = 125 μs). The line speed is calculated as follows:

$$\frac{24 \text{ channels}}{\text{frame}} \times \frac{8 \text{ bits}}{\text{channel}} = 192 \text{ bits per frame}$$

thus

$$\frac{192 \text{ bits}}{\text{frame}} \times \frac{8000 \text{ frames}}{\text{second}} = 1.536 \text{ Mbps}$$

Later, an additional bit (called the *framing bit*) is added to each frame. The framing bit occurs once per frame (8000-bps rate) and is recovered in the receiver, where it is used to maintain frame and sample synchronization between the TDM transmitter and receiver. As a result, each frame contains 193 bits, and the line speed for a T1 digital carrier system is

$$\frac{193 \text{ bits}}{\text{frame}} \times \frac{8000 \text{ frames}}{\text{second}} = 1.544 \text{ Mbps}$$

11-3-1 D-Type Channel Banks

Early T1 carrier systems used D1 *digital channel banks* (PCM encoders and decoders) with a seven-bit magnitude-only PCM code, analog companding, and $\mu = 100$. A later version of the D1 digital channel bank added an eighth bit (the signaling bit) to each PCM code for performing interoffice *signaling* (supervision between telephone offices, such as on hook, off hook, dial pulsing, and so forth). Since a signaling bit was added to each sample in every frame, the signaling rate was 8 kbps. In the early digital channel banks, the framing bit sequence was simply an alternating 1/0 pattern. Figure 11-4 shows the frame and bit alignment for T1-carrier systems that used D1 channel banks.

Over the years, T1 carrier systems have generically progressed through D2, D3, D4, D5, and D6 channel banks. D4, D5, and D6 channel banks use digital companding and eight-bit sign-magnitude-compressed PCM codes with $\mu = 255$.

Because the early D1 channel banks used a magnitude-only PCM code, an error in the most significant bit of a PCM sample produced a decoded error equal to one-half the total quantization range. Newer version digital channel banks used sign-magnitude codes, and an error in the sign bit causes a decoded error equal to twice the sample magnitude ($+V$ to $-V$ or vice versa) with a worst-case error equal to twice the total quantization range. However, in practice, maximum amplitude samples occur rarely; therefore, most errors have a magnitude less than half the coding range. On average, performance with sign-magnitude PCM codes is much better than with magnitude-only codes.

11-3-2 Superframe TDM Format

The 8-kbps signaling rate used with the early digital channel banks was excessive for signaling on standard telephone voice circuits. Therefore, with modern channel banks, a signaling bit is substituted only into the least significant bit of every sixth frame. Hence, five of every six frames have eight-bit resolution, while one in every six frames (the signaling frame) has only seven-bit resolution. Consequently, the signaling rate on each channel is only 1.333 kbps (8000 bps/6), and the average number of bits per sample is actually $7^{5/6}$ bits.

Because only every sixth frame includes a signaling bit, it is necessary that all the frames be numbered so that the receiver knows when to extract the signaling bit. Also, because the signaling is accomplished with a two-bit binary word, it is necessary to identify the most and least significant bits of the signaling word. Consequently, the superframe format shown in

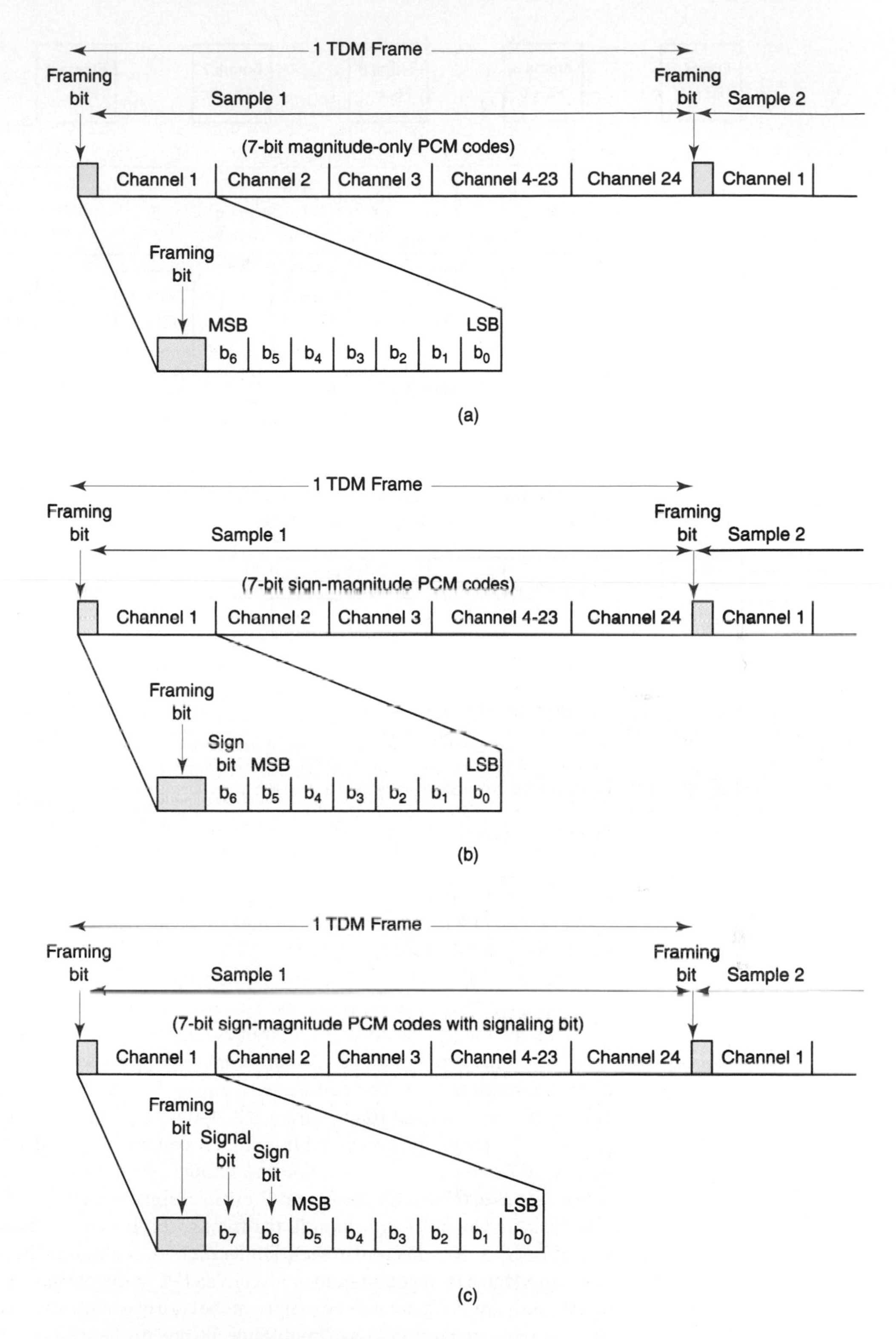

FIGURE 11-4 Early T1 Carrier system frame and sample alignment: (a) seven-bit magnitude-only PCM code; (b) seven-bit sign-magnitude code; (c) seven-bit sign-magnitude PCM code with signaling bit

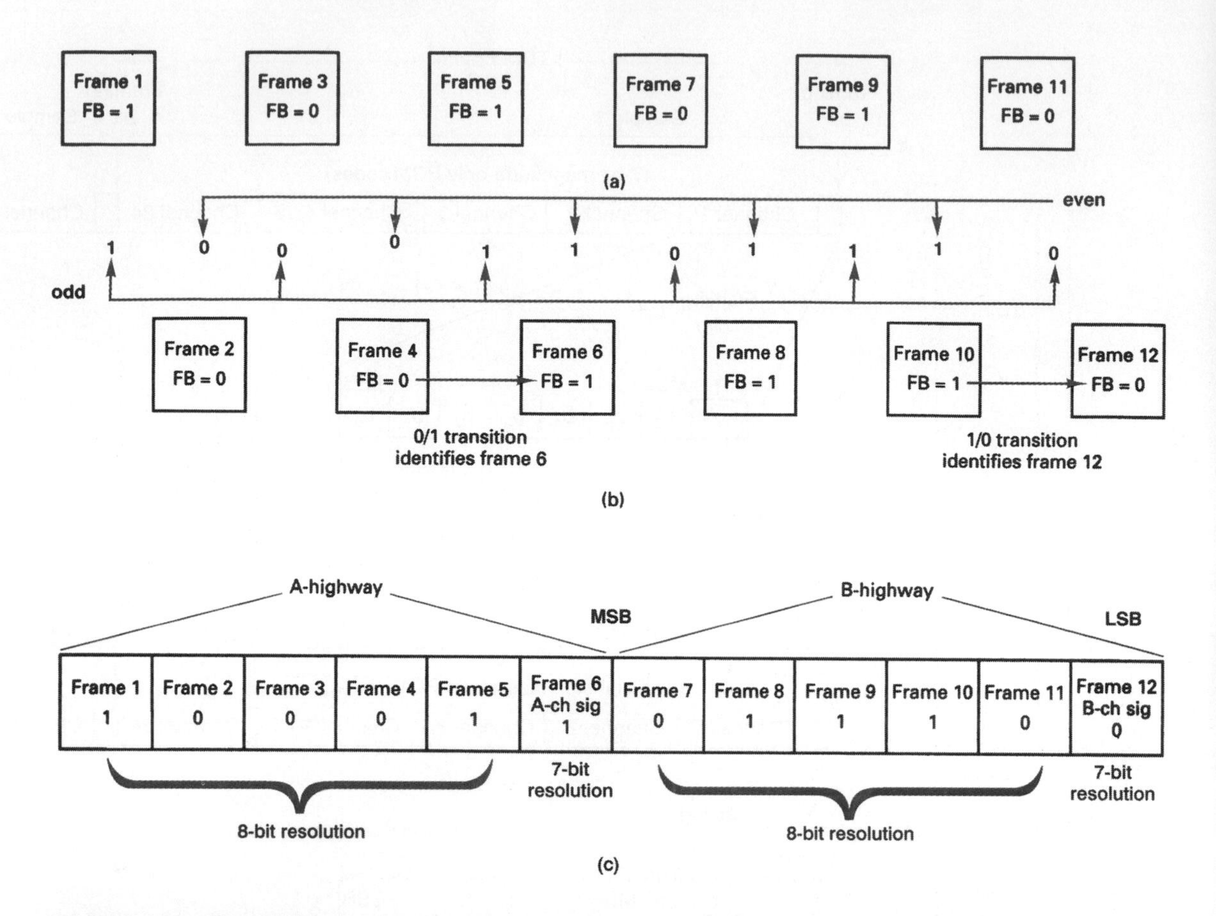

FIGURE 11-5 Framing bit sequence for the T1 superframe format using D2 or D3 channel banks: (a) frame synchronizing bits (odd-numbered frames); (b) signaling frame alignment bits (even-numbered frames); (c) composite frame alignment

Figure 11-5 was devised. Within each superframe, there are 12 consecutively numbered frames (1 to 12). The signaling bits are substituted in frames 6 and 12, the most significant bit into frame 6, and the least significant bit into frame 12. Frames 1 to 6 are called the A-highway, with frame 6 designated the A-channel signaling frame. Frames 7 to 12 are called the B-highway, with frame 12 designated the B-channel signaling frame. Therefore, in addition to identifying the signaling frames, the 6th and 12th frames must also be positively identified.

To identify frames 6 and 12, a different framing bit sequence is used for the odd- and even-numbered frames. The odd frames (frames 1, 3, 5, 7, 9, and 11) have an alternating 1/0 pattern, and the even frames (frames 2, 4, 6, 8, 10, and 12) have a 0 0 1 1 1 0 repetitive pattern. As a result, the combined framing bit pattern is 1 0 0 0 1 1 0 1 1 1 0 0. The odd-numbered frames are used for frame and sample synchronization, and the even-numbered frames are used to identify the A- and B-channel signaling frames (frames 6 and 12). Frame 6 is identified by a 0/1 transition in the framing bit between frames 4 and 6. Frame 12 is identified by a 1/0 transition in the framing bit between frames 10 and 12.

In addition to multiframe alignment bits and PCM sample bits, specific time slots are used to indicate alarm conditions. For example, in the case of a transmit power supply failure, a common equipment failure, or loss of multiframe alignment, the second bit in each channel is made a logic 0 until the alarm condition has cleared. Also, the framing bit in frame 12 is complemented whenever multiframe alignment is lost, which is assumed whenever frame alignment is lost. In addition, there are special framing conditions that must be avoided to maintain clock and bit synchronization in the receive demultiplexing equipment. Figure 11-6 shows the frame, sample, and signaling alignment for the T1 carrier system using D2 or D3 channel banks.

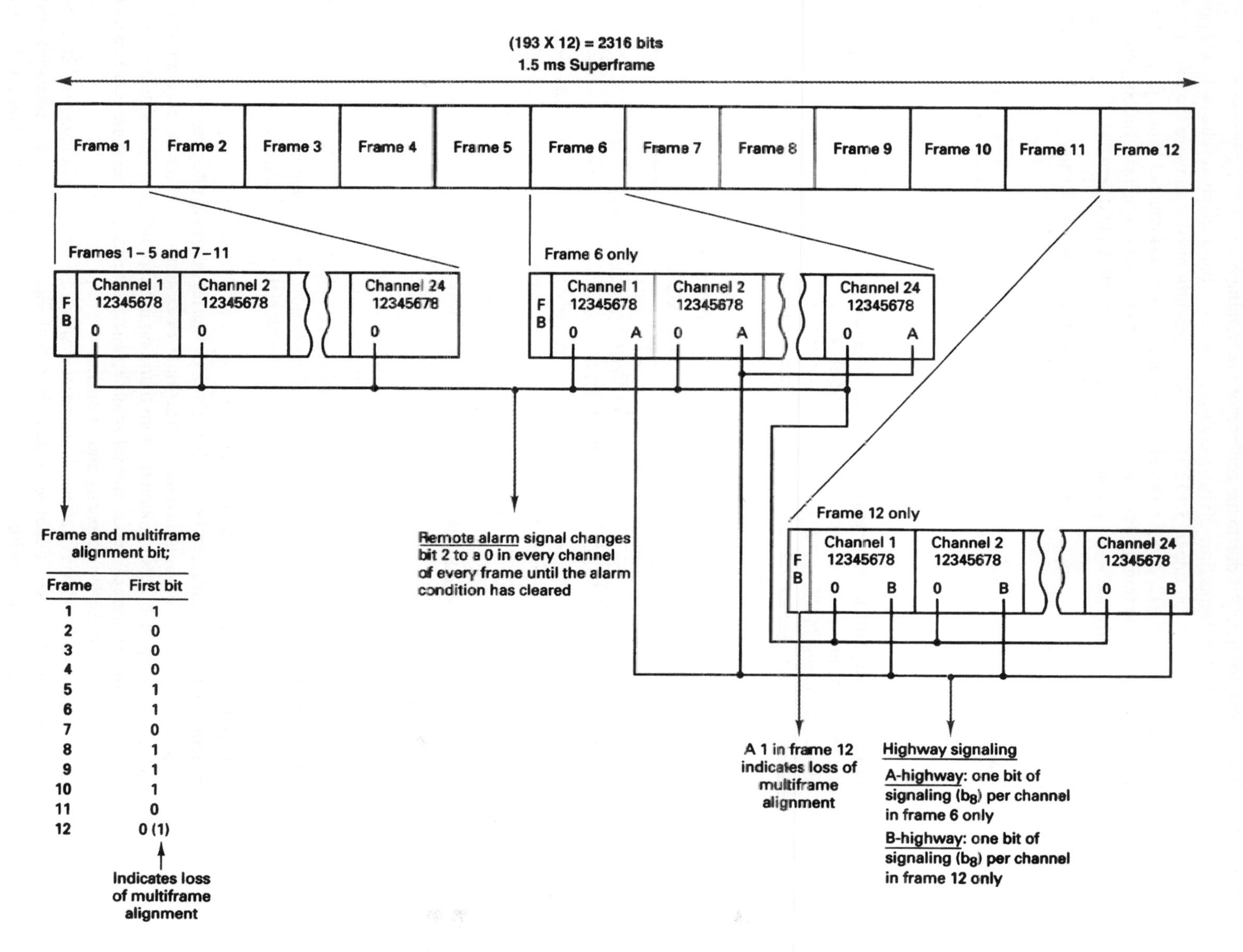

FIGURE 11-6 T1 carrier frame, sample, and signaling alignment for D2 and D3 channel banks

Figure 11-7a shows the framing bit circuitry for the 24-channel T1 carrier system using D2 or D3 channel banks. Note that the bit rate at the output of the TDM multiplexer is 1.536 Mbps and that the bit rate at the output of the 193-bit shift register is 1.544 Mbps. The 8-kHz difference is due to the addition of the framing bit.

D4 channel banks time-division multiplex 48 voice-band telephone channels and operate at a transmission rate of 3.152 Mbps. This is slightly more than twice the line speed for 24-channel D1, D2, or D3 channel banks because with D4 channel banks, rather than transmitting a single framing bit with each frame, a 10-bit frame synchronization pattern is used. Consequently, the total number of bits in a D4 (DS-1C) TDM frame is

$$\frac{8 \text{ bits}}{\text{channel}} \times \frac{48 \text{ channels}}{\text{frame}} = \frac{384 \text{ bits}}{\text{frame}} + \frac{10 \text{ framing bits}}{\text{frame}} = \frac{394 \text{ bits}}{\text{frame}}$$

and the line speed for DS-1C systems is

$$\frac{394 \text{ bits}}{\text{frame}} \times \frac{8000 \text{ frames}}{\text{second}} = 3.152 \text{ Mbps}$$

The framing for DS-1 (T1) PCM-TDM system or the framing pattern for the DS-1C (T1C) time-division multiplexed carrier system is added to the multiplexed digital signal at the output of the multiplexer. The framing bit circuitry used for the 48-channel DS-1C is shown in Figure 11-7b.

11-3-3 Extended Superframe Format

Another framing format recently developed for new designs of T1 carrier systems is the *extended superframe format.* The extended superframe format consists of 24 193-bit frames, totaling 4632 bits, of which 24 are framing bits. One extended superframe occupies 3 ms:

$$\left(\frac{1}{1.544 \text{ Mbits/s}}\right)\left(\frac{193 \text{ bits}}{\text{frame}}\right)(24 \text{ frames}) = 3 \text{ ms}$$

A framing bit occurs once every 193 bits; however, only 6 of the 24 framing bits are used for frame synchronization. Frame synchronization bits occur in frames 4, 8, 12, 16, 20, and 24 and have a bit sequence of 0 0 1 0 1 1. Six additional framing bits in frames 1, 5, 9, 13, 17, and 21 are used for an error detection code called CRC-6 (*cyclic redundancy checking*). The 12 remaining framing bits provide for a management channel called the *facilities data link* (FDL). FDL bits occur in frames 2, 3, 6, 7, 10, 11, 14, 15, 18, 19, 22, and 23.

The extended superframe format supports a four-bit signaling word with signaling bits provided in the second least significant bit of each channel during every sixth frame. The signaling bit in frame 6 is called the A bit, the signaling bit in frame 12 is called the B bit, the signaling bit in frame 18 is called the C bit, and the signaling bit in frame 24 is called the D bit. These signaling bit streams are sometimes called the A, B, C, and D *signaling channels* (or *signaling highways*). The extended superframe framing bit pattern is summarized in Table 11-1.

11-3-4 Fractional T Carrier Service

Fractional T carrier emerged because standard T1 carriers provide a higher capacity (i.e., higher bit rate) than most users require. Fractional T1 systems distribute the channels (i.e., bits) in a standard T1 system among more than one user, allowing several subscribers to share one T1 line. For example, several small businesses located in the same building can share one T1 line (both its capacity and its cost).

Bit rates offered with fractional T1 carrier systems are 64 kbps (1 channel), 128 kbps (2 channels), 256 kbps (4 channels), 384 kbps (6 channels), 512 kbps (8 channels), and 768 kbps (12 channels) with 384 kbps (1/4 T1) and 768 kbps (1/2 T1) being the most common. The minimum data rate necessary to propagate video information is 384 kbps.

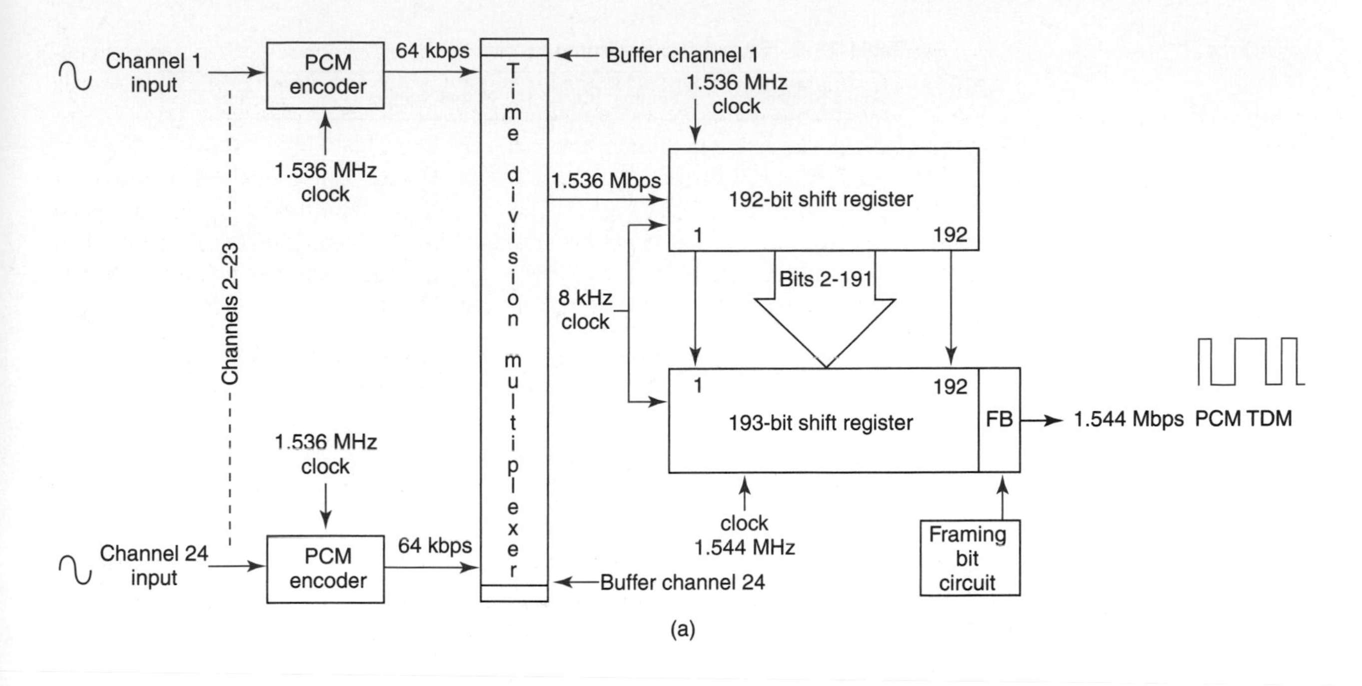

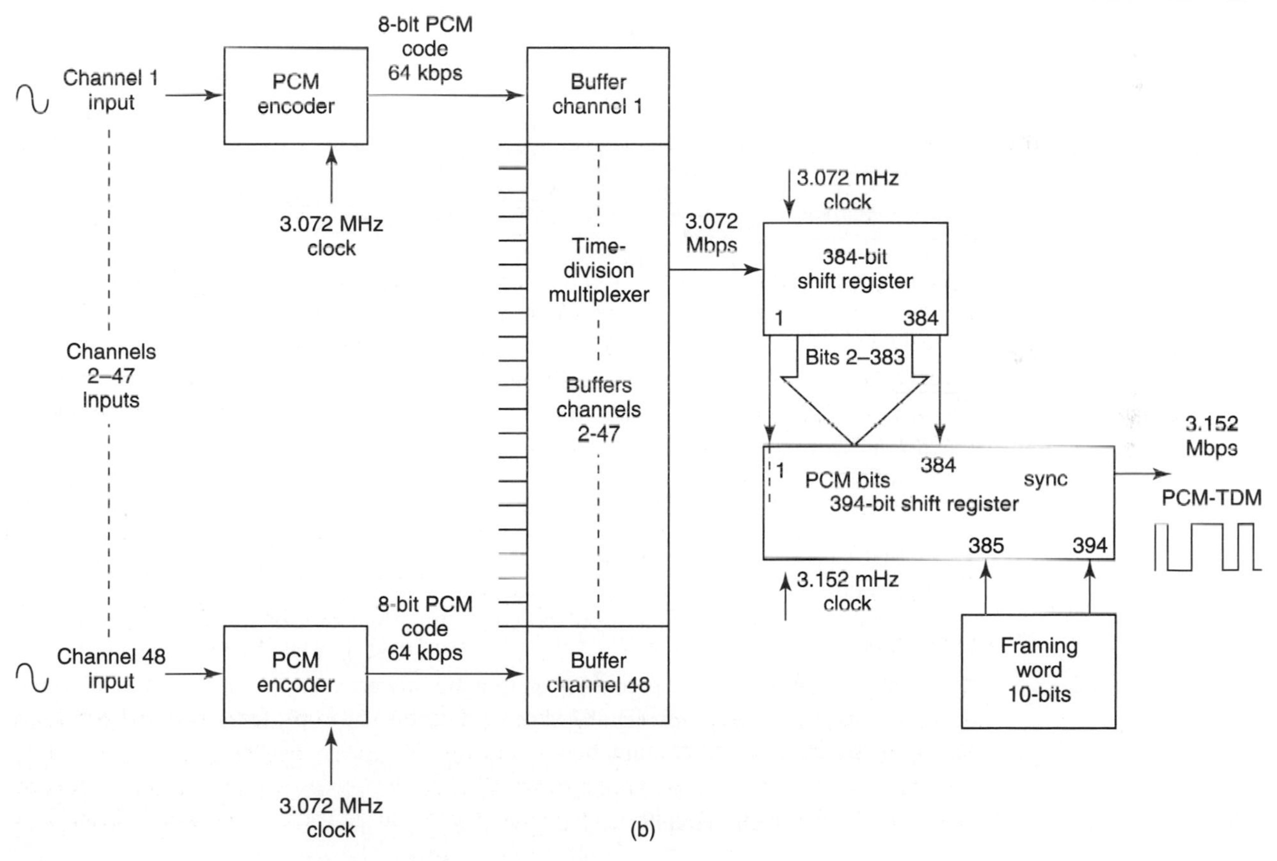

FIGURE 11-7 Framing bit circuitry T1 carrier system: (a) DS-1; (b) DS-1C

Table 11-1 Extended Superframe Format

Frame Number	Framing Bit	Frame Number	Framing Bit
1	C	13	C
2	F	14	F
3	F	15	F
4	S = 0	16	S = 0
5	C	17	C
6	F	18	F
7	F	19	F
8	S = 0	20	S = 1
9	C	21	C
10	F	22	F
11	F	23	F
12	S = 1	24	S = 1

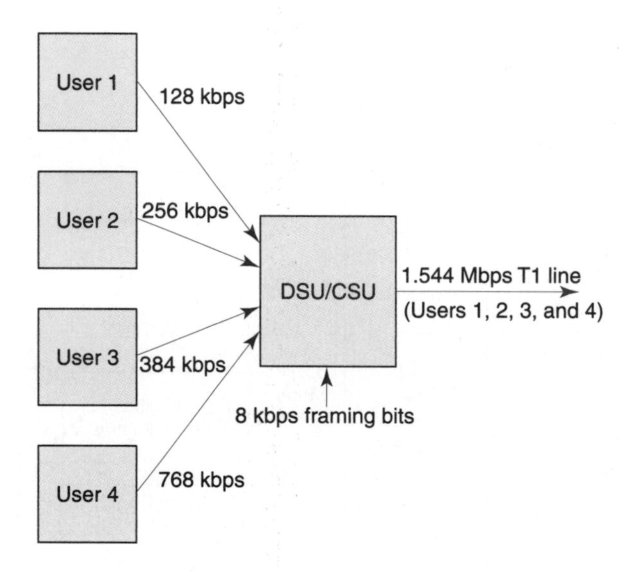

FIGURE 11-8 Fractional T1 carrier service

Fractional T3 is essentially the same as fractional T1 except with higher channel capacities, higher bit rates, and more customer options.

Figure 11-8 shows four subscribers combining their transmissions in a special unit called a *data service unit/channel service unit* (DSU/CSU). A DSU/CSU is a digital interface that provides the physical connection to a digital carrier network. User 1 is allocated 128 kbps, user 2 256 kbps, user 3 384 kbps, and user 4 768 kbps, for a total of 1.536 kbps (8 kbps is reserved for the framing bit).

11-4 NORTH AMERICAN DIGITAL HIERARCHY

Multiplexing signals in digital form lends itself easily to interconnecting digital transmission facilities with different transmission bit rates. Figure 11-9 shows the American Telephone and Telegraph Company (AT&T) North American Digital Hierarchy for multiplexing digital signals from multiple sources into a single higher-speed pulse stream suitable for transmission on the next higher level of the hierarchy. To upgrade from one level in the

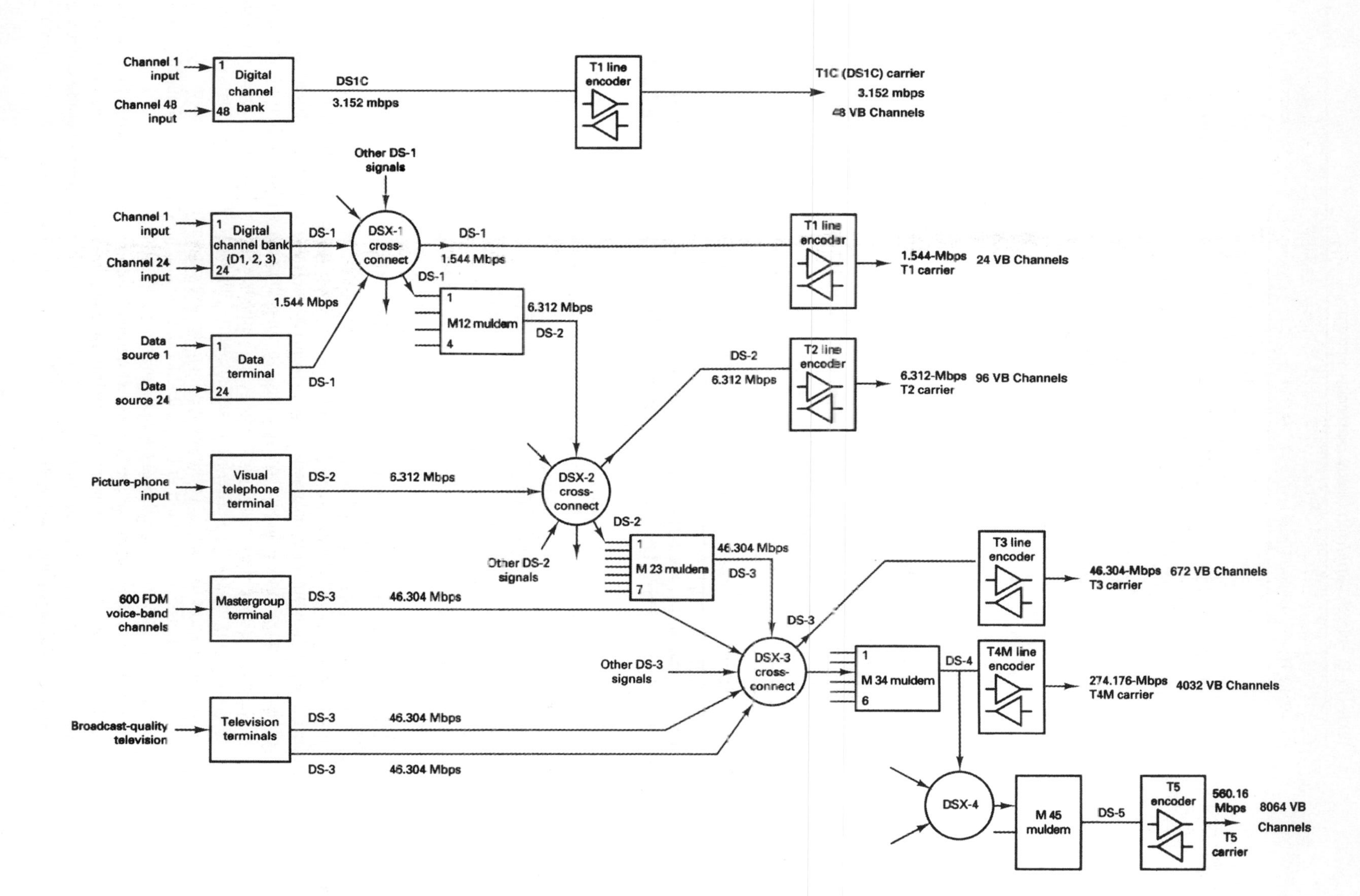

FIGURE 11-9 North American Digital Hierarchy

Table 11-2 North American Digital Hierarchy Summary

Line Type	Digital Signal	Bit Rate	Channel Capacities	Services Offered
T1	DS-1	1.544 Mbps	24	Voice-band telephone or data
Fractional T1	DS-1	64 kbps to 1.536 Mbps	24	Voice-band telephone or data
T1C	DS-1C	3.152 Mbps	48	Voice-band telephone or data
T2	DS-2	6.312 Mbps	96	Voice-band telephone, data, or picture phone
T3	DS-3	44.736 Mbps	672	Voice-band telephone, data, picture phone, and broadcast-quality television
Fractional T3	DS-3	64 kbps to 23.152 Mbps	672	Voice-band telephone, data, picture phone, and broadcast-quality television
T4M	DS-4	274.176 Mbps	4032	Same as T3 except more capacity
T5	DS-5	560.160 Mbps	8064	Same as T3 except more capacity

hierarchy to the next higher level, a special device called *muldem* (*mul*tiplexers/*dem*ultiplexer) is required. Muldems can handle bit-rate conversions in both directions. The muldem designations (M112, M23, and so on) identify the input and output digital signals associated with that muldem. For instance, an M12 muldem interfaces DS-1 and DS-2 digital signals. An M23 muldem interfaces DS-2 and DS-3 digital signals. As the figure shows, DS-1 signals may be further multiplexed or line encoded and placed on specially conditioned cables called T1 lines. DS-2, DS-3, DS-4, and DS-5 digital signals may be placed on T2, T3, T4M, or T5 lines, respectively.

Digital signals are routed at central locations called *digital cross-connects.* A digital cross-connect (DSX) provides a convenient place to make patchable interconnects and perform routine maintenance and troubleshooting. Each type of digital signal (DS-1, DS-2, and so on) has its own digital switch (DSX-1, DSX-2, and so on). The output from a digital switch may be upgraded to the next higher level of multiplexing or line encoded and placed on its respective T lines (T1, T2, and so on).

Table 11-2 lists the digital signals, their bit rates, channel capacities, and services offered for the line types included in the North American Digital Hierarchy.

11-4-1 Mastergroup and Commercial Television Terminals

Figure 11-10 shows the block diagram of a mastergroup and commercial television terminal. The mastergroup terminal receives voice-band channels that have already been frequency-division multiplexed (a topic covered later in this chapter) without requiring that each voice-band channel be demultiplexed to voice frequencies. The signal processor provides frequency shifting for the mastergroup signals (shifts it from a 564-kHz to 3084-kHz bandwidth to a 0-kHz to 2520-kHz bandwidth) and dc restoration for the television signal. By shifting the mastergroup band, it is possible to sample at a 5.1-MHz rate. Sampling of the commercial television signal is at twice that rate or 10.2 MHz.

When the bandwidth of the signals to be transmitted is such that after digital conversion it occupies the entire capacity of a digital transmission line, a single-channel terminal is provided. Examples of such single-channel terminals are mastergroup, commercial television, and picturephone terminals.

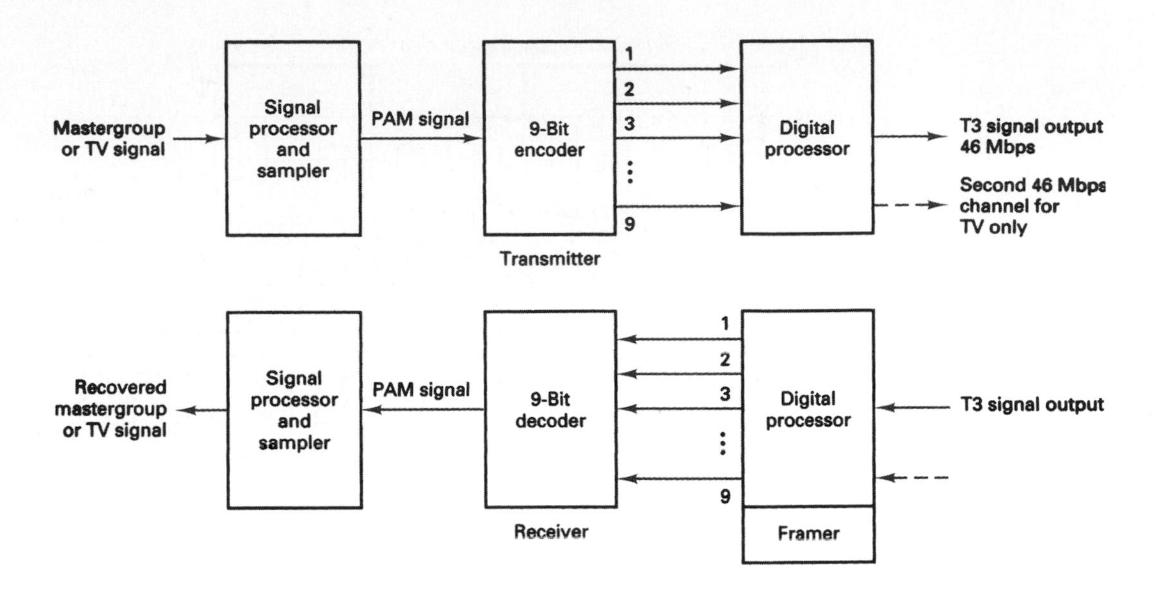

FIGURE 11-10 Block diagram of a mastergroup or commercial television digital terminal

To meet the transmission requirements, a nine-bit PCM code is used to digitize each sample of the mastergroup or television signal. The digital output from the terminal is, therefore, approximately 46 Mbps for the mastergroup and twice that much (92 Mbps) for the television signal.

The digital terminal shown in Figure 11-10 has three specific functions: (1) It converts the parallel data from the output of the encoder to serial data, (2) it inserts frame synchronizing bits, and (3) it converts the serial binary signal to a form more suitable for transmission. In addition, for the commercial television terminal, the 92-Mbps digital signal must be split into two 46-Mbps digital signals because there is no 92-Mbps line speed in the digital hierarchy.

11-4-2 Picturephone Terminal

Essentially, *picturephone* is a low-quality video transmission for use between nondedicated subscribers. For economic reasons, it is desirable to encode a picturephone signal into the T2 capacity of 6.312 Mbps, which is substantially less than that for commercial network broadcast signals. This substantially reduces the cost and makes the service affordable. At the same time, it permits the transmission of adequate detail and contrast resolution to satisfy the average picturephone subscriber. Picturephone service is ideally suited to a differential PCM code. Differential PCM is similar to conventional PCM except that the exact magnitude of a sample is not transmitted. Instead, only the difference between that sample and the previous sample is encoded and transmitted. To encode the difference between samples requires substantially fewer bits than encoding the actual sample.

11-4-3 Data Terminal

The portion of communications traffic that involves data (signals other than voice) is increasing exponentially. Also, in most cases the data rates generated by each individual subscriber are substantially less than the data rate capacities of digital lines. Therefore, it seems only logical that terminals be designed that transmit data signals from several sources over the same digital line.

Data signals could be sampled directly; however, this would require excessively high sample rates, resulting in excessively high transmission bit rates, especially for sequences

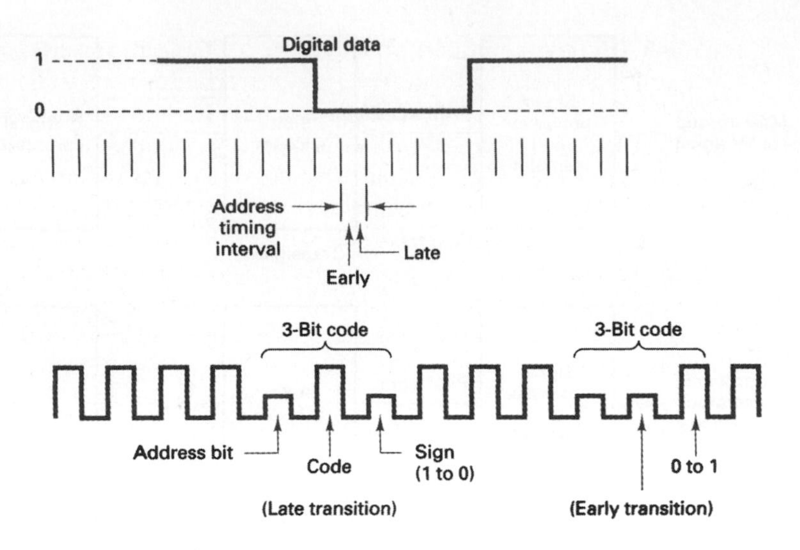

FIGURE 11-11 Data coding format

of data with few or no transitions. A more efficient method is one that codes the transition times. Such a method is shown in Figure 11-11. With the coding format shown, a three-bit code is used to identify when transitions occur in the data and whether that transition is from a 1 to a 0 or vice versa. The first bit of the code is called the address bit. When this bit is a logic 1, this indicates that no transition occurred; a logic 0 indicates that a transition did occur. The second bit indicates whether the transition occurred during the first half (0) or during the second half (1) of the sample interval. The third bit indicates the sign or direction of the transition; a 1 for this bit indicates a 0-to-1 transition, and a 0 indicates a 1-to-0 transition. Consequently, when there are no transitions in the data, a signal of all 1s is transmitted. Transmission of only the address bit would be sufficient; however, the sign bit provides a degree of error protection and limits error propagation (when one error leads to a second error and so on). The efficiency of this format is approximately 33%; there are three code bits for each data bit. The advantage of using a coded format rather than the original data is that coded data are more efficiently substituted for voice in analog systems. Without this coding format, transmitting a 250-kbps data signal requires the same bandwidth as would be required to transmit 60 voice channels with analog multiplexing. With this coded format, a 50-kbps data signal displaces three 64-kbps PCM-encoded channels, and a 250-kbps data stream displaces only 12 voice-band channels.

11-5 DIGITAL CARRIER LINE ENCODING

Digital line encoding involves converting standard logic levels (TTL, CMOS, and the like) to a form more suitable to telephone line transmission. Essentially, six primary factors must be considered when selecting a line-encoding format:

1. Transmission voltages and DC component
2. Duty cycle
3. Bandwidth considerations
4. Clock and framing bit recovery
5. Error detection
6. Ease of detection and decoding

11-5-1 Transmission Voltages and DC Component

Transmission voltages or levels can be categorized as being either *unipolar* (UP) or *bipolar* (BP). Unipolar transmission of binary data involves the transmission of only a single nonzero voltage level (e.g., either a positive or a negative voltage for a logic 1 and 0 V [ground] for a logic 0). In bipolar transmission, two nonzero voltages are involved (e.g., a positive voltage for a logic 1 and an equal-magnitude negative voltage for a logic 0 or vice versa).

Over a digital transmission line, it is more power efficient to encode binary data with voltages that are equal in magnitude but opposite in polarity and symmetrically balanced about 0 V. For example, assuming a 1-ohm resistance and a logic 1 level of +5 V and a logic 0 level of 0 V, the average power required is 12.5 W, assuming an equal probability of the occurrence of a logic 1 or a logic 0. With a logic 1 level of +2.5 V and a logic 0 level of −2.5 V, the average power is only 6.25 W. Thus, by using bipolar symmetrical voltages, the average power is reduced by a factor of 50%.

11-5-2 Duty Cycle

The *duty cycle* of a binary pulse can be used to categorize the type of transmission. If the binary pulse is maintained for the entire bit time, this is called *nonreturn to zero* (NRZ). If the active time of the binary pulse is less than 100% of the bit time, this is called *return to zero* (RZ).

Unipolar and bipolar transmission voltages can be combined with either RZ or NRZ in several ways to achieve a particular line-encoding scheme. Figure 11-12 shows five line-encoding possibilities.

In Figure 11-12a, there is only one nonzero voltage level (+V − logic 1); a zero voltage indicates a logic 0. Also, each logic 1 condition maintains the positive voltage for the entire bit time (100% duty cycle). Consequently, Figure 11-12a represents a unipolar nonreturn-to-zero signal (UPNRZ). Assuming an equal number of 1s and 0s, the average dc voltage of a UPNRZ waveform is equal to half the nonzero voltage (V/2).

In Figure 11-12b, there are two nonzero voltages (+V = logic 1 and −V = logic 0) and a 100% duty cycle is used. Therefore, Figure 11-12b represents a bipolar nonreturn-to-zero signal (BPNRZ). When equal-magnitude voltages are used for logic 1s and logic 0s, and assuming an equal probability of logic 1s and logic 0s occurring, the average dc voltage of a BPNRZ waveform is 0 V.

In Figure 11-12c, only one nonzero voltage is used, but each pulse is active for only 50% of a bit time ($t_b/2$). Consequently, the waveform shown in Figure 11-12c represents a unipolar return-to-zero signal (UPRZ). Assuming an equal probability of 1s and 0s occurring, the average dc voltage of a UPRZ waveform is one-fourth the nonzero voltage (V/4).

Figure 11-12d shows a waveform where there are two nonzero voltages (+V = logic 1 and −V = logic 0). Also, each pulse is active only 50% of a bit time. Consequently, the waveform shown in Figure 11-8d represents a bipolar return-to-zero (BPRZ) signal. Assuming equal-magnitude voltages for logic 1s and logic 0s and an equal probability of 1s and 0s occurring, the average dc voltage of a BPRZ waveform is 0 V.

In Figure 11-12e, there are again two nonzero voltage levels (−V and +V), but now both polarities represent logic 1s, and 0 V represents a logic 0. This method of line encoding is called *alternate mark inversion* (AMI). With AMI transmissions, successive logic 1s are inverted in polarity from the previous logic 1. Because return to zero is used, the encoding technique is called *bipolar-return-to-zero alternate mark inversion* (BPRZ-AMI). The average dc voltage of a BPRZ-AMI waveform is approximately 0 V regardless of the bit sequence.

With NRZ encoding, a long string of either logic 1s or logic 0s produces a condition in which a receive may lose its amplitude reference for optimum discrimination between received 1s and 0s. This condition is called *dc wandering*. The problem may also arise when there is a significant imbalance in the number of 1s and 0s transmitted. Figure 11-13 shows how dc wandering is produced from a long string of successive logic 1s. It can be seen that after a long string of 1s, 1-to-0 errors are more likely than 0-to-1 errors. Similarly, long strings of logic 0s increase the probability of 0-to-1 errors.

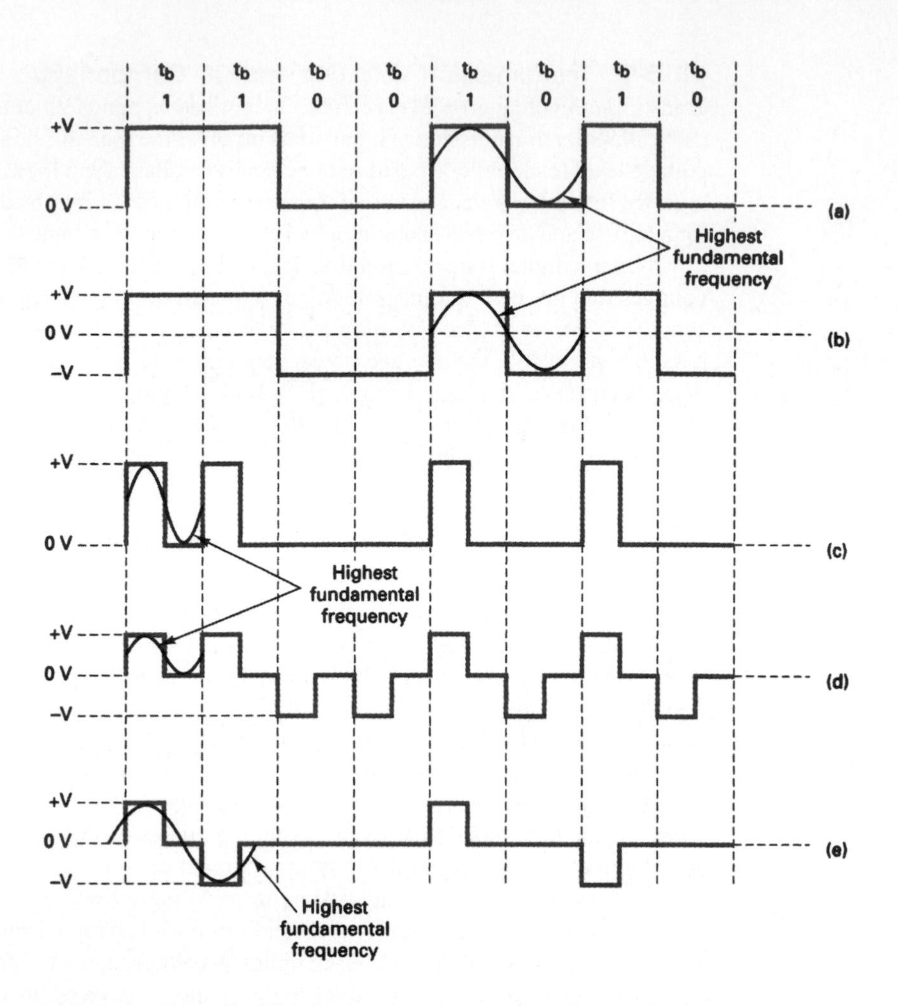

FIGURE 11-12 Line-encoding formats: (a) UPNRZ; (b) BPNRZ; (c) UPRZ; (d) BPRZ; (e) BPRZ-AMI

FIGURE 11-13 DC wandering

The method of line encoding used determines the minimum bandwidth required for transmission, how easily a clock may be extracted from it, how easily it may be decoded, the average dc voltage level, and whether it offers a convenient means of detecting errors.

11-5-3 Bandwidth Requirements

To determine the minimum bandwidth required to propagate a line-encoded digital signal, you must determine the highest fundamental frequency associated with the signal (see Figure 11-12). The highest fundamental frequency is determined from the worst-case (fastest transition) binary bit sequence. With UPNRZ, the worst-case condition is an alternating 1/0 sequence; the period of the highest fundamental frequency takes the time of two bits and, therefore, is equal to one-half the bit rate ($f_b/2$). With BPNRZ, again the worst-case condition is an alternating 1/0 sequence, and the highest fundamental frequency is one-half the bit rate ($f_b/2$). With UPRZ, the worst-case condition occurs when two successive logic 1s occur. Therefore, the minimum bandwidth is equal to the bit rate (f_b). With BPRZ encoding, the worst-case condition occurs for successive logic 1s or successive logic 0s, and the minimum bandwidth is again equal to the bit rate (f_b). With BPRZ-AMI, the worst-case condition is two or more consecutive logic 1s, and the minimum bandwidth is equal to one-half the bit rate ($f_b/2$).

11-5-4 Clock and Framing Bit Recovery

To recover and maintain clock and framing bit synchronization from the received data, there must be sufficient transitions in the data waveform. With UPNRZ and BPNRZ encoding, a long string of 1s or 0s generates a data signal void of transitions and, therefore, is inadequate for clock recovery. With UPRZ and BPRZ-AMI encoding, a long string of 0s also generates a data signal void of transitions. With BPRZ, a transition occurs in each bit position regardless of whether the bit is a 1 or a 0. Thus, BPRZ is the best encoding scheme for clock recovery. If long sequences of 0s are prevented from occurring, BPRZ-AMI encoding provides sufficient transitions to ensure clock synchronization.

11-5-5 Error Detection

With UPNRZ, BPNRZ, UPRZ, and BPRZ encoding, there is no way to determine if the received data have errors. However, with BPRZ-AMI encoding, an error in any bit will cause a bipolar violation (BPV, or the reception of two or more consecutive logic 1s with the same polarity). Therefore, BPRZ-AMI has a built-in error-detection mechanism. T carriers use BPRZ-AMI with +3 V and −3 V representing a logic 1 and 0 V representing a logic 0.

Table 11-3 summarizes the bandwidth, average voltage, clock recovery, and error-detection capabilities of the line-encoding formats shown in Figure 11-12. From Table 11-3, it can be seen that BPRZ-AMI encoding has the best overall characteristics and is, therefore, the most commonly used encoding format.

11-5-6 Digital Biphase

Digital biphase (sometimes called the *Manchester code* or *diphase*) is a popular type of line encoding that produces a strong timing component for clock recovery and does not cause dc wandering. Biphase is a form of BPRZ encoding that uses one cycle of a square wave at 0° phase to represent a logic 1 and one cycle of a square wave at 180° phase to represent a logic 0. Digital biphase encoding is shown in Figure 11-14. Note that a transition occurs in the center of every signaling element regardless of its logic condition or phase. Thus, biphase produces a strong timing component for clock recovery. In addition, assuming an equal probability of 1s and 0s, the average dc voltage is 0 V, and there is no dc wandering. A disadvantage of biphase is that it contains no means of error detection.

Biphase encoding schemes have several variations, including *biphase M, biphase L,* and *biphase S.* Biphase M is used for encoding SMPTE (Society of Motion Picture and Television Engineers) time-code data for recording on videotapes. Biphase M is well suited for this application because it has no dc component, and the code is self-synchronizing (self-clocking). Self-synchronization is an import feature because it allows clock recovery from the

Table 11-3 Line-Encoding Summary

Encoding Format	Minimum BW	Average DC	Clock Recovery	Error Detection
UPNRZ	$f_b/2^*$	+V/2	Poor	No
BPNRZ	$f_b/2^*$	0 V*	Poor	No
UPRZ	f_b	+V/4	Good	No
BPRZ	f_b	0 V*	Best*	No
BPRZ-AMI	$f_b/2^*$	0 V*	Good	Yes*

*Denotes best performance or quality.

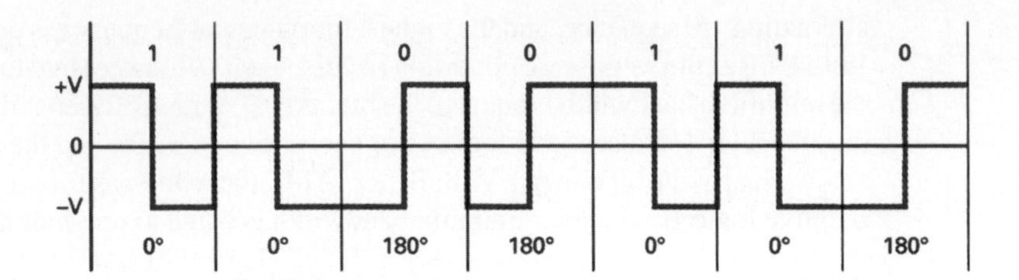

FIGURE 11-14 Digital biphase

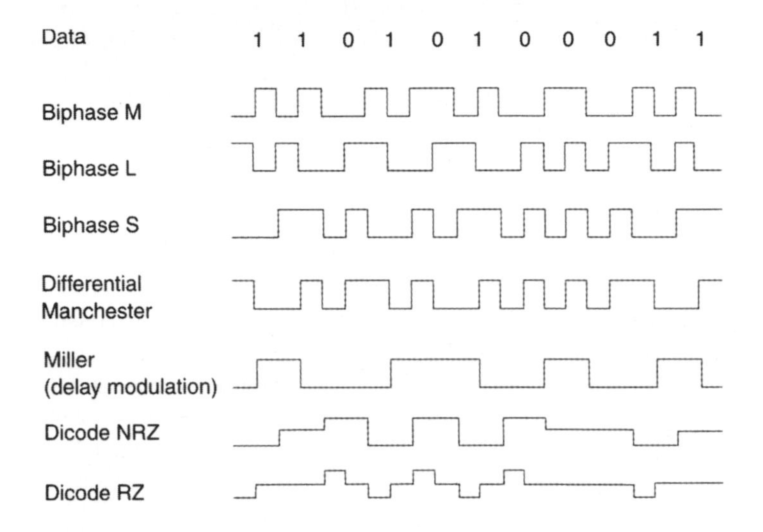

FIGURE 11-15 Biphase, Miller, and dicode encoding formats

data stream even when the speed varies with tape speed, such as when searching through a tape in either the fast or the slow mode. Biphase L is commonly called the Manchester code. Biphase L is specified in IEEE standard 802.3 for Ethernet local area networks (see Chapter 23).

Miller codes are forms of *delay-modulated codes* where a logic 1 condition produces a transition in the middle of the clock pulse, and a logic 0 produces no transition at the end of the clock intervals unless followed by another logic 0.

Dicodes are multilevel binary codes that use more than two voltage levels to represent the data. Bipolar RZ and RZ-AMI are two dicode encoding formats already discussed. Dicode NRZ and dicode RZ are two more commonly used dicode formats.

Figure 11-15 shows several variations of biphase, Miller, and dicode encoding, and Table 11-4 summarizes their characteristics.

11-6 T CARRIER SYSTEMS

T carriers are used for the transmission of PCM-encoded time-division multiplexed digital signals. In addition, T carriers utilize special line-encoded signals and metallic cables that have been conditioned to meet the relatively high bandwidths required for high-speed digital transmission. Digital signals deteriorate as they propagate along a cable because of power losses in the metallic conductors and the low-pass filtering inherent in parallel-wire transmission lines. Consequently, *regenerative repeaters* must be placed at periodic intervals. The distance between repeaters depends on the transmission bit rate and the line-encoding technique used.

Table 11-4 Summary of Biphase, Miller, and Dicode Encoding Formats

Biphase M (biphase-mark)
  1 (hi)—transition in the middle of the clock interval
  0 (low)—no transition in the middle of the clock interval
  Note: There is always a transition at the beginning of the clock interval.
Biphase L (biphase-level/Manchester)
  1 (hi)—transition from high to low in the middle of the clock interval
  0 (low)—transition from low to high in the middle of the clock interval
Biphase S (biphase-space)
  1 (hi)—no transition in the middle of the clock interval
  0 (low)—transition in the middle of the clock interval
  Note: There is always a transition at the beginning of the clock interval.
Differential Manchester
  1 (hi)—transition in the middle of the clock interval
  0 (low)—transition at the beginning of the clock interval
Miller/delay modulation
  1 (hi)—transition in the middle of the clock interval
  0 (low)—no transition at the end of the clock interval unless followed by a zero
Dicode NRZ
  One-to-zero and zero-to-one data transitions change the signal polarity. If the data remain constant, then a zero-voltage level is output.
Dicode RZ
  One-to-zero and zero-to-one data transitions change the signal polarity in half-step voltage increments. If the data do not change, then a zero-voltage level is output.

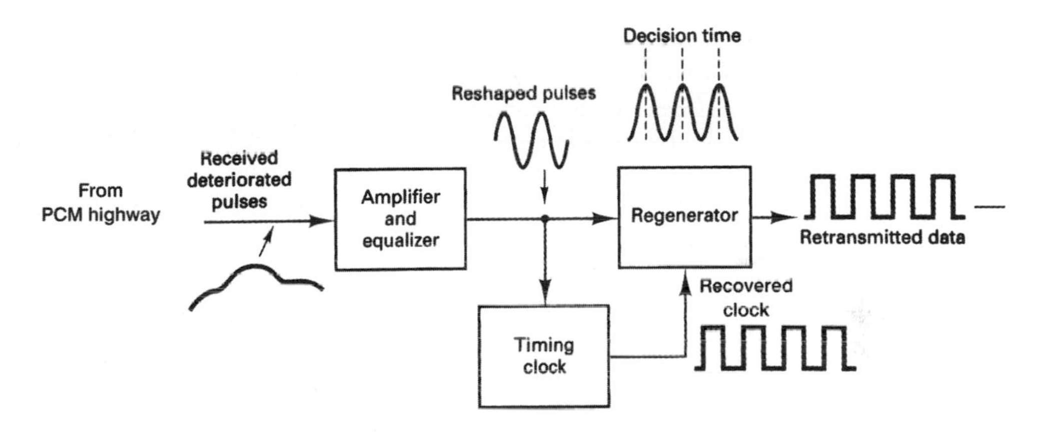

FIGURE 11-16 Regenerative repeater block diagram

Figure 11-16 shows the block diagram for a regenerative repeater. Essentially, there are three functional blocks: an *amplifier/equalizer,* a *timing clock recovery circuit,* and the *regenerator* itself. The amplifier/equalizer filters and shapes the incoming digital signal and raises its power level so that the regenerator circuit can make a pulse–no pulse decision. The timing clock recovery circuit reproduces the clocking information from the received data and provides the proper timing information to the regenerator so that samples can be made at the optimum time, minimizing the chance of an error occurring. A regenerative repeater is simply a threshold detector that compares the sampled voltage received to a reference level and determines whether the bit is a logic 1 or a logic 0.

Spacing of repeaters is designed to maintain an adequate signal-to-noise ratio for error-free performance. The signal-to-noise ratio at the output of a regenerative repeater is exactly what it was at the output of the transmit terminal or at the output of the previous regenerator (i.e., the signal-to-noise ratio does not deteriorate as a digital signal propagates through a regenerator; in fact, a regenerator reconstructs the original pulses with the original signal-to-noise ratio).

11-6-1 T1 Carrier Systems

T1 carrier systems were designed to combine PCM and TDM techniques for short-haul transmission of 24 64-kbps channels with each channel capable of carrying digitally encoded voice-band telephone signals or data. The transmission bit rate (*line speed*) for a T1 carrier is 1.544 Mbps, including an 8-kbps framing bit. The lengths of T1 carrier systems typically range from about 1 mile to over 50 miles.

T1 carriers use BPRZ-AMI encoding with regenerative repeaters placed every 3000, 6000, or 9000 feet. These distances were selected because they were the distances between telephone company manholes where regenerative repeaters are placed. The transmission medium for T1 carriers is generally 19- to 22-gauge twisted-pair metallic cable.

Because T1 carriers use BPRZ-AMI encoding, they are susceptible to losing clock synchronization on long strings of consecutive logic 0s. With a folded binary PCM code, the possibility of generating a long string of contiguous logic 0s is high. When a channel is idle, it generates a 0-V code, which is either seven or eight consecutive logic zeros. Therefore, whenever two or more adjacent channels are idle, there is a high likelihood that a long string of consecutive logic 0s will be transmitted. To reduce the possibility of transmitting a long string of consecutive logic 0s, the PCM data were complemented prior to transmission and then complemented again in the receiver before decoding. Consequently, the only time a long string of consecutive logic 0s are transmitted is when two or more adjacent channels each encode the maximum possible positive sample voltage, which is unlikely to happen.

Ensuring that sufficient transitions occur in the data stream is sometimes called *ones density.* Early T1 and T1C carrier systems provided measures to ensure that no single eight-bit byte was transmitted without at least one bit being a logic 1 or that 15 or more consecutive logic 0s were not transmitted. The transmissions from each frame are monitored for the presence of either 15 consecutive logic 0s or any one PCM sample (eight bits) without at least one nonzero bit. If either of these conditions occurred, a logic 1 is substituted into the appropriate bit position. The worst-case conditions were

	MSB	LSB	MSB	LSB	
Original DS-1 signal	1000	0000	0000	0001	14 consecutive 0s (no substitution)
	MSB	LSB	MSB	LSB	
Original DS-1 signal	1000	0000	0000	0000	15 consecutive 0s
Substituted DS-1 signal	1000	0000	0000	0010 (↑ Substituted bit)	

A 1 is substituted into the second least significant bit, which introduces an encoding error equal to twice the amplitude resolution. This bit is selected rather than the least significant bit because, with the superframe format, during every sixth frame the LSB is the signaling bit, and to alter it would alter the signaling word.

If at any time 32 consecutive logic 0s are received, it is assumed that the system is not generating pulses and is, therefore, out of service.

With modern T1 carriers, a technique called *binary eight zero substitution* (B8ZS) is used to ensure that sufficient transitions occur in the data to maintain clock synchronization. With

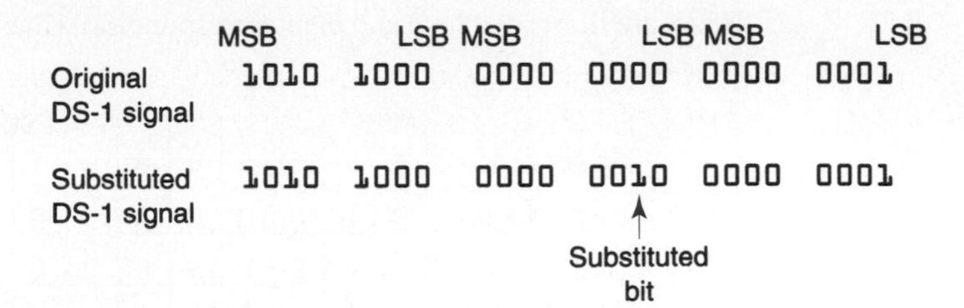

B8ZS, whenever eight consecutive 0s are encountered, one of two special patterns is substituted for the eight 0s, either + − 0 − + 0 0 0 or − + 0 + − 0 0 0. The + (plus) and − (minus) represent bipolar logic 1 conditions, and a 0 (zero) indicates a logic 0 condition. The eight-bit pattern substituted for the eight consecutive 0s is the one that purposely induces bipolar violations in the fourth and seventh bit positions. Ideally, the receiver will detect the bipolar violations and the substituted pattern and then substitute the eight 0s back into the data signal. During periods of low usage, eight logic 1s are substituted into idle channels. Two examples of

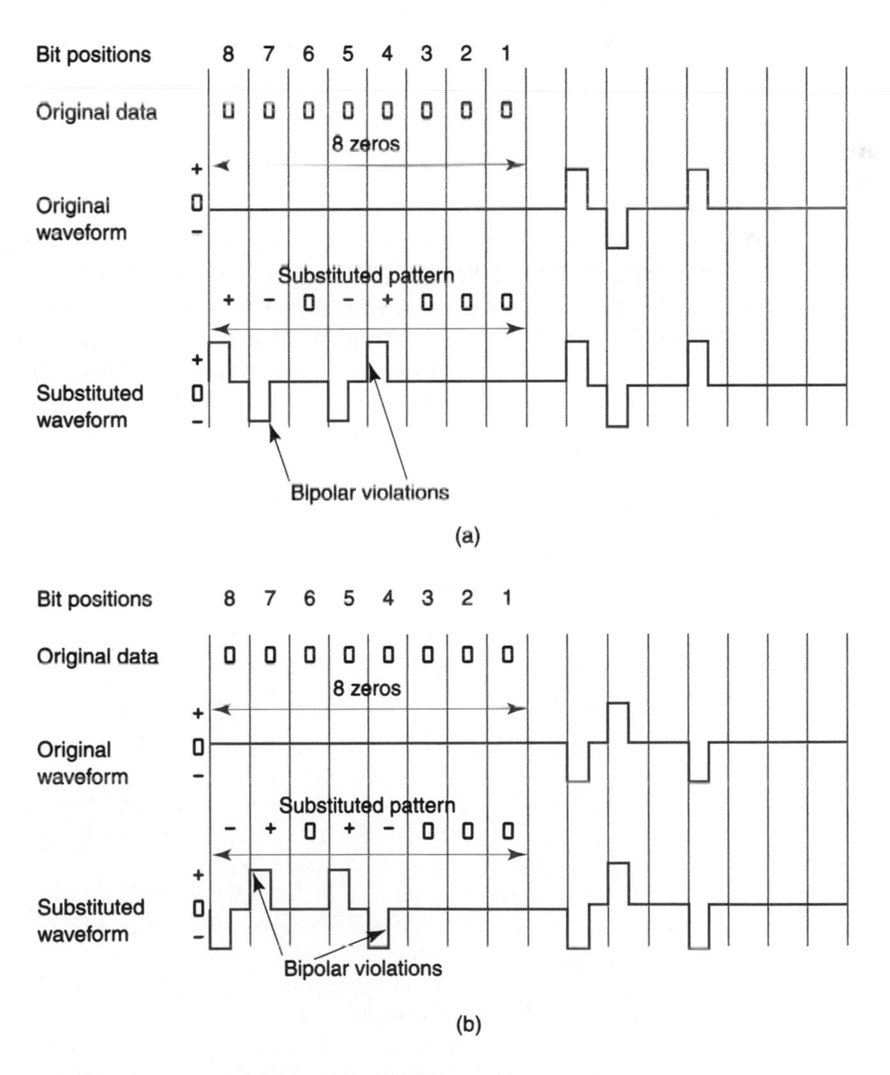

FIGURE 11-17 Waveforms for B8ZS example: (a) substitution pattern 1; (b) substitution pattern 2

B8ZS are illustrated here and their corresponding waveforms shown in Figures 11-17a and b, respectively:

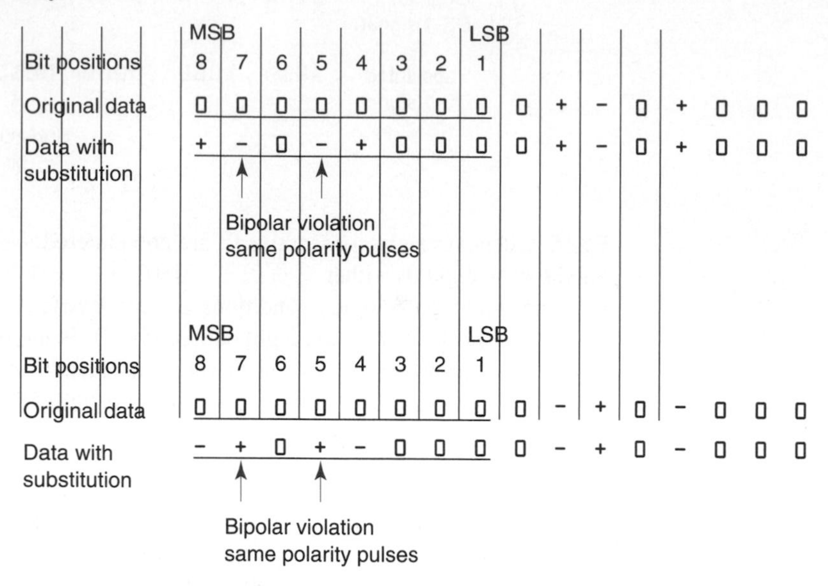

11-6-2 T2 Carrier System

T2 carriers time-division multiplex 96 64-kbps voice or data channels into a single 6.312-Mbps data signal for transmission over twisted-pair copper wire up to 500 miles over a special LOCAP (low capacitance) metallic cable. T2 carriers also use BPRZ-AMI encoding; however, because of the higher transmission rate, clock synchronization is even more critical than with a T1 carrier. A sequence of six consecutive logic 0s could be sufficient to cause loss of clock synchronization. Therefore, T2 carrier systems use an alternative method of ensuring that ample transitions occur in the data. This method is called *binary six zero substitution* (B6ZS).

With B6ZS, whenever six consecutive logic 0s occur, one of the following binary codes is substituted in its place: $0 - + 0 + -$ or $0 + - 0 - +$. Again $+$ and $-$ represent logic 1s, and 0 represents a logic 0. The six-bit code substituted for the six consecutive 0s is selected to purposely cause a bipolar violation. If the violation is detected in the receiver, the original six 0s can be substituted back into the data signal. The substituted patterns produce bipolar violations (i.e., consecutive pulses with the same polarity) in the second and fourth bits of the substituted patterns. If DS-2 signals are multiplexed to form DS-3 signals, the B6ZS code must be detected and removed from the DS-2 signal prior to DS-3 multiplexing. An example of B6ZS is illustrated here and its corresponding waveform shown in Figure 11-18.

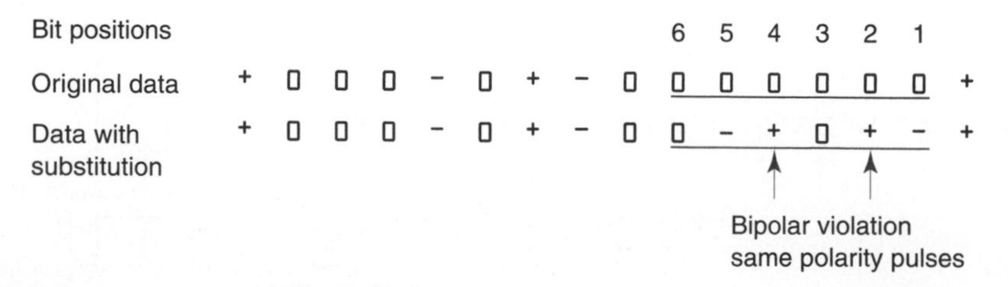

11-6-3 T3 Carrier System

T3 carriers time-division multiplex 672 64-kbps voice or data channels for transmission over a single 3A-RDS coaxial cable. The transmission bit rate for T3 signals is 44.736 Mbps. The coding technique used with T3 carriers is *binary three zero substitution* (B3ZS).

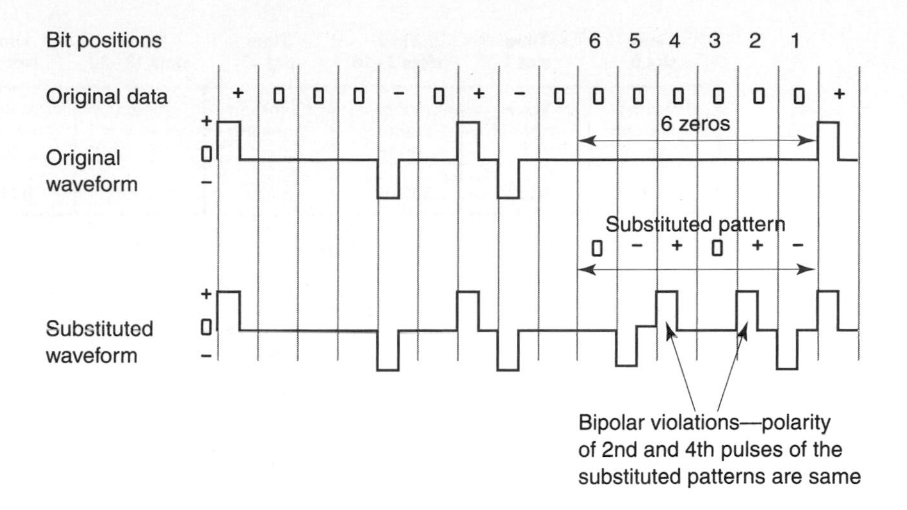

FIGURE 11-18 Waveform for B6ZS example

Substitutions are made for any occurrence of three consecutive 0s. There are four substitution patterns used: 00−, −0−, 00+, and +0+. The pattern chosen should cause a bipolar error in the third substituted bit. An example of B3ZS is shown here:

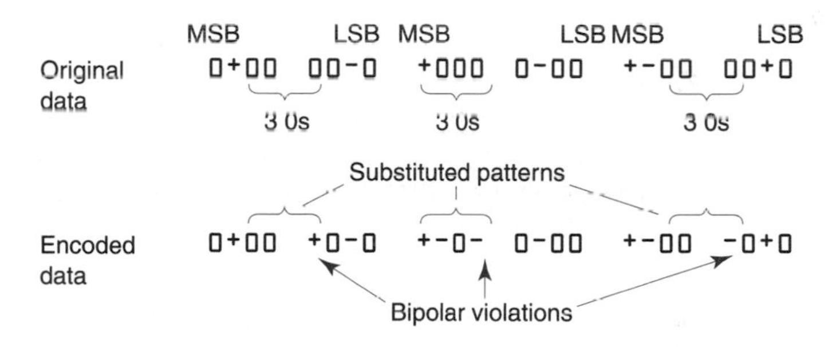

11-6-4 T4M and T5 Carrier Systems

T4M carriers time-division multiplex 4032 64-kbps voice or data channels for transmission over a single T4M coaxial cable up to 500 miles. The transmission rate is sufficiently high that substitute patterns are impractical. Instead, T4M carriers transmit scrambled unipolar NRZ digital signals; the scrambling and descrambling functions are performed in the subscriber's terminal equipment.

T5 carriers time-division multiplex 8064 64-kbps voice or data channels and transmit them at a 560.16 Mbps rate over a single coaxial cable.

11-7 EUROPEAN DIGITAL CARRIER SYSTEM

In Europe, a different version of T carrier lines is used, called *E-lines.* Although the two systems are conceptually the same, they have different capabilities. Figure 11-19 shows the frame alignment for the E1 European standard PCM-TDM system. With the basic E1 system, a 125-μs frame is divided into 32 equal time slots. Time slot 0 is used for a frame alignment pattern and for an alarm channel. Time slot 17 is used for a *common signaling channel* (CSC). The signaling for all 30 voice-band channels is accomplished on the common signaling channel. Consequently, 30 voice-band channels are time-division multiplexed into each E1 frame.

Time slot 0	Time slot 1	Time slots 2–16	Time slot 17	Time slots 18–30	Time slot 31
Framing and alarm channel 8 bits	Voice channel 1 8 bits	Voice channels 2–15 112 bits	Common signaling channel 8 bits	Voice channels 16–29 112 bits	Voice channel 30 8 bits

(a)

Time slot 17

Frame	Bits 1234	Bits 5678
0	0000	xyxx
1	ch 1	ch 16
2	ch 2	ch 17
3	ch 3	ch 18
4	ch 4	ch 19
5	ch 5	ch 20
6	ch 6	ch 21
7	ch 7	ch 22
8	ch 8	ch 23
9	ch 9	ch 24
10	ch 10	ch 25
11	ch 11	ch 26
12	ch 12	ch 27
13	ch 13	ch 28
14	ch 14	ch 29
15	ch 15	ch 30

16 frames equal one multiframe; 500 multiframes are transmitted each second

x = spare
y = loss of multiframe alignment if a 1

4 bits per channel are transmitted once every 16 frames, resulting in a 500 words per second (2000 bps) signaling rate for each channel

(b)

FIGURE 11-19 CCITT TDM frame alignment and common signaling channel alignment: (a) CCITT TDM frame (125 μs, 256 bits, 2.048 Mbps); (b) common signaling channel

Table 11-5 European Transmission Rates and Capacities

Line	Transmission Bit Rate (Mbps)	Channel Capacity
E1	2.048	30
E2	8.448	120
E3	34.368	480
E4	139.264	1920

With the European E1 standard, each time slot has eight bits. Consequently, the total number of bits per frame is

$$\frac{8 \text{ bits}}{\text{time slot}} \times \frac{32 \text{ time slots}}{\text{frame}} = \frac{256 \text{ bits}}{\text{frame}}$$

and the line speed for an E-1 TDM system is

$$\frac{256 \text{ bits}}{\text{frame}} \times \frac{8000 \text{ frames}}{\text{second}} = 2.048 \text{ Mbps}$$

The European digital transmission system has a TDM multiplexing hierarchy similar to the North American hierarchy except the European system is based on the 32-time-slot (30-voice-channel) E1 system. The *European Digital Multiplexing Hierarchy* is shown in Table 11-5. Interconnecting T carriers with E carriers is not generally a problem because most multiplexers and demultiplexers are designed to perform the necessary bit rate conversions.

11-8 DIGITAL CARRIER FRAME SYNCHRONIZATION

With TDM systems, it is imperative not only that a frame be identified but also that individual time slots (samples) within the frame be identified. To acquire frame synchronization, a certain amount of overhead must be added to the transmission. There are several methods used to establish and maintain frame synchronization, including added-digit, robbed-digit, added-channel, statistical, and unique-line code framing.

11-8-1 Added-Digit Framing

T1 carriers using D1, D2, or D3 channel banks use *added-digit framing.* A special *framing digit* (framing pulse) is added to each frame. Consequently, for an 8-kHz sample rate, 8000 digits are added each second. With T1 carriers, an alternating 1/0 frame-synchronizing pattern is used.

To acquire frame synchronization, the digital terminal in the receiver searches through the incoming data until it finds the framing bit pattern. This encompasses testing a bit, counting off 193 more bits, and then testing again for the opposite logic condition. This process continues until a repetitive alternating 1/0 pattern is found. Initial frame synchronization depends on the total frame time, the number of bits per frame, and the period of each bit. Searching through all possible bit positions requires N tests, where N is the number of bit positions in the frame. On average, the receiving terminal dwells at a false framing position for two frame periods during a search; therefore, the maximum average synchronization time is

$$\text{synchronization time} = 2NT = 2N^2 t_b \tag{11-1}$$

where N = number of bits per frame
T = frame period of Nt_b
t_b = bit time

For the T1 carrier, $N = 193$, $T = 125\ \mu s$, and $t_b = 0.648\ \mu s$; therefore, a maximum of 74,498 bits must be tested, and the maximum average synchronization time is 48.25 ms.

11-8-2 Robbed-Digit Framing

When a short frame is used, added-digit framing is inefficient. This occurs with single-channel PCM systems. An alternative solution is to replace the least significant bit of every nth frame with a framing bit. This process is called *robbed-digit framing.* The parameter n is chosen as a compromise between reframe time and signal impairment. For $n = 10$, the SQR is impaired by only 1 dB. Robbed-digit framing does not interrupt transmission but instead periodically replaces information bits with forced data errors to maintain frame synchronization.

11-8-3 Added-Channel Framing

Essentially, *added-channel framing* is the same as added-digit framing except that digits are added in groups or words instead of as individual bits. The European time-division multiplexing scheme previously discussed uses added-channel framing. One of the 32 time slots in each frame is dedicated to a unique synchronizing bit sequence. The average number of bits to acquire frame synchronization using added-channel framing is

$$\frac{N^2}{2(2^K - 1)} \tag{11-2}$$

where N = number of bits per frame
K = number of bits in the synchronizing word

For the European E1 32-channel system, $N = 256$ and $K = 8$. Therefore, the average number of bits needed to acquire frame synchronization is 128.5. At 2.048 Mbps, the synchronization time is approximately 62.7 μs.

11-8-4 Statistical Framing

With *statistical framing,* it is not necessary to either rob or add digits. With the gray code, the second bit is a logic 1 in the central half of the code range and a logic 0 at the extremes. Therefore, a signal that has a centrally peaked amplitude distribution generates a high probability of a logic 1 in the second digit. Hence, the second digit of a given channel can be used for the framing bit.

11-8-5 Unique-Line Code Framing

With *unique-line code framing,* some property of the framing bit is different from the data bits. The framing bit is made either higher or lower in amplitude or with a different time duration. The earliest PCM-TDM systems used unique-line code framing. D1 channel banks used framing pulses that were twice the amplitude of normal data bits. With unique-line code framing, either added-digit or added-word framing can be used, or specified data bits can be used to simultaneously convey information and carry synchronizing signals. The advantage of unique-line code framing is that synchronization is immediate and automatic. The disadvantage is the additional processing requirements necessary to generate and recognize the unique bit.

11-9 BIT VERSUS WORD INTERLEAVING

When time-division multiplexing two or more PCM systems, it is necessary to interleave the transmissions from the various terminals in the time domain. Figure 11-20 shows two methods of interleaving PCM transmissions: *bit interleaving* and *word interleaving.*

T1 carrier systems use word interleaving; eight-bit samples from each channel are interleaved into a single 24-channel TDM frame. Higher-speed TDM systems and delta mod-

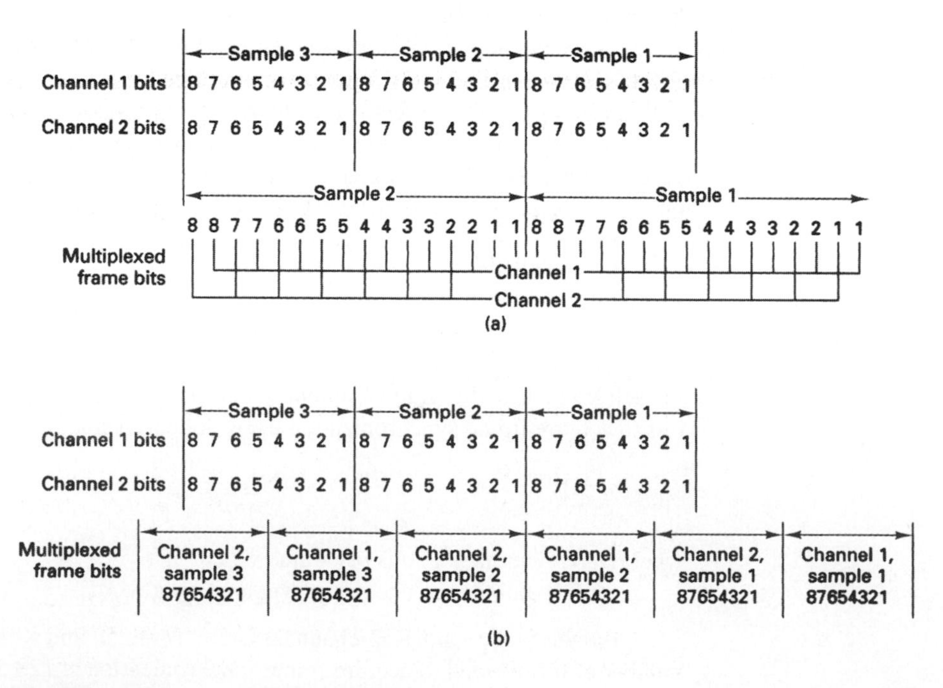

FIGURE 11-20 Interleaving: (a) bit; (b) word

ulation systems use bit interleaving. The decision as to which type of interleaving to use is usually determined by the nature of the signals to be multiplexed.

11-10 STATISTICAL TIME-DIVISION MULTIPLEXING

Digital transmissions over a synchronous TDM system often contain an abundance of time slots within each frame that contain no information (i.e., at any given instant, several of the channels may be idle). For example, TDM is commonly used to link remote data terminals or PCs to a common server or mainframe computer. A majority of the time, however, there are no data being transferred in either direction, even if all the terminals are active. The same is true for PCM-TDM systems carrying digital-encoded voice-grade telephone conversations. Normal telephone conversations generally involve information being transferred in only one direction at a time with significant pauses embedded in typical speech patterns. Consequently, there is a lot of time wasted within each TDM frame. There is an efficient alternative to synchronous TDM called *statistical time-division multiplexing*. Statistical time division multiplexing is generally not used for carrying standard telephone circuits but are used more often for the transmission of data when they are called *asynchronous* TDM, *intelligent* TDM, or simply *stat muxs.*

A statistical TDM multiplexer exploits the natural breaks in transmissions by dynamically allocating time slots on a demand basis. Just as with the multiplexer in a synchronous TDM system, a statistical multiplexer has a finite number of low-speed data input lines with one high-speed multiplexed data output line, and each input line has its own digital encoder and buffer. With the statistical multiplexer, there are n input lines but only k time slots available within the TDM frame (where $k > n$). The multiplexer scans the input buffers, collecting data until a frame is filled, at which time the frame is transmitted. On the receive end, the same holds true, as there are more output lines than time slots within the TDM frame. The demultiplexer removes the data from the time slots and distributes them to their appropriate output buffers.

Statistical TDM takes advantage of the fact that the devices attached to the inputs and outputs are not all transmitting or receiving all the time and that the data rate on the multiplexed line is lower than the combined data rates of the attached devices. In other words, statistical TDM multiplexers require a lower data rate than synchronous multiplexers need to support the same number of inputs. Alternately, a statistical TDM multiplexer operating at the same transmission rate as a synchronous TDM multiplexer can support more users.

Figure 11-21 shows a comparison between statistical and synchronous TDM. Four data sources (A, B, C, and D) and four time slots, or epochs (t_1, t_2, t_3, and t_4). The synchronous multiplexer has an output data rate equal to four times the data rate of each of the input channels. During each sample time, data are collected from all four sources and transmitted regardless of whether there is any input. As the figure shows, during sample time t_1, channels C and D have no input data, resulting in a transmitted TDM frame void of information in time slots C_1 and D_1. With a statistical multiplexer, however, the empty time slots are not transmitted. A disadvantage of the statistical format, however, is that the length of a frame varies and the positional significance of each time slot is lost. There is no way of knowing beforehand which channel's data will be in which time slot or how many time slots are included in each frame. Because data arrive and are distributed to receive buffers unpredictably, address information is required to ensure proper delivery. This necessitates more overhead per time slot for statistical TDM because each slot must carry an address as well as data.

The frame format used by a statistical TDM multiplexer has a direct impact on system performance. Obviously, it is desirable to minimize overhead to improve data throughput. Normally, a statistical TDM system will use a synchronous protocol such as HDLC (described in detail in a later chapter). With statistical multiplexing, control bits must be included

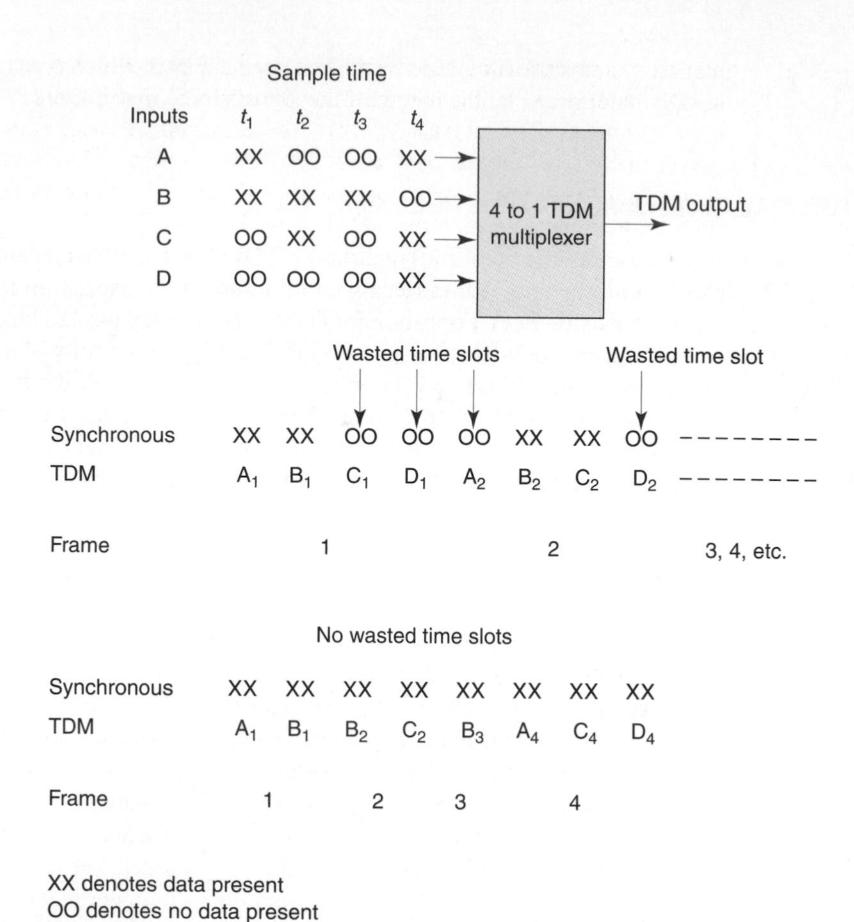

FIGURE 11-21 Comparison between synchronous and statistical TDM

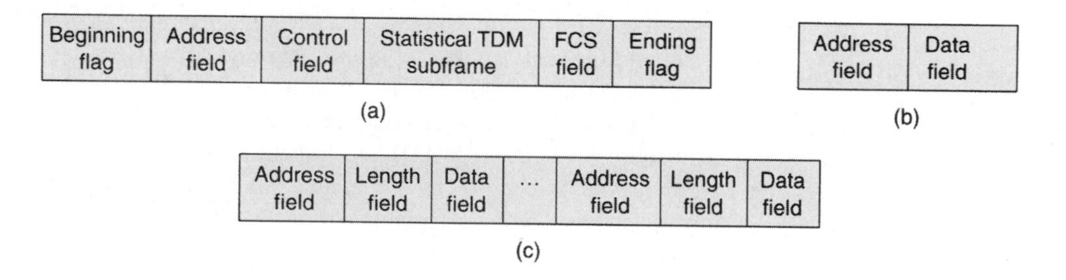

FIGURE 11-22 Statistical TDM frame format: (a) overall statistical TDM frame; (b) one-source per frame; (c) multiple sources per frame

within the frame. Figure 11-22a shows the overall frame format for a statistical TDM multiplexer. The frame includes beginning and ending flags that indicate the beginning and end of the frame, an address field that identifies the transmitting device, a control field, a statistical TDM subframe, and a frame check sequence field (FCS) that provides error detection.

Figure 11-22b shows the frame when only one data source is transmitting. The transmitting device is identified in the address field. The data field length is variable and limited only by the maximum length of the frame. Such a scheme works well in times of light loads but rather inefficiently under heavy loads. Figure 11-14c shows one way to improve the efficiency by allowing more than one data source to be included within a single frame. With multiple sources, however, some means is necessary to specify the length of the data stream from each source. Hence, the statistical frame consists of sequences of data fields labeled with an address and a bit count. There are several techniques that can be used to further improve efficiency. The address field can be shortened by using relative addressing where each address specifies the position of the current source relative to the previously transmitted source and the total number of sources. With relative addressing, an eight-bit address field can be replaced with a four-bit address field.

Another method of refining the frame is to use a two-bit label with the length field. The binary values 01, 10, and 11 correspond to a data field of 1, 2, or 3 bytes, respectively, and no length field necessary is indicated by the code 00.

11-11 CODECS AND COMBO CHIPS

11-11-1 Codec

A *codec* is a large-scale integration (LSI) chip designed for use in the telecommunications industry for private branch exchanges (PBXs), central office switches, digital handsets, voice store-and-forward systems, and digital echo suppressors. Essentially, the codec is applicable for any purpose that requires the digitizing of analog signals, such as in a PCM-TDM carrier system.

Codec is a generic term that refers to the coding functions performed by a device that converts analog signals to digital codes and digital codes to analog signals. Recently developed codecs are called *combo* chips because they combine codec and filter functions in the same LSI package. The input/output filter performs the following functions: bandlimiting, noise rejection, antialiasing, and reconstruction of analog audio waveforms after decoding. The codec performs the following functions: analog sampling, encoding/decoding (analog-to-digital and digital-to-analog conversions), and digital companding.

11-11-2 Combo Chips

A combo chip can provide the analog-to-digital and the digital-to-analog conversions and the transmit and receive filtering necessary to interface a full-duplex (four-wire) voice telephone circuit to the PCM highway of a TDM carrier system. Essentially, a combo chip replaces the older codec and filter chip combination.

Table 11-6 lists several of the combo chips available and their prominent features.

Table 11-6 Features of Several Codec/Filter Combo Chips

2916 (16-Pin)	2917 (16-Pin)	2913 (20-Pin)	2914 (24 Pin)
μ-law companding only	A-law companding only	μ/A-law companding	μ/A-law companding
Master clock, 2.048 MHz only	Master clock, 2.048 MHz only	Master clock, 1.536 MHz, 1.544 MHz, or 2.048 MHz	Master clock, 1.536 MHz, 1.544 MHz, or 2.048 MHz
Fixed data rate	Fixed data rate	Fixed data rate	Fixed data rate
Variable data rate, 64 kbps–2.048 Mbps	Variable data rate, 64 kbps–4.096 Mbps	Variable data rate, 64 kbps–4.096 Mbps	Variable data rate, 64 kbps–4.096 Mbps
78-dB dynamic range	78-dB dynamic range	78-dB dynamic range	78-dB dynamic range
ATT D3/4 compatible	ATT D3/4 compatible	ATT D3/4 compatible	ATT D3/4 compatible
Single-ended input	Single-ended input	Differential input	Differential input
Single-ended output	Single-ended output	Differential output	Differential output
Gain adjust transmit only	Gain adjust transmit only	Gain adjust transmit and receive	Gain adjust transmit and receive
Synchronous clocks	Synchronous clocks	Synchronous clocks	Synchronous clocks
			Asynchronous clocks
			Analog loopback
			Signaling

11-11-2-1 General operation. The following major functions are provided by a combo chip:

1. Bandpass filtering of the analog signals prior to encoding and after decoding
2. Encoding and decoding of voice and call progress signals
3. Encoding and decoding of signaling and supervision information
4. Digital companding

Figure 11-23a shows the block diagram of a typical combo chip. Figure 11-23b shows the frequency response curve for the transmit bandpass filter, and Figure 11-23c shows the frequency response for the receive low-pass filter.

11-11-2-2 Fixed-data-rate mode. In the *fixed-data-rate mode,* the master *transmit* and *receive clocks* on a combo chip (CLKX and CLKR) perform the following functions:

1. Provide the master clock for the on-board switched capacitor filter
2. Provide the clock for the analog-to-digital and digital-to-analog converters
3. Determine the input and output data rates between the codec and the PCM highway

Therefore, in the fixed-data-rate mode, the transmit and receive data rates must be either 1.536 Mbps, 1.544 Mbps, or 2.048 Mbps—the same as the master clock rate.

Transmit and receive frame synchronizing pulses (FSX and FSR) are 8-kHz inputs that set the transmit and receive sampling rates and distinguish between *signaling* and *nonsignaling* frames. $\overline{\text{TSX}}$ is a *time-slot strobe buffer enable* output that is used to gate the PCM word onto the PCM highway when an external buffer is used to drive the line. $\overline{\text{TSX}}$ is also used as an external gating pulse for a time-division multiplexer (see Figure 11-24a).

Data are transmitted to the PCM highway from DX on the first eight positive transitions of CLKX following the rising edge of FSX. On the receive channel, data are received from the PCM highway from DR on the first eight falling edges of CLKR after the occurrence of FSR. Therefore, the occurrence of FSX and FSR must be synchronized between codecs in a multiple-channel system to ensure that only one codec is transmitting to or receiving from the PCM highway at any given time.

Figures 11-24a and b show the block diagram and timing sequence for a single-channel PCM system using a combo chip in the fixed-data-rate mode and operating with a master clock frequency of 1.536 MHz. In the fixed-data-rate mode, data are input and output for a single channel in short bursts. (This mode of operation is sometimes called the *burst mode.*) With only a single channel, the PCM highway is active only 1/24 of the total frame time. Additional channels can be added to the system provided that their transmissions are synchronized so that they do not occur at the same time as transmissions from any other channel.

From Figure 11-24, the following observations can be made:

1. The input and output bit rates from the codec are equal to the master clock frequency, 1.536 Mbps.
2. The codec inputs and outputs 64,000 PCM bits per second.
3. The data output (DX) and data input (DR) are enabled only 1/24 of the total frame time (125 μs).

To add channels to the system shown in Figure 11-24, the occurrence of the FSX, FSR, and $\overline{\text{TSX}}$ signals for each additional channel must be synchronized so that they follow a timely sequence and do not allow more than one codec to transmit or receive at the

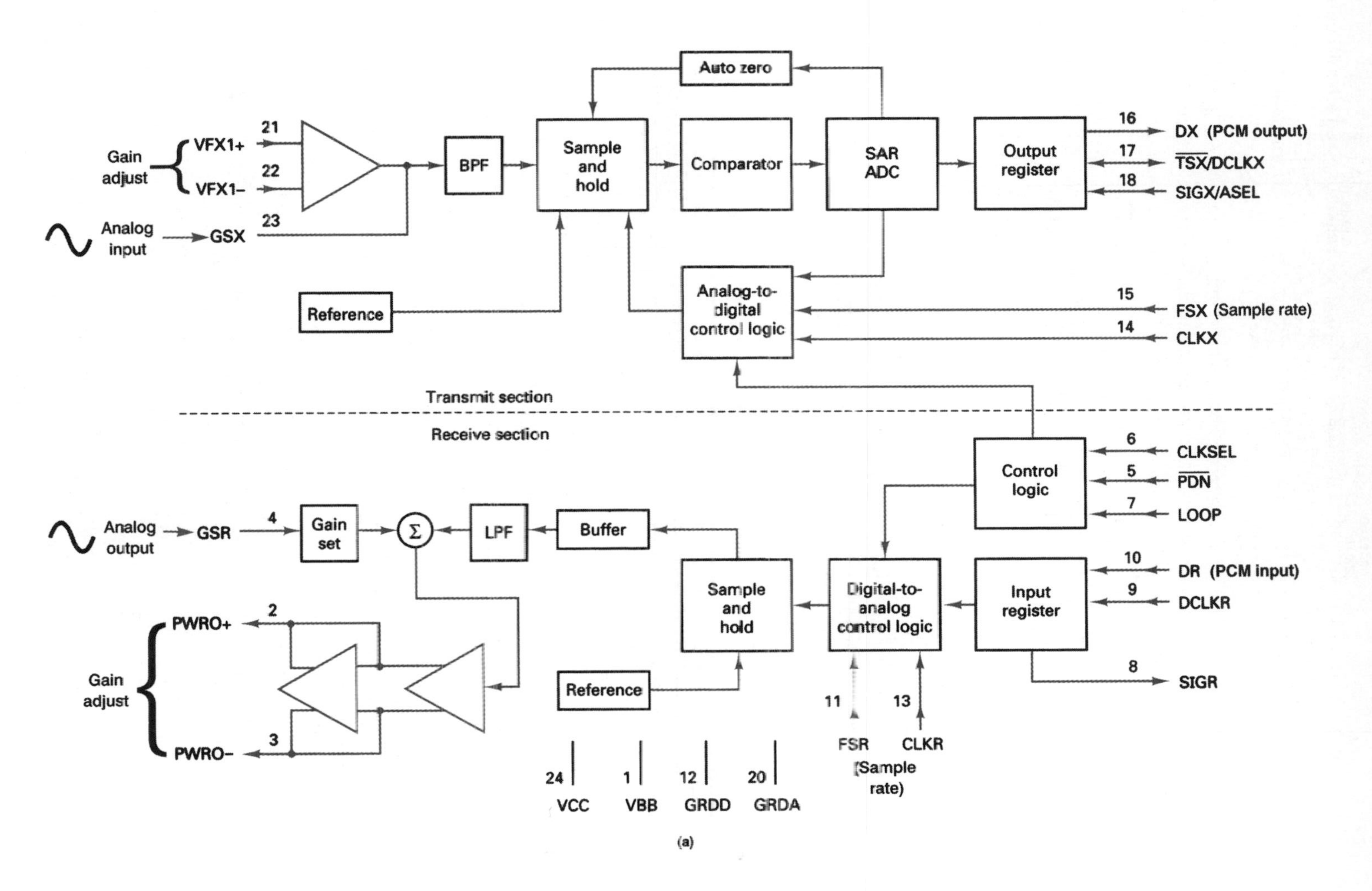

FIGURE 11-23 Combo chip: (a) block diagram; (b) transmit BPF response curve; (c) receive LPF response curve (*Continued*)

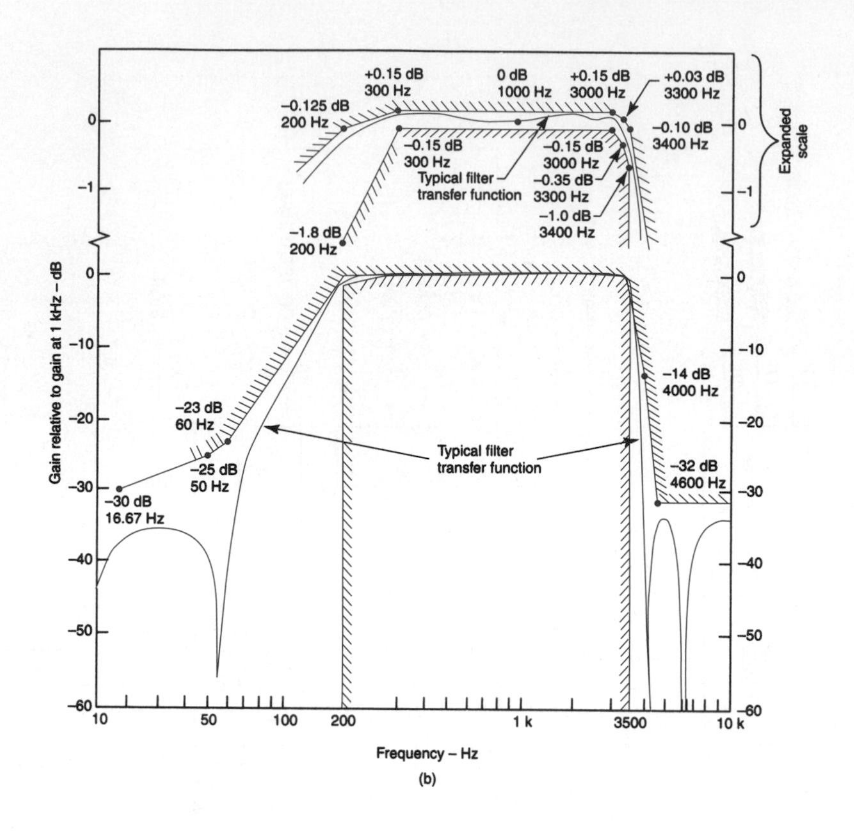

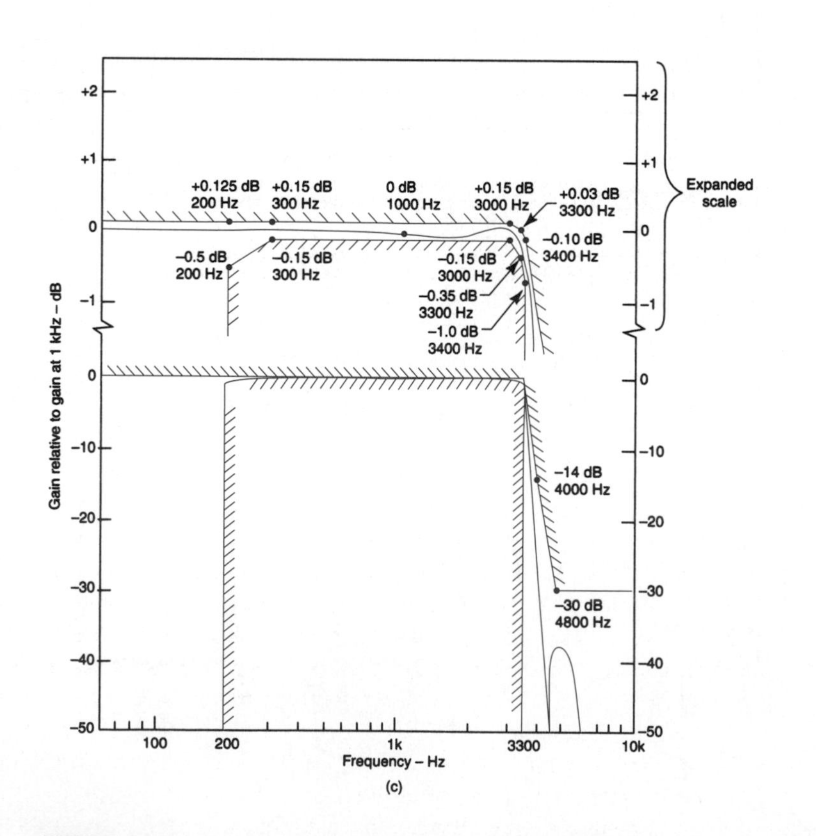

FIGURE 11-23 (*Continued*) Combo chip: (b) transmit BPF response curve; (c) receive LPF response curve

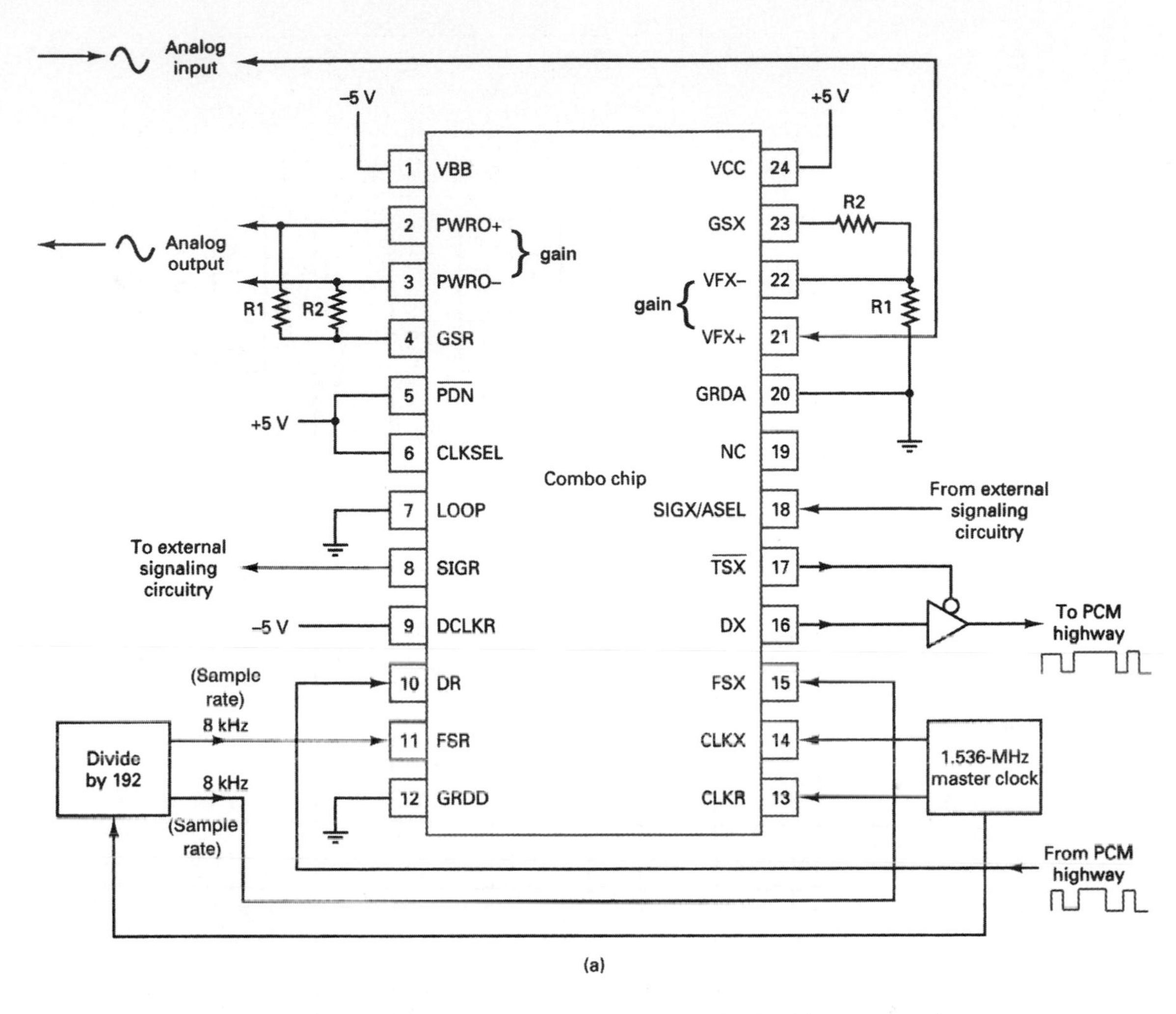

FIGURE 11-24 Single-channel PCM system using a combo chip in the fixed-data-rate mode: (a) block diagram; (*Continued*)

same time. Figures 11-25a and b show the block diagram and timing sequence for a 24-channel PCM-TDM system operating with a master clock frequency of 1.536 MHz.

11-11-2-3 Variable-data-rate mode. The *variable-data-rate mode* allows for a flexible data input and output clock frequency. It provides the ability to vary the frequency of the transmit and receive bit clocks. In the variable-data-rate mode, a master clock frequency of 1.536 MHz, 1.544 MHz, or 2.048 MHz is still required for proper operation of the onboard bandpass filters and the analog-to-digital and digital-to-analog converters. However, in the variable-data-rate mode, DCLKR and DCLKX become the data clocks for the receive and transmit PCM highways, respectively. When FSX is high, data are transmitted onto the PCM highway on the next eight consecutive positive transitions of DCLKX. Similarly, while FSR is high, data from the PCM highway are clocked into the codec on the next eight consecutive negative transitions of DCLKR. This mode of operation is sometimes called the *shift register mode.*

On the transmit channel, the last transmitted PCM word is repeated in all remaining time slots in the 125-μs frame as long as DCLKX is pulsed and FSX is held active high.

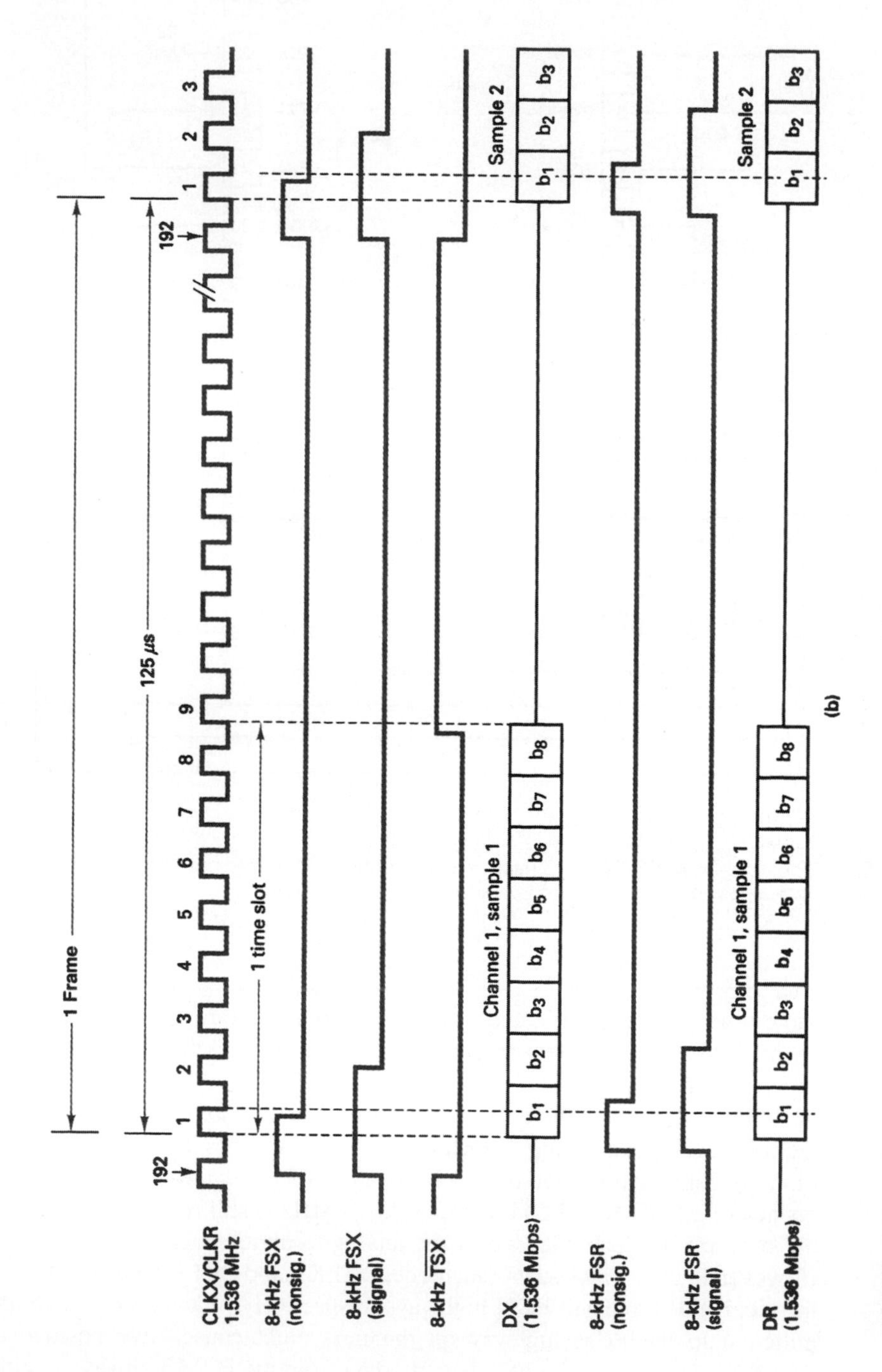

FIGURE 11-24 *(Continued)* (b) timing sequence

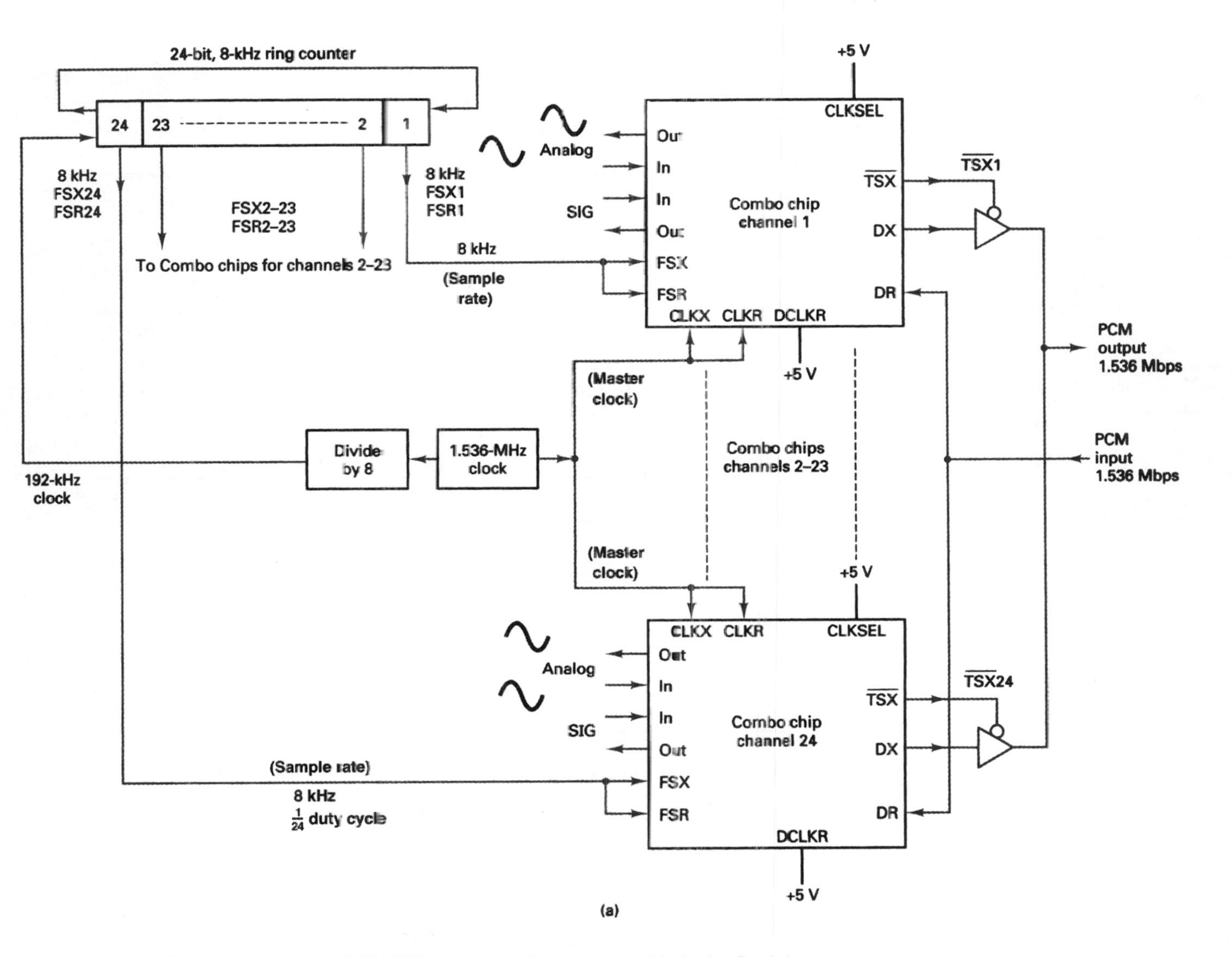

FIGURE 11-25 Twenty-four channel PCM-TDM system using a combo chip in the fixed-data-rate mode and operating with a master clock frequency of 1.536 MHz: (a) block diagram; (*Continued*)

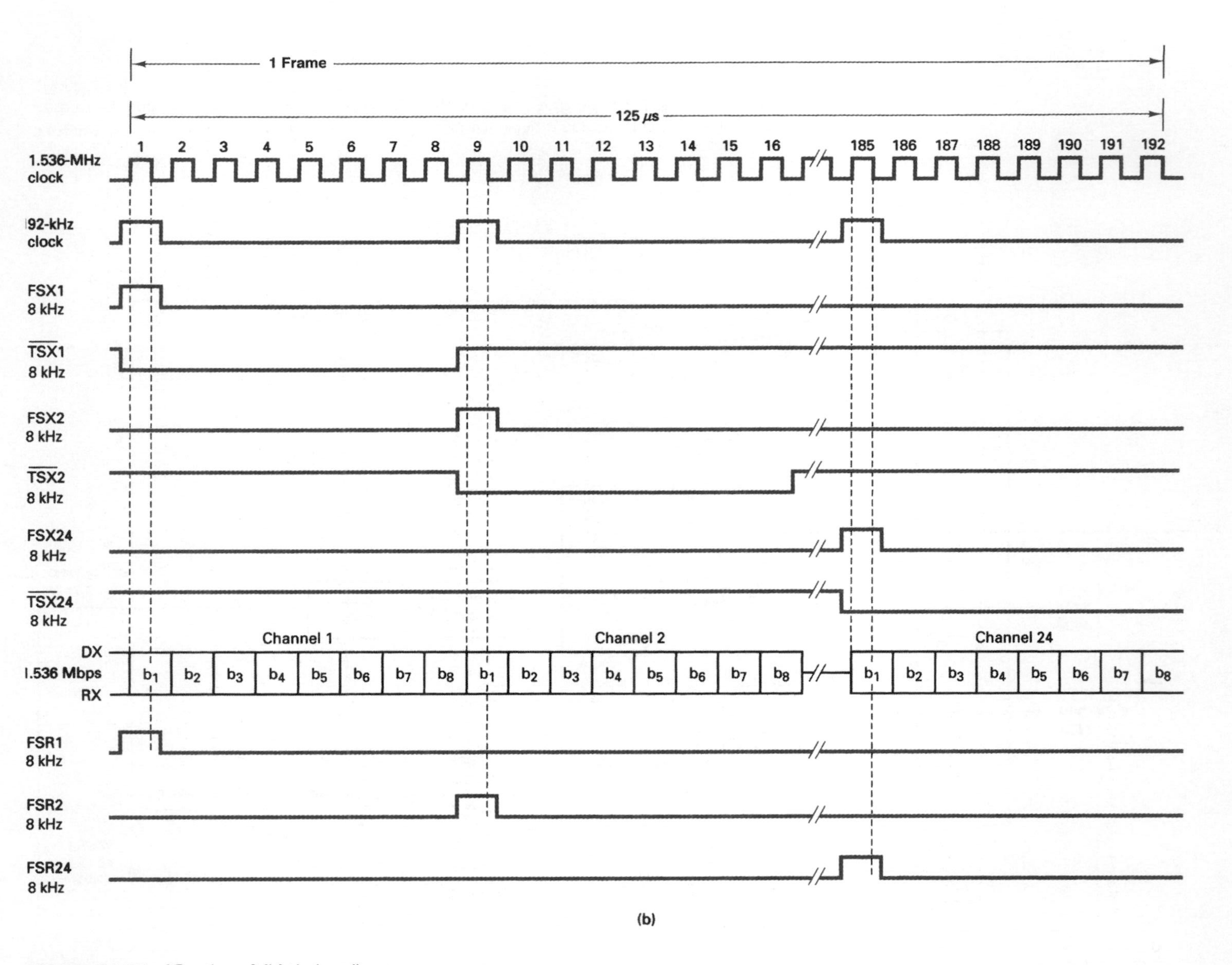

FIGURE 11-25 (*Continued*) (b) timing diagram

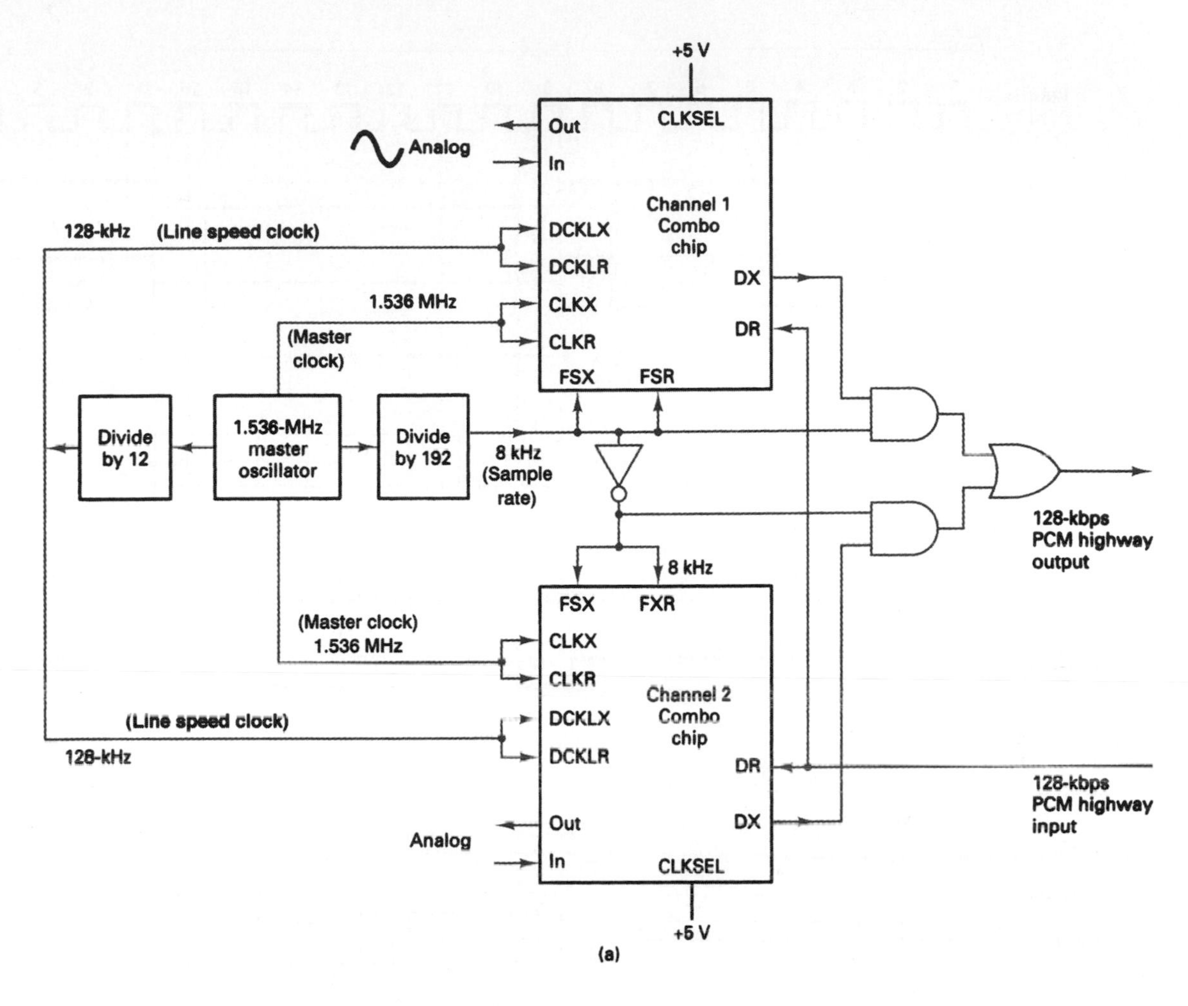

FIGURE 11-26 Two-channel PCM-TDM system using a combo chip in the variable-data-rate mode with a master clock frequency of 1.536 MHz: (a) block diagram; (*Continued*)

This feature allows the PCM word to be transmitted to the PCM highway more than once per frame. Signaling is not allowed in the variable-data-rate mode because this mode provides no means to specify a signaling frame.

Figures 11-26a and b shows the block diagram and timing sequence for a two-channel PCM-TDM system using a combo chip in the variable-data-rate mode with a master clock frequency of 1.536 MHz, a sample rate of 8 kHz, and a transmit and receive data rate of 128 kbps.

With a sample rate of 8 kHz, the frame time is 125 μs. Therefore, one eight-bit PCM word from each channel is transmitted and/or received during each 125-μs frame. For 16 bits to occur in 125 μs, a 128-kHz transmit and receive data clock is required:

$$t_b = \frac{1 \text{ channel}}{8 \text{ bits}} \times \frac{1 \text{ frame}}{2 \text{ channels}} \times \frac{125\ \mu\text{s}}{\text{frame}} = \frac{125\ \mu\text{s}}{16 \text{ bits}} = \frac{7.8125\ \mu\text{s}}{\text{bit}}$$

$$\text{bit rate} = \frac{1}{t_b} = \frac{1}{7.8125\ \mu\text{s}} = 128 \text{ kbps}$$

or

$$\frac{8 \text{ bits}}{\text{channel}} \times \frac{2 \text{ channels}}{\text{frame}} \times \frac{8000 \text{ frames}}{\text{second}} = 128 \text{ kbps}$$

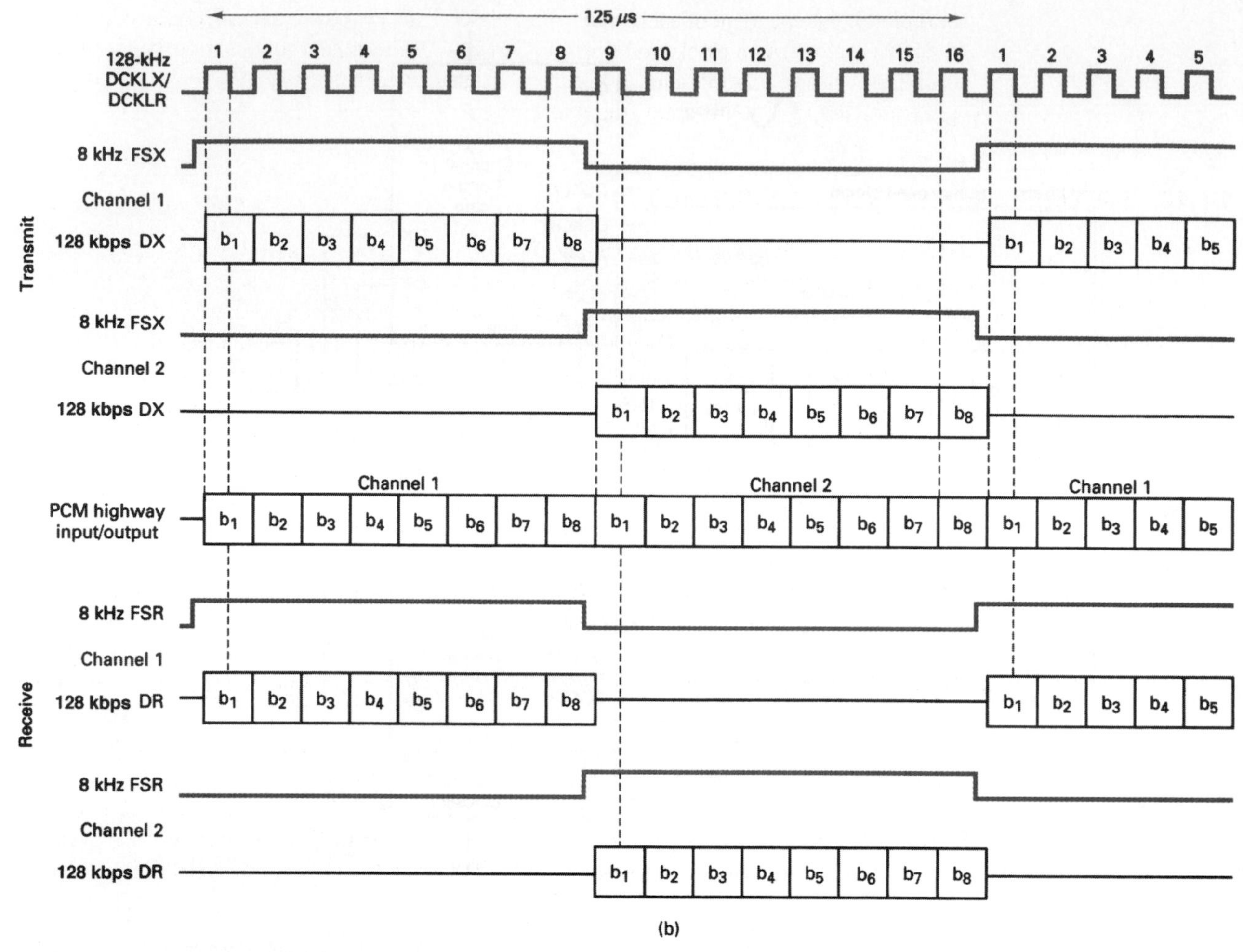

FIGURE 11-26 (*Continued*) (b) timing diagram

The transmit and receive enable signals (FSX and FSR) for each codec are active for one-half of the total frame time. Consequently, 8-kHz, 50% duty cycle transmit and receive data enable signals (FSX and FXR) are fed directly to one codec and fed to the other codec 180° out of phase (inverted), thereby enabling only one codec at a time.

To expand to a four-channel system, simply increase the transmit and receive data clock rates to 256 kHz and change the enable signals to 8-kHz, 25% duty cycle pulses.

11-11-2-4 Supervisory signaling. With a combo chip, *supervisory signaling* can be used only in the fixed-data-rate mode. A transmit signaling frame is identified by making the FSX and FSR pulses twice their normal width. During a transmit signaling frame, the signal present on input SIGX is substituted into the least significant bit position (b_1) of the encoded PCM word. At the receive end, the signaling bit is extracted from the PCM word prior to decoding and placed on output SIGR until updated by reception of another signaling frame.

Asynchronous operation occurs when the master transmit and receive clocks are derived from separate independent sources. A combo chip can be operated in either the synchronous or the asynchronous mode using separate digital-to-analog converters and voltage references in the transmit and receive channels, which allows them to be operated completely independent of each other. With either synchronous or asynchronous operation, the master clock, data clock, and time-slot strobe must be synchronized at the beginning of each frame. In the variable-data-rate mode, CLKX and DCLKX must be synchronized once per frame but may be different frequencies.

11-12 FREQUENCY-DIVISION MULTIPLEXING

With *frequency-division multiplexing* (FDM), multiple sources that originally occupied the same frequency spectrum are each converted to a different frequency band and transmitted simultaneously over a single transmission medium, which can be a physical cable or the Earth's atmosphere (i.e., wireless). Thus, many relatively narrow-bandwidth channels can be transmitted over a single wide-bandwidth transmission system without interfering with each other. FDM is used for combining many relatively narrowband sources into a single wideband channel, such as in public telephone systems. Essentially, FDM is taking a given bandwidth and subdividing it into narrower segments with each segment carrying different information.

FDM is an analog multiplexing scheme; the information entering an FDM system must be analog, and it remains analog throughout transmission. If the original source information is digital, it must be converted to analog before being frequency-division multiplexed.

A familiar example of FDM is the commercial AM broadcast band, which occupies a frequency spectrum from 535 kHz to 1605 kHz. Each broadcast station carries an information signal (voice and music) that occupies a bandwidth between 0 Hz and 5 kHz. If the information from each station were transmitted with the original frequency spectrum, it would be impossible to differentiate or separate one station's transmissions from another. Instead, each station amplitude modulates a different carrier frequency and produces a 10-kHz signal. Because the carrier frequencies of adjacent stations are separated by 10 kHz, the total commercial AM broadcast band is divided into 107 10-kHz frequency slots stacked next to each other in the frequency domain. To receive a particular station, a receiver is simply tuned to the frequency band associated with that station's transmissions. Figure 11-27 shows how commercial AM broadcast station signals are frequency-division multiplexed and transmitted over a common transmission medium (Earth's atmosphere).

With FDM, each narrowband channel is converted to a different location in the total frequency spectrum. The channels are stacked on top of one another in the frequency domain. Figure 11-28a shows a simple FDM system where four 5-kHz channels are frequency-division multiplexed into a single 20-kHz combined channel. As the figure shows,

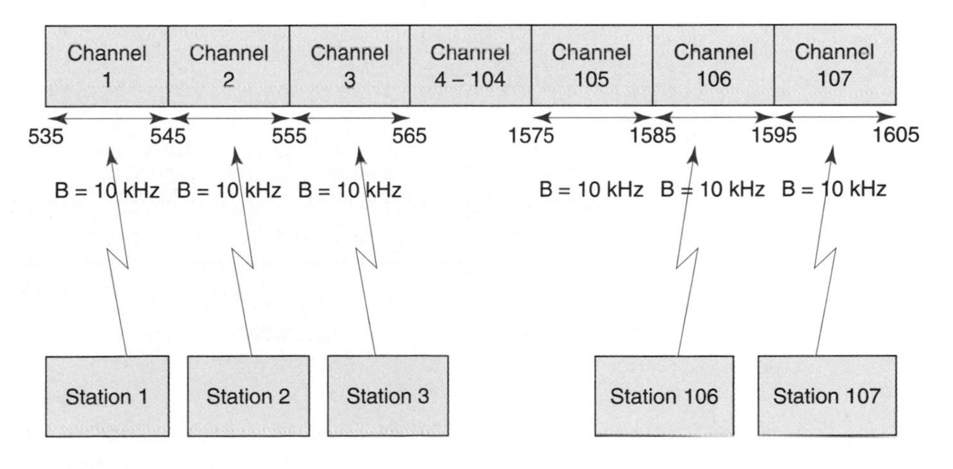

FIGURE 11-27 Frequency-division multiplexing of the commercial AM broadcast band

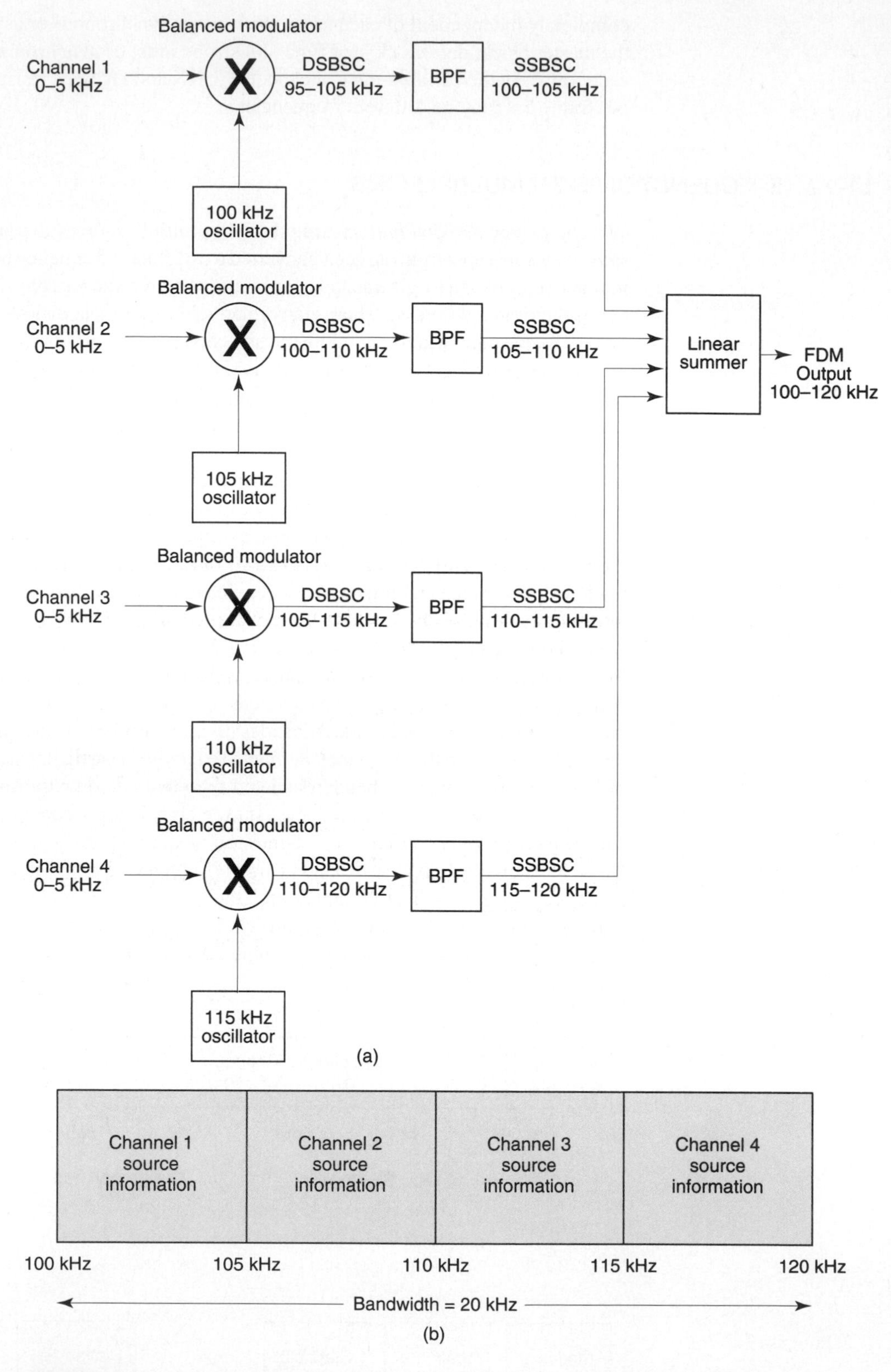

FIGURE 11-28 Frequency-division multiplexing: (a) block diagram; (b) frequency spectrum

channel 1 signals amplitude modulate a 100-kHz carrier in a balanced modulator, which inherently suppresses the 100-kHz carrier. The output of the balanced modulator is a double-sideband suppressed-carrier waveform with a bandwidth of 10 kHz. The double sideband waveform passes through a bandpass filter (BPF) where it is converted to a single sideband signal. For this example, the lower sideband is blocked; thus, the output of the BPF occupies the frequency band between 100 kHz and 105 kHz (a bandwidth of 5 kHz).

Channel 2 signals amplitude modulate a 105-kHz carrier in a balanced modulator, again producing a double sideband signal that is converted to single sideband by passing it through a bandpass filter tuned to pass only the upper sideband. Thus, the output from the BPF occupies a frequency band between 105 kHz and 110 kHz. The same process is used to convert signals from channels 3 and 4 to the frequency bands 110 kHz to 115 kHz and 115 kHz to 120 kHz, respectively. The combined frequency spectrum produced by combining the outputs from the four bandpass filters is shown in Figure 11-28b. As the figure shows, the total combined bandwidth is equal to 20 kHz, and each channel occupies a different 5-kHz portion of the total 20-kHz bandwidth.

There are many other applications for FDM, such as commercial FM and television broadcasting, high-volume telephone and data communications systems, and cable television and data distribution networks. Within any of the commercial broadcast frequency bands, each station's transmissions are independent of all the other stations' transmissions. Consequently, the multiplexing (stacking) process is accomplished without synchronization between stations. With a high-volume telephone communications system, many voice-band telephone channels may originate from a common source and terminate in a common destination. The source and destination terminal equipment is most likely a high-capacity electronic switching system (ESS). Because of the possibility of a large number of narrowband channels originating and terminating at the same location, all multiplexing and demultiplexing operations must be synchronized.

11-13 AT&T'S FDM HIERARCHY

Although AT&T is no longer the only long-distance common carrier in the United States, it still provides the vast majority of the long-distance services and, if for no other reason than its overwhelming size, has essentially become the standards organization for the telephone industry in North America.

AT&T's nationwide communications network is subdivided into two classifications: *short haul* (short distance) and *long haul* (long distance). The T1 carrier explained earlier in this chapter is an example of a short-haul communications system.

Figure 11-29 shows AT&T's long-haul FDM hierarchy. Only the transmit terminal is shown, although a complete set of inverse functions must be performed at the receiving terminal. As the figure shows, voice channels are combined to form groups, groups are combined to form supergroups, and supergroups are combined to form mastergroups.

11-13-1 Message Channel

The *message channel* is the basic building block of the FDM hierarchy. The basic message channel was originally intended for the analog voice transmission, although it now includes any transmissions that utilize voice-band frequencies (0 kHz to 4 kHz), such as data transmission using voice-band data modems. The basic voice-band (VB) circuit is called a basic 3002 channel and is actually bandlimited to approximately a 300-Hz to 3000-Hz frequency band, although for practical design considerations it is considered a 4-kHz channel. The basic 3002 channel can be subdivided and frequency-division multiplexed into 24 narrower-band 3001 (telegraph) channels.

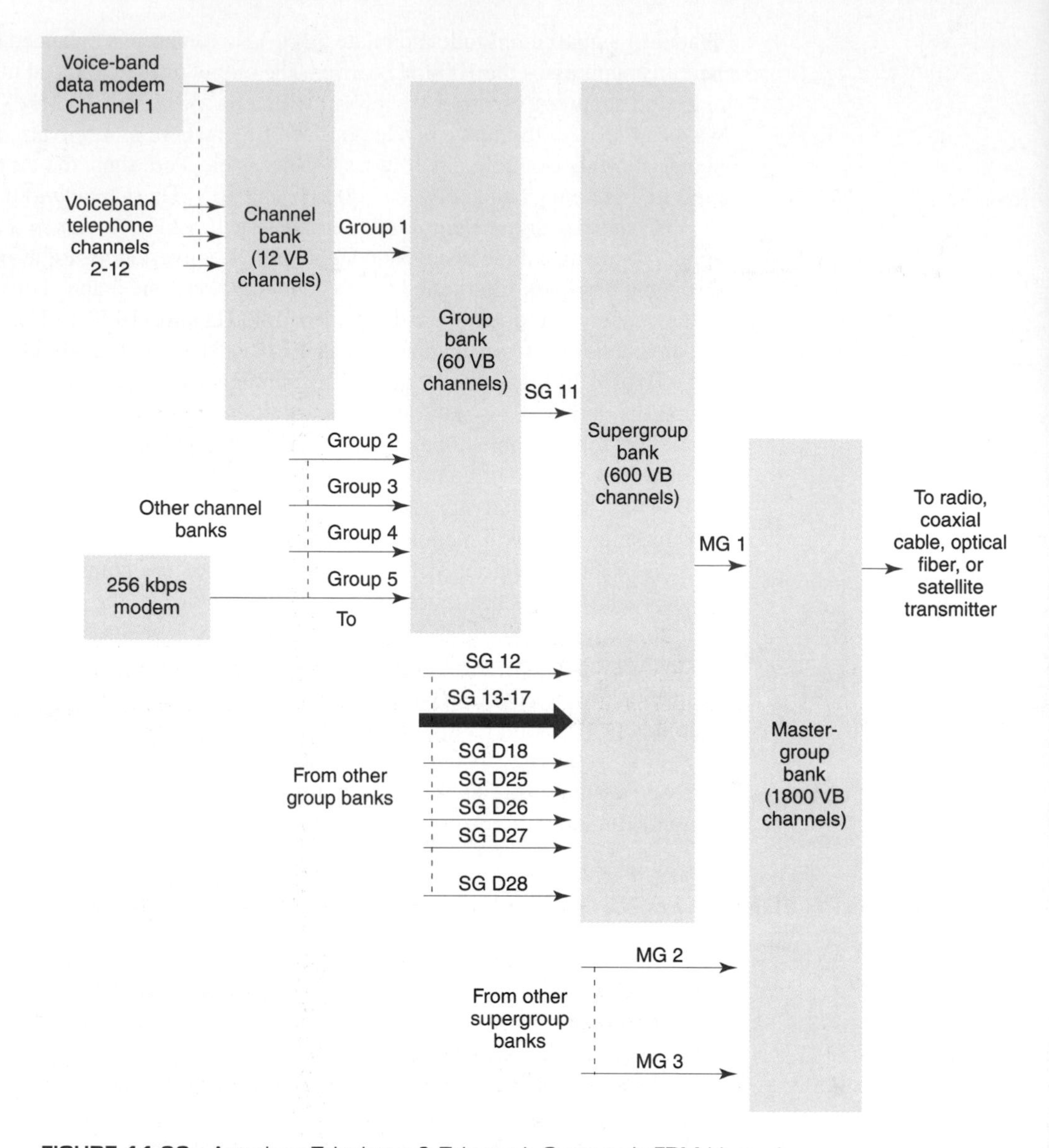

FIGURE 11-29 American Telephone & Telegraph Company's FDM hierarchy

11-13-2 Basic Group

A *group* is the next higher level in the FDM hierarchy above the basic message channel and, consequently, is the first multiplexing step for combining message channels. A basic group consists of 12 voice-band message channels multiplexed together by stacking them next to each other in the frequency domain. Twelve 4-kHz voice-band channels occupy a combined bandwidth of 48 kHz (4 × 12). The 12-channel modulating block is called an A-type (analog) channel bank. The 12-channel group output from an A-type channel bank is the standard building block for most long-haul broadband telecommunications systems.

11-13-3 Basic Supergroup

The next higher level in the FDM hierarchy shown in Figure 11-29 is the *supergroup*, which is formed by frequency-division multiplexing five groups containing 12 channels each for a combined bandwidth of 240 kHz (5 groups × 48 kHz/group or 5 groups × 12 channels/group × 4 kHz/channel).

11-13-4 Basic Mastergroup

The next highest level of multiplexing shown in Figure 11-29 is the *mastergroup*, which is formed by frequency-division multiplexing 10 supergroups together for a combined capacity of 600 voice-band message channels occupying a bandwidth of 2.4 MHz (600 channels × 4 kHz/channel or 5 groups × 12/channels/group × 10 groups/supergroup). Typically, three mastergroups are frequency-division multiplexed together and placed on a single microwave or satellite radio channel. The capacity is 1800 VB channels (3 mastergroups × 600 channels/mastergroup) utilizing a combined bandwidth of 7.2 MHz.

11-13-5 Larger Groupings

Mastergroups can be further multiplexed in mastergroup banks to form *jumbogroups* (3600 VB channels), *multijumbogroups* (7200 VB channels), and *superjumbogroups* (10,800 VB channels).

11-14 COMPOSITE BASEBAND SIGNAL

Baseband describes the modulating signal (intelligence) in a communications system. A single message channel is baseband. A group, supergroup, or mastergroup is also baseband. The composite baseband signal is the total intelligence signal prior to modulation of the final carrier. In Figure 11-29, the output of a channel bank is baseband. Also, the output of a group or supergroup bank is baseband. The final output of the FDM multiplexer is the *composite* (total) baseband. The formation of the composite baseband signal can include channel, group, supergroup, and mastergroup banks, depending on the capacity of the system.

11-14-1 Formation of Groups and Supergroups

Figure 11-30 shows how a group is formed with an A-type channel bank. Each voice-band channel is bandlimited with an antialiasing filter prior to modulating the channel carrier. FDM uses single-sideband suppressed-carrier (SSBSC) modulation. The combination of the balanced modulator and the bandpass filter makes up the SSBSC modulator. A balanced modulator is a double-sideband suppressed-carrier modulator, and the bandpass filter is tuned to the difference between the carrier and the input voice-band frequencies (LSB). The ideal input frequency range for a single voice-band channel is 0 kHz to 4 kHz. The carrier frequencies for the channel banks are determined from the following expression:

$$f_c = 112 - 4n \text{ kHz} \tag{11-3}$$

where n is the channel number. Table 11-7 lists the carrier frequencies for channels 1 through 12. Therefore, for channel 1, a 0-kHz to 4-kHz band of frequencies modulates a 108-kHz carrier. Mathematically, the output of a channel bandpass filter is

$$f_{\text{out}} = (f_c - 4 \text{ kHz}) \text{ to } f_c \tag{11-4}$$

where f_c = channel carrier frequency ($112 - 4n$ kHz) and each voice-band channel has a 4-kHz bandwidth.

For channel 1, $f_{\text{out}} = 108 \text{ kHz} - 4 \text{ kHz} = 104 \text{ kHz to } 108 \text{ kHz}$
For channel 2, $f_{\text{out}} = 104 \text{ kHz} - 4 \text{ kHz} = 100 \text{ kHz to } 104 \text{ kHz}$
For channel 12, $f_{\text{out}} = 64 \text{ kHz} - 4 \text{ kHz} = 60 \text{ kHz to } 64 \text{ kHz}$

The outputs from the 12 A-type channel modulators are summed in the *linear* combiner to produce the total group spectrum shown in Figure 11-30b (60 kHz to 108 kHz). Note that the total group bandwidth is equal to 48 kHz (12 channels × 4 kHz).

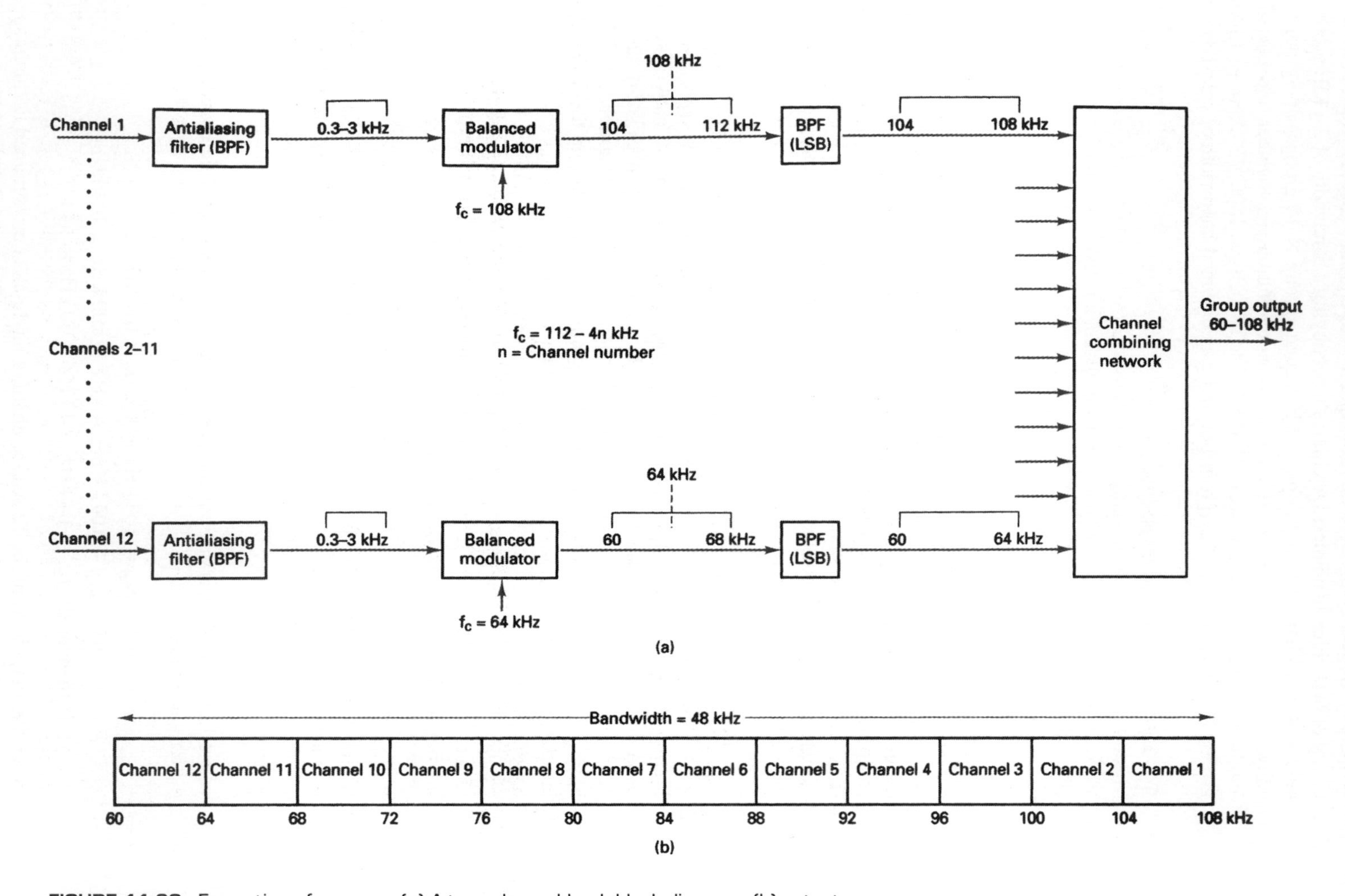

FIGURE 11-30 Formation of a group: (a) A-type channel bank block diagram; (b) output spectrum

Table 11-7 Channel Carrier Frequencies

Channel	Carrier Frequency (kHz)
1	108
2	104
3	100
4	96
5	92
6	88
7	84
8	80
9	76
10	72
11	68
12	64

Table 11-8 Group Carrier Frequencies

Group	Carrier Frequency (kHz)
1	420
2	468
3	516
4	564
5	612

Figure 11-31a shows how a supergroup is formed with a group bank and combining network. Five groups are combined to form a supergroup. The frequency spectrum for each group is 60 kHz to 108 kHz. Each group is mixed with a different group carrier frequency in a balanced modulator and then bandlimited with a bandpass filter tuned to the difference frequency band (LSB) to produce a SSBSC signal. The group carrier frequencies are derived from the following expression:

$$f_c = 372 + 48n \text{ kHz}$$

where n is the group number. Table 11-8 lists the carrier frequencies for groups 1 through 5. For group 1, a 60-kHz to 80-kHz group signal modulates a 420-kHz group carrier frequency. Mathematically, the output of a group bandpass filter is

$$f_{out} = (f_c - 108 \text{ kHz}) \text{ to } (f_c - 60 \text{ kHz})$$

where f_c = group carrier frequency (372 + 48n kHz) and for a group frequency spectrum of 60 KHz to 108 KHz

$$\text{Group 1}, f_{out} = 420 \text{ kHz} - (60 \text{ kHz to } 108 \text{ kHz}) = 312 \text{ kHz to } 360 \text{ kHz}$$
$$\text{Group 2}, f_{out} = 468 \text{ kHz} - (60 \text{ kHz to } 108 \text{ kHz}) = 360 \text{ kHz to } 408 \text{ kHz}$$
$$\text{Group 5}, f_{out} = 612 \text{ kHz} - (60 \text{ kHz to } 108 \text{ kHz}) = 504 \text{ kHz to } 552 \text{ kHz}$$

The outputs from the five group modulators are summed in the linear combiner to produce the total supergroup spectrum shown in Figure 11-31b (312 kHz to 552 kHz). Note that the total supergroup bandwidth is equal to 240 kHz (60 channels × 4 kHz).

11-15 FORMATION OF A MASTERGROUP

There are two types of mastergroups: L600 and U600 types. The L600 mastergroup is used for low-capacity microwave systems, and the U600 mastergroup may be further multiplexed and used for higher-capacity microwave radio systems.

11-15-1 U600 Mastergroup

Figure 11-32a shows how a U600 mastergroup is formed with a supergroup bank and combining network. Ten supergroups are combined to form a mastergroup. The frequency spectrum for each supergroup is 312 kHz to 552 kHz. Each supergroup is mixed with a different supergroup carrier frequency in a balanced modulator. The output is then bandlimited to the difference frequency band (LSB) to form a SSBSC signal. The 10 supergroup carrier

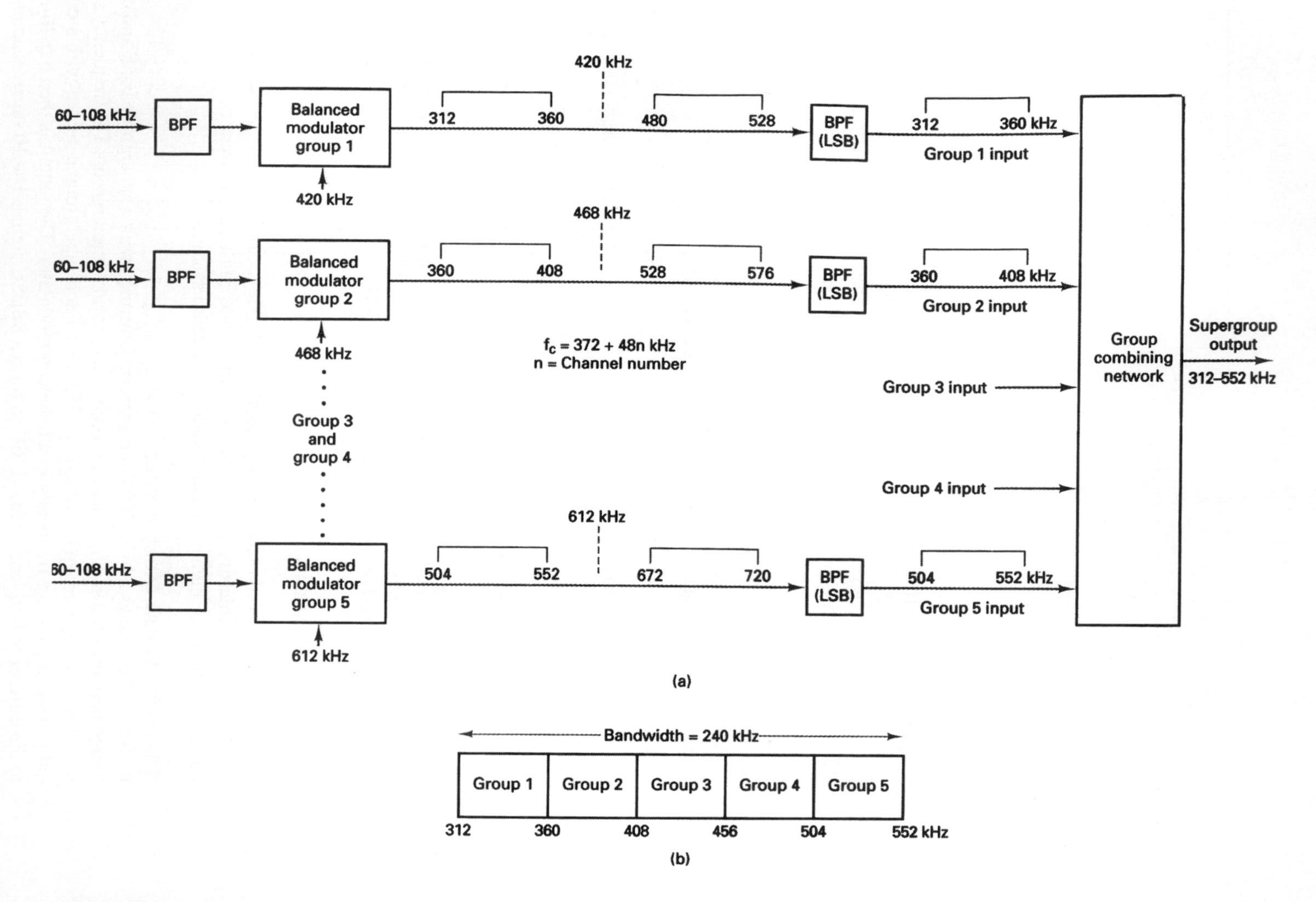

FIGURE 11-31 Formation of a supergroup: (a) group bank and combining network block diagram; (b) output spectrum

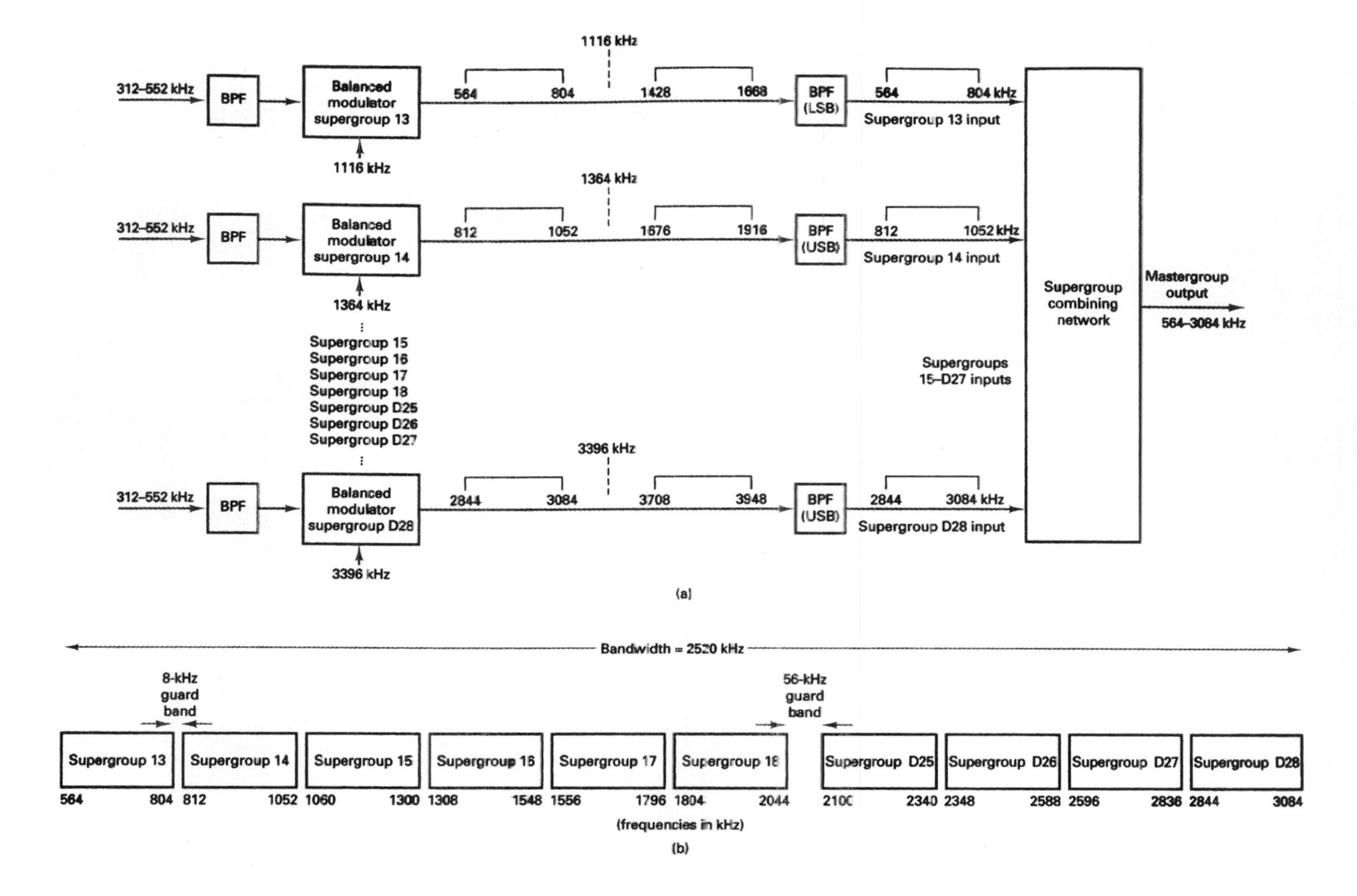

FIGURE 11-32 Formation of a U600 mastergroup: (a) supergroup bank and combining network block diagram; (b) output spectrum

Table 11-9 Supergroup Carrier Frequencies for a U600 Mastergroup

Supergroup	Carrier Frequency (kHz)
13	1116
14	1364
15	1612
16	1860
17	2108
18	2356
D25	2652
D26	2900
D27	3148
D28	3396

frequencies are listed in Table 11-9. For supergroup 13, a 312-kHz to 552-kHz supergroup band of frequencies modulates a 1116-kHz carrier frequency. Mathematically, the output from a supergroup bandpass filter is

$$f_{out} = f_c - f_s \text{ to } f_c$$

where f_c = supergroup carrier frequency
f_s = supergroup frequency spectrum (312 kHz to 552 kHz)

For supergroup 13, f_{out} = 1116 kHz − (312 kHz to 552 kHz) = 564 kHz to 804 kHz
For supergroup 14, f_{out} = 1364 kHz − (312 kHz to 552 kHz) = 812 kHz to 1052 kHz
For supergroup D28, f_{out} = 3396 kHz − (312 kHz to 552 kHz) = 2844 kHz to 3084 kHz

The outputs from the 10 supergroup modulators are summed in the linear summer to produce the total mastergroup spectrum shown in Figure 11-32b (564 kHz to 3084 kHz). Note that between any two adjacent supergroups, there is a void band of frequencies that is not included within any supergroup band. These voids are called *guard bands*. The guard bands are necessary because the demultiplexing process is accomplished through filtering and down-converting. Without the guard bands, it would be difficult to separate one supergroup from an adjacent supergroup. The guard bands reduce the *quality factor* (Q) required to perform the necessary filtering. The guard band is 8 kHz between all supergroups except 18 and D25, where it is 56 kHz. Consequently, the bandwidth of a U600 mastergroup is 2520 kHz (564 kHz to 3084 kHz), which is greater than is necessary to stack 600 voice-band channels (600 × 4 kHz = 2400 kHz).

Guard bands were not necessary between adjacent groups because the group frequencies are sufficiently low, and it is relatively easy to build bandpass filters to separate one group from another.

In the channel bank, the antialiasing filter at the channel input passes a 0.3-kHz to 3-kHz band. The separation between adjacent channel carrier frequencies is 4 kHz. Therefore, there is a 1300-Hz guard band between adjacent channels. This is shown in Figure 11-33.

11-15-2 L600 Mastergroup

With an L600 mastergroup, 10 supergroups are combined as with the U600 mastergroup, except that the supergroup carrier frequencies are lower. Table 11-10 lists the supergroup carrier frequencies for an L600 mastergroup. With an L600 mastergroup, the composite baseband spectrum occupies a lower-frequency band than the U-type mastergroup (Figure 11-34). An L600 mastergroup is not further multiplexed. Therefore, the maximum channel capacity for a microwave or coaxial cable system using a single L600 mastergroup is 600 voice-band channels.

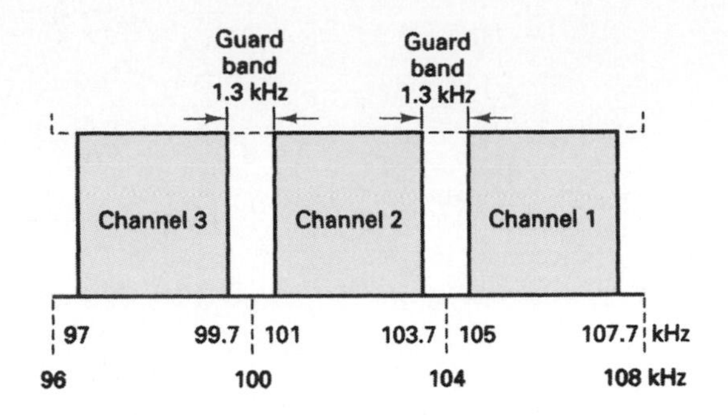

FIGURE 11-33 Channel guard bands

Table 11-10 Supergroup Carrier Frequencies for a L600 Mastergroup

Supergroup	Carrier Frequency (kHz)
1	612
2	Direct
3	1116
4	1364
5	1612
6	1860
7	2108
8	2356
9	2724
10	3100

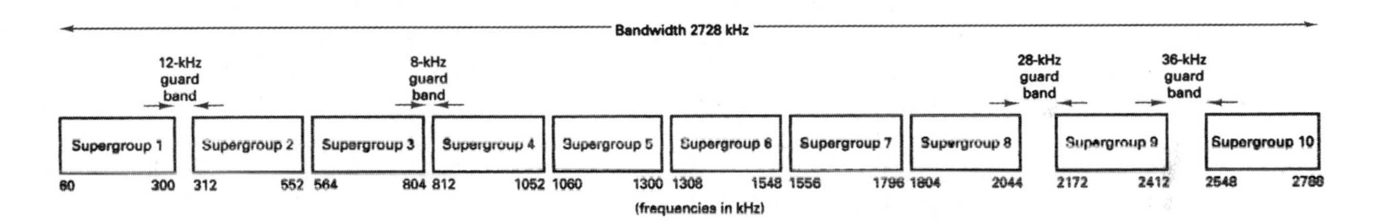

FIGURE 11-34 L600 mastergroup

11-15-3 Formation of a Radio Channel

A *radio channel* comprises either a single L600 mastergroup or up to three U600 mastergroups (1800 voice-band channels). Figure 11-35a shows how an 1800-channel composite FDM baseband signal is formed for transmission over a single microwave radio channel. Mastergroup 1 is transmitted directly as is, while mastergroups 2 and 3 undergo an additional multiplexing step. The three mastergroups are summed in a mastergroup combining network to produce the output spectrum shown in Figure 11-35b. Note the 80-kHz guard band between adjacent mastergroups.

The system shown in Figure 11-35 can be increased from 1800 voice-band channels to 1860 by adding an additional supergroup (supergroup 12) directly to mastergroup 1. The additional 312-kHz to 552-kHz supergroup extends the composite output spectrum from 312 kHz to 8284 kHz.

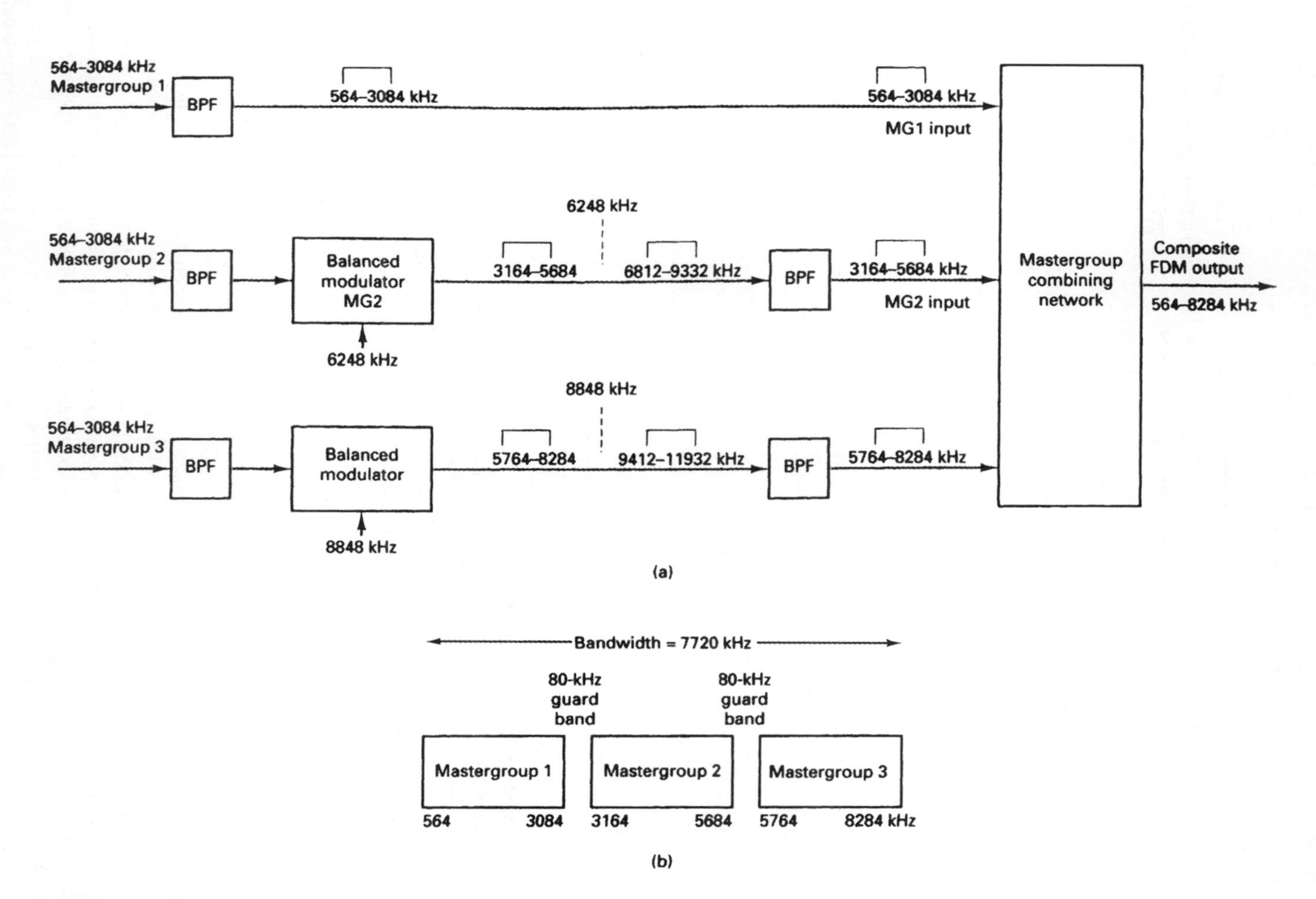

FIGURE 11-35 Three-mastergroup radio channel: (a) block diagram; (b) output spectrum

11-16 WAVELENGTH-DIVISION MULTIPLEXING

During the last two decades of the 20th century, the telecommunications industry witnessed an unprecedented growth in data traffic and the need for computer networking. The possibility of using *wavelength-division multiplexing* (WDM) as a networking mechanism for telecommunications routing, switching, and selection based on wavelength begins a new era in optical communications.

WDM promises to vastly increase the bandwidth capacity of optical transmission media. The basic principle behind WDM involves the transmission of multiple digital signals using several wavelengths without their interfering with one another. Digital transmission equipment currently being deployed utilizes optical fibers to carry only one digital signal per fiber per propagation direction. WDM is a technology that enables many optical signals to be transmitted simultaneously by a single fiber cable.

WDM is sometimes referred to as simply *wave-division multiplexing*. Since wavelength and frequency are closely related, WDM is similar to frequency-division multiplexing (FDM) in that the idea is to send information signals that originally occupied the same band of frequencies through the same fiber at the same time without their interfering with each other. This is accomplished by modulating injection laser diodes that are transmitting highly concentrated light waves at different wavelengths (i.e., at different optical frequencies). Therefore, WDM is coupling light at two or more discrete wavelengths into and out of an optical fiber. Each wavelength is capable of carrying vast amounts of information in either analog or digital form, and the information can already be time- or frequency-division multiplexed. Although the information used with lasers is almost always time-division multiplexed digital signals, the wavelength separation used with WDM is analogous to analog radio channels operating at different carrier frequencies. However, the carrier with WDM is in essence a wavelength rather than a frequency.

11-16-1 WDM versus FDM

The basic principle of WDM is essentially the same as FDM, where several signals are transmitted using different carriers, occupying nonoverlapping bands of a frequency or wavelength spectrum. In the case of WDM, the wavelength spectrum used is in the region of 1300 or 1500 nm, which are the two wavelength bands at which optical fibers have the least amount of signal loss. In the past, each window transmitted a signal digital signal. With the advance of optical components, each transmitting window can be used to propagate several optical signals, each occupying a small fraction of the total wavelength window. The number of optical signals multiplexed with a window is limited only by the precision of the components used. Current technology allows over 100 optical channels to be multiplexed into a single optical fiber.

Although FDM and WDM share similar principles, they are not the same. The most obvious difference is that optical frequencies (in THz) are much higher than radio frequencies (in MHz and GHz). Probably the most significant difference, however, is in the way the two signals propagate through their respective transmission media. With FDM, signals propagate at the same time and through the same medium and follow the same transmission path. The basic principle of WDM, however, is somewhat different. Different wavelengths in a light pulse travel through an optical fiber at different speeds (e.g., blue light propagates slower than red light). In standard optical fiber communications systems, as the light propagates down the cable, wavelength dispersion causes the light waves to spread out and distribute their energy over a longer period of time. Thus, in standard optical fiber systems, wavelength dispersion creates problems, imposing limitations on the system's performance. With WDM, however, wavelength dispersion is the essence of how the system operates.

With WDM, information signals from multiple sources modulate lasers operating at different wavelengths. Hence, the signals enter the fiber at the same time and travel through the same medium. However, they do not take the same path down the fiber. Since each

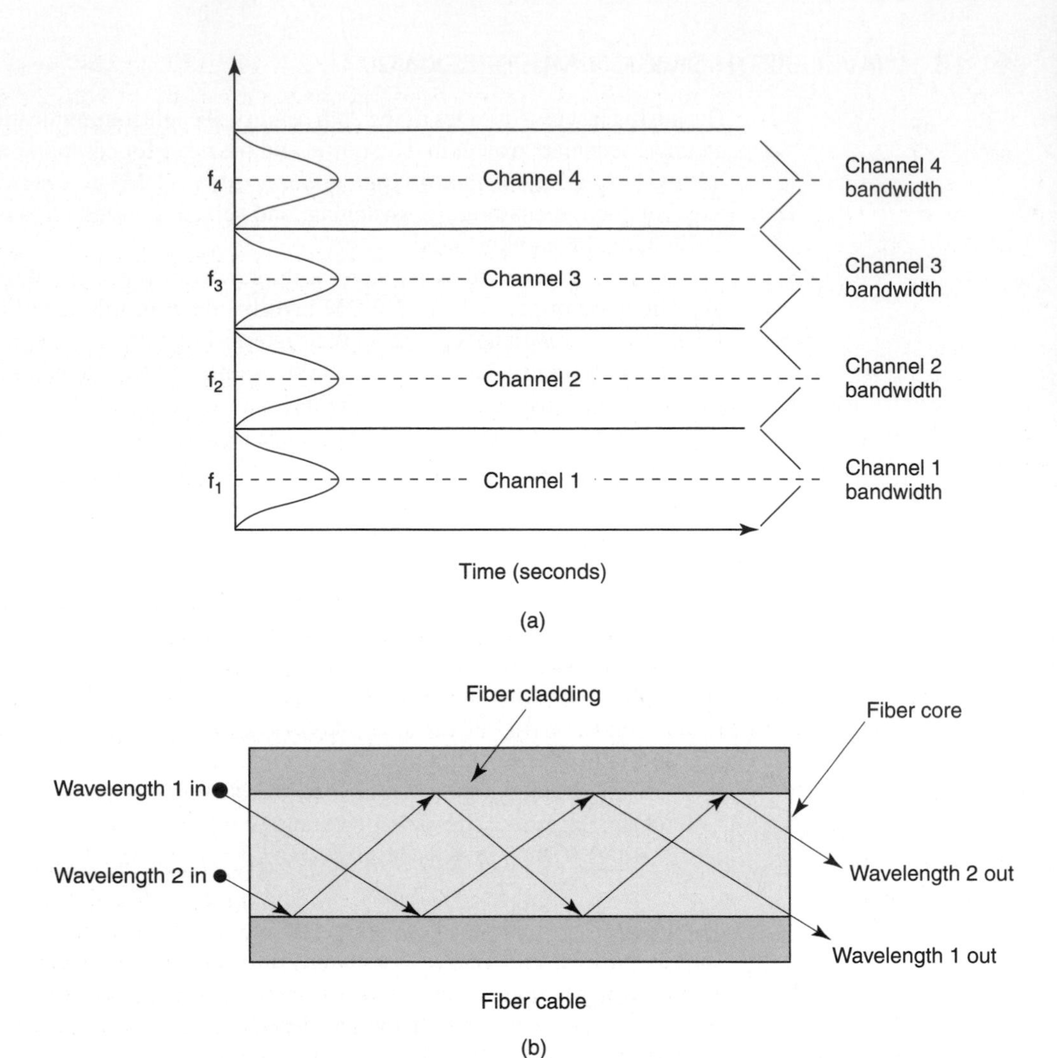

FIGURE 11-36 (a) Frequency-division multiplexing; (b) wave-length-division multiplexing

wavelength takes a different transmission path, they each arrive at the receive end at slightly different times. The result is a series of rainbows made of different colors (wavelengths) each about 20 billionths of a second long, simultaneously propagating down the cable.

Figure 11-36 illustrates the basic principles of FDM and WDM signals propagating through their respective transmission media. As shown in Figure 11-36a, FDM channels all propagate at the same time and over the same transmission medium and take the same transmission path, but they occupy different bandwidths. In Figure 11-36b, it can be seen that with WDM, each channel propagates down the same transmission medium at the same time, but each channel occupies a different bandwidth (wavelength), and each wavelength takes a different transmission path.

11-16-2 Dense-Wave-Division Multiplexing, Wavelengths, and Wavelength Channels

WDM is generally accomplished at approximate wavelengths of 1550 nm (1.55 μm) with successive frequencies spaced in multiples of 100 GHz (e.g., 100 GHz, 200 GHz, 300 GHz, and so on). At 1550-nm and 100-GHz frequency separation, the wavelength separation is approximately 0.8 nm. For example, three adjacent wavelengths each separated by 100 GHz correspond to wavelengths of 1550.0 nm, 1549.2 nm, and 1548.4 nm. Using a multiplexing technique called dense-wave-division multiplexing (D-WDM), the spacing between adjacent frequencies is considerably less. Unfortunately, there does not seem to be a standard definition of exactly what D-WDM means. Generally, optical systems carrying multiple optical signals spaced more than 200 GHz or 1.6 nm apart in the vicinity of 1550 nm are considered standard WDM. WDM systems carrying multiple optical signals in the vicinity of 1550 nm with less than 200-GHz separation are considered D-WDM. Obviously, the more wavelengths used in a WDM system, the closer they are to each other and the denser the wavelength spectrum.

Light waves are comprised of many frequencies (wavelengths), and each frequency corresponds to a different color. Transmitters and receivers for optical fiber systems have been developed that transmit and receive only a specific color (i.e., a specific wavelength at a specific frequency with a fixed bandwidth). WDM is a process in which different sources of information (channels) are propagated down an optical fiber on different wavelengths where the different wavelengths do not interfere with each other. In essence, each wavelength adds an optical lane to the transmission superhighway, and the more lanes there are, the more traffic (voice, data, video, and so on) can be carried on a single optical fiber cable. In contrast, conventional optical fiber systems have only one channel per cable, which is used to carry information over a relatively narrow bandwidth. A Bell Laboratories research team recently constructed a D-WDM transmitter using a single femtosecond, erbium-doped fiber-ring laser that can simultaneously carry 206 digitally modulated wavelengths of color over a single optical fiber cable. Each wavelength (channel) has a bit rate of 36.7 Mbps with a channel spacing of approximately 36 GHz.

Figure 11-37a shows the wavelength spectrum for a WDM system using six wavelengths, each modulated with equal-bandwidth information signals. Figure 11-37b shows how the output wavelengths from six lasers are combined (multiplexed) and then propagated over a single optical cable before being separated (demultiplexed) at the receiver with wavelength selective couplers. Although it has been proven that a single, ultrafast light source can generate hundreds of individual communications channels, standard WDM communications systems are generally limited to between 2 and 16 channels.

WDM enhances optical fiber performance by adding channels to existing cables. Each wavelength added corresponds to adding a different channel with its own information source and transmission bit rate. Thus, WDM can extend the information-carrying capacity of a fiber to hundreds of gigabits per second or higher.

11-16-3 Advantages and Disadvantages of WDM

An obvious advantage of WDM is enhanced capacity and, with WDM, full-duplex transmission is also possible with a single fiber. In addition, optical communications networks use optical components that are simpler, more reliable, and often less costly than their electronic counterparts. WDM has the advantage of being inherently easier to reconfigure (i.e., adding or removing channels). For example, WDM local area networks have been constructed that allow users to access the network simply by tuning to a certain wavelength.

There are also limitations to WDM. Signals cannot be placed so close in the wavelength spectrum that they interfere with each other. Their proximity depends on system design parameters, such as whether optical amplification is used and what optical technique is used to combine and separate signals at different wavelengths. The International Telecommunications Union adopted a standard frequency grid for D-WDM with a spacing of 100 GHz or integer multiples of 100 GHz, which at 1550 nm corresponds to a wavelength spacing of approximately 0.8 nm.

With WDM, the overall signal strength should be approximately the same for each wavelength. Signal strength is affected by fiber attenuation characteristics and the degree of amplification, both of which are wavelength dependent. Under normal conditions, the

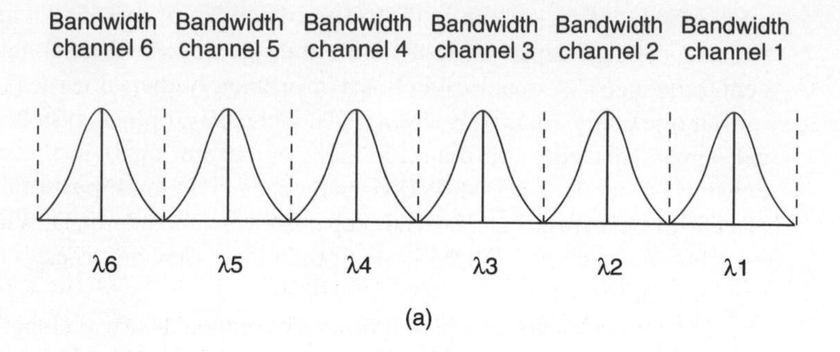

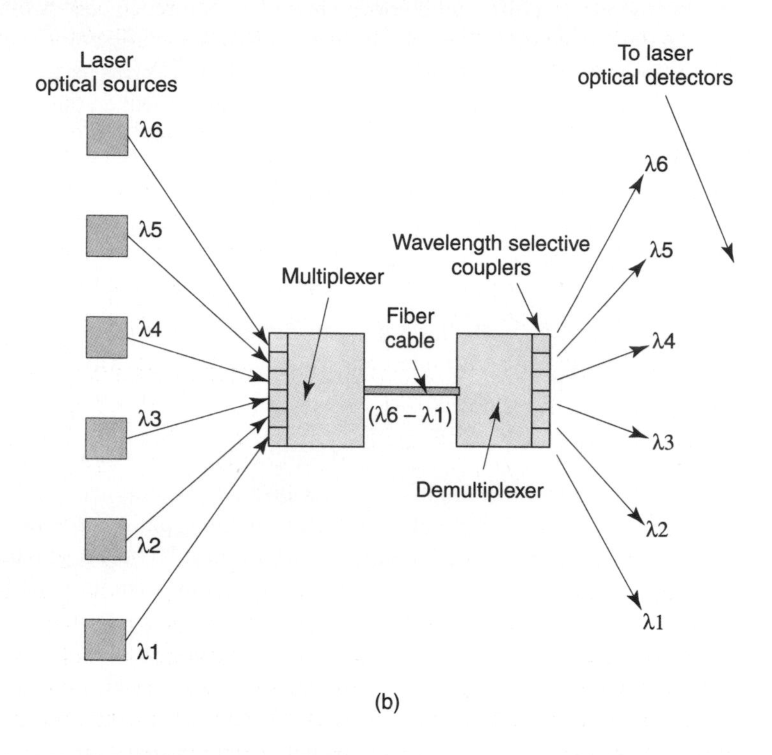

FIGURE 11-37 (a) Wavelength spectrum for a WDM system using six wavelengths; (b) multiplexing and demultiplexing six lasers

wavelengths chosen for a system are spaced so close to one another that attenuation differs very little among them.

One difference between FDM and WDM is that WDM multiplexing is performed at extremely high optical frequencies, whereas FDM is performed at relatively low radio and baseband frequencies. Therefore, radio signals carrying FDM are not limited to propagating through a contained physical transmission medium, such as an optical cable. Radio signals can be propagated through virtually any transmission medium, including free space. Therefore, radio signals can be transmitted simultaneously to many destinations, whereas light waves carrying WDM are limited to a two-point circuit or a combination of many two-point circuits that can go only where the cable goes.

The information capacity of a single optical cable can be increased n-fold, where n represents how many different wavelengths the fiber is propagating at the same time. Each wavelength in a WDM system is modulated by information signals from different sources. Therefore, an optical communications system using a single optical cable propagating n separate wavelengths must utilize n modulators and n demodulators.

11-16-4 WDM Circuit Components

The circuit components used with WDM are similar to those used with conventional radio-wave and metallic-wire transmission systems; however, some of the names used for WDM couplers are sometimes confusing.

11-16-4-1 Wavelength-division multiplexers and demultiplexers. *Multiplexers* or *combiners* mix or combine optical signals with different wavelengths in a way that allows them to all pass through a single optical fiber without interfering with one another. *Demultiplexers* or *splitters* separate signals with different wavelengths in a manner similar to the way filters separate electrical signals of different frequencies. Wavelength demultiplexers have as many outputs as there are wavelengths, with each output (wavelength) going to a different destination. Multiplexers and demultiplexers are at the terminal ends of optical fiber communications systems.

11-16-4-2 Wavelength-division add/drop multiplexer/demultiplexers. *Add/drop multiplexer/demultiplexers* are similar to regular multiplexers and demultiplexers except they are located at intermediate points in the system. Add/drop multiplexers and demultiplexers are devices that separate a wavelength from a fiber cable and reroute it on a different fiber going in a different direction. Once a wavelength has been removed, it can be replaced with a new signal at the same wavelength. In essence, add/drop multiplexers and demultiplexers are used to reconfigure optical fiber cables.

11-16-4-3 Wavelength-division routers. WDM *routers* direct signals of a particular wavelength to a specific destination while not separating all the wavelengths present on the cable. Thus, a router can be used to direct or redirect a particular wavelength (or wavelengths) in a different direction from that followed by the other wavelengths on the fiber.

11-16-4-4 Wavelength-division couplers. WDM *couplers* enable more efficient utilization of the transmission capabilities of optical fibers by permitting different wavelengths to be combined and separated. There are three basic types of WDM couplers: *diffraction grating, prism,* and *dichroic filter.* With diffraction gratings or prisms, specific wavelengths are separated from the other optic signal by reflecting them at different angles. Once a wavelength has been separated, it can be coupled into a different fiber. A dichroic filter is a mirror with a surface that has been coated with a material that permits light of only one wavelength to pass through while reflecting all other wavelengths. Therefore, the dichroic filter can allow two wavelengths to be coupled in different optical fibers.

11-16-4-5 WDM and the synchronous optical network. The *synchronous optical network* (SONET) is a multiplexing system similar to conventional time-division multiplexing except SONET was developed to be used with optical fibers. The initial SONET standard is OC-1. This level is referred to as *synchronous transport level 1* (STS-1). STS-1 has a 51.84-Mbps synchronous frame structure made of 28 DS-1 signals. Each DS-1 signal is equivalent to a single 24-channel T1 digital carrier system. Thus, one STS-1 system can carry 672 individual voice channels (24×28). With STS-1, it is possible to extract or add individual DS-1 signals without completely disassembling the entire frame.

OC-48 is the second level of SONET multiplexing. It combines 48 OC-1 systems for a total capacity of 32,256 voice channels. OC-48 has a transmission bit rate of 2.48332 Gbps (2.48332 billion bits per second). A single optical fiber can carry an OC-48 system. As many as 16 OC-48 systems can be combined using wave-division multiplexing. The light spectrum is divided into 16 different wavelengths with an OC-48 system attached to each transmitter for a combined capacity of 516,096 voice channels ($16 \times 32,256$).

QUESTIONS

11-1. Define *multiplexing.*

11-2. Describe time-division multiplexing.

11-3. Describe the Bell System T1 carrier system.

11-4. What is the purpose of the signaling bit?

11-5. What is frame synchronization? How is it achieved in a PCM-TDM system?

11-6. Describe the superframe format. Why is it used?

11-7. What is a codec? A combo chip?

11-8. What is a fixed-data-rate mode?

11-9. What is a variable-data-rate mode?

11-10. What is a DSX? What is it used for?

11-11. Explain *line coding.*

11-12. Briefly explain unipolar and bipolar transmission.

11-13. Briefly explain return-to-zero and nonreturn-to-zero transmission.

11-14. Contrast the bandwidth considerations of return-to-zero and nonreturn-to-zero transmission.

11-15. Contrast the clock recovery capabilities with return-to-zero and nonreturn-to-zero transmission.

11-16. Contrast the error detection and decoding capabilities of return-to-zero and nonreturn-to-zero transmission.

11-17. What is a regenerative repeater?

11-18. Explain B6ZS and B3ZS. When or why would you use one rather than the other?

11-19. Briefly explain the following framing techniques: added-digit framing, robbed-digit framing, added-channel framing, statistical framing, and unique-line code framing.

11-20. Contrast *bit* and *word interleaving.*

11-21. Describe frequency-division multiplexing.

11-22. Describe a message channel.

11-23. Describe the formation of a group, a supergroup, and a mastergroup.

11-24. Define *baseband* and *composite baseband.*

11-25. What is a guard band? When is a guard band used?

11-26. Describe the basic concepts of wave-division multiplexing.

11-27. What is the difference between WDM and D-WDM?

11-28. List the advantages and disadvantages of WDM.

11-29. Give a brief description of the following components: wavelength-division multiplexer/demultiplexers, wavelength-division add/drop multiplexers, and wavelength-division routers.

11-30. Describe the three types of wavelength-division couplers.

11-31. Briefly describe the SONET standard, including OC-1 and OC-48 levels.

PROBLEMS

11-1. A PCM-TDM system multiplexes 24 voice-band channels. Each sample is encoded into seven bits, and a framing bit is added to each frame. The sampling rate is 9000 samples per second. BPRZ-AMI encoding is the line format. Determine

a. Line speed in bits per second.

b. Minimum Nyquist bandwidth.

11-2. A PCM-TDM system multiplexes 32 voice-band channels each with a bandwidth of 0 kHz to 4 kHz. Each sample is encoded with an 8-bit PCM code. UPNRZ encoding is used. Determine

a. Minimum sample rate.

b. Line speed in bits per second.

c. Minimum Nyquist bandwidth.

11-3. For the following bit sequence, draw the timing diagram for UPRZ, UPNRZ, BPRZ, BPNRZ, and BPRZ-AMI encoding:

bit stream: 1 1 1 0 0 1 0 1 0 1 1 0 0

11-4. Encode the following BPRZ-AMI data stream with B6ZS and B3ZS:

+−0000+−+0−00000+−00+

11-5. Calculate the 12 channel carrier frequencies for the U600 FDM system.

11-6. Calculate the five group carrier frequencies for the U600 FDM system.

11-7. A PCM-TDM system multiplexes 20 voice-band channels. Each sample is encoded into eight bits, and a framing bit is added to each frame. The sampling rate is 10,000 samples per second. BPRZ-AMI encoding is the line format. Determine

a. The maximum analog input frequency.
b. The line speed in bps.
c. The minimum Nyquist bandwidth.

11-8. A PCM-TDM system multiplexes 30 voice-band channels each with a bandwidth of 0 kHz to 3 kHz. Each sample is encoded with a nine-bit PCM code. UPNRZ encoding is used. Determine

a. The minimum sample rate.
b. The line speed in bps.
c. The minimum Nyquist bandwidth.

11-9. For the following bit sequence, draw the timing diagram for UPRZ, UPNRZ, BPRZ, BPNRZ, and BPRZ-AMI encoding:

bit stream: 1 1 0 0 0 1 0 1 0 1

11-10. Encode the following BPRZ-AMI data stream with B6ZS and B3ZS:

+000000+ 000+00−

11-11. Calculate the frequency range for a single FDM channel at the output of the channel, group, supergroup, and mastergroup combining networks for the following assignments:

CH	GP	SG	MG
2	2	13	1
6	3	18	2
4	5	D25	2
9	4	D28	3

11-12. Determine the frequency that a single 1-kHz test tone will translate to at the output of the channel, group, supergroup, and mastergroup combining networks for the following assignments:

CH	GP	SG	MG
4	4	13	2
6	4	16	1
1	2	17	3
11	5	D26	3

11-13. Calculate the frequency range at the output of the mastergroup combining network for the following assignments:

GP	SG	MG
3	13	2
5	D25	3
1	15	1
2	17	2

11-14. Calculate the frequency range at the output of the mastergroup combining network for the following assignments:

SG	MG
18	2
13	3
D26	1
14	1

CHAPTER 12

Metallic Cable Transmission Media

CHAPTER OUTLINE

OBJECTIVES

- Define the general categories of transmission media
- Define *guided* and *unguided* transmission media
- Define *metallic transmission lines*
- Explain the difference between transverse and longitudinal waves
- Define and describe transverse electromagnetic waves
- Describe the following characteristics of electromagnetic waves: wave velocity, frequency, and wavelength
- Describe balanced and unbalanced transmission lines
- Explain baluns
- Describe the following parallel-conductor transmission lines: open-wire, twin-lead, twisted-pair, unshielded twisted-pair, shielded twisted-pair, and coaxial transmission lines
- Describe plenum cable
- Describe a transmission-line equivalent circuit
- Define the following transmission characteristics: *characteristic impedance* and *propagation constant*
- Describe transmission-line propagation and velocity factor
- Explain what is meant by the electrical length of a transmission line

- Describe the following transmission-line losses: conductor loss, dielectric heating loss, radiation loss, coupling loss, and corona
- Define and describe incident and reflected waves
- Explain the difference between resonant and nonresonant transmission lines
- Describe the term *reflection coefficient*
- Describe standing waves and standing wave ratio
- Analyze standing waves on open and shorted transmission lines
- Define *transmission line input impedance*
- Describe how to match impedances on a transmission line
- Describe time-domain reflectometry
- Describe microstrip and stripline transmission lines

12-1 INTRODUCTION

Transmission media can be generally categorized as either *unguided* or *guided.* Guided transmission media are those with some form of conductor that provides a conduit in which electromagnetic signals are contained. In essence, the conductor directs the signal that is propagating down it. Only devices physically connected to the medium can receive signals propagating down a guided transmission medium. Examples of guided transmission media are copper wire and optical fiber. Copper wires transport signals using electrical current, whereas optical fibers transport signals by propagating electromagnetic waves through a nonconductive material. Unguided transmission media are wireless systems (i.e., those without a physical conductor). Unguided signals are emitted then radiated through air or a vacuum (or sometimes water). The direction of propagation in an unguided transmission medium depends on the direction in which the signal was emitted and any obstacles the signal may encounter while propagating. Signals propagating down an unguided transmission medium are available to anyone who has a device capable of receiving them. Examples of unguided transmission media are air (Earth's atmosphere) and free space (a vacuum).

A *cable transmission medium* is a guided transmission medium and can be any physical facility used to propagate electromagnetic signals between two locations in a communications system. A physical facility is one that occupies space and has weight (i.e., one that you can touch and feel as opposed to a wireless transmission medium, such as Earth's atmosphere or a vacuum). Physical transmission media include metallic cables (transmission lines) and optical cables (fibers). *Metallic transmission lines* include *open wire, twin lead,* and *twisted-pair* copper wire as well as *coaxial cable,* and optical fibers include plastic- and glass-core fibers encapsulated in a wide assortment of cladding materials. *Cable transmission systems* are the most common means of interconnecting devices in local area networks because cable transmission systems are the only transmission medium suitable for the transmission of digital signals. Cable transmission systems are also the only acceptable transmission media for digital carrier systems such as T carriers.

12-2 METALLIC TRANSMISSION LINES

A *transmission line* is a *metallic conductor system* used to transfer electrical energy from one point to another using electrical current flow. More specifically, a transmission line is two or more electrical conductors separated by a nonconductive insulator (dielectric), such as a pair of wires or a system of wire pairs. A transmission line can be as short as a few inches or span several thousand miles. Transmission lines can be used to propagate dc or low-frequency ac (such as 60-cycle electrical power and audio signals) or to propagate very high frequencies (such as microwave radio-frequency signals). When propagating low-frequency signals, transmission line behavior is rather simple and quite predictable. However, when propagating high-frequency signals, the characteristics of transmission lines become more involved, and their behavior is somewhat peculiar to a student of lumped-constant circuits and systems.

12-3 TRANSVERSE ELECTROMAGNETIC WAVES

There are two basic kinds of waves: *longitudinal* and *transverse.* With longitudinal waves, the displacement (amplitude) is in the direction of propagation. A surface wave of water is a longitudinal wave. Sound waves are also longitudinal. With transverse waves, the direction of displacement is perpendicular to the direction of propagation. Electromagnetic waves are transverse waves.

Propagation of electrical power along a transmission line occurs in the form of *transverse electromagnetic* (TEM) *waves.* A wave is an *oscillatory motion.* The vibration of a particle excites similar vibrations in nearby particles. A TEM wave propagates primarily in the nonconductor (dielectric) that separates the two conductors of the transmission line. Therefore, a wave travels or propagates itself through a medium. Electromagnetic waves are produced by the acceleration of an electric charge. In conductors, current and voltage are always accompanied by an electric (E) field and a magnetic (H) field in the adjoining region of space. Figure 12-1 shows the spatial relationships between the E and H fields of an electromagnetic wave. From the figure, it can be seen that the E and H fields are perpendicular to each other (at 90° angles) at all points. This is referred to as *space quadrature.* Electromagnetic waves that travel along a transmission line from the source to the load are called *incident waves,* and those that travel from the load back toward the source are called *reflected waves.*

12-4 CHARACTERISTICS OF ELECTROMAGNETIC WAVES

Three of the primary characteristics of electromagnetic waves are *wave velocity, frequency,* and *wavelength.*

12-4-1 Wave Velocity

Waves travel at various speeds, depending on the type of wave and the characteristics of the propagation medium. Sound waves travel at approximately 1100 feet per second in the normal atmosphere. Electromagnetic waves travel much faster. In free space (a vacuum), TEM waves travel at the speed of light: $c = 186{,}283$ statute miles per second, or 299,793,000

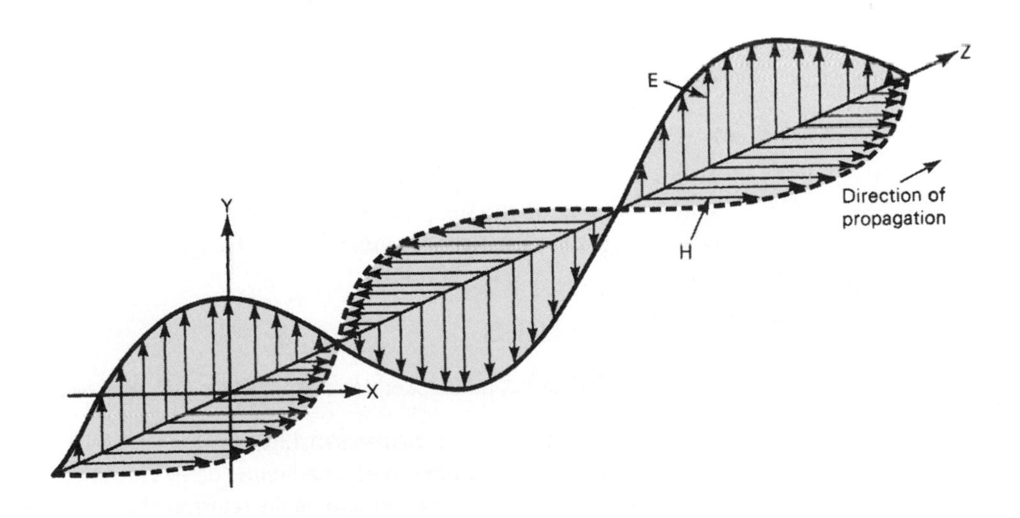

FIGURE 12-1 Transverse electromagnetic wave

meters per second, rounded off to 186,000 mi/s and 3×10^8 m/s. However, in air (such as Earth's atmosphere), TEM waves travel slightly slower, and along a transmission line, electromagnetic waves travel considerably slower.

12-4-2 Frequency and Wavelength

The oscillations of an electromagnetic wave are periodic and repetitive. Therefore, they are characterized by a frequency. The rate at which the periodic wave repeats is its *frequency.* The distance of one cycle occurring in space is called the *wavelength* and is determined from the following equation:

$$\text{distance} = \text{velocity} \times \text{time} \tag{12-1}$$

If the time for one cycle is substituted into Equation 12-1, we get the length of one cycle, which is called the wavelength and represented by the Greek lowercase lambda (λ).

$$\lambda = \text{velocity} \times \text{period}$$

$$= v \times T \tag{12-2}$$

where λ = wavelength
v = velocity
T = period

And because $T = 1/f$

$$\lambda = \frac{v}{f} \tag{12-3}$$

For free-space propagation, $v = c$; therefore, the length of one cycle is

$$\lambda = \frac{c}{f} = \frac{3 \times 10^8 \text{ m/s}}{f \text{ cycles/s}} = \frac{\text{meters}}{\text{cycle}} \tag{12-4}$$

To solve for wavelength in feet or inches, Equation 12-4 can be rewritten as

$$\lambda = \frac{1.18 \times 10^9 \text{ in/s}}{f \text{ cycles/s}} = \frac{\text{inches}}{\text{cycle}} \tag{12-5}$$

$$\lambda = \frac{9.83 \times 10^8 \text{ ft/s}}{f \text{ cycles/s}} = \frac{\text{feet}}{\text{cycle}} \tag{12-6}$$

Figure 12-2 shows a graph of the displacement and direction of propagation of a transverse electromagnetic wave as it travels along a transmission line from a source to a load. The horizontal (X) axis represents distance and the vertical (Y) axis displacement (voltage). One wavelength is the distance covered by one cycle of the wave. It can be seen that the wave moves to the right or propagates down the line with time. If a voltmeter were placed at any stationary point on the line, the voltage measured would fluctuate from zero to maximum positive, back to zero, to maximum negative, back to zero again, and then the cycle repeats.

12-5 TYPES OF TRANSMISSION LINES

Transmission lines can be generally classified as *balanced* or *unbalanced.*

12-5-1 Balanced Transmission Lines

With two-wire balanced lines, both conductors carry current; however, one conductor carries the signal, and the other conductor is the return path. This type of transmission is called *differential,* or *balanced,* signal transmission. The signal propagating down the wire is measured as the potential difference between the two wires. Figure 12-3 shows a balanced

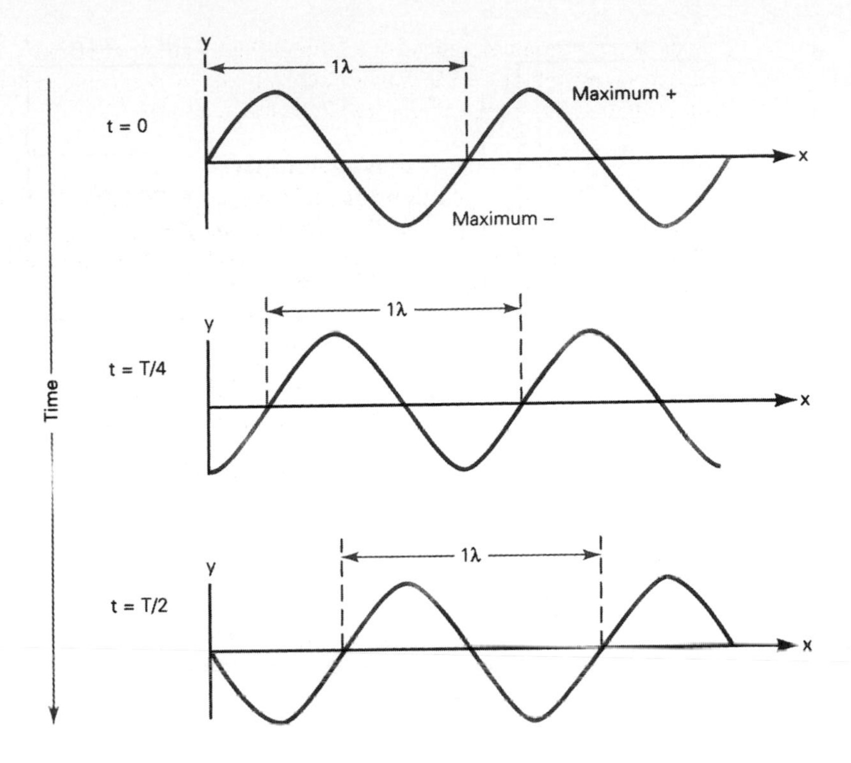

FIGURE 12-2 Displacement and velocity of a transverse wave as it propagates down a transmission line

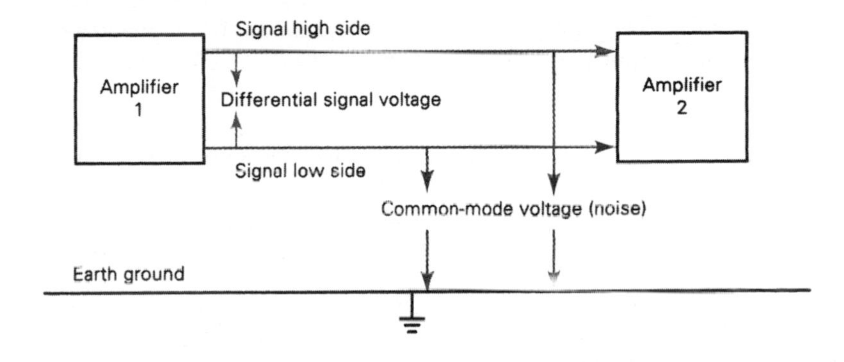

FIGURE 12-3 Differential, or balanced, transmission system

transmission line system. Both conductors in a balanced line carry signal currents. The two currents are equal in magnitude with respect to electrical ground but travel in opposite directions. Currents that flow in opposite directions in a balanced wire pair are called *metallic circuit currents.* Currents that flow in the same direction are called *longitudinal currents.* A balanced wire pair has the advantage that most *noise interference* (sometimes called *common-mode interference*) is induced equally in both wires, producing longitudinal currents that cancel in the load. The cancellation of common mode signals is called *common-mode rejection.* Common-mode rejection ratios of 40 dB to 70 dB are common in balanced transmission lines. Any pair of wires can operate in the balanced mode, provided that neither wire is at ground potential. This includes coaxial cable that has two center conductors and a shield. The shield is generally connected to ground to prevent static interference from penetrating the center conductor.

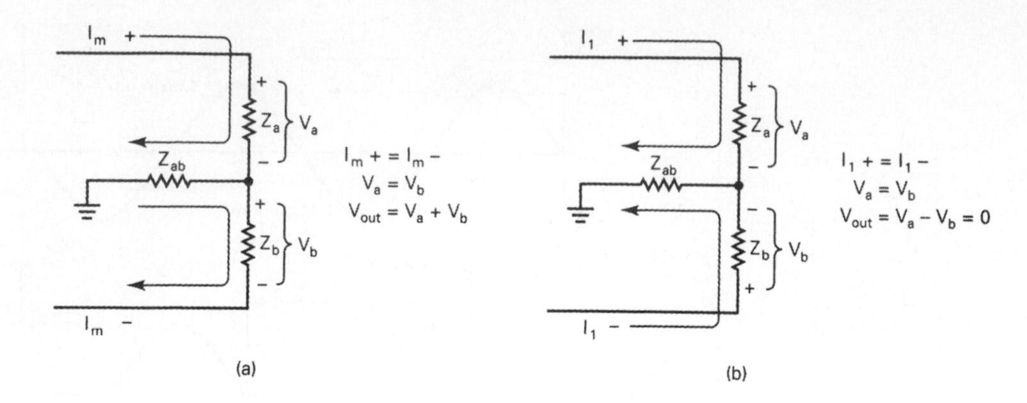

FIGURE 12-4 Results of metallic and longitudinal currents on a balanced transmission line: (a) metallic currents due to signal voltages; (b) longitudinal currents due to noise voltages

Figure 12-4 shows the result of metallic and longitudinal currents on a two-wire balanced transmission line. From the figure, it can be seen that the longitudinal currents (often produced by static interference) cancel in the load.

12-5-2 Unbalanced Transmission Lines

With an unbalanced transmission line, one wire is at ground potential, whereas the other wire is at signal potential. This type of transmission line is called *single-ended,* or *unbalanced,* signal transmission. With unbalanced signal transmission, the ground wire may also be the reference for other signal-carrying wires. If this is the case, the ground wire must go wherever any of the signal wires go. Sometimes this creates a problem because a length of wire has resistance, inductance, and capacitance and, therefore, a small potential difference may exist between any two points on the ground wire. Consequently, the ground wire is not a perfect reference point and is capable of having noise induced into it.

Unbalanced transmission lines have the advantage of requiring only one wire for each signal, and only one ground line is required no matter how many signals are grouped into one conductor. The primary disadvantage of unbalanced transmission lines is reduced immunity to common-mode signals, such as noise and other interference.

Figure 12-5 shows two unbalanced transmission systems. The potential difference on each signal wire is measured from that wire to a common ground reference. Balanced transmission lines can be connected to unbalanced lines and vice versa with special transformers called baluns.

12-5-3 Baluns

A circuit device used to connect a balanced transmission line to an unbalanced load is called a *balun* (balanced to unbalanced). Or more commonly, an unbalanced transmission line, such as a coaxial cable, can be connected to a balanced load, such as an antenna, using a special transformer with an unbalanced primary and a center-tapped secondary winding. The outer conductor (*shield*) of an unbalanced coaxial transmission line is generally connected to ground. At relatively low frequencies, an ordinary transformer can be used to isolate the ground from the load, as shown in Figure 12-6a. The balun must have an electrostatic shield connected to earth ground to minimize the effects of stray capacitances.

For relatively high frequencies, several different kinds of transmission-line baluns exist. The most common type is a *narrowband* balun, sometimes called a *choke, sleeve,* or *bazooka* balun, which is shown in Figure 12-6b. A quarter-wavelength sleeve is placed around and connected to the outer conductor of a coaxial cable. Consequently, the impedance seen looking back into the transmission line is formed by the sleeve, and the outer conductor and is equal to infinity (i.e., the outer conductor no longer has a zero impedance to

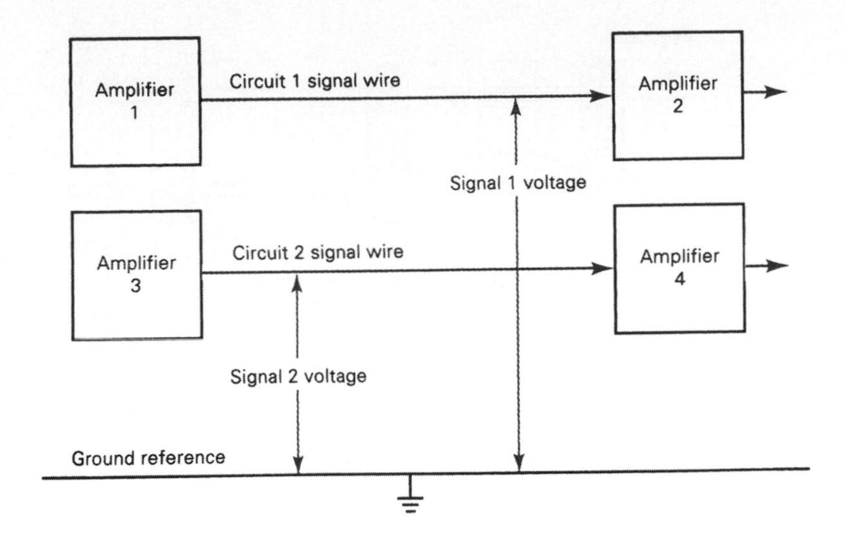

FIGURE 12-5 Single-ended, or unbalanced, transmission system

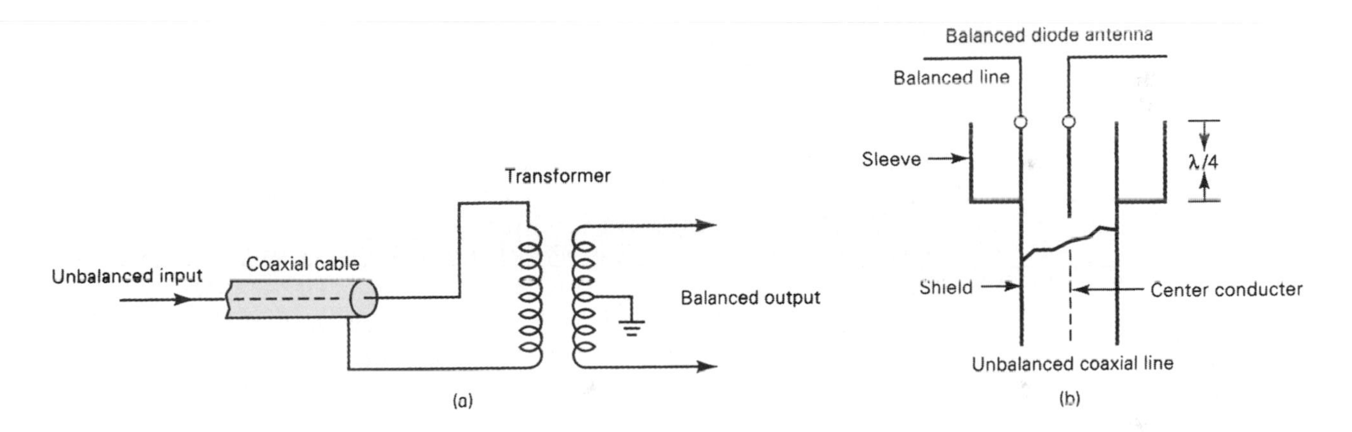

FIGURE 12-6 Baluns: (a) transformer balun; (b) bazooka balun

ground). Thus, one wire of the balanced pair can be connected to the sleeve without short-circuiting the signal. The second conductor is connected to the inner conductor of the coaxial cable.

12-6 METALLIC TRANSMISSION LINES

All data communications systems and computer networks are interconnected to some degree or another with cables, which are all or part of the transmission medium transporting signals between computers. Although there is an enormous variety of cables manufactured today, only a handful of them are commonly used for data communications circuits and computer networks. Belden, which is a leading cable manufacturer, lists more than 2000 different types of cables in its catalog. The most common metallic cables used to interconnect data communications systems and computer networks today are *parallel-conductor transmission lines* and *coaxial transmission lines.*

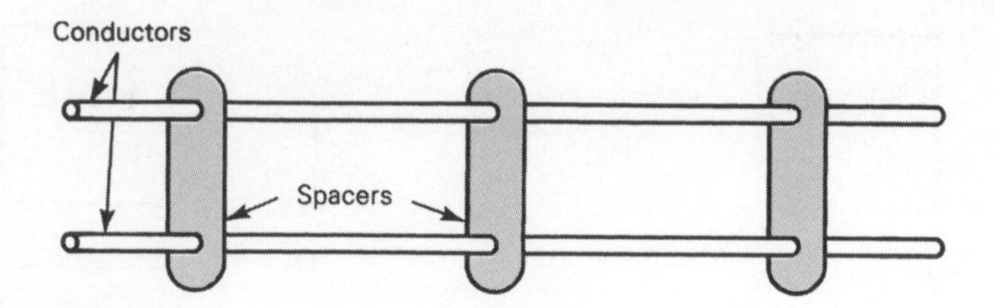

FIGURE 12-7 Open-wire transmission line

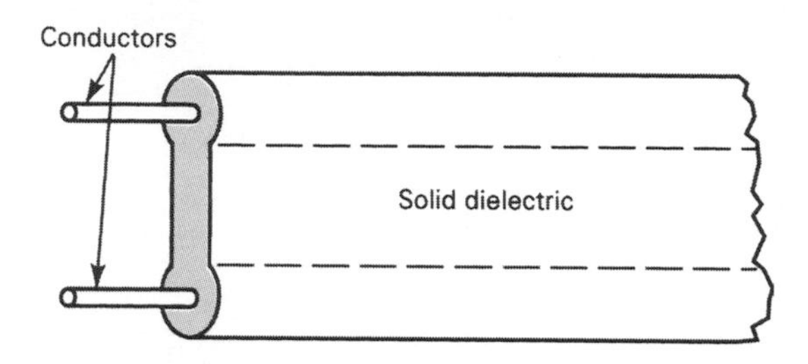

FIGURE 12-8 Twin lead two-wire transmission line

12-6-1 Parallel-Conductor Transmission Lines

Parallel-wire transmission lines are comprised of two or more metallic conductors (usually copper) separated by a nonconductive insulating material called a *dielectric.* Common dielectric materials include air, rubber, polyethylene, paper, mica, glass, and Teflon.

The most common parallel-conductor transmission lines are *open wire, twin lead,* and *twisted pair* [including *unshielded twisted-pair* (UTP) and *shielded twisted-pair* (STP)].

12-6-1-1 Open-wire transmission line. *Open-wire transmission lines* are two-wire parallel conductors (see Figure 12-7). Open-wire transmission lines consist simply of two parallel wires, closely spaced and separated by air. Nonconductive spacers are placed at periodic intervals not only for support but also to keep the distance between the conductors constant. The distance between the two conductors is generally between 2 inches and 6 inches. The dielectric is simply the air between and around the two conductors in which the TEM wave propagates. The only real advantage of this type of transmission line is its simple construction. Because there is no shielding, radiation losses are high, and the cable is susceptible to picking up signals through mutual induction, which produces crosstalk. Crosstalk occurs when a signal on one cable interferes with a signal on an adjacent cable. The primary use of open-wire transmission lines is in standard voice-grade telephone applications.

12-6-1-2 Twin lead. *Twin lead* is another form of two-wire parallel-conductor transmission line and is shown in Figure 12-8. Twin lead is essentially the same as open-wire transmission line except that the spacers between the two conductors are replaced with a continuous solid dielectric that ensures uniform spacing along the entire cable. Uniform spacing is a desirable characteristic for reasons that are explained later in this chapter. Twin-lead transmission line is the flat, brown cable typically used to connect televisions to rooftop antennas. Common dielectric materials used with twin-lead cable are Teflon and polyethylene.

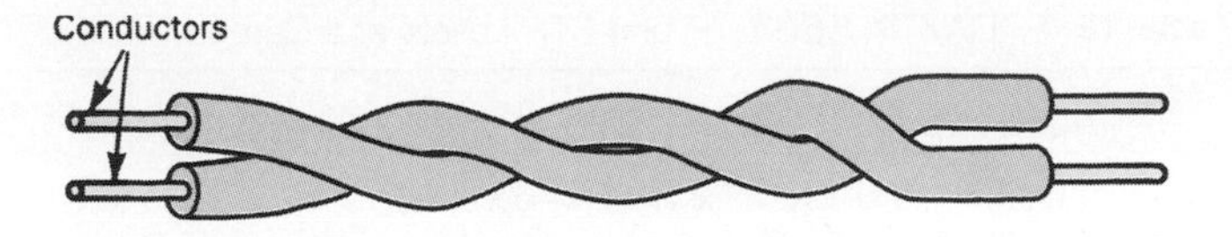

FIGURE 12-9 Twisted-pair two-wire transmission line

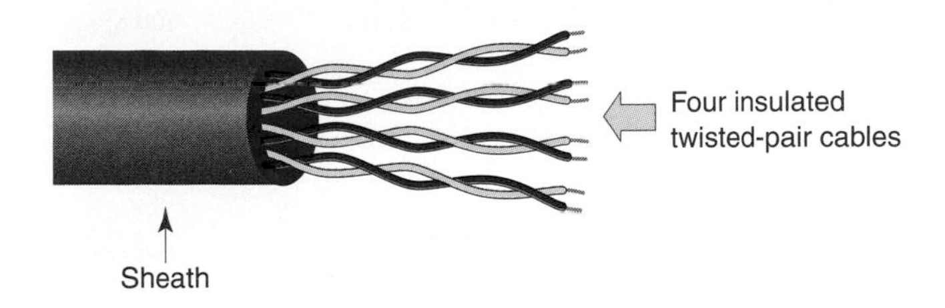

FIGURE 12-10 Unshielded twisted-pair (UTP) cable

12-6-1-3 Twisted-pair transmission lines. A *twisted-pair* transmission line (shown in Figure 12-9) is formed by twisting two insulated conductors around each other. Twisted pairs are often stranded in *units,* and the units are then cabled into *cores* containing up to 3000 pairs of wire. The cores are then covered with various types of *sheaths* forming cables. Neighboring pairs are sometimes twisted with different pitches (twist length) to reduce the effects of *electromagnetic interference* (EMI) and *radio frequency interference* (RFI) from external sources (usually man-made), such as fluorescent lights, power cables, motors, relays, and transformers. Twisting the wires also reduces crosstalk between cable pairs.

The size of twisted-pair wire varies from 16-gauge (16 AWG [American Wire Gauge]) to 26 gauge. The higher the wire gauge, the smaller the diameter and the higher the resistance. Twisted-pair cable is used for both analog and digital signals and is the most commonly used transmission medium for telephone networks and building cabling systems. Twisted-pair transmission lines are also the transmission medium of choice for most local area networks because twisted-pair cable is simple to install and relatively inexpensive when compared to coaxial and optical fiber cables.

There are two basic types of twisted-pair transmission lines specified by the EIA/TIA 568 Commercial Building Telecommunications Cabling Standard for local area networks: 100-ohm *unshielded twisted pair* and 150-ohm *shielded twisted pair.* A typical network utilizes a variety of cabling technologies, depending on the network's size, its topology, and the protocol used. The 568 standard provides guidelines for interconnecting various cabling technologies by dividing network-wiring systems into six subsystems: *horizontal cabling, backbone cabling, work area, telecommunications closet, equipment room,* and *building entrance.* The six subsystems specified in the 568 standard are described in more detail in a later chapter.

Unshielded twisted-pair (UTP) cable consists of two copper wires where each wire is separately encapsulated in PVC (polyvinyl chloride) insulation (see Figure 12-10). Because a wire can act like an antenna, the wires are twisted two or more times at varying lengths to reduce crosstalk and interference. By carefully controlling the number of twists per foot and the manner in which multiple pairs are twisted around each other, manufacturers can improve the bandwidth (i.e., bit rate) of the cable pair significantly. The minimum number of twists for UTP cable is two per foot.

Most telephone systems use UTP cable, and the majority of new buildings are prewired with UTP cable. Generally, more cable is installed than is initially needed, providing room

Table 12-1 EIA/TIA 568 UTP and STP Levels and Categories

Cable Type	Intended Use	Data Rate	Distance
Level 1 (UTP)	Standard voice and low-speed data	2400 bps	18,000 feet
Level 2 (UTP)	Standard voice and low-speed data	4 Mbps	18,000 feet
Category 3 (UTP/STP)	Low-speed local area networks	16 Mbps and all level 2 applications	100 meters
Category 4 (UTP/STP)	Low-speed local area networks	20 Mbps and all category 3 applications	100 meters
Category 5 (UTP/STP)	High-speed local area networks	100 Mbps	100 meters
Enhanced category 5 (UTP/STP)	High-speed local area networks and asynchronous transfer mode (ATM)	350 Mbps	100 meters or more
Proposed New Categories			
Category 6 (UTP/STP)	Very high-speed local area networks and asynchronous transfer mode (ATM)	550 Mbps	100 meters or more
Category 7 shielded screen twisted pair (STP)	Ultra-high-speed local area networks and asynchronous transfer mode (ATM)	1 Gbps	100 meters or more
Foil twisted pair (STP)	Ultra-high-speed local area networks and asynchronous transfer mode (ATM); designed to minimize EMI susceptibility and maximize EMI immunity	>1 Gbps	?
Shielded foil twisted pair (STP)	Ultra-high-speed local area networks and asynchronous transfer mode (ATM); designed to minimize EMI susceptibility and maximize EMI immunity	>1 Gbps	?

for orderly growth. This is one of the primary reasons why UTP cable is so popular. UTP cable is inexpensive, flexible, and easy to install. It is the least expensive transmission medium, but it is also the most susceptible to external electromagnetic interference.

To meet the operational requirements for local area networks, the EIA/TIA 568 standard classifies UTP twisted-pair cables into levels and categories that certify maximum data rates and recommended transmission distances for both UTP and STP cables (see Table 12-1). Standard UTP cable for local area networks is comprised of four pairs of 22- or 24-gauge copper wire where each pair of wires is twisted around each other.

There are seven primary unshielded twisted-pair cables classified by the EIA/TIA 568 standard: level 1, level 2, category 3, category 4, category 5, enhanced category 5, and category 6.

1. *Level 1.* Level 1 cable (sometimes called category 1) is ordinary thin-copper, voice-grade telephone wire typically installed before the establishment of the 568 standard. Many of these cables are insulated with paper, cord, or rubber and are, therefore, highly susceptible to interference caused by insulation breakdown. Level 1 cable is suitable only for voice-grade telephone signals and very low-speed data applications (typically under 2400 bps).

2. *Level 2.* Level 2 cable (sometimes called category 2) is only marginally better than level 1 cable but well below the standard's minimum level of acceptance. Level 2 cables are also typically old, leftover voice-grade telephone wires installed prior to the establishment of the 568 standard. Level 2 cables comply with IBM's Type 3 specification GA27-3773-1, which was developed for IEEE 802.5 Token ring local area networks operating at transmission rates of 4 Mbps.

3. *Category 3.* Category 3 (CAT-3) cable has more stringent requirements than level 1 or level 2 cables and must have at least three turns per inch, and no two pairs within the same cable can have the same number of turns per inch. This specification provides the cable more immunity to crosstalk. CAT-3 cable was designed to accommodate the requirements for two local area networks: IEEE 802.5 Token Ring (16 Mbps) and IEEE 802.3 10Base-T Ethernet (10 Mbps). In essence, CAT-3 cable is used for virtually any voice or data transmission rate up to 16 Mbps and, if four wire pairs are used, can accommodate transmission rates up to 100 Mbps.

4. *Category 4.* Category 4 (CAT-4) cable is little more than an upgraded version of CAT-3 cable designed to meet tighter constraints for attenuation (loss) and crosstalk. CAT-4 cable was designed for data transmission rates up to 20 Mbps. CAT-4 cables can also handle transmission rates up to 100 Mbps using cables containing four pairs of wires.

5. *Category 5.* Category 5 (CAT-5) cable is manufactured with more stringent design specifications than either CAT-3 or CAT-4 cables, including cable uniformity, insulation type, and number of turns per inch (12 turns per inch for CAT-5). Consequently, CAT-5 cable has better attenuation and crosstalk characteristics than the lower cable classifications. Attenuation in simple terms is simply the reduction of signal strength with distance, and crosstalk is the coupling of signals from one pair of wires to another pair. *Near-end crosstalk* refers to coupling that takes place when a transmitted signal is coupled into the receive signal at the same end of the cable.

CAT-5 cable is the cable of choice for most modern-day local area networks. CAT-5 cable was designed for data transmission rates up to 100 Mbps; however, data rates in excess of 500 Mbps are sometimes achieved. CAT-5 cable is UTP cable comprised of four pairs of wires, although only two (pairs 2 and 3) were intended to be used for connectivity. The other two wire pairs are reserved spares. The following standard color code is specified by the EIA for CAT-5 cable:

Pair 1: blue/white stripe and blue
Pair 2: orange/white stripe and orange
Pair 3: green/white stripe and green
Pair 4: brown/white stripe and brown

Each wire in a CAT-5 cable can be a single conductor or a bundle of stranded wires referred to as *Cat-5 solid* or *CAT-5 flex*, respectively. When both cable types are used in the same application, the solid cable is used for backbones and whenever the cable passes through walls or ceilings. The stranded cable is typically used for patch cables between hubs and patch panels and for drop cables that are connected directly between hubs and computers.

6. *Enhanced category 5.* Enhanced category 5 (CAT-5E) cables are intended for data transmission rates up to 250 Mbps, although they often operate successfully at rates up to 350 Mbps and higher.

7. *Category 6.* Category 6 (CAT-6) cable is a recently proposed cable type comprised of four pairs of wire capable of operating at transmission data rates up to 550 Mbps. CAT-6 cable is very similar to CAT-5 cable except CAT-6 cable is designed and fabricated with closer tolerances and uses more advanced connectors.

Shielded twisted-pair (STP) cable is a parallel two-wire transmission line consisting of two copper conductors separated by a solid dielectric material. The wires and dielectric are enclosed in a conductive metal sleeve called a *foil.* If the sleeve is woven into a mesh, it is called a *braid.* The sleeve is connected to ground and acts as a shield, preventing signals from radiating beyond their boundaries (see Figure 12-11). The sleeve also keeps electromagnetic noise and radio interference produced in external sources from reaching the signal conductors. STP cable is thicker and less flexible than UTP cable, making it more difficult and expensive to install. In addition, STP cable requires an additional grounding connector and is more expensive to manufacture. However, STP cable offers greater security and greater immunity to interference.

FIGURE 12-11 Shielded twisted-pair (STP) cable

Table 12-2 Attenuation and Crosstalk Characteristics of Twisted-Pair Cable

Frequency (MHz)	CAT-3 UTP	CAT-5 UTP	150-ohm STP
Attenuation (dB per 100 meters)			
1	2.6	2.0	1.1
4	5.6	4.1	2.2
16	13.1	8.2	4.4
25	—	10.4	6.2
100	—	22.0	12.3
300	—	—	21.4
Near-End Crosstalk (dB)			
1	41	62	58
4	32	53	58
16	23	44	50.4
25	—	41	47.5
100	—	32	38.5
300	—	—	31.3

There are seven primary STP cables classified by the EIA/TIA 568 standard: category 3, category 4, category 5, enhanced category 5, category 7, foil twisted pair, and shielded-foil twisted pair.

1. *Category 7.* Category 7 *shielded-screen twisted-pair* cable (SSTP) is also called PiMF (pairs in metal foil) cable. SSTP cable is comprised of four pairs of 22 or 23 AWG copper wire surrounded by a common metallic foil shield followed by a braided metallic shield.

2. *Foil twisted pair. Foil twisted-pair* cable is comprised of four pairs of 24 AWG copper wire encapsulated in a common metallic-foil shield with a PVC outer sheath. Foil twisted-pair cable has been deliberately designed to minimize EMI susceptibility while maximizing EMI immunity.

3. *Shielded-foil twisted pair. Shielded-foil twisted-pair* cable is comprised of four pairs of 24 AWG copper wires surrounded by a common metallic foil shield encapsulated in a braided metallic shield. Shielded-foil twisted-pair cables offer superior EMI protection.

12-6-2 Attenuation and Crosstalk Comparison

Table 12-2 shows a comparison of the attenuation and near-end crosstalk characteristics of three of the most popular types of twisted-pair cable. Attenuation is given in dB of loss per 100 meters of cable with respect to frequency. Lower dB values indicate a higher-quality cable, and the smaller the differences in the dB value for the various frequencies, the better the frequency response. Crosstalk is given in dB of attenuation between the transmit signal and the signal returned due to crosstalk with higher dB values indicating less crosstalk.

12-6-3 Plenum Cable

Plenum is the name given to the area between the ceiling and the roof in a single-story building or between the ceiling and the floor of the next higher level in a multistory building. Metal rectangular-shaped air ducts were traditionally placed in the plenum to control airflow in the building (both for heating and for cooling). In more modern buildings, the ceiling itself is used to control airflow because the plenum goes virtually everywhere in the building. This presents an interesting situation if a fire should occur in the plenum because the airflow would not be contained in a fire-resistant duct system. For ease of installation, networking cables are typically distributed throughout the building in the plenum. Traditional (*nonplenum*) cables use standard PVC sheathing for the outside insulator; such sheathing is highly toxic when ignited. Therefore, if a fire should occur, the toxic chemicals produced from the burning PVC would propagate through the plenum, possibly contaminating the entire building.

The National Electric Code (NEC) requires plenum cable to have special fire-resistant insulation. Plenum cables are coated with Teflon, which does not emit noxious chemicals when ignited, or special fire-resistant PVC, which is called *plenum-grade PVC.* Therefore, plenum cables are used to route signals throughout the building in the air ducts. Because plenum cables are considerably more expensive than nonplenum cables, traditional PVC coated cables are used everywhere else.

12-6-4 Coaxial (Concentric) Transmission Lines

In the recent past, parallel-conductor transmission lines were suited only for low data transmission rates. At higher transmission rates, their radiation and dielectric losses as well as their susceptibility to external interference were excessive. Therefore, *coaxial* cables were often used for high data transmission rates to reduce losses and isolate transmission paths. However, modern UTP and STP twisted-pair cables operate at bit rates in excess of 1 Gbps at much lower costs than coaxial cable. Twisted-pair cables are also cheaper, lighter, and easier to work with than coaxial cables. In addition, many extremely high-speed computer networks prefer optical fiber cables to coaxial cables. Therefore, coaxial cable is seeing less and less use in computer networks, although they are still a very popular transmission line for analog systems, such as cable television distribution networks.

The basic coaxial cable consists of a center conductor surrounded by a dielectric material (insulation), then a *concentric* (uniform distance from the center) shielding, and finally a rubber environmental protection outer jacket. *Shielding* refers to the woven or stranded mesh (or braid) that surrounds some types of coaxial cables. A coaxial cable with one layer of foil insulation and one layer of braided shielding is referred to as *dual shielded.* Environments that are subject to exceptionally high interference use *quad shielding,* which consists of two layers of foil insulation and two layers of braided metal shielding.

The center conductor of a coaxial cable is the signal wire and the braider outer conductor is the signal return (ground). The center conductor is separated from the shield by a solid dielectric material or insulated spacers when air is used for the dielectric. For relatively high bit rates, the braided outer conductor provides excellent shielding against external interference. However, at lower bit rates, the use of shielding is usually not cost effective.

Essentially, there are two basic types of coaxial cables: *rigid air filled* and *solid flexible.* Figure 12-12a shows a rigid air coaxial line. It can be seen that a tubular outer conductor surrounds the center conductor coaxially and that the insulating material is air. The outer conductor is physically isolated and separated from the center conductor by space, which is generally filled with Pyrex, polystyrene, or some other nonconductive material. Figure 12-12b shows a solid flexible coaxial cable. The outer conductor is braided, flexible, and coaxial to the center conductor. The insulating material is a solid nonconductive polyethylene material that provides both support and electrical isolation between the inner and outer conductors. The inner conductor is a flexible copper wire that can be either solid or hollow (cellular).

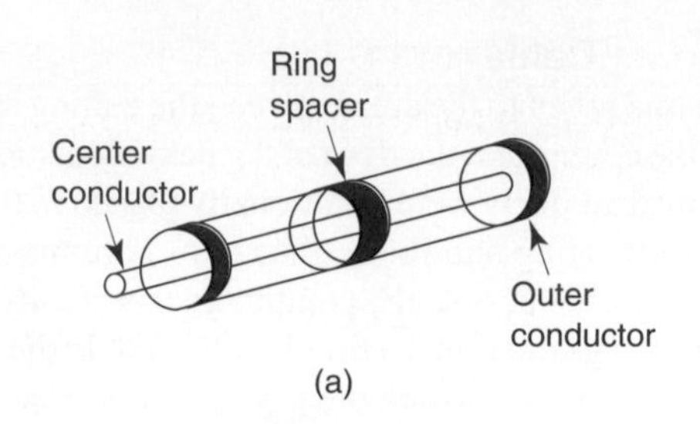

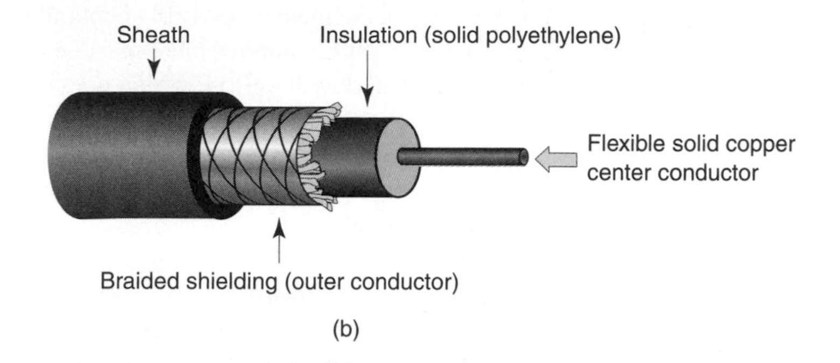

FIGURE 12-12 Coaxial or concentric transmission line: (a) rigid air-filled; (b) solid flexible

Rigid air-filled coaxial cables are relatively expensive to manufacture, and to minimize losses, the air insulator must be relatively free of moisture. Solid coaxial cables have lower losses than hollow cables and are easier to construct, install, and maintain. Both types of coaxial cables are relatively immune to external radiation, radiate little themselves, and are capable of operating at higher bit rates than their parallel-wire counterparts. For these reasons, coaxial cable is more secure than twisted-pair cable. Coaxial cables can also be used over longer distances and support more stations on a shared-media network than twisted-pair cable. The primary disadvantage of coaxial transmission lines is their poor cost-to-performance ratio, low reliability, and high maintenance.

The RG numbering system typically used with coaxial cables refers to cables approved by the U.S. Department of Defense (DoD). The DoD numbering system is used by most cable manufacturers for generic names; however, most manufacturers have developed several variations of each cable using their own product designations. For example, the Belden product number 1426A cross-references to one of several variations of the RG 58/U solid copper-core coaxial cable manufactured by Belden.

Table 12-3 lists several common coaxial cables and several of their parameters. Keep in mind that these values may vary slightly from manufacturer to manufacturer and for different variations of the same cable manufactured by the same company.

12-6-4-1 Coaxial cable connectors. There are essentially two types of coaxial cable connectors: standard *BNC connectors* and *type-N connectors.* BNC connectors are sometimes referred to as "bayonet mount," as they can be easily twisted on or off. N-type connectors are threaded and must be screwed on and off. Several BNC and N-type connectors are shown in Figure 12-13.

Table 12-3 Coaxial Cable Characteristics

DoD Reference No.	Characteristic Impedance (ohms)	Velocity Factor	10-MHz Attenuation (dB/p 100 ft)	Conductor Size	Capacitance (pF/ft)
RG-8/A-AU	52	0.66	0.585	12	29.5
RG-8/U foam	50	0.80	0.405	12	25.4
RG-58/A-AU	53	0.66	1.250	20	28.5
RG-58 foam	50	0.79	1.000	20	25.4
RG-59/A-AU	73	0.84	0.800	20	16.5
RG-59 foam	75	0.79	0.880	20	16.9

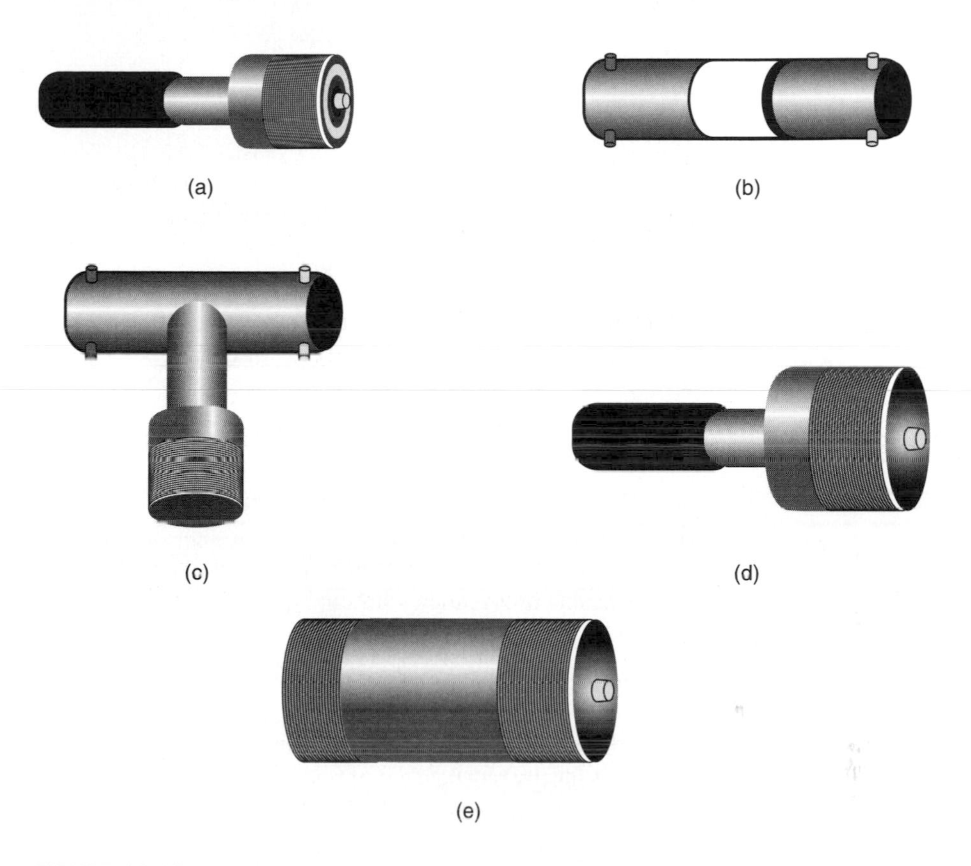

FIGURE 12-13 Coaxial cable connectors: (a) BNC connector; (b) BNC barrel; (c) BNC T; (d) Type-N; and (e) Type-N barrel

12-7 METALLIC TRANSMISSION LINE EQUIVALENT CIRCUIT

12-7-1 Uniformly Distributed Transmission Lines

The characteristics of a transmission line are determined by its electrical properties, such as wire *conductivity* and insulator *dielectric constant,* and its physical properties, such as wire diameter and conductor spacing. These properties, in turn, determine the *primary electrical constants: series dc resistance* (*R*), *series inductance* (*L*), *shunt capacitance* (*C*), and *shunt conductance* (*G*). Resistance and inductance occur along the line, whereas capacitance and conductance occur between the conductors. The primary constants are uniformly distributed throughout the length of the line and, therefore, are commonly called *distributed parameters.* To simplify analysis, distributed parameters are commonly given for a unit length of cable to form an artificial electrical model of the line. The combined parameters

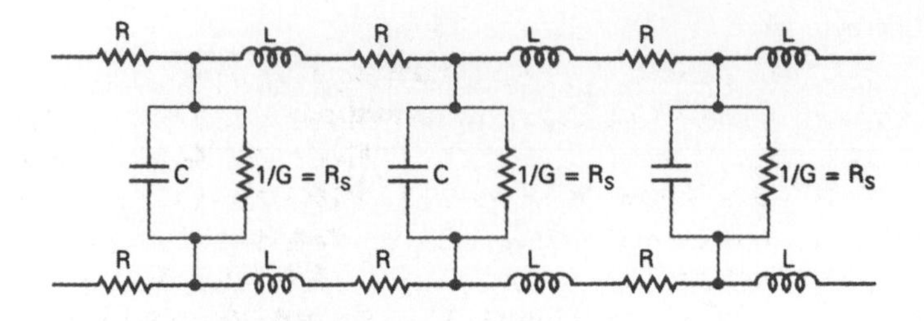

FIGURE 12-14 Two-wire parallel transmission line, electrical equivalent circuit

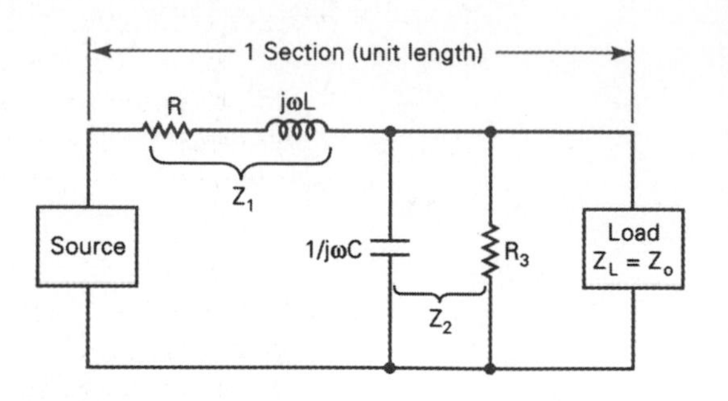

FIGURE 12-15 Equivalent circuit for a single section of transmission line terminated in a load equal to Z_o

are called *lumped parameters.* For example, series resistance is generally given in ohms per unit length (i.e., ohms per meter or ohms per foot).

Figure 12-14 shows the electrical equivalent circuit for a metallic two-wire parallel transmission line showing the relative placement of the various lumped parameters. For convenience, the conductance between the two wires is shown in reciprocal form and given as a shunt *leakage resistance* (R_s).

12-7-2 Transmission Characteristics

The transmission characteristics of a transmission line are called *secondary constants* and are determined from the four primary constants. The secondary constants are characteristic impedance and propagation constant.

12-7-2-1 Characteristic impedance. For maximum power transfer from the source to the load (i.e., no reflected power), a transmission line must be terminated in a purely resistive load equal to the *characteristic impedance* of the transmission line. The characteristic impedance (Z_o) of a transmission line is a complex quantity that is expressed in ohms, is ideally independent of line length, and cannot be directly measured. Characteristic impedance (sometimes called *surge impedance*) is defined as the impedance seen looking into an infinitely long line or the impedance seen looking into a finite length of line that is terminated in a purely resistive load with a resistance equal to the characteristic impedance of the line. A transmission line stores energy in its distributed inductance and capacitance. If a transmission line is infinitely long, it can store energy indefinitely; energy from the source enters the line, and none of it is returned. Therefore, the line acts as a resistor that dissipates all the energy. An infinitely long line can be simulated if a finite line is terminated in a purely resistive load equal to Z_o; all the energy that enters the line from the source is dissipated in the load (this assumes a totally lossless line).

Figure 12-15 shows a single section of a transmission line terminated in a load Z_L equal to Z_o. The impedance seen looking into a line of n such sections is determined from the following equation:

$$Z_o^2 = Z_1 Z_2 + \frac{Z_L^2}{n} \tag{12-7}$$

where n is the number of sections. For an infinite number of sections, Z_L^2/n approaches 0 if

$$\lim \frac{Z_L^2}{n}\bigg|_{n\to\infty} = 0$$

Then, $$Z_o = \sqrt{Z_1 Z_2} \tag{12-8}$$

where $$Z_1 = R + j\omega L$$

$$Y_2 = \frac{1}{Z_2} = \frac{1}{R_s} + \frac{1}{1/j\omega C}$$

$$= G + j\omega C$$

$$Z_2 = \frac{1}{G + j\omega C}$$

Therefore, $$Z_o = \sqrt{(R + j\omega L)\frac{1}{G + j\omega C}} \tag{12-9}$$

or $$Z_o = \sqrt{\frac{R + j\omega L}{G + j\omega C}} \tag{12-10}$$

For extremely low frequencies, the resistances dominate and Equation 12-10 simplifies to

$$Z_o = \sqrt{\frac{R}{G}} \tag{12-11}$$

For extremely high frequencies, the inductance and capacitance dominate and Equation 12-10 simplifies to

$$Z_o = \sqrt{\frac{j\omega L}{j\omega C}} = \sqrt{\frac{L}{C}} \tag{12-12}$$

From Equation 12-12, it can be seen that for high frequencies, the characteristic impedance of a transmission line approaches a constant, is independent of both frequency and length, and is determined solely by the distributed inductance and capacitance. It can also be seen that the phase angle is 0°. Therefore, Z_o looks purely resistive, and all the incident energy is absorbed by the line.

From a purely resistive approach, it can easily be seen that the impedance seen looking into a transmission line made up of an infinite number of sections approaches the characteristic impedance. This is shown in Figure 12-16. Again, for simplicity, only the series resistance R and the shunt resistance R_s are considered. The impedance seen looking into the last section of the line is simply the sum of R and R_s. Mathematically, Z_1 is

$$Z_1 = R + R_s = 10 + 100 = 110$$

Adding a second section, Z_2, gives

$$Z_2 = R + \frac{R_s Z_1}{R_s + Z_1} = 10 + \frac{100 \times 110}{100 + 110} = 10 + 52.38 = 62.38$$

and a third section, Z_3, is

$$Z_3 = R + \frac{R_s Z_2}{R_s + Z_2}$$

$$= 10 + \frac{100 \times 62.38}{100 + 62.38} = 10 + 38.42 = 48.32$$

A fourth section, Z_4, is

$$Z_4 = 10 + \frac{100 \times 48.32}{100 + 48.32} = 10 + 32.62 = 42.62$$

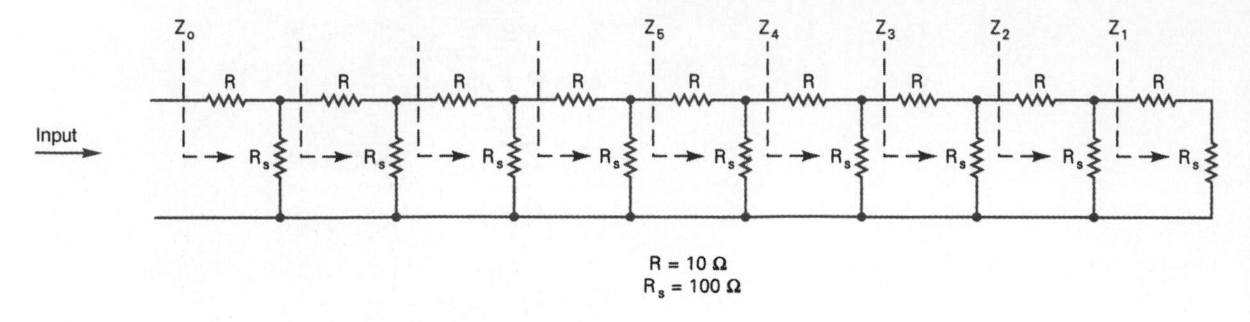

FIGURE 12-16 Characteristic impedance of a transmission line of infinite sections or terminated in load equal to Z_o

It can be seen that after each additional section, the total impedance seen looking into the line decreases from its previous value; however, each time the magnitude of the decrease is less than the previous value. If the process shown above were continued, the impedance seen looking into the line will decrease asymptotically toward 37 Ω, which is the characteristic impedance of the line.

If the transmission line shown in Figure 12-16 were terminated in a load resistance $Z_L = 37\ \Omega$, the impedance seen looking into any number of sections would equal 37 Ω, the characteristic impedance. For a single section of line, Z_o is

$$Z_o = Z_1 = R + \frac{R_s \times Z_L}{R_s + Z_L} = 10 + \frac{100 \times 37}{100 + 37} = 10 + \frac{3700}{137} = 37\ \Omega$$

Adding a second section, Z_2, is

$$Z_o = Z_2 = R + \frac{R_s \times Z_1}{R_s + Z_1} = 10 + \frac{100 \times 37}{100 + 37} = 10 + \frac{3700}{137} = 37\ \Omega$$

Therefore, if this line were terminated into a load resistance $Z_L = 37\ \Omega$, $Z_o = 37\ \Omega$ no matter how many sections are included.

The characteristic impedance of a transmission cannot be measured directly, but it can be calculated using Ohm's law. When a source is connected to an infinitely long line and a voltage is applied, a current flows. Even though the load is open, the circuit is complete through the distributed constants of the line. The characteristic impedance is simply the ratio of the source voltage (E_o) to the line current (I_o). Mathematically, Z_o is

$$Z_o = \frac{E_o}{I_o} \qquad \textbf{(12-13)}$$

where Z_o = characteristic impedance (ohms)
E_o = source voltage (volts)
I_o = transmission line current (amps)

The characteristic impedance of a two-wire parallel transmission line with an air dielectric can be determined from its physical dimensions (see Figure 12-17a) and the formula

$$Z_o = 276 \log \frac{D}{r} \qquad \textbf{(12-14)}$$

where Z_o = characteristic impedance (ohms)
D = distance between the centers of the two conductors (inches)
r = radius of the conductor (inches)

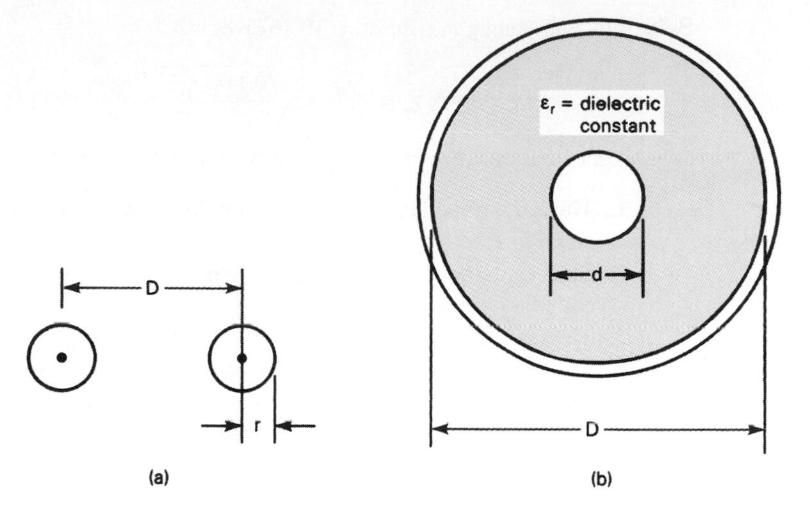

FIGURE 12-17 Physical dimensions of transmission lines: (a) two-wire parallel transmission line; (b) coaxial-cable transmission line

Example 12-1

Determine the characteristic impedance for an air dielectric two-wire parallel transmission line with a D/r ratio of 12.22.

Solution Substituting into Equation 12-14, we obtain

$$Z_o = 276 \log(12.22) = 300 \text{ ohms}$$

The characteristic impedance of a concentric coaxial cable can also be determined from its physical dimensions (see Figure 12-17b) and the formula

$$Z_o = \frac{138}{\sqrt{\epsilon_r}}\left(\log\frac{D}{d}\right) \tag{12-15}$$

where Z_o = characteristic impedance (ohms)
D = inside diameter of the outer conductor (inches)
ϵ_r = relative dielectric constant of the insulating material (unitless)

Example 12-2

Determine the characteristic impedance for an RG-59A coaxial cable with the following specifications: $d = 0.025$ inches, $D = 0.15$ inches, and $\epsilon_r = 2.23$.

Solution Substituting into Equation 12-15, we obtain

$$Z_o = \frac{138}{\sqrt{2.23}}\left(\log\frac{0.15 \text{ in.}}{0.25 \text{ in.}}\right) = 71.9 \text{ ohms}$$

For extremely high frequencies, characteristic impedance can be determined from the inductance and capacitance of the cable using the following formula:

$$Z_o = \sqrt{\frac{L}{C}} \tag{12-16}$$

Example 12-3

Determine the characteristic impedance for an RG-59A coaxial cable with the following specifications: $L = 0.118$ μH/ft and $C = 21$ pF/ft.

Solution Substituting into Equation 12-16 gives us

$$Z_o = \sqrt{\frac{L}{C}} = \sqrt{\frac{0.118 \times 10^{-6}\text{H/ft}}{21 \times 10^{-12}\text{F/ft}}} = 75\ \Omega$$

Transmission lines can be summarized thus far as follows:

1. The input impedance of an infinitely long line at radio frequencies is resistive and equal to Z_o.
2. Electromagnetic waves travel down the line without reflections; such a line is called nonresonant.
3. The ratio of voltage to current at any point along the line is equal to Z_o.
4. The incident voltage and current at any point along the line are in phase.
5. Line losses on a nonresonant line are minimum per unit length.
6. Any transmission line that is terminated in a purely resistive load equal to Z_o acts as if it were an infinite line.
   a. $Z_i = Z_o$.
   b. There are no reflected waves.
   c. V and I are in phase.
   d. There is maximum transfer of power from source to load.

12-7-2-2 Propagation constant. *Propagation constant* (sometimes called *propagation coefficient*) is used to express the attenuation (signal loss) and the phase shift per unit length of a transmission line. As a signal propagates down a transmission line, its amplitude decreases with distance traveled. The propagation constant is used to determine the reduction in voltage or current with distance as a TEM wave propagates down a transmission line. For an infinitely long line, all the incident power is dissipated in the resistance of the wire as the wave propagates down the line. Therefore, with an infinitely long line or a line that looks infinitely long, such as a finite line terminated in a matched load ($Z_o = Z_L$), no energy is returned or reflected back toward the source. Mathematically, the propagation constant is

$$\gamma = \alpha + j\beta \qquad \textbf{(12-17)}$$

where γ = propagation constant (unitless)
α = attenuation coefficient (nepers per unit length)
β = phase shift coefficient (radians per unit length)

The propagation constant is a complex quantity defined by

$$\gamma = \sqrt{(R + j\omega L)(G + j\omega C)} \qquad \textbf{(12-18)}$$

Because a phase shift of 2π rad occurs over a distance of one wavelength,

$$\beta = \frac{2\pi}{\lambda} \qquad \textbf{(12-19)}$$

At intermediate and radio frequencies, $\omega L > R$ and $\omega C > G$; thus,

$$\alpha = \frac{R}{2Z_o} + \frac{GZ_o}{2} \qquad \textbf{(12-20)}$$

and

$$\beta = \omega\sqrt{LC} \qquad \textbf{(12-21)}$$

The current and voltage distribution along a transmission line that is terminated in a load equal to its characteristic impedance (a matched line) are determined from the formulas

$$I = I_s e^{-l\gamma} \qquad \textbf{(12-22)}$$

$$V = V_s e^{-l\gamma} \qquad \textbf{(12-23)}$$

where I_s = current at the source end of the line (amps)
V_s = voltage at the source end of the line (volts)
γ = propagation constant
l = distance from the source at which the current or voltage is determined

For a matched load $Z_L = Z_o$ and for a given length of cable l, the loss in signal voltage or current is the real part of γl, and the phase shift is the imaginary part.

12-8 WAVE PROPAGATION ON A METALLIC TRANSMISSION LINE

Electromagnetic waves travel at the speed of light when propagating through a vacuum and nearly at the speed of light when propagating through air. However, in metallic transmission lines, where the conductor is generally copper and the dielectric materials vary considerably with cable type, an electromagnetic wave travels much more slowly.

12-8-1 Velocity Factor and Dielectric Constant

Velocity factor (sometimes called *velocity constant*) is defined simply as the ratio of the actual velocity of propagation of an electromagnetic wave through a given medium to the velocity of propagation through a vacuum (free space). Mathematically, velocity factor is

$$V_f = \frac{V_p}{c} \tag{12-24}$$

where V_f = velocity factor (unitless)
V_p = actual velocity of propagation (meters per second)
c = velocity of propagation through a vacuum (3×10^8 m/s)

and rearranging Equation 12-24 gives

$$V_f \times c = V_p \tag{12-25}$$

The velocity at which an electromagnetic wave travels through a transmission line depends on the dielectric constant of the insulating material separating the two conductors. The velocity factor is closely approximated with the formula

$$V_p = \frac{1}{\sqrt{\epsilon_r}} \tag{12-26}$$

where ϵ_r is the dielectric constant of a given material (the permittivity of the material relative to the permittivity of a vacuum − the ratio ϵ/ϵ_o, where ϵ is the permittivity of the dielectric and ϵ_o is the permittivity of air).

Velocity factor is sometimes given as a percentage, which is simply the absolute velocity factor multiplied by 100. For example, an absolute velocity factor $V_f = 0.62$ may be stated as 62%.

Dielectric constant is simply the relative permittivity of a material. The relative dielectric constant of air is 1.0006. However, the dielectric constant of materials commonly used in transmission lines ranges from 1.4872 to 7.5, giving velocity factors from 0.3651 to 0.82. The velocity factors and relative dielectric constants of several insulating materials are listed in Table 12-4.

Dielectric constant depends on the type of insulating material used. Inductors store magnetic energy, and capacitors store electric energy. It takes a finite amount of time for an inductor or a capacitor to take on or give up energy. Therefore, the velocity at which an electromagnetic wave propagates along a transmission line varies with the inductance and

Table 12-4 Velocity Factor and Dielectric Constant

Material	Velocity Factor (V_f)	Relative Dielectric Constant (ϵ_r)
Vacuum	1.0000	1.0000
Air	0.9997	1.0006
Teflon foam	0.8200	1.4872
Teflon	0.6901	2.1000
Polyethylene	0.6637	2.2700
Paper, paraffined	0.6325	2.5000
Polystyrene	0.6325	2.5000
Polyvinyl chloride	0.5505	3.3000
Rubber	0.5774	3.0000
Mica	0.4472	5.0000
Glass	0.3651	7.5000

capacitance of the cable. It can be shown that time $T = \sqrt{LC}$. Therefore, inductance, capacitance, and velocity of propagation are mathematically related by the formula

$$\text{velocity} \times \text{time} = \text{distance}$$

Therefore

$$V_p = \frac{\text{distance}}{\text{time}} = \frac{D}{T} \tag{12-27}$$

Substituting $\sqrt{LC}$ for time yields

$$V_p = \frac{D}{\sqrt{LC}} \tag{12-28}$$

If distance is normalized to 1 meter, the velocity of propagation for a lossless transmission line is

$$V_p = \frac{1_m}{\sqrt{LC}_{\text{sec}}} = \frac{1}{\sqrt{LC}}\frac{\text{meters}}{\text{second}} \tag{12-29}$$

where V_p = velocity of propagation (meters per second)
$\sqrt{LC}$ = seconds
L = inductance per unit length (H/m)
C = capacitance per unit length (F/m)

Example 12-4

For a given length of RG 8A/U coaxial cable with a distributed capacitance $C = 96.6$ pF/m, a distributed inductance $L = 241.56$ nH/m, and a relative dielectric constant $\epsilon_r = 2.3$, determine the velocity of propagation and the velocity factor.

Solution From Equation 12-16,

$$V_p = \frac{1}{\sqrt{(96.6 \times 10^{-12})(241.56 \times 10^{-9})}} = 2.07 \times 10^8 \text{ m/s}$$

From Equation 12-24,

$$V_f = \frac{2.07 \times 10^8 \text{ m/s}}{3 \times 10^8 \text{ m/s}} = 0.69$$

From Equation 12-26,

$$V_f = \frac{1}{\sqrt{2.3}} \approx 0.66$$

Because wavelength is directly proportional to velocity and the velocity of propagation of a TEM wave varies with dielectric constant, the wavelength of a TEM wave also varies with dielectric constant. Therefore, for transmission media other than free space, Equation 12-3 can be rewritten as

$$\lambda = \frac{V_p}{f} = \frac{cV_f}{f} = \frac{c}{f\sqrt{\epsilon_r}} \tag{12-30}$$

12-8-2 Electrical Length of a Transmission Line

The length of a transmission line relative to the length of the wave propagating down it is an important consideration when analyzing transmission-line behavior. At low frequencies (long wavelengths), the voltage along the line remains relatively constant. However, for high frequencies, several wavelengths of the signal may be present on the line at the same time. Therefore, the voltage along the line may vary appreciably. Consequently, the length of a transmission line is often given in wavelengths rather than in linear dimensions. Transmission-line phenomena apply to long lines. Generally, a transmission line is defined as long if its length exceeds 1/16th of a wavelength; otherwise, it is considered short. A given length of transmission line may appear short at one frequency and long at another frequency. For example, a 10-m length of transmission line at 1000 Hz is short ($\lambda = 300{,}000$ m; 10 m is only a small fraction of a wavelength). However, the same line at 6 GHz is long ($\lambda = 5$ cm; the line is 200 wavelengths long). It will be apparent later in this chapter, in Chapter 9, and in Appendix A that electrical length is used extensively for transmission-line calculations and antenna design.

12-8-3 Delay Lines

In the previous section, it was shown that the velocity of propagation of an electromagnetic wave depends on the media in which it is traveling. The velocity of an electromagnetic wave in free space (i.e., a vacuum) is the speed of light (3×10^8 m/s), and the velocity is slightly slower through the Earth's atmosphere (i.e., air). The velocity of propagation through a metallic transmission line is effected by the cable's electrical constants: inductance and capacitance. The velocity of propagation of a metallic transmission line is somewhat less than the velocity of propagation through either free space or the Earth's atmosphere.

Delay lines are transmission lines designed to intentionally introduce a time delay in the path of an electromagnetic wave. The amount of time delay is a function of the transmission line's inductance and capacitance. The inductance provides an opposition to changes in current, as does the charge and discharge times of the capacitance. Delay time is calculated as follows:

$$t_d = LC \text{ (seconds)} \tag{12-31}$$

where t_d = time delay (seconds)
L = inductance (henrys)
C = capacitance (farads)

If inductance and capacitance are given per unit length of transmission line (such as per foot or per meter), the time delay will also be per unit length (i.e., 1.5 ns/meter).

The time delay introduced by a length of coaxial cable is calculated with the following formula:

$$t_d = 1.016\,\epsilon \tag{12-32}$$

where ϵ is the dielectric constant of cable.

12-9 TRANSMISSION LINE LOSSES

For analysis purposes, metallic transmission lines are often considered to be totally lossless. In reality, however, there are several ways in which signal power is lost in a transmission line. They include *conductor loss, radiation loss, dielectric heating loss,*

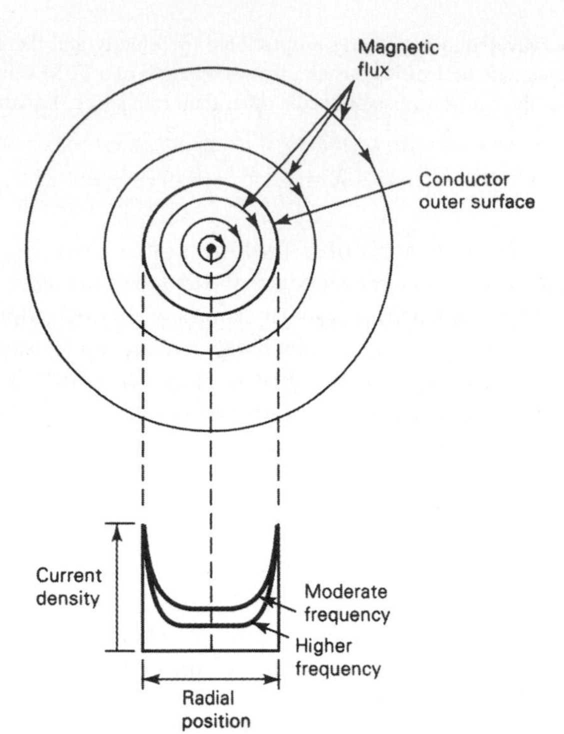

FIGURE 12-18 Isolated round conductor showing magnetic lines of flux, current distributions, and the skin effect

coupling loss, and *corona.* Cable manufacturers generally lump all cable losses together and specify them as attenuation loss in decibels per unit length (e.g., dB/m, dB/ft, and so on).

12-9-1 Conductor Losses

Because electrical current flows through a metallic transmission line and the line has a finite resistance, there is an inherent and unavoidable power loss. This is sometimes called *conductor loss* or *conductor heating loss* and is simply an I^2R power loss. Because resistance is distributed throughout a transmission line, conductor loss is directly proportional to the square of the line length. Also, because power dissipation is directly proportional to the square of the current, conductor loss is inversely proportional to characteristic impedance. To reduce conductor loss, simply shorten the transmission line or use a larger-diameter wire (i.e., one with less resistance). Keep in mind, however, that changing the wire diameter also changes the characteristic impedance and, consequently, the current.

Conductor loss depends somewhat on frequency because of a phenomenon called the *skin effect.* When current flows through an isolated round wire, the magnetic flux associated with it is in the form of concentric circles surrounding the wire core. This is shown in Figure 12-18. From the figure, it can be seen that the flux density near the center of the conductor is greater than it is near the surface. Consequently, the lines of flux near the center of the conductor encircle the current and reduce the mobility of the encircled electrons. This is a form of self-inductance and causes the inductance near the center of the conductor to be greater than at the surface. Therefore, at high frequencies, most of the current flows along the surface (outer skin) of the conduct rather than near its center. This is equivalent to reducing the cross-sectional area of the conductor and increasing the opposition to current flow (i.e., resistance). The additional opposition has a 0° phase angle and is, therefore, a resistance and not a reactance.

Therefore, the ac resistance of the conductor is proportional to the square root of the frequency. The ratio of the ac resistance to the dc resistance of a conductor is called the *resistance ratio.* Above approximately 100 MHz, the center of a conductor can be completely removed and have absolutely no effect on the total conductor loss or EM wave propagation.

Conductor loss in metallic transmission lines varies from as low as a fraction of a decibel per 100 meters for rigid air dielectric coaxial cable to as high as 200 dB per 100 meters for a solid dielectric flexible coaxial cable. Because both I^2R losses and dielectric losses are proportional to length, they are generally lumped together and expressed in decibels of loss per unit length (i.e., dB/m).

12-9-2 Dielectric Heating Losses

A difference of potential between two conductors of a metallic transmission line causes *dielectric heating.* Heat is a form of energy and must be taken from the energy propagating down the line. For air dielectric transmission lines, the heating loss is negligible. However, for solid-core transmission lines, dielectric heating loss increases with frequency.

12-9-3 Radiation Losses

If the separation between conductors in a metallic transmission line is an appreciable fraction of a wavelength, the electrostatic and electromagnetic fields that surround the conductor cause the line to act as if it were an antenna and transfer energy to any nearby conductive material. The energy radiated is called *radiation loss* and depends on dielectric material, conductor spacing, and length of the transmission line. Radiation losses are reduced by properly shielding the cable. Therefore, shielded cables (such as STP and coaxial cable) have less radiation loss than unshielded cables (such as twin lead, open wire, and UTP). Radiation loss is also directly proportional to frequency.

12-9-4 Coupling Losses

Coupling loss occurs whenever a connection is made to or from a transmission line or when two sections of transmission line are connected together. Mechanical connections are discontinuities, which are locations where dissimilar materials meet. Discontinuities tend to heat up, radiate energy, and dissipate power.

12-9-5 Corona

Corona is a luminous discharge that occurs between the two conductors of a transmission line when the difference of potential between them exceeds the breakdown voltage of the dielectric insulator. Generally, when corona occurs, the transmission line is destroyed.

12-10 INCIDENT AND REFLECTED WAVES

An ordinary transmission line is bidirectional; power can propagate equally well in both directions. Voltage that propagates from the source toward the load is called *incident voltage,* and voltage that propagates from the load toward the source is called *reflected voltage.* Similarly, there are incident and reflected currents. Consequently, incident power propagates toward the load, and reflected power propagates toward the source. Incident voltage and current are always in phase for a resistive characteristic impedance. For an infinitely long line, all the incident power is stored by the line, and there is no reflected power. Also, if the line is terminated in a purely resistive load equal to the characteristic impedance of the line, the load absorbs all the incident power (this assumes a lossless line). For a more practical definition, reflected power is the portion of the incident power that was not absorbed by the load. Therefore, the reflected power can never exceed the incident power.

12-10-1 Resonant and Nonresonant Transmission Lines

A transmission line with no reflected power is called a *flat* or *nonresonant* line. A transmission line is nonresonant if it is of infinite length or if it is terminated with a resistive

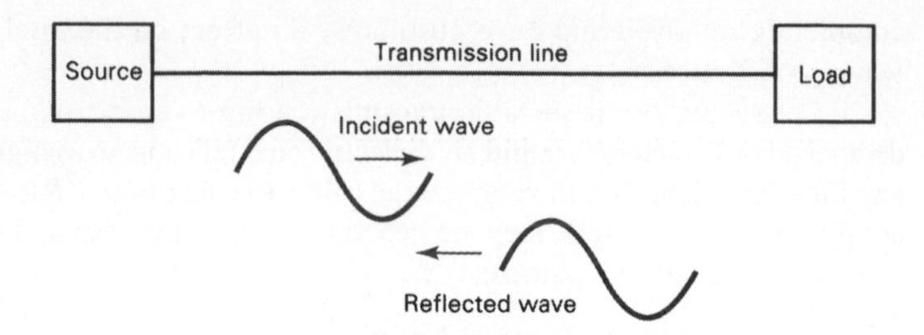

FIGURE 12-19 Source, load, transmission line, and their corresponding incident and reflected waves

load equal in ohmic value to the characteristic impedance of the transmission line. On a flat line, the voltage and current are constant throughout its length, assuming no losses. When the load is not equal to the characteristic impedance of the line, some of the incident power is reflected back toward the source. If the load is either a short or an open circuit, all the incident power is reflected back toward the source. If the source were replaced with an open or a short and the line were lossless, energy present on the line would reflect back and forth (oscillate) between the source and load ends similar to the way energy is transferred back and forth between the capacitor and inductor in an *LC* tank circuit. This is called a *resonant* transmission line. In a resonant line, energy is alternately transferred between the magnetic and electric fields of the distributed inductance and capacitance of the line. Figure 12-19 shows a source, transmission line, and load with their corresponding incident and reflected waves.

12-10-2 Reflection Coefficient

The *reflection coefficient* (sometimes called the *coefficient of reflection*) is a vector quantity that represents the ratio of reflected voltage to incident voltage or reflected current to incident current. Mathematically, the reflection coefficient is gamma, Γ, defined by

$$\Gamma = \frac{E_r}{E_i} \text{ or } \frac{I_r}{I_i} \tag{12-33}$$

where Γ = reflection coefficient (unitless)
E_i = incident voltage (volts)
E_r = reflected voltage (volts)
I_i = incident current (amps)
I_r = reflected current (amps)

From Equation 12-33 it can be seen that the maximum and worst-case value for Γ is 1 ($E_r = E_i$), and the minimum value and ideal condition occur when $\Gamma = 0$ ($E_r = 0$).

12-11 STANDING WAVES

When $Z_o = Z_L$, all the incident power is absorbed by the load. This is called a *matched line.* When $Z_o \neq Z_L$, some of the incident power is absorbed by the load, and some is returned (reflected) to the source. This is called an *unmatched* or *mismatched line.* With a mismatched line, there are two electromagnetic waves, traveling in opposite directions, present on the line at the same time (these waves are in fact called *traveling waves*). The two traveling waves set up an interference pattern known as a *standing wave.* This is shown in Figure 12-20. As the incident and reflected waves pass each other, stationary patterns of voltage and current are produced on the line. These sta-

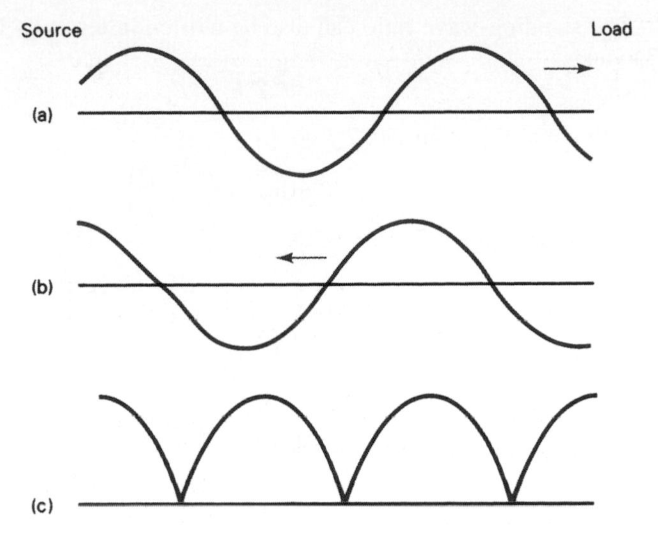

FIGURE 12-20 Developing a standing wave on a transmission line: (a) incident wave; (b) reflected wave; (c) standing wave

tionary waves are called standing waves because they appear to remain in a fixed position on the line, varying only in amplitude. The standing wave has minima (nodes) separated by a half wavelength of the traveling waves and maxima (antinodes) also separated by a half wavelength.

12-11-1 Standing-Wave Ratio

The *standing-wave ratio* (SWR) is defined as the ratio of the maximum voltage to the minimum voltage or the maximum current to the minimum current of a standing wave on a transmission line. SWR is often called the *voltage standing-wave ratio* (VSWR). Essentially, SWR is a measure of the mismatch between the load impedance and the characteristic impedance of the transmission line. Mathematically, SWR is

$$\text{SWR} = \frac{V_{\text{max}}}{V_{\text{min}}} \text{(unitless)} \tag{12-34}$$

The voltage maxima (V_{max}) occur when the incident and reflected waves are in phase (i.e., their maximum peaks pass the same point on the line with the same polarity), and the voltage minima (V_{min}) occur when the incident and reflected waves are 180° out of phase. Mathematically, V_{max} and V_{min} are

$$V_{\text{max}} = E_i + E_r \tag{12-35}$$

$$V_{\text{min}} = E_i - E_r \tag{12-36}$$

Therefore, Equation 12-34 can be rewritten as

$$\text{SWR} = \frac{V_{\text{max}}}{V_{\text{min}}} = \frac{E_i + E_r}{E_i - E_r} \tag{12-37}$$

From Equation 12-37, it can be seen that when the incident and reflected waves are equal in amplitude (a total mismatch), SWR = infinity. This is the worst-case condition. Also, from Equation 12-37, it can be seen that when there is no reflected wave ($E_r = 0$), SWR = E_i/E_i, or 1. This condition occurs when $Z_o = Z_L$ and is the ideal situation.

The standing-wave ratio can also be written in terms of Γ. Rearranging Equation 12-34 yields

$$\Gamma E_i = E_r \tag{12-38}$$

Substituting into Equation 12-37 gives us

$$\text{SWR} = \frac{E_i + E_i\Gamma}{E_i - E_i\Gamma} \tag{12-39}$$

Factoring out E_i yields

$$\text{SWR} = \frac{E_i(1 + \Gamma)}{E_i(1 - \Gamma)} = \frac{1 + \Gamma}{1 - \Gamma} \tag{12-40}$$

Cross multiplying gives

$$\text{SWR}(1 - \Gamma) = 1 + \Gamma \tag{12-41}$$

$$\text{SWR} - \text{SWR}\,\Gamma = 1 + \Gamma \tag{12-42}$$

$$\text{SWR} = 1 + \Gamma + (\text{SWR})\Gamma \tag{12-43}$$

$$\text{SWR} - 1 = \Gamma(1 + \text{SWR}) \tag{12-44}$$

$$\Gamma = \frac{\text{SWR} - 1}{\text{SWR} + 1} \tag{12-45}$$

Example 12-5

For a transmission line with incident voltage $E_i = 5$ V and reflected voltage $E_r = 3$ V, determine

a. Reflection coefficient.
b. SWR.

Solution **a.** Substituting into Equation 12-33 yields

$$\Gamma = \frac{E_r}{E_i} = \frac{3}{5} = 0.6$$

b. Substituting into Equation 12-37 gives us

$$\text{SWR} = \frac{E_i + E_r}{E_i - E_r} = \frac{5 + 3}{5 - 3} = \frac{8}{2} = 4$$

Substituting into Equation 12-45, we obtain

$$\Gamma = \frac{4 - 1}{4 + 1} = \frac{3}{5} = 0.6$$

When the load is purely resistive, SWR can also be expressed as a ratio of the characteristic impedance to the load impedance or vice versa. Mathematically, SWR is

$$\text{SWR} = \frac{Z_o}{Z_L} \text{ or } \frac{Z_L}{Z_o} \text{ (whichever gives an SWR greater than 1)} \tag{12-46}$$

The numerator and denominator for Equation 12-46 are chosen such that the SWR is always a number greater than 1 to avoid confusion and comply with the convention established in Equation 12-37. From Equation 12-46, it can be seen that a load resistance $Z_L = 2Z_o$ gives the same SWR as a load resistance $Z_L = Z_o/2$; the degree of mismatch is the same.

The disadvantages of not having a matched (flat) transmission line can be summarized as follows:

1. One hundred percent of the source incident power is not absorbed by the load.
2. The dielectric separating the two conductors can break down and cause corona as a result of the high-voltage standing-wave ratio.
3. Reflections and re-reflections cause more power loss.

4. Reflections cause ghost images.
5. Mismatches cause noise interference.

Although it is highly unlikely that a transmission line will be terminated in a load that is either an open or a short circuit, these conditions are examined because they illustrate the worst-possible conditions that could occur and produce standing waves that are representative of less severe conditions.

12-11-2 Standing Waves on an Open Line

When incident waves of voltage and current reach an open termination, none of the power is absorbed; it is all reflected back toward the source. The incident voltage wave is reflected in exactly the same manner as if it were to continue down an infinitely long line. However, the incident current is reflected 180° reversed from how it would have continued if the line were not open. As the incident and reflected waves pass, standing waves are produced on the line. Figure 12-20 shows the voltage and current standing waves on a transmission line that is terminated in an open circuit. It can be seen that the voltage standing wave has a maximum value at the open end and a minimum value one-quarter wavelength from the open. The current standing wave has a minimum value at the open end and a maximum value one-quarter wavelength from the open. It stands to reason that maximum voltage occurs across an open and there is minimum current.

The characteristics of a transmission line terminated in an open can be summarized as follows:

1. The voltage incident wave is reflected back just as if it were to continue (i.e., no phase reversal).
2. The current incident wave is reflected back 180° from how it would have continued.
3. The sum of the incident and reflected current waveforms is minimum at the open.
4. The sum of the incident and reflected voltage waveforms is maximum at the open.

From Figure 12-21, it can also be seen that the voltage and current standing waves repeat every one-half wavelength. The impedance at the open end $Z = V_{max}/I_{min}$ and is maximum. The impedance one-quarter wavelength from the open $Z = V_{min}/I_{max}$ and is minimum. Therefore, one-quarter wavelength from the open an impedance inversion occurs, and additional impedance inversions occur each one-quarter wavelength.

Figure 12-21 shows the development of a voltage standing wave on a transmission line that is terminated in an open circuit. Figure 12-22 shows an incident wave propagating down a transmission line toward the load. The wave is traveling at approximately the speed of light; however, for illustration purposes, the wave has been frozen at eighth-wavelength intervals. In Figure 12-22a, it can be seen that the incident wave has not reached the open. Figure 12-22b shows the wave one time unit later (for this example, the wave travels one-eighth wavelength per time unit). As you can see, the wave has moved

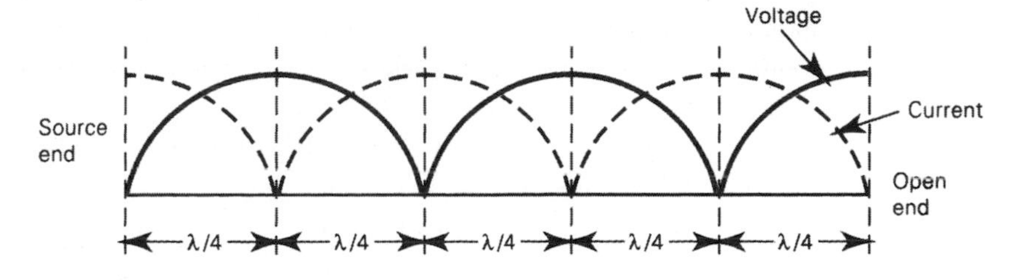

FIGURE 12-21 Voltage and current standing waves on a transmission line that is terminated in an open circuit

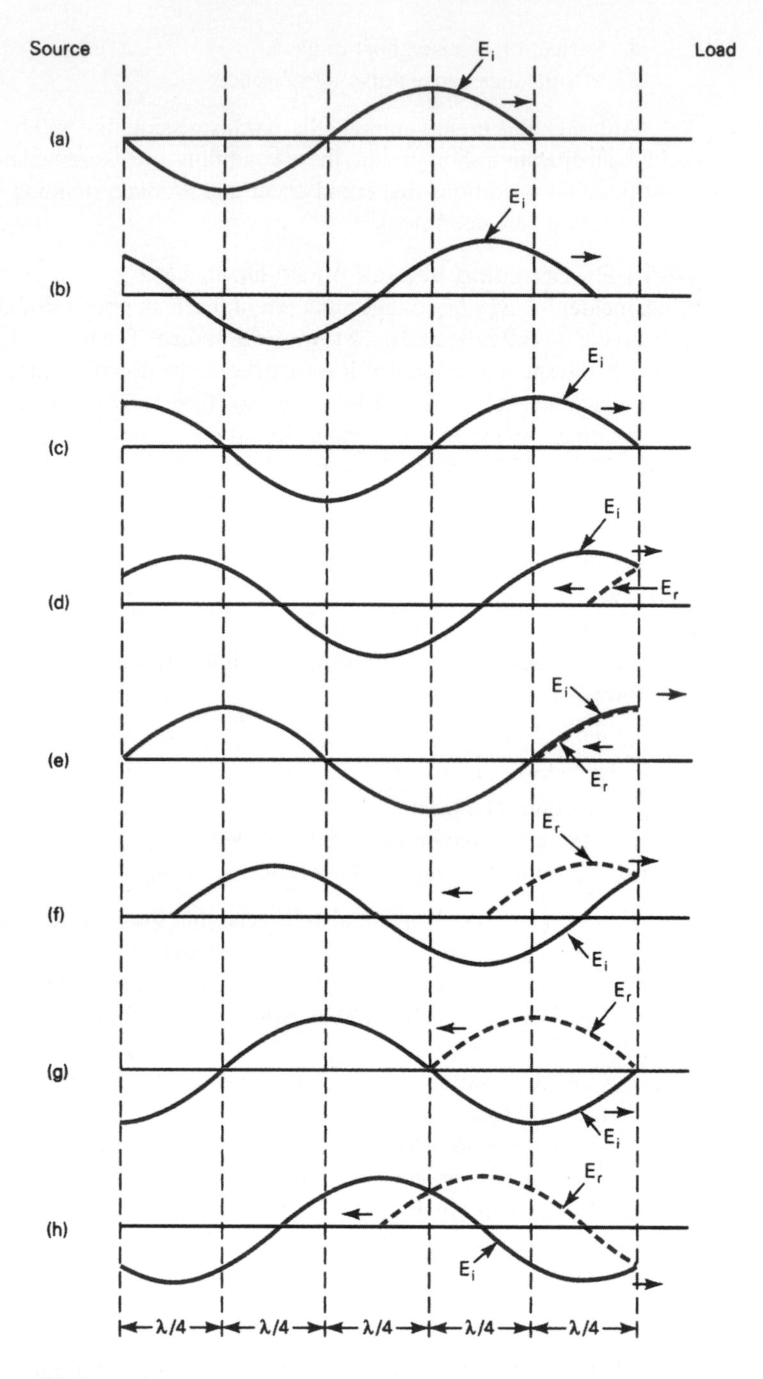

FIGURE 12-22 Incident and reflected waves on a transmission line terminated in an open circuit (*Continued*)

one-quarter wavelength closer to the open. Figure 12-22c shows the wave just as it arrives at the open. Thus far, there has been no reflected wave and, consequently, no standing wave. Figure 12-22d shows the incident and reflected waves one time unit after the incident wave has reached the open; the reflected wave is propagating away from the open. Figures 12-22e, f, and g show the incident and reflected waves for the next three time units. In Figure 12-22e, it can be seen that the incident and reflected waves are at their maximum

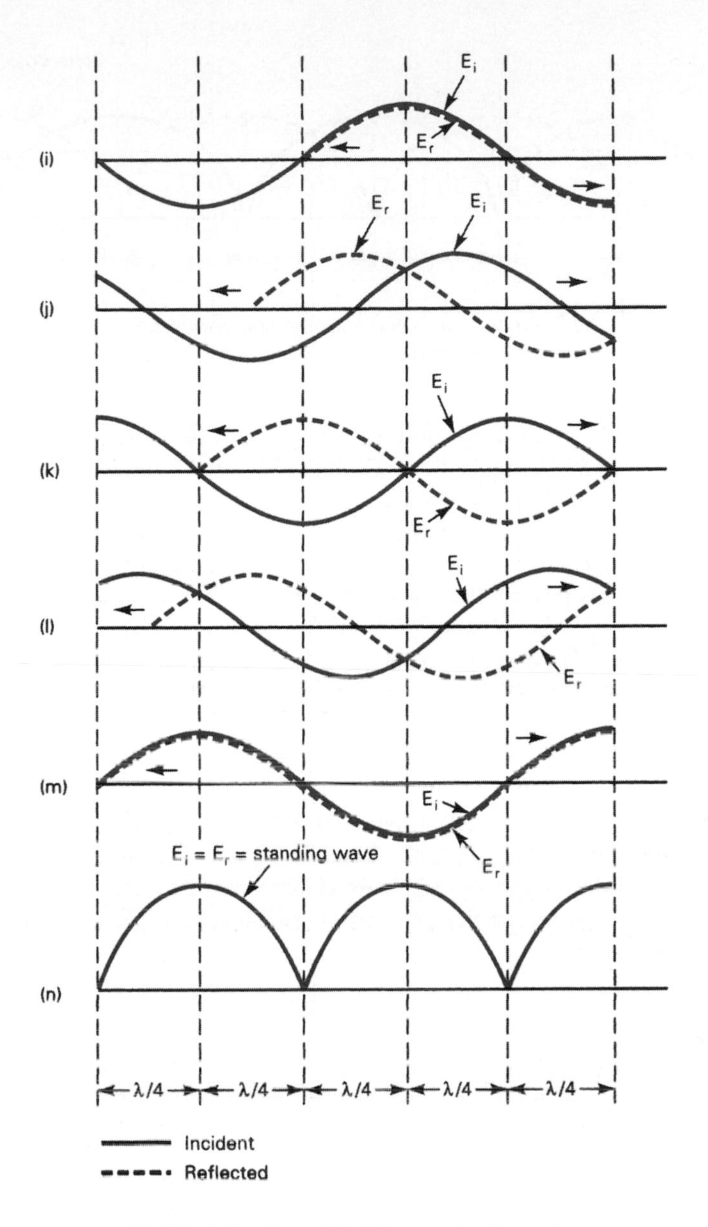

FIGURE 12-22 (*Continued*) Incident and reflected waves on a transmission line terminated in an open circuit

positive values at the same time, thus producing a voltage maximum at the open. It can also be seen that one-quarter wavelength from the open end of the transmission line the sum of the incident and reflected waves (the standing wave) is always equal to 0 V (a minimum). Figures 12-22h through m show propagation of the incident and reflected waves until the reflected wave reaches the source, and Figure 12-22n shows the resulting standing wave. It can be seen that the standing wave remains stationary (the voltage nodes and antinodes remain at the same points); however, the amplitude of the antinodes varies from maximum positive to zero to maximum negative and then repeats. For an open load, all the incident voltage is reflected ($E_r = E_i$); therefore, $V_{max} = E_i + E_r$, or $2E_i$. A similar illustration can be shown for a current standing wave (however, remember that the current reflects back with a 180° phase inversion).

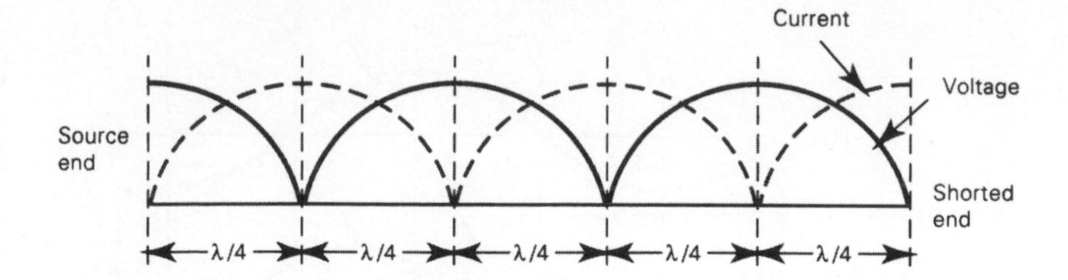

FIGURE 12-23 Voltage and current standing waves on a transmission line that is terminated in a short circuit

12-11-3 Standing Waves on a Shorted Line

As with an open line, none of the incident power is absorbed by the load when a transmission line is terminated in a short circuit. However, with a shorted line, the incident voltage and current waves are reflected back in the opposite manner. The voltage wave is reflected 180° reversed from how it would have continued down an infinitely long line, and the current wave is reflected in exactly the same manner as if there were no short.

Figure 12-23 shows the voltage and current standing waves on a transmission line that is terminated in a short circuit. It can be seen that the voltage standing wave has a minimum value at the shorted end and a maximum value one-quarter wavelength from the short. The current standing wave has a maximum value at the short and a minimum value one-quarter wavelength back. The voltage and current standing waves repeat every one-quarter wavelength. Therefore, there is an impedance inversion every quarter-wavelength interval. The impedance at the short $Z = V_{min}/I_{max}$ = minimum, and one-quarter wavelength back $Z = V_{max}/I_{min}$ = maximum. Again, it stands to reason that a voltage minimum will occur across a short and there is maximum current.

The characteristics of a transmission line terminated in a short can be summarized as follows:

1. The voltage incident wave is reflected back 180° reversed from how it would have continued.
2. The current incident wave is reflected back the same as if it had continued.
3. The sum of the incident and reflected current waveforms is maximum at the short.
4. The sum of the incident and reflected voltage waveforms is zero at the short.

For a transmission line terminated in either a short or an open circuit, the reflection coefficient is 1 (the worst case), and the SWR is infinity (also the worst-case condition).

12-12 TRANSMISSION-LINE INPUT IMPEDANCE

In the preceding section, it was shown that, when a transmission line is terminated in either a short or an open circuit, there is an *impedance inversion* every quarter-wavelength. For a loss-less line, the impedance varies from infinity to zero. However, in a more practical situation where power losses occur, the amplitude of the reflected wave is always less than that of the incident wave except at the termination. Therefore, the impedance varies from some maximum to some minimum value or vice versa, depending on whether the line is terminated in a short or an open. The input impedance for a lossless line seen looking into a transmission line that is terminated in a short or an open can be resistive, inductive, or capacitive, depending on the distance from the termination.

12-12-1 Phasor Analysis of Input Impedance: Open Line

Phasor diagrams are generally used to analyze the input impedance of a transmission line because they are relatively simple and give a pictorial representation of the voltage and cur-

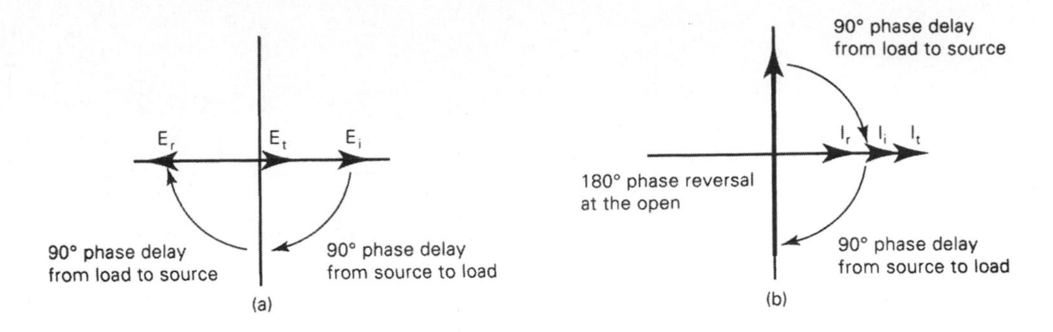

FIGURE 12-24 Voltage and current phase relationships for a quarter-wave line terminated in an open circuit: (a) voltage phase relationships; (b) current phase relationships

rent phase relationships. Voltage and current phase relations refer to variations in time. Figures 12-21 to 12-23 show standing waves of voltage and current plotted versus distance and, therefore, are not indicative of true phase relationships. The succeeding sections use phasor diagrams to analyze the input impedance of several transmission-line configurations.

12-12-1-1 Quarter-wavelength transmission line. Figure 12-24a shows the phasor diagram for the voltage, and Figure 12-24b shows the phasor diagram for the current at the input to a quarter-wave section of a transmission line terminated in an open circuit. I_i and V_i are the in-phase incident current and voltage waveforms, respectively, at the input (source) end of the line at a given instant in time. Any reflected voltage (E_r) present at the input of the line has traveled one-half wavelength (from the source to the open and back) and is, consequently, 180° behind the incident voltage. Therefore, the total voltage (E_t) at the input end is the sum of E_i and E_r. $E_t = E_i + E_r \angle 180°$, and, assuming a small line loss, $E_t = E_i - E_r$. The reflected current is delayed 90° propagating from the source to the load and another 90° from the load back to the source. Also, the reflected current undergoes a 180° phase reversal at the open. The reflected current has effectively been delayed 360°. Therefore, when the reflected current reaches the source end, it is in phase with the incident current, and the total current $I_t = I_i + I_r$. By examining Figure 12-24, it can be seen that E_t and I_t are in phase. Therefore, the input impedance seen looking into a transmission line one-quarter wave-length long that is terminated in an open circuit $Z_{\text{in}} = E_t \angle 0° / I_t \angle 0° = Z_{\text{in}} \angle 0°$. Z_{in} has a 0° phase angle, is resistive, and is minimum. Therefore, a quarter-wavelength transmission line terminated in an open circuit is equivalent to a series resonant *LC* circuit.

Figure 12-25 shows several voltage phasors for the incident and reflected waves on a transmission line that is terminated in an open circuit and how they produce a voltage standing wave.

12-12-1-2 Transmission line less than one-quarter wavelength long. Figure 12-26a shows the voltage phasor diagram, and Figure 12-26b shows the current phasor diagram for a transmission line that is less than one-quarter wavelength long (λ/4) and terminated in an open circuit. Again, the incident current (I_i) and voltage (E_i) are in phase. The reflected voltage wave is delayed 45° traveling from the source to the load (a distance of one-eighth wavelength) and another 45° traveling from the load back to the source (an additional one-eighth wavelength). Therefore, when the reflected wave reaches the source end, it lags the incident wave by 90°. The total voltage at the source end is the vector sum of the incident and reflected waves. Thus, $E_t = \sqrt{E_i^2 + E_r^2} = E_t \angle -45°$. The reflected current wave is delayed 45° traveling from the source to the load and another 45° from the load back to the source (a total distance of one-quarter wavelength). In addition, the reflected current wave has undergone a 180° phase reversal at the open prior to being reflected. The reflected current wave has been delayed a total of 270°. Therefore, the reflected wave effectively leads

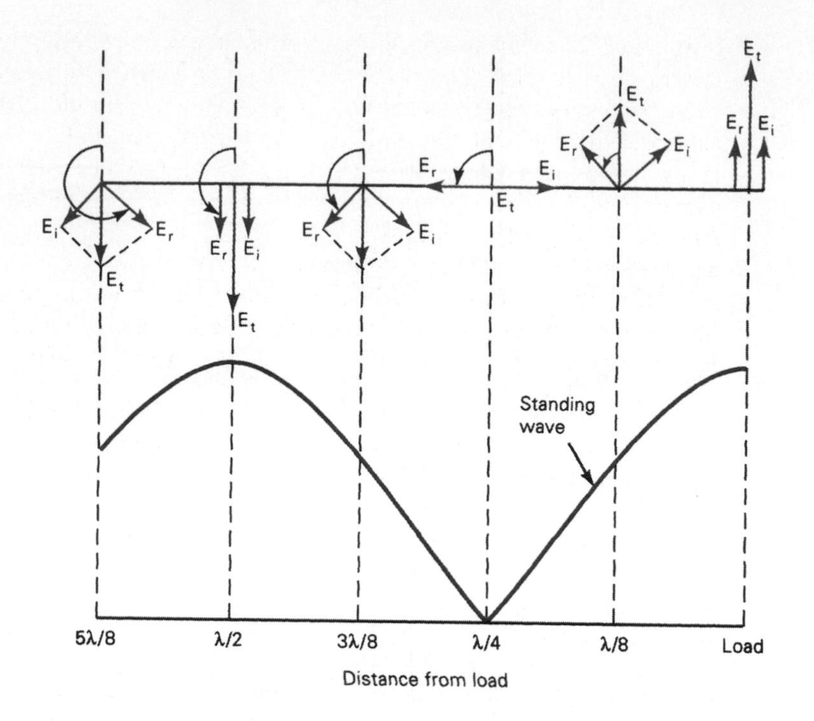

FIGURE 12-25 Vector addition of incident and reflected waves producing a standing wave

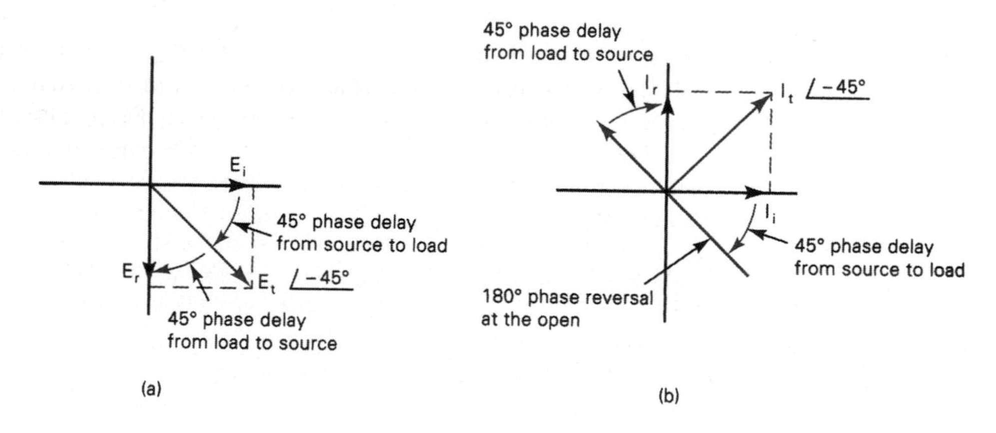

FIGURE 12-26 Voltage and current phase relationships for a transmission line less than one-quarter wavelength terminated in an open circuit: (a) voltage phase relationships; (b) current phase relationships

the incident wave by 90°. The total current at the source end is the vector sum of the present and reflected waves. Thus, $I_t = \sqrt{I_i^2 + I_r^2} = I_t \angle{+45°}$. By examining Figure 12-26, it can be seen that E_t lags I_t by 90°. Therefore, $Z_{in} = E_t \angle{-45°} / I_t \angle{+45°} = Z_{in} \angle{-90°}$. Z_{in} has a $-90°$ phase angle and, therefore, is capacitive. Any transmission line that is less than one-quarter wavelength and terminated in an open circuit is equivalent to a capacitor. The amount of capacitance depends on the exact electrical length of the line.

12-12-1-3 Transmission line more than one-quarter wavelength long. Figure 12-27a shows the voltage phasor diagram and Figure 12-27b shows the current phasor diagram for a transmission line that is more than one-quarter wavelength long and terminated in an open circuit. For this example, a three-eighths wavelength transmission line is used. The reflected voltage is delayed three-quarters wavelength or 270°. Therefore, the reflected voltage effectively leads the incident voltage by 90°. Consequently, the total voltage $E_t = \sqrt{E_i^2 + E_r^2}\,\underline{/+45°} = E_t\,\underline{/+45°}$. The reflected current wave has been delayed 270° and undergone an 180° phase reversal. Therefore, the reflected current effectively lags the incident current by 90°. Consequently, the total current $I_t = \sqrt{I_i^2 + I_r^2}\,\underline{/-45°} = I_t\underline{/-45°}$. Therefore, $Z_{in} = E_t\underline{/+45°}/I_t\underline{/-45°} = Z_{in}\,\underline{/+90°}$. Z_{in} has a +90° phase angle and is therefore inductive. The magnitude of the input impedance equals the characteristic impedance at eighth wavelength points. A transmission line between one-quarter and one-half wavelength that is terminated in an open circuit is equivalent to an inductor. The amount of inductance depends on the exact electrical length of the line.

12-12-1-4 Open transmission line as a circuit element. From the preceding discussion and Figures 12-24 through 12-27, it is obvious that an open transmission line can behave as a resistor, an inductor, or a capacitor, depending on its electrical length. Because standing-wave patterns on an open line repeat every half-wavelength interval, the input impedance also repeats. Figure 12-28 shows the variations in input impedance for an open transmission line of various electrical lengths. It can be seen that an open line is resistive and maximum at the open and at each successive half-wavelength interval and resistive and minimum one-quarter wavelength from the open and at each successive half-wavelength interval. For electrical lengths less than one-quarter wavelength, the input impedance is capacitive and decreases with length. For electrical lengths between one-quarter and one-half wavelength, the input impedance is inductive and increases with length. The capacitance and inductance patterns also repeat every half-wavelength.

12-12-2 Phasor Analysis of Input Impedance: Shorted Line

The following explanations use phasor diagrams to analyze shorted transmission lines in the same manner as with open lines. The difference is that with shorted transmission lines, the voltage waveform is reflected back with a 180° phase reversal, and the current waveform is reflected back as if there were no short.

12-12-2-1 Quarter-wavelength transmission line. The voltage and current phasor diagrams for a quarter-wavelength transmission line terminated in a short circuit are identical to those shown in Figure 12-24 except reversed. The incident and reflected voltages are in phase; therefore, $E_t = E_i + E_r$ and maximum. The incident and reflected currents are 180° out of phase; therefore, $I_t = I_i - I_r$ and minimum. $Z_{in} = E_i\,\underline{/0°}/I_t\,\underline{/0°}$ and maximum.

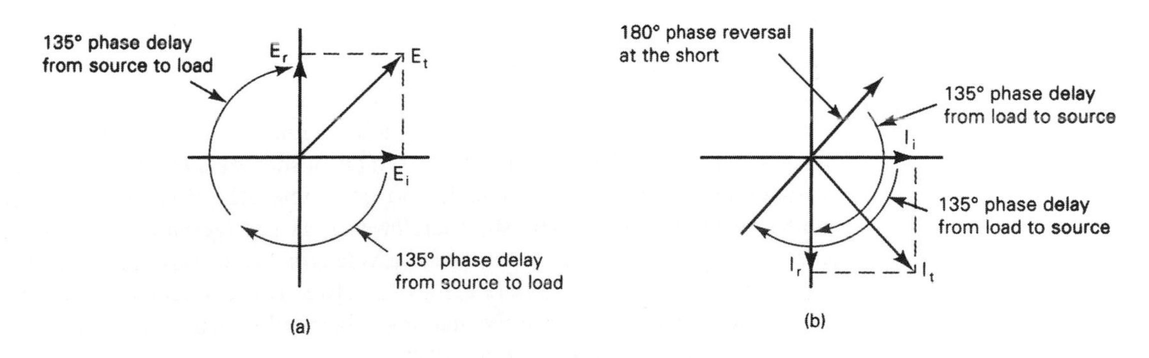

FIGURE 12-27 Voltage and current phase relationships for a transmission line more than one-quarter wavelength terminated in an open circuit: (a) voltage phase relationships; (b) current phase relationships

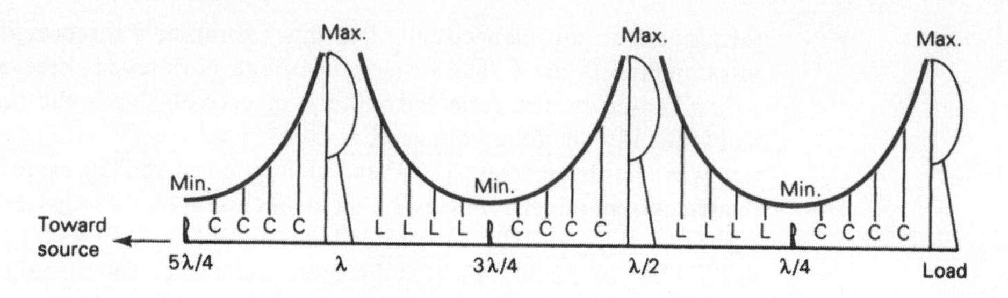

FIGURE 12-28 Input impedance variations for an open-circuited transmission line

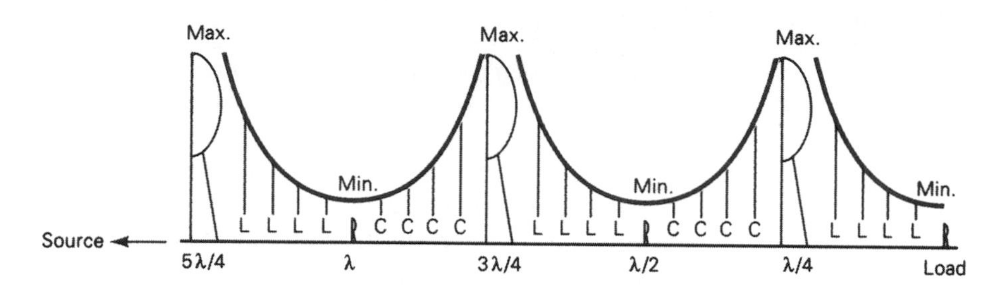

FIGURE 12-29 Input impedance variations for a short-circuited transmission line

Z_{in} has a 0° phase angle, is resistive, and is maximum. Therefore, a quarter-wavelength transmission line terminated in a short circuit is equivalent to a parallel *LC* circuit.

12-12-2-2 Transmission line less than one-quarter wavelength long. The voltage and current phasor diagrams for a transmission line less than one-quarter wavelength long and terminated in a short circuit are identical to those shown in Figure 12-26 except reversed. The voltage is reversed 180° at the short, and the current is reflected with the same phase as if it had continued. Therefore, the total voltage at the source end of the line leads the current by 90°, and the line looks inductive.

12-12-2-3 Transmission line more than one-quarter wavelength long. The voltage and current phasor diagrams for a transmission line more than one-quarter wavelength long and terminated in a short circuit are identical to those shown in Figure 12-27 except reversed. The total voltage at the source end of the line lags the current by 90°, and the line looks capacitive.

12-12-2-4 Shorted transmission line as a circuit element. From the preceding discussion, it is obvious that a shorted transmission line can behave as if it were a resistor, an inductor, or a capacitor, depending on its electrical length. On a shorted transmission line, standing waves repeat every half-wavelength; therefore, the input impedance also repeats. Figure 12-29 shows the variations in input impedance of a shorted transmission line for various electrical lengths. It can be seen that a shorted line is resistive and minimum at the short and at each successive half-wavelength interval and resistive and maximum one-quarter wavelength from the short and at each successive half-wavelength interval. For electrical lengths less than one-quarter wavelength, the input impedance is inductive and increases with length. For electrical lengths between one-quarter and one-half wavelength, the input impedance is capacitive and decreases with length. The inductance and capacitance patterns also repeat every half-wavelength interval.

12-12-2-5 Transmission-line input impedance summary. Figure 12-30 summarizes the transmission-line configurations described in the preceding sections, their input

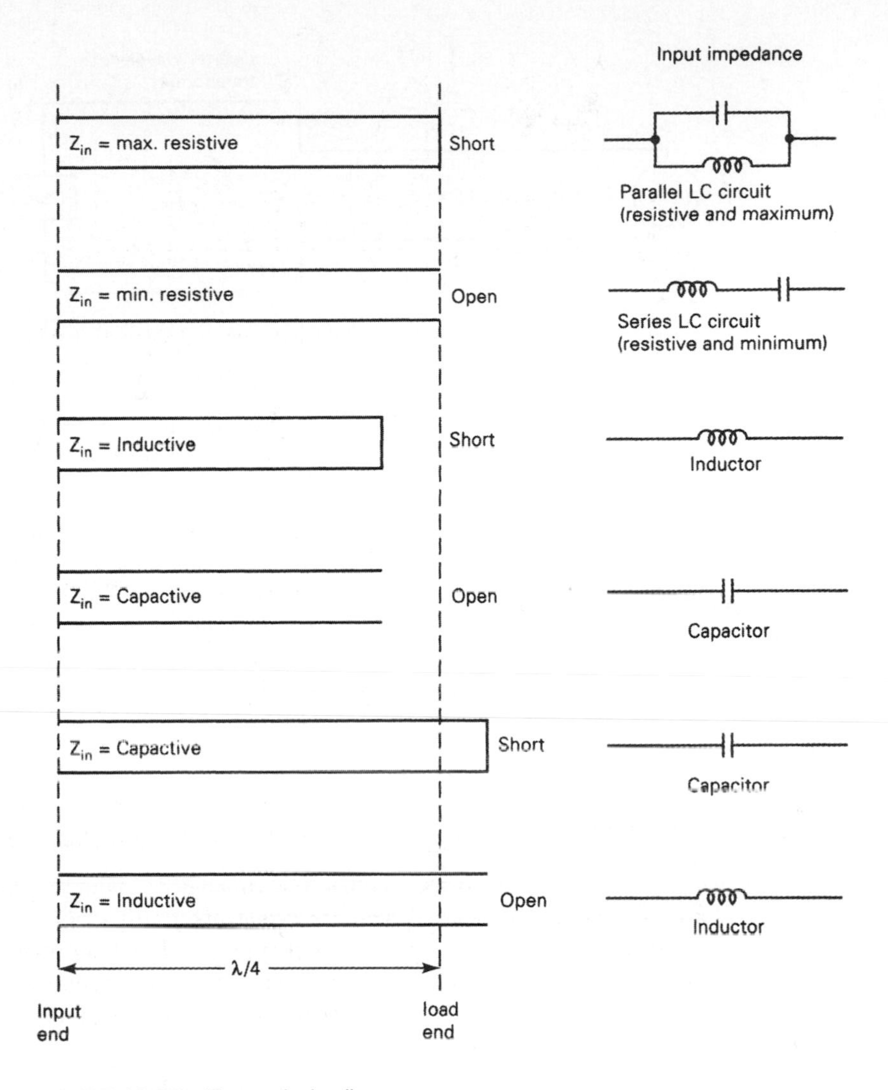

FIGURE 12-30 Transmission-line summary

impedance characteristics, and their equivalent *LC* circuits. It can be seen that both shorted and open sections of transmission lines can behave as resistors, inductors, or capacitors, depending on their electrical length.

12-12-3 Transmission-Line Impedance Matching

Power is transferred most efficiently to a load when there are no reflected waves, that is, when the load is purely resistive and equal to Z_o. Whenever the characteristic impedance of a transmission line and its load are not matched (equal), standing waves are present on the line, and maximum power is not transferred to the load. Standing waves cause power loss, dielectric breakdown, noise, radiation, and *ghost signals.* Therefore, whenever possible, a transmission line should be matched to its load. Two common transmission-line techniques are used to match a transmission line to a load having an impedance that is not equal to Z_o. They are quarter-wavelength transformer matching and stub matching.

12-12-3-1 Quarter-wavelength transformer matching. *Quarter-wavelength transformers* are used to match transmission lines to purely resistive loads whose resistance is not equal to the characteristic impedance of the line. Keep in mind that a quarter-wavelength transformer is not actually a transformer but rather a quarter-wavelength section of

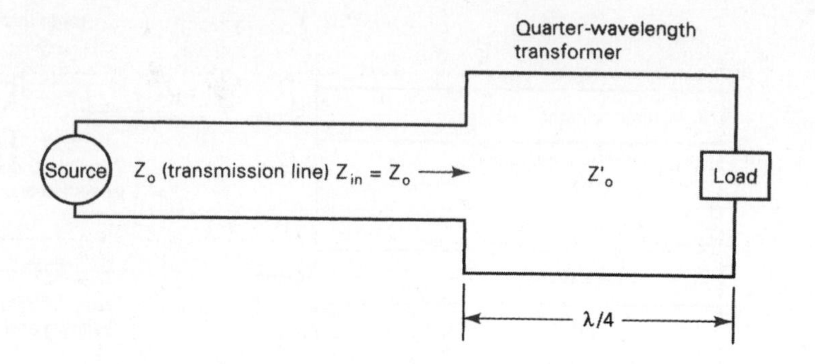

FIGURE 12-31 Quarter-wavelength transformer

transmission line that acts as if it were a transformer. The input impedance to a transmission line varies from some maximum value to some minimum value or vice versa every quarter-wavelength. Therefore, a transmission line one-quarter wavelength long acts as a *step-up* or *step-down transformer,* depending on whether Z_L is greater than or less than Z_o. A quarter-wavelength transformer is not a broadband impedance-matching device; it is a quarter-wavelength at only a single frequency. The impedance transformations for a quarter-wavelength transmission line are as follows:

1. $R_L = Z_o$: The quarter-wavelength line acts as a transformer with a 1:1 turns ratio.
2. $R_L > Z_o$: The quarter-wavelength line acts as a step-down transformer.
3. $R_L < Z_o$: The quarter-wavelength line acts as a step-up transformer.

As with a transformer, a quarter-wavelength transformer is placed between a transmission line and its load. A quarter-wavelength transformer is simply a length of transmission line one-quarter wavelength long. Figure 12-31 shows how a quarter-wavelength transformer is used to match a transmission line to a purely resistive load. The characteristic impedance of the quarter-wavelength section is determined mathematically from the formula

$$Z_0' = \sqrt{Z_o Z_L} \qquad \textbf{(12-47)}$$

where Z'_0 = characteristic impedance of a quarter-wavelength transformer
Z_o = characteristic impedance of the transmission line that is being matched
Z_L = load impedance

Example 12-6

Determine the physical length and characteristic impedance for a quarter-wavelength transformer that is used to match a section of RG-8A/U ($Z_o = 50\ \Omega$) to a 150-Ω resistive load. The frequency of operation is 150 MHz, and the velocity factor $V_f = 1$.

Solution The physical length of the transformer depends on the wavelength of the signal. Substituting into Equation 12-3 yields

$$\lambda = \frac{c}{f} = \frac{3 \times 10^8 \text{ m/s}}{150 \text{ MHz}} = 2 \text{ m}$$

$$\frac{\lambda}{4} = \frac{2 \text{ m}}{4} = 0.5 \text{ m}$$

The characteristic impedance of the 0.5-m transformer is determined from Equation 12-47:

$$Z_o' = \sqrt{Z_o Z_L} = \sqrt{(50)(150)} = 86.6\ \Omega$$

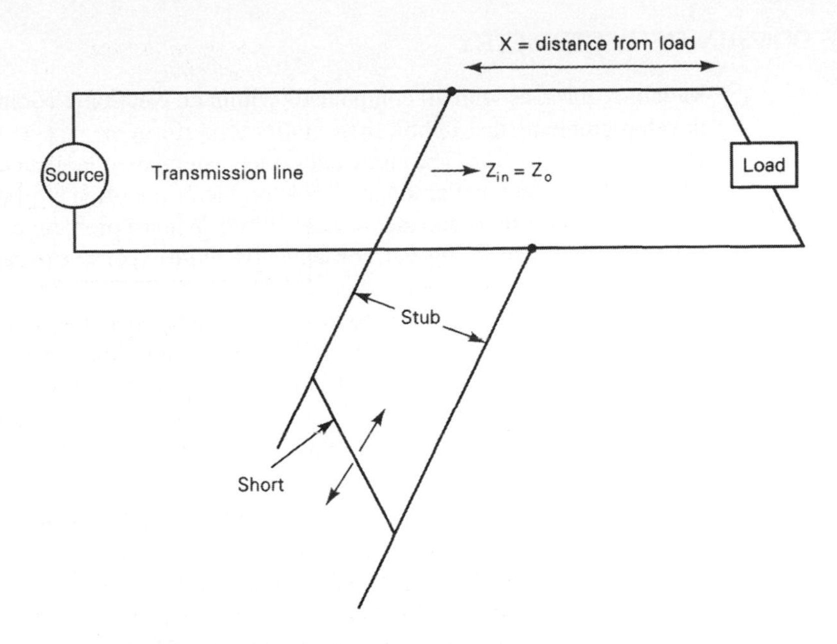

FIGURE 12-32 Shorted stub impedance matching

12-12-3-2 Stub matching. When a load is purely inductive or purely capacitive, it absorbs no energy. The reflection coefficient is 1, and the SWR is infinity. When the load is a complex impedance (which is usually the case), it is necessary to remove the reactive component to match the transmission line to the load. Transmission-line *stubs* are commonly used for this purpose. A transmission-line stub is simply a piece of additional transmission line that is placed across the primary line as close to the load as possible. The susceptance of the stub is used to tune out the susceptance of the load. With *stub matching,* either a shorted or an open stub can be used. However, shorted stubs are preferred because open stubs have a tendency to radiate, especially at the higher frequencies.

Figure 12-32 shows how a shorted stub is used to cancel the susceptance of the load and match the load resistance to the characteristic impedance of the transmission line. It has been shown how a shorted section of transmission line can look resistive, inductive, or capacitive, depending on its electrical length. A transmission line that is one-half wavelength or shorter can be used to tune out the reactive component of a load.

The process of matching a load to a transmission line with a shorted stub is as follows:

1. Locate a point as close to the load as possible where the conductive component of the input admittance is equal to the characteristic admittance of the transmission line:

$$Y_{\text{in}} = G - j\text{B}, \text{ where } G = \frac{1}{Z_o}$$

2. Attach the shorted stub to the point on the transmission line identified in step 1.
3. Depending on whether the reactive component at the point identified in step 1 is inductive or capacitive, the stub length is adjusted accordingly:

$$Y_{\text{in}} = G_o - j\text{B} + j\text{B}_{\text{stub}}$$
$$= G_o$$

if

$$\text{B} = \text{B}_{\text{stub}}$$

For a more complete explanation of stub matching using the Smith chart, refer to Appendix A.

12-13 TIME-DOMAIN REFLECTOMETRY

Metallic cables, as with all components within an electronic communications system, can develop problems that inhibit their ability to perform as expected. Cable problems often create unique situations because cables often extend over large distances, sometimes as far as several thousand feet or longer. Cable problems are often attributed to chemical erosion at cross-connect points and mechanical failure. When a problem occurs in a cable, it can be extremely time consuming and, consequently, quite expensive to determine the type and exact location of the problem.

A technique that can be used to locate an impairment in a metallic cable is called *time-domain reflectometry* (TDR). With TDR, transmission-line impairments can be pinpointed within several feet at distances of 10 miles. TDR makes use of the well-established theory that transmission-line impairments, such as shorts and opens, cause a portion of the incident signal to return to the source. How much of the transmitted signal returns depends on the type and magnitude of the impairment. The point in the line where the impairment is located represents a discontinuity to the signal. This discontinuity causes a portion of the transmitted signal to be reflected rather than continuing down the cable. If no energy is returned (i.e., the transmission line and load are perfectly matched), the line is either infinitely long or terminated in a resistive load with an impedance equal to the characteristic impedance of the line. TDR operates in a fashion similar to *radar.* A short duration pulse with a fast rise time is propagated down a cable, then the time for a portion of that signal to return to the source is measured. This return signal is sometimes called an *echo.* Knowing the velocity of propagation on the cable, the exact distance between the impairment and the source can be determined using the following mathematical relationships:

$$d = \frac{v \times t}{2} \tag{12-48}$$

where d = distance to the discontinuity (meters)
v = actual velocity (meters per second)
$v = k \times c$ (meters per second)
k = velocity factor (v/c) (unitless)
c = velocity in a vacuum (3×10^8 meters per second)
t = elapsed time (seconds)

The elapsed time is measured from the leading edge of the transmitted pulse to the reception of the reflected signal as shown in Figure 12-33a. It is important that the transmitted pulse be as narrow as possible. Otherwise, when the impairment is located close to the source, the reflected signal could return while the pulse is still being transmitted (Figure 12-33b), making it difficult to detect. For signals traveling at the speed of light (c), the velocity of

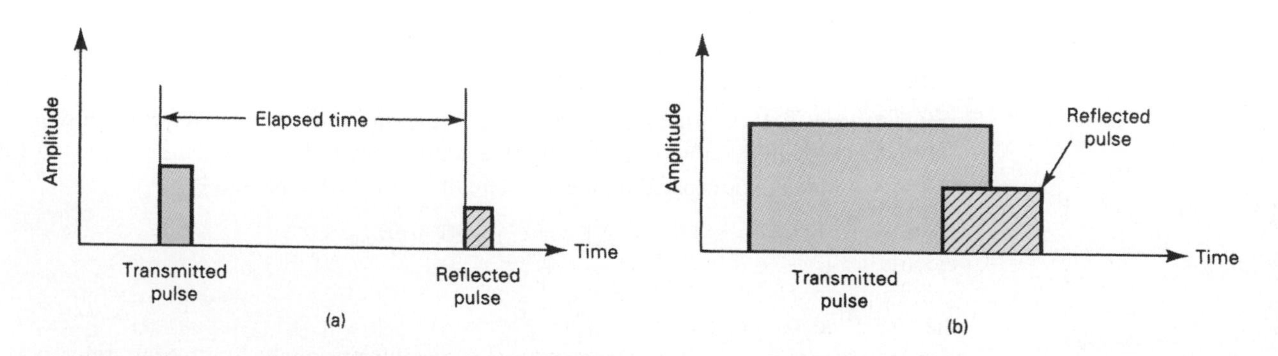

FIGURE 12-33 Time-domain reflectometry: (a) elapsed time; (b) transmitted pulse too long

propagation is 3×10^8 m/s, or approximately 1 ns/ft. Consequently, a pulse width of several microseconds would limit the usefulness of TDR only to cable impairments that occurred several thousand feet or farther away. Producing an extremely narrow pulse was one of the limiting factors in the development of TDR for locating cable faults on short cables.

Example 12-7

A pulse is transmitted down a cable that has a velocity of propagation of 0.8 c. The reflected signal is received 1 μs later. How far down the cable is the impairment?

Solution Substituting into Equation 12-48,

$$d = \frac{(0.8c) \times 1\ \mu s}{2}$$

$$= \frac{0.8 \times (3 \times 10^8\ m/s) \times 1 \times 10^6\ s}{2} = 120\ m$$

Example 12-8

Using TDR, a transmission-line impairment is located 3000 m from the source. For a velocity of propagation of 0.9c, determine the time elapsed from the beginning of the pulse to the reception of the echo.

Solution Rearranging Equation 12-48 gives

$$t = \frac{2d}{v} = \frac{2d}{k \times c}$$

$$= \frac{2(3000\ m)}{0.9(3 \times 10^8\ m/s)} = 22.22\ \mu s$$

12-14 MICROSTRIP AND STRIPLINE TRANSMISSION LINES

At frequencies below about 300 MHz, the characteristics of open and shorted transmission lines, such as those described earlier in this chapter, have little relevance. Therefore, at low frequencies, standard transmission lines would be too long for practical use as reactive components or tuned circuits. For high-frequency (300 MHz to 3000 MHz) applications, however, special transmission lines constructed with copper patterns on a *printed circuit* (PC) board called *microstrip* and *stripline* have been developed to interconnect components on PC boards. Also, when the distance between the source and load ends of a transmission line is a few inches or less, standard coaxial cable transmission lines are impractical because the connectors, terminations, and cables themselves are simply too large.

Both microstrip and stripline use the traces (sometimes called *tracks*) on the PC board itself. The traces can be etched using the same processes as the other traces on the board; thus, they do not require any additional manufacturing processes. If the lines are etched onto the surface of the PC board only, they are called microstrip lines. When the lines are etched in the middle layer of a multilayer PC board, they are called striplines. Microstrip and stripline can be used to construct transmission lines, inductors, capacitors, tuned circuits, filters, phase shifters, and impedance matching devices.

12-14-1 Microstrip

Microstrip is simply a flat conductor separated from a ground plane by an insulating dielectric material. A simple single-track microstrip line is shown in Figure 12-34a. The ground plane serves as the circuit common point and must be at least 10 times wider than the top conductor and must be connected to ground. The microstrip is generally either one-quarter or one-half wavelength long at the frequency of operation and equivalent to an unbalanced transmission line. Shorted lines are usually preferred to open lines because open

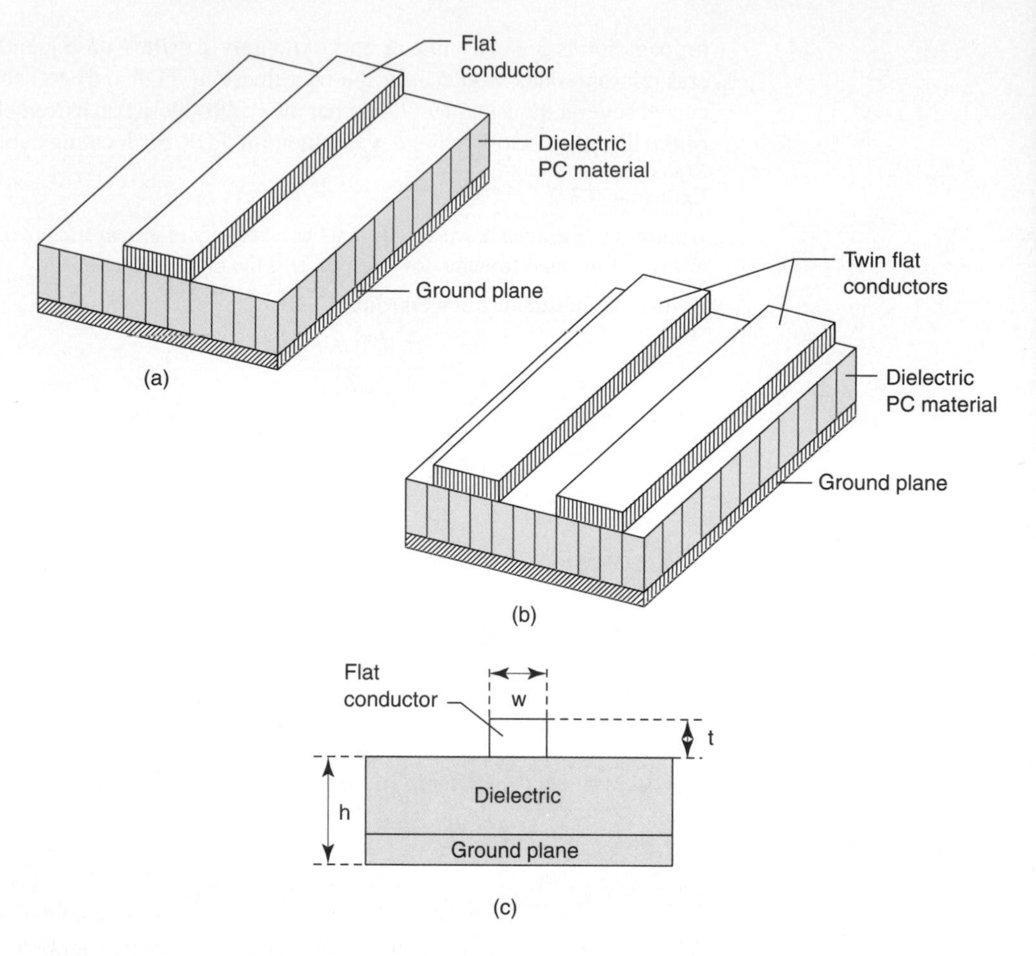

FIGURE 12-34 Microstrip transmission line: (a) unbalanced; (b) balanced; and (c) dimensions

lines have a greater tendency to radiate. Figure 12-34a shows a two-wire balanced microstrip transmission line.

As with any transmission line, the characteristic impedance of a microstrip line is dependent on its physical characteristics. Therefore, any characteristic impedance between 50 ohms and 200 ohms can be achieved with microstrip lines by simply changing its dimensions. The same is true for stripline. Unfortunately, every configuration of microstrip has its own unique formula. The formula for calculating the characteristic impedance of an unbalanced microstrip line such as the one shown in Figure 12-34c is

$$Z_o = \frac{87}{\sqrt{\epsilon + 1.41}} \ln\left(\frac{5.98h}{0.8w + t}\right) \tag{12-49}$$

where Z_o = characteristic impedance (ohms)
ϵ = dielectric constant (FR-4 fiberglass $\epsilon = 4.5$ and Teflon $\epsilon = 3$)
w = width of copper trace*
t = thickness of copper trace*
h = distance between copper trace and the ground plane (i.e., thickness of dielectric)*

*The dimensions for *w, t,* and *h* can be any linear unit (inches, millimeters, and so on) as long as they all use the same unit.

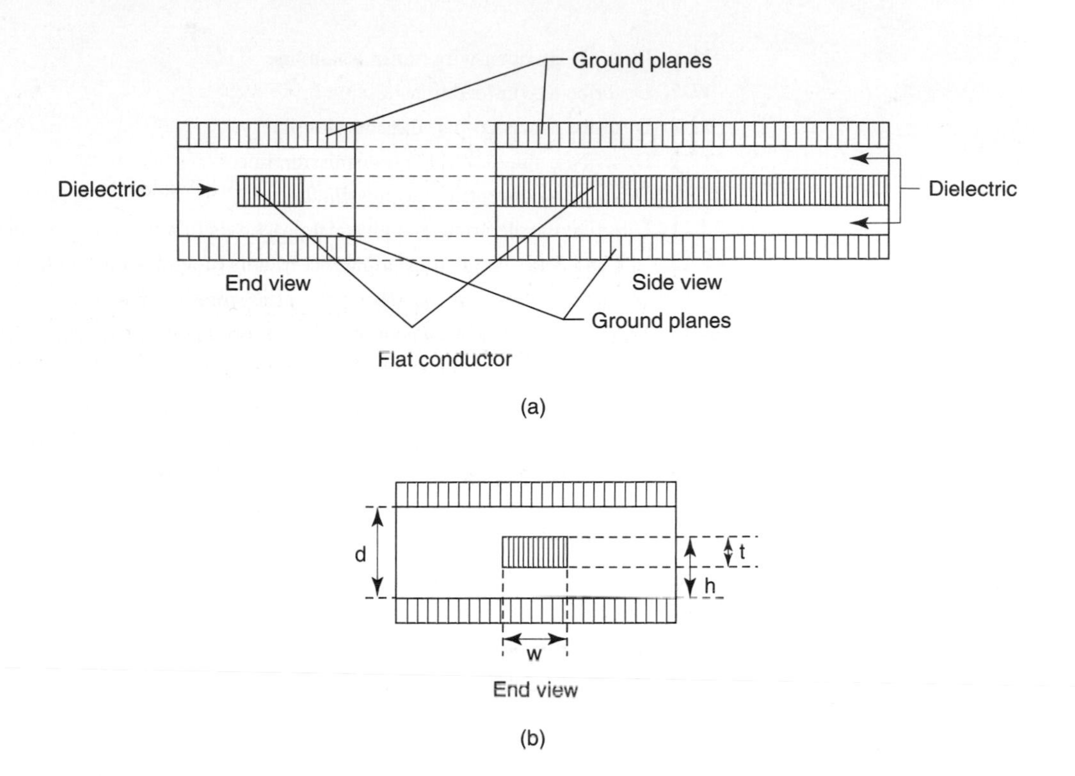

FIGURE 12-35 Stripline transmission line: (a) end and side views; (b) dimensions

12-14-2 Stripline

Stripline is simply a flat conductor sandwiched between two ground planes, as shown in Figure 12-35. Although stripline is more difficult to manufacture than microstrip, it is less likely to radiate; thus, losses in stripline are lower than with microstrip. Again, the length of a stripline is either one-quarter or one-half wavelength, and shorted lines are used more often than open lines. The characteristic impedance of a stripline configured as shown in Figure 12-35 is

$$Z_o = \frac{60}{\epsilon}\ln\left[\frac{4d}{0.67\pi w(0.8 + t/h)}\right] \tag{12-50}$$

where Z_o = characteristic impedance (ohms)
ϵ = dielectric constant (FR-4 fiberglass $\epsilon = 4.5$ and Teflon $\epsilon = 3$)
d = dielectric thickness*
w = width of conducting copper trace*
t = thickness of conducting copper trace*
h = distance between copper trace and the ground plane*

*The dimensions for *d, w, t,* and *h* can be any linear unit (inches, millimeters, and so on) as long as they all use the same unit.

QUESTIONS

12-1. Define *transmission line.*

12-2. Describe a transverse electromagnetic wave.

12-3. Define *wave velocity.*

12-4. Define *frequency* and *wavelength* for a transverse electromagnetic wave.

12-5. Describe balanced and unbalanced transmission lines.

12-6. Describe an open-wire transmission line.

12-7. Describe a twin-lead transmission line.

12-8. Describe a twisted-pair transmission line.

12-9. Describe a shielded-cable transmission line.

12-10. Describe a concentric transmission line.

12-11. Describe the electrical and physical properties of a transmission line.

12-12. List and describe the four primary constants of a transmission line.

12-13. Define *characteristic impedance* for a transmission line.

12-14. What properties of a transmission line determine its characteristic impedance?

12-15. Define *propagation constant* for a transmission line.

12-16. Define *velocity factor* for a transmission line.

12-17. What properties of a transmission line determine its velocity factor?

12-18. What properties of a transmission line determine its dielectric constant?

12-19. Define *electrical length* for a transmission line.

12-20. List and describe five types of transmission-line losses.

12-21. Describe an incident wave; a reflected wave.

12-22. Describe a resonant transmission line; a nonresonant transmission line.

12-23. Define *reflection coefficient.*

12-24. Describe standing waves; standing-wave ratio.

12-25. Describe the standing waves present on an open transmission line.

12-26. Describe the standing waves present on a shorted transmission line.

12-27. Define *input impedance* for a transmission line.

12-28. Describe the behavior of a transmission line that is terminated in a short circuit that is greater than one-quarter wavelength long; less than one-quarter wavelength.

12-29. Describe the behavior of a transmission line that is terminated in an open circuit that is greater than one-quarter wavelength long; less than one-quarter wavelength long.

12-30. Describe the behavior of an open transmission line as a circuit element.

12-31. Describe the behavior of a shorted transmission line as a circuit element.

12-32. Describe the input impedance characteristics of a quarter-wavelength transmission line.

12-33. Describe the input impedance characteristics of a transmission line that is less than one-quarter wavelength long; greater than one-quarter wavelength long.

12-34. Describe quarter-wavelength transformer matching.

12-35. Describe how stub matching is accomplished.

12-36. Describe time-domain reflectometry.

PROBLEMS

12-1. Determine the wavelengths for electromagnetic waves in free space with the following frequencies: 1 kHz, 100 kHz, 1 MHz, and 1 GHz.

12-2. Determine the frequencies for electromagnetic waves in free space with the following wavelengths: 1 cm, 1 m, 10 m, 100 m, and 1000 m.

12-3. Determine the characteristic impedance for an air-dielectric transmission line with D/r ratio of 8.8.

12-4. Determine the characteristic impedance for an air-filled concentric transmission line with D/d ratio of 4.

12-5. Determine the characteristic impedance for a coaxial cable with inductance $L = 0.2\ \mu\text{H/ft}$ and capacitance $C = 16\ \text{pF/ft}$.

12-6. For a given length of coaxial cable with distributed capacitance $C = 48.3\ \text{pF/m}$ and distributed inductance $L = 241.56\ \text{nH/m}$, determine the velocity factor and velocity of propagation.

12-7. Determine the reflection coefficient for a transmission line with incident voltage $E_i = 0.2$ V and reflected voltage $E_r = 0.01$ V.

12-8. Determine the standing-wave ratio for the transmission line described in problem 12-7.

12-9. Determine the SWR for a transmission line with maximum voltage standing-wave amplitude $V_{max} = 6$ V and minimum voltage standing-wave amplitude $V_{min} = 0.5$.

12-10. Determine the SWR for a 50-Ω transmission line that is terminated in a load resistance $Z_L = 75\ \Omega$.

12-11. Determine the SWR for a 75-Ω transmission line that is terminated in a load resistance $Z_L = 50\ \Omega$.

12-12. Determine the characteristic impedance for a quarter-wavelength transformer that is used to match a section of 75-Ω transmission line to a 100-Ω resistive load.

12-13. Using TDR, a pulse is transmitted down a cable with a velocity of propagation of 0.7*c*. The reflected signal is received 1.2 μs later. How far down the cable is the impairment?

12-14. Using TDR, a transmission-line impairment is located 2500 m from the source. For a velocity of propagation of 0.95*c*, determine the elapsed time from the beginning of the pulse to the reception of the echo.

12-15. Using TDR, a transmission-line impairment is located 100 m from the source. If the elapsed time from the beginning of the pulse to the reception of the echo is 833 ns, determine the velocity factor.

12-16. Determine the wavelengths for electromagnetic waves with the following frequencies: 5 kHz, 50 kHz, 500 kHz, and 5 MHz.

12-17. Determine the frequencies for electromagnetic waves with the following wavelengths: 5 cm, 50 cm, 5 m, and 50 m.

12-18. Determine the characteristic impedance for an air-dielectric transmission line with a *D/r* ratio of 6.8.

12-19. Determine the characteristic impedance for an air-filled concentric transmission line with a *D/d* ratio of 6.

12-20. Determine the characteristic impedance for a coaxial cable with inductance $L = 0.15$ μH/ft and capacitance $C = 20$ pF/ft.

12-21. For a given length of coaxial cable with distributed capacitance $C = 24.15$ pF/m and distributed inductance $L = 483.12$ nH/m, determine the velocity factor and velocity of propagation.

12-22. Determine the reflection coefficient for a transmission line with incident voltage $E_i = 0.4$ V and reflected voltage $E_r = 0.002$ V.

12-23. Determine the standing-wave ratio for the transmission line described in problem 12-22.

12-24. Determine the SWR for a transmission line with a maximum voltage standing-wave amplitude $V_{max} = 8$ V and a minimum voltage standing-wave amplitude $V_{min} = 0.8$ V.

12-25. Determine the SWR for a 50-Ω transmission line that is terminated in a load resistance $Z_L = 60\ \Omega$.

12-26. Determine the SWR for a 60-Ω transmission line that is terminated in a load resistance $Z_L = 50\ \Omega$.

12-27. Determine the characteristic impedance for a quarter-wave transformer that is used to match a section of 50-Ω transmission line to a 60-Ω resistive load.

CHAPTER 13

Optical Fiber Transmission Media

CHAPTER OUTLINE

OBJECTIVES

- Define *optical communications*
- Present an overview of the history of optical fibers and optical fiber communications
- Compare the advantages and disadvantages of optical fibers over metallic cables
- Define *electromagnetic frequency* and *wavelength spectrum*
- Describe several types of optical fiber construction
- Explain the physics of light and the following terms: velocity of propagation, refraction, refractive index, critical angle, acceptance angle, acceptance cone, and numerical aperture
- Describe how light waves propagate through an optical fiber cable
- Define *modes of propagation* and *index profile*
- Describe the three types of optical fiber configurations: single-mode step index, multimode step index, and multimode graded index
- Describe the various losses incurred in optical fiber cables
- Define *light source* and *optical power*
- Describe the following light sources: light-emitting diodes and injection diodes
- Describe the following light detectors: PIN diodes and avalanche photodiodes
- Describe the operation of a laser
- Explain how to calculate a link budget for an optical fiber system

13-1 INTRODUCTION

Optical fiber cables are the newest and probably the most promising type of guided transmission medium for virtually all forms of digital and data communications applications, including local, metropolitan, and wide area networks. With optical fibers, electromagnetic waves are guided through a media composed of a transparent material without using electrical current flow. With optical fibers, electromagnetic light waves propagate through the media in much the same way that radio signals propagate through Earth's atmosphere.

In essence, an *optical communications system* is one that uses light as the carrier of information. Propagating light waves through Earth's atmosphere is difficult and often impractical. Consequently, optical fiber communications systems use glass or plastic fiber cables to "*contain*" the light waves and guide them in a manner similar to the way electromagnetic waves are guided through a metallic transmission medium.

The *information-carrying capacity* of any electronic communications system is directly proportional to bandwidth. Optical fiber cables have, for all practical purposes, an infinite bandwidth. Therefore, they have the capacity to carry much more information than their metallic counterparts or, for that matter, even the most sophisticated wireless communications systems.

For comparison purposes, it is common to express the bandwidth of an analog communications system as a percentage of its carrier frequency. This is sometimes called the *bandwidth utilization ratio.* For instance, a VHF radio communications system operating at a carrier frequency of 100 MHz with 10-MHz bandwidth has a bandwidth utilization ratio of 10%. A microwave radio system operating at a carrier frequency of 10 GHz with a 10% bandwidth utilization ratio would have 1 GHz of bandwidth available. Obviously, the higher the carrier frequency, the more bandwidth available, and the greater the information-carrying capacity. Light frequencies used in optical fiber communications systems are between 1×10^{14} Hz and 4×10^{14} Hz (100,000 GHz to 400,000 GHz). A bandwidth utilization ratio of 10% would be a bandwidth between 10,000 GHz and 40,000 GHz.

13-2 HISTORY OF OPTICAL FIBER COMMUNICATIONS

In 1880, Alexander Graham Bell experimented with an apparatus he called a *photophone.* The photophone was a device constructed from mirrors and selenium detectors that transmitted sound waves over a beam of light. The photophone was awkward and unreliable and had no real practical application. Actually, visual light was a primary means of communicating long before electronic communications came about. Smoke signals and mirrors were used ages ago to convey short, simple messages. Bell's contraption, however, was the first attempt at using a beam of light for carrying information.

Transmission of light waves for any useful distance through Earth's atmosphere is impractical because water vapor, oxygen, and particulates in the air absorb and attenuate the signals at light frequencies. Consequently, the only practical type of optical communications system is one that uses a fiber guide. In 1930, J. L. Baird, an English scientist, and C. W. Hansell, a scientist from the United States, were granted patents for scanning and transmitting television images through uncoated fiber cables. A few years later, a German scientist named H. Lamm successfully transmitted images through a single glass fiber. At that time, most people considered fiber optics more of a toy or a laboratory stunt and, consequently, it was not until the early 1950s that any substantial breakthrough was made in the field of fiber optics.

In 1951, A. C. S. van Heel of Holland and H. H. Hopkins and N. S. Kapany of England experimented with light transmission through *bundles* of fibers. Their studies led to the development of the *flexible fiberscope,* which is used extensively in the medical field. It was Kapany who coined the term "fiber optics" in 1956.

In 1958, Charles H. Townes, an American, and Arthur L. Schawlow, a Canadian, wrote a paper describing how it was possible to use stimulated emission for amplifying light waves (laser) as well as microwaves (maser). Two years later, Theodore H. Maiman, a scientist with Hughes Aircraft Company, built the first optical maser.

The *laser* (*l*ight *a*mplification by *s*timulated *e*mission of *r*adiation) was invented in 1960. The laser's relatively high output power, high frequency of operation, and capability of carrying an extremely wide bandwidth signal make it ideally suited for high-capacity communications systems. The invention of the laser greatly accelerated research efforts in fiber-optic communications, although it was not until 1967 that K. C. Kao and G. A. Bockham of the Standard Telecommunications Laboratory in England proposed a new communications medium using *cladded* fiber cables.

The fiber cables available in the 1960s were extremely *lossy* (more than 1000 dB/km), which limited optical transmissions to short distances. In 1970, Kapron, Keck, and Maurer of Corning Glass Works in Corning, New York, developed an optical fiber with losses less than 2 dB/km. That was the "big" breakthrough needed to permit practical fiber optics communications systems. Since 1970, fiber optics technology has grown exponentially. Recently, Bell Laboratories successfully transmitted 1 billion bps through a fiber cable for 600 miles without a regenerator.

In the late 1970s and early 1980s, the refinement of optical cables and the development of high-quality, affordable light sources and detectors opened the door to the development of high-quality, high-capacity, efficient, and affordable optical fiber communications systems. By the late 1980s, losses in optical fibers were reduced to as low as 0.16 dB/km, and in 1988 NEC Corporation set a new long-haul transmission record by transmitting 10 gigabytes per second over 80.1 kilometers of optical fiber. Also in 1988, the American National Standards Institute (ANSI) published the *Synchronous Optical Network* (*SONET*). By the mid-1990s, optical voice and data networks were commonplace throughout the United States and much of the world.

13-3 OPTICAL FIBERS VERSUS METALLIC CABLE FACILITIES

Communications through glass or plastic fibers has several advantages over conventional metallic transmission media for both telecommunication and computer networking applications.

13-3-1 Advantages of Optical Fiber Cables

The advantages of using optical fibers include the following:

1. *Wider bandwidth and greater information capacity.* Optical fibers have greater information capacity than metallic cables because of the inherently wider bandwidths available with optical frequencies. Optical fibers are available with bandwidths up to several thousand gigahertz. The *primary electrical constants* (resistance, inductance, and capacitance) in metallic cables cause them to act like low-pass filters, which limit their transmission frequencies, bandwidth, bit rate, and information-carrying capacity. Modern optical fiber communications systems are capable of transmitting several gigabits per second over hundreds of miles, allowing literally millions of individual voice and data channels to be combined and propagated over one optical fiber cable.

2. *Immunity to crosstalk.* Optical fiber cables are immune to crosstalk because glass and plastic fibers are nonconductors of electrical current. Therefore, fiber cables are not surrounded by a changing magnetic field, which is the primary cause of crosstalk between metallic conductors located physically close to each other.

3. *Immunity to static interference.* Because optical fiber cables are nonconductors of electrical current, they are immune to static noise due to electromagnetic interference (EMI) caused by lightning, electric motors, relays, fluorescent lights, and other electrical noise sources (most of which are man-made). For the same reason, fiber cables do not radiate electromagnetic energy.

4. *Environmental immunity.* Optical fiber cables are more resistant to environmental extremes (including weather variations) than metallic cables. Optical cables also operate over a wider temperature range and are less affected by corrosive liquids and gases.

5. *Safety and convenience.* Optical fiber cables are safer and easier to install and maintain than metallic cables. Because glass and plastic fibers are nonconductors, there are no electrical currents or voltages associated with them. Optical fibers can be used around volatile liquids and gasses without worrying about their causing explosions or fires. Optical fibers are also smaller and much more lightweight and compact than metallic cables. Consequently, they are more flexible, easier to work with, require less storage space, cheaper to transport, and easier to install and maintain.

6. *Lower transmission loss.* Optical fibers have considerably less signal loss than their metallic counterparts. Optical fibers are currently being manufactured with as little as a few-tenths-of-a-decibel loss per kilometer. Consequently, optical regenerators and amplifiers can be spaced considerably farther apart than with metallic transmission lines.

7. *Security.* Optical fiber cables are more secure than metallic cables. It is virtually impossible to tap into a fiber cable without the user's knowledge, and optical cables cannot be detected with metal detectors unless they are reinforced with steel for strength.

8. *Durability and reliability.* Optical fiber cables last longer and are more reliable than metallic facilities because fiber cables have a higher tolerance to changes in environmental conditions and are immune to corrosive materials.

9. *Economics.* The cost of optical fiber cables is approximately the same as metallic cables. Fiber cables have less loss and require fewer repeaters, which equates to lower installation and overall system costs and improved reliability.

13-3-2 Disadvantages of Optical Fiber Cables

Although the advantages of optical fiber cables far exceed the disadvantages, it is important to know the limitations of the fiber. The disadvantages of optical fibers include the following:

1. *Interfacing costs.* Optical fiber cable systems are virtually useless by themselves. To be practical and useful, they must be connected to standard electronic facilities, which often require expensive interfaces.

2. *Strength.* Optical fibers by themselves have a significantly lower tensile strength than coaxial cable. This can be improved by coating the fiber with standard *Kevlar* and a protective jacket of PVC. In addition, glass fiber is much more fragile than copper wire, making fiber less attractive where hardware portability is required.

3. *Remote electrical power.* Occasionally, it is necessary to provide electrical power to remote interface or regenerating equipment. This cannot be accomplished with the optical cable, so additional metallic cables must be included in the cable assembly.

4. *Optical fiber cables are more susceptible to losses introduced by bending the cable.* Electromagnetic waves propagate through an optical cable by either refraction or reflection. Therefore, bending the cable causes irregularities in the cable dimensions, resulting in a loss of signal power. Optical fibers are also more prone to manufacturing defects, as even the most minor defect can cause excessive loss of signal power.

5. *Specialized tools, equipment, and training.* Optical fiber cables require special tools to splice and repair cables and special test equipment to make routine measurements. Not only is repairing fiber cables difficult and expensive, but technicians working on optical cables also require special skills and training. In addition, sometimes it is difficult to locate faults in optical cables because there is no electrical continuity.

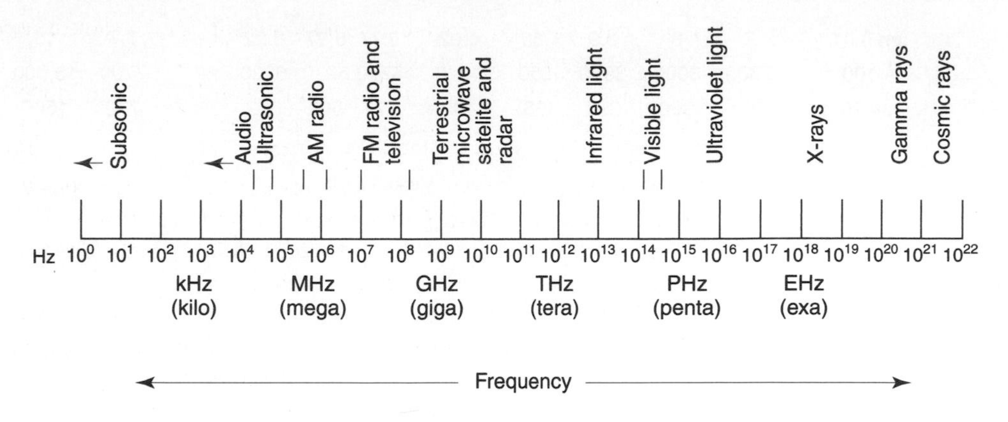

FIGURE 13-1 Electromagnetic frequency spectrum

13-4 ELECTROMAGNETIC SPECTRUM

The total electromagnetic frequency spectrum is shown in Figure 13-1. From the figure, it can be seen that the frequency spectrum extends from the subsonic frequencies (a few hertz) to cosmic rays (10^{22} Hz). The light frequency spectrum can be divided into three general bands:

1. *Infrared.* The band of light frequencies that is too high to be seen by the human eye with wavelengths ranging between 770 nm and 10^6 nm. Optical fiber systems generally operate in the infrared band.
2. *Visible.* The band of light frequencies to which the human eye will respond with wavelengths ranging between 390 nm and 770 nm. This band is visible to the human eye.
3. *Ultraviolet.* The band of light frequencies that are too low to be seen by the human eye with wavelengths ranging between 10 nm and 390 nm.

When dealing with ultra-high-frequency electromagnetic waves, such as light, it is common to use units of wavelength rather than frequency. Wavelength is the length that one cycle of an electromagnetic wave occupies in space. The length of a wavelength depends on the frequency of the wave and the velocity of light. Mathematically, wavelength is

$$\lambda = \frac{c}{f} \tag{13-1}$$

where λ = wavelength (meters/cycle)
c = velocity of light (300,000,000 meters per second)
f = frequency (hertz)

With light frequencies, wavelength is often stated in microns, where 1 micron = 10^{-6} meter (1 μm), or in nanometers (nm), where 1 nm = 10^{-9} meter. However, when describing the optical spectrum, the unit angstrom is sometimes used to express wavelength, where 1 angstrom = 10^{-10} meter, or 0.0001 micron. Figure 13-2 shows the total electromagnetic wavelength spectrum.

13-5 BLOCK DIAGRAM OF AN OPTICAL FIBER COMMUNICATIONS SYSTEM

Figure 13-3 shows a simplified block diagram of a simplex optical fiber communications link. The three primary building blocks are the transmitter, the receiver, and the optical fiber cable. The transmitter is comprised of a voltage-to-current converter, a light source, and a source-to-fiber interface (light coupler). The fiber guide is the transmission medium, which

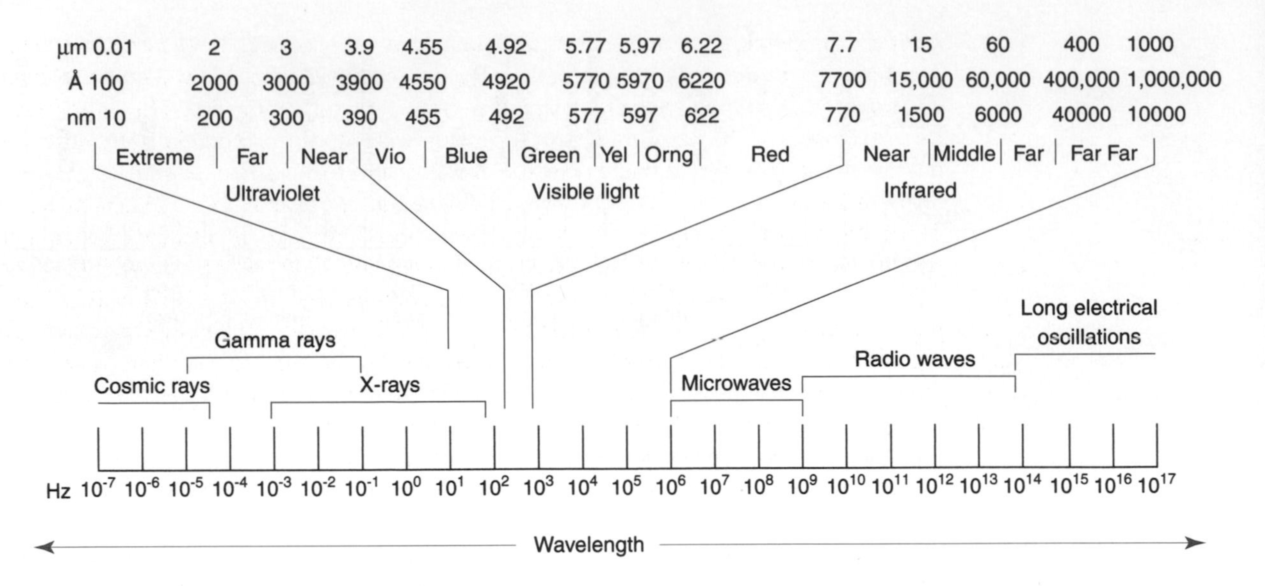

FIGURE 13-2 Electromagnetic wavelength spectrum

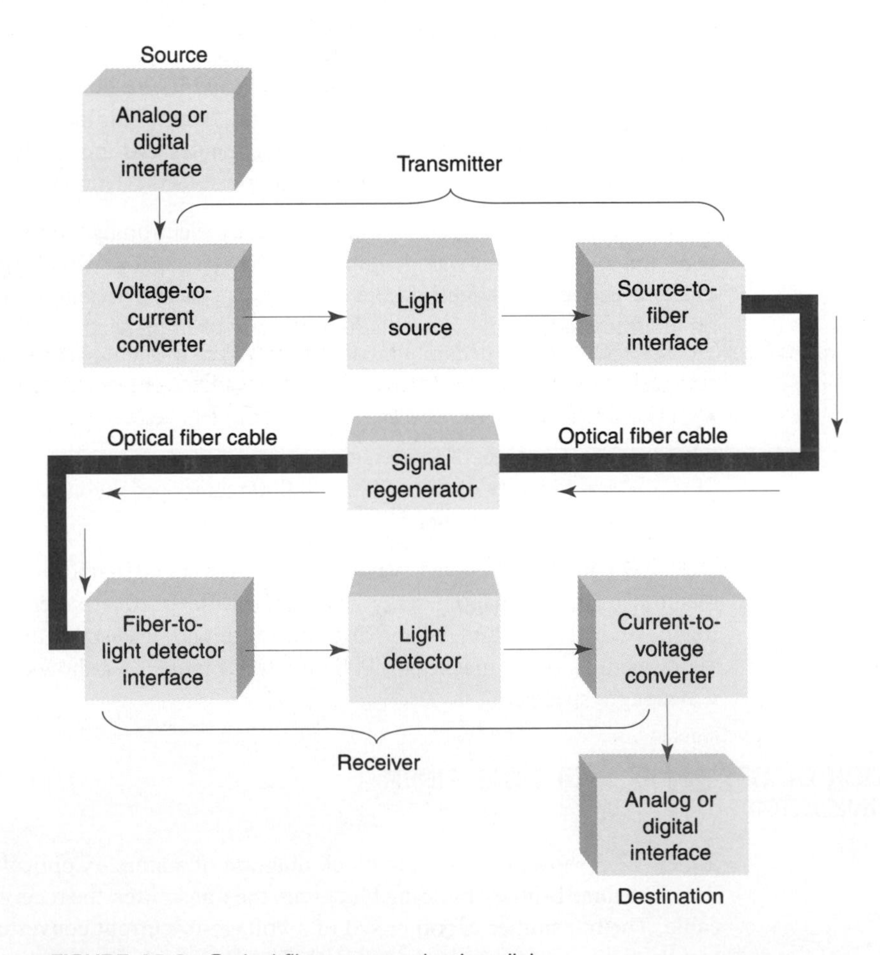

FIGURE 13-3 Optical fiber communications link

is either an ultrapure glass or a plastic cable. It may be necessary to add one or more regenerators to the transmission medium, depending on the distance between the transmitter and receiver. Functionally, the regenerator performs light amplification. However, in reality the signal is not actually amplified; it is reconstructed. The receiver includes a fiber-to-interface (light coupler), a photo detector, and a current-to-voltage converter.

In the transmitter, the light source can be modulated by a digital or an analog signal. The voltage-to-current converter serves as an electrical interface between the input circuitry and the light source. The light source is either an *infrared light-emitting diode* (LED) or an *injection laser diode* (ILD). The amount of light emitted by either an LED or ILD is proportional to the amount of drive current. Thus, the voltage-to-current converter converts an input signal voltage to a current that is used to drive the light source. The light outputted by the light source is directly proportional to the magnitude of the input voltage. In essence, the light intensity is modulated by the input signal.

The source-to-fiber coupler (such as an optical lens) is a mechanical interface. Its function is to couple light emitted by the light source into the optical fiber cable. The optical fiber consists of a glass or plastic fiber core surrounded by a cladding and then encapsulated in a protective jacket. The fiber-to-light detector-coupling device is also a mechanical coupler. Its function is to couple as much light as possible from the fiber cable into the light detector.

The light detector is generally a PIN (*p*-type-*i*ntrinsic-*n*-type) diode, an APD (avalanche photodiode), or a *phototransistor*. All three of these devices convert light energy to current. Consequently, a current-to-voltage converter is required to produce an output voltage proportional to the original source information. The current-to-voltage converter transforms changes in detector current to changes in voltage.

The analog or digital interfaces are electrical interfaces that match impedances and signal levels between the information source and destination to the input and output circuitry of the optical system.

13-6 OPTICAL FIBER TYPES

13-6-1 Optical Fiber Construction

The actual fiber portion of an optical cable is generally considered to include both the fiber *core* and its *cladding* (see Figure 13-4). A special lacquer, silicone, or acrylate coating is generally applied to the outside of the cladding to seal and preserve the fiber's strength,

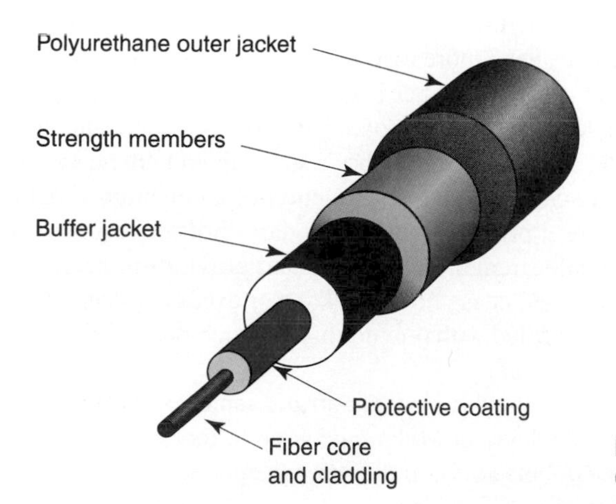

FIGURE 13-4 Optical fiber cable construction

helping maintain the cables attenuation characteristics. The coating also helps protect the fiber from moisture, which reduces the possibility of the occurrence of a detrimental phenomenon called *stress corrosion* (sometimes called *static fatigue*) caused by high humidity. Moisture causes silicon dioxide crystals to interact, causing bonds to break down and spontaneous fractures to form over a prolonged period of time. The protective coating is surrounded by a *buffer jacket,* which provides the cable additional protection against abrasion and shock. Materials commonly used for the buffer jacket include steel, fiberglass, plastic, flame-retardant polyvinyl chloride (FR-PVC), Kevlar yarn, and paper. The buffer jacket is encapsulated in a *strength member,* which increases the tensile strength of the overall cable assembly. Finally, the entire cable assembly is contained in an outer polyurethane jacket.

There are three essential types of optical fibers commonly used today. All three varieties are constructed of either glass, plastic, or a combination of glass and plastic:

Plastic core and cladding

Glass core with plastic cladding (called PCS fiber [plastic-clad silica])

Glass core and glass cladding (called SCS [silica-clad silica])

Plastic fibers are more flexible and, consequently, more rugged than glass. Therefore, plastic cables are easier to install, can better withstand stress, are less expensive, and weigh approximately 60% less than glass. However, plastic fibers have higher attenuation characteristics and do not propagate light as efficiently as glass. Therefore, plastic fibers are limited to relatively short cable runs, such as within a single building.

Fibers with glass cores have less attenuation than plastic fibers, with PCS being slightly better than SCS. PCS fibers are also less affected by radiation and, therefore, are more immune to external interference. SCS fibers have the best propagation characteristics and are easier to terminate than PCS fibers. Unfortunately, SCS fibers are the least rugged, and they are more susceptible to increases in attenuation when exposed to radiation.

The selection of a fiber for a given application is a function of the specific system requirements. There are always trade-offs based on the economics and logistics of a particular application.

13-6-1-1 Cable configurations. There are many different cable designs available today. Figure 13-5 shows examples of several optical fiber cable configurations. With loose tube construction (Figure 13-5a), each fiber is contained in a protective tube. Inside the tube, a polyurethane compound encapsules the fiber and prevents the intrusion of water. A phenomenon called *stress corrosion* or *static fatigue* can result if the glass fiber is exposed to long periods of high humidity. Silicon dioxide crystals interact with the moisture and cause bonds to break down, causing spontaneous fractures to form over a prolonged period. Some fiber cables have more than one protective coating to ensure that the fiber's characteristics do not alter if the fiber is exposed to extreme temperature changes. Surrounding the fiber's cladding is usually a coating of either lacquer, silicon, or acrylate that is typically applied to seal and preserve the fiber's strength and attenuation characteristics.

Figure 13-5b shows the construction of a constrained optical fiber cable. Surrounding the fiber are a primary and a secondary buffer comprised of Kevlar yarn, which increases the tensile strength of the cable and provides protection from external mechanical influences that could cause fiber breakage or excessive optical attenuation. Again, an outer protective tube is filled with polyurethane, which prevents moisture from coming into contact with the fiber core.

Figure 13-5c shows a *multiple-strand* cable configuration, which includes a steel central member and a layer of Mylar tape wrap to increase the cable's tensile strength. Figure 13-5d shows a ribbon configuration for a telephone cable, and Figure 13-5e shows both end and side views of a PCS cable.

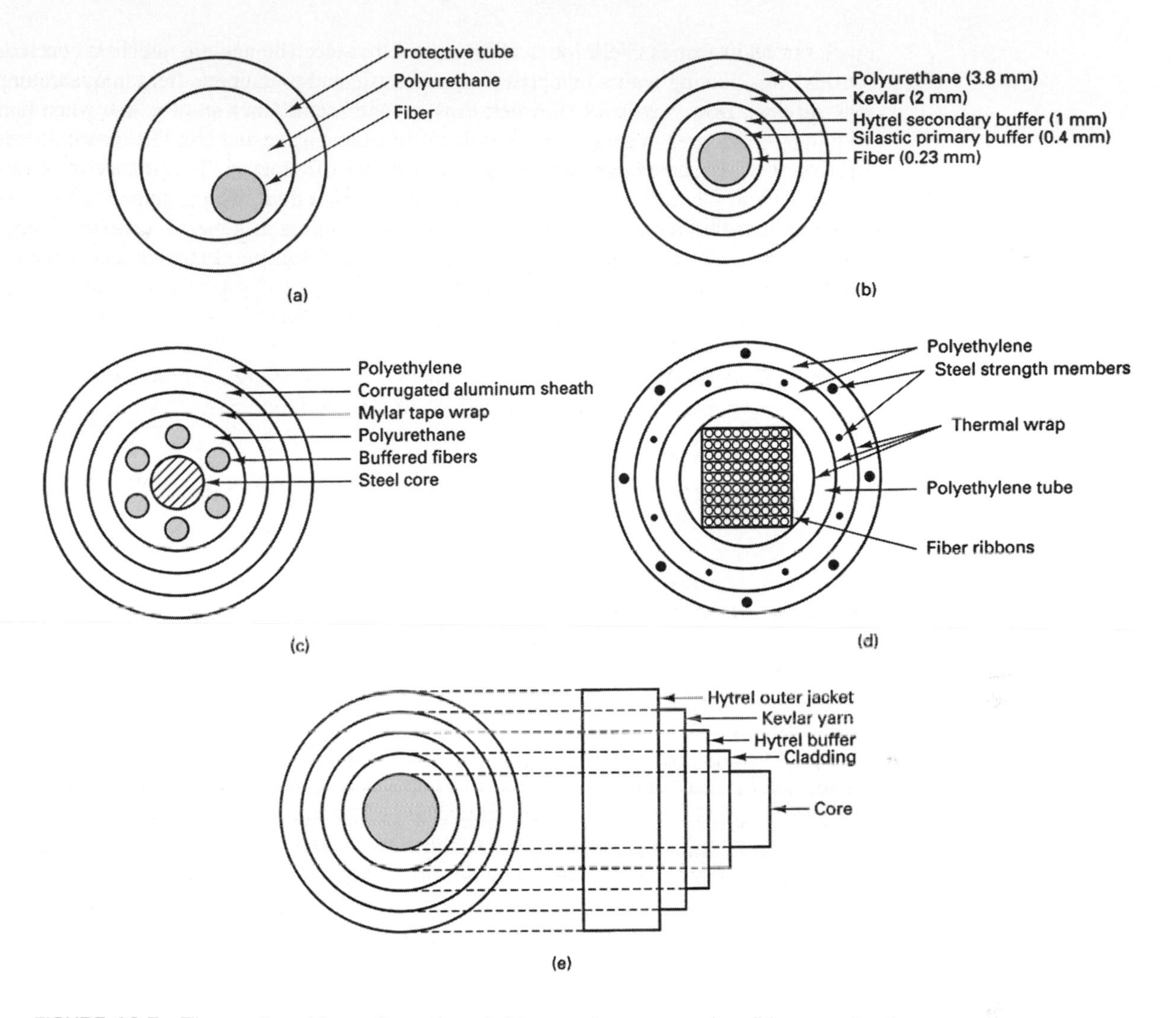

FIGURE 13-5 Fiber optic cable configurations: (a) loose tube construction; (b) constrained fiber; (c) multiple strands; (d) telephone cable; (e) plastic-silica cable

As mentioned, one disadvantage of optical fiber cables is their lack of tensile (pulling) strength, which can be as low as a pound. For this reason, the fiber must be reinforced with strengthening material so that it can withstand mechanical stresses it will typically undergo when being pulled and jerked through underground and overhead ducts and hung on poles. Materials commonly used to strengthen and protect fibers from abrasion and environmental stress are steel, fiberglass, plastic, FR-PVC (flame-retardant polyvinyl chloride), Kevlar yarn, and paper. The type of cable construction used depends on the performance requirements of the system and both economic and environmental constraints.

13-7 LIGHT PROPAGATION

13-7-1 The Physics of Light

Although the performance of optical fibers can be analyzed completely by application of Maxwell's equations, this is necessarily complex. For most practical applications, geometric wave tracing may be used instead.

In 1860, James Clerk Maxwell theorized that electromagnetic radiation contained a series of oscillating waves comprised of an electric and a magnetic field in quadrature (at 90° angles). However, in 1905, Albert Einstein and Max Planck showed that when light is emitted or absorbed, it behaves like an electromagnetic wave and also like a particle, called a *photon,* which possesses energy proportional to its frequency. This theory is known as *Planck's law.* Planck's law describes the photoelectric effect, which states, "When visible light or high-frequency electromagnetic radiation illuminates a metallic surface, electrons are emitted." The emitted electrons produce an electric current. Planck's law is expressed mathematically as

$$E_p = hf \tag{13-2}$$

where E_p = energy of the photon (joules)
h = Planck's constant = $6.625 \times 10^{-34}\ J - s$
f = frequency of light (photon) emitted (hertz)

Photon energy may also be expressed in terms of wavelength. Substituting Equation 13-1 into Equation 13-2 yields

$$E_p = hf \tag{13-3a}$$

or

$$E_p = \frac{hc}{\lambda} \tag{13-3b}$$

An atom has several energy levels or states, the lowest of which is the ground state. Any energy level above the ground state is called an *excited state.* If an atom in one energy level decays to a lower energy level, the loss of energy (in electron volts) is emitted as a photon of light. The energy of the photon is equal to the difference between the energy of the two energy levels. The process of decaying from one energy level to another energy level is called *spontaneous decay* or *spontaneous emission.*

Atoms can be irradiated by a light source whose energy is equal to the difference between ground level and an energy level. This can cause an electron to change from one energy level to another by absorbing light energy. The process of moving from one energy level to another is called *absorption.* When making the transition from one energy level to another, the atom absorbs a packet of energy (a photon). This process is similar to that of emission.

The energy absorbed or emitted (photon) is equal to the difference between the two energy levels. Mathematically,

$$E_p = E_2 - E_1 \tag{13-4}$$

where E_p is the energy of the photon (joules).

13-7-2 Optical Power

Light intensity is a rather complex concept that can be expressed in either *photometric* or *radiometric* terms. *Photometry* is the science of measuring only light waves that are visible to the human eye. Radiometry, on the other hand, measures light throughout the entire electromagnetic spectrum. In photometric terms, light intensity is generally described in terms of luminous *flux density* and measured in lumens per unit area. Radiometric terms, however, are often more useful to engineers and technologists. In radiometric terms, *optical power* measures the rate at which electromagnetic waves transfer light energy. In simple terms, optical power is described as the flow of light energy past a given point in a specified time. Optical power is expressed mathematically as

$$P = \frac{d(\text{energy})}{d(\text{time})} \tag{13-5a}$$

or $$= \frac{dQ}{dt} \quad \textbf{(13-5b)}$$

where P = optical power (watts)
dQ = instantaneous charge (joules)
dt = instantaneous change in time (seconds)

Optical power is sometimes called *radiant flux* (ϕ), which is equivalent to joules per second and is the same power that is measured electrically or thermally in watts. Radiometric terms are generally used with light sources with output powers ranging from tens of microwatts to more than 100 milliwatts. Optical power is generally stated in decibels relative to a defined power level, such as 1 mW (dBm) or 1 μW (dBμ). Mathematically stated,

$$\text{dBm} = 10 \log\left[\frac{P\,(\text{watts})}{0.001\,(\text{watts})}\right] \quad \textbf{(13-6)}$$

and

$$\text{db}\mu = 10 \log\left[\frac{P\,(\text{watts})}{0.000001\,(\text{watts})}\right] \quad \textbf{(13-7)}$$

Example 13-1

Determine the optical power in dBm and dBμ for power levels of

a. 10 mW
b. 20 μW

Solution

a. Substituting into Equations 13-6 and 13-7 gives

$$\text{dBm} = 10 \log\frac{10\text{ mW}}{1\text{ mW}} = 10\text{ dBm}$$

$$\text{dB}\mu = 10 \log\frac{10\text{ mW}}{1\,\mu\text{W}} = 40\text{ dB}\mu$$

b. Substituting into Equations 13-6 and 13-7 gives

$$\text{dBm} = 10 \log\frac{20\,\mu\text{W}}{1\text{ mW}} = -17\text{ dBm}$$

$$\text{dB}\mu = 10 \log\frac{20\,\mu\text{W}}{1\mu\text{W}} = 13\text{ dB}\mu$$

13-7-3 Velocity of Propagation

In free space (a vacuum), electromagnetic energy, such as light waves, travels at approximately 300,000,000 meters per second (186,000 mi/s). Also, in free space the velocity of propagation is the same for all light frequencies. However, it has been demonstrated that electromagnetic waves travel slower in materials more dense than free space and that all light frequencies do not propagate at the same velocity. When the velocity of an electromagnetic wave is reduced as it passes from one medium to another medium of denser material, the light ray changes direction or refracts (bends) toward the normal. When an electromagnetic wave passes from a more dense material into a less dense material, the light ray is refracted away from the normal. The *normal* is simply an imaginary line drawn perpendicular to the interface of the two materials at the point of incidence.

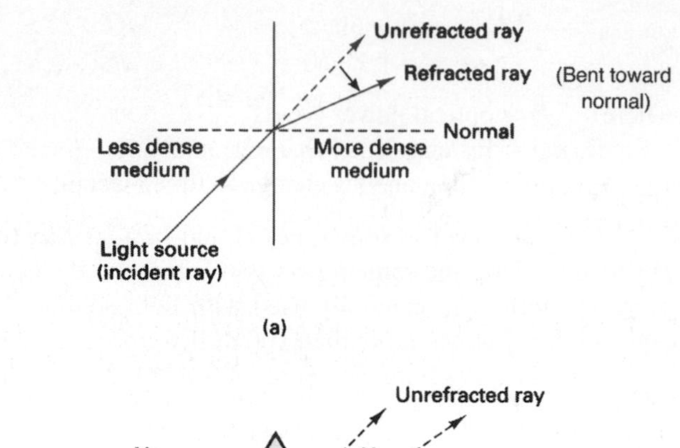

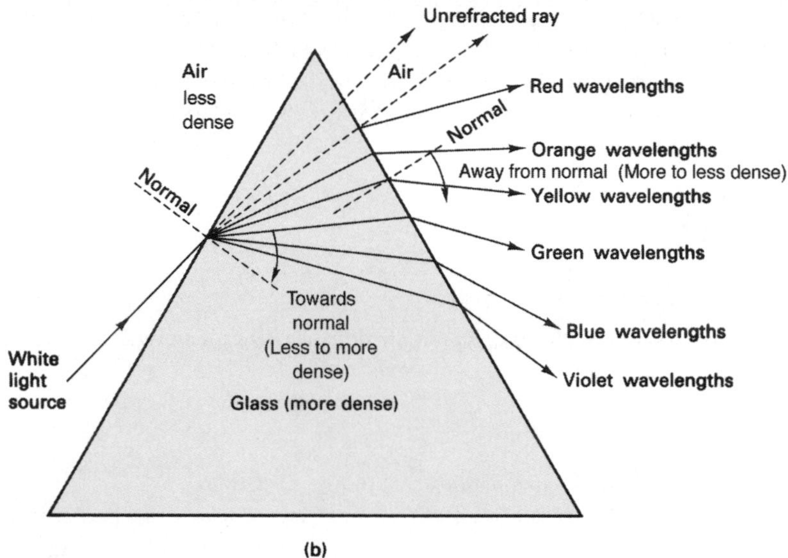

FIGURE 13-6 Refraction of light: (a) light refraction; (b) prismatic refraction

13-7-3-1 Refraction. For light-wave frequencies, electromagnetic waves travel through Earth's atmosphere (air) at approximately the same velocity as through a vacuum (i.e., the speed of light). Figure 13-6a shows how a light ray is refracted (bent) as it passes from a less dense material into a more dense material. (Actually, the light ray is not bent; rather, it changes direction at the interface.) Figure 13-6b shows how sunlight, which contains all light frequencies (*white light*), is affected as it passes through a material that is more dense than air. Refraction occurs at both air/glass interfaces. The violet wavelengths are refracted the most, whereas the red wavelengths are refracted the least. The spectral separation of white light in this manner is called *prismatic refraction.* It is this phenomenon that causes rainbows, where water droplets in the atmosphere act as small prisms that split the white sunlight into the various wavelengths, creating a visible spectrum of color.

13-7-3-2 Refractive index. The amount of bending or refraction that occurs at the interface of two materials of different densities is quite predictable and depends on the *refractive indexes* of the two materials. Refractive index is simply the ratio of the velocity of propagation of a light ray in free space to the velocity of propagation of a light ray in a given material. Mathematically, refractive index is

$$n = \frac{c}{v} \qquad \text{(13-8)}$$

where n = refractive index (unitless)
c = speed of light in free space (3×10^8 meters per second)
v = speed of light in a given material (meters per second)

Although the refractive index is also a function of frequency, the variation in most light wave applications is insignificant and, thus, omitted from this discussion. The indexes of refraction of several common materials are given in Table 13-1.

13-7-3-3 Snell's law. How a light ray reacts when it meets the interface of two transmissive materials that have different indexes of refraction can be explained with *Snell's law.* A refractive index model for Snell's law is shown in Figure 13-7. The *angle of incidence* is the angle at which the propagating ray strikes the interface with respect to the normal, and the *angle of refraction* is the angle formed between the propagating ray and the normal after the ray has entered the second medium. At the interface of medium 1 and medium 2, the incident ray may be refracted toward the normal or away from it, depending on whether n_1 is greater than or less than n_2. Hence, the angle of refraction can be larger or

Table 13-1 Typical Indexes of Refraction

Material	Index of Refraction[a]
Vacuum	1.0
Air	1.0003 (≈1)
Water	1.33
Ethyl alcohol	1.36
Fused quartz	1.46
Glass fiber	1.5–1.9
Diamond	2.0–2.42
Silicon	3.4
Gallium-arsenide	2.6

[a]Index of refraction is based on a wavelength of light emitted from a sodium flame (589 nm).

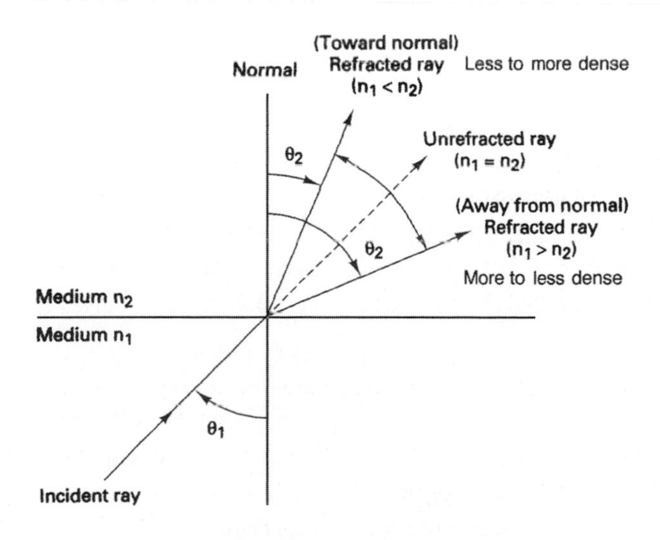

FIGURE 13-7 Refractive model for Snell's law

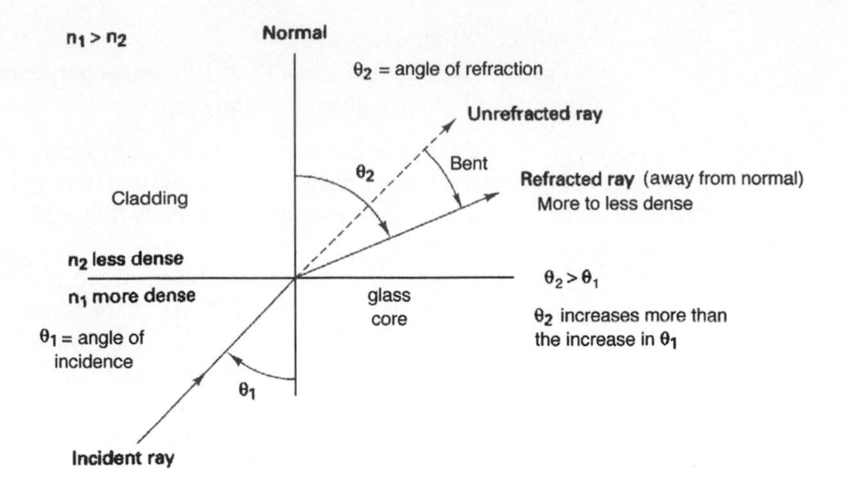

FIGURE 13-8 Light ray refracted away from the normal

smaller than the angle of incidence, depending on the refractive indexes of the two materials. Snell's law stated mathematically is

$$n_1 \sin \theta_1 = n_2 \sin \theta_2 \tag{13-9}$$

where n_1 = refractive index of material 1 (unitless)
n_2 = refractive index of material 2 (unitless)
θ_1 = angle of incidence (degrees)
θ_2 = angle of refraction (degrees)

Figure 13-8 shows how a light ray is refracted as it travels from a more dense (higher refractive index) material into a less dense (lower refractive index) material. It can be seen that the light ray changes direction at the interface, and the angle of refraction is greater than the angle of incidence. Consequently, when a light ray enters a less dense material, the ray bends away from the normal. The normal is simply a line drawn perpendicular to the interface at the point where the incident ray strikes the interface. Similarly, when a light ray enters a more dense material, the ray bends toward the normal.

Example 13-2

In Figure 13-8, let medium 1 be glass and medium 2 be ethyl alcohol. For an angle of incidence of 30°, determine the angle of refraction.

Solution From Table 13-1,

$$n_1 \text{ (glass)} = 1.5$$
$$n_2 \text{ (ethyl alcohol)} = 1.36$$

Rearranging Equation 13-9 and substituting for n_1, n_2, and θ_1 gives us

$$\frac{n_1}{n_2} \sin \theta_1 = \sin \theta_2$$

$$\frac{1.5}{1.36} \sin 30 = 0.5514 = \sin \theta_2$$

$$\theta_2 = \sin^{-1} 0.5514 = 33.47°$$

The result indicates that the light ray refracted (bent) or changed direction by 33.47° at the interface. Because the light was traveling from a more dense material into a less dense material, the ray bent away from the normal.

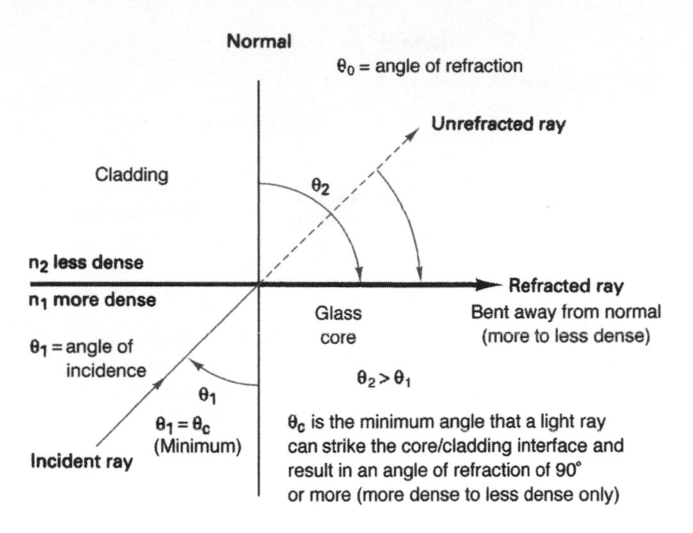

FIGURE 13-9 Critical angle refraction

13-7-3-4 Critical angle. Figure 13-9 shows a condition in which an incident ray is striking the glass/cladding interface at an angle ($_1$) such that the angle of refraction (θ_2) is 90° and the refracted ray is along the interface. This angle of incidence is called the *critical angle* (θ_c), which is defined as the minimum angle of incidence at which a light ray may strike the interface of two media and result in an angle of refraction of 90° or greater. It is important to note that the light ray must be traveling from a medium of higher refractive index to a medium with a lower refractive index (i.e., glass into cladding). If the angle of refraction is 90° or greater, the light ray is not allowed to penetrate the less dense material. Consequently, total reflection takes place at the interface, and the angle of reflection is equal to the angle of incidence. Critical angle can be represented mathematically by rearranging Equation 13-9 as

$$\sin \theta_1 - \frac{n_2}{n_1} \sin \theta_2$$

With $\theta_2 = 90°$, θ_1 becomes the critical angle (θ_c), and

$$\sin \theta_c = \frac{n_2}{n_1}(1) = \qquad \sin \theta_c = \frac{n_2}{n_1}$$

and

$$\theta_c = \sin^{-1} \frac{n_2}{n_1} \tag{13-10}$$

where θ_c is the critical angle.

From Equation 13-10, it can be seen that the critical angle is dependent on the ratio of the refractive indexes of the core and cladding. For example a ratio $n_2/n_1 = 0.77$ produces a critical angle of 50.4°, whereas a ratio $n_2/n_1 = 0.625$ yields a critical angle of 38.7°.

Figure 13-10 shows a comparison of the angle of refraction and the angle of reflection when the angle of incidence is less than or more than the critical angle.

13-7-3-5 Acceptance angle, acceptance cone, and numerical aperture. In previous discussions, the source-to-fiber aperture was mentioned several times, and the critical and acceptance angles at the point where a light ray strikes the core/cladding interface were explained. The following discussion addresses the light-gathering ability of a fiber, which is the ability to couple light from the source into the fiber.

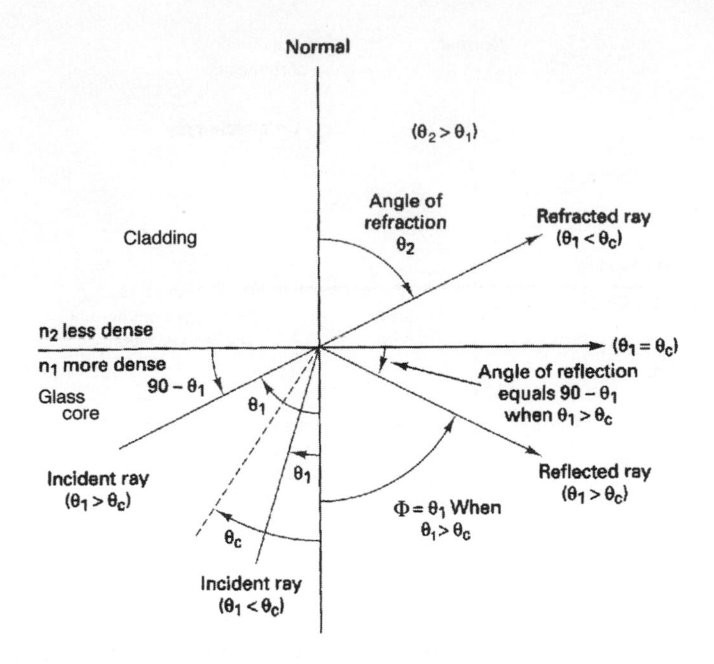

FIGURE 13-10 Angle of reflection and refraction

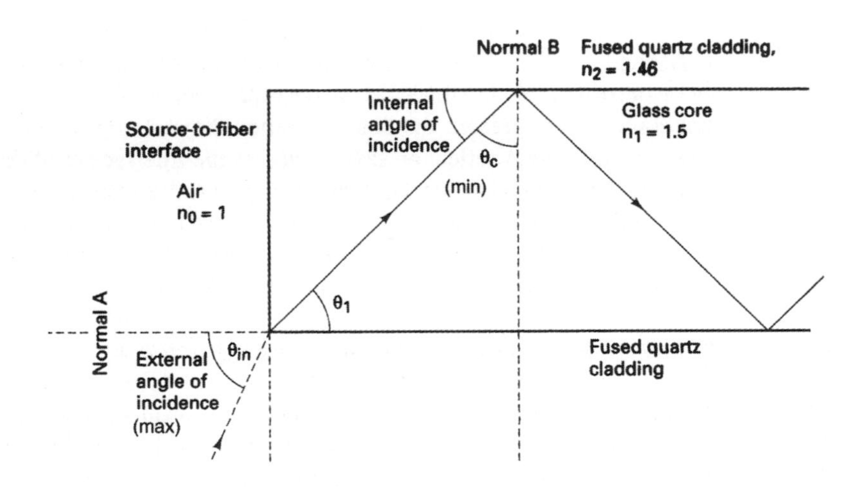

FIGURE 13-11 Ray propagation into and down an optical fiber cable

Figure 13-11 shows the source end of a fiber cable and a light ray propagating into and then down the fiber. When light rays enter the core of the fiber, they strike the air/glass interface at normal A. The refractive index of air is approximately 1, and the refractive index of the glass core is 1.5. Consequently, the light enters the cable traveling from a less dense to a more dense medium, causing the ray to refract toward the normal. This causes the light rays to change direction and propagate diagonally down the core at an angle that is less than the external angle of incidence (θ_{in}). For a ray of light to propagate down the cable, it must strike the internal core/cladding interface at an angle that is greater than the critical angle (θ_c). Using Figure 13-12 and Snell's law, it can be shown that the maximum angle that external light rays may strike the air/glass interface and still enter the core and propagate down the fiber is

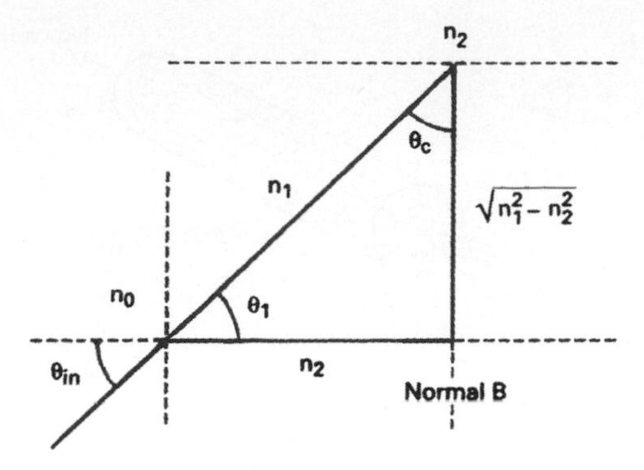

FIGURE 13-12 Geometric relationship of Equations 13-11a and b

$$\theta_{in(max)} = \sin^{-1}\frac{\sqrt{n_1^2 - n_2^2}}{n_o} \quad \textbf{(13-11a)}$$

where $\theta_{in(max)}$ = acceptance angle (degrees)
n_o = refractive index of air (1)
n_1 = refractive index of glass fiber core (1.5)
n_2 = refractive index of quartz fiber cladding (1.46)

Since the refractive index of air is 1, Equation 13-11a reduces to

$$\theta_{in(max)} = \sin^{-1}\sqrt{n_1^2 - n_2^2} \quad \textbf{(13-11b)}$$

$\theta_{in(max)}$ is called the *acceptance angle* or *acceptance cone half-angle*. $\theta_{in(max)}$ defines the maximum angle in which external light rays may strike the air/glass interface and still propagate down the fiber. Rotating the acceptance angle around the fiber core axis describes the acceptance cone of the fiber input. Acceptance cone is shown in Figure 13-13a, and the relationship between acceptance angle and critical angle is shown in Figure 13-13b. Note that the critical angle is defined as a minimum value and that the acceptance angle is defined as a maximum value. Light rays striking the air/glass interface at an angle greater than the acceptance angle will enter the cladding and, therefore, will not propagate down the cable.

Numerical aperture (NA) is closely related to acceptance angle and is the figure of merit commonly used to measure the magnitude of the acceptance angle. In essence, numerical aperture is used to describe the light-gathering or light-collecting ability of an optical fiber (i.e., the ability to couple light into the cable from an external source). The larger the magnitude of the numerical aperture, the greater the amount of external light the fiber will accept. The numerical aperture for light entering the glass fiber from an air medium is described mathematically as

$$NA = \sin\theta_{in} \quad \textbf{(13-12a)}$$

and

$$NA = \sqrt{n_1^2 - n_2^2} \quad \textbf{(13-12b)}$$

Therefore

$$\theta_{in} = \sin^{-1} NA \quad \textbf{(13-12c)}$$

where θ_{in} = acceptance angle (degrees)
NA = numerical aperture (unitless)
n_1 = refractive index of glass fiber core (unitless)
n_2 = refractive index of quartz fiber cladding (unitless)

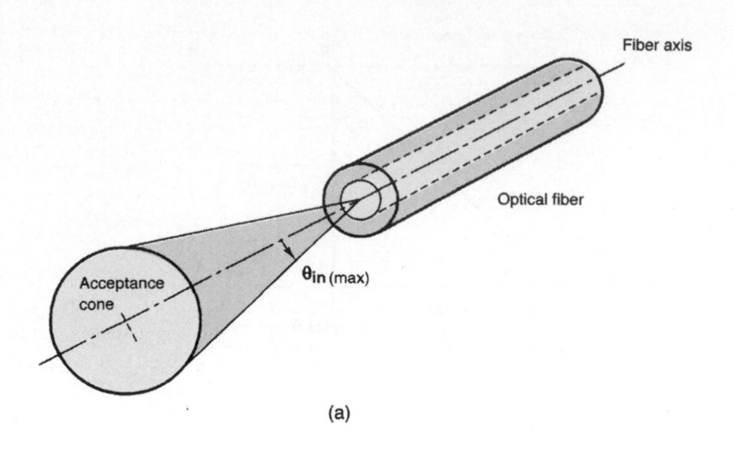

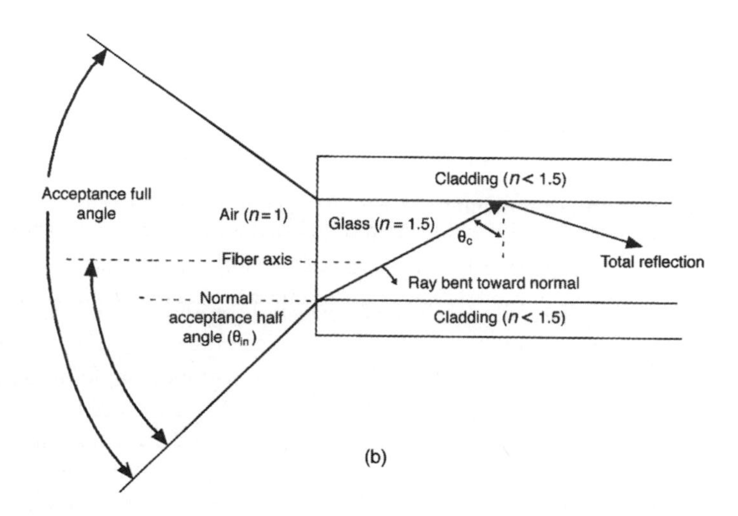

FIGURE 13-13 (a) Acceptance angle; (b) acceptance cone

A larger-diameter core does not necessarily produce a larger numerical aperture, although in practice larger-core fibers tend to have larger numerical apertures. Numerical aperture can be calculated using Equations 13-12a or b, but in practice it is generally measured by looking at the output of a fiber because the light-guiding properties of a fiber cable are symmetrical. Therefore, light leaves a cable and spreads out over an angle equal to the acceptance angle.

13-8 OPTICAL FIBER CONFIGURATIONS

Light can be propagated down an optical fiber cable using either reflection or refraction. How the light propagates depends on the *mode of propagation* and the *index profile* of the fiber.

13-8-1 Mode of Propagation

In fiber optics terminology, the word *mode* simply means path. If there is only one path for light rays to take down a cable, it is called *single mode.* If there is more than one path, it is called *multimode.* Figure 13-14 shows single and multimode propagation of light rays down an optical fiber. As shown in Figure 13-14a, with single-mode propagation, there is only one

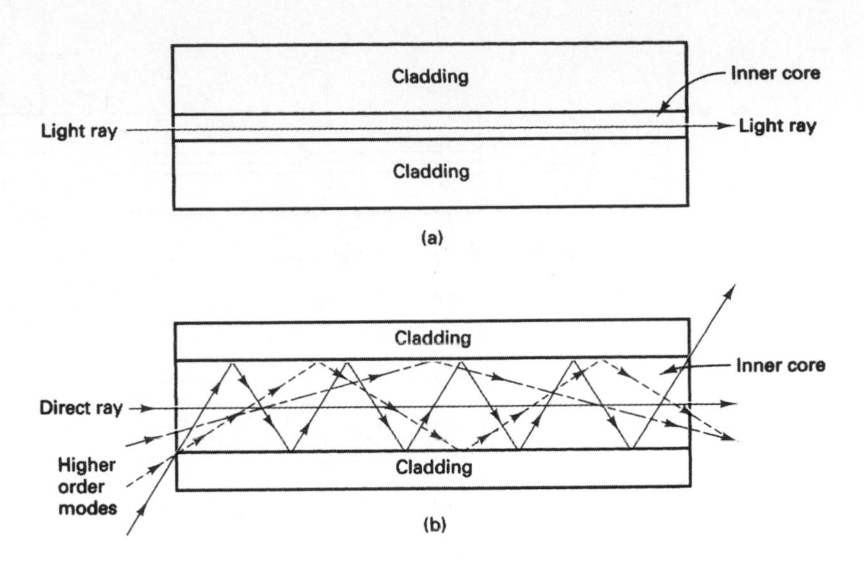

FIGURE 13-14 Modes of propagation: (a) single mode; (b) multimode

path for light rays to take, which is directly down the center of the cable. However, as Figure 13-14b shows, with multimode propagation there are many higher-order modes possible, and light rays propagate down the cable in a zigzag fashion following several paths.

The number of paths (modes) possible for a multimode fiber cable depends on the frequency (wavelength) of the light signal, the refractive indexes of the core and cladding, and the core diameter. Mathematically, the number of modes possible for a given cable can be approximated by the following formula:

$$N \approx \left(\frac{\pi d}{\lambda}\sqrt{n_1^2 - n_2^2}\right)^2 \qquad \text{(13-13)}$$

where N = number of propagating modes
d = core diameter (meters)
λ = wavelength (meters)
n_1 = refractive index of core
n_2 = refractive index of cladding

A multimode step-index fiber with a core diameter of 50 μm, a core refractive index of 1.6, a cladding refractive index of 1.584, and a wavelength of 1300 nm has approximately 372 possible modes.

13-8-2 Index Profile

The index profile of an optical fiber is a graphical representation of the magnitude of the refractive index across the fiber. The refractive index is plotted on the horizontal axis, and the radial distance from the core axis is plotted on the vertical axis. Figure 13-15 shows the core index profiles for the three types of optical fiber cables.

There are two basic types of index profiles: step and graded. A *step-index* fiber has a central core with a uniform refractive index (i.e., constant density throughout). An outside cladding that also has a uniform refractive index surrounds the core; however, the refractive index of the cladding is less than that of the central core. From Figures 13-15a and b, it can be seen that in step-index fibers, there is an abrupt change in the refractive index at the core/cladding interface. This is true for both single and multimode step-index fibers.

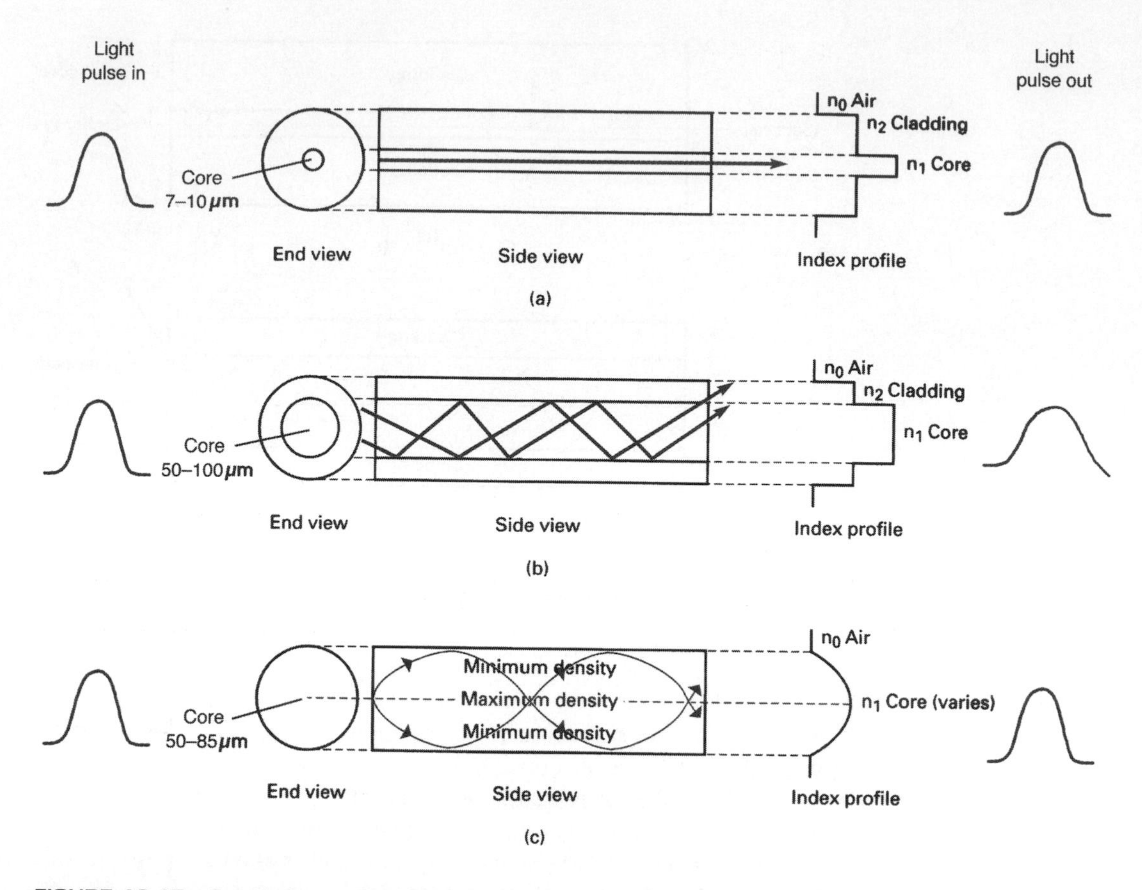

FIGURE 13-15 Core index profiles: (a) single-mode step index; (b) multimode step index; (c) multimode graded index

In the *graded-index* fiber, shown in Figure 13-15c, it can be see that there is no cladding, and the refractive index of the core is nonuniform; it is highest in the center of the core and decreases gradually with distance toward the outer edge. The index profile shows a core density that is maximum in the center and decreases symmetrically with distance from the center.

13-9 OPTICAL FIBER CLASSIFICATIONS

Propagation modes can be categorized as either multimode or single mode, and then multimode can be further subdivided into step index or graded index. Although there are a wide variety of combinations of modes and indexes, there are only three practical types of optical fiber configurations: *single-mode step-index, multimode step index,* and *multimode graded index.*

13-9-1 Single-Mode Step-Index Optical Fiber

Single-mode step-index fibers are the dominant fibers used in today's telecommunications and data networking industries. A single-mode step-index fiber has a central core that is significantly smaller in diameter than any of the multimode cables. In fact, the diameter is sufficiently small that there is essentially only one path that light may take as it propagates down the cable. This type of fiber is shown in Figure 13-16a. In the simplest form of single-mode step-index fiber, the outside cladding is simply air. The refractive index of the glass core (n_1) is approximately 1.5, and the refractive index of the air cladding (n_2) is 1. The large

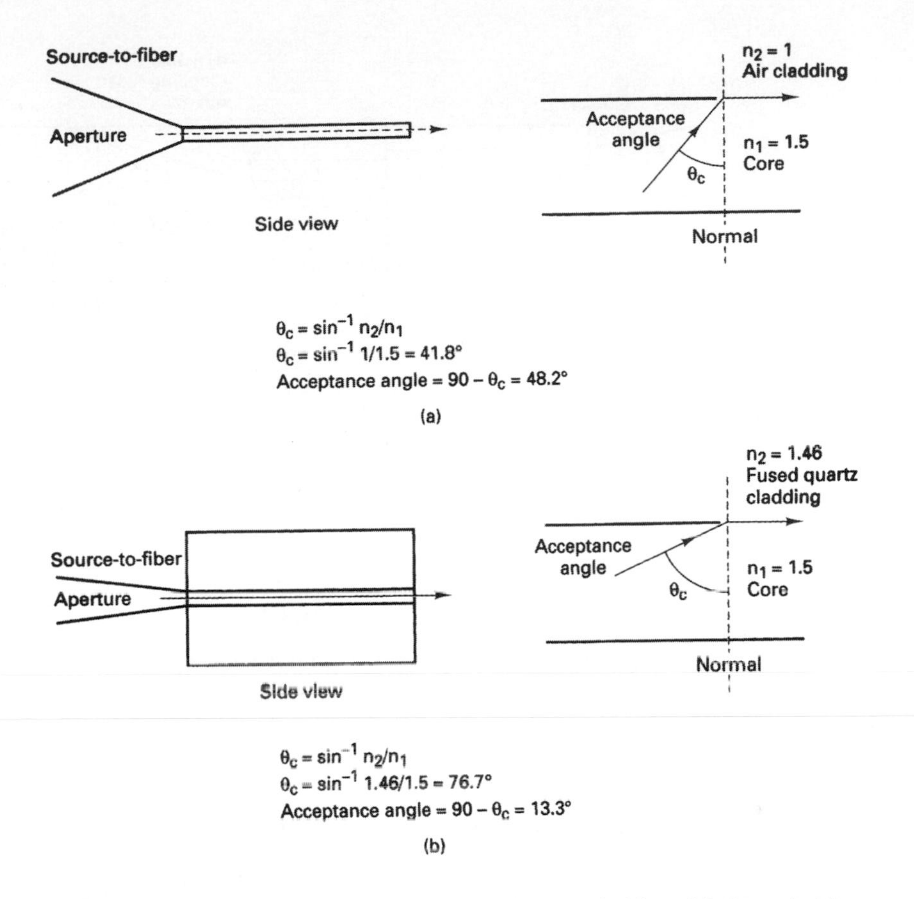

FIGURE 13-16 Single-mode step-index fibers: (a) air cladding; (b) glass cladding

difference in the refractive indexes results in a small critical angle (approximately 42°) at the glass/air interface. Consequently, a single-mode step-index fiber has a wide external acceptance angle, which makes it relatively easy to couple light into the cable from an external source. However, this type of fiber is very weak and difficult to splice or terminate.

A more practical type of single-mode step-index fiber is one that has a cladding other than air, such as the cable shown in Figure 13-16b. The refractive index of the cladding (n_2) is slightly less than that of the central core (n_1) and is uniform throughout the cladding. This type of cable is physically stronger than the air-clad fiber, but the critical angle is also much higher (approximately 77°). This results in a small acceptance angle and a narrow source-to-fiber aperture, making it much more difficult to couple light into the fiber from a light source.

With both types of single-mode step-index fibers, light is propagated down the fiber through reflection. Light rays that enter the fiber either propagate straight down the core or, perhaps, are reflected only a few times. Consequently, all light rays follow approximately the same path down the cable and take approximately the same amount of time to travel the length of the cable. This is one overwhelming advantage of single-mode step-index fibers, as explained in more detail in a later section of this chapter.

13-9-2 Multimode Step-Index Optical Fiber

A *multimode step-index* optical fiber is shown in Figure 13-17. Multimode step-index fibers are similar to the single-mode step-index fibers except the center core is much larger with the multimode configuration. This type of fiber has a large light-to-fiber aperture and, consequently, allows more external light to enter the cable. The light rays that strike the core/cladding interface at an angle greater than the critical angle (ray A) are propagated down the core in a zigzag fashion, continuously reflecting off the interface boundary. Light

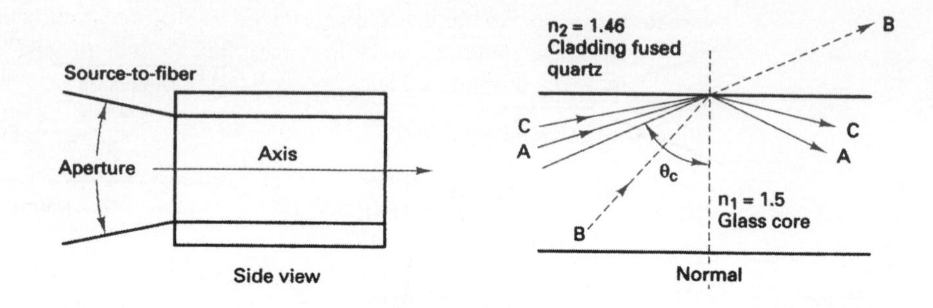

FIGURE 13-17 Multimode step-index fiber

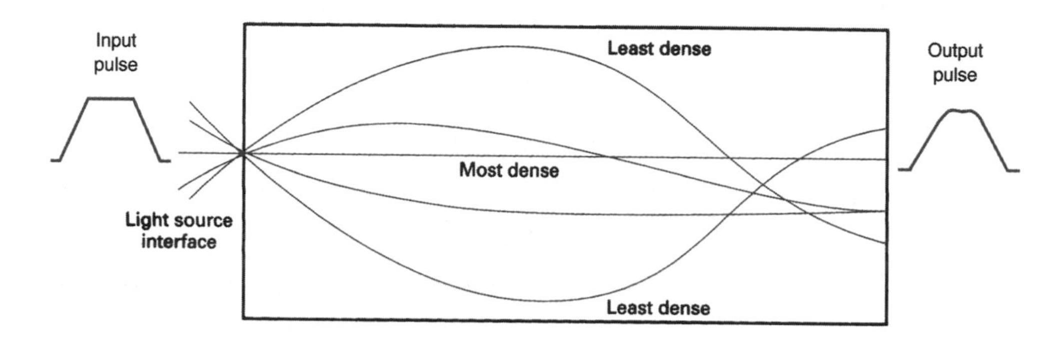

FIGURE 13-18 Multimode graded-index fiber

rays that strike the core/cladding interface at an angle less than the critical angle (ray B) enter the cladding and are lost. It can be seen that there are many paths that a light ray may follow as it propagates down the fiber. As a result, all light rays do not follow the same path and, consequently, do not take the same amount of time to travel the length of the cable.

13-9-3 Multimode Graded-Index Optical Fiber

A multimode graded-index optical fiber is shown in Figure 13-18. Graded-index fibers are characterized by a central core with a nonuniform refractive index. Thus, the cable's density is maximum at the center and decreases gradually toward the outer edge. Light rays propagate down this type of fiber through refraction rather than reflection. As a light ray propagates diagonally across the core toward the center, it is continually intersecting a less dense to more dense interface. Consequently, the light rays are constantly being refracted, which results in a continuous bending of the light rays. Light enters the fiber at many different angles. As the light rays propagate down the fiber, the rays traveling in the outermost area of the fiber travel a greater distance than the rays traveling near the center. Because the refractive index decreases with distance from the center and the velocity is inversely proportional to refractive index, the light rays traveling farthest from the center propagate at a higher velocity. Consequently, they take approximately the same amount of time to travel the length of the fiber.

13-9-4 Optical Fiber Comparison

13-9-4-1 Single-mode step-index fiber. Advantages include the following:

1. Minimum dispersion: All rays propagating down the fiber take approximately the same path; thus, they take approximately the same length of time to travel down the cable. Consequently, a pulse of light entering the cable can be reproduced at the receiving end very accurately.

2. Because of the high accuracy in reproducing transmitted pulses at the receive end, wider bandwidths and higher information transmission rates (bps) are possible with single-mode step-index fibers than with the other types of fibers.

Disadvantages include the following:

1. Because the central core is very small, it is difficult to couple light into and out of this type of fiber. The source-to-fiber aperture is the smallest of all the fiber types.
2. Again, because of the small central core, a highly directive light source, such as a laser, is required to couple light into a single-mode step-index fiber.
3. Single-mode step-index fibers are expensive and difficult to manufacture.

13-9-4-2 Multimode step-index fiber. Advantages include the following:

1. Multimode step-index fibers are relatively inexpensive and simple to manufacture.
2. It is easier to couple light into and out of multimode step-index fibers because they have a relatively large source-to-fiber aperture.

Disadvantages include the following:

1. Light rays take many different paths down the fiber, which results in large differences in propagation times. Because of this, rays traveling down this type of fiber have a tendency to spread out. Consequently, a pulse of light propagating down a multimode step-index fiber is distorted more than with the other types of fibers.
2. The bandwidths and rate of information transfer rates possible with this type of cable are less than that possible with the other types of fiber cables.

13-9-4-3 Multimode graded-index fiber. Essentially, there are no outstanding advantages or disadvantages of this type of fiber. Multimode graded-index fibers are easier to couple light into and out of than single-mode step-index fibers but are more difficult than multimode step-index fibers. Distortion due to multiple propagation paths is greater than in single-mode step-index fibers but less than in multimode step-index fibers. This multimode graded-index fiber is considered an intermediate fiber compared to the other fiber types.

13-10 LOSSES IN OPTICAL FIBER CABLES

Power loss in an optical fiber cable is probably the most important characteristic of the cable. Power loss is often called *attenuation* and results in a reduction in the power of the light wave as it travels down the cable. Attenuation has several adverse effects on performance, including reducing the system's bandwidth, information transmission rate, efficiency, and overall system capacity.

The standard formula for expressing the total power loss in an optical fiber cable is

$$A_{(\text{dB})} = 10 \log\left(\frac{P_{\text{out}}}{P_{\text{in}}}\right) \quad \textbf{(13-14)}$$

where $A_{(\text{dB})}$ = total reduction in power level, attenuation (unitless)
P_{out} = cable output power (watts)
P_{in} = cable input power (watts)

In general, multimode fibers tend to have more attenuation than single-mode cables, primarily because of the increased scattering of the light wave produced from the dopants in the glass. Table 13-2 shows output power as a percentage of input power for an optical

Table 13-2 % Output Power versus Loss in dB

Loss (dB)	Output Power (%)
1	79
3	50
6	25
9	12.5
10	10
13	5
20	1
30	0.1
40	0.01
50	0.001

Table 13-3 Fiber Cable Attenuation

Cable Type	Core Diameter (μm)	Cladding Diameter (μm)	NA (unitless)	Attenuation (dB/km)
Single mode	8	125	—	0.5 at 1300 nm
	5	125	—	0.4 at 1300 nm
Graded index	50	125	0.2	4 at 850 nm
	100	140	0.3	5 at 850 nm
Step index	200	380	0.27	6 at 850 nm
	300	440	0.27	6 at 850 nm
PCS	200	350	0.3	10 at 790 nm
	400	550	0.3	10 at 790 nm
Plastic	—	750	0.5	400 at 650 nm
	—	1000	0.5	400 at 650 nm

fiber cable with several values of decibel loss. A 13-dB cable loss reduces the output power to 50% of the input power.

Attenuation of light propagating through glass depends on wavelength. The three wavelength bands typically used for optical fiber communications systems are centered around 0.85 microns, 1.30 microns, and 1.55 microns. For the kind of glass typically used for optical communications systems, the 1.30-micron and 1.55-micron bands have less than 5% loss per kilometer, while the 0.85-micron band experiences almost 20% loss per kilometer.

Although total power loss is of primary importance in an optical fiber cable, attenuation is generally expressed in decibels of loss per unit length. Attenuation is expressed as a positive dB value because by definition it is a loss. Table 13-3 lists attenuation in dB/km for several types of optical fiber cables.

The optical power in watts measured at a given distance from a power source can be determined mathematically as

$$P = P_t \times 10^{-Al/10} \tag{13-15}$$

where P = measured power level (watts)
P_t = transmitted power level (watts)
A = cable power loss (dB/km)
l = cable length (km)

Likewise, the optical power in decibel units is

$$P(\text{dBm}) = P_{\text{in}}(\text{dBm}) - Al(\text{dB}) \tag{13-16}$$

where P = measured power level (dBm)
P_{in} = transmit power (dBm)
Al = cable power loss, attenuation (dB)

Example 13-3

For a single-mode optical cable with 0.25-dB/km loss, determine the optical power 100 km from a 0.1-mW light source.

Solution Substituting into Equation 13-15 gives

$$\begin{aligned} P &= 0.1\text{mW} \times 10^{-\{[(0.25)(100)]/(10)\}} \\ &= 1 \times 10^{-4} \times 10^{\{[(0.25)(100)]/(10)\}} \\ &= (1 \times 10^{-4})(1 \times 10^{-2.5}) \\ &= 0.316\ \mu\text{W} \end{aligned}$$

and

$$\begin{aligned} P(\text{dBm}) &= 10 \log\left(\frac{0.316\ \mu\text{W}}{0.001}\right) \\ &= -35\ \text{dBm} \end{aligned}$$

or by substituting into Equation 13-16

$$\begin{aligned} P(\text{dBm}) &= 10 \log\left(\frac{0.1\ \text{mW}}{0.001\ \text{W}}\right) - [(100\ \text{km})(0.25\ \text{dB/km})] \\ &= -10\ \text{dBm} - 25\ \text{dB} \\ &= -35\ \text{dBm} \end{aligned}$$

Transmission losses in optical fiber cables are one of the most important characteristics of the fibers. Losses in the fiber result in a reduction in the light power, thus reducing the system bandwidth, information transmission rate, efficiency, and overall system capacity. The predominant losses in optical fiber cables are the following:

- Absorption loss
- Material, or Rayleigh, scattering losses
- Chromatic, or wavelength, dispersion
- Radiation losses
- Modal dispersion
- Coupling losses

13-10-1 Absorption Losses

Absorption losses in optical fibers is analogous to power dissipation in copper cables; impurities in the fiber absorb the light and convert it to heat. The ultrapure glass used to manufacture optical fibers is approximately 99.9999% pure. Still, absorption losses between 1 dB/km and 1000 dB/km are typical. Essentially, there are three factors that contribute to the absorption losses in optical fibers: *ultraviolet absorption, infrared absorption,* and *ion resonance absorption.*

13-10-1-1 Ultraviolet absorption. Ultraviolet absorption is caused by valence electrons in the silica material from which fibers are manufactured. Light ionizes the valence electrons into conduction. The ionization is equivalent to a loss in the total light field and, consequently, contributes to the transmission losses of the fiber.

13-10-1-2 Infrared absorption. Infrared absorption is a result of photons of light that are absorbed by the atoms of the glass core molecules. The absorbed photons are converted to random mechanical vibrations typical of heating.

13-10-1-3 Ion resonance absorption. Ion resonance absorption is caused by OH^- ions in the material. The source of the OH^- ions is water molecules that have been trapped in the glass during the manufacturing process. Iron, copper, and chromium molecules also cause ion absorption.

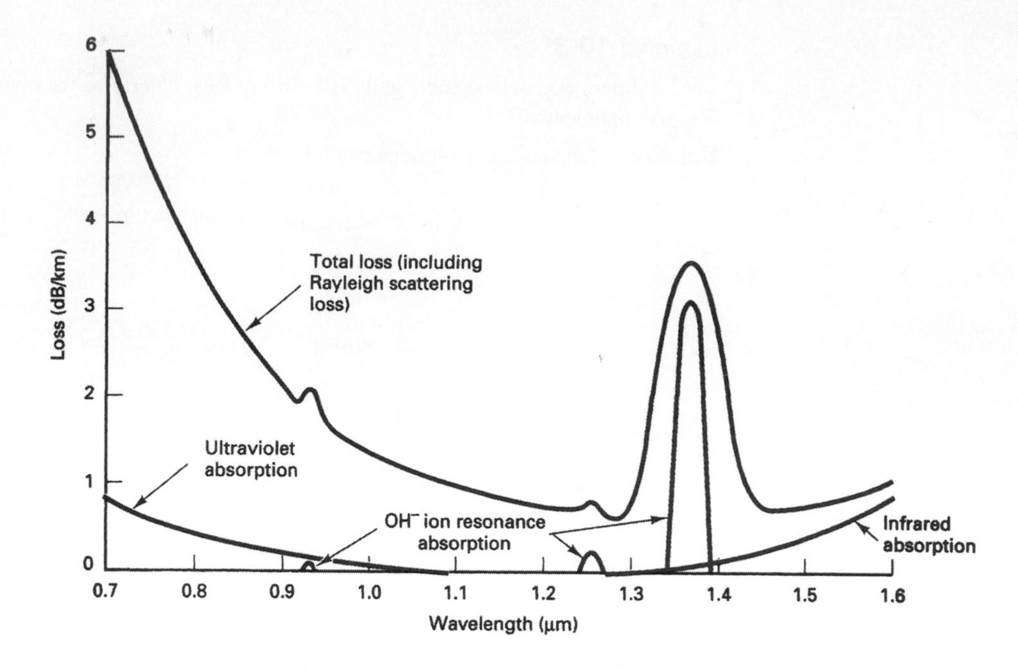

FIGURE 13-19 Absorption losses in optical fibers

Figure 13-19 shows typical losses in optical fiber cables due to ultraviolet, infrared, and ion resonance absorption.

13-10-2 Material, or Rayleigh, Scattering Losses

During manufacturing, glass is drawn into long fibers of very small diameter. During this process, the glass is in a plastic state (not liquid and not solid). The tension applied to the glass causes the cooling glass to develop permanent submicroscopic irregularities. When light rays propagating down a fiber strike one of these impurities, they are diffracted. Diffraction causes the light to disperse or spread out in many directions. Some of the diffracted light continues down the fiber, and some of it escapes through the cladding. The light rays that escape represent a loss in light power. This is called *Rayleigh scattering loss*. Figure 13-20 graphically shows the relationship between wavelength and Rayleigh scattering loss.

13-10-3 Chromatic, or Wavelength, Dispersion

Light-emitting diodes (LEDs) emit light containing many wavelengths. Each wavelength within the composite light signal travels at a different velocity when propagating through glass. Consequently, light rays that are simultaneously emitted from an LED and propagated down an optical fiber do not arrive at the far end of the fiber at the same time, resulting in an impairment called *chromatic distortion* (sometimes called *wavelength dispersion*). Chromatic distortion can be eliminated by using a monochromatic light source such as an injection laser diode (ILD). Chromatic distortion occurs only in fibers with a single mode of transmission.

13-10-4 Radiation Losses

Radiation losses are caused mainly by small bends and kinks in the fiber. Essentially, there are two types of bends: microbends and constant-radius bends. *Microbending* occurs as a result of differences in the thermal contraction rates between the core and the cladding material. A microbend is a miniature bend or geometric imperfection along the axis of the fiber and represents a discontinuity in the fiber where Rayleigh scattering can occur. Microbending losses generally contribute less than 20% of the total attenuation in a fiber.

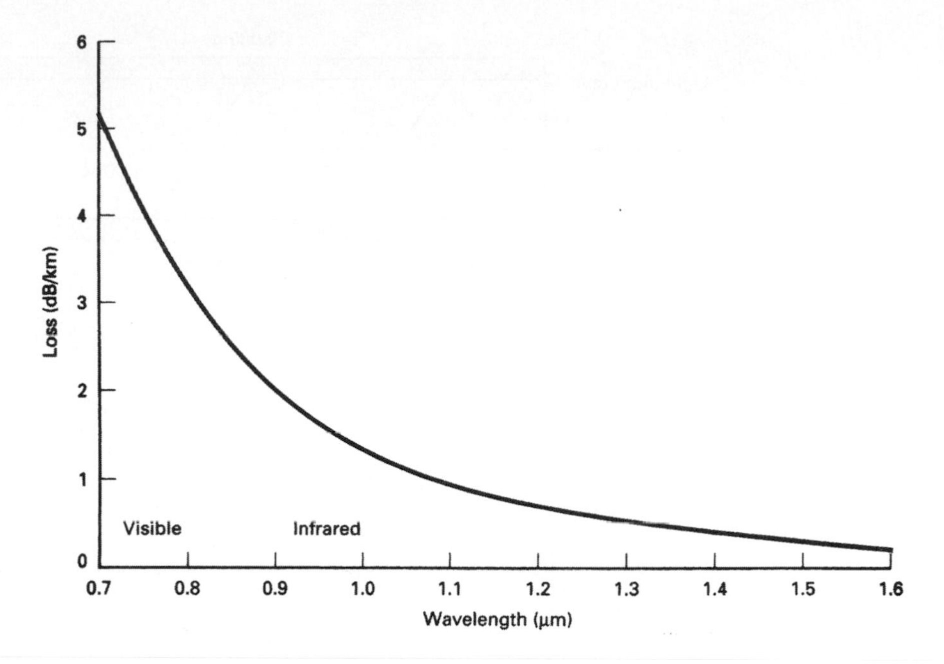

FIGURE 13-20 Rayleigh scattering loss as a function of wavelength

Constant-radius bends are caused by excessive pressure and tension and generally occur when fibers are bent during handling or installation.

13-10-5 Modal Dispersion

Modal dispersion (sometimes called *pulse spreading*) is caused by the difference in the propagation times of light rays that take different paths down a fiber. Obviously, modal dispersion can occur only in multimode fibers. It can be reduced considerably by using graded-index fibers and almost entirely eliminated by using single-mode step-index fibers.

Modal dispersion can cause a pulse of light energy to spread out in time as it propagates down a fiber. If the pulse spreading is sufficiently severe, one pulse may interfere with another. In multimode step-index fibers, a light ray propagating straight down the axis of the fiber takes the least amount of time to travel the length of the fiber. A light ray that strikes the core/cladding interface at the critical angle will undergo the largest number of internal reflections and, consequently, take the longest time to travel the length of the cable.

For multimode propagation, dispersion is often expressed as a *bandwidth length product* (BLP) or *bandwidth distance product* (BDP). BLP indicates what signal frequencies can be propagated through a given distance of fiber cable and is expressed mathematically as the product of distance and bandwidth (sometimes called *linewidth*). Bandwidth length products are often expressed in MHz − km units. As the length of an optical cable increases, the bandwidth (and thus the bit rate) decreases in proportion.

Example 13-4

For a 300-meter optical fiber cable with a BLP of 600 MHz−km, determine the bandwidth.

Solution $B = \dfrac{600\text{ MHz} - \text{km}}{0.3\text{ km}}$

$B = 2\text{ GHz}$

Figure 13-21 shows three light rays propagating down a multimode step-index optical fiber. The lowest-order mode (ray 1) travels in a path parallel to the axis of the fiber. The middle-order mode (ray 2) bounces several times at the interface before traveling the length

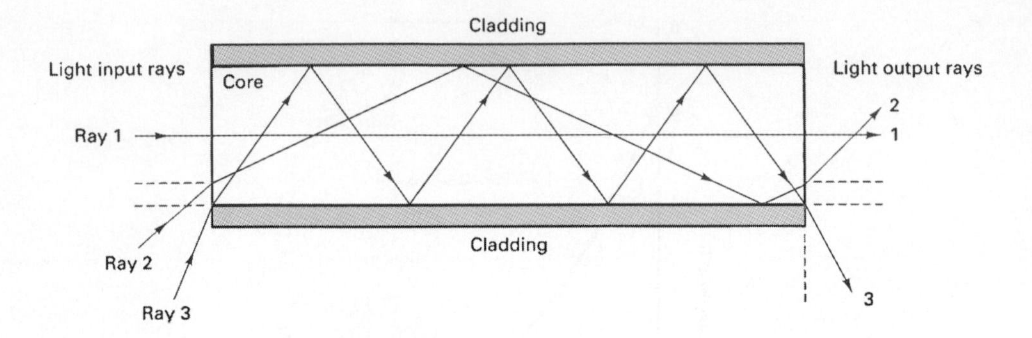

FIGURE 13-21 Light propagation down a multimode step-index fiber

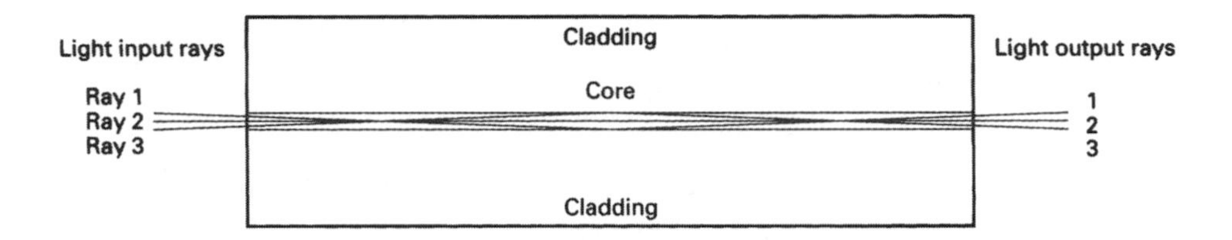

FIGURE 13-22 Light propagation down a single-mode step-index fiber

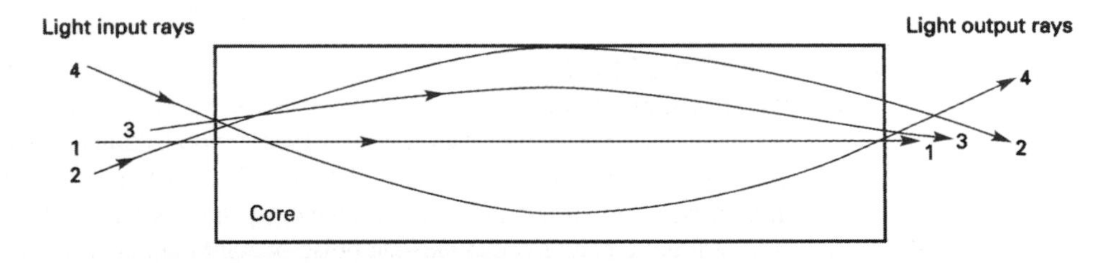

FIGURE 13-23 Light propagation down a multimode graded-index fiber

of the fiber. The highest-order mode (ray 3) makes many trips back and forth across the fiber as it propagates the entire length. It can be seen that ray 3 travels a considerably longer distance than ray 1 over the length of the cable. Consequently, if the three rays of light were emitted into the fiber at the same time, each ray would reach the far end at a different time, resulting in a spreading out of the light energy with respect to time. This is called modal dispersion and results in a stretched pulse that is also reduced in amplitude at the output of the fiber.

Figure 13-22 shows light rays propagating down a single-mode step-index cable. Because the radial dimension of the fiber is sufficiently small, there is only a single transmission path that all rays must follow as they propagate down the length of the fiber. Consequently, each ray of light travels the same distance in a given period of time, and modal dispersion is virtually eliminated.

Figure 13-23 shows light propagating down a multimode graded-index fiber. Three rays are shown traveling in three different modes. Although the three rays travel different paths, they all take approximately the same amount of time to propagate the length of the fiber. This is because the refractive index decreases with distance from the center, and the velocity at which a ray travels is inversely proportional to the refractive in-

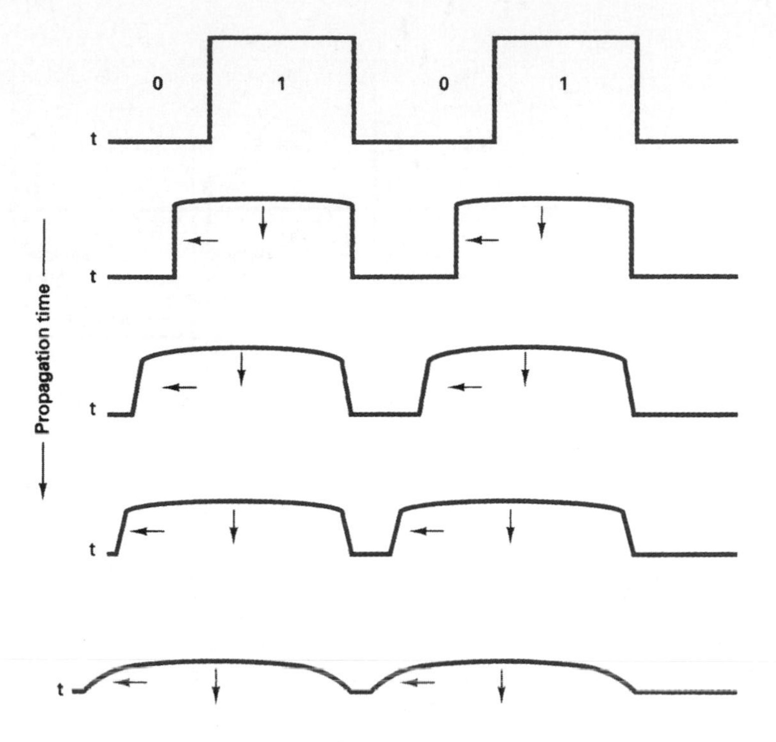

FIGURE 13-24 Pulse-width dispersion in an optical fiber cable

dex. Consequently, the farther rays 2 and 3 travel from the center of the cable, the faster they propagate.

Figure 13-24 shows the relative time/energy relationship of a pulse of light as it propagates down an optical fiber cable. From the figure, it can be seen that as the pulse propagates down the cable, the light rays that make up the pulse spread out in time, causing a corresponding reduction in the pulse amplitude and stretching of the pulse width. This is called *pulse spreading* or *pulse-width dispersion* and causes errors in digital transmission. It can also be seen that as light energy from one pulse falls back in time, it will interfere with the next pulse, causing intersymbol interference.

Figure 13-25a shows a unipolar return-to-zero (UPRZ) digital transmission. With UPRZ transmission (assuming a very narrow pulse), if light energy from pulse A were to fall back (*spread*) one bit time (t_b), it would interfere with pulse B and change what was a logic 0 to a logic 1. Figure 13-25b shows a unipolar nonreturn-to-zero (UPNRZ) digital transmission where each pulse is equal to the bit time. With UPNRZ transmission, if energy from pulse A were to fall back one-half of a bit time, it would interfere with pulse B. Consequently, UPRZ transmissions can tolerate twice as much delay or spread as UPNRZ transmissions.

The difference between the absolute delay times of the fastest and slowest rays of light propagating down a fiber of unit length is called the *pulse-spreading constant* (Δt) and is generally expressed in nanoseconds per kilometer (ns/km). The total pulse spread (ΔT) is then equal to the pulse-spreading constant (Δt) times the total fiber length (L). Mathematically, ΔT is

$$\Delta T_{(\text{ns})} = \Delta t_{(\text{ns/km})} \times L_{(\text{km})} \quad \textbf{(13-17)}$$

For UPRZ transmissions, the maximum data transmission rate in bits per second (bps) is expressed as

$$f_{b(\text{bps})} = \frac{1}{\Delta t \times L} \quad \textbf{(13-18)}$$

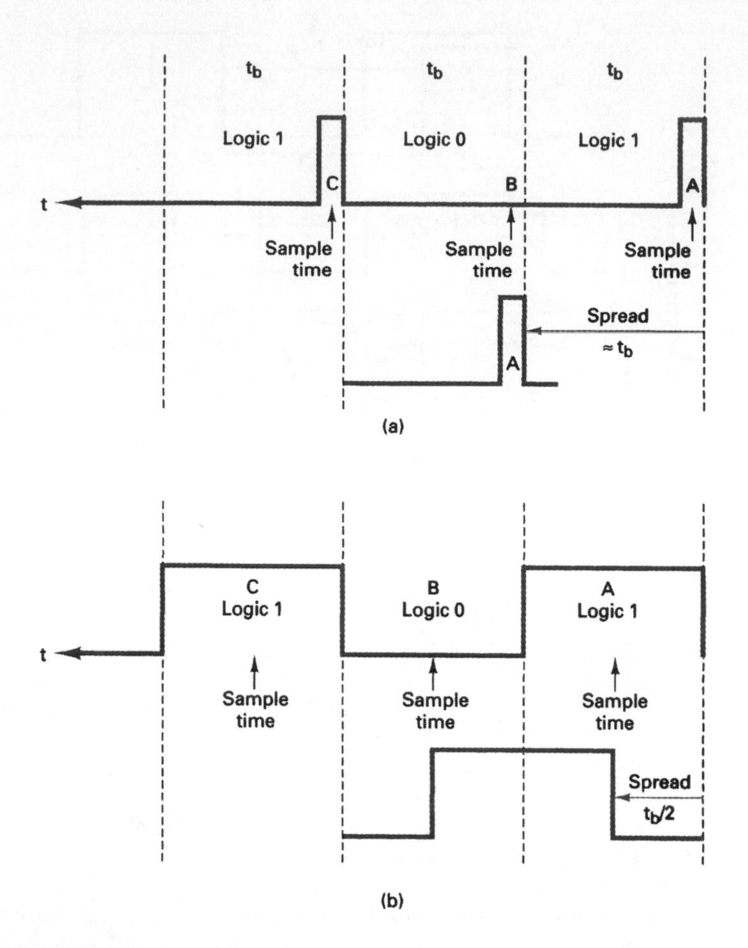

FIGURE 13-25 Pulse spreading of digital transmissions: (a) UPRZ; (b) UPNRZ

and for UPNRZ transmissions, the maximum transmission rate is

$$f_{b(\text{bps})} = \frac{1}{2\Delta t \times L} \tag{13-19}$$

Example 13-5

For an optical fiber 10 km long with a pulse-spreading constant of 5 ns/km, determine the maximum digital transmission rates for

a. Return-to-zero.

b. Nonreturn-to-zero transmissions.

Solution

a. Substituting into Equation 13-18 yields

$$f_b = \frac{1}{5 \text{ ns/km} \times 10 \text{ km}} = 20 \text{ Mbps}$$

b. Substituting into Equation 13-19 yields

$$f_b = \frac{1}{(2 \times 5 \text{ ns/km}) \times 10 \text{ km}} = 10 \text{ Mbps}$$

The results indicate that the digital transmission rate possible for this optical fiber is twice as high (20 Mbps versus 10 Mbps) for UPRZ as for UPNRZ transmission.

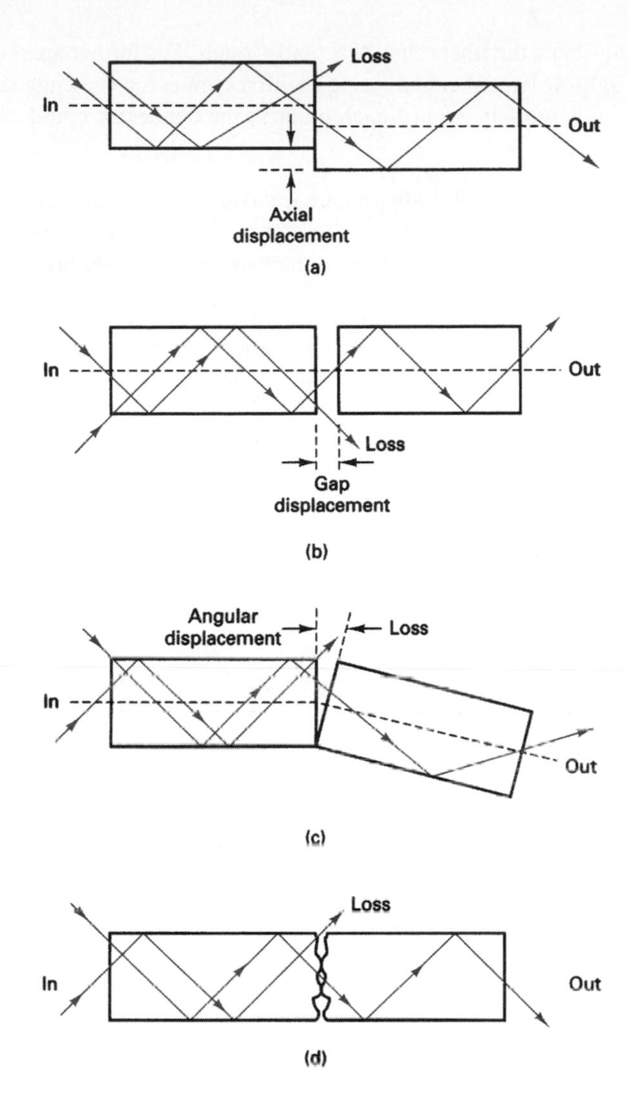

FIGURE 13-26 Fiber alignment impairments: (a) lateral misalignment; (b) gap displacement; (c) angular misalignment; (d) surface finish

13-10-6 Coupling Losses

Coupling losses are caused by imperfect physical connections. In fiber cables, coupling losses can occur at any of the following three types of optical junctions: light source-to-fiber connections, fiber-to-fiber connections, and fiber-to-photodetector connections. Junction losses are most often caused by one of the following alignment problems: lateral misalignment, gap misalignment, angular misalignment, and imperfect surface finishes.

13-10-6-1 Lateral displacement. *Lateral displacement* (*misalignment*) is shown in Figure 13-26a and is the lateral or axial displacement between two pieces of adjoining fiber cables. The amount of loss can be from a couple tenths of a decibel to several decibels. This loss is generally negligible if the fiber axes are aligned to within 5% of the smaller fiber's diameter.

13-10-6-2 Gap displacement (misalignment). *Gap displacement* (*misalignment*) is shown in Figure 13-26b and is sometimes called *end separation.* When splices are made in optical fibers, the fibers should actually touch. The farther apart the fibers, the greater the loss of light. If two fibers are joined with a connector, the ends should not touch because the two ends rubbing against each other in the connector could cause damage to either or both fibers.

13-10-6-3 Angular displacement (misalignment). Angular displacement (misalignment) is shown in Figure 13-26c and is sometimes called *angular displacement.* If the angular displacement is less than 2°, the loss will typically be less than 0.5 dB.

13-10-6-4 Imperfect surface finish. *Imperfect surface finish* is shown in Figure 13-26d. The ends of the two adjoining fibers should be highly polished and fit together squarely. If the fiber ends are less than 3° off from perpendicular, the losses will typically be less than 0.5 dB.

13-11 LIGHT SOURCES

The range of light frequencies detectable by the human eye occupies a very narrow segment of the total electromagnetic frequency spectrum. For example, blue light occupies the higher frequencies (shorter wavelengths) of visible light, and red hues occupy the lower frequencies (longer wavelengths). Figure 13-27 shows the light wavelength distribution produced from a tungsten lamp and the range of wavelengths perceivable by the human eye. As the figure shows, the human eye can detect only those lightwaves between approximately 380 nm and 780 nm. Furthermore, light consists of many shades of colors that are directly related to the heat of the energy being radiated. Figure 13-27 also shows that more visible light is produced as the temperature of the lamp is increased.

Light sources used for optical fiber systems must be at wavelengths efficiently propagated by the optical fiber. In addition, the range of wavelengths must be considered because the wider the range, the more likely the chance that chromatic dispersion will occur. Light

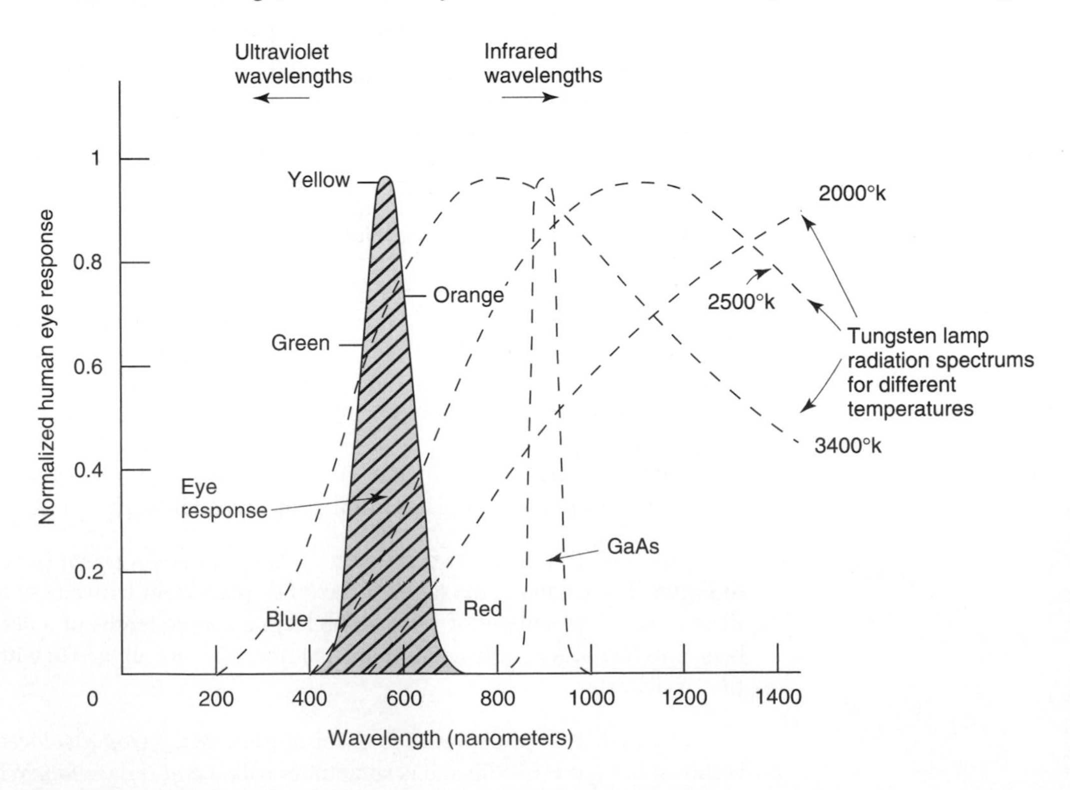

FIGURE 13-27 Tungsten lamp radiation and human eye response

sources must also produce sufficient power to allow the light to propagate through the fiber without causing distortion in the cable itself or in the receiver. Lastly, light sources must be constructed so that their outputs can be efficiently coupled into and out of the optical cable.

13-12 OPTICAL SOURCES

There are essentially only two types of practical light sources used to generate light for optical fiber communications systems: LEDs and ILDs. Both devices are constructed from semiconductor materials and have advantages and disadvantages. Standard LEDs have spectral widths of 30 nm to 50 nm, while injection lasers have spectral widths of only 1 nm to 3 nm (1 nm corresponds to a frequency of about 178 GHz). Therefore, a 1320-nm light source with a spectral linewidth of 0.0056 nm has a frequency bandwidth of approximately 1 GHz. Linewidth is the wavelength equivalent of bandwidth.

Selection of one light-emitting device over the other is determined by system economic and performance requirements. The higher cost of laser diodes is offset by higher performance. LEDs typically have a lower cost and a corresponding lower performance. However, LEDs are typically more reliable.

13-12-1 LEDs

An LED is a *p-n* junction diode, usually made from a semiconductor material such as aluminum-gallium-arsenide (AlGaAs) or gallium-arsenide-phosphide (GaAsP). LEDs emit light by spontaneous emission—light is emitted as a result of the recombination of electrons and holes.

When forward biased, minority carriers are injected across the *p-n* junction. Once across the junction, these minority carriers recombine with majority carriers and give up energy in the form of light. This process is essentially the same as in a conventional semiconductor diode except that in LEDs certain semiconductor materials and dopants are chosen such that the process is radiative; that is, a photon is produced. A photon is a quantum of electromagnetic wave energy. Photons are particles that travel at the speed of light but at rest have no mass. In conventional semiconductor diodes (germanium and silicon, for example), the process is primarily nonradiative, and no photons are generated. The energy gap of the material used to construct an LED determines the color of light it emits and whether the light emitted by it is visible to the human eye.

To produce LEDs, semiconductors are formed from materials with atoms having either three or five valence electrons (known as Group III and Group IV atoms, respectively, because of their location in the periodic table of elements). To produce light wavelengths in the 800-nm range, LEDs are constructed from Group III atoms, such as gallium (Ga) and aluminum (Al), and a Group IV atom, such as arsenide (As). The junction formed is commonly abbreviated GaAlAs for gallium-aluminum-arsenide. For longer wavelengths, gallium is combined with the Group III atom indium (In), and arsenide is combined with the Group V atom phosphate (P), which forms a gallium-indium-arsenide-phosphate (GaInAsP) junction. Table 13-4 lists some of the common semiconductor materials used in LED construction and their respective output wavelengths.

Table 13-4 Semiconductor Material Wavelengths

Material	Wavelength (nm)
AlGaInP	630–680
GaInP	670
GaAlAs	620–895
GaAs	904
InGaAs	980
InGaAsP	1100–1650
InGaAsSb	1700–4400

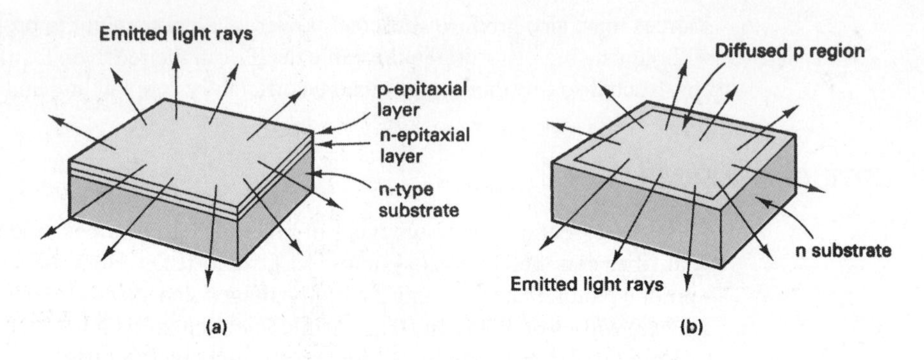

FIGURE 13-28 Homojunction LED structures: (a) silicon-doped gallium arsenide; (b) planar diffused

13-12-2 Homojunction LEDs

A *p-n* junction made from two different mixtures of the same types of atoms is called a homojunction structure. The simplest LED structures are homojunction and epitaxially grown, or they are single-diffused semiconductor devices, such as the two shown in Figure 13-28. *Epitaxially grown* LEDs are generally constructed of silicon-doped gallium-arsenide (Figure 13-28a). A typical wavelength of light emitted from this construction is 940 nm, and a typical output power is approximately 2 mW (3 dBm) at 100 mA of forward current. Light waves from homojunction sources do not produce a very useful light for an optical fiber. Light is emitted in all directions equally; therefore, only a small amount of the total light produced is coupled into the fiber. In addition, the ratio of electricity converted to light is very low. Homojunction devices are often called *surface emitters.*

Planar diffused homojunction LEDs (Figure 13-28b) output approximately 500 μW at a wavelength of 900 nm. The primary disadvantage of homojunction LEDs is the nondirectionality of their light emission, which makes them a poor choice as a light source for optical fiber systems.

13-12-3 Heterojunction LEDs

Heterojunction LEDs are made from a *p*-type semiconductor material of one set of atoms and an *n*-type semiconductor material from another set. Heterojunction devices are layered (usually two) such that the concentration effect is enhanced. This produces a device that confines the electron and hole carriers and the light to a much smaller area. The junction is generally manufactured on a substrate backing material and then sandwiched between metal contacts that are used to connect the device to a source of electricity.

With heterojunction devices, light is emitted from the edge of the material and are therefore often called *edge emitters.* A *planar heterojunction LED* (Figure 13-29) is quite similar to the epitaxially grown LED except that the geometry is designed such that the forward current is concentrated to a very small area of the active layer.

Heterojunction devices have the following advantages over homojunction devices:

The increase in current density generates a more brilliant light spot.

The smaller emitting area makes it easier to couple its emitted light into a fiber.

The small effective area has a smaller capacitance, which allows the planar heterojunction LED to be used at higher speeds.

Figure 13-30 shows the typical electrical characteristics for a low-cost infrared light-emitting diode. Figure 13-30a shows the output power versus forward current. From the figure, it can be seen that the output power varies linearly over a wide range of input cur-

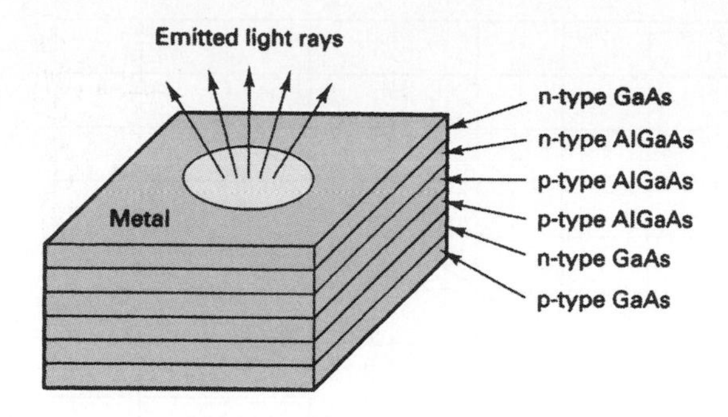

FIGURE 13-29 Planar heterojunction LED

rent (0.5 mW [−3 dBm] at 20 mA to 3.4 mW [5.3 dBm] at 140 mA). Figure 13-30b shows output power versus temperature. It can be seen that the output power varies inversely with temperature between a temperature range of −40°C to 80°C. Figure 13-30c shows relative output power in respect to output wavelength. For this particular example, the maximum output power is achieved at an output wavelength of 825 nm.

13-12-4 Burrus Etched-Well Surface-Emitting LED

For the more practical applications, such as telecommunications, data rates in excess of 100 Mbps are required. For these applications, the etched-well LED was developed. Burrus and Dawson of Bell Laboratories developed the etched-well LED. It is a surface-emitting LED and is shown in Figure 13-31. The Burrus etched-well LED emits light in many directions. The etched well helps concentrate the emitted light to a very small area. Also, domed lenses can be placed over the emitting surface to direct the light into a smaller area. These devices are more efficient than the standard surface emitters, and they allow more power to be coupled into the optical fiber, but they are also more difficult and expensive to manufacture.

13-12-5 Edge-Emitting LED

The edge-emitting LED, which was developed by RCA, is shown in Figure 13-32. These LEDs emit a more directional light pattern than do the surface-emitting LEDs. The construction is similar to the planar and Burrus diodes except that the emitting surface is a stripe rather than a confined circular area. The light is emitted from an active stripe and forms an elliptical beam. Surface-emitting LEDs are more commonly used than edge emitters because they emit more light. However, the coupling losses with surface emitters are greater, and they have narrower bandwidths.

The *radiant* light power emitted from an LED is a linear function of the forward current passing through the device (Figure 13-33). It can also be seen that the optical output power of an LED is, in part, a function of the operating temperature.

13-12-6 ILD

Lasers are constructed from many different materials, including gases, liquids, and solids, although the type of laser used most often for fiber-optic communications is the semiconductor laser.

The ILD is similar to the LED. In fact, below a certain threshold current, an ILD acts similarly to an LED. Above the threshold current, an ILD oscillates; lasing occurs. As current passes through a forward-biased *p-n* junction diode, light is emitted by spontaneous emission at a frequency determined by the energy gap of the semiconductor material. When a particular current level is reached, the number of minority carriers and photons produced on either side of the *p-n* junction reaches a level where they begin to collide with already excited minority carriers. This causes an increase in the ionization energy level and makes the carriers unstable. When this happens, a typical carrier recombines with an opposite type

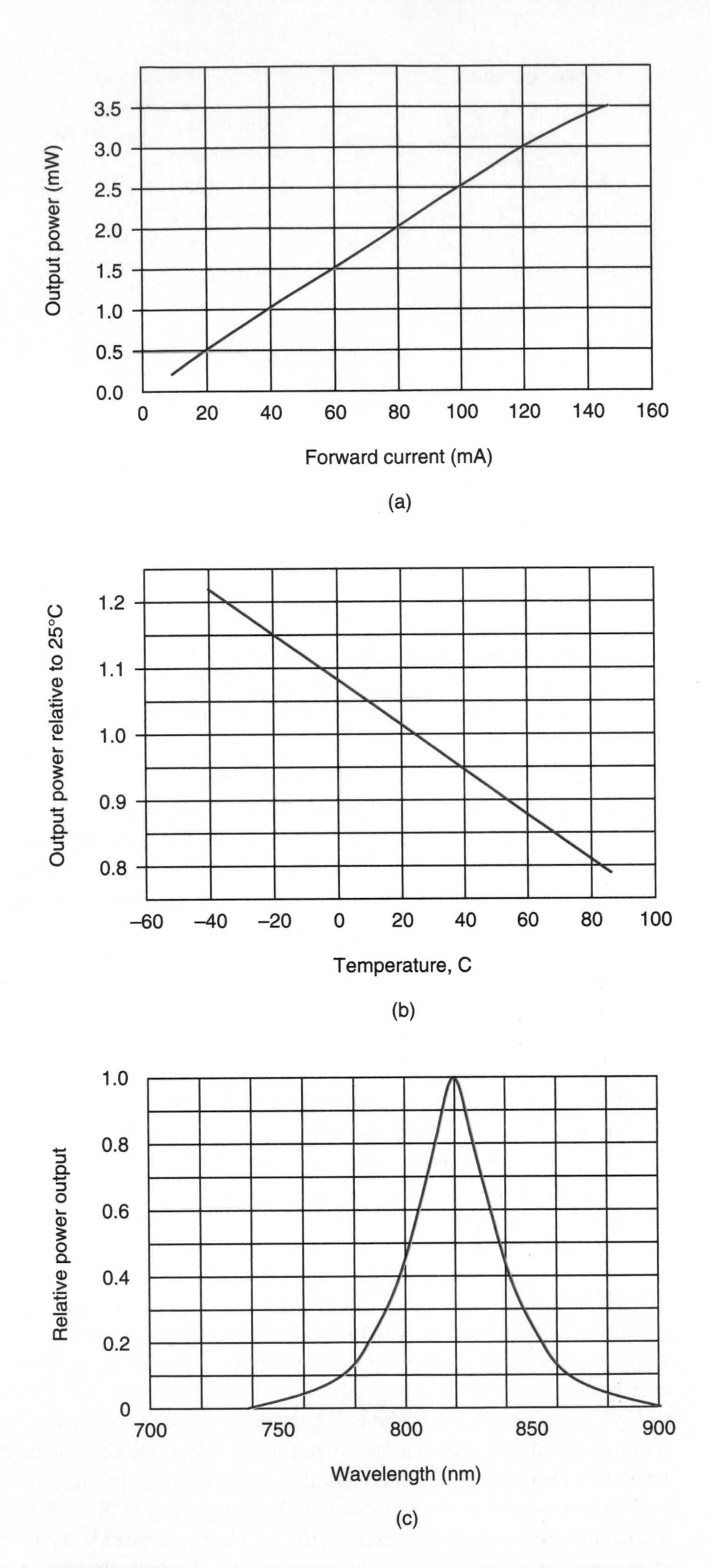

FIGURE 13-30 Typical LED electrical characteristics: (a) output power-versus-forward current; (b) output power-versus-temperature; and (c) output power-versus-output wavelength

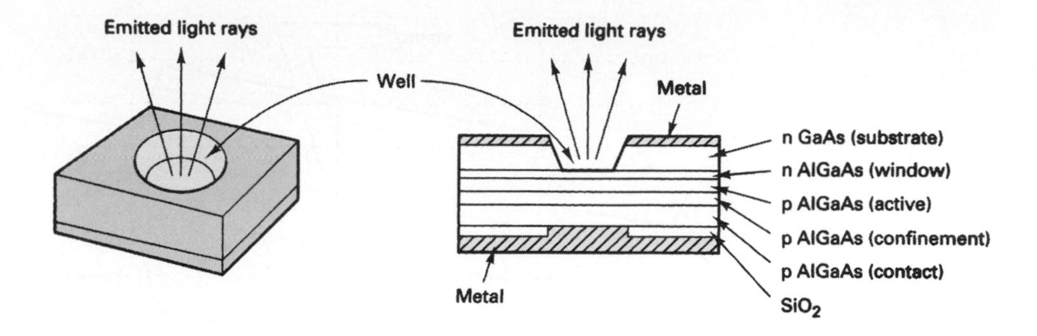

FIGURE 13-31 Burrus etched-well surface-emitting LED

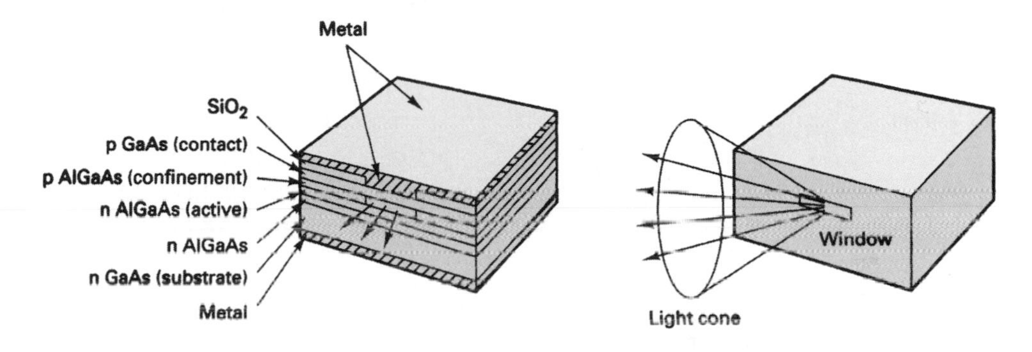

FIGURE 13-32 Edge-emitting LED

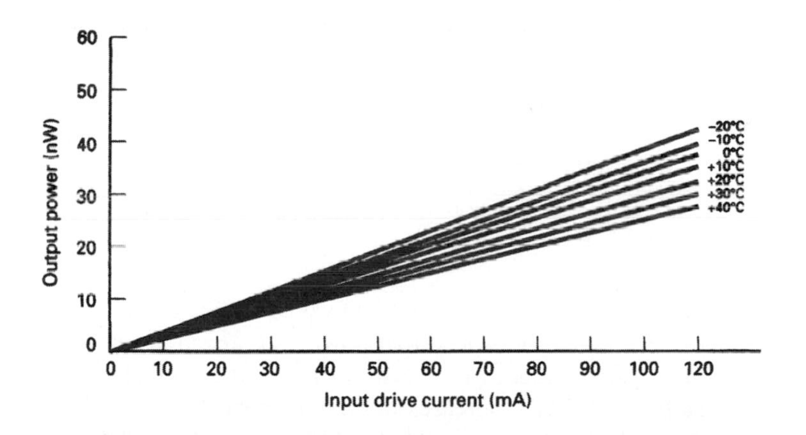

FIGURE 13-33 Output power versus forward current and operating temperature for an LED

of carrier at an energy level that is above its normal before-collision value. In the process, two photons are created; one is stimulated by another. Essentially, a gain in the number of photons is realized. For this to happen, a large forward current that can provide many carriers (holes and electrons) is required.

The construction of an ILD is similar to that of an LED (Figure 13-34) except that the ends are highly polished. The mirrorlike ends trap the photons in the active region and, as they reflect back and forth, stimulate free electrons to recombine with holes at a higher-than-normal energy level. This process is called *lasing*.

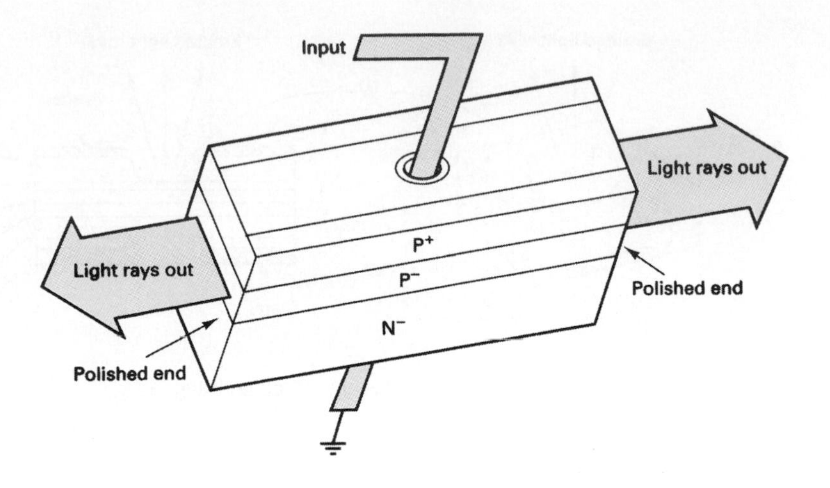

FIGURE 13-34 Injection laser diode construction

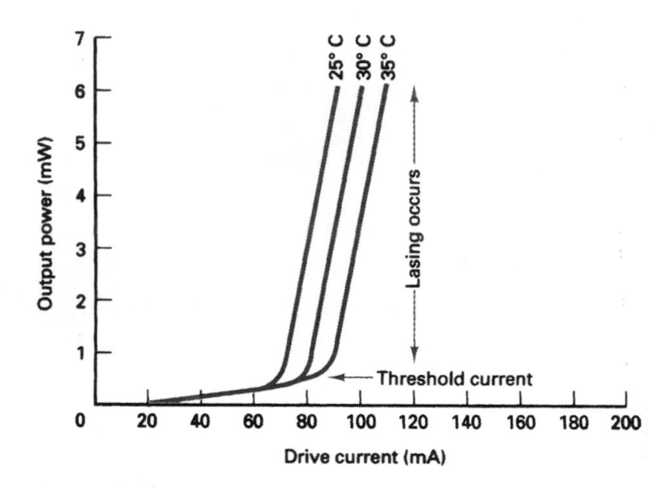

FIGURE 13-35 Output power versus forward current and temperature for an ILD

The radiant output light power of a typical ILD is shown in Figure 13-35. It can be seen that very little output power is realized until the threshold current is reached; then lasing occurs. After lasing begins, the optical output power increases dramatically, with small increases in drive current. It can also be seen that the magnitude of the optical output power of the ILD is more dependent on operating temperature than is the LED.

Figure 13-36 shows the light radiation patterns typical of an LED and an ILD. Because light is radiated out the end of an ILD in a narrow concentrated beam, it has a more direct radiation pattern.

ILDs have several advantages over LEDs and some disadvantages. Advantages include the following:

ILDs emit coherent (orderly) light, whereas LEDs emit incoherent (disorderly) light. Therefore, ILDs have a more direct radian pattern, making it easier to couple light emitted by the ILD into an optical fiber cable. This reduces the coupling losses and allows smaller fibers to be used.

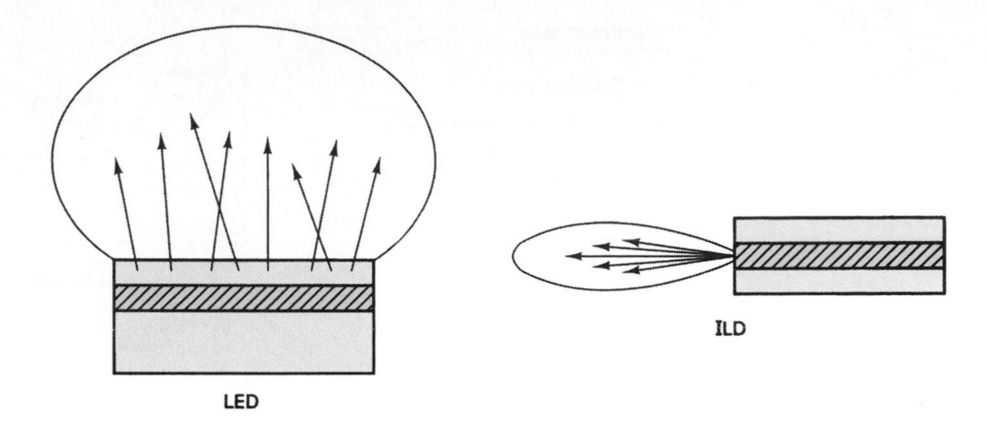

FIGURE 13-36 LED and ILD radiation patterns

The radiant output power from an ILD is greater than that for an LED. A typical output power for an ILD is 5 mW (7 dBm) and only 0.5 mW (−3 dBm) for LEDs. This allows ILDs to provide a higher drive power and to be used for systems that operate over longer distances.

ILDs can be used at higher bit rates than LEDs.

ILDs generate monochromatic light, which reduces chromatic or wavelength dispersion.

Disadvantages include the following:

ILDs are typically 10 times more expensive than LEDs.

Because ILDs operate at higher powers, they typically have a much shorter lifetime than LEDs.

ILDs are more temperature dependent than LEDs.

13-13 LIGHT DETECTORS

There are two devices commonly used to detect light energy in fiber-optic communications receivers: PIN diodes and APDs.

13-13-1 PIN Diodes

A *PIN diode* is a *depletion-layer photodiode* and is probably the most common device used as a light detector in fiber-optic communications systems. Figure 13-37 shows the basic construction of a PIN diode. A very lightly doped (almost pure or intrinsic) layer of *n*-type semiconductor material is sandwiched between the junction of the two heavily doped *n*- and *p*-type contact areas. Light enters the device through a very small window and falls on the carrier-void intrinsic material. The intrinsic material is made thick enough so that most of the photons that enter the device are absorbed by this layer. Essentially, the PIN photodiode operates just the opposite of an LED. Most of the photons are absorbed by electrons in the valence band of the intrinsic material. When the photons are absorbed, they add sufficient energy to generate carriers in the depletion region and allow current to flow through the device.

13-13-1-1 Photoelectric effect. Light entering through the window of a PIN diode is absorbed by the intrinsic material and adds enough energy to cause electronics to move from the valence band into the conduction band. The increase in the number of electrons that move into the conduction band is matched by an increase in the number of holes in the

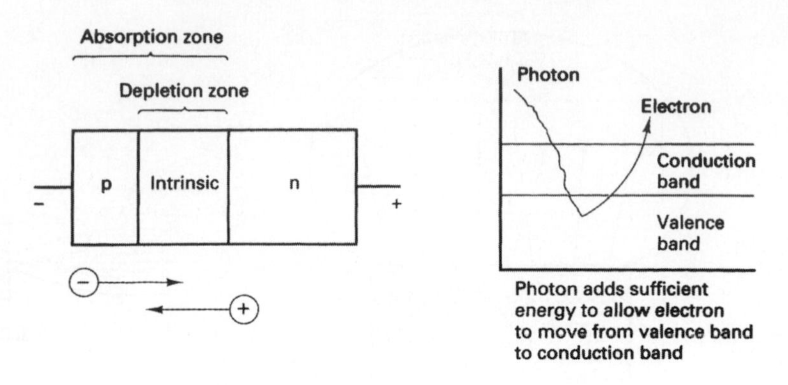

FIGURE 13-37 PIN photodiode construction

valence band. To cause current to flow in a photodiode, light of sufficient energy must be absorbed to give valence electrons enough energy to jump the energy gap. The energy gap for silicon is 1.12 eV (electron volts). Mathematically, the operation is as follows:

For silicon, the energy gap (E_g) equals 1.12 eV:

$$1 \text{ eV} = 1.6 \times 10^{-19} \text{ J}$$

Thus, the energy gap for silicon is

$$E_g = (1.12 \text{ eV})\left(1.6 \times 10^{-19} \frac{\text{J}}{\text{eV}}\right) = 1.792 \times 10^{-19} \text{ J}$$

and

$$\text{energy } (E) = hf \tag{13-20}$$

where h = Planck's constant = 6.6256×10^{-34} J/Hz
f = frequency (hertz)

Rearranging and solving for f yields

$$f = \frac{E}{h} \tag{13-21}$$

For a silicon photodiode,

$$f = \frac{1.792 \times 10^{-19} \text{ J}}{6.6256 \times 10^{-34} \text{ J/Hz}} = 2.705 \times 10^{14} \text{ Hz}$$

Converting to wavelength yields

$$\lambda = \frac{c}{f} = \frac{3 \times 10^8 \text{ m/s}}{2.705 \times 10^{14} \text{ Hz}} = 1109 \text{ nm/cycle}$$

13-13-2 APDs

Figure 13-38 shows the basic construction of an APD. An APD is a *pipn* structure. Light enters the diode and is absorbed by the thin, heavily doped *n*-layer. A high electric field intensity developed across the *i-p-n* junction by reverse bias causes impact ionization to occur. During impact ionization, a carrier can gain sufficient energy to ionize other bound electrons. These ionized carriers, in turn, cause more ionizations to occur. The process continues as in an avalanche and is, effectively, equivalent to an internal gain or carrier multiplication. Consequently, APDs are more sensitive than PIN diodes and require less additional amplification. The disadvantages of APDs are relatively long transit times and additional internally generated noise due to the avalanche multiplication factor.

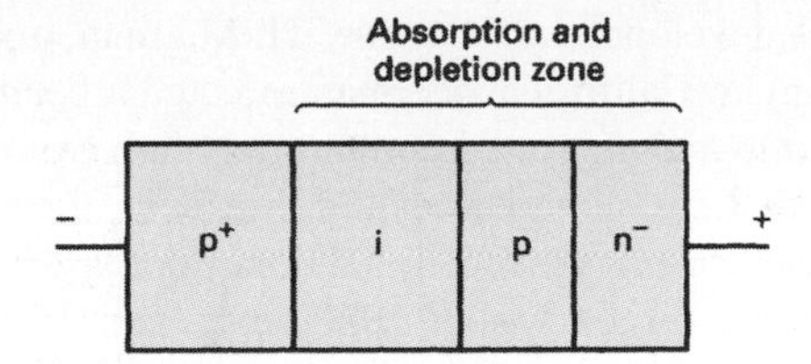

FIGURE 13-38 Avalanche photodiode construction

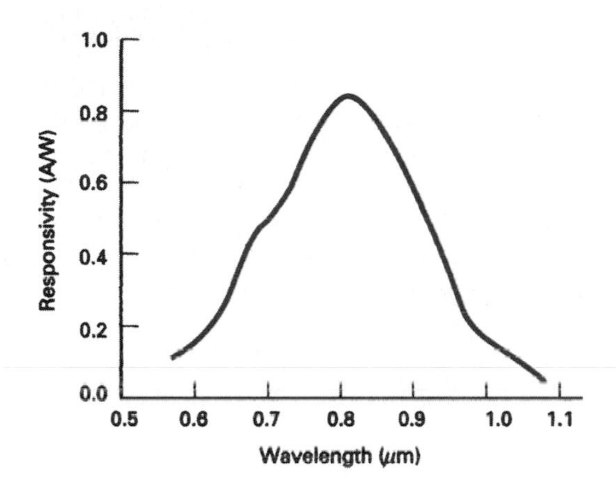

FIGURE 13-39 Spectral response curve

13-13-3 Characteristics of Light Detectors

The most important characteristics of light detectors are the following:

1. *Responsivity.* A measure of the conversion efficiency of a photodetector. It is the ratio of the output current of a photodiode to the input optical power and has the unit of amperes per watt. Responsivity is generally given for a particular wavelength or frequency.
2. *Dark current.* The leakage current that flows through a photodiode with no light input. Thermally generated carriers in the diode cause dark current.
3. *Transit time.* The time it takes a light induced carrier to travel across the depletion region of a semiconductor. This parameter determines the maximum bit rate possible with a particular photodiode.
4. *Spectral response.* The range of wavelength values that a given photodiode will respond. Generally, relative spectral response is graphed as a function of wavelength or frequency, as shown in Figure 13-39.
5. *Light sensitivity.* The minimum optical power a light detector can receive and still produce a usable electrical output signal. Light sensitivity is generally given for a particular wavelength in either dBm or dBμ.

13-14 LASERS

Laser is an acronym for *l*ight *a*mplification *s*timulated by the *e*mission of *r*adiation. Laser technology deals with the concentration of light into a very small, powerful beam. The acronym was chosen when technology shifted from microwaves to light waves. Basically, there are four types of lasers: gas, liquid, solid, and semiconductor.

The first laser was developed by Theodore H. Maiman, a scientist who worked for Hughes Aircraft Company in California. Maiman directed a beam of light into ruby crystals with a xenon flashlamp and measured emitted radiation from the ruby. He discovered that when the emitted radiation increased beyond threshold, it caused emitted radiation to become extremely intense and highly directional. Uranium lasers were developed in 1960 along with other rare-earth materials. Also in 1960, A. Javin of Bell Laboratories developed the helium laser. Semiconductor lasers (injection laser diodes) were manufactured in 1962 by General Electric, IBM, and Lincoln Laboratories.

13-14-1 Laser Types

Basically, there are four types of lasers: gas, liquid, solid, and semiconductor.

1. *Gas lasers.* Gas lasers use a mixture of helium and neon enclosed in a glass tube. A flow of coherent (one frequency) light waves is emitted through the output coupler when an electric current is discharged into the gas. The continuous light-wave output is monochromatic (one color).
2. *Liquid lasers.* Liquid lasers use organic dyes enclosed in a glass tube for an active medium. Dye is circulated into the tube with a pump. A powerful pulse of light excites the organic dye.
3. *Solid lasers.* Solid lasers use a solid, cylindrical crystal, such as ruby, for the active medium. Each end of the ruby is polished and parallel. The ruby is excited by a tungsten lamp tied to an ac power supply. The output from the laser is a continuous wave.
4. *Semiconductor lasers.* Semiconductor lasers are made from semiconductor *p-n* junctions and are commonly called ILDs. The excitation mechanism is a dc power supply that controls the amount of current to the active medium. The output light from an ILD is easily modulated, making it very useful in many electronic communications applications.

13-14-2 Laser Characteristics

All types of lasers have several common characteristics. They all use (1) an active material to convert energy into laser light, (2) a pumping source to provide power or energy, (3) optics to direct the beam through the active material to be amplified, (4) optics to direct the beam into a narrow powerful cone of divergence, (5) a feedback mechanism to provide continuous operation, and (6) an output coupler to transmit power out of the laser.

The radiation of a laser is extremely intense and directional. When focused into a fine hairlike beam, it can concentrate all its power into the narrow beam. If the beam of light were allowed to diverge, it would lose most of its power.

13-14-3 Laser Construction

Figure 13-40 shows the construction of a basic laser. A power source is connected to a flashtube that is coiled around a glass tube that holds the active medium. One end of the glass tube is a polished mirror face for 100% internal reflection. The flashtube is energized by a trigger pulse and produces a high-level burst of light (similar to a flashbulb). The flash causes the chromium atoms within the active crystalline structure to become excited. The process of pumping raises the level of the chromium atoms from ground state to an excited energy state. The ions then decay, falling to an intermediate energy level. When the population of ions in the intermediate level is greater than the ground state, a population inversion occurs. The population inversion causes laser action (lasing) to occur. After a period of time, the excited chromium atoms will fall to the ground energy level. At this time, photons are emitted. A photon is a packet of radiant energy. The emitted photons strike atoms and two other photons are emitted (hence the term "stimulated emission"). The frequency of the energy determines the strength of the photons; higher frequencies cause greater-strength photons.

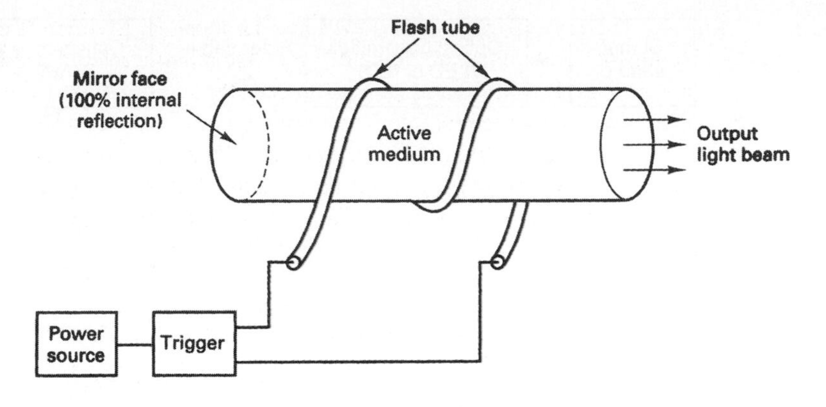

FIGURE 13-40 Laser construction

13-15 OPTICAL FIBER SYSTEM LINK BUDGET

As with any communications system, optical fiber systems consist of a source and a destination that are separated by numerous components and devices that introduce various amounts of loss or gain to the signal as it propagates through the system. Figure 13-41 shows two typical optical fiber communications system configurations. Figure 13-41a shows a repeaterless system where the source and destination are interconnected through one or more sections of optical cable. With a repeaterless system, there are no amplifiers or regenerators between the source and destination.

Figure 13-41b shows an optical fiber system that includes a repeater that either amplifies or regenerates the signal. Repeatered systems are obviously used when the source and destination are separated by great distances.

Link budgets are generally calculated between a light source and a light detector; therefore, for our example, we look at a link budget for a repeaterless system. A repeaterless system consists of a light source, such as an LED or ILD, and a light detector, such as an APD connected by optical fiber and connectors. Therefore, the link budget consists of a light power source, a light detector, and various cable and connector losses. Losses typical to optical fiber links include the following:

1. *Cable losses.* Cable losses depend on cable length, material, and material purity. They are generally given in dB/km and can vary between a few tenths of a dB to several dB per kilometer.
2. *Connector losses.* Mechanical connectors are sometimes used to connect two sections of cable. If the mechanical connection is not perfect, light energy can escape, resulting in a reduction in optical power. Connector losses typically vary between a few tenths of a dB to as much as 2 dB for each connector.
3. *Source-to-cable interface loss.* The mechanical interface used to house the light source and attach it to the cable is seldom perfect. Therefore, a small percentage of optical power is not coupled into the cable, representing a power loss to the system of several tenths of a dB.
4. *Cable-to-light detector interface loss.* The mechanical interface used to house the light detector and attach it to the cable is also not perfect and, therefore, prevents a small percentage of the power leaving the cable from entering the light detector. This, of course, represents a loss to the system usually of a few tenths of a dB.
5. *Splicing loss.* If more than one continuous section of cable is required, cable sections can be fused together (spliced). Because the splices are not perfect, losses ranging from a couple tenths of a dB to several dB can be introduced to the signal.

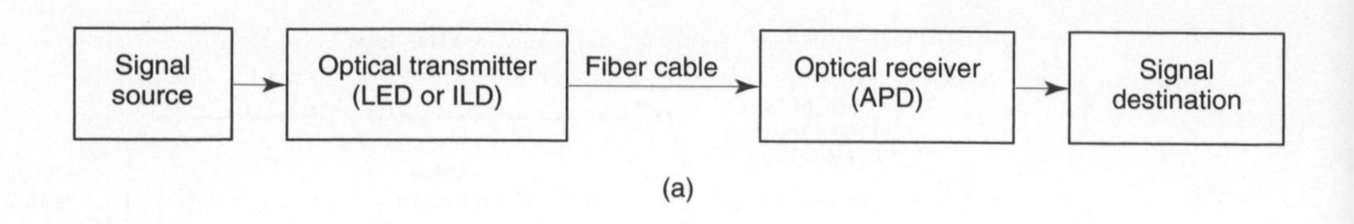

(a)

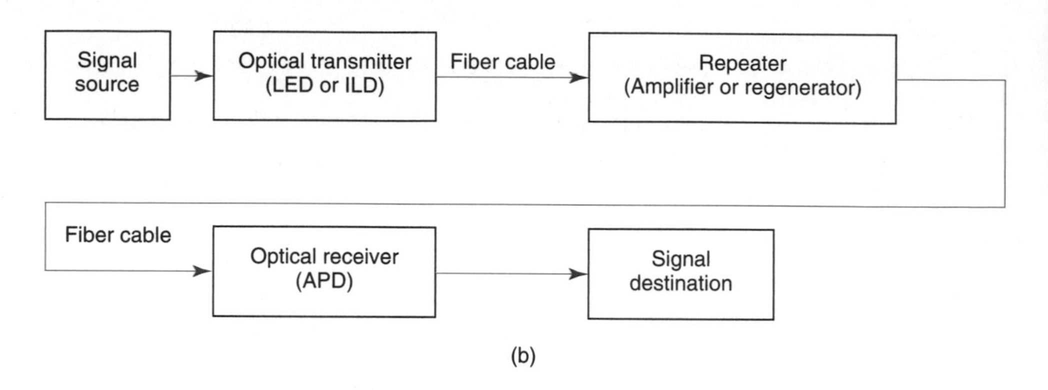

(b)

FIGURE 13-41 Optical fiber communications systems: (a) without repeaters; (b) with repeaters

6. *Cable bends.* When an optical cable is bent at too large an angle, the internal characteristics of the cable can change dramatically. If the changes are severe, total reflections for some of the light rays may no longer be achieved, resulting in refraction. Light refracted at the core/cladding interface enters the cladding, resulting in a net loss to the signal of a few tenths of a dB to several dB.

As with any link or system budget, the useful power available in the receiver depends on transmit power and link losses. Mathematically, receive power is represented as

$$P_r = P_t - \text{losses} \quad \textbf{(13-22)}$$

where P_r = power received (dBm)
P_t = power transmitted (dBm)
losses = sum of all losses (dB)

Example 13-6

Determine the optical power received in dBm and watts for a 20-km optical fiber link with the following parameters:

LED output power of 30 mW
Four 5-km sections of optical cable each with a loss of 0.5 dB/km
Three cable-to-cable connectors with a loss of 2 dB each
No cable splices
Light source-to-fiber interface loss of 1.9 dB
Fiber-to-light detector loss of 2.1 dB
No losses due to cable bends

Solution The LED output power is converted to dBm using Equation 13-6:

$$P_{\text{out}} = 10 \log \frac{30 \text{ mW}}{1 \text{ mW}}$$
$$= 14.8 \text{ dBm}$$

The cable loss is simply the product of the total cable length in km and the loss in dB/km. Four 5-km sections of cable is a total cable length of 20 km; therefore,

$$\text{total cable loss} = 20 \text{ km} \times 0.5 \text{ dB/km}$$
$$= 10 \text{ dB}$$

Cable connector loss is simply the product of the loss in dB per connector and the number of connectors. The maximum number of connectors is always one less than the number of sections of cable. Four sections of cable would then require three connectors; therefore,

$$\text{total connector loss} = 3 \text{ connectors} \times 2 \text{ dB/connector}$$
$$= 6 \text{ dB}$$

The light source-to-cable and cable-to-light detector losses were given as 1.9 dB and 2.1 dB, respectively. Therefore,

$$\text{total loss} = \text{cable loss} + \text{connector loss} + \text{light source-to-cable loss} + \text{cable-to-light detector loss}$$
$$= 10 \text{ dB} + 6 \text{ dB} + 1.9 \text{ dB} + 2.1 \text{ dB}$$
$$= 20 \text{ dB}$$

The receive power is determined by substituting into Equation 13-22:

$$P_r = 14.8 \text{ dBm} - 20 \text{ dB}$$
$$= -5.2 \text{ dBm}$$
$$= 0.302 \text{ mW}$$

QUESTIONS

13-1. Define a fiber-optic system.

13-2. What is the relationship between information capacity and bandwidth?

13-3. What development in 1951 was a substantial breakthrough in the field of fiber optics? In 1960? In 1970?

13-4. Contrast the advantages and disadvantages of fiber-optic cables and metallic cables.

13-5. Outline the primary building blocks of a fiber-optic system.

13-6. Contrast glass and plastic fiber cables.

13-7. Briefly describe the construction of a fiber-optic cable.

13-8. Define the following terms: *velocity of propagation, refraction,* and *refractive index.*

13-9. State Snell's law for refraction and outline its significance in fiber-optic cables.

13-10. Define *critical angle.*

13-11. Describe what is meant by *mode of operation;* by *index profile.*

13-12. Describe a step-index fiber cable; a graded-index cable.

13-13. Contrast the advantages and disadvantages of step-index, graded-index, single-mode, and multimode propagation.

13-14. Why is single-mode propagation impossible with graded-index fibers?

13-15. Describe the source-to-fiber aperture.

13-16. What are the *acceptance angle* and the *acceptance cone* for a fiber cable?

13-17. Define *numerical aperture.*

13-18. List and briefly describe the losses associated with fiber cables.

13-19. What is *pulse spreading?*

13-20. Define *pulse spreading constant.*

13-21. List and briefly describe the various coupling losses.

13-22. Briefly describe the operation of a light-emitting diode.

13-23. What are the two primary types of LEDs?

13-24. Briefly describe the operation of an injection laser diode.

13-25. What is lasing?

13-26. Contrast the advantages and disadvantages of ILDs and LEDs.

13-27. Briefly describe the function of a photodiode.

13-28. Describe the photoelectric effect.

13-29. Explain the difference between a PIN diode and an APD.

13-30. List and describe the primary characteristics of light detectors.

PROBLEMS

13-1. Determine the wavelengths in nanometers and angstroms for the following light frequencies:

a. 3.45×10^{14} Hz
b. 3.62×10^{14} Hz
c. 3.21×10^{14} Hz

13-2. Determine the light frequency for the following wavelengths:

a. 670 nm
b. 7800 Å
c. 710 nm

13-3. For a glass ($n = 1.5$)/quartz ($n = 1.38$) interface and an angle of incidence of 35°, determine the angle of refraction.

13-4. Determine the critical angle for the fiber described in problem 13-3.

13-5. Determine the acceptance angle for the cable described in problem 13-3.

13-6. Determine the numerical aperture for the cable described in problem 13-3.

13-7. Determine the maximum bit rate for RZ and NRZ encoding for the following pulse-spreading constants and cable lengths:

a. $\Delta t = 10$ ns/m, $L = 100$ m
b. $\Delta t = 20$ ns/m, $L = 1000$ m
c. $\Delta t = 2000$ ns/km, $L = 2$ km

13-8. Determine the lowest light frequency that can be detected by a photodiode with an energy gap = 1.2 eV.

13-9. Determine the wavelengths in nanometers and angstroms for the following light frequencies:

a. 3.8×10^{14} Hz
b. 3.2×10^{14} Hz
c. 3.5×10^{14} Hz

13-10. Determine the light frequencies for the following wavelengths:

a. 650 nm
b. 7200 Å
c. 690 nm

13-11. For a glass ($n = 1.5$)/quartz ($n = 1.41$) interface and an angle of incidence of 38°, determine the angle of refraction.

13-12. Determine the critical angle for the fiber described in problem 13-11.

13-13. Determine the acceptance angle for the cable described in problem 13-11.

13-14. Determine the numerical aperture for the cable described in problem 13-11.

13-15. Determine the maximum bit rate for RZ and NRZ encoding for the following pulse-spreading constants and cable lengths:

a. $\Delta t = 14$ ns/m, $L = 200$ m
b. $\Delta t = 10$ ns/m, $L = 50$ m
c. $\Delta t = 20$ ns/m, $L = 200$ m

13-16. Determine the lowest light frequency that can be detected by a photodiode with an energy gap = 1.25 eV.

13-17. Determine the optical power received in dBm and watts for a 24-km optical fiber link with the following parameters:

LED output power of 20 mW
Six 4-km sections of optical cable each with a loss of 0.6 dB/km
Three cable-to-cable connectors with a loss of 2.1 dB each
No cable splices
Light source-to-fiber interface loss of 2.2 dB
Fiber-to-light detector loss of 1.8 dB
No losses due to cable bends

CHAPTER 14

Electromagnetic Wave Propagation

CHAPTER OUTLINE

OBJECTIVES

- Define *free-space electromagnetic wave propagation*
- Define *electromagnetic wave polarization*
- Define and describe the terms *rays* and *wavefronts*
- Define *electromagnetic radiation* and describe the terms *power density* and *field intensity*
- Explain characteristic impedance of free space
- Describe a spherical wavefront
- Explain the inverse square law
- Define and explain wave attenuation and absorption
- Define and describe the optical properties of radio waves: refraction, reflection, and diffraction
- Define the modes of terrestrial propagation: ground waves, space waves, and sky waves
- Define and explain the following terms: critical frequency, critical angle, virtual height, and maximum usable frequency
- Define *skip* and *skip distance*
- Define *free-space* path loss and describe the variables that constitute it
- Define *fade* and *fade margin*

14-1 INTRODUCTION

In Chapter 13, we explained transverse electromagnetic (TEM) waves and also described how metallic wires can be used as a transmission medium to transfer TEM waves from one point to another. However, very often in electronic communications systems, it is impractical or impossible to interconnect two pieces of equipment with a physical facility such as a metallic wire or cable. This is especially true when the equipment is separated by large spans of water, rugged mountains, or harsh desert terrain or when communicating with satellite transponders orbiting 22,300 miles above Earth. Also, when the transmitters and receivers are mobile, as with two-way radio communications and cellular telephone, providing connections with metallic facilities is impossible. Therefore, Earth's atmosphere is often used as a transmission medium.

Free-space propagation of electromagnetic waves is often called *radio-frequency* (RF) *propagation* or simply *radio propagation.* Although free space implies a vacuum, propagation through Earth's atmosphere is often referred to as free-space propagation and can often be treated as just that, the primary difference being that Earth's atmosphere introduces losses and impairments to the signal that are not encountered in a vacuum. TEM waves will propagate through any dielectric material, including air. TEM waves, however, do not propagate well through lossy conductors, such as seawater, because the electric fields cause currents to flow in the material that rapidly dissipate the wave's energy.

Radio waves are electromagnetic waves and, like light, propagate through free space in a straight line with a velocity of approximately 300,000,000 meters per second. Other forms of electromagnetic waves include infrared, ultraviolet, X rays, and gamma rays. To propagate radio waves through Earth's atmosphere, it is necessary that the energy be radiated from the source, then the energy must be captured at the receive end. Radiating and capturing energy are antenna functions and are explained in Chapter 15, and the properties of electromagnetic waves were explained in Chapter 13. The purpose of this chapter is to describe the nature, behavior, and optical properties of radio waves propagating through Earth's atmosphere.

14-2 ELECTROMAGNETIC WAVES AND POLARIZATION

In essence, an *electromagnetic wave* is electrical energy that has escaped into free space. Electromagnetic waves travel in a straight line at approximately the speed of light and are made up of magnetic and electric fields that are at right angles to each other and at right angles to the direction of propagation. The essential properties of radio waves are frequency, intensity, direction of travel, and plane of polarization.

Radio waves are a form of electromagnetic radiation similar to light and heat. They differ from these other radiations in the manner in which they are generated and detected and in their frequency range. A radio wave consists of traveling electric and magnetic fields, with the energy evenly divided between the two types of fields.

The *polarization* of a plane electromagnetic wave is simply the orientation of the electric field vector in respect to the surface of the Earth (i.e., looking at the horizon). If the polarization remains constant, it is described as *linear polarization. Horizontal polarization* and *vertical polarization* are two forms of linear polarization. If the electric field is propagating parallel to the Earth's surface, the wave is said to be horizontally polarized. If the electric field is propagating perpendicular to the Earth's surface, the wave is said to be vertically polarized. If the polarization vector rotates 360° as the wave moves one wavelength through space and the field strength is equal at all angles of polarization, the wave is described as having *circular polarization.* When the field strength varies with changes in polarization, this is described as *elliptical polarization.* A rotating wave can turn in either direction. If the vector rotates in a clockwise direction, it is right handed, and if the vector rotates in a counterclockwise direction, it is considered left handed.

14-3 RAYS AND WAVEFRONTS

Electromagnetic waves are invisible; therefore, they must be analyzed by indirect methods using schematic diagrams. The concepts of *rays* and *wavefronts* are aids to illustrating the effects of electromagnetic wave propagation through free space. A ray is a line drawn along the direction of propagation of an electromagnetic wave. Rays are used to show the relative direction of electromagnetic wave propagation; however, it does not necessarily represent the propagation of a single electromagnetic wave. Several rays are shown in Figure 14-1 (R_a, R_b, R_c, and so on). A wavefront shows a surface of constant phase of electromagnetic waves. A wavefront is formed when points of equal phase on rays propagated from the same source are joined together. Figure 14-1 shows a wavefront with a surface that is perpendicular to the direction of propagation (rectangle *ABCD*). When a surface is plane, its wavefront is perpendicular to the direction of propagation. The closer to the source, the more complicated the wavefront becomes.

Most wavefronts are more complicated than a simple plane wave. Figure 14-2 shows a point source, several rays propagating from it, and the corresponding wavefront. A *point source* is a single location from which rays propagate equally in all directions (an *isotropic source*). The wavefront generated from a point source is simply a sphere with

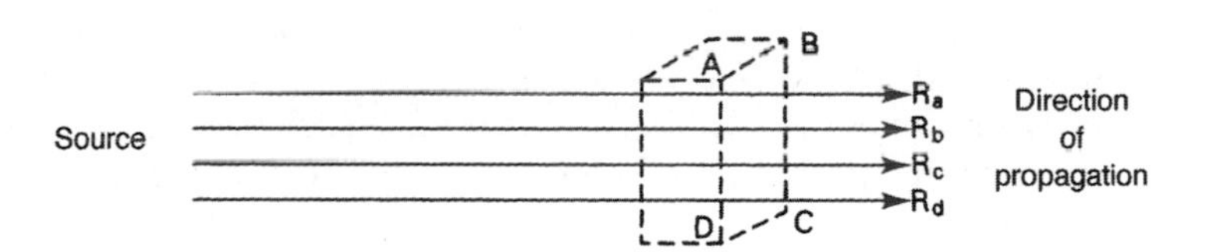

FIGURE 14-1 Plane wave

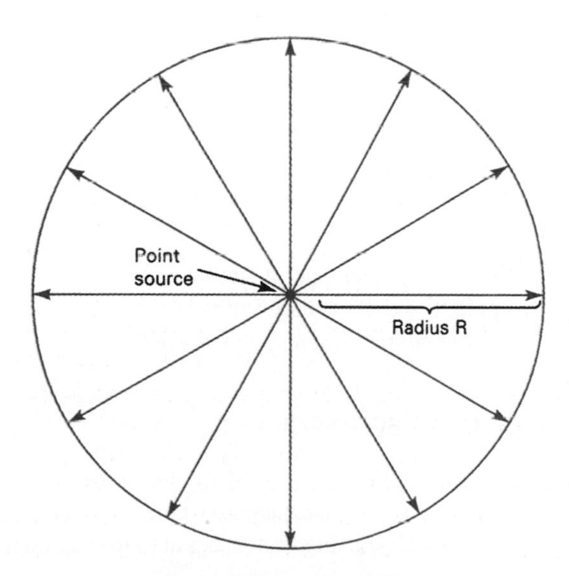

FIGURE 14-2 Wavefront from a point source

radius R and its center located at the point of origin of the waves. In free space and a sufficient distance from the source, the rays within a small area of a spherical wavefront are nearly parallel. Therefore, the farther from a source, the more wave propagation appears as a plane wavefront.

14-4 ELECTROMAGNETIC RADIATION

Radio waves are electromagnetic waves simply because they are made up of an electric and a magnetic field. The *magnetic field* is an invisible force field produced by a magnet, such as a conductor when current is flowing through it. Magnetic fields are continuous; however, it is standard for performing calculations and measurements to represent a magnetic field with individual lines of force. The strength of a magnetic field (H) produced around a conductor (such as a wire or an antenna) is expressed mathematically as

$$H = \frac{1}{2\pi d} \tag{14-1}$$

where H = magnetic field (ampere turns per meter)
d = distance from wire (meters)

Electric fields are also invisible force fields produced by a difference in voltage potential between two conductors. Electric field strength (E) is expressed mathematically as

$$E = \frac{q}{4\pi\epsilon d^2} \tag{14-2}$$

where E = electric field strength (volts per meter)
q = charge between conductors (coulombs)
ϵ = permittivity (farads per meter)
d = distance between conductors (meters)

Permittivity is the dielectric constant of the material separating the two conductors (i.e., the dielectric insulator). The permittivity of air or free space is approximately 8.85×10^{-12} F/m.

14-4-1 Power Density and Field Intensity

Electromagnetic waves represent the flow of energy in the direction of propagation. The rate at which energy passes through a given surface area in free space is called *power density.* Therefore, power density is energy per unit time per unit of area and is usually given in watts per square meter. *Field intensity* is the intensity of the electric and magnetic fields of an electromagnetic wave propagating in free space. Electric field intensity is usually given in volts per meter and magnetic field intensity in ampere turns per meter (At/m). Mathematically, power density is

$$\mathcal{P} = \mathcal{E}\mathcal{H} \quad \text{W/m}^2 \tag{14-3}$$

where $\mathcal{P}$ = power density (watts per meter squared)
$\mathcal{E}$ = rms electric field intensity (volts per meter)
$\mathcal{H}$ = rms magnetic field intensity (ampere turns per meter)

14-5 CHARACTERISTIC IMPEDANCE OF FREE SPACE

The electric and magnetic field intensities of an electromagnetic wave in free space are related through the characteristic impedance (resistance) of free space. The characteristic impedance of a lossless transmission medium is equal to the square root of the ratio of its magnetic permeability to its electric permittivity. Mathematically, the characteristic impedance of free space (Z_s) is

$$Z_s = \sqrt{\frac{\mu_0}{\epsilon_0}} \tag{14-4}$$

where Z_s = characteristic impedance of free space (ohms)
μ_0 = magnetic permeability of free space (1.26×10^{-6} H/m)
ϵ_0 = electric permittivity of free space (8.85×10^{-12} F/m)

Substituting into Equation 14-4, we have

$$Z_s = \sqrt{\frac{1.26 \times 10^{-6}}{8.85 \times 10^{-12}}} = 377\ \Omega$$

Therefore, using Ohm's law, we obtain

$$\mathcal{P} = \frac{\mathcal{E}^2}{377} = 377\mathcal{H}^2\ \text{W/m}^2 \tag{14-5}$$

$$\mathcal{H} = \frac{\mathcal{E}}{377}\ \text{At/m} \tag{14-6}$$

14-6 SPHERICAL WAVEFRONT AND THE INVERSE SQUARE LAW

14-6-1 Spherical Wavefront

Figure 14-3 shows a point source that radiates power at a constant rate uniformly in all directions. Such a source is called an *isotropic radiator.* A true isotropic radiator does not exist. However, it is closely approximated by an *omnidirectional antenna.* An isotropic radiator produces a spherical wavefront with radius R. All points distance R from the source lie on the surface of the sphere and have equal power densities. For example, in Figure 14-3, points A and B are an equal distance from the source. Therefore, the power densities at

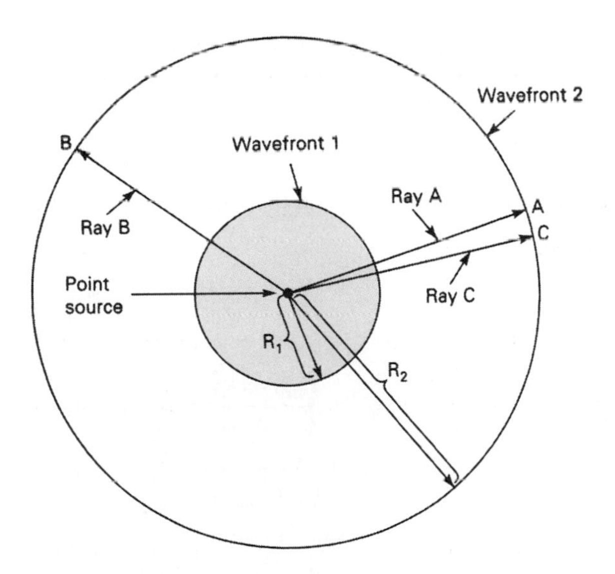

FIGURE 14-3 Spherical wavefront from an isotropic source

points A and B are equal. At any instant of time, the total power radiated, P_r watts, is uniformly distributed over the total surface of the sphere (this assumes a lossless transmission medium). Therefore, the power density at any point on the sphere is the total radiated power divided by the total area of the sphere. Mathematically, the power density at any point on the surface of a spherical wavefront is

$$\mathcal{P} = \frac{P_{\text{rad}}}{4\pi R^2} \tag{14-7}$$

where P_{rad} = total power radiated (watts)
R = radius of the sphere (which is equal to the distance from any point on the surface of the sphere to the source)
$4\pi R^2$ = area of the sphere

and for a distance R_a meters from the source, the power density is

$$\mathcal{P}_a = \frac{P_{\text{rad}}}{4\pi R_a^2}$$

Equating Equations 14-5 and 14-7 gives

$$\frac{P_{\text{rad}}}{4\pi R^2} = \frac{\mathscr{E}^2}{377}$$

Therefore,

$$\mathscr{E}^2 = \frac{377 P_{\text{rad}}}{4\pi R^2} \quad \text{and} \quad \mathscr{E} = \frac{\sqrt{30 P_{\text{rad}}}}{R} \tag{14-8}$$

14-6-2 Inverse Square Law

From Equation 14-7, it can be seen that the farther the wavefront moves from the source, the smaller the power density (R_a and R_c move farther apart). The total power distributed over the surface of the sphere remains the same. However, because the area of the sphere increases in direct proportion to the distance from the source squared (i.e., the radius of the sphere squared), the power density is inversely proportional to the square of the distance from the source. This relationship is called the *inverse square law.* Therefore, the power density at any point on the surface of the outer sphere is

$$\mathcal{P}_2 = \frac{P_{\text{rad}}}{4\pi R_2^2}$$

and the power density at any point on the inner sphere is

$$\mathcal{P}_1 = \frac{P_{\text{rad}}}{4\pi R_1^2}$$

Therefore,

$$\frac{\mathcal{P}_2}{\mathcal{P}_1} = \frac{P_{\text{rad}}/4\pi R_2^2}{P_{\text{rad}}/4\pi R_1^2} = \frac{R_1^2}{R_2^2} = \left(\frac{R_1}{R_2}\right)^2 \tag{14-9}$$

From Equation 14-9, it can be seen that as the distance from the source doubles the power density decreases by a factor of 2^2, or 4. When deriving the inverse square law of radiation (Equation 14-9), it was assumed that the source radiates isotropically, although it is not necessary; however, it is necessary that the velocity of propagation in all directions be uniform. Such a propagation medium is called an *isotropic medium.*

Example 14-1

For an isotropic antenna radiating 100 W of power, determine

a. Power density 1000 m from the source.
b. Power density 2000 m from the source.

Solution **a.** Substituting into Equation 14-7 yields

$$\mathcal{P}_1 = \frac{100}{4\pi 1000^2} = 7.96\ \mu\text{W/m}^2$$

b. Again, substituting into Equation 14-7 gives

$$\mathcal{P}_2 = \frac{100}{4\pi 2000^2} = 1.99\ \mu\text{W/m}^2$$

or, substituting into Equation 14-9, we have

$$\frac{\mathcal{P}_2}{\mathcal{P}_1} = \frac{1000^2}{2000^2} = 0.25$$

or

$$\mathcal{P}_2 = 7.96\ \mu\text{W/m}^2\,(0.25) = 1.99\ \mu\text{W/m}^2$$

14-7 WAVE ATTENUATION AND ABSORPTION

Free space is a vacuum, so no loss of energy occurs as a wave propagates through it. As waves propagate through free space, however, they spread out, resulting in a reduction in power density. This is called *attenuation* and occurs in free space as well as the Earth's atmosphere. Since Earth's atmosphere is not a vacuum, it contains particles that can absorb electromagnetic energy. This type of reduction of power is called *absorption loss* and does not occur in waves traveling outside Earth's atmosphere.

14-7-1 Attenuation

The inverse square law for radiation mathematically describes the reduction in power density with distance from the source. As a wavefront moves away from the source, the continuous electromagnetic field that is radiated from that source spreads out. That is, the waves move farther away from each other, and, consequently, the number of waves per unit area decreases. None of the radiated power is lost or dissipated because the wavefront is moving away from the source; the wave simply spreads out or disperses over a larger area, decreasing the power density. The reduction in power density with distance is equivalent to a power loss and is commonly called *wave attenuation.* Because the attenuation is due to the spherical spreading of the wave, it is sometimes called the *space attenuation* of the wave. Wave attenuation is generally expressed in terms of the common logarithm of the power density ratio (dB loss). Mathematically, wave attenuation (γ_a) is

$$\gamma_a = 10 \log \frac{\mathcal{P}_1}{\mathcal{P}_2} \qquad \textbf{(14-10)}$$

The reduction in power density due to the inverse square law presumes free-space propagation (a vacuum or nearly a vacuum) and is called wave attenuation. The reduction in power density due to nonfree-space propagation is called *absorption.*

14-7-2 Absorption

Earth's atmosphere is not a vacuum. It is made up of atoms and molecules of various substances, such as gases, liquids, and solids, and some of these materials are capable of absorbing electromagnetic waves. As an electromagnetic wave propagates through Earth's atmosphere, energy is transferred from the wave to the atoms and molecules of the atmosphere. As an electromagnetic wave passes through the atmosphere, it interchanges energy with free electrons and ions. If the ions do not collide with gas molecules or other ions, all the energy is converted back into electromagnetic energy, and the wave continues propagating with no loss of intensity. However, if the ions collide with other particles, they dissipate the energy that they have acquired from the electromagnetic wave, resulting in absorption of the energy. Since absorption of energy is dependent on the collision of particles, the greater the particle density, the greater the probability of collisions and the greater the absorption.

Wave absorption by the atmosphere is analogous to an I2R power loss. Once absorbed, the energy is lost forever and causes an attenuation (reduction) in the voltage and magnetic field intensities and a corresponding reduction in power density. Absorption of radio frequencies in a normal atmosphere depends on frequency and is relatively insignificant below approximately 10 GHz. Figure 14-4a shows atmospheric absorption in decibels per kilometer due to oxygen and water vapor for radio frequencies above 10 GHz. It can be seen that certain frequencies are affected more or less by absorption, creating peaks and valleys in the curves. Water vapor causes significant attenuation of electromagnetic waves at the higher frequencies. The first absorption band due to water vapor peaks at approximately 22 GHz, and the first absorption band caused by oxygen peaks at approximately 60 GHz.

Figure 14-4b shows atmospheric absorption in decibels per kilometer due to rainfall. It can be seen that the effect of rain on electromagnetic wave propagation is insignificant below approximately 6 GHz. At higher frequencies, however, rain attenuates radio transmission much more severely. The electromagnetic energy is absorbed and scattered by the raindrops, and this effect becomes even more pronounced when the length of the wave (wavelength) approaches the size of the raindrop.

Wave attenuation due to absorption depends not on the distance from the radiating source but, rather, on the total distance that the wave propagates through the atmosphere. In other words, for a *homogeneous medium* (one with uniform properties throughout), the absorption experienced during the first mile of propagation is the same as for the last mile. Also, abnormal atmospheric conditions, such as heavy rain or dense fog, absorb more energy than a normal atmosphere. Atmospheric absorption (η) for a wave propagating from R_1 to R_2 is $\gamma(R_2 - R_1)$, where γ is the absorption coefficient. Therefore, wave attenuation depends on the ratio R_2/R_1, and wave absorption depends on the distance between R_1 and R_2. In a more practical situation (i.e., an *inhomogeneous medium*), the absorption coefficient varies considerably with location, thus creating a difficult problem for radio systems engineers.

14-8 OPTICAL PROPERTIES OF RADIO WAVES

In Earth's atmosphere, ray–wavefront propagation may be altered from free-space behavior by *optical* effects such as *refraction, reflection, diffraction,* and *interference.* Using rather unscientific terminology, refraction can be thought of as *bending,* reflection as *bouncing,* diffraction as *scattering,* and interference as *colliding.* Refraction, reflection, diffraction, and interference are called optical properties because they were first observed in the science of optics, which is the behavior of light waves. Because light waves are high-frequency electromagnetic waves, it stands to reason that optical properties will also apply to radio-wave propagation. Although optical principles can be analyzed completely by application of Maxwell's equations, this is necessarily complex. For most applications, *geometric ray tracing* can be substituted for analysis by Maxwell's equations.

14-8-1 Refraction

Refraction is sometimes referred to as the bending of the radio-wave path. However, the ray does not actually bend. Electromagnetic refraction is actually the changing of direction of an electromagnetic ray as it passes obliquely from one medium into another with different

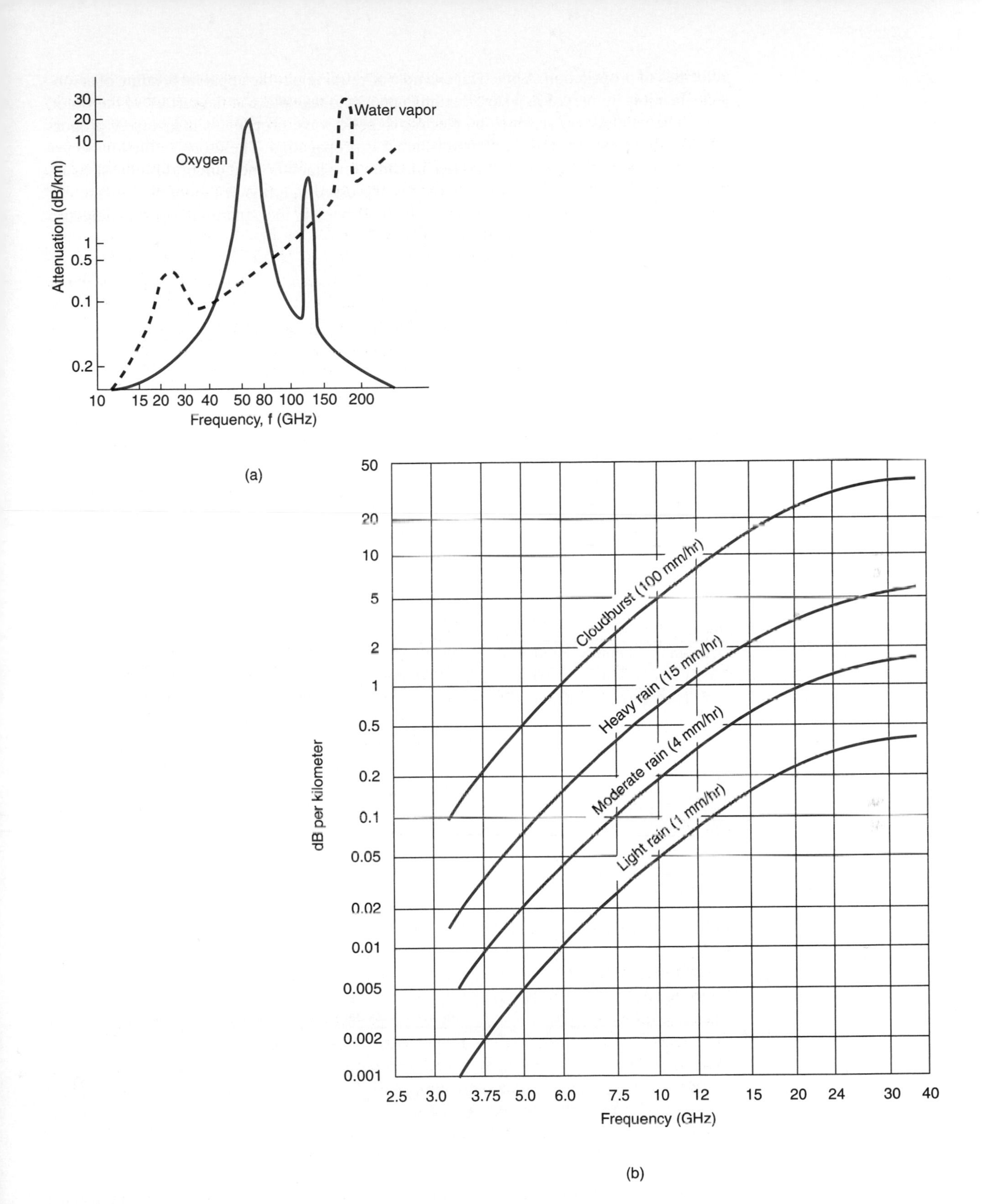

FIGURE 14-4 Atmospheric absorption of electromagnetic waves: (a) water vapor and oxygen; and (b) rainfall

velocities of propagation. A practical example of refraction is the apparent bending of an object when it is immersed in water. Bending appears to take place at the surface of the water.

The velocity (v) at which an electromagnetic wave propagates is inversely proportional to the density of the medium in which it is propagating. Therefore, refraction occurs whenever a radio wave passes from one medium into another medium of different density. Refraction of electromagnetic waves can be expressed in terms of the refractive index of the atmosphere it is passing through. Refractive index is the square root of the dielectric constant and is expressed mathematically as

$$n = \sqrt{k} \tag{14-11}$$

where n = the refractive index (unitless)
k = equivalent dielectric constant relative to free space (vacuum)

and

$$k = \sqrt{1 - \frac{81N}{f^2}} \tag{14-12}$$

where N = number of electrons per cubic centimeter
f = frequency (kHz)

Examination of Equation 14-12 shows that real values of the refractive index will always be less than unity. It can also be seen that the deviation of the refractive index from unity becomes greater the lower the frequency and the higher the electron density. When $f^2 < 81N$, the refractive index is imaginary, and the atmosphere is unable to propagate the electromagnetic wave without attenuation.

Figure 14-5 shows refraction of a wavefront at a *plane* boundary between two media with different densities. For this example, medium 1 is less dense than medium 2 ($v_1 > v_2$). It can be seen that ray A enters the more dense medium before ray B. Therefore, ray B propagates more rapidly than ray A and travels distance B–B' during the same time that ray A travels distance A–A'. Therefore, wavefront ($A'B'$) is *tilted* or bent in a downward direction. Because a ray is defined as being perpendicular to the wavefront at all points, the rays in Figure 14-5 have changed direction at the interface of the two media. Whenever a ray passes from a less dense to a

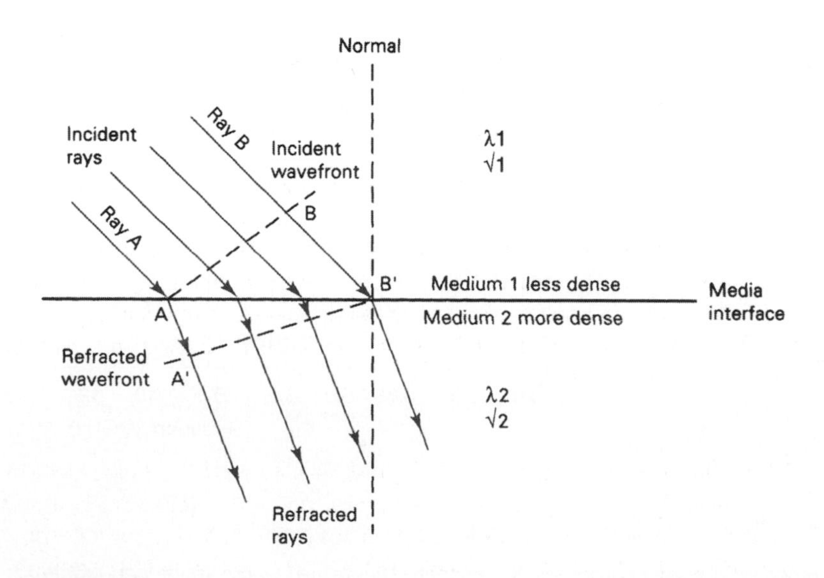

FIGURE 14-5 Refraction at a plane boundary between two media

more dense medium, it is effectively bent toward the *normal.* (The normal is simply an imaginary line drawn perpendicular to the interface at the point of incidence.) Conversely, whenever a ray passes from a more dense to a less dense medium, it is effectively bent away from the normal. The *angle of incidence* is the angle formed between the incident wave and the normal, and the *angle of refraction* is the angle formed between the refracted wave and the normal.

The amount of bending or refraction that occurs at the interface of two materials of different densities is quite predictable and depends on the *refractive index* (also called the *index of refraction*) of the two materials. The refractive index is simply the ratio of the velocity of propagation of a light ray in free space to the velocity of propagation of a light ray in a given material. Mathematically, the refractive index is

$$n = \frac{c}{v} \tag{14-13}$$

where n = refractive index (unitless)
c = speed of light in free space (3×10^8 m/s)
v = speed of light in a given material (meters per second)

The refractive index is also a function of frequency. However, the variation in most applications is insignificant and, therefore, is omitted from this discussion. How an electromagnetic wave reacts when it meets the interface of two transmissive materials that have different indexes of refraction can be explained with *Snell's law,* which simply states that

$$n_1 \sin \theta_1 = n_2 \sin \theta_2 \tag{14-14}$$

and

$$\frac{\sin \theta_1}{\sin \theta_2} = \frac{n_2}{n_1} \tag{14-15}$$

where n_1 = refractive index of material 1
n_2 = refractive index of material 2
θ_1 = angle of incidence (degrees)
θ_2 = angle of refraction (degrees)

and because the refractive index of a material is equal to the square root of its dielectric constant,

$$\frac{\sin \theta_1}{\sin \theta_2} = \sqrt{\frac{\epsilon_{r2}}{\epsilon_{r1}}} \tag{14-16}$$

where ϵ_{r1} = dielectric constant of medium 1
ϵ_{r2} = dielectric constant of medium 2

Refraction also occurs when a wavefront propagates in a medium that has a *density gradient* that is perpendicular to the direction of propagation (i.e., parallel to the wavefront). Figure 14-6 shows wavefront refraction in a transmission medium that has a gradual variation in its refractive index. The medium is more dense near the bottom and less dense at the top. Therefore, rays traveling near the top travel faster than rays near the bottom and, consequently, the wavefront tilts downward. The tilting occurs in a gradual fashion as the wave progresses, as shown.

14-8-2 Reflection

Reflect means to cast or turn back, and *reflection* is the act of reflecting. Electromagnetic reflection occurs when an incident wave strikes a boundary of two media and some or all of the incident power does not enter the second material. The waves that do not penetrate the second medium are reflected. Figure 14-7 shows electromagnetic wave reflection at a plane boundary between two media. Because all the reflected waves remain in medium 1,

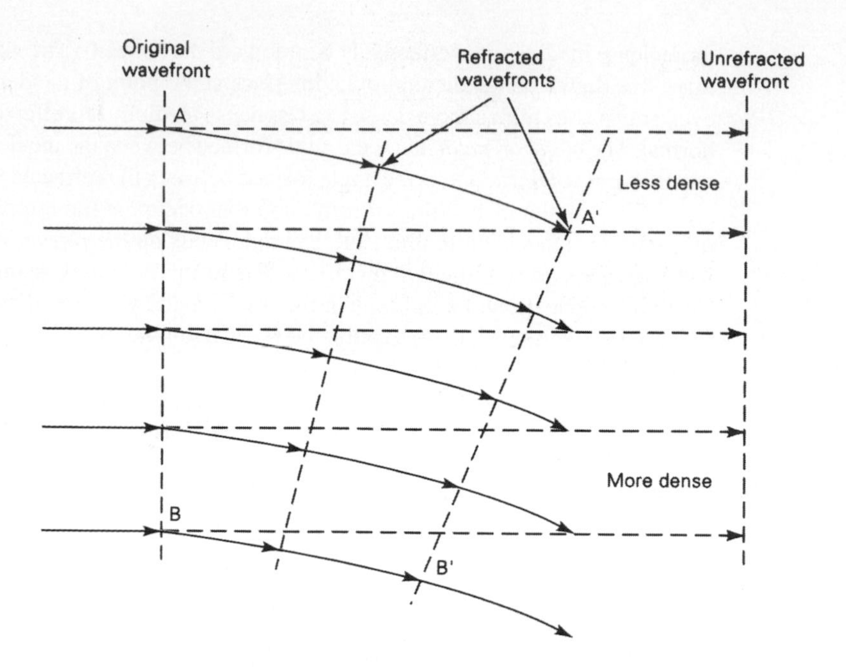

FIGURE 14-6 Wavefront refraction in a gradient medium

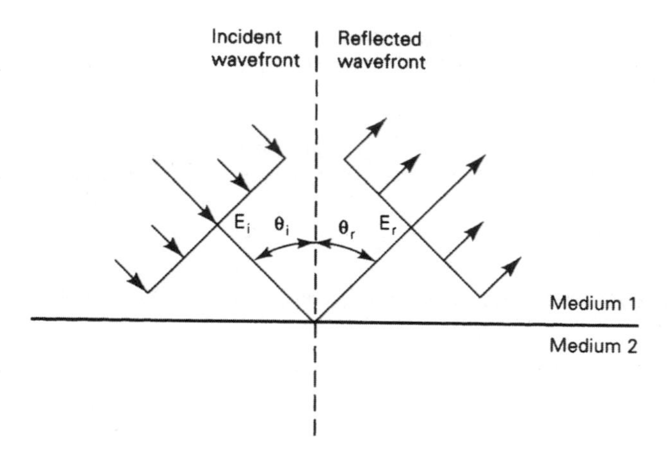

FIGURE 14-7 Electromagnetic reflection at a plane boundary of two media

the velocities of the reflected and incident waves are equal. Consequently, the *angle of reflection* equals the *angle of incidence* ($\theta_i = \theta_r$). However the reflected voltage field intensity is less than the incident voltage field intensity. The ratio of the reflected to the incident voltage intensities is called the *reflection coefficient,* Γ (sometimes called the *coefficient of reflection*). For a perfect conductor, $\Gamma = 1$. Γ is used to indicate both the relative amplitude of the incident and reflected fields and the phase shift that occurs at the point of reflection. Mathematically, the reflection coefficient is

$$\Gamma = \frac{E_r e^{j\theta_r}}{E_i e^{j\theta_i}} \tag{14-17}$$

$$\Gamma = \frac{E_r}{E_i} e^{j(\theta_r - \theta_i)} \tag{14-18}$$

where Γ = reflection coefficient (unitless)
E_i = incident voltage intensity (volts)
E_r = reflected voltage intensity (volts)
θ_i = incident phase (degrees)
θ_r = reflected phase (degrees)

The ratio of the reflected and incident power densities is Γ. The portion of the total incident power that is not reflected is called the *power transmission coefficient* (T) (or simply the *transmission coefficient*). For a perfect conductor, $T = 0$. The *law of conservation of energy* states that for a perfect reflective surface, the total reflected power must equal the total incident power. Therefore,

$$T + |\Gamma| = 1 \quad \textbf{(14-19)}$$

For imperfect conductors, both $|\Gamma|$ and T are functions of the angle of incidence, the electric field polarization, and the dielectric constants of the two materials. If medium 2 is not a perfect conductor, some of the incident waves penetrate it and are absorbed. The absorbed waves set up currents in the resistance of the material, and the energy is converted to heat. The fraction of power that penetrates medium 2 is called the *absorption coefficient* (or sometimes the *coefficient of absorption*).

When the reflecting surface is not plane (i.e., it is curved), the curvature of the reflected wave is different from that of the incident wave. When the wavefront of the incident wave is curved and the reflective surface is plane, the curvature of the reflected wavefront is the same as that of the incident wavefront.

Reflection also occurs when the reflective surface is *irregular* or *rough;* however, such a surface may destroy the shape of the wavefront. When an incident wavefront strikes an irregular surface, it is randomly scattered in many directions. Such a condition is called *diffuse reflection,* whereas reflection from a perfectly smooth surface is called *specular* (mirrorlike) *reflection.* Surfaces that fall between smooth and irregular are called *semirough surfaces.* Semirough surfaces cause a combination of diffuse and specular reflection. A semirough surface will not totally destroy the shape of the reflected wavefront. However, there is a reduction in the total power. The *Rayleigh criterion* states that a semirough surface will reflect as if it were a smooth surface whenever the cosine of the angle of incidence is greater than $\lambda/8d$, where d is the depth of the surface irregularity and λ is the wavelength of the incident wave. Reflection from a semirough surface is shown in Figure 14-8. Mathematically, Rayleigh's criterion is

$$\cos \theta_i > \frac{\lambda}{8d} \quad \textbf{(14-20)}$$

14-8-3 Diffraction

Diffraction is defined as the modulation or redistribution of energy within a wavefront when it passes near the edge of an *opaque* object. Diffraction is the phenomenon that allows light or radio waves to propagate (*peek*) around corners. The previous discussions of refraction and reflection assumed that the dimensions of the refracting and reflecting surfaces were large with respect to a wavelength of the signal. However, when a wavefront passes near an obstacle or discontinuity with dimensions comparable in size to a wavelength, simple geometric analysis cannot be used to explain the results, and *Huygens's principle* (which is deduced from Maxwell's equations) is necessary.

Huygens's principle states that every point on a given spherical wavefront can be considered as a secondary point source of electromagnetic waves from which other secondary waves (wavelets) are radiated outward. Huygens's principle is illustrated in Figure 14-9. Normal wave propagation considering an infinite plane is shown in Figure 14-9a. Each second-

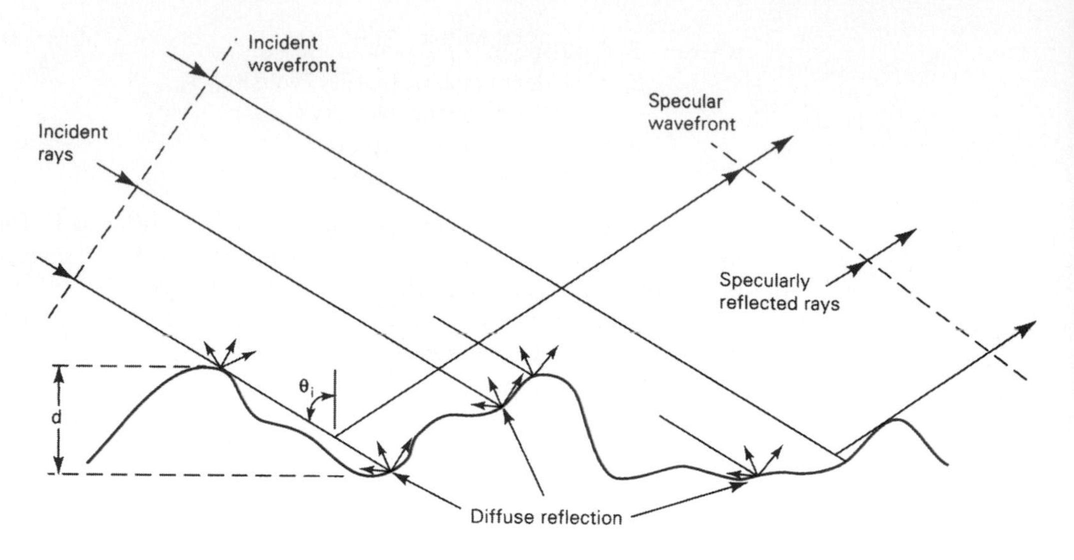

FIGURE 14-8 Reflection from a semirough surface

ary point source (p_1, p_2, and so on) radiates energy outward in all directions. However, the wavefront continues in its original direction rather than spreading out because cancellation of the secondary wavelets occurs in all directions except straight forward. Therefore, the wavefront remains plane.

When a finite plane wavefront is considered, as in Figure 14-9b, cancellation in random directions is incomplete. Consequently, the wavefront spreads out, or *scatters.* This scattering effect is called *diffraction.* Figure 14-9c shows diffraction around the edge of an obstacle. It can be seen that wavelet cancellation occurs only partially. Diffraction occurs around the edge of the obstacle, which allows secondary waves to "sneak" around the corner of the obstacle into what is called the *shadow zone.* This phenomenon can be observed when a door is opened into a dark room. Light rays diffract around the edge of the door and illuminate the area behind the door.

14-8-4 Interference

Interfere means to come into opposition, and *interference* is the act of interfering. Radio-wave interference occurs when two or more electromagnetic waves combine in such a way that system performance is degraded. Refraction, reflection, and diffraction are categorized as geometric optics, which means that their behavior is analyzed primarily in terms of rays and wavefronts. Interference, on the other hand, is subject to the principle of *linear superposition* of electromagnetic waves and occurs whenever two or more waves simultaneously occupy the same point in space. The principle of linear superposition states that the total voltage intensity at a given point in space is the sum of the individual wave vectors. Certain types of propagation media have nonlinear properties; however, in an ordinary medium (such as air or Earth's atmosphere), linear superposition holds true.

Figure 14-10 shows the linear addition of two instantaneous voltage vectors whose phase angles differ by angle θ. It can be seen that the total voltage is not simply the sum of the two vector magnitudes but rather the phasor addition of the two. With free-space propagation, a phase difference may exist simply because the *electromagnetic polarizations* of two waves differ. Depending on the phase angles of the two vectors, either addition or subtraction can occur. (This implies simply that the result may be more or less than either vector because the two electromagnetic waves can reinforce or cancel.)

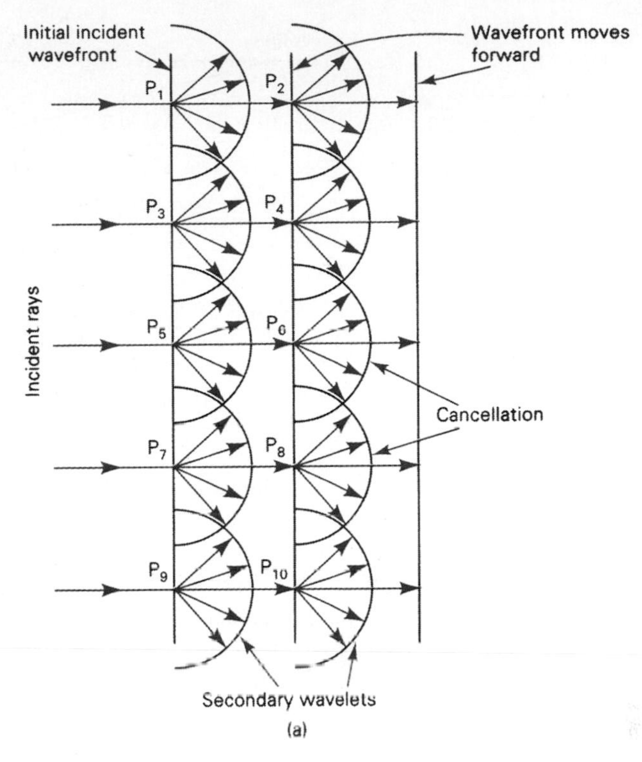

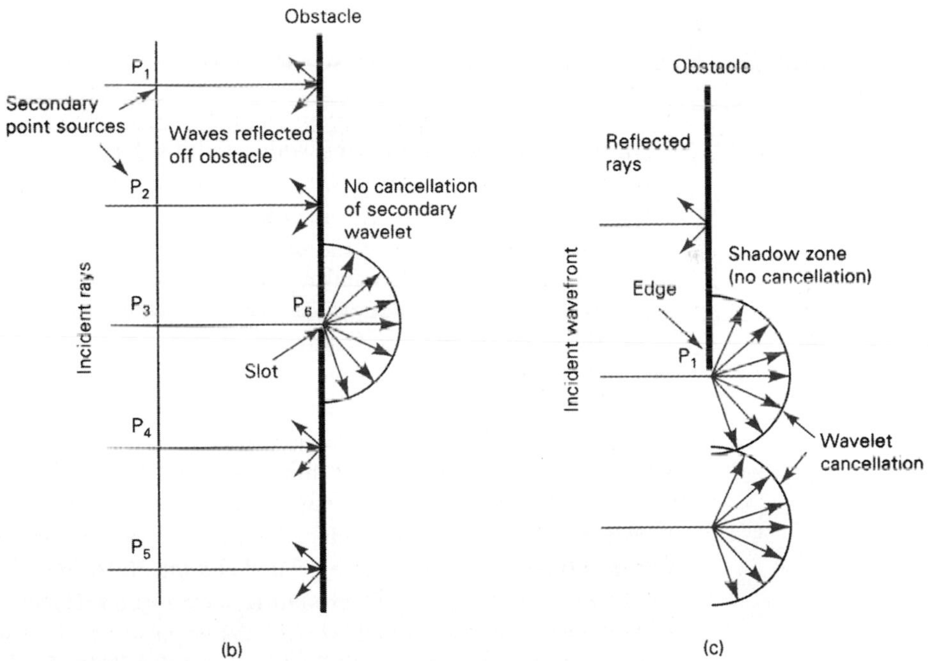

FIGURE 14-9 Electromagnetic wave diffraction: (a) Huygens's principle for a plane wavefront; (b) finite wavefront through a slot; (c) around an edge

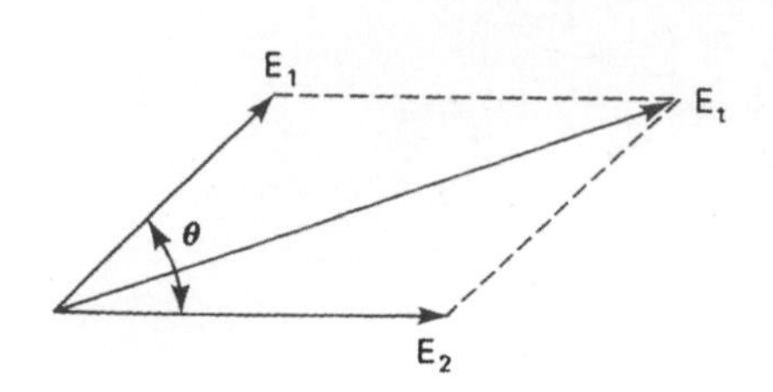

FIGURE 14-10 Linear addition of two vectors with differing phase angles

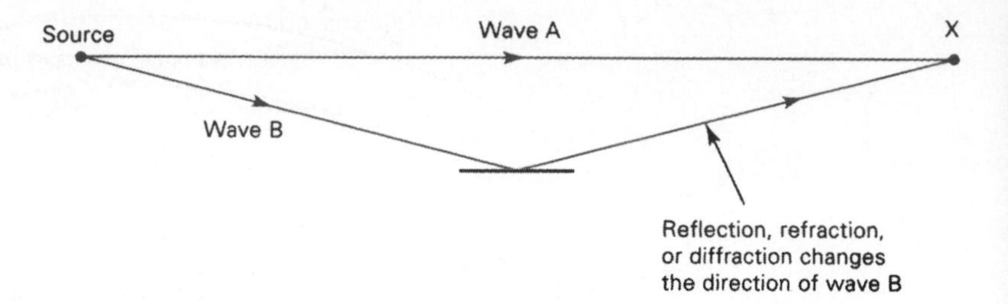

FIGURE 14-11 Electromagnetic wave interference

Figure 14-11 shows interference between two electromagnetic waves in free space. It can be seen that at point *X,* the two waves occupy the same area of space. However, wave *B* has traveled a different path than wave *A* and, therefore, their relative phase angles may be different. If the difference in distance traveled is an odd integral multiple of one-half wavelength, reinforcement takes place. If the difference is an even integral multiple of one-half wavelength, total cancellation occurs. More likely, the difference in distance falls somewhere between the two, and partial cancellation occurs. For frequencies below VHF, the relatively large wavelengths prevent interference from being a significant problem. However, with UHF and above, wave interference can be severe.

14-9 TERRESTRIAL PROPAGATION OF ELECTROMAGNETIC WAVES

Electromagnetic waves traveling within Earth's atmosphere are called *terrestrial waves,* and communications between two or more points on Earth is called *terrestrial radio communications.* Terrestrial waves are influenced by the atmosphere and Earth itself. In terrestrial radio communications, electromagnetic waves can be propagated in several ways, depending on the type of system and the environment. As previously explained, electromagnetic waves travel in straight lines except when Earth and its atmosphere alter their path. Essentially, there are three ways of propagating electromagnetic waves within Earth's atmosphere: ground wave, space wave, and sky wave propagation.

Figure 14-12 shows the three modes of propagation possible between two radio terrestrial antennas. Path 1 shows a direct or free-space wave, and path 2 shows a ground-reflected wave. Direct and ground-reflected waves together are called *space waves.* Path 3 shows a surface wave, which consists of the electric and magnetic fields associated with the currents induced in the ground. The magnitude of the ground current depends on the constants of the ground and the electromagnetic wave polarization. The cumulative sum of the direct, ground-reflected, and surface waves is sometimes referred to as the *ground wave,* which is confusing because a surface wave by itself is also sometimes called a ground wave. Path 4 is called the *sky wave,* which depends on the presence of the ionized layers above Earth that return some of the energy that otherwise would be lost in outer space.

Each of the four propagation modes exists in every radio system; however, some are negligible in certain frequency ranges or over a particular type of terrain. At frequencies below approximately 2 MHz, surface waves provide the best coverage because ground losses increase rapidly with frequency. Sky waves are used for high-frequency applications, and space waves are used for very high frequencies and above.

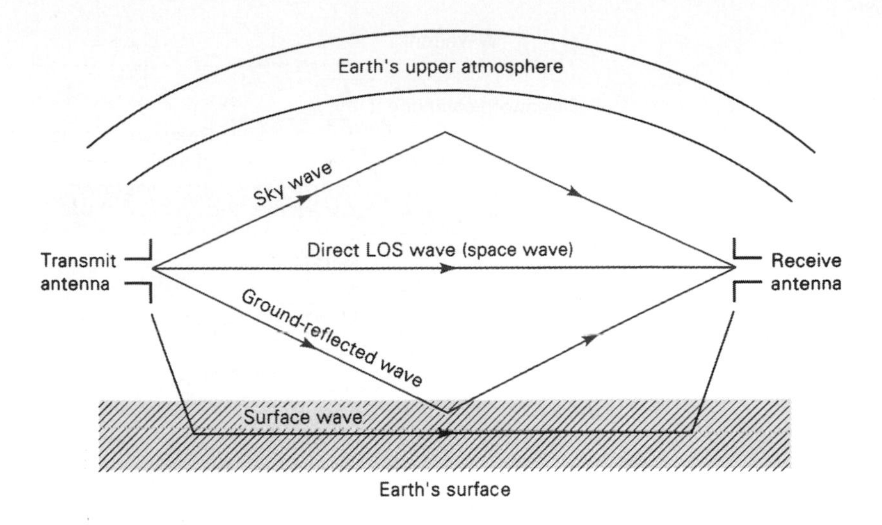

FIGURE 14-12 Normal modes of wave propagation

14-9-1 Surface Wave Propagation

A *surface wave* is an Earth-guided electromagnetic wave that travels over the surface of Earth. As a surface wave moves over Earth's surface, it is accompanied by charges induced in the Earth. The charges move with the wave, producing a current. Since the Earth offers resistance to the flow of current, energy is dissipated in a manner very similar to those in a transmission line. Earth's surface also has dielectric losses. Therefore, surface waves are attenuated as they propagate. Because energy is absorbed from the surface wave, the portion of the wave in contact with Earth's surface is continuously wiped out. The energy is replenished by diffraction of energy downward from the portions of the ground wave immediately above Earth's surface. This phenomenon produces a slight forward tilt in the wavefront as shown in Figure 14-13a.

Attenuation of the surface wave due to absorption depends on the conductivity of Earth's surface and the frequency of the electromagnetic wave. Surface waves propagate best over a good conductor. Table 14-1 lists the relative conductivity of various Earth surfaces. From the table, it is apparent that the best surface wave transmission occurs over seawater and that the highest degree of attenuation is over jungle areas. Attenuation over all types of terrain increases rapidly with frequency. Extremely high losses make it impractical to use surface waves for long-distance transmission of high-frequency electromagnetic waves.

Ground waves must be vertically polarized because the electric field in a horizontally polarized wave would be parallel to Earth's surface, and such waves would be short-circuited by the conductivity of the ground.

Figure 13-9b shows surface wave propagation. Earth's atmosphere has a gradient density (i.e., the density decreases gradually with distance from Earth's surface), which also causes the wavefront to tilt progressively forward. Therefore, the wave propagates around the Earth, remaining close to its surface, and if enough power is transmitted, the wavefront could propagate beyond the horizon or even around the entire circumference of the Earth. However, care must be taken when selecting the frequency and terrain over which surface waves will propagate to ensure that the wavefront does not tilt excessively and simply turn over, lie flat on the ground, and cease to propagate.

Surface wave propagation is commonly used for ship-to-ship and ship-to-shore communications, for radio navigation, and for maritime mobile communications. Surface waves are used at frequencies as low as 15 kHz.

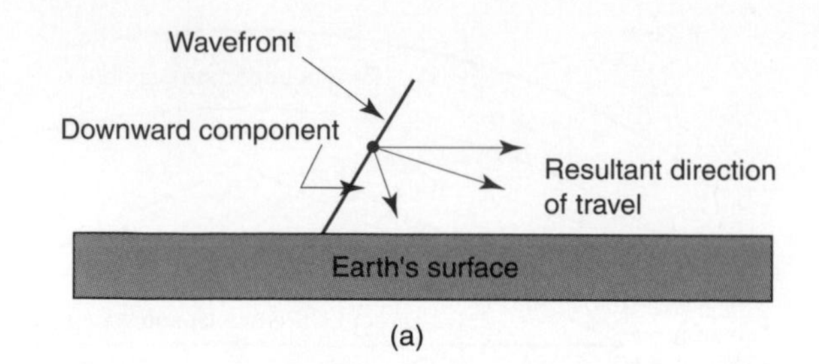

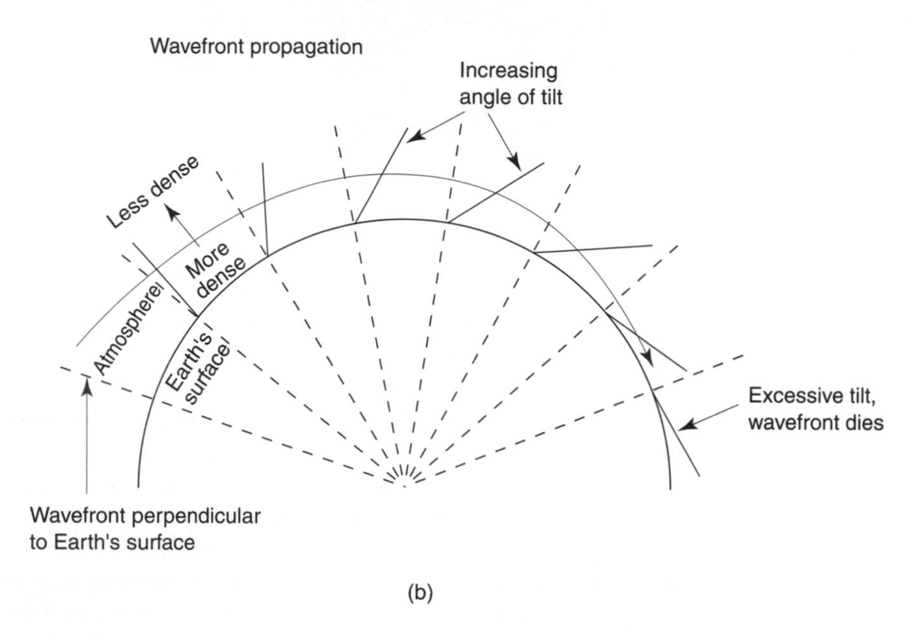

FIGURE 14-13 Surface waves: (a) movement of surface wave over Earth's surface; (b) surface wave propagation

Table 14-1 Relative Conductivity of Earth Surfaces

Surface	Relative Conductivity
Seawater	Good
Flat, loamy soil	Fair
Large bodies of freshwater	Fair
Rocky terrain	Poor
Desert	Poor
Jungle	Unusable

The disadvantages of surface waves are as follows:

Ground waves require a relatively high transmission power.
Ground waves are limited to very low, low, and medium frequencies (VLF, LF, and MF) requiring large antennas (the reason for this is explained in Chapter 11).
Ground losses vary considerably with surface material and composition.

The advantages of ground wave propagation are as follows:

Given enough transmit power, ground waves can be used to communicate between any two locations in the world.
Ground waves are relatively unaffected by changing atmospheric conditions.

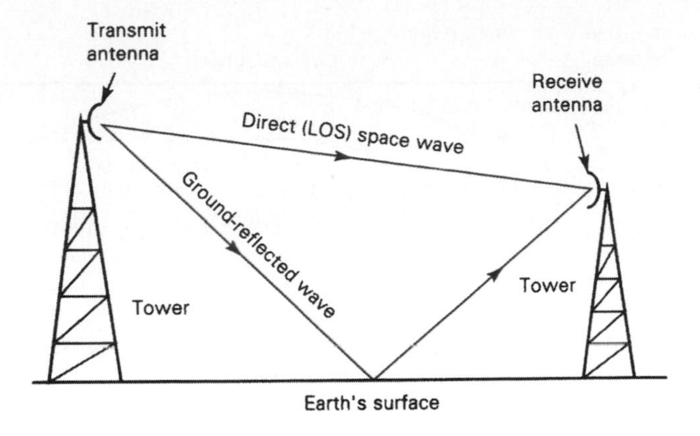

FIGURE 14-14 Space-wave propagation

14-9-2 Space Wave Propagation

Space wave propagation of electromagnetic energy includes radiated energy that travels in the lower few miles of Earth's atmosphere. Space waves include both direct and ground-reflected waves (see Figure 14-14). *Direct waves* travel essentially in a straight line between the transmit and receive antennas. Space wave propagation with direct waves is commonly called *line-of-sight* (LOS) *transmission.* Therefore, direct space wave propagation is limited by the curvature of the Earth. Ground-reflected waves are waves reflected by Earth's surface as they propagate between the transmit and receive antennas.

Figure 14-14 shows space wave propagation between two antennas. It can be seen that the field intensity at the receive antenna depends on the distance between the two antennas (attenuation and absorption) and whether the direct and ground-reflected waves are in phase (interference).

The curvature of Earth presents a horizon to space wave propagation commonly called the *radio horizon.* Because of atmospheric refraction, the radio horizon extends beyond the *optical horizon* for the common *standard atmosphere.* The radio horizon is approximately four-thirds that of the optical horizon. Refraction is caused by the troposphere because of changes in its density, temperature, water vapor content, and relative conductivity. The radio horizon can be lengthened simply by elevating the transmit or receive antennas (or both) above Earth's surface with towers or by placing them on top of mountains or high buildings.

Figure 14-15 shows the effect of antenna height on the radio horizon. The line-of-sight radio horizon for a single antenna at sea level is given as

$$d = \sqrt{2h} \tag{14-21}$$

where d = distance to radio horizon (miles)
h = antenna height above sea level (feet)

Therefore, for a transmit and receive antenna, the distance between the two antennas at sea level is

$$d = d_t + d_r$$

or

$$d = \sqrt{2h_t} + \sqrt{2h_r} \tag{14-22}$$

where d = total distance (miles)
d_t = radio horizon for transmit antenna (miles)
d_r = radio horizon for receive antenna (miles)
h_t = transmit antenna height (feet)
h_r = receive antenna height (feet)

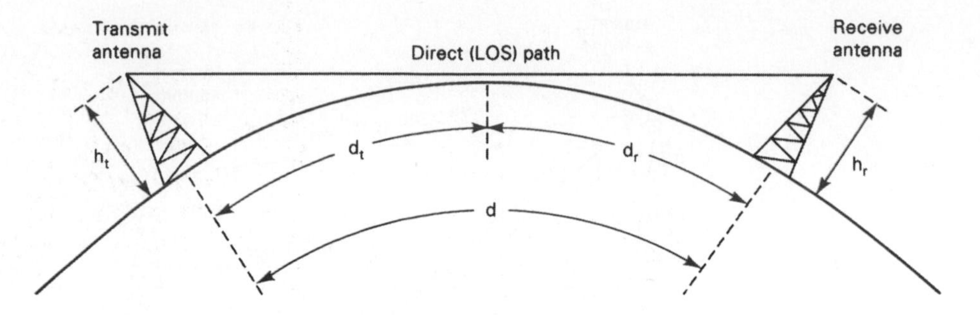

FIGURE 14-15 Space waves and radio horizon

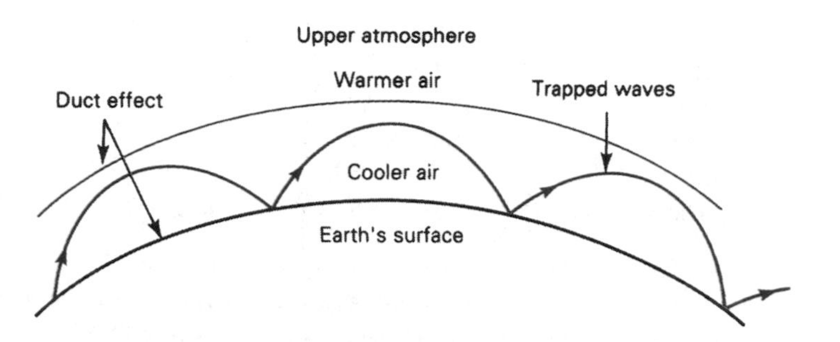

FIGURE 14-16 Duct propagation

The maximum distance between a transmitter and a receiver over average terrain can be approximated in metric units by the following equation:

$$d_{(max)} = \sqrt{17h_t} + \sqrt{17h_r} \quad \textbf{(14-23)}$$

where $d_{(max)}$ = maximum distance between transmitter and receiver (kilometers)
h_t = height of transmit antenna above sea level (meters)
h_r = height of receive antenna above sea level (meters)

From Equations 14-22 and 14-23, it can be seen that the space wave propagation distance can be extended simply by increasing either the transmit or the receive antenna height or both.

Because the conditions in Earth's lower atmosphere are subject to change, the degree of refraction can vary with time. A special condition called *duct propagation* occurs when the density of the lower atmosphere is such that electromagnetic waves are trapped between it and Earth's surface. The layers of the atmosphere act as a duct, and an electromagnetic wave can propagate for great distances around the curvature of Earth within this duct. Duct propagation is shown in Figure 14-16.

14-9-3 Sky Wave Propagation

Electromagnetic waves that are directed above the horizon level are called *sky waves*. Typically, sky waves are radiated in a direction that produces a relatively large angle with reference to Earth. Sky waves are radiated toward the sky, where they are either reflected or refracted back to Earth by the ionosphere. Because of this, sky wave propagation is sometimes called *ionospheric propagation*. The ionosphere is the region of space located approximately 50 km to 400 km (30 mi to 250 mi) above Earth's surface. The ionosphere is the upper portion of Earth's atmosphere. Therefore, it absorbs large quantities of the sun's radiant energy, which ionizes the air molecules, creating free electrons. When a radio wave

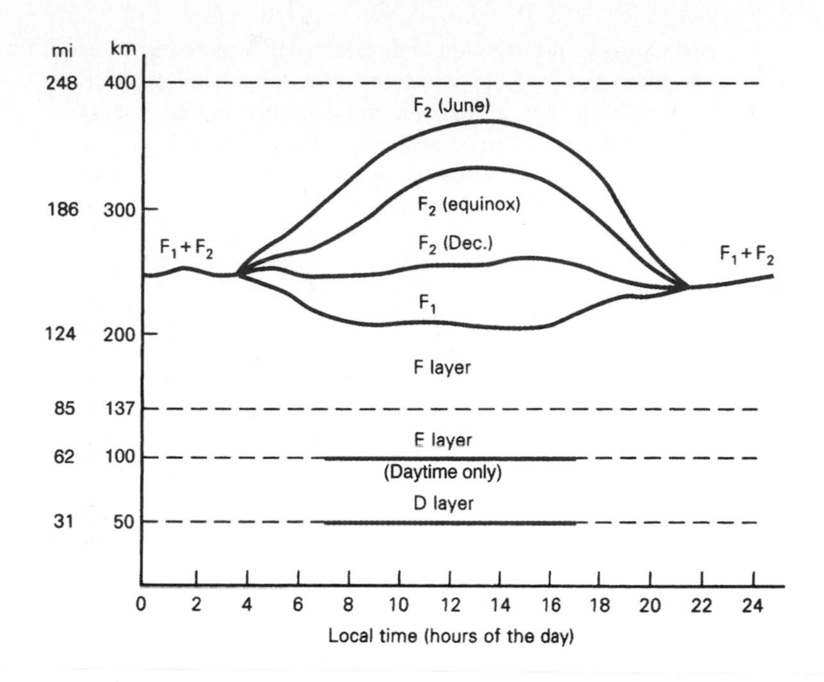

FIGURE 14-17 Ionospheric layers

passes through the ionosphere, the electric field of the wave exerts a force on the free electrons, causing them to vibrate. The vibrating electrons decrease current, which is equivalent to reducing the dielectric constant. Reducing the dielectric constant increases the velocity of propagation and causes electromagnetic waves to bend away from the regions of high electron density toward regions of low electron density (i.e., increasing refraction). As the wave moves farther from Earth, ionization increases; however, there are fewer air molecules to ionize. Therefore, the upper atmosphere has a higher percentage of ionized molecules than the lower atmosphere. The higher the ion density, the more refraction. Also, because of the ionosphere's nonuniform composition and its temperature and density variations, it is *stratified.* Essentially, three layers make up the ionosphere (the D, E, and F layers) and are shown in Figure 14-17. It can be seen that all three layers of the ionosphere vary in location and in *ionization density* with the time of day. They also fluctuate in a cyclic pattern throughout the year and according to the 11-year *sunspot cycle.* The ionosphere is most dense during times of maximum sunlight (during the daylight hours and in the summer).

14-9-3-1 D layer. The *D layer* is the lowest layer of the ionosphere and is located approximately between 30 miles and 60 miles (50 km to 100 km) above Earth's surface. Because it is the layer farthest from the sun, there is little ionization. Therefore, the D layer has very little effect on the direction of propagation of radio waves. However, the ions in the D layer can absorb appreciable amounts of electromagnetic energy. The amount of ionization in the D layer depends on the altitude of the sun above the horizon. Therefore, it disappears at night. The D layer reflects VLF and LF waves and absorbs MF and HF waves. (See Table 1-6 for VLF, LF, MF, and HF frequency regions.)

14-9-3-2 E layer. The *E layer* is located approximately between 60 miles and 85 miles (100 km to 140 km) above Earth's surface. The E layer is sometimes called the *Kennelly-Heaviside layer* after the two scientists who discovered it. The E layer has its maximum density at approximately 70 miles at noon, when the sun is at its highest point. As with the D layer, the E layer almost totally disappears at night. The E layer aids MF surface wave propagation and reflects HF waves somewhat during the daytime. The upper portion of the E layer is sometimes considered separately and is called the *sporadic* E layer because it seems to come and go rather unpredictably. The sporadic E layer is caused by *solar flares* and *sunspot activity.* The sporadic E layer is a thin layer with a very high ionization density. When it appears, there generally is an unexpected improvement in long-distance radio transmission.

14-9-3-3 F layer. The *F layer* is actually made up of two layers, the F_1 and F_2 layers. During the daytime, the F_1 layer is located between 85 miles and 155 miles (140 km to 250 km) above Earth's surface; the F_2 layer is located 85 miles to 185 miles (140 km to 300 km) abovc Earth's surface during the winter and 155 miles to 220 miles (250 km to 350 km) in the summer. During the night, the F_1 layer combines with the F_2 layer to form a single layer. The F_1 layer absorbs and attenuates some HF waves, although most of the waves pass through to the F_2 layer, where they are refracted back to Earth.

14-10 PROPAGATION TERMS AND DEFINITIONS

14-10-1 Critical Frequency and Critical Angle

Frequencies above the UHF range are virtually unaffected by the ionosphere because of their extremely short wavelengths. At these frequencies, the distances between ions are appreciably large and, consequently, the electromagnetic waves pass through them with little noticeable effect. Therefore, it stands to reason that there must be an upper frequency limit for sky wave propagation. *Critical frequency* (f_c) is defined as the highest frequency that can be propagated directly upward and still be returned to Earth by the ionosphere. The critical frequency depends on the ionization density and, therefore, varies with the time of day and the season. If the vertical angle of radiation is decreased, frequencies at or above the critical frequency can still be refracted back to Earth's surface because they will travel a longer distance in the ionosphere and, thus, have a longer time to be refracted. Therefore, critical frequency is used only as a point of reference for comparison purposes. However, every frequency has a maximum vertical angle at which it can be propagated and still be refracted back by the ionosphere. This angle is called the *critical angle.* The critical angle θ_c is shown in Figure 14-18.

A measurement technique called *ionospheric sounding* is sometimes used to determine the critical frequency. A signal is propagated straight up from the Earth's surface and gradually increased in frequency. At the lower frequencies, the signal will be completely absorbed by the atmosphere. As the frequency is increased, however, it (or some portion of it) will be returned to Earth. At some frequency, however, the signal will pass through the Earth's atmosphere into outer space and not return to Earth. The highest frequency that will be returned to Earth in the vertical direction is the critical frequency.

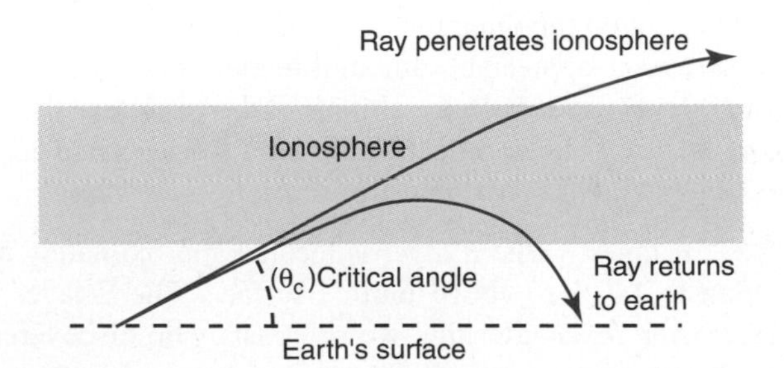

FIGURE 14-18 Critical angle

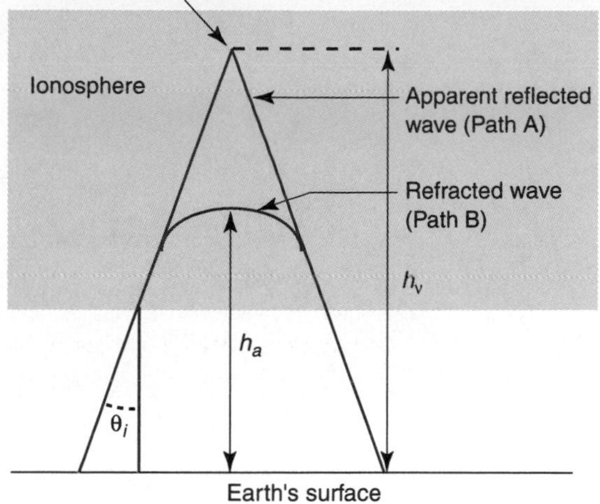

FIGURE 14-19 Virtual and actual height

14-10-2 Virtual Height

Virtual height is the height above Earth's surface from which a refracted wave appears to have been reflected. Figure 14-19 shows a wave that has been radiated from Earth's surface toward the ionosphere. The radiated wave is refracted back to Earth and follows path B. The actual maximum height that the wave reached is height h_a. However, path A shows the projected path that a reflected wave could have taken and still been returned to Earth at the same location. The maximum height that this hypothetical reflected wave would have reached is the virtual height (h_v).

14-10-3 Maximum Usable Frequency

The *maximum usable frequency* (MUF) is the highest frequency that can be used for sky wave propagation between two specific points on Earth's surface. It stands to reason, then, that there are as many values possible for MUF as there are points on Earth and frequencies—an infinite number. MUF, as with the critical frequency, is a limiting frequency for sky wave propagation. However, the maximum usable frequency is for a specific angle of incidence (shown in Figure 14-19). Mathematically, MUF is

$$\text{MUF} = \frac{\text{critical frequency}}{\cos \theta_i} \tag{14-24}$$

$$= \text{critical frequency} \times \sec \theta_i \tag{14-25}$$

where θ_i is the angle of incidence.

Equation 14-24 is called the *secant law*. The secant law assumes a flat Earth and a flat reflecting layer which, of course, can never exist. Therefore, MUF is used only for making preliminary calculations.

Because of the general instability of the ionosphere, the highest frequency used between two points is often selected lower than the MUF. It has been proven that operating at a frequency 85% of the MUF provides more reliable communications. This frequency is sometimes called the *optimum working frequency* (OWF).

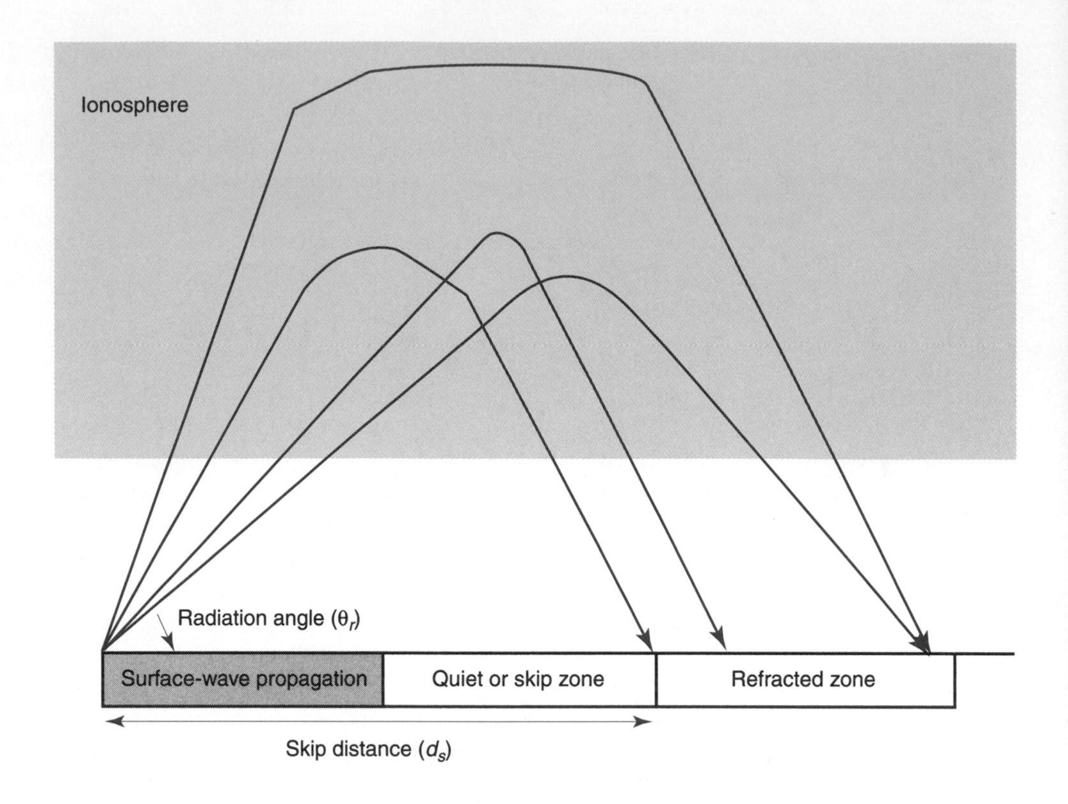

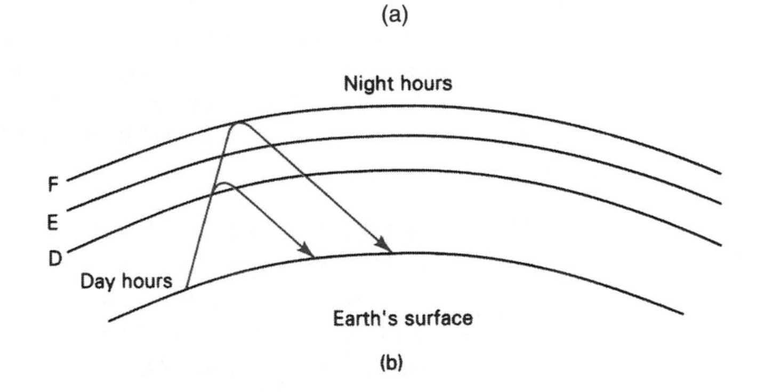

FIGURE 14-20 (a) Skip distance; (b) daytime-versus-nighttime propagation

14-10-4 Skip Distance and Skip Zone

Skip distance (d_s) is defined as the minimum distance from a transmit antenna that a sky wave at a given frequency will be returned to Earth. The frequency must be less than the maximum usable frequency and propagated at its critical angle. Figure 14-20a shows several rays with different radiation angles being radiated from the same point on Earth. It can be seen that the distance to where the wave returns to Earth moves closer to the transmit antenna as the radiation angle increases. When the radiation angle (θ_r) exceeds the critical angle (θ_c), the wave penetrates the ionosphere and escapes Earth's atmosphere.

At distances greater than the skip distance, two rays can take different paths and still be returned to the same point on Earth. The two rays are called the *lower ray* and the *upper,* or *Pedersen, ray.* The Pedersen ray is usually of little significance, as it tends to be much weaker than the lower ray because it spreads over a much larger area than the lower ray. The Pedersen ray becomes important when circumstances prevent the lower ray from reaching a particular point.

The area between where the surface waves are completely dissipated and the point where the first sky wave returns to Earth is called the *quiet,* or *skip, zone* because in this area there is no reception. Each frequency may have a different skip distance and skip zone.

Figure 14-20b shows the effect on the skip distance of the disappearance of the D and E layers during nighttime. Effectively, the *ceiling* formed by the ionosphere is raised, allowing sky waves to travel higher before being returned to Earth. This effect explains how faraway radio stations are sometimes heard during nighttime hours that cannot be heard during daylight hours.

14-11 FREE-SPACE PATH LOSS

Free-space path loss is often defined as the loss incurred by an electromagnetic wave as it propagates in a straight line through a vacuum with no absorption or reflection of energy from nearby objects. Free-space path loss is a misstated and often misleading definition because no energy is actually dissipated. Free-space path loss is a fabricated engineering quantity that evolved from manipulating communications system link budget equations into a particular format (link budget equations are covered in Chapter 23). The link equations include transmit antenna gain, free-space path loss, and the effective area of the receiving antenna (i.e., the receiving antenna gain). The manipulation of antenna gain terms results is a distance- and frequency-dependent term called free-space path loss (the relationship between antenna gain and path loss is shown in Chapter 15).

Free-space path loss assumes ideal atmospheric conditions so that no electromagnetic energy is actually lost or dissipated—it merely spreads out as it propagates away from the source, resulting in lower relative power densities. A more appropriate term for the phenomena is *spreading loss.* Spreading loss occurs simply because of the inverse square law. The mathematical expression for free-space path loss is

$$L_p = \left(\frac{4\pi D}{\lambda}\right)^2 \quad \textbf{(14-26)}$$

and because $\lambda = \frac{c}{f}$, Equation 14-26 can be written as

$$L_p = \left(\frac{4\pi f D}{c}\right)^2 \quad \textbf{(14-27)}$$

where L_p = free-space path loss (unitless)
D = distance (kilometers)
f = frequency (hertz)
λ = wavelength (meters)
c = velocity of light in free space (3 × 108 meters per second)

Converting to dB yields

$$L_p(\text{dB}) = 10 \log\left(\frac{4\pi f D}{c}\right)^2 \quad \textbf{(14-28)}$$

or

$$L_p(\text{dB}) = 20 \log\left(\frac{4\pi f D}{c}\right) \quad \textbf{(14-29)}$$

Separating the constants from the variables gives

$$L_p = 20 \log\left(\frac{4\pi}{c}\right) + 20 \log f + 20 \log D \quad \textbf{(14-30)}$$

For frequencies in MHz and distances in kilometers,

$$L_p = \left(\frac{4\pi(10^6)(10^3)}{3 \times 10^8}\right) + 20 \log f_{(\text{MHz})} + 20 \log D_{(\text{km})} \quad \textbf{(14-31)}$$

or $$L_p = 32.4 + 20 \log f_{(\mathrm{MHz})} + 20 \log D_{(\mathrm{km})} \qquad \textbf{(14-32)}$$

When the frequency is given in GHz and the distance in km,

$$L_p = 92.4 + 20 \log f_{(\mathrm{GHz})} + 20 \log D_{(\mathrm{km})} \qquad \textbf{(14-33a)}$$

When the frequency is given in GHz and the distance in miles,

$$L_p = 96.6 + 20 \log f_{(\mathrm{GHz})} + 20 \log D_{(\mathrm{m})} \qquad \textbf{(14-33b)}$$

Example 14-2

For a carrier frequency of 6 GHz and a distance of 50 km, determine the free-space path loss.

Solution

$$L_p = 32.4 + 20 \log 6000 + 20 \log 50$$
$$= 32.4 + 75.6 + 34 = 142 \text{ dB}$$

or

$$L_p = 92.4 + 20 \log 6 + 20 \log 50$$
$$= 92.4 + 15.6 + 34 = 142 \text{ dB}$$

14-12 FADING AND FADE MARGIN

Radio communications between remote locations, whether earth to earth or earth to satellite, require propagating electromagnetic signals through free space. As an electromagnetic wave propagates through Earth's atmosphere, the signal may experience intermittent losses in signal strength beyond the normal path loss. This loss is attributed to several different phenomena and can include both short- and long-term effects. This variation in signal loss is called *fading* and can be caused by natural weather disturbances, such as rainfall, snowfall, fog, hail and extremely cold air over a warm Earth. Fading can also be caused by man-made disturbances, such as irrigation, or from multiple transmission paths, irregular Earth surfaces, and varying terrains. The types and causes of fading are covered in more detail in Chapter 24. To accommodate temporary fading, an additional loss is added to the normal path loss. This loss is called *fade margin.*

Solving the Barnett-Vignant reliability equations for a specified annual system availability for an unprotected, nondiversity system yields the following expression:

$$F_m = \underbrace{30 \log D}_{\substack{\text{multipath}\\ \text{effect}}} + \underbrace{10 \log (6ABf)}_{\substack{\text{terrain}\\ \text{sensitivity}}} - \underbrace{10 \log (1 - R)}_{\substack{\text{reliability}\\ \text{objectives}}} - \underbrace{70}_{\text{constant}} \qquad \textbf{(14-34)}$$

where F_m = fade margin (decibels)
D = distance (kilometers)
f = frequency (gigahertz)
R = reliability expressed as a decimal (i.e., 99.99% = 0.9999 reliability)
$1 - R$ = reliability objective for a one-way 400-km route
A = roughness factor
= 4 over water or a very smooth terrain
= 1 over an average terrain
= 0.25 over a very rough, mountainous terrain
B = factor to convert a worst-month probability to an annual probability
= 1 to convert an annual availability to a worst-month basis
= 0.5 for hot humid areas
= 0.25 for average inland areas
= 0.125 for very dry or mountainous areas

Example 14-3

Determine the fade margin for the following conditions: distance between sites, $D = 40$ km; frequency, $f = 1.8$ GHz; smooth terrain; humid climate; and a reliability objective 99.99%.

Solution Substituting into Equation 14-34 yields

$$F_m = 30 \log 40 + 10 \log[(6)(4)(0.5)(1.8)] - 10 \log(1 - 0.9999) - 70$$
$$= 48.06 + 13.34 - (-40) - 70 = 31.4 \text{ dB}$$

QUESTIONS

14-1. Describe an electromagnetic ray; a wavefront.

14-2. Describe power density; voltage intensity.

14-3. Describe a spherical wavefront.

14-4. Explain the inverse square law.

14-5. Describe wave attenuation.

14-6. Describe wave absorption.

14-7. Describe refraction. Explain Snell's law for refraction.

14-8. Describe reflection.

14-9. Describe diffraction. Explain Huygens's principle.

14-10. Describe the composition of a good reflector.

14-11. Describe the atmospheric conditions that cause electromagnetic refraction.

14-12. Define *electromagnetic wave interference.*

14-13. Describe ground wave propagation. List its advantages and disadvantages.

14-14. Describe space wave propagation.

14-15. Explain why the radio horizon is at a greater distance than the optical horizon.

14-16. Describe the various layers of the ionosphere.

14-17. Describe sky wave propagation.

14-18. Explain why ionospheric conditions vary with time of day, month of year, and so on.

14-19. Define *critical frequency; critical angle.*

14-20. Describe virtual height.

14-21. Define *maximum usable frequency.*

14-22. Define *skip distance* and give the reasons that it varies.

14-23. Describe path loss.

14-24. Describe fade margin.

14-25. Describe fading.

PROBLEMS

14-1. Determine the power density for a radiated power of 1000 W at a distance 20 km from an isotropic antenna.

14-2. Determine the power density for problem 14-1 for a point 30 km from the antenna.

14-3. Describe the effects on power density if the distance from a transmit antenna is tripled.

14-4. Determine the radio horizon for a transmit antenna that is 100 ft high and a receiving antenna that is 50 ft high and for antennas at 100 m and 50 m.

14-5. Determine the maximum usable frequency for a critical frequency of 10 MHz and an angle of incidence of 45°.

14-6. Determine the electric field intensity for the same point in problem 14-1.

14-7. Determine the electric field intensity for the same point in problem 14-2.

14-8. For a radiated power $P_{rad} = 10$ kW, determine the voltage intensity at a distance 20 km from the source.

14-9. Determine the change in power density when the distance from the source increases by a factor of 4.

14-10. If the distance from the source is reduced to one-half its value, what effect does this have on the power density?

14-11. The power density at a point from a source is 0.001 μW, and the power density at another point is 0.00001 μW; determine the attenuation in decibels.

14-12. For a dielectric ratio $\sqrt{\epsilon_{r2}/\epsilon_{r1}} = 0.8$ and an angle of incidence $\theta_i = 26°$, determine the angle of refraction, θ_r.

14-13. Determine the distance to the radio horizon for an antenna located 40 ft above sea level.

14-14. Determine the distance to the radio horizon for an antenna that is 40 ft above the top of a 4000-ft mountain peak.

14-15. Determine the maximum distance between identical antennas equidistant above sea level for problem 14-13.

14-16. Determine the power density for a radiated power of 1200 W at distance of 50 km from an isotropic antenna.

14-17. Determine the power density for problem 14-16 for a point 100 km from the same antenna.

14-18. Describe the effects on power density if the distance from a transmit antenna is reduced by a factor of 3.

14-19. Determine the radio horizon for a transmit antenna that is 200 ft high and a receiving antenna that is 100 ft high and for antennas at 200 m and 100 m.

14-20. Determine the maximum usable frequency for a critical frequency of 20 MHz and an angle of incidence of 35°.

14-21. Determine the voltage intensity for the same point in problem 14-16.

14-22. Determine the voltage intensity for the same point in problem 14-17.

14-23. Determine the change in power density when the distance from the source decreases by a factor of 8.

14-24. Determine the change in power density when the distance from the source increases by a factor of 8.

14-25. If the distance from the source is reduced to one-quarter its value, what effect does this have on the power density?

14-26. The power density at a point from a source is 0.002 μW, and the power density at another point is 0.00002 μW; determine the attenuation in dB.

14-27. For a dielectric ratio of 0.4 and an angle of incidence $\theta_i = 18$, determine the angle of refraction θ_r.

14-28. Determine the distance to the radio horizon for an antenna located 80 ft above sea level.

14-29. Determine the distance to the radio horizon for an antenna that is 80 ft above the top of a 5000-ft mountain.

14-30. Determine the maximum distance between identical antennas equidistant above sea level for problem 14-28.

14-31. Determine the path loss for the following frequencies and distances:

f (MHz)	*D* (km)
400	0.5
800	0.6
3000	10
5000	5
8000	20
18,000	15

14-32. Determine the fade margin for a 30-km microwave hop. The RF frequency is 10 GHz, the terrain is water, and the reliability objective is 99.995%.

CHAPTER 15

Antennas and Waveguides

CHAPTER OUTLINE

OBJECTIVES

- Define *antennas* and *waveguide*
- Describe basic antenna operation
- Define the term *antenna reciprocity*
- Define and describe antenna radiation patterns
- Define and describe the differences between antenna directivity and gain
- Explain effective isotropic radiated power
- Define and describe capture area, captured power density, and capture power
- Define *antenna polarization*
- Define *antenna beamwidth, bandwidth,* and *input impedance*
- Define and describe the operation of an elementary doublet
- Describe the operation of a half-wave dipole
- Describe the operation of a quarter-wave dipole

- Explain antenna loading
- Define and describe the operation of several antenna arrays
- Describe the operation of the following special-purpose antennas: Yagi-Uda, turnstile, log periodic, loop, phased array, and helical
- Describe the operation of a parabolic reflector
- Describe the basic operation of waveguides

15-1 INTRODUCTION

An *antenna* is a metallic conductor system capable of radiating and capturing electromagnetic energy. Antennas are used to interface transmission lines to the atmosphere, the atmosphere to transmission lines, or both. In essence, a transmission line couples energy from a transmitter to an antenna or from an antenna to a receiver. The antenna, in turn, couples energy received from a transmission line to the atmosphere and energy received from the atmosphere to a transmission line. At the transmit end of a free-space radio communications system, an antenna converts electrical energy traveling along a transmission line into electromagnetic waves that are emitted into space. At the receive end, an antenna converts electromagnetic waves in space into electrical energy on a transmission line.

A *waveguide* is a special type of transmission line that consists of a conducting metallic tube through which high-frequency electromagnetic energy is propagated. A waveguide is used to efficiently interconnect high-frequency electromagnetic waves between an antenna and a transceiver.

Radio waves are electrical energy that has escaped into free space in the form of transverse electromagnetic waves. The escaped radio waves travel at approximately the velocity of light and are comprised of magnetic and electric fields that are at right angles to each other and at right angles to the direction of travel. The plane parallel to the mutually perpendicular lines of the electric and magnetic fields is called the *wavefront.* The wave always travels in a direction at right angles to the wavefront and may go forward or backward, depending on the relative direction of the lines of magnetic and electric flux. If the direction of either the magnetic or the electric flux reverses, the direction of travel is reversed. However, reversing both sets of flux has no effect on the direction of propagation.

All electrical circuits that carry alternating current radiate a certain amount of electrical energy in the form of electromagnetic waves. However, the amount of energy radiated is negligible unless the physical dimensions of the circuit approach the dimensions of a wavelength of the wave. For example, a power line carrying 60-Hz current with 20 feet of separation between conductors radiates virtually no energy because a wavelength at 60 Hz is over 3000 miles long, and 20 feet is insignificant in comparison. In comparison, an inductor (coil) 1 cm long carrying a 6-GHz signal will radiate a considerable amount of energy because 1 cm is comparable with the 5-cm wavelength.

15-2 BASIC ANTENNA OPERATION

From the previous discussion, it is apparent that the size of an antenna is inversely proportional to frequency. A relatively small antenna can efficiently radiate high-frequency electromagnetic waves, while low-frequency waves require relatively large antennas. Every antenna has directional characteristics and radiate more energy in certain directions relative to other directions. Directional characteristics of antennas are used

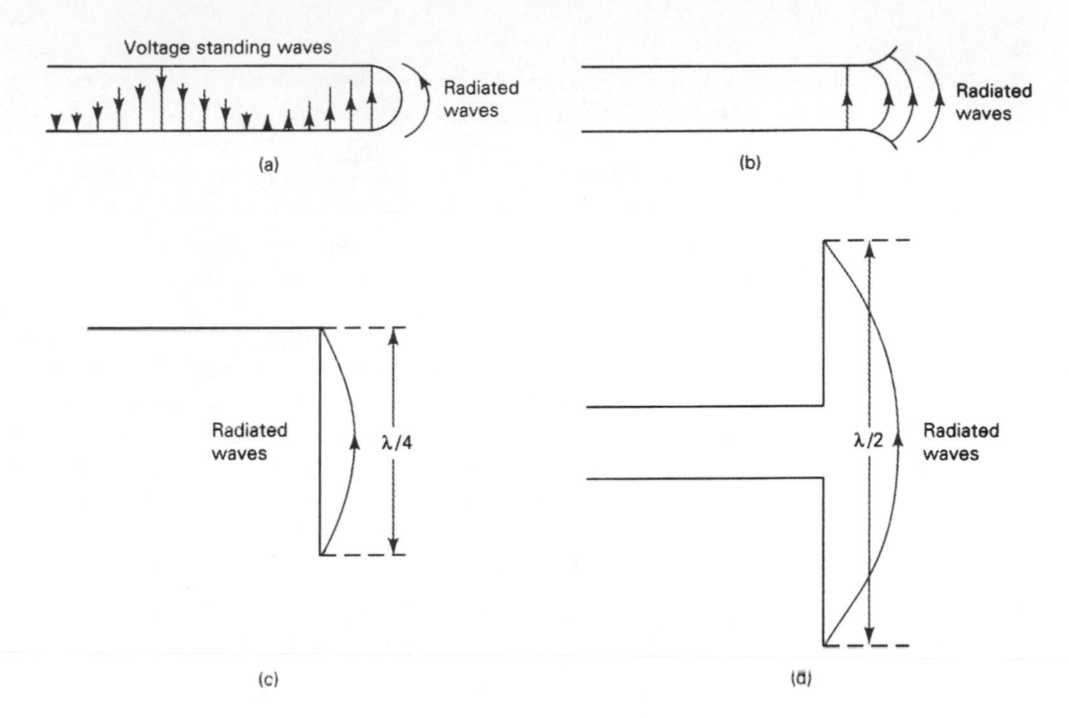

FIGURE 15-1 Radiation from a transmission line: (a) transmission-line radiation; (b) spreading conductors; (c) Marconi antenna; (d) Hertz antenna

to concentrate radiation in a desired direction or capture energy arriving from a particular direction.

For an antenna to efficiently receive radio signals, it must abstract energy from the radio wave as it passes by the receiving point. Electromagnetic wave reception occurs in an antenna because the electromagnetic flux of the wave cuts across the antenna conductor, inducing a voltage into the conductor that varies with time in exactly the same manner as the current flowing in the antenna that radiated the wave. The induced voltage, along with the current it produces, represents energy that the antenna absorbs from the passing wave.

Basic antenna operation is best understood by looking at the voltage standing-wave patterns on a transmission line, which are shown in Figure 15-1a. The transmission line is terminated in an open circuit, which represents an abrupt discontinuity to the incident voltage wave in the form of a phase reversal. The phase reversal results in some of the incident voltage being radiated, not reflected back toward the source. The radiated energy propagates away from the antenna in the form of transverse electromagnetic waves. The *radiation efficiency* of an open transmission line is extremely low. Radiation efficiency is the ratio of radiated to reflected energy. To radiate more energy, simply spread the conductors farther apart. Such an antenna is called a *dipole* (meaning two poles) and is shown in Figure 15-1b.

In Figure 15-1c, the conductors are spread out in a straight line to a total length of one-quarter wavelength. Such an antenna is called a basic *quarter-wave antenna* or a *vertical monopole* (sometimes called a Marconi antenna). A half-wave dipole is called a *Hertz antenna* and is shown in Figure 15-1d.

15-2-1 Antenna Equivalent Circuit

In radio communications systems, transmitters are connected to receivers through transmission lines, antennas, and free space. Electromagnetic waves are coupled from transmit

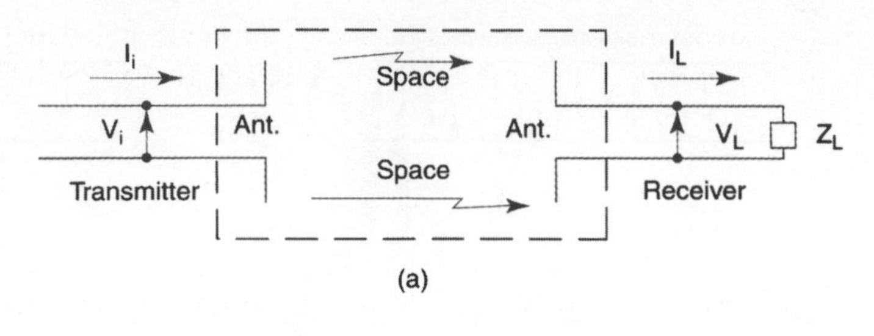

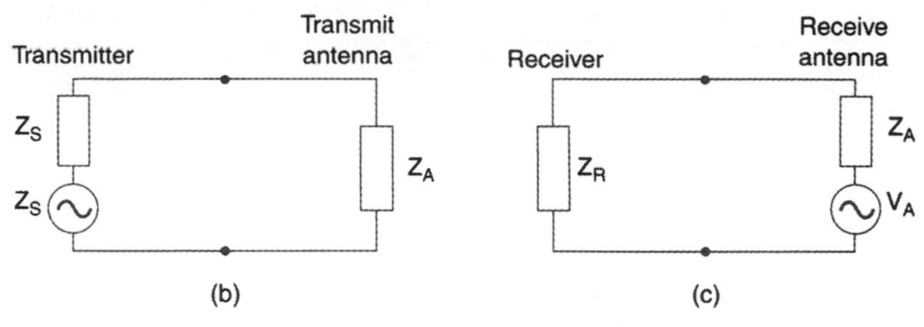

FIGURE 15-2 (a) Antenna as a four-terminal network; (b) transmit antenna equivalent circuit; (c) receive antenna equivalent circuit

to receive antennas through free space in a manner similar to the way energy is coupled from the primary to the secondary of a transformer. With antennas, however, the degree of coupling is much lower than with a transformer, and an electromagnetic wave is involved rather than just a magnetic wave. An antenna coupling system can be represented with a four-terminal network as shown in Figure 15-2a. Electromagnetic energy must be transferred from the transmitting antenna to free space and then from free space to the receiving antenna. Figure 15-2c shows the equivalent circuit for a transmit antenna, and Figure 15-2c shows the equivalent circuit for a receive antenna.

15-3 ANTENNA RECIPROCITY

A basic antenna is a *passive reciprocal device*—passive in that it cannot actually amplify a signal, at least not in the true sense of the word (however, you will see later in this chapter that an antenna can have gain). An antenna is a reciprocal device in that the transmit and receive characteristics and performance are identical (i.e., gain, directivity, frequency of operation, bandwidth, radiation resistance, efficiency, and so on).

Transmit antennas must be capable of handling high powers and, therefore, must be constructed with materials that can withstand high voltages and currents, such as metal tubing. Receive antennas, however, produce very small voltages and currents and can be constructed from small-diameter wire. In many radio communications systems, however, the same antenna is used for transmitting and receiving. If this is the case, the antenna must be constructed from heavy-duty materials. If one antenna is used for both transmitting and receiving, some means must be used to prevent the high-power transmit signals from being coupled into the relatively sensitive receiver. A special coupling device called a *diplexer* can be used to direct the transmit and receive signals and provide the necessary isolation.

Standard antennas have no active components (diodes, transistors, FETs, and so on); therefore, they are passive and reciprocal. In practice an active antenna does not exist. What is commonly called an active antenna is actually the combination of a passive antenna and a low-noise amplifier (LNA). Active antennas are nonreciprocal (i.e., they either transmit or receive but not both). It is important to note that active as well as passive antennas introduce power losses regardless of whether they are used for transmitting or receiving signals. Antenna gain is a misleading term that is explained in detail later in this chapter.

15-4 ANTENNA COORDINATE SYSTEM AND RADIATION PATTERNS

15-4-1 Antenna Coordinate System

The directional characteristics of an electromagnetic wave radiated or received by an antenna are generally described in terms of spherical coordinates as shown in Figure 15-3. Imagine the antenna placed in the center of the sphere; the distance to any point on the surface of the sphere can be defined in respect to the antenna by using the radius of the sphere d and angles θ and Φ. The x-y plane shown in the figure is referred to as the equatorial plane, and any plane at right angles to it is defined as a meridian plane.

15-4-2 Radiation Pattern

A *radiation pattern* is a *polar* diagram or graph representing field strengths or power densities at various angular positions relative to an antenna. If the radiation pattern is plotted in terms of electric field strength ($\mathscr{E}$) or power density ($\mathscr{P}$), it is called an *absolute* radiation pattern (i.e., variable distance, fixed power). If it plots field strength or power density with respect to the value at a reference point, it is called a *relative* radiation pattern (i.e., variable power, fixed distance). Figure 15-4a shows an absolute radiation pattern for an unspecified antenna. The pattern is plotted on *polar* coordinate paper with the heavy solid line representing points of equal power density (10 μW/m^2). The circular gradients indicate distance in 2-km steps. It can be seen that maximum radiation is in a direction 90° from the reference. The power density 10 km from the antenna in a 90° direction is 10 μW/m^2. In a 45° direction, the point of equal power density is 5 km from the antenna; at 180°, only 4 km; and in a −90° direction, there is essentially no radiation.

In Figure 15-4a, the primary beam is in a 90° direction and is called the *major lobe.* There can be more than one major lobe. There is also a *secondary* beam or *minor* lobe in a direction other than that of the major lobe. Normally, minor lobes represent undesired radiation or reception. Because the major lobe propagates and receives the most energy, that lobe is called the *front* lobe (the front of the antenna). Lobes adjacent to the front lobe are called *side* lobes (the 180° minor lobe is a side lobe), and lobes in a direction exactly opposite the front lobe are called *back* lobes (there is no back lobe shown on this pattern). The

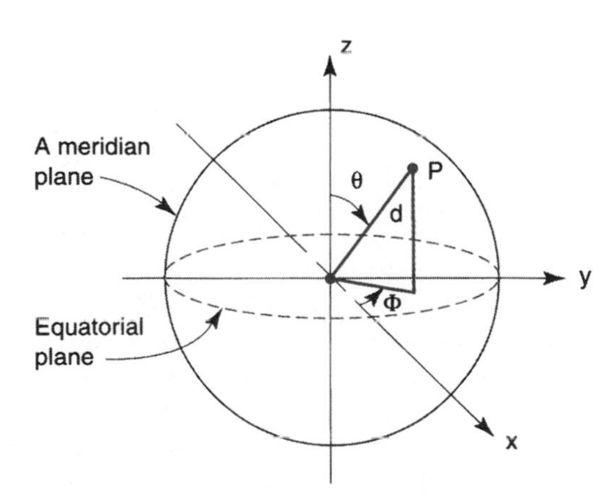

FIGURE 15-3 Spherical coordinates

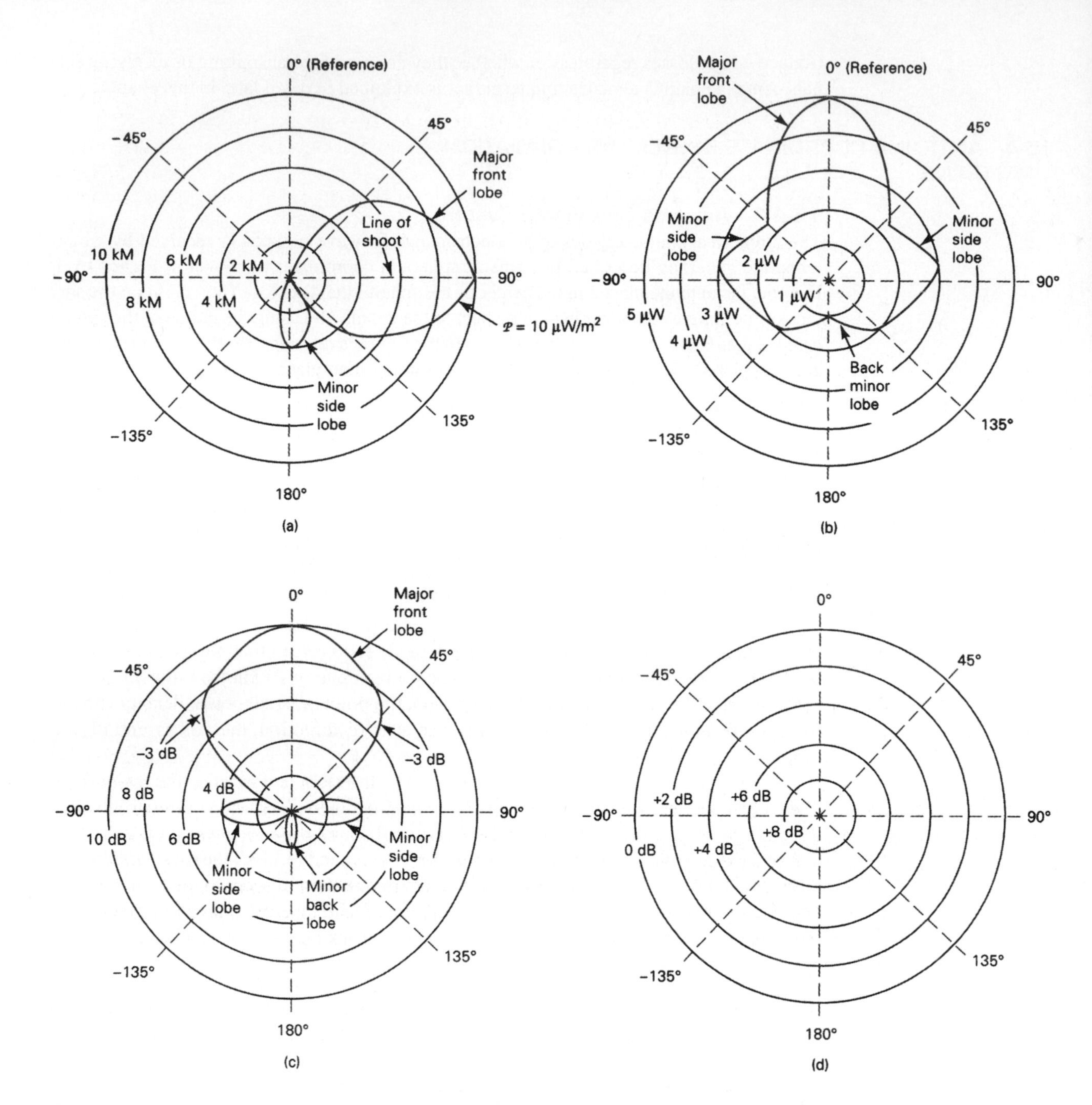

FIGURE 15-4 Radiation patterns: (a) absolute (fixed power) radiation pattern; (b) relative (fixed distance) radiation pattern; (c) relative (fixed distance) radiation pattern in decibels; and (d) relative (fixed distance) radiation pattern in decibels for an omnidirectional (point source) antenna

ratio of the front lobe power to the back lobe power is simply called the *front-to-back ratio,* and the ratio of the front lobe to a side lobe is called the *front-to-side ratio.* The line bisecting the major lobe, or pointing from the center of the antenna in the direction of maximum radiation, is called the *line of shoot,* or sometimes *point of shoot.*

Figure 15-4b shows a relative radiation pattern for an unspecified antenna. The heavy solid line represents points of equal distance from the antenna (10 km), and the circular gradients indicate power density in 1-μW/m^2 divisions. It can be seen that maximum radiation (5 μW/m^2) is in the direction of the reference (0°), and the antenna radiates the least power (1 μW/m^2) in a direction 180° from the reference. Consequently, the front-to-back ratio is 5:1 = 5. Generally, relative field strength and power density are plotted in decibels (dB), where dB = 20 log ($\mathscr{E}/\mathscr{E}_{max}$) or 10 log ($\mathscr{P}/\mathscr{P}_{max}$). Figure 15-4c shows a relative radiation pattern for power density in decibels. In a direction ±45° from the reference, the power density is −3 dB (half-power) relative to the power density in the direction of maximum radiation (0°). Figure 15-4d shows a relative radiation pattern for power density for an omnidirectional antenna. An omnidirectional antenna radiates energy equally in all directions; therefore, the radiation pattern is simply a circle (actually, a sphere). Also, with an omnidirectional antenna, there are no front, back, or side lobes because radiation is equal in all directions.

The radiation patterns shown in Figure 15-4 are two dimensional. However, radiation from an actual antenna is three dimensional. Therefore, radiation patterns are taken in both the horizontal (from the top) and the vertical (from the side) planes. For the omnidirectional antenna shown in Figure 15-4d, the radiation patterns in the horizontal and vertical planes are circular and equal because the actual radiation pattern for an isotropic radiator is a sphere.

Recall from Chapter 9 that a true isotropic radiator radiates power at a constant rate uniformly in all directions. An ideal isotropic antenna also radiates all the power supplied to it. Isotropic radiators do not exist, however, and they are used only for analytical descriptions and comparisons.

15-4-3 Near and Far Fields

The radiation field that is close to an antenna is not the same as the radiation field that is at a great distance. The term *near field* refers to the field pattern that is close to the antenna, and the term *far field* refers to the field pattern that is at great distance. During one-half of a cycle, power is radiated from an antenna where some of the power is stored temporarily in the near field. During the second half of the cycle, power in the near field is returned to the antenna. This action is similar to the way in which an inductor stores and releases energy. Therefore, the near field is sometimes called the *induction field.* Power that reaches the far field continues to radiate outward and is never returned to the antenna. Therefore, the far field is sometimes called the *radiation field.* Radiated power is usually the more important of the two; therefore, antenna radiation patterns are generally given for the far field. The near field is defined as the area within a distance D^2/λ from the antenna, where λ is the wavelength and D the antenna diameter in the same units.

15-4-4 Radiation Resistance and Antenna Efficiency

All the power supplied to an antenna is not radiated. Some of it is converted to heat and dissipated. *Radiation resistance* is somewhat "unreal" in that it cannot be measured directly. Radiation resistance is an ac antenna resistance and is equal to the ratio of the power radiated by the antenna to the square of the current at its feedpoint. Mathematically, radiation resistance is

$$R_r = \frac{P_{rad}}{i^2} \tag{15-1}$$

where R_r = radiation resistance (ohms)
P_{rad} = power radiated by the antenna (watts)
i = antenna current at the feedpoint (ampere)

Radiation resistance is the resistance that, if it replaced the antenna, would dissipate exactly the same amount of power that the antenna radiates. The radiation resistance of an antenna as described in Equation 15-1 is in a sense a fictitious quantity because it is referenced to an arbitrary point on the antenna that would have different current values for different reference points. It is common practice, however, to refer the radiation resistance

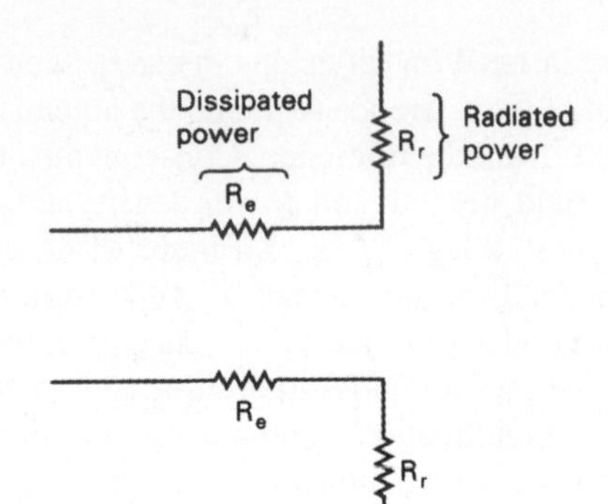

FIGURE 15-5 Simplified equivalent circuit of an antenna

to the current maximum point or sometimes the current at the feed point, although in many cases the two points are one in the same. When referenced to the current maximum point, radiation resistance is sometimes called *loop radiation resistance* because a current maximum is also called a current loop.

It seems apparent that radiation resistance is at times a rather nebulous concept as it is not always easily measured. It is a useful concept only when it is readily measurable and has no meaning for antennas in which there is no clearly defined current value to which it can be referenced.

Antenna efficiency is the ratio of the power radiated by an antenna to the sum of the power radiated and the power dissipated or the ratio of the power radiated by the antenna to the total input power. Mathematically, antenna efficiency is

$$\eta = \frac{P_{\text{rad}}}{P_{\text{in}}} \times 100 \tag{15-2a}$$

where η = antenna efficiency (percentage)
P_{rad} = radiated power (watts)
P_{in} = input power (watts)
$P_{\text{in}} = P_{\text{rad}} + P_d$

or

$$\eta = \frac{P_{\text{rad}}}{P_{\text{rad}} + P_d} \times 100 \tag{15-2b}$$

where P_{rad} = power radiated by antenna (watts)
P_d = power dissipated in antenna (watts)

Figure 15-5 shows a simplified electrical equivalent circuit for an antenna. Some of the input power is dissipated in the effective resistance (ground resistance, corona, imperfect dielectrics, eddy currents, and so on), and the remainder is radiated. The total antenna power is the sum of the dissipated and radiated powers. Therefore, in terms of resistance and current, antenna efficiency is

$$\eta = \frac{i^2 R_r}{i^2(R_r + R_e)} = \frac{R_r}{R_r + R_e} \tag{15-3}$$

where η = antenna efficiency
i = antenna current (ampere)
R_r = radiation resistance (ohms)
R_e = effective antenna resistance (ohms)

15-5 ANTENNA GAIN

The terms *directive gain* and *power gain* are often misunderstood and, consequently, misused. Directive gain is the ratio of the power density radiated in a particular direction to the power density radiated to the same point by a reference antenna, assuming both antennas are radiating the same amount of power. The relative power density radiation pattern for an antenna is actually a directive gain pattern if the power density reference is taken from a standard reference antenna, which is generally an isotropic antenna. The maximum directive gain is called *directivity.* Mathematically, directive gain is

$$D = \frac{\mathcal{P}}{\mathcal{P}_{\text{ref}}} \tag{15-4}$$

where D = directive gain (unitless)
$\mathcal{P}$ = power density at some point with a given antenna (watts per meter squared)
$\mathcal{P}_{\text{ref}}$ = power density at the same point with a reference antenna (watts per meter squared)

Power gain is the same as directive gain except that the total power fed to the antenna is used (i.e., antenna efficiency is taken into account). It is assumed that the given antenna and the reference antenna have the same input power and that the reference antenna is lossless ($\eta = 100\%$). Mathematically, power gain (A_p) is

$$A_p = D\eta \tag{15-5}$$

If an antenna is lossless, it radiates 100% of the input power, and the power gain is equal to the directive gain. The power gain for an antenna is given in decibels as

$$A_{P(\text{dB})} = 10 \log (D\eta) \tag{15-6}$$

For an isotropic reference, the power gain (dB) of a half-wave dipole is approximately 1.64 (2.15 dB). It is usual to state the power gain in decibels when referring to a $\lambda/2$ dipole (dBd). However, if reference is made to an isotropic radiator, the decibel figure is stated as dBi, or dB/isotropic radiator, and is 2.15 dB greater than if a half-wave dipole were used for the reference. It is important to note that the power radiated from an antenna can never exceed the input power. Therefore, the antenna does not actually amplify the input power. An antenna simply concentrates its radiated power in a particular direction. Therefore, points that are located in areas where the radiated power is concentrated realize an apparent gain relative to the power density at the same points had an isotropic antenna been used. If gain is realized in one direction, a corresponding reduction in power density (a loss) must be realized in another direction. The direction in which an antenna is "pointing" is always the direction of maximum radiation. Because an antenna is a reciprocal device, its radiation pattern is also its reception pattern. For maximum *captured* power, a receive antenna must be pointing in the direction from which reception is desired. Therefore, receive antennas have directivity and power gain just as transmit antennas do.

15-5-1 Effective Isotropic Radiated Power

Effective isotropic radiated power (EIRP) is defined as an equivalent transmit power and is expressed mathematically as

$$\text{EIRP} = P_{\text{rad}} D_t \text{ (watts)} \tag{15-7a}$$

where P_{rad} = total radiated power (watts)
D_t = transmit antenna directive gain (unitless)

or

$$\text{EIRP}_{(\text{dBm})} = 10 \log \frac{P_{\text{rad}}}{0.001} + 10 \log D_t \tag{15-7b}$$

or

$$\text{EIRP}_{(\text{dBW})} = 10 \log (P_{\text{rad}} D_t) \tag{15-7c}$$

Equation 15-7a can be rewritten using antenna input power and power gain as

$$\text{EIRP} = P_{\text{in}}A_t \tag{15-7d}$$

where P_{in} = total antenna input power (watts)
A_t = transmit antenna power gain (unitless)

or

$$\text{EIRP}_{(\text{dBm})} = 10 \log \left(\frac{P_{\text{in}}A_t}{0.001}\right) \tag{15-7e}$$

$$\text{EIRP}_{(\text{dBW})} = 10 \log (P_{\text{in}}A_t) \tag{15-7f}$$

EIRP or simply ERP (effective radiated power) is the equivalent power that an isotropic antenna would have to radiate to achieve the same power density in the chosen direction at a given point as another antenna. For instance, if a given transit antenna has a power gain of 10, the power density a given distance from the antenna is 10 times greater than it would have been had the antenna been an isotropic radiator. An isotropic antenna would have to radiate 10 times as much power to achieve the same power density. Therefore, the given antenna effectively radiates 10 times as much power as an isotropic antenna with the same input power and efficiency.

To determine the power density at a given point distance R from a transmit antenna, Equation 15-5 can be expanded to include the transmit antenna gain and rewritten as

$$\mathcal{P} = \frac{P_{\text{in}}A_t}{4\pi R^2} \tag{15-8a}$$

or in terms of directive gain

$$\mathcal{P} = \frac{P_{\text{rad}}D_t}{4\pi R^2} \tag{15-8b}$$

where $\mathcal{P}$ = power density (watts per meter squared)
P_{in} = transmit antenna input power (watts)
P_{rad} = power radiated from transmit antenna (watts)
A_t = transmit antenna power gain (unitless)
D_t = transmit antenna directive power gain (unitless)
R = distance from transmit antenna (meters)

Example 15-1

For a transmit antenna with a power gain $A_t = 10$ and an input power $P_{\text{in}} = 100$ W, determine

a. EIRP in watts, dBm, and dBW.
b. Power density at a point 10 km from the transmit antenna.
c. Power density had an isotropic antenna been used with the same input power and efficiency.

Solution **a.** Substituting into Equations 15-7d, e, and f yields

$$\begin{aligned}\text{EIRP} &= (100\text{ W})(10)\\ &= 1000\text{ W}\end{aligned}$$

$$\begin{aligned}\text{EIRP}_{(\text{dBm})} &= 10 \log \frac{1000}{0.001}\\ &= 60\text{ dBm}\end{aligned}$$

$$\begin{aligned}\text{EIRP}_{(\text{dBW})} &= 10 \log 1000\\ &= 30\text{ dBW}\end{aligned}$$

b. Substituting into Equation 15-8a gives

$$\begin{aligned}\mathcal{P} &= \frac{(100\text{ W})(10)}{4\pi(10{,}000\text{ m})^2}\\ &= 0.796\ \mu\text{W/m}^2\end{aligned}$$

c. Substituting into Equation 15-5, we obtain

$$\mathcal{P} = \frac{(100\text{ W})}{4\pi(10{,}000\text{ m})^2}$$
$$= 0.0796\ \mu\text{W/m}^2$$

It can be seen from Example 15-1 that the power density at a point 10 km from the transmit antenna is 10 times greater with the given antenna than it would be had an isotropic radiator been used. To achieve the same power density, the isotropic antenna would require an input power 10 times greater or 1000 W. The transmit antenna in the example effectively radiates the equivalent of 1000 W.

Example 15-2

For a transmit antenna with a radiation resistance $R_r = 72$ ohms, an effective antenna resistance $R_e = 8$ ohms, a directive gain $D = 20$, and an input power $P_{\text{in}} = 100$ W, determine

a. Antenna efficiency.
b. Antenna gain (absolute and dB).
c. Radiated power in watts, dBm, and dBW.
d. EIRP in watts, dBm, and dBW.

Solution **a.** Antenna efficiency is found by substituting into Equation 15-3:

$$\eta = \frac{72}{72 + 8} \times 100$$
$$= 90\%$$

b. Antenna gain is simply the product of the antenna's efficiency and directive gain:

$$A = (0.9)(20)$$
$$= 18$$

and

$$A(\text{dB}) = 10 \log 18$$
$$= 12.55\text{ dB}$$

c. Radiated power is found by rearranging Equation 15-2a:

$$P_{\text{rad}} = \eta P_{\text{in}}$$
$$= (0.9)(100\text{ W})$$
$$= 90\text{ W}$$

$$P_{\text{rad(dBm)}} = 10 \log \frac{90}{0.001}$$
$$= 49.54\text{ dBm}$$

$$P_{\text{rad(dBW)}} = 10 \log 90$$
$$= 19.54\text{ dBW}$$

d. EIRP is found by substituting into Equations 15-7d, e, and f:

$$\text{EIRP} = (100\text{ W})(18)$$
$$= 1800\text{ W}$$

$$\text{EIRP}_{\text{(dBm)}} = 10 \log \frac{1800}{0.001}$$
$$= 62.55\text{ dBm}$$

$$\text{EIRP}_{\text{(dBW)}} = 10 \log 1800$$
$$= 32.55\text{ dBW}$$

The relationships among directivity, power gain, power radiated, EIRP, and receive power density are best shown with an example.

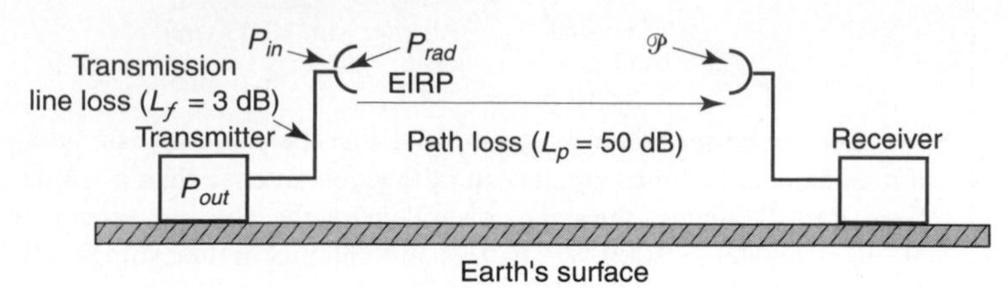

FIGURE 15-6 Figure for Example 15-3

Example 15-3

Given the free-space radio transmission system shown in Figure 15-6 with the following transmission characteristics:

transmitter power out = 40 dBm
transmission line loss $L_f = 3$ dB
free-space path loss $L_p = 50$ dB

a. Determine the antenna input power (P_{in}), radiated power (P_{rad}), EIRP, and receive power density ($\mathcal{P}$) for an isotropic transmit antenna with a directivity of unity $D_t = 1$ and an efficiency $\eta = 100\%$ (a gain of 1).
b. Determine the antenna input power (P_{in}), radiated power (P_{rad}), EIRP, and receive power density ($\mathcal{P}$) for a transmit antenna with a directivity $D_t = 10$ and an efficiency $\eta = 50\%$.
c. Determine the antenna input power (P_{in}), radiated power (P_{rad}), EIRP, and receive power density ($\mathcal{P}$) for a transmit antenna with a power gain $A_t = 5$.

Solution a. The antenna input power in dBm is

$$P_{\text{in}} = P_{\text{out}} - L_f$$
$$P_{\text{in}} = 40 \text{ dBm} - 3 \text{ dB}$$
$$= 37 \text{ dBm}$$

Radiated power in dBm is

$$P_{\text{rad}} = P_{\text{in}} + \eta \qquad \text{where } P_{\text{in}} \text{ is in dBm and } \eta_{(\text{dB})} = 10\log(1) = 0 \text{ dB}$$
$$= 37 \text{ dBm} + 0 \text{ dB}$$
$$= 37 \text{ dBm}$$

Effective isotropic radiated power in dBm is determined by expanding Equation 15-7c

$$\text{EIRP} = P_{\text{rad}} + 10\log(A_t) \qquad \text{where } P_{\text{rad}} \text{ is in dBm and } A_t = D_t\eta \text{ and is an absolute value}$$
$$= 37 \text{ dBm} + 10\log(1 \times 1)$$
$$= 37 \text{ dBm} + 0 \text{ dB}$$
$$= 37 \text{ dBm}$$

Receive power density ($\mathcal{P}$) is simply the EIRP minus the free-space path loss

$$\mathcal{P} = 37 \text{ dBm} - 50 \text{ dB}$$
$$= -13 \text{ dBm}$$

b. The antenna input power in dBm is

$$P_{\text{in}} = P_{\text{out}} - L_f$$
$$P_{\text{in}} = 40 \text{ dBm} - 3 \text{ dB}$$
$$= 37 \text{ dBm}$$

Radiated power in dBm is

$$P_{\text{rad}} = P_{\text{in}} + \eta \qquad \text{where } P_{\text{in}} \text{ is in dBm and } \eta_{(\text{dB})} = 10\log(0.5) = -3 \text{ dB}$$
$$= 37 \text{ dBm} + (-3 \text{ dB})$$
$$= 34 \text{ dBm}$$

Effective isotropic radiated power in dBm is determined by expanding Equation 15-7c

$$\begin{aligned} \text{EIRP} &= P_{rad} + 10\log(D_t) \quad \text{where } P_{rad} \text{ is in dBm and } D_t \text{ is an absolute value} \\ &= 34\text{ dBm} + 10\log(10) \\ &= 34\text{ dBm} + 10\text{ dB} \\ &= 44\text{ dBm} \end{aligned}$$

Receive power density ($\mathcal{P}$) is simply the EIRP minus the free-space path loss

$$\begin{aligned} \mathcal{P} &= 44\text{ dBm} - 50\text{ dB} \\ &= -6\text{ dBm} \end{aligned}$$

c. The antenna input power in dBm is

$$\begin{aligned} P_{in} &= P_{out} - L_f \\ P_{in} &= 40\text{ dBm} - 3\text{ dB} \\ &= 37\text{ dBm} \end{aligned}$$

Radiated power in dBm is

$$\begin{aligned} P_{rad} &= P_{in} + \eta \quad \text{where } P_{in} \text{ is in dBm and } \eta_{(dB)} = 10\log(0.5) = -3\text{ dB} \\ &= 37\text{ dBm} + (-3\text{ dB}) \\ &= 34\text{ dBm} \end{aligned}$$

Effective isotropic radiated power in dBm is determined by expanding Equation 15-7c

$$\begin{aligned} \text{EIRP} &= P_{rad} + A_t \quad \text{where } P_{rad} \text{ is in dBm and } A_t = 10\log(5) = 7\text{ dB} \\ &= 37\text{ dBm} + 7\text{ dB} \\ &= 44\text{ dBm} \end{aligned}$$

Receive power density ($\mathcal{P}$) is simply the EIRP minus the free-space path loss

$$\begin{aligned} \mathcal{P} &= 44\text{ dBm} - 50\text{ dB} \\ &= -6\text{ dBm} \end{aligned}$$

From Example 15-3, it can be seen that the receive power density is the same for the antennas specified in sections b and c. This is because an antenna with a directivity of 10 dB and an efficiency of 50% (0.5) is identical to an antenna with a power gain of 7 dB ([10 log (10 × 0.5)]. It can also be seen that the receive power density increases over that of an omnidirectional antenna shown in section a by a factor equal to the transmit antenna gain (7 dB).

15-6 CAPTURED POWER DENSITY, ANTENNA CAPTURE AREA, AND CAPTURED POWER

15-6-1 Captured Power Density

Antennas are reciprocal devices; thus, they have the same radiation resistance, efficiency, power gain, and directivity when used to receive electromagnetic waves as they have when transmitting electromagnetic waves. Therefore, Equation 15-8a can be expanded to include the power gain of the receiver antenna and rewritten as

$$C = \frac{(P_{in})(A_t)(A_r)}{4\pi R^2} \qquad \textbf{(15-9)}$$

where C = captured power density (watts per meter squared)
P_{in} = transmit antenna input power (watts)
A_t = transmit antenna power gain (unitless)
A_r = receive antenna power gain (unitless)
R = distance between transmit and receive antennas (meters)

Captured power density is the power density (W/m^2) in space and a somewhat misleading quantity. What is more important is the actual power (in watts) that a receive antenna produces at its output terminals which, of course, depends on how much power is captured by the receive antenna and the antenna's efficiency.

15-6-2 Antenna Capture Area and Captured Power

Although a reciprocal relationship exists between transmitting and receiving antenna properties, it is often more useful to describe receiving properties in a slightly different way. Whereas power gain is the natural parameter for describing the increased power density of a transmitted signal due to the directional properties of the transmiting antenna, a related quantity called *capture area* is a more natural parameter for describing the reception properties of an antenna.

The capture area of an antenna is an *effective area* and can be described as follows. A transmit antenna radiates an electromagnetic wave that has a power density at the receive antenna's location in W/m^2. This is not the actual power received but rather the amount of power incident on, or passing through, each unit area of any imaginary surface that is perpendicular to the direction of propagation of the electromagnetic waves. A receiving antenna exposed to the electromagnetic field will have radio-frequency voltage and current induced in it, producing a corresponding radio-frequency power at the antenna's output terminals. In principle, the power available at the antenna's output terminals (in watts) is the *captured power.* The captured power can be delivered to a load such as a transmission line or a receiver's input circuitry. For the captured power to appear at the antenna's output terminals, the antenna must have captured power from a surface in space immediately surrounding the antenna.

Captured power is directly proportional to the received power density and the effective capture area of the receive antenna. As one might expect, the physical cross-sectional area of an antenna and its effective capture area are not necessarily equal. In fact, sometimes antennas with physically small cross-sectional areas may have effective capture areas that are considerably larger than their physical areas. In these instances, it is as though the antenna is able to reach out and capture or absorb power from an area larger than its physical size.

There is an obvious relationship between an antenna's size and its ability to capture electromagnetic energy. This suggests that there must also be a connection between antenna gain and the antenna's receiving cross-sectional area. Mathematically, the two quantities are related as follows:

$$A_c = \frac{A_r \lambda^2}{4\pi} \qquad \text{(15-10)}$$

where A_c = effective capture area (meters squared)
λ = wavelength of receive signal (meters)
A_r = receive antenna power gain (unitless)

Rearranging Equation 15-10 and solving for antenna gain gives us

$$A_r = \frac{A_c 4\pi}{\lambda^2} \qquad \text{(15-11)}$$

Because antennas are reciprocal devices, the power received or captured by an antenna is the product of the power density in the space immediately surrounding the receive antenna and the antenna's effective area. Mathematically, captured power is

$$P_{cap} = P\,A_c \qquad \text{(15-12)}$$

where P_{cap} = captured power (watts)
$P = \frac{P_{in} A_t}{4\pi R^2}$ power density (watts per meter squared)
A_c = capture area (square meters)
P_{in} = power input to transmit antenna (watts)
A_t = transmit antenna power gain (unitless)
R = distance between transmit and receive antennas (meters)

Substituting Equation 15-10 into Equation 15-12 gives

$$P_{\text{cap}} = P\left(\frac{A_r\lambda^2}{4\pi}\right) \tag{15-13}$$

Substituting $\frac{P_{\text{in}} A_t}{4\pi R^2}$ for P (Equation 15-8a) in Equation 15-13 yields

$$P_{\text{cap}} = \left(\frac{P_{\text{in}} A_t}{4\pi R^2}\right)\left(\frac{A_r\lambda^2}{4\pi}\right) \tag{15-14}$$

$$= \frac{P_{\text{in}} A_t A_r \lambda^2}{4\pi R^2(4\pi)} \tag{15-15}$$

$$= \frac{P_{\text{in}} A_t A_r \lambda^2}{16\pi^2 R^2} \tag{15-16}$$

or

$$= (P_{\text{in}} A_t A_r)\left(\frac{\lambda^2}{16\pi^2 R^2}\right) \tag{15-17}$$

and since $P_{\text{in}} A_t = \text{EIRP}$ and $\frac{\lambda^2}{16\pi^2 R^2}$ = path loss (L_p), Equation 15-17 can be written as

$$P_{\text{cap}} = (\text{EIRP})(A_r)(L_p) \tag{15-18}$$

Converted to decibel units, Equation 15-17 becomes

$$P_{\text{cap}}(\text{dBm}) = 10\log\left(\frac{P_{\text{in}} A_t A_r}{0.001}\right) - 10\log\left(\frac{\lambda^2}{16\pi^2 R^2}\right)$$

captured power (dBm) − path loss (L_p) (dB)

Note: The formula for path loss was derived by expanding Equation 14-26.

Example 15-4

For a receive power density of 10 μW/m² and a receive antenna with a capture area of 0.2 m², determine

a. Captured power in watts.
b. Captured power in dBm.

Solution **a.** Substituting into Equation 15-12 yields

$$P_{\text{cap}} = (10\ \mu\text{W/m}^2)(0.2\ \text{m}^2)$$
$$= 2\ \mu\text{W}$$

b.

$$P_{\text{cap}}(\text{dBm}) = 10\log\frac{2\ \mu\text{W}}{0.001\ \text{W}}$$
$$= -27\ \text{dBm}$$

15-7 ANTENNA POLARIZATION

The *polarization* of an antenna refers simply to the orientation of the electric field radiated from it. An antenna may be *linearly* (generally either horizontally or vertically polarized, assuming that the antenna elements lie in a horizontal or vertical plane), *elliptically,* or *circularly polarized.* If an antenna radiates a vertically polarized electromagnetic wave, the antenna is defined as vertically polarized; if an antenna radiates a horizontally polarized electromagnetic wave, the antenna is said to be horizontally polarized; if the radiated electric field rotates in

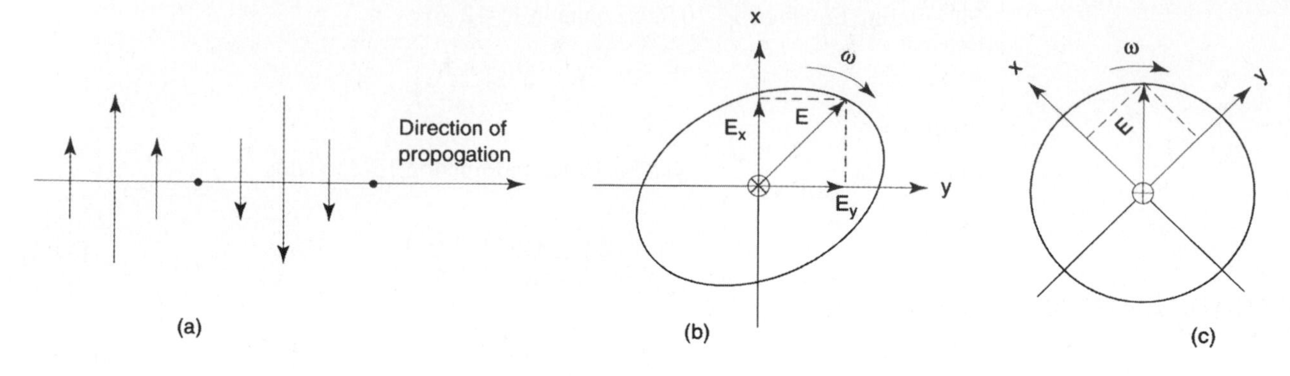

FIGURE 15-7 Antenna polarizations: (a) linear; (b) elliptical polarization; (c) circular polarization

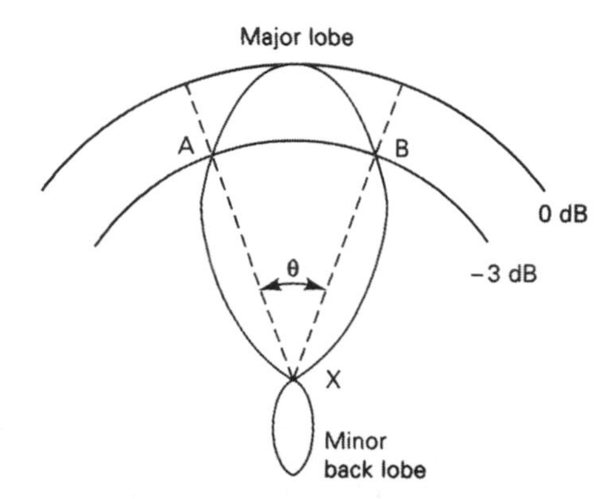

FIGURE 15-8 Antenna beamwidth

an elliptical pattern, it is elliptically polarized; and if the electric field rotates in a circular pattern, it is circularly polarized. Figure 15-7 shows the various polarizations described.

15-8 ANTENNA BEAMWIDTH

Antenna *beamwidth* is simply the angular separation between the two half-power (−3 dB) points on the major lobe of an antenna's plane radiation pattern, usually taken in one of the "principal" planes. The beamwidth for the antenna whose radiation pattern is shown in Figure 15-8 is the angle formed between points A, X, and B (angle θ). Points A and B are the half-power points (the power density at these points is one-half of what it is an equal distance from the antenna in the direction of maximum radiation). Antenna beamwidth is sometimes called −3-dB beamwidth or half-power beamwidth.

Antenna gain is inversely proportional to beamwidth (i.e., the higher the gain of an antenna, the narrower the beamwidth). An omnidirectional (isotropic) antenna radiates equally well in all directions. Thus, it has a gain of unity and a beamwidth of 360°. Typical antennas have beamwidths between 30° and 60°, and it is not uncommon for high-gain microwave antennas to have a beamwidth as low as 1°.

15-9 ANTENNA BANDWIDTH

Antenna *bandwidth* is vaguely defined as the frequency range over which antenna operation is "satisfactory." Bandwidth is normally taken as the difference between the half-power frequencies (difference between the highest and lowest frequencies of operation) but sometimes refers to variations in the antenna's input impedance. Antenna bandwidth is often expressed as a percentage of the antenna's optimum frequency of operation.

Example 15-5

Determine the percent bandwidth for an antenna with an optimum frequency of operation of 400 MHz and −3-dB frequencies of 380 MHz and 420 MHz.

Solution

$$\text{Bandwidth} = \frac{420 - 380}{400} \times 100$$
$$= 10\%$$

15-10 ANTENNA INPUT IMPEDANCE

Radiation from an antenna is a direct result of the flow of RF current. The current flows to the antenna through a transmission line, which is connected to a small gap between the conductors that make up the antenna. The point on the antenna where the transmission line is connected is called the antenna input terminal or simply the *feedpoint.* The feedpoint presents an ac load to the transmission line called the *antenna input impedance.* If the transmitter's output impedance and the antenna's input impedance are equal to the characteristic impedance of the transmission line, there will be no standing waves on the line, and maximum power is transferred to the antenna and radiated.

Antenna input impedance is simply the ratio of the antenna's input voltage to input current. Mathematically, input impedance is

$$Z_{\text{in}} = \frac{E_i}{I_i} \tag{15-19}$$

where Z_{in} = antenna input impedance (ohms)
E_i = antenna input voltage (volts)
I_i = antenna input current (ampere)

Antenna input impedance is generally complex; however, if the feedpoint is at a current maximum and there is no reactive component, the input impedance is equal to the sum of the radiation resistance and the effective resistance.

15-11 BASIC ANTENNA

15-11-1 Elementary Doublet

The simplest type of antenna is the *elementary doublet.* The elementary doublet is an electrically short dipole and is often referred to simply as a *short dipole, elementary dipole,* or *Hertzian dipole.* Electrically short means short compared with one-half wavelength but not necessarily one with a uniform current (generally, any dipole that is less than one-tenth wavelength long is considered electrically short). In reality, an elementary doublet cannot be achieved; however, the concept of a short dipole is useful in understanding more practical antennas.

An elementary doublet has uniform current throughout its length. However, the current is assumed to vary sinusoidally in time and at any instant is

$$i(t) = I \sin(2\pi ft + \theta)$$

where $i(t)$ = instantaneous current (amperes)
I = peak amplitude of the RF current (amperes)
f = frequency (hertz)
t = instantaneous time (seconds)
θ = phase angle (radians)

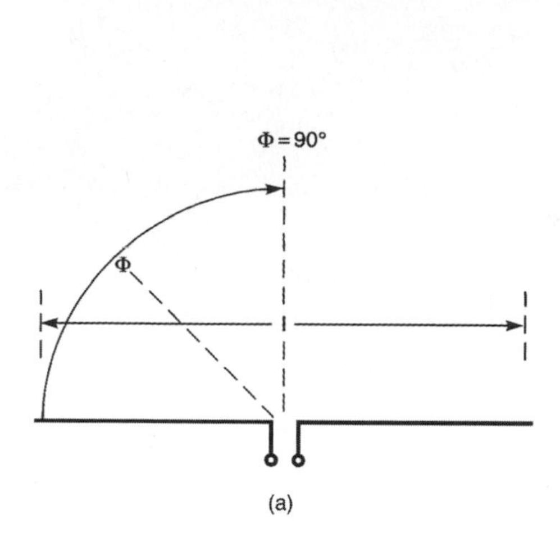

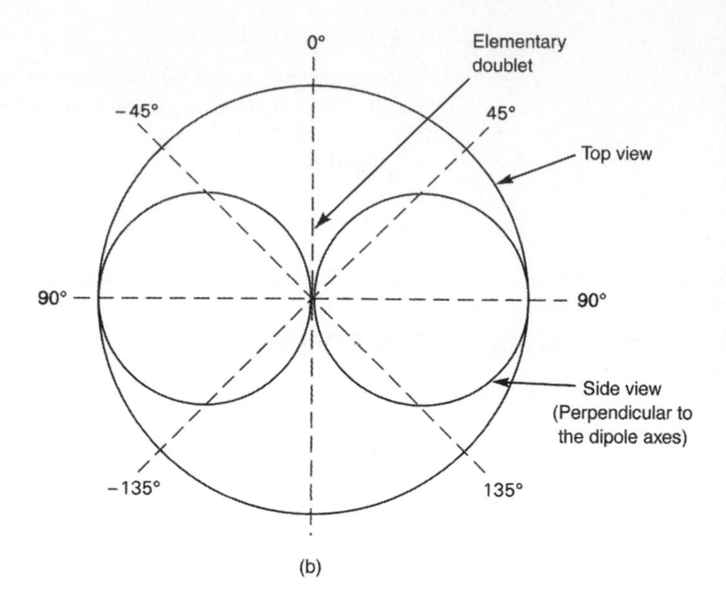

FIGURE 15-9 (a) Elementary doublet; (b) relative radiation pattern (top view)

With the aid of Maxwell's equations, it can be shown that the far (radiation) field is

$$\mathscr{E} = \frac{60\pi Il \sin \phi}{\lambda R} \tag{15-20}$$

where $\mathscr{E}$ = electric field intensity (volts per meter)
I = dipole current (amperes rms)
l = end-to-end length of the dipole (meters)
R = distance from the dipole (meters)
λ = wavelength (meters)
ϕ = angle between the axis of the antenna and the direction of radiation as shown in Figure 15-9a

Plotting Equation 15-20 gives the relative electric field intensity pattern for an elementary dipole, which is shown in Figure 15-9b. It can be seen that radiation is maximum at right angles to the dipole and falls off to zero at the ends.

The relative power density pattern can be derived from Equation 15-10 by substituting $\mathscr{P} = \mathscr{E}^2/120\pi$. Mathematically, we have

$$\mathscr{P} = \frac{30\pi I^2 l^2 \sin^2\phi}{\lambda^2 R^2} \tag{15-21}$$

15-12 HALF-WAVE DIPOLE

The linear half-wave dipole is one of the most widely used antennas at frequencies above 2 MHz. At frequencies below 2 MHz, the physical length of a half-wavelength antenna is prohibitive. The half-wave dipole is generally referred to as a *Hertz antenna* after Heinrich Hertz, who was the first to demonstrate the existence of electromagnetic waves.

A Hertz antenna is a *resonant* antenna. That is, it is a multiple of quarter-wavelengths long and open circuited at the far end. Standing waves of voltage and current exist along a resonant antenna. Figure 15-10 shows the idealized voltage and current distributions along a half-wave dipole. Each pole of the antenna looks as if it were an open quarter-wavelength section of transmission line. Thus, there is a voltage maximum and current minimum at the

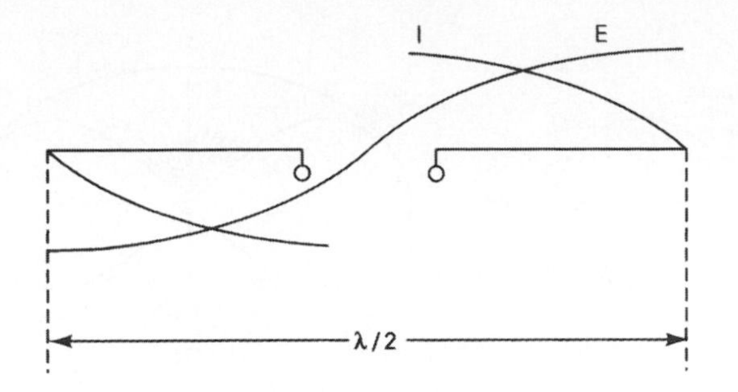

FIGURE 15-10 Idealized voltage and current distributions along a half-wave dipole

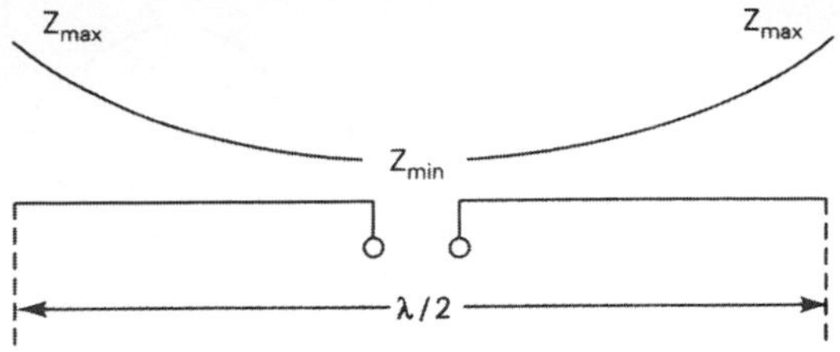

FIGURE 15-11 Impedance curve for a center-fed half-wave dipole

ends and a voltage minimum and current maximum in the middle. Consequently, assuming that the feedpoint is in the center of the antenna, the input impedance is E_{min}/I_{max} and a minimum value. The impedance at the ends of the antenna is E_{max}/I_{min} and a maximum value. Figure 15-11 shows the impedance curve for a center-fed half-wave dipole. The impedance varies from a maximum value at the ends of approximately 2500 Ω to a minimum value at the feedpoint of approximately 73 Ω (of which between 68 Ω and 70 Ω is the radiation resistance).

For an ideal antenna, the efficiency is 100%, directivity equals power gain, and the radiation resistance equals the input impedance (73); thus

$$\begin{aligned} D = A &= \frac{120}{\text{radiation resistance}} \\ & \frac{120}{72} \\ &= 1.667 \end{aligned} \qquad \textbf{(15-22)}$$

and $$10 \log 1.64 = 2.18 \text{ dB}$$

A wire radiator such as a half-wave dipole can be thought of as an infinite number of elementary doublets placed end to end. Therefore, the radiation pattern can be obtained by integrating Equation 15-20 over the length of the antenna. The free-space radiation pattern for a half-wave dipole depends on whether the antenna is placed horizontally or vertically with respect to Earth's surface. Figure 15-12a shows the vertical (from the side) radiation pattern for a vertically mounted half-wave dipole. Note that two major lobes radiate in opposite directions that are at right angles to the antenna. Also note that the lobes are not circles. Circular lobes are obtained only for the ideal case when the current is constant throughout the antenna's length, and this is unachievable in a practical antenna. Figure 15-12b shows the cross-sectional view. Note that the radiation pattern has a figure-eight pattern and resembles the shape of a doughnut. Maximum radiation is in a plane parallel to Earth's surface. The higher the angle of elevation is, the less the radiation, and for 90° there is no radiation. Figure 15-12c shows the horizontal (from the top) radiation pattern for a vertically mounted half-wave dipole. The pattern is circular because radiation is uniform in all directions perpendicular to the antenna.

15-12-1 Ground Effects on a Half-Wave Dipole

The radiation patterns shown in Figure 15-12 are for free-space conditions. In Earth's atmosphere, wave propagation is affected by antenna orientation, atmospheric absorption, and ground effects, such as reflection. The effect of ground reflection for an ungrounded half-wave dipole is shown in Figure 15-13. The antenna is mounted an appreciable number of wavelengths (height h) above the surface of Earth. The field strength at any given point in space is the sum of the direct and ground-reflected waves. The ground-reflected wave

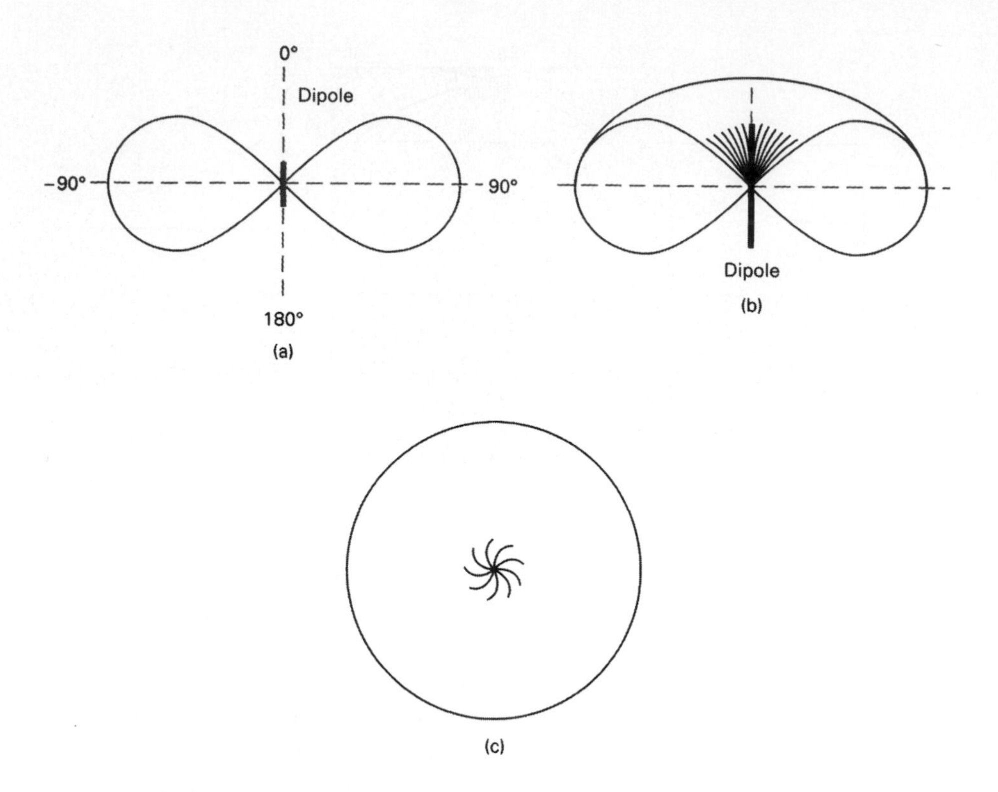

FIGURE 15-12 Half-wave dipole radiation patterns: (a) vertical (side) view of a vertically mounted dipole; (b) cross-sectional view; (c) horizontal (top) view

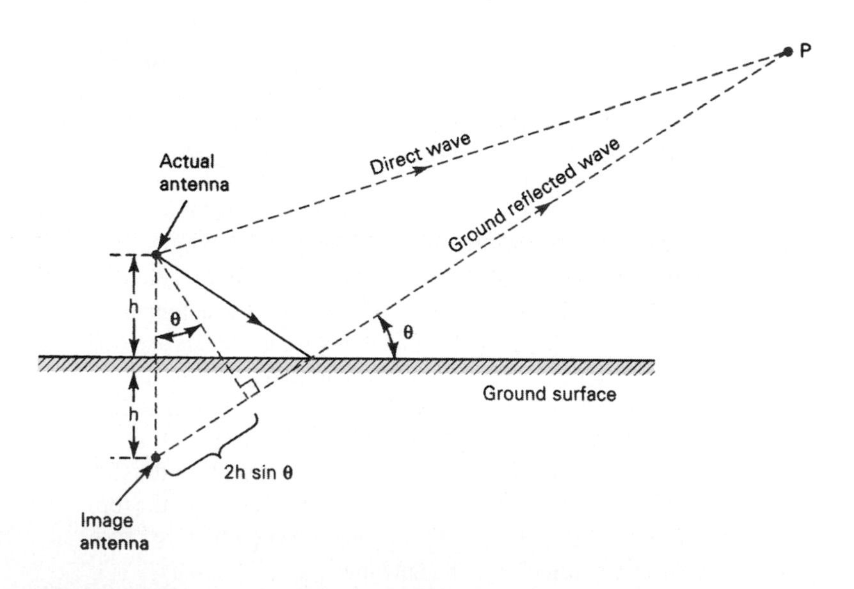

FIGURE 15-13 Ground effects on a half-wave dipole

appears to be radiating from an image antenna distance h below Earth's surface. This apparent antenna is a mirror image of the actual antenna. The ground-reflected wave is inverted 180° and travels a distance $2h \sin \theta$ farther than the direct wave to reach the same point in space (point P). The resulting radiation pattern is a summation of the radiations from the actual antenna and the mirror antenna. Note that this is the classical ray-tracing technique used in optics.

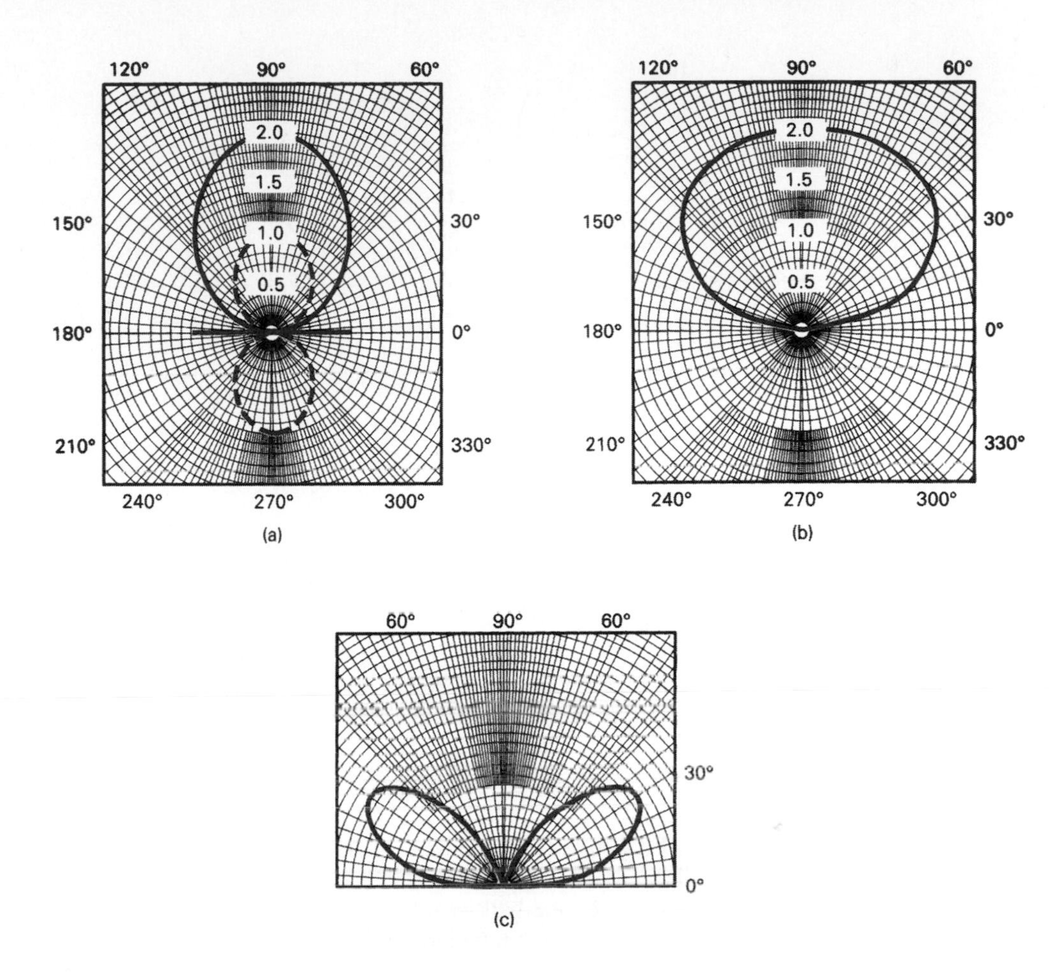

FIGURE 15-14 Vertical radiation pattern for a half-wave dipole: (a) in a plane through antenna; (b) in a plane at right angles to antenna; (c) horizontal dipole one-half wavelength above ground

Figure 15-14 shows the vertical radiation patterns for a horizontally mounted half-wave dipole one-quarter and one-half wavelength above the ground. For an antenna mounted one-quarter wavelength above the ground, the lower lobe is completely gone, and the field strength directly upward is doubled. Figure 15-14a shows in the dotted line the free-space pattern and in the solid line the vertical distribution in a plane through the antenna, and Figure 15-14b shows the vertical distribution in a plane at right angles to the antenna. Figure 15-14c shows the vertical radiation pattern for a horizontal dipole one-half wavelength above the ground. The figure shows that the pattern is now broken into two lobes, and the direction of maximum radiation (end view) is now at 30° to the horizontal instead of directly upward. There is no component along the ground for horizontal polarization because of the phase shift of the reflected component. Ground-reflected waves have similar effects on all antennas. The best way to eliminate or reduce the effect of ground reflected waves is to mount the antenna far enough above Earth's surface to obtain free-space conditions. However, in many applications, this is impossible. Ground reflections are sometimes desirable to get the desired elevation angle for the major lobe's maximum response.

The height of an ungrounded antenna above Earth's surface also affects the antenna's radiation resistance. This is due to the reflected waves cutting through or intercepting the antenna and altering its current. Depending on the phase of the ground-reflected wave, the antenna current can increase or decrease, causing a corresponding increase or decrease in the input impedance.

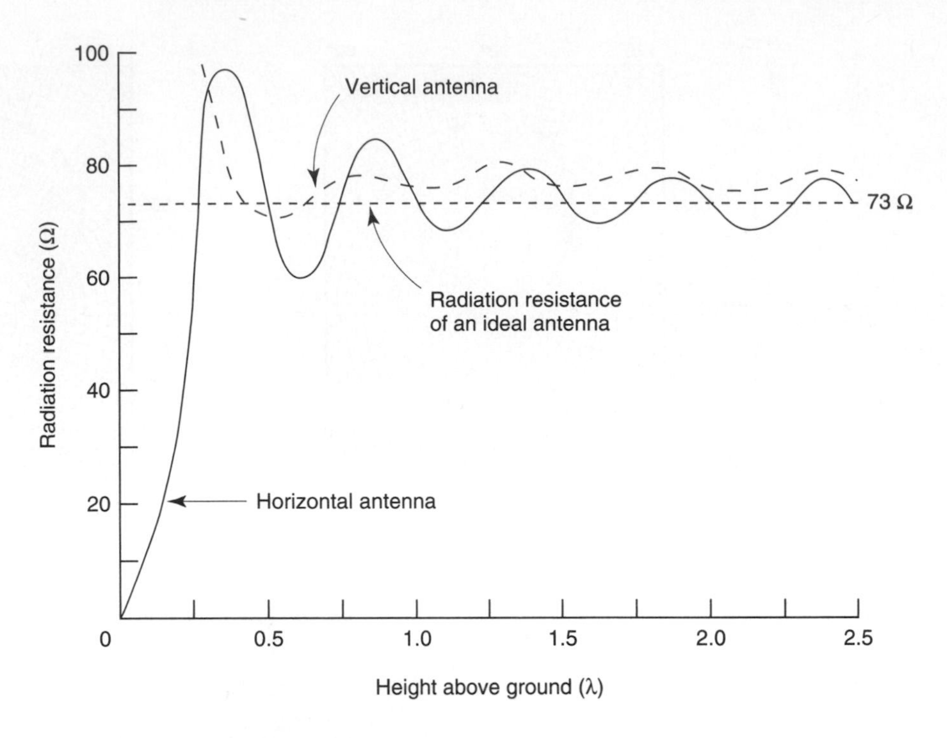

FIGURE 15-15 Radiation resistance versus height above ground

Figure 15-15 shows how the radiation resistance of vertical and horizontal half-wave dipoles varies with distance above the Earth's surface. As the figure shows, once beyond approximately one-half wavelength above ground, the effect of reflections is reduced dramatically, and the radiation resistance remains relatively constant.

15-13 GROUNDED ANTENNA

A *monopole* (single pole) antenna one-quarter wavelength long, mounted vertically with the lower end either connected directly to ground or grounded through the antenna coupling network, is called a *Marconi antenna.* The characteristics of a Marconi antenna are similar to those of the Hertz antenna because of the ground-reflected waves. Figure 15-16a shows the voltage and current standing waves for a quarter-wave grounded antenna. It can be seen that if the Marconi antenna is mounted directly on Earth's surface, the actual antenna and its *image* combine and produce exactly the same standing-wave patterns as those of the half-wave ungrounded (Hertz) antenna. Current maxima occur at the grounded ends, which causes high current flow through ground. To reduce power losses, the ground should be a good conductor, such as rich, loamy soil. If the ground is a poor conductor, such as sandy or rocky terrain, an artificial *ground plane* system made of heavy copper wires spread out radially below the antenna may be required. Another way of artificially improving the conductivity of the ground area below the antenna is with a *counterpoise.* A counterpoise is a wire structure placed below the antenna and erected above the ground. The counterpoise should be insulated from earth ground. A counterpoise is a form of capacitive ground system; capacitance is formed between the counterpoise and Earth's surface.

Figure 15-16b shows the radiation pattern for a quarter-wave grounded (Marconi) antenna. It can be seen that the lower half of each lobe is canceled by the ground-reflected waves. This is generally of no consequence because radiation in the horizontal direction is increased, thus increasing radiation along Earth's surface (ground waves) and improving area coverage. It can also be seen that increasing the antenna length improves horizontal ra-

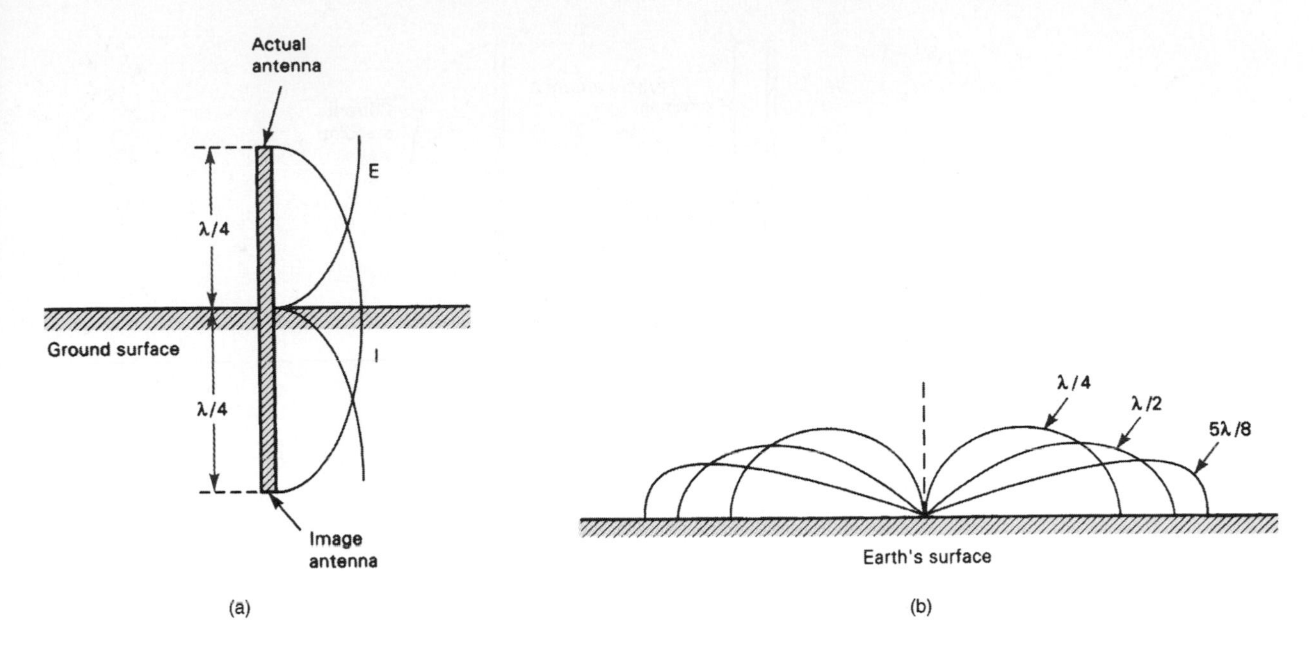

FIGURE 15-16 Quarter-wave grounded antenna: (a) voltage and current standing waves; (b) radiation pattern

diation at the expense of sky-wave propagation. This is also shown in Figure 15-16b. Optimum horizontal radiation occurs for an antenna that is approximately five-eighths wavelength long. For a one-wavelength antenna, there is no ground wave propagation.

A Marconi antenna has the obvious advantage over a Hertz antenna of being only half as long. The disadvantage of a Marconi antenna is that it must be located close to the ground.

15-14 ANTENNA LOADING

Thus far, we have considered antenna length in terms of wavelengths rather than physical dimensions. By the way, how long is a quarter-wavelength antenna? For a transmit frequency of 1 GHz, one-quarter wavelength is 0.075 m (2.95 in). However, for a transmit frequency of 1 MHz, one-quarter wavelength is 75 m, and at 100 kHz, one-quarter wavelength is 750 m. It is obvious that the physical dimensions for low-frequency antennas are not practical, especially for mobile radio applications. However, it is possible to increase the electrical length of an antenna by a technique called *loading.* When an antenna is loaded, its physical length remains unchanged, although its effective electrical length is increased. Several techniques are used for loading antennas.

15-14-1 Loading Coils

Figure 15-17a shows how a coil (inductor) added in series with a dipole antenna effectively increases the antenna's electrical length. Such a coil is appropriately called a *loading coil.* The loading coil effectively cancels out the capacitance component of the antenna input impedance. Thus, the antenna looks as if it were a resonant circuit, is resistive, and can now absorb 100% of the incident power. Figure 15-17b shows the current standing-wave patterns on an antenna with a loading coil. The loading coil is generally placed at the bottom of the antenna, allowing the antenna to be easily tuned to resonance. A loading coil effectively increases the radiation resistance of the antenna by approximately 5 Ω. Note also that the current standing wave has a maximum value at the coil,

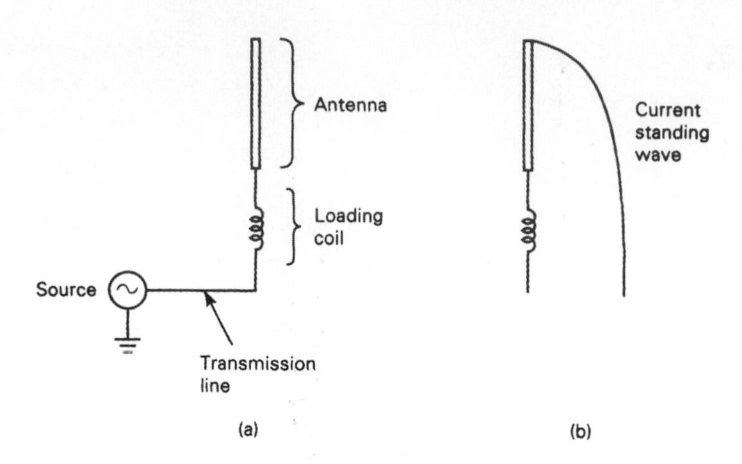

FIGURE 15-17 Loading coil: (a) antenna with loading coil; (b) current standing wave with loading coil

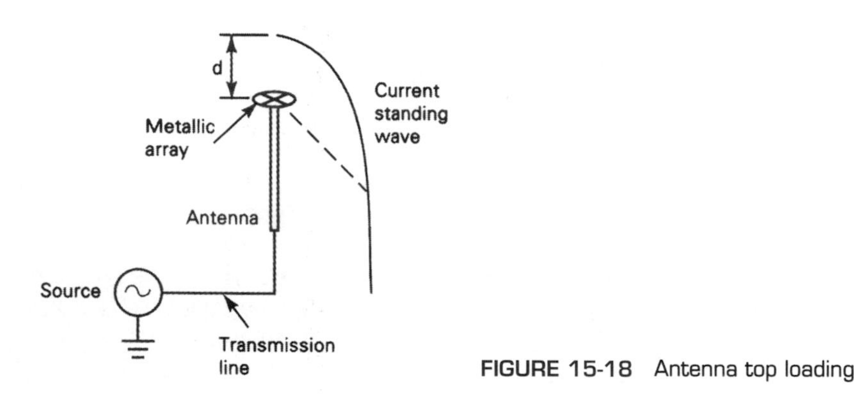

FIGURE 15-18 Antenna top loading

increasing power losses, creating a situation of possible corona, and effectively reducing the radiation efficiency of the antenna.

15-14-2 Top Loading

Loading coils have several shortcomings that can be avoided by using a technique called antenna *top loading.* With top loading, a metallic array that resembles a spoked wheel is placed on top of the antenna. The wheel increases the shunt capacitance to ground, reducing the overall antenna capacitance. Antenna top loading is shown in Figure 15-18. Notice that the current standing-wave pattern is pulled up along the antenna as though the antenna length had been increased distance *d,* placing the current maximum at the base. Top loading results in a considerable increase in the radiation resistance and radiation efficiency. It also reduces the voltage of the standing wave at the antenna base. Unfortunately, top loading is awkward for mobile applications.

The current loop of the standing wave can be raised even further (improving the radiation efficiency even more) if a *flat top* is added to the antenna. If a vertical antenna is folded over on top to form an L or T, as shown in Figure 15-19, the current loop will occur nearer the top of the radiator. If the flat top and vertical portions are each one-quarter wavelength long, the current maximum will occur at the top of the vertical radiator.

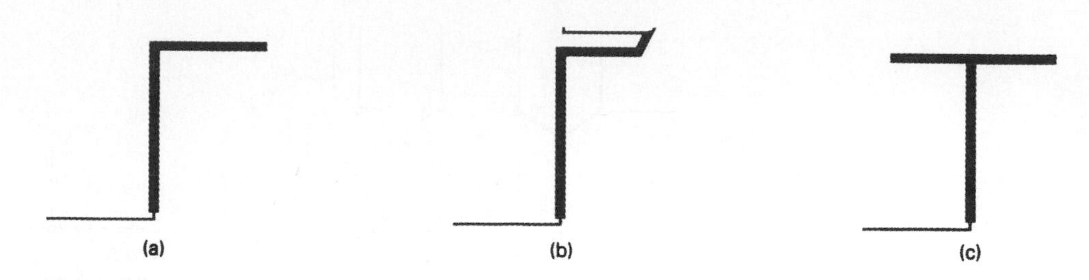

FIGURE 15-19 Flat-top antenna loading

15-15 ANTENNA ARRAYS

An antenna *array* is formed when two or more antenna elements are combined to form a single antenna. An antenna element is an individual radiator, such as a half- or quarter-wave dipole. The elements are physically placed in such a way that their radiation fields interact with each other, producing a total radiation pattern that is the vector sum of the individual fields. The purpose of an array is to increase the directivity of an antenna system and concentrate the radiated power within a smaller geographic area.

In essence, there are two types of antenna elements: *driven* and *parasitic* (nondriven). Driven elements are directly connected to the transmission line and receive power from or are driven by the source. Parasitic elements are not connected to the transmission line; they receive energy only through mutual induction with a driven element or another parasitic element. A parasitic element that is longer than the driven element from which it receives energy is called a *reflector.* A reflector effectively reduces the signal strength in its direction and increases it in the opposite direction. Therefore, it acts as if it were a concave mirror. This action occurs because the wave passing through the parasitic element induces a voltage that is reversed 180° with respect to the wave that induced it. The induced voltage produces an in-phase current, and the element radiates (it actually reradiates the energy it just received). The reradiated energy sets up a field that cancels in one direction and reinforces in the other. A parasitic element that is shorter than its associated driven element is called a *director.* A director increases field strength in its direction and reduces it in the opposite direction. Therefore, it acts as if it were a convergent convex lens. This is shown in Figure 15-20.

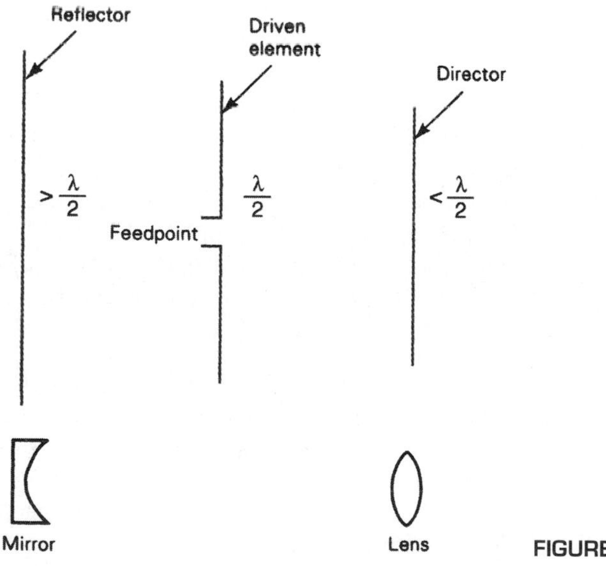

FIGURE 15-20 Antenna array

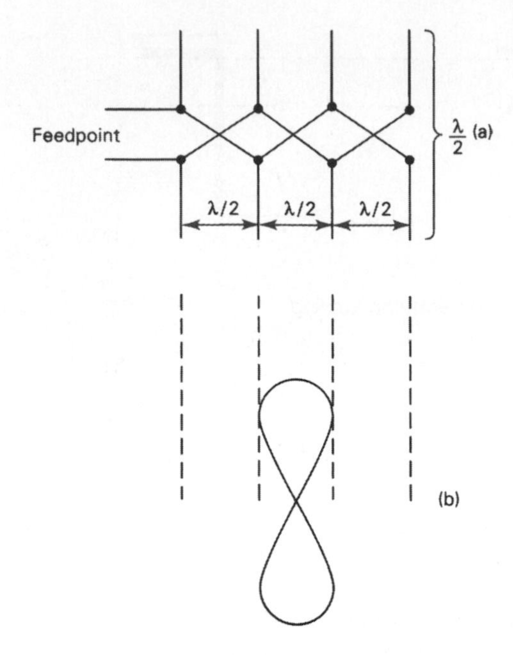

FIGURE 15-21 Broadside antenna: (a) broadside array; (b) radiation pattern

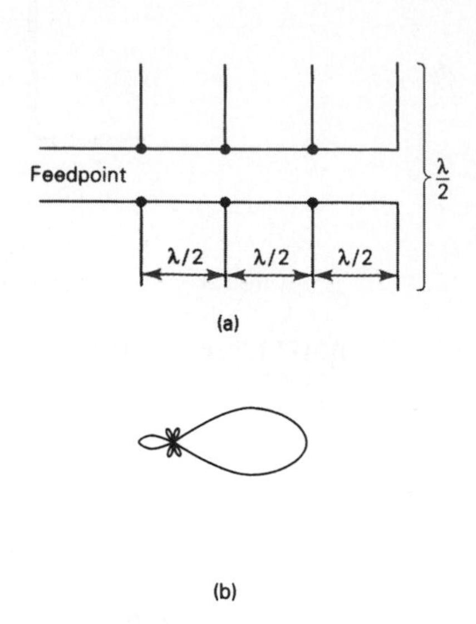

FIGURE 15-22 End-fire antenna: (a) end-fire array; (b) radiation pattern (side view)

Radiation directivity can be increased in either the horizontal or vertical plane, depending on the placement of the elements and whether they are driven. If not driven, the pattern depends on whether the elements are directors or reflectors. If driven, the pattern depends on the relative phase of the feeds.

15-15-1 Broadside Array

A *broadside array* is one of the simplest types of antenna arrays. It is made by simply placing several resonant dipoles of equal size (both length and diameter) in parallel with each other and in a straight line (collinear). All elements are fed in phase from the same source. As the name implies, a broadside array radiates at right angles to the plane of the array and radiates very little in the direction of the plane. Figure 15-21a shows a broadside array that is comprised of four driven half-wave elements separated by one-half wavelength. Therefore, the signal that is radiated from element 2 has traveled one-half wavelength farther than the signal radiated from element 1 (i.e., they are radiated 180° out of phase). Crisscrossing the transmission line produces an additional 180° phase shift. Therefore, the currents in all the elements are in phase, and the radiated signals are in phase and additive in a plane at right angles to the plane of the array. Although the horizontal radiation pattern for each element by itself is omnidirectional, when combined their fields produce a highly directive bidirectional radiation pattern (Figure 15-21b). Directivity can be increased even further by increasing the length of the array by adding more elements.

15-15-2 End-Fire Array

An *end-fire array* is essentially the same element configuration as the broadside array except that the transmission line is not crisscrossed between elements. As a result, the fields are additive in line with the plane of the array. Figure 15-22 shows an end-fire array and its resulting radiation pattern.

Table 15-1 shows the effects on the gain, beamwidth, and front-to-back ratio of a half-wave dipole by adding directors and reflectors.

Table 15-1

Type	Power Gain (dB)	Beamwidth (°)	Front-to-Back Ratio (dB)
Isotropic radiator	0	360	0
Half-wave dipole	2.16	80	0
1 element (director or reflector)	7.16	52	15 or 20
2 elements (1 reflector + 1 director or 2 directors)	10.16	40	23
3 elements (1 reflector + 2 directors)	12.06	36	24
4 elements (1 reflector + 3 directors)	13.26	34	25
6 elements (1 reflector + 5 directors)	14.56	30	27
10 elements (1 reflector + 9 directors)	15.66	24	29.5

Note. 0.1 λ = director spacing, 0.15 λ = reflector spacing, and 0.05 λ = length change per element.

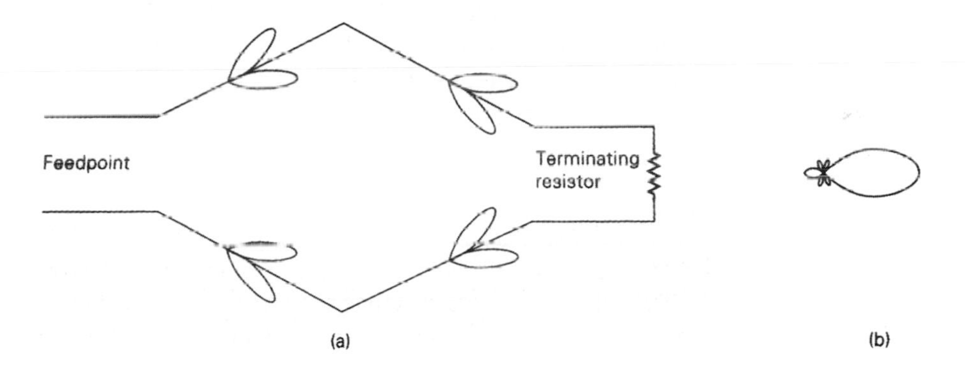

FIGURE 15-23 Rhombic antenna: (a) rhombic array; (b) radiation pattern

15-15-3 Nonresonant Array: The Rhombic Antenna

The *rhombic antenna* is a nonresonant antenna that is capable of operating satisfactorily over a relatively wide bandwidth, making it ideally suited for HF transmission (range 3 MHz to 30 MHz). The rhombic antenna is made up of four nonresonant elements each several wavelengths long. The entire array is terminated in a resistor if unidirectional operation is desired. The most widely used arrangement for the rhombic antenna resembles a transmission line that has been pinched out in the middle; it is shown in Figure 15-23. The antenna is mounted horizontally and placed one-half wavelength or more above the ground. The exact height depends on the precise radiation pattern desired. Each set of elements acts as a transmission line terminated in its characteristic impedance; thus, waves are radiated only in the forward direction. The terminating resistor absorbs approximately one-third of the total antenna input power. Therefore, a rhombic antenna has a maximum efficiency of 67%. Gains of over 40 (16 dB) have been achieved with rhombic antennas.

15-16 SPECIAL-PURPOSE ANTENNAS

15-16-1 Folded Dipole

A two-wire *folded dipole* and its associated voltage standing-wave pattern are shown in Figure 15-24a. The folded dipole is essentially a single antenna made up of two elements.

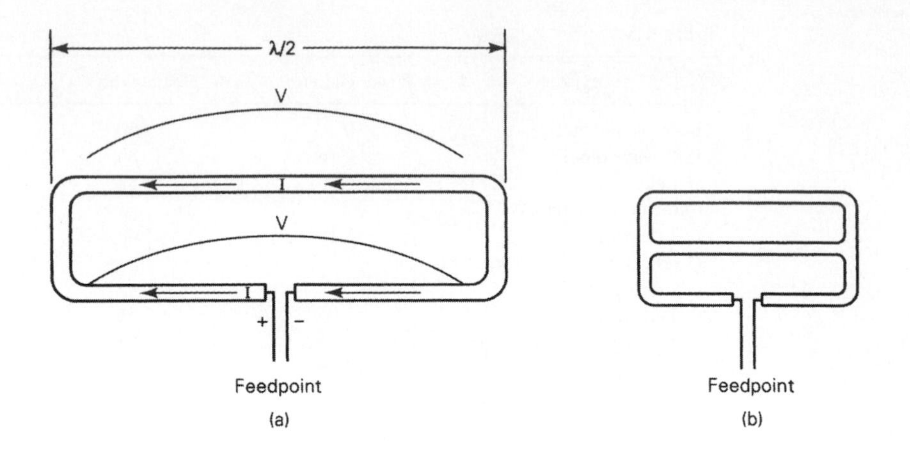

FIGURE 15-24 (a) Folded dipole; (b) three-element folded dipole

One element is fed directly, whereas the other is conductively coupled at the ends. Each element is one-half wavelength long. However, because current can flow around corners, there is a full wavelength of current on the antenna. Therefore, for the same input power, the input current will be one-half that of the basic half-wave dipole, and the input impedance is four times higher ($4 \times 72 = 288$). The input impedance of a folded dipole is equal to the half-wave impedance (72 Ω) times the number of folded wires squared. For example, if there are three dipoles, as shown in Figure 15-24b, the input impedance is $3^2 \times 72 = 648\ \Omega$. Another advantage of a folded dipole over a basic half-wave dipole is wider bandwidth. The bandwidth can be increased even further by making the dipole elements larger in diameter (such an antenna is appropriately called a *fat dipole*). However, fat dipoles have slightly different current distributions and input impedance characteristics than thin ones.

15-16-1-1 Yagi-Uda antenna. A widely used antenna that commonly uses a folded dipole as the driven element is the *Yagi-Uda antenna,* named after two Japanese scientists who invented it and described its operation. (The Yagi-Uda generally is called simply Yagi.) A Yagi antenna is a linear array consisting of a dipole and two or more parasitic elements: one reflector and one or more directors. A simple three-element Yagi is shown in Figure 15-25a. The driven element is a half-wavelength folded dipole. (This element is referred to as the driven element because it is connected to the transmission line; however, it is generally used for receiving only.) The reflector is a straight aluminum rod approximately 5% longer than the dipole, and the director is cut approximately 5% shorter than the driven element. The spacing between elements is generally between 0.1 and 0.2 wavelength. Figure 15-25b shows the radiation pattern for a Yagi antenna. The typical directivity for a Yagi is between 7 dB and 9 dB. The bandwidth of the Yagi can be increased by using more than one folded dipole, each cut to a slightly different length. Therefore, the Yagi antenna is commonly used for VHF television reception because of its wide bandwidth (the VHF TV band extends from 54 MHz to 216 MHz). Table 15-2 lists the element spacings for Yagi arrays with from two to eight elements.

15-16-2 Turnstile Antenna

A turnstile antenna is formed by placing two dipoles at right angles to each other, 90° out of phase, as shown in Figure 15-26a. The radiation pattern shown in Figure 15-26b is the sum of the radiation patterns from the two dipoles, which produces a nearly omnidirectional pattern. Turnstile antenna gains of 10 or more dB are common.

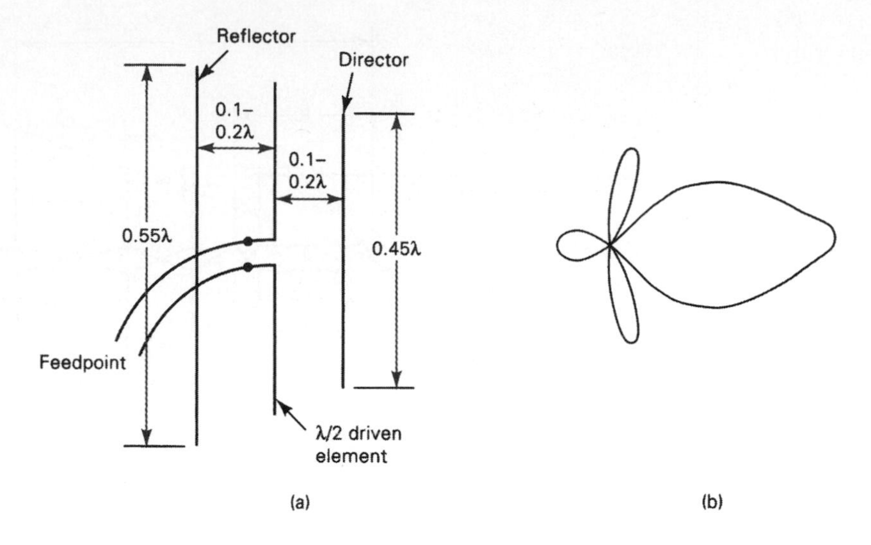

FIGURE 15-25 Yagi-Uda antenna: (a) three-element Yagi; (b) radiation pattern

Table 15-2 Element Spacing for Yagi Arrays (All Units in Wavelength, λ)

Element Spacing	Number of Elements 2	3	4	5	6	7	8
Reflector from driven element	0.19	0.19	0.19	0.18	0.18	0.18	0.18
Director 1 from driven element	—	0.17	0.16	0.16	0.16	0.16	0.15
Director 2 from director 1	—	—	0.16	0.18	0.20	0.21	0.22
Director 3 from director 2	—	—	—	0.20	0.25	0.30	0.30
Director 4 from director 3	—	—	—	—	0.28	0.28	0.29
Director 5 from director 4	—	—	—	—	—	0.30	0.30
Director 6 from director 5	—	—	—	—	—	—	0.35
Director 7 from director 6	—	—	—	—	—	—	—

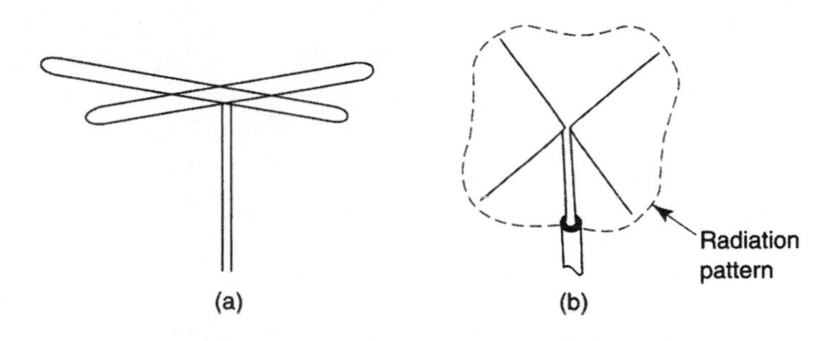

FIGURE 15-26 (a) Turnstile antenna; (b) radiation pattern

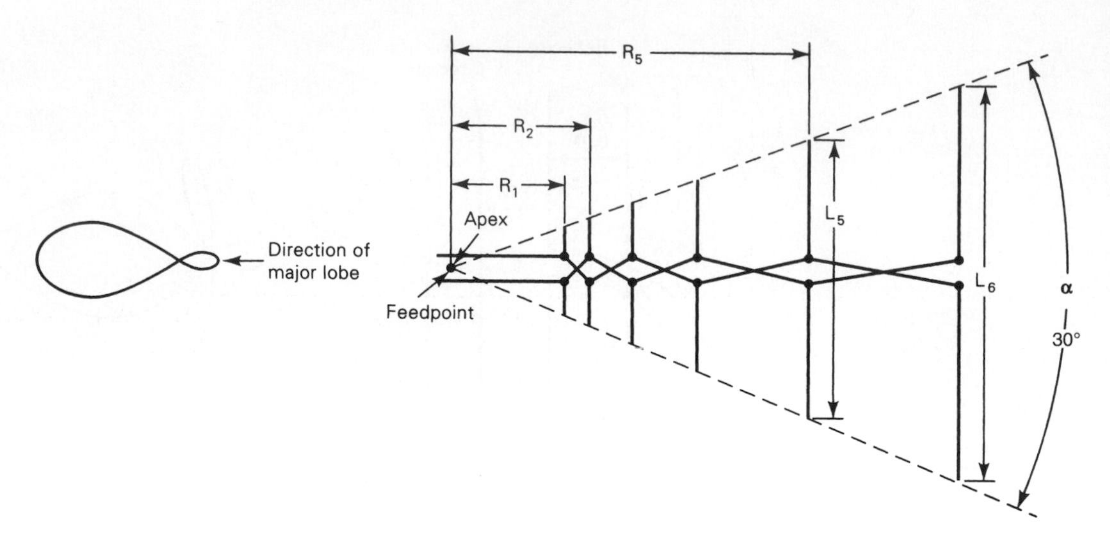

FIGURE 15-27 Log-periodic antenna

15-16-3 Log-Periodic Antenna

A class of frequency-independent antennas called *log periodics* evolved from the initial work of V. H. Rumsey, J. D. Dyson, R. H. DuHamel, and D. E. Isbell at the University of Illinois in 1957. The primary advantage of log-periodic antennas is the independence of their radiation resistance and radiation pattern to frequency. Log-periodic antennas have bandwidth ratios of 10:1 or greater. The bandwidth ratio is the ratio of the highest to the lowest frequency over which an antenna will satisfactorily operate. The bandwidth ratio is often used rather than simply stating the percentage of the bandwidth to the center frequency. Log periodics are not simply a type of antenna but rather a class of antenna because there are many different types, some that are quite unusual. Log-periodic antennas can be unidirectional or bidirectional and have a low to moderate directive gain. High gains may also be achieved by using them as an element in a more complicated array.

The physical structure of a log-periodic antenna is repetitive, which results in repetitive behavior in its electrical characteristics. In other words, the design of a log-periodic antenna consists of a basic geometric pattern that repeats, except with a different size pattern. A basic log-periodic dipole array is probably the closest that a log period comes to a conventional antenna; it is shown in Figure 15-27. It consists of several dipoles of different length and spacing that are fed from a single source at the small end. The transmission line is crisscrossed between the feedpoints of adjacent pairs of dipoles. The radiation pattern for a basic log-period antenna has maximum radiation outward from the small end. The lengths of the dipoles and their spacing are related in such a way that adjacent elements have a constant ratio to each other. Dipole lengths and spacings are related by the formula

$$\frac{R_2}{R_1} = \frac{R_3}{R_2} = \frac{R_4}{R_3} = \frac{1}{\tau} = \frac{L_2}{L_1} = \frac{L_3}{L_2} = \frac{L_4}{L_3} \tag{15-23}$$

or

$$\frac{1}{\tau} = \frac{R_n}{R_{n-1}} = \frac{L_n}{L_{n-1}}$$

where R = dipole spacing (inches)
L = dipole length (inches)
τ = design ratio (number less than 1)

The ends of the dipoles lie along a straight line, and the angle where they meet is designated α. For a typical design, $\tau = 0.7$ and $\alpha = 30°$. With the preceding structural stipu-

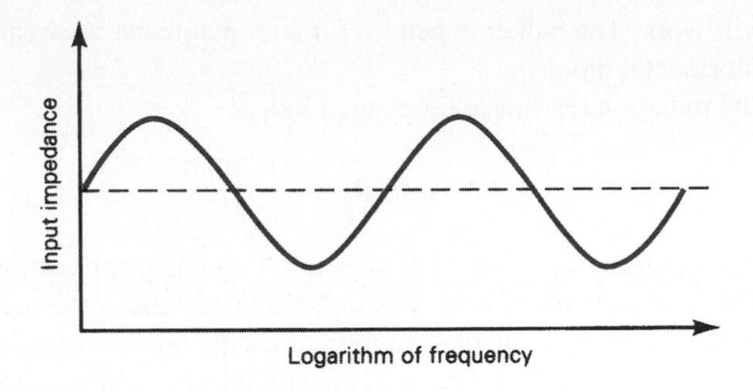

FIGURE 15-28 Log-periodic input impedance versus frequency

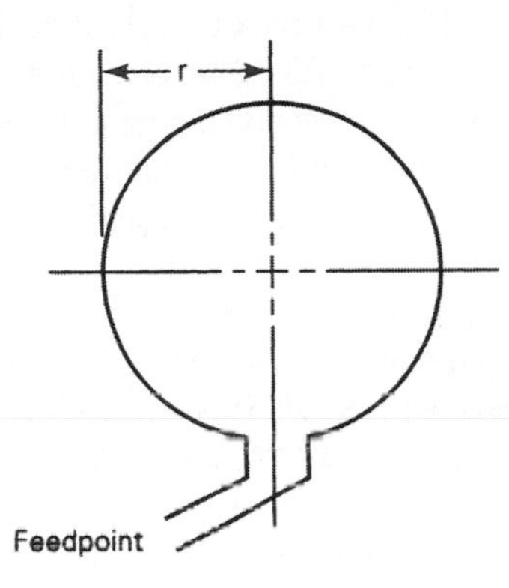

FIGURE 15-29 Loop antenna

lations, the antenna input impedance varies repetitively when plotted as a function of frequency and periodically when plotted against the log of the frequency (hence the name "log periodic"). A typical plot of the input impedance is shown in Figure 15-28. Although the input impedance varies periodically, the variations are not necessarily sinusoidal. Also, the radiation pattern, directivity, power gain, and beamwidth undergo a similar variation with frequency.

The magnitude of a log-frequency period depends on the design ratio and, if two successive maxima occur at frequencies f_1 and f_2, they are related by the formula

$$\log f_2 - \log f_1 = \log\frac{f_2}{f_1} = \log\frac{1}{\tau} \qquad \textbf{(15-24)}$$

Therefore, the measured properties of a log-periodic antenna at frequency f will have identical properties at frequency τf, $\tau^2 f$, $\tau^3 f$, and so on. Log-periodic antennas, like rhombic antennas, are used mainly for HF and VHF communications. However, log-periodic antennas do not have a terminating resistor and are, therefore, more efficient. Very often, TV antennas advertised as "high-gain" or "high-performance" antennas are log-period antennas.

15-16-4 Loop Antenna

The most fundamental *loop antenna* is simply a single-turn coil of wire that is significantly shorter than one wavelength and carries RF current. Such a loop is shown in Figure 15-29. If the radius (r) is small compared with a wavelength, current is essentially in phase throughout the loop. A loop can be thought of as many elemental dipoles connected together. Dipoles are straight; therefore, the loop is actually a polygon rather than circular. However, a circle can be approximated if the dipoles are assumed to be sufficiently short. The loop is surrounded by a magnetic field that is at right angles to the wire, and the directional pattern is independent of its exact shape. Generally, loops are circular; however, any shape will work. The radiation pattern for a loop antenna is essentially the same as that of a short horizontal dipole.

The radiation resistance for a small loop is

$$R_r = \frac{31{,}200\,A^2}{\lambda^4} \tag{15-25}$$

where A is the area of the loop. For very low frequency applications, loops are often made with more than one turn of wire. The radiation resistance of a multiturn loop is simply the radiation resistance for a single-turn loop times the number of turns squared. The polarization of a loop antenna, as that of an elemental dipole, is linear. However, a vertical loop is vertically polarized, and a horizontal loop is horizontally polarized.

Small vertically polarized loops are very often used as direction-finding antennas. The direction of the received signal can be found by orienting the loop until a null or zero value is found. This is the direction of the received signal. Loops have an advantage over most other types of antennas in direction finding in that loops are generally much smaller and, therefore, more easily adapted to mobile communications applications.

15-16-5 Phased Array Antennas

A *phased array antenna* is a group of antennas or a group of antenna arrays that, when connected together, function as a single antenna whose beamwidth and direction (i.e., radiation pattern) can be changed electronically without having to physically move any of the individual antennas or antenna elements within the array. The primary advantage of phased array antennas is that they eliminate the need for mechanically rotating antenna elements. In essence, a phased array is an antenna whose radiation pattern can be electronically adjusted or changed. The primary application of phased arrays is in radar when radiation patterns must be capable of being rapidly changed to follow a moving object. However, governmental agencies that transmit extremely high-power signals to select remote locations all over the world, such as Voice of America, also use adjustable phased antenna arrays to direct their transmissions.

The basic principle of phased arrays is based on interference among electromagnetic waves in free space. When electromagnetic energies from different sources occupy the same space at the same time, they combine, sometimes constructively (aiding each other) and sometimes destructively (opposing each other).

There are two basic kinds of phased antenna arrays. In the first type, a single relatively high-power output device supplies transmit power to a large number of antennas through a set of power splitters and phase shifters. How much of the total transmit power goes to each antenna and the phase of the signal are determined by an intricate combination of adjustable attenuators and time delays. The amount of loss in the attenuators and the phase shift introduced in the time delays is controlled by a computer. The time delays pass the RF signal without distorting it, other than to provide a specific amount of time delay (phase shift). The second kind of phased antenna arrays uses approximately as many low-power variable output devices as there are radiating elements, and the phase relationship among the output signals is controlled with phase shifters. In both types of phased arrays, the radiation pattern is selected by changing the phase delay introduced by each phase shifter. Figure 15-30 shows a phased antenna array that uses several identical antenna elements, each with its own adjustable phase delay.

15-16-6 Helical Antenna

A *helical antenna* is a broadband VHF or UHF antenna that is ideally suited for applications for which radiating circular rather than horizontal or vertical polarized electromagnetic waves are required. A helical antenna can be used as a single-element antenna or

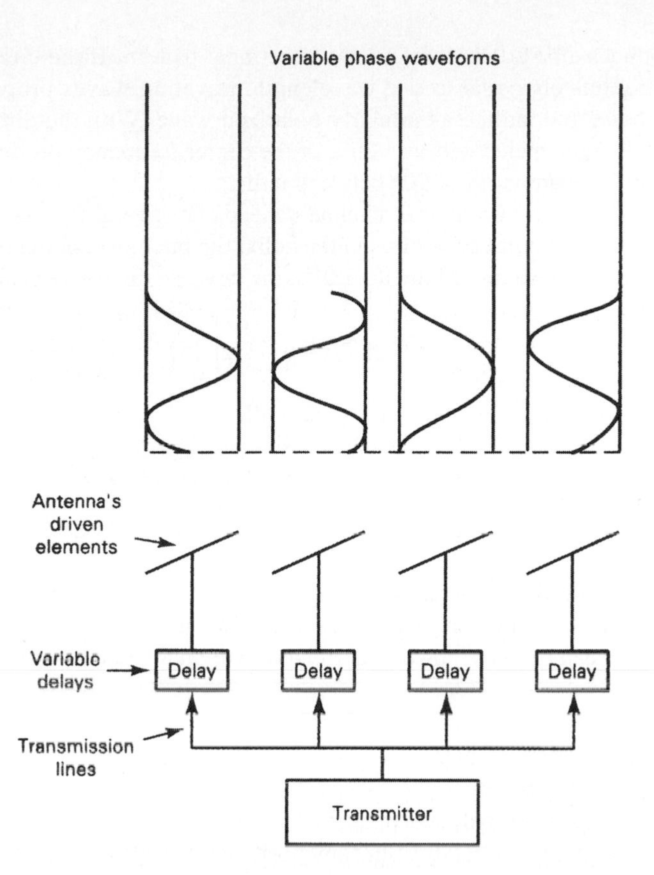

FIGURE 15-30 Phased array antenna

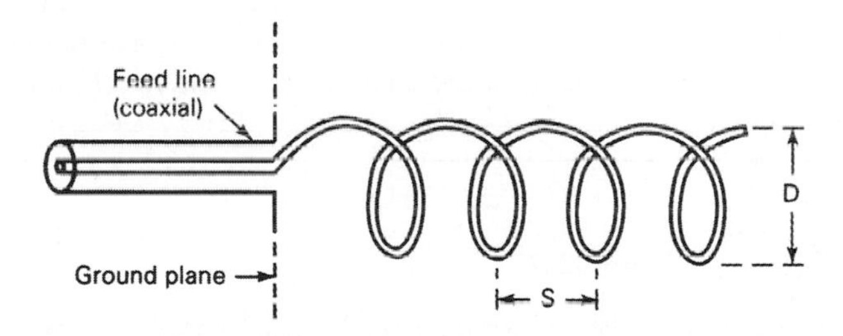

FIGURE 15-31 End-fire helical antenna

stacked horizontally or vertically in an array to modify its radiation pattern by increasing the gain and decreasing the beamwidth of the primary lobe.

A basic end-fire helical antenna is shown in Figure 15-31. The driven element of the antenna consists of a loosely wound rigid helix with an axis length approximately equal to the product of the number of turns and the distance between turns (pitch). A helical antenna is mounted on a ground plane made up of either solid metal or a metal screen that resembles chicken wire. With a helical antenna, there are two modes of propagation: *normal* and *axial*. In the normal mode, electromagnetic radiation is in a direction at right angles to the axis of the helix. In the axial mode, radiation is in the axial direction and produces a broadband, relatively directional pattern. If the circumference of the helix is approximately equal to one wavelength, traveling waves propagate around the turns of the helix and radiate a circularly polarized wave. With the dimensions shown in Figure 15-31, frequencies within ±20% of the center frequency produce a directivity of almost 25 and a beamwidth of 90° between nulls.

The gain of a helical antenna depends on several factors, including the diameter of the helix, the number of turns in the helix, the pitch or spacing between turns, and the frequency of operation. Mathematically, the power gain of a helical antenna is

$$A_{p(\mathrm{dB})} = 10 \log\left[15\left(\frac{\pi D}{\lambda}\right)^2 \frac{(NS)}{\lambda}\right] \tag{15-26}$$

where $A_{p(\mathrm{dB})}$ = antenna power gain (dB)
D = helix diameter (meters)
N = number of turns (any positive integer)
S = pitch (meters)
λ = wavelength (meters per cycle)

Typically, a helical antenna will have between a minimum of 3 or 4 and a maximum of 20 turns and power gains between 15 dB and 20 dB. The 3-dB beamwidth of a helical antenna can be determined with the following mathematical expression:

$$\theta = \frac{52}{(\pi D/\lambda)(\sqrt{NS/\lambda})} \tag{15-27}$$

where θ = beamwidth (degrees)
D = helix diameter (meters)
N = number of turns (any positive integer)
S = pitch (meters)
λ = wavelength (meters per cycle)

From Equations 15-26 and 15-27, it can be seen that, for a given helix diameter and pitch, the power gain increases proportionally to the number of turns and the beamwidth decreases. Helical antennas provide bandwidths anywhere between ±20% of the center frequency up to as much as a 2:1 span between the maximum and minimum operating frequencies.

15-17 UHF AND MICROWAVE ANTENNAS

Antennas used for UHF (0.3 GHz to 3 GHz) and microwave (1 GHz to 100 GHz) must be highly directive. An antenna has an apparent gain because it concentrates the radiated power in a narrow beam rather than sending it uniformly in all directions, and the beamwidth decreases with increases in antenna gain. The relationship among antenna area, gain, and beamwidth are shown in Figure 15-32. Microwave antennas ordinarily have half-power beamwidths on the order of 1° or less. A narrow beamwidth minimizes the effects of interference from outside sources and adjacent antennas. However, for line-of-site transmission, such as used with microwave radio, a narrow beamwidth imposes several limitations, such as mechanical stability and fading, which can lead to problems in antenna lineup.

All the electromagnetic energy emitted by a microwave antenna is not radiated in the direction of the *main lobe* (beam); some of it is concentrated in *minor lobes* called *side lobes,* which can be sources of interference into or from other microwave signal paths.

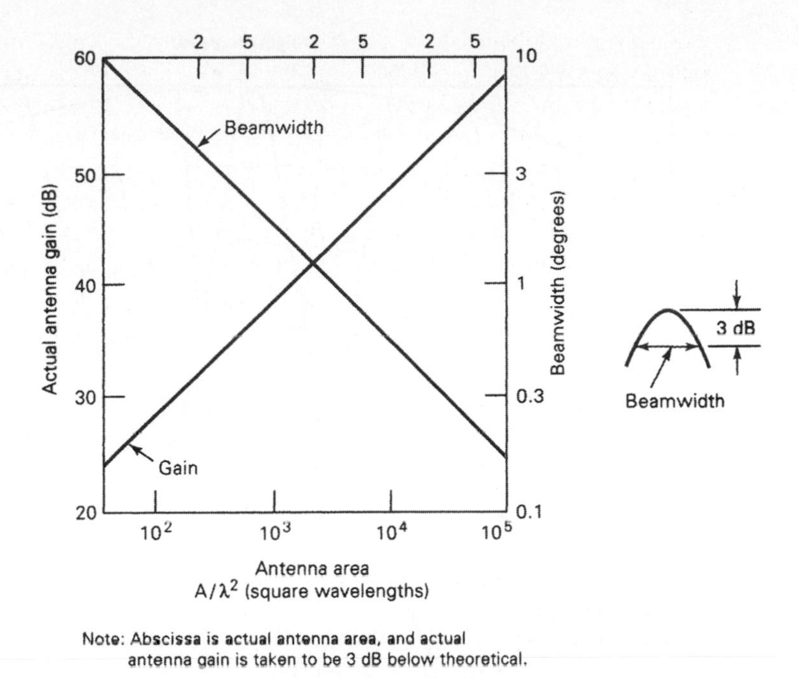

FIGURE 15-32 Antenna power gain and beamwidth relationship

Figure 15-33 shows the relationship between the main beam and the side lobes for a typical microwave antenna, such as a parabolic reflector.

Three important characteristics of microwave antennas are the front-to-back ratio, side-to-side coupling, and back-to-back coupling. The *front-to-back ratio* of an antenna is defined as the ratio of its maximum gain in the forward direction to its maximum gain in its backward direction. The front-to-back ratio of an antenna in an actual installation may be 20 dB or more below its isolated or free-space value because of foreground reflections from objects in or near the main transmission lobe. The front-to-back ratio of a microwave antenna is critical in radio system design because the transmit and receive antennas at repeater stations are often located opposite each other on the same structure (microwave radio systems and repeaters are discussed in more detail in Chapter 23). *Side-to-side* and *back-to-back coupling* express in decibels the coupling loss between antennas carrying transmitter output signals and nearby antennas carrying receiver input signals. Typically, transmitter output powers are 60 dB or higher in signal level than receiver input levels; accordingly, the coupling losses must be high to prevent a transmit signal from one antenna interfering with a receive signal of another antenna.

Highly directional (high gain) antennas are used with *point-to-point* microwave systems. By focusing the radio energy into a narrow beam that can be directed toward the receiving antenna, the transmitting antenna can increase the effective radiated power by several orders of magnitude over that of a nondirectional antenna. The receiving antenna, in a manner analogous to that of a telescope, can also increase the effective received power by a similar amount. The most common type of antenna used for microwave transmission and reception is the parabolic reflector.

15-17-1 Parabolic Reflector Antenna

Parabolic reflector antennas provide extremely high gain and directivity and are very popular for microwave radio and satellite communications links. A parabolic antenna consists of two main parts: a *parabolic reflector* and the active element called the *feed mechanism.*

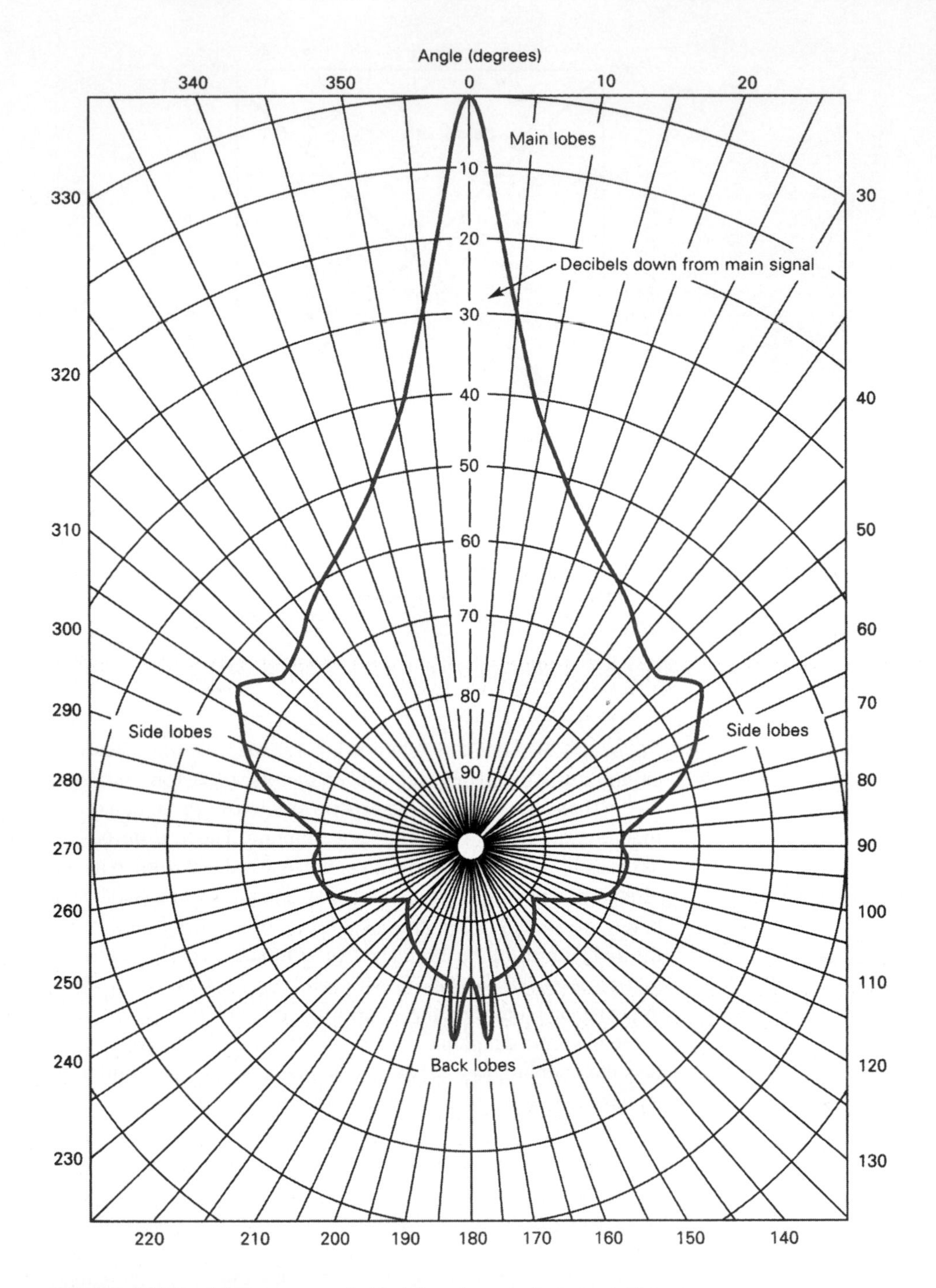

FIGURE 15-33 Main beam and side lobes for a typical parabolic antenna

In essence, the feed mechanism houses the primary antenna (usually a dipole or a dipole array), which radiates electromagnetic waves toward the reflector. The reflector is a passive device that simply reflects the energy radiated by the feed mechanism into a concentrated, highly directional emission in which the individual waves are all in phase with each other (an in-phase wavefront).

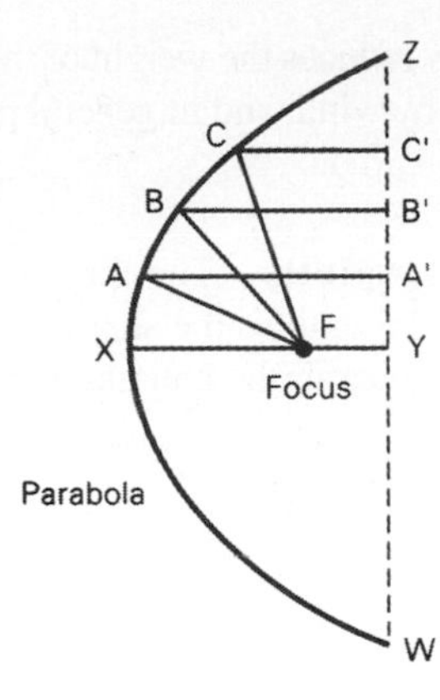

FIGURE 15-34 Geometry of a parabola

15-17-1-1 Parabolic reflectors. The parabolic reflector is probably the most basic component of a parabolic antenna. Parabolic reflectors resemble the shape of a plate or dish; therefore, they are sometimes called *parabolic dish* antennas or simply *dish* antennas. To understand how a parabolic reflector works, it is necessary to first understand the geometry of a *parabola.* A parabola is a plane curve that is expressed mathematically as $y = ax^2$ and defined as the locus of a point that moves so that its distance from another point (called the *focus*) added to its distance from a straight line (called the *directrix*) is of constant length. Figure 15-34 shows the geometry of a parabola whose focus is at point F and whose axis is line XY.

For the parabola shown in Figure 15-34, the following relationships exist:

$$FA + AA' = FB + BB' = FC + CC' = k \quad \text{(a constant length)}$$

and FX = focal length of the parabola (meters)
k = a constant for a given parabola (meters)
WZ = directrix length (meters)

The ratio of the focal length to the diameter of the mouth of the parabola (FX/WZ) is called the *aperture ratio* or simply *aperture* of the parabola; the same term is used to describe camera lenses. A parabolic reflector is obtained when the parabola is revolved around the XY axis. The resulting curved surface dish is called a *paraboloid.* The reflector behind the bulb of a flashlight or the headlamp of an automobile has a paraboloid shape to concentrate the light in a particular direction.

A parabolic antenna consists of a paraboloid reflector illuminated with microwave energy radiated by a feed system located at the focus point. If electromagnetic energy is radiating toward the parabolic reflector from the focus, all radiated waves will travel the same distance by the time they reach the directrix, regardless from which point on the parabola they are reflected. Thus, all waves radiated toward the parabola from the focus will be in phase when they reach the directrix (line WZ). Consequently, radiation is concentrated along the XY axis, and cancellation takes place in all other directions. A paraboloid reflector used to receive electromagnetic energy exhibits exactly the same behavior. Thus, a parabolic antenna exhibits the *principle of reciprocity* and works equally well as a receive antenna for waves arriving from the XY direction (normal to the directrix). Rays received from all other directions are canceled at that point.

It is not necessary that the dish have a solid metal surface to efficiently reflect or receive the signals. The surface can be a mesh and still reflect or receive almost as much energy as a solid surface, provided the width of the openings is less than 0.1 wavelength. Using a mesh rather than a solid conductor considerably reduces the weight of the reflector. Mesh reflectors are also easier to adjust, are affected less by wind, and in general provide a much more stable structure.

15-17-1-2 Parabolic antenna beamwidth. The three-dimensional radiation from a parabolic reflector has a main lobe that resembles the shape of a fat cigar in direction *XY*. The approximate −3-dB beamwidth for a parabolic antenna in degrees is given as

$$\theta = \frac{70\lambda}{D} \tag{15-28}$$

or

$$\theta = \frac{70c}{fD} \tag{15-29}$$

where θ = beamwidth between half-power points (degrees)
λ = wavelength (meters)
$c = 3 \times 10^8$ meters per second
D = antenna mouth diameter (meters)
f = frequency (hertz)

and

$$\phi_0 = 2\theta \tag{15-30}$$

where ϕ_0 equals the beamwidth between nulls in the radiation pattern (degrees). Equations 15-28 through 15-30 are accurate when used for antennas with large apertures (i.e., narrow beamwidths).

15-17-1-3 Parabolic antenna efficiency (η). In a parabolic reflector, reflectance from the surface of the dish is not perfect. Therefore, a small portion of the signal radiated from the feed mechanism is absorbed at the dish surface. In addition, energy near the edge of the dish does not reflect but rather is diffracted around the edge of the dish. This is called *spillover* or *leakage*. Because of dimensional imperfections, only about 50% to 75% of the energy emitted from the feed mechanism is actually reflected by the paraboloid. Also, in a real antenna the feed mechanism is not a point source; it occupies a finite area in front of the reflector and actually obscures a small area in the center of the dish and causes a shadow area in front of the antenna that is incapable of either gathering or focusing energy. These imperfections contribute to a typical efficiency for a parabolic antenna of only about 55% ($\eta = 0.55$). That is, only 55% of the energy radiated by the feed mechanism actually propagates forward in a concentrated beam.

15-17-1-4 Parabolic antenna power gain. For a transmit parabolic antenna, the power gain is approximated as

$$A_p = \eta\left(\frac{\pi D}{\lambda}\right)^2 \tag{15-31a}$$

where A_p = power gain with respect to an isotropic antenna (unitless)
D = mouth diameter of parabolic reflector (meters)
η = antenna efficiency (antenna radiated power relative to the power radiated by the feed mechanism) (unitless)
λ = wavelength (meters per cycle)

For a typical antenna efficiency of 55% ($\eta = 0.55$), Equation 15-31a reduces to

$$A_p = \frac{5.4D^2f^2}{c^2} \tag{15-31b}$$

where c is the velocity of propagation (3×10^8 m/s). In decibel form,

$$A_{p(\mathrm{dB})} = 20 \log f_{(\mathrm{MHz})} + 20 \log D_{(\mathrm{m})} - 42.2 \tag{15-31c}$$

where A_p = power gain with respect to an isotropic antenna (decibels)
D = mouth diameter of parabolic reflector (meters)
f = frequency (megahertz)
42.2 = constant (decibels)

For an antenna efficiency of 100%, add 2.66 dB to the value computed with Equation 15-31c.

From Equations 15-31a, b, and c, it can be seen that the power gain of a parabolic antenna is inversely proportional to the wavelength squared. Consequently, the area (size) of the dish is an important factor when designing parabolic antennas. Very often, the area of the reflector itself is given in square wavelengths (sometimes called the *electrical* or *effective* area of the reflector). The larger the area, the larger the ratio of the area to a wavelength and the higher the power gain.

For a receive parabolic antenna, the surface of the reflector is again not completely illuminated, effectively reducing the area of the antenna. In a receiving parabolic antenna, the effective area is called the *capture area* and is always less than the actual mouth area. The capture area can be calculated by comparing the power received with the power density of the signal being received. Capture area is expressed mathematically as

$$A_c = kA \tag{15-32}$$

where A_c = capture area (square meters)
A = actual area (square meters)
k = aperture efficiency, a constant that is dependent on the type of antenna used and configuration (approximately 0.55 for a paraboloid fed by a half-wave dipole)

Therefore, the power gain for a receive parabolic antenna is

$$A_p = \frac{4\pi A_c}{\lambda^2} = \frac{4\pi k A}{\lambda^2} \tag{15-33a}$$

Substituting the area of the mouth of a paraboloid into Equation 15-33a, the power gain of a parabolic receive antenna with an efficiency $\eta = 0.55$ can be closely approximated as

$$A_p = 5.4\left(\frac{D}{\lambda}\right)^2 \tag{15-33b}$$

where D = dish diameter (meters)
λ = wavelength (meters per cycle)

In decibel form,

$$A_{p(\mathrm{dB})} = 10 \log\left[5.4\left(\frac{D}{\lambda}\right)^2\right] \tag{15-33c}$$

The term k in Equation 15-32 is called *aperture efficiency* (or sometimes *illumination efficiency*). Aperture efficiency considers both the radiation pattern of the primary radiator and the effect introduced by the ratio of the focal length of the antenna to the reflector diameter (*f/D*). This ratio is called *aperture number.* Aperture number determines the angular aperture of the reflector, which indirectly determines how much of the primary radiation is reflected by the parabolic dish. Figure 15-35 illustrates radiation directions for parabolic reflectors (a) when the focal point is outside the reflector and (b) when the focal point is inside the reflector.

The transmit power gain calculated using Equation 15-31c and the receive antenna power gain calculated using Equation 15-33c will yield approximately the same results for a given antenna, thus proving the reciprocity of parabolic antennas.

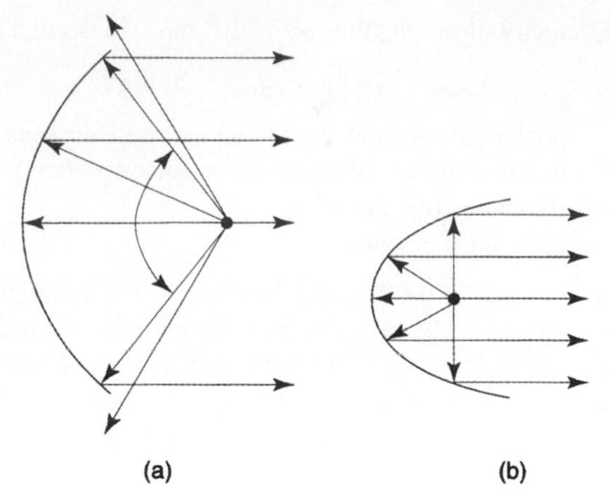

FIGURE 15-35 Radiation directions for parabolic reflectors: (a) focal point outside the reflector; (b) focal point inside the reflector

The radiation pattern shown in Figure 15-33 is typical for both transmit and receive parabolic antennas. The power gain within the main lobe is approximately 75 dB more than in the backward direction and almost 65 dB more than the maximum side lobe gain.

Example 15-6

For a 2-m-diameter parabolic reflector with 10 W of power radiated by the feed mechanism operating at 6 GHz with a transmit antenna efficiency of 55% and an aperture efficiency of 55%, determine

a. Beamwidth.
b. Transmit power gain.
c. Receive power gain.
d. EIRP.

Solution a. The beamwidth is found by substituting into Equation 15-29:

$$\theta = \frac{70(3 \times 10^8)}{(6 \times 10^9)(2)} = 1.75°$$

b. The transmit power gain is found by substituting into Equation 15-31c:

$$A_{p(\text{dB})} = 20 \log 6000 + 20 \log 2 - 42.2 = 39.4 \text{ dB}$$

c. The receive power gain is found by substituting into Equation 15-33c:

$$\lambda = \frac{c(\text{m/s})}{\text{frequency (Hz)}} = \frac{3 \times 10^8}{6 \times 10^9} = 0.05 \text{ m/cycle}$$

$$A_{p(\text{dB})} = 10 \log\left[5.4\left(\frac{2}{0.05}\right)^2\right] = 39.4 \text{ dB}$$

d. The EIRP is the product of the radiated power times the transmit antenna gain or, in decibels,

$$\text{EIRP} = A_{p(\text{dB})} + P_{\text{radiated(dBm)}}$$

$$= 39.4 + 10 \log \frac{10}{0.001}$$

$$= 39.4 \text{ dB} + 40 \text{ dBm} = 79.4 \text{ dBm}$$

15-17-2 Feed Mechanisms

The feed mechanism in a parabolic antenna actually radiates the electromagnetic energy and, therefore, is often called the *primary antenna.* The feed mechanism is of primary importance because its function is to radiate the energy toward the reflector. An ideal feed

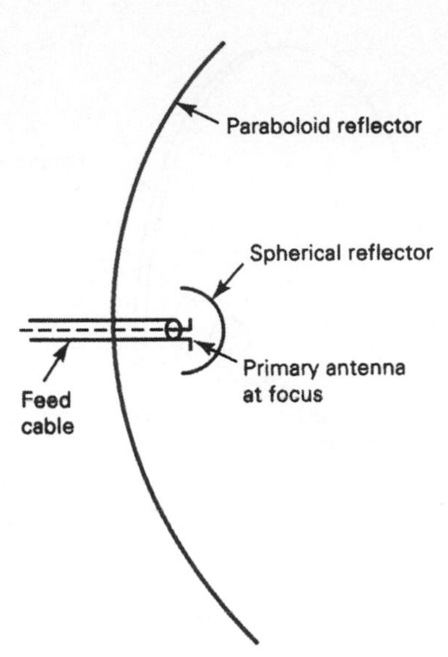

FIGURE 15-36 Parabolic antenna with a center feed

mechanism should direct all the energy toward the parabolic reflector and have no shadow effect. In practice, this is impossible to accomplish, although if care is taken when designing the feed mechanism, most of the energy can be radiated in the proper direction, and the shadow effect can be minimized. There are three primary types of feed mechanisms for parabolic antennas: center feed, horn feed, and Cassegrain feed.

15-17-2-1 Center feed. Figure 15-36 shows a diagram for a center-fed paraboloid reflector with an additional *spherical reflector*. The primary antenna is placed at the focus. Energy radiated toward the reflector is reflected outward in a concentrated beam. However, energy not reflected by the paraboloid spreads in all directions and has the tendency of disrupting the overall radiation pattern. The spherical reflector redirects such emissions back toward the parabolic reflector, where they are rereflected in the proper direction. Although the additional spherical reflector helps concentrate more energy in the desired direction, it also has a tendency to block some of the initial reflections. Consequently, the good it accomplishes is somewhat offset by its own shadow effect, and its overall performance is only marginally better than without the additional spherical reflector.

15-17-2-2 Horn feed. Figure 15-37a shows a diagram for a parabolic reflector using a horn feed. With a horn-feed mechanism, the primary antenna is a small horn antenna rather than a simple dipole or dipole array. The horn is simply a flared piece of waveguide material that is placed at the focus and radiates a somewhat directional pattern toward the parabolic reflector. When a propagating electromagnetic field reaches the mouth of the horn, it continues to propagate in the same general direction, except that, in accordance with Huygens's principle, it spreads laterally, and the wavefront eventually becomes spherical. The horn structure can have several different shapes, as shown in Figure 15-37b: sectoral (flaring only in one direction), pyramidal, or conical. As with the center feed, a horn feed presents somewhat of an obstruction to waves reflected from the parabolic dish.

The beamwidth of a horn in a plane containing the guide axis is inversely proportional to the horn mouth dimension in that plane. Approximate formulas for the half-power beamwidths of optimum-flare horns in the E and H planes are

$$\theta_E = \frac{56\lambda}{d_E} \qquad \text{(15-34a)}$$

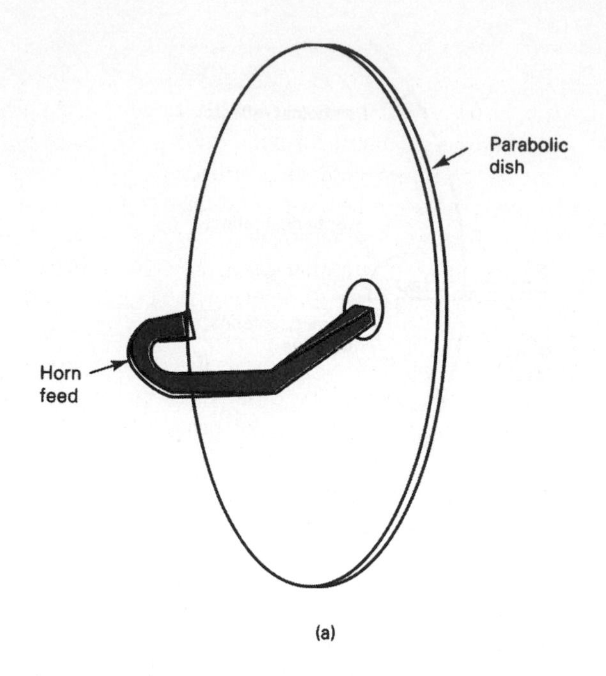

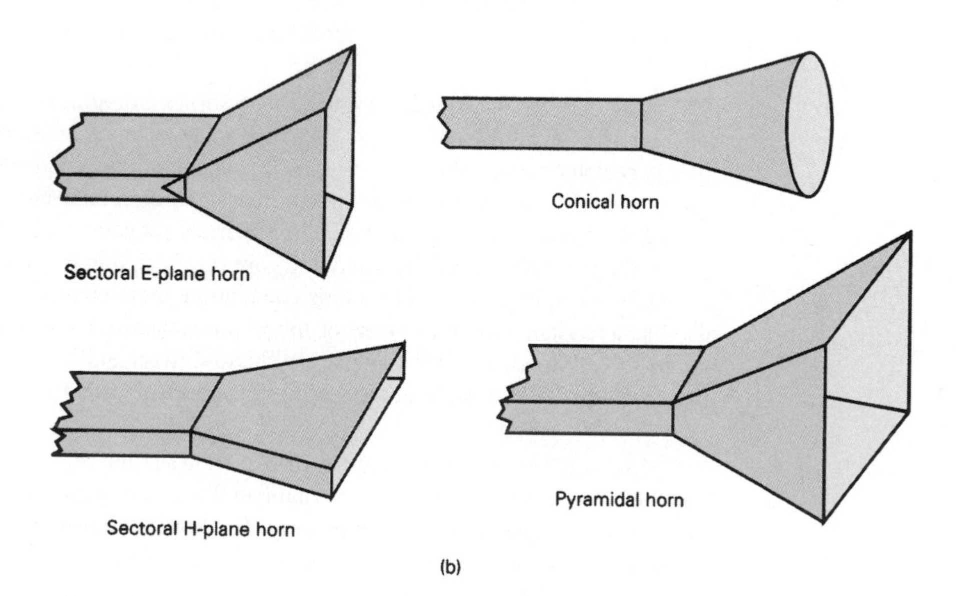

FIGURE 15-37 Parabolic antenna with a horn feed: (a) horn feed; (b) waveguide horn types

$$\theta_H = \frac{56\lambda}{d_H} \tag{15-34b}$$

where θ_E = half-power E-plane beamwidth (degrees)
θ_H = half-power H-plane beamwidth (degrees)
λ = wavelength (meters)
d_E = E-plane mouth dimension (meters)
d_H = H-plane mouth dimension (meters)

15-17-2-3 Cassegrain feed. The Cassegrain feed is named after an 18th-century astronomer and evolved directly from astronomical optical telescopes. Figure 15-38 shows the basic geometry of a Cassegrain-feed mechanism. The primary radiating source is located in or just behind a small opening at the vertex of the paraboloid rather than at the focus. The primary antenna is aimed at a small secondary reflector (*Cassegrain subreflector*) located between the vertex and the focus.

The rays emitted from the primary antenna are reflected from the Cassegrain subreflector and then illuminate the main parabolic reflector just as if they had originated at the focus. The rays are collimated by the parabolic reflector in the same way as with the center- and horn-feed mechanisms. The subreflector must have a hyperboloidal curvature to reflect the rays from the primary antenna in such a way as to function as a *virtual source* at the paraboloidal focus. The Cassegrain feed is commonly used for receiving extremely weak signals or when extremely long transmission lines or waveguide runs are required and it is necessary to place low-noise preamplifiers as close to the antenna as possible. With the Cassegrain feed, preamplifiers can be placed just before the feed mechanism and not be an obstruction to the reflected waves.

15-17-3 Conical Horn Antenna

A *conical horn* antenna consists of a cone that is truncated in a piece of circular waveguide as shown in Figure 15-39. The waveguide in turn connects the antenna to either the transmitter or the receiver. If the horn itself is used as the antenna, the *cone angle* θ (sometimes

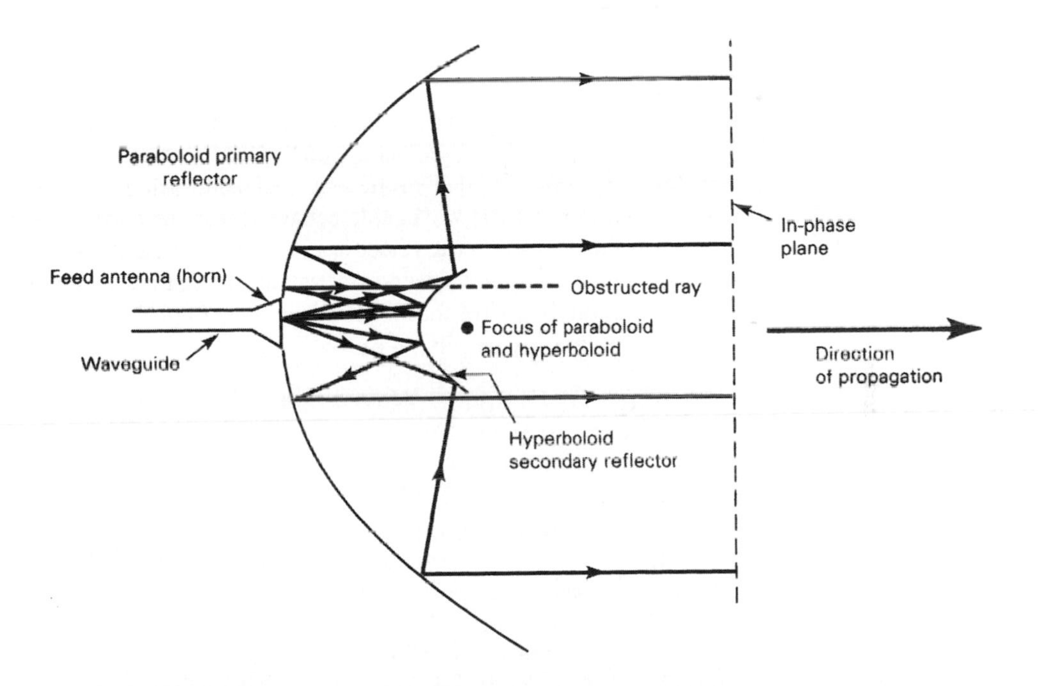

FIGURE 15-38 Parabolic antenna with a Cassegrain feed

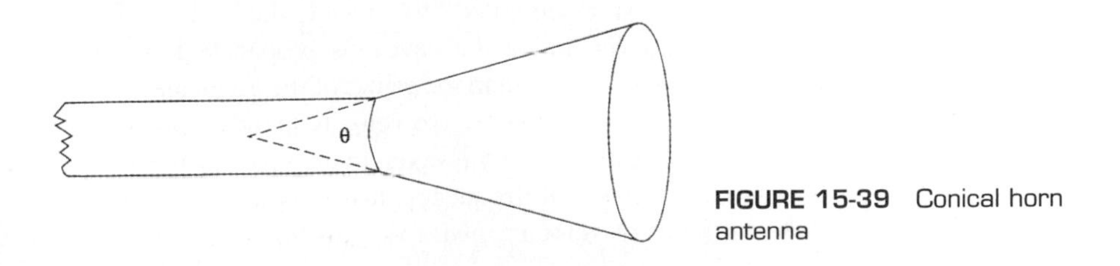

FIGURE 15-39 Conical horn antenna

called the flare angle) is made approximately 50°. In this case, the length of the truncated cone determines the antenna gain. When a conical horn is used as the feed mechanism for a parabolic dish, the flare angle and length are adjusted for optimum illumination of the reflector. The simplest feed mechanism is when the mouth of the conical horn is located at the focal point of the reflector.

15-18 WAVEGUIDES

Parallel-wire transmission lines, including coaxial cables, cannot effectively propagate electromagnetic energy above approximately 20 GHz because of the attenuation caused by skin effect and radiation losses. In addition, parallel-wire transmission lines cannot be used to propagate signals with high powers because the high voltages associated with them cause the dielectric separating the two conductors to break down. Consequently, parallel-wire transmission lines are impractical for many UHF and microwave applications. There are several alternatives, including optical fiber cables and waveguides.

In its simplest form, a *waveguide* is a hollow conductive tube, usually rectangular in cross section but sometimes circular or elliptical. The dimensions of the cross section are selected such that electromagnetic waves can propagate within the interior of the guide (hence the name "waveguide"). A waveguide does not conduct current in the true sense but rather serves as a boundary that confines electromagnetic energy. The walls of the waveguide are conductors and, therefore, reflect electromagnetic energy from their surface. If the wall of the waveguide is a good conductor and very thin, little current flows in the interior walls, and, consequently, very little power is dissipated. In a waveguide, conduction of energy occurs not in the walls of the waveguide but rather through the dielectric within the waveguide, which is usually dehydrated air or inert gas. In essence, a waveguide is analogous to a metallic wire conductor with its interior removed. Electromagnetic energy propagates down a waveguide by reflecting back and forth in a zigzag pattern.

When discussing waveguide behavior, it is necessary to speak in terms of electromagnetic field concepts (i.e., electric and magnetic fields) rather than currents and voltages as for transmission lines. The cross-sectional area of a waveguide must be on the same order as the wavelength of the signal it is propagating. Therefore, waveguides are generally restricted to frequencies above 1 GHz.

15-18-1 Rectangular Waveguide

Rectangular waveguides are the most common form of waveguide. To understand how rectangular waveguides work, it is necessary to understand the basic behavior of waves reflecting from a conducting surface.

Electromagnetic energy is propagated through free space as transverse electromagnetic (TEM) waves with a magnetic field, an electric field, and a direction of propagation that are mutually perpendicular. For an electromagnetic wave to exist in a waveguide, it must satisfy Maxwell's equations through the guide. Maxwell's equations are necessarily complex and beyond the intent of this book. However, a limiting factor of Maxwell's equations is that a TEM wave cannot have a tangential component of the electric field at the walls of the waveguide. The wave cannot travel straight down a waveguide without reflecting off the sides because the electric field would have to exist next to a conductive wall. If that happened, the electric field would be short-circuited by the walls themselves. To successfully propagate a TEM wave through a waveguide, the wave must propagate down the guide in a zigzag manner, with the electric field maximum in the center of the guide and zero at the surface of the walls.

In transmission lines, wave velocity is independent of frequency, and for air or vacuum dielectrics, the velocity is equal to the velocity in free space. However, in waveguides the velocity varies with frequency. In addition, it is necessary to distinguish between two different kinds of velocity: *phase velocity* and *group velocity.* Group velocity is the velocity at which a wave propagates, and phase velocity is the velocity at which the wave changes phase.

15-18-1-1 Phase velocity and group velocity. Phase velocity is the apparent velocity of a particular phase of the wave (e.g., the crest or maximum electric intensity point). Phase velocity is the velocity with which a wave changes phase in a direction parallel to a conducting surface, such as the walls of a waveguide. Phase velocity is determined by measuring the wavelength of a particular frequency wave and then substituting it into the following formula:

$$v_{ph} = f\lambda \tag{15-35}$$

where v_{ph} = phase velocity (meters per second)
f = frequency (hertz)
λ = wavelength (meters per cycle)

Group velocity is the velocity of a group of waves (i.e., a pulse). Group velocity is the velocity at which information signals of any kind are propagated. It is also the velocity at which energy is propagated. Group velocity can be measured by determining the time it takes for a pulse to propagate a given length of waveguide. Group and phase velocities have the same value in free space and in parallel-wire transmission lines. However, if these two velocities are measured at the same frequency in a waveguide, it will be found that, in general, the two velocities are not the same. At some frequencies they will be nearly equal, and at other frequencies they can be considerably different.

The phase velocity is always equal to or greater than the group velocity, and their product is equal to the square of the free-space propagation velocity. Thus,

$$v_g v_{ph} = c^2 \tag{15-36}$$

where v_{ph} = phase velocity (meters per second)
v_g = group velocity (meters per second)
$c = 3 \times 10^8$ meters per second

Phase velocity may exceed the velocity of light. A basic principle of physics states that no form of energy can travel at a greater velocity than light (electromagnetic waves) in free space. This principle is not violated because it is group velocity, not phase velocity, that represents the velocity of propagation of energy.

Because the phase velocity in a waveguide is greater than its velocity in free space, the wavelength for a given frequency will be greater in the waveguide than in free space. The relationship among free-space wavelength, guide wavelength, and the free-space velocity of electromagnetic waves is

$$\lambda_g = \lambda_o \frac{v_{ph}}{c} \tag{15-37}$$

where λ_g = guide wavelength (meters per cycle)
λ_o = free-space wavelength (meters per cycle)
v_{ph} = phase velocity (meters per second)
c = free-space velocity of light (3×10^8 meters per second)

15-18-1-2 Cutoff frequency and cutoff wavelength. Unlike transmission lines that have a maximum frequency of operation, waveguides have a minimum frequency of operation called the *cutoff frequency.* The cutoff frequency is an absolute limiting frequency; frequencies above the cutoff frequency will not be propagated by the waveguide. Conversely, waveguides have a maximum wavelength that they can propagate, called the *cutoff wavelength.* The cutoff wavelength is defined as the smallest free-space wavelength that is just unable to propagate in the waveguide. In other words, only frequencies with

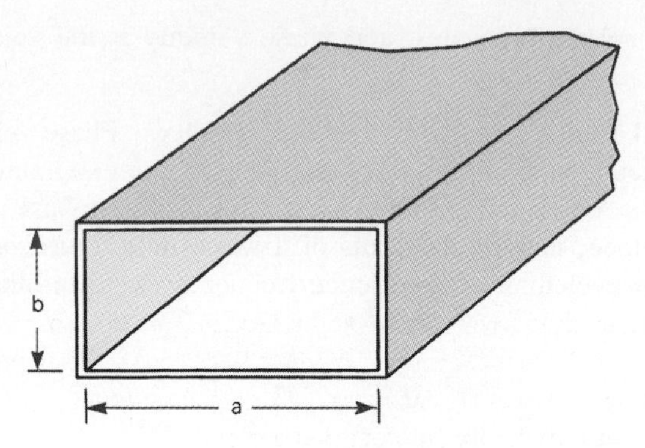

FIGURE 15-40 Cross-sectional view of a rectangular waveguide

wavelengths less than the cutoff wavelength can propagate down the waveguide. The cutoff wavelength and frequency are determined by the cross-sectional dimensions of the waveguide.

The mathematical relationship between the guide wavelength at a particular frequency and the cutoff frequency is

$$\lambda_g = \frac{c}{\sqrt{f^2 - f_c^2}} \tag{15-38}$$

where λ_g = guide wavelength (meters per cycle)
f = frequency of operation (hertz)
f_c = cutoff frequency (hertz)
c = free-space propagation velocity (3×10^8 meters per second)

Equation 15-38 can be rewritten in terms of the free-space wavelength as

$$\lambda_g = \frac{\lambda_o}{\sqrt{1 - (f_c/f)^2}} \tag{15-39}$$

where λ_g = guide wavelength (meters per cycle)
λ_o = free-space wavelength (meters per cycle)
f = frequency of operation (hertz)
f_c = cutoff frequency (hertz)

Combining Equations 15-37 and 15-39 and rearranging gives

$$v_{ph} = \frac{c(\lambda_g)}{\lambda_o} = \frac{c}{\sqrt{1 - (f_c/f)^2}} \tag{15-40}$$

It is evident from Equation 15-40 that if f becomes less than f_c, the phase velocity becomes imaginary, which means that the wave is not propagated. Also, it can be seen that as the frequency of operation approaches the cutoff frequency, the phase velocity and the guide wavelength become infinite, and the group velocity goes to zero.

Figure 15-40 shows a cross-sectional view of a piece of rectangular waveguide with dimensions a and b (a is normally designated the wider of the two dimensions). Dimension a determines the cutoff frequency of the waveguide according to the following mathematical relationship:

$$f_c = \frac{c}{2a} = \frac{c}{\lambda_c} \tag{15-41}$$

where f_c = cutoff frequency (hertz)
a = cross-sectional length (meters)

or, in terms of wavelength,

$$\lambda_c = 2a \tag{15-42}$$

where λ_c = cutoff wavelength (meters per cycle)
a = cross-sectional length (meters)

Equations 15-41 and 15-42 indicate that cutoff occurs at the frequency for which the largest transverse dimension of the guide is exactly one-half of the cutoff wavelength.

Figure 15-41 shows the top view of a section of rectangular waveguide and illustrates how electromagnetic waves propagate down the guide. For frequencies above the cutoff frequency (Figures 15-41a, b, and c), the waves propagate down the guide by reflecting back and forth between the wall at various angles. Figure 15-41d shows what happens to the electromagnetic wave at the cutoff frequency.

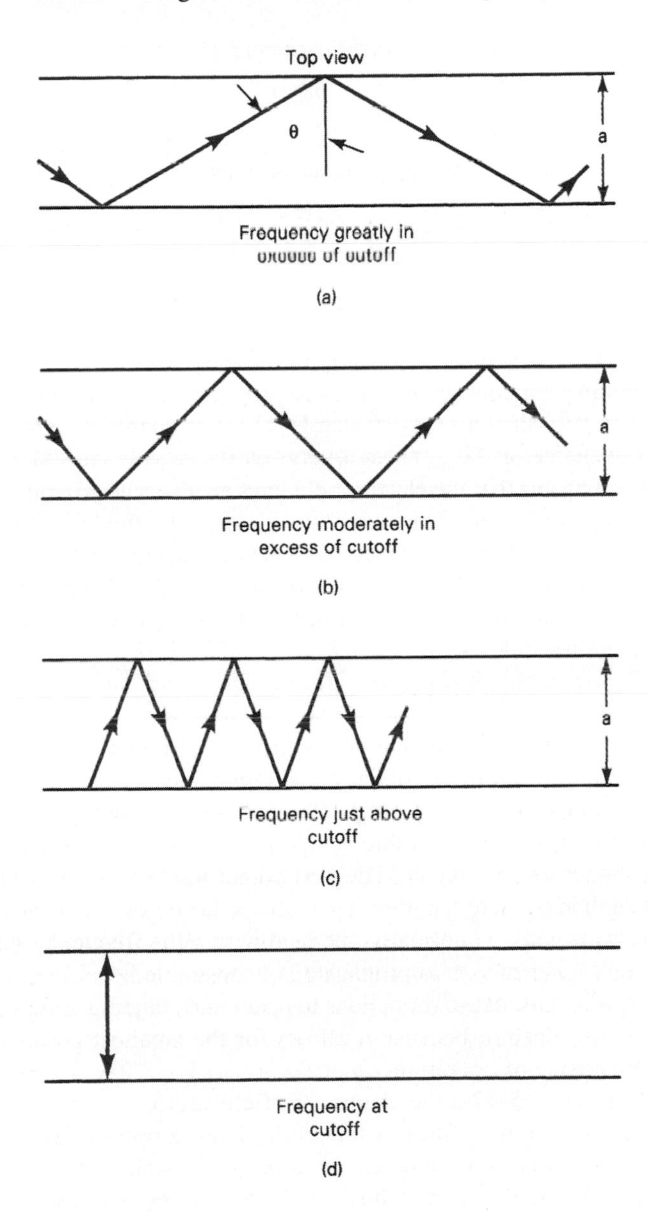

FIGURE 15-41 Electromagnetic wave propagation in a rectangular waveguide

Example 15-7

For a rectangular waveguide with a wall separation of 3 cm and a desired frequency of operation of 6 GHz, determine

a. Cutoff frequency.
b. Cutoff wavelength.
c. Group velocity.
d. Phase velocity.

Solution **a.** The cutoff frequency is determined by substituting into Equation 15-41:

$$f_c = \frac{3 \times 10^8 \text{ m/s}}{2(0.03 \text{ m})} = 5 \text{ GHz}$$

b. The cutoff wavelength is determined by substituting into Equation 15-42:

$$\lambda_c = 2(3 \text{ cm}) = 6 \text{ cm}$$

c. The phase velocity is found using Equation 15-41:

$$v_{ph} = \frac{3 \times 10^8}{\sqrt{1 - (5 \text{ GHz}/6 \text{ GHz})^2}} = 5.43 \times 10^8 \text{ m/s}$$

d. The group velocity is found by rearranging Equation 15-36:

$$v_g = \frac{c^2}{v_{ph}} = \frac{(3 \times 10^8)^2}{5.43 \times 10^8} = 1.66 \times 10^8 \text{ m/s}$$

15-18-1-3 Modes of propagation. Electromagnetic waves travel down a waveguide in different configurations called propagation *modes.* In 1955, the Institute of Radio Engineers published a set of standards. These standards designated the modes for rectangular waveguides as $TE_{m,n}$ for transverse-electric waves and $TM_{m,n}$ for transverse-magnetic waves. TE means that the electric field lines are everywhere transverse (i.e., perpendicular to the guide walls), and TM means that the magnetic field lines are everywhere transverse. In both cases, m and n are integers designating the number of half-wavelengths of intensity (electric or magnetic) that exist between each pair of walls. m is measured along the X-axis of the waveguide (the same axis the dimension a is measured on), and n is measured along the Y-axis (the same as dimension b).

Figure 15-42 shows the electromagnetic field pattern for a $TE_{1,0}$ mode wave. The $TE_{1,0}$ mode is sometimes called the dominant mode because it is the most "natural" mode. A waveguide acts as a high-pass filter in that it passes only those frequencies above the minimum or cutoff frequency. At frequencies above the cutoff frequency, higher-order TE modes of propagation, with more complicated field configurations, are possible. However, it is undesirable to operate a waveguide at a frequency at which these higher modes can propagate. The next higher mode possible occurs when the free-space wavelength is equal to length a (i.e., at twice the cutoff frequency). Consequently, a rectangular waveguide is normally operated within the frequency range between f_c and $2f_c$. Allowing higher modes to propagate is undesirable because they do not couple well to the load and, thus, cause reflections to occur and standing waves to be created. The $TE_{1,0}$ mode is also desired because it allows for the smallest possible size waveguide for a given frequency of operation.

In Figure 15-42a, the electric (E) field vectors are parallel to each other and perpendicular to the wide face of the guide. Their amplitude is greatest midway between the narrow walls and decreases to zero at the walls, in a cosinusoidal fashion. The magnetic (H) field vectors (shown by dashed lines) are also parallel to each other and perpendicular to the electric vectors. The magnetic intensity is constant in the vertical direction across the guide section. The wave is propagating in the longitudinal direc-

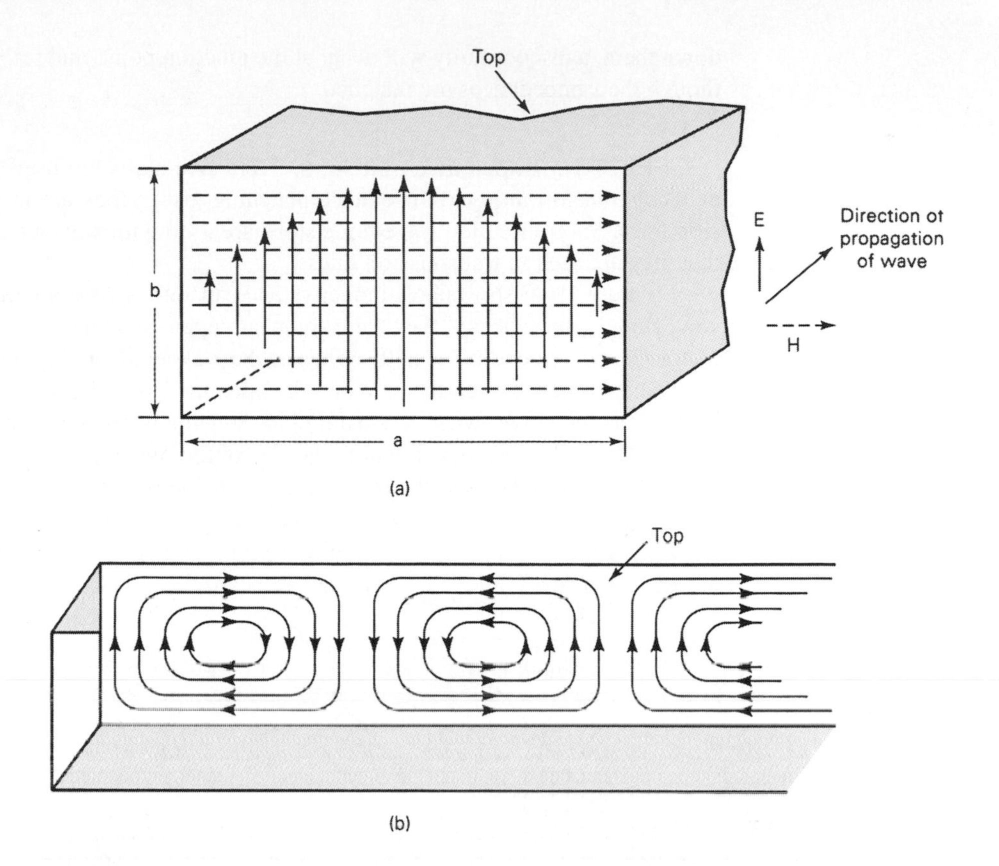

FIGURE 15-42 Electric and magnetic field vectors in a rectangular waveguide: (a) end view; (b) magnetic field configuration in a longitudinal section

tion of the guide, perpendicular to the E and H vectors. Figure 15-42b shows the magnetic field configuration in a longitudinal section of waveguide for the $TE_{1,0}$ propagation mode.

15-18-1-4 Characteristic impedance. Waveguides have a characteristic impedance that is analogous to the characteristic impedance of parallel-wire transmission lines and closely related to the characteristic impedance of free space. The characteristic impedance of a waveguide has the same significance as the characteristic impedance of a transmission line with respect to load matching, signal reflections, and standing waves. The characteristic impedance of a waveguide is expressed mathematically as

$$Z_o = \frac{377}{\sqrt{1 - (f_c/f)^2}} = 377\frac{\lambda_g}{\lambda_o} \quad \textbf{(15-43)}$$

where Z_o = characteristic impedance (ohms)
f_c = cutoff frequency (hertz)
f = frequency of operation (hertz)

Z_o is generally greater than 377 Ω. In fact, at the cutoff frequency, Z_o becomes infinite, and at a frequency equal to twice the cutoff frequency ($2f_c$), Z_o = 435 Ω. Two waveguides with the same length a dimension but different length b dimensions will have the same value of cutoff frequency and the same value of characteristic impedance. However, if these two waveguides are connected together end to end, and an electromagnetic wave is propagated down them, a discontinuity will occur at the junction point, and reflections will occur even though their impedances are matched.

15-18-1-5 Impedance matching. Reactive stubs are used in waveguides for impedance transforming and impedance matching just as they are in parallel-wire transmission lines. Short-circuited waveguide stubs are used with waveguides in the same manner that they are used in transmission lines.

Figure 15-43 shows how inductive and capacitive irises are installed in a rectangular waveguide to behave as if they were shunt susceptances. The irises consist of thin metallic plates placed perpendicular to the walls of the waveguide and joined to them at the edges, with an opening between them. When the opening is parallel to the narrow walls, the susceptance is inductive; when it is parallel to the wide walls, it is capacitive. The magnitude of the susceptance is proportional to the size of the opening.

A post placed across the narrowest dimension of the waveguide, as shown in Figure 15-44a, acts as an inductive shunt susceptance whose value depends on its diameter and its position in the transverse plane. Tuning screws, shown in Figure 15-44b, project partway across the narrow guide dimension, act as a capacitance, and may be adjusted.

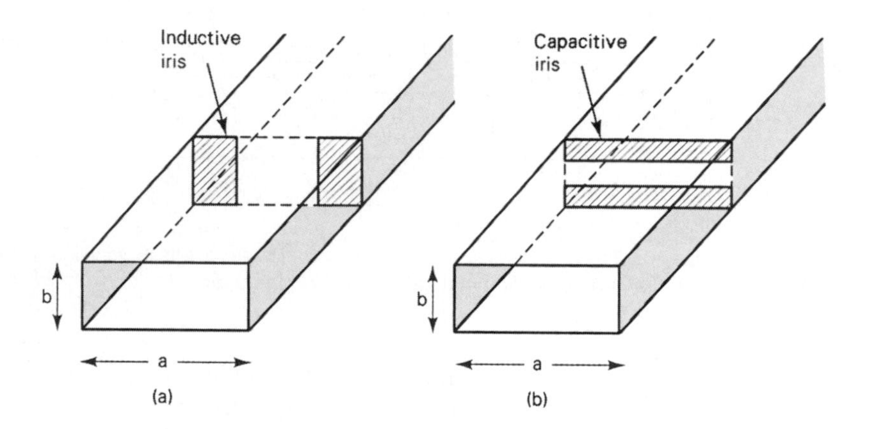

FIGURE 15-43 Waveguide impedance matching: (a) inductive iris; (b) capacitive iris

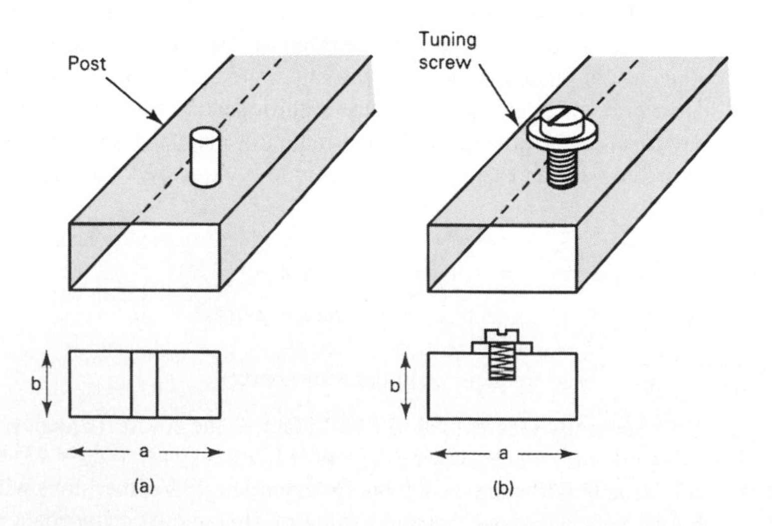

FIGURE 15-44 Waveguide impedance matching: (a) post; (b) tuning screw

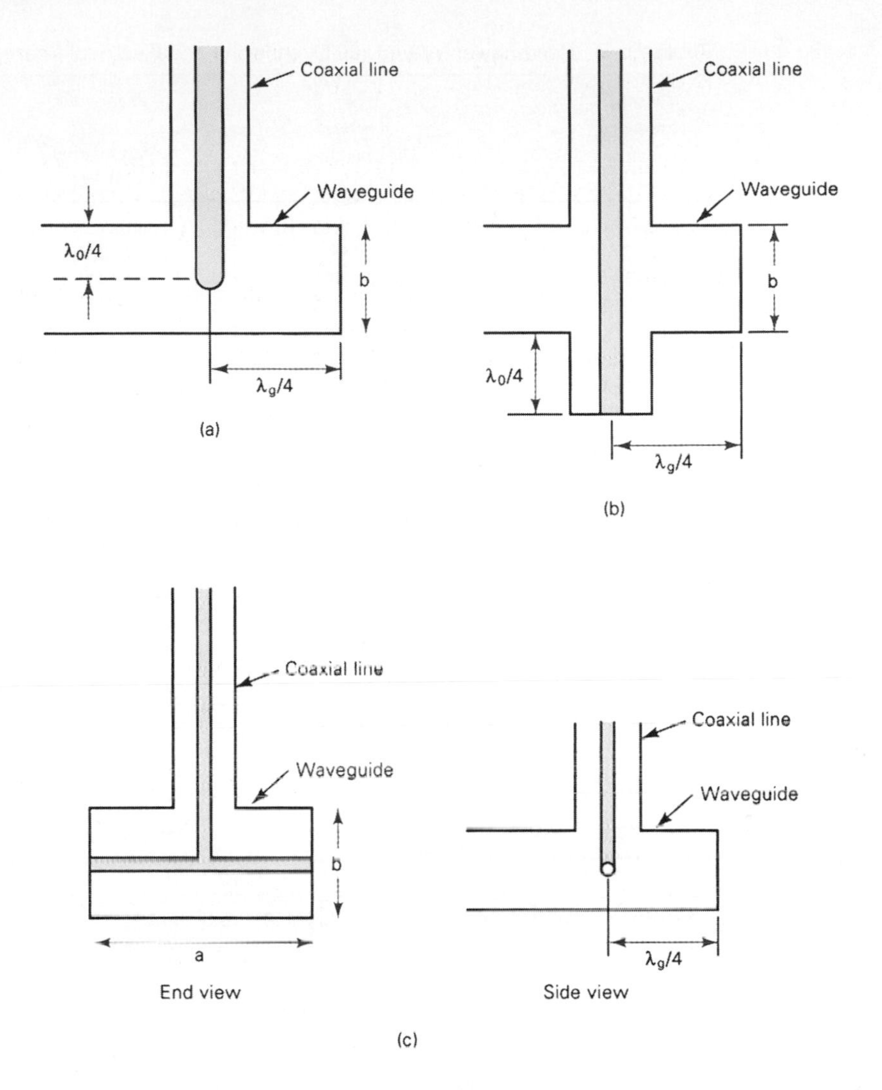

FIGURE 15-45 Transmission line-to-waveguide coupling: (a) quarter-wave probe coupler; (b) straight-through coupler; (c) cross-bar coupler

15-18-1-6 Transmission line-to-waveguide coupling. Figure 15-45 shows several ways in which a waveguide and transmission line can be joined together. The couplers shown can be used as wave launchers at the input end of a waveguide or as wave receptors at the load end of the guide. The dimension labeled $\lambda_o/4$ and $\lambda_g/4$ are approximate. In practice, they are experimentally adjusted for best results.

Table 15-3 lists the frequency range, dimensions, and electrical characteristics for several common types of rectangular waveguide.

15-18-2 Circular Waveguide

Rectangular waveguides are by far the most common; however, circular waveguides are used in radar and microwave applications when it is necessary or advantageous to propagate both vertically and horizontally polarized waves in the same waveguide. Figure 15-46 shows two pieces of circular waveguide joined together by a rotation joint.

The behavior of electromagnetic waves in circular waveguides is the same as it is in rectangular waveguides. However, because of the different geometry, some of the calculations are performed in a slightly different manner.

Table 15-3 Rectangular Waveguide Dimensions and Electrical Characteristics

Useful Frequency Range (GHz)	Outside Dimensions (mm)	Theoretical Average Attenuation (dB/m)	Theoretical Average (CW) Power Rating (kW)
1.12–1.70	169 × 86.6	0.0052	14,600
1.70–2.60	113 × 58.7	0.0097	6400
2.60–3.95	76.2 × 38.1	0.019	2700
3.95–5.85	50.8 × 25.4	0.036	1700
5.85–8.20	38.1 × 19.1	0.058	635
8.20–12.40	25.4 × 12.7	0.110	245
12.40–18.0	17.8 × 9.9	0.176	140
18.0–26.5	12.7 × 6.4	0.37	51
26.5–40.0	9.1 × 5.6	0.58	27
40.0–60.0	6.8 × 4.4	0.95	13
60.0–90.0	5.1 × 3.6	1.50	5.1
90.0–140	4.0 (diam.)	2.60	2.2
140–220	4.0 (diam.)	5.20	0.9
220–325	4.0 (diam.)	8.80	0.4

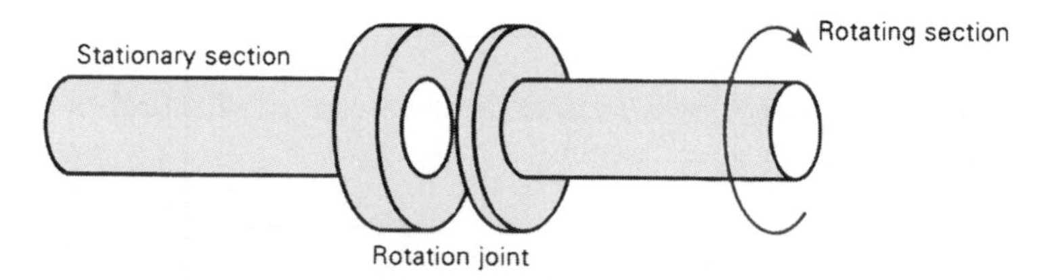

FIGURE 15-46 Circular waveguide with rotational joint

The cutoff wavelength for circular waveguides is given as

$$\lambda_o = \frac{2\pi r}{kr} \tag{15-44}$$

where λ_o = cutoff wavelength (meters per cycle)
r = internal radius of the waveguide (meters)
kr = solution of a Bessel function equation

Because the propagation mode with the largest cutoff wavelength is the one with the smallest value for kr (1.84), the $TE_{1,1}$ mode is dominant for circular waveguides. The cutoff wavelength for this mode reduces to

$$\lambda_o = 1.7d \tag{15-45}$$

where d is the waveguide diameter (meters).

Circular waveguides are easier to manufacture than rectangular waveguides and easier to join together. However, circular waveguides have a much larger area than a corresponding rectangular waveguide used to carry the same signal. Another disadvantage of circular waveguides is that the plane of polarization may rotate while the wave is propagating down it (i.e., a horizontally polarized wave may become vertically polarized and vice versa).

15-18-3 Ridged Waveguide

Figure 15-47 shows two types of ridged waveguide. A ridged waveguide is more expensive to manufacture than a standard rectangular waveguide; however, it also allows operation at lower frequencies for a given size. Consequently, smaller overall waveguide dimensions

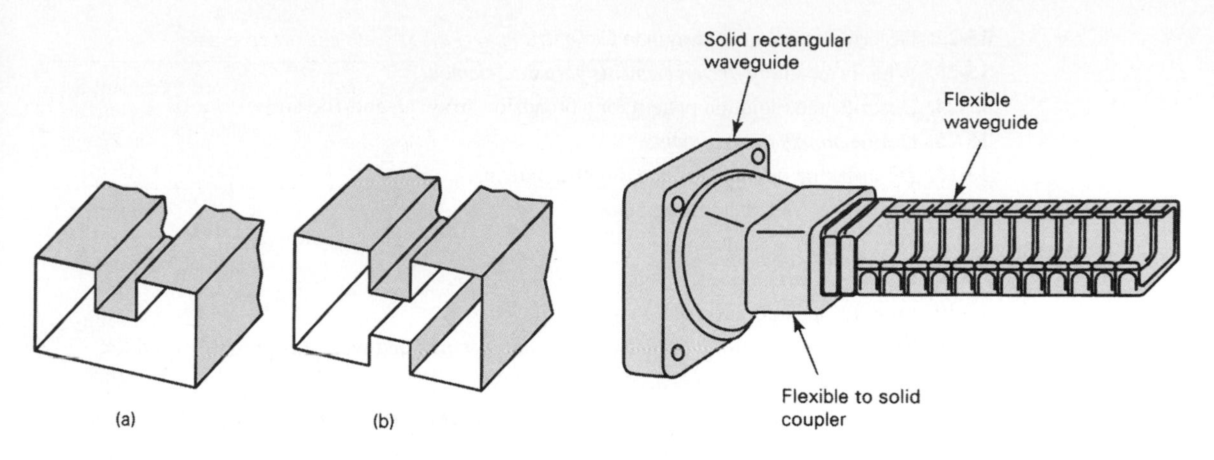

FIGURE 15-47 Ridged waveguide: (a) single ridge; (b) double ridge

FIGURE 15-48 Flexible waveguide

are possible using a ridged waveguide. A ridged waveguide has more loss per unit length than a rectangular waveguide. This characteristic, combined with its increased cost, limits its usefulness to specialized applications.

15-18-4 Flexible Waveguide

Figure 15-48 shows a length of flexible rectangular waveguide. A flexible waveguide consists of spiral-wound ribbons of brass or copper. The outside is covered with a soft dielectric coating (often rubber) to keep the waveguide air- and watertight. Short pieces of flexible waveguide are used in microwave systems when several transmitters and receivers are interconnected to a complex combining or separating unit. A flexible waveguide is also used extensively in microwave test equipment.

QUESTIONS

15-1. Define *antenna.*

15-2. Describe basic antenna operation using standing waves.

15-3. Describe a relative radiation pattern; an absolute radiation pattern.

15-4. Define *front-to-back ratio.*

15-5. Describe an omnidirectional antenna.

15-6. Define *near field* and *far field.*

15-7. Define *radiation resistance* and *antenna efficiency.*

15-8. Define and contrast *directive gain* and *power gain.*

15-9. What is the directivity for an isotropic antenna?

15-10. Define *effective isotropic radiated power.*

15-11. Define *antenna polarization.*

15-12. Define *antenna beamwidth.*

15-13. Define *antenna bandwidth.*

15-14. Define *antenna input impedance.* What factors contribute to an antenna's input impedance?

15-15. Describe the operation of an elementary doublet.

15-16. Describe the operation of a half-wave dipole.

15-17. Describe the effects of ground on a half-wave dipole.

15-18. Describe the operation of a grounded antenna.

15-19. What is meant by *antenna loading?*

15-20. Describe an antenna loading coil.

15-21. Describe antenna top loading.

15-22. Describe an antenna array.

15-23. What is meant by *driven element; parasitic element?*

15-24. Describe the radiation pattern for a broadside array; an end-fire array.

15-25. Define *nonresonant antenna.*

15-26. Describe the operation of the rhombic antenna.

15-27. Describe a folded dipole antenna.

15-28. Describe a Yagi-Uda antenna.

15-29. Describe a log-periodic antenna.

15-30. Describe the operation of a loop antenna.

15-31. Describe briefly how a *phased array antenna* works and what it is primarily used for.

15-32. Describe briefly how a *helical* antenna works.

15-33. Define the following terms: *main lobe, side lobes, side-to-side coupling,* and *back-to-back coupling.*

15-34. What are the two main parts of a *parabolic antenna?*

15-35. Describe briefly how a *parabolic reflector* works.

15-36. What is the purpose of the *feed mechanism* in a parabolic reflector antenna?

15-37. What is meant by the *capture area* of a parabolic antenna?

15-38. Describe how a *center-feed* mechanism works with a parabolic reflector.

15-39. Describe how a *horn-feed* mechanism works with a parabolic reflector.

15-40. Describe how a *Cassegrain feed* works with a parabolic reflector.

15-41. In its simplest form, what is a *waveguide?*

15-42. Describe *phase velocity; group velocity.*

15-43. Describe the *cutoff frequency* for a waveguide; *cutoff wavelength.*

15-44. What is meant by the TE mode of propagation? TM mode of propagation?

15-45. When is it advantageous to use a circular waveguide?

PROBLEMS

15-1. For an antenna with input power $P_{rad} = 100$ W, rms current $I = 2$ A, and effective resistance $R_e = 2\ \Omega$, determine

a. Antenna's radiation resistance.

b. Antenna's efficiency.

c. Power radiated from the antenna, P_{rad}.

15-2. Determine the directivity in decibels for an antenna that produces power density $\mathcal{P} = 2\ \mu W/m^2$ at a point when a reference antenna produces $0.5\ \mu W/m^2$ at the same point.

15-3. Determine the power gain in decibels for an antenna with directive gain $\mathcal{D} = 46$ dB and efficiency $\eta = 65\%$.

15-4. Determine the effective isotropic radiated power for an antenna with power gain $A_p = 43$ dB and radiated power $P_{in} = 200$ W.

15-5. Determine the effective isotropic radiated power for an antenna with directivity $\mathcal{D} = 33$ dB, efficiency $\eta = 82\%$, and input power $P_{in} = 100$ W.

15-6. Determine the power density at a point 20 km from an antenna without input power of 1000 W and a power gain $A_p = 23$ dB.

15-7. Determine the power density at a point 30 km from an antenna that has input power $P_{in} = 40$ W, efficiency $\eta = 75\%$, and directivity $\mathcal{D} = 16$ dB.

15-8. Determine the power density captured by a receiving antenna for the following parameters: transmit antenna input, $P_{in} = 50$ W; transmit antenna gain, $A_p = 30$ dB; distance between transmit and receive antennas, $d = 20$ km; and receive antenna directive gain, $A_p = 26$ dB.

15-9. Determine the directivity (in decibels) for an antenna that produces a power density at a point that is 40 times greater than the power density at the same point when the reference antenna is used.

15-10. Determine the effective radiated power for an antenna with directivity $\mathcal{D} = 400$ efficiency $\eta = 0.60$ and input power $P_{in} = 50$ W.

15-11. Determine the efficiency for an antenna with radiation resistance $R_r = 18.8\ \Omega$, effective resistance $R_e = 0.4\ \Omega$, and directive gain $\mathcal{D} = 200$.

15-12. Determine the power gain A_p for problem 15-11.

15-13. Determine the efficiency for an antenna with radiated power $P_{rad} = 44$ W, dissipated power $P_d = 0.8$ W, and directive gain $\mathcal{D} = 400$.

15-14. Determine power gain A_p for problem 15-13.

15-15. Determine the power gain and beamwidth for an end-fire helical antenna with the following parameters: helix diameter = 0.1 m, number of turns = 10, pitch = 0.05 m, and frequency of operation = 500 MHz.

15-16. Determine the beamwidth and transmit and receive power gains of a parabolic antenna with the following parameters: dish diameter of 2.5 m, a frequency of operation of 4 GHz, and an efficiency of 55%.

15-17. For a rectangular waveguide with a wall separation of 2.5 cm and a desired frequency of operation of 7 GHz, determine

a. Cutoff frequency.
b. Cutoff wavelength.
c. Group velocity.
d. Phase velocity.

15-18. For an antenna with input power $P_{rad} = 400$ W, rms current $i = 4$ A, and dc resistance $R_e = 4\ \Omega$, determine

a. Antenna's radiation resistance.
b. Antenna's efficiency.
c. Power radiated from the antenna, P_{rad}.

15-19. Determine the directivity in decibels for an antenna that produces a power density $\mathcal{P} = 4\ \mu W/m^2$ at a point in space when a reference antenna produces $0.4\ \mu W/m^2$ at the same point.

15-20. Determine the power gain in decibels for an antenna with a directive gain $\mathcal{D} = 50$ dB and an efficiency of 75%.

15-21. Determine the effective isotropic radiated power for an antenna with a power gain $A_p = 26$ dB and a radiated power $P_{in} = 400$ W.

15-22. Determine the effective isotropic radiated power for an antenna with a directivity $\mathcal{D} = 43$ dB, an efficiency of 75%, and an input power $P_{in} = 50$ W.

15-23. Determine the power density at a point 20 km from an antenna with an input power of 1200 W and a power gain $A_p = 46$ dB.

15-24. Determine the captured power density at a point 50 km from an antenna that has an input power $P_{in} = 100$ W, an efficiency of 55%, and a directivity $\mathcal{D} = 23$ dB.

15-25. Determine the power captured by a receiving antenna for the following parameters:

Power radiated $P_{in} = 100$ W

Transmit antenna directive gain $A_t = 40$ dB

Distance between transmit and receive antenna $d = 40$ km

Receive antenna directive gain $A_r = 23$ dB

15-26. Determine the directivity (in dB) for an antenna that produces a power density at a point that is 100 times greater than the power density at the same point when a reference antenna is used.

15-27. Determine the effective radiated power for an antenna with a directivity $\mathcal{D} = 300$, an efficiency = 80%, and an input power $P_{in} = 2500$ W.

15-28. Determine the efficiency for an antenna with radiation resistance $R_r = 22.2\ \Omega$, a dc resistance $R_e = 2.8\ \Omega$, and a directive gain $\mathcal{D} = 40$ dB.

15-29. Determine the power gain, G, for problem 15-28.

15-30. Determine the efficiency for an antenna with radiated power $P_{rad} = 65$ W, power dissipated $P_d = 5$ W, and a directive gain $\mathcal{D} = 200$.

15-31. Determine the power gain for problem 15-30.

CHAPTER 16

Telephone Instruments and Signals

CHAPTER OUTLINE

OBJECTIVES

- Define *communications* and *telecommunications*
- Define and describe *subscriber loop*
- Describe the operation and basic functions of a standard telephone set
- Explain the relationship among telephone sets, local loops, and central office switching machines
- Describe the block diagram of a telephone set
- Explain the function and basic operation of the following telephone set components: ringer circuit, on/off-hook circuit, equalizer circuit, speaker, microphone, hybrid network, and dialing circuit
- Describe basic telephone call procedures
- Define *call progress tones* and *signals*
- Describe the following terms: *dial tone, dual-tone multifrequency, multifrequency, dial pulses, station busy, equipment busy, ringing, ring back,* and *receiver on/off hook*
- Describe the basic operation of a cordless telephone
- Define and explain the basic format of caller ID
- Describe the operation of electronic telephones
- Describe the basic principles of paging systems

16-1 INTRODUCTION

Communications is the process of conveying information from one place to another. Communications requires a source of information, a transmitter, a receiver, a destination, and some form of transmission medium (connecting path) between the transmitter and the receiver. The transmission path may be quite short, as when two people are talking face to face with each other or when a computer is outputting information to a printer located in the same room. *Telecommunications* is long-distance communications (from the Greek word *tele* meaning "distant" or "afar"). Although the word "long" is an arbitrary term, it generally indicates that communications is taking place between a transmitter and a receiver that are too far apart to communicate effectively using only sound waves.

Although often taken for granted, the telephone is one of the most remarkable devices ever invented. To talk to someone, you simply pick up the phone and dial a few digits, and you are almost instantly connected with them. The telephone is one of the simplest devices ever developed, and the telephone connection has not changed in nearly a century. Therefore, a telephone manufactured in the 1920s will still work with today's intricate telephone system.

Although telephone systems were originally developed for conveying human speech information (voice), they are now also used extensively to transport data. This is accomplished using modems that operate within the same frequency band as human voice. Anyone who uses a telephone or a data modem on a telephone circuit is part of a global communications network called the *public telephone network* (PTN). Because the PTN interconnects subscribers through one or more switches, it is sometimes called the *public switched telephone network* (PSTN). The PTN is comprised of several very large corporations and hundreds of smaller independent companies jointly referred to as *Telco.*

The telephone system as we know it today began as an unlikely collaboration of two men with widely disparate personalities: Alexander Graham Bell and Thomas A. Watson. Bell, born in 1847 in Edinburgh, Scotland, emigrated to Ontario, Canada, in 1870, where he lived for only six months before moving to Boston, Massachusetts. Watson was born in a livery stable owned by his father in Salem, Massachusetts. The two met characteristically in 1874 and invented the telephone in 1876. On March 10, 1876, one week after his patent was allowed, Bell first succeeded in transmitting speech in his lab at 5 Exeter Place in Boston. At the time, Bell was 29 years old and Watson only 22. Bell's patent, number 174,465, has been called the most valuable ever issued.

The telephone system developed rapidly. In 1877, there were only six telephones in the world. By 1881, 3,000 telephones were producing revenues, and in 1883, there were over 133,000 telephones in the United States alone. Bell and Watson left the telephone business in 1881, as Watson put it, "in better hands." This proved to be a financial mistake, as the telephone company they left evolved into the telecommunications giant known officially as the American Telephone and Telegraph Company (AT&T). Because at one time AT&T owned most of the local operating companies, it was often referred to as the *Bell Telephone System* and sometimes simply as "*Ma Bell.*" By 1982, the Bell System grew to an unbelievable $155 billion in assets ($256 billion in today's dollars), with over one million employees and 100,000 vehicles. By comparison, in 1998, Microsoft's assets were approximately $10 billion.

AT&T once described the Bell System as "the world's most complicated machine." A telephone call could be made from any telephone in the United States to virtually any other telephone in the world using this machine. Although AT&T officially divested the Bell System on January 1, 1983, the telecommunications industry continued to grow at an unbelievable rate. Some estimate that more than 1.5 billion telephone sets are operating in the world today.

16-2 THE SUBSCRIBER LOOP

The simplest and most straightforward form of telephone service is called *plain old telephone service* (POTS), which involves subscribers accessing the public telephone network through a pair of wires called the *local subscriber loop* (or simply *local loop*). The local loop is the most fundamental component of a telephone circuit. A local loop is simply an unshielded twisted-pair transmission line (cable pair), consisting of two insulated conductors twisted together. The insulating material is generally a polyethylene plastic coating, and the conductor is most likely a pair of 116- to 26-gauge copper wire. A subscriber loop is generally comprised of several lengths of copper wire interconnected at junction and cross-connect boxes located in manholes, back alleys, or telephone equipment rooms within large buildings and building complexes.

The subscriber loop provides the means to connect a telephone set at a subscriber's location to the closest telephone office, which is commonly called an *end office, local exchange office,* or *central office.* Once in the central office, the subscriber loop is connected to an *electronic switching system* (ESS), which enables the subscriber to access the public telephone network. The local subscriber loop is described in greater detail in Chapter 17.

16-3 STANDARD TELEPHONE SET

The word *telephone* comes from the Greek words *tele,* meaning "from afar," and *phone,* meaning "sound," "voice," or "voiced sound." The standard dictionary defines a telephone as follows:

> *An apparatus for reproducing sound, especially that of the human voice (speech), at a great distance, by means of electricity; consisting of transmitting and receiving instruments connected by a line or wire which conveys the electric current.*

In essence, *speech* is sound in motion. However, sound waves are acoustic waves and have no electrical component. The basic telephone set is a simple analog transceiver designed with the primary purpose of converting speech or acoustical signals to electrical signals. However, in recent years, new features such as multiple-line selection, hold, caller ID, and speakerphone have been incorporated into telephone sets, creating a more elaborate and complicated device. However, their primary purpose is still the same, and the basic functions they perform are accomplished in much the same way as they have always been.

The first telephone set that combined a transmitter and receiver into a single hand-held unit was introduced in 1878 and called the Butterstamp telephone. You talked into one end and then turned the instrument around and listened with the other end. In 1951, Western Electric Company introduced a telephone set that was the industry standard for nearly four decades (the rotary dial telephone used by your grandparents). This telephone set is called the Bell System 500-type telephone and is shown in Figure 16-1a. The 500-type telephone set replaced the earlier 302-type telephone set (the telephone with the hand-crank magneto, fixed microphone, hand-held earphone, and no dialing mechanism). Although there are very few 500-type telephone sets in use in the United States today, the basic functions and operation of modern telephones are essentially the same. In modern-day telephone sets, the rotary dial mechanism is replaced with a Touch-Tone keypad. The modern Touch-Tone telephone is called a 2500-type telephone set and is shown in Figure 16-1b.

The quality of transmission over a telephone connection depends on the received volume, the relative frequency response of the telephone circuit, and the degree of interference. In a typical connection, the ratio of the acoustic pressure at the transmitter input to the corresponding pressure at the receiver depends on the following:

The translation of acoustic pressure into an electrical signal

The losses of the two customer local loops, the central telephone office equipment, and the cables between central telephone offices

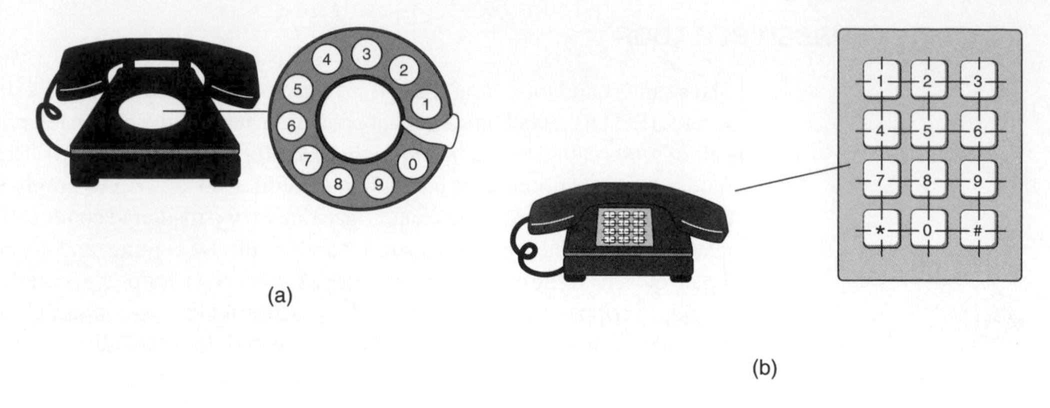

FIGURE 16-1 (a) 500-type telephone set; (b) 2500-type telephone set

The translation of the electrical signal at the receiving telephone set to acoustic pressure at the speaker output

16-3-1 Functions of the Telephone Set

The basic functions of a telephone set are as follows:

1. Notify the subscriber when there is an incoming call with an audible signal, such as a bell, or with a visible signal, such as a flashing light. This signal is analogous to an interrupt signal on a microprocessor, as its intent is to interrupt what you are doing. These signals are purposely made annoying enough to make people want to answer the telephone as soon as possible.
2. Provide a signal to the telephone network verifying when the incoming call has been acknowledged and answered (i.e., the receiver is lifted off hook).
3. Convert speech (acoustical) energy to electrical energy in the transmitter and vice versa in the receiver. Actually, the microphone converts the acoustical energy to mechanical energy, which is then converted to electrical energy. The speaker performs the opposite conversions.
4. Incorporate some method of inputting and sending destination telephone numbers (either mechanically or electrically) from the telephone set to the central office switch over the local loop. This is accomplished using either rotary dialers (pulses) or Touch-Tone pads (frequency tones).
5. Regulate the amplitude of the speech signal the calling person outputs onto the telephone line. This prevents speakers from producing signals high enough in amplitude to interfere with other people's conversations taking place on nearby cable pairs (crosstalk).
6. Incorporate some means of notifying the telephone office when a subscriber wishes to place an outgoing call (i.e., handset lifted off hook). Subscribers cannot dial out until they receive a dial tone from the switching machine.
7. Ensure that a small amount of the transmit signal is fed back to the speaker, enabling talkers to hear themselves speaking. This feedback signal is sometimes called *sidetone* or *talkback.* Sidetone helps prevent the speaker from talking too loudly.
8. Provide an open circuit (idle condition) to the local loop when the telephone is not in use (i.e., on hook) and a closed circuit (busy condition) to the local loop when the telephone is in use (off hook).
9. Provide a means of transmitting and receiving call progress signals between the central office switch and the subscriber, such as on and off hook, busy, ringing, dial pulses, Touch-Tone signals, and dial tone.

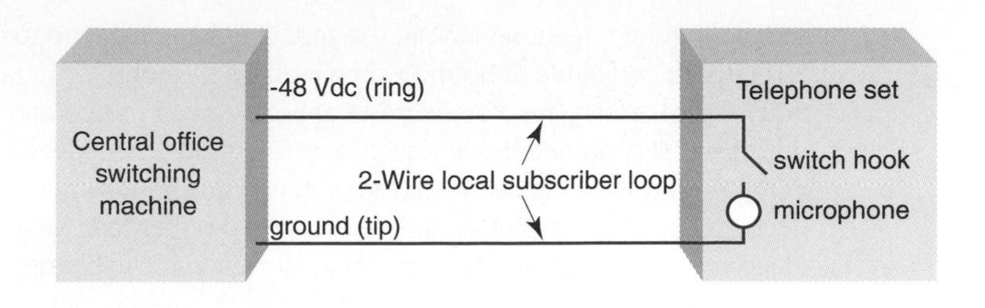

(a)

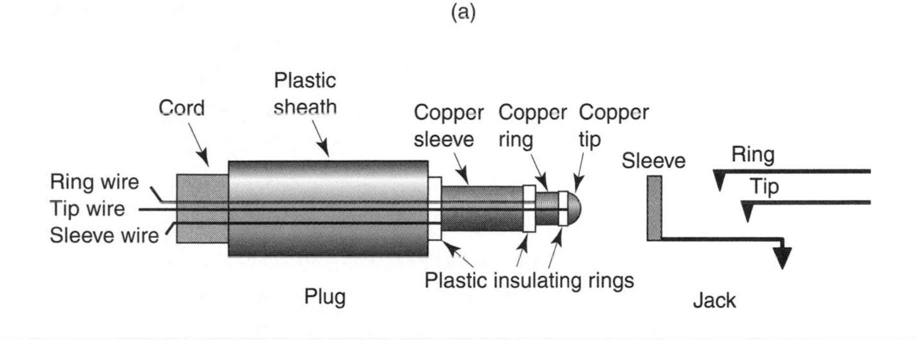

(b)

FIGURE 16-2 (a) Simplified two-wire loop showing telephone set hookup to a local switching machine; (b) plug and jack configurations showing tip, ring, and sleeve

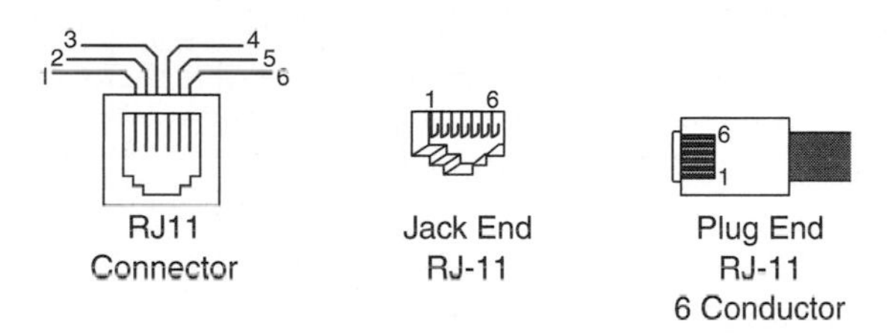

FIGURE 16-3 RJ-11 Connector

16-3-2 Telephone Set, Local Loop, and Central Office Switching Machines

Figure 16-2a shows how a telephone set is connected to a central office switching machine (local switch). As shown in the figure, a basic telephone set requires only two wires (one pair) from the telephone company to operate. Again, the pair of wires connecting a subscriber to the closest telephone office is called the *local loop.* One wire on the local loop is called the *tip,* and the other is called the *ring.* The names *tip* and *ring* come from the ¼-inch-diameter two-conductor phone plugs and patch cords used at telephone company switchboards to interconnect and test circuits. The tip and ring for a standard plug and jack are shown in Figure 16-2b. When a third wire is used, it is called the *sleeve.*

Since the 1960s, phone plugs and jacks have gradually been replaced in the home with a miniaturized plastic plug known as RJ-11 and a matching plastic receptacle (shown in Figure 16-3). *RJ* stands for *registered jacks* and is sometimes described as RJ-XX. RJ is a series of telephone connection interfaces (receptacle and plug) that are registered with the U.S. Federal Communications Commission (FCC). The term *jack* sometimes describes both the receptacle and the plug and sometimes specifies only the receptacle. RJ-11 is the most common telephone jack in use today and can have up to six conductors. Although an RJ-11 plug is capable of holding six wires in a 3/16-inch-by-3/16-inch body, only two wires (one pair) are necessary for a standard telephone circuit to operate. The other four wires can be used for a second telephone line and/or for some other special function.

As shown in Figure 16-2a, the switching machine outputs −48 Vdc on the ring and connects the tip to ground. A dc voltage was used rather than an ac voltage for several reasons: (1) to prevent power supply hum, (2) to allow service to continue in the event of a power outage, and (3) because people were afraid of ac. Minus 48 volts was selected to minimize electrolytic corrosion on the loop wires. The −48 Vdc is used for supervisory signaling and to provide talk battery for the microphone in the telephone set. On-hook, off-hook, and dial pulsing are examples of supervisory signals and are described in a later section of this chapter. It should be noted that −48 Vdc is the only voltage required for the operation of a standard telephone. However, most modern telephones are equipped with nonstandard (and often nonessential) features and enhancements and may require an additional source of ac power.

16-3-3 Block Diagram of a Telephone Set

A standard telephone set is comprised of a transmitter, a receiver, an electrical network for equalization, associated circuitry to control sidetone levels and to regulate signal power, and necessary signaling circuitry. In essence, a telephone set is an apparatus that creates an exact likeness of sound waves with an electric current. Figure 16-4 shows the functional block diagram of a *telephone set.* The essential components of a telephone set are the ringer circuit, on/off hook circuit, equalizer circuit, hybrid circuit, speaker, microphone, and a dialing circuit.

16-3-3-1 Ringer circuit. The telephone *ringer* has been around since August 1, 1878, when Thomas Watson filed for the first ringer patent. The *ringer circuit,* which was originally an electromagnetic bell, is placed directly across the tip and ring of the local loop. The purpose of the ringer is to alert the destination party of incoming calls. The audible tone from the ringer must be loud enough to be heard from a reasonable distance and offensive enough to make a person want to answer the telephone as soon as possible. In modern telephones, the bell has been replaced with an electronic oscillator connected to the speaker. Today, ringing signals can be any imaginable sound, including buzzing, a beeping, a chiming, or your favorite melody.

16-3-3-2 On/off hook circuit. The *on/off hook circuit* (sometimes called a *switch hook*) is nothing more than a simple single-throw, double-pole (STDP) switch placed across

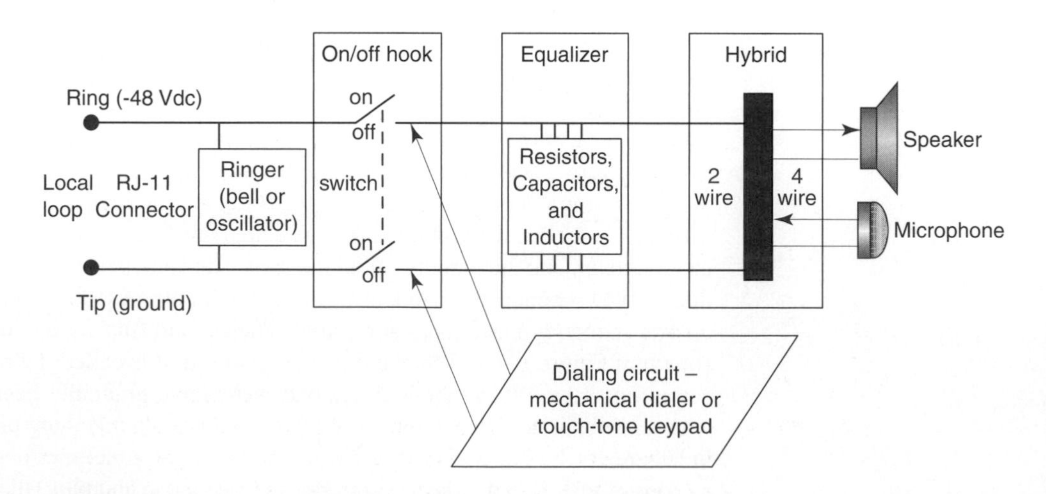

FIGURE 16-4 Functional block diagram of a standard telephone set

the tip and ring. The switch is mechanically connected to the telephone handset so that when the telephone is idle (on hook), the switch is open. When the telephone is in use (off hook), the switch is closed completing an electrical path through the microphone between the tip and ring of the local loop.

16-3-3-3 Equalizer circuit. *Equalizers* are combinations of passive components (resistors, capacitors, and so on) that are used to regulate the amplitude and frequency response of the voice signals. The equalizer helps solve an important transmission problem in telephone set design, namely, the interdependence of the transmitting and receiving efficiencies and the wide range of transmitter currents caused by a variety of local loop cables with different dc resistances.

16-3-3-4 Speaker. In essence, the *speaker* is the receiver for the telephone. The speaker converts electrical signals received from the local loop to acoustical signals (sound waves) that can be heard and understood by a human being. The speaker is connected to the local loop through the hybrid network. The speaker is typically enclosed in the *handset* of the telephone along with the microphone.

16-3-3-5 Microphone. For all practical purposes, the *microphone* is the transmitter for the telephone. The microphone converts acoustical signals in the form of sound pressure waves from the caller to electrical signals that are transmitted into the telephone network through the local subscriber loop. The microphone is also connected to the local loop through the hybrid network. Both the microphone and the speaker are transducers, as they convert one form of energy into another form of energy. A microphone converts acoustical energy first to mechanical energy and then to electrical energy, while the speaker performs the exact opposite sequence of conversions.

16-3-3-6 Hybrid network. The *hybrid network* (sometimes called a *hybrid coil* or *duplex coil*) in a telephone set is a special balanced transformer used to convert a two-wire circuit (the local loop) into a four-wire circuit (the telephone set) and vice versa, thus enabling full duplex operation over a two-wire circuit. In essence, the hybrid network separates the transmitted signals from the received signals. Outgoing voice signals are typically in the 1-V to 2-V range, while incoming voice signals are typically half that value. Another function of the hybrid network is to allow a small portion of the transmit signal to be returned to the receiver in the form of a *sidetone*. Insufficient sidetone causes the speaker to raise his voice, making the telephone conversation seem unnatural. Too much sidetone causes the speaker to talk too softly, thereby reducing the volume that the listener receives.

16-3-3-7 Dialing circuit. The *dialing circuit* enables the subscriber to output signals representing digits, and this enables the caller to enter the destination telephone number. The dialing circuit could be a rotary dialer, which is nothing more than a switch connected to a mechanical rotating mechanism that controls the number and duration of the on/off condition of the switch. However, more than likely, the dialing circuit is either an electronic dial-pulsing circuit or a Touch-Tone keypad, which sends various combinations of tones representing the called digits.

16-4 BASIC TELEPHONE CALL PROCEDURES

Figure 16-5 shows a simplified diagram illustrating how two telephone sets (subscribers) are interconnected through a central office dial switch. Each subscriber is connected to the switch through a local loop. The switch is most likely some sort of an electronic switching system (*ESS machine*). The local loops are terminated at the calling and called stations in telephone sets and at the central office ends to switching machines.

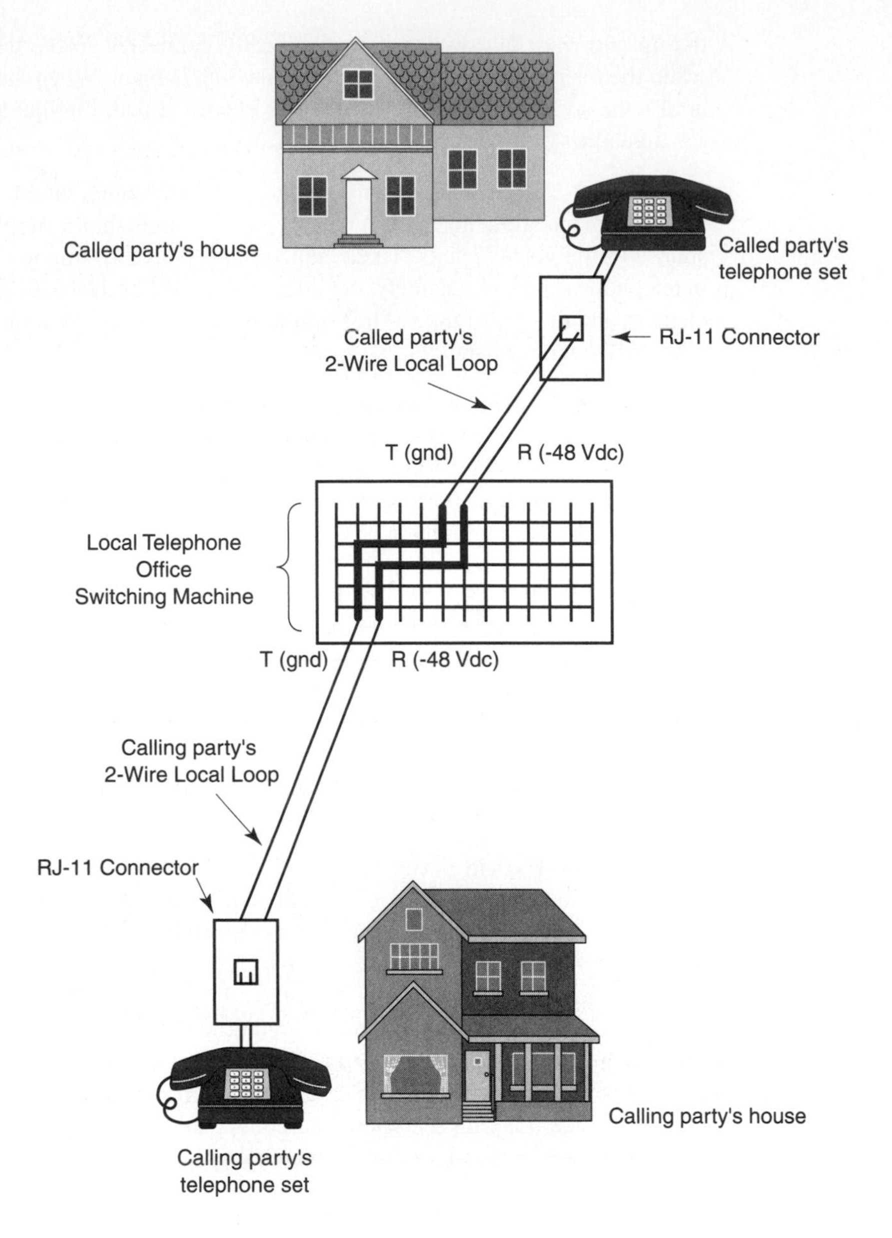

FIGURE 16-5 Telephone call procedures

When the calling party's telephone set goes off hook (i.e., lifting the handset off the cradle), the switch hook in the telephone set is released, completing a dc path between the tip and the ring of the loop through the microphone. The ESS machine senses a dc current in the loop and recognizes this as an off-hook condition. This procedure is referred to as *loop start operation* since the loop is completed through the telephone set. The amount of dc current produced depends on the wire resistance, which varies with loop length, wire gauge, type of wire, and the impedance of the subscriber's telephone. Typical loop resistance ranges from a few ohms up to approximately 1300 ohms, and typical telephone set impedances range from 500 ohms to 1000 ohms.

Completing a local telephone call between two subscribers connected to the same telephone switch is accomplished through a standard set of procedures that includes the 10 steps listed next. Accessing the telephone system in this manner is known as POTS (plain old telephone service):

Step 1 Calling station goes off hook.
Step 2 After detecting a dc current flow on the loop, the switching machine returns an audible dial tone to the calling station, acknowledging that the caller has access to the switching machine.
Step 3 The caller dials the destination telephone number using one of two methods: mechanical dial pulsing or, more likely, electronic dual-tone multifrequency (Touch-Tone) signals.
Step 4 When the switching machine detects the first dialed number, it removes the dial tone from the loop.
Step 5 The switch interprets the telephone number and then locates the local loop for the destination telephone number.
Step 6 Before ringing the destination telephone, the switching machine tests the destination loop for dc current to see if it is idle (on hook) or in use (off hook). At the same time, the switching machine locates a signal path through the switch between the two local loops.
Step 7a If the destination telephone is off hook, the switching machine sends a station busy signal back to the calling station.
Step 7b If the destination telephone is on hook, the switching machine sends a ringing signal to the destination telephone on the local loop and at the same time sends a ring back signal to the calling station to give the caller some assurance that something is happening.
Step 8 When the destination answers the telephone, it completes the loop, causing dc current to flow.
Step 9 The switch recognizes the dc current as the station answering the telephone. At this time, the switch removes the ringing and ring-back signals and completes the path through the switch, allowing the calling and called parties to begin their conversation.
Step 10 When either end goes on hook, the switching machine detects an open circuit on that loop and then drops the connections through the switch.

Placing telephone calls between parties connected to different switching machines or between parties separated by long distances is somewhat more complicated and is described in Chapter 17.

16-5 CALL PROGRESS TONES AND SIGNALS

Call progress tones and *call progress signals* are acknowledgment and status signals that ensure the processes necessary to set up and terminate a telephone call are completed in an orderly and timely manner. Call progress tones and signals can be sent from machines to machines, machines to people, and people to machines. The people are the subscribers (i.e., the calling and the called party), and the machines are the electronic switching systems in the telephone offices and the telephone sets themselves. When a switching machine outputs a call progress tone to a subscriber, it must be audible and clearly identifiable.

Signaling can be broadly divided into two major categories: *station signaling* and *interoffice signaling.* Station signaling is the exchange of signaling messages over local loops between stations (telephones) and telephone company switching machines. On the other hand, interoffice signaling is the exchange of signaling messages between switching machines. Signaling messages can be subdivided further into one of four categories: *alerting, supervising, controlling,* and *addressing.* Alerting signals indicate a request for service, such as going off hook or ringing the destination telephone. Supervising signals provide call status information, such as busy or ring-back signals. Controlling signals provide information in the form of announcements, such as number changed to another number, a number no longer in service, and so on. Addressing signals provide the routing information, such as calling and called numbers.

Examples of essential call progress signals are dial tone, dual tone multifrequency tones, multifrequency tones, dial pulses, station busy, equipment busy, ringing, ring-back, receiver on hook, and receiver off hook. Tables 16-1 and 16-2 summarize the most important call progress tones and their direction of propagation, respectively.

16-5-1 Dial Tone

Siemens Company first introduced *dial tone* to the public switched telephone network in Germany in 1908. However, it took several decades before being accepted in the United

Table 16-1 Call Progress Tone Summary

Tone or Signal	Frequency	Duration/Range
Dial tone	350 Hz plus 440 Hz	Continuous
DTMF	697 Hz, 770 Hz, 852 Hz, 941 Hz, 1209 Hz, 1336 Hz, 1477 Hz, 1633 Hz	Two of eight tones On, 50-ms minimum Off, 45-ms minimum, 3-s maximum
MF	700 Hz, 900 Hz, 1100 Hz, 1300 Hz, 1500 Hz, 1700 Hz	Two of six tones On, 90-ms minimum, 120-ms maximum
Dial pulses	Open/closed switch	On, 39 ms Off, 61 ms
Station busy	480 Hz plus 620 Hz	On, 0.5 s Off, 0.5 s
Equipment busy	480 Hz plus 620 Hz	On, 0.2 s Off, 0.3 s
Ringing	20 Hz, 90 vrms (nominal)	On, 2 s Off, 4 s
Ring-back	440 Hz plus 480 Hz	On, 2 s Off, 4 s
Receiver on hook	Open loop	Indefinite
Receiver off hook	dc current	20-mA minimum, 80-mA maximum,
Receiver-left-off-hook alert	1440 Hz, 2060 Hz, 2450 Hz, 2600 Hz	On, 0.1 s Off, 0.1 s

Table 16-2 Call Progress Tone Direction of Propagation

Tone or Signal	Direction
Dial tone	Telephone office to calling station
DTMF	Calling station to telephone office
MF	Telephone office to telephone office
Dial pulses	Calling station to telephone office
Station busy	Telephone office to calling subscriber
Equipment busy	Telephone office to calling subscriber
Ringing	Telephone office to called subscriber
Ring-back	Telephone office to calling subscriber
Receiver on hook	Calling subscriber to telephone office
Receiver off hook	Calling subscriber to telephone office
Receiver-left-off-hook alert	Telephone office to calling subscriber

States. Dial tone is an audible signal comprised of two frequencies: 350 Hz and 440 Hz. The two tones are linearly combined and transmitted simultaneously from the central office switching machine to the subscriber in response to the subscriber going off hook. In essence, dial tone informs subscribers that they have acquired access to the electronic switching machine and can now dial or use Touch-Tone in a destination telephone number. After a subscriber hears the dial tone and begins dialing, the dial tone is removed from the line (this is called *breaking dial tone*). On rare occasions, a subscriber may go off hook and not receive dial tone. This condition is appropriately called *no dial tone* and occurs when there are more subscribers requesting access to the switching machine than the switching machine can handle at one time.

16-5-2 Dual-Tone MultiFrequency

Dual-tone multifrequency (DTMF) was first introduced in 1963 with 10 buttons in Western Electric 1500-type telephones. DTMF was originally called *Touch-Tone.* DTMF is a more efficient means than dial pulsing for transferring telephone numbers from a subscriber's location to the central office switching machine. DTMF is a simple two-of-eight encoding scheme where each digit is represented by the linear addition of two frequencies. DTMF is strictly for signaling between a subscriber's location and the nearest telephone office or message switching center. DTMF is sometimes confused with another two-tone signaling system called *multifrequency signaling* (MF), which is a two-of-six code designed to be used only to convey information between two electronic switching machines.

Figure 16-6 shows the four-row-by-four-column keypad matrix used with a DTMF keypad. As the figure shows, the keypad is comprised of 16 keys and eight frequencies. Most household telephones, however, are not equipped with the special-purpose keys located in the fourth column (i.e., the A, B, C, and D keys). Therefore, most household telephones actually use two of seven tone encoding scheme. The four vertical frequencies (called the *low group frequencies*) are 697 Hz, 770 Hz, 852 Hz, and 941 Hz, and the four horizontal frequencies (called the *high group frequencies*) are 1209 Hz, 1336 Hz, 1477 Hz, and 1633 Hz. The frequency tolerance of the oscillators is ±.5%. As shown in Figure 16-6, the digits 2 through 9 can also be used to represent 24 of the 26 letters (Q and Z are omitted). The letters were originally used to identify one local telephone exchange from another,

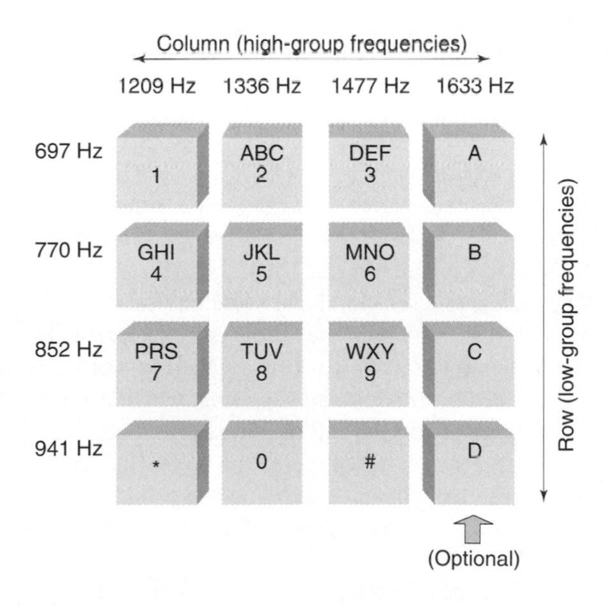

FIGURE 16-6 DTMF keypad layout and frequency allocation

Table 16-3 DTMF Specifications

Transmitter (Subscriber)	Parameter	Receiver (Local Office)
−10 dBm	Minimum power level (single frequency)	−25 dBm
+2 dBm	Maximum power level (two tones)	0 dBm
+4 dB	Maximum power difference between two tones	+4 dB
50 ms	Minimum digit duration	40 ms
45 ms	Minimum interdigit duration	40 ms
3 s	Maximum interdigit time period	3 s
Maximum echo level relative to transmit frequency level (−10 dB)		
Maximum echo delay (<20 ms)		

such as BR for Bronx, MA for Manhattan, and so on. Today, the letters are used to personalize telephone numbers. For example; 1-800-UPS-MAIL equates to the telephone number 1-800-877-6245. When a digit (or letter) is selected, two of the eight frequencies (or seven for most home telephones) are transmitted (one from the low group and one from the high group). For example, when the digit 5 is depressed, 770 Hz and 1336 Hz are transmitted simultaneously. The eight frequencies were purposely chosen so that there is absolutely no harmonic relationship between any of them, thus eliminating the possibility of one frequency producing a harmonic that might be misinterpreted as another frequency.

The major advantages for the subscriber in using Touch-Tone signaling over dial pulsing is speed and control. With Touch-Tone signaling, all digits (and thus telephone numbers) take the same length of time to produce and transmit. Touch-Tone signaling also eliminates the impulse noise produced from the mechanical switches necessary to produce dial pulses. Probably the most important advantage of DTMF over dial pulsing is the way in which the telephone company processes them. Dial pulses cannot pass through a central office exchange (local switching machine), whereas DTMF tones will pass through an exchange to the switching system attached to the called number.

Table 16-3 lists the specifications for DTME. The transmit specifications are at the subscriber's location, and the receive specifications are at the local switch. Minimum power levels are given for a single frequency, and maximum power levels are given for two tones. The minimum duration is the minimum time two tones from a given digit must remain on. The interdigit time specifies the minimum and maximum time between the transmissions of any two successive digits. An echo occurs when a pair of tones is not totally absorbed by the local switch and a portion of the power is returned to the subscriber. The maximum power level of an echo is 10 dB below the level transmitted by the subscriber and must be delayed less than 20 ms.

16-5-3 Multifrequency

Multifrequency tones (codes) are similar to DTMF signals in that they involve the simultaneous transmission of two tones. MF tones are used to transfer digits and control signals between switching machines, whereas DTMF signals are used to transfer digits and control signals between telephone sets and local switching machines. MF tones are combinations of two frequencies that fall within the normal speech bandwidth so that they can be propagated over the same circuits as voice. This is called *in-band signaling.* In-band signaling is rapidly being replaced by *out-of-band signaling,* which is discussed in Chapter 17.

MF codes are used to send information between the control equipment that sets up connections through a switch when more than one switch is involved in completing a call. MF codes are also used to transmit the calling and called numbers from the originating telephone office to the destination telephone office. The calling number is sent first, followed by the called number.

Table 16-4 lists the two-tone MF combinations and the digits or control information they represent. As the table shows, MF tones involve the transmission of two-of-six possi-

Table 16-4 Multifrequency Codes

Frequencies (Hz)	Digit or Control
700 + 900	1
700 + 1100	2
700 + 1300	4
700 + 1500	7
900 + 1100	3
900 + 1300	5
900 + 1500	8
1100 + 1300	6
1100 + 1500	9
1100 + 1700	Key pulse (KP)
1300 + 1500	0
1500 + 1700	Start (ST)
2600 Hz	IDLE

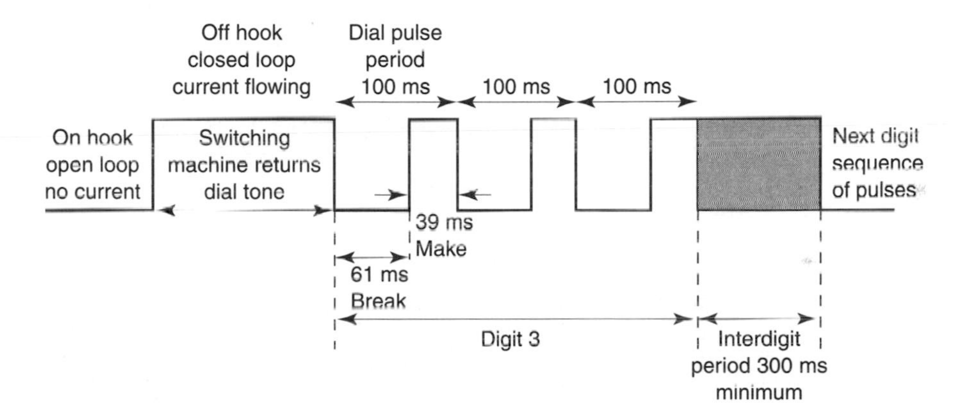

FIGURE 16-7 Dial pulsing sequence

ble frequencies representing the 10 digits plus two control signals. The six frequencies are 700 Hz, 900 Hz, 1100 Hz, 1300 Hz, 1500 Hz, and 1700 Hz. Digits are transmitted at a rate of seven per second, and each digit is transmitted as a 68-ms burst. The *key pulse* (KP) signal is a multifrequency control tone comprised of 1100 Hz plus 1700 Hz, ranging from 90 ms to 120 ms. The KP signal is used to indicate the beginning of a sequence of MF digits. The *start* (ST) signal is a multifrequency control tone used to indicate the end of a sequence of dialed digits. From the perspective of the telephone circuit, the ST control signal indicates the beginning of the processing of the signal. The IDLE signal is a 2600-Hz single-frequency tone placed on a circuit to indicate the circuit is not currently in use. For example, KP 3 1 5 7 3 6 1 0 5 3 ST is the sequence transmitted for the telephone number 315-736-1053.

16-5-4 Dial Pulses

Dial pulsing (sometimes called *rotary dial pulsing*) is the method originally used to transfer digits from a telephone set to the local switch. Pulsing digits from a rotary switch began soon after the invention of the automatic switching machine. The concept of dial pulsing is quite simple and is depicted in Figure 16-7. The process begins when the telephone set is lifted off hook, completing a path for current through the local loop. When the switching machine detects the off-hook condition, it responds with dial tone. After hearing the dial tone, the subscriber begins dial pulsing digits by rotating a mechanical dialing mechanism and then letting it return to its rest position. As the rotary switch returns to its rest position, it outputs a series of dial pulses corresponding to the digit dialed.

When a digit is dialed, the loop circuit alternately opens (breaks) and closes (makes) a prescribed number of times. The number of switch make/break sequences corresponds to the digit dialed (i.e., the digit 3 produces three switch openings and three switch closures). Dial pulses occur at 10 make/break cycles per second (i.e., a period of 100 ms per pulse cycle). For example, the digit 5 corresponds to five make/break cycles lasting a total of 500 ms. The switching machine senses and counts the number of make/break pairs in the sequence. The break time is nominally 61 ms, and the make time is nominally 39 ms. Digits are separated by an idle period of 300 ms called the *interdigit time.* It is essential that the switching machine recognize the interdigit time so that it can separate the pulses from successive digits. The central office switch incorporates a special *time-out circuit* to ensure that the break part of the dialing pulse is not misinterpreted as the phone being returned to its on-hook (idle) condition.

All digits do not take the same length of time to dial. For example, the digit 1 requires only one make/break cycle, whereas the digit 0 requires 10 cycles. Therefore, all telephone numbers do not require the same amount of time to dial or to transmit. The minimum time to dial pulse out the seven-digit telephone number 987-1234 is as follows:

digit	9	ID	8	ID	7	ID	1	ID	2	ID	3	ID	4
time (ms)	900	300	800	300	700	300	100	300	200	300	300	300	400

where ID is the interdigit time (300 ms) and the total minimum time is 5200 ms, or 5.2 seconds.

16-5-5 Station Busy

In telephone terminology, a *station* is a telephone set. A *station busy signal* is sent from the switching machine back to the calling station whenever the called telephone number is off hook (i.e., the station is in use). The station busy signal is a two-tone signal comprised of 480 Hz and 620 Hz. The two tones are on for 0.5 seconds, then off for 0.5 seconds. Thus, a busy signal repeats at a 60-pulse-per-minute (ppm) rate.

16-5-6 Equipment Busy

The *equipment busy signal* is sometimes called a *congestion tone* or a *no-circuits-available tone.* The equipment busy signal is sent from the switching machine back to the calling station whenever the system cannot complete the call because of equipment unavailability (i.e., all the circuits, switches, or switching paths are already in use). This condition is called *blocking* and occurs whenever the system is overloaded and more calls are being placed than can be completed. The equipment busy signal uses the same two frequencies as the station busy signal, except the equipment busy signal is on for 0.2 seconds and off for 0.3 seconds (120 ppm). Because an equipment busy signal repeats at twice the rate as a station busy signal, an equipment busy is sometimes called a *fast busy,* and a station busy is sometimes called a *slow busy.* The telephone company refers to an equipment busy condition as a *can't complete.*

16-5-7 Ringing

The *ringing signal* is sent from a central office to a subscriber whenever there is an incoming call. The purpose of the ringing signal is to ring the bell in the telephone set to alert the subscriber that there is an incoming call. If there is no bell in the telephone set, the ringing signal is used to trigger another audible mechanism, which is usually a tone oscillator circuit. The ringing signal is nominally a 20-Hz, 90-Vrms signal that is on for 2 seconds and then off for 4 seconds. The ringing signal should not be confused with the actual ringing sound the bell makes. The audible ring produced by the bell was originally made as annoying as possible so that the called end would answer the telephone as soon as possible, thus tying up common usage telephone equipment in the central office for the minimum length of time.

16-5-8 Ring-Back

The *ring-back signal* is sent back to the calling party at the same time the ringing signal is sent to the called party. However, the ring and ring-back signals are two distinctively different signals. The purpose of the ring-back signal is to give some assurance to the calling party that the destination telephone number has been accepted, processed, and is being rung. The ring-back signal is an audible combination of two tones at 440 Hz and 480 Hz that are on for 2 seconds and then off for 4 seconds.

16-5-9 Receiver On/Off Hook

When a telephone is *on hook,* it is not being used, and the circuit is in the *idle* (or *open*) *state.* The term *on hook* was derived in the early days of telephone when the telephone handset was literally placed on a hook (the hook eventually evolved into a cradle). When the telephone set is on hook, the local loop is open, and there is no current flowing on the loop. An on-hook signal is also used to terminate a call and initiate a disconnect.

When the telephone set is taken *off hook,* a switch closes in the telephone that completes a dc path between the two wires of the local loop. The switch closure causes a dc current to flow on the loop (nominally between 20 mA and 80 mA, depending on loop length and wire gauge). The switching machine in the central office detects the dc current and recognizes it as a receiver off-hook condition (sometimes called a *seizure* or *request for service*). The receiver off-hook condition is the first step to completing a telephone call. The switching machine will respond to the off-hook condition by placing an audible dial tone on the loop. The off-hook signal is also used at the destination end as an *answer signal* to indicate that the called party has answered the telephone. This is sometimes referred to as a *ring trip* because when the switching machine detects the off-hook condition, it removes (or trips) the ringing signal.

16-5-10 Other Nonessential Signaling and Call Progress Tones

There are numerous additional signals relating to initiating, establishing, completing, and terminating a telephone call that are nonessential, such as *call waiting tones, caller waiting tones, calling card service tones, comfort tones, hold tones, intrusion tones, stutter dial tone* (for voice mail), and *receiver off-hook tones* (also called *howler tones*).

16-6 CORDLESS TELEPHONES

Cordless telephones are simply telephones that operate without cords attached to the handset. Cordless telephones originated around 1980 and were quite primitive by today's standards. They originally occupied a narrow band of frequencies near 1.7 MHz, just above the AM broadcast band, and used the 117-vac, 60-Hz household power line for an antenna. These early units used frequency modulation (FM) and were poor quality and susceptible to interference from fluorescent lights and automobile ignition systems. In 1984, the FCC reallocated cordless telephone service to the 46-MHz to 49-MHz band. In 1990, the FCC extended cordless telephone service to the 902-MHz to 928-MHz band, which appreciated a superior signal-to-noise ratio. Cordless telephone sets transmit and receive over narrowband FM (NBFM) channels spaced 30 kHz to 100 kHz apart, depending on the modulation and frequency band used. In 1998, the FCC expanded service again to the 2.4-GHz band. Adaptive differential pulse code modulation and spread spectrum technology (SST) are used exclusively in the 2.4-GHz band, while FM and SST digital modulation are used in the 902-MHz to 928-MHz band. Digitally modulated SST telephones offer higher quality and more security than FM telephones.

In essence, a cordless telephone is a full-duplex, battery-operated, portable radio transceiver that communicates directly with a stationary transceiver located somewhere in

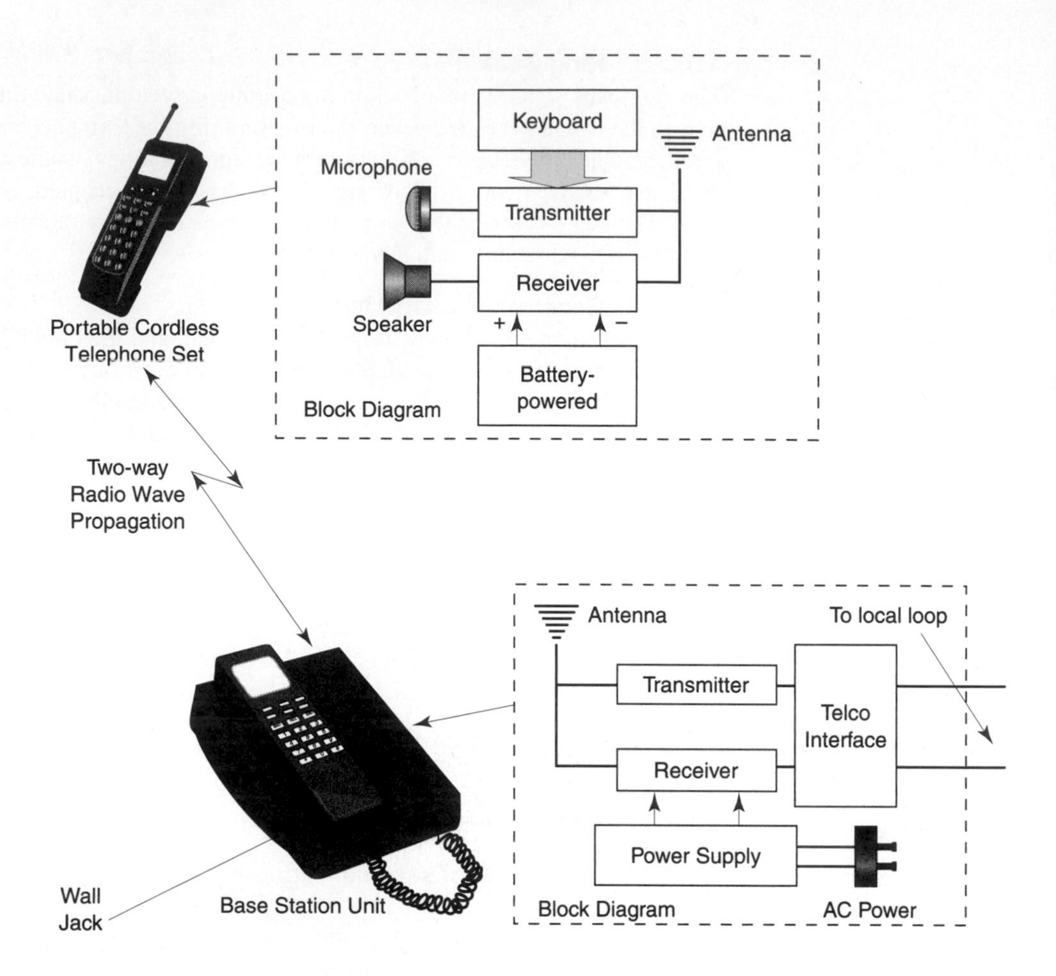

FIGURE 16-8 Cordless telephone system

the subscriber's home or office. The basic layout for a cordless telephone is shown in Figure 16-8. The base station is an ac-powered stationary radio transceiver (transmitter and receiver) connected to the local loop through a cord and telephone company interface unit. The interface unit functions in much the same way as a standard telephone set in that its primary function is to interface the cordless telephone with the local loop while being transparent to the user. Therefore, the base station is capable of transmitting and receiving both supervisory and voice signals over the subscriber loop in the same manner as a standard telephone. The base station must also be capable of relaying voice and control signals to and from the portable telephone set through the wireless transceiver. In essence, the portable telephone set is a battery-powered, two-way radio capable of operating in the full-duplex mode.

Because a portable telephone must be capable of communicating with the base station in the full-duplex mode, it must transmit and receive at different frequencies. In 1984, the FCC allocated 10 full-duplex channels for 46-MHz to 49-MHz units. In 1995 to help relieve congestion, the FCC added 15 additional full-duplex channels and extended the frequency band to include frequencies in the 43-MHz to 44-MHz band. Base stations transmit on high-band frequencies and receive on low-band frequencies, while the portable unit transmits on low-band frequencies and receives on high-band frequencies. The frequency assignments are listed in Table 16-5. Channels 16 through 25 are the original 10 full-duplex carrier frequencies. The maximum transmit power for both the portable unit and the base station is 500 mW. This stipulation limits the useful range of a cordless telephone to within 100 feet or less of the base station.

Table 16-5 43-MHz- to 49-MHz-Band Cordless Telephone Frequencies

Portable Unit		
Channel	Transmit Frequency (MHz)	Receive Frequency (MHz)
1	43.720	48.760
2	43.740	48.840
3	43.820	48.860
4	43.840	48.920
5	43.920	49.920
6	43.960	49.080
7	44.120	49.100
8	44.160	49.160
9	44.180	49.200
10	44.200	49.240
11	44.320	49.280
12	44.360	49.360
13	44.400	49.400
14	44.460	49.460
15	44.480	49.500
16	46.610	49.670
17	46.630	49.845
18	46.670	49.860
19	46.710	49.770
20	46.730	49.875
21	46.770	49.830
22	46.830	49.890
23	46.870	49.930
24	46.930	49.970
25	46.970	49.990

Note. Base stations transmit on the 49-MHz band and receive on the 46-MHz band.

Cordless telephones using the 2.4-GHz band offer excellent sound quality utilizing digital modulation and twin-band transmission to extend their range. With twin-band transmission, base stations transmit in the 2.4-GHz band, while portable units transmit in the 902-MHz to 928-MHz band.

16-7 CALLER ID

Caller ID (identification) is a service originally envisioned by AT&T in the early 1970s, although local telephone companies have only recently offered it. The basic concept of caller ID is quite simple. Caller ID enables the destination station of a telephone call to display the name and telephone number of the calling party before the telephone is answered (i.e., while the telephone is ringing). This allows subscribers to screen incoming calls and decide whether they want to answer the telephone.

The caller ID message is a simplex transmission sent from the central office switch over the local loop to a caller ID display unit at the destination station (no response is provided). The caller ID information is transmitted and received using Bell System 202-compatible modems (ITU V.23 standard). This standard specifies a 1200-bps FSK (frequency shift keying) signal with a 1200-Hz mark frequency (f_m) and a 2200-Hz space frequency (f_m). The FSK signal is transmitted in a burst between the first and second 20-Hz, 90-Vrms ringing signals, as shown in Figure 16-9a. Therefore, to ensure detection of the caller ID signal, the telephone must ring at least twice before being answering. The caller ID signal does not begin until 500 ms after the end of the first ring and must end 500 ms before the beginning of the second ring. Therefore, the caller ID signal has a 3-second window in which it must be transmitted.

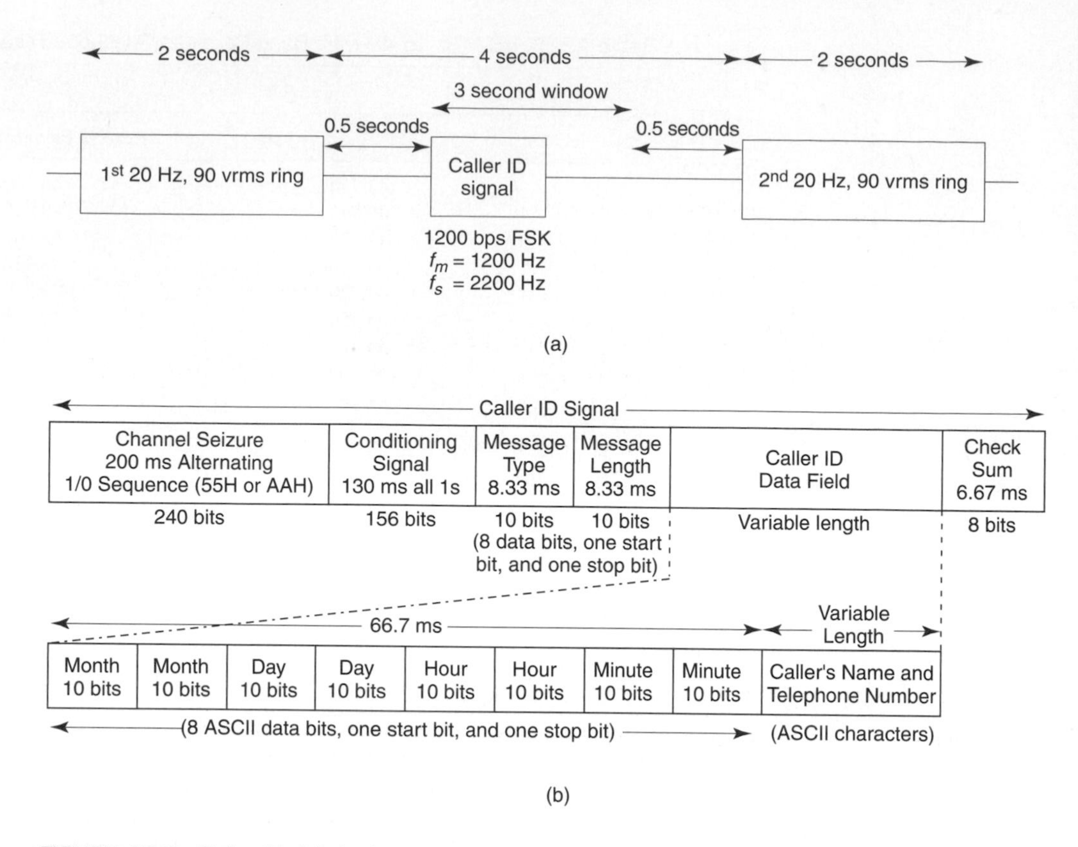

FIGURE 16-9 Caller ID: (a) ringing cycle; (b) frame format

The format for a caller ID signal is shown in Figure 16-9b. The 500-ms delay after the first ringing signal is immediately followed by the *channel seizure field,* which is a 200-ms-long sequence of alternating logic 1s and logic 0s (240 bits comprised of 120 pairs of alternating 1/0 bits, either 55 hex or AA hex). A *conditioning signal field* immediately follows the channel seizure field. The conditioning signal is a continuous 1200-Hz tone lasting for 130 ms, which equates to 156 consecutive logic 1 bits.

The protocol used for the next three fields—*message type field, message length field,* and *caller ID data field*—specifies asynchronous transmission of 16-bit characters (without parity) framed by one start bit (logic 0) and one stop bit (logic 1) for a total of 10 bits per character. The message type field is comprised of a 16-bit hex code, indicating the type of service and capability of the data message. There is only one message type field currently used with caller ID (04 hex). The message type field is followed by a 16-bit message length field, which specifies the total number of characters (in binary) included in the caller ID data field. For example, a message length code of 15 hex (0001 0101) equates to the number 21 in decimal. Therefore, a message length code of 15 hex specifies 21 characters in the caller ID data field.

The caller ID data field uses extended ASCII coded characters to represent a month code (01 through 12), a two-character day code (01 through 31), a two-character hour code in local military time (00 through 23), a two-character minute code (00 through 59), and a variable-length code, representing the caller's name and telephone number. ASCII coded digits are comprised of two independent hex characters (eight bits each). The first hex character is always 3 (0011 binary), and the second hex character represents a digit between 0 and 9 (0000 to 1001 binary). For example, 30 hex (0011 0000 binary) equates to the digit 0, 31 hex (0011 0001 binary) equates to the digit 1, 39 hex (0011 1001) equates to the digit

9, and so on. The caller ID data field is followed by a checksum for error detection, which is the 2's complement of the module 256 sum of the other words in the data message (message type, message length, and data words).

Example 16-1

Interpret the following hex code for a caller ID message (start and stop bits are not included in the hex codes):

04 12 31 31 32 37 31 35 35 37 33 31 35 37 33 36 31 30 35 33 xx

Solution 04—message type word

12—18 decimal (18 characters in the caller ID data field)

31, 31—ASCII code for 11 (the month of November)

32, 37—ASCII code for 27 (the 27th day of the month)

31, 35—ASCII code for 15 (the 15th hour—3:00 P.M.)

35, 37—ASCII code for 57 (57 minutes after the hour—3:57 P.M.)

33, 31, 35, 37, 33, 36, 31, 30, 35, 33—10-digit ASCII-coded telephone number (315 736–1053)

xx—checksum (00 hex to FF hex)

16-8 ELECTRONIC TELEPHONES

Although 500- and 2500-type telephone sets still work with the public telephone network, they are becoming increasingly more difficult to find. Most modern-day telephones have replaced many of the mechanical functions performed in the old telephone sets with electronic circuits. Electronic telephones use integrated-circuit technology to perform many of the basic telephone functions as well as a myriad of new and, and in many cases, nonessential functions. The refinement of microprocessors has also led to the development of multiple-line, full-feature telephones that permit automatic control of the telephone set's features, including telephone number storage, automatic dialing, redialing, and caller ID. However, no matter how many new gadgets are included in the new telephone sets, they still have to interface with the telephone network in much the same manner as telephones did a century ago.

Figure 16-10 shows the block diagram for a typical electronic telephone comprised of one multifunctional integrated-circuit chip, a microprocessor chip, a Touch-Tone keypad, a speaker, a microphone, and a handful of discrete devices. The major components included in the multifunctional integrated circuit chip are DTMF tone generator, MPU (microprocessor unit) interface circuitry, random access memory (RAM), tone ringer circuit, speech network, and a line voltage regulator.

The Touch-Tone keyboard provides a means for the operator of the telephone to access the DTMF tone generator inside the multifunction integrated-circuit chip. The external crystal provides a stable and accurate frequency reference for producing the dual-tone multifrequency signaling tones.

The tone ringer circuit is activated by the reception of a 20-Hz ringing signal. Once the ringing signal is detected, the tone ringer drives a piezoelectric sound element that produces an electronic ring (without a bell).

The voltage regulator converts the dc voltage received from the local loop and converts it to a constant-level dc supply voltage to operate the electronic components in the telephone. The internal speech network contains several amplifiers and associated components that perform the same functions as the hybrid did in a standard telephone.

The microprocessor interface circuit interfaces the MPU to the multifunction chip. The MPU, with its internal RAM, controls many of the functions of the telephone, such as

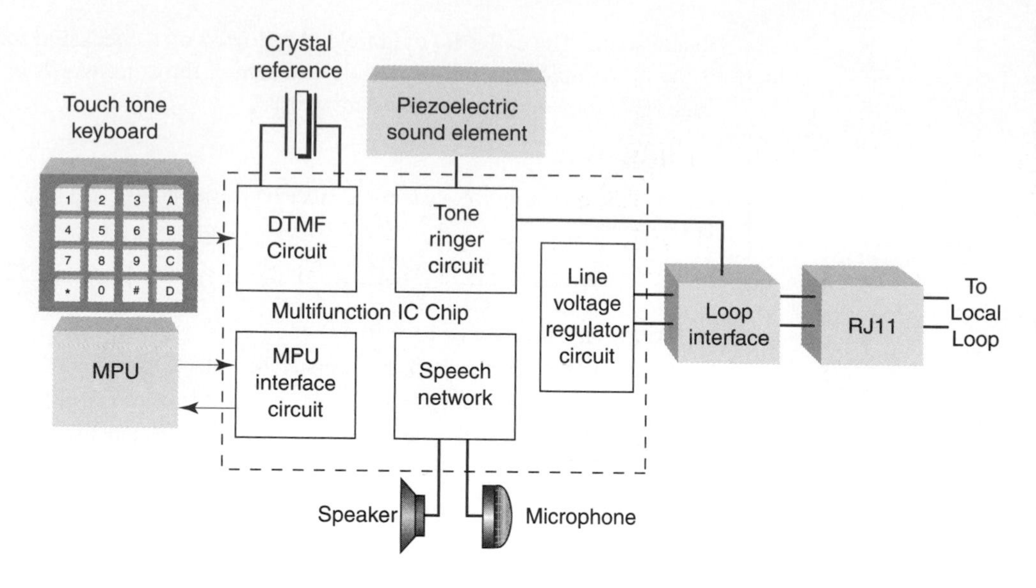

FIGURE 16-10 Electronic telephone set

number storage, speed dialing, redialing, and autodialing. The bridge rectifier protects the telephone from the relatively high-voltage ac ringing signal, and the switch hook is a mechanical switch that performs the same functions as the switch hook on a standard telephone set.

16-9 PAGING SYSTEMS

Most *paging syste*ms are simplex wireless communications system designed to alert subscribers of awaiting messages. Paging transmitters relay radio signals and messages from wire-line and cellular telephones to subscribers carrying portable receivers. The simplified block diagram of a paging system is shown in Figure 16-11. The infrastructure used with paging systems is somewhat different than the one used for cellular telephone system. This is because standard paging systems are one way, with signals transmitted from the paging system to portable pager and never in the reverse direction. There are narrow-, mid-, and wide-area pagers (sometimes called local, regional, and national). Narrow-area paging systems operate only within a building or building complex, mid-area pagers cover an area of several square miles, and wide-area pagers operate worldwide. Most pagers are mid-area where one centrally located high-power transmitter can cover a relatively large geographic area, typically between 6 and 10 miles in diameter.

To contact a person carrying a pager, simply dial the telephone number assigned that person's portable pager. The paging company receives the call and responds with a query requesting the telephone number you wish the paged person to call. After the number is entered, a *terminating signal* is appended to the number, which is usually the # sign. The caller then hangs up. The paging system converts the telephone number to a digital code and transmits it in the form of a digitally encoded signal over a wireless communications system. The signal may be simultaneously sent from more than one radio transmitter (sometimes called *simulcasting* or *broadcasting*), as is necessary in a wide-area paging system. If the paged person is within range of a broadcast transmitter, the targeted pager will receive the message. The message includes a notification signal, which either produces an audible beep or causes the pager to vibrate and the number the paged unit should call shown on an alphanumeric display. Some newer paging units are also capable of displaying messages as well as the telephone number of the paging party.

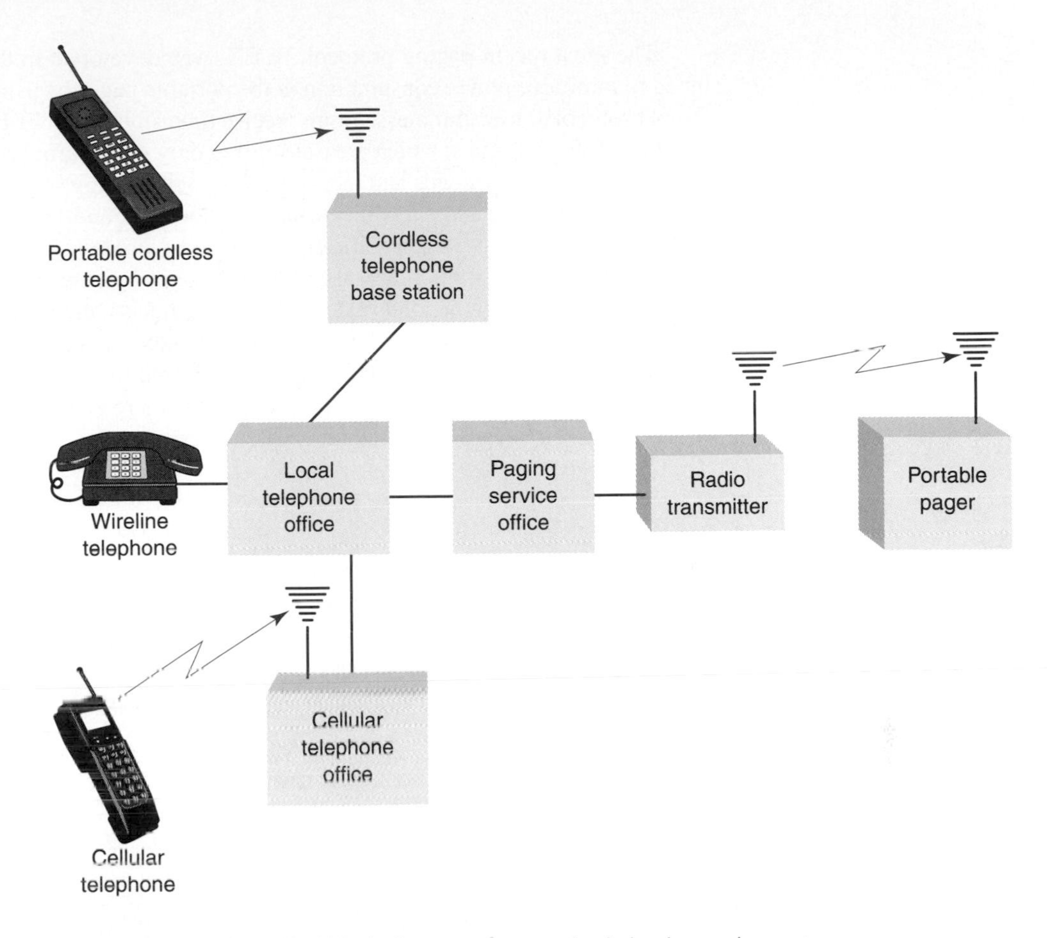

FIGURE 16-11 Simplified block diagram of a standard simplex paging system

Early paging systems used FM; however, most modern paging systems use FSK or PSK. Pagers typically transmit bit rates between 200 bps and 6400 bps with the following carrier frequency bands: 138 MHz to 175 MHz, 267 MHz to 284 MHz, 310 MHz to 330 MHz, 420 MHz to 470 MHz, and several frequency slots within the 900-MHz band.

Each portable pager is assigned a special code, called a *cap code*, which includes a sequence of digits or a combination of digits and letters. The cap code is broadcasted along with the paging party's telephone number. If the portable paging unit is within range of the broadcasting transmitter, it will receive the signal, demodulate it, and recognize its cap code. Once the portable pager recognizes its cap code, the callback number and perhaps a message will be displayed on the unit. Alphanumeric messages are generally limited to between 20 and 40 characters in length.

Early paging systems, such as one developed by the British Post Office called Post Office Code Standardization Advisory Group (POCSAG), transmitted a two-level FSK signal. POCSAG used an asynchronous protocol, which required a long preamble for synchronization. The preamble begins with a long *dotting sequence* (sometimes called a *dotting comma*) to establish clock synchronization. Data rates for POCSAG are 512 bps, 1200 bps, and 2400 bps. With POCSAG, portable pagers must operate in the *always-on mode* all the time, which means the pager wastes much of its power resources on nondata preamble bits.

In the early 1980s, the European Telecommunications Standards Institute (ETSI) developed the ERMES protocol. ERMES transmitted data at a 6250 bps rate using four-level FSK (3125 baud). ERMES is a synchronous protocol, which requires less time to synchronize. ERMES supports 16 25-kHz paging channels in each of its frequency bands.

The most recent paging protocol, FLEX, was developed in the 1990s. FLEX is designed to minimize power consumption in the portable pager by using a synchronous time-slotted protocol to transmit messages in precise time slots. With FLEX, each frame is comprised of 128 data frames, which are transmitted only once during a 4-minute period. Each frame lasts for 1.875 seconds and includes two synchronizing sequences, a header containing frame information and pager identification addresses, and 11 discrete data blocks. Each portable pager is assigned a specific frame (called a *home frame*) within the frame cycle that it checks for transmitted messages. Thus, a pager operates in the high-power standby condition for only a few seconds every 4 minutes (this is called the *wakeup time*). The rest of the time, the pager is in an ultra-low power standby condition. When a pager is in the wakeup mode, it synchronizes to the frame header and then adjusts itself to the bit rate of the received signal. When the pager determines that there is no message waiting, it puts itself back to sleep, leaving only the timer circuit active.

QUESTIONS

16-1. Define the terms *communications* and *telecommunications.*

16-2. Define *plain old telephone service.*

16-3. Describe a *local subscriber loop.*

16-4. Where in a telephone system is the *local loop?*

16-5. Briefly describe the basic functions of a standard *telephone set.*

16-6. What is the purpose of the *RJ-11 connector?*

16-7. What is meant by the terms *tip* and *ring?*

16-8. List and briefly describe the essential components of a standard telephone set.

16-9. Briefly describe the steps involved in completing a local telephone call.

16-10. Explain the basic purpose of *call progress tones* and *signals.*

16-11. List and describe the two primary categories of *signaling.*

16-12. Describe the following signaling messages: *alerting, supervising, controlling,* and *addressing.*

16-13. What is the purpose of *dial tone,* and when is it applied to a telephone circuit?

16-14. Briefly describe *dual-tone multifrequency* and *multifrequency* signaling and tell where they are used.

16-15. Describe *dial pulsing.*

16-16. What is the difference between a *station busy* signal and an *equipment busy* signal?

16-17. What is the difference between a *ringing* and a *ring-back* signal?

16-18. Briefly describe what happens when a telephone set is taken *off hook.*

16-19. Describe the differences between the operation of a *cordless telephone* and a *standard telephone.*

16-20. Explain how *caller ID* operates and when it is used.

16-21. Briefly describe how a paging system operates.

CHAPTER 17

The Telephone Circuit

CHAPTER OUTLINE

17-1 Introduction
17-2 The Local Subscriber Loop
17-3 Telephone Message–Channel Noise and Noise Weighting
17-4 Units of Power Measurement
17-5 Transmission Parameters and Private-Line Circuits
17-6 Voice-Frequency Circuit Arrangements
17-7 Crosstalk

OBJECTIVES

- Define *telephone circuit, message,* and *message channel*
- Describe the transmission characteristics a local subscriber loop
- Describe loading coils and bridge taps
- Describe loop resistance and how it is calculated
- Explain telephone message–channel noise and C-message noise weighting
- Describe the following units of power measurement: db, dBm, dBmO, rn, dBrn, dBrnc, dBrn 3-kHz flat, and dBrncO
- Define *psophometric noise weighting*
- Define and describe transmission parameters
- Define *private-line circuit*
- Explain bandwidth, interface, and facilities parameters
- Define *line conditioning* and describe C- and D-type conditioning
- Describe two-wire and four-wire circuit arrangements
- Explain hybrids, echo suppressors, and echo cancelers
- Define *crosstalk*
- Describe nonlinear, transmittance, and coupling crosstalk

17-1 INTRODUCTION

A *telephone circuit* is comprised of two or more facilities, interconnected in tandem, to provide a transmission path between a source and a destination. The interconnected facilities may be temporary, as in a standard telephone call, or permanent, as in a dedicated private-line telephone circuit. The facilities may be metallic cable pairs, optical fibers, or wireless carrier systems. The information transferred is called the *message,* and the circuit used is called the *message channel.*

Telephone companies offer a wide assortment of message channels ranging from a basic 4-kHz voice-band circuit to wideband microwave, satellite, or optical fiber transmission systems capable of transferring high-resolution video or wideband data. The following discussion is limited to basic voice-band circuits. In telephone terminology, the word *message* originally denoted speech information. However, this definition has been extended to include any signal that occupies the same bandwidth as a standard voice channel. Thus, a message channel may include the transmission of ordinary speech, supervisory signals, or data in the form of digitally modulated carriers (FSK, PSK, QAM, and so on). The network bandwidth for a standard voice-band message channel is 4 kHz; however, a portion of that bandwidth is used for *guard bands* and signaling. Guard bands are unused frequency bands located between information signals. Consequently, the effective channel bandwidth for a voice-band message signal (whether it be voice or data) is approximately 300 Hz to 3000 Hz.

17-2 THE LOCAL SUBSCRIBER LOOP

The *local subscriber loop* is the only facility required by all voice-band circuits, as it is the means by which subscriber locations are connected to the local telephone company. In essence, the sole purpose of a local loop is to provide subscribers access to the public telephone network. The local loop is a metallic transmission line comprised of two insulated copper wires (a pair) twisted together. The local loop is the primary cause of *attenuation* and *phase distortion* on a telephone circuit. Attenuation is an actual loss of signal strength, and phase distortion occurs when two or more frequencies undergo different amounts of phase shift.

The *transmission characteristics* of a cable pair depend on the wire diameter, conductor spacing, dielectric constant of the insulator separating the wires, and the conductivity of the wire. These physical properties, in turn, determine the inductance, resistance, capacitance, and conductance of the cable. The resistance and inductance are distributed along the length of the wire, whereas the conductance and capacitance exist between the two wires. When the insulation is sufficient, the effects of conductance are generally negligible. Figure 17-1a shows the electrical model for a copper-wire transmission line.

The electrical characteristics of a cable (such as inductance, capacitance, and resistance) are uniformly distributed along its length and are appropriately referred to as *distributed parameters.* Because it is cumbersome working with distributed parameters, it is common practice to lump them into discrete values per unit length (i.e., millihenrys per mile, microfarads per kilometer, or ohms per 1000 feet). The amount of attenuation and phase delay experienced by a signal propagating down a metallic transmission line is a function of the frequency of the signal and the electrical characteristics of the cable pair.

There are seven main component parts that make up a traditional local loop:

Feeder cable (*F1*). The largest cable used in a local loop, usually 3600 pair of copper wire placed underground or in conduit.

Serving area interface (*SAI*). A cross-connect point used to distribute the larger feeder cable into smaller distribution cables.

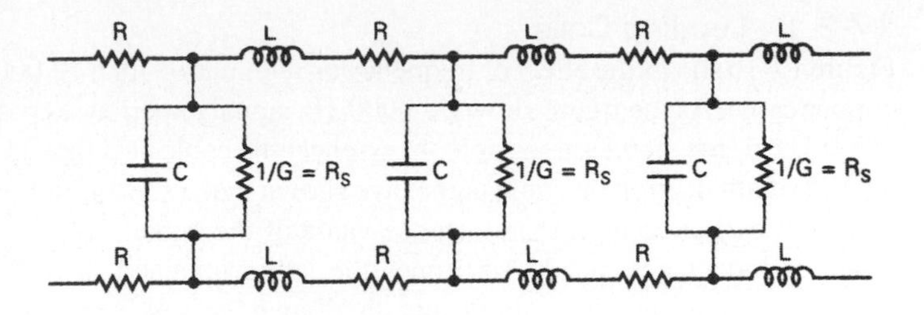

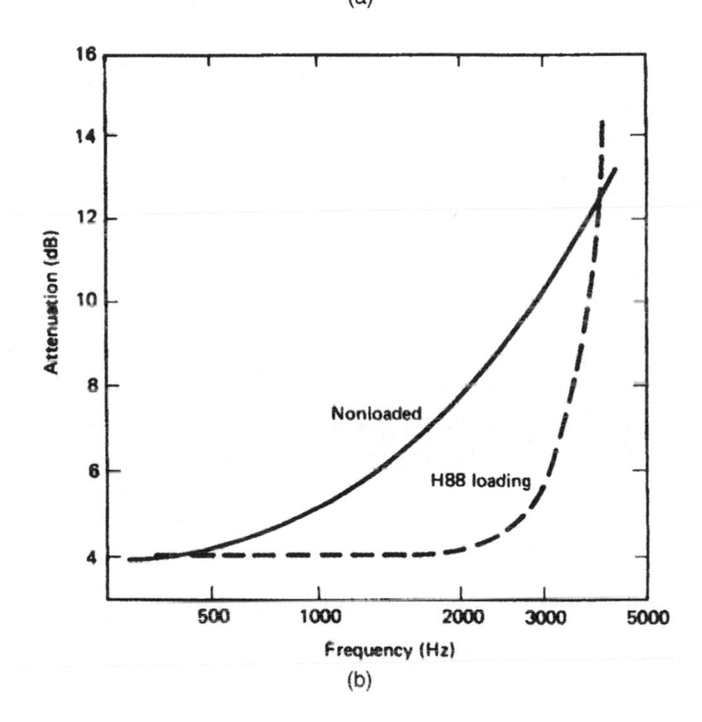

FIGURE 17-1 (a) Electrical model of a copper-wire transmission line, (b) frequency-versus-attenuation characteristics for unloaded and loaded cables

Distribution cable (F2). A smaller version of a feeder cable containing less wire pairs.

Subscriber or standard network interface (SNI). A device that serves as the demarcation point between local telephone company responsibility and subscriber responsibility for telephone service.

Drop wire. The final length of cable pair that terminates at the SNI.

Aerial. That portion of the local loop that is strung between poles.

Distribution cable and drop-wire cross-connect point. The location where individual cable pairs within a distribution cable are separated and extended to the subscriber's location on a drop wire.

Two components often found on local loops are loading coils and bridge taps.

17-2-1 Loading Coils

Figure 17-1b shows the effect of frequency on attenuation for a 12,000-foot length of 26-gauge copper cable. As the figure shows, a 3000-Hz signal experiences 6 dB more attenuation than a 500-Hz signal on the same cable. In essence, the cable acts like a low-pass filter. Extensive studies of attenuation on cable pairs have shown that a substantial reduction in attenuation is achieved by increasing the inductance value of the cable. Minimum attenuation requires a value of inductance nearly 100 times the value obtained in ordinary twisted-wire cable. Achieving such values on a uniformly distributed basis is impractical. Instead, the desired effect can be obtained by adding inductors periodically in series with the wire. This practice is called *loading,* and the inductors are called *loading coils.* Loading coils placed in a cable decrease the attenuation, increase the line impedance, and improve transmission levels for circuits longer than 18,000 feet. Loading coils allowed local loops to extend three to four times their previous length. A loading coil is simply a passive conductor wrapped around a core and placed in series with a cable creating a small electromagnet. Loading coils can be placed on telephone poles, in manholes, or on cross-connect boxes. Loading coils increase the effective distance that a signal must travel between two locations and cancels the capacitance that inherently builds up between wires with distance. Loading coils first came into use in 1900.

Loaded cables are specified by the addition of the letter codes A, B, C, D, E, F, H, X, or Y, which designate the distance between loading coils and by numbers, which indicate the inductance value of the wire gauge. The letters indicate that loading coils are separated by 700, 3000, 929, 4500, 5575, 2787, 6000, 680, or 2130 feet, respectively. B-, D-, and H-type loading coils are the most common because their separations are representative of the distances between manholes. The amount of series inductance added is generally 44 mH, 88 mH, or 135 mH. Thus, a cable pair designated 26H88 is made from 26-gauge wire with 88 mH of series inductance added every 6000 feet. The loss-versus-frequency characteristics for a loaded cable are relatively flat up to approximately 2000 Hz, as shown in Figure 17-1b. From the figure, it can be seen that a 3000-Hz signal will suffer only 1.5 dB more loss than a 500-Hz signal on 26-gauge wire when 88-mH loading coils are spaced every 6000 feet.

Loading coils cause a sharp drop in frequency response at approximately 3400 Hz, which is undesirable for high-speed data transmission. Therefore, for high-performance data transmission, loading coils should be removed from the cables. The low-pass characteristics of a cable also affect the phase distortion-versus-frequency characteristics of a signal. The amount of *phase distortion* is proportional to the length and gauge of the wire. Loading a cable also affects the phase characteristics of a cable. The telephone company must often add gain and delay equalizers to a circuit to achieve the minimum requirements. Equalizers introduce discontinuities or ripples in the bandpass characteristics of a circuit. Automatic equalizers in data modems are sensitive to this condition, and very often an overequalized circuit causes as many problems to a data signal as an underequalized circuit.

17-2-2 Bridge Taps

A *bridge tap* is an irregularity frequently found in cables serving subscriber locations. Bridge taps are unused sections of cable that are connected in shunt to a working cable pair, such as a local loop. Bridge taps can be placed at any point along a cable's length. Bridge taps were used for party lines to connect more than one subscriber to the same local loop. Bridge taps also increase the flexibility of a local loop by allowing the cable to go to more than one junction box, although it is unlikely that more than one of the cable pairs leaving a bridging point will be used at any given time. Bridge taps may or may not be used at some future time, depending on service demands. Bridge taps increase the flexibility of a cable by making it easier to reassign a cable to a different subscriber without requiring a person working in the field to cross connect sections of cable.

Bridge taps introduce a loss called *bridging loss.* They also allow signals to split and propagate down more than one wire. Signals that propagate down unterminated (open-circuited) cables reflect back from the open end of the cable, often causing interference with the original signal. Bridge taps that are short and closer to the originating or terminating ends often produce the most interference.

Bridge taps and loading coils are not generally harmful to voice transmissions, but if improperly used, they can literally destroy the integrity of a data signal. Therefore, bridge taps and loading coils should be removed from a cable pair that is used for data transmission. This can be a problem because it is sometimes difficult to locate a bridge tap. It is estimated that the average local loop can have as many as 16 bridge taps.

17-2-3 Loop Resistance

The dc resistance of a local loop depends primarily on the type of wire and wire size. Most local loops use 18- to 26-gauge, twisted-pair copper wire. The lower the wire gauge, the larger the diameter, the less resistance, and the lower the attenuation. For example, 26-gauge unloaded copper wire has an attenuation of 2.67 dB per mile, whereas the same length of 19-gauge copper wire has only 1.12 dB per mile. Therefore, the maximum length of a local loop using 19-gauge wire is twice as long as a local loop using 26-gauge wire.

The total attenuation of a local loop is generally limited to a maximum value of 7.5 dB with a maximum dc resistance of 1300 Ω, which includes the resistance of the telephone (approximately 120 Ω). The dc resistance of 26-gauge copper wire is approximately 41 Ω per 1000 feet, which limits the round-trip loop length to approximately 5.6 miles. The maximum distance for lower-gauge wire is longer of course.

The dc loop resistance for copper conductors is approximated by

$$R_{dc} = \frac{0.1095}{d^2} \qquad \textbf{(17-1)}$$

where R_{dc} = dc loop resistance (ohms per mile)
d = wire diameter (inches)

17-3 TELEPHONE MESSAGE–CHANNEL NOISE AND NOISE WEIGHTING

The *noise* that reaches a listener's ears affects the degree of annoyance to the listener and, to some extent, the intelligibility of the received speech. The total noise is comprised of room *background noise* and noise introduced in the circuit. Room background noise on the listening subscriber's premises reaches the ear directly through leakage around the receiver and indirectly by way of the sidetone path through the telephone set. Room noise from the talking subscriber's premises also reaches the listener over the communications channel. Circuit noise is comprised mainly of thermal noise, nonlinear distortion, and impulse noise, which are described in a later section of this chapter.

The measurement of interference (noise), like the measurement of volume, is an effort to characterize a complex signal. Noise measurements on a telephone message channel are characterized by how annoying the noise is to the subscriber rather than by the absolute magnitude of the average noise power. Noise interference is comprised of two components: annoyance and the effect of noise on intelligibility, both of which are functions of frequency. Noise signals with equal interfering effects are assigned equal magnitudes. To accomplish this effect, the American Telephone and Telegraph Company (AT&T) developed a weighting network called *C-message* weighting.

When designing the C-message weighting network, groups of observers were asked to adjust the loudness of 14 different frequencies between 180 Hz and 3500 Hz until the sound of each tone was judged to be equally annoying as a 1000-Hz reference tone in the absence of speech. A 1000-Hz tone was selected for the reference because empirical data indicated that 1000 Hz is the most annoying frequency (i.e., the best frequency response)

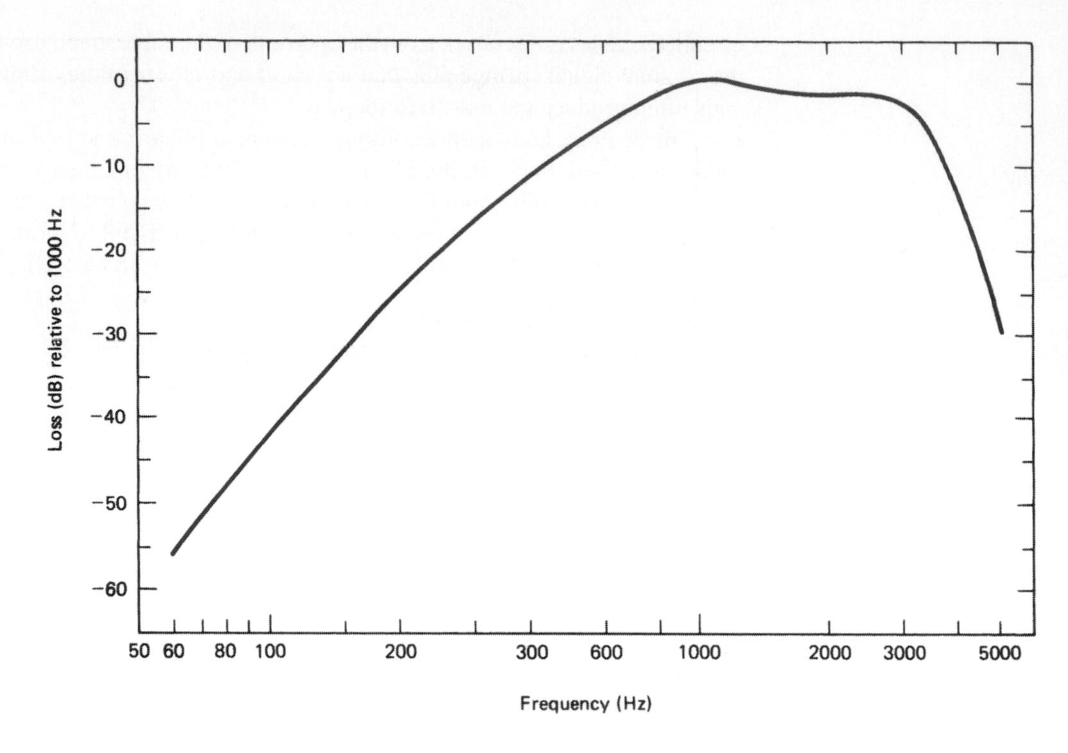

FIGURE 17-2 C-message weighting curve

to humans. The same people were then asked to adjust the amplitude of the tones in the presence of speech until the effect of noise on articulation (annoyance) was equal to that of the 1000-Hz reference tone. The results of the two experiments were combined, smoothed, and plotted, resulting in the C-message weighting curve shown in Figure 17-2. A 500-type telephone set was used for these tests; therefore, the C-message weighting curve includes the frequency response characteristics of a standard telephone set receiver as well as the hearing response of an average listener.

The significance of the C-message weighting curve is best illustrated with an example. From Figure 17-2, it can be seen that a 200-Hz test tone of a given power is 25 dB less disturbing than a 1000-Hz test tone of the same power. Therefore, the C-message weighting network will introduce 25 dB more loss for 200 Hz than it will for 1000 Hz.

When designing the C-message network, it was found that the additive effect of several noise sources combine on a root-sum-square (RSS) basis. From these design considerations, it was determined that a telephone message–channel noise measuring set should be a voltmeter with the following characteristics:

Readings should take into consideration that the interfering effect of noise is a function of frequency as well as magnitude.

When dissimilar noise signals are present simultaneously, the meter should combine them to properly measure the overall interfering effect.

It should have a transient response resembling that of the human ear. For sounds shorter than 200 ms, the human ear does not fully appreciate the true power of the sound. Therefore, noise-measuring sets are designed to give full-power indication only for bursts of noise lasting 200 ms or longer.

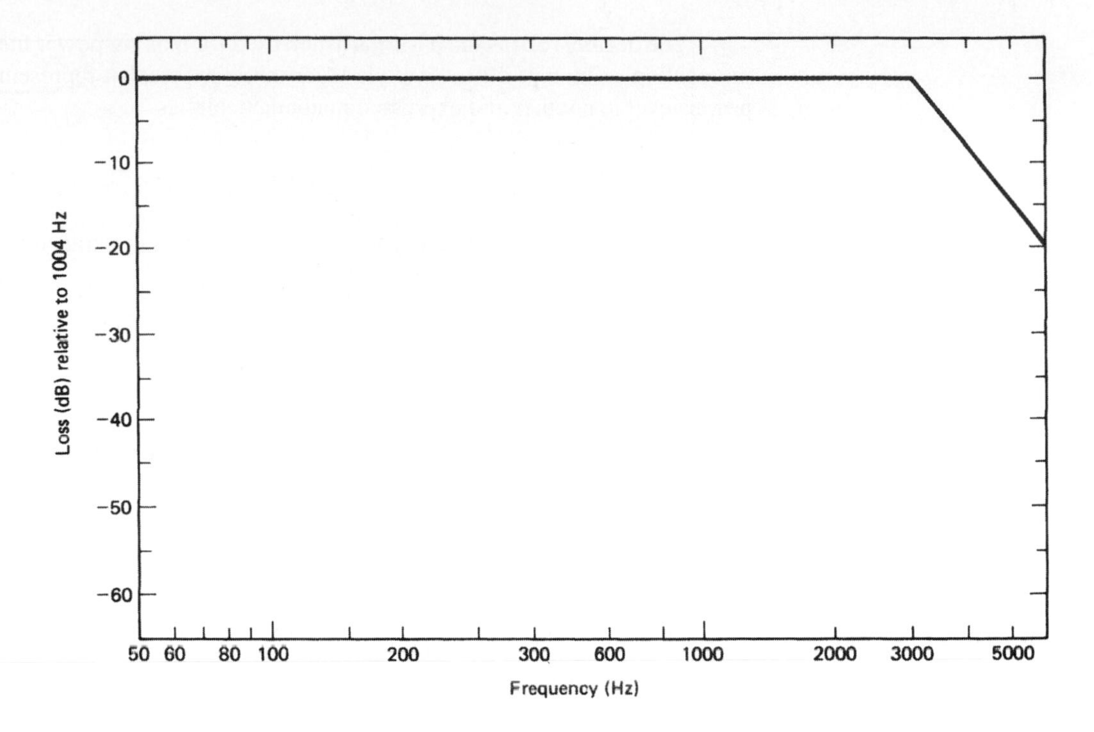

FIGURE 17-3 3-kHz flat response curve

When different types of noise cause equal interference as determined in subjective tests, use of the meter should give equal readings.

The reference established for performing message-channel noise measurements is −90 dBm (10^{-12} watts). The power level of −90 dBm was selected because, at the time, power levels could not measure levels below −90 dBm and, therefore, it would not be necessary to deal with negative values when reading noise levels. Thus, a 1000-Hz tone with a power level of −90 dBm is equal to a noise reading of 0 dBrn. Conversely, a 1000-Hz tone with a power level of 0 dBm is equal to a noise reading of 90 dBrn, and a 1000-Hz tone with a power level of −40 dBm is equal to a noise reading of 50 dBrn.

When appropriate, other weighting networks can be substituted for C-message. For example, a *3-kHz flat* network is used to measure power density of white noise. This network has a nominal low-pass frequency response down 3 dB at 3 kHz and rolls off at 12 dB per octave. A 3-kHz flat network is often used for measuring high levels of low-frequency noise, such as power supply hum. The frequency response for a 3-kHz flat network is shown in Figure 17-3.

17-4 UNITS OF POWER MEASUREMENT

17-4-1 dB and dBm

To specify the amplitudes of signals and interference, it is often convenient to define them at some reference point in the system. The amplitudes at any other physical location can then be related to this reference point if the loss or gain between the two points is known. For example, sea level is generally used as the reference point when comparing elevations. By referencing two mountains to sea level, we can compare the two elevations, regardless of where the mountains are located. A mountain peak in Colorado 12,000 feet above sea level is 4000 feet higher than a mountain peak in France 8000 feet above sea level.

The decibel (dB) is the basic yardstick used for making power measurements in communications. The unit dB is simply a logarithmic expression representing the ratio of one power level to another and expressed mathematically as

$$dB = 10 \log\left(\frac{P_1}{P_2}\right) \tag{17-2}$$

where P_1 and P_2 are power levels at two different points in a transmission system.

From Equation 17-2, it can be seen when $P_1 = P_2$, the power ratio is 0 dB; when $P_1 > P_2$, the power ratio in dB is positive; and when $P_1 < P_2$, the power ratio in dB is negative. In telephone and telecommunications circuits, power levels are given in dBm and differences between power levels in dB.

Equation 17-2 is essentially dimensionless since neither power is referenced to a base. The unit dBm is often used to reference the power level at a given point to 1 milliwatt. One milliwatt is the level from which all dBm measurements are referenced. The unit dBm is an indirect measure of absolute power and expressed mathematically as

$$dBm = 10 \log\left(\frac{P}{1\text{ mW}}\right) \tag{17-3}$$

where P is the power at any point in a transmission system. From Equation 17-3, it can be seen that a power level of 1 mW equates to 0 dBm, power levels above 1 mW have positive dBm values, and power levels less than 1 mW have negative dBm values.

Example 17-1

Determine

a. The power levels in dBm for signal levels of 10 mW and 0.5 mW.

b. The difference between the two power levels in dB.

Solution **a.** The power levels in dBm are determined by substituting into Equation 17-3:

$$dBm = 10 \log\left(\frac{10\text{ mW}}{1\text{ mW}}\right) = 10\text{ dBm}$$

$$dBm = 10 \log\left(\frac{0.5\text{ mW}}{1\text{ mW}}\right) = -3\text{ dBm}$$

b. The difference between the two power levels in dB is determined by substituting into Equation 17-2:

$$dB = 10 \log\left(\frac{10\text{ mW}}{0.5\text{ mW}}\right) = 13\text{ dB}$$

or

$$10\text{ dBm} - (-3\text{ dBm}) = 13\text{ dB}$$

The 10-mW power level is 13 dB higher than a 0.5-mW power level.

Experiments indicate that a listener cannot give a reliable estimate of the loudness of a sound but can distinguish the difference in loudness between two sounds. The ear's sensitivity to a change in sound power follows a logarithmic rather than a linear scale, and the dB has become the unit of this change.

17-4-2 Transmission Level Point, Transmission Level, and Data Level Point

Transmission level point (TLP) is defined as the optimum level of a test tone on a channel at some point in a communications system. The numerical value of the TLP does not describe the total signal power present at that point—it merely defines what the ideal level should be. The *transmission level* (TL) at any point in a transmission system is the ratio in dB of the power of a signal at that point to the power the same signal would be at a 0-dBm transmission level point. For example, a signal at a particular point in a transmission system measures −13 dBm. Is this good or bad? This could be answered only if it is known what the signal strength should be at that point. TLP does just that. The reference for TLP is 0 dBm. A −15-dBm TLP indicates that, at this specific point in the transmission system, the signal should measure −15 dBm. Therefore, the transmission level for a signal that measures −13 dBm at a −15-dBm point is −2 dB. A 0 TLP is a TLP where the signal power should be 0 dBm. TLP says nothing about the actual signal level itself.

Data level point (DLP) is a parameter equivalent to TLP except TLP is used for voice circuits, whereas DLP is used as a reference for data transmission. The DLP is always 13 dB below the voice level for the same point. If the TLP is −15 dBm, the DLP at the same point is −28 dBm. Because a data signal is more sensitive to nonlinear distortion (harmonic and intermodulation distortion), data signals are transmitted at a lower level than voice signals.

17-4-3 Units of Measurement

Common units for signal and noise power measurements in the telephone industry include dBmO, rn, dBrn, dBrnc, dBrn 3-kHz flat, and dBrncO.

17-4-3-1 dBmO. *dBmO* is dBm referenced to a zero transmission level point (0 TLP). dBmO is a power measurement adjusted to 0 dBm that indicates what the power would be if the signal were measured at a 0 TLP. dBmO compares the actual signal level at a point with what that signal level should be at that point. For example, a signal measuring −17 dBm at a −16-dBm transmission level point is −1 dBmO (i.e., the signal is 1 dB below what it should be, or if it were measured at a 0 TLP, it would measure −1 dBm).

17-4-3-2 rn (reference noise). *rn* is the dB value used as the reference for noise readings. Reference noise equals −90 dBm or 1 pW (1×10^{-12} W). This value was selected for two reasons: (1) Early noise measuring sets could not accurately measure noise levels lower than −90 dBm, and (2) noise readings are typically higher than −90 dBm, resulting in positive dB readings in respect to reference noise.

17-4-3-3 dBrn. *dBrn* is the dB level of noise with respect to reference noise (−90 dBm). dBrn is seldom used by itself since it does not specify a weighting. A noise reading of −50 dBm equates to 40 dBrn, which is 40 dB above reference noise $(-50 - [-90]) = 40$ dBrn.

17-4-3-4 dBrnc. *dBrnc* is similar to dBrn except dBrnc is the dB value of noise with respect to reference noise using C-message weighting. Noise measurements obtained with a C-message filter are meaningful, as they relate the noise measured to the combined frequency response of a standard telephone and the human ear.

17-4-3-5 dBrn 3-kHz flat. *dBrn 3-kHz flat* noise measurements are noise readings taken with a filter that has a flat frequency response from 30 Hz to 3 kHz. Noise readings taken with a 3-kHz flat filter are especially useful for detecting low-frequency noise, such as power supply hum. dBrn 3-kHz flat readings are typically 1.5 dB higher than dBrnc readings for equal noise power levels.

17-4-3-6 dBrncO. *dBrncO* is the amount of noise in dBrnc corrected to a 0 TLP. A noise reading of 34 dBrnc at a +7-dBm TLP equates to 27 dBrncO. dBrncO relates noise power readings (dBrnc) to a 0 TLP. This unit establishes a common reference point throughout the transmission system.

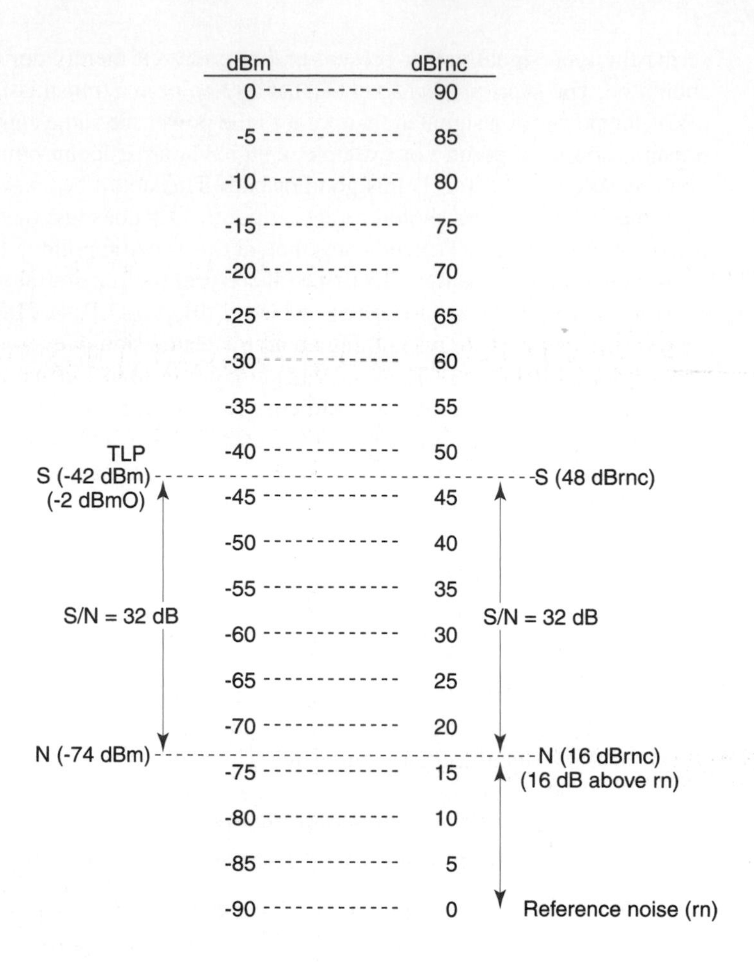

FIGURE 17-4 Figure for Example 17-2

Example 17-2

For a signal measurement of −42 dBm, a noise measurement of 16 dBrnc, and a −40-dBm TLP, determine

a. Signal level in dBrnc.
b. Noise level in dBm.
c. Signal level in dBmO.
d. signal-to-noise ratio in dB. (For the solutions, refer to Figure 17-4.)

Solution **a.** The signal level in dBrnc can be read directly from the chart shown in Figure 17-4 as 48 dBrnc. The signal level in dBrnc can also be computed mathematically as follows:

$$-42 \text{ dBm} - (-90 \text{ dBrn}) = 48 \text{ dBrnc}$$

b. The noise level in dBm can be read directly from the chart shown in Figure 17-4 as −74 dBm. The noise level in dBm can also be calculated as follows:

$$-90 + 16 = -74 \text{ dBm}$$

c. The signal level in dBmO is simply the difference between the actual signal level in dBm and the TLP or 2 dBmO as shown in Figure 17-4. The signal level in dBmO can also be computed mathematically as follows:

$$-42 \text{ dBm} - (-40 \text{ dBm}) = -2 \text{ dBmO}$$

d. The signal-to-noise ratio is simply the difference in the signal power in dBm and the noise power in dBm or the signal level in dBrnc and the noise power in dBrnc as shown in Figure 17-4 as 32 dB. The signal-to-noise ratio is computed mathematically as

$$-42 \text{ dBm} - (-74 \text{ dBm}) = 32 \text{ dB}$$

or

$$48 \text{ dBrnc} - 16 \text{ dBrnc} = 32 \text{ dB}$$

17-4-4 Psophometric Noise Weighting

Psophometric noise weighting is used primarily in Europe. Psophometric weighting assumes a perfect receiver; therefore, its weighting curve corresponds to the frequency response of the human ear only. The difference between C-message weighting and psophometric weighting is so small that the same conversion factor may be used for both.

17-5 TRANSMISSION PARAMETERS AND PRIVATE-LINE CIRCUITS

Transmission parameters apply to dedicated *private-line data circuits* that utilize the private sector of the public telephone network—circuits with bandwidths comparable to those of standard voice-grade telephone channels that do not utilize the public switched telephone network. Private-line circuits are direct connections between two or more locations. On private-line circuits, transmission facilities and other telephone company–provided equipment are hardwired and available only to a specific subscriber. Most private-line data circuits use four-wire, full-duplex facilities. Signal paths established through switched lines are inconsistent and may differ greatly from one call to another. In addition, telephone lines provided through the public switched telephone network are two wire, which limits high-speed data transmission to half-duplex operation. Private-line data circuits have several advantages over using the switched public telephone network:

Transmission characteristics are more consistent because the same facilities are used with every transmission.

The facilities are less prone to noise produced in telephone company switches.

Line conditioning is available only on private-line facilities.

Higher transmission bit rates and better performance is appreciated with private-line data circuits.

Private-line data circuits are more economical for high-volume circuits.

Transmission parameters are divided into three broad categories: bandwidth parameters, which include attenuation distortion and envelope delay distortion; interface parameters, which include terminal impedance, in-band and out-of-band signal power, test signal power, and ground isolation; and facility parameters, which include noise measurements, frequency distortion, phase distortion, amplitude distortion, and nonlinear distortion.

17-5-1 Bandwidth Parameters

The only transmission parameters with limits specified by the FCC are attenuation distortion and envelope delay distortion. *Attenuation distortion* is the difference in circuit gain experienced at a particular frequency with respect to the circuit gain of a reference frequency. This characteristic is sometimes referred to as *frequency response, differential gain,* and *1004-Hz deviation. Envelope delay distortion* is an indirect method of evaluating the phase delay characteristics of a circuit. FCC tariffs specify the limits for attenuation distortion and envelope delay distortion. To reduce attenuation and envelope delay distortion and improve the performance of data modems operating over standard message channels, it is often necessary to improve the quality of the channel. The process used to improve a basic telephone channel is called *line conditioning.* Line conditioning improves the high-frequency response of a message channel and reduces power loss.

The attenuation and delay characteristics of a circuit are artificially altered to meet limits prescribed by the *line conditioning* requirements. Line conditioning is available only to private-line subscribers at an additional charge. The *basic voice-band channel* (sometimes called a *basic 3002 channel*) satisfies the minimum line conditioning requirements. Telephone companies offer two types of special line conditioning for subscriber loops: C-type and D-type.

17-5-1-1 C-type line conditioning. *C-type conditioning* specifies the maximum limits for attenuation distortion and envelope delay distortion. C-type conditioning pertains to line impairments for which compensation can be made with filters and equalizers. This is accomplished with telephone company–provided equipment. When a circuit is initially turned up for service with a specific C-type conditioning, it must meet the requirements for that type of conditioning. The subscriber may include devices within the station equipment that compensate for minor long-term variations in the bandwidth requirements.

There are five classifications or levels of C-type conditioning available. The grade of conditioning a subscriber selects depends on the bit rate, modulation technique, and desired performance of the data modems used on the line. The five classifications of C-type conditioning are the following:

C1 and C2 conditioning pertain to two-point and multipoint circuits.

C3 conditioning is for access lines and trunk circuits associated with private switched networks.

C4 conditioning pertains to two-point and multipoint circuits with a maximum of four stations.

C5 conditioning pertains only to two-point circuits.

Private switched networks are telephone systems provided by local telephone companies dedicated to a single customer, usually with a large number of stations. An example is a large corporation with offices and complexes at two or more geographical locations, sometimes separated by great distances. Each location generally has an on-premise *private branch exchange* (PBX). A PBX is a relatively low-capacity switching machine where the subscribers are generally limited to stations within the same building or building complex. *Common-usage access lines* and *trunk circuits* are required to interconnect two or more PBXs. They are common only to the subscribers of the private network and not to the general public telephone network. Table 17-1 lists the limits prescribed by C-type conditioning for attenuation distortion. As the table shows, the higher the classification of conditioning imposed on a circuit, the flatter the frequency response and, therefore, a better-quality circuit.

Attenuation distortion is simply the frequency response of a transmission medium referenced to a 1004-Hz test tone. The attenuation for voice-band frequencies on a typical cable pair is directly proportional to the square root of the frequency. From Table 17-1, the attenuation distortion limits for a basic (unconditioned) circuit specify the circuit gain at any frequency between 500 Hz and 2500 Hz to be not more than 2 dB more than the circuit gain at 1004 Hz and not more than 3 dB below the circuit gain at 1004 Hz. For attenuation distortion, the circuit gain for 1004 Hz is always the reference. Also, within the frequency bands from 300 Hz and 499 Hz and from 2501 Hz to 3000 Hz, the circuit gain cannot be

Table 17-1 Basic and C-Type Conditioning Requirements

Attenuation Distortion (Frequency Response Relative to 1004 Hz)			Envelope Delay Distortion	
Channel Conditioning	Frequency Range (Hz)	Variation (dB)	Frequency Range (Hz)	Variation (μs)
Basic	300–499	+3 to −12	800–2600	1750
	500–2500	+2 to −8		
	2501–3000	+3 to −12		
C1	300–999	+2 to −6	800–999	1750
	1000–2400	+1 to −3	1000–2400	1000
	2401–2700	+3 to −6	2401–2600	1750
	2701–3000	+3 to −12		
C2	300–499	+2 to −6	500–600	3000
	500–2800	+1 to −3	601–999	1500
	2801–3000	+2 to −6	1000–2600	500
			2601–2800	3000
C3 (access line)	300–499	+0.8 to −3	500–599	650
	500–2800	+0.5 to −1.5	600–999	300
	2801–3000	+0.8 to −3	1000–2600	110
			2601–2800	650
C3 (trunk)	300–499	+0.8 to −2	500–599	500
	500–2800	+0.5 to −1	600–999	260
	2801–3000	+0.8 to −2	1000–2600	80
			2601–3000	500
C4	300–499	+2 to −6	500–599	3000
	500–3000	+2 to −3	600–799	1500
	3001–3200	+2 to −6	800–999	500
			1000–2600	300
			2601–2800	500
			2801–3000	1500
C5	300–499	+1 to −3	500–599	600
	500–2800	+0.5 to −1.5	600–999	300
	2801–3000	+1 to −3	1000–2600	100
			2601–2800	600

greater than 3 dB above or more than 12 dB below the gain at 1004 Hz. Figure 17-5 shows a graphical presentation of basic line conditioning requirements.

Figure 17-6 shows a graphical presentation of the attenuation distortion requirements specified in Table 17-1 for C2 conditioning, and Figure 17-7 shows the graph for C2 conditioning superimposed over the graph for basic conditioning. From Figure 17-7, it can be seen that the requirements for C2 conditioning are much more stringent than those for a basic circuit.

Example 17-3

A 1004 Hz test tone is transmitted over a telephone circuit at 0 dBm and received at −16 dBm. Determine

a. The 1004-Hz circuit gain.
b. The attenuation distortion requirements for a basic circuit.
c. The attenuation distortion requirements for a C2 conditioned circuit.

Solution **a.** The circuit gain is determined mathematically as

$$0 \text{ dBm} - (-16 \text{ dB}) = -16 \text{ dB (which equates to a loss of 16 dB)}$$

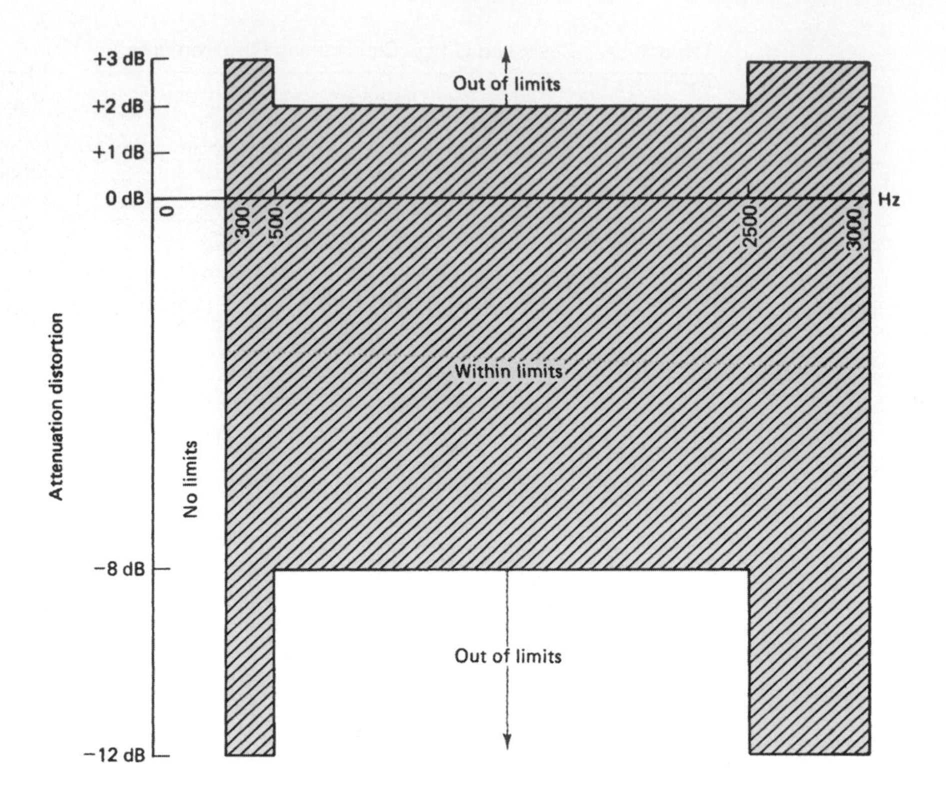

FIGURE 17-5 Graphical presentation of the limits for attenuation distortion for a basic 3002 telephone circuit

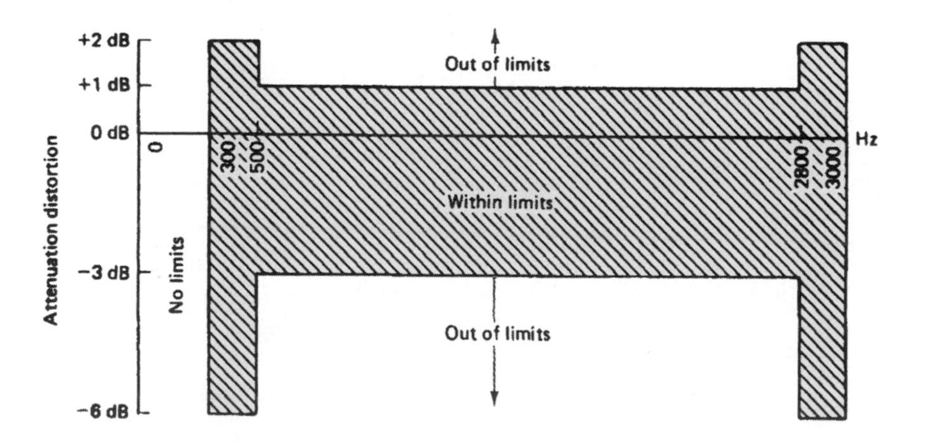

FIGURE 17-6 Graphical presentation of the limits for attenuation distortion for a C2 conditioned telephone circuit

b. Circuit gain requirements for a basic circuit can be determined from Table 17-1:

Frequency Band	*Requirements*	*Minimum Level*	*Maximum Level*
500 Hz and 2500 Hz	+2 dB and −8 dB	−24 dBm	−14 dBm
300 Hz and 499 Hz	+3 dB and −12 dB	−28 dBm	−13 dBm
2501 Hz and 3000 Hz	+3 dB and −12 dB	−28 dBm	−13 dBm

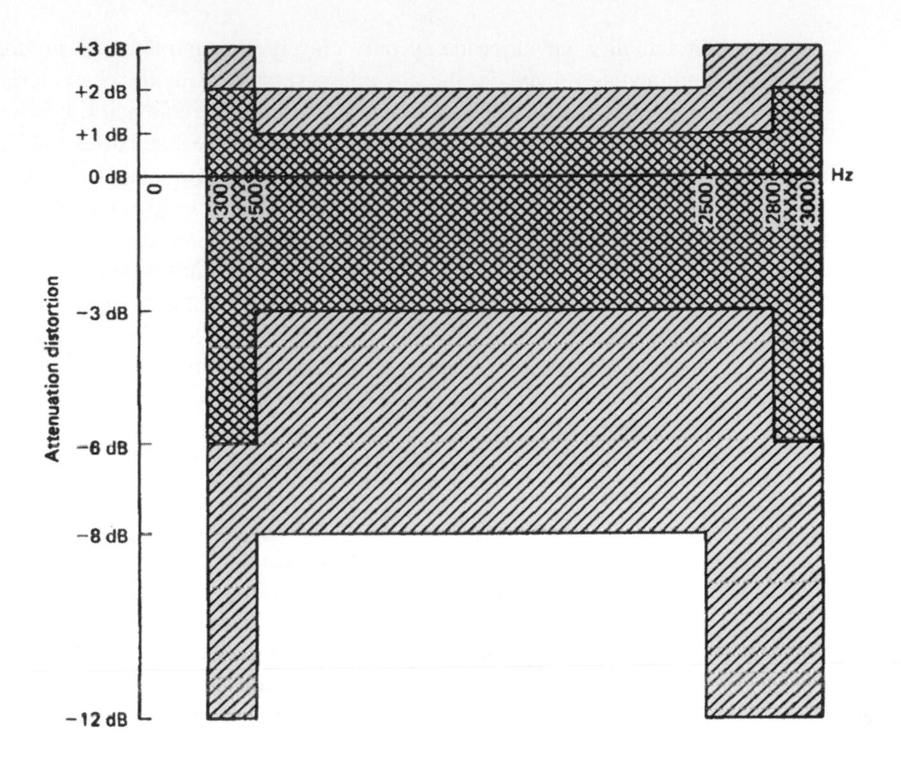

FIGURE 17-7 Overlay of Figure 9-5 over Figure 9-6 to demonstrate the more stringent requirements imposed by C2 conditioning compared to a basic (unconditioned) circuit.

c. Circuit gain requirements for a C2 conditioned circuit can be determined from Table 17-1:

Frequency Band	*Requirements*	*Minimum Level*	*Maximum Level*
500 Hz and 2500 Hz	+1 dB and −3 dB	−19 dBm	−15 dBm
300 Hz and 499 Hz	+2 dB and −6 dB	−22 dBm	−14 dBm
2801 Hz and 3000 Hz	+2 dB and −6 dB	−22 dBm	−14 dBm

A linear phase-versus-frequency relationship is a requirement for error-free data transmission—signals are delayed more at some frequencies than others. Delay distortion is the difference in phase shifts with respect to frequency that signals experience as they propagate through a transmission medium. This relationship is difficult to measure because of the difficulty in establishing a phase (time) reference. Envelope delay is an alternate method of evaluating the phase-versus-frequency relationship of a circuit.

The time delay encountered by a signal as it propagates from a source to a destination is called *propagation time,* and the delay measured in angular units, such as degrees or radians, is called *phase delay.* All frequencies in the usable voice band (300 Hz to 3000 Hz) do not experience the same time delay in a circuit. Therefore, a complex waveform, such as the output of a data modem, does not possess the same phase-versus-frequency relationship when received as it possessed when it was transmitted. This condition represents a possible impairment to a data signal. The *absolute phase delay* is the actual time required for a particular frequency to propagate from a source to a destination through a communications channel. The difference between the absolute delays of all the frequencies is phase distortion. A graph of phase delay-versus-frequency for a typical circuit is nonlinear.

By definition, envelope delay is the first derivative (slope) of phase with respect to frequency:

$$\text{envelope delay} = \frac{d\theta(\omega)}{d\omega} \tag{17-4}$$

In actuality, envelope delay only closely approximates $d\theta(\omega)/d\omega$. Envelope delay measurements evaluate not the true phase-versus-frequency characteristics but rather the phase of a wave that is the result of a narrow band of frequencies. It is a common misconception to confuse true phase distortion (also called delay distortion) with envelope delay distortion (EDD). *Envelope delay* is the time required to propagate a change in an AM envelope (the actual information-bearing part of the signal) through a transmission medium. To measure envelope delay, a narrowband amplitude-modulated carrier, whose frequency is varied over the usable voice band, is transmitted (the amplitude-modulated rate is typically between 25 Hz and 100 Hz). At the receiver, phase variations of the low-frequency envelope are measured. The phase difference at the different carrier frequencies is *envelope delay distortion.* The carrier frequency that produces the minimum envelope delay is established as the reference and is normalized to zero. Therefore, EDD measurements are typically given in microseconds and yield only positive values. EDD indicates the relative envelope delays of the various carrier frequencies with respect to the reference frequency. The reference frequency of a typical voice-band circuit is typically around 1800 Hz.

EDD measurements do not yield true phase delays, nor do they determine the relative relationships between true phase delays. EDD measurements are used to determine a close approximation of the relative phase delay characteristics of a circuit. Propagation time cannot be increased. Therefore, to correct delay distortion, equalizers are placed in a circuit to slow down the frequencies that travel the fastest more than frequencies that travel the slowest. This reduces the difference between the fastest and slowest frequencies, reducing the phase distortion.

The EDD limits for basic and conditioned telephone channels are listed in Table 17-1. Figure 17-8 shows a graphical representation of the EDD limits for a basic telephone chan-

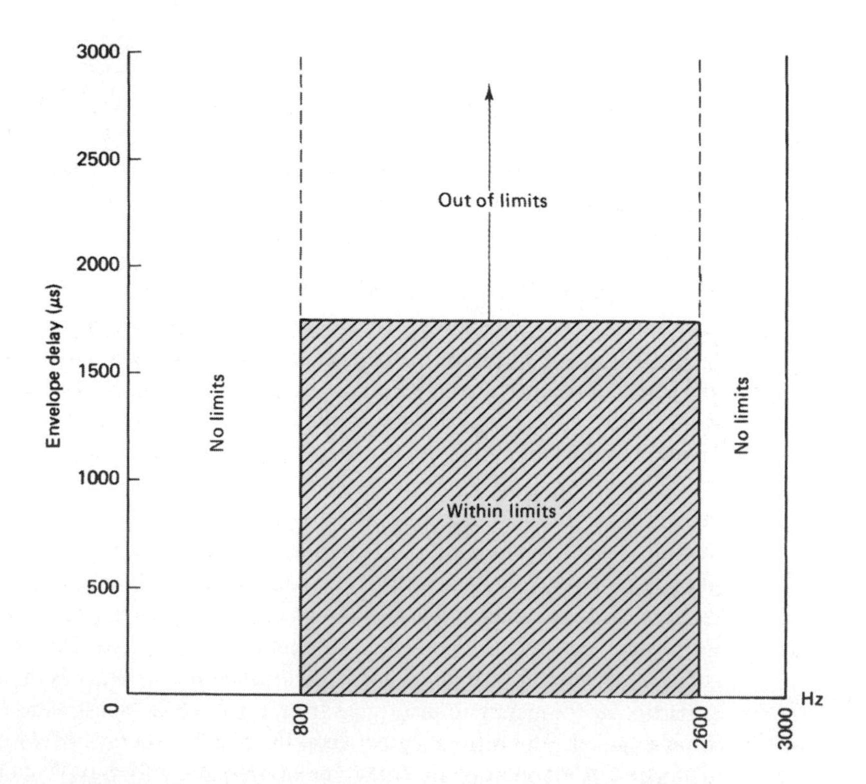

FIGURE 17-8 Graphical presentation of the limits for envelope delay in a basic telephone channel

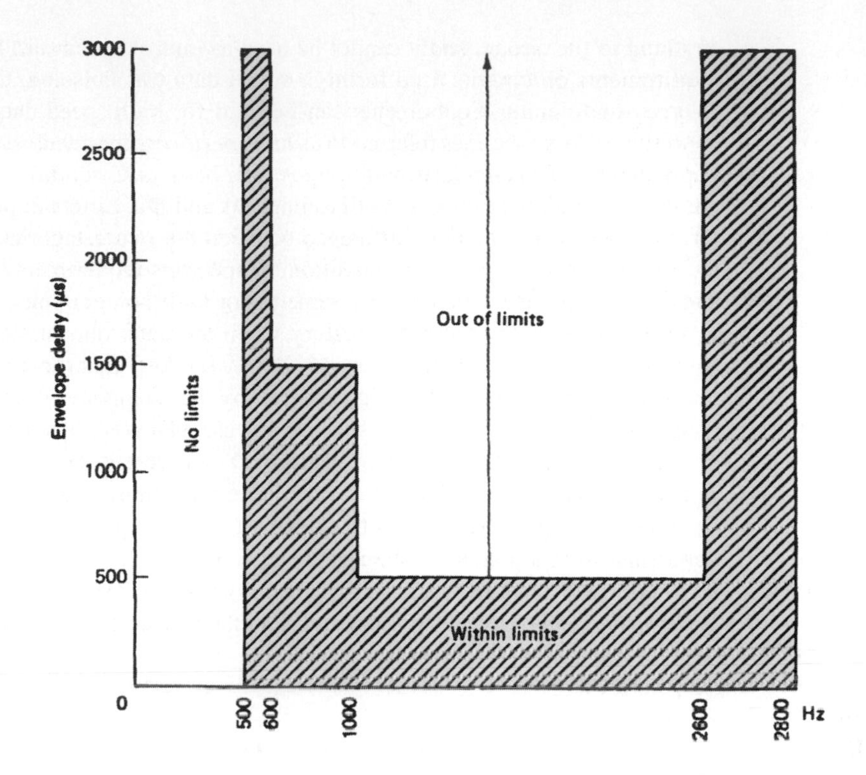

FIGURE 17-9 Graphical presentation of the limits for envelope delay in a telephone channel with C2 conditioning

nel, and Figure 17-9 shows a graphical representation of the EDD limits for a channel meeting the requirements for C2 conditioning. From Table 17-1, the EDD limit of a basic telephone channel is 1750 μs between 800 Hz and 2600 Hz. This indicates that the maximum difference in envelope delay between any two carrier frequencies (the fastest and slowest frequencies) within this range cannot exceed 1750 μs.

Example 17-4

An EDD test on a basic telephone channel indicated that an 1800-Hz carrier experienced the minimum absolute delay of 400 μs. Therefore, it is the reference frequency. Determine the maximum absolute envelope delay that any frequency within the 800-Hz to 2600-Hz range can experience.

Solution The maximum envelope delay for a basic telephone channel is 1750 μs within the frequency range of 800 Hz to 2600 Hz. Therefore, the maximum envelope delay is 2150 μs (400 μs + 1750 μs).

The absolute time delay encountered by a signal between any two points in the continental United States should never exceed 100 ms, which is not sufficient to cause any problems. Consequently, relative rather than absolute values of envelope delay are measured. For the previous example, as long as EDD tests yield relative values less than +1750 μs, the circuit is within limits.

17-5-1-2 D-type line conditioning. *D-type conditioning* neither reduces the noise on a circuit nor improves the signal-to-noise ratio. It simply sets the minimum requirements for *signal-to-noise* (S/N) *ratio* and *nonlinear distortion.* If a subscriber requests D-type conditioning and the facilities assigned to the circuit do not meet the requirements, a different facility is assigned. D-type conditioning is simply a requirement and does not add anything to the circuit, and it cannot be used to improve a circuit. It simply places higher requirements on circuits used for high-speed data transmission. Only circuits that meet D-type conditioning requirements can be used for high-speed data transmission. D-type conditioning is sometimes referred to as *high-performance conditioning* and can be applied to private-line data circuits in addition to either basic or C-conditioned requirements. There are two categories for D-type conditioning: D1 and D2. Limits imposed by D1 and D2 are virtually identical. The only difference between the two categories is the circuit arrangement to which they apply. D1 conditioning specifies requirements for two-point circuits, and D2 conditioning specifies requirements for multipoint circuits.

D-type conditioning is mandatory when the data transmission rate is 9600 bps because without D-type conditioning, it is highly unlikely that the circuit can meet the minimum performance requirements guaranteed by the telephone company. When a telephone company assigns a circuit to a subscriber for use as a 9600-bps data circuit and the circuit does not meet the minimum requirements of D-type conditioning, a new circuit is assigned. This is because a circuit cannot generally be upgraded to meet D-type conditioning specifications by simply adding corrective devices, such as equalizers and amplifiers. Telephone companies do not guarantee the performance of data modems operating at bit rates above 9600 bps over standard voice-grade circuits.

D-type conditioned circuits must meet the following specifications:

Signal-to-C-notched noise ratio: ≥28 dB

Nonlinear distortion

Signal-to-second order distortion: ≥35 dB

Signal-to-third order distortion: ≥40 dB

The signal-to-notched noise ratio requirement for standard circuits is only 24 dB, and they have no requirements for nonlinear distortion.

Nonlinear distortion is an example of correlated noise and is produced from nonlinear amplification. When an amplifier is driven into a nonlinear operating region, the signal is distorted, producing multiples and sums and differences (cross products) the original signal frequencies. The noise caused by nonlinear distortion is in the form of additional frequencies produced from nonlinear amplification of a signal. In other words, no signal, no noise. Nonlinear distortion produces distorted waveforms that are detrimental to digitally modulated carriers used with voice-band data modems, such as FSK, PSK, and QAM. Two classifications of nonlinear distortion are *harmonic distortion* (unwanted multiples of the transmitted frequencies) and *intermodulation distortion* (cross products [sums and differences] of the transmitted frequencies, sometimes called *fluctuation noise* or *cross-modulation noise*). Harmonic and intermodulation distortion, if of sufficient magnitude, can destroy the integrity of a data signal. The degree of circuit nonlinearity can be measured using either harmonic or intermodulation distortion tests.

Harmonic distortion is measured by applying a single-frequency test tone to a telephone channel. At the receive end, the power of the fundamental, second, and third harmonic frequencies is measured. Harmonic distortion is classified as second, third, *n*th order, or as total harmonic distortion. The actual amount of nonlinearity in a circuit is determined by comparing the power of the fundamental with the combined powers of the second and third harmonics. Harmonic distortion tests use a single-frequency (704-Hz) source (see Figure 17-10); therefore, no cross-product frequencies are produced.

Although simple harmonic distortion tests provide an accurate measurement of the nonlinear characteristics of analog telephone channel, they are inadequate for digital (T carrier) facilities. For this reason, a more refined method was developed that uses a multifrequency test-tone signal. Four test frequencies are used (see Figure 17-11): two designated

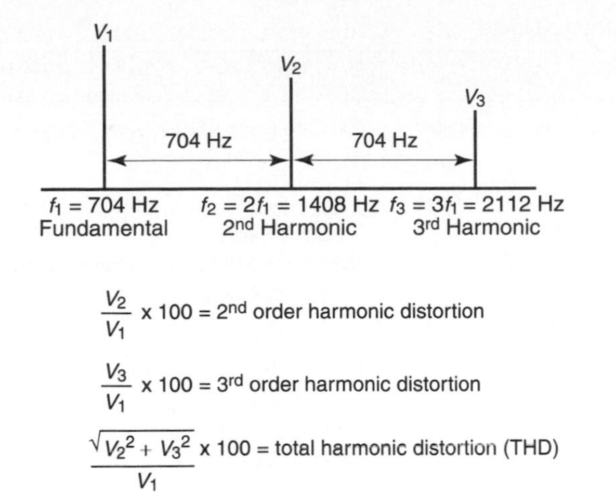

FIGURE 17-10 Harmonic distortion

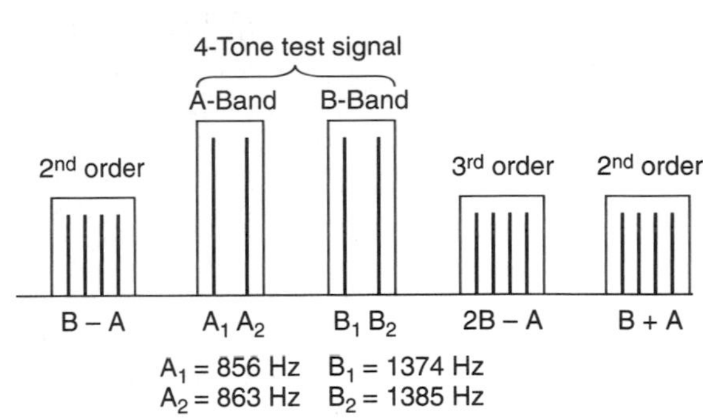

FIGURE 17-11 Intermodulation distortion

the A band (A1 = 856 Hz, A2 = 863 Hz) and two designated the B band (B1 = 1374 Hz and B2 = 1385 Hz). The four frequencies are transmitted with equal power levels, and the total combined power is equal to that of a normal data signal. The nonlinear amplification of the circuit produces multiples of each frequency (harmonics) and their cross-product frequencies (sum and difference frequencies). For reasons beyond the scope of this text, the following second- and third-order products were selected for measurement: B + A, B − A, and 2B − A. The combined signal power of the four A and B band frequencies is compared with the second-order cross products and then compared with the third order cross products. The results are converted to dB values and then compared to the requirements of D-type conditioning.

Harmonic and intermodulation distortion tests do not directly determine the amount of interference caused by nonlinear circuit gain. They serve as a figure of merit only when evaluating circuit parameters.

17-5-2 Interface Parameters

The two primary considerations of the interface parameters are electrical protection of the telephone network and its personnel and standardization of design arrangements. The interface parameters include the following:

Station equipment impedances should be 600 Ω resistive over the usable voice band.

Station equipment should be isolated from ground by a minimum of 20 MΩ dc and 50 kΩ ac.

The basic voice-grade telephone circuit is a 3002 channel; it has an ideal bandwidth of 0 Hz to 4 kHz and a usable bandwidth of 300 Hz to 3000 Hz.

The circuit gain at 3000 Hz is 3 dB below the specified in-band signal power.

The gain at 4 kHz must be at least 15 dB below the gain at 3 kHz.

The maximum transmitted signal power for a private-line circuit is 0 dBm.

The transmitted signal power for dial-up circuits using the public switched telephone network is established for each loop so that the signal is received at the telephone central office at −12 dBm.

Table 17-2 summarizes interface parameter limits.

Table 17-2 Interface Parameter Limits

Parameter	Limit
1. Recommended impedance of terminal equipment	600 Ω resistive ± 10%
2. Recommended isolation to ground of terminal equipment	At least 20 MΩ dc At least 50 kΩ ac At least 1500 V rms breakdown voltage at 60 Hz
3. Data transmit signal power	0 dBm (3-s average)
4. In-band transmitted signal power	2450-Hz to 2750-Hz band should not exceed signal power in 800-Hz to 2450-Hz band
5. Out-of-band transmitted signal power	
Above voice band:	
(a) 3995 Hz–4005 Hz	At least 18 dB below maximum allowed in-band signal power
(b) 4-kHz–10-kHz band	Less than –16 dBm
(c) 10-kHz–25-kHz band	Less than –24 dBm
(d) 25-kHz–40-kHz band	Less than –36 dBm
(e) Above 40 kHz	Less than –50 dBm
Below voice band:	
(f) rms current per conductor as specified by Telco but never greater than 0.35 A.	
(g) Magnitude of peak conductor-to-ground voltage not to exceed 70 V.	
(h) Conductor-to-conductor voltage shall be such that conductor-to-ground voltage is not exceeded. For an underground signal source, the conductor-to-conductor limit is the same as the conductor-to-ground limit.	
(i) Total weighted rms voltage in band from 50 Hz to 300 Hz, not to exceed 100 V. Weighting factors for each frequency component (f) are $f^2/10^4$ for f between 50 Hz and 100 Hz and $f^{3.3}/10^{6.6}$ for f between 101 Hz and 300 Hz.	
6. Maximum test signal power: same as transmitted data power.	

17-5-3 Facility Parameters

Facility parameters represent potential impairments to a data signal. These impairments are caused by telephone company equipment and the limits specified pertain to all private-line data circuits using voice-band facilities, regardless of line conditioning. Facility parameters include 1004-Hz variation, C-message noise, impulse noise, gain hits and dropouts, phase hits, phase jitter, single-frequency interference, frequency shift, phase intercept distortion, and peak-to-average ratio.

17-5-3-1 1004-Hz variation. The telephone industry has established 1004 Hz as the standard test-tone frequency; 1000 Hz was originally selected because of its relative location in the passband of a standard voice-band circuit. The frequency was changed to 1004 Hz with the advent of digital carriers because 1000 Hz is an exact submultiple of the 8-kHz sample rate used with T carriers. Sampling a continuous 1000-Hz signal at an 8000-Hz rate produced repetitive patterns in the PCM codes, which could cause the system to lose frame synchronization.

The purpose of the 1004-Hz test tone is to simulate the combined signal power of a standard voice-band data transmission. The 1004-Hz channel loss for a private-line data circuit is typically 16 dB. A 1004-Hz test tone applied at the transmit end of a circuit should be received at the output of the circuit at −16 dBm. Long-term variations in the gain of the transmission facility are called *1004-Hz variation* and should not exceed ±4 dB. Thus, the received signal power must be within the limits of −12 dBm to −20 dBm.

17-5-3-2 C-message noise. C-message noise measurements determine the average weighted rms noise power. Unwanted electrical signals are produced from the random movement of electrons in conductors. This type of noise is commonly called *thermal noise* because its magnitude is directly proportional to temperature. Because the electron movement is completely random and travels in all directions, thermal noise is also called *random noise,* and because it contains all frequencies, it is sometimes referred to as *white noise.* Thermal noise is inherently present in a circuit because of its electrical makeup. Because thermal noise is additive, its magnitude is dependent, in part, on the electrical length of the circuit.

C-message noise measurements are the terminated rms power readings at the receive end of a circuit with the transmit end terminated in the characteristic impedance of the telephone line. Figure 17-12 shows the test setup for conducting terminated C-message noise readings. As shown in the figure, a C-message filter is placed between the circuit and the power meter in the noise measuring set so that the noise measurement evaluates the noise with a response similar to that of a human listening to the noise through a standard telephone set speaker.

There is a disadvantage to measuring noise this way. The overall circuit characteristics, in the absence of a signal, are not necessarily the same as when a signal is present. Using compressors, expanders, and automatic gain devices in a circuit causes this difference. For this reason, *C-notched noise* measurements were developed. C-notched noise measurements differ from standard C-message noise measurements only in the fact that a *holding tone* (usually 1004 Hz or 2804 Hz) is applied to the transmit end of the circuit while the noise measurement is taken. The holding tone ensures that the circuit operation simulates a loaded voice or data transmission. *Loaded* is a communications term that indicates the presence of a signal power comparable to the power of an actual message transmission. A narrowband notch filter removes the holding tone before the noise power is measured. The test setup for making C-notched noise measurements is shown in Figure 17-13. As the figure shows, the notch filter is placed in front of the C-message filter, thus blocking the holding tone from reaching the power meter.

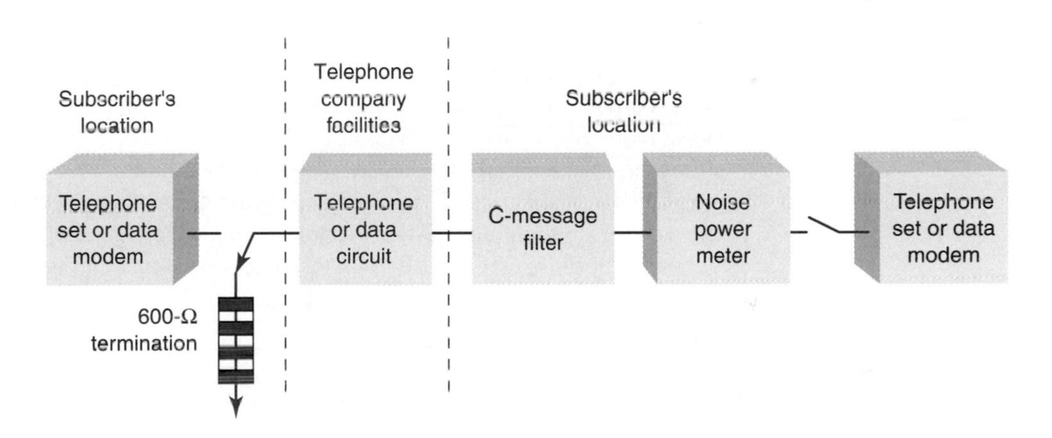

FIGURE 17-12 Terminated C-message noise test setup

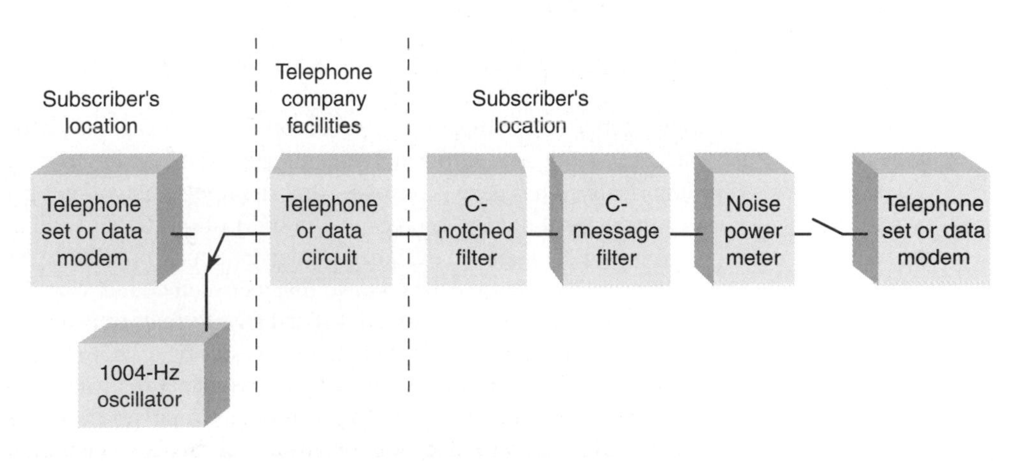

FIGURE 17-13 C-notched noise test setup

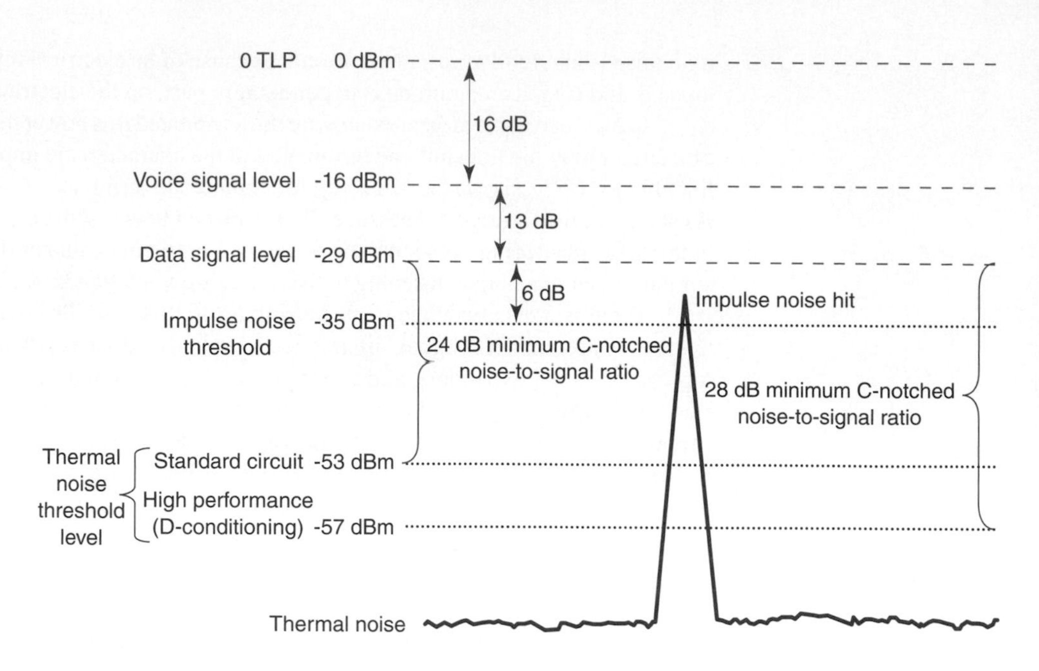

FIGURE 17-14 C-notched noise and impulse noise

The physical makeup of a private-line data circuit may require using several carrier facilities and cable arrangements in tandem. Each facility may be analog, digital, or some combination of analog and digital. Telephone companies have established realistic C-notched noise requirements for each type of facility for various circuit lengths. Telephone companies guarantee standard private-line data circuits a minimum signal-to-C-notched noise ratio of 24 dB. A standard circuit is one operating at less than 9600 bps. Data circuits operating at 9600 bps require D-type conditioning, which guarantees a minimum signal-to-C-notched noise ratio of 28 dB. C-notched noise is shown in Figure 17-14. Telephone companies do not guarantee the performance of voice-band circuits operating at bit rates in excess of 9600 bps.

17-5-3-3 Impulse noise. *Impulse noise* is characterized by high-amplitude peaks (impulses) of short duration having an approximately flat frequency spectrum. Impulse noise can saturate a message channel. Impulse noise is the primary source of transmission errors in data circuits. There are numerous sources of impulse noise—some are controllable, but most are not. The primary cause of impulse noise is man-made sources, such as interference from ac power lines, transients from switching machines, motors, solenoids, relays, electric trains, and so on. Impulse noise can also result from lightning and other adverse atmospheric conditions.

The significance of impulse noise hits on data transmission has been a controversial topic. Telephone companies have accepted the fact that the absolute magnitude of the impulse hit is not as significant as the magnitude of the hit relative to the signal amplitude. Empirically, it has been determined that an impulse hit will not produce transmission errors in a data signal unless it comes within 6 dB of the signal level as shown in Figure 17-14. Impulse hit counters are designed to register a maximum of seven counts per second. This leaves a 143-ms lapse called a *dead time* between counts when additional impulse hits are not registered. Contemporary high-speed data formats transfer data in a block or frame format, and whether one hit or many hits occur during a single transmission is unimportant, as any error within a message generally necessitates retransmission of the entire message. It has been determined that counting additional impulses during the time of a single transmission does not correlate well with data transmission performance.

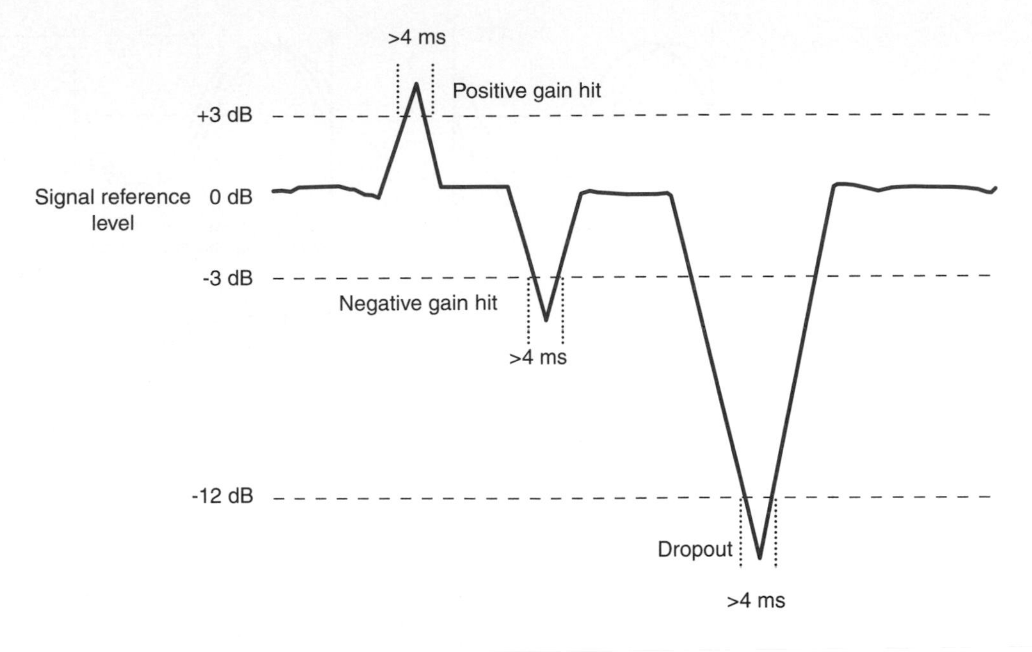

FIGURE 17-15 Gain hits and dropouts

Impulse noise objectives are based primarily on the error susceptibility of data signals, which depends on the type of modem used and the characteristics of the transmission medium. It is impractical to measure the exact peak amplitudes of each noise pulse or to count the number that occur. Studies have shown that expected error rates in the absence of other impairments are approximately proportional to the number of impulse hits that exceed the rms signal power level by approximately 2 dB. When impulse noise tests are performed, a 2802-Hz holding tone is placed on a circuit to ensure loaded circuit conditions. The counter records the number of hits in a prescribed time interval (usually 15 minutes). An impulse hit is typically less than 4 ms in duration and never more than 10 ms. Telephone company limits for recordable impulse hits is 15 hits within a 15-minute time interval. This does not limit the number of hits to one per minute but, rather, the average occurrence to one per minute.

17-5-3-4 Gain hits and dropouts. A *gain hit* is a sudden, random change in the gain of a circuit resulting in a temporary change in the signal level. Gain hits are classified as temporary variations in circuit gain exceeding ±3 dB, lasting more than 4 ms, and returning to the original value within 200 ms. The primary cause of gain hits is noise transients (impulses) on transmission facilities during the normal course of a day.

A *dropout* is a decrease in circuit gain (i.e., signal level) of more than 12 dB lasting longer than 4 ms. Dropouts are characteristics of temporary open-circuit conditions and are generally caused by deep fades on radio facilities or by switching delays. Gain hits and dropouts are depicted in Figure 17-15.

17-5-3-5 Phase hits. *Phase hits* (slips) are sudden, random changes in the phase of a signal. Phase hits are classified as temporary variations in the phase of a signal lasting longer than 4 ms. Generally, phase hits are not recorded unless they exceed ±20C° peak. Phase hits, like gain hits, are caused by transients produced when transmission facilities are switched. Phase hits are shown in Figure 17-16.

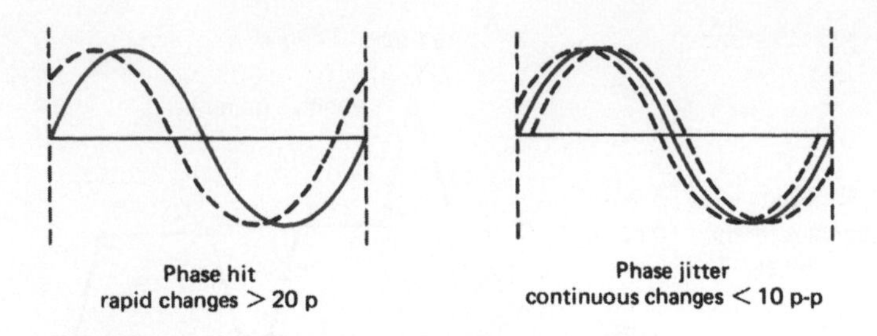

FIGURE 17-16 Phase hits and phase jitter

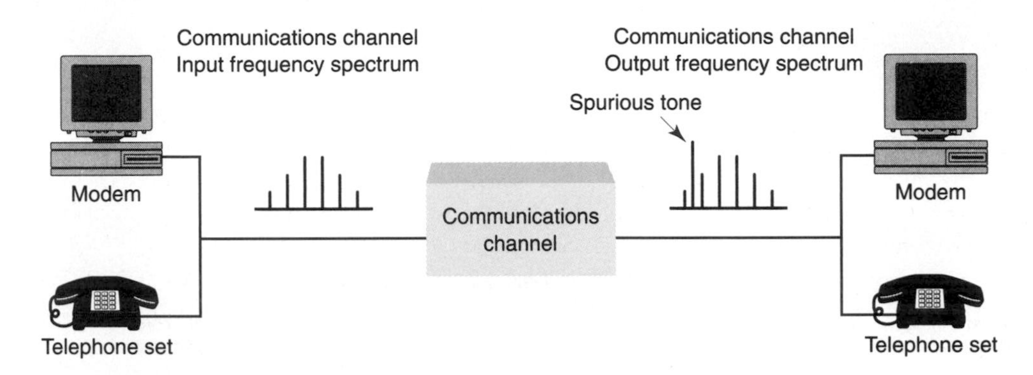

FIGURE 17-17 Single-frequency interference (spurious tone)

17-5-3-6 Phase jitter. *Phase jitter* is a form of incidental phase modulation—a continuous, uncontrolled variation in the zero crossings of a signal. Generally, phase jitter occurs at a 300-Hz rate or lower, and its primary cause is low-frequency ac ripple in power supplies. The number of power supplies required in a circuit is directly proportional to the number of transmission facilities and telephone offices that make up the message channel. Each facility has a separate phase jitter requirement; however, the maximum acceptable end-to-end phase jitter is 10° peak to peak regardless of how many transmission facilities or telephone offices are used in the circuit. Phase jitter is shown in Figure 17-16.

17-5-3-7 Single-frequency interference. *Single-frequency interference* is the presence of one or more continuous, unwanted tones within a message channel. The tones are called *spurious tones* and are often caused by crosstalk or cross modulation between adjacent channels in a transmission system due to system nonlinearities. Spurious tones are measured by terminating the transmit end of a circuit and then observing the channel frequency band. Spurious tones can cause the same undesired circuit behavior as thermal noise. Single-frequency interference is shown in Figure 17-17.

17-5-3-8 Frequency shift. *Frequency shift* is when the frequency of a signal changes during transmission. For example, a tone transmitted at 1004 Hz is received at 1005 Hz. Analog transmission systems used by telephone companies operate single-sideband suppressed carrier (SSBSC) and, therefore, require coherent demodulation. With coherent demodulation, carriers must be synchronous—the frequency must be reproduced exactly in the receiver. If this is not accomplished, the demodulated signal will be offset in frequency by the difference between transmit and receive carrier frequencies. The longer a circuit, the more analog transmission systems and the more likely frequency shift will occur. Frequency shift is shown in Figure 17-18.

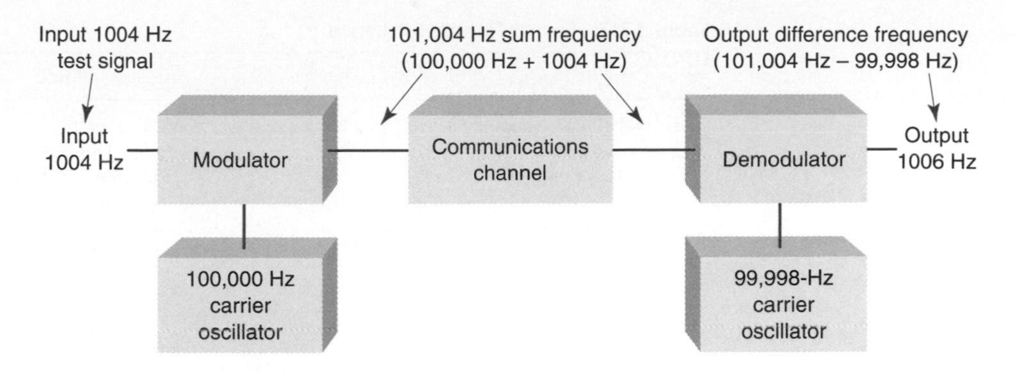

FIGURE 17-18 Frequency shift

17-5-3-9 Phase intercept distortion. *Phase intercept distortion* occurs in coherent SSBSC systems, such as those using frequency-division multiplexing when the received carrier is not reinserted with the exact phase relationship to the received signal as the transmit carrier possessed. This impairment causes a constant phase shift to all frequencies, which is of little concern for data modems using FSK, PSK, or QAM. Because these are practically the only techniques used today with voice-band data modems, no limits have been set for phase intercept distortion.

17-5-3-10 Peak-to-average ratio. The difficulties encountered in measuring true phase distortion or envelope delay distortion led to the development of peak-to-average ratio (PAR) tests. A signal containing a series of distinctly shaped pulses with a high peak voltage-to-average voltage ratio is transmitted. Differential delay distortion in a circuit has a tendency to spread the pulses, thus reducing the peak voltage-to-average voltage ratio. Low peak-to-average ratios indicate the presence of differential delay distortion. PAR measurements are less sensitive to attenuation distortion than EDD tests and are easier to perform.

17-5-3-11 Facility parameter summary. Table 17-3 summarizes facility parameter limits.

17-6 VOICE-FREQUENCY CIRCUIT ARRANGEMENTS

Electronic communications circuits can be configured in several ways. Telephone instruments and the voice-frequency facilities to which they are connected may be either *two wire* or *four wire.* Two-wire circuits have an obvious economic advantage, as they use only half as much copper wire. This is why most local subscriber loops connected to the public switched telephone network are two wire. However, most private-line data circuits are configured four wire.

17-6-1 Two-Wire Voice-Frequency Circuits

As the name implies, *two-wire transmission* involves two wires (one for the signal and one for a reference or ground) or a circuit configuration that is equivalent to using only two wires. Two-wire circuits are ideally suited to simplex transmission, although they are often used for half- and full-duplex transmission.

Figure 17-19 shows the block diagrams for four possible two-wire circuit configurations. Figure 17-19a shows the simplest two-wire configuration, which is a passive circuit consisting of two copper wires connecting a telephone or voice-band modem at one station through a telephone company interface to a telephone or voice-band modem at the

Table 17-3 Facility Parameter Limits

Parameter	Limit	
1. 1004-Hz loss variation	Not more than ±4 dB long term	
2. C-message noise	Maximum rms noise at modem receiver (nominal − 16 dBm point)	
Facility miles	*dBm*	*dBrncO*
0–50	−61	32
51–100	−59	34
101–400	−58	35
401–1000	−55	38
1001–1500	−54	39
1501–2500	−52	41
2501–4000	−50	43
4001–8000	−47	46
8001–16,000	−44	49
3. C-notched noise	(minimum values)	
(a) Standard voice-band channel	24-dB signal to C-notched noise	
(b) High-performance line	28-dB signal to C-notched noise	
4. Single-frequency interference	At least 3 dB below C-message noise limits	
5. Impulse noise		
Threshold with respect to 1004-Hz holding tone	*Maximum counts above threshold allowed in 15 minutes*	
0 dB	15	
+4 dB	9	
+8 dB	5	
6. Frequency shift	±5 Hz end to end	
7. Phase intercept distortion	No limits	
8. Phase jitter	No more than 10° peak to peak (end-to-end requirement)	
9. Nonlinear distortion (D-conditioned circuits only)		
Signal to second order	At least 35 dB	
Signal to third order	At least 40 dB	
10. Peak-to-average ratio	Reading of 50 minimum end to end with standard PAR meter	
11. Phase hits	8 or less in any 15-minute period greater than ±20 peak	
12. Gain hits	8 or less in any 15-minute period greater than ±3 dB	
13. Dropouts	2 or less in any 15-minute period greater than 12 dB	

destination station. The modem, telephone, and circuit configuration are capable of two-way transmission in either the half- or the full-duplex mode.

Figure 17-19b shows an active two-wire transmission system (i.e., one that provides gain). The only difference between this circuit and the one shown in Figure 17-19a is the addition of an amplifier to compensate for transmission line losses. The amplifier is unidirectional and, thus, limits transmission to one direction only (simplex).

Figure 17-19c shows a two-wire circuit using a digital T carrier for the transmission medium. This circuit requires a T carrier transmitter at one end and a T carrier receiver at the other end. The digital T carrier transmission line is capable of two-way transmission; however, the transmitter and receiver in the T carrier are not. The transmitter encodes the analog voice or modem signals into a PCM code, and the decoder in the receiver performs the opposite operation, converting PCM codes back to analog. The digital transmission medium is a pair of copper wire.

Figures 17-19a, b, and c are examples of *physical two-wire circuits,* as the two stations are physically interconnected with a two-wire metallic transmission line. Figure 17-19d shows an *equivalent two-wire circuit.* The transmission medium is Earth's atmosphere, and

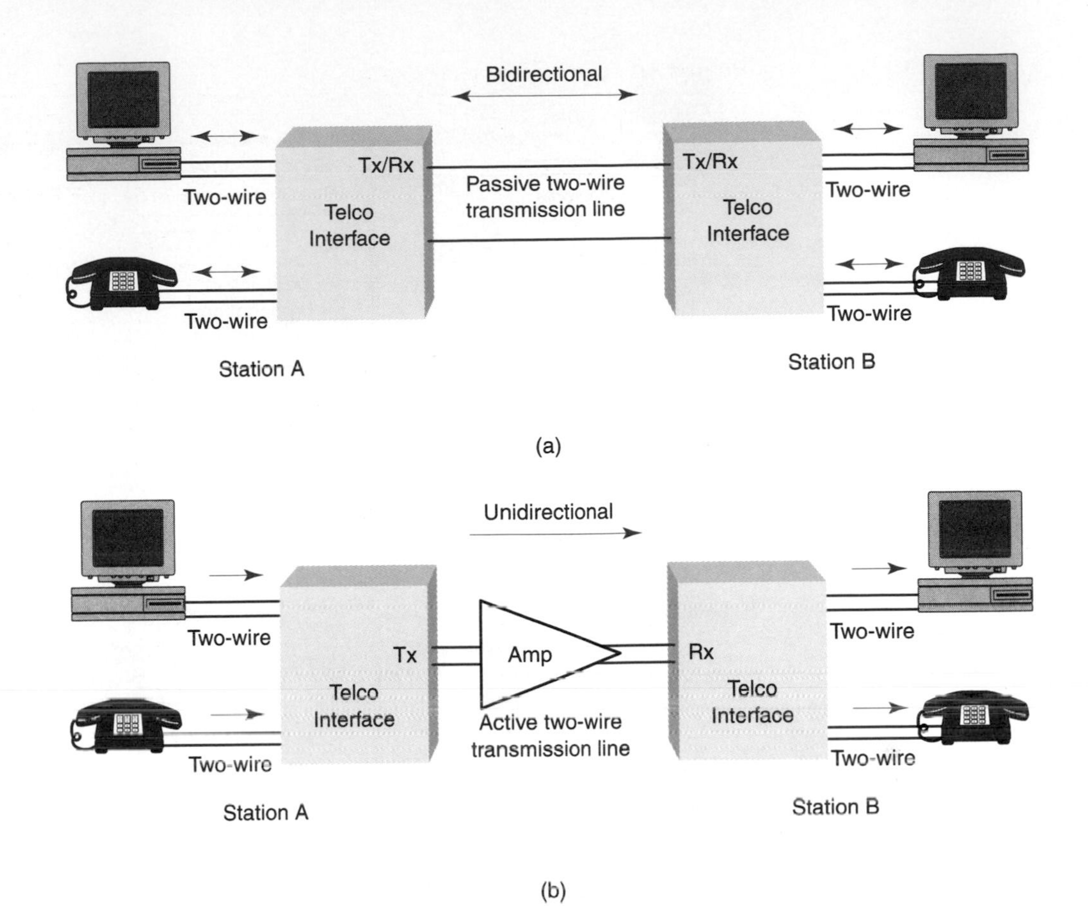

FIGURE 17-19 Two-wire configurations: (a) passive cable circuit; (b) active cable circuit (*Continued*)

there are no copper wires between the two stations. Although Earth's atmosphere is capable of two-way simultaneous transmission, the radio transmitter and receiver are not. Therefore, this is considered an equivalent two-wire circuit.

17-6-2 Four-Wire Voice-Frequency Circuits

As the name implies, *four-wire transmission* involves four wires (two for each direction—a signal and a reference) or a circuit configuration that is equivalent to using four wires. Four-wire circuits are ideally suited to full-duplex transmission, although they can (and very often do) operate in the half-duplex mode. As with two-wire transmission, there are two forms of four-wire transmission systems: *physical four wire* and *equivalent four wire.*

Figure 17-20 shows the block diagrams for four possible four-wire circuit configurations. As the figures show, a four-wire circuit is equivalent to two two-wire circuits, one for each direction of transmission. The circuits shown in Figures 17-20a, b, and c are physical four-wire circuits, as the transmitter at one station is hardwired to the receiver at the other station. Therefore, each two-wire pair is unidirectional (simplex), but the combined four-wire circuit is bidirectional (full duplex).

The circuit shown in Figure 17-20d is an equivalent four-wire circuit that uses Earth's atmosphere for the transmission medium. Station A transmits on one frequency (f_1) and receives on a different frequency (f_2), while station B transmits on frequency f_2 and receives on frequency f_1. Therefore, the two radio signals do not interfere with one another, and simultaneous bidirectional transmission is possible.

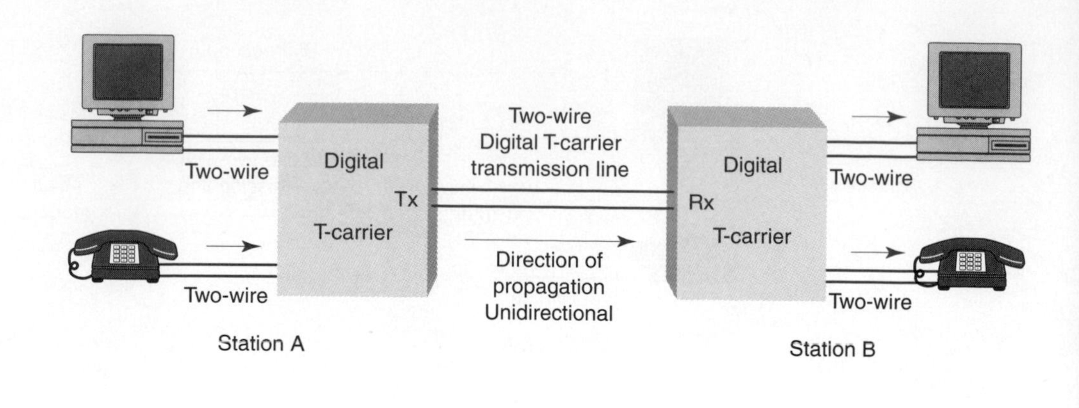

(c)

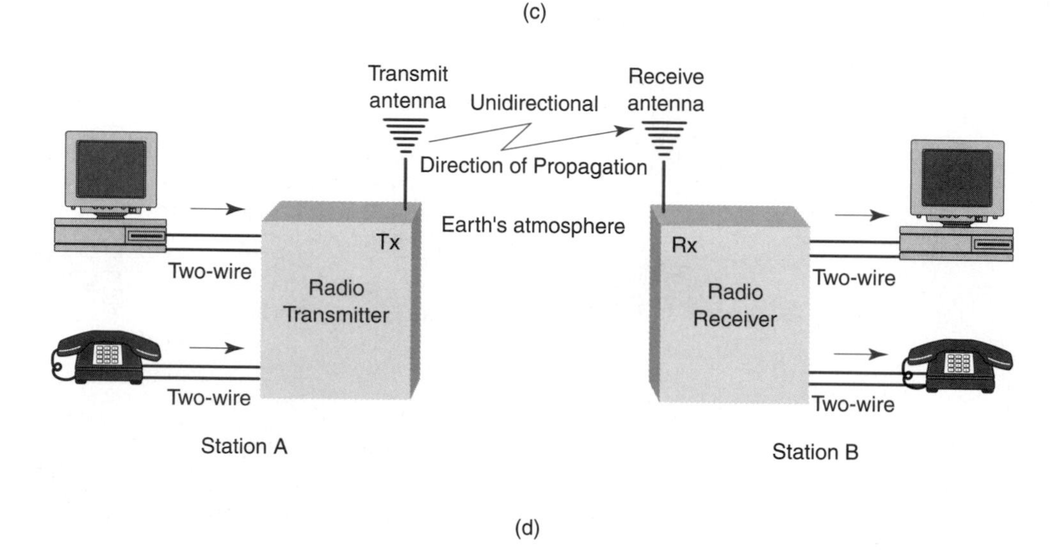

(d)

FIGURE 17-19 (Continued) (c) digital T-carrier system; (d) wireless radio carrier system

17-6-3 Two Wire versus Four Wire

There are several inherent advantages of four-wire circuits over two-wire circuits. For instance, four-wire circuits are considerably less noisy, have less crosstalk, and provide more isolation between the two directions of transmission when operating in either the half- or the full-duplex mode. However, two-wire circuits require less wire, less circuitry and, thus, less money than their four-wire counterparts.

Providing amplification is another disadvantage of four-wire operation. Telephone or modem signals propagated more than a few miles require amplification. A bidirectional amplifier on a two-wire circuit is not practical. It is much easier to separate the two directions of propagation with a four-wire circuit and install separate amplifiers in each direction.

17-6-4 Hybrids, Echo Suppressors, and Echo Cancelers

When a two-wire circuit is connected to a four-wire circuit, as in a long-distance telephone call, an interface circuit called a *hybrid,* or *terminating, set* is used to affect the interface. The hybrid set is used to match impedances and to provide isolation between the two directions of signal flow. The hybrid circuit used to convert two-wire circuits to four-wire circuits is similar to the hybrid coil found in standard telephone sets.

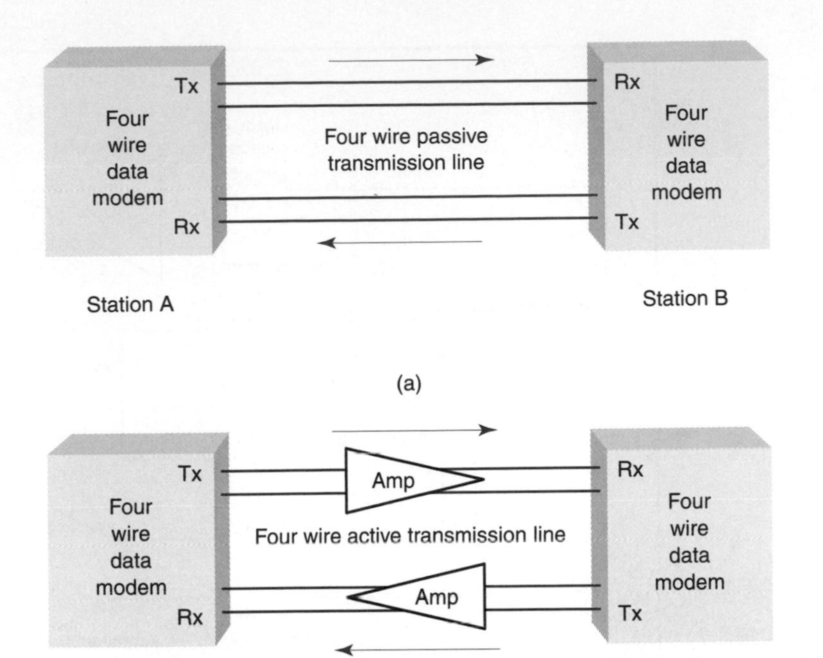

(b)

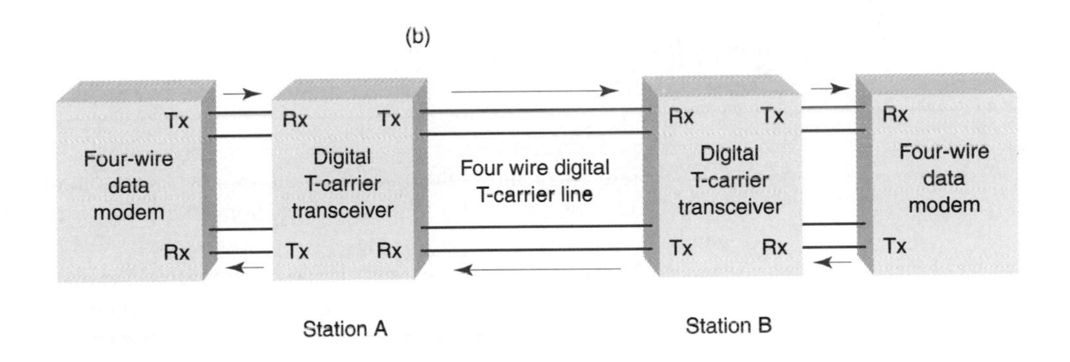

(c)

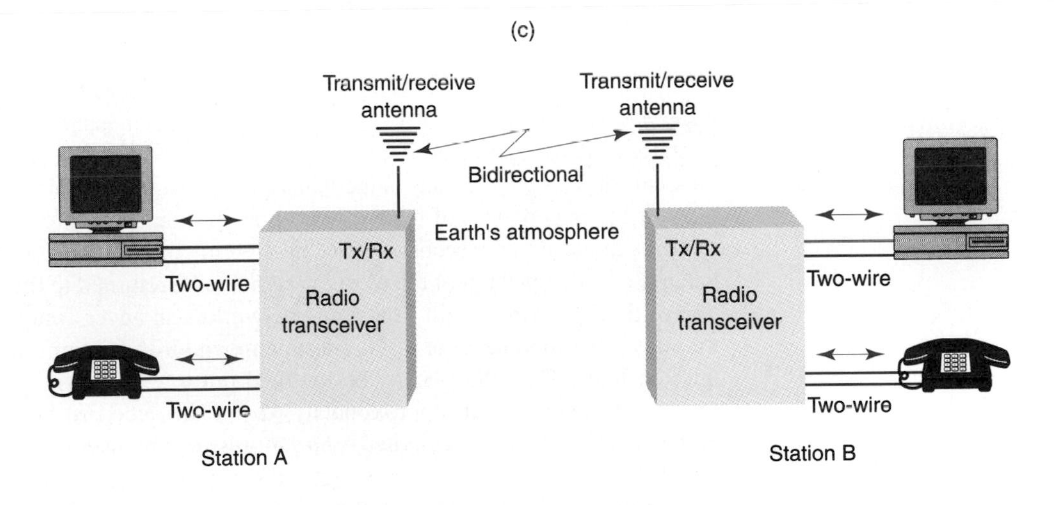

(d)

FIGURE 17-20 Four-wire configurations: (a) passive cable circuit; (b) active cable circuit; (c) digital T-carrier system; (d) wireless radio carrier system

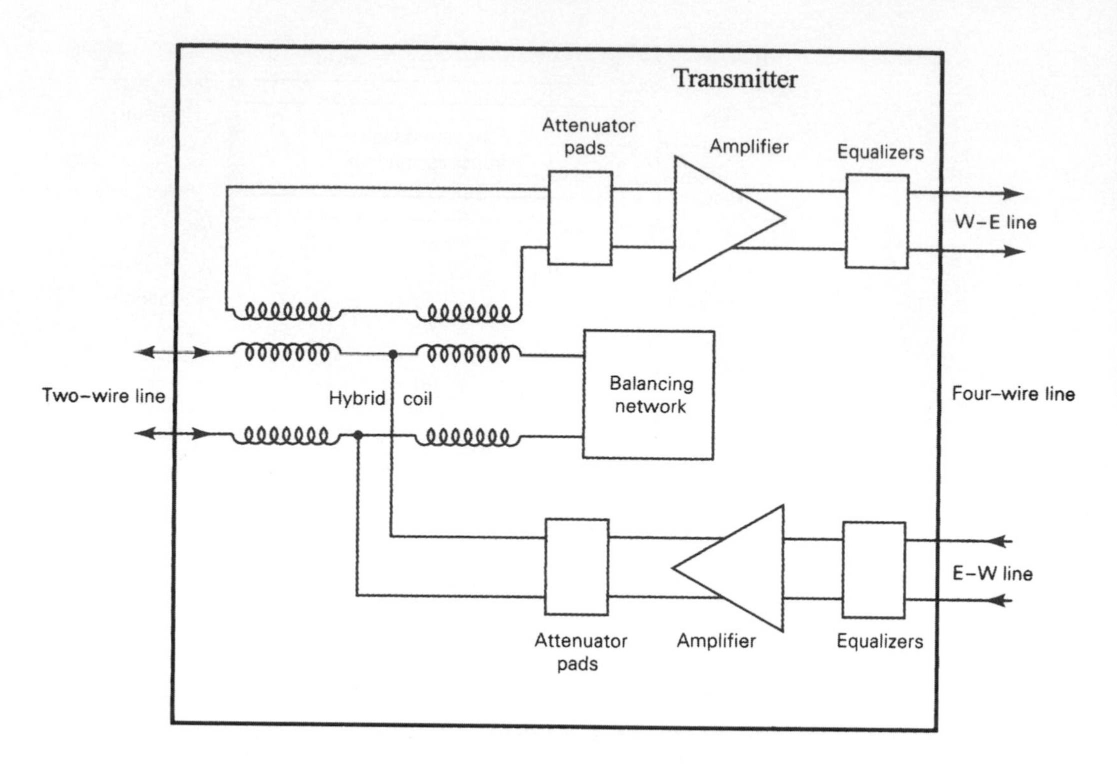

FIGURE 17-21 Hybrid (terminating) sets

Figure 17-21 shows the block diagram for a two-wire to four-wire hybrid network. The hybrid coil compensates for impedance variations in the two-wire portion of the circuit. The amplifiers and attenuators adjust the signal power to required levels, and the equalizers compensate for impairments in the transmission line that affect the frequency response of the transmitted signal, such as line inductance, capacitance, and resistance. Signals traveling west to east (W-E) enter the terminating set from the two-wire line, where they are inductively coupled into the west-to-east transmitter section of the four-wire circuit. Signals received from the four-wire side of the hybrid propagate through the receiver in the east-to-west (E-W) section of the four-wire circuit, where they are applied to the center taps of the hybrid coils. If the impedances of the two-wire line and the balancing network are properly matched, all currents produced in the upper half of the hybrid by the E-W signals will be equal in magnitude but opposite in polarity. Therefore, the voltages induced in the secondaries will be 180° out of phase with each other and, thus, cancel. This prevents any of the signals from being retransmitted to the sender as an echo.

If the impedances of the two-wire line and the balancing network are not matched, voltages induced in the secondaries of the hybrid coil will not completely cancel. This imbalance causes a portion of the received signal to be returned to the sender on the W-E portion of the four-wire circuit. Balancing networks can never completely match a hybrid to the subscriber loop because of long-term temperature variations and degradation of transmission lines. The talker hears the returned portion of the signal as an echo, and if the round-trip delay exceeds approximately 45 ms, the echo can become quite annoying. To eliminate this echo, devices called *echo suppressors* are inserted at one end of the four-wire circuit.

Figure 17-22 shows a simplified block diagram of an echo suppressor. The speech detector senses the presence and direction of the signal. It then enables the amplifier in the appropriate direction and disables the amplifier in the opposite direction, thus preventing the echo

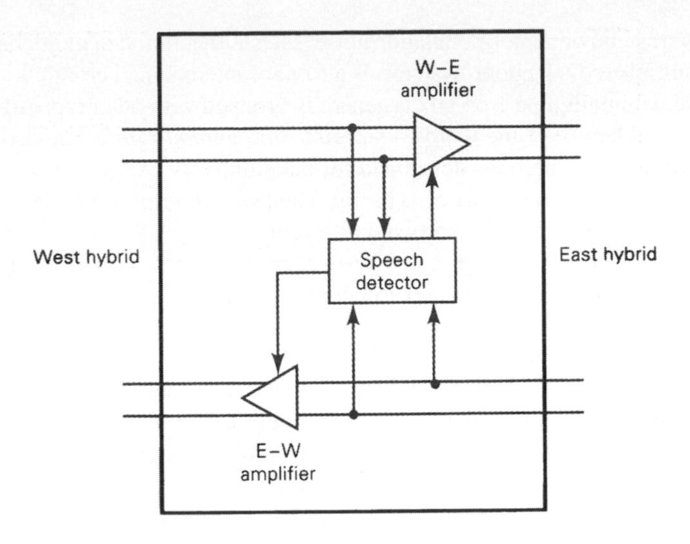

FIGURE 17-22 Echo suppressor

from returning to the speaker. A typical echo suppressor suppresses the returned echo by as much as 60 dB. If the conversation is changing direction rapidly, the people listening may be able to hear the echo suppressors turning on and off (every time an echo suppressor detects speech and is activated, the first instant of sound is removed from the message, giving the speech a choppy sound). If both parties talk at the same time, neither person is heard by the other.

With an echo suppressor in the circuit, transmissions cannot occur in both directions at the same time, thus limiting the circuit to half-duplex operation. Long-distance carriers, such as AT&T, generally place echo suppressors in four-wire circuits that exceed 1500 electrical miles in length (the longer the circuit, the longer the round-trip delay time). Echo suppressors are automatically disabled when they receive a tone between 2020 Hz and 2240 Hz, thus allowing full-duplex data transmission over a circuit with an echo suppressor. Full-duplex operation can also be achieved by replacing the echo suppressors with *echo cancelers.* Echo cancelers eliminate the echo by electrically subtracting it from the original signal rather than disabling the amplifier in the return circuit.

17-7 CROSSTALK

Crosstalk can be defined as any disturbance created in a communications channel by signals in other communications channels (i.e., unwanted coupling from one signal path into another). Crosstalk is a potential problem whenever two metallic conductors carrying different signals are located in close proximity to each other. Crosstalk can originate in telephone offices, at a subscriber's location, or on the facilities used to interconnect subscriber locations to telephone offices. Crosstalk is a subdivision of the general subject of interference. The term *crosstalk* was originally coined to indicate the presence of unwanted speech sounds in a telephone receiver caused by conversations on another telephone circuit.

The nature of crosstalk is often described as either *intelligible* or *unintelligible.* Intelligible (or near intelligible) crosstalk is particularly annoying and objectionable because the listener senses a real or fancied loss of privacy. Unintelligible crosstalk does not violate privacy, although it can still be annoying. Crosstalk between unlike channels, such as different types of carrier facilities, is usually unintelligible because of frequency inversion, frequency displacement, or digital encoding. However, such crosstalk often retains the syllabic pattern of speech and is more annoying than steady-state noise (such as thermal noise) with the same average power. Intermodulation noise, such as that found in multichannel frequency-division-multiplexed telephone systems, is a form of interchannel crosstalk that is usually unintelligible. Unintelligible crosstalk is generally grouped with other types of noise interferences.

The use of the words *intelligible* and *unintelligible* can also be applied to non-voice circuits. The methods developed for quantitatively computing and measuring crosstalk between voice circuits are also useful when studying interference between voice circuits and data circuits and between two data circuits.

There are three primary types of crosstalk in telephone systems: nonlinear crosstalk, transmittance crosstalk, and coupling crosstalk.

17-7-1 Nonlinear Crosstalk

Nonlinear crosstalk is a direct result of nonlinear amplification (hence the name) in analog communications systems. Nonlinear amplification produces harmonics and cross products (sum and difference frequencies). If the nonlinear frequency components fall into the passband of another channel, they are considered crosstalk. Nonlinear crosstalk can be distinguished from other types of crosstalk because the ratio of the signal power in the disturbing channel to the interference power in the disturbed channel is a function of the signal level in the disturbing channel.

17-7-2 Transmittance Crosstalk

Crosstalk can also be caused by inadequate control of the frequency response of a transmission system, poor filter design, or poor filter performance. This type of crosstalk is most prevalent when filters do not adequately reject undesired products from other channels. Because this type of interference is caused by inadequate control of the transfer characteristics or transmittance of networks, it is called *transmittance crosstalk.*

17-7-3 Coupling Crosstalk

Electromagnetic coupling between two or more physically isolated transmission media is called *coupling crosstalk.* The most common coupling is due to the effects of near-field mutual induction between cables from physically isolated circuits (i.e., when energy radiates from a wire in one circuit to a wire in a different circuit). To reduce coupling crosstalk due to mutual induction, wires are twisted together (hence the name *twisted pair*). Twisting the wires causes a canceling effect that helps eliminate crosstalk. Standard telephone cable pairs have 20 twists per foot, whereas data circuits generally require more twists per foot. Direct capacitive coupling between adjacent cables is another means in which signals from one cable can be coupled into another cable. The probability of coupling crosstalk occurring increases with cable length, signal power, and frequency.

There are two types of coupling crosstalk: near end and far end. *Near-end crosstalk* (NEXT) is crosstalk that occurs at the transmit end of a circuit and travels in the opposite direction as the signal in the disturbing channel. *Far-end crosstalk* (FEXT) occurs at the far-end receiver and is energy that travels in the same direction as the signal in the disturbing channel.

17-7-4 Unit of Measurement

Crosstalk interference is often expressed in its own special decibel unit of measurement, dBx. Unlike dBm, where the reference is a fixed power level, dBx is referenced to the level on the cable that is being interfered with (whatever the level may be). Mathematically, dBx is

$$\text{dBx} = 90 - (\text{crosstalk loss in decibels}) \tag{17-5}$$

where 90 dB is considered the ideal isolation between adjacent lines. For example, the magnitude of the crosstalk on a circuit is 70 dB lower than the power of the signal on the same circuit. The crosstalk is then 90 dB − 70 dBx = 20 dBx.

QUESTIONS

17-1. Briefly describe a *local subscriber loop.*

17-2. Explain what *loading coils* and *bridge taps* are and when they can be detrimental to the performance of a telephone circuit.

17-3. What are the designations used with *loading coils*?

17-4. What is meant by the term *loop resistance*?

17-5. Briefly describe *C-message noise weighting* and state its significance.

17-6. What is the difference between dB and dBm?

17-7. What is the difference between a TLP and a DLP?

17-8. What is meant by the following terms: dBmO, rn, dBrn, dBrnc, and dBrncO?

17-9. What is the difference between *psophometric noise weighting* and C-message weighting?

17-10. What are the three categories of *transmission parameters*?

17-11. Describe *attenuation distortion; envelope delay distortion.*

17-12. What is the reference frequency for attenuation distortion? Envelope delay distortion?

17-13. What is meant by *line conditioning*? What types of line conditioning are available?

17-14. What kind of circuits can have C-type line conditioning; D-type line conditioning?

17-15. When is D-type conditioning mandatory?

17-16. What limitations are imposed with D-type conditioning?

17-17. What is meant by *nonlinear distortion*? What are two kinds of nonlinear distortion?

17-18. What considerations are addressed by the *interface parameters*?

17-19. What considerations are addressed by *facility parameters*?

17-20. Briefly describe the following parameters: 1004-Hz variation, C-message noise, impulse noise, gain hits and dropouts, phase hits, phase jitter, single-frequency interference, frequency shift, phase intercept distortion, and peak-to-average ratio.

17-21. Describe what is meant by a *two-wire circuit; four-wire circuit*?

17-22. Briefly describe the function of a two-wire-to-four-wire *hybrid set.*

17-23. What is the purpose of an *echo suppressor; echo canceler*?

17-24. Briefly describe *crosstalk.*

17-25. What is the difference between *intelligible* and *unintelligible* crosstalk?

17-26. List and describe three types of crosstalk.

17-27. What is meant by near-end crosstalk; far-end crosstalk?

PROBLEMS

17-1. Describe what the following loading coil designations mean:

a. 22B44
b. 19H88
c. 24B44
d. 16B135

17-2. Frequencies of 250 Hz and 1 kHz are applied to the input of a C-message filter. Would their difference in amplitude be (greater, the same, or less) at the output of the filter?

17-3. A C-message noise measurement taken at a −22-dBm TLP indicates −72 dBm of noise. A test tone is measured at the same TLP at −25 dBm. Determine the following levels:

a. Signal power relative to TLP (dBmO)
b. C-message noise relative to reference noise (dBrn)
c. C-message noise relative to reference noise adjusted to a 0 TLP (dBrncO)
d. Signal-to-noise ratio

17-4. A C-message noise measurement taken at a −20-dBm TLP indicates a corrected noise reading of 43 dBrncO. A test tone at data level (0 DLP) is used to determine a signal-to-noise ratio of 30 dB. Determine the following levels:

a. Signal power relative to TLP (dBmO)
b. C-message noise relative to reference noise (dBrnc)
c. Actual test-tone signal power (dBm)
d. Actual C-message noise (dBm)

17-5. A test-tone signal power of −62 dBm is measured at a −61-dBm TLP. The C-message noise is measured at the same TLP at −10 dBrnc. Determine the following levels:

a. C-message noise relative to reference noise at a O TLP (dBrncO)
b. Actual C-message noise power level (dBm)
c. Signal power level relative to TLP (dBmO)
d. Signal-to-noise ratio (dB)

17-6. Sketch the graph for attenuation distortion and envelope delay distortion for a channel with C4 conditioning.

17-7. An EDD test on a basic telephone channel indicated that a 1600-Hz carrier experienced the minimum absolute delay of 550 μs. Determine the maximum absolute envelope delay that any frequency within the range of 800 Hz to 2600 Hz can experience.

17-8. The magnitude of the crosstalk on a circuit is 66 dB lower than the power of the signal on the same circuit. Determine the crosstalk in dBx.

CHAPTER 18

The Public Telephone Network

CHAPTER OUTLINE

OBJECTIVES

- Define *public telephone company*
- Explain the differences between the public and private sectors of the public telephone network
- Define *telephone instruments, local loops, trunk circuits,* and *exchanges*
- Describe the necessity for central office telephone exchanges
- Briefly describe the history of the telephone industry
- Describe operator-assisted local exchanges
- Describe automated central office switches and exchanges and their advantages over operator-assisted local exchanges
- Define *circuits, circuit switches,* and *circuit switching*
- Describe the relationship between local telephone exchanges and exchange areas
- Define *interoffice trunks, tandem trunks,* and *tandem switches*
- Define *toll-connecting trunks, intertoll trunks,* and *toll offices*
- Describe the North American Telephone Numbering Plan Areas
- Describe the predivestiture North American Telephone Switching Hierarchy
- Define the five classes of telephone switching centers
- Explain switching routes

- Describe the postdivestiture North American Telephone Switching Hierarchy
- Define *Common Channel Signaling System No. 7 (SS7)*
- Describe the basic functions of SS7
- Define and describe SS7 signaling points

18-1 INTRODUCTION

The telecommunications industry is the largest industry in the world. There are over 1400 independent telephone companies in the United States, jointly referred to as the *public telephone network* (PTN). The PTN uses the largest computer network in the world to interconnect millions of subscribers in such a way that the myriad of companies function as a single entity. The mere size of the PTN makes it unique and truly a modern-day wonder of the world. Virtually any subscriber to the network can be connected to virtually any other subscriber to the network within a few seconds by simply dialing a telephone number. One characteristic of the PTN that makes it unique from other industries is that every piece of equipment, technique, or procedure, new or old, is capable of working with the rest of the system. In addition, using the PTN does not require any special skills or knowledge.

18-2 TELEPHONE TRANSMISSION SYSTEM ENVIRONMENT

In its simplest form, a telephone transmission system is a pair of wires connecting two telephones or data modems together. A more practical transmission system is comprised of a complex aggregate of electronic equipment and associated transmission medium, which together provide a multiplicity of channels over which many subscriber's messages and control signals are propagated.

In general, a telephone call between two points is handled by interconnecting a number of different transmission systems in tandem to form an overall *transmission path* (*connection*) between the two points. The manner in which transmission systems are chosen and interconnected has a strong bearing on the characteristics required of each system because each element in the connection degrades the message to some extent. Consequently, the relationship between the performance and the cost of a transmission system cannot be considered only in terms of that system. Instead, a transmission system must be viewed with respect to its relationship to the complete system.

To provide a service that permits people or data modems to talk to each other at a distance, the communications system (telephone network) must supply the means and facilities for connecting the subscribers at the beginning of a call and disconnecting them at the completion of the call. Therefore, switching, signaling, and transmission functions must be involved in the service. The *switching function* identifies and connects the subscribers to a suitable transmission path. *Signaling functions* supply and interpret control and supervisory signals needed to perform the operation. Finally, transmission functions involve the actual transmission of a subscriber's messages and any necessary control signals. New transmission systems are inhibited by the fact that they must be compatible with an existing multi-trillion-dollar infrastructure.

18-3 THE PUBLIC TELEPHONE NETWORK

The public telephone network (PTN) accommodates two types of subscribers: *public* and *private.* Subscribers to the private sector are customers who lease equipment, transmission media (*facilities*), and services from telephone companies on a permanent basis. The leased circuits are designed and configured for their use only and are often referred to as *private-line* circuits or *dedicated* circuits. For example, large banks do not wish to share their communications network with other users, but it is not cost effective for them to construct their own networks. Therefore, banks lease equipment and facilities from public telephone companies and essentially operate a private telephone or data network within the PTN. The public telephone companies are sometimes called *service providers,* as they lease equipment and provide services to other private companies, organizations, and government agencies. Most metropolitan area networks (MANs) and wide area networks (WANs) utilize private-line data circuits and one or more service provider.

Subscribers to the public sector of the PTN share equipment and facilities that are available to all the public subscribers to the network. This equipment is appropriately called *common usage equipment,* which includes transmission facilities and telephone switches. Anyone with a telephone number is a subscriber to the public sector of the PTN. Since subscribers to the public network are interconnected only temporarily through switches, the network is often appropriately called the *public switched telephone network* (PSTN) and sometimes simply as the *dial-up network.* It is possible to interconnect telephones and modems with one another over great distances in fractions of a second by means of an elaborate network comprised of central offices, switches, cables (optical and metallic), and wireless radio systems that are connected by routing *nodes* (a node is a *switching point*). When someone talks about the public switched telephone network, they are referring to the combination of lines and switches that form a system of electrical routes through the network.

In its simplest form, data communications is the transmittal of digital information between two pieces of digital equipment, which includes computers. Several thousand miles may separate the equipment, which necessitates using some form of transmission medium to interconnect them. There is an insufficient number of transmission media capable of carrying digital information in digital form. Therefore, the most convenient (and least expensive) alternative to constructing an all-new all-digital network is to use the existing PTN for the transmission medium. Unfortunately, much of the PTN was designed (and much of it constructed) before the advent of large-scale data communications. The PTN was intended for transferring voice, not digital data. Therefore, to use the PTN for data communications, it is necessary to use a modem to convert the data to a form more suitable for transmission over the wireless carrier systems and conventional transmission media so prevalent in the PTN.

There are as many network configurations as there are subscribers in the private sector of the PTN, making it impossible to describe them all. Therefore, the intent of this chapter is to describe the public sector of the PTN (i.e., the public switched telephone network). Private-line data networks are described in later chapters of this book.

18-4 INSTRUMENTS, LOCAL LOOPS, TRUNK CIRCUITS, AND EXCHANGES

Telephone network equipment can be broadly divided into four primary classifications: instruments, local loops, exchanges, and trunk circuits.

18-4-1 Instruments

An *instrument* is any device used to originate and terminate calls and to transmit and receive signals into and out of the telephone network, such as a 2500-type telephone set, a cordless telephone, or a data modem. The instrument is often referred to as *station equipment* and the location of the instrument as the *station.* A *subscriber* is the operator or user of the instrument. If you have a home telephone, you are a subscriber.

18-4-2 Local Loops

As described in Chapters 8 and 9, the *local loop* is simply the dedicated cable facility used to connect an instrument at a subscriber's station to the closest telephone office. In the United States alone, there are several hundred million miles of cable used for local subscriber loops. Everyone who subscribes to the PTN is connected to the closest telephone office through a local loop. Local loops connected to the public switched telephone network are two-wire metallic cable pairs. However, local loops used with private-line data circuits are generally four-wire configurations.

18-4-3 Trunk Circuits

A *trunk circuit* is similar to a local loop except trunk circuits are used to interconnect two telephone offices. The primary difference between a local loop and a trunk is that a local loop is permanently associated with a particular station, whereas a trunk is a common-usage connection. A trunk circuit can be as simple as a pair of copper wires twisted together or as sophisticated as an optical fiber cable. A trunk circuit could also be a wireless communications channel. Although all trunk circuits perform the same basic function, there are different names given to them, depending on what types of telephone offices they interconnect and for what reason. Trunk circuits can be two wire or four wire, depending on what type of facility is used. Trunks are described in more detail in a later section of this chapter.

18-4-4 Exchanges

An *exchange* is a central location where subscribers are interconnected, either temporarily or on a permanent basis. Telephone company switching machines are located in exchanges. Switching machines are programmable matrices that provide temporary signal paths between two subscribers. Telephone sets and data modems are connected through local loops to switching machines located in exchanges. Exchanges connected directly to local loops are often called *local exchanges* or sometimes *dial switches* or *local dial switches.* The first telephone exchange was installed in 1878, only two years after the invention of the telephone. A central exchange is also called a *central telephone exchange, central office* (CO), *central wire center, central exchange, central office exchange,* or simply *central.*

The purpose of a telephone exchange is to provide a path for a call to be completed between two parties. To process a call, a switch must provide three primary functions:

Identify the subscribers
Set up or establish a communications path
Supervise the calling processes

18-5 LOCAL CENTRAL OFFICE TELEPHONE EXCHANGES

The first telephone sets were self-contained, as they were equipped with their own battery, microphone, speaker, bell, and ringing circuit. Telephone sets were originally connected directly to each other with heavy-gauge iron wire strung between poles, requiring a dedicated cable pair and telephone set for each subscriber you wished to be connected to. Figure 18-1a shows two telephones interconnected with a single cable pair. This is simple enough; however, if more than a few subscribers wished to be directly connected together, it became cumbersome, expensive, and very impractical. For example, to interconnect one subscriber to five other subscribers, five telephone sets and five cable pairs are needed, as shown in Figure 18-1b. To completely interconnect four subscribers, it would require six cable pairs, and each subscriber would need three telephone sets. This is shown in Figure 18-1c.

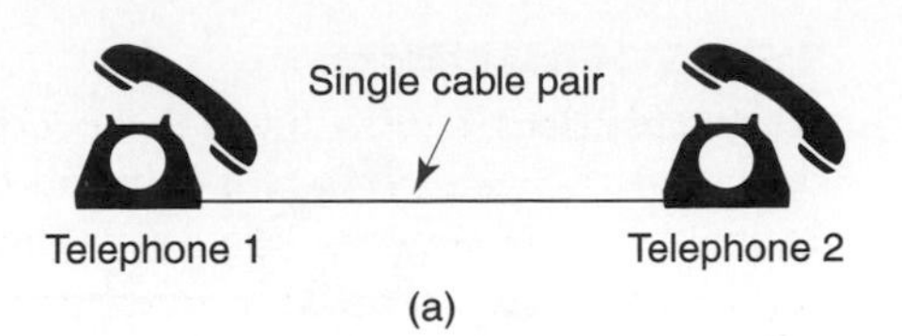

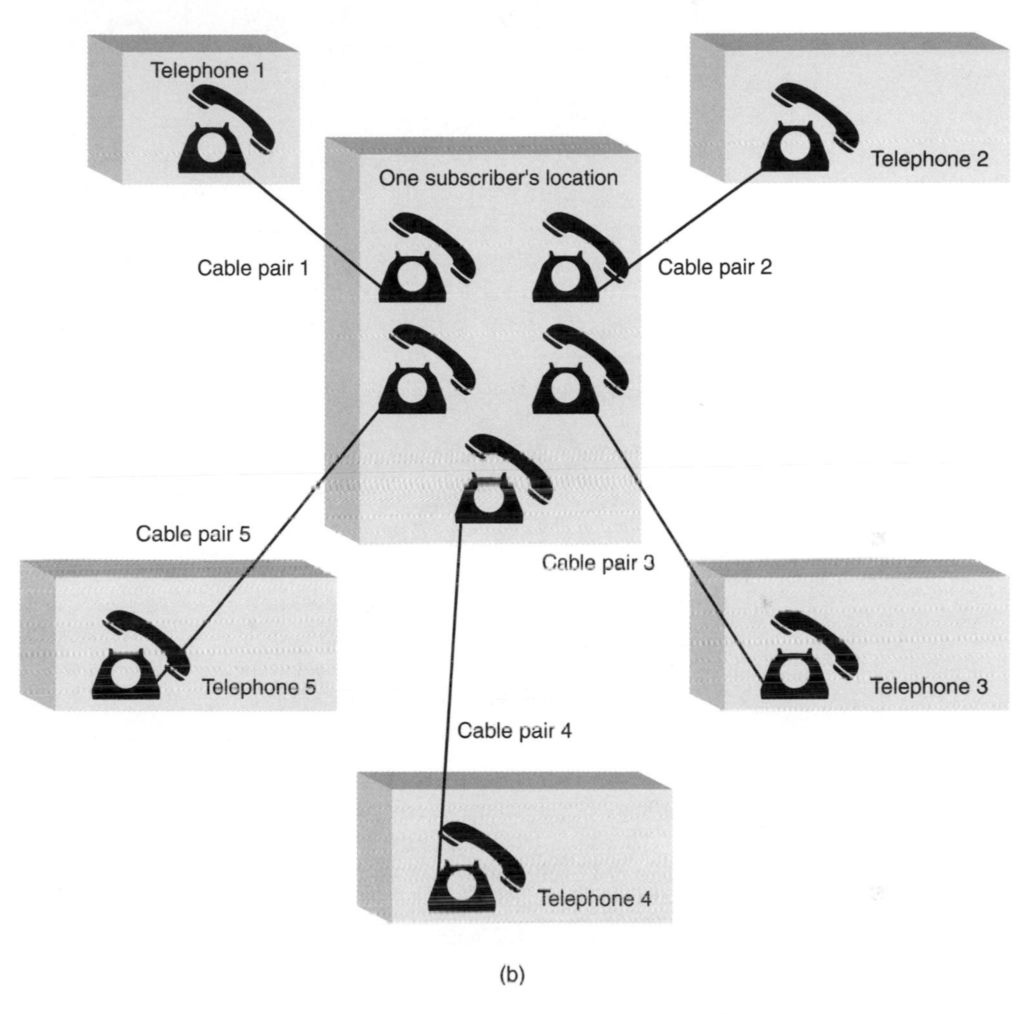

FIGURE 18-1 Dedicated telephone interconnections: (a) Interconnecting two subscribers; (b) Interconnecting one subscriber to five other telephone sets; (*Continued*)

The number of lines required to interconnect any number of stations is determined by the following equation:

$$N = \frac{n(n-1)}{2} \tag{18-1}$$

where n = number of stations (parties)
N = number of interconnecting lines

The number of dedicated lines necessary to interconnect 100 parties is

$$N = \frac{100(100-1)}{2} = 4950$$

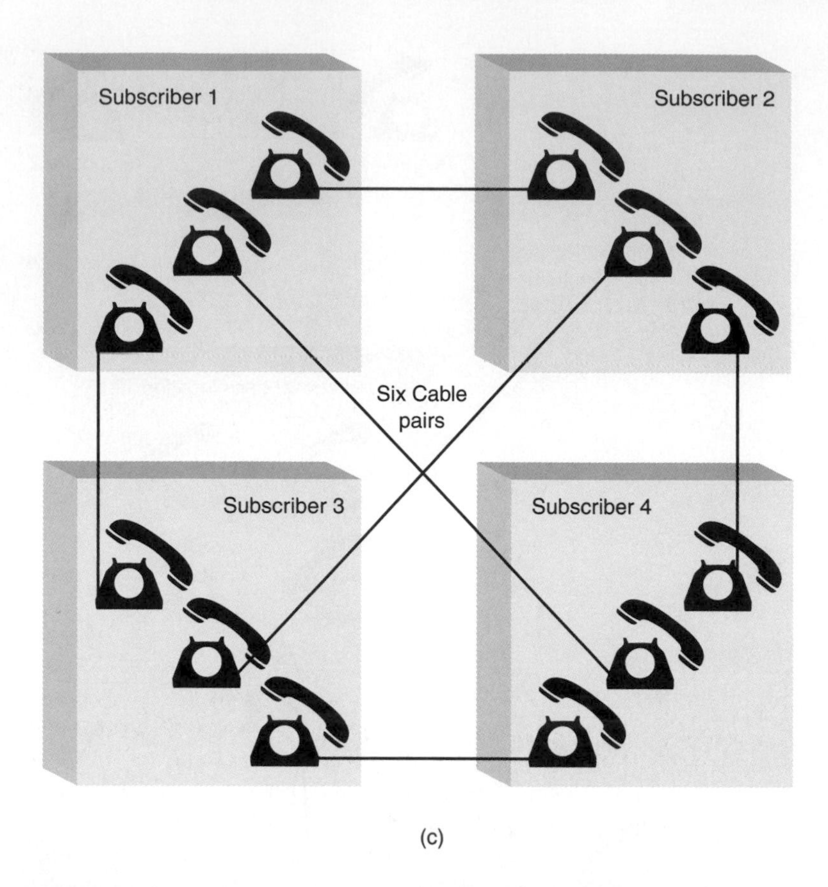

FIGURE 18-1 (*Continued*) (c) Interconnecting four subscribers

In addition, each station would require either 100 separate telephones or the capability of switching one telephone to any of 99 lines.

These limitations rapidly led to the development of the *central telephone exchange.* A telephone exchange allows any telephone connected to it to be interconnected to any of the other telephones connected to the exchange without requiring separate cable pairs and telephones for each connection. Generally, a community is served by only one telephone company. The community is divided into zones, and each zone is served by a different central telephone exchange. The number of stations served and the density determine the number of zones established in a given community. If a subscriber in one zone wishes to call a station in another zone, a minimum of two local exchanges is required.

18-6 OPERATOR-ASSISTED LOCAL EXCHANGES

The first commercial telephone switchboard began operation in New Haven, Connecticut, on January 28, 1878, marking the birth of the public switched telephone network. The switchboard served 21 telephones attached to only eight lines (obviously, some were party lines). On February 17 of the same year, Western Union opened the first large-city exchange in San Francisco, California, and on February 21, the New Haven District Telephone Company published the world's first telephone directory comprised of a single page listing only 50 names. The directory was immediately followed by a more comprehensive listing by the Boston Telephone Dispatch Company.

The first local telephone exchanges were *switchboards* (sometimes called *patch panels* or *patch boards*) where manual interconnects were accomplished using *patchcords* and

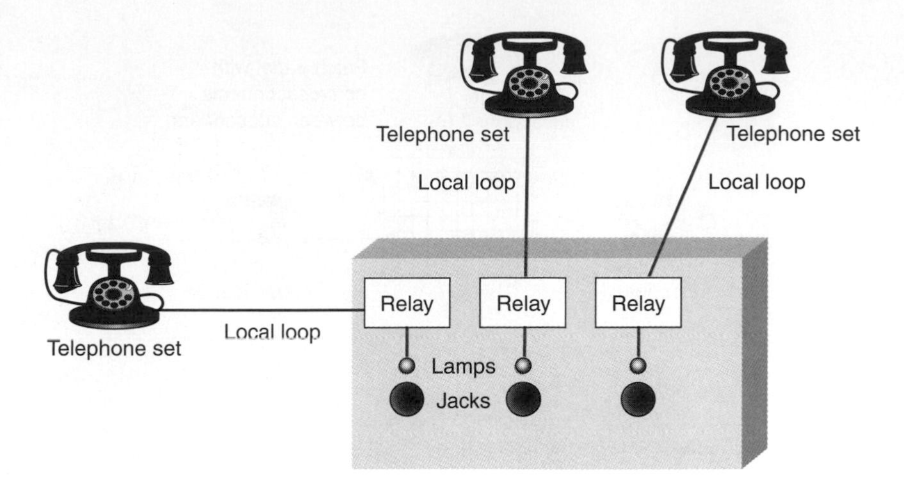

FIGURE 18-2 Patch panel configuration

jacks. All subscriber stations were connected through local loops to jacks on the switchboard. Whenever someone wished to initiate a call, they sent a ringing signal to the switchboard by manually turning a crank on their telephone. The ringing signal operated a relay at the switchboard, which in turn illuminated a supervisory lamp located above the jack for that line, as shown in Figure 18-2. Manual switchboards remained in operation until 1978, when the Bell System replaced their last cord switchboard on Santa Catalina Island off the coast of California near Los Angeles.

In the early days of telephone exchanges, each telephone line could have 10 or more subscribers (residents) connected to the central office exchange using the same local loop. This is called a *party line,* although only one subscriber could use their telephone at a time. Party lines are less expensive than private lines, but they are also less convenient. A private telephone line is more expensive because only telephones from one residence or business are connected to a local loop.

Connecting 100 private telephone lines to a single exchange required 100 local loops and a switchboard equipped with 100 relays, jacks, and lamps. When someone wished to initiate a telephone call, they rang the switchboard. An operator answered the call by saying, "Central." The calling party told the operator whom they wished to be connected to. The operator would then ring the destination, and when someone answered the telephone, the operator would remove her plug from the jack and connect the calling and called parties together with a special patchcord equipped with plugs on both ends. This type of system was called a *ringdown* system. If only a few subscribers were connected to a switchboard, the operator had little trouble keeping track of which jacks were for which subscriber (usually by name). However, as the popularity of the telephone grew, it soon became necessary to assign each subscriber line a unique telephone number. A switchboard using four digits could accommodate 10,000 telephone numbers (0000 to 9999).

Figure 18-3a shows a central office patch panel connected to four idle subscriber lines. Note that none of the telephone lines is connected to any of the other telephone lines. Figure 18-3b shows how subscriber 1 can be connected to subscriber 2 using a temporary connection provided by placing a patchcord between the jack for line 1 and the jack for line 2. Any subscriber can be connected to any other subscriber using patchcords.

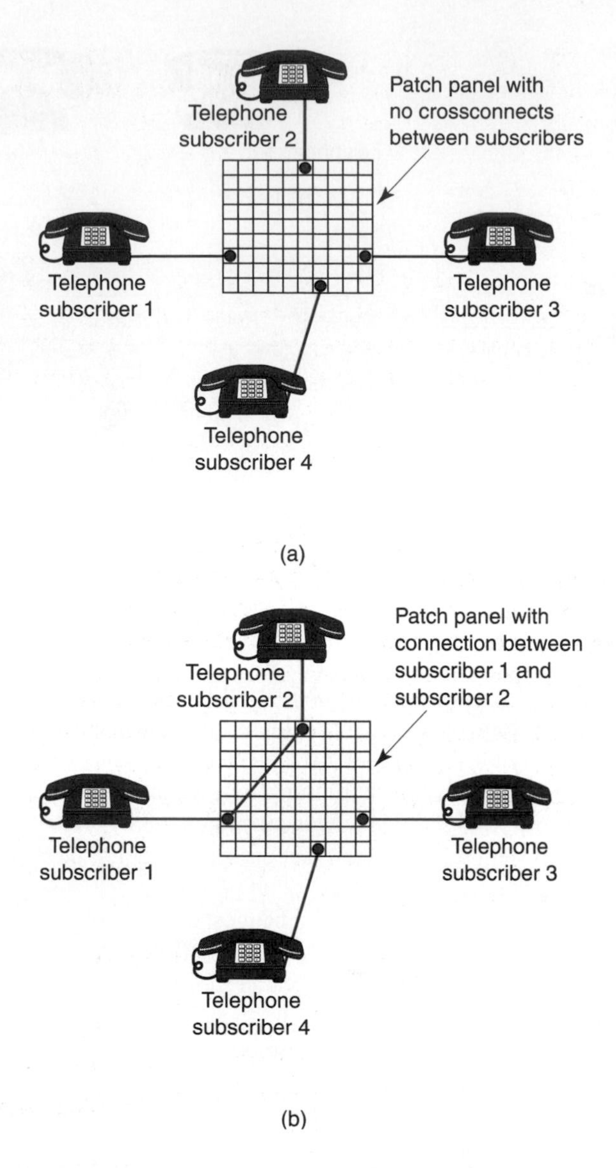

FIGURE 18-3 Central office exchange: (a) without interconnects; (b) with an interconnect

18-7 AUTOMATED CENTRAL OFFICE SWITCHES AND EXCHANGES

As the number of telephones in the United States grew, it quickly became obvious that operator-assisted calls and manual patch panels could not meet the high demand for service. Thus, automated switching machines and exchange systems were developed.

An *automated switching system* is a system of sensors, switches, and other electrical and electronic devices that allows subscribers to give instructions directly to the switch without having to go through an operator. In addition, automated switches performed interconnections between subscribers without the assistance of a human and without using patchcords.

In 1890 an undertaker in Kansas City, Kansas, named Alman Brown Strowger was concerned that telephone company operators were diverting his business to his competitors. Consequently, he invented the first automated switching system using electromechanical relays. It is said that Strowger worked out his original design using a cardboard box, straight pins, and a pencil.

With the advent of the Strowger switch, mechanical dialing mechanisms were added to the basic telephone set. The mechanical dialer allowed subscribers to manually dial the telephone number of the party they wished to call. After a digit was entered (dialed), a relay in the switching machine connected the caller to another relay. The relays were called *stepping relays* because the system stepped through a series of relays as the digits were entered. The stepping process continued until all the digits of the telephone number were entered. This type of switching machine was called a *step-by-step* (SXS) switch, *stepper,* or, perhaps more commonly, a *Strowger* switch. A step-by-step switch is an example of a progressive switching machine, meaning that the connection between the calling and called parties was accomplished through a series of steps.

Between the early 1900s and the mid-1960s, the Strowger switch gradually replaced manual switchboards. The Bell System began using steppers in 1919 and continued using them until the early 1960s. In 1938, the Bell System began replacing the steppers with another electromechanical switching machine called the *crossbar* (XBAR) *switch.* The first No. 1 crossbar was cut into service at the Troy Avenue central office in Brooklyn, New York, on February 14, 1938. The crossbar switch used sets of contact points (called *crosspoints*) mounted on horizontal and vertical bars. Electromagnets were used to cause a vertical bar to cross a horizontal bar and make contact at a coordinate determined by the called number. The most versatile and popular crossbar switch was the #5XB. Although crossbar switches were an improvement over step-by-step switches, they were short lived, and most of them have been replaced with *electronic switching systems* (ESS).

In 1965, AT&T introduced the No. 1 ESS, which was the first computer-controlled central office switching system used on the PSTN. ESS switches differed from their predecessors in that they incorporate *stored program control* (SPC), which uses software to control practically all the switching functions. SPC increases the flexibility of the switch, dramatically increases its reliability, and allows for automatic monitoring of maintenance capabilities from a remote location. Virtually all the switching machines in use today are electronic stored program control switching machines. SPC systems require little maintenance and require considerably less space than their electromechanical predecessors. SPC systems make it possible for telephone companies to offer the myriad of services available today, such as three-way calling, call waiting, caller identification, call forwarding, call within, speed dialing, return call, automatic redial, and call tracing. Electronic switching systems evolved from the No. 1 ESS to the No. 5 ESS, which is the most advanced digital switching machine developed by the Bell System.

Automated central office switches paved the way for totally *automated central office exchanges,* which allow a caller located virtually anywhere in the world to direct dial virtually anyone else in the world. Automated central office exchanges interpret telephone numbers as an address on the PSTN. The network automatically locates the called number, tests its availability, and then completes the call.

18-7-1 Circuits, Circuit Switches, and Circuit Switching

A *circuit* is simply the path over which voice, data, or video signals propagate. In telecommunications terminology, a circuit is the path between a source and a destination (i.e., between a calling and a called party). Circuits are sometimes called *lines* (as in telephone lines). A *circuit switch* is a programmable matrix that allows circuits to be connected to one another. Telephone company circuit switches interconnect input loop or trunk circuits to output loop or trunk circuits. The switches are capable of interconnecting any circuit connected to it to any other circuit connected to it. For this reason, the switching process is called *circuit switching* and, therefore, the public telephone network is considered a *circuit-switched network.* Circuit switches are *transparent.* That is, they interconnect circuits without altering the information on them. Once a circuit switching operation has been performed, a transparent switch simply provides continuity between two circuits.

18-7-2 Local Telephone Exchanges and Exchange Areas

Telephone exchanges are strategically placed around a city to minimize the distance between a subscriber's location and the exchange and also to optimize the number of stations connected to any one exchange. The size of the service area covered by an exchange depends on subscriber density and subscriber calling patterns. Today, there are over 20,000 local exchanges in the United States.

Exchanges connected directly to local loops are appropriately called *local exchanges.* Because local exchanges are centrally located within the area they serve, they are often called *central offices* (CO). Local exchanges can directly interconnect any two subscribers whose local loops are connected to the same local exchange. Figure 18-4a shows a local exchange with six telephones connected to it. Note that all six telephone numbers begin with 87. One subscriber of the local exchange can call another subscriber by simply dialing their seven-digit telephone number. The switching machine performs all tests and switching operations necessary to complete the call. A telephone call completed within a single local exchange is called an *intraoffice call* (sometimes called an *intraswitch* call). Figure 18-4b shows how two stations serviced by the same exchange (874-3333 to 874-4444) are interconnected through a common local switch.

In the days of manual patch panels, to differentiate telephone numbers in one local exchange from telephone numbers in another local exchange and to make it easier for people to

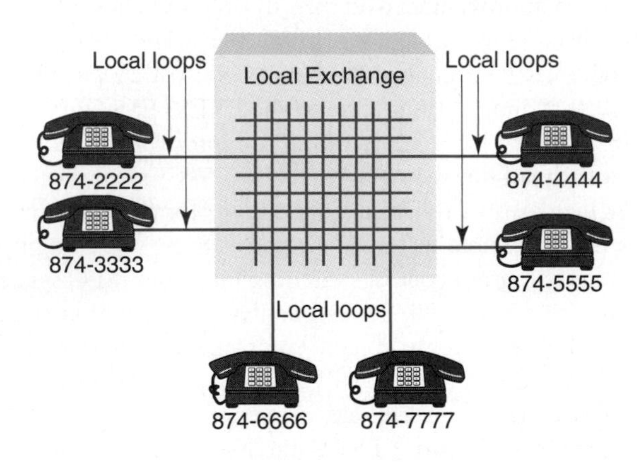

(a)

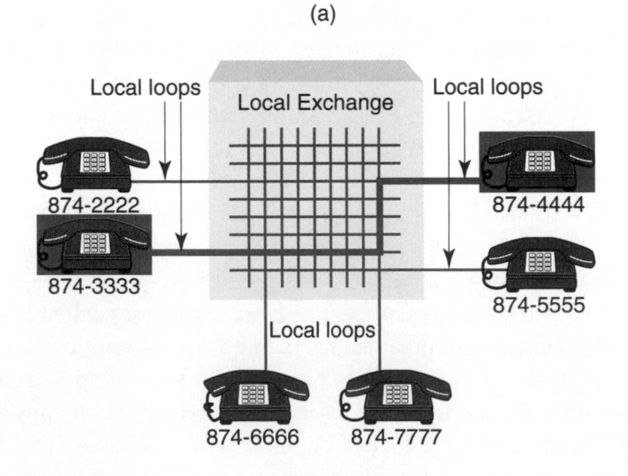

(b)

FIGURE 18-4 Local exchange: (a) no interconnections; (b) 874-3333 connected to 874-4444

remember telephone numbers, each exchange was given a name, such as Bronx, Redwood, Swift, Downtown, Main, and so on. The first two digits of a telephone number were derived from the first two letters of the exchange name. To accommodate the names with dial telephones, the digits 2 through 9 were each assigned three letters. Originally, only 24 of the 26 letters were assigned (Q and Z were omitted); however, modern telephones assign all 26 letters to oblige personalizing telephone numbers (the digits 7 and 9 are now assigned four letters each). As an example, telephone numbers in the Bronx exchange begin with 27 (B on a telephone dial equates to the digit 2, and R on a telephone dial equates to the digit 7). Using this system, a seven-digit telephone number can accommodate 100,000 telephone numbers. For example, the Bronx exchange was assigned telephone numbers between 270-0000 and 279-9999 inclusive. The same 100,000 numbers could also be assigned to the Redwood exchange (730-0000 to 739-9999).

18-7-3 Interoffice Trunks, Tandem Trunks, and Tandem Switches

Interoffice calls are calls placed between two stations that are connected to different local exchanges. Interoffice calls are sometimes called *interswitch* calls. Interoffice calls were originally accomplished by placing special plugs on the switchboards that were connected to cable pairs going to local exchange offices in other locations around the city or in nearby towns. Today telephone-switching machines in local exchanges are interconnected to other local exchange offices on special facilities called *trunks* or, more specifically, *interoffice trunks*. A subscriber in one local exchange can call a subscriber connected to another local exchange over an interoffice trunk circuit in much the same manner that they would call a subscriber connected to the same exchange. When a subscriber on one local exchange dials the telephone number of a subscriber on another local exchange, the two local exchanges are interconnected with an interoffice trunk for the duration of the call. After either party terminates the call, the interoffice trunk is disconnected from the two local loops and made available for another interoffice call. Figure 18-5 shows three exchange offices with

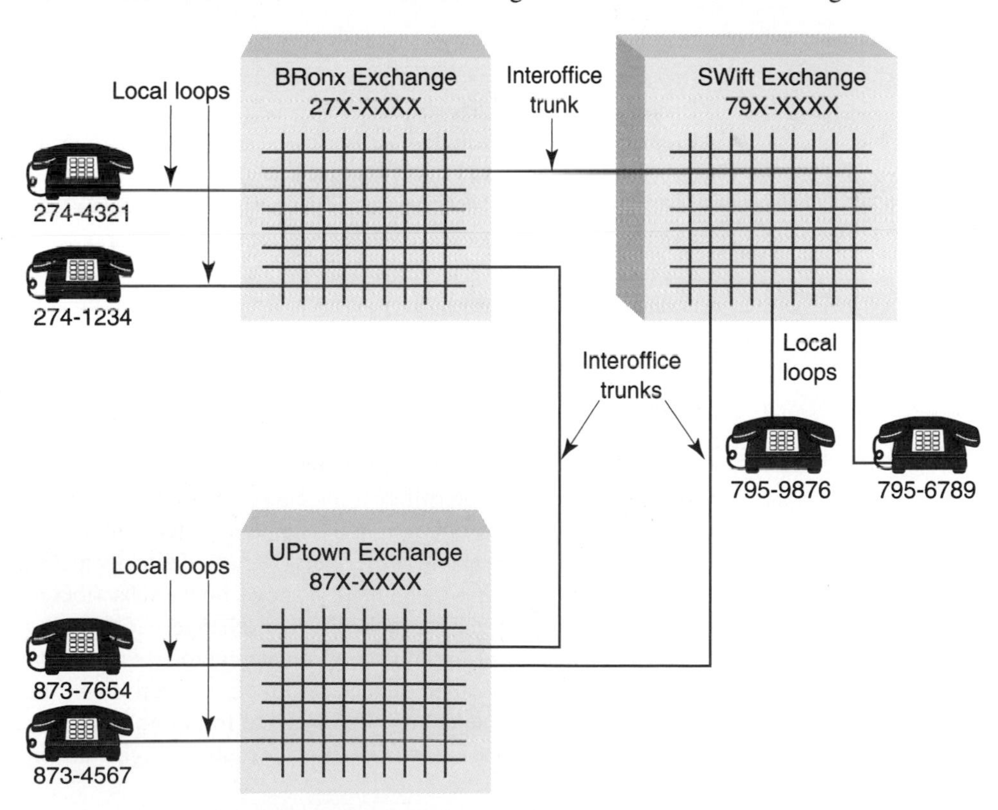

FIGURE 18-5 Interoffice exchange system

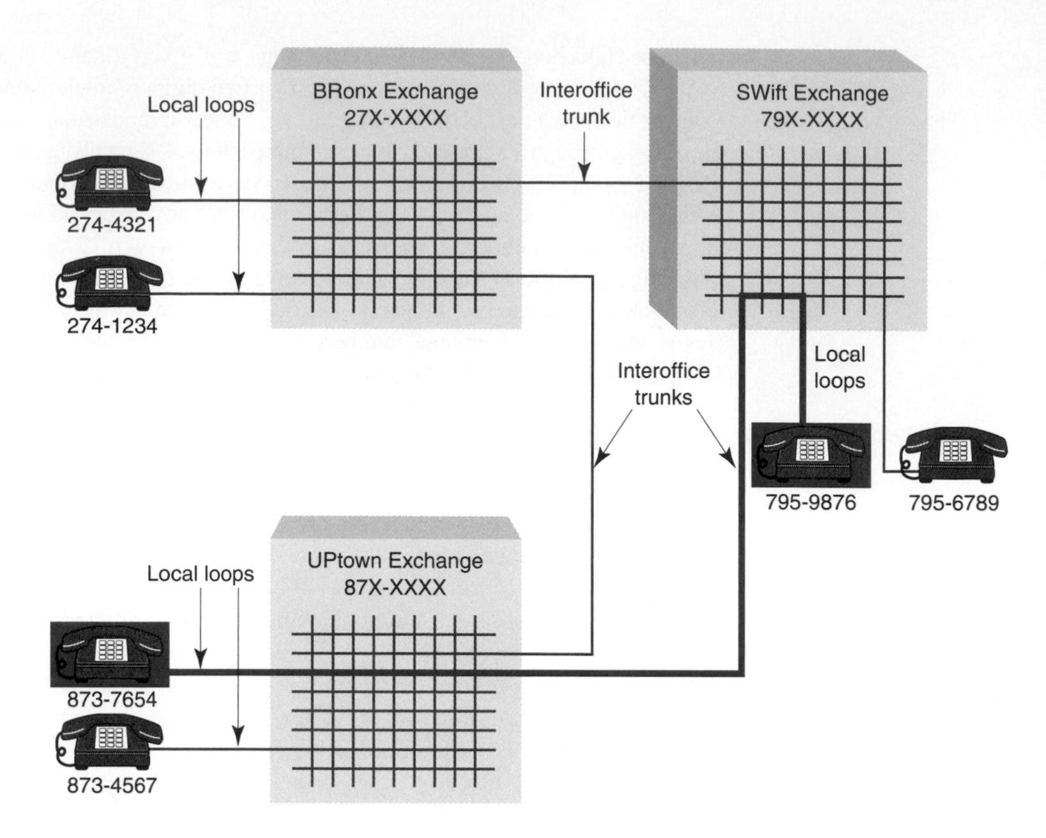

FIGURE 18-6 Interoffice call between subscribers serviced by two different exchanges

two subscribers connected to each. The telephone numbers for subscribers connected to the Bronx, Swift, and Uptown exchanges begin with the digits 27, 79, and 87, respectively. Figure 18-6 shows how two subscribers connected to different local exchanges can be interconnected using an interoffice trunk.

In larger metropolitan areas, it is virtually impossible to provide interoffice trunk circuits between all the local exchange offices. To interconnect local offices that do not have interoffice trunks directly between them, tandem offices are used. A *tandem office* is an exchange without any local loops connected to it (tandem meaning "in conjunction with" or "associated with"). The only facilities connected to the switching machine in a tandem office are trunks. Therefore, tandem switches interconnect local offices only. A *tandem switch* is called a *switcher's switch,* and trunk circuits that terminate in tandem switches are appropriately called *tandem trunks* or sometimes *intermediate trunks.*

Figure 18-7 shows two exchange areas that can be interconnected either with a tandem switch or through an interoffice trunk circuit. Note that tandem trunks are used to connect the Bronx and Uptown exchanges to the tandem switch. There is no name given to the tandem switch because there are no subscribers connected directly to it (i.e., no one receives dial tone from the tandem switch). Figure 18-8 shows how a subscriber in the Uptown exchange area is connected to a subscriber in the Bronx exchange area through a tandem switch. As the figure shows, tandem offices do not eliminate interoffice trunks. Very often, local offices have the capabilities to be interconnected with direct interoffice trunks as well as through a tandem office. When a telephone call is made from one local office to another, an interoffice trunk is selected if one is available. If not, a route through a tandem office is the second choice.

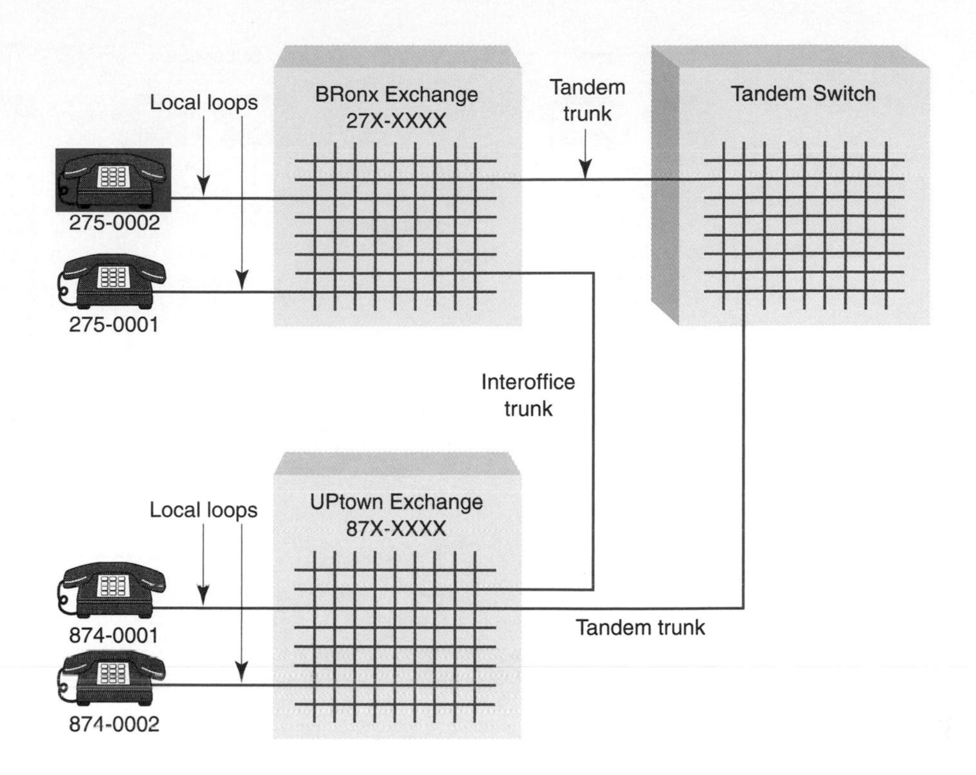

FIGURE 18-7 Interoffice switching between two local exchanges using tandem trunks and a tandem switch

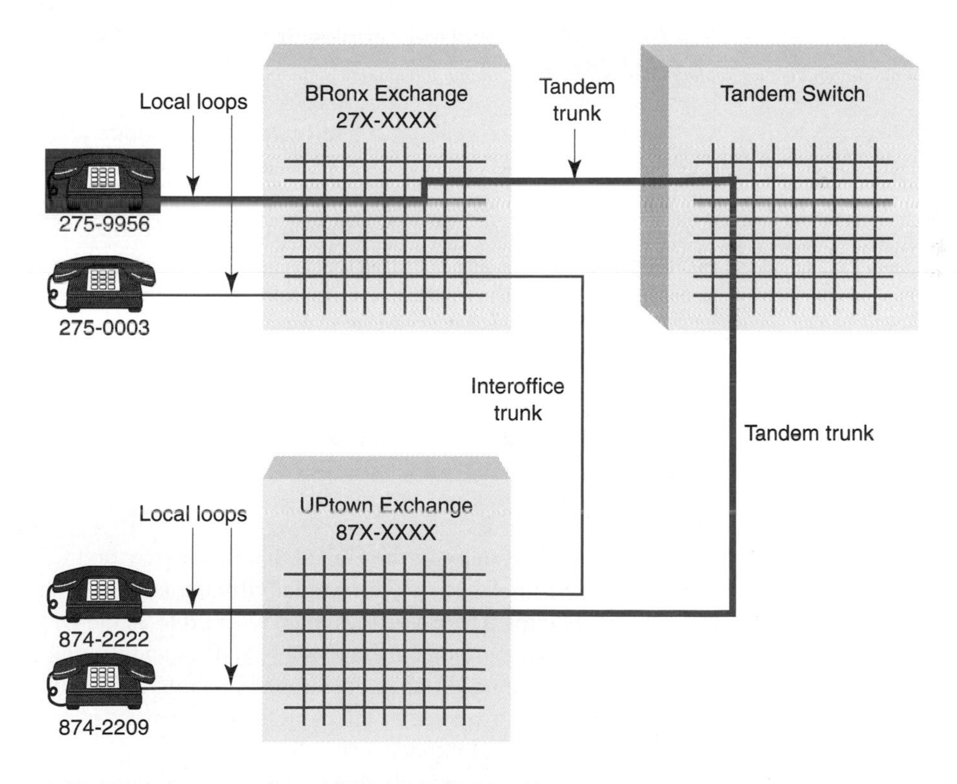

FIGURE 18-8 Interoffice call between two local exchanges through a tandem switch

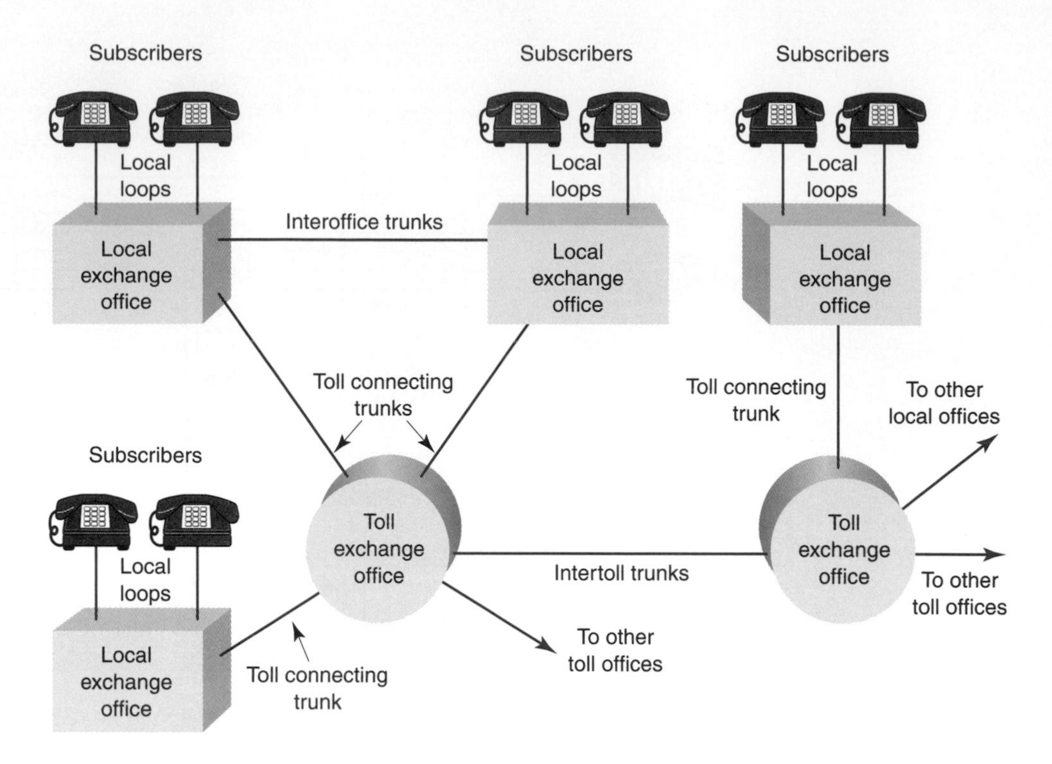

FIGURE 18-9 Relationship between local exchange offices and toll offices

18-7-4 Toll-Connecting Trunks, Intertoll Trunks, and Toll Offices

Interstate long-distance telephone calls require a special telephone office called a *toll office.* There are approximately 1200 toll offices in the United States. When a subscriber initiates a long-distance call, the local exchange connects the caller to a toll office through a facility called a *toll-connecting trunk* (sometimes called an *interoffice toll trunk*). Toll offices are connected to other toll offices with *intertoll trunks.* Figure 18-9 shows how local exchanges are connected to toll offices and how toll offices are connected to other toll offices. Figure 18-10 shows the network relationship between local exchange offices, tandem offices, toll offices, and their respective trunk circuits.

18-8 NORTH AMERICAN TELEPHONE NUMBERING PLAN AREAS

The *North American Telephone Numbering Plan* (NANP) was established to provide a telephone numbering system for the United States, Mexico, and Canada that would allow any subscriber in North America to direct dial virtually any other subscriber without the assistance of an operator. The network is often referred to as the DDD (*direct distance dialing*) network. Prior to the establishment of the NANP, placing a long-distance telephone call began by calling the long-distance operator and having her manually connect you to a trunk circuit to the city you wished to call. Any telephone number outside the caller's immediate area was considered a long-distance call.

North America is now divided into *numbering plan areas* (NPAs) with each NPA assigned a unique three-digit number called an *area code.* Each NPA is further subdivided into smaller service areas each with its own three-digit number called an *exchange code* (or

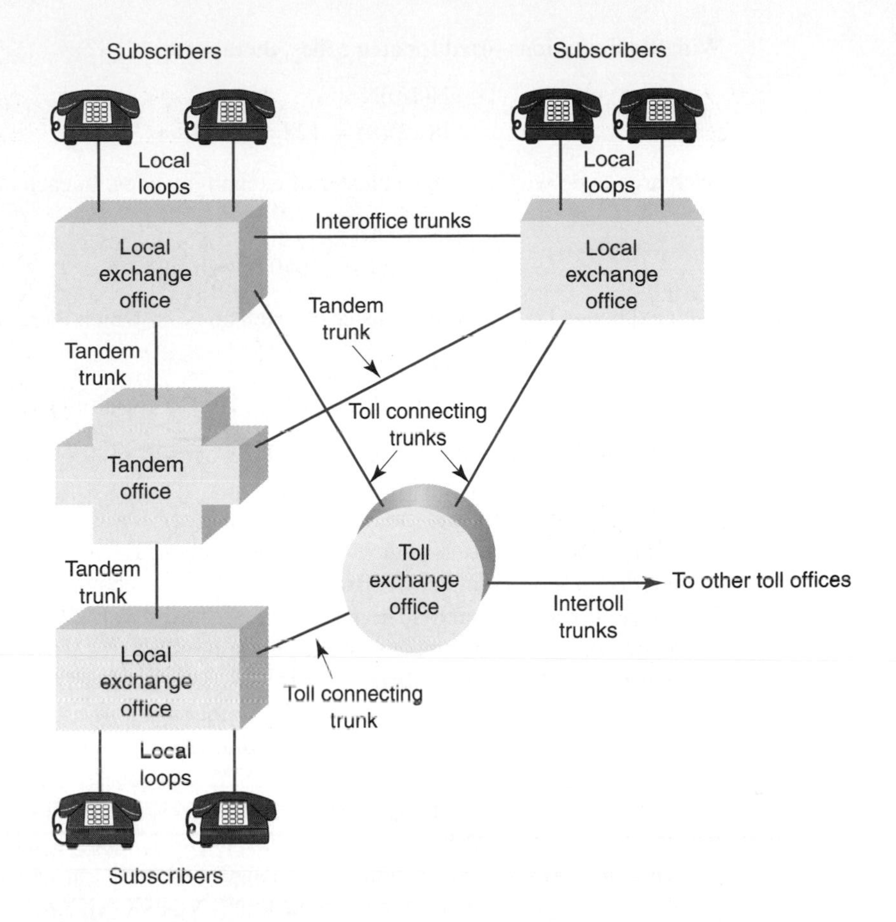

FIGURE 18-10 Relationship between local exchanges, tandem offices, and toll offices

prefix). Initially, each service area had only one central telephone switching office and one prefix. However, today a switching office can be assigned several exchange codes, depending on user density and the size of the area the office services. Each subscriber to a central office prefix is assigned a four-digit *extension number.* The three-digit area code represents the first three digits of a 10-digit telephone number, the three-digit prefix represents the next three digits, and the four-digit extension represents the last four digits of the telephone number.

Initially, within the North American Telephone Numbering Plan Area, if a digit could be any value from 0 through 9, the variable X designated it. If a digit could be any value from 2 through 9, the variable N designated it. If a digit could be only a 1 or a 0, it was designated by the variable 1/0 (one or zero). Area codes were expressed as N(1/0)N and exchange codes as NNX. Therefore, area codes could not begin or end with the digit 0 or 1, and the middle digit had to be either a 0 or a 1. Because of limitations imposed by electromechanical switching machines, the first two digits of exchange codes could not be 0 or 1, although the third digit could be any digit from 0 to 9. The four digits in the extension could be any digit value from 0 through 9. In addition, each NPA or area code could not have more than one local exchange with the same exchange code, and no two extension numbers within any exchanges codes could have the same four-digit number. The 18-digit telephone number was expressed as

$$\underbrace{\text{N(1/0)N}}_{\text{area code}} - \underbrace{\text{NNX}}_{\text{prefix}} - \underbrace{\text{XXXX}}_{\text{extension}}$$

With the limitations listed for area codes, there were

N(1/0)N
(8)(2)(8) = 128 possibilities

Each area code was assigned a cluster of exchange codes. In each cluster, there were

(N)(N)(X)
(8)(8)(10) = 640 possibilities

Each exchange code served a cluster of extensions, in which there were

(X)(X)(X)(X)
(10)(10)(10)(10) = 10,000 possibilities

With this numbering scheme, there were a total of (128)(640)(10,000) = 819,200,000 telephone numbers possible in North America.

When the NANP was initially placed into service, local exchange offices dropped their names and converted to their exchange number. Each exchange had 10 possible exchange codes. For example, the Bronx exchange was changed to 27 (B = 2 and r = 7). Therefore, it could accommodate the prefixes 270 through 279. Although most people do not realize it, telephone company employees still refer to local exchanges by their name. In January 1958, Wichita Falls, Texas, became the first American city to incorporate a true all-number calling system using a seven-digit number without attaching letters or names.

The popularity of cellular telephone has dramatically increased the demand for telephone numbers. By 1995, North America ran out of NPA area codes, so the requirement that the second digit be a 1 or a 0 was dropped. This was made possible because by 1995 there were very few electromagnetic switching machines in use in North America, and with the advent of SS7 signaling networks, telephone numbers no longer had to be transported over voice switching paths. This changed the numbering scheme to NXN-NNX-XXXX, which increased the number of area codes to 640 and the total number of telephones to 4,096,000,000. Figure 18-11 shows the North American Telephone Numbering Plan Areas as of January 2002.

The International Telecommunications Union has adopted an international numbering plan that adds a prefix in front of the area code, which outside North America is called a *city code.* The city code is one, two, or three digits long. For example, to call London, England, from the United States, one must dial 011-44-491-222-111. The 011 indicates an international call, 44 is the country code for England, 491 is the city code for London, 222 is the prefix for Piccadilly, and 111 is the three-digit extension number of the party you wish to call.

18-9 TELEPHONE SERVICE

A telephone connection may be as simple as two telephones and a single local switching office, or it may involve a multiplicity of communications links including several switching offices, transmission facilities, and telephone companies.

Telephone sets convert acoustic energy to electrical signals and vice versa. In addition, they also generate supervisory signals and address information. The subscriber loop provides a two-way path for conveying speech and data information and for exchanging ringing, switching, and supervisory signals. Since the telephone set and the subscriber loop are permanently associated with a particular subscriber, their combined transmission properties can be adjusted to meet their share of the total message channel objectives. For example, the higher efficiencies of new telephone sets and modems compensate for increased loop loss, permitting longer loop lengths or using smaller-gauge wire.

The small percentage of time (approximately 10% during busy hours) that a subscriber loop is utilized led to the development of line concentrators between subscribers and central offices. A concentrator allows many subscribers to share a limited number of lines

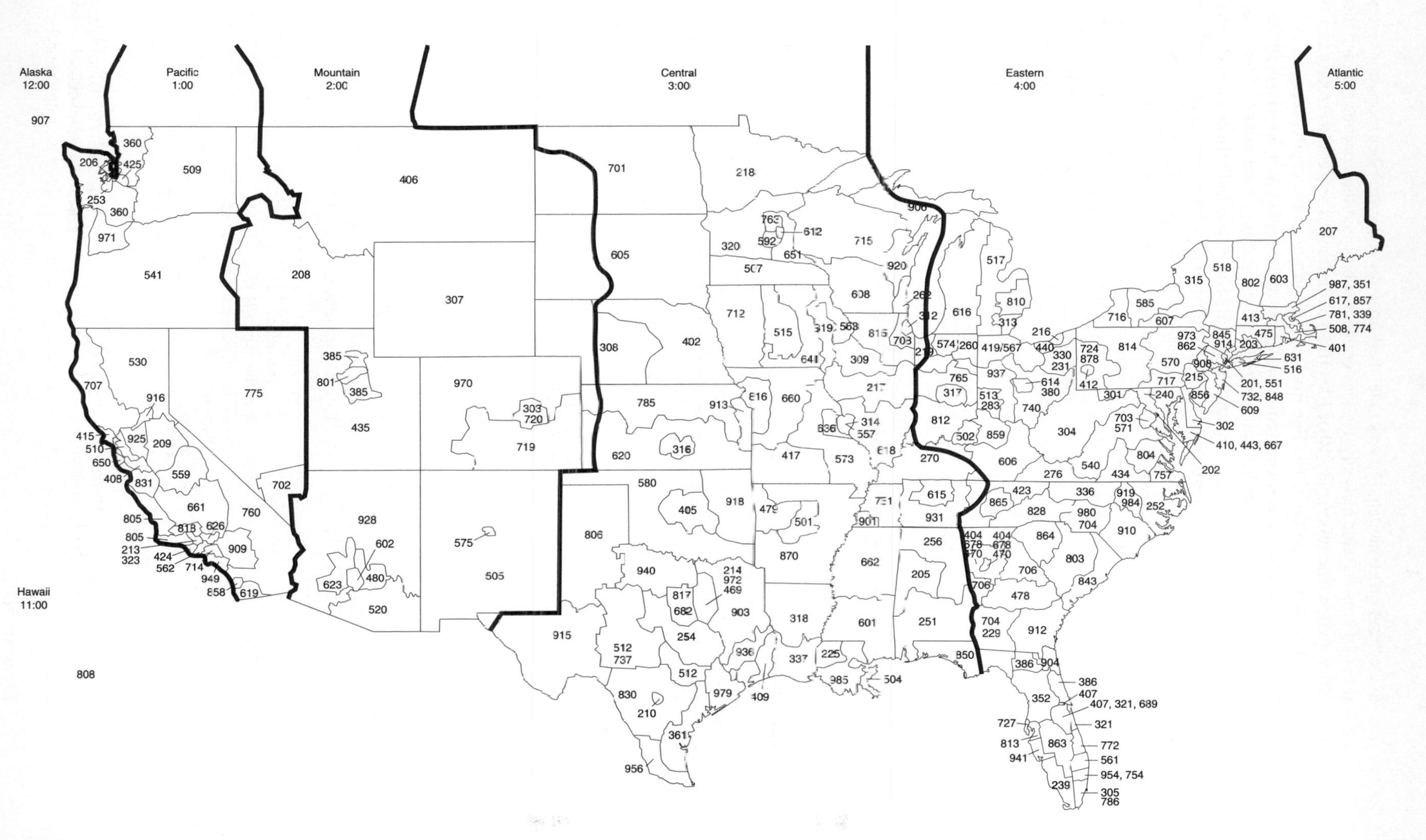

FIGURE 18-11 North American Telephone Numbering Plan Areas

to a central office switch. For example, there may be 100 subscriber loops connected to one side of a concentrator and only 10 lines connected between the concentrator and the central office switch. Therefore, only 10 of 100 (10%) of the subscribers could actually access the local office at any one time. The line from a concentrator to the central office is essentially a trunk circuit because it is shared (common usage) among many subscribers on an "as needed" basis. As previously described, trunk circuits of various types are used to interconnect local offices to other local offices, local offices to toll offices, and toll offices to other toll offices.

When subscribers are connected to a toll office through toll-connecting trunks, the message signal is generally handled on a two-wire basis (both directions of transmission on the same pair of wires). After appropriate switching and routing functions are performed at toll offices, messages are generally connected to intertoll trunks by means of a two-wire-to-four-wire *terminating set* (*term set* or *hybrid*), which splits the two directions of signal propagation so that the actual long-distance segment of the route can be accomplished on a four-wire basis (separate cable pairs for each direction). Signals are connected through intertoll trunks to remote toll-switching centers, which may in turn be connected by intertoll trunks to other toll-switching centers and ultimately reach the recipient of the call through a toll-connecting trunk, a local office, another four-wire-to-two-wire term set, a local switching office, and a final subscriber loop as shown in Figure 18-12. A normal two-point

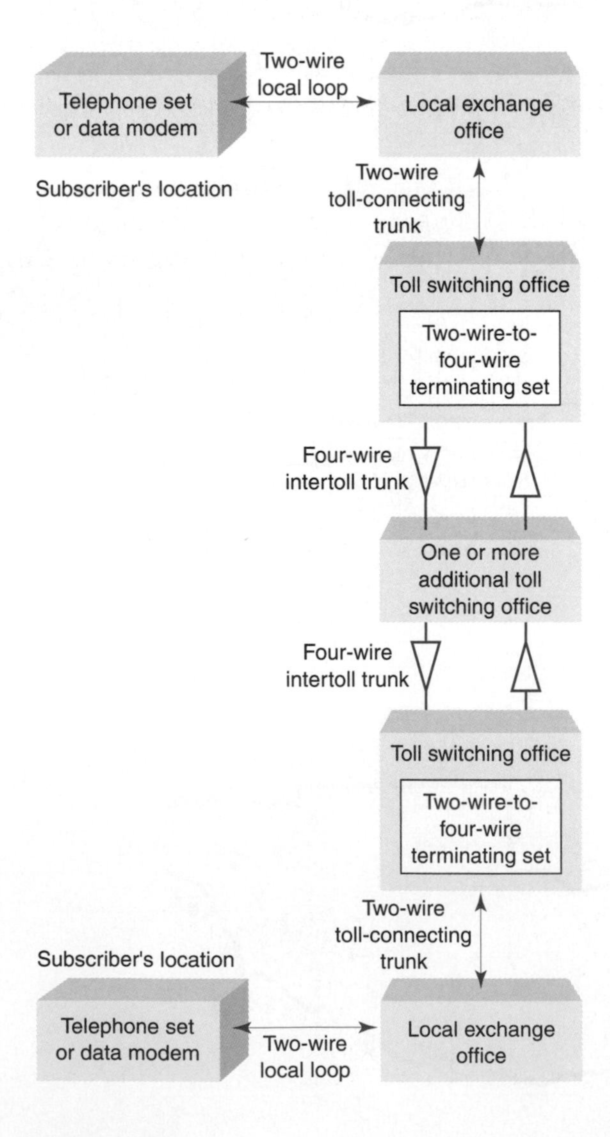

FIGURE 18-12 Long-distance telephone connection

telephone connection never requires more than two local exchange offices; however, there may be several toll-switching offices required, depending on the location of the originating and destination stations.

18-10 NORTH AMERICAN TELEPHONE SWITCHING HIERARCHY

With the advent of automated switching centers, a hierarchy of switching exchanges evolved in North America to accommodate the rapid increase in demand for long-distance calling. Thus, telephone company switching plans include a *switching hierarchy* that allows a certain degree of route selection when establishing a telephone call. A *route* is simply a path between two subscribers and is comprised of one or more switches, two local loops, and possibly one or more trunk circuits. The choice of routes is not offered to subscribers.

Telephone company switches, using software translation, select the best route available at the time a call is placed. The best route is not necessarily the shortest route. The best route is most likely the route requiring the fewest number of switches and trunk circuits. If a call cannot be completed because the necessary trunk circuits or switching paths are not available, the calling party receives an equipment (fast) busy signal. This is called *blocking.* Based on telephone company statistics, the likelihood that a call be blocked is approximately 1 in 100,000. Because software translations in automatic switching machines permit the use of alternate routes and each route may include several trunk circuits, the probability of using the same facilities on identical calls is unlikely. This is an obvious disadvantage of using the PSTN for data transmission because inconsistencies in transmission parameters occur from call to call.

18-10-1 Classes of Switching Offices

Before the divestiture of AT&T in 1984, the Bell System North American Switching Hierarchy consisted of five ranks or classes of switching centers as shown in Figure 18-13. The

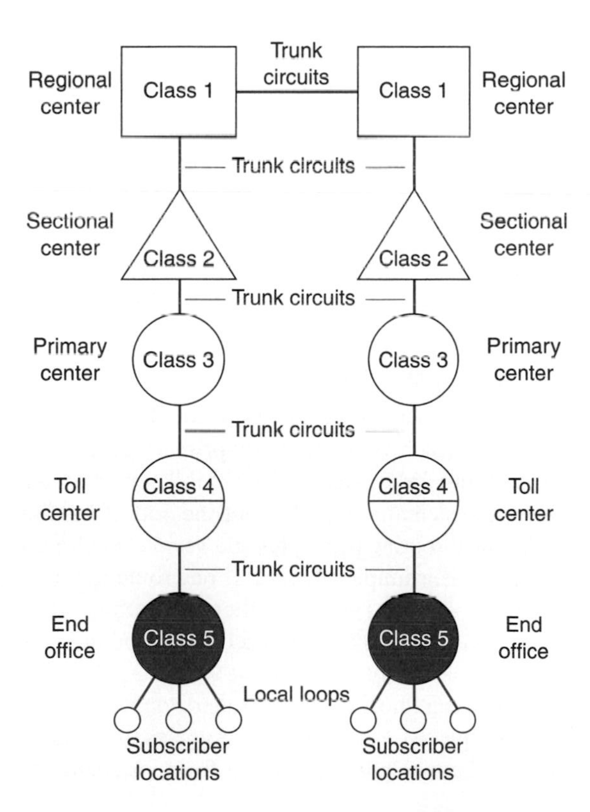

FIGURE 18-13 AT&T switching hierarchy prior to the 1984 divestiture

highest-ranking office was the regional center, and the lowest-ranking office was the end office. The five classifications of switching offices were as follows.

18-10-1-1 Class 5 end office. A class 5 office is a local exchange where subscriber loops terminated and received dial tone. End offices interconnected subscriber loops to other subscriber loops and subscriber loops to tandem trunks, interoffice trunks, and toll-connecting trunks. Subscribers received unlimited local call service in return for payment of a fixed charge each month, usually referred to as a *flat rate.* Some class 5 offices were classified as class 4/5. This type of office was called an *access tandem office,* as it was located in rural, low-volume areas and served as a dedicated class 5 office for local subscribers and also performed some of the functions of a class 4 toll office for long-distance calls.

18-10-1-2 Class 4 toll center. There were two types of class 4 offices. The class 4C toll centers provided human operators for both outward and inward calling service. Class 4P offices usually had only outward operator service or perhaps no operator service at all. Examples of operator-assisted services are person-to-person calls, collect calls, and credit card calls. Class 4 offices concentrated traffic in one switching center to direct outward traffic to the proper end office. Class 4 offices also provided centralized billing, provided toll customers with operator assistance, processed toll and intertoll traffic through its switching system, and converted signals from one trunk to another.

Class 3, 2, and 1 offices were responsible for switching intertoll-type calls efficiently and economically; to concentrate, collect, and distribute intertoll traffic; and to interconnect intertoll calls to all points of the direct distance dialing (DDD) network.

18-10-1-3 Class 3 primary center. This office provided service to small groups of class 4 offices within a small area of a state. Class 3 offices provided no operator assistance; however, they could serve the same switching functions as class 4 offices. A class 3 office generally had direct trunks to either a sectional or regional center.

18-10-1-4 Class 2 sectional center. Sectional centers could provide service to geographical regions varying in size from part of a state to all of several states, depending on population density. No operator services were provided; however, a class 2 office could serve the same switching functions as class 3 and class 4 offices.

18-10-1-5 Class 1 regional center. Regional centers were the highest-ranking office in the DDD network in terms of the size of the geographical area served and the trunking options available. Ten regional centers were located in the United States and two in Canada. Class 1 offices provided no operator services; however, they could serve the same switching functions as class 2, 3, or 4 offices. Class 1 offices had direct trunks to all the other regional centers.

18-10-2 Switching Routes

Regional centers served a large area called a *region.* Each region was subdivided into smaller areas called *sections,* which were served by primary centers. All remaining switching centers that did not fall into these categories were toll centers or end offices. The switching hierarchy provided a systematic and efficient method of handling long-distance telephone calls using hierarchical routing principles and various methods of automatic *alternate routing.* Alternate routing is a simple concept: If one route (path) is not available, select an alternate route that is available. Therefore, alternate routing often caused many toll offices to be interconnected in tandem to complete a call. When alternate routing is used, the actual path a telephone call takes may not resemble what the subscriber actually dialed. For example, a call placed between Phoenix, Arizona, and San Diego, California, may be routed from Phoenix to Albuquerque to Las Vegas to Los Angeles to San Diego. Common switching control equipment may have to add to, subtract from, or change the dialed information when

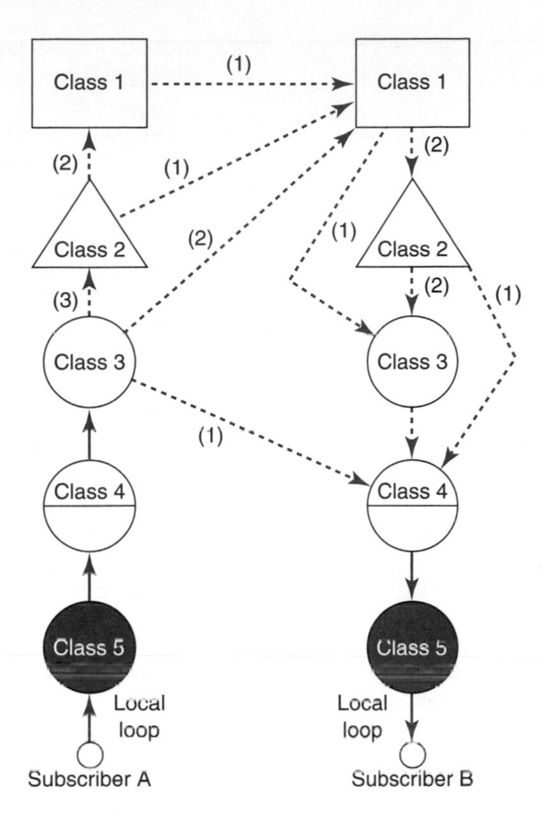

FIGURE 18-14 Choices of switching routes

routing a call to its destination. For example, an exchange office may have to add a prefix to a call with one, two, or three routing digits just to advance the call through an alternate route.

The five-class switching hierarchy is a *progressive switching scheme* that establishes an end-to-end route mainly through trial and error. Progressive switching is slow and unreliable by today's standards, as signaling messages are transported over the same facilities as subscriber's conversations using analog signals, such as multifrequency (MF) tones. Figure 18-14 shows examples of several choices for routes between subscriber A and subscriber B. For this example, there are 10 routes to choose from, of which only one requires the maximum of seven *intermediate links*. Intermediate links are toll trunks in tandem, excluding the two terminating links at the ends of the connection. In Figure 18-14, the first-choice route requires two intermediate links. Intermediate links are not always required, as in many cases a single *direct link,* which would be the first choice, exists between the originating and destination toll centers.

For the telephone office layout shown in Figure 18-15, the simplest connection would be a call between subscribers 1 and 2 in city A who are connected to the same end office. In this case, no trunk circuits are required. An interoffice call between stations 1 and 3 in city A would require using two tandem trunk circuits with an interconnection made in a tandem office. Consider a call originating from subscriber 1 in city A intended for subscriber 4 in city B. The route begins with subscriber 1 connected to end office 1 through a local loop. From the end office, the route uses a toll-connecting trunk to the toll center in city A. Between city A and city B, there are several route choices available. Because there is a high community of interest between the two cities, there is a direct intertoll trunk between City A and City B, which would be the first choice. However, there is an alternate route between city A and city B through the primary center in city C, which would probably be the second choice. From the primary center, there is a direct, high-usage intertoll trunk to both city A

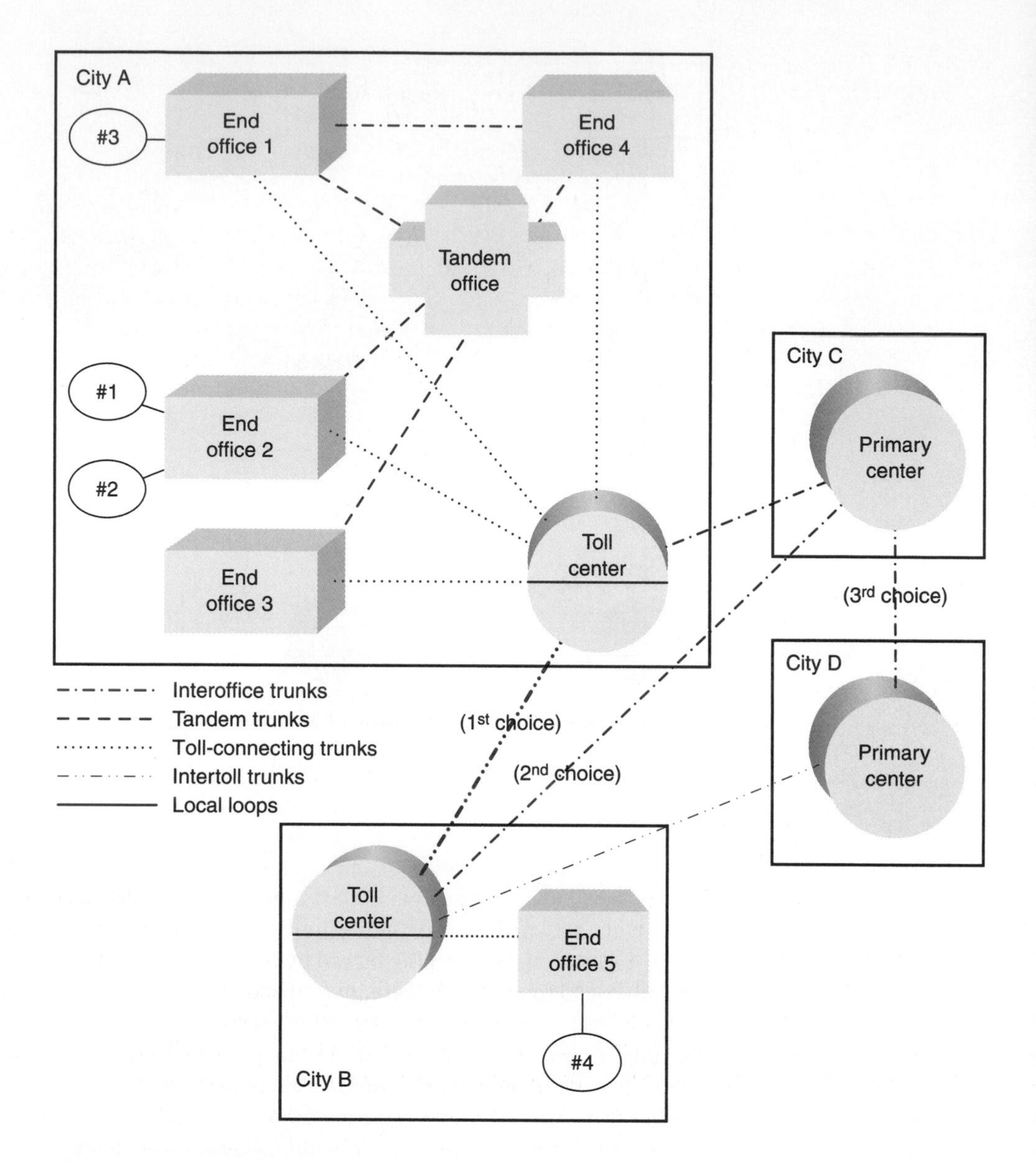

FIGURE 18-15 Typical switching routes

and city B (possibly the third choice), or, as a last resort, the toll centers in city A and city B could be interconnected using the primary centers in city C and city D (fourth choice).

The probability that a telephone call would require more than n links in tandem to reach the final destination decreases rapidly as n increases from 2 to 7. This is primarily because a large majority of long-distance toll calls are made between end offices associated with the same regional switching center, which of course would require fewer than seven toll trunks. Although the maximum number of trunks is seven, the average number for a typical toll call is only three. In addition, even when a telephone call was between telephones associated with different regional centers, the call was routed over the maximum of seven intermediate trunks only when all the normally available high-usage trunks are busy. The probability of this happening is only ρ^5, where ρ is the probability that all trunks in any one high-usage group are busy. Finally, many calls do not originate all the way down the hierarchy since each higher class of office will usually have class 5 offices homing on it that will act as class 4 offices for them.

18-11 COMMON CHANNEL SIGNALING SYSTEM NO. 7 (SS7) AND THE POSTDIVESTITURE NORTH AMERICAN SWITCHING HIERARCHY

Common Channel Signaling System No. 7 (i.e., SS7 or C7) is a global standard for telecommunications defined by the International Telecommunications Union (ITU) Telecommunications Sector (ITU-T). SS7 was developed as an alternate and much improved means of transporting signaling information through the public telephone network. The SS7 standard defines the procedures and protocol necessary to exchange information over the PSTN using a separate digital signaling network to provide wireless (cellular) and wireline telephone call setup, routing, and control. SS7 determines the switching path before any switches are actually enabled, which is a much faster and more reliable switching method than the old five-class progressive switching scheme. The SS7 signaling network performs its functions by exchanging telephone control messages between the SS7 components that support completing the subscribers' connection.

The functions of the SS7 network and protocol are as follows:

1. Basic call setup, management, and tear-down procedures
2. Wireless services, such as personal communications services (PCS), wireless roaming, and mobile subscriber authentication
3. Local number portability (LNP)
4. Toll-free (800/888) and toll (900) wire-line services
5. Enhanced call features, such as call forwarding, calling party name/number display, and three-way calling
6. Efficient and secure worldwide telecommunications service

18-11-1 Evolution of SS7

When telephone networks and network switching hierarchies were first engineered, their creators gave little thought about future technological advancements. Early telephone systems were based on transferring analog voice signals using analog equipment over analog transmission media. As a result, early telephone systems were not well suited for modern-day digital services, such as data, digitized voice, or digitized video transmission. Therefore, when digital services were first offered in the early 1960s, the telephone networks were ill prepared to handle them, and the need for an intelligent all-digital network rapidly became evident.

The ITU commissioned the Comiteé Consultatif International Téléphonique et Télégraphique (CCITT) to study the possibility of developing an intelligent all-digital telecommunications network. In the mid-1960s, the ITU-TS (International Telecommunications Union Telecommunications—Standardization Sector) developed a digital signaling standard known as *Signaling System No. 6* (SS6) that modernized the telephone industry. *Signaling* refers to the exchange of information between call components required to provide and maintain service. SS6, based on a proprietary, high-speed data communications network, evolved into *Signaling System No. 7* (SS7), which is now the telephone industry standard for most of the civilized world ("civilized world" because it was estimated that in 2002, more than half the people in the world had never used a telephone). High-speed packet data and out-of-band signaling characterize SS7. Out-of-band signaling is signaling that does not take place over the same path as the conversation. Out-of-band signaling establishes a separate digital channel for exchanging signaling information. This channel is called a *signaling link.*

The protocol used with SS7 uses a message structure, similar to X.25 and other message-based protocols, to request services from other networks. The messages propagate from one network to another in small bundles of data called *packets* that are independent of the subscriber voice or data signals they pertain to. In the early 1960s, the ITU-TS developed a *common channel signaling* (CCS) known as *Common Channel Interoffice Signaling System No. 6* (SS6). The basic concept of the common channel signaling is to use a facility (separate from the voice facilities) for transferring control and signaling information between telephone offices.

When first deployed in the United States, SS6 used a packet switching network with 2.4-kbps data links, which were later upgraded to 4.8 kbps. Signaling messages were sent as part of a data packet and used to request connections on voice trunks between switching offices. SS6 was the first system to use packet switching in the PSTN. Packets consisted of a block of data comprised of 12 signal units of 28 bits each, which is similar to the method used today with SS7.

SS7 is an architecture for performing out-of-band signaling in support of common telephone system functions, such as call establishment, billing, call routing, and information exchange functions of the PSTN. SS7 identifies functions and enables protocols performed by a telephone signaling network. The major advantages of SS7 include better monitoring, maintenance, and network administration. The major disadvantage is its complex coding.

Because SS7 evolved from SS6, there are many similarities between the two systems. SS7 uses variable-length signal units with a maximum length, therefore making it more versatile and flexible than SS6. In addition, SS7 uses 56-kbps data links (64 kbps for international links), which provide a much faster and efficient signaling network. In the future, data rates of 1.544 Mbps nationally and 2.048 Mbps internationally are expected.

In 1983 (just prior to the AT&T divestiture), SS6 was still widely used in the United States. When SS7 came into use in the mid-1980s, SS6 began to be phased out of the system, although SS6 was still used in local switching offices for several more years. SS7 was originally used for accessing remote databases rather than for call setup and termination. In the 1980s, AT&T began offering Wide Area Telephone Service (WATS), which uses a common 800 area code regardless of the location of the destination. Because of the common area code, telephone switching systems had a problem dealing with WATS numbers. This is because telephone switches used the area code to route a call through the public switched network. The solution involved adding a second number to every 800 number that is used by the switching equipment to actually route a call through the voice network. The second number is placed in a common, centralized database accessible to all central offices. When an 800 number is called, switching equipment uses a data link to access the database and retrieve the actual routing number. This process is, of course, transparent to the user. Once the routing number is known, the switching equipment can route the call using standard signaling methods.

Shortly after implementing the WATS network, the SS7 network was expanded to provide other services, such as call setup and termination. However, the database concept has proven to be the biggest advantage of SS7, as it can also be used to provide routing and billing information for all telephone services, including 800 and 900 numbers, 911 services, custom calling features, caller identifications, and a host of other services not yet invented.

In 1996, the FCC mandated *local number portability* (LNP), which requires all telephone companies to support the *porting* of a telephone number. Porting allows customers to change to a different service and still keep the same telephone number. For example, a subscriber may wish to change from *plain old telephone service* (POTS) to ISDN, which would have required changing telephone numbers. With LNP, the telephone number would remain the same because the SS7 database can be used to determine which network switch is assigned to a particular telephone number.

Today, SS7 is being used throughout the Bell Operating Companies telephone network and most of the independent telephone companies. This in itself makes SS7 the world's largest data communications network, as it links wireline telephone companies, cellular telephone companies, and long-distance telephone companies together with a common signaling system. Because SS7 has the ability to transfer all types of digital information, it supports most of the new telephone features and applications and is used with ATM, ISDN, and cellular telephone.

18-11-2 Postdivestiture North American Switching Hierarchy

Today, the North American telephone system is divided into two distinct functional areas: signaling and switching. The signaling network for the telephone system is SS7, which is used to determine how subscriber's voice and data signals are routed through the network. The switching network is the portion of the telephone network that actually transports the voice and data from one subscriber to another. The signaling part of the network establishes and disconnects the circuits that actually carry the subscriber's information.

After the divestiture of AT&T, technological advances allowed many of the functions distributed among the five classes of telephone offices to be combined. In addition, switching equipment was improved, giving them the capability to act as local switches, tandem switches, or toll switches. The new North American Switching Hierarchy consolidated many of the functions of the old hierarchy into two layers, with many of the functions once performed by the higher layers being located in the end office. Therefore, the postdivestiture telephone network can no longer be described as a hierarchy of five classes of offices. It is now seen as a system involving two decision points. The postdivestiture North American Switching Hierarchy is shown in Figure 18-16. Long-distance access is now accomplished through an access point called the *point-of-presence* (POP). The term *point-of-presence* is a telecommunications term that describes the legal boundaries for the responsibility of maintaining equipment and transmission lines. In essence, it is a demarcation point separating two companies.

After the divestiture of AT&T, calling areas were redefined and changed to *Local Access and Transport Areas* (LATAs) with each LATA having its own three-level hierarchy. Although the United States was originally divided into only 160 local access and transport areas, there are presently over 300 LATA dispersed throughout the United States. Within these areas, local telephone companies provide the facilities and equipment to interconnect

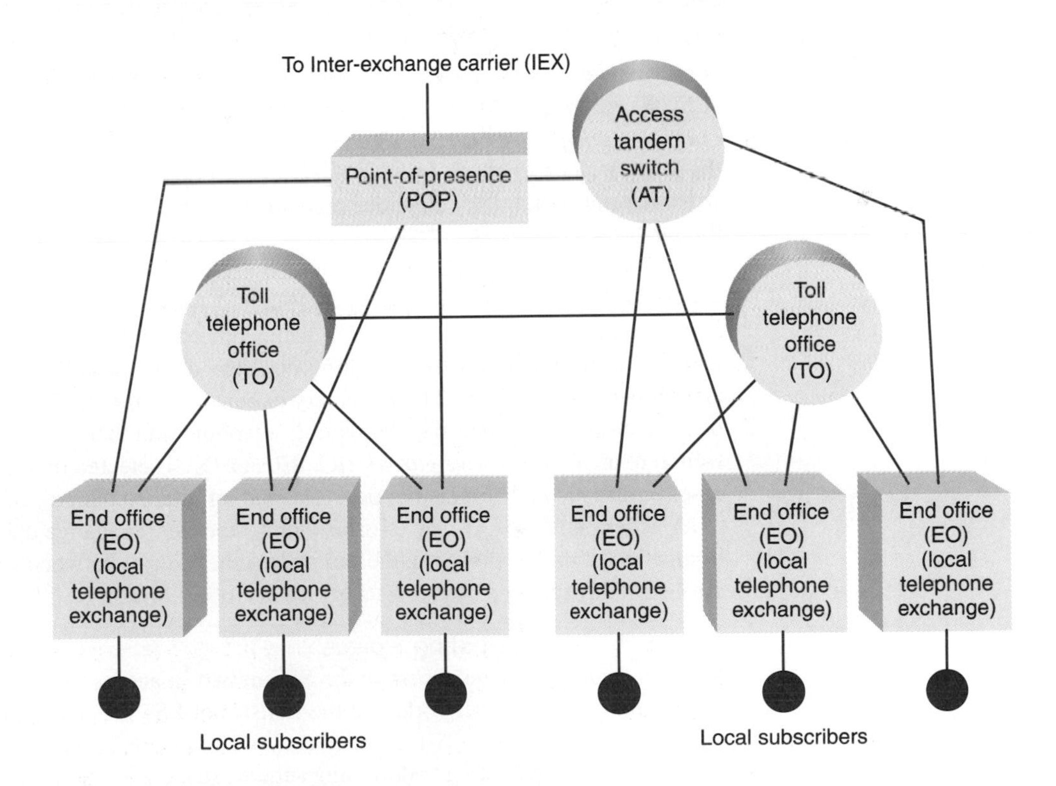

FIGURE 18-16 Postdivestiture North American Switching Hierarchy

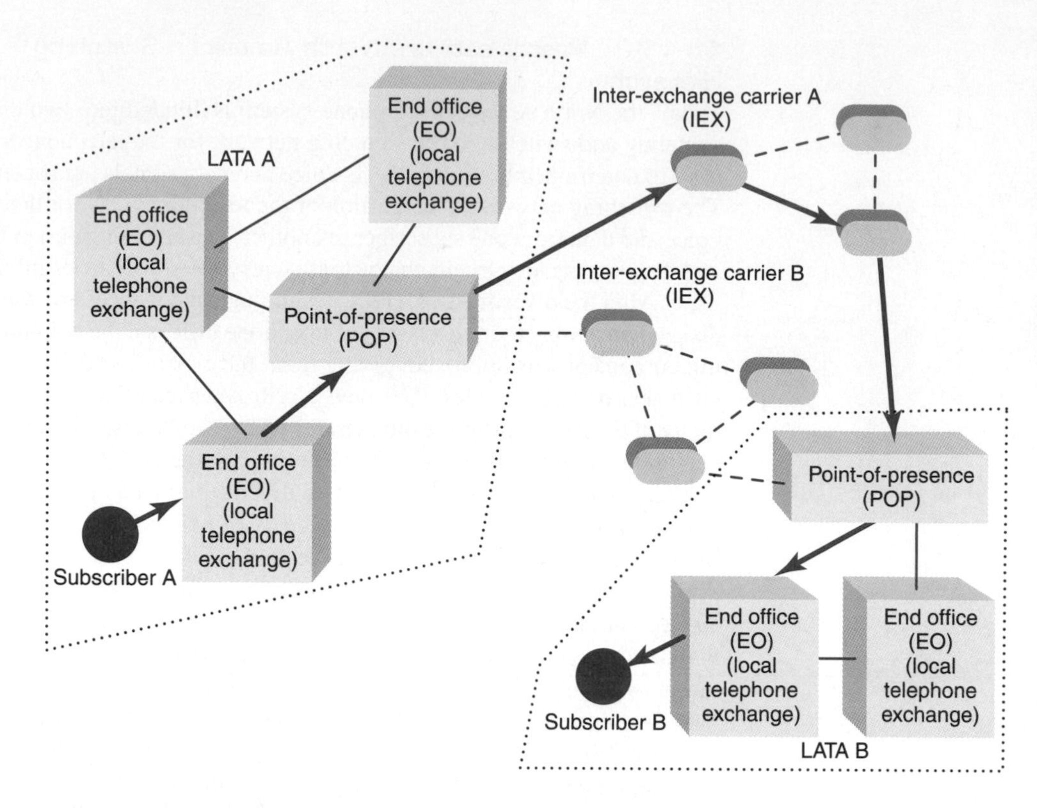

FIGURE 18-17 Example of an interexchange call between subscriber A in LATA A to subscriber B in LATA B

subscribers within the LATA. The telephone companies are called *local exchange carriers* (LECs), *exchange carriers* (ECs), *operating telephone companies* (OTCs), and *telephone operating companies* (TOCs). The areas serviced by local exchanges were redistributed by the Justice Department to provide telephone companies a more evenly divided area with equal revenue potential. Telephone calls made within a LATA are considered a function of the *intra-LATA network.*

Telephone companies further divided each LATA into a *local market* and a *toll market.* The toll market for a company is within its LATA but is still considered a long-distance call because it involves a substantial distance between the two local offices handling the call. These are essentially the only long-distance telephone calls local operating companies are allowed to provide, and they are very expensive. If the destination telephone number is in a different LATA than the originating telephone number, the operating company must switch to an *interexchange carrier* (IC, IEC, or IXC) selected by the calling party. In many cases, a direct connection is not possible, and an interexchange call must be switched first to an access tandem (AT) switch and then to the interexchange carrier point-of-presence. Figure 18-17 shows an example of an interexchange call between subscriber A in LATA A through interexchange carrier A to subscriber B in LATA B.

18-11-3 SS7 Signaling Points

Signaling points provide access to the SS7 network, access to databases used by switches inside and outside the network, and the transfer of SS7 messages to other signaling points within the network.

Every network has an addressing scheme to enable a node within the network to exchange signaling information with nodes it is not connected to by a physical link. Each node

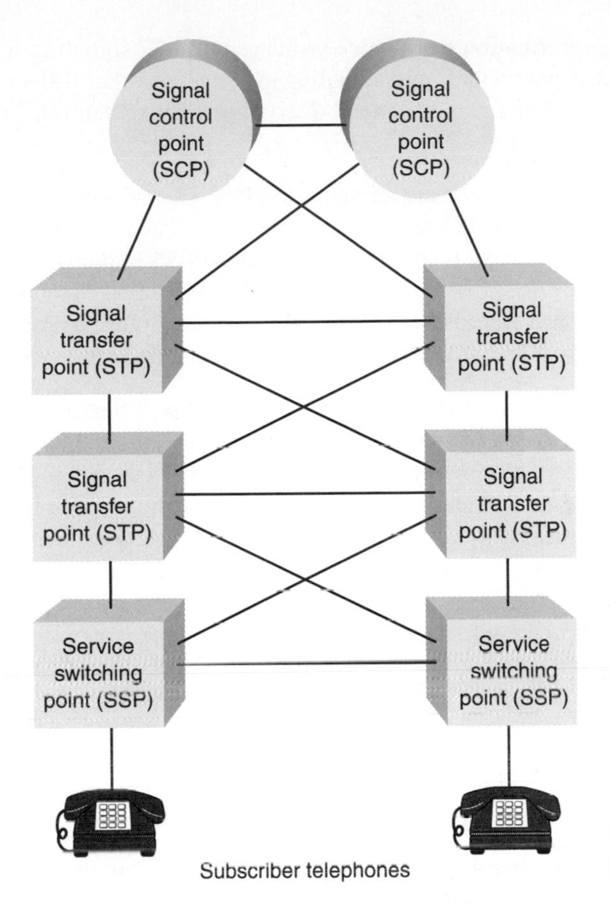

FIGURE 18-18 SS7 signaling point topology

is uniquely identified by a numeric *point code.* Point codes are carried in signaling messages exchanged between signaling points to identify the source and destination of each message (i.e., an *originating point code* and a *destination point code*). Each signaling point is identified as a *member* part of a cluster of signaling points. Similarly, a cluster is defined as being part of a complete network. Therefore, every node in the American SS7 network can be addressed with a three-level code that is defined by its network, cluster, and member numbers. Each number is an eight-bit binary number between 0 and 255. This three-level address is called the *point code.* A point code uniquely identifies a signaling point within the SS7 network and is used whenever it is addressed. A neutral party assigns network codes on a nationwide basis. Because there are a limited number of network numbers, networks must meet a certain size requirement to receive one. Smaller networks may be assigned one or more cluster numbers within network numbers 1, 2, 3, and 4. The smallest networks are assigned point codes within network number 5. The cluster they are assigned to is determined by the state where they are located. Network number 0 is not available for assignment, and network number 255 is reserved for future use.

The three types of signaling points are listed here, and a typical SS7 topology is shown in Figure 18-18.

18-11-3-1 Service switching points (SSPs). Service switching points (sometimes called *signal switching points*) are local telephone switches (in either end or tandem offices) equipped with SS7-compatible software and terminating signal links. The SSP provides the functionality of communicating with the voice switch by creating the packets or signal units necessary for transmission over the SS7 network. An SSP must convert signaling information from voice switches into SS7 signaling format. SSPs are basically local access points that send signaling messages to other SSPs to originate, terminate, or switch calls. SSPs may also send query messages to centralized databases to determine how to route a call.

18-11-3-2 Signal transfer points (STPs). Signal transfer points are the packet switches of the SS7 network. STPs serve as routers in the SS7 network, as they receive and route incoming signaling messages to the proper destination. STPs seldom originate a message. STPs route each incoming message to an outgoing signaling link based on routing information contained in the SS7 message. Because an STP acts like a network router, it provides improved utilization of the SS7 network by eliminating the need for direct links between all signaling points.

18-11-3-3 Service control points (SCPs). Service control points (sometimes called *signal control points*) serve as an interface to telephone company databases. The databases store information about subscriber's services, routing of special service numbers (such as 800 and 900 numbers), and calling card validation for fraud protection and provide information necessary for advanced call-processing capabilities. SCPs also perform protocol conversion from SS7 to X.25, or they can provide the capability of communicating with the database directly using an interface called a *primitive,* which provides access from one level of the protocol to another level. SCPs also send responses to SSPs containing a routing number(s) associated with the called number.

SSPs, STPs, and SCPs are interconnected with digital carriers, such as T1 or DS-0 links, which carry the signaling messages between the SS7 network devices.

18-11-4 SS7 Call Setup Example

A typical call setup procedure using the SS7 signaling network is as follows:

1. Subscriber A goes off hook and touch tones out the destination telephone number of subscriber B.
2. The local telephone translates the tones to binary digits.
3. The local telephone exchange compares the digits to numbers stored in a routing table to determine whether subscriber B resides in the same local switch as subscriber A. If not, the call must be transferred onto an outgoing trunk circuit to another local exchange.
4. After the switch determines that subscriber B is served by a different local exchange, an SS7 message is sent onto the SS7 network. The purposes of the message are as follows:
   - **i.** To find out if the destination number is idle.
   - **ii.** If the destination number is idle, the SS7 network makes sure a connection between the two telephone numbers is also available.
   - **iii.** The SS7 network instructs the destination switch to ring subscriber B.
5. When subscriber B answers the telephone, the switching path is completed.
6. When either subscriber A or subscriber B terminates the call by hanging up, the SS7 network releases the switching path, making the trunk circuits and switching paths available to other subscribers of the network.

QUESTIONS

18-1. What are the purposes of telephone network *signaling functions?*

18-2. What are the two types of subscribers to the public telephone network? Briefly describe them.

18-3. What is the difference between *dedicated* and *switched* facilities?

18-4. Describe the term *service provider.*

18-5. Briefly describe the following terms: *instruments, local loops, trunk circuits,* and *exchanges.*

18-6. What is a *local office telephone exchange?*

18-7. What is an *automated central office switch?*

18-8. Briefly describe the following terms: *circuits, circuit switches,* and *circuit switching.*

18-9. What is the difference between a *local telephone exchange* and an *exchange area?*

18-10. Briefly describe *interoffice trunks, tandem trunks,* and *tandem switches.*

18-11. Briefly describe *toll-connecting trunks, intertoll trunks,* and *toll offices.*

18-12. Briefly describe the *North American Telephone Numbering Plan.*

18-13. What is the difference between an *area code,* a *prefix,* and an *extension?*

18-14. What is meant by the term *common usage?*

18-15. What does *blocking* mean? When does it occur?

18-16. Briefly describe the *predivestiture North American Telephone Switching Hierarchy.*

18-17. Briefly describe the five *classes* of the predivestiture North American Switching Hierarchy.

18-18. What is meant by the term *switching route?*

18-19. What is meant by the term *progressive switching scheme?*

18-20. What is *SS7?*

18-21. What is *common channel signaling?*

18-22. What is meant by the term *local number portability?*

18-23. What is meant by the term *plain old telephone service?*

18-24. Briefly describe the *postdivestiture North American Switching Hierarchy.*

18-25. What is a LATA?

18-26. What is meant by the term *point-of-presence?*

18-27. Describe what is meant by the term *local exchange carrier.*

18-28. Briefly describe what is meant by SS7 *signaling points.*

18-29. List and describe the three SS7 signaling points.

18-30. What is meant by the term *point code?*

CHAPTER 19

Cellular Telephone Concepts

CHAPTER OUTLINE

OBJECTIVES

- Give a brief history of mobile telephone service
- Define *cellular telephone*
- Define *cell* and explain why it has a honeycomb shape
- Describe the following types of cells: macrocell, microcell, and minicell
- Describe edge-excited, center-excited, and corner-excited cells
- Define *service areas, clusters,* and *cells*
- Define *frequency reuse*
- Explain frequency reuse factor
- Define *interference*
- Describe co-channel and adjacent channel interference
- Describe the processes of cell splitting, sectoring, segmentation, and dualization
- Explain the differences between cell-site controllers and mobile telephone switching offices
- Define *base stations*
- Define and explain roaming and handoffs
- Briefly describe the purpose of the IS-41 protocol standard
- Define and describe the following cellular telephone network components: electronic switching center, cell-site controller, system interconnects, mobile and portable telephone units, and communications protocols
- Describe the cellular call procedures involved in making the following types of calls: mobile to wireline, mobile to mobile, and wireline to mobile

19-1 INTRODUCTION

The basic concepts of *two-way mobile telephone* are quite simple; however, mobile telephone systems involve intricate and rather complex communications networks comprised of analog and digital communications methodologies, sophisticated computer-controlled switching centers, and involved protocols and procedures. Cellular telephone evolved from two-way mobile FM radio. The purpose of this chapter is to present the fundamental concepts of cellular telephone service. Cellular services include standard *cellular telephone service* (CTS), *personal communications systems* (PCS), and *personal communications satellite systems* (PCSS).

19-2 MOBILE TELEPHONE SERVICE

Mobile telephone services began in the 1940s and were called MTSs (*mobile telephone systems* or sometimes *manual telephone systems,* as all calls were handled by an operator). MTS systems utilized frequency modulation and were generally assigned a single carrier frequency in the 35-MHz to 45-MHz range that was used by both the mobile unit and the base station. The mobile unit used a push-to-talk (PTT) switch to activate the transceiver. Depressing the PTT button turned the transmitter on and the receiver off, whereas releasing the PTT turned the receiver on and the transmitter off. Placing a call from a MTS mobile telephone was similar to making a call through a manual switchboard in the public telephone network. When the PTT switch was depressed, the transmitter turned on and sent a carrier frequency to the base station, illuminating a lamp on a switchboard. An operator answered the call by plugging a headset into a jack on the switchboard. After the calling party verbally told the operator the telephone number they wished to call, the operator connected the mobile unit with a patchcord to a trunk circuit connected to the appropriate public telephone network destination office. Because there was only one carrier frequency, the conversation was limited to half-duplex operation, and only one conversation could take place at a time. The MTS system was comparable to a party line, as all subscribers with their mobile telephones turned on could hear any conversation. Mobile units called other mobile units by signaling the operator who rang the destination mobile unit. Once the destination mobile unit answered, the operator disconnected from the conversation, and the two mobile units communicated directly with one another through the airways using a single carrier frequency.

MTS mobile identification numbers had no relationship to the telephone numbering system used by the public telephone network. Local telephone companies in each state, which were generally Bell System Operating Companies, kept a record of the numbers assigned to MTS subscribers in that state. MTS numbers were generally five digits long and could not be accessed directly through the public switched telephone network (PSTN).

In 1964, the Improved Mobile Telephone System (IMTS) was introduced, which used several carrier frequencies and could, therefore, handle several simultaneous mobile conversations at the same time. IMTS subscribers were assigned a regular PSTN telephone number; therefore, callers could reach an IMTS mobile phone by dialing the PSTN directly, eliminating the need for an operator. IMTS and MTS base station transmitters outputted powers in the 100-W to 200-W range, and mobile units transmitted between 5 W and 25 W. Therefore, IMTS and MTS mobile telephone systems typically covered a wide area using only one base station transmitter.

Because of their high cost, limited availability, and narrow frequency allocation, early mobile telephone systems were not widely used. However, in recent years, factors such as technological advancements, wider frequency spectrum, increased availability, and improved reliability have stimulated a phenomenal increase in people's desire to talk on the telephone from virtually anywhere, at any time, regardless of whether it is necessary, safe, or productive.

Today, mobile telephone stations are small handsets, easily carried by a person in their pocket or purse. In early radio terminology, the term *mobile* suggested any radio transmitter, receiver, or transceiver that could be moved while in operation. The term *portable* described a relatively small radio unit that was handheld, battery powered, and easily carried by a person moving at walking speed. The contemporary definition of mobile has come to mean moving at high speed, such as in a boat, airplane, or automobile, or at low speed, such as in the pocket of a pedestrian. Hence, the modern, all-inclusive definition of mobile telephone is any wireless telephone capable of operating while moving at any speed, battery powered, and small enough to be easily carried by a person.

Cellular telephone is similar to two-way mobile radio in that most communications occurs between base stations and mobile units. Base stations are fixed-position transceivers with relatively high-power transmitters and sensitive receivers. Cellular telephones communicate directly with base stations. Cellular telephone is best described by pointing out the primary difference between it and two-way mobile radio. Two-way mobile radio systems operate half-duplex and use PTT transceivers. With PTT transceivers, depressing the PTT button turns on the transmitter and turns off the receiver, whereas releasing the PTT button turns on the receiver and turns off the transmitter. With two-way mobile telephone, all transmissions (unless scrambled) can be heard by any listener with a receiver tuned to that channel. Hence, two-way mobile radio is a *one-to-many* radio communications system. Examples of two-way mobile radio are *citizens band* (CB), which is an AM system, and *public land mobile radio,* which is a two-way FM system such as those used by police and fire departments. Most two-way mobile radio systems can access the public telephone network only through a special arrangement called an *autopatch,* and then they are limited to half-duplex operation where neither party can interrupt the other. Another limitation of two-way mobile radio is that transmissions are limited to relatively small geographic areas unless they utilize complicated and expensive repeater networks.

On the other hand, cellular telephone offers full-duplex transmissions and operates much the same way as the standard wireline telephone service provided to homes and businesses by local telephone companies. Mobile telephone is a *one-to-one* system that permits two-way simultaneous transmissions and, for privacy, each cellular telephone is assigned a unique telephone number. Coded transmissions from base stations activate only the intended receiver. With mobile telephone, a person can virtually call anyone with a telephone number, whether it be through a cellular or a wireline service.

Cellular telephone systems offer a relatively high user capacity within a limited frequency spectrum providing a significant innovation in solving inherent mobile telephone communications problems, such as spectral congestion and user capacity. Cellular telephone systems replaced mobile systems serving large areas (cells) operating with a single base station and a single high-power transmitter with many smaller areas (cells), each with its own base station and low-power transmitter. Each base station is allocated a fraction of the total channels available to the system, and adjacent cells are assigned different groups of channels to minimize interference between cells. When demand for service increases in a given area, the number of base stations can be increased, providing an increase in mobile-unit capacity without increasing the radio-frequency spectrum.

19-3 EVOLUTION OF CELLULAR TELEPHONE

In the July 28, 1945, *Saturday Evening Post,* E. K. Jett, then the commissioner of the FCC, hinted of a cellular telephone scheme that he referred to as simply a *small-zone* radio-telephone system. On June 17, 1946, in St. Louis, Missouri, AT&T and Southwestern Bell introduced the first American commercial mobile radio-telephone service to private customers. In the same year, similar services were offered to 25 major cities throughout the United States. Each city utilized one base station consisting of a high-powered transmitter and a sensitive receiver that were centrally located on a hilltop or tower that covered an area within a 30- to 50-mile radius of the base station. In 1947, AT&T introduced a radio-telephone service they called *highway service* between New York and Boston. The system operated in the 35-MHz to 45-MHz band.

The first half-duplex, PTT FM mobile telephone systems introduced in the 1940s operated in the 35-MHz to 45-MHz band and required 120-kHz bandwidth per channel. In the early 1950s, the FCC doubled the number of mobile telephone channels by reducing the bandwidth to 60 kHz per channel. In 1960, AT&T introduced direct-dialing, full-duplex mobile telephone service with other performance enhancements, and in 1968, AT&T proposed the concept of a cellular mobile system to the FCC with the intent of alleviating the problem of spectrum congestion in the existing mobile telephone systems. Cellular mobile telephone systems, such as the *Improved Mobile Telephone System* (IMTS), were developed, and recently developed miniature integrated circuits enabled management of the necessarily complex algorithms needed to control network switching and control operations. Channel bandwidth was again halved to 30 kHz, increasing the number of mobile telephone channels by twofold.

In 1966, Don Adams, in a television show called *Get Smart,* unveiled the most famous mobile telephone to date: the fully mobile shoe phone. Some argue that the 1966 *Batphone supra* was even more remarkable, but it remained firmly anchored to the Batmobile, limiting Batman and Robin to vehicle-based telephone communications.

In 1974, the FCC allocated an additional 40-MHz bandwidth for cellular telephone service (825 MHz to 845 MHz and 870 MHz to 890 MHz). These frequency bands were previously allocated to UHF television channels 70 to 83. In 1975, the FCC granted AT&T the first license to operate a developmental cellular telephone service in Chicago. By 1976, the Bell Mobile Phone service for metropolitan New York City (approximately 10 million people) offered only 12 channels that could serve a maximum of 543 subscribers. In 1976, the FCC granted authorization to the American Radio Telephone Service (ARTS) to install a second developmental system in the Baltimore–Washington, D.C., area. In 1983, the FCC allocated 666 30-kHz half-duplex mobile telephone channels to AT&T to form the first U.S. cellular telephone system called Advanced Mobile Phone System (AMPS).

In 1991, the first digital cellular services were introduced in several major U.S. cities, enabling a more efficient utilization of the available bandwidth using voice compression. The calling capacity specified in the U.S. Digital Cellular (USDC) standard (EIA IS-54) accommodates three times the user capacity of AMPS, which used conventional frequency modulation (FM) and frequency-division multiple accessing (FDMA). The USDC standard specifies digital modulation, speech coding, and time-division multiple accessing (TDMA). Qualcomm developed the first cellular telephone system based on code-division multiple accessing (CDMA). The Telecommunications Industry Association (TIA) standardized Qualcomm's system as Interim Standard 95 (IS-95). On November 17, 1998, a subsidiary of Motorola Corporation implemented Iridium, a satellite-based wireless personal communications satellite system (PCSS).

19-4 CELLULAR TELEPHONE

The key principles of *cellular telephone* (sometimes called *cellular radio*) were uncovered in 1947 by researchers at Bell Telephone Laboratories and other telecommunications companies throughout the world when they developed the basic concepts and theory of cellular telephone. It was determined that by subdividing a relatively large geographic market area, called a *coverage zone,* into small sections, called *cells,* the concept of *frequency reuse* could be employed to dramatically increase the capacity of a mobile telephone channel. Frequency reuse is described in a later section of this chapter. In essence, cellular telephone systems allow a large number of users to share the limited number of *common-usage* radio

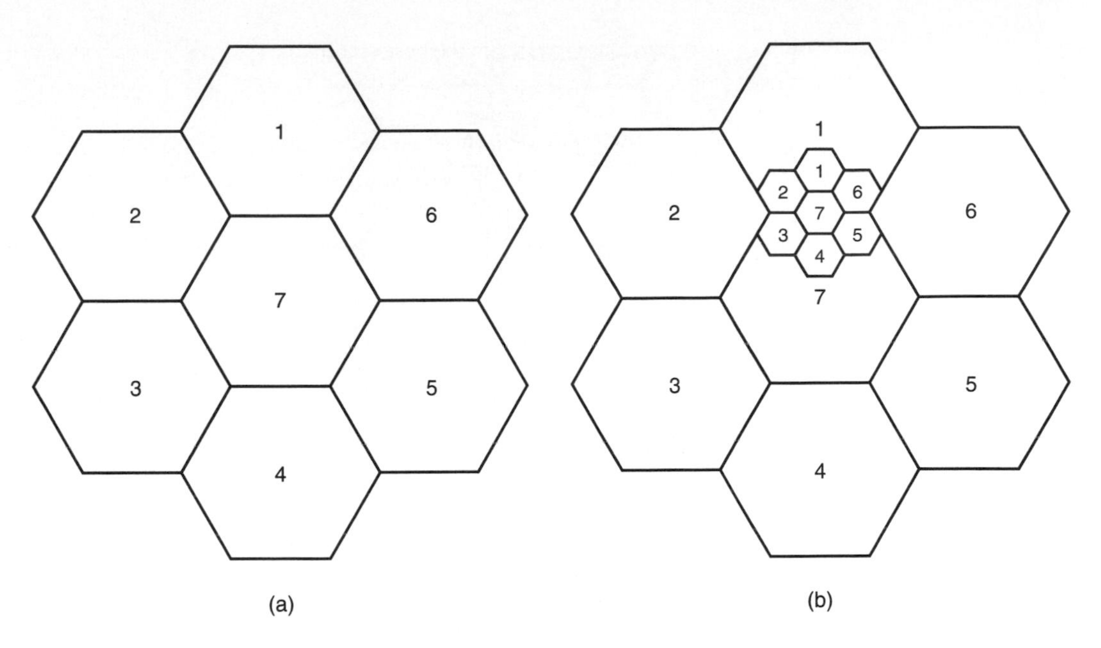

FIGURE 19-1 (a) Honeycomb cell pattern; (b) honeycomb pattern with two sizes of cells

channels available in a region. In addition, integrated-circuit technology, microprocessors and microcontroller chips, and the implementation of Signaling System No. 7 (SS7) have recently enabled complex radio and logic circuits to be used in electronic switching machines to store programs that provide faster and more efficient call processing.

19-4-1 Fundamental Concepts of Cellular Telephone

The fundamental concepts of cellular telephone are quite simple. The FCC originally defined geographic cellular radio coverage areas on the basis of modified 1980 census figures. With the cellular concept, each area is further divided into hexagonal-shaped cells that fit together to form a *honeycomb* pattern as shown in Figure 19-1a. The hexagon shape was chosen because it provides the most effective transmission by approximating a circular pattern while eliminating gaps inherently present between adjacent circles. A cell is defined by its physical size and, more importantly, by the size of its population and traffic patterns. The number of cells per system and the size of the cells are not specifically defined by the FCC and has been left to the providers to establish in accordance with anticipated traffic patterns. Each geographical area is allocated a fixed number of cellular voice channels. The physical size of a cell varies, depending on user density and calling patterns. For example, large cells (called *macrocells*) typically have a radius between 1 mile and 15 miles with base station transmit powers between 1 W and 6 W. The smallest cells (called *microcells*) typically have a radius of 1500 feet or less with base station transmit powers between 0.1 W and 1 W. Figure 19-1b shows a cellular configuration with two sizes of cell.

Microcells are used most often in high-density areas such as found in large cities and inside buildings. By virtue of their low effective working radius, microcells exhibit milder propagation impairments, such as reflections and signal delays. Macrocells may overlay clusters of microcells with slow-moving mobile units using the microcells and faster-moving units using the macrocells. The mobile unit is able to identify itself as either fast or slow moving, thus allowing it to do fewer cell transfers and location updates. Cell transfer algorithms can be modified to allow for the small distances between a mobile unit and the

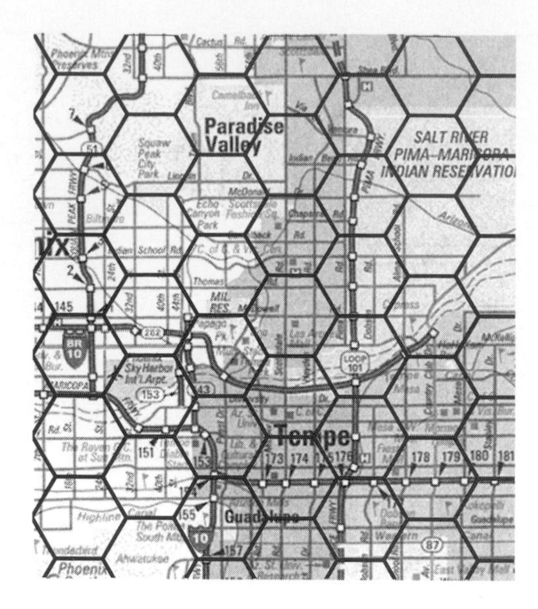

FIGURE 19-2 Hexagonal cell grid superimposed over a metropolitan area

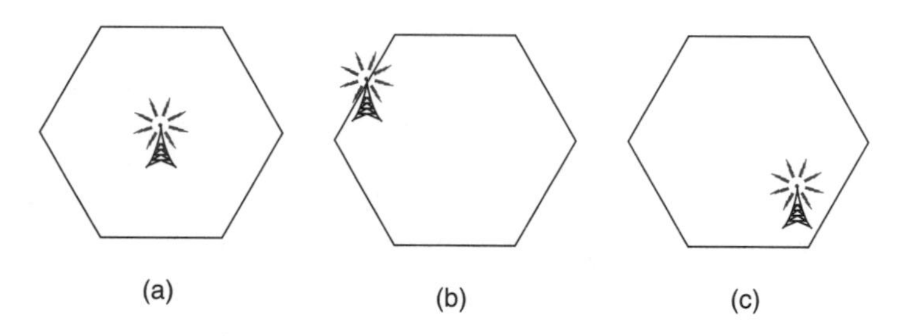

FIGURE 19-3 (a) Center excited cell; (b) edge excited cell; (c) corner excited cell

microcellular base station it is communicating with. Figure 19-2 shows what a hexagonal cell grid might look like when superimposed over a metropolitan area.

Occasionally, cellular radio signals are too weak to provide reliable communications indoors. This is especially true in well-shielded areas or areas with high levels of interference. In these circumstances, very small cells, called *picocells,* are used. Indoor picocells can use the same frequencies as regular cells in the same areas if the surrounding infrastructure is conducive, such as in underground malls.

When designing a system using hexagonal-shaped cells, base station transmitters can be located in the center of a cell (*center-excited cell* shown in Figure 19-3a), or on three of the cells' six vertices (*edge-* or *corner-excited cells* shown in Figures 19-3b and c). *Omnidirectional* antennas are normally used in center-excited cells, and sectored directional antennas are used in edge- and corner-excited cells (omnidirectional antennas radiate and receive signals equally well in all directions).

Cellular telephone is an intriguing mobile radio concept that calls for replacing a single, high-powered fixed base station located high above the center of a city with multiple, low-powered duplicates of the fixed infrastructure distributed over the coverage area on sites placed closer to the ground. The cellular concept adds a spatial dimension to the simple cable-trunking model found in typical wireline telephone systems.

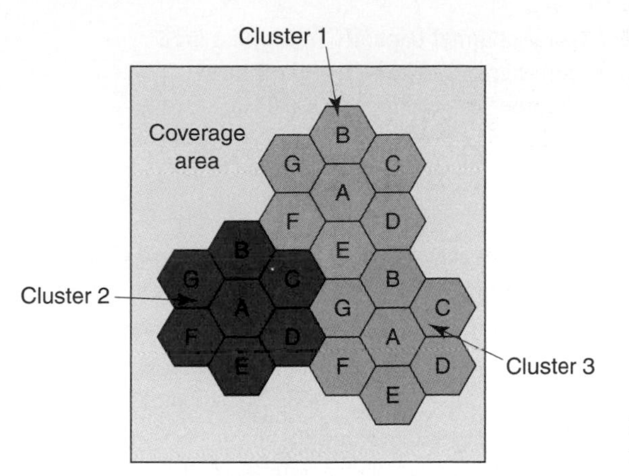

FIGURE 19-4 Cellular frequency reuse concept

19-5 FREQUENCY REUSE

Frequency reuse is the process in which the same set of frequencies (channels) can be allocated to more than one cell, provided the cells are separated by sufficient distance. Reducing each cell's coverage area invites frequency reuse. Cells using the same set of radio channels can avoid mutual interference, provided they are properly separated. Each cell base station is allocated a group of channel frequencies that are different from those of neighboring cells, and base station antennas are chosen to achieve a desired coverage pattern within its cell. However, as long as a coverage area is limited to within a cell's boundaries, the same group of channel frequencies may be used in different cells without interfering with each other, provided the two cells are sufficient distance from one another.

Figure 19-4 illustrates the concept of frequency reuse in a cellular telephone system. The figure shows a geographic cellular radio coverage area containing three groups of cells called *clusters*. Each cluster has seven cells in it, and all cells are assigned the same number of full-duplex cellular telephone channels. Cells with the same letter use the same set of channel frequencies. As the figure shows, the same sets of frequencies are used in all three clusters, which essentially increases the number of usable cellular channels available threefold. The letters A, B, C, D, E, F, and G denote the seven sets of frequencies.

The frequency reuse concept can be illustrated mathematically by considering a system with a fixed number of full-duplex channels available in a given area. Each service area is divided into clusters and allocated a group of channels, which is divided among N cells in a unique and disjoint channel grouping where all cells have the same number of channels but do not necessarily cover the same size area. Thus, the total number of cellular channels available in a cluster can be expressed mathematically as

$$F = GN \qquad \textbf{(19-1)}$$

where F = number of full-duplex cellular channels available in a cluster
G = number of channels in a cell
N = number of cells in a cluster

The cells that collectively use the complete set of available channel frequencies make up the cluster. When a cluster is duplicated m times within a given service area, the total number of full-duplex channels can be expressed mathematically as

$$C = mGN$$

or

$$= mF \qquad \textbf{(19-2)}$$

where C = total channel capacity in a given area
m = number of clusters in a given area
G = number of channels in a cell
N = number of cells in a cluster

Example 19-1

Determine the number of channels per cluster and the total channel capacity for a cellular telephone area comprised of 10 clusters with seven cells in each cluster and 10 channels in each cell.

Solution Substituting into Equation 19-1, the total number of full-duplex channels is

$$F = (10)(7)$$
$$= 70 \text{ channels per cluster}$$

Substituting into Equation 19-3, the total channel capacity is

$$C = (10)(7)(10)$$
$$= 700 \text{ channels total}$$

From Example 19-1, it can be seen that through frequency reuse, 70 channels (frequencies), reused in 10 clusters, produce 700 usable channels within a single cellular telephone area.

From Equations 19-1 and 19-2, it can be seen that the channel capacity of a cellular telephone system is directly proportional to the number of times a cluster is duplicated in a given service area. The factor N is called the *cluster size* and is typically equal to 3, 7, or 12. When the cluster size is reduced and the cell size held constant, more clusters are required to cover a given area, and the total channel capacity increases. The frequency reuse factor of a cellular telephone system is inversely proportional to the number of cells in a cluster (i.e., $1/N$). Therefore, each cell within a cluster is assigned $1/N$th of the total available channels in the cluster.

The number of subscribers who can use the same set of frequencies (channels) in non-adjacent cells at the same time in a small area, such as a city, is dependent on the total number of cells in the area. The number of simultaneous users is generally four, but in densely populated areas, that number may be significantly higher. The number of users is called the frequency reuse factor (FRF). The frequency reuse factor is defined mathematically as

$$FRF = \frac{N}{C} \qquad \textbf{(19-3)}$$

where FRF = frequency reuse factor (unitless)
N = total number of full-duplex channels in an area
C = total number of full-duplex channels in a cell

Meeting the needs of projected growth in cellular traffic is accomplished by reducing the size of a cell by splitting it into several cells, each with its own base station. Splitting cells effectively allows more calls to be handled by the system, provided the cells do not become too small. If a cell becomes smaller than 1500 feet in diameter, the base stations in adjacent cells would most likely interfere with one another. The relationship between frequency reuse and cluster size determines how cellular telephone systems can be rescaled when subscriber density increases. As the number of cells per cluster decreases, the possibility that one channel will interfere with another channel increases.

Cells use a hexagonal shape, which provides exactly six equidistant neighboring cells, and the lines joining the centers of any cell with its neighboring cell are separated by multiples of 60. Therefore, a limited number of cluster sizes and cell layouts is possible. To connect cells without gaps in between (*tessellate*), the geometry of a hexagon is such that the number of cells per cluster can have only values that satisfy the equation

$$N = i^2 + ij + j^2 \qquad \textbf{(19-4)}$$

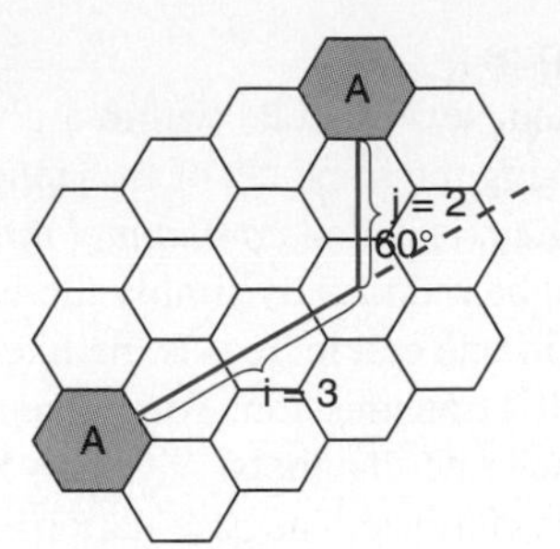

FIGURE 19-5 Locating first tier co-channel cells

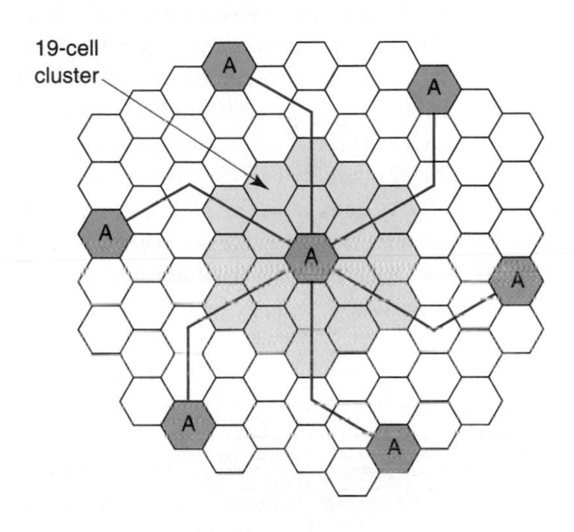

FIGURE 19-6 Determining first tier co-channel cells for Example 11-2

where N = number of cells per cluster
i and j = nonnegative integer values

The process of finding the tier with the nearest co-channel cells (called the *first tier*) is as follows and shown in Figure 19-5:

1. Move i cells through the center of successive cells.
2. Turn 60° in a counterclockwise direction.
3. Move j cells forward through the center of successive cells.

Example 19-2

Determine the number of cells in a cluster and locate the first-tier co-channel cells for the following values: $j = 2$ and $i = 3$.

Solution The number of cells in the cluster is determined from Equation 19-4:

$$N = 3^2 + (3)(2) + 2^2$$
$$N = 19$$

Figure 19-6 shows the six nearest first-tier 1 co-channel cells for cell A.

19-6 INTERFERENCE

The two major kinds of interferences produced within a cellular telephone system are *co-channel interference* and *adjacent-channel interference.*

19-6-1 Co-channel Interference

When frequency reuse is implemented, several cells within a given coverage area use the same set of frequencies. Two cells using the same set of frequencies are called *co-channel cells,* and the interference between them is called *co-channel interference.* Unlike thermal noise, co-channel interference cannot be reduced by simply increasing transmit powers because increasing the transmit power in one cell increases the likelihood of that cell's transmissions interfering with another cell's transmission. To reduce co-channel interference, a certain minimum distance must separate co-channels.

Figure 19-7 shows co-channel interference. The base station in cell A of cluster 1 is transmitting on frequency f_1, and at the same time, the base station in cell A of cluster 2 is transmitting on the same frequency. Although the two cells are in different clusters, they both use the A-group of frequencies. The mobile unit in cluster 2 is receiving the same frequency from two different base stations. Although the mobile unit is under the control of the base station in cluster 2, the signal from cluster 1 is received at a lower power level as co-channel interference.

Interference between cells is proportional not to the distance between the two cells but rather to the ratio of the distance to the cell's radius. Since a cell's radius is proportional to transmit power, more radio channels can be added to a system by either (1) decreasing the transmit power per cell, (2) making cells smaller, or (3) filling vacated coverage areas with new cells. In a cellular system where all cells are approximately the same size, co-channel interference is dependent on the radius (R) of the cells and the distance to the center of the nearest co-channel cell (D) as shown in Figure 19-8. Increasing the D/R ratio (sometimes called *co-channel reuse ratio*) increases the spatial separation between co-channel cells

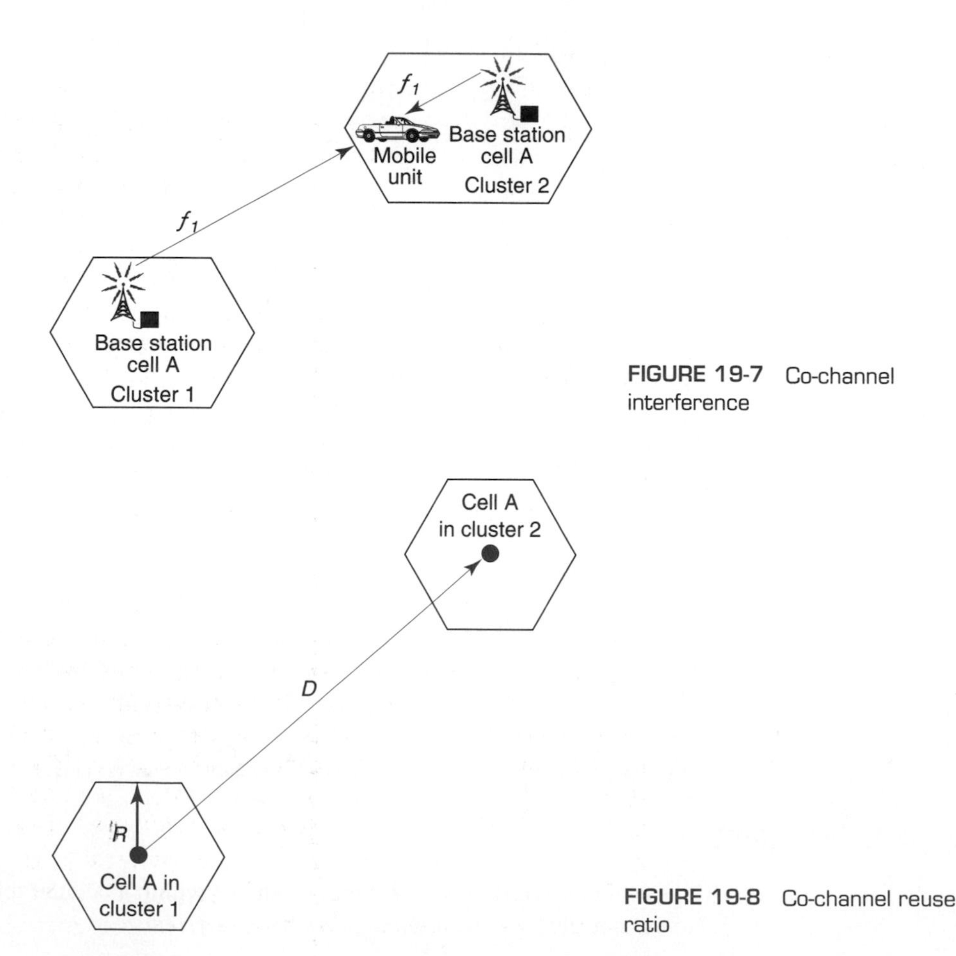

FIGURE 19-7 Co-channel interference

FIGURE 19-8 Co-channel reuse ratio

relative to the coverage distance. Therefore, increasing the co-channel reuse ratio (Q) can reduce co-channel interference. For a hexagonal geometry,

$$Q = \frac{D}{R} \tag{19-5}$$

where Q = co-channel reuse ratio (unitless)
D = distance to center of the nearest co-channel cell (kilometers)
R = cell radius (kilometers)

The smaller the value of Q, the larger the channel capacity since the cluster size is also smaller. However, a large value of Q improves the co-channel interference and, thus, the overall transmission quality. Obviously, in actual cellular system design, a trade-off must be made between the two conflicting objectives.

19-6-2 Adjacent-Channel Interference

Adjacent-channel interference occurs when transmissions from *adjacent channels* (channels next to one another in the frequency domain) interfere with each other. Adjacent-channel interference results from imperfect filters in receivers that allow nearby frequencies to enter the receiver. Adjacent-channel interference is most prevalent when an adjacent channel is transmitting very close to a mobile unit's receiver at the same time the mobile unit is trying to receive transmissions from the base station on an adjacent frequency. This is called the *near-far effect* and is most prevalent when a mobile unit is receiving a weak signal from the base station.

Adjacent-channel interference is depicted in Figure 19-9. Mobile unit 1 is receiving frequency f_1 from base station A. At the same time, base station A is transmitting frequency f_2 to mobile unit 2. Because mobile unit 2 is much farther from the base station than mobile unit 1, f_2 is transmitted at a much higher power level than f_1. Mobile unit 1 is located very close to the base station, and f_2 is located next to f_1 in the frequency spectrum (i.e., the adjacent channel); therefore, mobile unit 1 is receiving f_2 at a much higher power level than f_1. Because of the high power level, the filters in mobile unit 1 cannot block all the energy from f_2, and the signal intended for mobile unit 2 interferes with mobile unit 1's reception

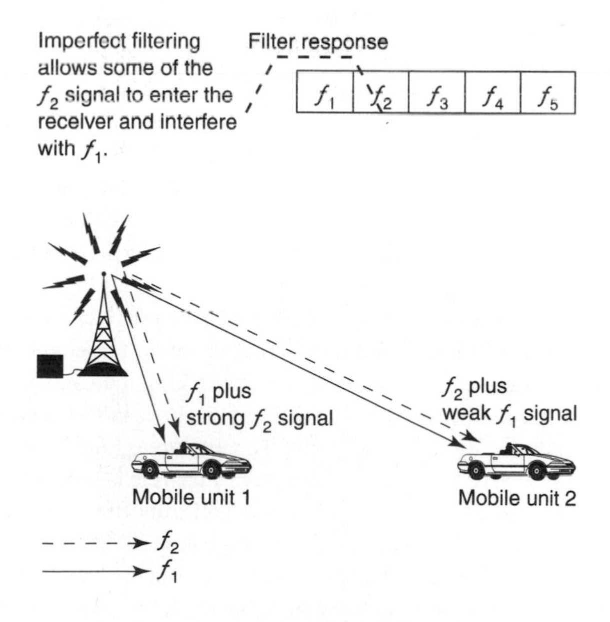

FIGURE 19-9 Adjacent-channel interference

of f_1. f_1 does not interfere with mobile unit 2's reception because f_1 is received at a much lower power level than f_2.

Using precise filtering and making careful channel assignments can minimize adjacent-channel interference in receivers. Maintaining a reasonable frequency separation between channels in a given cell can also reduce adjacent-channel interference. However, if the reuse factor is small, the separation between adjacent channels may not be sufficient to maintain an adequate adjacent-channel interference level.

19-7 CELL SPLITTING, SECTORING, SEGMENTATION, AND DUALIZATION

The Bell System proposed cellular telephone systems in the early 1960s as a means of alleviating congested frequency spectrums indigenous to wide-area mobile telephone systems using line-of-sight, high-powered transmitters. These early systems offered reasonable coverage over large areas, but the available channels were rapidly used up. For example, in the early 1970s, the Bell System could handle only 12 simultaneous mobile telephone calls at a time in New York City. Modern-day cellular telephone systems use relatively low-power transmitters and generally serve a much smaller geographical area.

Increases in demand for cellular service in a given area rapidly consume the cellular channels assigned the area. Two methods of increasing the capacity of a cellular telephone system are cell splitting and sectoring. Cell splitting provides for an orderly growth of a cellular system, whereas sectoring utilizes directional antennas to reduce co-channel and adjacent-channel interference and allow channel frequencies to be reassigned (reused).

19-7-1 Cell Splitting

Cell splitting is when the area of a cell, or independent component coverage areas of a cellular system, is further divided, thus creating more cell areas. The purpose of cell splitting is to increase the channel capacity and improve the availability and reliability of a cellular telephone network. The point when a cell reaches maximum capacity occurs when the number of subscribers wishing to place a call at any given time equals the number of channels in the cell. This is called the *maximum traffic load* of the cell. Splitting cell areas creates new cells, providing an increase in the degree of frequency reuse, thus increasing the channel capacity of a cellular network. Cell splitting provides for orderly growth in a cellular system. The major drawback of cell splitting is that it results in more *base station transfers* (handoffs) per call and a higher processing load per subscriber. It has been proven that a reduction of a cell radius by a factor of 4 produces a 10-fold increase in the handoff rate per subscriber.

Cell splitting is the resizing or redistribution of cell areas. In essence, cell splitting is the process of subdividing highly congested cells into smaller cells each with their own base station and set of channel frequencies. With cell splitting, a large number of low-power transmitters take over an area previously served by a single, higher-powered transmitter. Cell splitting occurs when traffic levels in a cell reach the point where channel availability is jeopardized. If a new call is initiated in an area where all the channels are in use, a condition called *blocking* occurs. A high occurrence of blocking indicates that a system is overloaded.

Providing wide-area coverage with small cells is indeed a costly operation. Therefore, cells are initially set up to cover relatively large areas, and then the cells are divided into smaller areas when the need arises. The area of a circle is proportional to its radius squared. Therefore, if the radius of a cell is divided in half, four times as many smaller cells could be created to provide service to the same coverage area. If each new cell has the same number of channels as the original cell, the capacity is also increased by a factor of 4. Cell splitting allows a system's capacity to increase by replacing large cells with several smaller cells while not disturbing the channel allocation scheme required to prevent interference between cells.

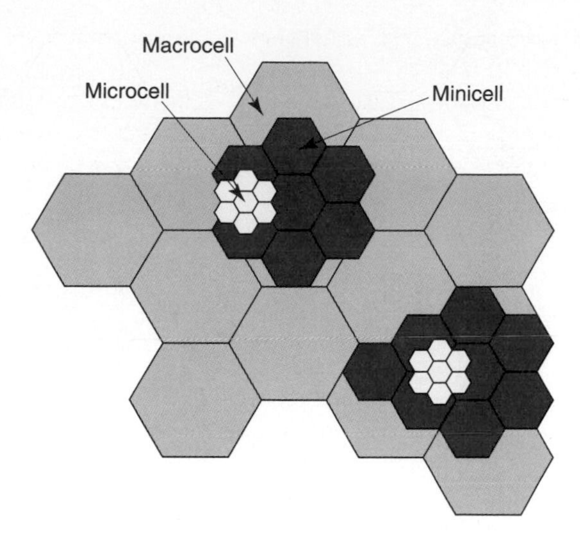

FIGURE 19-10 Cell splitting

Figure 19-10 illustrates the concept of cell splitting. Macrocells are divided into minicells, which are then further divided into microcells as traffic density increases. Each time a cell is split, its transmit power is reduced. As Figure 19-10 shows, cell splitting increases the channel capacity of a cellular telephone system by rescaling the system and increasing the number of channels per unit area (channel density). Hence, cell splitting decreases the cell radius while maintaining the same co-channel reuse ratio (D/R).

Example 19-3

Determine

a. The channel capacity for a cellular telephone area comprised of seven macrocells with 10 channels per cell.
b. Channel capacity if each macrocell is split into four minicells.
c. Channel capacity if each minicell is further split into four microcells.

Solution **a.**

$$\frac{10 \text{ channels}}{\text{cell}} \times \frac{7 \text{ cells}}{\text{area}} = 70 \text{ channels/area}$$

b. Splitting each macrocell into four minicells increases the total number of cells in the area to $4 \times 7 = 28$. Therefore,

$$\frac{10 \text{ channels}}{\text{cell}} \times \frac{28 \text{ cells}}{\text{area}} = 280 \text{ channels/area}$$

c. Further splitting each minicell into four microcells increases the total number of cells in the area to $4 \times 28 = 112$. Therefore,

$$\frac{10 \text{ channels}}{\text{cell}} \times \frac{112 \text{ cells}}{\text{area}} = 1120 \text{ channels/area}$$

From Example 19-3, it can be seen that each time the cells were split, the coverage area appreciated a fourfold increase in channel capacity. For the situation described in Example 19-3, splitting the cells twice increased the total capacity by a factor of 16 from 70 channels to 1120 channels.

19-7-2 Sectoring

Another means of increasing the channel capacity of a cellular telephone system is to decrease the D/R ratio while maintaining the same cell radius. Capacity improvement can be achieved by reducing the number of cells in a cluster, thus increasing the frequency reuse. To accomplish this, the relative interference must be reduced without decreasing transmit power.

In a cellular telephone system, co-channel interference can be decreased by replacing a single omnidirectional antenna with several directional antennas, each radiating within a

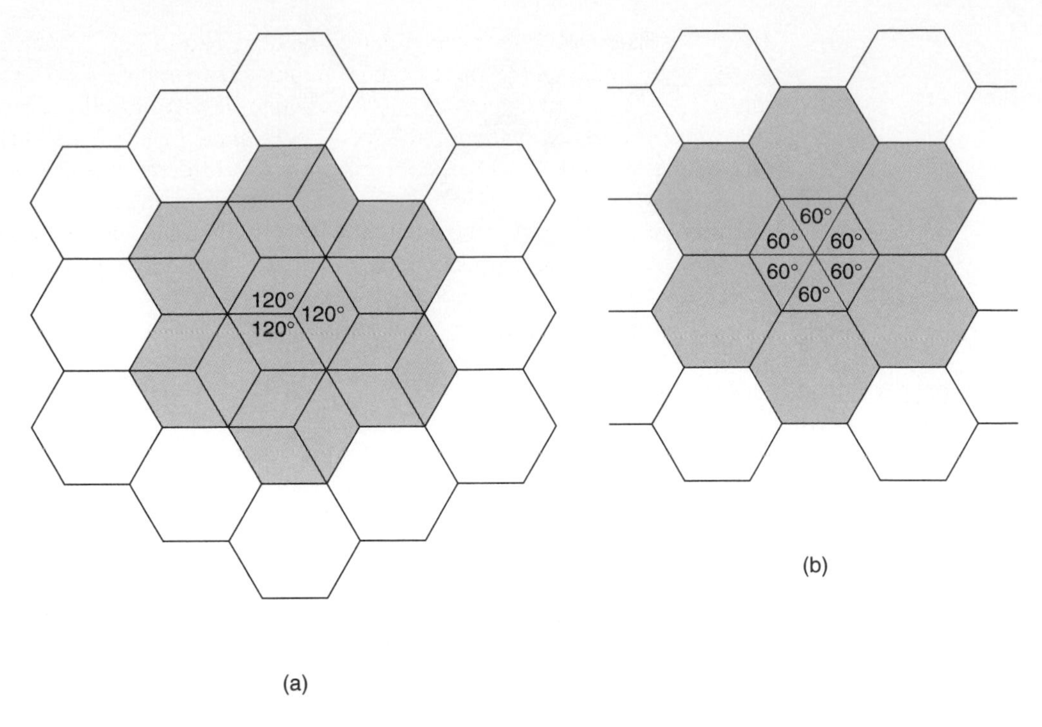

FIGURE 19-11 Sectoring: (a) 120-degree sectors; (b) 60-degree sectors

smaller area. These smaller areas are called *sectors,* and decreasing co-channel interference while increasing capacity by using directional antennas is called *sectoring*. The degree in which co-channel interference is reduced is dependent on the amount of sectoring used. A cell is normally partitioned either into three 60° or six 120° sectors as shown in Figure 19-11. In the three-sector configuration shown in Figure 19-11b, three antennas would be placed in each 120° sector—one transmit antenna and two receive antennas. Placing two receive antennas (one above the other) is called *space diversity.* Space diversity improves reception by effectively providing a larger target for signals radiated from mobile units. The separation between the two receive antennas depends on the height of the antennas above the ground. This height is generally taken to be the height of the tower holding the antenna. As a rule, antennas located 30 meters above the ground require a separation of eight wavelengths, and antennas located 50 meters above the ground require a separation of 11 wavelengths.

When sectoring is used, the channels utilized in a particular sector are broken down into sectored groups that are used only within a particular sector. With seven-cell reuse and 120° sectors, the number of interfering cells in the closest tier is reduced from six to two. Sectoring improves the signal-to-interference ratio, thus increasing the system's capacity.

19-7-3 Segmentation and Dualization

Segmentation and *dualization* are techniques incorporated when additional cells are required within the reuse distance. Segmentation divides a group of channels into smaller groupings or segments of mutually exclusive frequencies; cell sites, which are within the reuse distance, are assigned their own segment of the channel group. Segmentation is a means of avoiding co-channel interference, although it lowers the capacity of a cell by enabling reuse inside the reuse distance, which is normally prohibited.

Dualization is a means of avoiding full-cell splitting where the entire area would otherwise need to be segmented into smaller cells. When a new cell is set up requiring the same channel group as an existing cell (cell 1) and a second cell (cell 2) is not sufficiently far from cell 1 for normal reuse, the busy part of cell 1 (the center) is converted to a primary cell, and the same channel frequencies can be assigned to the new competing cell (cell 2). If all available channels need to be used in cell 2, a problem would arise because the larger secondary cell in cell 1 uses some of these, and there would be interference. In practice, however, cells are assigned different channels, so this is generally not a problem. A drawback of dualization is that it requires an extra base station in the middle of cell 1. There are now two base stations in cell 1: one a high-power station that covers the entire secondary cell and one a low-power station that covers the smaller primary cell.

19-8 CELLULAR SYSTEM TOPOLOGY

Figure 19-12 shows a simplified cellular telephone system that includes all the basic components necessary for cellular telephone communications. The figure shows a wireless radio network covering a set of geographical areas (cells) inside of which mobile two-way radio units, such as cellular or PCS telephones, can communicate. The radio network is defined by a set of radio-frequency transceivers located within each of the cells. The locations of these radio-frequency transceivers are called *base stations.* A base station serves as central control for all users within that cell. Mobile units (such as automobiles and pedestrians) communicate directly with the base stations, and the base stations communicate

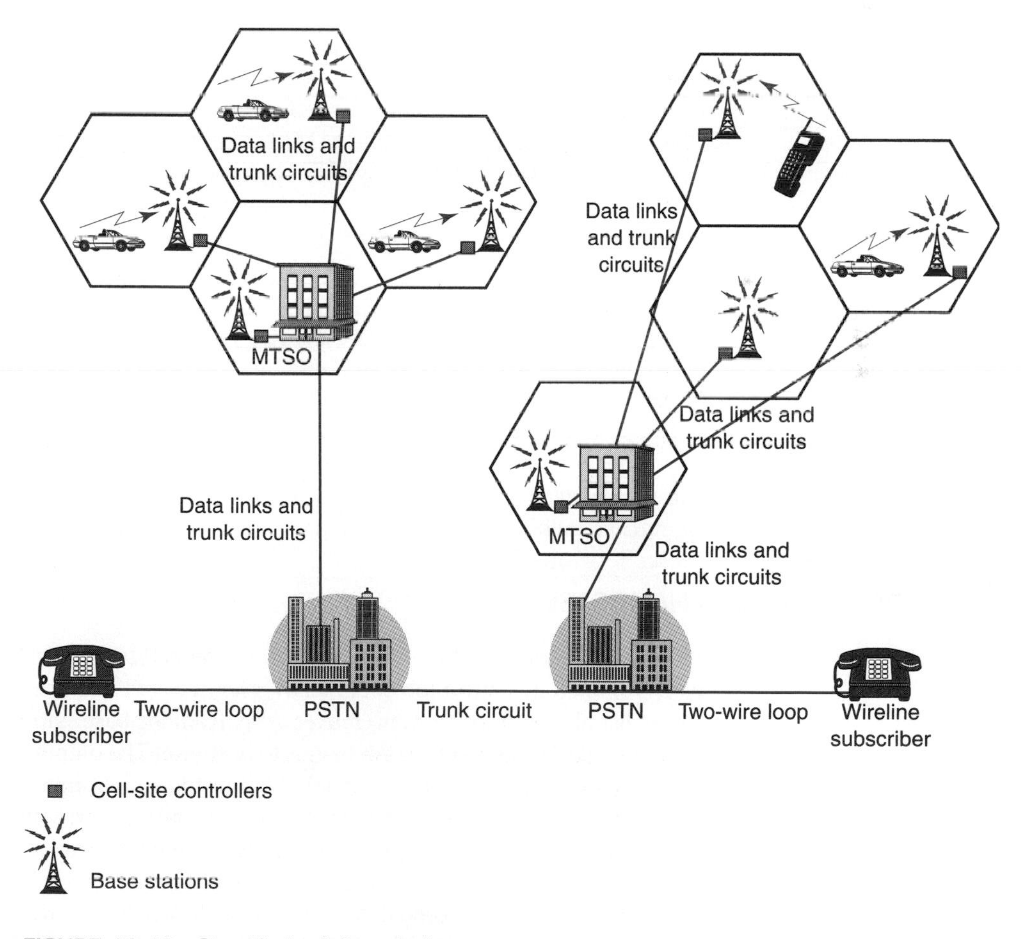

FIGURE 19-12 Simplified cellular telephone system

directly with a *Mobile Telephone Switching Office* (MTSO). An MTSO controls channel assignment, call processing, call setup, and call termination, which includes signaling, switching, supervision, and allocating radio-frequency channels. The MTSO provides a centralized administration and maintenance point for the entire network and interfaces with the public telephone network over wireline voice trunks and data links. MTSOs are equivalent to class 4 toll offices, except smaller. Local loops (or the cellular equivalent) do not terminate in MTSOs. The only facilities that connect to an MTSO are trunk circuits. Most MTSOs are connected to the SS7 signaling network, which allows cellular telephones to operate outside their service area.

Base stations can improve the transmission quality, but they cannot increase the channel capacity within the fixed bandwidth of the network. Base stations are distributed over the area of system coverage and are managed and controlled by an on-site computerized *cell-site controller* that handles all cell-site control and switching functions. Base stations communicate not only directly with mobile units through the airways using control channels but also directly with the MTSO over dedicated data control links (usually four wire, full duplex). Figure 19-12 shows how trunk circuits interconnect cell-site controllers to MTSOs and MTSOs with exchange offices within the PSTN.

The base station consists of a low-power radio transceiver, power amplifiers, a control unit (computer), and other hardware, depending on the system configuration. Cellular and PCS telephones use several moderately powered transceivers over a relatively wide service area. The function of the base station is to provide an interface between mobile telephone sets and the MTSO. Base stations communicate with the MTSO over dedicated data links, both metallic and nonmetallic facilities, and with mobile units over the airwaves using control channels. The MTSO provides a centralized administration and maintenance point for the entire network, and it interfaces with the PSTN over wireline voice trunks to honor services from conventional wireline telephone subscribers.

To complicate the issue, an MTSO is known by several different names, depending on the manufacturer and the system configuration. *Mobile Telephone Switching Office* (MTSO) is the name given by Bell Telephone Laboratories, *Electronic Mobile Xchange* (EMX) by Motorola, *AEX* by Ericcson, *NEAX* by NEC, and *Switching Mobile Center* (SMC) and *Master Mobile Center* (MMC) by Novatel. In PCS networks, the mobile switching center is called the MCS.

Each geographic area or cell can generally accommodate many different user channels simultaneously. The number of user channels depends on the accessing technique used. Within a cell, each radio-frequency channel can support up to 20 mobile telephone users at one time. Channels may be statically or dynamically assigned. Statically assigned channels are assigned a mobile unit for the duration of a call, whereas dynamically assigned channels are assigned a mobile unit only when it is being used. With both static and dynamic assignments, mobile units can be assigned any available radio channel.

19-9 ROAMING AND HANDOFFS

Roaming is when a mobile unit moves from one cell to another—possibly from one company's service area into another company's service area (requiring *roaming agreements*). As a mobile unit (car or pedestrian) moves away from the base station transceiver it is communicating with, the signal strength begins to decrease. The output power of the mobile unit is controlled by the base station through the transmission of up/down commands, which depends on the signal strength the base station is currently receiving from the mobile unit. When the signal strength drops below a predetermined threshold level, the electronic switching center locates the cell in the honeycomb pattern that is receiving the strongest signal from the particular mobile unit and then transfers the mobile unit to the base station in the new cell.

One of the most important features of a cellular system is its ability to transfer calls that are already in progress from one cell-site controller to another as the mobile unit moves

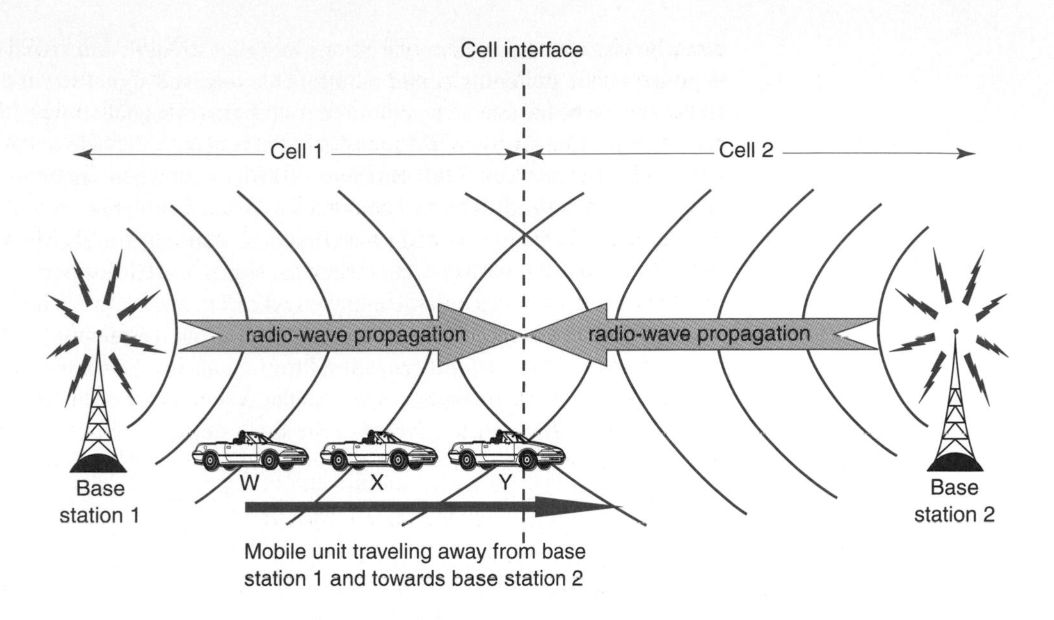

FIGURE 19-13 Handoff

from cell to cell within the cellular network. The base station transfer includes converting the call to an available channel within the new cell's allocated frequency subset. The transfer of a mobile unit from one base station's control to another base station's control is called a *handoff* (or *handover*). Handoffs should be performed as infrequently as possible and be completely *transparent* (*seamless*) to the subscriber (i.e., the subscribers cannot perceive that their facility has been switched). A handoff consists of four stages: (1) initiation, (2) resource reservation, (3) execution, and (4) completion. A connection that is momentarily broken during the cell-to-cell transfer is called a *hard handoff.* A hard handoff is a *break-before-make process.* With a hard handoff, the mobile unit breaks its connection with one base station before establishing voice communications with a new base station. Hard handoffs generally occur when a mobile unit is passed between disjointed systems with different frequency assignments, air interface characteristics, or technologies. A flawless handoff (i.e., no perceivable interruption of service) is called a *soft handoff* and normally takes approximately 200 ms, which is imperceptible to voice telephone users, although the delay may be disruptive when transmitting data. With a soft handoff, a mobile unit establishes contact with a new base station before giving up its current radio channel by transmitting coded speech signals to two base stations simultaneously. Both base stations send their received signals to the MTSO, which estimates the quality of the two signals and determines when the transfer should occur. A complementary process occurs in the opposite direction. A soft handoff requires that the two base stations operate synchronously with one another.

Figure 19-13 shows how a base station transfer is accomplished when a mobile unit moves from one cell into another (the figure shows a soft handoff). The mobile unit is moving away from base station 1 (i.e., toward base station 2). When the mobile unit is at positions W and X, it is well within the range of base station 1 and very distant from base station 2. However, when the mobile unit reaches position Y, it receives signals from base station 1 and base station 2 at approximately the same power level, and the two base stations should be setting up for a handoff (i.e., initiation and resource reservation). When the mobile unit crosses from cell 1 into cell 2, the handoff should be executed and completed.

Computers at cell-site controllers should transfer calls from cell to cell with minimal disruption and no degradation in the quality of transmission. The computers use *handoff decision algorithms* based on variations in signal strength and signal quality. When a call is in progress, the switching center monitors the received signal strength of each user channel. Handoffs can be initiated when the signal strength (or signal-to-interference ratio), measured by either the base station or the mobile unit's receiver, falls below a predetermined threshold level (typically between −90 dBm and −100 dBm) or when a network resource management needs to force a handoff to free resources to place an emergency call. During a handoff, information about the user stored in the first base station is transferred to the new base station. A condition called *blocking* occurs when the signal level drops below a usable level and there are no usable channels available in the target cell to switch to. To help avoid blocking or loss of a call during a handoff, the system employs a load-balancing scheme that frees channels for handoffs and sets handoff priorities. Programmers at the central switching site continually update the switching algorithm to amend the system to accommodate changing traffic loads.

The handoff process involves four basic steps:

1. *Initiation.* Either the mobile unit or the network determines the need for a handoff and initiates the necessary network procedures.
2. *Resource reservation.* Appropriate network procedures reserve the resources needed to support the handoff (i.e., a voice and a control channel).
3. *Execution.* The actual transfer of control from one base station to another base station takes place.
4. *Completion.* Unnecessary network resources are relinquished and made available to other mobile units.

19-9-1 IS-41 Standard

In the United States, roaming from one company's calling area into another company's calling area is called *interoperator roaming* and requires prior agreements between the two service providers. To provide seamless roaming between calling areas served by different companies, the Electronics Industries Association/Telecommunications Industry Association (EIA/TIA) developed the IS-41 protocol, which was endorsed by the Cellular Telecommunication Industry Association (CITA). IS-41 aligns with a subprotocol of the SS7 protocol stack that facilitates communications among databases and other network entities. The IS-41 standard is separated into a series of recommendations.

The principal purposes of IS-41 are to allow mobile units to roam and to perform handoffs of calls already in progress when a mobile unit moves from one cellular system into another without subscriber intervention. Before deployment of SS7, X.25 provided the carrier services for data messages traveling from one cell (the *home location register* [HLR]) to another cell (the *visitor location register* [VLR]). IS-41 provides the information and exchanges necessary to establish and cancel registration in various databases. IS-41 aligns with the ANSI version of SS7 to communicate with databases and other network functional entities.

IS-41 relies on a feature called *autonomous registration,* the process where a mobile unit notifies a serving MTSO of its presence and location through a base station controller. The mobile unit accomplishes autonomous registration by periodically transmitting its identity information, thus allowing the serving MTSO to continuously update its customer list. IS-41 allows MTSOs in neighboring systems to automatically register and validate locations of roaming mobile units so that users no longer need to manually register as they travel.

19-10 CELLULAR TELEPHONE NETWORK COMPONENTS

There are six essential components of a cellular telephone system: (1) an electronic switching center, (2) a cell-site controller, (3) radio transceivers, (4) system interconnections, (5) mobile telephone units, and (6) a common communications protocol.

19-10-1 Electronic Switching Centers

The *electronic switching center* is a digital telephone exchange located in the MTSO that is the heart of a cellular telephone system. The electronic switch performs two essential functions: (1) It controls switching between the public wireline telephone network and the cell-site base stations for wireline-to-mobile, mobile-to-wireline, and mobile-to-mobile calls, and (2) it processes data received from the cell-site controllers concerning mobile unit status, diagnostic data, and bill-compiling information. Electronic switches communicate with cell-site controllers using a data link protocol, such as X.25, at a transmission rate of 9.6 kbps or higher.

19-10-2 Cell-Site Controllers

Each cell contains one *cell-site controller* (sometimes called *base station controller*) that operates under the direction of the switching center (MTSO). Cell-site controllers manage each of the radio channels at each site, supervises calls, turns the radio transmitter and receiver on and off, injects data onto the control and voice channels, and performs diagnostic tests on the cell-site equipment. Base station controllers make up one part of the *base station subsystem.* The second part is the *base transceiver station* (BTS).

19-10-3 Radio Transceivers

Radio transceivers are also part of the base station subsystem. The radio transceivers (combination transmitter/receiver) used with cellular telephone system voice channels can be either narrowband FM for analog systems or either PSK or QAM for digital systems with an effective audio-frequency band comparable to a standard telephone circuit (approximately 300 Hz to 3000 Hz). The control channels use either FSK or PSK. The maximum output power of a cellular transmitter depends on the type of cellular system. Each cell base station typically contains one radio transmitter and two radio receivers tuned to the same channel (frequency). The radio receiver that detects the strongest signal is selected. This arrangement is called *receiver diversity.* The radio transceivers in base stations include the antennas (both transmit and receive). Modern cellular base station antennas are more aesthetically appealing than most antennas and can resemble anything from a window shutter to a palm tree to an architectural feature on a building.

19-10-4 System Interconnects

Four-wire leased lines are generally used to connect switching centers to cell sites and to the public telephone network. There is one dedicated four-wire trunk circuit for each of the cell's voice channels. There must also be at least one four-wire trunk circuit to connect switching centers to each cell-site controller for transferring control signals.

19-10-5 Mobile and Portable Telephone Units

Mobile and *portable telephone units* are essentially identical. The only differences are that portable units have a lower output power, have a less efficient antenna, and operate exclusively on batteries. Each mobile telephone unit consists of a control unit, a multiple-frequency radio transceiver (i.e., multiple channel), a logic unit, and a mobile antenna. The control unit houses all the user interfaces, including a built-in handset. The transceiver uses a frequency synthesizer to tune into any designated cellular system channel. The logic unit interrupts subscriber actions and system commands while managing the operation of the transceiver (including transmit power) and control units.

19-10-6 Communications Protocol

The last constituent of a cellular telephone system is the *communications protocol,* which governs the way telephone calls are established and disconnected. There are several layers of protocols used with cellular telephone systems, and these protocols differ between cellular networks. The protocol implemented depends on whether the voice and control channels are analog or digital and what method subscribers use to access the network. Examples of cellular communications protocols are IS-54, IS-136.2, and IS-95.

19-11 CELLULAR TELEPHONE CALL PROCESSING

Telephone calls over cellular networks require using two full-duplex radio-frequency channels simultaneously, one called the *user channel* and one called the *control channel.* The user channel is the actual voice channel where mobile users communicate directly with other mobile and wireline subscribers through a base station. The control channel is used for transferring control and diagnostic information between mobile users and a central cellular telephone switch through a base station. Base stations transmit on the *forward control channel* and *forward voice channel* and receive on the *reverse control channel* and *reverse voice channel.* Mobile units transmit on the reverse control channel and reverse voice channel and receive on the forward control channel and forward voice channel.

Completing a call within a cellular telephone system is similar to completing a call using the wireline PSTN. When a mobile unit is first turned on, it performs a series of start-up procedures and then samples the receive signal strength on all user channels. The mobile unit automatically tunes to the control channel with the strongest receive signal strength and synchronizes to the control data transmitted by the cell-site controller. The mobile unit interprets the data and continues monitoring the control channel(s). The mobile unit automatically rescans periodically to ensure that it is using the best control channel.

19-11-1 Call Procedures

Within a cellular telephone system, three types of calls can take place involving mobile cellular telephones: (1) mobile (cellular)-to-wireline (PSTN), (2) mobile (cellular)-to-mobile (cellular), and (3) wireline (PSTN)-to-mobile (cellular). A general description is given for the procedures involved in completing each of the three types of calls involving mobile cellular telephones.

19-11-1-1 Mobile (cellular)-to-wireline (PSTN) call procedures.

1. Calls from mobile telephones to wireline telephones can be initiated in one of two ways:
   a. The mobile unit is equivalently taken off hook (usually by depressing a talk button). After the mobile unit receives a dial tone, the subscriber enters the wireline telephone number using either a standard Touch-Tone keypad or with speed dialing. After the last digit is depressed, the number is transmitted through a reverse control channel to the base station controller along with the mobile unit's unique identification number (which is not the mobile unit's telephone number).
   b. The mobile subscriber enters the wireline telephone number into the unit's memory using a standard Touch-Tone keypad. The subscriber then depresses a send key, which transmits the called number as well as the mobile unit's identification number over a reverse control channel to the base station switch.
2. If the mobile unit's ID number is valid, the cell-site controller routes the called number over a wireline trunk circuit to the MTSO.
3. The MTSO uses either standard call progress signals or the SS7 signaling network to locate a switching path through the PSTN to the destination party.
4. Using the cell-site controller, the MTSO assigns the mobile unit a nonbusy user channel and instructs the mobile unit to tune to that channel.
5. After the cell-site controller receives verification that the mobile unit has tuned to the selected channel and it has been determined that the called number is on hook, the mobile unit receives an audible call progress tone (ring-back) while the wireline caller receives a standard ringing signal.
6. If a suitable switching path is available to the wireline telephone number, the call is completed when the wireline party goes off hook (answers the telephone).

19-11-1-2 Mobile (cellular)-to-mobile (cellular) call procedures.

1. The originating mobile unit initiates the call in the same manner as it would for a mobile-to-wireline call.
2. The cell-site controller receives the caller's identification number and the destination telephone number through a reverse control channel, which are then forwarded to the MTSO.
3. The MTSO sends a page command to all cell-site controllers to locate the destination party (which may be anywhere in or out of the service area).
4. Once the destination mobile unit is located, the destination cell-site controller sends a page request through a control channel to the destination party to determine if the unit is on or off hook.
5. After receiving a positive response to the page, idle user channels are assigned to both mobile units.
6. Call progress tones are applied in both directions (ring and ring-back).
7. When the system receives notice that the called party has answered the telephone, the switches terminate the call progress tones, and the conversation begins.
8. If a mobile subscriber wishes to initiate a call and all user channels are busy, the switch sends a directed retry command, instructing the subscriber's unit to reattempt the call through a neighboring cell.
9. If the system cannot allocate user channels through a neighboring cell, the switch transmits an intercept message to the calling mobile unit over the control channel.
10. If the called party is off hook, the calling party receives a busy signal.
11. If the called number is invalid, the calling party receives a recorded message announcing that the call cannot be processed.

19-11-1-3 Wireline (PSTN)-to-mobile (cellular) call procedures.

1. The wireline telephone goes off hook to complete the loop, receives a dial tone, and then inputs the mobile unit's telephone number.
2. The telephone number is transferred from the PSTN switch to the cellular network switch (MTSO) that services the destination mobile number.
3. The cellular network MTSO receives the incoming call from the PSTN, translates the received digits, and locates the base station nearest the mobile unit, which determines if the mobile unit is on or off hook (i.e., available).
4. If the mobile unit is available, a positive page response is sent over a reverse control channel to the cell-site controller, which is forwarded to the network switch (MTSO).
5. The cell-site controller assigns an idle user channel to the mobile unit and then instructs the mobile unit to tune to the selected channel.
6. The mobile unit sends verification of channel tuning through the cell-site controller.
7. The cell-site controller sends an audible call progress tone to the subscriber's mobile telephone, causing it to ring. At the same time, a ring-back signal is sent back to the wireline calling party.
8. The mobile answers (goes off hook), the switch terminates the call progress tones, and the conversation begins.

QUESTIONS

19-1. What is the contemporary mobile telephone meaning for the term *mobile?*

19-2. Contrast the similarities and differences between *two-way mobile radio* and *cellular telephone.*

19-3. Describe the differences between a cellular telephone *service area,* a *cluster,* and a *cell.*

19-4. Why was a *honeycomb pattern* selected for a cell area?

19-5. What are the differences between *macrocells, minicells,* and *microcells?*

19-6. What is meant by a *center-excited* cell? *Edge-excited* cell? *Corner-excited* cell?

19-7. Describe *frequency reuse.* Why is it useful in cellular telephone systems?

19-8. What is meant by *frequency reuse factor?*

19-8. Name and describe the two most prevalent types of interference in cellular telephone systems.

19-9. What significance does the *co-channel reuse factor* have on cellular telephone systems?

19-10. What is meant by the *near-far effect?*

19-11. Describe the concept of *cell splitting.* Why is it used?

19-12. Describe the term *blocking.*

19-13. Describe what is meant by *channel density.*

19-14. Describe *sectoring* and state why it is used.

19-15. What is the difference between a *MTSO* and a *cell-site controller?*

19-16. Explain what the term *roaming* means.

19-17. Explain the term *handoff.*

19-18. What is the difference between a *soft* and a *hard* handoff?

19-19. Explain the term *break before make* and state when it applies.

19-20. Briefly describe the purpose of the IS-41 standard.

19-21. Describe the following terms: *home location register, visitor location register,* and *autonomous registration.*

19-22. List and describe the six essential components of a *cellular telephone network.*

19-23. Describe what is meant by the term *forward channel; reverse channel.*

PROBLEMS

19-1. Determine the number of channels per cluster and the total channel capacity for a cellular telephone area comprised of 12 clusters with seven cells in each cluster and 16 channels in each cell.

19-2. Determine the number of cells in clusters for the following values: $j = 4$ and $i = 2$ and $j = 3$ and $i = 3$.

19-3. Determine the co-channel reuse ratio for a cell radius of 0.5 miles separated from the nearest co-channel cell by a distance of 4 miles.

19-4. Determine the distance from the nearest co-channel cell for a cell radius of 0.4 miles and a co-channel reuse factor of 12.

19-5. Determine

a. The channel capacity for a cellular telephone area comprised of seven macrocells with 16 channels per cell.

b. Channel capacity if each macrocell is split into four minicells.

c. Channel capacity if each minicell is further split into four microcells.

19-6. A cellular telephone company has acquired 150 full-duplex channels for a given service area. The company decided to divide the service area into 15 clusters and use a seven-cell reuse pattern and use the same number of channels in each cell. Determine the total number of channels the company has available for its subscribers at any one time.

CHAPTER 20

Cellular Telephone Systems

CHAPTER OUTLINE

OBJECTIVES

- Define *first-generation analog cellular telephone systems*
- Describe and outline the frequency allocation for the Advanced Mobile Telephone System (AMPS)
- Explain frequency-division multiple accessing (FDMA)
- Describe the operation of AMPS control channels
- Explain the AMPS classification of cellular telephones
- Describe the concepts of personal communications systems (PCS)
- Outline the advantages and disadvantages of PCS compared to standard cellular telephone
- Describe second-generation cellular telephone systems
- Explain the operation of N-AMPS cellular telephone systems
- Define *digital cellular telephone*
- Describe the advantages and disadvantages of digital cellular telephone compared to analog cellular telephone
- Describe time-division multiple accessing (TDMA)
- Describe the purpose of IS-54 and what is meant by dual-mode operation
- Describe IS-136 and explain its relationship to IS-54
- Describe the format for a USDC digital voice channel
- Explain the classifications of USDC radiated power
- Describe the basic concepts and outline the specifications of IS-95
- Describe code-division multiple accessing (CDMA)

- Outline the CDMA frequency and channel allocation for cellular telephone
- Explain the classifications of CDMA radiated power
- Summarize North American cellular and PCS systems
- Describe global system for mobile communications (GSM)
- Describe the services provided by GSM
- Explain GSM system architecture
- Describe the GSM radio subsystem
- Describe the basic concepts of a Personal Communications Satellite System (PCSS)
- Outline the advantages and disadvantages of PCSS over terrestrial cellular telephone systems

20-1 INTRODUCTION

Like nearly everything in the modern world of electronic communications, cellular telephone began as a relatively simple concept. However, the increased demand for cellular services has caused cellular telephone systems to evolve into complicated networks and internetworks comprised of several types of cellular communications systems. New systems have evoked new terms, such as *standard cellular telephone service* (CTS), *personal communications systems* (PCS), and *Personal Communications Satellite System* (PCSS), all of which are full-duplex mobile telephone systems that utilize the cellular concept.

Cellular telephone began as a relatively simple two-way analog communications system using frequency modulation (FM) for voice and frequency-shift keying (FSK) for transporting control and signaling information. The most recent cellular telephone systems use higher-level digital modulation schemes for conveying both voice and control information. In addition, the Federal Communications Commission has recently assigned new frequency bands for cellular telephone. The following sections are intended to give the reader a basic understanding of the fundamental meaning of the common cellular telephone systems and the terminology used to describe them.

20-2 FIRST-GENERATION ANALOG CELLULAR TELEPHONE

In 1971, Bell Telephone Laboratories in Murry Hill, New Jersey, proposed the cellular telephone concept as the *Advanced Mobile Telephone System* (AMPS). The cellular telephone concept was an intriguing idea that added a depth or spatial dimension to the conventional wireline trunking model used by the public telephone company at the time. The cellular plan called for using many low-profile, low-power cell-site transceivers linked through a central computer-controlled switching and control center. AMPS is a standard cellular telephone service (CTS) initially placed into operation on October 13, 1983, by Illinois Bell that incorporated several large cell areas to cover approximately 2100 square miles in the Chicago area. The original system used omnidirectional antennas to minimize initial equipment costs and employed low-power (7-watt) transmitters in both base stations and mobile units. Voice-channel radio transceivers with AMPS cellular telephones use narrowband frequency modulation (NBFM) with a usable audio-frequency band from 300 Hz to 3 kHz and a maximum frequency deviation of ± 12 kHz for 100% modulation. Using Carson's rule, this corresponds to an approximate bandwidth of 30 kHz. Empirical information determined that an AMPS 30-kHz telephone channel requires a minimum signal-to-interference ratio (SIR) of 18 dB for satisfactory performance. The smallest reuse factor that satisfied this requirement utilizing 120° directional antennas was 7. Consequently, the AMPS system uses a seven-cell reuse pattern with provisions for cell splitting and sectoring to increase channel capacity when needed.

20-2-1 AMPS Frequency Allocation

In 1980, the Federal Communications Commission decided to license two common carriers per cellular service area. The idea was to eliminate the possibility of a monopoly and

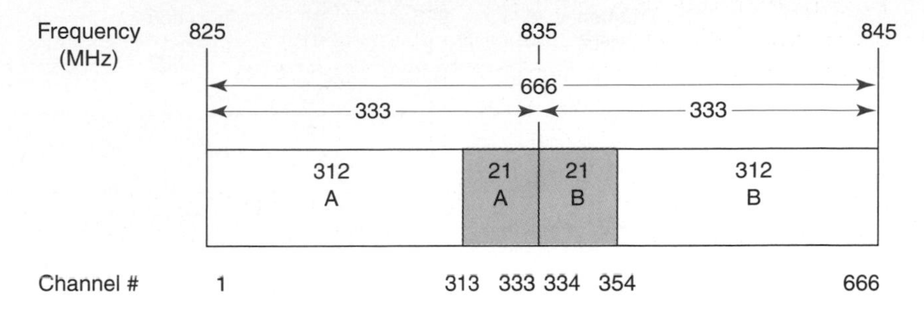

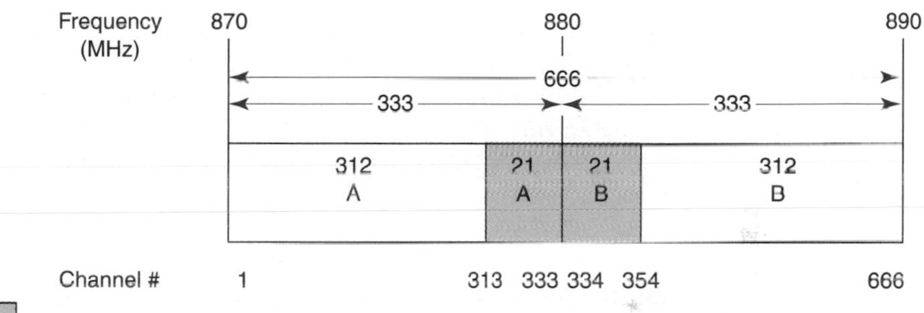

FIGURE 20-1 Original Advanced Mobile Phone Service (AMPS) frequency spectrum

provide the advantages that generally accompany a competitive environment. Subsequently, two frequency allocation plans emerged—system A and system B—each with its own group of channels that shared the allocated frequency spectrum. System A is defined for the non-wireline companies (i.e., cellular telephone companies) and system B for existing wireline companies (i.e., local telephone companies). The Federal Communications Commission initially assigned the AMPS system a 40-MHz frequency band consisting of 666 two-way channels per service area with 30-kHz spacing between adjacent channels.

Figure 20-1 shows the original frequency management system for the AMPS cellular telephone system. The A channels are designated 1 to 333, and the B channels are designated 334 to 666. For mobile units, channel 1 has a transmit frequency of 825.03 MHz, and channel 666 has a transmit frequency of 844.98 MHz. For base stations, channel 1 has a transmit frequency of 870.03 MHz, and channel 666 has a transmit frequency of 889.98 MHz. The receive frequencies are, of course, just the opposite.

Simultaneous transmission in both directions is a transmission mode called *full duplex* (FDX) or simply *duplexing.* Duplexing can be accomplished using frequency- or time-domain methods. *Frequency-division duplexing* (FDD) is used with AMPS and occurs when two distinct frequency bands are provided to each user. In FDD, each duplex channel actually consists of two simplex (one-way) channels (base station to mobile and mobile to base station). A special device called a duplexer is used in each mobile unit and base station to allow simultaneous transmission and reception on duplex channels.

Transmissions from base stations to mobile units are called *forward links,* whereas transmission from mobile units to base stations are called *reverse links.* (Forward links are

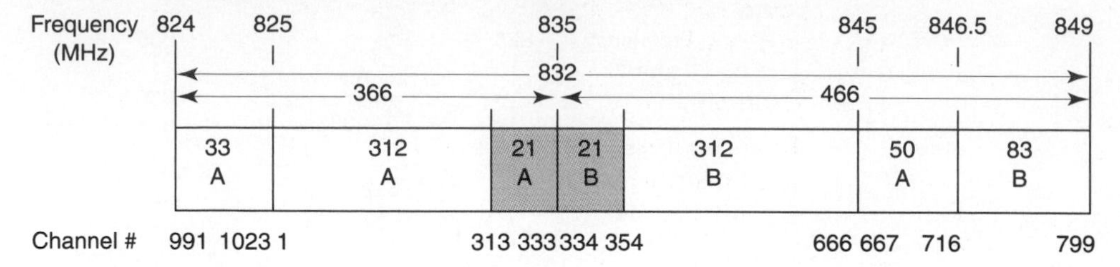

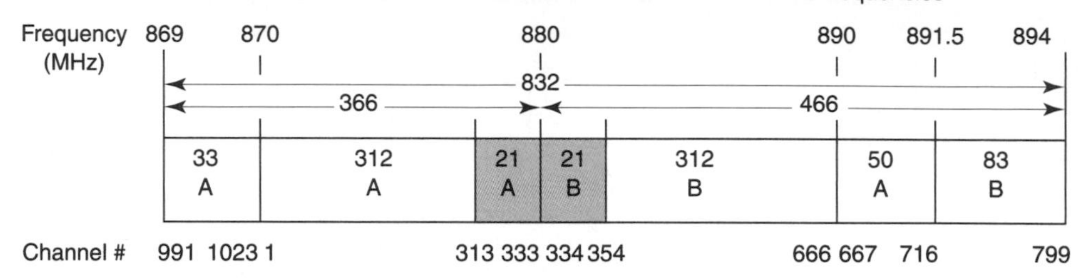

*Shaded areas denote control channels (A-system: 313 to 333 and B-system: 334 to 354)

FIGURE 20-2 Complete Advanced Mobile Phone Service (AMPS) frequency spectrum

sometimes called *downlinks* and reverse links are sometimes called *uplinks.*) The receiver for each channel operates 45 MHz above the transmit frequency. Consequently, every two-way AMPS radio channel consists of a pair of simplex channels separated by 45 MHz. The 45-MHz separation between transmit and receive frequencies was chosen to make use of inexpensive but highly selective duplexers in the mobile units.

In 1989, the Federal Communications Commission added an additional 10-MHz frequency spectrum to the original 40-MHz band, which increased the number of simplex channels by 166 for a total of 832 (416 full duplex). The additional frequencies are called the *expanded spectrum* and include channels 667 to 799 and 991 to 1023. The complete AMPS frequency assignment is shown in Figure 20-2. Note that 33 of the new channels were added below the original frequency spectrum and that the remaining 133 were added above the original frequency spectrum. With AMPS, a maximum of 128 channels could be used in each cell.

The mobile unit's transmit carrier frequency in MHz for any channel is calculated as follows:

$$f_t = 0.03\,N + 825 \qquad \text{for } 1 \le N \le 866 \tag{20-1}$$

$$f_t = 0.03(N - 1023) + 825 \qquad \text{for } 990 \le N \le 1023 \tag{20-2}$$

where f_t = transmit carrier frequency (MHz)
N = channel number

The mobile unit's receive carrier frequency is obtained by simply adding 45 MHz to the transmit frequency:

$$f_r = f_t + 45\text{ MHz} \tag{20-3}$$

The base station's transmit frequency for any channel is simply the mobile unit's receive frequency, and the base station's receive frequency is simply the mobile unit's transmit frequency.

Example 20-1

Determine the transmit and receive carrier frequencies for

a. AMPS channel 3.
b. AMPS channel 991.

Solution

a. The transmit and receive carrier frequencies for channel 3 can be determined from Equations 20-1 and 20-3:

$$\text{transmit} \qquad f_t = 0.03N + 825$$
$$= 0.03(3) + 825$$
$$= 825.09 \text{ MHz}$$
$$\text{receive} \qquad f_r = 825.09 \text{ MHz} + 45 \text{ MHz}$$
$$= 870.09 \text{ MHz}$$

b. The transmit and receive carrier frequencies for channel 991 can be determined from Equations 20-2 and 20-3:

$$\text{transmit} \qquad f_t = 0.03(991 - 1023) + 825$$
$$= 824.04 \text{ MHz}$$
$$\text{receive} \qquad f_r = 824.04 \text{ MHz} + 45 \text{ MHz}$$
$$= 869.04 \text{ MHz}$$

Table 20-1 summarizes the frequency assignments for AMPS. The set of control channels may be split by the system operator into subsets of dedicated control channels, paging channels, or access channels.

The Federal Communications Commission controls the allocation of cellular telephone frequencies (channels) and also issues licenses to cellular telephone companies to operate specified frequencies in geographic areas called *cellular geographic serving areas* (CGSA). CGSAs are generally designed to lie within the borders of a standard metropolitan statistical area (SMSA), which defines geographic areas used by marketing agencies that generally correspond to the area covered by a specific wireline LATA (local access and transport area).

20-2-2 Frequency-Division Multiple Accessing

Standard cellular telephone subscribers access the AMPS system using a technique called *frequency-division multiple accessing* (FDMA). With FDMA, transmissions are separated in the frequency domain—each channel is allocated a carrier frequency and channel bandwidth within the total system frequency spectrum. Subscribers are assigned a pair of voice channels (forward and reverse) for the duration of their call. Once assigned a voice channel, a subscriber is the only mobile unit using that channel within a given cell. Simultaneous transmissions from multiple subscribers can occur at the same time without interfering with one another because their transmissions are on different channels and occupy different frequency bands.

20-2-3 AMPS Identification Codes

The AMPS system specifies several identification codes for each mobile unit (see Table 20-2). The *mobile identification number* (MIN) is a 34-bit binary code, which in the United States represents the standard 10-digit telephone number. The MIN is comprised of a three-digit area code, a three-digit prefix (exchange number), and a four-digit subscriber (extension) number. The exchange number is assigned to the cellular operating company. If a subscriber changes service from one cellular company to another, the subscriber must be assigned a new cellular telephone number.

Table 20-1 AMPS Frequency Allocation

	AMPS
Channel spacing	30 kHz
Spectrum allocation	40 MHz
Additional spectrum	10 MHz
Total number of channels	832

System A Frequency Allocation

AMPS

Channel Number	Mobile TX, MHz	Mobile RX, MHz
1	825.030	870.030
313[a]	834.390	879.390
333[b]	843.990	879.990
667	845.010	890.010
716	846.480	891.480
991	824.040	869.040
1023	825.000	870.000

System B Frequency Allocation

Channel Number	Mobile TX, MHz	Mobile RX, MHz
334[c]	835.020	880.020
354[d]	835.620	880.620
666	844.980	890.000
717	846.510	891.000
799	848.970	894.000

[a]First dedicated control channel for system A.
[b]Last dedicated control channel for system A.
[c]First dedicated control channel for system B.
[d]Last dedicated control channel for system B.

Table 20-2 AMPS Identification Codes

Notation	Name	Length (Bits)	Description
MIN	Mobile identifier	34	Directory number assigned by operating company to a subscriber (telephone number)
ESN	Electronic serial number	32	Assigned by manufacturer to a mobile station (telephone)
SID	System identifier	15	Assigned by regulators to a geographical service area
SCM	Station class mark	4	Indicates capabilities of a mobile station
SAT	Supervisory audio tone	*	Assigned by operating company to each base station
DCC	Digital color code	2	Assigned by operating company to each base station

Another identification code used with AMPS is the *electronic serial number* (ESN), which is a 32-bit binary code permanently assigned to each mobile unit. The ESN are similar to the VIN (vehicle identification number) assigned to an automobile or the MAC address on a network interface card (NIC) in that the number is unique and positively identifies a specific unit.

Table 20-3 AMPS Mobile Phone Power Levels

	Class I		Class II		Class III	
Power Level	dBm	mW	dBm	mW	dBm	mW
0	36	4000	32	1600	28	640
1	32	1600	32	1600	28	640
2	28	640	28	640	28	640
3	24	256	24	256	24	256
4	20	102	20	102	20	102
5	16	41	16	41	16	41
6	12	16	12	16	12	16
7	8	6.6	8	6.6	8	6.6

The third identification code used with AMPS is the four-bit *station class mark* (SCM), which indicates whether the terminal has access to all 832 AMPS channels or only 666. The SCM also specifies the maximum radiated power for the unit (Table 20-3).

The *system identifier* (SID) is a 15-bit binary code issued by the FCC to an operating company when it issues it a license to provide AMPS cellular service to an area. The SID is stored in all base stations and all mobile units to identify the operating company and MTSO and any additional shared MTSO. Every mobile unit knows the SID of the system it is subscribed to, which is the mobile unit's *home system.* Whenever a mobile unit initializes, it compares its SID to the SID broadcast by the local base station. If the SIDs are the same, the mobile unit is communicating with its home system. If the SIDs are different, the mobile unit is roaming.

Local operating companies assign a two-bit *digital color code* (DCC) and a *supervisory audio tone* (SAT) to each of their base stations. The DCC and SAT help the mobile units distinguish one base station from a neighboring base station. The SAT is one of three analog frequencies (5970 Hz, 6000 Hz, or 6030 Hz), and the DCC is one of four binary codes (00, 01, 10, or 11). Neighboring base stations transmit different SAT frequencies and DCCs.

20-2-4 AMPS Control Channels

The AMPS channel spectrums are divided into two basic sets or groups. One set of channels is dedicated for exchanging control information between mobile units and base stations and is appropriately termed *control channels* (shaded areas in Figures 20-1 and 20-2). Control channels cannot carry voice information; they are used exclusively to carry service information. There are 21 control channels in the A system and 21 control channels in the B system. The remaining 790 channels make up the second group, termed *voice* or *user channels.* User channels are used for propagating actual voice conversations or subscriber data.

Control channels are used in cellular telephone systems to enable mobile units to communicate with the cellular network through base stations without interfering with normal voice traffic occurring on the normal voice or user channels. Control channels are used for call origination, for call termination, and to obtain system information. With the AMPS system, voice channels are analog FM, while control channels are digital and employ FSK. Therefore, voice channels cannot carry control signals, and control channels cannot carry voice information. Control channels are used exclusively to carry service information. With AMPS, base stations broadcast on the *forward control channel* (FCC) and listen on the *reverse control channel* (RCC). The control channels are sometimes called *setup* or *paging channels.* All AMPS base stations continuously transmit FSK data on the FCC so that idle cellular telephones can maintain lock on the strongest FCC regardless of their location. A subscriber's unit must be *locked* (sometimes called *camped*) on an FCC before it can originate or receive calls.

Each base station uses a control channel to simultaneously page mobile units to alert them of the presence of incoming calls and to move established calls to a vacant voice channel. The forward control channel transmits a 10-kbps data signal using FSK. Forward

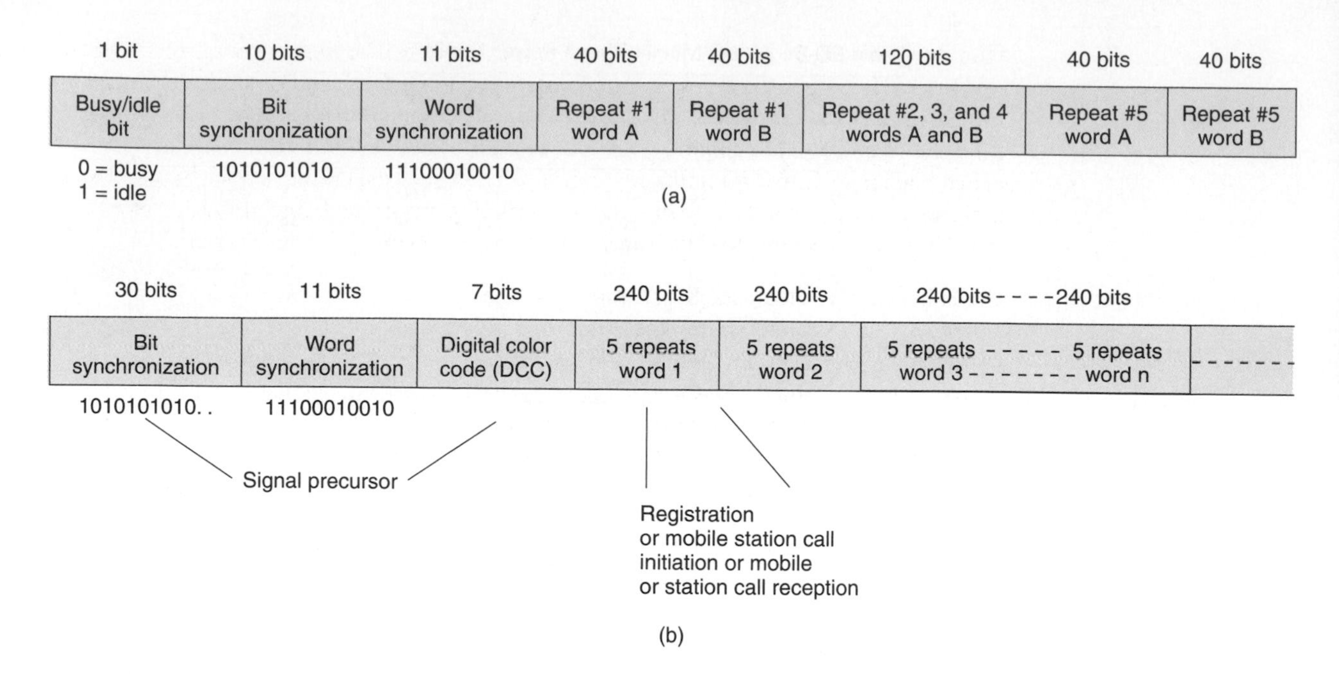

FIGURE 20-3 Control channel format: (a) forward control channel; (b) reverse control channel

control channels from base stations may contain overhead data, mobile station control information, or control file information.

Figure 20-3a shows the format for an AMPS forward control channel. As the figure shows, the control channel message is preceded by a 10-bit *dotting scheme,* which is a sequence of alternating 1s and 0s. The dotting scheme is followed by an 11-bit *synchronization word* with a unique sequence of 1s and 0s that enables a receiver to instantly acquire synchronization. The sync word is immediately followed by the message repeated five times. The redundancy helps compensate for the ill effects of fading. If three of the five words are identical, the receiver assumes that as the message.

Forward control channel data formats consist of three discrete information streams: stream A, stream B, and the busy-idle stream. The three data streams are multiplexed together. Messages to the mobile unit with the least-significant bit of their 32-bit *mobile identification number* (MIN) equal to 0 are transmitted on stream A, and MINs with the least-significant bit equal to 1 are transmitted on stream B. The busy-idle data stream contains *busy-idle bits,* which are used to indicate the current status of the reverse control channel (0 = busy and 1 = idle). There is a busy-idle bit at the beginning of each dotting sequence, at the beginning of each synchronization word, at the beginning of the first repeat of word A, and after every 10 message bits thereafter. Each message word contains 40 bits, and forward control channels can contain one or more words.

The types of messages transmitted over the FCC are the *mobile station control message* and the *overhead message train.* Mobile station control messages control or command mobile units to do a particular task when the mobile unit has not been assigned a voice channel. Overhead message trains contain *system parameter overhead messages, global action overhead messages,* and *control filler messages.* Typical mobile-unit control messages are *initial voice channel designation messages, directed retry messages, alert messages,* and *change power messages.*

Figure 20-3b shows the format for the reverse control channel that is transmitted from the mobile unit to the base station. The control data are transmitted at a 10-kbps rate and include *page responses, access requests,* and *registration requests.* All RCC messages begin with the RCC seizure precursor, which consists of a 30-bit *dotting sequence,* an 11-bit *synchronization word,* and the coded *digital color code* (DCC), which is added so that the control channel is not confused with a control channel from a nonadjacent cell that is reusing the same frequency. The mobile telephone reads the base station's DCC and then returns a coded version of it, verifying that the unit is locked onto the correct signal. When the call is finished, a 1.8-second *signaling time-out signal* is transmitted. Each message word contains 40 bits and is repeated five times for a total of 200 bits.

20-2-5 Voice-Channel Signaling

Analog cellular channels carry both voice using FM and digital signaling information using binary FSK. When transmitting digital signaling information, voice transmissions are inhibited. This is called *blank and burst:* The voice is blanked, and the data are transmitted in a short burst. The bit rate of the digital information is 10 kbps. Figure 20-4a shows the voice-channel signaling format for a forward voice channel, and Figure 20-4b shows the format for the reverse channel. The digital signaling sequence begins with a 101-bit dotting sequence that readies the receiver to receive digital information. After the dotting sequence, a synchronization word is sent to indicate the start of the message. On the forward voice channel, digital signaling messages are repeated 11 times to ensure the integrity of the message, and on the receive channel they are repeated 5 times. The forward channel uses 40-bit words, and the reverse channel uses 48-bit words.

20-3 PERSONAL COMMUNICATIONS SYSTEM

The Personal Communications System (PCS) is a relatively new class of cellular telephony based on the same basic philosophies as standard cellular telephone systems (CTSs), such as AMPS. However, PCS systems are a combination of cellular telephone networks and the *Intelligent Network,* which is the entity of the SS7 interoffice protocol that distinguishes the physical components of the switching network, such as the signal service point (SSP), signal control point (SCP), and signal transfer point (STP), from the services provided by the SS7 network. The services provided are distinctly different from the switching systems and protocols that promote and support them. PCS was initially considered a new service, although different companies have different visions of exactly what PCS is and what services it should provide. The Federal Communications Commission defines PCS mobile telephone as "a family of mobile or portable radio communications services, which provides services to individuals and business and is integrated with a variety of competing networks." In essence, PCS is the North American implementation of the European GSM standard.

Existing cellular telephone companies want PCS to provide broad coverage areas and fill in service gaps between their current service areas. In other words, they want PCS to be an extension of the current first- and second-generation cellular system to the 1850-MHz to 2200-MHz band using identical standards for both frequency bands. Other companies would like PCS to compete with standard cellular telephone systems but offer enhanced services and better quality using extensions of existing standards or entirely new standards. Therefore, some cellular system engineers describe PCS as a third-generation cellular telephone system, although the U.S. implementation of PCS uses modifications of existing cellular protocols, such as IS-54 and IS-95. Most cellular telephone companies reserve the designation third-generation PCS to those systems designed for transporting data as well as voice.

Although PCS systems share many similarities with first-generation cellular telephone systems, PCS has several significant differences that, most agree, warrant the use of a different name. Many of the differences are transparent (or at least not obvious) to the users of the networks. Probably the primary reason for establishing a new PCS cellular telephone system was because first-generation cellular systems were already overcrowded, and it was obvious that they would not be able to handle the projected demand

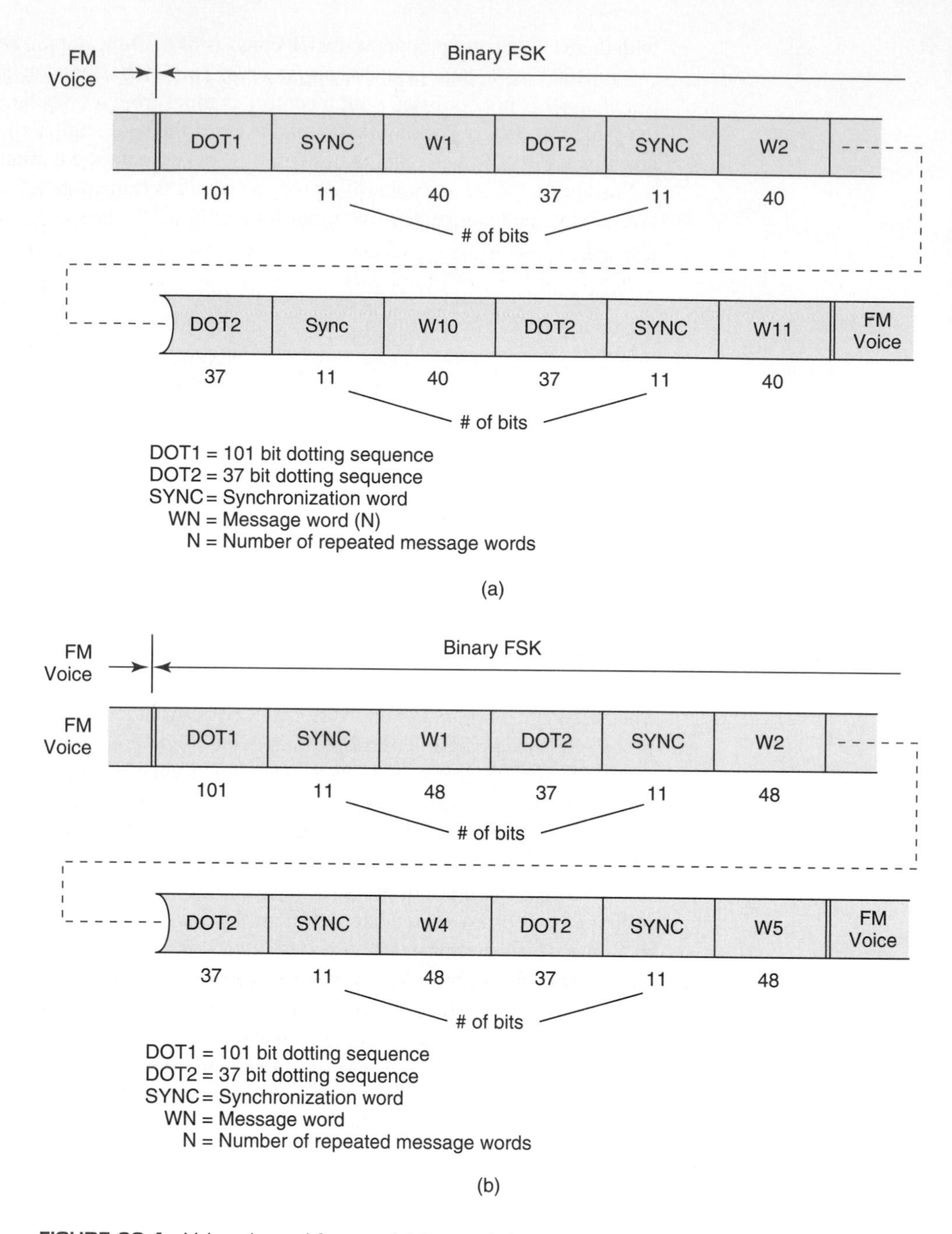

FIGURE 20-4 Voice channel format: (a) forward channel; (b) reverse channel

for future cellular telephone services. In essence, PCS services were conceived to provide subscribers with a low-cost, feature-rich wireless telephone service.

Differences between PCS systems and standard cellular telephone systems generally include but are certainly not limited to the following: (1) smaller cell size, (2) all digital, and (3) additional features. Cellular systems generally classified as PCS include IS-136 TDMA, GSM, and IS-95 CDMA.

The concept of *personal communications services* (also PCS) originated in the United Kingdom when three companies were allocated a band of frequencies in the 1.8-GHz band to develop a *personal communications network* (PCN) throughout Great Britain. The terms *PCS* and *PCN* are often used interchangeably. However, PCN refers to a wireless networking concept where any user can initiate or receive calls regardless of where they are using a portable, personalized transceiver. PCS refers to a new wireless system that incorporates enhanced network features and is more personalized than existing standard cellular telephone systems but does not offer all the features of an ideal PCN.

In 1990, the Federal Communications Commission adopted the term PCS to mean *personal communications services,* which is the North American implementation of the *global system for mobile communications.* However, to some people, PCS means *personal communications system,* which specifies a category or type of cellular telephone system. The exact nature of the services provided by PCS is not completely defined by the cellular telephone industry. However, the intention of PCS systems is to provide enhanced features to first- and second-generation cellular telephone systems, such as messaging, paging, and data services.

PCS is more of a concept than a technology. The concept being to assign everyone a *personal telephone number* (PTN) that is stored in a database on the SS7 network. This database keeps track of where each mobile unit can be reached. When a call is placed from a mobile unit, an *artificial intelligence network* (AIN) in SS7 determines where and how the call should be directed. The PCS network is similar to the D-AMPS system in that the MTSO stores three essential databases: *home location register, visitor location register,* and *equipment identification registry.*

Home location register (HLR). The HLR is a database that stores information about the user, including home subscription information and what supplementary services the user is subscribed to, such as call waiting, call hold, call forwarding, and call conferencing (three-way calling). There is generally only one HLR per mobile network. Data stored on the HLR are semipermanent, as they do not usually change from call to call.

Visitor location register (VLR). The VLR is a database that stores information about subscribers in a particular MTSO serving area, such as whether the unit is on or off and whether any of the supplementary services are activated or deactivated. There is generally only one VLR per mobile switch. The VLR stores permanent data, such as that found in the HLR, plus temporary data, such as the subscriber's current location.

Equipment identification registry (EIR). The EIR is a database that stores information pertaining to the identification and type of equipment that exists in the mobile unit. The EIR also helps the network identify stolen or fraudulent mobile units.

Many of the services offered by PCS systems are not currently available with standard cellular telephone systems, such as available mode, screen, private, and unavailable.

Available mode. The available mode allows all calls to pass through the network to the subscriber except for a minimal number of telephone numbers that can be blocked. The available mode relies on the delivery of the calling party number, which is checked against a database to ensure that it is not a blocked number. Subscribers can update or make changes in the database through the dial pad on their PCS handset.

Screen mode. The screen mode is the PCS equivalent to caller ID. With the screen mode, the name of the calling party appears on the mobile unit's display, which allows PCS users to screen calls. Unanswered calls are automatically forwarded to a *forwarding destination* specified by the subscriber, such as voice mail or another telephone number.

Private mode. With the private mode, all calls except those specified by the subscriber are automatically forwarded to a forwarding destination without ringing the subscriber's handset. Subscribers can make changes in the list of allowed calling numbers through the dial pad on their handset.

Unavailable mode. With the unavailable mode, no calls are allowed to pass through to the subscriber. Hence, all incoming calls are automatically forwarded to a forwarding destination.

PCS telephones are intended to be small enough to fit into a shirt pocket and use digital technology, which is quieter than analog. Their transmit power is relatively low; therefore, PCS systems utilize smaller cells and require more base stations than standard cellular systems for a given service area. PCS systems are sometimes called *microcellular* systems. The fundamental concept of PCS is to assign each mobile unit a PTN that is stored in a database on the SS7 common signaling network. The database keeps track of where mobile units are. When a call is placed for a mobile unit, the SS7 artificial intelligence network determines where the call should be directed.

The primary disadvantage of PCS is network cost. Employing small cells requires using more base stations, which equates to more transceivers, antennas, and trunk circuits. Antenna placement is critical with PCS. Large towers typically used with standard cellular systems are unacceptable in neighborhoods, which is where a large majority of PCS antennas must be placed.

PCS base stations communicate with other networks (cellular, PCS, and wireline) through a PCS switching center (PSC). The PSC is connected directly to the SS7 signaling network with a link to a signaling transfer point. PCS networks rely extensively on the SS7 signaling network for interconnecting to other telephone networks and databases.

PCS systems generally operate in a higher frequency band than standard cellular telephone systems. The FCC recently allocated an additional 160-MHz band in the 1850-MHz to 2200-MHz range. PCS systems operating in the 1900-MHz range are often referred to as *personal communications system 1900* (PCS 1900).

20-4 SECOND-GENERATION CELLULAR TELEPHONE SYSTEMS

First-generation cellular telephone systems were designed primarily for a limited customer base, such as business customers and a limited number of affluent residential customers. When the demand for cellular service increased, manufacturers searched for new technologies to improve the inherent problems with the existing cellular telephones, such as poor battery performance and channel unavailability. Improved batteries were also needed to reduce the size and cost of mobile units, especially those that were designed to be handheld. Weak signal strengths resulted in poor performance and a high rate of falsely initiated handoffs (*false handoffs*).

It was determined that improved battery performance and higher signal quality were possible only by employing digital technologies. In the United States, the shortcomings of the first-generation cellular systems led to the development of several second-generation cellular telephone systems, such as narrowband AMPS (N-AMPS) and systems employing the IS-54, IS-136, and IS-95 standards. A second-generation standard, known as *Global System for Mobile Communications* (GSM), emerged in Europe.

20-5 N-AMPS

Because of uncertainties about the practicality and cost effectiveness of implementing digital cellular telephone systems, Motorola developed a narrowband AMPS system called N-AMPS to increase the capacity of the AMPS system in large cellular markets. N-AMPS was originally intended to provide a short-term solution to the traffic congestion problem in the AMPS system. N-AMPS allows as many as three mobile units to use a single 30-kHz cellular channel at the same time. With N-AMPS, the maximum frequency deviation is reduced, reducing the required bandwidth to 10 kHz and thus providing a threefold increase in user capacity. One N-AMPS channel uses the carrier frequency for the existing AMPS channel and, with the other two channels, the carrier frequencies are offset by ±10 kHz. Each 10-kHz subchannel is capable of handling its own calls. Reducing the bandwidth degrades speech quality by lowering the signal-to-interference ratio. With narrower bandwidths, voice channels are more vulnerable to interference than standard AMPS channels and would generally require a higher frequency reuse factor. This is compensated for with the addition of an *interference avoidance scheme* called *Mobile Reported Interference* (MRI), which uses voice companding to provide synthetic voice channel quieting.

N-AMPS systems are *dual mode* in that mobile units are capable of operating with 30-kHz channels or with 10-kHz channels. N-AMPS systems use standard AMPS control channels for call setup and termination. N-AMPS mobile units are capable of utilizing four types of handoffs: *wide channel to wide channel* (30 kHz to 30 kHz), *wide channel to narrow channel* (30 kHz to 10 kHz), *narrow channel to narrow channel* (10 kHz to 10 kHz), and *narrow channel to wide channel* (10 kHz to 30 kHz).

20-6 DIGITAL CELLULAR TELEPHONE

Cellular telephone companies were faced with the problem of a rapidly expanding customer base while at the same time the allocated frequency spectrum remained unchanged. As is evident with N-AMPS, user capacity can be expanded by subdividing existing channels (band splitting), partitioning cells into smaller subcells (cell splitting), and modifying antenna radiation patterns (sectoring). However, the degree of subdivision and redirection is limited by the complexity and amount of overhead required to process handoffs between cells. Another serious restriction is the availability and cost of purchasing or leasing property for cell sites in the higher-density traffic areas.

Digital cellular telephone systems have several inherent advantages over analog cellular telephone systems, including better utilization of bandwidth, more privacy, and incorporation of error detection and correction.

AMPS is a first-generation analog cellular telephone system that was not designed to support the high-capacity demands of the modern world, especially in high-density metropolitan areas. In the late 1980s, several major manufacturers of cellular equipment determined that digital cellular telephone systems could provide substantial improvements in both capacity and performance. Consequently, the *United States Digital Cellular* (USDC) system was designed and developed with the intent of supporting a higher user density within a fixed-bandwidth frequency spectrum. Cellular telephone systems that use digital modulation, such as USDC, are called *digital cellular.*

The USDC cellular telephone system was originally designed to utilize the AMPS frequency allocation scheme. USDC systems comply with IS-54, which specifies dual-mode operation and backward compatibility with standard AMPS. USDC was originally designed to use the same carrier frequencies, frequency reuse plan, and base stations. Therefore, base stations and mobile units can be equipped with both AMPS and USDC channels within the same telephone equipment. In supporting both systems, cellular carriers are able to provide new customers with digital USDC telephones while still providing service to existing customers with analog AMPS telephones. Because the USDC system maintains compatibility with AMPS systems in several ways, it is also known as *Digital AMPS* (D-AMPS or sometimes DAMPS).

The USDC cellular telephone system has an additional frequency band in the 1.9-GHz range that is not compatible with the AMPS frequency allocation. Figure 20-5 shows the frequency spectrum and channel assignments for the 1.9-GHz band (sometimes called the PCS band). The total usable spectrum is subdivided into subbands (A through F); however, the individual channel bandwidth is limited to 30 kHz (the same as AMPS).

Reverse channel transmit frequency (GHz)

1.85 – 1.865	1.865 – 1.87	1.87 – 1.885	1.885 – 1.89	1.89 – 1.895	1.895 – 1.91
A-band (15 MHz) 449 30-kHz channels	D-band (15 MHz) 449 30-kHz channels	B-band (15 MHz) 449 30-kHz channels	E-band (15 MHz) 449 30-kHz channels	F-band (15 MHz) 449 30-kHz channels	C-band (15 MHz) 449 30-kHz channels

20 MHz separation between forward and reverse transmit frequency bands

Reverse channel transmit frequency (GHz)

1.93 – 1.945	1.945 – 1.95	1.95 – 1.965	1.965 – 1.97	1.97 – 1.975	1.975 – 1.99
A-band (15 MHz) 449 30-kHz channels	D-band (15 MHz) 449 30-kHz channels	B-band (15 MHz) 449 30-kHz channels	E-band (15 MHz) 449 30-kHz channels	F-band (15 MHz) 449 30-kHz channels	C-band (15 MHz) 449 30-kHz channels

FIGURE 20-5 1.9-GHz cellular frequency band

20-6-1 Time-Division Multiple Accessing

USDC uses *time-division multiple accessing* (TDMA) as well as frequency-division multiple accessing (FDMA). USDC, like AMPS, divides the total available radio-frequency spectrum into individual 30-kHz cellular channels (i.e., FDMA). However, TDMA allows more than one mobile unit to use a channel at the same time by further dividing transmissions within each cellular channel into time slots, one for each mobile unit using that channel. In addition, with AMPS FDMA systems, subscribers are assigned a channel for the duration of their call. However, with USDC TDMA systems, mobile-unit subscribers can only *hold* a channel while they are actually talking on it. During pauses or other normal breaks in a conversation, users must relinquish their channel so that other mobile units can use it. This technique of *time-sharing* channels significantly increases the capacity of a system, allowing more mobile-unit subscribers to use a system at virtually the same time within a given geographical area.

A USDC TDMA transmission frame consists of six equal-duration time slots enabling each 30-kHz AMPS channel to support three full-rate or six half-rate users. Hence, USDC offers as much as six times the channel capacity as AMPS. The original USDC standard also utilizes the same 50-MHz frequency spectrum and frequency-division duplexing scheme as AMPS.

The advantages of digital TDMA multiple-accessing systems over analog AMPS FDMA multiple-accessing systems are as follows:

1. Interleaving transmissions in the time domain allows for a threefold to sixfold increase in the number of mobile subscribers using a single cellular channel. Time-sharing is realized because of digital compression techniques that produce bit rates approximately one-tenth that of the initial digital sample rate and about one-fifth the initial rate when error detection/correction (EDC) bits are included.
2. Digital signals are much easier to process than analog signals. Many of the more advanced modulation schemes and information processing techniques were developed for use in a digital environment.
3. Digital signals (bits) can be easily encrypted and decrypted, safeguarding against eavesdropping.
4. The entire telephone system is compatible with other digital formats, such as those used in computers and computer networks.
5. Digital systems inherently provide a quieter (less noisy) environment than their analog counterparts.

20-6-2 EIA/TIA Interim Standard 54 (IS-54)

In 1990, the Electronics Industries Association and Telecommunications Industry Association (EIA/TIA) standardized the dual-mode USDC/AMPS system as Interim Standard 54 (IS-54), *Cellular Dual Mode Subscriber Equipment. Dual mode* specifies that a mobile station complying with the IS-54 standard must be capable of operating in either the analog AMPS or the digital (USDC) mode for voice transmissions. Using IS-54, a cellular telephone carrier could convert any or all of its existing analog channels to digital. The key criterion for achieving dual-mode operation is that IS-54 digital channels cannot interfere with transmissions from existing analog AMPS base and mobile stations. This goal is achieved with IS-54 by providing digital control channels and both analog and digital voice channels. Dual-mode mobile units can operate in either the digital or the analog mode for voice and access the system with the standard AMPS digital control channel. Before a voice channel is assigned, IS-54 mobile units use AMPS forward and reverse control channels to carry out user authentications and call management operations. When a dual-mode mobile unit transmits an access request, it indicates that it is capable of operating in the digital mode; then the base station will allocate a digital voice channel, provided one is available. The allocation procedure indicates the channel number (frequency) and the specific time slot (or slots) within that particular channel's TDMA frame. IS-54 specifies a 48.6-kbps rate per 30-kHz voice channel divided among three simultaneous users. Each user is allocated 13 kbps, and the remaining 9.6 kbps is used for timing and control overhead.

In many rural areas of the United States, analog cellular telephone systems use only the original 666 AMPS channels (1 through 666). In these areas, USDC channels can be added in the extended frequency spectrum (channels 667 through 799 and 991 through 1023) to support USDC telephones that roam into the system from other areas. In high-density urban areas, selected frequency bands are gradually being converted one at a time to the USDC digital standard to help alleviate traffic congestion. Unfortunately, this gradual changeover from AMPS to USDC often results in an increase in the interference and number of dropped calls experienced by subscribers of the AMPS system.

The successful and graceful transition from analog cellular systems to digital cellular systems using the same frequency band was a primary consideration in the development of the USDC standard. The introduction of N-AMPS and a new digital spread-spectrum standard has delayed the widespread deployment of the USDC standard throughout the United States.

20-6-3 USDC Control Channels and IS-136.2

The IS-54 USDC standard specifies the same 42 *primary control channels* as AMPS and 42 additional control channels called *secondary control channels.* Thus, USDC offers twice as many control channels as AMPS and is, therefore, capable of providing twice the capacity of control traffic within a given market area. Carriers are allowed to dedicate the secondary control channels for USDC-only use since AMPS mobile users do not monitor and cannot decode the new secondary control channels. In addition, to maintain compatibility with existing AMPS cellular telephone systems, the primary forward and reverse control channels in USDC cellular systems use the same signaling techniques and modulation scheme (FSK) as AMPS. However, a new standard, IS-136.2 (formerly IS-54, Rev.C), replaces FSK with $\pi/4$ DQPSK modulation for the 42 dedicated USDC secondary control channels, allowing digital mobile units to operate entirely in the digital domain. The IS-136.2 standard is often called North American-Time Division Multiple Accessing (NA-TDMA). IS-54 Rev.C was introduced to provide PSK (phase-shift keying) rather than FSK on dedicated USDC control channels to increase the control data rates and provide additional specialized services, such as paging and short messaging between private mobile user groups. *Short message service* allows for brief paging-type messages and short e-mail messages (up to 239 characters) that can be read on the mobile phone's display and entered using the keypad.

IS-136 was developed to provide a host of new features and services, positioning itself in a competitive market with the newer PCS systems. Because IS-136 specifies short messaging capabilities and private user-group features, it is well suited as a wireless paging system. IS-136 also provides an additional *sleep mode,* which conserves power in the mobile units. IS-136 mobile units are not compatible with IS-54 units, as FSK control channels are not supported.

The digital control channel is necessarily complex, and a complete description is beyond the scope of this book. Therefore, the following discussion is meant to present a general overview of the operation of a USDC digital control channel.

The IS-54 standard specifies three types of channels: analog control channels, analog voice channels, and a 10-kbps binary FSK digital control channel (DCCH). The IS-54 Rev.C standard (IS-136) provides for the same three types of channels plus a fourth—a digital control channel with a signaling rate of 48.6 kbps on USDC-only control channels. The new digital control channel is meant to eventually replace the analog control channel. With the addition of a digital control channel, a mobile unit is able to operate entirely in the digital domain, using the digital control channel for system and cell selection and channel accessing and the digital voice channel for digitized voice transmissions.

IS-136 details the exact functionality of the USDC digital control channel. The initial version of IS-136 was version 0, which has since been updated by revision A. Version 0 added numerous new services and features to the USDC digital cellular telephone system, including enhanced user services, such as short messaging and displaying the telephone number of the incoming call; sleep mode, which gives the telephone set a longer battery life when in the standby mode; private or residential system service; and enhanced security and validation against fraud. The newest version of IS-136, revision A, was developed to provide numerous new features and services by introducing an enhanced vocoder, over-the-air activation where the network operators are allowed to program information into telephones directly over the air, calling name and number ID, and enhanced hands-off and priority access to control channels.

IS-136 specifies several private user-group features, making it well adapted for wireless PBX and paging applications. However, IS-136 user terminals operate at 48.6 kbps and are, therefore, not compatible with IS-54 FSK terminals. Thus, IS-136 modems are more cost effective, as it is necessary to include only the 48.6-kbps modem in the terminal equipment.

20-6-3-1 Logical channels. The new digital control channel includes several *logical channels* with different functions, including the *random access channel* (RACH); the *SMS point-to-point, paging, and access response channel* (SPACH); the *broadcast control channel* (BCCH); and the *shared channel feedback* (SCF) channel. Figure 20-6 shows the logical control channels for the IS-136 standard.

20-6-3-2 Random access channel (RACH). RACH is used by mobile units to request access to the cellular telephone system. RACH is a unidirectional channel specified for transmissions from mobile-to-base units only. Access messages, such as origination, registration, page responses, audit confirmation, serial number, and message confirmation, are transmitted on the RACH. It also transmits messages that provide information on authentication, security parameter updates, and *short message service* (SMS) point-to-point messages. RACH is capable of operating in two modes using contention resolution similar to voice channels. RACH can also operate in a *reservation mode* for replying to a base-station command.

20-6-3-3 SMS point-to-point, paging, and access response channel (SPACH). SPACH is used to transmit information from base stations to specific mobile stations. RACH is a unidirectional channel specified for transmission from base stations to mobile units only and is shared by all mobile units. Information transmitted on the SPACH channel includes three separate logical subchannels: *SMS point-to-point messages, paging messages,* and *access response messages.* SPACH can carry messages related to a single mo-

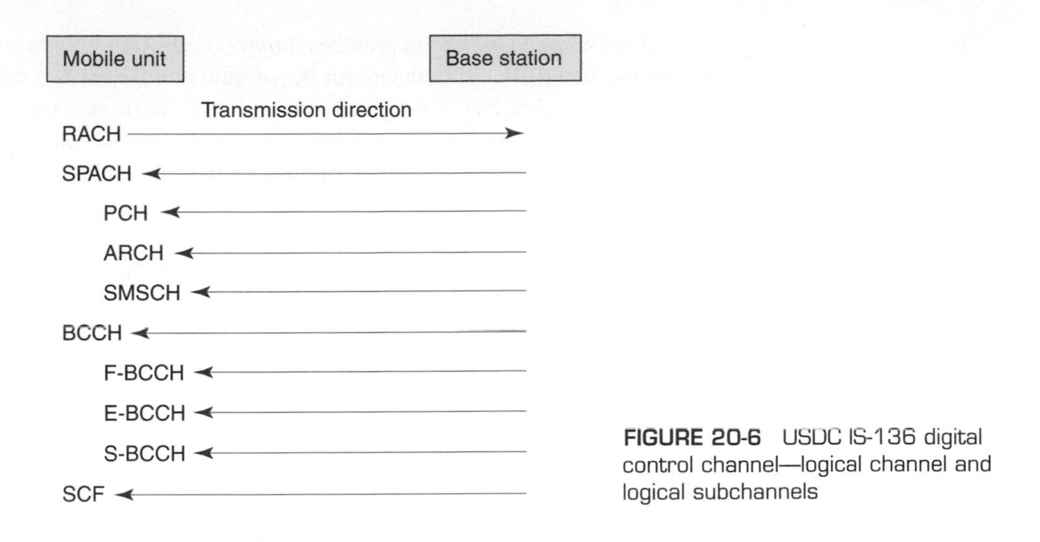

FIGURE 20-6 USDC IS-136 digital control channel—logical channel and logical subchannels

bile unit or to a small group of mobile units and allows larger messages to be broken down into smaller blocks for transmission.

The paging channel (PCH) is a subchannel of the logical channel of SPACH. PCH is dedicated to delivering pages and orders. The PCH transmits *paging messages, message-waiting messages,* and *user-alerting messages.* Each PCH message can carry up to five mobile identifiers. Page messages are always transmitted and then repeated a second time. Messages such as *call history count updates* and *shared secret data updates* used for the authentication and encryption process also are sent on the PCH.

The access response channel (ARCH) is also a logical subchannel of SPACH. A mobile unit automatically moves to an ARCH immediately after successful completion of contention- or reservation-based access on a RACH. ARCH can be used to carry assignments to another resource or other responses to the mobile station's access attempt. Messages assigning a mobile unit to an analog voice channel or a digital voice channel or redirecting the mobile to a different cell are also sent on the ARCH along with registration access (accept, reject, or release) messages.

The SMS channel (SMSCH) is used to deliver short point-to-point messages to a specific mobile station. Each message is limited to a maximum of 200 characters of text. Mobile-originated SMS is also supported; however, SMS where a base station can broadcast a short message designated for several mobile units is not supported in IS-136.

20-6-3-4 Broadcast control channel (BCCH). BCCH is an acronym referring to the F-BCCH, E-BCCH, and S-BCCH logical subchannels. These channels are used to carry generic, system-related information. BCCH is a unidirectional base station-to-mobile unit transmission shared by all mobile units.

The *fast broadcast control channel* (F-BCCH) broadcasts digital control channel (DCCH) structure parameters, including information about the number of F-BCCH, E-BCCH, and S-BCCH time slots in the DCCH frame. Mobile units use F-BCCH information when initially accessing the system to determine the beginning and ending of each logical channel in the DCCH frame. F-BCCH also includes information pertaining to access parameters, including information necessary for authentication and encryptions and information for mobile access attempts, such as the number of access retries, access burst size, initial access power level, and indication of whether the cell is barred. Information addressing the different types of registration, registration periods, and system identification information, including network type, mobile country code, and protocol revision, is also provided by the F-BCCH channel.

The *extended broadcast control channel* (E-BCCH) carries less critical broadcast information than F-BCCH intended for the mobile units. E-BCCH carries information about neighboring analog and TDMA cells and optional messages, such as emergency information, time and date messaging, and the types of services supported by neighboring cells.

The *SMS broadcast control channel* (S-BCCH) is a logical channel used for sending short messages to individual mobile units.

20-6-3-5 Shared channel feedback (SCF) channel. SCF is used to support random access channel operation by providing information about which time slots the mobile unit can use for access attempts and also if a mobile unit's previous RACH transmission was successfully received.

20-6-4 USDC Digital Voice Channel

Like AMPS, each USDC voice channel is assigned a 30-kHz bandwidth on both the forward and the reverse link. With USDC, however, each voice channel can support as many as three full-rate mobile users simultaneously by using digital modulation and a TDMA format called *North American Digital Cellular* (NADC). Each radio-frequency voice channel in the total AMPS FDMA frequency band consists of one 40-ms TDMA frame comprised of six time slots containing 324 bits each, as shown in Figure 20-7. For full-speech rate, three users

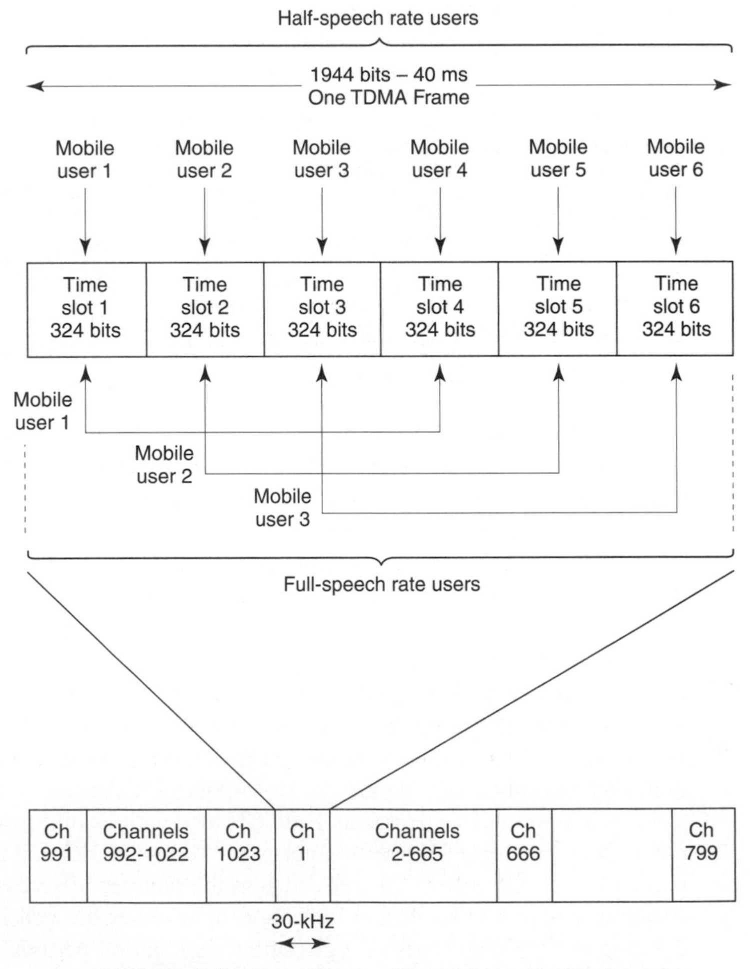

FIGURE 20-7 North American Digital Cellular TDMA frame format

share the six time slots in an equally spaced manner. For example, mobile user 1 occupies time slots 1 and 4, mobile user 2 occupies time slots 2 and 5, and mobile user 3 occupies time slots 3 and 6. For half-rate speech, each user occupies one time slot per frame. During their respective time slots, mobile units transmit short bursts (6.67 ms) of a digital-modulated carrier to the base station (i.e., *uplink transmissions*). Hence, full-rate users transmit two bursts during each TDMA frame. In the *downlink* path (i.e., from base stations to mobile units), base stations generally transmit continuously. However, mobile units only listen during their assigned time slot. The average cost per subscriber per base station equipment is lower with TDMA since each base station transceiver can be shared by up to six users at a time.

General Motors Corporation implemented a TDMA scheme called E-TDMA, which incorporates six half-rate users transmitting at half the bit rate of standard USDC TDMA systems. E-TDMA systems also incorporate *digital speech interpolation* (DSI) to dynamically assign more than one user to a time slot, deleting silence on calls. Consequently, E-TDMA can handle approximately 12 times the user traffic as standard AMPS systems and four times that of systems complying with IS-54.

Each time slot in every USDC voice-channel frame contains four data channels—three for control and one for digitized voice and user data. The full-duplex *digital traffic channel* (DTC) carries digitized voice information and consists of a *reverse digital traffic channel* (RDTC) and a *forward digital traffic channel* (FDTC) that carry digitized speech information or user data. The RDTC carries speech data from the mobile unit to the base station, and the FDTC carries user speech data from the base station to the mobile unit. The three supervisory channels are the *coded digital verification color code* (CDVCC), the *slow associated control channel* (SACCH), and the *fast associated control channel* (FACCH).

20-6-4-1 Coded digital verification color code. The purpose of the CDVCC color code is to provide co-channel identification similar to the SAT signal transmitted in the AMPS system. The CDVCC is a 12-bit message transmitted in every time slot. The CDVCC consists of an eight-bit digital voice color code number between 1 and 255 appended with four additional coding bits derived from a shortened Hamming code. The base station transmits a CDVCC number on the forward voice channel, and each mobile unit using the TDMA channel must receive, decode, and retransmit the same CDVCC code (handshake) back to the base station on the reverse voice channel. If the two CDVCC values are not the same, the time slot is relinquished for other users, and the mobile unit's transmitter will be automatically turned off.

20-6-4-2 Slow associated control channel. The SACCH is a signaling channel for transmission of control and supervision messages between the digital mobile unit and the base station while the mobile unit is involved with a call. The SACCH uses 12 coded bits per TDMA burst and is transmitted in every time slot, thus providing a signaling channel in parallel with the digitized speech information. Therefore, SACCH messages can be transmitted without interfering with the processing of digitized speech signals. Because the SACCH consists of only 12 bits per frame, it can take up to 22 frames for a single SACCH message to be transmitted. The SACCH carries various control and supervisory information between the mobile unit and the base station, such as communicating power-level changes and hand-off requests. The SACCH is also used by the mobile unit to report signal-strength measurements of neighboring base stations so, when necessary, the base station can initiate a *mobile-assisted handoff* (MAHO).

20-6-4-3 Fast associated control channel. The FACCH is a second signaling channel for transmission of control and specialized supervision and traffic messages between the base station and the mobile units. Unlike the CDVCC and SACCH, the FACCH does not have a dedicated time slot. The FACCH is a *blank-and-burst* type of transmission that, when transmitted, replaces digitized speech information with control and supervision messages within a subscriber's time slot. There is no limit on the number of speech frames that can be replaced with FACCH data. However, the digitized voice information is somewhat

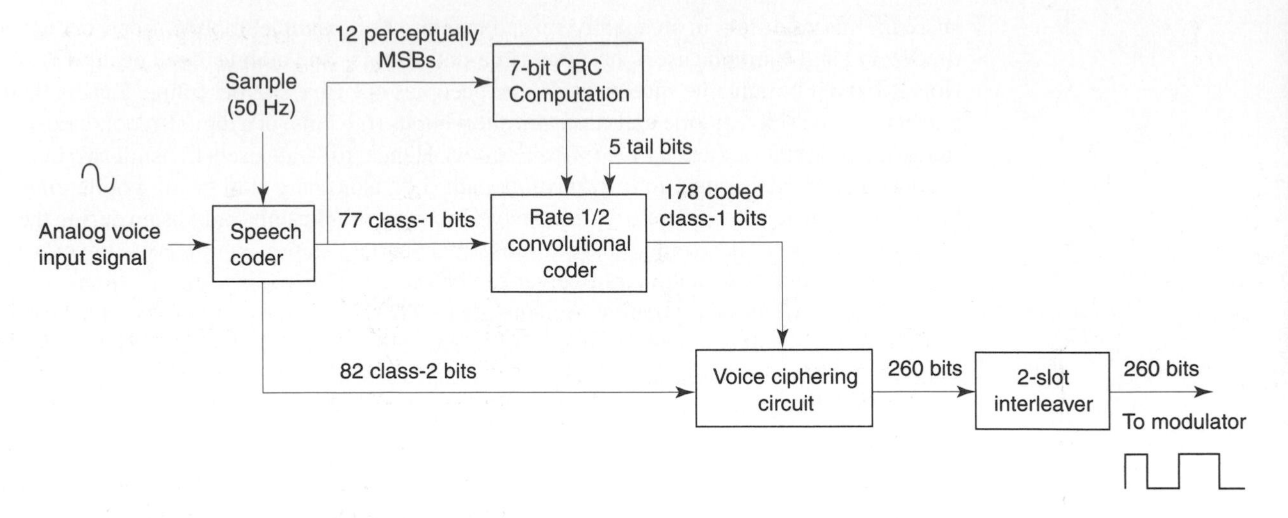

FIGURE 20-8 USDC digital voice-channel speech coder

protected by preventing an entire digitized voice transmission from being replaced by FACCH data. The 13-kbps net digitized voice transmission rate cannot be reduced below 3250 bps in any given time slot. There are no fields within a standard time slot to identify it as digitized speech or an FACCH message. To determine if an FACCH message is being received, the mobile unit must attempt to decode the data as speech. If it decodes in error, it then decodes the data as an FACCH message. If the cyclic redundancy character (CRC) calculates correctly, the message is assumed to be an FACCH message. The FACCH supports transmission of dual-tone multiple-frequency (DTMF) Touch-Tones, call release instruction, flash hook instructions, and mobile-assisted handoff or mobile-unit status requests. The FACCH data are packaged and interleaved to fit in a time slot similar to the way digitized speech is handled.

20-6-5 Speech Coding

Figure 20-8 shows the block diagram for a USDC digital voice-channel speech encoder. Channel error control for the digitized speech data uses three mechanisms for minimizing channel errors: (1) A rate one-half convolutional code is used to protect the more vulnerable bits of the speech coder data stream; (2) transmitted data are interleaved for each speech coder frame over two time slots to reduce the effects of Rayleigh fading; and (3) a cyclic redundancy check is performed on the most perceptually significant bits of the digitized speech data.

With USDC, incoming analog voice signals are sampled first and then converted to a binary PCM in a special *speech coder* (*vocoder*) called a *vector sum exciter linear predictive* (VSELP) *coder* or a *stochastically excited linear predictive* (SELP) *coder.* Linear predictive coders are time-domain types of vocoders that attempt to extract the most significant characteristics from the time-varying speech waveform. With linear predictive coders, it is possible to transmit good-quality voice at 4.8 kbps and acceptable, although poorer-quality, voice at lower bit rates.

Because there are many predictable orders in spoken word patterns, it is possible, using advanced algorithms, to compress the binary samples and transmit the resulting bit stream at a 13-kbps rate. A consortium of companies, including Motorola, developed the VSELP algorithm, which was subsequently adopted for the IS-54 standard. Error-detection and -correction (EDC) bits are added to the digitally compressed voice signals to reduce the effects of interference, bringing the final voice data rate to 48.6 kbps. Compression/expansion and error-detection/correction functions are implemented in the telephone handset by a special microprocessor called a digital signal processor (DSP).

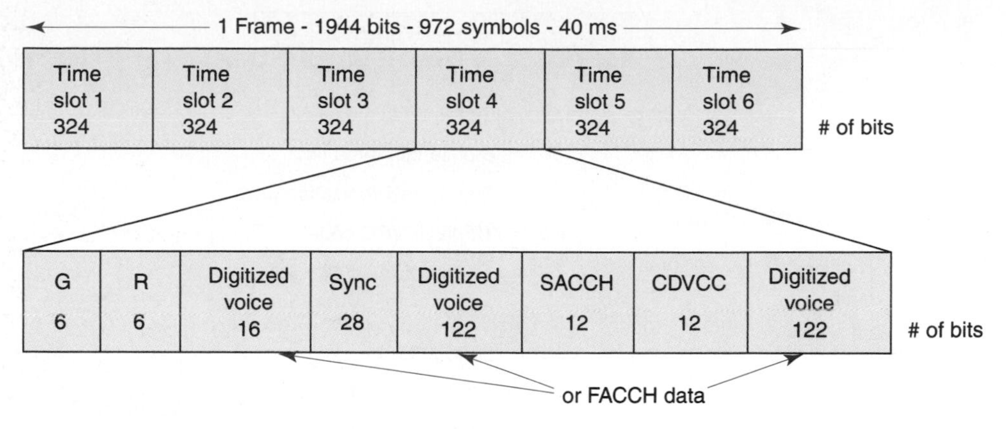

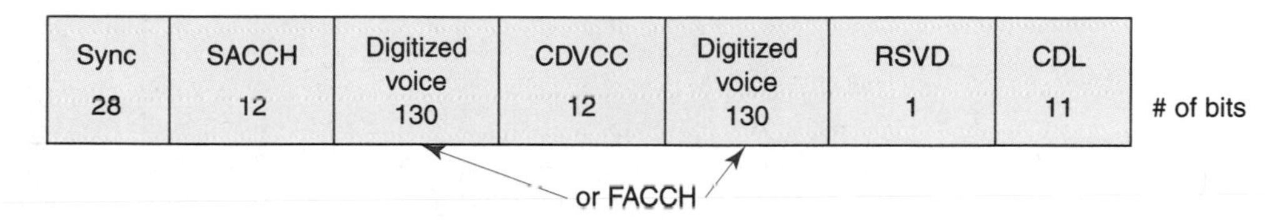

FIGURE 20-9 USDC digital voice channel slot and frame format

The VSELP coders output 7950 bps and produce a speech frame every 20 ms, or

$$\frac{7950 \text{ bits}}{\text{second}} \times \frac{20 \text{ ms}}{\text{frame}} = 159 \text{ bits-per-frame}$$

Fifty speech frames are outputted each second containing 159 bits each, or

$$\frac{50 \text{ frames}}{\text{second}} \times \frac{159 \text{ bits}}{\text{frame}} = 7950 \text{ bps}$$

The 159 bits included in each speech coder frame are divided into two classes according to the significance in which they are perceived. There are 77 class 1 bits and 82 class 2 bits. The class 1 bits are the most significant and are, therefore, error protected. The 12 most significant class 1 bits are block coded using a seven-bit CRC error-detection code to ensure that the most significant speech coder bits are decoded with a low probability of error. The less significant class 2 bits have no means of error protection.

After coding the 159 bits, each speech code frame is converted in a 1/2 convolution coder to 260 channel-coded bits per frame, and 50 frames are transmitted each second. Hence, the transmission bit rate is increased from 7950 bps for each digital voice channel to 13 kbps:

$$\frac{260 \text{ bits}}{\text{frame}} \times \frac{50 \text{ frames}}{\text{second}} = 13 \text{ kbps}$$

Figure 20-9 shows the time slot and frame format for the forward (base station to mobile unit) and reverse (mobile unit to base station) links of a USDC digital voice channel. USDC voice channels use frequency-division duplexing; thus, forward and reverse channel time slots operate on different frequencies at the same time. Each time slot carries interleaved digital voice data from the two adjacent frames outputted from the speech coder.

G1	RSDSDVSDWSDXSDYS	G2

Where G1 = 6-bit guard time
R = 6-bit length ramp time
S = 28-bit synchronization word
D = 12-bit CDVCC code
G2 = 44-bit guard time
V = 0000
W = 00000000
X = 000000000000
Y = 0000000000000000

FIGURE 20-10 USDC shortened burst digital voice channel format

In the reverse channel, each time slot contains two bursts of 122 digitized voice bits and one burst of 16 bits for a total of 260 digitized voice bits per frame. In addition, each time slot contains 28 synchronization bits, 12 bits of SACCH data, 12 bits of CDVCC bits, and six guard bits to compensate for differences in the distances between mobile units and base stations. The guard time is present in only the reverse channel time slots to prevent overlapping of received bursts due to radio signal transit time. The ramp-up time consists of six bits that allow gradual rising and falling of the RF signal energy within the time slot. Thus, a reverse channel time slot consists of 324 bits. If an FACCH is sent instead of speech data, one time slot of speech coding data is replaced with a 260-bit block of FACCH data.

In the forward channel, each time slot contains two 130-bit bursts of digitized voice data (or FACCH data if digitized speech is not being sent) for a total of 260 bits per frame. In addition, each forward channel frame contains 28 synchronization bits, 12 bits of SACCH data, 12 CDVCC bits, and 12 reserved bits for a total of 324 bits per time slot. Therefore, both forward and reverse voice channels have a data transmission rate of

$$\frac{324 \text{ bits}}{\text{time slot}} \times \frac{6 \text{ time slots}}{40 \text{ ms}} = 48.6 \text{ kbps}$$

A third frame format, called a *shortened burst,* is shown in Figure 20-10. Shortened bursts are transmitted when a mobile unit begins operating in a larger-diameter cell because the propagation time between the mobile and base is unknown. A mobile unit transmits shortened burst slots until the base station determines the required time offset. The default delay between the receive and transmit slots in the mobile is 44 symbols, which results in a maximum distance at which a mobile station can operate in a cell to 72 miles for an IS-54 cell.

20-6-6 USDC Digital Modulation Scheme

To achieve a transmission bit rate of 48.6 kbps in a 30-kHz AMPS voice channel, a *bandwidth (spectral) efficiency* of 1.62 bps/Hz is required, which is well beyond the capabilities of binary FSK. The spectral efficiency requirements can be met by using conventional pulse-shaped, four-phase modulation schemes, such as QPSK and OQPSK. However, USDC voice and control channels use a *symmetrical differential, phase-shift keying* technique known as π/4 DQPSK, or π/4 *differential quadriphase shift keying* (DQPSK), which offers several advantages in a mobile radio environment, such as improved co-channel rejection and bandwidth efficiency.

A 48.6-kbps data rate requires a symbol (baud) rate of 24.3 kbps (24.3 kilobaud per second) with a symbol duration of 41.1523 μs. The use of pulse shaping and π/4 DQPSK supports the transmission of three different 48.6-kbps digitized speech signals in a 30-kHz

Table 20-4 NA-TDMA Mobile Phone Power Levels

	Class I		Class II		Class III		Class IV
Power Level	dBm	mW	dBm	mW	dBm	mW	dBm
0	36	4000	32	1600	28	640	28
1	32	1600	32	1600	28	640	28
2	28	640	28	640	28	640	28
3	24	256	24	256	24	256	24
4	20	102	20	102	20	102	20
5	16	41	16	41	16	41	16
6	12	16	12	16	12	16	12
7	8	6.6	8	6.6	8	6.6	8
8	—	—	Dual mode only		—	—	4 dBm ± 3 dB
9	—	—	Dual mode only		—	—	0 dBm ± 6 dB
10	—	—	Dual mode only		—	—	− 4 dBm ± 9 dB

bandwidth with as much as 50 dB of adjacent-channel isolation. Thus, the bandwidth efficiency using $\pi/4$ DQPSK is

$$\eta = \frac{3 \times 48.6 \text{ kbps}}{30 \text{ kHz}}$$

$$= 4.86 \text{ bps/Hz}$$

where η is the bandwidth efficiency.

In a $\pi/4$ DQPSK modulator, data bits are split into two parallel channels that produce a specific phase shift in the analog carrier and, since there are four possible bit pairs, there are four possible phase shifts using a quadrature I/Q modulator. The four possible differential phase changes, $\pi/4$, $-\pi/4$, $3\pi/4$, and $-3\pi/4$, define eight possible carrier phases. Pulse shaping is used to minimize the bandwidth while limiting the intersymbol interference. In the transmitter, the PSK signal is filtered using a square-root raised cosine filter with a roll-off factor of 0.35. PSK signals, after pulse shaping, become a linear modulation technique, requiring linear amplification to preserve the pulse shape. Using pulse shaping with $\pi/4$ DQPSK allows for the simultaneous transmission of three separate 48.6-kbps speech signals in a 30-kHz bandwidth.

20-6-7 USDC Radiated Power

NA-TDMA specifies 11 radiated power levels for four classifications of mobile units, including the eight power levels used by standard AMPS transmitters. The fourth classification is for dual-mode TDMA/analog cellular telephones. The NA-TDMA power classifications are listed in Table 20-4. The highest power level is 4 W (36 dBm), and successive levels differ by 4 dB, with the lowest level for classes I through III being 8 dBm (6.6 mW). The lowest transmit power level for dual-mode mobile units is -4 dBm (0.4 mW) $\pm$ 9 dB. In a dual-mode system, the three lowest power levels can be assigned only to digital voice channels and digital control channels. Analog voice channels and FSK control channels transmitting in the standard AMPS format are confined to the eight power levels in the AMPS specification. Transmitters in the TDMA mode are active only one-third of the time; therefore, the average transmitted power is 4.8 dB below specifications.

20-7 INTERIM STANDARD 95 (IS-95)

FDMA is an access method used with standard analog AMPS, and both FDMA and TDMA are used with USDC. Both FDMA and TDMA use a frequency channelization approach to frequency spectrum management; however, TDMA also utilizes a time-division accessing approach. With FDMA and TDMA cellular telephones, the entire available cellular radio-frequency spectrum is subdivided into narrowband radio channels to be used for one-way communications links between cellular mobile units and base stations.

In 1984, Qualcomm Inc. proposed a cellular telephone system and standard based on spread-spectrum technology with the primary goal of increasing capacity. Qualcomm's new system enabled a totally digital mobile telephone system to be made available in the United States based on *code-division multiple accessing* (CDMA). The U.S. Telecommunications Industry Association recently standardized the CDMA system as Interim Standard 95 (IS-95), which is a mobile-to-base station compatibility standard for dual-mode wideband spread-spectrum communications. CDMA allows users to differentiate from one another by a unique code rather than a frequency or time assignment and, therefore, offers several advantages over cellular telephone systems using TDMA and FDMA, such as increased capacity and improved performance and reliability. IS-95, like IS-54, was designed to be compatible with existing analog cellular telephone system (AMPS) frequency band; therefore, mobile units and base stations can easily be designed for dual-mode operation. Pilot CDMA systems developed by Qualcomm were first made available in 1994.

NA-TDMA channels occupy exactly the same bandwidth as standard analog AMPS signals. Therefore, individual AMPS channel units can be directly replaced with TDMA channels, which are capable of carrying three times the user capacity as AMPS channels. Because of the wide bandwidths associated with CDMA transmissions, IS-95 specifies an entirely different channel frequency allocation plan than AMPS.

The IS-95 standard specifies the following:

1. Modulation—digital OQPSK (uplink) and digital QPSK (downlink)
2. 800-MHz band (IS-95A)
   45-MHz forward and reverse separation
   50-MHz spectral allocation
3. 1900-MHz band (IS-95B)
   90-MHz forward and reverse separation
   120-MHz spectral allocation
4. 2.46-MHz total bandwidth
   1.23-MHz reverse CDMA channel bandwidth
   1.23-MHz forward CDMA channel bandwidth
5. Direct-sequence CDMA accessing
6. 8-kHz voice bandwidth
7. 64 total channels per CDMA channel bandwidth
8. 55 voice channels per CDMA channel bandwidth

20-7-1 CDMA

With IS-95, each mobile user within a given cell, and mobile subscribers in adjacent cells use the same radio-frequency channels. In essence, frequency reuse is available in all cells. This is made possible because IS-95 specifies a direct-sequence, spread-spectrum CDMA system and does not follow the channelization principles of traditional cellular radio communications systems. Rather than dividing the allocated frequency spectrum into narrow-bandwidth channels, one for each user, information is transmitted (spread) over a very wide frequency spectrum with as many as 20 mobile subscriber units simultaneously using the same carrier frequency within the same frequency band. Interference is incorporated into the system so that there is no limit to the number of subscribers that CDMA can support. As more mobile subscribers are added to the system, there is a *graceful degradation* of communications quality.

With CDMA, unlike other cellular telephone standards, subscriber data change in real time, depending on the voice activity and requirements of the network and other users of the network. IS-95 also specifies a different modulation and spreading technique for the forward and reverse channels. On the forward channel, the base station simultaneously transmits user data from all current mobile units in that cell by using different spreading sequences (codes) for each user's transmissions. A pilot code is transmitted with the user data at a higher power level, thus allowing all mobile units to use coherent detection. On the reverse link, all mobile units respond in an asynchronous manner (i.e., no time or duration limitations) with a constant signal level controlled by the base station.

The speech coder used with IS-95 is the Qualcomm 9600-bps *Code-Excited Linear Predictive* (QCELP) coder. The vocoder converts an 8-kbps compressed data stream to a 9.6-kbps data stream. The vocoder's original design detects voice activity and automatically reduces the data rate to 1200 bps during silent periods. Intermediate mobile user data rates of 2400 bps and 4800 bps are also used for special purposes. In 1995, Qualcomm introduced a 14,400-bps vocoder that transmits 13.4 kbps of compressed digital voice information.

20-7-1-1 CDMA frequency and channel allocations. CDMA reduces the importance of frequency planning within a given cellular market. The AMPS U.S. cellular telephone system is allocated a 50-MHz frequency spectrum (25 MHz for each direction of propagation), and each service provider (system A and system B) is assigned half the available spectrum (12.5 MHz). AMPS common carriers must provide a 270-kHz guard band (approximately nine AMPS channels) on either side of the CDMA frequency spectrum. To facilitate a graceful transition from AMPS to CDMA, each IS-95 channel is allocated a 1.25-MHz frequency spectrum for each one-way CDMA communications channel. This equates to 10% of the total available frequency spectrum of each U.S. cellular telephone provider. CDMA channels can coexist within the AMPS frequency spectrum by having a wireless operator clear a 1.25-MHz band of frequencies to accommodate transmissions on the CDMA channel. A single CDMA radio channel takes up the same bandwidth as approximately 42 30-kHz AMPS voice channels. However, because of the frequency reuse advantage of CDMA, CDMA offers approximately a 10-to-1 channel advantage over standard analog AMPS and a 3-to-1 advantage over USDC digital AMPS.

For reverse (downlink) operation, IS-95 specifies the 824-MHz to 849-MHz band and forward (uplink) channels the 869-MHz to 894-MHz band. CDMA cellular systems also use a modified frequency allocation plan in the 1900-MHz band. As with AMPS, the transmit and receive carrier frequencies used by CDMA are separated by 45 MHz. Figure 20-11a shows the frequency spacing for two adjacent CDMA channels in the AMPS frequency band. As the figure shows, each CDMA channel is 1.23 MHz wide with a 1.25-MHz frequency separation between adjacent carriers, producing a 200-kHz guard band between CDMA channels. Guard bands are necessary to ensure that the CDMA carriers do not interfere with one another. Figure 20-11b shows the CDMA channel location within the AMPS frequency spectrum. The lowest CDMA carrier frequency in the A band is at AMPS channel 283, and the lowest CDMA carrier frequency in the B band is at AMPS channel 384. Because the band available between 667 and 716 is only 1.5 MHz in the A band, A band operators have to acquire permission from B band carriers to use a CDMA carrier in that portion of the frequency spectrum. When a CDMA carrier is being used next to a non-CDMA carrier, the carrier spacing must be 1.77 MHz. There are as many as nine CDMA carriers available for the A and B band operator in the AMPS frequency spectrum. However, the A and B band operators have 30-MHz bandwidth in the 1900-MHz frequency band, where they can facilitate up to 11 CDMA channels.

With CDMA, many users can share common transmit and receive channels with a transmission data rate of 9.6 kbps. Using several techniques, however, subscriber information is spread by a factor of 128 to a channel chip rate of 1.2288 Mchips/s, and transmit and receive channels use different spreading processes.

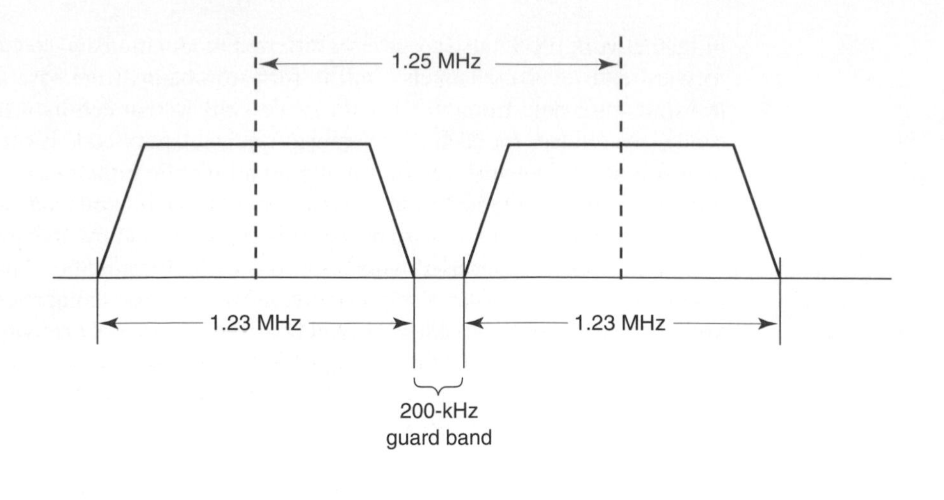

(a)

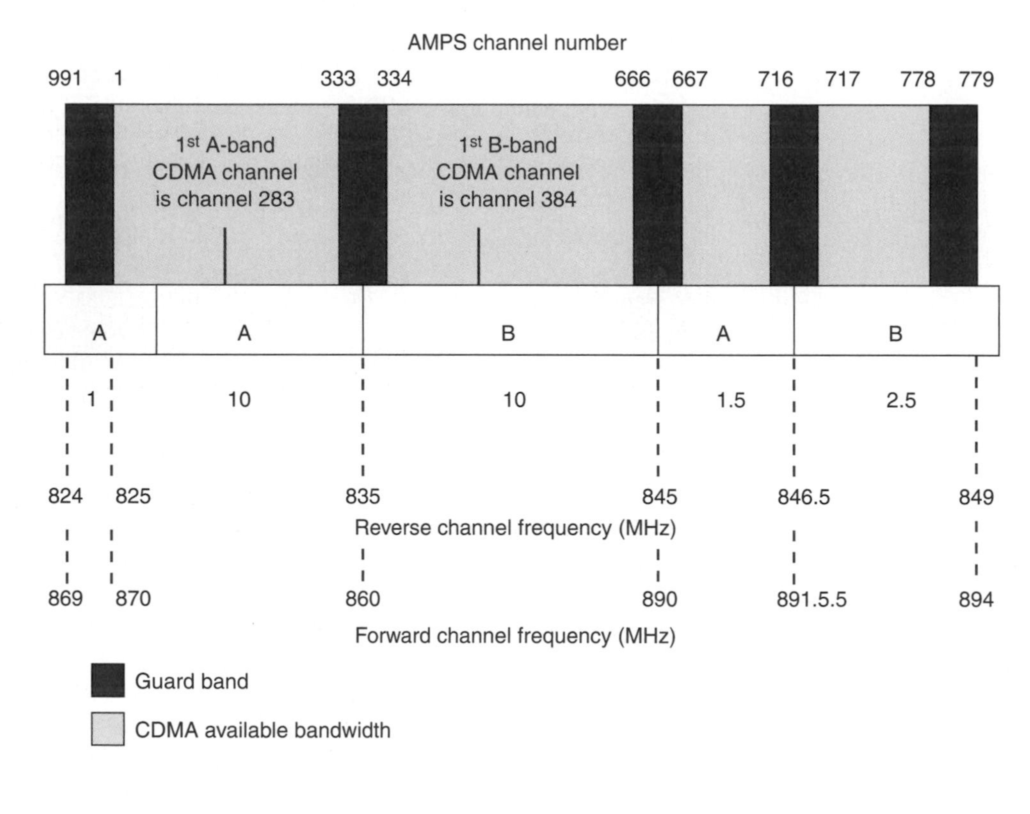

(b)

FIGURE 20-11 (a) CDMA channel bandwidth, guard band, and frequency separation; (b) CDMA channel location within the AMPS frequency spectrum

In the uplink channel, subscriber data are encoded using a rate 1/2 convolutional code, interleaved, and spread by one of 64 orthogonal spreading sequences using Walsh functions. Orthogonality among all uplink cellular channel subscribers within a given cell is maintained because all the cell signals are scrambled synchronously.

Downlink channels use a different spreading strategy since each mobile unit's received signal takes a different transmission path and, therefore, arrives at the base station at a different time. Downlink channel data streams are first convolutional encoded with a rate

1/3 convolution code. After interleaving, each block of six encoded symbols is mapped to one of the available orthogonal Walsh functions, ensuring 64-ary orthogonal signaling. An additional fourfold spreading is performed by subscriber-specified and base station-specific codes having periods of 2^{14} chips and 2^{15} chips, respectively, increasing the transmission rate to 1.2288 Mchips/s. Stringent requirements are enforced in the downlink channel's transmit power to avoid the near-far problem caused by varied receive power levels.

Each mobile unit in a given cell is assigned a unique spreading sequence, which ensures near perfect separation among the signals from different subscriber units and allows transmission differentiation between users. All signals in a particular cell are scrambled using a pseudorandom sequence of length 2^{15} chips. This reduces radio-frequency interference between mobiles in neighboring cells that may be using the same spreading sequence and provides the desired wideband spectral characteristics even though all Walsh codes do not yield a wideband power spectrum.

Two commonly used techniques for spreading the spectrum are *frequency hopping* and *direct sequencing*. Both of these techniques are characteristic of transmissions over a bandwidth much wider than that normally used in narrowband FDMA/TDMA cellular telephone systems, such as AMPS and USDC. For a more detailed description of frequency hopping and direct sequencing, refer to Chapter 25.

20-7-1-2 Frequency-hopping spread spectrum. Frequency-hopping spread spectrum was first used by the military to ensure reliable antijam and to secure communications in a battlefield environment. The fundamental concept of frequency hopping is to break a message into fixed-size blocks of data with each block transmitted in sequence except on a different carrier frequency. With frequency hopping, a pseudorandom code is used to generate a unique frequency-hopping sequence. The sequence in which the frequencies are selected must be known by both the transmitter and the receiver prior to the beginning of the transmission. The transmitter sends one block on a radio-frequency carrier and then switches (hops) to the next frequency in the sequence and so on. After reception of a block of data on one frequency, the receiver switches to the next frequency in the sequence. Each transmitter in the system has a different hopping sequence to prevent one subscriber from interfering with transmissions from other subscribers using the same radio channel frequency.

20-7-1-3 Direct-sequence spread spectrum. In direct-sequence systems, a high-bit-rate pseudorandom code is added to a low-bit-rate information signal to generate a high-bit-rate pseudorandom signal closely resembling noise that contains both the original data signal and the pseudorandom code. Again, before successful transmission, the pseudorandom code must be known to both the transmitter and the intended receiver. When a receiver detects a direct-sequence transmission, it simply subtracts the pseudorandom signal from the composite receive signal to extract the information data. In CDMA cellular telephone systems, the total radio-frequency bandwidth is divided into a few broadband radio channels that have a much higher bandwidth than the digitized voice signal. The digitized voice signal is added to the generated high-bit-rate signal and transmitted in such a way that it occupies the entire broadband radio channel. Adding a high-bit-rate pseudorandom signal to the voice information makes the signal more dominant and less susceptible to interference, allowing lower-power transmission and, hence, a lower number of transmitters and less expensive receivers.

20-7-2 CDMA Traffic Channels

CDMA traffic channels consist of a downlink (base station to mobile unit) channel and an uplink (mobile station to base station) channel. A CDMA downlink traffic channel is shown in Figure 20-12a. As the figure shows, the downlink traffic channel consists of up to 64 channels, including a broadcast channel used for control and traffic channels used to carry subscriber information. The broadcast channel consists of a pilot channel, a synchronization channel, up to seven paging channels, and up to 63 traffic channels. All these channels share the same 1.25-MHz CDMA frequency assignment. The traffic channel is identified

Pilot Channel	Synchronization Channel	Paging Channel 1 -------- Paging Channel 7	Traffic Channel 1 -------- Traffic Channel 62
		Optional and Variable (0–7)	Variable (55–62)

(a)

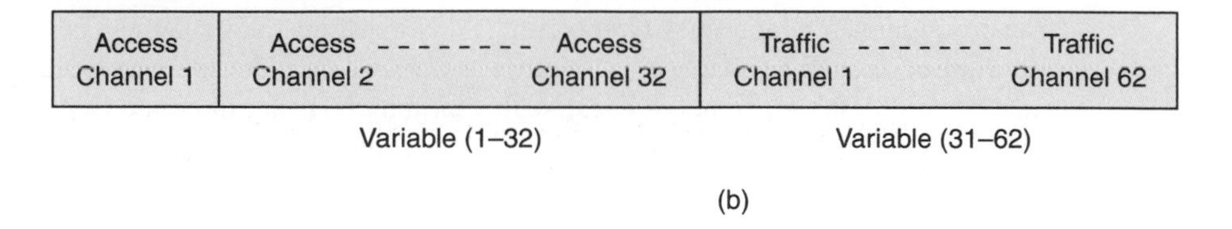

FIGURE 20-12 IS-95 traffic channels: (a) down-link; (b) up-link

by a distinct user-specific long-code sequence, and each access channel is identified by a distinct access channel long-code sequence.

The pilot channel is included in every cell with the purpose of providing a signal for the receiver to use to acquire timing and provide a phase reference for coherent demodulation. The pilot channel is also used by mobile units to compare signal strengths between base stations to determine when a handoff should be initiated. The synchronization channel uses a Walsh W32 code and the same pseudorandom sequence and phase offset as the pilot channel, allowing it to be demodulated by any receiver that can acquire the pilot signal. The synchronization channel broadcasts synchronization messages to mobile units and operates at 1200 bps. Paging channels convey information from the base station to the mobile station, such as system parameter messages, access parameter messages, CDMA channel list messages, and channel assignment messages. Paging channels are optional and can range in number between zero and seven. The paging channel is used to transmit control information and paging messages from the base station to the mobile units and operates at either 9600 bps, 4800 bps, or 2400 bps. A single 9600-bps pilot channel can typically support about 180 pages per second for a total capacity of 1260 pages per second.

Data on the downlink traffic channel are grouped into 20-ms frames. The data are first convolutionally coded and then formatted and interleaved to compensate for differences in the actual user data rates, which vary. The resulting signal is spread with a Walsh code and a long pseudorandom sequence at a rate of 1.2288 Mchips/s.

The uplink radio channel transmitter is shown in Figure 20-12b and consists of access channels and up to 62 uplink traffic channels. The access and uplink traffic channels use the same frequency assignment using direct-sequence CDMA techniques. The access channels are uplink only, shared, point-to-point channels that provide communications from mobile units to base stations when the mobile unit is not using a traffic channel. Access channels are used by the mobile unit to initiate communications with a base station and to respond to paging channel messages. Typical access channel messages include acknowledgements and sequence number, mobile identification parameter messages, and authentication parameters. The access channel is a random access channel with each channel subscriber uniquely identified by their pseudorandom codes. The uplink CDMA channel can contain up to a maximum of 32 access channels per supported paging channel. The uplink traffic channel operates at a variable data rate mode, and the access channels operate at a fixed 4800-bps rate. Access channel messages consist of registration, order, data burst, origination, page response, authentication challenge response, status response, and assignment completion messages.

Table 20-5 CDMA Power Levels

Class	Minimum EIRP	Maximum EIRP
I	−2 dBW (630 mW)	3 dBW (2.0 W)
II	−7 dBW (200 mW)	0 dBW (1.0 W)
III	−12 dBW (63 mW)	−3 dBW (500 mW)
IV	−17 dBW (20 mW)	−6 dBW (250 mW)
V	−22 dBW (6.3 mW)	−9 dBW (130 mW)

Subscriber data on the uplink radio channel transmitter are also grouped into 20-ms frames, convolutionally encoded, block interleaved, modulated by a 64-ary orthogonal modulation, and spread prior to transmission.

20-7-2-1 CDMA radiated power. IS-95 specifies complex procedures for regulating the power transmitted by each mobile unit. The goal is to make all reverse-direction signals within a single CDMA channel arrive at the base station with approximately the same signal strength (±1 dB), which is essential for CDMA operation. Because signal paths change continuously with moving units, mobile units perform power adjustments as many as 800 times per second (once every 1.25 ms) under control of the base station. Base stations instruct mobile units to increase or decrease their transmitted power in 1-dB increments (±0.5 dB).

When a mobile unit is first turned on, it measures the power of the signal received from the base station. The mobile unit assumes that the signal loss is the same in each direction (forward and reverse) and adjusts its transmit power on the basis of the power level of the signal it receives from the base station. This process is called *open-loop power setting.* A typical formula used by mobile units for determining their transmit power is

$$P_t \text{ dBm} = -76 \text{ dB} - P_r \qquad \textbf{(20-4)}$$

where P_t = transmit power (dBm)
P_r = received power (dBm)

Example 20-2

Determine the transmit power for a CDMA mobile unit that is receiving a signal from the base station at −100 dBm.

Solution Substituting into Equation 20-4 gives

$$P_t = -76 - (-100)$$
$$P_t = 24 \text{ dBm, or } 250 \text{ mW}$$

With CDMA, rather than limit the maximum transmit power, the minimum and maximum effective isotropic radiated power (EIRP) is specified (EIRP is the power radiated by an antenna times the gain of the antenna). Table 20-5 lists the maximum EIRPs for five classes of CDMA mobile units. The maximum radiated power of base stations is limited to 100 W per 1.23-MHz CDMA channel.

20-8 NORTH AMERICAN CELLULAR AND PCS SUMMARY

Table 20-6 summarizes several of the parameters common to North American cellular and PCS telephone systems (AMPS, USDC, and PCS).

Table 20-6 Cellular and PCS Telephone Summary

	Cellular System		
Parameter	AMPS	USDC (IS-54)	IS-95
Access method	FDMA	FDMA/TDMA	CDMA/FDMA
Modulation	FM	π/4 DQPSK	BPSK/QPSK
Frequency band			
Base station	869–894 MHz	869–894 MHz	869–894 MHz
Mobile unit	824–849 MHz	824–849 MHz	824–849 MHz
Base station	—	1.85–1.91 GHz	1.85–1.91 GHz
Mobile unit	—	1.93–1.99 GHz	1.93–1.99 GHz
RF channel bandwidth	30 kHz	30 kHz	1.25 MHz
Maximum radiated power	4 W	4 W	2 W
Control channel	FSK	PSK	PSK
Voice channels per carrier	1	3 or 6	Up to 20
Frequency assignment	Fixed	Fixed	Dynamic

20-9 GLOBAL SYSTEM FOR MOBILE COMMUNICATIONS

In the early 1980s, analog cellular telephone systems were experiencing a period of rapid growth in western Europe, particularly in Scandinavia and the United Kingdom and to a lesser extent in France and Germany. Each country subsequently developed its own cellular telephone system, which was incompatible with everyone else's system from both an equipment and an operational standpoint. Most of the existing systems operated at different frequencies, and all were analog. In 1982, the *Conference of European Posts and Telegraphs* (CEPT) formed a study group called *Groupe Spécial Mobile* (GSM) to study the development of a pan-European (*pan* meaning "all") public land mobile telephone system using ISDN. In 1989, the responsibility of GSM was transferred to the *European Telecommunications Standards Institute* (ETSI), and phase I of the GSM specifications was published in 1990. GSM had the advantage of being designed from scratch with little or no concern for being backward compatible with any existing analog cellular telephone system. GSM provides its subscribers with good quality, privacy, and security. GSM is sometimes referred to as the *Pan-European cellular system.*

Commercial GSM service began in Germany in 1991, and by 1993 there were 36 GSM networks in 22 countries. GSM networks are now either operational or planned in over 80 countries around the world. North America made a late entry into the GSM market with a derivative of GSM called *PCS-1900.* GSM systems now exist on every continent, and the acronym GSM now stands for *Global System for Mobile Communications.* The first GSM system developed was GSM-900 (phase I), which operates in the 900-MHz band for voice only. Phase 2 was introduced in 1995, which included facsimile, video, and data communications services. After implementing PCS frequencies (1800 MHz in Europe and 1900 MHz in North America) in 1997, GSM-1800 and GSM-1900 were created.

GSM is a second-generation cellular telephone system initially developed to solve the fragmentation problems inherent in first-generation cellular telephone systems in Europe. Before implementing GSM, all European countries used different cellular telephone standards; thus, it was impossible for a subscriber to use a single telephone set throughout Europe. GSM was the world's first totally digital cellular telephone system designed to use the services of SS7 signaling and an all-digital data network called *integrated services digital network* (ISDN) to provide a wide range of network services. With between 20 and 50 million subscribers, GSM is now the world's most popular standard for new cellular telephone and personal communications equipment.

20-9-1 GSM Services

The original intention was to make GSM compatible with ISDN in terms of services offered and control signaling. Unfortunately, radio-channel bandwidth limitations and cost prohibited GSM from operating at the 64-kbps ISDN basic data rate.

GSM telephone services can be broadly classified into three categories: *bearer services, teleservices,* and *supplementary services.* Probably the most basic bearer service provided by GSM is telephony. With GSM, analog speech signals are digitally encoded and then transmitted through the network as a digital data stream. There is also an emergency service where the closest emergency service provider is notified by dialing three digits similar to 911 services in the United States. A wide variety of data services is offered through GSM, where users can send and receive data at rates up to 9600 bps to subscribers in POTS (plain old telephone service), ISDN networks, Packet Switched Public Data Networks (PSPDN), and Circuit Switched Public Data Networks (CSPDN) using a wide variety of access methods and protocols, such as X.25. In addition, since GSM is a digital network, a modem is not required between the user and the GSM network.

Other GSM data services include Group 3 facsimile per ITU-T recommendation T.30. One unique feature of GSM that is not found in older analog systems is the *Short Message Service* (SMS), which is a bidirectional service for sending alphanumeric messages up to 160 bytes in length. SMS can be transported through the system in a store-and-forward fashion. SMS can also be used in a *cell-broadcast mode* for sending messages simultaneously to multiple receivers. Several supplemental services, such as *call forwarding* and *call barring,* are also offered with GSM.

20-9-2 GSM System Architecture

The system architecture for GSM as shown in Figure 20-13 consists of three major interconnected subsystems that interact among one another and with subscribers through

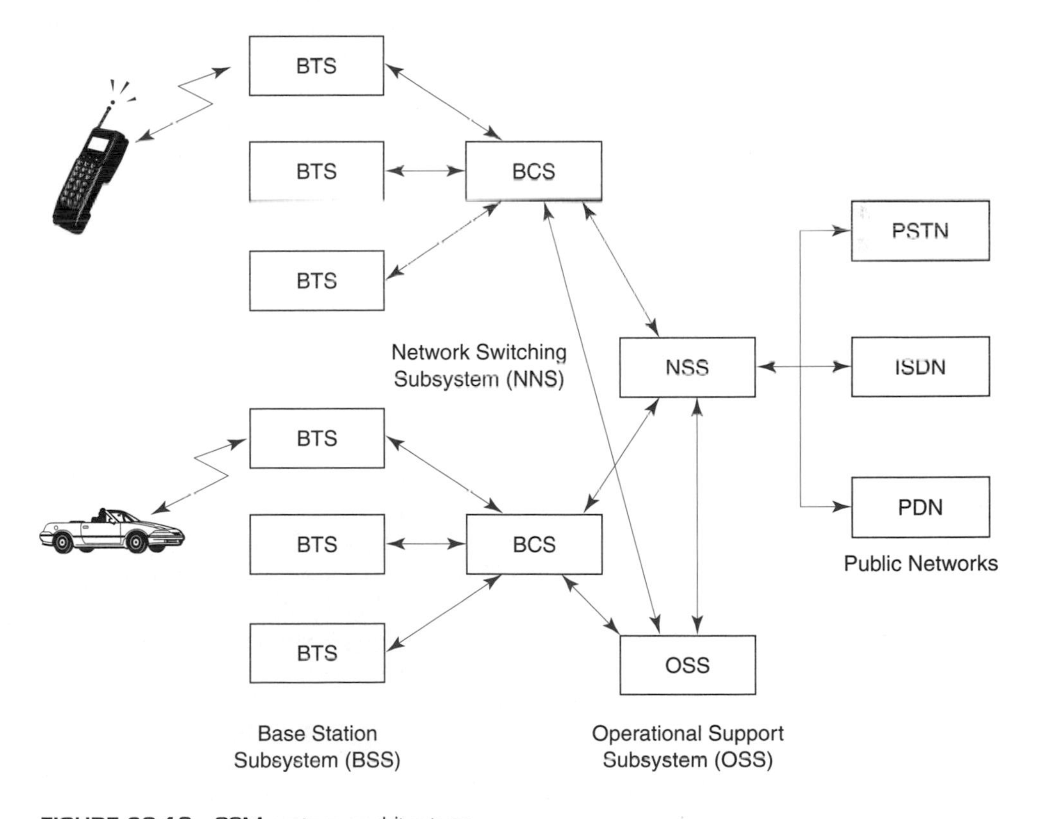

FIGURE 20-13 GSM system architecture

specified network interfaces. The three primary subsystems of GSM are *Base Station Subsystem* (BSS), *Network Switching Subsystem* (NSS), and *Operational Support Subsystem* (OSS). Although the mobile station is technically another subsystem, it is generally considered part of the base station subsystem.

The BSS is sometimes known as the radio *subsystem* because it provides and manages radio-frequency transmission paths between mobile units and the mobile switching center (MSC). The BSS also manages the radio interface between mobile units and all other GSM subsystems. Each BSS consists of many base station controllers (BSCs), which are used to connect the MCS to the NSS through one or more MSCs. The NSS manages switching functions for the system and allows the MSCs to communicate with other telephone networks, such as the public switched telephone network and ISDN. The OSS supports operation and maintenance of the system and allows engineers to monitor, diagnose, and troubleshoot every aspect of the GSM network.

20-9-3 GSM Radio Subsystem

GSM was originally designed for 200 full-duplex channels per cell with transmission frequencies in the 900-MHz band; however, frequencies were later allocated at 1800 MHz. A second system, called DSC-1800, was established that closely resembles GSM. GSM uses two 25-MHz frequency bands that have been set aside for system use in all member companies. The 890-MHz to 915-MHz band is used for mobile unit-to-base station transmissions (reverse-link transmissions), and the 935-MHz to 960-MHz frequency band is used for base station–to–mobile unit transmission (forward-link transmissions). GSM uses frequency-division duplexing and a combination of TDMA and FDMA techniques to provide base stations simultaneous access to multiple mobile units. The available forward and reverse frequency bands are subdivided into 200-kHz wide voice channels called *absolute radio-frequency channel numbers* (ARFCN). The ARFCN number designates a forward/reverse channel pair with 45-MHz separation between them. Each voice channel is shared among as many as eight mobile units using TDMA.

Each of the ARFCN channel subscribers occupies a unique time slot within the TDMA frame. Radio transmission in both directions is at a 270.833-kbps rate using binary Gaussian minimum shift keying (GMSK) modulation with an effective channel transmission rate of 33.833 kbps per user.

The basic parameters of GSM are the following:

1. GMSK modulation (Gaussian MSK)
2. 50-MHz bandwidth:
   890-MHz to 915-MHz mobile transmit band (reverse channel)
   935-MHz to 960-MHz base station transmit band (forward channel)
3. FDMA/TDMA accessing
4. Eight 25-kHz channels within each 200-kHz traffic channel
5. 200-kHz traffic channel
6. 992 full-duplex channels
7. Supplementary ISDN services, such as call diversion, closed user groups, caller identification, and *short messaging service* (SMS), which restricts GSM users and base stations to transmitting alphanumeric pages limited to a maximum of 160 seven-bit ASCII characters while simultaneously carrying normal voice messages.

20-10 PERSONAL SATELLITE COMMUNICATIONS SYSTEM

Mobile Satellite Systems (MSS) provide the vehicle for a new generation of wireless telephone services called *personal communications satellite systems* (PCSS). Universal wireless telephone coverage is a developing MSS service that promises to deliver mobile subscribers both traditional and enhanced telephone features while providing wide-area global coverage.

MSS satellites are, in essence, radio repeaters in the sky, and their usefulness for mobile communications depends on several factors, such as the space-vehicle altitude, orbital pattern, transmit power, receiver sensitivity, modulation technique, antenna radiation pattern (footprint), and number of satellites in its constellation. Satellite communications systems have traditionally provided both narrowband and wideband voice, data, video, facsimile, and networking services using large and very expensive, high-powered earth station transmitters communicating via high-altitude, geosynchronous earth-orbit (GEO) satellites. Personal communications satellite services, however, use low earth-orbit (LEO) and medium earth-orbit (MEO) satellites that communicate directly with small, low-power mobile telephone units. The intention of PCSS mobile telephone is to provide the same features and services offered by traditional, terrestrial cellular telephone providers. However, PCSS telephones will be able to make or receive calls anytime, anywhere in the world. A simplified diagram of a PCSS system is shown in Figure 20-14.

The key providers in the PCSS market include American Mobile Satellite Corporation (AMSC), Celsat, Comsat, Constellation Communications (Aries), Ellipsat (Ellipso), INMARSAT, LEOSAT, Loral/Qualcomm (Globalstar), TMI communications, TWR (Odysse), and Iridium LLC.

20-10-1 PCSS Advantages and Disadvantages

The primary and probably most obvious advantage of PCSS mobile telephone is that it provides mobile telephone coverage and a host of other integrated services virtually anywhere in the world to a truly global customer base. PCSS can fill the vacancies between land-based cellular and PCS telephone systems and provide wide-area coverage on a regional or global basis.

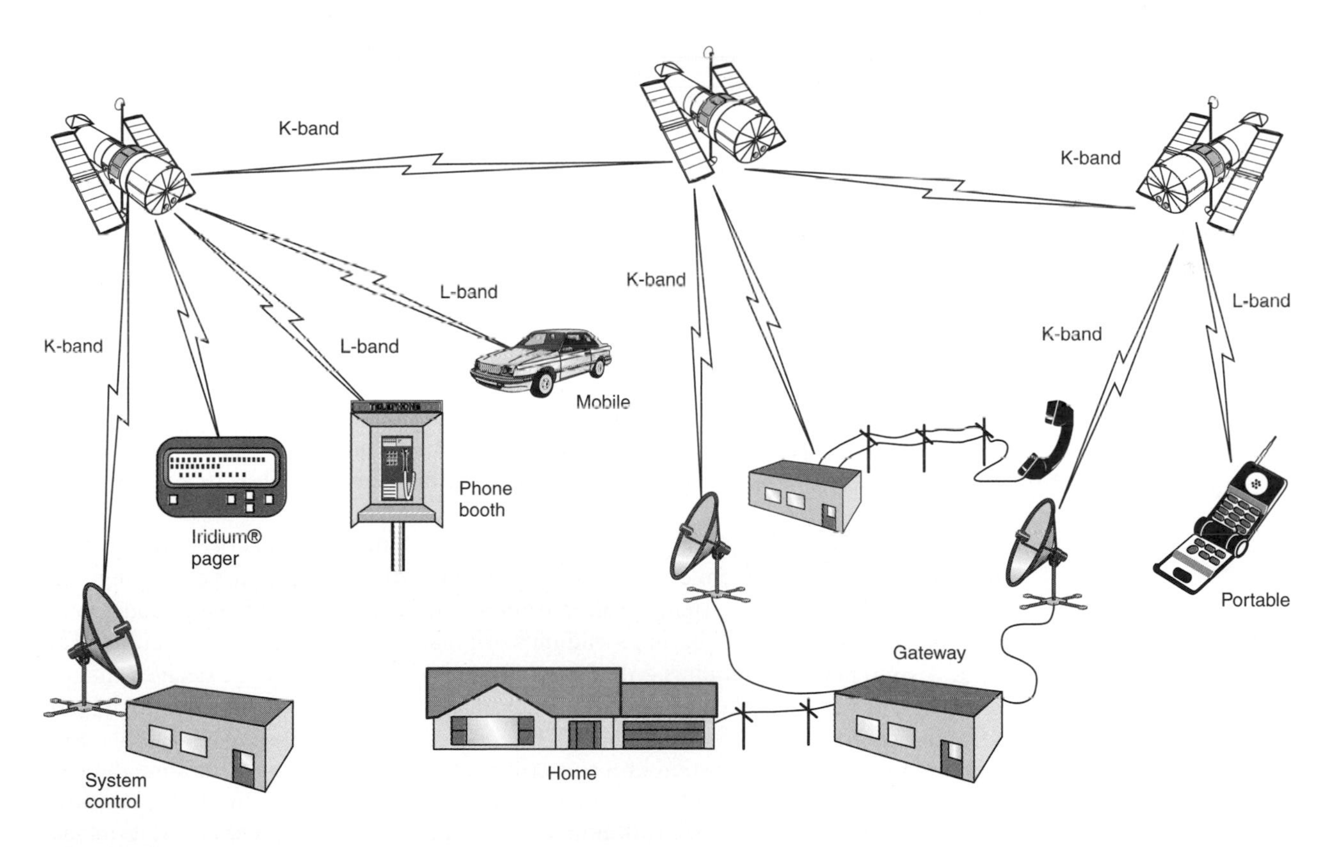

FIGURE 20-14 Overview of Iridium PCSS mobile telephone system

PCSS is ideally suited to fixed cellular telephone applications, as it can provide a full complement of telephone services to places where cables can never go because of economical, technical, or physical constraints. PCSS can also provide complementary and backup telephone services to large companies and organizations with multiple operations in diverse locations, such as retail, manufacturing, finance, transportation, government, military, and insurance.

Most of the disadvantages of PCSS are closely related to economics, with the primary disadvantage being the high risk associated with the high costs of designing, building, and launching satellites. There is also a high cost for the terrestrial-based networking and interface infrastructure necessary to maintain, coordinate, and manage the network once it is in operation. In addition, the intricate low-power, dual-mode transceivers are more cumbersome and expensive than most mobile telephone units used with terrestrial cellular and PCS systems.

20-10-2 PCSS Industry Requirements

PCSS mobile telephone systems require transparent interfaces and feature sets among the multitude of terrestrial networks currently providing mobile and wireline telephone services. In addition, the interfaces must be capable of operating with both ANSI and CCITT network constraints and be able to provide interpretability with AMPS, USDC, GMS, and PCS cellular telephone systems. PCSS must also be capable of operating dual-mode with *air-access protocols,* such as FDMA, TDMA, or CDMA. PCSS should also provide unique MSS feature sets and characteristics, such as inter-/intrasatellite handoffs, land based-to-satellite handoffs, and land-based/PCSS dual registration.

20-10-3 Iridium Satellite System

Iridium LLC is an international consortium owned by a host of prominent companies, agencies, and governments, including the following: Motorola, General Electric, Lockheed, Raytheon, McDonnell Douglas, Scientific Atlanta, Sony, Kyocera, Mitsubishi, DDI, Kruchinew Enterprises, Mawarid Group of Saudi Arabia, STET of Italy, Nippon Iridium Corporation of Japan, the government of Brazil, Muidiri Investments BVI, LTD of Venezuela, Great Wall Industry of China, United Communications of Thailand, the U.S. Department of Defense, Sprint, and BCE Siemens.

The *Iridium project,* which even sounds like something out of *Star Wars,* is undoubtedly the largest commercial venture undertaken in the history of the world. It is the system with the most satellites, the highest price tag, the largest public relations team, and the most peculiar design. The $5 billion, gold-plated *Iridium* mobile telephone system is undoubtedly (or at least intended to be) the Cadillac of mobile telephone systems. Unfortunately (and somewhat ironically), in August 1999, on Friday the 13th, Iridium LLC, the beleaguered satellite-telephone system spawned by Motorola's Satellite Communications Group in Chandler, Arizona, filed for bankruptcy under protection of Chapter 11. However, Motorola Inc., the largest stockholder in Iridium, says it will continue to support the company and its customers and does not expect any interruption in service while reorganization is under way.

Iridium is a satellite-based wireless personal communications network designed to permit a wide range of mobile telephone services, including voice, data, networking, facsimile, and paging. The system is called Iridium after the element on the periodic table with the atomic number 77 because Iridium's original design called for 77 satellites. The final design, however, requires only 66 satellites. Apparently, someone decided that element 66, dysprosium, did not have the same charismatic appeal as Iridium, and the root meaning of the word is "bad approach." The 66-vehicle LEO interlinked satellite constellation can track the location of a subscriber's telephone handset, determine the best routing through a network of ground-based gateways and intersatellite links, establish the best path for the telephone call, initiate all the necessary connections, and terminate the call on completion. The system also provides applicable revenue tracking.

With Iridium, two-way global communications is possible even when the destination subscriber's location is unknown to the caller. In essence, the intent of the Iridium system is to provide the best service in the telephone world, allowing telecommunication anywhere, anytime, and any place. The FCC granted the Iridium program a full license in January 1995 for construction and operation in the United States.

Iridium uses a GSM-based telephony architecture to provide a digitally switched telephone network and global dial tone to call and receive calls from any place in the world. This global roaming feature is designed into the system. Each subscriber is assigned a personal phone number and will receive only one bill, no matter in what country or area they use the telephone.

The Iridium project has a satellite network control facility in Landsdowne, Virginia, with a backup facility in Italy. A third engineering control complex is located at Motorola's SATCOM location in Chandler, Arizona.

20-10-3-1 System layout. Figure 20-14 shows an overview of the Iridium system. Subscriber telephone sets used in the Iridium system transmit and receive L-band frequencies and utilize both frequency- and time-division multiplexing to make the most efficient use of a limited frequency spectrum. Other communications links used in Iridium include EHF and SHF bands between satellites for telemetry, command, and control as well as routing digital voice packets to and from gateways. An Iridium telephone enables the subscriber to connect either to the local cellular telephone infrastructure or to the space constellation using its *dual-mode* feature.

Iridium gateways are prime examples of the advances in satellite infrastructures that are responsible for the delivery of a host of new satellite services. The purpose of the gateways is to support and manage roaming subscribers as well as to interconnect Iridium subscribers to the public switched telephone network. Gateway functions include the following:

1. Set up and maintain basic and supplementary telephony services
2. Provide an interface for two-way telephone communications between two Iridium subscribers and Iridium subscribers to subscribers of the public switched telephone network
3. Provide Iridium subscribers with messaging, facsimile, and data services
4. Facilitate the business activities of the Iridium system through a set of cooperative mutual agreements

20-10-3-2 Satellite constellation. Providing full-earth coverage is the underlying basis of the Iridium satellite system. Iridium uses 66 operational satellites (there are also some spares) configured at a mean elevation of 420 miles above Earth in six nearly-polar orbital planes (86.4° tilt), in which 11 satellites revolve around Earth in each orbit with an orbital time of 100 minutes, 28 seconds. This allows Iridium to cover the entire surface area of Earth and, whenever one satellite goes out of view of a subscriber, a different one replaces it. The satellites are phased appropriately in north–south necklaces forming *corotating planes* up one side of Earth, across the poles, and down the other side. The first and last planes rotate in opposite directions, creating a virtual *seam.* The corotating planes are separated by 31.6°, and the seam planes are 22° apart.

Each satellite is equipped with three L-band antennas forming a honeycomb pattern that consists of 48 individual spot beams with a total of 1628 cells aimed directly below the satellite, as shown in Figure 20-15. As the satellite moves in its orbit, the footprints move across Earth's surface, and subscriber signals are switched from one beam to the next or from one satellite to the next in a handoff process. When satellites approach the North or South Pole, their footprints converge, and the beams overlap. Outer beams are then turned off to eliminate this overlap and conserve power on the spacecraft. Each cell has 174 full-duplex voice channels for a total of 283,272 channels worldwide.

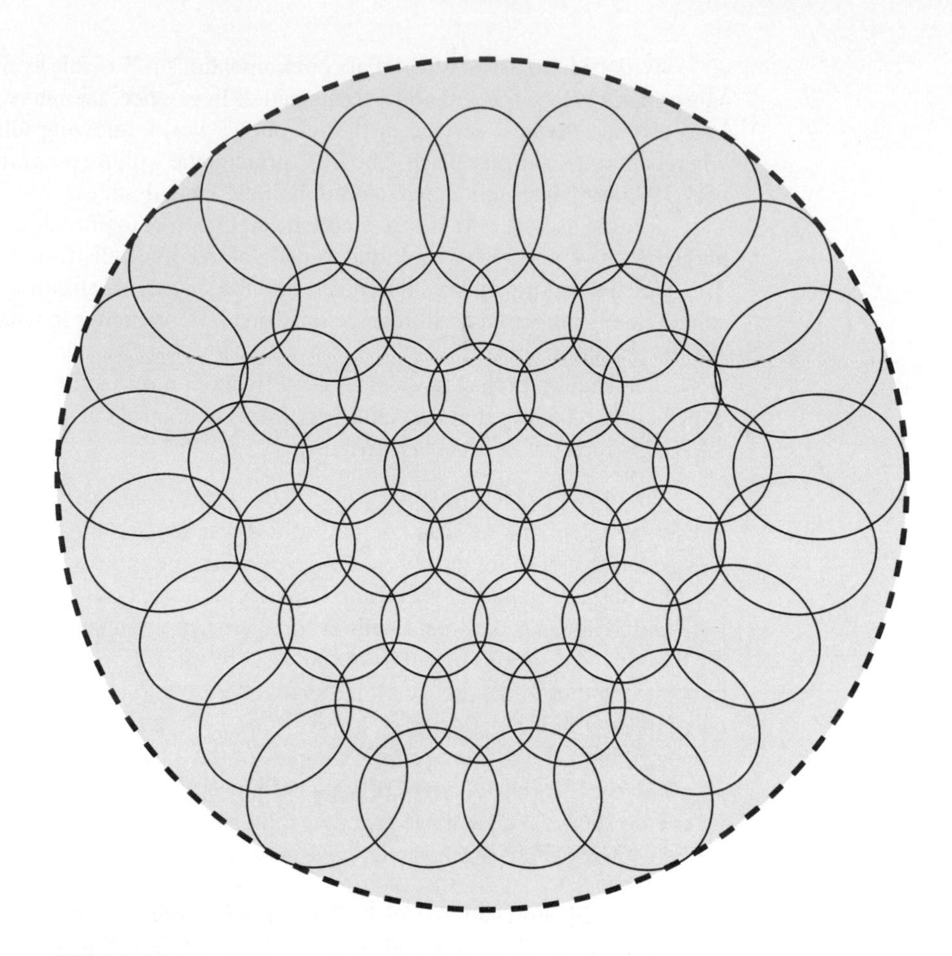

FIGURE 20-15 Iridium system spot beam footprint pattern

Using satellite *cross-links* is the unique key to the Iridium system and the primary differentiation between Iridium and the traditional satellite *bent-pipe system* where all transmissions follow a path from Earth to satellite to Earth. Iridium is the first mobile satellite to incorporate sophisticated, onboard digital processing on each satellite and cross-link capability between satellites.

Each satellite is equipped with four satellite-to-satellite cross-links to relay digital information around the globe. The cross-link antennas point toward the closest spacecraft orbiting in the same plane and the two adjacent corotating planes. *Feeder link* antennas relay information to the terrestrial gateways and the system control segment located at the earth stations.

20-10-3-3 Frequency plan and modulation. On October 14, 1994, the Federal Communication Commission issued a report and order Dockett #92-166 defining L-band frequency sharing for subscriber units in the 1616-MHz to 1626.5-MHz band. Mobile satellite system cellular communications are assigned 5.15 MHz at the upper end of this spectrum for TDMA/FDMA service. CDMA access is assigned the remaining 11.35 MHz for their service uplinks and a proportionate amount of the S-band frequency spectrum at 2483.5 MHz to 2500 MHz for their downlinks. When a CDMA system is placed into operation, the CDMA L-band frequency spectrum will be reduced to 8.25 MHz. The remaining 3.1 MHz of the frequency spectrum will then be assigned to either the Iridium system or another TDMA/FDMA system.

All Ka-band uplinks, downlinks, and cross-links are packetized TDM/FDMA using quadrature phase-shift keying (QPSK) and FEC 1/2 rate convolutional coding with Viterbi decoding. Coded data rates are 6.25 Mbps for gateways and satellite control facility links and 25 Mbps for satellite cross-links. Both uplink and downlink transmissions occupy 100 MHz of bandwidth, and intersatellite links use 200 MHz of bandwidth. The frequency bands are as follows:

L-band subscriber-to-satellite voice links = 1.616 GHz to 1.6265 GHz
Ka-band gateway downlinks = 19.4 GHz to 19.6 GHz
Ka-band gateway uplinks = 29.1 GHz to 29.3 GHz
Ka-intersatellite cross-links = 23.18 GHz to 23.38 GHz

QUESTIONS

20-1. What is meant by a *first-generation* cellular telephone system?

20-2. Briefly describe the AMPS system.

20-3. Outline the AMPS *frequency allocation.*

20-4. What is meant by the term *frequency-division duplexing*?

20-5. What is the difference between a *wireline* and *nonwireline* company?

20-6. Describe a *cellular geographic serving area.*

20-7. List and describe the three *classifications* of AMPS cellular telephones.

20-8. What is meant by the *discontinuous transmission mode*?

20-9. List the features of a *personal communications system* that differentiate it from a *standard cellular telephone network.*

20-10. What is the difference between a *personal communications network* and *personal communications services*?

20-11. Briefly describe the functions of a *home location register.*

20-12. Briefly describe the functions of a *visitor location register.*

20-13. Briefly describe the functions of an *equipment identification registry.*

20-14. Describe the following services: *available mode, screen mode, private mode,* and *unavailable mode.*

20-15. What is meant by a *microcellular system*?

20-16. List the advantages of a *PCS cellular system* compared to a *standard cellular system.*

20-17. List the disadvantage of a PCS cellular system.

20-18. What is meant by the term *false handoff*?

20-19. Briefly describe the N-AMPS cellular telephone system.

20-20. What is an *interference avoidance scheme*?

20-21. What are the four types of *handoffs* possible with N-AMPS?

20-22. List the advantages of a *digital cellular system.*

20-23. Describe the *United States Digital Cellular* system.

20-24. Describe the *TDMA scheme* used with USDC.

20-25. List the advantages of *digital* TDMA over *analog* AMPS FDMA.

20-26. Briefly describe the EIA/TIA *Interim Standard IS-54.*

20-27. What is meant by the term *dual mode*?

20-28. Briefly describe the EIA/TIA *Interim Standard IS-136.*

20-29. What is meant by the term *sleep mode*?

20-30. Briefly describe the *North American Digital Cellular* format.

20-31. Briefly describe the E-TDMA scheme.

20-32. Describe the differences between the radiated power classifications for USDC and AMPS.

20-33. List the IS-95 specifications.

20-34. Describe the *CDMA format* used with IS-95.

20-35. Describe the differences between the *CDMA radiated power procedures* and AMPS.

20-36. Briefly describe the GSM cellular telephone system.

20-37. Outline and describe the *services* offered by GSM.

20-38. Briefly describe the GSM *system architecture.*

What are the three *primary subsystems* of GSM?

Briefly describe the GSM *radio subsystem.*

List the *basic parameters* of GSM.

Briefly describe the *architecture* of a PCSS.

List the *advantages* and *disadvantages* of PCSS.

Outline the *industry requirements* of PCSS.

CHAPTER 21

Introduction to Data Communications and Networking

CHAPTER OUTLINE

OBJECTIVES

- Define the following terms: *data, data communications, data communications circuit,* and *data communications network*
- Give a brief description of the evolution of data communications
- Define *data communications network architecture*
- Describe data communications protocols
- Describe the basic concepts of connection-oriented and connectionless protocols
- Describe syntax and semantics and how they relate to data communications
- Define *data communications standards* and explain why they are necessary
- Describe the following standards organizations: ISO, ITU-T, IEEE, ANSI, EIA, TIA, IAB, ETF, and IRTF
- Define *open systems interconnection*
- Name and explain the functions of each of the layers of the seven-layer OSI model
- Define *station* and *node*
- Describe the fundamental block diagram of a two-station data communications circuit and explain how the following terms relate to it: *source, transmitter, transmission medium, receiver,* and *destination*
- Describe serial and parallel data transmission and explain the advantages and disadvantages of both types of transmissions

- Define *data communications circuit arrangements*
- Describe the following transmission modes: simplex, half duplex, full duplex, and full/full duplex
- Define *data communications network*
- Describe the following network components, functions, and features: servers, clients, transmission media, shared data, shared printers, and network interface card
- Define *local operating system*
- Define *network operating system*
- Describe peer-to-peer client/server and dedicated client/server networks
- Define *network topology* and describe the following: star, bus, ring, mesh, and hybrid
- Describe the following classifications of networks: LAN, MAN, WAN, GAN, building backbone, campus backbone, and enterprise network
- Briefly describe the TCP/IP hierarchical model
- Briefly describe the Cisco three-layer hierarchical model

21-1 INTRODUCTION

Since the early 1970s, technological advances around the world have occurred at a phenomenal rate, transforming the *telecommunications industry* into a highly sophisticated and extremely dynamic field. Where previously telecommunications systems had only voice to accommodate, the advent of very large-scale integration chips and the accompanying low-cost microprocessors, computers, and peripheral equipment has dramatically increased the need for the exchange of digital information. This, of course, necessitated the development and implementation of higher-capacity and much faster means of communicating.

In the data communications world, *data* generally are defined as information that is stored in digital form. The word *data* is plural; a single unit of data is a *datum.* Data communications is the process of transferring digital information (usually in binary form) between two or more points. *Information* is defined as knowledge or intelligence. Information that has been processed, organized, and stored is called data.

The fundamental purpose of a *data communications circuit* is to transfer digital information from one place to another. Thus, *data communications* can be summarized as the transmission, reception, and processing of digital information. The original source information can be in analog form, such as the human voice or music, or in digital form, such as binary-coded numbers or alphanumeric codes. If the source information is in analog form, it must be converted to digital form at the source and then converted back to analog form at the destination.

A *network* is a set of *devices* (sometimes called *nodes* or *stations*) interconnected by media links. *Data communications networks* are systems of interrelated computers and computer equipment and can be as simple as a personal computer connected to a printer or two personal computers connected together through the *public telephone network.* On the other hand, a data communications network can be a complex communications system comprised of one or more mainframe computers and hundreds, thousands, or even millions of remote terminals, personal computers, and workstations. In essence, there is virtually no limit to the capacity or size of a data communications network.

Years ago, a single computer serviced virtually every computing need. Today, the single-computer concept has been replaced by the networking concept, where a large number of separate but interconnected computers share their resources. Data communications networks and systems of networks are used to interconnect virtually all kinds of digital computing equipment, from *automatic teller machines* (ATMs) to bank computers; personal computers to information highways, such as the *Internet;* and workstations to mainframe computers. Data communications networks can also be used for airline and hotel reservation systems, mass media and news networks, and electronic mail delivery systems. The list of applications for data communications networks is virtually endless.

21-2 HISTORY OF DATA COMMUNICATIONS

It is highly likely that data communications began long before recorded time in the form of smoke signals or tom-tom drums, although they surely did not involve electricity or an electronic apparatus, and it is highly unlikely that they were binary coded. One of the earliest means of communicating electrically coded information occurred in 1753, when a proposal submitted to a Scottish magazine suggested running a communications line between villages comprised of 26 parallel wires, each wire for one letter of the alphabet. A Swiss inventor constructed a prototype of the 26-wire system, but current wire-making technology proved the idea impractical.

In 1833, Carl Friedrich Gauss developed an unusual system based on a five-by-five matrix representing 25 letters (I and J were combined). The idea was to send messages over a single wire by deflecting a needle to the right or left between one and five times. The initial set of deflections indicated a row, and the second set indicated a column. Consequently, it could take as many as 10 deflections to convey a single character through the system.

If we limit the scope of data communications to methods that use *binary coded* electri cal signals to transmit information, then the first successful (and practical) data communications system was invented by Samuel F. B. Morse in 1832 and called the *telegraph.* Morse also developed the first practical *data communications code,* which he called the *Morse code.* With telegraph, dots and dashes (analogous to logic 1s and 0s) are transmitted across a wire using electromechanical induction. Various combinations of dots, dashes, and pauses represented binary codes for letters, numbers, and punctuation marks. Because all codes did not contain the same number of dots and dashes, Morse's system combined human intelligence with electronics, as decoding was dependent on the hearing and reasoning ability of the person receiving the message. (Sir Charles Wheatstone and Sir William Cooke allegedly invented the first telegraph in England, but their contraption required six different wires for a single telegraph line.)

In 1840, Morse secured an American patent for the telegraph, and in 1844 the first telegraph line was established between Baltimore and Washington, D.C., with the first message conveyed over this system being "What hath God wrought!" In 1849, the first slow-speed telegraph printer was invented, but it was not until 1860 that high-speed (15-bps) printers were available. In 1850, Western Union Telegraph Company was formed in Rochester, New York, for the purpose of carrying coded messages from one person to another.

In 1874, Emile Baudot invented a telegraph *multiplexer,* which allowed signals from up to six different telegraph machines to be transmitted simultaneously over a single wire. The *telephone* was invented in 1875 by Alexander Graham Bell and, unfortunately, very little new evolved in telegraph until 1899, when Guglielmo Marconi succeeded in sending radio (wireless) telegraph messages. Telegraph was the only means of sending information across large spans of water until 1920, when the first commercial radio stations carrying voice information were installed.

It is unclear exactly when the first electrical computer was developed. Konrad Zuis, a German engineer, demonstrated a computing machine sometime in the late 1930s; however, at the time, Hitler was preoccupied trying to conquer the rest of the world, so the project fizzled out. Bell Telephone Laboratories is given credit for developing the first special-purpose computer in 1940 using electromechanical relays for performing logical operations. However, J. Presper Eckert and John Mauchley at the University of Pennsylvania are given credit by some for beginning modern-day computing when they developed the ENIAC computer on February 14, 1946.

In 1949, the U.S. National Bureau of Standards developed the first all-electronic diode-based computer capable of executing stored programs. The U.S. Census Bureau installed the machine, which is considered the first commercially produced American computer. In the 1950s, computers used punch cards for inputting information, printers for outputting information, and magnetic tape reels for permanently storing information. These early computers could process only one job at a time using a technique called *batch processing.*

The first general-purpose computer was an automatic sequence-controlled calculator developed jointly by Harvard University and International Business Machines (IBM) Corporation. The UNIVAC computer, built in 1951 by Remington Rand Corporation, was the first mass-produced electronic computer.

In the 1960s, batch-processing systems were replaced by on-line processing systems with terminals connected directly to the computer through serial or parallel communications lines. The 1970s introduced microprocessor-controlled microcomputers, and by the 1980s personal computers became an essential item in the home and workplace. Since then, the number of mainframe computers, small business computers, personal computers, and computer terminals has increased exponentially, creating a situation where more and more people have the need (or at least think they have the need) to exchange digital information with each other. Consequently, the need for data communications circuits, networks, and systems has also increased exponentially.

Soon after the invention of the telephone, the American Telephone and Telegraph Company (AT&T) emerged, providing both long-distance and local telephone service and data communications service throughout the United States. The vast AT&T system was referred to by some as the "Bell System" and by others as "Ma Bell." During this time, Western Union Corporation provided telegraph service. Until 1968, the AT&T operating tariff allowed only equipment furnished by AT&T to be connected to AT&T lines. In 1968, a landmark Supreme Court decision, the *Carterfone* decision, allowed non-Bell companies to interconnect to the vast AT&T communications network. This decision started the interconnect industry, which has led to competitive data communications offerings by a large number of independent companies. In 1983, as a direct result of an antitrust suit filed by the federal government, AT&T agreed in a court settlement to divest itself of operating companies that provide basic local telephone service to the various geographic regions of the United States. Since the divestiture, the complexity of the public telephone system in the United States has grown even more involved and complicated.

Recent developments in data communications networking, such as the *Internet, intranets,* and the *World Wide Web* (WWW), have created a virtual explosion in the data communications industry. A seemingly infinite number of people, from homemaker to chief executive officer, now feel a need to communicate over a finite number of facilities. Thus, the demand for higher-capacity and higher-speed data communications systems is increasing daily with no end in sight.

The *Internet* is a public data communications network used by millions of people all over the world to exchange business and personal information. The Internet began to evolve in 1969 at the *Advanced Research Projects Agency* (ARPA). ARPANET was formed in the late 1970s to connect sites around the United States. From the mid-1980s to April 30, 1995, the *National Science Foundation* (NSF) funded a high-speed backbone called NSFNET.

Intranets are private data communications networks used by many companies to exchange information among employees and resources. Intranets normally are used for security reasons or to satisfy specific connectivity requirements. Company intranets are generally connected to the public Internet through a *firewall,* which converts the intranet addressing system to the public Internet addressing system and provides security functionality by filtering incoming and outgoing traffic based on addressing and protocols.

The *World Wide Web* (WWW) is a server-based application that allows subscribers to access the services offered by the Web. Browsers, such as Netscape Communicator and Microsoft Internet Explorer, are commonly used for accessing data over the WWW.

21-3 DATA COMMUNICATIONS NETWORK ARCHITECTURE, PROTOCOLS, AND STANDARDS

21-3-1 Data Communications Network Architecture

A *data communications network* is any system of computers, computer terminals, or computer peripheral equipment used to transmit and/or receive information between two or more locations. *Network architectures* outline the products and services necessary for the individual components within a data communications network to operate together.

In essence, network architecture is a set of equipment, transmission media, and procedures that ensures that a specific sequence of events occurs in a network in the proper order to produce the intended results. Network architecture must include sufficient information to allow a program or a piece of hardware to perform its intended function. The primary goal of network architecture is to give the users of the network the tools necessary for setting up the network and performing data flow control. A network architecture outlines the way in which a data communications network is arranged or structured and generally includes the concept of *levels* or *layers* of functional responsibility within the architecture. The *functional responsibilities* include electrical specifications, hardware arrangements, and software procedures.

Networks and network protocols fall into three general classifications: *current, legacy,* and *legendary.* Current networks include the most modern and sophisticated networks and protocols available. If a network or protocol becomes a legacy, no one really wants to use it, but for some reason it just will not go away. When an antiquated network or protocol finally disappears, it becomes legendary.

In general terms, computer networks can be classified in two different ways: *broadcast* and *point to point.* With broadcast networks, all stations and devices on the network share a single communications channel. Data are propagated through the network in relatively short messages sometimes called *frames, blocks,* or *packets.* Many or all subscribers of the network receive transmitted messages, and each message contains an address that identifies specifically which subscriber (or subscribers) is intended to receive the message. When messages are intended for all subscribers on the network, it is called *broadcasting,* and when messages are intended for a specific group of subscribers, it is called *multicasting.*

Point-to-point networks have only two stations. Therefore, no addresses are needed. All transmissions from one station are intended for and received by the other station. With point-to-point networks, data are often transmitted in long, continuous messages, sometimes requiring several hours to send.

In more specific terms, point-to-point and broadcast networks can be subdivided into many categories in which one type of network is often included as a subnetwork of another.

21-3-2 Data Communications Protocols

Computer networks communicate using *protocols,* which define the procedures that the systems involved in the communications process will use. Numerous protocols are used today to provide networking capabilities, such as how much data can be sent, how it will be sent, how it will be addressed, and what procedure will be used to ensure that there are no undetected errors.

Protocols are arrangements between people or processes. In essence, a protocol is a set of customs, rules, or regulations dealing with formality or precedence, such as diplomatic or military protocol. Each functional layer of a network is responsible for providing a specific service to the data being transported through the network by providing a set of rules, called protocols, that perform a specific function (or functions) within the network. *Data communications protocols* are sets of rules governing the orderly exchange of data within the network or a portion of the network, whereas network architecture is a set of layers and protocols that govern the operation of the network. The list of protocols used by a system is called a *protocol stack,* which generally includes only one protocol per layer. *Layered network architectures* consist of two or more independent levels. Each level has a specific set of responsibilities and functions, including data transfer, flow control, data segmentation and reassembly, sequence control, error detection and correction, and notification.

21-3-2-1 Connection-oriented and connectionless protocols. Protocols can be generally classified as either *connection oriented* or *connectionless.* With a connection-oriented protocol, a logical connection is established between the endpoints (e.g., a *virtual circuit*) prior to the transmission of data. Connection-oriented protocols operate in a manner similar to making a standard telephone call where there is a sequence of actions and acknowledgments, such as setting up the call, establishing the connection, and then disconnecting. The actions and acknowledgments include dial tone, Touch-Tone signaling, ringing and ring-back signals, and busy signals.

Connection-oriented protocols are designed to provide a high degree of reliability for data moving through the network. This is accomplished by using a rigid set of procedures for establishing the connection, transferring the data, acknowledging the data, and then clearing the connection. In a connection-oriented system, each packet of data is assigned a unique sequence number and an associated acknowledgement number to track the data as they travel through a network. If data are lost or damaged, the destination station requests that they be re-sent. A connection-oriented protocol is depicted in Figure 21-1a. Characteristics of connection-oriented protocols include the following:

1. A connection process called a *handshake* occurs between two stations before any data are actually transmitted. Connections are sometimes referred to as *sessions, virtual circuits,* or *logical connections.*
2. Most connection-oriented protocols require some means of acknowledging the data as they are being transmitted. Protocols that use acknowledgment procedures provide a high level of network reliability.
3. Connection-oriented protocols often provide some means of error control (i.e., error detection and error correction). Whenever data are found to be in error, the receiving station requests a retransmission.
4. When a connection is no longer needed, a specific handshake drops the connection.

Connectionless protocols are protocols where data are exchanged in an unplanned fashion without prior coordination between endpoints (e.g., a datagram). Connectionless protocols do not provide the same high degree of reliability as connection-oriented protocols; however, connectionless protocols offer a significant advantage in transmission speed. Connectionless protocols operate in a manner similar to the U.S. Postal Service, where information is formatted, placed in an envelope with source and destination addresses, and then mailed. You can only hope the letter arrives at its destination. A connectionless protocol is depicted in Figure 21-1b. Characteristics of connectionless protocols are as follow:

1. Connectionless protocols send data with a source and destination address without a handshake to ensure that the destination is ready to receive the data.
2. Connectionless protocols usually do not support error control or acknowledgment procedures, making them a relatively unreliable method of data transmission.
3. Connectionless protocols are used because they are often more efficient, as the data being transmitted usually do not justify the extra overhead required by connection-oriented protocols.

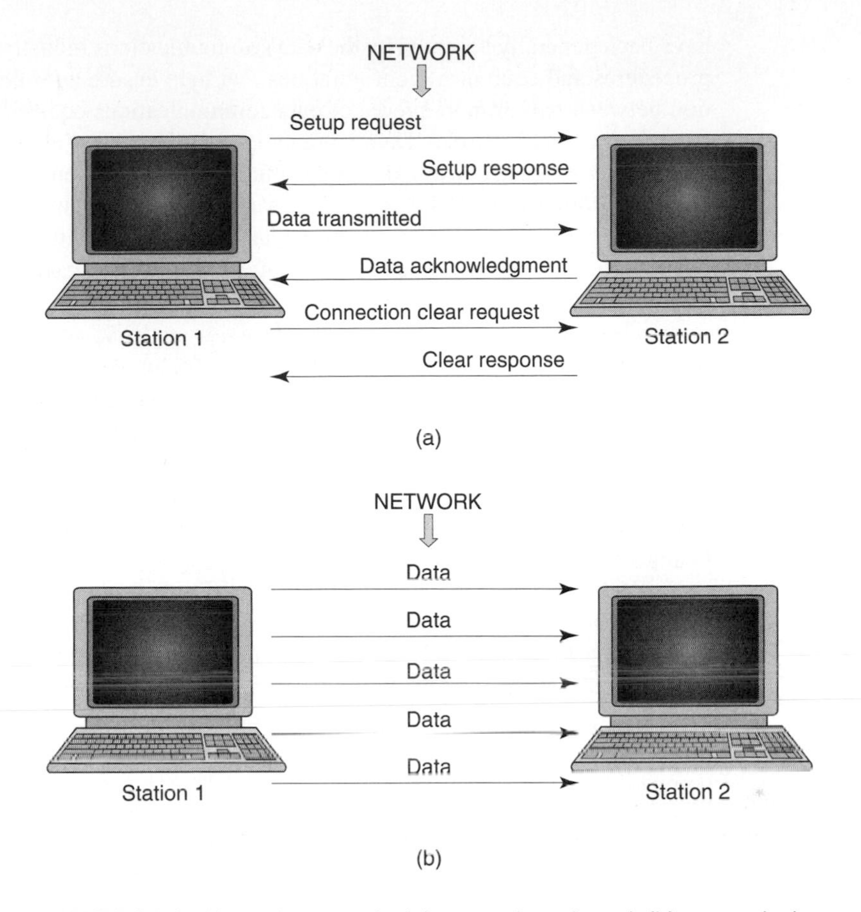

FIGURE 21-1 Network protocols: (a) connection-oriented; (b) connectionless

21-3-2-2 Syntax and semantics. Protocols include the concepts of *syntax* and *semantics.* Syntax refers to the structure or format of the data within the message, which includes the sequence in which the data are sent. For example, the first byte of a message might be the address of the source and the second byte the address of the destination. Semantics refers to the meaning of each section of data. For example, does a destination address identify only the location of the final destination, or does it also identify the route the data takes between the sending and receiving locations?

21-3-3 Data Communications Standards

During the past several decades, the data communications industry has grown at an astronomical rate. Consequently, the need to provide communications between dissimilar computer equipment and systems has also increased. A major issue facing the data communications industry today is worldwide compatibility. Major areas of interest are software and programming language, electrical and cable interface, transmission media, communications signal, and format compatibility. Thus, to ensure an orderly transfer of information, it has been necessary to establish standard means of governing the physical, electrical, and procedural arrangements of a data communications system.

A standard is an object or procedure considered by an authority or by general consent as a basis of comparison. Standards are authoritative principles or rules that imply a model or pattern for guidance by comparison. *Data communications standards* are guidelines that have been generally accepted by the data communications industry. The guidelines outline procedures and equipment configurations that help ensure an orderly transfer of information between two or more pieces of data communications equipment or two or more data communications networks. Data communications standards are not laws, however—they are simply suggested ways of implementing procedures and accomplishing results. If everyone complies with the standards, everyone's equipment, procedures, and processes will be compatible with everyone else's, and there will be little difficulty communicating information through the system. Today, most companies make their products to comply with standards.

There are two basic types of standards: *proprietary* (closed) system and *open* system. Proprietary standards are generally manufactured and controlled by one company. Other companies are not allowed to manufacture equipment or write software using this standard. An example of a proprietary standard is Apple Macintosh computers. Advantages of proprietary standards are tighter control, easier consensus, and a monopoly. Disadvantages include lack of choice for the customers, higher financial investment, overpricing, and reduced customer protection against the manufacturer going out of business.

With open system standards, any company can produce compatible equipment or software; however, often a royalty must be paid to the original company. An example of an open system standard is IBM's personal computer. Advantages of open system standards are customer choice, compatibility between venders, and competition by smaller companies. Disadvantages include less product control and increased difficulty acquiring agreement between vendors for changes or updates. In addition, standard items are not always as compatible as we would like them to be.

21-4 STANDARDS ORGANIZATIONS FOR DATA COMMUNICATIONS

A consortium of organizations, governments, manufacturers, and users meet on a regular basis to ensure an orderly flow of information within data communications networks and systems by establishing guidelines and standards. The intent is that all data communications equipment manufacturers and users comply with these standards. Standards organizations generate, control, and administer standards. Often, competing companies will form a joint committee to create a compromised standard that is acceptable to everyone. The most prominent organizations relied on in North America to publish standards and make recommendations for the data, telecommunications, and networking industries are shown in Figure 21-2.

21-4-1 International Standards Organization (ISO)

Created in 1946, the *International Standards Organization* (ISO) is the international organization for standardization on a wide range of subjects. The ISO is a voluntary, nontreaty organization whose membership is comprised mainly of members from the standards committees of various governments throughout the world. The ISO creates the sets of rules and standards for graphics and document exchange and provides models for equipment and system compatibility, quality enhancement, improved productivity, and reduced costs. The ISO is responsible for endorsing and coordinating the work of the other standards organizations. The member body of the ISO from the United States is the American National Standards Institute (ANSI).

21-4-2 International Telecommunications Union—Telecommunications Sector

The *International Telecommunications Union—Telecommunications Sector* (ITU-T), formerly the Comité Consultatif Internationale de Télégraphie et Téléphonie (CCITT), is one of four permanent parts of the International Telecommunications Union based in Geneva, Switzerland.

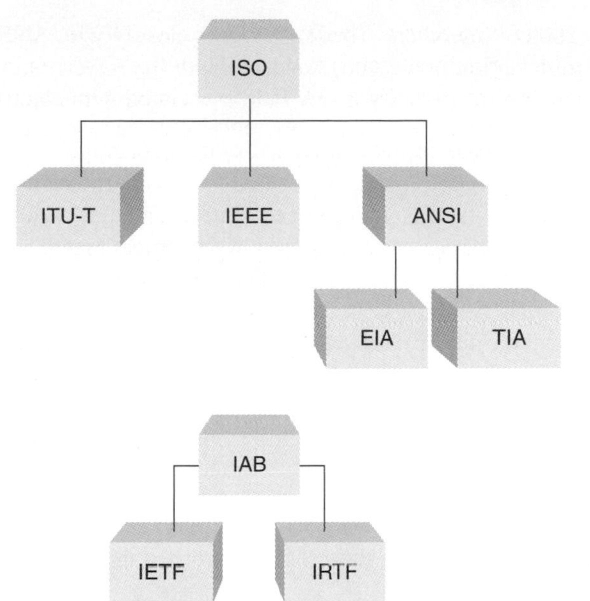

FIGURE 21-2 Standards organizations for data and network communications

Membership in the ITU-T consists of government authorities and representatives from many countries. The ITU-T is now the standards organization for the United Nations and develops the recommended sets of rules and standards for telephone and data communications. The ITU-T has developed three sets of specifications: the V series for modem interfacing and data transmission over telephone lines; the X series for data transmission over public digital networks, e-mail, and directory services; and the I and Q series for Integrated Services Digital Network (ISDN) and its extension Broadband ISDN (sometimes called the Information Superhighway).

The ITU-T is separated into 14 study groups that prepare recommendations on the following topics:

Network and service operation
Tariff and accounting principles
Telecommunications management network and network maintenance
Protection against electromagnetic environment effects
Outside plant
Data networks and open system communications
Characteristics of telematic systems
Television and sound transmission
Language and general software aspects for telecommunications systems
Signaling requirements and protocols
End-to-end transmission performance of networks and terminals
General network aspects
Transport networks, systems, and equipment
Multimedia services and systems

21-4-3 Institute of Electrical and Electronics Engineers

The *Institute of Electrical and Electronics Engineers* (IEEE) is an international professional organization founded in the United States and is comprised of electronics, computer, and communications engineers. The IEEE is currently the world's largest professional society with over 200,000 members. The IEEE works closely with ANSI to develop communications and information processing standards with the underlying goal of advancing theory, creativity, and product quality in any field associated with electrical engineering.

21-4-4 American National Standards Institute

The *American National Standards Institute* (ANSI) is the official standards agency for the United States and is the U.S. voting representative for the ISO. However, ANSI is a completely private, nonprofit organization comprised of equipment manufacturers and users of data processing equipment and services. Although ANSI has no affiliations with the federal government of the United States, it serves as the national coordinating institution for voluntary standardization in the United States. ANSI membership is comprised of people from professional societies, industry associations, governmental and regulatory bodies, and consumer groups.

21-4-5 Electronics Industry Association

The *Electronics Industries Associations* (EIA) is a nonprofit U.S. trade association that establishes and recommends industrial standards. EIA activities include standards development, increasing public awareness, and lobbying. The EIA is responsible for developing the RS (recommended standard) series of standards for data and telecommunications.

21-4-6 Telecommunications Industry Association

The *Telecommunications Industry Association* (TIA) is the leading trade association in the communications and information technology industry. The TIA facilitates business development opportunities and a competitive marketplace through market development, trade promotion, trade shows, domestic and international advocacy, and standards development. The TIA represents manufacturers of communications and information technology products and services providers for the global marketplace through its core competencies. The TIA also facilitates the convergence of new communications networks while working for a competitive and innovative market environment.

21-4-7 Internet Architecture Board

In 1957, the Advanced Research Projects Agency (ARPA), the research arm of the Department of Defense, was created in response to the Soviet Union's launching of *Sputnik.* The original purpose of ARPA was to accelerate the advancement of technologies that could possibly be useful to the U.S. military. When ARPANET was initiated in the late 1970s, ARPA formed a committee to oversee it. In 1983, the name of the committee was changed to the *Internet Activities Board* (IAB). The meaning of the acronym was later changed to the *Internet Architecture Board.*

Today the IAB is a technical advisory group of the Internet Society with the following responsibilities:

1. Oversees the architecture protocols and procedures used by the Internet
2. Manages the processes used to create Internet standards and serves as an appeal board for complaints of improper execution of the standardization processes
3. Is responsible for the administration of the various Internet assigned numbers
4. Acts as representative for Internet Society interests in liaison relationships with other organizations concerned with standards and other technical and organizational issues relevant to the worldwide Internet
5. Acts as a source of advice and guidance to the board of trustees and officers of the Internet Society concerning technical, architectural, procedural, and policy matters pertaining to the Internet and its enabling technologies

21-4-8 Internet Engineering Task Force

The *Internet Engineering Task Force* (IETF) is a large international community of network designers, operators, venders, and researchers concerned with the evolution of the Internet architecture and the smooth operation of the Internet.

21-4-9 Internet Research Task Force

The *Internet Research Task Force* (IRTF) promotes research of importance to the evolution of the future Internet by creating focused, long-term and small research groups working on topics related to Internet protocols, applications, architecture, and technology.

21-5 LAYERED NETWORK ARCHITECTURE

The basic concept of *layering* network responsibilities is that each layer adds value to services provided by sets of lower layers. In this way, the highest level is offered the full set of services needed to run a distributed data application. There are several advantages to using a *layered architecture.* A layered architecture facilitates *peer-to-peer communications protocols* where a given layer in one system can logically communicate with its corresponding layer in another system. This allows different computers to communicate at different levels. Figure 21-3 shows a layered architecture where layer N at the source logically (but not necessarily physically) communicates with layer N at the destination and layer N of any intermediate nodes.

21-5-1 Protocol Data Unit

When technological advances occur in a layered architecture, it is easier to modify one layer's protocol without having to modify all the other layers. Each layer is essentially independent of every other layer. Therefore, many of the functions found in lower layers have been removed entirely from software tasks and replaced with hardware. The primary disadvantage of layered architectures is the tremendous amount of overhead required. With layered architectures, communications between two corresponding layers requires a unit of data called a *protocol data unit* (PDU). As shown in Figure 21-4, a PDU can be a *header* added at the beginning of a message or a *trailer* appended to the end of a message. In a layered architecture, communications occurs between similar layers; however, data must flow through the other layers. Data flows downward through the layers in the source system and upward through the layers in the destination system. In intermediate systems, data flows upward first and then downward. As data passes from one layer into another, headers and trailers are added and removed from the PDU. The process of adding or removing PDU information is called *encapsulation/decapsulation* because it appears as though the PDU from the upper layer is encapsulated in the PDU from the lower layer during the downward

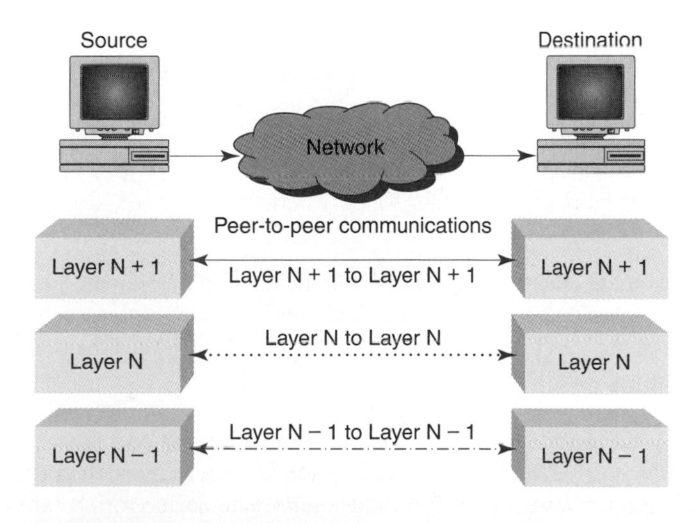

FIGURE 21-3 Peer-to-peer data communications

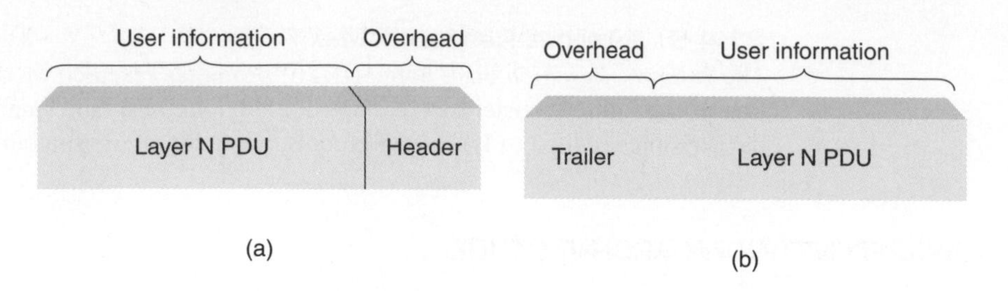

FIGURE 21-4 Protocol data unit: (a) header; (b) trailer

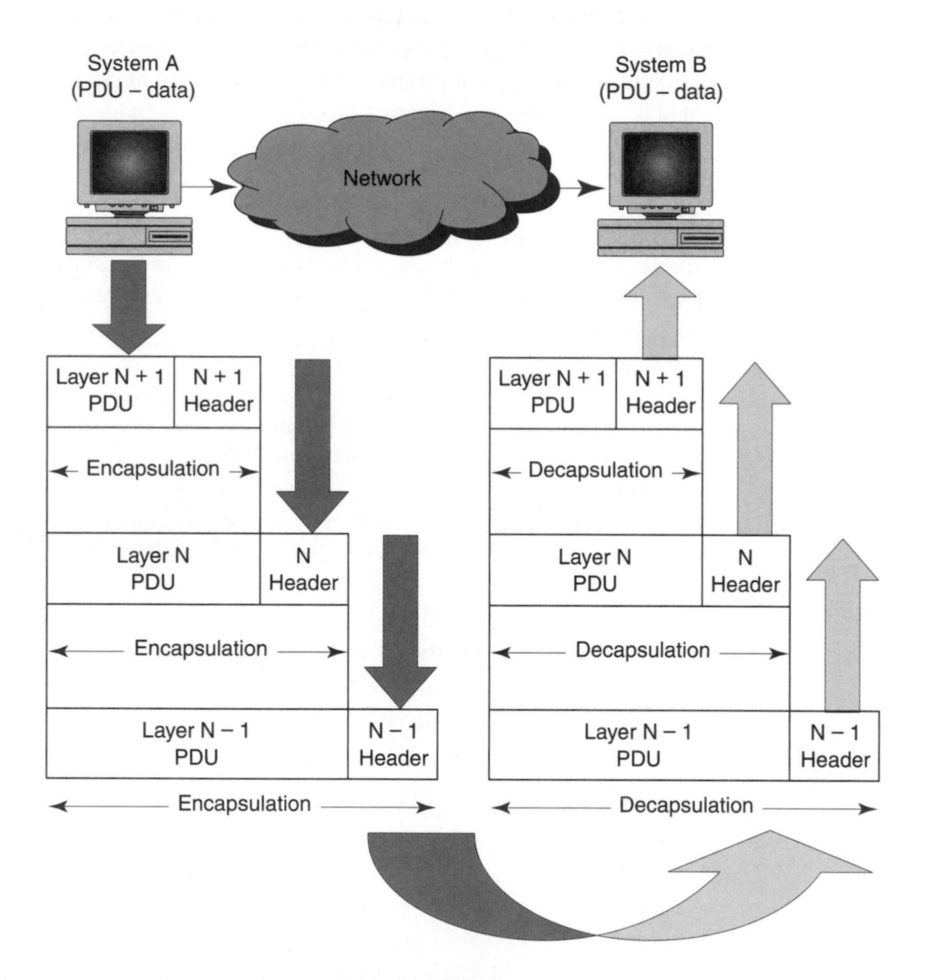

FIGURE 21-5 Encapsulation and decapsulation

movement and decapsulated during the upward movement. *Encapsulate* means to place in a capsule or other protected environment, and *decapsulate* means to remove from a capsule or other protected environment. Figure 21-5 illustrates the concepts of encapsulation and decapsulation.

In a layered protocol such as the one shown in Figure 21-3, layer N receive services from the layer immediately below it (N − 1) and provides services to the layer directly above it (N + 1). Layer N can provide service to more than one entity in layer N + 1 by using a *service access point* (SAP) *address* to define which entity the service is intended.

Information and network information passes from one layer of a multilayered architecture to another layer through a layer-to-layer *interface.* A layer-to-layer interface defines what information and services the lower layer must provide to the upper layer. A well-defined layer and layer-to-layer interface provide modularity to a network.

21-6 OPEN SYSTEMS INTERCONNECTION

Open systems interconnection (OSI) is the name for a set of standards for communicating among computers. The primary purpose of OSI standards is to serve as a structural guideline for exchanging information between computers, workstations, and networks. The OSI is endorsed by both the ISO and ITU-T, which have worked together to establish a set of ISO standards and ITU-T recommendations that are essentially identical. In 1983, the ISO and ITU-T (CCITT) adopted a seven-layer communications architecture reference model. Each layer consists of specific protocols for communicating.

The ISO seven-layer open systems interconnection model is shown in Figure 21-6. This hierarchy was developed to facilitate the intercommunications of data processing equipment by separating network responsibilities into seven distinct layers. As with any layered architecture, overhead information is added to a PDU in the form of headers and trailers. In fact, if all seven levels of the OSI model are addressed, as little as 15% of the transmitted message is actually source information, and the rest is overhead. The result of adding headers to each layer is illustrated in Figure 21-7.

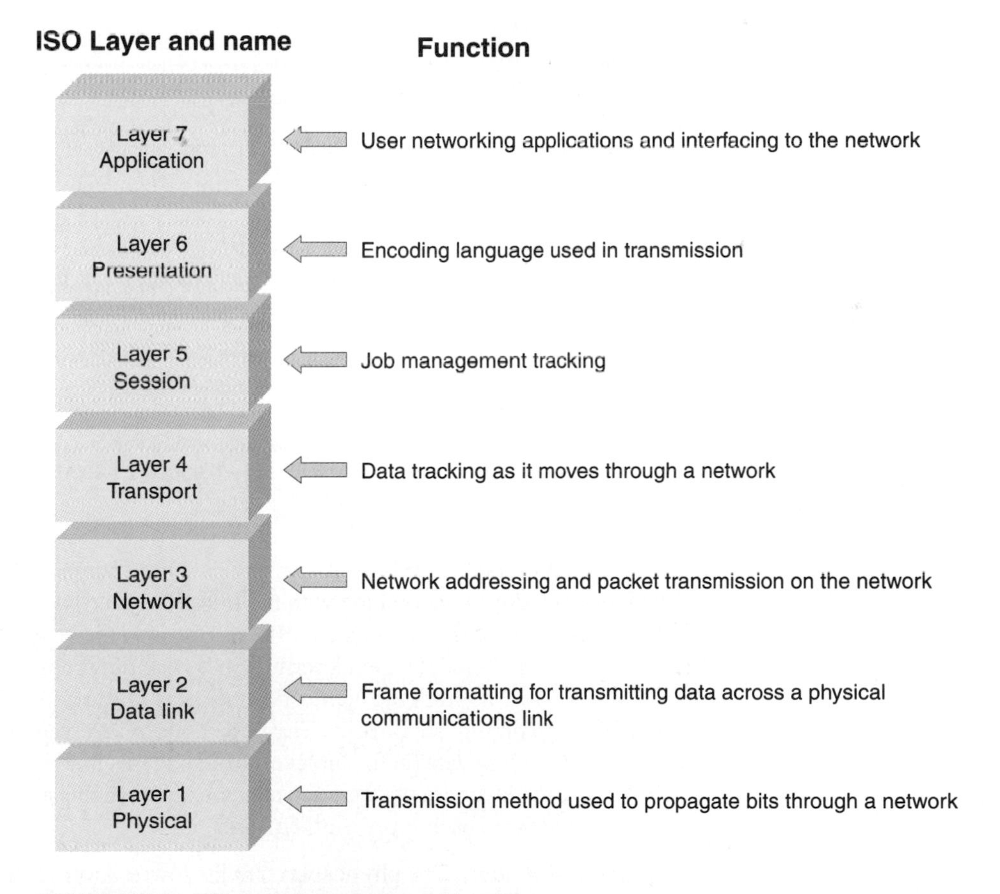

FIGURE 21-6 OSI seven-layer protocol hierarchy

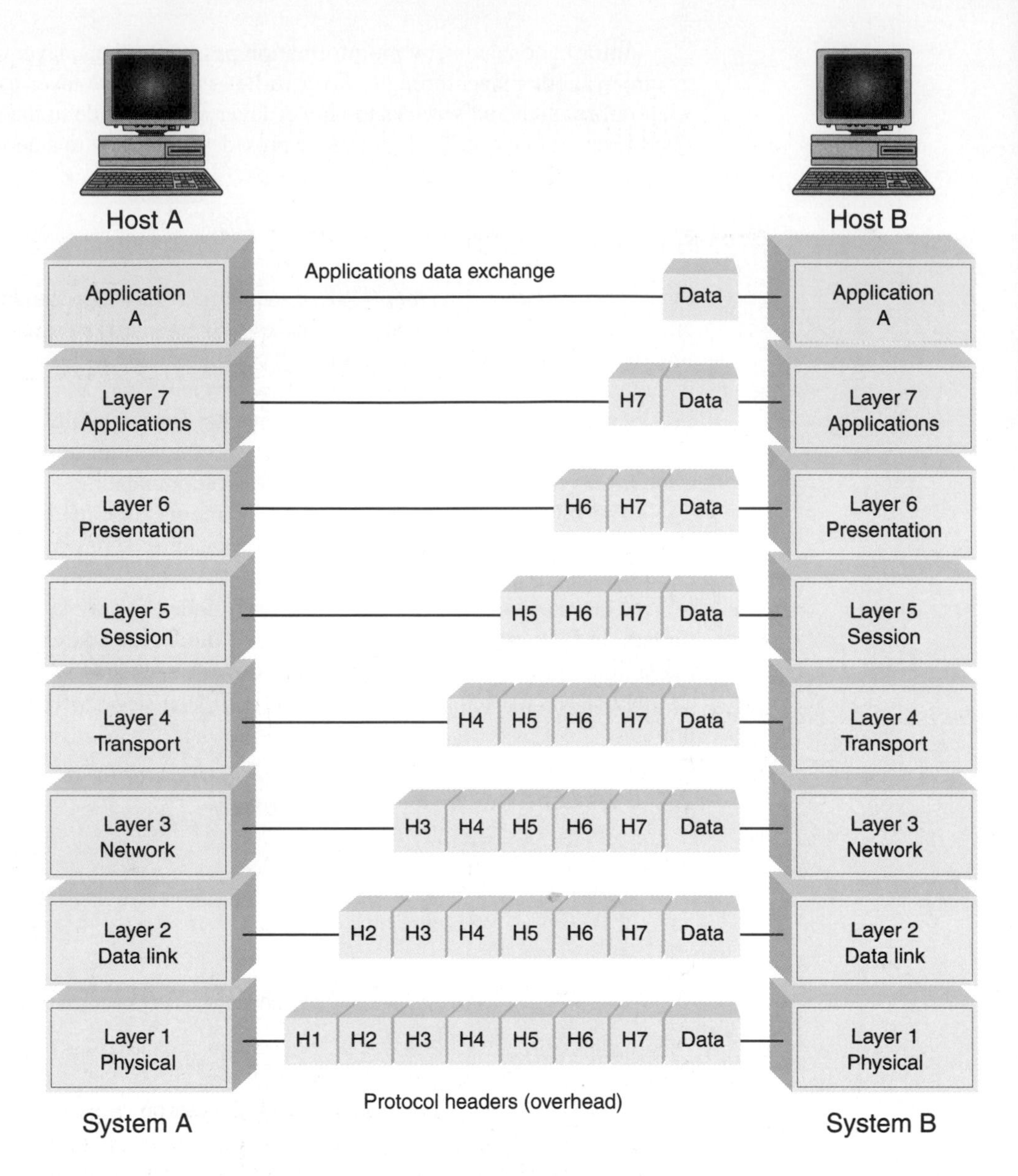

FIGURE 21-7 OSI seven-layer international protocol hierarchy. H7—applications header, H6—presentation header, H5—session header, H4—transport header, H3—network header, H2—data-link header, H1—physical header

In recent years, the OSI seven-layer model has become more academic than standard, as the hierarchy does not coincide with the Internet's four-layer protocol model. However, the basic functions of the layers are still performed, so the seven-layer model continues to serve as a reference model when describing network functions.

Levels 4 to 7 address the applications aspects of the network that allow for two host computers to communicate directly. The three bottom layers are concerned with the actual mechanics of moving data (at the bit level) from one machine to another. The basic services provided by each layer are discussed in a later chapter of this book. A brief summary of the services provided by each layer is given here.

1. *Physical layer.* The physical layer is the lowest level of the OSI hierarchy and is responsible for the actual propagation of unstructured data bits (1s and 0s) through a transmis-

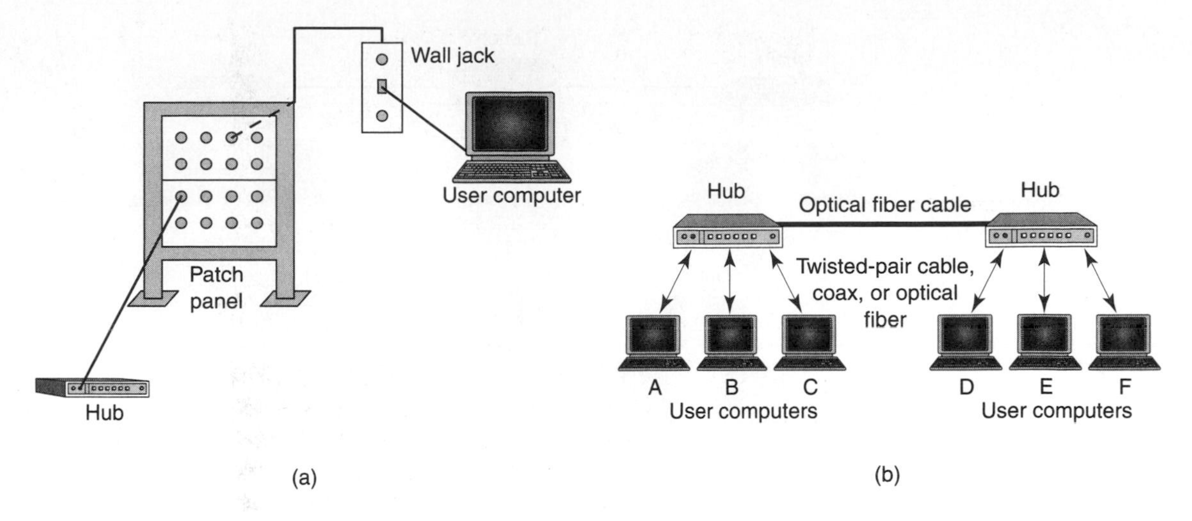

FIGURE 21-8 OSI layer 1—physical: (a) computer-to-hub; (b) connectivity devices

sion medium, which includes how bits are represented, the bit rate, and how bit synchronization is achieved. The physical layer specifies the type of transmission medium and the transmission mode (simplex, half duplex, or full duplex) and the physical, electrical, functional, and procedural standards for accessing data communications networks. Definitions such as connections, pin assignments, interface parameters, timing, maximum and minimum voltage levels, and circuit impedances are made at the physical level. Transmission media defined by the physical layer include metallic cable, optical fiber cable, or wireless radio-wave propagation. The physical layer for a cable connection is depicted in Figure 21-8a.

Connectivity devices connect devices on cabled networks. An example of a connectivity device is a hub. A hub is a transparent device that samples the incoming bit stream and simply repeats it to the other devices connected to the hub. The hub does not examine the data to determine what the destination is; therefore, it is classified as a layer 1 component. Physical layer connectivity for a cabled network is shown in Figure 21-8b.

The physical layer also includes the *carrier system* used to propagate the data signals between points in the network. Carrier systems are simply communications systems that carry data through a system using either metallic or optical fiber cables or wireless arrangements, such as microwave, satellites, and cellular radio systems. The carrier can use analog or digital signals that are somehow converted to a different form (encoded or modulated) by the data and then propagated through the system.

2. *Data-link layer.* The data-link layer is responsible for providing error-free communications across the physical link connecting primary and secondary stations (nodes) within a network (sometimes referred to as *hop-to-hop* delivery). The data-link layer packages data from the physical layer into groups called blocks, frames, or packets and provides a means to activate, maintain, and deactivate the data communications link between nodes. The data-link layer provides the final framing of the information signal, provides synchronization, facilitates the orderly flow of data between nodes, outlines procedures for error detection and correction, and provides the physical addressing information. A block diagram of a network showing data transferred between two computers (A and E) at the data-link level is illustrated in Figure 21-9. Note that the hubs are transparent but that the switch passes the transmission on to only the hub serving the intended destination.

3. *Network layer.* The network layer provides details that enable data to be routed between devices in an environment using multiple networks, subnetworks, or both. Networking components that operate at the network layer include routers and their software. The

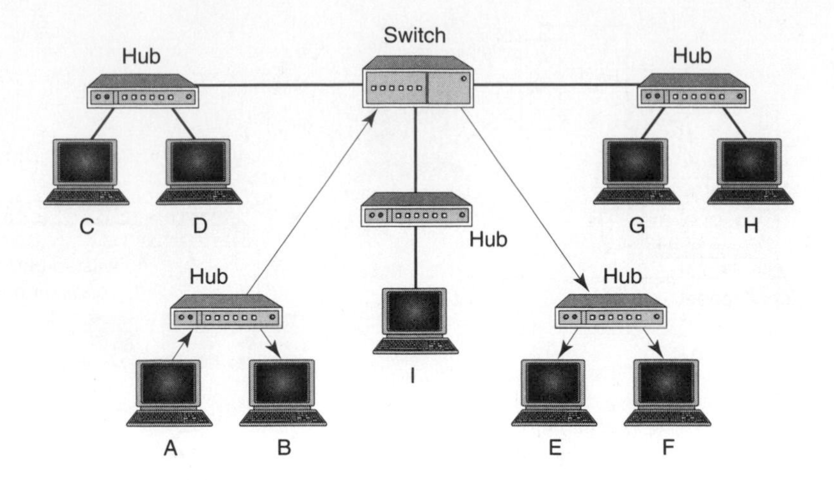

FIGURE 21-9 OSI layer 2—data link

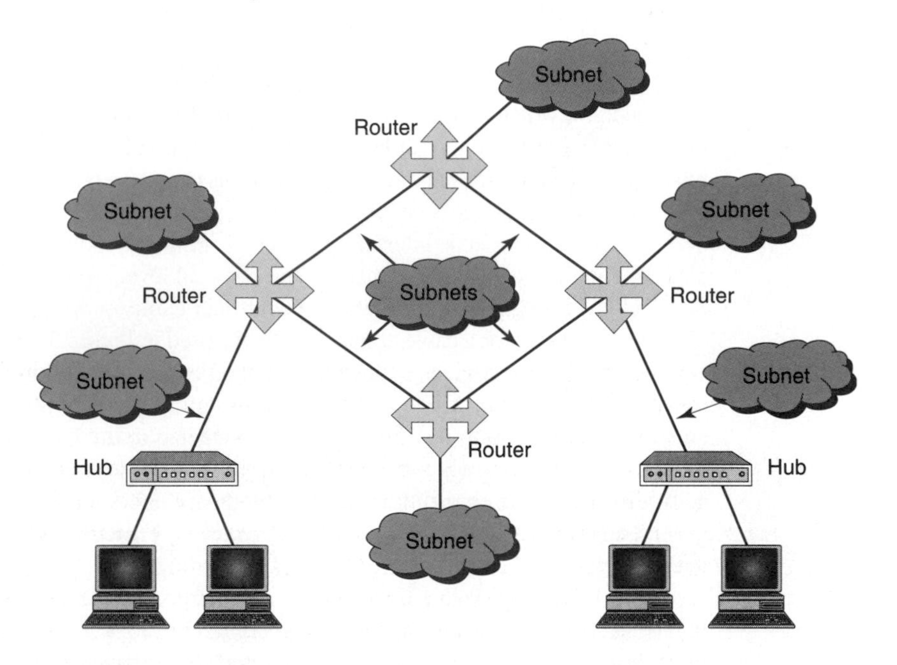

FIGURE 21-10 OSI layer 3—network

network layer determines which network configuration is most appropriate for the function provided by the network and addresses and routes data within networks by establishing, maintaining, and terminating connections between them. The network layer provides the upper layers of the hierarchy independence from the data transmission and switching technologies used to interconnect systems. It accomplishes this by defining the mechanism in which messages are broken into smaller data packets and routed from a sending node to a receiving node within a data communications network. The network layer also typically provides the source and destination network addresses (logical addresses), subnet information, and source and destination node addresses. Figure 21-10 illustrates the network layer of the OSI protocol hierarchy. Note that the network is subdivided into subnetworks that are separated by routers.

4. *Transport layer.* The transport layer controls and ensures the end-to-end integrity of the data message propagated through the network between two devices, which provides

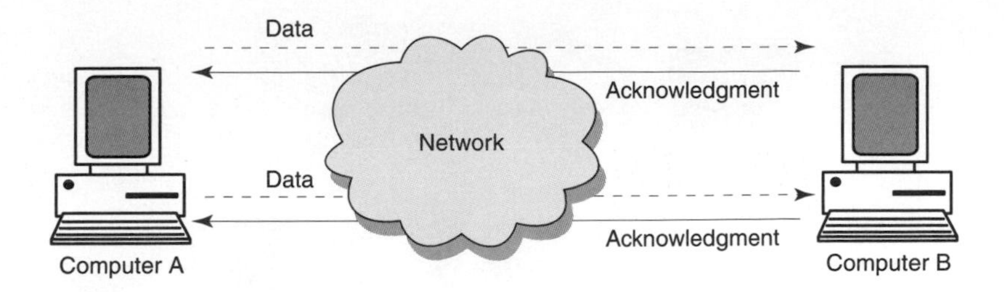

FIGURE 21-11 OSI layer 4—transport

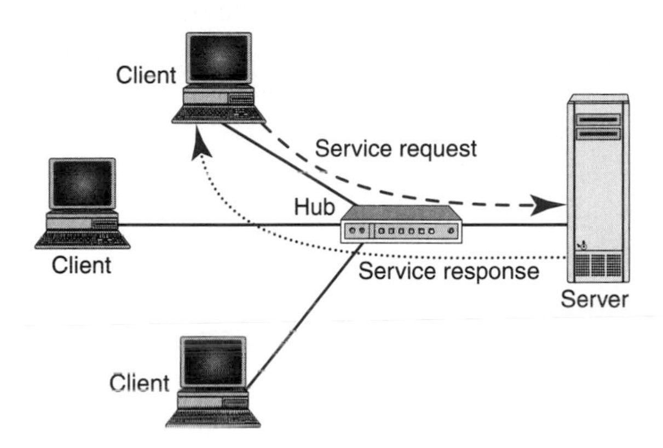

FIGURE 21-12 OSI layer 5—session

for the reliable, transparent transfer of data between two endpoints. Transport layer responsibilities includes message routing, segmenting, error recovery, and two types of basic services to an upper-layer protocol: connectionless oriented and connectionless. The transport layer is the highest layer in the OSI hierarchy in terms of communications and may provide data tracking, connection flow control, sequencing of data, error checking, and application addressing and identification. Figure 21-11 depicts data transmission at the transport layer.

5. *Session layer.* The session layer is responsible for network availability (i.e., data storage and processor capacity). Session layer protocols provide the logical connection entities at the application layer. These applications include file transfer protocols and sending e-mail. Session responsibilities include network log-on and log-off procedures and user authentication. A session is a temporary condition that exists when data are actually in the process of being transferred and does not include procedures such as call establishment, setup, or disconnect. The session layer determines the type of dialogue available (i.e., simplex, half duplex, or full duplex). Session layer characteristics include virtual connections between applications entities, synchronization of data flow for recovery purposes, creation of dialogue units and activity units, connection parameter negotiation, and partitioning services into functional groups. Figure 21-12 illustrates the establishment of a session on a data network.

6. *Presentation layer.* The presentation layer provides independence to the application processes by addressing any code or syntax conversion necessary to present the data to the network in a common communications format. The presentation layer specifies how end-user applications should format the data. This layer provides for translation between local representations of data and the representation of data that will be used for transfer between end users. The results of encryption, data compression, and virtual terminals are examples of the translation service.

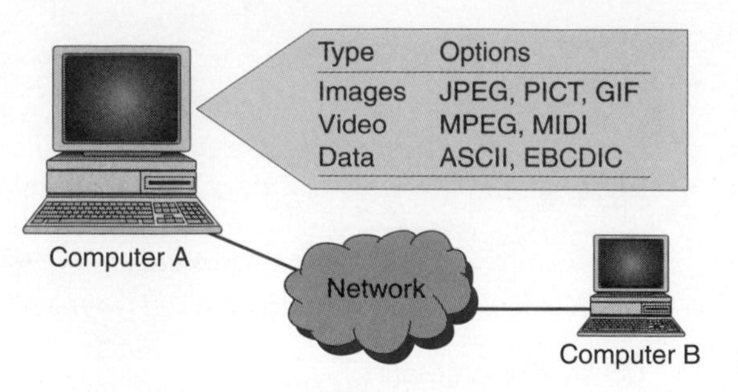

FIGURE 21-13 OSI layer 6—presentation

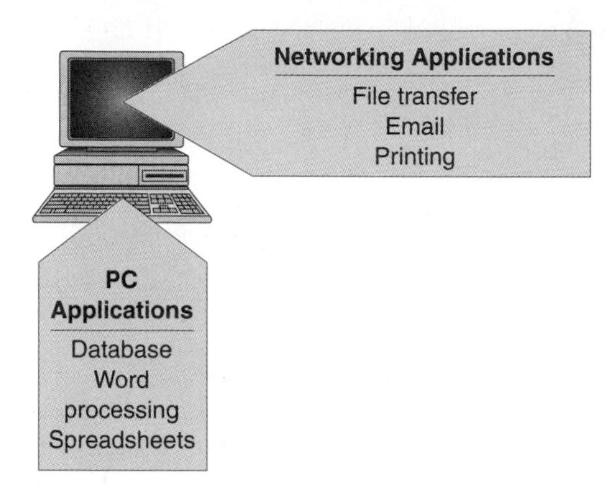

FIGURE 21-14 OSI layer 7—applications

The presentation layer translates between different data formats and protocols. Presentation functions include data file formatting, encoding, encryption and decryption of data messages, dialogue procedures, data compression algorithms, synchronization, interruption, and termination. The presentation layer performs code and character set translation (including ASCII and EBCDIC) and formatting information and determines the display mechanism for messages. Figure 21-13 shows an illustration of the presentation layer.

7. *Application layer.* The application layer is the highest layer in the hierarchy and is analogous to the general manager of the network by providing access to the OSI environment. The applications layer provides distributed information services and controls the sequence of activities within an application and also the sequence of events between the computer application and the user of another application. The application layer (shown in Figure 21-14) communicates directly with the user's application program.

User application processes require application layer service elements to access the networking environment. There are two types of service elements: CASEs (*common application service elements*), which are generally useful to a variety of application processes and SASEs (*specific application service elements*), which generally satisfy particular needs of application processes. CASE examples include association control that establishes, maintains, and terminates connections with a peer application entity and commitment, concurrence, and recovery that ensure the integrity of distributed transactions. SASE examples involve the TCP/IP protocol stack and include FTP (*file transfer protocol*), SNMP (*simple network management protocol*), Telnet (*virtual terminal protocol*), and SMTP (*simple mail transfer protocol*).

21-7 DATA COMMUNICATIONS CIRCUITS

The underlying purpose of a data communications circuit is to provide a transmission path between locations and to transfer digital information from one station to another using electronic circuits. A *station* is simply an endpoint where subscribers gain access to the circuit. A station is sometimes called a *node,* which is the location of computers, computer terminals, workstations, and other digital computing equipment. There are almost as many types of data communications circuits as there are types of data communications equipment.

Data communications circuits utilize electronic communications equipment and *facilities* to interconnect digital computer equipment. Communications facilities are physical means of interconnecting stations within a data communications system and can include virtually any type of physical transmission media or wireless radio system in existence. Communications facilities are provided to data communications users through public telephone networks (PTN), public data networks (PDN), and a multitude of private data communications systems.

Figure 21-15 shows a simplified block diagram of a two-station data communications circuit. The fundamental components of the circuit are source of digital information, transmitter, transmission medium, receiver, and destination for the digital information. Although the figure shows transmission in only one direction, bidirectional transmission is possible by providing a duplicate set of circuit components in the opposite direction.

Source. The information source generates data and could be a mainframe computer, personal computer, workstation, or virtually any other piece of digital equipment. The source equipment provides a means for humans to enter data into the system.

Transmitter. Source data is seldom in a form suitable to propagate through the transmission medium. For example, digital signals (pulses) cannot be propagated through a wireless radio system without being converted to analog first. The transmitter encodes the source information and converts it to a different form, allowing it to be more efficiently propagated through the transmission medium. In essence, the transmitter acts as an interface between the source equipment and the transmission medium.

Transmission medium. The transmission medium carries the encoded signals from the transmitter to the receiver. There are many different types of transmission media, such as free-space radio transmission (including all forms of wireless transmission, such as terrestrial microwave, satellite radio, and cellular telephone) and physical facilities, such as metallic and optical fiber cables. Very often, the transmission path is comprised of several different types of transmission facilities.

Receiver. The receiver converts the encoded signals received from the transmission medium back to their original form (i.e., decodes them) or whatever form is used in the destination equipment. The receiver acts as an interface between the transmission medium and the destination equipment.

Destination. Like the source, the destination could be a mainframe computer, personal computer, workstation, or virtually any other piece of digital equipment.

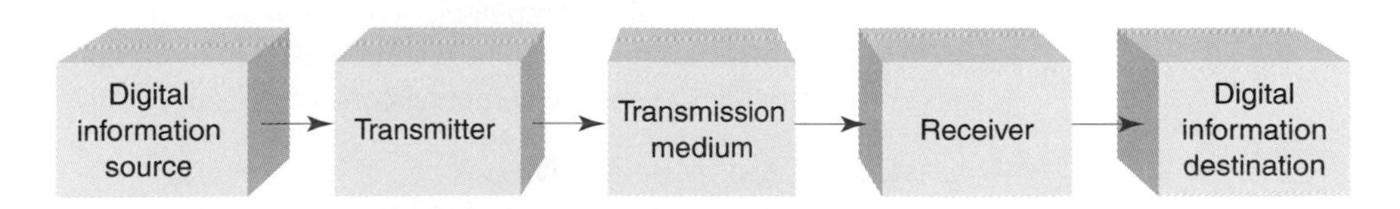

FIGURE 21-15 Simplified block diagram of a two-station data communications circuit

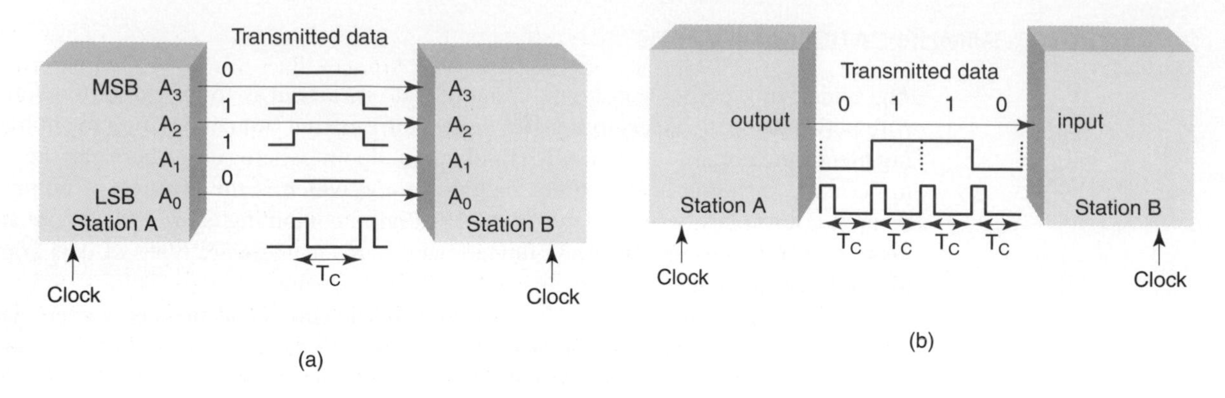

FIGURE 21-16 Data transmission: (a) parallel; (b) serial

21-8 SERIAL AND PARALLEL DATA TRANSMISSION

Binary information can be transmitted either in parallel or serially. Figure 21-16a shows how the binary code 0110 is transmitted from station A to station B in parallel. As the figure shows, each bit position (A_0 to A_3) has its own transmission line. Consequently, all four bits can be transmitted simultaneously during the time of a single clock pulse (T_C). This type of transmission is called *parallel by bit* or *serial by character.*

Figure 21-16b shows the same binary code transmitted serially. As the figure shows, there is a single transmission line and, thus, only one bit can be transmitted at a time. Consequently, it requires four clock pulses ($4T_C$) to transmit the entire four-bit code. This type of transmission is called *serial by bit.*

Obviously, the principal trade-off between parallel and serial data transmission is speed versus simplicity. Data transmission can be accomplished much more rapidly using parallel transmission; however, parallel transmission requires more data lines. As a general rule, parallel transmission is used for short-distance data communications and within a computer, and serial transmission is used for long-distance data communications.

21-9 DATA COMMUNICATIONS CIRCUIT ARRANGEMENTS

Data communications circuits can be configured in a multitude of arrangements depending on the specifics of the circuit, such as how many stations are on the circuit, type of transmission facility, distance between stations, and how many users are at each station. A data communications circuit can be described in terms of circuit configuration and transmission mode.

21-9-1 Circuit Configurations

Data communications networks can be generally categorized as either two point or multipoint. A *two-point* configuration involves only two locations or stations, whereas a *multipoint* configuration involves three or more stations. Regardless of the configuration, each station can have one or more computers, computer terminals, or workstations. A two-point circuit involves the transfer of digital information between a mainframe computer and a personal computer, two mainframe computers, two personal computers, or two data communications networks. A multipoint network is generally used to interconnect a single mainframe computer (host) to many personal computers or to interconnect many personal computers.

21-9-2 Transmission Modes

Essentially, there are four modes of transmission for data communications circuits: *simplex, half duplex, full duplex,* and *full/full duplex.*

21-9-2-1 Simplex. In the simplex (SX) mode, data transmission is unidirectional; information can be sent in only one direction. Simplex lines are also called *receive-only, transmit-only,* or *one-way-only* lines. Commercial radio broadcasting is an example of simplex transmission, as information is propagated in only one direction—from the broadcasting station to the listener.

21-9-2-2 Half duplex. In the half-duplex (HDX) mode, data transmission is possible in both directions but not at the same time. Half-duplex communications lines are also called *two-way-alternate* or *either-way* lines. Citizens band (CB) radio is an example of half-duplex transmission because to send a message, the *push-to-talk* (PTT) switch must be depressed, which turns on the transmitter and shuts off the receiver. To receive a message, the PTT switch must be off, which shuts off the transmitter and turns on the receiver.

21-9-2-3 Full duplex. In the full-duplex (FDX) mode, transmissions are possible in both directions simultaneously, but they must be between the same two stations. Full-duplex lines are also called *two-way simultaneous, duplex,* or *both-way* lines. A local telephone call is an example of full-duplex transmission. Although it is unlikely that both parties would be talking at the same time, they could if they wanted to.

21-9-2-4 Full/full duplex. In the full/full duplex (F/FDX) mode, transmission is possible in both directions at the same time but not between the same two stations (i.e., one station is transmitting to a second station and receiving from a third station at the same time). Full/full duplex is possible only on multipoint circuits. The U.S. postal system is an example of full/full duplex transmission because a person can send a letter to one address and receive a letter from another address at the same time.

21-10 DATA COMMUNICATIONS NETWORKS

Any group of computers connected together can be called a *data communications network,* and the process of sharing resources between computers over a data communications network is called *networking.* In its simplest form, networking is two or more computers connected together through a common transmission medium for the purpose of sharing data. The concept of networking began when someone determined that there was a need to share software and data resources and that there was a better way to do it than storing data on a disk and literally running from one computer to another. By the way, this manual technique of moving data on disks is sometimes referred to as *sneaker net.* The most important considerations of a data communications network are performance, transmission rate, reliability, and security.

Applications running on modern computer networks vary greatly from company to company. A network must be designed with the intended application in mind. A general categorization of networking applications is listed in Table 21-1. The specific application affects how well a network will perform. Each network has a finite capacity. Therefore, network designers and engineers must be aware of the type and frequency of information traffic on the network.

Table 21-1 Networking Applications

Application	Examples
Standard office applications	E-mail, file transfers, and printing
High-end office applications	Video imaging, computer-aided drafting, computer-aided design, and software development
Manufacturing automation	Process and numerical control
Mainframe connectivity	Personal computers, workstations, and terminal support
Multimedia applications	Live interactive video

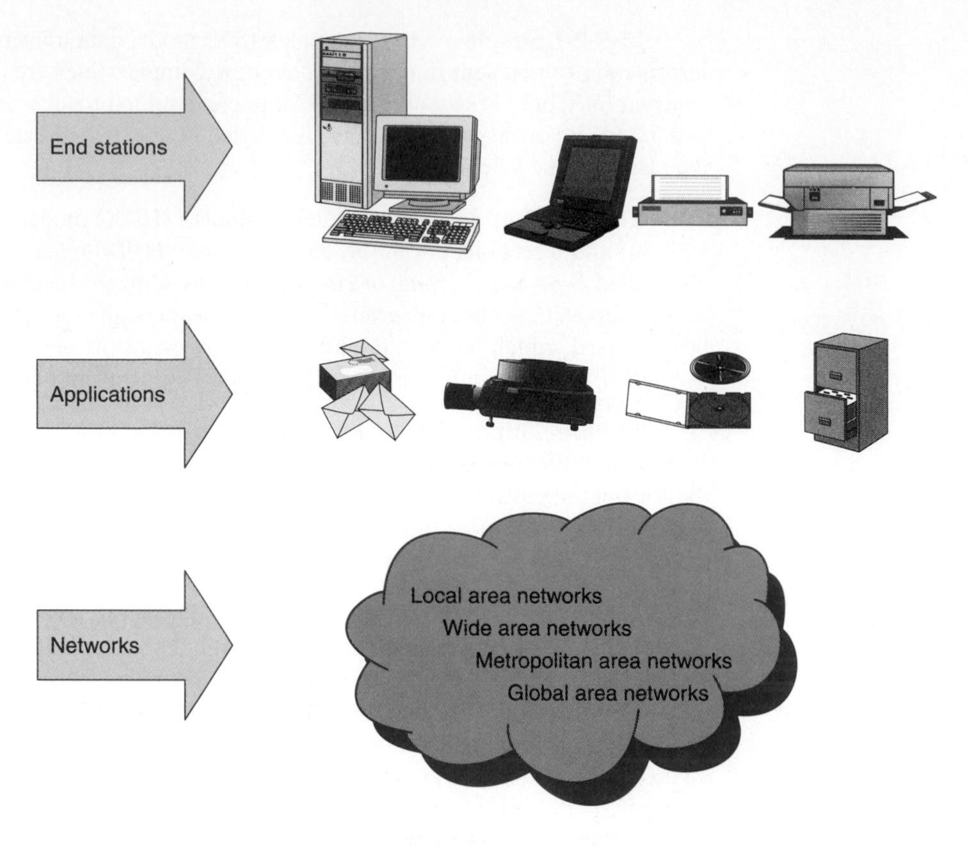

FIGURE 21-17 Basic network components

There are many factors involved when designing a computer network, including the following:

1. Network goals as defined by organizational management
2. Network security
3. Network uptime requirements
4. Network response-time requirements
5. Network and resource costs

The primary balancing act in computer networking is speed versus reliability. Too often, network performance is severely degraded by using error checking procedures, data encryption, and handshaking (acknowledgments). However, these features are often required and are incorporated into protocols.

Some networking protocols are very reliable but require a significant amount of overhead to provide the desired high level of service. These protocols are examples of connection-oriented protocols. Other protocols are designed with speed as the primary parameter and, therefore, forgo some of the reliability features of the connection-oriented protocols. These *quick protocols* are examples of connectionless protocols.

21-10-1 Network Components, Functions, and Features

Computer networks are like snowflakes—no two are the same. The basic components of computer networks are shown in Figure 21-17. All computer networks include some combination of the following: end stations, applications, and a network that will support the data traffic between the end stations. A computer network designed three years ago to support the basic networking applications of the time may have a difficult time supporting recently

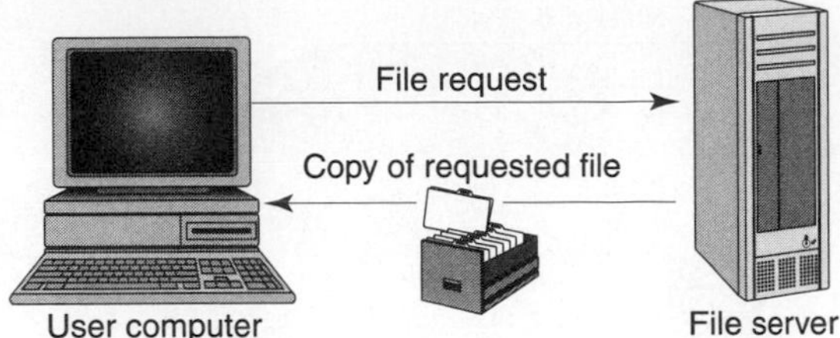

FIGURE 21-18 File server operation

developed high-end applications, such as medical imaging and live video teleconferencing. Network designers, administrators, and managers must understand and monitor the most recent types and frequency of networked applications.

Computer networks all share common devices, functions, and features, including servers, clients, transmission media, shared data, shared printers and other peripherals, hardware and software resources, network interface card (NIC), local operating system (LOS), and the network operating system (NOS).

21-10-1-1 Servers. *Servers* are computers that hold shared files, programs, and the network operating system. Servers provide access to network resources to all the users of the network. There are many different kinds of servers, and one server can provide several functions. For example, there are file servers, print servers, mail servers, communications servers, database servers, directory/security servers, fax servers, and Web servers, to name a few.

Figure 21-18 shows the operation of a *file server.* A user (client) requests a file from the file server. The file server sends a copy of the file to the requesting user. File servers allow users to access and manipulate disk resources stored on other computers. An example of a file server application is when two or more users edit a shared spreadsheet file that is stored on a server. File servers have the following characteristics:

1. File servers are loaded with files, accounts, and a record of the access rights of users or groups of users on the network.
2. The server provides a shareable virtual disk to the users (clients).
3. File mapping schemes are implemented to provide the virtualness of the files (i.e., the files are made to look like they are on the user's computer).
4. Security systems are installed and configured to provide the server with the required security and protection for the files.
5. Redirector or shell software programs located on the users' computers transparently activate the client's software on the file server.

21-10-1-2 Clients. *Clients* are computers that access and use the network and shared network resources. Client computers are basically the customers (users) of the network, as they request and receive services from the servers.

21-10-1-3 Transmission media. *Transmission media* are the facilities used to interconnect computers in a network, such as twisted-pair wire, coaxial cable, and optical fiber cable. Transmission media are sometimes called channels, links, or lines.

21-10-1-4 Shared data. *Shared data* are data that file servers provide to clients, such as data files, printer access programs, and e-mail.

21-10-1-5 Shared printers and other peripherals. *Shared printers* and *peripherals* are hardware resources provided to the users of the network by servers. Resources provided include data files, printers, software, or any other items used by clients on the network.

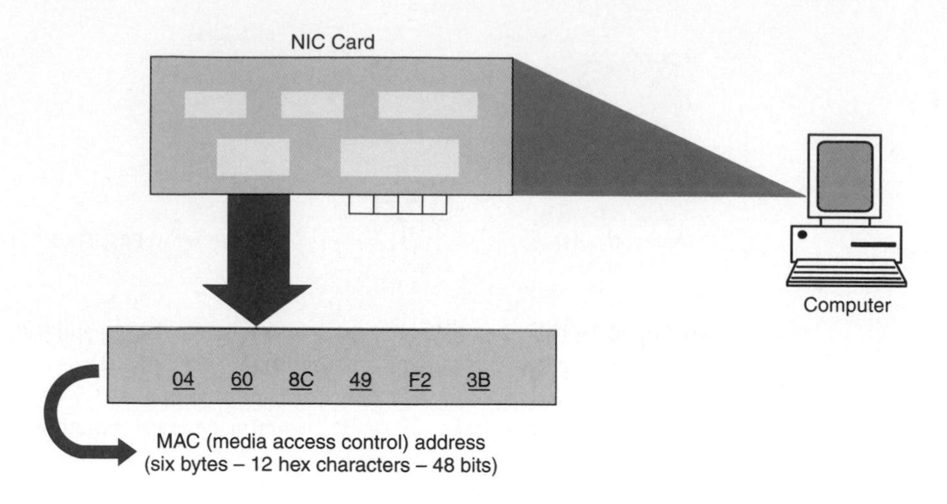

FIGURE 21-19 Network interface card (NIC)

21-10-1-6 Network interface card. Each computer in a network has a special expansion card called a *network interface card* (NIC). The NIC prepares (formats) and sends data, receives data, and controls data flow between the computer and the network. On the transmit side, the NIC passes frames of data on to the physical layer, which transmits the data to the physical link. On the receive side, the NIC processes bits received from the physical layer and processes the message based on its contents. A network interface card is shown in Figure 21-19. Characteristics of NICs include the following:

1. The NIC constructs, transmits, receives, and processes data to and from a PC and the connected network.
2. Each device connected to a network must have a NIC installed.
3. A NIC is generally installed in a computer as a daughterboard, although some computer manufacturers incorporate the NIC into the motherboard during manufacturing.
4. Each NIC has a unique six-byte media access control (MAC) address, which is typically permanently burned into the NIC when it is manufactured. The MAC address is sometimes called the physical, hardware, node, Ethernet, or LAN address.
5. The NIC must be compatible with the network (i.e., Ethernet—10baseT or token ring) to operate properly.
6. NICs manufactured by different vendors vary in speed, complexity, manageability, and cost.
7. The NIC requires drivers to operate on the network.

21-10-1-7 Local operating system. A *local operating system* (LOS) allows personal computers to access files, print to a local printer, and have and use one or more disk and CD drives that are located on the computer. Examples of LOSs are MS-DOS, PC-DOS, Unix, Macintosh, OS/2, Windows 3.11, Windows 95, Windows 98, Windows 2000, and Linux. Figure 21-20 illustrates the relationship between a personal computer and its LOS.

21-10-1-8 Network operating system. The *network operating system* (NOS) is a program that runs on computers and servers that allows the computers to communicate over a network. The NOS provides services to clients such as log-in features, password authenti-

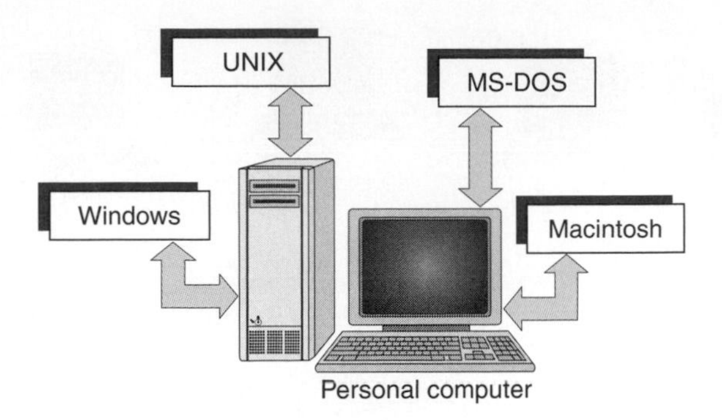

FIGURE 21-20 Local operating system (LOS)

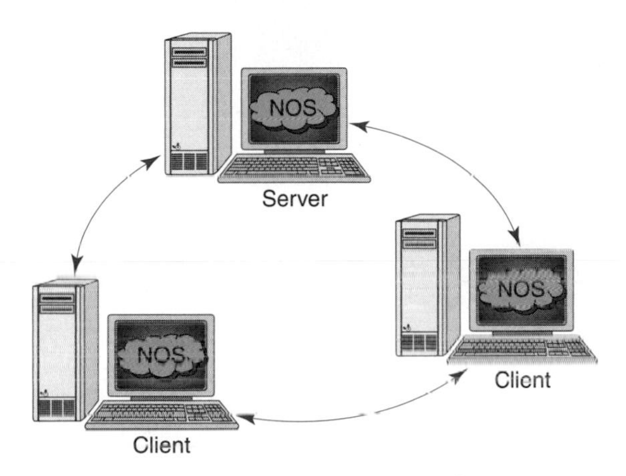

FIGURE 21-21 Network operating system (NOS)

cation, printer access, network administration functions, and data file sharing. Some of the more popular network operating systems are Unix, Novell NetWare, AppleShare, Macintosh System 7, IBM LAN Server, Compaq Open VMS, and Microsoft Windows NT Server. The NOS is software that makes communications over a network more manageable. The relationship between clients, servers, and the NOS is shown in Figure 21-21, and the layout of a local network operating system is depicted in Figure 21-22. Characteristics of NOSs include the following:

1. A NOS allows users of a network to interface with the network transparently.
2. A NOS commonly offers the following services: file service, print service, mail service, communications service, database service, and directory and security services.
3. The NOS determines whether data are intended for the user's computer or whether the data needs to be redirected out onto the network.
4. The NOS implements client software for the user, which allows them to access servers on the network.

21-10-2 Network Models

Computer networks can be represented with two basic network models: *peer-to-peer client/server* and *dedicated client/server.* The client/server method specifies the way in which two computers can communicate with software over a network. Although clients and servers are generally shown as separate units, they are often active in a single computer but not at the same time. With the client/server concept, a computer acting as a client initiates a software request from another computer acting as a server. The server computer responds and attempts

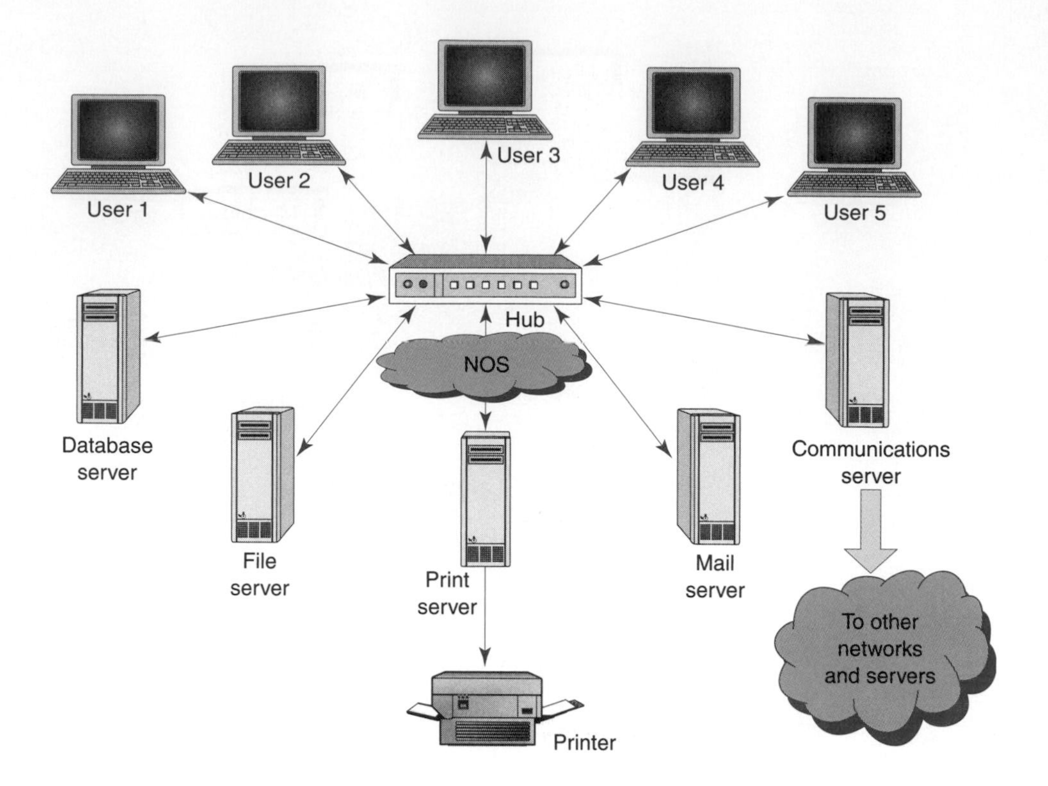

FIGURE 21-22 Network layout using a network operating system (NOS)

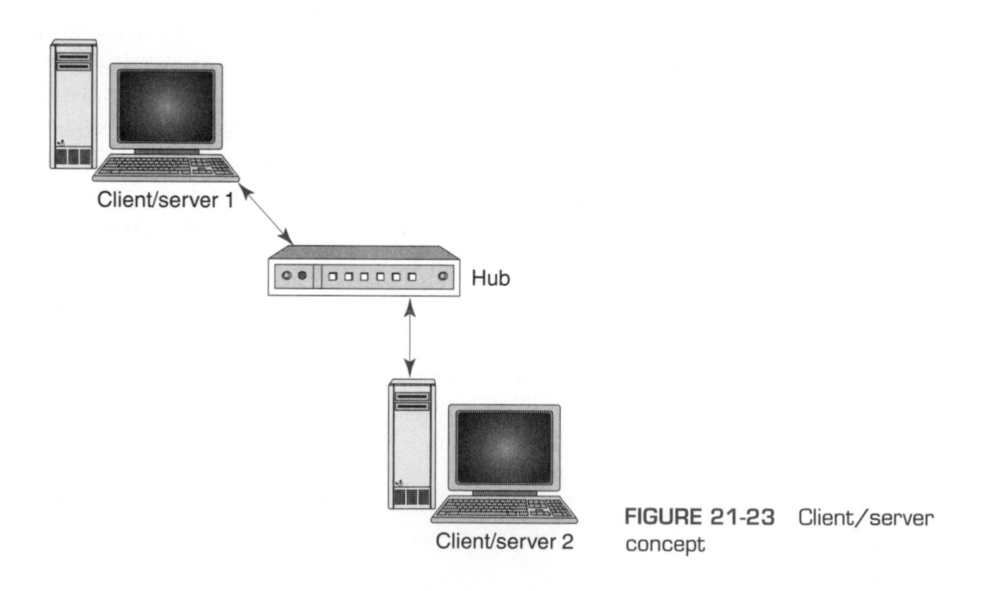

FIGURE 21-23 Client/server concept

to satisfy the request from the client. The server computer might then act as a client and request services from another computer. The client/server concept is illustrated in Figure 21-23.

21-10-2-1 Peer-to-peer client/server network. A *peer-to-peer client/server network* is one in which all computers share their resources, such as hard drives, printers, and so on, with all the other computers on the network. Therefore, the peer-to-peer operating system divides its time between servicing the computer on which it is loaded and servicing

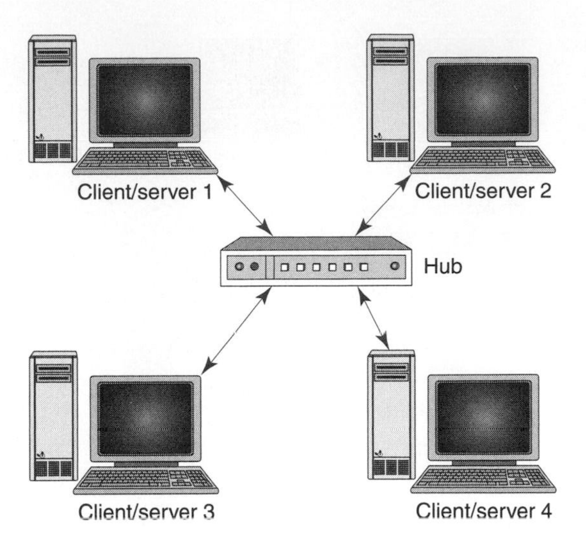

FIGURE 21-24 Peer-to-peer client/server network

requests from other computers. In a peer-to-peer network (sometimes called a *workgroup*), there are no dedicated servers or hierarchy among the computers.

Figure 21-24 shows a peer-to-peer client/server network with four clients/servers (users) connected together through a hub. All computers are equal, hence the name *peer.* Each computer in the network can function as a client and/or a server, and no single computer holds the network operating system or shared files. Also, no one computer is assigned network administrative tasks. The users at each computer determine which data on their computer are shared with the other computers on the network. Individual users are also responsible for installing and upgrading the software on their computer.

Because there is no central controlling computer, a peer-to-peer network is an appropriate choice when there are fewer than 10 users on the network, when all computers are located in the same general area, when security is not an issue, or when there is limited growth projected for the network in the immediate future. Peer-to-peer computer networks should be small for the following reasons:

1. When operating in the server role, the operating system is not optimized to efficiently handle multiple simultaneous requests.
2. The end user's performance as a client would be degraded.
3. Administrative issues such as security, data backups, and data ownership may be compromised in a large peer-to-peer network.

21-10-2-2 Dedicated client/server network. In a *dedicated client/server network,* one computer is designated the server, and the rest of the computers are clients. As the network grows, additional computers can be designated servers. Generally, the designated servers function only as servers and are not used as a client or workstation. The servers store all the network's shared files and applications programs, such as word processor documents, compilers, database applications, spreadsheets, and the network operating system. Client computers can access the servers and have shared files transferred to them over the transmission medium.

Figure 21-25 shows a dedicated client/server-based network with three servers and three clients (users). Each client can access the resources on any of the servers and also the resources on other client computers. The dedicated client/server-based network is probably

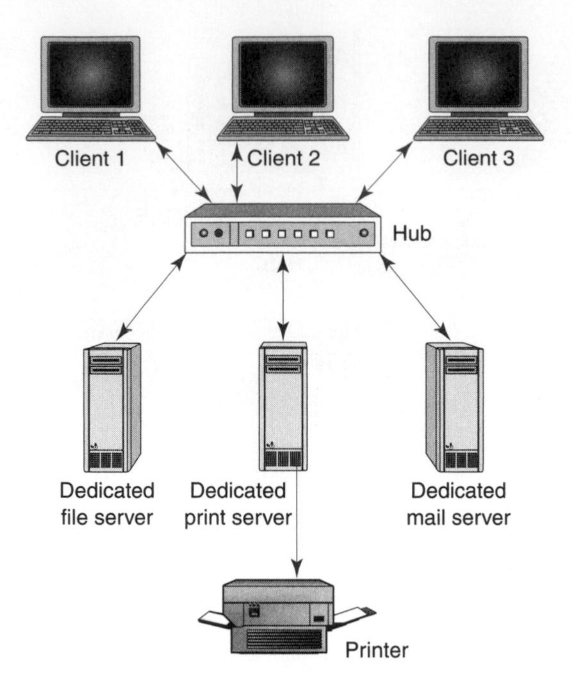

FIGURE 21-25 Dedicated client/server network

the most commonly used computer networking model. There can be a separate dedicated server for each function (i.e., file server, print server, mail server, etc.) or one single general-purpose server responsible for all services.

In some client/server networks, client computers submit jobs to one of the servers. The server runs the software and completes the job and then sends the results back to the client computer. In this type of client/server network, less information propagates through the network than with the file server configuration because only data and not applications programs are transferred between computers.

In general, the dedicated client/server model is preferable to the peer-to-peer client/server model for general-purpose data networks. The peer-to-peer model client/server model is usually preferable for special purposes, such as a small group of users sharing resources.

21-10-3 Network Topologies

Network topology describes the layout or appearance of a network—that is, how the computers, cables, and other components within a data communications network are interconnected, both physically and logically. The *physical topology* describes how the network is actually laid out, and the *logical topology* describes how data actually flow through the network.

In a data communications network, two or more stations connect to a link, and one or more links form a topology. Topology is a major consideration for capacity, cost, and reliability when designing a data communications network. The most basic topologies are *point to point* and *multipoint*. A point-to-point topology is used in data communications networks that transfer high-speed digital information between only two stations. Very often, point-to-point data circuits involve communications between a mainframe computer and another mainframe computer or some other type of high-capacity digital device. A two-point circuit is shown in Figure 21-26a.

A multipoint topology connects three or more stations through a single transmission medium. Examples of multipoint topologies are *star, bus, ring, mesh,* and *hybrid.*

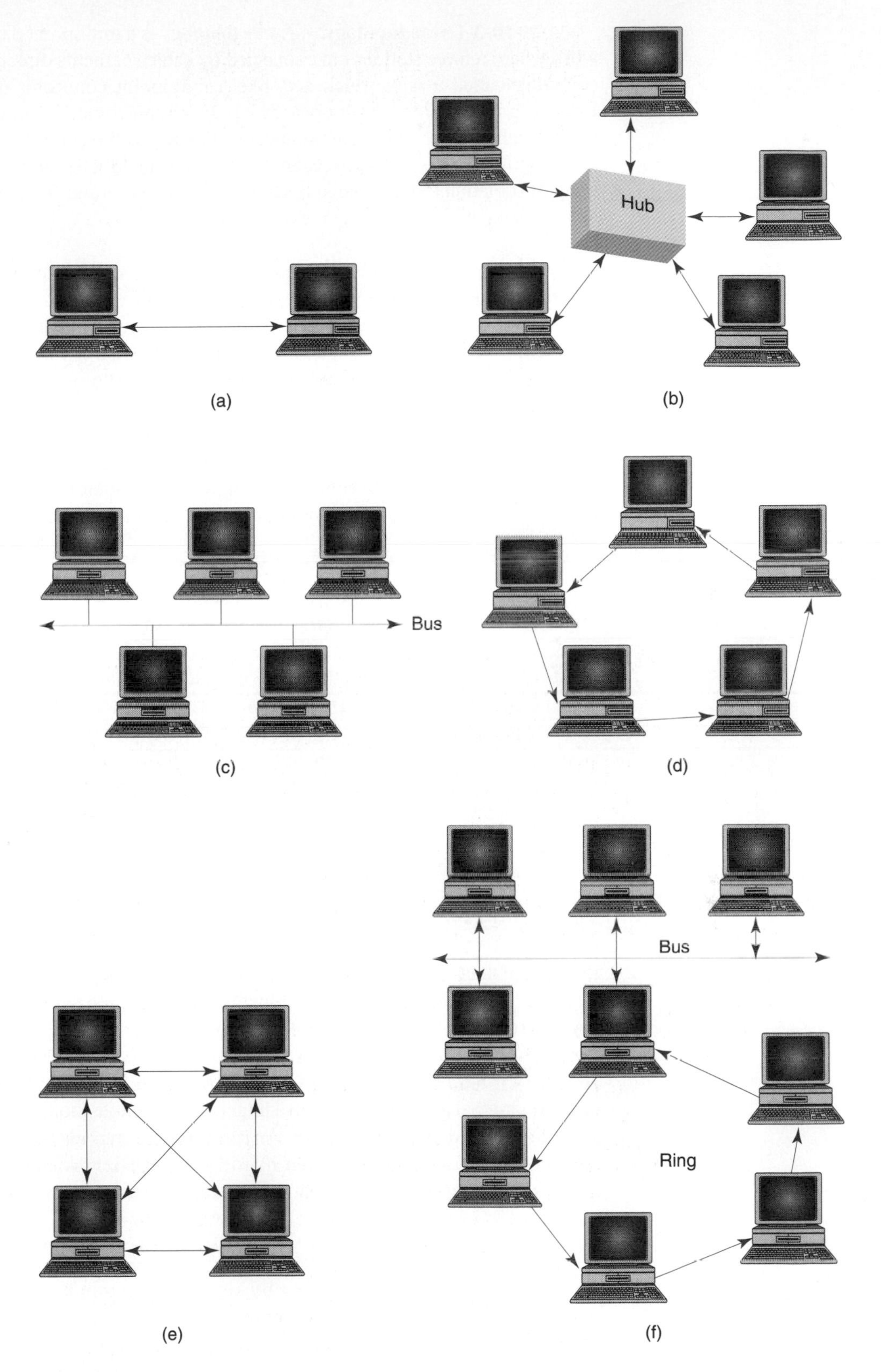

FIGURE 21-26 Network topologies: (a) point-to-point; (b) star; (c) bus; (d) ring; (e) mesh; (f) hybrid

21-10-3-1 Star topology. A *star topology* is a multipoint data communications network where remote stations are connected by cable segments directly to a centrally located computer called a *hub,* which acts like a multipoint connector (see Figure 21-26b). In essence, a star topology is simply a multipoint circuit comprised of many two-point circuits where each remote station communicates directly with a centrally located computer. With a star topology, remote stations cannot communicate directly with one another, so they must relay information through the hub. Hubs also have store-and-forward capabilities, enabling them to handle more than one message at a time.

21-10-3-2 Bus topology. A *bus topology* is a multipoint data communications circuit that makes it relatively simple to control data flow between and among the computers because this configuration allows all stations to receive every transmission over the network. With a bus topology, all the remote stations are physically or logically connected to a single transmission line called a *bus.* The bus topology is the simplest and most common method of interconnecting computers. The two ends of the transmission line never touch to form a complete loop. A bus topology is sometimes called *multidrop* or *linear bus,* and all stations share a common transmission medium. Data networks using the bus topology generally involve one centrally located host computer that controls data flow to and from the other stations. The bus topology is sometimes called a *horizontal bus* and is shown in Figure 21-26c.

21-10-3-3 Ring topology. A *ring topology* is a multipoint data communications network where all stations are interconnected in tandem (series) to form a closed loop or circle. A ring topology is sometimes called a *loop.* Each station in the loop is joined by point-to-point links to two other stations (the transmitter of one and the receiver of the other) (see Figure 21-26d). Transmissions are unidirectional and must propagate through all the stations in the loop. Each computer acts like a repeater in that it receives signals from down-line computers then retransmits them to up-line computers. The ring topology is similar to the bus and star topologies, as it generally involves one centrally located host computer that controls data flow to and from the other stations.

21-10-3-4 Mesh topology. In a *mesh topology,* every station has a direct two-point communications link to every other station on the circuit as shown in Figure 21-26e. The mesh topology is sometimes called *fully connected.* A disadvantage of a mesh topology is a fully connected circuit requires $n(n - 1)/2$ physical transmission paths to interconnect n stations and each station must have $n - 1$ input/output ports. Advantages of a mesh topology are reduced traffic problems, increased reliability, and enhanced security.

21-10-3-5 Hybrid topology. A *hybrid topology* is simply combining two or more of the traditional topologies to form a larger, more complex topology. Hybrid topologies are sometimes called *mixed topologies.* An example of a hybrid topology is the *bus star* topology shown in Figure 21-26f. Other hybrid configurations include the *star ring, bus ring,* and virtually every other combination you can think of.

21-10-4 Network Classifications

Networks are generally classified by size, which includes geographic area, distance between stations, number of computers, transmission speed (bps), transmission media, and the network's physical architecture. The four primary classifications of networks are *local area networks* (LANs), *metropolitan area networks* (MANs), *wide*

Table 21-2 Primary Network Types

Network Type	Characteristics
LAN (local area network)	Interconnects computer users within a department, company, or group
MAN (metropolitan area network)	Interconnects computers in and around a large city
WAN (wide area network)	Interconnects computers in and around an entire country
GAN (global area network)	Interconnects computers from around the entire globe
Building backbone	Interconnects LANs within a building
Campus backbone	Interconnects building LANs
Enterprise network	Interconnects many or all of the above
PAN (personal area network)	Interconnects memory cards carried by people and in computers that are in close proximity to each other
PAN (power line area network, sometimes called PLAN)	Virtually no limit on how many computers it can interconnect and covers an area limited only by the availability of power distribution lines

area networks (WANs), and *global area networks* (GANs). In addition, there are three primary types of interconnecting networks: *building backbone, campus backbone,* and *enterprise network.* Two promising computer networks of the future share the same acronym: the PAN (*personal area network*) and PAN (*power line area network,* sometimes called PLAN). The idea behind a personal area network is to allow people to transfer data through the human body simply by touching each other. Power line area networks use existing ac distribution networks to carry data wherever power lines go, which is virtually everywhere.

When two or more networks are connected together, they constitute an *internetwork* or *internet.* An internet (lowercase *i*) is sometimes confused with the *Internet* (uppercase *I*). The term *internet* is a generic term that simply means to interconnect two or more networks, whereas *Internet* is the name of a specific worldwide data communications network. Table 21-2 summarizes the characteristics of the primary types of networks, and Figure 21-27 illustrates the geographic relationship among computers and the different types of networks.

21-10-4-1 Local area network. *Local area networks* (LANs) are typically privately owned data communications networks in which 10 to 40 computer users share data resources with one or more file servers. LANs use a network operating system to provide two-way communications at bit rates typically in the range of 10 Mbps to 100 Mbps and higher between a large variety of data communications equipment within a relatively small geographical area, such as in the same room, building, or building complex (see Figure 21-28). A LAN can be as simple as two personal computers and a printer or could contain dozens of computers, workstations, and peripheral devices. Most LANs link equipment that are within a few miles of each other or closer. Because the size of most LANs is limited, the longest (or worst-case) transmission time is bounded and known by everyone using the network. Therefore, LANs can utilize configurations that otherwise would not be possible.

LANs were designed for sharing resources between a wide range of digital equipment, including personal computers, workstations, and printers. The resources shared can be software as well as hardware. Most LANs are owned by the company or organization

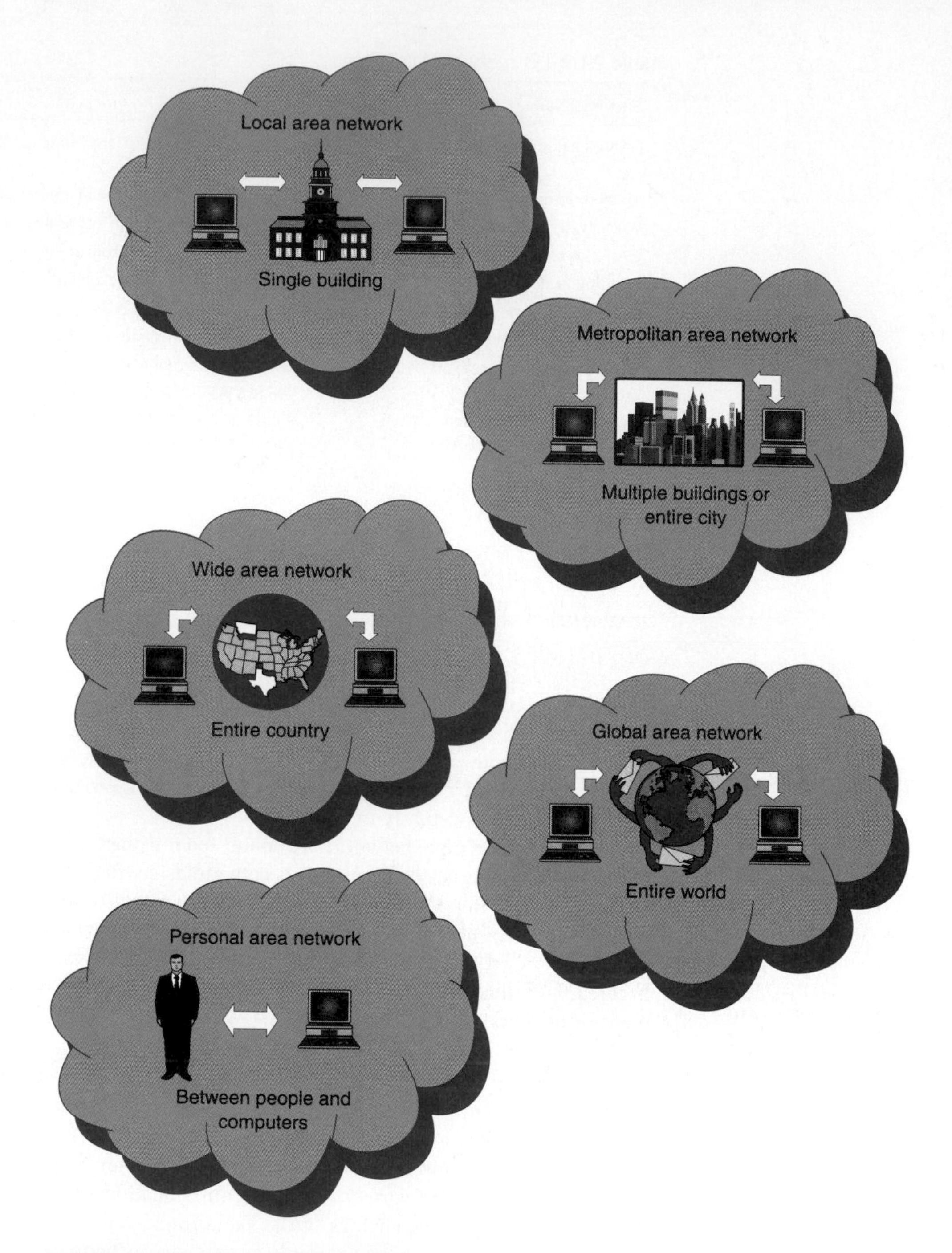

FIGURE 21-27 Computer network types

that uses it and have a connection to a building backbone for access to other departmental LANs, MANs, WANs, and GANs.

21-10-4-2 Metropolitan area network. A *metropolitan area network* (MAN) is a high-speed network similar to a LAN except MANs are designed to encompass larger areas, usually that of an entire city (see Figure 21-29). Most MANs support the transmission of both data and voice and in some cases video. MANs typically operate at

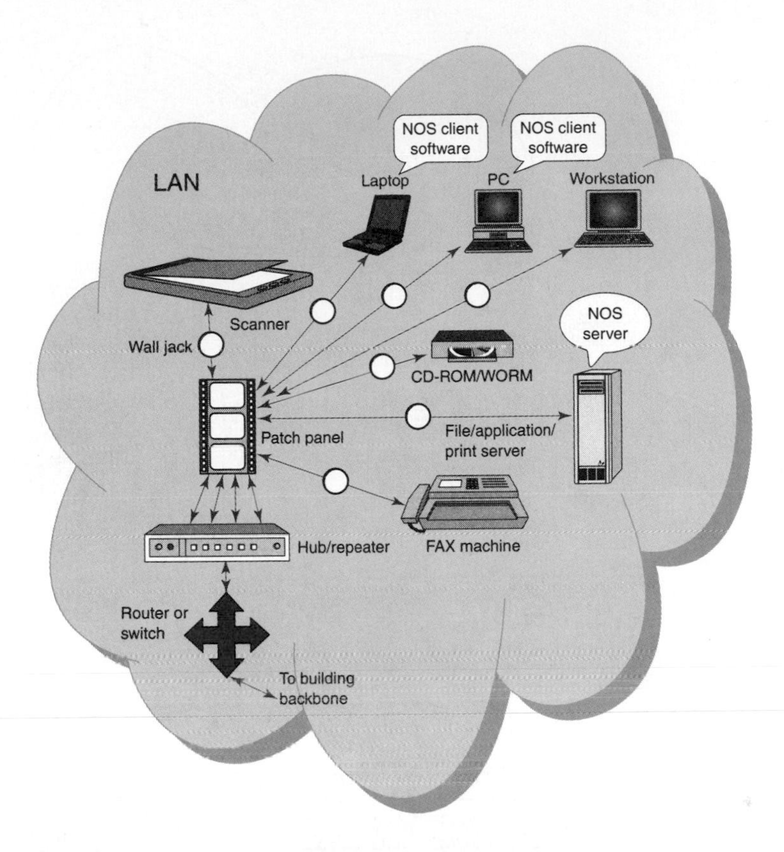

FIGURE 21-28 Local area network (LAN) layout

speeds of 1.5 Mbps to 10 Mbps and range from five miles to a few hundred miles in length. A MAN generally uses only one or two transmission cables and requires no switches. A MAN could be a single network, such as a cable television distribution network, or it could be a means of interconnecting two or more LANs into a single, larger network, enabling data resources to be shared LAN to LAN as well as from station to station or computer to computer. Large companies often use MANS to interconnect all their LANs.

A MAN can be owned and operated entirely by a single, private company, or it could lease services and facilities on a monthly basis from the local cable or telephone company. Switched Multimegabit Data Services (SMDS) is an example of a service offered by local telephone companies for handling high-speed data communications for MANs. Other examples of MANs are FDDI (fiber distributed data interface) and ATM (asynchronous transfer mode).

21-10-4-3 Wide area network. *Wide area networks* (WANs) are the oldest type of data communications network that provide relatively slow-speed, long-distance transmission of data, voice, and video information over relatively large and widely dispersed geographical areas, such as a country or an entire continent (see Figure 21-30). WANs typically interconnect cities and states. WANs typically operate at bit rates from 1.5 Mbps to 2.4 Gbps and cover a distance of 100 to 1000 miles.

WANs may utilize both public and private communications systems to provide service over an area that is virtually unlimited; however, WANs are generally obtained through service providers and normally come in the form of leased-line or circuit-switching technology. Often WANs interconnect routers in different locations. Examples of WANs are

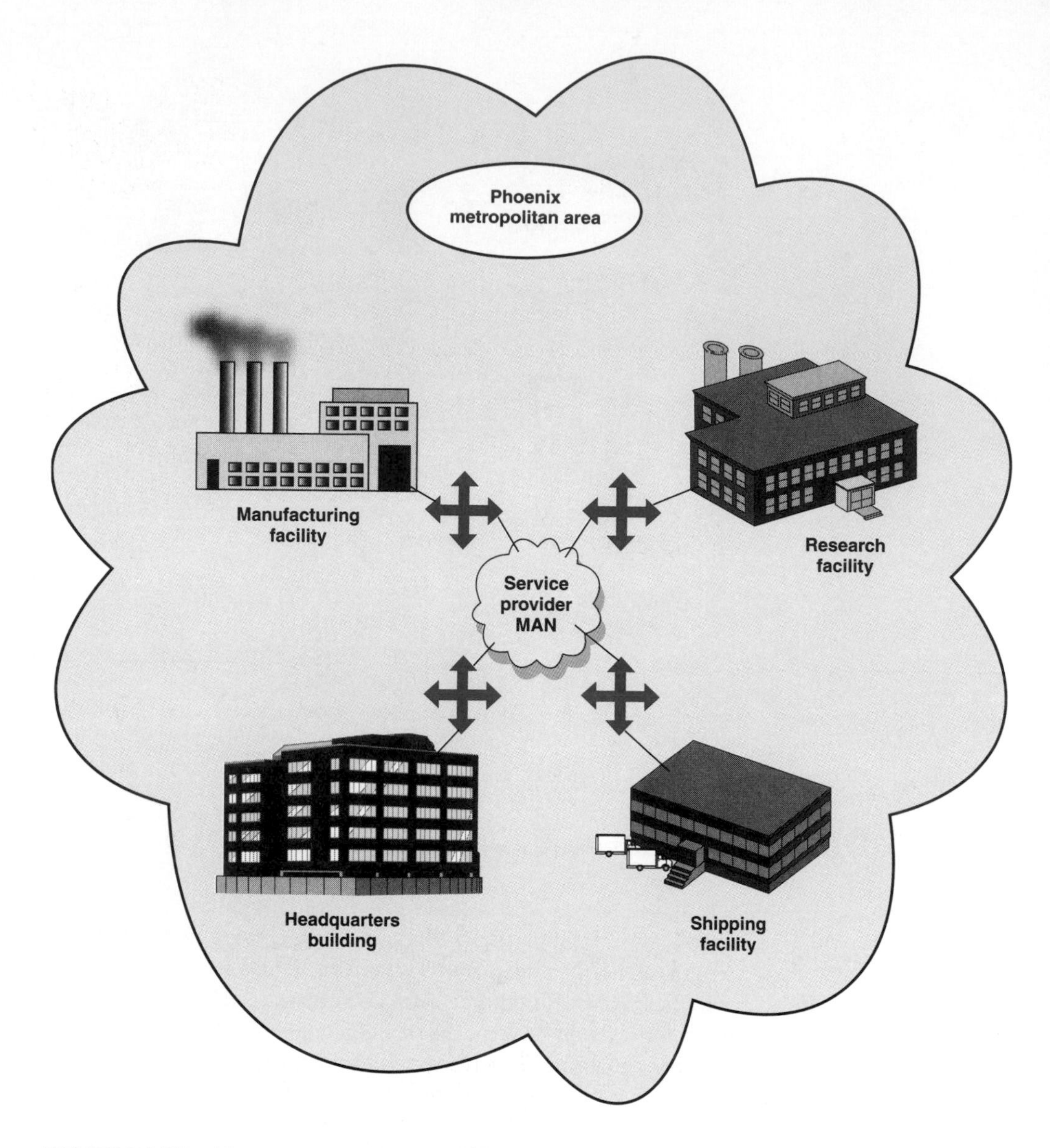

FIGURE 21-29 Metropolitan area network (MAN)

ISDN (integrated services digital network), T1 and T3 digital carrier systems, frame relay, X.25, ATM, and using data modems over standard telephone lines.

21-10-4-4 Global area network. *Global area networks* (GANs) provide connects between countries around the entire globe (see Figure 21-31). The Internet is a good example of a GAN, as it is essentially a network comprised of other networks that interconnects virtually every country in the world. GANs operate from 1.5 Mbps to 100 Gbps and cover thousands of miles

21-10-4-5 Building backbone. A *building backbone* is a network connection that normally carries traffic between departmental LANs within a single company. A building backbone generally consists of a switch or a router (see Figure 21-32) that can provide connectivity to other networks, such as campus backbones, enterprise backbones, MANs, WANs, or GANs.

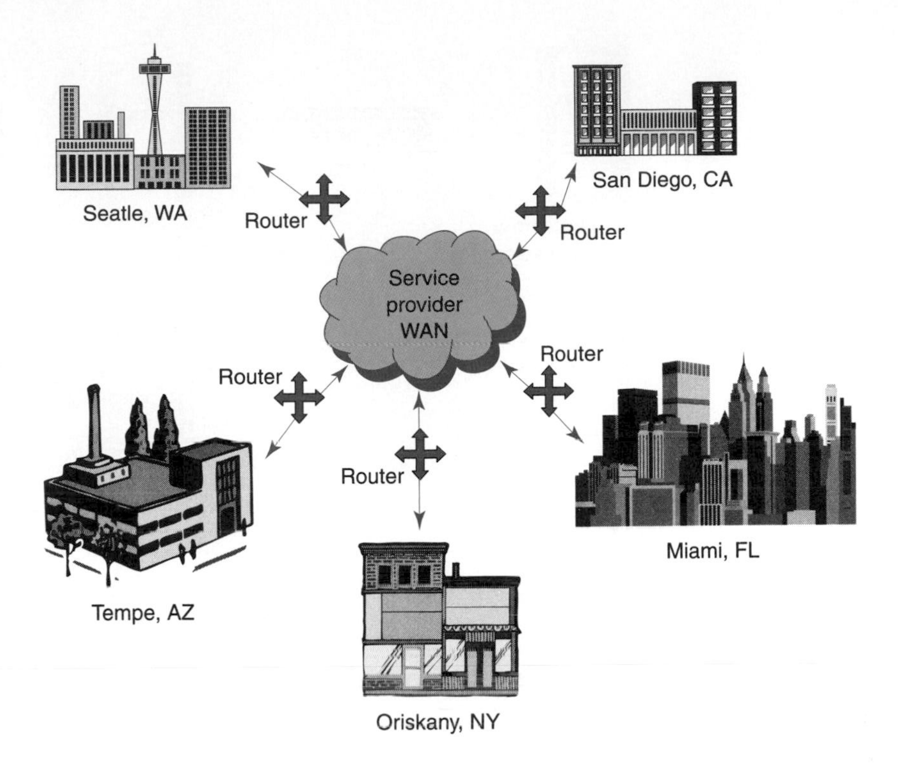

FIGURE 21-30 Wide area network (WAN)

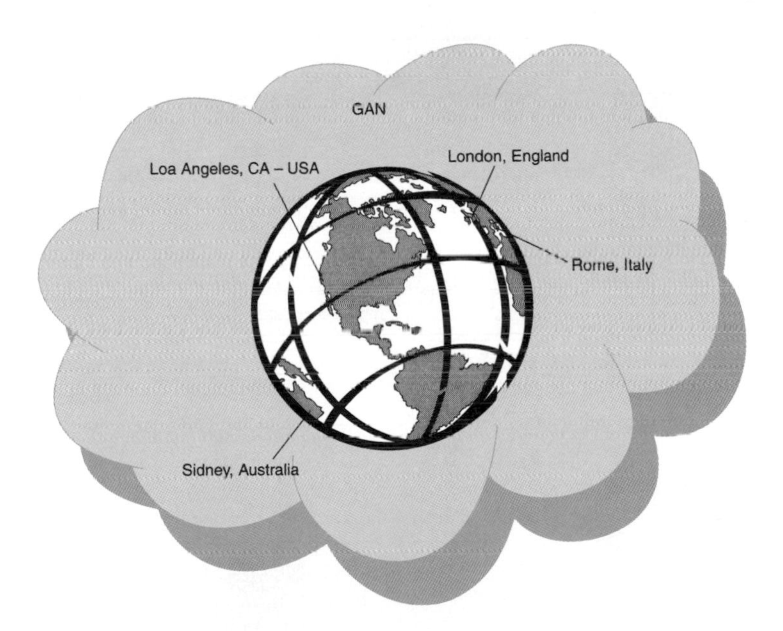

FIGURE 21-31 Global area network (GAN)

21-10-4-6 Campus backbone. A *campus backbone* is a network connection used to carry traffic to and from LANs located in various buildings on campus (see Figure 21-33). A campus backbone is designed for sites that have a group of buildings at a single location, such as corporate headquarters, universities, airports, and research parks.

A campus backbone normally uses optical fiber cables for the transmission media between buildings. The optical fiber cable is used to connect interconnecting devices, such as

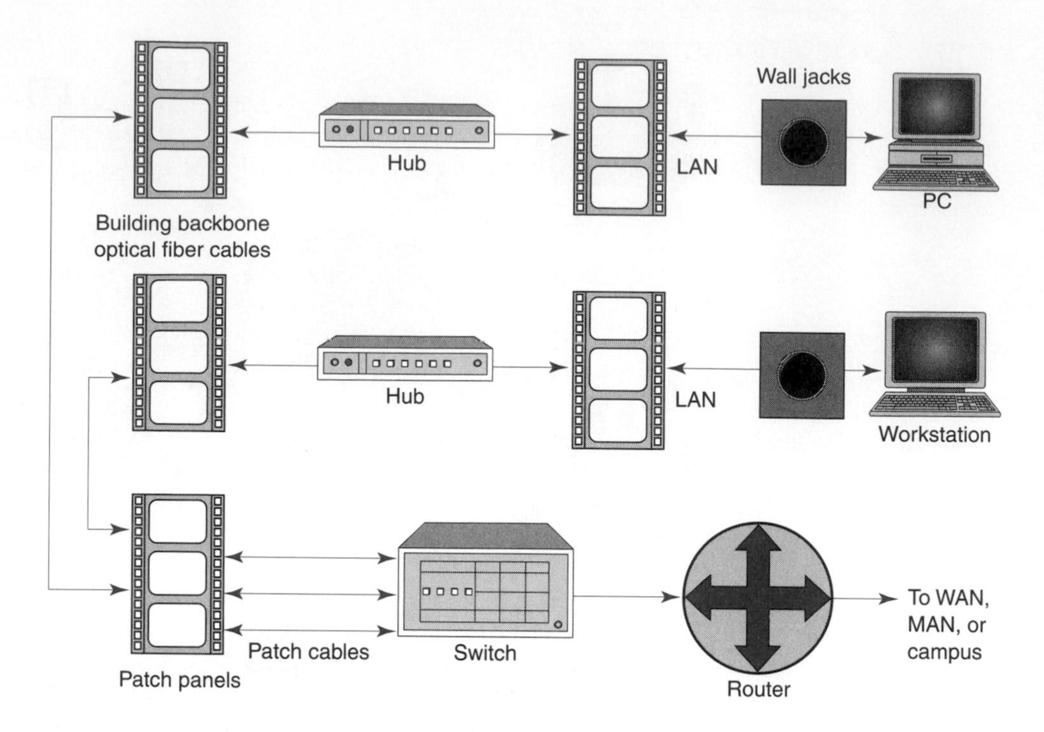

FIGURE 21-32 Building backbone

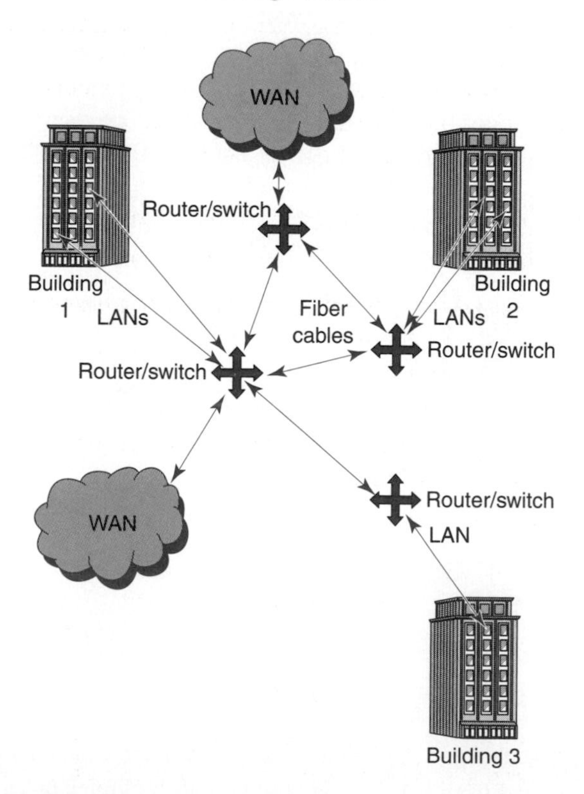

FIGURE 21-33 Campus backbone

bridges, routers, and switches. Campus backbones must operate at relatively high transmission rates to handle the large volumes of traffic between sites.

21-10-4-7 Enterprise networks. An *enterprise network* includes some or all of the previously mentioned networks and components connected in a cohesive and manageable fashion.

21-11 ALTERNATE PROTOCOL SUITES

The functional layers of the OSI seven-layer protocol hierarchy do not line up well with certain data communications applications, such as the Internet. Because of this, there are several other protocols that see widespread use, such as TCP/IP and the Cisco three-layer hierarchical model.

21-11-1 TCP/IP Protocol Suite

The *TCP/IP protocol suite* (*transmission control protocol/Internet protocol*) was actually developed by the Department of Defense before the inception of the seven-layer OSI model. TCP/IP is comprised of several interactive modules that provide specific functionality without necessarily operating independent of one another. The OSI seven-layer model specifies exactly which function each layer performs, whereas TCP/IP is comprised of several relatively independent protocols that can be combined in many ways, depending on system needs. The term *hierarchical* simply means that the upper-level protocols are supported by one or more lower-level protocols. Depending on whose definition you use, TCP/IP is a hierarchical protocol comprised of either three or four layers.

The three-layer version of TCP/IP contains the *network, transport,* and *application* layers that reside above two lower-layer protocols that are not specified by TCP/IP (the physical and data link layers). The network layer of TCP/IP provides internetworking functions similar to those provided by the network layer of the OSI network model. The network layer is sometimes called the *internetwork layer* or *internet layer.*

The transport layer of TCP/IP contains two protocols: TCP (transmission control protocol) and UDP (user datagram protocol). TCP functions go beyond those specified by the transport layer of the OSI model, as they define several tasks defined for the session layer. In essence, TCP allows two application layers to communicate with each other.

The applications layer of TCP/IP contains several other protocols that users and programs utilize to perform the functions of the three uppermost layers of the OSI hierarchy (i.e., the applications, presentation, and session layers).

The four-layer version of TCP/IP specifies the network access, Internet, host-to-host, and process layers:

Network access layer. Provides a means of physically delivering data packets using frames or cells

Internet layer. Contains information that pertains to how data can be routed through the *network*

Host-to-host layer. Services the process and Internet layers to handle the reliability and session aspects of data transmission

Process layer. Provides applications support

TCP/IP is probably the dominant communications protocol in use today. It provides a common denominator, allowing many different types of devices to communicate over a network or system of networks while supporting a wide variety of applications.

21-11-2 Cisco Three-Layer Model

Cisco defines a three-layer logical hierarchy that specifies where things belong, how they fit together, and what functions go where. The three layers are the core, distribution, and access:

Core layer. The core layer is literally the core of the network, as it resides at the top of the hierarchy and is responsible for transporting large amounts of data traffic reliably and quickly. The only purpose of the core layer is to switch traffic as quickly as possible.

Distribution layer. The distribution layer is sometimes called the *workgroup layer.* The distribution layer is the communications point between the access and the core layers that provides routing, filtering, WAN access, and how many data packets are allowed to access the core layer. The distribution layer determines the fastest way to handle service requests, for example, the fastest way to forward a file request to a server. Several functions are performed at the distribution level:

1. Implementation of tools such as access lists, packet filtering, and queuing
2. Implementation of security and network policies, including firewalls and address translation
3. Redistribution between routing protocols
4. Routing between virtual LANS and other workgroup support functions
5. Define broadcast and multicast domains

Access layer. The access layer controls workgroup and individual user access to internetworking resources, most of which are available locally. The access layer is sometimes called the *desktop layer.* Several functions are performed at the access layer level:

1. Access control
2. Creation of separate collision domains (segmentation)
3. Workgroup connectivity into the distribution layer

QUESTIONS

21-1. Define the following terms: *data, information,* and *data communications network.*

21-2. What was the first data communications system that used binary-coded electrical signals?

21-3. Discuss the relationship between network architecture and protocol.

21-4. Briefly describe broadcast and point-to-point computer networks.

21-5. Define the following terms: *protocol, connection-oriented protocols, connectionless protocols,* and *protocol stacks.*

21-6. What is the difference between syntax and semantics?

21-7. What are data communications standards, and why are they needed?

21-8. Name and briefly describe the differences between the two kinds of data communications standards.

21-9. List and describe the eight primary standards organizations for data communications.

21-10. Define the open systems interconnection.

21-11. Briefly describe the seven layers of the OSI protocol hierarchy.

21-12. List and briefly describe the basic functions of the five components of a data communications circuit.

21-13. Briefly describe the differences between serial and parallel data transmission.

21-14. What are the two basic kinds of data communications circuit configurations?

21-15. List and briefly describe the four transmission modes.

21-16. List and describe the functions of the most common components of a computer network.

21-17. What are the differences between servers and clients on a data communications network?

21-18. Describe a peer-to-peer data communications network.

21-19. What are the differences between peer-to-peer client/server networks and dedicated client/server networks?

21-20. What is a data communications network topology?

21-21. List and briefly describe the five basic data communications network topologies.

21-22. List and briefly describe the major network classifications.

21-23. Briefly describe the TCP/IP protocol model.

21-24. Briefly describe the Cisco three-layer protocol model.

C H A P T E R 22

Fundamental Concepts of Data Communications

CHAPTER OUTLINE

OBJECTIVES

- Define *data communication code*
- Describe the following data communications codes: Baudot, ASCII, and EBCDIC
- Explain bar code formats
- Define *error control, error detection,* and *error correction*
- Describe the following error-detection mechanisms: redundancy, checksum, LRC, VRC, and CRC
- Describe the following error-correction mechanisms: FEC, ARQ, and Hamming code
- Describe character synchronization and explain the differences between asynchronous and synchronous data formats
- Define the term *data communications hardware*
- Describe data terminal equipment
- Describe data communications equipment
- List and describe the seven components that make up a two-point data communications circuit
- Describe the terms *line control unit* and *front-end processor* and explain the differences between the two
- Describe the basic operation of a UART and outline the differences between UARTs, USRTs, and USARTs
- Describe the functions of a serial interface
- Explain the physical, electrical, and functional characteristics of the RS-232 serial interface
- Compare and contrast the RS-232, RS-449, and RS-530 serial interfaces

- Describe data communications modems
- Explain the block diagram of a modem
- Explain what is meant by Bell System–compatible modems
- Describe modem synchronization and modem equalization
- Describe the ITU-T modem recommendations

22-1 INTRODUCTION

To understand how a data communications network works as an entity, it is necessary first to understand the fundamental concepts and components that make up the network. The fundamental concepts of data communications include data communications code, error control (error detection and correction), and character synchronization, and fundamental hardware includes various pieces of computer and networking equipment, such as line control units, serial interfaces, and data communications modems.

22-2 DATA COMMUNICATIONS CODES

Data communications codes are often used to represent *characters* and *symbols,* such as letters, digits, and punctuation marks. Therefore, data communications codes are called *character codes, character sets, symbol codes,* or *character languages.*

22-2-1 Baudot Code

The *Baudot code* (sometimes called the *Telex code*) was the first *fixed-length character code* developed for machines rather than for people. A French postal engineer named Thomas Murray developed the Baudot code in 1875 and named the code after Emile Baudot, an early pioneer in telegraph printing. The Baudot code (pronounced *baw-dough*) is a *fixed-length source code* (sometimes called a *fixed-length block code*). With fixed-length source codes, all characters are represented in binary and have the same number of symbols (bits). The Baudot code is a five-bit character code that was used primarily for low-speed teletype equipment, such as the TWX/Telex system and radio teletype (RTTY). The latest version of the Baudot code is recommended by the CCITT as the International Alphabet No. 2 and is shown in Table 22-1.

22-2-2 ASCII Code

In 1963, in an effort to standardize data communications codes, the United States adopted the Bell System model 33 teletype code as the *United States of America Standard Code for Information Exchange* (USASCII), better known as ASCII-63. Since its adoption, ASCII (pronounced *as-key*) has progressed through the 1965, 1967, and 1977 versions, with the 1977 version being recommended by the ITU as International Alphabet No. 5, in the United States as ANSI standard X3.4-1986 (R1997), and by the International Standards Organization as ISO-14962 (1997).

ASCII is the standard character set for source coding the alphanumeric character set that humans understand but computers do not (computers only understand 1s and 0s). ASCII is a seven-bit fixed-length character set. With the ASCII code, the least-significant bit (LSB) is designated b_0 and the most-significant bit (MSB) is designated b_7 as shown here:

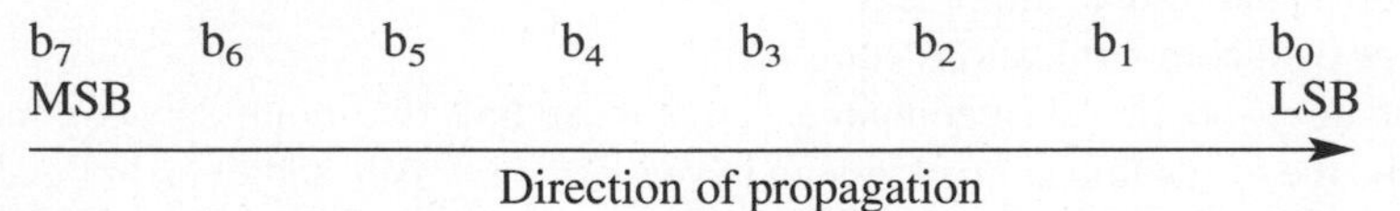

The terms *least* and *most significant* are somewhat of a misnomer because character codes do not represent weighted binary numbers and, therefore, all bits are equally sig-

Table 22-1 Baudot Code

		Bit				
Letter	Figure	Bit: 4	3	2	1	0
A	—	1	1	0	0	0
B	?	1	0	0	1	1
C	:	0	1	1	1	0
D	$	1	0	0	1	0
E	3	1	0	0	0	0
F	!	1	0	1	1	0
G	&	0	1	0	1	1
H	#	0	0	1	0	1
I	8	0	1	1	0	0
J	'	1	1	0	1	0
K	(	1	1	1	1	0
L	)	0	1	0	0	1
M	.	0	0	1	1	1
N	,	0	0	1	1	0
O	9	0	0	0	1	1
P	0	0	1	1	0	1
Q	1	1	1	1	0	1
R	4	0	1	0	1	0
S	bel	1	0	1	0	0
T	5	0	0	0	0	1
U	7	1	1	1	0	0
V	;	0	1	1	1	1
W	2	1	1	0	0	1
X	/	1	0	1	1	1
Y	6	1	0	1	0	1
Z	"	1	0	0	0	1
Figure shift		1	1	1	1	1
Letter shift		1	1	0	1	1
Space		0	0	1	0	0
Line feed (LF)		0	1	0	0	0
Blank (null)		0	0	0	0	0

nificant. Bit b_7 is not part of the ASCII code but is generally reserved for an error detection bit called the parity bit, which is explained later in this chapter. With character codes, it is more meaningful to refer to bits by their order than by their position; b_0 is the zero-order bit, b_1 the first-order bit, b_7 the seventh-order bit, and so on. However, with serial data transmission, the bit transmitted first is generally called the LSB. With ASCII, the low-order bit (b_0) is transmitted first. ASCII is probably the code most often used in data communications networks today. The 1977 version of the ASCII code with odd parity is shown in Table 22-2 (note that the parity bit is not included in the hex code).

22-2-3 EBCDIC Code

The *extended binary-coded decimal interchange code* (EBCDIC) is an eight-bit fixed-length character set developed in 1962 by the International Business Machines Corporation (IBM). EBCDIC is used almost exclusively with IBM mainframe computers and peripheral equipment. With eight bits, 2^8, or 256, codes are possible, although only 139 of the 256 codes are actually assigned characters. Unspecified codes can be assigned to specialized characters and functions. The name *binary coded decimal* was selected because the second hex character for all letter and digit codes contains only the hex values from 0 to 9, which have the same binary sequence as BCD codes. The EBCDIC code is shown in Table 22-3.

Table 22-2 ASCII-77: Odd Parity

	Binary Code										Binary Code								
Bit	7	6	5	4	3	2	1	0	Hex	Bit	7	6	5	4	3	2	1	0	Hex
NUL	1	0	0	0	0	0	0	0	00	@	0	1	0	0	0	0	0	0	40
SOH	0	0	0	0	0	0	0	1	01	A	1	1	0	0	0	0	0	1	41
STX	0	0	0	0	0	0	1	0	02	B	1	1	0	0	0	0	1	0	42
ETX	1	0	0	0	0	0	1	1	03	C	0	1	0	0	0	0	1	1	43
EOT	0	0	0	0	0	1	0	0	04	D	1	1	0	0	0	1	0	0	44
ENQ	1	0	0	0	0	1	0	1	05	E	0	1	0	0	0	1	0	1	45
ACK	1	0	0	0	0	1	1	0	06	F	0	1	0	0	0	1	1	0	46
BEL	0	0	0	0	0	1	1	1	07	G	1	1	0	0	0	1	1	1	47
BS	0	0	0	0	1	0	0	0	08	H	1	1	0	0	1	0	0	0	48
HT	1	0	0	0	1	0	0	1	09	I	0	1	0	0	1	0	0	1	49
NL	1	0	0	0	1	0	1	0	0A	J	0	1	0	0	1	0	1	0	4A
VT	0	0	0	0	1	0	1	1	0B	K	1	1	0	0	1	0	1	1	4B
FF	1	0	0	0	1	1	0	0	0C	L	0	1	0	0	1	1	0	0	4C
CR	0	0	0	0	1	1	0	1	0D	M	1	1	0	0	1	1	0	1	4D
SO	0	0	0	0	1	1	1	0	0E	N	1	1	0	0	1	1	1	0	4E
SI	1	0	0	0	1	1	1	1	0F	O	0	1	0	0	1	1	1	1	4F
DLE	0	0	0	1	0	0	0	0	10	P	1	1	0	1	0	0	0	0	50
DC1	0	0	0	1	0	0	0	1	11	Q	0	1	0	1	0	0	0	1	51
DC2	1	0	0	1	0	0	1	0	12	R	0	1	0	1	0	0	1	0	52
DC3	0	0	0	1	0	0	1	1	13	S	1	1	0	1	0	0	1	1	53
DC4	1	0	0	1	0	1	0	0	14	T	0	1	0	1	0	1	0	0	54
NAK	0	0	0	1	0	1	0	1	15	U	1	1	0	1	0	1	0	1	55
SYN	0	0	0	1	0	1	1	0	16	V	1	1	0	1	0	1	1	0	56
ETB	1	0	0	1	0	1	1	1	17	W	0	1	0	1	0	1	1	1	57
CAN	1	0	0	1	1	0	0	0	18	X	0	1	0	1	1	0	0	0	58
EM	0	0	0	1	1	0	0	1	19	Y	1	1	0	1	1	0	0	1	59
SUB	0	0	0	1	1	0	1	0	1A	Z	1	1	0	1	1	0	1	0	5A
ESC	1	0	0	1	1	0	1	1	1B	[	0	1	0	1	1	0	1	1	5B
FS	0	0	0	1	1	1	0	0	1C	\	1	1	0	1	1	1	0	0	5C
GS	1	0	0	1	1	1	0	1	1D	]	0	1	0	1	1	1	0	1	5D
RS	1	0	0	1	1	1	1	0	1E	∧	0	1	0	1	1	1	1	0	5E
US	0	0	0	1	1	1	1	1	1F	-	1	1	0	1	1	1	1	1	5F
SP	0	0	1	0	0	0	0	0	20	`	1	1	1	0	0	0	0	0	60
!	1	0	1	0	0	0	0	1	21	a	0	1	1	0	0	0	0	1	61
″	1	0	1	0	0	0	1	0	22	b	0	1	1	0	0	0	1	0	62
#	0	0	1	0	0	0	1	1	23	c	1	1	1	0	0	0	1	1	63
$	1	0	1	0	0	1	0	0	24	d	0	1	1	0	0	1	0	0	64
%	0	0	1	0	0	1	0	1	25	e	1	1	1	0	0	1	0	1	65
&	0	0	1	0	0	1	1	0	26	f	1	1	1	0	0	1	1	0	66
′	1	0	1	0	0	1	1	1	27	g	0	1	1	0	0	1	1	1	67
(	1	0	1	0	1	0	0	0	28	h	0	1	1	0	1	0	0	0	68
)	0	0	1	0	1	0	0	1	29	i	1	1	1	0	1	0	0	1	69
*	0	0	1	0	1	0	1	0	2A	j	1	1	1	0	1	0	1	0	6A
+	1	0	1	0	1	0	1	1	2B	k	0	1	1	0	1	0	1	1	6B
,	0	0	1	0	1	1	0	0	2C	l	1	1	1	0	1	1	0	0	6C
-	1	0	1	0	1	1	0	1	2D	m	0	1	1	0	1	1	0	1	6D
.	1	0	1	0	1	1	1	0	2E	n	0	1	1	0	1	1	1	0	6E
/	0	0	1	0	1	1	1	1	2F	o	1	1	1	0	1	1	1	1	6F
0	1	0	1	1	0	0	0	0	30	p	0	1	1	1	0	0	0	0	70
1	0	0	1	1	0	0	0	1	31	q	1	1	1	1	0	0	0	1	71
2	0	0	1	1	0	0	1	0	32	r	1	1	1	1	0	0	1	0	72
3	1	0	1	1	0	0	1	1	33	s	0	1	1	1	0	0	1	1	73
4	0	0	1	1	0	1	0	0	34	t	1	1	1	1	0	1	0	0	74
5	1	0	1	1	0	1	0	1	35	u	0	1	1	1	0	1	0	1	75
6	1	0	1	1	0	1	1	0	36	v	0	1	1	1	0	1	1	0	76
7	0	0	1	1	0	1	1	1	37	w	1	1	1	1	0	1	1	1	77
8	0	0	1	1	1	0	0	0	38	x	1	1	1	1	1	0	0	0	78

(Continued)

Table 22-2 *(Continued)*

	Binary Code										Binary Code								
Bit	7	6	5	4	3	2	1	0	Hex	Bit	7	6	5	4	3	2	1	0	Hex
9	1	0	1	1	1	0	0	1	39	y	0	1	1	1	1	0	0	1	79
:	1	0	1	1	1	0	1	0	3A	z	0	1	1	1	1	0	1	0	7A
;	0	0	1	1	1	0	1	1	3B	{	1	1	1	1	1	0	1	1	7B
<	1	0	1	1	1	1	0	0	3C	|	0	1	1	1	1	1	0	0	7C
=	0	0	1	1	1	1	0	1	3D	}	1	1	1	1	1	1	0	1	7D
>	0	0	1	1	1	1	1	0	3E	~	1	1	1	1	1	1	1	0	7E
?	1	0	1	1	1	1	1	1	3F	DEL	0	1	1	1	1	1	1	1	7F

NUL = null
SOH = start of heading
STX = start of text
ETX = end of text
EOT = end of transmission
ENQ = enquiry
ACK = acknowledge
BEL = bell
BS = back space
HT = horizontal tab
NL = new line
VT = vertical tab
FF = form feed
CR = carriage return
SO = shift-out
SI = shift-in
DLE = data link escape
DC1 = device control 1
DC2 = device control 2
DC3 = device control 3
DC4 = device control 4
NAK = negative acknowledge
SYN = synchronous
ETB = end of transmission block
CAN = cancel
SUB = substitute
ESC = escape
FS = field separator
GS = group separator
RS = record separator
US = unit separator
SP = space
DEL = delete

Table 22-3 EBCDIC Code

	Binary Code										Binary Code								
Bit	0	1	2	3	4	5	6	7	Hex	Bit	0	1	2	3	4	5	6	7	Hex
NUL	0	0	0	0	0	0	0	0	00		1	0	0	0	0	0	0	0	80
SOH	0	0	0	0	0	0	0	1	01	a	1	0	0	0	0	0	0	1	81
STX	0	0	0	0	0	0	1	0	02	b	1	0	0	0	0	0	1	0	82
ETX	0	0	0	0	0	0	1	1	03	c	1	0	0	0	0	0	1	1	83
	0	0	0	0	0	1	0	0	04	d	1	0	0	0	0	1	0	0	84
PT	0	0	0	0	0	1	0	1	05	e	1	0	0	0	0	1	0	1	85
	0	0	0	0	0	1	1	0	06	f	1	0	0	0	0	1	1	0	86
	0	0	0	0	0	1	1	1	07	g	1	0	0	0	0	1	1	1	87
	0	0	0	0	1	0	0	0	08	h	1	0	0	0	1	0	0	0	88
	0	0	0	0	1	0	0	1	09	i	1	0	0	0	1	0	0	1	89
	0	0	0	0	1	0	1	0	0A		1	0	0	0	1	0	1	0	8A
	0	0	0	0	1	0	1	1	0B		1	0	0	0	1	0	1	1	8B
FF	0	0	0	0	1	1	0	0	0C		1	0	0	0	1	1	0	0	8C
	0	0	0	0	1	1	0	1	0D		1	0	0	0	1	1	0	1	8D
	0	0	0	0	1	1	1	0	0E		1	0	0	0	1	1	1	0	8E
	0	0	0	0	1	1	1	1	0F		1	0	0	0	1	1	1	1	8F
DLE	0	0	0	1	0	0	0	0	10		1	0	0	1	0	0	0	0	90
SBA	0	0	0	1	0	0	0	1	11	j	1	0	0	1	0	0	0	1	91
EUA	0	0	0	1	0	0	1	0	12	k	1	0	0	1	0	0	1	0	92
IC	0	0	0	1	0	0	1	1	13	l	1	0	0	1	0	0	1	1	93
	0	0	0	1	0	1	0	0	14	m	1	0	0	1	0	1	0	0	94
NL	0	0	0	1	0	1	0	1	15	n	1	0	0	1	0	1	0	1	95
	0	0	0	1	0	1	1	0	16	o	1	0	0	1	0	1	1	0	96
	0	0	0	1	0	1	1	1	17	p	1	0	0	1	0	1	1	1	97
	0	0	0	1	1	0	0	0	18	q	1	0	0	1	1	0	0	0	98
EM	0	0	0	1	1	0	0	1	19	r	1	0	0	1	1	0	0	1	99
	0	0	0	1	1	0	1	0	1A		1	0	0	1	1	0	1	0	9A
	0	0	0	1	1	0	1	1	1B		1	0	0	1	1	0	1	1	9B
DUP	0	0	0	1	1	1	0	0	1C		1	0	0	1	1	1	0	0	9C
SF	0	0	0	1	1	1	0	1	1D		1	0	0	1	1	1	0	1	9D
FM	0	0	0	1	1	1	1	0	1E		1	0	0	1	1	1	1	0	9E

(Continued)

Table 22-3 *(Continued)*

Bit	Binary Code								Hex	Bit	Binary Code								Hex
	0	1	2	3	4	5	6	7			0	1	2	3	4	5	6	7	
ITB	0	0	0	1	1	1	1	1	1F		1	0	0	1	1	1	1	1	9F
	0	0	1	0	0	0	0	0	20		1	0	1	0	0	0	0	0	A0
	0	0	1	0	0	0	0	1	21	~	1	0	1	0	0	0	0	1	A1
	0	0	1	0	0	0	1	0	22	s	1	0	1	0	0	0	1	0	A2
	0	0	1	0	0	0	1	1	23	t	1	0	1	0	0	0	1	1	A3
	0	0	1	0	0	1	0	0	24	u	1	0	1	0	0	1	0	0	A4
	0	0	1	0	0	1	0	1	25	v	1	0	1	0	0	1	0	1	A5
ETB	0	0	1	0	0	1	1	0	26	w	1	0	1	0	0	1	1	0	A6
ESC	0	0	1	0	0	1	1	1	27	x	1	0	1	0	0	1	1	1	A7
	0	0	1	0	1	0	0	0	28	y	1	0	1	0	1	0	0	0	A8
	0	0	1	0	1	0	0	1	29	z	1	0	1	0	1	0	0	1	A9
	0	0	1	0	1	0	1	0	2A		1	0	1	0	1	0	1	0	AA
	0	0	1	0	1	0	1	1	2B		1	0	1	0	1	0	1	1	AB
	0	0	1	0	1	1	0	0	2C		1	0	1	0	1	1	0	0	AC
ENQ	0	0	1	0	1	1	0	1	2D		1	0	1	0	1	1	0	1	AD
	0	0	1	0	1	1	1	0	2E		1	0	1	0	1	1	1	0	AE
	0	0	1	0	1	1	1	1	2F		1	0	1	0	1	1	1	1	AF
	0	0	1	1	0	0	0	0	30		1	0	1	1	0	0	0	0	B0
	0	0	1	1	0	0	0	1	31		1	0	1	1	0	0	0	1	B1
SYN	0	0	1	1	0	0	1	0	32		1	0	1	1	0	0	1	0	B2
	0	0	1	1	0	0	1	1	33		1	0	1	1	0	0	1	1	B3
	0	0	1	1	0	1	0	0	34		1	0	1	1	0	1	0	0	B4
	0	0	1	1	0	1	0	1	35		1	0	1	1	0	1	0	1	B5
	0	0	1	1	0	1	1	0	36		1	0	1	1	0	1	1	0	B6
BOT	0	0	1	1	0	1	1	1	37		1	0	1	1	0	1	1	1	B7
	0	0	1	1	1	0	0	0	38		1	0	1	1	1	0	0	0	B8
	0	0	1	1	1	0	0	1	39		1	0	1	1	1	0	0	1	B9
	0	0	1	1	1	0	1	0	3A		1	0	1	1	1	0	1	0	BA
	0	0	1	1	1	0	1	1	3B		1	0	1	1	1	0	1	1	BB
RA	0	0	1	1	1	1	0	0	3C		1	0	1	1	1	1	0	0	BC
NAK	0	0	1	1	1	1	0	1	3D		1	0	1	1	1	1	0	1	BD
	0	0	1	1	1	1	1	0	3E		1	0	1	1	1	1	1	0	BE
SUB	0	0	1	1	1	1	1	1	3F		1	0	1	1	1	1	1	1	BF
SP	0	1	0	0	0	0	0	0	40	{	1	1	0	0	0	0	0	0	C0
	0	1	0	0	0	0	0	1	41	A	1	1	0	0	0	0	0	1	C1
	0	1	0	0	0	0	1	0	42	B	1	1	0	0	0	0	1	0	C2
	0	1	0	0	0	0	1	1	43	C	1	1	0	0	0	0	1	1	C3
	0	1	0	0	0	1	0	0	44	D	1	1	0	0	0	1	0	0	C4
	0	1	0	0	0	1	0	1	45	E	1	1	0	0	0	1	0	1	C5
	0	1	0	0	0	1	1	0	46	F	1	1	0	0	0	1	1	0	C6
	0	1	0	0	0	1	1	1	47	G	1	1	0	0	0	1	1	1	C7
	0	1	0	0	1	0	0	0	48	H	1	1	0	0	1	0	0	0	C8
	0	1	0	0	1	0	0	1	49	I	1	1	0	0	1	0	0	1	C9
¢	0	1	0	0	1	0	1	0	4A		1	1	0	0	1	0	1	0	CA
-	0	1	0	0	1	0	1	1	4B		1	1	0	0	1	0	1	1	CB
<	0	1	0	0	1	1	0	0	4C		1	1	0	0	1	1	0	0	CC
(	0	1	0	0	1	1	0	1	4D		1	1	0	0	1	1	0	1	CD
+	0	1	0	0	1	1	1	0	4E		1	1	0	0	1	1	1	0	CE
|	0	1	0	0	1	1	1	1	4F		1	1	0	0	1	1	1	1	CF
&	0	1	0	1	0	0	0	0	50	}	1	1	0	1	0	0	0	0	D0
	0	1	0	1	0	0	0	1	51	J	1	1	0	1	0	0	0	1	D1
	0	1	0	1	0	0	1	0	52	K	1	1	0	1	0	0	1	0	D2
	0	1	0	1	0	0	1	1	53	L	1	1	0	1	0	0	1	1	D3
	0	1	0	1	0	1	0	0	54	M	1	1	0	1	0	1	0	0	D4
	0	1	0	1	0	1	0	1	55	N	1	1	0	1	0	1	0	1	D5
	0	1	0	1	0	1	1	0	56	O	1	1	0	1	0	1	1	0	D6

(Continued)

Table 22-3 (*Continued*)

	Binary Code										Binary Code								
Bit	0	1	2	3	4	5	6	7	Hex	Bit	0	1	2	3	4	5	6	7	Hex
	0	1	0	1	0	1	1	1	57	P	1	1	0	1	0	1	1	1	D7
	0	1	0	1	1	0	0	0	58	Q	1	1	0	1	1	0	0	0	D8
	0	1	0	1	1	0	0	1	59	R	1	1	0	1	1	0	0	1	D9
!	0	1	0	1	1	0	1	0	5A		1	1	0	1	1	0	1	0	DA
$	0	1	0	1	1	0	1	1	5B		1	1	0	1	1	0	1	1	DB
*	0	1	0	1	1	1	0	0	5C		1	1	0	1	1	1	0	0	DC
)	0	1	0	1	1	1	0	1	5D		1	1	0	1	1	1	0	1	DD
:	0	1	0	1	1	1	1	0	5E		1	1	0	1	1	1	1	0	DE
¬	0	1	0	1	1	1	1	1	5F		1	1	0	1	1	1	1	1	DF
–	0	1	1	0	0	0	0	0	60	\	1	1	1	0	0	0	0	0	E0
/	0	1	1	0	0	0	0	1	61		1	1	1	0	0	0	0	1	E1
–	0	1	1	0	0	0	1	0	62	S	1	1	1	0	0	0	1	0	E2
	0	1	1	0	0	0	1	1	63	T	1	1	1	0	0	0	1	1	E3
	0	1	1	0	0	1	0	0	64	U	1	1	1	0	0	1	0	0	E4
	0	1	1	0	0	1	0	1	65	V	1	1	1	0	0	1	0	1	E5
	0	1	1	0	0	1	1	0	66	W	1	1	1	0	0	1	1	0	E6
	0	1	1	0	0	1	1	1	67	X	1	1	1	0	0	1	1	1	E7
	0	1	1	0	1	0	0	0	68	Y	1	1	1	0	1	0	0	0	E8
	0	1	1	0	1	0	0	1	69	Z	1	1	1	0	1	0	0	1	E9
	0	1	1	0	1	0	1	0	6A		1	1	1	0	1	0	1	0	EA
	0	1	1	0	1	0	1	1	6B		1	1	1	0	1	0	1	1	EB
%	0	1	1	0	1	1	0	0	6C		1	1	1	0	1	1	0	0	EC
	0	1	1	0	1	1	0	1	6D		1	1	1	0	1	1	0	1	ED
>	0	1	1	0	1	1	1	0	6E		1	1	1	0	1	1	1	0	EE
?	0	1	1	0	1	1	1	1	6F		1	1	1	0	1	1	1	1	EF
	0	1	1	1	0	0	0	0	70	0	1	1	1	1	0	0	0	0	F0
	0	1	1	1	0	0	0	1	71	1	1	1	1	1	0	0	0	1	F1
	0	1	1	1	0	0	1	0	72	2	1	1	1	1	0	0	1	0	F2
	0	1	1	1	0	0	1	1	73	3	1	1	1	1	0	0	1	1	F3
	0	1	1	1	0	1	0	0	74	4	1	1	1	1	0	1	0	0	F4
	0	1	1	1	0	1	0	1	75	5	1	1	1	1	0	1	0	1	F5
	0	1	1	1	0	1	1	0	76	6	1	1	1	1	0	1	1	0	F6
	0	1	1	1	0	1	1	1	77	7	1	1	1	1	0	1	1	1	F7
	0	1	1	1	1	0	0	0	78	8	1	1	1	1	1	0	0	0	F8
▲	0	1	1	1	1	0	0	1	79	9	1	1	1	1	1	0	0	1	F9
:	0	1	1	1	1	0	1	0	7A		1	1	1	1	1	0	1	0	FA
#	0	1	1	1	1	0	1	1	7B		1	1	1	1	1	0	1	1	FB
@	0	1	1	1	1	1	0	0	7C		1	1	1	1	1	1	0	0	FC
▲	0	1	1	1	1	1	0	1	7D		1	1	1	1	1	1	0	1	FD
–	0	1	1	1	1	1	1	0	7E		1	1	1	1	1	1	1	0	FE
”	0	1	1	1	1	1	1	1	7F		1	1	1	1	1	1	1	1	FF

DLE = data-link escape
DUP = duplicate
EM = end of medium
ENQ = enquiry
EOT = end of transmission
ESC = escape
ETB = end of transmission block
ETX = end of text
EUA = erase unprotected to address
FF = form feed
FM = field mark
IC = insert cursor
ITB = end of intermediate transmission block
NUL = null
PT = program tab
RA = repeat to address
SBA = set buffer address
SF = start field
SOH = start of heading
SP = space
STX = start of text
SUB = substitute
SYN = synchronous
NAK = negative acknowledge

FIGURE 22-1 Typical bar code

22-3 BAR CODES

Bar codes are those omnipresent black-and-white striped stickers that seem to appear on virtually every consumer item in the United States and most of the rest of the world. Although bar codes were developed in the early 1970s, they were not used extensively until the mid-1980s. A bar code is a series of vertical black bars separated by vertical white bars (called spaces). The widths of the bars and spaces along with their reflective abilities represent binary 1s and 0s, and combinations of bits identify specific items. In addition, bar codes may contain information regarding cost, inventory management and control, security access, shipping and receiving, production counting, document and order processing, automatic billing, and many other applications. A typical bar code is shown in Figure 22-1.

There are several standard bar code formats. The format selected depends on what types of data are being stored, how the data are being stored, system performance, and which format is most popular with business and industry. Bar codes are generally classified as being discrete, continuous, or two-dimensional (2D).

Discrete code. A discrete bar code has spaces or gaps between characters. Therefore, each character within the bar code is independent of every other character. Code 39 is an example of a discrete bar code.

Continuous code. A continuous bar code does not include spaces between characters. An example of a continuous bar code is the Universal Product Code (UPC).

2D code. A 2D bar code stores data in two dimensions in contrast with a conventional linear bar code, which stores data along only one axis. 2D bar codes have a larger storage capacity than one-dimensional bar codes (typically 1 kilobyte or more per data symbol).

22-3-1 Code 39

One of the most popular bar codes was developed in 1974 and called *Code 39* (also called *Code 3 of 9* and *3 of 9 Code*). Code 39 uses an alphanumeric code similar to the ASCII code. Code 39 is shown in Table 22-4. Code 39 consists of 36 unique codes representing the 10 digits and 26 uppercase letters. There are seven additional codes used for special characters, and an exclusive start/stop character coded as an asterisk (*). Code 39 bar codes are ideally suited for making labels, such as name badges.

Each Code 39 character contains nine vertical elements (five bars and four spaces). The logic condition (1 or 0) of each element is encoded in the width of the bar or space (i.e., *width modulation*). A wide element, whether it be a bar or a space, represents a logic 1, and a narrow element represents a logic 0. Three of the nine elements in each Code 39 character must be logic 1s, and the rest must be logic 0s. In addition, of the three logic 1s, two must be bars and one a space. Each character begins and ends with a black bar with alternating white bars in between. Since Code 39 is a discrete code, all characters are separated with an intercharacter gap, which is usually one character wide. The asterisks at the beginning and end of the bar code are start and stop characters, respectively.

Table 22-4 Code 39 Character Set

Character	Binary Code									Bars	Spaces	Check Sum Value
	b_8	b_7	b_6	b_5	b_4	b_3	b_2	b_1	b_0	$b_8b_6b_4b_2b_0$	$b_7b_5b_3b_1$	
0	0	0	0	1	1	0	1	0	0	00110	0100	0
1	1	0	0	1	0	0	0	0	1	10001	0100	1
2	0	0	1	1	0	0	0	0	1	01001	0100	2
3	1	0	1	1	0	0	0	0	0	11000	0100	3
4	0	0	0	1	1	0	0	0	1	00101	0100	4
5	1	0	0	1	1	0	0	0	0	10100	0100	5
6	0	0	1	1	1	0	0	0	0	01100	0100	6
7	0	0	0	1	0	0	1	0	1	00011	0100	7
8	1	0	0	1	0	0	1	0	0	10010	0100	8
9	0	0	1	1	0	0	1	0	0	01010	0100	9
A	1	0	0	0	0	1	0	0	1	10001	0010	10
B	0	0	1	0	0	1	0	0	1	01001	0010	11
C	1	0	1	0	0	1	0	0	0	11000	0010	12
D	0	0	0	0	1	1	0	0	1	00101	0010	13
E	1	0	0	0	1	1	0	0	0	10100	0010	14
F	0	0	1	0	1	1	0	0	0	01100	0010	15
G	0	0	0	0	0	1	1	0	1	00011	0010	16
H	1	0	0	0	0	1	1	0	0	10010	0010	17
I	0	0	1	0	0	1	1	0	0	01010	0010	18
J	0	0	0	0	1	1	1	0	0	00110	0010	19
K	1	0	0	0	0	0	0	1	1	10001	0001	20
L	0	0	1	0	0	0	0	1	1	01001	0001	21
M	1	0	1	0	0	0	0	1	0	11000	0001	22
N	0	0	0	0	1	0	0	1	1	00101	0001	23
O	1	0	0	0	1	0	0	1	0	10100	0001	24
P	0	0	1	0	1	0	0	1	0	01100	0001	25
Q	0	0	0	0	0	0	1	1	1	00011	0001	26
R	1	0	0	0	0	0	1	1	0	10010	0001	27
S	0	0	1	0	0	0	1	1	0	01010	0001	28
T	0	0	0	0	1	0	1	1	0	00110	0001	29
U	1	1	0	0	0	0	0	0	1	10001	1000	30
V	0	1	1	0	0	0	0	0	1	01001	1000	31
W	1	1	1	0	0	0	0	0	0	11000	1000	32
X	0	1	0	0	1	0	0	0	1	00101	1000	33
Y	1	1	0	0	1	0	0	0	0	10100	1000	34
Z	0	1	1	0	1	0	0	0	0	01100	1000	35
—	0	1	0	0	0	0	1	0	1	00011	1000	36
.	1	1	0	0	0	0	1	0	0	10010	1000	37
spacc	0	1	1	0	0	0	1	0	0	01010	1000	38
*	0	1	0	0	1	0	1	0	0	00110	1000	—
$	0	1	0	1	0	1	0	0	0	00000	1110	39
/	0	1	0	1	0	0	0	1	0	00000	1101	40
+	0	1	0	0	0	1	0	1	0	00000	1011	41
%	0	0	0	1	0	1	0	1	0	00000	0111	42

Figure 22-2 shows the Code 39 representation of the start/stop code (*) followed by an intercharacter gap and then the Code 39 representation of the letter A.

22-3-2 Universal Product Code

The grocery industry developed the *Universal Product Code* (UPC) sometime in the early 1970s to identify their products. The *National Association of Food Chains* officially adopted the UPC code in 1974. Today UPC codes are found on virtually every grocery item from a candy bar to a can of beans.

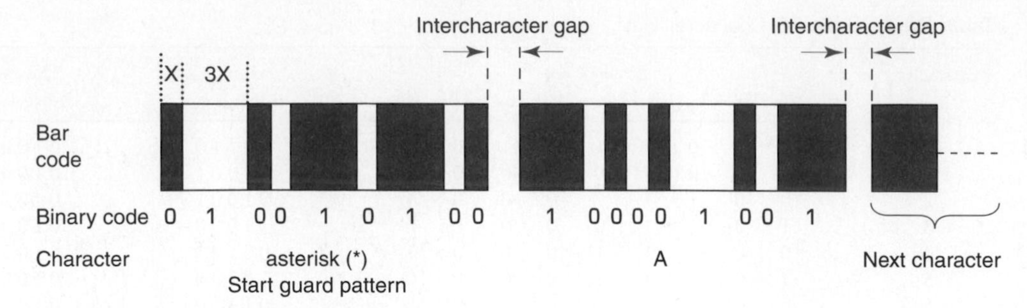

FIGURE 22-2 Code 39 bar code

Figures 22-3a, b, and c show the character set, label format, and sample bit patterns for the standard UPC code. Unlike Code 39, the UPC code is a continuous code since there are no intercharacter spaces. Each UPC label contains a 12-digit number. The two long bars shown in Figure 22-3b on the outermost left- and right-hand sides of the label are called the *start guard pattern* and the *stop guard pattern,* respectively. The start and stop guard patterns consist of a 101 (bar-space-bar) sequence, which is used to frame the 12-digit UPC number. The left and right halves of the label are separated by a *center guard pattern,* which consists of two *long bars* in the center of the label (they are called long bars because they are physically longer than the other bars on the label). The two long bars are separated with a space between them and have spaces on both sides of the bars. Therefore, the UPC center guard pattern is 01010 as shown in Figure 22-3b. The first six digits of the UPC code are encoded on the left half of the label (called the *left-hand characters*), and the last six digits of the UPC code are encoded on the right half (called the *right-hand characters*). Note in Figure 22-3a that there are two binary codes for each character. When a character appears in one of the first six digits of the code, it uses a left-hand code, and when a character appears in one of the last six digits, it uses a right-hand code. Note that the right-hand code is simply the complement of the left-hand code. For example, if the second and ninth digits of a 12-digit code UPC are both 4s, the digit is encoded as a 0100011 in position 2 and as a 1011100 in position 9. The UPC code for the 12-digit code 012345 543210 is

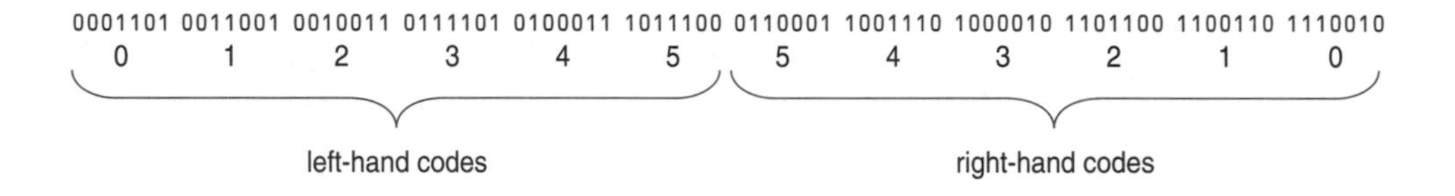

The first left-hand digit in the UPC code is called the *UPC number system character,* as it identifies how the UPC symbol is used. Table 22-5 lists the 10 UPC number system characters. For example, the UPC number system character 5 indicates that the item is intended to be used with a coupon. The other five left-hand characters are data characters. The first five right-hand characters are data characters, and the sixth right-hand character is a check character, which is used for error detection. The decimal value of the number system character is always printed to the left of the UPC label, and on most UPC labels the decimal value of the check character is printed on the right side of the UPC label.

With UPC codes, the width of the bars and spaces does not correspond to logic 1s and 0s. Instead, the digits 0 through 9 are encoded into a combination of two variable-

UPC Character Set

Left-hand character	Decimal digit	Right-hand character
0001101	0	1110010
0011001	1	1100110
0010011	2	1101100
0111101	3	1000010
0100011	4	1011100
0110001	5	1001110
0101111	6	1010000
0111011	7	1000100
0110111	8	1001000
0001011	9	1110100

(a)

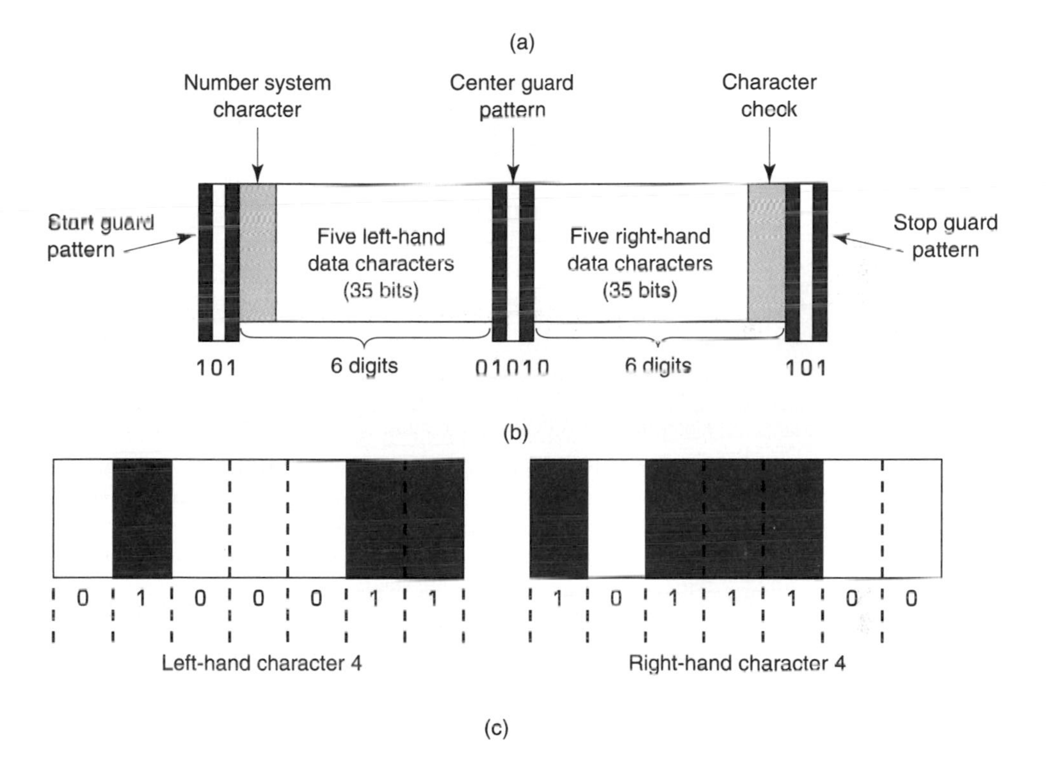

FIGURE 22-3 (a) UPC version A character set; (b) UPC label format; (c) left- and right-hand bit sequence for the digit 4

width bars and two variable-width spaces that occupy the equivalent of seven bit positions. Figure 22-3c shows the variable-width code for the UPC character 4 when used in one of the first six digit positions of the code (i.e., left-hand bit sequence) and when used in one of the last six digit positions of the code (i.e., right-hand bit sequence). A single bar (one bit position) represents a logic 1, and a single space represents a logic 0. However, close examination of the UPC character set in Table 22-5 will reveal that all UPC digits are comprised of bit patterns that yield two variable-width bars and two variable-width spaces, with the bar and space widths ranging from one to four bits. For the UPC character 4 shown in Figure 22-3c, the left-hand character is comprised of a one-bit space followed in order by a one-bit bar, a three-bit space, and a two-bit bar. The right-hand character is comprised of a one-bit bar followed in order by a one-bit space, a three-bit bar, and a two-bit space.

Table 22-5 UPC Number System Characters

Character	Intended Use
0	Regular UPC codes
1	Reserved for future use
2	Random-weight items that are symbol marked at the store
3	National Drug Code and National Health Related Items Code
4	Intended to be used without code format restrictions and with check digit protection for in-store marking of nonfood items
5	For use with coupons
6	Regular UPC codes
7	Regular UPC codes
8	Reserved for future use
9	Reserved for future use

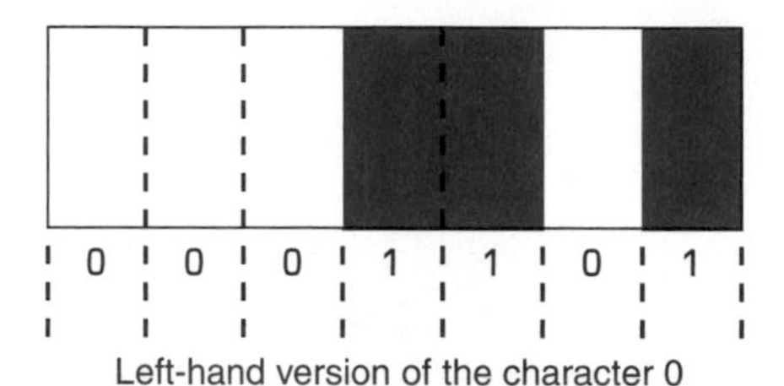

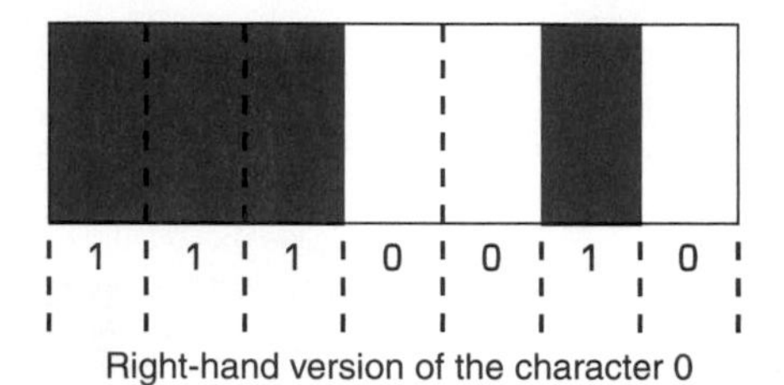

FIGURE 22-4 UPC character 0

Example 22-1

Determine the UPC label structure for the digit 0.

Solution From Figure 22-3a, the binary sequence for the digit 0 in the left-hand character field is 0001101, and the binary sequence for the digit 0 in the right-hand character field is 1110010.

The left-hand sequence is comprised of three successive 0s, followed by two 1s, one 0, and one 1. The three successive 0s are equivalent to a space three bits long. The two 1s are equivalent to a bar two bits long. The single 0 and single 1 are equivalent to a space and a bar, each one bit long.

The right-hand sequence is comprised of three 1s followed by two 0s, a 1, and a 0. The three 1s are equivalent to a bar three bits long. The two 0s are equivalent to a space two bits long. The single 1 and single 0 are equivalent to a bar and a space, each one bit long each. The UPC pattern for the digit 0 is shown in Figure 22-4.

22-4 ERROR CONTROL

A data communications circuit can be as short as a few feet or as long as several thousand miles, and the transmission medium can be as simple as a pair of wires or as complex as a microwave, satellite, or optical fiber communications system. Therefore, it is inevitable that errors will occur, and it is necessary to develop and implement error-control procedures. Transmission errors are caused by electrical interference from natural sources, such as lightning, as well as from man-made sources, such as motors, generators, power lines, and fluorescent lights.

Data communications errors can be generally classified as *single bit, multiple bit,* or *burst.* Single-bit errors are when only one bit within a given data string is in error. Single-bit errors affect only one character within a message. A multiple-bit error is when two or more nonconsecutive bits within a given data string are in error. Multiple-bit errors can affect one or more characters within a message. A burst error is when two or more consecutive bits within a given data string are in error. Burst errors can affect one or more characters within a message.

Error performance is the rate in which errors occur, which can be described as either an expected or an empirical value. The theoretical (mathematical) expectation of the rate at which errors will occur is called *probability of error* ($P[e]$), whereas the actual historical record of a system's error performance is called *bit error rate* (BER). For example, if a system has a $P(e)$ of 10^{-5}, this means that mathematically the system can expect to experience one bit error for every 100,000 bits transported through the system ($10^{-5} = 1/10^5 = 1/100{,}000$). If a system has a BER of 10^{-5}, this means that in the past there was one bit error for every 100,000 bits transported. Typically, a BER is measured and then compared with the probability of error to evaluate system performance. Error control can be divided into two general categories: *error detection* and *error correction.*

22-5 ERROR DETECTION

Error detection is the process of monitoring data transmission and determining when errors have occurred. Error-detection techniques neither correct errors nor identify which bits are in error—they indicate only when an error has occurred. The purpose of error detection is not to prevent errors from occurring but to prevent undetected errors from occurring.

The most common error-detection techniques are redundancy checking, which includes vertical redundancy checking, checksum, longitudinal redundancy checking, and cyclic redundancy checking.

22-5-1 Redundancy Checking

Duplicating each data unit for the purpose of detecting errors is a form of error detection called *redundancy.* Redundancy is an effective but rather costly means of detecting errors, especially with long messages. It is much more efficient to add bits to data units that check for transmission errors. Adding bits for the sole purpose of detecting errors is called *redundancy checking.* There are four basic types of redundancy checks: vertical redundancy checking, checksums, longitudinal redundancy checking, and cyclic redundancy checking.

22-5-1-1 Vertical redundancy checking. *Vertical redundancy checking* (VRC) is probably the simplest error-detection scheme and is generally referred to as *character parity* or simply *parity.* With character parity, each character has its own error-detection bit called the *parity bit.* Since the parity bit is not actually part of the character, it is considered a redundant bit. An *n*-character message would have *n* redundant parity bits. Therefore, the number of error-detection bits is directly proportional to the length of the message.

With character parity, a single parity bit is added to each character to force the total number of logic 1s in the character, including the parity bit, to be either an odd number (*odd parity*) or an even number (*even parity*). For example, the ASCII code for the letter C is 43 hex, or P1000011 binary, where the P bit is the parity bit. There are three logic 1s in the code, not counting the parity bit. If odd parity is used, the P bit is made a logic 0, keeping the total number of logic 1s at three, which is an odd number. If even parity is used, the P bit is made a logic 1, making the total number of logic 1s four, which is an even number.

The primary advantage of parity is its simplicity. The disadvantage is that when an even number of bits are received in error, the parity checker will not detect them because when the logic condition of an even number of bits is changed, the parity of the character remains the same. Consequently, over a long time, parity will theoretically detect only 50% of the transmission errors (this assumes an equal probability that an even or an odd number of bits could be in error).

Example 22-2

Determine the odd and even parity bits for the ASCII character R.

Solution The hex code for the ASCII character R is 52, which is P1010010 in binary, where P designates the parity bit.

For odd parity, the parity bit is a 0 because 52 hex contains three logic 1s, which is an odd number. Therefore, the odd-parity bit sequence for the ASCII character R is 01010010.

For even parity, the parity bit is 1, making the total number of logic 1s in the eight-bit sequence four, which is an even number. Therefore, the even-parity bit sequence for the ASCII character R is 11010010.

Other forms of parity include *marking parity* (the parity bit is always a 1), *no parity* (the parity bit is not sent or checked), and *ignored parity* (the parity bit is always a 0 bit if it is ignored). Marking parity is useful only when errors occur in a large number of bits. Ignored parity allows receivers that are incapable of checking parity to communicate with devices that use parity.

22-5-1-2 Checksum. *Checksum* is another relatively simple form of redundancy error checking where each character has a numerical value assigned to it. The characters within a message are combined together to produce an error-checking character (checksum), which can be as simple as the arithmetic sum of the numerical values of all the characters in the message. The checksum is appended to the end of the message. The receiver replicates the combining operation and determines its own checksum. The receiver's checksum is compared to the checksum appended to the message, and if they are the same, it is assumed that no transmission errors have occurred. If the two checksums are different, a transmission error has definitely occurred.

22-5-1-3 Longitudinal redundancy checking. *Longitudinal redundancy checking* (LRC) is a redundancy error detection scheme that uses parity to determine if a transmission error has occurred within a message and is therefore sometimes called *message parity.* With LRC, each bit position has a parity bit. In other words, b_0 from each character in the message is XORed with b_0 from all the other characters in the message. Similarly, b_1, b_2, and so on are XORed with their respective bits from all the characters in the message. Essentially, LRC is the result of XORing the "character codes" that make up the message, whereas VRC is the XORing of the bits within a single character. With LRC, even parity is generally used, whereas with VRC, odd parity is generally used.

The LRC bits are computed in the transmitter while the data are being sent and then appended to the end of the message as a redundant character. In the receiver, the LRC is recomputed from the data, and the recomputed LRC is compared to the LRC appended to the message. If the two LRC characters are the same, most likely no transmission errors have occurred. If they are different, one or more transmission errors have occurred.

Example 22-3 shows how are VRC and LRC are calculated and how they can be used together.

Example 22-3

Determine the VRCs and LRC for the following ASCII-encoded message: THE CAT. Use odd parity for the VRCs and even parity for the LRC.

Solution

Character		T	H	E	sp	C	A	T	LRC
Hex		54	48	45	20	43	41	54	2F
ASCII code	b_0	0	0	1	0	1	1	0	1
	b_1	0	0	0	0	1	0	0	1
	b_2	1	0	1	0	0	0	1	1
	b_3	0	1	0	0	0	0	0	1
	b_4	1	0	0	0	0	0	1	0
	b_5	0	0	0	1	0	0	0	1
	b_6	1	1	1	0	1	1	1	0
Parity bit (VRC)	b_7	0	1	0	0	0	1	0	0

The LRC is 00101111 binary (2F hex), which is the character "/" in ASCII. Therefore, after the LRC character is appended to the message, it would read "THE CAT/."

The group of characters that comprise a message (i.e., THE CAT) is often called a *block* or *frame* of data. Therefore, the bit sequence for the LRC is often called a *block check sequence* (BCS) or *frame check sequence* (FCS).

With longitudinal redundancy checking, all messages (regardless of their length) have the same number of error-detection characters. This characteristic alone makes LRC a better choice for systems that typically send long messages.

Historically, LRC detects between 95% and 98% of all transmission errors. LRC will not detect transmission errors when an even number of characters has an error in the same bit position. For example, if b_4 in an even number of characters is in error, the LRC is still valid even though multiple transmission errors have occurred.

22-5-1-4 Cyclic redundancy checking. Probably the most reliable redundancy checking technique for error detection is a convolutional coding scheme called *cyclic redundancy checking* (CRC). With CRC, approximately 99.999% of all transmission errors are detected. In the United States, the most common CRC code is CRC-16. With CRC-16, 16 bits are used for the block check sequence. With CRC, the entire data stream is treated as a long continuous binary number. Because the BCS is separate from the message but transported within the same transmission, CRC is considered a *systematic code.* Cyclic block codes are often written as (n, k) cyclic codes where n = bit length of transmission and k = bit length of message. Therefore, the length of the BCC in bits is

$$\text{BCC} = n - k$$

A CRC-16 block check character is the remainder of a binary division process. A data message polynominal $G(x)$ is divided by a unique generator polynominal function $P(x)$, the quotient is discarded, and the remainder is truncated to 16 bits and appended to the message as a BCS. The generator polynominal must be a prime number (i.e., a number divisible by only itself and 1). CRC-16 detects all single-bit errors, all double-bit errors (provided the divisor contains at least three logic 1s), all odd number of bit errors (provided the division contains a factor 11), all error bursts of 16 bits or less, and 99.9% of error bursts greater than 16 bits long. For randomly distributed errors, it is estimated that the likelihood of CRC-16 not detecting an error is 10^{-14}, which equates to one undetected error every two years of continuous data transmission at a rate of 1.544 Mbps.

With CRC generation, the division is not accomplished with standard arithmetic division. Instead, modulo-2 division is used, where the remainder is derived from an exclusive OR (XOR) operation. In the receiver, the data stream, including the CRC code, is divided by the same generating function $P(x)$. If no transmission errors have occurred, the remainder will be zero. In the receiver, the message and CRC character pass through a block check register. After the entire message has passed through the register, its contents should be zero if the receive message contains no errors.

Mathematically, CRC can be expressed as

$$\frac{G(x)}{P(x)} = Q(x) + R(x) \qquad \textbf{(22-1)}$$

where $G(x)$ = message polynominal
$P(x)$ = generator polynominal
$Q(x)$ = quotient
$R(x)$ = remainder

The generator polynomial for CRC-16 is

$$P(x) = x^{16} + x^{15} + x^2 + x^0$$

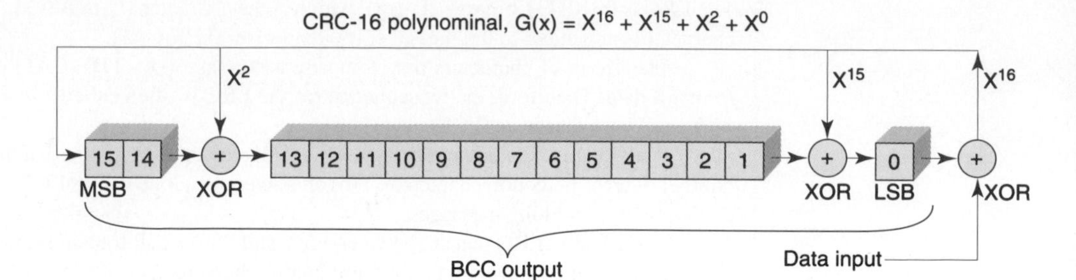

FIGURE 22-5 CRC-16 generating circuit

The number of bits in the CRC code is equal to the highest exponent of the generating polynomial. The exponents identify the bit positions in the generating polynomial that contain a logic 1. Therefore, for CRC-16, b_{16}, b_{15}, b_2, and b_0 are logic 1s, and all other bits are logic 0s. The number of bits in a CRC character is always twice the number of bits in a data character (i.e., eight-bit characters use CRC-16, six-bit characters use CRC-12, and so on).

Figure 22-5 shows the block diagram for a circuit that will generate a CRC-16 BCC. A CRC generating circuit requires one shift register for each bit in the BCC. Note that there are 16 shift registers in Figure 22-5. Also note that an XOR gate is placed at the output of the shift registers for each bit position of the generating polynomial that contains a logic 1, except for x^0. The BCC is the content of the 16 registers after the entire message has passed through the CRC generating circuit.

Example 22-4

Determine the BCS for the following data and CRC generating polynomials:

$$\text{Data } G(x) = x^7 + x^5 + x^4 + x^2 + x^1 + x^0$$
$$= 10110111$$

$$\text{CRC } P(x) = x^5 + x^4 + x^1 + x^0$$
$$= 110011$$

Solution First, $G(x)$ is multiplied by the number of bits in the CRC code, which is 5:

$$x^5(x^7 + x^5 + x^4 + x^2 + x^1 + x^0) = x^{12} + x^{10} + x^9 + x^7 + x^6 + x^5 = 1011011100000$$

Then the result is divided by $P(x)$:

```
                   1 1 0 1 0 1 1 1
1 1 0 0 1 1  | 1 0 1 1 0 1 1 1 0 0 0 0 0
               1 1 0 0 1 1
                 1 1 1 1 0 1
                 1 1 0 0 1 1
                   1 1 1 0 1 0
                   1 1 0 0 1 1
                     1 0 0 1 0 0
                     1 1 0 0 1 1
                       1 0 1 1 1 0
                       1 1 0 0 1 1
                         1 1 1 0 1 0
                         1 1 0 0 1 1
                           0 1 0 0 1 = CRC
```

The CRC is appended to the data to give the following data stream:

$$\overbrace{1\ 0\ 1\ 1\ 0\ 1\ 1\ 1}^{G(x)}\ \overbrace{0\ 1\ 0\ 0\ 1}^{CRC}$$

At the receiver, the data are again divided by $P(x)$ =:

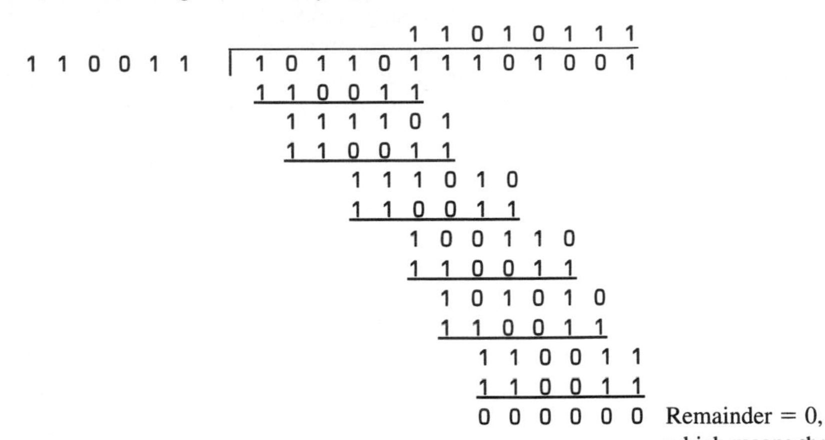

Remainder = 0, which means there were no transmission errors

22-6 ERROR CORRECTION

Although detecting errors is an important aspect of data communications, determining what to do with data that contain errors is another consideration. There are two basic types of error messages: *lost message* and *damaged message.* A lost message is one that never arrives at the destination or one that arrives but is damaged to the extent that it is unrecognizable. A damaged message is one that is recognized at the destination but contains one or more transmission errors.

Data communications network designers have developed two basic strategies for handling transmission errors: *error-detecting codes* and *error-correcting codes.* Error-detecting codes include enough redundant information with each transmitted message to enable the receiver to determine when an error has occurred. Parity bits, block and frame check characters, and cyclic redundancy characters are examples of error-detecting codes. Error-correcting codes include sufficient extraneous information along with each message to enable the receiver to determine when an error has occurred and which bit is in error.

Transmission errors can occur as single-bit errors or as bursts of errors, depending on the physical processes that caused them. Having errors occur in bursts is an advantage when data are transmitted in blocks or frames containing many bits. For example, if a typical frame size is 10,000 bits and the system has a probability of error of 10^{-4} (one bit error in every 10,000 bits transmitted), independent bit errors would most likely produce an error in every block. However, if errors occur in bursts of 1000, only one or two blocks out of every 1000 transmitted would contain errors. The disadvantage of bursts of errors is they are more difficult to detect and even more difficult to correct than isolated single-bit errors. In the modern world of data communications, there are two primary methods used for error correction: *retransmission* and *forward error correction.*

22-6-1 Retransmission

Retransmission, as the name implies, is when a receive station requests the transmit station to resend a message (or a portion of a message) when the message is received in error. Because the receive terminal automatically calls for a retransmission of the entire message, retransmission is often called ARQ, which is an old two-way radio term that means *automatic repeat request* or *automatic retransmission request.* ARQ is probably the most reliable method of error correction, although it is not necessarily the most efficient. Impairments on transmission media often occur in bursts. If short messages are used, the likelihood that impairments will occur during a transmission is small. However, short messages require more *acknowledgments* and *line turnarounds* than do long messages. Acknowledgments are when the recipient of data sends a short message back to the sender acknowledging receipt of the last transmission. The acknowledgment can indicate a successful transmission (positive acknowledgment) or an unsuccessful transmission (negative acknowledgment). Line turnarounds are when a receive station becomes the transmit station, such as when acknowledgments are sent or when retransmissions are sent in response to a negative acknowledgment. Acknowledgments and line turnarounds for error control are forms of overhead (data other than user information that must be transmitted). With long messages, less turnaround time is needed, although the likelihood that a transmission error will occur is higher than for short messages. It can be shown statistically that messages between 256 and 512 characters long are the optimum size for ARQ error correction.

There are two basic types of ARQ: discrete and continuous. *Discrete ARQ* uses *acknowledgments* to indicate the successful or unsuccessful reception of data. There are two basic types of acknowledgments: positive and negative. The destination station responds with a *positive acknowledgment* when it receives an error-free message. The destination station responds with a negative *acknowledgment* when it receives a message containing errors to call for a retransmission. If the sending station does not receive an acknowledgment after a predetermined length of time (called a *time-out*), it retransmits the message. This is called *retransmission after time-out.*

Another type of ARQ, called *continuous ARQ,* can be used when messages are divided into smaller blocks or frames that are sequentially numbered and transmitted in succession, without waiting for acknowledgments between blocks. Continuous ARQ allows the destination station to asynchronously request the retransmission of a specific frame (or frames) of data and still be able to reconstruct the entire message once all frames have been successfully transported through the system. This technique is sometimes called *selective repeat,* as it can be used to call for a retransmission of an entire message or only a portion of a message.

22-6-2 Forward Error Correction

Forward error correction (FEC) is the only error-correction scheme that actually detects and corrects transmission errors when they are received without requiring a retransmission. With FEC, redundant bits are added to the message before transmission. When an error is detected, the redundant bits are used to determine which bit is in error. Correcting the bit is a simple matter of complementing it. The number of redundant bits necessary to correct errors is much greater than the number of bits needed to simply detect errors. Therefore, FEC is generally limited to one-, two-, or three-bit errors.

FEC is ideally suited for data communications systems when acknowledgments are impractical or impossible, such as when simplex transmissions are used to transmit messages to many receivers or when the transmission, acknowledgment, and retransmission time is excessive, for example when communicating to far away places, such as deep-space vehicles. The purpose of FEC codes is to eliminate the time wasted for retransmissions. However, the addition of the FEC bits to each message wastes time itself. Obviously, a trade-off is made between ARQ and FEC, and system requirements determine which method is best suited to a particular application. Probably the most popular error-correction code is the Hamming code.

22-6-2-1 Hamming code. A mathematician named Richard W. Hamming, who was an early pioneer in the development of error-detection and -correction procedures, de-

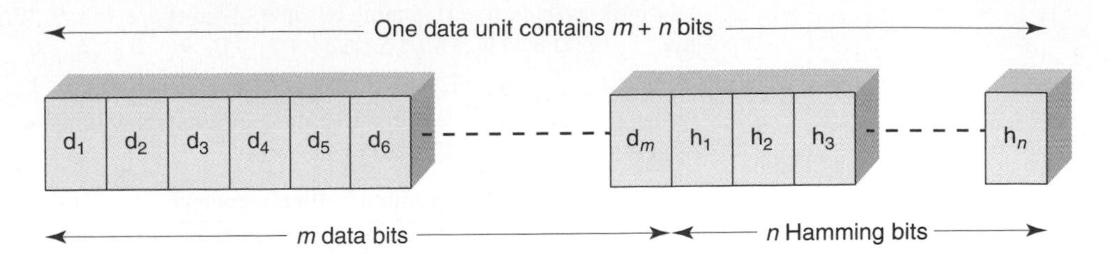

FIGURE 22-6 Data unit comprised of *m* character bits and *n* Hamming bits

veloped the *Hamming code* while working at Bell Telephone Laboratories. The Hamming code is an *error-correcting code* used for correcting transmission errors in synchronous data streams. However, the Hamming code will correct only single-bit errors. It cannot correct multiple-bit errors or burst errors, and it cannot identify errors that occur in the Hamming bits themselves. The Hamming code, as with all FEC codes, requires the addition of overhead to the message, consequently increasing the length of a transmission.

Hamming bits (sometimes called *error bits*) are inserted into a character at random locations. The combination of the data bits and the Hamming bits is called the Hamming code. The only stipulation on the placement of the Hamming bits is that both the sender and the receiver must agree on where they are placed. To calculate the number of redundant Hamming bits necessary for a given character length, a relationship between the character bits and the Hamming bits must be established. As shown in Figure 22-6, a data unit contains m character bits and n Hamming bits. Therefore, the total number of bits in one data unit is $m + n$. Since the Hamming bits must be able to identify which bit is in error, n Hamming bits must be able to indicate at least $m + n + 1$ different codes. Of the $m + n$ codes, one code indicates that no errors have occurred, and the remaining $m + n$ codes indicate the bit position where an error has occurred. Therefore, $m + n$ bit positions must be identified with n bits. Since n bits can produce 2^n different codes, 2^n must be equal to or greater than $m + n + 1$. Therefore, the number of Hamming bits is determined by the following expression:

$$2^n \geq m + n + 1 \qquad \textbf{(22-2)}$$

where n = number of Hamming bits
m = number of bits in each data character

A seven-bit ASCII character requires four Hamming bits ($2^4 > 7 + 4 + 1$), which could be placed at the end of the character bits, at the beginning of the character bits, or interspersed throughout the character bits. Therefore, including the Hamming bits with ASCII-coded data requires transmitting 11 bits per ASCII character, which equates to a 57% increase in the message length.

Example 22-5

For a 12-bit data string of 101100010010, determine the number of Hamming bits required, arbitrarily place the Hamming bits into the data string, determine the logic condition of each Hamming bit, assume an arbitrary single-bit transmission error, and prove that the Hamming code will successfully detect the error.

Solution Substituting $m = 12$ into Equation 22-2, the number of Hamming bits is

$$\text{for } n = 4 \qquad 2^4 = 16 \geq 12 + 4 + 1 = 17$$

Because 16 < 17, four Hamming bits are insufficient:

$$\text{for } n = 5 \qquad 2^5 = 32 \geq 12 + 5 + 1 = 18$$

Because 32 > 18, five Hamming bits are sufficient, and a total of 17 bits make up the data stream (12 data plus five Hamming).

Arbitrarily placing five Hamming bits into bit positions 4, 8, 9, 13, and 17 yields

bit position	17	16	15	14	13	12	11	10	9	8	7	6	5	4	3	2	1
	H	1	0	1	H	1	0	0	H	H	0	1	0	H	0	1	0

To determine the logic condition of the Hamming bits, express all bit positions that contain a logic 1 as a five-bit binary number and XOR them together:

Bit position	Binary number	
2	00010	
6	00110	
XOR	00100	
12	01100	
XOR	01000	
14	01110	
XOR	00110	
16	10000	
XOR	10110	= Hamming bits

$$b_{17} = 1, \quad b_{13} = 0, \quad b_9 = 1, \quad b_8 = 1, \quad b_4 = 0$$

The 17-bit Hamming code is

H				H				H	H				H			
1	1	0	1	0	1	0	0	1	1	0	1	0	0	0	1	0

Assume that during transmission, an error occurs in bit position 14. The received data stream is

1 1 0 0 0 1 0 0 1 1 0 1 0 0 0 1 0

(error: the fourth bit, 0)

At the receiver, to determine the bit position in error, extract the Hamming bits and XOR them with the binary code for each data bit position that contains a logic 1:

Bit position	Binary number	
Hamming bits	10110	
2	00010	
XOR	10100	
6	00110	
XOR	10010	
12	01100	
XOR	11110	
16	10000	
XOR	01110	= 14

Therefore, bit position 14 contains an error.

22-7 CHARACTER SYNCHRONIZATION

In essence, *synchronize* means to harmonize, coincide, or agree in time. *Character synchronization* involves identifying the beginning and end of a character within a message. When a continuous string of data is received, it is necessary to identify which bits belong to which characters and which bits are the MSBS and LSBS of the character. In essence, this is *character synchronization:* identifying the beginning and end of a character code. In data communications circuits, there are two formats commonly used to achieve character synchronization: asynchronous and synchronous.

22-7-1 Asynchronous Serial Data

The term asynchronous literally means "without synchronism," which in data communications terminology means "without a specific time reference." Asynchronous data transmis-

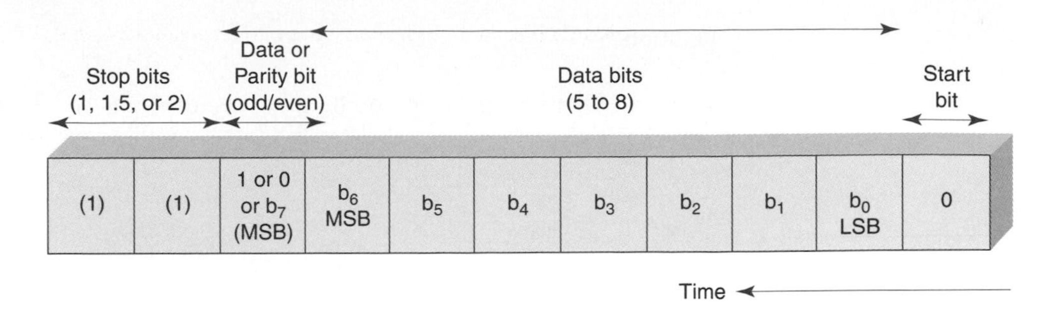

FIGURE 22-7 Asynchronous data format

sion is sometimes called *start-stop transmission* because each data character is framed between *start* and *stop bits.* The start and stop bits identify the beginning and end of the character, so the time gaps between characters do not present a problem. For asynchronously transmitted serial data, framing characters individually with start and stop bits is sometimes said to occur on a *character-by-character* basis.

Figure 22-7 shows the format used to frame a character for asynchronous serial data transmission. The first bit transmitted is the start bit, which is always a logic 0. The character bits are transmitted next, beginning with the LSB and ending with the MSB. The data character can contain between five and eight bits. The parity bit (if used) is transmitted directly after the MSB of the character. The last bit transmitted is the stop bit, which is always a logic 1, and there can be either one, one and a half, or two stop bits. Therefore, a data character may be comprised of between seven and 11 bits.

A logic 0 is used for the start bit because an idle line condition (no data transmission) on a data communications circuit is identified by the transmission of continuous logic 1s (called *idle line 1s*). Therefore, the start bit of a character is identified by a high-to-low transition in the received data, and the bit that immediately follows the start bit is the LSB of the character code. All stop bits are logic 1s, which guarantees a high-to-low transition at the beginning of each character. After the start bit is detected, the data and parity bits are clocked into the receiver. If data are transmitted in real time (i.e., as the operator types data into the computer terminal), the number of idle line 1s between each character will vary. During this *dead time,* the receive will simply wait for the occurrence of another start bit (i.e., high-to-low transition) before clocking in the next character. Obviously, both slipping over and slipping under produce errors. However, the errors are somewhat self-inflicted, as they occur in the receiver and are not a result of an impairment that occurred during transmission.

With asynchronous data, it is not necessary that the transmit and receive clocks be continuously synchronized; however, their frequencies should be close, and they should be synchronized at the beginning of each character. When the transmit and receive clocks are substantially different, a condition called *clock slippage* may occur. If the transmit clock is substantially lower than the receive clock, *underslipping* occurs. If the transmit clock is substantially higher than the receive clock, a condition called *overslipping* occurs. With overslipping, the receive clock samples the receive data slower than the bit rate. Consequently, each successive sample occurs later in the bit time until finally a bit is completely skipped.

Example 22-6

For the following sequence of bits, identify the ASCII-encoded character, the start and stop bits, and the parity bits (assume even parity and two stop bits):

Solution

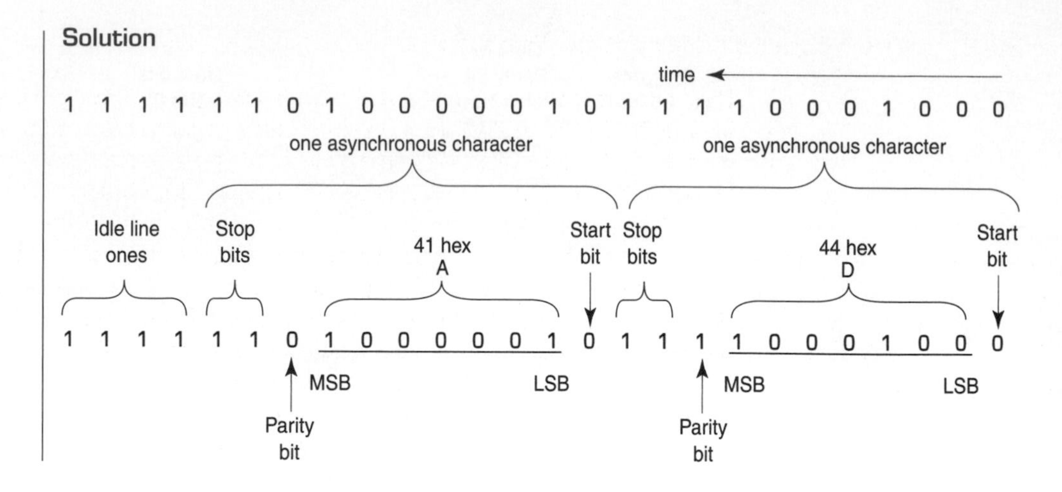

22-7-2 Synchronous Serial Data

Synchronous data generally involves transporting serial data at relatively high speeds in groups of characters called blocks or frames. Therefore, synchronous data are not sent in real time. Instead, a message is composed or formulated and then the entire message is transmitted as a single entity with no time lapses between characters. With synchronous data, rather than frame each character independently with start and stop bits, a unique sequence of bits, sometimes called a synchronizing (SYN) character, is transmitted at the beginning of each message. For synchronously transmitted serial data, framing characters in blocks is sometimes said to occur on a *block-by-block* basis. For example, with ASCII code, the SYN character is 16 hex. The receiver disregards incoming data until it receives one or more SYN characters. Once the synchronizing sequence is detected, the receiver clocks in the next eight bits and interprets them as the first character of the message. The receiver continues clocking in bits, interpreting them in groups of eight until it receives another unique character that signifies the end of the message. The end-of-message character varies with the type of protocol being used and what type of message it is associated with. With synchronous data, the transmit and receive clocks must be synchronized because character synchronization occurs only once at the beginning of a message.

With synchronous data, each character has two or three bits added to each character (one start and either one, one and a half, or two stop bits). These bits are additional overhead and, thus, reduce the efficiency of the transmission (i.e., the ratio of information bits to total transmitted bits). Synchronous data generally has two SYN characters (16 bits of overhead) added to each message. Therefore, asynchronous data are more efficient for short messages, and synchronous data are more efficient for long messages.

Example 22-7

For the following string of ASCII-encoded characters, identify each character (assume odd parity):

Solution

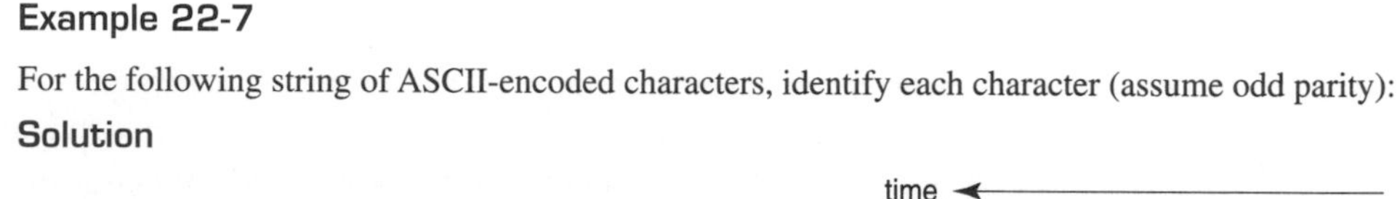

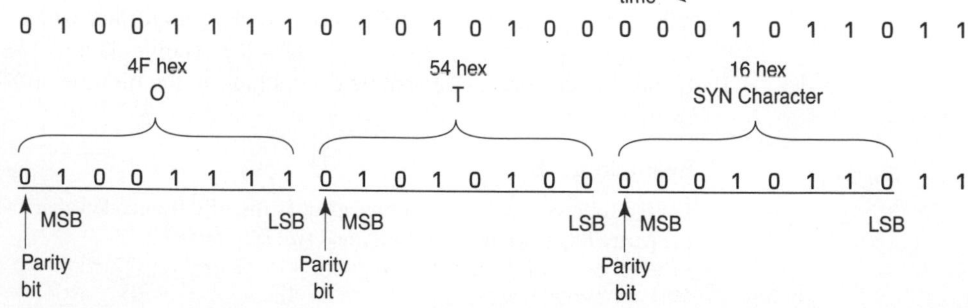

22-8 DATA COMMUNICATIONS HARDWARE

Digital information sources, such as personal computers, communicate with each other using the POTS (plain old telephone system) telephone network in a manner very similar to the way analog information sources, such as human conversations, communicate with each other using the POTS telephone network. With both digital and analog information sources, special devices are necessary to interface the sources to the telephone network.

Figure 22-8 shows a comparison between human speech (analog) communications and computer data (digital) communications using the POTS telephone network. Figure 22-8a shows how two humans communicate over the telephone network using standard analog telephone sets. The telephone sets interface human speech signals to the telephone network and vice versa. At the transmit end, the telephone set converts acoustical energy (information) to electrical energy and, at the receive end, the telephone set converts electrical energy back to acoustical energy. Figure 22-8b shows how digital data are transported over the telephone network. At the transmitting end, a telco interface converts digital data from the transceiver to analog electrical energy, which is transported through the telephone network. At the receiving end, a telco interface converts the analog electrical energy received from the telephone network back to digital data.

In simplified terms, a data communications system is comprised of three basic elements: a *transmitter* (source), a *transmission path* (data channel), and a *receiver* (destination). For two-way communications, the transmission path would be bidirectional and the source and destination interchangeable. Therefore, it is usually more appropriate to describe a data communications system as connecting two *endpoints* (sometimes called *nodes*) through a common communications channel. The two endpoints may not possess the same computing capabilities; however, they must be configured with the same basic components. Both endpoints must be equipped with special devices that perform unique functions, make the physical connection to the data channel, and process the data before they are transmitted and after they have been received. Although the special devices are

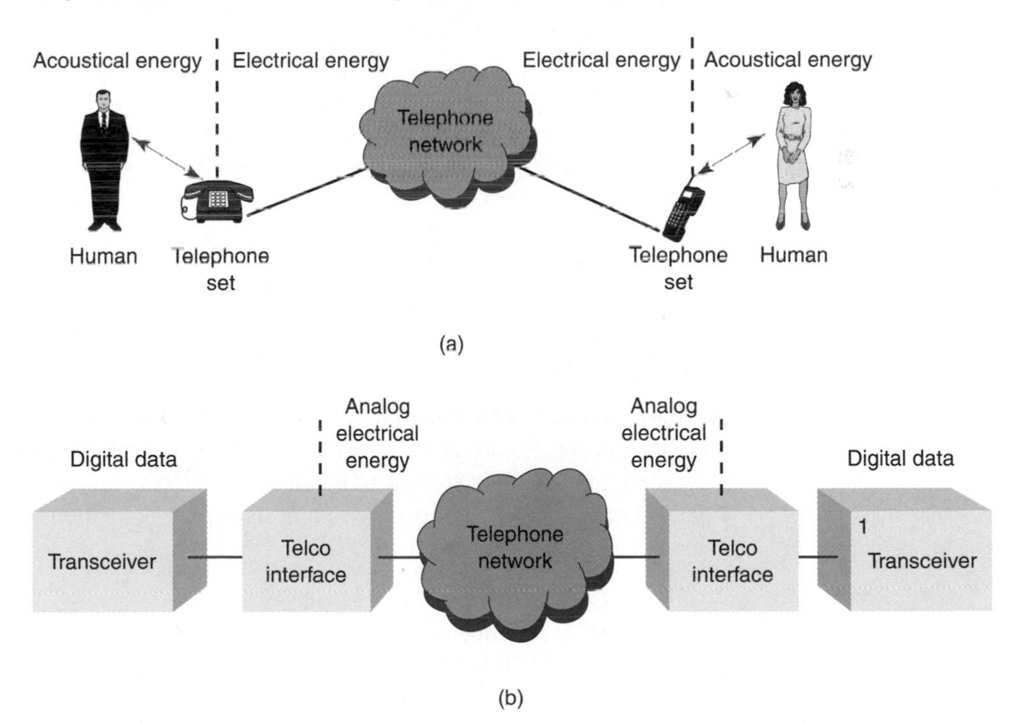

FIGURE 22-8 Telephone communications network: (a) human communications; (b) digital data communications

sometimes implemented as a single unit, it is generally easier to describe them as separate entities. In essence, all endpoints must have three fundamental components: *data terminal equipment, data communications equipment,* and a *serial interface.*

22-8-1 Data Terminal Equipment

Data terminal equipment (DTE) can be virtually any binary digital device that generates, transmits, receives, or interprets data messages. In essence, a DTE is where information originates or terminates. DTEs are the data communications equivalent to the person in a telephone conversation. DTEs contain the hardware and software necessary to establish and control communications between endpoints in a data communications system; however, DTEs seldom communicate directly with other DTEs. Examples of DTEs include video display terminals, printers, and personal computers.

Over the past 50 years, data terminal equipment has evolved from simple on-line printers to sophisticated high-level computers. Data terminal equipment includes the concept of *terminals, clients, hosts,* and *servers.* Terminals are devices used to input, output, and display information, such as keyboards, printers, and monitors. A client is basically a modern-day terminal with enhanced computing capabilities. Hosts are high-powered, high-capacity mainframe computers that support terminals. Servers function as modern-day hosts except with lower storage capacity and less computing capability. Servers and hosts maintain local databases and programs and distribute information to clients and terminals.

22-8-2 Data Communications Equipment

Data communications equipment (DCE) is a general term used to describe equipment that interfaces data terminal equipment to a transmission channel, such as a digital T1 carrier or an analog telephone circuit. The output of a DTE can be digital or analog, depending on the application. In essence, a DCE is a *signal conversion device,* as it converts signals from a DTE to a form more suitable to be transported over a transmission channel. A DCE also converts those signals back to their original form at the receive end of a circuit. DCEs are transparent devices responsible for transporting bits (1s and 0s) between DTEs through a data communications channel. The DCEs neither know nor do they care about the content of the data.

There are several types of DCEs, depending on the type of transmission channel used. Common DCEs are *channel service units* (CSUs), *digital service units* (DSUs), and *data modems.* CSUs and DSUs are used to interface DTEs to digital transmission channels. Data modems are used to interface DTEs to analog telephone networks. Because data communications channels are terminated at each end in a DCE, DCEs are sometimes called *data circuit-terminating equipment* (DCTE). CSUs and DSUs are described in a later chapter of this book, whereas data modems are described in subsequent sections of this chapter.

22-9 DATA COMMUNICATIONS CIRCUITS

A data modem is a DCE used to interface a DTE to an analog telephone circuit commonly called a POTS. Figure 22-9a shows a simplified diagram for a two-point data communications circuit using a POTS link to interconnect the two endpoints (endpoint A and endpoint B). As shown in the figure, a two-point data communications circuit is comprised of the seven basic components:

1. DTE at endpoint A
2. DCE at endpoint A
3. DTE/DCE interface at endpoint A
4. Transmission path between endpoint A and endpoint B
5. DCE at endpoint B
6. DTE at endpoint B
7. DTE/DCE interface at endpoint B

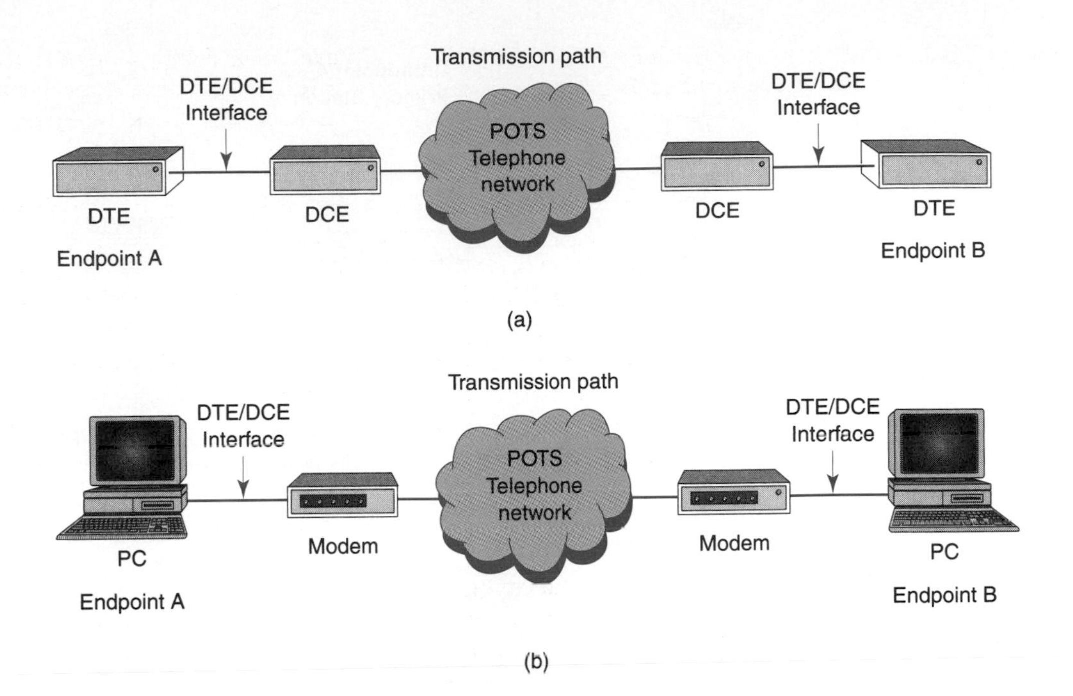

FIGURE 22-9 Two point data communications circuit: (a) DTE/DCE representation; (b) device representation

The DTEs can be terminal devices, personal computers, mainframe computers, front-end processors, printers, or virtually any other piece of digital equipment. If a digital communications channel were used, the DCE would be a CSU or a DSU. However, because the communications channel is a POTS link, the DCE is a data modem.

Figure 22-9b shows the same equivalent circuit as is shown in Figure 22-9a, except the DTE and DCE have been replaced with the actual devices they represent—the DTE is a personal computer, and the DCE is a modem. In most modern-day personal computers for home use, the modem is simply a card installed inside the computer.

Figure 22-10 shows the block diagram for a centralized multipoint data communications circuit using several POTS data communications links to interconnect three endpoints. The circuit is arranged in a bus topology with central control provided by a mainframe computer (host) at endpoint A. The host station is sometimes called the *primary station*. Endpoints B and C are called *secondary stations*. The primary station is responsible for establishing and maintaining the data link and for ensuring an orderly flow of data between itself and each of the secondary stations. Data flow is controlled by an applications program stored in the mainframe computer at the primary station.

At the primary station, there is a mainframe computer, a front-end processor (DTE), and a data modem (DCE). At each secondary station, there is a modem (DCE), a line control unit (DTE), and a *cluster* of terminal devices (personal computers, printers, and so on). The line control unit at the secondary stations is referred to as a *cluster controller,* as it controls data flow between several terminal devices and the data communications channel. Line control units at secondary stations are sometimes called station controllers (STACOs), as they control data flow to and from all the data communications equipment located at that station.

For simplicity, Figure 22-10 only shows one data circuit served by the mainframe computer at the primary station. However, there can be dozens of different circuits served by one mainframe computer. Therefore, the primary station line control unit (i.e., the front-end processor) must have enhanced capabilities for storing, processing, and retransmitting data it receives from all secondary stations on all the circuits it serves. The primary station stores software for database management of all the circuits it serves. Obviously, the duties

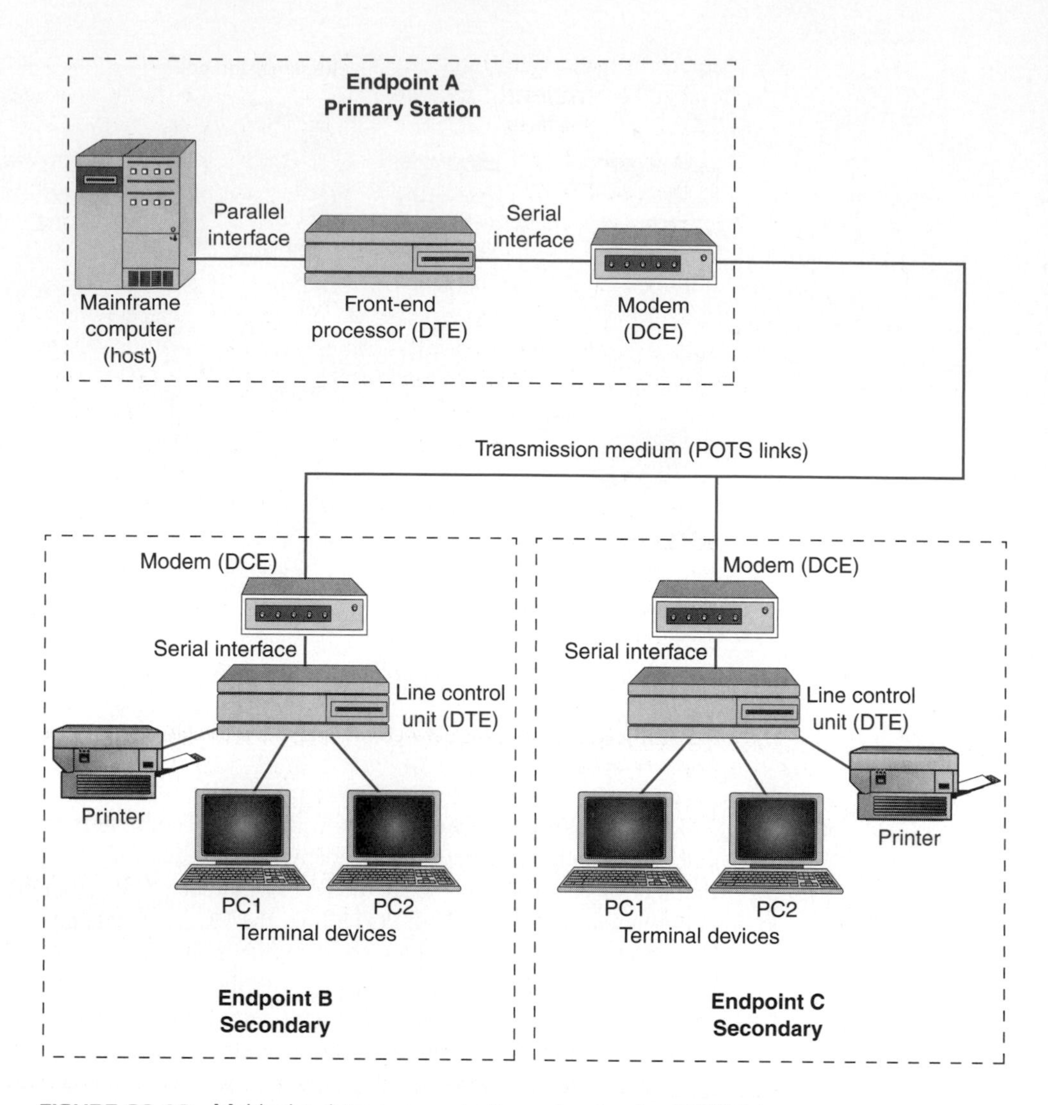

FIGURE 22-10 Multipoint data communications circuit using POTS links

performed by the front-end processor at the primary station are much more involved than the duties performed by the line control units at the secondary stations. The FEP directs data traffic to and from many different circuits, which could all have different parameters (i.e., different bit rates, character codes, data formats, protocols, and so on). The LCU at the secondary stations directs data traffic between one data communications link and a relative few terminal devices, which all transmit and receive data at the same speed and use the same data-link protocol, character code, data format, and so on.

22-10 LINE CONTROL UNIT

As previously stated, a line control unit (LCU) is a DTE, and DTEs have several important functions. At the primary station, the LCU is often called a FEP because it processes information and serves as an interface between the host computer and all the data communications circuits it serves. Each circuit served is connected to a different port on the FEP. The FEP directs the flow of input and output data between data communications circuits and their respective application programs. The data interface between the mainframe computer and the FEP transfers data in parallel at relatively high bit rates. However, data transfers between the modem and the FEP are accomplished in serial and at a much lower bit rate. The FEP at the primary station and the LCU at the secondary stations perform parallel-to-serial and serial-to-parallel conversions. They also house the circuitry that performs error detection and correction. In addition, data-link control characters are inserted and deleted in the FEP and LCUs (data-link control characters are described in Chapter 23).

Within the FEP and LCUs, a single special-purpose integrated circuit performs many of the fundamental data communications functions. This integrated circuit is called a *universal asynchronous receiver/transmitter* (UART) if it is designed for asynchronous data transmission, a *universal synchronous receiver/transmitter* (USRT) if it is designed for synchronous data transmission, and a *universal synchronous/asynchronous receiver/transmitter* (USART) if it is designed for either asynchronous or synchronous data transmission. All three types of circuits specify general-purpose integrated-circuit chips located in an LCU or FEP that allow DTEs to interface with DCEs. In modern-day integrated circuits, UARTs and USRTs are often combined into a single USART chip that is probably more popular today simply because it can be adapted to either asynchronous or synchronous data transmission. USARTs are available in 24- to 64-pin dual in-line packages (DIPs).

UARTS, USRTS, and USARTS are devices that operate external to the central processor unit (CPU) in a DTE that allow the DTE to communicate serially with other data communications equipment, such as DCEs. They are also essential data communications components in terminals, workstations, PCs, and many other types of serial data communications devices. In most modern computers, USARTs are normally included on the motherboard and connected directly to the serial port. UARTs, USRTs, and USARTs designed to interface to specific microprocessors often have unique manufacturer-specific names. For example, Motorola manufactures a special purpose UART chip it calls an *asynchronous communications interface adapter* (ACIA).

22-10-1 UART

A UART is used for asynchronous transmission of serial data between a DTE and a DCE. Asynchronous data transmission means that an asynchronous data format is used, and there is no clocking information transferred between the DTE and the DCE. The primary functions performed by a UART are the following:

1. Parallel-to-serial data conversion in the transmitter and serial-to-parallel data conversion in the receiver
2. Error detection by inserting parity bits in the transmitter and checking parity bits in the receiver
3. Insert start and stop bits in the transmitter and detect and remove start and stop bits in the receiver
4. Formatting data in the transmitter and receiver (i.e., combining items 1 through 3 in a meaningful sequence)
5. Provide transmit and receive status information to the CPU
6. Voltage level conversion between the DTE and the serial interface and vice versa
7. Provide a means of achieving bit and character synchronization

Transmit and receive functions can be performed by a UART simultaneously because the transmitter and receiver have separate control signals and clock signals and share a bidirectional data bus, which allows them to operate virtually independently of one another. In addition, input and output data are double buffered, which allows for continuous data transmission and reception.

Figure 22-11 shows a simplified block diagram of a line control unit showing the relationship between the UART and the CPU that controls the operation of the UART. The CPU coordinates data transfer between the line-control unit (or FEP) and the modem. The CPU is responsible for programming the UART's control register, reading the UART's status register, transferring parallel data to and from the UART transmit and receive buffer registers, providing clocking information to the UART, and facilitating the transfer of serial data between the UART and the modem.

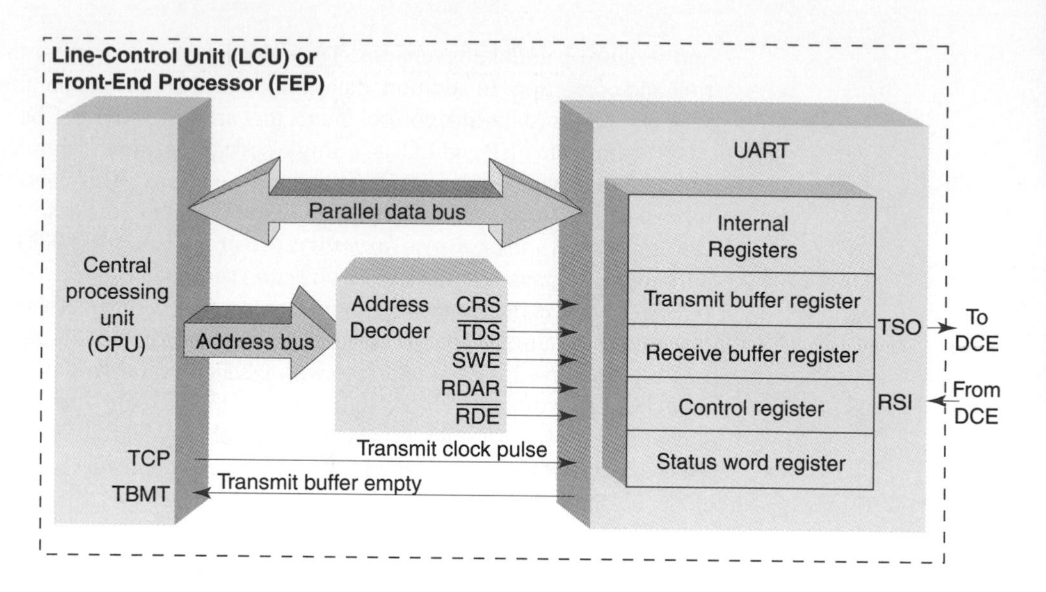

FIGURE 22-11 Line control unit UART interface

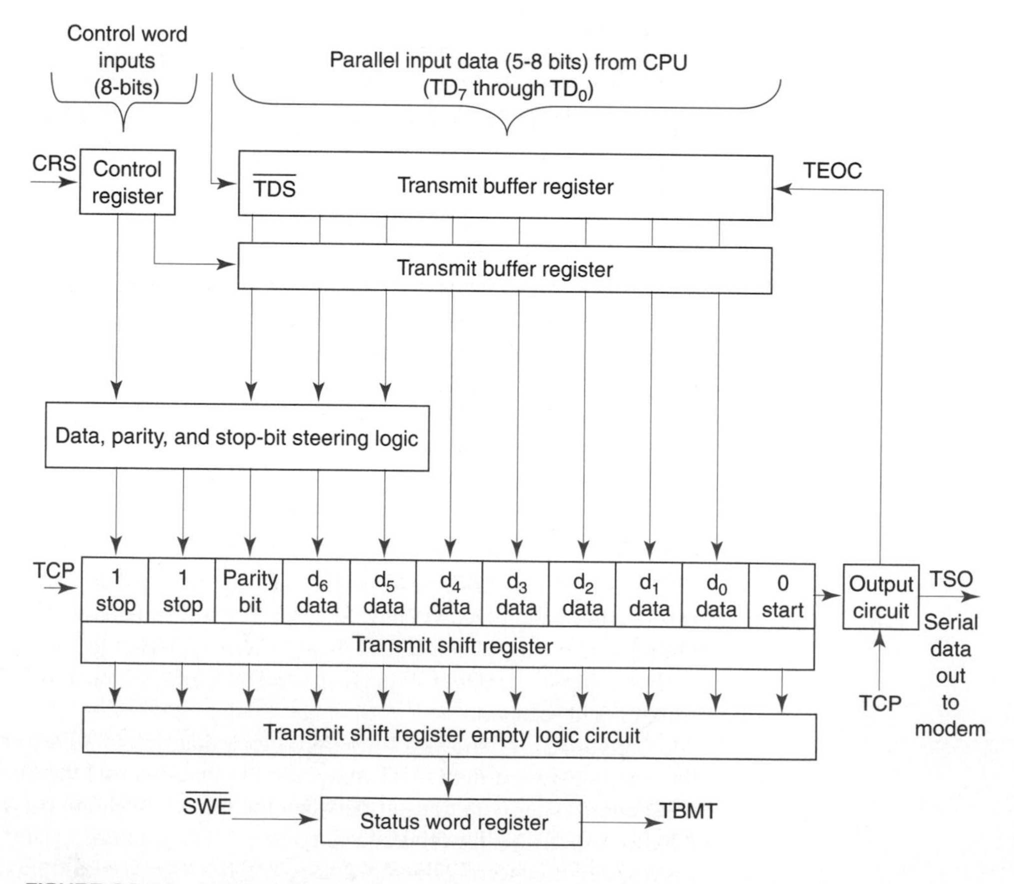

FIGURE 22-12 UART transmitter block diagram

Table 22-6 UART Control Register Inputs

D_7 and D_6

Number of stop bits

NSB1	NSB2	No. of Bits
0	0	Invalid
0	1	1
1	0	1.5
1	1	2

D_5 and D_4

NPB (parity or no parity)	
1	No parity bit (RPE disabled in receiver)
0	Insert parity bits in transmitter and check parity bits in receiver

POE (parity odd or even)	
1	Even parity
0	Odd parity

D_3 and D_2

Character length

NDB1	NDB2	Bits per Word
0	0	5
0	1	6
1	0	7
1	1	8

D_1 and D_0

Receive clock (baud rate factor)

RC1	RC2	Clock Rate
0	0	Synchronous mode
0	1	1X
1	0	16X
1	1	32X

A UART can be divided into two functional sections: the transmitter and the receiver. Figure 22-12 shows a simplified block diagram of a UART transmitter. Before transferring data in either direction, an eight-bit control word must be programmed into the UART control register to specify the nature of the data. The control word specifies the number of data bits per character; whether a parity bit is included with each character and, if so, whether it is odd or even parity; the number of stop bits inserted at the end of each character; and the receive clock frequency relative to the transmit clock frequency. Essentially, the start bit is the only bit in the UART that is not optional or programmable, as there is always one start bit, and it is always a logic 0. Table 22-6 shows the control-register coding format for a typical UART.

As specified in Table 22-6, the parity bit is optional and, if used, can be either odd or even. To select parity, NPB is cleared (logic 0), and to exclude the parity bit, NBP is set (logic 1). Odd parity is selected by clearing POE (logic 0), and even parity is selected by setting POE (logic 1). The number of stop bits is established with the NSB1 and NSB2 bits and can be one, one and a half, or two. The character length is determined by NDB1 and NDB2 and can be five, six, seven, or eight bits long. The maximum character length is 11 bits (i.e., one start bit, eight data bits, and two stop bits or one start bit, seven data bits, one parity bit, and two stop bits). Using a 22-bit character format with ASCII encoding is sometimes called *full ASCII.*

Figure 22-13 shows three of the character formats possible with a UART. Figure 22-13a shows an 11-bit data character comprised of one start bit, seven ASCII data bits, one odd-parity bit, and two stop bits (i.e., full ASCII). Figure 22-13b shows a nine-bit data character comprised of one start bit, seven ARQ data bits, and one stop bit, and Figure 22-13c shows another nine-bit data character comprised of one start bit, five Baudot data bits, one odd parity bit, and two stop bits.

A UART also contains a *status word register,* which is an *n*-bit data register that keeps track of the status of the UART's transmit and receive buffer registers. Typical status

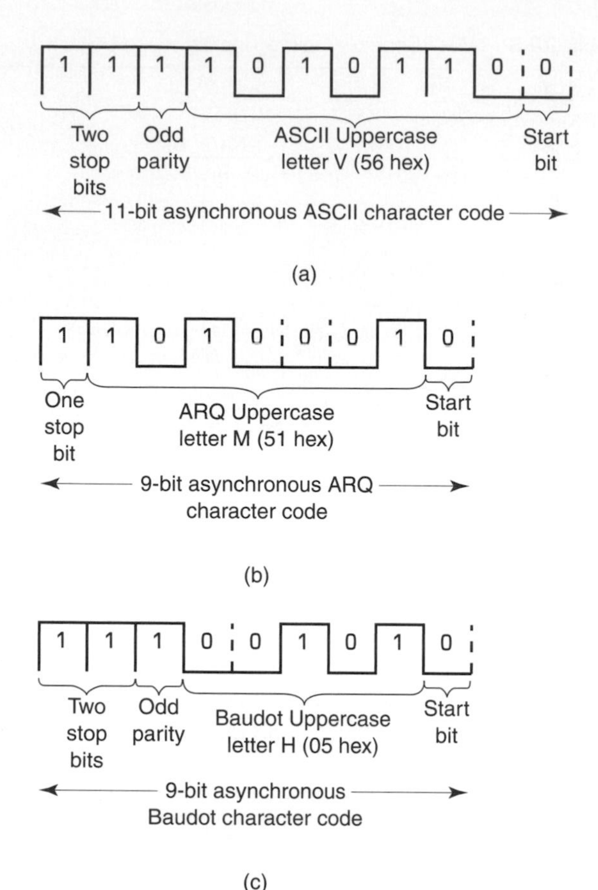

FIGURE 22-13 Asynchronous characters: (a) ASCII character; (b) ARQ character; (c) Baudot character

conditions compiled by the status word register for the UART transmitter include the following conditions:

TBMT: transmit buffer empty. Transmit shift register has completed transmission of a data character

RPE: receive parity error. Set when a received character has a parity error in it

RFE: receive framing error. Set when a character is received without any or with an improper number of stop bits

ROR: receiver overrun. Set when a character in the receive buffer register is written over by another receive character because the CPU failed to service an active condition on REA before the next character was received from the receive shift register

RDA: receive data available. A data character has been received and loaded into the receive data register

22-10-1-1 UART transmitter. The operation of the typical UART transmitter shown in Figure 22-12a is quite logical. However, before the UART can send or receive data, the UART control register must be loaded with the desired mode instruction word. This is accomplished by the CPU in the DTE, which applies the mode instruction word to the control word bus and then activates the control-register strobe (CRS).

Figure 22-14 shows the signaling sequence that occurs between the CPU and the UART transmitter. On receipt of an active *status word enable* ($\overline{\text{SWE}}$) signal, the UART sends a *transmit buffer empty* (TBMT) signal from the status word register to the CPU to indicate that the transmit buffer register is empty and the UART is ready to receive more

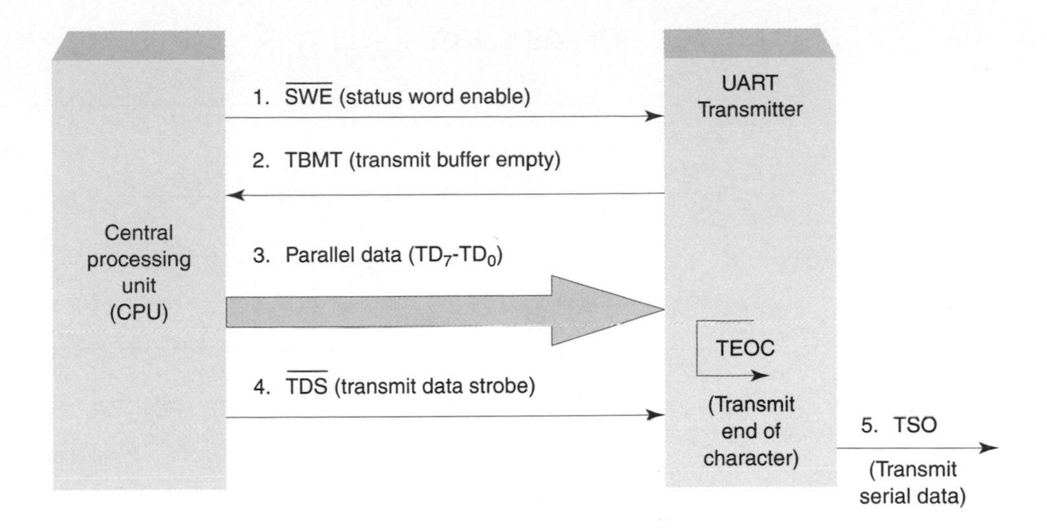

FIGURE 22-14 UART transmitter signal sequence

data. When the CPU senses an active condition of TBMT, it applies a parallel data character to the transmit data lines (TD_7 through TD_0) and strobes them into the transmit buffer register with an active signal on the *transmit data strobe* ($\overline{TDS}$) signal. The contents of the transmit buffer register are transferred to the transmit shift register when the *transmit end-of-character* (TEOC) signal goes active (the TEOC signal is internal to the UART and simply tells the transmit buffer register when the transmit shift register is empty and available to receive data). The data pass through the steering logic circuit, where it picks up the appropriate start, stop, and parity bits. After data have been loaded into the transmit shift register, they are serially outputted on the *transmit serial output* (TSO) pin at a bit rate equal to the transmit clock (TCP) frequency. While the data in the transmit shift register are serially clocked out of the UART, the CPU applies the next character to the input of the transmit buffer register. The process repeats until the CPU has transferred all its data.

22-10-1-2 UART receiver. A simplified block diagram for a UART receiver is shown in Figure 22-15. The number of stop bits and data bits and the parity bit parameters specified for the UART receiver must be the same as those of the UART transmitter. The UART receiver ignores the reception of idle line 1s. When a valid start bit is detected by the start bit verification circuit, the data character is clocked into the receive shift register. If parity is used, the parity bit is checked in the parity checker circuit. After one complete data character is loaded into the shift register, the character is transferred in parallel into the receive buffer register, and the *receive data available* (RDA) flag is set in the status word register. The CPU reads the status register by activating the *status word enable* ($\overline{SWE}$) signal and, if RDA is active, the CPU reads the character from the receive buffer register by placing an active signal on the receive data enable (RDE) pin. After reading the data, the CPU places an active signal on *the receive data available reset* ($\overline{RDAR}$) pin, which resets the RDA pin. Meanwhile, the next character is received and clocked into the receive shift register, and the process repeats until all the data have been received. Figure 22-16 shows the receive signaling sequence that occurs between the CPU and the UART.

22-10-1-3 Start-bit verification circuit. With asynchronous data transmission, precise timing is less important than following an agreed-on format or pattern for the data. Each transmitted data character must be preceded by a start bit and end with one or more

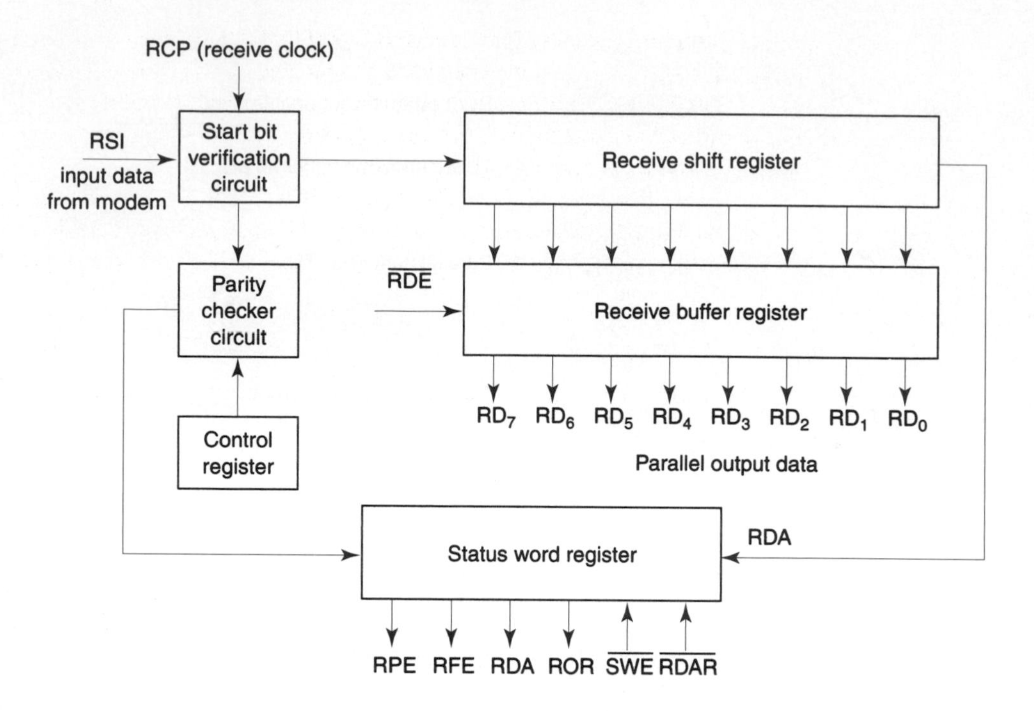

FIGURE 22-15 UART receiver block diagram

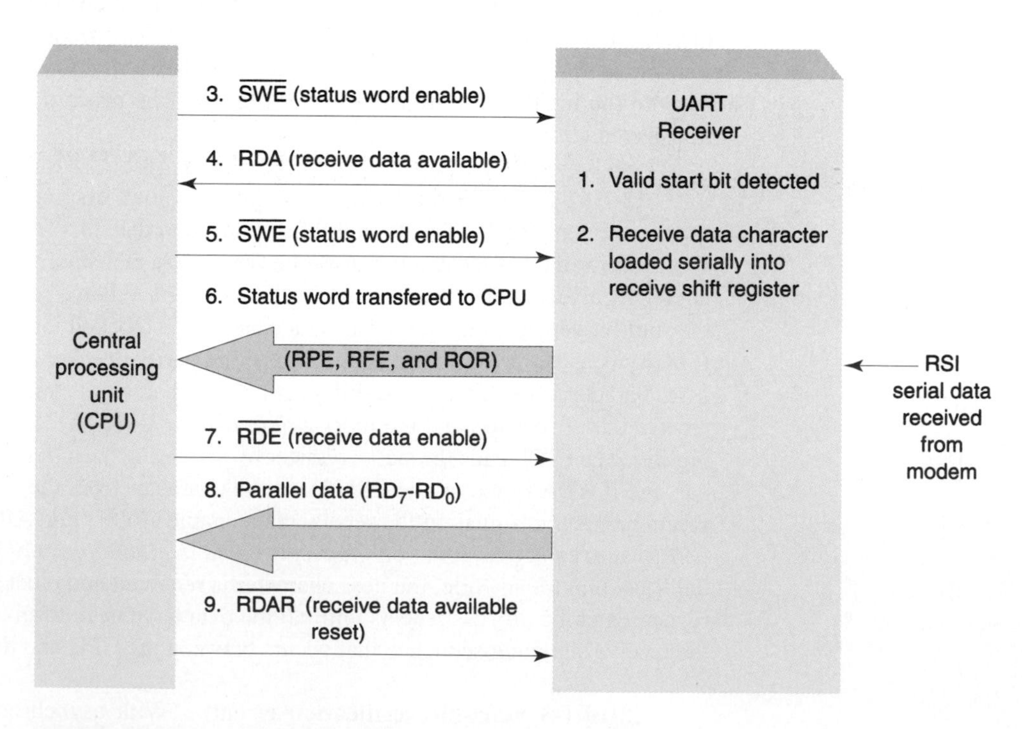

FIGURE 22-16 UART receive signal sequence

stop bits. Because data received by a UART have been transmitted from a distant UART whose clock is asynchronous to the receive UART, bit synchronization is achieved by establishing a timing reference at the center of each start bit. Therefore, it is imperative that a UART detect the occurrence of a valid start bit early in the bit cell and establish a timing reference before it begins to accept data.

The primary function of the start bit verification circuit is to detect valid start bits, which indicate the beginning of a data character. Figure 22-17a shows an example of how a noise hit can be misinterpreted as a start bit. The input data consist of a continuous string

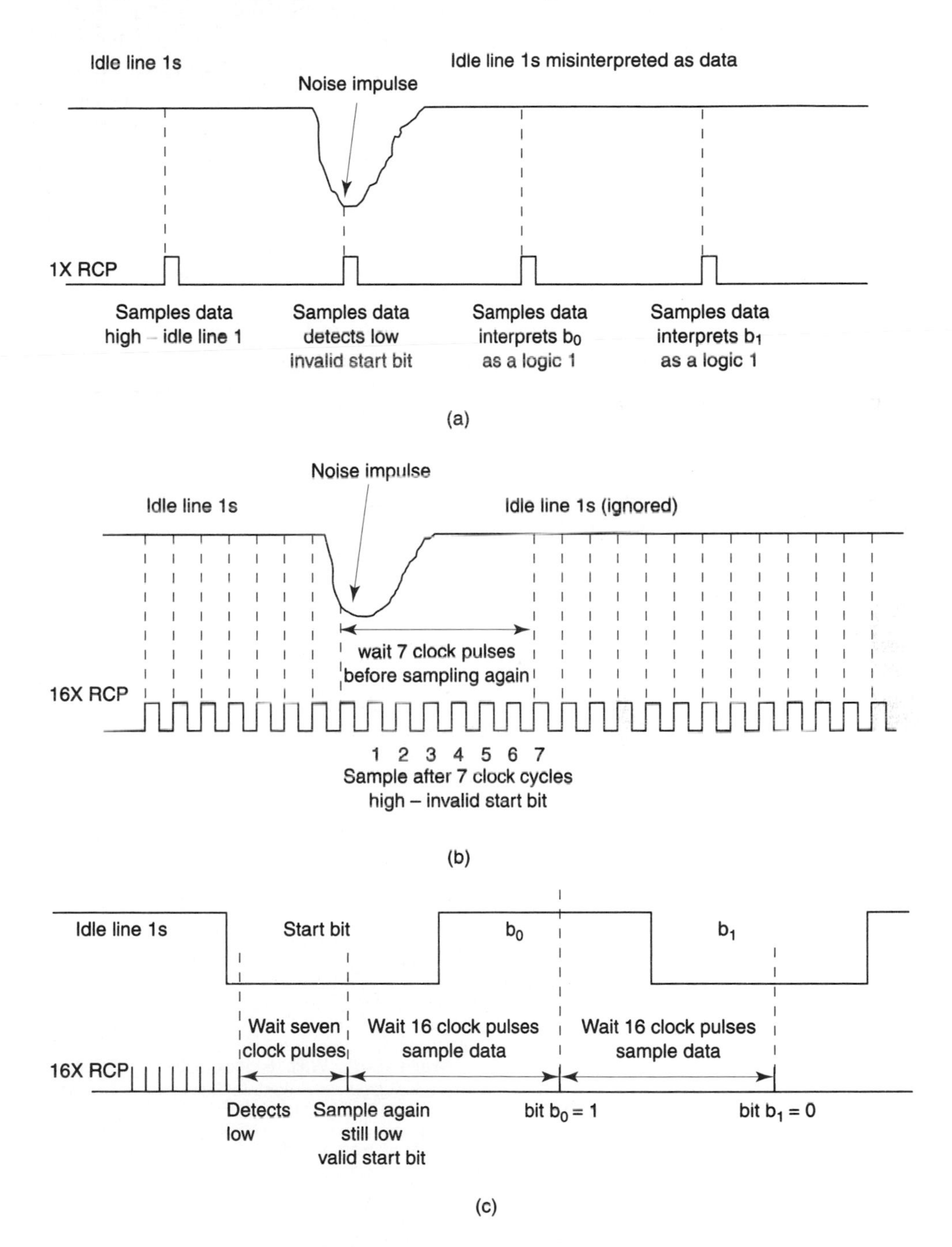

FIGURE 22-17 Start bit verification: (a) 1X RCP; (b) 16X RCP; (c) valid start bit

of idle line 1s, which are typically transmitted when there is no information. Idle line 1s are interpreted by a receiver as continuous stop bits (i.e., no data). If a noise impulse occurs that causes the receive data to go low at the same time the receiver clock is active, the receiver will interpret the noise impulse as a start bit. If this happens, the receiver will misinterpret the logic condition present during the next clock as the first data bit (b_0) and the following clock cycles as the remaining data bits (b_1, b_2, and so on). The likelihood of misinterpreting noise hits as start bits can be reduced substantially by clocking the UART receiver at a rate higher than the incoming data. Figure 22-17b shows the same situation as shown in Figure 22-17a, except the receive clock pulse (RCP) is 16 times (16×) higher than the receive serial data input (RSI). Once a low is detected, the UART waits seven clock cycles before resampling the input data. Waiting seven clock cycles places the next sample very near the center of the start bit. If the next sample detects a low, it assumes that a valid start bit has been detected. If the data have reverted to the high condition, it is assumed that the high-to-low transition was simply a noise pulse and, therefore, is ignored. Once a valid start bit has been detected and verified (Figure 22-17c), the start bit verification circuit samples the incoming data once every 16 clock cycles, which essentially makes the sample rate equal to the receive data rate (i.e., 16 RCP/16 = RCP). The UART continues sampling the data once every 16 clock cycles until the stop bits are detected, at which time the start bit verification circuit begins searching for another valid start bit. UARTs are generally programmed for receive clock rates of 16, 32, or 64 times the receive data rate (i.e., 16×, 32×, and 64×).

Another advantage of clocking a UART receiver at a rate higher than the actual receive data is to ensure that a high-to-low transition (valid start bit) is detected as soon as possible. This ensures that once the start bit is detected, subsequent samples will occur very near the center of each data bit. The difference in time between when a sample is taken (i.e., when a data bit is clocked into the receive shift register) and the actual center of a data bit is called the sampling error. Figure 22-18 shows a receive data stream sampled at a rate 16 times higher (16 RCP) than the actual data rate (RCP). As the figure shows, the start bit is not immediately detected. The difference in time between the beginning of a start bit and when it is detected is called the *detection error.* The maximum detection error is equal to the time of one receive clock cycle ($t_{cl} = 1/R_{cl}$). If the receive clock rate equaled the receive data rate, the maximum detection error would approach the time of one bit, which would mean that a start bit would not be detected until the very end of the bit time. Obviously, the higher the receive clock rate, the earlier a start bit would be detected.

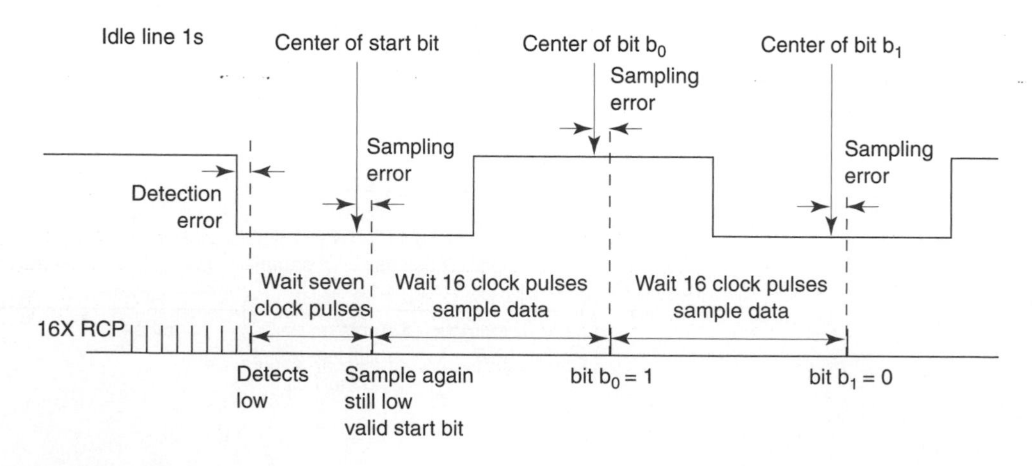

FIGURE 22-18 16X receive clock rate

Because of the detection error, successive samples occur slightly off from the center of the data bit. This would not present a problem with synchronous clocks, as the sampling error would remain constant from one sample to the next. However, with asynchronous clocks, the magnitude of the sampling error for each successive sample would increase (the clock would slip over or slip under the data), eventually causing a data bit to be either sampled twice or not sampled at all, depending on whether the receive clock is higher or lower than the transmit clock.

Figure 22-19 illustrates how sampling at a higher rate reduces the sampling error. Figures 22-19a and b show data sampled at a rate eight times the data rate (8×) and 16 times the data rate (16×), respectively. It can be seen that increasing the sample rate moves the sample time closer to the center of the data bit, thus decreasing the sampling error.

Placing stop bits at the end of each data character also helps reduce the *clock slippage* (sometimes called *clock skew*) problem inherent when using asynchronous transmit and receive clocks. Start and stop bits force a high-to-low transition at the beginning of each character, which essentially allows the receiver to resynchronize to the start bit at the beginning of each data character. It should probably be mentioned that with UARTs the data rates do not have to be the same in each direction of propagation (e.g., you could transmit data at 1200 bps and receive at 600 bps). However, the rate at which data leave a transmitter must be the same as the rate at which data enter the receiver at the other end of the circuit. If you transmit at 1200 bps, it must be received at the other end at 1200 bps.

22-10-2 Universal Synchronous Receiver/Transmitter

A *universal synchronous receiver/transmitter* (USRT) is used for synchronous transmission of data between a DTE and a DCE. Synchronous data transmission means that a synchronous data format is used, and clocking information is generally transferred between the DTE and the DCE. A USRT performs the same basic functions as a UART, except for

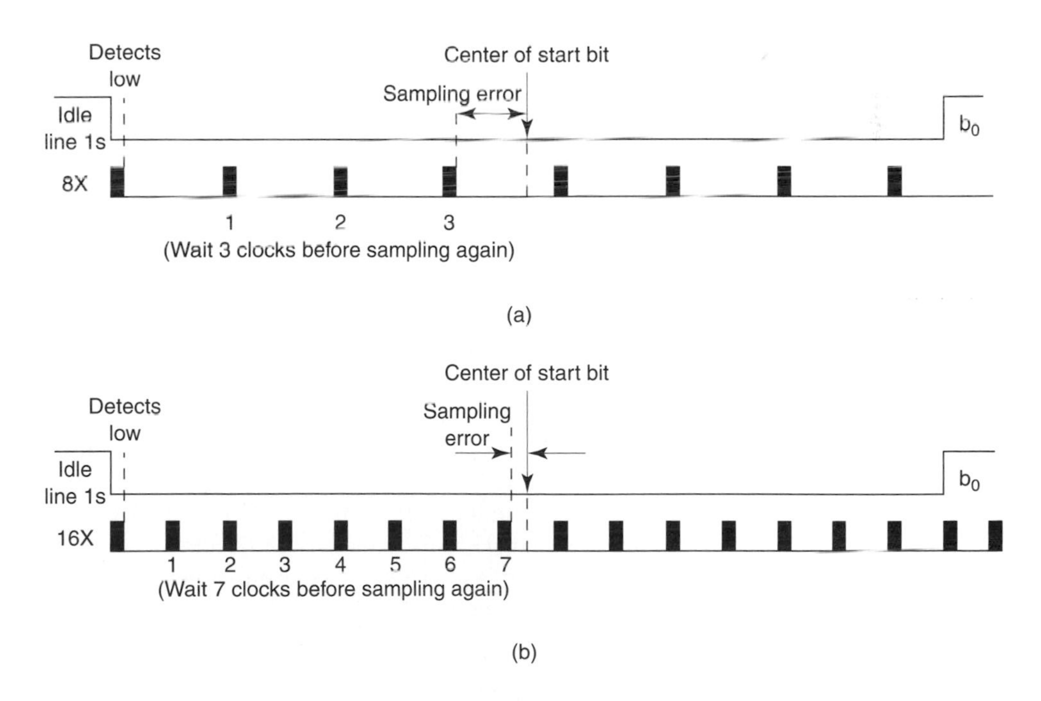

FIGURE 22-19 Sampling error: (a) 8X RCP; (b) 16X RCP

synchronous data (i.e., the start and stop bits are omitted and replaced by unique synchronizing characters). The primary functions performed by a USRT are the following:

1. Serial-to-parallel and parallel-to-serial data conversions
2. Error detection by inserting parity bits in the transmitter and checking parity bits in the receiver.
3. Insert and detect unique data synchronization (SYN) characters
4. Formatting data in the transmitter and receiver (i.e., combining items 1 through 3 in a meaningful sequence)
5. Provide transmit and receive status information to the CPU
6. Voltage-level conversion between the DTE and the serial interface and vice versa
7. Provide a means of achieving bit and character synchronization

22-11 SERIAL INTERFACES

To ensure an orderly flow of data between a DTE and a DCE, a standard serial interface is used to interconnect them. The serial interface coordinates the flow of data, control signals, and timing information between the DTE and the DCE.

Before serial interfaces were standardized, every company that manufactured data communications equipment used a different interface configuration. More specifically, the cable arrangement between the DTE and the DCE, the type and size of the connectors, and the voltage levels varied considerably from vender to vender. To interconnect equipment manufactured by different companies, special level converters, cables, and connectors had to be designed, constructed, and implemented for each application. A serial interface standard should provide the following:

1. A specific range of voltages for transmit and receive signal levels
2. Limitations for the electrical parameters of the transmission line, including source and load impedance, cable capacitance, and other electrical characteristics outlined later in this chapter
3. Standard cable and cable connectors
4. Functional description of each signal on the interface

In 1962, the Electronics Industries Association (EIA), in an effort to standardize interface equipment between data terminal equipment and data communications equipment, agreed on a set of standards called the *RS-232 specifications* (RS meaning "recommended standard"). The official name of the RS-232 interface is *Interface Between Data Terminal Equipment and Data Communications Equipment Employing Serial Binary Data Interchange.* In 1969, the third revision, RS-232C, was published and remained the industrial standard until 1987, when the RS-232D was introduced, which was followed by the RS-232E in the early 1990s. The RS-232D standard is sometimes referred to as the EIA-232 standard. Versions D and E of the RS-232 standard changed some of the pin designations. For example, data set ready was changed to DCE ready, and data terminal ready was changed to DTE ready.

The RS-232 specifications identify the mechanical, electrical, functional, and procedural descriptions for the interface between DTEs and DCEs. The RS-232 interface is similar to the combined ITU-T standards V.28 (electrical specifications) and V.24 (functional description) and is designed for serial transmission up to 20 kbps over a maximum distance of 50 feet (approximately 15 meters).

22-11-1 RS-232 Serial Interface Standard

The mechanical specification for the RS-232 interface specifies a cable with two connectors. The standard RS-232 cable is a sheath containing 25 wires with a DB25P-compatible

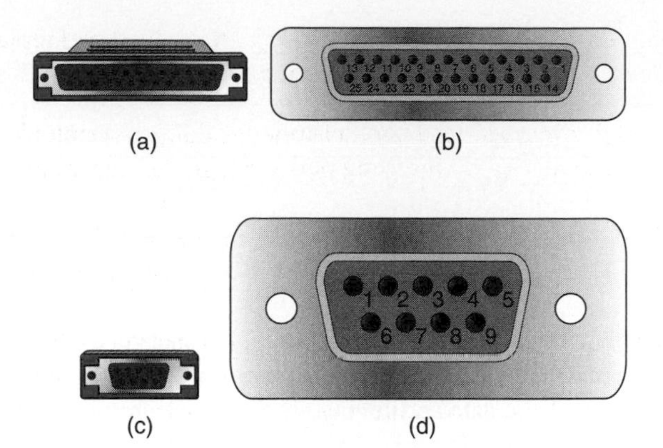

FIGURE 22-20 RS-232 serial interface connector: (a) DB25P; (b) DB25S; (c) DB9P; (d) DB9S

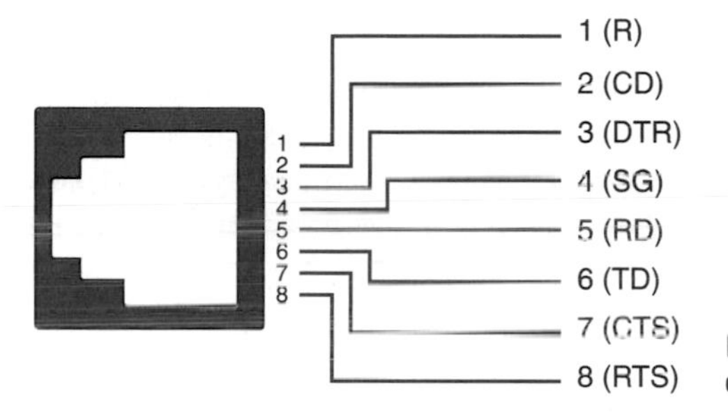

FIGURE 22-21 EIA-561 modular connector

male connector (plug) on one end and a DB25S-compatible female connector (receptacle) on the other end. The DB25P-compatible and DB25S-compatible connectors are shown in Figures 22-20a and b, respectively. The cable must have a plug on one end that connects to the DTE and a receptacle on the other end that connects to the DCE. There is also a special PC nine-pin version of the RS-232 interface cable with a DB9P-compatible male connector on one end and a DB9S-compatible connector at the other end. The DB9P-compatible and DB9S-compatible connectors are shown in Figures 22-20c and d, respectively (note that there is no correlation between the pin assignments for the two connectors). The nine-pin version of the RS-232 interface is designed for transporting asynchronous data between a DTE and a DCE or between two DTEs, whereas the 25-pin version is designed for transporting either synchronous or asynchronous data between a DTE and a DCE. Figure 22-21 shows the eight-pin EIA-561 modular connector, which is used for transporting asynchronous data between a DTE and a DCE when the DCE is connected directly to a standard two-wire telephone line attached to the public switched telephone network. The EIA-561 modular connector is designed exclusively for dial-up telephone connections.

Although the RS-232 interface is simply a cable and two connectors, the standard also specifies limitations on the voltage levels that the DTE and DCE can output onto or receive from the cable. The DTE and DCE must provide circuits that convert their internal logic levels to RS-232-compatible values. For example, a DTE using TTL logic interfaced to a DCE using CMOS logic is not compatible. *Voltage-leveling circuits* convert the internal voltage levels from the DTE and DCE to RS-232 values. If both the DCE and the DTE output and accept RS-232 levels, they are electrically compatible regardless of which logic family they use internally. A voltage leveler is called a *driver* if it outputs signals onto the cable and a

Table 22-7 RS-232 Voltage Specifications

	Data Signals		Control Signals	
	Logic 1	Logic 0	Enable (On)	Disable (Off)
Driver (output)	−5 V to −15 V	+5 V to +15 V	+5 V to +15 V	−5 V to −15 V
Terminator (input)	−3 V to −25 V	+3 V to +25 V	+3 V to +25 V	−3 V to −25 V

terminator if it accepts signals from the cable. In essence, a driver is a transmitter, and a terminator is a receiver. Table 22-7 lists the voltage limits for RS-232-compatible drivers and terminators. Note that the data and control lines use *non–return to zero, level* (NRZ-L) bipolar encoding. However, the data lines use negative logic, while the control lines use positive logic.

From examining Table 22-7, it can be seen that the voltage limits for a driver are more inclusive than the voltage limits for a terminator. The output voltage range for a driver is between +5 V and +15 V or between −5 V and −15 V, depending on the logic level. However, the voltage range in which a terminator will accept is between +3 V and +25 V or between −3 V and −25 V. Voltages between ±3 V are undefined and may be interpreted by a terminator as a high or a low. The difference in the voltage levels between the driver output and the terminator input is called *noise margin* (NM). The noise margin reduces the susceptibility to interface caused by noise transients induced into the cable. Figure 22-22a shows the relationship between the driver and terminator voltage ranges. As shown in Figure 22-22a, the noise margin for the minimum driver output voltage is 2 V (5 − 3), and the noise margin for the maximum driver output voltage is 10 V (25 – 15). (The minimum noise margin of 2 V is called the *implied noise margin.*) Noise margins will vary, of course, depending on what specific voltages are used for highs and lows. When the noise margin of a circuit is a high value, it is said to have *high noise immunity,* and when the noise margin is a low value, it has *low noise immunity.* Typical RS-232 voltage levels are +10 V for a high and −10 V for a low, which produces a noise margin of 7 V in one direction and 15 V in the other direction. The noise margin is generally stated as the minimum value. This relationship is shown in Figure 22-22b. Figure 22-22c illustrates the immunity of the RS-232 interface to noise signals for logic levels of +10 V and −10 V.

The RS-232 interface specifies single-end (unbalanced) operation with a common ground between the DTE and DCE. A common ground is reasonable when a short cable is used. However, with longer cables and when the DTE and DCE are powered from different electrical buses, this may not be true.

Example 22-8

Determine the noise margins for an RS-232 interface with driver signal voltages of ±6 V.

Solution The noise margin is the difference between the driver signal voltage and the terminator receive voltage, or

$$NM = 6 - 3 = 3\text{ V or } NM = 25 - 6 = 19\text{ V}$$

The minimum noise margin is 3 V.

22-11-1-1 RS-232 electrical equivalent circuit. Figure 22-23 shows the equivalent electrical circuit for the RS-232 interface, including the driver and terminator. With these electrical specifications and for a bit rate of 20 kbps, the nominal maximum length of the RS-232 interface cable is approximately 50 feet.

22-11-1-2 RS-232 functional description. The pins on the RS-232 interface cable are functionally categorized as either ground (signal and chassis), data (transmit and re-

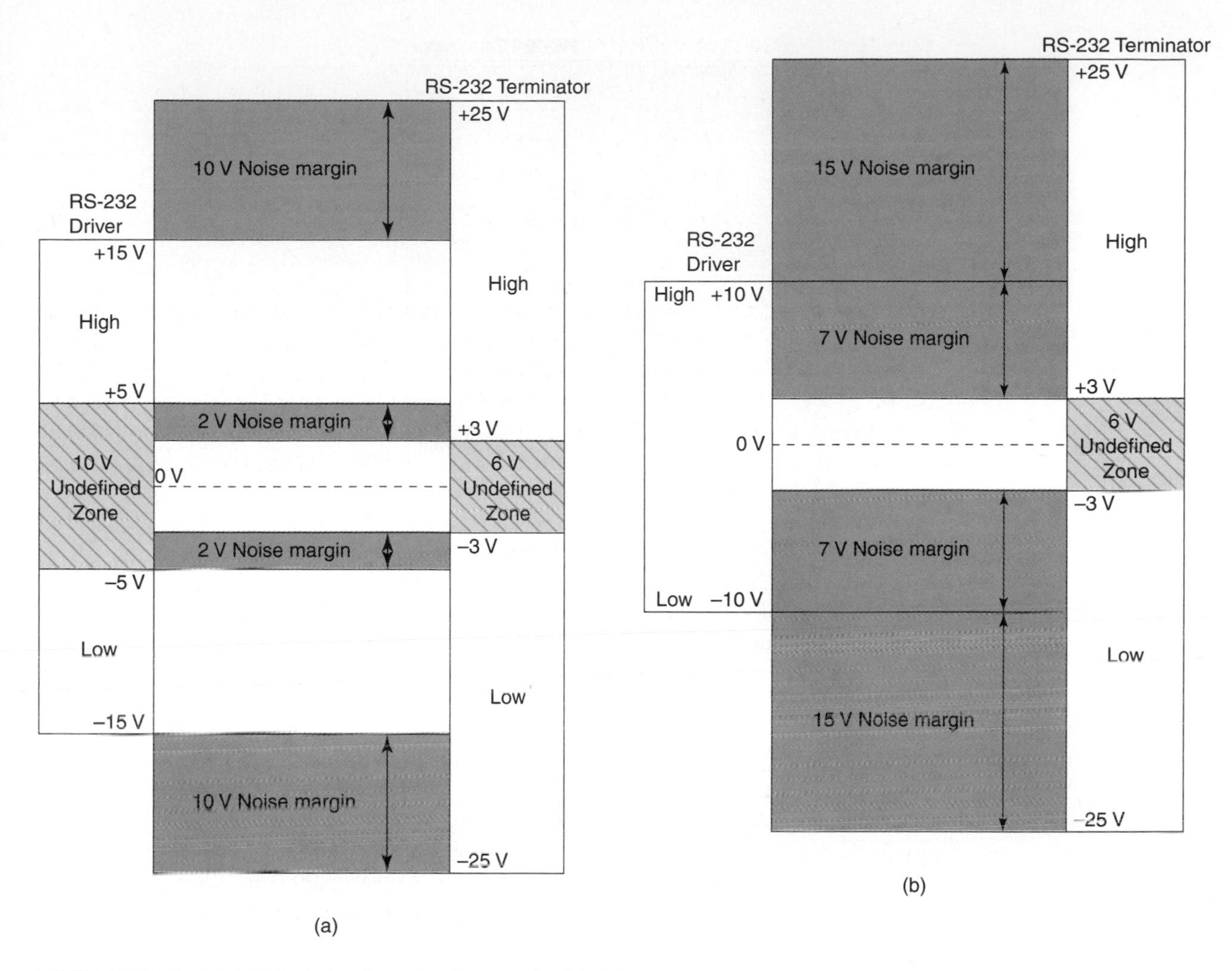

FIGURE 22-22 RS-232 logic levels and noise margin: (a) driver and terminator voltage ranges; (b) noise margin with a +10 V high and −10 V low (*Continued*)

ceive), control (handshaking and diagnostic), or timing (clocking signals). Although the RS-232 interface as a unit is bidirectional (signals propagate in both directions), each individual wire or pin is unidirectional. That is, signals on any given wire are propagated either from the DTE to the DCE or from the DCE to the DTE but never in both directions. Table 22-8 lists the 25 pins (wires) of the RS-232 interface and gives the direction of signal propagation (i.e., either from the DTE toward the DCE or from the DCE toward the DTE). The RS-232 specification designates the first letter of each pin with the letters A, B, C, D, or S. The letter categorizes the signal into one of five groups, each representing a different type of circuit. The five groups are as follows:

A—ground
B—data
C—control
D—timing (clocking)
S—secondary channel

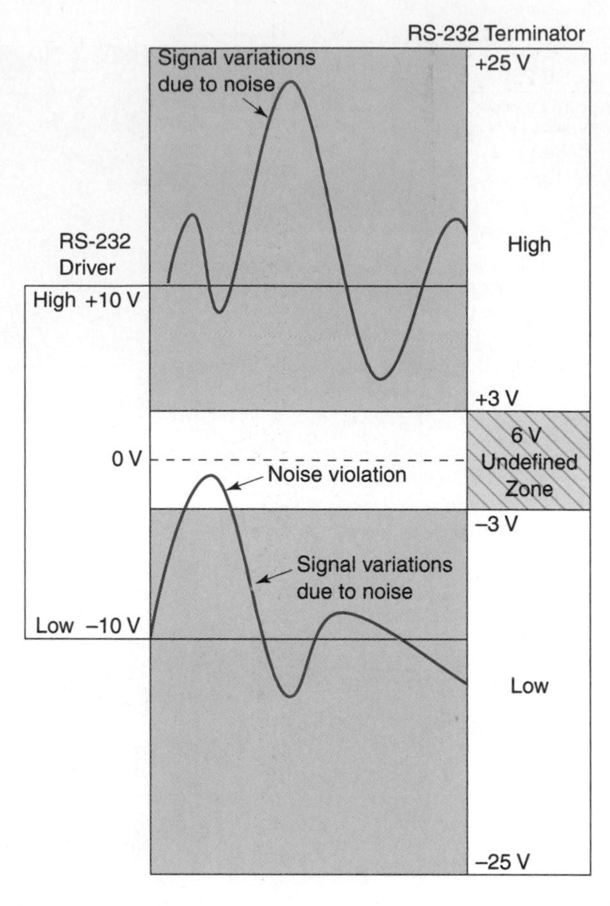

FIGURE 22-22 (*Continued*)
(c) noise violation

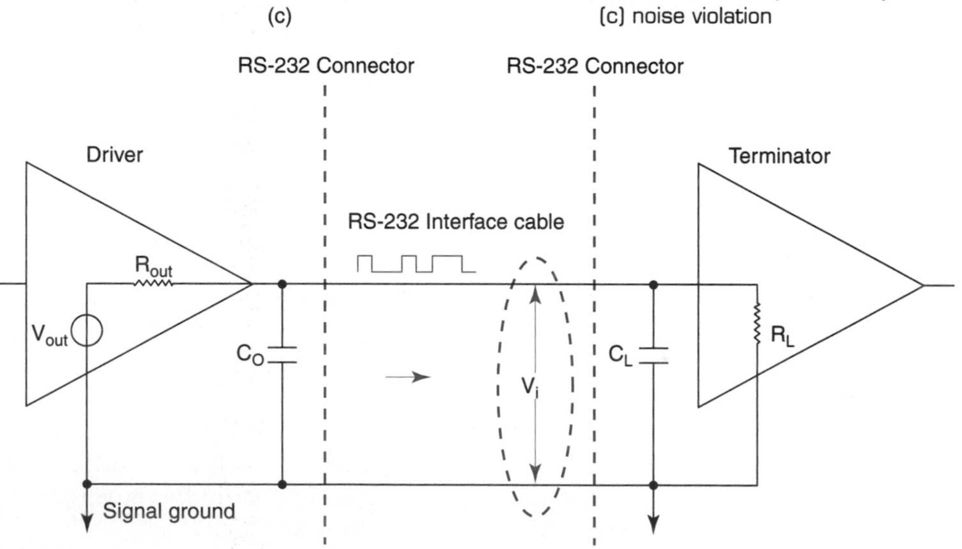

V_{out} — open-circuit voltage at the output of a driver (±5 V to ±15 V)
V_i — terminated voltage at the input to a terminator (±3 V to ±25 V)
C_L — load capacitance associated with the terminator, including the cable (2500 pF maximum)
C_O — capacitance seen by the driver including the cable (2500 pF maximum)
R_L — terminator input resistance (3000 Ω to 7000 Ω)
R_{out} — driver output resistance (300 Ω maximum)

FIGURE 22-23 RS-232 Electrical specifications

Table 22-8 EIA RS-232 Pin Designations and Direction of Propagation

Pin Number	Pin Name	Direction of Propagation
1	Protective ground (frame ground or chassis ground)	None
2	Transmit data (send data)	DTE to DCE
3	Receive data	DCE to DTE
4	Request to send	DTE to DCE
5	Clear to send	DCE to DTE
6	Data set ready (modem ready)	DCE to DTE
7	Signal ground (reference ground)	None
8	Receive line signal detect (carrier detect or data carrier detect)	DCE to DTE
9	Unassigned	None
10	Unassigned	None
11	Unassigned	None
12	Secondary receive line signal detect (secondary carrier detect or secondary data carrier detect)	DCE to DTE
13	Secondary clear to send	DCE to DTE
14	Secondary transmit data (secondary send data)	DTE to DCE
15	Transmit signal element timing—DCE (serial clock transmit—DCE)	DCE to DTE
16	Secondary receive data	DCE to DTE
17	Receive signal element timing (serial clock receive)	DCE to DTE
18	Unassigned	None
19	Secondary request to send	DTE to DCE
20	Data terminal ready	DTE to DCE
21	Signal quality detect	DCE to DTE
22	Ring indicator	DCE to DTE
23	Data signal rate selector	DTE to DCE
24	Transmit signal element timing—DTE (serial clock transmit—DTE)	DTE to DCE
25	Unassigned	None

Because the letters are nondescriptive designations, it is more practical and useful to use acronyms to designate the pins that reflect the functions of the pins. Table 22-9 lists the EIA signal designations plus the nomenclature more commonly used by industry in the United States to designate the pins.

Twenty of the 25 pins on the RS-232 interface are designated for specific purposes or functions. Pins 9, 10, 11, 18, and 25 are unassigned (unassigned does not necessarily imply unused). Pins 1 and 7 are grounds; pins 2, 3, 14, and 16 are data pins; pins 15, 17, and 24 are timing pins; and all the other pins are used for control or handshaking signals. Pins 1 through 8 are used with both asynchronous and synchronous modems. Pins 15, 17, and 24 are used only with synchronous modems. Pins 12, 13, 14, 16, and 19 are used only when the DCE is equipped with a secondary data channel. Pins 20 and 22 are used exclusively when interfacing a DTE to a modem that is connected to standard dial-up telephone circuits on the public switched telephone network.

There are two full-duplex data channels available with the RS-232 interface; one channel is for *primary data* (actual information), and the second channel is for *secondary data* (diagnostic information and handshaking signals). The secondary channel is sometimes used as a reverse or backward channel, allowing the receive DCE to communicate with the transmit DCE while data are being transmitted on the primary data channel.

Table 22-9 EIA RS-232 Pin Designations and Designations

Pin Number	Pin Name	EIA Nomenclature	Common U.S. Acronyms
1	Protective ground (frame ground or chassis ground)	AA	GWG, FG, or CG
2	Transmit data (send data)	BA	TD, SD, TxD
3	Receive data	BB	RD, RxD
4	Request to send	CA	RS, RTS
5	Clear to send	CB	CS, CTS
6	Data set ready (modem ready)	CC	DSR, MR
7	Signal ground (reference ground)	AB	SG, GND
8	Receive line signal detect (carrier detect or data carrier detect)	CF	RLSD, CD, DCD
9	Unassigned	—	—
10	Unassigned	—	—
11	Unassigned	—	—
12	Secondary receive line signal detect (secondary carrier detect or secondary data carrier detect)	SCF	SRLSD, SCD, SDCD
13	Secondary clear to send	SCB	SCS, SCTS
14	Secondary transmit data (secondary send data)	SBA	STD, SSD, STxD
15	Transmit signal element timing—DCE (serial clock transmit—DCE)	DB	TSET, SCT-DCE
16	Secondary receive data	SBB	SRD, SRxD
17	Receive signal element timing (serial clock receive)	DD	RSET, SCR
18	Unassigned	—	—
19	Secondary request to send	SCA	SRS, SRTS
20	Data terminal ready	CD	DTR
21	Signal quality detect	CG	SQD
22	Ring indicator	CE	RI
23	Data signal rate selector	CH	DSRS
24	Transmit signal element timing—DTE (serial clock transmit—DTE)	DA	TSET, SCT-DTE
25	Unassigned	—	—

The functions of the 25 RS-232 pins are summarized here for a DTE interfacing with a DCE where the DCE is a data communications modem:

Pin 1—protective ground, frame ground, or *chassis ground* (GWG, FG, or CG). Pin 1 is connected to the chassis and used for protection against accidental electrical shock. Pin 1 is generally connected to signal ground (pin 7).

Pin 2—transmit data or *send data* (TD, SD, or TxD). Serial data on the primary data channel are transported from the DTE to the DCE on pin 2. Primary data are the actual source information transported over the interface. The transmit data line is a transmit line for the DTE but a receive line for the DCE. The DTE may hold the TD line at a logic 1 voltage level when no data are being transmitted and between characters when asynchronous data are being transmitted. Otherwise, the TD driver is enabled by an active condition on pin 5 (clear to send).

Pin 3—receive data (RD or RxD). Pin 3 is the second primary data pin. Serial data are transported from the DCE to the DTE on pin 3. Pin 3 is the receive data pin for the DTE and the transmit data pin for the DCE. The DCE may hold the TD line at a logic 1 voltage level when no data are being transmitted or when pin 8 (RLSD) is inactive. Otherwise, the RD driver is enabled by an active condition on pin 8.

Pin 4—request to send (RS or RTS). For half-duplex data transmission, the DTE uses pin 4 to request permission from the DCE to transmit data on the primary data channel. When the DCE is a modem, an active condition on RTS turns on the modem's analog carrier. The RTS and CTS signals are used together to coordinate half-duplex data transmission between the DTE and DCE. For full-duplex data transmission, RTS can be held active permanently. The RTS driver is enabled by an active condition on pin 6 (data set ready).

Pin 5—clear to send (CS or CTS). The CTS signal is a handshake from the DCE to the DTE (i.e., modem to LCU) in response to an active condition on RTS. An active condition on CTS enables the TD driver in the DTE. There is a predetermined time delay between when the DCE receives an active condition on the RTS signal and when the DCE responds with an active condition on the CTS signal.

Pin 6—data set ready or *modem ready* (DSR or MR). DSR is a signal sent from the DCE to the DTE to indicate the availability of the communications channel. DSR is active only when the DCE and the communications channel are available. Under normal operation, the modem and the communications channel are always available. However, there are five situations when the modem or the communications channel are not available:

1. The modem is shut off (i.e., has no power).
2. The modem is disconnected from the communications line so that the line can be used for normal telephone voice traffic (i.e., in the voice rather than the data mode).
3. The modem is in one of the self-test modes (i.e., analog or digital loopback).
4. The telephone company is testing the communications channel.
5. On dial-up circuits, DSR is held inactive while the telephone switching system is establishing a call and when the modem is transmitting a specific response (answer) signal to the calling station's modem.

An active condition on the DSR lead enables the request to send driver in the DTE, thus giving the DSR lead the highest priority of the RS 232 control leads.

Pin 7—signal ground or *reference ground* (SG or GND). Pin 7 is the signal reference (return line) for all data, control, and timing signals (i.e., all pins except pin 1, chassis ground).

Pin 8—receive line signal detect, carrier detect, or *data carrier detect* (RLSD, CD, or DCD). The DCE uses this pin to signal the DTE when it determines that it is receiving a valid analog carrier (data carrier). An active RLSD signal enables the RD terminator in the DTE, allowing it to accept data from the DCE. An inactive RLSD signal disables the terminator for the DTE's receive data pin, preventing it from accepting invalid data. On half-duplex data circuits, RLSD is held inactive whenever RTS is active.

Pin 9—unassigned. Pin 9 is non–EIA specified; however, it is often held at +12 Vdc for test purposes (+P).

Pin 10—unassigned. Pin 10 is non–EIA specified; however, it is often held at −12 Vdc for test purposes (−P).

Pin 11—unassigned. Pin 11 is non–EIA specified; however, it is often designated as *equalizer mode* (EM) and used by the modem to signal the DTE when the modem is self-adjusting its internal equalizers because error performance is suspected to be poor. When the carrier detect signal is active and the circuit is inactive, the modem is retraining (resynchronizing), and the probability of error is high. When the receive line signal detect (pin 8) is active and EM is inactive, the modem is trained, and the probability of error is low.

Pin 12—secondary receive line signal detect, secondary carrier detect, or *secondary data carrier detect* (SRLSD, SCD, or SDCD). Pin 12 is the same as RLSD (pin 8), except for the secondary data channel.

Pin 13—secondary clear to send. The SCTS signal is sent from DCE to the DTE as a response (handshake) to the secondary request to send signal (pin 19).

Pin 14—secondary transmit data or *secondary send data* (STD, STD, or STxD). Diagnostic data are transmitted from the DTE to the DCE on this pin. STD is enabled by an active condition on SCTS.

Pin 15—transmission signal element timing or (*serial clock transmit*) *DCE* (TSET, SCT-DCE). With synchronous modems, the transmit clocking signal is sent from the DCE to the DTE on this pin.

Pin 16—secondary received data (SRD or SRxD). Diagnostic data are transmitted from the DCE to the DTE on this pin. The SRD driver is enabled by an active condition on secondary receive line signal detect (SRLSD).

Pin 17—receiver signal element timing or *serial clock receive* (RSET or SCR). When synchronous modems are used, clocking information recovered by the DCE is sent to the DTE on this pin. The receive clock is used to clock data out of the DCE and into the DTE on the receive data line. The clock frequency is equal to the bit rate on the primary data channel.

Pin 18—unassigned. Pin 11 is non–EIA specified; however, it is often used for the *local loopback* (LL) signal. Local loopback is a control signal sent from the DTE to the DCE placing the DCE (modem) into an analog loopback condition. Analog and digital loopbacks are described in a later section of this chapter.

Pin 19—secondary request to send (SRS or SRTS). SRTS is used by the DTE to bid for the secondary data channel from the DCE. SRTS and SCTS coordinate the flow of data on the secondary data channel.

Pin 20—data terminal ready (DTR). The DTE sends signals to the DCE on the DTR line concerning the availability of the data terminal equipment. DTR is used primarily with dial-up circuits to handshake with ring indicator (pin 22). The DTE disables DTR when it is unavailable, thus instructing the DCE not to answer an incoming call.

Pin 21—signal quality detector (SQD). The DCE sends signals to the DTE on this line indicating the quality of the received analog carrier. An inactive (low) signal on SQD tells the DTE that the incoming signal is marginal and that there is a high likelihood that errors are occurring.

Pin 22—ring indicator (RI). The RI line is used primarily on dial-up data circuits for the DCE to inform the DTE that there is an incoming call. If the DTE is ready to receive data, it responds to an active condition on RI with an active condition on DTR. DTR is a handshaking signal in response to an active condition on RI.

Pin 23—data signal rate selector (DSRS). The DTE used this line to select one of two transmission bit rates when the DCE is equipped to offer two rates. (The data rate selector line can be used to change the transmit clock frequency.)

Pin 24—transmit signal element timing or *serial clock transmit–DTE* (TSET, SCT-DTE). When synchronous modems are used, the transmit clocking signal is sent from the DTE to the DCE on this pin. Pin 24 is used only when the master clock is located in the DTE.

Pin 25—unassigned. Pin 5 is non–EIA specified; however, it is sometimes used as a control signal from the DCE to the DTE to indicate that the DCE is in either the remote or local loopback mode.

For asynchronous transmission using either the DB9P/S-modular connector, only the following nine pins are provided:

1. Receive line signal detect
2. Receive data

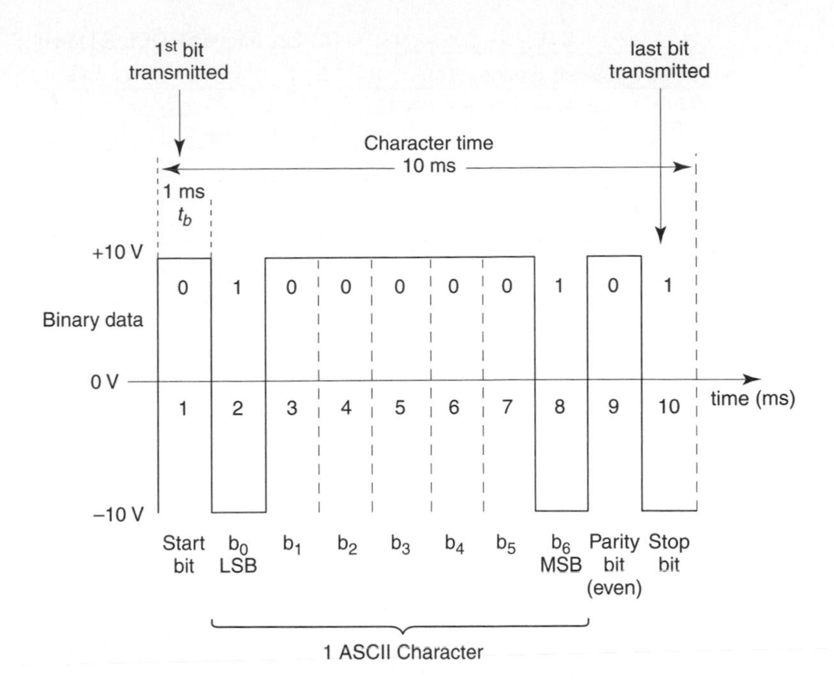

FIGURE 22-24 RS-232 data timing diagram—ASCII upper case letter A, 1 start bit, even parity, and one stop bit

3. Transmit data
4. Data terminal ready
5. Signal ground
6. Data set ready
7. Request to send
8. Clear to send
9. Ring indicator

22-11-1-3 RS-232 signals. Figure 22-24 shows the timing diagram for the transmission of one asynchronous data character over the RS-232 interface. The character is comprised of one start bit, one stop bit, seven ASCII character bits, and one even-parity bit. The transmission rate is 1000 bps, and the voltage level for a logic 1 is −10 V and for a logic 0 is +10 V. The time of one bit is 1 ms; therefore, the total time to transmit one ASCII character is 10 ms.

22-11-1-4 RS-232. Asynchronous Data Transmission. Figures 22-25a and b show the functional block diagram for the drivers and terminators necessary for transmission of asynchronous data over the RS-232 interface between a DTE and a DCE that is a modem. As shown in the figure, only the first eight pins of the interface are required, which includes the following signals: signal ground and chassis ground, transmit data and receive data, request to send, clear to send, data set ready, and receive line signal detect.

Figure 22-26a shows the transmitter timing diagram for control and data signals for a typical asynchronous data transmission over an RS-232 interface with the following parameters:

Modem RTS-CTS delay = 50 ms

DTE primary data message length = 100 ms

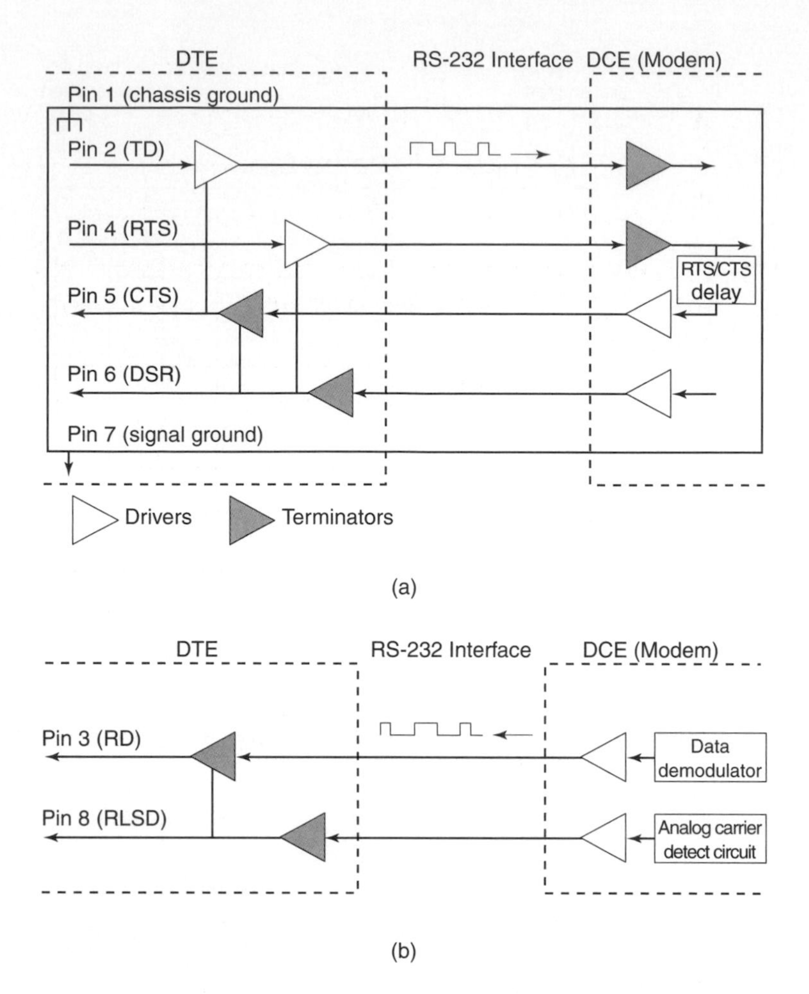

FIGURE 22-25 Functional block diagram for the drivers and terminators necessary for transmission of asynchronous data over the RS-232 interface between a DTE and a DCE (modem): (a) transmit circuits; (b) receive circuits

Modem training sequence = 50 ms
Propagation time = 10 ms
Modem RLSD turn-off delay time = 5 ms

When the DTE wishes to transmit data on the primary channel, it enables request to send ($t = 0$). After a predetermined RTS/CTS time delay time, which is determined by the modem (50 ms for this example), CTS goes active. During the 50-ms RTS/CTS delay, the modem outputs an analog carrier that is modulated by a unique bit pattern called a *training sequence.* The training sequence for asynchronous modems is generally nothing more than a series of logic 1s that produce 50 ms of continuous mark frequency. The analog carrier is used to initialize the communications channel and the distant receive modem (with synchronous modems, the training sequence is more involved, as it would also synchronize the carrier and clock recovery circuits in the distant modem). After the RTS/CTS delay, the transmit data (TD) line is enabled, and the DTE begins transmitting user data. When the transmission is

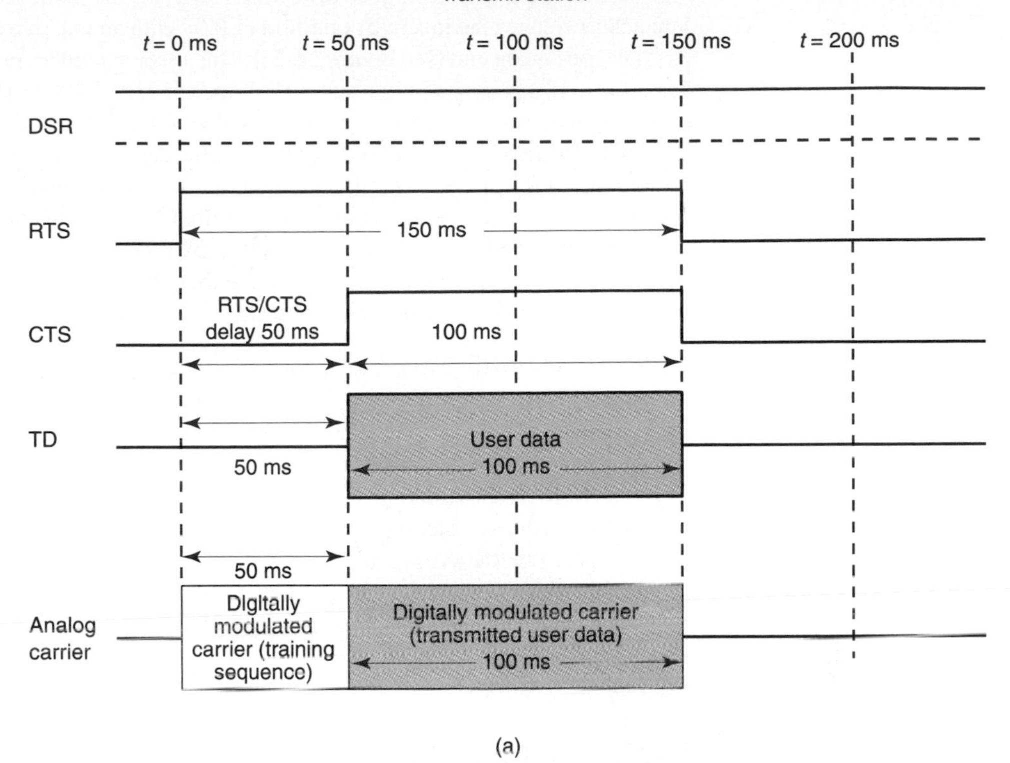

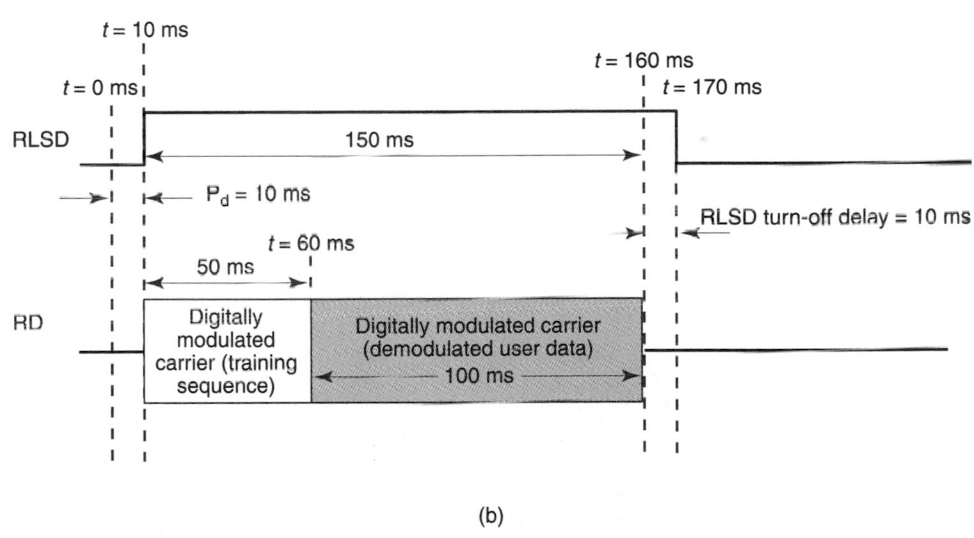

FIGURE 22-26 Typical timing diagram for control and data signals for asynchronous data transmission over the RS-232 interface between a DTE and a DCE (modem): (a) transmit timing diagram; (b) receive timing diagram

complete ($t = 150$ ms), RTS goes low, which turns off the modem's analog carrier. The modem acknowledges the inactive condition of RTS with an inactive condition on CTS.

At the distant end (see Figure 22-26b), the receive modem receives a valid analog carrier after a 10-ms propagation delay (P_d) and enables RLSD. The DCE sends an active RLSD signal across the RS-232 interface cable to the DT, which enables the receive data line (RD). However, the first 50 ms of the receive data is the training sequence, which is ignored by the DTE, as it is simply a continuous stream of logic 1s. The DTE identifies the beginning of the user data by recognizing the high-to-low transition caused by the first start bit ($t = 60$ ms). At the end of the message, the DCE holds RLSD active for a predetermined RLSD turn-off delay time (10 ms) to ensure that all the data received have been demodulated and outputted onto the RS-232 interface.

22-11-2 RS-449 Serial Interface Standards

In the mid-1970s, it appeared that data rates had exceeded the capabilities of the RS-232 interface. Consequently, in 1977, the EIA introduced the RS-449 serial interface with the intention of replacing the RS-232 interface. The RS-449 interface specifies a 37-pin primary connector (DB37) and a nine-pin secondary connector (DB9) for a total of 46 pins, which provide more functions, faster data transmission rates, and spans greater distances than the RS-232 interface. The RS-449 is essentially an updated version of the RS-232 interface except the RS-449 standard outlines only the mechanical and functional specifications of the interface.

The RS-449 primary cable is for serial data transmission, while the secondary cable is for diagnostic information. Table 22-10a lists the 37 pins of the RS-449 primary cable and their designations, and Table 22-10b lists the nine pins of the diagnostic cable and their designations. Note that the acronyms used with the RS-449 interface are more descriptive than those recommended by the EIA for the RS-232 interface. The functions specified by the RS-449 are very similar to the functions specified by the RS-232. The major difference

Table 22-10a RS-449 Pin Designations (37-Pin Connector)

Pin Number	Pin Name	EIA Nomenclature
1	Shield	None
19	Signal	SG
37	Send common	SC
20	Receive common	RC
28	Terminal in service	IS
15	Incoming call	IC
12, 30	Terminal ready	TR
11, 29	Data mode	DM
4, 22	Send data	SD
6, 24	Receive data	RD
17, 35	Terminal timing	TT
5, 23	Send timing	ST
8, 26	Receive timing	RT
7, 25	Request to send	RS
9, 27	Clear to send	CS
13, 31	Receiver ready	RR
33	Signal quality	SQ
34	New signal	NS
16	Select frequency	SF
2	Signal rate indicator	SI
10	Local loopback	LL
14	Remote loopback	RL
18	Test mode	TM
32	Select standby	SS
36	Standby indicator	SB

Table 22-10b RS-449 Pin Designations (Nine-Pin Connector)

Pin Number	Pin Name	EIA Nomenclature
1	Shield	None
5	Signal ground	SG
9	Send common	SC
2	Receive common	RC
3	Secondary send data	SSD
4	Secondary receive data	SRD
7	Secondary request to send	SRS
8	Secondary clear to send	SCS
6	Secondary receiver ready	SRF

between the two standards is the separation of the primary data and secondary diagnostic channels onto two separate cables.

The electrical specifications for the RS-449 were specified by the EIA in 1978 as either the RS-422 or the RS-423 standard. The RS-449 standard, when combined with RS-422A or RS-423A, were intended to replace the RS-232 interface. The primary goals of the new specifications are listed here:

1. Compatibility with the RS-232 interface standard
2. Replace the set of circuit names and mnemonics used with the RS-232 interface with more meaningful and descriptive names
3. Provide separate cables and connectors for the primary and secondary data channels
4. Provide single-ended or balanced transmission
5. Reduce crosstalk between signal wires
6. Offer higher data transmission rates
7. Offer longer distances over twisted-pair cable
8. Provide loopback capabilities
9. Improve performance and reliability
10. Specify a standard connector

The RS-422A standard specifies a balanced interface cable capable of operating up to 10 Mbps and span distances up to 1200 meters. However, this does not mean that 10 Mbps can be transmitted 1200 meters. At 10 Mbps, the maximum distance is approximately 15 meters, and 90 kbps is the maximum bit rate that can be transmitted 1200 meters.

The RS-423A standard specifies an unbalanced interface cable capable of operating at a maximum transmission rate of 100 kbps and span a maximum distance of 90 meters. The RS-442A and RS-443A standards are similar to ITU-T V.11 and V.10, respectively. With a bidirectional unbalanced line, one wire is at ground potential, and the currents in the two wires may be different. With an unbalanced line, interference is induced into only one signal path and, therefore, does not cancel in the terminator.

The primary objective of establishing the RS-449 interface standard was to maintain compatibility with the RS-232 interface standard. To achieve this goal, the EIA divided the RS-449 into two categories: *category I* and *category II* circuits. Category I circuits include only circuits that are compatible with the RS-232 standard. The remaining circuits are classified as category II. Category I and category II circuits are listed in Table 22-11.

Category I circuits can function with either the RS-422A (balanced) or the RS-423A (unbalanced) specifications. Category I circuits are allotted two adjacent wires for each RS-232-compatible signal, which facilitates either balanced or unbalanced operation. Category II circuits are assigned only one wire and, therefore, can facilitate only unbalanced (RS-423A) specifications.

Table 22-11 RS-449 Category I and Category II Circuits

Category I	
SD	Send data (4, 22)
RD	Receive data (6, 24)
TT	Terminal timing (17, 35)
ST	Send timing (5, 23)
RT	Receive timing (8, 26)
RS	Request to send (7, 25)
CS	Clear to send (9, 27)
RR	Receiver ready (13, 31)
TR	Terminal ready (12, 30)
DM	Data mode (11, 29)
Category II	
SC	Send common (37)
RC	Receive common (20)
IS	Terminal in service (28)
NS	New signal (34)
SF	Select frequency (16)
LL	Local loopback (10)
RL	Remote loopback (14)
TM	Test mode (18)
SS	Select standby (32)
SB	Standby indicator (36)

The RS-449 interface provides 10 circuits not specified in the RS-232 standard:

1. *Local loopback* (LL, pin 10). Used by the DTE to request a local (analog) loopback from the DCE
2. *Remote loopback* (RL, pin 14). Used by the DTE to request a remote (digital) loopback from the distant DCE
3. *Select frequency* (SF, pin 16). Allows the DTE to select the DCE's transmit and receive frequencies
4. *Test mode* (TM, pin 18). Used by the DTE to signal the DCE that a test is in progress
5. *Receive common* (RC, pin 20). Common return wire for unbalanced signals propagating from the DCE to the DTE
6. *Terminal in service* (IS, pin 28). Used by the DTE to signal the DCE whether it is operational
7. *Select standby* (SS, pin 32). Used by the DTE to request that the DCE switch to standby equipment in the event of a failure on the primary equipment
8. *New signal* (NS, pin 34). Used with a modem at the primary location of a multipoint data circuit so that the primary can resynchronize to whichever secondary is transmitting at the time
9. *Standby indicator* (SB, pin 36). Intended to be by the DCE as a response to the SS signal to notify the DTE that standby equipment has replaced the primary equipment
10. *Send common* (SC, pin 37). Common return wire for unbalanced signals propagating from the DTE to the DCE

22-11-3 RS-530 Serial Interface Standards

Since the data communications industry did not readily adopt the RS-449 interface, it came and went virtually unnoticed by most of industry. Consequently, in 1987 the EIA introduced another new standard, the RS-530 serial interface, which was intended to operate at data rates

Table 22-12 RS-530 Pin Designations

Signal Name	Pin Number(s)
Shield	1
Transmit data[a]	2, 14
Receive data[a]	3, 16
Request to send[a]	4, 19
Clear to send[a]	5, 13
DCE ready[a]	6, 22
DTE ready[a]	20, 23
Signal ground	7
Receive line signal detect[a]	8, 10
Transmit signal element timing (DCE source)[a]	15, 12
Receive signal element timing (DCE source)[a]	17, 9
Local loopback[b]	18
Remote loopback[b]	21
Transmit signal element timing (DTE source)[a]	24, 11
Test mode[b]	25

[a]Category I circuits (RS-422A).
[b]Category II circuits (RS-423A).

between 20 kbps and 2 Mbps using the same 25-pin DB-25 connector used by the RS-232 interface. The pin functions of the RS-530 interface are essentially the same as the RS-449 category I pins with the addition of three category II pins: local loopback, remote loopback, and test mode. Table 22-12 lists the 25 pins for the RS-530 interface and their designations.

Like the RS-449 standard, the RS-530 interface standard does not specify electrical parameters. The electrical specifications for the RS-530 are outlined by either the RS-422A or the RS-423A standard. The RS-232, RS-449, and RS-530 interface standards provide specifications for answering calls, but do not provide specifications for initiating calls (i.e., dialing). The EIA has a different standard, RS-366, for automatic calling units. The principal use of the RS-366 is for dial backup of private-line data circuits and for automatic dialing of remote terminals.

22-12 DATA COMMUNICATIONS MODEMS

The most common type of data communications equipment (DEC) is the *data communications modem.* Alternate names include *datasets, dataphones,* or simply *modems.* The word *modem* is a contraction derived from the words *modulator* and *demodulator.*

In the 1960s, the business world recognized a rapidly increasing need to exchange digital information between computers, computer terminals, and other computer-controlled equipment separated by substantial distances. The only transmission facilities available at the time were analog voice-band telephone circuits. Telephone circuits were designed for transporting analog voice signals within a bandwidth of approximately 300 Hz to 3000 Hz. In addition, telephone circuits often included amplifiers and other analog devices that could not propagate digital signals. Therefore, voice-band data modems were designed to communicate with each other using analog signals that occupied the same bandwidth used for standard voice telephone communications. Data communications modems designed to operate over the limited bandwidth of the public telephone network are called *voice-band modems.*

Because digital information cannot be transported directly over analog transmission media (at least not in digital form), the primary purpose of a *data communications modem* is to interface computers, computer networks, and other digital terminal equipment to analog communications facilities. Modems are also used when computers are too far apart to be directly interconnected using standard computer cables. In the transmitter (modulator) section of a modem, digital signals are encoded onto an analog carrier. The digital signals modulate the carrier, producing digitally modulated analog signals that are capable of being transported through the analog communications media. Therefore, the output of a modem is an analog signal that is carrying digital information. In the receiver section of a modem, digitally modulated analog signals are demodulated. Demodulation is the reverse process of modulation. Therefore, modem receivers (demodulators) simply extract digital information from digitally modulated analog carriers.

The most common (and simplest) modems available are ones intended to be used to interface DTEs through a serial interface to standard voice-band telephone lines and provide reliable data transmission rates from 300 bps to 56 kbps. These types of modems are sometimes called *telephone-loop modems* or *POTS modems,* as they are connected to the telephone company through the same local loops that are used for voice telephone circuits. More sophisticated modems (sometimes called *broadband modems*) are also available that are capable of transporting data at much higher bit rates over wideband communications channels, such as those available with optical fiber, coaxial cable, microwave radio, and satellite communications systems. Broadband modems can operate using a different set of standards and protocols than telephone loop modems.

A modem is, in essence, a transparent repeater that converts electrical signals received in digital form to electrical signals in analog form and vice versa. A modem is transparent, as it does not interpret or change the information contained in the data. It is a repeater, as it is not a destination for data—it simply repeats or retransmits data. A modem is physically located between digital terminal equipment (DTE) and the analog communications channel. Modems work in pairs with one located at each end of a data communications circuit. The two modems do not need to be manufactured by the same company; however, they must use compatible modulation schemes, data encoding formats, and transmission rates.

Figure 22-27 shows how a typical modem is used to facilitate the transmission of digital data between DTEs over a POTS telephone circuit. At the transmit end, a modem receives discrete digital pulses (which are usually in binary form) from a DTE through a serial digital interface (such as the RS-232). The DCE converts the digital pulses to analog signals. In essence, a modem transmitter is a *digital-to-analog converter* (DAC). The analog signals are then outputted onto an analog communications channel where they are transported through the system to a distant receiver. The equalizers and bandpass filters shape and band-limit the signal. At the destination end of a data communications system, a modem receives analog signals from the communications channel and converts them to digital pulses. In essence, a modem receiver is an *analog-to-digital converter* (ADC). The demodulated digital pulses are then outputted onto a serial digital interface and transported to the DTE.

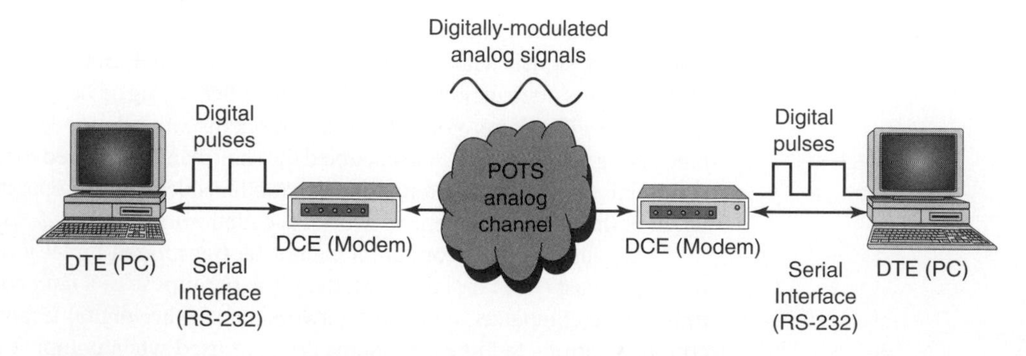

FIGURE 22-27 Data communications modems - POTS analog channel

22-12-1 Bits per Second versus Baud

The parameters *bits per second* (bps) and *baud* are often misunderstood and, consequently, misused. Baud, like bit rate, is a rate of change; however, baud refers to the rate of change of the signal on the transmission medium after encoding and modulation have occurred. Bit rate refers to the rate of change of a digital information signal, which is usually binary. Baud is the reciprocal of the time of one output *signaling element,* and a signaling element may represent several information bits. A signaling element is sometimes called a *symbol* and could be encoded as a change in the amplitude, frequency, or phase. For example, binary signals are generally encoded and transmitted one bit at a time in the form of discrete voltage levels representing logic 1s (highs) and logic 0s (lows). A baud is also transmitted one at a time; however, a baud may represent more than one information bit. Thus, the baud of a data communications system may be considerably less than the bit rate.

22-12-2 Bell System–Compatible Modems

At one time, Bell System modems were virtually the only modems in existence. This is because AT&T operating companies once owned 90% of the telephone companies in the United States, and the AT&T operating tariff allowed only equipment manufactured by Western Electric Company (WECO) and furnished by Bell System operating companies to be connected to AT&T telephone lines. However, in 1968, AT&T lost a landmark Supreme Court decision, the *Carterfone decision,* which allowed equipment manufactured by non-Bell companies to interconnect to the vast AT&T communications network, provided that the equipment met Bell System specifications. The Carterfone decision began the *interconnect industry,* which has led to competitive data communications offerings by a large number of independent companies.

The operating parameters for Bell System modems are the models from which the international standards specified by the ITU-T evolved. Bell System modem specifications apply only to modems that existed in 1968; therefore, their specifications pertain only to modems operating at data transmission rate of 9600 bps or less.

22-12-3 Modem Block Diagram

Figure 22-28 shows a simplified block diagram for a data communications modem. For simplicity, only the primary functional blocks of the transmitter and receiver are shown.

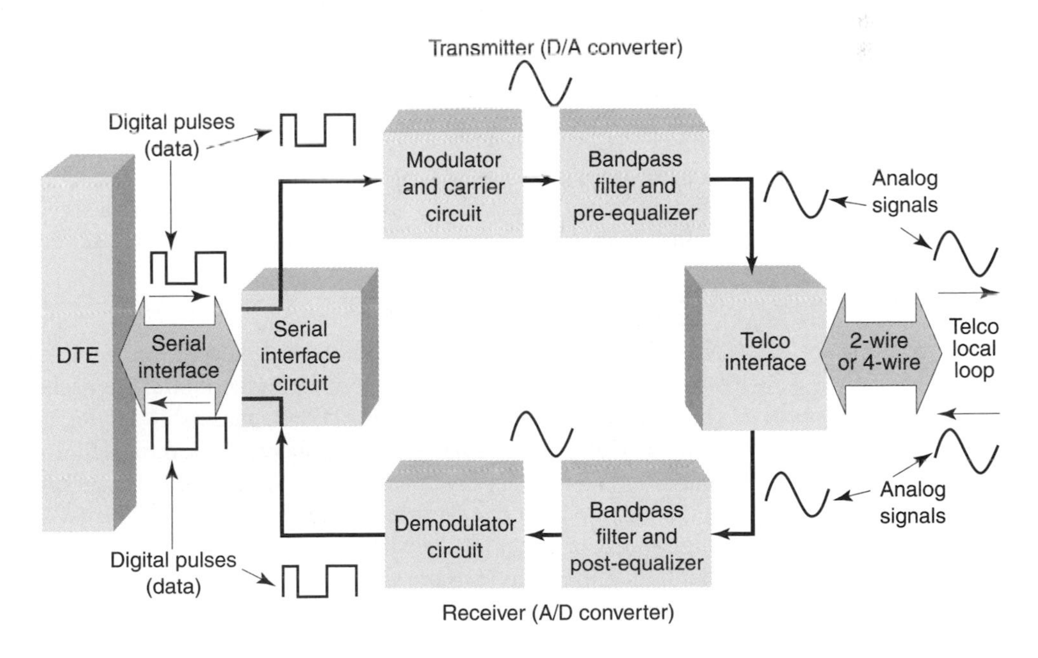

FIGURE 22-28 Simplified block diagram for an asynchronous FSK modem

The basic principle behind a modem transmitter is to convert information received from the DTE in the form of binary digits (bits) to digitally modulated analog signals. The reverse process is accomplished in the modem receiver.

The primary blocks of a modem are described here:

1. *Serial interface circuit.* Interfaces the modem transmitter and receiver to the serial interface. The transmit section accepts digital information from the serial interface, converts it to the appropriate voltage levels, and then directs the information to the modulator. The receive section receives digital information from the demodulator circuit, converts it to the appropriate voltage levels, and then directs the information to the serial interface. In addition, the serial interface circuit manages the flow of control, timing, and data information transferred between the DTE and the modem, which includes handshaking signals and clocking information.

2. *Modulator circuit.* Receives digital information from the serial interface circuit. The digital information modulates an analog carrier, producing a digitally modulated analog signal. In essence, the modulator converts digital changes in the information to analog changes in the carrier. The output from the modulator is directed to the transmit bandpass filter and equalizer circuit.

3. *Bandpass filter and equalizer circuit.* There are bandpass filter and equalizer circuits in both the transmitter and receiver sections of the modem. The transmit bandpass filter limits the bandwidth of the digitally modulated analog signals to a bandwidth appropriate for transmission over a standard telephone circuit. The receive bandpass filter limits the bandwidth of the signals allowed to reach the demodulator circuit, thus reducing noise and improving system performance. Equalizer circuits compensate for bandwidth and gain imperfections typically experienced on voiceband telephone lines.

4. *Telco interface circuit.* The primary functions of the telco interface circuit are to match the impedance of the modem to the impedance of the telephone line and regulate the amplitude of the transmit signal. The interface also provides electrical isolation and protection and serves as the demarcation (separation) point between subscriber equipment and telephone company–provided equipment. The telco line can be two-wire or four-wire, and the modem can operate half or full duplex. When the telephone line is two wire, the telco interface circuit would have to perform four-wire-to-two-wire and two-wire-to-four-wire conversions.

5. *Demodulator circuit.* Receives modulated signals from the bandpass filter and equalizer circuit and converts the digitally modulated analog signals to digital signals. The output from the demodulator is directed to the serial interface circuit, where it is passed on to the serial interface.

6. *Carrier and clock generation circuit.* The carrier generation circuit produces the analog carriers necessary for the modulation and demodulation processes. The clock generation circuit generates the appropriate clock and timing signals required for performing transmit and receive functions in an orderly and timely fashion.

22-12-4 Modem Classifications

Data communications modems can be generally classified as either *asynchronous* or *synchronous* and use one of the following digital modulation schemes: amplitude-shift keying (ASK), frequency-shift keying (FSK), phase-shift keying (PSK), or quadrature amplitude modulation (QAM). However, there are several additional ways modems can be classified, depending on which features or capabilities you are trying to distinguish. For example, modems can be categorized as internal or external; low speed, medium speed, high speed, or very high speed; wide band or voice band; and personal or commercial. Regardless of how modems are classified, they all share a common goal, namely, to convert digital pulses to analog signals in the transmitter and analog signals to digital pulses in the receiver.

Some of the common features provided data communications modems are listed here:

1. Automatic dialing, answering, and redialing
2. Error control (detection and correction)
3. Caller ID recognition
4. Self-test capabilities, including analog and digital loopback tests
5. Fax capabilities (transmit and receive)
6. Data compression and expansion
7. Telephone directory (telephone number storage)
8. Adaptive transmit and receive data transmission rates (300 bps to 56 kbps)
9. Automatic equalization
10. Synchronous or asynchronous operation

22-12-5 Asynchronous Voice-Band Modems

Asynchronous modems can be generally classified as low-speed voice-band modems, as they are typically used to transport asynchronous data (i.e., data framed with start and stop bits). Synchronous data are sometimes used with an asynchronous modem; however, it is not particularly practical or economical. Synchronous data transported by asynchronous modems is called *isochronous transmission.* Asynchronous modems use relatively simple modulation schemes, such as ASK or FSK, and are restricted to relatively low-speed applications (generally less than 2400 bps), such as telemetry and caller ID.

There are several standard asynchronous modems designed for low-speed data applications using the switched public telephone network. To operate full duplex with a two-wire dial-up circuit, it is necessary to divide the usable bandwidth of a voice-band circuit in half, creating two equal-capacity data channels. A popular modem that does this is the Bell System 103–compatible modem.

22-12-5-1 Bell system 103–compatible modem. The 103 modem is capable of full-duplex operation over a two-wire telephone line at bit rates up to 300 bps. With the 103 modem, there are two data channels, each with their own mark and space frequencies. One data channel is called the *low-band channel* and occupies a bandwidth from 300 Hz to 1650 Hz (i.e., the lower half of the usable voice band). A second data channel, called the *high-band channel,* occupies a bandwidth from 1650 Hz to 3000 Hz (i.e., the upper half of the usable voice band). The mark and space frequencies for the low-band channel are 1270 Hz and 1070 Hz, respectively. The mark and space frequencies for the high-band channel are 2225 Hz and 2025 Hz, respectively. Separating the usable bandwidth into two narrower bands is called *frequency-division multiplexing* (FDM). FDM allows full-duplex (FDX) transmission over a two-wire circuit, as signals can propagate in both directions at the same time without interfering with each other because the frequencies for the two directions of propagation are different. FDM allows full-duplex operation over a two-wire telephone circuit. Because FDM reduces the effective bandwidth in each direction, it also reduces the maximum data transmission rates. A 103 modem operates at 300 baud and is capable of simultaneous transmission and reception of 300 bps.

22-12-5-2 Bell system 202T/S modem. The 202T and 202S modem are identical except the 202T modem specifies four-wire, full-duplex operation, and the 202S modem specifies two-wire, half-duplex operation. Therefore, the 202T is utilized on four-wire private-line data circuits, and the 202S modem is designed for the two-wire switched public telephone network. Probably the most common application of the 202 modem today is caller ID, which is a simplex system with the transmitter in the telephone office and the receiver at the subscriber's location. The 202 modem is an asynchronous 1200-baud transceiver utilizing FSK with a transmission bit rate of 1200 bps over a standard voice-grade telephone line.

22-12-6 Synchronous Voice-Band Modems

Synchronous modems use PSK or quadrature amplitude modulation (QAM) to transport synchronous data (i.e., data preceded by unique SYN characters) at transmission rates between 2400 bps and 56,000 bps over standard voice-grade telephone lines. The modulated carrier is transmitted to the distant modem, where a coherent carrier is recovered and used to demodulate the data. The transmit clock is recovered from the data and used to clock the received data into the DTE. Because of the addition of clock and carrier recovery circuits, synchronous modems are more complicated and, thus, more expensive than asynchronous modems.

PSK is commonly used in medium speed synchronous voice-band modems, typically operating between 2400 bps and 4800 bps. More specifically, QPSK is generally used with 2400-bps modems and 8-PSK with 4800-bps modems. QPSK has a bandwidth efficiency of 2 bps/Hz; therefore, the baud rate and minimum bandwidth for a 2400-bps synchronous modem are 1200 baud and 1200 Hz, respectively. The standard 2400-bps synchronous modem is the Bell System 201C or equivalent. The 201C modem uses a 1600-Hz carrier frequency and has an output spectrum that extends from approximately 1000 Hz to 2200 Hz. Because 8-PSK has a bandwidth efficiency of 3 bps/Hz, the baud rate and minimum bandwidth for 4800-bps synchronous modems are 1600 baud and 1600 Hz, respectively. The standard 4800-bps synchronous modem is the Bell System 208A. The 208A modem also uses a 1600-Hz carrier frequency but has an output spectrum that extends from approximately 800 Hz to 2400 Hz. Both the 201C and the 208A are full-duplex modems designed to be used with four-wire private-line circuits. The 201C and 208A modems can operate over two-wire dial-up circuits but only in the simplex mode. There are also half-duplex two-wire versions of both modems: the 201B and 208B.

High-speed synchronous voice-band modems operate at 9600 bps and use 16-QAM modulation. 16-QAM has a bandwidth efficiency of 4 bps/Hz; therefore, the baud and minimum bandwidth for 9600-bps synchronous modems is 2400 baud and 2400 Hz, respectively. The standard 9600-bps modem is the Bell System 209A or equivalent. The 209A uses a 1650-Hz carrier frequency and has an output spectrum that extends from approximately 450 Hz to 2850 Hz. The Bell System 209A is a four-wire synchronous voice-band modem designed to be used on full-duplex private-line circuits. The 209B is the two-wire version designed for half-duplex operation on dial-up circuits.

Table 22-13 summarizes the Bell System voice-band modem specifications. The modems listed in the table are all relatively low speed by modern standards. Today, the Bell System–compatible modems are used primarily on relatively simple telemetry circuits, such as remote alarm systems and on metropolitan and wide-area private-line data networks, such as those used by department stores to keep track of sales and inventory. The more advanced, higher-speed data modems are described in a later section of this chapter.

22-12-7 Modem Synchronization

During the request-to-send/clear-to-send (RTS/CTS) delay, a transmit modem outputs a special, internally generated bit pattern called a *training sequence.* This bit pattern is used to synchronize (train) the receive modem at the distant end of the communications channel. Depending on the type of modulation, transmission bit rate, and modem complexity, the training sequence accomplishes one or more of the following functions:

1. Initializes the communications channel, which includes disabling echo and establishing the gain of automatic gain control (AGC) devices
2. Verifies continuity (activates RLSD in the receive modem)
3. Initialize descrambler circuits in receive modem
4. Initialize automatic equalizers in receive modem
5. Synchronize the receive modem's carrier to the transmit modem's carrier
6. Synchronize the receive modem's clock to the transmit modem's clock

Table 22-13 Bell System Modem Specifications

Bell System Designation	Transmission Facility	Operating Mode	Circuit Arrangement	Synchronization Mode	Modulation	Transmission Rate
103	Dial-up	FDM/FDX	Two wire	Asynchronous	FSK	300 bps
113A/B	Dial-up	FDM/FDX	Two wire	Asynchronous	FSK	300 bps
201B	Dial-up	HDX	Two wire	Synchronous	QPSK	2400 bps
201C	Private line	FDX	Four wire	Synchronous	QPSK	2400 bps
202S	Dial-up	HDX	Two wire	Asynchronous	FSK	1200 bps
202T	Private line	FDX	Four wire	Asynchronous	FSK	1800 bps
208A	Private line	FDX	Four wire	Synchronous	8-PSK	4800 bps
208B	Dial-up	HDX	Two wire	Synchronous	8-PSK	4800 bps
209A	Private line	FDX	Four wire	Synchronous	16-QAM	9600 bps
209B	Dial-up	HDX	Two wire	Synchronous	16-QAM	9600 bps
212A	Dial up	HDX	Two wire	Asynchronous	FSK	600 bps
212B	Private line	FDX	Four wire	Synchronous	QPSK	1200 bps

Dial-up = switched telephone network
Private line = dedicated circuit
FDM = frequency-division multiplexing
HDX = half duplex
FDX = full duplex
FSK = frequency-shift keying
QPSK = four-phase PSK
8-PSK = eight-phase PSK
16-QAM = 16-state QAM

22-12-8 Modem Equalizers

Equalization is the compensation for phase delay distortion and amplitude distortion inherently present on telephone communications channels. One form of equalization provided by the telephone company is C-type conditioning, which is available only on private-line circuits. Additional equalization may be performed by the modems themselves. *Compromise equalizers* are located in the transmit section of a modem and provide *preequalization*—they shape the transmitted signal by altering its delay and gain characteristics before the signal reaches the telephone line. It is an attempt by the modem to compensate for impairments anticipated in the bandwidth parameters of the communications line. When a modem is installed, the compromise equalizers are manually adjusted to provide the best error performance. Typically, compromise equalizers affect the following:

1. Amplitude only
2. Delay only
3. Amplitude and delay
4. Neither amplitude nor delay

Compromise equalizer settings may be applied to either the high- or low-voice-band frequencies or symmetrically to both at the same time. Once a compromise equalizer setting has been selected, it can be changed only manually. The setting that achieves the best error performance is dependent on the electrical length of the circuit and the type of facilities that comprise it (i.e., one or more of the following: twisted-pair cable, coaxial cable, optical fiber cable, microwave, digital T-carriers, and satellite).

Adaptive equalizers are located in the receiver section of a modem, where they provide postequalization to the received signals. Adaptive equalizers automatically adjust their gain and delay characteristics to compensate for phase and amplitude impairments encountered on the communications channel. Adaptive equalizers may determine the quality of the received signal within its own circuitry, or they may acquire this information from the demodulator or descrambler circuits. Whatever the case, the adaptive equalizer may continuously vary its settings to achieve the best overall bandwidth characteristics for the circuit.

22-13 ITU-T MODEM RECOMMENDATIONS

Since the late 1980s, the International Telecommunications Union (ITU-T, formerly CCITT), which is headquartered in Geneva, Switzerland, has developed transmission standards for data modems outside the United States. The ITU-T specifications are known as the V-series, which include a number indicating the standard (V.21, V.23, and so on). Sometimes the V-series is followed by the French word *bis,* meaning "second," which indicates that the standard is a revision of an earlier standard. If the standard includes the French word *terbo,* meaning "third," the bis standard also has been modified. Table 22-14 lists some of the ITU-T modem recommendations.

22-13-1 ITU-T Modem Recommendation V.29

The ITU-T V.29 specification is the first internationally accepted standard for a 9600-bps data transmission rate. The V.29 standard is intended to provide synchronous data transmission over four-wire leased lines. V.29 uses 16-QAM modulation of a 1700-Hz carrier frequency. Data are clocked into the modem in groups of four bits called quadbits, resulting in a 2400-baud transmission rate. Occasionally, V.29-compatible modems are used in the half-duplex mode over two-wire switched telephone lines. Pseudo full-duplex operation can be achieved over the two-wire lines using a method called *ping-pong.* With ping-pong, data sent to the modem at each end of the circuit by their respective DTE are buffered and automatically exchanged over the data link by rapidly turning the carriers on and off in succession.

Pseudo full-duplex operation over a two-wire line can also be accomplished using *statistical duplexing.* Statistical duplexing utilizes a 300-bps reverse data channel. The reverse channel allows a data operator to enter keyboard data while simultaneously receiving a file from the distant modem. By monitoring the data buffers inside the modem, the direction of data transmission can be determined, and the high- and low-speed channels can be reversed.

22-13-2 ITU-T Modem Recommendation V.32

The ITU-T V.32 specification provides for a 9600-bps data transmission rate with true full-duplex operation over four-wire leased private lines or two-wire switched telephone lines. V.32 also provides for data rates of 2400 bps and 4800 bps. V.32 specifies QAM with a carrier frequency of 1800 Hz. V.32 is similar to V.29, except with V.32 the advanced coding technique *trellis encoding* is specified (see Chapter 12). Trellis encoding produces a superior signal-to-noise ratio by dividing the incoming data stream into groups of five bits called *quintbits* (M-ary, where $M = 2^5 = 32$). The constellation diagram for 32-state trellis encoding was developed by Dr. Ungerboeck at the IBM Zuerich Research Laboratory and combines coding and modulation to improve bit error performance. The basic idea behind trellis encoding is to introduce controlled redundancy, which reduces channel error rates by doubling the number of signal points on the QAM constellation. The trellis encoding constellation used with V.32 is shown in Figure 22-29.

Full-duplex operation over two-wire switched telephone lines is achieved with V.32 using a technique called *echo cancellation.* Echo cancellation involves adding an inverted replica of the transmitted signal to the received signal. This allows the data transmitted from each modem to simultaneously use the same carrier frequency, modulation scheme, and bandwidth.

22-13-3 ITU-T Modem Recommendation V.32bis and V.32terbo

ITU-T recommendation V.32bis was introduced in 1991 and created a new benchmark for the data modem industry by allowing transmission bit rates of 14.4 kbps over standard

Table 22-14 ITU-T V-Series Modem Standards

ITU-T Designation	Specification
V.1	Defines binary 0/1 data bits as space/mark line conditions
V.2	Limits output power levels of modems used on telephone lines
V.4	Sequence of bits within a transmitted character
V.5	Standard synchronous signaling rates for dial-up telephone lines
V.6	Standard synchronous signaling rates for private leased communications lines
V.7	List of modem terminology in English, Spanish, and French
V.10	Unbalanced high-speed electrical interface specifications (similar to RS-423)
V.11	Balanced high-speed electrical interface specifications (similar to RS-422)
V.13	Simulated carrier control for full-duplex modem operating in the half-duplex mode
V.14	Asynchronous-to-synchronous conversion
V.15	Acoustical couplers
V.16	Electrocardiogram transmission over telephone lines
V.17	Application-specific modulation scheme for Group III fax (provides two-wire, half-duplex trellis-coded transmission at 7.2 kbps, 9.6 kbps, 12 kbps, and 14.4 kbps)
V.19	Low-speed parallel data transmission using DTMF modems
V.20	Parallel data transmission modems
V.21	0-to-300 bps full-duplex two-wire modems similar to Bell System 103
V.22	1200/600 bps full-duplex modems for switched or dedicated lines
V.22bis	1200/2400 bps two-wire modems for switched or dedicated lines
V.23	1200/75 bps modems (host transmits 1200 bps and terminal transmits 75 bps). V.23 also supports 600 bps in the high channel speed. V.23 is similar to Bell System 202. V.23 is used in Europe to support some videotext applications.
V.24	Known in the United States as RS-232. V.24 defines only the functions of the interface circuits, whereas RS-232 also defines the electrical characteristics of the connectors.
V.25	Automatic answering equipment and parallel automatic dialing similar to Bell System 801 (defines the 2100-Hz answer tone that modems send)
V.25bis	Serial automatic calling and answering—CCITT equivalent to the Hayes AT command set used in the United States
V.26	2400-bps four-wire modems identical to Bell System 201 for four-wire leased lines
V.26bis	2400/1200 bps half-duplex modems similar to Bell System 201 for two-wire switched lines
V.26terbo	2400/1200 bps full-duplex modems for switched lines using echo canceling
V.27	4800 bps four-wire modems for four-wire leased lines similar to Bell System 208 with manual equalization
V.27bis	4800/2400 bps four-wire modems same as V.27 except with automatic equalization
V.28	Electrical characteristics for V.24
V.29	9600-bps four-wire full-duplex modems similar to Bell System 209 for leased lines
V.31	Older electrical characteristics rarely used today
V.31bis	V.31 using optocouplers
V.32	9600/4800 bps full-duplex modems for switched or leased facilities
V.32bis	4.8-kbps, 7.2-kbps, 9.6-kbps, 12-kbps, and 14.4-kbps modems and rapid rate regeneration for full-duplex leased lines
V.32terbo	Same as V.32bis except with the addition of adaptive speed leveling, which boosts transmission rates to as high as 21.6 kbps
V.33	12.2 kbps and 14.4 kbps for four-wire leased communications lines
V.34	(V. fast) 28.8-kbps data rates without compression
V.34+	Enhanced specifications of V.34
V.35	48-kbps four-wire modems (no longer used)
V.36	48-kbps four-wire full-duplex modems
V.37	72-kbps four-wire full-duplex modems
V.40	Method teletypes use to indicate parity errors
V.41	An older obsolete error-control scheme
V.42	Error-correcting procedures for modems using asynchronous-to-synchronous conversion (V.22, B.22bis, V.26terbo, V.32, and V.32bis, and LAP M protocol)
V.42bis	Lempel-Ziv-based data compression scheme used with V.42 LAP M
V.50	Standard limits for transmission quality for modems
V.51	Maintenance of international data circuits

(Continued)

Table 22-14 (*Continued*)

ITU-T Designation	Specification
V.52	Apparatus for measuring distortion and error rates for data transmission
V.53	Impairment limits for data circuits
V.54	Loop test devices of modems
V.55	Impulse noise-measuring equipment
V.56	Comparative testing of modems
V.57	Comprehensive tests set for high-speed data transmission
V.90	Asymmetrical data transmission—receive data rates up to 56 kbps but restricts transmission bit rates to 33.6 kbps
V.92	Asymmetrical data transmission—receive data rates up to 56 kbps but restricts transmission bit rates to 48 kbps
V.100	Interconnection between public data networks and public switched telephone networks
V.110	ISDN terminal adaptation
V.120	ISDN terminal adaptation with statistical multiplexing
V.230	General data communications interface, ISO layer 1

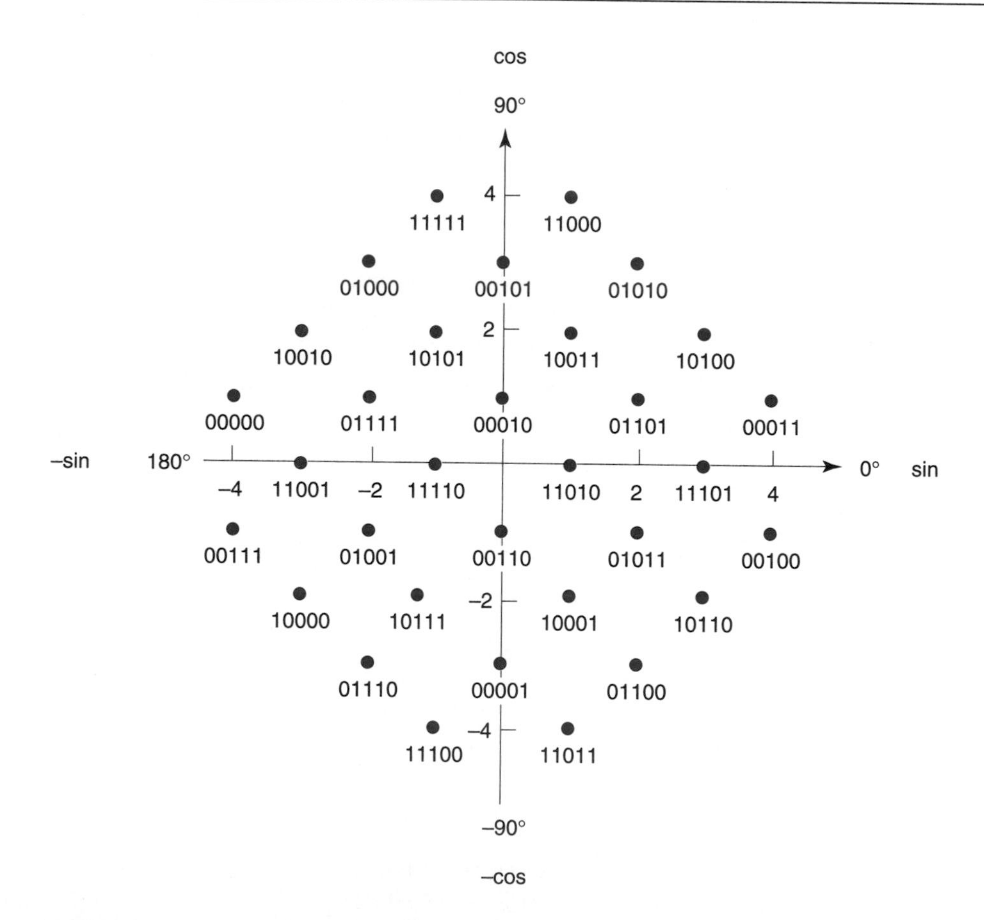

FIGURE 22-29 V.32 constellation diagram using Trellis encoding

voice-band telephone channels. V.32bis uses a 64-point signal constellation with each signaling condition representing six bits of data. The constellation diagram for V.32 is shown in Figure 22-30. The transmission bit rate for V.32 is six bits/code × 2400 codes/second = 14,400 bps. The signaling rate (baud) is 2400.

V.32bis also includes automatic *fall-forward* and *fall-back* features that allow the modem to change its transmission rate to accommodate changes in the quality of the commu-

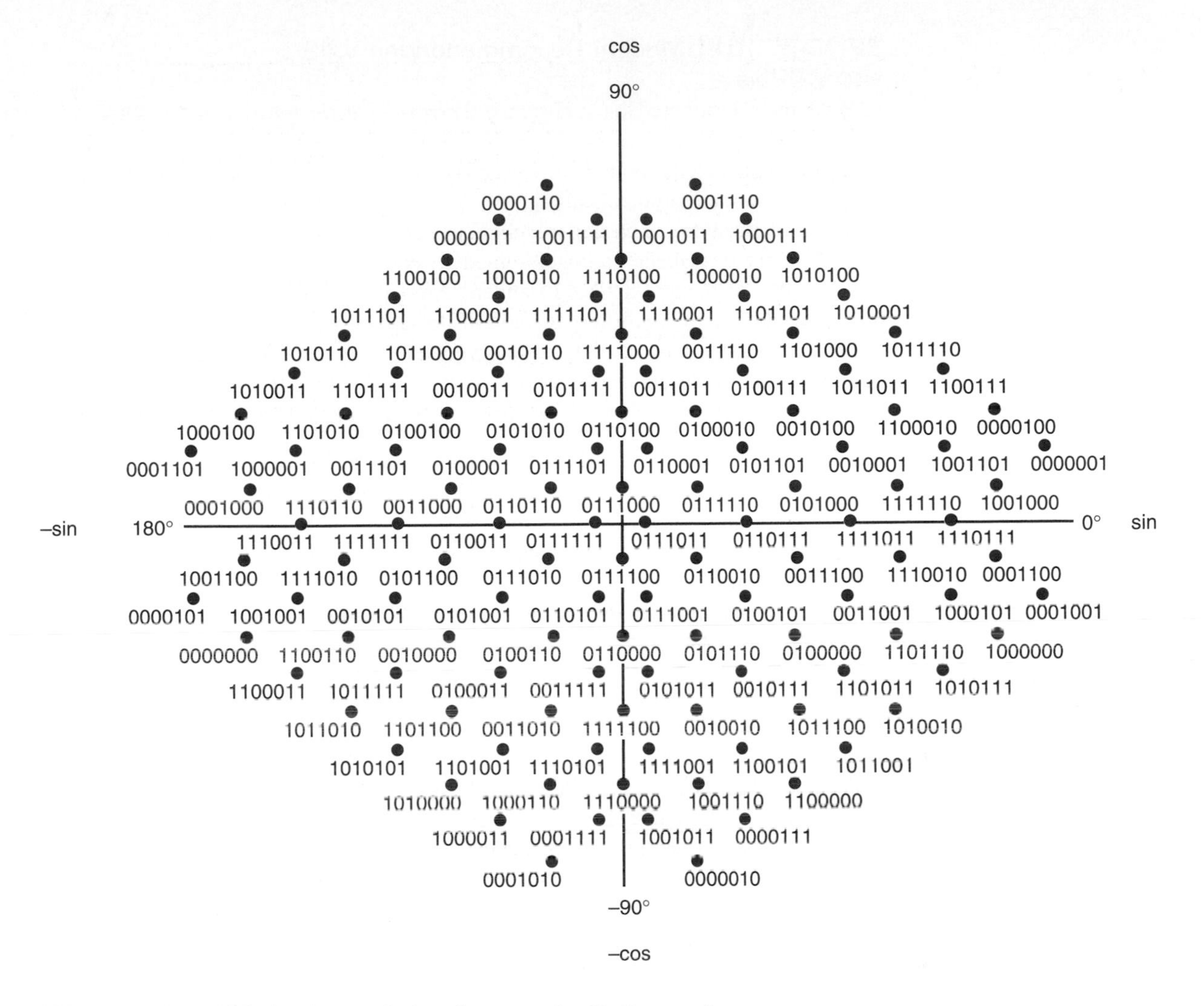

FIGURE 22-30 V.33 signal constellation diagram using Trellis encoding

nications line. The fall-back feature slowly reduces the transmission bit rate to 12.2 kbps, 9.6 kbps, or 4.8 kbps if the quality of the communications line degrades. The fall-forward feature gives the modem the ability to return to the higher transmission rate when the quality of the communications channel improves. V.32bis support Group III fax, which is the transmission standard that outlines the connection procedures used between two fax machines or fax modems. V.32bis also specifies the data compression procedure used during transmissions.

In August 1993, U.S. Robotics introduced V.32terbo. V.32terbo includes all the features of V.32bis plus a proprietary technology called *adaptive speed leveling*. V.32terbo includes two categories of new features: increased data rates and enhanced fax abilities. V.32terbo also outlines the new 19.2-kbps data transmission rate developed by AT&T.

22-13-4 ITU-T Modem Recommendation V.33

ITU-T specification V.33 is intended for modems that operate over dedicated two-point, private-line four-wire circuits. V.33 uses trellis coding and is similar to V.32 except a V.33 signaling element includes six information bits and one redundant bit, resulting in a data transmission rate of 14.4 kbps, 2400 baud, and an 1800-Hz carrier. The 128-point constellation used with V.33 is shown in Figure 22-30.

22-13-5 ITU-T Modem Recommendation V.42 and V.42bis

In 1988, the ITU adopted the V.42 standard *error-correcting procedures for DCEs* (modems). V.42 specifications address asynchronous-to-synchronous transmission conversions and error control that includes both detection and correction. V.42's primary purpose specifies a relatively new modem protocol called Link Access Procedures for Modems (LAP M). LAP M is almost identical to the packet-switching protocol used with the X.25 standard.

V.42bis is a specification designed to enhance the error-correcting capabilities of modems that implement the V.42 standard. Modems employing data compression schemes have proven to significantly surpass the data throughput performance of the predecessors. The V.42bis standard is capable of achieving somewhere between 3-to-1 and 4-to-1 compression ratios for ASCII-coded text. The compression algorithm specified is British Telecom's BTLZ. Throughput rates of up to 56 kbps can be achieved using V.42bis data compression.

22-13-6 ITU-T Modem Recommendation V.32 (V.fast)

Officially adopted in 1994, V.fast is considered the next generation in data transmission. Data rates of 28.8 kbps without compression are possible using V.34. Using current data compression techniques, V.fast modems will be able to transmit data at two to three times current data rates. V.32 automatically adapts to changes in transmission-line characteristics and dynamically adjusts data rates either up or down, depending on the quality of the communication channel.

V.34 innovations include the following:

1. Nonlinear coding, which offsets the adverse effects of system nonlinearities that produce harmonic and intermodulation distortion and amplitude proportional noise
2. Multidimensional coding and constellation shaping, which enhance data immunity to channel noise
3. Reduced complexity in decoders found in receivers
4. Precoding of data for more of the available bandwidth of the communications channel to be used by improving transmission of data in the outer limits of the channel where amplitude, frequency, and phase distortion are at their worst
5. Line probing, which is a technique that receive modems to rapidly determine the best correction to compensate for transmission-line impairments

22-13-7 ITU-T Modem Recommendation V.34+

V.34+ is an enhanced standard adopted by the ITU in 1996. V.34+ adds 31.2 kbps and 33.6 kbps to the V.34 specification. Theoretically, V.34+ adds 17% to the transmission rate; however, it is not significant enough to warrant serious consideration at this time.

22-13-8 ITU-T Modem Recommendation V.90

The ITU-T developed the V.90 specification in February 1998 during a meeting in Geneva, Switzerland. The V.90 recommendation is similar to 3COM's x2 and Lucent's K56flex in that it defines an asymmetrical data transmission technology where the upstream and downstream data rates are not the same. V.90 allows modem downstream (receive) data rates up to 56 kbps and upstream (transmit) data rates up to 33.6 kbps. These data rates are inappropriate in the United States and Canada, as the FCC and CRTC limit transmission rates offered by telephone companies to no more than 53 kbps.

22-13-9 ITU-T Modem Recommendation V.92

In 2000, the ITU approved a new modem standard called V.92. V.92 offers three improvements over V.90 that can be achieved only if both the transmit and receive modems and the Internet Service Provider (ISP) have V.92 compliant modems. V.92 offers (1) upstream transmission rate of 48 kbps, (2) faster call setup capabilities, and (3) incorporation of a hold option.

QUESTIONS

22-1. Define *data communications code.*

22-2. Give some of the alternate names for data communications codes.

22-3. Briefly describe the following data communications codes: Baudot, ASCII, and EBCDIC.

22-4. Describe the basic concepts of *bar codes.*

22-5. Describe a *discrete bar code; continuous bar code; 2D bar code.*

22-6. Explain the encoding formats used with *Code 39* and *UPC* bar codes.

22-7. Describe what is meant by *error control.*

22-8. Explain the difference between *error detection* and *error correction.*

22-9. Describe the difference between *redundancy* and *redundancy checking.*

22-10. Explain *vertical redundancy checking.*

22-11. Define *odd parity; even parity; marking parity.*

22-12. Explain the difference between *no parity* and *ignored parity.*

22-13. Describe how checksums are used for error detection.

22-14. Explain *longitudinal redundancy checking.*

22-15. Describe the difference between *character* and *message parity.*

22-16. Describe *cyclic redundancy checking.*

22-17. Define *forward error correction.*

22-18. Explain the difference between using *ARQ* and a *Hamming code.*

22-19. What is meant by *character synchronization*?

22-20. Compare and contrast *asynchronous* and *synchronous serial data formats.*

22-21. Describe the basic format used with asynchronous data.

22-22. Define the *start* and *stop bits.*

22-23. Describe *synchronous data.*

22-24. What is a *SYN character*?

22-25. Define and give some examples of *data terminal equipment.*

22-26. Define and give examples of *data communications equipment.*

22-27. List and describe the basic components that make up a *data communications circuit.*

22-28. Define *line control unit* and describe its basic functions in a data communications circuit.

22-29. Describe the basic functions performed by a *UART.*

22-30. Describe the operation of a UART transmitter and receiver.

22-31. Explain the operation of a *start bit verification circuit.*

22-32. Explain *clock slippage* and describe the effects of *slipping over* and *slipping under.*

22-33. Describe the differences between UARTs, USRTs, and USARTs.

22-34. List the features provided by *serial interfaces.*

22-35. Describe the purpose of a serial interface.

22-36. Describe the physical, electrical, and functional characteristics of the *RS-232 interface.*

22-37. Describe the *RS-449 interface* and give the primary differences between it and the RS-232 interface.

22-38. Describe *data communications modems* and explain where they are used in data communications circuits.

22-39. What is meant by a *Bell System–compatible* modem?

22-40. What is the difference between *asynchronous* and *synchronous* modems?

22-41. Define *modem synchronization* and list its functions.

22-42. Describe modem *equalization.*

22-43. Briefly describe the following ITU-T modem recommendations: V.29, V.32, V.32bis, V.32terbo, V.33, V.42, V.42bis, V.32fast, and V.34+.

PROBLEMS

22-1. Determine the hex codes for the following Baudot codes: C, J, 4, and /.

22-2. Determine the hex codes for the following ASCII codes: C, J, 4, and /.

22-3. Determine the hex codes for the following EBCDIC codes: C, J, 4, and /.

22-4. Determine the left- and right-hand UPC label format for the digit 4.

22-5. Determine the LRC and VRC for the following message (use even parity for LRC and odd parity for VCR):

```
D A T A sp C O M M U N I C A T I O N S
```

22-6. Determine the LRC and VRC for the following message (use even parity for LRC and odd parity for VCR):

```
A S C I I sp C O D E
```

22-7. Determine the BCS for the following data- and CRC-generating polynomials:

$$G(x) = x^7 + x^4 + x^2 + x^0 = 1\,0\,0\,1\,0\,1\,0\,1$$
$$P(x) = x^5 + x^4 + x^1 + x^0 = 1\,1\,0\,0\,1\,1$$

22-8. Determine the BCC for the following data- and CRC-generating polynomials:

$$G(x) = x^8 + x^5 + x^2 + x^0$$
$$P(x) = x^5 + x^4 + x^1 + x^0$$

22-9. How many Hamming bits are required for a single EBCDIC character?

22-10. Determine the Hamming bits for the ASCII character "B." Insert the hamming bits into every other bit location starting from the left.

22-11. Determine the Hamming bits for the ASCII character "C" (use odd parity and two stop bits). Insert the Hamming bits into every other location starting at the right.

22-12. Determine the noise margins for an RS-232 interface with driver output signal voltages of ±12 V.

22-13. Determine the noise margins for an RS-232 interface with driver output signal voltages of ±11 V.

CHAPTER 23

Data-Link Protocols and Data Communications Networks

CHAPTER OUTLINE

OBJECTIVES

- Define *data-link protocol*
- Define and describe the following data-link protocol functions: line discipline, flow control, and error control
- Define *character-* and *bit-oriented protocols*
- Describe asynchronous data-link protocols
- Describe synchronous data-link protocols
- Explain binary synchronous communications
- Define and describe *synchronous data-link control*
- Define and describe *high-level data-link control*
- Describe the concept of a public data network
- Describe the X.25 protocol
- Define and describe the basic concepts of asynchronous transfer mode
- Explain the basic concepts of integrated services digital network
- Define and describe the fundamental concepts of local area networks
- Describe the fundamental concepts of Ethernet
- Describe the differences between the various types of Ethernet
- Describe the Ethernet II and IEEE 802.3 frame formats

23-1 INTRODUCTION

The primary goal of *network architecture* is to give users of a network the tools necessary for setting up the network and performing data flow control. A network architecture outlines the way in which a data communications network is arranged or structured and generally includes the concepts of levels or layers within the architecture. Each layer within the network consists of specific *protocols* or rules for communicating that perform a given set of functions.

Protocols are arrangements between people or processes. A *data-link protocol* is a set of rules implementing and governing an orderly exchange of data between layer two devices, such as line control units and front-end processors.

23-2 DATA-LINK PROTOCOL FUNCTIONS

For communications to occur over a data network, there must be at least two devices working together (one transmitting and one receiving). In addition, there must be some means of controlling the exchange of data. For example, most communication between computers on networks is conducted half duplex even though the circuits that interconnect them may be capable of operating full duplex. Most data communications networks, especially local area networks, transfer data half duplex where only one device can transmit at a time. Half-duplex operation requires coordination between stations. Data-link protocols perform certain network functions that ensure a coordinated transfer of data. Some data networks designate one station as the *control station* (sometimes called the *primary station*). This is sometimes referred to as *primary-secondary* communications. In centrally controlled networks, the primary station enacts procedures that determine which station is transmitting and which is receiving. The transmitting station is sometimes called the *master station,* whereas the receiving station is called the *slave station.* In primary-secondary networks, there can never be more than one master at a time; however, there may be any number of slave stations. In another type of network, all stations are equal, and any station can transmit at any time. This type of network is sometimes called a *peer-to-peer network.* In a peer-to-peer network, all stations have equal access to the network, but when they have a message to transmit, they must contend with the other stations on the network for access to the transmission medium.

Data-link protocol *functions* include line discipline, flow control, and error control. *Line discipline* coordinates hop-to-hop data delivery where a hop may be a computer, a network controller, or some type of network-connecting device, such as a router. *Line discipline* determines which device is transmitting and which is receiving at any point in time. *Flow control* coordinates the rate at which data are transported over a link and generally provides an acknowledgment mechanism that ensures that data are received at the destination. *Error control* specifies means of detecting and correcting transmission errors.

23-2-1 Line Discipline

In essence, line discipline is coordinating half-duplex transmission on a data communications network. There are two fundamental ways that line discipline is accomplished in a data communications network: *enquiry/acknowledgment* (ENQ/ACK) and *poll/select.*

23-2-1-1 ENQ/ACK. Enquiry/acknowledgment (ENQ/ACK) is a relatively simple data-link-layer line discipline that works best in simple network environments where there is no doubt as to which station is the intended receiver. An example is a network comprised of only two stations (i.e., a two-point network) where the stations may be interconnected permanently (hardwired) or on a temporary basis through a switched network, such as the public telephone network.

Before data can be transferred between stations, procedures must be invoked that establish logical continuity between the source and destination stations and ensure that the destination station is ready and capable of receiving data. These are the primary purposes of line discipline procedures. ENQ/ACK line discipline procedures determine which device on a network can initiate a transmission and whether the intended receiver is available and ready to receive a message. Assuming all stations on the network have equal access to the transmission medium, a data session can be initiated by any station using ENQ/ACK. An exception would be a receive-only device, such as most printers, which cannot initiate a session with a computer.

The initiating station begins a session by transmitting a *frame, block,* or *packet* of data called an *enquiry* (ENQ), which identifies the receiving station. There does not seem to be any universally accepted standard definition of frames, blocks, and packets other than by size. Typically, packets are smaller than frames or blocks, although sometimes the term *packet* means only the information and not any overhead that may be included with the message. The terms *block* and *frame,* however, can usually be used interchangeably.

In essence, the ENQ sequence solicits the receiving station to determine if it is ready to receive a message. With half-duplex operation, after the initiating station sends an ENQ, it waits for a response from the destination station indicating its readiness to receive a message. If the destination station is ready to receive, it responds with a *positive acknowledgment* (ACK), and if it is not ready to receive, it responds with a *negative acknowledgment* (NAK). If the destination station does not respond with an ACK or a NAK within a specified period of time, the initiating station retransmits the ENQ. How many enquiries are made varies from network to network, but generally after three unsuccessful attempts to establish communications, the initiating station gives up (this is sometimes called a *time-out*). The initiating station may attempt to establish a session later, however, after several unsuccessful attempts; the problem is generally referred to a higher level of authority (such as a human).

A NAK transmitted by the destination station in response to an ENQ generally indicates a temporary unavailability, and the initiating station will simply attempt to establish a session later. An ACK from the destination station indicates that it is ready to receive data and tells the initiating station it is free to transmit its message. All transmitted message frames end with a unique terminating sequence, such as *end of transmission* (EOT), which indicates the end of the message frame. The destination station acknowledges all message frames received with either an ACK or a NAK. An ACK transmitted in response to a received message indicates the message was received without errors, and a NAK indicates that the message was received containing errors. A NAK transmitted in response to a message is usually interpreted as an automatic request for retransmission of the rejected message.

Figure 23-1 shows how a session is established and how data are transferred using ENQ/ACK procedures. Station A initiates the session by sending an ENQ to station B. Station B responds with an ACK indicating that it is ready to receive a message. Station A transmits message frame 1, which is acknowledged by station B with an ACK. Station A then transmits message frame 2, which is rejected by station B with a NAK, indicating that the message was received with errors. Station A then retransmits message frame 2, which is received without errors and acknowledged by station B with an ACK.

23-2-1-2 Poll/select. The poll/select line discipline is best suited to *centrally controlled* data communications networks using a multipoint topology, such as a bus, where one station or device is designated as the primary or host station and all other stations are designated as secondaries. Multipoint data communications networks using a single transmission medium must coordinate access to the transmission medium to prevent more than one station from attempting to transmit data at the same time. In addition, all exchanges of data must occur through the primary station. Therefore, if a secondary station wishes to transmit data to another secondary station, it must do so through the primary station. This is analogous to transferring data between memory devices in a computer using a central

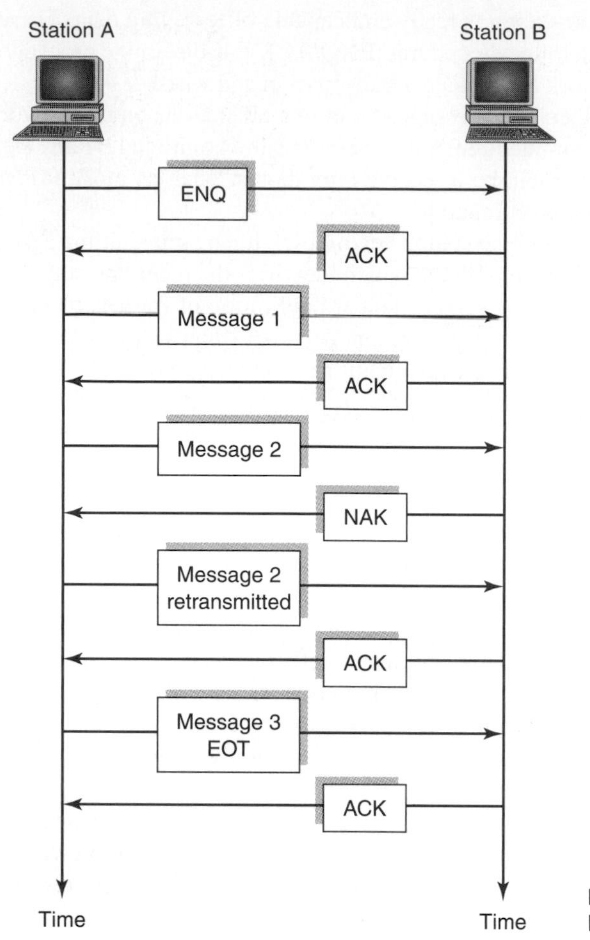

FIGURE 23-1 Example of ENQ/ACK line discipline

processing unit (CPU) where all data are read into the CPU from the source memory and then written to the destination memory.

In a poll/select environment, the primary station controls the data link, while secondary stations simply respond to instructions from the primary. The primary determines which device or station has access to the transmission channel (medium) at any given time. Hence, the primary initiates all data transmissions on the network with polls and selections.

A *poll* is a solicitation sent from the primary to a secondary to determine if the secondary has data to transmit. In essence, the primary designates a secondary as a transmitter (i.e., the master) with a poll. A *selection* is how the primary designates a secondary as a destination or recipient of data. A selection is also a query from the primary to determine if the secondary is ready to receive data. With two-point networks using ENQ/ACK procedures, there was no need for addresses because transmissions from one station were obviously intended for the other station. On multipoint networks, however, addresses are necessary because all transmissions from the primary go to all secondaries, and addresses are necessary to identify which secondary is being polled or selected. All secondary stations receive all polls and selections transmitted from the primary. With poll/select procedures, each secondary station is assigned one or more address for identification. It is the secondaries' responsibility to examine the address and determine if the poll or selection is intended for them. The primary has no address because transmissions from all secondary stations go only to the primary. A primary can poll only one station at a time; however, it can select more than one secondary at a time using *group* (more than one station) or *broadcast* (all stations) addresses.

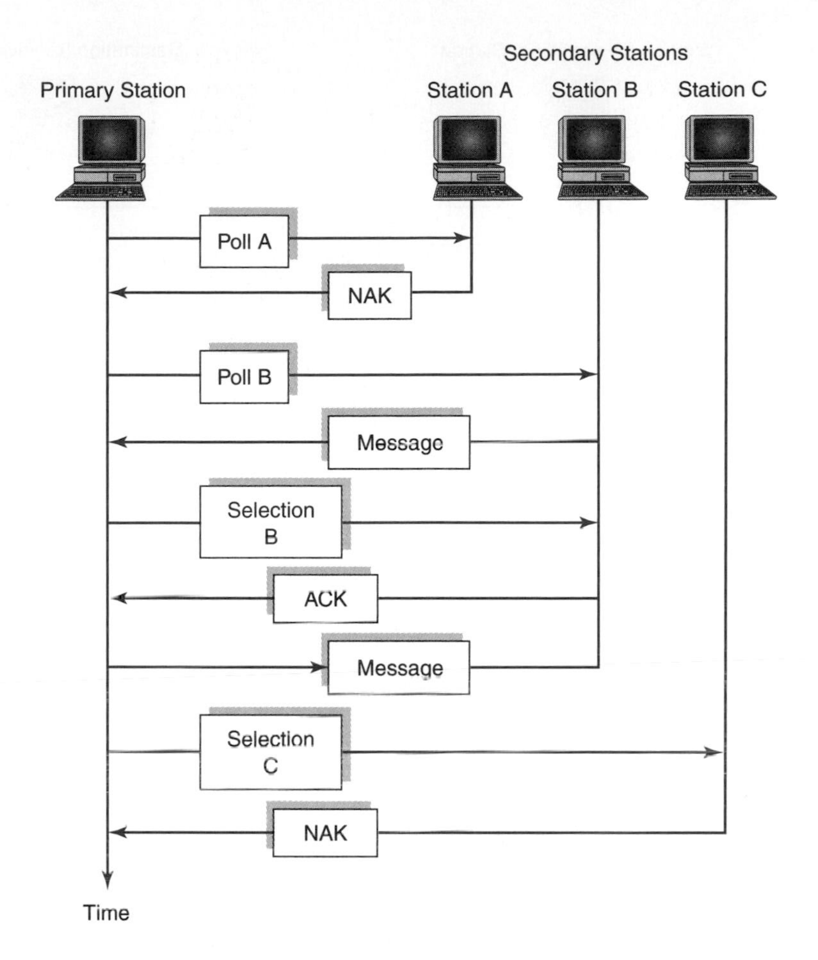

FIGURE 23-2 Example of poll/select line discipline

When a primary polls a secondary, it is soliciting the secondary for a message. If the secondary has a message to send, it responds to the poll with the message. This is called a positive acknowledgment to a poll. If the secondary has no message to send, it responds with a negative acknowledgment to the poll, which confirms that it received the poll but indicates that it has no messages to send at that time. This is called a negative acknowledgment to a poll.

When a primary selects a secondary, it is identifying the secondary as a receiver. If the secondary is available and ready to receive data, it responds with an ACK. If it is not available or ready to receive data, it responds with a NAK. These are called, respectively, positive and negative acknowledgments to a selection.

Figure 23-2 shows how polling and selections are accomplished using poll/select procedures. The primary polls station A, which responds with a negative acknowledgment to a poll (NAK) indicating it received the poll but has no message to send. Then the primary polls station B, which responds with a positive acknowledgment to a poll (i.e., a message). The primary then selects station B to see if it ready to receive a message. Station B responds with a positive acknowledgment to the selection (ACK), indicating that it is ready to receive a message. The primary transmits the message to station B. The primary then selects station C, which responds with a negative acknowledgment to the selection (NAK), indicating it is not ready to receive a message.

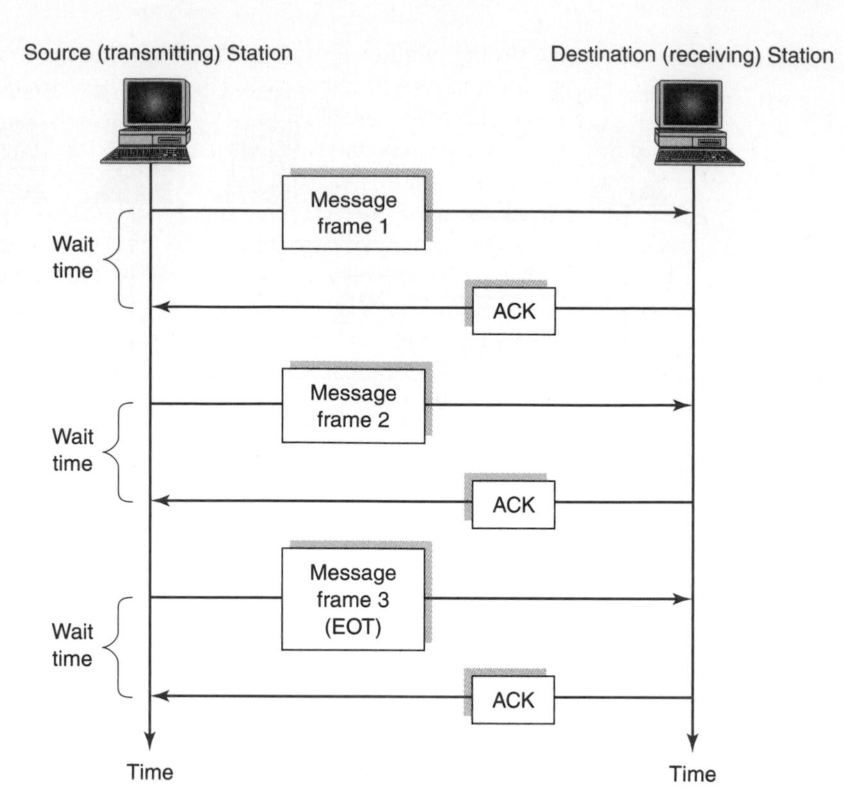

FIGURE 23-3 Example of stop-and-wait flow control

23-2-2 Flow Control

Flow control defines a set of procedures that tells the transmitting station how much data it can send before it must stop transmitting and wait for an acknowledgment from the destination station. The amount of data transmitted must not exceed the storage capacity of the destination station's buffer. Therefore, the destination station must have some means of informing the transmitting station when its buffers are nearly at capacity and telling it to temporarily stop sending data or to send data at a slower rate. There are two common methods of flow control: stop-and-wait and sliding window.

23-2-2-1 Stop-and-wait flow control. With *stop-and-wait* flow control, the transmitting station sends one message frame and then waits for an acknowledgment before sending the next message frame. After it receives an acknowledgment, it transmits the next frame. The transmit/acknowledgment sequence continues until the source station sends an end-of-transmission sequence. The primary advantage of stop-and-wait flow control is simplicity. The primary disadvantage is speed, as the time lapse between each frame is wasted time. Each frame takes essentially twice as long to transmit as necessary because both the message and the acknowledgment must traverse the entire length of the data link before the next frame can be sent.

Figure 23-3 shows an example of stop-and-wait flow control. The source station sends message frame 1, which is acknowledged by the destination station. After stopping transmission and waiting for the acknowledgment, the source station transmits the next frame (message frame 2). After sending the second frame, there is another lapse in time while the destination station acknowledges reception of frame 2. The time it takes the source station to transport three frames equates to at least three times as long as it would have taken to send the message in one long frame.

23-2-2-2 Sliding window flow control. With *sliding window* flow control, a source station can transmit several frames in succession before receiving an acknowledgment. There is only one acknowledgment for several transmitted frames, thus reducing the number of acknowledgments and considerably reducing the total elapsed transmission time as compared to stop-and-wait flow control.

The term *sliding window* refers to imaginary receptacles at the source and destination stations with the capacity of holding several frames of data. Message frames can be acknowledged any time before the window is filled with data. To keep track of which frames have been acknowledged and which have not, sliding window procedures require a modulo-n numbering system where each transmitted frame is identified with a unique sequence number between 0 and $n - 1$. n is any integer value equal to 2^x, where x equals the number of bits in the numbering system. With a three-bit binary numbering system, there are 2^3, or eight, possible numbers (0, 1, 2, 3, 4, 5, 6, and 7), and therefore the windows must have the capacity of holding $n - 1$ (seven) frames of data. The reason for limiting the number of frames to $n - 1$ is explained in Section 23-6-1-3.

The primary advantage of sliding window flow control is network utilization. With fewer acknowledgments (i.e., fewer line turnarounds), considerably less network time is wasted acknowledging messages, and more time can be spent actually sending messages. The primary disadvantages of sliding window flow control are complexity and hardware capacity. Each secondary station on a network must have sufficient buffer space to hold $2(n - 1)$ frames of data ($n - 1$ transmit frames and $n - 1$ receive frames), and the primary station must have sufficient buffer space to hold $m(2[n - 1])$, where m equals the number of secondary stations on the network. In addition, each secondary must store each unacknowledged frame it has transmitted and keep track of the number of each unacknowledged frame it transmits and receives. The primary station must store and keep track of all unacknowledged frames it has transmitted and received for each secondary station on the network.

23-2-3 Error Control

Error control includes both error detection and error correction. However, with the data-link layer, error control is concerned primarily with error detection and message retransmission, which is the most common method of error correction.

With poll/select line disciplines, all polls, selections, and message transmissions end with some type of end-of-transmission sequence. In addition, all messages transported from the primary to a secondary or from a secondary to the primary are acknowledged with ACK or NAK sequences to verify the validity of the message. An ACK means the message was received with no transmission errors, and a NAK means there were errors in the received message. A NAK is an automatic call for retransmission of the last message.

Error detection at the data-link layer can be accomplished with any of the methods described in Chapter 22, such as VRC, LRC, or CRC. Error correction is generally accomplished with a type of retransmission called *automatic repeat request* (ARQ) (sometimes called *automatic request for retransmission*). With ARQ, when a transmission error is detected, the destination station sends a NAK back to the source station requesting retransmission of the last message frame or frames. ARQ also calls for retransmission of missing or lost frames, which are frames that either never reach the secondary or are damaged so severely that the destination station does not recognize them. ARQ also calls for retransmission of frames where the acknowledgments (either ACKs or NAKs) are lost or damaged.

There are two types of ARQ: stop-and-wait and sliding window. Stop-and-wait flow control generally incorporates *stop-and-wait ARQ,* and sliding window flow control can implement ARQ in one of two variants: *go-back-n* frames or *selective reject* (SREJ). With go-back-n frames, the destination station tells the source station to go back n frames and retransmit all of them, even if all the frames did not contain errors. Go-back-n requests retransmission of the damaged frame plus any other frames that were transmitted after it. If the second frame in a six-frame message were received in error, five frames would be retransmitted. With selective reject, the destination station tells the source station to retransmit only the frame (or frames) received in error. Go-back-*n* is easier to implement, but it also wastes more time, as most of the frames retransmitted were not received in error. Selective reject is more complicated to implement but saves transmission time, as only those frames actually damaged are retransmitted.

23-3 CHARACTER- AND BIT-ORIENTED DATA-LINK PROTOCOLS

All data-link protocols transmit control information either in separate control frames or in the form of overhead that is added to the data and included in the same frame. Data-link protocols can be generally classified as either character or bit oriented.

23-3-1 Character-Oriented Protocols

Character-oriented protocols interpret a frame of data as a group of successive bits combined into predefined patterns of fixed length, usually eight bits each. Each group of bits represents a unique character. Control information is included in the frame in the form of standard characters from an existing character set, such as ASCII. Control characters convey important information pertaining to line discipline, flow control, and error control.

With character-oriented protocols, unique data-link control characters, such as start of text (STX) and end of text (ETX), no matter where they occur in a transmission, warrant the same action or perform the same function. For example, the ASCII code 02 hex represents the STX character. Start of text, no matter where 02 hex occurs within a data transmission, indicates that the next character is the first character of the text or information portion of the message. Care must be taken to ensure that the bit sequences for data-link control characters do not occur within a message unless they are intended to perform their designated data-link functions.

Character-oriented protocols are sometimes called *byte-oriented* protocols. Examples of character-oriented protocols are XMODEM, YMODEM, ZMODEM, KERMIT, BLAST, IBM 83B Asynchronous Data Link Protocol, and IBM Binary Synchronous Communications (BSC—bisync). Bit-oriented protocols are more efficient than character-oriented protocols.

23-3-2 Bit-Oriented Protocols

A *bit-oriented protocol* is a discipline for serial-by-bit information transfer over a data communications channel. With bit-oriented protocols, data-link control information is transferred as a series of successive bits that may be interpreted individually on a bit-by-bit basis or in groups of several bits rather than in a fixed-length group of *n* bits where *n* is usually the number of bits in a data character. In a bit-oriented protocol, there are no dedicated data-link control characters. With bit-oriented protocols, the control field within a frame may convey more than one control function.

Bit-oriented typically convey more information into shorter frames than character-oriented protocols. The most popular bit-oriented protocol are Synchronous Data Link Communications (SDLC) and High-Level Data Link Communications (HDLC).

23-4 ASYNCHRONOUS DATA-LINK PROTOCOLS

Asynchronous data-link protocols are relatively simple, character-oriented protocols generally used on two-point networks using asynchronous data and asynchronous modems. Asynchronous protocols, such as XMODEM and YMODEM, are commonly used to facilitate communications between two personal computers over the public switched telephone network.

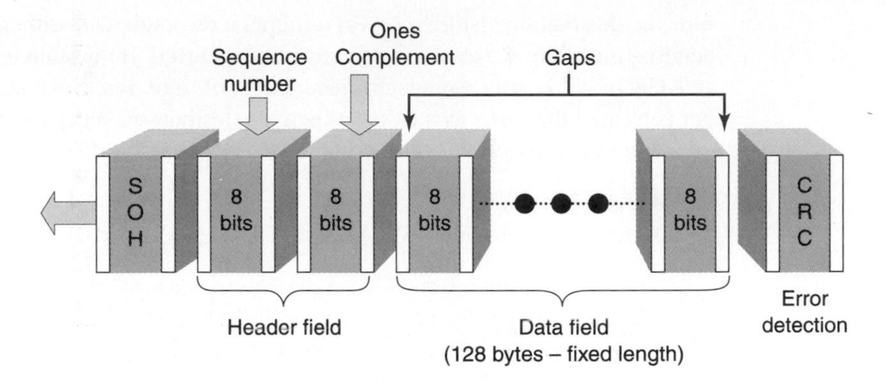

FIGURE 23-4 XMODEM frame format

23-4-1 XMODEM

In 1979, a man named Ward Christiansen developed the first *file transfer protocol* designed to facilitate transferring data between two personal computers (PCs) over the public switched telephone network. Christiansen's protocol is now called *XMODEM*. XMODEM is a relatively simple data-link protocol intended for low-speed applications. Although XMODEM was designed to provide communications between two PCs, it can also be used between a PC and a mainframe or host computer.

XMODEM specifies a half-duplex stop-and-wait protocol using a data frame comprised of four fields. The frame format for XMODEM contains four fields as shown in Figure 23-4. The four fields for XMODEM are the SOH field, header field, data field, and error-detection field. The first field of an XMODEM frame is simply a one-byte start of heading (SOH) field. SOH is a data-link control character that is used to indicate the beginning of a header. Headers are used for conveying system information, such as the message number. SOH simply indicates that the next byte is the first byte of the header. The second field is a two-byte sequence that is the actual header for the frame. The first header byte is called the sequence number, as it contains the number of the current frame being transmitted. The second header byte is simply the 2-s complement of the first byte, which is used to verify the validity of the first header byte (this is sometimes called *complementary redundancy*). The next field is the information field, which contains the actual user data. The information field has a maximum capacity of 128 bytes (e.g., 128 ASCII characters). The last field of the frame is an eight-bit CRC frame check sequence, which is used for error detection.

Data transmission and control are quite simple with the XMODEM protocol—too simple for most modern-day data communications networks. The process of transferring data begins when the destination station sends a NAK character to the source station. Although NAK is the acronym for a negative acknowledgment, when transmitted by the destination station at the beginning of an XMODEM data transfer, it simply indicates that the destination station is ready to receive data. After the source station receives the initial NAK character, it sends the first data frame and then waits for an acknowledgment from the destination station. If the data are received without errors, the destination station responds with an ACK character (positive acknowledgment). If the data is received with errors, the destination station responds with a NAK character, which calls for a retransmission of the data. After the originate station receives the NAK character, it retransmits the same frame. Each time the destination station receives a frame, it responds with either a NAK or an ACK, depending on whether a transmission error has occurred. If the source station does not receive an ACK or NAK after a predetermined length of time, it retransmits the last frame. A time-out is treated the same as a NAK. When the destination station wishes to prematurely terminate a transmission, it inserts a cancel (CAN) character.

23-4-2 YMODEM

YMODEM is a protocol similar to XMODEM except with the following exceptions:

1. The information field has a maximum capacity of 1024 bytes.
2. Two CAN characters are required to abort a transmission.
3. ITU-T-CRC 16 is used to calculate the frame check sequence.
4. Multiple frames can be sent in succession and then acknowledged with a single ACK or NAK character.

23-5 SYNCHRONOUS DATA-LINK PROTOCOLS

With *synchronous data-link protocols,* remote stations can have more than one PC or printer. A group of computers, printers, and other digital devices is sometimes called a *cluster.* A single line control unit (LCU) can serve a cluster with as many as 50 devices. Synchronous data-link protocols are generally used with synchronous data and synchronous modems and can be either character or bit oriented. One of the most common synchronous data-link protocols is IBM's binary synchronous communications (BSC).

23-5-1 Binary Synchronous Communications

Binary synchronous communications (BSC) is a synchronous character-oriented data-link protocol developed by IBM. BSC is sometimes called *bisync* or *bisynchronous communications.* With BSC, each data transmission is preceded by a unique synchronization (SYN) character as shown here:

S	S	
Y	Y	message
N	N	

The message can be a poll, a selection, an acknowledgment, or a message containing user information.

The SYN character with ASCII is 16 hex and with EBCDIC 32 hex. The SYN character places the receiver in the character (byte) mode and prepares it to receive data in eight-bit groupings. With BSC, SYN characters are always transmitted in pairs (hence the name *bisync* or *bisynchronous communications*). Received data are shifted serially one bit at a time through the detection circuit, where they are monitored in groups of 16 bits looking for two SYN characters. Two SYN characters are used to avoid misinterpreting a random eight-bit sequence in the middle of a message with the same bit sequence as a SYN character. For example, if the ASCII characters A and b were received in succession, the following bit sequence would occur:

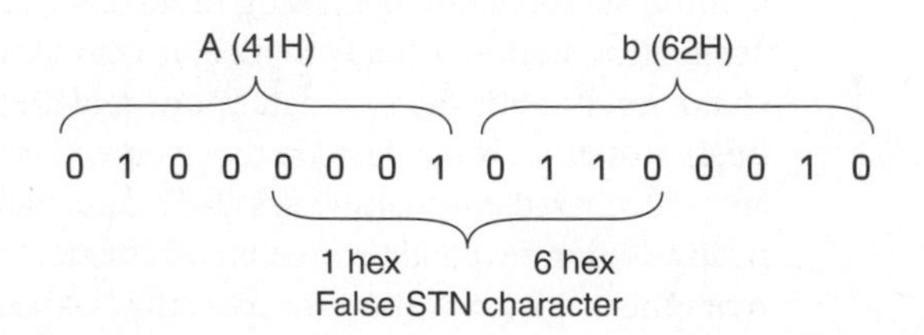

As you can see, it appears that a SYN character has been received when in fact it has not. To avoid this situation, SYN characters are always transmitted in pairs and, consequently, if only one is detected, it is ignored. The likelihood of two false SYN characters occurring one immediately after the other is remote.

23-5-1-1 BSC polling sequences. BSC uses a poll/select format to control data transmission. There are two polling formats used with bisync: general and specific. The format for a *general poll* is

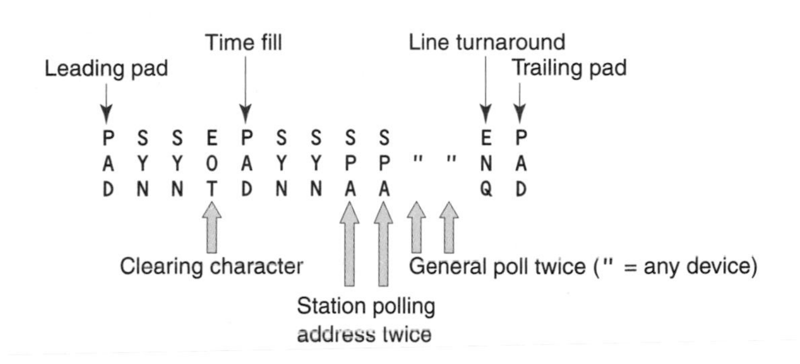

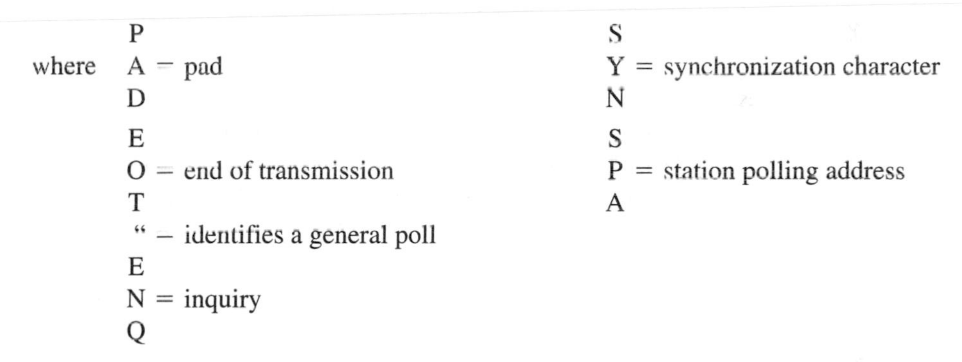

where PAD = pad
SYN = synchronization character
EOT = end of transmission
SPA = station polling address
" – identifies a general poll
ENQ = inquiry

The PAD character at the beginning of the sequence is called a *leading pad* and is either 55 hex or AA hex (01010101 or 10101010). A leading pad is simply a string of alternating 1s and 0s for clock synchronization. Immediately following the leading pad are two SYN characters that establish character synchronization. The EOT character is a *clearing character* that places all secondary stations into the line monitor mode. The PAD character immediately following the second SYN character is simply a string of successive logic 1s that serves as a time fill, giving each of the secondary stations time to clear. The number of logic 1s transmitted during this time fill is not necessarily a multiple of eight bits. Consequently, two more SYN characters are transmitted to reestablish character synchronization. Two *station polling address* (SPA) characters are transmitted for error detection (character redundancy). A secondary will not recognize or respond to a poll unless its SPA appears twice in succession. The two quotation marks signify that the poll is a general poll for any device at that station that has a formatted message to send. The enquiry (ENQ) character is sometimes called a *format* or *line turnaround* character because it simply completes the polling sequence and initiates a line turnaround.

The PAD character at the end of the polling sequence is a trailing pad (FF). The purpose of the trailing pad is to ensure that the RLSD signal in the receive modem is held active long enough for the entire message to be demodulated.

Table 23-1 Station and Device Addresses

Station or Device Number	SPA	SSA	DA	Station or Device Number	SPA	SSA	DA
0	sp	—	sp	16	&	0	&
1	A	/	A	17	J	1	J
2	B	S	B	18	K	2	K
3	C	T	C	19	L	3	L
4	D	U	D	20	M	4	M
5	E	V	E	21	N	5	N
6	F	W	F	22	O	6	O
7	G	X	G	23	P	7	P
8	H	Y	H	24	Q	8	Q
9	I	Z	I	25	R	9	R
10	[	-	[	26	]	:	]
11	.	,	.	27	$	#	$
12	<	%	<	28	*	@	*
13	(	—	(	29	)	‘	)
14	+	>	+	30	;	=	;
15	!	?	!	31	^	”	^

With BSC, there is a second polling sequence called a *specific poll.* The format for a specific poll is

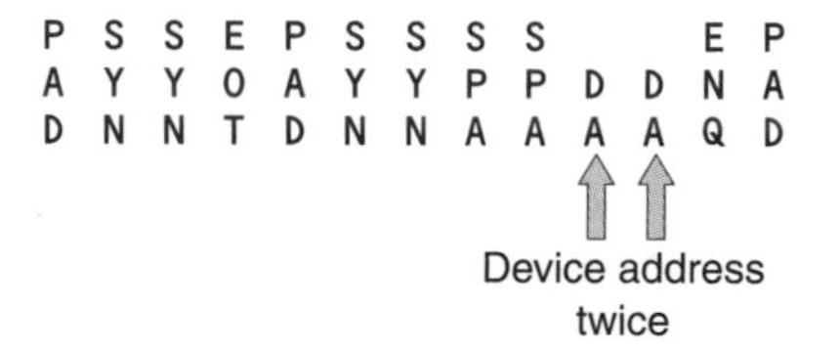

The character sequence for a specific poll is identical to a general poll except two device address (DA) characters are substituted for the two quotation marks. With a specific poll, both the station and the device address are included. Therefore, a specific poll is an invitation for only one specific device at a given secondary station to transmit its message. Again, two DA characters are transmitted for redundancy error detection.

Table 23-1 lists the station polling addresses, station selection addresses, and device addresses for a BSC system with a maximum of 32 stations and 32 devices.

With bisync, there are only two ways in which a secondary station can respond to a poll: with a formatted message or with an ACK. The character sequence for a ACK is

P S S E P
A Y Y O A
D N N T D

23-5-1-2 BSC selection sequence. The format for a selection with BSC is

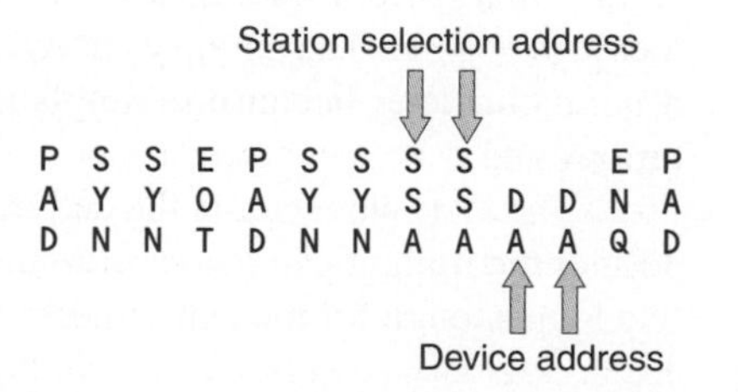

The sequence for a selection is very identical to a specific poll except two SSA characters are substituted for the two SPA characters. SSA stands for *station selection address.* All selections are specific, as they are for a specific device at the selected station.

A secondary station can respond to a selection with either a positive or a negative acknowledgment. A positive acknowledgment to a selection indicates that the device selected is ready to receive. The character sequence for a positive acknowledgment is

P	S	S	D		P
A	Y	Y	L	0	A
D	N	N	E		D

A negative acknowledgment to a selection indicates that the selected device is not ready to receive. A negative acknowledgment is called a reverse interrupt (RVI). The character sequence for a negative acknowledgment to a selection is

P	S	S	D		P
A	Y	Y	L	6	A
D	N	N	E		D

23-5-1-3 BSC message sequence. Bisync uses stop-and-wait flow control and stop-and-wait ARQ. Formatted messages are sent from secondary stations to the primary station in response to a poll and sent from primary stations to secondary stations after the secondary has been selected. Formatted messages use the following format:

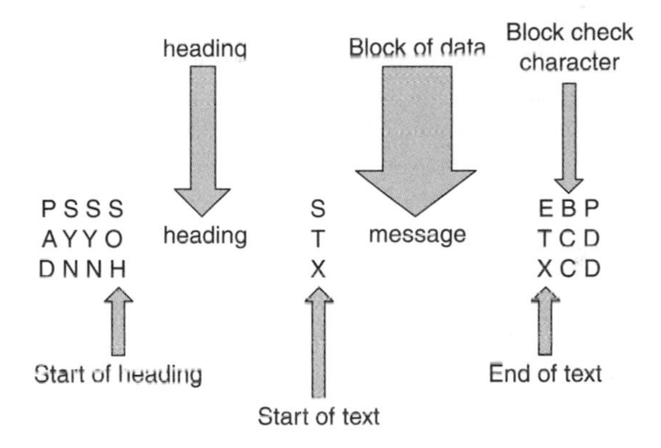

The *block check character* (BCC) uses longitudinal redundancy checking (LRC) with ASCII-coded messages and cyclic redundancy checking (CRC-16) for EBCDIC-coded messages (when CRC-16 is used, there are two BCCs). The BCC is sometimes called a block check sequence (BCS) because it does not represent a character; it is simply a sequence of bits used for error detection.

The BCC is computed beginning with the first character after SOH and continues through and includes the *end of text* (ETX) character. (If there is no heading, the BCC is computed beginning with the first character after start of text.) Data are transmitted in blocks or frames that are generally between 256 and 1500 bytes long. ETX is used to terminate the last block of a message. *End of block* (ETB) is used for multiple block messages to terminate all message blocks except the last one. The last block of a message is always terminated with ETX. The receiving station must acknowledge all BCCs.

A positive acknowledgment to a BCC indicates that the BCC was good, and a negative acknowledgment to a BCC indicates that the BCC was bad. A negative acknowledgment is an automatic request for retransmission (ARQ). The character sequences for positive and negative acknowledgments are the following:

Positive responses to BCCs (messages):

P	S	S	D		P	
A	Y	Y	L	0	A	(even-numbered blocks)
D	N	N	E		D	

or

P	S	S	D		P	
A	Y	Y	L	1	A	(odd-numbered blocks)
D	N	N	E		D	

Negative response to BCCs (messages):

P	S	S	N	P
A	Y	Y	A	A
D	N	N	K	D

where NAK = negative acknowledgment

23-5-1-4 BSC transparency. It is possible that a device attached to one or more of the ports of a station controller is not a computer terminal or printer. For example, a microprocessor-controlled system used to monitor environmental conditions (temperature, humidity, and so on) or a security alarm system. If so, the data transferred between it and the primary are not ASCII- or EBCDIC-encoded characters. Instead, they could be microprocessor op-codes or binary-encoded data. Consequently, it would be possible that an eight-bit sequence could occur within the message that is equivalent to a data-link control character. For example, if the binary code 00000011 (03 hex) occurred in a message, the controller would misinterpret it as the ASCII code for the ETX. If this happened, the controller would terminate the message and interpret the next sequence of bits as the BCC. To prevent this from occurring, the controller is made *transparent* to the data. With bisync, a *data-link escape* (DLE) character is used to achieve transparency. To place a controller into the transparent mode, STX is preceded by a DLE. This causes the controller to transfer the data to the selected device without searching through the message looking for data-link control characters. To come out of the transparent mode, DLE ETX is transmitted. To transmit a bit sequence equivalent to DLE as part of the text, it must be preceded by a DLE character (i.e., DLE DLE). There are only three additional circumstances with transparent data when it is necessary to precede a character with DLE:

1. *DLE ETB.* Used to terminate all blocks of data except the final block.
2. *DLE ITB.* Used to terminate blocks of transparent text other than the final block when ITB (end of intermittent block) is used for a block-terminating character.
3. *DLE SYN.* With bisync, two SYN characters are inserted in the text in messages lasting longer than 1 second to ensure that the receive controller maintains character synchronization.

23-6 SYNCHRONOUS DATA-LINK CONTROL

Synchronous data-link control (SDLC) is a synchronous bit-oriented protocol developed in the 1970s by IBM for use in *system network architecture* (SNA) environments. SDLC was the first link-layer protocol based on synchronous, bit-oriented operation. The Inter-

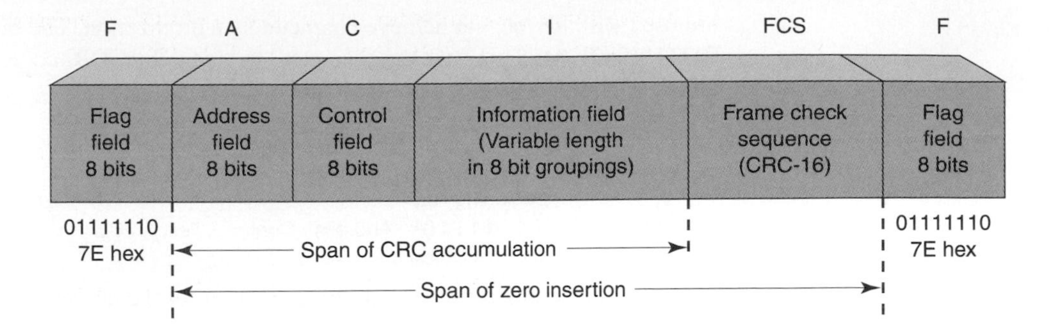

FIGURE 23-5 SDLC frame format

national Organization for Standardization modified SDLC and created high-level data-link control (HDLC) and the International Telecommunications Union—Telecommunications Standardization Sector (ITU-T) subsequently modified HDLC to create Link Access Procedures (LAP). The Institute of Electrical and Electronic Engineers (IEEE) modified HDLC and created IEEE 802.2. Although each of these protocol variations is important in its own domain, SDLC remains the primary SNA link-layer protocol for wide area data networks.

SDLC can transfer data simplex, half duplex, or full duplex and can support a variety of link types and topologies. SDLC can be used on point-to-point or multipoint networks over both circuit- and packet-switched networks. SDLC is a bit-oriented protocol (BOP) where there is a single control field within a message *frame* that performs essentially all the data-link control functions. SDLC frames are generally limited to 256 characters in length. EBCDIC was the original character language used with SDLC.

There are two types of network nodes defined by SDLC: *primary stations* and *secondary stations.* There is only one primary station in an SDLC circuit, which controls data exchange on the communications channel and issues *commands.* All other stations on an SDLC network are secondary stations, which receive commands from the primary and return (transmit) *responses* to the primary station.

There are three transmission states with SDLC: transient, idle, and active. The *transient state* exists before and after an initial transmission and after each line turnaround. A secondary station assumes the circuit is in an idle state after receiving 15 or more consecutive logic 1s. The *active state* exists whenever either the primary or one of the secondary stations is transmitting information or control signals.

23-6-1 SDLC Frame Format

Figure 23-5 shows an SDLC frame format. Frames transmitted from the primary and secondary stations use exactly the same format. There are five *fields* included in an SDLC frame:

1. Flag field (beginning and ending)
2. Address field
3. Control field
4. Information (or text) field
5. Frame check sequence field

23-6-1-1 SDLC flag field. There are two *flag fields* per frame, each with a minimum length of one byte. The two flag fields are the beginning flag and ending flag. Flags are used for the *delimiting sequence* for the frame and to achieve *frame and character synchronization.* The delimiting sequence sets the limits of the frame (i.e., when the frame begins and when it ends). The flag is used with SDLC in a manner similar to the way SYN characters are used with bisync—to achieve character synchronization. The bit sequence for a flag is 01111110 (7E hex), which is the character "=" in the EBCDIC code. There are several variations to how flags are transmitted with SDLC:

1. One beginning and one ending flag for each frame:

Beginning flag					Ending flag
01111110	Address	Control	Text	FCC	01111110

2. The ending flag from one frame is used for the beginning flag for the next frame:

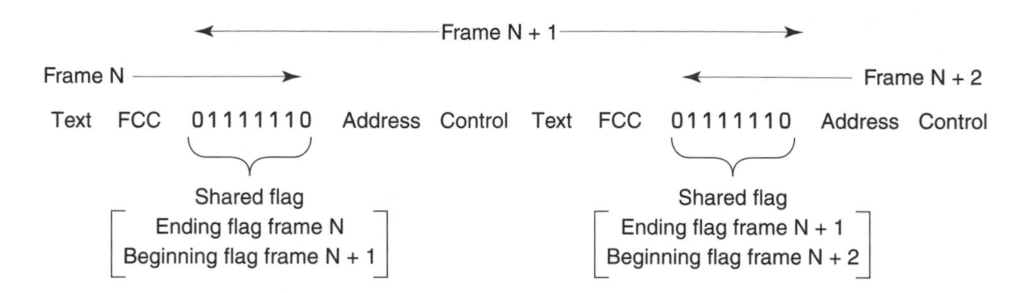

3. The last zero of an ending flag can be the first zero of the beginning flag of the next frame:

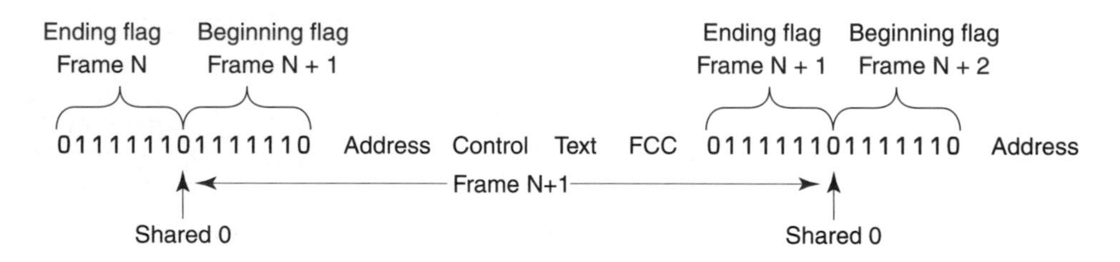

4. Flags are transmitted continuously during the time between frames in lieu of idle line 1s:

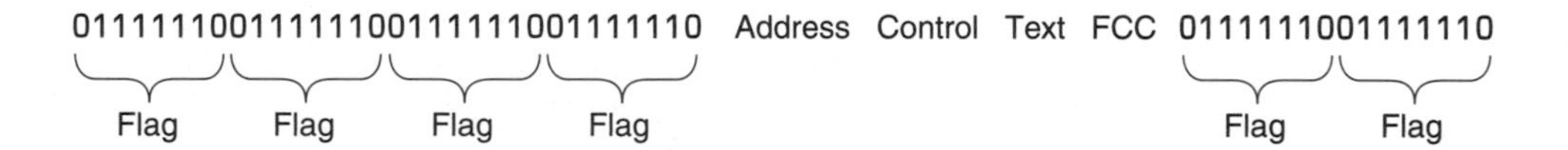

23-6-1-2 SDLC address field. An SDLC address field contains eight bits; therefore, 256 addresses are possible. The address 00 hex (00000000) is called the *null* address and is never assigned to a secondary station. The null address is used for network testing. The address FF hex (11111111) is called the *broadcast* address. The primary station is the only station that can transmit the broadcast address. When a frame is transmitted with the broadcast address, it is intended for all secondary stations. The remaining 254 addresses can be used as unique station addresses intended for one secondary station only or as group addresses that are intended for more than one secondary station but not all of them.

In frames sent by the primary station, the address field contains the address of the secondary station (i.e., the address of the destination). In frames sent from a secondary station, the address field contains the address of the secondary (i.e., the address of the station sending the message). The primary station has no address because all transmissions from secondary stations go to the primary.

23-6-1-3 SDLC control field. The control field is an eight-bit field that identifies the type of frame being transmitted. The control field is used for polling, confirming previously received frames, and several other data-link management functions. There are three frame formats with SDLC: information, supervisory, and unnumbered.

Information frame. With an *information frame,* there must be an information field, which must contain user data. Information frames are used for transmitting sequenced information that must be acknowledged by the destination station. The bit pattern for the control field of an information frame is

Bit:	b_0	b_1	b_2	b_3	b_4	b_5	b_6	b_7
Function:		*nr*		P or F $\overline{P}$ or $\overline{F}$		*ns*		0 Indicates information frame

A logic 0 in the high-order bit position identifies an information frame (I-frame). With information frames, the primary can select a secondary station, send formatted information, confirm previously received information frames, and poll a secondary station—with a single transmission.

Bit b_3 of an information frame is called a *poll* (P) or *not-a-poll* ($\overline{P}$) bit when sent by the primary and a *final* (F) or *not-a-final* ($\overline{F}$) bit when sent by a secondary. In frames sent from the primary, if the primary desires to poll the secondary (i.e., solicit it for information), the P bit in the control field is set (logic 1). If the primary does not wish to poll the secondary, the P bit is reset (logic 0). With SDLC, a secondary cannot transmit frames unless it receives a frame addressed to it with the P bit set. This is called the *synchronous response mode.* When the primary is transmitting multiple frames to the same secondary, b_3 is a logic 0 in all but the last frame. In the last frame, b_3 is set, which demands a response from the secondary. When a secondary is transmitting multiple frames to the primary, b_3 in the control field is a logic 0 in all frames except the last frame. In the last frame, b_3 is set, which simply indicates that the frame is the last frame in the message sequence.

In information frames, bits b_4, b_5, and b_6 of the control field are *ns* bits, which are used in the transmit sequence number (*ns* stands for "number sent"). All information frames must be numbered. With three bits, the binary numbers 000 through 111 (0 through 7) can be represented. The first frame transmitted by each station is designated frame 000, the second frame 001, and so on up to frame 111 (the eighth frame), at which time the count cycles back to 000 and repeats.

SDLC uses a sliding window ARQ for error correction. In information frames, bits b_0, b_1, and b_2 in the control field are the *nr* bits, which are the receive numbering sequence used to indicate the status of previously received information frames (*nr* stands for "number received"). The *nr* bits are used to confirm frames received without errors and to automatically request retransmission of information frames received with errors. The *nr* is the number of the next information frame the transmitting station expects to receive or the number of the next information frame the receiving station will transmit. The *nr* confirms received frames through $nr - 1$. Frame $nr - 1$ is the last information frame received without a transmission error. For example, when a station transmits $nr = 5$, it is confirming successful reception of previously unconfirmed frames up through frame 4. Together, the *ns* and *nr* bits are used for error correction (ARQ). The primary station must keep track of the *ns* and *nr* for each secondary station. Each secondary station must keep track of only its *ns* and *nr.* After all frames have been confirmed, the primary station's *ns* must agree with the secondary station's *nr* and vice versa.

For the following example, both the primary and secondary stations begin with their *nr* and *ns* counters reset to 000. The primary begins the information exchange by sending three information frames numbered 0, 1, and 2 (i.e., the *ns* bits in the control character for the three frames are 000, 001, and 010). In the control character for the three frames, the primary transmits an $nr = 0$ (i.e., 000). An $nr = 0$ is transmitted because the next frame the primary expects to receive from the secondary is frame 0, which is the secondary's present *ns*. The secondary responds with two information frames ($ns = 0$ and 1). The secondary received all three frames from the primary without any errors; therefore, the *nr* transmitted in the secondary's control field is 3, which is the number of the next frame the primary will send. The primary now sends information frames 3 and 4 with an $nr = 2$, which confirms the correct reception of frames 0 and 1 from the secondary. The secondary responds with frames $ns = 2$, 3, and 4 with an $nr = 4$. The $nr = 4$ confirms reception of only frame 3 from the primary ($nr - 1$). Consequently, the primary retransmits frame 4. Frame 4 is transmitted together with four additional frames ($ns = 5$, 6, 7, and 0). The primary's $nr = 5$, which confirms frames 2, 3, and 4 from the secondary. Finally, the secondary sends information frame 5 with an $nr = 1$, which confirms frames 4, 5, 6, 7, and 0 from the primary. At this point, all frames transmitted have been confirmed except frame 5 from the secondary. The preceding exchange of information frames is shown in Figure 23-6.

With SDLC, neither the primary nor the secondary station can send more than seven numbered information frames in succession without receiving a confirmation. For example, if the primary sent eight frames ($ns = 0$, 1, 2, 3, 4, 5, 6, and 7) and the secondary responded with an $nr = 0$, it is ambiguous which frames are being confirmed. Does $nr = 0$

Primary Station																	
Status	ns	0	1	2			3	4				4	5	6	7	0	
	nr:	0	0	0			2	2				5	5	5	5	5	
	P/$\overline{P}$	0	0	1			0	1				0	0	0	0	1	
Control Field	b_0	0	0	0			0	0				1	1	1	1	1	
	b_1	0	0	0			1	1				0	0	0	0	0	
	b_2	0	0	0			0	0				1	1	1	1	1	
	b_3	0	0	1			0	1				0	0	0	0	1	
	b_4	0	0	0			0	1				1	1	1	1	0	
	b_5	0	0	1			1	0				0	0	1	1	0	
	b_6	0	1	0			1	0				0	1	0	1	0	
	b_7	0	0	0			0	0				0	0	0	0	0	
hex code		00	02	14			46	58				A8	AA	AC	AE	B0	
Secondary Station																	
Status	ns:				0	1			2	3	4						5
	nr:				3	3			4	4	4						1
	F/$\overline{F}$				0	1			0	0	1						1
Control Field	b_0				0	0			1	1	1						0
	b_1				1	1			0	0	0						0
	b_2				1	1			0	0	0						1
	b_3				0	1			0	0	1						1
	b_4				0	0			0	0	1						1
	b_5				0	0			1	1	0						0
	b_6				0	1			0	1	0						1
	b_7				0	0			0	0	0						0
hex code					60	72			84	86	98						3A

FIGURE 23-6 SDLC exchange of information frames

mean that all eight frames were received correctly, or does it mean that frame 0 had an error in it and all eight frames must be retransmitted? (All frames beginning with $nr - 1$ must be retransmitted.)

Example 23-1

Determine the bit pattern for the control field of an information frame sent from the primary to a secondary station for the following conditions:

a. Primary is sending information frame 3 ($ns = 3$)
b. Primary is polling the secondary (P = 1)
c. Primary is confirming correct reception of frames 2, 3, and 4 from the secondary ($nr = 5$)

Solution

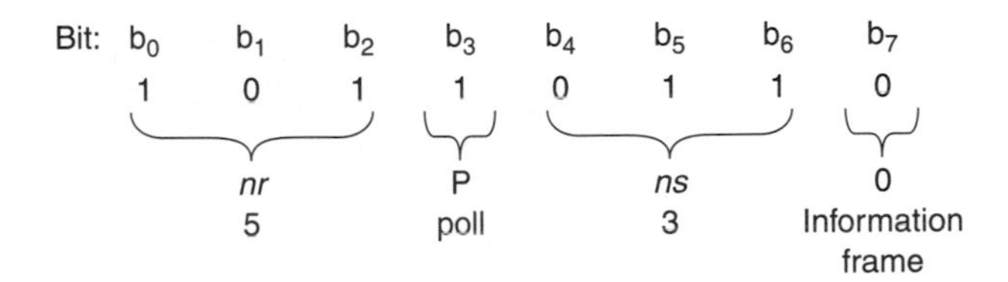

Example 23-2

Determine the bit pattern for the control field of an information frame sent from a secondary station to the primary for the following conditions:

a. Secondary is sending information frame 7 ($ns = 7$)
b. Secondary is not sending its final frame (F = 0)
c. Secondary is confirming correct reception of frames 2 and 3 from the primary ($nr = 4$)

Solution

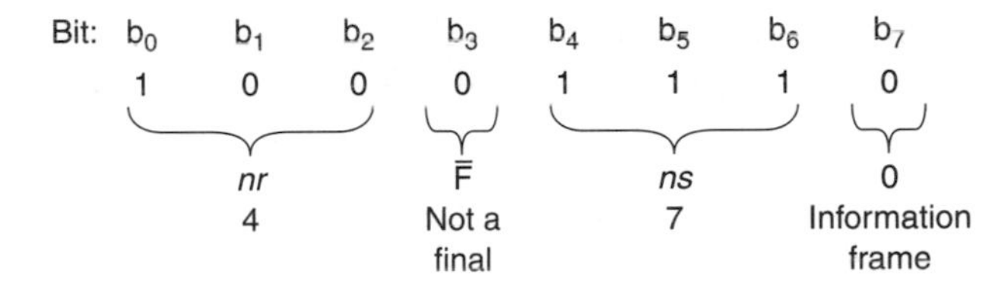

Supervisory frame. With *supervisory frames,* an information field is not allowed. Consequently, supervisory frames cannot be used to transfer numbered information; however, they can be used to assist in the transfer of information. Supervisory frames can be used to confirm previously received information frames, convey ready or busy conditions, and for a primary to poll a secondary when the primary does not have any numbered information to send to the secondary. The bit pattern for the control field of a supervisory frame is

Bit: b_0 b_1 b_2 b_3 b_4 b_5 b_6 b_7

Function: *nr* | P or F, $\bar{P}$ or $\bar{F}$ | X X Function code | 0 1 Indicates supervisory frame

A supervisory frame is identified with a 01 in bit positions b_6 and b_7, respectively, of the control field. With the supervisory format, bit b_3 is again the poll/not-a-poll or final/not-a-final bit, and b_0, b_1, and b_2 are the *nr* bits. Therefore, supervisory frames can be used by a primary to poll a secondary, and both the primary and the secondary stations can use supervisory frames to confirm previously received information frames. Bits b_4 and b_5 in a supervisory are the function code that either indicate the receive status of the station transmitting the frame or request transmission or retransmission of sequenced information frames. With two bits, there are four combinations possible. The four combinations and their functions are the following:

b_4	b_5	Receive Status
0	0	Ready to receive (RR)
0	1	Ready not to receive (RNR)
1	0	Reject (REJ)
1	1	Not used with SDLC

When a primary station sends a supervisory frame with the P bit set and a status of ready to receive, it is equivalent to a general poll. Primary stations can use supervisory frames for polling and also to confirm previously received information frames without sending any information. A secondary uses the supervisory format for confirming previously received information frames and for reporting its receive status to the primary. If a secondary sends a supervisory frame with RNR status, the primary cannot send it numbered information frames until that status is cleared. RNR is cleared when a secondary sends an information frame with the F bit = 1 or a supervisory frame indicating RR or REJ with the F bit = 0. The REJ command/response is used to confirm information frames through $nr - 1$ and to request transmission of numbered information frames beginning with the frame number identified in the REJ frame. An information field is prohibited with a supervisory frame, and the REJ command/response is used only with full-duplex operation.

Example 23-3

Determine the bit pattern for the control field of a supervisory frame sent from a secondary station to the primary for the following conditions:

a. Secondary is ready to receive (RR)
b. It is a final frame
c. Secondary station is confirming correct reception of frames 3, 4, and 5 ($nr = 6$)

Solution

Bit:	b_0	b_1	b_2	b_3	b_4	b_5	b_6	b_7
	1	1	0	1	0	0	0	1
		nr 6		F final	RR		Supervisory frame	

Unnumbered frame. An *unnumbered frame* is identified by making bits b_6 and b_7 in the control field both logic 1s. The bit pattern for the control field of an unnumbered frame is

Bit:	b_0	b_1	b_2	b_3	b_4	b_5	b_6	b_7
Function:	X	X	X	P or F $\bar{P}$ or $\bar{F}$	X	X	1	1
		Function code			Function code		Indicates unnumbered frame	

With an unnumbered frame, bit b_3 is again either the poll/not-a-poll or final/not-a-final bit. There are five X bits (b_0, b_1, b_2, b_4, and b_5) included in the control field of an unnumbered frame that contain the function code, which is used for various unnumbered commands and responses. With five bits available, there are 32 unnumbered commands/

Table 23-2 Unnumbered Commands and Responses

Binary Configuration								
b_0		b_7	Acronym	Command	Response	I Field Prohibited	Resets *ns* and *nr*
000	P/F	0011	UI	Yes	Yes	No	No
000	F	0111	RIM	No	Yes	Yes	No
000	P	0111	SIM	Yes	No	Yes	Yes
100	P	0011	SNRM	Yes	No	Yes	Yes
000	F	1111	DM	No	Yes	Yes	No
010	P	0011	DISC	Yes	No	Yes	No
011	F	0011	UA	No	Yes	Yes	No
100	F	0111	FRMR	No	Yes	No	No
111	F	1111	BCN	No	Yes	Yes	No
110	P/F	0111	CFGR	Yes	Yes	No	No
010	F	0011	RD	No	Yes	Yes	No
101	P/F	1111	XID	Yes	Yes	No	No
111	P/F	0011	TEST	Yes	Yes	No	No

responses possible. The control field in an unnumbered frame sent from a primary station is called a command, and the control field in an unnumbered frame sent from a secondary station is called a response. With unnumbered frames, there are neither *ns* nor *nr* bits included in the control field. Therefore, numbered information frames cannot be sent or confirmed with the unnumbered format. Unnumbered frames are used to send network control and status information. Two examples of control functions are placing a secondary station on- or off-line and initializing a secondary station's line control unit (LCU).

Table 23-2 lists several of the more common unnumbered commands and responses. Numbered information frames are prohibited with all unnumbered frames. Therefore, user information cannot be transported with unnumbered frames and, thus, the control field in unnumbered frames does not include *nr* and *ns* bits. However, information fields containing control information are allowed with the following unnumbered commands and responses: UI, FRMR, CFGR, TEST, and XID.

A secondary station must be in one of three modes: *initialization mode, normal response mode,* or *normal disconnect mode.* The procedures for placing a secondary station into the initialization mode are system specified and vary considerably. A secondary in the normal response mode cannot initiate unsolicited transmissions; it can transmit only in response to a frame received with the P bit set. When in the normal disconnect mode, a secondary is off-line. In this mode, a secondary station will accept only the TEST, XID, CFGR, SNRM, or SIM commands from the primary station and can respond only if the P bit is set.

The unnumbered commands and responses are summarized here:

1. *Unnumbered information* (*UI*). UI can be a command or a response that is used to send unnumbered information. Unnumbered information transmitted in the I-field is not acknowledged.
2. *Set initialization mode* (*SIM*). SIM is a command that places a secondary station into the initialization mode. The initialization procedure is system specified and varies from a simple self-test of the station controller to executing a complete IPL (initial program logic) program. SIM resets the *ns* and *nr* counters at the primary and secondary stations. A secondary is expected to respond to a SIM command with an unnumbered acknowledgment (UA) response.
3. *Request initialization mode* (*RIM*). RIM is a response sent by a secondary station to request the primary to send a SIM command.

4. *Set normal response mode* (*SNRM*). SNRM is a command that places a secondary into the normal response mode (NRM). A secondary station cannot send or receive numbered information frames unless it is in the normal response mode. Essentially, SNRM places a secondary station on-line. SNRM resets the *ns* and *nr* counters at both the primary and secondary stations. UA is the normal response to a SNRM command. Unsolicited responses are not allowed when a secondary is in the NRM. A secondary station remains in the NRM until it receives a disconnect (DISC) or SIM command.
5. *Disconnect mode* (*DM*). DM is a response transmitted from a secondary station if the primary attempts to send numbered information frames to it when the secondary is in the normal disconnect mode.
6. *Request disconnect* (*RD*). RD is a response sent by a secondary when it wants the primary to place it in the disconnect mode.
7. *Disconnect* (*DISC*). DISC is a command that places a secondary station in the normal disconnect mode (NDM). A secondary cannot send or receive numbered information frames when it is in the normal disconnect mode. When in the NDM, a secondary can receive only SIM or SNRM commands and can transmit only a DM response. The expected response to a DISC is UA.
8. *Unnumbered acknowledgment* (*UA*). UA is an affirmative response that indicates compliance to SIM, SNRM, or DISC commands. UA is also used to acknowledge unnumbered information frames.
9. *Frame reject* (*FRMR*). FRMR is for reporting procedural errors. The FRMR response is an answer transmitted by a secondary after it has received an invalid frame from the primary. A received frame may be invalid for any one of the following reasons:
   a. The control field contains an invalid or unassigned command.
   b. The amount of data in the information field exceeds the buffer space in the secondary station's controller.
   c. An information field is received in a frame that does not allow information fields.
   d. The *nr* received is incongruous with the secondary's *ns,* for example, if the secondary transmitted *ns* frames 2, 3, and 4 and then the primary responded with an *nr* = 7.

   A secondary station cannot release itself from the FRMR condition, nor does it act on the frame that caused the condition. The secondary repeats the FRMR response until it receives one of the following mode-setting commands: SNRM, DISC, or SIM. The information field for a FRMR response must contain three bytes (24 bits) and has the following format:

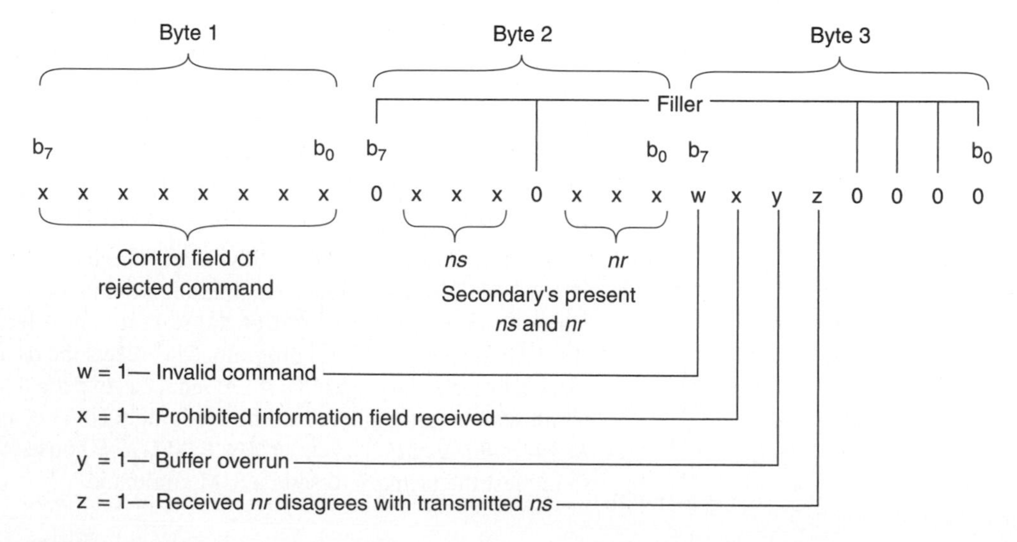

10. *TEST.* The TEST command/response is an exchange of frames between the primary station and a secondary station. An information field may be included with the TEST command; however, it cannot be sequenced (numbered). The primary sends a TEST command to a secondary in any mode to solicit a TEST response. If an information field is included with the command, the secondary returns it with its response. The TEST command/response is exchanged for link testing purposes.
11. *Exchange station identification (XID).* XID can be a command or a response. As a command, XID solicits the identification of a secondary station. An information field can be included in the frame to convey the identification data of either the primary or the secondary. For dial-up circuits, it is often necessary that the secondary station identify itself before the primary will exchange information frames with it, although XID is not restricted only to dial-up circuits.

23-6-1-4 SDLC information field. All information transmitted in an SDLC frame must be in the information field (I-field), and the number of bits in the information field must be a multiple of eight. An information field is not allowed with all SDLC frames; however, the data within an information field can be user information or control information.

23-6-1-5 Frame Check Character (FCC) field. The FCC field contains the error detection mechanism for SDLC. The FCC is equivalent to the BCC used with binary synchronous communications (BSC). SDLC uses CRC-16 and the following generating polynomial: $x^{16} + x^{12} + x^5 + x^1$. Frame check characters are computed on the data in the address, control and information fields.

23-6-2 SDLC Loop Operation

An SDLC loop operates half-duplex. The primary difference between the loop and bus configurations is that in a loop, all transmissions travel in the same direction on the communications channel. In a loop configuration, only one station transmits at a time. The primary station transmits first, then each secondary station responds sequentially. In an SDLC loop, the transmit port of the primary station controller is connected to the receive port of the controller in the first down-line secondary station. Each successive secondary station is connected in series with the transmission path with the transmit port of the last secondary station's controller on the loop connected to the receive port of the primary station's controller. Figure 23-7 shows the physical layout for an SDLC loop.

In an SDLC loop, the primary transmits sequential frames where each frame may be addressed to any or all of the secondary stations. Each frame transmitted by the primary station contains an address of the secondary station to which that frame is directed. Each secondary station, in turn, decodes the address field of every frame and then serves as a repeater for all stations that are down-loop from it. When a secondary station detects a frame with its address, it copies the frame, then passes it on to the next down-loop station. All frames transmitted by the primary are returned to the primary. When the primary has completed transmitting, it follows the last flag with eight consecutive logic 0s. A flag followed by eight consecutive logic 0s is called a *turnaround sequence*, which signals the end of the primary's transmissions. Immediately following the turnaround sequence, the primary transmits continuous logic 1s, which is called the go-ahead sequence. A secondary station cannot transmit until it receives a frame address to it with the P bit set, a turnaround sequence, and then a go-ahead sequence. Once the primary has begun transmitting continuous logic 1s, it goes into the receive mode.

The first down-loop secondary station that receives a frame addressed to it with the P bit set changes the go-ahead sequence to a flag, which becomes the beginning flag of that secondary station's response frame or frames. After the secondary station has transmitted its last frame, it again becomes a repeater for the idle line 1s from the primary, which become the go-ahead sequence for the next down-loop secondary station. The next secondary station that receives a frame addressed to it with the P bit set detects the turnaround sequence, any frames transmitted from up-loop

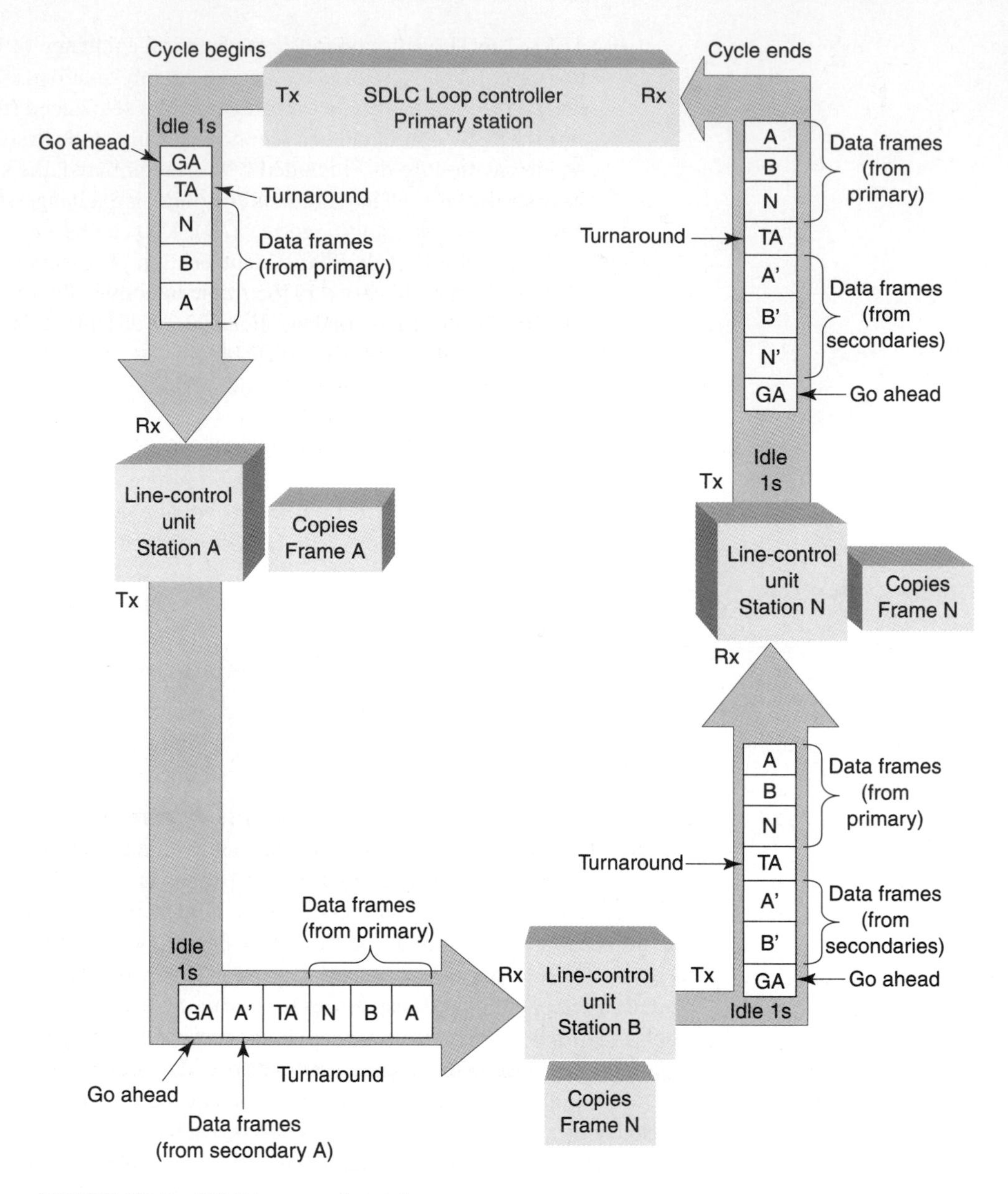

FIGURE 23-7 SDLC loop configuration

secondary stations, and then the go-ahead sequence. Each secondary station inserts its response frames immediately after the last frame transmitted by an up-loop secondary. Frames transmitted from the primary are separated from frames transmitted by the secondaries by the turnaround sequence. Without the separation, it would be impossible to tell which frames were from the primary and which were from a secondary, as their frame formats (including the address field) are identical. The cycle is completed when the primary station receives its own turnaround sequence, a series of response frames, and then the go-ahead sequence.

The previously described sequence is summarized here:

1. Primary transmits sequential frames to one or more secondary stations.
2. Each transmitted frame contains a secondary station's address.
3. After a primary has completed transmitting, it follows the last flag of the last frame with eight consecutive logic 0s (turnaround sequence) followed by continuous logic 1s (go-ahead sequence -0111111111111 - - - -).

4. The turnaround sequence alerts secondary stations of the end of the primary's transmissions.
5. Each secondary, in turn, decodes the address field of each frame and removes frames addressed to them.
6. Secondary stations serve as repeaters for any down-line secondary stations.
7. Secondary stations cannot transmit frames of their own unless they receive a frame with the P bit set.
8. The first secondary station that receives a frame addressed to it with the P bit set changes the seventh logic 1 in the go-ahead sequence to a logic 0, thus creating a flag. The flag becomes the beginning flag for the secondary station's response frames.
9. The next down-loop secondary station that receives a frame addressed to it with the P bit set detects the turnaround sequence, any frames transmitted by other up-loop secondary stations, and then the go-ahead sequence.
10. Each secondary station's response frames are inserted immediately after the last repeated frame.
11. The cycle is completed when the primary receives its own turnaround sequence, a series of response frames, and the go-ahead sequence.

23-6-2-1 SDLC loop configure command/response. The configure command/response (CFGR) is an unnumbered command/response that is used only in SDLC loop configurations. CFGR contains a one-byte *function descriptor* (essentially a subcommand) in the information field. A CFGR command is acknowledged with a CFGR response. If the low-order bit of the function descriptor is set, a specified function is initiated. If it is reset, the specified function is cleared. There are six subcommands that can appear in the configure command/response function field:

1. *Clear* (00000000). A *clear* subcommand causes all previously set functions to be cleared by the secondary. The secondary's response to a clear subcommand is another clear subcommand, 00000000.
2. *Beacon test* (*BCN*) (0000000X). The *beacon test* subcommand causes the secondary receiving it to turn on (00000001) or turn off (00000000) its carrier. The beacon response is called a carrier, although it is not a carrier in the true sense of the word. The beacon test command causes a secondary station to begin transmitting a beacon response, which is not a carrier. However, if modems were used in the circuit, the beacon response would cause the modem's carrier to turn on. The beacon test is used to isolate open-loop continuity problems. In addition, whenever a secondary station detects a loss of signal (either data or idle line ones), it automatically begins to transmit its beacon response. The secondary will continue transmitting the beacon until the loop resumes normal status.
3. *Monitor mode* (0000010X). The *monitor* command (00000101) causes the addressed secondary station to place itself into the monitor (receive-only) mode. Once in the monitor mode, a secondary cannot transmit until it receives either a monitor mode clear (00000100) or a clear (00000000) subcommand.
4. *Wrap* (0000100X). The *wrap* command (00001001) causes a secondary station to loop its transmissions directly to its receiver input. The wrap command places the secondary effectively off-line for the duration of the test. A secondary station takes itself out of the wrap mode when it receives a wrap clear (00001000) or clear (00000000) subcommand.
5. *Self-test* (0000101X). The self-test subcommand (00001011) causes the addressed secondary to initiate a series of internal diagnostic tests. When the tests are completed, the secondary will respond. If the P bit in the configure command is set, the secondary will respond following completion of the self-test or at its earliest opportunity. If the P bit is reset, the secondary will respond following completion of the test to the next poll-type frame it receives from the primary.

All other transmissions are ignored by the secondary while it is performing a self-test; however, the secondary will repeat all frames received to the next down-loop station. The secondary reports the results of the self-test by setting or clearing the low-order bit (X) of its self-test response. A logic 1 means that the tests were unsuccessful, and a logic 0 means that they were successful.

6. *Modified link test* (0000110X). If the *modified link test* function is set (X bit set), the secondary station will respond to a TEST command with a TEST response that has an information field containing the first byte of the TEST command information field repeated n times. The number n is system specified. If the X bit is reset, the secondary station will respond with a zero-length information field. The modified link test is an optional subcommand and is used only to provide an alternative form of link test to that previously described for the TEST command.

23-6-2-2 SDLC transparency. With SDLC, the flag bit sequence (01111110) can occur within a frame where it is not intended to be a flag. For instance, within the address, control, or information fields, a combination of one or more bits from one character combined with one or more bits from an adjacent character could produce a 01111110 pattern. If this were to happen, the receive controller would misinterpret the sequence for a flag, thus destroying the frame. Therefore, the pattern 01111110 must be prohibited from occurring except when it is intended to be a flag.

One solution to the problem would be to prohibit certain sequences of characters from occurring, which would be difficult to do. A more practical solution would be to make a receiver transparent to all data located between beginning and ending flags. This is called *transparency*. The *transparency mechanism* used with SDLC is called *zero-bit insertion* or *zero stuffing*. With zero-bit insertion, a logic 0 is automatically inserted after any occurrence of five consecutive logic 1s except in a designated flag sequence (i.e., flags are not zero inserted). When five consecutive logic 1s are received and the next bit is a 0, the 0 is automatically deleted or removed. If the next bit is a logic 1, it must be a valid flag. An example of zero insertion/deletion is shown here:

Original frame bits at the transmit station:

01111110	01101111	11010011	1110001100110101	01111110
Beginning flag	Address	Control	Frame check character	Ending flag

After zero insertion but prior to transmission:

01111110	01101111	101010011	11100001100110101	01111110
Beginning flag	Address	Control	Frame check character	Ending flag

Inserted zeros

After zero deletion at the receive end:

01111110	01101111	11010011	1110001100110101	01111110
Beginning flag	Address	Control	Frame check character	Ending flag

23-6-3 Message Abort

Message abort is used to prematurely terminate an SDLC frame. Generally, this is done only to accommodate high-priority messages, such as emergency link recovery procedures. A message abort is any occurrence of seven to 14 consecutive logic 1s. Zeros are not inserted in an abort sequence. A message abort terminates an existing frame and immediately begins the higher-

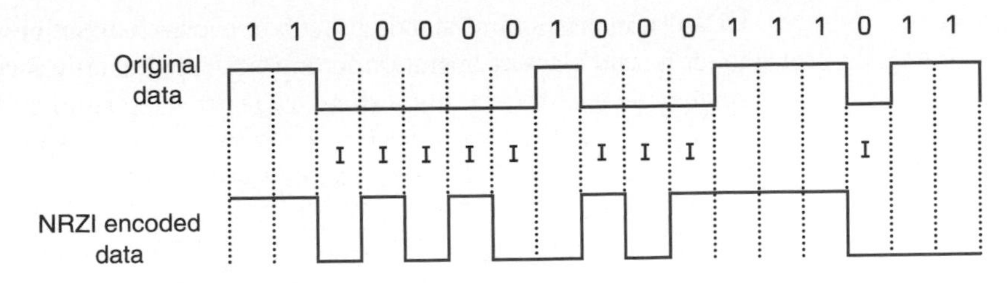

*I indicates to invert

FIGURE 23-8 NRZI encoding

priority frame. If more than 14 consecutive logic 1s occur in succession, it is considered an idle line condition. Therefore, 15 or more contiguous logic 1s places the circuit into the idle state.

23-6-4 Invert-on-Zero Encoding

With binary synchronous transmission such as SDLC, transmission and reception of data must be time synchronized to enable identification of sequential binary digits. Synchronous data communications assumes that bit or time synchronization is provided by either the DCE or the DTE. The master transmit clock can come from the DTE or, more likely, the DCE. However, the receive clock must be recovered from the data by the DCE and then transferred to the DTE. With synchronous data transmission, the DTE receiver must sample the incoming data at the same rate that it was outputted from the transmit DTE. Although minor variations in timing can exist, the receiver in a synchronous modem provides data clock recovery and dynamically adjusted sample timing to keep sample times midway between bits. For a DCE to recover the data clock, it is necessary that transitions occur in the data. Traditional unipolar (UP) logic levels, such as TTL (0 V and +5 V), do not provide transitions for long strings of logic 0s or logic 1s. Therefore, they are inadequate for clock recovery without placing restrictions on the data. *Invert-on-zero coding* is the encoding scheme used with SDLC because it guarantees at least one transition in the data for every seven bits transmitted. Invert-on-zero coding is also called NRZI (*nonreturn-to-zero inverted*).

With NRZI encoding, the data are encoded in the controller at the transmit end and then decoded in the controller at the receive end. Figure 23-8 shows examples of NRZI encoding and decoding. The encoded waveform is unchanged by 1s in the NRZI encoder. However, logic 0s cause the encoded transmission level to invert from its previous state (i.e., either from a high to a low or from a low to a high). Consequently, consecutive logic 0s are converted to an alternating high/low sequence. With SDLC, there can never be more than six logic 1s in succession (a flag). Therefore, a high-to-low transition is guaranteed to occur at least once every seven bits transmitted except during a message abort or an idle line condition. In a NRZI decoder, whenever a high/low or low/high transition occurs in the received data, a logic 0 is generated. The absence of a transition simply generates a logic 1. In Figure 23-8, a high level is assumed prior to encoding the incoming data.

NRZI encoding was originally intended for asynchronous modems that do not have clock recovery capabilities. Consequently, the receive DTE must provide time synchronization, which is aided by using NRZI-encoded data. Synchronous modems have built-in scrambler and descrambler circuits that ensure transitions in the data and, thus, NRZI encoding is unnecessary. The NRZI encoder/decoder is placed in between the DTE and the DCE.

23-7 HIGH-LEVEL DATA-LINK CONTROL

In 1975, the International Organization for Standardization (ISO) defined several sets of substandards that, when combined, are called *high-level data-link control* (HDLC). HDLC is a superset of SDLC; therefore, only the added capabilities are explained.

HDLC comprises three standards (subdivisions) that outline the frame structure, control standards, and class of operation for a bit-oriented data-link control (DLC):

1. ISO 3309
2. ISO 4335
3. ISO 7809

23-7-1 ISO 3309

The ISO 3309 standard defines the frame structure, delimiting sequence, transparency mechanism, and error-detection method used with HDLC. With HDLC, the frame structure and delimiting sequence are essentially the same as with SDLC. An HDLC frame includes a beginning flag field, an address field, a control field, an information field, a frame check character field, and an ending flag field. The delimiting sequence with HDLC is a binary 01111110, which is the same flag sequence used with SDLC. However, HDLC computes the frame check characters in a slightly different manner. HDLC can use either CRC-16 with a generating polynomial specified by CCITT V.41 for error detection. At the transmit station, the CRC characters are computed such that when included in the FCC computations at the receive end, the remainder for an errorless transmission is always FOB8. HDLC also offers an optional 32-bit CRC checksum.

HDLC has extended addressing capabilities. HDLC can use an eight-bit address field or an *extended addressing* format, which is virtually limitless. With extended addressing, the address field may be extended recursively. If b_0 in the address field is a logic 1, the seven remaining bits are the secondary's address (the ISO defines the low-order bit as b_0, whereas SDLC designates the high-order bit as b_0). If b_0 is a logic 0, the next byte is also part of the address. If b_0 of the second byte is a logic 0, a third address byte follows and so on until an address byte with a logic 1 for the low-order bit is encountered. Essentially, there are seven bits available in each address byte for address encoding. An example of a three-byte extended addressing scheme is shown in the following. Bit b_0 in the first two bytes of the address field are logic 0s, indicating that one or more additional address bytes follow. However, b_0 in the third address byte is a logic 1, which terminates the address field. There are a total of 21 address bits (seven in each byte):

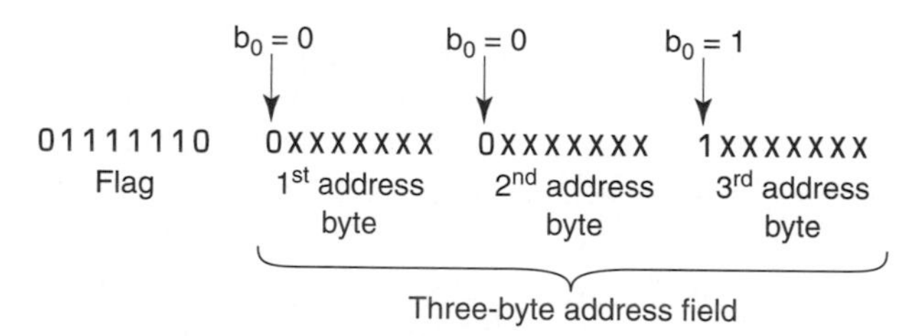

23-7-2 Information Field

HDLC permits any number of bits in the information field of an information command or response. With HDLC, any number of bits may be used for a character in the I field as long as all characters have the same number of bits.

23-7-3 Elements of Procedure

The ISO 3309 standard defines the elements of procedure for HDLC. The control and information fields have increased capabilities over SDLC, and there are two additional operational modes allowed with HDLC.

23-7-3-1 Control field. With HDLC, the control field can be extended to 16 bits. Seven bits are for the *ns,* and seven bits are for the *nr.* Therefore, with the extended control format, there can be a maximum of 127 outstanding (unconfirmed) frames at any given time. In essence, a primary station can send 126 successive information frames to a secondary station with the P bit = 0 before it would have to send a frame with the P bit = 1.

With HDLC, the supervisory format includes a fourth status condition: selective reject (SREJ). SREJ is identified by two logic 1s in bit positions b_4 and b_5 of a supervisory control field. With SREJ, a single frame can be rejected. A SREJ calls for the retransmission of only one frame identified by the three-bit *nr* code. A REJ calls for the retransmission of all frames beginning with frame identified by the three-bit *nr* code. For example, the primary sends I frames *ns* = 2, 3, 4, and 5. If frame 3 were received in error, a REJ with an *nr* of 3 would call for a retransmission of frames 3, 4, and 5. However, a SREJ with an *nr* of 3 would call for the retransmission of only frame 3. SREJ can be used to call for the retransmission of any number of frames except only one at a time.

23-7-3-2 HDLC operational modes. SDLC specifies only one operational mode, called the *normal response mode* (NRM), which allows secondaries to communicate with the primary only after the primary has given the secondary permission to transmit. With SDLC, when a station is logically disconnected from the network, it is said to be in the *normal disconnect mode.*

HDLC has two additional operational modes: *asynchronous response mode* (ARM) and *asynchronous balanced mode* (ABM). With ARM, secondary stations are allowed to send unsolicited responses (i.e., communicate with the primary without permission). To transmit, a secondary does not need to have received a frame from the primary with the P bit set. However, if a secondary receives a frame with the P bit set, it must respond with a frame with the F bit set. HDLC also specifies an *asynchronous disconnect mode,* which is identical to the normal disconnect mode except that the secondary can initiate an asynchronous DM or RIM response at any time.

The ISO 7809 standard combines previous standards 6159 (unbalanced) and 6256 (balanced) and outlines the class of operation necessary to establish the link-level protocol. Unbalanced operation is a class of operation logically equivalent to a multipoint private-line circuit with a polling environment. There is a single primary station responsible for central control of the network. Data transmission may be either half or full duplex.

Asynchronous balanced mode is a mode of operation logically equivalent to a two-point private-line circuit where each station has equal data-link responsibilities (a station can operate as a primary or as a secondary), which enables a station to initiate data transmission without receiving permission from any other station. Channel access is accomplished through contention on a two-wire circuit using the asynchronous response mode. Data transmission is half duplex on a two-wire circuit or full duplex over a four-wire circuit.

23-8 PUBLIC SWITCHED DATA NETWORKS

A *public switched data network* (PDN or PSDN) is a switched data communications network similar to the public telephone network except a PDN is designed for transferring data only. A public switched data network is comprised of one or more wide-area data networks designed to provide access to a large number of subscribers with a wide variety of computer equipment.

The basic principle behind a PDN is to transport data from a source to a destination through a network of intermediate *switching nodes* and transmission media. The switching nodes are not concerned with the content of the data, as their purpose is to provide *end stations* access to transmission media and other switching nodes that will transport data from node to node until it reaches its final destination. Figure 23-9 shows a public switched data network comprised of several switching nodes interconnected with *transmission links* (channels). The end-station devices can be personal computers, servers, mainframe computers, or any other piece of computer hardware capable of sending or receiving data. End stations are connected to the network through switching nodes. Data enter the network where they are routed through one or more intermediate switching nodes until reaching their destination.

Some switching nodes connect only to other switching nodes (sometimes called *tandem switching nodes* or *switchers switches*), while some switching nodes are connected to end stations as well. Node-to-node communications links generally carry multiplexed data

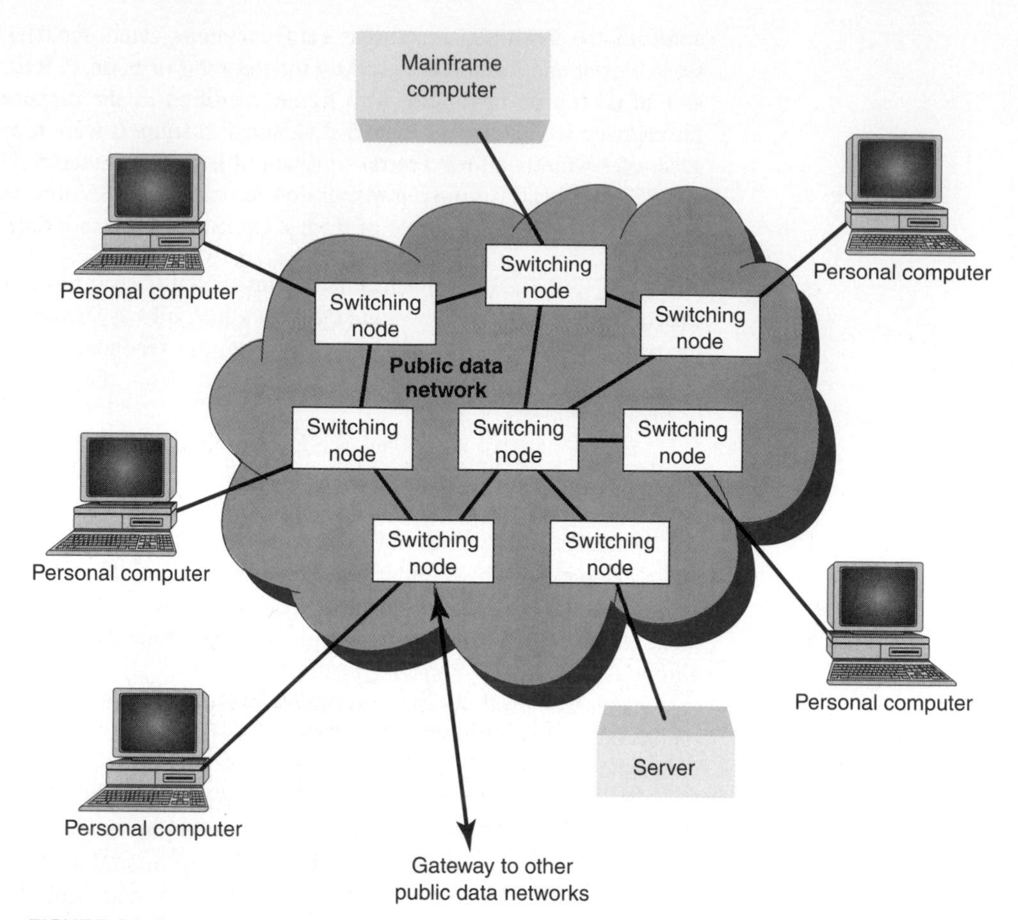

FIGURE 23-9 Public switched data network

(usually time-division multiplexing). Public data networks are not *direct connected;* that is, they do not provide direct communications links between every possible pair of nodes.

Public switched data networks combine the concepts of *value-added networks* (VANs) and *packet switching networks.*

23-8-1 Value-Added Network

A value-added network "adds value" to the services or facilities provided by a common carrier to provide new types of communication services. Examples of added values are error control, enhanced connection reliability, dynamic routing, failure protection, logical multiplexing, and data format conversions. A VAN comprises an organization that leases communications lines from common carriers such as AT&T and MCI and adds new types of communications services to those lines. Examples of value-added networks are GTE Telnet, DATAPAC, TRANSPAC, and Tymnet Inc.

23-8-2 Packet Switching Network

Packet switching involves dividing data messages into small bundles of information and transmitting them through communications networks to their intended destinations using computer-controlled switches. Three common switching techniques are used with public data networks: *circuit switching, message switching,* and *packet switching.*

23-8-2-1 Circuit switching. Circuit switching is used for making a standard telephone call on the public telephone network. The call is established, information is transferred, and then the call is disconnected. The time required to establish the call is called the *setup* time. Once the call has been established, the circuits interconnected by the network switches are allocated to a single user for the duration of the call. After a call has been established, information is transferred in *real time.* When a call is terminated, the circuits and switches are once again available for another user. Because there are a limited number of circuits and switching paths available, *blocking* can occur. Blocking is the inability to complete a call because there are no facilities or switching paths available between the source and destination locations. When circuit switching is used for data transfer, the terminal equipment at the source and destination must be compatible; they must use compatible modems and the same bit rate, character set, and protocol.

A circuit switch is a *transparent* switch. The switch is transparent to the data; it does nothing more than interconnect the source and destination terminal equipment. A circuit switch adds no value to the circuit.

23-8-2-2 Message switching. Message switching is a form of *store-and-forward* network. Data, including source and destination identification codes, are transmitted into the network and stored in a switch. Each switch within the network has message storage capabilities. The network transfers the data from switch to switch when it is convenient to do so. Consequently, data are not transferred in real time; there can be a delay at each switch. With message switching, blocking cannot occur. However, the delay time from message transmission to reception varies from call to call and can be quite long (possibly as long as 24 hours). With message switching, once the information has entered the network, it is converted to a more suitable format for transmission through the network. At the receive end, the data are converted to a format compatible with the receiving data terminal equipment. Therefore, with message switching, the source and destination data terminal equipment do not need to be compatible. Message switching is more efficient than circuit switching because data that enter the network during busy times can be held and transmitted later when the load has decreased.

A message switch is a *transactional* switch because it does more than simply transfer the data from the source to the destination. A message switch can store data or change its format and bit rate, then convert the data back to their original form or an entirely different form at the receive end. Message switching multiplexes data from different sources onto a common facility.

23-8-2-3 Packet switching. With packet switching, data are divided into smaller segments, called *packets,* prior to transmission through the network. Because a packet can be held in memory at a switch for a short period of time, packet switching is sometimes called a *hold-and-forward* network. With packet switching, a message is divided into packets, and each packet can take a different path through the network. Consequently, all packets do not necessarily arrive at the receive end at the same time or in the same order in which they were transmitted. Because packets are small, the hold time is generally quite short, message transfer is near real time, and blocking cannot occur. However, packet switching networks require complex and expensive switching arrangements and complicated protocols. A packet switch is also a transactional switch. Circuit, message, and packet switching techniques are summarized in Table 23-3.

23-9 CCITT X.25 USER-TO-NETWORK INTERFACE PROTOCOL

In 1976, the CCITT designated the X.25 user interface as the international standard for packet network access. Keep in mind that X.25 addresses only the physical, data-link, and network layers in the ISO seven-layer model. X.25 uses existing standards when possible. For example, X.25 specifies X.21, X.26, and X.27 standards as the physical interface, which correspond to EIA RS-232, RS-423A, and RS-422A standards, respectively. X.25 defines HDLC as the international standard for the data-link layer and the American National Standards Institute (ANSI) 3.66 *Advanced Data Communications Control Procedures*

Table 23-3 Switching Summary

Circuit Switching	Message Switching	Packet Switching
Dedicated transmission path	No dedicated transmission path	No dedicated transmission path
Continuous transmission of data	Transmission of messages	Transmission of packets
Operates in real time	Not real time	Near real time
Messages not stored	Messages stored	Messages held for short time
Path established for entire message	Route established for each message	Route established for each packet
Call setup delay	Message transmission delay	Packet transmission delay
Busy signal if called party busy	No busy signal	No busy signal
Blocking may occur	Blocking cannot occur	Blocking cannot occur
User responsible for message-loss protection	Network responsible for lost messages	Network may be responsible for each packet but not for entire message
No speed or code conversion	Speed and code conversion	Speed and code conversion
Fixed bandwidth transmission (i.e., fixed information capacity)	Dynamic use of bandwidth	Dynamic use of bandwidth
No overhead bits after initial setup delay	Overhead bits in each message	Overhead bits in each packet

Table 23-4 LAPB Commands

	Bit Number			
Command	8 7 6	5	4 3 2	1
1 (information)	*nr*	P	*ns*	0
RR (receiver ready)	*nr*	P	0 0 0	1
RNR (receiver not ready)	*nr*	P	0 1 0	1
REJ (reject)	*nr*	P	1 0 0	1
SABM (set asynchronous balanced mode)	0 0 1	P	1 1 1	1
DISC (disconnect)	0 1 0	P	0 0 1	1

(ADCCP) as the U.S. standard. ANSI 3.66 and ISO HDLC were designed for private-line data circuits with a polling environment. Consequently, the addressing and control procedures outlined by them are not appropriate for packet data networks. ANSI 3.66 and HDLC were selected for the data-link layer because of their frame format, delimiting sequence, transparency mechanism, and error-detection method.

At the link level, the protocol specified by X.25 is a subset of HDLC, referred to as *Link Access Procedure Balanced* (LAPB). LAPB provides for two-way, full-duplex communications between DTE and DCE at the packet network gateway. Only the address of the DTE or DCE may appear in the address field of a LAPB frame. The address field refers to a link address, not a network address. The network address of the destination terminal is embedded in the packet header, which is part of the information field.

Tables 23-4 and 23-5 show the commands and responses, respectively, for an LAPB frame. During LAPB operation, most frames are commands. A response frame is compelled only when a command frame is received containing a poll (P bit) = 1. SABM/UA is a command/response pair used to initialize all counters and timers at the beginning of a session. Similarly, DISC/DM is a command/response pair used at the end of a session. FRMR is a response to any illegal command for which there is no indication of transmission errors according to the frame check sequence field.

Information (I) commands are used to transmit packets. Packets are never sent as responses. Packets are acknowledged using *ns* and *nr* just as they were in SDLC. RR is sent by a station when it needs to respond (acknowledge) something but has no information packets to send. A response to an information command could be RR with F = 1. This procedure is called *checkpointing*.

Table 23-5 LAPB Responses

Response	Bit Number			
	8 7 6	5	4 3 2	1
RR (receiver ready)	*nr*	F	0 0 0	1
RNR (receiver not ready)	*nr*	F	0 1 0	1
REJ (reject)	*nr*	F	1 0 0	1
UA (unnumbered acknowledgment)	0 1 1	F	0 0 1	1
DM (disconnect mode)	0 0 0	F	1 1 1	1
FRMR (frame rejected)	1 0 0	F	0 1 1	1

REJ is another way of requesting transmission of frames. RNR is used for the flow control to indicate a busy condition and prevents further transmissions until cleared with an RR.

The network layer of X.25 specifies three switching services offered in a switched data network: permanent virtual circuit, virtual call, and datagram.

23-9-1 Permanent Virtual Circuit

A *permanent virtual circuit* (PVC) is logically equivalent to a two-point dedicated private-line circuit except slower. A PVC is slower because a hardwired, end-to-end connection is not provided. The first time a connection is requested, the appropriate switches and circuits must be established through the network to provide the interconnection. A PVC identifies the routing between two predetermined subscribers of the network that is used for all subsequent messages. With a PVC, a source and destination address are unnecessary because the two users are fixed.

23-9-2 Virtual Call

A *virtual call* (VC) is logically equivalent to making a telephone call through the DDD network except no direct end-to-end connection is made. A VC is a one-to-many arrangement. Any VC subscriber can access any other VC subscriber through a network of switches and communication channels. Virtual calls are temporary virtual connections that use common usage equipment and circuits. The source must provide its address and the address of the destination before a VC can be completed.

23-9-3 Datagram

A *datagram* (DG) is, at best, vaguely defined by X.25 and, until it is completely outlined, has very limited usefulness. With a DG, users send small packets of data into the network. Each packet is self-contained and travels through the network independent of other packets of the same message by whatever means available. The network does not acknowledge packets, nor does it guarantee successful transmission. However, if a message will fit into a single packet, a DG is somewhat reliable. This is called a *single-packet-per-segment* protocol.

23-9-4 X.25 Packet Format

A virtual call is the most efficient service offered for a packet network. There are two packet formats used with virtual calls: a call request packet and a data transfer packet.

23-9-4-1 Call request packet. Figure 23-10 shows the field format for a call request packet. The delimiting sequence is 01111110 (an HDLC flag), and the error-detection/correction mechanism is CRC-16 with ARQ. The link address field and the control field have little use and, therefore, are seldom used with packet networks. The rest of the fields are defined in sequence.

Format identifier. The format identifier identifies whether the packet is a new call request or a previously established call. The format identifier also identifies the packet numbering sequence (either 0–7 or 0–127).

Logical channel identifier (LCI). The LCI is a 12-bit binary number that identifies the source and destination users for a given virtual call. After a source user has

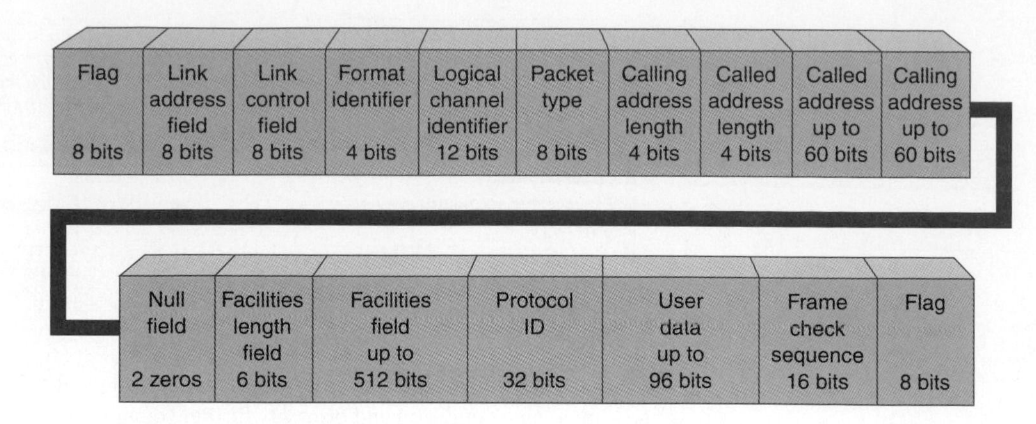

FIGURE 23-10 X.25 call request packet format

gained access to the network and has identified the destination user, they are assigned an LCI. In subsequent packets, the source and destination addresses are unnecessary; only the LCI is needed. When two users disconnect, the LCI is relinquished and can be reassigned to new users. There are 4096 LCIs available. Therefore, there may be as many as 4096 virtual calls established at any given time.

Packet type. This field is used to identify the function and the content of the packet (new request, call clear, call reset, and so on).

Calling address length. This four-bit field gives the number of digits (in binary) that appear in the calling address field. With four bits, up to 15 digits can be specified.

Called address length. This field is the same as the calling address field except that it identifies the number of digits that appear in the called address field.

Called address. This field contains the destination address. Up to 15 BCD digits (60 bits) can be assigned to a destination user.

Calling address. This field is the same as the called address field except that it contains up to 15 BCD digits that can be assigned to a source user.

Facilities length field. This field identifies (in binary) the number of eight-bit octets present in the facilities field.

Facilities field. This field contains up to 512 bits of optional network facility information, such as reverse billing information, closed user groups, and whether it is a simplex transmit or simplex receive connection.

Protocol identifier. This 32-bit field is reserved for the subscriber to insert user-level protocol functions such as log-on procedures and user identification practices.

User data field. Up to 96 bits of user data can be transmitted with a call request packet. These are unnumbered data that are not confirmed. This field is generally used for user passwords.

23-9-4-2 Data transfer packet. Figure 23-11 shows the field format for a data transfer packet. A data transfer packet is similar to a call request packet except that a data transfer packet has considerably less overhead and can accommodate a much larger user data field. The data transfer packet contains a send-and-receive packet sequence field that was not included with the call request format.

The flag, link address, link control, format identifier, LCI, and FCS fields are identical to those used with the call request packet. The send and receive packet sequence fields are described as follows:

Send packet sequence field. The P(s) field is used in the same manner that the *ns* and *nr* sequences are used with SDLC and HDLC. P(s) is analogous to *ns,* and P(r) is analogous

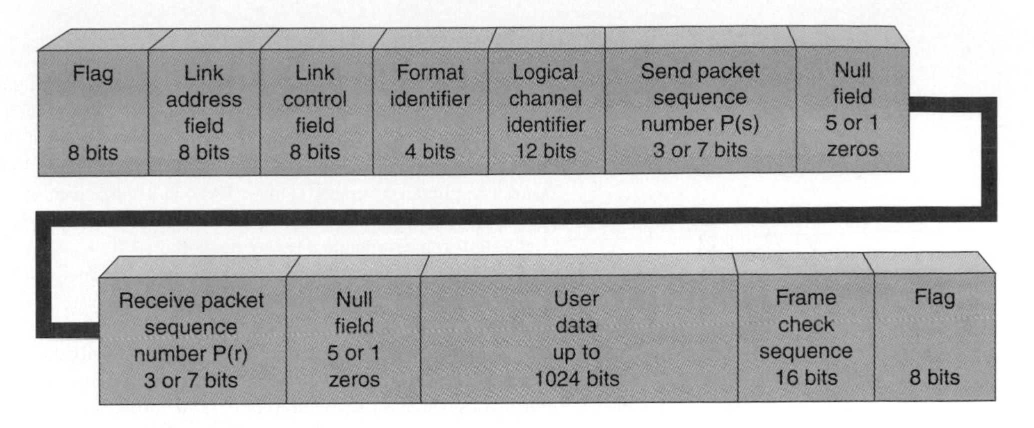

FIGURE 23-11 X.25 data transfer packet format

to *nr.* Each successive data transfer packet is assigned the next P(s) number in sequence. The P(s) can be a 14- or seven-bit binary number and, thus, number packets from either 0–7 or 0–127. The numbering sequence is identified in the format identifier. The send packet field always contains eight bits, and the unused bits are reset.

Receive packet sequence field. P(r) is used to confirm received packets and call for retransmission of packets received in error (ARQ). The I field in a data transfer packet can have considerably more source information than an I field in a call request packet.

23-9-5 The X Series of Recommended Standards

X.25 is part of the X series of ITU-T-recommended standards for public data networks. The X series is classified into two categories: X.1 through X.39, which deal with services and facilities, terminals, and interfaces, and X.40 through X.199, which deal with network architecture, transmission, signaling, switching, maintenance, and administrative arrangements. Table 23-6 lists the most important X standards with their titles and descriptions.

23-10 INTEGRATED SERVICES DIGITAL NETWORK

The data and telephone communications industry is continually changing to meet the demands of contemporary telephone, video, and computer communications systems. Today, more and more people have a need to communicate with each other than ever before. In order to meet these needs, old standards are being updated and new standards developed and implemented almost on a daily basis.

The *Integrated Services Digital Network* (ISDN) is a proposed network designed by the major telephone companies in conjunction with the ITU-T with the intent of providing worldwide telecommunications support of voice, data, video, and facsimile information within the same network (in essence, ISDN is the integrating of a wide range of services into a single multipurpose network). ISDN is a network that proposes to interconnect an unlimited number of independent users through a common communications network.

To date, only a small number of ISDN facilities have been developed; however, the telephone industry is presently implementing an ISDN system so that in the near future, subscribers will access the ISDN system using existing public telephone and data networks. The basic principles and evolution of ISDN have been outlined by the International Telecommunication Union-Telephony (ITU-T) in its recommendation ITU-T I.120 (1984). ITU-T I.120 lists the following principles and evolution of ISDN.

Table 23-6 ITU-T X Series Standards

X.1	International user classes of service in public data networks. Assigns numerical class designations to different terminal speeds and types.
X.2	International user services and facilities in public data networks. Specifies essential and additional services and facilities.
X.3	Packet assembly/disassembly facility (PAD) in a public data network. Describes the packet assembler/disassembler, which normally is used at a network gateway to allow connection of a start/stop terminal to a packet network.
X.20bis	Use on public data networks of DTE designed for interfacing to asynchronous full-duplex V-series modems. Allows use of V.24/V.28 (essentially the same as EIA RS-232).
X.21bis	Use on public data networks of DTE designed for interfacing to synchronous full-duplex V-series modems. Allows use of V.24/V.28 (essentially the same as EIA RS-232) or V.35.
X.25	Interface between DTE and DCE for terminals operating in the packet mode on public data networks. Defines the architecture of three levels of protocols existing in the serial interface cable between a packet mode terminal and a gateway to a packet network.
X.28	DTE/DCE interface for a start/stop mode DTE accessing the PAD in a public data network situated in the same country. Defines the architecture of protocols existing in a serial interface cable between a start/stop terminal and an X.3 PAD.
X.29	Procedures for the exchange of control information and user data between a PAD and a packet mode DTE or another PAD. Defines the architecture of protocols behind the X.3 PAD, either between two PADs or between a PAD and a packet mode terminal on the other side of the network.
X.75	Terminal and transit call control procedures and data transfer system on international circuits between packet-switched data networks. Defines the architecture of protocols between two public packet networks.
X.121	International numbering plan for public data networks. Defines a numbering plan including code assignments for each nation.

23-10-1 Principles of ISDN

The main feature of the ISDN concept is to support a wide range of voice (telephone) and nonvoice (digital data) applications in the same network using a limited number of standardized facilities. ISDNs support a wide variety of applications, including both switched and nonswitched (dedicated) connections. Switched connections include both circuit- and packet-switched connections and their concatenations. Whenever practical, new services introduced into an ISDN should be compatible with 64-kbps switched digital connections. The 64-kbps digital connection is the basic building block of ISDN.

An ISDN will contain intelligence for the purpose of providing service features, maintenance, and network management functions. In other words, ISDN is expected to provide services beyond the simple setting up of switched circuit calls.

A layered protocol structure should be used to specify the access procedures to an ISDN and can be mapped into the open system interconnection (OSI) model. Standards already developed for OSI-related applications can be used for ISDN, such as X.25 level 3 for access to packet switching services.

It is recognized that ISDNs may be implemented in a variety of configurations according to specific national situations. This accommodates both single-source or competitive national policy.

23-10-2 Evolution of ISDN

ISDNs will be based on the concepts developed for telephone ISDNs and may evolve by progressively incorporating additional functions and network features including those of any other dedicated networks such as circuit and packet switching for data so as to provide for existing and new services.

The transition from an existing network to a comprehensive ISDN may require a period of time extending over one or more decades. During this period, arrangements must be developed for the internetworking of services on ISDNs and services on other networks.

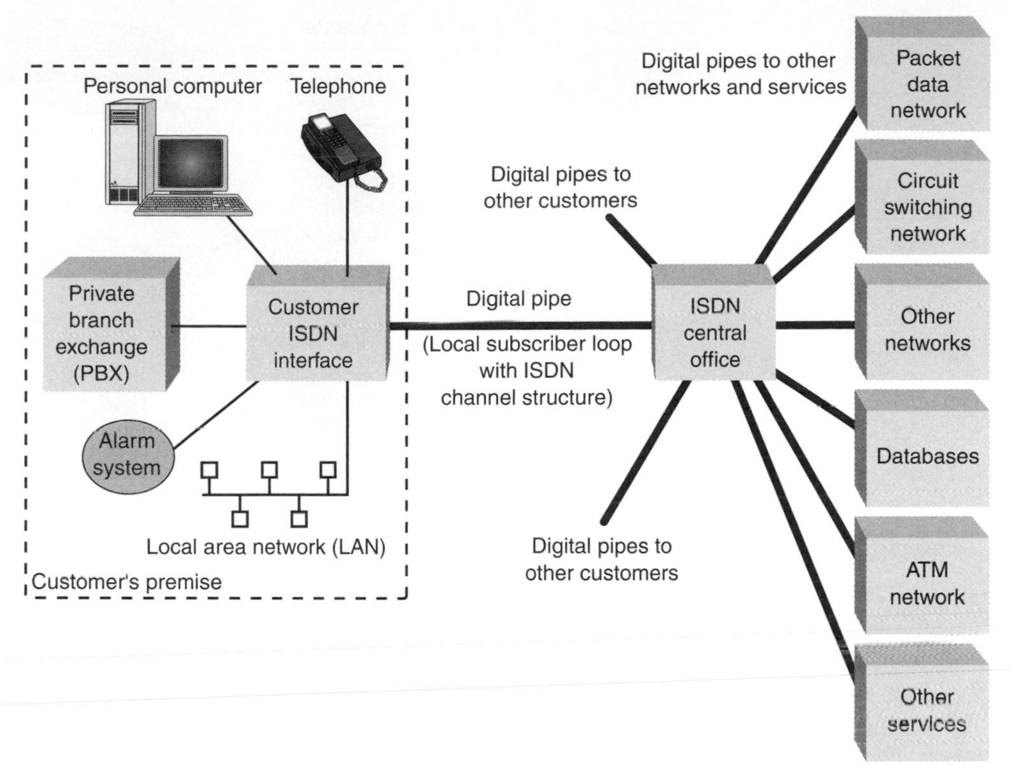

FIGURE 23-12 Subscriber's conceptual view of ISDN

In the evolution toward an ISDN, digital end-to-end connectivity will be obtained via plant and equipment used in existing networks, such as digital transmission, time-division multiplex, and/or space-division multiplex switching. Existing relevant recommendations for these constituent elements of an ISDN are contained in the appropriate series of recommendations of ITU-T and CCIR.

In the early stages of the evolution of ISDNs, some interim user-network arrangements may need to be adopted in certain countries to facilitate early penetration of digital service capabilities. An evolving ISDN may also include at later stages switched connections at bit rates higher and lower than 64 kbps.

23-10-3 Conceptual View of ISDN

Figure 23-12 shows a view of how ISDN can be conceptually viewed by a subscriber (customer) of the system. Customers gain access to the ISDN system through a local interface connected to a digital transmission medium called a *digital pipe*. There are several sizes of pipe available with varying capacities (i.e., bit rates), depending on customer need. For example, a residential customer may require only enough capacity to accommodate a telephone and a personal computer. However, an office complex may require a pipe with sufficient capacity to handle a large number of digital telephones interconnected through an on-premise private branch exchange (PBX) or a large number of computers on a local area network (LAN).

Figure 23-13 shows the ISDN user network, which illustrates the variety of network users and the need for more than one capacity pipe. A single residential telephone is at the low end of the ISDN demand curve, followed by a multiple-drop arrangement serving a telephone, a personal computer, and a home alarm system. Industrial complexes would be at the high end of the demand curve, as they require sufficient capacity to handle hundreds of telephones and several LANs. Although a pipe has a fixed capacity, the traffic on the pipe can be comprised of data from a dynamic variety of sources with varying signal types and bit rates that have been multiplexed into a single high-capacity pipe. Therefore, a customer can gain access to both circuit- and packet-switched services through the same

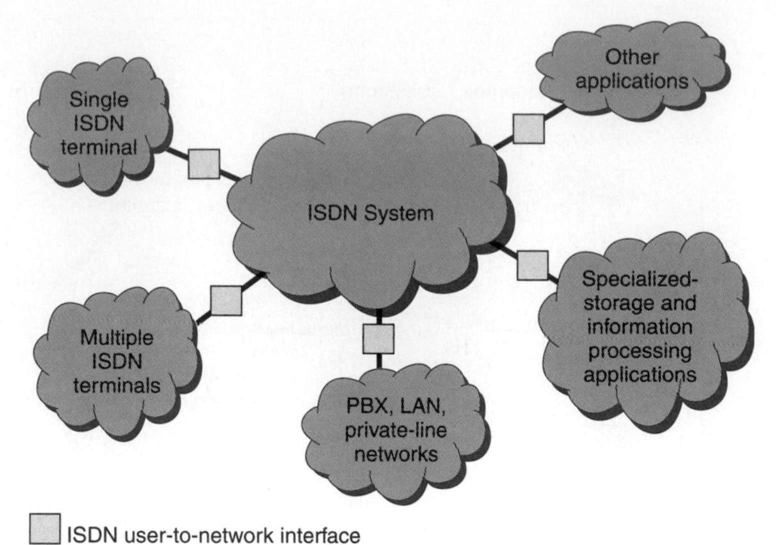

FIGURE 23-13 ISDN user network

pipe. Because of the obvious complexity of ISDN, it requires a rather complex control system to facilitate multiplexing and demultiplexing data to provide the required services.

23-10-4 ISDN Objectives

The key objectives of developing a worldwide ISDN system are the following:

1. *System standardization.* Ensure universal access to the network.
2. *Achieving transparency.* Allow customers to use a variety of protocols and applications.
3. *Separating functions.* ISDN should not provide services that preclude competitiveness.
4. *Variety of configurations.* Provide private-line (leased) and switched services.
5. *Addressing cost-related tariffs.* ISDN service should be directly related to cost and independent of the nature of the data.
6. *Migration.* Provide a smooth transition while evolving.
7. *Multiplexed support.* Provide service to low-capacity personal subscribers as well as to large companies.

23-10-5 ISDN Architecture

Figure 23-14 shows a block diagram of the architecture for ISDN functions. The ISDN network is designed to support an entirely new physical connector for the customer, a digital subscriber loop, and a variety of transmission services.

A common physical is defined to provide a standard interface connection. A single interface will be used for telephones, computer terminals, and video equipment. Therefore, various protocols are provided that allow the exchange of control information between the customer's device and the ISDN network. There are three basic types of ISDN channels:

1. B channel: 64 kbps
2. D channel: 16 kbps or 64 kbps
3. H channel: 384 kbps (H_0), 1536 kbps (H_{11}), or 1920 kbps (H_{12})

ISDN standards specify that residential users of the network (i.e., the subscribers) be provided a *basic access* consisting of three full-duplex, time-division multiplexed digital channels, two operating at 64 kbps (designated the B channels, for *bearer*) and one at 16 kbps (designated the D channel, for *data*). The B and D bit rates were selected to be compatible with existing DS1–DS4 digital carrier systems. The D channel is used for carrying signaling

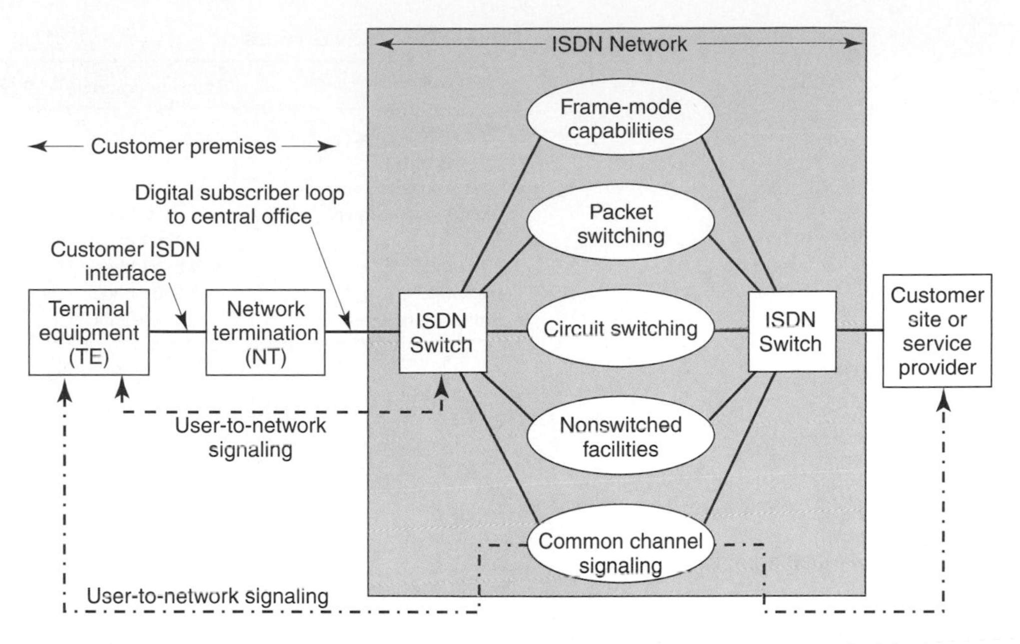

FIGURE 23-14 ISDN architecture

information and for exchanging network control information. One B channel is used for digitally encoded voice and the other for applications such as data transmission, PCM-encoded digitized voice, and videotex. The 2B + D service is sometimes called the *basic rate interface* (BRI). BRI systems require bandwidths that can accommodate two 64-kbps B channels and one 16-kbps D channel plus framing, synchronization, and other overhead bits for a total bit rate of 192 kbps. The H channels are used to provide higher bit rates for special services such as fast facsimile, video, high-speed data, and high-quality audio.

There is another service called the *primary service, primary access,* or *primary rate interface* (PRI) that will provide multiple 64-kbps channels intended to be used by the higher-volume subscribers to the network. In the United States, Canada, Japan, and Korea, the primary rate interface consists of 23 64-kbps B channels and one 64-kbps D channel (23B + D) for a combined bit rate of 1.544 Mbps. In Europe, the primary rate interface uses 30 64-kbps B channels and one 64-kbps D channel for a combined bit rate of 2.048 Mbps.

It is intended that ISDN provide a circuit-switched B channel with the existing telephone system; however, packet-switched B channels for data transmission at nonstandard rates would have to be created.

The subscriber's loop, as with the twisted-pair cable used with a common telephone, provides the physical signal path from the subscriber's equipment to the ISDN central office. The subscriber loop must be capable of supporting full-duplex digital transmission for both basic and primary data rates. Ideally, as the network grows, optical fiber cables will replace the metallic cables.

Table 23-7 lists the services provided to ISDN subscribers. BC designates a circuit-switched B channel, BP designates a packet-switched B channel, and D designates a D channel.

23-10-6 ISDN System Connections and Interface Units

ISDN subscriber units and interfaces are defined by their function and reference within the network. Figure 23-15 shows how users may be connected to an ISDN. As the figure shows, subscribers must access the network through one of two different types of entry devices: *terminal equipment type 1* (TE1) or *terminal equipment type 2* (TE2). TE1 equipment supports standard ISDN interfaces and, therefore, requires no protocol translation. Data enter

Table 23-7 ISDN Services

Service	Transmission Rate	Channel
Telephone	64 kbps	BC
System alarms	100 bps	D
Utility company metering	100 bps	D
Energy management	100 bps	D
Video	2.4–64 kbps	BP
Electronic mail	4.8–64 kbps	BP
Facsimile	4.8–64 kbps	BC
Slow-scan television	64 kbps	BC

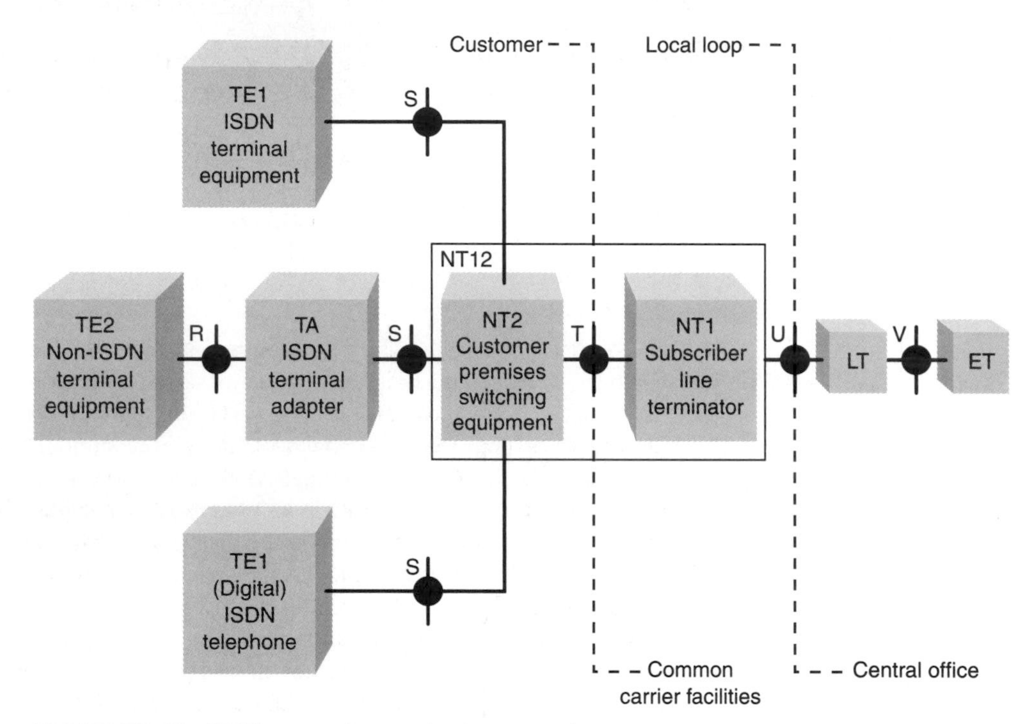

FIGURE 23-15 ISDN connections and reference points

the network and are immediately configured into ISDN protocol format. TE2 equipment is classified as non-ISDN; thus, computer terminals are connected to the system through physical interfaces such as the RS-232 and host computers with X.25. Translation between non-ISDN data protocol and ISDN protocol is performed in a device called a *terminal adapter* (TA), which converts the user's data into the 64-kbps ISDN channel B or the 16-kbps channel D format and X.25 packets into ISDN packet formats. If any additional signaling is required, it is added by the terminal adapter. The terminal adapters can also support traditional analog telephones and facsimile signals by using a 3.1-kHz audio service channel. The analog signals are digitized and put into ISDN format before entering the network.

User data at points designated as *reference point S* (*system*) are presently in ISDN format and provide the 2B + D data at 192 kbps. These reference points separate user terminal equipment from network-related system functions. *Reference point T* (*terminal*) locations correspond to a minimal ISDN network termination at the user's location. These reference points separate the network provider's equipment from the user's equipment. *Reference point R* (*rate*) provides an interface between non-ISDN-compatible user equipment and the terminal adapters.

Network termination 1 (NT1) provides the functions associated with the physical interface between the user and the common carrier and are designated by the letter *T* (these functions correspond to OSI layer 1). The NT1 is a boundary to the network and may be controlled by the ISDN provider. The NT1 performs line maintenance functions and supports multiple channels at the physical level (e.g., 2B + D). Data from these channels are time-division multiplexed together. Network terminal 2 devices are intelligent and can perform *concentration* and switching functions (functionally up through OSI level 3). NT2 terminations can also be used to terminate several S-point connections and provide local switching functions and two-wire-to-four-wire and four-wire-to-two-wire conversions. *U-reference points* refer to interfaces between the common carrier subscriber loop and the *central office switch.* A *U loop* is the media interface point between an NT1 and the central office.

Network termination 1,2 (NT12) constitutes one piece of equipment that combines the functions of NT1 and NT2. U loops are terminated at the central office by a *line termination* (LT) unit, which provides physical layer interface functions between the central office and the loop lines. The LT unit is connected to an *exchange termination* (ET) at *reference point* V. An ET routes data to an outgoing channel or central office user.

There are several types of transmission channels in addition to the B and D types described in the previous section. They include the following:

HO channel. This interface supports multiple 384-kbps HO channels. These structures are 3HO + D and 4HO + D for the 1.544 Mbps interface and 5HO + D for the 2.048-Mbps interface.

H11 channel. This interface consists of one 1.536-Mbps H11 channel (24 64-kbps channels).

H12 channel. European version of H11 that uses 30 channels for a combined data rate of 1.92 Mbps.

E channel. Packet switched using 64 kbps (similar to the standard D channel).

23-10-7 Broadband ISDN

Broadband ISDN (BISDN) is defined by the ITU-T as a service that provides transmission channels capable of supporting transmission rates greater than the primary data rate. With BISDN, services requiring data rates of a magnitude beyond those provided by ISDN, such as video transmission, will become available. With the advent of BISDN, the original concept of ISDN is being referred to as *narrowband* ISDN.

In 1988, the ITU-T first recommended as part of its I-series recommendations relating to BISDN: I.113, *Vocabulary of terms for broadband aspects of ISDN,* and I.121, *Broadband aspects of ISDN.* These two documents are a consensus concerning the aspects of the future of BISDN. They outline preliminary descriptions of future standards and development work.

The new BISDN standards are based on the concept of an *asynchronous transfer mode* (ATM), which will incorporate optical fiber cable as the transmission medium for data transmission. The BISDN specifications set a maximum length of 1 km per cable length but are making provisions for repeated interface extensions. The expected data rates on the optical fiber cables will be either 11 Mbps, 155 Mbps, or 600 Mbps, depending on the specific application and the location of the fiber cable within the network.

ITU-T classifies the services that could be provided by BISDN as interactive and distribution services. *Interactive services* include those in which there is a two-way exchange

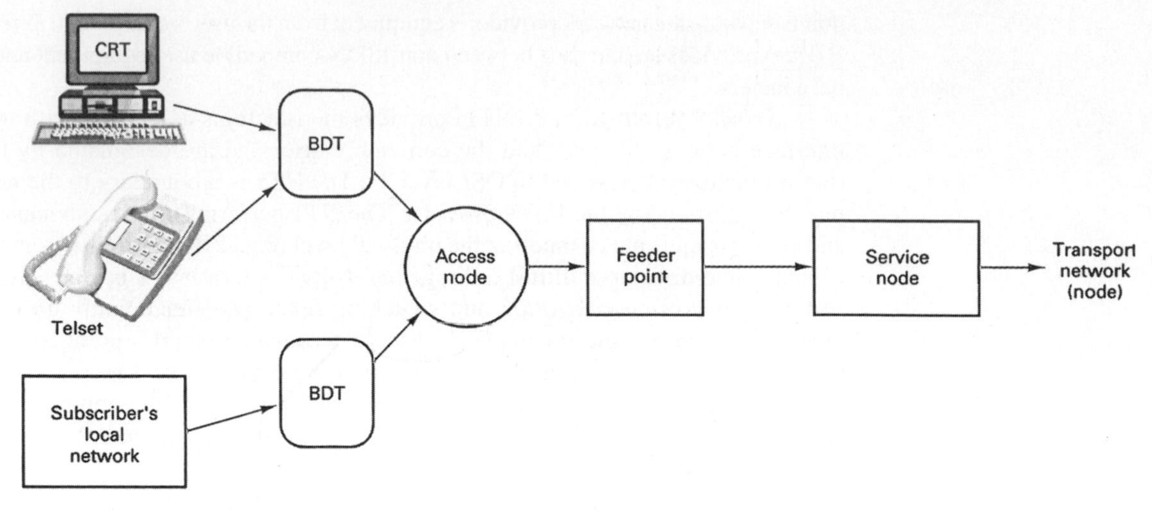

FIGURE 23-16 BISDN access

of information (excluding control signaling) between two subscribers or between a subscriber and a service provider. *Distribution services* are those in which information transfer is primarily from service provider to subscriber. On the other hand, *conversational services* will provide a means for bidirectional end-to-end data transmission, in real time, between two subscribers or between a subscriber and a service provider.

The authors of BISDN composed specifications that require the new services meet both existing ISDN interface specifications and the new BISDN needs. A standard ISDN terminal and a *broadband terminal interface* (BTI) will be serviced by the *subscriber's premise network* (SPN), which will multiplex incoming data and transfer them to the *broadband node.* The broadband node is called a *broadband network termination* (BNT), which codes the data information into smaller packets used by the BISDN network. Data transmission within the BISDN network can be asymmetric (i.e., access on to and off of the network may be accomplished at different transmission rates, depending on system requirements).

23-10-7-1 BISDN configuration. Figure 23-16 shows how access to the BISDN network is accomplished. Each peripheral device is interfaced to the *access node* of a BISDN network through a *broadband distant terminal* (BDT). The BDT is responsible for the electrical-to-optical conversion, multiplexing of peripherals, and maintenance of the subscriber's local system. Access nodes concentrate several BDTs into high-speed optical fiber lines directed through a *feeder point* into a *service node.* Most of the control functions for system access are managed by the service node, such as call processing, administrative functions, and switching and maintenance functions. The functional modules are interconnected in a star configuration and include switching, administrative, gateway, and maintenance modules. The interconnection of the function modules is shown in Figure 23-17. The central control hub acts as the end user interface for control signaling and data traffic maintenance. In essence, it oversees the operation of the modules.

Subscriber terminals near the central office may bypass the access nodes entirely and be directly connected to the BISDN network through a service node. BISDN networks that use optical fiber cables can utilize much wider bandwidths and, consequently, have higher transmission rates and offer more channel-handling capacity than ISDN systems.

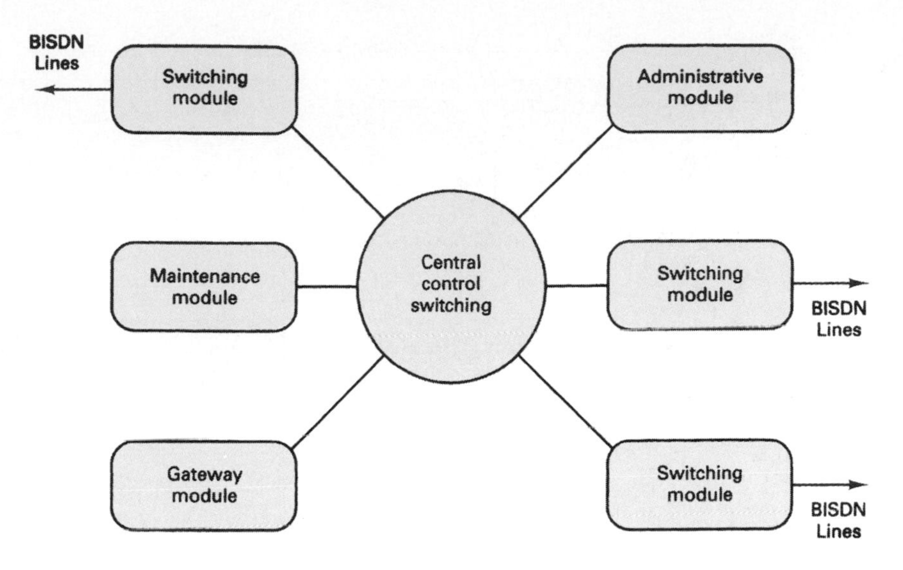

FIGURE 23-17 BISDN functional module interconnections

23-10-7-2 Broadband channel rates. The CCITT has published preliminary definitions of new broadband channel rates that will be added to the existing ISDN narrowband channel rates:

1. H21: 32.768 Mbps
2. H22: 43 Mbps to 45 Mbps
3. H4: 132 Mbps to 138.24 Mbps

The H21 and H22 data rates are intended to be used for full-motion video transmission for videoconferencing, video telephone, and video messaging. The H4 data rate is intended for bulk data transfer of text, facsimile, and enhanced video information. The H21 data rate is equivalent to 512 64-kbps channels. The H22 and H4 data rates must be multiples of the basic 64-kbps transmission rate.

23-11 ASYNCHRONOUS TRANSFER MODE

Asynchronous transfer mode (ATM) is a relatively new data communications technology that uses a high-speed form of packet switching network for the transmission media. ATM was developed in 1988 by the ITU-T as part of the BISDN. ATM is one means by which data can enter and exit the BISDN network in an asynchronous (time-independent) fashion. ATM is intended to be a carrier service that provides an integrated, high-speed communications network for corporate private networks. ATM can handle all kinds of communications traffic, including voice, data, image, video, high-quality music, and multimedia. In addition, ATM can be used in both LAN and WAN network environments, providing seamless internetworking between the two. Some experts claim that ATM may eventually replace both private leased T1 digital carrier systems and on-premise switching equipment.

Conventional electronic switching (ESS) machines currently utilize a central processor to establish switching paths and route traffic through a network. ATM switches, in contrast, will include self-routing procedures where individual cells (short, fixed-length packets of data) containing subscriber data will route their own way through the ATM switching

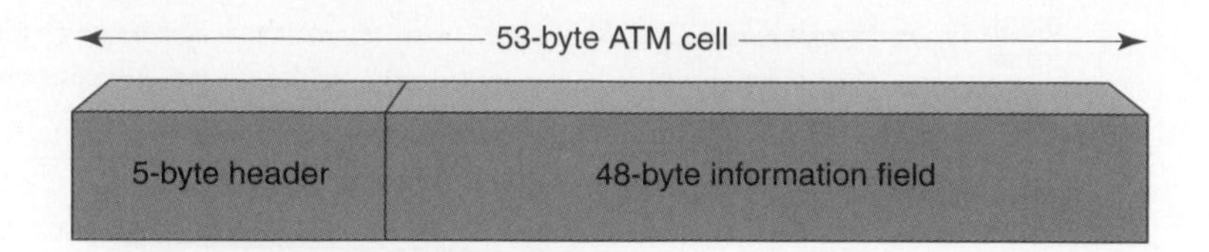

FIGURE 23-18 ATM cell structure

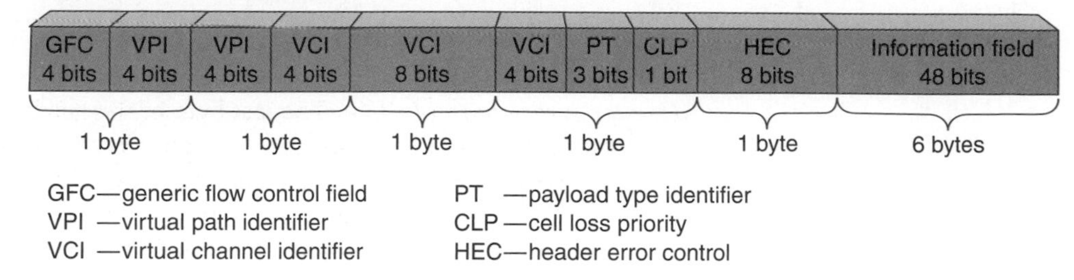

FIGURE 23-19 ATM five-byte header field structure

network in real time using their own address instead of relying on an external process to establish the switching path.

ATM uses *virtual channels* (VCs) and *virtual paths* (VPs) to route cells through a network. In essence, a virtual channel is merely a connection between a source and a destination, which may entail establishing several ATM links between local switching centers. With ATM, all communications occur on the virtual channel, which preserves cell sequence. On the other hand, a virtual path is a group of virtual channels connected between two points that could compromise several ATM links.

ATM incorporates *labeled channels* that are transferable at fixed data rates anywhere from 16 kbps up to the maximum rate of the carrier system. Once data have entered the network, they are transferred into fixed time slots called *cells.* An ATM cell contains all the network information needed to relay individual cells from node to node over a preestablished ATM connection. Figure 23-18 shows the ATM cell structure, which is a fixed-length data packet only 53 bytes long, including a five-byte header and a 48-byte information field. Fixed-length cells provide the following advantages:

1. A uniform transmission time per cell ensures a more uniform transit-time characteristic for the network as a whole.
2. A short cell requires a shorter time to assemble and, thus, shorter delay characteristics for voice.
3. Short cells are easier to transfer over fixed-width processor buses, it is easier to buffer the data in link queues, and they require less processor logic.

23-11-1 ATM Header Field

Figure 23-19 shows the five-byte ATM header field, which includes the following fields: generic flow control field, virtual path identifier, virtual channel identifier, payload type identifier, cell loss priority, and header error control.

Generic flow control field (*GFC*). The GFC field uses the first four bits of the first byte of the header field. The GFC controls the flow of traffic across the user network interface (UNI) and into the network.

Virtual path identifier (*VPI*) *and virtual channel identifier* (*VCI*). The 24 bits immediately following the GFC are used for the ATM address.

Payload type identifier (*PT*). The first three bits of the second half of byte 4 specify the type of message (payload) in cell. With three bits, there are eight different types of payloads possible. At the present time, however, types 0 to 3 are used for identifying the type of user data, types 4 and 5 indicate management information, and types 6 and 7 are reserved for future use.

Cell loss priority (*CLP*). The last bit of byte 4 is used to indicate whether a cell is eligible to be discarded by the network during congested traffic periods. The CLP bit is set by the user or cleared by the user. If set, the network may discard the cell during times of heavy use.

Header error control (*HEC*). The last byte of the header field is for error control and is used to detect and correct single-bit errors that occur in the header field only; the HEC does not serve as an entire cell check character. The value placed in the HEC is computed from the four previous bytes of the header field. The HEC provides some protection against the delivery of cells to the wrong destination address.

23-11-1-1 ATM information field. The 48-byte information field is reserved for user data. Insertion of data into the information field of a cell is a function of the upper half of layer 2 of the ISO-OSI seven-layer protocol hierarchy. This layer is specifically called the ATM Adaptation Layer (AAL). The AAL gives ATM the versatility necessary to facilitate, in a single format, a wide variety of different types of services ranging from continuous processes signals, such as voice transmission, to messages carrying highly fragmented bursts of data such as those produced from LANs. Because most user data occupy more than 48 bytes, the AAL divides information into 48-byte segments and places them into a series of segments. The five types of AALs are the following:

1. *Constant bit rate* (*CBR*). CBR information fields are designed to accommodate PCM-TDM traffic, which allows the ATM network to emulate voice or DSN services.
2. *Variable bit rate (VBR) timing-sensitive services.* This type of AAL is currently undefined; however, it is reserved for future data services requiring transfer of timing information between terminal points as well as data (i.e., packet video).
3. *Connection-oriented VBR data transfer.* Type 3 information fields transfer VBR data such as impulsive data generated at irregular intervals between two subscribers over a preestablished data link. The data link is established by network signaling procedures that are very similar to those used by the public switched telephone network. This type of service is intended for large, long-duration data transfers, such as file transfers or file backups.
4. *Connectionless VBR data transfer.* This AAL type provides for transmission of VBR data that does not have a preestablished connection. Type 4 information fields are intended to be used for short, highly bursty types of transmissions, such as those generated from a LAN.

23-11-2 ATM Network Components

Figure 23-20 shows an ATM network, which is comprised of three primary components: ATM endpoints, ATM switches, and transmission paths.

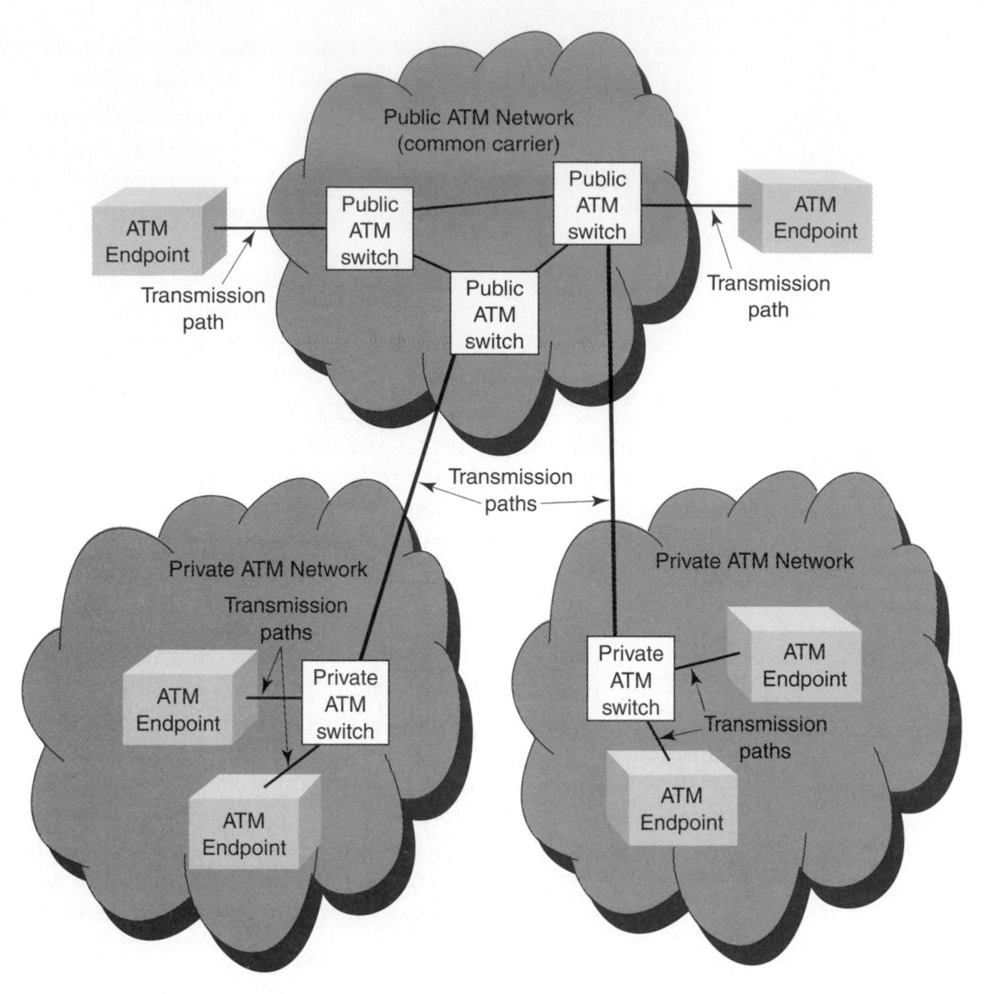

FIGURE 23-20 ATM network components

23-11-2-1 ATM endpoints. *ATM endpoints* are shown in Figure 23-21. As shown in the figure, endpoints are the source and destination of subscriber data and, therefore, they are sometimes called *end systems.* Endpoints can be connected directly to either a public or a private ATM switch. An ATM endpoint can be as simple as an ordinary personal computer equipped with an ATM network interface card. An ATM endpoint could also be a special-purpose network component that services several ordinary personal computers, such as an Ethernet LAN.

23-11-2-2 ATM switches. The primary function of an *ATM switch* is to route information from a source endpoint to a destination endpoint. ATM switches are sometimes called *intermediate systems,* as they are located between two endpoints. ATM switches fall into two general categories: public and private.

Public ATM switches. A *public ATM switch* is simply a portion of a public service provider's switching system where the service provider could be a local telephone company or a long-distance carrier, such as AT&T. An ATM switch is sometimes called a *network node.*

Private ATM switches. Private ATM switches are owned and maintained by a private company and sometimes called *customer premise nodes.* Private ATM switches are

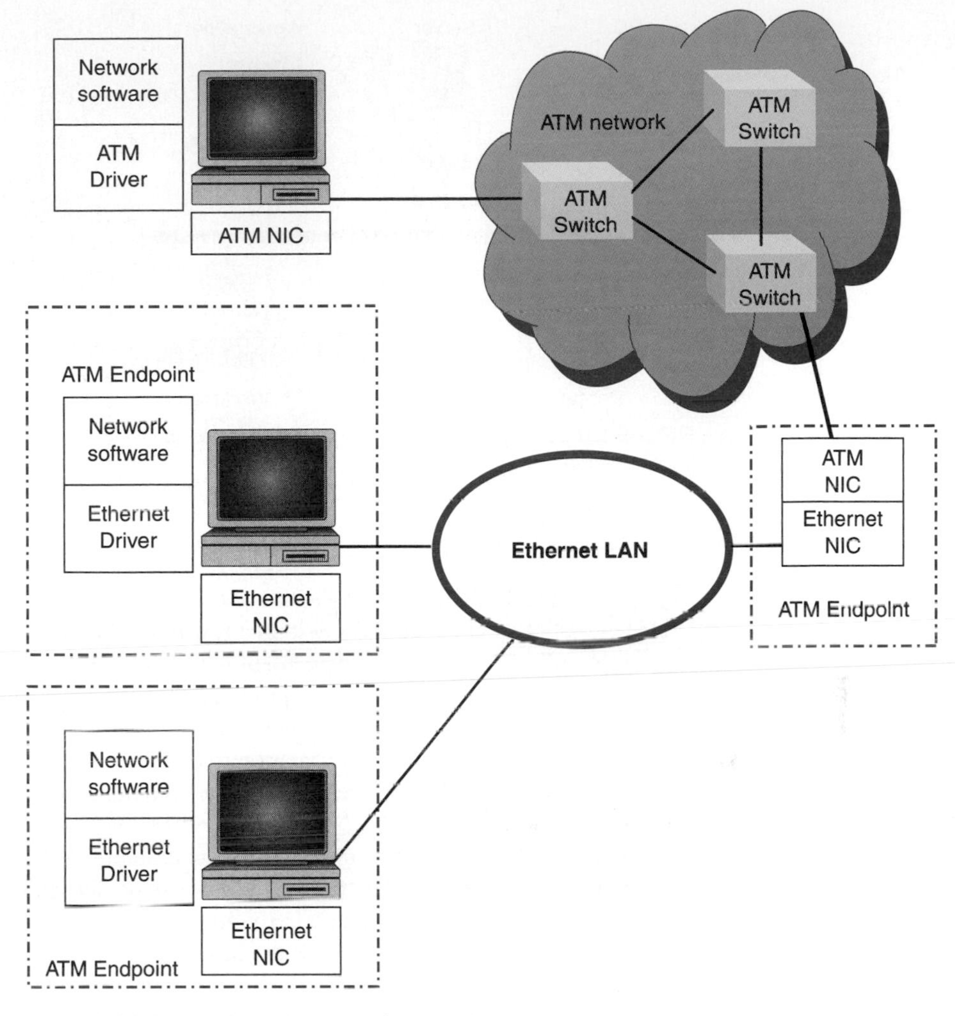

FIGURE 23-21 ATM endpoint implementations

sold to ATM customers by many of the same computer networking infrastructure vendors who provide ATM customers with network interface cards and connectivity devices, such as repeaters, hubs, bridges, switches, and routers.

23-11-2-3 Transmission paths. ATM switches and ATM endpoints are interconnected with physical communications paths called transmission paths. A transmission path can be any of the common transmission media, such as twisted-pair cable or optical fiber cable.

23-12 LOCAL AREA NETWORKS

Studies have indicated that most (80%) of the communications among data terminals and other data equipment occurs within a relatively small local environment. A *local area network* (LAN) provides the most economical and effective means of handling local data communication needs. A LAN is typically a privately owned data communications system in which the users share resources, including software. LANs provide two-way communications between a large variety of data communications terminals within a

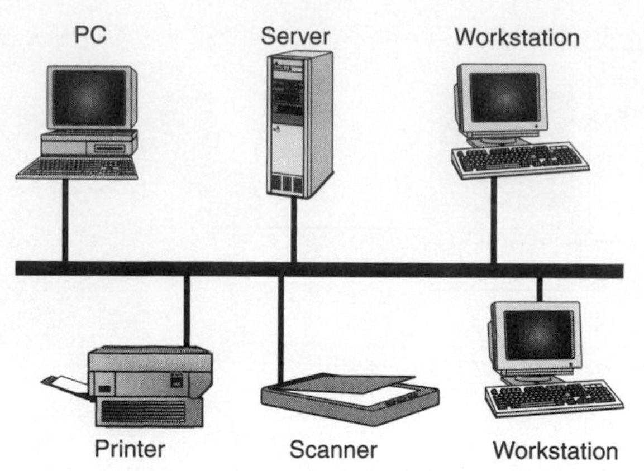

FIGURE 23-22 Typical local area network component configuration

limited geographical area such as within the same room, building, or building complex. Most LANs link equipment that is within a few miles of each other.

Figure 23-22 shows several personal computers (PCs) connected to a LAN to share common resources such as a modem, printer, or server. The server may be a more powerful computer than the other PCs sharing the network, or it may simply have more disk storage space. The server "serves" information to the other PCs on the network in the form of software and data information files. A PC server is analogous to a mainframe computer except on a much smaller scale.

LANs allow for a room full or more of computers to share common resources such as printers and modems. The average PC uses these devices only a small percentage of the time, so there is no need to dedicate individual printers and modems to each PC. To print a document or file, a PC simply sends the information over the network to the server. The server organizes and prioritizes the documents and then sends them, one document at a time, to the common usage printer. Meanwhile, the PCs are free to continue performing other useful tasks. When a PC needs a modem, the network establishes a *virtual connection* between the modem and the PC. The network is transparent to the virtual connection, which allows the PC to communicate with the modem as if they were connected directly to each other.

LANS allow people to send and receive messages and documents through the network much quicker than they could be sent through a paper mail system. *Electronic mail* (e-mail) is a communications system that allows users to send messages to each other through their computers. E-mail enables any PC on the network to send or receive information from any other PC on the network as long as the PCs and the server use the same or compatible software. E-mail can also be used to interconnect users on different networks in different cities, states, countries, or even continents. To send an e-mail message, a user at one PC sends its address and message along with the destination address to the server. The server effectively "relays" the message to the destination PC if they are subscribers to the same network. If the destination PC is busy or not available for whatever reason, the server stores the message and resends it later. The server is the only computer that has to keep track of the location and address of all the other PCs on the network. To send e-mail to subscribers of other networks, the server relays the message to the server on the destination user's network, which in turn relays the mail to the destination PC. E-mail can be used to send text information (letters) as well as program files, graphics, audio, and even video. This is referred to as multimedia communications.

LANs are used extensively to interconnect a wide range of data services, including the following:

- Data terminals
- Laser printers
- Graphic plotters
- Large-volume disk and tape storage devices
- Facsimile machines
- Personal computers
- Mainframe computers
- Data modems
- Databases
- Word processors
- Public switched telephone networks
- Digital carrier systems (T carriers)
- E-mail servers

23-12-1 LAN System Considerations

The capabilities of a LAN are established primarily by three factors: *topology, transmission medium,* and *access control protocol.* Together these three factors determine the type of data, rate of transmission, efficiency, and applications that a network can effectively support.

23-12-1-1 LAN topologies. The topology or physical architecture of a LAN identifies how the stations (terminals, printers, modems, and so on) are interconnected. The transmission media used with LANs include metallic *twisted-wire pairs, coaxial cable,* and *optical fiber cables.* Presently, most LANs use coaxial cable; however, optical fiber cable systems are being installed in many new networks. Fiber systems can operate at higher transmission bit rates and have a larger capacity to transfer information than coaxial cables.

The most common LAN topologies are the star, bus, bus tree, and ring, which are illustrated in Figure 23-23.

23-12-1-2 Star topology. The preeminent feature of the star topology is that each station is radially linked to a *central node* through a direct point-to-point connection as shown in Figure 23-23a. With a star configuration, a transmission from one station enters the central node, where it is retransmitted on all the outgoing links. Therefore, although the circuit arrangement physically resembles a star, it is logically configured as a bus (i.e., transmissions from any station are received by all other stations).

Central nodes offer a convenient location for system or station troubleshooting because all traffic between outlying nodes must flow through the central node. The central node is sometimes referred to as *central control, star coupler,* or *central switch* and typically is a computer. The star configuration is best adapted to applications where most of the communications occur between the central node and outlying nodes. The star arrangement is also well suited to systems where there is a large demand to communicate with only a few of the remote terminals. Time-sharing systems are generally configured with a star topology. A star configuration is also well suited for word processing and database management applications.

Star couplers can be implemented either passively or actively. When passive couplers are used with a metallic transmission medium, transformers in the coupler provide an electromagnetic linkage through the coupler, which passes incoming signals on to outgoing links. If optical fiber cables are used for the transmission media, coupling can be achieved by fusing fibers together. With active couplers, digital circuitry in the central node acts as a repeater. Incoming data are simply regenerated and repeated on to all outgoing lines.

One disadvantage of a star topology is that the network is only as reliable as the central node. When the central node fails, the system fails. If one or more outlying nodes fail, however, the rest of the users can continue to use the remainder of the network. When failure of any single entity within a network is critical to the point that it will disrupt service on the entire network, that entity is referred to as a *critical resource.* Thus, the central node in a star configuration is a critical resource.

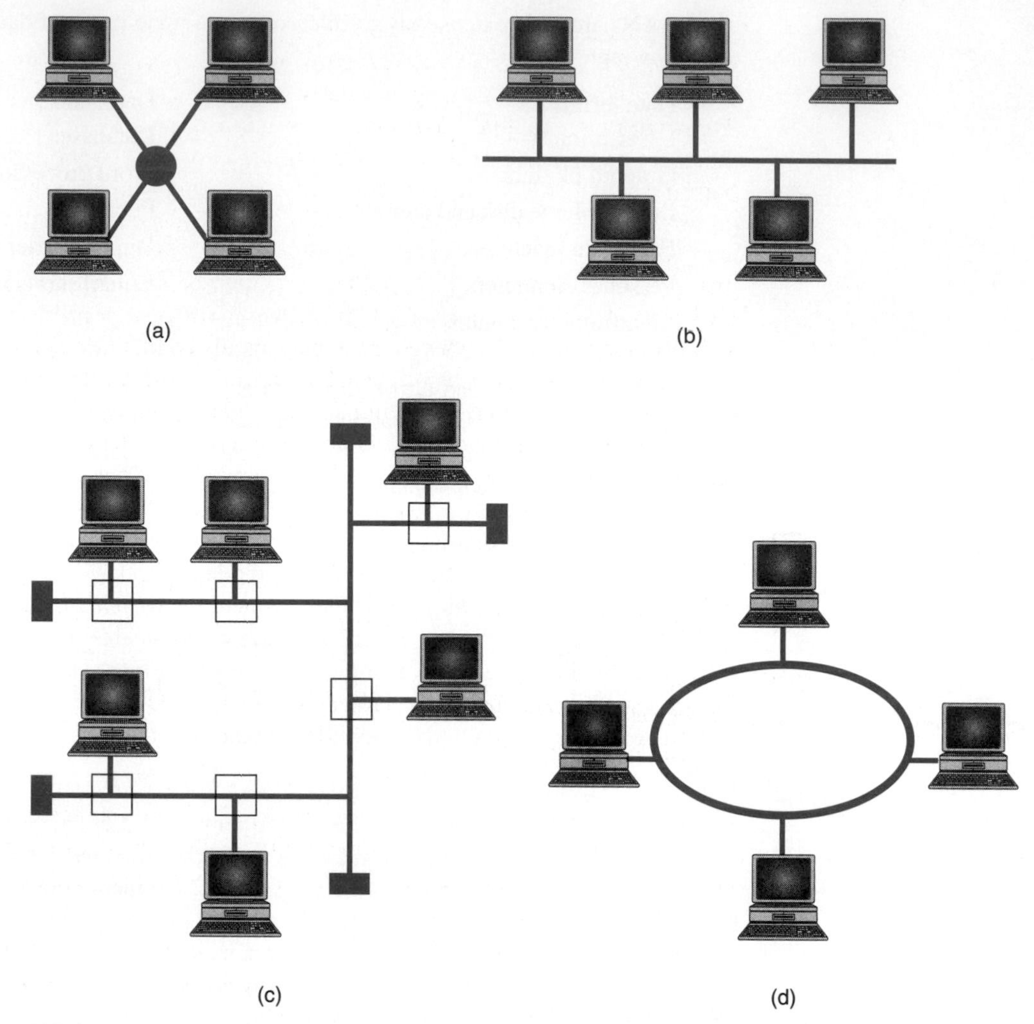

FIGURE 23-23 LAN Topologies: (a) star; (b) bus; (c) tree bus; (d) ring or loop

23-12-1-3 Bus topology. In essence, the bus topology is a multipoint or multidrop circuit configuration where individual nodes are interconnected by a common, shared communications channel as shown in Figure 23-23b. With the bus topology, all stations connect, using appropriate interfacing hardware, directly to a common linear transmission medium, generally referred to as a *bus*. In a bus configuration, network control is not centralized to a particular node. In fact, the most distinguishing feature of a bus LAN is that control is distributed among all the nodes connected to the LAN. Data transmissions on a bus network are usually in the form of small packets containing user addresses and data. When one station desires to transmit data to another station, it monitors the bus first to determine if it is currently being used. If no other stations are communicating over the network (i.e., the network is clear), the monitoring station can commence to transmit its data. When one station begins transmitting, all other stations become receivers. Each receiver must monitor all transmission on the network and determine which are intended for them. When a station identifies its address on a received data message, it acts on it or it ignores that transmission.

One advantage of a bus topology is that no special routing or circuit switching is required and, therefore, it is not necessary to store and retransmit messages intended for other nodes. This advantage eliminates a considerable amount of message identification overhead and processing time. However, with heavy-usage systems, there is a high likelihood that more than one station may desire to transmit at the same time. When transmissions from two or more stations occur simultaneously, a data collision occurs, disrupting data communications on the entire network. Obviously, a priority contention scheme is necessary to handle data collision. Such a priority scheme is called *carrier sense, multiple access with collision detect* (CSMA/CD), which is discussed in a later section of this chapter.

Because network control is not centralized in a bus configuration, a node failure will not disrupt data flow on the entire LAN. The critical resource in this case is not a node but instead the bus itself. A failure anywhere along the bus opens the network and, depending on the versatility of the communications channel, may disrupt communication on the entire network.

The addition of new nodes on a bus can sometimes be a problem because gaining access to the bus cable may be a cumbersome task, especially if it is enclosed within a wall, floor, or ceiling. One means of reducing installation problems is to add secondary buses to the primary communications channel. By branching off into other bases, a multiple bus structure called a *tree bus* is formed. Figure 23-23c shows a tree bus configuration.

23-12-1-4 Ring topology. With a ring topology, adjacent stations are interconnected by repeaters in a closed-loop configuration as shown in Figure 23-23d. Each node participates as a repeater between two adjacent links within the ring. The repeaters are relatively simple devices capable of receiving data from one link and retransmitting them on a second link. Messages, usually in packet form, are propagated in the simplex mode (one-way only) from node to node around the ring until it has circled the entire loop and returned to the originating node, where it is verified that the data in the returned message are identical to the data originally transmitted. Hence, the network configuration serves as an inherent error-detection mechanism. The destination station(s) can acknowledge reception of the data by setting or clearing appropriate bits within the control segment of the message packet. Packets contain both source and destination address fields as well as additional network control information and user data. Each node examines incoming data packets, copying packets designated for them and acting as a repeater for all data packets by retransmitting them (bit by bit) to the next down-line repeater. A repeater should neither alter the content of received packets nor change the transmission rate.

Virtually any physical transmission medium can be used with the ring topology. Twisted-wire pairs offer low cost but severely limited transmission rates. Coaxial cables provide greater capacity than twisted-wire pairs at practically the same cost. The highest data rates, however, are achieved with optical fiber cables, except at a substantially higher installation cost.

23-12-2 LAN Transmission Formats

Two transmission techniques or formats are used with LANs, baseband and broadband, to multiplex transmissions from a multitude of stations onto a single transmission medium.

23-12-2-1 Baseband transmission format. Baseband transmission formats are defined as transmission formats that use digital signaling. In addition, baseband formats use the transmission medium as a single-channel device. Only one station can transmit at a time, and all stations must transmit and receive the same types of signals (encoding schemes, bit rates, and so on). Baseband transmission formats time-division multiplex signals onto the transmission medium. All stations can use the media but only one at a time. The entire frequency spectrum (bandwidth) is used by (or at least made available to) whichever station is presently transmitting. With a baseband format, transmissions are bidirectional. A signal inserted at any point on the transmission medium propagates in both directions to the ends, where it is absorbed. Digital signaling requires a bus topology because digital signals cannot be easily propagated through the splitters and joiners necessary in a tree bus topology. Because of transmission line losses, baseband LANs are limited to a distance of no more than a couple miles.

23-12-2-2 Broadband transmission formats. Broadband transmission formats use the connecting media as a multichannel device. Each channel occupies a different frequency band within the total allocated bandwidth (i.e., frequency-division multiplexing). Consequently, each channel can contain different modulation and encoding schemes and operate at different transmission rates. A broadband network permits voice, digital data, and video to be transmitted simultaneously over the same transmission medium. However, broadband systems are unidirectional and require RF modems, amplifiers, and more complicated transceivers than baseband systems. For this reason, baseband systems are more prevalent. Circuit components used with broadband LANs easily facilitate splitting and joining operations; consequently, both bus and tree bus topologies are allowed. Broadband systems can span much greater distances than baseband systems. Distances of up to tens of miles are possible.

The layout for a baseband system is much less complex than a broadband system and, therefore, easier and less expensive to implement. The primary disadvantages of baseband are its limited capacity and length. Broadband systems can carry a wide variety of different kinds of signals on a number of channels. By incorporating amplifiers, broadband can span much greater distances than baseband. Table 23-8 summarizes baseband and broadband transmission formats.

23-12-3 LAN Access Control Methodologies

In a practical LAN, it is very likely that more than one user may wish to use the network media at any given time. For a medium to be shared by various users, a means of controlling access is necessary. Media-sharing methods are known as *access methodologies.* Network access methodologies describe how users access the communications channel in a LAN. The first LANs were developed by computer manufacturers; they were expensive and worked only with certain types of computers with a limited number of software programs. LANs also required a high degree of technical knowledge and expertise to install and maintain. In 1980, the IEEE, in an effort to resolve problems with LANs, formed the 802 Local Area Network Standards Committee. In 1983, the committee established several recommended standards for LANs. The two most prominent standards are IEEE Standard 802.3, which addresses an access method for bus topologies called *carrier sense, multiple access with collision detection* (CSMA/CD), and IEEE Standard 802.5, which describes an access method for ring topologies called *token passing.*

Table 23-8 Baseband versus Broadband Transmission Formats

Baseband	Broadband
Uses digital signaling	Analog signaling requiring RF modems and amplifiers
Entire bandwidth used by each transmission—no FDM	FDM possible, i.e., multiple data channels (video, audio, data, etc.)
Bidirectional	Unidirectional
Bus topology	Bus or tree bus topology
Maximum length approximately 1500 m	Maximum length up to tens of kilometers
Advantages	
Less expensive	High capacity
Simpler technology	Multiple traffic types
Easier and quicker to install	More flexible circuit configurations, larger area covered
Disadvantages	
Single channel	RF modem and amplifiers required
Limited capacity	Complex installation and maintenance
Grounding problems	Double propagation delay
Limited distance	

23-12-3-1 Carrier sense, multiple access with collision detection. CSMA/CD is an access method used primarily with LANs configured in a bus topology. CSMA/CD uses the basic philosophy that, "If you have something to say, say it. If there's a problem, we'll work it out later." With CSMA/CD, any station (node) can send a message to any other station (or stations) as long as the transmission medium is free of transmissions from other stations. Stations monitor (listen to) the line to determine if the line is busy. If a station has a message to transmit but the line is busy, it waits for an idle condition before transmitting its message. If two stations transmit at the same time, a *collision* occurs. When this happens, the station first sensing the collision sends a special jamming signal to all other stations on the network. All stations then cease transmitting (*back off*) and wait a random period of time before attempting a retransmission. The random delay time for each station is different and, therefore, allows for prioritizing the stations on the network. If successive collisions occur, the back-off period for each station is doubled.

With CSMA/CD, stations must contend for the network. A station is not guaranteed access to the network. To detect the occurrence of a collision, a station must be capable of transmitting and receiving simultaneously. CSMA/CD is used by most LANs configured in a bus topology. *Ethernet* is an example of a LAN that uses CSMA/CD and is described later in this chapter.

Another factor that could possibly cause collisions with CSMA/CD is *propagation delay.* Propagation delay is the time it takes a signal to travel from a source to a destination. Because of propagation delay, it is possible for the line to appear idle when, in fact, another station is transmitting a signal that has not yet reached the monitoring station.

23-12-3-2 Token passing. *Token passing* is a network access method used primarily with LANs configured in a ring topology using either baseband or broadband transmission formats. When using *token passing* access, nodes do not contend for the right to transmit data. With token passing, a specific packet of data, called a *token,* is circulated around the ring from station to station, always in the same direction. The token is generated by a designated station known as the *active monitor.* Before a station is allowed to transmit, it must first possess the token. Each station, in turn, acquires the token and examines the data frame to determine if it is carrying a packet addressed to it. If the frame contains a packet with the receiving station's address, it copies the packet into memory, appends any messages it has to send to the token, and then relinquishes the token by retransmitting all data packets and the token to the next node on the network. With token passing, each station has equal access to the transmission medium. As with CSMA/CD, each transmitted packet contains source and destination address fields. Successful delivery of a data frame is confirmed by the destination station by setting *frame status flags,* then forwarding the frame around the ring to the original transmitting station. The packet then is removed from the frame before transmitting the token. A token cannot be used twice, and there is a time limitation on how long a token can be held. This prevents one station from disrupting data transmissions on the network by holding the token until it has a packet to transmit. When a station does not possess the token, it can only receive and transfer other packets destined to other stations.

Some 16-Mbps token ring networks use a modified form of token passing methodology where the token is relinquished as soon as a data frame has been transmitted instead of waiting until the transmitted data frame has been returned. This is known as an *early token release mechanism.*

23-13 ETHERNET

Ethernet is a baseband transmission system designed in 1972 by Robert Metcalfe and David Boggs of the Xerox Palo Alto Research Center (PARC). Metcalfe, who later founded 3COM Corporation, and his colleagues at Xerox developed the first experimental Ethernet system to interconnect a Xerox Alto personal workstation to a graphical user interface. The first experimental Ethernet system was later used to link Altos workstations to each other and to link the workstations to servers and laser printers. The signal clock for the experimental Ethernet interface was derived from the Alto's system clock, which produced a data transmission rate of 2.94 Mbps.

Metcalfe's first Ethernet was called the Alto Aloha Network; however, in 1973 Metcalfe changed the name to Ethernet to emphasize the point that the system could support any computer, not just Altos, and to stress the fact that the capabilities of his new network had evolved well beyond the original Aloha system. Metcalfe chose the name based on the word *ether,* meaning "air," "atmosphere," or "heavens," as an indirect means of describing a vital feature of the system: the physical medium (i.e., a cable). The physical medium carries data bits to all stations in much the same way that *luminiferous ether* was once believed to transport electromagnetic waves through space.

In July 1976, Metcalfe and Boggs published a landmark paper titled "Ethernet: Distributed Packet Switching for Local Computer." On December 13, 1977, Xerox Corporation received patent number 4,063,220 titled "Multipoint Data Communications System with Collision Detection." In 1979, Xerox joined forces with Intel and Digital Equipment Corporation (DEC) in an attempt to make Ethernet an industry standard. In September 1980, the three companies jointly released the first version of the first Ethernet specification called the Ethernet Blue Book, DIX 1.0 (after the initials of the three companies), or Ethernet I.

Ethernet I was replaced in November 1982 by the second version, called Ethernet II (DIX 2.0), which remains the current standard. In 1983, the 802 Working Group of the IEEE released their first standard for Ethernet technology. The formal title of the standard was *IEEE 802.3 Carrier Sense Multiple Access with Collision Detection (CSMA/CD) Access Method and Physical Layer Specifications.* The IEEE subsequently reworked several sections of the original standard, especially in the area of the frame format definition, and in 1985 they released the 802.3a standard, which was called *thin Ethernet, cheapernet,* or *10Base-2 Ethernet.* In 1985, the IEEE also released the IEEE 802.3b 10Broad36 standard, which defined a transmission rate of 10 Mbps over a coaxial cable system.

In 1987, two additional standards were released: IEEE 802.3d and IEEE 802.3e. The 802.3d standard defined the *Fiber Optic Inter-Repeater Link* (FOIRL) that used two fiber optic cables to extend the maximum distance between 10 Mbps repeaters to 1000 meters. The IEEE 802.3e standard defined a 1-Mbps standard based on twisted-pair cable, which was never widely accepted. In 1990, the IEEE introduced a major advance in Ethernet standards: IEEE 802.3i. The 802.3i standard defined 10Base-T, which permitted a 10-Mbps transmission rate over simple category 3 unshielded twisted-pair (UTP) cable. The widespread use of UTP cabling in existing buildings created a high demand for 10Base-T technology. 10Base-T also facilitated a star topology, which made it much easier to install, manage, and troubleshoot. These advantages led to a vast expansion in the use of Ethernet.

In 1993, the IEEE released the 802.3j standard for 10Base-F (FP, FB, and FL), which permitted attachment over longer distances (2000 meters) through two optical fiber cables. This standard updated and expanded the earlier FOIRL standard. In 1995, the IEEE improved the performance of Ethernet technology by a factor of 10 when it released the 100-Mbps 802.3u 100Base-T standard. This version of Ethernet is commonly known as *fast Ethernet.* Fast Ethernet supported three media types: 100Base-TX, which operates over two pairs of category 5 twisted-pair cable; 100Base-T4, which operates over four pairs of category 3 twisted-pair cable; and 100Base-FX, which operates over two multimode fibers.

In 1997, the IEEE released the 802.3x standard, which defined full-duplex Ethernet operation. Full-duplex Ethernet bypasses the normal CSMA/CD protocol and allows two stations to communicate over a point-to-point link, which effectively doubles the transfer rate by allowing each station to simultaneously transmit and receive separate data streams. In 1997, the IEEE also released the IEEE 802.3y 100Base-T2 standard for 100-Mbps operation over two pairs of category 3 balanced transmission line.

In 1998, IEEE once again improved the performance of Ethernet technology by a factor of 10 when it released the 1-Gbps 802.3z 1000Base-X standard, which is commonly called *gigabit Ethernet.* Gigabit Ethernet supports three media types: 1000Base-SX, which operates with an 850-nm laser over multimode fiber; 1000Base-LX, which operates with a 1300-nm laser over single and multimode fiber; and 1000Base-CX, which operates over short-haul copper-shielded twisted-pair (STP) cable. In 1998, the IEEE also released the 802.3ac standard, which defines extensions to support virtual LAN (VLAN) tagging on Ethernet networks. In 1999, the release of the 802.3ab 1000Base-T standard defined 1-Gbps operation over four pairs of category 5 UTP cabling.

The topology of choice for Ethernet LANs is either a linear bus or a star, and all Ethernet systems employ carrier sense, multiple access with collision detect (CSMA/CD) for the accessing method.

23-13-1 IEEE Ethernet Standard Notation

To distinguish the various implementations of Ethernet available, the IEEE 802.3 committee has developed a concise notation format that contains information about the Ethernet system, including such items as bit rate, transmission mode, transmission medium, and segment length. The IEEE 802.3 format is

<data rate in Mbps><transmission mode><maximum segment length in hundreds of meters>

or

<data rate in Mbps><transmission mode><transmission media>

The transmission rates specified for Ethernet are 10 Mbps, 100 Mbps, and 1 Gbps. There are only two transmission modes: baseband (base) or broadband (broad). The segment length can vary, depending on the type of transmission medium, which could be coaxial cable (no designation), twisted-pair cable (T), or optical fiber (F). For example, the notation 10Base-5 means 10-Mbps transmission rate, baseband mode of transmission, with a maximum segment length of 500 meters. The notation 100Base-T specifies 100-Mbps transmission rate, baseband mode of transmission, with a twisted-pair transmission medium. The notation 100Base-F means 100-Mbps transmission rate, baseband transmission mode, with an optical fiber transmission medium.

The IEEE currently supports nine 10-Mbps standards, six 100-Mbps standards, and five 1-Gbps standards. Table 23-9 lists several of the more common types of Ethernet, their cabling options, distances supported, and topology.

Table 23-9 Current IEEE Ethernet Standards

Transmission Rate	Ethernet System	Transmission Medium	Maximum Segment Length
10 Mbps	10Base-5	Coaxial cable (RG-8 or RG-11)	500 meters
	10Base-2	Coaxial cable (RG-58)	185 meters
	10Base-T	UTP/STP category 3 or better	100 meters
	10Broad-36	Coaxial cable (75 ohm)	Varies
	10Base-FL	Optical fiber	2000 meters
	10Base-FB	Optical fiber	2000 meters
	10Base-FP	Optical fiber	2000 meters
100 Mbps	100Base-T	UTP/STP category 5 or better	100 meters
	100Base-TX	UTP/STP category 5 or better	100 meters
	100Base-FX	Optical fiber	400–2000 meters
	100Base-T4	UTP/STP category 5 or better	100 meters
1000 Mbps	1000Base-LX	Long-wave optical fiber	Varies
	1000Base-SX	Short-wave optical fiber	Varies
	1000Base-CX	Short copper jumper	Varies
	1000Base-T	UTP/STP category 5 or better	Varies

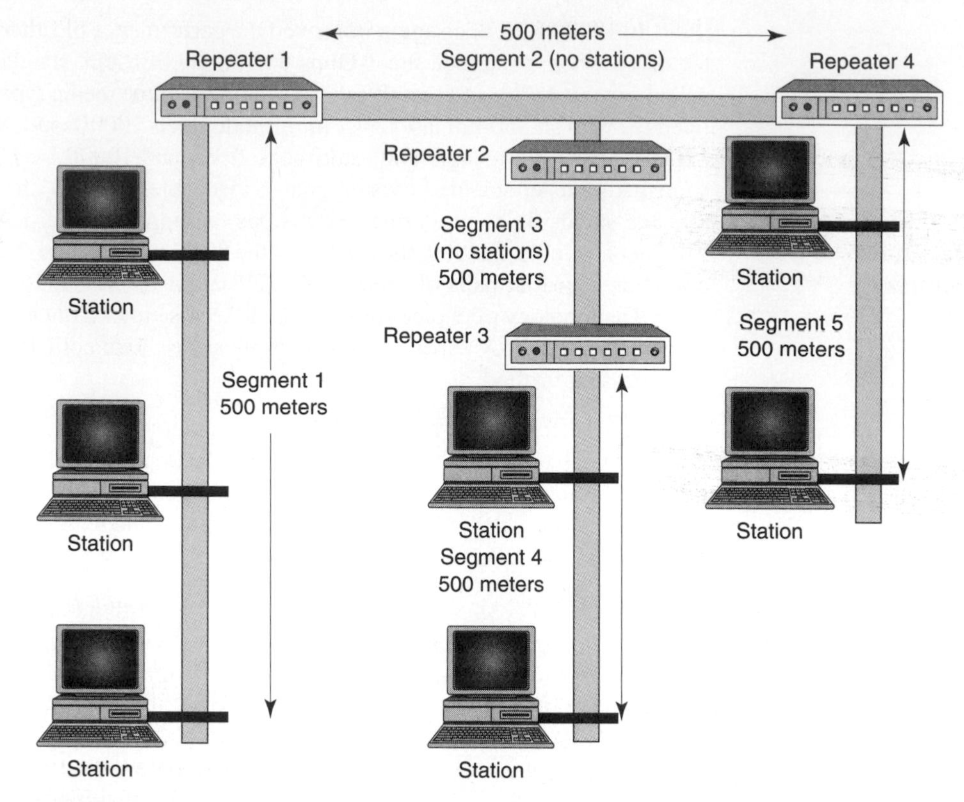

FIGURE 23-24 10 Mbps 5-4-3 Ethernet configuration

23-13-2 10-Mbps Ethernet

Figure 23-24 shows the physical layout for a 10Base-5 Ethernet system. The maximum number of cable segments supported with 10Base-5 Ethernet is five, interconnected with four repeaters or hubs. However, only three of the segments can be populated with nodes (computers). This is called the *5-4-3 rule:* five segments joined by four repeaters, but only three segments can be populated. The maximum segment length for 10Base-5 is 500 meters. Imposing maximum segment lengths are required for the CSMA/CD to operate properly. The limitations take into account Ethernet frame size, velocity of propagation on a given transmission medium, and repeater delay time to ensure collisions that occur on the network are detected.

On 10Base-5 Ethernet, the maximum segment length is 500 meters with a maximum of five segments. Therefore, the maximum distance between any two nodes (computers) is $5 \times 500 = 2500$ meters. The worst-case scenario for collision detection is when the station at one end of the network completes a transmission at the same instant the station at the far end of the network begins a transmission. In this case, the station that transmitted first would not know that a collision had occurred. To prevent this from happening, minimum frame lengths are imposed on Ethernet.

The minimum frame length for 10Base-5 is computed as follows. The velocity of propagation along the cable is assumed to be approximately two-thirds the speed of light, or

$$v_p = \frac{2}{3}v_c$$

$$v_p = \left(\frac{2}{3}\right)(3 \times 10^8 \text{ m/s})$$

$$v_p = 2 \times 10^8 \text{ m/s}$$

Thus, the length of a bit along a cable for a bit rate of 10 Mbps is

$$\text{bit length} = \frac{2 \times 10^8 \text{ m/s}}{10 \text{ mbps}} = 20 \text{ meters/bit}$$

and the maximum number of bits on a cable with a maximum length of 2500 meters is

$$\frac{2500 \text{ m}}{20 \text{ m/b}} = 125 \text{ bits}$$

Therefore, the maximum time for a bit to propagate end to end is

$$\frac{2500 \text{ m}}{2 \times 10^8 \text{ m/s}} = 12.5 \ \mu\text{s}$$

and the round-trip delay equals

$$2 \times 12.5 \ \mu\text{s} = 25 \ \mu\text{s}$$

Therefore, the minimum length of an Ethernet message for a 10-Mbps transmission rate is

$$\frac{\text{round-trip delay}}{\text{bit time}} = \frac{25 \ \mu\text{s}}{0.1 \ \mu\text{s}} = 250 \text{ bits}$$

where the time of a bit (t_b = 1/bit rate or 1/10 Mbps = 0.1 μs). However, the minimum number of bits is doubled and rounded up to 512 bits (64 eight-bit bytes).

10Base-5 is the original Ethernet that specifies a *thick* 50-Ω double-shielded RG-11 coaxial cable for the transmission medium. Hence, this version is sometimes called *thicknet* or *thick Ethernet.* Because of its inflexible nature, 10Base-5 is sometimes called *frozen yellow garden hose.* 10Base-5 Ethernet uses a bus topology with an external device called a *media access unit* (MAU) to connect terminals to the cable. The MAU is sometimes called a *vampire tap* because it connects to the cable by simply puncturing the cable with a sharp prong that extends into the cable until it makes contact with the center conductor. Each connection is called a *tap,* and the cable that connects the MAU to its terminal is called an *attachment unit interface* (AUI) or sometimes simply a *drop.* Within each MAU, a digital transceiver transfers electrical signals between the drop and the coaxial transmission medium. 10Base-5 supports a maximum of 100 nodes per segment. Repeaters are counted as nodes; therefore, the maximum capacity of a 10Base-5 Ethernet is 297 nodes. With 10Base-5, unused taps must be terminated in a 50-Ω resistive load. A drop left unterminated or any break in the cable will cause total LAN failure.

23-13-2-1 10Base-2 Ethernet. 10Base-5 Ethernet uses a 50-ΩRG-11 coaxial cable, which is thick enough to give it high noise immunity, thus making it well suited to laboratory and industrial applications. The RG-11 cable, however, is expensive to install. Consequently, the initial costs of implementing a 10Base-5 Ethernet system are too high for many small businesses. In an effort to reduce the cost, International Computer Ltd, Hewlett-Packard, and 3COM Corporation developed an Ethernet variation that uses thinner, less expensive 50-ΩRG-58 coaxial cable. RG-58 is less expensive to purchase and install than RG-11. In 1985, the IEEE 802.3 Standards Committee adopted a new version of Ethernet and gave it the name 10Base-2, which is sometimes called *cheapernet* or *thinwire* Ethernet.

10Base-2 Ethernet uses a bus topology and allows a maximum of five segments; however, only three can be populated. Each segment has a maximum length of 185 meters with no more than 30 nodes per segment. This limits the capacity of a 10Base-2 network to 96 nodes. 10Base-2 eliminates the MAU, as the digital transceiver is located inside the terminal and a simple BNC-T connector connects the network interface card (NIC) directly to the coaxial cable. This eliminates the expensive cable and the need to tap or drill into it.

With 10Base-2 Ethernet, unused taps must be terminated in a 50-Ω resistive load and a drop left unterminated, or any break in the cable will cause total LAN failure.

23-13-2-2 10Base-T Ethernet. 10Base-T Ethernet is the most popular 10-Mbps Ethernet commonly used with PC-based LAN environments utilizing a star topology. Because stations can be connected to a network hub through an internal transceiver, there is no need for an AUI. The "T" indicates unshielded twisted-pair cable. 10Base-T was developed to allow Ethernet to utilize existing voice-grade telephone wiring to carry Ethernet signals. Standard modular RJ-45 telephone jacks and four-pair UTP telephone wire are specified in the standard for interconnecting nodes directly to the LAN without an AUI. The RJ-45 connector plugs direction into the network interface card in the PC. 10Base-T operates at a transmission rate of 10 Mbps and uses CSMA/CD; however, it uses a multiport hub at the center of network to interconnect devices. This essentially converts each segment to a point-to-point connection into the LAN. The maximum segment length is 100 meters with no more than two nodes on each segment.

Nodes are added to the network through a port on the hub. When a node is turned on, its transceiver sends a DC current over the twisted-pair cable to the hub. The hub senses the current and enables the port, thus connecting the node to the network. The port remains connected as long as the node continues to supply DC current to the hub. If the node is turned off or if an open- or short-circuit condition occurs in the cable between the node and the hub, DC current stops flowing, and the hub disconnects the port from the network, allowing the remainder of the LAN to continue operating status quo. With 10Base-T Ethernet, a cable break affects only the nodes on that segment.

23-13-2-3 10Base-FL Ethernet. 10Base-FL (fiber link) is the most common 10-Mbps Ethernet that uses optical fiber for the transmission medium. 10Base-FL is arranged in a star topology where stations are connected to the network through an external AUI cable and an external transceiver called a fiber-optic MAU. The transceiver is connected to the hub with two pairs of optical fiber cable. The cable specified is graded-index multimode cable with a 62.5-μm-diameter core.

23-13-3 100-Mbps Ethernet

Over the past few years, it has become quite common for bandwidth-starved LANs to upgrade 10Base-T Ethernet LANs to 100Base-T (sometimes called fast Ethernet). The 100Base-T Ethernet includes a family of fast Ethernet standards offering 100-Mbps data transmission rates using CSMA/CD access methodology. 100-Mbps Ethernet installations do not have the same design rules as 10-Mbps Ethernet. 10-Mbps Ethernet allows several connections between hubs within the same segment (collision domain). 100-Mbps Ethernet does not allow this flexibility. Essentially, the hub must be connected to an internetworking device, such as a switch or a router. This is called the *2-1 rule*—two hubs minimum for each switch. The reason for this requirement is for collision detection within a domain. The transmission rate increased by a factor of 10; therefore, frame size, cable propagation, and hub delay are more critical.

IEEE standard 802.3u details operation of the 100Base-T network. There are three media-specific physical layer standards for 100Base-TX: 100Base-T, 100Base-T4, and 100Base-FX.

23-13-3-1 100Base-TX Ethernet. 100Base-TX Ethernet is the most common of the 100-Mbps Ethernet standards and the system with the most technology available. 100Base-TX specifies a 100-Mbps data transmission rate over two pairs of category 5 UTP or STP cables with a maximum segment length of 100 meters. 100Base-TX uses a physical star topology (half duplex) or bus (full duplex) with the same media access method (CSMA/CD) and frame structures as 10Base-T; however, 100Base-TX requires a hub port and NIC, both of which must be 100Base-TX compliant. 100Base-TX can operate full duplex in certain situations, such as from a switch to a server.

23-13-3-2 100Base-T4 Ethernet. 100Base-T4 is a physical layer standard specifying 100-Mbps data rates using two pairs of category 3, 4, or 5 UTP or STP cable. 100Base-T4 was devised to allow installations that do not comply with category 5 UTP cabling specifications. 100Base-T4 will operate using category 3 UTP installation or better; however, there are some significant differences in the signaling.

23-13-3-3 100Base-FX Ethernet. 100Base-FX is a physical layer standard specifying 100-Mbps data rates over two optical fiber cables using a physical star topology. The logical topology for 100Base-FX can be either a star or a bus. 100Base-FX is often used to interconnect 100Base-TX LANs to a switch or router. 100Base-FX uses a duplex optical fiber connection with multimode cable that supports a variety of distances, depending on circumstances.

23-13-4 1000-Mbps Ethernet

One-gigabit Ethernet (1 GbE) is the latest implementation of Ethernet that operates at a transmission rate of one billion bits per second and higher. The IEEE 802.3z Working Group is currently preparing standards for implementing gigabit Ethernet. Early deployments of gigabit Ethernet were used to interconnect 100-Mbps and gigabit Ethernet switches, and gigabit Ethernet is used to provide a *fat pipe* for high-density backbone connectivity. Gigabit Ethernet can use one of two approaches to medium access: half-duplex mode using CSMA/CD or full-duplex mode, where there is no need for multiple accessing.

Gigabit Ethernet can be generally categorized as either two-wire 1000Base-X or four-wire 1000Base-T. Two-wire gigabit Ethernet can be either 1000Base-SX for short-wave optical fiber, 1000Base-LX for long-wave optical fiber, or 1000Base-CX for short copper jumpers. The four-wire version of gigabit Ethernet is 1000Base-T. 1000Base-SX and 1000Base-LX use two optical fiber cables where the only difference between them is the wavelength (color) of the light waves propagated through the cable. 1000Base-T Ethernet was designed to be used with four twisted pairs of Category 5 UTP cables.

23-13-5 Ethernet Frame Formats

Over the years, four different Ethernet frame formats have emerged where network environment dictates which format is implemented for a particular configuration. The four formats are the following:

Ethernet II. The original format used with DIX.

IEEE 802.3. The first generation of the IEEE Standards Committee, often referred to as a raw IEEE 802.3 frame. Novell was the only software vendor to use this format.

IEEE 802.3 with 802.2 LLC. Provides support for IEEE 802.2 LLC.

IEEE 802.3 with SNAP similar to IEEE 802.3 but provides backward compatibility for 802.2 to Ethernet II formats and protocols.

Ethernet II and IEEE 802.3 are the two most popular frame formats used with Ethernet. Although they are sometimes thought of as the same thing, in actuality Ethernet II and IEEE 802.3 are not identical, although the term *Ethernet* is generally used to refer to any IEEE 802.3-compliant network. Ethernet II and IEEE 802.3 both specify that data be transmitted from one station to another in groups of data called *frames.*

23-13-5-1 Ethernet II frame format. The frame format for Ethernet II is shown in Figure 23-25 and is comprised of a preamble, start frame delimiter, destination address, source address, type field, data field, and frame check sequence field.

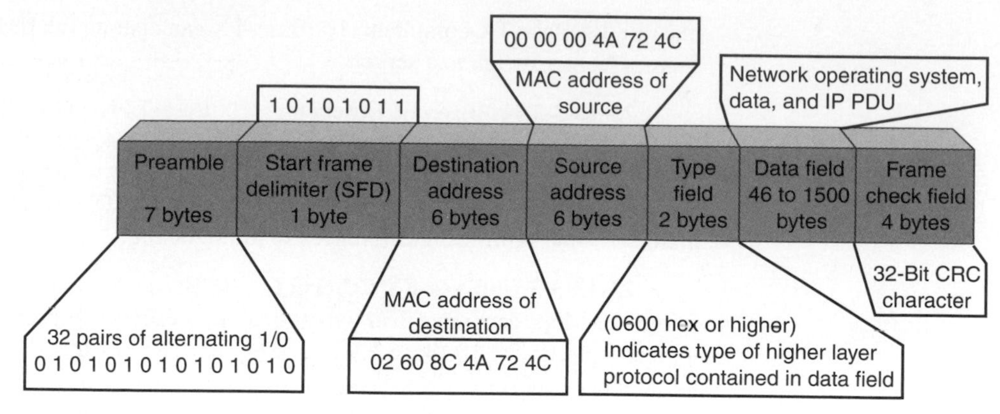

FIGURE 23-25 Ethernet II frame format

Preamble. The preamble consists of eight bytes (64 bits) of alternating 1s and 0s. The purpose of the preamble is to establish clock synchronization. The last two bits of the preamble are reserved for the start frame delimiter.

Start frame delimiter. The start frame delimiter is simply a series of two logic 1s appended to the end of the preamble, whose purpose is to mark the end of the preamble and the beginning of the data frame.

Destination address. The source and destination addresses and the field type make up the frame header. The destination address consists of six bytes (48 bits) and is the address of the node or nodes that have been designated to receive the frame. The address can be a unique, group, or broadcast address and is determined by the following bit combinations:

bit 0 = 0	If bit 0 is a 0, the address is interpreted as a unique address intended for only one station.
bit 0 = 1	If bit 0 is a 1, the address is interpreted as a multicast (group) address. All stations that have been preassigned with this group address will accept the frame.
bit 0–47	If all bits in the destination field are 1s, this identifies a broadcast address, and all nodes have been identified as receivers of this frame.

Source address. The source address consists of six bytes (48 bits) that correspond to the address of the station sending the frame.

Type field. Ethernet does not use the 16-bit type field. It is placed in the frame so it can be used for higher layers of the OSI protocol hierarchy.

Data field. The data field contains the information and can be between 46 bytes and 1500 bytes long. The data field is transparent. Data-link control characters and zero-bit stuffing are not used. Transparency is achieved by counting back from the FCS character.

Frame check sequence field. The CRC field contains 32 bits for error detection and is computed from the header and data fields.

23-13-5-2 IEEE 802.3 frame format. The frame format for IEEE 802.3 is shown in Figure 23-26 and consists of the following:

Preamble. The preamble consists of seven bytes to establish clock synchronization. The last byte of the preamble is used for the start frame delimiter.

Start frame delimiter. The start frame delimiter is simply a series of two logic 1s appended to the end of the preamble, whose purpose is to mark the end of the preamble and the beginning of the data frame.

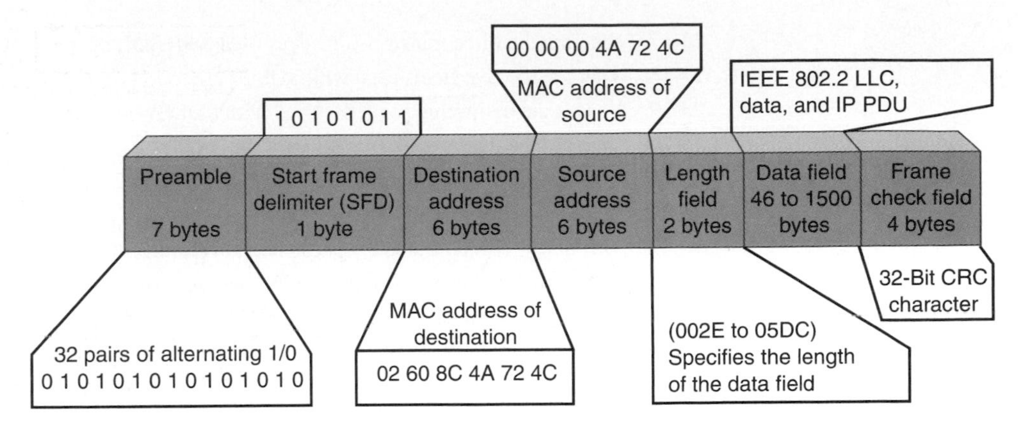

FIGURE 23-26 IEEE 802.3 frame format

Destination and source addresses. The destination and source addresses are defined the same as with Ethernet II.

Length field. The two-byte length field in the IEEE 802.3 frame replaces the type field in the Ethernet frame. The length field indicates the length of the variable-length logical link control (LLC) data field, which contains all upper-layered embedded protocols.

Logical link control (LLC). The LLC field contains the information and can be between 46 bytes and 1500 bytes long. The LLC field defined in IEEE 802.3 is identical to the LLC field defined for token ring networks.

Frame check sequence field. The CRC field is defined the same as with Ethernet II.

End-of-frame delimiter. The end-of-frame delimiter is a period of time (9.6 μs) in which no bits are transmitted. With Manchester encoding, a void in transitions longer than one-bit time indicates the end of the frame.

QUESTIONS

23-1. Define *data-link protocol.*

23-2. What is meant by a primary station? Secondary station?

23-3. What is a master station? Slave station?

23-4. List and describe the three data-link protocol functions.

23-5. Briefly describe the ENQ/ACK line discipline.

23-6. Briefly describe the poll/select line discipline.

23-7. Briefly describe the stop-and-wait method of flow control.

23-8. Briefly describe the sliding window method of flow control.

23-9. What is the difference between character- and bit-oriented protocols?

23-10. Describe the difference between asynchronous and synchronous protocols.

23-11. Briefly describe how the XMODEM protocol works.

23-12. Why is IBM's 3270 protocol called "bisync"?

23-13. Briefly describe the polling sequence for BSC, including the difference between a general and specific poll.

23-14. Briefly describe the selection sequence for BSC.

23-15. How does BSC achieve transparency?

23-16. What is the difference between a command and a response with SDLC?

23-17. What are the three transmission states used with SDLC?

23-18. What are the five fields used with SDLC?

23-19. What is the delimiting sequence used with SDLC?

23-20. What are the three frame formats used with SDLC?

23-21. What are the purposes of the *ns* and *nr* bit sequences?

23-22. What is the difference between P and F bits?

23-23. With SDLC, which frame types can contain an information field?

23-24. With SDLC, which frame types can be used for error correction?

23-25. What SDLC command/response is used for reporting procedural errors?

23-26. When is the configure command/response used with SDLC?

23-27. What is the go-ahead sequence? The turnaround sequence?

23-28. What is the transparency mechanism used with SDLC?

23-29. What supervisory condition exists with HDLC that is not included in SDLC?

23-30. What are the transparency mechanism and delimiting sequence for HDLC?

23-31. Briefly describe invert-on-zero encoding.

23-32. List and describe the HDLC operational modes.

23-33. Briefly describe the layout for a public switched data network.

23-34. What is a value-added network?

23-35. Briefly describe circuit, message, and packet switching.

23-36. What is a transactional switch? A transparent switch?

23-37. Explain the following terms: *permanent virtual circuit, virtual call,* and *datagram.*

23-38. Briefly describe an X.25 call request packet.

23-39. Briefly describe an X.25 data transfer packet.

23-40. Define *ISDN.*

23-41. List and describe the principles of ISDN.

23-42. List and describe the evolution of ISDN.

23-43. Describe the conceptual view of ISDN and what is meant by the term *digital pipe.*

23-44. List the objectives of ISDN.

23-45. Briefly describe the architecture of ISDN.

23-46. List and describe the ISDN system connections and interface units.

23-47. Briefly describe BISDN.

23-48. Briefly describe asynchronous transfer mode.

23-49. Describe the differences between virtual channels and virtual paths.

23-50. Briefly describe the ATM header field; ATM information field.

23-51. Describe the following ATM network components: ATM endpoints, ATM switches, ATM transmission paths.

23-52. Briefly describe a local area network.

23-53. List and describe the most common LAN topologies.

23-54. Describe the following LAN transmission formats: baseband and broadband.

23-55. Describe the two most common LAN access methodologies.

23-56. Briefly describe the history of Ethernet.

23-57. Describe the Ethernet standard notation.

23-58. List and briefly describe the 10-Mbps Ethernet systems.

23-59. List and briefly describe the 100-Mbps Ethernet systems.

23-60. List and briefly describe the 1000-Mbps Ethernet systems.

23-61. Describe the two most common Ethernet frame formats.

PROBLEMS

23-1. Determine the hex code for the control field in an SDLC frame for the following conditions: information frame, poll, transmitting frame 4, and confirming reception of frames 2, 3, and 4.

23-2. Determine the hex code for the control field in an SDLC frame for the following conditions: supervisory frame, ready to receive, final, and confirming reception of frames 6, 7, and 0.

23-3. Insert 0s into the following SDLC data stream:

111 001 000 011 111 111 100 111 110 100 111 101 011 111 111 111 001 011

23-4. Delete 0s from the following SDLC data stream:

010 111 110 100 011 011 111 011 101 110 101 111 101 011 100 011 111 00

23-5. Sketch the NRZI waveform for the following data stream (start with a high condition):

1 0 0 1 1 1 0 0 1 0 1 0

23-6. Determine the hex code for the control field in an SDLC frame for the following conditions: information frame, not a poll, transmitting frame number 5, and confirming the reception of frames 0, 1, 2, and 3.

23-7. Determine the hex code for the control field in an SDLC frame for the following conditions: supervisory frame, not ready to receive, not a final, and confirming reception of frames 7, 0, 1, and 2.

23-8. Insert 0s into the following SDLC data stream:

0110111111101100001111100101110001011111111011111001

23-9. Delete 0s from the following SDLC data stream:

0010111110011111011111011000100011111011101011000101

23-10. Sketch the NRZI levels for the following data stream (start with a high condition):

1 1 0 1 0 0 0 1 1 0 1

C H A P T E R 24

Microwave Radio Communications and System Gain

CHAPTER OUTLINE

OBJECTIVES

- Define *microwave*
- Describe microwave frequencies and microwave frequency bands
- Contrast the advantages and disadvantages of microwave
- Contrast analog versus digital microwave
- Contrast frequency modulation with amplitude modulation microwave
- Describe the block diagram for a microwave radio system
- Describe the different types of microwave repeaters
- Define *diversity* and describe several diversity systems
- Define *protection switching arrangements* and describe several switching system configurations
- Describe the operation of the various components that make up microwave radio terminal and repeater stations
- Identify the free-space path characteristics and describe how they affect microwave system performance
- Define *system gain* and describe how it is calculated for FM microwave radio systems

24-1 INTRODUCTION

Microwaves are generally described as electromagnetic waves with frequencies that range from approximately 500 MHz to 300 GHz or more. Therefore, microwave signals, because of their inherently high frequencies, have relatively short wavelengths, hence the name "micro" waves. For example, a 100-GHz microwave signal has a wavelength of 0.3 cm, whereas a 100-MHz commercial broadcast-band FM signal has a wavelength of 3 m. The wavelengths for microwave frequencies fall between 1 cm and 60 cm, slightly longer than infrared energy. Table 24-1 lists some of the microwave radio-frequency bands available in the United States. For full-duplex (two-way) operation as is generally required of microwave communications systems, each frequency band is divided in half with the lower half identified as the *low band* and the upper half as the *high band.* At any given radio station, transmitters are normally operating on either the low or the high band, while receivers are operating on the other band.

On August 17, 1951, the first transcontinental microwave radio system began operation. The system was comprised of 107 relay stations spaced an average of 30 miles apart to form a continuous radio link between New York and San Francisco that cost the Bell System approximately $40 million. By 1954, there were over 400 microwave stations scattered across the United States and, by 1958, microwave carriers were the dominant means of long-distance communications as they transported the equivalent of 13 million miles of telephone circuits.

The vast majority of electronic communications systems established since the mid-1980s have been digital in nature and, thus, carry voice, video, and data information in digital form.

However, terrestrial (earth-based) *microwave radio relay* systems using frequency (FM) or digitally modulated carriers (PSK or QAM) still provide approximately 35% of

Table 24-1 Microwave Radio-Frequency Assignments

Service	Frequency (MHz)	Band
Military	1710–1850	L
Operational fixed	1850–1990	L
Studio transmitter link	1990–2110	L
Common carrier	2110–2130	S
Operational fixed	2130–2150	S
Operational carrier	2160–2180	S
Operational fixed	2180–2200	S
Operational fixed television	2500–2690	S
Common carrier and satellite downlink	3700–4200	S
Military	4400–4990	C
Military	5250–5350	C
Common carrier and satellite uplink	5925–6425	C
Operational fixed	6575–6875	C
Studio transmitter link	6875–7125	C
Common carrier and satellite downlink	7250–7750	C
Common carrier and satellite uplink	7900–8400	X
Common carrier	10,700–11,700	X
Operational fixed	12,200–12,700	X
Cable television (CATV) studio link	12,700–12,950	Ku
Studio transmitter link	12,950–13,200	Ku
Military	14,400–15,250	Ka
Common carrier	17,700–19,300	Ka
Satellite uplink	26,000–32,000	K
Satellite downlink	39,000–42,000	Q
Satellite crosslink	50,000–51,000	V
Satellite crosslink	54,000–62,000	V

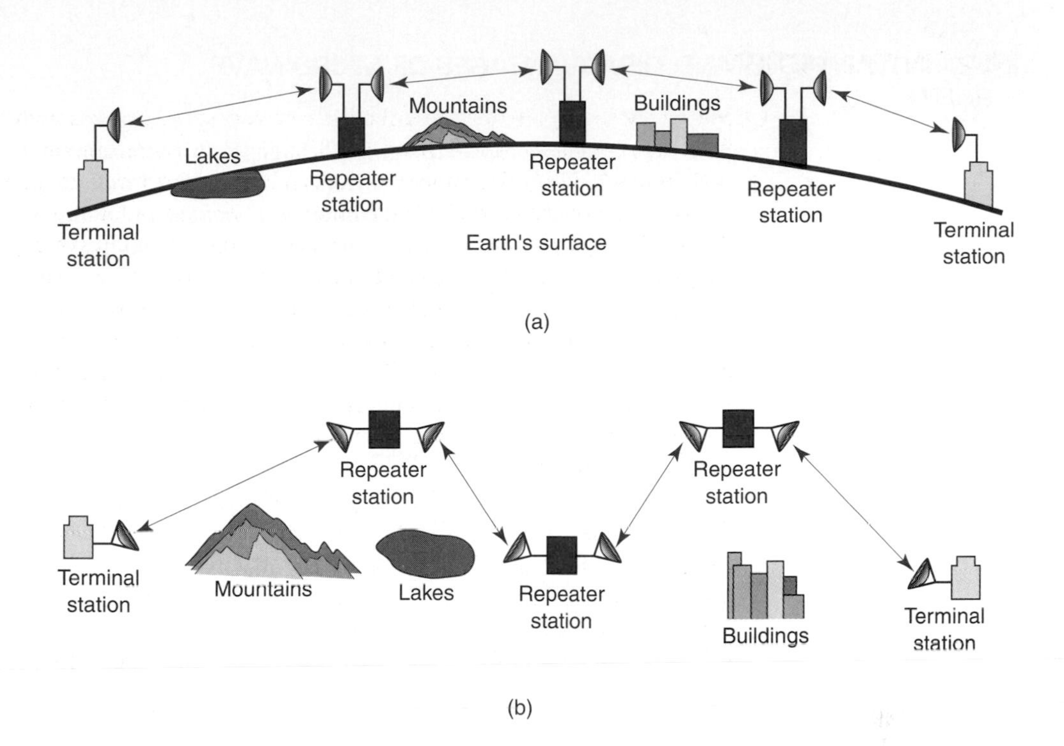

FIGURE 24-1 Microwave radio communications link: (a) side view; (b) top view

the total information-carrying circuit mileage in the United States. There are many different types of microwave systems operating over distances that vary from 15 miles to 4000 miles in length. *Intrastate* or *feeder service* microwave systems are generally categorized as *short haul* because they are used to carry information for relatively short distances, such as between cities within the same state. *Long-haul* microwave systems are those used to carry information for relatively long distances, such as *interstate* and *backbone* route applications. Microwave radio system capacities range from less than 12 voice-band channels to more than 22,000 channels. Early microwave systems carried frequency-division-multiplexed voice-band circuits and used conventional, noncoherent frequency-modulation techniques. More recently developed microwave systems carry pulse-code-modulated time-division-multiplexed voice-band circuits and use more modern digital modulation techniques, such as phase-shift keying (PSK) or quadrature amplitude modulation (QAM).

Figure 24-1 shows a typical layout for a microwave radio link. Information originates and terminates at the terminal stations, whereas the repeaters simply relay the information to the next downlink microwave station. Figure 24-1a shows a microwave radio link comprised of two terminal stations (one at each end) that are interconnected by three repeater stations. As the figure shows, the microwave stations must be geographically placed in such a way that the terrain (lakes, mountains, buildings, and so on) do not interfere with transmissions between stations. This sometimes necessitates placing the stations on top of hills, mountains, or tall buildings. Figure 24-1b shows how a microwave radio link appears from above. Again, the geographic location of the stations must be carefully selected such that natural and man-made barriers do not interfere with propagation between stations. Again, sometimes it is necessary to construct a microwave link around obstacles, such as large bodies of water, mountains, and tall buildings.

24-2 ADVANTAGES AND DISADVANTAGES OF MICROWAVE RADIO

Microwave radios propagate signals through Earth's atmosphere between transmitters and receivers often located on top of towers spaced about 15 miles to 30 miles apart. Therefore, microwave radio systems have the obvious advantage of having the capacity to carry thousands of individual information channels between two points without the need for physical facilities such as coaxial cables or optical fibers. This, of course, avoids the need for acquiring rights-of-way through private property. In addition, radio waves are better suited for spanning large bodies of water, going over high mountains, or going through heavily wooded terrain that impose formidable barriers to cable systems. The advantages of microwave radio include the following:

24-2-1 Advantages of Microwave Radio

1. Radio systems do not require a right-of-way acquisition between stations.
2. Each station requires the purchase or lease of only a small area of land.
3. Because of their high operating frequencies, microwave radio systems can carry large quantities of information.
4. High frequencies mean short wavelengths, which require relatively small antennas.
5. Radio signals are more easily propagated around physical obstacles such as water and high mountains.
6. Fewer repeaters are necessary for amplification.
7. Distances between switching centers are less.
8. Underground facilities are minimized.
9. Minimum delay times are introduced.
10. Minimal crosstalk exists between voice channels.
11. Increased reliability and less maintenance are important factors.

24-2-2 Disadvantages of Microwave Radio

1. It is more difficult to analyze and design circuits at microwave frequencies.
2. Measuring techniques are more difficult to perfect and implement at microwave frequencies.
3. It is difficult to implement conventional circuit components (resistors, capacitors, inductors, and so on) at microwave frequencies.
4. Transient time is more critical at microwave frequencies.
5. It is often necessary to use specialized components for microwave frequencies.
6. Microwave frequencies propagate in a straight line, which limits their use to line-of-sight applications.

24-3 ANALOG VERSUS DIGITAL MICROWAVE

A vast majority of the existing microwave radio systems are frequency modulation, which of course is analog. Recently, however, systems have been developed that use either phase-shift keying or quadrature amplitude modulation, which are forms of digital modulation. This chapter deals primarily with conventional FDM/FM microwave radio systems. Although many of the system concepts are the same, the performance of digital signals are evaluated quite differently. Chapter 25 deals with satellite systems that utilize PCM/PSK. Satellite radio systems are similar to terrestrial microwave radio systems; in fact, the two systems share many of the same frequencies. The primary difference between satellite and terrestrial radio systems is that satellite systems propagate signals outside Earth's atmosphere and, thus, are capable of carrying signals much farther while utilizing fewer transmitters and receivers.

24-4 FREQUENCY VERSUS AMPLITUDE MODULATION

Frequency modulation (FM) is used in microwave radio systems rather than amplitude modulation (AM) because AM signals are more sensitive to amplitude nonlinearities inherent in *wideband microwave amplifiers*. FM signals are relatively insensitive to this type of nonlinear distortion and can be transmitted through amplifiers that have compression or amplitude nonlinearity with little penalty. In addition, FM signals are less sensitive to random noise and can be propagated with lower transmit powers.

Intermodulation noise is a major factor when designing FM radio systems. In AM systems, intermodulation noise is caused by repeater amplitude nonlinearity. In FM systems, intermodulation noise is caused primarily by transmission gain and delay distortion. Consequently, in AM systems, intermodulation noise is a function of signal amplitude, but in FM systems, it is a function of signal amplitude and the magnitude of the frequency deviation. Thus, the characteristics of FM signals are more suitable than AM signals for microwave transmission.

24-5 FREQUENCY-MODULATED MICROWAVE RADIO SYSTEM

Microwave radio systems using FM are widely recognized as providing flexible, reliable, and economical point-to-point communications using Earth's atmosphere for the transmission medium. FM microwave systems used with the appropriate multiplexing equipment are capable of simultaneously carrying from a few narrowband voice circuits up to thousands of voice and data circuits. Microwave radios can also be configured to carry high-speed data, facsimile, broadcast-quality audio, and commercial television signals. Comparative cost studies have proven that FM microwave radio is very often the most economical means for providing communications circuits where there are no existing metallic cables or optical fibers or where severe terrain or weather conditions exist. FM microwave systems are also easily expandable.

A simplified block diagram of an FM microwave radio is shown in Figure 24-2. The *baseband* is the composite signal that modulates the FM carrier and may comprise one or more of the following:

1. Frequency-division-multiplexed voice-band channels
2. Time-division-multiplexed voice-band channels
3. Broadcast-quality composite video or picturephone
4. Wideband data

24-5-1 FM Microwave Radio Transmitter

In the FM *microwave transmitter* shown in Figure 24-2a, a *preemphasis* network precedes the FM deviator. The preemphasis network provides an artificial boost in amplitude to the higher baseband frequencies. This allows the lower baseband frequencies to frequency modulate the IF carrier and the higher baseband frequencies to phase modulate it. This scheme ensures a more uniform signal-to-noise ratio throughout the entire baseband spectrum. An FM deviator provides the modulation of the IF carrier that eventually becomes the main microwave carrier. Typically, IF carrier frequencies are between 60 MHz and 80 MHz, with 70 MHz the most common. *Low-index* frequency modulation is used in the FM deviator. Typically, modulation indices are kept between 0.5 and 1. This produces a *narrowband* FM signal at the output of the deviator. Consequently, the IF bandwidth resembles conventional AM and is approximately equal to twice the highest baseband frequency.

The IF and its associated sidebands are up-converted to the microwave region by the mixer, microwave oscillator, and bandpass filter. Mixing, rather than multiplying, is used to translate the IF frequencies to RF frequencies because the modulation index is unchanged

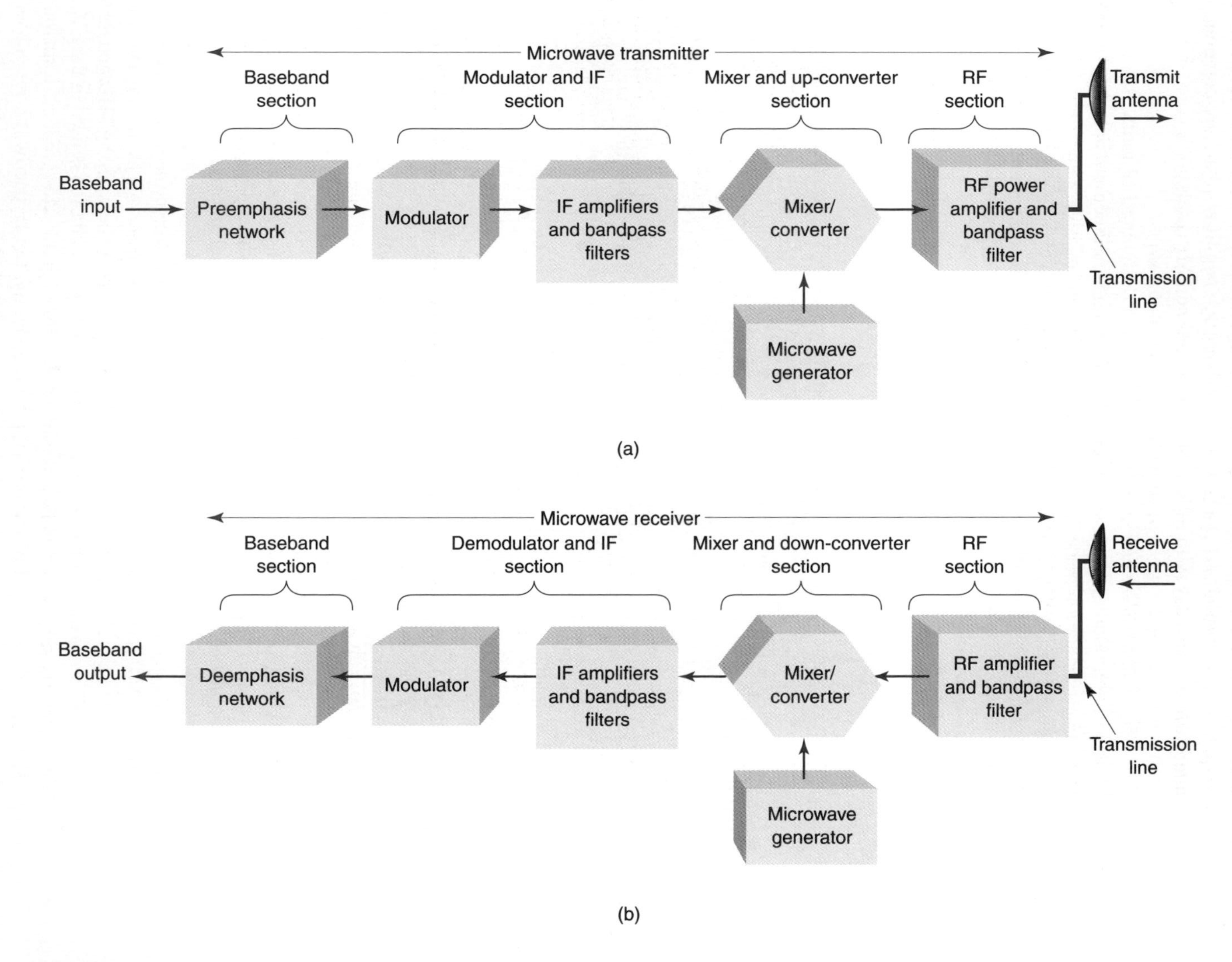

FIGURE 24-2 Simplified block diagram of a microwave radio: (a) transmitter; (b) receiver

by the heterodyning process. Multiplying the IF carrier would also multiply the frequency deviation and the modulation index, thus increasing the bandwidth.

Microwave generators consist of a crystal oscillator followed by a series of frequency multipliers. For example, a 125-MHz crystal oscillator followed by a series of multipliers with a combined multiplication factor of 48 could be used to a 6-GHz microwave carrier frequency. The channel-combining network provides a means of connecting more than one microwave transmitter to a single transmission line feeding the antenna.

24-5-2 FM Microwave Radio Receiver

In the FM microwave receiver shown in Figure 24-2b, the channel separation network provides the isolation and filtering necessary to separate individual microwave channels and direct them to their respective receivers. The bandpass filter, AM mixer, and microwave oscillator down-convert the RF microwave frequencies to IF frequencies and pass them on to the FM demodulator. The FM demodulator is a conventional, *noncoherent* FM detector (i.e., a discriminator or a PLL demodulator). At the output of the FM detector, a deemphasis network restores the baseband signal to its original amplitude-versus-frequency characteristics.

24-6 FM MICROWAVE RADIO REPEATERS

The permissible distance between an FM microwave transmitter and its associated microwave receiver depends on several system variables, such as transmitter output power, receiver noise threshold, terrain, atmospheric conditions, system capacity, reliability objectives, and performance expectations. Typically, this distance is between 15 miles and 40 miles. Long-haul microwave systems span distances considerably longer than this. Consequently, a single-hop microwave system, such as the one shown in Figure 24-2, is inadequate for most practical system applications. With systems that are longer than 40 miles or when geographical obstructions, such as a mountain, block the transmission path, *repeaters* are needed. A microwave repeater is a receiver and a transmitter placed back to back or in tandem with the system. A simplified block diagram of a microwave repeater is shown in Figure 24-3. The repeater station receives a signal, amplifies and reshapes it, and then retransmits the signal to the next repeater or terminal station down line from it.

The location of intermediate repeater sites is greatly influenced by the nature of the terrain between and surrounding the sites. Preliminary route planning generally assumes relatively flat areas, and path (hop) lengths will average between 25 miles and 35 miles between stations. In relatively flat terrain, increasing path length will dictate increasing the antenna tower heights. Transmitter output power and antenna gain will similarly enter into the selection process. The exact distance is determined primarily by line of-site path clearance and received signal strength. For frequencies above 10 GHz, local rainfall patterns could also have a large bearing on path length. In all cases, however, paths should be as level as possible. In addition, the possibility of interference, either internal or external, must be considered.

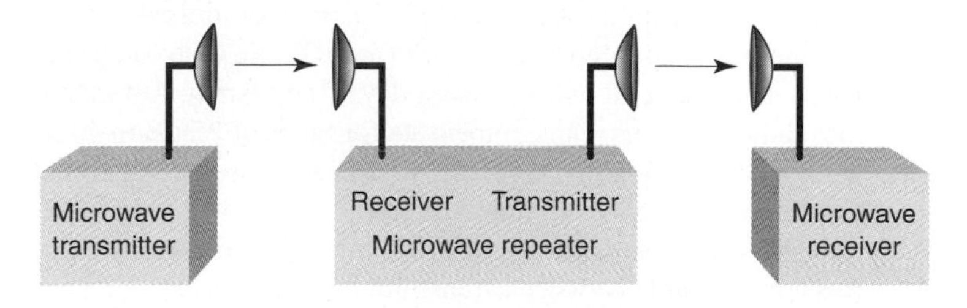

FIGURE 24-3 Microwave repeater

Basically, there are three types of microwave repeaters: IF, baseband, and RF (see Figure 24-4). IF repeaters are also called *heterodyne* repeaters. With an IF repeater (Figure 24-4a), the received RF carrier is down-converted to an IF frequency, amplified, reshaped, up-converted to an RF frequency, and then retransmitted. The signal is never demodulated below IF. Consequently, the baseband intelligence is unmodified by the repeater. With a baseband repeater (Figure 24-4b), the received RF carrier is down-converted to an IF frequency, amplified, filtered, and then further demodulated to baseband. The baseband signal, which is typically frequency-division-multiplexed voice-band channels, is further demodulated to a mastergroup, supergroup, group, or even channel level. This allows the baseband signal to be reconfigured to meet the routing needs of the overall communications network. Once the baseband signal has been reconfigured, it FM modulates an IF carrier, which is up-converted to an RF carrier and then retransmitted.

Figure 24-4c shows another baseband repeater configuration. The repeater demodulates the RF to baseband, amplifies and reshapes it, and then modulates the FM carrier. With this technique, the baseband is not reconfigured. Essentially, this configuration accomplishes the same thing that an IF repeater accomplishes. The difference is that in a baseband configuration, the amplifier and equalizer act on baseband frequencies rather than IF frequencies. The baseband frequencies are generally less than 9 MHz, whereas the IF frequencies are in the range 60 MHz to 80 MHz. Consequently, the filters and amplifiers necessary for baseband repeaters are simpler to design and less expensive than the ones required for IF repeaters. The disadvantage of a baseband configuration is the addition of the FM terminal equipment.

Figure 24-4d shows an RF-to-RF repeater. With RF-to-RF repeaters, the received microwave signal is not down-converted to IF or baseband; it is simply mixed (heterodyned) with a local oscillator frequency in a nonlinear mixer. The output of the mixer is tuned to either the sum or the difference between the incoming RF and the local oscillator frequency, depending on whether frequency up- or down-conversion is desired. The local oscillator is sometimes called a *shift oscillator* and is considerably lower in frequency than either the received or the transmitted radio frequencies. For example, an incoming RF of 6.2 GHz is mixed with a 0.2-GHz local oscillator frequency producing sum and difference frequencies of 6.4 GHz and 6.0 GHz. For frequency up-conversion, the output of the mixer would be tuned to 6.4 GHz, and for frequency down-conversion, the output of the mixer would be tuned to 6.0 GHz. With RF-to-RF repeaters, the radio signal is simply converted in frequency and then reamplified and transmitted to the next down-line repeater or terminal station. Reconfiguring and reshaping are not possible with RF-to-RF repeaters.

24-7 DIVERSITY

Microwave systems use *line-of-site* transmission; therefore a direct signal path must exist between the transmit and the receive antennas. Consequently, if that signal path undergoes a severe degradation, a service interruption will occur. Over time, radio path losses vary with atmospheric conditions that can vary significantly, causing a corresponding reduction in the received signal strength of 20, 30, or 40 or more dB. This reduction in signal strength is temporary and referred to as *radio fade*. Radio fade can last for a few milliseconds (short term) or for several hours or even days (long term). Automatic gain control circuits, built into radio receivers, can compensate for fades of 25 dB to 40 dB, depending on system design; however, fades in excess of 40 dB can cause a total loss of the received signal. When this happens, service continuity is lost.

Diversity suggests that there is more than one transmission path or method of transmission available between a transmitter and a receiver. In a microwave system, the purpose of using diversity is to increase the reliability of the system by increasing its availability.

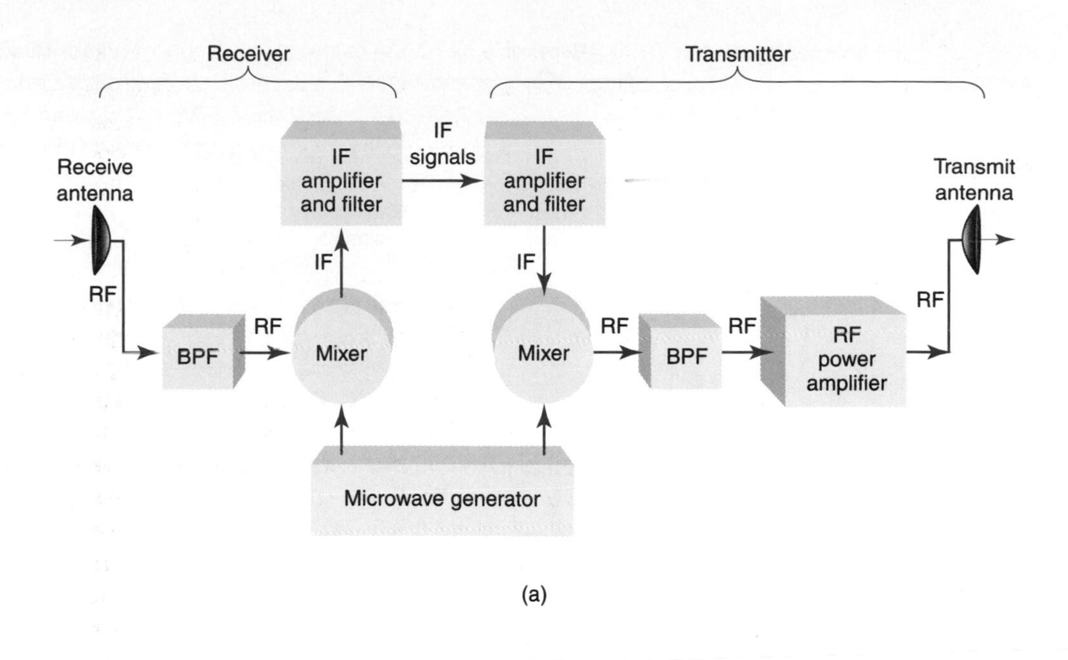

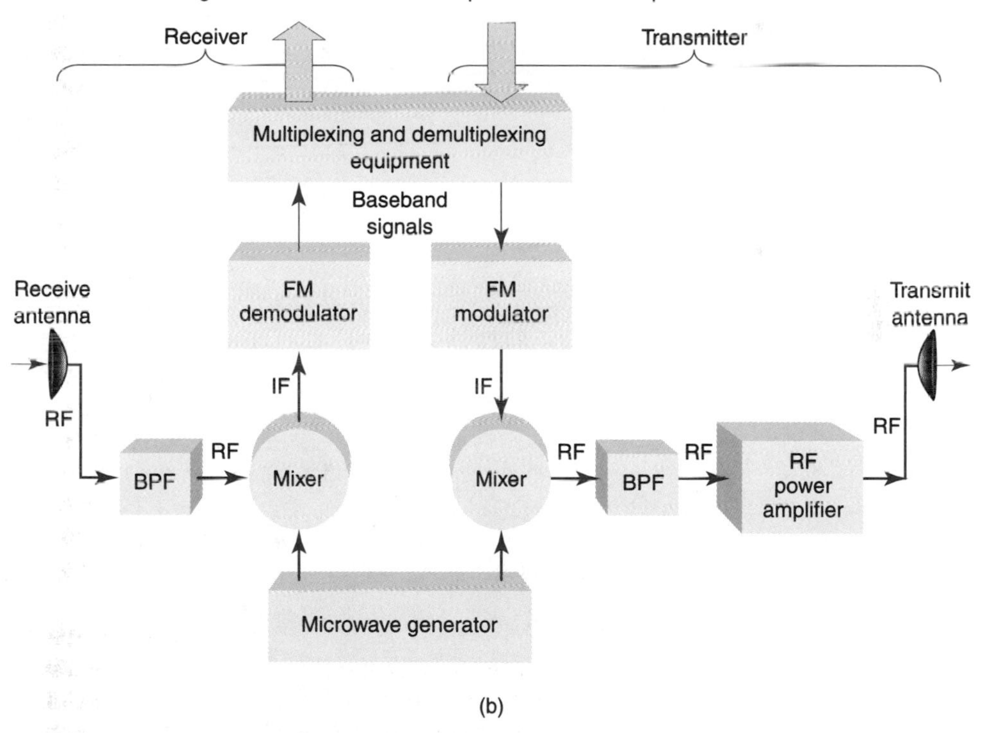

FIGURE 24-4 Microwave repeaters: (a) IF; (b) baseband; (*Continued*)

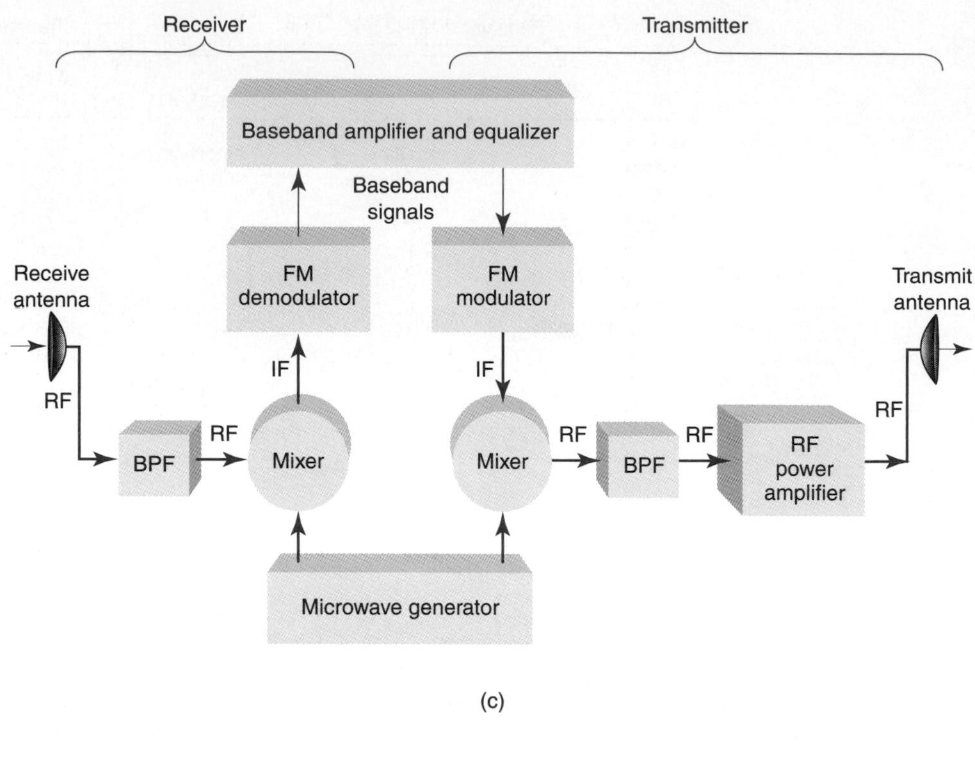

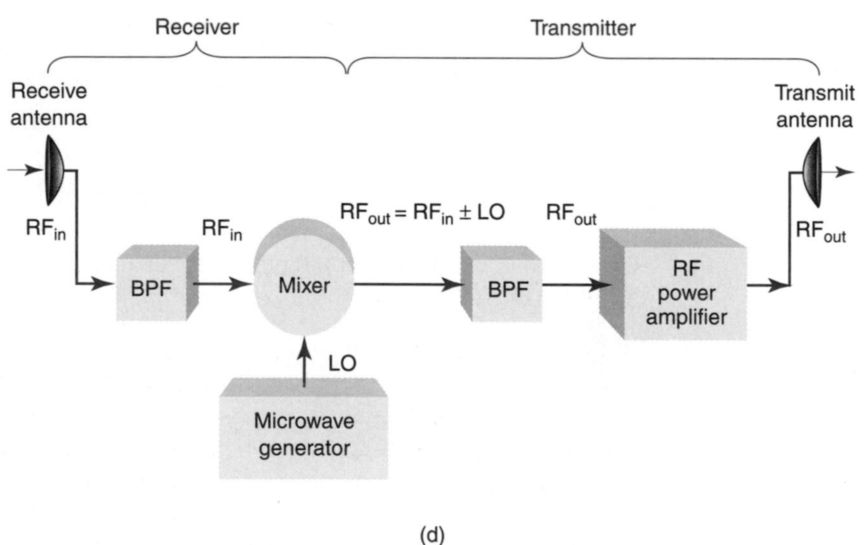

FIGURE 24-4 (*Continued*) Microwave repeaters: (c) baseband; (d) RF

Table 24-2 shows a relatively simple means of translating a given system reliability percentage into terms that are more easily related to experience. For example, a reliability percentage of 99.99% corresponds to about 53 minutes of outage time per year, while a reliability percentage of 99.9999% amounts to only about 32 seconds of outage time per year.

When there is more than one transmission path or method of transmission available, the system can select the path or method that produces the highest-quality received signal. Generally, the highest quality is determined by evaluating the carrier-to-noise (C/N) ratio at the receiver input or by simply measuring the received carrier power. Although there are

Table 24-2 Reliability and Outage Time

Reliability (%)	Outage Time (%)	Year (Hours)	Outage Time per Month (Hours)	Day (Hours)
0	100	8760	720	24
50	50	4380	360	12
80	20	1752	144	4.8
90	10	876	72	2.4
95	5	438	36	1.2
98	2	175	14	29 minutes
99	1	88	7	14.4 minutes
99.9	0.1	8.8	43 minutes	1.44 minutes
99.99	0.01	53 minutes	4.3 minutes	8.6 seconds
99.999	0.001	5.3 minutes	26 seconds	0.86 seconds
99.9999	0.0001	32 seconds	2.6 seconds	0.086 seconds

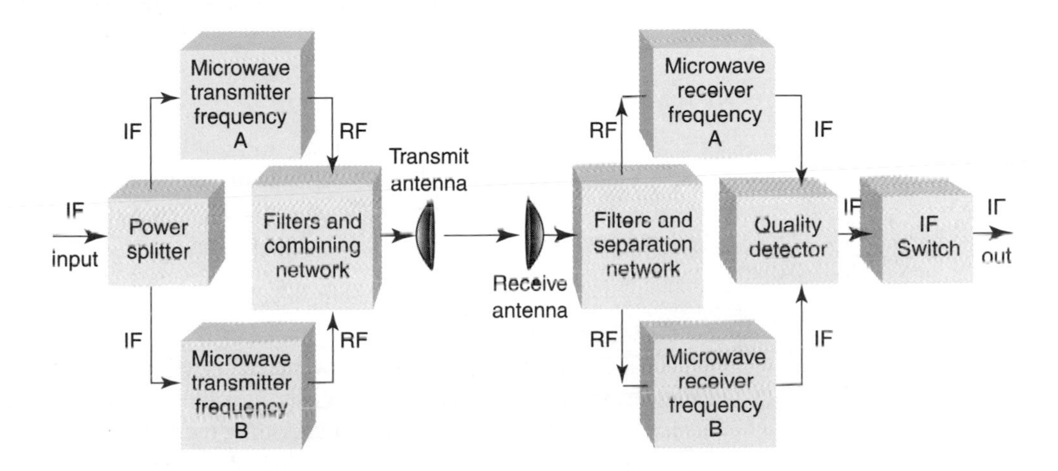

FIGURE 24-5 Frequency diversity microwave system

many ways of achieving diversity, the most common methods used are frequency, space, polarization, receiver, quad, or hybrid.

24-7-1 Frequency Diversity

Frequency diversity is simply modulating two different RF carrier frequencies with the same IF intelligence, then transmitting both RF signals to a given destination. At the destination, both carriers are demodulated, and the one that yields the better-quality IF signal is selected. Figure 24-5 shows a single-channel frequency-diversity microwave system.

In Figure 24-5a, the IF input signal is fed to a power splitter, which directs it to microwave transmitters A and B. The RF outputs from the two transmitters are combined in the channel-combining network and fed to the transmit antenna. At the receive end (Figure 24-5b), the channel separator directs the A and B RF carriers to their respective microwave receivers, where they are down-converted to IF. The quality detector circuit determines which channel, A or B, is the higher quality and directs that channel through the IF switch to be further demodulated to baseband. Many of the temporary, adverse atmospheric conditions that degrade an RF signal are frequency selective; they may degrade one frequency more than another. Therefore, over a given period of time, the IF switch may switch back and forth from receiver A to receiver B and vice versa many times.

Frequency-diversity arrangements provide complete and simple equipment redundance and have the additional advantage of providing two complete transmitter-to-receiver electrical paths. Its obvious disadvantage is that it doubles the amount of frequency spectrum and equipment necessary.

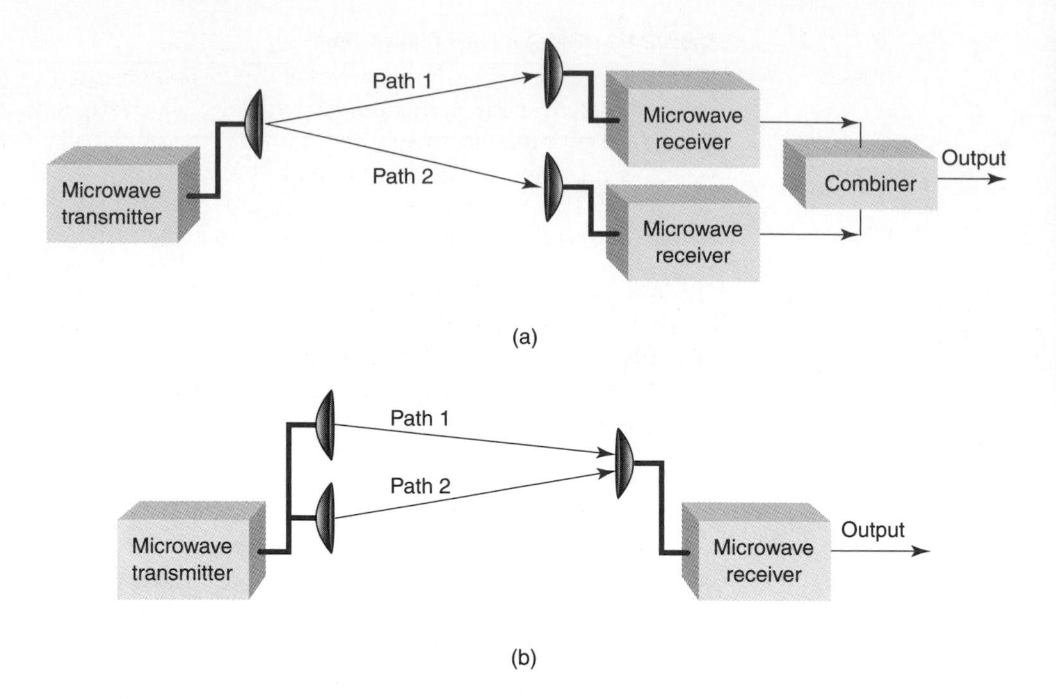

FIGURE 24-6 Space diversity: (a) two receive antennas; (b) two transmit antennas

24-7-2 Space Diversity

With space diversity, the output of a transmitter is fed to two or more antennas that are physically separated by an appreciable number of wavelengths. Similarly, at the receiving end, there may be more than one antenna providing the input signal to the receiver. If multiple receiving antennas are used, they must also be separated by an appreciable number of wavelengths. Figure 24-6 shows two ways to implement space diversity. Figure 24-6a shows a space diversity system using two transmit antennas, whereas Figure 24-6b shows a space diversity system using two receive antennas. The rule is to use two transmit antennas or two receive antennas but never two of each.

When space diversity is used, it is important that the electrical distance from a transmitter to each of its antennas and to a receiver from each of its antennas is an equal multiple of wavelengths long. This is to ensure that when two or more signals of the same frequency arrive at the input to a receiver, they are in phase and additive. If received out of phase, they will cancel and, consequently, result in less received signal power than if simply one antenna system were used. Adverse atmospheric conditions are often isolated to a very small geographical area. With space diversity, there is more than one transmission path between a transmitter and a receiver. When adverse atmospheric conditions exist in one of the paths, it is unlikely that the alternate path is experiencing the same degradation. Consequently, the probability of receiving an acceptable signal is higher when space diversity is used than when no diversity is used. An alternate method of space diversity uses a single transmitting antenna and two receiving antennas separated vertically. Depending on the atmospheric conditions at a particular time, one of the receiving antennas should be receiving an adequate signal. Again, there are two transmission paths that are unlikely to be affected simultaneously by fading.

Space-diversity arrangements provide for path redundancy but not equipment redundancy. Space diversity is more expensive than frequency diversity because of the additional antennas and waveguide. Space diversity, however, provides efficient frequency spectrum usage and a substantially greater protection than frequency diversity.

24-7-3 Polarization Diversity

With *polarization diversity,* a single RF carrier is propagated with two different electromagnetic polarizations (vertical and horizontal). Electromagnetic waves of different polarizations do not necessarily experience the same transmission impairments. Polarization diversity is generally used in conjunction with space diversity. One transmit/receive antenna pair is vertically polarized, and the other is horizontally polarized. It is also possible to use frequency, space, and polarization diversity simultaneously.

24-7-4 Receiver Diversity

Receiver diversity is using more than one receiver for a single radio-frequency channel. With frequency diversity, it is necessary to also use receiver diversity because each transmitted frequency requires its own receiver. However, sometimes two receivers are used for a single transmitted frequency.

24-7-5 Quad Diversity

Quad diversity is another form of hybrid diversity and undoubtedly provides the most reliable transmission; however, it is also the most expensive. The basic concept of quad diversity is quite simple: It combines frequency, space, polarization, and receiver diversity into one system. Its obvious disadvantage is providing redundant electronic equipment, frequencies, antennas, and waveguide, which are economical burdens.

24-7-6 Hybrid Diversity

Hybrid diversity is a somewhat specialized form of diversity that consists of a standard frequency-diversity path where the two transmitter/receiver pairs at one end of the path are separated from each other and connected to different antennas that are vertically separated as in space diversity. This arrangement provides a space-diversity effect in both directions—in one direction because the receivers are vertically spaced and in the other direction because the transmitters are vertically spaced. This arrangement combines the operational advantages of frequency diversity with the improved diversity protection of space diversity. Hybrid diversity has the disadvantage, however, of requiring two radio frequencies to obtain one working channel.

24-8 PROTECTION SWITCHING ARRANGEMENTS

To avoid a service interruption during periods of deep fades or equipment failures, alternate facilities are temporarily made available in a *protection switching* arrangement. The general concepts of protection switching and diversity are quite similar: Both provide protection against equipment failures and atmospheric fades. The primary difference between them is, simply, that diversity systems provide an alternate transmission path for only a single microwave link (i.e., between one transmitter and one receiver) within the overall communications system. Protection switching arrangements, on the other hand, provide protection for a much larger section of the communications system that generally includes several repeaters spanning a distance of 100 miles or more. Diversity systems also generally provide 100% protection to a single radio channel, whereas protection switching arrangements are usually shared between several radio channels.

Essentially, there are two types of protection switching arrangements: *hot standby* and *diversity.* With hot standby protection, each working radio channel has a dedicated backup or spare channel. With diversity protection, a single backup channel is made available to as many as 11 working channels. Hot standby systems offer 100% protection for each working radio channel. A diversity system offers 100% protection only to the first working channel to fail. If two radio channels fail at the same time, a service interruption will occur.

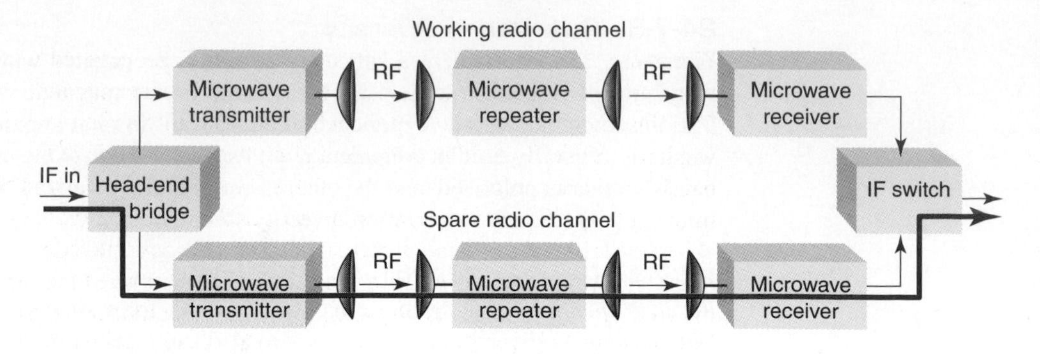

(a)

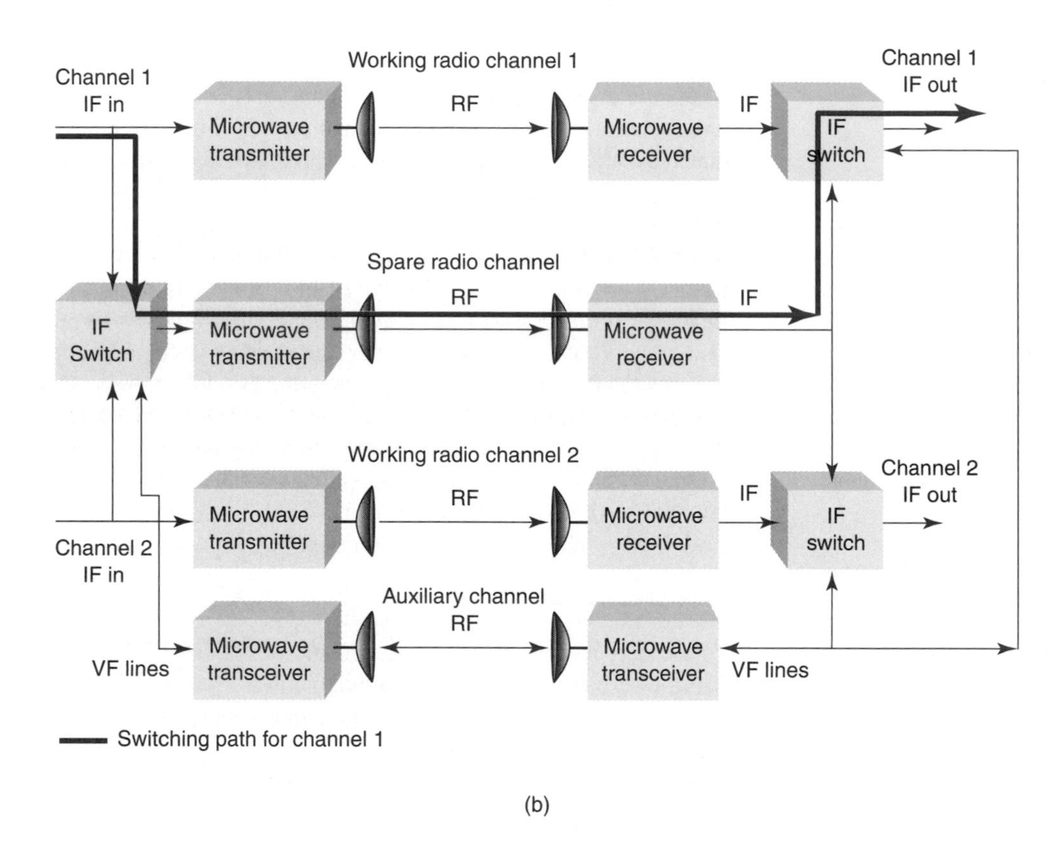

(b)

FIGURE 24-7 Microwave protection switching arrangements: (a) hot standby; (b) diversity

24-8-1 Hot Standby

Figure 24-7a shows a single-channel hot standby protection switching arrangement. At the transmitting end, the IF goes into a *head-end bridge,* which splits the signal power and directs it to the working and the spare (standby) microwave channels simultaneously. Consequently, both the working and the standby channels are carrying the same baseband information. At the receiving end, the IF switch passes the IF signal from the working channel to the FM terminal equipment. The IF switch continuously monitors the received signal power on the working channel and, if it fails, switches to the standby channel. When the IF signal on the working channel is restored, the IF switch resumes its normal position.

24-8-2 Diversity

Figure 24-7b shows a diversity protection switching arrangement. This system has two working channels (channel 1 and channel 2), one spare channel, and an *auxiliary* channel. The IF switch at the receive end continuously monitors the receive signal strength of both working channels. If either one should fail, the IF switch detects a loss of carrier and sends back to the transmitting station IF switch a VF (*voice frequency*) tone-encoded signal that directs it to switch the IF signal from the failed channel onto the spare microwave channel. When the failed channel is restored, the IF switches resume their normal positions. The auxiliary channel simply provides a transmission path between the two IF switches. Typically, the auxiliary channel is a low-capacity low-power microwave radio that is designed to be used for a maintenance channel only.

24-8-3 Reliability

The number of repeater stations between protection switches depends on the *reliability objectives* of the system. Typically, there are between two and six repeaters between switching stations.

As you can see, diversity systems and protection switching arrangements are quite similar. The primary difference between the two is that diversity systems are permanent arrangements and are intended only to compensate for temporary, abnormal atmospheric conditions between only two selected stations in a system. Protection switching arrangements, on the other hand, compensate for both radio fades and equipment failures and may include from six to eight repeater stations between switches. Protection channels also may be used as temporary communication facilities while routine maintenance is performed on a regular working channel. With a protection switching arrangement, all signal paths and radio equipment are protected. Diversity is used selectively—that is, only between stations that historically experience severe fading a high percentage of the time.

A statistical study of outage time (i.e., service interruptions) caused by radio fades, equipment failures, and maintenance is important in the design of a microwave radio system. From such a study, engineering decisions can be made on which type of diversity system and protection switching arrangement is best suited for a particular application.

Figure 24-8 shows a comparison between diversity and protection switching. As shown in the figure, protection switching arrangements protect against equipment failures

Protection Switching protects against adverse atmospheric conditions and equipment failures between switching stations.

Switching station — Tx Rx — Repeater station — Tx Rx — Repeater station — Tx Rx — Repeater station — Tx Rx — Switching station

Diversity Diversity Diversity Diversity

Protection Switching

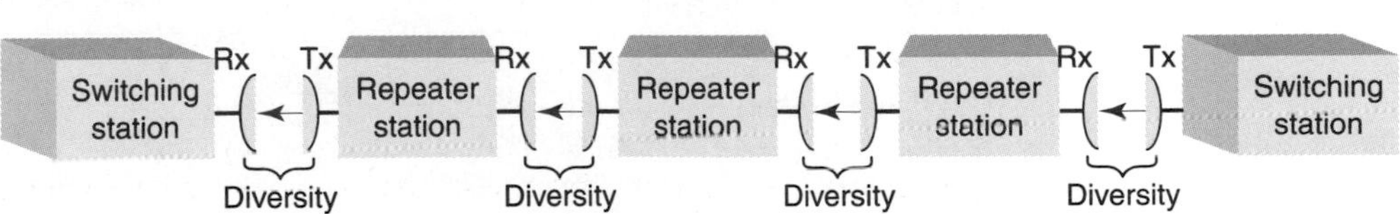

Diversity protects against adverse atmospheric conditions between transmit and receive antennas.

FIGURE 24-8 Comparison between diversity and protection switching

in any of the electronic equipment (transmitters, receivers, and so on) in any of the microwave stations between the two switching stations. Diversity, however, protects only against adverse atmospheric conditions between a transmit antenna and a receive antenna.

24-9 FM MICROWAVE RADIO STATIONS

Basically, there are two types of FM microwave stations: terminals and repeaters. *Terminal stations* are points in the system where baseband signals either originate or terminate. *Repeater stations* are points in a system where baseband signals may be reconfigured or where RF carriers are simply "repeated" or amplified.

24-9-1 Terminal Station

Essentially, a terminal station consists of four major sections: the baseband, wireline entrance link (WLEL), FM-IF, and RF sections. Figure 24-9 shows a block diagram of the baseband, WLEL, and FM-IF sections. As mentioned, the baseband may be one of several different types of signals. For our example, frequency-division-multiplexed voice-band channels are used.

24-9-1-1 Wireline entrance link. Often in large communications networks, such as the American Telephone and Telegraph Company (AT&T), the building that houses the radio station is quite large. Consequently, it is desirable that similar equipment be physically placed at a common location (i.e., all frequency-division-multiplexed [FDM] equipment in the same room). This simplifies alarm systems, providing dc power to the equipment, maintenance, and other general cabling requirements. Dissimilar equipment may be separated by a considerable distance. For example, the distance between the FDM equipment and the FM-IF section is typically several hundred feet and in some cases several miles. For this reason, a wireline entrance link (WLEL) is required. A WLEL serves as the interface between the multiplex terminal equipment and the FM-IF equipment. A WLEL generally consists of an amplifier and an equalizer (which together compensate for cable transmission losses) and level-shaping devices commonly called pre- and deemphasis networks.

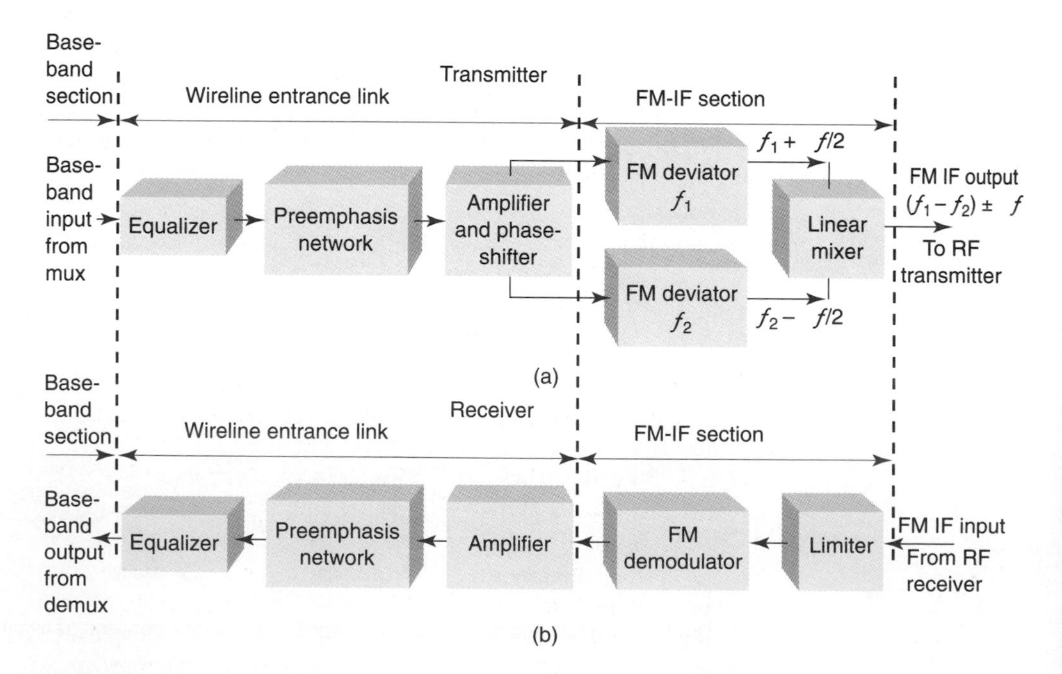

FIGURE 24-9 Microwave terminal station: (a) transmitter; (b) receiver

24-9-1-2 IF section. The FM terminal equipment shown in Figure 24-9 generates a frequency-modulated IF carrier. This is accomplished by mixing the outputs of two deviated oscillators that differ in frequency by the desired IF carrier. The oscillators are deviated in phase opposition, which reduces the magnitude of phase deviation required of a single deviator by a factor of 2. This technique also reduces the deviation linearity requirements for the oscillators and provides for the partial cancellation of unwanted modulation products. Again, the receiver is a conventional noncoherent FM detector.

24-9-1-3 RF section. A block diagram of the RF section of a microwave terminal station is shown in Figure 24-10. The IF signal enters the transmitter (Figure 24-10a) through a protection switch. The IF and compression amplifiers help keep the IF signal power constant and at approximately the required input level to the transmit modulator (*transmod*). A transmod is a balanced modulator that, when used in conjunction with a microwave generator, power amplifier, and bandpass filter, up-converts the IF carrier to an RF carrier and amplifies the RF to the desired output power. Power amplifiers for microwave radios must be capable of amplifying very high frequencies and passing very wide bandwidth signals. *Klystron tubes, traveling-wave tubes* (TWTs), and *IMPATT* (impact/avalanche and transit time) diodes are several of the devices currently being used in microwave power amplifiers. Because high-gain antennas are used and the distance between microwave stations is relatively short, it is not necessary to develop a high output power from the transmitter output amplifiers. Typical gains for microwave antennas range from 10 dB to 40 dB, and typical transmitter output powers are between 0.5 W and 10 W.

A *microwave generator* provides the RF carrier input to the up-converter. It is called a microwave generator rather than an oscillator because it is difficult to construct a stable circuit that will oscillate in the gigahertz range. Instead, a crystal-controlled oscillator operating in the range 5 MHz to 25 MHz is used to provide a base frequency that is multiplied up to the desired RF carrier frequency.

An *isolator* is a unidirectional device often made from a ferrite material. The isolator is used in conjunction with a channel-combining network to prevent the output of one transmitter from interfering with the output of another transmitter.

The RF receiver (Figure 24-10b) is essentially the same as the transmitter except that it works in the opposite direction. However, one difference is the presence of an IF amplifier in the receiver. This IF amplifier has an *automatic gain control* (AGC) circuit. Also, very often, there are no RF amplifiers in the receiver. Typically, a highly sensitive, low-noise balanced demodulator is used for the receive demodulator (receive mod). This eliminates the need for an RF amplifier and improves the overall signal-to-noise ratio. When RF amplifiers are required, high-quality, *low-noise amplifiers* (LNAs) are used. Examples of commonly used LNAs are tunnel diodes and parametric amplifiers.

24-10 MICROWAVE REPEATER STATION

Figure 24-11 shows the block diagram of a microwave IF repeater station. The received RF signal enters the receiver through the channel separation network and bandpass filter. The receive mod down-converts the RF carrier to IF. The IF AMP/AGC and equalizer circuits amplify and reshape the IF. The equalizer compensates for *gain-versus-frequency nonlinearities* and *envelope delay distortion* introduced in the system. Again, the transmod up-converts the IF to RF for retransmission. However, in a repeater station, the method used to generate the RF microwave carrier frequencies is slightly different from the method used in a terminal station. In the IF repeater, only one microwave generator is required to supply both the transmod and the receive mod with an RF carrier signal. The microwave generator, shift oscillator, and shift modulator allow the repeater to receive one RF carrier frequency, down-convert it to IF, and then up-convert the IF to a different RF carrier

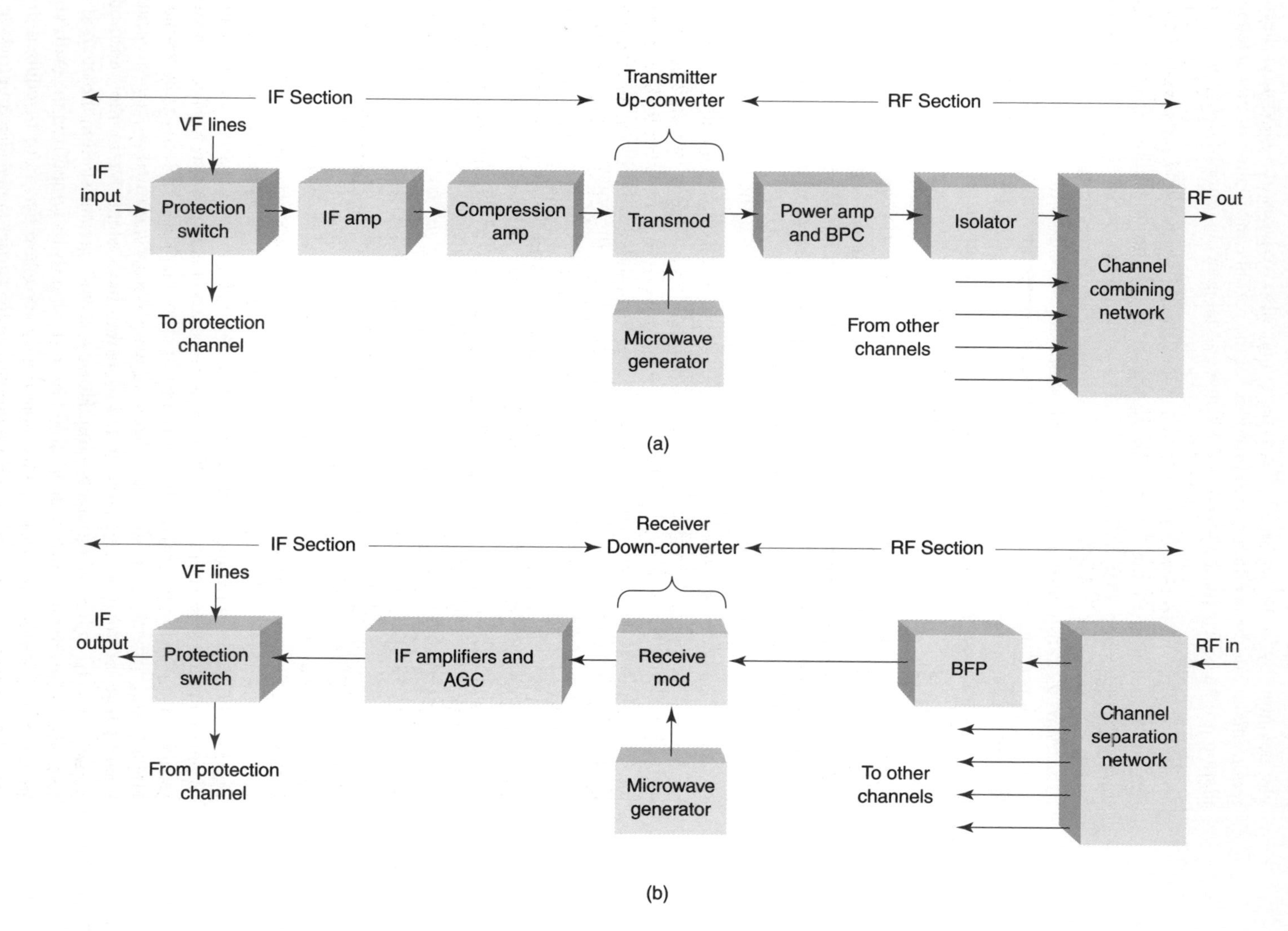

FIGURE 24-10 Microwave terminal station: (a) transmitter; (b) receiver

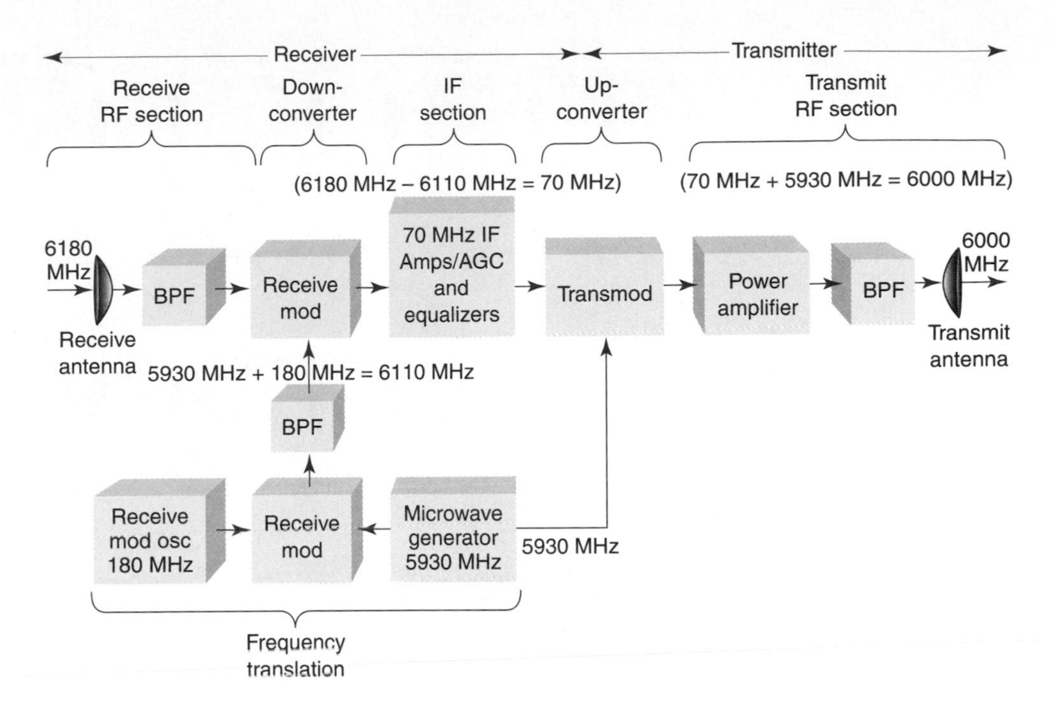

FIGURE 24-11 Microwave radio IF repeater block diagram

frequency. It is possible for station D to receive the transmissions from both station B and station C simultaneously (this is called *multihop interference* and is shown in Figure 24-12a). This can occur only when three stations are placed in a geographical straight line in the system. To prevent this from occurring, the allocated bandwidth for the system is divided in half, creating a low-frequency and a high-frequency band. Each station, in turn, alternates from a low-band to a high-band transmit carrier frequency (Figure 24-12b). If a transmission from station B is received by station D, it will be rejected in the channel separation network and cause no interference. This arrangement is called a high/low microwave repeater system. The rules are simple: If a repeater station receives a low-band RF carrier, then it retransmits a high-band RF carrier and vice versa. The only time that multiple carriers of the same frequency can be received is when a transmission from one station is received from another station that is three hops away. This is unlikely to happen.

Another reason for using a high/low-frequency scheme is to prevent the power that "leaks" out the back and sides of a transmit antenna from interfering with the signal entering the input of a nearby receive antenna. This is called *ringaround.* All antennas, no matter how high their gain or how directive their radiation pattern, radiate a small percentage of their power out the back and sides, giving a finite *front-to-back* ratio for the antenna. Although the front-to-back ratio of a typical microwave antenna is quite high, the relatively small amount of power that is radiated out the back of the antenna may be quite substantial compared with the normal received carrier power in the system. If the transmit and receive carrier frequencies are different, filters in the receiver separation network will prevent ringaround from occurring.

A high/low microwave repeater station (Figure 24-12b) needs two microwave carrier supplies for the down- and up-converting process. Rather than use two microwave generators, a single generator with a shift oscillator, a shift modulator, and a bandpass filter can generate the two required signals. One output from the microwave generator is fed directly into the transmod, and another output (from the same microwave generator) is mixed with the shift oscillator signal in the shift modulator to produce a second microwave carrier

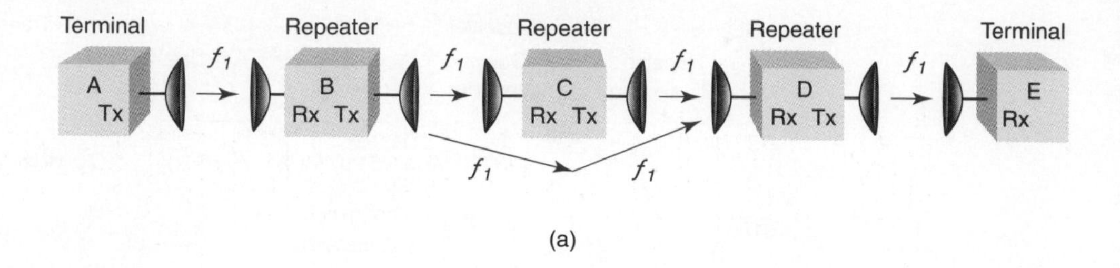

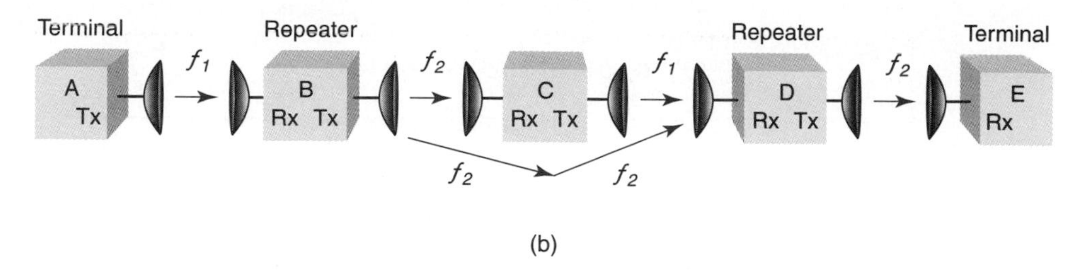

FIGURE 24-12 (a) Multihop interference; (b) high/low microwave system

frequency. The second microwave carrier frequency is offset from the first by the shift oscillator frequency. The second microwave carrier frequency is fed into the receive modulator.

Example 24-1

In Figure 24-11, the received RF carrier frequency is 6180 MHz, and the transmitted RF carrier frequency is 6000 MHz. With a 70-MHz IF frequency, a 5930-MHz microwave generator frequency, and a 180-MHz shift oscillator frequency, the output filter of the shift mod must be tuned to 6110 MHz. This is the sum of the microwave generator and the shift oscillator frequencies (5930 MHz + 180 MHz = 6110 MHz).

This process does not reduce the number of oscillators required, but it is simpler and cheaper to build one microwave generator and one relatively low-frequency shift oscillator than to build two microwave generators. This arrangement also provides a certain degree of synchronization between repeaters. The obvious disadvantage of the high/low scheme is that the number of channels available in a given bandwidth is cut in half.

Figure 24-13 shows a high/low-frequency plan with eight channels (four high band and four low band). Each channel occupies a 29.7-MHz bandwidth. The west terminal transmits the low-band frequencies and receives the high-band frequencies. Channels 1 and 3 (Figure 24-13a) are designated as *V channels.* This means that they are propagated with vertical polarization. Channels 2 and 4 are designated as H, or horizontally polarized, channels. This is not a polarization diversity system. Channels 1 through 4 are totally independent of each other; they carry different baseband information. The transmission of *orthogonally* polarized carriers (90° out of phase) further enhances the isolation between the transmit and receive signals. In the west-to-east direction, the repeater receives the low-band frequencies and transmits the high-band frequencies. After channel 1 is received and down-converted to IF, it is up-converted to a different RF frequency and a different polarization for retransmission. The low-band channel 1 corresponds to the high-band channel 11, channel 2 to channel 12, and so on. The east-to-west direction (Figure 24-13b) propagates the high- and low-band carriers in the sequence opposite to the west-to-east system. The polarizations are also reversed. If some of the power from channel 1 of the west terminal were to propagate directly to the east terminal receiver, it would have a different frequency and polarization than channel 11's transmissions. Consequently, it would not interfere with

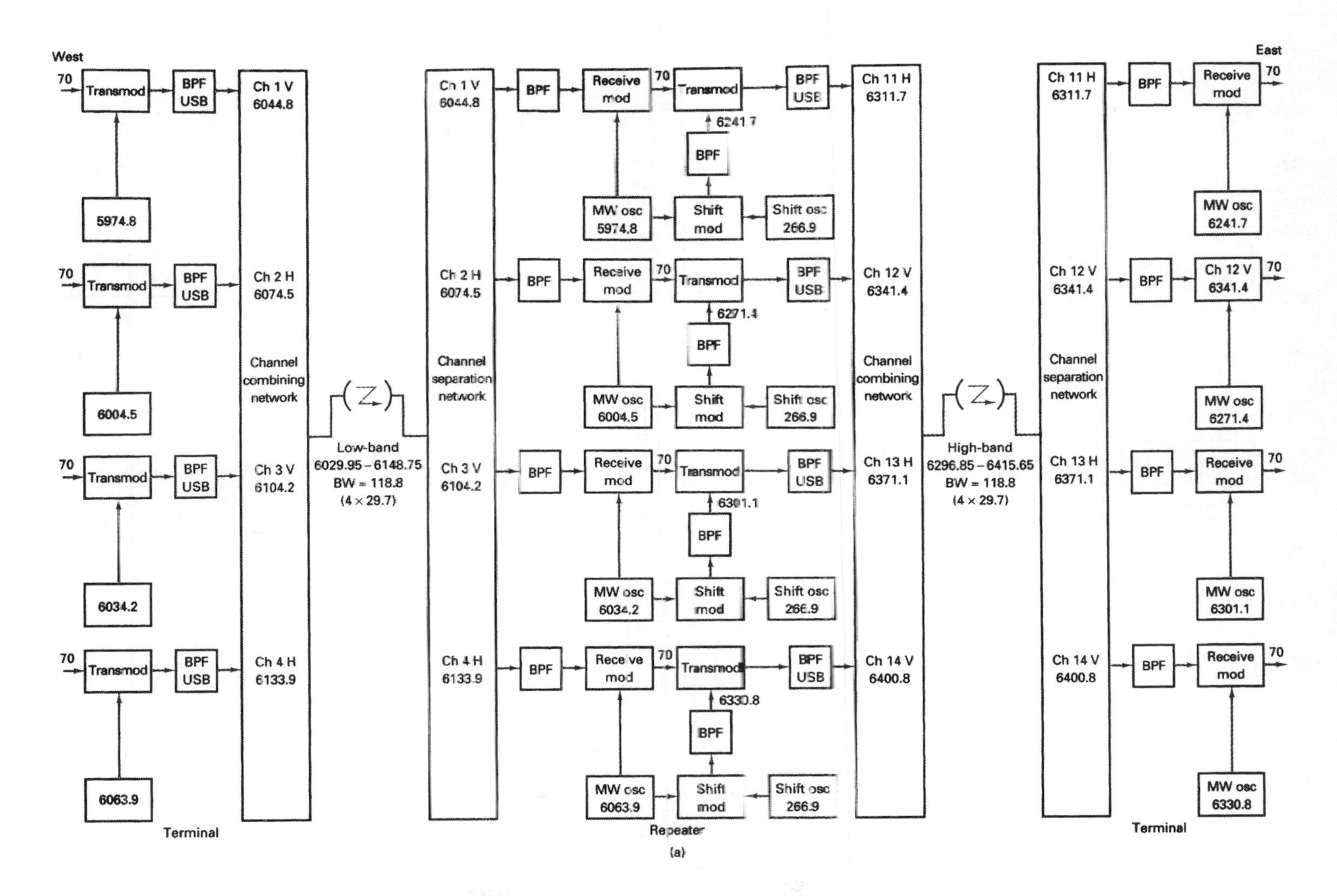

FIGURE 24-13 Eight-channel high/low frequency plan: (a) west to east; (*Continued*)

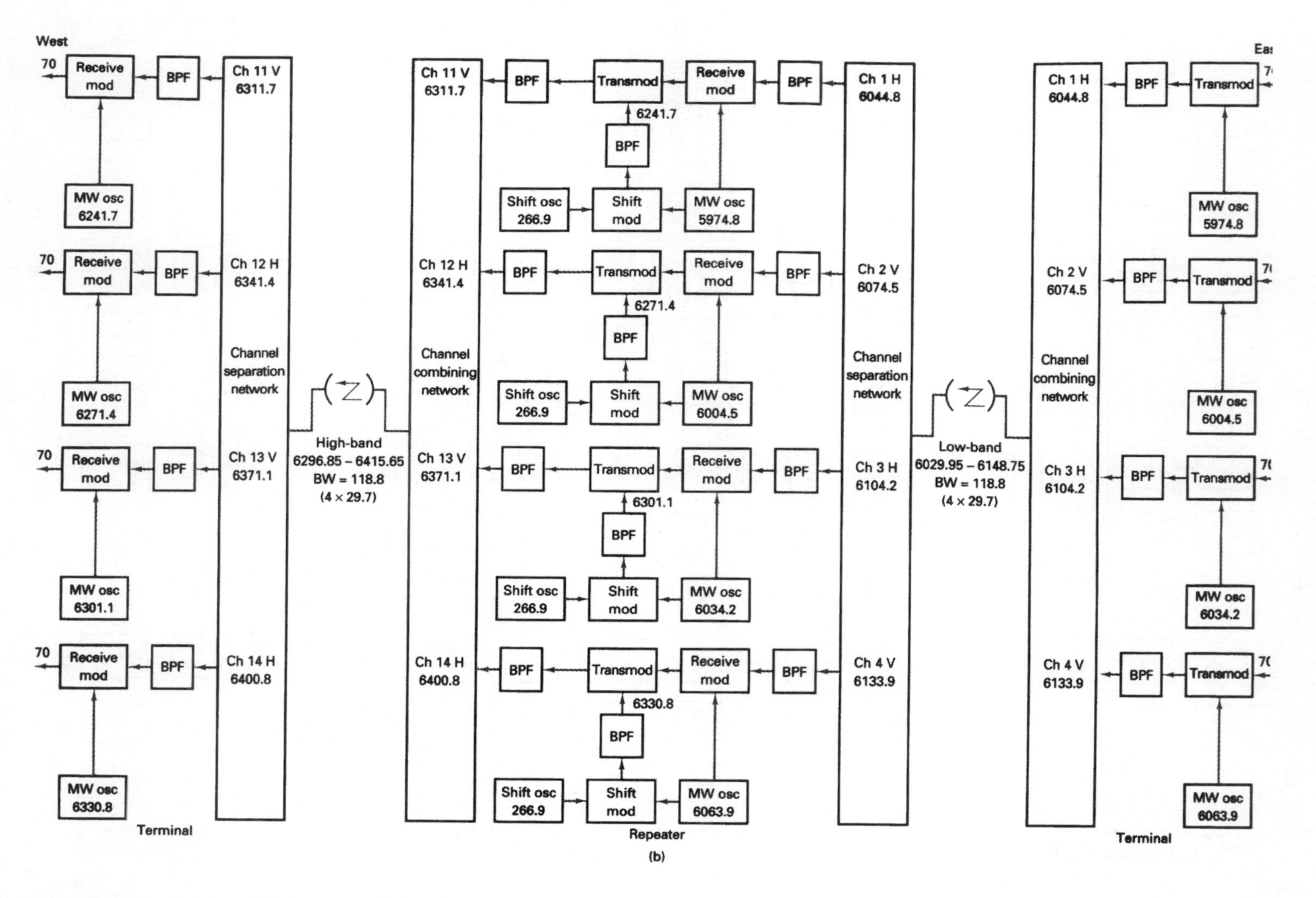

FIGURE 24-13 (*Continued*) Eight-channel high/low frequency plan: (b) east to west

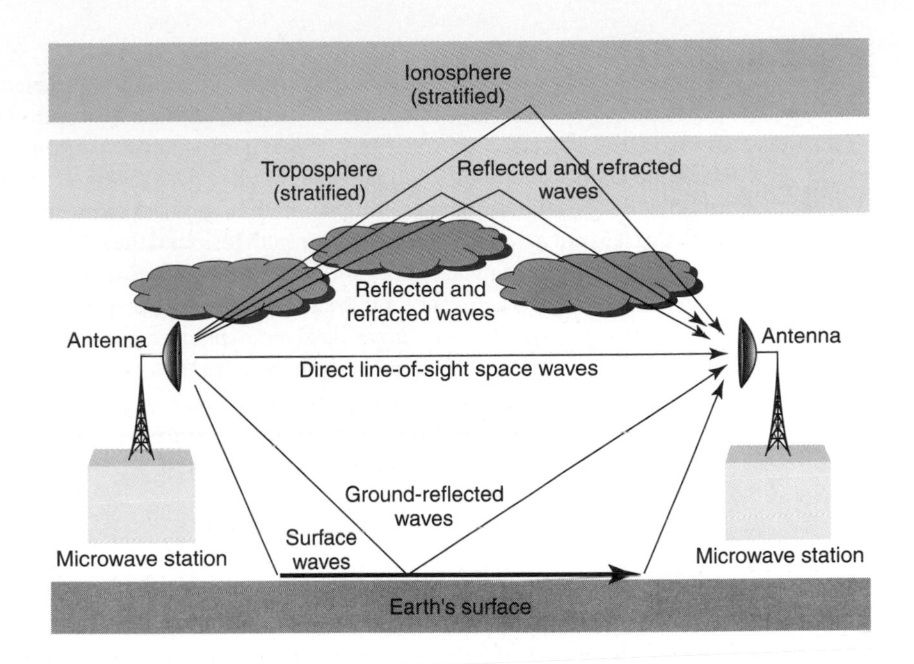

FIGURE 24-14 Microwave propagation paths

the reception of channel 11 (no multihop interference). Also, note that none of the transmit or receive channels at the repeater station has both the same frequency and polarization. Consequently, the interference from the transmitters to the receivers due to ringaround is insignificant.

24-11 LINE-OF-SIGHT PATH CHARACTERISTICS

The normal *propagation paths* between two radio antennas in a microwave radio system are shown in Figure 24-14. The *free-space path* is the *line-of-sight path* directly between the transmit and receive antennas (this is also called the *direct wave*). The *ground-reflected wave* is the portion of the transmit signal that is reflected off Earth's surface and captured by the receive antenna. The *surface wave* consists of the electric and magnetic fields associated with the currents induced in Earth's surface. The magnitude of the surface wave depends on the characteristics of Earth's surface and the electromagnetic polarization of the wave. The sum of these three paths (taking into account their amplitude and phase) is called the *ground wave*. The *sky wave* is the portion of the transmit signal that is returned (reflected) back to Earth's surface by the ionized layers of Earth's atmosphere.

All paths shown in Figure 24-14 exist in any microwave radio system, but some are negligible in certain frequency ranges. At frequencies below 1.5 MHz, the surface wave provides the primary coverage, and the sky wave helps extend this coverage at night when the absorption of the ionosphere is at a minimum. For frequencies above about 30 MHz to 50 MHz, the free-space and ground-reflected paths are generally the only paths of importance. The surface wave can also be neglected at these frequencies, provided that the antenna heights are not too low. The sky wave is only a source of occasional long-distance interference and not a reliable signal for microwave communications purposes. In this chapter, the surface and sky wave propagations are neglected, and attention is focused on those phenomena that affect the direct and reflected waves.

24-11-1 Free-Space Path Loss

Free-space path loss is often defined as the loss incurred by an electromagnetic wave as it propagates in a straight line through a vacuum with no absorption or reflection of energy from nearby objects. Free-space path loss is a misstated and often misleading definition because no energy is actually dissipated. Free-space path loss is a fabricated engineering quantity that evolved from manipulating communications system link budget equations, which include transmit antenna gain, free-space path loss, and the effective area of the receiving antenna (i.e., the receiving antenna gain) into a particular format. The manipulation of antenna gain terms results is a distance and frequency-dependent term called *free-space path loss.*

Free-space path loss assumes ideal atmospheric conditions, so no electromagnetic energy is actually lost or dissipated—it merely spreads out as it propagates away from the source, resulting in lower relative power densities. A more appropriate term for the phenomena is *spreading loss.* Spreading loss occurs simply because of the inverse square law. The mathematical expression for free-space path loss is

$$L_p = \left(\frac{4\pi D}{\lambda}\right)^2 \tag{24-1}$$

and because $\lambda = \frac{c}{f}$, Equation 14-26 can be written as

$$L_p = \left(\frac{4\pi f D}{c}\right)^2 \tag{24-2}$$

where L_p = free-space path loss (unitless)
D = distance (kilometers)
f = frequency (hertz)
λ = wavelength (meters)
c = velocity of light in free space (3 × 108 meters per second)

Converting to dB yields

$$L_p(\text{dB}) = 10 \log\left(\frac{4\pi f D}{c}\right)^2 \tag{24-3}$$

or

$$L_p(\text{dB}) = 20 \log\left(\frac{4\pi f D}{c}\right) \tag{24-4}$$

Separating the constants from the variables gives

$$L_p = 20 \log\left(\frac{4\pi}{c}\right) + 20 \log f + 20 \log D \tag{24-5}$$

For frequencies in MHz and distances in kilometers,

$$L_p = \left[\frac{4\pi(10^6)(10^3)}{3 \times 10^8}\right] + 20 \log f_{(\text{MHz})} + 20 \log D_{(\text{km})} \tag{24-6}$$

or

$$L_p = 32.4 + 20 \log f_{(\text{MHz})} + 20 \log D_{(\text{km})} \tag{24-7}$$

When the frequency is given in GHz and the distance in km,

$$L_p = 92.4 + 20 \log f_{(\text{GHz})} + 20 \log D_{(\text{km})} \tag{24-8}$$

When the frequency is given in GHz and the distance in miles,

$$L_p = 96.6 + 20 \log f_{(\text{GHz})} + 20 \log D_{(\text{miles})} \tag{24-9}$$

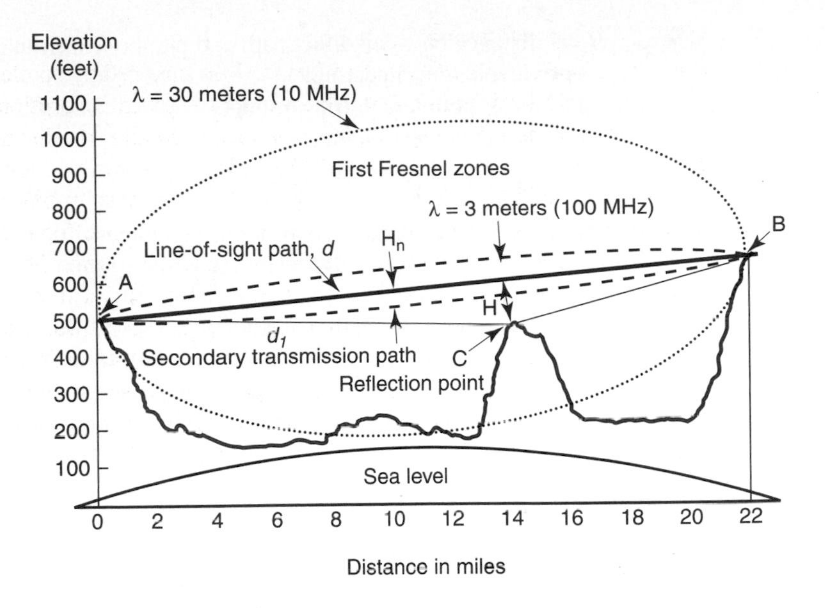

FIGURE 24-15 Microwave line-of-sight path showing first Fresnel zones

24-11-2 Path Clearance and Antenna Heights

The presence and topography of Earth's surface and the nonuniformity of the atmosphere above it can markedly affect the operating conditions of a microwave radio communications link. A majority of the time, the path loss of a typical microwave link can be approximated by the calculated free-space path loss. This is accomplished by engineering the path between transmit and receive antennas to provide an optical line-of-sight transmission path that should have adequate clearance with respect to surrounding objects. This clearance is necessary to ensure that the path loss under normal atmospheric conditions does not deviate from its nominal free-space value and to reduce the effects of severe fading that could occur during abnormal conditions.

The importance of providing an adequate path clearance is shown in Figure 24-15, which shows the profile of the path between the antennas of two microwave stations. For the antenna heights shown, the distance H represents the clearance of the line-of-sight path, AB, and the intervening terrain. Path ACB represents a secondary transmission path via reflection from the projection at location C. With no phase reversal at the point of reflection, the signal from the two paths would partially cancel whenever AB and ACB differed by an odd multiple of a half wavelength. When the grazing angle of the secondary wave is small, which is typically the case, a phase reversal will normally occur at the point of reflection (C). Therefore, whenever the distances AB and ACB differ by an odd multiple of a half wavelength, the energies of the received signals add rather than cancel. Conversely, if the lengths of the two paths differ by a whole number of half wavelengths, the signals from the two paths will tend to cancel.

The amount of clearance is generally described in terms of Fresnel (pronounced "fra-nell") zones. All points from which a wave could be reflected with an additional path length of one-half wavelength form an ellipse that defines the first Fresnel zone. Similarly, the boundary of the *n*th Fresnel zone consists of all points in which the propagation delay is $n/2$ wavelengths. For any distance, d_1, from antenna A, the distance H_n from the line-of-sight path to the boundary of the *n*th Fresnel zone is approximated by a parabola described as

$$H_n = \sqrt{\frac{n\lambda d_1(d - d_1)}{d}} \tag{24-10}$$

where H_n = distance between direct path and parabola surrounding it
λ = wavelength (linear unit)
d = direct path length (linear unit)
d_1 = reflected path length (linear unit)

and all linear units must be the same (feet, meters, cm, and so on).

The boundaries of the first Fresnel zones for $\lambda = 3$ meters (100 MHz) in the vertical plane through AB are shown in Figure 24-15. In any plane normal to AB, the Fresnel zones are concentric circles.

Measurements have shown that to achieve a normal transmission loss approximately equal to the free-space path loss, the transmission path should pass over all obstacles with a clearance of at least 0.6 times the distance of the first Fresnel zone and preferably by a distance equal to or greater than the first Fresnel zone distance. However, because of the effects of refraction, greater clearance is generally provided to reduce deep fading under adverse atmospheric conditions.

When determining the height of a microwave tower, a profile plot is made of the terrain between the proposed antenna sites, and the worst obstacle in the path, such as a mountain peak or ridge, is identified. The obstacle is used for a leverage point to determine the minimum path clearance between two locations from which the most suitable antenna heights are determined. Portable antennas, transmitters, and receivers are used to test the location to determine the optimum antenna heights.

24-11-3 Fading

The previous sections illustrated how free-space path loss is calculated. Path loss is a fixed loss, which remains constant over time. With very short path lengths at below 10 GHz, the signal level at the distant antenna can be calculated to within ± 1 dB. Provided that the transmit power remains constant, receive signal level (RSL) should remain uniform and constant over long periods of time. As the path length is extended, however, the measured receive signal level can vary around a nominal median value and remain in that range for minutes or hours and then suddenly drop below the median range and then return to the median level again. At other times and/or on other radio paths, the variation in signal level can be continuous for varying periods. Drops in receive signal level can be as much as 30 dB or more. This reduction in receive signal level is called *fading.*

Fading is a general term applied to the reduction in signal strength at the input to a receiver. It applies to propagation variables in the physical radio path that affect changes in the path loss between transmit and receive antennas. The changes in the characteristics of a radio path are associated with both atmospheric conditions and the geometry of the path itself (i.e., the relative position of the antenna with respect to the ground and surrounding terrain and obstacles). Substantial atmospheric conditions can transform an otherwise adequate line-of-sight path into an obstructed path because the effective path clearance approaches zero or becomes negative. Fading can occur under conditions of heavy ground fog or when extremely cold air moves over warm ground. The result in either case is a substantial increase in path loss over a wide frequency band. The magnitude and rapidity of occurrence of slow, flat fading of this type can generally be reduced only by using greater antenna heights.

A more common form of fading is a relatively rapid, frequency selective type of fading caused by interference between two or more rays in the atmosphere. The separate paths between transmit and receive antennas are caused by irregularities in the dielectric permittivity of the air, which varies with height. The transmission margins that must be provided against both types of fading are important considerations in determining overall system parameters and reliability objectives.

An interference type of fade may occur to any depth but, fortunately, the deeper the fade, the less frequently it occurs and the shorter its duration. Both the number of fades and the percentage of time a received signal is below a given level tend to increase as either the

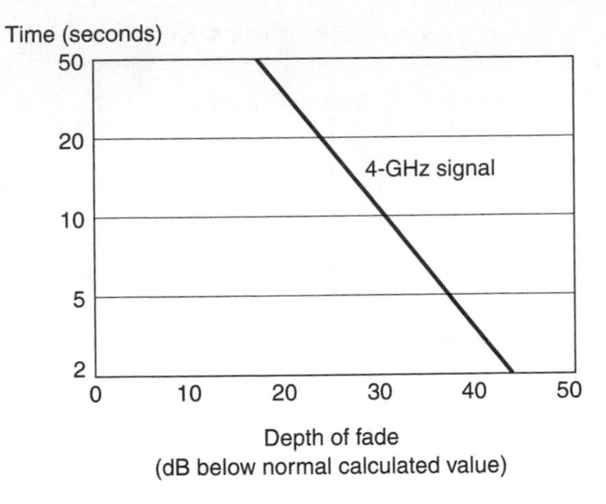

FIGURE 24-16 Median duration of fast fading

repeater spacing or the frequency of operation increases. Multiple paths are usually overhead, although ground reflections can occasionally be a factor. Using frequency or space diversity can generally minimize the effects of multipath fading.

Figure 24-16 shows the median duration of radio fades on a 4-GHz signal for various depths with an average repeater spacing of 30 miles. As shown in the figure, a median duration of a 20-dB fade is about 30 seconds, and the median duration of a 40-dB fade is about 3 seconds. At any given depth of fade, the duration of 1% of the fades may be as much as 10 times or as little as one-tenth of the median duration.

Multipath fading occurs primarily during nighttime hours on typical microwave links operating between 2 GHz and 6 GHz. During daytime hours or whenever the lower atmosphere is thoroughly mixed by rising convection currents and winds, the signals on a line-of-sight path are normally steady and at or near the calculated free-space values. On clear nights with little or no wind, however, sizable irregularities or layers can form at random elevations, and these irregularities in refraction result in multiple transmission path lengths on the order of a million wavelengths or longer. Multipath fading has a tendency to build up during nighttime hours with a peak in the early morning and then disappear as convection currents caused by heat produced during the early daylight hours break up the layers. Both the occurrence of fades and the percentage of time below a given receive signal level tend to increase with increases in repeater spacing or frequency.

24-12 MICROWAVE RADIO SYSTEM GAIN

In its simplest form, *system gain* (G_s) is the difference between the nominal output power of a transmitter (P_t) and the minimum input power to a receiver ($C_{\min}$) necessary to achieve satisfactory performance. System gain must be greater than or equal to the sum of all gains and losses incurred by a signal as it propagates from a transmitter to a receiver. In essence, system gain represents the net loss of a radio system, which is used to predict the reliability of a system for a given set of system parameters.

Ironically, system gain is actually a loss, as the losses a signal experiences as it propagates from a transmitter to a receiver are much higher than the gains. Therefore, the net system gain always equates to a negative dB value (i.e., a loss). Because system gain is defined as a net loss, individual losses are represented with positive dB values, while individual gains are represented with negative dB values. Figure 24-17 shows the diagram for a microwave system indicating where losses and gains typically occur.

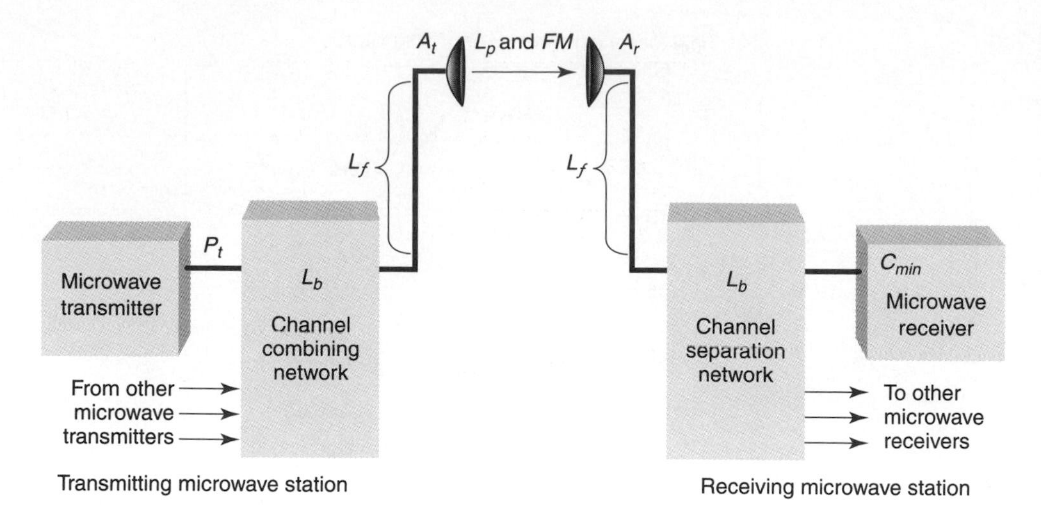

FIGURE 24-17 System gains and losses

Mathematically, system gain in its simplest form is

$$G_s = P_t - C_{\min} \tag{24-11}$$

where G_s = system gain (dB)
P_t = transmitter output power (dBm or dBW)
$C_{\min}$ = minimum receiver input power necessary to achieve a given reliability and quality objective

and where

$$P_t - C_{\min} \geq \text{losses} - \text{gains} \tag{24-12}$$

Gains
A_t = transmit antenna gain relative to an isotropic radiator (dB)
A_r = receive antenna gain relative to an isotropic radiator (dB)

Losses
L_p = free-space path loss incurred as a signal propagates from the transmit antenna to the receive antenna through Earth's atmosphere (dB)
L_f = transmission line loss between the distribution network (channel-combining network at the transmit station or channel separation network at the receive station) and its respective antenna (dB)
L_b = total coupling or branching loss in the channel-combining network between the output of a transmitter and the transmission line or from the output of a channel separation network and the receiver (dB)
FM = fade margin for a given reliability objective (dB)

A more useful expression for system gain is

$$G_{s(\text{dB})} = P_t - C_{\min} \geq FM_{(\text{dB})} + L_{p(\text{dB})} + L_{f(\text{dB})} + L_{b(\text{dB})} - A_{t(\text{dB})} - A_{r(\text{dB})} \tag{24-13}$$

Path loss can be determined from either Equation 24-8 or Equation 24-9, while feeder and branching losses depend on individual component specifications and diversity arrangements. Table 24-3 lists component specifications for several types of transmission lines for both space- and frequency-diversity systems. Antenna gain depends on the antenna's physical dimensions and the frequency of operation. Table 24-3 lists approximate antenna gains for parabolic antennas with several different diameters. The magnitude of the fade margin depends on several factors relating to the distance between transmit and receive antennas and the type of terrain the signal propagates over. Fade margin calculations are described in the next section of this chapter.

Table 24-3 System Gain Parameters

Frequency (GHz)	Feeder Loss (L_f) Type	Loss (dB/100 Meters)	Branching Loss (L_b) (dB) Diversity Frequency	Space	Antenna Gain (A_t or A_r) Diameter (Meters)	Gain (dB)
1.8	Air-filled coaxial cable	5.4	4	2	1.2	25.2
					2.4	31.2
					3.0	33.2
					3.7	34.7
					4.8	37.2
7.4	EWP 64 elliptical waveguide	4.7	3	2	1.2	37.1
					1.5	38.8
					2.4	43.1
					3.0	44.8
					3.7	46.5
8.0	EWP 69 elliptical waveguide	6.5	3	2	1.2	37.8
					2.4	43.8
					3.0	45.6
					3.7	47.3
					4.8	49.8

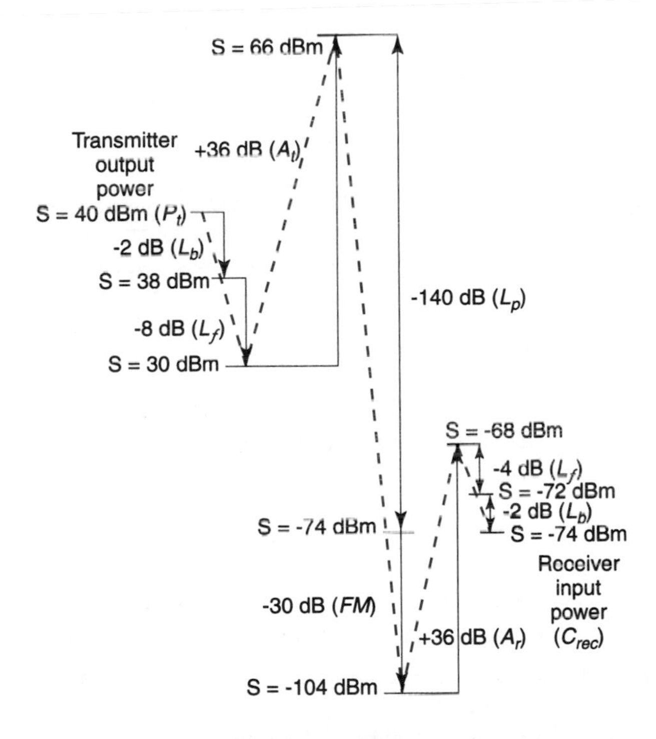

FIGURE 24-18 Microwave radio link signal levels relative to system gains and losses

Figure 24-18 shows a simplified diagram illustrating how the level of a signal changes as it propagates from a microwave transmitter to a microwave receiver for the following system parameters:

Transmit station $P_t = 40$ dBm
$L_b = 2$ dB
$L_f = 8$ dB
$A_t = 36$ dB

Atmosphere	$L_p = 140$ dB
	$FM = 30$ dB
Receive station	$A_r = 36$ dB
	$L_f = 4$ dB
	$L_b = 2$ dB

System gain is determined from Equation 24-13:

$$\begin{aligned} G_{s(\text{dB})} &= FM_{(\text{dB})} + L_{p(\text{dB})} + L_{f(\text{dB})} + L_{b(\text{dB})} - A_{(\text{dB})t} - A_{r(\text{dB})} \\ &= 30\text{ dB} + 140\text{ dB} + 12\text{ dB} + 4\text{ dB} - 36\text{ dB} - 36\text{ dB} \\ &= 114\text{ dB} \end{aligned}$$

and the receive signal level (C_{rec}) is simply the transmit power (P_t) minus system gain (G_s) or

$$\begin{aligned} C_{\text{rec}} &= 40\text{ dBm} - 114\text{ dB} \\ &= -74\text{ dBm} \end{aligned}$$

24-12-1 Fade Margin

Fade margin (sometimes called *link margin*) is essentially a "fudge factor" included in system gain equations that considers the nonideal and less predictable characteristics of radiowave propagation, such as multipath propagation (multipath loss) and terrain sensitivity. These characteristics cause temporary, abnormal atmospheric conditions that alter the free-space loss and are usually detrimental to the overall system performance. Fade margin also considers system reliability objectives. Thus, fade margin is included in system gain equations as a loss.

In April 1969, W. T. Barnett of Bell Telephone Laboratories described ways of calculating outage time due to fading on a nondiversity path as a function of terrain, climate, path length, and fade margin. In June 1970, Arvids Vignant (also of Bell Laboratories) derived formulas for calculating the effective improvement achievable by vertical space diversity as a function of the spacing distance, path length, and frequency.

Solving the Barnett-Vignant reliability equations for a specified annual system availability for an unprotected, nondiversity system yields the following expression:

$$F_m = \underbrace{30 \log D}_{\substack{\text{multipath}\\\text{effect}}} + \underbrace{10 \log (6ABf)}_{\substack{\text{terrain}\\\text{sensitivity}}} - \underbrace{10 \log (1 - R)}_{\substack{\text{reliability}\\\text{objectives}}} - \underbrace{70}_{\text{constant}} \qquad \textbf{(24-14)}$$

where F_m = fade margin (dB)
D = distance (kilometers)
f = frequency (gigahertz)
R = reliability expressed as a decimal (i.e., 99.99% = 0.9999 reliability)
$1 - R$ = reliability objective for a one-way 400-km route
A = roughness factor
= 4 over water or a very smooth terrain
= 1 over an average terrain
= 0.25 over a very rough, mountainous terrain
B = factor to convert a worst-month probability to an annual probability
= 1 to convert an annual availability to a worst-month basis
= 0.5 for hot humid areas
= 0.25 for average inland areas
= 0.125 for very dry or mountainous areas

Example 24-2

Consider a space-diversity microwave radio system operating at an RF carrier frequency of 1.8 GHz. Each station has a 2.4-m-diameter parabolic antenna that is fed by 100 m of air-filled coaxial cable. The terrain is smooth, and the area has a humid climate. The distance between stations is 40 km. A reliability objective of 99.99% is desired. Determine the system gain.

Solution Substituting into Equation 24-14, we find that the fade margin is

$$F_m = 30 \log 40 + 10 \log[(6)(4)(0.5)(1.8)] - 10 \log(1 - 0.9999) - 70$$
$$= 48.06 + 13.34 - (-40) - 70$$
$$= 48.06 + 13.34 + 40 - 70 = 31.4 \text{ dB}$$

Substituting into Equation 24-8, we obtain path loss:

$$L_p = 92.4 + 20 \log 1.8 + 20 \log 40$$
$$= 92.4 + 5.11 + 32.04 = 129.55 \text{ dB}$$

From Table 24-3,

$$L_b = 4 \text{ dB}(2 + 2 = 4)$$
$$L_f = 10.8 \text{ dB}(100 \text{ m} + 100 \text{ m} = 200 \text{ m})$$
$$A_t = A_r = 31.2 \text{ dB}$$

Substituting into Equation 24-13 gives us system gain:

$$G_s = 31.4 + 129.55 + 10.8 + 4 - 31.2 - 31.2 = 113.35 \text{ dB}$$

The results indicate that for this system to perform at 99.99% reliability with the given terrain, distribution networks, transmission lines, and antennas, the transmitter output power must be at least 113.35 dB more than the minimum receive signal level.

24-12-2 Receiver Threshold

Carrier-to-noise (C/N) ratio is probably the most important parameter considered when evaluating the performance of a microwave communications system. The minimum wideband carrier power (C_{min}) at the input to a receiver that will provide a usable baseband output is called the receiver *threshold* or, sometimes, receiver *sensitivity.* The receiver threshold is dependent on the wideband noise power present at the input of a receiver, the noise introduced within the receiver, and the noise sensitivity of the baseband detector. Before C_{min} can be calculated, the input noise power must be determined. The input noise power is expressed mathematically as

$$N = KTB \qquad (24\text{-}15)$$

where N = noise power (watts)
K = Boltzmann's constant (1.38×10^{-23} J/K)
T = equivalent noise temperature of the receiver (kelvin) (room temperature = 290 kelvin)
B = noise bandwidth (hertz)

Expressed in dBm,

$$N_{(\text{dBm})} = 10 \log \frac{KTB}{0.001} = 10 \log \frac{KT}{0.001} + 10 \log B$$

For a 1-Hz bandwidth at room temperature,

$$N = 10 \log \frac{(1.38 \times 10^{-23})(290)}{0.001} + 10 \log 1$$
$$= -174 \text{ dBm}$$

Thus,

$$N_{(\text{dBm})} = -174 \text{ dBm} + 10 \log B \qquad (24\text{-}16)$$

Example 24-3

For an equivalent noise bandwidth of 10 MHz, determine the noise power.

Solution Substituting into Equation 24-16 yields

$$N = -174 \text{ dBm} + 10 \log(10 \times 10^6)$$
$$= -174 \text{ dBm} + 70 \text{ dB} = -104 \text{ dBm}$$

If the minimum C/N requirement for a receiver with a 10-MHz noise bandwidth is 24 dB, the minimum receive carrier power is

$$C_{min} = \frac{C}{N} + N = 24 \text{ dB} + (-104 \text{ dBm}) = -80 \text{ dBm}$$

For a system gain of 113.35 dB, it would require a minimum transmit carrier power (P_t) of

$$P_t = G_s + C_{min} = 113.35 \text{ dB} + (-80 \text{ dBm}) = 33.35 \text{ dBm}$$

This indicates that a minimum transmit power of 33.35 dBm (2.16 W) is required to achieve a carrier-to-noise ratio of 24 dB with a system gain of 113.35 dB and a bandwidth of 10 MHz.

24-12-3 Carrier-to-Noise versus Signal-to-Noise Ratio

Carrier-to-noise (C/N) is the ratio of the wideband "carrier" (actually, not just the carrier but rather the carrier and its associated sidebands) to the wideband noise power (the noise bandwidth of the receiver). C/N can be determined at an RF or an IF point in the receiver. Essentially, C/N is a *predetection* (before the FM demodulator) signal-to-noise ratio. Signal-to-noise (S/N) is a *postdetection* (after the FM demodulator) ratio. At a baseband point in the receiver, a single voice-band channel can be separated from the rest of the baseband and measured independently. At an RF or IF point in the receiver, it is impossible to separate a single voice-band channel from the composite FM signal. For example, a typical bandwidth for a single microwave channel is 30 MHz. The bandwidth of a voice-band channel is 4 kHz. C/N is the ratio of the power of the composite RF signal to the total noise power in the 30-MHz bandwidth. S/N is the ratio of the signal power of a single voice-band channel to the noise power in a 4-kHz bandwidth.

24-12-4 Noise Factor and Noise Figure

Noise factor (F) and *noise figure* (NF) are figures of merit used to indicate how much the signal-to-noise ratio deteriorates as a signal passes through a circuit or series of circuits. Noise factor is simply a ratio of input signal-to-noise ratio to output signal-to-noise ratio. In other words, a ratio of ratios. Mathematically, noise factor is

$$F = \frac{\text{input signal-to-noise ratio}}{\text{output signal-to-noise ratio}} \text{ (unitless ratio)} \quad \textbf{(24-17)}$$

Noise figure is simply the noise factor stated in dB and is a parameter commonly used to indicate the quality of a receiver. Mathematically, noise figure is

$$NF = 10 \log \frac{\text{input signal-to-noise ratio}}{\text{output signal-to-noise ratio}} \text{ (dB)} \quad \textbf{(24-18)}$$

or

$$NF = 10 \log F \quad \textbf{(24-19)}$$

In essence, noise figure indicates how much the signal-to-noise ratio deteriorates as a waveform propagates from the input to the output of a circuit. For example, an amplifier with a noise figure of 6 dB means that the signal-to-noise ratio at the output is 6 dB less than it was at the input. If a circuit is perfectly noiseless and adds no additional noise to the signal, the signal-to-noise ratio at the output will equal the signal-to-noise ratio at the input. For a perfect, noiseless circuit, the noise factor is 1, and the noise figure is 0 dB.

An electronic circuit amplifies signals and noise within its passband equally well. Therefore, if the amplifier is ideal and noiseless, the input signal and noise are amplified the same, and the signal-to-noise ratio at the output will equal the signal-to-noise ratio at

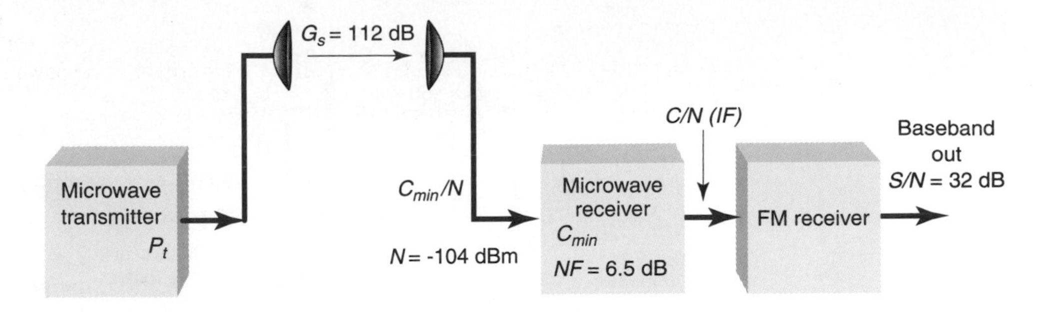

FIGURE 24-19 System gain diagram for Example 24-4

the input. In reality, however, amplifiers are not ideal. Therefore, the amplifier adds internally generated noise to the waveform, reducing the overall signal-to-noise ratio. The most predominant noise is thermal noise, which is generated in all electrical components. Therefore, all networks, amplifiers, and systems add noise to the signal and, thus, reduce the overall signal-to-noise ratio as the signal passes through them.

Example 24-4

Refer to Figure 24-19. For a system gain of 112 dB, a total noise figure of 6.5 dB, an input noise power of −104 dBm, and a minimum $(S/N)_{out}$ of the FM demodulator of 32 dB, determine the minimum receive carrier power and the minimum transmit power.

Solution To achieve a S/N ratio of 32 dB out of the FM demodulator, an input C/N of 15 dB is required (17 dB of improvement due to FM quieting). Solving for the receiver input carrier-to-noise ratio gives

$$\frac{C_{min}}{N} = \frac{C}{N} + NF_T = 15 \text{ dB} + 6.5 \text{ dB} = 21.5 \text{ dB}$$

Thus,

$$C_{min} = \frac{C_{min}}{N} + N = 21.5 \text{ dB} + (-104 \text{ dBm}) = -82.5 \text{ dBm}$$

$$P_t = G_s + C_{min} = 112 \text{ dB} + (-82.5 \text{ dBm}) = 29.5 \text{ dBm}$$

Example 24-5

For the system shown in Figure 24-20, determine the following: G_s, C_{min}/N, C_{min}, N, G_s, and P_t.

Solution The minimum C/N at the input to the FM receiver is 23 dB:

$$\frac{C_{min}}{N} = \frac{C}{N} + NF_T = 23 \text{ dB} + 4.24 \text{ dB} = 27.24 \text{ dB}$$

Substituting into Equation 24-16 yields

$$N = -174 \text{ dBm} + 10 \log B = -174 \text{ dBm} + 68 \text{ dB} = -106 \text{ dBm}$$

$$C_{min} = \frac{C_{min}}{N} + N = 27.24 \text{ dB} + (-106 \text{ dBm}) = -78.76 \text{ dBm}$$

Substituting into Equation 24-14 gives us

$$F_m = 30 \log 50 + 10 \log[(6)(0.25)(0.125)(8)] - 10 \log(1 - 0.99999) - 70 = 32.76 \text{ dB}$$

Substituting into Equation 24-8, we have

$$L_p = 92.4 \text{ dB} + 20 \log 8 + 20 \log 50 = 92.4 \text{ dB} + 18.06 \text{ dB} + 33.98 \text{ dB} = 144.44 \text{ dB}$$

From Table 24-3,

$$L_b = 4 \text{ dB}$$

$$L_f = 0.75(6.5 \text{ dB}) = 4.875 \text{ dB}$$

$$A_t = A_r = 37.8 \text{ dB}$$

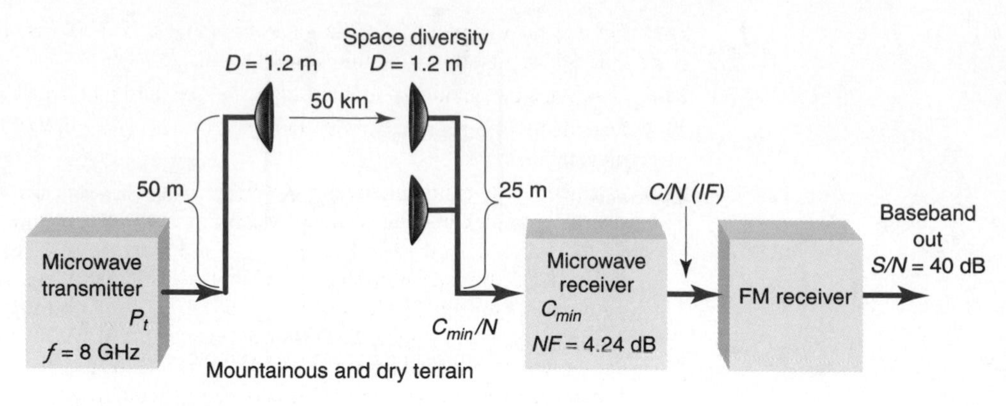

FIGURE 24-20 System gain diagram for Example 24-5

Note: The gain of an antenna increases or decreases proportional to the square of its diameter (i.e., if its diameter changes by a factor of 2, its gain changes by a factor of 4, which is 6 dB).

Substituting into Equation 24-13 yields

$$G_s = 32.76 + 144.44 + 4.875 + 4 - 37.8 - 37.8 = 110.475 \text{ dB}$$

$$P_t = G_s + C_{\min} = 110.475 \text{ dB} + (-78.76 \text{ dBm}) = 31.715 \text{ dBm}$$

QUESTIONS

24-1. What constitutes a short-haul microwave system? A long-haul microwave system?

24-2. Describe the baseband signal for a microwave system.

24-3. Why do FDM/FM microwave systems use low-index FM?

24-4. Describe a microwave repeater. Contrast baseband and IF repeaters.

24-5. Define *diversity.* Describe the three most commonly used diversity schemes.

24-6. Describe a protection switching arrangement. Contrast the two types of protection switching arrangements.

24-7. Briefly describe the four major sections of a microwave terminal station.

24-8. Define *ringaround.*

24-9. Briefly describe a high/low microwave system.

24-10. Define *system gain.*

24-11. Define the following terms: *free-space path loss, branching loss,* and *feeder loss.*

24-12. Define *fade margin.* Describe multipath losses, terrain sensitivity, and reliability objectives and how they affect fade margin.

24-13. Define *receiver threshold.*

24-14. Contrast carrier-to-noise ratio with signal-to-noise ratio.

24-15. Define *noise figure.*

PROBLEMS

24-1. Calculate the noise power at the input to a receiver that has a radio carrier frequency of 4 GHz and a bandwidth of 30 MHz (assume room temperature).

24-2. Determine the path loss for a 3.4-GHz signal propagating 20,000 m.

24-3. Determine the fade margin for a 60-km microwave hop. The RF carrier frequency is 6 GHz, the terrain is very smooth and dry, and the reliability objective is 99.95%.

24-4. Determine the noise power for a 20-MHz bandwidth at the input to a receiver with an input noise temperature of 290°C.

24-5. For a system gain of 120 dB, a minimum input C/N of 30 dB, and an input noise power of −115 dBm, determine the minimum transmit power (P_t).

24-6. Determine the amount of loss attributed to a reliability objective of 99.98%.

24-7. Determine the terrain sensitivity loss for a 4-GHz carrier that is propagating over a very dry, mountainous area.

24-8. A frequency-diversity microwave system operates at an RF carrier frequency of 7.4 GHz. The IF is a low-index frequency-modulated subcarrier. The baseband signal is the 1800-channel FDM system described in Chapter 11 (564 kHz to 8284 kHz). The antennas are 4.8-m-diameter parabolic dishes. The feeder lengths are 150 m at one station and 50 m at the other station. The reliability objective is 99.999%. The system propagates over an average terrain that has a very dry climate. The distance between stations is 50 km. The minimum carrier-to-noise ratio at the receiver input is 30 dB. Determine the following: fade margin, antenna gain, free-space path loss, total branching and feeder losses, receiver input noise power (C_{min}), minimum transmit power, and system gain.

24-9. Determine the overall noise figure for a receiver that has two RF amplifiers each with a noise figure of 6 dB and a gain of 10 dB, a mixer down-converter with a noise figure of 10 dB, and a conversion gain of −6 dB, and 40 dB of IF gain with a noise figure of 6 dB.

24-10. A microwave receiver has a total input noise power of −102 dBm and an overall noise figure of 4 dB. For a minimum C/N ratio of 20 dB at the input to the FM detector, determine the minimum receive carrier power.

24-11. Determine the path loss for the following frequencies and distances:

f (MHz)	D (km)
200	0.5
800	0.8
3000	5
5000	10
8000	25
18000	10

24-12. Determine the fade margin for a 30-km microwave hop. The RF frequency is 4 GHz, the terrain is water, and the reliability objective is 99.995%.

24-13. Determine the noise power for a 40-MHz bandwidth at the input to a receiver with an input temperature $T = 400°C$.

24-14. For a system gain of 114 dB, a minimum input C/N = 34 dB, and an input noise power of −111 dBm, determine the minimum transmit power (P_t).

24-15. Determine the amount of loss contributed to a reliability objective of 99.9995%.

24-16. Determine the terrain sensitivity loss for an 8-GHz carrier that is propagating over a very smooth and dry terrain.

24-17. A frequency-diversity microwave system operates at an RF = 7.4 GHz. The IF is a low-index frequency-modulated subcarrier. The baseband signal is a single mastergroup FDM system. The antennas are 2.4-m parabolic dishes. The feeder lengths are 120 m at one station and 80 m at the other station. The reliability objective is 99.995%. The system propagates over an average terrain that has a very dry climate. The distance between stations is 40 km. The minimum carrier-to-noise ratio at the receiver input is 28 dB. Determine the following: fade margin, antenna gain, free-space path loss, total branching and feeder losses, receiver input power (C_{min}), minimum transmit power, and system gain.

24-18. Determine the overall noise figure for a receiver that has two RF amplifiers each with a noise figure of 8 dB and a gain of 13 dB, a mixer down-converter with a noise figure of 6 dB, and a conversion gain of −6 dB, and 36 dB of IF gain with a noise figure of 10 dB.

24-19. A microwave receiver has a total input noise power of −108 dBm and an overall noise figure of 5 dB. For a minimum C/N ratio of 18 dB at the input to the FM detector, determine the minimum receive carrier power.

CHAPTER 25

Satellite Communications

CHAPTER OUTLINE

OBJECTIVES

- Define *satellite communications*
- Describe the history of satellite communications
- Explain Kepler's laws and how they relate to satellite communications
- Define and describe satellite orbital patterns and elevation categories
- Describe geosynchronous satellite systems and their advantages and disadvantages over other types of satellite systems
- Explain satellite look angles
- List and describe satellite classifications, spacing, and frequency allocation
- Describe the different types of satellite antenna radiation patterns
- Explain satellite system up- and downlink models
- Define and describe satellite system parameters
- Explain satellite system link equations
- Describe the significance of satellite link budgets and how they are calculated

25-1 INTRODUCTION

In astronomical terms, a *satellite* is a celestial body that orbits around a planet (e.g., the moon is a satellite of Earth). In aerospace terms, however, a satellite is a space vehicle launched by humans and orbits Earth or another celestial body. Communications satellites are man-made satellites that orbit Earth, providing a multitude of communication functions to a wide variety of consumers, including military, governmental, private, and commercial subscribers.

In essence, a *communications satellite* is a microwave repeater in the sky that consists of a diverse combination of one or more of the following: receiver, transmitter, amplifier, regenerator, filter, onboard computer, multiplexer, demultiplexer, antenna, waveguide, and about any other electronic communications circuit ever developed. A satellite radio repeater is called a *transponder,* of which a satellite may have many. A *satellite system* consists of one or more satellite space vehicles, a ground-based station to control the operation of the system, and a user network of earth stations that provides the interface facilities for the transmission and reception of terrestrial communications traffic through the satellite system.

Transmissions to and from satellites are categorized as either *bus* or *payload.* The bus includes control mechanisms that support the payload operation. The payload is the actual user information conveyed through the system. Although in recent years new data services and television broadcasting are more and more in demand, the transmission of conventional speech telephone signals (in analog or digital form) is still the bulk of satellite payloads.

In the early 1960s, AT&T released studies indicating that a few powerful satellites of advanced design could handle more telephone traffic than the entire existing AT&T long-distance communications network. The cost of these satellites was estimated to be only a fraction of the cost of equivalent terrestrial microwave or underground cable facilities. Unfortunately, because AT&T was a utility and government regulations prevented them from developing the satellite systems, smaller and much less lucrative companies were left to develop the satellite systems, and AT&T continued for several more years investing billions of dollars each year in conventional terrestrial microwave and metallic cable systems. Because of this, early developments in satellite technology were slow in coming.

25-2 HISTORY OF SATELLITES

The simplest type of satellite is a *passive reflector,* which is a device that simply "bounces" signals from one place to another. A passive satellite reflects signals back to Earth, as there are no gain devices on board to amplify or modify the signals. The moon is a natural satellite of Earth, visible by reflection of sunlight and having a slightly elliptical orbit. Consequently, the moon became the first passive satellite in 1954, when the U.S. Navy successfully transmitted the first message over this Earth-to-moon-to-Earth communications system. In 1956, a relay service was established between Washington, D.C. and Hawaii and, until 1962, offered reliable long-distance radio communications service limited only by the availability of the moon. Over time, however, the moon proved to be an inconvenient and unreliable communications satellite, as it is above the horizon only half the time and its position relative to Earth is constantly changing.

An obvious advantage of passive satellites is that they do not require sophisticated electronic equipment on board, although they are not necessarily void of power. Some passive satellites require *radio beacon transmitters* for tracking and ranging purposes. A beacon is a continuously transmitted unmodulated carrier that an earth station can lock on to and use to determine the exact location of a satellite so the earth station can align its antennas. Another disadvantage of passive satellites is their inefficient use of transmitted power. For example, as little as 1 part in every 10^{18} of an earth station's transmitted power is actually returned to earth station receiving antennas.

In 1957, Russia launched *Sputnik I,* the first *active* earth satellite. An active satellite is capable of receiving, amplifying, reshaping, regenerating, and retransmitting information. *Sputnik I* transmitted telemetry information for 21 days. Later in the same year, the United States launched *Explorer I,* which transmitted telemetry information for nearly five months.

In 1958, NASA launched *Score,* a 150-pound conical-shaped satellite. With an on-board tape recording, *Score* rebroadcast President Eisenhower's 1958 Christmas message. *Score* was the first artificial satellite used for relaying terrestrial communications. *Score* was a *delayed repeater* satellite as it received transmissions from earth stations, stored them on magnetic tape, and then rebroadcast them later to ground stations farther along in its orbit.

In 1960, NASA in conjunction with Bell Telephone Laboratories and the Jet Propulsion Laboratory launched *Echo,* a 100-foot-diameter plastic balloon with an aluminum coating. *Echo* passively reflected radio signals it received from large earth station antennas. *Echo* was simple and reliable but required extremely high-power transmitters at the earth stations. The first transatlantic transmission using a satellite was accomplished using *Echo.* Also in 1960, the Department of Defense launched *Courier,* which was the first transponder-type satellite. *Courier* transmitted 3 W of power and lasted only 17 days.

In 1962, AT&T launched *Telstar I,* the first active satellite to simultaneously receive and transmit radio signals. The electronic equipment in *Telstar I* was damaged by radiation from the newly discovered Van Allen belts and, consequently, lasted for only a few weeks. *Telstar II* was successfully launched in 1963 and was electronically identical to *Telstar I* except more radiation resistant. *Telstar II* was used for telephone, television, facsimile, and data transmissions and accomplished the first successful transatlantic video transmission.

Syncom I, launched in February 1963, was the first attempt to place a geosynchronous satellite into orbit. Unfortunately, *Syncom I* was lost during orbit injection; however, *Syncom II* and *Syncom III* were successfully launched in February 1963 and August 1964, respectively. The *Syncom III* satellite was used to broadcast the 1964 Olympic Games from Tokyo. The *Syncom* satellites demonstrated the feasibility of using geosynchronous satellites.

Since the *Syncom* projects, a number of nations and private corporations have successfully launched satellites that are currently being used to provide national as well as regional and international global communications. Today, there are several hundred satellite communications systems operating in virtually every corner of the world. These companies provide worldwide, fixed common-carrier telephone and data circuits; point-to-point television broadcasting; network television distribution; music broadcasting; mobile telephone service; navigation service; and private communications networks for large corporations, government agencies, and military applications.

Intelsat I (called *Early Bird*) was the first commercial telecommunications satellite. It was launched from Cape Kennedy in 1965 and used two transponders and a 25-MHz bandwidth to simultaneously carry one television signal and 480 voice channels. Intelsat stands for *Int*ernational *Tel*ecommunications *Sat*ellite Organization. Intelsat is a commercial global satellite network that manifested in 1964 from within the United Nations. Intelsat is a consortium of over 120 nations with the commitment to provide worldwide, nondiscriminatory satellite communications using four basic service categories: international public switched telephony, broadcasting, private-line/business networks, and domestic/regional communications. Between 1966 and 1987, Intelsat launched a series of satellites designated *Intelsat II, III, IV, V,* and *VI. Intelsat VI* has a capacity of 80,000 voice channels. Intelsat's most recent satellite launches include the 500, 600, 700, and 800 series space vehicles.

The former Soviet Union launched the first set of *domestic satellites* (Domsats) in 1966 and called them *Molniya,* meaning "lightning." Domsats are satellites that are owned, operated, and used by a single country. In 1972, Canada launched its first commercial satellite designated *Anik,* which is an Inuit word meaning "little brother." Western Union launched their first Westar satellite in 1974, and Radio Corporation of America (RCA) launched its first Satcom (*Sat*ellite *Com*munications) satellites in 1975. In the United States today, a publicly owned company called *Com*munications *Sat*ellite Corporation (Comsat) regulates the use and operation of U.S. satellites and also sets their tariffs. Although a company or government may own a satellite, its utilities are generally made available to anyone willing to pay for them. The United States currently utilizes the largest share of available worldwide satellite time (24%); Great Britain is second with 13%, followed by France with 6%.

25-3 KEPLER'S LAWS

A satellite remains in orbit because the centrifugal force caused by its rotation around Earth is counterbalanced by Earth's gravitational pull. In the early seventeenth century while investigating the laws of planetary motion (i.e., motion of planets and their heavenly bodies called moons), German astronomer Johannes Kepler (1571–1630) discovered the laws that govern satellite motion. The laws of planetary motion describe the shape of the orbit, the velocities of the planet, and the distance a planet is with respect to the sun. *Kepler's laws* may be simply stated as (1) the planets move in ellipses with the sun at one focus, (2) the line joining the sun and a planet sweeps out equal areas in equal intervals of time, and (3) the square of the time of revolution of a planet divided by the cube of its mean distance from the sun gives a number that is the same for all planets. Kepler's laws can be applied to any two bodies in space that interact through gravitation. The larger of the two bodies is called the *primary,* and the smaller is called the *secondary* or *satellite.*

Kepler's first law states that a satellite will orbit a primary body (like Earth) following an elliptical path. An ellipse has two *focal points* (*foci*) as shown in Figure 25-1a (F_1 and F_2), and the center of mass (called the barycenter) of a two-body system is always centered on one of the foci. Because the mass of Earth is substantially greater than that of the satellite, the center of mass will always coincide with the center of Earth. The geometric properties of the ellipse are normally referenced to one of the foci which is logically selected to be the one at the center of Earth.

For the semimajor axis (α) and the semiminor axis (β) shown in Figure 25-1a, the *eccentricity* (abnormality) of the ellipse can be defined as

$$\epsilon = \frac{\sqrt{\alpha^2 - \beta^2}}{\alpha} \tag{25-1}$$

where ϵ is eccentricity.

Kepler's second law, enunciated with the first law in 1609, is known as the *law of areas.* Kepler's second law states that for equal intervals of time a satellite will sweep out equal areas in the orbital plane, focused at the barycenter. As shown in Figure 25-1b, for a satellite traveling distances D_1 and D_2 meters in 1 second, areas A_1 and A_2 will be equal. Because of the equal area law, distance D_1 must be greater than distance D_2, and, therefore, velocity V_1 must be greater than velocity V_2. The velocity will be greatest at the point of closest approach to Earth (known as the *perigee*), and the velocity will be least at the farthest point from Earth (known as the *apogee*). Kepler's second law is illustrated in Figure 25-1b.

Kepler's third law, announced in 1619, is sometimes known as the *harmonic law.* The third law states that the square of the periodic time of orbit is proportional to the cube of the mean distance between the primary and the satellite. This mean distance is equal to the semimajor axis; thus, Kepler's third law can be stated mathematically as

$$\alpha = AP^{2/3} \tag{25-2}$$

where A = constant (unitless)
α = semimajor axis (kilometers)
P = mean solar earth days

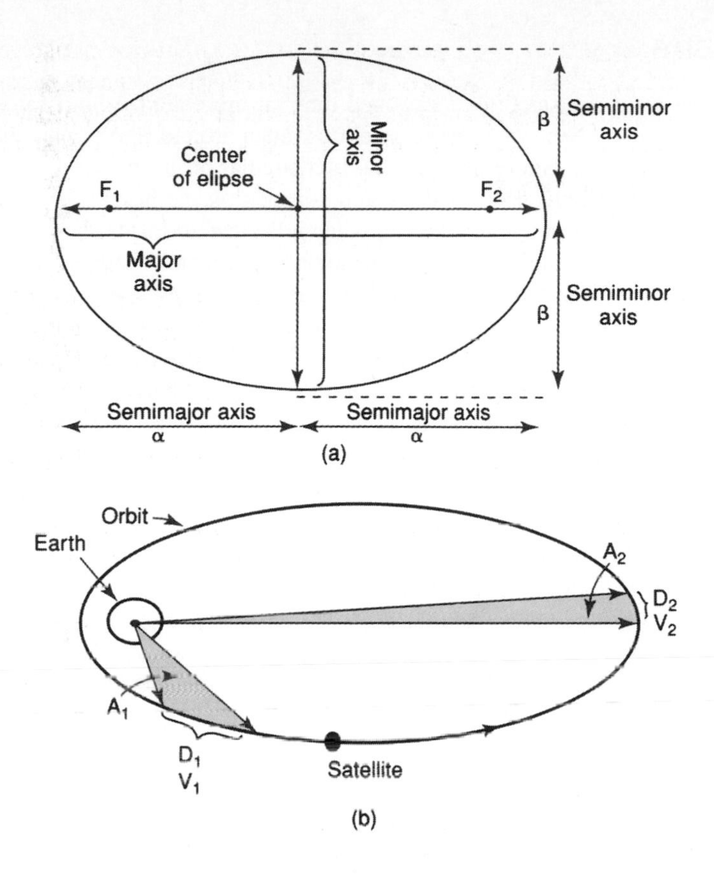

FIGURE 25-1 [a] Focal points F_1 and F_2, semimajor axis a, and semiminor axis b of an ellipse; [b] Kepler's second law

and P is the ratio of the time of one sidereal day (t_s = 23 hours and 56 minutes) to the time of one revolution of Earth on its own axis (t_e = 24 hours).

thus,

$$
\begin{aligned}
P &= \frac{t_s}{t_e} \\
&= \frac{1436 \text{ minutes}}{1440 \text{ minutes}} \\
&= 0.9972
\end{aligned}
$$

Rearranging Equation 25-2 and solving the constant A for earth yields

$$A = 42241.0979$$

Equations 25-1 and 25-2 apply for the ideal case when a satellite is orbiting around a perfectly spherical body with no outside forces. In actuality, Earth's equatorial bulge and external disturbing forces result in deviations in the satellite's ideal motion. Fortunately, however, the major deviations can be calculated and compensated for. Satellites orbiting close to Earth will be affected by atmospheric drag and by Earth's magnetic field. For more distant satellites, however, the primary disturbing forces are from the gravitational fields of the sun and moon.

25-4 SATELLITE ORBITS

Most of the satellites mentioned thus far are called *orbital* satellites, which are *nonsynchronous.* Nonsynchronous satellites rotate around Earth in an elliptical or circular pattern as shown in Figure 25-2a and b. In a circular orbit, the speed or rotation is constant; however, in elliptical orbits the speed depends on the height the satellite is above Earth. The speed of the satellite is greater when it is close to Earth than when it is farther away.

If the satellite is orbiting in the same direction as Earth's rotation (counterclockwise) and at an angular velocity greater than that of Earth ($\omega_s > \omega_e$), the orbit is called a *prograde* or *posigrade* orbit. If the satellite is orbiting in the opposite direction as Earth's rotation or in the same direction with an angular velocity less than that of Earth ($\omega_s < \omega_e$), the orbit is called a *retrograde* orbit. Most nonsynchronous satellites revolve around Earth in a prograde orbit. Therefore, the position of satellites in nonsynchronous orbits is continuously changing in respect to a fixed position on Earth. Consequently, nonsynchronous satellites have to be used when available, which may be as little as 15 minutes per orbit. Another disadvantage of orbital satellites is the need for complicated and expensive tracking equipment at the earth stations so they can locate the satellite as it comes into view on each orbit and then lock its antenna onto the satellite and track it as it passes overhead. A major advantage of orbital satellites, however, is that propulsion rockets are not required on board the satellites to keep them in their respective orbits.

25-4-1 Satellite Elevation Categories

Satellites are generally classified as having either a *low earth orbit* (LEO), *medium earth orbit* (MEO), or *geosynchronous earth orbit* (GEO). Most LEO satellites operate in the 1.0-GHz to 2.5-GHz frequency range. Motorola's satellite-based mobile-telephone system, *Iridium,*

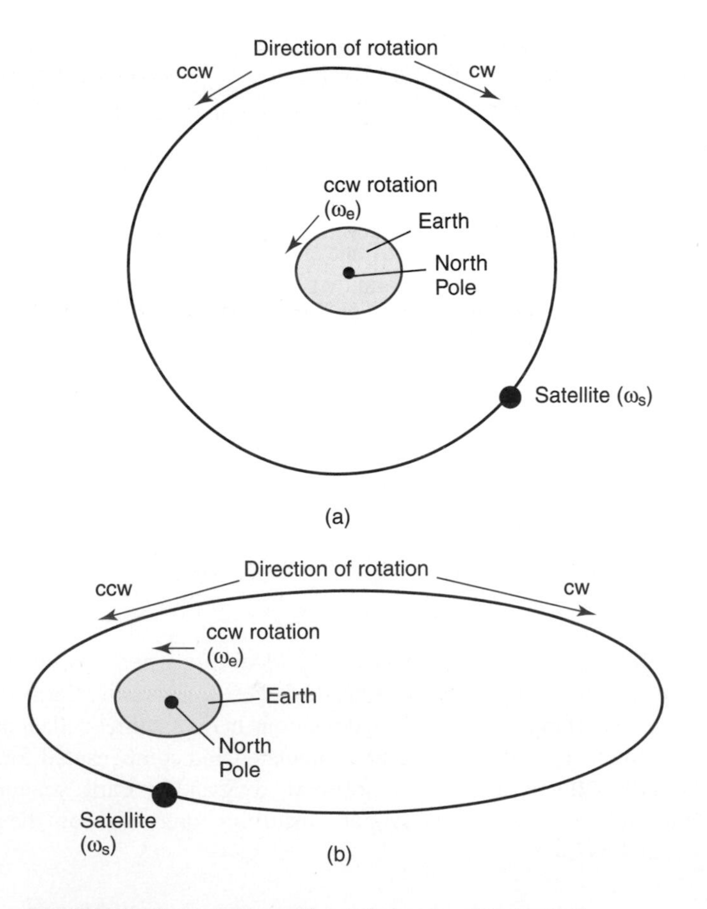

FIGURE 25-2 Satellite orbits: [a] circular; [b] elliptical

is a LEO system utilizing a 66-satellite constellation orbiting approximately 480 miles above Earth's surface. The main advantage of LEO satellites is that the path loss between earth stations and space vehicles is much lower than for satellites revolving in medium- or high-altitude orbits. Less path loss equates to lower transmit powers, smaller antennas, and less weight.

MEO satellites operate in the 1.2-GHz to 1.66-GHz frequency band and orbit between 6000 miles and 12,000 miles above Earth. The Department of Defense's satellite-based global positioning system, *NAVSTAR,* is a MEO system with a constellation of 21 working satellites and six spares orbiting approximately 9500 miles above Earth.

Geosynchronous satellites are high-altitude earth-orbit satellites operating primarily in the 2-GHz to 18-GHz frequency spectrum with orbits 22,300 miles above Earth's surface. Most commercial communications satellites are in geosynchronous orbit. Geosynchronous or *geostationary* satellites are those that orbit in a circular pattern with an angular velocity equal to that of Earth. Geostationary satellites have an orbital time of approximately 24 hours, the same as Earth; thus, geosynchronous satellites appear to be stationary, as they remain in a fixed position in respect to a given point on Earth.

Satellites in high-elevation, nonsynchronous circular orbits between 19,000 miles and 25,000 miles above Earth are said to be in *near-synchronous* orbit. When the near-synchronous orbit is slightly lower than 22,300 miles above Earth, the satellite's orbital time is lower than Earth's rotational period. Therefore, the satellite is moving slowly around Earth in a west-to-east direction. This type of near-synchronous orbit is called *subsynchronous.* If the orbit is higher than 22,300 miles above Earth, the satellite's orbital time is longer than Earth's rotational period, and the satellite will appear to have a reverse (retrograde) motion from east to west.

25-4-2 Satellite Orbital Patterns

Before examining satellite orbital paths, a basic understanding of some terms used to describe orbits is necessary. For the following definitions, refer to Figure 25-3:

Apogee. The point in an orbit that is located farthest from Earth

Perigee. The point in an orbit that is located closest to Earth

Major axis. The line joining the perigee and apogee through the center of Earth; sometimes called *line of apsides*

Minor axis. The line perpendicular to the major axis and halfway between the perigee and apogee (Half the distance of the minor axis is called the semiminor axis.)

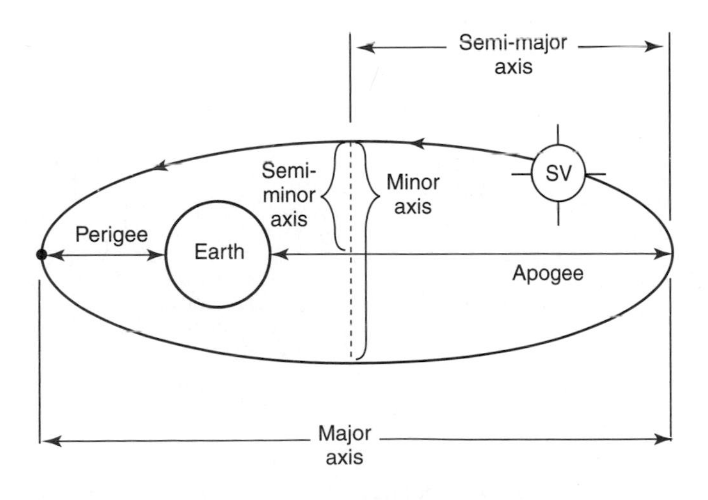

FIGURE 25-3 Satellite orbital terms

Although there is an infinite number of orbital paths, only three are useful for communications satellites. Figure 25-4 shows three paths that a satellite can follow as it rotates around Earth: inclined, equatorial, or polar. All satellites rotate around Earth in an orbit that forms a plane that passes through the center of gravity of Earth called the *geocenter.*

Inclined orbits are virtually all orbits except those that travel directly above the equator or directly over the North and South Poles. Figure 25-5a shows the *angle of inclination* of a satellite orbit. The angle of inclination is the angle between the Earth's equatorial plane and the

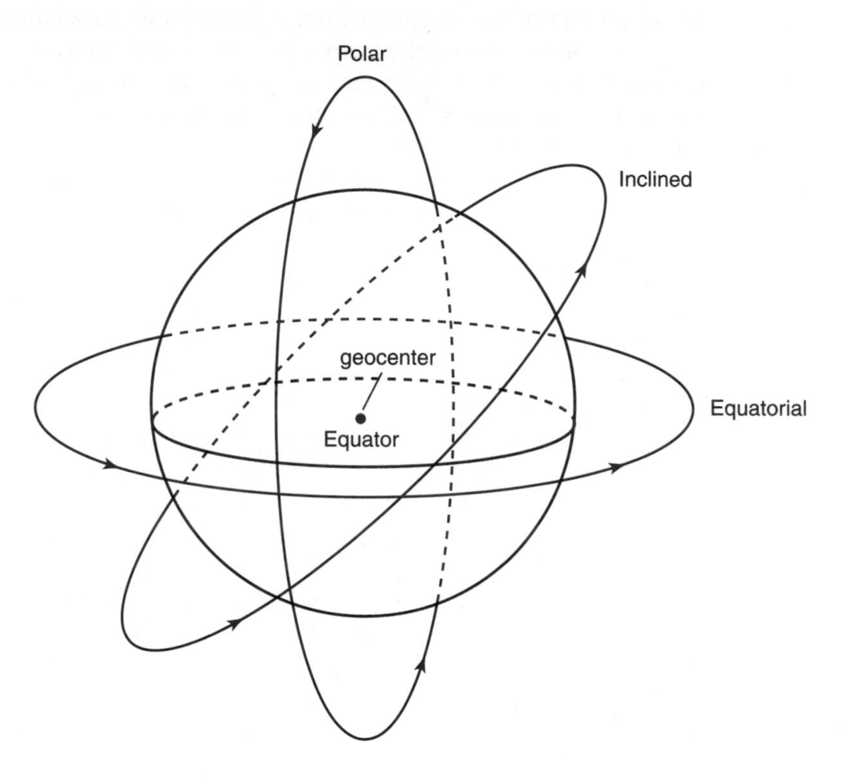

FIGURE 25-4 Satellite orbital patterns

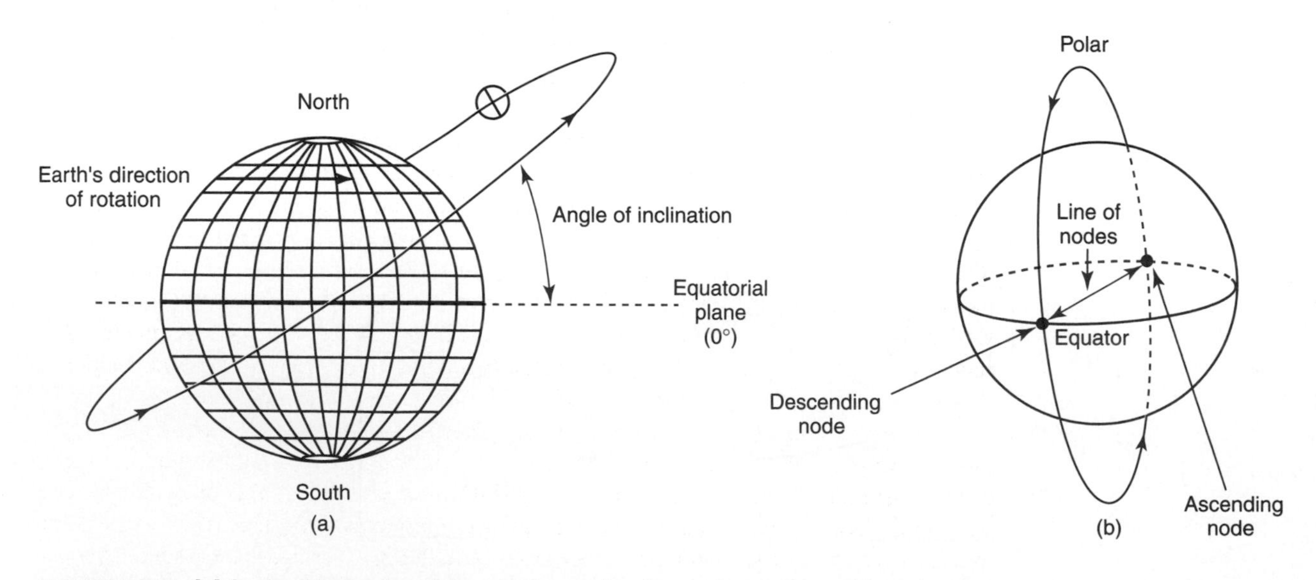

FIGURE 25-5 [a] Angle of inclination; [b] ascending node, descending node, and line of nodes

orbital plane of a satellite measured counterclockwise at the point in the orbit where it crosses the equatorial plane traveling from south to north. This point is called the *ascending node* and is shown in Figure 25-5b. The point where a polar or inclined orbit crosses the equatorial plane traveling from north to south is called the *descending node,* and the line joining the ascending and descending nodes through the center of Earth is called the *line of nodes.* Angles of inclination vary between 0° and 180°. To provide coverage to regions of high latitudes, inclined orbits are generally elliptical. Kepler's second law shows that the angular velocity of the satellite is slowest at its apogee. Therefore, the satellite remains visible for a longer period of time to the higher latitude regions if the apogee is placed above the high-latitude region.

An *equatorial orbit* is when the satellite rotates in an orbit directly above the equator, usually in a circular path. With an equatorial orbit, the angle of inclination is 0°, and there are no ascending or descending nodes and, hence, no line of nodes. All geosynchronous satellites are in equatorial orbits.

A *polar orbit* is when the satellite rotates in a path that takes it over the North and South Poles in an orbit perpendicular to the equatorial plane. Polar orbiting satellites follow a low-altitude path that is close to Earth and passes over and very close to both the North and South Poles. The angle of inclination of a satellite in a polar orbit is nearly 90°. It is interesting to note that 100% of Earth's surface can be covered with a single satellite in a polar orbit. Satellites in polar orbits rotate around Earth in a longitudinal orbit while Earth is rotating on its axis in a latitudinal rotation. Consequently, the satellite's radiation pattern is a diagonal line that forms a spiral around the surface of Earth that resembles a barber pole. As a result, every location on Earth lies within the radiation pattern of a satellite in a polar orbit twice each day.

Earth is not a perfect sphere, as it bulges at the equator. In fact, until the early 1800s, a 20,700-foot mountain in Ecuador called Volcan Chimborazo was erroneously thought to be the highest point on the planet. However, because of equatorial bulge, Volcan Chimborazo proved to be the farthest point from the center of the Earth. An important effect of the Earth's equatorial bulge is causing elliptical orbits to rotate in a manner that causes the apogee and perigee to move around the Earth. This phenomena is called *rotation of the line of apsides;* however, for an angle of inclination of 63.4°, the rotation of the line of apsides is zero. Thus, satellites required to have an apogee over a particular location are launched into orbit with an angle of inclination of 63.4°, which is referred to as the 63° slot.

One of the more interesting orbital satellite systems currently in use is the Commonwealth of Independent States (CIS) *Molniya* system of satellites, which is shown in Figure 25-6. The CIS is the former Soviet Union. Molniya can also be spelled *Molnya* and *Molnia,* which means "lightning" in Russian (in colloquial Russian, *Molniya* means "news flash"). *Molniya* satellites are used for government communications, telephone, television, and video.

The *Molniya* series of satellites use highly inclined elliptical orbits to provide service to the more northerly regions where antennas would have to be aimed too close to the horizon to detect signals from geostationary space vehicles rotating in an equatorial orbit. *Molniya* satellites have an apogee at about 40,000 km and a perigee at about 400 km. The apogee is reached while over the Northern Hemisphere and the perigee while over the Southern Hemisphere. The size of the ellipse was chosen to make its period exactly one-half a *sidereal day.* One sidereal day is the time it takes Earth to rotate back to the same constellation. The sidereal day for Earth is 23 hours and 56 minutes, slightly less than the time required for Earth to make one complete rotation around its own axis—24 hours. A sidereal day is sometimes called the *period* or *sidereal period.*

Because of its unique orbital pattern, the *Molniya* satellite is synchronous with the rotation of Earth. During a satellite's 12-hour orbit, it spends about 11 hours over the Northern Hemisphere. Three or more space vehicles follow each other in this orbit and *pass off* communications to each other so that continuous communications is possible while minimal earth station antenna tracking is necessary. Satellites with orbital patterns like *Molniya* are sometimes classified as having a highly elliptical orbit (HEO).

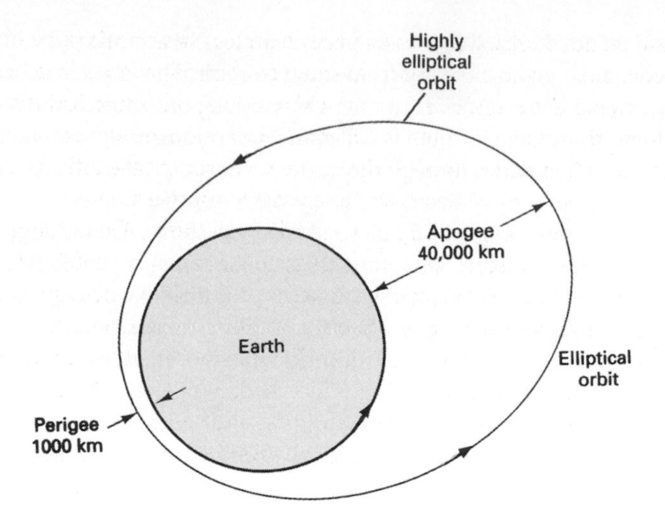

FIGURE 25-6 Soviet *Molniya* satellite orbit

25-5 GEOSYNCHRONOUS SATELLITES

As stated, geosynchronous satellites orbit Earth above the equator with the same angular velocity as Earth. Hence, geosynchronous (sometimes called *stationary* or *geostationary*) satellites appear to remain in a fixed location above one spot on Earth's surface. Since a geosynchronous satellite appears to remain in a fixed location, no special antenna tracking equipment is necessary—earth station antennas are simply pointed at the satellite. A single high-altitude geosynchronous satellite can provide reliable communications to approximately 40% of the earth's surface.

Satellites remain in orbit as a result of a balance between centrifugal and gravitational forces. If a satellite is traveling at too high a velocity, its centrifugal force will overcome Earth's gravitational pull, and the satellite will break out of orbit and escape into space. At lower velocities, the satellite's centrifugal force is insufficient, and gravity tends to pull the vehicle toward Earth. Obviously, there is a delicate balance between acceleration, speed, and distance that will exactly balance the effects of centrifugal and gravitational forces.

The closer to Earth a satellite rotates, the greater the gravitational pull and the greater the velocity required to keep it from being pulled to Earth. Low-altitude satellites orbiting 100 miles above Earth travel at approximately 17,500 mph. At this speed, it takes approximately 1.5 hours to rotate around Earth. Consequently, the time that a satellite is in line of sight of a particular earth station is 0.25 hour or less per orbit. Medium-altitude Earth-orbit satellites have a rotation period of between 5 and 12 hours and remain in line of sight of a particular earth station for between 2 and 4 hours per orbit. High-altitude earth-orbit satellites in geosynchronous orbits travel at approximately 6840 mph and complete one revolution of Earth in approximately 24 hours.

Geosynchronous orbits are circular; therefore, the speed of rotation is constant throughout the orbit. There is only one geosynchronous earth orbit; however, it is occupied by a large number of satellites. In fact, the geosynchronous orbit is the most widely used earth orbit for the obvious reason that satellites in a geosynchronous orbit remain in a fixed position relative to Earth and, therefore, do not have to be tracked by earth station antennas.

Ideally, geosynchronous satellites should remain stationary above a chosen location over the equator in an equatorial orbit; however, the sun and the moon exert gravitational forces, solar winds sweep past Earth, and Earth is not perfectly spherical. Therefore, these unbalanced forces cause geosynchronous satellites to drift slowly away from their assigned locations in a figure-eight excursion with a 24-hour period that follows a wandering path slightly above and below the equatorial plane. In essence, it occurs in a special type of inclined orbit sometimes called a *stationary inclined orbit.* Ground controllers must periodically adjust satellite positions to counteract these forces. If not, the excursion above and below the equator would build up at a rate of between 0.6° and 0.9° per year. In addition, geosynchronous satellites in an elliptical orbit also rift in an east or west direction as viewed from Earth. The process of maneuvering a satellite within a preassigned window is called *station keeping.*

There are several requirements for satellites in geostationary orbits. The first and most obvious is that geosynchronous satellites must have a 0° angle of inclination (i.e., the satellite vehicle must be orbiting directly above Earth's equatorial plane). The satellite must also be orbiting in the same direction as Earth's rotation (eastward—toward the morning sun) with the same angular (rotational) velocity—one revolution per day.

The semimajor axis of a geosynchronous earth orbit is the distance from a satellite revolving in the geosynchronous orbit to the center of Earth (i.e., the radius of the orbit measured from Earth's geocenter to the satellite vehicle). Using Kepler's third law as stated in Equation 25-2 with $A = 42241.0979$ and $P = 0.9972$, the semimajor axis α is

$$\begin{aligned}\alpha &= AP^{2/3}\\ &= (42241.0979)(0.9972)^{2/3}\\ &= 42{,}164 \text{ km}\end{aligned} \tag{25-3}$$

Hence, geosynchronous earth-orbit satellites revolve around Earth in a circular pattern directly above the equator 42,164 km from the center of Earth. Because Earth's equatorial radius is approximately 6378 km, the height above mean sea level (h) of a satellite in a geosynchronous orbit around Earth is

$$\begin{aligned}h &= 42{,}164 \text{ km} - 6378 \text{ km}\\ &= 35{,}786 \text{ km}\end{aligned}$$

or approximately 22,300 miles above Earth's surface.

25-5-1 Geosynchronous Satellite Orbital Velocity

The circumference (C) of a geosynchronous orbit is

$$\begin{aligned}C &= 2\pi(42{,}164 \text{ km})\\ &= 264{,}790 \text{ km}\end{aligned}$$

Therefore, the velocity (v) of a geosynchronous satellite is

$$\begin{aligned}v &= \frac{264{,}790 \text{ km}}{24 \text{ hr}}\\ &= 11{,}033 \text{ km/hr}\end{aligned}$$

or

$$v \approx 6840 \text{ mph}$$

25-5-2 Round-Trip Time Delay of Geosynchronous Satellites

The round-trip propagation delay between a satellite and an earth station located directly below it is

$$\begin{aligned}t &= \frac{d}{c}\\ &= \frac{2(35{,}768 \text{ km})}{3 \times 10^5 \text{ km/s}}\\ &= 238 \text{ ms}\end{aligned}$$

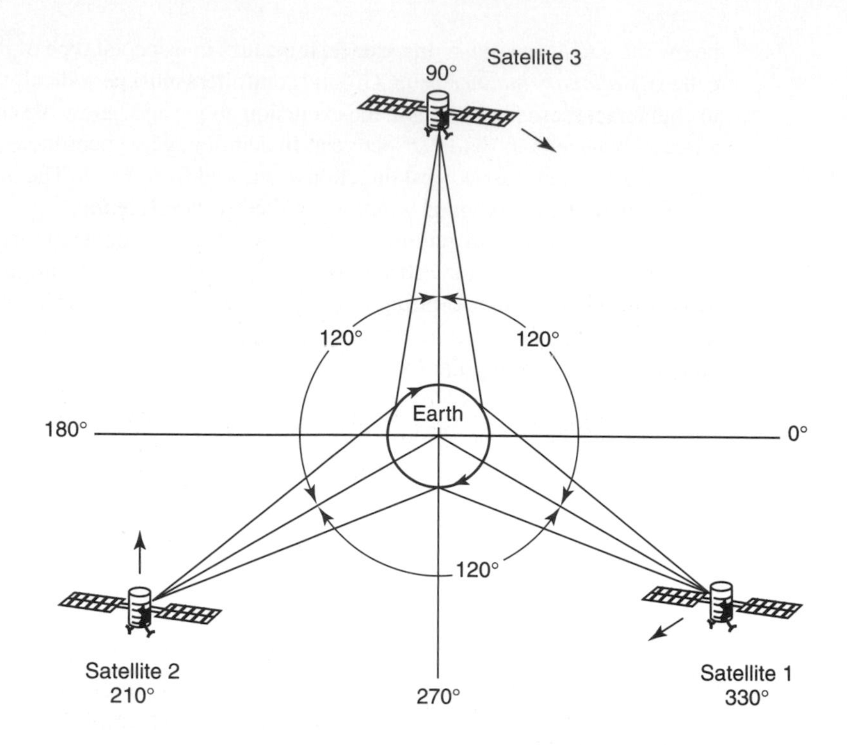

FIGURE 25-7 Three geosynchronous satellites in Clarke orbits

Including the time delay within the earth station and satellite equipment, it takes more than a quarter of a second for an electromagnetic wave to travel from an earth station to a satellite and back when the earth station is located at a point on Earth directly below the satellite. For earth stations located at more distant locations, the propagation delay is even more substantial and can be significant with two-way telephone conversations or data transmissions.

25-5-3 Clarke Orbit

A geosynchronous earth orbit is sometimes referred to as the *Clarke orbit* or *Clarke belt,* after Arthur C. Clarke, who first suggested its existence in 1945 and proposed its use for communications satellites. Clarke was an engineer, a scientist, and a science fiction author who wrote several books including *2001: A Space Odyssey.* The Clarke orbit meets the concise set of specifications for geosynchronous satellite orbits: (1) be located directly above the equator, (2) travel in the same direction as Earth's rotation at 6840 mph, (3) have an altitude of 22,300 miles above Earth, and (4) complete one revolution in 24 hours. As shown in Figure 25-7, three satellites in Clarke orbits separated by 120° in longitude can provide communications over the entire globe except the polar regions.

An international agreement initially mandated that all satellites placed in the Clarke orbit must be separated by at least 1833 miles. This stipulation equates to an angular separation of 4° or more, which limits the number of satellite vehicles in a geosynchronous earth orbit to less than 100. Today, however, international agreements allow satellites to be placed much closer together. Figure 25-8 shows the locations of several satellites in geosynchronous orbit around Earth.

25-5-4 Advantages and Disadvantages of Geosynchronous Satellites

The advantages of geosynchronous satellites are as follows:

1. Geosynchronous satellites remain almost stationary in respect to a given earth station. Consequently, expensive tracking equipment is not required at the earth stations.

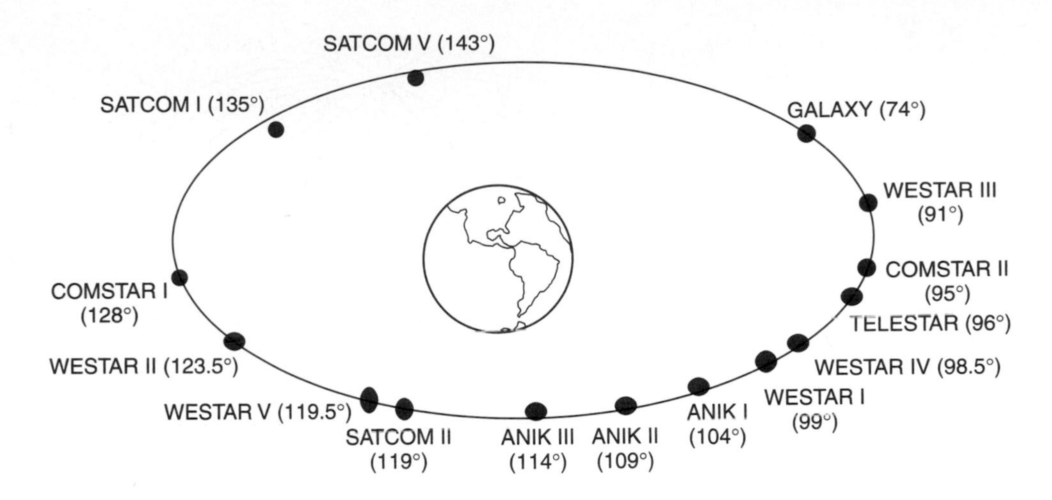

FIGURE 25-8 Satellites in geosynchronous earth orbits

2. Geosynchronous satellites are available to all earth stations within their *shadow* 100% of the time. The shadow of a satellite includes all the earth stations that have a line-of-sight path to it and lie within the radiation pattern of the satellite's antennas.
3. There is no need to switch from one geosynchronous satellite to another as they orbit overhead. Consequently, there are no transmission breaks due to switching times.
4. The effects of Doppler shift are negligible.

The disadvantages of geosynchronous satellites are as follows:

1. Geosynchronous satellites require sophisticated and heavy propulsion devices onboard to keep them in a fixed orbit.
2. High-altitude geosynchronous satellites introduce much longer propagation delays. The round-trip propagation delay between two earth stations through a geosynchronous satellite is between 500 ms and 600 ms.
3. Geosynchronous satellites require higher transmit powers and more sensitive receivers because of the longer distances and greater path losses.
4. High-precision spacemanship is required to place a geosynchronous satellite into orbit and to keep it there.

25-6 ANTENNA LOOK ANGLES

To optimize the performance of a satellite communications system, the direction of maximum gain of an earth station antenna (sometimes referred to as the *boresight*) must be pointed directly at the satellite. To ensure that the earth station antenna is aligned, two angles must be determined: the *azimuth* and the *elevation angle.* Azimuth angle and elevation angle are jointly referred to as the antenna *look angles.* With geosynchronous satellites, the look angles of earth station antennas need to be adjusted only once, as the satellite will remain in a given position permanently, except for occasional minor variations.

The location of a satellite is generally specified in terms of latitude and longitude similar to the way the location of a point on Earth is described; however, because a satellite is orbiting many miles above the Earth's surface, it has no latitude or longitude. Therefore, its location is identified by a point on the surface of earth directly below the satellite. This point is called the *subsatellite point* (SSP), and for geosynchronous satellites the SSP must fall on the equator. Subsatellite points and earth station locations are specified using standard

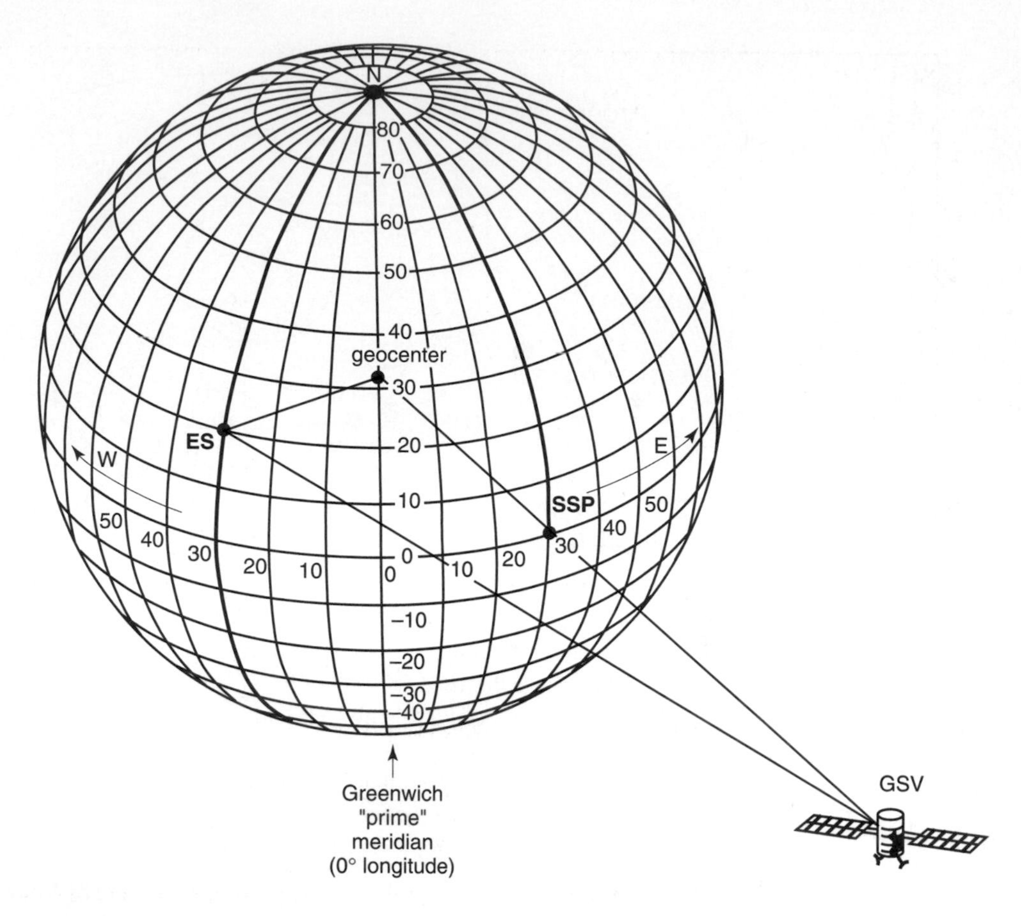

FIGURE 25-9 Geosynchronous satellite position, subsatellite point, and Earth longitude and latitude coordinate system

latitude and longitude coordinates. The standard convention specifies angles of longitude between 0° and 180° either east or west of the Greenwich prime meridian. Latitudes in the Northern Hemisphere are angles between 0° and 90°N and latitudes in the Southern Hemisphere are angles between 0° and 90°S. Since geosynchronous satellites are located directly above the equator, they all have a 0° latitude. Hence, geosynchronous satellite locations are normally given in degrees longitude east or west of the Greenwich meridian (for example, 122°W or 78°E). Figure 25-9 shows the position of a hypothetical geosynchronous satellite vehicle (GSV), its respective subsatellite point (SSP), and an arbitrarily selected earth station (ES) all relative to Earth's geocenter. The SSP for the satellite shown in the figure is 30°E longitude and 0° latitude. The earth station has a location of 30°W longitude and 20°N latitude.

25-6-1 Angle of Elevation

Angle of elevation (sometimes called *elevation angle*) is the vertical angle formed between the direction of travel of an electromagnetic wave radiated from an earth station antenna pointing directly toward a satellite and the horizontal plane. The smaller the angle of elevation, the greater the distance a propagated wave must pass through Earth's atmosphere. As with any wave propagated through Earth's atmosphere, it suffers absorption and may also be severely contaminated by noise. Consequently, if the angle of elevation is too small and the distance the wave travels through Earth's atmosphere is too long, the wave may

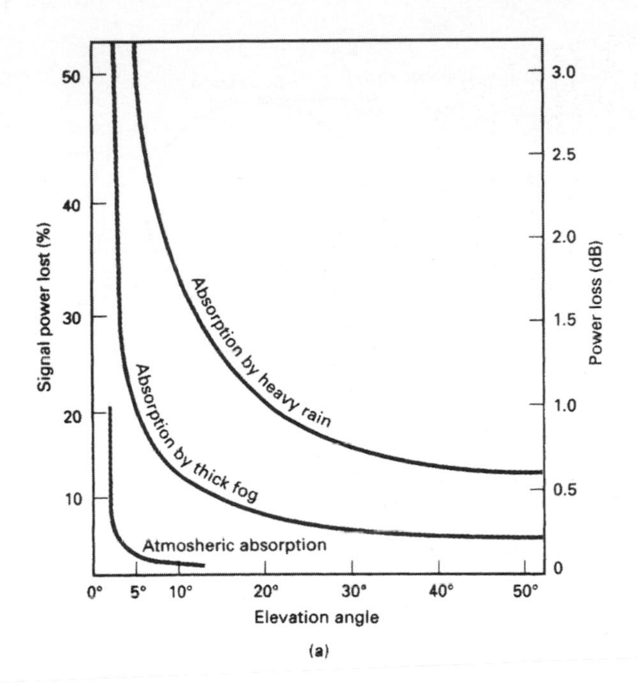

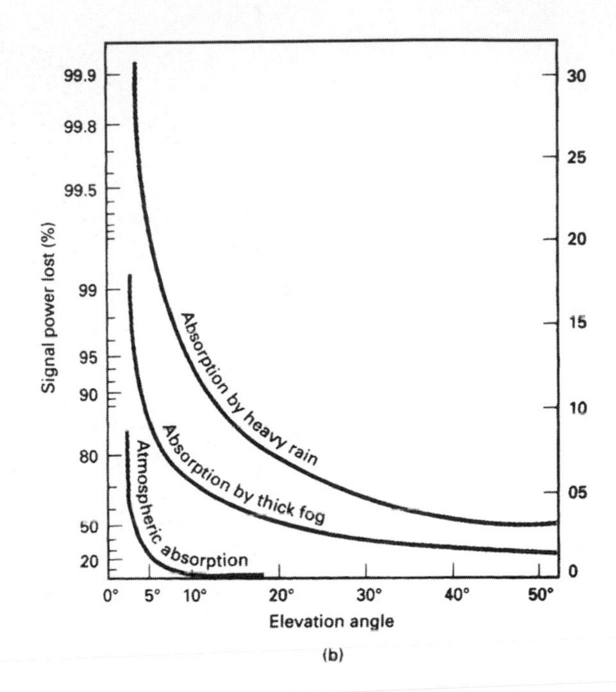

FIGURE 25-10 Attenuation due to atmospheric absorption: [a] 6/4-GHz band; [b] 14/12-GHz band

deteriorate to the extent that it no longer provides acceptable transmission quality. Generally, 5° is considered as the minimum acceptable angle of elevation. Figure 25-10 shows how the angle of elevation affects the signal strength of a propagated electromagnetic wave due to normal atmospheric absorption, absorption due to thick fog, and absorption due to heavy rainfall. It can be seen that the 14/12-GHz band shown in Figure 25-10b is more severely affected than the 6/4-GHz band shown in Figure 25-10a because of the smaller wavelengths associated with the higher frequencies. The figure also shows that at elevation angles less than 5°, the amount of signal power lost increases significantly. Figure 25-10b illustrates angle of elevation of an earth station antenna with respect to a horizontal plane.

25-6-2 Azimuth Angle

Azimuth is the horizontal angular distance from a reference direction, either the southern or northern most point of the horizon. *Azimuth angle* is defined as the horizontal pointing angle of an earth station antenna. For navigation purposes, azimuth angle is usually measured in a clockwise direction in degrees from true north. However, for satellite earth stations in the Northern Hemisphere and satellite vehicles in geosynchronous orbits, azimuth angle is generally referenced to true south (i.e., 180°). Figure 25-11a illustrates the azimuth angle referenced to due north (0°) and due south (180°), and Figure 25-11c shows elevation angle and azimuth of an earth station antenna relative to a satellite.

Angle of elevation and azimuth angle both depend on the latitude of the earth station and the longitude of both the earth station and the orbiting satellite. For a geosynchronous satellite in an equatorial orbit, the procedure for determining angle of elevation and azimuth is as follows: From a good map, determine the longitude and latitude of the earth station.

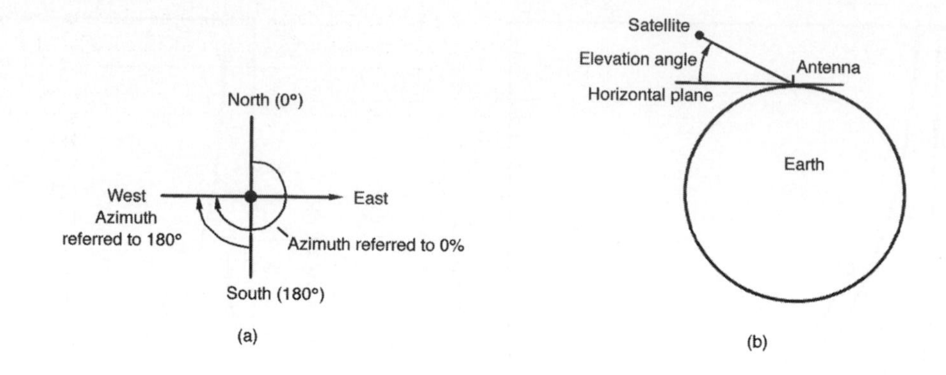

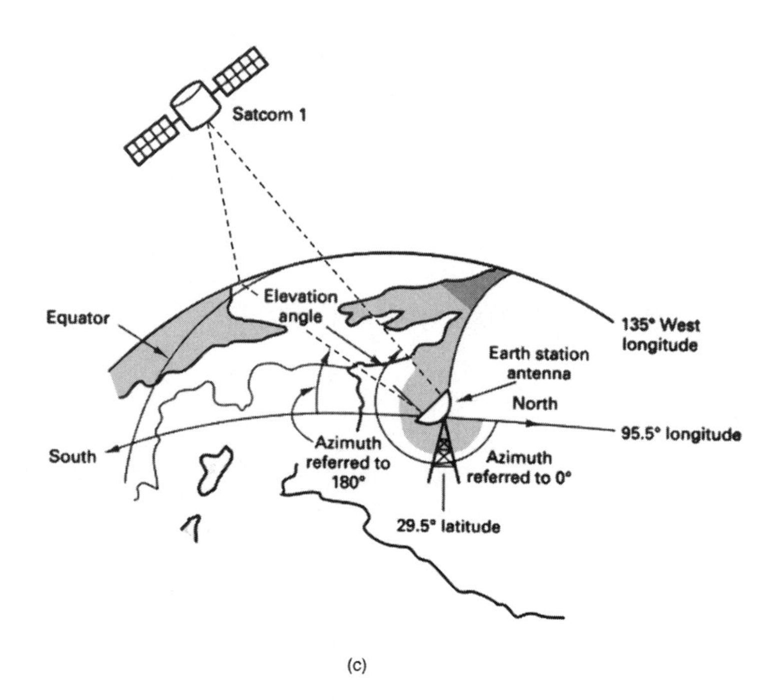

FIGURE 25-11 Azimuth and angle of elevation, "lookangles"

From Table 25-1, determine the longitude of the satellite of interest. Calculate the difference, in degrees (ΔL), between the longitude of the satellite and the longitude of the earth station. Then from Figure 25-12 determine the azimuth angle, and from Figure 25-13 determine the elevation angle. Figures 25-12 and 25-13 are for geosynchronous satellites in equatorial orbits.

Example 25-1

An earth station is located in Houston, Texas, which has a longitude of 95.5°W and a latitude of 29.5°N. The satellite of interest is RCA's *Satcom 1,* which has a longitude of 135°W. Determine the azimuth angle and elevation angle for the earth station.

Solution First determine the difference between the longitude of the earth station and the satellite vehicle:

$$\begin{aligned} \Delta L &= 135° - 95.5° \\ &= 39.5° \end{aligned}$$

Table 25-1 Longitudinal Position of Several Current Synchronous Satellites Parked in an Equatorial Arc[a]

Satellite	Longitude (°W)
Satcom I	135
Satcom II	119
Satcom V	143
Satcom C1	137
Satcom C3	131
Anik 1	104
Anik 2	109
Anik 3	114
Anik C1	109.25
Anik C2	109.15
Anik C3	114.9
Anik E1	111.1
Anik E2	107.3
Westar I	99
Westar II	123.5
Westar III	91
Westar IV	98.5
Westar V	119.5
Mexico	116.5
Galaxy III	93.5
Galaxy IV	99
Galaxy V	125
Galaxy VI	74
Telstar	96
Comstar I	128
Comstar II	95
Comstar D2	76.6
Comstar D4	75.4
Intelsat 501	268.5
Intelsat 601	27.5
Intelsat 701	186

[a]0° latitude.

Locate the intersection of ΔL and the earth station's latitude on Figure 25-12. From the figure, the azimuth angle is approximately 59° west of south (i.e., west of 180°). On Figure 25-13, locate the intersection of ΔL and the earth station's latitude. The angle of elevation is approximately 35°.

25-6-3 Limits of Visibility

For an earth station in any given location, the Earth's curvature establishes the *limits of visibility* (i.e., *line-of-sight limits*), which determine the farthest satellite away that can be seen looking east or west of the earth station's longitude. Theoretically, the maximum line-of-sight distance is achieved when the earth station's antenna is pointing along the horizontal (zero elevation angle) plane. In practice, however, the noise picked up from Earth and the signal attenuation from Earth's atmosphere at zero elevation angle is excessive. Therefore, an elevation angle of 5° is generally accepted as being the minimum usable elevation angle. The limits of visibility depend in part on the antenna's elevation and the earth station's longitude and latitude.

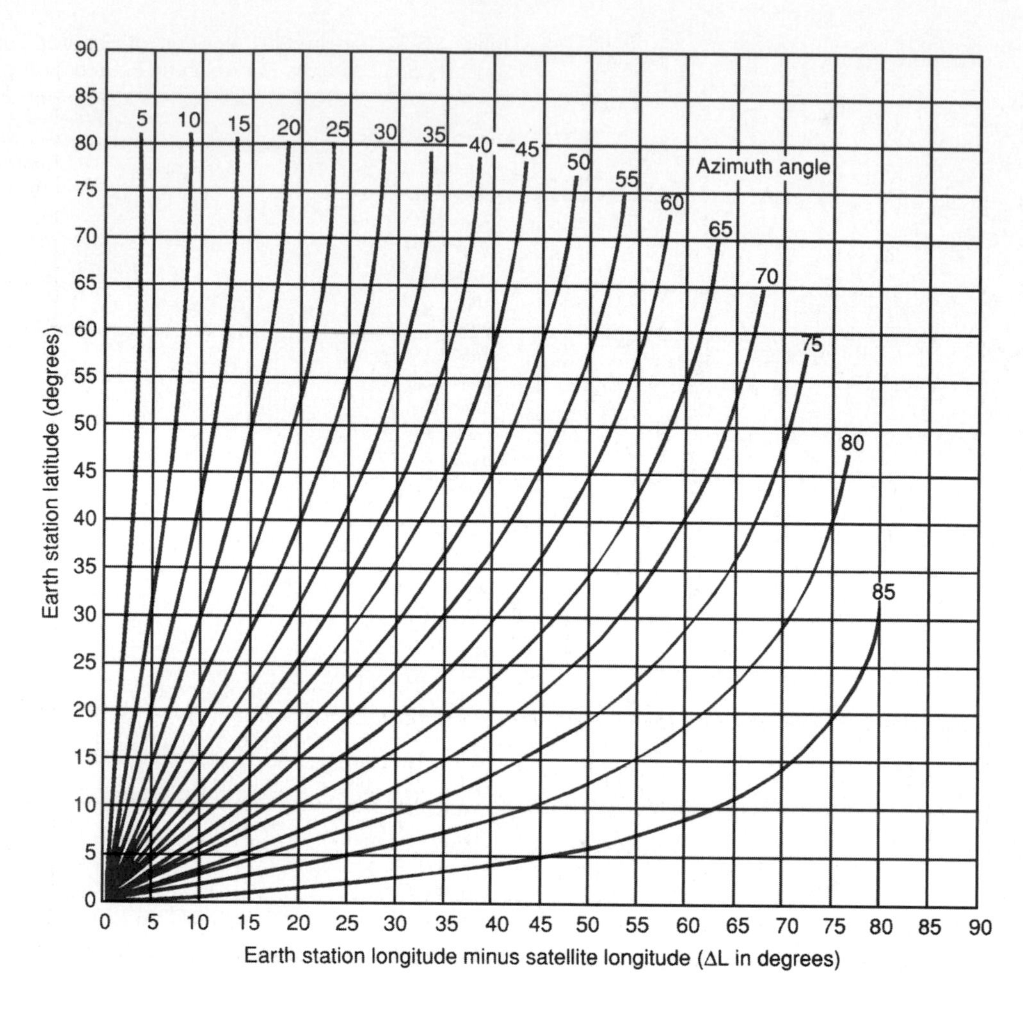

FIGURE 25-12 Azimuth angles for earth stations located in the northern hemisphere referenced to 180 degrees

25-7 SATELLITE CLASSIFICATIONS, SPACING, AND FREQUENCY ALLOCATION

The two primary classifications for communications satellites are *spinners* and *three-axis stabilizer satellites.* A spinner satellite uses the angular momentum of its spinning body to provide roll and yaw stabilization. With a three-axis stabilizer, the body remains fixed relative to Earth's surface, while an internal subsystem provides roll and yaw stabilization. Figure 25-14 shows the two main classifications of communications satellites.

Geosynchronous satellites must share a limited space and frequency spectrum within a given arc of a geostationary orbit. Each communications satellite is assigned a longitude in the geostationary arc approximately 22,300 miles above the equator. The position in the slot depends on the communications frequency band used. Satellites operating at or near the same frequency must be sufficiently separated in space to avoid interfering with each other (Figure 25-15). There is a realistic limit to the number of satellite structures that can be

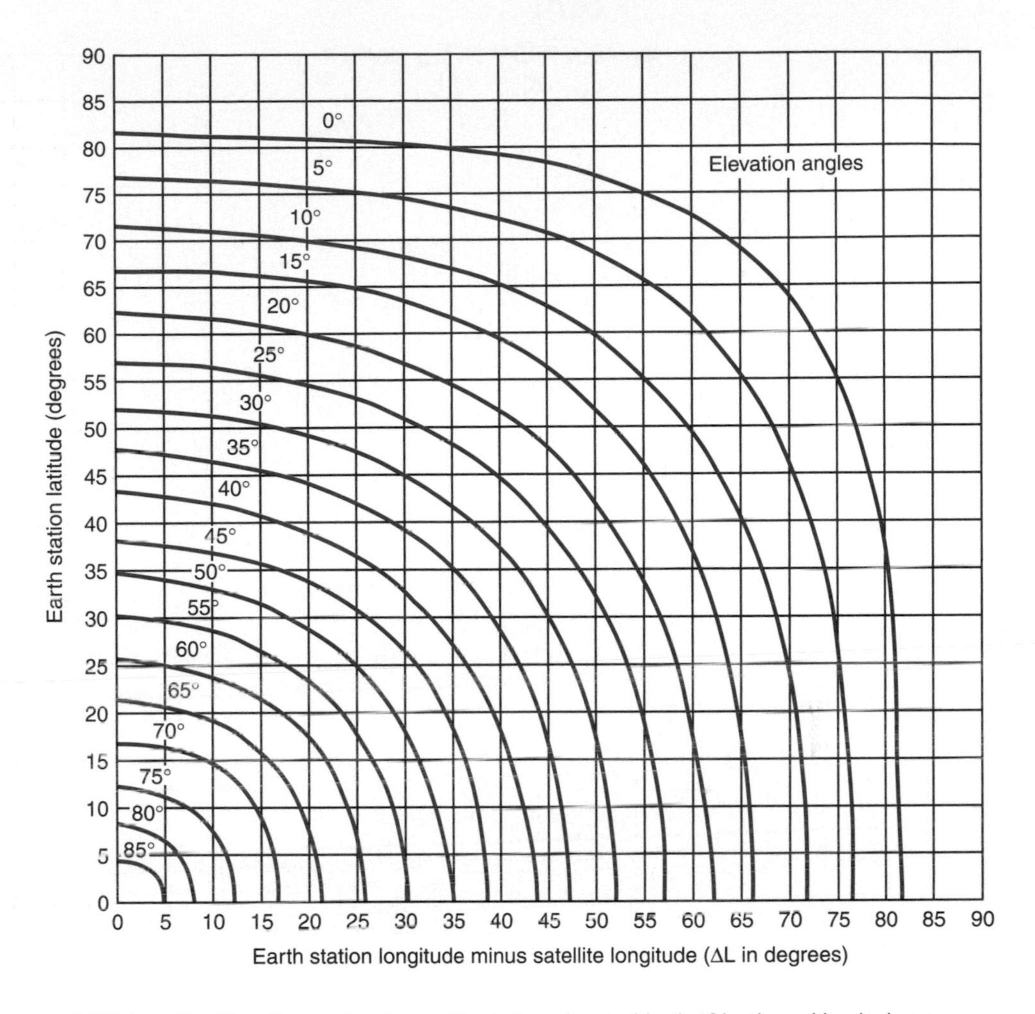

FIGURE 25-13 Elevation angles for earth stations located in the Northern Hemisphere

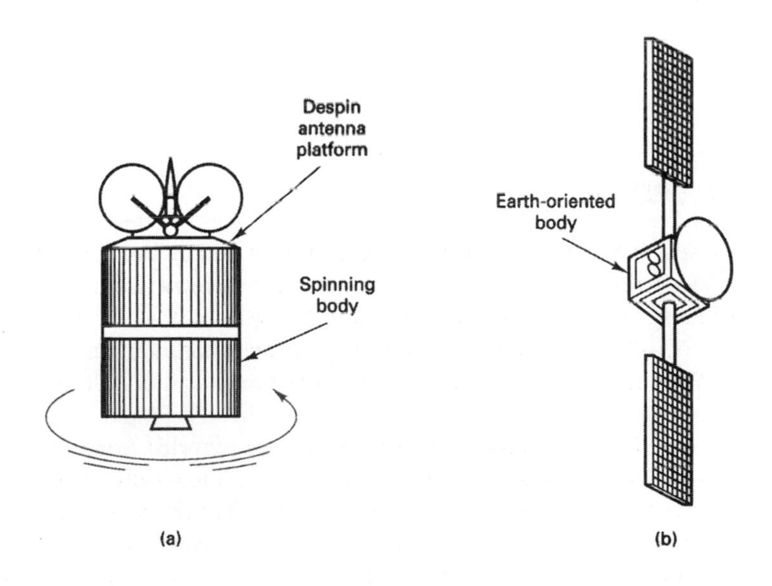

FIGURE 25-14 Satellite classes: [a] spinner; [b] three-axis stabilizer

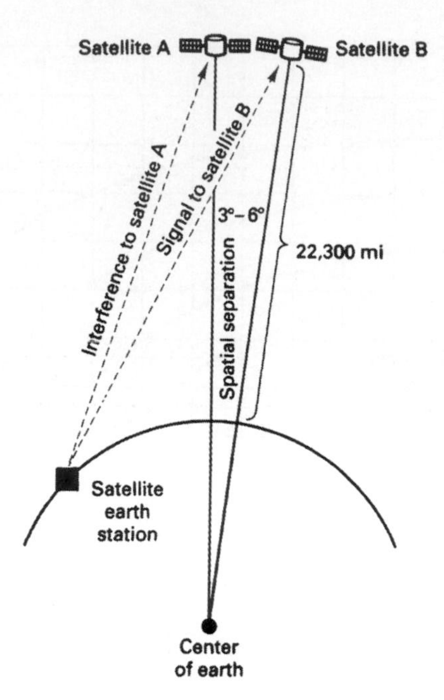

FIGURE 25-15 Spatial separation of satellites in geosynchronous orbit

stationed (*parked*) within a given area in space. The required *spatial separation* is dependent on the following variables:

1. Beamwidths and side lobe radiation of both the earth station and satellite antennas
2. RF carrier frequency
3. Encoding or modulation technique used
4. Acceptable limits of interference
5. Transmit carrier power

Generally, 1° to 4° of spatial separation is required, depending on the variables stated previously.

The most common carrier frequencies used for satellite communications are the 6/4-GHz and 14/12-GHz bands. The first number is the uplink (earth station-to-transponder) frequency, and the second number is the downlink (transponder-to-earth station) frequency. Different uplink and downlink frequencies are used to prevent ringaround from occurring (Chapter 24). The higher the carrier frequency, the smaller the diameter required of an antenna for a given gain. Most domestic satellites use the 6/4-GHz band. Unfortunately, this band is also used extensively for terrestrial microwave systems. Care must be taken when designing a satellite network to avoid interference from or with established microwave links.

Certain positions in the geosynchronous orbit are in higher demand than the others. For example, the mid-Atlantic position, which is used to interconnect North America and Europe, is in exceptionally high demand; the mid-Pacific position is another.

The frequencies allocated by the World Administrative Radio Conference (WARC) are summarized in Figure 25-16. Table 25-2 shows the bandwidths available for various services in the United States. These services include *fixed point* (between earth stations located at fixed geographical points on Earth), *broadcast* (wide-area coverage), *mobile* (ground-to-aircraft, ships, or land vehicles), and *intersatellite* (satellite-to-satellite cross-links).

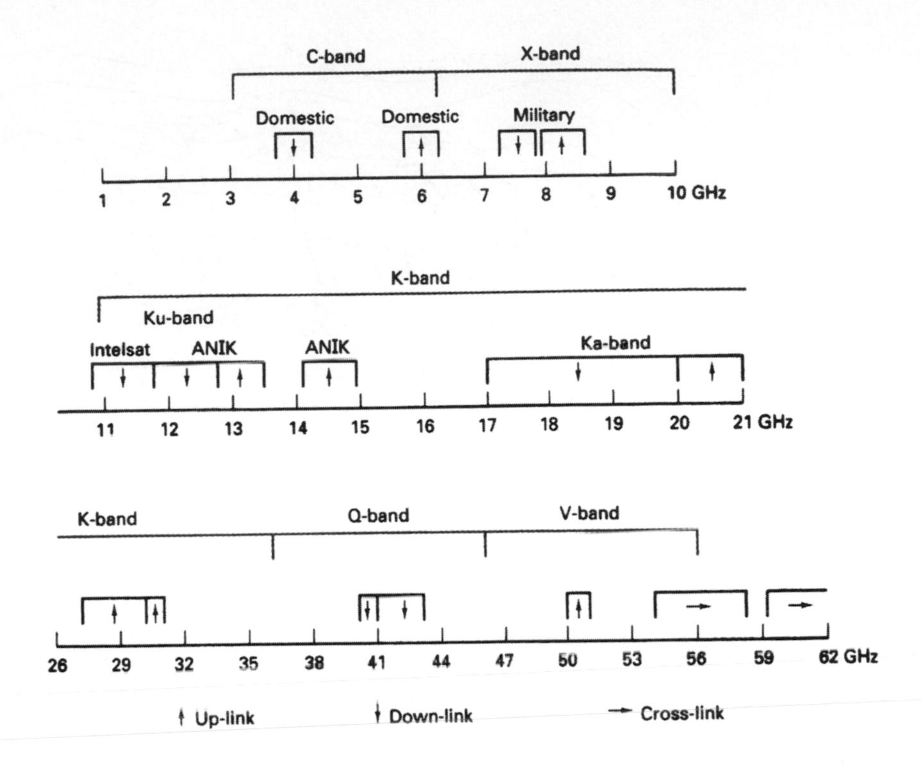

FIGURE 25-16 WARC satellite frequency assignments

Table 25-2 Satellite Bandwidths Available in the United States

	Frequency Band (GHz)			
Band	Uplink	Cross-Link	Downlink	Bandwidth (MHz)
C	5.9–6.4		3.7–4.2	500
X	7.9–8.4		7.25–7.75	500
Ku	14–14.5		11.7–12.2	500
Ka	27–30		17–20	—
	30–31		20–21	—
Q	—		40–41	1000
	—		41–43	2000
V	50–51		—	1000
(ISL)		54–58		3900
		59–64		5000

25-8 SATELLITE ANTENNA RADIATION PATTERNS: FOOTPRINTS

The area on Earth covered by a satellite depends on the location of the satellite in its orbit, its carrier frequency, and the gain of its antenna. Satellite engineers select the antenna and carrier frequency for a particular spacecraft to concentrate the limited transmitted power on a specific area of Earth's surface. The geographical representation of a satellite antenna's radiation pattern is called a *footprint* or sometimes a *footprint map.* In essence, a footprint of a satellite is the area on Earth's surface that the satellite can receive from or transmit to.

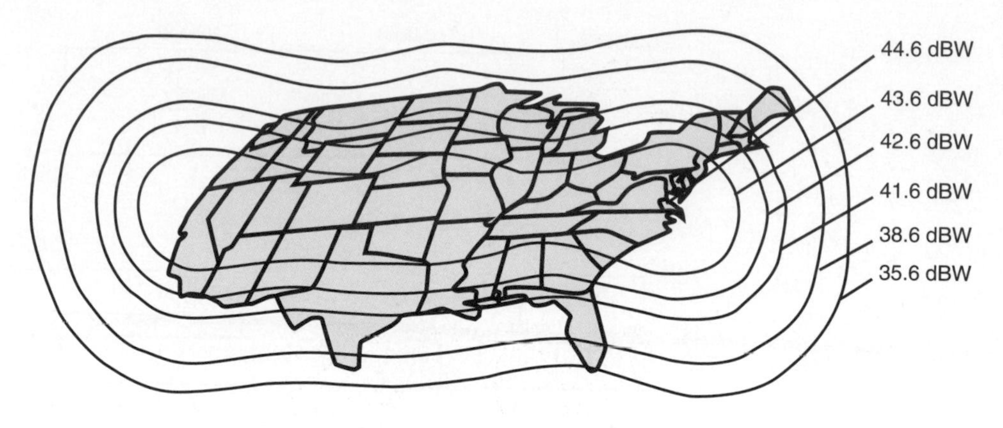

FIGURE 25-17 Satellite antenna radiation patterns (footprints)

The shape of a satellite's footprint depends on the satellite orbital path, height, and the type of antenna used. The higher the satellite, the more of the Earth's surface it can cover. A typical satellite footprint is shown in Figure 25-17. The contour lines represent limits of equal receive power density.

Downlink satellite antennas broadcast microwave-frequency signals to a selected geographic region within view (line of sight) of the spacecraft. The effective power transmitted is called *effective isotropic radiated power* (EIRP) and is generally expressed in dBm or dBW. A footprint map is constructed by drawing continuous lines between all points on a map with equal EIRPs. A distinctive footprint map is essentially a series of contour lines superimposed on a geographical map of the region served. A different footprint could exist for each beam from each communications satellite.

The pattern of the contour lines and power levels of a footprint are determined by precise details of the downlink antenna design as well as by the level of microwave power generated by each onboard channel. Although each transponder is a physically separate electronic circuit, signals from multiple transponders are typically downlinked through the same antenna. As might be expected, receive power levels are higher in areas targeted by the downlink antenna boresight and weaker in off-target areas. A receive antenna dish near the edge of a satellite coverage area must be larger than those located at or near the center of the footprint map. Extremely large-diameter earth station antennas are necessary for reception of satellite broadcasts in geographic areas located great distances from the downlink antenna boresight.

Characteristically, there are variations in footprint maps among satellites. For example, European Ku-band spacecraft generally have footprint radiation patterns that are circularly symmetric with power levels that decrease linearly in areas removed progressively further from the center of the satellite's boresight. American C-band satellites typically have relatively flat power levels over the region of coverage with fairly sharp drop-offs in power beyond the edges. Recently launched satellites such as the American DBS-1 (direct-broadcast satellites) have employed more sophisticated beam-shaping downlink antennas that permit designers to shape footprints to reach only specified targeted areas, hence not wasting power in nontargeted areas.

It is possible to design satellite downlink antennas that can broadcast microwave signals to cover areas on Earth ranging in size from extremely small cities to as much as 42% of the Earth's surface. The size, shape, and orientation of a satellite downlink antenna and the power generated by each transponder determine geographic coverage and EIRPs. Radiation patterns from a satellite antenna are generally categorized as either *spot, zonal, hemispherical,* or *earth* (global). The radiation patterns are shown in Figure 25-18.

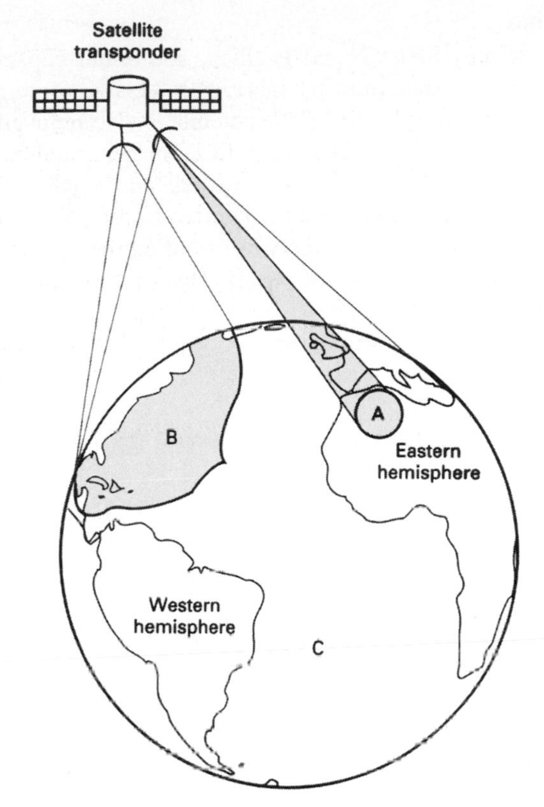

FIGURE 25-18 Beams: [a] spot; [b] zonal; [c] earth

25-8-1 Spot and Zonal Beams

The smallest beams are *spot beams* followed by *zonal beams.* Spot beams concentrate their power to very small geographical areas and, therefore, typically have proportionately higher EIRPs than those targeting much larger areas because a given output power can be more concentrated. Spot and zonal beams blanket less than 10% of the Earth's surface. The higher the downlink frequency, the more easily a beam can be focused into a smaller spot pattern. For example, the new breed of high-power Ku-band satellites can have multiple spot beams that relay the same frequencies by transmitting different signals to areas within a given country. In general, most Ku-band footprints do not blanket entire continental areas and have a more limited geographic coverage than their C-band counterparts. Therefore, a more detailed knowledge of the local EIRP is important when attempting to receive broadcasts from Ku-band satellite transmissions.

25-8-2 Hemispherical Beams

Hemispherical downlink antennas typically target up to 20% of the Earth's surface and, therefore, have EIRPs that are 3 dB or 50% lower than those transmitted by spot beams that typically cover only 10% of the Earth's surface.

25-8-3 Earth [Global] Beams

The radiation patterns of *earth coverage* antennas have a beamwidth of approximately 17° and are capable of covering approximately 42% of Earth's surface, which is the maximum view of any one geosynchronous satellite. Power levels are considerably lower with earth beams than with spot, zonal, or hemispherical beams, and large receive dishes are necessary to adequately detect video, audio, and data broadcasts.

25-8-4 Reuse

When an allocated frequency band is filled, additional capacity can be achieved by *reuse* of the frequency spectrum. By increasing the size of an antenna (i.e., increasing the antenna gain), the beamwidth of the antenna is also reduced. Thus, different beams of the same frequency can be directed to different geographical areas of Earth. This is called *frequency reuse.* Another method of frequency reuse is to use *dual polarization.* Different information signals can be transmitted to different earth station receivers using the same band of frequencies simply by orienting their electromagnetic polarizations in an orthogonal manner (90° out of phase). Dual polarization is less effective because Earth's atmosphere has a tendency to reorient or repolarize an electromagnetic wave as it passes through. Reuse is simply another way to increase the capacity of a limited bandwidth.

25-9 SATELLITE SYSTEM LINK MODELS

Essentially, a satellite system consists of three basic sections: an uplink, a satellite transponder, and a downlink.

25-9-1 Uplink Model

The primary component within the *uplink* section of a satellite system is the earth station transmitter. A typical earth station transmitter consists of an IF modulator, an IF-to-RF microwave up-converter, a high-power amplifier (HPA), and some means of bandlimiting the final output spectrum (i.e., an output bandpass filter). Figure 25-19 shows the block diagram of a satellite earth station transmitter. The IF modulator converts the input baseband signals to either an FM-, a PSK-, or a QAM-modulated intermediate frequency. The up-converter (mixer and bandpass filter) converts the IF to an appropriate RF carrier frequency. The HPA provides adequate gain and output power to propagate the signal to the satellite transponder. HPAs commonly used are klystons and traveling-wave tubes.

25-9-2 Transponder

A typical *satellite transponder* consists of an input bandlimiting device (BPF), an input *low-noise amplifier* (LNA), a *frequency translator,* a low-level power amplifier, and an output bandpass filter. Figure 25-20 shows a simplified block diagram of a satellite transponder. This transponder is an RF-to-RF repeater. Other transponder configurations are IF and baseband repeaters similar to those used in microwave repeaters. In Figure 25-20,

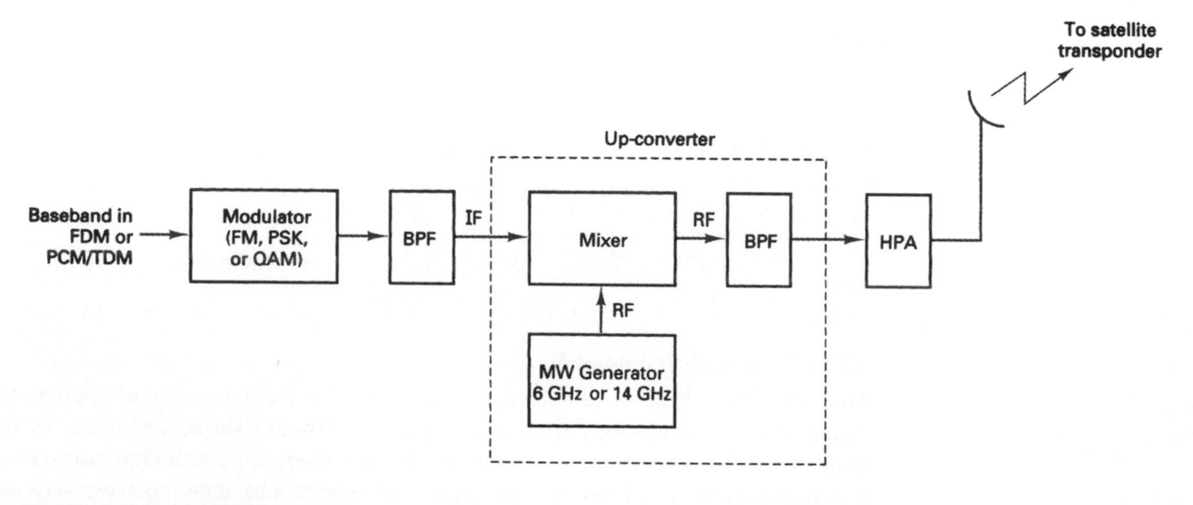

FIGURE 25-19 Satellite uplink model

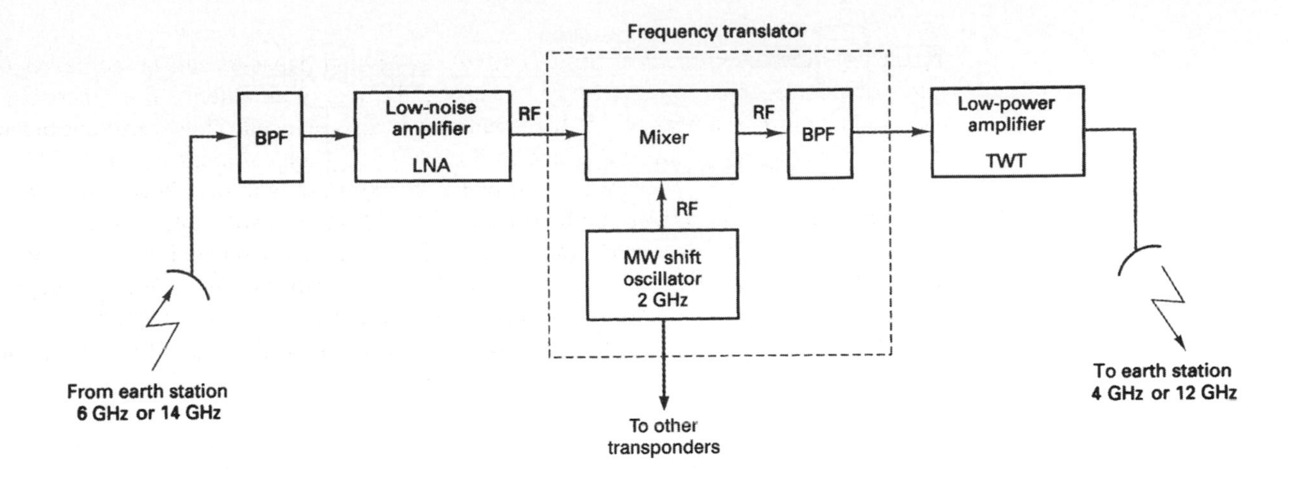

FIGURE 25-20 Satellite transponder

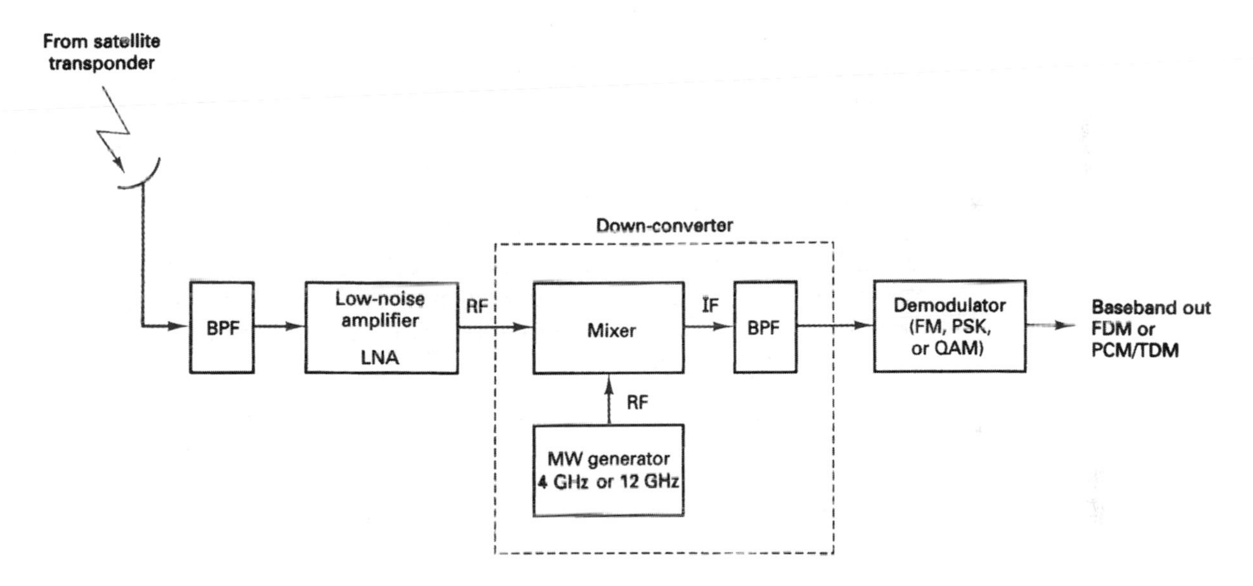

FIGURE 25-21 Satellite downlink model

the input BPF limits the total noise applied to the input of the LNA. (A common device used as an LNA is a tunnel diode.) The output of the LNA is fed to a frequency translator (a shift oscillator and a BPF), which converts the high-band uplink frequency to the low-band downlink frequency. The low-level power amplifier, which is commonly a traveling-wave tube, amplifies the RF signal for transmission through the downlink to earth station receivers. Each RF satellite channel requires a separate transponder.

25-9-3 Downlink Model

An earth station receiver includes an input BPF, an LNA, and an RF-to-IF down-converter. Figure 25-21 shows a block diagram of a typical earth station receiver. Again, the BPF limits the input noise power to the LNA. The LNA is a highly sensitive, low-noise device, such as a tunnel diode amplifier or a parametric amplifier. The RF-to-IF down-converter is a mixer/bandpass filter combination that converts the received RF signal to an IF frequency.

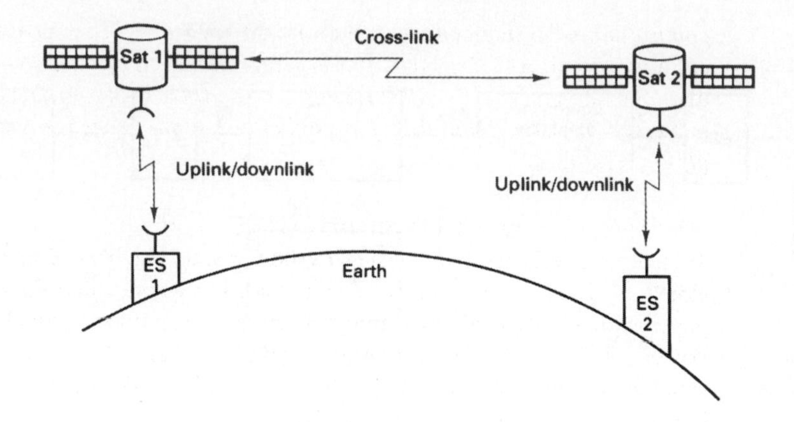

FIGURE 25-22 Intersatellite link

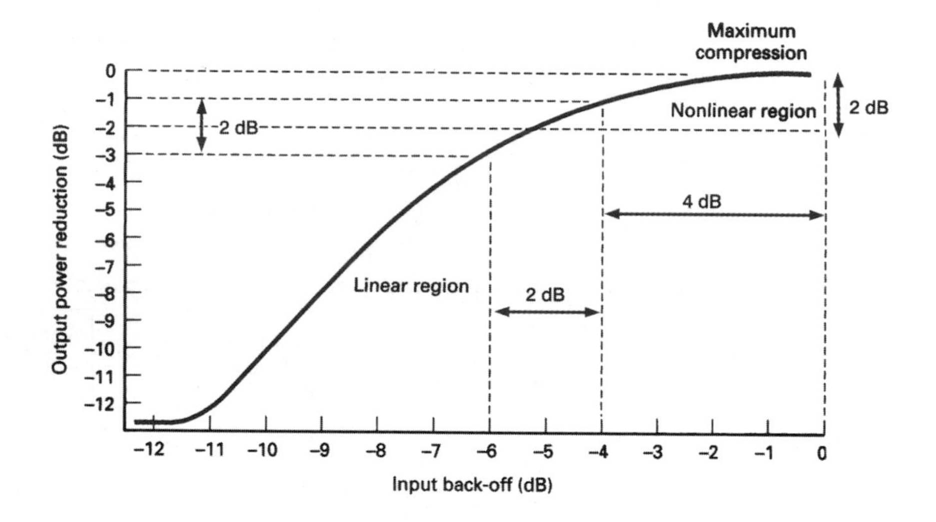

FIGURE 25-23 HPA input/output characteristic curve

25-9-4 Cross-Links

Occasionally, there is an application where it is necessary to communicate between satellites. This is done using *satellite cross-links* or *intersatellite links* (ISLs), shown in Figure 25-22. A disadvantage of using an ISL is that both the transmitter and the receiver are *space bound.* Consequently, both the transmitter's output power and the receiver's input sensitivity are limited.

25-10 SATELLITE SYSTEM PARAMETERS

25-10-1 Back-Off Loss

High-power amplifiers used in earth station transmitters and the traveling-wave tubes typically used in satellite transponders are *nonlinear devices;* their gain (output power versus input power) is dependent on input signal level. A typical input/output power characteristic curve is shown in Figure 25-23. It can be seen that as the input power is reduced by 4 dB, the output power is reduced by only 1 dB. There is an obvious *power compression.* To reduce the amount of intermodulation distortion caused by the nonlinear amplification of the HPA, the input power must be reduced (*backed off*) by several dB. This allows the HPA to operate in a more *linear* region. The amount the output level is backed off from rated levels is equivalent to a loss and is appropriately called *back-off loss* (L_{bo}).

25-10-2 Transmit Power and Bit Energy

To operate as efficiently as possible, a power amplifier should be operated as close as possible to saturation. The *saturated output power* is designated $P_{o(sat)}$ or simply P_t. The output power of a typical satellite earth station transmitter is much higher than the output power from a terrestrial microwave power amplifier. Consequently, when dealing with satellite systems, P_t is generally expressed in dBW (decibels in respect to 1 W) rather than in dBm (decibels in respect to 1 mW).

Most modern satellite systems use either phase-shift keying (PSK) or quadrature amplitude modulation (QAM) rather than conventional frequency modulation (FM). With PSK and QAM, the input baseband is generally a PCM-encoded, time-division-multiplexed signal that is digital in nature. Also, with PSK and QAM, several bits may be encoded in a single transmit signaling element. Consequently, a parameter more meaningful than carrier power is *energy per bit* (E_b). Mathematically, E_b is

$$E_b = P_t T_b \tag{25-4}$$

where E_b = energy of a single bit (joules per bit)
P_t = total saturated output power (watts or joules per second)
T_b = time of a single bit (seconds)

or, because $T_b = 1/f_b$, where f_b is the bit rate in bits per second,

$$E_b = \frac{P_t}{f_b} = \frac{\text{J/s}}{\text{b/s}} = \frac{\text{joules}}{\text{bit}} \tag{25-5}$$

Example 25-2

For a total transmit power (P_t) of 1000 W, determine the energy per bit (E_b) for a transmission rate of 50 Mbps.

Solution

$$T_b = \frac{1}{f_b} = \frac{1}{50 \times 10^6 \text{ bps}} = 0.02 \times 10^{-6} \text{ s}$$

(It appears that the units for T_b should be s/bit, but the per bit is implied in the definition of T_b, time of bit.)

Substituting into Equation 25-4 yields

$$E_b = 1000 \text{ J/s } (0.02 \times 10^{-6} \text{ s/bit}) = 20 \text{ µJ}$$

(Again the units appear to be J/bit, but the per bit is implied in the definition of E_b, energy per bit.)

$$E_b = \frac{1000 \text{ J/s}}{50 \times 10^6 \text{ bps}} = 20 \text{ µJ}$$

Expressed as a log with 1 joule as the reference,

$$E_b = 10 \log(20 \times 10^{-6}) = -47 \text{ dBJ}$$

It is common to express P_t in dBW and E_b in dBW/bps. Thus,

$$P_t = 10 \log 1000 = 30 \text{ dBW}$$

$$E_b = P_t - 10 \log f_b = P_t - 10 \log (50 \times 10^6)$$

$$= 30 \text{ dBW} - 77 \text{ dB} = -47 \text{ dBW/bps}$$

or simply −47 dBJ.

25-10-3 Effective Isotropic Radiated Power

Effective isotropic radiated power (EIRP) is defined as an equivalent transmit power and is expressed mathematically as

$$\text{EIRP} = P_{\text{in}} A_t \tag{25-6}$$

where EIRP = effective isotropic radiated power (watts)
P_{in} = antenna input power (watts)
A_t = transmit antenna gain (unitless ratio)

Expressed as a log,

$$\text{EIRP}_{(\text{dBW})} = P_{\text{in(dBW)}} + A_{t(\text{dB})} \tag{25-7}$$

In respect to the transmitter output,

$$P_{\text{in}} = P_t - L_{\text{bo}} - L_{bf}$$

Thus,

$$\text{EIRP} = P_t - L_{\text{bo}} - L_{bf} + A_t \tag{25-8}$$

where P_{in} = antenna input power (dBW per watt)
L_{bo} = back-off losses of HPA (decibels)
L_{bf} = total branching and feeder loss (decibels)
A_t = transmit antenna gain (decibels)
P_t = saturated amplifier output power (dBW per watt)

Example 25-3

For an earth station transmitter with an antenna output power of 40 dBW (10,000 W), a back-off loss of 3 dB, a total branching and feeder loss of 3 dB, and a transmit antenna gain of 40 dB, determine the EIRP.

Solution Substituting into Equation 25-6 yields

$$\begin{aligned}\text{EIRP} &= P_t - L_{\text{bo}} - L_{bf} + A_t \\ &= 40\text{ dBW} - 3\text{ dB} - 3\text{ dB} + 40\text{ dB} = 74\text{ dBW}\end{aligned}$$

25-10-4 Equivalent Noise Temperature

With terrestrial microwave systems, the noise introduced in a receiver or a component within a receiver was commonly specified by the parameter noise figure. In satellite communications systems, it is often necessary to differentiate or measure noise in increments as small as a tenth or a hundredth of a decibel. Noise figure, in its standard form, is inadequate for such precise calculations. Consequently, it is common to use *environmental temperature* (T) and *equivalent noise temperature* (T_e) when evaluating the performance of a satellite system. In Chapter 24, total noise power was expressed mathematically as

$$N = KTB \tag{25-9}$$

Rearranging and solving for T gives us

$$T = \frac{N}{KB} \tag{25-10}$$

where N = total noise power (watts)
K = Boltzmann's constant (joules per kelvin)
B = bandwidth (hertz)
T = temperature of the environment (kelvin)

Again from Chapter 24,

$$\text{F} = 1 + \frac{T_e}{T} \tag{25-11}$$

Table 25-3 Noise Unit Comparison

Noise Factor (F) (unitless)	Noise Figure (NF) (dB)	Equivalent Temperature (T_e) (°K)	dBK
1.2	0.79	60	17.78
1.3	1.14	90	19.54
1.4	1.46	120	20.79
2.5	4	450	26.53
10	10	2700	34.31

where T_e = equivalent noise temperature (kelvin)
F = noise factor (unitless)
T = temperature of the environment (kelvin)

Rearranging Equation 25-9, we have

$$T_e = T(\mathrm{F} - 1) \tag{25-12}$$

Typically, equivalent noise temperatures of the receivers used in satellite transponders are about 1000 K. For earth station receivers, T_e values are between 20 K and 1000 K. Equivalent noise temperature is generally more useful when expressed logarithmically referenced to 1 K with the unit of dBK, as follows:

$$T_{e(\mathrm{dBK})} = 10 \log T_e \tag{25-13}$$

For an equivalent noise temperature of 100 K, $T_{e(\mathrm{dBK})}$ is

$$T_e = 10 \log 100 \text{ or } 20 \text{ dBK}$$

Equivalent noise temperature is a hypothetical value that can be calculated but cannot be measured. Equivalent noise temperature is often used rather than noise figure because it is a more accurate method of expressing the noise contributed by a device or a receiver when evaluating its performance. Essentially, equivalent noise temperature (T_e) represents the noise power present at the input to a device plus the noise added internally by that device. This allows us to analyze the noise characteristics of a device by simply evaluating an equivalent input noise temperature. As you will see in subsequent discussions, T_e is a very useful parameter when evaluating the performance of a satellite system.

Noise factor, noise figure, equivalent noise temperature, and dBK are summarized in Table 25-3.

Example 25-4

Convert noise figures of 4 dB and 4.1 dB to equivalent noise temperatures. Use 300 K for the environmental temperature.

Solution Converting the noise figures to noise factors yields

$$\mathrm{NF} = 4 \text{ dB}, \mathrm{F} = 2.512$$
$$\mathrm{NF} = 4.1 \text{ dB}, \mathrm{F} = 2.57$$

Substituting into Equation 25-10 yields

$$T_e = 300(2.512 - 1)$$
$$= 453.6 \text{ K}$$
$$T_e = 300(2.57 - 1)$$
$$= 471 \text{ K}$$

From Example 25-4, it can be seen that a 0.1-dB difference in the two noise figures equated to a 17.4° difference in the two equivalent noise temperatures. Hence, equivalent noise temperature is a more accurate method of comparing the noise performances of two receivers or devices.

25-10-5 Noise Density

Simply stated, *noise density* (N_0) is the noise power normalized to a 1-Hz bandwidth, or the noise power present in a 1-Hz bandwidth. Mathematically, noise density is

$$N_0 = \frac{N}{B} = \frac{KT_eB}{B} = KT_e \qquad \textbf{(25-14)}$$

where N_0 = noise density (watts/per hertz) (N_0 is generally expressed as simply watts; the per hertz is implied in the definition of N_0),

$$1\ \text{W/Hz} = \frac{1\ \text{joule/sec}}{1\ \text{cycle/sec}} = \frac{1\ \text{joule}}{\text{cycle}}$$

N = total noise power (watts)
B = bandwidth (hertz)
K = Boltzmann's constant (joules/per kelvin)
T_e = equivalent noise temperature (kelvin)

Expressed as a log with 1 W/Hz as the reference,

$$N_{0(\text{dBW/Hz})} = 10 \log N - 10 \log B \qquad \textbf{(25-15)}$$

$$= 10 \log K + 10 \log T_e \qquad \textbf{(25-16)}$$

Example 25-5

For an equivalent noise bandwidth of 10 MHz and a total noise power of 0.0276 pW, determine the noise density and equivalent noise temperature.

Solution Substituting into Equation 25-12, we have

$$N_0 = \frac{N}{B} = \frac{276 \times 10^{-16}\ \text{W}}{10 \times 10^6\ \text{Hz}} = 276 \times 10^{-23}\ \text{W/Hz}$$

or simply 276×10^{-23} W.

$$N_0 = 10 \log(276 \times 10^{-23}) = -205.6\ \text{dBW/Hz}$$

or simply -205.6 dBW. Substituting into Equation 25-13 gives us

$$N_0 = 10 \log 276 \times 10^{-16} - 10 \log 10\ \text{MHz}$$
$$= -135.6\ \text{dBW} - 70\ \text{dB} = -205.6\ \text{dBW}$$

Rearranging Equation 25-12 and solving for equivalent noise temperature yields

$$T_e = \frac{N_0}{K}$$
$$= \frac{276 \times 10^{-23}\ \text{J/cycle}}{1.38 \times 10^{-23}\ \text{J/K}} = 200\ \text{K/cycle}$$

Expressed as a log, $T_e = 10 \log 200 = 23$ dBK

$$= N_0 - 10 \log K = N_0 - 10 \log 1.38 \times 10^{-23}$$
$$= -205.6\ \text{dBW} - (-228.6\ \text{dBWK}) = 23\ \text{dBK}$$

25-10-6 Carrier-to-Noise Density Ratio

C/N_0 is the average wideband carrier power-to-noise density ratio. The *wideband carrier power* is the combined power of the carrier and its associated sidebands. The noise density is the thermal noise present in a normalized 1-Hz bandwidth. The carrier-to-noise density ratio may also be written as a function of noise temperature. Mathematically, C/N_0 is

$$\frac{C}{N_0} = \frac{C}{KT_e} \qquad \textbf{(25-17)}$$

Expressed as a log,

$$\frac{C}{N_0}(\text{dB}) = C_{(\text{dBW})} - N_{0(\text{dBW})} \tag{25-18}$$

25-10-7 Energy of Bit-to-Noise Density Ratio

E_b/N_0 is one of the most important and most often used parameters when evaluating a digital radio system. The E_b/N_0 ratio is a convenient way to compare digital systems that use different transmission rates, modulation schemes, or encoding techniques. Mathematically, E_b/N_0 is

$$\frac{E_b}{N_0} = \frac{C/f_b}{N/B} = \frac{CB}{Nf_b} \tag{25-19}$$

E_b/N_0 is a convenient term used for digital system calculations and performance comparisons, but in the real world, it is more convenient to measure the wideband carrier power-to-noise density ratio and convert it to E_b/N_0. Rearranging Equation 25-18 yields the following expression:

$$\frac{E_b}{N_0} = \frac{C}{N} \times \frac{B}{f_b} \tag{25-20}$$

The E_b/N_0 ratio is the product of the carrier-to-noise ratio (C/N) and the noise bandwidth–to–bit rate ratio (B/f_b). Expressed as a log,

$$\frac{E_b}{N_0}(\text{dB}) = \frac{C}{N}(\text{dB}) + \frac{B}{f_b}(\text{dB}) \tag{25-21}$$

The energy per bit (E_b) will remain constant as long as the total wideband carrier power (C) and the transmission rate (bps) remain unchanged. Also, the noise density (N_0) will remain constant as long as the noise temperature remains constant. The following conclusion can be made: For a given carrier power, bit rate, and noise temperature, the E_b/N_0 ratio will remain constant regardless of the encoding technique, modulation scheme, or bandwidth.

Figure 25-24 graphically illustrates the relationship between an expected probability of error $P(e)$ and the minimum C/N ratio required to achieve the $P(e)$. The C/N specified is for the minimum double-sided Nyquist bandwidth. Figure 25-25 graphically illustrates the relationship between an expected $P(e)$ and the minimum E_b/N_0 ratio required to achieve that $P(e)$.

A $P(e)$ of $10^{-5}(1/10^5)$ indicates a probability that one bit will be in error for every 100,000 bits transmitted. $P(e)$ is analogous to the bit error rate (BER).

Example 25-6

A coherent binary phase-shift-keyed (BPSK) transmitter operates at a bit rate of 20 Mbps. For a probability of error $P(e)$ of 10^{-4},

a. Determine the minimum theoretical C/N and E_b/N_0 ratios for a receiver bandwidth equal to the minimum double-sided Nyquist bandwidth.

b. Determine the C/N if the noise is measured at a point prior to the bandpass filter where the bandwidth is equal to twice the Nyquist bandwidth.

c. Determine the C/N if the noise is measured at a point prior to the bandpass filter where the bandwidth is equal to three times the Nyquist bandwidth.

Solution **a.** With BPSK, the minimum bandwidth is equal to the bit rate, 20 MHz. From Figure 25-24, the minimum C/N is 8.8 dB. Substituting into Equation 25-20 gives us

$$\frac{E_b}{N_0} = \frac{C}{N} + \frac{B}{f_b}$$

$$= 8.8 \text{ dB} + 10 \log \frac{20 \times 10^6}{20 \times 10^6}$$

$$= 8.8 \text{ dB} + 0 \text{ dB} = 8.8 \text{ dB}$$

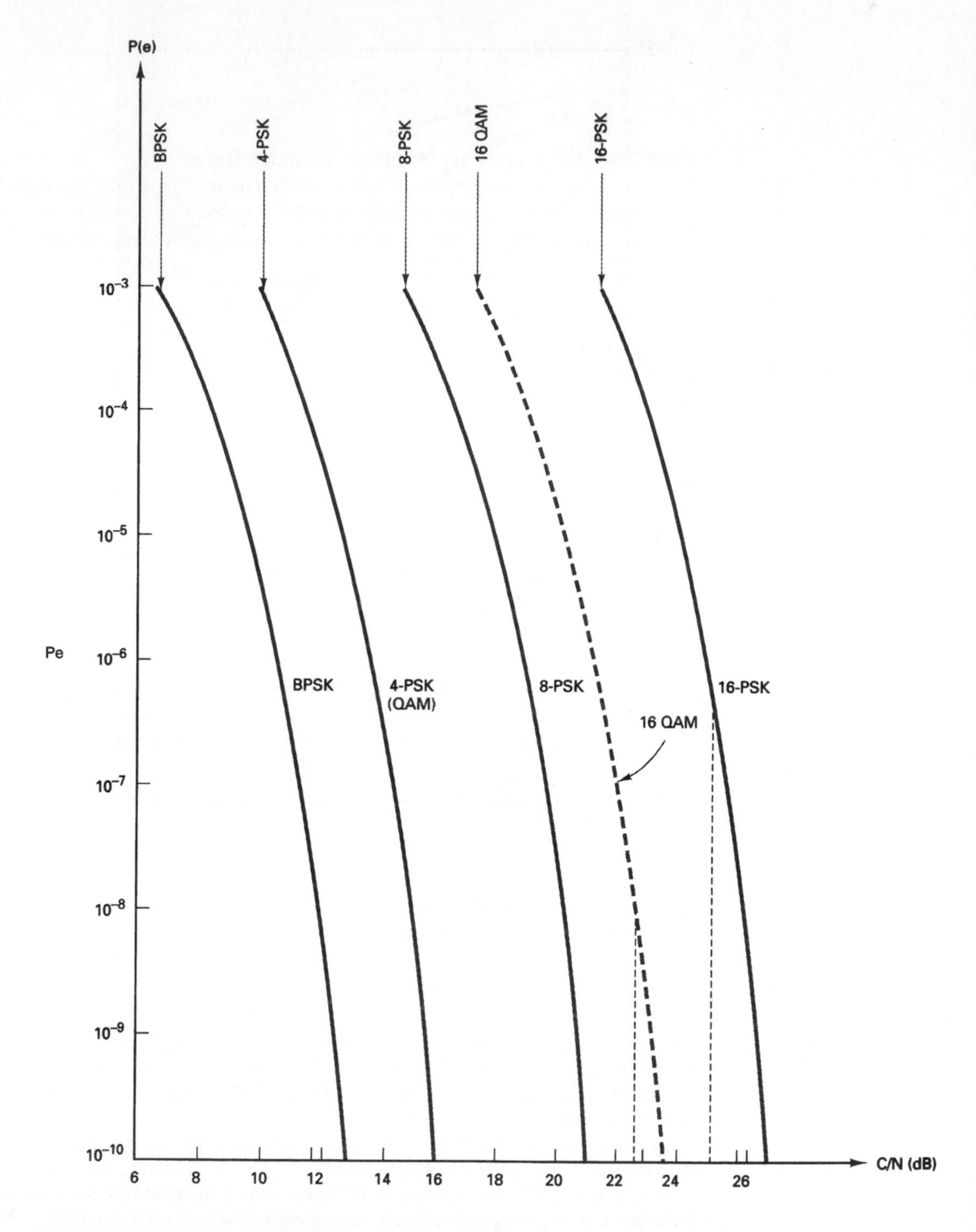

FIGURE 25-24 $P(e)$ performance of M-ary PSK, QAM, QPR, and M-ary APK coherent systems. The rms C/N is specified in the double-sided Nyquist bandwidth

Note: The minimum E_b/N_0 equals the minimum C/N when the receiver noise bandwidth equals the bit rate which for BPSK also equals the minimum Nyquist bandwidth. The minimum E_b/N_0 of 8.8 can be verified from Figure 25-25.

What effect does increasing the noise bandwidth have on the minimum C/N and E_b/N_0 ratios? The wideband carrier power is totally independent of the noise bandwidth. However, an increase in the bandwidth causes a corresponding increase in the noise power. Consequently, a decrease in C/N

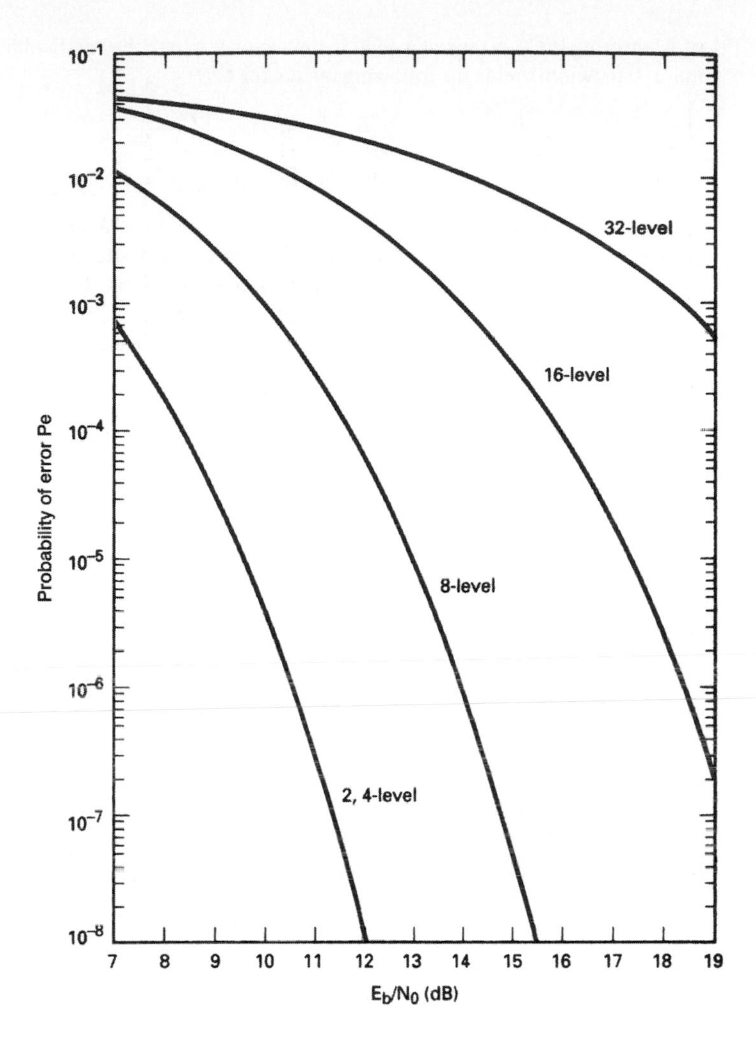

FIGURE 25-25 Probability of error $P(e)$ versus E_b/N_0 ratio for various digital modulation schemes

is realized that is directly proportional to the increase in the noise bandwidth. E_b is dependent on the wideband carrier power and the bit rate only. Therefore, E_b is unaffected by an increase in the noise bandwidth. N_0 is the noise power normalized to a 1-Hz bandwidth and, consequently, is also unaffected by an increase in the noise bandwidth.

b. Because E_b/N_0 is independent of bandwidth, measuring the C/N at a point in the receiver where the bandwidth is equal to twice the minimum Nyquist bandwidth has absolutely no effect on E_b/N_0. Therefore, E_b/N_0 becomes the constant in Equation 25-20 and is used to solve for the new value of C/N. Rearranging Equation 25-20 and using the calculated E_b/N_0 ratio, we have

$$\frac{C}{N} = \frac{E_b}{N_0} - \frac{B}{f_b}$$

$$= 8.8 \text{ dB} - 10 \log \frac{40 \times 10^6}{20 \times 10^6}$$

$$= 8.8 \text{ dB} - 10 \log 2$$

$$= 8.8 \text{ dB} - 3 \text{ dB} = 5.8 \text{ dB}$$

c. Measuring the C/N ratio at a point in the receiver where the bandwidth equals three times the minimum bandwidth yields the following results for C/N:

$$\frac{C}{N} = \frac{E_b}{N_0} - 10 \log \frac{60 \times 10^6}{20 \times 10^6}$$
$$= 8.8 \text{ dB} - 10 \log 3 = 4.03 \text{ dB}$$

The C/N ratios of 8.8, 5.8, and 4.03 dB indicate the C/N ratios that could be measured at the three specified points in the receiver and still achieve the desired minimum E_b/N_0 and $P(e)$.

Because E_b/N_0 cannot be directly measured to determine the E_b/N_0 ratio, the wideband carrier-to-noise ratio is measured and, then, substituted into Equation 25-20. Consequently, to accurately determine the E_b/N_0 ratio, the noise bandwidth of the receiver must be known.

Example 25-7

A coherent 8-PSK transmitter operates at a bit rate of 90 Mbps. For a probability of error of 10^{-5},

a. Determine the minimum theoretical C/N and E_b/N_0 ratios for a receiver bandwidth equal to the minimum double-sided Nyquist bandwidth.

b. Determine the C/N if the noise is measured at a point prior to the bandpass filter where the bandwidth is equal to twice the Nyquist bandwidth.

c. Determine the C/N if the noise is measured at a point prior to the bandpass filter where the bandwidth is equal to three times the Nyquist bandwidth.

Solution **a.** 8-PSK has a bandwidth efficiency of 3 bps/Hz and, consequently, requires a minimum bandwidth of one-third the bit rate, or 30 MHz. From Figure 25-24, the minimum C/N is 18.5 dB. Substituting into Equation 25-20, we obtain

$$\frac{E_b}{N_0} = 18.5 \text{ dB} + 10 \log \frac{30 \text{ MHz}}{90 \text{ Mbps}}$$
$$= 18.5 \text{ dB} + (-4.8 \text{ dB}) = 13.7 \text{ dB}$$

b. Rearranging Equation 25-20 and substituting for E_b/N_0 yields

$$\frac{C}{N} = 13.7 \text{ dB} - 10 \log \frac{60 \text{ MHz}}{90 \text{ Mbps}}$$
$$= 13.7 \text{ dB} - (-1.77 \text{ dB}) = 15.47 \text{ dB}$$

c. Again, rearranging Equation 25-20 and substituting for E_b/N_0 gives us

$$\frac{C}{N} = 13.7 \text{ dB} - 10 \log \frac{90 \text{ MHz}}{90 \text{ Mbps}}$$
$$= 13.7 \text{ dB (dB)} = 13.7 \text{ dB}$$

It should be evident from Examples 25-6 and 25-7 that the E_b/N_0 and C/N ratios are equal only when the noise bandwidth is equal to the bit rate. Also, as the bandwidth at the point of measurement increases, the C/N decreases.

When the modulation scheme, bit rate, bandwidth, and C/N ratios of two digital radio systems are different, it is often difficult to determine which system has the lower probability of error. E_b/N_0 is independent of bandwidth and modulation scheme, so it is a convenient common denominator to use for comparing the probability of error performance of two digital radio systems.

25-10-8 Gain-To-Equivalent Noise Temperature Ratio

Gain-to-equivalent noise temperature ratio (G/T_e) is a figure of merit used to represent the quality of a satellite or earth station receiver. The G/T_e ratio is the ratio of the receive antenna gain (G) to the equivalent system noise temperature (T_e) of the receiver. G/T_e is expressed mathematically as

$$\frac{G}{T_e} = G - 10 \log(T_s) \qquad (25\text{-}22)$$

where G = receive antenna gain (dB)
T_s = operating or system temperature (degrees Kelvin)

and $\quad T_s = T_a + T_r$

where $\quad T_a$ = antenna temperature (degrees Kelvin)
T_r = receiver effective input noise temperature (degrees Kelvin)

It should be noted that the ratio of G to T_e involves two different quantities. Antenna gain is a unitless value, whereas temperature has the unit of degrees Kelvin. The reference temperature is 1 K; therefore, the decibel notation for G/T_e is dBK^{-1} or dB/K, which should not be interpreted as decibels per degree Kelvin.

25-11 SATELLITE SYSTEM LINK EQUATIONS

The error performance of a digital satellite system is quite predictable. Figure 25-26 shows a simplified block diagram of a digital satellite system and identifies the various gains and losses that may affect the system performance. When evaluating the performance of a digital satellite system, the uplink and downlink parameters are first considered separately, then the overall performance is determined by combining them in the appropriate manner. Keep in mind, a digital microwave or satellite radio simply means that the original and demodulated baseband signals are digital in nature. The RF portion of the radio is analog, that is, FSK, PSK, QAM, or some other higher-level modulation riding on an analog microwave carrier.

25-11-1 Link Equations

The following *link equations* are used to separately analyze the uplink and downlink sections of a single radio-frequency carrier satellite system. These equations consider only the ideal gains and losses and effects of thermal noise associated with the earth station transmitter, earth station receiver, and the satellite transponder.

Uplink Equation

$$\frac{C}{N_0} = \frac{A_t P_{\text{in}}(L_p L_u) A_r}{K T_e} = \frac{A_t P_{\text{in}}(L_p L_u)}{K} \times \frac{G}{T_e} \tag{25-23}$$

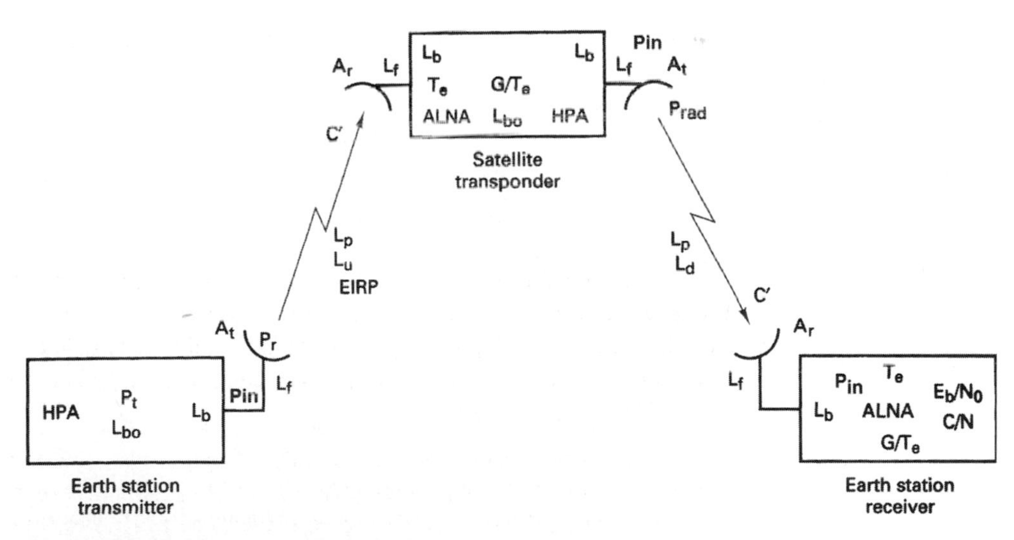

FIGURE 25-26 Overall satellite system showing the gains and losses incurred in both the uplink and downlink sections. HPA, high-power amplifier; P_t, HPA output power; L_{bo}, back-off loss; L_f, feeder loss; L_b, branching loss; A_t, transmit antenna gain; P_r total radiated power = $P_t - L_{bo} - L_b - L_f$; EIRP, effective isotropic radiated power $P_{rad} A_t$; L_u, additional uplink losses due to atmosphere; L_p, path loss; A_r, receive antenna gain; G/T_e, gain-to-equivalent noise ratio; L_d, additional downlink losses due to atmosphere; LNA, low-noise amplifier; C/T_e, carrier-to-equivalent noise ratio; C/N_0, carrier-to-noise density ratio; E_b/N_0, energy of bit-to-noise density ratio; C/N, carrier-to-noise ratio

where L_d and L_u are the additional uplink and downlink atmospheric losses, respectively. The uplink and downlink signals must pass through Earth's atmosphere, where they are partially absorbed by the moisture, oxygen, and particulates in the air. Depending on the elevation angle, the distance the RF signal travels through the atmosphere varies from one earth station to another. Because L_p, L_u, and L_d represent losses, they are decimal values less than 1. G/T_e is the receive antenna gain plus the gain of the LNA divided by the equivalent input noise temperature.

Expressed as a log,

$$\frac{C}{N_0} = \underbrace{10 \log A_t P_{\text{in}}}_{\substack{\text{EIRP} \\ \text{earth} \\ \text{station}}} - \underbrace{20 \log\left(\frac{4\pi D}{\lambda}\right)}_{\substack{\text{free-space} \\ \text{path loss} \\ L_p}} + \underbrace{10 \log\left(\frac{G}{T_e}\right)}_{\substack{\text{satellite} \\ G/T_e}} - \underbrace{10 \log L_u}_{\substack{\text{additional} \\ \text{atmospheric} \\ \text{losses}}} - \underbrace{10 \log K}_{\substack{\text{Boltzmann's} \\ \text{constant}}} \quad \textbf{(25-24)}$$

$$= \text{EIRP (dBW)} - L_p(\text{dB}) + \frac{G}{T_e}(\text{dBK}^{-1}) - L_u(\text{dB}) - K(\text{dBWK}) \quad \textbf{(25-25)}$$

Downlink Equation

$$\frac{C}{N_0} = \frac{A_t P_{\text{in}}(L_p L_d) A_r}{K T_e} = \frac{A_t P_{\text{in}}(L_p L_d)}{K} \times \frac{G}{T_e} \quad \textbf{(25-26)}$$

Expressed as a log,

$$\frac{C}{N_0} = \underbrace{10 \log A_t P_{in}}_{\substack{\text{EIRP} \\ \text{satellite}}} - \underbrace{20 \log\left(\frac{4\pi D}{\lambda}\right)}_{\substack{\text{free-space} \\ \text{path loss} \\ L_p}} + \underbrace{10 \log\left(\frac{G}{T_e}\right)}_{\substack{\text{earth} \\ \text{station} \\ G/T_e}} - \underbrace{10 \log L_d}_{\substack{\text{additional} \\ \text{atmospheric} \\ \text{losses}}} - \underbrace{10 \log K}_{\substack{\text{Boltzmann's} \\ \text{constant}}} \quad \textbf{(25-27)}$$

$$= \text{EIRP(dBW)} - L_p(\text{dB}) + \frac{G}{T_e}(\text{dBK}^{-1}) - L_d\,(\text{dB}) - K\,(\text{dBWK})$$

25-12 LINK BUDGET

Table 25-4 lists the system parameters for three typical satellite communication systems. The systems and their parameters are not necessarily for an existing or future system; they are hypothetical examples only. The system parameters are used to construct a *link budget*. A link budget identifies the system parameters and is used to determine the projected *C/N* and E_b/N_0 ratios at both the satellite and earth station receivers for a given modulation scheme and desired *P*(*e*).

Example 25-8

Complete the link budget for a satellite system with the following parameters.

Uplink

1. Earth station transmitter output power at saturation, 2000 W	33 dBW
2. Earth station back-off loss	3 dB
3. Earth station branching and feeder losses	4 dB
4. Earth station transmit antenna gain (from Figure 25-27, 15 m at 14 GHz)	64 dB
5. Additional uplink atmospheric losses	0.6 dB
6. Free-space path loss (from Figure 25-28, at 14 GHz)	206.5 dB
7. Satellite receiver G/T_e ratio	-5.3 dBK^{-1}
8. Satellite branching and feeder losses	0 dB
9. Bit rate	120 Mbps
10. Modulation scheme	8-PSK

Table 25-4 System Parameters for Three Hypothetical Satellite Systems

	System A: 6/4 GHz, earth coverage QPSK modulation, 60 Mbps	System B: 14/12 GHz, earth coverage 8-PSK modulation, 90 Mbps	System C: 14/12 GHz, earth coverage 8-PSK modulation, 120 Mbps
Uplink			
Transmitter output power (saturation, dBW)	35	25	33
Earth station back-off loss (dB)	2	2	3
Earth station branching and feeder loss (dB)	3	3	4
Additional atmospheric (dB)	0.6	0.4	0.6
Earth station antenna gain (dB)	55	45	64
Free-space path loss (dB)	200	208	206.5
Satellite receive antenna gain (dB)	20	45	23.7
Satellite branching and feeder loss (dB)	1	1	0
Satellite equivalent noise temperature (K)	1000	800	800
Satellite G/T_e (dBK^{-1})	−10	16	−5.3
Downlink			
Transmitter output power (saturation, dBW)	18	20	10
Satellite back-off loss (dB)	0.5	0.2	0.1
Satellite branching and feeder loss (dB)	1	1	0.5
Additional atmospheric loss (dB)	0.8	1.4	0.4
Satellite antenna gain (dB)	16	44	30.8
Free-space path loss (dB)	197	206	205.6
Earth station receive antenna gain (dB)	51	44	62
Earth station branching and feeder loss (dB)	3	3	0
Earth station equivalent noise temperature (K)	250	1000	270
Earth station G/T_e (dBK^{-1})	27	14	37.7

Downlink

1. Satellite transmitter output power at saturation, 10 W	10 dBW
2. Satellite back-off loss	0.1 dB
3. Satellite branching and feeder losses	0.5 dB
4. Satellite transmit antenna gain (from Figure 25-27, 0.37 m at 12 GHz)	30.8 dB
5. Additional downlink atmospheric losses	0.4 dB
6. Free-space path loss (from Figure 25-28, at 12 GHz)	205.6 dB
7. Earth station receive antenna gain (15 m, 12 GHz)	62 dB
8. Earth station branching and feeder losses	0 dB
9. Earth station equivalent noise temperature	270 K
10. Earth station G/T_e ratio	37.7 dBK^{-1}
11. Bit rate	120 Mbps
12. Modulation scheme	8 -PSK

Solution *Uplink budget:* Expressed as a log,

$$\text{EIRP (earth station)} = P_t + A_t - L_{bo} - L_{bf}$$
$$= 33 \text{ dBW} + 64 \text{ dB} - 3 \text{ dB} - 4 \text{ dB} = 90 \text{ dBW}$$

Carrier power density at the satellite antenna:

$$C' = \text{EIRP (earth station)} - L_p - L_u$$
$$= 90 \text{ dBW} - 206.5 \text{ dB} - 0.6 \text{ dB} = -117.1 \text{ dBW}$$

C/N_0 at the satellite:

$$\frac{C}{N_0} = \frac{C}{KT_e} = \frac{C}{T_e} \times \frac{1}{K} \qquad \text{where } \frac{C}{T_e} = C' \times \frac{G}{T_e}$$

Thus,
$$\frac{C}{N_0} = C' \times \frac{G}{T_e} \times \frac{1}{K}$$

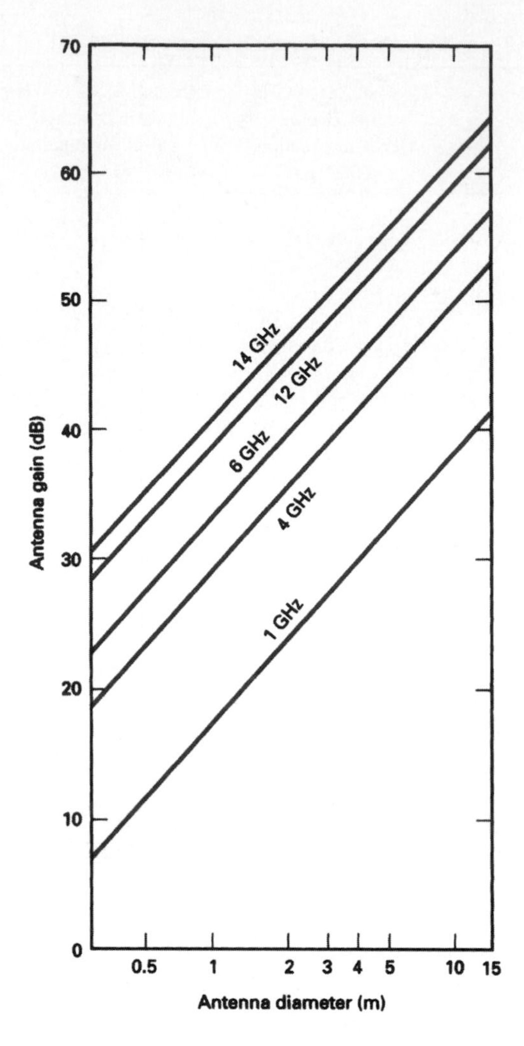

FIGURE 25-27 Antenna gain based on the gain equation for a parabolic antenna:

$$A\ [\text{db}] = 10 \log \eta\ [\pi\ D/\lambda]^2$$

where D is the antenna diameter, λ = the wavelength, and η = the antenna efficiency. Here $\eta = 0.55$. To correct for a 100% efficient antenna, add 2.66 dB to the value.

Expressed as a log,

$$\frac{C}{N_0} = C' + \frac{G}{T_e} - 10 \log(1.38 \times 10^{-23})$$

$$= -117.1 \text{ dBW} + (-5.3 \text{ dBK}^{-1}) - (-228.6 \text{ dBWK}) = 106.2 \text{ dB}$$

Thus,

$$\frac{E_b}{N_0} = \frac{C/f_b}{N_0} = \frac{C}{N_0} - 10 \log f_b$$

$$= 106.2 \text{ dB} - 10\,(\log 120 \times 10^6) = 25.4 \text{ dB}$$

and for a minimum bandwidth system,

$$\frac{C}{N} = \frac{E_b}{N_0} - \frac{B}{f_b} = 25.4 - 10 \log \frac{40 \times 10^6}{120 \times 10^6} = 30.2 \text{ dB}$$

Downlink budget: Expressed as a log,

$$\text{EIRP (satellite transponder)} = P_t + A_t - L_{bo} - L_{bf}$$

$$= 10 \text{ dBW} + 30.8 \text{ dB} - 0.1 \text{ dB} - 0.5 \text{ dB}$$

$$= 40.2 \text{ dBW}$$

Carrier power density at earth station antenna:

$$C' = \text{EIRP} - L_p - L_d$$

$$= 40.2 \text{ dBW} - 205.6 \text{ dB} - 0.4 \text{ dB} = -165.8 \text{ dBW}$$

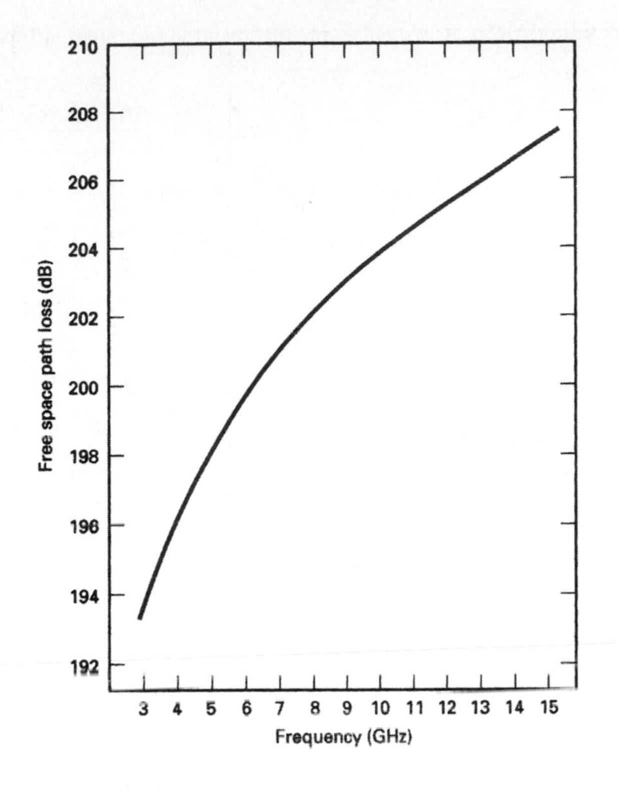

Elevation angle correction:

Angle	+dB
90°	0
45°	0.44
0°	1.33

FIGURE 25-28 Free-space path loss [L_p] determined from $L_p = 183.5 + 20 \log f$ [GHz], elevation angle = 90°, and distance = 35,930 km

C/N_0 at the earth station receiver:

$$\frac{C}{N_0} = \frac{C}{KT_e} = \frac{C}{T_e} \times \frac{1}{K} \quad \text{where } \frac{C}{T_e} = C' \times \frac{G}{T_e}$$

Thus,

$$\frac{C}{N_0} = C' \times \frac{G}{T_e} \times \frac{1}{K}$$

Expressed as a log,

$$\frac{C}{N_0} = C' + \frac{G}{T_e} - 10 \log(1.38 \times 10^{-23})$$

$$= -165.8 \text{ dBW} + (37.7 \text{ dBK}^{-1}) - (-228.6 \text{ dBWK}) = 100.5 \text{ dB}$$

An alternative method of solving for C/N_0 is

$$\frac{C}{N_0} = C' + A_r - T_e - K$$

$$= -165.8 \text{ dBW} + 62 \text{ dB} - 10 \log 270 - (-228.6 \text{ dBWK})$$

$$= -165.8 \text{ dBW} + 62 \text{ dB} - 24.3 \text{ dBK}^{-1} + 228.6 \text{ dBWK} = 100.5 \text{ dB}$$

$$\frac{E_b}{N_0} = \frac{C}{N_0} - 10 \log f_b$$

$$= 100.5 \text{ dB} - 10 \log(120 \times 10^6)$$

$$= 100.5 \text{ dB} - 80.8 \text{ dB} = 19.7 \text{ dB}$$

and for a minimum bandwidth system,

$$\frac{C}{N} = \frac{E_b}{N_0} - \frac{B}{f_b} = 19.7 - 10 \log \frac{40 \times 10^6}{120 \times 10^6} = 24.5 \text{ dB}$$

With careful analysis and a little algebra, it can be shown that the overall energy of bit-to-noise density ratio (E_b/N_0), which includes the combined effects of the uplink ratio $(E_b/N_0)_u$ and the downlink ratio $(E_b/N_0)_d$, is a standard product over the sum relationship and is expressed mathematically as

$$\frac{E_b}{N_0}(\text{overall}) = \frac{(E_b/N_0)_u(E_b/N_0)_d}{(E_b/N_0)_u + (E_b/N_0)_d} \tag{25-28}$$

where all E_b/N_0 ratios are in absolute values. For Example 25-25, the overall E_b/N_0 ratio is

$$\frac{E_b}{N_0}(\text{overall}) = \frac{(346.7)(93.3)}{346.7 + 93.3} = 73.5$$

$$= 10 \log 73.5 = 18.7 \text{ dB}$$

As with all product-over-sum relationships, the smaller of the two numbers dominates. If one number is substantially smaller than the other, the overall result is approximately equal to the smaller of the two numbers.

The system parameters used for Example 25-9 were taken from system C in Table 25-4. A complete link budget for the system is shown in Table 25-5.

Table 25-5 Link Budget for Example 25-10

Uplink	
1. Earth station transmitter output power at saturation, 2000 W	33 dBW
2. Earth station back-off loss	3 dB
3. Earth station branching and feeder losses	4 dB
4. Earth station transmit antenna gain	64 dB
5. Earth station EIRP	90 dBW
6. Additional uplink atmospheric losses	0.6 dB
7. Free-space path loss	206.5 dB
8. Carrier power density at satellite	−117.1 dBW
9. Satellite branching and feeder losses	0 dB
10. Satellite G/T_e ratio	-5.3 dBK^{-1}
11. Satellite C/T_e ratio	-122.4 dBWK^{-1}
12. Satellite C/N_0 ratio	106.2 dB
13. Satellite C/N ratio	30.2 dB
14. Satellite E_b/N_0 ratio	25.4 dB
15. Bit rate	120 Mbps
16. Modulation scheme	8-PSK
Downlink	
1. Satellite transmitter output power at saturation, 10 W	10 dBW
2. Satellite back-off loss	0.1 dB
3. Satellite branching and feeder losses	0.5 dB
4. Satellite transmit antenna gain	30.8 dB
5. Satellite EIRP	40.2 dBW
6. Additional downlink atmospheric losses	0.4 dB
7. Free-space path loss	205.6 dB
8. Earth station receive antenna gain	62 dB
9. Earth station equivalent noise temperature	270 K
10. Earth station branching and feeder losses	0 dB
11. Earth station G/T_e ratio	37.7 dBK^{-1}
12. Carrier power density at earth station	−165.8 dBW
13. Earth station C/T_e ratio	-128.1 dBWK^{-1}
14. Earth station C/N_0 ratio	100.5 dB
15. Earth station C/N ratio	24.5 dB
16. Earth station E_b/N_0 ratio	19.7 dB
17. Bit rate	120 Mbps
18. Modulation scheme	8-PSK

QUESTIONS

25-1. Briefly describe a satellite.

25-2. What is a passive satellite? An active satellite?

25-3. Contrast nonsynchronous and synchronous satellites.

25-4. Define *prograde* and *retrograde.*

25-5. Define *apogee* and *perigee.*

25-6. Briefly explain the characteristics of low-, medium-, and high-altitude satellite orbits.

25-7. Explain equatorial, polar, and inclined orbits.

25-8. Contrast the advantages and disadvantages of geosynchronous satellites.

25-9. Define *look angles, angle of elevation,* and *azimuth.*

25-10. Define *satellite spatial separation* and list its restrictions.

25-11. Describe a "footprint."

25-12. Describe spot, zonal, and earth coverage radiation patterns.

25-13. Explain *reuse.*

25-14. Briefly describe the functional characteristics of an uplink, a transponder, and a downlink model for a satellite system.

25-15. Define *back-off loss* and its relationship to saturated and transmit power.

25-16. Define *bit energy.*

25-17. Define *effective isotropic radiated power.*

25-18. Define *equivalent noise temperature.*

25-19. Define *noise density.*

25-20. Define *carrier-to-noise density ratio* and *energy of bit-to-noise density ratio.*

25-21. Define *gain-to-equivalent noise temperature ratio.*

25-22. Describe what a satellite link budget is and how it is used.

PROBLEMS

25-1. An earth station is located at Houston, Texas, that has a longitude of 99.5° and a latitude of 29.5° north. The satellite of interest is *Satcom V.* Determine the look angles for the earth station antenna.

25-2. A satellite system operates at 14-GHz uplink and 11-GHz downlink and has a projected $P(e)$ of 10^{-7}. The modulation scheme is 8-PSK, and the system will carry 120 Mbps. The equivalent noise temperature of the receiver is 400 K, and the receiver noise bandwidth is equal to the minimum Nyquist frequency. Determine the following parameters: minimum theoretical C/N ratio, minimum theoretical E_b/N_0 ratio, noise density, total receiver input noise, minimum receive carrier power, and the minimum energy per bit at the receiver input.

25-3. A satellite system operates at 6-GHz uplink and 4-GHz downlink and has a projected $P(e)$ of 10^{-6}. The modulation scheme is QPSK and the system will carry 100 Mbps. The equivalent receiver noise temperature is 290 K, and the receiver noise bandwidth is equal to the minimum Nyquist frequency. Determine the C/N ratio that would be measured at a point in the receiver prior to the BPF where the bandwidth is equal to (a) 1½ times the minimum Nyquist frequency and (b) 3 times the minimum Nyquist frequency.

25-4. Which system has the best projected BER?

a. 8-QAM, C/N = 15 dB, $B = 2f_N$, $f_b = 60$ Mbps

b. QPSK, C/N = 16 dB, $B = f_N$, $f_b = 40$ Mbps

25-5. An earth station satellite transmitter has an HPA with a rated saturated output power of 10,000 W. The back-off ratio is 6 dB, the branching loss is 2 dB, the feeder loss is 4 dB, and the antenna gain is 40 dB. Determine the actual radiated power and the EIRP.

25-6. Determine the total noise power for a receiver with an input bandwidth of 20 MHz and an equivalent noise temperature of 600 K.

25-7. Determine the noise density for Problem 25-6.

25-8. Determine the minimum C/N ratio required to achieve a $P(e)$ of 10^{-5} for an 8-PSK receiver with a bandwidth equal to f_N.

25-9. Determine the energy per bit-to-noise density ratio when the receiver input carrier power is -100 dBW, the receiver input noise temperature is 290 K, and a 60-Mbps transmission rate is used.

25-10. Determine the carrier-to-noise density ratio for a receiver with a -70-dBW input carrier power, an equivalent noise temperature of 180 K, and a bandwidth of 20 MHz.

25-11. Determine the minimum C/N ratio for an 8-PSK system when the transmission rate is 60 Mbps, the minimum energy of bit-to-noise density ratio is 15 dB, and the receiver bandwidth is equal to the minimum Nyquist frequency.

25-12. For an earth station receiver with an equivalent input temperature of 200 K, a noise bandwidth of 20 MHz, a receive antenna gain of 50 dB, and a carrier frequency of 12 GHz, determine the following: G/T_e, N_0, and N.

25-13. For a satellite with an uplink E_b/N_0 of 14 dB and a downlink E_b/N_0 of 18 dB, determine the overall E_b/N_0 ratio.

25-14. Complete the following link budget:

Uplink parameters

1. Earth station transmitter output power at saturation, 1 kW
2. Earth station back-off loss, 3 dB
3. Earth station total branching and feeder losses, 3 dB
4. Earth station transmit antenna gain for a 10-m parabolic dish at 14 GHz
5. Free-space path loss for 14 GHz
6. Additional uplink losses due to the Earth's atmosphere, 0.8 dB
7. Satellite transponder G/T_e, -4.6 dBK^{-1}
8. Transmission bit rate, 90 Mbps, 8-PSK

Downlink parameters

1. Satellite transmitter output power at saturation, 10 W
2. Satellite transmit antenna gain for a 0.5-m parabolic dish at 12 GHz
3. Satellite modulation back-off loss, 0.8 dB
4. Free-space path loss for 12 GHz
5. Additional downlink losses due to Earth's atmosphere, 0.6 dB
6. Earth station receive antenna gain for a 10-m parabolic dish at 12 GHz
7. Earth station equivalent noise temperature, 200 K
8. Earth station branching and feeder losses, 0 dB
9. Transmission bit rate, 90 Mbps, 8-PSK

25-15. An earth station is located at Houston, Texas, that has a longitude of 99.5° and a latitude of 29.5° north. The satellite of interest is *Westar III*. Determine the look angles from the earth station antenna.

25-16. A satellite system operates at 14 GHz uplink and 11 GHz downlink and has a projected $P(e)$ of one bit error in every 1 million bits transmitted. The modulation scheme is 8-PSK, and the system will carry 90 Mbps. The equivalent noise temperature of the receiver is 350 K, and the receiver noise bandwidth is equal to the minimum Nyquist frequency. Determine the following parameters: minimum theoretical C/N ratio, minimum theoretical E_b/N_0 ratio, noise density, total receiver input noise, minimum receive carrier power, and the minimum energy per bit at the receiver input.

25-17. A satellite system operates a 6-GHz uplink and 4-GHz downlink and has a projected $P(e)$ of one bit error in every 100,000 bits transmitted. The modulation scheme is 4-PSK, and the system will carry 80 Mbps. The equivalent receiver noise temperature is 120 K, and the receiver noise bandwidth is equal to the minimum Nyquist frequency. Determine the following:

a. The C/N ratio that would be measured at a point in the receiver prior to the BPF where the bandwidth is equal to two times the minimum Nyquist frequency.

b. The C/N ratio that would be measured at a point in the receiver prior to the BPF where the bandwidth is equal to three times the minimum Nyquist frequency.

25-18. Which system has the best projected BER?

a. QPSK, C/N = 16 dB, $B = 2f_N$, $f_b = 40$ Mbps

b. 8-PSK, C/N = 18 dB, $B = f_N$, $f_b = 60$ Mbps

25-19. An earth station satellite transmitter has an HPA with a rated saturated output power of 12,000 W. The back-off ratio of 4 dB, the branching loss is 1.5 dB, the feeder loss is 5 dB, and the antenna gain is 38 dB. Determine the actual radiated power and the EIRP.

25-20. Determine the total noise power for a receiver with an input bandwidth of 40 MHz and an equivalent noise temperature of 800 K.

25-21. Determine the noise density for Problem 25-20.

25-22. Determine the minimum *C/N* ratio required to achieve a *P*(*e*) of one bit error for every 1 million bits transmitted for a QPSK receiver with a bandwidth equal to the minimum Nyquist frequency.

25-23. Determine the energy of bit-to-noise density ratio when the receiver input carrier power is −85 dBW, the receiver input noise temperature is 400 K, and a 50-Mbps transmission rate.

25-24. Determine the carrier-to-noise density ratio for a receiver with a −80-dBW carrier input power, equivalent noise temperature of 240 K, and a bandwidth of 10 MHz.

25-25. Determine the minimum *C/N* ratio for a QPSK system when the transmission rate is 80 Mbps, the minimum energy of bit-to-noise density ratio is 16 dB, and the receiver bandwidth is equal to the Nyquist frequency.

25-26. For an earth station receiver with an equivalent input temperature of 400 K, a noise bandwidth of 30 MHz, a receive antenna gain of 44 dB, and a carrier frequency of 12 GHz, determine the following: G/T_e, N_0, and N.

25-27. For a satellite with an uplink E_b/N_0 of 16 dB and a downlink E_b/N_0 of 13 dB, determine the overall E_b/N_0.

25-28. Complete the following link budget:

Uplink parameters

1. Earth station output power at saturation, 12 kW
2. Earth station back-off loss, 4 dB
3. Earth station branching and feeder losses, 2 dB
4. Earth station antenna gain for a 10-m parabolic dish at 14 GHz
5. Free-space path loss for 14 GHz
6. Additional uplink losses due to Earth's atmosphere, 1 dB
7. Satellite transponder G/T_e, −3 dBk
8. Transmission bit rate, 80 Mbps
9. Modulation scheme, 4-PSK

Downlink parameters

1. Satellite transmitter output power at saturation, 5 W
2. Satellite station transmit antenna gain for a 0.5-m parabolic dish at 12 GHz
3. Satellite modulation back-off loss, 1 dB
4. Free-space path loss for 12 GHz
5. Additional downlink losses due to Earth's atmosphere, 1 dB
6. Earth station receive antenna gain for a 10-m parabolic dish at 12 GHz
7. Earth station equivalent noise temperature, 300 K
8. Transmission bit rate, 80 Mbps
9. Modulation scheme, 4-PSK

CHAPTER 26

Satellite Multiple Accessing Arrangements

CHAPTER OUTLINE

OBJECTIVES

- Define *multiple accessing*
- Describe the operation of FDM/FM satellite systems
- Describe frequency-division multiple accessing
- Describe time-division multiple accessing
- Describe code-division multiple accessing
- Define *channel capacity*
- Explain the basic concepts of satellite radio navigation
- Describe the operation of the NAVSTAR Global Positioning System

26-1 INTRODUCTION

In Chapter 25 we analyzed the link parameters of *single-channel satellite transponders.* In this chapter, we will extend the discussion of satellite communications to systems designed for *multiple carriers.* Whenever multiple carriers are utilized in satellite communications, it is necessary that a *multiple-accessing format* be established over the system. This format allows for a distinct separation between the uplink and downlink transmissions to and from a multitude of different earth stations. Each format has its own specific characteristics, advantages, and disadvantages.

26-2 FDM/FM SATELLITE SYSTEMS

Figure 26-1a shows a single-link (two earth stations) *fixed-frequency* FDM/FM system using a single satellite transponder. With earth coverage antennas and for full-duplex operation, each link requires two RF satellite channels (i.e., four RF carrier frequencies, two uplink and two downlink). In Figure 26-1a, earth station 1 transmits on a high-band carrier (f11, f12, f13, and so on) and receives on a low-band carrier (f1, f2, f3, and so on). To avoid interfering with earth station 1, earth station 2 must transmit and receive on different RF carrier frequencies. The RF carrier frequencies are fixed and the satellite transponder is simply an RF-to-RF repeater that provides the uplink/downlink frequency translation. This arrangement is economically impractical and also extremely inefficient. Additional earth stations can communicate through different transponders within the same satellite structure (Figure 26-1b), but each additional link requires four more RF carrier frequencies. It is unlikely that any two-point link would require the capacity available in an entire RF satellite channel. Consequently, most of the available bandwidth is wasted. Also, with this arrangement, each earth station can communicate with only one other earth station. The RF satellite channels are fixed between any two earth stations; thus, the voice-band channels from each earth station are committed to a single destination.

In a system where three or more earth stations wish to communicate with each other, fixed-frequency or *dedicated-channel* systems such as those shown in Figure 26-1, are inadequate; a method of *multiple accessing* is required. That is, each earth station using the satellite system has a means of communicating with each of the other earth stations in the

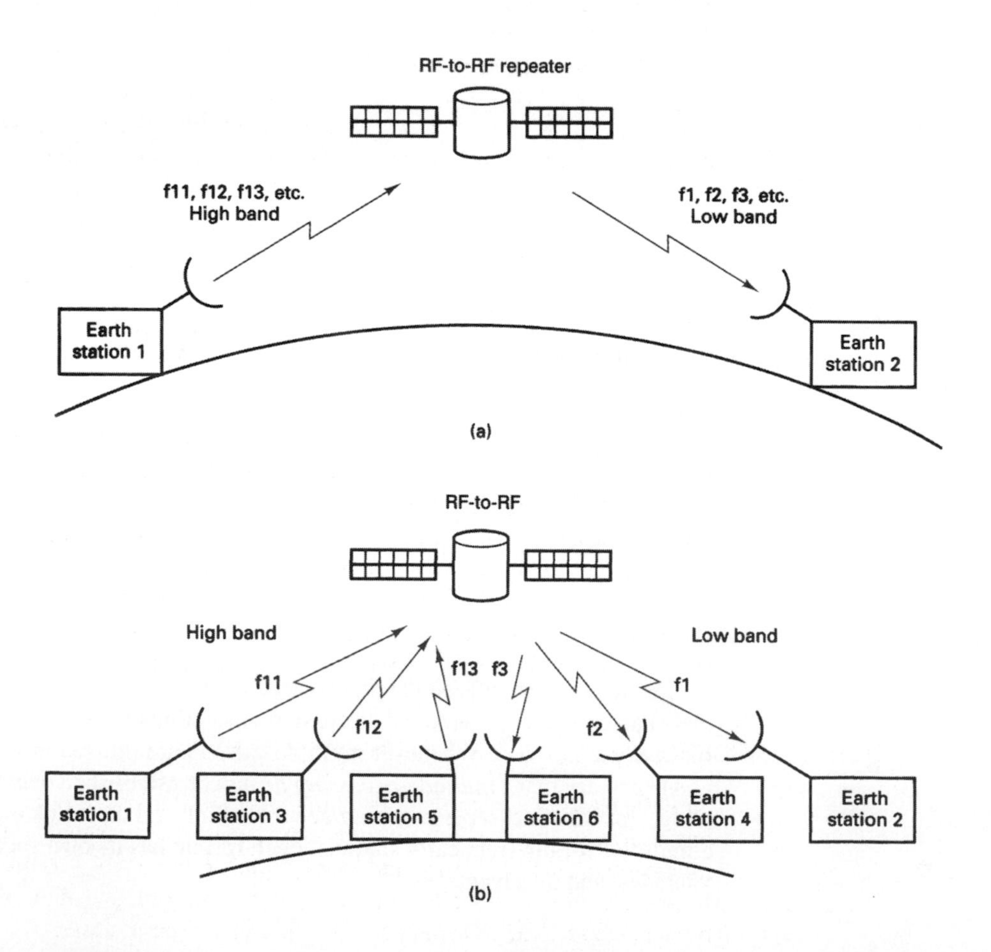

FIGURE 26-1 Fixed-frequency earth station satellite system: (a) single link; (b) multiple link

system through a common satellite transponder. Multiple accessing is sometimes called *multiple destination* because the transmissions from each earth station are received by all the other earth stations in the system. The voice-band channels between any two earth stations may be *preassigned* (*dedicated*) or *demand-assigned* (*switched*). When preassignment is used, a given number of the available voice-band channels from each earth station are assigned a dedicated destination. With demand assignment, voice-band channels are assigned on an as-needed basis. Demand assignment provides more versatility and more efficient use of the available frequency spectrum. On the other hand, demand assignment requires a control mechanism that is common to all the earth stations to keep track of channel routing and the availability of each voice-band channel.

Remember, in an FDM/FM satellite system, each RF channel requires a separate transponder. Also, with FDM/FM transmissions, it is impossible to differentiate (separate) multiple transmissions that occupy the same bandwidth. Fixed-frequency systems may be used in a multiple-access configuration by switching the RF carriers at the satellite, reconfiguring the baseband signals with multiplexing/demultiplexing equipment on board the satellite, or using multiple spot beam antennas (reuse). All three of these methods require relatively complicated, expensive, and heavy hardware on the spacecraft.

Communications satellites operating in the C-band are allocated a total bandwidth of 500 MHz symmetrical around the satellite's center frequency. This is often referred to as one satellite channel, which is further divided into radio channels. Most communications satellites carry 12 transponders (radio channel transmitter/receiver pairs), each with 36 MHz of bandwidth. The carriers of the 12 transponders are frequency-division multiplexed with a 4-MHz guard band between each of them and a 10-MHz guard band on both ends of the 500-MHz assigned frequency spectrum.

If adjacent transponders in the 500-MHz spectrum are fed from a quadrature-polarized antenna, the number of transponders (radio channels) available in one satellite channel can be doubled to 24. Twelve odd-numbered transponders transmit and receive with a vertically polarized antenna, and 12 even-numbered transponders transmit and receive on a horizontally polarized antenna. The carrier frequencies of the even channels are offset 20 MHz from the carrier frequencies of the odd-numbered transponders to reduce crosstalk between adjacent transponders. This method of assigning adjacent channels different electromagnetic polarizations is called *frequency reuse* and is possible by using orthogonal polarization and spacing adjacent channels 20 MHz apart. Frequency reuse is a technique for achieving better utilization of the available frequency spectrum.

26-2-1 Anik-E Communications Satellite

Anik is an Eskimo word meaning "little brother." The Anik-E communications satellites are Domsats (domestic satellites) operated by Telsat Canada. Figure 26-2 shows the frequency and polarization plan for the Anik-E satellite system. One group of 12 radio channels (group A) uses horizontal polarization, and one group of 12 radio channels (group B) uses vertical polarization for 24 total radio channels, each with 36 MHz of bandwidth. There is a 4-MHz bandwidth between adjacent radio channels and a 10-MHz bandwidth at each end of the spectrum for a total satellite channel bandwidth of 500 MHz. There are 12 primary radio channels and 12 spare or preemptible radio channels.

26-3 MULTIPLE ACCESSING

Satellite *multiple accessing* (sometimes called *multiple destination*) implies that more than one user has access to one or more radio channels (transponders) within a satellite communications channel. Transponders are typically leased by a company or a common carrier for the purpose of providing voice or data transmission to a multitude of users. The method by which a satellite transponder's bandwidth is used or accessed depends on the multiple-accessing method utilized.

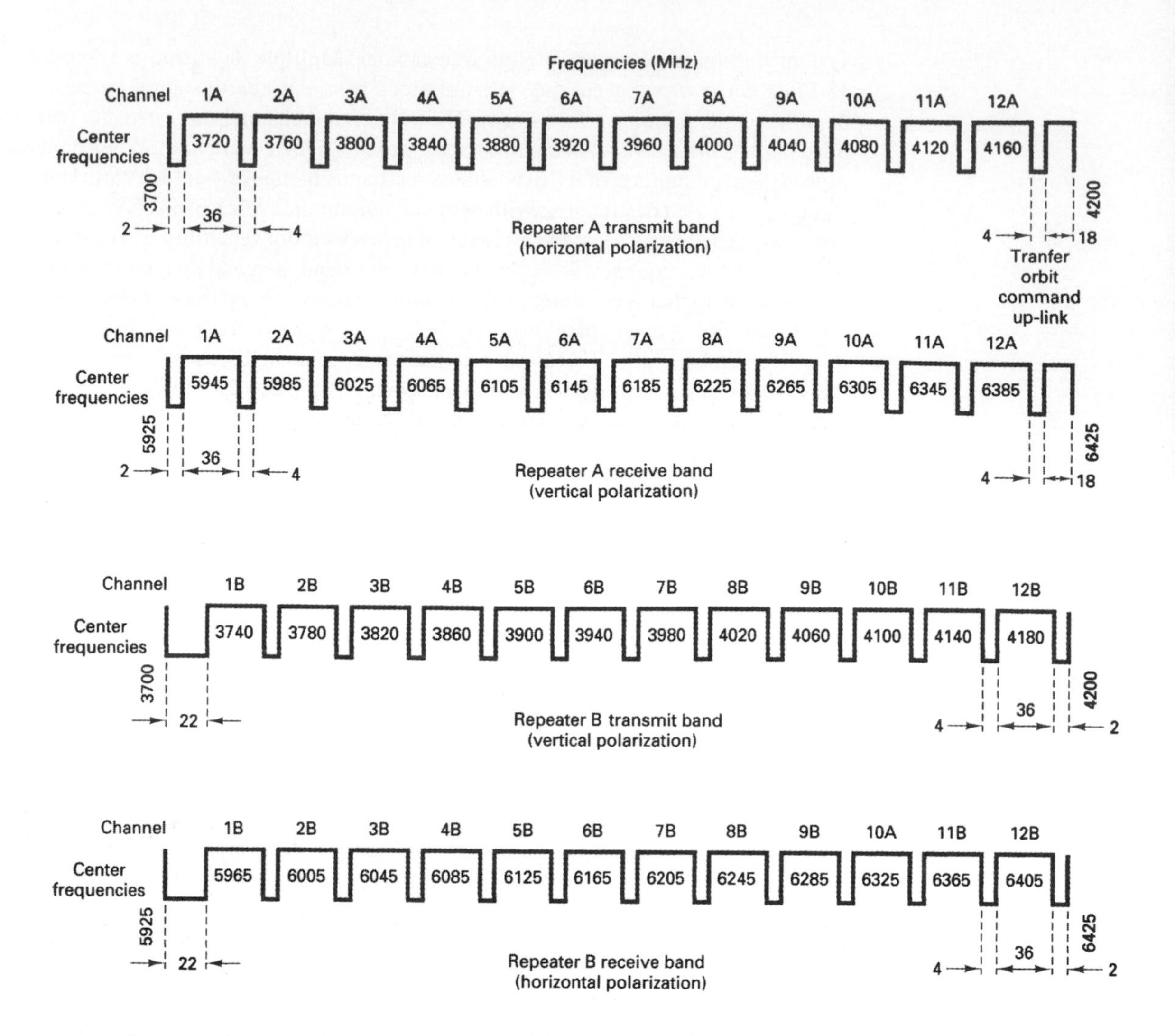

FIGURE 26-2 *Anik-E* frequency and polarization plan

Figure 26-3 illustrates the three most commonly used *multiple-accessing arrangements:* frequency-division multiple accessing (FDMA), time-division multiple accessing (TDMA), and code-division multiple accessing (CDMA). With FDMA, each earth station's transmissions are assigned specific uplink and downlink frequency bands within an allotted satellite bandwidth; they may be *preassigned* or *demand assigned.* Consequently, FDMA transmissions are separated in the frequency domain and, therefore, must share the total available transponder bandwidth as well as the total transponder power. With TDMA, each earth station transmits a short burst of information during a specific time slot (*epoch*) within a TDMA frame. The bursts must be synchronized so that each station's burst arrives at the satellite at a different time. Consequently, TDMA transmissions are separated in the time domain, and with TDMA the entire transponder bandwidth and power are used for each transmission but for only a prescribed interval of time. With CDMA, all earth stations transmit within the same frequency band and, for all practical purposes, have no limitations on when they may transmit or on which carrier frequency. Thus, with CDMA, the entire satellite transponder bandwidth is used by all stations on a continuous basis. Signal separation is accomplished with envelope encryption/decryption techniques.

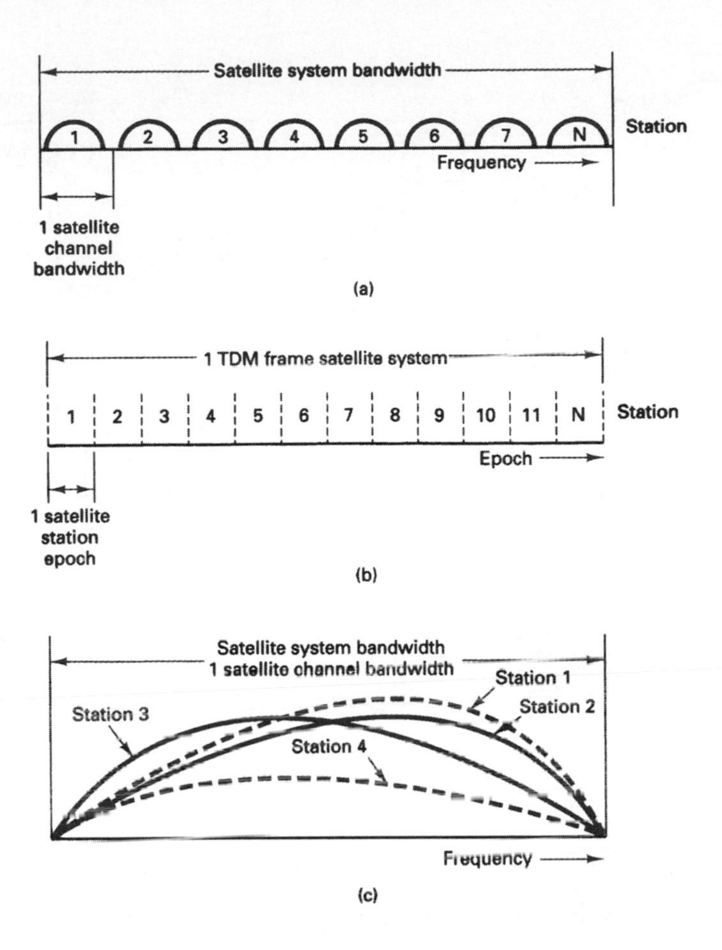

FIGURE 26-3 Multiple-accessing arrangements: (a) FDMA; (b) TDMA; (c) CDMA

26-3-1 Frequency-Division Multiple Access

Frequency-division multiple access (FDMA) is a method of multiple accessing where a given RF bandwidth is divided into smaller frequency bands called *subdivisions.* Each subdivision has its own IF carrier frequency. A control mechanism is used to ensure that two or more earth stations do not transmit in the same subdivision at the same time. Essentially, the control mechanism designates a receive station for each of the subdivisions. In demand-assignment systems, the control mechanism is also used to establish or terminate the voice-band links between the source and destination earth stations. Consequently, any of the subdivisions may be used by any of the participating earth stations at any given time. If each subdivision carries only one 4-kHz voice-band channel, this is known as a *single-channel per carrier* (SCPC) system. When several voice-band channels are frequency-division multiplexed together to form a composite baseband signal comprised of groups, supergroups, or even mastergroups, a wider subdivision is assigned. This is referred to as *multiple-channel per carrier* (MCPC).

Carrier frequencies and bandwidths for FDM/FM satellite systems using multiple-channel-per-carrier formats are generally assigned and remain fixed for a long period of time. This is referred to as *fixed-assignment, multiple access* (FDM/FM/FAMA). An alternate channel allocation scheme is *demand-assignment, multiple access* (DAMA). Demand assignment allows all users continuous and equal access of the entire transponder bandwidth by assigning carrier frequencies on a temporary basis using a statistical assignment process. The first FDMA demand-assignment system for satellites was developed by Comsat for use on the *Intelsat* series *IVA* and *V* satellites.

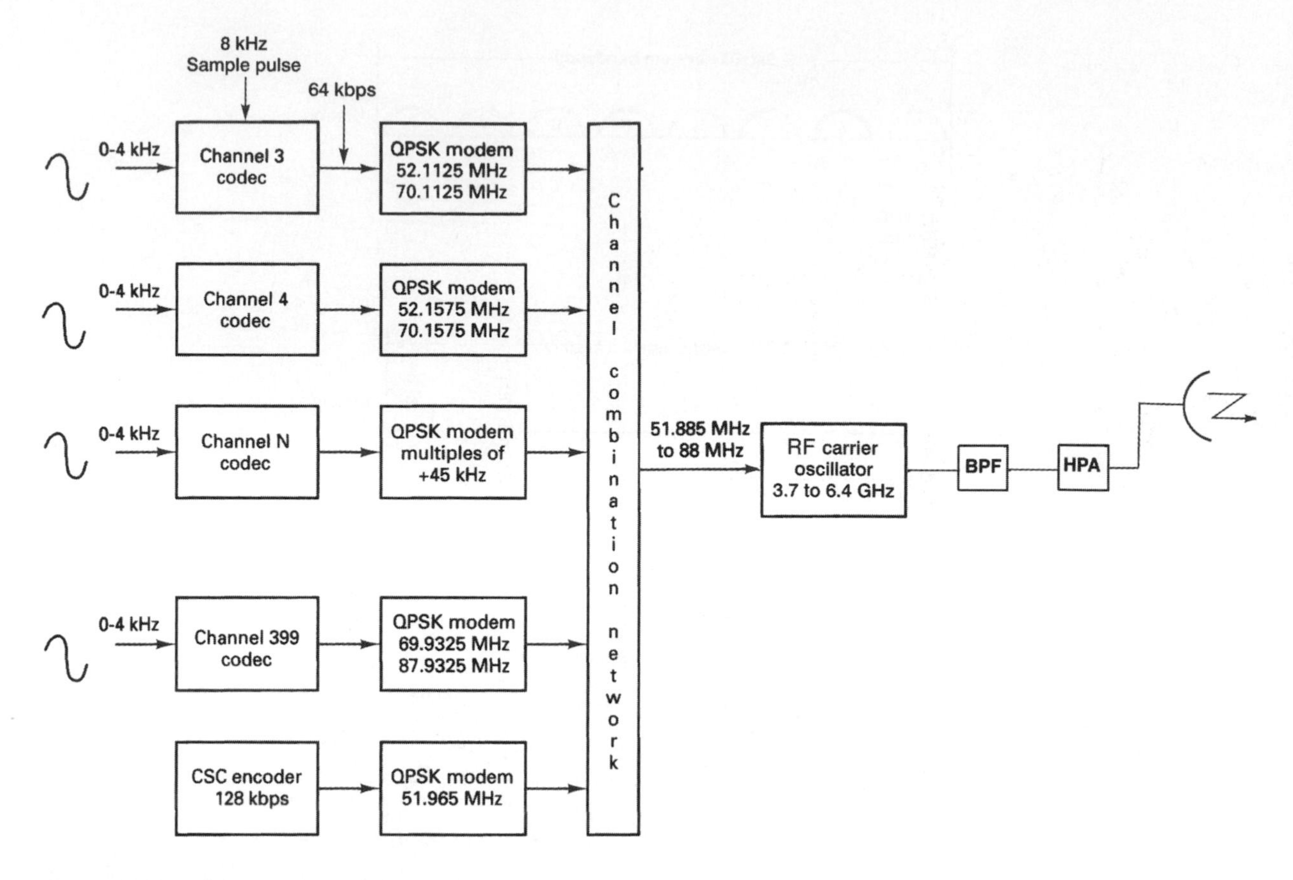

FIGURE 26-4 FDMA, SPADE earth station transmitter

26-3-1-1 SPADE DAMA satellite system. SPADE is an acronym for *s*ingle-channel-per-carrier *P*CM multiple-*a*ccess *d*emand-assignment *e*quipment. Figures 26-4 and 26-5 show the block diagram and IF frequency assignments, respectively, for SPADE.

With SPADE, 800 PCM-encoded voice-band channels separately QPSK modulate an IF carrier signal (hence the name *single-carrier per channel,* SCPC). Each 4-kHz voice-band channel is sampled at an 8-kHz rate and converted to an eight-bit PCM code. This produces a 64-kbps PCM code for each voice-band channel. The PCM code from each voice-band channel QPSK modulates a different IF carrier frequency. With QPSK, the minimum required bandwidth is equal to one-half the input bit rate. Consequently, the output of each QPSK modulator requires a minimum bandwidth of 32 kHz. Each channel is allocated a 45-kHz bandwidth, allowing for a 13-kHz guard band between pairs of frequency-division-multiplexed channels. The IF carrier frequencies begin at 52.0225 MHz (low-band channel 1) and increase in 45-kHz steps to 87.9775 MHz (high-band channel 400). The entire 36-MHz band (52 MHz to 88 MHz) is divided in half, producing two 400-channel bands (a low band and a high band). For full-duplex operation, 400 45-kHz channels are used for one direction of transmission, and 400 are used for the opposite direction. Also, channels 1, 2, and 400 from each band are left permanently vacant. This reduces the number of usable full-duplex voice-band channels to 397. The 6-GHz C-band extends from 5.725 GHz to 6.425 GHz (700 MHz). This allows for approximately 19 36-MHz RF channels per system. Each RF channel has a capacity of 397 full-duplex voice-band channels.

Each RF channel (Figure 26-5) has a 160-kHz *common signaling channel* (CSC). The CSC is a time-division-multiplexed transmission that is frequency-division multiplexed

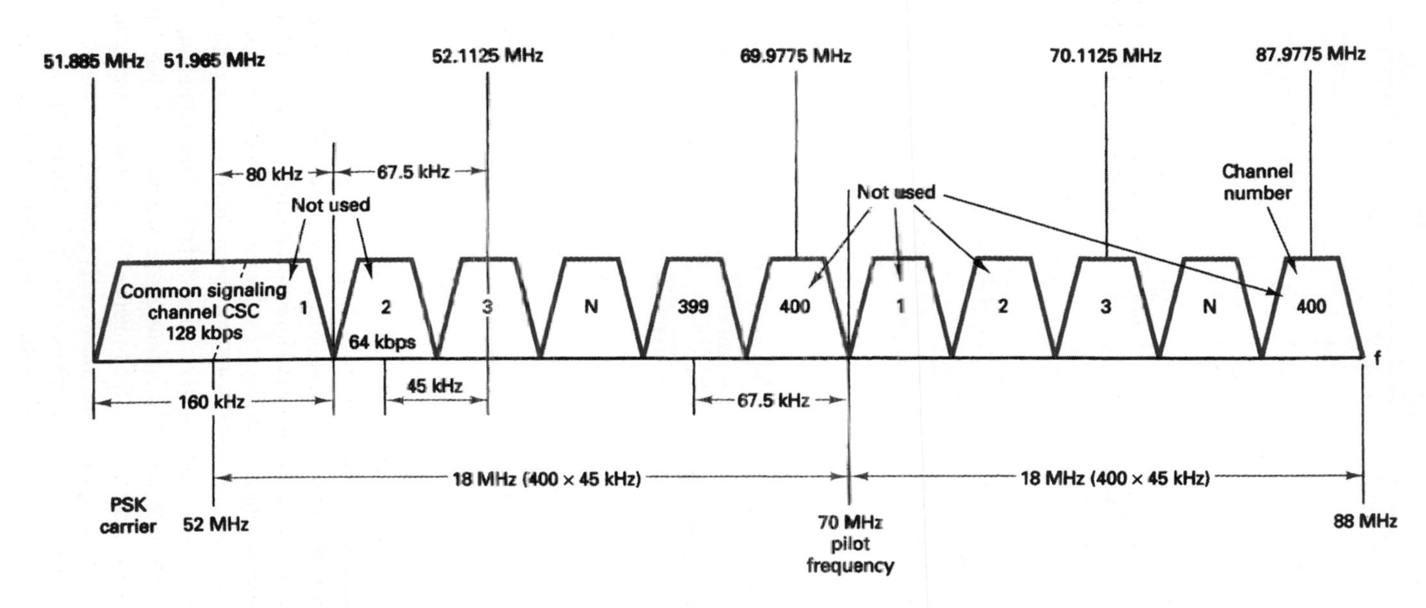

FIGURE 26-5 Carrier frequency assignments for the *Intelsat* single-channel-per-carrier PCM multiple-access demand-assignment equipment (SPADE)

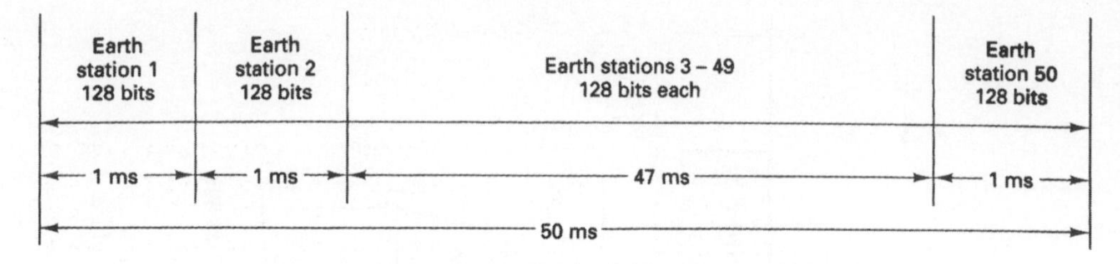

FIGURE 26-6 FDMA, SPADE common signaling channel (CSC)

into the IF spectrum below the QPSK-encoded voice-band channels. Figure 26-6 shows the TDM frame structure for the CSC. The total frame time is 50 ms, which is subdivided into 50 1-ms epochs. Each earth station transmits on the CSC channel only during its preassigned 1-ms time slot. The CSC signal is a 128-bit binary code. To transmit a 128-bit code in 1 ms, a transmission rate of 128 kbps is required. The CSC code is used for establishing and disconnecting voice-band links between two earth station users when demand-assignment channel allocation is used.

The CSC channel occupies a 160-kHz bandwidth, which includes the 45 kHz for low-band channel 1. Consequently, the CSC channel extends from 51.885 MHz to 52.045 MHz. The 128-kbps CSC binary code QPSK modulates a 51.965-MHz carrier. The minimum bandwidth required for the CSC channel is 64 kHz; this results in a 48-kHz guard band on either side of the CSC signal.

With FDMA, each earth station may transmit simultaneously within the same 36-MHz RF spectrum but on different voice-band channels. Consequently, simultaneous transmissions of voice-band channels from all earth stations within the satellite network are interleaved in the frequency domain in the satellite transponder. Transmissions of CSC signals are interleaved in the time domain.

An obvious disadvantage of FDMA is that carriers from multiple earth stations may be present in a satellite transponder at the same time. This results in cross-modulation distortion between the various earth station transmissions. This is alleviated somewhat by shutting off the IF subcarriers on all unused 45-kHz voice-band channels. Because balanced modulators are used in the generation of QPSK, carrier suppression is inherent. This also reduces the power load on a system and increases its capacity by reducing the idle channel power.

26-3-2 Time-Division Multiple Access

Time-division multiple access (TDMA) is the predominant multiple-access method used today. It provides the most efficient method of transmitting digitally modulated carriers (PSK). TDMA is a method of time-division multiplexing digitally modulated carriers between participating earth stations within a satellite network through a common satellite transponder. With TDMA, each earth station transmits a short *burst* of a digitally modulated carrier during a precise time slot (epoch) within a TDMA frame. Each station's burst is synchronized so that it arrives at the satellite transponder at a different time. Consequently, only one earth station's carrier is present in the transponder at any given time, thus avoiding a collision with another station's carrier. The transponder is an RF-to-RF repeater that simply receives the earth station transmissions, amplifies them, and then retransmits them in a downlink beam that is received by all the participating earth stations. Each earth station receives the bursts from all other earth stations and must select from them the traffic destined only for itself.

Figure 26-7 shows a basic TDMA frame. Transmissions from all earth stations are synchronized to a *reference burst.* Figure 26-7 shows the reference burst as a separate transmission, but it may be the *preamble* that precedes a reference station's transmission of data. Also, there may be more than one synchronizing reference burst.

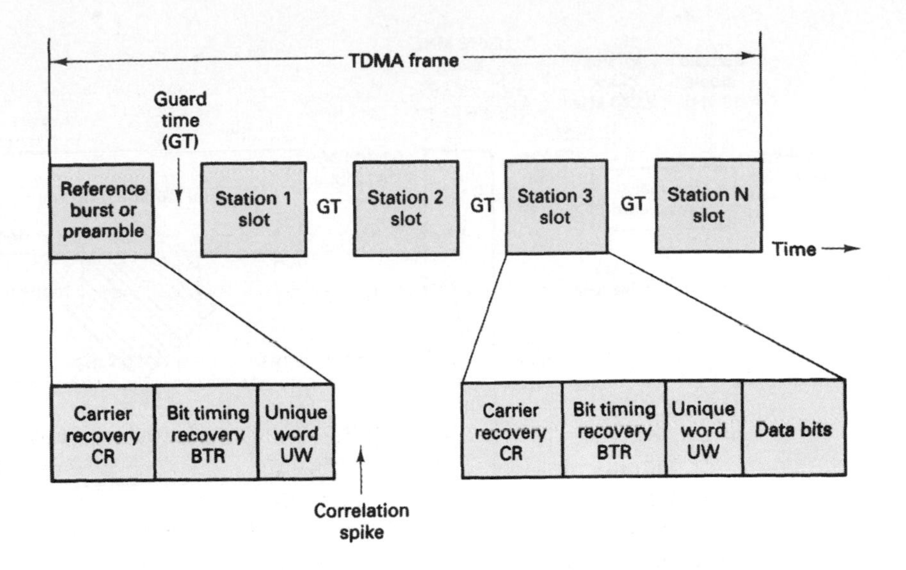

FIGURE 26-7 Basic time-division multiple-accessing (TDMA) frame

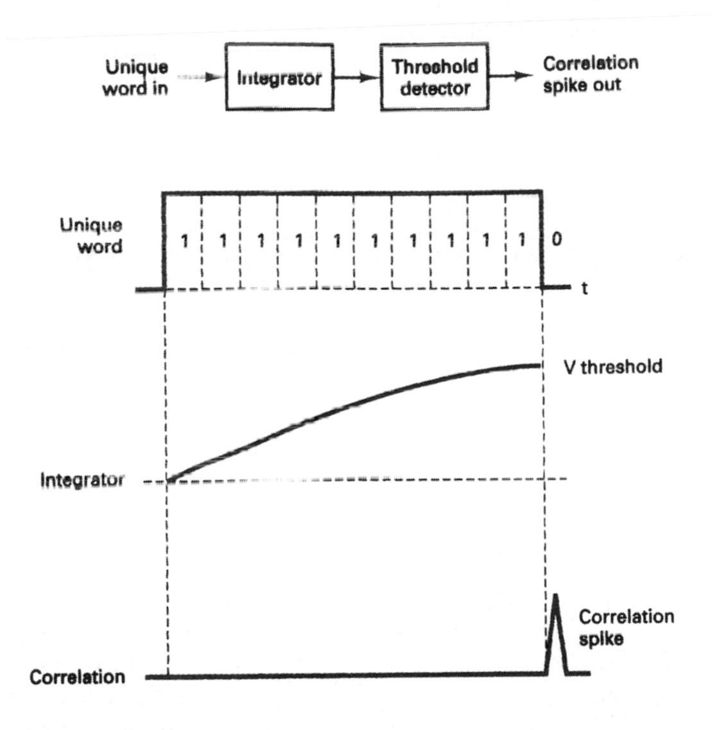

FIGURE 26-8 Unique word correlator

The reference burst contains a *carrier recovery sequence* (CRS), from which all receiving stations recover a frequency and phase coherent carrier for PSK demodulation. Also included in the reference burst is a binary sequence for *bit timing recovery* (BTR, i.e., clock recovery). At the end of each reference burst, a *unique word* (UW) is transmitted. The UW sequence is used to establish a precise time reference that each of the earth stations uses to synchronize the transmission of its burst. The UW is typically a string of 20 successive binary 1s terminated with a binary 0. Each earth station receiver demodulates and integrates the UW sequence. Figure 26-8 shows the result of the integration process. The integrator and threshold detector are designed so that the threshold voltage is reached

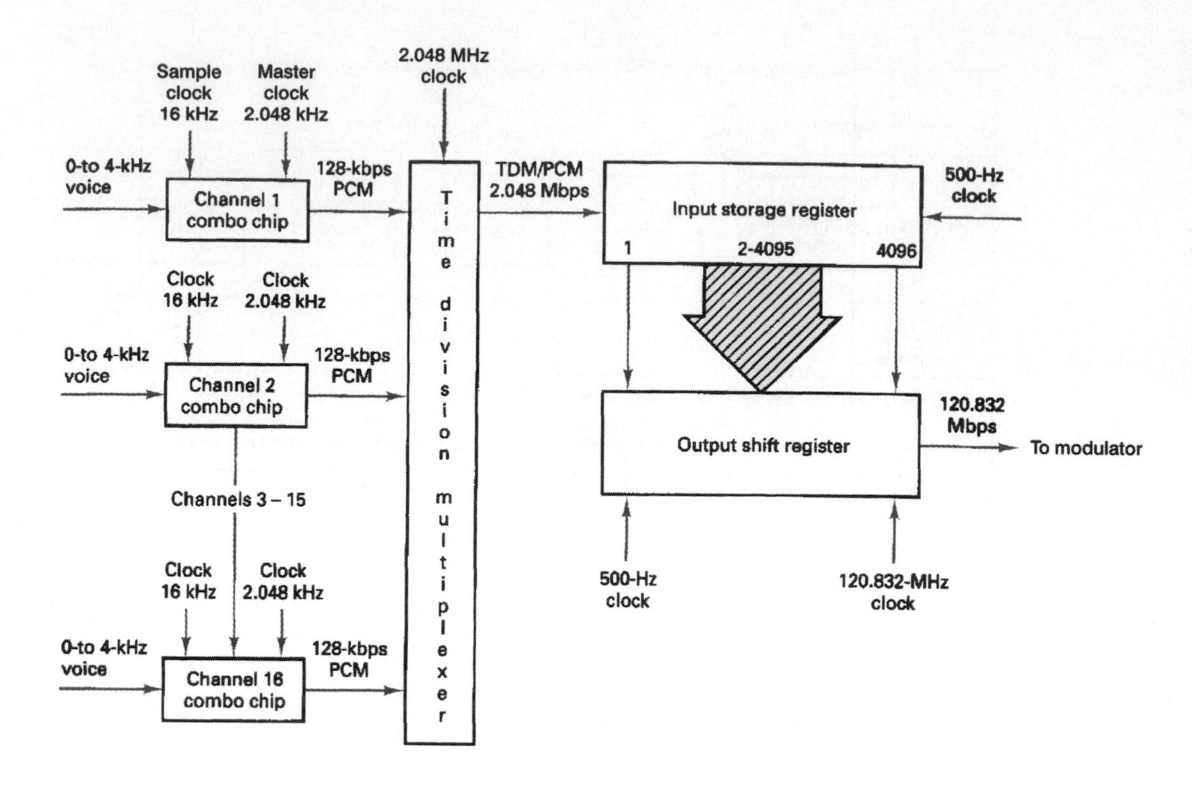

FIGURE 26-9 TDMA, CEPT primary multiplex frame transmitter

precisely when the last bit of the UW sequence is integrated. This generates a *correlation spike* at the output of the threshold detector at the exact time the UW sequence ends.

Each earth station synchronizes the transmission of its carrier to the occurrence of the UW correlation spike. Each station waits a different length of time before it begins transmitting. Consequently, no two stations will transmit the carrier at the same time. Note the *guard time* (GT) between transmissions from successive stations. This is analogous to a guard band in a frequency-division-multiplexed system. Each station precedes the transmission of data with a *preamble.* The preamble is logically equivalent to the reference burst. Because each station's transmissions must be received by all other earth stations, all stations must recover carrier and clocking information prior to demodulating the data. If demand assignment is used, a common signaling channel also must be included in the preamble.

26-3-2-1 CEPT primary multiplex frame. Figures 26-9 and 26-10 show the block diagram and timing sequence, respectively, for the CEPT primary multiplex frame. (CEPT is the Conference of European Postal and Telecommunications Administrations; the CEPT sets many of the European telecommunications standards.) This is a commonly used TDMA frame format for digital satellite systems.

Essentially, TDMA is a *store-and-forward* system. Earth stations can transmit only during their specified time slot, although the incoming voice-band signals are continuous. Consequently, it is necessary to sample and store the voice-band signals prior to transmission. The CEPT frame is made up of eight-bit PCM-encoded samples from 16 independent

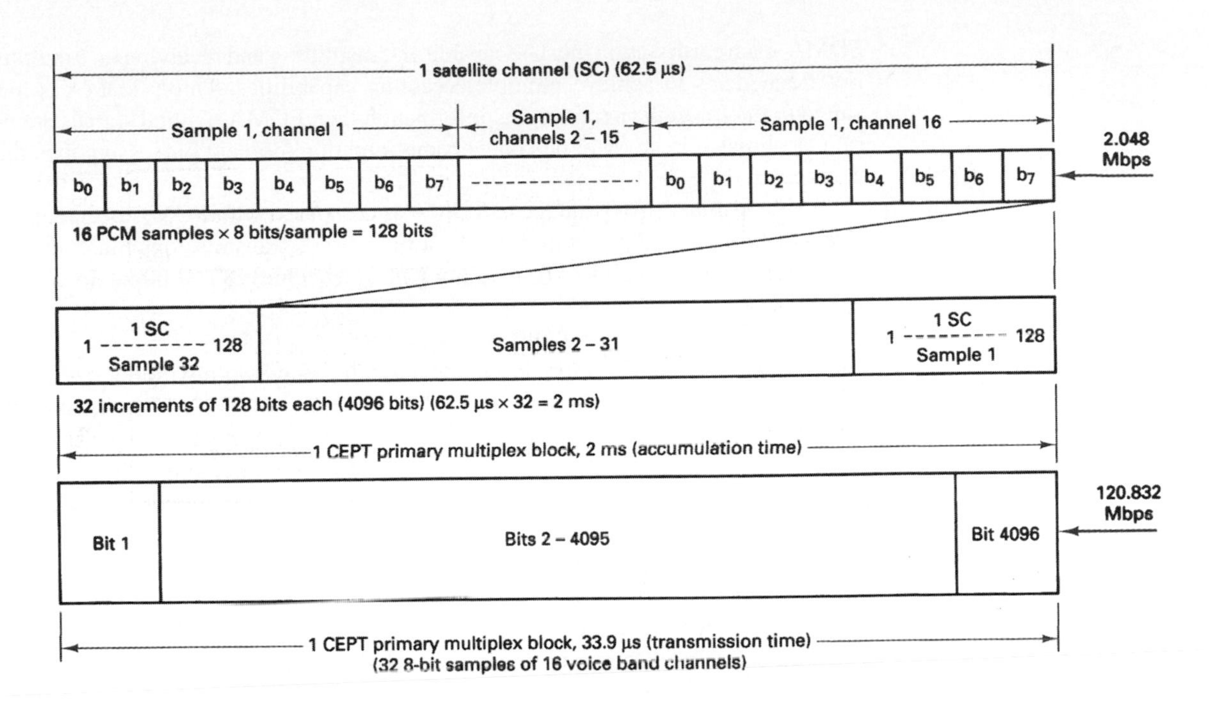

FIGURE 26-10 TDMA, CEPT primary multiplex frame

voice-band channels. Each channel has a separate codec that samples the incoming voice signals at a 16-kHz rate and converts those samples to eight-bit binary codes. This results in 128 kbps transmitted at a 2.048-MHz rate from each voice channel codec. The 16 128-kbps transmissions are time-division multiplexed into a subframe that contains one eight-bit sample from each of the 16 channels (128 bits). It requires only 62.5 μs to accumulate the 128 bits (2.048-Mbps transmission rate). The CEPT multiplex format specifies a 2-ms frame time. Consequently, each earth station can transmit only once every 2 ms and, therefore, must store the PCM-encoded samples. The 128 bits accumulated during the first sample of each voice-band channel are stored in a holding register, while a second sample is taken from each channel and converted into another 128-bit *subframe.* This 128-bit sequence is stored in the holding register behind the first 128 bits. The process continues for 32 subframes (32×62.5 μs = 2 ms). After 2 ms, 32 eight-bit samples have been taken from each of 16 voice-band channels for a total of 4096 bits ($32 \times 8 \times 16 = 4096$). At this time, the 4096 bits are transferred to an output shift register for transmission. Because the total TDMA frame is 2 ms long and during this 2-ms period each of the participating earth stations must transmit at different times, the individual transmissions from each station must occur in a significantly shorter time period. In the CEPT frame, a transmission rate of 120.832 Mbps is used. This rate is the 59th multiple of 2.048 Mbps. Consequently, the actual transmission of the 4096 accumulated bits takes approximately 33.9 μs. At the earth station receivers, the 4096 bits are stored in a holding register and shifted at a 2.048-Mbps rate. Because all the clock rates (500 Hz, 16 kHz, 128 kHz, 2.048 MHz, and 120.832 MHz) are synchronized, the PCM codes are accumulated, stored, transmitted, received, and then decoded in perfect synchronization. To the users, the voice transmission appears to be a continuous process.

There are several advantages of TDMA over FDMA. The first, and probably the most significant, is that with TDMA only the carrier from one earth station is present in the satellite transponder at any given time, thus reducing intermodulation distortion. Second, with

FDMA, each earth station must be capable of transmitting and receiving on a multitude of carrier frequencies to achieve multiple-accessing capabilities. Third, TDMA is much better suited to the transmission of digital information than FDMA. Digital signals are more naturally acclimated to storage, rate conversions, and time-domain processing than their analog counterparts.

The primary disadvantage of TDMA as compared with FDMA is that in TDMA precise synchronization is required. Each earth station's transmissions must occur during an exact time slot. Also, bit and frame timing must be achieved and maintained with TDMA.

26-3-3 Code-Division Multiple Access

With FDMA, earth stations are limited to a specific bandwidth within a satellite channel or system but have no restriction on when they can transmit. With TDMA, an earth station's transmissions are restricted to a precise time slot but have no restriction on what frequency or bandwidth it may use within a specified satellite system or channel allocation. With *code-division multiple access* (CDMA), there are no restrictions on time or bandwidth. Each earth station transmitter may transmit whenever it wishes and can use any or all the bandwidth allocated a particular satellite system or channel. Because there is no limitation on the bandwidth, CDMA is sometimes referred to as *spread-spectrum multiple access;* transmissions can spread throughout the entire allocated bandwidth. Transmissions are separated through envelope encryption/decryption techniques. That is, each earth station's transmissions are encoded with a unique binary word called a *chip code.* Each station has a unique chip code. To receive a particular earth station's transmission, a receive station must know the chip code for that station.

Figure 26-11 shows the block diagram of a CDMA encoder and decoder. In the encoder (Figure 26-11a), the input data (which may be PCM-encoded voice-band signals or raw digital data) is multiplied by a unique chip code. The product code PSK modulates an IF carrier, which is up-converted to RF for transmission. At the receiver (Figure 26-11b), the RF is down-converted to IF. From the IF, a coherent PSK carrier is recovered. Also, the chip code is acquired and used to synchronize the receive station's code generator. Keep in mind, the receiving station knows the chip code but must generate a chip code that is synchronous in time with the receive code. The recovered synchronous chip code multiplies the recovered PSK carrier and generates a PSK-modulated signal that contains the PSK carrier plus the chip code. The received IF signal that contains the chip code, the PSK carrier, and the data information is compared with the received IF signal in the *correlator.* The function of the correlator is to compare the two signals and recover the original data. Essentially, the correlator subtracts the recovered PSK carrier + chip code from the received PSK carrier + chip code + data. The resultant is the data.

The correlation is accomplished on the analog signals. Figure 26-12 shows how the encoding and decoding is accomplished. Figure 26-12a shows the correlation of the correctly received chip code. A $+1$ indicates an in-phase carrier, and a -1 indicates an out-of-phase carrier. The chip code is multiplied by the data (either $+1$ or -1). The product is either an in-phase code or one that is 180° out of phase with the chip code. In the receiver, the recovered synchronous chip code is compared in the correlator with the received signaling elements. If the phases are the same, a $+1$ is produced; if they are 180° out of phase, a -1 is produced. It can be seen that if all the recovered chips correlate favorably with the incoming chip code, the output of the correlator will be a $+6$ (which is the case when a logic 1 is received). If all the code chips correlate 180° out of phase, a -6 is generated (which is the case when a logic 0 is received). The bit decision circuit is simply a threshold detector. Depending on whether a $+6$ or -6 is generated, the threshold detector will output a logic 1 or a logic 0, respectively.

As the name implies, the correlator looks for a correlation (similarity) between the incoming coded signal and the recovered chip code. When a correlation occurs, the bit decision circuit generates the corresponding logic condition.

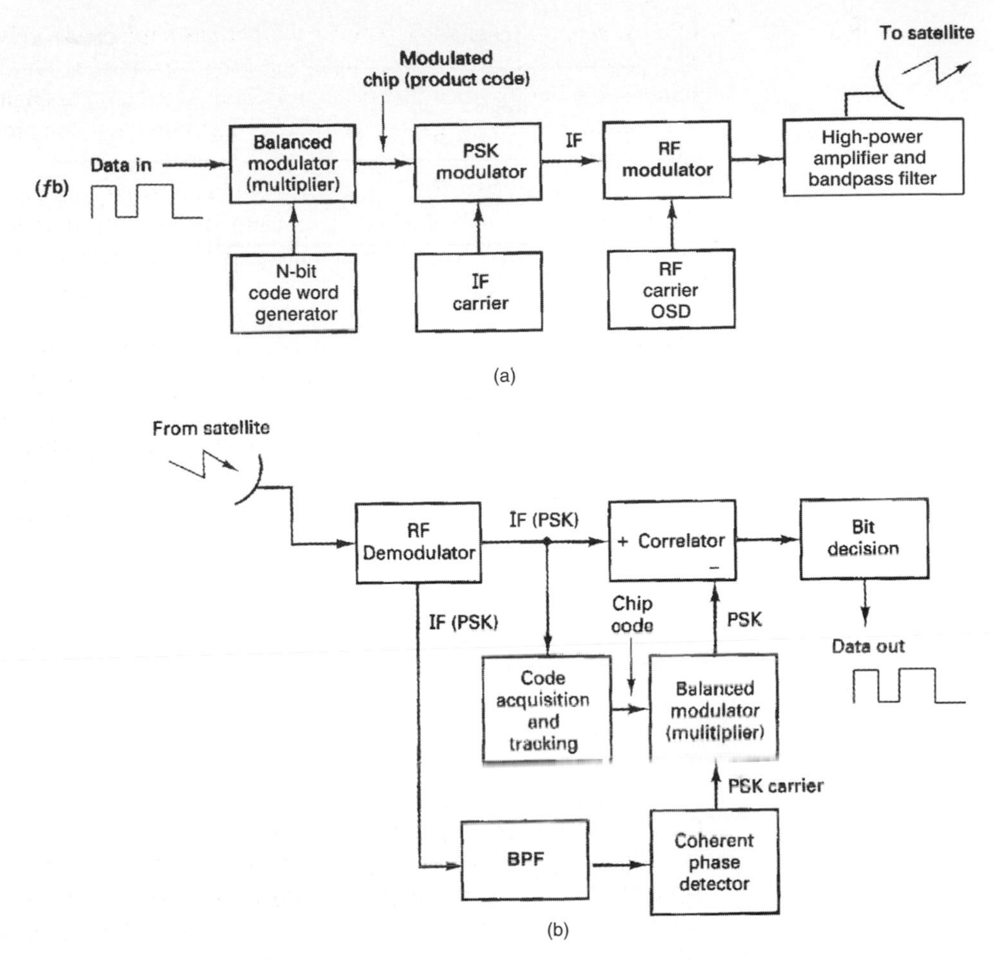

FIGURE 26-11 Code-division multiple access (CDMA): (a) encoder; (b) decoder

With CDMA, all earth stations within the system may transmit on the same frequency at the same time. Consequently, an earth station receiver may be receiving coded PSK signals simultaneously from more than one transmitter. When this is the case, the job of the correlator becomes considerably more difficult. The correlator must compare the recovered chip code with the entire received spectrum and separate from it only the chip code from the desired earth station transmitter. Consequently, the chip code from one earth station must not correlate with the chip codes from any of the other earth stations.

Figure 26-12b shows how such a coding scheme is achieved. If half the bits within a code were made the same and half were made exactly the opposite, the resultant would be zero cross correlation between chip codes. Such a code is called an *orthogonal code.* In Figure 26-12b, it can be seen that when the orthogonal code is compared with the original chip code, there is no correlation (i.e., the sum of the comparison is zero). Consequently, the orthogonal code, although received simultaneously with the desired chip code, had absolutely no effect on the correlation process. For this example, the orthogonal code is received in exact time synchronization with the desired chip code; this is not always the case. For systems that do not have time-synchronous transmissions, codes must be developed where there is no correlation between one station's code and any phase of another station's code.

The primary difference between spread-spectrum PSK transmitters and other types of PSK transmitters is the additional modulator where the code word is multiplied by the incoming data. Because of the pseudorandom nature of the code word, it is often referred

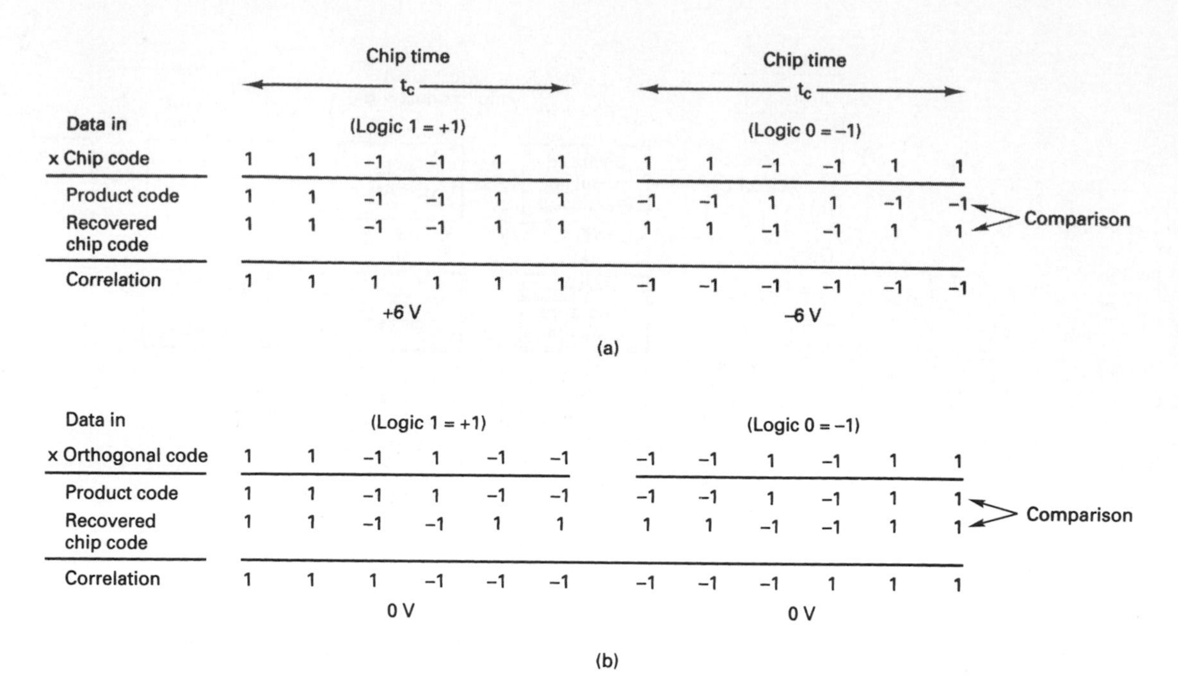

FIGURE 26-12 CDMA code/data alignment: (a) correct code; (b) orthogonal code

to as *pseudorandom noise* (PRN). The PRN must have a high autocorrelation property with itself and a low correlation property with other transmitter's pseudorandom codes. The code word rate (R_{cw}) must exceed the incoming data rate (R_d) by several orders of magnitude. In addition, the code rate must be statistically independent of the data signal. When these two conditions are satisfied, the final output signal spectrum will be increased (spread) by a factor called the *processing gain.* Processing gain is expressed mathematically as

$$G = \frac{R_{cw}}{R_d} \qquad \textbf{(26-1)}$$

where G is processing gain and $R_{cw} >> R_d$.

A spread-spectrum signal cannot be demodulated accurately if the receiver does not possess a despreading circuit that matches the code word generator in the transmitter. Three of the most popular techniques used to produce the spreading function are *direct sequence, frequency hopping,* and a combination of direct sequence and frequency hopping called *hybrid direct-sequence frequency hopping* (hybrid-DS/FH).

26-3-3-1 Direct sequence. Direct-sequence spread spectrum (DS-SS) is produced when a bipolar data-modulated signal is linearly multiplied by the spreading signal in a special balanced modulator called a *spreading correlator.* The spreading code rate $R_{cw} = 1/T_c$, where T_c is the duration of a single bipolar pulse (i.e., the chip). Chip rates are 100 to 1000 times faster than the data message; therefore, chip times are 100 to 1000 times shorter in duration than the time of a single data bit. As a result, the transmitted output frequency spectrum using spread spectrum is 100 to 1000 times wider than the bandwidth of the initial PSK data-modulated signal. The block diagram for a direct-sequence spread-spectrum system is shown in Figure 26-13. As the figure shows, the data source directly modulates the carrier signal, which is then further modulated in the spreading correlator by the spreading code word.

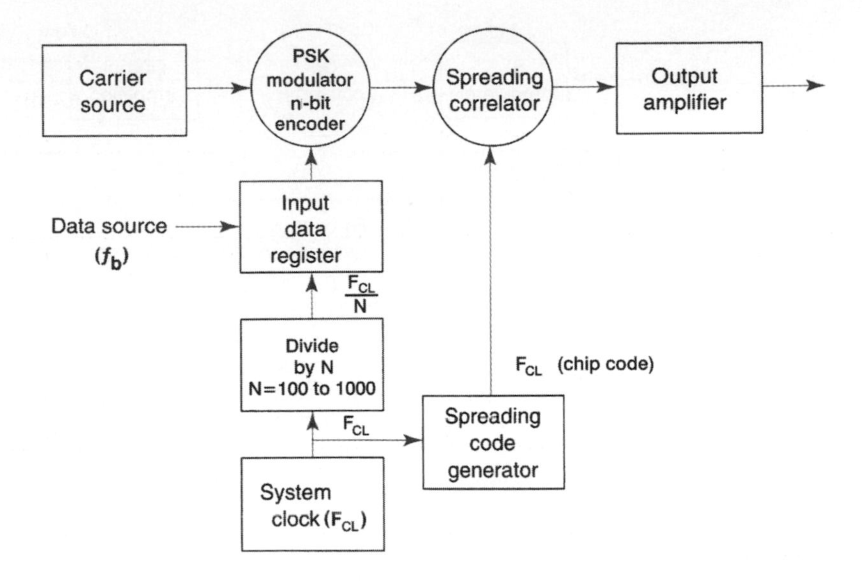

FIGURE 26-13 Simplified block diagram for a direct-sequence spread-spectrum transmitter

The spreading (chip) codes used in spread-spectrum systems are either *maximal-length sequence codes,* sometimes called *m-sequence codes* or *Gold codes.* Gold codes are combinations of maximal-length codes invented by Magnavox Corporation in 1967, especially for multiple-access CDMA applications. There is a relatively large set of Gold codes available with minimal correlation between chip codes. For a reasonable number of satellite users, it is impossible to achieve perfectly orthogonal codes. You can only design for a minimum cross correlation among chips.

One of the advantages of CDMA was that the entire bandwidth of a satellite channel or system may be used for each transmission from every earth station. For our example, the chip rate was six times the original bit rate. Consequently, the actual transmission rate of information was one-sixth of the PSK modulation rate, and the bandwidth required is six times that required to simply transmit the original data as binary. Because of the coding inefficiency resulting from transmitting chips for bits, the advantage of more bandwidth is partially offset and is, thus, less of an advantage. Also, if the transmission of chips from the various earth stations must be synchronized, precise timing is required for the system to work. Therefore, the disadvantage of requiring time synchronization in TDMA systems is also present with CDMA. In short, CDMA is not all that it is cracked up to be. The most significant advantage of CDMA is immunity to interference (jamming), which makes CDMA ideally suited for military applications.

26-3-3-2 Frequency-hopping spread spectrum. Frequency hopping is a form of CDMA where a digital code is used to continually change the frequency of the carrier. The carrier is first modulated by the data message and then up-converted using a frequency-synthesized local oscillator whose output frequency is determined by an *n*-bit pseudorandom noise code produced in a spreading code generator. The simplified block diagram for a frequency-hopping spread-spectrum transmitter is shown in Figure 26-14.

With frequency hopping, the total available bandwidth is partitioned into smaller frequency bands, and the total transmission time is subdivided into smaller time slots. The idea is to transmit within a limited frequency band for only a short time, then switch to another

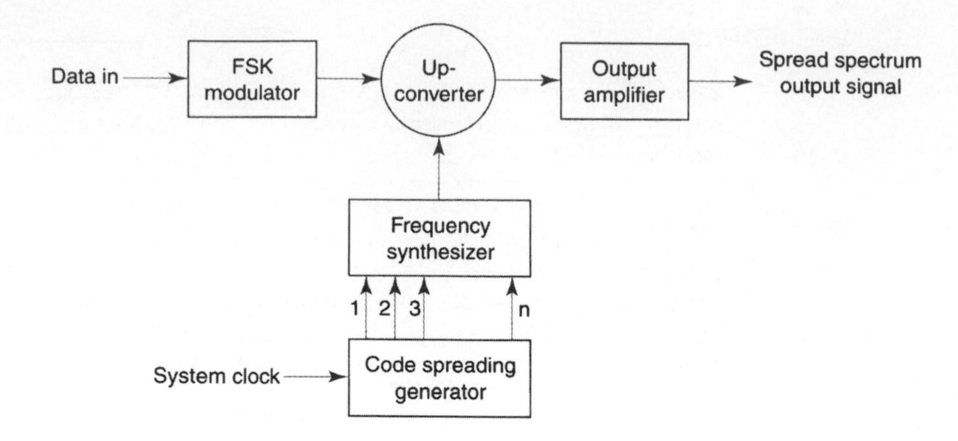

FIGURE 26-14 Simplified block diagram of a frequency-hopping spread-spectrum transmitter

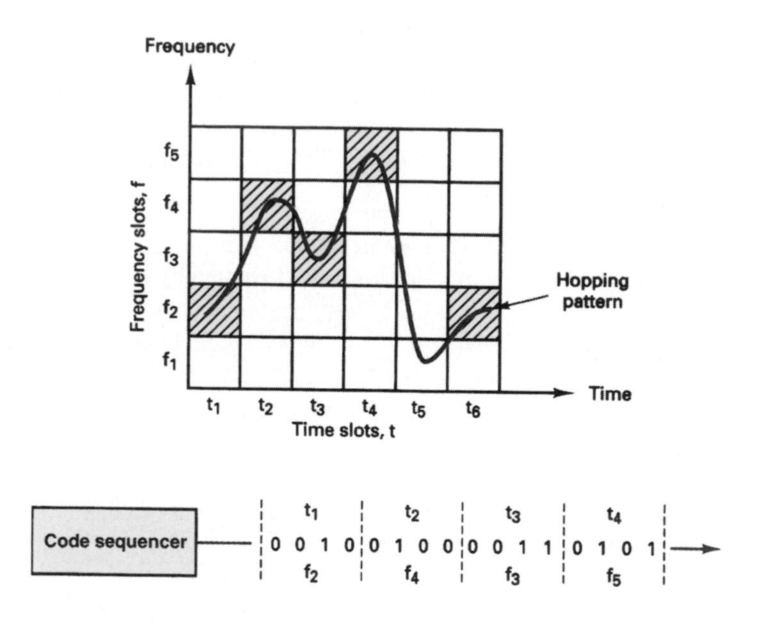

FIGURE 26-15 Frequency time-hopping matrix

frequency band and so on. This process continues indefinitely. The frequency-hopping pattern is determined by a binary spreading code. Each station uses a different code sequence. A typical hopping pattern (frequency-time matrix) is shown in Figure 26-15.

With frequency hopping, each earth station within a CDMA network is assigned a different frequency-hopping pattern. Each transmitter switches (hops) from one frequency band to the next according to their assigned pattern. With frequency hopping, each station uses the entire RF spectrum but never occupies more than a small portion of that spectrum at any one time.

FSK is the modulation scheme most commonly used with frequency hopping. When it is a given station's turn to transmit, it sends one of the two frequencies (either mark or space) for the particular band in which it is transmitting. The number of stations in a given frequency-hopping system is limited by the number of unique hopping patterns that can be generated.

26-4 CHANNEL CAPACITY

Essentially, there are two methods used to interface terrestrial voice-band channels with satellite channels: digital noninterpolated interfaces (DNI) and digital speech interpolated interfaces (DSI).

26-4-1 Digital Noninterpolated Interfaces

A *digital noninterpolated interface* assigns an individual terrestrial channel (TC) to a particular satellite channel (SC) for the duration of the call. A DNI system can carry no more traffic than the number of satellite channels it has. Once a TC has been assigned an SC, the SC is unavailable to the other TCs for the duration of the call. DNI is a form of preassignment; each TC has a permanent dedicated SC.

26-4-2 Digital Speech Interpolated Interfaces

A *digital speech interpolated interface* assigns a terrestrial channel to a satellite channel only when speech energy is present on the TC. DSI interfaces have *speech detectors* that are similar to *echo suppressors;* they sense speech energy, then seize an SC. Whenever a speech detector senses energy on a TC, the TC is assigned to an SC. The SC assigned is randomly selected from the idle SCs. On a given TC, each time speech energy is detected, the TC could be assigned to a different SC. Therefore, a single TC can use several SCs for a single call. For demultiplexing purposes, the TC/SC assignment information must be conveyed to the receive terminal. This is done on a common signaling channel similar to the one used on the SPADE system. DSI is a form of demand assignment; SCs are randomly assigned on an as-needed basis.

With DSI it is apparent that there is a *channel compression;* there can be more TCs assigned than there are SCs. Generally, a TC:SC ratio of 2:1 is used. For a full-duplex (two-way simultaneous) communication circuit, there is speech in each direction 40% of the time, and for 20% of the time the circuit is idle in both directions. Therefore, a DSI gain of slightly more than 2 is realized. The DSI gain is affected by a phenomenon called *competitive clipping.* Competitive clipping is when speech energy is detected on a TC and there is no SC to assign it to. During the *wait* time, speech information is lost. Competitive clipping is not noticed by a subscriber if its duration is less than 50 ms.

To further enhance the channel capacity, a technique called *bit stealing* is used. With bit stealing, channels can be added to fully loaded systems by stealing bits from the in-use channels. Generally, an overload channel is generated by stealing the least-significant bit from seven other satellite channels. Bit stealing results in eight channels with seven-bit resolution for the time that the *overload channel* is in use. Consequently, bit stealing results in a lower SQR than normal.

26-4-3 Time-Assignment Speech Interpolation

Time-assignment speech interpolation (TASI) is a form of analog channel compression that has been used for suboceanic cables for many years. TASI is very similar to DSI except that the signals interpolated are analog rather than digital. TASI also uses a 2:1 compression ratio. TASI was also the first means used to scramble voice for military security. TASI is similar to a packet data network; the voice message is chopped up into smaller segments made of sounds or portions of sounds. The sounds are sent through the network as separate bundles of energy, then put back together at the receive end to reform the original voice message.

26-5 SATELLITE RADIO NAVIGATION

Navigation can be defined as the art or science of plotting, ascertaining, or directing the course of movements, in other words, *knowing where you are and being able to find your way around.* The most ancient and rudimentary method of navigation is *wandering.* Wandering is simply continuing to travel about until you reach your destination, assuming of course that you have one. My good friend and worldwide traveler, Todd Ferguson, once said, "True travel has no destination." Wandering is the popular navigation technique used by many students during their first week of classes at all colleges and universities. Probably the earliest effective or useful means of navigation is *celestial navigation.* With celestial navigation, direction and distance are determined from precisely timed sightings of celestial bodies, including the stars and moon. This is a primitive technique that dates back thousands of years. An obvious disadvantage of celestial navigation is that it works best at night, preferably with clear skies.

Another rather rudimentary method of navigation is *piloting.* Piloting is fixing a position and direction with respect to familiar, significant landmarks, such as railroad tracks, water towers, barns, mountain peaks, and bodies of water. Piloting derived its name from early aircraft pilots who used this method of navigation.

Dead (ded) *reckoning* is a navigation technique that determines position by extrapolating a series of measured velocity increments. The term *dead* is derived from the word "*ded*uced" and not necessarily from the fate of the people who used the technique. Dead reckoning was used quite successfully by Charles Lindbergh in 1927 during his historic 33-hour transatlantic journey and quite unsuccessfully by Amelia Earhart in 1937 during her attempt to make the first around-the-world flight.

Although each of the navigation methods described thus far had its place in time, undoubtedly the most accurate navigation technique to date is *radio* or *electronic navigation.* With radio navigation, position is determined by measuring the travel time of an electromagnetic wave as it moves from a transmitter to a receiver. There are approximately 100 different types of domestic radio navigation systems currently being used. Some use terrestrial (land-based) broadcast transmitters and others use satellite (space-based) broadcast transmitters. The most accurate and useful radio navigation systems include the following:

Decca (terrestrial surface broadcast)
Omega (terrestrial surface broadcast; provides global coverage)
Loran (terrestrial surface broadcast)
Navy Transit GPS (low-orbit satellite broadcast; provides global coverage)
Navstar GPS (medium-orbit satellite broadcast; provides global coverage)

Loran and Navstar are the two most often used radio navigation systems today.

26-5-1 Loran Navigation

Until recently, *Loran* (*Lo*ng *Ra*nge *N*avigation) was the most effective, reliable, and accurate means of radio navigation. Loran-A was developed during World War II, and the most recent version, Loran-C, surfaced in 1980. Today, Loran is used primarily for recreational aircraft and ships.

With Loran, receivers acquire specially coded signals from two pairs of high-powered, land-based transmitters whose locations are precisely known. The elapsed time between reception of the coded signals is precisely measured and converted in the receiver to distance using the propagation speed of electromagnetic waves. Using basic geometry and the relationship between distance (d), speed (v), and time (t) ($d = vt$), the location of the receiver can be determined with a high degree of accuracy. There is only one set of coordinates that possess a particular time (distance) relationship from four sources.

Loran is only as accurate as the preciseness of the transmission times of the coded signals. System errors are due primarily to propagation problems, such as the fact that Earth's surface is not smooth or perfectly round. Atmospheric conditions and multiple transmission paths can also adversely affect the performance of Loran. However, probably the most prominent disadvantage of Loran is the fact that it does not provide continuous worldwide coverage. Land-based transmitters can be located only where there is land, which is a relatively small proportion of Earth's surface. Consequently, there are locations where Loran signals simply cannot be received (dead spots). However good or bad Loran may have been or could be is unimportant because a newer, better technique of radio navigation called Navstar GPS has emerged that utilizes satellite-based transmitters.

26-5-2 Navstar GPS

Navstar is an acronym for *Nav*igation *S*ystem with *T*ime *A*nd *R*anging, and *GPS* is an abbreviation of Global Positioning System. Navstar GPS is the newest and most accurate system of radio navigation available. Navstar GPS is a satellite-based open navigation system, which simply means that it is available to anyone equipped with a GPS receiver. The United States Department of Defense (DoD) developed Navstar to provide continuous, highly precise position, velocity, and time information to land-, sea-, air-, and space-based users. In essence, Navstar GPS is a space-based navigation, three-dimensional positioning, and time-distribution system. The intent of the system is to use a combination of ground stations, orbiting satellites, and special receivers to provide navigation capabilities to virtually everyone, at any time, anywhere in the world, regardless of weather conditions. The Navstar Satellite System was completed in 1994 and is maintained by the United States Air Force.

26-5-3 GPS Services

GPS provides two levels of service or accuracy: standard positioning service and precise positioning service.

26-5-3-1 Standard positioning service. The *standard positioning service* (SPS) is a positioning and timing service that is available to all GPS users (military, private, and commercial) on a continuous, worldwide basis with no direct charge. SPS will provide a predictable positioning accuracy that 95% of the time is to within 100 m horizontally, 156 m vertically, and 185 m 3-D, with a time transfer accuracy to UTC (*Universal Transverse Mercator Grid*) within 340 nanoseconds. The accuracy of the SPS service can be downgraded during times of national emergencies. For security reasons, the accuracy of the SPS service is intentionally degraded by the DoD through the use of a technique called *selective availability*. Selective availability (SA) is accomplished by manipulating navigation message orbit data (epsilon) and/or the satellite clock frequency (dither).

26-5-3-2 Precise positioning service. The *precise positioning service* (PPS) is a highly accurate military positioning, velocity, and timing service that is available on a continuous, worldwide basis to users authorized by the DoD. PPS user equipment provides a predictable positioning accuracy 95% of the time of at least 22 m horizontally, 27.7 m vertically, and 35.4 m 3-D, and a time transfer accuracy to UTC within 200 nanoseconds. Only authorized users with cryptographic equipment and keys and specially equipped receivers can use the precise positioning service. PPS was designed primarily to be used by the U.S. and allied military, certain U.S. government agencies, and selected civil users specifically approved by the U.S. government.

26-5-4 Navstar Segments

Navstar GPS consists of three segments: a *space segment,* a *ground control segment,* and a *user segment.*

26-5-4-1 Satellite segment. The U.S. Air Force Space Command (AFSC) formally declared the Navstar GPS satellite system as being fully operational as of April 27, 1995. The satellite segment, sometimes called the *space segment,* consists of 24 operational satellites

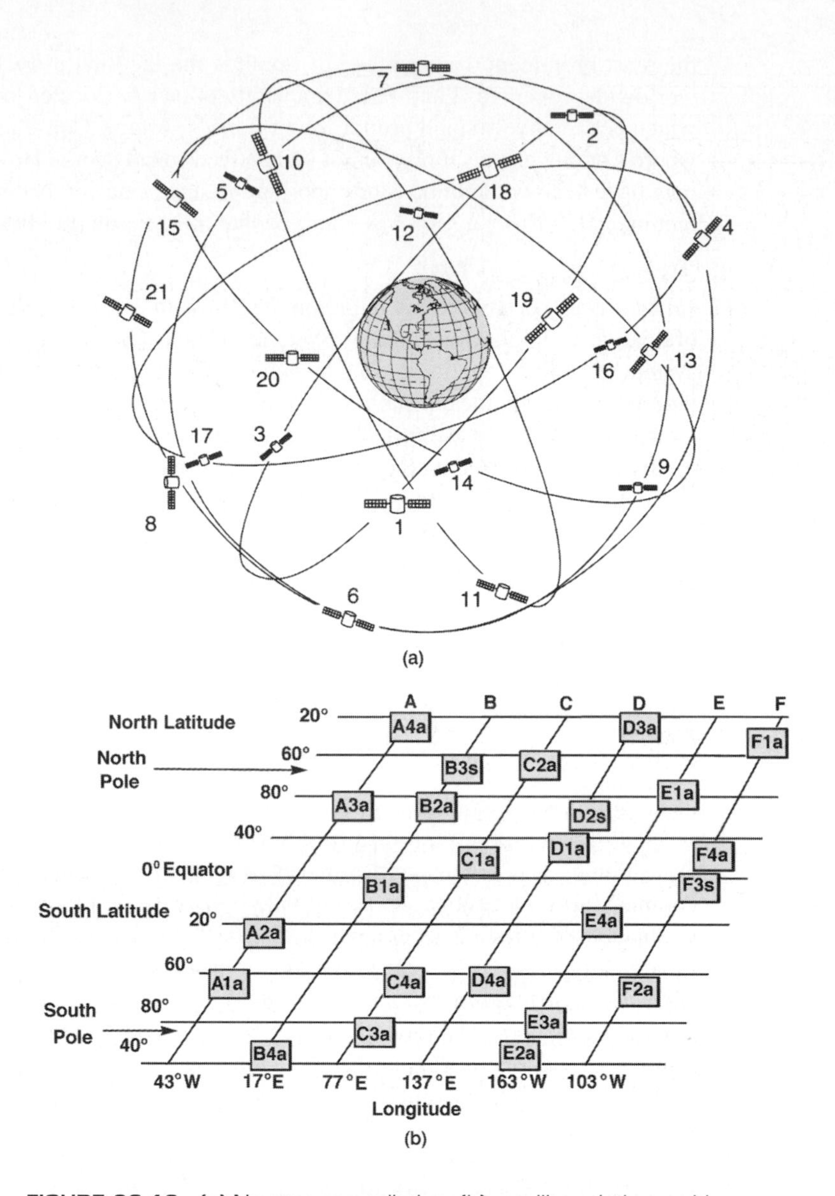

FIGURE 26-16 (a) Navstar constellation; (b) satellite relative positions

revolving around Earth in six orbital planes approximately 60° apart with four satellites in each plane. There are 21 working satellites and three satellites reserved as spaces. In the event of a satellite failure, one of the spare space vehicles can be moved into its place. (There are actually more than 24 satellites now, as some of the older space vehicles have been replaced with newer satellites with more modern propulsion and guidance systems.) Figure 26-16a shows orbital patterns for the 21 working satellites in the Navstar constellation, and Figure 26-16b shows the relative positions of the 24 satellites in respect to each other.

Navstar satellites are not geosynchronous. The satellites revolve around Earth in a circular pattern with an inclined orbit. The angle of elevation at the ascending node is 55° with respect to the equatorial plane. The average elevation of a Navstar satellite is 9476 statute miles (approximately 20,200 km) above Earth. Navstar satellites take approximately 12 hours to orbit Earth. Therefore, their position is approximately the same at the same sidereal time each day (the satellites actually appear 4 minutes earlier each day). Figure 26-17 shows the orbits of several Navstar satellites superimposed over a Mercator projection of

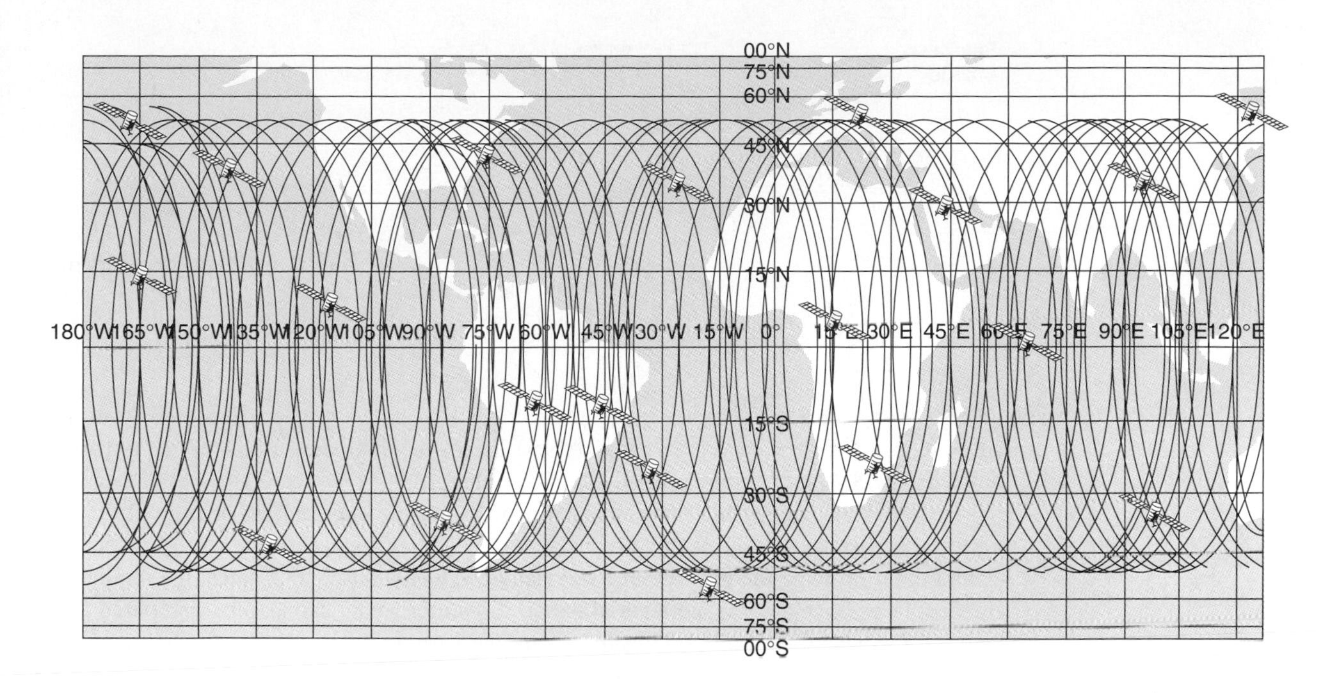

FIGURE 26-17 Mercator projection of Navstar satellite orbits

the world. As the figure shows, the satellites spiral around Earth in six planes virtually covering the surface of the entire globe.

The position of the Navstar satellites in orbit is arranged such that between five and eight satellites will be in view of any user at all times, thereby ensuring continuous worldwide coverage. Information from three satellites is needed to calculate a navigational unit's horizontal location on Earth's surface (*two-dimensional reporting*), but information from four satellites enables a receiver to also determine its altitude (*three-dimensional reporting*). Three-dimensional reporting is obviously more crucial on land because ground surfaces are not constant, whereas the surface of a large body of water is. Navstar satellites broadcast navigation and system data, atmospheric propagation correction data, and satellite clock bias information.

26-5-4-2 Navstar satellite groupings. There have been three distinct groups plus one subgroup of Navstar satellites. The groups are designated as *blocks.* The 11 block I prototype satellites were intended to be used only for system testing. Block II satellites were the first fully functional satellites that included onboard *cesium atomic clocks* for producing highly accurate timing signals. Block II satellites are capable of detecting certain error conditions, then automatically transmitting a coded message indicating that it is out of service. Block II satellites can operate for approximately 3.5 days between receiving updates and corrections from the control segment of the system. Block IIa satellites are identical to the standard block II versions except they can operate continuously for 180 days between uploads from the ground. The latest satellites, block IIR versions, can operate for 180 days between uploads and possess autonomous navigation capabilities by generating their own navigation information. Thus, the accuracy of the system using block IIR satellites can be maintained longer between uploads.

26-5-4-3 Navstar satellite identification. Each satellite has three identifying numbers. The first number is the Navstar number that identifies the specific satellite onboard hardware. The second number is the space vehicle (SV) number, which is assigned according to the order of the vehicle's launch. The third number is a *pseudorandom noise*

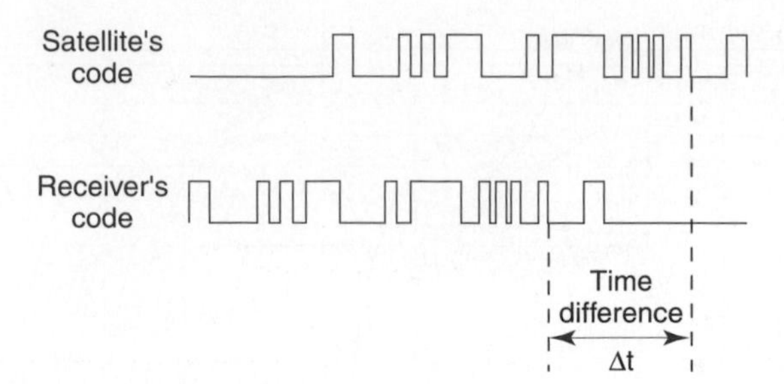

FIGURE 26-18 GPS pseudorandom timing code

(PRN) *code* number. This unique integer number is used to encrypt the signal from that satellite. Some GPS receivers identify the satellite from which they are receiving transmissions by the SV numbers; others use the PRN number.

Each Navstar satellite continually transmits a daily updated set of digitally coded *ephemeris data* that describes its precise orbit. *Ephemeris* is a term generally associated with a table showing the position of a heavenly body on a number of dates in a regular sequence, in essence, an astronomical almanac. Ephemeris data can also be computed for a satellite that specifies where on Earth the satellite is directly above at any given instant in terms of latitude and longitude coordinates.

26-5-4-4 Satellite ranging. The GPS system works by determining how long it takes a radio signal transmitted from a satellite to reach a land-based receiver and then using that time to calculate the distance between the satellite and the earth station receiver. Radio waves travel at approximately the speed of light, or 3×10^8 m/s. If a receiver can determine exactly when a satellite began sending a radio message and exactly when the message was received, then it can determine the propagation (delay) time. From the propagation time, the receiver can determine the distance between it and the satellite using the simple mathematical relationship

$$d = v \times t \qquad \textbf{(26-2)}$$

where d = distance between satellite and receiver (meters)
v = velocity (3×10^8 m/s)
t = propagation time (seconds)

The trick, of course, is in determining exactly when the synchronizing signal left the satellite. To determine this, the satellite transmitter and earth station receiver produce identical synchronizing (pseudorandom) codes at exactly the same time, as shown in Figure 26-18. Each satellite continuously transmits its precise synchronizing code. After a synchronizing code is acquired, a receiver simply compares the received code with its own locally produced code to determine the propagation time. The time difference multiplied by the velocity of the radio signal gives the distance to the satellite.

Figure 26-19 illustrates how an aircraft can determine the range (distance) it is from four different satellites by simply measuring the propagation (delay) times and multiplying them by the speed of light. Again, simultaneous equations can be used to determine the aircraft's longitude and latitude.

For a receiver on Earth to determine its longitude and latitude, it must receive signals from three or more satellites identifying the satellite vehicle number or their pseudorandom timing code (PRN) and each satellite's location. The location of a satellite is described using a three-dimensional coordinate relative to Earth's center as shown in Figure 26-20. Earth's center is the reference point with coordinates 0, 0, 0. Thus, each satellite space vehicle has an X_s, Y_s, Z_s coordinate that pinpoints its location in respect to Earth's geocenter.

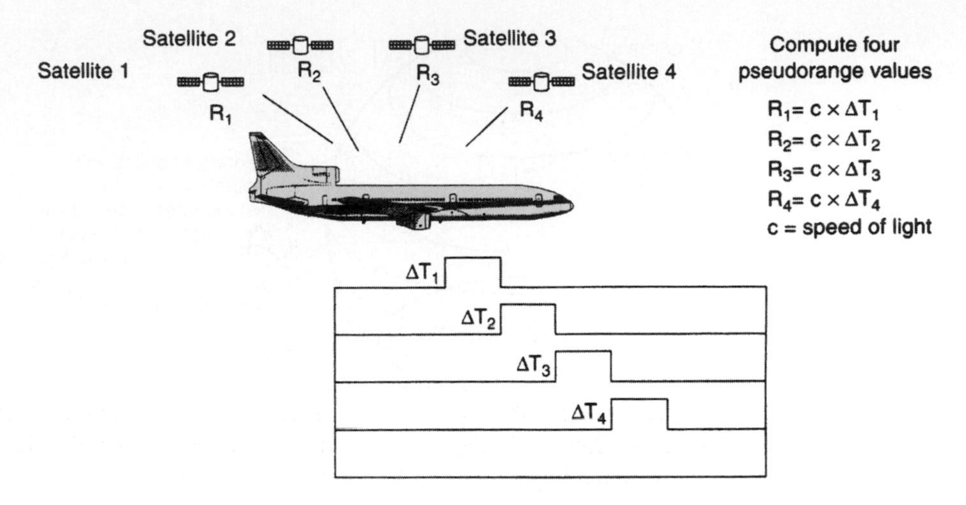

FIGURE 26-19 GPS ranging solution

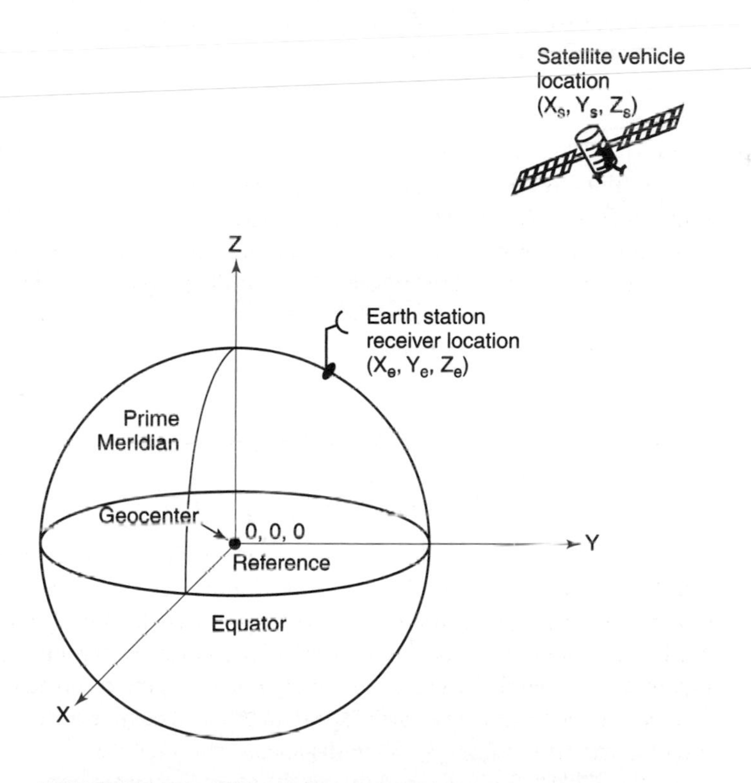

FIGURE 26-20 GPS satellite and earth station receiver coordinate system

The coordinates of the satellites, however, must be updated continually because they vary slightly as the satellite orbits Earth. The location of an earth station also has a three-dimensional coordinate, X_e, Y_e, Z_e, referenced to Earth's center as shown in Figure 26-20.

If an earth station receiver knows the location of a single satellite and the distance the satellite is from the receiver, it knows that it must be located somewhere on an imaginary sphere centered on the satellite with a radius equal to the distance the satellite is from the receiver. This is shown in Figure 26-21a. If the receiver knows the location of two satellites

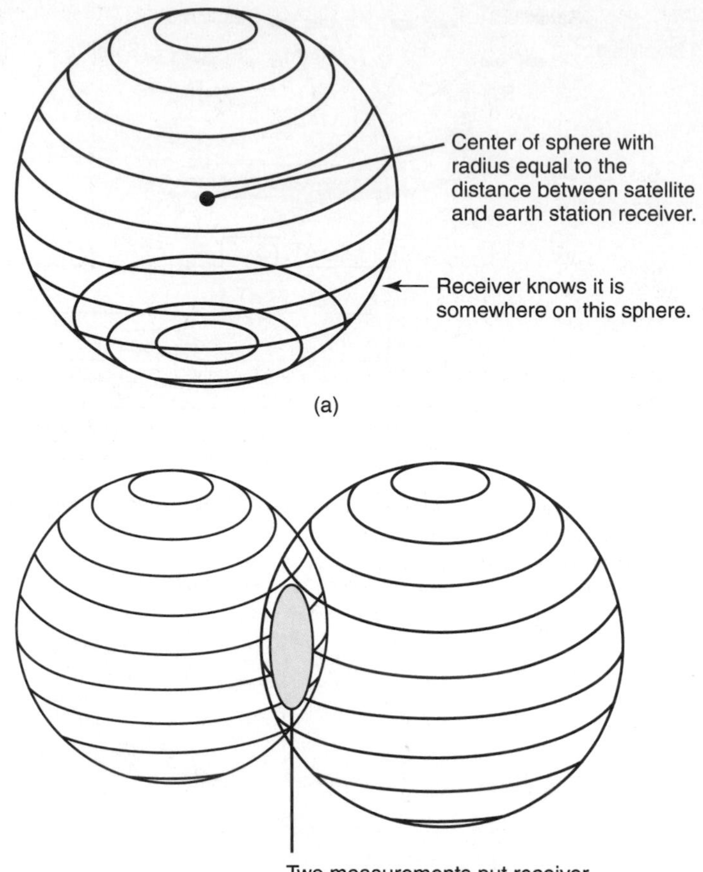
Center of sphere with
radius equal to the
distance between satellite
and earth station receiver.
Receiver knows it is
somewhere on this sphere.
(a)
Two measurements put receiver
somewhere on this circle.
(b)

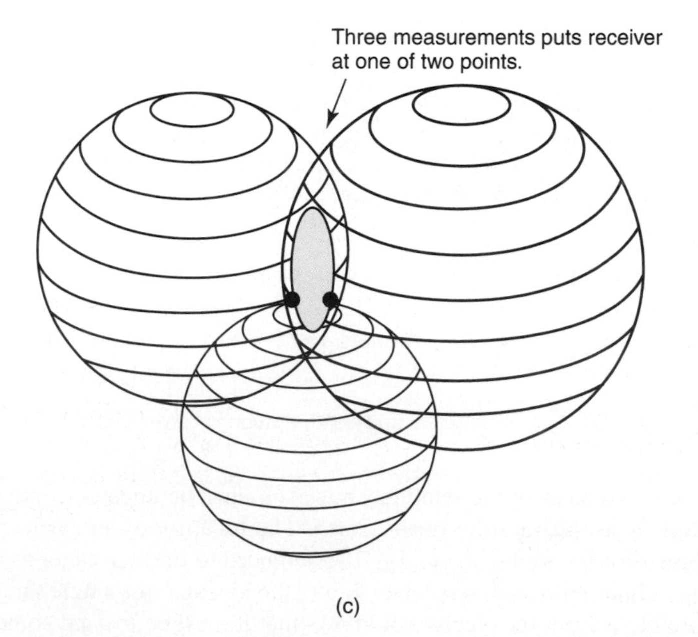
Three measurements puts receiver
at one of two points.
(c)

FIGURE 26-21 Earth station receiver location relative to the distance from (a) one satellite, (b) two satellites, and (c) three satellites

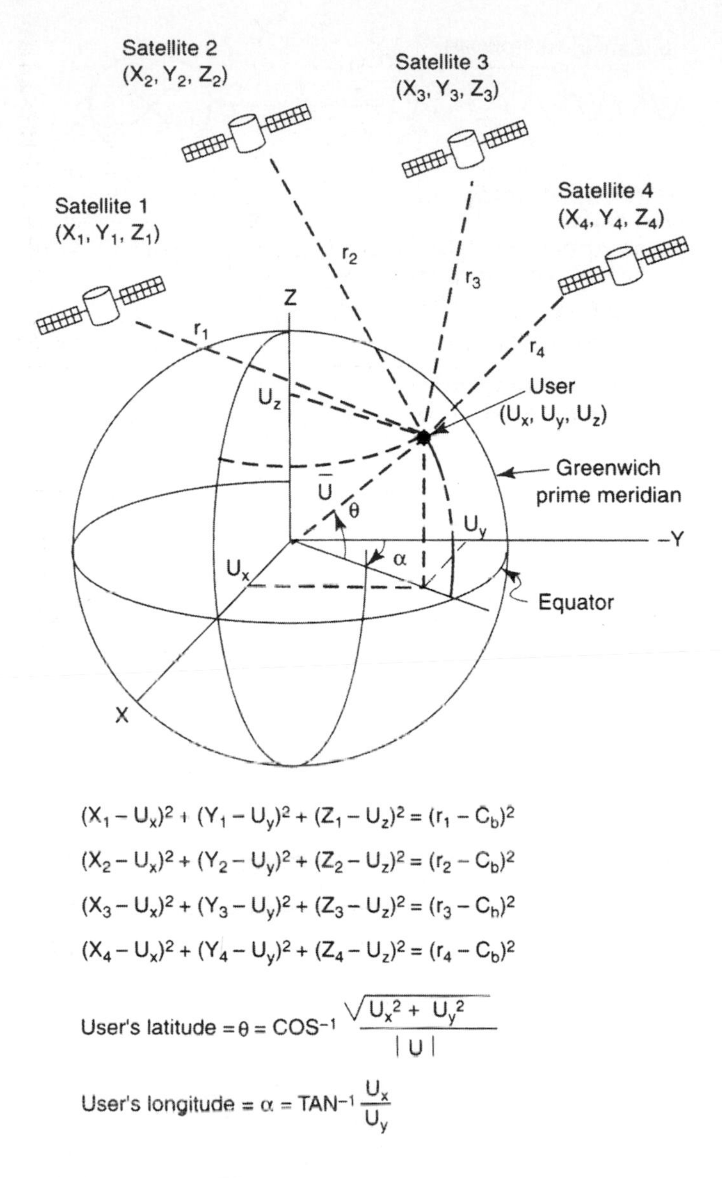

FIGURE 26-22 GPS satellite position calculations

and their distances from the receiver, it can narrow its location to somewhere on the circle formed where the two spheres intersect as shown in Figure 26-21b. If the location and distance to a third satellite is known, a receiver can pinpoint its location to one of two possible locations in space as shown in Figure 26-21c. The GPS receivers can usually determine which point is the correct location as one location is generally a ridiculous value. If the location and distance from a fourth satellite is known, the altitude of the earth station can also be determined.

Figure 26-22 shows there are three unknown position coordinates (x, y, and z). Therefore, three equations from three satellites are required to solve for the three unknown coordinates. A fourth unknown is the error in the receiver's clock, which affects the accuracy of the time-difference measurement. To eliminate the *clock bias error* (C_b), a fourth satellite is needed to produce the fourth equation necessary to solve for four unknowns using simultaneous equations. The solutions to the simultaneous equations for determining latitude and longitude are given in Figure 26-22.

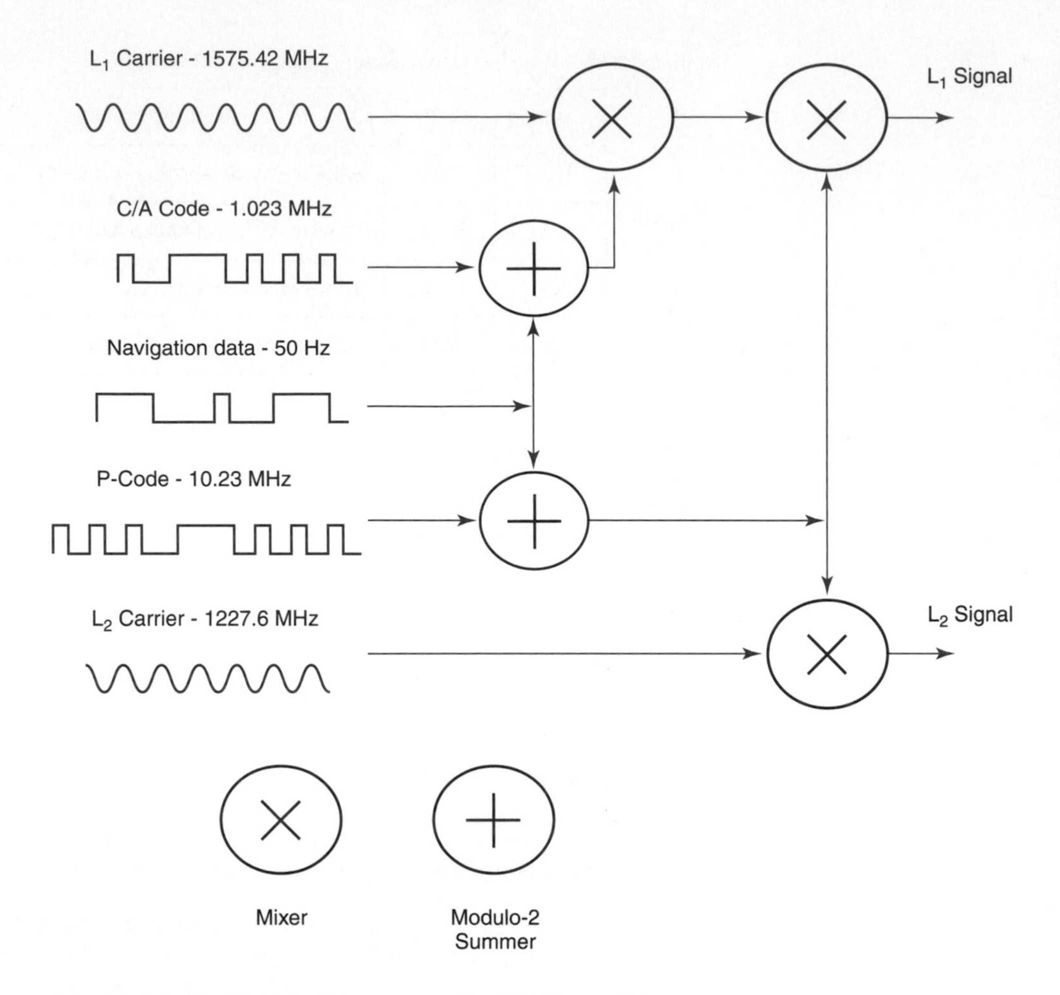

FIGURE 26-23 Simplified Navstar satellite CDMA transmitter

26-5-4-5 GPS satellite signals. All Navstar satellites transmit on the same two *L-band* microwave carrier frequencies: L_1 = 1575.42 MHz and L_2 = 1227.6 MHz. The L_1 signal carries the navigation message and the standard positioning service (SPS) code signals. The L_2 signal is used by the precise positioning service (PPS) equipped receivers to measure the ionospheric delay. GPS satellites use code division multiple accessing (CDMA spread *spectrum*), which allows all 24 satellites to transmit simultaneously on both carriers without interfering with each other. Three pseudorandom binary codes modulate the L_1 and L_2 carriers as shown in Figure 26-23.

1. The coarse/acquisition (C/A) code is a repeating pseudorandom noise (PRN) code with a 1.023-MHz chip rate and a period of 1 ms. This noiselike code modulates the L_1 carrier signal, spreading its spectrum over approximately a 1-MHz bandwidth. Each satellite vehicle has a different C/A PRN code which is used primarily to acquire the P-code. Satellites are often identified by their unique PRN number.
2. The precision (P) code has a 10.23-MHz rate, lasts a period of seven days, and contains the principle navigation ranging code. In the *antispoofing* (AS) mode, the P-code is encrypted into the Y-code which requires a classified AS module in each receiver channel and is used only by authorized users. The P(Y) code is the basis for the precision positioning system.

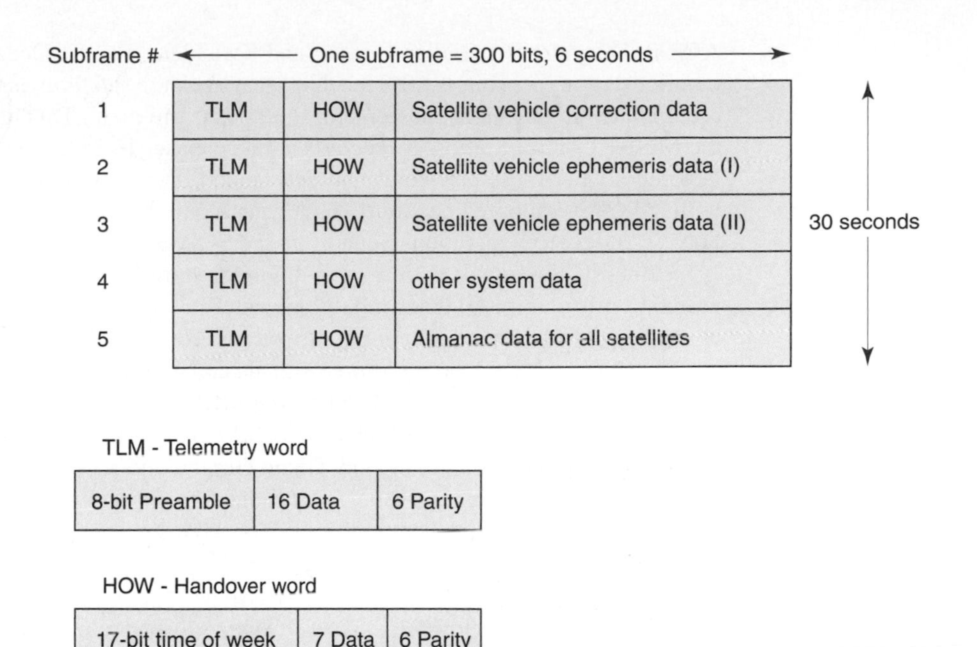

FIGURE 26-24 Navigation data frame format

3. The Y-code is used in place of the P-code whenever the antispoofing (AS) mode of operation is activated. Antispoofing guards against fake transmissions of satellite data by encrypting the P-code to form the Y-code.

Because of the spread-spectrum characteristics of the modulated carriers, the Navstar system provides a large margin of resistance to interference. Each satellite transmits a navigation message containing its orbital elements, clock behavior, system time, and status messages. In addition, an almanac is provided which gives the approximate ephemeris data for each active satellite and, thus, allows the users to find all the satellites once the first one has been acquired.

The navigation message modulates the L_1-C/A code signal as shown in Figure 26-23. The navigation message is a 50-Hz signal made of data bits that describe the GPS satellite orbits, clock corrections, and other system parameters. The data format used for the navigation message is shown in Figure 26-24. The navigation data frame consists of 1500 data bits divided into five 300-bit subframes. A navigation frame is transmitted once every 30 s (6 s for each subframe) for a transmission rate of 50 bps. Each subframe is preceded by a telemetry word (TLM) and a handover word (HOW). The TLM contains an eight-bit preamble, 24 data bits, and six parity bits; the HOW contains a 17-bit code identifying the time of week, seven data bits, and six parity bits.

The first subframe contains satellite vehicle correction data and the second and third subframes contain ephemeris parameter data. The fourth and fifth subframes are used to transmit different pages of system data including almanac data for all systems. Twenty-five frames (125 subframes) constitute the complete navigation message that is sent over a 12.5-minute period.

26-5-4-6 Control segment. The Navstar control segment, called the *operational control system* (OCS), includes all the fixed-location ground-based *monitor stations* located throughout the world, a *Master Control Station* (MCS), and uplink transmitters. There are passive monitor stations located in California, Hawaii, Alaska, Ascencion Island (off West

Africa), Diego Garcia (Indian Ocean), and Kwajalein (Pacific Ocean), among others. The monitor stations are simply GPS receivers that track the satellites as they pass overhead and accumulate ranging and ephemeris (orbital) data from them. This information is relayed to the Master Control Station at Falcon Air Force Base, 12 miles east of Colorado Springs, Colorado, where it is processed and determined if the actual satellite position compares with the GPS-computed position. The Master Control Station is managed by the U.S. Air Force's 2nd Space Operations Squadron (2nd SOPS).

The MCS receives data from the monitor stations in real time, 24 hours a day, and uses that information to determine if any satellites are experiencing clock or ephemeris changes and to detect equipment malfunctions. New navigation and ephemeris information is calculated from the monitored signals and uploaded to the satellites once or twice per day. The information calculated by the MCS, along with routine maintenance commands, is relayed to the satellites by ground-based uplink antennas. The ground antennas are located at Ascencion Island, Diego Garcia, and Kwajalein. The antenna facilities transmit to the satellites via an *S-band* radio link. In addition to its main function, the MCS maintains a 24-hour computer bulletin board system with the latest system news and status.

26-5-4-7 User segment. The GPS user segment consists of all the GPS receivers and the user community. GPS receivers convert signals received from space vehicles into position, velocity, and time estimates. Four satellites are required to compute the four dimensions of *x, y, z* (position) and time. GPS receivers are used for navigation, positioning, time dissemination, mapping, guidance systems, surveying, photogrammetry, public safety, archaeology, geology, geophysics, wildlife, aviation, marine, and numerous other research applications. Navigation in three dimensions, however, is the primary function of GPS.

GPS navigation receivers are made for aircraft, ships, ground vehicles, and handheld units for individuals. GPS navigation precise positioning is possible using GPS receivers at reference locations providing corrections and relative positioning data for remote receivers. Surveying, geodetic control, and plate tectonic studies are examples. Time and frequency dissemination, based on the precise clocks on board the space vehicles and controlled by the monitor stations, is another use for GPS. Astronomical observatories, telecommunications facilities, and laboratory standards can be set to precise time signals or controlled to accurate frequencies by special-purpose GPS receivers.

26-5-5 Differential GPS

Differential GPS makes standard GPS even more accurate. Differential GPS works by canceling out most of the natural and man-made errors that creep into normal GPS measurements. Inaccuracies in GPS signals come from a variety of sources, such as satellite clock drift, imperfect orbits, and variations in Earth's atmosphere. These imperfections are variable and difficult if not impossible to predict. Therefore, what is needed is a means to measure the actual errors as they are occurring.

With differential GPS, a second receiver is placed in a location whose exact position is known. It calculates its position from the satellite data and then compares it with its known position. The difference between the calculated and known positions is the error in the GPS signal. Differential GPS is only practical in locations where you can leave a receiver permanently, such as near an airport.

Sources of errors in GPS are satellite clocks, selective availability, ephemeris, atmospheric delays, multipath, receiver clocks, and so on. Table 26-1 summarizes GPS error sources for both standard and differential GPS.

Table 26-1 Summary of GPS Error Sources

Per Satellite Accuracy	Standard GPS	Differential GPS
Satellite clocks	1.5	0
Orbit errors	2.5	0
Ionosphere	5.0	0.4
Troposphere	0.5	0.2
Receiver noise	0.3	0.3
Multipath reception	0.6	0.6
Selective availability	30	0
Typical position accuracy (meters)		
Horizontal	50	1.3
Vertical	78	2.0
3-D	93	2.4

QUESTIONS

26-1. Discuss the drawbacks of using FDM/FM modulation for satellite multiple-accessing systems.

26-2. Contrast *preassignment* and *demand assignment*.

26-3. What are the three most common multiple-accessing arrangements used with satellite systems?

26-4. Briefly describe the multiple-accessing arrangements listed in Question 26-3.

26-5. Briefly describe the operation of Comsat's SPADE system.

26-6. What is meant by *single carrier per channel?*

26-7. What is a common-signaling channel and how is it used?

26-8. Describe what a reference burst is for TDMA and explain the following terms: *preamble, carrier recovery sequence, bit timing recovery, unique word,* and *correlation spike.*

26-9. Describe guard time.

26-10. Briefly describe the operation of the CEPT primary multiplex frame.

26-11. What is a store-and-forward system?

26-12. What is the primary advantage of TDMA as compared with FDMA?

26-13. What is the primary advantage of FDMA as compared with TDMA?

26-14. Briefly describe the operation of a CDMA multiple-accessing system.

26-15. Describe a chip code.

26-16. Describe what is meant by an orthogonal code.

26-17. Describe cross correlation.

26-18. What are the advantages of CDMA as compared with TDMA and FDMA?

26-19. What are the disadvantages of CDMA?

26-20. What is a Gold code?

26-21. Describe frequency hopping.

26-22. What is a frequency-time matrix?

26-23. Describe digital noninterpolated interfaces.

26-24. Describe digital speech interpolated interfaces.

26-25. What is channel compression, and how is it accomplished with a DSI system?

26-26. Describe competitive clipping.

26-27. What is meant by *bit stealing?*

26-28. Describe time-assignment speech interpolation.

PROBLEMS

26-1. How many satellite transponders are required to interlink six earth stations with FDM/FM modulation?

26-2. For the SPADE system, what are the carrier frequencies for channel 7? What are the allocated passbands for channel 7? What are the actual passband frequencies (excluding guard bands) required?

26-3. If a 512-bit preamble precedes each CEPT station's transmission, what is the maximum number of earth stations that can be linked together with a single satellite transponder?

26-4. Determine an orthogonal code for the following chip code (101010). Prove that your selection will not produce any cross correlation for an in-phase comparison. Determine the cross correlation for each out-of-phase condition that is possible.

26-5. How many satellite transponders are required to interlink five earth stations with FDM/FM modulation?

26-6. For the SPADE system, what are the carrier frequencies for channel 9? What are the allocated passbands for channel 10? What are the actual passband frequencies for channel 12 (excluding the guard bands)?

26-7. If a 256-bit preamble precedes each CEPT station's transmission, what is the maximum number of earth stations that can be linked together with a single satellite transponder?

26-8. Determine an orthogonal code for the following chip code (010101). Prove that your selection will not produce any cross correlation for an in-phase comparison. Determine the cross correlation for each out-of-phase condition that is possible.

APPENDIX A

The Smith Chart

INTRODUCTION

Mathematical solutions for transmission-line impedances are laborious. Therefore, it is common practice to use charts to graphically solve transmission-line impedance problems. Equation A-1 is the formula for determining the impedance at a given point on a transmission line:

$$Z = Z_O\left[\frac{Z_L + jZ_O \tan \beta S}{Z_O + jZ_L \tan \beta S}\right] \quad \textbf{(A-1)}$$

where Z = line impedance at a given point
Z_L = load impedance
Z_O = line characteristic impedance
βS = distance from the load to the point where the impedance value is to be calculated

Several charts are available on which the properties of transmission lines are graphically presented. However, the most useful graphical representations are those that give the impedance relations that exist along a lossless transmission line for varying load conditions. The *Smith chart* is the most widely used transmission-line calculator of this type. The Smith chart is a special kind of impedance coordinate system that portrays the relationship of impedance at any point along a uniform transmission line to the impedance at any other point on the line.

The Smith chart was developed by Philip H. Smith at Bell Telephone Laboratories and was originally described in an article entitled "Transmission Line Calculator" (*Electronics,* January 1939). A Smith chart is shown in Figure A-1. This chart is based on two sets of *orthogonal* circles. One set represents the ratio of the resistive component of the line impedance (R) to the characteristic impedance of the line (Z_O), which for a lossless line is also purely resistive. The second set of circles represents the ratio of the reactive component of

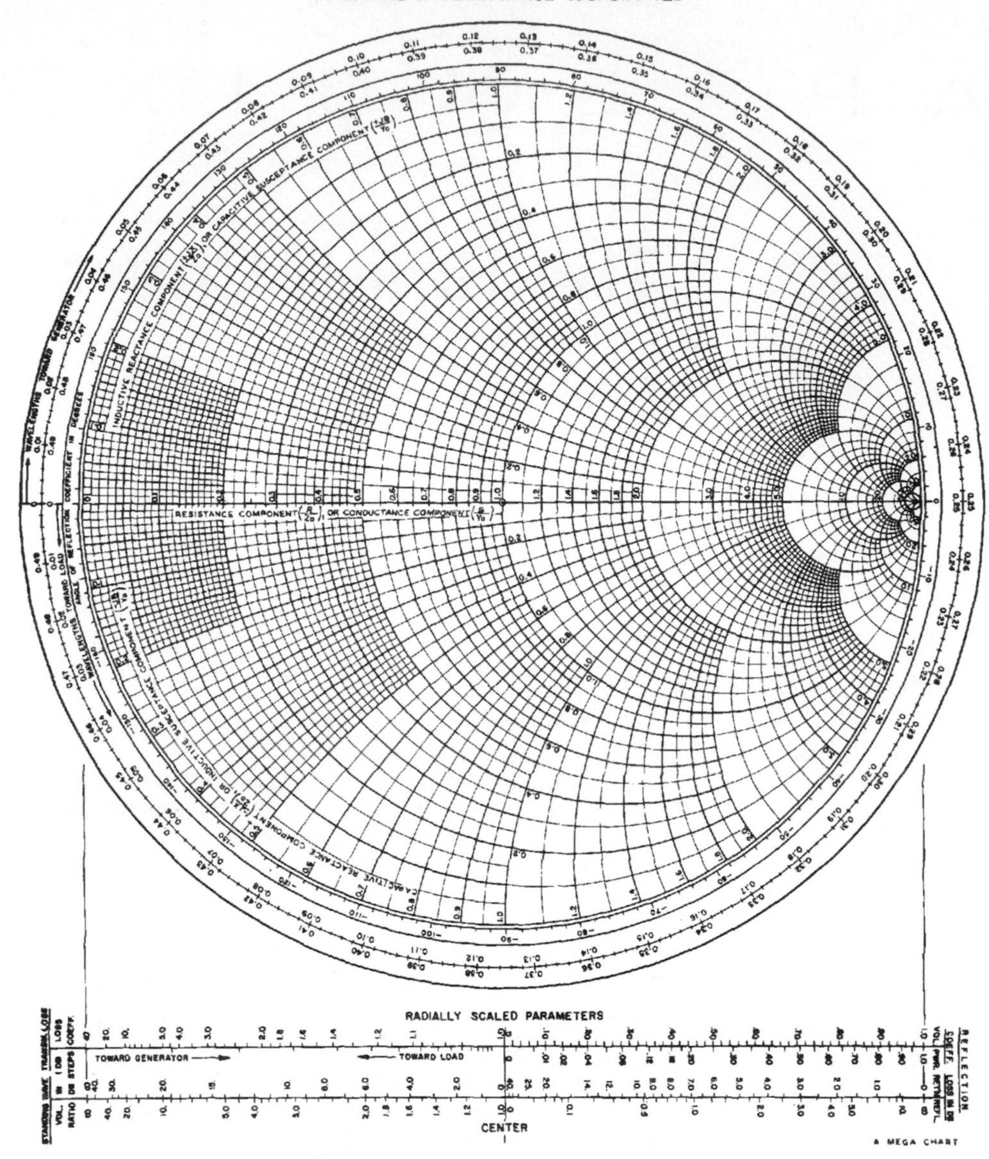

FIGURE A-1 Smith chart, transmission-line calculator

the line impedance ($\pm jX$) to the characteristic impedance of the line (Z_O). Parameters plotted on the Smith chart include the following:

1. Impedance (or admittance) at any point along a transmission line
   a. Reflection coefficient magnitude (Γ)
   b. Reflection coefficient angle in degrees
2. Length of transmission line between any two points in wavelengths
3. Attenuation between any two points
   a. Standing-wave loss coefficient
   b. Reflection loss

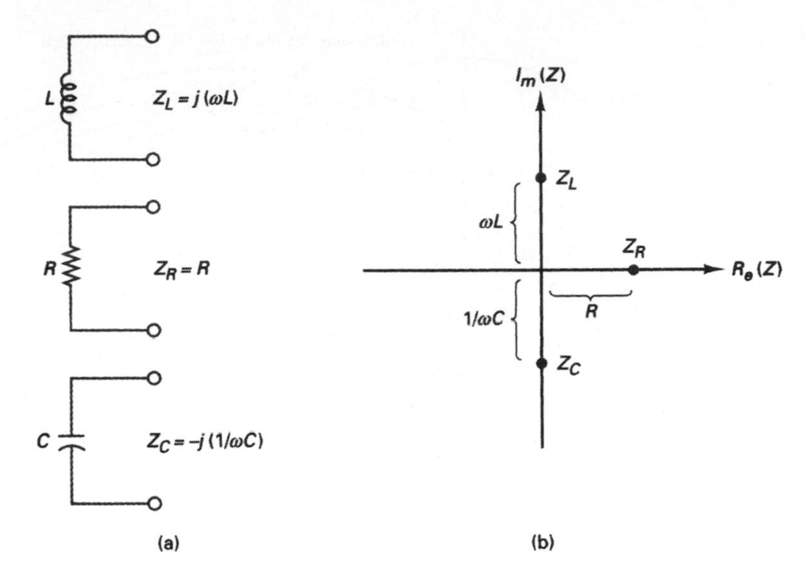

FIGURE A-2 (a) Typical circuit elements; (b) impedances graphed on rectangular coordinate plane (Note: ω is the angular frequency at which Z is measured.)

4. Voltage or current standing-wave ratio
   a. Standing-wave ratio
   b. Limits of voltage and current due to standing waves

SMITH CHART DERIVATION

The impedance of a transmission line, Z, is made up of both *real* and *imaginary* components of either sign ($Z = R \pm jX$). Figure A-2a shows three typical circuit elements, and Figure A-2b shows their impedance graphed on a *rectangular* coordinate plane. All values of Z that correspond to passive networks must be plotted on or to the right of the imaginary axis of the Z plane (this is because a negative real component implies that the network is capable of supplying energy). To display the impedance of all possible passive networks on a rectangular plot, the plot must extend to infinity in three directions ($+R$, $+jX$, and $-jX$). The Smith chart overcomes this limitation by plotting the complex *reflection coefficient*,

$$\Gamma = \frac{z - 1}{z + 1} \qquad \textbf{(A-2)}$$

where z equals the impedance normalized to the characteristic impedance (i.e., $z = Z/Z_O$).

Equation A-2 shows that for all passive impedance values, z, the magnitude of Γ is between 0 and 1. Also, because $|\Gamma| \leq 1$, the entire right side of the z plane can be mapped onto a circular area on the Γ plane. The resulting circle has a radius $r = 1$ and a center at $\Gamma = 0$, which corresponds to $z = 1$ or $Z = Z_O$.

Lines of Constant $R_e(z)$

Figure A-3a shows the rectangular plot of four lines of constant resistance $R_e(z) = 0, 0.5, 1$, and 2. For example, any impedance with a real part $R_e = 1$ will lie on the $R = 1$ line. Impedances with a positive reactive component (X_L) will fall above the real axis, whereas

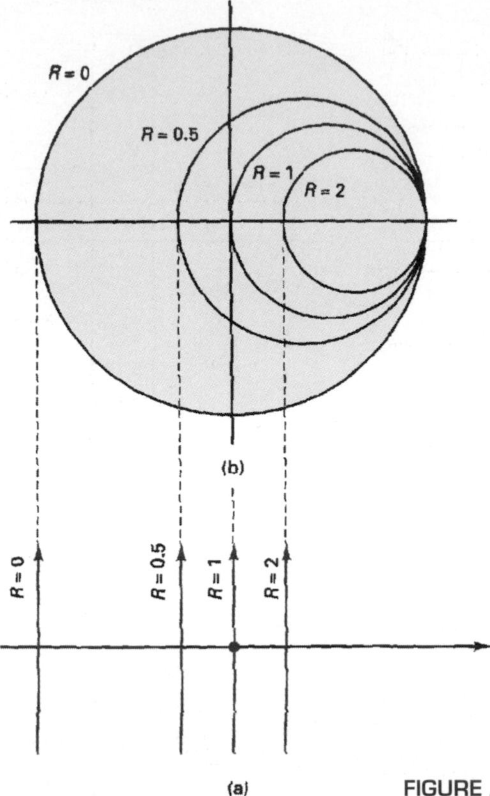

FIGURE A-3 (a) Rectangular plot; (b) Γ plane

impedances with a negative reactive component (X_C) will fall below the real axis. Figure A-3b shows the same four values of R mapped onto the Γ plane. $R_e(z)$ are now circles of $R_e(\Gamma)$. However, inductive impedances are still transferred to the area above the horizontal axis, and capacitive impedances are transferred to the area below the horizontal axis. The primary difference between the two graphs is that with the circular plot the lines no longer extend to infinity. The infinity points all meet on the plane at a distance of 1 to the right of the origin. This implies that for $z = \infty$ (whether real, inductive, or capacitive), $\Gamma = 1$.

Lines of Constant $X(z)$

Figure A-4a shows the rectangular plot of three lines of constant inductive reactance ($X =$ 0.5, 1, and 2), three lines of constant capacitive reactance ($X = -0.5$, -1, and -2), and a line of zero reactance ($X = 0$). Figure A-4b shows the same seven values of jX plotted onto the Γ plane. It can be seen that all values of infinite magnitude again meet at $\Gamma = 1$. The entire rectangular z plane curls to the right, and its three axes (which previously extended infinitely) meet at the intersection of the $\Gamma = 1$ circle and the horizontal axis.

Impedance Inversion (Admittance)

Admittance (Y) is the mathematical inverse of Z (i.e., $Y = 1/Z$). Y, or for that matter any complex number, can be found graphically using the Smith chart by simply plotting z on the complex Γ plane and then rotating this point 180° about $\Gamma = 0$. By rotating every point on the chart by 180°, a second set of coordinates (the y coordinates) can be developed that is an inverted mirror image of the original chart. See Figure A-5a. Occasionally, the admittance coordinates are superimposed on the same chart as the impedance coordinates. See Figure A-5b. Using the combination chart, both impedance and admittance values can be read directly by using the proper set of coordinates.

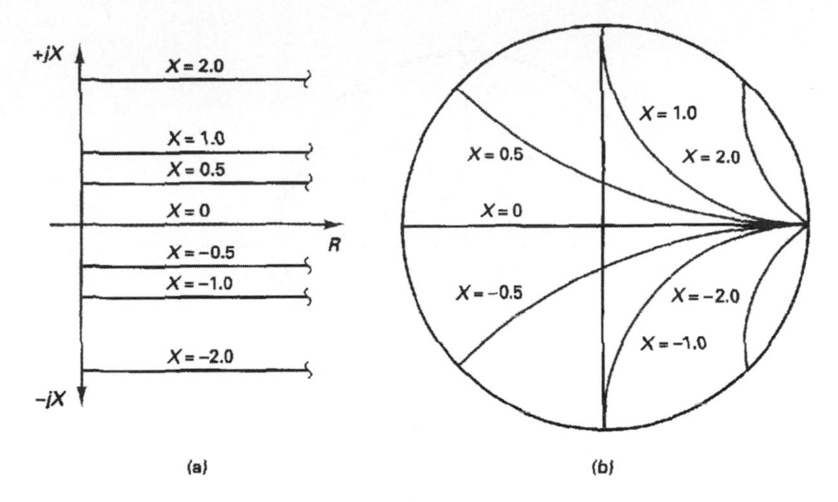

FIGURE A-4 (a) Rectangular plot; (b) Γ plane

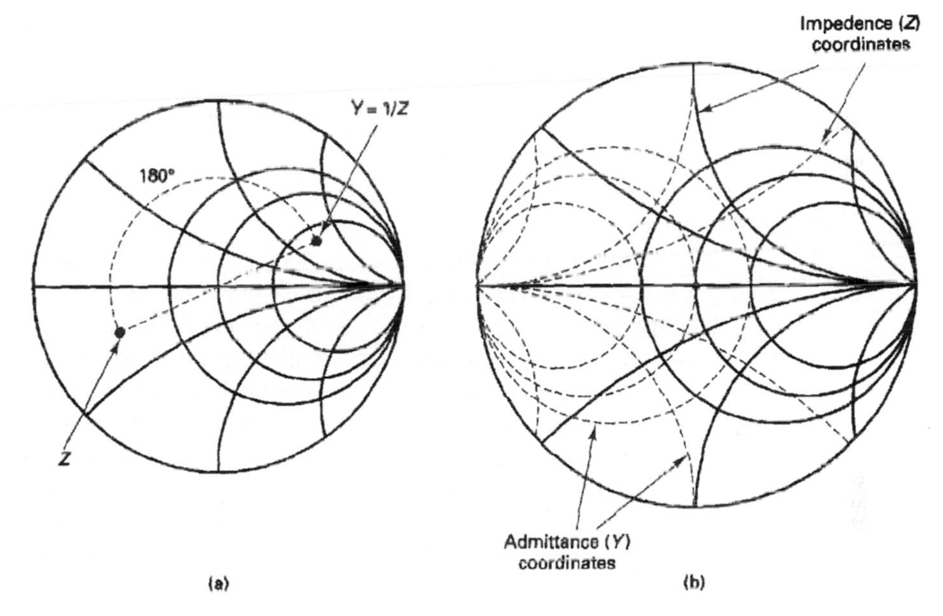

FIGURE A-5 Impedance inversion

Complex Conjugate

The *complex conjugate* can be determined easily using the Smith chart by simply reversing the sign of the angle of Γ. On the Smith chart, Γ is usually written in polar form and angles become more negative (phase lagging) when rotated in a clockwise direction around the chart. Hence, 0° is on the right end of the real axis and ±180° is on the left end. For example, let $\Gamma = 0.5\underline{/+150°}$. The complex conjugate, Γ*, is $0.5\ \underline{/-150°}$. In Figure A-6, it is shown that Γ* is found by mirroring Γ about the real axis.

PLOTTING IMPEDANCE, ADMITTANCE, AND SWR ON THE SMITH CHART

Any impedance *Z* can be plotted on the Smith chart by simply *normalizing* the impedance value to the characteristic impedance (i.e., $z = Z/Z_O$) and plotting the real and imaginary

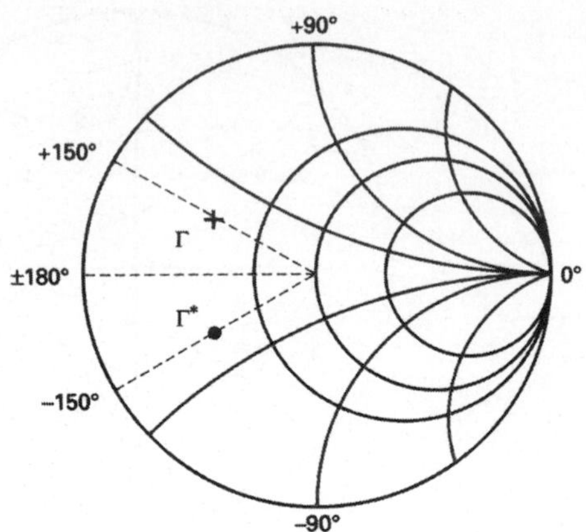

FIGURE A-6 Complex conjugate

parts. For example, for a characteristic impedance $Z_O = 50\ \Omega$ and an impedance $Z = 25\ \Omega$ resistive, the normalized impedance z is determined as follows:

$$z = \frac{Z}{Z_O} = \frac{25}{50} = 0.5$$

Because z is purely resistive, its plot must fall directly on the horizontal axis ($\pm jX = 0$). $Z = 25$ is plotted on Figure A-7 at point A (i.e., $z = 0.5$). Rotating 180° around the chart gives a normalized admittance value $y = 2$ (where $y = Y/Y_O$). y is plotted on Figure A-7 at point B.

As previously stated, a very important characteristic of the Smith chart is that any lossless line can be represented by a circle having its origin at $1 \pm j0$ (the center of the chart) and radius equal to the distance between the origin and the impedance plot. Therefore, the *standing-wave ratio* (SWR) corresponding to any particular circle is equal to the value of Z/Z_O at which the circle crosses the horizontal axis on the right side of the chart. Therefore, for this example, SWR = 0.5 ($Z/Z_O = 25/50 = 0.5$). It should also be noted that any impedance or admittance point can be rotated 180° by simply drawing a straight line from the point through the center of the chart to where the line intersects the circle on the opposite side.

For a characteristic impedance $Z_O = 50$ and an inductive load $Z = +j25$, the normalized impedance z is determined as follows:

$$z = \frac{Z}{Z_O} = \frac{1jX}{Z_O} = \frac{1j25}{50} = 1j0.5$$

Because z is purely inductive, its plot must fall on the $R = 0$ axis, which is the outer circle on the chart. $z = +j0.5$ is plotted on Figure A-8 at point A, and its admittance $y = -j2$ is graphically found by simply rotating 180° around the chart (point B). SWR for this example must lie on the far right end of the horizontal axis, which is plotted at point C and corresponds to SWR = ∞, which is inevitable for a purely reactive load. SWR is plotted at point C.

For a complex impedance $Z = 25 + j25$, z is determined as follows:

$$z = \frac{25\ 1\ j25}{50} = 0.5\ 1\ j0.5$$

$$= 0.707\angle 45^\circ$$

Therefore,

$$Z = 0.707\ \angle 45^\circ \times 50 = 35.35\ \angle 45^\circ$$

and

$$Y = \frac{1}{35.35\angle 45^\circ} = 0.02829\angle -45^\circ$$

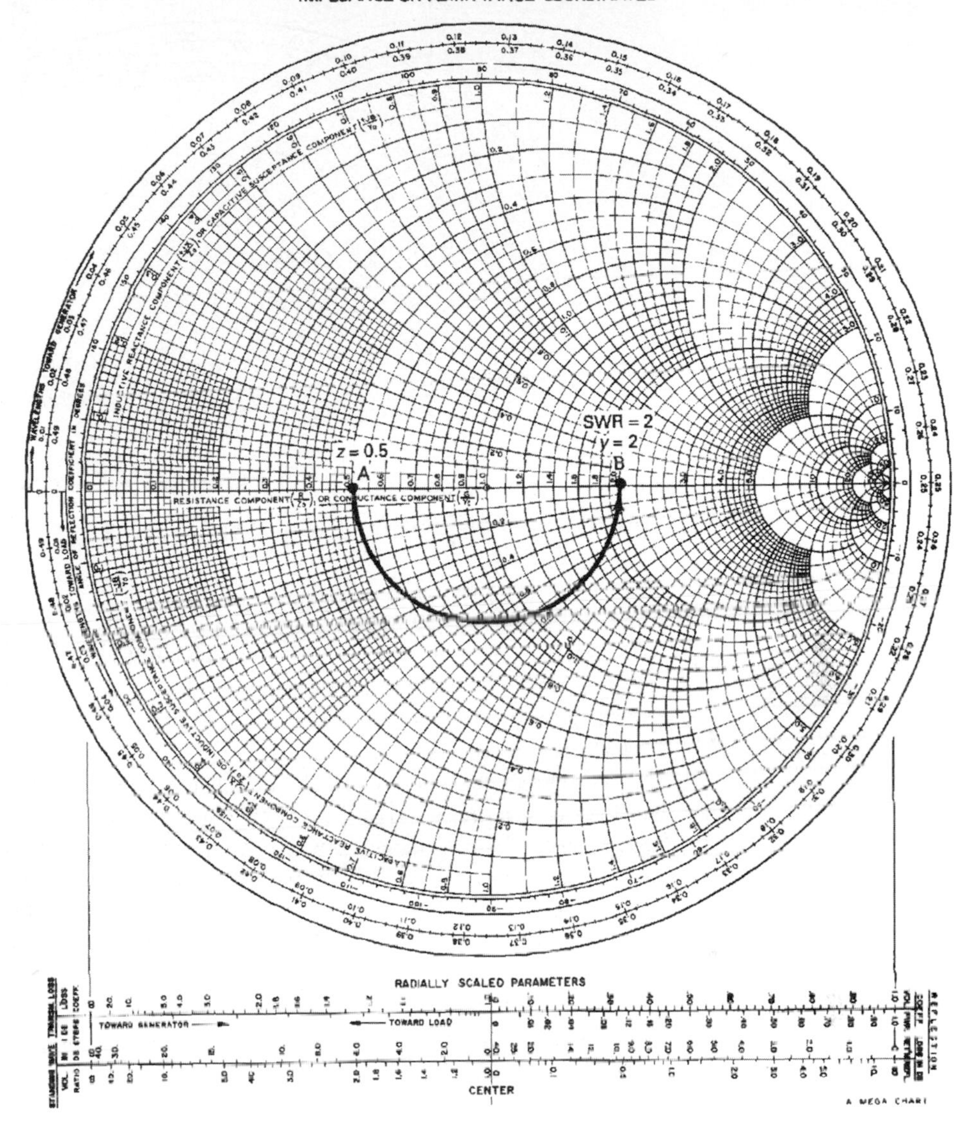

FIGURE A-7 Resistive impedance

Thus, $$y = \frac{Y}{Y_O} = \frac{0.02829}{0.02}\underline{/-45^\circ} = 1.414\underline{/-45^\circ}$$

and $$y = 1 - j1$$

z is plotted on the Smith chart by locating the point where the $R = 0.5$ arc intersects the $X = 0.5$ arc on the top half of the chart. $z = 0.5 + j0.5$ is plotted on Figure A-9 at point A, and y is plotted at point B $(1 - j1)$. From the chart, SWR is approximately 2.6 (point C).

INPUT IMPEDANCE AND THE SMITH CHART

The Smith chart can be used to determine the input impedance of a transmission line at any distance from the load. The two outermost scales on the Smith chart indicate distance in

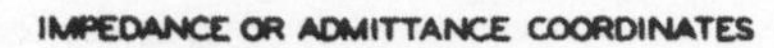

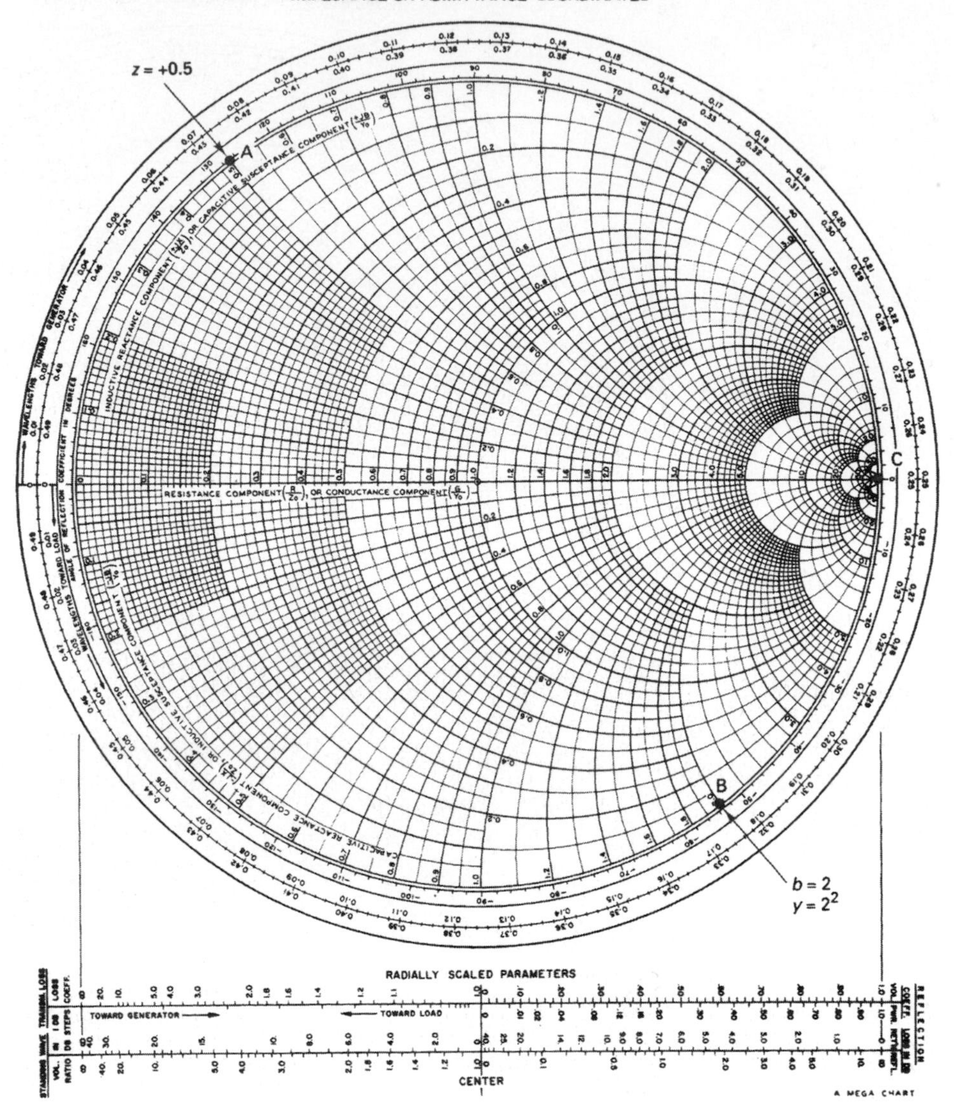

FIGURE A-8 Inductive load

wavelengths (see Figure A-1). The outside scale gives distance from the load toward the generator and increases in a clockwise direction, and the second scale gives distance from the source toward the load and increases in a counterclockwise direction. However, neither scale necessarily indicates the position of either the source or the load. One complete revolution (360°) represents a distance of one-half wavelength (0.5λ), half of a revolution (180°) represents a distance of one-quarter wavelength (0.25λ), and so on.

A transmission line that is terminated in an open circuit has an impedance at the open end that is purely resistive and equal to infinity (Chapter 12). On the Smith chart, this point is plotted on the right end of the $X = 0$ line (point *A* on Figure A-10). As you move toward the source (generator), the input impedance is found by rotating around the chart in a clockwise direction. It can be seen that input impedance immediately becomes capacitive and maximum. As you rotate farther around the circle (move toward the generator), the capac-

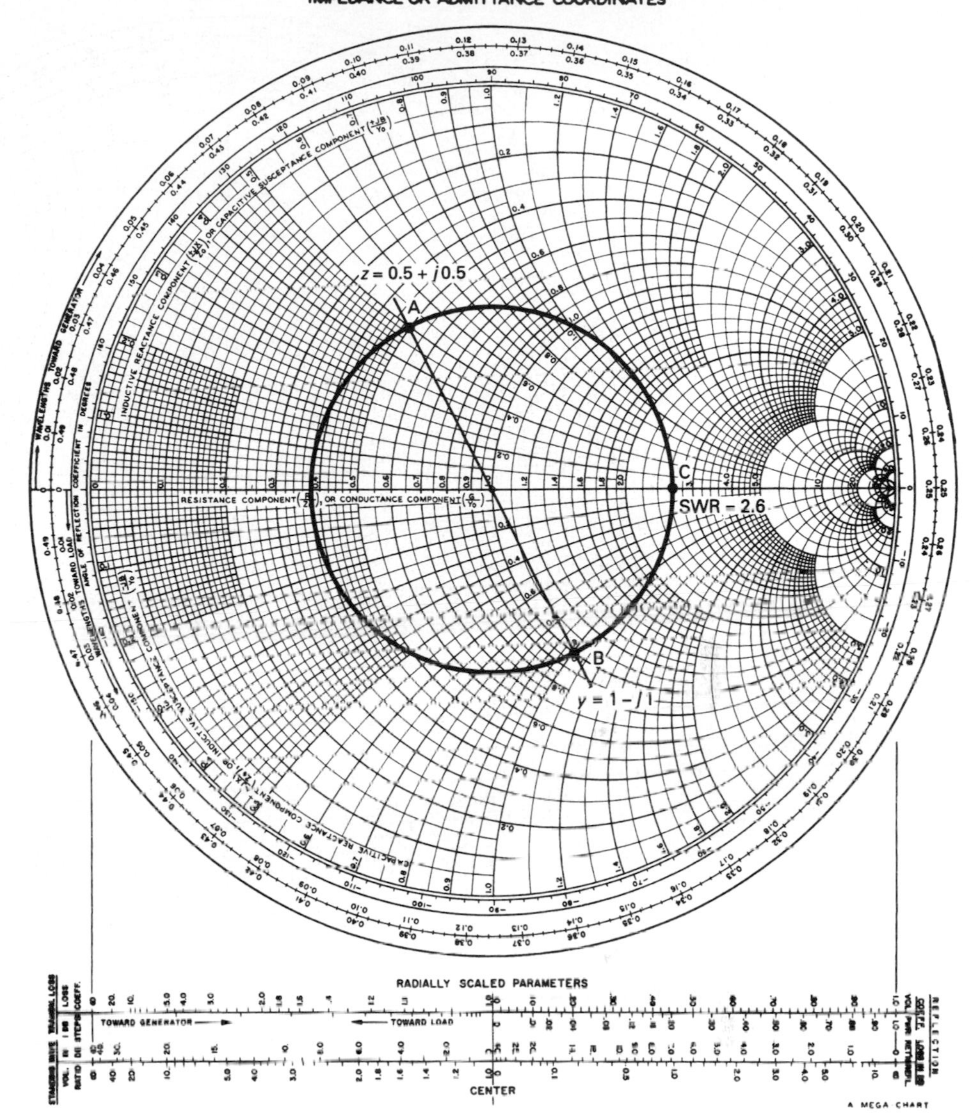

FIGURE A-9 Complex impedance

itance decreases to a normalized value of unity (i.e., $z = -j1$) at a distance of one-eighth wavelength from the load (point C on Figure A-10) and a minimum value just short of one-quarter wavelength. At a distance of one-quarter wavelength, the input impedance is purely resistive and equal to 0 Ω (point B on Figure A-10). As described in Chapter 12, there is an impedance inversion every one-quarter wavelength on a transmission line. Moving just past one-quarter wavelength, the impedance becomes inductive and minimum; then the inductance increases to a normalized value of unity (i.e., $z = +j1$) at a distance of three-eighths wavelength from the load (point D on Figure A-10) and a maximum value just short of one-half wavelength. At a distance of one-half wavelength, the input impedance is again purely resistive and equal to infinity (return to point A on Figure A-10). The results of the preceding analysis are identical to those achieved with phase analysis in Chapter 12 and plotted in Figure 12-25.

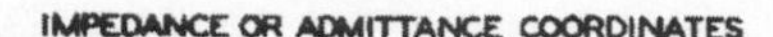

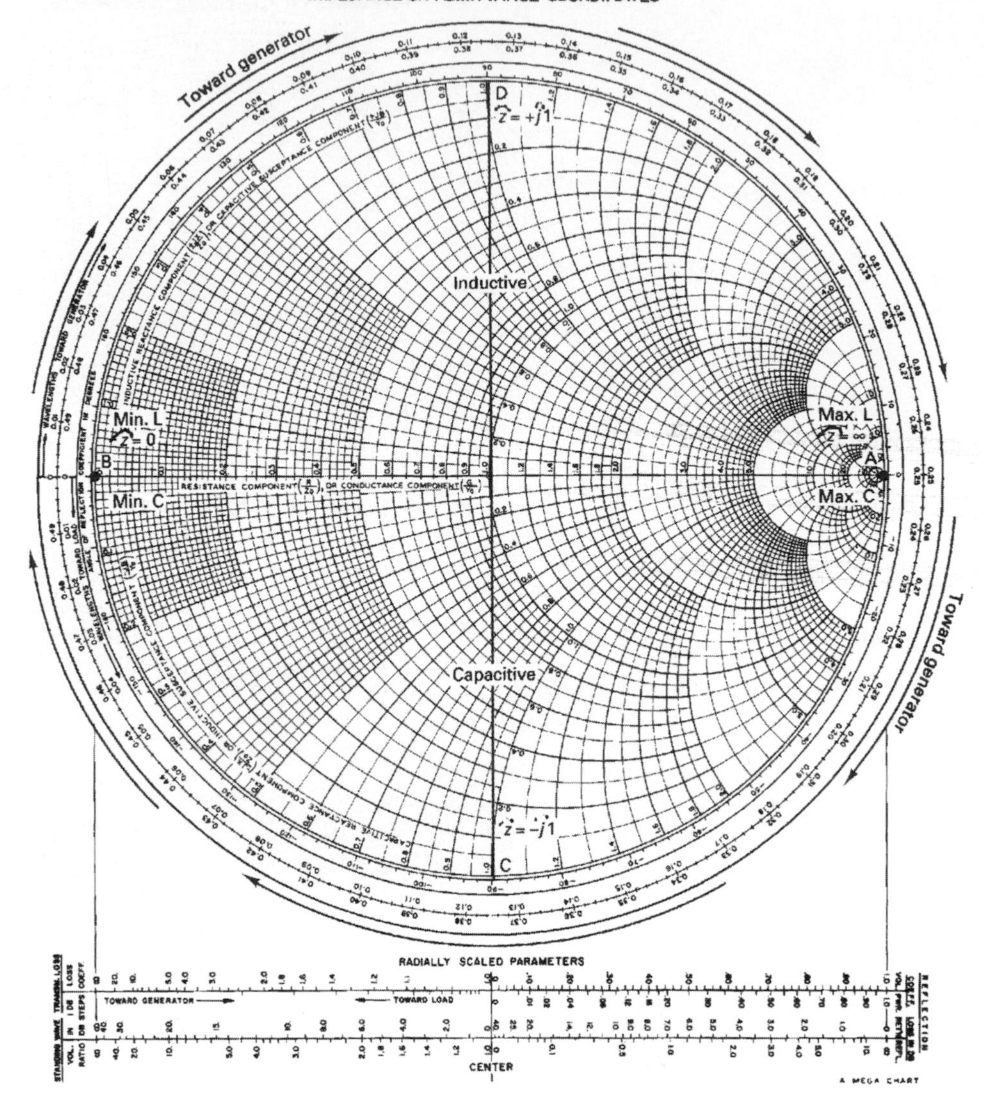

FIGURE A-10 Transmission-line input impedance for shorted and open line

A similar analysis can be done with a transmission line that is terminated in a short circuit, although the opposite impedance variations are achieved as with an open load. At the load, the input impedance is purely resistive and equal to 0. Therefore, the load is located at point *B* on Figure A-10, and point *A* represents a distance one-quarter wavelength from the load. Point *D* is a distance of one-eighth wavelength from the load and point *C*, a distance of three-eighths wavelength. The results of such an analysis are identical to those achieved with phasors in Chapter 12 and plotted in Figure 12-26.

For a transmission line terminated in a purely resistive load not equal to Z_O, Smith chart analysis is very similar to the process described in the preceding section. For example, for a load impedance $Z_L = 37.5\ \Omega$ resistive and a transmission-line characteristic impedance $Z_O = 75\ \Omega$, the input impedance at various distances from the load is determined as follows:

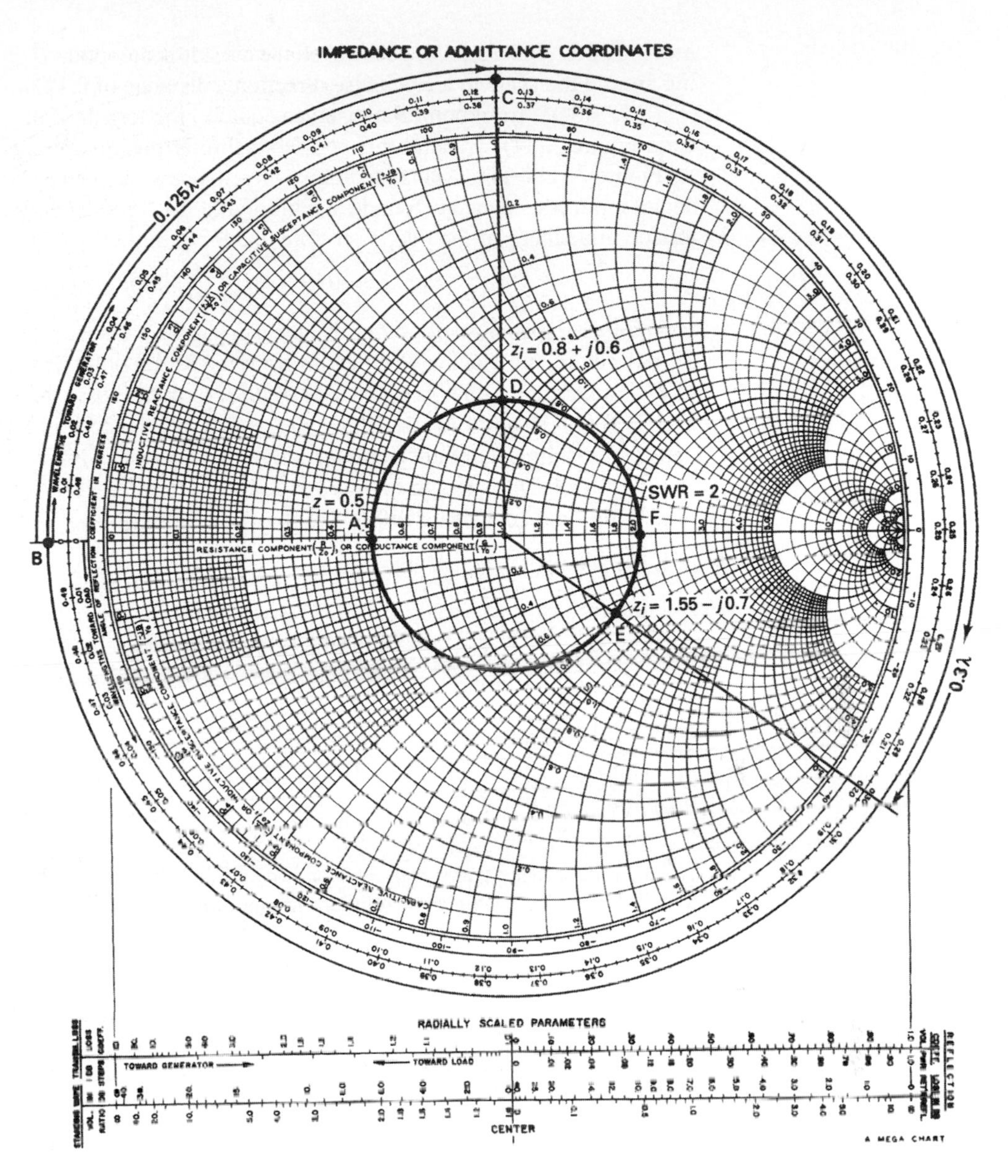

FIGURE A-11 Input impedance calculations

1. The normalized load impedance z is

$$z = \frac{Z_L}{Z_O} = \frac{37.5}{75} = 0.5$$

2. $z = 0.5$ is plotted on the Smith chart (point A on Figure A-11). A circle is drawn that passes through point A with its center located at the intersection of the $R = 1$ circle and the $x = 0$ arc.

3. SWR is read directly from the intersection of the $z = 0.5$ circle and the $X = 0$ line on the right side (point F), SWR = 2. The impedance circle can be used to describe all impedances along the transmission line. Therefore, the input impedance (Z_i) at a distance of 0.125λ from the load is determined by extending the z circle to the outside of the chart, moving point A to a similar position on the outside scale (point B on Figure A-11) and moving around the scale in a clockwise direction a distance of 0.125λ.

4. Rotate from point B a distance equal to the length of the transmission line (point C on Figure A-11). Transfer this point to a similar position on the $z = 0.5$ circle (point D on Figure A-11). The normalized input impedance is located at point D ($0.8 + j0.6$). The actual input impedance is found by multiplying the normalized impedance by the characteristic impedance of the line. Therefore, the input impedance Z_i is

$$Z_i = (0.8 + j0.6)75 = 60 + j45$$

Input impedances for other distances from the load are determined in the same way. Simply rotate in a clockwise direction from the initial point a distance equal to the length of the transmission line. At a distance of 0.3λ from the load, the normalized input impedance is found at point $E(z = 1.55 - j0.7$ and $Z_i = 116.25 - j52.5)$. For distances greater than 0.5λ, simply continue rotating around the circle, with each complete rotation accounting for 0.5λ. A length of 1.125 is found by rotating around the circle two complete revolutions and an additional 0.125λ.

Example A-1

Determine the input impedance and SWR for a transmission line 1.25λ long with a characteristic impedance $Z_O = 50\ \Omega$ and a load impedance $Z_L = 30 + j40\ \Omega$.

Solution The normalized load impedance z is

$$z = \frac{30 + j40}{50} = 0.6 + j0.8$$

z is plotted on Figure A-12 at point A and the impedance circle is drawn. SWR is read off the Smith chart from point B.

$$\text{SWR} = 2.9$$

The input impedance 1.25λ from the load is determined by rotating from point C 1.25λ in a clockwise direction. Two complete revolutions account for 1λ. Therefore, the additional 0.25λ is simply added to point C.

$$0.125\lambda + 0.25\lambda = 0.37\lambda \text{ (point } D\text{)}$$

Point D is moved to a similar position on the $z = 0.6 + j0.8$ circle (point E), and the input impedance is read directly from the chart.

$$z_i = 0.63 - j0.77$$
$$Z_i = 50(0.63 - j0.77) = 31.5 - j38.5$$

Quarter-Wave Transformer Matching with the Smith Chart

As described in Chapter 12, a length of transmission line acts as a transformer (i.e., there is an impedance inversion every one-quarter wavelength). Therefore, a transmission line with the proper length located the correct distance from the load can be used to match a load to the impedance of the transmission line. The procedure for matching a load to a transmission line with a quarter-wave transformer using the Smith chart is outlined in the following steps.

1. A load $Z_L = 75 + j50\ \Omega$ can be matched to a 50-Ω source with a quarter-wave transformer. The normalized load impedance z is

$$z = \frac{75 + j50}{50} = 1.5 + j1$$

2. $z = 1.5 + j1$ is plotted on the Smith chart (point A, Figure A-13) and the impedance circle is drawn.

3. Extend point A to the outermost scale (point B). The characteristic impedance of an ideal transmission line is purely resistive. Therefore, if a quarter-wave transformer is lo-

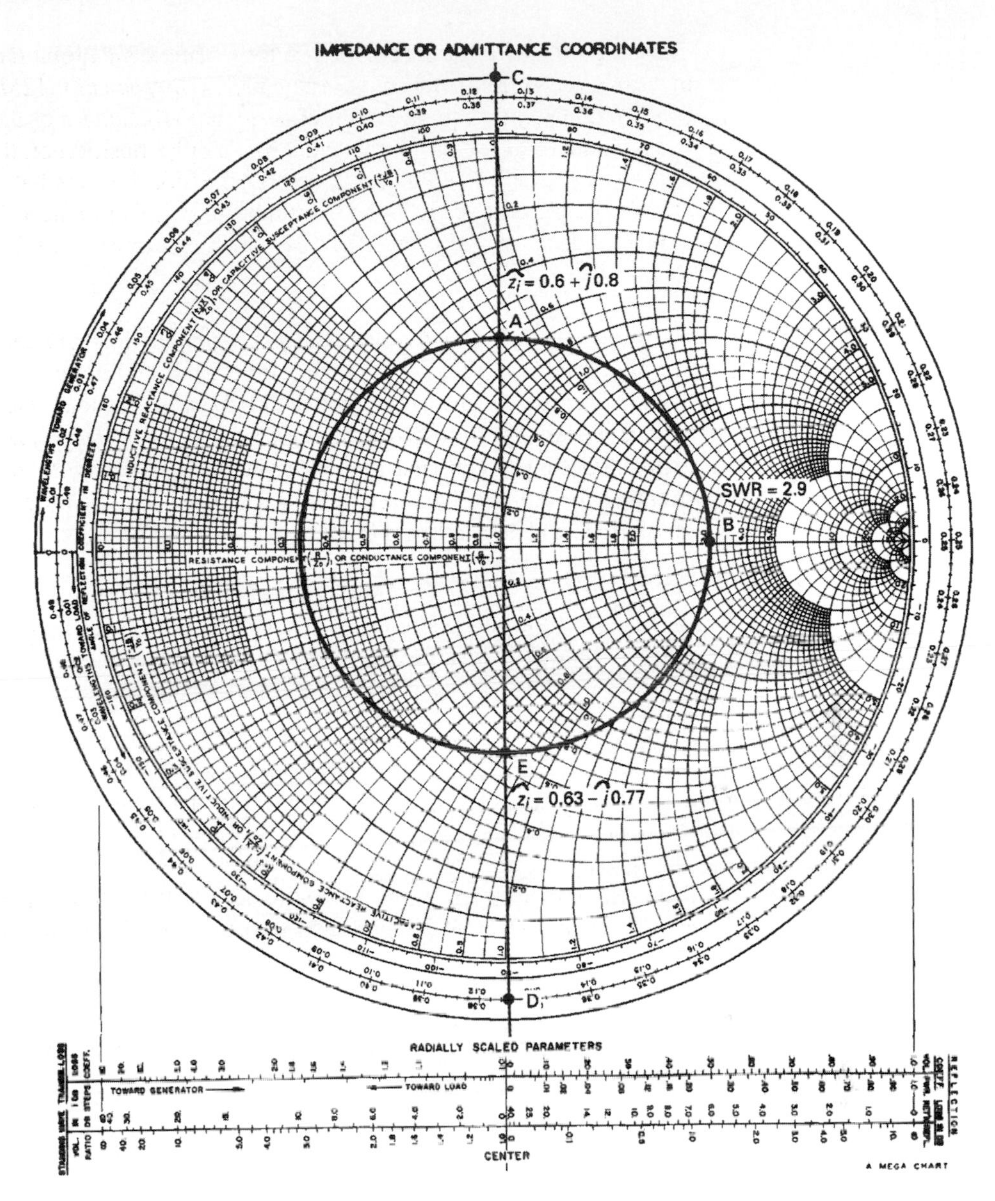

FIGURE A-12 Smith chart for Example A-1

cated at a distance from the load where the input impedance is purely resistive, the transformer can match the transmission line to the load. There are two points on the impedance circle where the input impedance is purely resistive: where the circle intersects the $X = 0$ line (points C and D on Figure A-13). Therefore, the distance from the load to a point where the input impedance is purely resistive is determined by simply calculating the distance in wavelengths from point B on Figure A-13 to either point C or D, whichever is the shortest. The distance from point B to point C is

point C	0.250λ
− point B	$-\ 0.192\lambda$
distance	0.058λ

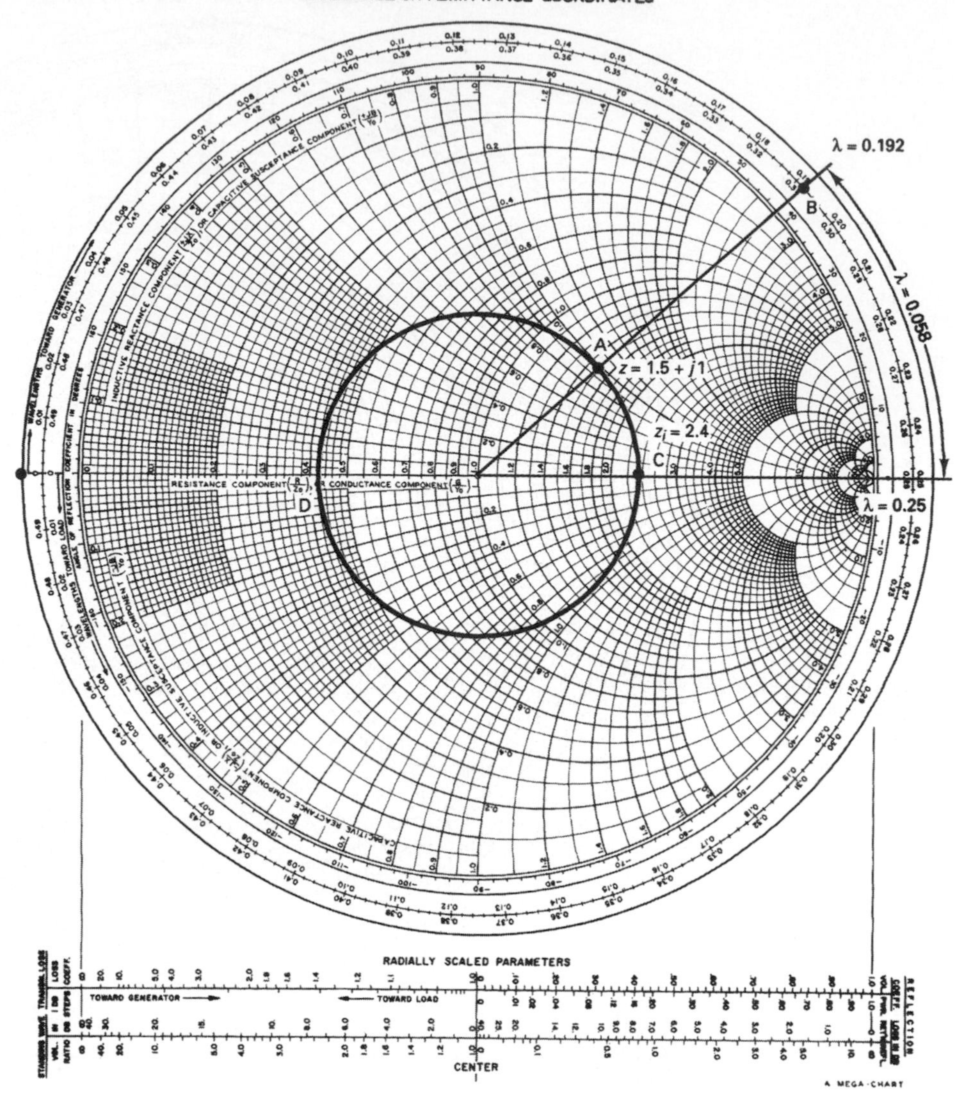

FIGURE A-13 Smith chart, quarter-wave transformer

If a quarter-wave transformer is placed 0.058λ from the load, the input impedance is read directly from Figure A-13, $z_i = 2.4$ (point C).

4. Note that 2.4 is also the SWR of the mismatched line and is read directly from the chart.

5. The actual input impedance $Z_i = 50(2.4) = 120\ \Omega$. The characteristic impedance of the quarter-wave transformer is determined from Equation 12-47.

$$Z'_O = \sqrt{Z_O Z_1} = \sqrt{50 \times 120} = 77.5\ \Omega$$

Thus, if a quarter-wavelength of a 77.5-Ω transmission line is inserted 0.058λ from the load, the line is matched. It should be noted that a quarter-wave transformer does not totally eliminate standing waves on the transmission line. It simply eliminates them from the transformer back to the source. Standing waves are still present on the line between the transformer and the load.

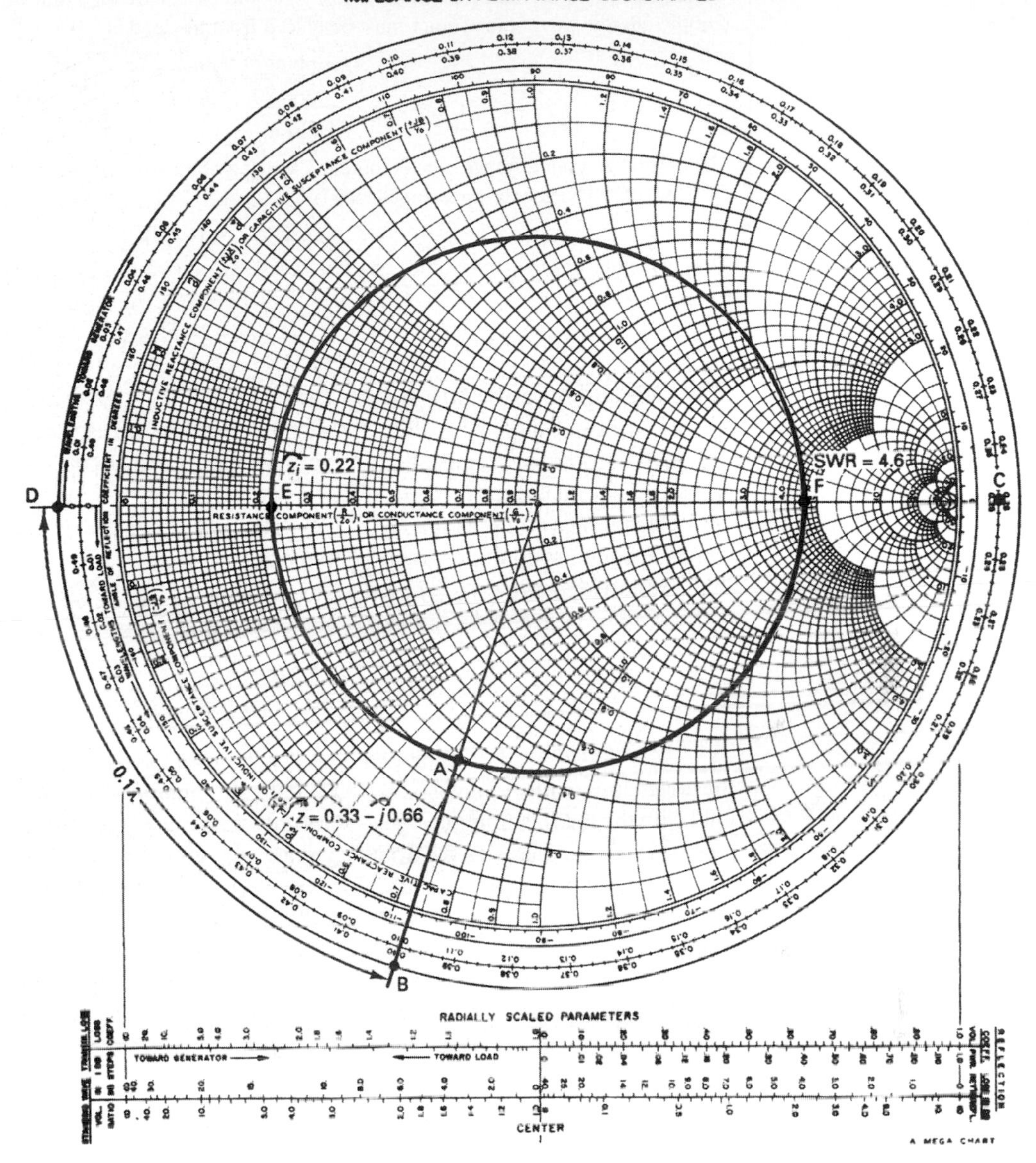

FIGURE A-14 Smith chart for Example A-2

Example A-2

Determine the SWR, characteristic impedance of a quarter-wave transformer, and the distance the transformer must be placed from the load to match a 75-Ω transmission line to a load $Z_L = 25 - j50$.

Solution The normalized load impedance z is

$$z = \frac{25 - j50}{75} = 0.33 - j0.66$$

z is plotted at point A on Figure A-14 and the corresponding impedance circle is drawn. The SWR is read directly from point F:

$$\text{SWR} = 4.6$$

The closest point on the Smith chart where Z_i is purely resistive is point D. Therefore, the distance that the quarter-wave transformer must be placed from the load is

$$\begin{array}{ll} \text{point } D & 0.5\lambda \\ -\text{ point } B & 0.4\lambda \\ \hline \text{distance} & 0.1\lambda \end{array}$$

The normalized input impedance is found by moving point D to a similar point on the z circle (point E), $z_i = 0.22$. The actual input impedance is

$$Z_i = 0.22(75) = 16.5\ \Omega$$

The characteristic impedance of the quarter-wavelength transformer is again found from Equation 12-47.

$$Z'_O = \sqrt{75 \times 16.5} = 35.2\ \Omega$$

Stub Matching with the Smith Chart

As described in Chapter 12, shorted and open stubs can be used to cancel the reactive portion of a complex load impedance and, thus, match the load to the transmission line. Shorted stubs are preferred because open stubs have a greater tendency to radiate.

Matching a complex load $Z_L = 50 - j100$ to a 75-Ω transmission line using a shorted stub is accomplished quite simply with the aid of a Smith chart. The procedure is outlined in the following steps:

1. The normalized load impedance z is

$$z = \frac{50 - j100}{75} = 0.67 - j1.33$$

2. $z = 0.67 - j1.33$ is plotted on the Smith chart shown in Figure A-15 at point A and the impedance circle is drawn in. Because stubs are shunted across the load (i.e., placed in parallel with the load), admittances are used rather than impedances to simplify the calculations, and the circles and arcs on the Smith chart are now used for conductance and susceptance.

3. The normalized admittance y is determined from the Smith chart by simply rotating the impedance plot, z, 180°. This is done on the Smith chart by simply drawing a line from point A through the center of the chart to the opposite side of the circle (point B).

4. Rotate the admittance point clockwise to a point on the impedance circle where it intersects the $R = 1$ circle (point C). The real component of the input impedance at this point is equal to the characteristic impedance Z_O, $Z_{in} = R \pm jX$, where $R = Z_O$. At point C, the admittance $y = 1 + j1.7$.

5. The distance from point B to point C is how far from the load the stub must be placed. For this example, the distance is $0.18\lambda - 0.09\lambda = 0.09\lambda$. The stub must have an impedance with a zero resistive component and a susceptance that has the opposite polarity (i.e., $y_s = 0 - j1.7$).

6. To find the length of the stub with an admittance $y_s = 0 - j1.7$, move around the outside circle of the Smith chart (the circle where $R = 0$), having a wavelength identified at point D, until an admittance $y = 1.7$ is found (wavelength value identified at point E). You begin at point D because a shorted stub has minimum resistance ($R = 0$) and, consequently, a susceptance $B = \infty$. Point D is such a point. (If an open stub were used, you would begin your rotation at the opposite side of the $X = 0$ line, point F.)

7. The distance from point D to point E is the length of the stub. For this example, the length of the stub is $0.334\lambda - 0.25\lambda = 0.084\lambda$.

Example A-3

For a transmission line with a characteristic impedance $Z_O = 300\ \Omega$ and a load with a complex impedance $Z_L = 450 + j600$, determine SWR, the distance a shorted stub must be placed from the load to match the load to the line, and the length of the stub.

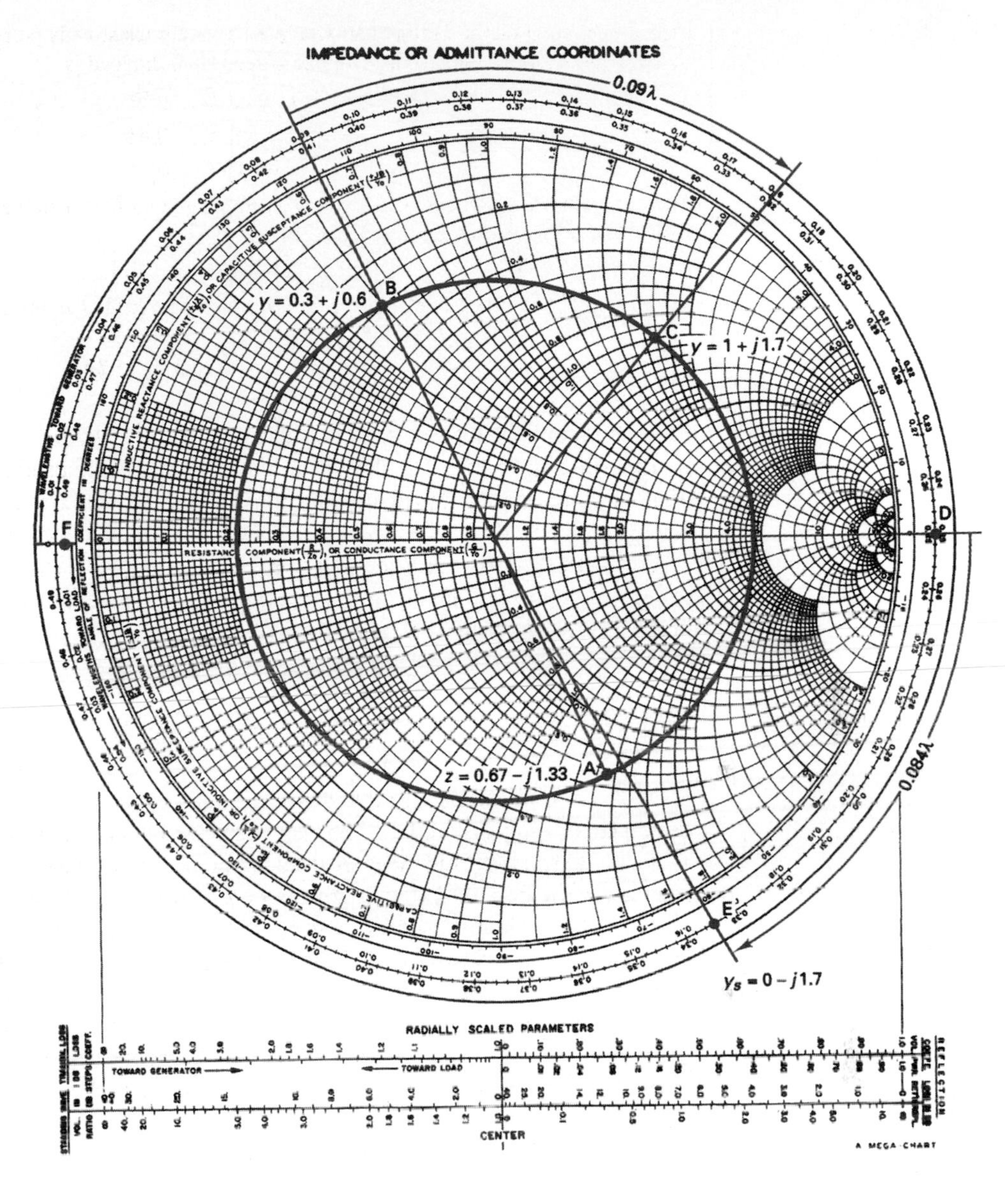

FIGURE A-15 Stub matching, Smith chart

Solution The normalized load impedance z is

$$z = \frac{450 + j600}{300} = 1.5 + j2$$

$z = 1.5 + j2$ is plotted on Figure A-16 at point A and the corresponding impedance circle is drawn. SWR is read directly from the chart at point B.

$$\text{SWR} = 4.7$$

Rotate point A 180° around the impedance circle to determine the normalized admittance.

$$y = 0.24 - j0.325 \text{ (point } C\text{)}$$

To determine the distance from the load to the stub, rotate clockwise around the outer scale beginning from point C until the circle intersects the $R = 1$ circle (point D).

$$y = 1 + j1.7$$

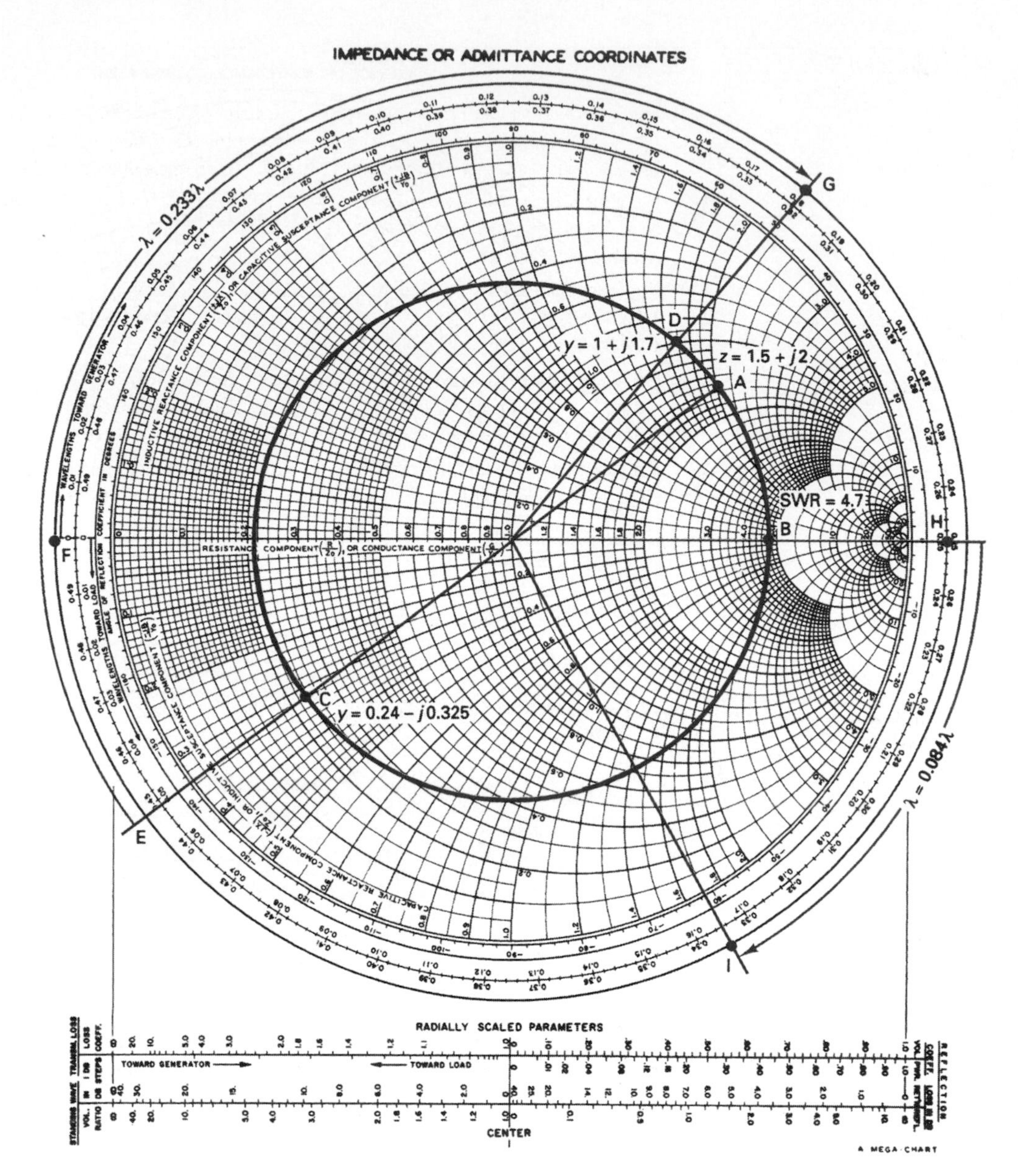

FIGURE A-16 Smith chart for Example A-3

The distance from point C to point D is the sum of the distances from point E to point F and point F to point G.

$$
\begin{aligned}
E \text{ to } F &= 0.5\lambda - 0.449\lambda = 0.051\lambda \\
+\, F \text{ to } G &= \underline{0.18\lambda - 0\lambda} \quad = \underline{0.18\lambda} \\
&\text{total distance} \quad = 0.231\lambda
\end{aligned}
$$

To determine the length of the shorted stub, calculate the distance from the $y = \infty$ point (point H) to the $y_s = 0 - j1.7$ point (point I).

$$\text{stub length} = 0.334\lambda - 0.25\lambda = 0.084\lambda$$

PROBLEMS

A-1. For a coaxial transmission line with the following characteristics: $Z_O = 72$ ohms, $\varepsilon_r = 2.02$, $f = 4.2$ GHz, and $Z_L = 30 - j60$ ohms; determine the following: (a) VSWR, (b) impedance 6.6 inches from load, and (c) minimum purely resistive impedance on the line (Z_{min}).

A-2. For a terminated twin-lead transmission line with the following characteristics: $Z_O = 300$ ohms, $\varepsilon_r = 2.56, f = 48$ MHz, and $Z_L = 73$ ohms; determine the following: (a) length of the stub and (b) shortest distance from the load that the stub can be placed.

A-3. For a terminated coaxial transmission line with the following characteristics: VSWR $= 3.0, f =$ 1.6 GHz, $\varepsilon_r = 2.02$, $Z_O = 50$ ohms, and distance from a VSWR null (minimum) to the load of 11.0 inches; determine the load impedance (Z_L).

A-4. For a terminated coaxial transmission line with the following characteristics: $Z_L = 80 - j120$ ohms, $Z_O = 50$ ohms, $\varepsilon_r = 1.0, f = 9.6$ GHz; determine the following: (a) the shortest distance from the load to the transformer, (b) transformer length, and (c) transformer impedance.

A-5. For a shorted coaxial transmission line with the following characteristics: $Z_O = 50$ ohms, $\varepsilon_r =$ $1.0, f = 3.2$ GHz, and $Z_{in} = -j80$ ohms; determine the length of the transmission line.

Answers to Selected Problems

CHAPTER 1

1-1. **a.** 7 dB
b. 11.8 dB
c. 14 dB
d. 21 dB
e. 33 dB
f. 40 dB
g. 50 dB
1-3. **a.** 400
b. 1.6
c. 19,952
d. 398,000
1-5. **a.** −60 dBm
b. −90 dBm
c. −128.5 dBm
1-7. **a.** −20 dBm
b. 0.2 W, 23 dBm
c. 23 dB, −10 dB, 30 dB
d. 43 dB
1-9. 12.1 dBm
1-11. **a.** 6 meters
b. 0.75 meters
c. 75 cm
d. 3 cm
1-13. Halved, doubled
1-15. **a.** 290 K

b. 256 K

c. 300 K

d. 223 K

1-17. 20 kHz

1-19. 3 kHz, 6 kHz, and 9 kHz
5 kHz, 10 kHz, and 15 kHz

1-21. a. 20 dB

b. 26 dB

c. −6 dB

d. 12 dB

1-23. F = 11.725 and NF = 10.7 dB

1-25. a. N = 8.28×10^{-17} W or −130.8 dBm

b. $V_n = 1.287 \times 10^{-7}$ V

1-27. 6.48 dB

1-29. 3 dB

1-31. 21.4 dB

1-33. a. 2.1×10^{-7} V

b. 1.4×10^{-7} V

c. 8×10^{-7} V

CHAPTER 2

2-1. 10 kHz

2-3. 100 kHz, 200 kHz, and 300 kHz

2-5. 1st harmonic = 500 Hz, $V_1 = 10.19\ V_p$
3rd harmonic = 1500 Hz, $V_3 = 3.4\ V_p$
5th harmonic = 2500 Hz, $V_5 = 2.04\ V_p$

2-7. Nonlinear amplifier with a single input frequency = 1 kHz

2-9. 7 kHz, 14 kHz, and 21 kHz; 4 kHz, 8 kHz, and 12 kHz

CHAPTER 3

3-1. a. 19.9984 MHz

b. 19.0068 MHz

c. 20.0032 MHz

3-3. $L = 100\ \mu H, f = 159.2$ kHz

3-5. $C_d = 0.0022\ \mu F$

3-7. $V_{out} = 1.5$ V, $V_{out} = 0.5$ V, $V_{out} = 1.85$ V

3-9. 20 kHz/rad, 125.6 krad/s or 102 dB

3-11. $V_d = 0.373$ V

3-13. $\theta_e = 0.830.25$ rad

3-15. $f_{out} = 647{,}126$ Hz

3-17. a. 9.9976 MHz

b. 10.0024 MHz

c. 9.9988 MHz

3-19. ±26°

3-21. $f = 65$ kHz

3-23. a. $f = 80$ kHz

b. $f = 40$ kHz

c. $f = 20$ kHz/volt

3-25. **a.** 12 kHz/rad
b. 5 kHz
c. 0.0833 V
d. 0.4167 V
e. 0.4167 rad
f. ±18.84 kHz
3-27. 4.8 kHz
3-29. ±39.15 kHz

CHAPTER 4

4-1. **a.** LSB = 95 kHz to 100 kHz, USB = 100 kHz to 105 kHz
b. $B = 10$ kHz
c. USF = 103 kHz, LSF = 97 kHz
4-3. $V_{max} = 25\ V_p$, $V_{min} = 15\ V_p$, m = 0.25
4-5. 15 V_p each
4-9. $V_c = 16\ V_p$, $V_{usf} = V_{lsf} = 3.2\ V_p$
4-11. **a.** 4 V_p
b. 12 V_p
c. 8 V_p
d. 0.667
e. 66.7%
4-13. **a.** 20 W
b. 10 W
c. 1000 W
d. 1020 W
4-15. 3000 W
4-17. **a.** $A_{max} = 112$, $A_{min} = 32$
b. $V_{out(max)} = 0.224\ V_p$, $V_{out(min)} = 0.064\ V_p$
4-21. **a.** 0.4
b. 40%
4-23. **a.** 20 V_p
b. 20 V_p
c. 10 V_p
d. 0.5
e. 50%
4-25. **a.** 50%, 3 V_p
b. 12 V_p
c. 18 V_p and 6 V_p
4-27. **a.** 36.8%
b. 38 V_p and 12 V_p
4-29. **a.** 64 W
b. 464 W
4-31. **a.** $A_{max} = 162$, $A_{min} = 18$
b. 1.62 V_p, 0.18 V_p

CHAPTER 5

5-1. 8 kHz
5-3. 900 degrees
5-5. $f_{if} = 600$ kHz, $f_{usf} = 604$ kHz, and $f_{lsf} = 596$ kHz

5-7. **a.** 1810 kHz
b. 122
5-9. **a.** 20 kHz
b. 14,276 Hz
5-11. 6 dB
5-13. Low-end bandwidth = 9 kHz, high-end bandwidth = 26.7 kHz
5-15. 13.6 kHz
5-17. **a.** 37.685 MHz
b. 48.33 MHz
5-19. 40.539 kHz and 23.405 kHz
5-21. 82 dB
5-23. 81.5 dB

CHAPTER 6

6-1. **a.** 396 kHz to 404 kHz
b. 397.2 kHz to 402.8 kHz
6-3. **a.** 34.1 MHz to 34.104 MHz
b. 34.1015 MHz
6-5. **a.** 28 MHz to 28.003 MHz
b. 28.0022 MHz
6-7. **a.** 496 kHz to 500 kHz
b. 497 kHz
6-9. **a.** 36.196 MHz to 36.204 MHz and 36.2 MHz carrier
b. 36.1975 MHz, 36.2 MHz, and 36.203 MHz carrier
6-11. 10.602 MHz, BFO = 10.6 MHz
6-13. **a.** 202 kHz and 203 kHz
b. 5.57 W and 2.278 W
6-15. **a.** 294 kHz to 306 kHz
b. 295.5 kHz to 304.5 kHz
6-17. **a.** 331.5 MHz to 31.505 MHz
b. 31.5025 MHz
6-19. **a.** 27.0003 MHz to 27.004 MHz
b. 27.0018 MHz
6-21. **a.** 594 kHz to 599.7 kHz
b. 597.5 kHz

CHAPTER 7

7-1. 0.5 kHz/V and 1 kHz
7-3. **a.** 40 kHz
b. 80 kHz
c. 20
7-5. 80%
7-7. **a.** 4
b.

J_n	V_n	f
0.22	1.76	carrier
0.58	4.64	1st sideband
0.35	2.80	2nd sideband
0.13	1.04	3rd sideband
0.03	0.24	4th sideband

7-9. 50 kHz

7-11. 2

7-13. 11.858 W, 3.87 W, 0.247 W, and 0.008 W

7-15. 0.0377 radians

7-17. a. 7.2 radians

b. 14.4 Hz

c. 90 MHz

7-19. 0.3 rads/V

7-21. 8 radians

7-23. 1, 2, 3, 4, 8, 14

7-25. 40 kHz, 10 kHz

7-27. 20 kHz

7-29. 144 kHz and 116 kHz

7-31. a. 15 kHz

b. 0.03535 rad

c. 20 dB

7-33. a. 1 kHz and 18 kHz

b. 0.1, 1.8

c. 56 kHz

CHAPTER 8

8-1. 18 dB, 22 dB

8-3. 114.15 MHz, 103.445 MHz

8-5. 20.4665 MHz and 20.3335 MHz

8-7. 0.4

8-9. 3.3 dB, 13.3 dB

8-11. −88.97 dBm

8-13. 1 V

8-15. 1.2 V

8-17. 50 dB

CHAPTER 9

9-1. 16 kHz, 4000 baud

9-3. 22 kHz, 10 kbaud

9-5. 5 MHz, 5 Mbaud

9-7. I = 1, Q = 0

9-9.

Q	I	C	Phase
0	0	0	−22.5°
0	0	1	−67.5°
0	1	0	22.5°
0	1	1	67.5°
1	0	0	−157.5°
1	0	1	−112.5°
1	1	0	157.5°
1	1	1	112.5°

9-11.

Q	Q'	I	I'	Phase
0	0	0	0	−45°
1	1	1	1	135°
1	0	1	0	135°
0	1	0	1	−45°

9-13. Input 00110011010101
XNOR 101110111001100

9-15. 40 MHz

9-17.

Q	I	Phase
0	0	−135°
0	1	−45°
1	0	135°
1	1	45°

9-19. 3.33 MHz, 3.33 Mbaud

9-21. 2.5 MHz, 2.5 Mbaud

9-23. **a.** 2 bps/Hz
b. 3 bps/Hz
c. 4 bps/Hz

CHAPTER 10

10-1. **a.** 8 kHz
b. 20 kHz

10-3. 10 kHz

10-5. 6 kHz

10-7. **a.** −2.12 V
b. −0.12 V
c. +0.04 V
d. −2.12 V
e. 0 V

10-9. 1200 or 30.8 dB

10-11. **a.** +0.01 to +0.03 V
b. −0.01 to −0.03 V
c. +20.47 to +20.49 V
d. −20.47 to −20.49 V
e. +5.13 to +5.15 V
f. +13.63 to +13.65 V

10-13.

f_{in}	f_s
2 kHz	4 kHz
5 kHz	10 kHz
12 kHz	24 kHz
20 kHz	40 kHz

10-15.

f_{in}	f_s
2.5 kHz	1.25 kHz
4 kHz	2 kHz
9 kHz	4.5 kHz
11 kHz	5.5 kHz

10-17.

N	DR	db
7	63	6
8	127	12
12	2047	66
14	8191	78

10-19.

μ	gain
255	0.948
100	0.938
255	1.504

10-21. 50.8 dB, 50.8 dB, 60.34 dB, 36.82 dB

10-23.

V_{in}	12-bit	8-bit	12-bit	decoded V	% Error
−6.592	100110011100	11011001	100110011000	−6.528	0.98
+12.992	001100101100	01100010	001100101000	+12.929	0.495
−3.36	100011010010	11001010	100011010100	−3.392	0.94

10-25. 11001110, 00110000, 01011111, 00100000

CHAPTER 11

11-1. **a.** 1.521 Mbps

b. 760.5 kHz

11-3. +-0000+-+0-00000+-00+-+0
00-　　　00-

11-5.

channel	f(kHz)
1	108
2	104
3	100
4	96
5	92
6	88
7	84
8	80
9	76
10	72
11	68
12	64

11-7. **a.** 5 kHz

b. 1.61 Mbps

c. 805 kHz

11-11.

CH	GP	SG	MG
100–104	364–370	746–750	746–750
84–88	428–432	1924–1928	4320–4324
92–96	526–520	2132–2136	4112–4116
72–76	488–492	2904–2908	5940–5944

11-13.

GP	SG	MG	MG out (kHz)
3	13	2	5540–5598
5	25	3	6704–6752
1	15	1	1248–1296
2	17	2	4504–4552

CHAPTER 12

12-1.

f	λ
1 kHz	300 km
100 kHz	3 km
1 MHz	300 m
1 GHz	0.3 m

12-3. 261 ohms

12-5. 111.8 ohms

12-7. 0.05

12-9. 12

12-11. 1.5

12-13. 252 m

12-15. 0.8

12-17.

λ	f
5 cm	6 GHz
50 cm	600 MHz
5 m	60 MHz
50 m	6 MHz

12-19. 107.4 ohms

12-21. 86.6 ohms

12-23. 1.01

12-25. 1.2

12-27. 54.77 ohms

CHAPTER 13

13-1. **a.** 869 nm, 8690 A°
b. 828 nm, 8280 A°
c. 935 nm, 9350 A°

13-3. 38.57°

13-5. 56°

13-7. **a.** RZ = 1 Mbps, NRZ = 500 kbps
b. RZ = 50 kbps, NRZ = 25 kbps
c. RZ = 250 kbps, NRZ = 125 kbps

13-9. **a.** 789 nm, 7890 A°
b. 937 nm, 9370 A°
c. 857 nm, 8570 A°

13-11. 42°

13-13. 36°

13-15. **a.** RZ = 357 kbps, NRZ = 179 kbps
b. RZ = 2 Mbps, NRZ = 1 Mbps
c. RZ = 250 kbps, NRZ = 125 kbps

CHAPTER 14

14-1. 0.2 $\mu w/m^2$

14-3. Increase by a factor of 9

14-5. 14.14 MHz

14-7. 0.0057 V/m

14-9. Increase by a factor of 16

14-11. 20 dB

14-13. 8.94 mi

14-15. 17.89 mi

14-17. 0.0095 $\mu W/m^2$

14-19. 34.14 mi or 96.57 km
14-21. 3.79 mV/m
14-23. Increase by a factor of 64
14-25. Increase by a factor of 16
14-27. 50.58°
14-29. 12.65 mi

CHAPTER 15

15-1. **a.** 25 ohms
b. 92.6%
c. 92.6 W
15-3. 38.13 dB
15-5. 82 W, 163611.5 W, 82.1 dBm, 52.1 dBW
15-7. 30 W, 0.106 $\mu W/m^2$
15-9. 16 dB
15-11. 97.9%
15-13. 99.2%
15-15. 5.25 dB, 108.7°
15-17. **a.** 6 GHz
b. 5 cm
c. 1.54×10^8 m/s
d. 5.82×10^8 m/s
15-19. 10 dB
15-21. 82 dBm
15-23. 0.5 $\mu W/m^2$
15-25. 9.947 $\mu W/m^2$
15-27. 87.78 dBm
15-29. 39.5 dB

CHAPTER 16—NO PROBLEMS

CHAPTER 17

17-1. **a.** 22-gauge wire with 44 mH inductance every 3000 feet
b. 19-gauge wire with 88 mH inductance every 6000 feet
c. 24-gauge wire with 44 mH inductance every 3000 feet
d. 16-gauge wire with 135 mH inductance every 3000 feet
17-3. **a.** −3 dBrnO
b. 18 dBrnc
c. 40 dBrnO
d. 47 dB
17-5. **a.** 51 dBrncO
b. −100 dBm
c. −1 dBmO
d. 36 dB
17-7. 2300 μs

CHAPTER 18—NO PROBLEMS

CHAPTER 19

19-1. 112 channels per cluster, 1344 total channel capacity

19-3. 8

19-5. **a.** 112

b. 448

c. 1792

CHAPTER 20—NO PROBLEMS

CHAPTER 21—NO PROBLEMS

CHAPTER 22

22-1. C = 0E, J = 1A, 4 = 0A, / = 17

22-3. C = C3, J = D1, 4 = F4, / = 61

22-5. 10100000 binary, A0 hex

22-7. 1000010010100000 binary, 84A0 hex

22-9. 4

22-11. Hamming bits = 0010 in positions 8, 6, 4, and 2

22-13. 8 V

CHAPTER 23

23-1. B8 hex

23-3. 4 inserted zeros

23-5. 8A hex

23-7. 65 hex

23-9. 4 deleted zeros

CHAPTER 24

24-1. −99.23 dBm

24-3. 28.9 dB

24-5. −85 dBm

24-7. 1.25 dB

24-9. 6.39 dB

24-11. 72.4 dB, 84.4 dB, 115.9 dB, 126.8 dBm, 138.8 dB, 137.5 dB

24-13. −94.3 dBm

24-15. 53 dB

24-17. $FM = 31.6$ dB, $A_t = A_r = 43.1$ dB, $L_p = 141.8$ dB, $L_b = 6$ dB, $L_f = 9.4$ dB, $N = -106$ dBm, $C_{min} = -78$ dBm, $P_t = 21.6$ dBm

24-19. −81 dBm

CHAPTER 25

25-1. Elevation angle = 51°, azimuth = 33°

25-3. **a.** 11.74 dB

b. 8.5 dB

25-5. P_{rad} = 28 dBW, EIRP = 68 dBW

25-7. −200.8 dBW

25-9. 26.18 dB

25-11. 19.77 dB

25-13. 12.5 dB

25-15. Elevation angle = 55°, azimuth = 15° west of south

25-17. **a.** C/N = 10.2 dB

b. C/N = 8.4 dB

25-19. 68.3 dBW

25-21. −199.6 dBj

25-23. N_o = −202.5 dBW, E_b = −162 dBj, E_b/N_o = 40.5 dB

25-25. 19 dB

25-27. 11.25 dB

CHAPTER 26

26-1. 15 transponders

26-3. 44 stations

26-5. 10 transponders

Index

Entries in **bold** indicate table or figure entry